Usability Evaluation and Interface Design:

Cognitive Engineering, Intelligent Agents and Virtual Reality

HUMAN FACTORS AND ERGONOMICS

Gavriel Salvendy, Series Editor

Bullinger, H.-J., and Ziegler, J. (Eds.) : *Human–Computer Interaction: Ergonomics and User Interfaces*

Bullinger, H.-J., and Ziegler, J. (Eds.) : *Human–Computer Interaction: Communication, Cooperation, and Application Design*

Stephanidis, C. (Ed.) : *User Interfaces for All: Concepts, Methods, and Tools*

Smith, M. J., Salvendy, G., Harris, D., and Koubeck, R. J. (Eds.) : *Usability Evaluation and Interface Design: Cognitive Engineering, Intelligent Agents and Virtual Reality*

Smith, M. J., and Salvendy, G. (Eds.) : *Systems, Social and Internationalization Design Aspects of Human–Computer Interaction*

Stephanidis, C. (Ed.) : *Universal Access in HCI: Towards an Information Society for All*

Stanney, K. (Ed.) : *Handbook of Virtual Environments Technology: Design, Implementation, and Applications*

Meister, D., and Enderwick, T. : *Human Factors in System Design, Development, and Testing*

Usability Evaluation and Interface Design:

Cognitive Engineering, Intelligent Agents and Virtual Reality

Volume 1 of the Proceedings of HCI International 2001

9th International Conference on Human–Computer Interaction Symposium on Human Interface (Japan) 2001

4th International Conference on Engineering Psychology and Cognitive Ergonomics

1st International Conference on Universal Access in Human–Computer Interaction August 5–10, 2001 New Orleans, Louisiana, USA

Edited by

Michael J. Smith
University of Wisconsin–Madison

Gavriel Salvendy
Purdue University

Don Harris
Cranfield University

Richard J. Koubek
Wright State University

2001

LAWRENCE ERLBAUM ASSOCIATES, PUBLISHERS
Mahwah, New Jersey   London

Copyright © 2001 by Lawrence Erlbaum Associates, Inc.
All rights reserved. No part of this book may be reproduced in any form, by photostat, microform, retrieval system, or any other means, without prior written permission of the publisher.

Lawrence Erlbaum Associates, Inc., Publishers
10 Industrial Avenue
Mahwah, NJ 07430

Library of Congress Cataloging-in-Publication Data

Usability Evaluation and Interface Design: Cognitive Engineering, Intelligent Agents and Virtual Reality / edited by Michael J. Smith, Gavriel Salvendy, Don Harris, and Richard J. Koubek
     p.   cm.
Includes bibliographical references and index.
ISBN 0-8058-3607-1 (cloth : alk. paper) Volume 1

ISBN 0-8058-3608-X (Volume 2)
ISBN 0-8058-3609-8 (Volume 3)
ISBN 0-8058-3610-1 (Set)

2001

Books published by Lawrence Erlbaum Associates are printed on acid-free paper, and their bindings are chosen for strength and durability.

Printed in the United States of America
10 9 8 7 6 5 4 3 2 1

USABILITY EVALUATION AND INTERFACE DESIGN: COGNITIVE ENGINEERING, INTELLIGENT AGENTS AND VIRTUAL REALITY
M.J. Smith, G. Salvendy, D. Harris, R.J. Koubek (Editors)

PREFACE

A total of 2,738 individuals from industry, academia, research institutes, and governmental agencies from 37 countries submitted their work for presentation at the Ninth International Conference on Human-Computer Interaction held in New Orleans, Louisiana U.S.A., 5-10 August 2001. Only those submittals that were judged to be of high scientific quality were included in the program. These papers address the latest research and application in the human aspects of design and use of computing systems. The papers accepted for presentation thoroughly cover the entire field of human-computer interaction, including the cognitive, social, ergonomic, and health aspects of work with computers. These papers address major advances in knowledge and effective use of computers in a variety of diversified application areas, including offices, financial institutions, manufacturing, electronic publishing, construction, and health care.

We are most grateful to the following cooperating organizations:

Chinese Academy of Sciences   International Ergonomics Association
Human Factors and Ergonomics Society   Japan Ergonomics Society
Institute of Industrial Engineers   Japan Management Association

The 317 papers contributing to this book cover the following areas:

Interface design, interactive systems, agents, virtual reality, mixed reality, sensory augmentation, web applications, information retrieval, physiological aspects of HCI, and cognitive ergonomics

The select papers on a complementary part of human-computer interaction are presented in two companion books: *Systems, Social and Internationalization Design Aspects of Human-Computer Interaction*, edited by M.J. Smith and G. Salvendy, and *Universal Access in HCI: Towards an Information Society for All*, edited by C. Stephanidis.

We wish to thank the following Board members who so diligently contributed to the success of the conference and to the direction of the content of this book. The conference Board members include:

Organizational Board

Hans-Jorg Bullinger, *German*   Jean-Claude Sperandio, *France*
Bengt Knave, *Sweden*   Hiroshi Tamura, *Japan*
Takao Ohkubo, *Japan*   Kong-shi Xu, *China*
Susumu Saito, *Japan*   Kan Ahang, *China*
Ben Shneiderman, *USA*

Ergonomics and Health Aspects of Work with Computers

Arne Aaras, *Norway*   Peter Kern, *Germany*
Pascale Carayon, *France*   Danuta Koradecka, *Poland*
Barbara Cohen, *USA*   Helmut Krueger, *Switzerland*
Marvin J. Dainoff, *USA*   Aura Matias, *Philippines*
Martin Helander, *Sweden*   Steven L. Sauter, *USA*
Waldemar Karwowski, *USA*   Gunnella Westlander, *Sweden*

Human Interface and the Management of Information

Yuichiro Anzai, *Japan*	Shogo Nishida, *Japan*
Lajos Balint, *Hungary*	Leszek Pacholski, *Poland*
Gunilla Bradley, *Sweden*	Robert W. Proctor, *USA*
Vincent Duffy, *Hong Kong*	Lawrence M. Schleifer, *USA*
Michitaka Hirose, *Japan*	E. Eugene Schultz, *USA*
Yasufumi Kume, *Japan*	Tsutomu Tabe, *Japan*
Mark R. Lehto, *USA*	Eberhard Ulich, *Switzerland*
Kee Yong Lim, *Singapore*	Tjerk W. van der Schaaf, *Netherlands*
Ann Majchrzak, *USA*	John R. Wilson, *UK*
Fiona Nah, *USA*	Sakae Yamamoto, *Japan*
Nobuto Nakamura, *Japan*	Bernhard Zimolong, *Germany*

Human-Computer Interaction

Albert G. Arnold, *Netherlands*	Aaron Marcus, *USA*
Nuray M. Aykin, *USA*	Nicolas Marmaras, *Greece*
Sebastiano Bagnara, *Italy*	Masaki Nakagawa, *Japan*
Nigel Bevan, *UK*	Celestine A. Ntuen, *USA*
George J. Boggs, *USA*	Katsuhiko Ogawa, *Japan*
Klaus-Peter Faehnrich, *Germany*	Kjell Ohlsson, *Sweden*
Pierre Falzon, *France*	Michelle Robertson, *USA*
Xiaowen Fang, *USA*	Dominique L. Scapin, *France*
Ephraim Glinert, *USA*	Andrew Sears, *USA*
Nancy Lightner, *USA*	Pentti K. Seppala, *Finland*
Sheue-Ling Hwang, *Taiwan*	Kay Stanney, *USA*
Julie A. Jacko, *USA*	Tomio Watanabe, *Japan*
Kari Lindstrom, *Finland*	Nong Ye, *USA*
Yili Liu, *USA*	Juergen Ziegler, *IGermany*
John Long, *UK*	Wenli Zhu, *USA*

Engineering Psychology and Cognitive Ergononmics

Chris Babar, *UK*	Peter Jorna, *Netherlands*
Ken Boff, *USA*	Masaaki Kurosu, *Japan*
Guy Boy, *France*	Kenneth R. Laughery, Sr., *USA*
Carlo Cacciabue, *Italy*	David Morrison, *Australia*
Judy Edworthy, *UK*	Sundaram Narayanan, *USA*
Arthur Fisk, *USA*	Reiner Onken, *Germany*
Margie Galer-Flyte, *UK*	Eduardo Salas, *USA*
Michael W. Haas, *USA*	Neville Stanton, *UK*
Craig Harvey, *USA*	Eric Tang, *ROC*
Erik Hollnagel, *Sweden*	Christopher D. Wickens, *USA*
Kenji Itoh, *Japan*	

Universal Access in Human-Computer Interaction

Demosthenes Akoumianakis, *Greece*	Alfred Kobsa, *Germany*
Elisabeth Andre, *Germany*	Harry Murphy, *USA*
David Benyon, *UK*	Michael Pieper, *Germany*
Noelle Carbonell, *France*	Christian Stary, *Austria*
Jan Ekberg, *Finland*	Hirotada Ueda, *Japan*
Pier Luigi Emiliani, *Italy*	Jean Vanderdonckt, *Belgium*
Jon Gunderson, *USA*	Gregg Vanderheiden, *USA*
Seppo Haataza, *Finland*	Annika Waern, *Sweden*
Ilias Iakovidis, *Belgium*	Gerhard Weber, *Germany*
Julie Jacko, *USA*	Michael Wilson, *UK*
Arthur Karshmer, *USA*	Toshiki Yamoka, *Japan*

This conference could not have been held without the diligent work of Kim Gilbert, the conference administrator and Myrna Kasdorf, the program administrator, who were both invaluable in the completion of this book. Also, a special thanks goes to Dr. Xiaowen Fang, the student liaison, and Nancy Lightner, the registration chair, for all of their outstanding efforts.

Michael J. Smith
University of Wisconsin-Madison
Madison, Wisconsin 53706 USA

Gavriel Salvendy
Purdue University
West Lafayette, Indiana 47907 USA

Constantine Stephanidis
ICS Forth
Crete, Greece GR71110

Richard J. Koubek
Wright State University
Dayton, Ohio 45435 USA

Don Harris
Cranfield University
Cranfield, Bedford MK43 OAL UK

June 2001

HCI International 2003

The Tenth International Conference on Human-Computer Interaction, HCI International 2003, will take place jointly with the Symposium on Human Interface (Japan) 2003, the 5th International Conference on Engineering Psychology and Cognitive Ergonomics, and the 2nd International Conference on Universal Access in Human-Computer Interaction, on the island of Crete, Greece, 22-27 June 2003. The conference will cover a broad spectrum of HCI-related themes, including theoretical issues, methods, tools and processes for HCI design, new interface techniques and applications. The conference will offer a pre-conference program with tutorials and workshops, parallel paper sessions panels, and exhibitions. For more information please visit the URL address: http://hcii2003.ics.forth.gr

General Chair:

Professor Constantine Stephanidis
ICS-FORTH and University of Crete
Heraklion, Crete, Greece
Telephone: +30-81-391741
Fax: +30-81-391740
Email: cs@ics.forth.gr

The proceedings will be published by Lawrence Erlbaum and Associates.

USABILITY EVALUATION AND INTERFACE DESIGN: COGNITIVE ENGINEERING, INTELLIGENT AGENTS AND VIRTUAL REALITY
Table of Contents

ix

SECTION 2. VIRTUAL REALITY

SECTION 3. THE WEB AND APPLICATIONS

xiii

xiv

SECTION 4. PHYSIOLOGICAL ASPECTS OF HUMAN-TECHNOLOGY INTERACTION

SECTION 5. COGNITIVE ERGONOMICS

An evaluation of gesture recognition for PDAs

A. Sears and R. Arora[1]

Laboratory for Interactive Systems Design, Information Systems Department, UMBC, 1000 Hilltop Circle, Baltimore, MD 21250

Handheld computing devices are being used for an ever-increasing variety of tasks. While current devices typically provide a stylus-activated soft keyboard, small physical keyboard, gesture recognition, or a telephone keypad-based technique to support data entry, little is known about the relative merits of these alternatives. In this article, we report on an empirical comparison of two common gesture recognition techniques: Jot and Graffiti. Our results demonstrate that, for novice users, Jot allows for more rapid text entry. Our results also indicate that data entry rates reported elsewhere may be overly optimistic if used to predict novice performance under realistic conditions.

1. INTRODUCTION

Mobile, handheld, computing devices are becoming increasingly common. Personal Digital Assistants (PDAs), cellular phones, and pagers support an increasing array of activities, which is making data entry increasingly important. Individuals no longer simply select phone numbers from a list, instead they schedule appointments, reply to text-pages and email, and browse the WWW. Various text-entry techniques can support these tasks, including stylus-activated soft keyboards, small physical keyboards, gesture recognition, and telephone keypad-based techniques. To date, few empirical studies have evaluated the relative merits of these techniques. In this article, we begin this process by evaluating the relative usability of the two most common gesture recognition techniques: Jot and Graffiti. The experiment reported below investigates the effectiveness of these technique when used by novice users to perform realistic tasks.

2. RELATED RESEARCH

Many researchers have investigated stylus-based data entry for computers. Some researchers focused on the relationship between recognition accuracy and user acceptance (e.g., LaLomia 1994; Frankish, Hull and Morgan 1995). Others reported data entry rates ranging from 14 to 18 words per minute (wpm), for various gesture recognition systems (e.g., Chang and MacKenzie 1994; MacKenzie, Nonnecke, McQueen, Riddersma and Meltz 1994; MacKenzie, Nonnecke, Riddersma, McQueen and Meltz 1994; MacKenzie and Chang 1999; McQueen, MacKenzie, Nonnecke, Riddersma and Meltz 1994; McQueen, MacKenzie and Zhang 1995). However, all of these studies involved data entry using a tablet connected to a PC, not a PDA. Three studies focused on numeric characters while a fourth focused on lowercase letters. One study did explore tasks involving mixed upper- and lower-case letters, but no numbers or symbols were included. Further, for all of these studies participants were instructed to aim for both speed and accuracy, but to ignore any errors and continue with their task. Restricted character sets, the use of a

---

[1] This research was funded by Motorola. The authors gratefully acknowledge their generous support.

desktop tablet rather than a PDA, and instructions to ignore errors make it difficult to apply these results to realistic tasks where users would correct errors while using mobile, handheld, computing devices.

While Lewis does describe a study where users did interact with a handheld device, the study was a simulated perfect handwriting recognition where any attempt to enter a letter was considered correct (Lewis 1999). Using this simulated 100% accurate handwriting recognition, participants entered text at 21-24wpm depending on the task.

3. RESEARCH OBJECTIVES AND HYPOTHESES

The current experiment investigates the effectiveness of Jot and Graffiti for realistic tasks users will encounter when using Internet-enabled mobile handheld computing devices. The focus is on novice performance, thereby providing insights regarding the adoption of the technology (since such decisions are often based on limited interactions) and the effectiveness of these techniques for infrequent users. Our hypothesis is that novices will be able to enter text more quickly using Jot than they will using Graffiti. This hypothesis is motivated by the observation that the strokes required by Jot more closely approximate the resulting characters than the strokes required by Graffiti. It is further motivated by the belief that this similarity will allow novices to learn and use Jot more quickly. For example, when using Jot, most strokes resemble the resulting character (e.g., A, B, 1, 2, ?) while a few are less intuitive (e.g., "). In contrast, when using Graffiti, many strokes still resemble the resulting character (e.g., B, C, D), but others may remind users of the resulting character only after the connection is established (e.g., A, F, T), and others appear to have little in common with the character they generate (e.g., [, ", %).

4. METHOD

4.1. Subjects

Thirty-one UMBC students volunteered to participate in the study. Participants received a payment of $10.00 as compensation for their time. To better represent the potential users of these techniques, participation was limited to individuals who were not in an information technology or engineering oriented major (e.g., Information Systems, Computer Science, or Computer Engineering). Further, participants could not have any prior experience using Jot or Graffiti. Informed consent was obtained prior to participation.

Participants were randomly assigned to use either Jot or Graffiti. Fifteen participants used Jot, sixteen used Graffiti. Sixteen participants were female. The average age of the participants was 23.4. All participants were regular computer users. Ten participants used cellular phones, pagers, or PDAs an average of 2.2 times per day while 21 did not use any of these devices.

4.2. Apparatus

Jot and Graffiti can be used on a single PDA, but the techniques have been optimized for different platforms. For example, Jot is available for the Palm OS, but is substantially slower than Graffiti running on the Palm OS. To ensure that the techniques would perform as intended, we chose to use two PDAs: a Palm III for Graffiti and a Casio Cassiopia E100 for Jot. These devices were selected because they are similar in size, weight, and display area. The devices were configured such that users interacted with a monochrome display, there was no audio feedback, and the ink-trace feature available on the Casio was disabled. Participants used the built-in memo application when using Graffiti and the Note Taker application when using Jot.

2

4.3. Tasks

Six tasks that varied in length, content, and complexity were utilized. Task one involved entering a name and address as may be done when recording contact information. Tasks two and three involved entering relatively simple URLs based on the belief that more complex URLs would be accessed via bookmarks, searches, or browsing from simpler URLs. Tasks four through six involved entering varying amounts of basic alphanumeric data as may be done when recording information about an upcoming appointment or providing a quick response to an email message. Appendix A contains the exact text participants were asked to enter for each task.

4.4. Procedure

Input technique was treated as a between-group variable with each participant using either Jot or Graffiti. Task was treated as a within-subject variable. After reading and signing the consent form, participants were given a brief orientation to the technique they would use. This introduction included a demonstration of how to write several lowercase letters, uppercase letters, numbers, and symbols. Next, participants were allowed 10 minutes to practice using the technique. Participants were free to practice using either sample tasks we provided or any other text. Finally, each participant completed the six tasks in a unique random order. Participants were given one task at a time on a single sheet of paper to ensure that the tasks were completed in the appropriate order. They were allowed to review the text and begin the task when they were ready. Participants were instructed to complete the task as they would under realistic usage conditions, balancing speed and accuracy. As a result, participants typically corrected some, but not all errors. Participants were allowed to take a break before beginning a new task. These breaks averaged approximately 30 seconds. After completing all six tasks, participants completed a questionnaire that investigated their perceptions of the technique used as well as various demographics. Participant interactions were videotaped to allow for a detailed analysis of the results.

5. RESULTS

5.1. Data entry rates

To allow for comparisons between tasks and with other studies, all results are reported based upon a standard of 5 characters per word. Table 1 includes the means and standard deviations for data entry rates. A one-way analysis of variance (ANOVA) with repeated measures for task was used to assess the effect of input technique. Significant main effects were found for both technique and task ($F(1,29)=11.5$, $p<0.005$; $F(5,145)=18.6$, $p<0.001$). A significant interaction between technique and task was also identified ($F(5,145)=4.3$, $p<0.002$). Overall, Jot allowed users to complete these tasks more quickly. The interaction indicates that the benefits of Jot varied among tasks. For example, data entry rates were approximately 20% faster for Jot on task one whereas they were over 90% faster for task three.

Table 1: Means and standard deviations (in parentheses) for data entry rates (in wpm).

	Task						
	1	2	3	4	5	6	Total
Jot	5.10	7.91	7.32	7.35	8.79	7.74	7.37
	(2.13)	(2.61)	(2.52)	(3.51)	(2.85)	(2.55)	(2.88)
Graffiti	4.30	5.01	3.81	4.99	6.14	5.44	4.95
	(1.10)	(2.20)	(1.40)	(2.04)	(1.89)	(1.64)	(1.87)

5.2. Error rates

Uncorrected errors were identified by comparing the desired result to the text produced by each participant. Each missing, extra, or incorrect word was counted as a single error. Table 2 includes the means and standard deviations for error rates (percentage of words containing an error). A one-way ANOVA with repeated measures for task was used to assess the effect of input technique on errors. A significant main effect was not found for either technique or task ($F(1,29)=0.2$, n.s.; $F(5,145)=0.6$, n.s.). No significant interaction between technique and task was found ($F(5,145)=1.7$, n.s.). Overall, there were no significant differences between Jot and Graffiti: on average approximately 10% of the words in the resulting text contained an error.

Table 2: Means and standard deviations (in parentheses) for error rates.

	Task						
	1	**2**	**3**	**4**	**5**	**6**	**Total**
Jot	0.09	0.06	0.06	0.06	0.11	0.12	0.08
	(0.20)	(0.12)	(0.12)	(0.16)	(0.29)	(0.31)	(0.21)
Graffiti	0.14	0.24	0.11	0.10	0.11	0.05	0.12
	(0.45)	(0.53)	(0.17)	(0.28)	(0.34)	(0.14)	(0.34)

5.4. Subjective satisfaction

A satisfaction questionnaire was administered after all six tasks were completed. Reliability was assessed using Cronbach's alpha, yielding a value of $\alpha=0.94$ which indicates that the questionnaire is highly reliable. A one-way ANOVA with repeated measures for question was used to assess the effect of input technique on responses with planned post-hoc comparisons to identify any significant differences for individual questions. Participants using Jot were more interested in using the device in the future (Graffiti ($F(1,28)=8.8$, $p<0.01$). Further, participants using Jot felt entering URLs was easier and faster than participants using Graffiti ($F(1,28)=5.8$, $p<0.05$; $F(1,28)=4.6$, $p<0.05$ respectively).

6. CONCLUSIONS

Using Jot, novices entered text at 7.37wpm as compared to only 4.95wpm when using Graffiti. This represents a 49% increase in productivity without any increase in uncorrected errors. While Jot did allow for more rapid data entry, both techniques resulted in data entry rates that are much lower than those reported elsewhere. Earlier studies reported data entry rates ranging of 14 to 18wpm for gesture recognition, but even our fastest participant never reached these speeds. A review of the different experimental methodologies and tasks employed may help explain the faster data entry rates reported elsewhere. Earlier studies did limit practice, but did not use PDAs and focused on tasks that may not accurately represent the activities PDA users will engage in. More importantly these earlier studies had users ignore errors and employed restricted alphabets. As a result, 14-18wpm may better represent a theoretical optimal level of performance (i.e., when recognition accuracy approaches 100%) for new users after limited practice. Lewis' result of 24wpm was obtained using a simulated 100% accurate handwriting recognition system and therefore may represent a target for expert performance after extended practice (Lewis 1999). In contrast, we believe that our results of 5-8wpm more accurately represent performance for new users after limited practice using currently available state-of-the-art technologies.

REFERENCES

Chang, L. and MacKenzie, I. S. (1994). A comparison of two handwriting recognizers for pen-based computers. *Proceedings of CASCON '94*, Toronto: IBM Canada, 364-371.

Frankish, C., Hull, R., and Morgan, P. (1995). Recognition accuracy and user acceptance of pen interfaces. Proceedings of CHI '95. New York: ACM Press, pp. 503-510.

LaLomia, M. J. (1994). User acceptance of handwritten recognition accuracy. Conference Companion CHI '94. New York: ACM Press, pp. 107.

Lewis, J. R. (1999). Input rates and user preference for three small-screen input methods: Standard keyboard, predictive keyboard, and handwriting. *Proceedings of the Human Factors and Ergonomics Society 43rd Annual Meeting*, 425-428.

MacKenzie, I. S., Nonnecke, B., McQueen, C., Riddersma, S. and Meltz, M. (1994). A comparison of three methods of character entry on pen-based computers. *Proceedings of the Human Factors and Ergonomics Society 38th Annual Meeting*, Santa Monica, CA: Human Factors Society, 330-334.

MacKenzie, I. S., Nonnecke, B., Riddersma, S., McQueen, C. and Meltz, M. (1994). Alphanumeric entry on pen-based computers. *International Journal of Human-Computer Studies*, 41, 775-792.

MacKenzie, I. S., and Chang, L. (1999). A performance comparison of two handwriting recognizers. *Interacting with Computers*, 11, 283-297.

McQueen, C., MacKenzie, I. S., Nonnecke, B., Riddersma, S. and Meltz, M. (1994). A comparison of four methods of numeric entry on pen-based computers. *Proceedings of Graphics Interface '94*. Toronto, Ontario: Canadian Information Processing Society, 75-82.

McQueen, C., MacKenzie, I. S. and Zhang, S. X. (1995). An extended study of numeric entry on pen-based computers. *Proceedings of Graphics Interface '95*, Toronto: Canadian Information Processing Society, 215-222.

APPENDIX A: EXPERIMENTAL TASKS

Task One
John Doe
8374 Maple Dr.
Apt. 36-C
Baltimore, MD 21250
(410) 391-4398
jdoe@gl.umbc.edu

Task Two
www.giraffe837.com

Task Three
www.travelocity.com/vaca23

Task Four
Department Meeting

Task Five
Meeting with Bob and Sue about annual budget

Task Six
The meeting this Tuesday has been changed to 2 pm. Please notify me if there is a conflict in your schedule. Bring all materials regarding Alpha project with you to this meeting. I will send more details later in the week.

Investigating PDA web browsing through eye movement analysis

Holly S. Bautsch-Vitense[1], Gottlieb J. Marmet[1], and Julie A. Jacko[2]

[1]Department of Industrial Engineering
University of Wisconsin-Madison
1513 University Avenue
Madison, WI 53706

[2]School of Industrial & Systems Engineering
Georgia Institute of Technology
765 Ferst Drive
Atlanta, GA 30332-0205

ABSTRACT

With the growing popularity of PDAs, and the growing requirements for wireless communication, the use of Internet browsers on mobile devices is rapidly increasing. Analysis of eye movement can provide a novel approach for evaluating the effectiveness of browsing the Internet on Personal Digital Assistants (PDA)s. In this paper, we investigate the interaction styles of individuals using a PDA to browse Web pages. Two distinct format styles for viewing the Web pages were studied. The first style presented the pages on the PDA in a standard format as would appear on the Internet, and the second style presented the pages in a specialized format specifically designed for a PDA. These two styles were investigated to determine which more clearly presented information to the users. The results indicate that the specialized Web channel style is a more efficient method for conveying Internet content.

1   INTRODUCTION

With the rapid increase in the use of mobile computing devices, many questions have surfaced regarding efficiency, accuracy, reliability, and overall use of these devices. Some research has looked into issues such as, the comparison between display sizes of a PDA and typical computer monitor [4], development of specialized Web browsers from academia to rival commercially available products [2,3], discussions of the proper method to transform normally available Internet content to fit onto PDA devices [1], and guidelines established at the W3C on mobile content authoring [8]. However, research in the field of PDAs (e.g., Visor, Cassiopeia, and Palm) concerned with interface design have not yet been thoroughly identified or investigated in the current literature. More specifically, much needs to be learned to understand PDA use and the various interaction styles of its users. The interaction style of individuals using a PDA does not mimic the interaction style used with traditional computer interfaces, due to differences in display size, physical size, mobility and functionality [4]. It also cannot be assumed that traditional computer display interface guidelines and design recommendations can be applied to PDA displays.

The trend in PDA devices is to provide similar information to what is available on a personal computer, however this information has to be presented on hardware that has a smaller display screen, lower resolution, and no color or a limited range of colors. Additionally PDA devices have small memory capacities, less powerful CPUs, limited data input methods and a limited source of power [8]. As a result unique standards and guidelines are needed to fully utilize this growing technology. However, before standards and guidelines can be developed, a knowledge base of empirical research must be constructed to define the user population and common usage patterns employed by current users.

As a result of the growing trend to stay informed through wireless communication, the latest versions of PDA devices have wireless connection capabilities. Through these capabilities, users are able to connect to the Internet at any time from any location. Several applications have been developed to enable PDAs and other handheld devices to view Internet Web pages. Although effects of small PDA displays have been documented [4], there is very little known about how these new PDA Web browsers can best present information. As a result, this investigation evaluated the use of two PDA Web page formats. Eye movement analysis was used to capture and evaluate the use of these Web page formats. Eye gaze performance and task completion time, conveyed the users' ability to perform a task using a Web page presented on a PDA.

Because of its availability and popularity at the time of the study, a 3Com Palm III™ was selected as the PDA device to be investigated. The Web pages of interest were downloaded onto the Palm so as to avoid any delays resulting from real-time connectivity to the Internet.

The objective of the study was to determine the best method for presenting Web pages on a Palm. The format in which the Web pages were presented was the focus of this study, not the type of browser. Two different format styles were investigated, one presented the Web page as it would appear on a personal computer, and the other presented

6

the page in a format specifically designed from PDAs. The use of the two styles of Web pages were captured for novice and expert users. Although the use of eye movement analysis is not a novel approach for evaluating the effectiveness of a Web browser, its use in evaluating PDA Web browsers is so far uncharted territory.

2   PDA WEB BROWSING

Two styles of viewing Web pages were investigated. The first style was of standard Web page format. Web pages presented in this format appeared similarly as they would on a typical computer monitor viewing the Internet. The pages were displayed in full form. Due to the difference in display sizes of a computer monitor and Palm, the content of a standard Web page appeared differently on the Palm display. Graphics and text that would be displayed in full color and high resolution on a computer monitor were resized to fit the Palm and were displayed in a lower resolution. Text also wrapped to fit the Palm display. Because there was no horizontal scrolling capability, the page length was longer than observed for the specialized Web channel.

The second Web page style investigated displayed the Web pages in a specialized Web channel format. Specialized Web channels are Web pages specifically designed and formatted for use on PDAs. Web channels were developed to allow users of mobile devices access to Internet information in much more concise manner. For instance, there were relatively no graphics and no advertisements, and the formatting of the text was more conducive to a small display (smaller font and conservative spacing). There are several Web browsers available for use on a PDA; ProxiWeb [3], HandWeb [7], PalmScape [5], Power Browser and AvantGo, to name a few. These browsers are PDA versions of an Internet Web browser, such as Microsoft Explorer™ or Netscape Navigator™. AvantGo was selected to display the Web pages for this study because it provided the capability to view both a standard Web page style and a specialized Web channel style of the same content.

3   METHODS

3.1 Participants: Sixteen participants were recruited from the University of Wisconsin College of Engineering to participate in the study. PDA expertise was gathered by a pre-task, user profile questionnaire. If a participant reported that they had extensive experience with a PDA and considered themselves experts, they were classified as an expert user in this study. Conversely, if a participant reported using a PDA for a very limited time or never at all, they were classified as a novice for the study. Seven participants were classified as self-reported expert PDA users due to their extensive experience with the hardware. Their ages ranged from 23 to 40 years (mean = 31). The nine remaining participants were classified as self-reported novice PDA users. Their ages ranged from 20 to 38 years (mean = 24).

3.2 Experimental Setting: Due to eye tracking technology requirements and novel research goals, unique methods were employed in the experimental setup. Applying an eye tracking system (Applied Science Laboratories Model (ASL) 501™ eye tracker) to the unique application of tracking participants' use of a PDA, posed several challenges.

In order to understand these challenges, several aspects of eye tracking technology must be first explained. Calibration of the eye tracking system is vital for assuring the accurate collection of eye movement. To accurately capture movements, the system must be calibrated and used within a fixed scene space. A scene space is defined by a set of surfaces, consisting of one calibration surface and multiple other bounded surfaces. In this study the display screen served as the calibration surface and the surface from which data were collected. Another aspect of the eye tracking system that posed a challenge related to the fact that a fairly constant visual angle must be maintained while data are collected. It is critical that participants maintain a constant visual angle while viewing the display screen. In order to ensure that this is the case, the participant must maintain a fixed distance from the viewing surface (e.g., display screen) throughout the duration of the study. However inherently, a PDA is designed to be used while being held in a variety of positions at arms reach or an approximate distance from the eye of two feet. Unfortunately, this typical use pattern could not be replicated for this study if an eye tracking system was to capture eye movement. As already discussed, the eye tracking system also required a static surface from which to capture eye position. Another challenge imposed by the eye tracking system was the minimum size of the PDA's display. Typically, eye tracking systems are used to investigate design characteristics of computer interfaces, aircraft cockpits, controls systems of unmanned aircraft vehicles, and virtual environments, to name a few. However, in this study the display surface under investigation was much smaller than usual. The Palm III used in the study had a screen size of 6 x 6 cm (160 x 160 pixels). Typically the eye tracking system works best when the calibration points subtend a visual angle of 10 degrees. When using the Palm in a fairly reasonable manner (i.e., held 2 feet in front of the user), the calibration points only subtend a visual angle of 4 degrees.

To meet the first challenge of maintaining a fixed surface, an adjustable stand was built to both secure and support the Palm. This allowed for an image of the Palm to be captured by an overhead mounted camera and relayed to a 27" display screen for viewing by the participant. This approach to the experimental setup also met the second challenge of viewing surface size. The projected image of the Palm on the display screen was over three times the size of the original Palm. As a result, the minimum visual angle of the projected image of the Palm was 11 degrees. Thus through incorporating an overhead camera and adjustable stand, all of the challenges posed by the eye tracking system were met.

3.3 Experimental Design: The study used a 2x2 nested factorial repeated measures design. The independent variables were Palm expertise (novice and expert), and Web page style (standard Web page and specialized Web channel). The dependent variables consisted of two performance measures: task completion time, and eye movement fixation dwell time. All participants were presented with both Web page styles. The order in which they were presented was counterbalanced to minimize order effects.

Because each participant evaluated both the standard Web page style and specialized Web channel style twice, two Web pages, USA Today and Fox Sports, were used as content material, both of equivalent complexity. Both styles (standard Web page and specialized Web channel) of the USA Today and Fox Sports pages contained the same article.

3.4 Procedure: All participants were provided training on how to use the Palm in our experimental setting. Once the participant performed the training tasks to a sufficient level and felt comfortable with the operation of the Palm in our experimental setup, they were fitted with the head-mounted optics of the eye tracking system.

Calibration of the equipment was then performed. Once calibration was verified, the trials began. The participants completed two sets of two trials. Each trial presented one of the two Web page styles, standard or specialized channel. The participants were first instructed to turn on the Palm and then locate the Web viewing application, AvantGo. Once AvantGo was loaded, one of the two Web page styles displayed either the USA Today or Fox Sports Web pages. The participants were instructed to read the entire article and answer a question. The questions used in the study were designed to motivate thorough review of the displayed information. Once the participant found the answer to the question the trial ended. This procedure was repeated a total of four times per participant with varying Web page styles.

3.5 Analyses: Task completion times along with fixation dwell times were collected for each participant. The participants were analyzed by category, either novice or expert, to determine if there was an effect of experience. Task completion time was defined as the time from when the Web page loaded on the Palm until the participant completed the task of identifying the answer to the question. The fixation dwell time metric was captured in the context of predefined areas of interest. Figure 1 shows the six areas of interest used in the analyses; (1) title bar, (2) region 1, (3) region 2, (4) region 3, (5) region 4, (6) scroll bar.

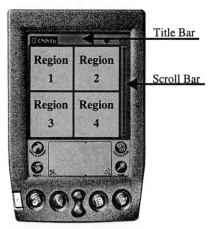

Figure 1. Six areas of interest.

8

Eye movements falling outside these predefined regions were represented in an "all-other" region. These areas of interest permitted more detailed analyses of the eye movement. The fixation algorithm criteria used in this study defined the start of a fixation after an eye gaze remained within approximately 1 degree visual area for at least 200 msec. Eye blinks of up to 300 msec. could occur during a fixation without terminating the fixation. Fixation dwell time defines the total time spent fixating on a given area of interest.

4   RESULTS

In summary, analyses of the two performance measures revealed differences with respect to where participants viewed information on the display. The total completion time and dwell time metrics demonstrated that eye gaze pattern varied with respect to which Web page style was presenting the information. No significant effect of experience was found for the dependent variables, the following results were analyzed without respect to expertise categories.

4.1 Task Completion Time: The repeated measures analysis performed on task completion time revealed a main effect of Web page style ($F_{(1,14)}$ = 18.986; p<.05). Figure 2 shows the mean task completion time for each of the Web styles across trials 1 and 2.

A significant interaction was not found, but the Scheffe post-hoc analysis did reveal significant differences between trial 1 for the standard Web style (mean task completion time = 1.728) and trials 1 (mean task completion time = .5531) and 2 (mean task completion time = .7656) for the specialized Web channel style. Since no significant differences were found between the trials for one particular Web style, no learning effect was observed for either the standard Web or specialized Web channel styles.

4.2 Dwell Time: Analyses performed on dwell time revealed significant main effects of Web style ($F_{(1,14)}$ = 21.104; p<.05) and area of interest ($F_{(6,84)}$ = 9.977; p<.05). Figure 3 shows these main effects across all areas of interest. On average, the participants spent 28.7 seconds less time fixating on a given area of interest while in the specialized Web channel style than the standard style. Scheffe post-hoc analysis revealed a significant difference with respect to dwell time for the all-other regions between the standard Web and specialized Web channel styles.

A significant interaction was observed between Web style and area of interest on dwell time ($F_{(6,84)}$ = 3.998; p<.05). The interaction occurred in regions 1 and 4 between the two Web styles. Region 1 recorded a higher dwell time than region 4 for the standard style. However, this trend reversed for the specialized Web channel style.

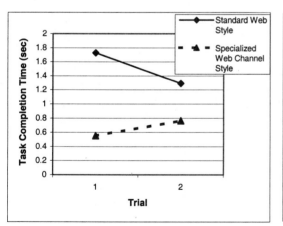

Figure 2. Mean Task Completion Time _x_ Trial.

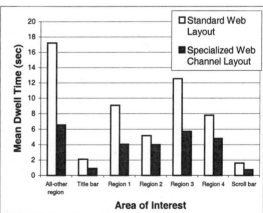

Figure 3. Mean Dwell Time _x_ Area of Interest.

5 DISCUSSION AND CONCLUSIONS

The study showed that the specialized Web channel was a more efficient style for presenting Web pages on the Palm. The experimental task was completed 1.7 seconds faster when Web pages were shown in the specialized channel style than the standard style.

In interpreting the meaning of the fixations' dwell times, the depth at which information was processed is conveyed. The longer the dwell time, generally the more substantial information processing is required. The standard style was found to have a longer mean dwell time than the channel style. This is an indication that the standard style required more in-depth cognitive processing. This metric characterizes how information on the Palm was detected and comprehended [6], and that the specialized channel presented the Web pages in a much more efficient manner.

Results indicated that there was no difference in performance between expert and non-expert participants. This can be attributed to the specific expertise of Palm application use of our participants. Although our experts were well versed in the standard Palm applications, none had viewed Web pages on the Palm previously and therefore their interaction was as novel as those classified as non-experts. Additionally, we found no evidence of improvement in task performance as participants encountered the second trial. Again, this can be attributed to the task. The total time to review material presented on the Palm was brief and the task of answering a question was simple. Thus, significant changes in performance as participants encountered the second trial were not realized.

Significant results were obtained for the all-other regions for the standard and specialized channel styles. For trials involving the specialized Web channels, a mean of 6.55 seconds was spent on the all-other region. However, for the standard Web pages, a mean of 17.23 seconds was spent on the all-other region. This difference may be attributed to the fact that the participants looked at the question (located on the left side of the Palm on the adjustable stand) more often due to the more complex format of the article. It is possible that the channel style afforded the participants to read the question less frequently due to the speed at which they could locate the answer. The article was found to be presented much more concisely through the channel style.

Although this research confirms previous results found when evaluating web browsing used on personal computers and this is a first step towards establishing empirical research focused on PDAs. This research provides empirical data to support the creation of PDA specific Web content. Our findings suggest that simply wrapping content to fit on a smaller display is not a sufficient method of transformation.

In addition, eye movement analysis as applied for a PDA, has been shown to be a useful tool in the evaluation of such interfaces

6 ACKNOWLEDGMENTS

The National Science Foundation in a grant awarded to Dr. Julie Jacko (BES-9896304) funded this research.

7 REFERENCES

1. Bickmore, W.T., & Schilitt, N.B. (1997). Digestor: Device-independent Access to the World Wide Web. In *Proceedings of the 6th International WWW Conference* (pp. 511-517).

2. Buyukkokten, O., Garcia-Molina, H., Paepcke, A., & Winograd, T. (2000). Power Browser: Efficient Web Browsing for PDAs. In *Human Factor in Computing Systems: CHI 00 Conference Proceedings* (pp. 430-437). New York: ACM Press.

3. Fox, A., Goldberg, I., Gribble, S.D., Lee, D.C., Polito, A., & Brewe, E.A. (1998). Experience with Top Gun Wingman: A Proxy-Based Graphical Web Browser for the 3Com PalmPilot. In *Conference Reports of Middleware '98, Lake District, England.*

4. Jones, M., Marsden, G., Mohd-Nasir, N., Boone, K., & Buchanan, G. (1999). Improving Web Interaction on Small Displays. In *Proceedings of the 8th International WWW Conference* (pp. 51-59).

5. Kazuho, O., PalmScape: http://palmscape.ilinx.co.jp/

6. Kotval & Goldberg. (1998). Eye movements and interface component grouping: an evaluation method. *Proceeding of the 42nd Annual Meeting of the Human Factors Society.* Santa Monica, CA: Human factors Society, pp. 486-490.

7. Smartcode Software, HandWeb: http://www.smartcodesoft.com/.

8. W3C, HTML 4.0 Guidelines for Mobile Access. http://www.w3.org/TR/NOTE-html40-mobile/

Development of a Gesture-Based Interface for Mobile Computing

Jack Maxwell Vice, B.S., Corinna E. Lathan, Ph.D., and James B. Sampson, Ph.D.

AnthroTronix, Inc., 387 Technology Drive, College Park, MD 20742
&
U.S. Army Soldier Systems Center, Natick, MA 01760

ABSTRACT

The goal of this project was to develop a robust, gesture-based human computer interface for a mobile computing system. Specifically, a "dataglove" system was designed and tested for application to a wearable computer system. The focus to date has been on capturing gestures to communicate voiceless messages among team members in a DARPA funded Digital Military Police (Digital MP) program developed through the U.S. Army Soldier Systems Center. However, the technologies developed here have potential applications in many disciplines outside of the military.

1. INTRODUCTION

The Defense Advanced Research Projects Agency (DARPA) and the US Army Natick Soldier Center (NSC) are supporting the development of a robust, portable, small, lightweight, and low-cost wearable communications and information management system that will assist highly mobile users by providing complete PC and communications capabilities. This system is known as the "Digital MP". Its planned demonstration is being tailored to the operational capability requirements of the MP soldier. The "Digital MP's" support features include a hands-free, voice-based interface tailored to specific users and situational needs and it will also provide interface support for required peripherals (Figure 1). The system is being designed to be easy to use, to be compatible with legacy systems (software and hardware), to be comfortable to wear, and have a battery solution that is easily recharged, exchanged and provides power for a complete day's activities on a single charge.

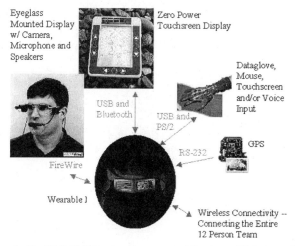

Figure 1: The Digital MP System Components (Courtesy of U.S. Army Soldier Systems Center, Natick, MA 01760.)

Gestures, or movements related primarily, but not exclusively, to the hands, are electronically captured through a "dataglove" to communicate a variety of command and control functions such as mouse emulation, robot control, team communications, and alerts. A vocabulary of commands allows the operator to operate in silent mode or in a physically limiting environment. The gestures are easily reconfigured to meet a variety of user specifications and environments.

2. SCENARIO DESCRIPTIONS

The following scenarios provide examples of how a gesture recognition communication system can provide functionality that didn't previously exist and consequently, increase combat effectiveness.

2.1 Jungle Patrol
One example of the utility of the instrumented glove for Digital MP is a jungle patrol. During dismounted patrols, communications between operators is accomplished using arm and hand signals. Line of sight visibility is necessary for unassisted hand signals and in very dense vegetation, such as in the jungle, or at night, line of sight is difficult to maintain. Our system can recognize the MP's hand signal and communicate a predefined message to other operators, without violating noise discipline and without line of sight. For instance, the point man could, upon encountering a minefield, give the hand signal for freeze and the entire platoon would see and hear the message. Normally the hand signal would have to be relayed from the point man through the platoon and some of the men could still be stepping through the minefield as the message is relayed.

2.2 Checkpoint Monitoring
The military checkpoint is another example of how a gesture recognition system can increase combat effectiveness. A military checkpoint is a security choke point where personnel and vehicles are identified and searched or detained. At a military checkpoint, a number of hazardous situations can develop. One such situation is that of duress. For security personnel, a situation of duress occurs when an aggressor intimidates the security personnel with the threat of deadly force in hopes of preventing the security personnel from communicating the situation to others. In a bank, bank personnel commonly use a secret button to sound a silent alarm for this type of situation. Military personnel usually have a duress code word or phrase to communicate the situation. The problem with a duress code word is that it still requires audible communication. With gesture recognition, the user could give a covert, unique duress gesture that wouldn't cause any suspicion in the aggressor. Additionally the MP could, unbeknownst to the aggressor, use gestures to communicate further information about the situation such as the appropriate response action.

3. SYSTEM DESCRIPTION

A primary measure of good interface design is how well the interface is matched to tasks it will support. As a large number of tasks have been identified in the Digital MP program, design of a single, optimal interface is nearly impossible. Appropriate feedback to the soldier is highly dependent upon the operational environment in which the system is being used. In addition, use of the system may be passed from one user to another throughout a mission. As a result, we have focused on developing an easily configurable interface for a mobile computing environment.

3.1 Configurable User Interface
A common method of non-verbal communication in combat situations is gesture-based signals. Hand gestures provide an intuitive method of communication between MPs. As each individual military squad has unique, continually changing gestures, the ability to configure gestures will enhance the usefulness of a gesture-based communication method.

The glove has six bend sensors, a two-axis tilt sensor, and two pressure sensors. There is one bend sensor for each finger and one for the wrist. The tilt sensor is on the back of the glove, between the wrist and fingers to provide orientation data about the hand. The two pressure sensors are located at the tip of the index and middle fingers.

A software package, NeatTools, was installed to collect and store motion data from an operator. A 16-bit Analog to Digital converter (TNG-3B) provided 8 analog and 8 digital signals. TNG-3B (MindTel, LLC, Syracuse, NY) is a serial port interface device. The PC through the serial port connection provides all power to the A/D interface.

3.2 System Capabilities
- Fingertip command and control: The ability of the operator to interact with the system is independent of physical surroundings and limitations. The embedded sensors place control at the fingertips of the

operator independent of hand position in space. All 9 sensors can be used as analog input for things like mouse pointer, virtual keyboard or scrolling a map or menu.

- The ability to recognize static and dynamic gestures: Accurate gesture recognition is achieved by utilizing seven of the nine sensors built into the glove.
- Interoperability with all wearable devices: The gloved interface systems architecture maintains maximum flexibility by adhering to personal computer I/O standards. Integration with advanced tactical helmet and body worn systems, provides compelling user input options previously unavailable in those systems. The software prototyping component of the system allows gesture data to be communicated over the Internet via standard socket communications.

4. EVALUATION

Nine Military Police enlisted soldiers (MOS 95B) participated in the evaluation of the Digital-MP system for four days in November 2000 at Ft. Polk, LA. The climatic conditions were cool to comfortable (45-70F) with an occasional light rain. All the soldiers had at least two years experience with their MP unit. The mean age was 23.1 years with a range of 20-32. The first day started with an orientation class on the features and capabilities of the system followed by individual training on each of the capabilities; the menu system, voice control, mouse-wheel control, face recognition software, map and GPS operations, and camera aiming of the wearable system. Language translation and the data glove capabilities were demonstrated on lap-top systems. Each soldier learned to use each feature of the system until they felt comfortable performing the related tasks.

On the following days, small teams of three soldiers navigated outside with the wearable systems around the test area, taking GPS readings of themselves and team members, communicating to each other on their "push-to-talk" computer radios and conducting face recognition of each other, sharing results of face recognitions, and entering information into their computer logs. All activities were done in the company of one or more of the system design team technicians. When the system or a computer function did not work the technician fixed the problem on the spot then allowing the soldier to continue to operate the system.

4.1 Subjective Ratings of the Dataglove

On the last day of testing, each soldier was given a human factors questionnaire for rating and evaluating each feature of the system and the two separate demo systems; language translation and the dataglove. The rating scale used was an 11-point scale ranging from −5 to +5 with the middle score being 0, representing a borderline good or bad rating for the feature or subsystem. Soldiers were also asked to write in comments on each feature and capability. Following the individual questionnaire the team was then asked to discuss and comment as a group on the system and its features. Except for some comparison rating other components, only the dataglove results are presented here for discussion.

Key questions asked about each of the features or functions were perceived: Ease of Use, Speed of Operation, Reliability, and Potential Utility. Although the dataglove was evaluated separately from the wearable system the soldiers were able to consider its application relative to the tasks performed during the test. The dataglove ratings for each of the key questions were: 2.0 Ease of Use, 2.5 Speed of Operation, 1.7 Reliability, and 2.3 Potential Utility. These scores were among the highest of all the other features where the average rating was a positive 1.04 with a standard deviation of 1.48. Figure 2 shows how the dataglove compares to the other components in terms of Potential Utility. Nearly all the main features of the system had a high mean rating (+2.0 and above) except for the glasses-mounted display and camera, which both received strong negative ratings.

The following comments on the dataglove were written or recorded during the debriefing session:
The glove will be useful for the infantry.
Digital glove is good for hand and arm signals.
It will be useful in the field where you can't talk. You can warn of a threat or call a halt.
I recommend the glove be made more durable but need to also maintain dexterity.
It's potentially a good alternative to the mouse I/O.
Should design it so it can be transferred to another soldier if necessary.

The following is a summary of the evaluation of the dataglove for this demonstration test:

a. Ease of use: Very easy to use. Minimal training required.
b. Reliability: As demonstrated, highly reliable
c. Function and Utility: Users saw great potential and utility.
d. Comments and Recommendations: System needs to be integrated into DMP system and tested with the full system in the field.

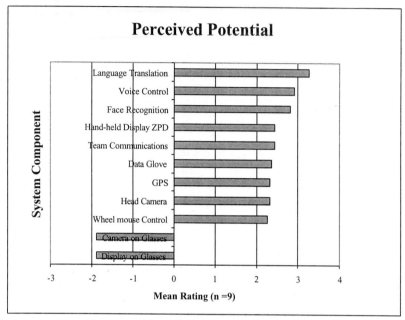

Figure 2: Perceived Potential of the Digital MP System Components

4.1   Gesture Recognition Resolution

In addition to the subjective questionnaire data collected at Ft. Polk, the gesture recognition resolution was tested. After demonstrating the gesture to voice-text capabilities of 10 pre-defined gestures, our focus was - on the fly gesture creation and portability between users without re-calibration.

A size large data glove and software were tested with five soldiers using ten gestures. The sensors and gestures were calibrated for the investigator only. The accuracy setting in the pattern recognition software was adjusted between 88% and 95% in order to determine the accuracy required for 100% recognition. The accuracy setting was also adjusted up to see at what point the gestures would not be recognized. As shown in Table 1, at a 90% accuracy setting, all the gestures were properly recognized. Not all subjects were tested at all accuracy ratings.

Table 1: Number of non-recognitions at % accuracy setting:

Subject #	Glove size	88%	90%	91%	95%
1	L	-	0	-	4
2	M	-	-	0	3
3	S	-	0	-	1
4	L	-	0	-	2
5	M	0	0	0	4

5.   CONCLUSIONS AND FUTURE WORK

A dataglove, capturing gestural movements, was designed, tested, and evaluated as an alternate interface for a portable, lightweight, and low-cost wearable digital MP system. The dataglove interface was highly rated as a means of interacting during highly mobile activities. Further tests will include operational environment testing.

In addition, through initial testing, feedback indicates the following improvements to optimize the system performance:

- High-resolution finger and hand position information and six degree-of-freedom orientation information. High resolution will maximize gesture recognition precision. Increased degrees-of-freedom will maximize gesture recognition vocabulary and flexibility. Both of these depend on concurrent advanced sensor development.

- Haptic (force) and vibrotactile feedback will enhance visualization and situational awareness for both local and remote environments. For example, tactile signals could be used to communicate alerts to the operator.

- As with all wearable components, our glove requirements will include miniaturization, increased durability and reliability, and low power. Ideally, sensors will be wireless, unobtrusive sensors and have high hazardous environment reliability. Hardened systems will ensure combat-effectiveness of individual soldiers and crews on the battlefield under worldwide environmental extremes.

ACKNOWLEDGMENTS

This research was supported by the DARPA funded Digital Military Police Program administered through the U.S. Army Soldier Systems Center in Natick, MA.

REFERENCES FOR DATA-GLOVE PAPER

Stedmon, A.W., Moore, P.M., Kalawsky, R.S., Aung, M., Purcell, J., Reeh, C. &York, T. "It's not what you wear, it's how you wear it: Human factors of wearable computers", Proceedings of the Human Factors and Ergonomics Society 43rd Annual meeting, 1999, 1050-1054.

Yun, M.H., Han, S.H., Kim, K.J., & Han, S. "Measuring customer perceptions on product usability: Development of image and impression attributes of consumer electronic products", Proceedings of the Human Factors and Ergonomics Society 43rd Annual meeting, 1999, 486-490.

Miller, M.A. & Stimart, R. P. "The user interface design process: The good, the bad, & we did what we could in two weeks", In Perlman, G.P., Green, G.K. & Wogalter (Eds) Human Factors Perspectives on Human-Computer Interaction 1983-1994, 1994. Human Factors & Ergonomics Society Publication, 1995.

Data entry for mobile devices using soft keyboards: Understanding the effect of keyboard size

Y. Zha and A. Sears[1]

Laboratory for Interactive Systems Design, Information Systems Department, UMBC, 1000 Hilltop Circle, Baltimore, MD 21250

Personal Digital Assistants (PDAs) often provide a small stylus-activated soft keyboard to support data entry activities, as do some cellular phones. However, there is little data regarding the effect keyboard size has on performance or user satisfaction. In this article, we report on an experiment designed to investigate these issues in the context of a Palm-style QWERTY keyboard. Our results demonstrate that keyboard size does not affect data entry rates or error rates. Further, effects on subjective satisfaction ratings are minimal and do not suggest that users prefer larger keyboards. These results have important implications as new, even smaller, devices are developed.

1. INTRODUCTION

Mobile, handheld, computing devices are becoming increasingly common with PDAs, cell phones, and pagers support an ever expanding array of activities. As the range of applications continues to expand, data entry is becoming more important. To support data entry activities, many mobile devices employ small stylus-activated soft keyboards. At the same time, limited data exists regarding the efficacy of this common data entry technique. In this article, we begin the process of investigating the efficacy of small stylus-activated keyboards for use with mobile devices. We focus on the effect of keyboard size when novice users perform realistic tasks.

2. RELATED RESEARCH

Numerous studies have investigated the efficacy of stylus-activated QWERTY-style keyboards. For example, various studies have reported data entry rates ranging from 11-40 words per minute (wpm) depending on the amount of practice, the tasks performed, the technology used, and the instructions provided to participants (Lewis 1999; Lewis, LaLomia, and Kennedy 1999; MacKenzie and Zhang 1999; MacKenzie, Zhang, and Soukoreff 1999; MacKenzie, Nonnecke, McQueen, Riddersma, and Meltz 1994). These studies were all motivated by the fundamental problem of entering data on small, handheld, mobile computing devices. Each study included a QWERTY-style keyboard as one of the alternatives explored. Data entry rates ranged from 11 to 28wpm for initial performance and as high as 40wpm after extensive practice. Each of these studies does provide useful insights, but the methodologies and tasks employed make it difficult to generalize these results to situations where individuals are completing realistic tasks using a PDA. Some studies used paper mockups of the keyboards, others used desktop tablets, and the one study that did use a PDA utilized a soft keyboard almost twice as large as those found on the most PDAs used today. Several studies had users address errors in unrealistic ways

---

[1] This research was funded by Motorola. The authors gratefully acknowledge their generous support.

by either ignoring errors or typing the correct letter without deleting the incorrect letter and other studies used tasks that are not representative of those that users would actually encounter.

3. RESEARCH OBJECTIVES

Soft keyboards are readily available on PDAs, yet we have little knowledge regarding their effectiveness for novice users. Given the experimental designs and tasks utilized in earlier studies, our first objective is to gain insight into the effectiveness of a soft keyboard under realistic usage conditions. At this time, there are no published reports documenting the effect keyboard size has on performance, errors, or satisfaction in the context of soft keyboards as small as those that will be required for use on future mobile devices. Therefore, our second objective is to determine whether or not keyboard size affects performance or preferences in the context of data entry activities using a stylus-activated soft keyboard.

4. METHODS

4.1. Subjects

Thirty UMBC students volunteered to participate in the study. Participants received a payment of $10.00 as compensation for their time. To better represent the potential users of these techniques, participation was limited to individuals who were not in an information technology or engineering oriented major (e.g., Information Systems, Computer Science, or Computer Engineering). Informed consent was obtained prior to participation. Participants were randomly assigned to use one of three different sized keyboards. Ten participants used each keyboard size. Seventeen participants were female. The average age of the participants was 25.6. All participants were regular computer users. Fourteen participants used cell phones, pagers, or PDAs an average of 2.5 times per day while 16 did not use any of these devices.

4.2. Apparatus

A Casio Cassiopia E100 was used in this study. The E100 provides a display/input region measuring approximately 6.1x7.9cm. To maximize the size of each key, a two-screen QWERTY-style keyboard similar to that used in Palm devices was employed (see Figure 1). This design provides access to all letters and some symbols from the primary keyboard (on the left), but requires users to access a secondary keyboard (on the right) to type numbers and some other symbols. The primary keyboard (on the left) was 3.2cm, 4.3cm, or 5.4cm wide. Throughout the remainder of this paper, these keyboards will be referred to as small, medium, and large. Alphabetic keys were square, measuring approximately 2.6mm, 3.5mm, and 4.4mm per side for the small, medium, and large keyboards respectively. The small keyboard was the smallest keyboard that could be created and displayed given the screen resolution and fonts available on this device. This keyboard could fit on many, but not all, cellular phones. The large keyboard represents a typical size for a soft keyboard on a PDA, using virtually the entire width of the display. The medium keyboard provided a midpoint for comparison purposes. The top of the keyboard was located at the same vertical position on the screen regardless of the keyboard size. The keyboards were centered horizontally.

4.3. Tasks

Six tasks that varied in length, content, and complexity were utilized. Task one involved entering a name and address as may be done when recording contact information. Tasks two and three involved entering relatively simple URLs based on the belief that more complex URLs would be accessed via bookmarks, searches, or browsing from simpler URLs. Tasks four through six involved entering varying amounts of basic alphanumeric data as may be done when recording information about an upcoming appointment or providing a quick response to an email message. Appendix A contains the exact text participants were asked to enter for each task.

Figure 1: The two-screen keyboard used in the current study. The "abc" and "123" keys allow users to switch between the keyboards.

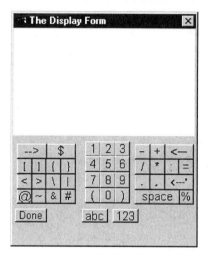

4.4. Procedure

Keyboard size was treated as a between-group variable with each participant interacting with either the small, medium, or large keyboard. Task was treated as a within-subject variable. After reading and signing the consent form, participants were given a brief orientation to the system they would use. This introduction included a demonstration of how to enter lowercase and uppercase letters, numbers, and symbols. Next, participants were allowed 10 minutes to practice using either sample tasks we provided or any other text. Finally, each participant completed the six tasks in a unique random order. Participants were given one task at a time on a single sheet of paper to ensure that the tasks were completed in the appropriate order. They were allowed to review the text and begin the task when they were ready. Participants were instructed to complete the task as they would under realistic usage conditions, balancing speed and accuracy. As a result, participants typically corrected some, but not all errors. Participants were allowed to take a break before beginning a new task. These breaks averaged approximately 30 seconds. After completing all six tasks, participants completed a questionnaire that investigated their perceptions of the keyboard used as well as various demographics. All participant interactions were recorded automatically by the software being utilized.

5. RESULTS

5.1. Data entry rates

To allow for comparisons between tasks and with other studies, all results are reported based upon a standard of 5 characters per word. Means and standard deviations for data entry rates, measured in wpm, are reported in Table 1. A one-way analysis of variance (ANOVA) with repeated measures for task was used to assess the effect of keyboard size. There was no significant main effect for keyboard size (F(2,27)=0.6, n.s.), but there was a significant effect for task (F(5,135)=50.2, p<0.001).

Table 1: Mean data entry rates (in wpm) and standard deviations (in parentheses) for each of six tasks.

| | | Task | | | | | |
		1	2	3	4	5	6
	Small	7.49	7.60	9.33	9.66	10.39	12.26
		(0.99)	(1.99)	(1.82)	(2.98)	(3.20)	(2.81)
Keyboard	**Medium**	7.25	6.77	7.56	7.67	10.20	12.56
Size		(1.40)	(1.01)	(1.18)	(2.73)	(2.84)	(2.71)
	Large	6.98	7.31	9.33	8.53	10.49	12.62
		(0.85)	(1.59)	(2.07)	(2.31)	(2.74)	(1.58)

5.2.   Error rates

Uncorrected errors were identified by comparing the desired result to the text produced by each participant. Each missing, extra, or incorrect word was counted as a single error. Means and standard deviations for uncorrected error rates are reported in Table 2. A one-way ANOVA with repeated measures for task was used to assess the effect of keyboard size. There was no significant main effect for keyboard size or task (F(2,27)=1.0, n.s.; F(5,135)=0.7, n.s.).

Table 2: Means and standard deviations for uncorrected error rates for each of six tasks.

| | | Task | | | | | |
		1	2	3	4	5	6
	Small	0.02	0.03	0.02	0.06	0.03	0.05
		(0.03)	(0.09)	(0.06)	(0.18)	(0.08)	(0.04)
Keyboard	**Medium**	0.00	0.03	0.00	0.00	0.00	0.04
Size		(0.00)	(0.09)	(0.00)	(0.00)	(0.00)	(0.04)
	Large	0.05	0.06	0.04	0.03	0.01	0.04
		(0.12)	(0.12)	(0.12)	(0.09)	(0.04)	(0.08)

5.3.   Subjective satisfaction

A satisfaction questionnaire was administered after all six tasks were completed. Reliability was assessed using Cronbach's alpha, yielding a value of $\alpha=0.83$, which indicates that the questionnaire is highly reliable. A one-way ANOVA with repeated measures for question was used to assess the effect of input technique on responses with planned post-hoc comparisons to identify any significant differences for individual questions. Keyboard size only had a significant effect on responses to one question regarding the ease of entering URLs (F(2,27)=8.8, p<0.002), but no consistent pattern could be identified between keyboard size and participant responses indicating that participants did not consistently prefer larger or smaller keyboards.

6. CONCLUSIONS

Since many individuals make decisions regarding the adoption of these technologies after minimal use, and many others use these technologies only occasionally, the current study focused on novice use. In this context, keyboard size had no effect on data entry rates or the number of uncorrected errors in the resulting text. Responses to only one of fifteen questions on number of uncorrected errors in the resulting text. Responses to only one of fifteen questions on the satisfaction questionnaire were affected by keyboard size and even those results did not show a consistent pattern. Our results confirm that small soft keyboards that could fit on a cellular phone are not only feasible, but that they result in data entry rates, error rates, and preference ratings that are comparable to larger keyboards currently used on PDAs. However, the task in which the users engaged does have a significant effect on data entry rates with data entry rates ranging from 7.2 to 12.5wpm. Overall, it appears that tasks that require users to switch between the two keyboards result in slower data entry. While confirming that small soft keyboards are feasible in the context of cellular phones, our results raise questions about the generalizability of the faster data entry rates reported elsewhere. As reported above, earlier studies had users ignore errors, focused on simple tasks, or had users interact with paper mockups or a desktop tablet. We suggest that it is critical to consider the underlying experimental methodology and the tasks users completed when interpreting the results of these studies.

REFERENCES

Lewis, J. R. (1999). Input rates and user preference for three small-screen input methods: Standard keyboard, predictive keyboard, and handwriting. *Proceedings of the Human Factors and Ergonomics Society 43rd Annual Meeting.* Pp. 425-428.

Lewis, J. R., LaLomia, M. J. & Kennedy, P. J. (1999). Evaluation of typing key layouts for stylus input. *Proceedings of the Human Factors and Ergonomics Society 43rd Annual Meeting,* pp. 420-424. Santa Monica, CA: Human Factors Society.

MacKenzie, I. S., Zhang, S. X. (1999) The design and evaluation of a high-performance soft keyboard. *Proceedings of the ACM Conference on Human Factors in Computing Systems - CHI '99,* pp. 25-31. New York: ACM.

MacKenzie, I. S., Zhang, S. X. & Soukoreff, R. W. (1999). Text entry using soft keyboards. *Behaviour & Information Technology,* 18, 235-244.

MacKenzie, I. S., Nonnecke, B., McQueen, C., Riddersma, S. & Meltz, M. (1994). A comparison of three methods of character entry on pen-based computers. *Proceedings of the Human Factors and Ergonomics Society 38th Annual Meeting,* pp. 330-334. Santa Monica, CA: Human Factors Society.

APPENDIX A: EXPERIMENTAL TASKS

Task One
John Doe
8374 Maple Dr.
Apt. 36-C
Baltimore, MD 21250
(410) 391-4398
jdoe@gl.umbc.edu

Task Two
www.giraffe837.com

Task Three
www.travelocity.com/vaca23

Task Four
Department Meeting

Task Five
Meeting with Bob and Sue about annual budget

Task Six
The meeting this Tuesday has been changed to 2 pm. Please notify me if there is a conflict in your schedule. Bring all materials regarding Alpha project with you to this meeting. I will send more details later in the week.

VIBROTACTILE FEEDBACK FOR HANDLING VIRTUAL CONTACT IN IMMERSIVE VIRTUAL ENVIRONMENTS

Robert W. Lindeman[1] and James N. Templeman[2]

[1]Dept. of Computer Science, The George Washington University, 801 22nd St., NW, Washington, DC 20052 USA
[2]CODE 5513, Naval Research Laboratory, 4555 Overlook Ave, SW, Washington, DC 20375 USA

ABSTRACT

This paper addresses the issue of improving a user's perception of contacts they make with virtual objects in virtual environments. Because these objects have no physical component, the user's perceptual understanding of the material properties of the object, and of the nature of the contact, is hindered, often limited solely to visual feedback. Many techniques for providing haptic feedback to compensate for the lack of touch in virtual environments have been designed and implemented. These systems have increased our understanding of the nature of how humans perceive contact. However, providing effective, general-purpose haptic feedback solutions has proven illusive.

We propose a more-holistic approach, incorporating feedback to several modalities in concert. We describe a low-cost, prototype system we have developed for delivering vibrotactile feedback to the user, discuss different parameters that can be manipulated in order to provide different sensations, and propose ways in which this simple feedback can be combined with feedback to other modalities to create a better understanding of virtual contact.

1. INTRODUCTION

Virtual contact research addresses the problem of what feedback to provide when the user comes into contact with a purely virtual object (PVO) within a virtual environment (VE). We define virtual reality as fooling the senses into believing they are experiencing something they are not actually experiencing. In the visual domain, for instance, this means creating the scenery and viewpoint changes that induce the user into believing that what they are seeing is "real," or at least predictable.

As humans, we interact with our environment using multiple feedback channels, all coordinated to help us make sense of the world around us. The lack of multimodal feedback in current VE systems, however, hinders users from fully understanding the nature of contacts between the user and objects in these environments. It has been found that because real-world contact combines feedback spanning multiple channels (*e.g.*, tactile and visual), providing feedback to multiple channels in VEs can improve user performance (Kontarinis and Howe, 1995; Lindeman *et al.*, 1999). Grasping virtual controls, opening virtual doors, and using a probe to explore a volumetric data set can all be made more effective by providing additional, multimodal feedback. In essence, we are addressing the need for supporting effective user actions given a sensorially-deprived environment, because the only feedback the user receives is that which we provide.

2. PREVIOUS WORK

Next we review previous work into providing feedback to the different sensory channels, with particular focus on the haptic channel.

2.1 The Sensory Channels

For the most part, it is the visual sense that has received the most attention from VE researchers. Some work has been done on fooling the other senses. After visuals, the auditory sense has received the most attention (Wenzel, 1992). The area of haptics has received increasing attention from VE researchers over the past decade. The proprioceptive sense has also received some attention (Mine *et al.*, 1997). The senses of smell and taste have received less attention, because of their inherently intrusive nature (Hoffman *et al.*, 1998).

A common finding of researchers in these different sensing modalities is that it is not enough to address only one of the senses; that to give a deeper feeling of immersion, multiple senses need to be stimulated simultaneously. Furthermore, providing more than one type of stimuli allows researchers to achieve adequate results using lower "resolution" displays (Srinivasan, 1994). This becomes important for low-cost applications, like game consoles or aids for the hearing impaired (Tan *et al.*, 1999). For example, relatively-simple haptic feedback could be combined with high-quality visual images, which have become very inexpensive to produce, in order to create a similar sense of contact produced by more-expensive haptic displays.

2.2 The Human Haptic System

Though much work has been done in the study of the human haptic system, there is still no unequivocal evidence showing exactly how haptic input is transformed from external stimuli into internal understanding. Broadly, the human haptic system can be broken down into two major subsystems. The ***tactile*** system, referring to the sense of contact with an object, receives information mediated by the response of mechanoreceptors in the skin within and around the contact area. Experiments have been performed into defining the characteristics and thresholds of the different mechanoreceptors (Srinivasan, 1994; Kontarinis and Howe, 1995). The ***kinesthetic*** system (also called the *proprioceptive* system), referring to the position and motion of limbs along with the associated forces, receives information from sensory receptors in the skin around the joints, joint capsules, tendons, and muscles, as well as from motor-command signals (Srinivasan, 1994). The tactile and kinesthetic systems play roles of varying influence depending on the task we are performing. If we arrange their influence as axes on a graph, we can define a space for plotting tasks we perform in VEs as shown in Figure 1.

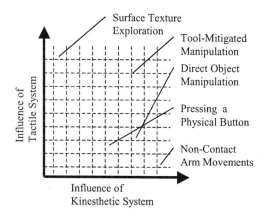

Figure 1: Influence of Tactile System vs. Influence of Kinesthetic System

Srinivasan (1994) uses the terms *exploration* and *manipulation* to describe a framework for studying biomedical, sensory, motor, and cognitive subsystems. The process of exploration is mainly concerned with extracting attributes of objects, and is therefore a tactile-dominated task. Manipulation is mainly concerned with modifying the environment, and is therefore dominated by kinesthetic feedback. Indeed, both subsystems play a role in almost every type of interaction; it is the strength of their influence that varies.

In terms of producing haptic feedback, passive approaches use properties inherent in physical props to convey stimuli (Hinckley *et al.*, 1994; Lindeman *et al.*, 1999). These systems do not require any computer control to provide a stimulus, and can provide high-fidelity, inexpensive (both computationally and monetarily) feedback. An example of this would be to register a physical railing with the representation of a railing in a VE, allowing the user to feel the real railing when reaching for the virtual one (Hodges *et al.*, 1995). Because these objects are passive, however, the range of feedback any given object can provide is limited in both type and strength. Therefore, a more common approach to stimulating the haptic sense is through the use of active-haptic feedback (Salisbury *et al.*, 1995). These approaches typically deliver stimuli using some sort of force-reflecting device, such as an arm-linkage employing sensors and actuators (Brooks *et al.*, 1990; Howe, 1992), force-feedback gloves (Burdea *et al.*, 1992), or master manipulators (Iwata, 1990).

Much of the empirical work into determining how we sense touch has focused on the hands, and, in particular, on the finger pad (Craig, 1998). Some approaches combine tactile and kinesthetic stimulation into a single system. Howe (1992) describes a teleoperation system for supporting a precision pinch grasp, using a two-fingered linkage. Wellman and Howe (1995) augmented this device by attaching inverted speakers, controlled by output from a PC sound card, to each of the master finger brackets, thereby adding a vibratory component to the feedback system. Kontarinis *et al.* (1995) describe the addition of arrays of shape memory allow (SMA) wire actuators to this system for providing feedback directly to the finger pad. Howe *et al.* (1995) describe methods for improving the responsiveness of SMA actuators by producing faster heating response using feed-forward derivative compensation, faster cooling response using pneumatic cooling jets, and reduced hysteresis through the use of position sensing LEDs.

Some researchers have conducted studies using SMA arrays (Bensmaïa and Hollins, 2000; Craig, 1998). In addition to SMA, other materials can also be used for tactile stimulation (see Fletcher, 1996, for a comparison of material properties). Some researchers have focused on our ability to discern combinations of sinusoidal wave forms at differing frequencies (Tan *et al.*, 1999; Bensmaïa and Hollins, 2000). Others have looked at our ability to discern patterns in the presence of temporal masking of pattern elements (Craig, 1998). In teleoperation experiments, researchers have looked at how vibratory feedback can be used to regulate the amount of pressure applied at the slave side (Amanat *et al.*, 1994).

Some researchers have begun to explore the use of vibrating motors, similar to those used in pagers and cell phones, as a means of providing inexpensive haptic feedback (Massimino and Sheridan, 1993; Rupert, 2000; Cheng *et al.*, 1996). We propose combining these low-cost vibrotactile (VT) feedback units with feedback through other channels to relay contact information to the user. In the absence of actual, physical walls, tactors mounted on the extremities of a user (*e.g.*, on the arm) could be triggered to simulate contact between the arm and a virtual wall. It is hoped that this integration of VT feedback into a VE system would thereby improve a user's sense of contact made with objects in the VE.

Commercial products have also emerged, both in the home entertainment and research areas. Simple vibrotactile attachments for game console controllers have been commercially available for some time now. Nintendo, Sega, and Sony all produce devices utilizing eccentric motors to add vibration to the gaming experience. Consumer-grade, force-feedback joysticks from Microsoft and Logitech, and commercial-grade force-feedback joysticks from Immersion Corporation (Okamura *et al.*, 1998), use actuators to control the resistance/force delivered to handle controllers. Virtual Technologies, Inc., produces a glove instrumented with six VT units, five on the fingers and one on the palm. This is a representative, rather than exhaustive, list of commercial force-feedback offerings, and underscores the attention currently being paid to the use of VT feedback by industry.

3. RESEARCH QUESTIONS

We see several important steps within VT feedback research where more work is required, and posit them as fertile areas of short- to medium-term research. These include:

1. A survey of technologies available for providing VT feedback,
2. Methodical studies of the parameter sensitivity of VT feedback, across devices, users, applications, *etc.*,
3. A framework for combining VT with other feedback modalities, and
4. Exploration of application areas (tasks) where this holistic feedback could be effective.

We are currently addressing these issues in our research.

3.1 Current Prototype

We have built a prototype VT feedback system using commercial, off-the-shelf tactors, and a proprietary control circuit. A simple protocol is used to send commands from a host computer to a microcontroller unit (MCU) via a standard serial connection (Figure 2). The MCU interprets the commands from the host, and through control circuitry, is able to set the vibration level of any tactor to one of 32 levels.

A high-level Application Programming Interface (API) allows the application to control the vibration level of each tactor individually. The tactors are inexpensive DC pager motors, each with an eccentric weight, and a maximum of 1,000 revolutions per second. By integrating this type of system into a traditional VE system, we can experiment with ways of combining visual, audio, and haptic feedback to produce effective contact feedback.

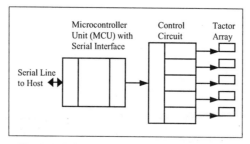

Figure 2: Prototype vibrotactile feedback system.

3.2 Varying VT Feedback

There are a number of parameters that can be manipulated to provide varying VT feedback. These can be divided into parameters which affect each tactor individually, and those that affect a group of tactors.

Parameters for Individual Tactors:

1. *Frequency*: Modulation can be used to vary the frequency of the feedback (Bensmaïa and Hollins, 2000; Tan *et al.*, 1999).
2. *Amplitude*: Modulation can be used to vary the intensity of the feedback (Tan *et al.*, 1999).
3. *Temporal Delay*: By varying the time-delay of a stimulus, we can aid the identification of spatial patterns (Massimino and Sheridan, 1993; Craig, 1998; Tan *et al.*, 1999).

Parameters for Groups of Tactors:

1. *Waveform*: Applying different waveforms to the propagation of a stimulus across the array will allow more complex stimulation to be displayed. These waveforms can be viewed as tactile "images," and their propagation can be viewed as "moving images."
2. *Tactor Placement*: It is well known that the concentration and sensitivity (*i.e.*, type) of haptic receptors in the human body vary with location. Feedback will need to vary with the location on the body being stimulated.
3. *Interpolation Method*: By varying the vibration of adjacent tactors over space and time, a relatively-sparse area of the skin can be fooled into believing that the tactor resolution is higher than it actually is (Rupert, 2000).

Our prototype allows us to rapidly configure, deploy, and experiment with a wide range of form factors. We are currently focusing on stylus, hand-controller, and glove form factors. Physical props could be outfitted with VT devices to provide additional feedback. For instance, a rifle prop could be outfitted to give the user a sense of bumping the barrel into something, or resting it on a support. As a test environment, we are experimenting with the use of a single tactor, mounted on a stylus, for exploring a volume data set. As the user moves the stylus through the data set, the vibration fed back through the stylus is proportional to the value of a particular variable in the data. Experiments could be done to compare how the user's perception of the data set varies when 1) the visual and VT channels are fed from the same variable, 2) the visual and VT channels are fed from different, but complementary variables, 3) the visual and VT channels are fed from different, but conflicting variables, 4) other senses are stimulated in addition to using the visual and VT channels.

4.   CONCLUSION

We have designed a prototype system for integrating VT feedback with visual and auditory feedback in VEs. This system is flexible enough for us to test different combinations of feedback, and measure the impact of these combinations on the ability of users to understand the nature of contact with purely virtual objects. After further research and testing, we believe vibrotactile feedback will have a significant impact on improving the user experience in virtual environments.

ACKNOWLEDGEMENTS

Funding for this work was provided by the Office of Naval Research under Grant No. N000140110107.

REFERENCES

Amanat, I.Z., Riviere, C.N., and Thakor, N.V. "Teleoperation of a dexterous robotic hand using vibrotactile feedback," Proc. of 16th Annual Int'l. Conf. of the IEEE Eng. in Med. and Bio. Soc., 1994, pp. 1059-1060.

Bensmaïa, S.J. and Hollins, M. "Complex tactile waveform discrimination," J. of the Acoustical Soc. of America, 2000, 108(3), pp. 1236-1245.

Brooks, F., Ouh-Young, M., Batter, J., and Kilpatrick, P. "Project GROPE - Haptic displays for scientific visualization," Proc. of ACM SIGGRAPH '90, 1990, pp. 177-185.

Burdea, G., Zhuang, J., Roskos, E., Silver, D., and Langrana, N. "A portable dexterous master with force feedback," Presence: Teleoperators and Virtual Environments, 1992, 1(1), pp. 18-28.

Cheng, L.-T., Kazman, R., and Robinson, J. "Vibrotactile feedback in delicate virtual reality operations," Proc. of the Fourth ACM Int'l. Conf. on Multimedia, 1996, pp. 243-251.

Craig, J.C. "Vibrotactile pattern isolation/integration," Perception & Psychophysics, 1998, 60(5), pp. 888-899.

Fletcher, R. "Force transduction materials for human-technology interfaces," IBM Sys. J., 1996, 35(3&4), pp. 630-638.

Hinckley, K., Pausch, R., Goble, J.C., and Kassell, N.F. "Passive real-world interface props for neurosurgical visualization," Proc. of ACM CHI '94, 1994, pp. 452-458.

Hodges, L.F., Rothbaum, B.O., Kooper, R., Opdyke, D., Meyer, T., North, M., de Graaff, J.J., and Williford, J. "Virtual environments for treating the fear of heights," IEEE Computer, 1995, 28(7), pp. 27-34.

Hoffman, H.G., Hollander, A., Schroder, K., Rousseau, S., and Furness, T. "Physically touching and tasting virtual objects enhances the realism of virtual environments," Virtual Reality: Research, Development, and Application, 1998, 3, pp. 226-234.

Howe, R.D. "A force-reflecting teleoperated hand system for the study of tactile sensing in precision manipulation," Proc. of the 1992 IEEE Int'l. Conf. on Robotics and Autom., 1992, vol. 2, pp. 1321-1326.

Howe, R.D., Kontarinis, D.A., and Peine, W.J. "Shape memory alloy actuator controller design for tactile displays," Proc. of the 34th IEEE Conf. on Decision and Control, 1995, vol. 4, pp. 3540-3544.

Iwata, H. "Artificial reality with force-feedback: Development of desktop virtual space with compact master manipulator," Computer Graphics, 1990, 24(4), pp. 165-170.

Kontarinis, D.A., and Howe, R.D. "Tactile display of vibratory information in teleoperation and virtual environments," Presence: Teleoperators and Virtual Environments, 1995, 4(4), pp. 387-402.

Kontarinis, D.A., Son, J.S., Peine, W.J., and Howe, R.D. "A tactile shape sensing and display system for teleoperated manipulation," Proc. of the Int'l. Conf. on Robotics and Autom., 1995, pp. 641-646.

Lindeman, R.W., Sibert, J.L., and Hahn, J.K. "Towards usable VR: An empirical study of user interfaces for immersive virtual environments," Proc. of ACM CHI '99, 1999, pp. 64-71.

Massimino, M.J., and Sheridan, T.B. "Sensory substitution for force feedback in teleoperation," Presence: Teleoperators and Virtual Environments, 1993, 2(4), pp. 344-352.

Mine, M., Brooks, F., and Séquin, C. "Moving objects in space: Exploiting proprioception in virtual-environment interaction," Proc. of ACM SIGGRAPH '97, 1997, pp. 19-26.

Okamura, A.M., Dennerlein, J.T., and Howe, R.D. "Vibration feedback models for virtual environments," Proc. of the IEEE Int'l. Conf. on Robotics and Autom., 1998, pp. 674-679.

Rupert, A. "An instrumentation solution for reducing spatial disorientation mishaps," IEEE Eng. in Med. and Bio., March/April, 2000, pp. 71-80.

Salisbury, K., Brock, D., Massie, T., Swarup, N., and Zilles, C. "Haptic rendering: programming touch interaction with virtual objects," Proc. of the 1995 Symp. on Interactive 3D Graphics, 1995, pp. 123-130.

Srinivasan, M.A. "Haptic Interfaces," Virtual Reality: Scientific and Technical Challenges," in N. Durlach and A. Mavor (Eds.) Report to the Committee on Virtual Reality Research and Development, National Research Council, National Academy Press, Washington, DC, 1994, pp. 161-187.

Tan, H.Z., Durlach, N.I., Reed, C.M., and Rabinowitz, W.M. "Information transmission with a multi-finger tactual display," Perception & Psychophysics, 1999, 61(6), pp. 993-1008.

Wellman, P. and Howe, R.D. "Towards realistic vibrotactile display in virtual environments," in T. Alberts (Ed.) Proceeding of the ASME Dynamics Sys. and Control Division, Symposium on Haptic Interfaces for Virtual Environment and Teleoperator Sys., 1995, DSC-Vol. 57-2, pp. 713-718.

Wenzel, E. "Localization in virtual acoustic displays," Presence: Teleoperators and Virtual Environments, 1992, 1(1), pp. 80-107.

25

Guided Design and Evaluation of Distributed, Collaborative 3D Interaction in Projection Based Virtual Environments

Gernot Goebbels[1], Vali Lalioti[2], Thomas Mack[1]

[1] German National Research Center for Information Technology
Department of Virtual Environments
[2] Computer Science Department
University of Pretoria, South Africa

ABSTRACT

In this paper we present a framework for the design and evaluation of distributed, collaborative 3D interaction focussing on projection based systems. We discuss the issues of collaborative 3D interaction using audio/video for face-to-face communication and the differences in using rear projection based Virtual Environments.

1. INTRODUCTION

The vision in Collaborative Virtual Environments (CVE) is to provide distributed teams with a virtual space where they can meet as if face-to-face, co-exist and collaborate while sharing and manipulating virtual data in real time. Therefore the environment needs to provide shared data representation, shared manipulation, integrate real-time video and audio communication and control between remote participants and at the same time provide a natural way of interacting with the shared data. For supporting the implementation and realization of such CVEs we report on our framework for the design and evaluation of distributed, collaborative 3D interaction focussing on projection based systems. The approach focuses on our CVE interaction taxonomy that supports the development of applications for small groups working together in rear projection-based VEs making use of video conferencing and 6DOF input devices. Design guidelines and the evaluation of different collaboration metaphors, operations, feedback components and user interfaces are also presented in the paper.

2. CVE INTERACTION FRAMEWORK

In order to find out how to support users we start with a very detailed *User's Task Description (UTD)*. A following *User's Task Analysis (UTA)* determines the so-called *User+Need Space (UNS)* which itself is the originator of the flow within our CVE taxonomy graph. The taxonomy can be found in earlier papers (Goebbels et al., 2000a; Goebbels et al., 2000b). This UNS relays the information extracted by the UTA of the UTD. We recommend to do an extensive, detailed description and analysis of the user's task in order to find out how the user's needs can be classified and addressed. Then the UNS deals with the following groups of issues:

- representation components, work mode, input/output device combinations, auxiliary tools as operations, metaphors and interaction techniques as well as actions and action feedback.

Representation Components are a very important part of Virtual Environments since they determine the representation of the visual parts of the application. The components are the representation of the user, the remote user, the environment, the virtual input device, the virtual tools and finally the representation of the data model and functionality.

The *Application+Interaction Space (AIS)* describes how users interact, with each other and collaboratively with the data set, in the virtual environment. In order to find the best interaction we first have to understand the low-level makeup of interaction. Therefore we have to split down interaction tasks and to find interaction templates which can be combined to form more complex interactions.

Awareness-Action-Feedback loops denote such interaction templates. These AAF loops allow us to understand and analyse very tiny steps in interactions. When analysing an interaction task of a single user with a data set we divide an autonomous AAF loop into four blocks. The first two blocks belong to the awareness phase where the user starts with proprioception (Mine et al., 1997). Proprioception allows the user to be aware of where s/he stands and looks

26

at, the position and orientation of body parts like arms, hands and fingers and everything that allows users to perceive themselves in relation to the environment. The next step is to be aware of the physical input devices held in the users hands and the virtual tool representations connected to them. The position and orientation of the virtual data set is perceived in this phase too. The user is then ready to perform an action. This action can for example simply be to move the hand together with the physical input device. After the action phase a feedback phase follows. In this phase the user perceives the feedback from the action without which it is impossible to analyse the result of the action. In this case the user perceives the movement of the virtual tool representations as s/he moved the input device together with the hand. After the perception of the status of the situation the user can decide if the task is completed and therefore break the loop or whether the task is not completed yet and therefore prepare for the next action starting with the first block again.

Collaborative Awareness-Action-Feedback loops are of the same structure as the autonomous AAF loops. In addition to the autonomous AAF loops, the user perceives the co-presence during the awareness phase. It is comparable to proprioception but now information about the remote partner is queried. An interesting component represents the perception of co-knowledge and co-status. We found out that knowing that your partner is aware of you is one of the most important steps in this awareness phase. The user can confirm this status check either by voice or with the help of a gesture like the "thumbs up". The action and the feedback phase are equal to the ones of the autonomous AAF loop. The Awareness-Action-Feedback loops are templates. With the help of operations, metaphors and interaction techniques described in (Goebbels et al., 2000b) it is now possible to give those templates a "face". Depending on the user's subtask appropriate operations, metaphors and interaction techniques are chosen for each action.

We designed and implemented a medical CVE application according to the taxonomy and collaborative and autonomous AAF loops. We chose the most appropriate metaphors, operations, interaction techniques and representation components for this application. Two 2-sided Responsive Workbenches were used as the displays systems. The technical setup and an example from a real-time collaborative session are shown in Figure 1 (Goebbels et al., 2000a).

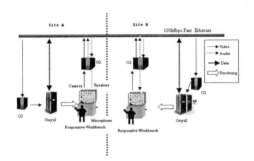 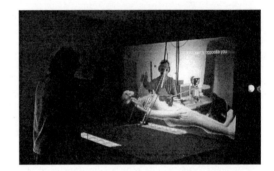

Figure 1. Built and used setup with two collaborative RWBs. Snap shot of a real-time collaborative session.

2.1 Evaluation

Three different evaluation methods are applicable when assessing Collaborative Virtual Environments (CVE). The expert heuristic, the formative and the summative evaluation (Hix and Hartson, 1993; Hix et al., 1999; Nielson, 1993). These evaluation methods make it possible to substantiate or refute realizations of a specific CVE. Assessing evaluators have to be no VE experts and not part of the developers team. Expert heuristic and formative evaluation are applied in alternating cycles in the early design state of the CVE. Based on the expert's knowledge, problems concerning usability can be solved following the expert's recommendations. After these recommendations are considered in a new and better design of the CVE the summative evaluation is applied. The objective of this evaluation method is to compare been different CVEs designed with the information obtained from the User+Need Space. Hence the output of the summative evaluation enables to statistically compare different realizations of interaction techniques, operations, representation components etc. and to choose the most appropriate one in terms of usability. However, important when planning an evaluation is to determine items which are assessable. This is often the most complex part. This collection of items is necessary to formulate specific questionnaires and hence to find and eliminate disturbance factors from the implementation of the CVE. For the assessment of the CVE the following factors are determined with respect to the User+Need Space defined by the User Task Analysis (see section 2.):

- menu representations
- virtual tool representations
- representation of data and its functionality
- environmental representations
- input devices
- physical equipment and cabling
- data processing and system reaction time
- graphical and acoustical resolution and quality
- network transfer rate
- perception of the own presence within the CVE
- perception of the partner's co-presence within the CVE
- perception of the collaboration in terms of equality of rights
- perception of the quality of collaboration
- frequency with which the user looked to the partner
- frequency with which the user spoke with the partner

Considering all these evaluation items in one session is almost impossible, since the items mentioned above evaluate too many different aspects of *Human-Computer-Human* interaction. In order to address this number of items special evaluation sessions are defined, namely the usability session, co-presence session and co-work session. An introduction is given prior to the evaluation sessions. During this introduction the evaluators are informed about the display system, the equipment and the environment they are going to work with. The objective is to create almost same conditions for all evaluators, since this is necessary for comparing numerical results of the formative and summative evaluations.

In the *usability session* the users (evaluators) interact autonomously within the VE for about five minutes.

During the interaction an external observer is taking notes and filling out a special observer questionnaire. This VE expert is observing the non-expert evaluator during the usability, the co-presence and the co-work sessions. Beside querying specific information about the time the user had to think and to debate before performing actions the questionnaire leaves space for informal observations. Especially this questionnaire helps to assess items which are difficult to be assessed by the evaluators themselves such like questions *"Did the user loose concentration during a session ?"* or *"How quickly could the user correct mistakes and continue the work ?"*. Information if the evaluator lost concentration during a session has an impact on the analysis and the way the numerical results have to be interpreted. However, this information can also imply the high cognitive load of interaction in the Collaborative Virtual Environment. Beside the overall ability to interact with the system critical incidents are very interesting to the observer.

In the *co-presence session* the user works again in the CVE but now with another data set. In contrast to the latter session an experienced user who has been involved in the development process is remotely present within the same environment through an audio/video connection. The experienced user explains the task, the data set, the input devices and the tools remotely to the evaluator. The remote partner who acts like a supervisor does not use any input devices or tools, but only gestures and verbal instructions. The task is to position three bones as precisely as possible to complement a human skeleton. These bones lay in front of the evaluator and look very similar to each other. If the evaluator does not know what to do the supervisor gives advise about the tools to be used, how to query information about the bones, how to change the viewpoint etc..

In the *co-work session* the task is slightly different. The task is to position three bones belonging to three different pairs to complement the human female skeleton collaboratively by both users. Each bone in a pair belongs to the left or the right side of the skeleton (i.e. the femur bone of the right and the left leg). A set of three of these bones lay in front of each user. As the users stand opposite each other, on different sides of the skeleton, they have to find out which bones belong to their side as the bones are mixed. Bones which belong to the partner's side can be exchanged by passing it over. To ensure further collaboration during the task the human female skeleton is covered by its skin. In order to position the bones, the particular part of the skeleton has to be made visible by cutting away the skin in this region. It is not possible to cut the skin permanently. This means that the cutting user holds the skin cutter while the other user positions the bone.

3. EVALUATION RESULTS

The expert heuristic, formative and summative evaluations for the different sessions delivered usability findings and recommendations.

The User+Need Space (UNS) for the considered evaluation scenario determines different representation forms for generic and content specific operations. For the generic operations a toolbar is designed whereas the content specific operations are grouped by a special ring menu (Goebbels et al., 2000b). In early designs of the CVE the generic toolbar was configurable in terms of its position by the user. The idea behind was that a dominant right-handed user might want to position the menu somewhere else in space than a dominant left-hander. Evaluation results showed that configuration of menus has a negative impact on the cognitive load. Additionally it is not really used in limited interaction spaces offered for example by the Responsive Workbench (RWB). Working with both hands at a RWB, the total viewing frustum is accessible in contrast to CAVE like display systems. Thus during the formative and summative evaluation the toolbar was positioned close to the users body within arm distance corresponding to the vendor's tray metaphor. Working at a RWB this toolbar is fixed whereas it is attached to the users body position when working in a CAVE or cylindrical and wall display systems.

Similar problems are encountered when using ring menus described in (Goebbels et al., 2000b). When a user intersects the data with the menu pick ray in the right hand the ring menu appears attached to the left hand and vice versa. This corresponds to the metaphor of handling a painter's palette with respect to dominant right and left-handers. The advantages were assumed to be the comfortable handling of this ring menu since it does not occlude any object being handled this way. For detaching the ring menu, over the shoulder deletion was integrated (Mine et al., 1997). Evaluation results showed that the handling making use of the painter's palette metaphor is not always as comfortable as assumed. The reason is that the user first has to recognize that the status of the hand changed as something is suddenly attached to it. Then the user has to look at the ring menu in order to select a content specific operation using the other hand. This is particularly annoying if the hand is busy with another task already. Additionally this metaphor makes it impossible to concentrate on the data set as the user is forced to turn the head towards the ring menu. In the improved design the ring menu is attached to the calling hand holding the menu pick ray. It follows the translation of the user's hand whereas the rotation of the user's wrist is used to intersect the ring pieces with the pick ray. The advantages are that the menu appears within the user's gaze and disappears as soon as the user releases the stylus button again. The menu is designed to be 70% transparent to avoid occlusion of data.

As already mentioned the menus group operations together. In order to apply operations tools are selected, e.g. the zoom operation requires a special zoom tool. The tools are represented by 3D icons which are attached to the buttons of the toolbar or to the choices of the ring menu. Usability findings showed that representations for the snap back tool, the information tool and the skin cutting tool were not appropriate in the early CVE design. Now the snap back tool is represented by a three dimensional hook icon, the information by a three dimensional "i" letter and the skin cutter tool by a three dimensional knife icon. These virtual tool representations increased the evaluator's tool recognition rate by almost 80%.

Evaluation results indicated also that early approaches using two pinch gloves as input devices were not really addressing the user's needs. Reasons are the uncomfortable usage when working stand-alone collaboratively and trying to hand over pinch gloves to another user. Another encountered problem using pinch gloves together with pick rays is that it is almost impossible to keep pointing somewhere and additionally snap with the middle finger and the thumb for selection. Similar problems using pinch gloves have been encountered in (Hix et al., 1999). Improvements are made by using a special three button tool in one hand and a stylus in the other. The reason for not using three button tools in both hands refers to the high cognitive load of their usage due to the many buttons. After modification evaluation showed that the stylus is rather used in the dominant and the three button tool in the non-dominant hand.

A *sharing viewpoint* metaphor is implemented for manipulating the users' viewpoint (Goebbels et al., 2000b). Evaluation results showed that an exo-centric viewpoint manipulation is better than an ego-centric when *standing almost beside the partner*. In this context exo-centric manipulation is based on how a user would act in real world by moving laterally. When sharing the same viewpoint *(looking through the partner's eyes)* or sharing the mirrored viewpoint *(looking from opposite the partner)* ego-centric viewpoint manipulation is implemented. This manipulation is realized by pressing and releasing a special button on the three button tool. These observations are valid working at a Responsive Workbench. Because of the limited interaction space it is possible to access the data set visually from all sides by manipulating the viewpoint as described above. However, other own evaluations showed that in the CVE implemented using a CAVE and a cylindrical display no ego-centric viewpoint manipulation is needed. Here users prefer exo-centric viewpoint manipulation due to the larger interaction space and the perception of entire immersion.

29

In the co-work session the evaluators complement a female skeleton by missing bones. There the task is aggravated as the skin of the body is cut in order to make the skeleton visible. Usability findings indicated that users prefer to get a quick overview of the situation. This leads to the implementation of a content specific wireframe operation. The users are able to only render the skin of the body in wireframe and thus have a direct view onto the underlying skeleton. With this strategies can be discussed and collaborative tasks can be planned more quickly. This content specific wireframe operation is only usable for getting an overview. For complementing the skeleton the skin has still to be cut.

In addition to that observations of critical incidents during the co-presence session are made. These critical incidents occur due to network drop outs, indicating that the perception of co-presence is interrelated with the video frame rate. Further experiments with the video frame rate as parameter showed that the perception of co-presence vanishes completely if the video frame rate sinks below 12 fps.

4. CONCLUSIONS

We presented our interaction taxonomy for designing and creating Collaborative Virtual Environments. They provide distributed collaborative teams with a virtual space where they could meet as if face-to-face, coexist and collaborate while sharing and manipulating virtual data. Further we discussed the issues involved in bringing together Human Computer Interaction and Human to Human Communication, focusing on projection-based Virtual Environment systems. Evaluation result derived from alternating cycles of expert heuristic and formative and summative evaluations are also discussed.

The work reported was supported by the Humboldt-University of Berlin and the German Ministry of Research and Technology (BMBF) under grant number 01KX9712/1.

REFERENCES

Goebbels, G., Aquino, P., Lalioti, V., and Goebel, M. (2000a). Supporting team work in collaborative virtual environments. In Proceedings of ICAT 2000 - The Tenth International Conference on Artificial Reality and Tele-existence.

Goebbels, G., Lalioti, V., and Goebel, M. (2000b). On collaboration in distributed virtual environments. In The Journal of Three Dimensional Images, Japan, 14(4) , pages 42-47.

Hix, D. and Hartson, H. R. (1993). User interface development: Ensuring usability through product and process. New York: John Wiley and Sons.

Hix, D., II, E. S., Gabbard, J. L., McGee, M., Durbin, J., and King, T. (1999). User-centered design and evaluation of a real-time battlefield visualization virtual environment. IEEE, pages 96-103.

Mine, M., Frederick, P., Brooks, J., and Sequin, C. (1997). Moving objects in space: Exploiting proprioception in virtual environment interaction. Proceedings of SIGGRAPH 97, Los Angeles, CA.

Nielson, J. (1993). Usability engineering. Academic Press.

3D Interaction and Visualization in the Industrial Environment

Stuart Goose, Ingo Gruber, Sandra Sudarsky, Ken Hampel, Brent Baxter, [†]Nassir Navab

Multimedia Department, [†]Imaging and Visualization Department

Siemens Corporate Research

755 College Road East, Princeton, NJ 08540, USA

Tel: 1-609-734-3391

E-mail: {sgoose, igruber, bbaxter, hampel, sudarsky, navab}@scr.siemens.com

ABSTRACT

Process visualization has come a long way from control panels with switches and lights. Contemporary factories are monitored and controlled by sophisticated software where graphics and animations have usurped gauges and knobs. At Siemens Corporate Research we have been focusing for some time now on applying 3D interaction and visualization techniques to the industrial automation domain. Described in this paper are several novel aspects of the Siemens Three-dimensional Automation Graphical Environment (STAGE).

1. INTRODUCTION

Dynamic process visualization technology has progressed significantly from previous generations of control panels equipped predominantly with switches and lights. Contemporary factories are monitored and controlled by sophisticated plant management software systems where graphics and animations have usurped gauges and knobs. The current generation of plant management software facilitates the hierarchical collections of 2D animated graphics for process visualization, as seen in figure 1 (a).

Siemens is the world's largest supplier of products, systems, solutions and services in the industrial and building technology sectors. The four main markets targeted are manufacturing, process, building and logistics automation. In order to help maintain this leading position, at Siemens Corporate Research one future trend that we have been focusing on is applying 3D interaction and visualization techniques to the industrial automation domain. Our goal has been to enable plant technicians not only view the live process in 3D, but also to interact with the objects in a convenient, intuitive and multi-modal manner.

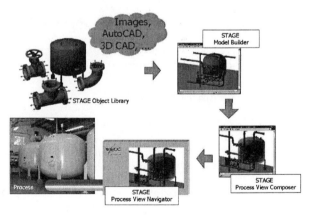

(a)   (b)

Figure 1: Process monitoring in (a) 2 dimensions and (b) 3 dimensions.

Our work on 3D for the factory floor is realized through a project called STAGE: Siemens Three-dimensional Automation Graphical Environment. This project is a multi-disciplinary effort to build a comprehensive solution

31

involving Siemens researchers from multimedia, image processing and automation divisions. Image processing techniques have been developed for expediting 3D scene construction from 2D models and/or representative images. An overview of the steps from creation of the scene to the eventual scene navigation in STAGE can be appreciated in figure 1 (b).

Reported in the remainder of the paper are several novel aspects of STAGE. Context-sensitive speech-driven interaction with 3D objects is presented in Section 2. In section 3, techniques are reported for achieving greater realism in modeling dynamics through the use of textures, audio and shadows. A brief discussion of related work is provided in Section 4. Section 5 proposes areas for further research and provides some concluding remarks.

2. SPEECH[3]: AUGMENTING 3D INTERACTION WITH SPEECH TECHNOLOGIES

We have developed a framework, called Speech[3], that utilizes a 3D VRML browser together with commodity speech recognition and speech synthesis engines to enable end-users to issue spoken commands to complex 3D objects in a scene and receive spoken feedback. The motivation for the Speech[3] framework was to ease the interaction in 3D environments and inherit the potential benefits of a multi-modal interface [4]. VRML browsers have no native support for speech technology, although most support playing static digital audio files in 3D while some support also streamed audio. Hence, we developed a framework for this purpose.

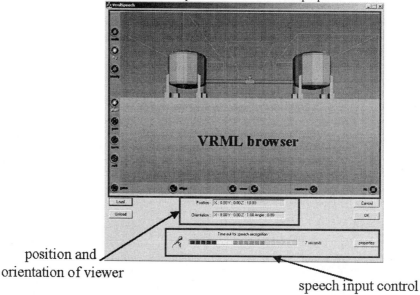

Figure 2: Annotated screen-shot of the interface. The proximity sensor bounding boxes can also be seen.

With Speech[3], VRML proximity sensors are employed to circumscribe and demarcate the 3D objects representing the industrial equipment in the 3D scene to be speech-enabled. As the user moves around the 3D scene, Speech[3] is able to detect via notification events from the proximity sensors within the VRML browser when the user is in the vicinity of a specific object. Speech[3] can then "load" a precise speech grammar into the recognition engine that specifies the speech commands to which the 3D object in the immediate vicinity can respond. Symmetrically, Speech[3] "unloads" the speech grammar when the user leaves the vicinity of the 3D object. This enables the framework to offer a context-sensitive speech-driven interface to an arbitrarily complex 3D scene.

In addition to speech recognition, Speech[3] can also generate multiple simultaneous 3D parameterized synthesized speech streams. This is particularly important for dynamic environments, such as industrial automation, for a variety of reasons: communicating the real-time process values of the industrial equipment, guidance, and the rich listening/browsing of HTML on-line documentation [3]. These services cannot be adequately fulfilled using static digital audio files.

32

The challenge of providing multiple simultaneous 3D parameterized synthesized speech streams was met by building a component that accepts a text string and a 3D coordinate and which wrapped an instantiation of the speech synthesis engine and passing the resulting output into Microsoft DirectSound3D. Empirical use proved that "speech occlusion" was annoying for the user. Often the users did not detect the spoken output, and on other occasions the users found it irritating to have to maneuver to a specific orientation before they could properly hear the speech of the 3D object. Hence a design decision was made to enable the 3D objects to emit speech always in the direction of the user. Whenever Speech[3] framework receives notifications from the VRML browser that the user has changed position, it immediately alters the geometry of all currently active DirectSound3D speech streams to modify dynamically the direction of the spoken output(s). This has the affect of making the currently active speech streams in the proximity follow the user.

The "cocktail party effect" [1], the ability of a person to focus their listening attention primarily upon a single talker while surrounded by others also holding noisy conversations, has been acknowledged for some time. This ability can be exploited judiciously in order to impart information to the user concurrently when within the bounds of a proximity sensor. Although this simultaneity worked, it was significantly improved upon by employing multiple distinct synthesized voices for different categories of information; increasing the volume of the most important synthesized voice and decreasing relatively the volume of other voices; ensuring that the multiple concurrent sources emanating the speech streams avoid impinging upon the trajectories of each other.

The Speech[3] framework takes as input an XML file that comprises the original VRML scene and multiple XML nodes that specify the speech interaction information for given VRML objects in the scene. This additional XML information specifies the speech recognition grammars and the corresponding text strings to be spoken. For example, if the user approaches a container tank he or she could enquire, "Current status?" To which the container tank might reply, "34% full of water at a temperature of 62 degrees Celsius." The XML file allows text strings containing parameters whose values must be retrieved from the underlying plant management software before the Speech[3] framework can speak the reply. In this example, the Speech[3] framework dynamically queried the management software to obtain the values of "34", "water" and "62".

In a noisy environment, such as a factory, it is necessary to avoid Speech[3] listening to spurious noise and misinterpreting it as speech commands. Hence, the user can configure Speech[3] "wake-up" in response to specified phrase. Once Speech[3] is "awake" it listens for a configurable period of time before being deactivitated and returning to "sleep". It can be seen from the speech input control in figure 2 that a line of colored LEDs indicate a countdown of the time remaining before Speech[3] returns to sleep.

Anecdotal experience learned from colleagues at Philips speech research indicated that when users are confronted with a speech recognition system and are not aware of the permitted vocabulary, they tend to avoid using the system. To avoid this situation in Speech[3], when a user enters the proximity sensor for a given 3D object the available speech commands can be spoken to the user, displayed using a transparent 3D graphic sign, or both. It is also feasible for two overlapping 3D objects in the scene to have the same valid speech command. Our solution is to detect the speech command conflict and to query the user further as to which 3D object the command should be applied.

It was considered that VoiceXML would be a suitable language for describing the voice interaction to complement the VRML, but unfortunately VoiceXML constructs proved too limiting for use in the Speech[3] framework. For example, VoiceXML provides no mechanisms for achieving the multiple concurrent 3D parameterized speech synthesis support described above. VoiceXML also does not have a standard mechanism for specifying and employing different speech synthesis voices in order to achieve a varied and captivating voice rendering.

3. INTRODUCING REALISM

3.1 Modeling Fluid Flow Using Texture and Sound

The objective of this work was to create a library of 3D objects to simulate water, oil, and gas flow within a VRML environment. With this objective in mind, a number of objects including, pipes, valves and tanks were designed. Each of these objects was defined in terms of a geometrical description as well as certain specified behaviors. The basic behavior associated with these objects, which corresponds to a fluid/gas flow simulation, was implemented based on moving textures and special sounds.

For instance, pipes were implemented as transparent cylindrical objects with varied lengths and widths. Different image textures mapped to the cylindrical objects were used to specify different kinds of liquids and gases (see Figure 3). A time sensor was incorporated to animate the texture coordinate transformations, producing the visual effect of flow along the pipes. This simple implementation had the problem that the rate of flow was not constant; the texture appeared to move slower in longer pipes than in shorter ones. In order to fix this problem, the cycle interval of the time sensor was defined proportional to the length of the pipe, forcing a constant rate of flow independent of the pipe length. Similarly, valves and tanks were created with their corresponding geometry and behaviors. Valves were designed with an associated cylinder sensor whose rotation could be adjusted to control the valve opening and consequently the rate of flow. Once the objects were connected, events were routed from one object to the next creating a cascade of events that ultimately simulated the required flow. For instance, Figure 3 shows a valve connected to a series of pipes, connected to a tank. As the user increases the opening of the valve, an event is triggered, sending the value of the new rate of flow to the pipes. From this value, the animation of the texture transform is updated and the intensity of the associated sound is elevated. This behavior propagates through neighboring objects until reaching the tank. When these objects are integrated into STAGE, the rate of flow is not longer controlled by a user, but mapped directly to the real values obtained from the plant management software.

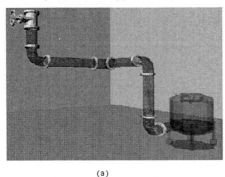 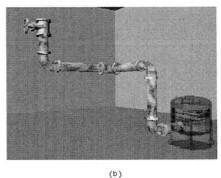

(a)   (b)

Figure 3: Use of textures and sounds to simulate water (a) and gas flow (b)

3.1 Introducing Realism Through Shadows

Shadows provide important perceptual cues that help disambiguate the location, shape and size of objects in a 3D scene. Without shadows it is difficult to determine the spatial relationship between objects; it is hard to distinguish whether an object is sitting directly on top of another object or floating in space. Shadows have been extensively studied in the field of computer graphics and a number of algorithms have been proposed that generate very realistic shadows for static scenes. However, shadow computation is a time consuming operation that cannot easily be performed at interactive rates. Because of this complexity, interactive virtual worlds are often created without shadows and often appear very artificial and ambiguous. In fact, the VRML specification does not require shadows to be generated. Standard VRML objects allow the light to pass through them without casting any shadows. We propose a technique to compute dynamic shadows by restricting the shadow generation problem to the case of polyhedral objects and planar surfaces [7]. These shadows are created as the projection of the underlying objects onto the given planes as seen from the light sources. Based on this technique we propose a new VRML node and demonstrate the dynamic generation of shadows as objects and light sources are interactively manipulated.

4. DISCUSSION OF RELATED WORK

The benefits of virtual environments within the industrial sector are clear [2]. The authors found little evidence of prior work conducted into the integration of speech technology within virtual environments. Ressler *et al* [6] describes a technique for integrating speech synthesis output within VRML. While their approach is satisfactory, in comparison with Speech[3] their solution has limitations including: the speech synthesis is not spatialized; only a single speech stream can be played at once; there is no means for parameterizing the speech output. In addition,

Ressler *et al* do not consider at all the integration of speech recognition. Mynatt *et al* [5] describe a system called Audio Aura for providing office workers with rich auditory cues (via wireless headphones) within the context of VRML that describes the current state of the physical objects that interest them. Although Audio Aura and Speech[3] have their roots in VRML, they differ in many ways. Speech[3] supports speech in and out for dialog, whereas Audio Aura is concerned with generating a continuous audio output that, by defining *ecologies*, is a metaphor for the local environment.

5. FUTURE WORK AND CONCLUSIONS

Current work includes integrating the Speech[3] framework with an augmented reality framework to assist a mobile maintenance technician, as seen in figure 4. With him is a tiny notebook computer equipped with a built-in camera and wireless connection to factory database server. He points the camera at a nearby visual marker -- one of many distributed around the plant. The coded visual markers are designed so each is not only uniquely coded but also provides features for computing the relative pose (translation and orientation) of the camera. The camera transmits the code and pose information to the plant's online database. The server then computes not only where the engineer is located, but also his exact orientation in space. The engineer can then interact with the malfunctioning pump using the Speech[3] features. Another avenue of investigation is in the application of noise reduction techniques and hardware to improve the performance of speech recognition in noise environments.

Described in this paper are several novel aspects of the Siemens Three-dimensional Automation Graphical Environment (STAGE). The Speech[3] framework augments a VRML browser to offer a context-sensitive speech-driven interface to an arbitrarily complex 3D scene. Techniques for achieving greater realism in modeling dynamics through the use of textures, audio and shadows were also described.

Figure 4: A mobile maintenance technician benefiting from augmented reality and Speech[3] technologies.

REFERENCES

1. Arons, B., A Review of the Cocktail Party Effect, *Journal of the American Voice I/O Society 12*, pages 35-50, July 1992.

2. Dai, F., Virtual Reality for Industrial Applications, Springer-Verlag, 1998.

3. Goose, S. and Möller, C., A 3D Audio Only Interactive Web Browser: Using Spatialization to Convey Hypermedia Document Structure, *Proceedings of the ACM International Conference on Multimedia*, Orlando, USA, pages 363-371, October 1999.

4. Grasso, M., Ebert, D. and Finin, T., The Integrality of Speech in Multimodal Interfaces, *ACM Transactions on Computer Human Interaction*, 5(4):303-325, December 1998.

5. Mynatt, E., Back, M., Want, R., Baer, M. and Ellis, J., Designing Audio Aura, *ACM International Conference on Computer Human Interaction*, Los Angeles, USA, pages 566-573, 1998.

6. Ressler, S. and Wang, Q., Making VRML Accessible for People with Disabilities, *ASSETS 98*, Marina del Rey, USA, pages 50-55, April 1998.

7. Sudarsky, S., Generating Dynamic Shadows for Virtual Reality Applications, submitted to *IEEE Information Visualisation 2001*.

Interfacing tangible input devices to a 3D virtual environment for users with special needs

S.V. Cobb[1], T. Starmer[2], R.C. Cobb[2], S. Tymms[2], T.P. Pridmore[3] and D. Webster[3]

[1]VIRART, University of Nottingham, UK
[2]School of Mechanical, Materials, Manufacturing Engineering and Management, University of Nottingham, UK
[3]School of Computer Science and Information Technology, University of Nottingham, UK
[1] sue.cobb@nottingham.ac.uk, [2] richard.cobb@nottingham.ac.uk, [3]tony.pridmore@nottingham.ac.uk

1   Abstract

Virtual environments have been used over the last ten years for teaching life skills to children with learning disabilities and/or physical disabilities. They are now being considered as a means of providing practical training of everyday tasks for adults rehabilitating following a stroke. However, current desktop virtual reality systems require the use of a mouse and keyboard or joystick to interact with the VE, and this excludes a high percentage of users in particular, those with special needs. Progressive research at the University of Nottingham is being carried out to examine and develop solutions for interfacing alternative input devices to facilitate greater accessibility to virtual environments. Preliminary studies have shown that users with special needs can more successfully interact with objects in a virtual environment using direct manipulation devices compared with a standard mouse input, and that real objects are more acceptable than toy replicas. However, the current devices are dedicated to performing a single task and so work is continuing to improve flexibility in the interface system to facilitate performance of a range of virtual environment tasks.

2   Introduction

Virtual Reality (VR) technology can be used to provide computer-simulated Virtual Environments (VEs) replicating everyday situations and tasks. Real-time interaction allows the user to explore and practise real world scenarios and tasks in their own time and, where possible, unaided. This provides the opportunity to make mistakes without embarrassment, to learn from these mistakes and practise the activities again and again until the user has sufficient confidence to try performing the task in real life.

A number of research teams have developed virtual environments for life skills training. A 'Virtual City' comprising a House, Café, Supermarket and Transport System has been created to allow children and adults with learning disabilities to learn about and practice tasks such as shopping, cooking, washing, crossing the road and catching a bus (Brown et al., 1999; Neale et al, 1999; Neale, Cobb and Wilson, 2001). A virtual environment designed for stroke rehabilitation patients to practice making a cup of coffee in a virtual kitchen has been developed (Davies et al., 1999; Lindén et al., 2000) and a virtual environment for patients with traumatic brain injury to practice preparation of a simple meal (Christiansen et al., 1998) demonstrate the potential of VR for training specific tasks.

With the use of desktop PCs, these environments are affordable and therefore potentially useful in schools and rehabilitation centres. During ten years of virtual environment development, VIRART have worked closely with the Shepherd School in Nottingham to identify and investigate the utility and usability of virtual environments for special needs education (Cobb, Neale and Stewart, 2001). However, current systems still require the use of a mouse and keyboard to interact with the VE, which excludes a high percentage of users in special needs schools; "Although 60% of our students can access the computer in some way, there is still a group of students whom still cannot and would benefit greatly by doing so. It is for this group that suitable solution needs to be found." (Stewart, 1999).

This current research aims to address this issue by improving accessibility to a virtual environment for students with special needs who are unable to use a standard PC independently. Progressive studies have been conducted to examine alternative methods of interfacing to a Virtual Coffee-making task. The ultimate objective is to use the virtual environment as a training medium supporting real task learning by using real objects interfaced to the computer. The research approach has required progressive development exploring issues relating to ease of interfacing to the virtual environment and user acceptance and usability of alternative input devices.

3 The Virtual Coffee-making task

The virtual coffee-making task is situated within a 3D virtual kitchen environment. This environment was built by VIRART in 1995 as part of a range of life skills training environments (Cobb, Neale and Stewart, 2001). The task comprises seven discrete task steps; turning on a tap, switching on a kettle, opening a coffee jar, spooning coffee into a mug, closing the coffee jar, pouring hot water into a mug, and pouring in milk. To help navigate around the virtual kitchen automatic viewpoints are used. Figure 1 shows the viewpoint at the virtual worktop and Figure 2 shows the viewpoint at the sink. The virtual object to be activated can be seen clearly in each viewpoint and instructions are given verbally at each step. If the task is completed successfully at each step feedback is given by performance of the required activity within the virtual environment and the verbal instruction for the next step is given. If the task is not completed successfully a spoken error message is given and the instruction is repeated. When the overall task is completed a verbal message congratulates the user and a cup of steaming coffee is displayed in the virtual environment.

4 Study 1: Keyboard-mounted tangible objects

4.1 Description of Device

A prototype device has been developed to support use of the Virtual-Coffee-making task (Starmer, 2000). The device is mounted upon the computer keyboard and consists of toy-sized replicas of real objects (e.g. a tap, switch, coffee jar, spoon, kettle, milk carton, etc.), shown in Figure 3. The student activates these objects in turn to move through the coffee-making sequence (e.g. turn the tap, press the switch, twist the lid on the coffee jar, tilt the spoon, tip the kettle and tip the milk carton). When the object is activated correctly, mechanical levers underneath the device operate to press down on one of the keys on the keyboard. This key input to the computer activates a response in the virtual environment and initiates the next instruction. If the key is not fully depressed or the wrong key is pressed, the verbal instruction is repeated.

The initial idea for developing an overlay keyboard-mounted device was prompted by existing keyboard overlay systems for interaction with computer games, which were very popular with the children at Shepherd School. Moreover, a system which activated key-pressing on the a standard computer keyboard was the simplest method for interfacing to the existing virtual environment software.

4.2 Evaluation of the keyboard-mounted device

User evaluation trials were conducted to assess the potential success of the prototype by comparing it against performance of the same task using the standard PC interface currently employed (keyboard and mouse). Five students at the Shepherd School participated in the evaluation study. All performed the coffee-making scenario twice; once with standard PC input devices and once with the keyboard-mounted device. The sequence order was randomised such that three students used the standard interface first and then the keyboard-mounted device and two students used the keyboard-mounted device first and then the standard interface. The virtual environment was run on a Pentium II computer and projected onto a large display screen.

Input of control action for each step differed between the two trials. In the standard interface condition, the user had to use the mouse to position the cursor over the required virtual object and press the left mouse button to confirm selection of that object. For the keyboard-mounted interface, the user had to manipulate the corresponding object on the device (e.g. turn the tap, tip the kettle).

A range of assessment measures derived from an earlier study evaluating use of input devices by students with special needs (Crosier, 1996) were applied:
1. Performance Time: Time taken to complete the task.
2. Errors: Four types of error were recorded:
   - Selecting the wrong gadget.
   - Failing to attempt to complete the task step
   - Failure to complete the task. Attempts to use the gadget but not completing the task step.
   - Device misuse. Using a gadget on the device or the mouse in an inappropriate manner.
3. Engagement with task: An engagement rating on a four-point scale was assigned for each task step.
4. Effort: An effort rating on a five-point scale was assigned for each task step.

The results showed that only two out of the five students completed the task using the mouse interface, whereas for the keyboard-mounted device all five students successfully completed all steps without any support. As would be expected, time taken to complete the task varied between students. For the two students who completed the task in both conditions, performance time was slightly faster using the keyboard-mounted device compared with the mouse. For the three students who did not complete the task using the mouse, their performance time was more than twice as long using the keyboard-mounted device compared with the other two students, indicating weaker manipulation skills. The important result here is that the keyboard-mounted device allowed these three students to complete the task, which had not been possible using a standard computer mouse. Levels of engagement varied between individuals but overall were higher for the keyboard-mounted device than the mouse. For all students, effort levels were much lower using the keyboard-mounted device than with the mouse.

4.3   Conclusions

Observation studies indicated that, in general, the students interacted very well with the keyboard-mounted device. One particular student had never before interacted with a computer unaided and staff at the school commented that they had considered this student to have poor task-attendance. However, it was noted that, when using the prototype keyboard-mounted interface, this student's attention significantly improved.

The keyboard-mounted system represents an improvement on standard input modalities for this user group. However, the requirement that all relevant movements result in the mechanical depression of some key or keys on a standard keyboard significantly restricts interaction. Manipulable objects are constrained to fit over a standard keyboard, producing toy-sized objects. As a result these objects must be small and closely spaced, with only limited lateral movement being possible.

A continuation project aimed to relax these restrictions by embedding simple sensors within real objects which, while still physically tethered to the computer, may be held and moved more naturally. Manipulation of tethered objects results not in the physical depression of keys on the keyboard, but in the transmission of equivalent signals to the computer, via hard-wiring directly into the keyboard port for the PC.

5   Study 2: Tethered tangible objects

5.1   Criteria for Design

A design specification was created based on the results of task analysis, technical research and pilot testing. It was important to involve users in the design process and so a user walkthrough method was applied to help establish the design criteria. Observations were made of the students using real objects to pretend to make coffee. This was used to determine any physical or cognitive restrictions that might affect their ability to use the device. Actions such as manipulation of objects and co-ordination when using more than one object were assessed so as to decide on criteria to include in the design.

General criteria included safety considerations, maintenance requirements and overall cost of the device. More specific criteria that were highlighted by the results of the pilot trials related to the particular strengths and weaknesses of the chosen students. These included the maximum reach distance required, which particularly important as one student was confined to a wheelchair. Another example was the need to avoid actions that would require the use of two hands as a certain candidate could only use one.

5.2   Description of the Device

A prototype of the new device has been developed (Tymms, 2001), shown in Figure 4. The objects are real objects that are used to make coffee in real life. Some of the objects are fixed and some unfixed in positions on a wooden platform. The users are required to lift the objects and move them in a manner replicating the way in which they would be used in real life when making coffee. The objects that can be moved are not tethered to the base or to any other objects. Instead a combination of reed switches and micro-switches is used to determine the location of the objects. The switches are soldered to the wires taken from a standard computer keyboard so that the device is remote from the keyboard. For example, the cup is fixed in position and contains a reed switch that detects a magnet in the milk carton when placed above it. When activated, the corresponding signal is relayed to the VE software resulting in the display of milk being poured into the virtual coffee cup.

5.3   Evaluation of the worktop device

Initial observations of two students using the device with the Virtual Coffee-making task indicate positive responses. Both seemed to enjoy the tasks and claimed to find the device fun when questioned. It was clear that the device was successful in allowing both candidates to interact with the environment. Both students had previously been unable to complete the task with a mouse and, although they could complete the task using the keyboard-mounted device, they made many mistakes. They also seemed to find the device relatively easy to use. The students were asked to complete the sequence twice and both students showed an improvement in their ability to use the device on the second attempt. This suggests that the inclusion of real objects and the requirement to mimic real life actions was effective in improving these students' ability to interact with the virtual environment and could therefore be successful as a teaching aid.

Several design issues were raised as a result of the observations. For example, the robustness of the device needs to be improved as one student in particular was very rough with the objects. These issues are to be addressed before the final testing and evaluation of the prototype.

6   *Study 3: Un-tethered tangible objects*

While the development of tangible interfaces to the virtual environment shows some promise in supporting training of real world tasks, the need for hard-wired interfacing to the VE software inevitably limits flexibility of the system. In a parallel investigation, we are examining the possibility of providing un-tethered interaction via the use of techniques in machine vision. Here, a camera positioned above the work surface acquires colour images of the student's hands as he/she manipulates real, free-standing cups, spoons, coffee jars, etc (see Figure 5). The colour of an image region can be represented in a number of ways, though RGB (red, green, blue) is perhaps the most widely known colour space. It has been shown that if an alternative hue, saturation, intensity (HSI) space is used, image regions depicting human skin exhibit very tightly clustered hue and saturation values which are largely independent of race, skin tone, etc. Skin may therefore be detected by computing hue and saturation at each image location and marking those locations whose HS values fall within the expected range. As the expected HS values vary with camera response and illumination conditions some initial calibration is required, but this merely requires a small number of images to be acquired and portions of skin identified using a mouse. The expected HS range can then be determined automatically. If the objects to be manipulated are selected to have significantly different, easily identifiable HS values, image segmentation software can be produced which identifies each item of interest. By computing simple measurements of the distribution of colours seen in each region, tracking software can identify detected objects in subsequent images. Estimates can then be made of the motion of each object relative to the camera.

The aim of this work is to identify the relative positions and movement of coloured regions corresponding to hands, spoons, etc and transmit signals equivalent to keyboard presses to the virtual environment as a result. In this approach, any motion or configuration of objects is potentially usable as long as the hands and/or items of interest remain in the field of view of the camera and appropriate vision software can be developed. To date, image segmentation and region tracking software has been developed and is being tested on images of a user working with the tethered tangible interface described above. Initial results are shown in figure 6, where image locations considered to depict skin are labelled black. The user's hand is clearly identifiable as the largest connected region of black pixels. The robustness of the approach will be tested by comparing signals generated by the vision software with those produced by the tethered system. User evaluation will then proceed as before.

7   *Discussion and Conclusions*

This preliminary research has shown the potential for improved accessibility to virtual environments aimed at teaching practical life skills by interfacing tangible objects to the computer. Students found it much easier to operate the virtual task using the keyboard-mounted device rather than standard PC input devices. Whilst this device offered some range of input actions replicating real-world actions (e.g. pressing the power switch, turning the tap, tipping the kettle), these were limited by the need to fix the objects in place. The consequence of this was that there was no opportunity to move the objects in relation to each other, as you would need to do to complete the task in real life. This resulted in a very unnatural action for the spoon. This device is useful if the learning objective is merely to identify and select the objects in the correct order, but has limited value in preparation for performing the task in real life.

The worktop device provided a more realistic representation of the actual tasks, using real objects and adding the requirement to move objects in relation to each other. Of course one obvious disadvantage of this approach is that it is still dedicated to one task only. In continuing work we intend to consider how we can provide greater versatility to increase the number and range of skills-based tasks which can be trained in this manner. This may include modularising the device perhaps by providing a range of interchangeable objects for performance of different tasks. Though development is currently at an early stage the machine vision approach may provide some of the required flexibility: calibration to new objects is simple, and many visual tracking systems automatically learn the appearance of the sequences of actions they are to identify.

8 *Acknowledgements*

These projects were conducted in part fulfilment of a BEng Honours degree in Manufacturing Engineering and Operations Management and a BSc Honours degree in Computer Science at the University of Nottingham. The authors would like to thank the teachers and students at the Shepherd who gave up their time and contributed many ideas to the development of these prototypes. Also to Steven Kerr of VIRART for extensive adaptations to the virtual environment to make it compatible with the prototypes. Special thanks are given to Barry Holdsworth from the Manufacturing Workshops who helped with the engineering development.

Figure 1. The Virtual Coffee-making task   **Figure 2. Viewpoint at the sink**   **Figure 3. The Keyboard-mounted device**

Figure 4. The Worktop device   **Figure 5. Aerial view of the task**   **Figure 6. Visual detection of skin**

9 *References*

Brown, D. J., Neale, H. R., Cobb, S. V. G., & Reynolds, H. (1999). Development and evaluation of the virtual city. International Journal of Virtual Reality, 4(1), 28-41.

Christiansen, C., Abreu, B., Ottenbacher, K., Huffman, K., Masel, B. and Culpepper, R. (1998) Task Performance in Virtual Environments Used for Cognitive Rehabilitation after Traumatic Brain Injury, Arch Phys Med Rehabil, Vol 79, 888-892.

Cobb, S., Neale, H. and Stewart, D. (2001) Virtual Environments - Improving accessibility to learning? Paper to be presented at the 1st International Conference on Universal Access in Human-Computer Interaction (UAHCI), New Orleans, 5th-10th August

Crosier, J.K. (1996) Experimental comparison of different input devices into virtual reality systems for use by children with SLD. BEng thesis, School of M3EM, University of Nottingham, UK.

Davies, R.C., Johansson, G., Boschian, K., Lindén, A., Minör, U. and Sonesson, B. (1998) A practical example using virtual reality in the assessment of brain injury. Paper presented at the 2nd European Conference on Disability, Virtual Reality and Associated Technologies, Skovde, Sweden.

Lindén, A., Davies, R.C., Boschian, K., Minör, U., Olsson, R., Sonesson, B., Wallergård, M. and Johansson, G. (2000) Special considerations for navigation and interaction in virtual environments for people with brain injury. Proc. 3rd Intl. Conf. Disability, Virtual Reality and assoc. Tech. (ICDVRAT), Alghero, Italy, 287-296. ISBN 0 7049 1142 6

Neale, H. R., Cobb, S. V., & Wilson, J. R. (2001). Involving users with learning disabilities in virtual environment design. Paper to be presented at the 1st International Conference on Universal Access in Human-Computer Interaction (UAHCI), New Orleans, 5th-10th August

Neale, H. R., Brown, D. J., Cobb, S. V. G., & Wilson, J. R. (1999). Structured evaluation of Virtual Environments for special needs education. Presence: teleoperators and virtual environments, 8(3), 264-282.

Starmer, T.J (2000) Design and development of an alternative computer input device for children with special needs. BEng thesis, School of M3EM, University of Nottingham, UK.

Stewart, D. (1999) Personal Communication. Head Teacher of Shepherd School, Nottingham, UK.

Tymms, S.J. (2001) Design of tangible input devices for special needs users. BEng thesis, School of M3EM, University of Nottingham, UK.

Webster, D. (2001) Hand tracking for HCI. BSc thesis, School of CS&IT, University of Nottingham, UK.

A Study of the Relative Importance of Visual Cues in Desktop Virtual Environments

Sonali S. Morar   Robert D. Macredie   Timothy Cribbin

Department of Information Systems and Computing
Brunel University, Uxbridge UB8 3PH, UK
{Sonali.Morar}, {Robert.Macredie}, {Timothy.Cribbin}@brunel.ac.uk

ABSTRACT

Visual Depth Cues integrate to produce the essential depth and dimensionality of Desktop Virtual Environments. This study discusses Desktop Virtual Environments in terms of visual search task objectives and presents the results of an investigation that identifies the effects of initial probe positions, texture and motion parallax visual depth cues on precise depth judgements made within a Desktop Virtual Environment that is viewed egocentrically. Results indicate that initial probe positions significantly effect precise depth judgements, texture is only significantly effective for certain conditions and motion parallax, in support of results from previous studies, does not improve depth judgements for egocentrically viewed Desktop Virtual Environments. The results help us to understand the effects of certain visual depth cues for egocentric views, suggesting that visual search task objectives may effect the visual depth cues used to emphasise depth within respective Desktop Virtual Environments.

1. INTRODUCTION

Desktop virtual environments (DVEs) are increasingly used to support a varied number of visual search tasks. As the use of these particular synthetic environments increases, so does the need to provide accurate and effective virtual representations (Wickens and Hollands 2000). This is particularly important for visual search tasks that involve precise depth judgements and are considered safety critical, such as those within the medical, vehicle or aviation industries. Extensive research has been undertaken to identify the relative dominance of visual depth cues that combine to create the inherent illusion of depth. The implications of visual depth cue studies broadly relate to perceiving the real world and perceiving DVEs under specific contexts (Dosher et al. 1986, Surdick et al. 1997). However, research identifying visual depth cues within particular DVEs that support different visual search task objectives, is limited (Hubona et al. 1999, Wickens 2000) even though recent studies have reflected the significance of task differences (Bradshaw et al. 2000). This paper discusses the perception of DVEs by highlighting the importance of the visual search tasks conducted within them. It is becoming apparent that perceiving depth within DVEs may be affected by visual search task objectives. Visual search tasks involving accurate spatial perceptions may require DVEs emphasising certain visual depth cues which may differ from DVEs that aim to control or facilitate specific user interactions. The study reported in this paper investigates the effects of initial probe placements, texture (a relatively weaker depth cue) and motion parallax (a relatively dominant visual depth cue) on conducting precise depth judgements within a DVE that is viewed egocentrically.

2. VISUAL SEARCH TASKS IN DESKTOP VIRTUAL ENVIRONMENTS

As the use of DVEs increases, so does the reliance on presenting virtual representations that can support precise depth judgements for visual search tasks. Extensive research has been undertaken within the aviation field in particular, since the challenge of representing air space accurately for pilots (Wickens and Hollands 2000) and understanding the operations within complex aircraft systems for engineers (Hubona et al. 1999) is considered of utmost importance. In these cases, research has acknowledged the inherent ambiguity of presenting the three dimensional environment on a two-dimensional display and has gone on to identify the specific visual cues or spatial display characteristics that impact on task performance. For instance, Hubona et al. (1997) found that controlling object motion effected the perceptual accuracy with the time taken as a trade-off. They suggested that further research was necessary to explore the effects of motion on other visual cues such as luminance, a visual depth cue that can provide information about the shape of an object (Ramachandran 1988). Eyles (1991) recognised the DVE as a 'spatial display' and a 'spatial instrument' where the former presented clear and precise representations of space and the latter supported controlled user interactions. Hendrix and Barfield (1995) investigated the visual depth cues that were effective for spatial instrument design and found that monocular and perspective cues were effective for depth judgement in azimuth but not altitude. Their study did not take in to account motion, which they accepted

41

might be an influential factor for spatial perception. Recent studies have identified the relative importance of task differences and viewing distances on depth perception in the real world (Bradshaw 2000). Likewise, Wickens (2000) proposed that different tasks involving depth perception might be effected by different viewpoints of DVEs. Accordingly, research has indicated the need to investigate broader factors effecting perception of visual depth cues when conducting visual search tasks within DVEs, since these depth cues provide the essential and inherent dimensionality (Hubona et al. 1999).

3. IMPACT OF VISUAL DEPTH CUES ON DEPTH JUDGEMENT
Recent studies have suggested that depth perception within DVEs may be effected by factors such as task differences, viewing distances and changing viewpoints (Wickens 2000). Many studies have used egocentric views for experimental depth judgement tasks (Surdick et al. 1997, Hubona et al. 1997, Westerman and Cribbin 1998). This is especially relevant when the DVE is considered a 'spatial instrument' that aims to control user interactions (Eyles 1991). In these instances, research has identified the relatively dominant and weaker visual cues and has attempted to relate findings to the most appropriate visual cue combination models (Wickens et al. 1989, Jacobs and Fine 1999). Broadly, results have shown that motion parallax is a dominant visual cue and relative brightness is, in contrast, very weak for depth perception on DVEs (Wickens et al. 1989). Surdick et al. (1997) found relative brightness reduced in effectiveness considerably as distances increased. Jacobs and Fine (1999) investigated cue combination strategies and concluded that visual cues were most effective when that particular cue was more informative in a given task. Further investigation of the egocentric view should reveal whether weaker cues, in terms of effectiveness for depth perception across distances, would be used for 'fine tunning' precise depth judgements within DVEs (Hubona et al. 1999). Combinations of visual depth cues are integrated to provide the illusion of depth essential in the display of DVEs and, according to research, the most dominant visual cue in a given scene would take precedence for depth perception (Landy et al. 1995). This has been emphasised for situations where visual depth cues are perceived in conflict and it is proposed that the dominant or least ambiguous visual depth cue be accordingly perceived (Dosher et al. 1986). The effective presentation of visual depth cues may impact on egocentric views of conditions requiring precise depth judgements. The impact of a dominant visual cue, motion parallax, has been investigated in respect of depth judgement (Hubona et al. 1997, Bradshaw et al. 2000). Hubona et al. (1997) found that for tasks involving the egocentric perception of computer generated objects, controlled object motion improved perceptual accuracy but time taken to complete this task increased. This suggests that observers spend time judging depth when the display is momentarily still. Bradshaw et al. (2000) present results indicating that the motion viewing condition had no effect on depth judgement tasks in the real world, which is accordingly discussed in terms of similar studies undertaken using simulated stimuli. This appears to suggest that although motion is a relatively dominant visual cue, it may not be entirely effective for egocentric views of precise depth judgements in DVEs.

4. HYPOTHESES
Three aspects of undertaking precise depth judgements in DVEs that are viewed egocentrically are emphasised through the discussion of three hypotheses in this section which form the basis for the experimental work reported in this paper. The impact of probe positioning, texture and motion are focussed on particularly in order to inform the empirical methodology.

4.1 Impact of Probe Positioning on Precise Depth Judgements
Much empirical work has used egocentric views for depth judgement tasks (Surdick et al. 1997, Hubona et al. 1997, Westerman and Cribbin 1998). However, research is limited in respect to identifying the visual depth cues that effectively support DVEs in the context of spatial instruments presenting various spatial displays (Eyles 1991). Since many of these empirical depth judgement tasks involve matching the depths of two equi-distance probes, the initial positioning of these probes may have an impact on depth judgement as much as the particular visual depth cues that are being investigated. Hendrix and Barfield (1995) proposed categories of monocular visual depth cues used to present DVEs; four visual cues; luminance, texture, height and motion, representing each category, have been selected to test the following hypothesis: *H1 Depth judgement accuracy is effected by the initial positions of the probes.*

4.2 Effect of Texture on Precise Depth Judgements in Different Visual Cue Conditions
For egocentric views of DVEs requiring precise depth judgements, certain visual depth cues may vary in their effectiveness (Surdick et al. 1997). Westerman and Cribbin (1998) found that there was no difference between texture and luminance which were both considered as providing qualitatively similar depth information (Landy et al.

1995). However texture has been identified as a visual cue that can also provide spatial orientation (Cutting and Millard 1984, Wanger et al. 1992). In addition to this, Nagata (1991) illustrated that texture was slightly more effective across viewing distances than luminance. It is therefore proposed that although texture is a relatively weak visual depth cue, it may still provide precise depth judgements for smaller distances where the dominant visual cues may be inaccurate. To identify the effectiveness of texture when combined with a qualitatively different cue, the following hypothesis is tested: *H2 Precise depth judgements are improved when texture is added.*

4.3 Effect of Motion Parallax on Precise Depth Judgement for Egocentric Views

Wanger et al. (1992) found that motion did not have a significant effect when positioning probes within a DVE that was viewed egocentrically. They believed that the motion confused participants and this effected their ability to perceive the cue accurately. However, it could also be suggested that participants did not need such a dominant cue for the egocentric view of the virtual objects. For instance, Bradshaw (2000) found that motion did not have any effect on the depth matching tasks that were undertaken in the real world and Delucia (1991) found that motion presented inaccurate distance perceptions of two virtual objects, if the ground intercept cue was not added to the objects. To identify the effectiveness of static representations of DVEs viewed egocentrically, the following hypothesis is tested: *H3 Depth judgement accuracy does not significantly improve with motion.*

5. METHOD

32 participants (16 male and 16 female) volunteered from the first year undergraduate and masters student population of Brunel University, UK. They completed a test of visual acuity to ensure that their acuity was normal or corrected-to-normal. Participants performed an experimental task that involved matching the z-axis distances of two probes. This was an interactive task that focussed on measuring participant's depth perception ability and has commonly been used in previous depth matching experiments (Delucia 1991, Westerman and Cribbin 1998). Two grey probes of varying sizes were presented against a red background. The probes were randomly separated in position in the x-axis and maintained the same y-axis positions. The probes differed in respect to their initial positions whereby the interactive, target probe was either in front of, very close to, or behind the static, reference probe. Participants were required to match the z-axis positions of the probes and then they pressed the space bar to record the measurement and were accordingly presented with the next trial. Luminance was illustrated by positioning a virtual light source near the participant which meant that as the probe was moved closer towards to the participant, it became lighter. Participants assessed the relative brightness of the probes in order to judge their respective depths. Texture was applied as a checkerboard bitmap to the probes. As the interactive, target probe was moved closer towards the participant, the checkerboard pattern increased in size. Relative height was illustrated by the target probe appearing to rise higher as it was moved backwards from the participant and descend as it was moved closer towards the participant. Both the probes swaying illustrated motion parallax, and as the participant moved the target probe forward, the swaying movement increased accordingly. The probes were viewed egocentrically and participants undertook 20 trials on each condition. The DVE graphics were created using Wild Tangent software and the graphics were presented using Viglen P3 400 hardware.

All participants completed sixteen conditions consisting firstly of the four cues in isolation, then six combinations of the cues in pairs, then four conditions of the cues in combinations of three and finally a combination with all four cues together. Each condition comprised of 20 trials. However, for the purpose of testing the three hypotheses, only certain conditions were analysed. The first set investigated the impact of probe positioning under four conditions: (1) with and without luminance; (2) with and without texture; (3) with and without height; and (4) with and without motion. The second set investigated the effect of texture on conducting precise depth judgements in three conditions: (1) with and without luminance; (2) with and without height; and (3) with and without motion. The third set investigated the effect of motion when perceiving four conditions: (1) with and without luminance; (2) with and without texture; (3) with and without height; and (4) with and without a combination of luminance, texture and height. The dependent measure was the z-axis error of distance between the two probes.

6. RESULTS AND ANALYSIS

The dependent measure was the mean (absolute) error of distance between the two probes. Three further dependent measures were singlemean, doublemean and combinationmean which were calculate by averaging the performance of the single cues, double cues and combination cues, respectively. Data for the repeated measures ANOVA failed to meet the assumption of sphericity ($p < 0.01$) and therefore the Huynh-Feldt value was used instead. The analysis indicated that there was a significant difference between the three means $F(2, 62) = 18.12$; $p < 0.01$. Figure 1 clearly illustrates the improvement of depth judgement as qualitatively different cues are added to a combination.

Figure 1. Difference between Single, Double
visual cue condition.

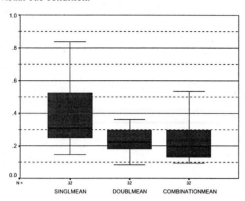

SINGLMEAN DOUBLMEAN COMBINATIONMEAN

Figure 2. Effect of Motion Parallax on combination
and Combination Depth Judgement Averages.

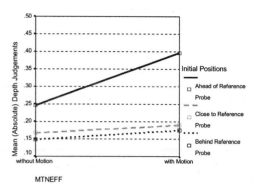

MTNEFF

For H1, there was a significant main effect for the three initial positions with and without luminance $F(2, 62) = 13.057$; $p < 0.001$. There was also a significant main effect with adding luminance $F(1, 31) = 82.83$; $p < 0.001$ and a significant interaction effect between the initial positions and adding luminance $F(2, 62) = 8.376$; $p < 0.05$. There was a significant main effect for the three initial positions with and without texture $F(2, 62) = 12.202$; $p < 0.001$. Main effects analysis was also significant for the addition of texture $F(1, 31) = 57.066$; $p < 0.001$ and the interaction effect was significant $F(2, 62) = 6.618$; $p < 0.05$. The analysis for main effects also showed that initial positions $F(2, 62) = 12.22$; $p < 0.001$ and the addition of relative height $F(1, 31) = 66.67$; $p < 0.001$ were significant. The interaction effect for initial positions and height was also significant $F(2, 62) = 6.853$; $p < 0.05$. Main effects for the initial positions in the motion parallax condition was significant $F(2, 62) = 16.58$; $p < 0.001$, as was the addition of motion cue $F(1, 31) = 56.21$: $p < 0.001$, and the interaction between the two $F(2, 62) = 4.04$; $p < 0.05$. Therefore, in all single cue conditions, results indicated that initial positions were significantly effective, supporting the hypothesis.

H2 examines the effects of adding texture to different cue conditions. ANOVA main effects indicate that there were no significant main effects for the luminance condition or for the height condition. However, there is significant main effects for the motion parallax condition $F(1, 31) = 8.447$; $p < 0.05$. For the combined cue condition of luminance, height and motion parallax, there were no significant main effects. Therefore results supported the hypothesis only for the motion parallax condition.

For H3, repeated measures ANOVA was used to test the effect of adding motion to single cues; luminance, texture and height and to a combination of the three. There were no significant main effects for motion parallax added to the luminance, texture and height cues respectively. However, when motion parallax was added to the condition with luminance, texture and height in combination, main effects analysis was significant $F(1, 31) = 4.492$; $p < 0.05$. Figure 2. suggests that motion parallax does not improve depth judgements in this instance, supporting H3.

7. DISCUSSION
Results from this study seem to suggest that, for egocentric views of precise depth judgements made within DVEs, the initial positioning of probes has a significant effect, texture only improves depth judgement when motion parallax is present and motion parallax only improves depth judgement in a combined cue condition. Since motion parallax had been acknowledged as a dominant visual cue (Wickens et al. 1989), this study attempted to determine the effect it had on precise depth judgements when viewing probes within DVEs, egocentrically. Previous studies had concluded that motion parallax did not make a significant difference, however their results did not distinguish between egocentric and exocentric views (Wanger et al. 1992). Certain studies have used the egocentric view for the depth matching tasks (Hendrix and Barfield 1995, Westerman and Cribbin 1998). Results in this study have shown that motion does not significantly improve depth judgement for egocentric views, broadly supporting previous results (Delucia 1991, Bradshaw et al. 2000). This suggests that static displays may be more effective for precise depth judgements that are made in DVEs that are egocentrically viewed. Hubona et al. (1997) found that

controlled motion was more effective and the results of this study may partially support that idea since continuous motion was used for the experimental task. Further investigations would reveal whether controlled motion improves precise depth judgements for egocentric views and whether texture, or other weaker visual cues, provide detail for virtual distances that may be lost through motion parallax.

This study concentrated on precise depth judgements made within DVEs that were viewed egocentrically. This was in an attempt to recognise the relation between visual depth cues to provide the appropriate illusion of depth necessary for various visual search tasks. Wickens (2000) conducted a study that investigated the visual perceptiveness of a DVE, which was viewed egocentrically and exocentrically. He concluded that due to the limitations of both types of views, visual search tasks in the context of flight simulation would benefit from having more than one view being presented simultaneously. It may however be possible to identify the visual search tasks in terms of their broad objectives such that they are 'spatial instruments' presenting 'spatial displays' that may require varying depth judgements. It may then be possible to define the characteristics that would support the visual search tasks, in terms of the ideal display views, the prospective level of interaction, the required amount of motion and most effective combination of visual depth cues. The salient feature of DVEs is that it is an interactive virtual environment and therefore views can be manipulated to perceive varying amounts of detail, at any given time.

REFERENCES

Bradshaw, M.F., Parton, A.D., and Glennerster, A. (2000). The task-dependent use of binocular disparity and motion parallax information. *Vision Research, 40*, 3725-3734.

Cutting J.E. and Millard R.T. (1984). Three gradients and the perception of flat and curved surfaces *Journal of Experimental Psychology: General, 113*, 198-216.

DeLucia, P.R. (1991). Pictorial and motion-based information for depth perception. *Journal of Experimental Psychology: Human Perception and Performance, 17*, 738-748.

Dosher, A.D., Sperling, G., and Wurst, S.A. (1986). Tradeoffs between stereopsis and Proximity Luminance Covariance as determinants of perceived 3D structure, *Vision Research, 26*, 973-990.

Eyles, D.E. (1991). A computer graphics systems for visualising spacecraft in orbit, *Pictorial Communication in Virtual and Real Environments*, (2nd ed.). Ellis, S.R., Kaiser, M., and Grunwald, A.J. (Eds.). Bristol, Pa: Taylor & Francis.

Hendrix, C. and Barfield, W. (1995). Relationship between monocular and binocular depth cues for judgements of spatial information and spatial instrument design, *Displays, 16*, 103-113.

Hubona, G.S., Shirah, G.W and Fout, D.G. (1997). The effects of motion and stereopsis on three-dimensional visualization, *International Journal of Human-Computer Studies, 47*, 609-627.

Hubona, G.S., Wheeler, P.N., Shirah, G.W and Brandt, M. (1999). The relative contributions of stereo, lighting, and scenes in promoting 3D depth visualization, *ACM Transactions on Computer-Human Interaction, 6* (3), 214-242.

Jacobs, R. A. & Fine, I (1999). Experience-dependent integration of texture and motion cues to depth, *Vision Research, 39*, 4062-4075.

Landy, M.S., Maloney, L.T., Johnston, E.B. and Young, M. (1995). Measurement and modelling of depth cue combination: in defense of weak fusion, *Vision Research, 35*, 389-412.

Nagata, S. (1991). How to reinforce perception of depth in single two-dimensional pictures, *Pictorial Communication in Virtual and Real Environments*, (2nd ed.). Ellis, S.R., Kaiser, M., and Grunwald, A.J. (Eds.). Bristol, Pa: Taylor & Francis.

Ramachandran, V. S. (1988). Perception of shape from shading. *Nature, 331*, 163-166.

Surdick, R.T., Davis, E.T. and Hodges, L.F. (1997). The perception of distance in simulated visual displays: A Comparison of the effectiveness and accuracy of multiple depth cues across viewing distances. *Presence, 6*, 513-531.

Wanger, L.R., Ferwerda, J.A. and Greenberg, D.P. (1992). Perceiving spatial relationships in computer-generated images. *IEEE Computer Graphics & Applications, 12*, 44-58.

Westerman, S.J. and Cribbin, T. (1998). Individual differences in the use of depth cues: implications for computer- and video-based tasks. *Acta Psychologica, 99*, 293-310.

Wickens, C.D. (2000). *The When and How of Using 2-D and 3-D Displays for Operational Tasks.* Human Factors and Ergonomics Society Annual Meeting, San Diego, CA.

Wickens, C.D. and Hollands J.G. (2000). *Engineering Psychology and Human Performance* (3rd ed.) New Jersey: Prentice Hall.

Wickens, C.D., Todd, S., & Seidler, K. (1989). *Three Dimensional Displays: Perception, Implementation, and Applications* (CSERIAC Rep: CSERIAC-SOAR-89-001). Ohio: Wright-Patterson Air Force Base.

HCI Challenges In Designing for Users with Disabilities

J.C. Muzio and M. Serra

Department of Computer Science
University of Victoria
Victoria, BC Canada
jmuzio@csr.uvic.ca, mserra@csr.uvic.ca

ABSTRACT. Human-computer interaction (HCI) specialists propose methodologies for building interfaces to software applications. HCI emphasizes developing a deep understanding of user characteristics and tasks. As important as HCI is for designing programs used by the general public, it is even more important if the target audience is disabled. We designed software to be used in the Traumatic Brain Injury Program at a local hospital to assist the cognitive rehabilitation of patients with brain injuries. We gathered feedback from therapists and patients in the hospital environment to refine our product. The reactions of all users determined how successful we were at designing a program with such stringent requirements.

1. INTRODUCTION

The rehabilitation of Acquired Brain Injured patients at the Gorge Road Hospital (Victoria, B.C., Canada) has involved computer software as one component in the process. Original software was developed many years ago and contained the basic functionality for a set of ten activities, used either as part of the rehabilitation process or for assessment. No graphical interface was present and the software, while considered to be effective, presented itself no better than paper-based exercises. A new program, Indigo, was created, which incorporates all the activities and presents them in an effective manner, together with tools for scoring and feedback. Therapists need the system to evaluate, monitor, rehabilitate and for measurement purposes. The software needs to be repeatedly used and enjoyed both by the patients and by the therapists. The challenges to writing software to be used by such disparate groups are significant, particularly given the wide range of severity of the injuries suffered by the patients.

The activities were analyzed and restructured; they were significantly enhanced and new features added. The principles of HCI, the study of cognitive psychology, the characteristics of the patient population, and strict software engineering guidelines were all incorporated into the design process of Indigo. The therapists involved in the rehabilitation program were questioned to determine the results that were most important and represent the patients' abilities. Finally three sequential versions of the software were delivered and have been tested as extensively as possible. At the moment the software is ready to be distributed to other similar centers.

We report here on the process of developing Indigo from the point of view of the HCI issues as they relate to this special set of users: the disabled patients and the supporting therapists. While many items can be considered mainstream in the literature, their importance was enhanced by the special needs of the user group.

2. REQUIREMENTS

HCI is user-centered and accentuates the need for designing the visual representation of the program before focusing on the computing itself. A good interface design is extremely important for this group of disabled patients and their supporting therapists. This hospital unit is for Acquired Brain Injuries, which are all types of brain injuries acquired as the result of disease processes, ranging from MS (Multiple Sclerosis), strokes, and aneurysms, to traumatic brain injuries (typically as a result of motor vehicle accidents). As the software is part of a patient's therapy, they have no choice but to use it. Since the goal is to assess and retrain cognitive functions which worked normally before the brain injury, the patients are prone to suffer from very high frustration levels and poor concentration, so simplicity of use is critical and a primary objective is to spare the cognitive functions of the patients as much as possible by designing an easy system. Patients also suffer from a lowered threshold of indignation, which means that, if they do not enjoy using the program, they will quit early. On the other hand, the therapists are prescribing the program and are responsible for setting it up and monitoring the patient's progress, thus they also have to feel comfortable. There are no alternative programs available, so we were compelled to develop an excellent product that could be used by other rehabilitation institutions as well as by patients working on their own.

These patients typically exhibit deficits in the following areas:
- Memory - specifically new learning is affected.
- Attention and concentration - sustaining attention during a task, shifting attention from one task to another and dividing attention.
- Executive functions - analyzing and synthesizing information, sequencing and goal directed activities.
- Perception - difficulties in auditory and visual functions, recognition of objects.
- Language abilities - quality and quantity of communication.

The rehabilitation of patients varies widely as each patient has different deficits and brings a wide range of previous skills and experiences to the process. In general, the rehabilitation process lasts from four to six months. A typical patient needs an assessment to determine the main problems and to chart the course of therapy. Therapy then includes a number of activities, of both remedial and learning types, to compensate for their deficits.

Issues that are common to almost all brain injuries are lessening of cognitive function, impaired accuracy of responses and a deterioration of the speed of responses. As well problems that are less universal, but occur frequently in the general population with disabilities, are loss of attention and concentration, learning and memory deficits, integrative thinking, planning and organizing, and lack of control over emotions and behaviour. This means that new information cannot be presented quickly, in a complex manner or in competition with other information. There are decreases in executive functions, reasoning and judgement. It is difficult for a brain-injured person to approach new situations. There are difficulties with sequencing and with operations that require multiple simultaneous decisions. Some patients are also severely challenged with physical effects. Reduced fine motor coordination and dexterity prevent some patients from using a mouse successfully so we have to allow a seamless combination of mouse and keyboard controls.

Study of the principles of cognitive psychology, HCI and software engineering helped to determine the overall principles of good program design for the general population. Those identified principles were then examined with respect to this specific patient population. There was an increased emphasis placed on ease of use over all other considerations, as it was determined that simplicity of design was of paramount importance.

The range of disabilities and deficits encountered by people with brain injuries were considered and taken into account during the design process. An assessment of the patients involved in the ward at that time was undertaken, the information collected included: age, time since injury, computer usage before injury, characteristics and education before the injury, cognitive problems, emotional and behavioral problems and physical problems.

3. DESIGN ISSUES

We tried to design software that had consistent and limited options. The limitations learned from the brain injured users gave us insights towards more general HCI paradigms for people with disabilities which overlap in some of the effects, whether they be psychological or physical.

Primary Design Issues That Were Identified:

- Use bigger graphic elements i.e. fonts, buttons, icons etc.
- Very few colours, clearly *distinct* from one another.
- Sound is used to reinforce the visual information, but used very sparingly.
- Minimize the quantity of information that must be remembered from one screen to the next.
- Use familiarity and imagery for things that must be remembered.
- Reduce the normally suggested number of maximum elements on a screen from 7 ± 2 to 4 ± 2; this was a crucial point.
- Direct users' attention by structuring and grouping elements.
- Avoid simultaneous tasks.

- Avoid lengthy written information.
- Offer a narrow and shallow decision structure with few choices for options.
- Account for patients who cannot use the mouse or part of the keyboard due to motor impairments.
- Minimize the number of gross motor movements e.g. back and forth between mouse and keyboard.
- Minimize the number of transitions between gross and fine motor movements.
- Avoid situations in which the user feels 'trapped' in a screen which can trigger severe frustration.
- Keep things simple.

4. METHODOLOGY

Our program is composed of ten activities. The main new strategy was to use a game metaphor, with an accompanying game board as visualization, to present the activities and to make it easy for the user to understand and navigate the software. A short summary of the activities can be found in the Appendix.

4.1 Analysis of the existing software

The ten activities that were in use at the hospital were analyzed in depth: their functionality (to derive well defined software engineering specifications); their interface characteristics (to derive the HCI specifications); their cognitive psychology content (to make sure that only enhancements to the original rehabilitation goals would be introduced). The instruction screens, activities and scoring were all recorded. The original activities lacked a GUI and were entirely text based. For each activity observed, problems were noted, such as:
- ° the patient having to remember the name of a saved file,
- ° feedback not being displayed long enough,
- ° activities that can't be stopped before completion,
- ° meaning of options not explained,
- ° instructions not clear or inaccurate.

Fixing these problems as well as enhancing the activities was given emphasis during the design process and provided a starting point for the design.

4.2 Survey of Therapists

The therapists asked for meaningful scores for each exercise. The scores reflected both a short term use for the assessment of the progress of the patients and also a long term goal of being able, in the future, to collect valid statistical data on a large group of patients. It should be remembered that the software is used both for continuous rehabilitation on a daily basis - thus the obvious need for incremental scores with real time feedback and reinforcement - as well as for monthly assessment of some skills (e.g. memory) – requiring comparison tables of scores. For each activity, the most important information on the patient's performance was identified in consultation with the therapists. In general, the system should compute and save precise scores for all activities and then store the data in a form readily available for consultation. The patient should be able to choose how many times the exercise is to be executed. In addition, the therapists were given a questionnaire regarding new features and the activity of the icons and buttons.

4.3 Evolution of the program

Once a beta version of the program was completed, it was installed on two of the computers at the Gorge Road Hospital. The therapists were given a demonstration and were walked through each activity. The therapists went through one activity at a time becoming familiar with it, and they started a notebook for comments and questions on each activity. If they had any desired changes, there were discussed to clarify what exactly was wanted, and then the Indigo program was updated. After significant changes were completed, a new build was installed at the hospital. Once the therapists were comfortable with an activity they would introduce patients to it. The therapists were able to get initial feedback from their patients that had the most patience and highest functioning level. The characteristics of the patient population unfortunately made it impossible to administer a formal survey. Other patients made comments on their own, or the therapists made note of the areas that gave them difficulty, the functions they did not understand and the patients response to the program. Over time, all of the improvements and clarifications were incorporated into Indigo, and it was ready for more extensive testing.

4.4 Navigating through the program

At the opening of the application, the users type their name and, as with all subsequent screen, the focus can be changed using either the mouse or the space bar on the keyboard. Once the Indigo game board is loaded, the ten activities are depicted in a very simple fashion by a game board with squares surrounding a central panel (similar to a Monopoly board). This game metaphor proved to be a winning strategy, as it combined the ease of use, the simplicity and the effectiveness of functionality which were deemed critically necessary (see above), and enabled the interface issues to maintain a consistency throughout, eliminating a major source of frustration. Moreover, the game board itself provided a connection of personal familiarity for patients and reduced the computer anxiety.

When a user chooses an activity, a screen of instructions appears at first - unique for each activity, yet always completely consistent in their layouts. The option choices, clearly shown on the game board for each activity, can be different, as some games are so intensive that they allow only one repetition, while others can be customized, together with the therapist, for speed, repetitions, length, difficulty. The "Back", "Help", "Exit", "Scores" buttons are always in the same locations on the board and directly accessible, and are integrated within the functionality. For example, if the Help button is pressed while an activity is running, the timing clock continues to count. This implies that the patient's time score will be higher, a perfectly valid case. As the patient progresses in the rehabilitation, the Help button will be used less often, improving the score, and avoiding the patient's attention from wandering during the pause in the activity.

4.5 Direct feedback

The therapists took extensive notes on the daily use of the system and some patients were also able to comment. The main changes made as a result of feedback were in the following categories:
- Color intensity – the contrast was fine, but the choices were changed slightly to less brighter shades.
- Fonts sizes and their ease of adjustment – extremely useful for vision problems.
- Range of difficulty levels – the ease of adjustment through the interface enables patients to make the changes as well as the therapists.
- Game metaphor – easily understood by all, even with the most varied background of users.
- Help and Instructions – their uniformity was enhanced even more.
- Scoring feedback – many changes were made here providing the easiest possible interface to both patients and therapists to check scores at any point in time, with a fine or coarse grain window, which was truly appreciated. For example, when an item was missed during a number search activity, the position as well as the item were emphasized in a non intrusive fashion, yet allowing the therapist to judge quickly whether patients were having trouble scanning from a particular side of their vision.

The main feedback notes were that patients actually had to be allowed, upon request, to continue activities beyond the allotted time, as they found the pleasant interface an enormous change from their regular expectation; most of all, many patients were able to use the software on their own after a few sessions, something never achieved before. It is hoped that this new level of enthusiasm will enhance their recovery process and make it rather less frustrating than formerly.

The program has been accepted with enthusiasm by both the therapists and patients. The program required alterations, but the design process was successful in creating a product that met the needs of the therapists and patients. The patients seem to enjoy using the program and are able to navigate through it easily for the most part. The figure in the Appendix shows 2 screen shots: the left one is of the main game board, while the right one is of the Number Search activity (they need color to look appropriate).

5. CONCLUSION

In conclusion, the process for the creation of Indigo, while long and involved, was necessary and effective. The careful study of the HCI interface design issues relative to the needs of the special group of users is what made the product successful, beyond what the simple functionality could ever have provided. The therapists who have used the program have responded positively to it. Several comments have highlighted the constructive and beneficial aspects of the program for the patients. It has required a considerable amount of work on the therapists' behalf to complete this process, from the beginning consultation and lengthy questionnaire to the discussions during the beta-testing phase, and their total involvement in the development of the enhancements.

6. REFERENCES

Collins, D. (1995) <u>Designing Object-Oriented User Interfaces.</u> Benjamin-Cummings.

Kapor, M. (1990) <u>A Software Design Manifesto,</u> [On-Line]. Available: www.kapor.com/homepages/mkapor/ Software_Design_Manifesto.html.

Laurel, B. (1990) <u>The Art of Human-Computer Interface Design,</u> Addison-Wesley.

Mullet, K. (1995) <u>Designing Visual Interfaces: Communication Oriented Techniques.</u> SunSoft Press.

Marion,, D.W. (1999) <u>Traumatic Brain Injury.</u> Thieme Medical.

Winograd, T. (1996) <u>Bringing Design to Software.</u> ACM Press, Reading, Mass. And Addison-Wesley.

7. APPENDIX: Indigo activities

1. **Logic Sequences:** pick the next logical step in a sequence.
2. **Circle Chase:** keep a smaller circle inside a bigger circle while the outside bigger circle moves.
3. **Mirror Image:** find objects on one side which are a rotation or reflection of objects on the other side.
4. **Vision Drill:** as two vertical bars move towards each from the outside towards the center of the screen, at different speeds, one must stop them at the precise moment when they meet.
5. **Count the Shapes:** in a grid with many shapes of different colors, one must count the objects of the same shape, or same color, or both.
6. **Missing Number:** numbers are scrolled in a set sequence with one missing which has to be identified.
7. **Number Search:** in 4 quadrants, numbers are scattered, and one must find and give the quadrant position of a given number.
8. **Reaction Time:** react as quickly as possible to an auditory or visual event by mouse or keyboard clicking.
9. **Towers of Hanoi:** the typical problem of moving some blocks of decreasing shapes from one peg to another, using only one extra temporary peg and maintaining the order of bigger shapes under smaller shapes.
10. **Memory Challenge:** letters or numbers are displayed on a grid, later covered; one is asked to remember where an item was located.

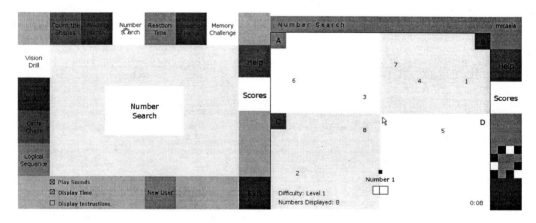

MATHAINO: SIMULTANEOUS LEGACY INTERFACE MIGRATION TO MULTIPLE PLATFORMS

Rohit V. Kapoor, Eleni Stroulia

Department of Computing Science, University of Alberta, Edmonton, Alberta, T6G 2E8, Canada
{kapoor, stroulia}@cs.ualberta.ca

ABSTRACT

With the proliferation of different computing devices and platforms, it is becoming increasingly important for organizations to migrate their existing software systems to new environments, possibly more than one, with minimum effort and risk. In this paper, we introduce Mathaino, a platform independent, non-invasive tool that can be used to migrate, in a semi-automated manner, text-based legacy interfaces to modern web-centric target platforms, by constructing wrappers as front-ends around the original legacy-system user interface.

1. INTRODUCTION

Migration of legacy systems to new computing platforms and paradigms is one of the most challenging problems in today's IT industry and it has been the subject of a lot of software engineering research. Several proposed approaches (Bisbal 1999, Merlo et. al. 1993, Quilici 1995) advocate the construction of simple wrappers to "call" modules of the legacy system or the complete migration of the legacy system to the new platform. However, both approaches have major drawbacks. With time, the new system, constructed as a result of the full-scale migration process, ages into another legacy system forcing the user to start another migration. As this task consumes immense time and human resources, the process of re-migration becomes quite expensive; the ever-increasing rate of innovation in the field of information technology has only made the situation worse. On the other hand, the major problem with the "module wrapping" solutions is risk mitigation. Often legacy systems are expensive systems that form the core of business model of their respective organizations. Thus, invasive solutions that require modifications to the legacy code, or even simpler wrapper solutions that require any modification to the existing system setup, are sometimes too risky to implement.

Mathaino is a platform independent, non-invasive tool that can be used to migrate text-based legacy interfaces to modern web-centric target platforms, by constructing wrappers as front-ends around the original legacy-system user interface. In addition, Mathaino aims to further automate the legacy migration process by supporting the simultaneous migration to multiple platforms. In the rest of this paper, we briefly describe the architecture of Mathaino, its design-time environment and the wrapper construction process, and its run-time components. We conclude with some of the advantages of this approach over other competing technologies.

2. THE OVERALL MATHAINO ARCHITECTURE

Mathaino has been designed according to a layered architectural pattern. Mathaino utilizes two Java Virtual Machines (Java 1 and Java 2) and the Java ODBC API in its operational platform. The Mathaino base system, which forms the next layer, provides a plugin framework, over which the rest of the system is built. The various Mathaino subsystems plug into this base system. This architecture makes Mathaino quite extendable, in that, additional capabilities can be added as plugins to the base system. Further, the use of the layered architecture pattern, coupled with the use of Java and XML makes the entire Mathaino system very portable. Currently, versions of Mathaino exist for Windows NT/98/95, LINUX, and Solaris platforms.

There are two types of Mathaino plugins, (a) design-time components and (b) run-time components. Mathaino's design-time components are integrated into an environment that is used by the developer to analyze and model the legacy interface. The produced model provides the basis on which the run-time environment re-implements for the legacy system a new user interface for each one of a variety of new platforms. This decomposition of the various Mathaino components enables Mathaino to port the legacy systems in a platform independent manner. While the run-time components of Mathaino are platform specific, to implement a user interface for the legacy system on a particular target platform, the design-time components are completely platform independent. The various components that comprise these two environments are discussed further in the following sections.

51

3. THE DESIGN-TIME SUBSYSTEM

The Mathaino IDE has been built around a component-based architecture. Thus, all the design-time components exist as separate objects that plug into the main IDE. Mathaino's design-time components include (a) a set of *template* plugins, (b) an *analyzer* plugin, (c) a *domain-modeling* plugin, and (d) an *abstract UI generation* plugin. Mathaino builds on our earlier work on the URGenT system and the first three types of plugins perform transformations similar to the URGenT task-analysis step, described earlier (Kong, Stroulia, Matichuk 1999). In this paper, we focus on the abstract user interface generation process at design time and the run-time interpretation of this abstract user interface by Mathaino's platform-specific run-time components.

The first three components of the design-time environment take as input a set of task-specific traces, i.e., traces recorded by users of the legacy system performing several different instances of the same task on the legacy interface. Mathaino analyzes these traces comparatively to construct an abstract model of (a) the navigation of the user through the legacy system screens during the performance of the task at hand, (b) the types of information input by the user to the legacy interface and displayed by the system interface to the user throughout this navigation, (c) the domain of values for these types of information, and (d) the interdependencies among these information values.

Mathaino improves on the analysis capabilities of URGenT in two respects. First, it is able to formulate patterns to describe the location of different information elements displayed on the legacy screen, relative to landmark labels on the screen, instead of relying solely on the coordinates of the displayed elements. It is, therefore, able to analyze an increased class of screens including static and dynamic – but well-structured – screens. Second, given sufficient examples, it is able to analyze non-deterministic tasks, that is, tasks whose instances may require navigation through any one of a number of alternative navigation paths through the system screens.

At the end of the analysis phase, the abstract user-interface generator is invoked. This component is responsible for defining the new, front-end, interface for the legacy system task, in terms of a set of abstract forms. Each abstract form corresponds to a set of screens of the legacy system. For each abstract form, the user-interface generator produces a plan for navigating at run-time through the legacy screens corresponding to the abstract form. The complete set of forms defines a new user interface, that can be used as a front-end to the original legacy interface and through which the user can accomplish the given task.

The abstract user-interface generator produces automatically the specification of a new user interface for the legacy system, based on model-based user-interface design heuristics (Vanderdonckt 1995). To that end, it uses the model produced during the earlier analysis phase. As mentioned previously, each abstract form is a logical grouping of legacy interface screens. Thus, the first issue for the abstract user interface generation process is to decompose the set of screens into logical groups, each one corresponding to a distinct abstract form. We have developed a set of heuristics based on the observed information flow, from the user to the legacy interface and from the legacy interface to the user, to produce this correspondence between legacy screens and abstract user interface forms.

A trace of a particular task on the legacy system consists of a linear sequence of screens. The first step is to identify the starting screen for each form. To calculate these division points of the screen navigation, the abstract user-interface generator assumes that the value of each output type of information that has been identified previously by the user is essential for inferring the set of values of the required input information occurring after it. Thus, Mathaino starts a new form as soon as it encounters any screen that contains a required input information field, preceded by a previously identified output information field.

After Mathaino has created the list of starting screens for the various forms according to the previous heuristic, the next step is to normalize this list. The factors that must be accounted for are:
   a.   The first screen of the trace must be the starting screen for the first form. In case it has not already been included into the list it is added now.
   b.   To cleanly exit the legacy system it is necessary that the last screen of the trace be the terminating screen for the last form. Thus, in case it has not already been included into the list, it is added now.
   c.   Also Mathaino makes sure that all fields identified as output fields in the analysis phase are displayed on some form in the new UI. Thus, if any screens containing output fields have been ignored, Mathaino creates a new form to accommodate their presentation.

After the original legacy screen navigation has been decomposed into a set of abstract forms, two more tasks have to be accomplished by the design-time environment of Mathaino: first, to select the appropriate widgets for receiving and display the information fields associated with each form, and second, to decide the relative layout of these widgets on the form.

Mathaino proposes an abstract widget for each input and output information field, based on a simple set of heuristics. For example, for fields with an enumerated range of values Mathaino chooses the combo box if the number of possible choices is more than two and a set of radio buttons otherwise. Similarly fields with values in open-ended domains are assigned a simple text box widget by default. However, the user can override these settings and choose another widget for the input field if he so desires. For example, a very common task is to replace the simple input box widget with a password widget for password fields. The run-time Mathaino components, responsible for interpreting the forms on a specific target platform, will supply a widget in the target platform that most closely matches the abstract input widget selected by the user.

Mathaino lays out each form in a tabular manner. The user is able to customize this layout pattern as follows:
a. The user can change the number of layout columns. Mathaino automatically rearranges the input widgets in accordance with the new layout columns.
b. For output fields, the user can change the wrap length for each output field. All the data for that field are then wrapped at the new wrap length. This is helpful as many text based legacy systems rely on a particular wrap length to generate a tabular layout of data. However, Mathaino might forsake this setting for devices whose physical screen is not capable of showing the requisite number for characters per line.
c. The user can customize the labels for both output fields and input widgets, and can also choose various alignments, relative to the input and output fields, for this text.

After the user is satisfied with the abstract user interface, i.e., the selected widgets and their layout, Mathaino generates an XML representation for its specification. The run-time interpreters of Mathaino are responsible for interpreting this XML specification of the abstract user interface into the actual platform-specific target user interface. As the basic Mathaino subsystems are platform independent, the interpreters are also responsible for interpreting platform-specific messages received via the new user interface back into a platform independent form that can be understood by the back-end Mathaino navigation subsystems. In this manner, the user can create a single, unified user-interface specification for a task in the legacy system, which can then be implemented simultaneously on multiple platforms.

4. THE RUN-TIME SUBSYSTEM

The Mathaino run-time environment consists of two subsystems. The front-end interpreters are responsible for interpreting the abstract user interface specification on the new platform and interacting with the user, displaying and receiving information. The back-end remote system navigator is responsible for enacting the navigation on the legacy interface, i.e., forwarding the user input, as received by the front-end interface, to the appropriate screens and fields of the legacy interface and retrieving the expected user output from the legacy interface to forward it to the front end.

4.1 The Remote System Navigator

The remote system navigator has been built over the Celware Legacy Connector, a sophisticated terminal emulator developed by Celcorp (Celcorp), the industrial sponsor of the CelLEST project (CelLEST, Stroulia et. al 1999) of which Mathaino is a subsystem. Celcorp has been involved in the process of legacy application integration for the past decade and is being recognized as a leader in the industry. They have developed a state-of-the-art suite of intelligent tools for legacy-application modeling, and terminal data stream automation and integration with other applications running on heterogeneous platforms. The Celware Legacy Connector used by Mathaino's run-time environment is capable of sending keyboard input and receiving terminal output from a variety of legacy systems. Currently it supports IBM 3270, VT100 and HLLAPI (High Level Language API) terminals. This makes the majority of legacy systems accessible through it.

The remote system navigator of Mathaino includes a basic navigation API that forms a higher layer of abstraction over the Celware Legacy Connector API. Further, it includes a form navigator component that has been built on top of this basic navigation API. The form navigator can execute the navigation plan for any abstract form that has been produced by the abstract user interface generator. Any component, such as the platform-specific abstract user interface interpreters, that uses the form navigator is required to communicate the plan data to and from it in a platform independent manner. Therefore, the user-interface interpreters need to retranslate the platform specific messages into a form understandable by the back-end form navigator. The following two sections discuss the two platform specific interpreters that currently exist for Mathaino.

4.2 The XHTML Interpreter

XHTML is the next generation of HTML. Currently XHTML 1.0 has been approved as a W3C recommendation (XHTML). Essentially XHTML is a reformulation of HTML 4.01 in XML. Thus, any XHTML document is also a valid XML document. As most Mathaino components produce content in XML, XHTML was a natural choice for implementing the abstract user interface on platforms with web browsers. Currently, both Microsoft's Internet Explorer (versions > 5.0) and Netscape's Navigator (Mozilla MI18 or Navigator 6 PR 3) can parse XHTML web pages. Thus, the XHTML user interface produced by this interpreter can be executed on any computer that is capable of running any one of these web browsers.

Mathaino's XHTML interpreter maps the abstract user interface forms to appropriate XHTML CGI forms. The interpreter has been built using the Java Servlet API and it executes as a servlet on the web server. It dynamically parses the abstract forms and translates them into XHTML at run-time. It also parses the CGI response produced by the client web browser into the platform independent form required by the form navigator. The XHTML interpreter maps the abstract user interface widgets into appropriate CGI widgets. For example, the range fields are mapped to either a combo box widget or a set of radio buttons. The interpreter also uses XHTML tables to layout the web page in a format that closely resembles the format chosen by the user for the abstract UI. In this way, the XHTML interpreter can execute the interface, specified by the design-time abstract user-interface generator, on web browsers in a manner transparent to the end user. Web pages produced by Mathaino's XHTML interpreter have been validated by W3C as being 100% XHTML 1.0 compliant.

4.3 The WML Interpreter

Like XHTML, WML too is an XML based markup language. However, unlike XHTML, WML is targeted towards WAP devices like cell phones, PDAs etc. The WAP Forum (WAPF) has approved WML as a standard markup language for displaying web pages on wireless devices. In its present version, the WML platform presents several constraints to the user-interface development process. For example, the size of any WML web page (this can also be loosely referred to as a WML Deck) is restricted to a maximum of 1200 bytes. Further, input widgets like combo boxes, radio buttons, edit boxes are simply not available on WAP devices. Thus, any WML based form can only include simple text fields for keyboard input.

Hence, to translate the abstract forms to their corresponding WML cards, Mathaino has to make several adjustments. For example, in case of range fields Mathaino uses a simple menu that numerically lists each possible choice and for fields with values in an open-ended domain, it uses a simple text input field. Further, each abstract form must be translated to several WML decks if it exceeds the maximum limit of 1200 characters. Even with these platform specific restrictions, Mathaino makes an attempt to render the abstract user interface as accurately as possible on the target platform. For example, the input decks on the WML device always follow the output decks so that the user has a chance to see the output before he encounters the input fields. Also even though the maximum possible length of a deck is 1200 characters long, Mathaino makes sure that the user sees the entire contents of an output field by splitting it over multiple WML decks. The WML interpreter internally caches the user responses to the multiple decks that have been generated for the single abstract user-interface form. This allows it translate the WML responses back into a unified form that is understood by the form navigator component.

Hence, by translating the abstract user interface at runtime into a target UI, Mathaino can implement the legacy system on even highly restrictive platforms like WML. Further, as Mathaino uses the native user-interface widgets to render the forms, the user gets the same look and feel as any other native application for that platform.

5. RELATED WORK, EVALUATION, AND CONCLUSIONS

In this paper, we discussed Mathaino, a prototype tool for simultaneous legacy-interface migration to multiple target platforms, based on a model of the task-specific interaction between the user and the system, as extracted from recorded traces of this interaction.

Mathaino's approach to legacy interface migration offers several advantages over the alternatives:

a. Mathaino is a highly automated tool for task-specific legacy interface migration. Its degree of automation significantly reduces the development effort and risk for migrating existing systems to new platforms.
b. Mathaino addresses the problem of simultaneous legacy interface migration to multiple platforms, such as XHTML for a variety of web browsers, and WML for WAP devices. These two platforms have widely varying requirements, and demonstrate the flexibility of Mathaino's migration processes.
c. Mathaino is non invasive, that is it does not require any modifications to the original legacy interface or code, and as such, it reduces the risk of the migration effort.

We have evaluated Mathaino on several legacy systems, on which we have developed interfaces for different tasks. Some of these systems include the book-search task in the Harvard on-line library, a claims-processing task in the legacy system of an insurance company, and the email-reading task with the PINE email reader as our legacy system. Figure 1 illustrates a legacy system screen containing output data being rendered on an XHTML compliant web browser and a cell phone. Our successful experience of dealing with these systems demonstrates the robustness of the solution being proposed in this paper. Mathaino demonstrates a lightweight and risk free technique of migrating legacy systems that certainly provides an interesting alternative to previously proposed solutions.

Figure 1: (Left) A snapshot from the Harvard on-line library, displaying books with a given title; (Right) Snapshots from the same interface, as rendered on XHTML and WML browsers after Mathaino-based migration.

ACKNOWLEDGEMENTS

This work was supported by a generous contribution from Celcorp and a Collaborative Research and Development grant by NSERC 215451-98, and ASERC, the Alberta Software Engineering Research Consortium.

REFERENCES

Bisbal J., Lawless D., Wu B., and Grimson J., "Legacy Information Systems: Issues and Directions", IEEE Software, 16(5) September/October 1999, pp 103-111.

Celcorp, Internal Technical Documentation, http://www.celcorp.com.

CelLEST, http://www.cs.ualberta.ca/~stroulia/CELLEST.

Kong, L., Stroulia E., Matichuk, B., "Legacy Interface Migration: A Task-Centered Approach", 8th International Conference on Human-Computer Interaction August 22-27, 1999 Munich, Germany, pp. 1167-1171.

Merlo, E., Girard, J., Kontogiannis, K., Panangaden, P., and Mori, R. D., "Reverse engineering of user interfaces", Proceeding of the Working Conference on Reverse Engineering, Baltimore, MD, May 1993, pp. 171-179.

Quilici A., Reverse engineering of legacy systems: a path toward success, Proceedings of the 17th International Conference on Software engineering, April 24 - 28, 1995, Seattle, WA USA, pp. 333-336.

Stroulia, E., El-Ramly, M., Kong, L., Sorenson P. and Matichuk. B., "Reverse Engineering Legacy Interfaces: An Interaction-Driven Approach", Proceedings of the 6th Working Conference on Reverse Engineering, October 1999, Atlanta, Georgia USA. pp. 292-302.

Vanderdonckt, J., "Knowledge-Based Systems for Automated User Interface Generation: the TRIDENT Expierence" RP-95-010. Namur: Facultés Universitaires Notre-Dame de la Paix, Institut d'Informatique 1995.

WAPF, The WAP Forum, WML 1.1 DTD, http://www.wapforum.org/DTD/wml_1_1.dtd.

XHTML™ 1.0: The Extensible HyperText Markup Language, A Reformulation of HTML 4 in XML 1.0, W3C Recommendation 26 January, http://www.w3.org/TR/xhtml1.

A User Interface for Accessing to Information Spaces by the Manipulation of Physical Objects

Masanori Sugimoto
University of Tokyo
7-3-1 Hongo, Bunkyo-ku,
Tokyo, 113-0033, Japan

Fusako Kusunoki
Tama Art University / JST
2-1723, Yarimizu, Hachioji
Tokyo, 192-0394, Japan

Hiromichi Hashizume
National Institute of Informatics
2-1-2 Hitotsubashi, Chiyoda-ku
Tokyo, 101-8430, Japan

Abstract

In this paper, a system for accessing to information spaces through the manipulation of physical objects is described. We have devised a new type of sensor-embedded board that can integrate physical and virtual worlds: users can search information by placing pieces on the surface of the board, and are given visual feedback projected on the board through a LCD projector. This immersive environment allows users to access to information in an intuitive and easy manner. Through several preliminary user studies, we have been evaluating the system.

1. INTRODUCTION

In this paper, a system for accessing to information spaces through the manipulation of physical objects is described. Due to the recent development of information technologies, such as the Internet, it has become possible for us to access to almost infinite number of information resources all over the world immediately. Although we sometimes have difficulties in connecting to Web sites (slow network etc.), the problem we feel serious now is how to discover useful information from a vast mixture of useful and useless information. In our daily activities, we use a computer for working, learning and playing: we write documents, draw figures, and enjoy web surfing with a computer. It may not be difficult for us to use a computer. Even if we first struggle with a keyboard and a mouse for a moment, we soon become accustom to use them. But, are our computers actually easy to use for us? How about people who have not had chances to use a computer, for example, the elderly or infants?

We think that current computers are not well designed based on natural human cognition: we live in a physical world and interact with real objects. This may cause a serious problem for people who are not good at thinking of how they interact with a virtual world with a computer.

In this paper, we propose a system that enables users to access to information spaces and search information by interacting with physical objects. We developed a novel board-type input device by embedding sensors. The feature of the board is that it can rapidly recognize multiple objects on its surface. Users of the system can construct a query by arranging physical pieces. It then submits the query to an index server to retrieve information on Internet resources. Search results are visualized and projected on the board through a LCD projector. Users can access to and explore information spaces in an immersive environment where a physical

56

world of the board is overlaid with a virtual world generated by a computer. Through some informal user studies, we confirmed that users enjoyed the system as they would when they played a game.

2. SYSTEM CONIFIGURATION

2.1 Overview

Figure 1 shows an overview of the system. It composed of a sensor-embedded board, a personal computer (or a workstation), and a LCD projector. When physical pieces are placed on the surface of the board, a software module called *board control module* starts the recognition of their arrangement. Then, *query formulator* submits a query constructed based on the piece arrangement to an index server, and retrieve the result. *Visualization module* visualizes the result on the board through a LCD projector. Data transmission between the board and a computer is done through their RS-232C interfaces.

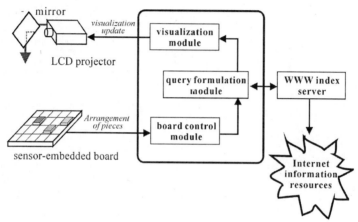

Figure 1. An overview of the system

2.2 A Board for Object Identification

In order to develop our multiple input device, we have so far investigated several approaches [6]. Through these investigations, we decided to use the RFID (Radio Frequency Identification) technology. RFID is a non-contact object identification and data transfer technology. The RFID system consists of two components: an antenna (with a transceiver and decoder) and a tag (Figure 2). An antenna emits radio signals to activate a tag, and writes data to the tag or reads data from it in the electromagnetic field produced by the antenna. An antenna combined with the transceiver and decoder is called a reader. It decodes data encoded in a tag's integrated circuit (IC) and passes the data to an attached computer. In our current implementation of the board system, 24 x 20 antennas are laid out like a checkerboard. When an object that embeds a tag is placed on the surface of the board, the board can recognize what objects exist and where they are on it, through the data transmission between the tag and an antenna.

Unlike using visual tags and a camera for object recognition, the RFID is a reliable and unobtrusive technology: the performance of data transmission between a tag and an antenna is stable, and a tag is small enough to make it invisible by embedding it in a piece. However, the data transmission speed between a reader and a tag is not very fast (10ms to 20ms). This may cause serious communication delays; for example, it will take 1 to 2 seconds to identify a hundred pieces. The authors devised a new method that reduces the processing time for data transmission. In the current implementation, one CPU is attached to a unit that comprises 4 × 4 squares as shown in Figure 4. When the board receives a command for detecting the arrangement of pieces on it from an attached computer, it first sends a read command to all the CPUs simultaneously through the interface CPU. Then, the CPU of each unit sequentially activates and controls 16 antennas in its unit, so that each of them activates a tag and reads its data. Finally, the board sends data about tags from the CPUs through the interface CPU. The architecture of the board enables parallel processing of communication between tags and readers, and makes the communication time theoretically independent of the number of units, or the size of the board. The time taken to acquire the arrangement of pieces on the current version of the board is less than 0.05 second [1][2][3][4][5].

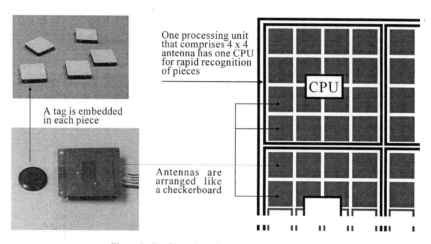

One processing unit that comprises 4 x 4 antenna has one CPU for rapid recognition of pieces

CPU

A tag is embedded in each piece

Antennas are arranged like a checkerboard

Figure 2. Configuration of a sensor-embedded board

2.2 A User Interface for Information Retrieval

A user interface for accessing to information spaces is visualized on the board, and composed of two parts: a palette and a workbench. On its palette, pieces to be used for users' query are placed. Data written in a tag which is embedded in a piece is projected on its surface. Users can pick up any pieces from the palette and move them to the workbench. When pieces are placed on the workbench, the system recognizes them as a part of a user's query. Based on the arrangement of pieces on the workbench, the system constructs a query formula, and submits it to a WWW index server as shown in Figure 1.

In order for users to construct a query formula in an intuitive manner, we implemented rules for expressing "AND" and "OR". As shown in Figure 3, when two pieces (data written in them are A and B, respectively) are neighboring with each other, the system interprets these pieces as "A AND B" (Figure 3(a)). When these pieces are isolated with each other, the system interprets them as "A OR B" (Figure 3(b)). In the case of Figure 3 (c), the system interprets the arrangement of pieces as "(A AND B AND C) OR (D AND E)".

When the system sends a query arranged on the board to an index server on the Internet, it then receives its search results, and visualizes the results around their own related pieces. We have implemented two visualization modes: (Mode-1) visualization based on scores, and (Mode-2) visualization based on Venn diagram. In Mode-1, scores given by the index server to each retrieved document are used for visualization: a document including a query term which is written in a piece is shown around the piece. When its score is high, an icon of the document is visualized close to the piece. In Mode-2, the system first calculates intersections and differences of a document set retrieved through a current query formula. For example, when the system accepts a query shown in Figure 3(c), it calculates an intersection and union of a document set retrieved through "(*A* AND *B* AND *C*) OR (*D* AND *E*)". Then, the system visualizes their icons around their related query pieces. A document itself appears when a new piece (not a query piece) is placed on an icon of the document.

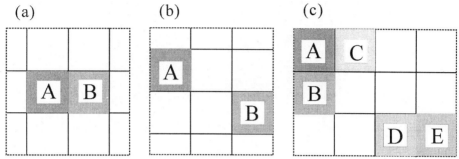

Figure 3: Query formulas that represent (a) "*A* AND *B*", (b)"*A* OR *B*", and
(c) "(*A* AND *B* AND *C*) OR (*D* AND *E*)", respectively

4. INFORMAL USER STUDIES

We have so far carried out informal user studies to evaluate the system (Figure 4), and received various feedback and comments. Most users told us that they could use and enjoy the system as they would play a game. Users repeatedly moved pieces and changed their queries. A query construction method proposed in this paper was acceptable to the users: they gave us a comment that the combination of manipulating physical pieces on the board and visual feedback of their manipulation was useful, and they could construct a query in an intuitive manner. The system was also used in a group setting: three or four users sit around the board and cooperatively search information. They simultaneously manipulated pieces and constructed a query. Each user interacted not only with the system but also with other users (discussing seriously or chatting). The fact that users can try the system in an easy manner and enjoy it is vital for people who are not accustomed to, or not well motivated to use computers.

5. CONCLUSIONS

This paper describes a system for accessing information spaces by manipulating physical objects. The system allows users to search information in an intuitive manner. We have devised a new type of sensor-embedded board that can quickly recognize different types of numerous objects. By projecting the visualization of information spaces on the board, the system can generate an immersive environment that integrates physical and virtual worlds. We are now evaluating the system and have received several feedback from users through preliminary experiments. Future research will be aimed at improving the system and carrying out more evaluations.

Figure 4. The system in use

Acknowledgement

This research project received support from Grand-in-Aid for Scientific Research funded by Japanese Society for the Promotion of Science, Precursory Research for Embryonic Science and Technology by Japan Science and Technology Corporation, and from the Telecommunications Advancement Foundation.

REFERENCES

[1] Kusunoki, F., Sugimoto, M., and Hashizume, H.: "Toward the Integration of Physical and Virtual Worlds for Supporting Group Learning", *Digital Cities: Technologies, Experiences, and Future Perspectives*, Springer-Verlag, pp. 224-235 (2000).

[2] Kusunoki, F., Sugimoto, M., and Hashizume, H.: "A Group Learning Support System that Enhances Externalization of Thinking", *Journal of Institute for Electronics, Information, and Communication Engineers*, Vol.J83-DI, No.6, pp.580-587 (2000) (In Japanese).

[3] Kusunoki, F., Sugimoto, M., and Hashizume, H.: "How Can CSCL Systems Support Pupils' Externalization?", In *Proc. of International Workshop on New Technologies for Collaborative Learning (NTCL2000)*, Hyogo, Japan, pp.109-118 (2000).

[4] Sugimoto, M., Kusunoki, F., and Hashizume, H.: "ePro: A System for Supproting Collaboration that Enhances Interactions", In *Proc. of IEEE Systems, Man, and Cybernetics 2000 Conference (SMC2000)*, Nashville, TN, pp.745-750 (2000).

[5] Sugimoto, M., Kusunoki, F., and Hashizume, H.: "Supporting Face-to-Face Group Activities with a Sensor-Embedded Board", In *Proc. of ACM CSCW2000 Workshop on Shared Environments to Support Face-to-Face Collaboration*, Philadelphia, PA, pp.46-49 (2000).

[6] Sugimoto, M., Kusunoki, F., and Hashizume, H.: "E^2board: An Electronically Enhanced Board for Games and Group Activity Support", In *Proc. of Affective Human Factor Design (CAHD2001)*, 2001 (to appear).

TOWARDS AN INTERACTIVE SYSTEMS DESIGN
BASED ON THE USER TASK WITH PETRI NETS

Dimitri Tabary[1], Rémi Bastide[2], Philippe Palanque[2], Mourad Abed[1]

[1]LAMIH
Université de Valenciennes et du Hainaut-Cambrésis
Le Mont Houy
F-59313 Valenciennes cedex 9, France
{dimitri.tabary; mourad.abed}@univ-valenciennes.fr

[2]LIHS
Université Toulouse 1
Place Anatole France
F-31042 Toulouse cedex, France
{bastide; palanque}@univ-tls1.fr
Tel: 33-561-633-588 Fax: 33-561-633-798
Tel: 33-327-511-461 Fax: 33-327-511-316

ABSTRACT

Currently, two points of view can be in opposition to develop an interactive system. The actors of software engineering think the IHM be a by-product of design software (for example, generable automatically starting from a model of class UML) and the community human factors think the software be a by-product of task and activity analysis, and who himself ignore methodological aspects of software engineering.
We propose a federator framework ensuring a design based on the tasks of the user. For that, we use the formalism of the Interactive Cooperative Objects (ICOs) to describe the different models (Task, Domain, User and Dialogue) of the method Task Object Oriented Design (TOOD).

1. INTRODUCTION

Currently, two points of view can be in opposition to develop an interactive system. The actors of software engineering think the IHM be a by-product of design software (for example, generable automatically starting from a model of class UML) and the community human factors think the software be a by-product of task and activity analysis, and who himself ignore methodological aspects of software engineering (Forbig, 1999).
In the same way, the state of the art reveals the need for developing methods of interactive systems design based on a formal approach. This state will not be developed in this article for lack of place; the interested readers can refer to (CADUI'96). These methods must integrate the user and his task in the design. Moreover, they must support the passage between the models and the stages of the cycle of development.
To bring closer these two currents, this paper presents a federator framework ensuring a design based on the tasks of the user. For that, we use the formalism of the Interactive Cooperative Objects (ICO) (Bastide et all, 1998) to describe the different models (Task, Domain, User and Dialogue) of the method TOOD (Task Object Oriented Design (Tabary & Al, 2000)). Due to lack of place, the example of the method and the discussion can't appear in this article. However, the example is available on request by email.

2. KNOWLEDGE REPRESENTATION

Many research on the specification and the generation of Human-Machine System (HMS) prototype are in the current of the Model-Based user interface Design (MBD) paradigm. This paradigm refers to the absolute, to an explicit, mostly declarative, description, capturing the semantics of the application and all knowledge necessary to the specification of the appearance as well as the behavior of the interactive system (Sukaviriya and al., 1994). It is part of the line of UIMSs (User Interface Management Systems) (Szekely, 1996). It provides a true alternative for the construction of interfaces. It recommends a top-down apprehension and requires the drawing up of formal specifications by the designers, rather than computer programming. This knowledge, described in a high level specialized language, is translated or interpreted for the total or partial generation of the applicative code.

The approaches based on this paradigm are based on the integration of a task model, of a user model, a domain model and a dialogue model (abstract and concrete). The design of the models is carried out without imposed order and requires mapping between the models. These approaches thus make it possible to freely begin the systems design with the modelling of the data or the tasks users.

Moreover, the need for formalizing the knowledge extracted the existent and need analysis is currently known in all the work devoted to the engineering of the systems interactive. Current work as point out it Palanque and Bastide (Palanque and Bastide, 1995), are based all on formal approaches, by generally using general formalisms rather than notations ad hoc.

Likewise, to control the design of a system, the paradigms of the approach with objects suggest the simultaneous taking into account of the static aspects (or structural) and of the dynamic (behaviour). The formal methods of design of IHM thus integrate generally these two components. They often try to use two formalisms dedicated to each one of these two aspects. However notations or formalisms different do not support transitions easy between the models. To fill, these obstacles we propose to use the ICOs to describe the objects of each model

The Interactive Cooperative Objects (ICOs, (Palanque, 1992)) formalism can be compared to an object-oriented language dedicated to the construction of man-machine interfaces, see figure 1. ICO use concepts borrowed from object-oriented approach (dynamic instantiation, classification, encapsulation, inheritance) to describe the structural aspects of systems and uses high-level Petri nets to describe their dynamic aspects. The objects interact by invocation of services. A class of ICO describes the structure of its instances by four components: the behaviour, the services, the state and the presentation. The behaviour of an ICO states how the object reacts to external stimuli according to its inner state. This behaviour is described by a high-level Petri net called the Object Control Structure (ObCS) of the object (Bastide, 1992). The dialect of Petri net used is called Object Petri net (Sibertin-Blanc, 1985). An ICO offers a set of services. They define the interface offered by the object to its environment. The state of an ICO is the distribution and the value of the tokens in the places of ObCS. This allows defining how the current state influences the availability of services, and conversely how the performance of a service influences the state of the object. Finally, a presentation could associate with ICO. It is a structured set of widgets organises in a set of windows. The formal definition of ICOs can be to consult in (Palanque and Bastide, 1995; Bastide et all, 1998).

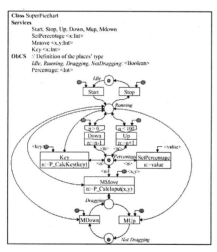

Figure 1: ICO sample
(extract from Palanque et all, 1998)

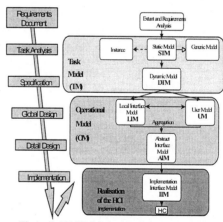

Figure 2: TOOD Method Development Cycle

3. METHODOLOGY

The TOOD design process can be divided into four major stages, Figure 2:
- The existent and requirement analysis is based on its user's activity and it forms the entry point and the basis for any new designs.
- The Task Model (TM) concerns the description of the user tasks of the system to be designed. It makes it possible to describe the user task in a coherent and complete way. Two models are created at this level: the Static Task Model (STM) and the Dynamic Task Model (DTM), in order to be able to use it for the HCI specification. We also identify the domain objects and the different users of the system in this part.
- The Operational Model (OM) makes it possible to specify the HCI objects in a Local Interface Model (LIM), as well as the user procedures in a User Model (UM) of the system to be designed. It uses the needs and the characteristics of the functional task model in order to result in an Abstract Interface Model (AIM) which is compatible with the user's objectives and procedures. The dynamic of domain object is described in this part.
- The Creation of the HCI concerns the computer implementation of the specifications resulting from the previous stage, supported by the multi-agent software architecture defined in the Interface Implementation Model (IIM).

The TOOD method is based on the transformation of a series of models to cover the together stages of the cycle of development, analysis of the task with the implementation. A task model, a user model and a domain model support the phase of specification and a dialogue model the phase of design, as Figure 2 shows it.

The designer is free to begin the specification with one of the models. However these three models are interdependent, indeed a task depends on the user who carries it out and the objects, which are available. On the another with dimensions roles of the users also depend on the tasks which they achieve. In the same way, the user has a point of view on the objects, which it handles in the system. At the end of the specification, it is necessary to carry out a first loop of evaluation checking and validating the coherence of the models by a confrontation with the customer and/or the users and the experts. It can be interesting when the formalism, for example the Petri nets, makes it possible to check generic properties.

3.1   Task Model

The task model, like MAD (Scapin and Pierret-Golbreich, 1990), ADEPT (Johnson, et al.), DIANE+ (Tarby, 1993), GRAAL (Van-Eylen and Hiraclides, 1996), GLADIS++ (Buisine, 1999) is designed as a means to take the user and his task into account as early as possible and to allow a more natural integration into the software development cycle. The aim is to abstract the user task information necessary for the formal design of interfaces. The task model makes it possible to make a rough sketch of the task in order to show up the "What?" that is to say the main treatments, the main data and the resources (human and system) which are necessary in order to accomplish the task. Two types of model result from this stage of specification, one a static model to express the structure of the task, the other a dynamic model in order to represent its behaviour as regards its environment. Based on an object-oriented approach, the Static Task Model (STM) shows up task classes to represent the treatments expected from the system. The Dynamic Task Model (DTM) uses ICOs to describe how the task manages the data and the resources it handles according to its internal state.

3.2   Domain Model

In TOOD, the data are called Domain Objects. They compose two sets: Functional Objects and the User Objects. The user objects form the conceptual representation that is made the user of work he achieves. The functional objects as for them intervene in the achievement of the tasks without the user being necessarily conscious of their presences. These domain objects are formalised in a similar way to the tasks (static and dynamic). In addition to the structure and dynamic of the objects, these models also give a synthetic and overall view of the composition and heritage links between the objects.

3.3   User Model

The user is modelled also by two models one static other dynamics. This modelling is carried out for each type of user. The static model defines the name of the user group, specifies their roles (operator, organic controller...). It is possible that certain user group has privileges (administrator, maker decision...). It indicates the privileged interaction methods of the various types of users. Authors also specify the experiment of the user in the task, the experiment of the system, the motivation, the experiment of a complex means of interaction as well as the cognitive profile and the preferences of user (Vanderdonckt, 1997). The dynamic part models the behaviour of the user in procedures of actions. These procedures contain all the elements necessary to describe precisely how to achieve a goal, i.e. mainly the operations, the constraints of precedence and the actions on the interactive objects. The user has, according to (Abed, 1990), three major statuses Perception, Cognition and Action. He/she perceives, is sensitive to a certain number of events of interlocking, of stimulation on behalf of the interface, he can read a certain number of data which he has the possibility of modifying. He defines a solution with respect to his observation and acts on the interface to achieve the goal which he laid down. The state of the ICO thus consists of three places Cognition (C), Perception (P) and Action (A), see figure 3.

3.4   Operational Model

The following design stage refines then each final task until elementary treatments applied to the clearly detailed domain objects. It contributes to the expression of the link between the applicative part and HCI part of the system. The elementary treatments are at the base of description of the behavior of the resources (human and system) of the tasks identified in the preceding stage, it acts of the model of dialogue of the task. This model describes the interactivity between resources system and the users (human resources). This description is performed in parallel in order to ensure that the system functions subjacent to the system resources comply with the user expectations in order to perform the task required of him, whilst leaving him a certain degree of freedom in his work. Thus the system resource model describes their behaviour; that is to say "how?" these resources react to outside stimuli (user and/or application events). The user model, which forms the second part of the operational model, models the behaviour of the user during procedure actions. These procedures contain all the elements necessary in order to describe exactly how to achieve an aim, that is to say primarily the operations, the orders of precedence and the actions on the interactive components.

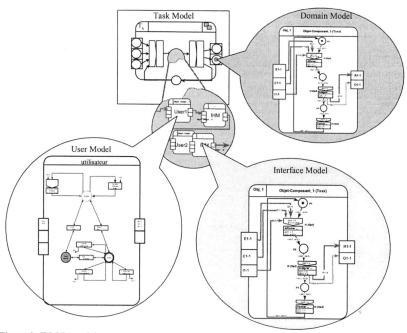

Figure 3: TOOD models

4. DISCUSSION

Thus all the models share the same representation (figure 3). A static part identifies the mapping between the models and a dynamic part indicates the behavior of the entities the ones compared to the others. The models thus defined make it possible to share the work of design and implementation between the specialists in the software engineering and human factors engineering. Thus a team working on the task and domain model can develop the functional core. One other team starting from the task model, the user model and the dialogue model can make the user interface. The two teams will be link by the need for respecting the requests for services between the definite objects in the different models. The use of ICO as formalism allows associating rendering specifications to any part of the Petri net. We consider that rendering deals with state: the very purpose of rendering is to make the inner state of the application visible to the user. It is therefore logical to associate rendering specification with the places of the Petri net[1].

REFERENCES

Abed, M. "Contribution à la modélisation de la tâche par des outils de spécification exploitant les mouvements oculaires: Application à la conception et à l'évaluation des interfaces homme-machine", September 1990 Ph.D Thesis University of Valenciennes, France.

Bastide, R. "Objets coopératifs: Un formalisme pour la modélisation des systèmes concurrents", February 1992, Ph.D. Thesis, University of Toulouse I, France.

Bastide, R., Palanque, P., Le D.H., Muñoz, J. " Integrating rendering specifications into a formalism for the design of interactive systems", In proceedings of Design Specification and verification of Interactive Systems '98 Springer Verlag, 1998, pp. 171-191.

Buisine, A. "Vers une démarche industrielle pour le développement d'interfaces Homme-Machine", 1999. Ph.D. Thesis, University of Rouen, France.

CADUI'96 "Computer-Aided Design of User Interface" Vanderdonckt J. (Ed.); Namur, Belgium Presses universitaires, 1996.

Forbig, P. "How Software Engineers deal with Task Models", In Bullinger, H. & Ziegler, J; (Eds.) Human-Computer Interaction: Ergonomics and User Interfaces. Hillsdale, New Jersey Lawrence Erlbaum, 1999, vol. 1, pp. 1322-1326.

Johnson, P., Wilson, S., Markopoulos, P & Pycock, J. (1993). "ADEPT-Advanced Design Environment for Prototyping with Task Models", In Proceeding InterCHI'93. ACM Press, Amsterdam, The Netherlands, 1993, pp. 56-57.

Palanque, P. and Bastide, R. Spécifications formelles pour l'ingénierie des interfaces homme-machine. Technique et Science Informatique, 1995, Vol. 14, No4, 473-500.

Palanque, P. "Modélisation par objets coopératifs interactifs d'interfaces homme-machine dirigées par l'utilisateur". September, 1992 Ph.D. Thesis, Université de Toulouse I.

Scapin, D.L. & Pierret-Golbreich, C. "Towards a method for task description: MAD", In Berlinguet, L. & Berchelette, D. (Eds.) Work with Display Unit' 89. Elsevier Science Publishers. Amsterdam, The Netherlands, 1990, pp. 371-380

Sibertin-Blanc, C. "High-level Petri nets with Data Structure", In proceedings 6th EWPNA, Espoo (Finland), 1985.

Sukaviriya, P., S. Kovacevic, J.D. Foley, B.A. Meyers, D.R. Olsen, D.R., Schneider-Hufschmidt M. "Model-Based Interface, What are they and Why Should We Care?", In proceedings UIST'94, November 2-4, 1994, pp. 133-135.

Szekely, P. "Retrospective and challenge for Model Based Interface Development", In CADUI'96, 1996 pp. xxi-xliv.

Tabary, D., Abed, M. & Kolski, C. "Object-Oriented Modelling of Manual, Automatic, Interactive Task in Mono or Multi-User Contexts using the TOOD method", In IFAC 2nd Conf on Management and Control of Production and Logistics (MCPL'2000) 5-8 July 2000, Grenoble – France.

Tarby, J.C. (1993) "Gestion Automatique du Dialogue Homme-Machine à partir de spécifications Conceptuelles", september 1993, PhD. Thesis, University of Toulouse I, France.

Vanderdonckt, J. "Conception assistée de la présentation d'une interface homme-machine ergonomique pour une application de gestion hautement interactive", July 1997, PhD of University Faculty of Notre-Dame de la Paix.

Van-Eylen, H. & Hiraclides, G. GRAAL: En quête d'une démarche de développement d'interface utilisateur. Angkor, Paris, 1996.

---

[1] See http://lis.univ-tlse1.fr/palanque/ICOs.htm

PROPOSITION FOR AN AUTOMATIC DESIGN OF SOHTCO INTERFACES FROM COOPERATIVE PROCESS SIMULATION

E. ADAM, D. TABARY, M. ABED

LAMIH - UMR CNRS 8530,
Université de Valenciennes et du Hainaut Cambrésis,
F - 59313 Valenciennes Cedex 9
{emmanuel.adam; dimitri.tabary, mourad.abed}@univ-valenciennes.fr

ABSTRACT

Automatic or semi-automatic specification of human-machine interface, on the basis of ergonomic rules and task analysis is a large field of research since about ten years. In this article, we propose to use a approved method (the TOOD method) in order to propose a automatic specification of a HOlonic Multi-Agent System for assistance with COoperative Work. This HOMASCOW is composed of intelligent agents structured following holonic principles and which interacts with actors of a complex administrative process.

1. INTRODUCTION

Automatic or semi-automatic specification of human-machine interface, on the basis of ergonomic rules and task analysis is a large field of research since about ten years (Abed, 1990).

In this article, we propose to go by this principle in order to propose an automatic specification of HOMASCOW interfaces (meaning HOlonic MultiAgent System for assistance with COoperative Work, or SOHTCO in french). This HOMASCOW is composed of intelligent agents structured following holonic principles.

This proposition pursues a first application of AMOMCASYS method (meaning Adaptable Modeling Method for Complex Administrative Systems) on complex administrative processes. This method proposes a modeling of cooperative activities of complex administrative organization on four axes: data axe, data flow axe, processing axe and dynamic axe.

This dynamic modeling is based on a particular Petri net: parameterized Petri net. These nets are part of high level Petri nets and allow to clearly represent complex processes thanks to their abstraction possibilities. Actually, in parameterized Petri net, it is possible to unify, in a parameter, a coherent set of objects or values. This allows to create an abstraction and to manipulate set of objects rather than individuals objects. So we use parameterized Petri net to model actor activities and interruption of activities of workflow process actors.

This Petri net allows to represent actors, documents on which they are working and activity interrupts. From this Petri net, we have built a simulator allowing to follow processes interactively. Moreover, simulation allows to interactively test new organizational solutions. In fact, principal function of this simulation is that it allows to get functioning rules of a cooperative process from which specification of a distributed information system could be envisaged.

These different rules constitute assistance rules used by our multiagent system for cooperative work (HOMASCOW) (Adam, 2001). Actually, consequently to the hypothesis that there is now, in any administrative system, at least one workstation (PC or other) per office or area, we have distributed the multiagent system around actors of these cooperative systems. Using assistance rules, multiagent system has to advice them in their activities and to make cooperation easier. HOMASCOW aim is to increase both autonomy of actors by an appropriate help, cooperation by an easier communication, actor stability by a watch on theirs activities.

So, the HOMASCOW that we propose, for a given process, is based on the structure of the human organization and on three levels of agent (a *process responsible agent*, which manage *user responsible agents*, which manage *executing agents*).

On the basis of our first specification of these agents, which follows the application of AMOMCASYS method on an actual case (processes of a patent department in a large company), we could propose an automatic specification of first and second level agents from simulation of workflow processes.

In this paper, we present first the TOOD method. Then the structure of HOlonic MultiAgent System for helping COoperative Work are described. And finally, we will show how to combine these two projects in order to obtain an automatic design of HOMASCOW interfaces from cooperative process simulation.

The TOOD method

In the aim of generating a prototype for the presentation and dialogue of an interactive system, we suggest a user task object oriented design method named TOOD (Task Object Oriented Design, (Tabary and Abed, 1998)). It is based on the transformation of a series of models in order to cover all the stages in the development cycle, from the specification to the implementation. In the specification phase, it relies on a task model and in the design phase it relies on an operational model.

Our models base itself on the use of Object Petri Nets (OPNs) and their extensions dedicated to the specification of the systems human-machine, notably the cooperative objects and the interactive cooperative objects (Palanque and Bastide, 1995)

The task model makes it possible to make a rough sketch of the task in order to show up the "What?" that is to say the main treatments, the main data and the resources (human and system) which are necessary in order to accomplish the task. Two types of model result from this stage of specification, one a static model to express the structure of the task, the other a dynamic model in order to represent its behaviour as regards its environment. Based on an object-oriented approach, the Static Task Model (STM) shows up task classes to represent the treatments expected from the system. The Dynamic Task Model (DTM) uses OPNs to describe how the task manages the data and the resources it handles according to its internal state.

In TOOD, the data are called Domain Objects. They are represented in TOOD as objects and are formalised in a similar way to the tasks (static and dynamic). In addition to the structure and dynamic of the objects, these models also give a synthetic and overall view of the composition and heritage links between the objects.

The following stage in design then refines each task until terminal treatment processes and clearly detailed objects are produced, along with the expression of the link between the applicative part and the HCI part of the system. The terminal treatment processes are at the basis of the description of the resource behaviour (human and system) for the tasks identified in the previous stage; this is the operational task model. This model describes, using OPNs, the mechanism for interactivity between the system resources and the user. This description is performed in parallel in order to ensure that the system functions subjacent to the system resources comply with the user expectations in order to perform the task required of him, whilst leaving him a certain degree of freedom in his work. Thus the system resource model describes their behaviour; that is to say "how?" these resources react to outside stimuli (user and/or application events). The user model, which forms the second part of the operational model, models the behaviour of the user during procedure actions. These procedures contain all the elements necessary in order to describe exactly how to achieve an aim, that is to say primarily the operations, the orders of precedence and the actions on the interactive components.

A design assistance environment developed in JAVA supports these stages. This environment makes it possible to input and simulate models and also allows the automatic generation of the HCI. The approach uses two different generators: one for the application (transforms the tasks into JAVA classes); the other for the interface (the system resources). A last model is responsible for automating their placing in relationship in such a way as to obtain a completely functional interactive application. The interface level includes the dialogue controller, as well as the links between the presentation level and the user data management level.

Automatic design of a HOlonic MultiAgent System for helping COoperative Work (HOMASCOW)

A MultiAgent System (MAS) can be defined as being a group of agents which interact between themselves directly or indirectly (through of a database). An agent may be defined as being an intelligent entity, which is part of a multiagent system. The space in which the system, and therefore the agent, evolves is called the world. The agent is capable of perceiving and modifying its environment (what surrounds it). An agent must be part of a multiagent system, it has therefore to have capacities for communication. As regards the notion of intelligence, the following principle should be underlined: an entity is intelligent if it is capable of learning, that is to say of adapting its knowledge. An agent can therefore be defined as follows: an agent is an adaptive, rational and autonomous entity, capable of communication and action. It may also be adaptive and its adaptivity degree may vary from an agent type to another.

An agent generally has acquaintances; these are agents with which it communicates or interacts. Each agent has elements of knowledge concerning its environment. These elements of knowledge are also called representations or beliefs. An agent has one or several objectives, which are also called goals or desires. To reach their goals, some agents are able to plan its actions.

More generally, according to their goal, MAS have to satisfy certain principles.

2.1 Characteristics of Holonic MultiAgent System (HOLOMAS)

A Holonic MultiAgent System is a MAS following principle of holonic system (an organizational concept proposed by (Koestler, 1969)): it has a tree architecture; its parts, that could be composed of one or more agents, are autonomous, cooperative and stable. One of the main advantages of a HOLOMAS is its recursive architecture (Gerber, 1999): an entity is considered both as a whole made up of other entities and as being part of a set. This characteristic makes the setting-up of HOLOMAS easier. Indeed, each part of the HOLOMAS (and so each holonic agent) should have the same characteristics as the whole, plus one or more specialties. So, it is necessary to identify what are the minimal characteristics that must have the holonic agents.

2.2 Characteristics of a HOlonic MultiAgent System for helping COoperative Work

The HOMASCOW that we propose aims at advice actors of complex administrative system in their activities and a make cooperation easier.

In order to keep coherence within a distributed system, some researches have shown that at least two kinds of rules are necessary: individual rules at each module level and social rules that specifies possible interactions between modules. In our case, we propose the use of the three levels of rules (personal rules, local rules, and global rules) distributed in the organization of the HOMASCOW:

- a first level constituted of a holonic agent responsible of the process, linked to the human actor responsible of the process. This agent contains global rules (data flow rules),
- a second level constituted of a set of holonic agent "responsible of workplace", each one being dedicated to an actor. These agents contain personal rules and local rules relative to users' activities,
- a third and last level constituted of a set of holonic agents that makes up HOMASCOW basis. They are responsible of interactions with actors and holonic agents, and responsible of the document management.

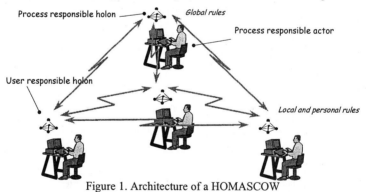

Figure 1. Architecture of a HOMASCOW

Number of holonic agents making up HOMASCOW basis is variable, it depends on the number of documents managed by actors. This data is not taken into account by AMOMCASYS method.

However, number of holonic agents responsible of workplace directly results of the modeling and the simulation. Moreover, activities of these agents, as well as activities of the holonic responsible of the process, are already specified. In addition to "classical" activities that has to own an agent (communicate, interact, protect), productive activities of these agents consist in following internal representation of the process and in guiding actor according to this representation.

Figure 1 represents an example of the general architecture of a HOMASCOW. In this figure, a sub-HOMASCOW is associated to each user. It contains local and personal rules. To the user responsible of the process is associated a sub-HOMASCOW that contains the global data flow rules.

2.3 Proposition for an automatic design of HOMASCOW

The AMOMCASYS method proposes the use of a simulator, which allows, for an administrative process, to obtain the three levels of rules (personal, local and global rules) (Adam, 1999).

Use of the simulator implies also to define actors of the process, used documents, workplaces and links between them. From the simulation, we obtain an initialization file whose structure is shown in figure 2a.

Thanks to the recursive architecture of the HOMASCOW, we can give an adapted file (fig. 2b) to a holonic agent which will behave like the process responsible agent and will create the different user agents. These user agents will create then the minimal basis of the HOMASCOW (that is to say, for each sub-HOMASCOW, an interface agent, a data responsible agent, an emission responsible agent and a reception responsible agent). Figure 2 shows the initialization files use by the simulator and the HOMASCOW. Except for the IP address of the agent (and some others characteristics such as the fact to be busy, free for instance), it is possible to transform the first file (fig. 2a) to the second one.

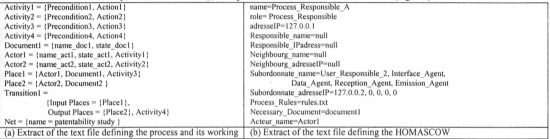

Activity1 = {Precondition1, Action1}	name=Process_Responsible_A
Activity2 = {Precondition2, Action2}	role= Process_Responsible
Activity3 = {Precondition3, Action3}	adresseIP=127.0.0.1
Activity4 = {Precondition4, Action4}	Responsible_name=null
Document1 = {name_doc1, state_doc1}	Responsible_IPadress=null
Actor1 = {name_act1, state_act1, Activity1}	Neighbourg_name=null
Actor2 = {name_act2, state_act2, Activity2}	Neighbourg_adresseIP=null
Place1 = {Actor1, Document1, Activity3}	Subordonnate_name=User_Responsible_2, Interface_Agent,
Place2 = {Actor2, Document2 }	Data_Agent, Reception_Agent, Emission_Agent
Transition1 =	Subordonnate_adresseIP=127.0.0.2, 0, 0, 0, 0
{Input Places = {Place1},	Process_Rules=rules.txt
Output Places = {Place2}, Activity4}	Necessary_Document=document1
Net = {name = patentability study }	Acteur_name=Actor1
(a) Extract of the text file defining the process and its working	(b) Extract of the text file defining the HOMASCOW

Figure 2. Initialization files use by the simulator and the HOMASCOW

Currently, we are able to create a sub-HOMASCOW from data issued from the initialization file. And, in order to be able to built a more complete architecture of the HOMASCOW, we apply the MAGIQUE architecture (Bensaid, 1999) particularly well suit for design hierarchic multi-agent architecture.

But, if the AMOMCASYS method allows to give sufficient information to design automatically some agent of a HOMASCOW system for a given process, it needs to simulate it, that is particularly long in term of time. That is why it is interesting to see what the TOOD method should bring to our problem.

4. Contribution of the TOOD method to the Automatic design of HOMASCOW

In TOOD, each activity is the realization of a task. A task is defined by an input interface, an output interface and a set of resources. The Input Interface specifies the list of describer objects on which the task is performed. These objects form the input parameters for the task, which means that the task needs to access the referenced user objects. It is made up of three types of data: the Triggers, the Controls and the Input Data. The Output Interface specifies the final state of the task. It is made up of two types of object: The Reactions and the Output Data. The Resources represent the agents carrying out the task. These agents can be the human operators and/or the interactive objects and/or application components of the system.

The TOOD environment makes it possible to input the activity, the pre-condition, the post-condition and the actors of the HOMASCOW architecture. This environment enables to simulate models and also allows the automatic generation of the HCI. The approach uses two different generators one for the application (transforms the tasks into JAVA classes), the other for the textual specification generation.

TOOD thus makes it possible to publish a great number of agents. It allows their simulation and the generation of their specifications.

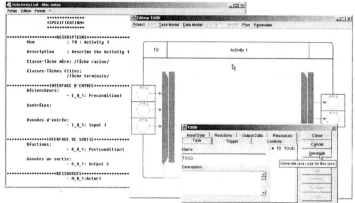

Figure 2. Screen copy of the TOOD environment

5. Conclusion

We have specified a multi-agent system assisting the actors in an administrative system, following the holonic approach. This approach enabled us to recursively design agents from rules describing the working mechanism of the administrative system. If the AMOMCASYS method allows to analyze and model human organization, it does not permit to automatically design the assistance system (HOMASCOW). That is why we turn to TOOD method in order to obtain a semi-automatic design of interactive multi-agent system from activities modeling.

REFERENCES

Abed M., Contribution à la modélisation de la tâche par des outils de spécification exploitant les mouvements oculaires: application à la conception et à l'évaluation des interfaces homme-machine. (in french), PhD Thesis, Université de Valenciennes et du Hainaut-Cambrésis, September, 1990

Adam E., "Specifications of intelligent human-machine interfaces for helping cooperation in human-organizations", In H.J. Bullinger, J. Ziegler (Ed.), Ergonomics and user interfaces, vol. 1, London: Lawrence Erlbaum Associates, 1999, pp. 311-315.

Adam E., Mandiau, R., Kolski, C. Application of a holonic multi-agent system for cooperative work to administrative processes. Journal of Applied Systems Studies, 2001, vol. 2, No 1.

Bensaid N.,Contribution à la réalisation d'une modèle d'architecture multi-agent hiérarchique. Thèse de doctorat, Université des Sciences et Technologies de Lille, France, Mai 1999.

Gerber C., Siekmann J., Vierke G., "Holonic Multi-Agent Systems". Research report, RR-99-03, March 1999, DFKI GmbH, Germany.

Koestler A., The Ghost in the Machine. Arkana Books, London, 1969.

Palanque P. and Bastide R. "Spécifications formelles pour l'ingénierie des interfaces homme-machine. Technique et Science Informatique", 1995, vol.14, No 4, pp. 473-500.

Tabary D. and Abed M. TOOD: An Object-oriented methodology for describing user tasks in interface design and specification. La Lettre de l'IA, 1998, No 134, pp. 107-114.

AGENTS ORIENTED SPECIFICATION OF INTERACTIVE SYSTEMS

Hacène MAOUDJI, Houcine EZZEDINE and André PENINOU

L.A.M.I.H., UMR CNRS 8530 - Le Mont Houy, University of Valenciennes,
F-59313 Valenciennes cedex 09, France
{hacene.maoudji, houcine.ezzedine, andre.peninou}@univ-valenciennes.fr

ABSTRACT

In this article, we propose an agent oriented specification for Human-Machine Interfaces design intended to process control and command. In most cases, models of interactive systems are much more concerned by user controlled applications and do not consider the special features of supervision applications. We propose an agent based model of interfaces where classical components of interactive systems (Application, Dialogue Controller, Interface) are modeled as multi-agent systems. In this model, agents are able to carry out both the control and the command functions of the process. Our approach has been applied in an industrial context of Passengers Information Systems (PIS) in urban transport systems. Some results of this large-scale application are discussed.

1 INTRODUCTION

Currently, the approaches used for interactive systems design promote a modular structuring with the aim of a better apprehension of reactivity, flexibility, maintainability and re-usability. However, in most cases, models are much more concerned with user controlled applications and do not consider the special features of supervision applications. Thus, we propose an agent based model of interfaces where classical components of interactive systems are modeled as multi-agent systems. The specialized agents proposed are able to carry out both the control function, that is enabling the communication from the process towards the user through output devices, and the command function enabling the communication from the user towards the process through input devices. In the context of supervision, each one of these two functions have particular characteristics related to the supervision specificities (real-time processes, ...). The paper is structured as follows. Section 2 highlights the potential benefits of agent based modeling of interactive systems. Section 3 describes the industrial context of experimentation for Passengers Information Systems (PIS) in urban transport systems. Sections 4 describes our approach. Finally, section 5 discusses the application of this model.

2 A MULTI-AGENT APPROACH FOR HUMAN-MACHINE INTERFACES

There is no single definition of the concept of agent common to all researchers (Ferber 1995; Wooldridge and Jennings 1995). In our context, we will comply with the definition of Wooldridge and Jennings (1995) who identified four key characteristics of an agent : 1) autonomy towards the user and its environment, 2) social ability to communicate with other agents, 3) reactivity towards the changes of its environment, 4) proactive behavior by the way of goal-oriented initiatives. The general principle of multi-agent systems is thus to distribute a system into a certain number of co-operating entities able to communicate and coordinate their behavior in order to achieve a common goal.

A key step of the engineering of Human-Machine Interfaces (HMI) is the choice and the specification of an architecture. An architecture supplies the designer with a generic structure from which an interactive application can be developed. Two main types of architectures can be distinguished : functional components architectures and structural components architectures. In the first category we can find the two most known models : Seeheim model and its evolution in ARCH model. In the second category, we can find : agents models and objects models.

Compared to the types of architectures previously mentioned, our approach is intermediate since it borrows, in its principles, from the two types of models : it is both functional and structural (Maoudji and al., 2001). We propose to use the three classical functional components : Interface, Dialogue Controller, Interface to the Application (or Application). These functional components are clearly identifiable and propose a decomposition according to three subproblems requiring a differentiated and relatively independent solution (Pfaff, 1985). We think that each one of these functional components can be break down according to a structural view, that is in the form of multi-agent systems. The long-term objective is to develop systems able to adapt themselves in a dynamic way to the complexity of data as well as the complexity of the user.

3 INDUSTRIAL CONTEXT OF THE EXPERIMENTATION

We have tried out our approach within the framework of a cooperative project including an industrial firm, the SEMURVAL company which exploits the tramway/bus transport system of the town of Valenciennes. Other french research laboratories also take part in this project : INRETS, LAIL, I3D. This project is supported by the Regional Council of the Nord/Pas de Calais (France). The part of project which we are concerned with is relating to the definition of a supervision interface for the controller of traveler information in the transport system of the city of Valenciennes. The transport system offers information displayed in stations and cars concerning the schedules of cars, cars stops, as well as connections. This information is automatically managed by an operating help system getting real-time data about cars positions and conditions. Nevertheless, the information has to be checked.

Supervision process purpose is to control constantly the state of a process and to submit commands for the control of this same process. The controlled process has its own behavior. This behavior depends on parameters that the operator can neither control nor modify (Millot, 1988). In our case for example, traffic jam may directly change the bus schedules, but the controller cannot manage the traffic jam. Within this framework, the application must enable the controller : a) to follow the progress of the automatic traveler information over the network (stations) and in the cars, b) to control this information when necessary according to company objectives (what information to give in the case of significant delays for example) or to exceptional situations (broken-down cars, etc).

4 AGENT-ORIENTED SPECIFICATION OF THE INTERFACE

4.1 Application component specification

The Application component is a multi-agent model which simulates the real process but also is directly coupled with this process. The role of this multi-agent system is to manage the traveler information displayed in cars and stations and to calculate automatically the information to display (delay of cars, scheduled changes of routes or timetables, etc.). Considering the natural distribution of the process, both geographically and structurally, the relevant agents are the cars and the stations. Cars and stations agents communicate with each other and so, information flows through the routes. Each agent uses its own rules in order to decide : 1) how to cooperate with the other agents, 2) what traveler information to display. Lastly, these agents can accept external commands, i.e. command submitted by the controller. The model of agent used in our case is thus reactive. According to information received from the environment, the traveler information is never calculated by a centralized process but only locally according to local rules.

The different behaviors of each agent enable to easily encompass the variety of the possible situations such as : line having frequent traffic jams and delays on some sections, connection stations, line with frequency or fixed schedules, etc. The behavioral rules of each agents are designed according to these characteristics. Lastly, in a medium-term, this approach enables to scale-up the information system depending on its evolution over the time : add-on of new lines, set up of temporary lines, changes of lines, etc.

4.2 Interface Component Specification

The Interface component is composed of agents which manage the display of the process monitoring and the commands submitted by the controller. The agents are composed of various basic interactive elements in a logical way in order to make up interface objects which are relevant to the user : The graphic line view, traveler information displayed in stations, traveler information displayed in cars, the function for modifying traveler information. Each agent manages the state of a user interface object in relation to input/output functions : requests to the Dialogue Controller for possible modifications of the process, submittings of the commands of the human controller to the Dialogue Controller. In particular, each agent manages the state of its current display (displayed or not). For example, in case of alert, an agent can behave in order to trigger off its display, even if the user does not prompt for explicitly. It can also trigger off the display of other agents in order to supply the user with a coherent display of the context. This point is directly bound to supervision.

4.3 Dialogue Controller component specification

The Dialogue Controller component is composed of agents representing the same elements of process as for the Application component (cars and stations). Their role is not any more to simulate the real process, but to : a) relay the modifications observed on the level of the application on the relevant interface agents, b) to relay the command actions towards the relevant application agents. This multi-agent system is used as an abstraction with regards to application agents. It also manages its own constraints related to the interaction such as the proposal for standardized solutions of information according to the situation. This element is however not developed yet to date.

This system keeps watch on the application agents and will have, in the long term, to detect abnormal situations in order to propose a help to the controllers

5   INFORMATION SUPPORT SYSTEM (ISS) DESIGN

5.1   Model description

The full model of the system is mainly built on three parts : 1) The simulator (Application), which deals information with system like the ESS does, 2) A multi-agent system(dialogue controller), composed of resources providing and interface control agents, 3) A graphical interface (Presentation), intended to bring information to regulator and seize his requests.

The Information Support System (ISS) goal is to permit visualization, edition and creation of traveler intended information. Such data is exchanged with the Exploitation Support System (ESS), real-time knowing all vehicles positions and states, and also with the Decision Support System (DSS) which helps regulators of the network. ESS is fully independant towards the ISS, which needs to be able to understand information he receives and produce understandable new ones. To this end, the intermediate agent is mainly intended for filtering and buffering data, translating and checking requests consistency.

5.2   The simulator

The simulator regroups the necessary information to the generation of messages given out by the EHS.
It is also able to adapt this information according to the human requests and updates (Translating and consistency checking abilities).
These data represent stops, lines, vehicles, lists of scheduled stops and arrivals, alerts and messages intended for regulation or travel information.

A module generates time-based events, in order to modify simulator's data.
Examples or these changes could be: new delay, vehicle delay modification, change in scheduled stops, broadcast of a new station alert, and so on.

5.3   The multi-agent dialogue controller system

The multi-agent dialogue controller system affords interface needs, i.e. data and services.
This architecture tends to make the HMI fully independent from the ESS. It keeps a local ESS state representation, built from messages sent by the simulator. In this way, the system only keeps user useful information. This multi-agents system determines if it is useful to do a new state interface notification from the current received message. So, if necessary, it asks the interface to update itself using its data. These entities standing for stops and vehicles are light weighted compared to simulator's ones. This information is static, especially in the scheduled stops and arrivals which are only ESS modifiable.

5.4   The graphical interface

The graphical interface depends on the intermediate agent in which it draws its resources (it is an agent belonging to the multi-agent system dialogue controller). Any interface modification reflects one of the intermediate agent. Interface seized actions and parameters are transmitted to the intermediate agent, which will notify back when update will be available.

5.4.1   View of a line

This view is composed of graphical elements, such as stops, sub-lines, vehicles... (fig. 1). These elements have a variable representation, according to the state of the entity they are representing : A class of problems = one color, a type of vehicle = a shape, ...

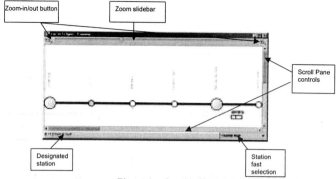

Figure 1 – Graphic line view

5.4.2 Detailed view of a station

The detailed view of a station is brought by acting on its associated representation in line view. The view is composed of several independant panels (Messages and alerts, state synthesis, and scheduled stops and arrivals information), each one linked to data that concerns itself by the way of the intermediate agent. It shows scheduled arrivals information to user, by the way of a set of tabs, depending of a selectable drop-down directions list (fig. 2a). Each view is instanciated by the intermediate agent and will be notified for a data update.

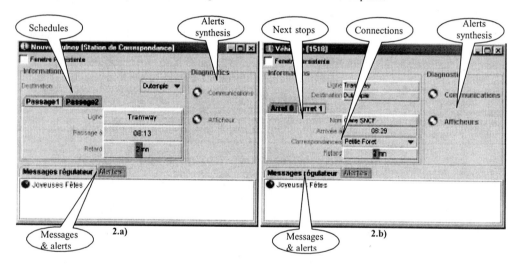

Figure 2 – Station view (2.a left) and vehicle (2.b right)

5.4.3 Detailed view of a vehicle

The detailed view of a vehicle (fig 2.b) brings the same basis as the station one, excepts of the representation of its associated scheduled stops information. A drop-down list box contains the arrival available correspondences, shown considering the delay time. These correspondences are determined by one of the agent services.

5.4.4 The message center

The message center is the component which brings regulator ability to view a synthetically view of all the messages sent towards the travelers. It is composed of two panels : 1) the first one manages a list of all messages (fig. 3a). This list is build from agent's data about the whole messages. 2) the second one, instantiated by the last one, is dedicated to create and edit messages (fig. 3.b). It is intended to validate and pass the message to be distributed the agent. The agent, then, will be in charge to deliver this message (updating the local state of each stop or vehicle), and communicate the resulting updates to the interface. According to recipients, the message type will be unicasted, multicasted or broadcasted. This characteristic leads to a specific representation in the interface (by an icon).

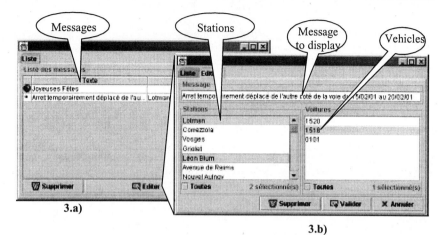

Figure 3 – Message center : message list (3.a left) and the message editing panel (3.b à right)

6   CONCLUSION

Our agent oriented interface specification approach has been applied in the setting of bringing a significative help for the traveler information. A model was built following the agent specification. The use of a intermediate agent (dialogue controller), besides the fact or permitting a ESS least load by filtering and keeping a local copy of its data, permitted a fully independent implementation of the interface towards the ESS (or the simulator). The modular decomposition of the interface brought possibilities of interface distribution. This distribution allowed to compose the final interface, without worrying about interactions between views.

In the same way, it permits us to consider a spatial distribution of the elements : On several regulation stations, or in a redundancy manner.

One of the perspectives of this work is to propose a more interactive applications design method in the domain of control/command process.

7   REFERENCES

Ferber, J. (1995). Multi-Agent Systems : Introduction to Distributed Artificial Intelligence. Addison-Wesley.

Maoudji, H., Ezzedine, E. and Péninou, A. (2001). Towards agent oriented specification of human-machine interface : application to the transport systems. Proceedings of the 8th IFAC/IFIP/IFORS/IEA Symposium on Human-Machine Systems, HMS'2001, Kassel, Germany, september.

Millot, P. (1988). Supervision des procédés automatisés et ergonomie. Hermes. Paris.

Bass L., Coutaz J., (1991). Developing software for the user interface, Addison-Wesley.

Wooldridge, M. and Jennings, N.R. (1995). Intelligent Agents: Theory and Practice. The Knowledge Engineering Review, 10 (2), pp. 115-152.

Contribution of the Petri nets and the Multi Agent System in HCI specification

Meriem RIAHI, Faouzi MOUSSA

LIP2-Département des Sciences de l'Informatique – Faculté des Sciences de Tunis
Campus universitaire 1060 le belvédère Tunis- Tunisia
Email: meriem.riahi@insat.rnu.tn, faouzi.moussa@ensi.rnu.tn,

ABSTRACT

This paper studies a formal technique based on Petri nets for modeling the human-machine system (HMS) behavior, in order to identify the user requirements. These user requirements allow the deduction of the necessary interface objects in terms of informational and control objects. A multi-agent system approach is proposed to generate the specifications of the appropriate displays deduced for the control and the monitoring of the system.

1. INTRODUCTION

Considering the crucial role of the User Interface (UI) in control rooms of industrial processes with a high degree of security such as chemical industries, nuclear processes and transport systems, the tendency consists in presenting data on graphical screens. Indeed, graphical interfaces represent the predominant way: (i) of informing the operators of the process evolution and (ii) of assisting them during their mental problem-solving tasks (Alty, 1985; Rasmussen, 1986; Sheridan, 1997). Thus, creating the interface, the designer must decide on what to present to the operators according to the functional context and the corresponding human tasks, and how to present the information graphically.

The main purpose of this research work, is the study of a global methodology for UI design and development (Riahi, 1998). We are especially interested in the identification of the user requirements from the HMS analysis and in the process of interface design taking these user requirements into consideration. In fact, we remind, here, the process of modeling the human-machine system (HMS) behavior, expressing the interaction of the operator within the interface according to different functioning contexts of the system. Then, we develop how to specify and generate the interface considering the user requirements deduced from this model and using a multi-agent system approach.

2. HMS BEHAVIOUR MODELLING

Controlling a process, the operator needs to know, instantly, the changes of the process state. This information will be transmitted to him through the different displays of the interface. Thus, any modification of the system's state will imply a changing in the graphical displays. This change can affect either the object's parameters (colors, shapes,...) or the display contents (appearance or disappearance of some graphical objects,..). To model this aspect, we have proposed a model based on interpreted Petri nets describing the system's evolution according to the different events (Moussa, 1999; Riahi, 2000). In fact, Petri nets can express with reliability the different aspects of concurrency, parallelism and iteration of the operator's actions and offers a helpful panel of techniques to validate the Graphical User Interface (GUI) specification before their generation (David, 1992; Palanque, 1997; Williem, 1988).

This model takes into account the results of the system analysis and its different contexts of functioning, on the first hand, and the operator's tasks analysis, on the other hand. The places model the operator's states according to a functioning state of the system and the transitions model the evolution between these states. For better understanding, we remind, first, in section 2.1 and 2.2 the major concepts of this approach, we detail, after, the multi agent system approach for specifying and generating the HCI.

2.1 Task modeling using Petri nets

An operator's task is composed of a well-organized set of elementary actions. The elementary structure is composed of 3 places and 2 transitions representing the operator's behavior while executing his action according to the system's evolution (figure 1). The places of this elementary structure model the operator's state before, during and after the execution of his action.

We model the elementary actions of the operator by elementary structures of Petri nets and then we proceed by typical compositions as sequential, parallel, choice, iteration, etc., to achieve the model of human task in case of any dysfunctioning of the system.

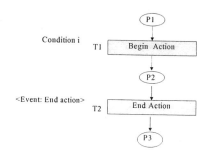

Figure 1: elementary structure

A global model is then constructed integrating the different evolutions of the functioning system's states performing, thus, the modeling of the human-machine system behavior and allowing the deduction of user requirements.

2.2 Deduction of the user requirements

The HMS behavior model expresses the different functioning states of the system and the operator's actions intervening in case of any dysfunctioning. Thus, deducing the user requirements consists in identifying the appropriate set of informational variables for each state and the set of control variables, the operator needs to intervene in order to correct an abnormal situation. In fact, we consider in the HMS behavior model the transitions «Begin Action», and we associate to them the adequate variables, either informational or control, which refer to the user's requirements.

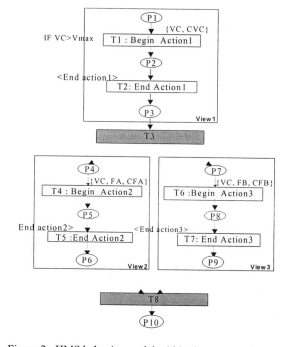

Figure 2: HMS behavior model within the user requirements

Thus, for example, at the state P2 of the model given in figure 2, the operator disposes of the relevant user requirements to perform his action well. For instance, according to the action A1, the informational variable VC informs the operator of the current level, the control variable CVC allows him to decrease this level by manipulating the valve. Once the informational and control variables have been identified, we have to deduce the necessary objects of the UI: to each informational variable, we associate a "graphical informational object " and to each control variable, we associate a " graphical control object ".

The graphical specification step of the HCI is now to achieve. It is presented in the following section.

3. UI SPECIFICATION

For UI presentation, we propose to take advantage of the research carried out since the eighties, by some researchers (Johannsen, 1995; Kolski, 1996; Kolski, 1997; Moussa, 2000) concerning knowledge-based approaches for automatic design and/or evaluation of UI used in process control. These systems are able to decide on the appropriate graphical displays, taking into consideration, on the one hand the characteristics of each sub-system (the list of its functioning states, the user requirements associated to each state,...) and on the other hand, specific formalized guidelines stored in its knowledge bases, in order to ensure the best ergonomic quality possible (O'Hara, 1999; Vanderdonckt, 1994).

However, the major problem often encountered when dealing with a knowledge based system consists in the huge quantity of formalized data to store and manipulate as a whole. Using meta knowledge did not improve so much the performance of the system. Thus, we explore, in this paper, a new way of UI specification using a Multi Agent System (MAS) based approach. After a brief presentation of MAS, we will focus more on the architecture of the proposed approach.

An agent may be defined as being an intelligent entity, which is part of a multi-agent system. It is capable of perceiving and modifying the environment what surrounds it. Generally, an agent has capacities for communication (direct or indirect through a database). Cognitive agents own knowledge on their environment, on the multi-agent system, and on themselves. They can be defined as adaptive, rational and autonomous entities, capable of communication and action. They may also be adaptive and theirs adaptivity degrees may vary from an agent type to another. On the contrary, reactive agents act by reaction to stimuli, without any knowledge or representation of their environment (Mandiau, 1999, Mandiau, 2000).

So, multi-agent systems allow to efficiently distribute the knowledge. They also permit to manage them when they are composed of different layers of agents. For example, in (Adam, 2001), use of holonic rules allow to design a multi-agent system whose parts are composed of sub-systems dedicated to a user.

In this way, the sub-systems are composed of a manager agent, which contain knowledge associated to the user and several executing agents owning knowledge relative to theirs specialties (for example, the interface agent owns knowledge about representation of the data).

Complex systems are powerful, flexible and give, thanks to the correlation, a good way to implement a collaborative task. The cognitive agents will decide, in a cooperative work, of the UI specification. As shown in figure 2, the gray boxes express the different functioning states of the HMS. Each state of functioning is supported by a GUI named view. A view states of one or more linked displays according to the intrinsic features of the system and to the functioning state of the process. For example, we associate to the box 1, which stands for a normal state of functioning, a supervision view and a historical view. while, the boxes 2 and 3 are relative to two states of dysfunctioning, we associate, so, to them control views.

An agent "view" models each graphical display. It defines the specification of the display in terms of graphical objects and deals with the dialogue control of those objects (actions as enable/disable, display/mask). The agent view interacts, in addition to the Petri net model of the HMS (figure 3), with other agents specialized in further area of knowledge as: graphical presentation of the objects, use of colors, choice of interaction modes, etc.

Each agent has an objective which allows it to state on the type of view to specify (supervision, control, etc.).

As knowledge on the environment, the agent possesses the set of the user requirements. Its perception is ensured by the variables states. Indeed, the Petri net affects the values of the logical variables according to the functioning state of the process. This is done thanks to a process of communication around a shared memory between the Petri net and the MAS. When perceiving a variation the agent view updates its environment and executes the adequate graphical commands (enable/disable, display/mask, change colors, etc).

The agent view collaborates with the other agents by invocating them to participate in the process of specifying the GI specification. They will answer questions about the interaction modes, the colors to adopt, etc.

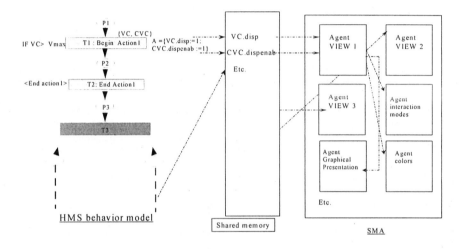

Figue 3 . Interaction Communication process between Petri net model and MAS

4. CONCLUSION

We centred in this paper our presentation on the study of a new promising way to specify the GUI. According to the capitalized experience in the field of manipulating knowledge based systems. We find very interesting to explore the MAS to try to take advantage of the power of collaborative work between agents, each one offering a field of specialization and a class of answers in response to specific questions rising along the process of HCI specification. We are validating this approach on a pedagogical example implemented with C++ and XML. The results will be presented further.

ACKNOWLEDGEMENT

The authors would like to thank Emmanuel ADAM for his contribution in this paper.

REFERENCES

Adam, E., Mandiau, R., Kolski, C. (2001). Application of a holonic multi-agent system for cooperative work to administrative processes. Journal of Applied Systems Studies, 2001, vol. 2, No 1.

Alty, J.L., Elzer, P., Holst, O., Johannsen, G., Savory, S. (1985). Literature and user survey of issues to man-machine interfaces for supervision and control systems. *Report of Esprit project P600*, Copenhagen: C.R.I., 1985.

David, R., ALLA, H. (1992). *Du GRAFCET aux Réseaux de Petri*, Editions HERMES, Paris, 1992.

Johannsen, G. (1995). Conceptual design of multi-human machine interfaces. *Proceedings 6th IFAC/IFIP/IFORS/IEA Symposium on Analysis, Design and Evaluation of Man-Machine Systems*, MIT, Cambridge, USA, June 27-29, 1995.

Kolski, C., Moussa, F. (1996). Two examples of tools for working with guidelines for process control field: Synop and ERGO-CONCEPTOR, *Third Annual Meeting of the International Special Interest Group on "Tools for working with guidelines"*, J. Vanderdonckt (Ed.), Namur, Belgium, 4 June 1996.

Kolski, C. (1997). Interfaces homme-machine, application aux systèmes industriels complexes. Editions HERMES, Paris, 1997.

Mandiau, R., Le Strugeon, E., Agimont, G. (1999). Study of the influence of organizational structure of a multi-agent system. International networking and information system journal, pp.153-181, vol. 2, n°2, 1999.

Mandiau, R. (2000). Modélisaton et Evaluation d'organisations multi-agents, habilitation à diriger des recherches, Université de Valenciennes et du Hainaut-Cambrésis, Décembre 2000.

Moussa, F., Riahi, M., Moalla, M. Kolski, C. (1999). A Petri net method for specification and automatic generation of user interface. In « Human-Computer Interaction: Ergonomics and user interfaces, volume 1, Proceedings HCI'99, 8th International Conference on Human-Computer Interaction, Munich, Germany, August », H.J. Bullinger, J. Ziegler (Eds.), Lawrence Erlbaum Associates, 1999, pp. 988-992.

Moussa, F., Kolski, C. Riahi, M. (2000). A Model Based Approach to Semi-Automated User Interface Generation for Process Control Interactive Applications. Interacting with Computers 12, 3 (2000) 245–279.

O'Hara, J., Brown, W. (1999). Software Tool for the Use of Human Factors Engineering Guidelines to Conduct Control Room Evaluations. In: Büllinger, H.J., Ziegler, J. (Eds.): Human-Computer Interaction: communication, cooperation and application design. Lawrence Erlbaum Associates, London, 973–977, 1999.

Palanque, P. Paterno F. (1997). Formal Methods in Human-Computer Interaction, Springer Verlag, 1997.

Rasmussen, J. (1986). Information processing and human-machine interaction, an approach to cognitive engineering. *Elsevier Science Publishing, 1986.*

Riahi, M., Moussa, F., Moalla, M. Kolski, C. (1998). Vers une spécification formelle des interfaces homme-machine basée sur l'utilisation des réseaux de Petri. In M.F. Barthet (Ed.), Actes du 6ème Colloque ERGO IA'98 Ergonomie et Informatique Avancée. (pp. 196-205). Bayonne: ESTIA/ILS, 1998.

Riahi, M., Moussa, F., Kolski, C. Moalla, M. (2000). Use of interpreted petri nets for human-machine dialogue specification in process control. Proceedings ACIDCA'2000 International Conference on Artificial and Computational Intelligence for Decision, Control and Automation in Engineering and Industrial Applications, 22-24 March, Monastir, Tunisia.

Sheridan, T.B. (1997). Task analysis, task allocation and supervisory control. In Handbook of Human-Computer Interaction, Helander M.G., Landauer T.K., Prabhu P. (Eds), North Holland, Amsterdam, pp. 87-105, 1997.

Vanderdonckt, J. (1994). Guide ergonomique des interfaces homme-machine. Presses Universitaires de Namur, Namur, Belgium, 1994.

Williem, R., Biljon, V. (1988). Extending Petri Nets for specifying Man-Machine dialogues, International Journal of Man-Machine Studies, vol. 28, 1988, pp. 437-45.

THE TASK-TO-PRESENTATION-DIALOG MAPPING PROBLEM

Quentin Limbourg and Jean Vanderdonckt

Université catholique de Louvain, Place des Doyens, 1 – B-1348 Louvain-la-Neuve, Belgium
{Limbourg, Vanderdonckt}@qant.ucl.ac.be – http://www.qant.ucl.ac.be/membres/jv

ABSTRACT
The model-based design research community has already produced several successful attempts to generate a target model from a source model. In this paper, we investigate to what extent a dialog model and a presentation model can be systematically derived from a task model. It is believed that these models can work better hand by hand. For this purpose, the ConcurTaskTrees notation has been exploited to provide a task model for a single-user interactive application. From the definition of operators involved in this notation, a set of systematic rules for deriving both dialog and presentation models at the same time are developed. For each operator, a decision tree presenting designers with design alternatives is proposed and based on dialog attributes (i.e., dialog control, dialog sequencing, dialog mode and function triggering mode) and dialog properties (e.g., visibility, observability, browsability). The refined version of the Windows Transition graphical notation is finally exploited to graphically represent the models that can be generated.

1. INTRODUCTION

In model-based approaches (Paterno, 2000), starting from a task model for model derivation is important:

1. It is naturally the definitive knowledge source where the task is expressed according to the user words, not in terms of system vocabularies.

2. Other models do not contain expressive information enough to derive something significant. The expressiveness power of the derivation can be maximized.

3. Other types of models, like data models or domain models are not especially appropriate as they induce a set of particular tasks (e.g., insert, delete, modify, list, check, search, print, transfer), which are not necessarily corresponding to the user task. Rather, to carry out another interactive task, which can be any combination of parts or whole of these tasks, the user has to switch from one predefined task to an-other, which is not user-centered. Exceptions can still occur: for instance, Teallach (Barclay et al., 1999) derives task, presentation and dialog models from a domain model provided by the underlying data base management system. In this case, UIs for data-intensive systems are the target applications, which is a specific domain in information systems.

4. Finally, derivation of other models from a task model is also allowed, e.g., an activity-chaining graph from a task model (Vanderdonckt & Bodart, 1993) to come closer to the designer's world, but there is a potential risk of information loss in the transformation.

To derive presentation and dialog models from a task model, two usual types of tasks relations can be exploited:

1. *Structural relations* according which a task is recursively decomposed into sub-tasks to end up with actions working on domain objects.
2. *Temporal relations* that provide constraints for ordering (sub-)tasks according to the task logic.

2. DERIVATION RULES FOR PRESENTATION AND DIALOG MODELS

Before defining and exploring derivation rules, working hypotheses need to be set up. In the rest of this paper, the ConcurTaskTrees notation (Pterno, Mancini, and Meniconi, 1997) is used as it already supports both structural and temporal relations that are commonly used in task analysis. The Window Transition refined notation is assumed to model the mid-level dialog, along with four dialog attributes.

Figure 1. Composition of presentation and dialog elements for Wi.

Firstly, a mono-window approach is adopted here, that is one window is generated for each (sub-)task element appearing in the decomposition. A multi-window approach where several windows are generated either for a single (sub-)task element or for grouped elements has already been analyzed. Let Ti denote the ith task or sub-task element in this task model and Wi the corresponding window (Figure 1). Only system tasks with or without user feedback (such as user confirmation for launching an automated process) and interactive tasks are here considered. No user tasks or other types of task (e.g., mechanical task on a machine) are covered as they are not linked to the interactive system. Let Pi denote the presentation part which is proprietary to Wi. Pi typically consists of control widgets like edit box, radio button, check box, list box. Let Di denote the dialog part which is proprietary to Wi. Di typically consists of dialog-oriented widgets like "Ok", "Cancel", "Close", "Search" push buttons, icons, command gestures, function keys. Let Ci,j denote a dialog widget allowing a transition from Wi to Wj. There can be several such widgets for similar or different transitions.

Secondly, for the purpose of this paper, the four dialog attributes affecting the selection of Wi, Pi, Di, Ci, and Ci,j are defined as: dialog control is explicit, dialog sequencing is multi-threaded multi-programmed, dialog mode is asynchronous, and function triggering mode is manual and explicitly displayed (thus, the graphical representation in figure 1.) Other values for these dialog attributes can be considered equally, but they here facilitate exemplification and are usually defined as such.

Thirdly, it is assumed that widget selection is performed independently as follows: each domain element manipulated by the task element Ti gives rise to one or several abstract interaction objects [4,16,17]. These objects are turned into concrete ones, belonging to Pi.

Fourthly, for the simplicity and the concision of this paper, and without any loss of generality, it is assumed that any parent task element (except the root element) is decomposed into two child task elements. From this simplification, a generalization can be easily applied by expanding the derivation rules for any combination of elements.

3. THE ENABLING OPERATOR WITH INFORMATION PASSING: T2 [I]>> T3

This operator is one of the most frequent ones as it allows the presentation of I, the information passed from sub-task T2 to sub-task T3. Let us denote by WI the window materializing I, PI the presentation of this information and by DI the dialog related to this information if needed. For instance, it could be a "Search" push button, a "Validate" icon. In order to explore design alternatives that this operator may engender, we introduce the *visibility* property. The information passed between task elements is said *visible* if and only if the user is able to access the information in some certain interaction. When no interaction is required, this information is said *observable*. When some interaction is required, this information is said *browsable*. The introduction of this property is motivated by the observation that not all information items should be displayed all together at a time. Critical information should definitely be observable, but unimportant information should not. Rather, the user should be able to access this information, but only on demand. Information browsable can be in this state *internally* or *externally* depending on the information is presented within or outside the scope of the initial window, respectively. In the last case, the dialog can be *modal*, respectively *modeless*, if the externally browsable information should be terminated, respectively not terminated, before returning to the initiating window. Information simultaneously visible in multiple task elements is said *shared*. The designer can specify these parameters either globally for a particular task or locally by redefinition for particular sub-tasks.

Figure 2 depicts a decision tree resulting from the examination of different configuration cases induced by visibility from sub-tasks T2 and T3. Each configuration ID will be referred to in the text. The A region as graphically defined is repeated for (11), (12, (13), and (14) configurations.

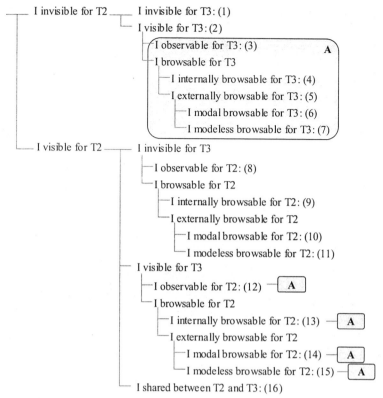

Figure 2. Decision tree of alternative specifications for the enabling operator with information passing.

Configuration (1) is similar to the simple enabling operator. Configuration (2) is similar to that operator too, but W3 is replaced by the window configuration reproduced in figure 3a. When I is not observable for T3, it is considered that I should be browsable in some way. When I is internally browsable (4), figure 3b reproduces the resulting expanding window with two dedicated dialog elements: CI>> expands the current window to let I appear (e.g., via a "More>>" button) while CI<< reduced the current window to remove I (e.g, via a "Less<<" button). When I is externally browsable (5), two cases between W3 and WI are possible: the dialog could be modal (6) or modeless (7) as represented with the marker on figure 3c.

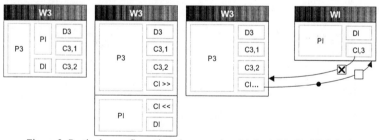

Figure 3. Particular configuration changes when I is invisible for T2 (a,b,c).

When I is observable for T2 (8), only W2 should be replaced by figure 4a. When I is internally observable for T2 (9), W2 becomes structured as represented in figure 4b. And, similarly to above, when I is externally observable for T2, figure 4c provides the change for modeless case (11). For modal case (10), the marker is removed.

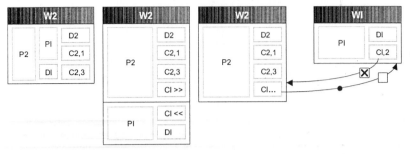

Figure 4. Particular configuration changes when I is visible for T2 (a,b,c).

The visibility of I for both T2 and T3 raises a potential problem of redundancy and/or consistency. When I is observable for T2, the pattern represented by the "A" rectangle on figure 1 is duplicated for each sub-case. In these configurations, there is no risk of redundancy of I presentation and dialog elements since the sequence between T2 and T3 imposes that W2 and W3 cannot be displayed simultaneously. Therefore, I is never displayed two times on the user screen. For instance, figure 5a graphically represents the I observability in both T2 and T3, which is a possible configuration in (12). Since opening W3 closes W2 and vice versa, I is only displayed once.

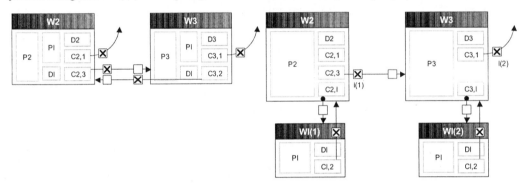

Figure 5. Particular configuration changes when I is visible for both T2 and T3 (a,b).

This reasoning also holds for subsequent categories (13), (14) and (15) except the case where the two instances of WI are modeless. They should be independent due to the enabling operator between the two sub-tasks: this is why closing W2, respectively W3, also closes the first, respectively the second, instance of WI as shown in figure 5b. Figure 6 graphically depicts the last special configuration (16) where I is shared among W2 and W3. In this case, WI can be created by both W2 and W3, but once. WI should stay visible until T3 is achieved in W3, not before. Therefore, W2 could be created from W2 and destroyed from W3 (i.e. when leaving W3).

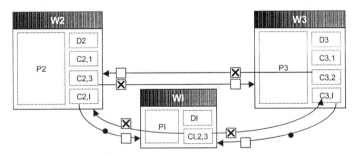

Figure 6. Particular configuration when I is shared among T2 and T3.

4. CONCLUSION

Although theoretically possible configurations are gathered in figure 1, it might be desirable to avoid some configurations for usability reasons. Here is an incomplete set of guidelines providing some assistance to the designer this way:

- The invisibility of I from both T2 and T3 (1) should be avoided since no user feedback of I is produced towards the user.
- Modeless browsability of I in T2 and/or T3 should be preferred to modal browsability as the window initiating WI and WI can be used simultaneously, thus not preventing the user to forget information in the initiating window.
- I should be visible in W3 in some way as it will be used to achieve T3.
- WI, as every window, should be controlled by one dialog element at a time, but several dialog elements concurrently. It can be updated according to the results produced in other related windows.

The presentation and dialog should foster information *persistence*, the faculty of I remaining visible is some way when the transition from W2 to W3 is operated.

REFERENCES

Ballardin, G., Mancini, C., Paternò, F.: Computer-Aided Analysis of Cooperative Applications. In Vanderdonckt, J., Puerta, A. (eds.): Proceedings of the 3rd Int. Conf. on Computer-Aided Design of User Interfaces CADUI'99 (Louvain-la-Neuve, 21-23 October 1999). Kluwer Academics Publishers, Dordrecht (1999) 257–270

Barclay, P.J., Griffiths, T., Mc Kirdy, J., Paton, N.W., Cooper, R., Kennedy, J.: The Teallach Tool : Using Models for Flexible User Interface Design. In Vanderdonckt, J., Puerta, A. (eds.) : Proc. of 3rd Int. Conf. on Computer-Aided Design of User Interfaces CADUI'99 (Louvain-la-Neuve, 21-23 October 1999). Kluwer Academics, Dordrecht (1999) 139–158. Accessible at http://img.cs.man.ac.uk/teallach/publications/Cadui99/CADUI99.ps

Johnson, P., Johnson, H., Wilson, S.: Rapid prototyping of user interfaces driven by task models. In Carroll, J.M. (ed.). Scenario-based design: envisioning work and technology in system development. John Wiley, New York (1995) 209–246

Paternò, F., Mancini, C., Meniconi, S.: ConcurTaskTrees: A Diagrammatic Notation for Specifying Task Models. In Proc. of IFIP Int. Conf. on Human-Computer Interaction Interact '97 (Sydney, July 1997). Chapman & Hall, London (1997) 362–369

Paternò, F.: Model-Based Design and Evaluation of Interactive Application. Springer Verlag, Berlin (1999).

Vanderdonckt, J., Bodart, F.: Encapsulating Knowledge for Intelligent Interaction Objects Selection. In Proc. of ACM Conf. on Human Aspects in Computing Systems InterCHI'93 (Amsterdam, 24-29 April 1993). ACM Press, New York (1993) 424–429. Accessible at http://www.qant.ucl.ac.be/membres/jv/publi/Inter CHI93-Encaps.pdf

Wilson, S., Johnson, P.: Bridging the Generation Gap: From Work Tasks to User Interface Designs. In Vanderdonckt, J. (ed.). Proceedings of the 2nd Int. Workshop on Computer-Aided Design of User Interfaces CADUI'96 (Namur-5-7 June 1996). Presses Universitaires de Namur, Namur (1996) 77–94.

Knowledge Discovery in Human-Machine Systems

John Murray, Ph.D.
Pacific Consultants LLC
Mountain View, CA

Thomas J. Ayres, Ph.D.
Exponent
Menlo Park, CA

Madeleine M. Gross, Ph.D.
EPRI
Palo Alto, CA

ABSTRACT

Various industrial process management centers and network control rooms gather and archive comprehensive numeric and textual documentation on their daily operations. These constitute valuable information sources on the circumstances surrounding incidents or mishaps associated with human errors. This paper examines some aspects of a process to apply advanced text analysis techniques to help discover information regarding the context of such events, with particular attention to domain concept identification and taxonomy development.

1. INTRODUCTION

Human performance has been noted as having critical importance for safety in many systems operations environments (e.g. Lees & Laundry, 1989; Gross et al, 2000). In some fields, such as nuclear power generation, a high level of government oversight and regulatory control applies to the industry. The data archives resulting from such oversight form a significant reservoir of information about the circumstances surrounding any human error-associated accident, incident, or near-miss. Additional value is provided when this information is considered in conjunction with the long-term plans and day-to-day operational activities around the time of such incidents. The very abundance and variety of such information, however, can overwhelm a human analyst who is attempting to discern trends and patterns in such data (Murray & Liu, 1997).

This paper briefly characterizes some typical sources of numeric data and text material available in these operational environments. Analysis of such documentation would be expected to help understand the circumstances associated with human performance deficiencies. Some research approaches to automating this type of text analysis are described.

Complex operations centers typically have several different activity-recording, data-gathering, and exception event notification procedures. In addition to exception event reports, a large array of routine telemetry, numerical or coded operational data is often gathered and archived on a daily basis. Figure 1 offers an illustration of this data environment. The challenge is to develop an analysis process which can combine chronologically-based data with the individual reports that capture current status at a given instant in time.

SumGen (Maybury, 1995) is an example of how the chronological component of this type of system can be designed. It is an object-oriented tool for analyzing chronological event messages occurring during a military engagement. It is designed to infer and summarize the overall process and activities in progress, for command and control purposes. A candidate application for this approach may be found in the U.S. Army's Land Warrior system (Murray, 2000).

The combined set of textual reports and coded or numeric data constitutes a valuable source of useful information about the circumstances leading up to and surrounding human performance failures. The challenge here is to tease out the sometimes-obscure features of the data that, in conjunction with key phrases and nuances of text, help to explain the context and situation surrounding particular types of errors and mishaps. It is particularly important to identify practical methods of integrating the analysis of textual material with the search for patterns in the chronological data.

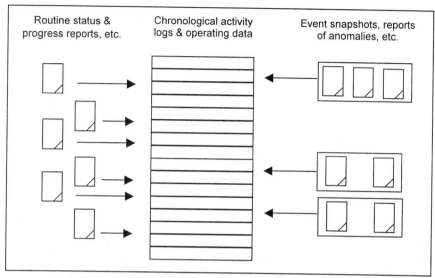

Figure 1: Data environment comprising instantaneous and chronological components (Murray et al, 1999).

2. IDENTIFYING DOMAIN CONCEPTS

Many text analysis systems incorporate some domain-dependent attributes, typically identified as a taxonomy of salient words and phrases. A thesaurus, which groups words and phrases into related concepts, may also be used. In the ideal case, a specialized dictionary or list of common terms used in the domain is already available and can be adapted for use by the analysis system. However, this is often not the case, and the task of manually generating such lists can be quite laborious. Thus, automating some portion of this activity is of interest. It is obviously not sufficient simply to search a text data corpus for the keyword entries in the taxonomy. An initial, syntactic parsing of the text can produce information about parts of speech, etc. There also needs to be a semantic-level analysis of the concepts extracted.

AutoSlog (Riloff, 1993) was a system that automatically built a domain-specific dictionary of concept nodes. The dictionary was intended for use by text analysis systems such as Circus (Lehnert et al, 1992), which extract selective concepts from a corpus of texts in structuring their information findings. Each entry in the domain-specific concept node dictionary had a set of data slots to be filled by lexical analysis of the texts. Concept node lists help to classify the contents of a large corpus of texts by developing them into augmented relevancy signatures. A signature represents the presence of some key phrase in a text, and is composed of the concept nodes, its triggering keyword, and some related semantic features. The semantic features are derived from the particular usage of the keyword, and provide information on the document context where the key phrase was found. In addition to the use of concept nodes in augmented relevancy signatures, the authors also describe their application in case-based text classification tasks.

Another approach to identifying concepts in documents for classification purposes is described by Mani & Bloedorn (1997). Concepts denoted by words, phrases, and proper names in a text are represented positionally as nodes in a graph. The edges linking them correspond to the semantic relations between them. These links may be whatever relationships can be feasibly identified in the context of the domain - for example, the role or job title that links a person to a business or organization.

SPARSER (McDonald, 2000) is a fully-implemented system for natural language understanding, which was designed for rapid analysis of short, single-subject text items. It conducts an in-depth, linguistically-principled analysis of those components of the text that it can interpret, and ignores the rest. The output from SPARSER is a domain-level semantic model made up of objects, properties, and relations that were detected in the text. However, since a domain-level model developed in this way is expressed in terminology drawn from the text, it must still be integrated in some useful way with any existing domain taxonomies.

Iwanska et al (2000) describe how automated development of a taxonomy is feasible by focusing on basic natural-language constructs in the text corpus. For example, the phrase "not only A, but B" implies the relationship sense that B is a stronger notion that A. Similarly, "X, such as Y...Z" offers the sense that Y...Z are examples or instances of X. This type of approach appears to be suited to descriptive or definitional documents, which discuss some aspects of the domain, rather than event-related reports. Some useful material for human performance investigations might be obtained in this manner from standard operating procedures or user manuals.

3. TAXONOMY DEVELOPMENT

A significant issue facing text data exploration projects is the development of a rich and useful taxonomy that covers the domain of operational performance, human error, and workplace safety. Various industries have standard technical glossaries and category lists that can help address this task. However, the challenge is to link the abstract or generic codes represented by these lists with the salient words that actually appear in a correspondingly-coded event report. The terminology used by individuals who write reports and document events does not necessarily conform to the structured categories provided by the glossaries and coding schemes.

The job of developing a usable text analysis dictionary is been addressed in some fields. For example, the LexiQuest suite of text analysis products includes specialized dictionaries that are tailored to the financial, medical, and legal professions, as well as several other domains. In the human factors arena, some similar work has been done in connection with the modeling activity using the Quorum text analysis system (McGreevy & Statler, 1998). The development of such a taxonomy relies upon detailed input from subject matter experts (SMEs). One way to achieve this is to use a set of source material that has already been classified or pre-coded in some fashion. Assuming that such a coded corpus of text material is available, we now discuss a process that can be used for exploring such a taxonomy development.

From the complete text data corpus of coded event reports, a subset is selected that have been coded with causal factors involving human performance lapses and errors. Reports that include other causal factors as well may need to be analyzed separately. This subset is used to build a semantic network, using an approach such as that referenced earlier (Mani & Bloedorn, 1996), to derive a salient common terminology for them. This will form the basis for the human factors-related vocabulary. The intent here is to identify the domain concepts that are distinctive to human performance.

Most vocabulary entries referring to system equipment and components should be excluded in this analysis, since these words can occur in reports that are coded with other causal factors unrelated to human performance. Typical exceptions to this rule are terms representing data displays, control devices, and other user interaction equipment. The edited vocabulary is then used to seed a cluster analysis, which is applied to a larger subset of the text data reports. This subset should be representative of the overall corpus, in terms of the mix of coded causal factors.

4: CLUSTER ANALYSIS

Cluster analysis is typically used to arrange a collection of documents so that similar documents are grouped together, and documents in different groups are clearly distinct from one another. An example of a hierarchical clustering model, which relies upon neural networks using adaptive resonance theory, is discussed by Muñoz (1997). The network used to generate compound keywords and create semantically-meaningful term associations. The strength of associations between terms is measured numerically, and is used in combination with co-occurrences to produce concept clusters.

Hierarchical clustering is provided by various commercially-available products, including IBM's Intelligent Miner for Text toolset. Such products generally offer a tree view of the document collection. Documents in one cluster share a high degree of similarity, which is determined on a document-by-document basis from the weight and overlap of the keywords derived from the edited vocabulary. The tool enables the user to adjust the grouping thresholds and identify when different patterns of tree structures and clusters begin to form. This clustering process is used to refine the vocabulary contents. The intent here is to shape the clustering mechanisms, in order to help reflect the coded causal factors that were assigned by the human analysts. A number of iterations of this process are likely to be required, to fine-tune the process by adjusting the weightings assigned to certain phrases, combinations of terms, etc. The resultant coding-vocabulary combination can be viewed as a basic taxonomy of features associated with the human factors issues identified in the corpus of event reports. When it has been adequately refined, a testing and validation phase is intended, where the clustering process is applied to the entire text data corpus.

Note that the reliability of this overall process hinges upon the accuracy of the original coding of causal factors. For this part of the project, the pre-coding of reports corresponds to the input from subject matter experts. As mentioned, SME input is required to properly prepare these type of analysis tools. Some industry organizations and regulatory bodies have established standardized coding schemes for characterizing failures and errors in their respective fields. For example, the topic is an area of interest in the areas of aviation (e.g. Gosling et al, 1998) and nuclear generation (US NRC, 2000). The US Dept. of Energy standard for root cause analysis procedures also includes some generic human performance causal factors (US DoE, 1992).

It is recognized that the work of developing a taxonomy of this nature is a complex task. For example, it is a significant challenge to deal sensibly with nicknames, site-specific acronyms, misspellings, etc. By narrowing the domain of interest to just the terminology that concerns human-system interaction, it may be possible to enhance the accuracy of the system. A reference lexical system can be used to help resolve the semantic inter-relationships among general language terms, such as synonyms, constituent parts, etc. WordNet is one such lexical system, the latest version of which is being enhanced to use contextual information derived from the texts under analysis (Harabagiu & Moldovan, 2000).

ACKNOWLEDGEMENT

Portions of the background work for this paper were carried out in conjunction with the Automated Text Analysis task of the Automated Human Performance Analysis Tools Program, under the direction of Madeleine Gross (*mgross@epri.com*), Program Manager for the Strategic Human Performance program at EPRI.

REFERENCES

Cox E (2000). Free-Form Text Data Mining: Integrating fuzzy systems, self-organizing neural nets, and rule-based knowledge bases. PC AI, 14 (5), p22-26.

Gosling G, Roberts K, & Jayaswal A, (1998). Improving the Representation of Human Error in the Use of the Flight Crew Integration Tool, UCB-ITS-RR-98-5, Institute of Transportation Studies, Berkeley, CA.

Gross, M. M., Ayres, T. J. & Murray, J. (2000). Analysis of human error at electric utilities. Proceedings of the 44th Annual Meeting of the Human Factors and Ergonomics Society, Vol. 3, 173-176.

Harabagiu S, & Moldovan D, (2000). Enriching the WordNet Taxonomy with Contextual Knowledge Acquired from Text, NLP & Knowledge Representation [Iwanska & Shapiro eds.], AAAI, 302-333.

Iwanska L, Mata N, & Kruger K, (2000). Fully Automatic Acquisition of Taxonomic Knowledge from Large Corpora of Texts [Iwanska & Shapiro eds.], AAAI, 335-345.

Lees R. & Laundry B. (1989). "Increasing the Understanding of Industrial Accidents: An Analysis of Potential Major Injury Records," Canadian I. of Public Health, Vol. 80, No. 6.

Lehnert W, Cardie C, Fisher D, McCarthy J, Riloff E, & Soderland S. (1992). Description of the Circus system as used for MUC-4. Proc. of the Fourth Message Understanding Conference, pp. 282-288.

Mani I & Bloedom E, (1997). Multi-document summarization by graph search and matching. Proceedings of the 14th National Conference on Artificial Intelligence, 622-628.

Maybury, M, (1995). Generating summaries from event data. Information Processing & Mgmnt, 31 (5), 735-752.

McDonald D, (2000). Issues in the Representation of Real Text: The Design of KRISP, Natural Language Processing and Knowledge Representation [Iwanska & Shapiro eds.], AAAI, 77-110.

McGreevy M & Statler I, (1998). Rating the Relevance of Quorum-selected ASRS Incident Narratives to a "Controlled Flight into Terrain" Accident, NASA Tech. Memo 1998-208749, NASA Ames Research Center, CA.

Muñoz A, (1997). Compound Key Word Generation from Document Databases Using A Hierarchical Clustering ART Model. Intelligent Data Analysis, 1 (1), http://www-east.elsevier.com/ida/browse/96-5/ida96-5.htm.

Murray J (2000). Wearable Computers in Battle: recent Advances in the Land Warrior System, International. Symposium on Wearable Computers 2000, Atlanta, GA.

Murray, J., Gross, M. M. & Ayres, T. J. (1999). Human error in power plants: A search for pattern and context. Proceedings of the Silicon Valley Ergonomics Conference & Exposition, 187-191.

Murray J & Liu Y, (1997). Hortatory Operations in Highway Traffic Management, IEEE Trans. on Systems, Man, and Cybernetics (A), Vol. 27(3).

Riloff E, (1993). Automatically constructing a dictionary for information extraction tasks. Proceedings of the 11th Conference on Artificial Intelligence, 811-816.

US NRC, (2000). Human Factors Information System (HFIS), http://www.nrc.gov/NRR/HFIS /index.htm, U.S. Nuclear Regulatory Commission, Washington, D.C.

US DoE, (1992). Root Cause Analysis Guidance Document, DOE-NE-STD-1004-92, U.S. Department of Energy, Washington, D.C.

Designing Mobile Applications: Challenges, Methodologies, and Lessons Learned

Roman Longoria, Ph.D.

Oracle Corporation; 500 Oracle Parkway; MS 2OP10; Redwood Shores, California 94065; USA
roman.longoria@oracle.com or roman.longoria@acm.org

Abstract

With mobile devices and wireless infrastructures becoming more powerful and ubiquitous, the corporate world is striving to use these technologies to keep their businesses competitive. One way is to provide access to enterprise applications via wireless, mobile devices. Designing mobile enterprise applications provides unique challenges. The obvious challenge is to design for small devices and usage in a mobile context; however, one may also need to design for integration with desktop products, diverse and highly vertical applications, novice users as well as domain experts, and a scalability of functionality that usually exceeds traditional consumer-oriented mobile applications and services.

Strategies for the design of these applications include a User Centered Design (UCD) process that focuses on users, tasks, and the contextual environment. Strategies for building these applications include either developing directly for a specific device, or using a technology stack that allows the developer to build a single application that can be accessed on a variety of mobile devices. The combination of these strategies is reinforced by a methodology that focuses on the realistic usage of a mobile application, and also reduces the development cost and time.

In following the UCD process, iterative design cycles and usability testing have yielded a wealth of design ideas and validation data. Lessons learned are provided in this paper as well as prototypical designs that tested well and those that tested poorly.

1. Challenges for the Designer

Designing usable software applications has never been an easy task. It seems that just as we are getting a handle on designing good interfaces, the technology changes as we are faced with a new set of challenges. Software user interfaces have evolved from character based interfaces to GUI desktop applications to the web and all of its facets. Modern wireless mobile applications are the newest wrinkle in the UI design landscape. Although applications running on mobile devices are by no means a new phenomenon, there is no question that recent technical advances have catapulted the mobile craze to a new, higher level. There are many challenges in designing wireless applications for mobile devices. Some of these involve parameters dictated by the hardware itself, the wireless transmission technology, and requirements specific to the applications themselves.

There are obvious hardware constraints that hinder the usability and utility of mobile applications. For example, the screen size to display and interact with information is very small. The usable display area varies within and between device types. At Oracle, we have been able to come up with device type specific estimates (in pixels) for safe design heuristics. For standard web-enabled phones, a typical width is around 96 pixels (or 12 characters). The height depends on how many lines the phone has. Our heuristic is basically 8 pixels in height per display line (e.g., 3 lines = 24 pixels, 6 lines = 48 pixels). Personal digital assistants (PDAs) also vary in screen size. For the Palm™ we design for 153 X 144. For Windows Pocket PC devices, we design for 240 X 268. For horizontally oriented "smart phones", basically a combination of phones and PDAs, we design for 240 X 120. Compare these numbers to typical desktop or web applications designed for either 1024 X 768 or 800 X 600. No matter the device, every pixel counts, and all effort is made to maximize the available display area.

Another hardware constraint is that text input methodologies are annoying at best. Entering data on a phone or PDA is tedious and requires a certain level of dexterity. Even new text input and hand writing recognition technologies do not really ameliorate the problem. Voice input may someday be a viable solution, but as of yet, it is not perfectly reliable. To put it simply, applications that require text input annoy users.

Low bandwidth transmissions are also a hindrance. Slow download time seems to be second only to text input as a user annoyance. Most devices can download between 9K to 14K per second (no faster than a 10 year old modem!). This makes every page count even more.

To make matters worse, designers and developers are faced with technology that is in constant flux, and users are faced with all of the idiosyncratic device specific differences in behavior. The devices themselves are often awkward

91

or clumsy to hold, much less use in a truly "mobile" context. In addition, there is no single industry platform, and there are no user interface standards or conventions that a designer can count on. Things become even more complicated when one considers the evolving combination of phones and PDAs.

Other challenges facing designers have to do with the applications themselves. Although the problems here are certainly related, or indeed imposed by, the hardware and technical constraints mentioned above, not all design challenges rest on the technology. There has been tremendous excitement and growth in consumer–oriented mobile services, but it seems that there still is a basic lack of understanding about what users want and need in mobile contexts. Just how much web-browsing or e-commerce do people want to do from their phone or PDA?

This problem becomes increasingly important as the corporate world seeks to have its enterprise-level business software become "mobilized". Additional challenges arise in this case due to migrating the functionality of desktop or web-based applications to mobile devices (for the purposes of this paper the term "desktop application" will refer to both traditional "thick-client" and web-based applications). Just what is the mobile context of your enterprise users? What assumptions can one make about the needs of the users in a mobile context? How much positive or negative transfer is there between desktop applications and mobile applications? These are all major questions that should be addressed in the design of the mobile applications.

Trying to defining *the* mobile context is a difficult endeavor. Are users in a car, in a train, at the airport, on the golf course, on a yacht, on a sales call, at home? Are they walking, talking, sitting, standing, leaning on bar, in well lit conditions or in the dark? You can never really know, and you cannot, and probably should not, design for all the possibilities.

There are a few things you can count on. By definition, in the mobile context users are not seated in front of a computer. They have a critical need for either information or to perform some task. This information or task is usually a very specific subset of what is available from the desktop. Users probably want to *get in* and *get out* as fast as possible.

One of the most challenging aspects of designing mobile applications is not the actual screen layout, but is the defining of functional requirements and information architecture. "Feature shedding" is a critical aspect of designing mobile applications. Trying to fit all of a desktop application's functionality or directly mapping its task/information hierarchy into a mobile application will cause far more usability and utility problems than you are trying to solve. Although you may be used to thinking what types of features you can add to your software product, when designing mobile applications, it is just as important to consider those features that you should leave out.

2. Tools and Methodologies for the Designer and Usability Engineer

The good news is that all of the traditional user-centered design tools of the trade are perfectly applicable in designing mobile applications. This should not be a big surprise. From needs analyses, prototyping, and evaluation, an iterative cycle of designing and data collection is just as important for mobile applications as it is for any other type of product. Just because mobile applications are small and lightweight (if you have done your job right) does not mean that the same level of validation via usability testing is not necessary. To the contrary, evaluation is just as critical because the cost of a user error or an inefficient navigation flow is so high due to long download times.

Ideally, the place to start is with needs and task analyses. Questions to ask include "what tasks are users doing outside of their offices". What information is critical in supporting their tasks? What limitations does being out of the office place on their work or on their role in the organization's work process? This will help identify and prioritize the features to keep and the features to shed in the mobile application.

After identifying the features for the application, the next steps are fairly straight forward. Creating task flow diagrams is relatively simple and will help articulate the UI architecture. Prototyping the application should follow and is a great way to visually and interactively demonstrate the flow of the application. Fortunately, prototyping mobile applications is not a difficult task. Wireless Markup Language for mobile phones (WML) is very similar to HTML, and easy to learn. Creating "web-clipping" applications for the Palm™ is also simple and consists mainly of writing HTML. Tools to help build applications and emulators to run them are available for free on the Internet (http://www.palmos.com/dev/tech/webclipping/; http://developer.phone.com/ ; http://forum.nokia.com/main.html).

Usability testing mobile applications with emulators has its pros and cons. Emulators run on a desktop, and provide the user with a high-fidelity simulation of the application. Testing can be run in a traditional usability lab, and screen capturing devices can be used to record the users' interactions. Testing with emulators allows for capturing data on the users' performance and understanding of the application's navigation structures, clarity, and efficiency. Subjective data can also be informative on the utility of the features and perceived usefulness of the application.

One should be cautious however in basing all usability assessments on emulator testing. Since the emulators run on a desktop and either access local files or utilize high speed Internet connections, one cannot capture the effects of the slow download time on users' performance and subjective ratings. For example, longer download time may adversely affect performance because the users may get distracted within their mobile contextual environment, or may simply forget information that was stored in their short term memory. Also, the environment may hinder performance. Low lighting conditions, noisy distractions, traffic, or whatever else the users may be faced with can all have effects on how well the user can interact with the mobile application. Subjective ratings can also be influenced by slow download time. Applications which are feature rich and aesthetically enhanced with graphics may test well on a fast emulator, but may annoy users in real-time environments when users are in a hurry to get the information they need or to perform some task. Subjective ratings can also be influenced by the hardware itself. The combination of text input, viewing the small displays, and software operation can reduce the users' overall satisfaction with the application.

Testing on live applications with the actual devices can provide performance, subjective and very realistic anecdotal information. These tests can be effective to see if users can find, recite, or recall information while using a mobile application. One can also obtain subjective ratings, such as user satisfaction and perceptions of the utility of the application. These subjective ratings are actually more realistic than those captured via emulator testing because any effects that a slow download time and hardware human factors would have on these ratings would be captured. There is also the possibility to capture anecdotal information that may be caused by some environmental factors. This type of information can be difficult to anticipate from testing inside a usability lab. An additional benefit is that mobile applications are relatively easier to develop, compared to a traditional desktop application, developed in C++ for example. Therefore, testing live code does not necessarily imply that you are testing a product late in the development cycle.

Testing on live applications also has its difficulties. Without the aid of a screen capturing system, observation of user interaction is problematic. Having the users stand or sit still and pointing a video camera at the device can provide some degree of viewing, but the video quality is poor, and can be obscured if the user moves or shakes. Anchoring the device in a cradle can help, but this reduces the fidelity of the environment. It is possible to develop server-side tools to record user navigation and even errors, but this is costly as development resources are needed.

One drawback, whether testing live applications or via an emulator, is that you are confined by the existing technology stack. For example, for phones, you are restricted to today's WML, HDML, or cHTML. For PDAs you are restricted to HTML 3.2 or the subset used in Palm™ "web clipping" applications. These restrictions will prevent prototyping conceptual, "blue sky" designs. For future oriented designs, one will be better off using some other prototyping tool.

An additional, more technical, tool that both helps and hinders product design is a technology stack that allows one to design and build a single application that can be accessed on a variety of mobile devices. This "device-agnostic" approach saves development resources. Instead of developing specifically for each device's native language, the application is developed using meta-data structures. The technology stack then correctly translates the application so that it is accessed on different devices. If done properly, the same application would maintain consistency in its overall task flow and hierarchy, its terminology, and its general concepts. However, one would be able to make the most out of the differences in the devices. For example, one could make use of the extra display area in PDAs by forcing a different layout than is used for phone applications. This type of mobile application design and development approach is relatively new and will be topic of a future paper. One question that would need to be addressed is the implications that this approach would have on the somewhat competing need to allocate functionality based on the different types of devices. In other words, does one really want to surface the same functionality and application structure on a phone as is surfaced on a PDA?

3. General Lessons Learned

For over two and a half years, the Mobile Architecture User Interface team at Oracle has been designing and testing consumer and enterprise applications and platforms. Through iterative design cycles, usability evaluations, and customer needs analyses, a variety of design ideas and validation data has been generated. The following discussion outlines very high level design heuristics gleaned from these finding.

There is a need. Through interviews with existing and potential customers, it is clear that there is a need for enterprise level mobile applications that exceeds a pure marketing drive. For Enterprise Resource Planning, Customer Relationship Management, and Business Intelligence software, customers have a vested interest in accessing key functionality and information. The need for data and to accomplish critical tasks still exists in the mobile context. Invest the resources to identify what subset of functionality is needed for the user population and to define requirements for each application.

Every pixel counts. Although this sounds like an obvious comment, it is meant to emphasize the need to rethink traditional screen layout principles when designing for PDAs and phones. The critical balance between a clean layout and presenting data to the user often results in rethinking how the data is presented. For example, instead of showing data in tabular format, concatenating the columnar data into a single string saves space and has only a minor impact on visual scanning speed, especially when combined with putting the key differential data at the beginning of the string.

Every round trip counts. There is a fine line between the need to keep pages simple and the need to put enough information on the page to decrease the number of round trips to the server. When designing and testing on an emulator, it is easy to forget that the real world user is burdened by the slow download times of today's wireless technology. What may seem acceptable navigation when there is no "performance hit" may be unbearable to the end user who has to wait several seconds for each page to download. It is therefore critical to surface key information immediately and allow the user to drill down if needed.

Don't try to shove a desktop application into a mobile device. One should not have to map a desktop application's hierarchy or full feature set into the mobile application. Remember that the mobile application should solve an immediate and critical need of the user. The trick is to find out what that need is and to ignore the rest. Define the use case scenario. Ask "is there an easier way to accomplish the goal". For example, if it is easier for the user to dial 411 to get a phone number and address, one should not expect the user to use an application which only provides that data. Also, for enterprise applications, there is a threshold for each unique user need at which the user feels comfortable plugging in a laptop to access the desktop application. The mobile application should focus on the functionality and information that is below that threshold.

Avoid data entry. No matter how much the user input technology improves for mobile devices, the need for data input will continue to annoy users. Mobile applications should build in convenient mechanisms that ameliorate the need for and frequency of data input. For example, using pre-set data, either defined by the user or intelligently by the system, is very helpful in allowing form input without requiring text input.

Allow for desktop-based customization. Leave the heavy lifting to the desktop. Users really do not expect to customize applications from the device itself. While there are many types of customizations that users may want for a mobile application (e.g., what customer or employee data to show in the directory; what business intelligence to show) our data shows that users would accept the need to customize the mobile application via some desktop user interface. Focus on identifying the key customizations needed, designing for the discovery of the customization functionality, and then designing the actual screen layout for the customization pages.

Keep your navigation model simple and clear. Allowing the user to go to too many places at once can lead to unclear expectations for navigation and the users getting lost and ultimately irritated. The application hierarchy should not go beyond 3 or 4 layers deep. In addition you should avoid excessive "intra-application" navigation. At Oracle, we have found that by limiting navigation points to a few constant anchors (e.g., "Home" and "Portal"), users were more likely to find their way around the application and mobile web site. Also, when there is a potential for lateral navigation in application hierarchies, allowing a constant model for "drilling-up" is imperative. Similarly, in web-based applications, users are prone to use the devices' native "back" navigation functionality as much as possible.

Think modular. Creating modular, "plug-in", functionality will allow one to provide added value to applications when contextually appropriate. For example, functionality which allows a user to get directions might work well as a stand-alone application, but could also be very useful when it is integrated inside of another application. For instance, a user in a mobile context, looking up the address to a customer's location, may very well be on the way to the customer and need directions. By designing and building the "Get Directions" functionality as a "module", one can embed it at the point when it is needed and avoid superfluous navigation. Many web applications are doing this now, so really it is not a new concept. It is important to remember that a thorough analysis of the mobile context should identify which modular functionality to emphasize.

Think details vs. action lists. In enterprise applications, the level of functionality often exceeds that of typical consumer-oriented applications. Likewise, mobile enterprise applications' functionality can scale up and pose interesting design dilemmas. One such dilemma is the distinction between detail-oriented and action-oriented designs. Detail-oriented refers to those applications where the primary objective of the user is to get some information (e.g., finding a customer's address, phone number, or account balance). Action-oriented refers to those applications which are more heavily loaded with tasks the user can perform (e.g., buying from a store, submitting an expense report).

For detail-oriented applications, the goal is to get the information to the user as quickly and efficiently as possible. The user may then want to do something with the information, but the information is the primary goal. To go back to the customer directory example, the primary goal of the user may be to get the contact information of a customer. Once the user gets the information, the user would probably want to do something with the information (e.g., call, get directions, email). Thus one design option would be to immediately surface the customer details and then progressively disclose the added functionality.

For action-oriented applications, the user may need to choose from a variety of tasks to perform, only one of which may be to view detailed information. In these cases it is useful to design in a manner which surfaces the tasks that can be performed by presenting a menu to the user. If the user wants to drill down to get detailed information, he can, but he is not forced to. For example, in a mobile expense reporting application, users have the ability to select an existing expense report and review its details. But users can also edit it, submit it for approval, or delete it. In usability testing, users wanted to be able to perform these actions without "drilling" into the details. By not taking the user directly to the details page, all possible functionality is surfaced and more easily discovered. In general, find the most likely, primary action(s) and surface it.

Don't get too caught up in the devices. There are basically two schools of thought on which wireless mobile devices are used to access a company's enterprise mobile applications. Either the individual company is going to dictate which devices are going to be allowed, or the framework will allow access to any mobile device with a wireless connection. It seems unreasonable that many large companies would dictate specific device usage (at least for wireless connectivity). So if one is focusing on a potentially broad spectrum of devices, one may wish to focus design efforts on those with the largest market share. The specifics of the devices, however, are likely to change over time. Focus on those attributes of the device which seem more stable. For example, the pixel size of the display area is probably less stable as a device property as compared to the markup language used to deliver applications to that device.

Design for today but plan for tomorrow. Although it may not be wise to get too caught up in the specifics of the devices, one should be aware of them, for they are more likely to impact immediate design decisions. A benefit of working on mobile applications, is that the development cycle tends to be relatively short, so it is likely that one will be able to revise an application's design when new technology emerges. However, whether it is the promise of more bandwidth or better devices, one will never really know when the technology or device companies are going to deliver.

Write Corporate UI Standards. When producing multiple enterprise applications, it is important to standardize the look and feel. Providing design heuristics, although very useful, is not enough to ensure this. UI standards help provide a consistent layout, terminology, navigation model, as well as the opportunity for corporate branding. Even though mobile applications are smaller and have a lot of restrictions in their designs, there are more degrees of freedom than one might think. Without a standard approach to the design of the application suite, one stands the risk of different applications diverging on critical aspects of their user models, which in turn, can hinder the transfer of learning between the products, hinder users' performance, and be very irritating to users.

In addition, UI standards help reduce the workload of designers and usability engineers by reducing some "lower level" design decisions, such as terminology. This frees them up to tackle more complex product specific design and architectural issues, which standards may not be able to address.

4. Conclusions
Mobile applications seem to be the next evolutionary step in consumer and enterprise software. While designing these applications has its challenges, traditional usability and UI design methodologies are still applicable. The design heuristics provided in this paper should transcend continually changing mobile technologies and hardware. The future will no doubt bring new challenges and much more research will be needed. With new technology comes the chance for design innovation, and ultimately the ability to provide users with what they want and need.

Lessons Learned in Developing Human-Computer Interfaces for Infantry Wearable Computer Systems

Theodore V. Hromadka, III

Exponent – Failure Analysis Associates, Inc, Menlo Park, CA
tvhromadka@exponent.com

Abstract

The U.S. Army's Land Warrior program is adapting commercially available technologies for use by dismounted infantry in a "digital battlefield". The Land Warrior system focuses on providing real-time situational awareness and communication abilities to the soldier. Given the environment and consequences of using this system, the human-computer interface becomes critically important. This paper reviews the evolution of the software interface and the hardware used to operate it.

1. Background

In order to understand the challenges facing the user interface, one must first understand the underlying system, its functions, and the expected use cases. The U.S. Army Land Warrior system is a fighting system that is being developed to be worn by infantry soldiers. It includes advanced communication abilities to allow soldiers to more effectively perform their mission. It is meant to provide complete "situational awareness" to each member of the group. The system hardware includes geographic position sensors, a daylight video sight, a thermal weapon sight, a laser rangefinder/digital compass, and a radio, all of which are linked to and controlled by a wearable computer. The Land Warrior system provides each soldier's position and heading. All soldiers are wirelessly linked through both voice and data communications. Digital images can be captured and sent, as can text reports, laser targeting data, etc. Soldier positions, aerial photos, targets, and other useful information can be overlaid on top of a digitized map that is kept synchronized across each Land Warrior system, thus accomplishing the mechanics of providing situational awareness.

Bridging the gap between this information machinery and the soldier himself is the human-computer interface (HCI). The interface hardware consists of a monocular head-mounted display (HMD), voice headset, and two devices that act essentially as computer mice. One mouse is worn on the user's chest or belt, and the second is mounted on the soldier's rifle in an accessible location. In some configurations, a wearable keyboard and an optional handheld touchscreen are also included.

Most of the HCI effort went into the software interface, as the hardware could not be altered as easily in the constricted timeframe. The Land Warrior prototype system studied in this paper was developed under extreme time pressure in order to participate in a large warfighting experiment. It should also be noted that many software interface decisions were made in response to the characteristics of the hardware, proving the axiom of "software follows hardware". As such, the software interface, and especially its evolution, will be the main focus of this paper. It is also important to understand the typical use case of the Land Warrior system. The system is used by an infantry soldier, often while he is engaged in other activities, and who often must stay awake and alert for up to 96 hours without sleeping. The system must be simple enough to be learned quickly, yet must also be efficient and complex enough to be useful to expert users, all without interfering with the accomplishment of non-system tasks, as discussed in Nielsen (1993). The user is often operating in the harshest environments possible, while under severe strain and exhaustion, and requiring rapid, efficient task accomplishment. Above all, the system is used in situations that are literally life-or-death critical.

2. First Version of the Software Interface

With the infantry use case in mind, the first software user interface was designed around the principles of simplicity and efficiency. This version was designated version 0.5, under the theory that it was a halfway point on the road to a fully developed interface (Brooks, 1995). The version 0.5 system was never to be used in the field by soldiers; it was an experimental test system used to improve the design of the later version 0.6 system. Much of the interface architecture was based on research and prototyping performed by a previous contractor as described in the Soldier Computer Interface Design Document (1997), modified to reflect advances in technology.

Hardware requirements imposed several restrictions on the UI from the beginning. The non-illuminated screen is occasionally viewed by night-vision devices, which are monochromatic. The head-mounted display supported only a limited screen size and resolution, reducing the amount of information visible at any one time. In fact, the initial HMD, while having excellent resolution and screen size parameters, was yellowscale, which had obvious impact on the UI color scheme. The cursor controller, which was essentially a mouse, was mounted on the soldier's rifle. Operating it was similar to using a one-fingered joystick held up sideways, making drag-and-drop basically impossible. It was initially very difficult to adapt to the change in left-right orientation, although after a few weeks of use the soldiers became surprisingly adept.

The subsequent versions of the UI software share many characteristics of the first. Therefore, v0.5 will be described in greater detail, and subsequent version descriptions will consist primarily of evolutionary changes. The key features of the v0.5 UI were permanent status bars running along the top and bottom of the screen, a permanent menu bar along the left side that also functioned as a "new arrival" indicator, and a workspace area to the right that took up the majority of the screen. The workspace held the main information of interest to the user, sub-toolbars along the top for controlling that information, and often acted as a control panel in its own right. Despite research to the contrary (Nielsen, 1993), basing the UI around modes was found to be extremely useful, as the information was so easily classified as belonging to a unique UI mode.

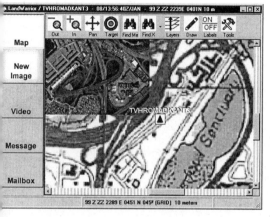

Figure 1. v0.5 Map Mode

The permanent status bars (Figure 1) were chosen to provide constant, hands-free access to the most frequently needed information, as recommended by Norman (1990). The top status bar listed user network identification, date and time, and position coordinates. The bottom status bar was broken in three sections and was used for data connection status (e.g. n-percent of a message has been sent), miscellaneous workspace information (e.g. the map coordinates of the map point under the mouse cursor), and notification of a new position update from another Land Warrior, respectively. In this way, the user was constantly streamed data that had rapid expiration times.

Early tests showed that the map was clearly the most important workspace, so it was designated as the default mode. Looking at different parts of the map presented the most difficult challenge. Sometimes the user wished to see a specific part of the terrain, sometimes the user wanted to find another user or position. The UI was constructed accordingly; the map could be set to jump to, and center around another user, track the user himself, shuffle multiple layers of maps (i.e., display an aerial photo on top of a military map), zoom in and out, or simply pan around the map using scroll bars or by clicking on edges of the map. Map functions were selected using the workspace toolbar. Buttons were marked with large images and small text captions to aid quick selection. If the mouse cursor was over a button, the normally grayscale button was illuminated in orange, and a tooltip (bubble caption) hovered near the button temporarily, explaining and describing the button (McKay, 1999). The most commonly used functions were given the most accessible locations; rarely-used functions were located under a submenu, as is typical in most software today.

Icons and map overlays were given careful evaluation. Maps are produced using high-contrast colors and in accordance with the four-color-theorem; as such, it was nearly impossible to find icon colors that would never blend in with the map itself on occasion. A solution to the color-contrast issue was not found until the next version of software. Icon captioning issues were resolved by providing a captioning on/off toggle, greatly reducing screen clutter while still allowing for icon identifcation as needed. Several factors were resolved by military regulations (Pacific Northwest National Laboratory [PNNL], 1999; U. S. Department of Defense [DoD], 1999a; U. S. DoD, 1999b). Overlays and units were given colors according to their relationship to the user, icons were positioned such that their upper left corner was the true geographic position, and icons were chosen in accordance with what type of unit it represented. Changing between modes was accomplished through the main menu toolbar located on the left side of the screen. Like the status bars, this toolbar was displayed at all times. The loss of screen real estate was justified by ease of jumping between modes.

This toolbar doubled as an indicator of newly arrived information. When a new map, image, text message, etc arrived over the data network, the appropriate mode button was illuminated and the word "New" was added (Figure 1). Clicking on that button shifted the UI to that mode and opened up the newly arrived information (Figure 2). The use of the "New" indicator took up the real estate that would have been used for graphic icons on these buttons, but in this prototype the tradeoff was acceptable.

The Image Mode interface was similar to those of common commercial image software tools such as Microsoft Paint. Image mode was used for viewing, transmitting, or freehand drawing on maps, image stills, diagrams, or other graphics. Priority was placed on minimizing mouse action, due to the awkwardness of the hardware. Control buttons were large and constantly available at the top of the screen; line drawing could be performed by connecting dots in a manner similar to CAD software.

Here again the benefits of continuous user feedback were realized when sergeants from the TSM-Soldier office in Fort Benning, Georgia suggested implementing the ability to store and select pre-drawn military icon symbols, rather than attempting to draw the symbols by hand. A map symbol toolbar was created accordingly. Users could create groups of symbols ahead of time, and symbols

97

were automatically colored by unit relationship (enemy, friendly, etc). As another concession to awkward mouse hardware, traditional drag-and-drop interface was abandoned in favor of "click-and-carry", where the user would click on the map symbol at the toolbar, release the mouse button, move the cursor to the symbol destination, and click again, thus carrying and dropping the symbol without needing to hold the mouse button down during the entire cursor movement process.

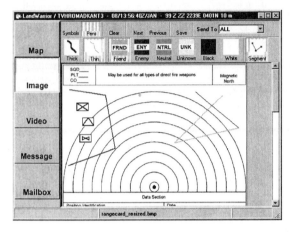 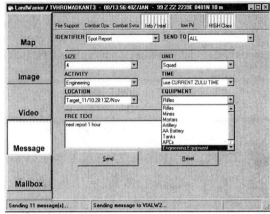

Figure 2. v0.5 Image Mode   Figure 3. v0.5 Message Creation Screen

The interface for creating and sending messages (Figure 3) was also adjusted to accommodate the hardware. Military messages tend to be very structured in order to enable rapid dissemination and good compression. Land Warrior messages are composed by selecting items from drop-down menus. Ideally, the user can pre-select which items appear as the default values. A free text box is also provided, with a soft keyboard option, to accommodate unanticipated text. Messages are read in a separate "Mailbox" interface, which closely parallels common commercial e-mail utilities. The mailbox interface remained basically the same across all versions of the UI and is discussed more fully in the second interface section.

3. Chief Lessons Learned from the First Software Iteration

During the extensive write-and-test, high user-feedback development cycle, many significant results about wearable and infantry-specific user interfaces were obtained.

The most important lesson learned was that developers must use the hardware and adjust the UI accordingly. Testing the software UI using real hardware is key. While it is true that the hardware is often unavailable during the software development process, the importance of keeping hardware limitations in mind while developing software cannot be overemphasized. In an unscientific experiment, the software was modified such that each mouse click was timestamped and recorded in a log file. Using this method, time to accomplish a sequence of tasks was measured. A group of 5 people, considered expert users of the software and hardware, were asked to change from Image Mode to Map Mode, place two map symbols in the upper left corner, draw a default-setting arrow from upper left to lower right, and then draw a large red arrow from lower right to upper right. This task set was performed 3 times each, using both v0.5 and v0.6 software. The test was performed indoors in a low-stress environment. Each version of software was run on desktop hardware (large monitor, commercial mouse) and compared

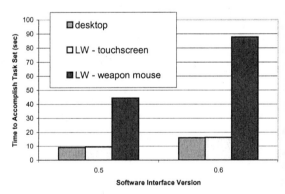

Figure 4. Comparison of Desktop Interface Hardware vs. Wearable Interface Hardware

against runs performed using (and wearing) the appropriate Land Warrior system. To ensure that computer characteristics were not a variable, the same computers were used with each set of I/O devices. The times were then averaged and are charted in Figure 4. Time to accomplish the task set were about the same on the desktop and the touchscreen, but were typically five times longer when attempted on the wearable system. This experiment clearly demonstrates the need to test software using the actual interface hardware.

everal software interface adjustments were made after testing on actual Land Warrior hardware. Buttons were enlarged to facilitate gloved fingers on the touchscreen and unwieldy mice. Related buttons were clustered close together to minimize mouse travel. It was generally found that the cursor control functioned best as an omnidirectional, pen-like instrument when used on a map or image, but as more efficient if it snapped or tabbed between buttons on menus and toolbars. This dual-mode interface is being investigated in 1.0 software.

Second Generation Software

The second version of the software interface, v0.6, was constructed in response to changes in the system hardware. The chest-mounted body mouse was added, the HMD was improved, and the weapon-mounted mouse was moved from its location above the trigger to a symmetric spot on both sides of the magazine well. The chest-mounted mouse also included buttons for operating the voice radio and toggling between computer screen and video view modes directly, without going through the software interface. The software interface resembled the previous version's in functionality, with these important differences:

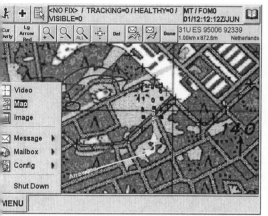

Figure 5. v0.6 Map Mode (full contrast map)

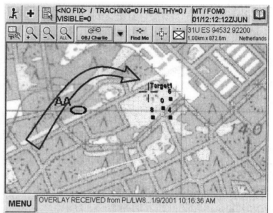

Figure 6. v0.6 Map Mode (dimmed map)

A flyout menu-driven architecture was instituted in place of many of the fixed toolbars (Figure 5) This change was made in order to increase screen real estate, and in order to provide for more functions, but came at the cost of reducing efficiency. On the menus, care was taken to consistently place graphic icons next to the text captions, with icons of related functions sharing common graphic details (Paap & Cooke, 1997).

The problem of icon/map contrast was solved by having a "dimming" feature, where the map was partially grayed out, bringing the icons into better relief (compare Figures 5 and 6).

In Mailbox mode, messages are listed in a grid at the top, and the text is displayed in a window below. In this new version, double-clicking on the bottom window text moved the software into Create Message mode, with that earlier message's data automatically filled into the drop down menus and free text box (Figure 3).

The drawing interface was reworked to allowed for complex arrows, lines, and other symbols to be drawn in CAD fashion, with minimal mouse action.

Observations from Usage Experiments

platoon of approximately 40 Land Warrior systems participated in a week-long, ultra-realistic Advanced Warfighting Experiment in September 2000 at Fort Polk, LA. Due to the fact that only the participants were allowed into the war game, detailed data about the interface usage is not available. However, anecdotal evidence and user comments were freely given.

- The new mouse location on the magazine well was not entirely successful. While it did eliminate the interference issues with the trigger, it was too far away from the weapon grip for many users' fingers to reach without letting go of the grip. Additionally, it is possible that the chest-mounted mouse was used more often during training because of its additional radio and display control buttons, and so became the natural mouse to use in the field. Almost all of the users preferred the chest-mounted mouse, despite its initially confusing left-right orientation.
- The soldiers frequently were seen to be twisting their heads around, trying to look at the corners or sides of the HMD screen. It was very difficult to adapt to the screen moving with the head. The soldiers typically flipped the HMD up and out of the

way, folding it down only to look at the screen for a moment and then returning it to its non-interfering position. Some users detached the HMD and kept it in a vest pouch, like a pocket watch. The general lessons learned from observing HMD usage was the soldiers adjusted the equipment to suit their needs, and typically preferred to have the display in an accessible but non-intrusive location when not in use.

- Cognitive load could become an important factor as the user grows fatigued or is engaged in combat (Marshall, 2000). In situations of high stress, and generally after staying awake for long periods of time, the soldiers relied less on the complicated features of Land Warrior and more on their years of non-Land Warrior training. Part of this is attributable to the intensity of military training, where behavior is learned to the point of instinct as a matter of survival. This is clearly an area requiring further study.
- Reducing screen clutter, especially in Map mode, will be a major focus of the next interface. The map image shown in Figures 5 and 6 use examples of smaller, cleaner symbols than those typically used in military maps. Current military graphics are oriented more towards units and groups, and the Land Warrior map is scaled to individuals.
- Voice communication was greatly preferred over text messaging for short messages, for obvious reasons. The on-screen keyboard was found to be too slow for word processing, and is being replaced in the next version.

6. Conclusions

This paper reviewed two versions of the User Interface (UI) developed for the Land Warrior system. We found that, in general, soldier started up the Land Warrior system, set the map to stay centered on their own position, and relied on the direct-to-hardware voice radio and video camera buttons included on the chest-mounted mouse device. Many features, especially text messaging, were seldom used, mostly because their implementation was slow or unavailable. The usefulness of the voice communication and the map showing everyone's positions is without question. The map could benefit from a reduction in screen clutter, starting with smaller map icons. The UI to date has been purely visual; auditory cues should be investigated, especially for navigation or incoming messages. Mouse movement might be more effective if the cursor control sometimes functions as a normal omnidirectional mouse in certain workspaces (e.g., drawing on a map), and sometimes functions like a "tab" key when navigating among menus or buttons. Automatically moving the mouse to the most frequently chosen following button in a sequence was proposed, but dropped due to time constraints, and should be explored. It is difficult to determine if users neglected certain functions of the system due to the user interface, the utility of the function itself, simple psychological inertia resulting from years of being trained to operate without the system, or still other reasons. As the hardware improves and more is learned about how best to utilize a completely novel system like Land Warrior, the user interface will adapt and evolve in answer to those questions. The lessons learned from this initial development period will be invaluable in this and other wearable or military projects.

Acknowledgements

The author wishes to thank the members of the 3rd Battalion, 75th Ranger Regiment and also the 82nd Airborne Division for their incalculable patience, feedback, and assistance in developing these systems. John Murray and Lawrence Stallman were the principal architects of the user interfaces, and their efforts under severe time pressure are greatly admired.

References

Brooks, Frederick P., Jr. (1995). The Mythical Man-Month (Rev. ed.). Chapel Hill, NC: Addison-Wesley.

Hughes Aircraft Company (1997). Soldier Computer Interface Design Document, Rev. B. Unpublished report, contract DAAB07-95-C-H302.

Marshall, S. L. A. (2000). Men Against Fire. Norman, OK: University of Oklahoma Press.

McKay, Everett N. (1999). Developing User Interfaces for Microsoft Windows. Redmond, WA: Microsoft Press.

Nielsen, Jakob (1993). Usability Engineering. San Francisco: Morgan Kaufmann Publishers, Inc.

Norman, Donald A. (1990). The Design of Everyday Things. New York, NY: Doubleday.

Paap, K., & Cooke, N. (1997). Design of Menus. In M. G. Helander, T. K. Landauer, & P. V. Prabhu (Eds.), Handbook of Human-Computer Interaction (2nd ed.) (pp. 533-572). Amsterdam: Elsevier Science B.V.

Pacific Northwest National Laboratory (1999). U.S. Army Weapon Systems Human-Computer Interface (WSHCI) Style Guide, Version 3. Defense Information Systems Agency.

U.S. Department of Defense (U.S. DoD) (1999a). User Interface Specifications for the Defense Information Infrastructure (DII), Version 4.0. Defense Information Systems Agency.

U.S. Department of Defense (U.S. DoD) (1999b). Military Standard 2525B, Common Warfighting Symbology Version 15. Defense Information Systems Agency.

Recent Investigations of User Interface Improvements for a Military Wearable Computer System

Lawrence E. Stallman
lstallman@exponent.com

Exponent® Failure Analysis Associates®, Inc., Menlo Park, CA

ABSTRACT

The development of an effective military wearable computer system involves many complex issues, including efficiently providing functionality to the user in stressful and life-critical situations. An iterative collaboration involving subject matter experts and human factors specialists was conducted to rapidly develop human-computer interaction improvements to the system. This paper addresses selected key issues that were explored during the development.

In an effort to provide the user with functionality characteristics appropriate for certain environmental or task situations, audible presentation of information was studied. The effectiveness and attention load required of the user to perform land navigation using either visual or audible information in varying conditions was evaluated.

During the evaluation, the user performed a primary land navigation task in a simulated environment with the aid of either a visual map or an auditory cue while responding to a secondary identification of friend or foe task. An automated data collection method was used to document the subject performance during the land navigation evaluation. Preliminary results from the tests indicate that the audible presentation of land navigation information is an effective supplement to visual presentation of land navigation information.

1. INTRODUCTION

Traditionally, land navigation has been an inherently complex task, involving terrain feature orientation, map reading, and the use of a compass. The manual performance of this task can require several minutes or more, can be inaccurate, and often requires frequent stops to verify the correct route. Further, it is a particularly difficult task to perform at night.

With the introduction of global positioning system (GPS) equipment, soldiers are able to navigate terrain with much greater accuracy and with far less effort. Coupled with a wearable computer, display system, and related equipment, the soldier can instantly see his location and compass heading on the actual map, as well as his destination. This capability allows the individual soldier to proceed without stopping for manual position checks and without risk of getting lost, even at night. In demonstration field tests of prototype equipment, soldiers have been able to navigate to a specific location in half the time it normally takes without such equipment (Bianks, 2000).

However, this capability must be provided in a way that does not impede the soldier's situational awareness – his ability to detect and react to enemy or environmental information. The soldier should be able to navigate and still be aware of his environment, particularly while traveling across dangerous terrain at night. This capability should also not induce a burden on the soldier in time-critical or otherwise stressful situations, such as approaching or exiting areas where enemy contact is planned or likely.

There are also system engineering and logistical issues to consider. Currently available computers and display devices, necessary to display maps and the soldier's position, consume significant amounts of power. The batteries required to provide adequate power for a typical mission are large and cumbersome additions to the soldier's load, and also create significant logistical difficulties concerning handling and resupply in the field. Reductions in power consumption of such a system could have significant benefits.

2. RESEARCH

While there are clear benefits to providing soldiers with a visual map display, it is also reasonable to conclude that there are problems with this approach. Interpreting the complex visual information on the map requires concentration, preventing the soldier from both maintaining continual awareness of the environment and moving effectively while viewing the map. Further, the use of a display at night can induce light adaptation issues, where the soldier's eyes become accustomed to the brightness of the display and require time to readjust to the darker surroundings.

The use of audible cues to provide navigation information appears to be a reasonable solution. The audible cue would announce the distance and direction to the destination. The use of audible information doesn't induce any light adaptation issues, and is also appropriate for situations where the soldier would want to keep moving while the information is presented (Weinschenk & Barker, 2000, p. 194).

Certainly, audible cues could mask important environmental sounds, but it is reasonable to expect that the soldier would be able to adjust the volume to a desired level. This complements the noise discipline often required of soldiers, in that voice communications and other emitted sounds cannot allow an enemy to detect the soldier. The audible cue could also be suppressed and initiated, on demand, when in a situation amenable to the subtle distraction of the audible information.

If a networked capability is provided to allow the soldier to receive other soldiers' positions, it is probable that the most efficient means of interpreting this information is through a visual representation. This complex, critically valuable information would be difficult to quickly communicate via audible cues. As such, it is proposed that audible information would supplement the visual information. Specifically, a typical usage scenario might be that the soldier checks the visual display periodically, perhaps every twenty or thirty minutes, confirming the location of other soldiers and the terrain features through which he will navigate. The soldier would then turn the display off and rely on audible cues to navigate between visual checks. Immediate power savings can be realized while the display and associated equipment is turned off or placed in a low-power state. Certainly, the operational scenario might call for more or less frequent use of the visual display.

3. EXPERIMENTATION

An experiment was designed to investigate the merit of both visual and auditory augmentation, as well as the effect both approaches had on the subject's ability to respond to external stimuli. The experiment involved the performance of a primary land navigation task, where the subject would navigate through a course of waypoint locations and a secondary task, and a secondary identification of friend or foe (IFF) task. The subject would acknowledge the presentation of enemy information, and would simply ignore friendly information.

The primary task consisted of maneuvering through a three-dimensional simulated environment. Four waypoints, situated generally at north, south, east, and west locations in the simulated environment, were used to construct various courses of approximately equal distance. The environment included multiple buildings, fences, and other obstacles, through which the subject would maneuver to reach the waypoints. This prevented the subject from simply determining the necessary heading and traveling in a straight line to each waypoint.

The secondary task IFF information was presented by visual and auditory means at random intervals while the subject performed the primary land navigation task. Visual IFF information was presented as an image of a man with a blue (friend) or red (foe) armband displayed on top of the simulated environment image. The image would appear for a random interval of time and then disappear. Auditory IFF information was presented as a pre-recorded male voice saying the word 'Friend' or the word 'Foe' at random intervals. The subject would have to interpret the visual or audible friend or foe indications, and acknowledge the foe indications by pressing a button.

These tasks were performed for five different test conditions. Two involved visual augmentation via the use of a map display that showed a map of the simulated environment, the subject's current location and compass heading, and the locations of the four waypoints. Another two conditions involved auditory augmentation, consisting of a synthesized female voice providing the subject with the distance and heading to the current waypoint. The synthesized female voice was selected to provide distinction from the male voice and prevent unrealistic expectations of the system (Weinschenk & Barker, 2000, p. 109). The last test condition lacked any augmentation via visual display or auditory cues.

For both the visual and auditory augmentation test conditions, each type of augmentation was provided under optimal and poor conditions. The poor conditions were intended to represent situations where the augmentation's effectiveness was reduced by reasonable environmental factors. The optimal visual test displayed the map with adequate brightness and contrast, while the poor visual test map display was adjusted so that the map was difficult to see. The poor visual test condition was intended to mimic the effect of using the map display in bright sunlight, an observed worst-case with current display systems. The optimal audio test condition provided the auditory cues without background noise, other than the random auditory IFF utterances. The poor audio test condition introduced loud background noise – again, a plausible worst-case scenario.

3.1 Experimental Design

The experimental environment was controlled between test conditions and subjects, such that the ambient and experimental equipment light and sound levels were measured with light and sound meters and repeated for each test condition. The test condition was the only independent variable for the experiment, and the dependent variables included the time required to complete the primary land navigation task and the time required to acknowledge the secondary IFF task foe indication.

The order of the test conditions was randomized among subjects. The only exception is the unaugmented baseline test condition, which was always the last test condition. This was done to investigate whether or not the subject had learned the land navigation course waypoint locations. If the subject navigated directly to the waypoint locations without augmentation, then the performance for the earlier test conditions would necessarily be suspect.

The land navigation course for each test condition was also randomized between subjects, although the land navigation course was the same for each of the three repetitions for each test condition. The course was presented to the subject prior to beginning each test condition, and was available for review by the subject during the test condition. This relieved the subject from having to remember the land navigation course and therefore removed an undesired variable from the experiment.

For the visual augmentation test conditions, the map was provided during the entire test condition. Each waypoint was reached when the boundary of the subject's symbol on the map display intersected the boundary of the waypoint symbol on the map. For auditory augmentation test conditions, the navigation information was provided every 10 seconds, and each waypoint was reached when the subject was within 50 meters of the waypoint. The waypoint-reached decision parameters were determined from the map display coordinate resolution during the development of the experiment, and are considered to be equivalent.

The poor augmentation test conditions were not designed to directly compare the effectiveness of the augmentations, but to investigate the utility of the augmentation in conceivable, severe conditions. If the poor augmentation test conditions showed no benefit compared to the unaugmented test condition, then the viability of the augmentation for conditions between optimal and poor would be suspect.

The secondary task IFF information was randomized for all repetitions. Both the visual and auditory IFF information would randomly present either the friend or the foe indication. The visual IFF indication would appear at random intervals between 5 and 10 seconds, at random locations on the simulated environment display, and would be displayed for random lengths of time between 0.5 and 5 seconds. The auditory IFF indication would be played at random intervals, between 5 and 10 seconds. The IFF indication timing parameters were determined experimentally from trial runs of the experiment, and are intended to provide meaningful experimental conclusions rather than mimic real-world conditions.

3.2 Experimental System Configuration

A single personal computer generated the simulated environment, map display, auditory cue, and IFF indications. The simulated environment and visual IFF indications were displayed on a wall by a video projector, and the map display was presented on a standard 21-inch monitor. The auditory cues and IFF indications were played through stereo speakers located in front of the subject. The background noise for the poor auditory augmentation test condition was provided by a small radio tuned to an unused frequency. The experimental system configuration is shown in Figure 1.

The map display monitor's distance from the subject's seat was intended to provide an image size similar to available wearable displays. The monitor was also rotated such that the subject was required to noticeably shift focus between the simulated display and the map display. This was done in an effort to introduce the effect of focal distance shifting from a wearable display close to the eye and the real, three-dimensional environment.

The subject moved through the simulated environment by pressing cursor keys on a standard personal computer keyboard and moving a standard mouse. The cursor keys allowed the subject to move forward and backward, as well as sliding to the left and right. The mouse allowed the subject to rotate to the left or right, as well as look up or down. The subject used his dominant hand to control the mouse and his non-dominant hand to control the cursor keys.

Figure 1. Experimental System Configuration

The subject acknowledged the secondary task foe indications by pressing any of several buttons on the front face of a standard game controller with his dominant hand. The game controller was placed in his lap below his dominant hand.

3.3 Procedure
A general overview of the experiment was read to all of the subjects before the testing began. All subjects were instructed to refrain from discussing the experiment with other subjects until the testing was completed.

Each subject was then individually tested with no other subjects present. Before the formal testing began, the subject was allowed to practice under optimal conditions with both auditory and visual augmentation. The subject was also presented with examples of the secondary IFF task. The subject was allowed to continue practicing until familiar with the experiment and the experimental equipment. Each subject completed the practice session in less than 5 minutes.

For each test condition, the subject was read a set of instructions describing the parameters of the test condition. The subject consecutively completed all three repetitions for each test condition, without stopping. Relevant experimental data was logged, including the subject's location in the simulated environment, the IFF information events, and the foe indication acknowledgement input time. The time required to complete all three repetitions was also manually recorded with a stopwatch.

When the third repetition was completed, the subject was asked to answer several questions and provide any general comments. The subject was allowed to take short breaks between test conditions.

3.4 Subjects
A total of 8 subjects were available for the experiment. The small sample size was not part of the experimental design, but a consequence of time constraints and subject availability. All subject participation was voluntary. The subjects were all junior enlisted Army soldiers, and none of the subjects had any vision or hearing difficulties.

4. EXPERIMENTAL DATA
Each subject performed three repetitions of each of the five test conditions, for a total of 120 individual trials. Four error trials were removed from the analysis of the experimental data. The average total time to complete all three repetitions for each test condition is shown in Figure 2.

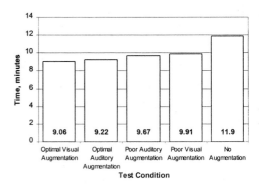

Figure 2. Land Navigation Total Course Times

The number of unacknowledged visual and auditory foe indications, per repetition, for the optimal augmentation test conditions is shown in Figure 3.

5. ANALYSIS

Given the limited number of subjects and trials, overly specific conclusions cannot be drawn from the experimental data. Despite the small sample size, useful observations can be made from this experiment (Nielsen, 1993, p. 166).

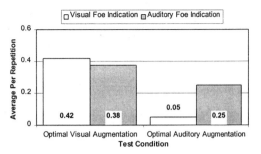

Figure 3. Unacknowledged Foe Indications

The unaugmented test condition yielded statistically significant differences from the augmented test conditions at $\alpha = 0.05$. We can therefore conclude that the augmentation improved performance in the land navigation task. Further, the poor augmentation condition results suggest that the augmentations provide benefit even in severe conditions.

The subjects did not memorize the waypoints. The subjects' movement during the unaugmented test condition clearly indicated that the waypoints were not known, and the augmented test condition results were not invalid for this reason. In fact, the unaugmented test condition data could not be analyzed because the individual test course repetitions could not be reliably identified.

The visual and auditory augmentations provided the same amount of land navigation benefit for optimal augmentation test conditions. There was no statistically significant difference between the optimal augmentation test conditions (p-value>0.1), indicating that the auditory cues were just as effective as the visual map display.

The optimal visual and auditory augmentation test conditions resulted in the same amount of unacknowledged auditory foe indications. There was no statistically significant difference between the optimal augmentation test conditions (p-value>0.1), indicating that subjects were equally distracted for both test conditions.

The optimal visual augmentation test condition exhibited more unacknowledged visual foe indications than the optimal auditory augmentation test condition. There was a statistically significant difference between the optimal augmentation test conditions (p-value<0.05). As expected, the subject's concentration on the visual map display impaired his ability to detect the visual foe indication on the simulated environment display.

6. CONCLUSION

This experiment attempted to quantify the utility of visual and auditory presentation of land navigation information. While the experimental data was not entirely conclusive, important observations were made that substantiate the use of auditory cues as a supplement to visual map displays for land navigation tasks.

Further research into this issue is warranted, both to investigate areas of statistical uncertainty and optimize implementation issues. In addition, more detailed analysis based on real land navigation tasks and operational considerations are strongly encouraged.

7. ACKNOWLEDGEMENTS

Sincere gratitude is offered to: the Army soldiers who participated in the experiment; Jeff Fernandez and Gary Riccio for their significant support in the design of this experiment; Lucas Vogel, Chuck Bonnett, Jean Dyer, and all of the folks at Fort Benning, GA who helped conduct the experiment; to everyone at the Army's Project Manager-Soldier Systems and TRADOC Systems Manager-Soldier offices for their dedication to soldiers; and to the gracious reviewers and those who have provided encouragement along the way.

REFERENCES

Biank, T. S. (2000, September 17). Soldiers testing computers in combat. Fayetteville Observer. Retrieved April 2, 2001, from the World Wide Web: http://www.fayettevilleobserver.com

Nielsen, Jakob (1993). Usability Engineering. San Francisco: Morgan Kaufmann Publishers, Inc.

Weinschenk, S., & Barker, D. (2000). Designing Effective Speech Interfaces. New York: John Wiley & Sons, Inc.

INTERNATIONALIZATION OF WEBSITES:
THE NEXT GREAT CHALLENGE IN INTERFACE DESIGN

Lawrence E. Burgee

Loyola College in Maryland
2034 Greenspring Drive
Timonium, MD 21093-4114
lburgee@loyola.edu

ABSTRACT

This paper discusses the importance of internationalization and localization for websites. Localization is a long-term commitment and is much more than just translation. There are many localization considerations when developing an internationally friendly website. The following areas of localization are addressed: language selection and translation; website name; computing environment; and, images, symbols, pictures, gestures, and color. As the Web truly becomes worldwide, internationalization and localization initiatives will continue to grow.

1. INTRODUCTION

This paper discusses the importance of internationalization and localization for websites. The need for website localization is examined followed by a look at specific areas where localization is critical for successful global eBusiness. Why bother? It is difficult enough to produce a website with good usability for the home market without worrying about possible users in other countries or cultures. The very nature of the *World Wide* Web allows for users from anywhere in the world to access any desired website.

The Web is rapidly changing from a U.S.-centric offering to a truly global platform for the exchange of ideas, information, products, and services. The concept of eCommerce or eBusiness has emerged over the past several years and is beginning to have profound impact on the ways in which we conduct business. CommerceNet (the industry consortium) states that "eCommerce is the use of inter-networked computers to create and transform business relationships. It is most commonly associated with buying and selling information, products, and services via the Internet, but it is also used to transfer and share information within organizations through intranets to improve decision-making and eliminate duplication of effort. The new paradigm of eCommerce is built not just on transactions but on building, sustaining and improving relationships, both existing and potential (CommerceNet)." This paper will refer to eCommerce and eBusiness and will consider both terms to be essentially the same.

International business occurs when organizations expand from the domestic or *home* market and seek opportunities in other countries or regions. For example, when a U.S.-based company begins to sell its products in Germany or hires a Chinese firm to manufacture components, it is conducting international business. In international business, face-to-face meetings are much less common due to the distance between participants and the time and expense required to meet in person. Meetings between potential business partners may be expensive but necessary in order to build a level of trust sufficient to consummate a transaction. Different languages and dialects can make it very difficult to communicate and often times prompts the need for translators. This also impacts the use of slang and abbreviations. Different cultures, business protocols, and infrastructures lead to a host of challenges for those entering the international arena.

Global eBusiness refers to the combination of international business and eBusiness. Users interact with each other through the use of computing devices and face-to-face communications between people are mostly non-existent (due to limited worldwide bandwidth). Global eBusiness has only been a reality for the past 5-7 years as the common World Wide Web protocol has paved the way for standard Internet usage. The challenges of international business are compounded as the *personal* element is removed from the business process and buyers and sellers (possibly in the form of electronic software agents) conduct business electronically.

2. RISE OF GLOBAL EBUSINESS

In 1993, there were only 90,000 Web users worldwide. By 1999, there were 171 million Web users and in 2000, 304 million users. By 2005, there will be more than 1 billion users, 700 million (70%) of which will be outside of North America. In 1996, more than 80% of Web users' first language was English. By November 2000, only 49.9% of Web users' first language was English (see Table 1). By 2004, only one-third of Web users will be native English speakers (ITTA Inc., 2000). The United States is reaching a saturation level of Web users, while most of the remainder of the world is just getting started. Most future gains in Internet usage will come from non-English-speaking users outside of the United States. Table 2 shows the top 10 nations in Internet use at year-end 2003.

As of September 2000, there are 4.5 million websites worldwide, 17.7 million registered domains, and more than 1 billion unique indexable webpages. 78% of all websites are published in English (only) and 96% of eCommerce sites (conducting transactions) are in English (ITTA Inc., 2000). Worldwide eCommerce revenues will increase more than 600% over the next four years from US$ 233 billion in 2000 to US$ 1.4 trillion in 2004 (CyberAtlas, 2000). The U.S. will lead in eCommerce revenues with a significant portion of revenues derived from international sales.

Table 1: Web Users First Language

Language	1996	2000
English	80%	49.9%
Chinese	1%	7.6%
Japanese	4%	7.2%
German	1%	5.9%
Spanish	1%	5%

3. ...ZATIONS

Internationalization is the process of creating an application (or product or service) that is technically capable of being adapted to operate in multiple locales. *Localization* is the task of taking an internationalized application (or product or service) and delivering a locale-specific variant of it. *Globalization* is the combination of internationalization and localization. *Personalization* is the process delivering an individual-specific (personal) variant of an application. Personalization is the natural evolution of localization but will most likely take decades to achieve. *Channelization* refers to the fact that applications must increasingly be architected to support multiple target devices such as personal digital assistants (Palm Pilots, Handspring Visors, etc.), cell phones, and TV set-top devices (WebTV) (Jones, 2000).

Table 2: Top 10 Nations in Internet Use at Year-End 2003

Rank	Nation	Internet Users	Share %
1	U.S.A.	135.7	36.2
2	Japan	26.9	7.18
3	Germany	19.1	5.10
4	United Kingdom	17.9	4.77
5	China	15.8	4.20
6	Canada	15.2	4.05
7	South Korea	14.8	3.95
8	Italy	11.6	3.08
9	Brazil	10.6	2.84
10	France	9.0	2.39
Total	Worldwide	374.9	100

4. LOCALIZATION SPECIFICS

Websites are no different from traditional products and services when it comes to the need for internationalization, localization, and in sum, globalization. As the Web evolves to a non-U.S.-centric communications and commerce medium, organizations conducting eBusiness must address the specific target markets in which they choose (or hope) to compete. Users are much more likely to buy online from sites that support their language of choice.

There are many localization considerations when developing an internationally friendly website. The following areas of localization will be addressed: language selection and translation; website name; computing environment; and, images, symbols, pictures, gestures, and color. There are many other areas of localization which are important to consider including: video, sound, and other multimedia options; search capabilities; sort order; formats (date, time, number, phone number, address, etc.); abbreviations; currency; printing - paper sizes; business culture, procedures, and protocol; fiscal law; social law (i.e. privacy); accounting requirements; taxation; weights; education styles; meals and eating times; distribution methods; and income levels.

4.1 Language Selection and Translation

It is important to carefully select the target market for a website. Just because a website is successful in the home market is no guarantee that it will be in a new culturally-different market. Language selection is a major consideration and leads into the translation process. In 2000, a website translated and published in just seven languages will reach nearly 84% of the Web population (see Table 3) (Global Reach, 2000). These languages are English, Chinese, Japanese, German, Spanish, Korean, and French. Four of these languages are Romance character-based languages (English, German, Spanish, and French), and three are Asian symbol-based languages (Japanese, Chinese, and Korean). Of course, this mix is changing as the English dominance wanes and dozens of other countries experience rapidly increasing numbers of Web users. Also, it is important to remember that many countries have multiple languages and dialects and many languages are spoken in more than one country.

Table 3: Online Language Populations

Language	2000	Language	2000
English	49.9%	Korean	4.1%
Chinese	7.6%	French	4.0%
Japanese	7.2%	Italian	3.2%
German	5.9%	Portuguese	2.5%
Spanish	5.0%	Russian	2.4%

A major task in preparing a website for international use is to translate the text into the language of each local market. This can be very expensive and time consuming but is imperative to attracting viewers in the target countries. It is essential that those tasked with internationalization prepare their material in such a way that it can be easily translated into more than one language. Human translators are the most effective at performing proper translations, however, they can be very costly and may not fully understand the Web page content or mindset of the intended viewers. Electronic (software) translators are becoming very popular but usually only achieve translation accuracy rates of 60-70% depending on the languages involved. SYSTRAN is one of the leading suppliers of translation software and offers free translation on the AltaVista Web portal at world.altavista.com.

Jargon, slang, and colloquialisms can be universally confusing and should be eliminated from the original website during the internationalization process. Acronyms, abbreviations, and mnemonics should likewise be avoided. Many terms lack abbreviations in the target language, and some languages do not allow abbreviations. Simple and plain typefaces should be used for all language translations. Stylized fonts may make characters harder for international users to recognize (Horton, 1993). Some languages are not read left-to-right, as in English. Arabic, for example, is read right-to-left. Chinese is read top-to-bottom, right-to-left. Asian languages can be very difficult to translate due to their symbolic character sets and present a major design challenge.

Information expands when it is translated. As a general rule, translated information has a length inversely proportional to the length of the Web page content in the original language. This means that shorter information units require a greater percentage of expansion space (approximately 100 to 200% for 10 characters or less) than do longer information units (approximately 30% for more than 70 characters), such as blocks of text (Merrill & Shanoski, 1992). Text buttons and menus can be much wider or lengthy and can cause major headaches if not accounted for during design. Consider how this menu bar expands as it is translated from English to French:

English | File   Edit   View   Print   Help   *French* | Fichier   Édition   Visualisation   Impression   Aide

4.2 Website Name

The website name is critical from both a usability and marketing standpoint. The top 50 websites (November 2000) were examined by the author and some interesting observations made. Many of the websites (see Table 6 for the top 25) names mean nothing and can be considered culturally-neutral (yahoo, geocities, amazon, ebay, altavista) and many are acronyms (aol, msn, nbci, cnn). All of the names end in .com (except #7 – America Online's proprietary site). The minimum length is 2 characters, the maximum is 17 characters, the mean is 6.5 characters, and the standard deviation is 2.87, yielding a majority of names in the 5-10 character range. According to Miller's famous article, *The Magic Number Seven, Plus or Minus Two*, this is to be expected (Miller, 1956). Websites named for a non-U.S. market should have the appropriate extension in place of .com. For example, the website for Amazon in the U.S. is www.amazon.com; the website for Amazon in Germany is www.amazon.de; and the website for Amazon in France is www.amazon.fr. Facilitate name guessing by localizing with the expected extension.

Table 4: Top Websites of November 2000

1	yahoo	6	passport	11	nbci	16	netscape	21	iwon
2	aol	7	aol proprietary	12	grab	17	angelfire	22	americangreetings
3	msn	8	amazon	13	excite	18	tripod	23	about
4	microsoft	9	ebay	14	altavista	19	speedyclick	24	cnn
5	geocities	10	lycos	15	ask	20	bluemountain	25	msnbc

Make it easy for international users to find the localized versions of your website. Consider the use of a splash page, a page the user will arrive at before navigating to the home page. For example, when users go to www.disneylandparis.com, they are presented with a page that lists six language options: Français, English, Deutsch, Nederlands, Espanõl, and Italiano. These are the available languages written in the native language complete with localized accent characters (ç, ð). Selecting the desired language takes the user to the appropriate localized home page. Avoid using flags for language selection as many users cannot identify their home flag and many languages could be represented by one flag. Also avoid forcing language selection in English. On the home page for United Parcel Service at www.ups.com, a Chinese-speaking visitor is expected to click on *Global Regions* and *Asia Pacific* in order to reach the selection page for China. A Chinese person should not have to understand English in order to find the Chinese version of a website!

4.3 Computing Environment

The computing environment varies in different regions of the world. The most common device for accessing the Web is the personal computer with an Internet browser. However, increasing channelization dictates the necessity to support multiple target devices such as personal digital assistants, cell phones, and TV set-top devices. It is predicted that in 2005, only 43% of European households will use PCs to hook up to the Web (and it will decline after 2005), compared with 74% in the United States. Most Europeans are expected to reach the Web in ways different from those of Americans (Echikson & Reinhardt, 2000). Localization specialists will have to adapt to the European preference for smaller, simpler, and cheaper interactive devices. Compared to Americans, Europeans reside in smaller living quarters, pay much higher rates for electricity, pay by the minutes for local phone calls, and are already accustomed to cellular phones that can operate seamlessly across more than seventeen countries in Europe. It is clear that the main Internet access device in Europe will be wireless and mobile. Japan also has a steadily increasing level of wireless Web access, primarily via cellular phones.

In the United States, Internet access is readily available at home through plentiful high-grade phone lines at speeds up to 53 kbs, and through newer very high speed cable modem and DSL services. Most large companies provide high-speed access and most public libraries even offer fast connections to visitors who may not have access at home or work. Internet Service Providers allow unlimited access for a flat monthly fee and some ISPs are even free. Local phone calls are flat fee and usually carry no additional surcharges. The connectivity convenience and affordability in the U.S. is not like the situation in many other parts of the world. Outside North America, many ISPs charge by the minute for Internet access, local telephone calls require a per-minute charge, energy costs (electric) are often much higher and less reliable, PCs are much more expensive to purchase and maintain, and connection speeds are much lower (in many areas less than 33 kbs). On the positive side, non-U.S. penetration of wireless digital phones is much higher in Europe and Asia. It is important to consider users computing environments in order to succeed with localization. There will likely be many American companies that pay millions of dollars to localize websites to be viewed on a PC browser only to find later that the target users are using alternative devices.

4.4 Images, Symbols, Pictures, Gestures, and Color

Images are the visual language of a culture and like words, don't always translate. What we recognize in our culture may have little or no meaning in another. Websites, all with a global audience, are particularly susceptible to this problem. For example, because a trash can in Thailand may look like a small wicker basket, a Thai user is likely to be confused by an image of an American-style trash can found in Apple interfaces (Sukaviriya & Moran, 1990). The American trash can also confused British users who indicated it looked much more like a postal box than a British trash can. A number of studies have concluded that many images do not convey the same meaning in all cultures (Russo & Boor, 1993). To succeed in an international market, images must be carefully selected and designed. Designers must be sufficiently aware of differences among cultures to recognize images that are culturally specific and isolate them during the internationalization process.

Commonly used symbols may also be subject to misinterpretation across cultures. Inappropriate use of symbols can destroy the meaning intended by a webpage. Using a picture of a stork, for example in Singapore is risky because it symbolizes maternal death. In Japan, the number 4 is like the unlucky number 13 in America, and 7 is unlucky in Ghana, Kenya, and Singapore (Russo & Boor, 1993). The wise old owl in the United States is seen to represent evil and witchcraft in many areas of Central and South America. Gestures are very culture-specific and should never be depicted on webpages intended for an international audience. Web of Culture at www.webofculture.com is an excellent resource for gesture information for many different countries.

Color associations also differ among cultures. In America mail boxes are blue, in England they are red, and in France and Greece they are yellow. In Medieval times, woman in Europe wore green at weddings to symbolize fertility. Over the centuries, white evolved as the appropriate color for brides. In the United States, people typically associate white with weddings and black with funerals. In China and India, the opposite is true. The color of text can also carry meaning. For example, Chinese text should not be in red unless a very personal message is being conveyed (and hence should not appear on most Chinese websites) (Galitz, 1997).

5. CONCLUSION

In summary, localization is a long-term commitment and is much more than just translation. The above analysis shows that there are many areas and issues to be considered for website localization. Internationalization from the start should lead to targeted localization that is consistent with the mission of the organization. Localization will require a great deal of effort and may dictate the need for partnerships, joint-ventures, and/or localization specialists or vendors. As the Web truly becomes worldwide, website globalization initiatives will continue to grow.

REFERENCES

CommerceNet. *eCommerce defined*. Available: www.commerce.net/resources/pw/chap1-9/pg2.html.

CyberAtlas. (2000). *120 million Web users shop online* (April 11). CyberAtlas. Available: http://cyberatlas.internet.com/big_picture/demographics/article/0,1323,6601_338561,00.html.

Echikson, W. D., & Reinhardt, A. (2000, December 11). Suddenly, PCs are falling flat. *Business Week,* 66.

Galitz, W. O. (1997). *The essential guide to user interface design.* New York: John Wiley & Sons, Inc.

Global Reach. (2000). *Global Internet statistics: online language populations* (September). Global Reach. Available: http://www.glreach.com/globstats.

Horton, W. (1993). The almost universal language: graphics for international documents. *Technical Communication, Fourth Quarter*, 683-693.

ITTA Inc. (2000). *State of the Internet 2000* (September 1). US Internet Council

ITTA Inc. Available: www.itta.com/internet2000.htm.

Jones, N. (2000). *Building globalized applications* (February 1). GartnerGroup. Available: www.gartner.com.

Merrill, C. K., & Shanoski, M. (1992, October 13-16). *Internationalizing online information.* Paper presented at the SIGDOC'92: Proceedings of the 10th annual international conference on systems documentation, Ottawa, Canada.

Miller, G. A. (1956). The magical number seven, plus or minus two: some limits on our capacity for processing information. *Psychological Review, 63*, 81-97.

Russo, P., & Boor, S. (1993). *How fluent is your interface? designing for international users.* Paper presented at the CHI'93: Conference proceedings on human factors in computing systems.

Sukaviriya, P., & Moran, L. (1990). User interfaces for Asia. In J. Nielsen (Ed.), *Designing user interfaces for international use* (pp. 189-218). New York: Elsevier Science Publishing.

An Exploratory Study of Situational Error on the Web

Dr. Jonathan Lazar
Department of Computer and Information Sciences
Towson University
8000 York Road
Towson, MD 21252
jlazar@towson.edu

Dr. Anthony Norcio
Department of Information Systems
UMBC
1000 Hilltop Circle
Baltimore, MD 21250
norcio@umbc.edu

Abstract

Users frequently make errors, and these errors can be frustrating, and can keep the users from reaching their task goals. In the networked environment of the Internet, there are an increased number of transactions and components involved, increasing the opportunity that the user will not be able to reach their task goal. Some of these errors are not due to the action of the users, but rather, are due to factors outside of their control, such as the inaccessibility of a network resource. These errors have previously been named **Situational Errors**, because while the user perceives that an error has occurred, the specific *situation* is possibly the cause of the error, rather than the user's actions. While theoretical examples of situational error have been presented, no experiments have been performed to examine situational errors and determine how often they occur, as well as what types of situational error can occur. This paper presents the methodology for an experiment on situational error.

1. Introduction

Errors can occur frequently when users are interacting with computers. This is especially true with novice users, who are not as familiar with the procedures required to successfully complete a task, and therefore, might not be able to reach their task goal (Lazar & Norcio, 1999; Lazar & Norcio, 2000b). Donald Norman's taxonomy of error describes, at the top level, two different types of error: mistakes and slips (Norman, 1983). A mistake is when the user does not choose the appropriate commands to reach their task goal (Norman, 1983). A slip is when the user chooses the correct commands to reach their task goal, but does not enter the commands correctly (due to incorrect spelling, etc.) (Norman, 1983). Both of these errors occur because of the incorrect actions of the user. What if a hard disk crashes, or there is a hardware failure? These errors are not due to the actions of the user, however, the user is in control of this situation, because they can replace any malfunctioning part of their computer. However, this situation is very different in the networked environment. Imagine a situation where the user has performed all actions in a correct manner, is not able to reach their task goal, and has no control of the situation and no chance to rectify the situation. This is the world of the situational error.

2. Situational Error

Novice users sometimes find the Internet and the Web to be very confusing. The users must first know the web location of the content that they are interested in, or be able to find the location of the content by use of a search engine. However, knowing the correct location (the URL) of the requested content is not enough; users may have problems actually accessing that content. When the user submits a request for a specific web page (by providing the URL), it is questionable whether they will be able to reach the web content, their task goal. With all of the networks and servers involved in accessing a web page, there is an increased likelihood of error (Lazar & Norcio, 1999; Lazar & Norcio, 2000b). A user may not be able to access the web site that they want, although the user has not performed any commands incorrectly. For instance, the user may correctly type in the URL for a web site, such as the Washington Post (http://www.washingtonpost.com), but is unable to access the web site due to problems with the Washington Post web server, the Internet Service Provider (of either the user or the Washington Post), or problems on the user's local network. (Lazar & Norcio, 1999; Lazar & Norcio, 2000b). In addition, a remote web site may have failed (Johnson, 1998). The error is not rooted in the user's own computer, so the user essentially has no control over correcting this type of error and reaching their task goal. This type of error has been named a **situational error**, because the error is due to situational factors which are out of the user's control (Lazar & Norcio, 1999; Lazar & Norcio, 2000b). Although the user has not performed any actions incorrectly, it has been postulated that novice users tend to view these occurrences as errors, and then blame themselves, when in fact the user has not caused the error (Lazar & Norcio, 1999; Lazar & Norcio, 2000b).

2.1 Download Time

If a web page takes a long time to load, the user may get frustrated and could possibly assume that there is an error (Lazar & Norcio, 1999; Lazar & Norcio, 2000b). In a distributed network environment, the delay between the time that a user makes a request and the time that the user is presented with the material requested can change the user's perception of the web content itself. A number of research studies on network delay of web pages have been previously published. In Ramsay, Barbesi, and Preece (1998), it was found that increased download time can change the perception of whether the material is interesting (Ramsay, Barbesi & Preece, 1998). Jacko, Sears, and Borella reported that the user's perception of the quality of the material was affected by download time (Jacko, Sears & Borella, 2000). In addition, Sears, Jacko, and Dubach report that increased download time increases the user's feelings of being lost (Sears, Jacko & Dubach, 2000). Sears and Jacko (2000) provide a good discussion of the causes of network delay (Sears & Jacko, 2000). Traditionally, when performing a most complex task, users expect a response from a computer within 10-15 seconds (Shneiderman, 1998). Jakob Nielsen suggests no more than a 10-second download is acceptable for the web environment (Nielsen, 2000; Nielsen, 2001). Of course, many users still have slow dial-up connections, and web designers have no control over this factor (Nielsen, 2001). A long download time can confuse or frustrate the user, possibly causing the user to perceive that an error has occurred.

2.2 Implications

If users can possibly perceive many different situations on the web as errors, it is important to learn more about these types of errors. When novice users browse the web, how often do situational errors occur, how do users perceive the errors, and what types of situational

112

errors are most frequent? Since there are not previously published studies relating to situational error, this study will be an exploratory study. This exploratory study could also serve as a pilot study for a larger study at a later time. There are numerous implications for learning more about situational error. When it is determined which situational errors are most troublesome, then related work can begin in three areas: system design, training design, and documentation design. Relating to system design, error messages continue to be some of the worst parts of the human-computer interface (Lazar & Norcio, 2000a). At the recent CHI 2001 conference, Bill Gates admitted that even he cannot understand many of the error messages provided by Microsoft operating systems. Training methods can be improved so that they incorporate appropriate responses to error. And documentation can be improved so that it better prepares users to deal with errors. These areas will all improve the user interaction experience.

3. Proposed Methodology

This research experiment is currently under progress, and it is expected that preliminary results will be available at the time of the conference. A total of ten subjects will take part in this exploratory study. Two separate groups of five novice subjects each will receive training on how to browse the web. The subjects will receive exploratory training, which was previously found to be an effective training method for the web (Lazar & Norcio, 2001). In exploratory training, users receive an exploration of how the web environment and the browser function, but they do not receive any direct instructions on what to type in. The actual training script for exploratory training has been previously tested for clarity. As the subjects receive the training on using the web, they will be encouraged to explore the web environment. Subjects will be asked to indicate when they feel that an error has occurred. Interaction logging would be inappropriate, since the goal is not to learn about system error, but the user **perception** of error. When the user indicates that an error has occurred, a description of the error will be recorded. After the experiment, the data will be analyzed to look for patterns, to determine how users perceive situational errors, and to determine which situational errors are most frequent.

4. Summary

It is hoped that by learning more about situational errors, the entire user interaction experience on the web can be improved. A new trend in the field of human-computer interaction is universal usability--designing systems that can be used by a range of different users in a range of different settings, using a range of different platforms (Shneiderman, 2000). By learning more about the errors that are perceived to occur, user interfaces can be improved to better assist these many different user groups and situations. Error messages in web browsers can be improved, and training methods for the web can more effectively address the issue of error. Most error messages do not provide helpful information for users, instead offering cryptic phrases such as "server could not connect" (Lazar & Norcio, 2000a). By knowing which situational errors occur frequently, error messages can be improved to more effectively assist the user in these situations. With a greater understanding of situational error, training methods and documentation for the web can be designed to specifically address situational error, so that users have a better understanding of the situational error and are therefore less frustrated.

113

5. References

Jacko, J., Sears, A. & Borella, M. (2000). The effect of network delay and media on user perceptions of web resources. Behaviour and Information Technology, 19(6), 427-439.

Johnson, C. (1998). Electronic gridlock, information saturation, and the unpredictability of information retrieval over the world wide web. In P. Palanque & F. Paterno (Eds.), Formal Methods in Human-Computer Interaction (pp. 261-282). London: Springer.

Lazar, J., & Norcio, A. (1999). To Err Or Not To Err, That Is The Question: Novice User Perception of Errors While Surfing The Web. Proceedings of the Information Resource Management Association 1999 International Conference, 321-325.

Lazar, J., & Norcio, A. (2000a). Intelligent Error Message Design for the Internet. Proceedings of the 2000 World Multiconference on Systemics, Cybernetics and Informatics, 532-535.

Lazar, J., & Norcio, A. (2000b). System and Training Design for End-User Error. In S. Clarke & B. Lehaney (Eds.), Human-Centered Methods in Information Systems: Current Research and Practice (pp. 76-90). Hershey, PA: Idea Group Publishing.

Lazar, J., & Norcio, A. (2001). Training Novice Users in Developing Strategies for Responding to Errors When Browsing the Web. paper under review.

Nielsen, J. (2000). Designing web usability: The practice of simplicity. Indianapolis: New Riders Publishing.

Nielsen, J. (2001). The need for speed (Available at: http://www.useit.com).

Norman, D. (1983). Design rules based on analyses of human error. Communications of the ACM, 26(4), 254-258.

Ramsay, J., Barbesi, A., & Preece, J. (1998). A psychological investigation of long retrieval times on the World Wide Web. Interacting with Computers, 10, 77-86.

Sears, A., & Jacko, J. (2000). Understanding the relation between network quality of service and the usability of distributed multimedia documents. Human-Computer Interaction, 15(1), 43-68.

Sears, A., Jacko, J., & Dubach, E. (2000). International aspects of WWW usability and the role of high-end graphical enhancements. International Journal of Human-Computer Interaction, 12 (2), 243-263.

Shneiderman, B. (1998). Designing the User Interface: Strategies for Effective Human-Computer Interaction. (3rd ed.). Reading, Masssachusetts: Addison-Wesley.

Shneiderman, B. (2000). Universal Usability: Pushing Human-Computer Interaction Research to Empower Every Citizen. Communications of the ACM, 43(5), 84-91.

Designing for Effective Information Presentation: The Effects of Cultural Differences on Speed, Accuracy, and Perceptions on Usability and Aesthetics

Jantawan Noiwan
Department of Business Administration
Faculty of Management Sciences
Prince of Songkla University
Hatyai, Songkla 90110 Thailand
E-mail: noiwan@umbc.edu

Anthony F. Norcio
Department of Information Systems
University of Maryland Baltimore County
1000 Hilltop Circle
Baltimore, Maryland 21250 USA
E-mail: norcio@umbc.edu

ABSTRACT

Despite the increasing popularity of animated on-line banner advertising on Web pages, Web users report difficulty in acquiring information on such Web pages. On-line banner can increase brand awareness and generate a "click-through" rate. However, little is known about the effect of animated on-line banner advertising on on-line human information processing. Moreover, as realizing the growth of current interests in cultural interface design, the study will be conducted in a culturally comparative basis. The major purpose of this study is to answer how color, a basic but powerful feature of graphic design, affects user performance and preference in seeking information on a Web page containing an animated banner advertising between American and Thai subjects.

1. INTRODUCTION

Among various types of on-line advertising media (e.g., banners, buttons, and text links), Web banners have become the most widely used on-line advertising media (Meland, 2000). Researchers and practitioners have responded enthusiastically to on-line banner advertising. However, empirical evidence, in terms of cognitive perspectives in human information processing on Web pages containing animated banner advertisings, are remarkably scarce. Still, on-line information seeking has quickly become a daily activity for humans and a number of Web publishers have quickly increased to provide current on-line information and archives for their customers through the Internet. Understanding strategies in acquiring desired information is essential and guides research and practice in Web page design. Among a variety of graphic components on screen, color is one of the powerful components of design.

More importantly, the Internet, the global communication channel, allows buyers and sellers who have different cultural backgrounds and speak different languages to interact with each other. The success of e-commerce could depend on the effectiveness of managing cultural differences of users from different parts of the world. To localize an interface by taking cultural factors into account must be considered with care. Similar to meanings of music, language, and image, perceptions of color differ from culture to culture. As cultural backgrounds could influence the learned responses and reactions to color (Eiseman, 2000), color preferences might be considered culturally dependent. Indeed, real-world evidence in differences of color preferences across cultures in various areas such as fashion and packaging designs could be found daily. Interface designers need to understand color appreciation and color response of people in different cultures and regions, since effective use of color can create several benefits, and these benefits must be given to all people throughout the world. This study attempts to explore effects of combinations of text and background colors of Web animated banner graphics on performance and preference in seeking information on Web pages containing animated banner graphics between American and Thai subjects.

2. RELATED STUDIES

Attention generally refers to a selectivity of processing (Eysenck & Keane, 1995), concentration effort on a stimulus, or the limited resources available to the cognitive system (Ashcraft, 1998). Humans always encounter many information sources simultaneously, but they cannot easily attend to more than one source of information at a time because of the limitation on attentional resource. When attempting to concentrate on one stimulus, a person ignores the surrounding stimuli or distractions. This process is called as filtering or selecting (Ashcraft, 1998). However, a study shows that objects in a visual peripheral system can draw human attention (Driver & Baylis, 1989). In the cognitive psychology arena, visual attention is categorized into various types. The number of attention categories varies. For instance, Wickens and Hollands (2000) categorize attention into three types: selective attention, focused attention, and divided attention. From a theoretical point of view, this study considers divided attention as the framework. Within this framework, it is assumed that attentional resource is limited in nature, and therefore, attention is a human capacity that can control human response. The purpose of divided attention in this study is to distribute limited resources between two tasks, including seeking target words among non-target words and attending also to the animated banner advertisings. Wickens & Hollands (2000) emphasize that two tasks can interfere with each other when they have the same stimulus modality either visual or auditory. In other words, tasks can be performed more easily when each of them is using a different modality.

A recent study by Zhang (1999) reports significant differences in different conditions of animated graphics in on-line information seeking. The study investigates the impact of animated graphics in searching target words on Web pages containing animated graphics in relationship to task difficulty, animated graphic color, animated graphic content, and instruction to ignore such animated graphics (Zhang, 1999). The results show that the animated graphics worsen user performance in searching for the target words. For example, an animated graphic that is similar but irrelevant to a task distracts a user's attention more than an animated graphic that is dissimilar to a task does. Such results become more negative when users are instructed not to ignore the animated graphics. Nevertheless, in terms of advertising banners, animated ones tend to be more effective in increasing a click-through rate than static ones.

Noiwan & Emurian (2001) investigate the effects of target word density (i.e., high, medium, and low) and Web page presentation styles (i.e., no graphics, static graphics, and animated graphics) on search time and user preferences. The results show a significant effect of target density on search time. Search time on low-density pages is significantly briefer than on high-density pages, an outcome that validated the experimental protocol. However, no significant effect is found for page presentation style, and the interaction between target density and presentation style is not significant. Self-report data showed that static graphics pages and animated graphics pages are sometimes perceived differently in terms of usability and aesthetics, and both styles are perceived as visually appealing to users. Several studies show the effects of color in human information processing. (Hoadley, 1989), for example, states that color, one of the attributes of a visual stimulus, can attract human attention. Moreover, Marcus (1992) expresses that color is the most complicated visual component. Furthermore, extensive studies of color in visual attention, particularly in visual search, show that color is one of the best ways to distinguish a stimulus from the surroundings (e.g., D'Zmura, 1991). To date, only one study has been investigated in this context (Zhang, 1999). In that study, bright color is the vital attribute of animated banner, which can greatly distract user attention. The brightness attribute is explored in two levels: bright color and dull color. The result shows that an animated graphic with bright color distracts a user's attention more than an animated graphic with dull color.

As cultural backgrounds could influence the learned responses and reactions to color (Eiseman, 2000), color preferences in usability perspectives might be considered culturally dependent. Several factors influencing human preferences of unique colors or color combinations have been investigated such as age, gender, emotion, personality, and nationality by rating separately or by comparing between each pair of colors on a subjective scale (Kreitler & Kreitler, 1972). In the investigation between age and color preference, for instance, Terwogt and Hoeksma (1995) state that preferences tend to change as a result of social and cultural influences. Kreitler and Kreitler (1972) suggest that an answer to color preferences depends on a context of a question. Grieve (1991) suggests that color preference must be assessed in regard to the objects or the contexts of the perceived colors. Therefore, like or dislike in unique colors or color combinations should be identified into specific aspects in regard to a specific task that a user performs. In interface design, such aspects might include the usability of an interface. For example, in interacting with an interface, one might think that yellow graphics could increase an interface's pleasantness, whereas gray menus could increase an interface's usefulness.

Usability might differ from one culture to the next. Cultural backgrounds could affect designers' ideas in designing interfaces. Designers develop systems based on their assumptions, while users perceive those systems with the users' assumptions (Kaplan, 2000). Due to different thinking styles, attitudes, feelings, and behaviors of users influenced by their cultures, users from different cultures might perceive usable elements or concepts differently. For instance, in Web design, most Web usability guidelines are developed with an American perspective. What Americans perceive as usable might not be usable in other countries or cultures, and vice versa. Differences between Thai and American cultures affect Web design (Noiwan & Norcio, 2000). The major differences between Thai and American Web sites are visual designs and designs of structure of information. However, no evidence has been reported that these so-called undesirable interface elements as mentioned above are also unusable among Thai users.

3. METHODOLOGY

3.1 Research Questions

a. Are there performance differences in target-word searching on different Web pages containing animated banner graphics with different combinations of text and background colors?

b. Are there differences in perceived usability on different Web pages containing animated banner graphics with different combinations of text and background colors?

c. What are the relationships between performance in information searching and perceived usability on different Web pages containing animated banner graphics with different combinations of text and background colors?

d. Does culture affect performance in information searching and perceived usability on different Web pages containing animated banner graphics with different combinations of text and background color?

3.2 Experimental Design

The experiment in this study uses 2 X 6, within-subjects, full-factorial design for both cultural groups: Thai and American. With respect to cross-cultural comparison, 2 x 6 x 2 mixed factorial design is utilized whereby cultural group (Thai and American) is the between-subjects factor. Each subject performs all 13 experimental tasks. Both Thai and American groups are presented with the same sequences of tasks. In each cultural group, the two independent variables include background color of an animated banner (red, yellow, blue, orange, violet, and green) and text color of an animated banner (black and white). Moreover, extensive studies on color preferences for hues utilize these six basic colors (i.e., Eysenck, 1981; Saito, 1994). With regard to the between-subjects design, cultural group, including Thai and American, is used as another independent variable. It is important to clarify that the study does not aim at exploring superiority of subjects between cultures. Rather, the study attempts to explore some possible cultural differences in interacting with a computer interface. The ultimate goal is to design an interface that could facilitate cognition and perception of users from different cultural groups. The dependent variables for the experiment include (1) total target-word search time, (2) total target-word accuracy, (3) banner-word selection accuracy, and (4) self-reports of usability.

Automatic recording of search time in seconds and target-word search accuracy in numbers of errors begins when a Web page is displayed to a subject, and it ends when a subject finishes the searching task. Banner-word selecting accuracy is assessed after a subject finishes searching a Web page by asking him or her to identify the word that is not shown in three animated banners. Four multiple choices are provided; one is the wrong answer, and the others are the right ones. Perceived usability of banner-color use is assessed using a self-report series of brief questions immediately after a subject finishes each task. Five statements are evaluated, namely, (1) a color use of banner draws my attention from seeking target words, (2) a color use of banner facilitates readability of the banners' words, (3) a color use of banner increases visual appeal of the Web page, (4) a color use of banner makes the Web page seems interesting, and (5) a color use of banner makes the Web page seems enjoyable. A subject rates each factor varied in seven scales, namely, extremely disagree, quite disagree, slightly disagree, neither agree nor disagree, slightly agree, quite agree, and extremely agree.

Two groups of participants join in this study: American subject group and Thai subject group. Each cultural group consists of 30 female and 30 male volunteers. To reduce undesired variations among subjects other than cultural differences, factors such as Internet experience, age, and gender are controlled. Wining monetary rewards is used as an incentive to motivate subjects to join the experiment and to perform the experimental tasks as fast and accurate as possible. The first prize costs $100.00 and the second prize costs $50.00.

3.3 Procedures

Each subject is seated in front of the laptop computer and is informed about the brief instructions by the experimenter. The subject uses a mouse to select target words from a task page and choices on a questionnaire page. The subject first signs the consent form after being informed about their rights. Then, the subject fills out the pre-experiment questionnaire page about demographic, Internet usage information, and color preferences. The subject is instructed to scan a Web page "as in reading", that is to read line by line and from left to right within a line, as fast and accurate as possible. The subject is also instructed not to ignore the words on animated banners. Then, the subject performs a practice session with an actual test program. For the actual experiment, each subject is presented with 13 different Web pages containing different animated graphics. A pre-task page is shown to present the subject with the target words. Then, a subject is required to search for 16 target words in a task Web page. When a target word is found, a subject must click on it one time. Time and error in word searching task is automatically recorded. When all target words are clicked, the subject is automatically taken to a post-task page, to rate usability of the corresponding experimental Web page and select the word that is not appeared on three animated banners. Then, the subject must take a break for 5 seconds by closing their eyes for relaxation. Afterward, the subject has to click a button on a screen to begin the next experiment.

3.4 Experimental Material

a. Target Words and Non-Target Words

With respect to words utilized in a target-word searching task, an early study by Zhang (1999) was conducted by using non-word strings as random combinations of one to four letters in her first experiment and words not controlled for numbers of letters in her second experiment. Extending on this early work, for American subjects in this study, all words are five-letter, which can be read with a single eye fixation. This experiment is controlled for word familiarity by taken from a corpus of American-English words. Allowing uncontrolled word familiarity might cause difficulty in interpretation of measures (Kreuz, 1987). For experimenting with Thai subjects, five-letter words, controlling for word familiarity, taken from a corpus of Thai words are used.

In Zhang's study, target words in each Web page were 8% of the total words displayed on a page (Zhang, 1999). Moreover, in the pilot study of this study, 16 target words (8% of the total words) were randomly distributed all across a Web page containing 200 total words. In each experimental Web page, 200 words, including 184 non-target words and 16 target words are randomly listed by not attempting to make sentences. Each word is separated by a space. No period is used. A target word is varied in each experimental Web page. Locations of non-target and target words are randomized by a computer program; however, target words are not allowed to be placed next to each other. Words shown in each Web page are selectable; when a subject clicks on a word, which is originally black, is changed to gray and then information of the word is recorded into a database.

b. Animated Banner Graphics

Extending from the early study by Zhang (1999), this study utilizes animated banner graphics. Each as with a word inside, which is considered as a task-similar graphic. The effects of varying combinations of text and background colors of animated banner graphics are investigated. Since the purpose of the study is to investigate how users share attention to two stimuli simultaneously, three animated banner graphics are used in each Web page in order to investigate how users perceive words on banner graphics while they are also searching for target words from textual information. Such three animated graphics are located at the right side of textual information across text lines, as commonly shown in current American and Thai Web sites. Relevant to this matter, for English text which is read from left to right, empirical evidence shows that the effective field of view or the perceptual span or the visual area that eyes can cover in each fixation includes about 3 or 4 characters to the left of fixation and about 14 to 15 to the right of fixation, because more informative text is lying to the right of fixation (Rayner & Pollatsek, 1989).

With regard to a banner graphic size, a common size of banners on the right side of screen is 125 by 125 pixel is used. Even though, a 468 by 60 pixel size has been the most widely used currently, a 125 by 125 pixel size is now more frequently used. Zhang (1999) also utilize the graphic that has almost the same size as 125 X 125 pixels, namely, the 110 X 110 pixel graphic. Words on banners are drawn from the non-target word database used for searching task. Such words are designed to change in size as used in Zhang (1999). With the purpose of readability, a San Serif font is used. The background colors are red, orange, yellow, green, blue, and violet. The text colors are white and black. All colors are commonly used in Web banner graphics. Moreover, such colors have also been used to investigate color preferences and color meanings across cultural groups. In particular, a moderate level of color brightness and saturation are chosen to minimize overly distracted effects.

c. Experimental Web Pages

In this study, 13 experimental Web pages are developed: 12 pages are the experimental conditions of cross-products from the two levels of text color of a banner and the six levels of background color of a banner, and one page is a based-line page with no banner graphic. With the objective of eliminating the potential role of practice effects in the interpretation of the results, the study utilizes a Latin Square design variation to distribute the order of page presentation events over subjects. An example of experimental Web pages for American subjects is shown in the figure 1.

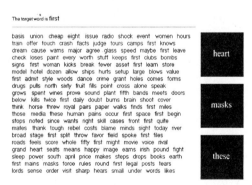

Figure 1. An example of experimental Web pages for American subjects

4. CONCLUSION

The paper presents the ongoing experimental study exploring the effects of combinations of text and background colors of Web animated banner graphics on performance and preference in seeking information on Web pages containing animated banner graphics between American and Thai subjects. The study will utilize a target-word searching task and explores variations in text and background color combinations of animated banners. Within-subjects, full-factorial design is utilized for each cultural group. Within-subjects factors include banner-text color and banner-background color and a between-subjects factor includes a cultural group. The dependent variables include total search time for the target words, target-word search accuracy, banner-word selection accuracy, and self-reports of usability. Currently, the study is under the subject recruitment process.

REFERENCES

Ashcraft, M. H. (1998). Fundamentals of Cognition. New York: Addison-Wesley.

Driver, J., & Baylis, G. (1989). Movement and Visual Attention: The Spotlight Metaphor Breaks down. Journal of Experimental Psychology: Human Perception and Performance, 15(3), 448-456.

D'Zmura, M. (1991). Color in Visual Search. Vision Research, 31(6), 951-966.

Eiseman, L. (2000). PANTONE all about Color, [Online]. Pantone. Available: http://www_pantone_com-allaboutcolor-allaboutcolor_aspID=43.htm [2000, September 18].

Eysenck, H. J. (1981). Aesthetic Preference and Individual Differences. In D. O'Hare (Ed.), Psychology and the Arts (pp. 76-101). Brighton, Sussex, UK: Harvester.

Eysenck, M. W., & Keane, M. T. (1995). Cognitive Psychology: A Student's Handbook. East Sussex, UK: Lawrence Erlbaum Associates.

Grieve, K. W. (1991). Traditional Beliefs and Color Perception. Perceptual and Motor Skills, 72, 1319-1323.

Hoadley, E. D. (1989). The Functions of Color in Human Information Processing. Baltimore, MD: Loyola College in Maryland.

Kaplan, B. (2000). Culture Counts: How Institutional Values Affect Computer Use. MD Computing, 17(1), 23-25.

Kreitler, H., & Kreitler, S. (1972). Psychology of the Arts. Durham, NC: Duke University Press.

Kreuz, R. J. (1987). The Subjective Familiarity of English Homophones. Memory & Cognition, 15, 154-168.

Marcus, A. (1992). Graphic Design for Electronic Documents and User Interfaces. New York: ACM Press.

Maxwell, S. E., & Delaney, H. D. (1990). Designing Experiments and Analyzing Data: A Model Comparison Perspective. CA: Brooks/Cole Publishing.

Meland, M. (2000). Banner Ads Get Sexy. Forbes, 2000, 28-29.

Noiwan, J., & Emurian, H. H. (2001). The Effects of Target Density and Graphics Presentation Style on Search Time and Reports of Usability and Preferences. Paper presented at the The 2001 Information Resources Management Association International Conference: IRMA 2001, Toronto, Ontario, Canada.

Noiwan, J., & Norcio, A. F. (2000, July 23 - 26). A Comparative Analysis on Web Heuristic Usability between Thai Academic Web Sites and US Academic Web Sites. Paper presented at the 4th World Multiconference on Systemics, Cybernetics and Informatics: SCI 2000, Orlando, Florida.

Rayner, K., & Pollatsek, A. (1989). The Psychology of Reading. Englewood Cliffs, NJ: Prentice-Hall.

Saito, M. (1994). Cross-cultural Study on Color Preference. Japanese Psychological Research, 36, 219-232.

Terwogt, M. M., & Hoeksma, J. B. (1995). Colors and Emotions: Preferences and Combinations. Journal of General Psychology, 122(1), 5-17.

Zhang, P. (1999). Will You Use Animation on Your Web Pages? Experiments of Animation Effects and Implications for Web User Interface Design and Online Advertising. In F. Sudweeks & C. T. Romm (Eds.), Doing Business on the Internet: Opportunities and Pitfalls (pp. 35-52). Great Britain: Springer-Verlag.

Abstract Interfaces in Three-Dimensional Space

Darniet K. Jennings

Department of Information Systems
University of Maryland Baltimore County
1000 Hilltop Circle, Baltimore, MD 21250
djenni1@umbc.edu

Abstract

Abstract Interfaces are key elements in linking objects in both physical and digitally-computerized environments. In computing, the relevant domains of interest when dealing with intersecting objects and information can be aggregated into three distinct classes of abstract interfaces: those of the user, the software, and physical hardware interfaces. In all cases, abstract interfaces serve as a means for placing disparate elements contained in separate objects or conceptual groupings into context with one another. The number of incongruous properties brought into alignment through the interface abstraction determines the effectiveness of the interface. Within three-dimensional virtual environments the visual, human cognitive, and internal representation aspects of the display design are equally important in determining the effectiveness of the highly visual interface. Four different types of context can be used as key indicators for how well an interface solution has mapped the incongruous elements onto one another. The visual, spatial, relational, and logical contexts related to three-dimensional environments and their effective use are discussed. The SLVR model, a methodology that can be applied to evaluate the effectiveness of a given interface along the four context dimensions is introduced.

1. Introduction

The spatial metaphor, the cognitive model employed by human beings to orient themselves in physical space, is the most powerful conceptual tool employed by human beings in their interpretation of the environment about them (Lakoff & Johnson, 1980; Maglio & Barrett, 1997). It is developed in youth, and becomes more expansive and refined as the individual forming and modifying the spatial cognitive model increases his or her interaction in physical space while living day to day (Lakoff & Johnson, 1980). Orientation, specifically the notion of "up" is perhaps the most powerful component within the spatial metaphor. Its dominance is due to the overriding effect of gravity on all terrestrial objects. Magnitude and direction are two other dominant indicators within the spatial metaphor. So powerful is this understanding of the physical world, that the spatial metaphor is the dominant schema—consciously or unconsciously—for adapting novel sensory data into a contextual framework (Lakoff & Johnson, 1980). Oftentimes, abstractions are used in complex systems or ideas to adapt these complexities into a more palatable form for those who must use the information presented from those sources. Some of these abstractions are created extemporaneously by the users themselves while viewing the sensory input, and other times these abstractions are created by content developers with the end user in mind. Recently, many content developers have begun to develop interfaces and interaction methods that appeal directly to the spatial metaphor as a way of maximizing usability and information transfer. How users create their own abstractions has been examined extensively in psychology and philosophy. How these abstractions are created by computer application content developers to cater to the spatial arena, particularly when dealing with three-dimensional environments where the spatial metaphor is most dominant, is open for exploration.

2. Abstract Interfaces in Software Development

When dealing with the digital phenomena native to a computerized system, abstractions are necessary because human beings' cognitive processes are ill-suited for communicating with a computer at a primitive (i.e. bit) level. The natural environment and the objects that exist within this environment are analog in nature. The continuous nature of this environment allows for a range of flexibility when dealing with sensory information in this form, and also allows information to be reconstructed based on patterns within the analog system. A digital state exists almost exclusively within a digital computing system, and does not directly translate to an analog understanding. The disconnect between points in an analog continuum and digital states is a fundamental problem in system usability. An illustration of the problem associated with a human's difficulty using digital computers, and how it is dealt with, can be seen in the use of higher level programming languages (and the reasons for their development). Today's high-level languages are much closer to what humans consider written communication than the early machine coding schemes performed using commands expressed in binary or assembly code. This was necessary because human beings do not think or speak in binary, however the use of high-level languages places a heavier burden on the system because ultimately the instructions the hardware executes must be expressed as a binary stream and the system must process the higher level instructions with the objective of converting those statements to machine code. This need to translate information moving between two objects with dissimilar communication, interaction, or transformation patterns is the domain of the abstract interface. An abstract layer is formal separator between two distinct objects that facilitates communication between the objects. Figure 1 illustrates the relationship between an abstract layer (the intersection of the two related property domains of objects A and B) and the two objects with dissimilar domains themselves. The difference in color tone between of the arrow portions on either side of the interface represents the idea that object A and object B communicate in different methods from one another. That the arrow shades in the figure on either side of the abstract layer appear in the native protocol color of the object is a direct result of the actions of the layer. When A views B, it is in reality seeing an abstracted form of B as dictated by the abstraction and vice versa. As computing itself has become ubiquitous, and computing power becoming a non-issue for traditional tasks, the most important emergent element is the discussion of how system components (including the user) fit together and how they communicate. Failures of computer systems and their interfaces (which to the user *is* the system) is oftentimes not in the developer's intent, the system logic, or the talent of the programmer(s)—it lies directly with whether or not the interface meets the communication needs of the objects on either side of the interface. Does A receive what it needs to from B in its necessary context and vice versa?

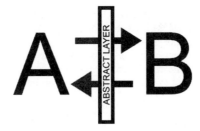

Figure 1: An Abstract Interface Conceptual Diagram

The emphasis here is that the abstract interface is the only point of contact between the two dissimilar objects on either side. It is not a transmission medium or a bridge because they merely exist as connectors between two points that could communicate through any other medium that provides connectivity between the two. In the case of an abstract layer, communication is *only* possible through the layer because it serves as a conversation manager as well as a facilitator. There are three primary classes of abstract interfaces of concern in computing and system and software development. Each form deals with a crucial element of the system model and serves a distinct purpose. They are in no particular order the abstract *software*, *hardware*, and *user* interface.

Abstract Interface Classes and Descriptions		
Abstract Interface Class	**Description**	**Example**
Software	A layer between two portions of software or an interface between two different systems. These are necessary to connect dissimilar software constructs, or to allow related constructs to interact effectively.	A portion of code written to merge two dissimilar software modules or a system interface to link two distinct computer systems.
Hardware	Enables two objects to be joined together physically. Can be software but is dedicated to the coordination of two physical devices.	Connecting a new hard drive to a personal computer.
User	A layer that exists between the user and the object of the user's attention	A common graphical user interface

Table 1: Three Classes of Abstract Interfaces in Computing

Table 1 details the properties of each class. There are several instances of sub-classes for each (accessors, encapsulators, manipulators, etc.), though a formal taxonomy of the various types of abstract interfaces has not been published. While the abstract layer clarifies the context of information passing through it for the recipient on the other side, it does not necessarily alter all of the information passing through it. In some cases, only portions of the information need be changed. In other instances, only a certain segment of the information need be transferred through the abstract layer at all. Just as important as determining how the information needs to be altered from one side of the abstraction to the other is the question of what is relevant to each side of the interface relationship.

This notion translates directly into how the development of interfaces for three-dimensional environments should be approached. In order to create the proper interface, the particulars of the users' needs and expectations must be established and validated. Because three-dimensional environments and interfaces are by definition associated with the spatial metaphor, they are fundamentally linked to the experiential metaphor employed by all human beings across all tasks and disciplines (Lakoff & Johnson, 1980). The key to success in creating abstractions in three-dimensional virtual environments is defining which portions of the spatial metaphor must be met to encourage user acceptance of the space as a true three-dimensional space. Any user should be able to sit down in front of the interface to a three-dimensional system and be aware that it is a three-dimensional environment and be able to navigate and interact with minimal instruction--and above all should not need to be told that the environment in fact mimics reality. An acceptable three-dimensional environment should appeal to a user's sense of space in a way to make that determination implicit and have obvious rules (should they deviate from a user's physical reality) that do not violate what users know to be true about physical space (Nielsen, 1993).

3. Complexities in Virtual Space

Virtual Environments, in a popular sense, are any environments designed for users that are created artificially. The discussion on virtual environments (VEs) here is limited to only those created for display and use within a computing medium. The underlying composition of a three-dimensional space is consistent whether speaking about an immersive or non-immersive environment. The perspective of the user does not change between the two, merely the way the perspective is displayed to the user. Since the display uses a three-dimensional Cartesian reference paradigm for the placement and movement of objects within the environment, the users should find the movement within the environment to be intuitive provided the physics of the environment have been implemented in a way that is familiar to the users' expectations (Kalawsky, 1993). Three-dimensional environments are laid out according to the Cartesian Coordinate System. There are three major axes, typically labeled X (horizontal), Y (vertical), and Z (diagonal). Elements in three-dimensional space are laid out using the combined values to determine where the surface points of the objects meet the environment. Objects are created with polygon primitives (spheres, cubes, etc.) or extruded polygonal regions. Six Degrees of Freedom (6DOF) in movement is allowed in the environment on the three major axes, with rotation around each axis an option in three dimensions. The freedom of movement gives users much more flexibility in their use of the interface, but can also increase complexity of system use by creating disorientation or the feeling of being lost (Kalawsky, 1993).

The technological issues with three-dimensional virtual environments are numerous, many of them having a direct impact of the viability of any system employing a three-dimensional interface. These issues deal with image latency and consequently motion sickness, disorientation, CPU processing limits, and a breakdown in immersion (Brooks Jr., 1999). However, as technology continues to improve these issues are becoming less of a factor, and the source of the shortcomings present in current three-dimensional interfaces may not be so much in how these environments appear as in how they behave. Early video games serve as a good reference for how the level of immersion and satisfaction a user derives from a virtual environment is driven to a large extent by how a program relays visual output to a user in terms of behavior rather than appearance (Weiss, 1998). The earliest games such as Pong by Atari Corporation are exemplars in that users of these games could be engaged for hours in a task that was designed to mimic a game of table tennis while looking nothing like the actual game and barely managing to mimic the real behavior of the game as witnessed in true space (Weiss, 1998). However, as Lakoff and Johnson and others have stated it is not always necessary to carry over all elements of an experiential domain to have a strong relationship between the two the source and target domain (Lakoff & Johnson, 1980).

Pong, though it did not have the visual and technical complexity of today's video games, had the critical, fundamental elements from table tennis in place (Weiss, 1998). There were present in the game a playing "surface", two "paddles", and a "ball". Beyond that it had the ability for a use to play alone or against a human opponent, varying angles and speeds for striking the "ball", a definite way to score and defeat an opponent, and the capacity for strategic thinking and ball placement. In reality, the surface was just the empty background portion of the screen, the paddles were two oblong rectangles on either side of the screen, and the ball was a square that bounced back and forth between the two rectangles depending on how it was struck by the opposing rectangle. It was not a high-fidelity visual representation of table tennis, nor was it a high-fidelity simulation of gaming behavior in table tennis, but it was effective because the subset of properties mapped from the real world game into the video game were consistent with the spatial schema people had conceptualized based on table tennis. The user schema was satisfied by the game's behavior, and the logic of the schema was not violated by the game's behavior. In terms of serving user expectation and providing a good interface between user and software Pong did well, and the success of the early games such as Pong paved the way for the multi-billion dollar industry that video gaming has become. Interestingly enough, even with the massive processing power present in today's gaming systems and the advanced graphics capability of these same systems, a common complaint among video game enthusiasts is that video games are no longer "fun". The games themselves, though they are more technologically complex and visually more true to life, oftentimes are missing the fundamental elements necessary for enjoyment that in the simpler games were the only elements possible. In any virtual environment either three-dimensional or no the fundamental elements are key to the success or failure of the environment in the eyes of the user. What, then, are the determinants of interface success and which abstract elements of spatial reality are essential?

4. Abstract User Interfaces in Three-Dimensional Applications

Given that technology does not currently exist to grant users a true-to-life, fully immersive three-dimensional environment, at best there exists only a partial, yet convincing, domain mapping between the task being attempted in the computerized environment and the users' perception of how that same task might be performed in physical space according to their internal judgment adapted from the spatial metaphor. As mentioned in the previous section, three-dimensional virtual spaces are compelled to match the spatial metaphor much more closely than their spatially devoid two-dimensional cousins. The key issue here is the interface provided for interacting with the environment must match the portions of the spatial metaphor relevant for the task being performed by the user at the appropriate moment for the interface (Norman, 1993), and ultimately the environment and application to be considered a success. Figure 2 displays a visual representation of a model developed to quantify the extent to which any given three-dimensional interface provides support to various key components that support the spatial metaphor for a given application. The model contains subjective and objective measures, and requires the input of an individual not involved in the development process of the software user interface being evaluated.

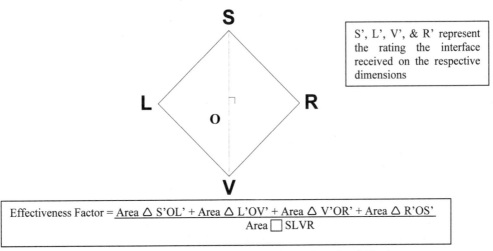

Figure 2: The SLVR Model of Three-Dimensional User Interface Effectiveness

The model strives to incorporate all meaningful elements for a given dimension, with the resultant *Effectiveness Factor* serving as a quantitative rating upon which a particular three-dimensional interface can be measured against other three-dimensional interfaces in terms of overall adherence to the spatial metaphor. As the factor approaches one the interface moves closer to optimal in its effectiveness based on the evaluation criteria. The four dimensions of the model refer to the four fundamental contextual components relevant to a three-dimensional representation of physical space. They are the Spatial, Logical, Visual, and Relational contexts present in the organization of the virtual environment. The properties of each are as follows:

Spatial (S):

- Each object in a three-dimensional virtual space has a spatial coordinate and spatial correspondence to all other objects in the space within the scope of the object
- Depending on the granularity of the coordinate scale, some objects may appear to inhabit the same coordinates as other objects though they are logically separated (i.e. virtual pages inside a virtual book)
- The consistency with which these objects maintain their relationships with one another with respects to a terrestrial based physics system (and thus the spatial metaphor) comprises the spatial context

Logical (L):

- The established rules must be consistent throughout the environment for successful user navigation
- It is of benefit to have behavioral rules mimic those of physical reality
- The extent to which the environment functions within its established rules is its logical context

Visual (V):

- Each user has particular view on events in virtual space
- These views may or may not match the spatial or relational views due to alteration, distortion, or filtering
- How the user perceives and is presented with the environment is the visual context

Relational (R):

- Objects are not stored in the way that they appear onscreen
- How objects are stored can have a substantial effect on the usability of a particular environment
- The underlying linkages between the virtual objects is the relational context

5. Conclusion

Context defines our existence. Useful information at its root is information that can be understood within a context relevant to the particular user at a particular point in time. The area of Human-Computer Interaction has placed itself at the vanguard of bringing computing power and its benefits to the general user. The focus is and must continue to be on creating systems and interfaces that fit into an individual's conceptual schema. The primary disconnect when dealing with a user interface presented in three dimensions is that a person brings to that application an extensive catalogue of experience working in three dimensions along with more refined expectations than they would have with a two-dimensional display that is less similar to their typical tasks in physical space. The tendency for people to complicate system use by forcing the spatial metaphor onto an application where its use is inappropriate is prevalent and documented (Maglio & Barrett, 1997). That these problems would be exacerbated in three-dimensional interfaces is understandable because by their very nature a three-dimensional interface is an invitation for a user to employ the spatial metaphor. The abstract evaluation framework presented here proposes an approach to identifying and quantifying the specific categories of properties that determine how well a virtual environment appeals to the spatial metaphor employed by humans—and thus ultimately how effective that interface may be. Additional research on this subject could include further refinement of the framework, and testing of the expanded model to determine its effectiveness in predicting user satisfaction and efficiency using an interface in a particular virtual space.

References

Brooks Jr., F. P. (1999). What's Real About Virtual Reality? *IEEE Computer Graphics And Applications, 19*(6), 16.

Kalawsky, R. S. (1993). *The Science of Virtual Reality and Virtual Environments : A Technical, Scientific and Engineering Reference on Virtual Environments*. Wokingham, England ; Reading, Mass.: Addison-Wesley.

Lakoff, G., & Johnson, M. (1980). *Metaphors We Live By*. Chicago: University of Chicago Press.

Maglio, P. P., & Barrett, R. (1997). *On the Trail of Information Searchers*. Paper presented at the Nineteenth Annual Conference of the Cognitive Science Society, Mahwah, NJ.

Nielsen, J. (1993). *Usability Engineering*. San Diego, California: Academic Press.

Norman, D. (1993). *Things That Make Us Smart*: Perseus Books.

Weiss, M. J. (1998). *Some words to live by: Avoid missing ball for high score*, [WWW]. Discovery Communications, Inc. Available: http://www.discovery.com/stories/history/toys/PONG/birthday1.html [2001, February 20].

Practical Approaches to Fuzzy Object-Oriented User Modeling

Thawatchai Piyawat

Department of Business Administration
Faculty of Management Sciences
Prince of Songkla University
Hatyai, Songkla 90110 Thailand
E-mail: piyawat@umbc.edu

Anthony F. Norcio

Department of Information Systems
University of Maryland Baltimore County
1000 Hilltop Circle
Baltimore, Maryland 21250 USA
E-mail: norcio@umbc.edu

ABSTRACT

Fuzzy Object-Oriented User Modeling (FOOUM) extends the original stereotype-based user modeling approach to overcome the limitations on individualization, and feature identification, and handing reasoning under uncertainty. FOOUM employs object orientation to structuralize user models and their stereotypes. Relationships among them are graded using fuzzy set methods to embed uncertainty of user information. FOOUM aims to provide finer user modeling results than does the original stereotypical approach. As a result, FOOUM shows potentials to be employed in several types of systems including cooperative systems, interest-based systems, and user observation and improvement systems. Especially, FOOUM aims to deliver results that are properly relevant to the arbitrary variance of user interests for interest-based systems.

1 INTRODUCTION

User modeling is the process of gathering goals, knowledge, and preferences of users and representing them in data structures called user models. Presently, stereotype-based user modeling is the most frequently used approach for building user-modeling systems. Nevertheless, the approach holds a number of limitations. To overcome the shortcomings of the conventional approach and to enhance the efficiency of user modeling, Fuzzy Object-Oriented User Modeling (FOOUM) is defined such that it is the extension of the stereotypical approach by employing object orientation to construct user model classes and objects and using fuzzy set methods as a way to encode the uncertainty of user information. The practical applications of FOOUM are discussed in this paper.

2 USER MODELING AND USER MODELS

Orwant defines user modeling as "… nothing more than a fancy term for automated personalization" (Orwant, 1996). From this perspective, user modeling is the activity of a system to provide active personalization to its users. In other words, a system automatically adapts itself according to user preferences to enhance users' performance. However, personalization is not the only purpose of user modeling. Indeed, user modeling is a system process of helping users to reach their goals of using the system with less effort and more effectiveness. In order to accomplish the objectives of user modeling, a system must maintain information about knowledge levels of its users. This constitutes user models such that they represent all the necessary information and assumptions required in the modeling processes for particular users (Kobsa, 1994). As a result, users' *goals*, *knowledge levels*, and *preferences* are vitally important elements of user modeling.

To structure user models, the *stereotype-based user modeling approach* proposed by Rich (Rich, 1979) is commonly used. A stereotype is a pre-defined user model in which each stereotype contains one or more name-value pairs of attribute elements called *facets*. Stereotypes are organized in a generalization hierarchy in which stereotypes inherit properties from its ancestors. A stereotype has one or more characteristic properties called *triggers* used to identify its applicability to a user who exposes information according to them.

Conclusively, user-modeling systems gather users' goals, knowledge levels, and preferences to construct user models. The user models are used to facilitate the user-system interaction in order to enhance the users' performance of using the systems. Presently, the stereotypical approach or its variations are typically employed to constitute user models in most user-modeling systems.

3 LIMITATIONS OF CONVENTIONAL STEREOTYPE-BASED USER MODELING APPROACH

Following Chen's discussion (Chen, 1995), the limitations of the traditional stereotype-based user modeling approach that reduce its representational ability are as follows:

3.1 Reasoning under Uncertainty

Uncertainty occurs due to a lack of sufficient information one needs to satisfy the truth of a statement (Smets, 1997). Uncertainty is primarily relevant to the aspect of belief, not knowledge (Smets, 1998). The degree of certainty for a statement depends on the degree of belief one has in the truth of the statement. Therefore, user information tends to be less certain because of its inconsistency. User information always contains conflict, incoherence, inconsistence, and confusion that are four causes of inconsistency (Smets, 1998). As a result, although users continuously reveal their information, the variety of information actually causes uncertainty of the information in the larger picture.

For the stereotype-based user-modeling systems, when conflicted evidence of a stereotype attribute is detected, the stereotype needs revision in order to maintain its truth and consistency (Chen, 1995). However, since stereotypes are organized in a generalization hierarchy, and most properties of stereotypes are based on default assumptions; the revision may cause dependency-directed backtracking effects. Although most stereotype-based user modeling techniques provide conflict resolutions, those methods are mostly ad hoc routines for making immediate modeling decisions. Conflict resolutions are generally inefficient and primarily achieved by either eliminating inconsistency or not considering inconsistency as a factor. Therefore, for user-modeling systems with an extended running period, their stereotype hierarchies eventually need revisions. Nevertheless, the revision process is normally operationally expensive. It consumes both time and processing power. In the real world environment in which conflicts frequently occur, this significantly degrades the stereotype-based user modeling approach.

3.2 Individualization

The generalization hierarchy of stereotype-based user modeling provides a simple method of stereotypical classification in which each stereotype inherits attributes from its ancestor. According to Rich's original approach (Rich, 1979), stereotype-based user modeling employs the single-inheritance feature, i.e., a stereotype has only one parent. Based on the original mechanism of stereotype-based user modeling, each individual user has only one stereotype to associate with him or her at any given time. The stereotype is chosen from the hierarchy when a user exposes his or her information that matches with the stereotype's trigger attributes. Nevertheless, if there is no appropriate stereotype existing in the hierarchy, a new stereotype is created as a child node of the most-possibly matched stereotype for the user. A required attribute is then added to the newly created stereotype.

This mechanism gives a possibility to implement a simple user modeling system for a closed and definable environment. However, when applying this method to a wider user space, it cannot effectively handle the diversity of users. Nonetheless, as human behavior is indeed distinct rather than similar, providing flexible individualization is an important goal of user modeling (Rich, 1999). In fact, human "traits" can be loosely categorized. However, when considering the traits in detail, each person is unique. Therefore, this uniqueness must be taken into account in order to model users effectively.

3.3 Feature Identification

The original method of organizing stereotypes into a generalization hierarchy limits the flexibility of the stereotype-based approach to adjust attributes in stereotypes. During interactions, users may demonstrate traits that cause adding or dropping attributes from the active stereotypes. However, the stereotype-based approach does not provide an effective way to update those attributes. When a conflict arises, most stereotype-based systems simply replace the active stereotype with its ancestor (Huang, Mccalla, Greer, & Neufeld, 1991). Using that method, it results in losing the attributes that still apply to users. In addition, the ancestor stereotype may not actually be applicable to users.

In another scenario, when users demonstrate their traits that do not concur to any stereotype, a new stereotype for the user is created as a child node of the most-possibly applicable stereotype. Attributes from the parent stereotype are then inherited to the child. Afterward, required attributes are added to the new stereotype. This method introduces inconsistency and ambiguity to the generalization hierarchy because the newly added attributes may already exist elsewhere in other stereotypes located at other paths of the generalization tree. Nonetheless, the stereotype hierarchical technique does not provide the ability to identify and reorganize the required attributes from stereotypes resided in different paths (Chen & Norcio, 1993).

4   FUZZY OBJECT-ORIENTED USER MODELING (FOOUM)

4.1 Problem Analysis

Chiu states that user models are incomplete, uncertain, ambiguous, unstructured, and unstable (Chiu, 1993). These characteristics pose a challenge to researchers on methods for capturing user information and representing it in systematic structures. Certainly, the structures require great flexibility in order to cope with the difficult nature of user models. Furthermore, the structures must be able to overcome the limitations of conventional user modeling approaches as discussed in Section 3 as well.

By analyzing Chiu's statement, the problems of user modeling can be divided into two categories.

1. *Problems of representational structure* Problems in this category are the results of the unstructured and unstable nature of user information.
2. *Problems of uncertainty and vagueness management* Problems in this category are the results of the incomplete, uncertain, and ambiguous characteristics of user information.

Although those problems target different domains, each problem needs solutions that also provide positive conditions to resolve the other. Therefore, in order to develop an effective approach for user modeling, one cannot solve each problem separately. Instead, both problems must be considered as a whole while attempting to overcome each issue. This paper considers that object orientation and fuzzy set theory has potential to overcome those problems as in the following discussions.

4.2 The Solution for Representational Structure Problems

As specified above, the representational structure for user modeling must be extensively flexible in order to cope with the unstructured and unstable nature of user information. Considering this requirement, the object-oriented approach shows its proficient adaptability to support varied structures of data modeling and manipulation concepts (Beeri, 1990). In addition, the structural framework of object orientation intends to handle representational activities in the way in which they are completely transparent to the user (Gyseghem, Caluwe, & Vandenberghe, 1993). Furthermore, the object-oriented approach is constructed after the real world system, which consists of objects that are composed of and collaborating with others. Therefore, employing object orientation to represent user models, which are the structures of natural information retrieved from users, is expected to promote effectiveness in the representational and utilization ability of user modeling. Consequently, the extensible structure of the object-oriented paradigm is expected to be able to override the unstructured nature of user information, and the flexibility of object orientation, in conjunction with supporting features from the fuzzy set approach discussed below, aims to overcome the unstable behavior of the information. Furthermore, as each object holds its uniqueness in its system, the object-oriented approach shows high potential to promote the individualization and the feature identification.

4.3 The Solution for Uncertainty and Vagueness Management Problems

Other than unstructured and unstable characteristics, additional properties of user models as Chiu (Chiu, 1993) pointed out are incomplete, uncertain, and ambiguous in which they add intense difficulty in processing and representing user models. Nonetheless, as these characteristics are primary dispositions of user information, an effective user modeling approach must define vigorous methods to handle them properly. Advantageously, for problems in this domain, fuzzy set theory establishes a sound foundation to handle vague data sets efficiently. Fuzzy logic, defined based on fuzzy set theory, addresses a paradigm of managing referential decisions on incomplete, uncertain, and ambiguous information. Noticeably, this type of information is exact as found in user models. Hence, to develop an effective user modeling approach in which it can appropriately handle the vague nature of user information, fuzzy set theory is a potential choice to employ in the development.

As a generalization of classical set theory, fuzzy set theory is applied to nearly every field that the applications of crisp sets has been involved. As a result, fuzzy set theory is extended in various prospects to compromise various areas (Zimmermann, 1996). For databases, Buckles and Petry (Buckles & Petry, 1982) introduce the use of *similarity relation* (Zadeh, 1971) as a foundation of fuzzy relational operations. A similarity relation is defined as a fuzzy relation that is reflexive, symmetrical, and max-min transitive (Zimmermann, 1996). Similarity, in other words, is a generalization of equality (Yazici, George, & Aksoy, 1998). For each attribute domain D, similarity relations are characterized over the domain elements:

$$s : D \times D \rightarrow [0,1]$$

In similarity relations for $x, y, z \in D$, the following properties hold:

$$s(x, x) = 1 \qquad \text{(reflexivity)}$$
$$s(x, y) = s(y, x) \qquad \text{(symmetry)}$$
$$s(x, z) \leq \max_{\forall y \in D}[\min(s(x, y), s(y, z))] \qquad \text{(transitivity)}$$

Although a number of papers propose several methods of applying fuzzy set theory to database arena, the similarity-based system appears to contain the most mathematical soundness (Ling, Yaacob, & Phang, 1997). George, et al. (George, Buckles, & Petry, 1993), later extend the similarity-based fuzzy database approach to the object-oriented data model. Their *similarity-based Fuzzy Object-Oriented Data model* (FOOD) is primarily adapted to Geographical Information Systems (GIS) as presented in (Morris, 1999) and (Cross & Firat, 2000). However, it also shows significant potential to apply in various fields that involve vague information (Morris, 1999). As a result, it is extended and used for eliciting effectiveness to stereotype-based user modeling as discussed in detail in (Piyawat, 2001).

4.4 FOOUM Methodology

Fuzzy Object-Oriented User Modeling (FOOUM) is primarily based on the stereotype-based user modeling. The stereotypical approach contains two primary elements, i.e., stereotypes and user models in which each user models belong to a stereotype, and stereotypes are hierarchically arranged. Similarly, FOOUM approach consists of two types of elements, viz., user model classes and user model objects. In comparison to the stereotype-based approach, user model classes are the stereotypes of user model objects, and user model classes are placed into a multiple-inheritance hierarchical structure. Classes and objects in FOOUM contain multi-valued attributes to represent collections of user traits. In the aspect of representing user information, the fundamental differences between FOOUM and the original stereotypical approach are that FOOUM allows multiple-valued attributes, multiple-inheritance of classes, and multiple-instance of objects from classes. Fuzzy set theory is employed to encode degrees of certainty of values to attributes, attributes to objects, objects to their classes, and classes to their superclasses. As a result, FOOUM is expected to be able to represent user information and resolve user-modeling decision more effectively than does the original stereotype-based approach. The effectiveness of FOOUM is studied in (Piyawat, 2001).

5 FOOUM IN PRACTICE

5.1 Overview

User modeling may be considered as an application area of knowledge representation in which it primarily involves representing the knowledge of a computer system about its users. The computer system uses the knowledge to accomplish its tasks in serving its users. For most computer applications, user modeling enhances their ability of delivering their services to their users. Notwithstanding, several applications, such as tutoring systems, are exclusively relied on their user modeling activities. In this section, application areas in which FOOUM shows its potentials are discussed.

5.2 Cooperative Systems

User modeling has potential to be employed in applications designed to facilitate cooperation among users, i.e., Computer-Supported Cooperate Work systems (CSCW), and between users and systems, i.e., Intelligent User Interfaces (IUI). The ability of applications to identify goals, preferences, and knowledge levels of users in detail as delivered by FOOUM is important for assisting the team cooperation.

5.3 Interested-based Systems

By knowing goals, knowledge levels, and preferences of users, applications can both deliver and filter out contents based on the interests of users. In this e-commerce age, this feature is appealing to both application developers and end users. Especially, FOOUM shows high potentials to use in interest-based systems since user interests are frequently changed and laid into inconsistency areas in which they are difficult to categorize beforehand into groups as required by the original stereotype-based user modeling approach.

5.4 User Observation and Improvement Systems

Systems in this type observe user behaviors. Using the retrieved information, the systems provide users help and guidance according to their purposes of using the systems. Those systems may further be discussed as in the following categories.

5.4.1 Tutoring Systems

Generally, in order to instruct learners successfully and effectively, a tutor must first understand their students. Therefore, when a system assumes the role of tutoring, it certainly needs to model its users in order to provide its services. Tutoring systems are designed to examine and improve some skills of students. Based on this goal, tutoring systems may be considered as user modeling systems that target on specific skills of users. They differ from general user modeling systems in the way that, besides modeling users, tutoring systems provide training for improving the skills as well. FOOUM aims to aid the complex tutoring systems where series of lessons cannot be determined in advance, as FOOUM is designed to open for discovering all possible sets of lessons with different degrees of needs.

5.4.2 Unsolicited Advising Systems

Unsolicited advising systems monitor users while they are interacting with the systems. When they find ineffective patterns of operations, they voluntarily provide advice to users. More advanced unsolicited advising systems may also detect users' goals and supply guidelines to users for achieving their goals effectively. In some circumstances, the systems may perform tasks of detecting goals and completing them on behalf of users as well. FOOUM is expected to outperform the stereotype-based user modeling systems in complex systems where user goals are difficult to identify specifically.

5.4.3 Operational Control Systems

User modeling may be used for the purpose of control. In a critical system that requires precise operational procedures, modeling the person who operates it may be useful to detect and prevent problems before they actually occur. Similar to the benefit of unsolicited advising systems, FOOUM aims to provide ability to determine goals of user in complex environments where operational procedures are unable to serialize to limited sets to be the base to build stereotypes.

5.4.4 User Performance Measurement Systems

Certainly, by modeling its users, a system can reveal the users' performance in various aspects. The result of measuring user performance can later be used in several purposes, e.g., system improvement or organizational management. While the stereotypical user-modeling systems categorize a user into a particular group, FOOUM systems show the belonging of a user to multiple groups with different degrees of certainty. As a result, FOOUM can retain the uniqueness of performance of each user in an assortment of points. This extended information surely helps achieving the purposes of measuring user performance in superior detail.

6 REFERENCES

Beeri, C. (1990). A Formal Approach to Object-oriented Databases. *Data and Knowledge Engineering, 5*, 353-382.

Buckles, B. P., & Petry, F. E. (1982). Fuzzy Databases and Their Applications. In M. M. Gupta & E. Sanchez (Eds.), *Fuzzy Information and Decision Processes*. New York, NY: North-Holland Publishing Company.

Chen, Q. (1995). *Pattern Recognition and Classification in User Modeling: A Neural Network Approach.* Unpublished Dissertation, University of Maryland Baltimore County, Baltimore, MD.

Chen, Q., & Norcio, A. F. (1993). *An Associative Approach in Dynamic User Modeling.* Paper presented at the 5th International Conference on Human-Computer Interaction.

Chiu, C. (1993). *Reasoning about domain knowledge level for dynamic user modeling in adaptive human-computer interfaces: A fuzzy logic/neural network approach.* Unpublished Dissertation, University of Maryland Baltimore County, Baltimore, MD.

Cross, V., & Firat, A. (2000). Fuzzy Objects for Geographical Information Systems. *Fuzzy Sets and Systems, 113*, 19-36.

George, R., Buckles, B. P., & Petry, F. E. (1993). Modelling Class Hierarchies in the Fuzzy Object-oriented Data Model. *Fuzzy Sets and Systems, 60*, 259-272.

Gyseghem, N. V., Caluwe, R. D., & Vandenberghe, R. (1993). *UFO: Uncertainty and Fuzziness in an Object-oriented Model.* Paper presented at the Second IEEE International Conference on Fuzzy Systems.

Huang, X., Mccalla, G. I., Greer, J. E., & Neufeld, E. (1991). Revising deductive knowledge and stereotype knowledge in a student model. *User modeling and user-adapted interaction, 1*(1), 87-115.

Kobsa, A. (1994). *User Modeling and User-Adapted Interaction.* Paper presented at the Proceedings of the CHI '94 conference companion on Human factors in computing systems.

Ling, T. C., Yaacob, M. H., & Phang, K. K. (1997). *Fuzzy Database Framework--Relational versus Object-oriented Model.* Paper presented at the Intelligent Information Systems.

Morris, H. A. (1999). *A Fuzzy Object Oriented Approach for Managing Spatial Data with Uncertainty.* Unpublished Doctoral Dissertation, Tulane University, New Orleans.

Orwant, J. (1996). For want of a bit the user was lost: Cheap user modeling. *IBM Systems Journal, 35*(3&4), 398-416.

Piyawat, T. (2001) *Fuzzy Object Oriented User Modeling: Theory, Representation, and Effectiveness.* Unpublished Dissertation Proposal, University of Maryland Baltimore County, Baltimore, MD.

Rich, E. (1979). User Modeling via Stereotypes. *Cognitive Science, 3*, 329-354.

Rich, E. (1999). Users are individuals: individualizing user models. *International Journal of Human-Computer Studies, 51*, 323-338.

Smets, P. (1997). Imperfect Information: Imprecision - Uncertainty. In A. Motro & P. Smets (Eds.), *Uncertainty management in information systems : from needs to solutions* (Vol. XVI, pp. 464). Boston: Kluwer Academic.

Smets, P. (1998). Theories of Uncertainty. In E. Ruspini & P. P. Bonissone & W. Pedrycz (Eds.), *Handbook of Fuzzy Computation*. Philadephia, PA: Institute of Physics Publishing.

Yazici, A., George, R., & Aksoy, D. (1998). Design and Implementation Issues in the Fuzzy Object-oriented Data Model. *Journal of Information Sciences, 108*, 241-260.

Zadeh, L. A. (1971). Similarity Relations and Fuzzy Orderings. *Information Sciences, 3*(2), 177-200.

Zimmermann, H.-J. (1996). *Fuzzy Set Theory and Its Applications* (3rd ed.). Boston, MA: Kluwer Academic Publishers.

Color as Communication:
Nontextual Methods for the Conveyance of Information

Darren Denenberg

Department of Information Systems
The University of Maryland, Baltimore County
1000 Hilltop Circle
Baltimore, Maryland
21250

One of the most important aspects of an information system is its ability to convey useful information to an end user. This is most commonly accomplished by presenting textual or numerical information on a screen or other output device that a user can then read and interpret, allowing them to determine the next appropriate course of action based on that information. However, as information technology becomes more commonplace and the amount of information continually increases, it becomes necessary to consider new ways of communicating that information to users. One way this can be accomplished is by using a non-textual means of communication to present an additional layer of information along with the base meaning that the text supplies, such as displaying a text message in red. This not only conveys the textual meaning of the message, it presents an additional layer of information that the user can immediately interpret and act upon. This paper will examine the sue of color as a non-textual method of information presentation as it pertains to the information technology domain, investigate the current application and usefulness of the method, and present possible areas for future research and applications in which these concepts could have significant impact.

1. A Brief Introduction to Non-Textual Communication

The use of non-textual means of communication is not a new concept. In fact, much of the information that we encounter during our working day is communicated through non-textual means, sometimes as a supplement to a textual message, and sometimes in and of itself. Most people in the United States are familiar with the stop sign, illustrated in figure 1.1 below.

Figure 1.1

This particular sign uses one textual and two non-textual layers of information to communicate three distinct, discrete messages. Upon first encountering this sign, one would think that it's purpose is to warn drivers to stop at an intersection or dangerous crossing. While this is true, it must be pointed out that the word "danger" appears nowhere on this sign. In fact, the important message of "danger" is communicated not with text at all, but in this case with the color red used for the background. The color red is recognized virtually universally as an indicator of danger and is used from highway signs to traffic lights to warning labels on potentially dangerous products. The final message that is communicated via this particular sign is that it is a "regulatory" sign, meaning it is used to give a command (as opposed to warning signs or marker signs, etc.) to drivers. This is communicated through the use of the octagonal shape, which is used exclusively for stop signs.

These may seem like inconsequential concerns, however the Federal Highway Administration, in their Manual on Uniform Traffic Control Devices (MUTCD), puts forth very strict guidelines concerning the text, color, and shape of not just stop signs, but all traffic signs. Concerning text, the MUTCD mandates that no other text can appear on the sign except the message the sign is attempting to communicate, any other textual communications have to be attached underneath the sign on a separate placard. Concerning color, it is mandated that the red background used on stop signs not only be visible whether in daylight or reflected in headlights, but that it appears the same hue regardless. Finally, regarding the shape, the octagon has been federally mandated to be used exclusively for stop signs, and over time the that particular shape, along with similar ones such as the hexagon, are commonly interpreted as representing potential danger.

2. Psychological and Cultural Issues Regarding Color as Non-Textual Communication

Most cultures communicate meanings without the use of text through color. However, it is important to realize that by no means are the meanings of those colors consistent among different cultures. For example, very often in Asian cultures the color white is used to communicate mourning and sadness, thus it is very common to see people wearing white at funerals. Conversely, black is a color used to indicate joy and celebration. This is in stark contrast to western cultures where those meanings are reversed. British culture interprets the color red as signifying royalty, whereas in India it is often the color representing purity and used in wedding outfits, yet in the United States it represents danger. The Dani tribe of New Zealand makes no distinction among colors at all, they classify all colors as either 'black' or 'white,' with lighter colors including blue and yellow falling into the 'white' category and darker colors such as brown and gray falling into the 'black' category. This categorization is maintained across shades as well, with a light blue such as cyan being classified as 'white' while the darker shade navy blue is classified as 'black.'

The significance of the cultural issues regarding color presented here is that studies of color as a method of communication, be they psychological, sociological, or technological studies, are by their nature not generalizable across cultures. That is, no study of color as a method of communication can be applied to any culture outside of the one in which it was conducted because of the cultural variations in the meanings of those colors.

Psychologically, the effects of various colors on humans have been studied extensively. Detailed discussions of these studies are beyond the scope of this paper, however special mention should be made of two general but significant color-related phenomena in order to illustrate its potential; The first of these is the ability of red to instantaneously increase a subject's galvanic skin response (Wilson, 1966), the second being the ability of dark colored clothing to impart an impression of aggressiveness and toughness on an individual, the latter being true regardless of whether one is examining sports teams that wear dark-colored uniforms or a job applicant who wears a dark colored suit to their interview (Damhorst & Reed, 1986).

3. A Condensed Review of the Literature Investigating Color as a Display Method

Having considered the brief examination of color presented here, this paper will now examine how can these constructs be applied to the domain of information systems, and review some of the literature that has examined the effects of these methods on information displays.

Most of the research involved in color as a method of communication for information displays has focused on the use of colors to assign levels of urgency to a communication such as a line in a graph or text presented on a display. To investigate the entire body of literature in this domain is not possible in this confined space, however a brief review of some of the more relevant literature will be presented. Additionally, this paper does not concern itself with any literature focusing on color as a design or aesthetic construct.

One study that directly examined the effects of a color display as opposed to a monochrome display was conducted by Benbasat, *et al* (Benbasat, Dexter, & Todd, 1986). In this study, graduate and undergraduate students were recruited from marketing courses taught through a business school and presented with what is known as the "Brand Manager's Allocation problem." This challenges participants to determine the best means of dividing up an advertising budget across various territories with the goal of reaping the most profit. The study examined the effects of tabular v. graphical presentations as well as color v. monochrome, and it was discovered that the benefits of color apply more strongly to graphical as opposed to tabular presentations. The study found several significant benefits of using color as a method of presentation: Color line graphs were more easily interpreted by subjects, the use of color graphs showed an improvement in task performance over repeated trials, color enhanced the accuracy of the decision making process when subjects were under a time limitation, color greatly enhanced the task performance of those subjects who demonstrated difficulty in determining individual presentations in a display containing a multitude of information. Additionally, color was rated by subjects as more easily interpreted and more helpful to the decision making process as opposed to monochrome.

Not all of the color presentations were beneficial, however. Subjects using multicolor graphs requested more reports in order to make their decision as opposed to those using monocolor graphs. Additionally, the subjects in the color groups took longer to arrive at a decision, which tends to be true when one is not constrained by a time limit. As was mentioned earlier, color appears to enhance the decision making process and decrease the time it takes a subject to arrive at a decision when a time constraint is present.

Christ (Christ, 1975) conducted a literature review investigating the research on the use of color in visual displays, the relevant results of which will be discussed here. This review separated the discussion into reviews relating to unidimensional displays and reviews relating to multidimensional displays, the former referring to a display where the targets are differentiated by only one stimulus dimension (such as color) and the latter referring to a display in which targets can differ among several stimulus dimensions (such as color and shape, among others).

Concerning unidimensional displays, or displays in which target and distractor stimuli vary across only one dimension such as color, Christ's review found that color is a superior identifying attribute when compared to size, brightness, and shape. In this case, an identifying characteristic is the characteristic a subjects searches for to identify a target. According to the author's literature review, the only identifying characteristic superior to color was the alphanumeric symbol. This effect was exacerbated as exposure time to the display decreased or as the number of distractors increased.

When considering unidimensional displays, that is, displays in which target and distractor stimuli vary among multiple dimensions such as color *and* shape and a subject is required to identify both, the author's findings were very similar to those of unidimensional displays; color was s superior identifying attribute when compared to size and shape. However, unlike unidimensional displays, the superior identifying nature of the alphanumeric symbols that was present with unidimensional displays disappeared when presented in a multidimensional display and were in fact equal.

When examining the superiority of alphanumeric symbols relative to colors as identifying attributes when presented in a multidimensional display, it should be noted that Christ's review revealed that as stimulus information increases, the gap between color and alphanumeric dimensions decreases as far as identification accuracy is concerned, and as the alphanumeric digits became more similar in structure (for example, E and F as opposed to X and O) color actually became a superior identifying attribute and became increasingly superior as the similarity of the alphanumeric characters increased.

Notwithstanding the previous discussion, it needs to be considered how an individual's own cognitive approach to interpreting visual displays effects their performance in a specific experimental setting. One study that examined this concept compared the effects of black and white displays as opposed to color-coded displays as far as their ability to improve the learning of field-dependent and field-independent subjects was concerned (Dwyer & Moore, 1991). Specifically, this study presented subjects with a booklet containing an essay about the various parts and functions of the human heart, with accompanying illustrations in color for one group and in black and white for the other. Subjects then had to draw a diagram of a heart with all the described parts indicated in their proper places, and an identification task in which subjects had to correctly identify the parts of a heart through a multiple choice questionnaire. What this study discovered was that while field-dependent subjects consistently scored higher on the drawing tests and the identification tests as was expected, this was only true when evaluating performances of those involved in the black and white group. When color coding was introduced, those differences were discovered to no longer be significant, thus indicating color coding could be used as means by which those whose learning skills are hampered by a specific cognitive circumstance could be brought up to a competitive level through the use of something as simple as the use of color.

4. Conclusion

While this discussion of color as a display construct and method of non-textual communication has been brief and was in no way all-inclusive, it makes clear the fact that what seems a mundane concept, color, can have significant effects on those who view it, and this is true whether one refers to increased decision-making ability, pedagogical effects on those with cognitive learning deficiencies, search and identification capabilities, or even the effect it has on one's interpretation of another. It can be used to focus attention to particular element in a display where the amount of distractors may be significant, such as the displays used in air traffic control towers or operating rooms. The need for accurate and fast decision making in these environments necessitates the ability to communicate as much information as possible while not interfering with the decision-making process.

To use color randomly and without any direction or purpose when designing the output of an information display would cause serious, yet perhaps not so obvious problems: It would confuse those using the display, it would reduce their ability to make appropriate decisions based on the information presented, it would impair the ability of some to learn from the information presented on the display, and it would actually require users of the display to take longer in order to comprehend and store the information cognitively, effectively defeating the purpose of presenting information. There is much more to be researched, but this review clearly illustrates that it is a significant facet of information systems design.

References

1. Benbasat, I., Dexter, A. S., & Todd, P. (1986). An Experimental Program Investigating Color-Enhanced and Graphical Information Presentation: An Integration of the Findings. *Communications of the ACM, 29*(11), 1094-1105.
2. Christ, R. E. (1975). Review and Analysis of Color Coding Research for Visual Displays. *Human Factors, 17*(6), 542-570.
3. Damhorst, M. L., & Reed, J. A. P. (1986). Clothing Color Value and Facial Expression: Effects on Evaluations of Female Job Applicants. *Social Behaviour and Personality, 14*(1), 89-98.
4. Dwyer, F. M., & Moore, D. M. (1991). Effect of Color Coding on Visually Oriented Tests With Students of Different Cognitive Styles. *The Journal of Psychology, 125*(6), 677-680.
5. Wilson, G. D. (1966). Arousal Properties of Red Versus Green. *Perceptual and Motor Skills, 23*(3), 947-949.

AUGMENTING PHYSICAL TOOLS: BRIDGING THE CHASM OF LATE ADOPTION WITH MULTIMODAL LANGUAGE

David R. McGee

Philip R. Cohen

Pacific Northwest National Laboratory, P.O. Box 999, Richland, WA 99352, david.mcgee@pnl.gov

Oregon Graduate Institute, 20000 NW Walker Rd., Beaverton, OR 97006, pcohen@cse.ogi.edu

ABSTRACT

Computing systems that attempt to automate safety-critical operations, such as those found in hospitals, traffic control centers, and military command posts, often do not account for the way that physical artifacts and human language constitute a medium for collaborative activity in those environments. Consequently, these approaches fail to win the favor of these naturally conservative end-users. In this paper we examine ethnographies of these three environments to understand their common characteristics and the artifacts used in them. Based on these comparisons, we argue that when automating these types of environments, designers must begin to deliver tools that: (1) support co-present collaboration, and (2) are reliable in the extreme. Otherwise, our solutions will continue to fall into the chasm that separates early and late-adopters of technology. Consequently, we propose a technique for augmenting physical artifacts by observing the use of multimodal language coincident with them. This approach blends the benefits of computation with those of physical artifacts, thereby bridging a technology chasm which separates early adopters from conservative users.

1. INTRODUCTION

Despite the breakthroughs that computing has offered in the past half-century, people still spend most of their time working and playing in the real world: meeting each other, sketching out diagrams together, writing and editing as a team, collaborating on designs, etc., all without the aid of computers. Due to the tangible nature of everyday physical tools, people frequently discard computational substitutes. Put simply, people often find physical tools more appropriate than computational ones.

For example, flight controllers across the globe still insist upon using pens and paper flight strips to negotiate each aircraft's route through their airspace (Mackay, 1999). They use shared symbols to annotate the strips, providing not only a record of change, but a tool for collaboration; they lay out the flight strips on boards as an indicator of relative height and distance. Similarly, military officers use a paper map, pens, and Post-it™ notes to both track the position and disposition of units in the field and to plan future conduct (McGee, Cohen & Wu, 2000). Officers draw symbols on Post-it notes that represent people and equipment then array these Post-its on a shared map board that is visible to all who enter the command center. Health care professionals create bundles of highly specific information on paper, consisting of preformatted charts, but also whatever happens to be at hand, such as the back of a gauze pad (Gorman et al., 2000; Heath & Luff, 2000).

These three domains represent a class of work environments in which computing automation has largely failed. Professionals in these domains remain technologically conservative, because choosing tools that are not an improvement over existing tools can result in tragic consequences. Our view is that new approaches to human computer interaction that *blend* the distinctive qualities of these environments and their existing tools with computation will bridge the chasm between these necessarily pragmatic work processes and the technological innovation they deserve.

In the next two sections, we will further characterize the three challenging environments introduced earlier in terms of the tasks being performed, the people who populate them, and the artifacts they cling to. We will examine the properties of decision environments where the chasm between early and late-term adopters is literally an abyss because the risk of adopting safety-critical systems that simply do not assist, or worse yet have the potential to fail, is immense. We will describe how designs to automate these tasks fail to account for a considerable part of their purpose; namely, as a means of robust problem-solving within teams—oftentimes collaborating without expressly "collaborating" (i.e. by observing others, overhearing, etc.). This failure is what ultimately led us to our previously reported set of constraints (McGee, Cohen & Wu, 2000) and our design for a system, Rasa (McGee & Cohen, 2001), which abides by them and allows users to retain the properties of their current tools yet still reap the benefits of digitization. This approach is one way to bridge the chasm that separates late-adopters of technology from potentially beneficial applications.

2. TASK PROPERTIES

The three tasks described in the introduction share a number of similar characteristics. (1) Users are building situational awareness: primarily formulating plans and creating mental representations for real-world processes, activities, and objects using uncertain information. (2) The objectives of each cannot be achieved by working alone; instead in order to reduce uncertainty, users rely upon expert judgments of, interaction with, and observation of their colleagues, whose specialties span the problem space. (3) Moreover, collaboration is at the heart of these tasks and is attained primarily using shared language, ensuring that the risks are mitigated. (4) What is more, this activity is distinct from traditional "collaboration" in that these users rely not only on direct communication but also on situational cues. These tasks are embedded in an environment in which the "end-user" is never just an individual at work, but in which "secondary" interactions, such as overhearing and directly or peripherally observing this work, are just as important, sometimes even more so. (5) Despite severe constraints in time, attention, and expertise, decisions must be made quickly to protect human life, often leading to satisficing (making a satisfactory, but perhaps not optimal, decision). (6) Each of these environments must operate almost continuously, requiring several shifts. Consequently, particular individuals may or may not be on hand when a crisis arises. The rapid transfer of prioritized and situation-specific knowledge is essential when shift changes occur or when new plans are developed. (7) Safe operation is a paramount concern in these tasks. In each, expert human decisions are relied upon to reduce uncertainty and safeguard and protect lives. In sum: these are human-centered, collaborative decision tasks that are mission-critical in nature and must be conducted in a safety-conscious fashion. Because of the potential loss of life due to mistakes or failures, robust procedures and tools have been adopted that diminish these risks as much as possible.

3. PHYSICAL ARTIFACT PROPERTIES

Let us examine why physical artifacts, as opposed to computational tools, are specifically chosen in these environments. First, the kinds of co-located, situational, collaborative activities that are the norm in these environments are difficult, if not impossible, to support with current computing systems. By their nature, these tasks often involve people interacting dynamically with shared tools, whether pointing at flight strips that represent flight paths in jeopardy or grabbing a Post-it note and arguing about "its" potential damaging effects on friendly unit positions. (See Figure 1 for one example.) Today's computers tend to reduce or eliminate the amount of human-human communication surrounding them. However, this is just the kind of behavior that collabora-

Figure 1. Officers engaged in several discussions of an evolving situation at a command post map.

tors, like the officers in command posts, rely on to assess a situation. Second, the ability to communicate efficiently and effectively using shared languages aids in the rapid evolution of problem solving. Forms of this, such as written symbology or shorthand, allow us to quickly perform actions such as naming and referring; however, this ability is missing from most computational interfaces. Moreover, humans interact *multimodally* (with speech, gesture, and written symbols) to enhance the bandwidth of their communication. Consequently, people choose artifacts that increase multimodal communication channels rather than detract from them. Third, because of the reliance on collaboration to arrive at rapid decisions, and the number of personnel required to execute these complex tasks, users are frequently interrupted, leading to a reliance on tools that serve as persistent memory aids, such as written notes, and robust placeholders for task state. Finally, the tools that they choose are common and fail-proof. They must be quickly replaceable, interchangeable with what can naturally be found at hand, and reliable. Flight strips can be made with a pen, a piece of paper, and a pair of scissors; maps are often simply sketched when a full-relief map is unavailable, inappropriate, or unnecessary. Therefore, these artifacts remain in use even as computing attempts to supplant them because they are reliable, support collaboration, and use symbols and language that serve to aid memory and cognition, whereas computing systems halt work upon failure, fail to make work visible to collaborators, and restrict the flow of human spoken, gestural, and symbolic communication.

4. DESIGN CONSTRAINTS

In previous work, we identified five key constraints for designing systems to retain the benefits of the artifacts identified in a military command post (McGee, Cohen & Wu, 2000). We assert that they apply as well to the two other domains depicted here. These constraints are provided for convenience sake in Table 1. In the table, we use the term *augmentation* to denote how users annotate physical artifacts in order to extend the objects' meaning: e.g., commanders write on Post-it notes so that they represent particular units in the field, and controllers write on flight strips so that they represent a different flight plan than the one scheduled.

Two derived constraints follow from these. A corollary of the minimality, human performance, and human understanding constraints is that in order to function in a given environment, human-machine interfaces must be based on the current work style, including those interfaces necessary to change the meaning of an object. Since the minimality constraint reminds us that the system should adopt the common tools in the environment, any computational aid should be as invisible as possible. Our solution to this dilemma is to add sufficient sensing mechanisms to the environment so that users can rely, as much as is feasible, on existing multimodal language to augment artifacts in the environment.

A second consequence, based on the minimality and human understanding constraints, is that the system must rely on the language of the work practice to establish the proper representational relationships between the augmented objects and the digital world. These denotational relationships should be analogous to the ones being created between the physical artifacts and the real world entities that they represent.

Table 1. Design constraints

Minimality Constraint	Changes to the work practice must be minimal. The system should work with user's current tools, language, and conventions.
Human Performance Constraint	Multiple end-users must be able to perform augmentations.
Malleability Constraint	The meaning of an augmentation should be changeable; at a minimum, it should be incrementally so.
Human Understanding Constraint	Users must be able to perceive and understand their own augmentations unaided by technology. Moreover, multiple users should be able to do likewise, even if some are not present physically or temporally.
Robustness Constraint	The work must be able to continue without interruption should the system fail.

5. HOW THE CONSTRAINTS ENSURE THAT THE PROPERTIES ARE SATISFIED

Together these constraints ensure that any design to automate environments accounts for the properties that the existing tools provide. The minimality constraint addresses the need for computational artifacts to be based on the existing common tools, procedures, and language. Consequently, users can continue to create and interact actively with the current tangible artifacts, regardless of failures or inconsistencies in the associated computing systems. The human performance constraint ensures that the common shared language becomes the means by which users interact with the computational component of the system. The malleability constraint guarantees that artifacts can be extended. The human understanding constraint requires that all users understand how each artifact is augmented, i.e., what it represents in the real world. The corollaries ensure two things: (1) that both human understanding and the ability to extend objects are possible even without computational aid, and (2) that the system uses an interface that is essentially invisible, relying on the natural spoken and written language of the task rather than contemporary approaches like graphical user interfaces. In essence, these constraints serve to ensure that the properties we identified earlier are retained.

6. RASA

Based on these constraints, we developed a system, Rasa, which takes advantage of the officers' use of task-oriented language in a command post by recognizing written symbols and spoken language as officers update a command post's map (McGee & Cohen, 2001). Rasa captures speech with close-talking microphones or microphone arrays placed strategically near the tools being augmented. These tools are a high-resolution contour map and Post-it notes.

136

The map and its plastic overlays are attached to a touch-sensitive digitizing tablet and the Post-its are on a radiosensitive digitizing pen tablet. The map, which could just as well be a blank sheet of paper, is first registered to a specific area of the Earth by touching two of its corners and speaking the coordinates for each. Once the map is registered, officers can (1) update the underlying digital system by pointing at Post-its already on the map, but not added to the digital system, (2) add new units to the system by drawing on a new Post-it, or (3) draw directly on the map's plastic overlay various symbols representing changes to the terrain (these are typically man-made such as berms, fortifications, etc.).

To add a new unit to the map, an officer draws a symbol on a Post-it representing a military unit. These symbols constitute a language with which an officer can express variables for each unit, such as its size, machinery, etc. While the user is drawing the symbol, they are also free to add information, such as the unit's nickname, to the object via speech. Both the spoken and written language are then captured and recognized—speech is recognized by an off-the-shelf recognizer, and the symbols are recognized by our own three-level hierarchy of neural network recognizers (Wu, Oviatt & Cohen, 1999). Rasa parses the output of the recognizers, producing meaning representations in the form of typed feature structures. These feature structures are fused in Rasa's multimodal integrator (Johnston, 1998), by firing declarative rules in its chart parser that match the feature structures of the input, unifying the inputs according to the rule, and evaluating the constraints declared there against features in the input. Once fusion is complete, the system projects the unit onto the map and prepares a spoken utterance to request a confirmation from the user.

By taking advantage of the language already present, Rasa is able to deliver a capability wherein its users can continue to use the physical tools and surrounding procedures that support the robust, collaborative, decision-support needed for safe conduct of this work.

7. TECHNOLOGY CHASMS

Physical tools are so familiar that mimicking them, e.g., building metaphors of physical tools and environments, has been and still is *the* major theme in human interface design. Norman, in his classic "The Design of Everyday Things" (Norman, 1988) examines the distance between (1) a person's intended actions and the actions actually required to affect the state of the world, and (2) actual changes brought about in the world and the readiness with which these changes are perceived. These distances Norman calls the gulfs of execution and evaluation respectively. The function and possible actions of common objects are perceived from their physical attributes and social standards relating to their use, leading to what Norman calls a *natural mapping*. An object's ability to express these functional mappings is commonly referred to as the object's *affordances* (Gibson, 1979). People simply perceive affordances for everyday objects. Chairs afford sitting and trees afford shade, for example. This principle—ensuring that mappings are natural—along with visibility, consistency and coherence, and feedback, comprise the basis of Norman's thesis on principled design.

Contemporary computational interfaces fail to take advantage of the affordances and natural mappings of physical objects; consequently, they create a technological chasm (Moore, 1991), which prevents more conservative workers, like those mentioned earlier, from using computational tools. According to Moore, this chasm, which represents the span of acceptance of technology products from early adopters to the much larger group of so-called "pragmatists and conservatives," is vast and difficult-to-cross. The pragmatists want evolution rather than revolution, the opposite of the early adopters. They want technology that enhances and integrates into existing work practices and systems, whereas early adopters are willing to expend energy on learning, understanding, and advocating technology that is only potentially fruitful. Moore's conservative users want *continuous* rather than *discontinuous innovations*, which force the customer to change his or her behavior, infrastructure (e.g., current hardware and software), processes, etc. For whatever reason, pragmatists and conservatives are unwilling to make these adjustments.

We contend that supporting natural mappings and the affordances of *physical objects* can eliminate these gulfs as they are typically introduced by desktop computing interfaces. Moreover, we believe that this represents a way to bridge the chasm between early to late-adopters.

8. CONCLUSIONS

In general, users in the work environments discussed in this paper have resisted attempts to computerize their tasks. We suggest a number of reasons for this. The spoken and written language of these and many other tasks makes users' collaboration possible—teamwork that is critical to their success and safety. Yet, designs often fail to account for the effects of the technology on the existing human-human collaboration resulting from the use of *language and shared artifacts* in a situated context. Moreover, physical tools are more efficient, reliable, convenient, and cheap—and they get the job done.

However, computing systems, most especially personal computers, do not exhibit these properties of physical artifacts. As a result, users in our target environments have chosen one of two options: (1) retain the "redundant" physical artifact-based work practice and use it in tandem with the computing system, thus doubling the users' effort or worse, or (2) dispose of the computing automation system altogether and continue to rely on the physical artifacts. We argue that users should have more options than these: options that blend useful physical tools with computation. We have presented one embodiment of this proposition in the form of Rasa, which augments a commander's map and paper tools to capture and understand the spoken and written language used naturally there.

ACKNOWLEDGEMENTS
This work was supported in part by the Command Post of the Future Program, DARPA under contract number N66001-99-D-8503. Pacific Northwest National Laboratory is operated by Battelle for the U.S. Department of Energy under contract DE-AC06-76RLO 1830. The authors wish to thank their colleagues at PNNL and OGI for their support, especially those who contributed to or reviewed this work.

REFERENCES
Gibson, J. J. (1979). *The Ecological Approach to Visual Perception*: Houghton-Mifflin.

Gorman, P., Ash, J., Lavelle, M., Lyman, J., Delcambre, L. & Maier, D. (2000). "Bundles in the Wild: Managing Information to Solve Problems and Maintain Situation Awareness." *Library Trends 49* (2).

Heath, C. & Luff, P. (2000). *Technology in Action*. Cambridge, UK: Cambridge University Press.

Johnston, M. (1998). Unification-based multimodal parsing. *Proceedings of the International Joint Conference of the Association for Computational Linguistics and the International Committee on Computational Linguistics*, 624-630. Montreal, Canada: Association for Computational Linguistics Press.

Mackay, W. E. (1999). "Is paper safer? The role of flight strips in air traffic control." *ACM Transactions on Computer-Human Interaction 6* (4), 311-340.

McGee, D. R. & Cohen, P. R. (2001). Creating tangible interfaces by transforming physical objects with multimodal language. *Proceedings of the International Conference on Intelligent User Interfaces*, 113-119. Santa Fe, NM: ACM Press.

McGee, D. R., Cohen, P. R. & Wu, L. (2000). Something from nothing: Augmenting a paper-based work practice with multimodal interaction. *Proceedings of the Conference on Designing Augmented Reality Environments*, 71-80. Helsingor, Denmark: ACM Press.

Moore, G. A. (1991). *Crossing the Chasm: Marketing and Selling High-Tech Goods to Mainstream Customers*. New York: Harper Business.

Norman, D. A. (1988). *The Design of Everyday Things*. New York, NY: Currency/Doubleday.

Wu, L., Oviatt, S. & Cohen, P. (1999). "Multimodal integration - A statistical view." *IEEE Transactions on Multimedia 1* (4), 334-341.

Capabilities and Limitations of Wizard of Oz Evaluations of Speech User Interfaces

Wallace J. Sadowski

IBM Voice Systems
1555 Palm Beach Lakes Blvd.
West Palm Beach, FL 33401
wjsadows@us.ibm.com

Abstract

This paper addresses the use of the Wizard of Oz (WOZ) technique to assist the development and evaluation of speech-enabled interactive voice response (IVR) systems. In a WOZ study, people who are representative of the target-product audience attempt to use speech to control a *simulation* of the application. Using the WOZ methodology, developers can test a system before building a functional prototype. Such studies can guide and influence system development when the most flexibility for changes exists. Discussed in this paper are capabilities and limitations to consider when using the WOZ methodology for the development and testing of speech-enabled systems (Sadowski, 2001). Briefly discussed are two recent studies (Sadowski and Lewis, 2000a; 2000b) that illustrate some of the capabilities and limitations of the WOZ methodology.

1. Introduction

The Wizard of Oz (WOZ) usability testing methodology enables developers to evaluate an application before it is functional. This approach uses a trained human tester, serving as the 'wizard,' to simulate the machine side of the human-computer interaction. John Kelley applied the term "Wizard of Oz" to this method in 1980, but the simulation of speech-based systems by humans emerged at least as early as the 1970's (Thomas, 1976).

By using the WOZ methodology, developers can test a system before expending the time and money required to build a functional prototype. WOZ tests allow an evaluator to make initial estimates of average call duration, average utterance length, prompt clarity, and other usability attributes of the system. Such studies can guide and influence system development when the most flexibility for changes exists. Developers can use the WOZ technique to explore proposed changes to existing applications, to evaluate prototypes, or to facilitate the development of prospective applications. However, to use WOZ simulations to assess any system one must be able to specify and simulate the systems' interface characteristics in detail. The wizard must understand and simulate the systems functional and operational capabilities and limitations accurately for an effective WOZ evaluation.

139

The WOZ approach to usability evaluation has advantages over traditional usability testing methods using fully developed and functional systems. Listed below are some of the benefits and advantages of this testing paradigm for evaluating speech-enabled interactive voice response (IVR) systems.

2. Benefits and Advantages of the WOZ testing paradigm

1. Enables developers to evaluate speech-enabled systems for usability issues before completion of the system development (e.g., coding, prompt recordings, etc.).

2. Allows a developer to modify call flow paths and test these various paths for effectiveness, efficiency, and ease of navigation without requiring software coding.

3. Enables the wizard to evaluate the effectiveness of the self-revealing help prompts in moving the user forward through the interface.

4. WOZ simulation is an efficient method to determine users responses to a system prompt. This information can aid in the identification of words or phrases to include in the grammar set[1].

5. Enables the developers to easily rearrange the breadth and depth of the menu structure for evaluating issues such as ease of navigation, time on task, user preference, and load on users' working memory.

6. Enables developers to generate task scenarios and observe the human-computer interaction to determine the most intuitive sequencing of prompts for user input and which prompts most effectively elicit that information.

7. Provides the means to evaluate existing applications regarding their suitability for a speech-based interface.

8. Allows the wizard to "simulate" various recognition rates.

9. Allows the simulation of various system response times (SRTs) to determine when users become frustrated or annoyed.

Because an operational system is not required, the Wizard of Oz (WOZ) technique permits initial usability evaluations and the ability to incorporate resulting recommended changes before software coding is complete. This permits developers to continue revising the system during testing and helps reduce the required changes to the functional system by identifying usability issues during the developmental stages. However, there are limits to the effectiveness and appropriateness of WOZ testing.

---

[1] Some researchers have suggested that the style and content of users' utterances vary as a function of whether users believe they are speaking to a computer or a human being (Hauptmann and Rudnicky, 1988; Richards and Underwood, 1984).

3. Limitations of the WOZ testing paradigm

1. The wizard must know what words and/or phrases are valid inputs. Additionally, developers cannot predict when substitution errors may occur unless user input is feed through a recognition engine.

2. The wizard must be very familiar with how the system should respond to out of grammar (OOG) inputs, silence timeout's (STOs), spoke-too-soon (STS) inputs, or spoke-way-too-soon (SWTS) inputs. The wizard must detect these errors consistently and present the appropriate response(s) when they do occur.

3. The wizard must be careful not to influence user responses through subtle changes in voice intonation or human-like reactions to responses. The wizard must ensure there are no subtle variations in the content of prompts or how they are presented.

4. Because the wizard serves as the systems 'voice,' subjective ratings of intelligibility and pleasantness are invalid unless the live system's voice is used during testing (e.g., recorded prompts or a text to speech engine).

5. The wizard cannot predict, and thus cannot simulate, the speech rate of the operational system.

6. The wizard may have difficulty in keeping the pace of the system and user interaction consistent and appropriate.

7. The WOZ methodology does not enable a valid measure of system response times.

8. When simulating a full duplex system the wizard may have difficulty hearing or understanding user input if the user "barges-in" while the wizard is concurrently speaking. The wizard also needs to know the prompt-stopping characteristics of the intended system (i.e., energy detection, speech detection, or word/phrase recognition).

9. The workload on the wizard required to simulate the system appropriately may necessitate the use of additional testers or equipment to accurately collect the data on user performance.

4. Applying the WOZ technique

The primary goal of the first study (Sadowski and Lewis, 2000a) was to evaluate the IBM WebSphere "WebVoice" demo. Its three component applications perform basic library, banking, and calendar functions. The purpose of this evaluation was to ensure that the applications were consistent with speech user interface usability guidelines (IBM, 2000) and to ensure that those guidelines produced a usable interface. Because a working prototype was not available for testing, we used the WOZ usability evaluation technique. This WOZ study uncovered six usability problems in the call flow including; remote placement of information related to the always-active commands, confusion regarding exiting the system, and ambiguous, confusing or inconsistent prompts.

For the second study, participants used a functional version of the same system. The primary purpose of the second study (Sadowski and Lewis, 2000b) was to determine if the recommended changes (based upon the earlier WOZ evaluation) alleviated the targeted usability problems and to identify remaining usability issues. As expected, the functional prototype had usability problems, but those problems were primarily associated with aspects of the system that were not capable of being predicted and/or simulated in the WOZ study. Specifically, there was only one problem identified in the call flow, but there were five problems associated with system attributes such as the speaking rate of the artificial voice, an ambiguous production of multi-digit numbers, and the failure to recognize dates spoken as all digits.

5. Discussion

Many of the limitations and benefits of the Wizard of Oz technique exist because testing occurs without an operational system. This allows developers and usability engineers to test design-hypotheses without involving substantial software coding. However, because a person is simulating the system, some characteristics of the system are not suitable for evaluation using this method. For example, developers should use a functional system to obtain accurate and reliable estimates of system characteristics such as system response time, recognition accuracy and ratings of the system's voice.

In general, the overall utility of the WOZ testing depends upon an accurate simulation of the system by the wizard. The wizard is responsible for ensuring user's interactions with the system and the responses they receive simulate those they would experience using an operational system. This poses potential problems regarding subtle voice cues, wizard workload, and the promptness of responses from the wizard to user input. The wizard must also be able to identify improper user input and provide the user with appropriate responses. Many of the difficulties regarding the accurate simulation of a speech-enabled IVR system can be overcome or minimized through careful and deliberate organization of the system responses (i.e., the scripted responses) and by using a "system savvy" wizard to ensure consistent and appropriate responses to user input. In summary, the WOZ methodology is a useful tool for aiding in the development and evaluation of speech-enabled IVR system design, but it is important to be aware of both its capabilities and limitations.

References

Hauptmann, A., & Rudnicky, A. (1988). Talking to computers; an empirical investigation. *International Journal of Man-Machine Studies, 28*, 583-604.

IBM. (2000). *Designing a speech user interface* (in IBM Voice Server Programmers Guide). West Palm Beach, FL: Author.

Richards, M. & Underwood, K. (1984). Talking to machines. How are people naturally inclined to speak? In E.D. Megaw (Ed.) *Contemporary Ergonomics*. London: Taylor and Francis.

Sadowski, W. J. (2001). *Benefits and Limitations of Applying the Wizard of Oz Usability Methodology to the Development and Evaluation of Speech-Enabled Interactive Voice Recognition (IVR) Systems* (IBM Tech Report 29.3402). Raleigh, NC: International Business Machines Corp.

Sadowski, W. J. & Lewis, J. R. (2000a). *Wizard of Oz usability evaluation of the IBM WebSphere "WebVoice" demo* (Tech. Report 29.3321) Raleigh, NC: International Business Machines Corp.

Sadowski, W. J. & Lewis, J. R. (2000b). *Usability Evaluation of the IBM WebSphere "WebVoice" Demo* (IBM Tech Report 29.3387). Raleigh, NC: International Business Machines Corp.

Thomas, J. (1976). *A method for studying natural language dialogue* (Research Report RC-5882). Yorktown, NY: International Business Machines Corp.

Intelligibility and Acceptability of Short Phrases Generated by Embedded Text-To-Speech Engines

Huifang Wang[1]
James R. Lewis[2]

[1]Cisco Systems, Inc.
255 W. Tasman Dr., SJC-J/2
San Jose, CA 95134
huifwang@cisco.com[1]

[2]IBM Voice Systems
1555 Palm Beach Lakes Blvd.
West Palm Beach, Florida
jimlewis@us.ibm.com

Abstract

We investigated the intelligibility and acceptability of three formant text-to-speech (TTS) engines suitable for use in devices with embedded speech recognition capability. Listeners transcribed and rated recordings of short phrases from four text domains (U.S. currency, dates, digits and proper names) produced by three commercially-available embedded TTS engines and a human speaker. The human voice received the best intelligibility and acceptability scores, and one of the TTS engines had superior intelligibility and acceptability relative to the two others. The results suggest that the ability to accurately produce names (the least constrained and least accurately transcribed text domain) was the system characteristic that best discriminated among the engines. The intelligibility and acceptability scores were generally consistent. Listeners transcribed shorter phrases more accurately than longer phrases, but acceptability ratings were independent of phrase length.

1. Introduction

Products with embedded speech capability typically have extreme resource constraints. For this reason, current text-to-speech engines for embedded products employ formant synthesis-by-rule to produce speech output. Because the competition in the embedded product space is fierce, we are always interested in understanding the strengths and weaknesses of our selected solution against the other solutions available in the marketplace.

The primary purpose of this experiment was to assess the intelligibility and acceptability of three formant text-to-speech (TTS) engines suitable for use in devices with embedded speech capability. Secondary purposes of the experiment were to:

- Benchmark the TTS engines against a recorded human voice
- Investigate any interaction among the engines and four different types of text commonly produced by embedded TTS engines (U.S. currency, dates, digits and proper names)
- Investigate the effects of phrase length on intelligibility and acceptability (measuring acceptability with the Mean Opinion Scale, or MOS)

We had an interest in the potential effects of phrase length on intelligibility because there are competing cognitive forces involved in the transcription of short phrases. One hypothesis is that transcription accuracy for shorter phrases should be better than for longer phrases because there is less material to remember. An alternative hypothesis is that longer phrases provide more acoustic and linguistic context, which should make it easier for listeners to compensate for any intelligibility problems in the production of artificial speech. It is likely that both hypotheses are true, but it wasn't clear which would exert the greater influence on transcription accuracy.

---

[1] At the time we designed this experiment, Huifang Wang was an employee of IBM Voice Systems in West Palm Beach, Florida.

With regard to acceptability rating with the MOS, Johnston (1996) had listeners judge the quality of natural speech degraded with time frequency warping. He found a significant relationship in the expected direction for judgements using the MOS Listening Effort item (greater degradation led to poorer ratings). He also found that using sentences as stimuli in an experiment of three TTS voices yielded results that were just as sensitive as those using longer paragraphs. The stimuli in the current experiment were shorter than those used by Johnston, ranging in length from 2 to 14 syllables. These stimuli provide an opportunity to extend the previous finding of the stability of MOS ratings to shorter speech samples and for the overall MOS (rather than a single item). Additional evidence that listener ratings of speech acceptability are reasonably independent of the length of the speech sample could have important implications for the design of more efficient experiments.

2. Method

2.1. Participants

Sixteen people participated in the experiment. Half of the participants were IBM employees and half were employees of a temporary employment agency, hired for the purpose of participating in this experiment. Each group had an equal number of males and females, with approximately equal age distribution across the groups (about half over and half under 40 years of age).

2.2. Materials

Each participant heard all four voices (human speech and three different commercially-available TTS voices, identified in this paper as TTS1, TTS2, and TTS3). The test texts were short phrases from four text domains: U.S. currency (e.g., "one dollar and twenty-three cents"), dates (e.g., "January first two thousand one"), numbers (e.g., "five five five one two three four") and proper names (e.g., "U.S. Forty Toll West"). The phrases in each domain had the same average number of syllables (6.375) to control for potential confounding due to phrase length differences. For purpose of controlling the distribution of phrase length across speech production conditions, we categorized the phrases by pairs as Short (1,2), Medium (3,4), Medium-Long (5,6) and Long (7,8).

2.3. Procedure

The experimental design used Latin squares to systematically counterbalance the presentation of test phrases among the various experimental conditions and to systematically counterbalance the order of presentation of the experimental conditions for both intelligibility and acceptability rating sessions for both groups of participants (IBM and agency employees). During an intelligibility session, participants listened to a test phrase from one of the text domains produced by one of the systems (or the human) and then transcribed it, with two trials per system (for a total of eight trials per session). Transcription accuracy scores were 1 for perfect transcription and 0 for incorrect transcription. After completing an intelligibility session, participants listened to both of the test phrases (from the assigned pair of phrases) produced by a given voice during the intelligibility session, and then rated the acceptability of that voice using the Mean Opinion Scale (MOS,), doing this for each voice. Participants continued in this fashion until they had completed the intelligibility and acceptability rating trials for all voices and for all text domains. (See the Appendix for the MOS items used in this experiment. See Lewis, 2001, for information about the psychometric properties of the MOS).

3. Results

The design of this experiment allowed two different arrangements of the data for the purpose of factorial analysis of variance: by Text Domain or by Phrase Length. Because our primary development interest was in the effects of Text Domain, that arrangement received the more detailed analyses.

3.1. Intelligibility

Arrangement by text domain. Analysis of variance on the intelligibility scores (averaged across the two trials) revealed significant main effects for both Voice ($F(3,36) = 3.98$, $p = .015$) and Text Domain ($F(3,36) = 29.3$, $p = .0000001$), and a marginally significant Voice by Text Domain interaction ($F(9,108) = 1.8$, $p = .08$). The between-subjects variables of Group and Gender were not significant ($p > .70$), and did not significantly interact with the other variables. A set of t-tests indicated that listening to both TTS2 and TTS3 resulted in significantly poorer transcription accuracy (77% and 70% respectively) than the 89% accuracy for the human voice (TTS2: $t(15) = 2.2$, $p = .04$; TTS3: $t(15) = 4.6$, $p = .0003$), and TTS1 (84%) had somewhat better accuracy than TTS3

$(t(15) = 2.0, p = .06)$. Transcription accuracies for all text domains were significantly different (all $p < .01$), with 95% accuracy for currency , 86% for times and dates, 78% for numbers and 63% for proper names. The nature of the Voice by Text Domain interaction on intelligibility was that all the voices showed similar patterns for the domains of currency, date and number, but were distinctly different for the domain of name, with accuracies ranging from 34% for TTS4 to 91% for the human voice (see Figure 1).

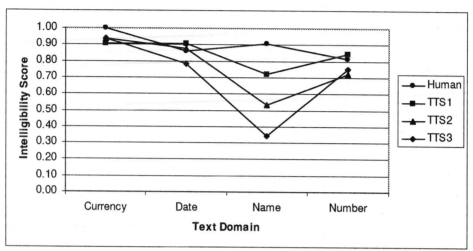

Figure 1. Voice by Text Domain Interaction (Intelligibility)

Arrangement by phrase length. An analysis of variance indicated a significant main effect of phrase length $(F(3,36) = 5.6, p = .003)$, but no additional significant effects other than the previously described significant main effect of Voice. A set of t-tests indicated more accurate transcription for short (89%) than medium-long (79%) and long phrases (70%) $(t(15) = 2.67, p = .02,$ and $t(15) = 3.83, p = .002$ respectively) and more accurate transcription for medium (83%) than long phrases $(t(15)=2.32, p=.04)$.

3.2. Acceptability

Arrangement by text domain. Using the MOS factor structure described by Lewis (2001), analysis of variance detected significant main effects of Voice $(F(3,36) = 30.3, p = .0000000006)$, Text Domain $(F(3,36) = 4.0, p = .015)$, and Factor $(F(2,24) = 8.9, p = .001)$. The main effects and interaction of Gender and Group were not significant (all $p > .50$). The Voice by Text Domain $(F(9,108) = 4.8, p < .00002)$, Voice by Factor $(F(6,72) = 12.2, p = .000000002)$, Text Domain by Factor $(F(6,72) = 2.8, p = .02)$, and Voice by Text Domain by Factor $(F(18,216) = 10.0, p < .0000000001)$ interactions were significant. There were also significant Factor by Gender by Group $(F(2,24) = 4.8, p = .02))$ and Voice by Factor by Gender by Group $(F(6,72) = 2.4, p = .04)$ interactions. In these analyses MOS scores can range from 0 to 4, with lower numbers indicating better ratings. A set of t-tests on the main effect of Voice showed that all differences were statistically significant ($p < .10$ for TTS2 vs. TTS3, $p < .0004$ for all other differences, with an overall mean rating of 0.27 for the human voice, 1.09 for TTS1, 1.54 for TTS2, and 1.76 for TTS3). A similar set of tests on Text Domain showed poorer ratings for Date (1.29) and Name (1.35) than Number (1.06) or Currency (0.96) (all $p < .03$). A set of t-tests on the main effect of Factor indicated significantly poorer ratings for Naturalness (1.49) than Intelligibility (1.01) $(t(15) = 5.33, p = .0001)$ and Speaking Rate (1.13) $(t(15) = 2.5, p = .02)$. There is not sufficient space in this paper to describe all interactions in detail, but the key attributes of the significant two-way interactions were:

- The Voice by Text Domain interaction illustrated the clear separation between the human and TTS voices. For currency and number domains, TTS1 and TTS2 voices had almost equal acceptance. The greatest difference between TTS1 and TTS2 occurred for the production of the times and dates.
- Figure 2 shows the Voice by Factor interaction. Decomposing the overall MOS into its component factors (Intelligibility, Naturalness, and Speaking Rate) revealed a difference in profile between the artificial and

146

human voices. All of the voices occupied different vertical locations on the graph, consistent with the finding of a significant main effect of Voice. The human voice had about equal ratings for Intelligibility and Naturalness, with a somewhat poorer rating for Speaking Rate. All of the artificial voices received about equal ratings for Intelligibility and Speaking Rate, with somewhat poorer ratings for Naturalness.

• The most salient characteristic of the Text Domain by Factor interaction was the relatively poor rating for the name domain on the Intelligibility scale, which was consistent with the measured intelligibility of names relative to the other text domains.

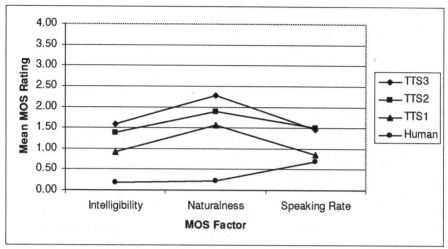

Figure 2. Voice by Factor Interaction (Acceptability)

Arrangement by phrase length. The most interesting finding from rearranging the data to investigate the effect of phrase length was that there was no significant effect of Phrase Length ($F(3,36) = .08$, $p = .97$, with mean ratings of 1.16 for Short, 1.18 for Medium, 1.15 for Medium-Long, and 1.18 for Long) and no Voice by Phrase Length interaction ($F(9,108) = .54$, $p = .85$), consistent with the findings of Johnston (1996).

4. Discussion

The key findings from the experiment were:

4.1. TTS1 was the Best Artificial Voice

Among the three TTS engines, TTS1 was the most intelligible and acceptable. The results of this experiment show no reason for us to consider using TTS2 or TTS3 in place of TTS1 (our current engine).

4.2. The Human Voice was Most Intelligible and Acceptable

TTS1 was the best artificial voice, but the recorded human voice had the best intelligibility scores and received the best acceptability ratings. Whenever possible (which could be rarely in embedded systems), developers should use a recorded human voice rather than an artificial voice.

4.3. The Least Intelligible Text Domain was that of Proper Names

The pattern of results by text domain for intelligibility scores was interesting in that the least constrained domain (names) had the poorest intelligibility overall and the greatest variability by voice. The significant Voice by Text Domain interaction suggests that the ability to accurately produce names was the system characteristic that best discriminated among the three TTS systems.

4.4. Memory Load Dominates Contextual Effects when Transcribing Short Phrases

The finding that longer phrases had poorer transcription accuracy than shorter phrases demonstrates that contextual effects are weaker than the effects of memory load when transcribing short phrases. The practical implication of this is that there is no reason to have systems produce longer phrases than necessary with the belief that this will enhance intelligibility. Consistent with the guidelines presented in Balentine and Morgan (1999), system messages should not be any longer than they need to be, whether produced by a human or artificial voice.

4.5. MOS Ratings are Stable over a Wide Variety of Phrase Lengths

The main effect of Phrase Length was not marginally nonsignificant -- it was dramatically nonsignificant ($p = .97$). The detection of other effects is evidence that the failure to detect effects of Phrase Length was not due to an insensitive experimental design. Listeners seem to arrive at their judgements of speech output quality quickly and produce consistent ratings even when the text sample is very short. The practical implication of this finding is that it should be possible to run very efficient studies comparing artificial voices when using the MOS as the instrument for assessing perception of voice quality.

4.6. MOS Ratings and Intelligibility Scores Tend to Lead to Similar Decisions

The results for intelligibility scores and acceptance ratings were highly similar, leading to identical conclusions regarding the relative quality of the voices. For example, the patterns of means for the significant main effects of Voice were virtually identical for both variables. Using data from the current experiment, Lewis (2001) found that intelligibility scores tend to correlate significantly with the MOS Intelligibility factor. This, along with the stability of the MOS over phrase lengths, supports the continued use of the MOS in experiments conducted for the purpose of evaluating the quality of TTS systems.

5. References

Balentine, B., & Morgan, D. P. (1999). *How to build a speech recognition application: A style guide for telephony dialogs.* San Ramon, CA: Enterprise Integration Group.

Johnston, R. D. (1996). Beyond intelligibility: The performance of text-to-speech synthesisers. *BT Technology Journal, 14*, 100-111.

Lewis, J. R. (2001). *Psychometric properties of the MOS* (In the proceedings of this conference).

6. Appendix. MOS Used in the Experiment

The MOS text shown here is the same as that used in the experiment, but with the format modified to fit.

1. Global Impression: Your answer must indicate how you rate the sound quality of the voice you have heard.
   [ ] Excellent [ ] Good [ ] Fair [ ] Poor [ ] Bad

2. Listening Effort: Your answer must indicate the degree of effort you had to make to understand the message.
   [ ] No effort required [ ] Slight effort required [ ] Effort required [ ] Major effort required
   [ ] Message not understood with any feasible effort

3. Comprehension Problems: Your answer must indicate if you found single words hard to understand.
   [ ] None [ ] Few [ ] Some [ ] Many [ ] Every word

4. Speech Sound Articulation: Your answer must indicate if the speech sounds are clearly distinguishable.
   [ ] Yes, very clearly [ ] Yes, clearly enough [ ] Fairly clear [ ] No, not very clear [ ] No, not at all

5. Pronunciation: Your answer must indicate if you noticed any anomalies in the naturalness of sentence pronunciation.
   [ ] No [ ] Yes, but not annoying [ ] Yes, slightly annoying [ ] Yes, annoying [ ] Yes, very annoying

6. Speaking Rate: Your answer must indicate if you found the speed of delivery of the message appropriate.
   [ ] Yes [ ] Yes, but slower than preferred [ ] Yes, but faster than preferred [ ] No, too slow
   [ ] No, too fast

7. Voice Pleasantness: Your answer must indicate if you found the voice you have heard pleasant.
   [ ] Very pleasant [ ] Pleasant [ ] Fair [ ] Unpleasant [ ] Very unpleasant

Psychometric Properties of the Mean Opinion Scale

James R. Lewis

IBM Voice Systems
1555 Palm Beach Lakes Blvd.
West Palm Beach, Florida
jimlewis@us.ibm.com

Abstract

The Mean Opinion Scale (MOS) is a seven-item questionnaire used to evaluate speech quality. Analysis of existing data revealed (1) two MOS factors (Intelligibility and Naturalness, plus a single independent Rate item), (2) good reliability for Overall MOS and for subscales based on the Intelligibility and Naturalness factors, (3) appropriate sensitivity of MOS factors, (4) validity of MOS factors related to paired comparisons, and (5) validity of MOS Intelligibility related to intelligibility scores. In conclusion, the current MOS has acceptable psychometric properties, but adding items to the Naturalness scale and increasing the number of scale steps from five to seven should improve its reliability.

1. Introduction

The Mean Opinion Scale (MOS) is the method for evaluating text-to-speech (TTS) quality recommended by the International Telecommunications Union (ITU). The MOS is a Likert-style questionnaire, typically with seven 5-point scale items addressing the following TTS characteristics: (1) Global Impression, (2) Listening Effort, (3) Comprehension Problems, (4) Speech Sound Articulation, (5) Pronunciation, (6) Speaking Rate, and (7) Voice Pleasantness.

It might seem that articulation tests that assess intelligibility (such as rhyme tests) would be more suitable for evaluating artificial speech than a subjective tool such as the MOS. Most modern text-to-speech systems, although more demanding on the listener than natural speech (Paris, Thomas, Gilson, & Kincaid, 2000), are quite intelligible (Johnston, 1996). "Once a speech signal has breached the 'intelligibility threshold', articulation tests lose their ability to discriminate. ... it is precisely because people's opinions are so sensitive, not just to the signal being heard, but also to norms and expectations, that opinion tests form the basis of all modern speech quality assessment methods." (Johnston, 1996, pp. 102, 103)

Developers of products that use artificial speech output need reliable and valid tools for evaluating the quality of TTS systems. When the tool is a questionnaire that collects subjective ratings (like the MOS), it is important to understand its psychometric properties. The goal of psychometrics is to establish the quality of psychological measures (Nunnally, 1978). Some of the metrics of psychometric quality are reliability (consistent measurement), validity (measurement of the intended attribute), and sensitivity (responds to specific experimental manipulations).

1.1. Brief Review of Psychometric Practice

Reliability. The most common measurement of a scale's reliability is coefficient alpha (Nunnally, 1978). Coefficient alpha can range from 0 (completely unreliable) to 1 (perfectly reliable). For purposes of research or evaluation in which the final score will be the average of ratings from more than one questionnaire, the minimally acceptable reliability is .70 (Landauer, 1988).

Validity. Researchers commonly use the correlation coefficient to assess criterion-related validity (the relationship between the measure of interest and a different concurrent or predictive measure). The magnitude of the correlation does not need to be large to provide evidence of validity, but the correlation should be significant.

Sensitivity. A measurement is sensitive if it responds to experimental manipulation. For a measurement to result in statistically significant differences in an experiment, it must be both reliable and valid.

Number of scale steps. All other things being equal, a greater number of scale steps will enhance scale reliability, but with rapidly diminishing returns. As the number of scale steps increases from two to twenty, there is an initially rapid increase in reliability that tends to level off at about seven steps (Nunnally, 1978). After eleven steps there is very little gain in reliability from increasing the number of steps. Lewis (1993) found that mean differences between experimental groups measured with questionnaire items having seven steps correlated more strongly with the observed significance level of statistical tests than did similar measurements using items that had five scale steps.

Factor analysis. Factor analysis is a statistical procedure that examines the correlations among variables to discover groups of related variables (Nunnally, 1978). Because summated (Likert) scales are more reliable than single item scores and it is easier to interpret and present a smaller number of scores, it is common to conduct a factor analysis to determine if there is a statistical basis for the formation of measurement scales based on factors. Generally, a factor analysis requires five participants per item to ensure stable factor estimates (Nunnally, 1978). There are a number of methods for estimating the number of factors in a set of scores, including discontinuity and parallel analysis (Coovert & McNelis, 1988).

1.2. Previous Research in MOS Psychometrics

Reliability. A literature review turned up no previous work reporting MOS reliability in any form.

Validity. Salza et al. (1996) measured the overall quality of three Italian TTS synthesis systems with a common prosodic control but different diphones and synthesizers using both paired comparisons and the MOS. Their results showed good agreement between the two measurement methods, providing evidence for the validity of the MOS. Johnston (1996) had listeners judge the quality of natural speech degraded with time frequency warping. He found a significant relationship in the expected direction for judgements using the MOS Listening Effort item (greater degradation led to poorer ratings).

Sensitivity. Johnston (1996) found that the MOS Listening Effort item showed statistically significant differences among the ratings of three TTS systems, and that this item was more sensitive than a more general item asking listeners to rate the overall quality of the system. He also found that using sentences as stimuli yielded results that were just as sensitive as those using longer paragraphs.

Yabuoka et al. (2000) investigated the relationship between five distortion scales (differential spectrum, phase, waveform, cepstrum distance, and amplitude) and MOS ratings. They were able to calculate statistically significant regression formulas for predicting MOS ratings from manipulations of the distortion scales. Unfortunately, they did not report the exact type of MOS that they used in the experiment.

Factor structure. The factor structure of the MOS is currently in question. Kraft and Portele (1995), using an eight-item version of the MOS (with an additional 'Naturalness' item), reported two factors – one interpreted as intelligibility (segmental attributes) and one as naturalness (suprasegmental, or prosodic attributes). The Speaking Rate (Speed) item did not fall in either of the two factors. More recently, Sonntag et al. (1999), using the same version of the MOS (but with 6-point rather than 5-point scales), reported only a single factor.

1.3. Goals of the Current Research

The goals of the current research were to (1) evaluate the factor structure of the 7-item 5-point-scale version of the MOS (the version reported by Salza et al., 1996, adapted for use in our lab), (2) estimate the reliability of the overall MOS score and any revealed factors, (3) investigate the sensitivity of the MOS scores, and (4) extend the work on validity of the MOS.

2. Method

2.1. Factor Analysis and Reliability Evaluation

Over the last two years we have conducted a number of experiments in which participants have completed the MOS. In some of these experiments we have also collected paired-comparison data and, in the most recent (Wang & Lewis, 2001), we also collected intelligibility scores. Participants in these experiments have included in approximately equal numbers, males and females, persons older and younger than 40 years old, and IBM and non-IBM employees. Drawing from six of these experiments I assembled a database of 73 independent completions of the version of the MOS that we have been using (taken from Salza et al, 1996). (Note: Using the guideline that the number of completed questionnaires required for factor analysis is five times the number of items in the questionnaire (Nunnally, 1978), the minimum required number of MOS questionnaires is 35, well below the 73 questionnaires in the database.) This database was the source for a factor analysis, reliability assessment (both of the overall MOS and the factors identified in the factor analysis) and sensitivity investigation using analysis of variance on the independent variable of System.

2.2. Validity Evaluations

Relationship to paired comparisons. Data from a classified IBM report provided an opportunity to replicate the finding of Salza et al. (1996) that MOS ratings correlate significantly with paired comparisons. In the experiment described in the report, listeners provided paired comparisons after listening to samples from each of two TTS voices, then provided MOS ratings for each voice after hearing them a second time.

Relationship to intelligibility scores. Data from Wang and Lewis (2001) provided an opportunity to investigate the correlation between MOS ratings and intelligibility scores. In that experiment, listeners heard a variety of types of short phrases produced by four TTS voices, with the task to write down what the voice was saying. After finishing that intelligibility task, listeners heard the samples for each voice a second time and provided MOS ratings after reviewing each voice.

3. Results

3.1. Factor Analysis

After conducting a factor analysis of the MOS database, a parallel analysis (Coovert & McNelis, 1988) on the resulting eigenvalues indicated a three-factor solution that accounted for about 71% of the variance. Table 1 shows the results of the three-factor varimax-rotated solution, with bolded text to highlight the factor on which each item had the highest load. Note that the third factor only contains a single item. In normal use of the term, a factor has more than one contributing item, so in this report the conclusion is that the MOS has two factors with one item (Speaking Rate) not associated with either factor. Labeling factors is always a subjective exercise, but the factors do appear to be consistent with the factors reported by Kraft and Portele (1995), with items 2-5 (Listening Effort, Comprehension Problems, Speech Sound Articulation, Pronunciation) forming an Intelligibility factor and items 1 and 7 (Global Impression, Voice Pleasantness) forming a Naturalness factor.

Table 1. Three-Factor Varimax-Rotated Solution

	FAC1	FAC2	FAC3
MOS1	0.327	**0.900**	0.194
MOS2	**0.629**	0.370	0.427
MOS3	**0.693**	0.104	0.358
MOS4	**0.672**	0.433	0.294
MOS5	**0.746**	0.437	0.139
MOS6	0.322	0.204	**0.754**
MOS7	0.182	**0.665**	0.139

3.2. Reliability

Coefficient alpha for the overall MOS was 0.89, with 0.88 for the Intelligibility factor and 0.81 for the Naturalness factor. (It isn't possible to compute coefficient alpha for a single item.) Thus, the reliability of the MOS was acceptable. The reliability of the Naturalness subscale was somewhat lower than the Intelligibility subscale, probably due to it only having two items.

3.3. Sensitivity

Overall MOS rating. A between-subjects one-way analysis of variance on the overall MOS rating was statistically significant ($F(4, 68) = 7.6, p = .00004$). As expected, the recorded human voice (Wave) received the best rating, followed by the concatenative and formant-synthesized voices respectively.

Analysis by factor. Figure 1 shows the relationship among the TTS systems in the database and the MOS factors (including Speaking Rate). A mixed-factors analysis of variance indicated a significant main effect of System ($F(4, 68) = 9.6, p = .000003$), a significant main effect of MOS Factor ($F(2, 136) = 14.7, p = .000002$), and a significant System by Factor interaction ($F(8, 136) = 3.1, p = .003$).

3.4. Validity

Correlation with paired comparisons. Correlations computed among the final preference votes (paired comparisons) of 16 listeners exposed to two distinctly different TTS systems and the mean difference scores for MOS ratings for both systems indicated that the validity coefficients for overall MOS, Naturalness and

Intelligibility were significant ($p < .10$, $r = .55$, .49, and .46 respectively). The correlation between paired comparisons and Speaking Rate was not significant ($r = .36$, $p = .172$).

Correlation with intelligibility scores from Wang and Lewis (2001). The only significant validity coefficient was that for Intelligibility ($r = -.43$, $p = .10$), which indicates evidence for both convergent and divergent validity. The evidence for convergent validity (having a significant relationship where expected) is the correlation between the MOS Intelligibility factor and the overall intelligibility score from Wang and Lewis. The evidence for divergent validity (failing to correlate significantly with scores hypothesized to tap into different constructs) is the non-significant correlations between the overall intelligibility score and the other MOS measurements (Overall MOS: $r = -.38$, $p = .15$; Naturalness: $r = -.19$, $p = .48$; Speaking Rate: $r = -.26$, $p = .33$).

Figure 1. Interaction of TTS System and MOS Factor

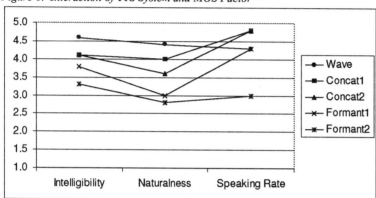

4. Discussion

The version of the MOS derived from Salza et al. (1996) seems to have reasonably good psychometric properties. The factor analysis of the current data resulted in a factor structure similar to that of Kraft and Portele (1995), specifically two factors (Intelligibility and Naturalness) and an unrelated item for Speaking Rate. The reliability of the overall MOS and its subscales is acceptable. Furthermore, the data replicated the validity result of Salza et al. by showing a significant correlation between paired comparison data and MOS data (Overall MOS, Naturalness, and Intelligibility). The data also indicated appropriate convergent and divergent validity for the intelligibility scores from Wang and Lewis (2001). Note that this result is similar to that reported by Johnston (1996), who found that the Listening Effort item (which is part of the Intelligibility factor) was more sensitive to degradation of speech intelligibility than the Global Effort item (which is part of the Naturalness factor).

Using principles from psychometrics (Nunnally, 1978), it should be possible to improve the reliability of the MOS. Rather than using 5-point scales with an anchor at each step, general reliability should improve slightly with a change to 7-point bipolar scales. Because the Naturalness factor had somewhat weaker reliability than the Intelligibility factor, it would be reasonable to add at least one more item to the MOS that is likely to tap into the construct of Naturalness.

The MOS Speaking Rate item failed to fall onto either the Intelligibility or Naturalness factor in both the current study and in Kraft and Portele (1995). This might have happened because Speaking Rate is truly independent of either of these constructs, or might have been an artifact due to the unique labeling of the scale points for this item. The other items have scales that have a clear ordinal pattern, such as "Excellent", "Good", "Fair", "Poor", and "Bad" for the Global Impression item. The labels for the Speaking Rate item are, in contrast, "Yes", "Yes, but slower than preferred", "Yes, but faster than preferred", "No, too slow", and "No, too fast", which do not have a clear top-to-bottom ordinal relationship. If the item assessing Speaking Rate had the same structure as the other items in the MOS, a future factor analysis could determine less ambiguously whether Speaking Rate is truly independent of Intelligibility and Naturalness.

5. A Proposed New Version of the MOS

The key proposals are to increase the number of scale steps per item from five to seven (using bipolar scales), to increase the number of items related to Naturalness, and to make the structure of the Speaking Rate item consistent with the other items. These modifications should improve the reliability of the MOS and, by extension, its other psychometric properties because reliability constrains the magnitude of validity coefficients (Nunnally, 1978) and limits a scale's sensitivity. The summary version of the revised MOS presented here shows the text of the item and its bipolar labels. (For the completely revised MOS and more details, see Lewis, 2001.)

1. *Global Impression:* Please rate the sound quality of the voice you heard. (Very Bad / Excellent)
2. *Listening Effort:* Please rate the degree of effort you had to make to understand the message. (Impossible Even with Much Effort / No Effort Required)
3. *Comprehension Problems:* Were single words hard to understand? (All Words Hard to Understand / All Words Easy to Understand)
4. *Speech Sound Articulation:* Were the speech sounds clearly distinguishable? (Not at All Clear / Very Clear)
5. *Pronunciation:* Did you notice any problems in the naturalness of sentence pronunciation? (Very Many Problems / Didn't Notice Any)
6. *Voice Pleasantness:* Was the voice you heard pleasant to listen to? (Very Unpleasant / Very Pleasant)
7. *Voice Naturalness:* Did the voice sound natural? (Very Unnatural / Very Natural)
8. *Ease of Listening:* Would it be easy to listen to this voice for long periods of time? (Very Difficult / Very Easy)
9. *Speaking Rate:* Was the speed of delivery of the message appropriate? (Poor Rate of Speech / Perfect Rate of Speech -- also include this additional sub-item: "If unsatisfactory, please circle one: Too Slow or Too Fast")

If the proposed changes work as expected, the revised MOS items 2-5 will continue to form an Intelligibility factor. Items 1 and 6-8 should form a Naturalness factor with substantially greater reliability (possibly in excess of .90) than the current Naturalness factor due to the additional items and the shift from five to seven scale steps. The change in the structure of item 9 (formerly item 6) should make it possible to determine whether Speaking Rate is truly independent of the other two factors without losing the ability to determine if a listener finds it too slow or fast.

6. References

Coovert, M. D., & McNelis, K. (1988). Determining the number of common factors in factor analysis: A review and program. *Educational and Psychological Measurement, 48*, 687-693.

Johnston, R. D. (1996). Beyond intelligibility: The performance of text-to-speech synthesisers. *BT Technology Journal, 14*, 100-111.

Kraft, V., & Portele, T. (1995). Quality evaluation of five German speech synthesis systems. *Acta Acustica, 3*, 351-365.

Landauer, T. K. (1988). Research methods in human-computer interaction. In M. Helander (Ed.), *Handbook of human-computer interaction.* New York: Elsevier.

Lewis, J. R. (1993). Multipoint scales: Mean and median differences and observed significance levels. *International Journal of Human-Computer Interaction, 5*, 383-392.

Lewis, J. R. (2001). *Psychometric properties of the Mean Opinion Scale* (Tech. Report in press -- will be available at http://sites.netscape.net/jrlewisinfl after publication).

Nunnally, J. C. (1978). *Psychometric theory.* New York: McGraw-Hill.

Paris, C. R., Thomas, M. H., Gilson, R. D., & Kincaid, J. P. (2000). Linguistic cues and memory for synthetic and natural speech. *Human Factors, 42*, 421-431.

Salza, P. L., Foti, E., Nebbia, L., & Oreglia, M. (1996). MOS and pair comparison combined methods for quality evaluation of text to speech systems. *Acta Acustica, 82*, 650-656.

Sonntag, G. P., Portele, T., Haas, F., & Kohler, J. (1999). Comparative evaluation of six German TTS systems. In *Eurospeech '99* (pp. 251-254). Budapest: Technical University of Budapest.

Wang, H., & Lewis, J. R. (2001). Intelligibility and acceptability of short phrases generated by text-to-speech (to appear in the conference proceedings for Human-Computer Interaction International '01).

Yabuoka, H., Nakayama, T., Kitabayashi, Y., & Asakawa, Y. (2000). Investigations of independence of distortion scales in objective evaluation of synthesized speech quality. *Electronics and Communications in Japan, Part 3, 83*, 14-22.

A PERSPECTIVE ON INTELLIGENT INFORMATION INTERFACES FOR MOBILE USERS

Thomas Rist
DFKI Saarbrücken, Germany. Email: rist@dfki.de

Abstract

Mobile access to information services can not be accomplished by simply carrying over WIMP style interfaces as know from the PC world. Rather, information delivery needs to be customized in order to benefit mobile users. This paper discusses in particular the customization of graphical representations for tiny displays.

1. INTRODUCTION

During the last few years, we have seen a variety of mobile appliances such as mobile phones, micro PDA's, and also first working prototypes of the next generation's wrist watches that - in addition to their original functionality - provide wireless access to the Internet and the World-Wide-Web. While it is debatable whether web-browsing is amongst the most useful applications for the users of such mobile devices, there is no doubt that these appliances provide a high potential for a broad range of new information services that can accommodate for the specific needs of mobile users. For example, think of the commuter who whishes to get the newest travel info, whereas the stock jobber may wish to inspect the development of shares and perform transactions while being on the move. It is our believe that many of these information services should be brought to the user by means of clever combinations of written text, voice and sound, graphics, and animation. Design concepts as known from the design of WIMP and multi-media interfaces running on a PC, however, cannot be simply carried over to the mobile domain. Firstly, mobile users are often situational disabled in contrast to users sitting in front of a stationary PC. For instance, while walking or driving, users might prefer to receive requested information verbally instead of graphical displays. Consequently, the design of services and user interfaces must take into account the specific conditions that result from non-stationary system usage. Secondly, the large variety of new portable computing and communication devices adds another level of complexity to information systems and systems which are to support telecommunication and collaborative work since one has to consider the fact that that different users may not be equipped equally in terms of output and input capabilities. Intelligent mechanisms for on-the-fly adaptation of input/output modalities offer a promising perspective for the support of mobile users who want to access information services or participate in multi-user applications. With a focus on information presentation, DFKI has explored automated adaptation mechanisms to serve mobile users in a number of different areas of application. This contribution briefly reports on the objectives and achievements of this work, and finally sketches some ideas towards future research.

2. ACCESS TO INFORMATION SOURCES AND SERVICES

For a number of years DFKI has been developing systems that search and gather information from the WWW in order to satisfy a user's particular information needs. As opposed to mere search engines, these systems deploy so-called information agents which embody knowledge where and how to find certain types of information. Taking as input a user's request for information, the agents visit a number of web sites, work through the web pages in order to identify and extract relevant bits and collect them in a common information pool. This information gathering phase is then followed by a presentation generation phase in which the collected contents is coherently presented. For the purpose of illustration, we refer to DFKI's AiA Travel Agent [1] and to DFKI's Personal Picture Finder [2].

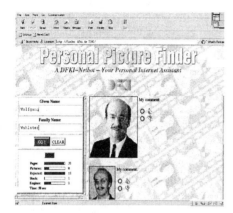

Figure 1. Left: WWW interface of DFKI's Personal Picture Finder. For a given name the system searches the web for a picture of the person. Right: A typical collection of web pages (weather, hotel information and route descriptions) consulted by the AiA travel agent(s) in response to a user query.

The travel agent assists a user in preparing for a journey. For a given travel destination a swarm of agents gather information that might be of interest for the travel. The resulting pool of gathered information typically contains local weather news/forecasts, recommendations where to stay and where to eat, and an overview of local cultural events and entertainments. In a number of cases, the discovery of certain bits of information will trigger further agents to acquire complementary bits. For instance, if a hotel or a restaurant is found relevant, a dedicated "map agent" is triggered to search for a map showing how to get to the place. The right hand side of Figure 1 shows a typical collection of different pieces of textual and graphical material retrieved by the information agents in response to a user query. The Personal Picture Finder is a service that takes as input the name of a person and searches the web for a portrait picture of the person (cf. left part of Fig.1). Such a service can be especially useful when preparing for a meeting with persons for the first time, or when one has to collect someone not met face-to-face before at the airport or the train station.

While in their original versions both services have been designed for PC users who are assumed to interface with the services via an ordinary WWW browser, it is quite obvious that users on the move could greatly benefit from these services as well if they could access the services by mobile devices, such as WAP-enabled mobile phones.

Currently, several attempts are made to develop transformation mechanisms that take as input arbitrary information sources (e.g. html pages) and deliver information presentations that can be displayed on mobile devices with limited display capabilities, such as lack of screen real estate and lack of colors. In case of a textual source, a straightforward approach is to fragment the text into yet displayable chunks. However, this strategy can easily result in huge stacks of WML pages which are too unpractical to read on a mobile device. Other approaches [3, 4] provide filtering and parsing technology that enables a user to specify which textual contents of a web page shall be extracted and converted into a format displayable on a mobile device. In the case of visual media, such as graphics, animation and video, however, partitioning is often not possible at all, and the transformation of graphics is yet an almost untouched challenge. At DFKI we started to tackle the problem of how to transform graphical representations so that they can be displayed on very small displays, such as a 100*60 pixel display of a mobile phone. In particular, we are currently investigating different approaches to solve the graphical transformation problem: *uniformed transformations, informed transformations,* and *re-generation of graphics.*

In the first and the second approach a transformation (usually comprising down-scaling and reduction of colors) is applied to a source graphics in order to obtain a target graphics that meets the requirements of the target display screen. The transformation can be uninformed (or blind) in so far, that only little information about the source graphics is taken into account when selecting and adjusting the transformation. Unfortunately, it is very difficult to find a general-purpose transformation that reliably produces suitable results for the large variety of graphics found on the WWW. Figure 2 illustrates the problem. While the applied transformation produces an acceptable result for the source graphics in the first case, the result is less acceptable when applying the same transformation to the source graphics in the second case.

Figure 2. Applying the same transformation T to two different source pictures in order to obtain a small black and white target picture that can be displayed on a mobile device.

A more promising approach starts with an analysis of the source picture in order to inform the selection and adjustment of transformation parameters.

Basically the analysis phase performs a classification of source pictures amongst syntactic or even semantic features. For instance, in our current work, the set of implemented semantic classifiers comprises classifiers that distinguish between portrait and non-portrait images, outdoor versus indoor images, outdoor images that show a scene with blue sky, clouds, sunset, water, forest or meadows, and snow-covered landscapes.

However, it is still difficult to make an assignment between recognized features of an image on the one hand, and available transformations and their parameter adjustments on the other hand. We are currently investigating in how far this problem can be solved by deploying machine learning techniques. That is, in a training phase, a graphics design expert manually

156

assigns images to transformations and thereby allow the system to recognize and generalize correspondences between image features and transformation parameters. In our current test setting we use some 430 features to characterize images. In contrast, our repertoire of transformations is yet quite small. By means of a software package for machine learning [5] we trained the system with a set of 130 images. So far, the achieved scores for adequate selections of transformations are in the range of 60-70 %. While this result is encouraging, further refinements of the approach are required.

The third approach, *re-generation of graphics,* does not aim at a modification of a source graphics at the picture level. Rather, the idea is to generate a new picture from a content description obtained from a deeper semantic analysis of the source graphics. We illustrate this approach by a transformation of a route description as it may be obtained from a routing service available on the WWW.

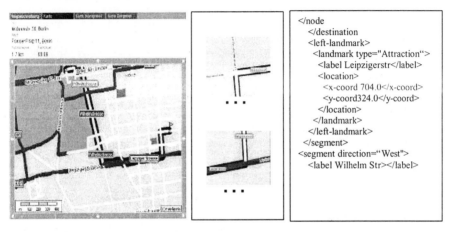

Figure 3. Analyzing a map in order to construct a symbolic description (here in XML format) of the core contents (here a route description).

Once a semantic representation of the picture content has been extracted, it is possible to generate a variety of new graphical representations which adopt different graphical styles in order to meet resource limitations of the output device and/or a user's personal style preferences. Figure 4 presents several variants of graphical displays showing a routing maneuver.

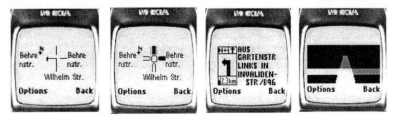

Figure 4. Display variants of a routing maneuver ("turn left on next crossing") shown on a WAP-enabled mobile phone (Nokia 7110).

3. FURTHER DIRECTIONS

In the introduction to this contribution we argued that mobile access to information services can not be accomplished by simply carrying over WIMP style interfaces as know from the PC world. Further, it has been argued that the information delivery needs to be customized in order to benefit the mobile users. The current contribution elaborated in more detail the customization of graphical representations (cf. Section 2). In our research we also address multimodality as a further important aspect of future mobile interfaces. Especially the combination of voice input together with graphical selections as well as the combined display of graphical and audio output is very likely to become a predominant interaction paradigm for users of tiny mobile devices. Envisaged scenarios include: (a) the user formulates information requests likewise: verbally, by a graphical selection, or by a combination of both; (b) the user can flexibly switch between several information display modes including the display of text and graphics, audio display, and effective combinations of multiple modes (see also [6]). While the concept of multimodal interfaces is not a new one [7], its instantiation for the case of users of mobile devices, however, reveals a number of new challenges to be dealt with including technical issues (e.g., how to perform robust speech analysis on a mobile device, how to flexibly translate contents given in one medium into another format), and usability issues (e.g., how to decide on which media combinations are most appropriate considering a mobile user's current task and situation).

ACKNOWLEDGEMENT

Thanks to Patrick Brandmeier who contributed to the work presented in this paper.

REFERENCES

[1] André, E., Rist, T. and Müller, J. (1999). Employing AI Methods to Control the Behavior of Animated Interface Agents. *Applied Artificial Intelligence* 13:415-448.

[2] Endres, C., Meyer, M., and Wahlster, W.: Personal Picture Finder: Ein Internet-Agent zur wissensbasierten Suche nach Personenphotos, Online'99 (in German). Service temporarily available: *http://finder.dfki.de:7000/*

[3] GoSMS.com Ltd., TIP Reference Manual. GoSMS.com Ltd, Tel-Aviv, Israel, Nov. 2000.

[4] Bergström, A., Jaksetic, P. and Nordin, P. Enhancing Information Retrieval by Automatic Acquisition of Textual Relations Using Genetic Programming. In: Proceedings of Intelligent User Interfaces (IUI) 2000, ACM Press, 2000.

[7] Witten, I.H. and Frank, E.: Data mining: Practical machine learning tools and techniques with Java implementations. Morgan Kaufmann, San Francisco. 2000. Software temporär verfügbar: http://www.cs.waikato.ac.nz/~ml/weka/

[6] Rist, T.: Towards Services that Enable Ubiquitous Access to Virtual Communication Spaces. In Proc. of UAHCI 2001, this issue.

[7] Maybury, M. and Wahlster, W. editors. 1998. Readings in Intelligent User Interfaces. Morgan Kaufmann Press.

POST-WIMP INTERACTIVE SYSTEMS:
MODELING VISUAL AND HAPTIC INTERACTION

Piero Mussio (1), Umberto Cugini (2) and Monica Bordegoni (2)

(1) Dipartimento di Elettronica per l'Automazione, Universita' di Brescia, Italy
(2) Dipartimento di Ingegneria Industriale, Universita' di Parma, Italy

ABSTRACT

Post Wimp Interactive Systems are end-user oriented systems characterized by new modalities of interaction. The new interaction modalities, based for example on gesture or speech for input and output, make available a stream of multimodal events to the user and to the system. Users and system must recognize patterns of interest in the stream of events and interpret them to achieve the goals of the interactive task. The goal of a successful design is to bring the system semantics to reflect the user's one, so that each event occurring during the interaction is properly understood by the user and adequately managed by the computer. This report describes some initial steps toward this goal.

1. INTRODUCTION

Post Wimp Interactive Systems are end-user oriented systems characterized by new modalities of interaction. End–users (users in the following) have different cultures, skills and dexterity levels and interact with the system in different working contexts to achieve different tasks. The new interaction modalities, based for example on gesture or speech for input and output, make available a stream of multimodal events to the user and to the system.

To achieve the goals of the task being executed, users and system must recognize patterns of interest for the interaction in the stream of events.

The patterns that users recognize in the stream depend on their culture, skill as well as on the working context. The system, on its side, is designed to recognize patterns that are meaningful for its designer.

The interpretation of the patterns in the input stream is uncertain because some patterns may be subject to different interpretation depending on the situation in which they occur and both the users and the system can make errors in the interpretation. Moreover, the user and the designer may detect different patterns or assign different meanings to a same pattern.

These features influence the way post-wimp interactive system are designed and developed and force designers to evolve their classic design techniques of interactive systems (Myers, 2000).

This paper discusses these problems from an engineering point of view, in that the final goal of our research, is to understand how to design interactive systems, which users may accept and use in the execution of their tasks, trusting them (Nielsen, 1993).

The discussion starts introducing a recently developed model of Human Computer Interaction recently proposed by the Pictorial Computing Laboratory (Bottoni, 1996), focusing on the communication aspects of WIMP interaction.

According to this model, human and computer communicate by materializing and interpreting visual patterns in a sequence of messages at successive time instants, the humans using human natural cognitive criteria, the computer using the criteria programmed in it. The goal of a successful design is to bring the system semantics to reflect the user's one, so that the messages exchanged during the interaction are properly understood by the user and adequately managed by the computer.

This paper proposes a first step in extending this model to post WIMP interactive systems, studying some problems arising in *visual and haptic interaction.*

In visual and haptic interaction, humans and system recognize patterns in the stream of multimodal events combining haptic and visual events. Some results of initial experiments demonstrate the potential power of such a combination and give indications on the modelling of these multimodal processes.

2. MODELLING THE COMMUNICATION ASPECTS

The Pictorial Computing Laboratory models WIMP interaction as a process based on the exchange of images between two participants, namely the human user and the computer (Bottoni, 1996). Both human and computer interpret every event occurring during interaction with reference to the whole image appearing on the computer display screen. This image is formed by text, graphs, pictures, icons, etc. and therefore represents a multimedia

159

message, and materializes and conveys the meaning intended by the sender and must be interpreted – i.e. associated with a possible different meaning – by the receiver.

The images on the screen are produced (materialized) by the human or by the system through actions that materialize the results of human reasoning and computer computations.

Human and computer communicate by materializing and interpreting a sequence of images at successive instants of time $t_1,....,t_n$. Humans interpret and materialize the images using human natural cognitive criteria, the computer using the criteria programmed in it. The images describe the state of the interaction, and their interpretation determines the next action of the human and the next computation of the computer.

Following Preece et al., the Pictorial Computing Laboratory approach to interactive system design can be classified as an *holistic* one (Preece, 1996). In the design process, the decisions about the way in which the interface should look and how it should behave are taken depending on how this will be physically communicated to users and attention is focused both on the appearance and the behaviour of the interface. However, differently from other holistic approaches, the PCL approach adopts a formal technique to specify the computational meaning of what is progressively defined (Bottoni, 1999). In this way the interactive system conceptual model is incrementally defined, which can be displayed as a concrete reality to users and in consequence can be validated by usability techniques while each refinement is also formally verified.

During the interaction the computer assumes the double role of the tool used by the human to materialize his messages and of the second participant to the interaction. The human changes the image on the screen acting on the input devices of the computer. Therefore, the human uses the computer as the tool used to materialize the message. However, the computer also computes the reactions to the human's actions.

In this model, the I/O devices through which the human performs his/her actions, in particular the haptic ones – the mouse and the keyboard – appears as secondary channels of communication and hence actions where treated only at a high level of abstraction.

Humans interpret the images on the screen by recognizing *characteristic structures* (*css* or structures for short), i.e. sets of image pixels that users recognize as functional or perceptual units. The *cs* recognition results into the association of a meaning with a structure. Humans express the meaning attributed to the *cs* by a verbal description. In an image a human may recognize several *css*: combining their meanings, humans derive the meaning of the whole image on the screen.

Conversely, the computer associates graphical entities (set of pixels) on the screen with computational constructs. Each computational construct -here denoted by u- represents the meaning associated with a graphical entity on the screen, as intended by the system designer.

It is exactly this association that makes the computer able to interpret the captured user actions (such as clicking on a button), with respect to the image on the screen, possibly firing computational activities whose results are materialized on the screen, via creation, deletion, or modification of *css*. The association between a cs and the corresponding u is called a *characteristic pattern* and can be formalized introducing two functions intcp and matcp the first mapping the cs into u and the second u into the cs. Hence a *characteristic pattern* (cp) is specified by cp=<cs, u, <intcp, matcp>>.

The image I on the screen is seen as a cs formed by a set of component css, each associated with its u. Hence the inage I can be associated to a description d, derived by the composition of the set of all descriptions, associated with the component css. Two functions, int and mat, can be defined on the basis of the individual functions in the cps, and the relationship between the image i and its semantics d is summarized by a *visual sentence*, the triple: vs=<i,d,<int, mat>>. A set of vss is a Visual Language (VL).

In any interaction process, two interpretation processes occur: one internal to the computer, in which each image is associated with a computational meaning, as defined by the designer and implemented in the computer; and one proper to the human performing the task, depending on his/her role in the task, as well as on his/her culture, experience and skill. Hence, two semantics must be taken into account in describing the interaction process and in the identification of the features characterizing it. As observed in (Chang, 1996), users can achieve their tasks if they associate with each cs and with the whole image a meaning similar to the one associated by the computer, i.e. that is if an *adequate communication* is reached.

From this point of view, the goal of a successful design is to bring the system semantics to reflect the user's one, so that the messages exchanged during the interaction are properly understood by the user and adequately managed by the computer.

This goal becomes difficult to be achieved when ambiguous messages occur and equivocal situations arise.

As an example, figure 1 has been generated by the computer using three well-defined geometrical models. The image is ambiguous for the user, who is not able to interpret univocally the three elements and may flip between the different interpretations.

Figure 1. Images of three 3D models: each image can be interpreted as a carved box, two boxes or a box in a corner.

3. HAPTIC INTERACTION

When haptic tools are used not only as operational tools, but become primary communication devices in that their signals become communication messages, one speaks of haptic interaction. By using a haptic device, human-computer interaction is augmented with the sense of touch: a user touches virtual objects and can explore a 3-D space.

Several kinds of devices have been proposed for haptic interaction, from one-actuator - for example, the PHANToM device (SensAble Technologies), shown in Figure 2, to multiple actuators - for example, the haptic display (Hirota, 1995). In any device, each actuator transmits a sampled, quantized signal, in terms of point applied forces reproducing the physical interaction with an element of a surface, having specific characteristics (roughness, softness, etc.). The human hand/finger receives this set of stimuli, and the person decodes all that as a local, tactile description of a 3-D surface.

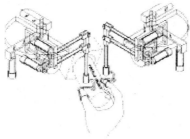

Figure 2. Touching a virtual object using two PHANToM haptic devices by SensAble Technologies, Inc.

In haptic interaction human and computer exchange:
   a) a set of impulses which are applied by the computer through some device to the human's hand and/or fingers and;
   b) a set of movements performed by the human through the hand and/or finger on an haptic device and mapped by the device int a time sequence of digital signals for the computer.

Due to current technological constraints, haptic devices are limited in resolution and dimensions. They send a high granularity set of signals that describe a limited part (or even a point) of the surface being explored. This limitation may induce some *aliasing* effects that may determine ambiguous or equivocal situations.

In order to understand an environment, the human has to integrate in time the set of data that receives from the haptic device. The elementary signals by which the haptic structures are recognized are the impulses that the computer is able to generate or detect and transmit. The identification of haptic structures can be modeled as a two stages process: some feature of the object being touched – such as its elasticity or hardness- can be recognized statically, by interpreting the instantaneous signal $s(t)$ that in general is only sufficient to start forming a mental model of the object. Other features, such as its geometrical or textural properties, can only be evaluated in time by perceiving continuity or discontinuity in a set of signals $s(t_1)$, $s(t_2)$ ....$s(t_n)$, each one representing an instantaneous signal, so as to complete a mental model of the surfaces present in the environment. These sets of perceptible haptic events that the user recognizes as functional or perceptual units and with which s/he can associate a meaning – and a name expressing the meaning - are called *haptic characteristic structures* (hcs).

The most elementary signal detectable (generable) is here called '*hapxel*', and is described by a quadruple $<x, y, z, s(t)>$, where x, y ,z are the coordinates describing the position of the haptic device in the virtual space and $s(t)$ the digitized signal.

A *hapxel* plays in haptic communication a role analogous to the role of the pixel in the visual communication. A set of one or more *hapxel*, which constitute a functional or perceptual unit for the human, constitutes haptic characteristic structures (hcs). In general, hcs*s* are recognized and materialized by human or by the computer integrating sequences of *hapxel* (signals in the 3-D space) in time: this fact reduces the analogy with visual css whose recognition and materialization requires integration in space.

The frequency at which the haptic device transmits impulses to the human hand/fingers determines the human perception and influences interpretation of the set of signals. Similarly to the visual refresh rate, some studies have shown how continuous perception is achieved only if the haptic refresh rate is not below 1KHz (Massie, 1998). If the refresh rate goes below this threshold, humans may feel some discontinuities that simulate non-existing patterns. These artifacts are the origin of possible *aliasing* effects, where information is distorted because of low-frequency sampling. Moreover, in currently available haptic devices, other disorienting features may exist, determined by the specific technology adopted.

To make the reasoning concrete, let us examine some simple experiment performed using one the exposition will be based on examples from some experiments performed using a PHANToM device.

4. EXPERIMENTING COMBINED VISUAL-HAPTIC INTERACTION

In order to verify the power of haptic interaction we have run some experiments with twenty subjects (Faconti, 2000). The experiments are based on the 3D objects shown in Figure 1, which generate according to a deterministic geometric model. Figure 3 shows another view of the models, in order to allow the reader to understand what the objects are. This view was not accessible to the subjects of the experiments.

In the first experiment, subjects were only allowed to exploit their visual capabilities. The three images were shown to the subjects, one at a time, one after the other. Users were asked to describe what they recognized. The images resulted ambiguous and equivoque classifications were generated in particular the second and third images, which were correctly interpreted by 14% of subjects.

In the second experiment, subjects were allowed to exploit only their haptic capabilities. They had to explore a 3D model of one object at a using one PHANToM device, i.e. only receiving a haptic stimulus corresponding to one *hapxel* at a time. Few subjects were able to say which object they were touching, regardless of their experience and skills. In the case of the two boxes object, none of the subjects was able to recognize the objects.

These very poor results may have explained in several ways. The device by itself has a very low bandwidth of information considered both in absolute terms and when compared to the sense of touch in reality. In real environment, humans do not use only one finger during shape recognition and, even in that case, the number of receptors conveying haptic information is more than a thousand order of magnitude of what the PHANToM provides.

In the third experiment the subjects were allowed to exploit their visual and haptic capabilities simultaneously, being permitted to relate visual, and haptic clues. In fact, the coupling of two modalities providing redundant information on the scene to be explored increases the correct interpretation of the objects. In the example of the two boxes object, 83% of the subjects recognized the correct object.

Figure 3. Objects of Figure 1 shown from a different point of view, not accessible to the subjects of the experiment.

The situation in which both haptic and visual modalities are present is more complex to be specified, from the point of view of the designer. In fact, we have set of pixels – i.e. vcss – and set of *hapxels* – i.e. hcss that are related and determine one only entity – a characteristic pattern - to be recognized or materialized. Hence, one internal description d of the entity of interest must store the properties related to the different sensory modalities as well as the type of the cp.

Hence, in this experiment, in which for the exploration of each object the visual representation of the virtual space is maintained constant, an instantaneous multimodal cp is specified as:

$$mcp(t)=<<hcs(t),vcs,>,<hu(t),vu>,<vint, vmat>,<hint(t), hmat(t)>>$$

in which haptic components (h-) are time-dependent, while the visual ones (v-) do not vary in time.
However, these instantaneous multimodal cp are the instantaneous projections of a unique multimodal cp:

$$mcp=<<chcs,vcs,>,<chd,vd>,<vint, vmat>,<hint, hmat>>.$$

Here, also the haptic components is time-independent, but must be expressed as a reactive feature of the characteristic pattern, in that it depends also on the user's action and not only on the physical characteristics of the objects being modeled.

5. CONCLUSIONS

The goal of a successful HCI design is always to bring the system semantics to reflect the user's one, so that the messages exchanged during the interaction are properly understood by the user and adequately managed by the computer. Indeed, achieving this goal requires the capability of managing reactive properties, coordinating the different modes and obtaining the different projections along the different modes.

This paper has outlined problems related to multimodal interaction combining haptic and visual events, and has presented some preliminary results of experiments that underline the power of visuo-haptic interaction. These issues appear to be the open problems to be explored in the design of post-wimp interactive systems.

REFERENCES

Bottoni, P., Costabile, M.F., Levialdi, S., Mussio, P. "A Visual Approach to HCI", In ACM SIGCHI Bulletin, Vol.28, No.3, July 1996, pp.50-55, also available at the URL: www.acm.org/sigchi/bulletin/1996.3/levialdi.html.

Bottoni, P., Costabile, M.F., Mussio, P. "Specification and Dialog Control of Visual Interaction through Visual Rewriting Systems", In ACM TOPLAS,Vol. 21, No.6, pp.1077-1136, 1999.

Chang, S.K. and Mussio, P. "Customized Visual Language Design". In International Conference on Software Engineering and Knowledge Engineering - SEKE'96, pp. 553-562, 1996.

Cugini, U., Bordegoni, M., Rizzi, C., De Angelis, F., and Prati, M. "Modelling and Haptic Interaction with non–rigid materials". In Eurographics'99 - State-of-the-Art Reports, Milano, 1999.

Faconti, G., Massink, M., Bordegoni, M., De Angelis, F. and Booth, S. "Haptic Cues for Image Disambiguation". In Computer Graphics Forum. Blackwell, 19(3), pp. 169-178, 2000.

Hirota, K. and M. Hirose, M. "Providing Force Feedback in Virtual Environments". In Proc. IEEE Computer Graphics and Applications, pp. 22-30, September 1995.

Massie, T. "Physical interaction: The Nuts and Bolts of Using Touch Interfaces", In Physical Interaction: The Nuts and Bolts of Using Interfaces with Computer Graphics Applications, SIGGRAPH 98, Tutorial 1, 1998.

Myers B., Hudson S.E. and Puasch R. " Past, Present, Future of User Interface Software Tools". *ACM Trans. On Computer-Human Interaction*, Vol.7, No.1, March 2000, pp.3-28..

Preece J., Rogers Y., Sharp H., Benion D., S. Holland S. and Carey T. "Human-Computer Interaction", Addison Wesley, 1996.

SensAble Technologies, Inc., URL: www.sensable.com.

Nielsen, J. "Usability Engineering", Academic Press, 1993.

163

The Sentient Map as a New Paradigm for Human-Computer Interface

Shi-Kuo Chang, Min Zhao and Xuan. Zou
Department of Computer Science
University of Pittsburgh, Pittsburgh, PA 15260 USA

Abstract

The *sentient map* is a map that can increase the user's awareness by sensing the user's input and generating appropriate reactions such as retrieving and presenting the relevant information to the user. We describe the sentient m an application as the user interface for the Growing Book that is a constantly expanding electronic book for e-learning, then discuss the design of the experimental sentient map system.

1. The Sentient Map

The *sentient map* is a map that can increase the user's awareness by sensing the user's input and generating appropriate reactions such as retrieving and presenting the relevant information to the user. Sentient maps can serve indexes and lead the user to more information. In practice a sentient map is a gesture-enhanced interface for an informat system. A companion paper [Chang01] describes the motivations for the sentient map paradigm and the gesture qu language in greater detail. Therefore in the present paper we will concentrate on illustrating the use of the sentient ma one important application area, i.e., e-learning.

First, we repeat here the formal definition of the sentient map [Chang01].

A sentient map *m=(type, profile, v, IC)* has a *type*, a *profile*, a *visual appearance*, and a set of *teleactivities*.

A sentient map's *type* can be geographical map, directory page, web page, document and so on.

A sentient map's *profile* consists of attributes specified by its creator for this map type.

A sentient map's *visual appearance* is defined by a visual sentence *v*, which is created according to a visual gram or a multidimensional grammar.

The *teleactivities* of a sentient map is defined by a collection of *index cells IC* [Chang95]. Each index cell *ic* may h a visual appearance, forming a *teleaction object*. These teleaction objects are usually overlaid on some background.

Ignoring the type and profile, a sentient map is basically a collection of teleaction objects including a backgro object.

A *composite sentient map* is defined recursively as a composition of several sentient maps.

Maps are widely used to present spatial/temporal information to serve as a guide, or an index, so that the viewer of map can obtain certain desired information. Often a map has embedded in it the creator's intended viewpoints an purposes. A web page can also be regarded as a map, with the URLs as indexes to other web pages. In fact, any docum can be regarded as a map in a multi-dimensional space. Moreover, with associated scripts, the maps can also be m active [Chang95]. These two notions, that data can be viewed as maps and that maps can be made active, led u propose a new paradigm for visual information retrieval - the sentient map. The natural way to interact with a sentient is by means of gestures. In the companion paper [Chang01] we describe how simple gestures called *c-gestures* consis of mouse clicks and keyboard strokes are used to interact with a sentient map.

This paper is organized as follows. For empirical study, Section 2 describes an e-learning environment called Growing Book that serves as a test bed. In Section 3, we discuss the design of the sentient map system.

2. The Sentient Map Interface for the Growing Book

A *Growing Book* is an electronic book co-developed by a group of teachers who are geographically dispe throughout the world and collaborate in teaching and research [Chang00]. Since the course materials are consta evolving, the Growing Book must be constantly updated and expanded. The Growing Book is used by each teacher in the local classroom as well as in the distance learning environment. Therefore the Growing Book must be accessibl multilingual students. The chapters of he Growing Book are owned by different instructors who may utilize an vide different tools for distance learning, self learning and assessment. In what follows, we describe a prototype [301] and its sentient map interface, to serve as a test bed to investigate how to design and manage the Growing Book so it can be accessed universally by people with different linguistic skills and cultural background for effective teaching research.

A sentient map *m* for the Growing Book typically has the type "chapter". Its *profile* has the following attributes: apter_No: the numerical id of the chapter, **Chapter_Title:** the title of the chapter, **Chapter_Password:** the password order to update the profile or perform privileged operations, **Chapter_URL:** the URL of this chapter, **Author:** the hor, or authors, of this chapter, **Teacher:** the teacher, or teachers, of this chapter, **Student:** the student, or students, of chapter, **Who_is_Who:** the known authorities on the subject matter of this chapter, **Center_of_Excellence:** the known ters of excellence on research related to this chapter, **Reference:** the references for this chapter, **Tool:** the software ls useful for this chapter, **Privacy:** a list of user names who want to keep their personal information off the awareness , and **Awareness:** the awareness list specifies what the user want to be informed of.

This profile is used to increase the awareness of the user. The commands supported by the Growing Book include the owing: ABSTRACT: Create an abstraction of a chapter at a certain level. SEND_AUTHOR: Send a message to the hor(s) of a chapter. SEND_TEACHER: Send a message to the teacher(s) of a chapter. SEND_STUDENT: Send a ssage to the student(s) of a chapter. ADD_AUTHOR: Add to the author(s) of a chapter. ADD_TEACHER: Add to the cher(s) of a chapter. ADD_STUDENT: Add to the student(s) of a chapter. DROP_AUTHOR: Drop from the author(s) a chapter. DROP_TEACHER: Drop from the teacher(s) of a chapter. DROP_STUDENT: Drop from the student(s) of a pter. SET_AWARENESS: Set the awareness profile. SET_PRIVACY: Add someone to the privacy list. ECK_AWARENESS: Check the awareness profile. AWARE: Display info one should be aware of. ADD_WHO-IS-IO: Add to who-is-who. ADD_CENTER: Add to centers of excellence. ADD_REFERENCE: Add to reference(s). D_TOOL: Add to tool(s). ADD_WATERMARK: Add watermarks to the html files of a chapter. SPLAY_WATERMARK: Display watermarks of the html files of a chapter. WEAVE: Weave pieces of the same type, h as sound bytes, into a presentation. MATCHPAR: Define Growing Book working directory and matching parameters.

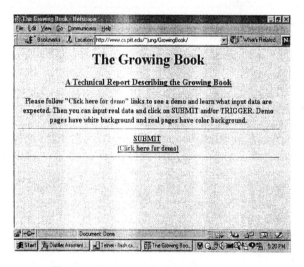

Figure 1. Top page of the Growing Book.

The Growing Book is designed using the Active Index System as the underlying coordination and control system. The nmands described above are message types for the Active Index System. In Figure 1, the top page of the Growing Book lustrated. The user can click on the red DEMO link to first see a demo and learn what input data are expected. Then user can input data and click on the SUBMIT link.

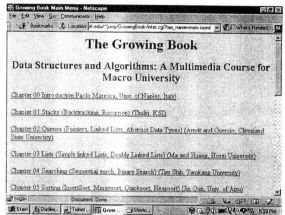

Figure 2. The contents of the Growing Book.

The contents of the Growing Book are illustrated in Figure 2. The user can enter a number of commands to manipul the Growing Book. For example, the *weave* command weaves the media objects of a certain type together to give streamed presentation. The *matchpar* command enables the user to set the parameters to search the Growing Book. On the parameters are set, the user can search for documents similar to a document.

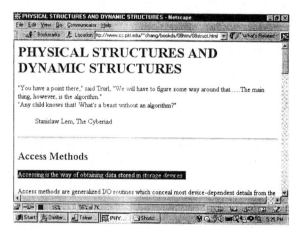

Figure 3. Finding similar documents by highlighting a paragraph in a document.

The words in the paragraph highlighted in Figure 3 are matched against the words of the documents in the Growi Book at different levels of abstraction, so that the matched documents can be identified. The Sentient Map can then used to visualize and retrieve the documents. Since more details can be found in the companion paper [Chang01], only brief discussion is given here. The Sentient Map can be integrated with the NetICE environment for virt teleconferencing developed by Dr. T. Chen at Carnegie Mellon University [NetICE99]. Students and instructors can u this environment as a virtual classroom. Several users represented by their *avatars* are engaged in a discussion, as sho in Figure 4. When they need information they may turn their attention to the sentient map (in bitmap-formatted mo displayed on the wall, and more information becomes available and is also visible to all the participants of the virt teleconference. In the current prototype system, three maps represent three point query types -- instructor, subject and ti as illustrated by Figure 5. The user can then use the point c-gesture (a double click) to query on the map. For example the user clicks on the subject map, an input window will pop up to let him/her input what subject he is interested in, su as "software engineering" or "data structure". At the same time, a world map will be displayed to let user go on w his/her gesture, as illustrated by Figure 6.

ure 4. A NetICE conference room.   Figure 5. A Sentient Map.   Figure 6. Sentient Map after point gesture.

The Design of the Sentient Map System

The system architecture of the sentient map system is illustrated in Figure 7. The sentient map system, whether in ap-formatted mode or html-formatted mode, accepts the user's gestures, pre-processes them, and sends messages to Gesture Analyzer. The Gesture Analyzer is an IC-based analyzer [Chang95]. The different messages drive the ICs to nge from one state to another and perform pre-defined actions. For example, commands for the Growing Book are the ently supported message types. The Gesture Analyzer produces SQL query for the database. After the result is eved from the database, it is visualized and the user can see the well-formatted results for the input gesture query.

The **Sentient Map Interface** in html-formatted mode will be explained here. It deals with web pages in html nat and has three components. In the first component, a function **getString** is written in JavaScript and embedded in l files. It will get the highlighted text in a web page and change it to the format that can be used as a parameter in a ST" message, for example, by changing "space" to "underscore". In the second component, a CGI function eline.cgi will accept that modified parameter, extract it to its original format by calling **getcgivars()** (i.e. changing derscore" back to "space"). Then this parameter will be analyzed by several methods, and the database most relevant : is selected by a majority vote of the results, and the results are visualized. In the third component, an applet called TQuery is activated. The functionality of this applet is to present the interface of the query input, perform that query he related database and show the final result. In this applet, we defined several functions, such as function **getQuery()** will write a query by using SQL (Structured Query Language), **getTable()** that will connect to the related database by g JDBC (Java Database Connectivity), and **displayResult()** that will show the final query results by popping up a ll window.

The **IC-based Gesture Analyzer** of the sentient map interface sends user's input to the IC-based Gesture lyzer. An index cell (IC) is an active object with states, transitions and actions. The IC development tools include the builder, the IC_compiler and the IC_manager. The designer uses the IC_builder to design the ICs' states and actions, then uses the IC_compiler to compile them. The IC_manager manages the ICs. The IC for the Gesture Analyzer is trated in Figure 8.

The **Similarity Matcher** performs matching of sentient maps. Currently it is based upon keyword matching, and a version will support matching of images and videos. Finally, the **Result Visualizer** supports the visualization of eved results.

erences:

ng95] S. K. Chang, "Towards a Theory of Active Index", Journal of Visual Languages and Computing, Vol. 6, No. 1, March 1995, 101-118.

ang00] S. K. Chang, T. Arndt, F. R. Guo, S. Levialdi, A. C. Liu, J. H. Ma, T. Shih, G. Tortora, "MACRO UNIVERSITY -- A Framework for a Federation of Virtual Universities", International Journal of Computer Processing of Oriental Languages, Vol. 13, No. 3, September 2000.

ang01] S. K. Chang, "Gesture Query for the Sentient Map", Proceedings of 1st International Conference on Universal Access in Human-Computer Interaction, New Orleans, Louisiana, USA - August 5-10, 2001.

01] www.cs.pitt.edu/~jung/GrowingBook/ (2001).

ICE99] http://amp.ece.cmu.edu/proj_NetICE.htm (1999).

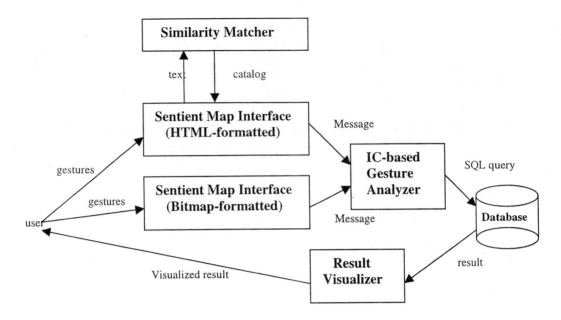

Figure 7. Sentient Map System Architecture.

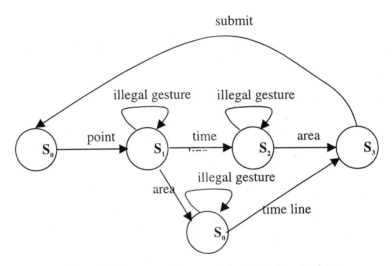

Figure 8. IC state transition graph for the Gesture Analyzer.

Internet-Based Electronic Data Exchange Interface for E-Business Integration[1]

Amy J.C. Trappey, Steven Chen and Pei-Hsun Ho
Department of Industrial Engineering and Engineering Management
National Tsing Hua University
Hsinchu 300, Taiwan
trappey@ie.nthu.edu.tw

ABSTRACT

This research focuses on the development of an Internet-based electronic trade exchange technology and the generation of neutral trading files in XML format. Traditional Electronic Data Interchange (EDI) transfers EDI messages via value added networks (VANs) or private lines, which require major EDI enabling expenses. When many trading exchange activities and corresponding data must integrate with the corporate ERP system, the related data may need tedious modification and conversion. These tasks often overburden small companies with a limited number of MIS staffs. Even large companies take major efforts and resources in EB enabling. Using the Internet-based trade exchange approach, the EDI implementation cost can be lowered and the implementation time can be shortened. An open database format for EDI-related data integration makes the internal and external data mapping processes feasible and easier. The research aims to develop an electronic trade exchange solution for enterprises that are pursuing B-to-B e-commerce with small to medium enterprises. Through business process reengineering (BPR), trading message inflows are modeled using an integrated information architecture. Thus, SMEs can transfer or receive electronic messages through the Internet-based trade exchange system adapting efficient inflow and outflow processes. Inside an enterprise, its ERP system can access the trading data via an open database. The system can further export the neutral files in XML/EDI format to external trading partners. With the standard format, the data can be easily integrated with other Electronic Commerce (EC) applications. Finally, a prototype system is implemented to demonstrate the Internet-based trade exchange concept at work.

1. INTRODUCTION

Enterprises are often required by customers to send business documents via Internet using EDI standards (Hinkkanen, Kalakota, Saengcharoenrat, Stallaert, & Whinston, 1999). Through electronic data interchanges, enterprises lower the transactional complexity, enhance the competitive advantage and reduce the transactional mistakes (Bergeron & Raymond, 1997). XML/EDI (Raymond & Blili, 1997) is a transactional standard that suits the needs of small and medium enterprises (SMEs) because documents presented using XML/EDI can be displayed in Web browsers and also conform to the EDI standard. SMEs can transfer electronic messages through Internet-based B2B trade exchange systems and adapt efficient data inflow and outflow processes (XEDI.ORG, 2000).

---

[1] This research is supported by ROC National Science Council Research Grant NSC89-2213-E-007-104.

2. FUNCTIONAL MODELING OF EDI PORTAL SYSTEM

This research divides the portal system into the front-end subsystem and the back-end subsystem. The front-end manages user login, transmission and reception of EDI messages, as well as other operational processes. The back-end sets up the format of database, processes information translation and controls limits of authority for users and groups. First, using the function view in the architecture of integrated information system (Davenport & Short, 1990), the two key functional requirements for the front-end and the back-end subsystems are defined in Figures 1 and 2.

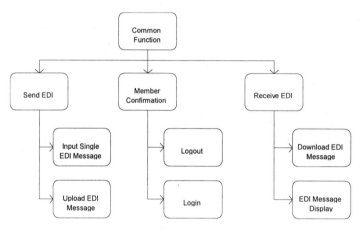

Figure 1. Functional tree for the front-end EDI subsystem.

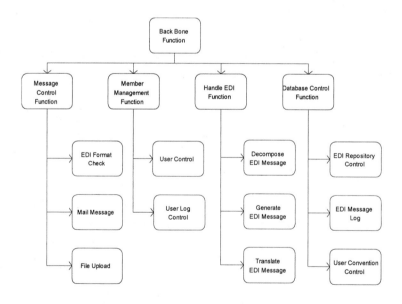

Figure 2. Functional tree for the back-end EDI subsystem.

2.1   The Processes of Front-end Operations

The EDI system focuses on the exchange of transactional data. The basic data processing included in the system are (1) send EDI messages, (2) member confirmation, and (3) receive EDI messages. All processes are performed using the Internet/Web interface. Users can send the transactional data by Web browser. The system will import the data into the member database after receiving the data and send an email confirmation to notify the member.

2.1.1   Send EDI messages. Remote user can fill in the request form in the Web page to send the order, order change, order response and invoice. The system will translate the data into XML/EDI format, prepare the EDI file for uploading to the server, and select the transactional target. After the server verifies the EDI format, the EDI message will be sent to the target partner.

2.1.2   Receive EDI messages. Remote user can get into his inbox, i.e., the message box, display the messages, and respond the messages. When user requests for downloading the transaction file, the system provides the function to download the transaction file with EDI and XML format.

2.2   The Processes of Back-end Operations

The back-end system divides into four major functions including message control, database control, membership management and EDI message handling. Processing tEDI messages is the key function of an EDI system. Because an EDI file always consists of many segments and components, the system must read, identify, interpret, and process the EDI information recursively as shown in Figure 3.

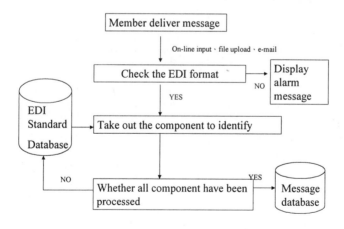

Figure 3. EDI message process workflow.

171

3. DEMONSTRATION OF WEB-BASED EDI PORTAL SYSTEM

Remote users access the EDI system using a common Web browser. The EDI system will ask for the user ID and password to check authorization status.   When the authorization procedure is completed, the system will list the functions that the specific user is permitted to perform as shown in Figure 4.   Figures 5 and 6 show remote users requesting to send XML/EDI file and on-line key-ins to trading partners respectively:

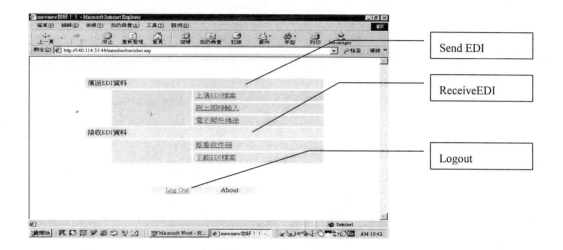

Figure 4. The main interface of the prototyped EDI portal system.

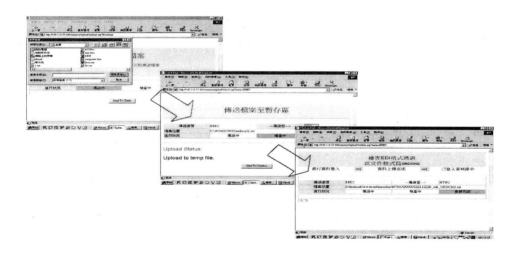

Figure 5. Send XML/EDI message to a trading partner with an existing document file.

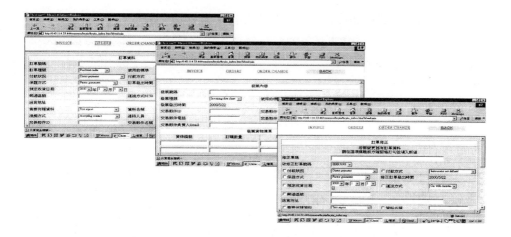

Figure 6. Send XML/EDI message to a trading partner using on-line input interface.

4. CONCLUSION

In this paper, a Web-based electronic data exchange system based on XML/EDI schema is developed to achieve the neutral data exchange between different computer-based systems. This paper presents the key functions and the standard features of an EDI portal system. There are further functions that can be developed and modified in details to better satisfy practical needs for industry. The cost and manpower spent in data exchanges can be significantly reduced using XML/EDI standard and Web-enabled EDI portal system as depicted in this research.

REFERENCES

Bergeron, F. & Raymond, L. (1997). Managing EDI for corporate advantage: A longitudinal study. Information Management, 31, 319-333.

Davenport, T. H. & Short. J. E. (1990). The new industrial engineering: Information technology and business process redesign. Sloan Management Review, 31(4), 11-27.

Hinkkanen, A., Kalakota, R., Saengcharoenrat, P., Stallaert, J., & Whinston, A. B. (1999, October). Open EDI concept. http://yama.bus.utexas.edu/ejou/articles/art_1.html.

Raymond, L. & Blili, S. (1997). Adopting EDI in a network enterprise: the case of subcontracting SMEs. European Journal of Purchasing & Supply Management, 3(3), 165-175.

Scheer, A. W. (2000). ARIS - Business Process Modeling, 2nd Revision. Berlin: Springer-Verlag.

XEDI.ORG (2000, May). XML and EDI:peaceful co-existence. http://www.xedi.org/whitepapers.html.

Storefront Interface Analysis and Evaluation for E-Commerce Applications[1]

Charles V. Trappey[2]

Department of Management Science

National Chiao Tung University

Hsinchu 300, Taiwan

Amy J.C. Trappey and Claire Ya-Ping Chang

Department of Industrial Engineering & Engineering Management

National Tsing Hua University

Hsinchu 300, Taiwan

ABSTRACT

The purpose of this research is to derive a means for evaluating websites, to define the elements underlying their success, and provide a standard for generalizing the content of Internet storefronts. The proposed schema uses HTML tags to categorize user interface elements and information elements presented via Internet storefronts. In order to explore the role that these elements play in characterizing popular and unpopular Internet storefronts, a content analysis of the distribution of elements within best-rated and worst-rated Internet storefronts was conducted. A code sheet was created to generalize the elements underlying Internet storefronts. Given a valid codesheet to categorize elements, the distribution of elements for popular and unpopular commercial websites was derived. The research demonstrates that once the critical elements underlying a successful site are derived, then these elements can be used to re-engineer new, more successful commercial sites.

1. INTRODUCTION

E-commerce on the World Wide Web is an emerging way of doing business and the principles underlying retail success are little understood. Firms are using the web as a distribution channel, a medium for marketing, and a place for transactions. However, among the growing number of commercial websites, many are failures in regard to profitability and user satisfaction (Creativegood.com, 2001). On the positive side, the abundance of commercial sites provides consumers with new ways of interacting with merchants and a means to control the information they receive and the information they send.

---

[1] This research is supported by ROC National Science Council Research Grant NSC 89-2416-H-009-043.

[2] Please send all correspondences to Dr. Charles Trappey, Department of Management Science, National Chiao Tung University, Hsinchu 300, Taiwan; trappey@cc.nctu.edu.tw.

2. METHODOLOGY

Researchers argue that to create a successful website, measurements and classifications are needed to define the elements underlying the human-computer interaction (Anton & Postmus, 1999; Donaton, 1995; Hoffman, Novak & Chatterjee, 1995). In order to analyze and evaluate websites, an element schema code sheet is designed. Since most websites are constructed using HTML tags and other scripting languages such as Vbscript, Javascript, and Active Server Pages, the element schema reflects this set of tags (World Wide Web Consortium, 1999a,b,c). Owing to various display styles and additional functions supported by different programming languages, the components of websites are classified using subjective assessments. The subjective assessments include category, usage/function, display styles, and status. Status is used to determine whether the element can link to other web resources. Dynamic status means that the page can link to other URL addresses or interact with the server, while static status means that the page cannot form other links. Furthermore, the functions are classified according to purpose and according to the way the elements are displayed (Table 1).

2.1 Sample

A popular site listing the Top10 websites (ZD Inc, 1999a-f) was used as the criterion to test the code sheet and to derive the re-engineering principles. Both the code sheet designer and a code sheet verifier evaluated websites and circled elements on the code sheet that corresponded to elements displayed at a sample of sites. Finally, the numbers of elements identified were compared between the two individuals. The validated code sheet was used to generalize the elements underlying best and worst business-to-consumer (B2C) Internet storefronts.

2.2 Proportion of Elements

A statistical test for differences between proportions of elements that distinguish best and worst Internet storefronts was conducted on data abstracted using the code sheet. Significant differences were found between the proportion of elements that define advertisements, links with word descriptions, links with images, buttons, selections, and checkboxes. Of these six elements, links with word descriptions and checkboxes are the most popular types of storefront elements.

2.3 Principal Components

A principal component analysis of eighteen best and eighteen worst websites (Anton & Postmus, 1999) was conducted to divide the elements into groups in terms of their characteristics:

(1) Transaction triggers – a set that includes advertisements, quick service by image, and quick service by text. Transaction triggers stimulate customers and possible increase shopping behavior.

(2) Information offers – a set that includes catalogues of classified information, catalogues with a list of subcategories, links with word descriptions, read only words, and read only pictures. This set of elements provides customers with information.

(3) Data exchange elements – a set that includes blocks for single-line input, radio choice, selection, and checkboxes. The data exchange set is essential for processing transactions and helps companies obtain customer information.

175

Table 1. Elements defined in the code sheet

No.	Name	Function	Exhibition	HTML Syntax	Status (dynamic/static)
1	Advertisement	Promote or conduct transactions	Images	Anchor, IMG	D
2	Trademark	Trademark	Images	IMG	S
3	Quick service by image	Links to URL address and triggers interactive services	Images or logos	Anchor, IMG	D
4	Quick service by words	The same as 3	Text	Anchor	D
5	Catalogue of classified information	Grouping information	Information classified into a list of items	Anchor, IMG, TABLE	D
6	Catalogue with a list of subcategories	Grouping information	Information classified with subcategories in detail	Anchor, IMG, TABLE	D
7	Links with word descriptions	Links to URL address	Text	Anchor	D
8	Links with images	Links to URL address	Pictures or icons	Anchor, IMG	D
9	Headline news	Notify headline news	Key words related to the headline news	Anchor	D
10	Statistical data	Commercial statistical data	Table with statistical data which can be linked to other information	Anchor, TABLE	D
11	Block for single-line input	Input a single-line word or asterisks with/without limited characters	Given a block	INPUT (word, password), ISINDEX	D
12	Button	Reset data or submit data to the server	A push button	INPUT (reset, submit)	D
13	Field for multi-line input	Input multi-line word	A field with multi-rows and multi-columns	TEXTAREA	D
14	Radio choice	Single choice	A list of radio buttons	INPUT (radio)	D
15	Selection	One or more than one choices	A drop-down menu or a scrolled list box	SELECT	D
16	Checkbox	One or more choices	A checkbox	INPUT (checkbox)	D
17	Read only words	Read only information	Text		S
18	Read only pictures	Read only picture	Pictures	IMG	S

2.4 Re-engineering Websites

The final objective of the research is to provide standards for re-engineering successful, customer driven Internet storefronts. The data show that in order to offer more information and introduce products efficiently, the

176

use of linkable and non-linkable text is essential. Quick services for triggering interaction is necessary since quick service gathers information from customers and continues the process flow of the transaction. Customers prefer multiple choices (i.e., checkboxes) rather than single choices and the drop-down menus. Groups of buttons are not suggested since some buttons may stop the transaction flow. Text is more favorable than graphic presentations, and the number of the advertisements should be reduced to eliminate page clutter.

3.   RESULTS

Using the statistical analysis results, a new Internet storefront prototype is re-engineered and demonstrated. First, text is used more than images; quick services are used for triggering interaction and to obtain personnel information.   Second, although checkboxes are favored, it is not necessary to use checkboxes on the first page of the storefront.   Third, the use of advertisements is limited on the prototype, and both linkable and non-linkable texts are used.   The end result is a re-engineered storefront that depicts the statistically derived principles.

4.   CONCLUSION

Principal components analysis indicates that crucial elements can be divided into three groups: transaction triggers, information offers and data exchange. Combined with the statistical results from the differences between two population proportions, it is concluded that quick services by text (element 4) are helpful during the stage of triggering transactions whereas during the stage of information gathering, text is better than pictures, so elements 7 and 17 are recommended.   Moreover, in the process of purchasing, multiple selections (element 16) are preferred by customers.   In conclusion, elements 4, 7, 16, and 17 are recommended for Internet storefronts.   Using these elements draws net surfers to the site and facilitates online purchasing (Figure 1).

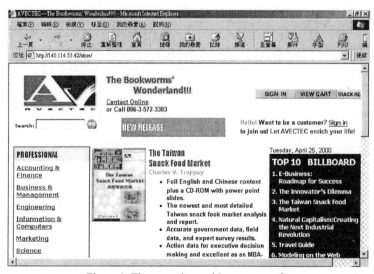

Figure 1. The re-engineered Internet storefront

REFERENCES

Anton, J.M. and Postmus, R. F. (1999). The CRM Performance Index for Web-Based Customer Relationship Management. Purdue University.

Csikszentmihalyi, M. (1990). Flow: The psychology of optimal experience (1st ed.). New York: Harper Perennial.

Creativegood.com (2001). http://www.creativegood.com/holiday2000/.

Donaton, S. (1995). Pathfinder blazes a trail to ads. Advertising Age, 66, 19.

Hoffman, D.L., Novak, T.P., & Chatterjee P. (1995). Commercial scenarios for the web: opportunities and challenges. [On-line]. Journal of Computer-Mediated Communication, Special Issue on Electronic Commerce, 1 (3). Available from: http://shum.huji.ac.il/jcmc/vol1/issue3/vol1no3.html.

ZD Inc. (1999a). Aspects of a good customer experience. Retrieved December 1, 1999 from World Wide Web: http://www.zdnet.com/enterprise/e-business/evaluations/0,7689,2277653-2,00.html.

ZD Inc. (1999b). Aspects of a bad customer experience. Retrieved December 1, 1999 from World Wide Web: http://www.zdnet.com/enterprise/e-business/evaluations/0,7689,2277653-3,00.html.

ZD Inc. (1999c). Complete list: worst examples of e-commerce. Retrieved December 1, 1999 from World Wide Web: http://www.zdnet.com/enterprise/e-business/evaluations/0,7689,2298718,00.html.

ZD Inc. (1999d). Our methodology. Retrieved December 1, 1999 from World Wide Web: http://www.zdnet.com/enterprise/e-business/evaluations/0,7689,2277653,00.html.

ZD Inc. (1999e). Top 10 best examples of e-commerce. Retrieved December 1, 1999 from World Wide Web: http://www.zdnet.com/enterprise/e-business/evaluations/0,7689,2282637,00.html.

ZD Inc. (1999f). Top 10 worst examples of e-commerce. Retrieved December 1, 1999 from World Wide Web: http://www.zdnet.com/enterprise/e-business/evaluations/0,7689,2282631,00.html.

World Wide Web Consortium (1999a). Forms in HTML documents. Retrieved November 26, 1999 from the World Wide Web: http://www.w3.org/TR/1999/PR-html40-19990824/interact/forms.html.

World Wide Web Consortium (1999b). Links in HTML documents. Retrieved November 26, 1999 from the World Wide Web: http://www.w3.org/TR/1999/PR-html40-19990824/struct/links.html.

World Wide Web Consortium (1999c). Tables in HTML documents. Retrieved November 26, 1999 from the World Wide Web: http://www.w3.org/TR/1999/PR-html40-19990824/struct/tables.html.

AN APPLICATION OF LESSONS LEARNED IN VIRTUAL REALITY TO THE SERVICES INDUSTRY

Vincent G. Duffy

Department of Industrial Engineering and Engineering Management
The Hong Kong University of Science and Technology
Clear Water Bay, Kowloon, Hong Kong
Tel: (852) 2358-8237, Fax: (852) 2358-0062; Email: vduffy@ust.hk

ABSTRACT

New technologies such as virtual reality can only be implemented effectively with knowledge and consideration of the context in which they are to be used. To enable applications, it appears necessary to develop domain knowledge regarding the new technologies while giving special consideration for previously developed understanding in other industrial settings such as manufacturing. This paper addresses some research questions recently studied regarding some human and organizational aspects in virtual reality and manufacturing. Current practical limitations are noted. A brief summary of lessons learned in manufacturing and their relevance in implementing the new virtual tools in risk reduction, and training for safe decision-making in the aircraft services industry is discussed. Suggestions for future work are intended to provide some insight for effective utilization of virtual reality in support of human resource planning in the services industry.

1. INTRODUCTION AND LITERATURE REVIEW

A number of organizational, human and social aspects that are important in concurrent product development and manufacturing (Duffy and Salvendy, 1997) are also important in human resource planning in the services industry (Duffy and Salvendy, 2000). Some of the following discussion will focus on how these are important, conceptually, when considering software tools developed for human resource planning in the services industry. As well, it is necessary to develop domain knowledge about technologies such as virtual reality based training in the context in which they may be used (Duffy and Salvendy, 2000).

2. ESTABLISHING DOMAIN KNOWLEDGE AND LESSONS LEARNED

Social aspects - Geographically distributed work and the trend toward global manufacturing are giving rise to the need for technologies that give users a greater sense of presence that have the potential for yielding results similar to those of co-located work (Duffy and Salvendy, 1997; 1999).

Organizational aspects – Models showing relative time for training and system development are presented in Duffy and Salvendy (2000) and Kan, Duffy and Su (2001). The models suggest that increased task complexity and system knowledge requirements will increase training time and system development time. A knowledge modelling methodology used for developing intelligent virtual reality-based industrial training systems is shown in Ye, *et al.* (2000). A Petri net-based knowledge elicitation scheme and reasoning mechanism, and an agent-based knowledge representation pattern are used to enable real-time interactive and immersive experience for the trainee. The difficulty of developing such a knowledge base system for intelligent virtual training systems is not only that the training domain knowledge involved varies a lot and is complex, but also that the interaction style between the virtual training system and trainees is quite different from traditional training approaches.

Human aspects – Perception and transfer of training - The impact of advanced technical and virtual training on hazard and risk perception is shown in Duffy (2000a) and Duffy, Wu and Goonetilleke (2000). When training is done through a computer simulated industrial task in the optimum cutting condition, the optimum cutting condition can be determined by the trained subject. Whereas the untrained subject perceives a greater hazard when traditional physiological stimulus such as frequency of the cutting sound is increased. This implies that sound cues in the virtual environment can help train operators to correctly perceive a more hazardous condition that is different from intuition or their physiology would predict.

Other studies were done at UST in Hong Kong in order to show the effects of virtual lighting on risk perception (Duffy, 2001) and eye fatigue (Duffy and Chan, 2000). In an Internet virtual factory, subjects viewing machine operation in two significantly different virtual lighting conditions perceived the operator to be at greater risk in the darker environment (Duffy, 2000b). These findings are consistent with those expected under different lighting conditions in a real factory. The most recent work, based on the RGC project of 1998 (Duffy, 1998, RGC/HKUST 6168/98E) shows a methodology for the study of risk perception in a computer simulated industrial environment (Wu, Duffy and Ng, 2000). This study focused on differences in perception based on different combinations of distance between machines, lighting level, and sound level. The methodology for comparing the sound in virtual and real is shown in Duffy, Wu and Goonetilleke (1999). Finally, a study of perceived inter-object and traversed distance in virtual and real environments is shown in Wu, Duffy and Leung (2000).

In the course of research on the RGC project of 1999 (Duffy, 1999b, RGC/HKUST 6211/99E), it was found that a 'simulated tool breakage' in a virtual training sequence did improve the 'decision making' performance of subjects (Duffy, 1999). The animation appeared to help subjects to more easily recognize differences in sound between the 'good' and 'poor' cutting condition. It was also found that 'mental workload' increased with a 'disincentive' in the training task. It was reported that subjects struggled with decisions regarding whether or not to shut the machine down during what they perceived to be an 'unsafe' cutting condition. This technique, when used in virtual training systems, appears to enable dual coding (Duffy, 2000c).

2. ONGOING WORK AND FUTURE WORK

Previous work has focused on identifying changes in perception, based on changing parameters in the virtual environment. As well, some aspects of the virtual training sequence that can influence decision-making have been identified. It is important to begin to integrate those findings to provide a methodology for improving safety and reducing risk in industrial environments. Given results of previous research, it is believed that a methodology can be developed with regard to improving safety and risk reduction that incorporates not only training for the correct procedure, but also how to gain compliance when improper operating conditions are correctly recognized. The ongoing and future research is intended to enable use of the proposed methodology to implement forms of computer-based training that will lead to lasting change with regard to safety recognition and behavior.

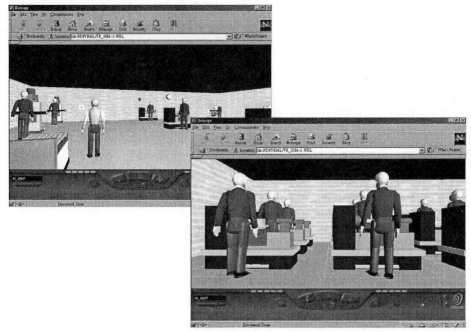

Figure 1. Current capabilities. The figure in the lower right, shown with a Netscape Internet browser, indicates a virtual scene with 0.5m distance between machines. A human (upper left) added in the scene can move along a specified path after the user presses a button. The AGV also is moving along a specified path in this Internet virtual environment.

The primary objectives of the ongoing and future work are as follows:

- Show that, under restricted hearing, subjects will underestimate the speed in virtual training.
- Develop ability to predict performance and retention based on training mode and training conditions.
- Reduce mental workload in the real task after virtual training.
- Gain operator compliance in speed selection after virtual training in some select conditions.
- Extend the lab findings into the service industry to improve safety and reduce risk.

The research will be implemented in two phases. The first phase will focus on further establishing empirical results related to perception and performance in a lab setting. Unique to this study is the Internet-based dynamic virtual environment that has many similarities to the real industrial setting that these results can be applied to (see Figure 1). The second phase will require re-creating a similar virtual environment under the guidance of an aircraft services provider so details of the real work conditions can be incorporated, trained and improved. Details of some of the preliminary discussions of the first phase can be found in Ng and Duffy, (2001).

ACKNOWLEDGEMENTS

This study is sponsored, in part, by the Competitive Earmarked Research Grants from the Research Grants Council of Hong Kong HKUST/CERG 6168/98E and HKUST/CERG 6211/99E. I would also express sincere appreciation to Tracy Cheung, Vivian Lau, Calvin Or, Parry Ng and Flora Wu for their assistance in the most recent software development efforts.

REFERENCES

Duffy, V.G. & Chan, H.S. (2000). Effects of virtual lighting on eye fatigue, *Human Factors and Ergonomics in Manufacturing*, accepted.

Duffy, V.G. & Salvendy, G. (1997). Prediction of effectiveness of concurrent engineering in electronics manufacturing in the U.S., *Human Factors and Ergonomics in Manufacturing*: 7 (4), 351-373.

Duffy, V.G. & Salvendy, G. (1999). Relating company performance to staff perceptions: the impact of concurrent engineering on time to market, *International Journal of Production Research*: 37 (4), 821-834.

Duffy, V.G. & Salvendy, G. (2000). Concurrent engineering and virtual reality for human resource planning, *Computers in Industry*, 42, 109-125.

Duffy, V.G., Wu, F. & Goonetilleke, R.S. (1999). Toward a comparison of risk perception in real and computer simulated industrial environments, *Proceedings of the 26th International Conference on Computers and Industrial Engineering*, Melbourne, Australia, December 15-17, pp.539-544.

Duffy, V.G., Wu, F. & Goonetilleke, R. (2000). Impact of advanced technical and virtual training on hazard and risk perception, working paper.

Duffy, V.G. (1998). Improving safety and reducing risk: Modeling the workplace through empirical studies of Stevens power law in virtual and real work environments. *Research Grants Council of Hong Kong*, Competitive Earmarked Research Grant, July 1998, Project HKUST 6168/98E.

Duffy, V.G. (1999). Impact of an intelligent tutor on risk and sound perception in CNC machining, *Proceedings of the IEEE Systems, Man and Cybernetics Conference*, SMC99, Tokyo, Japan, October 12-15, pp. 1091-1094.

Duffy, V.G. (1999b). Modeling risk reduction and industrial performance based on the impact of training in the virtual environment. Research Grants Council of Hong Kong, Competitive Earmarked Research Grant, Project HKUST 6211/99E.

Duffy, V.G. (2000a). Toward online virtual safety training in the manufacturing and service industries, IEA2000/HFES2000, *Proceedings of the XIVth Triennial Congress of the International Ergonomics Association and the 44th Annual Meeting of the Human Factors and Ergonomics Society*, San Diego, California, USA, July 30-August 4, 2000, vol 4, pp. 341-344.

Duffy, V.G. (2000b). Effects of virtual lighting on risk perception and eye fatigue, *7th HAAMAHA Conference, The 7th International Conference on Human Aspects of Advanced Manufacturing: Agility and Hybrid Automation*, Krakow, Poland, August 27-30, 2000, pp. 106-109.

Duffy, V.G. (2000c). Impact of a simulated accident on decision-making performance, working paper, submitted to *International Journal of Cognitive Ergonomics*.

Duffy, V.G. (2001). Effects of virtual lighting on risk perception, working paper, submitted to *Ergonomics*.

Kan, H.Y., Duffy, V.G. & Su, C.J. (2001). An Internet virtual reality collaborative environment for effective product design, *Computers in Industry*, in press.

Ng, P.P.W. & Duffy, V.G. (2001). Impact of dynamic virtual and real objects on perception of hazard and risk. *HCI International 2001, 9th International Conference on Human-Computer Interaction*, New Orleans, Louisiana, USA, August 5-10, 2001, accepted.

Wu, F., Duffy, V.G., & Leung, G.T.C. (2000). A study of perceived inter-object and traversed distance in virtual and real environments, working paper, submitted to *International Journal of Cognitive Ergonomics*.

Wu, F., Duffy, V.G. & Ng. P.P.W. (2000). A methodology for the study of risk perception in a computer simulated industrial environment, working paper, submitted to *Human Factors*.

Ye, L.L., Duffy, V.G., Yen, B.P.-C., Lin, F. and Su, C.J., 2000, Knowledge modelling methodology for intelligent virtual reality-based industrial training systems, *Asian Journal of Ergonomics*, in press.

A Survey of Design Guidelines for Usability of Web Sites

Wang, Chau-Hung

Department of Business Administration at Soochow University
56, Sec. 1, Kuei-Yang St., Taipei, Taiwan, R.O.C.

Abstract

In this paper, we will survey many web sites design guidelines to induce some suggestions for user-centered design web sites. Some of these suggestions are widely applied, others need to be promoted extensively. However, we hope these suggestions will help designers to develop user-friendly web sites. Finally, we will introduce web sites automatic evaluation system and the concept of universal usability.

1. The characteristic of the web sites

The World Wide Web (WWW) site is one type of the computer interface in the network. Its main purpose is to deliver and present information among users at different places. The special characteristics of the web sites are using hypertext and hypermedia to communicate multi-media information to different users in the world. By strengthening coherence and weakening cognitive overhead will assist readers in the construction of their mental models to increase the readability of a hyperdocument. Thuring [9] proposed eight design principles to support the construction of a mental model of a hyperdocument.

- Use typed link labels to represent semantic relations between information units.
- Indicate equivalencies between information units to help reduce the impression of fragmentation.
- Preserve the context of information units by showing neighboring nodes in the document structure and their relationships to increase the coherence.
- Use higher order information units to aggregate information thus lending more structure to the document.
- Visualize the structure of the document to provide a most useful overview and helps the reader to identify major topics and their relations.
- Include cues into the visualization of structure which show the reader's current

position, the way that led to this position and navigational options for moving on.

- Provide a set of complementary navigation facilities which cover aspects of direction and distance.
- Use a stable screen layout with windows of fixed position and default size to diminish cognitive overhead.

Three important factors must be considered in the process of design web sites: 1. Task support: the interface should meet user expectation and support the tasks users need to get done. 2. Usability: the interface should be easy to learn, easy to remember, pleasant to use and cause few errors. 3. Aesthetics: the interface should communicate visually, helping users absorb information and carry out the tasks they need to do while minimizing information overload.[2] To neglect any one of these factors will decrease capability of computer interface. From the viewpoint of web's content, we can breaks down sites into four categories: informational sites, search sites, multipurpose portal sites, and entertainment sites. However, Shneiderman [7] described that there are triple threats in the WWW: the users' environments are highly varied, quality control is shaky, and rugged independence limits common interface design patterns.

Present marketing efforts that view the Web as predominately electronic publishing can be thought of as first-generation. As technology progresses, and developers move from HTML and CGI (Common Gateway Interface) scripts to Java and object-oriented development, we are seeing the birth of second-generation systems. Second-generation systems may include elements of electronic publishing, but also take advantage of interfaces and gateways to databases and models to provide richer systems. Meanwhile, O'keefe [5] proposed a framework using Web-based CDSS (Customer Decision Support Systems) and generic Web and Internet facilities to support customer decision making in B2B or B2C.

2. Usability

The International Standards Organization (ISO) proposed the definition of usability as that the usability of a product is the degree to which specific users can achieve specific goals within a particular environment; effectively, efficiently, comfortably, and in an acceptable manner. Booth [1] thought that usefulness, effectiveness (or ease of use), learnability, and attitude (or likeability) as the four essential features for a usability system. Thus, a user can find an interface element to be problematic for many reasons: It might make the system harder to learn; it might make it slower for users to perform their tasks; it may cause usage errors; or it may simply be ugly or otherwise unpleasing. Usability problems can be defined as aspects of a user interface that may cause the resulting system to have reduced usability for the end user. [4] In addition, in the field of human-computer interaction many measures (such as time, error, and attitude) are used to evaluate the usability of a system qualitatively and quantitatively. Otherwise, Pearrow[6] defined the usability as following: Usability is the broad discipline of applying sound scientific observation, measurement, and design principles to creation and maintenance of Web sites in order to bring about the greatest ease of use, ease of learnability, amount of usefulness, and least amount of discomfort for the humans who have to use the system.

However, Head [2] thought common usability problems on the web are:

- Unclear labeling and vocabulary
- Users need to remember too much
- Graphics are cluttered and overused
- Inaccurate understanding of site design
- Poor match between site design and user's needs
- Navigation problems
- Designed without clear target-user populations in mind
- Design is not guided by users' goals
- Insufficient privacy and security

3. Design guidelines for web sites

Nielson [3,4] proposed 10 usability heuristics: visibility of system status, match the system to the real world, user control and freedom, consistency and standards, error prevention, recognition rather than recall, flexibility and efficiency of use, aesthetic and minimalist design, help users recognize, diagnose, and recover from errors, and help and documentation. Wang [10] provided 10 usability design guidelines for web sites from the viewpoint of cognitive limits: using familiarity language and word, consistency of system design, avoiding excessive memory, increase the effective and adaptive, brief and esthetic screen, clear and definite item names and contents, presenting hierarchical information, full navigation function, using CGI to increase interactive, avoiding to show many windows in one screen. Pearrow [6] provided other heuristics for web site especially: chunking, use the inverted pyramid style of writing, important information belongs "above the fold", avoid gratuitous use of feature, make your pages "scannable", and keep download and response times low. Head [2] proposed 7 universal design principles of web:

- Equitable use: This means that the design is useful and marketable to any group of users.
- Flexibility in use: This means that the design accommodates a wide range of individual preferences and abilities.
- Simple and intuitive use: Use of the design is easy to understand, regardless of the user's experience, knowledge, language skills, or current concentration level.
- Perceptible information: The design communicates necessary information effectively

185

to the user regardless of ambient conditions or the user's sensory abilities.

- Tolerance of error: The design minimizes hazards and the adverse consequences of accidental or unintended actions.
- Low physical effort: The design can be used efficiently and comfortably and with a minimum of fatigue.
- Size and space: These for approach, reach, manipulation, and use are appropriate regardless of the user's body size, posture, mobility.

Finally, Head [2] from three directions (task support, usability, and aesthetics) proposed 22 web site design evaluation checklist on the book entitled "Design Wise: a guide for evaluating the interface design of information resources."

Besides, it exists a kind of web site that can automatically evaluate the web sites after you key in the site address, it is called web site automatic evaluation system. It will give you the evaluated result according the following directions:

- Browser compatibility: This test determines how well your web page displays across different browsers.
- Link validity: This test locates any "dead" links that your page may have.
- Polularity: This test is an indicator of the strength of your online presence.
- Load time: This test determines how fast your web page loads at common connection speeds.
- Submission readiness: This test determines whether your web page is ready to be successfully indexed by directories and search engines.
- Spelling: This test isolates words that may not be spelled correctly.
- HTML validity: this test checks the accuracy of the HTML code for your web page(s).

Pearrow [6] as well proposed the measures about the web sites in a automatic evaluation system, such as link integrity (searching for broken links), load times, types of content on a site (HTML, Java, Javascript, other embedded content such as video or audio data), and the number of ways to traverse to particular page.

4. Universal usability

Universal usability can be defined as having more than 90% of all households as successful users of information and communications services at least once a week. Designers of old technologies as postal services, telephones, and television have reached the goal of universal usability, but computing technology is still too difficult to use for many people. Shneiderman [8] proposed three research agenda to study universal usability:

- Technology variety: supporting a broad range of hardware, software, and network access;

- User diversity: Accommodating users with different skill, knowledge, age, gender, disabilities, disabling conditions (mobility, sunlight, noise), literacy, culture, income, and so forth; and

- Gaps in user knowledge: bridging the gap between what users know and what they need to know.

5. Reference

1. Booth, P. A., " An Introduction to Human-Computer Interaction," LEA Ltd., London, UK, 1989.

2. Head, A. J., "Design Wise: A Guide for Evaluating the interface Design of Information Resources," Information Today, Inc. Medford, New Jersey, 1999.

3. Nielsen, J., "Ten usability heuristics," the article are presented at the web site www.useit.com/papers/heuristic/heuristic_list.html.

4. Nielsen, J. and Mack, R. L., "Usability Inspection Methods," John Wiley & Sons, Inc. New York, 1994.

5. O'keefe, R. M. and Mceachern, T., "Web-based customer decision support systems," Communications of the ACM, Vol. 41, No. 3, pp. 71-78, 1998.

6. Pearrow, M., "Web Site Usability Handbook," Charles River Media, Inc. Rockland, Massachusetts, 2000.

7. Shneiderman, B., "Is the web really different fro everything else?" in ACM conference proceedings, CHI 98: Human Factors in Computer System, New York, 1998.

8. Shneiderman, B., "Universal Usability," Communications of the ACM, Vol.43, No. 5, 2000.

9. Thuring, M., Hannemann, J., and Haake, J. M., "Hypermedia and cognition: designing for comprehension," Communications of the ACM, Vol. 38, No. 8, pp. 57-66, 1995.

10. Wang, C. H., "User interface design for the home page in WWW," Journal of Industrial Engineer, Vol.16, No. 2., pp. 265-276, 1999(Chinese).

The Influence of Usability Principles on Developer's Performance

Tzai-Zang Lee   Wan-Ya Lin
Department of Industrial Management Science
National Cheng Kung University
Tainan, Taiwan 701
Email: leetz@mail.ncku.edu.tw

ABSTRACT

To develop a usable and user friendly software is very important. The first step is to choose a good software development tool package which has good user interface usability. Good usability is one of the goals of user interface in software development.   Good usability facilitates the smooth interaction between human and computers.   Researchers have proposed a good number of usability principles.   Are those principles really existing and functioning in the process of software development?   How do they influence the software developer's performance? This research tries to get answers for these two questions.

According to Dix and colleagues (1998), usability consists of three important principles: learnability, flexibility, and robustness. Learnability facilitates users' learning of programming tool, consists of predictability, synthesizability, familiarity, generalizability, and consistency. Flexibility serves to enhance the exchange of information between users and system, including dialogue initiativeness, multi-threading, task migratability, substitutivity, and customizability. Robustness helps on goal attainment and evaluation, consists of observability, recoverability, responsiveness, and task conformance.

Software developers in the development of management information system from various companies were surveyed to make sure that principles of user interface design do exist. A questionnaire of 28 questions consisted of learnability, flexibility, and robustness principles was constructed and distributed to 246 programmers.   Their responses were analyzed by using a factor analysis.   Those components with eigenvalues above one were kept .

Based on the results of the factor analysis, an experiment was conducted to control the extent of usability principles as independent variables and measure speed of performance and error rate as dependent variables. There were seven conditions of usability: one with all of three usability principles, three with either one of usability principles, and three with either two of usability principles. Seven tasks were arranged for subjects to perform. Each subject has to perform all seven tasks, each under different usability condition.   A 7 by 7 Latin square was adopted to level off any carry over effects. A HyperCam was used to record completion time and error rate as dependent variables.

Results showed that tasks that with principle of learnability or flexibility in the development tool did shorten the completion time, but robustness did not have this effect. Tasks with either one of the three usability principles had significant effects on lowering error rate. The results suggested that usability principles do influence software developer's performance.

Keywords: Usability, learnability, flexibility, robustness.

1. INTRODUCTION

The advancement of technology and information system has made computer a must of modern life. A lot of jobs have to be done by using computers. To accomplish works with good user interface system not only enhances ease of use, but also increases productivity.

On the other hand, by one reason or another, some user interface tools are difficult to learn and use, that in turn lower task performance. Those negative factors such as complicated data entry, unclear error messages, weak debugging power, and cluttered screen. (Gelach and Kuo, 1991)

In order to improve usability of user interface system, some design principles have been proposed. A software with good usability can contribute to lower physiological and psychological stress, higher user-system interaction, higher productivity and quality (Reed, et al, 1999)

A computer program usually can be divided into user interface and application programs. User interface is the interaction media between human and computer system. This communication media serves to link between users and computers. Through user interface, intention of users can be transformed into machine language to be understood and to be carried out. According to Collins (1995), user interface including: conceptual model, presentation language, action language, and implementation model.

In another words, computer users enter their commands via input devices, get feedback from display terminals or other output devices. The usability of user interface system is to enhance the functionality and user friendliness of that system. For programmers, user interface is the communication media to accommodate the requirement of design with capability of system. There are five styles of user interface: command-oriented, natural language, menu-driven, form-oriented, and direct manipulation (Shneiderman, 1998).

Usability of interface system can ease the usage of users (Dix et al, 1998). On the other hand, a system without or with bad usability would lengthen the work time, increase errors, and cause frustration of users (Henneman, 1999). Usability is the system attributes collection that can help users to accomplish their jobs efficiently, effectively, and satisfactorily (Henneman, 1999).

According to Dix et al (1998), there are three main categories of principles to support usability of an interactive system: learnability, flexibility, and robustness. Each category of principle in turn can be divided into more specific principles. Learability of an interactive system allows novice users to know how to use it at the beginning and understand how to reach a maximal performance. There are five specific principles to support learnability: predictability, synthesizability, familiarity generabizability, and consistency. Flexibility of an interactive system allows the end-user and the system exchange information in multiple ways. There are also five principles to support flexibility: dialog initiative, multi-threading, task migratability, sensitivity, and customizability: Robustness of an interactive system assists the accomplishment and assessment of system goals. There are four specific principles to support robustness: observability, recoverability, responsiveness, and task conformance.

The current research was carried out to check the existence and extent of usability principles, and also to check how usability principles affect programming speed and error rate.

2. METHODOLOGY

A survey was conducted to make sure that usability principles do exist during software development. An experiment was done to check how the extent of usability principles affect the programming performance.

2.1   Usability survey

An HTML (Hyper Text Markup Language) questionnaire of 28 questions was sent to 246 programmers through email. The questionnaire was consisted of questions on learnability, flexibility, and robustness. A Cronbach's α analysis was done to check the reliability of the questionnaire.

By using factor analysis, those items with factor loading 0.6 or higher were chosen. Those components with eigenvalues above one were kept by using the same names.

2.2   Experimental design

A 7×7 Latin square design was employed to test the null hypothesis:

(1) Learnability has no effect on programming speed.
(2) Learnability has no effect on error rate.
(3) Flexibility has no effect on programming speed.
(4) Flexibility has no effect on error rate.
(5) Robustness has no effect on programming speed.
(6) Robustness has no effect on error rate.

Seven programmers with one year or longer programming experience served as subjects. They were asked to write up a program for each one of seven tasks, each under one of seven experiment conditions to prevent learning or carry over effects. Each subject has to perform all seven tasks, each under different usability condition. Seven experiment conditions were arranged as followed: one with all three usability principles, three with either two of the usability principles, and three with either one of the usability principles.

A HyperCam software was used to record the programming speed and error rate.

3. RESULTS

Performance data from the experiment were analyzed. There were significant differences on programming speed and error rate by comparing subjects and experiment conditions. Further analysis were done to check the effect of each usability principle on programming speed and error rate separately.

3.1 Programming speed

Results from analysis of variance by using programming speed as dependent variable as shown on Table 1. There are very significant effects (P<0.01) of learnability and flexibility principles on programming speed. Principle of robustness has a chose but not significant effect (p=0.057) on programming time.

Table 1 ANOVA of effects of usability principles on programming speed

ANOVA; Var.: Time; R-sqr = .51173; Adj: .45194					
3 factors at two levels; MS Residual= 9483.689					
DV: Time					
	SS	df	MS	F	P
(1) Learnability	198240.6	1	198240.6	23.06535	2.02E-05
(2) Flexibility	101040.1	1	101040.1	11.75605	0.001371
(3) Robustness	32914.29	1	32914.29	3.829587	0.057024
1 by 2	1.285714	1	1.285714	0.00015	0.990299
1 by 3	41.28571	1	41.28571	0.004804	0.945073
2 by 3	3087	1	3087	0.359173	0.552185
Error	360978.9	42	8594.735		
Total SS	742828	48			

3.2 Error rate

Results form analysis of variance by using error rate as dependent variable as shown on Table 2. There are very significant influences (p<0.01) of learnability and flexibility principles on error rate. Principle of robustness has significant effect (p<1015) on error rate.

Table 2 ANOVA of effects of usability principles on error rate

ANOVA; Var.: Error rate; R-sqr = .60794; Adj: .55994					
3 factors at two levels; MS Residual= 10.22131					
DV: Error rate					
	SS	Df	MS	F	P
(1) Learnability	429.9083	1	429.9083	38.31746	2.12E-07
(2) Flexibility	273.3117	1	273.3177	24.36062	1.31E-05
(3) Robustness	65.60825	1	65.60825	5.847618	0.020014
1 by 2	37.16767	1	37.16767	3.312729	0.075877
1 by 3	5.302277	1	5.302277	0.472588	0.495578
2 by 3	0.518696	1	0.518696	0.046231	0.830797
Error	471.2255	42	22.21965		

Total SS	1246.669	48			

4. CONCLUSIONS

This study was done in two stages. The first stage was to survey the existence of usability principles. Through factor analysis, principles of learnability, flexibility, and robustness were confirmed. Based on the results form the first stage, an experiment was conducted to check how principles affect programming performance.

With a minor exception, all the three usability principles have significant effects on programming speed an error rate.

When principle of learnability is functioning, it allows "users to understand how to use it initially and then how to attain a maximal level of performance"(P. 162, Dix et al, 1998). This principle helps to shorten the programming time and lower error rate.

Similarly, when the principle of flexibility is functioning, it allows the end-user to exchange information in multiple ways. The versatility does increase programming performance.

REFERENCES

Collins, D. (1995). *Designing Object-Oriented User Interface*. Redwood City: CA: Benjamin Cummings Publishing Company.

Dix, A., Finlay, J., Abowd, G., and Beale, R.(1998). *Human-Computer Interaction* (2nd ed.) London: Prentice Hall Europe.

Gerlach, J. H. and Kuo, F. Y. (1991). Understanding human-computer interaction for information system design. *MIS Quarterly*, 15, 527-549.

Henneman, R. L. (1999). Design for usability: process, skills and tools. *Information, Knowledge, & Systems Management,* 2, 133-145.

Reed, P., Holoaway, K., Isensee, S., Buie, E., Fox, J., and Williams, J. (1999). User interface guideline and standards: process, issues, and prospects. *Interacting with Computer,* 12, 119-142

Shneiderman, B. (1998). *Designing the User Interface-Strategies for Effective Human-Computer Interaction* (3rd ed.). Reading, Massachusetts: Addison-Wesley Publishing Company.

The Development of a Cognitive-Driven Expert System Interface

Kuo-Wei Su[1,*], Thu-Hua Liu[2], and Sheue-Ling Hwang[3]

[1,*]Department of Accounting & Statistics, Tak Ming College,
Taipei, Taiwan, R.O.C., 114.
Tel: 886-2-26585801 ext. 324
Fax: 886-2-25463916
E-mail: kwsu@mail.takming.edu.tw
[2]College of Management, Chang Gung University,
Taoyuan, Taiwan, R.O.C., 259.
[3]Institute of Industrial Engineering & Engineering Management,
National Tsing Hua University, Hsinchu, Taiwan, R.O.C, 300.

Abstract

Cognitive compatibility plays an important role in the development of the human-computer interface. A good user interface will lead to better user/expert system interaction and task performance. The study as reported here is to design the expert system (ES) interface, which is originally consistent with the cognitive types of the users that are seldom mentioned in the previous research. The knowledge base of the system is represented by a fault decision tree diagram, and is incorporated to communicate between the maintainers and the computer. Furthermore, we also evaluated and compared the effectiveness of expert system interface with traditional maintenance handbook. As a case study, a fault recovery expert system for the maintenance department of the diesel engine bus system of Taipei City has been developed. Findings of the study have important implications in expert system interface design, including the specific features of changeability, traceability, plenitude, and qualitative description in this maintenance area.

1. Introduction

It is well recognized that the nature and quality of users' interaction with expert systems (ES) has been an increasing concern today. Understanding of the user's cognitive structure and the users' task is a critical component in the development of user-centered interface design. A good interface will enhance user/expert system interaction and task performance.

The design of the user interface is influenced by many factors, amongst these the mental model of the user's thinking processes, aspects of usability and the explanation capabilities of the knowledge-based expert system (KBES), have been discussed in the psychological and artificial intelligence literature [2][6]. During recent years, some areas of psychological research have contributed to a better understanding of what goes on in the minds of workers and engineers working with computerized systems [3][6], which could be referred as "cognitive compatibility".

The study as reported here is to design the ES interface, which is originally consistent with the cognitive types of the users that are seldom mentioned in the previous research. The purpose for this study is to avoid or eliminate human cognitive errors and misjudgments in the maintenance operation when adopting expert system as a decision aid. The research takes the maintenance department of the diesel engine bus system of Taipei City as a case study.

2. A user/expert system interface development for maintenance tasks

2.1 Framework description of a developed model of expert system interface (DMESI)

This study constructed a developed model of expert system interface (DMESI) in a maintenance domain. Figure 1 formulates the scheme of DMESI.

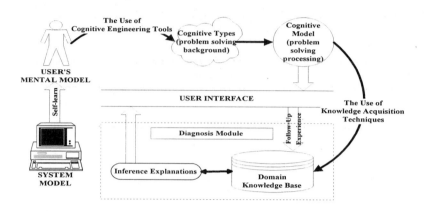

Figure 1. A developed model of expert system interface (DMESI)

The notions of users' (i.e. the experienced maintainers) mental model and system model are discussed and considered in the success of DMESI.   Users' mental model was regarded as a critical issue and was put in the forward phase of the interface design.   A *user mental model* is required by a KBES to help to identify what needs to be explained, to determine the depth and complexity of the explanation, and to establish the knowledge necessary to assist the user in achieving his/her diagnostic planning and understanding the diagnostic solution [2].   Different artificial intelligence researchers use different definitions of a *user mental model* [5][11], and thus they disagree in how to create the *user mental model*.   On the other hand, the *user mental model* can represent a certain aspect of the user, which is implicit within the system and used to adapt the system to suit the user.   It seems difficult to have a unified *user mental model,* which embodies all the user requirements because not all users, even in the same field, think or process information in the same mode.   Nevertheless, from the survey in the previous study [7], the maintainers' cognitive types (Individual's cognitive type reveals how one perceives, analyzes, and interprets the acquired information) were consistent in the maintenance field.   By the DMESI, one may go deeply into the theory of cognitive psychology to expose whether a unified *user mental model* exists or not.

The *system model*, on the contrary, is the model of the expert system as seen and made by the maintainers.   For a novice user, he/she will self-learn from the *system model*.   However, an expert user could share his/her opinion of maintenance with domain knowledge base through expert system interface.   The major symbol of the expert *system model* is a diagnosis module.   The diagnosis module could be characterized by ease of use, capabilities, friendliness, complexity, multi-representations, reasoning process, and rapid responses.   A domain knowledge base represented by a fault tree diagram and an inference explanation was incorporated into the diagnosis module.   Generally, a domain knowledge base is simply a set of rules, procedures and objects created in order to reproduce the expertise of a senior maintainer.   The knowledge, acquired through maintainer interviews, is then organized in a most possible transparent manner, in order to allow a maintainer to make modifications and corrections without training or knowledge of implementation details [9].   Here knowledge bases are constructed in the development process of a maintenance protocol [7] and set down by a specific knowledge acquisition technique.   Moreover, the inference explanation utilizes search strategies, reasoning direction, and evidence aggregation to simulate the intelligent behavior of a senior maintainer.   Through viewing the system's inference explanations, a user may support or modify an existing mental model or create a new one.

3. Representation and evaluation of the system

3.1 Representing knowledge-based expert system (KBES)

A graphical display of the user interface provides maintenance's users with a flexible mechanism for viewing knowledge base structure and is shown in Figure 2. As a pre-processing step, a forward flow arrangement of 'How to do it', the procedure knowledge was represented through a simple text-processing phase to be transformed into an expert system [7]. The forms of 'Decision Trees' representation chosen for the knowledge could match the structure of the tasks in hand. It captures analysis information in hierarchy. The decision tree represents failure modes, possible causes, and analysis processing of underlying problems in the given event.

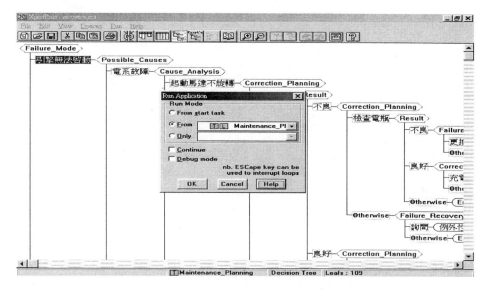

Figure 2. Decision tree and inference explanations

The interface explanations is part of the interface environment that utilizes production rules, procedures and data in order to simulate the intelligent behavior of a competent maintainers. It executes the following tasks:

- Forward and backward chaining of rules
- Start up of procedures
- Debug mode of rules

The three tasks are common to the majority of inference engines. The explanations offered by the user interface are not limited to the conventional 'why' or 'how' explanation; deep explanation about a specific task can be provided through manual or senior maintainers. The experts considered that the known quantity offering of possible clauses and the mile reference of preventive maintenance are not necessary. More detail contents (enough knowledge cues) but simple interface were suggested by assurances of experts aimed at users of *ISTJ* cognitive type. But if the users are in the *ENFP* cognitive type and skill-based behavior, then more keywords but few knowledge in detail appearance are needed for the interface design [7].

Direct editing of decision trees or attributes view is easy to be implemented in the interface. Specifically, because the users have been encouraged to comment on the KBES, and the feedback process through KBES interface is easy, it is much easier to expand the new knowledge, new features and capabilities. In real life applications, far more complex decision making paths will involve backward and forward 'Chaining' structures, where one tree invokes other trees. However, the complexity of the path remains transparent to the run time maintainers.

3.2 The effectiveness comparison with manual of maintenance

Hu et al. [4] suggested that interface design may have a significant effect on system-user concept communication, regardless of users' familiarity with the search task, and that a graphical user interface may be more effective in supporting such communication than a list-based design. Desire for effective knowledge access and sharing across the specified domain have made manual-based increasingly ineffective or obsolete. Through the KBES interface manipulation, users can search the problem solving strategy easily and flexibly or expand on their new knowledge into the KBES.

In summary, comparing with manual, specific features of changeability, traceability, and plenitude can serve as the benefits of KBES interface. The first one, changeability permits that the expert systems continuously evolve and therefore continuously expand ([8]. Traceability refers to design features that help users trace the causes of inadequate or erroneous actions [10] and the plenitude follows users' requirements to render further information.

3.3 Evaluating KBES

The prototype of the KBES (Figure 2) was presented to and tested by maintainers (novice and senior). The evaluation tests (walkthroughs with the user to view document and paper designs) of a diagnostic process through KBES were executed under several mechanical failure modes. They could communicate with the computer via a message board and find a diagnostic route quickly and validly after the trials. Generally, they appreciated the straightforward way of working with this system and expected that the human error and maintenance time will be reduced. And further, each time a KBES is exposed to a new case, or whenever there are changes in the environment, the system must be refined (new capabilities are often added at this time). Each time a substantial refinement is made, an evaluation should follow. Development and evaluation of a system continue as long as improvements are achieved [1].

Encouraged by this positive feedback, a prototype system was built and tested. Additionally, for making a good judgment on a failure problem, the diagnostic skill should take into account the following aspects suggested by senior maintainers:

1) For avoiding unnecessary maintenance, maintainers should sense the operational regular function of the system.
2) To consider the environmental factors.
3) To take the drivers' behavior in driving as a reference.
4) To understand the situation (5W: "Who", "Where", "Which", "When", and "What") of failure clearly.
5) To consider the problems (quality, quantity, and timing) of the mechanism, the oil route, and the circuitry separately, to think them in simplification, and to solve them in details.
6) To notice the failure alarm affected by other subsystems without ignore the major problem.
7) To run a trial with driver.

In the concept of cognitive compatibility, the test has proved its value in practice and the above suggestions can be used for the criteria of the knowledge base formation.

4. Conclusions

Findings of the study have important implications in expert system interface design, including the specific features of changeability, traceability, plenitude, and qualitative description in this maintenance area. More detail contents but simple interface has become the fundamental of KBES interface design.

According to the positive results of the KBES interface evaluation, this paper suggests two design guidelines for expert system developers, as listed below.

1) Based on the user's cognitive type, the inference explanations should be tailored closely toward the users' mental model.
2) Easy operation and logic of KBES interface will induce user to understand and upgrade the domain knowledge.

5. References

[1] Awad, E. M. (1996). *Building expert systems*. Minneapolis/St. Paul: West Publishing.

[2] Berrais, A. (1997). Knowledge-based expert systems: user interface implication. *Advances in Engineering Software, 28,* 31-41.

[3] Fuchs-Frohnhofen, P., Hartmann, E. A., Brandt, D., & Weydandt, D. (1996). Designing human-machine interfaces to match the user's mental models. *Control Eng. Practice, 4,* 13-18.

[4] Hu, P. J. H., Ma, P. C., & Chau, P. Y. K. (1999). Evaluation of user interface designs for information retrieval system: a computer-based experiment. *Decision Support System, 27,* 125-143.

[5] Mckeown, K. R. (1990). User modelling and user interfaces. In *AAAI-90, Proc. Eighth National Conf. Artificial Intelligence* (pp. 1138-1139). American Association for Artificial Intelligence.

[6] Rook, F. W., & Donnell, M. L. (1993). Human cognition and expert system interface: mental models and inference explanations. *IEEE Transaction on Systems, Man, and Cybernetics, 23,* 1649-1661.

[7] Su, K. W., Hwang, S. L., & Liu, T. H. (2000). Knowledge architecture and framework design for preventing human error in maintenance tasks. *Expert Systems with Applications, 19*(3), 219-228.

[8] Karimi, J., & Briggs, P. L. (1996). Software maintenance support for knowledge-based systems. *Journal of Systems and Software, 34*(3), 134-143.

[9] Kaszkurewicz, E., Bhaya, A., & Ebecken, N. F. F. (1997). A fault detection and diagnosis module for oil production plants in offshore platforms. *Expert Systems with Applications, 12*(2), 189-194.

[10] Kontogiannis, T. (1997). A framework for the analysis of cognitive reliability in complex systems: a recovery centered approach. *Reliability Engineering and System Safety, 58,* 233-250.

[11] Williges, R. C. (1987). *The use of models in human computer interfaces design*. Ergnomics Society Lecture, Swansea.

Speech Completion:
New Speech Interface with On-demand Completion Assistance

Masataka Goto, Katunobu Itou, Tomoyosi Akiba, and Satoru Hayamizu

National Institute of Advanced Industrial Science and Technology (former Electrotechnical Laboratory)
1-1-1 Umezono, Tsukuba, Ibaraki 305-8568, JAPAN.
m.goto@aist.go.jp

Abstract

This paper describes a novel speech interface function, called *speech completion*, that helps a user enter a word or phrase by *completing* (filling in the rest of) a phrase fragment uttered by the user. Although the concept of completion has been widely used in text-based interfaces, effective completion for speech has not been proposed. We enable a user to invoke the speech-completion function intentionally and effortlessly by building an interface that displays completion candidates when a filled pause is uttered (a vowel is lengthened) during a phrase. The filled pause can be considered a nonverbal modality that has not been used in speech input interfaces. In our experience with a system that includes a filled-pause detector and a speech recognizer capable of listing completion candidates, the effectiveness of speech completion was confirmed.

1 INTRODUCTION

Current speech-input interfaces have not fully exploited the potential of speech. Although human speech has two aspects, verbal information (e.g., words) and nonverbal information (e.g., hesitation), most speech recognizers utilize only the modality of verbal information. They are therefore, as it were, nothing more than a computer keyboard that sometimes makes key-recognition errors; even if the precision of speech recognizers could be improved, it is difficult to build an interface that is handier than a keyboard. The purpose of this study is to build a speech interface that makes full use of the role nonverbal speech information plays in human-human communication.

From among various nonverbal information, we focus on a filled pause (the lengthening of a vowel), which is a hesitation phenomenon and is apt to reflect the mental state of a speaker (e.g., trying to think of a subsequent word) (Takubo, 1995; Rose, 1998). When a speaker cannot remember an entire phrase and hesitates with a filled pause, the listener sometimes helps the speaker recall it: the listener suggests options obtained by *completing* the partially uttered fragment (i.e., by filling in the rest of it). For example, when a speaker cannot remember the Japanese phrase *"maikeru jakuson"* (in English, "Michael Jackson")[1] and stumbles, saying *"maikeru–"* ("Michael–") with a filled pause *"ru–"* ("l–"), a listener can help the speaker by asking whether the speaker intended to say *"maikeru jakuson"* ("Michael Jackson"). Although one of the reasons that speech communication is comfortable is that we can expect a listener to help us this way when we utter vague or incomplete information, this phenomenon has been given little attention in speech recognition research.[2]

The concept of *completing* a fragment has been widely used in text-based interfaces. For example, several text editors (e.g., Emacs) and UNIX shells (e.g., tcsh and bash) provide functions completing the names of files and commands. These functions fill in the rest of a partially typed fragment when a *completion-trigger key* (typically the Tab key) is pressed. Completion functions for pen-based interfaces, such as POBox (Masui, 1998), have also been proposed. Even though completion is so convenient that it becomes indispensable to those who have used it, effective completion functions for speech input interfaces have not been developed because there has been no way to trigger them during natural speech input.

In this paper we describe a completion function, called *speech completion*, that enables a user to enter a word or phrase by uttering a fragment of it with a filled pause. The following sections explain the basic concept of speech completion, describe the design and implementation of a speech recognition interface with the speech-completion function, show experimental results obtained with the interface, and discuss directions of future work.

---

[1] When a foreign name like "Michael Jackson" is written or pronounced in Japanese, the Japanese style is used: *"maikeru jakuson."*

[2] In most current speech recognizers, it is understood that the user prepares the content in advance and pronounces it carefully and precisely. Although hesitation phenomena such as filled pauses and restarts occur frequently in spontaneous speech, current recognizers accept only fluent speech without these phenomena because they tend to cause recognition errors. In addition, only few attempts have been made at making good use of the valuable roles these phenomena play in human-human communication.

198

2 SPEECH COMPLETION

Speech completion, the general term for interface functions that enable a user to invoke completion assistance during speech input, has the following three advantages:

1. It helps the user recall uncertain phrases.

2. It saves labor when the input word or phrase is a long one.

3. It reduces the psychological pressure of being forced to utter the whole content carefully and precisely, something that most current speech recognizers force users to do.

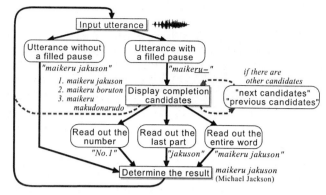

Figure 1: Flowchart for our speech input interface with speech completion assistance.

Since completion should not become annoying, it should be invoked only when the user wants to obtain completion candidates. For effective and practical speech completion, we therefore use an intentional filled pause to trigger a speech-completion function that fills in the rest of a partially uttered fragment. Since the filled pause is a very natural trigger, the user can invoke the speech-completion function intentionally and effortlessly. This is especially true for Japanese, a moraic language in which every mora ends with a vowel that can be lengthened. In fact, speakers typically use filled pauses to gain time to recall a word or to wait for a listener to help with word choice.

Although various completion levels — such as those of the word, phrase, clause, and sentence — can be considered, in this paper we concentrate on word-level and phrase-level completion (this completion can be naturally extended to the sentence level). We deal with words registered in the system vocabulary of a speech recognizer, and phrases such as the names of musicians and songs can be registered as single words.

3 SPEECH INPUT INTERFACE WITH SPEECH COMPLETION ASSISTANCE

We designed a speech-input-interface system (Figure 1) that can assist a user by completing any system vocabulary word (which can be either a single word or a phrase). It works as follows:

1. When the user does not remember the last part of a word or phrase and utters the first part while intentionally lengthening its last syllable (making a filled pause),[3] the system displays a numbered list of completion candidates whose beginnings acoustically resemble the uttered fragment.

   When, for example, the Japanese phrase *"maikeru jakuson"* (in English, "Michael Jackson") is registered as one word in the system vocabulary, a user uttering the fragment *"maikeru–"* ("Michael–") gets completion candidates such as *"maikeru jakuson"* ("Michael Jackson"), *"maikeru boruton"* ("Michael Bolton"), and *"maikeru makudonarudo"* ("Michael McDonald").

2. The user can invoke the display of other candidates by uttering the turning-the-page phrases, "next candidates" and "previous candidates," displayed whenever there are too many candidates to fit on the computer screen. If all the candidates are inappropriate or the user wants to enter another word, the user can simply ignore the displayed candidates and proceed with the next utterance.

3. When the user selects one of the candidates by uttering (reading out) either its number, the last part of the word, or the entire word, that word is highlighted and used as the speech input result.

4 IMPLEMENTATION

Figure 2 shows the architecture of our speech-completion system. The boxes in the figure represent different processes, and the four main processes are those of the filled-pause detector (Section 4.1), the speech recognizer (Section 4.2), the interface manager, and the graphics manager. Those processes can be distributed over a LAN (Ethernet) and connected by using a network protocol called *RVCP (Remote Voice Control Protocol)*, which is an extension of RMCP (Goto, Neyama, & Muraoka, 1997). RVCP supports timestamp-based synchronization for real-time information handling and supports efficient multicast-based information sharing without the overhead of multiple transmission.

---

[3]The user can insert a filled pause at an arbitrary position while uttering a word or phrase.

The filled-pause detector controls the two modes of the speech recognizer, the *normal mode* and the *completion mode*. In the completion mode triggered by a filled pause, the recognizer generates a numbered list of completion candidates that is sent to the interface manager managing the state transition of the interface (flowchart shown in Figure 1). The graphics manager manages a front-end GUI and displays on the screen the recognition results and a pop-up window containing the candidate list.

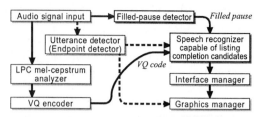

Figure 2: System architecture.

4.1 Filled-pause detector

Because speech completion is impractical without a real-time method for detecting filled pauses that is independent of vocabulary and language, we use a robust filled-pause detection method (Goto, Itou, & Hayamizu, 1999). It is a bottom-up method that can detect an intentionally lengthened vowel in any word without using top-down information (language model). It determines the beginning and end of each filled pause by finding two acoustical features of filled pauses, small fundamental frequency transitions and small spectral envelope deformations. Those features are evaluated by using a sophisticated instantaneous-frequency-based analysis (Goto et al., 1999). Figure 3 shows an example of a detected filled pause.

Note that, as shown in Figure 2, the processing of the real-time filled-pause detector that directly analyzes the input audio signal is executed in parallel with that of the following HMM-based speech recognizer.

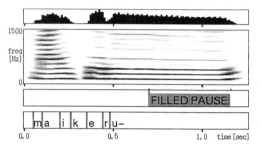

Figure 3: An example of a detected filled pause for "maike$\underline{ru}$–": the power and the spectrum (top), the filled pause (middle), and the phoneme sequence recognized by the speech recognizer that inhibits the transition from a vowel phoneme to the next phoneme during the detected filled pause (bottom).

4.2 Speech recognizer capable of listing completion candidates

To provide a list of completion candidates whenever a filled pause was detected, we extended an HMM-based speech recognizer, *niNja* (Itou, Hayamizu, & Tanaka, 1992). The extended recognizer receives VQ codes and the detected filled pause and sends the following results to the interface manager:

- The top N_{choice} completion candidates

  The completion candidates are generated at the beginning of each filled pause and are sent immediately (i.e., before the end of the current utterance). Each candidate includes the information of which parts (phonemes) of it have been uttered (how much of it has been uttered).

- The top N_{result} recognition results

  At the endpoint of each utterance, the recognition results of that utterance are sent.

The recognizer uses not only a vocabulary of words to be input and be completed (e.g., the names of musicians and songs) but also a vocabulary for operating the interface (e.g., the candidate numbers and the turning-the-page phrases). All these words are stored in a tree structure as shown in Figure 4, where the wedge marks represent multiple hypotheses maintained by a frame-synchronous Viterbi beam search decoder.[4]

The recognizer enters the completion mode when the beginning of a filled pause is detected (about 200 ms after a vowel is lengthened). It generates candidates by tracing from the top N_{seed} hypotheses (at the beginning of the pause) to the leaves (Figure 4): the candidates are obtained by deriving from the vocabulary tree those words that share the prefix corresponding to each incomplete word hypothesis of the uttered fragment. The top N_{choice} candidates (leaves) are sorted and numbered in order of likelihood and then sent to the interface manager. We call the nodes corresponding to the top N_{seed} hypotheses the *speech completion seeds*. For example, if the top black circle in Figure 4 is a seed, the completion candidates obtained are "Michael Bolton" and "Michael Jackson."

---

[4]Because single phonemes cannot be recognized accurately enough, most up-to-date speech recognizers do not determine the phoneme sequence of a word phoneme by phoneme but instead choose the maximum likelihood hypothesis while pursuing multiple hypotheses predicting the next phoneme.

To enable the user to select the correct candidate by reading out the last part of it, the speech-completion system must be able to recognize last-part fragments that are not registered as vocabulary words. We therefore use an *entry node table* in which are listed the roots (nodes) from which the decoder starts searching. In the normal mode, only the root of the vocabulary tree is listed. During the utterance just after the listing of completion candidates, speech completion seeds are temporarily added to the table as shown in Figure 4. Although a candidate can be selected by uttering just the last part, the recognition result sent to the interface manager is the entire word.

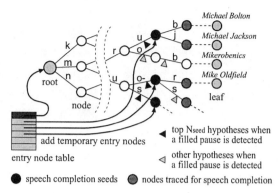

5 EXPERIMENTAL RESULTS

We tested the system with a system vocabulary comprising about 500 names of musicians and songs in Japanese. Our current implementation uses the following parameter values, which should be adjusted according to the vocabulary and the purpose of speech input interface: N_{choice} = 20, N_{result} = 5, and N_{seed} = 5. Figure 5 shows an example of the graphics output displayed during speech completion.

Figure 4: Obtaining completion candidates on the vocabulary tree at the beginning of a filled pause and adding speech completion seeds to the entry node table.

Experimental results showed that our system can provide a helpful list of completed full names when a filled pause is uttered. We have confirmed that the speech-completion function is intuitive enough to be used without any training and is effective for entering uncertain phrases. The completion candidates are especially useful when the input phrase is a long proper name.

6 DISCUSSION

Our speech completion research began with the intention of making speech-recognition technologies user-friendly. This approach suggests various directions for future work.

6.1 Autocompletion functions

While text-based (keyboard-based) completion functions manually invoked by a completion-trigger key were mentioned in Section 1, autocompletion functions have also been used in Reactive Keyboard (Darragh, Witten, & James, 1990) and for URL entry on web browsers. These autocompletion functions automatically list up completion candidates every time a user types a key. The interface with such functions is also called *predictive interface* and is considered effective.

Such autocompletion functions, however, are not suitable for use in speech input interfaces. In keyboard input, there is no ambiguity in recognizing which key is pressed and the boundaries between successive characters are obvious. In speech input, on the other hand, the recognition of each phoneme is ambiguous and the boundaries between successive phonemes are not easily determined.[5] It is therefore hard to determine when autocompletion candidates should be displayed. Even if the candidates are displayed at arbitrary regular intervals, it is very difficult to continually provide candidates as good as those obtained in text-based autocompletion. The autocompletion of speech input is thus likely to become annoying and not be practical since a predictive interface tends not to be used when it is less convenient than an interface that is not predictive. In our speech-completion function, by enabling a user to explicitly invoke the completion function by using an intentional filled pause, we have successfully built a practical interface that does not interfere with the user at all when the completion is not needed.

6.2 Multimodal interface using multiple modalities in speech audio signals

As stated in Section 1, most current speech interfaces use only the modality of verbal information that can be handled by speech recognition. Our speech-completion function, on the other hand, achieves a user-friendly interface because it exploits the modality of nonverbal information — the filled pauses. We regard our speech completion as a multimodal interface that makes use of multiple modalities contained in speech audio signals.

---

[5]Metaphorically speaking, speech is not block letters but cursive letters that are not easily segmented.

(1) Uttering *"maikeru–."*   (3) Uttering *"No. 2."*

(2) A pop-up window containing completion candidates appears.

(4) The second candidate is highlighted and bounces.

(5) The selected candidate *"maikeru jakuson"* is determined to be the recognition result.

Figure 5: Screen snapshots of speech completion when the phrase "maikeru jakuson" *("Michael Jackson") is entered.*

Starting from our speech-completion interface, interfaces that are more user-friendly could be built by introducing other nonverbal modalities. In contrast to a computer keyboard, the current speech recognizers have dealt with only a part of the normal letter keys.[6] In this study, on the other hand, the role of the special key "Tab" (the typical completion-trigger key in text-based interfaces such as UNIX shells and the Emacs editor) is triggered by the filled pause. This approach opens up new vistas for future research that assigns other nonverbal information (e.g., pitch and speech rate) to special keys.

Furthermore, speech interfaces can go beyond the limitations of computer keyboard functions because speech has both verbal and nonverbal modalities that can naturally and simultaneously provide different functions. In our speech-completion function, for example, the voice during a filled pause simultaneously conveys both phoneme information (verbal modality) and the user's mental state (nonverbal modality); this function can be considered more efficient and natural than text completion using the completion-trigger key. Typical nonverbal modalities can thus provide meta communicative functions that make ordinary verbal communication rich. By fully bringing out this kind of potential capabilities of speech, we will be able to build excellent interfaces that enhance well-known advantages of speech interfaces, such as hands-free and fast input speed.

7   CONCLUSION

We have described the new speech interface function *"speech completion,"* which fills in the missing part of a partially uttered fragment in order to help a user enter an uncertain phrase. While we have confirmed the effectiveness of this function for the task of inputting the names of musicians and songs, it can also be immediately applied to various other speech applications. We believe it will become as indispensable in speech interfaces as text completion is in good text-based interfaces.

REFERENCES

Darragh, J. J., Witten, I. H., & James, M. L. (1990). The Reactive Keyboard: A Predictive Typing Aid. *IEEE Computer*, *23*(11), 41–49.

Goto, M., Itou, K., & Hayamizu, S. (1999). A Real-time Filled Pause Detection System for Spontaneous Speech Recognition. In *Proceedings of Eurospeech '99*, pp. 227–230.

Goto, M., Neyama, R., & Muraoka, Y. (1997). RMCP: Remote Music Control Protocol — Design and Applications —. In *Proceedings of International Computer Music Conference*, pp. 446–449.

Itou, K., Hayamizu, S., & Tanaka, H. (1992). Continuous speech recognition by context-dependent phonetic HMM and an efficient algorithm for finding N-best sentence hypotheses. In *Proceedings of ICASSP 92*, pp. I–21–24.

Masui, T. (1998). An Efficient Text Input Method for Pen-based Computers. In *Proceedings of CHI'98*, pp. 328–335.

Rose, R. L. (1998). The communicative value of filled pauses in spontaneous speech. Master's thesis, University of Birmingham.

Takubo, Y. (1995). Towards a Linguistic Model of Speech Performance (*in Japanese*). *Journal of Information Processing Society of Japan*, *36*(11), 1020–1026.

---

[6] In this paper, we consider that a computer keyboard consists of two parts, the normal letter keys for typing alphabetical and numeric characters and the other special keys such as Tab, Delete, and Escape keys.

202

A Support System for Visually Impaired Persons Using Acoustic Interface – Recognition of 3-D Spatial Information –

Yoshihiro KAWAI and Fumiaki TOMITA

National Institute of Advanced Industrial Science and Technology (AIST)
AIST Tsukuba Central 2, 1-1-1, Umezono, Tsukuba, Ibaraki 305–8568, Japan.
y.kawai@aist.go.jp   f.tomita@aist.go.jp

Abstract

Without visual information, visually impaired people incur many inconveniences in their daily and social lives. It is important to prepare an infrastructure that enables visually impaired users to easily understand the circumstances of their surroundings. However, many problems cannot be solved by infrastructure maintenance and development alone. Therefore, we are developing an active support system – an intelligent support device – that would provide access to 3-D spatial information surrounding the user via an acoustic interface. The proposed system is expected to be useful in the situations where the infrastructure is incomplete and the situation changes in real-time. This paper describes a prototype system, processing of 3-D visual information, and some basic experiments on an acoustic interface using 3-D virtual sound.

1 Introduction

Much of the information that humans acquire from the outside world is obtained through sight. Without this facility visually impaired people incur many inconveniences in their daily and social lives. Much research has been done worldwide on support systems for the visually impaired [2, 5, 7]. However, there are still many problems in representing the real-time information that is changing around the user.

Canes and seeing-eye dogs are widely used as walking support devices in active operator support systems. However, the range that a user can sense with a cane is limited, and the use of seeing-eye dog, still entails problems of availability and practicability. Support devices using electronic technologies have been developed, but considerable training is needed to use them. It is important to prepare an infrastructure that enables visually impaired users to easily understand the circumstances in their surroundings. Surface bumps, braille panels, and audio traffic signals are in use. However, economic realities limit their installation and availability. Many problems cannot be solved by infrastructure maintenance and development alone. Therefore, we aimed to develop an active support system – an intelligent support device – that would provide access to 3-D spatial information surrounding the user via an acoustic interface.

Among other reported visual aid systems using sound are a system that uses ultrasonic waves, Kaspa (Sonic Vision), one that displays images by differences of frequency pitch and volume, one that utilizes a speaker array, and one that utilizes stereophonic effects [6]. However, the target of these systems is mainly a two-dimensional space. Three-dimensional sound can provide more real-world information because it includes an intuitive feeling of depth, as well as a feeling of front and rear. There is a GUI access system that uses 3-D sound, but as it shows only the relative position between windows, 3-D sound is not being fully utilized.

We are developing a support system that displays 3-D visual information using 3-D virtual sound. Our system is unique in that 3-D environment information is acquired for a task set by the user, and it is represented by 3-D virtual sound. Images captured by small stereo cameras are analyzed in the context of a given task to obtain a 3-D structure and object recognition is performed. The results are then conveyed to the user via 3-D virtual sound. This system is expected to be useful in situations where the infrastructure is incomplete and the situation surrounding the user is changing in real-time. In addition, this system can be used without much learning because it provides information via virtual sound superposed on actual environment sounds. This method dose not replace or impede the user's existing external auditory sense. We assume that it could be used, for example, to assist while playing sports or walking, as shown in Figure 1 and Figure 4 (a). While the importance to walking assistance needs no further explanation, assistance while playing sports is certainly also important. The sports that visually impaired persons can do are limited to those with two-dimensional movements, such as jogging, roller skating, floor volleyball, etc. Many visually

Figure 1: Application example (Assistance of playing sports).

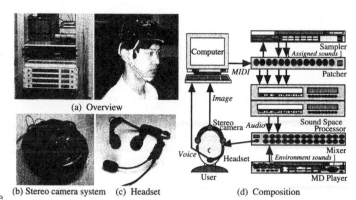

(a) Overview

(b) Stereo camera system   (c) Headset   (d) Composition

Figure 2: Overview and composition of the system.

impaired persons have voiced requests to be able to play sports that involve 3-D movements together with sighted people. So we believe that our proposed system is one answer that satisfies their needs.

This paper presents an outline of our prototype system, the 3-D visual information processing methods to acquire the 3-D information surrounding the user, and some basic experiments on an acoustic interface using 3-D virtual sound.

2 System

We have built the prototype system shown in Figure 2 (a). Figure 2 (b) is a stereo camera system with three small cameras, mounted on a helmet, and (c) is a headset with a microphone and headphones. Figure 2 (d) shows the configuration of this system.

The images captured by the stereo cameras are sent to a computer and analyzed. After the 3-D data is reconstructed, the user's targets are recognized and tracked. Both the status of the objects acquired and the sound assigned to each object from a sampler are input to the sound space processor. A sound image for each target is mapped in a 3-D virtual sound space and transmitted to the user. Although as yet not implemented, we may also have a microphone for the user to set a task by voice, which may be processed by a voice recognition engine.

2.1 Stereo camera system

It is desirable for the visual information input unit to be small and lightweight because these devices will be mounted on the user's head. We mounted an aluminum frame with three small cameras on a helmet (shown in Figure 2 (b)). Each camera is a 1/4-inch color CCD camera. The total weight of the helmet is about 650 g. We have set the focus of the lens at more than 2 m, a point beyond the reach of a visually impaired person's cane. The reason for using three cameras is to reduce the calculation of correspondence problems with horizontal lines during stereo image analysis.

The advantages of using cameras are that they are suited for object recognition and tracking, measurement of distant objects (e.g., discrimination of the red/green color of a traffic signal from far away) and for character information readability (e.g., characters on a sign). Although it is still difficult to analyze images to obtain 3-D visual information, there exists a potential use for recent pattern recognition techniques in our application. For example, we have been developing the Versatile Volumetric Vision (VVV) system, a general-purpose system which can be used for many purposes in many fields [9]. The details are described in section 3.1.

2.2 Acoustic system using three-dimensional virtual sound

With recent developments in virtual reality technologies, the technical progress of acoustics in virtual space has been remarkable. We can use 3-D virtual sound readily, since some 3-D sound equipment is already in the commercial market. We assembled our acoustic system using the sound space processor RSS-10 made by

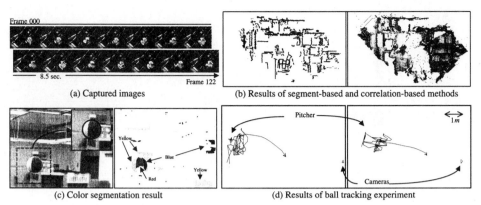

(a) Captured images

(b) Results of segment-based and correlation-based methods

(c) Color segmentation result

(d) Results of ball tracking experiment

Figure 3: Results of thrown ball tracking experiment.

Roland corporation (the left side in Figure 2 (a)). This is a device which calculates an arbitrary 3-D virtual sound space using Head Related Transfer Functions (HRTFs).

As an output device, we selected bone conduction headphones, which do not entirely cover the user's ears, and therefore do not impede the user's hearing and understanding of environment sounds. It is very important not to impede a user's auditory sense.

3 Three-dimensional visual information and auditory information

3.1 Three-dimensional visual information

We briefly explained that our method of obtaining 3-D information would use three captured images. A feature of our proposed method is that 3-D shape recognition and tracking are possible even though only range information and color information is obtained. The details of these algorithms have been reported [3, 8, 9]. In this paper, we will show the results of a 3-D visual information processing experiment in which the user plays catch ball as one application example in Figure 1.

The technology for tracking a moving object is necessary to help visually impaired persons play sports. Figure 3 (a) shows a captured stereo image using the stereo camera system in Figure 2. In this experiment, we performed a test of tracking a sponge ball. The ball is colored red, blue, and yellow. First, a pitcher held a beanbag in both hands, and then threw it to the system as shown in Figure 3 (a). The number of total images was 123 frames per 8.5 *sec.* (14.5 fps). The overall image size was 160×120 pixels, but at first, some images were captured at 640×480 pixels (4.5 fps) to detect the position of the target ball. The pitcher had to show the ball clearly to enable it to be recognized easily (to calculate initial parameters). After it was recognized, its positions were tracked using the algorithm. The reconstructed result for one scene (frame number 099) using the segment-based method is shown in the left figure of Figure 3 (b). Distance measurement using the correlation-based method was also done at the same time to detect obstacle objects (the right figure in Figure 3 (b)). A tracking process using only the color segmentation method was performed in this experiment to track a fast-moving object, the thrown ball. The ball included the primary colors (red, blue, and yellow), which are easily segmented in the RGB color space, as shown in Figure 3 (c). In each camera's image, the regions with each of the color parts were segmented, and only one region, the region boundary of each color space, was selected. For this region, a circle was fitted and the 2-D coordinates of its center were calculated. Then, the 3-D position of the moving ball was calculated using three positions. This process was iterated repeated to trace the ball. The time required to search the ball's new position was reduced by guessing it from obtained transformation parameters. The result of tracking the ball is shown in Figure 3 (d). Although the position resolution was low because small images were used, the tracked path was fairly accurate. These position parameters were sent to the sound space processor with an assigned sound to calculate the virtual sound based on HRTFs.

205

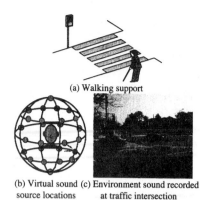

(a) Walking support

(b) Virtual sound (c) Environment sound recorded
source locations at traffic intersection

Figure 4: Experiment setup.

Table 1: Results of tests on auditory localization.

(a) Case A: without environment sound

Subject	T1	T2	Time
A3	11.5%	48.1%	2.8 sec.
B3	19.2%	53.8%	3.7 sec.
C3	11.5%	38.5%	9.4 sec.
Avg.	14.1%	46.8%	5.0 sec.

(b) Case B: with environment sound

Subject	T1	T2	Time
A3	13.5%	44.2%	2.9 sec.
B3	5.8%	34.6%	4.6 sec.
C3	9.6%	36.5%	9.5 sec.
Avg.	9.6%	38.4%	5.7 sec.

(c) Analysis of mis-recognized answers of subject A3 at T2

Subject A3	Error rate	Error type		
		Front&Rear	Up&Down	Other
Case A	51.9%	14.8%	59.3%	33.3%
Case B	55.8%	41.4%	51.7%	20.7%

Small fixed focus cameras mounted on a frame are used as the input device. They do not comprise an active camera system, so there is a problem in that the object information is not reconstructed closer than $2\ m$ (See Figure 3 (d)), because it was not captured in every camera. However, this camera setting was decided based on the reach accessible using a cane, as was described in section 2.1. Also, this experiment was carried out off-line, so the computation time was not in real time. We are now developing a real-time on-line system.

3.2 Three-dimensional auditory information

One method that uses the auditory sense for sensory substitution informs the user of the environment situation using a synthetic voice, however, playing sounds is a superior means for immediate understanding.

Not all visual information is converted to auditory information. The only sounds output are those representing the targets needed to do the task. However, for obstacles or in dangerous situations, such as the presence of stairs, walls, or cars, an alarm or voice alert is output to warn the user of dangerous objects.

Regarding the output sound, if it already exists in the real world, the same sound is used. Otherwise, a sound that the user can easily recognize is assigned. It is important to create and superimpose a sound space that is the same as the real environment as much as possible. The user can change these sound settings and parameters at will, and choose favorite sound sources from the sampler in Figure 2 (d).

4 Basic tests on auditory localization

We conducted some basic tests on auditory localization of 3-D sound in a virtual space to develop an acoustic interface. The results of some experiments have been described in [4]. In this paper, we explain sound localization tests to investigate the influence of environment sounds. These tests were performed on the assumption of an application of the walking support system shown in Figure 4 (a).

The subjects were two males and one female. The sound locations for all directions in the virtual sound space are shown in Figure 4 (b). Twenty-six sound sources were arrayed on a sphere with a radius of 3.0 m. They were located at both north and south poles, and at 8 points each on a horizontal plane and cross horizontal planes at ± 45 degrees, at 45-degree intervals. The height of the sphere center from the floor was 3.2 m. The sound was presented at random 52 times. Each sound source was assigned a unique number. After having been briefed as to the correspondence between numbers and directions in advance, the subjects responded with the number that they felt. For the sound source in these tests, a sound similar to that heard at audio traffic lights was used. The environment sounds included the sounds of automobiles which were recorded at a pedestrian crossing shown in Figure 4 (c). We performed two sound position recognition tests, one without the environment sounds (Case A), and one with the environment sounds (Case B). Table 1 (a) and (b) shows the rates of correct answers and recognition time for Case A and Case B.

Tolerance n (T_n) is the rate of correct answers according to the following:

T_1: rate when the answer exactly matched the actual direction.

T$_2$: rate when the answer was accepted as correct if its distance was within one position from side to side or up and down.

The average of completely correct answers at T$_1$ was very low (in Case A only 14.1% and in Case B 9.6%). This result shows that environment sounds degrade the sound image localization. There are two main reasons. One is the narrow range of the bone conduction headphones. The effective frequency band is only 30 to 3000 Hz at more than 80 dB. Another is the problem of the sound source. We used a simple sound, which was similar to the sound at an audio traffic signal, instead of pink noise, which is similar to environment sounds and easy to recognize. The frequency band is narrower than pink noise, so it was more difficult to localize sound images.

Table 1 (c) shows an analysis of subject A3's errors at T$_2$. He mis-recognized to almost the same degree in both Case A and Case B. However, the errors between the front and rear directions in Case A were larger than in Case B. We can assume that the influence of the environment sounds is large for sound position recognition between the front and rear directions. It is known that frequency is related to the recognition of front and rear directions [1], so the difference of frequencies should be emphasized to improve recognition between the front and rear directions.

5   Conclusions

We are developing a recognition support system using 3-D virtual sound to enable visually impaired persons to obtain reliable information about their environment. We described the design of our prototype system, 3-D visual and auditory information processing methods, and some basic experimental results of our acoustic tests. We showed that it is a usable system if tasks are limited, though there is still a problem of image processing time. Problems in sound image localization were clarified through the experiments. For example, we found that there were many mis-recognitions between the front and rear directions.

In the future, we will first complete this project as an on-line system and develop algorithms for computer vision to analyze 3-D visual information. We will take these experimental results into account and develop an acoustic interface so that the rates of sound image localization are improved, for example, by altering the sound source frequencies for specified locations and directions that are likely to be mis-recognized. We will also develop a user interface that allows task setting by voice, and investigate the influence of virtual sounds superimposed on environment sounds.

References

[1] J. Blauert, "Spatial Hearing (Revised Edition)", The MIT Press, 1996.

[2] T. Ifukube, T. Sasaki, C. Peng: "A blind mobility aid modeled after echolocation of bats", IEEE Trans. BME-38, 5, pp.461–465, 1991.

[3] Y. Kawai, T. Ueshiba, Y. Ishiyama, Y. Sumi, F. Tomita, "Stereo Correspondence Using Segment Connectivity", Proc. of ICPR'98, I, pp.648–651, 1998.

[4] Y. Kawai, M. Kobayashi, H. Minagawa, M. Miyakawa, F. Tomita, "A Support System for Visually Impaired Persons Using Three-Dimensional Virtual Sound", Proc. of ICCHP2000, pp.327–334, 2000.

[5] J. M. Loomis, R. G. Golledge, R. L Klatzky., J. M. Speige, J. Tietz, "Personal guidance system for the visually impaired", Proc. of ASSETS'94, pp.85–91, 1994.

[6] J. M. Loomis, C. Hebert, J. G. Cicinelli, "Active localization of virtual sounds," J. Acoust. Sot. Am., 88, 4, pp.1757–1764, 1990.

[7] P. B. L. Meijer, "An Experimental System for Auditory Image Representations", IEEE Trans. Biomed. Eng., 39, 2, pp.112–121, 1992.

[8] Y. Sumi, Y. Kawai, T. Yoshimi, F. Tomita, "Recognition of 3D free-form objects using segment-based stereo vision", Proc. of ICCV'98, pp.668–674, 1998.

[9] F. Tomita, T. Yoshimi, T. Ueshiba, Y. Kawai, Y.Sumi, "R&D of Versatile 3D Vision System VVV", Proc. of SMC'98, TP17-2, pp.4510–4516, 1998.

COMAP: A CONTENT MAPPER FOR AUDIO-MEDIATED COLLABORATIVE WRITING

Masood Masoodian, Saturnino Luz

Department of Computer Science, The University of Waikato, Hamilton, New Zealand
Department of Computer Science, Trinity College Dublin, Dublin, Ireland

ABSTRACT

Although textual communication is perhaps the most common form of human-to-human interaction over the Internet, other communication channels such as audio are also becoming more widely available as viable means of Internet-based group communication. This diversity of communication media has created a need for effective ways of recording, combining, and retrieving the contents of group interaction which takes place on the Internet in a multimedia fashion. This paper describes an approach to recording, extracting, organising and retrieving the audio and textual material generated during Internet-based group meetings. The aim of the research presented in this paper is to allow mapping and viewing the content of meetings based on the concept of temporal and contextual neighbourhoods. The temporal and contextual neighbourhood concept is currently being further developed through a prototype system called COMAP.

1. INTRODUCTION

Collaborative group work in face-to-face meeting environments often involves creation or discussion of shared documents. Even when documents are not the focus of a group activity, in most cases, some form of a text artefact, such as minutes of the meeting, is created which later on serves as means of sharing the contents of the meeting with those who may or may not have attended it. Such documents generally assist people in working on collaborative activities, for instance by reminding them of the responsibilities they may have undertaken during the meeting, or perhaps by providing them with information without which carrying out their group task may not be possible. It is therefore clear that in many group work scenarios, shared group documents play a central part in the whole group work and interaction process.

Another important part of group work, namely the verbal communication between collaborating team members, is however largely lost except for the parts which maybe recorded in the minutes of the group meetings. Even when minutes are recorded by a team member or a secretary, these end up being just the recollections or interpretations of how the minute-taker may have perceived the meeting, rather than the verbatim content of what was said or discussed during the meeting. However, it must be said that in some cases verbal communication of the meetings are recorded, either using audio recordings, or by means of full content minutes.

A more challenging problem which is faced by the collaborate teams is that even when all the meeting contents are recorded and kept, there are no links between the audio recordings and other shared group documents used, created, or changed during the meetings. Even though some computer applications allow audio annotation of documents, these are generally made only for parts of the documents, and the playback mechanisms are usually crude, requiring the playback of the whole recording to be able to find the necessary parts.

With the rapid growth of the Internet in terms of its popularity, as well as its speed, reliability, and the number of collaborative tools available, more and more people are using it to communicate and work with others across distance. Despite the ease with which the Internet can be used to allow effective group work between physically remote people, it also suffers from all the problems associated with face-to-face meetings in its inability in providing means of combining and retrieving the multimedia contents of meetings.

Development of the Internet tools for supporting group work has mainly been guided by existing research which have clearly shown the importance of shared workspaces in supporting collaborative work (Whittaker et al., 1993). This need has been met, though only partially, through the development of groupware products which allow document creation, editing, sharing, and retrieval synchronously and asynchronously over the Internet (Appelt, 1999; Roseman and Greenberg, 1996; Fitzpatrick et al., 1995). The critical role which the audio communication plays in the effectiveness of group work has also been identified in numerous empirical studies (Masoodian, 1996; Masoodian et al., 1995). As such, other Internet tools have also been developed to support audio communication across distance between collaborative team members. It could therefore be said that using one of the many shared document applications along with an audio link would be sufficient as means of carrying out effective group work over the Internet. Although in most collaborative work situations, specially those involving cooperative tasks (Masoodian, 1996), this is not far from the truth, the post-meeting problem of accessing and relating the contents of the meeting still remains, as is the case with the face-to-face meetings.

The aim of the research described in this paper is to create a computer supported collaborative work system which allows: (i) recording of audio communication between remote users, (ii) creation of shared documents, (iii) automatic indexing of audio recordings based on the contents of shared documents, and (iv) access to audio recordings of meetings through the contents of shared documents, and vice-versa. It should be pointed out that even though the focus of this research is to facilitate content retrieval for Internet-based meetings, many of the techniques discussed here could also be easily adopted to support face-to-face meeting scenarios.

2. ISSUES IN AUDIO AND TEXT INDEXING

Currently there are only a handful of systems that allow creation of links between audio recordings and the work produced during collaborative meetings, though it may be to a very limited degree. Moran et al. (1997) describe a system which supplements written records of a meeting by relying on browsing of hand-annotated audio recordings. Recent research on speech-mediated meeting tools has used spontaneous speech recognition technology to support audio scanning (Hirschberg et al., 1999; Whittaker et al., 1999), automated meeting transcription, and information retrieval from meeting records (Waibel et al., 2001). Most of these recent developments originated from topic detection tasks on broadcast news. While information retrieval from speech data in general constitutes a useful tool in meeting browsing, audio-mediated collaborative writing presents unique characteristics and challenges which are not optimally addressed by standard speech to text techniques. In spite of considerable advances in the last decade or so, spontaneous speech transcription is still far from perfect. Even in perfectly fluent speech, such as that found in news broadcasts, error rates tend to exceed 10%. In meeting dialogues, where false starts, dysfluencies and ungrammaticalities are rule rather than exception, one should expect much higher error rates. In addition, proper names and other "named entities", which are likely to constitute an important class of the information one would need to extract from the audio track, tend to be out of the vocabulary of the recognizer, posing extra problems for a full text-to-speech approach.

Despite such difficulties, it is argued here that collaborative scenarios where both speech and text are used by meeting participants have characteristics that may help produce more productive functionality than traditional information retrieval or meeting transcription methods. These characteristics are described below.

3. TEMPORAL AND CONTEXTUAL NEIGHBOURHOODS

In existing systems the indices for accessing different parts of the recorded audio contents of meetings are generated by using manually inserted timestamps in documents, or by using a speech recogniser on audio recordings. The major limitation in using these methods is that the former relies on manual entry of the timestamps, while the latter utilises the audio contents but ignores shared documents, which often serve as a complementary medium of collaboration.

A meeting support tool, COMAP (COntent MAPper), has been designed to overcome these limitations. COMAP (Luz and Masoodian, 2000) divides the contents of meeting documents into logical sections. These logical sections are then linked to different segments of the audio recording of the meeting using a number of techniques. After the meeting automatic processes run off-line which analyse the content of the shared document and the audio recordings. Therefore the use of COMAP tool can be divided into during-meeting and post-meeting scenarios:

- During a meeting the participants interact with each other using an audio channel and a shared document tool, both of which are provided by COMAP. The system acts as a listener, recording the audio and textual contents of the meeting, as well as a number of interaction events such as the various timestamps.
- After the meeting a user can view the contents of it using a COMAP meeting browser. In this mode however, the system acts as an assistant for browsing the meeting documents along with the audio communication.

209

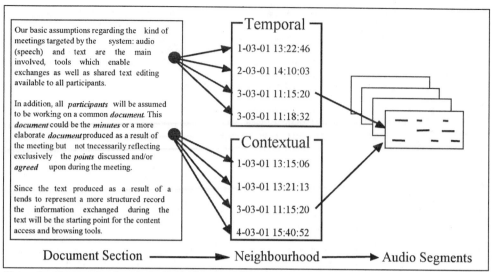

Our basic assumptions regarding the kind of meetings targeted by the system: audio (speech) and text are the main involved, tools which enable exchanges as well as shared text editing available to all participants.

In addition, all *participants* will be assumed to be working on a common *document*. This *document* could be the *minutes* or a more elaborate *document* produced as a result of the meeting but not tnecessarily reflecting exclusively the *points* discussed and/or *agreed* upon during the meeting.

Since the text produced as a result of a tends to represent a more structured record the information exchanged during the text will be the starting point for the content access and browsing tools.

Temporal
1-03-01 13:22:46
2-03-01 14:10:03
3-03-01 11:15:20
3-03-01 11:18:32

Contextual
1-03-01 13:15:06
1-03-01 13:21:13
3-03-01 11:15:20
4-03-01 15:40:52

Document Section ⟶ Neighbourhood ⟶ Audio Segments

Figure 1: Temporal and contextual neighbourhood links between audio and textual contents of meetings

COMAP automatically links the text sections to different segments of the audio recording using the concept of neighbourhood (Figure 1). COMAP neighbourhoods can be defined as follows:

- Temporal neighbourhood: a segment of the audio recording is in temporal neighbourhood of a document section if that audio segment was recorded while the section was being created, changed, or discussed by the participants. There can be multiple audio segments in temporal neighbourhood of a document section, each corresponding to different points in time when that section was active.
- Contextual neighbourhood: a segment of the audio recording is in contextual neighbourhood of a document section when it shares a number of keywords with that section. These keywords are identified automatically by analysing the contents of the audio recording as well as the document. Once again, a document section can have multiple audio segments in its neighbourhood.

4. FROM TEXT TO AUDIO INDEXING

At present COMAP assumes that the shared meeting documents are textual, and that they are created, edited or discussed during the meetings. Since the text produced as a result of a meeting represents a more structured record of the information exchanged during the meeting, text is the starting point for the content indexing and browsing tools of COMAP.

Accessing content is easier in written text than in recordings of spoken language. Browsing through a large text is an activity which takes advantage of the parallel nature of the written medium. Audio browsing, on the other hand, has traditionally relied upon tools which employ a tape recorder metaphor: the user goes through the content sequentially, skipping over parts presumed to be less important by fast-forwarding and rewinding. Attempts have been made at providing a more structured presentation of audio contents of meetings by displaying a graphical view of turn-taking patterns on an interface component based on a "musical score" metaphor (Luz and Roy, 1999). In meeting scenarios such as the one described above, however, the issue arises of how to "salvage" the discussion process that gave rise to a certain section, by indexing textual contents to speech recordings. COMAP supports this process by mapping the textual sections of a document to their temporal and contextual audio neighbourhoods (Figure 1).

Mapping temporal neighbourhoods can be done by timestamping production and manipulation of audio and text. After the meeting, while browsing through the meeting document, the user will be able to select a section and retrieve a stack of audio segments associated with that section. In order to single out text sections and audio segments the system needs to be able to define neighbourhoods in each media. Neighbourhoods in text can be determined via syntactic, punctuation and formatting cues, or explicitly marked up in the way described below. COMAP assumes that each audio segment corresponding to a text section will contain the interval delimited by the first and last continuous textual timestamps for that section up to the first detectable silence before and after these times.

Mapping contextual neighbourhoods requires each section of the text to be processed so that a set of keywords can be identified. Once this has been done, the audio recording of the meeting must be scanned for these keywords. This is an aspect where COMAP presents advantages over existing approaches to meeting browsing based on continuous speech recognition. Unlike full automatic transcription, word-spotting can effectively detect proper names and short phrases in audio recording. Once the named entities and other relevant keywords have been extracted from the collaborative text editor, the audio needs to be segmented according to frequency and time of occurrence of keywords and linked to the text. Identification of keywords is done by using a rule-based part-of-speech tagger and information extraction algorithms.

As mentioned, temporal neighbourhoods require timestamps in both audio and text. This is achieved in COMAP through the use of tools built on the Real-Time Protocol (RTP), in which timestamping is the native mechanism for data packet transmission and synchronization. Audio segments are naturally time stamped due to the sequential nature of the modality.

Keeping track of text creation and modification involves defining the smallest units of data for timestamping. In COMAP, smallest units of text are currently assumed to be paragraphs. The shared document editing tool of COMAP has been designed to use XML as the format for processing and storing the shared meeting documents. Figure 2 shows the XML definition of a COMAP document, along with a short example of how a section and paragraph are annotated with respect to author and modification times. Transmission of text does not require the finer level of synchronisation required by audio multicast packets, therefore larger abstract data units (ADU) have been used in the definition of the COMAP text tool.

```
<?xml version="1.0" standalone="no"?>
<!DOCTYPE COMAP-TEXT [
<!ELEMENT COMAP-TEXT (COMAP-SECTION,COMAP-PARGPH)+>
<!ATTLIST COMAP-TEXT CREATED CDATA #REQUIRED
        CREATOR CDATA #REQUIRED
        LASTOPEN CDATA #REQUIRED>
<!ELEMENT COMAP-SECTION (#PCDATA|COMAP-PARGPH)*>
<!ATTLIST COMAP-SECTION CREATED CDATA #REQUIRED
        CREATOR CDATA #REQUIRED
        LASTOPEN CDATA #REQUIRED>
<!ELEMENT COMAP-PARGPH (#PCDATA|COMAP-PARTICIPANTS)*>
<!ATTLIST COMAP-PARGPH CREATED CDATA #REQUIRED
        CREATOR CDATA #REQUIRED
        LASTOPEN CDATA #REQUIRED>
<!ELEMENT COMAP-PARTICIPANT (#PCDATA)*>
<!ATTLIST COMAP-PARTICIPANT UID CDATA #REQUIRED
        TIMESTAMP CDATA #REQUIRED>
]>
<COMAP-TEXT CREATED="01012001" CREATOR="luz" LASTOPEN="02032001">
<COMAP-SECTION CREATED="01032001" CREATOR="luz" LASTOPEN="02032001">
<COMAP-PARGPH CREATED="02032001" CREATOR="luz" LASTOPEN="02032001">
Discussing COMAP XML design and so on..........
<COMAP-PARTICIPANT UID=luz TIMESTAMP="02032001-10:53"/>
<COMAP-PARTICIPANT UID=luz TIMESTAMP="02032001-12:53"/>
<COMAP-PARTICIPANT UID=masood TIMESTAMP="02032001-12:01"/>
</COMAP-PARGPH></COMAP-SECTION></COMAP-TEXT>
```

Figure 2: XML definition of a COMAP document

211

5. CONCLUSIONS

Effective collaborative group work often relies on the availability of an audio communication channel as well as a shared workspace in which group documents can be viewed, edited, discussed and changed. Although these requirements are important in both face-to-face and remote interaction environments, they are much more critical when people are working together remotely. In recent years tools have been developed to support both of these requirements of group work over the Internet. However, what is still missing is the existence of tools which allow content of the shared documents to be linked to audio recordings of the conversation between group participants during remote meetings.

This paper has described a prototype tool called COMAP which has been designed to facilitate indexing and retrieval of the multimedia meeting contents by dividing meeting documents into logical sections and linking them into segments of audio recordings using the temporal and contextual neighbourhood relationships. These neighbourhood relationships are created and maintained through automatically inserted timestamps and keyword analysis of the contents of both textual and audio components of the meetings.

Further development of the COMAP system is currently underway. It is believed that the ultimate usability of the COMAP tool will be dependant on a number of factors, such as the ideal size of a logical document section (ranging from a word, to a line or paragraph). These factors can only be identified through testing of the system in real meetings.

REFERENCES

Appelt, W. "WWW Based Collaboration with the BSCW System", In proceedings of Current Trends in Theory and Practice of Informatics, SOFSEM'99, Springer Lecture Notes in Computer Science 1725, 1999, 66-78.

Fitzpatrick, G., Tolone, W. and Kaplan, S. "Work, Locales and Distributed Social Worlds", In proceedings of European Conference on Computer-Supported Cooperative Work, ECSCW '95, 1995, 1-16.

Hirschberg, J., Whittaker, S., Hindle, D., Pereira, F. and Sighal, A. "Finding Information in Audio: A New Paradigm for Audio Browsing and Retrieval", In proceedings of Accessing Information in Spoken Audio, ESCA ETRW '99, 1999.

Luz, S. and Masoodian, M. "Mapping Collaborative Text and Audio Communication over the Internet", In proceedings of World Conference on the WWW and Internet, WebNet'2000, 2000, 769-770.

Luz, S.F. and Roy, D.M. "Meeting Browser: A System for Visualising and Accessing Audio in Multicast Meetings", In proceedings of IEEE Multimedia and Signal Processing Workshop, 1999, 489-494.

Masoodian, M. "Human-to-Human Communication Support for Computer-Based Shared Workspace Collaboration", Ph.D. Thesis, The University of Waikato, Hamilton, New Zealand, 1996.

Masoodian, M., Apperley, M. and Frederikson, L. "Video Support for Shared Work-Space Interaction: An Empirical Study", Interacting with Computers, Vol. 7, No. 3, 1995, 237-253.

Moran, T.P., Palen, L., Harrison, S., Chiu, P., Kimber, D., Minneman, S., van Melle, W. and Zellweger, P. "'I'll get that off the audio': A Case Study of Salvaging Multimedia Meeting Records", In proceedings of Human Factors in Computing Systems, CHI '97, 1997, 202-209.

Roseman, M. and Greenberg, S. "TeamRooms: Network Places for Collaboration", In proceedings of ACM Conference on Computer Supported Cooperative Work, CSCW '96, 1996, 325-333.

Waibel, A., Bett, M., Metze, F., Ries, K., Schaaf, T., Schulz, T., Soltau, H., Yu, H. and Zechner, K. "Advances in Automatic Meeting Creation and Access", In proceedings of the International Conference on Acoustics, Speech and Signal Processing, 2001, in print.

Whittaker, S., Geelhoed, E. and Robinson, E. "Shared Workspaces: How Do They Work and When Are They Useful?", International Journal of Man-Machine Studies, Vol. 39, No. 5, 1993, 813-842.

Whittaker, S., Hirschberg, J., Choi, J., Hindle, D., Pereira, F. and Singhal, A. "SCAN: Designing and Evaluating User Interfaces to Support Retrieval from Speech Archives", In proceedings of Conference on Research and Development in Information Retrieval, SIGIR'99, 1999, 26-33.

An Experimental Study on Potentiality of Voice Input Device for Human-Machine Interface in Industrial Plants

Miwa NAKANISHI , Yusaku OKADA

KEIO University

The purpose of this study is to investigate the fundamental characteristics of the operation by voice input device (voice operation). In particular, we examined the way of information choosing in CRT operation experimentally.

Comparing touch operation with voice operation reveals the followings. The required time of voice operation is longer than touch operation. And, variations of subjects' performance of voice operation are also large. Moreover, in voice operation, a subject's emotion effects on his operation much directly. However a subject can memorize his operation and the state of the system by utterance. Thus, a subject's strategy of his performance in voice operation tends to depend on short-term memories.

Hence, the primary consideration in introducing voice input device into the industrial plants were obtained.

1) It should be avoided that voice input device is used alone as a Human-Machine Interface in CRT operation. However voice input device may work effective as a support of the other existing input devices.

2) Voice operation would be appropriate for not-well-trained operators to get skilled.

1. Introduction

Now, more and more advanced automation plants adopt key or touch panel as a human-machine interface. While human-machine interaction becomes more flexible by such interfaces, it is also a fact that the new burden is brought to an operator's cognition or process of thinking. For example, complicated structure of information is the noteworthy burden in CRT operation for the operator whose main task is surveillance. Such a burden is the potential factor of human error, in particular, for operators with low skill level. However, because of restrictions of a screen space, it is difficult to improve these problems under the existing interfaces. Therefore, strengthening operators by education and training is the present method to ease the problems.

Then, in this study, we focus on the potentiality of voice input device as an interface, which is not caught by restrictions of a screen space, and supports education and training. The operation by voice input device (voice operation) is examined by comparing with the operation by touch panel (touch operation). Voice input device is said to have some remarkable characteristics such as making remote operation possible and enabling the aimed information to turn out by only one call. In this study, further, it is clarified how voice operation influences an operator's cognition by constructing an information processing model. And we will consider the introduction of voice input devices into CRT operation in industrial plants.

2. Method

In the experiment, the simple simulator which has the functions of changing pictures and changing components was programmed. In order to compare touch operation and voice operation, the same task was performed by the two different interfaces. Voice input device 'Speachnavi' was used here. The picture on CRT was divided into four as shown in Fig. 1. The buttons for changing components are placed on left-hand side, which are constituted from every ten pieces each by A and B. The problem number of the experiment, and the button for the completion of an experiment are shown in the lower right. The window displaying logic pictures, and the buttons for change of pictures are placed on the upper right. The initial state of each component was either failure or off. Failure button was yellow and off button was blue. A subject's task is displaying logic pictures of all the given problem numbers, choosing necessary components, changing the components of off to on, and making only one course passable. The component turned to on was displayed in red. For one problem, a subject has to make courses of both A and B passable. When passage of all the courses satisfying each problem is attained, a subject judges one experiment to be completed. However, since some components are out of order, a subject may judge that no course can be passed. Problem numbers and unusable components are given in various numbers and combination for every experiment. Here, in order to carry out comparative evaluation, 30 kinds of patterns were prepared beforehand. 15 healthy students who are from 21 year-old to 24 year-old performed the experiment every 30 times each by touch operation and voice operation. In voice operation, some words which would be used in the experiment were registered into the

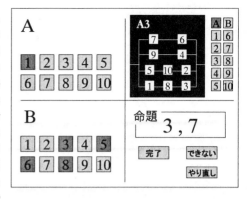

Fig.1 A picture design of the experiment system

system so that control can be performed with the words which don't have sense of incongruity for the subjects. Action of a subject during experiment was recorded with time as an operation log.

3. Result

3-1 The analysis of the required time

Fig.2 shows the distribution of the required time when all subjects were given the same problems. In the many case, while about 50 seconds were required in voice operation, about 30 seconds were required in touch operation. It is because utterance for every control takes several seconds in voice operation. Moreover, the variation of the required time in voice operation is also larger than that in touch operation. It explains that each subject has different speed for speech. Thus, in voice operation, it was found that everyone could not necessarily control it at the same pace.

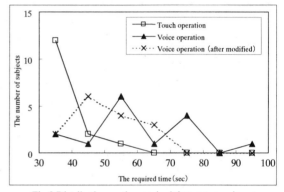

Fig.2 Distribution on the required time an experiment

The following peculiar phenomena were seen during the experiment in voice operation. One is non-recognition, that control is not completed because the system does not recognize the word. Another is incorrect-recognition, that the input different from a subject's intension is brought into the system because the system recognizes the word as one different from the word which a subject spoke actually. When non-recognition happened, the subject had to speak again and again until control was completed. When incorrect-recognition happened, another input is needed for recovery of the incorrect control. Such things are also the causes which enlarged the required time and the variation in voice operation. Although the present voice input device has high ability of recognition to some extent, it is still weak to delicate disorder of utterance, such as a break and rhythm of utterance, too loud/small voice, and a noise by breathe. Furthermore, since utterance is greatly dependent on an operator's mental elements, such as a lack of confidence, impatience, and anger. However, it is expectable enough that voice input device develops in respect of hardware further from now on. Therefore, if it can recognize the utterance in various states of mind of the human being of indefinite a large number more correctly, it will be used effectively. For example, the data after modified in Fig.2 shows the distribution of the required time which removed the time for non-recognition and incorrect recognition from the required time of the voice operation. If the technical fault with voice input device and disorder of the utterance is held down, it is expected that voice operation be attained in time near touch operation.

3-2 The analysis of human error

The process from looking at a logic picture to turning a certain component can be extracted as one with some cognitive activities. The human error in this process is observed and considered.

First, compared with touch operation, human error of a consciousness was not observed so much in voice operation. When a button is touched, a part of screen is covered by a subject's hand. On the other hand, since voice operation doesn't need to use hand, there is no factor which prevents a subject from looking at CRT. Therefore, there was no human error such as overlook of the covered information. Moreover, there is a tendency that an operator's eyes follow his finger in touch operation. Therefore, in case the picture design needs movement of a viewpoint in operation, short-term memory is required. Contrastively, since an operator's viewpoint doesn't concentrate on his finger in voice operation, there is little influence by the design of the picture.

On the contrary, typical human error of voice operation is the incorrect input such as a mistaken word and poor input, that causes non-recognizing and incorrect-recognition of the system. Speaking the wrong word is as of the same kind human error as touching the wrong button, and it is caused by incorrect judgment, misapprehension, and carelessness. On the other hand, poor input is the human error peculiar to voice operation, and it is caused by utterance beyond the range in which a system can recognizes the word correctly. The individual difference is not only large, but voice is influenced by the mind. Touch operation is indirectly affected through judgment or thinking rather than the input itself is directly affected, even if a subject touches a button under unusual mind. However, in voice operation, impatience or the lack of confidence appears in the strength of voice, a break, disorder of a breath, and so on. And, subject's mind changes the input. Under the situation, even when judgment and thinking are right, incorrect-recognition and non-recognition happen to the system. Thus a gap is produced between an intention of a subject and an output of the system. Therefore, a subject might think that the action of a system is contrary to his command, and the distrust over the system arises. In fact, the following comments were heard from subjects after the experiment; "Since the displayed logic picture was different from what I spoke, I suspected the equipment is out of order", "I was perplexed in the beginning and gradually got angry with the equipment which did not recognize the word correctly even if I speak repeatedly" and so on. In this way, once a subject's mind gets worse further, he may talk in loud voice, or may sigh, and may attach excessive prefix and suffix. It was also connected with incorrect-recognition and non-recognition, and a vicious circle of operation was produced. Besides it, there was a case that the subject spoke in loud voice relatively since the circumference was not quiet. It also caused the poor input. Furthermore, since the error by the poor input is thought for the subject to have carried out the right input

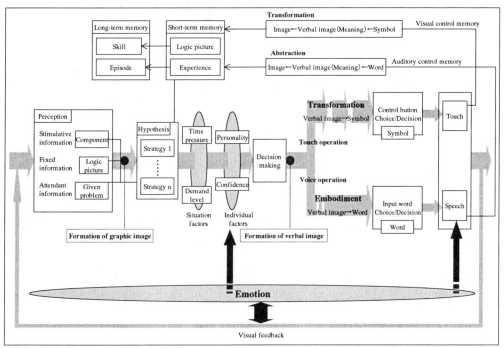

Fig.3 The information processing model

215

apparently, a subject seldom notices the produced incorrect operation. In the case, a subject went on the operation without recovering his mistake, and then, the whole experiment got confused further.

The followings were found as a result of examining human error. Voice operation has the advantage that operation can be continued without an operator's view being barred. On the other hand, it is the greatest difficulty that an operator's emotions influence his operation directly. Emotions are the state of the heart which cannot be described, i.e., impatience and displeasure, anger, and so on. It is difficult for man to be aware of it or to control it. Therefore, once the human error caused by emotions arises, the human error also transfigures emotions and becomes the potential factor of another human error. Especially, voice operation tends to lapse into the vicious circle which worsens the relation between a system and an operator.

From the above consideration, we constructed the information processing model as shown in Fig.3. It expresses the cognitive characteristics of touch operation and voice operation.

Actual plant operation is classified roughly into two scenes. One is the usual case in which operators are mainly engaged in monitoring. The other is the urgent case in which active intervention of operators is needed. Although mind of the operator is comparatively quiet in usual, it turns around to an unstable state in the urgent case. Furthermore, under the situation, a surrounding atmosphere may make noisy. Taking these things into consideration, especially voice operation is not desirable in the urgent case because it depends on an operator's emotions and surrounding situation. Moreover, even if it is the usual case, an operator's mind is not always stable and it is also difficult for an operator to know his own mind. Therefore, it is safe to use voice input device in the scenes which are not critical, such as change of pictures in monitoring. Moreover, it is also necessary to make an operator know that he can always recover it by other interfaces when unexpected error arises in voice operation.

3-3 The strategy in a recovery task

If a subject tries to end an experiment without solving the given problem, a system notifies of an error. In this case, a subject has to turn a component correctly with reference to some logic pictures corresponding to problem numbers. This is a recovery task. Then, our attention is directed to the process of information retrieval for the recovery task. Here, recovery tasks are classified into three cases according to how many times the logic pictures were referred to. It is shown in Fig.4. For example, when a

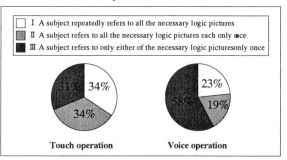

I A subject repeatedly refers to all the necessary logic pictures
II A subject refers to all the necessary logic pictures each only once
III A subject refers to only either of the necessary logic pictures only once

Touch operation   Voice operation

Fig.4 Classification of recovery task

subject receives the response "the state of the component of A is not right" from a system in the experiment which gives two problems of 3 and 7, the recovery tasks are classified as follows. I is the case where the subject repeatedly refers to the logic pictures of A-3 and A-7. II is the case where the subject refers to these each only once. III is the case where the subject refers to only either. In case I, the subject remembers little about the past control, and logic pictures. Then, since the subject tends to take much time in discovering his error, I is not desirable case. In case II, although an error can be discovered without retrieving too much information as a result, the subject didn't have enough the memory which helps him to think. On the other hand, in case III, the subject was able to omit a part of process of information retrieval since he had memorized the contents of the past control, and logic pictures. Thus, this is the desirable case. According to Fig.4, in touch operation, there is much number of case I and case II. On the other hand, in voice operation, case III occupies the large rate. From this result, the followings can be reasoned about a recovery task of voice operation. The subject may have immediately found out the doubtful logic picture without re-referring to all the corresponding logic pictures in order, since he had an idea about the scene where the error was committed. The human error such as trying to finish an experiment without solving all the given problems was caused by a subject's forgetting to refer to a logic picture, referring to an incorrect logic picture, forgetting to turn a component, or turning an incorrect component. In the voice operation performed by means of the words, he may have memorized something performed just before without his particular conscious. According to the

216

memory, the plan for discovering the error may have been decided. Moreover, since the subject had memorized some logic pictures seen before the recovery task, he may have omitted referring to a part of logic pictures. Of course, also in touch operation, while the subject repeated the experiment, he memorized logic pictures to some extent. However, since the subject feels it the same to touch any button, touch operation is monotonous. Thus, the subject feels weariness and the decline of his concentration. Contrastively, since the subject can get feed back of his operation by hearing in voice operation, he can escape the fall of his concentration.

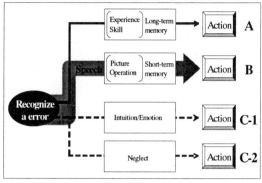

Fig.5 Strategies in operations for recovery task

The followings were obtained from the above. In voice operation, it is easy for a subject to memorize the performance which he did just before. Thus, the plan for the next task tends to be influenced by the last control.

Fig.5 shows the strategy in which an operator determines the following action after recognizing an error. As for the strategy of the recovery tasks in present plant operation, the tendency changes with an operator's skill level. The strategy of an enough skilled operator tends to be the pattern (A) depending on long-term memory, such as experience and knowledge, in many cases. On the other hand, the strategy of an operator with low skill level tends to be the pattern (C-1) depending on intuition or emotions, or the pattern (C-2) of giving priority to the ease of search in many cases. However, the strategy in voice operation may be strongly guided to the pattern (B) not depending on an operator's skill level but depending on the short-term memory of just before operation. Naturally, the pattern (A) is the optimum strategy for recovery tasks. Therefore, it is not necessarily desirable for the enough skilled operator that the strategy (A) is transposed to (B) depending on just before operation. Contrastively, it is desirable for the operator with low skill level to have clearer plan for the task by the strategy (B) based on short-term memory instead of the unstable strategy (C-1) or (C-2). In addition, the knowledge may be soon fixed as an operator's long-term memory by the repetition of the pattern (B) based on short-term memory.

From the result, it is obtained that voice operation has great potentiality for enhancing an operator's memorizing and studying. Therefore, voice operation may work effectively for the operator with low skill level. It is expected that the efficiency of training of operation be increased.

4. Conclusion

This research compared and examined voice operation and touch operation through the experiment. Furthermore, a potentiality of introducing a voice input device into practices was also considered. Consequently, followings were obtained as characteristics of voice operation.

First, it is difficult to control the system quickly. Besides, the control speed depends on an operator's individual characteristic greatly. Moreover, an operator's emotions are directly reflected in his operation. On the other hand, an operator tends to memorize his operation and the state of the system because he speaks his operation itself. To take such characteristics into consideration, followings can be suggested about operation in industrial plants.

In the first place, voice operation it should be avoided that voice input device alone is used as a human-machine interface like touch operation and keyboard operation. In particular, it is not desirable to use in the scene where the state of mind of operators is unusual, such the urgency. Therefore, it is recommended to introduce voice input device as assistance of the existing interface. In the second, the voice operation's effect of short-term memory and study is effective for the operator with low skill level. Therefore, it is recommended voice input device is utilized for the increase in efficiency of training and education.

References

Nakanishi,M., Okada,Y.: A Study on Control to Change over a State of a System by Voice Operation, Journal of the Society for Industrial Plant Human Factors of Japan, 6(1), in printing, 2001.

A Voice-Based Java Programming Environment

Angela Bowers, Clement Allen, Martina Davis

Florida A&M University
Computer and Information Sciences
Tallahassee, FL 32307
{abowers, allen, davism}@cis.famu.edu

ABSTRACT

Today's computer users want more instinctive interactions with their computers. Using speech within applications is one way of meeting this need. The tool we developed is a voice programming environment for creating Java programs. In addition to being able to directly type a program in the tool, it allows the user to speak common programming tasks. A user study was conducted as an introductory way to determine the feasibility and practicality of using speech during programming. We have shown that voice programming is a practical and feasible complement to programming in an integrated development environment (IDE).

1. INTRODUCTION

Savvy computer users of today want more intuitive and natural interactions with their computers. Spoken language dialogue systems provide one way to meet the wants of these users. Spoken dialogue systems allow the user to interact with the system using speech to accomplish some purpose. Some example systems that use spoken dialogue are: The Philips Automatic Train Timetable Information System described [2], and the ARISE project [4].

An important issue for today's computer professionals is the field of computer programming. Generally speaking programmers develop programs in one of two ways: a text editor, or integrated development environment (IDE). The editor provides no support for typical tasks when programming. For example, a program may not be compiled in the editor, but must be compiled elsewhere. The goal of an IDE is to package many functions associated with developing a program in one place. So, while some, if not most of the program is still typewritten, other services are also provided. Some example IDE functions include code compilation and templates for programming constructs. The purpose of these IDE functions is increased productivity for the programmer.

Even if an IDE is used to program, the programmer generally interacts with the programming environment in a limited number of ways. The majority of the program is typed using the keyboard, with the mouse being used occasionally (in some programming environments) to carry out some menu function. For those developers who are not proficient typists, the "hunt-and-peck" style of typing diverts attention away from the task at hand—designing and writing a correct program. Moreover, for those that have limited use of their hands (i.e. some physically challenged persons), programming can become a frustrating, if not impossible proposition. A natural resolution to this problem is to look at using speech during the programming process. A programming environment that would translate the thoughts (as evidenced by spoken words) of the programmer into a program would be a viable solution. While there is some research in voice programming, there is limited research dealing with multimodal IDEs that include speech.

There are several issues that must be considered when discussing voice programming as a solution to these concerns. Will multiple modes be beneficial to the programmer? Will the use of speech allow the programmer to more effectively think about the task of writing a program? Or will the speech commands be more things to remember, and detract from programming? Also, would the programmer primarily use one mode (i.e. do all programming in speech)? Answering questions such as these are necessary to determine the practicality of this system.

The remainder of the paper will be organized as follows. The proposed system, which will allow for voice-based programming, will be introduced in section 2. In section 3, a user study of the system will be presented. This study attempts to determine how this voice-based programming environment will affect the efficiency, accuracy, and effort expended of programmers using the system. Finally, section 4 will summarize the paper and explore possible future research work.

2. A VOICE-BASED JAVA PROGRAMMING ENVIRONMENT

As described in section 1, voice programming will be explored as it relates to all programmers. This section will list some goals of this system, and describe the system features in greater detail.

2.1 Goals

We have identified several goals that this system should meet in order to answer the concerns of those programmers who would use this system.

- Multi-modal: Our application will allow speech as input, as well as the keyboard and mouse. Allowing multiple modes as input gives the user flexibility, in that the user can choose exactly how she wishes to interact with the computer at any given time. The user chooses the best modality for the task.
- Independent of Level of Expertise: The system should have the ability to be used by novice and expert users, and it should be able to take into account the special needs of each type of user. This goal of usefulness to any user level can be viewed in two areas: level of expertise in programming, and level of expertise in spoken dialogue systems. As it relates to programming, the system should not unnecessarily slow down an expert programmer by being overly helpful for tasks in which the expert is already proficient. For novices, the system should not be unduly complex, and should provide help options. As it relates to spoken dialogue systems, the system should be largely intuitive to a novice user, using natural language. The user should not have to fumble about to choose a word to perform some task. For an expert user in spoken dialogue systems, the system should not be significantly different from other spoken dialogue systems in function.
- Intelligent: The system should be able to assist in the programming process, and not passively respond to user commands. Firstly, the system should allow for the identification of problems in programs before compile time. Secondly, the system should be able to help the programmer in translating her thoughts to program code. For example, if the programmer said, "initialize the values in the array to zero," a program segment would be generated which would perform this function.

2.2 Features

The tool we developed is a voice-based IDE for creating Java programs. The functions of the systems can be categorized into three areas: programming constructs, edit functions, and file functions.

- Programming constructs: Programming language constructs are the building blocks for a program. This tool will create a template for each of the main programming constructs in the Java programming language. The following Java programming constructs are supported in the tool: class, interface, method, field, loops, decisions, exception handling, and branching.
- Edit functions: Edit functions allow the user to manipulate pieces or sections of the program. The edit functions include: cut/copy/paste, undo/redo, find/replace, and move (moves to a given position in the program).
- File functions: File functions allow the user to manipulate the entire program. The file functions include: compile, run, new, save, open, and print.

2.3 Using the System

The system is very similar to a program text editor in that the program may be typed directly into the tool. However, speech commands allow for more efficient execution of the commands a user may wish to do. A representative of each type of system function (programming constructs, edit functions, file functions) will be described later.

Programming constructs: This includes all constructs in the Java programming language. When the user wishes to create a new programming construct, the user should speak the appropriate command. As an example, to add a new class, the user would say, "add class." A template for a new class would appear in the programming area as shown in Figure 1. She would then say, "add method." A template for a new method would then appear in the programming area as shown in Figure 2.

Edit commands: As stated earlier, edit commands allow a programmer to edit sections of the program by speaking the appropriate command. Given Figure 2 as a starting point, if the user, for example, wants to undo the last thing that was done in the program, she would say, "undo," and the program screen would look like Figure 1.

File commands: File commands allow a programmer to manipulate the entire program. If the user wants to open a Java program in the tool, she would say "open," then a dialog box would come up, and the user could select the file to open.

As stated before, each of the commands that can be done using speech can also be done without speech, that is, with the keyboard and mouse. For file and edit functions, the user can use the menu to do the same commands. For programming constructs, the user can use the menu, or she can type the programming construct directly into the programming area.

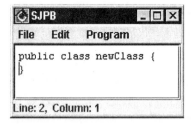

Figure 1: Screen after user speaks "add class."

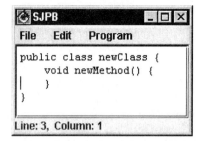

Figure 2: Screen after user speaks "add method."

2.4 Example

To illustrate how a person might use the system, we will go through a simple eight-line program that prints all the numbers from 1 to 100 and their squares to the standard output. The following are the steps the user would take to create the program.

1) Say "add class"; **2)** Move the cursor to the class name and change the name of the class to Squares (see Figure 3) **3)** Position cursor inside class and say "main"; **4)** Position cursor inside main method and say "for"; **5)** Position cursor inside for loop and say "add field"; **6)** Type "i<=100" and "i++"; **7)** Type "System.out.println("i, i*i")" (see Figure 4)

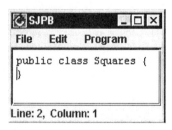

Figure 3: Screen after completion of step 2.

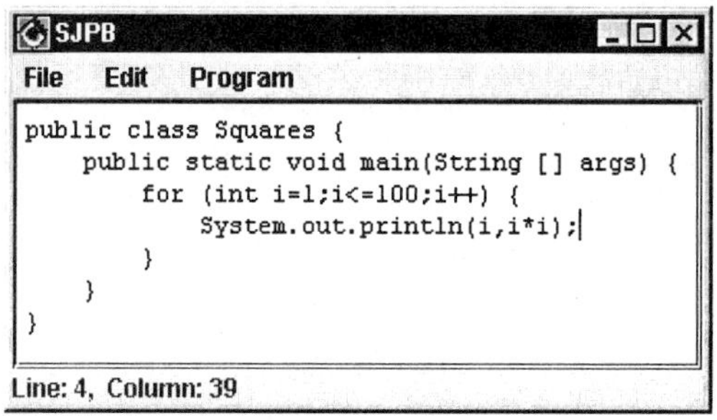

Figure 4: Screen after completion of step 7.

2.5 Architecture

Figure 5 gives a visual representation of the architecture of the system. The tool is built on JSBB [1]. JSBB is a visual tool for modeling spoken dialogue systems. In JSBB, a developer of a speech application can model the dialogue that an application will undergo in interacting with the user, and JSBB will generate the code to support speech in the application. JSBB, in turn, uses the Java Speech API classes in order to converse with the speech recognition engine.

3. USER STUDIES

A user study was conducted as an introductory way to determine feasibility and practicality of using speech during programming. It also examines the effects of speech on the programming process.

3.1 Procedures

There were 9 volunteers that served as test subjects for the study. All participants were university students with some level of Java experience. Each participant was tested separately. The procedures for the study were as follows: First, each test subject was explained the purpose of the experiment. Second, the subjects were given a demonstration of the tool. Third, the speech engine was trained to recognize the subject. In the fourth step, the actual testing was performed. There were two different test situations for the participants. Eight of the participants were given three Java programs and asked to use the tool to reproduce the code. One of the programs is shown in Figure 6. In the last step of the testing, a survey was given to each of the participants in order to ascertain their opinions about using the tool.

Programming Tool
JSBB
Java Speech API
Speech Engine

Figure 5: Voice Programming Tool System Architecture

```java
public class Prg1 {
    public static void main (String args[]) {
        int avg;
        int sum=0;

        for (int i=0;i<=100;i++) {
            sum = sum+i;
        }
        avg=sum/100;
        System.out.println("The average of the numbers"
            +"1 to 100 is: " + avg);
    }
}
```

Figure 6: User study program

4.2 Results

The results of the experiment are shown in Table 1. The following metrics were measured for each program.

- Average time elapsed: The amount of time (in minutes) from the first command, keystroke, or mouse click to the correct completion of the program.

Average number of failures: The average number of times the system failed to respond correctly to a spoken command. This includes the scenario of the user not recognizing a word the user spoke, as well as not responding at all to a spoken command.

- Average number of times menu used: The average number of times the menu was used instead of speech
- Average number of words spoken: The average number of words spoken for the program.
- Number possible commands spoken: The total number of programming language commands that could have been spoken.
- % possible commands spoken: The percentage of the total possible programming language commands that the participant actually spoke.

Program	Avg. time elapsed (min)	Avg. # failures	Avg. # times menu used	Avg. # words spoken	# possible commands spoken	% possible commands spoken
1	6.222	2.222	0.333	5.667	6	94.444
2	6.750	2.714	1.375	7.286	12	60.714
3	6.875	3.714	0.625	13.286	21	63.265
all programs	6.600	2.826	0.826	8.478		72.808

Table 1: User Study Metrics and Results.

The average time taken to input each program was between six and seven minutes. When looking at this data, it can be observed that on average, less than 3 words were misunderstood by the system. Also, there was an average of less than 3 failures for each program. The menu, which is an alternative to the speech commands, was used for an average of less than 1 time. As evidenced by the data, speech was used almost three-fourths of the time when it was possible to use voice commands.

After the study, each participant was given a survey to complete. Table 2 gives the questions and results of that survey. Questions 1-5 and question 7 were on a scale of 1 to 5, 1 = never, 5=always. Question 8 is on a scale of 1 to 5, 1 = beginner, 5=expert. Question 9 is on a scale of 1 to 3, 1 = no experience, 2 = used some spoken dialogue system, 3 = developed a spoken dialogue system. As seen by the results, the participants overwhelmingly found the tool intuitive and easy to understand. The participants indicated that the tool understood their speech commands most of the time. They also found it relatively easy to program using the tool. The tool was rarely slow to carry out their commands. All in all, 78% of the participants would use the tool to create Java programs. The average level of Java experience is 3, which means that the average participant has taken a Java class, and has done some additional programming. The average level of spoken dialogue systems experience is 1.33, which denotes someone who has no spoken dialogue systems experience.

Survey question	Avg..	Survey question	Avg.
1. Was the tool easy to understand?	4.77	6. Would you use this tool?	78% yes
2. Did the tool understand what you said?	4.22	7. How often was the tool slow?	1.77
3. Was it easy to program using the tool?	4	8. What type of Java Programming experience?	3
4. Did you know the commands to use?	4.55	9. What type of spoken dialogue experience?	1.33
5. Did the tool work the way you expected?	4.11		

Table 2: User Survey Results

3.3 Conclusions

Although the results of this study are preliminary because of the relatively few number of test users, there are several conclusions that can be reached from the user studies. Most importantly, voice programming is a practical and feasible alternative to programming in a text editor or IDE. The low amount of complete failures and misunderstandings of speech commands reinforces the feasibility of using voice as an input modality for programming. Also, the participants in the study used speech in a majority of the situations in which it was possible to use speech, even when presented with other options. Speech was seen as a viable alternative to typing the information directly into the tool. As evidenced by the comments made by the participants, even more common commands should be implemented. For example, many users looked for a speech command that would generate a print statement in the code (i.e. System.out.println("hello"). The participants found the tool more than adequate in terms of ease of use, intuitiveness, and response time. The majority of the participants would use a tool such as this for programming. Also, the majority of the participants were beginning users as it relates to spoken dialogue systems, and they had no problems understanding how to program using speech and the tool.

4. SUMMARY

We examined the ways programmers traditionally develop programs and observed the fact that inputting programs is traditionally limited to the keyboard and mouse. For any person that has problems with these methods of input, programming may be an exasperating task. Our approach to this situation was to explore voice programming as a complement an integrated development environment. In addition to typing or using the mouse to write a program, speech may be used to carry out tasks in the programming process. From our data, it is usable by both novices and experts to programming and speech technology.

Our future work will involve improving the system so that it understands more programming constructs. Other common commands can be implemented in the tool for even more efficient programming. Other ideas for future work will involve auto-tracking and extending the tool in order for programmers to logically group or associate different classes. In addition, we plan to add more intelligence to the tool. One goal is to allow a user to speak high level programming tasks, for example, "sort this array." We would also like to test the tool on physically challenged persons in order to observe the effectiveness for this population of users.

5. REFERENCES

[1] Allen, C., Stoecklin., S., Bobbie, P., Chen, Q., Wu, P., "An Architecture for Designing Distributed Spoken Dialogue Interfaces", Proceedings of the 11th IEEE International Conference of Tools with Artificial Intelligence, Chicago, IL, 1999.

[2] Aust, H., Oerder, M., Seide, F., and Steinbiss, V. "The Philips automatic train timetable information system". Speech Communication 17, 249-262, 1995.

[3] C. Bettinin, S. Chin., "Towards a speech oriented programming environment", IEEE Region 10 Conference on Computer and Communication Systems, Hong-Kong, September 1990.

[4] Sanderman, A, Cremers, A., Boves, L., and Strum, J. "Evaluation of the dutch train timetable information system developed in the ARISE project", Proceedings of IVTTA, pp, 91-96, 1998.

Feedback Requirements for a Direct Voice Input System

Philip S. E. Farrell, Mike Perlin, and Ghee W. Ho

Defence and Civil Institute of Environmental Medicine, 1133 Sheppard Avenue West, Toronto, Canada, M3M 3B9

ABSTRACT

Direct Voice Input (DVI) has been identified as an alternative control method for computer-based systems within military aircraft. Past studies revealed that DVI affords heads up time, which positively influences safety and situational awareness. DVI also requires new crew coordination procedures that effectively redistribute workload. In this study, DVI feedback requirements were investigated for helicopter operations. Perceptual Control Theory was used to determine information requirements, and an experiment was performed to determine the effectiveness of various feedback forms. Data trends show that the heads up and heads down visual displays perform similarly, and both are slightly better than audio feedback in terms of the speed and accuracy of assessing the feedback.

1. INTRODUCTION

Advances in cockpit technologies will continue to impact operations within military aircraft and affect procurement strategies. Moving map displays, helmet mounted symbology, and flight management systems are three examples of technologies that may require alternative interfaces in order to ensure safe and effective human-computer interaction. Direct Voice Input (DVI) is being considered as an alternate control interface in the Canadian Forces Griffon Helicopter cockpit (tactical and transport missions) to allow pilots to interact with new aircraft systems while keeping their hands on the cyclic and collective.

Voice recognition software converts sound patterns into a form that the computer recognizes. Kobierski and Swail (1997) resolved one technical challenge and reported 99% recognition rates in the noisy Griffon Helicopter. Integrating DVI in a busy workspace generates human factors challenges as well, such as grammar development, multi-task interference, attention demands, and crew coordination procedures, which have been addressed in several studies (Farrell et al., 2001).

This paper investigates the form and content of feedback for operating the communication functions of the Control Display Unit (CDU) with DVI in the Griffon multi-task environment. The CDU provides an interface to the navigation and communication systems in the Griffon Helicopter where radio frequencies and way-point details are entered. The CDU has a 5x5" CRT display with soft keys and a customized keyboard. Information is organized in a nested menu structure. This nested menu structure and the CDU location in the cockpit (aft of the center console) requires heads down time to read and enter data, causing attention to be shifted from outside to inside the cockpit. DVI may possibly reduce workload by minimizing head and eye movements, and increase situational awareness by maintaining the point of regard out of the window.

Feedback information requirements for the proposed DVI system were generated using a Perceptual Control Theory (PCT) framework of human-computer interaction. The effects of type of feedback were explored experimentally using Signal Detection Theory (SDT). If DVI were the only task, then the optimal form for the feedback could be predicted from what is known about how interface modalities interfere with each other (Herdman and Beckett, 1996), however, the DVI load is added to existing sensory, cognitive, and psychomotor loads in the helicopter environment.

2. FEEDBACK INFORMATION REQUIREMENTS

A PCT approach (Farrell & Chéry, 1998) was used to derive the potential feedback information requirements. In this case, the approach begins with a Griffon Helicopter composite scenario. A section of the scenario involves communication with an outside agent using the aircraft radios. The pilot sets up the communication link by choosing the proper radio and frequency.

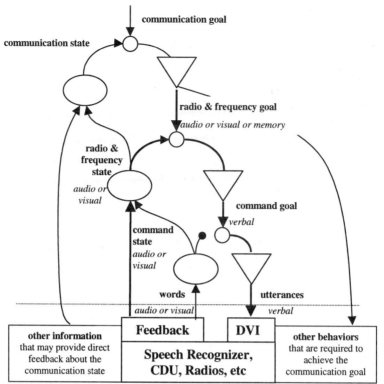

Figure 1. A PCT model of the proposed DVI system showing the feedback information requirements

Next, a goal decomposition of the scenario section is performed as illustrated in Figure 1. The highest level goal is to be in a desired communication state. One of many sub-tasks or behaviors needed to achieve the goal is to set the desired radio and frequency. Behaviors become sub-goals at the next level down in the PCT paradigm. Therefore, the desired radio and frequency becomes a sub-goal that is further decomposed into the command goal and utterance behavior. The world then changes state and sensory information is generated about the words, command, and radio and frequency, and this feedback is used to generate internal perceptions about the command, radio and frequency, and communication states, respectively. Each perception is compared to its goal, thus closing the control loop.

The PCT model provides design guidance. For example, all abstraction levels may be controlled when manual input is integrated with DVI, which may be necessary as the vocabulary increases. Also, the model suggests that explicit feedback for all levels will reduce the cognitive integration loads. The interface designer must decide whether to display each aspect of the feedback information explicitly, or leave it up to the human to integrate the information and generate the higher level perceptions. A design decision was made to provide feedback for the command state only for the following experiment. No direct control over the commanded words was possible (indicated by the broken line), unlike manual input with delete and insert capability. Thus, the entire command was repeated if an utterance error was made or was not recognized. Also, participants did not control for the communication goal, and so no other information was given to help generate a perception about the communication state. The proposed feedback display was based on a DVI Action-Target-Parameter structure (Churchill & Herdman, 2000). For example, if the command was "set radio 1 to 118.5" then the feedback display would be "radio 1 is set to 118.5", which contained information about the command and radio and frequency state.

3.   FEEDBACK FORM REQUIREMENTS

3.1 Purpose

The italicized text in Figure 1 indicates control and display modalities of the DVI system. Other aviation, navigation, and communication tasks will also use audio, visual, memory (cognitive), and verbal processes, and potentially interfere with the DVI task. The purpose of the experiment is to determine any performance differences due to interference effects between the DVI feedback forms in a Griffon Helicopter multi-task simulated environment. The results will be used as advice when integrating DVI in a fielded system.

3.2 Experimental Design

In the PCT model, goal achievement is possible only after the feedback information is perceived correctly. The user must pay attention in order to detect any errors in the feedback. If the user is occupied with other tasks then there is the potential to perceive *no error* when one exists (miss), or perceive an *error* when none exists (false alarm). "Signal Detection Theory (SDT) is applicable in any situation in which there are two discrete states of the world (call these signal and noise) that cannot easily be discriminated" (Wickens, 1992). In this case, an *error* and *no error* are the two discrete states of the feedback. Since error detection is the variable under investigation, the *error* has been defined as the signal and *no error* as the noise. Table 1 outlines the four possible outcomes using SDT terminology. The performance measures for comparing different feedback forms are the detection accuracy - the user's ability to detect an error - and the detection time.

Participants used DVI to set communication links, and were given correct or erroneous radio and frequency feedback. Probability data were generated for a Receiver Operating Characteristic (ROC) curve which is a plot of the Hit probability, $P_H = f_H/(f_H + f_M)$ (f - frequency), versus the False Alarm probability, $P_{FA} = f_{FA}/(f_{FA} + f_{CR})$ over a range of theoretical biases to the Signal-to-Noise Ratio (SNR). That is, if SNR is zero (no signal) then subjects will tend to respond with "no error" and $(P_H, P_{FA}) \rightarrow (0,0)$. Similarly, if SNR is infinite then $(P_H, P_{FA}) \rightarrow (1,1)$.

Table 1. SDT Terminology for experiment.

Participant's Assessment	Signal (error)	Noise (no error)
"error"	Hit (H)	False Alarm (FA)
"no error"	Miss (M)	Correct Rejection (CR)

Table 2. Feedback form conditions for each flight and SNR

	Condition 1	Condition 2	Condition 3	Condition 4
PNF	HUD/CDU	VO/CDU	HUD/VO	CDU
PF	VO	HUD	CDU	HUD/VO

The ROC curve shows whether there are feedback form sensitivity (d') differences. That is, a feedback form will be more sensitive to a correct response if its curve is closer the upper left corner of the ROC curve (i.e., high P_H and low P_{FA}) than another feedback form's curve. In addition to the SDT analysis, the detection time is used to determine performance differences. The detection time is defined as the time difference between the onset of the feedback and the participant's response.

The experimental design involved four manipulations: feedback form, SNR, crew position, and mission segment. Combinations of three feedback forms were investigated: a visual heads up display (HUD), an audio voice output (VO), and a heads down display (CDU). The CDU was included since it is likely to remain in the cockpit regardless of the study's conclusions. The SNR values of 1:3, 1:1, and 3:1 were chosen so that empirical data might be compared to the theoretical ROC curve. Since the Pilot Flying (PF) and the Pilot Not Flying (PNF) perform different tasks, the crew position may affect the feedback form performance. Similarly, different tasks were executed in different flight phases, and so hover and cruise were chosen for the mission segment manipulation.

The experimental conditions are given in Table 2. Note that the CDU appears once per condition due to computer processing limitations. The PNF was given the CDU whenever possible, which reflects operating procedures in the Griffon Helicopter. Participants worked as a crew and completed a total of 96 DVI commands during the hover and cruise segments of each flight. For each of the four display conditions, participants switched crew position and repeated the flight. The crew repeated these eight flights at the three different SNRs for a total of 24 flights.

3.3 Apparatus

The experiment is performed in the Aircraft Crewstation Demonstrator (ACD), which is a fixed based, low–fidelity, rapidly reconfigurable simulator for investigating new cockpit technologies with the human-in-the-loop. The ACD was configured to represent a Griffon Helicopter. Drivers for the DVI system, feedback displays, and desired command displays were written as well as a means for introducing errors and a data collection application. The PF occupied the right seat and PNF occupied the left seat in the ACD. Both positions had virtual instrument panels with a window containing the desired command. The CDU display was in the center floor console position, the HUD was a Liquid Crystal Display mounted on the simulator's glare shield, and VO was heard through the DVI headset.

3.4 Participants and Procedure

Participants were 18 years and older and were recruited from a pool of trained "simulator pilots" (> 10 hours simulator flying time) that included DCIEM employees, students, military personnel, and the general public. The results reported in this paper include data from three of the investigators, since the data from the expected 16 participants had not been fully collected. Participants were briefed on the experiment and individually trained the DVI recognizer with the grammar developed for this experiment. During a flight, a DVI command was displayed on the instrument panel and the participant uttered the command. The feedback forms were displayed, and the participant decided whether or not the feedback was correct by saying "right" or "wrong". Participants were asked to make the assessment as quickly and as accurately as possible. All utterances and displays were recorded and time stamped. The total time to conduct the experiment was 14 hours spread over six days.

4. RESULTS AND DISCUSSION

Eight subjects participated in this initial test, and 3 of them were investigators. Thus, the results presented below are data trends only. Up to sixteen subjects will participate in the full experiment. The ROC curves were plotted, and it was immediately observed that the data points were clustered in the upper left corner. This has two implications. First, the mean values of the calculated biases for each SNR condition (c_{25} = -.38, c_{50} = -.13, and c_{75} = .02) did not differ significantly from c = 0 (no bias). Thus, the data from the separate SNR conditions may be grouped together. Second, the mean accuracy for all manipulations was 90% compared to the baseline accuracy of 100%. Regardless of any statistical difference, any display combination would most likely suffice in practice.

Table 3 shows the sensitivity values for each of the conditions, however there are no significant differences. The visual displays performed equally well and better than VO, which moderated the performance. The feedback forms performed better in the hover than during the cruise, and there was little difference between crew positions.

Table 3. Mean sensitivity values, d'

PNF			PF		
Hover	Feedback	Cruise	Hover	Feedback	Cruise
3.37	HUD/CDU	2.96	3.04	HUD	2.92
3.43	CDU	2.96	3.43	CDU	2.73
3.22	VO/HUD	2.94	3.32	VO/HUD	2.81
3.36	VO/CDU	2.65	2.94	VO	2.54

Table 4. Mean probability ratio, κ

PNF			PF		
Hover	Feedback	Cruise	Hover	Feedback	Cruise
.86	HUD/CDU	.79	.81	HUD	.79
.86	CDU	.80	.87	CDU	.76
.85	VO/HUD	.79	.84	VO/HUD	.76
.86	VO/CDU	.75	.79	VO	.69

The data was further analyzed as a single probability space ($P_H + P_M + P_{CR} + P_{FA} = 1$) as opposed to the dual probability space of SDT ($P_H + P_M = 1$ and $P_{CR} + P_{FA} = 1$). The probability ratio, kappa (κ) is defined as $\kappa = (P_o - P_c)/(1 - P_c)$ where $P_o = (f_H + f_{CR})/(f_H + f_M + f_{FA} + f_{CR})$ is the observed probability of a correct detection and P_c is the expected chance probability (Reynold, 1977). When κ is zero, the observed probability is equal to chance, or $P_o = P_c$. For example, flipping a coin 100 times is likely to yield $P_o \approx P_c = 0.5$. If the participant predicts (detects) the coin face correctly every time, then $P_o = 1$ and $\kappa = 1$. $\kappa > 0$ means that the probability of a correct detection is greater than chance and must be weighted or influenced, in this case, by the four manipulations. The κ trends in Table 4 are identical to the d' trends. Moreover, κ provides a normalized number of the observed probability for a given feedback form.

Table 5. Baseline Detection Times (sec)

Feedback	Mean	Median	Std Dev	Skew
HUD	2.00	1.81	0.564	1.08
CDU	2.30	2.00	1.06	1.34
VO	3.65	3.57	0.41	1.02

Table 6. Experimental Median Detection Times (sec)

PNF			PF		
Hover	Feedback	Cruise	Hover	Feedback	Cruise
1.89	HUD/CDU	2.32	1.98	HUD	2.22
2.09	CDU	2.44	2.42	CDU	2.71
2.59	VO/HUD	3.01	2.75	VO/HUD	3.09
2.90	VO/CDU	3.18	3.72	VO	3.82

The detection time data was analyzed and the results are shown in Table 5. Baseline data was collected for the three separate feedback forms only since the baseline for a combined feedback form is likely to be the fastest individual time. Note that the data is skewed and so the median value best represents the distribution. The CDU takes up to 0.5 seconds before the feedback begins, and VO takes about 3 seconds to complete the command. When an error was heard the participant could respond as soon as possible, but for a correct response the entire feedback needed to be listened to before responding. The HUD and VO displays required little or no head movement, while the CDU display required some time to look down. These times are included in the detection time results. The detection times from fastest to slowest are HUD, CDU, and VO.

The experimental times in Table 6 were compared to the baseline. All times were greater than baseline as expected under increased load. Again, the visual displays were faster than the audio display. VO moderates the visual display times when in combination. PNF and hover were slightly faster than PF and cruise, respectively, in all conditions.

5. CONCLUSIONS

A PCT analysis provided feedback information requirements for the DVI system. An Action-Target-Parameter structure was used to develop the vocabulary and grammar, and integrate the feedback into a single display. This structure made the feedback salient and contributed to the high accuracy probabilities. The experimental trends showed that all feedback forms were 90% accurate on average and about 0.5 seconds slower while multi-tasking. The feedback forms performed better in the hover segment than in the cruise segment with little difference between the PNF and PF performance. The audio display performed consistently poorer than the visual displays although these differences are not statistically significant. Interestingly, the CDU is similar to the HUD, which would not have been predicted at the beginning of the study.

For implementation in a Griffon Helicopter, the feedback requirements will then be dictated by the operational impact of DVI. As the number of systems operated by DVI increase, a consistent design of the DVI vocabulary and grammar will become even more critical. It is likely that manual input would be integrated with DVI in order to ensure control at all abstraction levels. Also, explicit information for all levels would be desirable so to reduce the cognitive load of information integration. The sensitivity of the feedback will be degraded as threats, additional communications, poor visibility, and system failures are added to the scenario. The differences in feedback form most likely will be pronounced in a full operational environment.

REFERENCES

Churchill, L.L. & Herdman, C. (2000). *CH146 Griffon Direct Voice Input Control For The Canadian Forces Utility Tactical Transport Helicopter. Final Project Report.* Prepared for DCIEM under contract no. W8477-8-AC40/003/SV. BSC Document Number 1000-1029, BAE Systems Canada, Kanata, Ontario, Canada. August, 2000.

Farrell, P. S. E. Chéry S. (1998). PAT: Perceptual Control Theory Analysis Technique. *Proceedings of the Human Factors and Ergonomics Society 42nd Annual Meeting.* Chicago, Illinois. 5 – 9 October, 1998.

Farrell, P. S. E., Churchill, L. L., Wellwood, M. & Herdman, C. M. (2001). Integrating Direct Voice Input in a Helicopter Crew Environment. *Proceedings of the Eleventh International Symposium on Aviation Psychology.* Columbus, Ohio. 5 – 8 March, 2001.

Herdman C. M., & Beckett, L. (1996). Code-Specific Processes in Word Naming: Evidence Supporting a Dual-Route Model of Word Recognition. *Journal of Experimental Psychology Human Perception and Performance,* 1996 Vol. 22, No. 5, 1149-1165

Powers, W.T. Clark, R. K., & McFarland, R. L. (1960). A general feedback theory of human behavior: Part 1, *Perceptual and Motor* Skills, 1960, 11, 71-88.

Reynolds H.T. (1977). *The analysis of cross-correlations.* Freeions, Collier, Macmillan Publishers. New York, USA.

Swail, C. & Kobierski, R. (1997). Direct Voice Input for Control of an Avionics Management System. In *American Helicopter Society Avionics and Crew Systems Technical Specialists' Meeting.* Philadelphia.

Wickens, C (1992). *Engineering Psychology and Human Performance, Second Edition.* HarperCollins Publishers, New York, NY, 1992.

A COGNITIVE LINGUISTIC PERSPECTIVE ON THE USER INTERFACE

by Inger Lytje

Aalborg University, Department of Communication, Kroghstræde 3, DK-9220 Aalborg, Denmark
e-mail: inger@hum.auc.dk

ABSTRACT

The user interface is analyzed in cognitive linguistic terms as an interactive multimedia text in which different expression modalities, such as natural language, speech and visual representations, interact. At the expression plane the user interface is analyzed in terms of interface signs which make it possible for the user to interact with different kinds of information. At the content plane, the user interface is analyzed in terms of cognitive semantic categories comprising metaphor, cognitive blend, frame and script. Software quality is related to the semiotic model of the user interface, and some important quality criteria are suggested, stressing the enhancement of user competence, learning and understanding. The evaluation procedures are concived as text-based analysis of the user interface combined with experimental prototyping. It is suggested that quality be enhanced through an appropriate organization of the design process comprising conceptualization followed by experimental prototyping.

1. INTRODUCTION

In the software industry people have become aware that software has to be evaluated from a user point of view before it is implemented in the user environment. In this connection, one has to consider the quality criteria and the methods for evaluating software quality.

Usability engineering conceives software quality in terms of user effectiveness, as effective users reduce the cost of the employer and thereby raise his profits in the short term. Usability is measured in quantitative terms through usability testing in laboratories (Nielsen, 1993) and through heuristic evaluation, which is based on the idea that there are some basic characteristics of usable interfaces. Usability engineering can be seen as a kind of experience based debugging of the user interface through which a series of inappropriate features are eradicated. Meaning and understanding play a minor role as quality criteria, and learning is considered a cost that has to be minimized. Usability engineering only offers some experience-based guidelines of design through the notion of heuristic evaluation.

The Direct Manipulation idea about software (Hutchins, Hollan & Norman, 1986) conceives software quality in terms of minimizing the gulf between the user's knowledge and intentions and the information system. When the gulf is bridged effectively, the cognitive effort of the user is minimized, and in this case the software is considered to be high quality. Task oriented graphical interface languages are considered from the point of view that the user interface has to be articulated in terms of lexical signs refering to the task domain and to the user's

world. It means that the interface language should match with the user's conceptualization of his or her task. The direct manipulation paradigm offers guidelines for design of graphical user interfaces that are understood as a model world of objects and relations between objects in which the user acts and interacts. The quality criteria are ease-of-use and easy-to-learn.

Natural language interfaces in terms of speech and writing based on natural language understanding technologies (e.g., Allen, 1987) have for many years been considered the most sought after kind of interface from the point of view that natural language is the most natural way for humans to communicate. Until recently it was taken for granted that natural language was based on a very restricted channel of communication which only allows for a "teletype approach" (Stock, 1995). It has now been suggested, however, that natural language interfaces should be understood as multi-modal interfaces integrating natural language with visual, gesture, motion and other kinds of modalities (Waibel, Tue Vo, Duchnowski & Manke, 1996). Moreover, it has been suggested that combining several different human communication modalities is the most natural thing to do.

While the Direct Manipulation tradition talks about restricted task oriented interface languages, I will place myself within the natural language tradition, and consequently I will consider the interface language to be at the same level of generality as natural languages. Cognitive linguistics offers a framework for understanding language meaning as a combination of different modalities and consequently for understanding the interactive multimedia text that is implemented in the computer (Fauconnier, 1999; Lakoff & Johnson, 1980; Langacker, 1987; Schank & Abelson, 1977). Within this framework, I am going to develop an understanding of the user interface as an interplay between 1) the structural characteristics of the interface signs, 2) the meaning of the signs and 3) the implementation of the signs. In this way, the notion of meaning and understanding as a prerequisite of learning and acting become important reference points of quality criteria and quality measures, linking usability to the enhancement of user skills and competence. This notion of quality is suggested, as competent users are very productive in the long term. They are able to reason and solve problems in problematic situations, to adapt the system to other user environments and to participate in design processes. Furthermore, they are able to conceptualize their knowledge and consequently to share it with other people.

Adapting and applying cognitive linguistic theory to software has consequences for the design process. User oriented software design has been practised using participative and cooperative design strategies (Kyng & Mathiassen, 1997). The idea is that design has to be based on a full understanding of the social environment in which the software is used. This understanding is obtained through open design practices in which users participate, or through the use of social science methods such as surveys, questionnaires and ethnographic studies. As part of this strategy, the software is evaluated through a prototyping process that evaluates the software in a real user environment. The success of cooperative design, though, very much depends on the quality of the first prototype and on the dialog between designer and user (Lytje, 2000). Applying cognitive linguistic theory makes it possible to conceptualize the software in cognitive linguistic terms before the experimental prototyping process begins, and this enhances the quality of the first prototype.

2. SOFTWARE AS A SEMIOTIC PRODUCT

In order to establish quality criteria and methods for evaluating software quality from a user point of view, I will suggest an analysis of software in cognitive linguistic terms. The semiotic analysis draws the attention towards the interaction between the structural characteristics of the user interface and the user's cognition when using and understanding the interface. To that we add a third dimension, that is the implementation of the sign in the computer. The interface sign can be depicted in the following way:

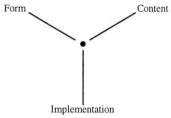

Figure 1. The interface sign

The node in the middle symbolizes the interface sign, and the figure tells us that the interface sign should be understood as a unity of form, content and implementation. The overall philosophy of cognitive linguistic theory cfocuses on the relationship between linguistic form and semantic content. The idea is that the form motivates the construction of meaning in the mind of the subject which experiences the form. Moreover it is assumed that meaning is constructed in terms of cognitive processes such as figure-ground, trajector-landmark, metaphor analogy, framing, mental space, prototypes, metonymy, polysemy, conceptual blending, fictive motion, force dynamics (Fauconnier, 1999). It is assumed that subjects engage in quite similar constructions on the basis of similar grammatical prompts. The adaptation of this semiotic model to the computer presupposes that meaning construction in general is based on the cognitive processes mentioned above. It seems reasonable to make this presupposition as cognitive linguistics is seen as a general science of meaning construction. It also presupposes that human-computer interaction is understood as a kind of language activity. I do believe that this is the case: When using and designing software we engage in language activity, and the language we use is symbolic, but also visual, spatial and iconic. This corresponds with the notion of imaginary, image schemas and spatiality that are some of the pervasive categories in cognitive linguistics (Langacker, 1987).

2.1 Analyzing software at the expression plane

The signs in the user interface can be seen as symbolic representations of input-out mechanisms through which the user interacts with different kinds of information. I suggest that we consider the following kinds of information:

229

- *Process.*
  Algorithmic processes such as spelling checking.
- *Object.*
  E.g. text, picture and music.
- *Relation.*
  Quality of object information such as typography or color.
- *Dynamic object.*
  E.g. the drawing module of a text processor.
- *Structure.*
  Links the signs, e.g. menu and frame.
- *Layout.*
  Facilitates the users reading and understanding of the sign, e.g. a caption.

The following is an account of the different ways the users can interact with different kinds of information through interface signs. The examples refer to Visual Basic.

- *Interacting with process information*:
  - Controlling the process, e.g. CommandButton and OptionButton.
  - Graphical simulation of the process, e.g. the hourglass.
- *Interacting with object information:*
  - Display, e.g. PictureBox and TextBox.
  - Transfer of information to and from external databases and archives, e.g. DataControl.
  - Inserting and editing of information, e.g. TextBox.
  - Navigation, e.g. ScrollBar.
  - Adding and editing relational information
- *Interacting with dynamic object*:
  - Activating the object, e.g. clicking the icons in the toolbar.
  - Embedding, e.g. embedding a drawing program into a text processor.
  - Display, e.g. the presentation of a dynamic object as a window.

Interface signs are considered the basic lexical categories in a design language that conceptualizes the computer as human computer interaction, e.g. Visual Basic. This is unlike to ordinary programming languages that conceive the computer as a computational logic and database languages that conceive the computer as a database for information storage and retrieval. The basic semantic structures, *process, thing and relation,* (Langacker, 1987) can be recognized in the list of information-types above. The different kinds of information refer to different kinds of semantic structure. In this way the computer based signs are linked to semantic structure and consequently to the level of meaning.

2.2 Analyzing software at the content plane

At the content level, the user interface is analyzed in terms of metaphor (Lakoff & Johnson, 1980), blending (Fauconnier, 1999), frame (Fillmore, 1982; Langacker, 1987) and script (Schank, 1977). The content plane is related to the expression plane through meaning construction that is motivated by signs and sign structures. The cognitive schemas are activated when understanding a semiotic product, such as a natural language text or a software product, which has been implemented in a multimedia computer.

As readers and users, as writers and designers, we go through metaphorical and other kinds of figurative reasoning processes, through which we understand the meaning of the signs and sign structures we have before us. Through these reasoning processes, different cognitive domains are blended. For instance, the knowledge domain of 'mail' and the knowledge domain of 'computer' have been separated until recently, but through the concept 'electronic mail' they are brought together in the following way: Through a generalization process, the concept 'mail' is related to the concept 'communication of addressed messages between a sender and a receiver', and subsequently this general concept is specified, using the computer and computer networks as the medium of implementation.

Activating a frame during understanding processes means the cognitive activation of a contextual background of understanding. In the user interface, the signs have to be understood on a contextual background, and this contextual background has to be made explicit. Otherwise the user will not be able to understand the meaning of the sign, unless the contextual background is already in the user's mind. It indicates that cognitive framing has to include both the background knowledge, which is made explicit in the user interface, and the background knowledge, which is already in the user's mind. The reason why graphical user interfaces are experienced to be more user friendly than command language user interfaces might be that the contextual background is made explicit in the graphical user interface, while the background knowledge in command language user interfaces has to reside in the user's mind.

Human computer interaction can be understood as an event sequence through which the user and the software exchange information. The user interface can be seen as a script for the possible event sequences in which the user can participate (Schank, 1977).

2.3 Implementation

The signs and the sign structures in the user interface can be implemented in various ways. There are some important requirements to consider in this connection, many of which are described by Jakob Nielsen (Nielsen, 1993). Generally speaking, the software should be implemented in such a way that the user feels comfortable when using the system. It means that the software engine should be robust, deadlocks should be prevented, the user's memory should not be overloaded and the navigation should be flexible allowing for regrets and corrections of misunderstandings. Error handling, response time and transparency of the software also depends on implementation strategies.

3. EVALUATION AND DESIGN

The semiotic model of the interface sign as depicted in figure 1, can be used as a reference point for evaluating software form a user point of view. Quality is related to the triad, form, content and implementation. Below, some important quality criteria (C) and evaluation methods (M) are listed:

- (C) Has the software been conceptualized in such a way that the end user can learn how to interact with the software in a meaningful way. (M) A textual analysis of the user interface will reveal whether the interface text is consistent, complete, coherent and focused in relation to the user domain.
- (C) Have the signs in the interface been formed, separated and linked in such a way that the reference to different kinds of information and different kinds of interaction styles are made clear. (M) A sign based analysis at the expression plane will reveal whether this is the case.
- (C) Is the user interface formed in such a way that misunderstandings are prevented. That is, does the user interface communicate the message clearly to the user. (M) Experiments with end users will reveal whether there are too many failures and whether end users misunderstand some of the signs.
- (C) Is the user interface implemented in such a way, that the user does not get into trouble. (M) Heuristic evaluation and usability test may reveal that.
- (C) Is the preference of media and the interaction between different media expressions suitable. (M) A sign based analysis will reveal that.

The semiotic model can also be used as a reference point for designing software from a user point of view.

First, the designer has to develop a conceptual model of the software (the user interface) in cognitive semantic terms: metaphor, cognitive blend, frame, script etc. In order to be able to do that in a meaningful way, the designer has to research the use domain. The conceptual

model has to be signified in terms of interface signs, and finally the software has to be implemented as a prototype.

Secondly, the designer has to organize a prototyping process with real users in a real user environment in order to evaluate the conceptualization in an experimental way. During this phase, the software is evaluated in relation to potential users' knowledge and background. The participants in the process reason and communicate about the semantic model and the way this model has been designed, and they discuss the issue of social change. During this phase the conceptual model might be overruled by practical and instrumental considerations, which are caused by the use situation and the conditions of use.

Thirdly, the designer has to organize experiments with end users in order to test implementation details.

4. CONCLUSION

It has been suggested that the user interface of a software product is conceived as a kind of text that can be analyzed in cognitive linguistic terms. The user interface is constructed by the means of interface signs and sign structures, which are implemented in the computer, and the user interacts with the software through these interface signs. They make up the lexical categories of a design language, which conceives the computer as human-computer interaction. The interface signs are conceived as a sign relation between form and content, and this sign relation is implemented in the computer. At the expression plane, the signs are defined according to user interaction with different kinds of information. At the content plane, the meaning of the signs and sign structures is conceived as cognitive schemas that are activated in the user's mind when reading and understanding the signs.

The semiotic model of the user interface is used as a reference point for evaluating software from a user point of view. Some important quality criteria and evaluation methods have been suggested comprising media preference, comprehensibility and learnability. The latter criteria contribute to the enhancement of user's skills and competence. This is of great value because competent users are more productive and more satisfied than unskilled users in the long term. The usability criteria that most software designers have become familiar with are related to the implementation of signs and sign structures. They concern the engineering and debugging aspects of the user interface while the other criteria are related to the notion of understanding.

The semiotic model also makes up a reference point of design. The designer is invited to open the design process with a cognitive semantic analysis of the user's (cognitive) domain. Subsequently, this cognitive semantic model has to be signified in terms of interface signs. In this connection, the designer has to consider the media of expression and the interaction between them in order to communicate to the user in an appropriate way. The user interface has to be voiced in such a way that it communicates the meaning of the interface clearly to the user.

REFERENCES

Allen, J. (1987). *Natural Language Understanding*. Menlo Park: Benjamin/Cumming.

Fauconnier, G. (1999). Methods and generalizations. In T. Janssen & G. Redeker (Eds.), *Cognitive Linguistics: foundations, scope and methodology* (pp. 95-127). Berlin: Mouton de Gruyter.

Fields, R. & Wright, P. (2000). Understanding work and designing artefacts. *International Journal of Human-Computer Studies*, 53, 1-4.

Fillmore, C. J. (1982). Frame Semantics. In The Linguistic Society of Korea (Ed.), *Linguistics in the Morning Calm* (pp. 111-137). Seoul: Hanshin.

Hutchins, E. L., Hollan, J. D. & Norman, D. A. (1986). Direct Manipulation Interfaces. In D. A. Norman & S. W. Draper (Eds.) *User Centered System Design* (pp. 87-124). Hillsdale, New Jersey: Lawrence Erlbaum.

Kyng, M. & Mathiassen; L. (1997). *Computers and Design in Context*. Cambridge, Massachusetts: The MIT press.

Lakoff, G. & Johnson, M. (1980). *Metaphors we live by*. Chicago: University of Chicago Press.

Langacker, R. W. (1987). *Foundations of Cognitive Grammar, vol. 1. Theoretical Prerequisites*. Stanford. California: Stanford University Press.

Lytje, I.(2000). *Software som tekst. En teori om systemudvikling*. Aalborg, Danmark: Aalborg University Press.

Nielsen, J. (1993). *Usability Engineering*. San Diego: Morgan Kaufman.

Pressman, R. S. (2000). *Software Engineering, A Practitioner's Approach, European Adaption*. London: McGraw-Hill

Schank, R. & Abelson, R. (1977). *Scripts, Plans, Goals and Understanding, An Inquiry into Human Knowledge Structures*. Hillsdale, New Jersey: Lawrence Erlbaum.

Stock, O. (1995). A Third Modality of Natural Language. *Artificial Intelligence Review*, 9, 129-146.

Waibel, A., Tue Vo, M., Duchnowski, P. & Manke, S. (1996). Multimodal Interfaces. *Artificial Intelligence Review*, 10, 299-319.

Simulating recognition errors in speech user interface prototyping

Matthias Peissner, Frank Heidmann, Jürgen Ziegler

Fraunhofer-Institute for Industrial Engineering (IAO), Nobelstr. 12, D-70569 Stuttgart, Germany

ABSTRACT

We have developed a Wizard of Oz simulation tool which allows scenario-based simulation of speech systems for the conduction of empirical studies with future users. This paper focuses on the adequate integration of recognition errors as they are an important feature of speech-based applications. The presented solution considers the aspects of reliability and validity. Both are necessary preconditions for the immediate transferability of simulation results to the real system.

1. SPEECH USER INTERFACE PROTOTYPING

In the field of GUI design it has become common practice to test usability in early development stages. By using paper prototypes important design decisions can be met on the empirical basis of tests with future users. In comparison to the vast amount of empirical studies and guidelines concerning the usability of GUIs, we know very little about how to design effective speech user interfaces (SUI). Moreover SUI designers face the essential difficulty of getting a sound feeling for the dialogue flow by merely inspecting a written dialogue specification. For these reasons it is even more important to include prototyping and usability testing early in the design process of user-friendly interactive voice response systems (IVR systems).

The speech equivalent to a paper prototype is a Wizard of Oz (WOZ) study (Weinschenk & Barker, 2000), where a human (the wizard) simulates the role of the computer during testing and starts different recorded system prompts dependent on what the user said. Usability testing with the WOZ technique can lead to valuable results regarding the following topics:

- *Designing a user-oriented grammar:*
  In very early development stages WOZ studies can pinpoint the utterances which are typically used in order to control the available functions. Given a sufficient number of subjects the transcriptions of the test sessions can give a representative image of how users would expect the system to understand. The most frequently recorded utterances can serve as a valid basis for a user-centred grammar. This way, the time-consuming procedure of pilot testing including iterative grammar modifications and recognition tuning can be shortened or even partially avoided (Pearl, 2000).

- *Comparison of different systems / system versions:*
  Alternative design decisions can quickly be acted out and tested with future users. Especially the different effects of alternative prompt versions on the users' performance and attitudes towards the system can be evaluated.

- *Overall ergonomic evaluation:*
  WOZ experiments can take the traditional role of usability tests in evaluation and troubleshooting. The detection of major problems of use in an early development stage enables iterative redesign and reconception without the otherwise necessary phases of implementation.

Necessary precondition for the validity of a WOZ study is that the interaction between user and "machine" (here the wizard) has to be as realistic as possible. Otherwise, the gained results cannot be transferred immediately to the real situation of system use. This means, that on one hand, the subject in a WOZ study must actually belief that she interacts with a real system, which is a matter of adequate instruction. On the other hand, the simulation must not differ from the specified system behaviour in essential aspects. Among others, this refers to the reliability of speech recognition which is treated in detail in the following section, and to the available complexity of the dialogue. With high complexity applications it is necessary to do scenario-based testing in order to reduce the amount of probable user utterances. This supports the wizard's decision by giving a situation specific pre-selection of probable options for "system" reactions.

2. PROBLEM

Speech technology is probabilistic in nature and therefore recognition errors are inherent in any speech-based application. Furthermore, situations of recognition errors are especially crucial to usability variables such as effectiveness and efficiency in task solving and user acceptance (Yankelovich, Levow & Marx, 1995).

233

Therefore, it will be indispensable in most cases to carefully simulate error situations in WOZ studies in order to achieve data about questions like: How frustrating do users experience recognition errors in the application in question? Do the mechanisms of error management actually assist the users in correction? Do the users recognise the occurrence of an error at all?

How should recognition errors be included into the simulation design? Even if you had in mind the whole grammar of the recognition system you would never be able to anticipate the system's behaviour. This unpredictability of recognition errors is still increased if the IVR system is used from a cellular phone. Obviously, a simple rule-based model for simulation of recognition errors is not applicable. On the other hand, bare arbitrariness or intuition as basis for the wizard's decisions will bias the test results. In order to ensure reliability and validity of a WOZ study the following aspects have to be taken into account:

- *Realistic probabilities for recognition errors:*
  A predefined and realistic probability for correct understandings, substitution errors and rejection errors is a precondition for a sound evaluation of the relevant usability criteria. And it allows controlled examination of the consequences of various confidence thresholds. The confidence threshold defines the minimal probability of correct classification needed to execute an action. Probabilities below the threshold lead to rejection usually accompanied by a prompt like 'Sorry, I could not understand you. Please repeat.' Necessary data stem from knowledge of the used recogniser and of relevant parameters of the used classification scheme.

- *Standardised simulation:*
  Without using automatic speech recognition it will never be possible to completely eliminate influences on simulation performance that arise from the wizard's decisions. These influences cannot be held constant over time, different persons and situations. It is an important goal to achieve a maximum level of objectivity by reducing the possible options, the consequences and the need of human decisions to a manageable minimum. Only under comparable test conditions different systems or system versions and the performance of different user groups can be compared adequately. For the comparison of different prototype versions the simulated recognition performance should be balanced in order to avoid undesired side effects.

- *Interactivity:*
  Although standardisation is an important feature, especially in within-subjects designs of system comparison, interactivity is essential for the validity of the results. That means that, despite standardisation of the simulation system, responses must depend on what the user says. Strict balancing (i.e. constant predefined sequences of correct recognition, substitution error and rejection in both conditions) and randomising (i.e. constant predefined frequencies of correct recognition, substitution error and rejection in both conditions) do not consider occurring training effects in the users' speech performance which are likely to support higher recognition rates in the version presented in the second position.

One method to support the simulation of recognition errors is the use of filters, e.g. vocoders which distort the spoken input, in order to help the wizard perform to the system's expected level (Bernsen, Dybkaer, and Dybkaer, 1998). Filters suffice the requirements of interactivity and standardisation. But it is questionable if they can support a realistic simulation of errors. Firstly, the relationship between the probability of recognition errors and the physical intensity of the filter is not straightforward and has to be investigated empirically before. Secondly, a deterministic filter that constantly distorts the input signal might be no appropriate model for a highly probabilistic process. Human speech performance is probabilistic. Even if two utterances sound completely identically for another person the acoustic signals will never be totally the same. Environmental noise, recording and transmission are also probabilistic factors that make it impossible to anticipate the acoustic quality of the system input signal. Finally, the procedure of recognition itself is probabilistic in nature as it follows a statistical classification scheme.

3. OUR APPROACH
We have developed a software tool that supports WOZ simulations of IVR systems (see figure 1 for the GUI).

3.1 The WOZ-GUI
Each button (except those for the scenario selection) on the simulation GUI stands for a set of user utterances, a specific subset of the grammar. For ease of use each button is labelled with the corresponding grammar or at least a part of it.

The scenario-based approach makes it possible to simulate even highly complex applications. Any scenario consists of one or more pairs of user utterance and system prompt. When a scenario is started the main frame displays a matrix of buttons each representing an expected user utterance. For illustration, let us take a scenario which include a call John Smith and after that to change his number entry in the telephone book. The first target user utterance is something like "I'd like to call John Smith" which is represented by the * button in the

234

first column. The other buttons in the first line represent expected variations form the target utterance in this first sub-task, e.g. "I want to place a call" or "Go to telephone book" or "John Smith". These utterances start other actions, e.g. feed-forward prompts that shall obtain the missing data in order to accomplish a transaction (e.g. "Whom would you like to call?"). When feed-forward prompts are played which are not part of the target path (the first column of the main frame) a child window is popped up displaying buttons representing possible user utterances in the actual sub-dialogue. When all data is captured which is needed to proceed to the next sub-task of the scenario the window is shut again. The following lines are built up the same way: the first button represents the next target user utterance, the other buttons of the same line stand for other expected utterances. Simultaneously displaying all stages of a scenario instead of only the actual one allows to appropriately react to users who choose another than the expected order of actions.

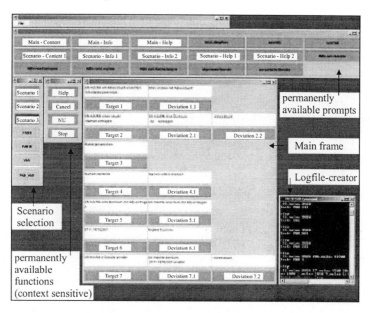

Figure 1: Graphical User Interface simulation tool

The frame on the top of the screen displays specific help prompts and other prompts which are permanently available independent of the actual scenario and context of use. Examples for corresponding user utterances are "Content" and "Help on <function x>".

On the left of the main frame the control frame displays buttons for functions which are context sensitive. The user utterances "help" and "cancel" or an out-of-grammar utterance (OOG) start different prompts dependent on the actual context of use. The same with barge-in which may be switched off in certain contexts.

The DOS-box in the right lower corner displays the text which is written to the log-file during interaction. The log-file includes information about what button was pressed at what time and what prompt was played. Log-file analysis can discover detours and wanderings and identify especially problematic areas in the dialogue flow. More over they contain information about the actual rate of correct understandings which is important for further evaluations.

3.2 Simulating speech recognition errors

The occurrence of recognition errors is integrated on the basis of the above mentioned considerations concerning realistic error probabilities, standardisation und interactivity. The imperfect performance of speech recognition is modelled by human speech understanding which is restricted to a predefined set of utterances (the grammar) and added a probabilistic element of uncertainty. The wizard only has to decide whether the user utterance *can* be understood given the grammar restrictions and then to press a corresponding button. She has *not* to assess if the real system would correctly understand. Each button – except those for scenario selection - starts a random function which triggers one of a certain set of possible actions. For example, pressing the target button is followed by the target action (target prompt), or another not desired action (simulating a substitution error), or rejection ('Sorry, I could not understand you. Please repeat.'). The probabilities for correct recognition and the

235

errors of substitution and rejection are customised according to experiences or expectations regarding the real system (see figure 2).

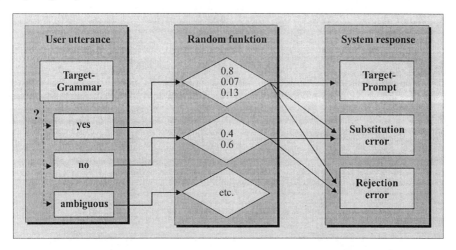

Figure 2: Probabilities for correct recognition and errors of substitution and rejection

3.3 Required definitions and wizard's skills

The following parameters are necessary input variables for running a WOZ simulation with our software. Any testing scenario must be decomposed into all necessary user utterances (the target utterances) and the corresponding target prompts. Expected variations should be identified for any target user utterance. For any target prompt a set of suitable prompts for simulating substitution errors must be given. And finally, any available user utterance (displayed button on the GUI) must be assigned the probabilities for correct recognition, substitution error and rejection error. These probabilities may vary between different classes of utterances, e.g. digits, utterances containing more than one token, etc.

The wizard holds a key role for the validity of the study. She must know exactly the dialogue flow, especially the permanent available functions and prompts, in order to be able to manage the complex GUI. Furthermore, deep knowledge of the grammar including permissible variations is needed. Extensive training will be indispensable in order to gain the needed speed and competence in the decision process.

3.4 First testing experiences

We conducted a study in order to compare two different prompt versions. We used exactly the same dialogue model and simulation parameters (grammar, error rates) with two different prompt versions. The questions were: Do users notice a difference? Do they establish different mental models, including hypothesis concerning the reliability of recognition? Do user utterances differ between both prompt versions? How do different prompt versions affect the acceptance and likeability of an IVR system? We used a within-subjects design with 20 subjects. Any subject worked on 10 scenarios, 5 scenarios with each version. Presentation order was balanced. Subjects were instructed that they were to test two different prototypes and to judge which version was more comfortable to use. It showed that the actual performance of the "speech recognition" which was a matter of random was a dominant aspect for the preference of one or the other version. Having only a small number of user utterances for each version and assuming a rather low probability of 80 % for correct recognition it was not rare that the actual performance differed noticeable between both versions. Even very small differences were discovered (e.g. "one misrecognition at the first two at the second version"). This effect would also have appeared if a real IVR system was used for testing with different prompt styles. Strict balancing is surely not the right way to avoid this effect. But it should be reduced by increasing the number of necessary user utterances. Taken this into account the conducted study provided valuable information, e.g. the transcripts of the interactions which indicate that one prompt version evokes user utterances which can be assumed to increase the performance of speech recognition compared to the other version. We will not discuss further results here in detail. Boyce (Boyce, 2000) reports a comparable study and results.

Another WOZ study with 10 subjects was conducted in order to check overall usability of a speech portal before the real system was completely implemented and tuned. Before our tests the specified grammar was rather sparse and so one valuable by-product was a collection of utterances used in order to accomplish the given tasks. The results indicated, among other things, that the provided grammar urgently needed to be extended in order to allow for efficient use of the IVR system. First clues for these extensions could be given but needed to be validated and completed by a more extensive study. Important recommendations on the basis of the results concerned modifications in prompting especially in cases of rejection as a consequence of an OOG utterance or as a consequence of recognition failure. As user utterances are strongly dependent on system prompting it is important that these changes are done *before* extensive pilot testing which aims at optimising speech recognition regarding recorded user utterances.

In both studies our wizard's decisions were not free of human error as she had to meet complex decisions in a very short time. Erroneous decisions can never be totally prevented but they are no problem as long as they do not arise systematically. However, it seems to be very difficult for a human "speech recogniser" to repeatedly reject a OOG utterance when the subject is trying so hard. Cases of transaction success as a consequence of a too tolerant wizard decision should be marked in the transcripts in order to be taken in to account in further evaluations.

4. CONCLUSIONS

Designing and developing an IVR system is a complex task due to close interdependencies between the essential aspects of dialogue architecture, prompting, grammar and reliability of speech recognition. In order to find application specific solutions for user-friendly SUIs, prototyping and WOZ studies will be indispensable as they help to save time and costs in an iterative development cycle. On the other hand the conduction of WOZ studies is a favourable approach for scientific investigations in order to collect empirical data as a sound basis for guiding principles of ergonomic SUI design.

Our approach to simulating recognition errors in WOZ studies allows for realistic simulation conditions. It allows to see how users cope with recognition errors. And it enables to evaluate the effectiveness of the mechanisms the system provides in order to support users in discovering and coping with recognition errors.

REFERENCES

Bernsen, N.O., Dybkaer, H., and Dybkaer, L., "Designing Interactive Speech Systems: From First Ideas to User Testing", London: Springer, 1998. c

Boyce, J. S., "Natural spoken dialogue systems for telephony applications", Communications of the ACM, 2000, Vol.43, No. 9, pp. 29-34.

Pearl, C., "A Prototyping Tool for Telephone Speech Applications." CHI 2000 Workshop on Natural Language Interfaces, The Hague, Netherlands, April 3, 2000, (http://www.cs.utep.edu/novick/nlchi/papers/Pearl.htm).

Weinschenk, S. and Barber, D., "Designing Effective Speech Interfaces". New York: John Wiley & Sons, 2000.

Yankelovich, N.; Levow, G.A. and Marx, M. "Designing SpeechActs: Issues in Speech User Interfaces." CHI '95 Proceedings. ACM Conference on Human Factors in Computing Systems. Denver, CO, May 7-11, 1995.

A Study on Real-time Gesture Classification Method

H.Shimoda, T.Sasai and H.Yoshikawa

Graduate School of Energy Science, Kyoto University, Gokasho, Uji, Kyoto 611-0011, JAPAN

ABSTRACT

A real-time and automatic gesture classification method from dynamic image of upper half of the body is proposed, as a basic study to recognize naturally expressed gesture during conversation. The method consists of (1)feature extraction, (2)feature analysis and (3)classification, and a prototype system has been developed based on the method, followed by the system evaluation by subject experiment. Comparing the classification by the system with that by human subjects, most of gestures were classified appropriately in real time, however, there were a few gestures which could not be segmented properly, especially when the gesture has only small head movement such as light nodding.

1. INTRODUCTION

As information technology has been improved quickly and computers has been introduced to everywhere in our daily life, the relationship between human and machine becomes more important. In order to realize smooth and natural communication between human and machine, not only humans access to machines, but also machines recognize human status and make suitable responses from the human status.

In case that humans communicate each other, they use not only verbal language but also lots of modalities such as facial expression, gesture, voice tone. They are called nonverbal messages, and Mehrabian said that 93% of information is communicated by nonverbal messages (Mehrabian, 1972).

In this study, the authors pay attention to the gesture among the nonverbal messages. The important gesture during conversation is not patterned sign gesture but motions such as "adaptor", which are unconsciously and naturally expressed during the conversation (Ekman and Friesen, 1969). Such natural motions (gestures), however, have many individual variations, it is therefore difficult for machine to recognize them. Almost all of the conventional studies recognize predefined patterned gestures (Gao *et al*, 2000, Wilson and Bobick, 1999).

The authors aim at recognizing the naturally expressed gestures, and have proposed an automatic and real-time classification method. The final goal of this study is to realize individual adaptive interface. If such interface is applied to the personal computer, the user can operate it more comfortably and naturally. The concept of the individual adaptive interface is to recognize individual habits and favors from not only user's operational information but also human signals such as gesture, facial expression, voice tone, view direction and eye blink, and can make adaptive response to the user. The facial expression, voice tone, view direction and eye blink has common tendencies corresponding to the user's internal status. The naturally expressed gesture, however, does not have such common tendency. The gesture classification, which is proposed in this study, is therefore positioned as the first step of recognizing such gestures.

2. AUTOMATIC CLASSIFICATION METHOD

In this study, the objective motion for classification is upper half of the body gesture when human makes conversation sitting on the chair. Concretely, the input is dynamic image of upper half of the body captured by a color CCD camera, and the gestures are extracted and classified from the image. The output is the classification results and the start time of each gesture.

The proposed method consists of (1)feature extraction, (2)feature analysis, and (3)classification. Figure 1 shows the outline of the method. Firstly, feature values are extracted, which expresses movement of face and both hands. Then, the distinctive motions are analyzed from the variations of the feature values, and each distinctive motion is expressed as a feature vector. Lastly, the feature vectors are classified by using real-time cluster analysis. The details of each process will be explained below.

2.1 Feature Extraction

As mentioned in the previous section, the objective gesture is body motion when sitting on the chair. Therefore, the gesture mainly consists of the movement of upper half of the body. The authors pay attention to the movements of face, right hand and left hand because they can well describe the feature of the gesture and they can be extracted

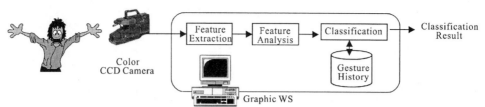

Fig.1. Outline of Automatic Classification Method

easily by simple image processing in real time. Therefore, the time variation of the position and shape information of these three anatomies are extracted in this process. Concretely, regions of these anatomies are recognized by image processing, and the center positions of the regions are calculated. As the shape information, the areas and aspect ratios (ratio of vertical versus horizontal length) are also calculated. In this method, the anatomies are basically extracted by using color information of the input image. However, there are some problems of only using color information such as (1) a hand can not be extracted when it is hidden behind the body, (2) anatomies are extracted as one region when they overlap each other. In order to solve these problems, this method employs special processing using the past frame information.

2.2 Calculation of Feature Values

Here, feature values are calculated from the extracted face, right and left hand region. Table 1 shows the feature values to be calculated. These values are calculated for each video frame and averaged in the last three frames in order to remove noise factors.

Table 1. Feature Values

Explanation of Feature Values
(1) (X,Y) Coordinate of face center
(2) Area of Face
(3) Aspect ratio of face region
(4) (X,Y) Coordinate of right hand center
(5) Area of right hand
(6) (X,Y) Coordinate of left hand center
(7) Area of left hand

2.3 Feature Analysis

Here, the gestures are detected by analyzing time series of the feature values, and then feature vector is made, which expresses the distinctive motion of the detected gesture.

First, the moving part of the time series is segmented as a gesture. This process is called gesture segmentation. In case that the time variation of feature values does not exceed the variations that are caused by the tiny sway of the body, it is defined as "no motion". And the period except the "no motion", it is defined as a gesture.

Then, the time series of the feature values in a gesture are analyzed and the distinctive information is extracted. The feature vector is organized from the information. The factors of the feature vector should be enough to express the distinctive motion of the gesture. Table 2 shows the concrete factors, which were decided from observing humans gestures during their conversation.

Table 2. Factors of Feature Vector

	Factor of Feature Values		Factor of Feature Values
1	Time period	13	Y coordinate of right hand when gesture starts
2	X coordinate of face when gesture starts	14	Y coordinate of right hand when gesture ends
3	X coordinate of face when gesture ends	15	Maximum variation of Y coordinate of right hand
4	Maximum variation of X coordinate of face	16	Maximum variation of right hand area
5	Y coordinate of face when gesture starts	17	X coordinate of left hand when gesture starts
6	Y coordinate of face when gesture ends	18	X coordinate of left hand when gesture ends
7	Maximum variation of Y coordinate of face	19	Maximum variation of X coordinate of left hand
8	Maximum variation of face area	20	Y coordinate of left hand when gesture starts
9	Maximum variation of face aspect ratio	21	Y coordinate of left hand when gesture ends
10	X coordinate of right hand when gesture starts	22	Maximum variation of Y coordinate of left hand
11	X coordinate of right hand when gesture ends	23	Maximum variation of left hand area
12	Maximum variation of X coordinate of right hand		

2.4 Classification

Here, the feature vectors are classified in real time. The classification result is expressed as a group of the feature vectors (a cluster). First, in order to reduce the varieties of each factor of the feature vector, they are standardized from their standard deviations. Then, the factors are weighted to be classified appropriately. The coefficients of standardizing and weighting were decided from the results of subject experiments.

And, the standardized and weighted feature vectors are classified by real-time cluster analysis. Averages of the factors in the feature vectors are calculated as the representative vectors of the clusters beforehand. And when a new gesture occurs, the feature vector of the gesture is compared with the representative vectors and classified into the nearest cluster. In case that the distances between the feature vector and all of the representative vectors are beyond a threshold, a new cluster is created and the feature vector is classified into the new cluster. In this process, the distance between vectors is calculated as Minkovsky distance.

3. PROTOTYPE SYSTEM DEVLOPMENT

The authors have constructed a prototype system based on the above method. In this section, the hardware and software of the system is explained.

3.1 Hardware Configuration

The hardware of the prototype system consists of the following components; (1)Graphic Workstation O2(SGI), (2)CCD Camera EVI-D30(SONY), (3)Video Cassette Recorder WV-D9000(SONY) and (4)21inch Display(SGI).

The dynamic image of the upper half of the body taken by the CCD camera is put into the multiport video processor(MVP) of the graphic workstation. The MVP converts the image into digital signal and stores it into video memory. The workstation processes the images based on the proposed method. The classification results are shown on the display in real time.

3.2 Software Configuration

The software of the prototype system consists of the following subsystems; (1)Feature Extraction Subsystem, (2)Feature Analysis Subsystem and (3)Classification Subsystem. Video Library(VL) and OpenGL is also employed for video signal input and classification result output, respectively.

4. EVALUATIONAL EXPERIMENT OF PROTOTYPE SYSTEM

4.1 Purpose of Experiment

The purpose of the experiment is to evaluate how appropriate the prototype system can classify human gestures, and to confirm that the system can classify them in real time.

In this experiment, the classification results by the system and those by three human subjects are compared. The dynamic image of the upper half of the body as an input of the system was taken and recorded beforehand. While processing gesture classification, the processing time of each subsystem is measured.

4.2 Prepared Dynamic Image

In order to prepare the dynamic image to be used in the experiment, the upper half of the body during conversation was taken by the CCD camera and recorded in the VCR. The situation when taking the dynamic image is shown in Figure 2.

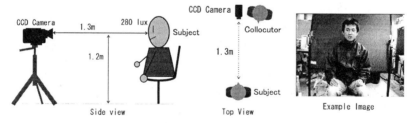

Fig.2. Situation When Taking Dynamic Image

The subject was sitting on the chair and talking with the collocutor. The length of the recorded image is about 1 hour. In the experiment, 15 minutes of the recorded image was used for the evaluation.

240

4.3 Experiment Procedure

The experiment procedure was as follows;
(1) The three subject (SK, OY, MT) were asked to watch 1 minute example video image and explained about the experiment.
(2) The recorded image was shown to the subject to classify the gestures. They were asked to find gestures and record (a)starting time of each gesture, (b)brief description of the gesture and (c)classification of the gesture. They could watch the recorded image back and forth by operating the VCR.
(3) The same image was put in the prototype system to classify the gestures by the system. When classifying the gestures, the processing time of each video frame was measured.

4.4 Experimental Result

Table 3 shows the classification result of the system and three subjects. In the table, frame number, classification result of the system, those of subject SK, OT and MT are shown from the left column. The alphabets indicated as the results by the subject are classification of each subject. There is no relationship of alphabets among the subjects. The "XX" means that the subject could not classify the gesture because it was too complicated. Table 4 shows the number of segmented gestures and clusters by the system and the subjects. Table 5 shows the detail comparison of classification result by the system and subject MT. Table 6 shows the average of the processing time of each subsystem.

Table 4. Segmented Gestures and Clusters by System and Subjects

Number of	System	Subject		
		SK	OY	MT
Segmented Gestures	75	62	53	77
Clusters	19	19	20	25

Table 6. Processing Time (Average)

Process	Time [msec]
Capturing dynamic image (H/W Process)	0.35
Feature Extraction Subsystem	25.99
Feature Analysis Subsystem	0.08
Classification Subsystem	0.20
Total	26.62

4.5 Discussion

Table 4 shows that the number of segmented gestures by the system is almost the same as or a little more than those by the subjects. It means that the system could pick up most of the movement without missing. There is however a little gestures, which are recognized by all the subjects but the system. They are small head movement such as light nodding. The reason why the system could not segment such gesture is because head movement was so small that the system did not recognize it as a gesture.

On the other hand, the number of clusters classified by the system is almost the same as or a little less than those by the subjects. Table 3 shows that the classifications by the subjects are not all the same but have individual varieties. According to the subjects' reports, it took more than two hours for the subjects to classify all the gestures in only 15 minute video image, so that their classification standard had not been stable from the start to the end of the video image. This means that it was difficult for the subjects to classify all the gestures appropriately.

Next, the classification result by the system is compared with that by the subject MT in detail. In Table 5, the cluster #1 by the system has 9 of both hand gestures among 11 gestures classified by the subject. The cluster #2 has 13 of right hand gestures among 18 gestures by the subject. The cluster #3 has 13 of left hand gestures among 16 gestures. This result shows that most of the gestures were classified appropriately by the system. In addition, it might be possible for the system to classify the gesture which human subjects classified as "XX" shown in Table 3. In Table 3, however, there are four gestures which all the subject recognized as a gesture but the system could not segment them.

The frame rate of the input image is 30 frames per second. It is, therefore, necessary that the processing time for each frame is less than 33.3 msec in order to classify the gestures without delay. Table 6 shows that the average of total processing time was 26.62 msec. Most of the processing time was spent in the feature extraction subsystem because it deals with two-dimensional data.

241

5. CONCLUSION

In this study, the authors have proposed an automatic gesture classification method as a basic study of recognition for naturally expressed gesture during conversation. The method consists of (1)feature extraction, (2)feature analysis and (3)classification. And a prototype system has been developed based on the method. Then, the system has been evaluated by comparing with human subject classification. The result shows that the system classified most of the gestures appropriately, however, the system could not segment a few gestures which has only small head movement such as light nodding.

Table 3. Classification Result

Frame	System	SK	OY	MT
182	1			A
650	1			A
723	1			
1050	1			
1793	2	A	A	B
1866	2			
2125		B	B	C
2792	1			D
2821	3	C	C	
2909	3			
3515	2	B	B	C
3938	2			A
4178				E
4456		D	D	F
4535	2	E	D	F
4590	3	B	E	G
4657			B	H
4763	2			
4818	3	E	F	I
4866		E	F	
5006	2	D	D	F
5053	2		D	F
5185	3	E	H	I
5408		F		A
5512	4	G		
5554	5			
5664	6		I	J
7150	2	H	G	K
7654	3	I	J	L
8004	3	I	J	L
8057	6	J	XX	XX
8254	7	J		
8485	3		J	L
8524	3	K		
8656	3		K	XX
8707	8	L	L	XX
8837	1			E
8904	2	M	L	M
9018	2		D	N
9047	2	A	D	N
9171	9	C	M	XX
9277	1			
9403	3	E	H	O
9810			G	N
9829	10	D	XX	J
10105	1	N		E
10388		J	N	XX
10719			N	
11121	3	E	H	O
11605	3	E	F	P

Frame	System	SK	OY	MT
12616	2	G	O	Q
12661	1			A
14033				A
14803	3	I	J	L
15005	11	G	O	Q
15370	2	A	A	N
15425		A	A	N
15770				A
15946	12		XX	XX
16000	2	G		
16450		F		A
16628	2	D	D	F
16988	3	J	M	XX
17071	13	H	P	R
17158	14	J	Q	S
17236	15			
17271	1	O	N	T
17591	2	A	R	XX
17730	2	M	R	B
17927	1			A
18344	2	D	D	F
18587			P	
18739	1	P		A
18763	16			A
19030			P	A
19995				A
20367	10	Q	R	U
21040	3	J	S	V
21113	3			W
21972		B	T	
22136	17	G	I	J
22211	2	G	R	J
22833			P	A
23230				X
23380				A
23577	1	R		A
23604	16			
23669	18	P		
23779	2	S		XX
23871	2			
23949	1			
24144	1	P		A
24393				A
25311	19	P		A
26136				A
26459		P		
26563	2			A
26829	3	E	H	G
26978	3	E	J	G
27004		E		

Table 5. Comparison of System and Subject MT

System	Frame	Subject MT	Description by subject MT
1	182	A	Move hands on the thighs
	650	A	Put hands between thighs
	723		
	1050		
	2792	D	Scratch head
	8837	E	Put hands between thighs
	9277		
	10105	E	Put hands between thighs
	12661	A	Move hands on the thighs
	17271	T	Fix up sleeve
	17927	A	Move hands on the thighs
	18739	A	Move hands on the thighs
	23577	A	Move hands on the thighs
	23949		
	24144	A	Move hands on the thighs
2	1793	B	Raise right hand
	1866		
	3515	C	Raise both hands a little
	4535	F	Point a finger of right hand
	4763		
	5006	F	Point a finger of right hand
	5053	F	Point a finger of right hand
	7150	K	Point a finger of right hand
	8904	M	Wave right hand
	9018	N	Point a finger of right hand
	9047	N	Raise right hand
	12616	Q	Move something by hands
	15370	N	Raise right hand
	16000		
	16628	F	Point a finger of right hand
	17591	XX	Phonecall gesture
	17730	B	Phonecall gesture
	18344	F	Point a finger of right hand
	22211	J	"Apart from that" gesture
	23779	XX	"Jumble" gesture
	23871		
3	26563	A	Move hands on the thighs
	2821		
	2909		
	4590	G	Wave left hand
	4818	I	Stroke right hand down
	5185	I	Raise left hand
	7654	L	Point a finger of left hand
	8004	L	Point a finger of left hand
	8485	L	Point a finger of left hand
	8524		
	8656	XX	Touch chin by left hand
	9403	O	Raise left hand
	11121	O	Raise left hand
	11605	P	Raise left hand
	14803	L	Point a finger of left hand
	16988	XX	Point a finger of left hand
	21040	V	Point a finger of left hand back
	21113	W	Turn over both hands
	26829	G	Wave left hand
	26978	G	Wave left hand
4	5512		
5	5554		
6	5664	J	"Apart from that" gesture
	8057	XX	Count number by hands

⋮ Continued

REFERENCES

Mehrabian, A., "Nonverbal Communication", Aldine-Atherton, Chicago, Illinois, 1972.

Ekman, P. and Friesen, V., "The Repertoire of Nonverbal Behavior: Categories, Origins, Usage, and Coding", Semiotica, 1969, Vol.1, pp.49-98.

Gao, W., Ma, J., Wu, J., Wang, C., "Sign Language Recognition Based on HMM/ANN/DP", International Journal of Pattern Recognition and Artificial Intelligence, 2000, Vol.14, No. 5, pp.587-602.

Wilson, A. D. and Bobick, A. F., "Parametric Hidden Markov Models for Gesture Recognition", IEEE Transactions on Pattern Analysis and Machine Intelligence, 1999, Vol.21, No. 9, pp.884-900.

Non-Verbal Communication System in Cyberspace

Shigeo MORISHIMA

Seikei University/ATR Spoken Language Translation Research Lab.
3-3-1 Kichijoji-kitamachi Musashino Tokyo 180-8633 JAPAN
shigeo@ee.seikei.ac.jp

Abstract:

In this paper, we describe a recent research results about how to generate an avatar's face on a real-time process exactly copying a real person's face to realize non-verbal communication in cyberspace. It is very important for synthesis of a real avatar to duplicate emotion and impression precisely included in original face image and voice. Face fitting tool from multi-angle camera images is introduced to make a real 3D face model with real texture and geometry very close to the original. When avatar is speaking something, voice signal is very essential to decide a mouth shape feature. So real-time mouth shape control mechanism is proposed by conversion from speech parameters to lip shape parameters using multilayered neural network. For dynamic modeling of facial expression, muscle structure constraint is introduced to generate a facial expression naturally with a few parameters. We also tried to get muscle parameters automatically to decide an expression from local motion vector on face calculated by optical flow in video sequence.

1 INTRODUCTION

Our final goal is to generate a virtual space close to the real communication environment between network users or between human and machine. There is an avatar in cyberspace projecting the feature of each user which has a real texture-mapped face to generate facial expression and action which is controlled by multi-modal input signal. Very important factor to make an avatar look believable or alive depends on how well an avatar can duplicate a real human's expression and impression on a face precisely.

In section 2, Face fitting tool from multi-angle camera images is introduced to make a real 3D face model with real texture and geometry very close to the original. This fitting tool is GUI based system and easy mouse operation picking up each feature point on face contour and face parts can help easy construction of 3D personal face model.

When avatar is speaking, voice signal is very essential to determine a mouth shape feature. So real-time mouth shape control mechanism is proposed by conversion from speech parameters to lip shape parameters using neural network. This neural network can realize an interpolation between specific mouth shapes given as learning data [1,2]. Emotional factor sometimes can be captured by speech parameters. This media conversion mechanism is described in section 3.

For dynamic modeling of facial expression, muscle structure constraint is introduced to make a facial expression naturally with a few parameters. We also tried to get muscle parameters automatically from local motion vector on face calculated by optical flow in video sequence. So 3D facial expression transition is duplicated from original people by analysis of 2D camera captured video image without landmarks on face. These are expressed in section 4.

2 FACE MODELING

To generate a realistic avatar's face, a generic face model is adjusted to user's face image. The generic face model has all of the control rules for facial expressions defined by FACS parameter as a 3D movement of grid points or a facial muscle structure physically simulated to modify model geometry.

Fig.1 shows a personal model both before and after fitting process for front view image by using our original GUI based face fitting tool. Synthesized face is coming out by mapping of blended texture generated by user's images onto the modified personal face model. And also to construct 3D model more accurately, we introduce multi-view face image fitting tool. Fig.2 shows fitting result to captured images from several angles. Fig.3 shows example of reconstructed face using 9 views images

3 VOICE DRIVEN TALKING AVATAR

To realize lip synchronization, spoken voice is analyzed and converted into mouth shape parameters with neural network on frame by frame basis.

Multiple users' communication system in cyberspace is constructed based on server-client system. In this system, only a few parameters and voice signal are transmitted through network. Avatar's face are synthesized by these parameters locally at client system.

3.1 Parameter conversion

At server system, voice from each client is phonetically analyzed and converted to mouth shape and expression parameters. LPC Cepstrum parameters are converted into mouth shape parameters by neural network trained by vowel features. Fig.4 shows neural network structure for parameter conversion. 20 dimensional Cepstrum parameter are calculated every 32ms with 32ms frame length. At client system, on-line captured voice of each user is digitized by 16KHz and 16bits, and is transmitted to server system frame-by-frame through network in communication system. And then mouth shape of each avatar in cyberspace is synthesized by this mouth shape parameters received at each client.

3.2 Mouth shape editor

Mouth shape can be easily edited by our mouth shape editor. We can change each mouth parameter to decide a specific mouth shape on preview window. Typical vowel mouth shapes are shown in Fig.5. Our special mouth model has polygons for mouth inside and teeth. For parameter conversion from LPC Cepstrum to mouth shape, only mouth shapes for 5 vowel and nasals are defined as training set. We have defined all of the mouth shapes for Japanese phoneme and English phoneme by using this mouth shape editor. Fig. 6 shows a synthesized avatar's face speaking phoneme /a/.

3.3 Multiple points communication system

Location information of each avatar, mouth shape parameters and emotion parameters are transmitted every 1/30 seconds to client system. Distance between every 2 users are calculated by the avatar location information, and voice from every user except himself is mixed and amplified with gain according to the distance. So the voice from the nearest avatar is very loud and one from far away is silent.

Based on facial expression parameters and mouth shape parameters, avatar face is synthesized frame by frame. And

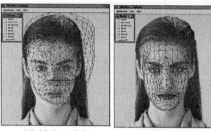

a) Initial model   b) Fitted model

Fig.1 Frontal model fitting by GUI tool

Fig.2 Multi-view fitting result

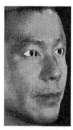   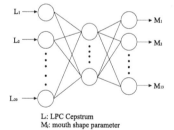

L: LPC Cepstrum
M: mouth shape parameter

Fig.3 Synthesized face   Fig.4 Network for parameter conversion

a) Shape for /a/   b) Shape for /i/   c) Shape for /u/   d) Shape for /o/

Fig.5 Typical mouth shapes

244

avatar body is located on cyberspace according to the location information. There are two modes for displaying, view from avatar's own eyes for eye contact and view from sky to search for other users in cyberspace and these views can be chosen by menu in window. Fig.7 shows a communication system in cyberspace with avatar.

3.4 User adaptation

When new user comes in, his face model and voice model have to be registered before operation. For voice adaptation, 75 persons' voice data including 5 vowels are pre-captured and database for weights of neural network and voice parameters are constructed. So speaker adaptation is performed by choosing the optimum weight from database. Training of neural network for every 75 persons' data has already finished before operation. When new nonregistered speaker comes in, he has to speak 5 vowels into microphone. LPC Cepstrum is calculated for every 5 vowels and this is given into the neural network. And then mouth shape is calculated by selected weight and error between true mouth shape and generated mouth shape is evaluated. This process is applied to all of the database one by one and the optimum weight is selected when the minimum error is detected.

3.5 Entertainment application

An interactive movie system we constructed is an image creating system in which user can control facial expression and lip motion of his face image inserted into movie scene. User gives voice by microphone and pushing keys which determine expression and special effect. His own video program can be generated on realtime.

At first, once a frontal face image of visitor is captured by camera. 3D generic wireframe model is fitted onto user's face image to generate personal 3D surface model. At interactive process, a famous movie scene is going on and face part of actor or actress is replaced with visitor's face. And also facial expression and lip shape are controlled synchronously by captured voice. Fig.8 shows the original movie clip and Fig.9 shows the result of fitting of face model into this scene. Fig.10 shows an user's face inserted into actor's face after color correction. Any expression can be appended and any scenario can be given by user independent to original story in this interactive movie system.

4 MUSCLE CONSTRAINT FACE MODEL

Muscle based face image synthesis is one of the most realistic approaches to the realization of a lifelike agent in computers. A facial muscle model is composed of facial tissue elements and simulated muscles. In this model, forces are calculated effecting a facial tissue element by contraction of each muscle string, so the combination of each muscle contracting force decides a specific facial expression. This muscle parameter is determined on a trial and error basis by comparing the sample photograph and a generated image using our Muscle-Editor to generate a specific face image. In

Fig.6 Synthesized face speaking /a/

Fig.7 Communication system in cyberspace

Fig.8 Original movie clip

Fig.9 Fitting result of face model

Fig.10 Reconstructed face by user's

this section, we propose the strategy of automatic estimation of facial muscle parameters from 2D optical flow using a neural network. This corresponds to the 3D facial motion capturing from 2D camera image under the physics based condition without markers.

We introduce multilayer backpropagation network for the estimation of muscle parameter. The face expression is then resynthesized from estimated muscle parameter to evaluate how well an impression of original expression can be recovered. We also tried to generate animation using the captured data from an image sequence. As a result, we can get and synthesize image sequence which give an impression very close to the original video.

4.1 Layered dynamic tissue model

The human skull is covered by a deformable tissue which has five distinct layers. Four layers (epidermis, dermis, subcutaneous connective tissue, and fascia) comprise the skin, and the fifth layer comprises the muscles responsible for facial expression. In accordance with the structure of real skin, we employ a synthetic tissue model constructed from the elements illustrated in Fig.11, consisting of nodes interconnected by deformable springs (the lines in the figure). The epidermal surface is defined by nodes 1, 2, and 3, which are connected by epidermal springs. The epidermal nodes are also connected by dermal-fatty layer springs to nodes 4, 5, and 6, which define the fascia surface. Fascia nodes are interconnected by fascia springs. They are also connected by muscle layer springs to skull surface nodes 7, 8, 9. The facial tissue model is implemented as a collection of node and spring data structures. The node data structure includes variables to represent the nodal mass, position, velocity, acceleration, and net force. The spring data structure comprises the spring stiffness, the natural length of the spring, and pointers to the data structures of the two nodes that are interconnected by the spring. Newton's laws of motion govern the response of the tissue model to force [7]. This leads to a system of coupled, second-order ordinary differential equations that relate the node positions, velocities, and accelerations to the nodal forces.

4.2 Facial muscle model

Fig.12 shows our simulated muscles. Black line means location of each facial muscle in a layered tissue face model. Each muscle is modeled by combination of separated muscles and left part and right part of face are symmetric. For example, frontalis has a triangularly spreaded feature, so 3 kinds of linear muscles model this frontalis on one side. Frontalis pulls up the eyebrows and makes wrinkles in the forehead. Corrugator, which is modeled by 4 springs located between eyes, pulls the eyebrow together and makes wrinkles between left and right eyebrows. However those muscles can't pull down the eyebrows and make the eyes thin. So Orbicularis oculi muscle is also appended in our model. Orbicularis oculi is separated into upper part, and lower part. To make muscle control simple around the eye area, the Orbicularis oculi is modeled with a single function in our model. Normally, muscles are located between a bone node and a fascia node. But the Orbicularis oculi has an irregular style, whereby it is attached between fascia nodes in a ring configuration; it has 8 linear muscles which approximate a ring muscle. Contraction of ring muscle makes the eye thin. The muscles around mouth are very important for the production of speaking scenes. Most of the Japanese speaking scenes are composed of vowels, so we mainly focused on the reproduction of vowel mouth shapes as a first step and relocated the muscles around mouth [8]. As a result, a final facial muscle model has 14 muscle springs in the forehead area and 27 muscle springs in the mouth area.

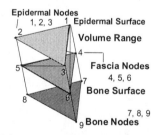

Fig.11 Layered tissue element

Fig.12 Facial muscle model

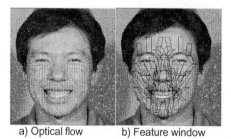

a) Optical flow   b) Feature window

Fig.13 Motion feature vector on face

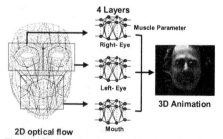

Fig.14 Conversion from motion to animation

4.3 Motion capture on face

A personal face model is constructed by fitting the generic control model to personal range data. Optical flow vector is calculated in a video sequence and accumulate motion in the sequence from neutral to each expression. And then motion vectors are averaged in each window shown in Fig.13. This window location is determined automatically in each video frame.

A layered neural network finds a mapping from the motion vector to the muscle parameter. A four-layer structure is chosen to effectively model the nonlinear performance. The first layer is the input layer, which corresponds to 2D motion vector. The second and third layers are hidden layers. Units of the second layer have a linear function and those of the third layer have a sigmoid function. The fourth layer is the output layer, corresponding to muscle parameter, and it has linear units. Linear functions in the input and output layers are introduced to maintain the range of input and output values.

A simpler neural network structure can help to speed up the convergence process in learning and reduce the calculation cost in parameter mapping, so the face area is divided into three subareas as shown in Fig. 14. They are the mouth area, left-eye area, and right-eye area, each giving on independent skin motion. Three independent neural networks are prepared for these three areas.

Learning patterns are composed of basic expressions and the combination of their muscle contraction forces. We create 6 basic facial expressions consisting of anger, disgust, fear, happiness, sadness and surprise. Basic mouth shapes for vowels "a", "i", "u", "e" and "o", and a closed mouth shape.

Motion vector is given to a neural network from each video frame and then a facial animation is generated from the output muscle parameter sequence. This is the test of the effect of interpolation on our parameter conversion method based on the generalization of the neural network. Fig.15 shows a result of expression regeneration for surprise from original video sequence.

5 CONCLUSION

To generate a realistic avatar's face for face-to-face communication, multi-modal signal is introduced to duplicate original face expression. Voice is used to realize lip synchronization. Video captured image is used to regenerate original face expression under facial muscle constraint. All kinds of methods are based on neural network to realize a parameter mapping.

Reference

[1] Morishima, S. and Harashima H. (1991). *A Media Conversion from Speech to Facial Image for Intelligent Man-Machine Interface*, IEEE JSAC, Vol.9, No.4, pp. 594-600.

[2] Morishima, S. (1997). *Virtual Face-to-Face Communication Driven by Voice Through Network*. Workshop on Perceptual User Interfaces, pp.85-86.

[3] Ekman, P. and Friesen, W.V. (1978). *Facial Action Coding System*. Consulting Psychologists Press Inc.

[4] Essa, I., Darrell, T. and Pentland, A. (1994). *Tracking Facial Motion. Proceedings of Workshop on Motion and Nonrigid and Articulated Objects*, pp.36-42.

[5] Mase, K. (1991). *Recognition of Facial Expression from Optical Flow*. IEICE Transactions, Vol E 74, No. 10.

[6] Morishima, S. (1996). *Modeling of Facial Expression and Emotion for Human Communication System*. Displays 17, pp.15-25, Elsevier.

[7] Lee, Y, Terzopoulos, D. and Waters, K. (1995). *Realistic modeling for facial animation*. Proceedings of SIGGRAPH'95, pp.55-62.

[8] Sera, H., Morishima, S.and Terzopoulos, D. (1996). *Physics-based muscle model for mouth shape control*. Proceedings of Robot and Human Communication, pp. 207-212.

[9] Schlosberg, H.(1954). *Three Dimension of Emotion*, Psychological Review, 61(2), 81-88.

[10] Smith, C.A. and Ellsworth, P.C.(1985). *Pattern of Cognitive Appraisal in Emotion*, Journal of Personality and Social Psychology, 48(4), pp813-838.

a) Original   b) Regenerated

Fig.15 Face expression regeneration

Embodied Interface for Emergence and Co-share of 'Ba'

Yoshiyuki MIWA[1], Shigeru WESUGI[2], Chikara Ishibiki[2], and Shirou ITAI[2]

[1]Waseda University, 3-4-1,Ohkubo, Shinjuku, Tokyo, Japan
miwa@mn.waseda.ac.jp

[2]Graduate School of Science and Engineering, Waseda University,

3-4-1,Ohkubo, Shinjuku, Tokyo, Japan
wesugi@computer.org, {ishibiki, itai}@miwa.mech.waseda.ac.jp

Abstract

This paper describes a conceptual idea of 'Emergence Reality' applying emergence and co-share of 'Ba' for creating a co-existing feeling, and development of two prototyped interface systems. Firstly, the interface is designed for support of three-dimensional modeling between remote locations by bridging between computer-generating virtual space and physically-existing real space subconsciously and by creating co-existing virtual space. Secondly, the interface is communication system applying face robot as bodily media and is expected to create a remote situation by combining robot with video image. Development and some experiments give expectations of 'Ba' communication creating actuality at remote location.

1.'Ba' communication

The message transmitted over cyberspace among remote locations is separated from a sender and just explicit information, and hardly handles implicit information such as conditions and fluctuations taking place in mind, emotional activity and bodily sense in general. On the other hand, during the face-to-face activities this implicit information is expressed through mutual bodily interaction. And mutual bodily-synchronized interaction, which is called as an entrainment phenomenon (Condon, 1974), is considered to cause the co-sharing of the semantic situation in the dialogue or the context (Miwa, 2000; Watanabe, 1997). 'Ba' ,as authors defined in this paper, is an implicit space accompanied with feeling requiring to define the context, and is generated within mind. And we believe a co-existing feeling or 'being together, here' emerges among people by co-sharing 'Ba'.

On the other hand, other studies of supporting telecommunication have been conducted these days: clear board for drawing the pictures and characters each other(Ishii,1994), hyper-mirror applying mirror metaphor(Morikawa,1998), virtual conference by computer graphic avatar (Greenhalgh, 1997), and collaborative work with robots between remote locations(Kotoku,2000). However above mentioned researches handle only explicit information, so that they leave a lot of problems, for examples it doesn't create a co-existing feeling well, adjust timing and spacing each other, nor help to sense mutual situation.

In order to investigate 'Ba' communication technique for emergence of co-existing feeling between remote locations, authors continuously have designed and developed based on bodily interaction (Miwa, 2001;Shimizu, 2000;WESUGI, 2000; MIWA, 1999). But it is a present situation that 'Ba' expression technique is still developing which evokes a co-existing feeling at remote locations within self. Then for approaching the problem we designed and developed two interface systems: co-creative modeling system bridging between real and virtual space, and communication system applying robot as bodily media. This paper describes in following chapters a conceptual idea of 'Emergence Reality' for evoking 'Ba', and those interface systems and results of experiments that they are expected to create 'Ba', which is different from present virtual reality produces.

Figure1: Co-creative telecommunication

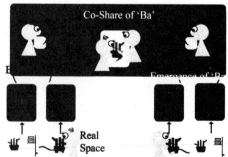

Figure2: Emergence and Co-share of 'Ba'

2.Emergence Reality for 'Ba' Technique

In communicating activity, explicit information transmitting, for example improving image and sound quality, is not always best way to handle 'Ba' for a co-existing feeling. It must be a good example that handwriting letter carries vividly feeling or emotional activities of the writer. In order to handle implicit information such as conditions, fluctuations taking place, bodily sense in mind, how much media includes bodily movements or embodied interaction is more important than how clear image is, considering an entrainment phenomenon as a clue. 'Emergence Reality ' proposed by authors' group is a conceptual idea of creating a co-existing feeling based on 'Ba' by spatial expression of embodied interaction. And this idea is also called 'Partial Reality ' from the view how to express a part for creating whole situation. The idea will be available for 'Ba' communication in the future, by which transmitting the message on co-sharing 'Ba' comparing to only transmitting explicit information.

3.Hands-on Modeling Interface

Authors aim at design and development of interface system supporting hands-on three-dimensional geometric modeling, for examples clay modeling, building blocks and sand playing, between remote locations. This system has two main features in design: modeling with physically real hand in virtual space and 'building together' in co-shared virtual space not one-way instruction (Wesugi, 2000). On the other hand, modeling interface systems so far are insufficient to satisfy the features as mentioned in previous chapter. In order to solve this, authors designed two-step approach as depicted figure 2. The first step is bridging between physically hands-on working real space and computer generating virtual space showing visualized modeling process through embodied interface subconsciously. The second step is creation of co-existing virtual space for a feeling of 'building together'.

3.1 Embodied Interface for Bridging between Real and Virtual

An embodied interface system was prototyped so far for bridging between real space and virtual space as depicted in figure 3. The system for flower arrangement-like enables users to actually put leaf-unit into /out of stem-unit and to see the virtual plant and the real hands with optical see-through Head-Mounted Display (HMD). And a clue to solve an occlusion problem is found by adjusting the difference of luminance among hands, virtual objects, and flower-like device respectively as depicted figure 4. Some experiments of free modeling by six male adults show a feeling of consistent integration of real and virtual occurs when they operate the flower-like devise during one part of body touching the device, and when they vibrate stem or leave units dynamically. These comments show that the subconscious integration between real and virtual occurs through emergence of 'Ba' caused by introducing embodied interface by bodily interaction, not only through accurate registering real image and virtual image geometrically.

3.2 Peripheral Representation of Embodied Interaction

Authors designed and developed spatial representation for sensing mutual situation through co-share of 'Ba' during hands-on modeling activities in the virtual space. The unique feature of the representation is situation-accompanied expression such as finger pointing and gaze not only explicit pointer such as arrow and characters. And we hope to get clues to design expression stimulating co-share of 'Ba' by synchronizing mutual bodily movements. These representations are three types: embodied avatar applying eye, blink and head movements, gaze direction vector applying head movement, and peripheral frame of which location, posture, and transparency changes by applying head movement signal, and are shown in figure 5.

The examination was carried out for investigating gaze effect of these representations as a clue to follow target operated by mouse among three couples. This results that the embodied avatar is available for gaze direction, and the direction vector is valuable to instruct one-way. Additionally peripheral frame for sensing mutual situation contributes to a feeling of gaze awareness in spite avatar face out of HMD view area, and to induce others gaze subconsciously, but not to direct indication. This peripheral frame is expected to create remote situation through emergence and co-share of 'Ba'.

Luminance of hands is low $\longrightarrow$ high

Figure3: Utilizing the flower arrangement system   **Figure4:** Solution of Occulusion

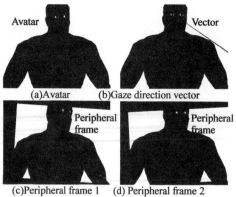

(a)Avatar   (b)Gaze direction vector

(c)Peripheral frame 1   (d) Peripheral frame 2

Figure5: Three representations

(a)Combpined Image   (b)Actual Modeling

Figure 6: Hands-on Modeling

Figure 7: Mulutiple Hands-on Modeling System

Authors based on above results so far and have prototyped interface system as shown in figure 6, which enables users to model structures with bricks both in real space and in virtual space by combining virtual objects on the real structures and their hands with HMD. Then this system is expected to multiple users between remote locations as shown in figure 7.

4.Eye Robot on Emergence Reality
4.1 Robot as Bodily Media

'Ba' is considered to be generated within self and to appear outside as facial expression and as bodily movement. Authors designed and developed Eye robot as 'Ba' expression media in real time based on gaze, blink and nod under this theory (Miwa, 1999). Additionally this robot changes its facial frame, interval between eyes, and look by applying physiological signals such as respiration and cardiac beat. Variable facial expression of this robot is shown in figure 8. Facial frame expands when breathing in, and contracts when breathing out. In non face-to-face communicating activities this robot participates in conversation as an agent at remote location, and moves by utilizing gaze and nod of operator. In the conversation experiment a phenomenon nearly like entrainment, by which bodily movement of observer got involved into robot movement, was often observed as shown in figure 9 comparing to video conversation. Additionally some experiments show that robot movement can express characteristics of operators. These results mean embodied robot can help emergence and co-sharing of 'Ba'. On the other hand, only an interaction with robot is insufficient to give observers a feeling of 'existing with the operator'.

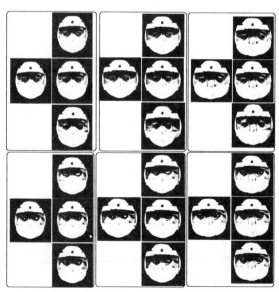

Figure 8: Variable facial expression

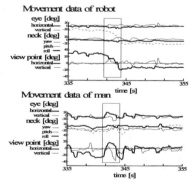

(a)An example showing the dynamic attention-shared

(b)Communication scene to many people
(right:action showing the attention-shared phenomenon)
Figure 9: Communicating people

(a) HMD   (b)Projector

Figure 10: Technique and utilization of Partial Reality

4.2 Expression by Combining Robot with Video Image

Embodied expression through only media robot hardly creates a co-existing feeling as mentioned above. In order to solve this problem, authors designed a method combining the robot with video image. And we investigated the technique caused that actuality of the operator emerged in the video image and coherency was generated between robot and situation where operator exists through means that eye and head movements of robot corresponded to those of operator. Then we investigated two ways of overlaying: optical see-through HMD and projector casting video image onto the robot. In each system the robot moves with video image overlapped on the face. The difference of those is background situation. Firstly in the HMD system background is where observer exists. Secondly in the projector system background is a video image where the operator exists. Each configuration and composite images in experiments are shown in figure 10. Some communication experiments present a feeling of 'operator is existing ' increases surely on this method when the gaze, blink and neck movement of robot synchronize with those of operator in video image.

Additionally an extremely interesting phenomenon is found that background itself in the video image have actuality, for example as if hands grew out of robot and appeared touching a pot plant when remote operator touches incidentally the pot plant around.

This result means actuality emerges in things as well as operator where the operator exists by partial movements of robot. But the phenomenon doesn't appear when an observer watches carefully an operator touching a pot plant.

5.Conclusions

Authors aim at design and development of telecommunication system based on emergence and co-share of 'Ba' for generating a co-existing feeling. And we have proposed a conceptual idea 'Emergence Reality' creating remote actuality from partially embodied interaction as positive measures of 'Ba' communication. In order to obtain a clue to establish the idea, two communication systems were designed and developed.

Firstly the system is for three-dimensional modeling between remote locations, and has two unique features: embodied interface bridging real and virtual, and representations for creating co-existing feeling in such virtual space. The system for flower arrangement-like activity enables participants to hands-on modeling in both real and virtual space. And it shows the embodied interface is valuable to stimulate bridging real and virtual subconsciously through emergence of 'Ba'. In the future work experiments should be investigated to measure emergence of the consistent integration by bodily activity. And in order to create co-sharing virtual space, three representations, embodied avatar, direction vector, and peripheral frame of which location, posture, and transparency changes, were developed, and their effects were investigated. Some experiments, following target between a couple, show these representations are available to express others gaze, and are expected to sense gaze awareness. Especially peripheral frame representation is expected to be helpful to realize 'Emergence Reality' fro creating a co-existing feeling through 'Ba' activity, because the representation might stimulate to synchronize mutual bodily interaction by dynamic background fluctuations.

Secondly eye robot was designed and developed as media representing facial expression and bodily movements in real time based on gaze, blink and nod. Additionally in order to increase actuality of operator and things in the video image combining the robot with video image is attempted by overlapping video image onto the robot. And this technique is different from AR, projecting virtual image onto real space, or MR, expanding AR idea, in principle, because AR or MR handles only explicit information on the basis of quantity. On the other hand, the purpose of this technique is to give actuality to the operator and things where it exists by partial movements of robot. Especially it is considered to relate to handle implicit information on the basis of quality or in semantic manner. An phenomenon as if arms of operator grew out of robot presents that a coherency emerges

251

between robot and actual situation around its operator, though it might be a kind of illusion. And this result propose that the lack of actuality in the video image can be complemented by means emergence of 'Ba' in mind based on robot movements partially synchronized with video image, not by complete human-mimetic robot as well as operator. This robot communication system has many problems which must be solved, especially what an accurate register of robot and video image requires in order to give actuality to the image, and how much partial information requires in order to create remote situation.

Authors' unique technique 'Emergence Reality' is expected to contribute to 'Ba' communication in the near future, which creates a co-existing feeling based on emergence and co-share of 'Ba' through subconscious embodied interface.

Acknowledgements

The authors would like to thank Shimizu Hiroshi for his helpful comments on this research.

References

Condon,W.S.,Sander,L. ,(1974),Neonate movement is synchronized with adult speech,*Science*,No.183,pp.99-101.

Greenhalgh,C.,Bullock,A.,Tromp,J.,Benford,.S.,(1997),Evaluating the network and usability characteristics of virtual reality conferencing, *BT Technol J*,Vol.15,No.4,pp.101-119.

Ishii,H.,Kobayashi,M.,Arita,K.,: Interactive Design of Seamless Collaboration Media, *COMMUNICATION OF THE ACM*, Vol.37, No.8, pp.83-97,(1994).

Shimizu,H., et al., Ba and Co-creation (Japanese), NTT publishing co., Ltd., pp.23-177, 2000.

Kotoku,T.,Chong,N.,et al.: Tele-collaboration system with on-line graphic simulator, *Proc. of SICE System Integration Division Annual Conference*,pp.127−128,(2000).

Miwa,Y.,Ishibiki,C.,WESUGI.S,(2001),'Ba'Expression Technique for Communication- Proposal for Emergence Reality-, *Japan Society of Physiological Anthropology 45th Annual conference* ,(Japanese) .

Miwa,Y., et al.,(2000), Measurement of Entrainment Emergence Process, *Journal of Human Interface Society*, vol. 2, no. 2, pp.185-191.

Miwa,Y., et al. , (1999),Performance robot serving as communication media: development of an 'eyeball robot' investigating bodily communication, *Advanced Robotics*, Vol.13, No.3, pp.279-281.

Morikawa,O.,Maesako,T. ,(1998).HyperMirror:Toward Pleasant-to-use Video Mediated Communication System, *CSCW98*, pp.149-158.

Watanabe,T., et al.,(1997),Evaluation of the Entrainment Between a Speaker's Burst-Pause of Speech and Respiration and a Listener's Respiration in Face-to-Face Communication, *Proc. of the 6th IEEE International Workshop on Robot and Human Communication*, pp.392-397.

Wesugi, S., Yoshiyuki, M., (2000),Embodied Interface for Emergence and Transmission of 'Ba', *Proc. of the Human Interface Symposium 2000*, pp.73-76, (Japanese).

E-COSMIC: Embodied Communication System for Mind Connection

Tomio Watanabe

Faculty of Computer Science and System Engineering, Okayama
Prefectural University 111 Kuboki, Soja, Okayama, JAPAN 719-1197
CREST of JST (Japan Science and Technology)

On the basis of the entrainment among verbal and nonverbal behavior in face-to-face communication, Embodied Communication System for Mind Connection (E-COSMIC) is developed for supporting essential human interaction in communication. E-COSMIC consists of an embodied virtual communication system for human interaction analysis by synthesis and a speech driven embodied interaction system. The system would form the foundation of mediated communication technologies as well as the methodology for the analysis and understanding of human interactions.

1. Introduction

Human communicates not only by verbal message but also by nonverbal behavior such as nodding and gesture. In particular, the coherently related synchrony of biorhythms among verbal and nonverbal behavior between talkers, which is called entrainment in communication, plays an important role in human interaction. The embodied communication closely related with behavioral entrainment is an essential form of communication in which the relation of interaction between talkers is formed through mutual bodies. Hence, the clarification of this mechanism is indispensable to the development of human interaction system.

In this present paper, focusing on the embodied entrainment, the concept of E-COSMIC is proposed for supporting and analyzing human interaction in communication. The prototype of the system is developed, and the sensory evaluation and behavioral analysis in human interaction demonstrate the effectiveness of the system.

2. Concept of E-COSMIC

The outline of E-COSMIC is shown in Fig.1. E-COSMIC mainly consists of an embodied virtual communication system and a speech driven embodied interaction system. The former is developed for human interaction analysis by synthesis, and the latter is for supporting human interaction on the basis of the analysis using the former.

In the embodied virtual communication system, two types of avatars for human interaction analysis by synthesis are developed (Watanabe and Okubo, 1999). One is a VirtualActor as a human avatar which represents interactive behavior such as the motions of head, arms and body; the other is a VirtualWave as an abstract avatar in which human behavior is simplified as the motion of wave to clarify an essential role of interaction. The system provides networked virtual communication environment in which two remote talkers can share embodied interaction through their VirtualActors / VirtualWaves including themselves in the same virtual space. The analysis by synthesis for interaction in communication is performed by processing the behavior of VirtualActors such as cutting or delaying the motion and voice of VirtualActors in various conditions of the spatial relation and positions of VirtualActors. Three VirtualActors analyze group communication systematically.

On the basis of the human interaction analysis by using the embodied virtual communication system, the speech driven embodied interaction system is developed for supporting human interaction by generating the communicative motions of robot called InterRobot coherently related to speech input (Watanabe et al, 2000). InterActor is the electronic media version of InterRobot as physical media, which sets free from the hardware restriction of InterRobot in GUI based network communication. SAKURA is a speech driven embodied entrainment communication system for activating group. The system creates virtual communication room where InterActors with both functions of speaker and listener are entrained one another as a teacher and some students by generating the whole body motion such as nodding, blinking and the actions of head, arms and waist coherently related speech input. By using SAKURA, talkers can communicate with the sense of unity through the entrained InterActors in the same virtual room by only speech input via network.

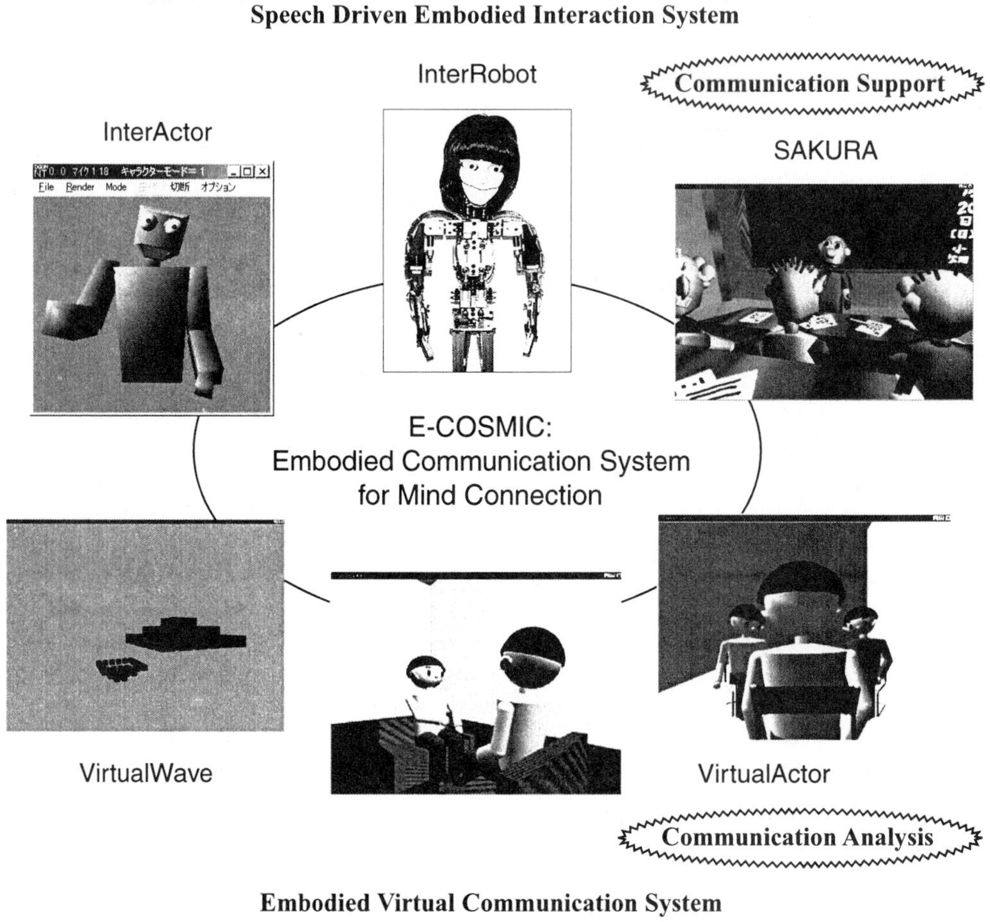

Speech Driven Embodied Interaction System

Figure 1: E-COSMIC

3. Embodied Interaction System

3.1 Concept of InterRobots System

The concept of InterRobots system for supporting human interaction is shown in Fig. 2. The system consists of two InterRobots which have both functions of speaker and listener. When talker 1 speaks to InterRobot 1, InterRobot 1 responses to the utterance by means of its entire body motions by nodding, blinking and gestures in the manner of a listener for talker 1. Thus, the talker can talk smoothly. Then, the speech is transmitted via a network to the remote InterRobot 2. InterRobot 2 can effectively transmit the talker 1's message to talker 2 by generating body motion in the manner of a speaker on the basis of a time series of the speech, and presenting both the speech and the synchronized body motion simultaneously. Talker 2 in the role of a speaker this time achieves communication by transmitting his speech via InterRobot 2 as listener and InterRobot 1 as speaker to the talker 1 in the same way. The only information transmitted or received by this system is voice. The system supports the sharing of mutual embodiment in communication by generating the related motions from the time series of speech on the basis of the relation between voice and body motion during the speech. InterRobot is replaced by InterActor in GUI based network communication.

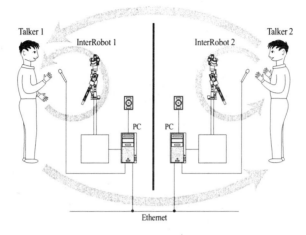

Figure 2: Concept of InterRobots system

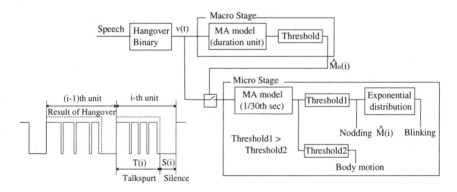

Figure 3: Listener's interaction model

3.2 Speech Driven Interaction Model

From the analysis of the relation between a speaker's voice and a listener's motions, the importance of the relation with the listener's head motion, which mainly included nodding, was pointed out (Watanabe et al, 2000). A speech driven interaction model was introduced as shown in Fig. 3. This model consists of two stages - macro and micro. The macro stage estimates the existence of nodding in one duration unit which consists of a talkspurt (ON) episode $T(i)$ and the following silence (OFF) episode $S(i)$ with a hangover value of 4/30s (Watanabe, 1990). We defined $Mu(i)$ in binary code

255

according to whether or not the nodding starts in the i-th unit. The estimator $\hat{Mu}(i)$ of Mu(i) at the macro stage is an MA model, expressed as the weighted sum of unit speech activity R;

$$\hat{M}_u(i) = \sum_{j=1}^{J} a(j)R(i-j) + u(i) \qquad (1)$$

$$R(i) = \frac{T(i)}{T(i) + S(i)} \qquad (2)$$

        a(j): linear prediction coefficient,   T(i): talkspurt duration in the i-th duration unit
        S(i): silence duration in the i-th duration unit,   u(i): noise

If $\hat{Mu}(i)$ is over a threshold value, the macro stage judges that nodding exists in that duration unit and transfers decision-making to the micro stage. The micro stage decides the onset of nodding. The estimator $\hat{M}(i)$ of M(i) is also an MA model, and expressed as the weighted sum of the binary voice signal V(i) in each video frame of 1/30 s over the past 2 (1/30 x 60) s.

$$\hat{M}(i) = \sum_{j=1}^{K} b(j)V(i-j) + w(i) \qquad (3)$$

        b(i): linear prediction coefficient,   w(i): noise

The estimator has two thresholds. If the estimator exceeds the first threshold value, the micro stage judges that the time, t, is the body motion onset. If the estimator is higher than the second, higher, threshold, the model confirms nodding. Because nodding follows the body motion, a listener's expressive action is realized. Figure 4 shows the result of estimation. This demonstrates the effectiveness of the model in comparison with the estimation by the only micro stage. The eye-blinks were generated with the timing of exponential intervals from each nodding onset in response to speech input, because the listener's eye-blinks took place during nodding simultaneously and the eye-blinking intervals were regarded as an exponential distribution. From the analysis of the communication experiment, the speaker's action model was made as its own MA model of the burst-pause of speech to the whole body motion in place of the

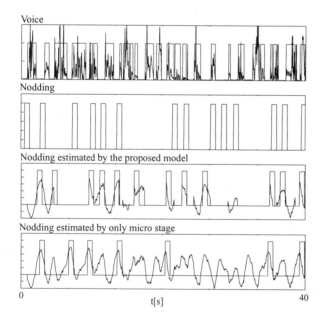

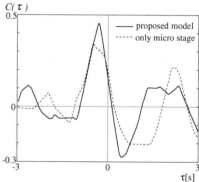

Figure 4: Time series of a speaker's voice, a listener's nodding, the estimated noddings, and cross-correlations $C(\tau)$ between the actual nodding and the estimated noddings.

listener's nodding in the same way as the listener's action model.

3.3 InterRobots System

By introducing both listener and speaker action models into a robot, InterRobot with the functions of both listener and speaker was realized. By using two InterRobots, a speech driven embodied interaction robots system was developed. A scene of remote communication by the system is shown in Fig. 5. Each subject carried out a verbal dialogue with the InterRobot in a separate room and his speech was transmitted or received via 100 MB Ethernet.

To confirm the effectiveness of InterRobots system for smooth interaction, we compared among (A) only voice, (B) projected picture of InterRobot and (C) InterRobot. In any communication mode, we used a microphone as input device and a loud speaker as output device. In the communication mode (B), we used the projected picture of the front view of the InterRobot to the screen installed at the front of the talker. The size of the picture was same as the real InterRobot. The subjects were five pairs of ten male students. After conversation via this system, the subjects filled out a questionnaire. The result of sensory evaluation is shown in Table 1. This demonstrates the effectiveness of InterRobot as the physical media of communication.

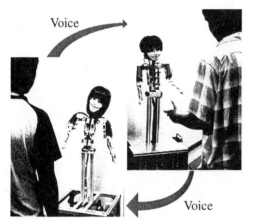

Voice

Voice

Figure 5: Remote communication

Table 1: Result of sensory evaluation.

	(A)	(B)	(C)	Total
(A) Voice		0	0	0
(B) Projection	10		1	11
(C) InterRobot	10	9		19

4. Conclusions

E-COSMIC is a robust and practical system for supporting essential human interaction in communication, in which talkers can share and entrain their embodiment in communication in actual or virtual space. The system would be expected to form the foundation of mediated communication technologies as well as the methodology for the analysis and understanding of various human interactions.

References

Watanabe, T. & Okubo, M. (1999), An Embodied Virtual Communication System for Human Interaction Sharing, Proc. of 1999 IEEE International Conference on Systems, Man, and Cybernetics, pp.1060-1065.

Watanabe, T., Okubo, M. & Ogawa, H. (2000), An Embodied Interaction Robots System Based on Speech, Journal of Robotics and Mechatronics, Vol.12, No.2, pp.126-134.

Watanabe, T. (1990), The Adaptation of Machine Conversational Speed to Speaker Utterance Speed in Human-Machine Communication, IEEE Trans. Systems, Man, and Cybernetics, Vol.SMC-20, No.2, pp.502-507.

Unconstrained Gesture Recognition

John Bellando and Ravi Kothari

Department of Electrical & Computer Engineering & Computer Science
University of Cincinnati, Cincinnati, OH 45221-0030

Abstract

Gesture plays a significant role in human communication, and the recognition of gestures is integral to natural Human Computer Interaction (HCI). In this paper, we propose a single camera based computational paradigm for the recognition of human gestures. Besides being inexpensive and relatively easy to implement, the proposed paradigm is scalable in that it can be extended to two cameras so as to recognize a broader set of gestures. The proposed method relies on modeling the hand at each instant of time as the union of a set of identically parameterized regions (disks). A tree structure is used to maintain relational structure between the disks. Edit distances between the models obtained at successive time instants are accumulated to generate a signature for each gesture. A feed-forward neural network with added memory neurons is then used to recognize the gestures. The network is able to provide accurate and robust gesture recognition of similar but previously unseen gestures in almost all cases.

1 Introduction

The integral role that gesture plays in human communication motivates gesture-based Human Computer Interaction (HCI) [1]. Inclusion of gestures often makes communication more natural and efficient. For example, some computer tasks such as the manipulation of objects in 3 dimensions, are more easily accomplished used gesture. In addition, accurate modeling of gestures in a virtual environment greatly adds to the immersive quality, providing human qualities to otherwise static computer avatars.

Much of the work directed towards using gesture in HCI can be categorized as *3D model-based* or *Appearance-based*. 3D model-based methods typically rely on complex sensors, such as hand trackers and datagloves, which provide highly accurate hand models at the expense of being intrusive and restricting the natural movements of a user. In some cases, the sensors are un-tethered; however additional restrictions are placed on the user. For example, the use of markers on the hand are partially intrusive, and have been used for recognizing sign language postures [2]. Applications in which *3D model-based* methods have been used include interfacing with robots [3], interacting with virtual environments [4], and accessing graphical editing software [5].

Appearance-based methods on the other hand are typically based on camera-based acquisition systems, typically involving multiple cameras [6], and allow for natural and unrestricted user movement. An entire range of techniques have been proposed to process each image; for example velocity and curvature for the detection of T'ai Chi movements [7], orientation histograms [8], pattern space trajectories [9], deformable 2D templates, eigenvectors and image moments.

Towards developing a gesture-based interface for HCI, we propose an inexpensive, non-intrusive system for the modeling and recognition of human gestures. Though the method as presented herein is based on a single camera (e.g. QuickCam) and can robustly recognize a subset of gestures, it is scalable in that it can be modified to incorporate two cameras, thereby recognizing a broader set of gestures. The proposed method relies on modeling the hand at each instant of time as the union of a set of identically parameterized regions (disks). A tree structure is used to maintain relational structure between the disks. Edit distances between the models obtained at successive time instants are then used accumulated to generate a signature for each gesture which are then recognized by a feed-forward neural network with added memory neurons. Our results show that the network is able to provide accurate and robust gesture recognition of similar but previously unseen gestures in almost all cases.

In the following we describe our methodology in greater detail and present some experimental results.

258

2   Modeling of the Hand

The model of the hand is derived from a two-dimensional image of the hand and consists of *spatial* modeling (representing the shape of the hand) and *relational* modeling (representing the relation between components describing the shape). Towards spatial modeling, we represent the approximation $\hat{S}$ of a hand shape S as,

$$\hat{S} = D_1 \cup D_2 \cup \cdots \cup D_n = \bigcup_{i=1}^{n} D_i \qquad ; \qquad D_i \cap D_j = \emptyset \qquad (1)$$

where D_i represents a compact region parameterized as a *disk*. Each disk requires three parameters to represent it (the center coordinates and the radius). To determine the placement of the disks we use a combination of the distance transform and a heuristic search strategy.

Distance Transform: The distance transform [10] is a spatial transformation that operates on a binary image (in which 1 represents the foreground or object of interest and 0 represents the background) to provide an image in which the value of a pixel is its smallest distance to the background in the binary image. Succinctly, $g(x,y) = T(f(x,y))$, where $f(x,y)$ represents the binary image, $T(\cdot)$ the distance transform and $g(x,y)$ represents the resulting image. A binary hand image and the corresponding distance transform is shown in Figure 1. Although not clearly visible along the fingers, the ridges tend to form a skeletal image of the hand. It is along these lines that we place the disks.

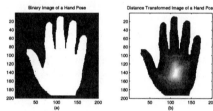

Figure 1: (a) The original binary hand image; (b) The Distance Transform image. The background has been shown as white only for clarity; it is value 0 (black).

Disk Generation: Let $\tilde{S}$ be the remaining hand shape which needs to be modeled. At the beginning of the process, $\tilde{S} = S$, and when the model is completed, $\tilde{S} = S - \hat{S}$. Using the results of the distance transform, the first disk is placed with its center at the location of the highest pixel value in the distance transform image (usually this is in the center of the palm of the hand). The radius of this initial disk is equal to the value at that pixel location. Recall that the value of the pixel in the distance transform image represents the smallest distance to the background (non-hand region). The disk region generated becomes D_1 in the set representation and is assured of being the largest region in the final model. In order to provide an ordered listing of the regions, it is necessary that $A(D_1) \geq A(D_2) \geq A(D_3) \geq \cdots \geq A(D_n)$, where $A(D_i)$ is the area of the disk, D_i. This ordering is assured using the distance transform as the basis for constructing the model of the hand, since with each pass, the highest point (or further pixel from the background) is used as the disks's center. After the initial disk is determined, the pixels which fall within the radius of the disk are removed. Succinctly, $\tilde{S}_i = \tilde{S}_{i-1} - A(D_i)$, where $\tilde{S}_i$ is the remaining hand shape to be modeled after the ith disk has been generated, $\tilde{S}_{i-1}$ is the hand shape before the addition of the disk D_i, and $A(D_i)$ is the area of the ith disk that was just generated. The process is repeated and progressively smaller disks are added until the *normalized approximation error*, $J_i = (S - \tilde{S}_i)/S$, falls below a predetermined threshold. Observe that $0 \leq J_i \leq 1$. In our simulations we found that J_i of approximately 0.8 provides a good compromise between the accuracy of representation and the cost of representation. The results for an outstretched hand are shown in the left panel of Figure 2.

Relational Modeling: Relational information is introduced based on two assumptions – (i) the palm of the hand is represented by the largest disk, and (ii) disks further from the largest disk are similar in size or smaller. Under these two assumptions, a hand can be viewed as a tree where the palm represents the root node and progressively smaller disks are successive children. Disks representing the tip of the fingers then form the leaf node. A modified breadth-first visitation scheme is used to incorporate relational information in the model. While the complete details are not presented here due to lack of space, the results of the tree generation for the outstretched hand is shown in the right panel of Figure 2.

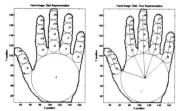

Figure 2: (Left panel) The completed model of the hand using disk regions with a *normalized approximation error* of 0.83. The numbering of the disks indicates the order in which the disks were placed. (Right Panel) The disk-tree model of an outstretched hand.

3   Generating the Gesture Signature

We generate the signature of a gesture based on the tree edit distance obtained from models constructed at successive instants of time. The edit distance encapsulates the changes that the disk-tree model of the hand has undergone from one instant to the other and these changes, and when accumulated over time, allow capturing the temporal characteristics of a gesture. The approach we have adopted is based on a modification of the one in [11] and the overall goal is to find the minimum change that would allow a tree model generated from the hand at some instant to be matched to the tree model generated from the next at the next instant. There are three different costs that need to be defined: the cost of a match of a node in one model to a node in the other model, the cost of deleting a node in the model and the cost of inserting a node in the model. The latter two costs are required as the number of disks in each model are not necessarily the same.

The cost to match one node (disk; say i_1) in one model (T_1) to a node (disk; say j_1) in another model (T_2) can be represented as $\gamma(T_1[i_1] \to T_2[j_1])$ and is based on the distance between the center's of the two disks being matched as well as the size of the disk. Consider two nodes (disks) from the algorithm, $D_1 = T_1[i_1]$ and $D_2 = T_2[j_1]$. The cost of matching these two disks is given by:

$$J = f\left(\|D_{1_{x,y}} - D_{2_{x,y}}\| + \|D_{1_r} - D_{2_r}\|\right) \tag{2}$$

where $D_{i_{x,y}}$ is the center—(x, y) pair—of disk D_i, D_{i_r} is the radius of disk D_i, and $f()$ is a threshold penalty function which is linear up to a threshold point and exponential beyond that point. The threshold and exponential value were empirically determined to be 20 and 1.5 respectively.

The cost of deleting a node ($\gamma(T_1[i_1] \to \Lambda)$) and the cost of adding a node ($\gamma(\Lambda \to T_2[j_1])$) were given constant values of 40. This value was twice the threshold for Equation (2) and allowed for the diameter of the disk being added or deleted to be roughly the same as the threshold for similarity between matched disks.

Once correspondence is established between nodes (disks) in a model obtained at an instant of time and a node (disk) in a model obtained from a subsequent instant of time, it becomes feasible to track the movements of the individual disks (see Figure 3).

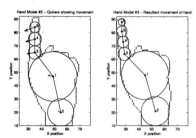

Figure 3: The left panel shows a model with the quivers indicating the motion necessary to reach the model in the next frame (right panel).

The complete signature of a gesture based on the tree edit distance obtained from models constructed at successive instants of time. To make the signature compact, we restricted the model at each instant of time to the largest 11 disks and generated a signature based on,

$$\mathcal{G} = [[P_1 P_2 \ldots P_{11}][P_{1_x} P_{1_y} P_{2_x} P_{2_y} \ldots P_{11_x} P_{11_y}][T_{1_x} T_{1_y} T_{2_x} T_{2_y} \ldots T_{11_x} T_{11_y}]] \tag{3}$$

where P_i represents the parent of Disk i, P_{i_x} and P_{i_y} are the relative x, y distances from Disk i to its parent, and T_{i_x} and T_{i_y} are the x, y movements of Disk i from the previous frame to its current position.

Each successive frame provides a 55 dimensional vector which, over time, provides information about the displacement of the largest disks in the hand model. We used a fixed number of frames for each gesture (50) though this should be varied. As there is some correlation between the 55 dimensions, we used dimensionality reduction to map the high dimensional space to a lower dimensional manifold. We used Principal Component Analysis (PCA) [12] and retained the eigenvectors corresponding to 36 largest eigenvalues. The original signatures are then represented in this 36 dimensional manifold.

4 Recognition of Gestures

To recognize the overall signature of a gesture (36 dimensional vector over 50 frames) we selected a neural network well suited for temporal sequence processing [13]. The network can be broadly defined as a *context model* in that it uses an explicit memory mechanism [13] to store past states, i.e., every neuron in the network, except the output layer neurons) has a corresponding memory neuron attached to it. This explicit mechanism allows "long term" memory to be more effectively retained. This explicit storage converts the overall network in to feed-forward network allowing for a simple training algorithm. Due to lack of space, we are omitting additional details.

5 Results of the Recognition System

The lexicon of gestures consisted of 12 examples of 9 distinct gestures ('*No*', '*Sign No*', '*Come*', '*Come Here*', '*Understand*', '*Why*', '*Angry*', '*GoodBye*', and '*Look*'). Each of the 9 representative gestures was acquired 12 times by an average of 4 to 5 different participants, resulting in 108 total gestures. After the 108 gestures were modeled and correspondence established between the 50 hand poses for each gesture, the memory network discussed in Section 4 was used for training and classification. For the purposes of performing the gesture recognition, the 12 examples of each gesture were divided into a training set of 8 samples for each gesture, resulting in a total of 72 training patterns. The remaining 4 samples of each gesture were used as 32 testing patterns to gauge the generalization of the network on gestures it had not been trained on.

We constructed a network with 36 input neurons, 180 hidden neurons, and 9 output neurons—one for each of the gestures in the lexicon (a bipolar 1-of-N coding scheme was used; a 1 at the location for a particular gesture and -1 for the remaining outputs). The network learning coefficients, η and γ were set at 0.01 and 0.05 respectively. The constants of the activation functions (c_1, c_2, k_1 and k_2) were all chosen as 1, and the desired outputs of the network were bi-polar vectors containing a 1 at the correct gesture output and -1 otherwise. Training was stopped after 1242 epochs. After training was complete, the 50 time frames for each of the 72 training signatures was then presented to the system for recognition. The output for a specific signature was taken to be the mean of its output vector over the course of 50 time frames.

Table 1 show the confusion matrix generated for the training and test pattern gestures. The training data shows excellent accuracy with the slight exception of the gesture '*Understand*' and one example of the gesture '*Come*'. The testing data shows a similar disparity for the gesture '*Understand*' and a single pattern for the gesture '*Come Here*' is misclassified. These anomalies are present because of the similarities between certain gestures over various time intervals. The recognition accuracy of 95.8% for the training data and 88.9% for the testing data are compared with that reported by other authors in Table 2. While results of this paper are comparable and often better, based upon both the number and types of gestures recognized as well as the overall accuracy of the system, these comparisons should bear in mind that the data used in each study is different.

6 Conclusion

In this paper, we presented a computational methodology to generate a model of the hand based on a single image frame. Tracking of the gesture is done by generating models for successive image frames and using the edit distance to find the smallest operation required to match successive models. The resulting edit operations are then be input to a context based neural network classifier for actual recognition of the gestures. In its present form, the proposed method provides robust recognition results for a small library

Table 1: The confusion matrix obtained during training (testing).

	No	SignNo	Come	Here	Under	Why	Angry	Bye	Look
No	8(4)	0(0)	0(0)	0(0)	0(0)	0(0)	0(0)	0(0)	0(0)
Sign No	0(0)	8(4)	0(0)	0(0)	0(0)	0(0)	0(0)	0(0)	0(0)
Come	0(0)	0(0)	7(4)	1(0)	0(0)	0(0)	0(0)	0(0)	0(0)
Come Here	0(0)	0(0)	0(1)	8(3)	0(0)	0(0)	0(0)	0(0)	0(0)
Understand	0(1)	0(0)	0(0)	0(0)	6(1)	0(0)	2(1)	0(0)	0(1)
Why	0(0)	0(0)	0(0)	0(0)	0(0)	8(4)	0(0)	0(0)	0(0)
Angry	0(0)	0(0)	0(0)	0(0)	0(0)	0(0)	8(4)	0(0)	0(0)
Bye	0(0)	0(0)	0(0)	0(0)	0(0)	0(0)	0(0)	8(4)	0(0)
Look	0(0)	0(0)	0(0)	0(0)	0(0)	0(0)	0(0)	0(0)	8(4)

Table 2: A comparison of results obtained. The number of gestures in the table is given as the number of unique gestures × the number of examples for each unique gesture.

Method	Gesture Type	Input to System	# of Gestures	Accuracy	
				Training	Testing
This paper	Dynamic	2D Image	9 × 12	95.8%	88.9%
HMM [5]	Dynamic	2D Image	12 × 1	85%	N/A
Recursive HMM [14]	Static	2D Image	6 × 100	N/A	96%
FFNN [15]	Static	Dataglove	5 × 1	100%	N/A

of gestures. However, the overall methodology is scalable in that it can be used when multiple cameras are available. In that setting, the modeling of the hand can be done using spheres as opposed to the disks.

References

[1] J. Preece, et al., *Human-Computer Interaction*, Addison-Wesley, New York, NY (1994).

[2] R. Cipolla, Y. Okamoto, and Y. Kuno, "Robust Structure from Motion Using Motion Parallax," *1993 IEEE 4th International Conference on Computer Vision*, pp. 374–383 (1993).

[3] C. Lee and Y. Xu, "Online, Interactive Learning of Gestures for Human/Robot Interfaces," *Proceedings of the 1996 IEEE International Conference on Robotics and Automation*, Vol. 4, pp. 2982–2987 (1996).

[4] R. Jacoby, M. Ferneau, and J. Humphries, "Gestural Interaction in a Virtual Environment," *Stereoscopic Displays and Virtual Reality Systems*, Vol. 2177, pp. 355–364 (1994).

[5] B.W. Min, et al., "Hand Gesture Recognition Using Hidden Markov Models," *Proceedings of the 1997 International Conference on Systems, Man and Cybernetics*, Vol. 5, pp. 4232–4235 (1997).

[6] A. Utsumi, et al., "Direct Manipulation Scene Creation in 3D: Estimating Hand Postures from Multiple-camera Images," *Visual Proceedings of SIGGRAPH '97*, (1996).

[7] L.W. Campbell, et. al., "Invariant Features for 3-D Gesture Recognition," *Proc. of the Second Intl. Conference on Automatic Face and Gesture Recognition*, pp. 157–162 (1996).

[8] W.T. Freeman and M. Roth, "Orientation Histograms for Hand Gesture Recognition," *International Workshop on Automatic Face and Gesture Recognition*, Zurich, Switzerland (1995).

[9] S. Nagaya, S. Seki, and R. Oka, "A Theoretical Consideration of Pattern Space Trajectory for Gesture Spotting Recognition," *Proc. of the Second Intl. Conference on Automatic Face and Gesture Recognition*, pp. 72–77 (1996).

[10] R. M. Haralick and L. G. Shapiro, *Computer and Robot Vision*, Addison-Wesley, Reading, MA (1992).

[11] K. Zhang and D. Shasha, "Simple Fast Algorithms for the Editing Distance Between Trees and Related Problems," *SIAM Journal of Computing*, Vol. 18, No. 6, pp. 1245-1262 (1989).

[12] I.T. Jolliffe, *Principal Component Analysis*, Springer-Verlag, NY (1986).

[13] P.S. Sastry, G. Santharam, and K.P. Unnikrishnan, "Memory Neuron Networks for Identification and Control of Dynamical Systems," *IEEE Transactions on Neural Networks*, Vol. 5, No. 2, pp. 306–319 (1994).

[14] J. Schlenzig, E. Hunter, and R. Jain, "Vision Based Hand Gesture Interpretation Using Recursive Estimation," *28th Asilomar Conference on Signals, Systems and Computers*, Vol. 2, pp. 1267–1271 (1994).

[15] R. Beale and A. D. N. Edwards, "Recognising Postures and Gestures Using Neural Networks," *Neural Networks and Pattern Recognition in Human-Computer Interaction*, Ellis Horwood, NY, pp. 163–172 (1992).

INTERMODAL DIFFERENCES IN DISTRACTION EFFECTS WHILE CONTROLLING AUTOMOTIVE USER INTERFACES

Michael Geiger[*], Martin Zobl[*], Klaus Bengler[+], Manfred Lang[*]

[*]Institute for Human-Machine Communication
Technical University of Munich, D-80290 Munich, Germany
[+] BMW AG, dep. EV-22, D-80788 Munich, Germany

geiger@ei.tum.de, zobl@ei.tum.de, klaus.bengler@bmw.de, lang@ei.tum.de

ABSTRACT

In this study, the haptical and gestural user input modes are compared with regard to distraction from a controlling task similar to steering a car. The examination is carried out in a driving-simulation laboratory. While performing the controlling task permanently, the test subject has to execute a variety of user input with a given modality. Haptical input is done by using buttons or a rotary knob while gestural input consists of dynamic right hand movements in a designated space. During the experiment, the controlling error, the object recognition performance, and the required duration for each user input are measured. The results show a significant benefit for gestural input.

1. INTRODUCTION

Due to the multiplicity of coexisting electronic devices in modern luxury and upper class automobiles, the drivers' task to control this huge functional range while driving has become increasingly complex. Some examples of these devices are navigation systems, audio and video components like CD player, radio, and television, mobile phone, car computer, air condition, and any additional conceivable unit such as internet applications. For the purpose of enabling the user to handle these car devices, an optimized man machine interface is desirable. Nowadays, user input in cars is mainly done haptically via buttons or rotary knobs. Our basic approach is to use additional input channels besides the tactile one. Humans mainly pass on information using speech and gestures. This study examines the potential of visual gesture input to control in-car devices with regard to distraction. We understand gestures as dynamic right hand movements performed in the field of vision of a camera mounted at the car's ceiling. The gesture vocabulary used is the outcome of foregoing studies and is reduced to an intuitive set of hand-gestures. Our assumption that gestural input might cause less distraction effects than haptical input bases on recent psychological experiments on selective visual processing [Deu98] [Pap99]. These studies have shown that goal-directed deictic movements are generally coupled with substantial mental distraction effects, which are verified by an almost nonexistent capability of object recognition performance. They conclude that it is not possible to attend to one object and, at the same time, to point to another. Transferring these results to our problem, the haptical actuation of a button is similar to a goal directed movement. Accordingly, the use of dynamic hand gestures, which are less directional than haptical interactions, promises an efficient reduction in loss of attention.

2. METHODOLOGY

2.1 Test Environment and Conditions

To ensure an authentic environment, this study is carried out in a driving-simulation laboratory (cf. Fig. 1). In order to avoid undesired side effects on the data, caused by the characteristics of the driving simulation application (implemented car model, graphics etc.), we have developed an abstract steering task (cf. Fig. 2) displayed on a projection area in front of the car. Thereby, the steering nominal is given as a marker following horizontal movements similar to a road course. The controlling task consists of following the nominal marker with a second marker, which is controlled by the subject's steering wheel, whereby the angle of the steering wheel is translated into horizontal deflection of the actual marker.

While steering, the subject has to perform a variety of user input (cf. input tasks, Tab. 1) with a given modality. Haptical input is done using buttons or a rotary knob placed in the midconsole (cf. Fig. 3a) and according dynamic right hand gestures in a designated space above the midconsole (cf. Fig. 3b).

To get comparable data, the sequential control of the experiment is automated, which means that all events happen exactly at the same time. The input tasks are read out to the subject via pre-recorded speech. After the test, the subject is asked by which kind of user input (haptical or gestural) it felt more distracted from the steering task, and which one was more comfortable.

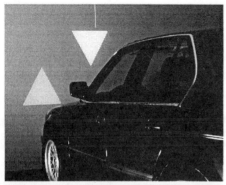

Figure 1: driving simulation laboratory

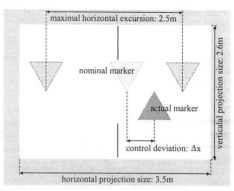

Figure 2: steering task

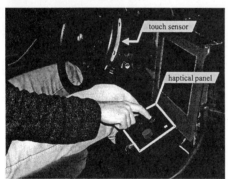

Figure 3a: haptical user input (select button)

Figure 3b: gestural user input (telephone gesture)

		Input Tasks			
		Select	Phone	Increase	Decrease
Input Method	Haptical	Press 'select' button	Press 'phone' button	Turn rotary knob clockwise	Turn rotary knob counter clockwise
	Gestural	Point to front	Lift virtual phone receiver	Up-movement	Down-movement

Table 1: input tasks

2.2 Data Logging

In order to estimate the amount of distraction while processing user input, the nominal and actual x-positions, the object recognition performance and the required duration are measured and stored to a log file.

2.2.1 Nominal x-Position and Actual x-Position

Both nominal x-position and actual x-position are recorded with a sampling rate of 50 Hz (cf. curves Fig. 5). During the test design, the motion of the nominal marker was set to an average speed that allows the controlling to be done with a very small deviation error if the subject exclusively concentrates on that main task. The x-position data is used to calculate the control error (cf. Results) which describes the subject's capability of following the road.

2.2.2 Required Duration

The required duration is the time interval while the steering wheel is released, measured by a touch sensor (cf. Fig. 3a) mounted to the steering wheel. Before the experiment starts, the subject is instructed to keep this sensor pressed all the time unless releasing is necessary to do an input task. Additionally, all input tasks have to be done as fast as possible and without interruption. To stimulate the subject to perform as fast as possible, an acoustical feedback is presented at the end of every task in form of a sine tone with a frequency that increases (and becomes more unpleasant) with elapsed time. The required duration is one measure for the efficiency of the respective input method.

2.2.3 Object recognition performance

Object recognition performance is measured as recognition error rate with 2AFC, in which the subject's task is to recognize the identity of a visual object presented for a short time. The object (seven-segment display, cf. Fig. 4) is presented at the actual position of the nominal marker within a varying time delay after the release of the steering wheel.

Figure 4: visual objects to identify and mask item

To assure the object's presentation coming along during the execution of the input task, this delay lasts at least 100 ms but does not exceed 500 ms. The object is then replaced by a mask item ('8') 70 ms later in order to avoid the imprint of the pattern to the retina. The object recognition performance indicates whether the subject's visual attention is directed to the front (traffic event) or not.

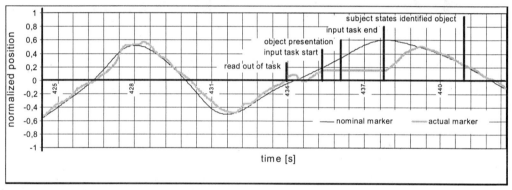

Figure 5: Timechart (excerpt): x-positions of nominal and actual marker as well as occurring events during input task; Remarkable: even the task's read out causes an attention loss; in this example, the subject stops the controlling task entirely while doing user input.

2.3 Problem of delayed input task performance

During the test design the following problem occurred. The subjects did not perform the input tasks - as demanded - as fast as possible, but waited for the object presentation before doing user input. The implementation of acoustic feedback (cf. Required Duration) could not solve this problem completely. Therefore, we invented additional events in order to prevent the subject attuning to the object presentation during an input task: On the one hand, now there are tasks without object presentation. On the other hand, random object occurrence outside an input task is implemented. Altogether, there are 32 input tasks: 16 haptical and 16 gestural user inputs, whereas only half of the tasks come along with a displayed object. Furthermore, there are 16 object presentations outside input tasks.

3. RESULTS

For data comparison, the differences of the matched observations $d_i=x_i^{hap}-x_i^{ges}$ for all measured factors are regarded. Since the differences are apparently normally distributed the t-test is used. The null hypothesis H_0: $\mu_d=0$ suggests that gestural and haptical values are identical and is tested against the alternative H_A: $\mu_d>0$ that haptical input produces higher values than gestural input. In our case, higher values stand for poorer suitability for in-car use concerning all measured data.

3.1 Controlling Error

The controlling error is calculated by the root mean square of the horizontal deviation Δx (cf. Fig. 2) between the nominal and actual value of the steering task [Joh93]. To obtain distance independent values, the pixel difference is converted to a deviation angle. The shape of the two distributions (cf. Fig. 6) is almost identical whereas the distribution for haptical input is shifted to higher deviation angles. The null hypothesis that the distribution of the differences has a mean of $\mu_d=0$ can be rejected on a significance level $\alpha<0.01$. The $1-\alpha$ interval for the true mean is $\mu_d\in[0.18\ 0.58]$, whereas a deviation angle of 0.58 deg corresponds to an offset of about 1m in a distance of 100m.

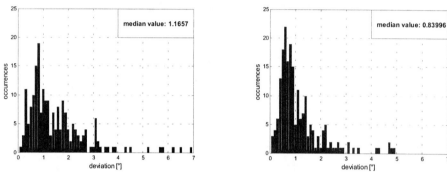

Figure 6: control error for haptical input (left) and for gestural input (right)

3.2 Required Duration

The data for the required duration (cf. Fig. 7) show the same results as for the controlling error, with the $1-\alpha$ interval for the true mean $\mu_d\in[0.36\ 0.57]$. This means that in average haptical input takes about 1.4 times longer than a gestural one.

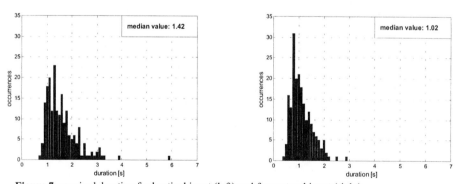

Figure 7: required duration for haptical input (left) and for gestural input (right)

266

3.3 Object Recognition Performance

The error rates (cf. Fig. 8) also show that gestural input is superior to haptical input regarding the recognition performance. The median recognition error rate for gestural input is about half the size of the haptical one. Using 2AFC, an error rate of 0.5 means pure guess. The null hypothesis can be rejected with $\alpha < 0.01$, too. The $1-\alpha$ interval for the true mean is $\mu_d \in [0.08 \ 0.32]$. The subject's attention is obviously not directed to the front with haptical input.

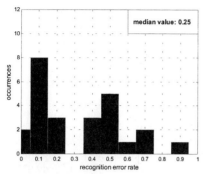 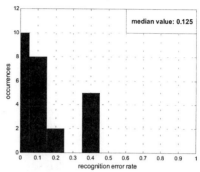

Figure 8: recognition error rate for haptical input (left) and for gestural input (right)

3.4 User Acceptance

After the experiment, the subjects complete a questionnaire. The questionnaire results are in correspondence to the objective outcomes. According to the subjects' statements, gestural user input distracts less (94% of the subjects) than haptical (6%) and is more pleasant (76% vs. 24%).

4. CONCLUSIONS

The findings argue for substantially reduced distractions when using gestural user input for controlling an automotive MMI. Therefore, we expect a significant gain in user acceptance by using gestural user input for certain tasks in an automotive environment. The next step in our work will be to implement the institutes gesture recognition system [Mor99] into the car environment.

REFERENCES

[Deu98]   Deubel, H., Schneider, W. X., Paprotta, I. (1998). Selective Dorsal and Ventral Processing: Evidence for a Common Attentional Mechanism in Reaching and Perception. *Visual Cognition*, pp.81-107

[Joh93]   Johannsen, G. (1993). *Mensch-Maschine-Systeme* Available: Springer Verlag, ISBN 3-540-56152-8, Berlin-Heidelberg-New York.

[Mor99]   Morguet, P., Lang, M. (1999). Comparison of Approaches to Continous Hand Gesture Recognition for a Visual Dialog System. *Proceedings ICASSP 99 (Phoenix, Arizona, USA), IEEE, Vol. 6*, pp. 3549-3552

[Pap99]   Paprotta, I. (1999). *Selektive Wahrnehmung und Handlungssteuerung, Die Kopplung von visueller Aufmerksamkeit und Bewegungszielselektion.* Available: Shaker Verlag Aachen, ISBN 3-8265-6588-6.

AN INTUITIVE PEN-GESTURAL INTERFACE FOR SYNTACTIC-SEMANTIC ANNOTATION OF NON-CURSIVE HANDWRITTEN INPUT

Jörg Hunsinger

Institute for Human-Machine Communication
Technical University of Munich, D-80290 Munich, Germany
hunsinger@ei.tum.de

ABSTRACT

Any system designed for handwriting or drawing recognition, especially as far as probabilistic decoding components are involved, must be extensively trained by means of manually acquired, segmented, and syntactic-semantically annotated reference material. In order to standardize and facilitate these error-prone operations, we developed a structured graphical interface which utilizes simple pen gesturing as a highly efficient technique for graphical input manipulation. In particular, we use this tool to acquire and annotate training corpora of handwritten mathematical formulas. By applying corresponding feature extraction and iterative training mechanisms, the resulting set of probabilistic parameters is then fed into our single-stage top-down semantic decoder for formula understanding. Nevertheless, the interface is suitable for any other non-cursive handwriting or drawing application. On the signal near levels, the derived feature vectors may be processed by any handwritten symbol recognition stage.

1. INTRODUCTION

Our interface was especially designed for a pen operated LCD digitizing tablet providing instantaneous visual feedback. For enhanced interactivity we set a high value on on-line visualization features as well as back and forth switching facilities between the different involved subtasks. All the necessary worksteps require merely simple pen gesturing and/or pen selection events. The software is platform independent due to the use of the Tcl/Tk scripting language [1]. In the following section we give a detailed description of the relevant interface functionality.

2. METHODOLOGY

The main menu serves to switch freely between four basic workspaces:

2.1 Acquisition Workspace

Any new samples may be written down with arbitrary positioning on a scrollable area. Pen movements are sampled at 60 Hz so that sequences of handwriting strokes (i.e. segments between a pen-down and a subsequent pen-up event) successively appear on screen. All the collected data are stored online in an internal binary format. To keep the temporal order intact, which is implicitly evaluated in our decoding approach, only strokewise backward deletions, i.e. repeated undo operations, are allowed. Fig. 1 shows a typical handwriting sample being entered in this workspace.

2.2 Segmentation Workspace

After one or more handwriting or drawing samples have been collected, their piecewise segmentation to syntactic units (i.e. handwritten symbols or drawing elements, respectively) takes place in a separate workspace. The purpose of this stage is to group strokes belonging together by performing erase- or encircle-type pen gestures. Special color coding methods and - optionally - surrounding rectangles around every segmented stroke group (or single stroke, respectively) indicate segmentation consistency at a glance. During pen gesture formation, all the captured strokes are promptly highlighted to ease gesture execution and to visualize segmentation success incrementally. Symbol fragmentation is possible via the definition of two or more symbolic components assigned to a single semantic unit (cf. Sec. 2.4). In the mathematics domain, such a strategy makes sense for supporting e.g. a radical sign drawn in two parts, interrupted by a radicand expression. The segmentation scenario is illustrated in Fig. 2.

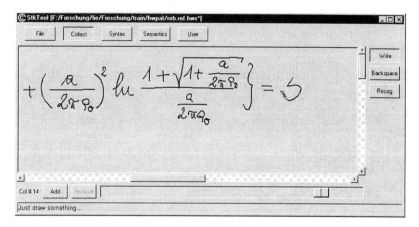

Figure 1: Acquisition Workspace. The lower horizontal scrollbar serves to browse the set of already acquired handwriting samples ("collections"). Via the Backspace button the user may delete one or more handwriting strokes in backward order.

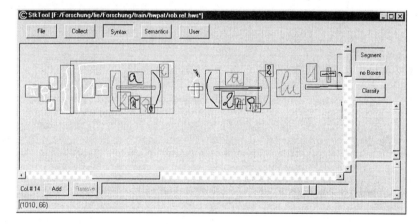

Figure 2: Segmentation Workspace. Initially, all the acquired handwriting strokes appear in alternating color according to writing order (here: black and dark grey in turn). During pen gesture formation, the captured strokes take on a different color (here: white), and will show a single bounding box around them afterwards. The pen gesture trajectory is displayed as a thin curve (here: white) during execution. Alternate coloring is also maintained between the final segmented stroke groups analogically. Stroke groups that were already syntactically and/or semantically annotated (cf. below) are indicated by another uniform color (here: light grey).

2.3 Syntactic Annotation Workspace

In the third workspace every segment of grouped handwriting strokes is assigned to a syntactic value, i.e. a symbol contained in the supported lexicon of the considered application domain. This is achieved by a) selecting the appropriate symbol entry from a structured list and b) performing a pen gesture to capture one or more segments from the sample being edited. The annotation process is also supported by color coding techniques and can be recontrolled by moving the pen on any annotated segment: The assigned syntactic value will then automatically be selected in the structured list mentioned above; additionally, all the segments assigned to the same entry will be highlighted. Fig. 3 gives an example of the syntactic annotation functionality. The acquired segmentation and syntactic annotation knowledge is postprocessed by our corresponding feature extraction methods and stored in a compressed database of reference patterns for every supported symbol class [2] [3].

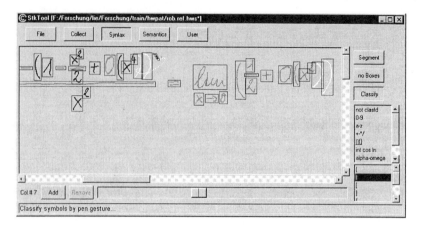

Figure 3: Syntactic Annotation Workspace. At first, one of the supported syntactic values (i.e. symbols) is selected – either by choosing it from the structured list on the right hand side, or by capturing a handwriting segment already assigned to that value (in this example: one of the white right braces on the right). Via pen gestures extra segments are then added on. Here, another two right braces are being assigned at a single blow. Different colors (here: dark grey or black, resp.) are used to distinguish between already or not yet annotated segments.

2.4   Semantic Annotation Workspace

Finally, the different syntactic constituents of a segmented and syntactically annotated sample must be allocated to semantic attributes. To this end, a so-called Semantic Structure, i.e. a compact hierarchical representation of the logical contents of a given handwritten expression, may be loaded into a separate window section inside the fourth workspace. In the special case of mathematical formulas, this object is automatically generated by the use of a conventional LaTeX compatible formula editor and subsequent data transformation due to text based chart parsing. The hierarchical structure of this representation is then - also by using intuitive pen gestures - mapped onto the collection of annotated symbols, so that successively any semantically closed subexpression of the given input (e.g. a fraction denominator) is correlated to a semantic unit of the corresponding semantic representation. Again color coding and (hierarchical) surrounding rectangle display are used to facilitate these steps and to quickly revise their consistency. Afterwards all the necessary (probabilistic) parameters which describe the semantic attributes contained in the choice and positional arrangement of the different symbols or symbol groups, respectively, are derived by the use of appropriate training procedures [4] [5]. The semantic annotation features are summarized in Fig. 4.

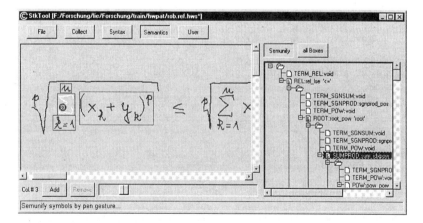

Figure 4: Semantic Annotation Workspace. In the right hand area the Semantic Structure of the given

sample formula is displayed as a hierarchical list. By selecting an entry the corresponding handwritten symbol(s) get highlighted (here: white), and bounding boxes appear around all semantic successor subexpressions.

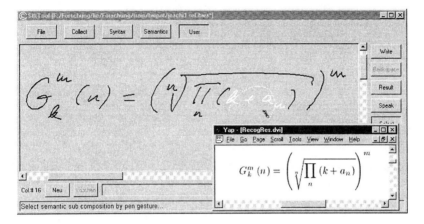

Figure 5: Interactive Formula Recognition. In this example, the given handwritten formula was correctly recognized and transformed to a typesetting format. By utilizing the recognized formula structure, the interface allows the user to perform "intelligent" pen gestures (i.e., only semantically self-contained subexpressions are selectable) in order to correct or modify formula parts.

3. RESULTS & CONCLUSIONS

We have demonstrated that intuitive pen gestures are well suited for a convenient and straightforward execution of segmentation and annotation tasks on handwritten or handdrawn input. By the use of our general pen-gestural interface the resulting sample data may systematically be evaluated on the different abstraction levels involving signal, syntactic, and semantic knowledge to supply a corresponding online recognition system with all the necessary probabilistic parameters. As an example, we implemented a robust probabilistic semantic decoder for handwritten mathematical formulas on the basis of the presented interface [5]. For interactive correction or modification tasks, our pen gesturing technique was also integrated in the end-user interface (cf. Fig. 5). In the future it should be examined to which extent the same approach succeeds for other application domains.

REFERENCES

[1] Welch B. B.: Practical Programming in Tcl and Tk, ISBN 0-13-022028-0, Prentice-Hall, New Jersey, USA, 1999.

[2] Intel Recognition Primitives Library Reference Manual, Order Number 637785-007, Intel Corporation, 1998.

[3] Hunsinger J. and Lang M.: *A Single-Stage Top-Down Probabilistic Approach towards Understanding Spoken and Handwritten Mathematical Formulas*, Proc. 6th International Conference on Spoken Language Processing (ICSLP 2000), Beijing, China, Vol. IV, pp. 386-389, October 2000.

[4] Hunsinger J. and Lang M.: *A Speech Understanding Module for a Multimodal Mathematical Formula Editor*, Proc. IEEE International Conference on Acoustics, Speech, and Signal Processing (ICASSP) 2000, Istanbul, Turkey, Vol. IV, pp. 2413-2416, June 2000.

[5] Hunsinger J., Lieb R., Lang M.: *Real-time Structural Analysis of Handwritten Mathematical Formulas by Probabilistic Context-free Geometry Modelling*, to appear in: Proc. 2001 International Symposium on Intelligent Multimedia, Video & Speech Processing (ISIMP 2001), Hong Kong, China, May 2-4, 2001.

AN AUTOMATIC, ADAPTIVE HELP SYSTEM TO SUPPORT GESTURAL OPERATION OF AN AUTOMOTIVE MMI

Ralf Nieschulz[*], Michael Geiger[*], Klaus Bengler[+], Manfred Lang[*]

[*]Institute for Human-Machine Communication
Technical University of Munich, D-80290 Munich, Germany
[+] BMW Group, dep. EV-22, D-80788 Munich, Germany

[nieschulz | geiger | lang]@ei.tum.de, klaus.bengler@bmw.de

ABSTRACT

In this study we surveyed an approach to determine an user's need of assistance while operating an automotive human-machine interface by means of gestures. In an extensive usability study we investigated three characteristic features that seemed to be suitable for this approach. The first stage of our 2-stage methodology is a neural network based classification that determines if a user needs help while performing a certain task. In the second stage a heuristic postprocess based on statistics determines which help this user actually needs in the given context. This approach was then evaluated in an offline application.

1. INTRODUCTION

Gesture controlled operation of in car devices can provide a comfortable way of interaction. But one can assume, that not many people are familiar to use it intuitively. Therefore we present an automatic, adaptive help system, which supports the novice user to operate the human-machine interface (MMI). It seems to be perspicuous that the user's need for assistance or his uncertainty does have an effect on his cognitive performance, i.e. the reflection time, as well as on the execution of the gestures concerning duration and quality. Based on this consideration, our goal is to provide an automatic, but unobtrusive assistance to minimize the amount of help requests.

2. METHODOLOGY

2.1. Experimental Environment

The experiment was conducted in the institute's driving simulation laboratory (cf. fig. 2.2a). It consisted of two parts: 1) user interaction in a parked car and 2) user interaction while driving (simulation). This paper will present the results of part 1. The study was carried out applying the so called 'Wizard of Oz' methodology. This means that an experimental manager ('wizard') telecontrols occurring events and is able to influence the system's behavior, while the test person is told to interact with an already implemented und functioning system. The test subject was seated inside the car and confronted with an automotive MMI. The experimental manager resided in the control room. He had visual connection to the subject from two points of view (cf. fig. 2.2b). Subject and wizard communicated via audio intercom.

The GUI was optimized for operation using right hand gestures and simulates devices such as radio, cassette-player, CD-player, CD-changer, telephone and navigation system. It is the result of a study about controlling automotive devices by gestures. Subjects had to interact with the MMI exclusively by gestures to avoid interfering side effects. The experimental manager observed the subject's right hand movements via monitor and interpreted them on the basis of a given vocabulary. This means that the 'wizard' acted like a gesture recognition system. He only accepted valid gestures that were part of the vocabulary corpus. The implemented gesture vocabulary consists of a set of intuitive right handed movements which resulted from former studies of our institute [Zob01]. The test person had no instructions at all, except to use only gestures for interaction. If the subject had any problems concerning the manner of performing gestures or the appliance of the MMI, he/she could push a 'help request button' (cf. fig. 2.2c) to get context sensitive help. This assistance is provided in several different audio-visual 'help packages' (cf. fig 2.1).

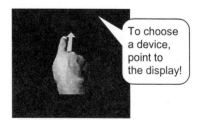

Figure 2.1: Example for a 'help package'

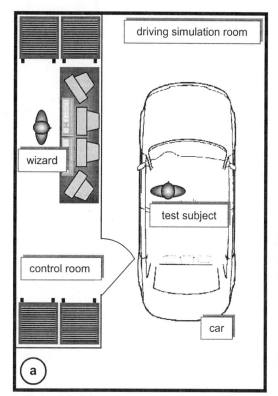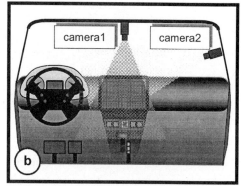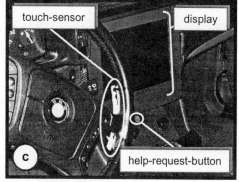

Figure 2.2: Schematic experimental setup: **a)** plan view of driving simulation laboratory; **b)** camera positions: camera 1 views the space where gestural input is performed, camera 2 shows a front view of the subject; **c)** layout of touch-sensor, help-request-button and display (also outlined in b)

During the experiment, the test subject was asked to perform a variety of typical tasks which were read out automatically by a virtual moderator. Such a task could be "Switch to CD number 4 and listen to track 5" or "Call Mr. Hunsinger". The experiment was divided into several task blocks of increasing difficulty. After each block, the subject was interviewed. The person was asked how he/she gets along with the gestures and the operation of the MMI. Furthermore the experimental subject should explain whether he/she was satisfied with the given context sensitive assistance.

Throughout the whole test procedure all relevant data were automatically logged to hard disk with time stamp. These contain besides the status of the MMI (i.e. active device, current CD title, volume setting, current destination, etc.) every request for assistance and all moments the right hand of the experimental subject left the steering wheel and returned to it. This information is given by a touch sensor mounted to the steering wheel (cf. fig. 2.2c). The subject was instructed to keep this sensor always pressed unless gestural interaction had to be done.

2.2. Preprocessing (Feature Extraction)

In order to infer the user's need of assistance, three features are determined as input for the help system. The first feature is the execution duration t_e of the gestures. This represents the time, while the right hand is off the steering wheel. The second feature maps the time the user has to think before executing a gesture. This feature is named 'cognition time' t_c. It is actually the time between the execution of separate gestures. Both were measured by the touch sensor located on the steering wheel (cf. fig. 2.2c). The third feature - execution quality of the gestures k - is estimated by the experimental manager and assigned to six categories: 'unknown', 'very bad', 'bad', 'acceptable', 'good' and 'very good'. Thereby he has to consider two aspects: how well each gesture is executed itself and how well it can be distinguished from other gestures. The execution quality corresponds to a confidence measure of a real gesture recognition system. Later on the 'Wizard of Oz' will be replaced by our real-time gesture recognition system [Mor99]. All three features are then preprocessed to get a standardized feature space.

The first two features t_e and t_c are averaged by using former measurements of the actual user, as well as corresponding measurements of all test subjects. They are calculated as follows:

$$t_{e_{norm}}[n] = \frac{t_e[n]}{\left(1-\frac{1}{w_e}\right)t_{e_{norm}}[n-1]+\frac{1}{w_e}t_{e_0}} \quad \text{and} \quad t_{c_{norm}}[n] = \frac{t_c[n]}{\left(1-\frac{1}{w_c}\right)t_{c_{norm}}[n-1]+\frac{1}{w_c}t_{c_0}}$$

with $t_e[n]$, $t_c[n]$: actual execution duration, resp. actual cognition time

$t_{e_{norm}}[n]$, $t_{c_{norm}}[n]$: execution duration, resp. cognition time, normalized to previous averages

t_{e_0}, t_{c_0} : each: average over all gestures and experimental subjects

w_e, w_c : each: weight, to adjust the ratio between users and overall average

Then both features are non-linearly scaled to a data-range of 0 to 1 regarding standard deviation of all test data (cf. fig. 2.3). This leads to the features p_1 and p_2. Thereby the average durations $t_{e_{norm}}[n]=1$ and $t_{c_{norm}}[n]=1$ are mapped to $p_1[n]=0.5$ and $p_2[n]=0.5$. These operations provide adaptation to the users gestural behavior concerning execution duration and cognition time.

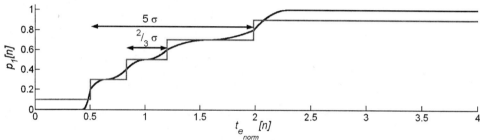

Figure 2.3: The blue line shows the non-linearly scaling function regarding standard deviation σ of all test data, the red line the five categories as described in sec. 2.3, both shown on $p_1[n]$ exemplary

The third feature - execution quality k - is averaged using a time interval, which is weighted with a memory function $w[n]$. The goal of $w[n]$ is to evaluate past gestures less than the current one. This is realized by a descending exponential characteristic: $w[n]=e^{-n}$. The feature is then scaled to a data-range of 0 to 1, too. This results in:

$$p_3[n] = \frac{1}{\sum\limits_{m=0}^{\infty} w[m]} \sum\limits_{m=0}^{\infty} w[m]p[n-m] = \frac{1}{1+e^{-1}}\sum\limits_{m=0}^{\infty}e^{-m}p[n-m]$$

Since $w[7]=e^{-7}<10^{-3}$ it is a good choice to only evaluate the past seven gestures, so that the time interval of $w[n]$ is seven gestures. To simplify matters p_3 can be determined recursively:

$$p_3[n] = (1-e^{-1})\,k[n-1]+e^{-1}p_3[n-1]$$

Feature p_3 can be called a current average of gesture quality. This is to avoid assistance for a user who performs a bad gesture only once. The bad quality of this gesture will be averaged in the context of the last gestures. That way this operation also provides an adaptation to the user's gestural behavior in consideration of the last gestures.

Furthermore the statement, whether the subject actually needs assistance ('yes' or 'no'), is determined as target feature $T[n]$.

2.3. Creating the underlying neural network

The feature vectors of the first two-thirds of the experimental subjects were then used to train a neural network, which supplies the statement, whether the subject needs help, or not. First the training data are categorized linearly: the first two features p_1 and p_2 in the five categories 'very short', 'short', 'normal', 'fast' and 'very fast' (cf. fig. 2.3) and the third feature p_3 into the same six categories as k above (cf. sec. 2.2). In consequence all feature vectors are mapped to a maximum of 150 training vectors. As the neural network is built as a probabilistic neural network (PNN) based on a radial basis network (RBN), the first layer contains a maximum of 150 neurons depending on the training material. As a positive side effect the memory usage of the neural network will be decreased as well as the system's speed will be increased. As the training data do not cover the whole feature space there are areas which are represented only by very few neurons. In order to avoid an over-weighting, resulting from this, the individual neurons are weighted by their normalized average distance to all other neurons in the feature space. This is done because in training material of present quantity (the training material consists of about 2000 gestures) there are always outliers and because there is much more data for the statement 'the user doesn't need assistance' than for the statement 'the user needs assistance'. By these modifications it can be avoided that the network will almost always determine the class which represents the first statement. This means to the PNN that its recognition performance increases in stability.

2.4. Application and postprocessing

In the current status the help system is applied off-line. The logged data sets of the remaining one-third of the test subjects are used as test data. The individual features are preprocessed the same way as above, but not categorized. Then the input feature vector is fed into the neural network, which supplies the result, whether the user needs assistance, or not. As the network does not take into account the context of the MMI, it is necessary to find out the accurate assistance as a function of the context and the user's operation history. Therefore the system searches the help database for the statistically most probable 'help package'. Statistics of the current user (e.g. which gestures have been used recently, in which context, were they used in the correct manner, which assistance was already presented to the user, etc.?) as well as of all test subjects (e.g. with which gestures / with which functionalities of the MMI did the users get along best / worst, etc.?) are considered. If necessary the postprocessing algorithm sometimes suppresses assistance, even if the neural network determines that the user needs help. Such possible cases are checked heuristically and lead to an improvement of the overall recognition result. If, for example, the user was thinking a long time before the latest gesture and does a very slow and uncertain right hand move, yet, this caused what he wanted, e.g. to accept a telephone call for the first time, the MMI should not provide information of how to accept a telephone call.

3. RESULTS

In sum 2935 gestures of 18 experimental subjects were evaluated. In 304 cases subjects needed assistance. 2013 gestures were used as training data. The remaining 922 data sets were used for recognition. As can be seen in fig. 3.1 the overall recognition rate of the neural network is 92.0 % and the error rate is 8.0 %. The error rate can be reduced by relatively 24 % doing the postprocessing (cf. fig. 3.2). The recognition rate of the case, that the user doesn't need

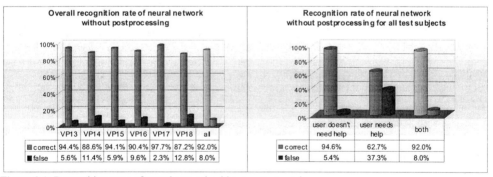

Figure 3.1: Recognition rates of neural network without postprocessing

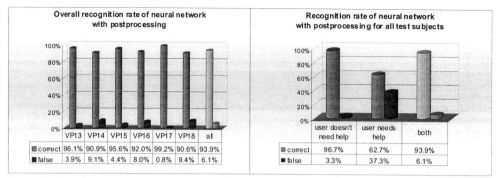

Figure 3.2: Recognition rates of neural network with postprocessing

help, is 94.6 %, while the rate of the other case, where the user needs assistance, is only 62.7 %. That is because of the fact that only about 10 % of the training material represents this target. The postprocessing has only effect in the first case and increases its recognition rate to 96.7 %.

Reclassification of the training material shows similar results (cf. fig. 3.3). But here the recognition rate for the target 'user needs assistance' is more than 10 % higher.

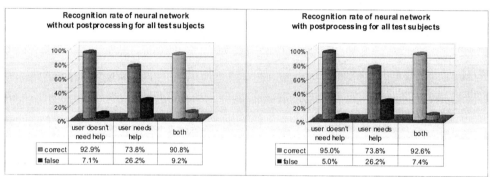

Figure 3.3: Reclassification with training data: Recognition rates of neural network for all experimental subjects

4.   CONCLUSION

The results of our study show that the system is able to recognize reliably, whether the user is in need of assistance, or not. The statistical postprocessing ensures that most of the time the accurate assistance is presented to the user, so that the operation of the MMI is facilitated. We are quite confident that in the future the help system will also work very well with the input of our institute's real-time gesture recognition system [Mor99]. Follow-up studies will have to take into account that different gestures need different execution durations and that subjects need different cognition times in order to plan different gestures. In addition one has to consider that different gestures typically show different confidence measures.

REFERENCES

[Mor99]   Morguet, P., Lang, M. (1999). Comparison of Approaches to Continuous Hand Gesture Recognition for a Visual Dialog System. *Proceedings ICASSP 99 (Phoenix, Arizona, USA), IEEE, Vol. 6*, 3549 (4 pp.).

[Zob01]   Zobl, M., Geiger, M., Bengler, K., Lang, M. (2001). A Usability Study on Hand Gesture Controlled Operation of In-Car Devices. *Poster Proceedings HCII 2001 (New Orleans, Louisiana, USA), (this conference).*

Interface Design of Video Scout:
A Selection, Recording, and Segmentation System for TVs

John Zimmerman[†], George Marmaropoulos*, Clive van Heerden*

Philips Research[†] and Philips Design*

345 Scarborough Road, Briarcliff Manor, NY, USA, 10510

{john.zimmerman, george.marmaropoulos, clive.vanheerden}@philips.com

1 Abstract

Video Scout is a prototype retrieval application that allows Personal Video Recorders to actually *watch* the TV programs they record. By analyzing the visual, audio, and transcript data, Scout can segment and index TV programs, finding and recording specific video clips that match requests in users' profiles. For example: if users request information on Philips, Scout will watch news programs and capture any stories it finds on Philips. The Scout interface offers a familiar TV environment where users can interact with whole TV programs and video clips organized by topic. Scout also provides users with tools for managing their profiles. This paper captures the Video Scout interface design process, from concept sketches to user testing to final prototype design.

2 Background

PVRs, like TiVo® [1], have changed the way people interact with television. The user's task has changed from (a) finding something to watch from 100+ channels to (b) finding something to record from 10,000+ weekly shows. By monitoring shows users watch and by providing an interface for rating shows, PVRs construct user profiles that allow them to recommend and automatically record programs. This is the first smart TV appliance.

Video Scout represents a future direction for PVRs in that it actually watches the programs it records. Scout analyzes the visual, audio, and transcript data and integrates the results in a Bayesian engine. By searching for matches between users' profiles and the indexed clips, Scout can provide users with the clips they are looking for. For example: if users want information on an earthquake that just happened, Scout can watch the news programs and extract just the stories on this earthquake. (For a more detailed description of the Scout engine, please see [2]). This paper details the Video Scout interface design process, from concept sketches to user testing to final prototype design. It illustrates how feedback from the user test influenced the look and feel of the final prototype.

3 Concept Sketches

Our design process started with three concept sketches. (i) Raindrops: offers users tools for finding events within whole TV programs. (ii) Seeker: displays a playback system for topic-based video clips that are automatically recorded according to users' profiles. (iii) TV Magnets: presents an interface for specifying the kinds of content users want to attract and repel. We implemented these concepts with simple interactivity in Macromedia Director and demonstrated them to test subjects using a video projector to create the illusion of a futuristic TV.

The Raindrops concept (figure 1) presents a TV program guide where current and future shows fall like "rain". Stored programs are displayed at the bottom of the screen. Over time the stored programs slowly sink into the drain, indicating to users that they are going to be deleted. The system provides tools for finding specific segments from whole TV shows. The example below displays all of the fieldgoals from a stored football game.

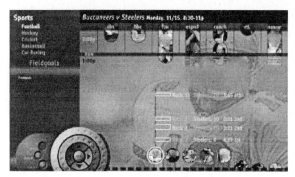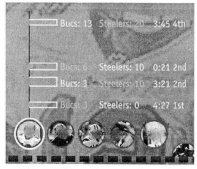

Figure 1. Raindrops sketch and detail of a stored and segmented football game.

The Seeker concept (figure 2) finds video clips on specific topics based on users' profiles. In the example below, the system employs users' TV viewing histories to infer which *celebrities* they like. It then records all video clips it finds with these celebrities. When users begin watching TV, they see all of the clips the system has collected. A graphical scroll bar located on the right of the list reveals the relative length of the individual clips. Users can build a playlist on the right from the list of all captured clips on the left.

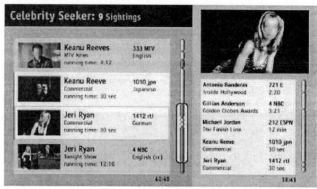

Figure 2. Celebrity Seeker Sketch

The TV Magnets concept (figure 3) lets users choose topics to be recorded. Users create different kinds of magnets—event magnet, news magnet, people magnet, etc.—that attract and repel specific story segments. Users can individually adjust the strength of attraction or repulsion for topics. In the example below the user has specified an event magnet that attracts stories about plane crashes, earthquakes, tornadoes, and floods. The same magnet repels TV commercials and stories on weather.

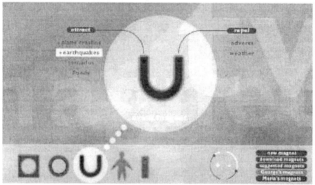

Figure 3. TV Magnet Sketch

3.1 Reactions

Of the three concepts, Raindrops was best received. Subjects liked selecting TV shows by image or face as well as by text. They also liked the new feature that let them identify specific events from within a program.

In general, they were more confused by Seeker and TV Magnets. *At first, almost all subjects had trouble understanding why they would want only a piece of a TV show instead of a whole TV show.* They often expressed anxiety that if they only had a video clip, they might be missing something better. They said that both systems were cumbersome, offering too many options that got in the way of trying to watch TV.

With Seeker, subjects indicated that they did not want narrative content cut into pieces unless these pieces worked as a preview, a task beyond the semantic ability of the Scout engine. Several also stated that this technology would be great for tracking financial news.

For TV Magnets, users liked specifying the kind of information to be recorded. They especially liked the idea of repelling advertisements. Their main concern, however, was the fear of a *need to constantly manage the magnets* to get the right content.

4 Prototype Design

We based the Scout interface on the look and feel of Raindrops because this design had received the best feedback, and *it allowed us to lead users into the idea of video clips* by initially presenting whole TV shows. Our main task was to take the best pieces of the other designs and incorporate them into Raindrops.

We divided the prototype into two main sections called Program Guide and TV Magnets. The Program Guide section works a lot like Raindrops, allowing users to interact with whole TV shows. The TV Magnets section offers users access to their profiles and video clips organized by topic.

4.1 Grid

Building on the Raindrops design, we divided the screen into three zones. The left side, called the tool zone, lets users adjust parameters, changing the results displayed on the right side of the screen. The right side was divided into the broadcast zone (top) and the stored zone (bottom). The broadcast zone displays live and future shows. The stored zone displays recorded TV shows and video clips. Each zone is flexible and can dynamically expand, allowing users more space. We designed these zones in order to free users from "tree" menu structures they found cumbersome in many TV products available today.

4.2 Program Guide

The Program Guide is the first screen users see (figure 4). Our guide differs from traditional TV guides in two ways. First, *we display live and stored content in a single interface*. Second, we changed the orientation of the time/channel grid to support the metaphor of *raining* content. This rotation also helps explain the relationship between broadcast and stored content. Live shows fall across the current time boundary—the dark line separating the broadcast and stored zones—and join other stored shows at the bottom of the screen.

We chose to display stored and broadcast content together because they represent similar values to users. When users select something to watch, they are indicating how important this item is in terms of their time.

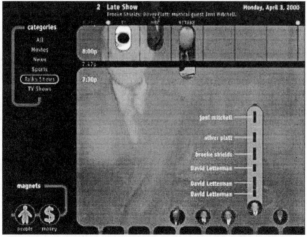

Figure 4. Program Guide showing a stored talk show's program and commercial segments.

The tools zone (left) contains category filters. By selecting a filter, users can reduce the number of broadcast and stored shows that appear. Below the categories are the TV magnet indicators. These feedback devices inform users when the system has stored new topic-based video clips.

The broadcast zone (top right) displays all shows being broadcast that match the highlighted category. The shows appear as pill-shaped images with their heights indicating the length of the program. Shows selected to be recorded have a circle in the upper left corner. When a recorded show crosses the current time barrier, it transforms into a circle and joins the other recorded shows at the bottom of the screen. Shows not selected to be recorded dissolve as they cross the current time barrier.

Users can expand stored shows to see an initial segmentation based on TV commercials. Dark segments indicate program content separated by TV commercials. In this example, a talkshow, summary labels indicate guest and host segments. Organization of segments matches the time axis for broadcast shows. The lowest segment labeled "David Letterman" represents the beginning of the show, and the top segment labeled "joni mitchell" represents the end of the show. For talk shows, users can select two additional segmentations. They can see all of the jokes in the program—the system listens for a laugh and searches back for the end of the previous laugh and calls this a joke—and they can see a segmentation based on musical performance. In the case of figure 4, this allows users to quickly play a video clip of Joni Mitchell singing. The initial segmentation by TV commercial works for all TV shows; however, the additional segmentations are specific to the recorded show's genre.

4.3 TV Magnets

The TV Magnets section displays video clips organized by topic. Figure 5 below shows the Financial TV Magnet screen. The tool zone lists all of the topics and includes company names, stock markets, and high-level financial topics. The black circles with the white Ns indicate topics that have attracted new content. As users navigate up and down this list, stored clips matching the highlighted topic appear on the right in the stored zone. Each clip represents an individual story on the topic.

TV shows to be analyzed in the future for financial information appear in the broadcast zone. However, *not all programs that appear will be recorded*. For example: Scout does not know if a financial news broadcast will contain information on Microsoft before it is actually broadcast. Instead, Scout selects shows that are likely to contain desired content. It then records these shows and searches for stories within the recorded content. Any segment that matches a topic is stored, and the rest is deleted.

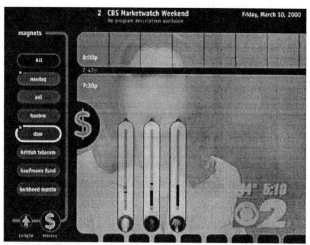

Figure 5. TV Magnet showing Financial Magnet results.

Figure 5 shows four clips that have been attracted on the topic "dow". Faces pulled from the video stream help indicate what shows the clips come from. In this example, one clip comes from *Marketwatch* and three clips come from two different episodes of *Wall Street Week*. We chose to display the video clip in relation to the whole program for three reasons. First, this display is consistent with the segmentations users see in the program guide section. Second, the length of the clip in relation to the whole show offers a quick preview as to how much detail is covered in the clip. Third, topics that appear at the beginning or appear several times within a program have been given a higher priority by the broadcaster and this information may be important to users who select one clip over another.

TV Magnets also provide users with tools for refining the content that gets recorded. We created magnets for financial news and celebrities. For celebrities, Scout automatically records video clips on the most watched celebrities that appeared in users' TV viewing histories. For financial news, Scout records video clips on companies and financial topics that match the users' investment portfolios. Adding the investment portfolio and viewing history to the user profile allows Scout to begin recording right out of the box.

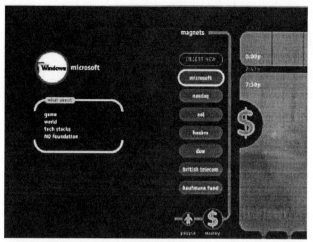

Figure 6. Modify TV Magnets

By default, Scout records all information on a listed topic. However, users can adjust what is recorded by adding keyword filters between the "what about" brackets (figure 6). In figure 6, only stories about Microsoft that mention "game", "world", or "tech stocks" will be recorded. Additionally, no stories that match these keywords but also mention Bill Gate's Foundation will be recorded. The "NO" indicates the repel feature of the magnet. By playing with the keywords, users can adjust the system over time, until it records exactly what they want.

4.4 Implementation

We built the Scout user interface using Macromedia Director, and we used Tabuleiro's DirectMedia Xtra for controlling MPEG files. We recorded approximately 30 hours of programming and analyzed it using our video analysis engine [2]. As users navigate the prototype, the Director executable reads in text files output by this engine and dynamically lays out screens.

In order to create a more TV-like experience, we used a Philips Pronto remote control for user input. We developed the prototype to respond to keyboard commands and trained the Pronto using a wireless keyboard.

5 Conclusion

We created the Video Scout prototype using a traditional user-centered design process of concept sketches, user testing, and design. Our main challenge was to create a new TV interface that comfortably allows users to interact with both whole TV programs and with video clips extracted from TV programs. Scout represents a future direction for PVRs because it gives the user much more control of the content that gets recorded. User profiles are automatically generated, but can be also be modified by the users themselves. In this way, Scout can quickly be trained to retrieve only the content users really want. And as the amount of video content people have to choose from increases with digital cable, digital satellite, and downloadable video from the Internet, the need for Scout to control information overload becomes clearer.

In the future we hope to use this prototype with users to better assess what applications they would like this technology to perform in their homes.

6 References

[1] http://www.tivo.com
[2] Dimitrova, N., McGee, T., Agnihotri, L., Dagtas, S., & Jasinschi, R. (2000). On Selective Video Content Analysis and Filtering, *Proceedings of SPIE Storage and Retrieval for Media Databases*, 2000, San Jose, CA, USA, 359-368.
[3] Jasinschi, R., Dimitrova, N., McGee, T., Agnihotri, L., Zimmerman, J., & Li, D. (2001). Integrated Multimedia Processing for Topic Segmentation and Classification, *Proceedings of International Conference on Image Processing*, 2001, Thessaloniki, Greece.

A Pattern-Supported Approach to the User Interface Design Process

Asa Granlund

Ericsson Radio Systems AB
Box 1248
S-581 12 LINKÖPING, Sweden
asa.granlund@era.ericsson.se

Daniel Lafrenière

GESPRO Technologies
1245 chemin Sainte-Foy, Édifice 1
Québec (QC), Canada G1S 4P2
lafrenid@gespro.com

David A. Carr

Institutionen för Systemteknik
Luleå University of Technology
S-971 87 LULEÅ, Sweden
david@sm.luth.se

ABSTRACT

Patterns describe generic solutions to common problems in context. Originating from the world of architecture, patterns have been used mostly in object-oriented programming and data analysis. The goal of HCI patterns is to create an inventory of solutions to help designers (and usability engineers) to resolve UI development problems that are common, difficult and frequently encountered. In this paper, we present our pattern-supported approach to user interface design in the context of information visualization. Using a concrete example from the telecommunications domain, we will focus on a task/subtask pattern to illustrate how knowledge about a task and an appropriate interaction design solution can be captured and communicated.

1   INTRODUCTION

1.1   What Is a Pattern?

A pattern is a formalized description of a proven concept that expresses non-trivial solutions to a UI design problem. The primary goal of patterns in general is to create an inventory of solutions to help UI designers resolve UI development problems that are common, difficult and frequently encountered. (adapted from Loureiro & Plummer, 1999)

A pattern is a format for describing a solution to a design problem. Patterns originate from architecture and were introduced by Christopher Alexander (Alexander, et. al., 1977; Alexander, 1979) in the mid-70's. Alexander noticed that certain solutions always apply to the same recurring problems and developed patterns as a design knowledge documentation method.

Software Engineering (Gamma, Helm, Johnson & Vlissides, 1995) adopted the pattern as a way to facilitate reuse of software. Early attempts at reuse of components often failed because the units were too small and did not mesh well together. Software patterns were adopted to allow sharing of larger units, and they specify in quite fine detail how components interact. As such they are much more prescriptive than patterns for architecture.

User interface designers also noticed that certain design problems occurred over and over. These problems generally have known good solutions. However, there has been a problem communicating them. Guidelines represent a possible solution, but they are generally seen as hard to interpret and as requiring excessive effort to find relevant material (Mahemoff & Johnston, 1998). For this reason, there has been an increasing interest in patterns to document user-interface design solutions. The SIGCHI'97 workshop on patterns (Bayle, et. al., 1998) saw patterns as a way of dealing with the increasing complexity and diversity of HCI design.

Tidwell (1999) describes patterns as "… possible good solutions to a common design problem within a certain context, by describing the invariant qualities of all those solutions." Simply put, patterns can be said to provide powerful and generic design guidance in a format that is consistent and easy to read and understand – they convey knowledge about good design.

Patterns are used implicitly by many skilled UI designers who have found solutions that have worked for them in the past. However, these designers usually keep little in the way of formal (documented) descriptions of these solutions. Thus, there are in fact both *implicit* patterns and formal (*explicit*) patterns. Explicit patterns can be used as a means of collecting and *formalizing* this knowledge. Using HCI patterns for capturing and documenting design knowledge is currently a hot topic, and there are many reasons for this interest (Erickson, 1998):

- Patterns provide a *lingua franca* that can be read and understood by all, regardless of background.

- The existing formal ways of documenting UI design knowledge are often weak – patterns offer a good way of capturing and transferring this knowledge. They are presented consistently, are easy to read, and provide background reasoning. The format provides information about the problem at hand, the context, a solution and also the rationale behind this solution.

- They promote reuse.
- Patterns are a valuable source of information, supporting both the analysis and the current situation and the design of the new system.

However, we believe that patterns **cannot** serve as a single source of design knowledge. They must still be complemented by traditional sources of information. But, they will point to information that is generally valid (for a specific domain) and also to designs that have proven good for similar projects. Since the description format provides reasoning and motivation, a pattern's relevance for the current project can be tested and evaluated.

Patterns must also be part of a language of interrelated patterns, participating in and supporting each other, in order to be truly useful. The pattern language works on different scales, and promotes the iterative growth of a design.

1.2   HCI Patterns Are Different

Patterns differ from design guidelines. Guidelines aim at coherence among user interfaces by documenting all the intricacies of a particular user interface (such as Windows or MacOS). However, guidelines focus mostly on window/widget issues while neglecting the knowledge required for proper UI design. In fact, the major forces influencing design: the user, the context, and the task, are missing from guidelines. Design rationale is missing, too. Patterns capture and document all of this important knowledge. They are also more invariant over time.

HCI patterns also differ from other pattern types. Software engineering patterns tend to specify a very strong interrelationship among component descriptions. The emphasis is on interface specification among components. However, user-interface designers see these patterns as too rigid and too detailed. User interface designers are concerned with esthetics and social aspects as well as function. They also want freedom to innovate and express themselves. These desires, combined with the fact that HCI is a young discipline where much is unknown, make a pure engineering approach inappropriate. Thus, HCI patterns are closer to architectural patterns. However, HCI patterns are also tied to software systems and as such must take software issues into consideration.

1.3   The PSA Approach

Up to this point, most of the work on patterns in HCI has focused on screen design issues. Our pattern-supported approach (PSA) to the user interface design process suggests a wider scope for the use of patterns by looking at the overall user-oriented interface design process. PSA addresses patterns not only at the design phase, but *before* design. (See Figure 1.)

Based on the fact that the usability of a system emerges as the product of the user, the task and the context of use, PSA integrates this knowledge in most of its patterns, dividing the forces in the pattern description correspondingly (i.e., describing Task, User, and Context forces). PSA provides a *double-linked chain* of patterns (parts of an emerging pattern language) that *support* each step of the design process (Figure 2). For example, *task* patterns point to *Structure and Navigation Patterns*, which in turn point to *GUI Design Patterns*, and vice-versa. These patterns offer a way to capture and communicate knowledge from previous designs (including the knowledge from system definition, task/user analysis and structure & navigation design). Given a mature language of patterns belonging to the described classes, the PSA approach provides an entry point to this pattern language, and suggests (without restricting the pattern usage) a chain of appropriate patterns at different levels of analysis and design.

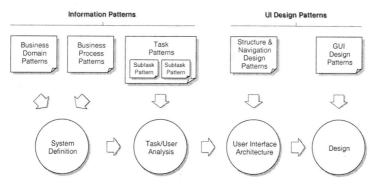

Figure 1 – The PSA Framework

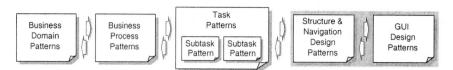

Figure 2 – Links between PSA Patterns

Here is a brief overview of the PSA patterns:

- **Business Domain Patterns** describe the type of business, its goals, plus the typical actors and business processes involved. They provide a starting point for initially defining the system design by pointing to relevant *Business Process* patterns and thereby to *Task* patterns. They help communicate the *System Vision*.

- **Business Process Patterns** describe typical processes and actors involved in the delivery of services/goods in compliance with the business goals. They narrow down the system definition and point to specific *Task* patterns to be considered.

- **Task Patterns** are used for capturing and passing on knowledge about the task, typical users, and their work context from previous similar projects, and for suggesting an appropriate interaction design solution. They point to *Structure & Navigation Design* patterns that describe solutions that have proven suitable for the task type in previous designs.

- **Structure & Navigation Design Patterns** describe ways to structure information and implement navigation in order to support the user's task. This design is based on the information described in the *Task* patterns.

- **GUI Design Patterns** document GUI design issues based upon information described in the *Task Patterns* and *Structure & Navigation Design Patterns*. They are based on the work of Tidwell (1999).

PSA is concerned with domain specific pattern support, that is: *Business Domain*, *Business Process* and *Task* patterns that describe a **specific** domain. We believe that this level of specificity is necessary – if the patterns are too abstract, they will not be useful in practice. However, it is desirable to make them as general and invariant as possible. Knowledge gained from this work will later be used to attempt the generalization that will also support transfer of design knowledge between domains.

Structure & Navigation Design patterns and *GUI Design* patterns, however, are **general** from the start, since they are not dependent on the domain. This means that the information in the *Business Domain* and *Business Process* and *Task* patterns leads up to suggestions for suitable *Structure & Navigation Design* patterns. *Structure & Navigation Design* patterns in turn point to *GUI Design* patterns, but the *Structure & Navigation Design* pattern in itself is **not** based on domain specific knowledge about task, user, and context.

2 FORMAT AND USE OF PSA PATTERNS

In order to demonstrate the use of PSA patterns, we will discuss the form of PSA patterns and illustrate their use with excerpts from a task pattern and a subtask pattern taken from our project to build a pattern language for system design for radio network management (mobile telephony). The patterns selected document part of the activity around optimizing the operation of the radio network. They also incorporate recommendations for the use of information visualization in this activity. The patterns incorporate guidance on design for visualization (Carr, 1999) into the context of managing cellular telephone networks.

We present only task and subtask patterns. In a design project, a set of relevant task patterns would have been selected during discussions with stakeholders, using Business Domain and Business Process patterns to guide the selection. The Task pattern can then be used for planning task and user analysis (and in some cases substitute for these activities, where for some reason, they cannot be carried out).

2.1 Task Patterns

Task patterns describe complex tasks, and point to subtask patterns that in turn describe parts of the task. They are used for capturing knowledge from previous similar projects, and use Description, Context and Forces to pass on the knowledge regarding: the task, typical users, their work context, and an appropriate interaction design solution. The Task pattern description consists of the following sections:

- **Name** – describes the task. A pattern called "Radio Network Optimization " would be a more specific instantiation of the more generic pattern "Network Optimization", which can apply to many different domains.

- **Context** – describes the goal of the design, specifying the user and the requirements of the task.

- **Problem** – describes the design problem at hand. Whenever there are competing concerns (forces) there will also be a problem – without the competing concerns, the solution is trivial, and there is no problem.

- **Example** – is used to clarify the task. This adds a feeling for the task at hand, and PSA tries to use a storyboard in order to make the example more vivid and easier to understand.

- **Forces** – describe all the, often conflicting, factors that influence design, directly or indirectly. PSA uses detailed forces, specific to the domain, but generic within it. The richness provides all the recurring information that would typically be gathered during task and user analysis. Information that is "project specific" does not fit into the Force description. Forces are divided into task, user, and context forces. The task forces describe task characteristics and special requirements. They build on general knowledge (e.g., regarding visualization tasks), but are made specific to the task (e.g., radio network optimization) in order to help the reader more clearly understand the problem. The reader avoids mental translation between general task characteristics and radio network optimization. The Force descriptions about the user are inspired by "personas" (Cooper, 1999), in that they paint a picture of a typical user. The information could easily be used as a basis for a full "persona" description in order to exemplify the user. Context forces are used to capture environmental and social factors that influence the work.

- **Design Solution** – communicates design considerations for the task and "guidelines" emerging from the forces. Since this pattern describes a complex task containing many subtasks, this interaction design solution (Figure 3) concerns overall interaction issues. For design on lower levels, each contained subtask pattern will contribute its own guidance.

- **Resulting Subtask Patterns** – describe the smaller tasks that are part of the complex task described in the pattern. These are generic, making up building blocks, while at the same time having their own pattern descriptions with forces, related Subtask patterns, and related Structure & Navigation patterns. For example, the Subtask patterns for our Optimize Radio Network pattern are: Graphic Overview, Zoom, Filter Details-on-demand, Relate, and History. They are derived from Shneiderman's visual information seeking tasks (Shneiderman, 1996).

- **Resulting Structure & Navigation Patterns** – suggest a way of navigating (and thereby implicitly structuring) the data/functions on the level of the complex task described in the pattern, in other words, the overall structure and navigation. Structure and navigation for parts of the task may well be different, and this is described in the resulting subtask patterns.

- **Resulting GUI Design Patterns** – suggest suitable patterns for GUI design solutions.

2.2 Subtask Patterns

All complex tasks are made up of smaller tasks – subtasks. Trying to describe a complex task in one pattern would be very difficult, and the description would quickly become large and unwieldy. Using the strength of a pattern language, we can let the task pattern point to subtask patterns that participate in the complex overall task. The subtask patterns are typically (at least from our experience so far) generic and not dependant on user task and context. This is due to the granularity of the description. They have the same components as the task pattern, describe solu-

---

The task forces emphasize the lack of structured workflow, the need to handle large amounts of data, and the need for displaying data based on the context of the specific network optimization problem. These forces and the specifics of the problem statement suggest information visualization as a solution. Information visualization is appropriate where the user has a task that is not easy to specify and involves large amounts of data (Carr, 1999). One can assume that the user will follow Shneiderman's information seeking mantra (Shneiderman, 1996): "Overview first, zoom and filter, then details on demand". This suggests the following organization for the design:

- Use a two-dimensional map image for overview. This map image should be as simple and uncluttered as possible, only providing the critical information at any given time.

- Provide zooming and panning of the map. Semantic zooming should be supported, where greater detail is revealed as the user zooms in.

- Provide flexible, direct-manipulation-based filtering. Dynamic queries (Ahlberg & Shneiderman, 1994) are an example design.

- Provide details-on-demand for all relevant objects in the map image. Make sure that the user doesn't lose spatial context when drilling-down.

- When using multiple views of the same data, make sure that their contents are synchronized. Consider a design similar to "snap-together visualizations" (North & Shneiderman, 2000) where users can quickly construct their own customized visualization consisting of multiple coordinated views.

Figure 3 – Design solution for the Optimize Radio Network pattern.

tions for interaction design and point to structure and navigation patterns at this lower level of detail. This should be quite intuitive – any complex system may have an overall structure, navigation model, and interaction design solution, and at the same time have subparts with varying solutions within it. The subtask pattern also refers to related subtask patterns.

There are two main differences between task and subtask pattern descriptions. Subtask forces are generally a subset of task forces or derived from them. They are specific to the subtask, and are not divided into Task, User, and Context Forces. At this level of detail the description is general for many complex tasks, regardless of the domain. From this point of view, forces are more general than for task patterns, and independent of task, user, or context. Also, subtask patterns have been found to be independent of a specific problem. Their design solution needs to be described in general terms

3   CONCLUSIONS AND FUTURE WORK

Up to this point, we have had only positive feedback from interaction designers presented with the approach. People appreciate the strength of the format, and believe it would really support them in their work. However, we have just started building a language of patterns, and many questions remain unsolved.

From a constructional point of view, we are currently working with the format of the structure and navigation patterns. Originally, the approach offered conceptual design patterns, but as these turned out to be too abstract to be useful, and we turned to the more practical subtask patterns. We are, however, striving to capture the more complex aspects of modeling.

We are also concerned with the robustness of the chain of patterns that we offer. What happens if some but not all forces apply? Can the link to the next level of patterns be trusted? We are thinking of having a "template" component for which pointers are always valid. But at the same time, we are concerned that the approach and related patterns will become too unwieldy. Our goal is that the user of the patterns should never be concerned with the construction of the patterns – they must be easy and intuitive.

But above all, the approach and the patterns need to be adapted and validated through practical usage. Today, the patterns do not fully supply a lingua franca, but are more or less targeting interaction designers. The descriptions, structure and level of detail must be adapted to fit actual design projects. This can only be done through iteration-based, practical use. In addition, whether or not the patterns can hold their promise of facilitating communication must be evaluated.

4   REFERENCES

Ahlberg, C. & Shneiderman, B. (1994) Visual information seeking: tight coupling of dynamic query filters with starfield displays, *Human Factors in Computing Systems Proceedings of CHI'94*, Boston, MA, 365-371.

Alexander, C. (1979) *The Timeless Way of Building*. Oxford University Press, New York, NY.

Alexander, C., Ishikawa, S., Silverstein, M., Jacobson, M., Fiskdahl-King, I. & Angel, S. (1977) *A Pattern Language: Towns, Buildings, Construction*. Oxford University Press, New York, NY.

Bayle, E., Bellamy R., Casaday, G., Erickson, T., Fincher, S., Grinter, B., Gross, B., Lehder, D., Marmolin, H., Moore, B., Potts, C., Skousen, G., & Thomas, J. (1998) Putting it all together: towards a pattern language for interaction design, *SGICHI Bulletin*, 30(1), 17-23.

Carr, D. (1999) Guidelines for designing information visualization applications, *Proceedings of ECUE'99*, Stockholm, Sweden.

Cooper, A. (1999). *The Inmates are Running the Asylum*. SAMS, Indianapolis, IN, ISBN 0-62-31649-8.

Erickson, T. (1998) Interaction Pattern Languages: A Lingua Franca for Interaction Design?, *UPA 98 Conference*, Washington, DC.

Gamma, E., Helm, R., Johnson, R., & Vlissides, R. (1995) *Design Patterns: Elements of Reusable Object-Oriented Software*. Addison-Wesley, Reading, MA, ISBN 0-201-63361-2.

Loureiro, K. & Plummer, D. (1999). *AD Patterns: Beyond Objects and Components*. Research Note # COM-08-0111, Gartner Group.

Mahemoff, M, & Johnston, L. (1998) Principles for a usability-oriented pattern language, *OZCHI '98 Proceedings*, Adelaide, Australia, 132-139.

North, C., & Shneiderman, B. (2000) Snap-together visualization: a user interface for coordinating visualizations via relational schemata, *Proceedings of Advanced Visual Interfaces 2000*, Palermo, Italy, 128-135.

Shneiderman, B (1996). The eyes have it: a task by data type taxonomy for information visualizations, *Proceedings of 1996 IEEE Visual Languages*, Boulder, CO, 336-343.

Tidwell, J. (1999). *Common Ground: A Pattern Language for Human-Computer Interface Design*. http://www.mit.edu/~jtidwell/interaction_patterns.html

Effects of Time Delay and Manipulator Speed on a Telerobotic Peg-in-Hole Task

J. Corde Lane, Craig R. Carignan, and David L. Akin

Space Systems Laboratory, University of Maryland, College Park, Maryland 20742

1. Background

The Space Systems Laboratory is currently developing the Ranger Telerobotic Shuttle Experiment (RTSX), an advanced telerobot testbed designed to demonstrate that robots can perform maintenance tasks on space stations and satellites. The Ranger TSX vehicle uses four manipulators, shown in Figure 1, to perform its required tasks. A six-degree of freedom (DOF) positioning leg connects the main body of the vehicle to the pallet, providing mobility for the vehicle to optimize its working position. Two 8 DOF dexterous arms are then used to perform the servicing task. A seven DOF video arm is used to provide the remote operator with the desired view of the work area.

Ranger will be controlled from a local control station from the aft flight deck of the Space Shuttle, and also remotely from the ground. Figure 2 shows the layout of an operator console for the ground control station and the two hand controllers used to directly control the robot. Video from different cameras on the robot can be viewed from three video monitors. A Silicon Graphics® workstation displays a user interface with graphical simulation of the robot. The flight control station has a functional equivalent of hand controllers, monitors, and computer, but it is packaged differently from the control station used aboard the Space Shuttle.

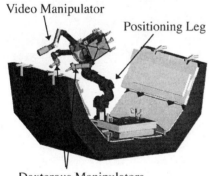

Figure 1: Ranger TSX Arm Configuration
Figure 2: Ground Control Station Layout Time Delay

Depending on the communication link between the ground control station and the robot, a 1.3 sec to 10 sec time delay will be experienced. Research has shown that time delay can significantly degrade operator performance[9]. Held, et al.[4] found that even at a time delay of 0.3 seconds, operators tend to decouple their command movements from the robotic system's response. Ferrell[3] showed that as time delay increased, the completion time for a 2DOF manipulator-positioning task increased proportionally. Black[2] had similar results illustrating the increase in task time with higher levels of delay for a 6DOF manipulator task.

To alleviate some of the negative effects of time delay, many researchers use some form of predictive display[1,6,8]. Operators use the predictive display, which is not affected by time delay, to help guide the actual manipulator to a location. This involves sending commands to a simulation, which then predicts how the arm will react without time delay. This prediction is typically overlaid on the display of the actual robot so that the operator will know the results of their commands.

The ability to control a remote device depends on the speed being commanded. To reduce the occurrence of collisions remote systems may be moved with slow velocities. For a particular task, an optimum velocity of the system might be determined to complete the task both quickly and without excessive error. This study evaluated controlling a simulated remote manipulator under changing conditions of time delay and use of a predictive display. The effects of those two may modify the optimum velocity that the manipulator should be controlled.

2. Experiment

In this investigation, four subjects performed a simulated peg-in-hole task. Using the ground control station, shown in Figure 2, subjects used the hand controllers to command the right dexterous manipulator tip in Cartesian rate. The velocity of the tip was controlled in the manipulator tip's frame of reference. As Figure 3 illustrates, if the tool was rotated 30 degrees, a rightward command could appear as if the manipulator was moving to the diagonally instead. This would cause the subjects to either correct or compensate for any angular differences in the manipulator's orientation.

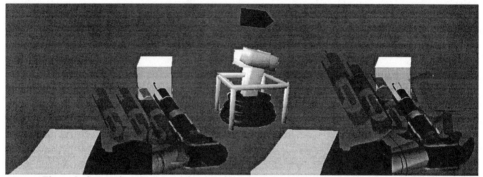

Figure 3: Comparison Between Visual (left) and Manipulator tip (right) Frame of Reference

The task was to insert the manipulator tip (~1cm diameter) into a 2cm diameter hole contained inside a cubic foot box. During the test the subject could switch between one of three views: an overall shot of the worksite and the arm (Figure 4), a close-up view from the right side (Figure 5), and a top view (Figure 6) of the hole. One of the subjects used the overall view to coarsely move the manipulator close to the hole, and then would use the close-up views for the final insertion. The other three subjects exclusively used the close views even during initial movements.

On half of the tests with time delay, a translucent predictive display of the manipulator was overlaid on the display of the actual manipulator. This arm could be controlled directly and had no delay associated with it. The actual arm would follow the same path as the predictive display as the time delay expired, except when a collision occurred. A collision would cause the predictive display, unaware of any obstacles, to move while the actual manipulator would remain stuck at the surface of the box. This difference became very helpful to the subjects in quickly observing and correcting for impacts. Without this visual cue, subjects controlling under time delay without a predictive display would take longer to realize an impact had occurred.

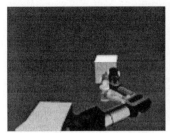  

Figure 4: Overview Camera View   Figure 5: Right Side Camera View   Figure 6: Top Camera View

The simulation model differed in several significant ways from reality to facilitate computer performance. If the manipulator tip touched the surface inside the hole or outside the box, it would remain stuck until the subject pulled the manipulator back. The simulation's low fidelity contact model would not allow any sliding. Since even the slightest contact with the inside of the hole would make the manipulator stop, the front face of the box was made translucent, allowing the subject to view the manipulator tip while inserting it. This unrealistic visual cue was given to offset the strict contact model. To help the simulation to run in real-time only kinematic motion was calculated. No dynamics or other perturbations were modeled in the simulation. This reduced the realistic feel of controlling the manipulator, which slightly overshoots and oscillates with every motion. These effects would cause deviations between the actual manipulator and the predictive display, which would reduce the effectiveness of the predictive display.

Subjects were instructed to insert the manipulator tip into the hole as fast as possible while avoiding impacts with the box. Both the time to complete the task and the number of times the manipulator impacted the box were tracked. A test would begin by the arm being randomly placed at one of 32 initial positions from the box. A pause for a few seconds allowed the operator to assess the manipulator's position and how to proceed. Then the hole would appear in the center of the box and the subject would command the manipulator tip for insertion. Once the manipulator tip touched the back of the hole, the time to complete that trial and number of impacts was calculated. Then the manipulator would be placed at a new initial position and the process would start again. When all 32 initial positions were performed that test was completed.

The subjects performed 32 insertions for each set of independent variables: time delay, use of predictive display, and manipulator speed. A total of 35 test cell combinations were tested. Three time delays were tested: 0 sec, 1.5 sec, and 3 sec. During time delay testing, subjects performed the experiment with and without a predictive display to investigate how much the predictive display helped the subject compensate for the time delay. Finally, a range of maximum manipulator tip speeds was tested: 1, 2, 3, 4, 5, 7.5, and 9 in/sec. This range was used to determine if there was an optimal manipulator speed for performing this task.

The hypothesis was that time delay would produce a strong negative effect on performance; a large increase in task completion time and number of impacts was expected. The use of the predictive display was expected to reduce the increase in time and number of impacts to be slightly higher than the no delay levels. Finally, it was anticipated a maximum manipulator speed of around 4 or 5 in/sec would produce the optimal performance. Speeding up or slowing down would likely have adverse affects. Slower speeds will unnecessarily increase completion times, while faster speeds may be too difficult for the subjects to control.

3. Results and Discussion

As expected, time delay had a pronounced affect on performance. Figures 6 and 7 show how time delay and use of a predictive display affected the task completion time and the number of impacts respectively. Each of the five bars are averaged between all four subjects and the seven different speeds tested. The lighter bars indicate when only the actual manipulator was shown; and the darker bars show the results when subjects used a predictive display overlaid on top of the actual. The letters within the bar indicate its grouping determined by ANOVA at 0.01 statistical significance level.

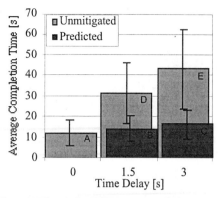

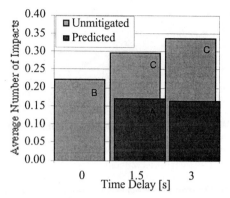

Figure 7: Time Delay Effects on Completion Time   Figure 8: Time Delay Effects on Number of Impacts

As Figure 7 indicates, an increase in time delay causes a statistically significant increase in completion time. At 1.5 sec delay an increase of over 150% occurred and a 3 sec delay resulted with an increase over 250%. This effect of time delay was dramatically reduced by the use of the predictive display; approximately 90% of the increased time was eliminated. The pronounced effect of the predictive display was partially due to the lack of any error between the predictive and actual display. Typical errors due to modeled dynamics, thermal variations, video aberrations, and sensor loss of calibration cause deviations between the predictive and actual displays. This deviation lessens the advantage of the predictive display[6, 7]. However, even with these errors a predictive display can still be very useful[5,6].

The predictive display also provided additional visual cues when an impact occurred, due to the predictive display having no knowledge of the environment. Although the predictive display gave no direct immediate feedback that an impact had occurred, a small deviation between the predictive and actual display would indicate that an obstacle probably held the actual manipulator back. If a predictive display were presented to the subjects when testing with no time delay, a small decrease in completion time would be expected due to this additional feedback.

The number of overall times the manipulator impacted the box was low. Figure 8 shows that the average number of impacts for one insertion trial was about 0.25, or one impact during four-insertion trials. Even though large standard deviation existed, the ANOVA still found statistical significance at the 0.01 level. As expected, when no predictive display was used, time delay increased the number of impacts by about 40%.

What was not anticipated was that when subjects had access to the predictive display, they made 25% fewer impacts when compared to no time delay. This decrease in impacts was attributed to increase attention to avoid impacts during time delay. Without time delay, an impact could be quickly diagnosed and corrected. This promoted the subjects to maneuver the arm faster accepting the risk of a quick impact. However, as the delay increased, subjects may not realize an impact occurred for several seconds. Therefore, an impact became more costly as time delay increased. Subjects probably used the predictive display more attentively to reduce the chance of an expensive impact. As mentioned before, the addition of realistic errors between the predictive and actual display can reduce this benefit of fewer impacts.

The wide range of maximum manipulator speeds caused a smaller variation in performance as was expected. Results are plotted in Figure 9 and Figure 10; the 0.01 level ANOVA groupings for each speed are listed below the graphs. Figure 9 shows only a 45% increase between the optimal speed (7.5 in/sec) and the slowest speed (1 in/sec), and speeds 4 in/s and above varied by less than 6%. Although the speed of 7.5 in/s was determined to be optimal by the ANOVA, any speed above 3 in/s would most likely prove satisfactory. A bias existed for the 7.5 in/s speed, since previous experimentation had used that speed as the baseline. Therefore three of the four subjects had experience controlling the manipulator at 7.5 in/s. This could have caused the minor performance improvement. Figure 10 shows that only the fastest manipulator speed of 9 in/s produced a statistically significant higher number of impacts. That speed facilitated recklessly fast commands causing a 40% increase of impacts when compared with other speeds.

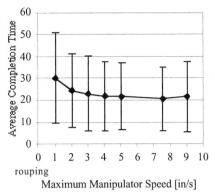

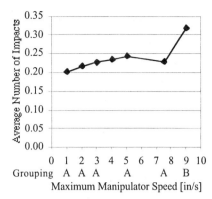

Figure 9: Manipulator Speed Effects on Completion Time

Figure 10: Manipulator Speed Effects on Number of Impact

In summary, the most influential affect on the subject's performance was time delay. Secondly, a predictive display played a vital role in reducing the negative affects of time delay. Finally, the control speed of the manipulator proved to be the least important in affecting performance. Some drawbacks of his study were that it did not simulate realistic dynamics of the manipulator or contact modeling between the tooltip and the box. More research is required to help define the usefulness of predictive displays under these more accurate predictive conditions. The effect of calibration errors between the predictive and actual displays would help define the benefit of using predictive displays for real-time control of robots.

4. References

1. Bejczy, Antal K. and Won S. Kim. "Predictive Displays and Shared Compliance Control for Time-Delayed Manipulation." Proceedings IEEE International Workshop on Intelligent Robots Systems, IROS. 1990
2. Black, J. H. Factorial Study Of Remote Manipulation With Transmission Time Delay. M.S. Thesis MIT, 1971.
3. Ferrell, William R. "Remote Manipulation With Transmission Delay." IEEE Transactions in Human Factors in Electronics,6(1). 1965.
4. Held R., A. Efstathiou, and M. Greene. "Adaptation to Displaced and DelayedVisual Feedback from the Hand." Journal of Experimental Psychology, 72. 1966.
5. Kelley, C. R. "Predictor Instruments Look into the Future" Control Engineering, 9(3) 1962.
6. Lane, J. Corde. Human Factors Optimization of Virtual Environment Attributes for a Space Telerobotic Control Station. Diss. Univ. of Maryland,College Park. 2000.
7. Mar, Linda E. Human Control Performance In Operation Of A Time-Delayed Master-Slave Telemanipulator. B.S. Thesis MIT, 1985.
8. Noyes, Mark V. Superposition Of Graphics On Low Bit Rate Video As An Aid In Teleoperation. M.S. Thesis MIT, 1982.
9. Sheridan, Thomas B. Telerobotics, Automation, And Human Supervisory Control. Cambridge: MIT Press, 1992.

Absolute Thresholds of Perception for Changes in a Graphical User Interface

James R. Dabrowski and Ethan V. Munson*

Department of EECS
University of Wisconsin-Milwaukee
Milwaukee, WI 53211 USA

Abstract

For years, software engineers have been told that applications must respond to user events within 100 milliseconds for them to seem instantaneous. Yet, despite this widespread acceptance of 100 milliseconds as the limit of human perception for changes in a graphical user interface, no one seems to have ever attempted to establish this limit through experimentation. In the present research, we attempt to establish thresholds of perception for changes in a graphical user interface using an adaptive tracking method. For keyboard interactions, subjects did not notice delays of approximately 140 milliseconds whereas, for mouse interactions, subjects did not notice delays of around 195 milliseconds. These delays are significantly greater than the 100 millisecond conventional wisdom in use today. Given these findings, further research is clearly needed to firmly establish lower bounds on application responsiveness so that software and operating system engineers can more precisely tune the interactive real-time responsiveness of their systems.

1 Introduction

Miller (Miller, 1968) was the first researcher to systematically address system response time and user satisfaction. He reasoned that expected or tolerable delays in conversations between persons would also be expected or tolerable when people interact with computers in a similar manner. He felt that a computer should take no more than 100 to 200 milliseconds to respond to a key-press or other similar action by the user. Since then, this 100 millisecond limit seems to have taken hold in the software engineering community even though Miller established it, not through experimentation, but rather through his beliefs about what acceptable delays would be.

After Miller, only two studies were performed that manipulated overall system response time on an action-by-action basis in a concerted effort to measure the effect of system response time on user performance and satisfaction. In 1981, Youmans (Youmans, 1981) found that preferred delays for some basic text-editing commands were tuned to a response time of 300 milliseconds — far greater than that limit proposed by Miller. Another study (Williges and Williges, 1982) had subjects perform standard database interactions with a computer and found that system delay and echo rate had the greatest impact on user productivity yet the levels of acceptable delay reported by them were also far greater than those suggested by Miller.

The remainder of the system response time studies focused on whether or not sub-second response times were truly needed and what effect system response time variability had on user satisfaction and ability to perform a given task. Some studies (Barber and Lucas, 1983; Lambert, 1984) found that decreases in system response time lead to more productivity, but others found that that increased productivity came at the cost of more errors or generally poorer work (Grossberg et al., 1976; Kosmatka, 1984). Others found that delays in system response time generally led to more frustration on the part of users, but did not necessarily decrease their satisfaction with the system (Goodman and

*The authors may be reached by e-mail at {jimd, munson}@cs.uwm.edu. This research was supported by an award from the USENIX Student Research Grants Program. Ethan Munson was also supported by the U. S. Department of Defense and by NSF CAREER award CCR-9734102.

Spence, 1978; Goodman and Spence, 1981; Goodman and Spence, 1982; Dannenbring, 1983; Dannenbring, 1984; Guynes, 1988). Finally, a group of researchers has systematically investigated the effect of system response time on user anxiety and task performance and have concluded that for any given task, there is an optimal delay that will minimize anxiety and maximize performance (Kuhmann et al., 1987; Kuhmann, 1989; Thum et al., 1995; Kohlisch and Schaefer, 1996; Kohlisch and Kuhmann, 1997). This early line of research into system response time essentially concluded with the findings that faster is generally better, but these increases in speed may come at the cost of accuracy and anxiety in the user.

None of the early studies, however, investigated users' abilities to perceive the delay associated with the system. It seems that as a baseline, before one can address the effect of delay on user satisfaction, one must first know the point at which they can physically perceive a delay. The most thorough investigation of this issue in computer science was by Card, Moran and Newell (Card et al., 1983). The main thrust of their work, however, was on reaction times — how long it takes people to respond to changes in an interface — rather than on people's abilities to perceive those changes. This difference is critical. Card, Moran and Newell did not specifically investigate the issue of human perception but instead relied on existing research to make estimates of the limits of human perception. They established 100 milliseconds as the average time it takes users to perceive a change on a computer screen, which corresponds to the limit suggested by Miller. However, the work on which they base this limit was a review of the psychological literature that measured the thresholds of perception for certain visual stimuli in which the subject was a passive observer under nearly ideal detection conditions. Since their analysis is based heavily on pure psychophysiological research, it is not clear whether these basic science results can be directly applied to general computer use.

Despite this widespread acceptance of 100 milliseconds as the limit of human perception for changes in a graphical user interface(Nielsen, 1994; Shneiderman, 1998), no one seems to have ever attempted to establish this limit through experimentation. The work cited by Card, Moran and Newell is perhaps the most applicable and most scientific. However, in those studies, all the subjects were passive observers whose task was to detect delay between limited, specific stimuli. They were not active users, interacting with an interface, causing the changes to occur and then making a decision as to whether they perceived a delay in the responsiveness of the system. It is possible, even likely, that when the subject is the agent of change rather than a passive observer of a visual stimulus, the subject will be more or less likely to notice the delays. Which is the case is unknown, but it bears investigation.

2 Methods

2.1 Participants

A total of 21 subjects were used in this study. The mean age of the subjects was 26.52 years with a standard deviation of 7.53 years. There were 18 males and 3 females tested. All the subjects were students at a mid-sized, urban university. Most would be classified as experienced, high frequency computer users. Most reported having three or more years of experience using Windows-based computers and that they used such systems ten or more hours per week. All subjects were paid ten dollars for their participation in the study. A typical testing session lasted from 45 minutes to one hour.

2.2 Apparatus

To determine thresholds of perception, an application was created using Visual Basic version 6.0 that was instrumented such that its responsiveness could be controlled at the millisecond level. This application was run on a standard PC running the WindowsNT operating system. The application consisted of five experimental tasks, a training task and a configuration window, each of which was accessed from a main-menu window. They are each described briefly below. For a more thorough description of the entire application see Dabrowski's thesis (Dabrowski, 2000).

For three of the tasks, the interaction with the application was with the mouse. For the button task, users were instructed to click on a button in a button-bar. The application systematically delayed the amount of time between the user click and the appearance of a colored box on the screen. For the dialog box task, users clicked an "OK" button in a dialog box to make it disappear. The application controlled the amount of time between the clicking of the "OK" button and the disappearance of the dialog box. For the menu task, users were instructed to click on specific menus. Again, the application controlled the amount of delay between the clicking on the menu by the user and the appearance of the menu.

	Menu	Button	Dialog	Typing	Form
Mean	171.98	197.56	192.98	156.07	146.81
Std. Dev.	64.68	68.26	54.74	52.63	46.22
Std. Error	14.11	14.89	11.95	11.48	10.09
Minimum	22.50	55.00	59.50	39.93	37.50
Maximum	307.08	366.67	267.92	241.31	215.45

Table 1: Means, standard deviations, standard errors, maximum and minimum values obtained for each task. Numbers shown are in milliseconds.

In the other two tasks, the interaction with the application was through the keyboard. In the form fill-in task, users were presented with a random word and were instructed to type that word into a text-box. In the sentence task, users were presented with a full sentence to type into a text box. For these typing tasks, we did not control the amount of time between the actual striking of the key on the keyboard with the drawing of that character on the screen. Instead, we controlled the rate of appearance of characters after the user began typing. Once the users began typing, the characters would appear only after the pre-specified delay and at regular intervals thereafter.

2.3 Procedure

Each subject was tested according to the procedure described below. There were no significant deviations from this procedure for any of the subjects tested in this experiment.

After completion of a training task, all subjects completed each of the five tasks described above. The tasks were presented in a random order — the only condition upon the randomization was that each task had to appear in each of the presentation orders an equal number of times. Within each task an adaptive tracking method with a standard staircase (Cornsweet, 1962) was used to determine the thresholds of perception for that particular task. For all tasks, an initial delay level of 200 milliseconds was selected. For each task, the subject was presented with the window as described above and told to follow the on-screen instructions. Once the subject performed the indicated action, a dialog box would appear on the screen asking the subject to indicate whether s/he noticed a delay between the time s/he performed the indicated action and the time the change occurred on the screen. If the subject did not notice a delay at that particular level, the application increased the delay in the system by a step size of five milliseconds for the next repetition of the task. If the subject did notice a delay, the application decreased the amount of delay in the system by the same five milliseconds. The dialog box then disappeared and the subject had to once again perform the same task.

This procedure was repeated until such time as the subject achieved fifteen reversals. A reversal occurred when a subject reported perceiving a delay on a trial at one delay level and then reported not seeing a delay on the next trial at a slightly smaller delay level, or vice versa. When the subject achieved fifteen reversals, the application automatically terminated that particular task. The use of this standard staircase with fifteen reversals is the same procedure described by Cornsweet (Cornsweet, 1962) and has been used extensively in determining thresholds in the field of psychology. After the completion of a particular task, the experimenter then indicated to the subject which task s/he should perform next until all tasks were completed.

3 Results

Table 1 shows the mean thresholds of detection obtained in this experiment along with other relevant statistics. The mean thresholds of perception for each task were calculated in the standard manner of averaging each subject's last twelve reversals for that particular task (Cornsweet, 1962).

A one-way, repeated measures ANOVA was performed to determine if there were any significant differences between the means obtained for the various tasks. The overall F-value obtained was 8.58 which was significant at the $p < .05$ level. Pair-wise Tukey-tests were performed to determine which tasks had significantly different thresholds of perception. Table 2 shows the differences in means between each of the tasks. A difference of 30.2 milliseconds or greater was needed for significance at $p < .05$. As can be seen from the table, both typing tasks had significantly lower

	Menu	Button	Dialog	Typing	Form
Menu	-	25.58	21.00	15.91	25.17
Button	-	-	4.48	41.49†	50.75†
Dialog	-	-	-	36.91†	46.17†
Typing	-	-	-	-	9.26
Form	-	-	-	-	-

†indicates $p < .05$.

Table 2: Results of pair-wise Tukey tests. Numbers shown are differences between threshold means. A difference greater than 30.2 milliseconds indicates significance at $p < .05$.

	Mean	df	t-value	P-value
Menu	171.98	20	5.10	<.0001
Button	197.56	20	6.55	<.0001
Dialog	192.98	20	7.78	<.0001
Typing	156.07	20	4.88	<.0001
Form	146.81	20	4.64	.0002

Table 3: Results of t-tests comparing obtained thresholds to 100 millisecond conventional wisdom.

thresholds of perception than two of the mouse tasks (the dialog task and the button task), but were not significantly shorter than the menu task.

Finally, to test whether the obtained thresholds of perception for the five tasks were significantly different from the 100 millisecond conventional wisdom, two-tailed t-tests were performed. As can be seen from Table 3, all the obtained thresholds were significantly greater than 100 milliseconds.

4 Conclusion

First, let us consider the conventional wisdom that applications must respond within 100 milliseconds for them to seem instantaneous to users. The present study found that significantly longer delays go unnoticed by most users. It would appear that a 140 millisecond response time is "fast enough" for virtually any type of interaction with a computer and that delays of up to 200 milliseconds for mouse events might go unnoticed.

Secondly, the results seem to show that for button-clicking events, users had significantly higher thresholds of perception than for keyboard events, though neither of these were significantly different from menu-clicking events. There is, however, a methodological problem with the current study that leaves that conclusion in question.

In the current study, all timer events were triggered by the mouse-down event. Yet, in the standard Windows interface, button objects execute their code on the "click" event. Therefore, the thresholds of perception for both mouse-related tasks involving buttons (the dialog task and the button task) may have been artificially inflated over the menu task to account for the amount of time it took the subjects to release the mouse button.

Despite this improper instrumentation, there were still differences in the thresholds of perception for mouse-related versus typing tasks even though individually these differences were not statistically significant. This potential difference needs further investigation to determine if there is a single threshold of perception for all changes on a computer monitor regardless of the type of user interaction or if indeed the interaction style causes differences in users' abilities to detect those changes. We are currently conducting additional studies in an attempt to determine which of the above explanations is more accurate.

References

Barber, R. E. and Lucas, H. C. (1983). System response time, operator productivity, and job satisfaction. *Communications of the ACM*, 26(11):972 – 986.

Card, S., Moran, T. P., and Newell, A. (1983). *The psychology of human computer interaction*. Lawrence Erlbaum Associates, Hillsdale, NJ.

Cornsweet, T. N. (1962). The staircase-method in psychophysics. *American Journal of Psychology*, 75:485–491.

Dabrowski, J. (2000). Absolute thresholds of perception for changes in a graphical user interface. Master's thesis, University of Wisconsin-Milwaukee, Milwaukee, WI.

Dannenbring, G. L. (1983). The effect of computer response time on user performance and satisfaction: a preliminary investigation. *Behavior Research Methods and Instrumentation*, 15(2):213 – 216.

Dannenbring, G. L. (1984). System response time and user performance. *IEEE Transactions on Systems, Man and Cybernetics*, SMC-14(3):473 – 478.

Goodman, T. and Spence, R. (1982). The effects of potentiometer dimensionality, system response time, and time of day on interactive graphical problem solving. *Human Factors*, 24(4):437 – 456. (Goodman and Spence, 1978).

Goodman, T. J. and Spence, R. (1978). The effect of computer system response time on interactive computer-aided. In *Proceedings of the SIGGRAPH '78 Conference*, New York. ACM.

Goodman, T. J. and Spence, R. (1981). The effect of computer system response time variability on interactive graphical problem solving. *IEEE Transactions on Systems, Man and Cybernetics*, SMC-11(3):207 – 216.

Grossberg, M., Wiesen, R. A., and Yntema, D. B. (1976). An experiment on problem solving with delayed computer responses. *IEEE Transactions on Systems, Man and Cybernetics*, SMC-6:219 – 222.

Guynes, J. L. (1988). Impact of system response time on state anxiety. *Communications of the ACM*, 31(3):342 – 347.

Kohlisch, O. and Kuhmann, W. (1997). System response time and readiness for task execution — the optimum duration of inter-task delays. *Ergonomics*, 40(4):265 – 280.

Kohlisch, O. and Schaefer, F. (1996). Physiological changes during computer tasks: Responses to mental load or to motor demands? *Ergonomics*, 39(2):213 – 224.

Kosmatka, L. J. (1984). A user challenges value of subsecond response time. *Computerworld*, 17:1 – 18.

Kuhmann, W. (1989). Experimental investigation of stress-inducing properties of system response time. *Ergonomics*, 32(3):271 – 280.

Kuhmann, W., Boucsein, W., Schaefer, F., and Alexander, J. (1987). Experimental investigation of psychophysiological stress-reactions induced by different system response times in human-compter interaction. *Ergonomics*, 30(6):933 – 943.

Lambert, G. N. (1984). A comparative study of system response time on program developer productivity. *IBM Systems Journal*, 23(1):36 – 43.

Miller, R. B. (1968). Response time in man-computer conversational transactions. In *Proceedings of the Spring Computer Conference*, pages 268 – 277, Montvale, NJ. AFIPS Press.

Nielsen, J. (1994). *Usability Engineering*. Morgan Kaufmann.

Shneiderman, B. (1998). *Designing the User Interface: Strategies for effective human-computer interaction*. Addison Wesley Longman, 3rd edition.

Thum, M., Boucsein, W., and Kuhmann, W. (1995). Standardized task strain and system response times in human-computer interaction. *Ergonomics*, 38(7):1342 – 1352.

Williges, R. C. and Williges, B. H. (1982). Modeling the human operator in computer-based data entry. *Human Factors*, 24(3):285 – 299.

Youmans, D. M. (1981). An experiment investigating the user requirements of future office workstations with emphasis on preferred response times. Technical Report HF058, IBM Hursley Human Factors Laboratory, Winchester.

EXTENDING KAINDL'S SCENARIO-BASED REQUIREMENTS DERIVATION TECHNIQUE FOR USER INTERFACE DESIGN

Gitesh K. Raikundalia

Centre for Internet Computing and E-Commerce
Swinburne University of Technology
P.O. Box 218
Hawthorn 3122 Australia
Email: gitesh@it.swin.edu.au
Ph: +61 3 9214 4383 Fax: +61 3 9819 0823

ABSTRACT

Kaindl's scenario-based, object oriented method for requirements capture has proven to be useful for addressing the purposes within a scenario. Scenarios have been highly useful in software development. Unfortunately, scenarios typically have been used without explicitly dealing with the particular purposes of specific interactions within the scenario, or the interactions have been neglected or left implicit. Kaindl's method complements scenarios with their purposes achieves a more complete and consistent definition of a task model and requirements.

This paper presents application of the method to the design of a Web electronic meeting document manager. It is shown how Kaindl's method has been extended further, including user interface design. The Unified Modelling Language (UML) is the current standard in object oriented analysis and design. The paper illustrates how UML can be used in an extension of Kaindl's method in software design.

1. INTRODUCTION

In software engineering, a scenario is used commonly to illustrate the use of a tool in a series of interactions with a user. Exploitation of scenarios has been growing recently for many reasons. For instance, scenarios are easy for end-users to understand and therefore are highly useful in agreement with them about functionality of software. Thus, scenarios have proven relevant in capturing formal software requirements.

Hermann Kaindl (1995) presents a method for requirements capture and task modelling based upon viewing scenarios as behavioural requirements connected with their *purposes*—the functional requirements. Unfortunately, scenarios have been used exclusively without explicitly dealing with the particular purposes of specific interactions of the scenario, e.g., Potts (1995), Hsia, Samuel, Gao, Kung, Toyoshima & Chen (1994). Such interactions have been neglected or left implicit. Viewing part of the behaviour of a tool in a scenario, even if highly descriptive, does not clearly provide justification why this behaviour is valid. Kaindl's method of complementing scenarios with their purposes achieves a more complete and consistent definition of a task model and requirements.

This paper presents an extension to Kaindl's method, including an extension to the extent of design of user interfaces. The extension arose from initial application of the method to development of a Web meeting document management system (*WMDMS*). The experience led to taking the method further from an object oriented specification of a tool to a full visual design of the tool and the use of relevant tabular representations. The visual depiction of a system's user interface is designed using the Unified Modelling Language (UML), the current standard in object oriented analysis and design (see www.rational.com).

2. WMDMS SCENARIO

The scenario created for development of the WMDMS illustrates interactions with the tool, such as viewing of an agenda by meeting participants. The scenario covers activities in a series of three meetings, where each meeting consists of five subphases. Five participants are involved in the meetings. A very small portion of the scenario is shown in Table 1. The scenario is organised within nine tables, one table for each of the three major meeting phases—pre-meeting, in-meeting and post-meeting phases. The scenario contains more than five hundred events like those in Table 1. The period of time in which an event occurs is represented by:

<meeting phase><meeting number>[<day1>,<time1>–<day2>,<time2>]

The <meeting phase> is one of the five pre-meeting phases (e.g., pre-meeting phase), the meeting number is the number of the current meeting in a series of related meetings, *day1* and *day2* are the day(s) over which the event occurs, *time1* is the time (in 24 hour time) when the event begins and *time2* is the time when the event ends.

Table 1 shows that three participants submit an agenda item and the fourth simply views the current version of the agenda. These events fall into the category of agenda development. The document used by participants for these events is the current version of the agenda. A hyperlink from "Agenda" takes the developer to the agenda document.

Time (days, hrs)	Event category	Participants				Documents
		Peter	**Emma**	**John**	**Graham**	
pre[10, 1453-1458]	Develop agenda	Submits agenda item				Agenda
pre[10, 1616-1627]			Submits agenda item			
pre[10, 1653-1659]					Submits agenda item	
pre[11, 0945-0952]				Views agenda		

Table 1: Portion of extensive scenario showing agenda-related activity

3. APPLYING KAINDL'S METHOD FOR FUNCTIONAL REQUIREMENTS DERIVATION

The first step of the method involves determination of the functionality of the system. This is done by determining semi-formalisms representing system functions. The method uses Functional Representation (FR) (Chandrasekaran, Goel & Iwasaki, 1993) used in other areas, such as device design. FRs need to be created for the structure of the entire electronic meeting environment and the functionality of all objects in the system. The lesson learnt from this stage of our research is the use of a Web table in representing the FRs. Figure 1 shows an FR created for a WMDMS function and the equivalent tabular representation of the FR. All functions for the entire system are collected together in HTML tables.

Function　　　(Provide participants with generation of meeting discussion analyses)
System　　　(WMDMS)
Given　　　(Meeting transcript)
Goal　　　(Provide specific documents for applying discussion analyses)
By-behaviour　(WMDMS generation of meeting derivative information)

Function	Given	Goal	By-Behaviour
Provide generation of meeting discussion analyses	Meeting transcript	Provide specific documents for applying discussion analyses	WMDMS generation of meeting derivative information

(a) FR for an electronic meeting environment function

(b) Tabulated FR for an electronic meeting environment function

Figure 1: Creation of FR for a WDMDS function

3.1 Complementing the Scenario with Purposes

This collection of functions represents all the purposes of the scenario. Kaindl's work connects the purposes to the scenario using By-Function hyperlinks. In our research, the scenario tabulation is extended by one column to direct the developer to the function. The scenario table in Table 1 now appears as shown as Figure 2(a). Each of the two

298

entries in the By-Function column in Figure 2 are hyperlinks to the FR tables corresponding to these functions. This situation is shown in the Figure using arrows to reflect the linking.

4. EXTENDING THE METHOD FOR USER INTERFACE REQUIREMENTS

After functional requirements were derived, it was perceived that the method could be applied to derivation of user interface requirements (another finding of the work). This extension was pursued as it was believed that a consistent definition and representation of *all* requirements for the WMDMS could be achieved by using Kaindl's method throughout. Web page elements are already treated as objects since they form the Document Object Model manipulated by JavaScript and JScript. A separation between the document type and its user interface is required, where each of these is a separate object. A document's user interface object (IO) is composed of all IOs contained in its Web page.

Event category	Participants				Docume nts	By-Function
	Peter	Emma	John	Grah.		
Develop agenda	Submits agenda item				Agenda	Allow agenda creation;
		Submits agenda item				Provide access to agenda

Function	Given	Goal	By-Behaviour
Allow agenda creation	Agenda details from participants	Provide agenda document	Internal WMDMS behaviour
Provide access to agenda	Agenda	Document usage for item coverage in meeting	Access agenda

(a) Scenario table extended with By-Function column   (b) Portion of FR table

Figure 2: Creation of FR for a WDMDS function

Presentation is separated from functionality. The appearance of an IO is not prescribed, and the designer can determine this by using techniques such as Cascading Style Sheets accounting for different browser types. Instead, interest lies in issues such as the information the IO represents or if the IO is or is not a hyperlink.

4.1 Object Design

The final extension to Kaindl's method is the visual design of object hierarchies representing the documents' IOs. Kaindl's method describes object decomposition, yet does not address the final step of visually modelling and implementing the final system.

Meeting: Home page details – 5 June, 1995

Agenda

Agenda item no.	Agenda item	Documents needed
1	Reasons for using Web	
1.1	increasing use	printed material containing articles
1.2	facilities/features	printed material about multimedia features
1.3	advantages over alternative forms of advertising	
1.4	competitor use?	known URLs of company Web pages
2	Cost	Cost description
2.1	monetary	
2.2	effort involved	

Figure 3: User interface design for agenda interface object

299

Use is made of a CASE modelling tool, such as the popular Rational Rose (www.rational.com), and the Unified Modelling Language (UML), which is the current standard in object oriented analysis and design (see the same Web site). Using these two, the user interface is constructed from the requirements derived from the scenario. Since the user interface is decomposable into other objects and objects are in relationship with each other, object oriented techniques such as aggregation and association are modelled easily for the user interface. Figure 3 shows the Web page design of the agenda document. Figure 4 shows the object model created in Rose for the Web page in Figure 3.

Figure 4 shows that the agenda page is represented using an agenda IO. Like Web pages these days are expected to have some associated metadata in addition to the usual page content. Metadata includes details such as date of page creation, date of page modification, author, etc. Thus the agenda IO can store such metadata and would have an corresponding *processMetadata()* method. The agenda IO is composed of *Heading* and *Item table* objects. The Heading object's *writeHeading()* method prints the meeting name and date to the page. *Item table* has a method for creating the HTML code for the item table. As the Figure shows, the *Table body* object contains 1 or more *Table row* objects, each object creating and printing its row.

Object-oriented design also involves creation of interaction diagrams for systems where dynamicity exists in the system. Therefore, in addition to the object model in Figure 4 representing a static snapshot of the page, a corresponding interaction diagram may also be created (not shown) to capture dynamic change of a page.

Once a Web page has been designed, in line with typical prototyping techniques, these designs can be presented to users to obtain their feedback. From iterative development, the designs generated directly from requirements derivation may be only the beginning of the entire development process.

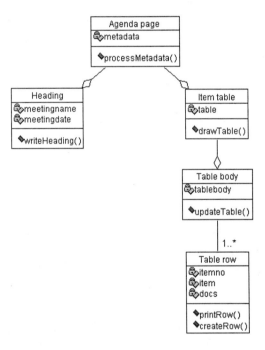

Figure 4: UML object diagram design for agenda interface object

300

5. IMPLEMENTING THE SYSTEM

Logan was implemented with Common Gateway Interface (CGI) scripts. CGI scripts may be written in almost any language, including C++. Rose generates C++ code corresponding to a given model drawn by the user. The generated code can be used as a skeleton for the developer to code the methods for each user interface. For instance, the signatures for *writeHeading()* method of the *Heading* object and the *updateTable()* method of the *Table body* object are automatically generated. As Java is a very popular Web language and is relevant for dynamic pages, modelling tools that generate Java code would be in greater demand for Web tool implementation.

6. CONCLUSION

This paper presented an extension to Kaindl's method of object oriented, scenario-based requirements derivation. The extensions included the use of Web tables and hypertext for storing, presenting and linking different aspects of the scenario and the use of the method to determine user interface requirements. Due to the object oriented nature of the design, the current standard in object oriented analysis and design, UML, is applied to the visual design of the system. A CASE tool like Rational Rose also assists in the implementation of the design by generating skeleton code. Kaindl (1995) does not describe how visual design and implementation may proceed from design, however, this paper does provide one possibility for doing so.

7. FUTURE WORK

Use cases are a relatively new method for determining functional requirements that make use of scenarios. A very interesting future work would be to examine the relationship between use case scenarios, as constructed easily using tools like Rose, and Kaindl's scenario-based method. This would assist in discovering how use cases can be applied to Kaindl's method.

REFERENCES

Chandrasekaran, B., Goel, A. & Iwasaki, Y. (1993). Functional representation as design rationale. *IEEE Computer*, *26* (1), 48–56.

Hsia, P., Samuel, J., Gao, J., Kung, D. Toyoshima, Y. & Chen, C. (1994). Formal approaches to scenario analysis. *IEEE Software*, *11* (2), 33–41.

Kaindl, H. (1995). An integration of scenarios with their purposes in task modeling. In: *Proceedings of the Symposium of Designing Interactive Systems*, *Michigan*, Aug. 23-25, 1995 (pp. 227–235).

Potts, C. (1995). Using schematic scenarios to understand user needs. In: *Proceedings of the Symposium of Designing Interactive Systems*, *Michigan*, Aug. 23-25, 1995 (pp. 247–256).

Assessing Head-Tracking in a Desktop Haptic Environment

Chris Raymaekers, Joan De Boeck, Karin Coninx

Expertise Centre for Digital Media
Limburg University Centre
Wetenschapspark 2, 3590 Diepenbeek
Belgium
Tel. +32 11 26 84 11 fax +32 11 26 84 00
{chris.raymaekers, joan.deboeck, karin.coninx}@luc.ac.be

ABSTRACT

A common problem in desktop haptic applications is a lack of depth perception. In this work, we have adopted the solution of head-tracking from immersive setups in a though-the-window environment. This both enlarges the user's workspace and allows the user to easily look at the virtual scene from different perspectives. In addition, we have assessed this setup in a formal user test. Although we did not measure a significant improvement in time or efficiency, the results of this test show us that users take up a different strategy in exploring the scene and are more involved in the virtual world when using head tracking.

1. INTRODUCTION

Over the past few years, haptic devices, such as the PHANToM device (Massie and Salisbury 1994) found their way to desktop environments. While Miller and Zelevnik have integrated the PHANToM device in the XWindows system (Miller. and Zelevnik 1998) and Oakley et al. have used this device in the Microsoft Windows GUI (Oakley et al. 2000), most researchers such as Chen (Chen 1999) and Gormann (Gormann et al. 1998) are concentrating on 3D applications. Although, in this kind of applications, the user is supposed to interact in three dimensions, a common annoyance is a lack of depth perception (Giess et al. 2000) and spatial awareness (Arenault and Ware 2000).

In order to assist the user exploring and understanding the virtual environment more thoroughly, we have developed and tested a solution in which the user's head movements are mapped to the virtual camera position. This method allows users to look at objects from a different point-of-view and enlarges the virtual workspace. This approach is comparable to the approaches used in JDCAD/JDCAD+ (Liang and Green 1994)(Green and Halliday 1996) and HoloSketch (Deering 1995)(Deering 1996). However, the value of this contribution is that haptic feedback, using a PHANToM device, and head tracking are combined in our experimental setup (which is not commonly found in literature) and that formal testing has been done.

This paper gives a short technical overview of our solution; afterwards, the results of a formal user experiment are presented and discussed.

2. HEAD TRACKING

In order to track the user's head movements, we have mounted a magnetic tracker onto a cap (see fig. 1). We call this setup a "head-tracking device". The movements of the user's head are tracked and superimposed to the virtual camera without affecting the position of the virtual pointer, which is represented by the position (and orientation) of the PHANToM device, relative to the user.

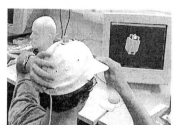

Fig. 1: Head-tracking device

Before being applied, head movements are processed by a transfer function that limits the response range, as depicted in figure 2. Small head movements are filtered out by a quadratic function to protect the user from motion sickness due to involuntary head movements. Large head movements are attenuated to prevent the virtual world to be wiped away when the user turns his head in the other direction. Finally, the transfer function parameters are adapted for each degree of freedom in such a manner that the transfer function amplifies the translations in the important directions (translations along the X-axis and the Y-axis) and attenuates distracting rotations around the Z-axis (tilting the head).

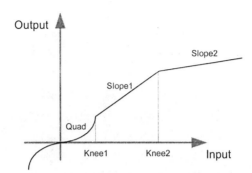

Fig. 2.: Transfer function for the head-tracking device

3. EXPERIMENTAL SETUP

To test the usefulness of our solution, we have set up a user experiment in which users had to evaluate this new tool: two targeting tasks in a through-the-window VR setup, using Sensable's PHANToM device were examined. The use of the head-tracking device was compared to a condition with a fixed viewpoint. The targeting task consisted of four virtual scenes, ranging in complexity (see fig. 3), in which the pointer had to be moved from a fixed position, around some obstacles to a target object: a red-and-white square, positioned on a cube. Each of these scenes had to be undertaken twice in both conditions. The test persons first had the chance to practice the task in another virtual room, which only contained the target. A person, who was able to reach the target twice in less than 30 seconds, was considered as a subject with enough experience for the test.

Twenty-two volunteers, twenty males and two females, all participated with their dominant hand in a counterbalanced repeated measures design. The dependent variables we measured were: elapsed time, distance covered by the virtual pointer and number of erroneous touches. After completing the test in each particular condition the test subjects were asked to answer a series of questions. When the second condition was completed, a comparing questionnaire was presented.

4. RESULTS

After statistical analysis of the data, using one-way ANOVA over *all* measurement data, we had to conclude that users perform better *without* the head-tracking device. However, if only *the very last trial* of each person is considered, no statistically significant difference can be found (see table 1). Similar results were found for the other independent variables, but these results are not so pronounced. As a first conclusion, we thus can state that people appear to be slower when they use the head-tracking. However, once the virtual room has been known well, performances in both cases show to be equal.

	All measurements	Last trial
With head-tracking device	4719	3231
Without head-tracking device	3987	3200
P-value	0.0252	0.9086

Table 1. Average timings in milliseconds

To go on with this conclusion, video recordings taken during the tests demonstrate that users behave in a more explorative manner when using head-tracking, which explains the apparent decrease in performance. This behaviour is in contrast with the more "trial-and-error" approach of the users who are not able to modify their viewpoint. Furthermore, most subjects raised the fact on the system that the head-tracking device would be of more importance when the scene becomes more complex, or when certain objects are (slightly) obscured by others. As can be seen from figure 3, the target is in all scenes (partially) visible. Only in the last scene the pointer is obscured in the start position. We used this kind of scenes intentionally, because otherwise the test condition without the head-tracking device would certainly be more difficult or impossible to perform, thus biasing the results.

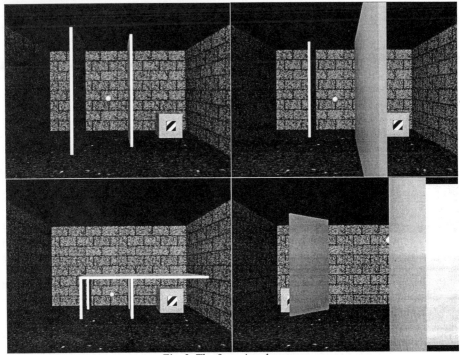

Fig. 3. The four virtual scenes

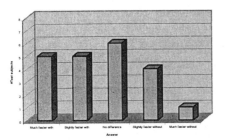

Fig. 4. Subjective speed in both conditions

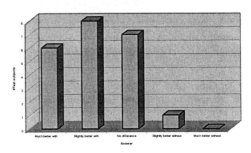

Fig. 5. Depth perception in both conditions

Figure 4 turns out that the users' subjective appreciation is not always confirmed by the objective measurements. In the last questionnaire, we asked the test subjects in which case they were able to reach their target in a faster manner. Almost half of the test persons reported that they were faster when they used the head-tracking device, while the other persons thought it did not make a difference, or that they were less efficient when using the head-tracking device. This can be explained by the fact that users seem to make a subjective difference between the time needed to explore the scene and the time needed to reach the target, which the measurements in the software do not. In each trial, the position of the PHANToM was fixed for ten seconds; after a beep signal, this restriction was released and the measurements started. We believed that the subjects would benefit from the head-tracking during this time by exploring the scene and start moving the pointer to the target after the beep. However, most persons remained passive during the waiting period and did not start to explore the scene until after the beep.

From the other answers of the comparing questionnaire, we can conclude that most persons report a better understanding of the virtual world: 63% assert having a better depth perception using the head-tracking device, while only one person found that the tracking degraded his depth perception (see fig. 5 and 6). Moreover, as can be seen from figure 7, almost 73% of our test subjects (16 out of 22) state that working with the head-tracking device is more comfortable. No single user is indifferent with regard to the head-tracking device.

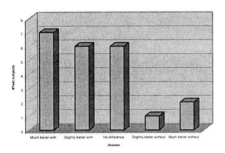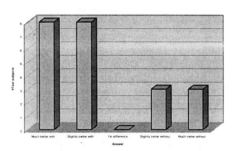

Fig. 6. Understanding of the scenes in both conditions   Fig. 7. Agreeability of both conditions

Besides this, we found that some test persons used the head-tracking device without noticing it: they turned their head in order to explore the scene, but afterwards reported not to have used it. One user even turned his head to look at something in his first trial, where he did not knew anything about the head-tracking device. We can thus conclude that head tracking presents a natural metaphor, even in a through-the-window VR setup.
We think these results support the use of head-tracking, since using haptic feedback adds an extra dimension to the user interface: a bad understanding of an object's depth, can lead to a diminished comfort due to the physical stress of bumping into objects.

5. CONCLUSIONS AND FUTURE WORK

We have investigated head tracking in a desktop haptic environment. The results of a formal user test assessing this combination have been presented in this paper.
Summarizing, we can state that the head-tracking device does not result in a direct increase of performance, but rather allows the user to explore the scene in a more intuitive and natural manner and to provide him with a better spatial awareness. This is appreciated by most of our users, as we are considering e.g. object-modelling applications (Raymaekers et al. 1999). We believe that the results of this experiment are transferable to other exploratory and dynamic virtual environments using similar haptic feedback.
Current research has looked at the possibility to use head tracking to observe objects. This method also allows for a larger field-of-view than is made possible by a standard computer monitor. This "virtual" field-of-view could be used to add extra user interface elements in the virtual world. For instance, a 3D menu could be placed just left of the normal view. If the user looked to the left, the menu would become visible and accessible. This allows user interface elements to be integrated in the virtual world, without cluttering the user's field-of-view.
Furthermore, our research looked at ways to improve the "visual" interaction. Further research is needed in order to assess the value of force feedback in various conditions.

6. ACKNOWLEDGMENTS

Part of the work presented in this paper is subsidised by the Flemish Government and EFRO (European Fund for Regional Development).
We would also like to thank Wouter Kempen (graduate student at the University of Maastricht who made his thesis at the EDM-LUC) for his help in conducting the user experiments.

REFERENCES

Arenault, R., Ware, C. (2000) Eye-Hand Co-ordination with Force Feedback. Proceedings of CHI 2000, Apr 1-6, Den Haag, NL, pp. 408-414

Chen, E. (1999) Six Degree-Of-Freedom Haptic System as a Desktop Virtual Prototyping Interface. Proceedings of First International Workshop on Virtual Reality and Prototyping", June 1999, Laval, FR, pp. 97-106

Deering, M.F. (1995) HoloSketch: A Virtual Reality Sketching/Animation Tool. ACM Transactions on Computer-Human Interaction, vol. 2(3), 1995, pp. 220-238

Deering, M.F. (1996) The HoloSketch VR Sketching System. Communications of the ACM, vol 39(5), 1996, pp.54-61

Giess, C., Töpfer, S., Meinzer, H. (2000) Can Shadows Improve Haptic Interaction in Virtual Environments? Proceedings of PURS 2000, Selected Readings in Vision and Graphics, 6-7 July 2000, Zurich, CH, pp. 49-54

Gormann, P., Lieser, J., Murray, W., Haluck, R., Krummel, T. (1998) Assessment and Validation of a Force Feedback Virtual Reality Based Surgical Simulator. Proceedings of the Third PHANToM Users Group Workshop, October 3-6 1998, Dedham, MA, USA

Green, M., Halliday, S. (1996) A Geometric Modeling and Animation System for Virtual Reality. Communications of the ACM, vol. 39(5), May 1996, pp. 46-53

Massie, T., Salisbury, J. (1994), The PHANToM Haptic Interface: A Device for Probing Virtual Objects. Proceedings of the ASME winter Annual Meeting, Interfaces for Virtual Environments and Teleoperator Systems, November 1994, Chicago, IL, USA, 1994

Miller, T., Zelevnik R. (1998) An Insidious Haptic Invasion: Adding Force to the X Desktop. Proceedings of the 11th ACM UIST'98 symposium on user interface software and technology, ACM, San Francisco, CA USA, 1-4 November 1998, pp. 59-66

Liang, J., Green, M. (1994) JDCAD: A Highly Interactive 3D Modeling System. Computer & Graphics, vol. 18(4), 1994, pp. 499-506

Oakley, I, McGee, M., Brewster, S., Gray, P. (2000) Putting the Feel in 'Look and Feel'. Proceedings of CHI 2000, April 1-6 2000, Den Haag, NL, pp. 415-422

Raymaekers, C., De Weyer, T., Coninx, K., Van Reeth, F., Flerackers, E. (1999) ICOME: an Immersive 3D Object Modelling Environment, Virtual Reality Journal, 1999, vol. 4(2), pp. 129-138

EASYCOM: DESIGNING AN INTUITIVE AND PERSONALIZED INTERFACE FOR UNIFIED REALTIME COMMUNICATION AND COLLABORATION

Hubertus Hohl and Manfred Burger

Siemens AG, Corporate Technology CT IC 5, Otto-Hahn-Ring 6, D-81730 München, Germany
E-Mail: hubertus.hohl@mchp.siemens.de, manfred.burger@mchp.siemens.de

ABSTRACT
In this paper, we report on an evolutionary approach to unified communications based on an easy-to-use visual interaction paradigm. Our objective is to enable intuitive and attractive interactions with screen-based telephony and telecommunications clients complementing ease and joy of use. We describe EasyCom, a client for IP-Telephony that seamlessly integrates handling of voice, video, and data communication processes into a unified interface metaphor. We focus on the evolution of the system from the original user interface conception towards a product solution guided by field trials and analysis of customer feedback.

1. INTRODUCTION

Although telephones are a ubiquitous means for simple everyday communication, they are not always simple to use. Especially enhanced-convenience phones operated in offices and call centers support lots of sophisticated features, e.g. call swapping, call transfer, call forwarding, or Conferencing. User experience shows that the more features are supported the more difficult phones are to use. With screen-based devices there is a great potential for improving the control of complex synchronous communication processes. Today, telephony applications are running on a variety of stationary and mobile devices, ranging from PCs, web tablets, and palmtop computers to cell phones. With the convergence of the plain old telephony world into Computer-Telephony-Integration (CTI) and Voice-over-IP based telecommunications solutions, screen-based devices are noticeably used to control or even replace phones in offices, call centers and public kiosk systems. However, existing applications still stick to mimic the use of conventional phones. Their standalone user interfaces are neither suited for nor extensible enough to manage realtime voice, video, and data communication processes appropriately. Our objective is to overcome these limitations by utilizing a new interaction paradigm for enriched computer-mediated communication.

2. A VISUAL APPROACH TO HANDLING COMPLEX COMMUNICATION FUNCTIONS

EasyCom's conceptual model is based on the *Communication Circle* metaphor, originally introduced in (Grundel, Schneider-Hufschmidt, 1999). The user is placed in the center of a circle visualizing communication partners inside dedicated sectors. Different sector colors and arrangements serve as indicators of partner's communication state (active, on hold, incoming call, outgoing call). With easy-to-learn and fun-to-use drag-and-drop actions on visualized partners and sectors, the current communication situation can be directly manipulated. Figure 1, for example, shows a series of interactions characteristic for business telephony.

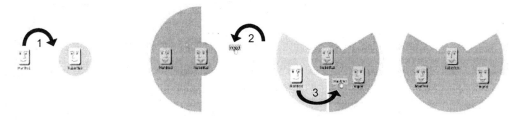

Figure 1: Setting up a conference with drag-and-drop interactions: (1) accept incoming call, (2) place consultation call (e.g., by dragging partner out of address book), (3) set up conference with both partners.

Multiple calls can be distinguished easily inside the Communication Circle by visual display of communication partners as user-configurable image icons. To further improve the visual awareness of the current communication state, we introduced so-called *Talking Heads* (cf. Figure 2), animated characters that symbolize a partner's activity (i.e. mouth opened and looking at each other when call is active, mouse closed when call is held). The user can choose to visualize partners by pictures, self-created icons, or by Talking Heads.

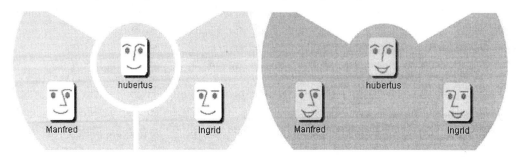

Figure 2: Talking Heads visualize communication states of partners: for held calls (left), for active calls (right).

Although the EasyCom user interface also offers conventional context-dependent function keys to manage telephony features (cf. Figure 4), the primary interactions are based on drag-and-drop. We observed that users rapidly internalize these gestures accepting them as natural once they have been learned. Drag-and-drop actions are convenient, but not self-explanatory for the first time. To overcome this problem, we consistently enhanced the user interface by advanced visual feedback techniques: drag-and-drop indicators and a graphical function preview.

DRAG-AND-DROP INDICATORS
Interface elements that are draggable, such as buttons, list items, partner icons, and circle sectors are consistently equipped with graphical indications (cf. Figure 3). For example, to hang up a conference as a whole, the user just drags the indicator of the conference sector out of the circle and drops it onto the background. To avoid cluttering the user interface with graphical clues, we use rollover indicators that only become visible when the mouse hovers over draggable elements. In this way, users are animated to explore the interface step by step or discover new functionality serendipitously.

Figure 3: Drag-and-drop indicators for address book entries (left), Talking Heads (middle), circle sectors (right).

GRAPHICAL FUNCTION PREVIEW
When dropping an item the action that will be performed is not always obvious beforehand. The preview circle provides a ready overview of the resulting call situation. Whenever an item is dragged over a valid drop target, the preview circle displays the final communication situation resulting from the intended action. The preview function is also utilized in combination with function keys, providing a form of tool tip that graphically describes the intended action. For example, Figure 4 shows the graphical function preview for initiating a conference out of a consultation either by clicking on the function key „Conference" or by performing a drag-and-drop action.

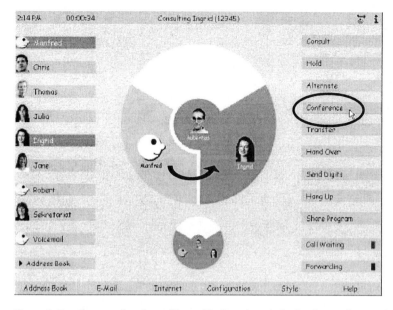

Figure 4: EasyCom user interface with graphical preview circle showing conference situation.

3. FROM TELEPHONY TO UNIFIED COMMUNICATIONS

Originally, the Communication Circle paradigm focused on telephony applications. Operated on desktop computers it is reasonable to extend the metaphor to support unified communications, adding value by integrating or connecting common PC-based communication and collaboration tools. To bridge the gap between synchronous communication and asynchronous messaging e-mail transfer to partners has been integrated into EasyCom. By simply dragging an e-mail icon into a partner's sector, the user's standard e-mail tool is launched and recipient's address filled in. By this means, it is very handy to send minutes to all participants in a conference.

To support teleworking in spacially distributed teams, we integrated video display, data sharing and transfer into the interaction paradigm. For example, the user simply drags a file from a file browser into the partner's sector to launch and work together on a document (cf. Figure 5). Similarily, participants can quickly exchange small pieces of personal data, such as business cards or multimedia information, in the context of the current voice call.

Figure 5: Teleworking with integrated video and shared documents.

4. PERSONALIZING THE USER INTERFACE

Not only for consumer products but also for professional use, the hedonic quality of a graphical user interface has an important influence on the system's appeal (Hassenzahl et al., 2000) (Tractinsky et al., 2000). The novelty and originality of a user interface perceived by the user constitutes one of the success factors for systems with custom look-and-feel and novel use of graphics. Customer feedback has shown that EasyCom users want to be empowered to tailor the user interface to their personal needs and preferences. We introduced end-user-customization by offering a set of predefined design variants with scaleable display formats (cf. Figure 6). Beyond it, we plan to provide a design tool that enables end users to add their personal touch to the user interface. They can interactively modify existing or create new graphic designs by configuring design properties such as colors, fonts, icons, talking heads, etc.

Figure 6: Some EasyCom design variants.

A different approach to personalization is to provide configurable communication feature sets on the user interface. In this way, the system can be tailored to specific usage contexts by site administrators or even individual end users. By enabling the use of personal interactive customization tools users will be empowered to perform communication tasks in a very specific individualized manner (Riecken, 2000). For instance, in a public kiosk system or in a personal mobile setting, restricted or individual feature sets might be appropriate. In call and contact centers, the user interface should provide fast access to specific functionality for routing incoming calls. In secretariates and offices, sophisticated supplementary telephony features are mandatory.

5. BEYOND THE DESKTOP – AND INTO HAND-HELDS

Originally, the graphic design of EasyCom's user interface targeted at screenphones and PCs with medium- or large-sized displays. The original version of the interface covered a display area of 800x600 pixels. On desktop systems which are usually operated using multiple overlapping windows, users complained that EasyCom consumed too much screen space needed for other applications that are necessary to perform concurrent business tasks. Our first approach was to provide scaled-down versions of the user interface covering sizes from 400x300 up to 800x600 pixels. In this way, the user interface suits well for full-screen mode used on screenphones and public kiosk systems as well as for window-based desktop applications.

To use screen space even more efficiently, we are currently following an alternative approach: The monolithic EasyCom user interface is unitized into separate logical building blocks, so-called *EasyBlocks*. An EasyBlock provides specific functionality that can be accessed independently or in combination with other blocks using the familiar drag-and-drop interactions. For example, the central Communication Circle, speed dials, address book, key pad, redial list, and call journal compose individual EasyBlocks. They can be easily and flexibly minimized, arranged, and grouped together on the screen using a special snap-to-fit interaction technique. In this way, each user can tailor his or her preferred screen configuration hidding unnecessary functionality while keeping the overall consistent interaction paradigm.

Another reason for scaling down and modularizing the user interface is to transfer the unique EasyCom user experience to mobile and hand-held communication devices with small-sized displays, such as cellular phones, smart phones, pagers, wearable computers, and PDAs (cf. Figure 7). The main idea here is about utilizing the EasyBlocks concept for device-adapted interaction.

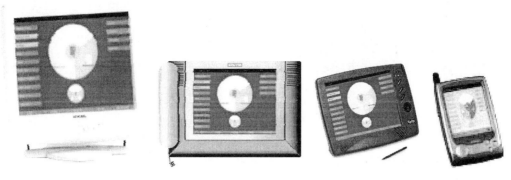

Figure 7: Scaling down the EasyCom user interface from the desktop to the palmtop.

6. RESULTS AND CONCLUSIONS

EasyCom covers a wide range of application scenarios for realtime communication and collaboration. Originally, EasyCom has been deployed as a PC client for Siemens HiPath 5000, a family of H.323-compliant Voice-over-IP based communications solutions for enterprises. Meanwhile, EasyCom has also been utilized successfully as a CTI user interface for various other telephony systems. Initial usability tests on an early prototype, as well as field trails and customer feedback on the product have shown that the EasyCom user interface approach is well-suited for unified communications breaking the barrier between traditional phones and modern telecommunications services. It appeared that the consistent and easy-to-use interaction paradigm allows simple and effective handling of complex multimedia communication functions and supplementary telephony features. Through the integration of telephony, data sharing and transfer, instant messaging, and web access on one user interface, the efficiency of user's personal workflow associated with day-to-day business tasks can be increased considerably. Besides, it turned out that the unique and attractive user interface philosophy offered across diverse telephony platforms and devices is highly recognized by customers.

Although the original approach primarily concentrated on professionals operating the system in office environments and call centers, we believe the interaction paradigm is also suited for public kiosk systems, mobile personal communication devices, and consumer products such as web tablets used at home. Equipped with touch-sensitive color displays (or even micro displays available in the near future) and pen- or gesture-based interaction the EasyCom approach could allow spontaneous communication anywhere, embodying a unified and device-adapted interaction paradigm.

REFERENCES

Grundel, C. and Schneider-Hufschmidt M. "A direct manipulation user interface for the control of communication processes – making call handling manageable -", In H.-J. Bullinger and J. Ziegler (Eds.) Human-Computer Interaction: Communication, Cooperation, and Application Design. Lawrence Erlbaum Assoc., 1999, 8-13.

Hassenzahl, M., Platz A., Burmester M., and Lehner K. "Hedonic and ergonomic quality aspects determine a software's appeal", In Proceedings of the CHI 2000, ACM, 201-208.

Riecken, D. "Personal End-User Tools", In Communications of the ACM, August 2000, Vol. 43, No. 8, pp. 89-91.

Tractinsky, T., Katz. A.S., and Ikar, D. "What is beautiful is usable", In Interacting with Computers: the Interdisciplinary Journal of Human-Computer Interaction, December 2000, Vol. 13, Issue 2, pp. 127-145.

USING A BRAIN-COMPUTER INTERFACE IN VIRTUAL AND REAL WORLDS

Jessica D. Bayliss[a][*] and Brent Auernheimer[b]

[a]Computer Science, University of Rochester, Rochester NY 14627 USA

[b]Computer Science, California State University, Fresno CA 93740 USA

A brain-computer interface (BCI) links an individual's EEG signals with a computer. This can enable locked-in patients to communicate, as well as to give them control over their environment. Due to the difficulty of processing the small and noisy EEG signals, little work has been done on making these interfaces usable. Since immersive virtual environments may be motivational for patients, we combined them with a BCI. Subjects' self-reported qualitative experience did not necessarily match their objective performance. We discuss this and future work in the light of the current results.

1. INTRODUCTION

Current BCI systems are slow in comparison with normal human actions. This is because of the low bandwidth of the signals as well as the time to recognize and process the signal. This complicates the creation of usable BCIs. As an example of usability difficulties encountered by BCI users, consider the Thought Translation Device. It was the only means of communication for several locked-in patients (Birbaumer et. al, 1999). The communication rate was two characters per minute, and only two of five locked-in patients were able and willing to use the system. Usability issues, including how a specific application may affect the EEG signals it uses for control, are especially important for BCIs and are discussed in Section 2.

Since an immersive virtual environment may provide motivation for locked-in patients, we looked at performance under three conditions: in an immersive virtual apartment while wearing a head-mounted display (VR condition), while looking at the same apartment scene on a computer monitor (MONITOR condition), and while looking at a virtual monitor in the head-mounted display (FIXED DISPLAY condition). We discuss future work in the context of obtained results.

2. USABILITY ISSUES

Developing usable BCIs is a challenge for system designers. The complexity of the hardware, real-time processing demands of the software, and small number of test users can leave little time for usability testing and tuning. This is changing. Moore and Kennedy (2000) overview human-computer interface and training issues for an implanted BCI. The user moved a

---

[*] This research was supported by NIH/PHS grant 1-P41-RR09283. It was also facilitated in part by a National Physical Science Consortium Fellowship and by stipend support from NASA Goddard Space Flight Center.

cursor by imaging movements in his left hand. This resulted in a communication rate of about three letters per minute.

In work with brain injury patients, Cole et al. (1994) note that system developers' design process must change. They note: "...our work with brain injury patients has shown that patient-system performance is extremely sensitive to ... minor design parameters: furthermore, that the brain injury survivor needs to be viewed as a relatively sensitive component, while the computer system design needs to be the most flexible." In fact, about two thirds of user interface changes and three quarters of the functionality were requested by patients or clinicians. They conclude "it is clear that at least some ... changes would not have been suggested by those with systems expertise [developers] because those changes were either counter-intuitive or violated accepted guidelines."

Table 1: Usability aspects of several BCI systems

System	Training Time	Number of Choices	Speed [+]	Errors	Retention	Subjective Satisfaction
Brain Response Interface	10-60 minutes	64	30	10%	Excellent	Considered
P3 Character Recognition	Minutes	36	4	5%	Excellent	Not Discussed
Mu Rhythm Training	15-20 sessions	2	20	10%	Not mentioned	Not Discussed
Thought Translation Device	Months*	27	2	10-30%	Not Good	Indirectly discussed
Implanted Device	Months*	N/A	2	Not reported	Excellent	Not Discussed

Table 1 is a comparison of existing systems. Many BCI papers do not discuss subjective satisfaction at all so the category for subjective satisfaction only includes whether or not it was considered in the papers about the system.

While the systems mentioned use different signals and applications, they have many things in common. Signals may be broken into three general categories: implant-based, evoked potentials, and operant conditioning. Both evoked potential and operant conditioning methods are normally externally-based BCIs, as the electrodes are located on top of the scalp.

Evoked potentials (EPs) are a reaction to stimuli. Exogenous components of an evoked potential, or those components influenced primarily by physical stimulus properties, generally take place within the first 200 milliseconds after stimulus onset. Visual evoked potentials (VEPs) fall into this category and are used in the Brain Response Interface (Sutter, 1992).

Cognitive evoked potentials take longer to occur and may occur in different sensory modalities. Chapman and Bragdon (1964) discovered a positive wave - the P3 - peaking around

300 msec after task-relevant stimuli. While the P3 is affected by several characteristics of stimuli, the common ones are stimulus frequency (less frequent stimuli produce a larger response) and task relevance. Thus, the more relevant and infrequent the stimulus is, the larger the P3. Farwell and Donchin (1988) first successfully used the P3 as a BCI with their character recognition application.

EPs are discrete are produced by subjects without training. They are inherent responses to stimuli. The P3 has the benefit of enabling both auditory and visual control. However, EPs require that the user wait for presentation of the appropriate stimulus. Often flashing lights over the item of interest are used to trigger EPs. These flashing lights may be annoying, and in the worst case may cause epilepsy if they occur at the wrong frequency. Visual EPs require that the user have good eye movement control in order to focus on the appropriate stimuli. Those who are completely locked-in do not have eye control.

Several methods use operant conditioning on spontaneous EEG signals for BCI control. The main feature of operant conditioning is that it enables continuous rather than discrete control. Biofeedback enables users to control the amplitude of a specific type of ongoing EEG activity (most often mu waves). This may be a drawback: continuous control is fatiguing for patients and fatigue may cause changes in performance. Also, operant conditioning methods using spontaneous EEG are not always easily learned as evidenced by the training times in Table 1. In some cases operant conditioning may not work and when it does work there may be retention problems. When looking at these systems, it is normal to see that only four out of six subjects mastered control. The mu rhythm training system (McFarland et. al, 1997) and the Thought Translation Device (Birbaumer et. al, 1999) use operant conditioning.

One example in the table above shows an internal BCI. Two patients used the implanted brain-computer interface system of Kennedy and Bakay (1998). Each patient learned to control a cursor. The velocity of the cursor was determined by the neural firing rate. Neural wave shapes were converted to triples of pulses corresponding to mouse input. The first and second pulses controlled the X and Y position of the cursor, and the third pulse was the mouse click. Several months were required for the neural tissue to grow into the implant and the general communication rate was not better than externally BCIs.

3. THE STUDY

Since users may be more motivated by an alternative display and since degree of motivation may change the nature of EEG signals, the experiment addressed the following: (1) Is there a difference in BCI control between being immersed in a virtual environment and staring at a computer monitor?, (2) Are there any signal differences between the two conditions?, and (3) Are there learning effects over time?

The particular BCI application allowed simple control of items in a virtual apartment. The signal chosen to control the BCI was the P3 evoked potential. This signal is discrete and appropriate for environmental control.

3.1 Experimental Setup

In two of the three conditions, the virtual apartment is presented to the subject through a head-mounted display (HMD). Bayliss and Ballard (2000) previously showed that evoked potentials were reliably obtained in a virtual environment while wearing an HMD.

A trigger pulse containing information about the presentation of a specific stimulus was sent to the EEG acquisition system whenever a stimulus was presented. While an epoch size from

(100 msec to 1.5 seconds) was specified, the data was recorded continuously. Seven electrode sites were arranged on the heads of nine subjects with a linked mastoid reference. Sites FZ, CZ, PZ, P3, P4, as well as a lower and an upper vertical electro-oculographic (EOG) channel were used.

3.1 The Experiment

Nine non-handicapped subjects experienced four conditions. The virtual apartment contained five items to be controlled: a television set, a floor lamp, a radio, and commands to say "HI" "BYE" to virtual person in the apartment. Controlling "physical" devices turned them on and off, and the verbal commands made the graphical person appear or disappear. All feedback was visual. For example, the radio displayed musical notes above it whenever it was turned on. In order to elicit a P3 response, a transparent button located on each item occasionally flashed red. The rate of button flashing was once per second.

In order to train the real-time recognition algorithm, subjects performed a simple task in the virtual apartment environment. For the first five minutes of the experiment, all subjects counted the number of red button flashes on the floor lamp. The next three conditions lasted for approximately five minutes each and were presented in a random order. The conditions were: VR condition (control the apartment items while fully immersed in VR using an HMD), MONITOR condition (control the apartment items while watching the apartment on a computer monitor), and FIXED SCREEN HMD condition (control the apartment items while watching a virtual monitor in the HMD).

Each of these conditions consisted of 250 trials. The FIXED SCREEN HMD condition was originally chosen because it used the HMD but unlike the VR condition, was not immersive. The FIXED SCREEN HMD condition is similar to that of a locked-in individual without voluntary head movement.

3.2 Results

Table 2 shows the throughput in items controlled per minute. There is a slight difference ($p<0.05$) between the MONITOR and FIXED SCREEN HMD conditions, but none of the other control differences are significant. The amplitude of the P3 signal does not vary significantly between conditions.

Individuals varied widely in their ability to control items and they tended to prefer the VR condition irrespective of their performance (six subjects preferred the VR condition to all other conditions). The FIXED SCREEN HMD condition was consistently rated the lowest. Subjects did not like that they could not move their head in order to concentrate on different items. This implies that even when subjects maintain eye control, they may not want to use a visual interface if they can't move their heads. Of the subjects who did not like the VR condition, all of them complained of the HMD fitting in an uncomfortable manner over their electrodes.

Table 2: Average number of tasks/minute accomplished by each subject.

Condition/Subjects	1	2	3	4	5	6	7	8	9
VR	3.8	3.7	4.5	3.3	1.9	1.7	2.6	0.9	3.1
MONITOR	4.5	5.6	6.1	3.5	1.2	1.2	3.3	1.3	2.9
FIXED SCREEN HMD	3.0	3.2	4.3	3.8	2.4	1.7	3.3	0.7	2.7

There are no statistically significant learning results over time, although six of the nine subjects performed better on their last task than the first one. When asked why they might have performed better, subjects made comments like, "I learned to relax and just let the system work", and, "I learned to make the blinking light more relevant to myself." Subjects could very well have improved by relaxing, as tightening muscles causes muscle-related noise and may decrease the performance of the signal recognition algorithm.

4. DISCUSSION

The goal of the experiment was to determine differences in the performance of individuals using a BCI in an immersive virtual environment compared to a computer monitor. The results show there are no significant differences between these conditions, but subjects performed significantly worse in the FIXED SCREEN HMD condition. The subjective experiences of subjects mirror this performance, but surprising most subjects liked the VR condition even though they did not perform the best on it. The FIXED SCREEN HMD condition is the most similar condition to the visual interfaces commonly used by handicapped individuals unable to move their heads. The poor results on this condition provide evidence that other types of visual interfaces should be explored, or even that other modalities should be tested.

REFERENCES

Bayliss, J.D. & Ballard. D.H. (2000). Recognizing evoked potentials in a virtual environment, *Advances in Neural Information Processing Systems 12*.

Birbaumer, N., Ghanayim, N., Hinterberger, T., Iversen, I., Kotchoubey, B., Kubler, A., Perelmouter, J., Taub, E. & Flor, H. (1999). A spelling device for the paralysed, *Nature 398*, 297—298.

Chapman, R. M. & Bragdon, H.R. (1964). Evoked responses to numerical and non-numerical visual stimuli while problem solving, *Nature 203*, 1155—1157.

Cole, E., Dehdashti, P., Petti, L. & Anger, M (1994). Participatory design for sensitive interface parameters: Contributions of traumatic brain injury patients to their prosthetic software, *CHI94 Conference Companion*, ACM.

Farwell, L.A., & Donchin, E. (1988). Talking off the top of your head: toward a mental prosthesis utilizing event-related brain potentials, *Electroenceph. Clin. Neurophysiol.*, 510-523.

Kennedy, P.R. & Bakay, R.A.E. (1988). Restoration of neural output from a paralyzed patient by a direct brain connection, *NeuroReport* 9, 1707—1711.

McFarland, D.J., Lefkowicz, A.T. & Wolpaw, J.R. (1997). Design and operation of an EEG-based brain-computer interface with digital signal processing technology, *Behav. Research Methods, Instruments, and Computers* 29(3), 337-345.

Moore, M.M. & Kennedy, P.R. (2000). Human factors issues in the Neural Signals direct brain-computer interface. *Proceedings of ASSETS'00*, ACM.

Sutter, E.E. (1992). The brain response interface: communication through visually-induced electrical brain responses, *Journal of Microcomputer Applications 15*, 31-45.

PODS: Interpreting Spatial and Temporal Environmental Information[1]

Edoardo Biagioni

esb@hawaii.edu
University of Hawai'i at Mānoa

Abstract

We are building wireless networks of environmental sensors to investigate why endangered species of plants will grow in one area but not in neighboring areas. We collect large amounts of data automatically and repeatedly. This paper describes requirements for presenting this real-time information in ways that allow scientists as well as more casual observers to detect significant patterns in time and space.

1   The Problem

Rare and endangered species of plants are rare and endangered because they only grow in few locations. These locations evidently have some special properties that support the growth of these endangered plants, properties that are not present in adjacent locations. We have started building devices to capture environmental information and transmit it in real-time over the internet. The environmental information we have started to collect includes rainfall, moisture, temperature, sunlight, and wind. Because these plants are in remote areas, we use wireless digital communication to relay the information back to the Internet, and from there to a database. We are also installing digital cameras that can help capture events, such as animal visits, that are not recorded by the other sensors. We call such a grouping of sensors and cameras a *pod*. Pods should be inexpensive so that many can be deployed. To make deployment easier, the network must be self-organizing: any *ad-hoc* layout of pods that allows wireless communication should be sufficient for data transfer back to one or more internet-connected base stations.

All of these sensor pods must be camouflaged in a way that they remain unnoticed by either humans or animals. We have camouflaged pods to look like either rocks or tree branches, and they have gone suitably undisturbed (and presumably unnoticed) over many months in areas much more crowded than our intended areas of deployment.

Ad-hoc wireless sensor networks present interesting issues of routing in wireless networks, of energy management, of data prediction and compression, and of the design of the sensors themselves – for example, we have not yet found a good design for unobtrusive instruments for measuring wind using only low power. In this paper we focus on the issue of presenting relevant data to users of this near-real-time information.

---

[1] This research was sponsored by the Defense Advanced Research Projects Agency, under the title "A Remote Ecological Micro-Sensor Network."

1.1 Information

One of the reasons for the near-real-time data collection is to allow scientists to take action when situations occur that are in some way interesting or unexpected. For example, a scientist may request pictures or request that special data be collected, or may even visit the site in person to observe for example flowering or the effect of herbivore browsing. The most important data we need to present is therefore the most recent data collected, but we must present it in the context of historical and geographical data.

1.2 Geographical Gradients and Temporal Predictability

In Hawai'i, weather gradients are very sharp, with rainforests and deserts sometimes less than 10 miles apart. We need to present our data to make obvious the geographic and temporal nature of weather gradients. We are therefore not surprised that endangered plants are limited to relatively small areas, but we suffer from a lack of reliable information about the weather in these small areas, since data from collection stations nearby may not provide reliable information about the weather at the location where the species of interest grows. By using a very dense network of relatively inaccurate sensors, we hope to be able to detect correlations among weather phenomena in nearby areas, giving us knowledge that may help protect plants that might need specific weather conditions to survive. Unlike conventional weather displays, we need to focus on the gradients rather than the peaks of phenomena, and on the two dimensions of what the "normal" weather is for a certain area, and on how much the weather differs from "normal" at the current time and in the recent past. If one area is always warmer than another, we want that to be clear in the data presentation. If one area is always warmer than another, but today is cooler, we need to flag this as a significant event likely to be interesting to a human.

We have also started to study the temporal correlation of sensor data. We have placed some sensors near our laboratories on the island of Oahu and have noticed, by repeated observation, that though sometimes the sky is clear in the morning (we observe smooth changes over time in the light measurements), it has never been clear in the afternoon. We need the ability to automatically deduct this information from the data and, to the degree possible, present high-level summaries as well as detailed information.

1.3 Images

Some of our pods are being designed to hold cameras. In contrast to conventional web cameras, these are high-resolution still cameras that can capture many of the details (such as insect visits) that are significant in the life of a plant. With a battery-powered wireless network, there is always a tradeoff between the amount of data transferred and the lifetime of the batteries. We can take more pictures and replace the batteries more frequently, or have fewer pictures and longer battery lifetime. There is sufficient processing power in a pod that, if it is possible to automatically identify phenomena of interest, it will be possible to automatically adjust the frame rate of cameras in different locations to cover a range of weather phenomena. Phenomena such as rain that might be meaningful to a human in the field may not be as significant as wind, for example, which might prevent pollinators from reaching the plants. We have placed preliminary cameras in the field (http://www.botany.hawaii.edu/pods/) to begin the process of learning what pictures are significant and worth the energy to transmit them.

2 Human Interfaces

Standard ways of displaying the data include tables, graphs, and maps. These give us the ability to express large amounts of both time-sequence and spatial information simultaneously. We have planned relatively large and dense network with approximately a hundred sensors. We can draw maps showing the position and instantaneous value for each sensor, but this does a poor job of emphasizing gradients and identifying and making visible significant geographical boundaries in the values of sensor measurements. One of our preliminary user interfaces (http://red2.ics.hawaii.edu/cgi-bin/location) uses colors to indicate temperature and rainfall on a map (blue for rain, cyan for dry cool, red for dry hot), and in case of rain, uses dot sizes to indicate how recently the rain was detected. We have also been generating daily plots of all our sensor measurements, showing variations during the course of the current day. Some of these variations are predictable – more light during the day, and more variation in light intensity during the afternoon – and some, such as rain, are less predictable.

We consider these interfaces preliminary because, although they do a good job of presenting small slices of the data, they fail to present the big picture. The remainder of this section sketches some characteristics of user interfaces that we have envisioned but have not yet realized.

The first characteristic is flexibility and configurability. A scientist interested in the solar radiation in a given area should be able to focus in on the data of interest, ignoring irrelevant data. This is a challenge of choosing which dimensions of the data to display. Not only do we have the three axes of space and the fourth dimension of time, we have one dimension for each type of sensor in our pods. Correlations between any set of dimensions may be of interest to scientists and ecologists.

In contrast, students doing a project for high school might want to focus on year-to-year data, ignoring more ephemeral phenomena. This is the second characteristic, and it may be described as the ability to "zoom" into specific sets or summaries of the data. Another description of this characteristic is the range and precision of a query on the sensor data. The precision might be spatial, temporal, and over ranges of sensor values.

In general, users might want to query the database using both different dimensions and different range and precision for each dimension. We are aware of much work having been done in many related ares, for example as described by Swan and others (1998), by Larson (2001), or even for the presentation of weather information, for example NOAA (2001). We look forward to adapting these ideas to our application as much as possible.

For our high-resolution camera images, we are planning to use large displays to provide specialists with close-up long-term study of the species (*phenology*), and non-specialists with immersive experiences of special environments such as rainforests and drylands and such weather phenomena as rainshowers and droughts.

3 Machine Learning

The ultimate user interface, for this and for other problems, is one that can respond to natural-language queries or can directly infer the user's interest and display the appropriate data in response. We do not expect useful interfaces of this kind to be available within the lifetime of this project.

We are aware of and intend to investigate the applicability of existing machine-learning algorithms such as presented by Mitchell (1997) to do data predictions. Successful data prediction would contribute to our project in two ways. By only sending data that does not match the prediction, we can substantially reduce the amount of data traffic and hence the power consumption of the pods. In addition, by having a model that predicts the expected set of readings from the sensors, it becomes much easier to identify unusual situations, whether it is a shift, weakening, or reversal in a gradient, or a long-term trend that can be evaluated by a human to determine its significance.

Some of the machine learning algorithms require substantial computing power to derive a model of the data. Since the data is available in the laboratory, these computations can be done on powerful computers. The computed models can then be distributed back to the pods, and used to both reduce the data flow and to clearly signal unexpected events. We are aware that for our application, small variations from the model are usually not significant, so the machine learning algorithms need only approximately predict the actual measurements.

4   Related Work

Many groups are working on distributed sensor networks for a variety of purposes. There is a conference track (Dasarathy, 2001) devoted to sensor fusion architectures, algorithms, and applications. Some of the characteristics that appear to be specific to our application include the overall predictability of the data (we are measuring weather and plants, rather than animals, troops, or industrial processes), and the strong interest in detecting phenomena (combinations of measurements and measurement patterns in space and time) for which we have no *a priori* knowledge.

We mentioned above some of the related work in human computer interfaces for information retrieval and for distributed sensor data and weather data. Weather data is usually presented with different symbols and sometimes colors on a two-dimensional map. These symbols usually represent different values for temperature, pressure, and wind at different locations at a given time. Recently animations have been used to display the evolution of weather systems over time. It appears to the author that in general, weather data is represented in such a way that a user searching for a specific item of data can easily locate that item, but a user looking for unknown correlations is unlikely to find unexpected patterns or unusual combinations of data. In addition, it is very hard to represent time sequences of data meaningfully using these techniques – the animations tend to provide relatively little information compared to the charts, in part because the movies focus more on the "big picture" and less on the details, and in part because it is hard for humans to focus on the details in a moving picture.

Information retrieval does accomodate searches for patterns and items that are not known *a priori*. Unlike in most of IR, where spatial and temporal relations are just one more dimension of the retrieval, we expect to have user interfaces designed to focus users' attention on significant temporal and spatial correlations.

5   Future Work

As is common in research generally and specifically in the design of human-computer interfaces, we expect to keep improving our interfaces through a succession of prototypes, and make gradual progress towards a satisfactory interface for our sensor data. The initial focus of our work is on deploying the sensor nets themselves, but once the sensors are in place, we will need to have suitable means of determining what is happening in the field, if anything, and for focusing our attention on areas which are likely to prove interesting. We are interested in making all our data available to both scientists and to individuals interested in self- or group-study, including students at the high school and university levels.

6   Acknowledgements

The PODS project would not be possible without my co-principal investigators, Kim Bridges and Brian Chee.

References

Belur Dasarathy (2001), program chair. Sensor Fusion: Architectures, Algorithms, and Applications V One of the tracks in the AeroSense conference.

Ray Larson (2001). `http://sherlock.berkeley.edu/geo_ir/PART1.html`, retrieved April 10, 2001.

Tom Mitchell (1997). Machine Learning, McGraw Hill, 1997, ISBN 0070428077.

NOAA (2001). `http://iwin.nws.noaa.gov/iwin/main.html`, retrieved April 12, 2001.

Russell Swan, James Allan, Don Byrd (1998). Evaluating a Visual Retrieval Interface: AspInquery at TREC-6 Position paper for the CHI '98 Workshop on Innovation and Evaluation in Information Exploration Interfaces (Los Angeles, April 1998).

ADAPT: Predicting User Action Planning

Stephanie M. Doane

Department of Psychology, Box 6161, Mississippi State University, Mississippi State, MS 39762-6161

ABSTRACT
A cognitive model of novice and expert aviation pilot action planning called ADAPT models human performance in a dynamically changing simulated flight environment. Rigorous tests of ADAPT's predictive validity compared performance of individual pilots to that of their respective models. Individual pilots were asked to execute a series of flight maneuvers using a flight simulator, and their eye fixations and control movements were recorded in a time-synched database. Models of 25 individual pilots were constructed, and used to simulate execution of the same flight maneuvers performed by the individual pilots. The time-synched eye fixations and control movements of individual pilots and their respective models were compared. The model explains and predicts a significant portion of pilot visual attention and control movements during flight as a function of piloting expertise. This paper focuses on measuring the fit between human and modeled pilot performance, and the possible implications for flight training and cockpit design.

1. INTRODUCTION
A theoretically based computational model of cognition was used to explain and predict expert and novice pilot behaviors during simulated flight maneuvers. In the present research, planning and action take place in tandem and the user (a pilot) works on multiple tasks at any given moment. Accomplishing tasks in the present research refers to the successful prioritization of multiple goals and actions in a dynamically changing task environment. The theoretical premise is that comprehension-based mechanisms identical to those used to understand a list of words, narrative prose, and algebraic word problems constrain problem solving episodes as well. This premise rests on Kintsch's (1988;1998) construction-integration theory of comprehension. Specifically, Kintsch's (1988) theory presumes that low-level associations between incoming contextual information (e.g., task instructions) and background knowledge (e.g., domain knowledge) are constructed and used to constrain knowledge activation via a constraint-based integration process. The resulting pattern of context-sensitive knowledge activations is referred to as a situation model and represents the current state of comprehension. The present research examines the predictive validity of the claim that comprehension-based processes play a central role in cognition (e.g., Doane, Sohn, Adams, & McNamara, 1994; Gernsbacher, 1990; Kintsch, 1988;1998; Schmalhofer & Tschaitschian, 1993; van Dijk & Kintsch, 1983). Proponents of this view have proposed detailed cognitive models of comprehension (e.g., Kintsch, 1988; Doane & Sohn, 2000; Doane, Sohn, McNamara & Adams, 2000), and provided evidence for the importance of comprehension for understanding cognition in general (e.g., Gernsbacher, 1990).

1.1 Research Goals
The present study extends the Construction-Integration theory of comprehension to account for user cognition in the complex and dynamically changing environment of airplane piloting. Specifically, it evaluates whether ADAPT, a construction-integration model of piloting skill can predict the focus of pilot visual attention and control manipulation during simulated flight maneuvers. This was accomplished by simulating the performance of twenty-five actual pilots on twelve segments of flight, and then comparing human and model performance data to determine ADAPT's predictive validity. This extension makes two important contributions. First, it tests the ability of a cognitive theory to predict visual information gathering and multiple task performance in an applied task environment. Such tests are necessary to support the centrality of comprehension-based processes beyond the controlled laboratory environment. Second, rigorous tests of the predictive rather than descriptive validity of the computational theory were performed. The present study represents a significant methodological contribution for evaluation of computational models of human cognition (e.g., Thagard, 1989; Doane & Sohn, 2000). Computational cognitive models of operator behavior in real-time environments have in the past been plagued by ad hoc explanations of how task-sensitive knowledge activation occurs. The present effort demonstrates that ADAPT can predict visual information gathering and action planning performance for multiple pilots during simulated flight maneuvers.

1.2 Construction/Integration Model
The construction-integration model (Kintsch, 1988) was initially developed to explain certain phenomena of text comprehension, such as word sense disambiguation. The model describes how we use contextual information to assign a single meaning to words that have multiple meanings. For example, the appropriate assignment of meaning

for the word "bank" is different in the context of conversations about paychecks (money "bank") and about swimming (river "bank"). In Kintsch's view, this can be explained by representing memory as an associative network where the nodes in the network contain propositional representations of knowledge about the current context or task, general (context-independent) declarative facts, and If/Then rules that represent possible plans of action (Mannes & Kintsch, 1991). The declarative facts and plan knowledge are similar to declarative and procedural knowledge contained in ACT-R (e.g., Anderson, 1993). When the model simulates comprehension in the context of a specific task (e.g., reading a paragraph for a later memory test), a set of weak symbolic production rules construct an associative network of knowledge interrelated on the basis of superficial similarities between propositional representations of knowledge without regard to task context. This associated network of knowledge is then integrated via a constraint-satisfaction algorithm that propagates activation throughout the network, strengthening connections between items relevant to the current task context and inhibiting or weakening connections between irrelevant items. This integration phase results in context-sensitive knowledge activation constrained by inter-item overlap and current task relevance.

The ability to simulate context-sensitive knowledge activation is most important for the present work. The present research examines the construction of adaptive, novel plans of action rather than retrieval of known routine procedures (e.g., Holyoak, 1991). Symbolic/connectionist architectures such as the one utilized in the present study use symbolic rules to interrelate knowledge in a network, and then spread activation throughout the network using connectionist constraint-satisfaction algorithms. This architecture has significant advantages over solely symbolic or connectionist forms for researchers interested in context-sensitive aspects of adaptive problem solving (e.g., Broadbent, 1993; Holyoak, 1991; Holyoak & Thagard, 1989; Mannes & Doane, 1991; Thagard, 1989). van Dijk and Kintsch (1983) and Kintsch (1988; 1994) suggest that comprehending text that describes a problem to be solved (e.g., an algebra story problem) involves retrieving relevant factual knowledge, utilizing appropriate procedural knowledge (e.g., knowledge of algebraic and arithmetic operations), and formulating a solution plan. Kintsch's framework has successfully modeled interactive problem solving tasks. Using a computational construction/ integration model called UNICOM, Doane et al. (1992; 2000) modeled command production performance and the skill acquisition of UNIX users while they interacted with an instructional tutor (UNIX is a command-line based computer operating system). In UNICOM, instructional text and the current state of the operating system serve as cues for activation of the relevant knowledge and for organizing this knowledge to produce an action sequence. The focus of the analysis was not so much on understanding the text per se, but on the way these instructions activated the UNIX knowledge relevant to the performance of the specified task.

2. METHODOLOGY

In the present research, the visual attention and flight performance of twenty-five individual pilots during simulated flight maneuvers was modeled. The eye and control movements of novice and expert private aviation pilots were tracked while they completed flight maneuvers during simulated flight. Results from this study were used to build and test the ADAPT model of pilot visual attention and flight performance. The eye-movement study was part of a larger project and a joint effort with additional collaborators. The results from this study are only summarized here. Further details are available in Fox et al. (1995) and in Doane and Sohn (2000).

2.1 Participants

25 participants, student pilots and instructor pilots, from the Institute of Aviation at the University of Illinois participated in the study. All participants who participated in the experiment did so voluntarily and received $5 per hour. A pre-experimental questionnaire was administered to determine the level of pilots' flight experience. Analysis of the questionnaire resulted in three expertise groups: Novice, intermediate, and expert groups contained eight, eleven, and six participants, respectively. Novices had an average total flight time of 49 hours; Intermediates had an average of 481 hours; Experts had an average of 1467 hours.

2.2 Flight Simulator

A Gateway 2000 computer with an SVGA graphics card produced the flight simulation. Pilots used a sidearm-mounted joystick to provide control inputs such as roll (lateral stick movement), pitch (fore-aft stick movement), and power (push or pull back movement of a button atop the stick).

2.3 Flight Simulation Procedure

Each segment was preceded by a 30 second straight and level lead-in. This allowed participants time to read the segment instructions displayed in a small instruction panel displayed on screen (See Figure 1). The changes for each flight segment were stipulated in an instruction panel and could be accomplished within 60 seconds in the first five

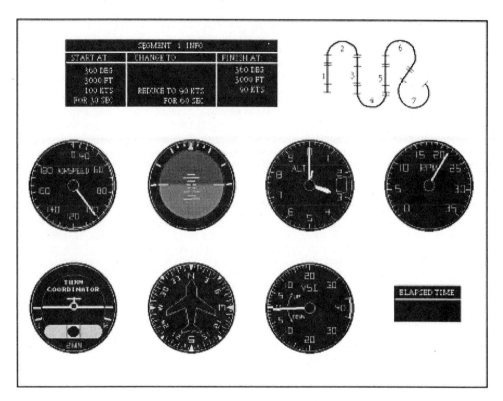

Figure 1. Example flight simulator cockpit display. Depicted are 7 flight instruments including the airspeed indicator, altitude indicator, altimeter, tachometer, turn coordinator, heading indicator, and vertical speed indicator (VSI). Also depicted are boxed tables showing the experimental instructions, flight path, and elapsed time in the segment.

segments and within 75 seconds in the last two segments. If the participants were not within +/-50 feet of altitude, +/-5 degrees of heading, and +/-5 knots of airspeed at the end of the allotted time frame, the program ended the segment and the participants were placed at the beginning of the next lead-in leg. This procedure enabled the recording of seven independent segments of flight performance for each subject. The time required to complete all the seven segments was about 11 minutes.

2.4 ADAPT Architecture
ADAPT represents human memory as an associative network where each proposition representing knowledge constitutes a single node. The discussion of ADAPT's architecture uses, for the most part, planning terms as defined by Allen and Perrault (1980). The main exceptions are the use of the terms "outcome" which in the present work refers to the cognitive consequence of performing a mental action and "in the world knowledge" which in the present work refers to a cognitive understanding of the transient state of the world. ADAPT represents the three major classes of knowledge proposed by Kintsch (1988); world knowledge (dynamically changing knowledge of the environment), general knowledge (e.g., declarative facts), and plan element knowledge (e.g., procedural knowledge represented as if/then rules). Twenty-five individual knowledge bases were constructed to represent the individual pilots that participated in the empirical study. The details of this procedure are provided in Doane & Sohn (2000). The contents of each pilot's knowledge base were determined through observations of a small portion of the pilot's eye-scan, control movement, and airplane performance data. Data were sampled from six different 7 to 15 second time blocks ("windows") for each individual pilot. Thus, 56 seconds of empirical performance data were sampled to build individual knowledge bases that were then used to "predict" approximately 11 minutes of pilot behavior during simulated flight maneuvers. The sampling windows were chosen to score pilot knowledge while initiating, maintaining, and finishing flight maneuvers. Missing knowledge was scored using an overlay method (see VanLehn, 1988), and explicit knowledge scoring rules were devised. Knowledge scoring rules were used to

324

determine if pilots possessed specific types of knowledge A total of 28 flight knowledge scoring rules were devised and applied to evaluate the performance observed in the performance sampling window for each pilot. Once an individual knowledge base for each pilot was constructed, the model was "trained" by using the initial knowledge base to simulate the first 10 seconds of performance in segment 1. Following training the model was "tested" by simulating pilot performance in each of the seven flight segments included in the empirical study without our intervention.

2.6 ADAPT Testing

ADAPT was tested by simulating each pilot's performance on the seven segments of flight. A given pilot's knowledge base was accessed by ADAPT, and the model was given desired flight goals and beginning plane status. This information matched what was given to individual pilots at the start of a given flight segment. The model executed a construction and integration cycle, and fired the most activated plan element whose preconditions were met in either world or general knowledge. The outcome(s) of the plan element fired were then added to the world knowledge and the need and trace propositions were updated to indicate the revised state of the world. Following knowledge base revisions, the model determined if working memory capacity or decay thresholds had been exceeded. If so, the model retained the most activated (capacity) and recent (decay) propositions that fell within the limits set for the individual model during the training phase. This procedure was repeated for each segment of flight until the model obtained the desired flight goals or exceeded arbitrary time (cycle) limits (approximately 150 cycles). The entire procedure was automated.

3. RESULTS

The main data of interest is the match between the sequence of actions observed for individual pilots and the corresponding sequence of plans fired by their ADAPT simulated performance. Using a computation method described in Doane and Sohn (2000), the mean percent matches between the individual pilots and models was obtained as a function of expertise. The mean percent matches on the cognitive plans were 89%, 88%, and 88% for novices, intermediates, and experts respectively. The same matches on the action plans were 80%, 79%, and 78% for the three expertise groups respectively. Given that 25 pilots were modeled over 7 segments of flight using a very small window of data from each pilot, Doane and Sohn (2000) obtained high matches between the pilots and models (see Table 1).

Table 1 Mean percent match between models and pilots as a function of expertise and maneuvering status, M = mean, SD = standard deviation of the mean.

	Novice		Intermediate		Expert	
Status	M	SD	M	SD	M	SD
	Cognitive Plans					
Initiating	89	18	88	10	94	5
Maintaining	87	7	85	14	82	14
Finishing	89	9	89	8	87	7
	Action Plans					
Initiating	85	14	82	10	87	9
Maintaining	74	13	72	15	68	12
Finishing	81	10	81	9	78	12

Overall the model showed a high ability to predict how the pilots would perform. The term "predict" is used because the knowledge base used in each simulation was based on a subset of the pilot performance data. Once simulations began, the models flew the flight segments in an automated fashion by ADAPT. It suggests that ADAPT, a comprehension-based model of adaptive planning is effective for understanding the constraint-based activation of pilot knowledge in flight. Given the extent of the modeling (25 pilots over 7 segments of flight) and the rigor of the evaluation procedure, this is a significant finding. It suggests that ADAPT, a constraint-based model of adaptive planning is effective for understanding the constraint-based activation of pilot knowledge in flight.

4. CONCLUSIONS

A comprehension-based model accounted for a significant amount of pilot visual attention and flight performance. Using a model based on the construction-integration theory of comprehension, individual knowledge bases were developed using a small subset of pilot performance data. The trained knowledge bases were used by ADAPT to "fly" flight segments flown by individual pilots, and this allowed the prediction of significant aspects of user performance. Methodologically, rigorous training and testing methods more commonly used by researchers in machine learning were applied to develop a predictive and descriptive knowledge-based model. Modeling 25 individual pilots flying seven segments of flight allows rigorous testing of qualitative and quantitative fits of the model to the actual performance data. That is, one can go beyond mere speculation that the model provides a "good description" of the data. ADAPT predicts many aspects of pilot performance during flight.

There is an enormous amount of interest in adaptive training systems, a component of which is the model of student knowledge or "student model." The present findings have clear implications for the acquisition and adaptation of student model components of intelligent instructional systems. Our results suggest that a significant portion of pilot visual attention and control movements can be anticipated during simulated flight. Given this, the anticipated pilot behavior could be used to customize instruction, or even to warn pilots about upcoming situations in an effort to modify their behavior.

REFERENCES

Allen, J. F. & Perrault, C. R. (1980). Analyzing intention in utterances. *Artificial Intelligence, 15*, 143-178.

Anderson, J. R. (1993). *Rules of the Mind*. Hillsdale, NJ: Erlbaum.

Broadbent, D. E. (1993). Comparison with human experiments. In D. E. Broadbent (Ed.), *The simulation of human intelligence*. (pp. 198-217). Cambridge, MA: Blackwell.

Doane, S. M., Mannes, S. M., Kintsch, W., & Polson, P. G. (1992). Modeling User Command Production: A Comprehension-Based Approach. *User Modeling and User Adapted Interaction, 2(3)*, 249-285.

Doane, S. M., McNamara, D. S., Kintsch, W., Polson, P. G., & Clawson, D. (1992). Prompt comprehension in UNIX command production. *Memory and Cognition, 20(4)*, 327-343.

Doane, S. M. & Sohn, Y.W. (1999). Instrument Scanning as a Function of Piloting Expertise. Manuscript submitted for publication.

Doane, S. M., Sohn, Y. W., Adams, D., & McNamara, D. S. (1994). Learning from instruction: A comprehension-based approach. *Proceedings of the 16th Annual Conference of the Cognitive Science Society*, 254-259. Atlanta, GA: Erlbaum.

Doane, S. M., Sohn, Y. W., McNamara, D. S., & Adams, D (2000). Comprehension-based skill acquisition. *Cognitive Science, 24*, 1-52.

Doane, S. M. & Sohn, Y.W. (2000). ADAPT: A predictive cognitive model of user visual attention and action planning. *User Modeling and User Adapted Interaction, 10*, 1-45.

Fox, J., Merwin, D., Marsh, R., Sohn, Y., Doane, S., Kramer, A., Lintern, G. & Wickens, C. (1995). Modeling pilot skill acquisition. *Paper presented at the Biennial Aviation Psychology Conference*. Columbus, Ohio.

Gernsbacher, M. A. (1990). *Language comprehension as structure building*. Hillsdale, NJ: Erlbaum.

Holyoak, K. J. (1991). Symbolic connectionism: toward third-generation theories of expertise. In K. A. Ericsson & J. Smith (Eds.), *Toward a general theory of expertise*. (pp. 301-336). Cambridge University Press.

Holyoak, K. J., & Thagard, P. (1989). Analogical mapping by constraint satisfaction. *Cognitive Science, 13*, 295-355.

Kintsch, W. (1998). *Comprehension: A Paradigm for Cognition*. Cambridge, MA: Cambridge University Press.

Kintsch, W. (1988). The use of knowledge in discourse processing: A construction-integration model. *Psychological Review, 95*, 163-182.

Kintsch, W. (1994). The psychology of discourse processing. In M. A. Gernsbacher (Ed.), *Handbook of Psycholinguistics*. (pp. 721-739). San Diego: Academic Press.

Mannes, S. M., & Doane, S. M. (1991). A hybrid model of script generation: Or getting the best of both worlds. *Connection Science, 3(1)*, 61-87.

Mannes, S. M., & Kintsch, W. (1991). Routine computing tasks; Planning as understanding. *Cognitive Science, 15(3)*, 305-342.

Schmalhofer, F., & Tschaitschian, B. (1993). The acquisition of a procedure schema from text and experiences. *Proceedings of the 15th Annual Conference of the Cognitive Science Society*. (pp. 883-888). Hillsdale, NJ: Erlbaum.

Sohn, Y. W., & Doane, S. M. (1997). Cognitive constraints on computer problem solving skills. *Journal of Experimental Psychology: Applied, 3*, 288-312.

Thagard, P. (1989). Explanatory coherence. *Brain and Behavioral Sciences, 12*, 435-467.

van Dijk, T. A., & Kintsch, W. (1983). *Strategies of discourse comprehension*. New York: Academic Press.

VanLehn, K. (1988). Student modeling. In M. C. Polson & J. J. Richardson (Eds.), *Foundations of intelligent tutoring systems*. (pp. 55-76). Hillsdale, NJ: Erlbaum.

Capturing Students' Note-Taking Strategy with Audio Recording Techniques

Martha E. Crosby

Department of Information
and Computer Sciences
University of Hawaii
Honolulu, HI 96822
crosby@hawaii.edu

Marie K. Iding

Department of Educational
Psychology
University of Hawaii
Honolulu, HI 96822
miding@hawaii.edu

Thomas Speitel

Curriculum Research and
Development Group
University of Hawaii
Honolulu, HI 96822
speitel@hawaii.edu

ABSTRACT

The ability now exists to instantly provide large amounts of information to students using modern technology. When too much information is presented at once, however, students are overwhelmed and their efficiency at processing this information drops dramatically. Taking notes is a common strategy used by students to cope with information overload. Yet, effective note-taking is a skill that must be learned. In this study, we ask whether the use of appropriate tools can help students use selection when they take notes. We first describe relevant points from the note-taking research. Next, we describe "Past record," a computer program designed to assist students in the note taking process. Then we describe how this technological innovation can be useful for note-taking activities. Finally, we present an experiment in which the use of "Past record" technology for note-taking was tested among seventh and eighth students learning information about marine science.

1. INTRODUCTION

Results from note-taking research suggest that students learn more effectively when they are required to use some selection criteria, yet researchers have found that students tend to take incomplete notes and record only a small percentage of the ideas presented (e.g., Hartley & Cameron, 1967). Researchers have also found that when students take notes they are more likely to record portions of the material verbatim rather than relate the presented material to information they already know or organize the material into a meaningful framework, (Bretzing & Kulhavy, 1981; Kiewra & Fletcher, 1984). Considering the cognitive overload of trying to listen to a presentation, record the information, and at the same time integrate and reformulate the material recording verbatim quotes may be all the students can realistically accomplish, especially if the note-taking is done with pencil and paper.

This cognitive overload associated with note-taking in response to spoken or oral inputs can be increased by situational task demands in addition to individual differences among listeners. For example, the note-taking task becomes increasingly difficult for students who are learning disabled, are not proficient in the target language, are not familiar with terminology used in the discipline (i.e., are novices) or have motor impairments. Other characteristics of learners can be developmental in nature, e.g., lack of familiarity with the note-taking process or conventions for note-taking. Similarly, there are developmental components associated with short-term memory capacity.

It is reasonable to hypothesize that content-novices as well as younger students would experience some difficulty with effective note-taking, and be more likely to take notes that consist of unrelated bits of information. Students with motor, sensory or cognitive impairments are also liable to experience difficulty. It appears likely that technological innovations in conjunction with effective research-based educational practices could improve the efficacy of the note-taking process for many students.

What, then constitutes effective note-taking? Ideally, effective note-taking involves the employment of selection criteria (Slavin, 2000), rather than indiscriminate recording of points. Furthermore, Slavin (2000) contends "Note-taking that requires some mental processing is more effective than simply writing down what was read (Kiewra, 1991; Kiewra et al., 1991)" (p. 204). Further, Marten and Saljo (1976a; 1976b) differentiated between students who take surface versus deep processing approaches in note-taking. McKeachie (1999) described their distinctions quite effectively:

> Some students process the material as little as possible, simply trying to remember the words the instructor says and doing little else. This would be...a "surface approach." Other students try to see the implications of what the lecturer is saying, relate what is currently being said to other information...and try to understand what the author intended. They elaborate and translate the instructor's words into their own. (p. 74).

This latter approach is associated with deeper processing. McKeachie (1999) argues that students would benefit from taking less notes and listening more attentively when instructors present "new, difficult material" (p. 73). He also provides a practical set of guidelines for instructors to keep in mind in considering the efficacy of their students' note-taking in response to lectures. These points are clearly tied to familiarity with short-term memory limitations. For example, McKeachie (1999) explains, "Note-taking thus is dependent on one's ability, derived from past experience (long term memory), to understand what is being said and to hold it in working memory long enough to write it down" (p. 72). He further explains,

> "When students are in an area of new concepts or when the instructor is using language that is not entirely familiar to the students, students may be processing the lecture word by word or phrase by phrase and lose the meaning of a sentence or of a paragraph before the end of the thought is reached" (p. 73).

In summary, it appears that effective note-taking involves at least two processes. First, students must be selective about the material they take notes on and second, they must process oral inputs at a deeper level. To the extent that technology can reduce much of the complexity of the this process, students' cognitive resources can be freed up to attend to selectivity and deeper processing. Therefore, in the present study, we investigated the effectiveness of a technological innovation called, "Past record" for facilitating novice students' note-taking in science.

3. AUDIO PAST RECORDING

"Past record," is the name of computer software designed to assist students in the note taking process. This audio past recording software places audio events in a small random access memory buffer, which is constantly being overwritten. The time duration of the buffer allows the user to "reach back into the past" when they start the recording process to hard drive. The "present" audio is sent immediately to hard drive. When the "present" recording process is stopped, the buffered "past" audio is inserted into the front of the audio file. This allows the user to selectively decide what they wish to record after they have heard it; hence the name past recording. See Speitel and Iding, (1997) for further information about this software.

4. PILOT STUDY

In a pilot study, Speitel and Iding, (1997) compared several techniques students often use in response to remembering lectured information. This study included the following experimental conditions: (1) listening (not taking notes but concentrating on listening to presented material); (2)

note taking (using pencil and paper to take notes from presented material); (3) normal recording (recording whatever they selected via an audio recorder) and (4) audio past recording (a newly developed tool to help selective note-taking). Results from this pilot study suggested that audio recording devices might help students be selective in their note-taking strategies. Since regular audio recording makes it difficult to be selective because the decision of when to start and stop recording must be made before or after audio occurrences, we wished to determine whether audio past recording would enable the students to be more selective in their note-taking than regular recording. The test data and student responses suggested insufficiencies in the processes of listening and written note taking as a way to facilitate selection of information. Straight listening is limited by the barrage of the stream and the capacity and latent period of short-term memory. Written note taking required "fast writing" and "good listening" which often do not go together. However, student comments indicated that both regular recording and past recording could provide a useful means of selecting important audio information in lecture and lecture-like settings.

5. EXPERIMENTAL METHOD

An further experiment was designed to investigate how each "Past record" mode (regular record and past record) influenced students' comprehension and selection strategies. The participants consisted of twenty-six seventh and twenty-six eighth grade students from the University of Hawaii Laboratory School. Each grade of students were further divided into two sections of 13 students each. A balanced repeated measures design was used for the experiment. Over a period of four days, half of each grade participated in the experiment while the other half went to their regular science class. The students listened to two lectures about marine animals. The lectures were on material that was new to the students. Both lectures were presented by the same member of the research team. The lecture material contained the following three types of information: (1) extraneous (2) important -cued and (3) important - incidental. An example of cued information would be "the most important point is..." The incidental important information was presented without the cue words. The students were given 5 minutes to listen to the material they recorded and were then given a quiz on the material. The students were told about the time limit in order to encourage them to be selective in the information they recorded. At the end of the experiment, the students were given a questionnaire concerning their preference

6. RESULTS

The data suggests that the past record condition has an advantage over the regular record condition. When student attitude data was examined concerning the "Past record" mode Only 11 of the 52 students preferred the regular record mode to the past record mode. In addition, figure 1 shows that the average percent correct on the comprehension exam for the eighth grade students was significantly better ($p<.001$) in the past recorder condition (80.8) than in the recorder condition (61.5) However, this was not the case for the seventh grade students where the mean scores for the past recorder condition and recorder condition were not significantly different. However, for the seventh grade students. the mean comprehension scores were significantly lower than they were for the eight grade students. This finding suggests that although past recording was able to provide a useful way to select important audio information in lecture settings for the eight grade students, neither past recording or regular recording helped the seventh grade in this situation. Possibly, more instruction in the note-taking process is needed for the seventh graders before technological innovation can be employed for successful note-taking activities.

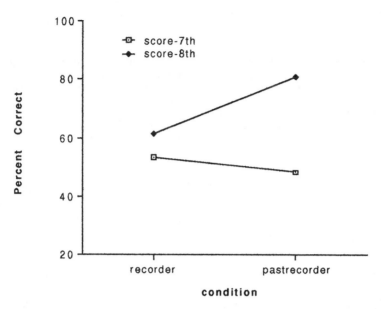

Figure 1. Lesson Comprehension for 7th and 8th Grade Students Using "Past recorder" Software

7. IMPLICATIONS

Audio past recording has a variety of potential applications, both for special needs and for very specific tasks. For example, Speitel and Iding (1997) described some of these tasks:

> Adaptations of this technology are also being tested with the visually impaired in meeting and lecture situations. It is anticipated that partial hearing impaired students will also be able to utilize Past recording. Learning disabled students, especially those who have difficulty processing and/or producing written language can also benefit. The general public as well as professionals who work with clients (teachers, doctors, lawyers, architects, etc.) or do research (social and natural scientists) can use this technology. Sources of audio and video can include phone, radio, TV, and all recordable natural and human phenomena (p. 347).

In a later paper, Iding, Speitel and Crosby (2001) described

> other possible uses for Past Record, including for persons with visual, auditory and other motor impairments that would affect note-taking skills. We also noted that Past Record could be useful for ESL and foreign language teachers and students, and for speech pathologists and linguists interested in recording utterances. We have not yet fully investigated its usefulness for other performance situations besides teaching, although we speculate that this technology would be valuable for those practicing artistic or musical performances, speeches, and story-telling (p. 10).

ACKNOWLEDGMENTS

This work was supported in part by ONR grant no. N00014970578 and DARPA Space and Naval Warfare Systems via ONR grant no. N660019818911. The authors wish to thank Thanhtruc Nguyen and Tyra Shimabuku for their assistance with the research design, data collection and data analysis.

REFERENCES

Bretzing, B. H., & Kulhavy, R. W. (1981). Note-taking and passage style. *Journal of Educational Psychology, 73*, 242-250.

Hartley, J. & Cameron, A. (1967). Some observations on the efficiency of lecturing. *Educational Review, 20*, 3-7.

Iding, M., Speitel, T., & Crosby, M. (2001, August). Machine Interviewing For Assessing Student Learning: Automatic Interviewer and Past Recorder Technology. Paper accepted for presentation at IFIP World Conference on Computers in Education WCCE2001, Copenhagen, Denmark.

Kiewra, K. A., DuBois, N. F., Christian, D., McShane, A., Meyerhoffer, M. & Roskelley, D. (1991). Note-taking functions and techniques. *Journal of Educational Psychology, 83*, 240-245.

Kiewra, K. A., & Fletcher, H. J. (1984). The relationship between levels of note-taking and achievement. *Human Learning, 3*, 273-280.

Marten, F. & Saljo, R. (1976a). On quantitative differences in learning: I-Outcome and process. *British Journal of Educational Psychology, 46*, 4-11.

Marten, F. & Saljo, R. (1976b). On quantitative differences in learning: II-Outcome as a function of the learner's conception of the task. *British Journal of Educational Psychology, 46*, 115-127.

McKeachie, W. J. (1999). Teaching Tips: Strategies, Research, and Theory for College and University Teachers (10th Ed.). Boston: Houghton Mifflin.

Slavin, R. E. (2000). Educational Psychology: Theory and Practice (6th Ed.). Needham Heights, MA: Allyn & Bacon.

Speitel, T., & Iding, M. (1997). Audio past recording software for computer-assisted learning. *Computer Assisted Language Learning, 10(4)*, 339-347.

Speitel, T., Iding, M., Crosby, M., & Shimabuku, T. (1999, October) Computer-aided learner electronic portfolio software development. *Conference Proceedings for WEBNET 99 World Conference on the WWW and Internet.*

Predicting User Task with a Neural Network

Laurel King

University of Hawaii, College of Business Administration, 2404 Maile Way, Honolulu, HI 96822
laurelk@hawaii.edu

ABSTRACT

This study investigates whether human eye fixations have a consistent and recognizable pattern for a particular task and scene. Neural networks are used to predict which scene a participant was searching based on the participant's eye movement data. A back-propagation neural network was chosen for its ability to make inferences based on consistent training data. The eye movement data was converted into facts and target patterns with which to train different versions of networks. Several back-propagation neural networks using different configurations and combinations of variables were designed and trained to determine the network with the highest trainability and accuracy. Variables based on each participant's eye fixations tracked during each of the scenes were used to train the networks to recognize the pattern for that particular scene. After training, each network was tested on new data to see if it was capable of correctly categorizing the scenes from the eye movement information. The performance of the resulting networks is evaluated and conclusions regarding the usefulness of the model are explored. A neural network was found to be capable of predicting the type of scene by knowing whether or not there was a fixation in certain locations, but not by the number of fixations in certain locations. Also, the fixation information was not able to predict the exact scene, but it was somewhat successful at predicting the type of scene.

1. INTRODUCTION

Eye-tracking technology has been used to understand how different individuals look at a particular scene, what they are looking at when they accomplish specific tasks, and how their eye movements adapt to varying conditions and tasks (e.g., Crosby and Stelovsky, 1990, Hegarty and Steinhoff, 1997, Viviani, 1990). Algorithms can be used to calculate the length and position of their fixations, pupil size, and the direction and length of saccades. As with many studies of perception, the results of eye tracking studies are complicated by individual differences, and also by the vast amount of recorded data that must be collected and analyzed. It is for this reason that a neural network was selected to generalize from this plethora of data to try to predict the participants' task from their eye movements. If this is possible, it would suggest that there is more similarity than difference in the way people view a particular scene. It would also support various interactive applications for eye tracking and better interface design. A back-propagation neural network was chosen as the tool because of its ability to infer from generalizations based on consistent training data. The study investigates whether there is a consistent difference that can be recognized by a neural network between the eye movements of subjects as they look at each scene in the data set.

Artificial neural networks are modeled after biological neural networks that use input from neurons to generate signals that activate other neurons that generate output signals based on the input signals received. Artificial neural networks are parallel processing programs that use an algorithm to adjust the connection weights between the input neurons, hidden neurons, and output neurons in the system so that the network can learn to recognize or find specific patterns in a set of data. Unlike rule-based systems, neural networks are able to make inferences based on the previous patterns they have learned. A back-propagation neural network program named *BrainMaker* of California Scientific Software was used to create the networks for this study. Back-propagation neural networks must be trained using input data and matching patterns. As the network is trained to recognize certain patterns, the difference between the target pattern and the network's output is propagated back through the network and used to adjust the connection weights to minimize the error and improve performance.

The eye movement data set was converted into facts and patterns with which to train the networks. Several back-propagation neural networks using different configurations and combinations of variables were designed and trained to determine the networks with the highest trainability and accuracy. Information on fixations from each subject's eye movements tracked during each of the scenes was used to train the networks to recognize the pattern for any particular scene. After training, each network was tested on new data to see if it was capable of correctly categorizing the scenes from eye movements. The successfulness of the resulting networks is compared and the usefulness of the model is explored.

2. METHODOLOGY

The data for this study consists of eye-tracking data from 30 participants who each looked at 73 different scenes to accomplish a task. Dr. Martha E. Crosby collected the data in a separate research project and kindly offered it for use in this study. The data included one scene with a textual explanation of the task, and the rest were pictures of arrows. The subjects were asked to count the number of vertical arrows that pointed down in each display. There were 12 variations of scenes, two each with one to six arrows. Each of the twelve scenes had three versions, one without distracters, one with oblique arrows as distracters, and one with vertical arrows pointing up as distracters. There were small and large arrow versions of the 36 scenes to total 72 scenes. For the purposes of this study, the small and large arrows were combined for 36 patterns plus one text pattern. Small examples of the full-screen scenes that were used are shown below in Figure 1.

Figure 1. Sample Scenes

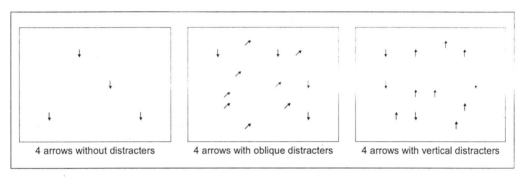

| 4 arrows without distracters | 4 arrows with oblique distracters | 4 arrows with vertical distracters |

A neural network requires that each example of a pattern have the same number of inputs. By dividing the screen into twelve sections it was possible to compare the data for different individuals and scenes where the number of fixations varied from subject to subject and scene to scene. Each of the scenes was divided into a grid with twelve squares in order to count the number of fixations per square of the grid. Even though each individual looked at the objects in the target scene in a different order, it was hypothesized that there would be some pattern as to where one looked at the scene, despite different abilities in pattern recognition and peripheral vision.

2.1 Inputs

Creating neural networks is a process of trial and error to find a combination of inputs, outputs, and network size that is most effective in determining the desired output pattern. The first step is to determine which variables should be used as inputs to produce the correct output. Initially, ten varieties of neural networks were created using combinations of the inputs using fixation counts in each of the squares of the grid, whether the subject had answered correctly, and the total number of fixations. The networks did not train well and the original random weights were often as good as the trained networks. Other factors such as gender and the subject symbol were added as inputs to explore whether the fixations correlated more with the subject or with the scene. The subject symbol worsened performance greatly, but gender did not seem to affect it. This implies that the neural network did not weight the two gender inputs very highly. However, adding the 30 subject symbols to the inputs appeared to have confused the system. At this point in the study, it seemed that there would not be enough similarity among subjects to allow a neural network to determine the task.

Next, the fixation inputs were normalized by using the percentage of total fixations in each square, rather than the actual counts. However, this did not result in much better trainability, although it did test better. Generally, neural networks perform better with normalized data. The algorithm looks for similarities and differences between the inputs of the various patterns, so it was surprising that percentages did not improve trainability. In the third phase, the fixation input was made into a binary that indicated whether or not there was a fixation in each of the twelve squares. This was more successful than the other networks had been, presumably because the number of fixations per square was an individual difference as opposed to a pattern from which to predict the scene.

2.2 Output Patterns

A neural network's performance is measured by how much better it can predict than would be expected by chance. It is for this reason that the network should be trained and tested on examples that have equal numbers of all patterns, otherwise it will adjust its weights to the most common patterns. In this sense, the data for this study was ideal for training a neural network because the networks were presented with equal number of examples of the different patterns other than the text page. Since the number of patterns was 37, there was only a 1 in 37 or a 2.7% chance of any one of them being selected randomly. However, the probability of a particular arrow pattern appearing was 2 out of 73 and 1 out of 73 for the text pattern, because there were fewer text patterns in the data set.

Since the first network with all 37 scenes as the target patterns did not train well, several simplified versions of output patterns were tested. They included all the original 37 scenes and the combined scenes with 1-3 arrows and 4-6 arrows with and without distracters. Also, the 13 scene types were used as target patterns with the distracter type entered as one of the inputs. The subject symbol (with scene category and distracter types entered as inputs) and the type of distracter were also used as target patterns in order to see whether the individual differences were consistent enough to be predicted by the network. Also, it was important to see if the distracter type was a determining factor in eye movement patterns.

2.3 Network Size

Each of the networks tested was tested using one hidden layer and then two hidden layers. Also, the number of hidden neurons was increased to see if the performance improved. None of the networks improved from an increase in the number of hidden neurons or layers beyond having the number of inputs and hidden neurons equal. This implies that there is a continuous relationship between the inputs and the outputs.

Smaller data sets with fewer target patterns were also tested to see if there was too much data for the networks. These networks did train better in that the RMS fell faster and more dramatically to indicate that the weights were being adjusted to successfully minimize the RMS. However, the testing results did not differ significantly from the networks based on the full data set, so the results have not been included in the next section.

3. RESULTS

Initial results seemed to indicate that it would be impossible to determine the task from the participant's eye fixations using a neural network. However, some positive results were found by manipulating the inputs and outputs of the networks. Table 1 shows a comparison of the training and testing statistics for the best performing networks created to predict the seven scene types in each phase of the study. Network C2 which used the presence of fixations in the grid, the total number of fixations, and whether or not the subject made an error counting the arrows predicted 28.7% of the 7 categories with distracter type in testing. The differences in the inputs, network size, and the change in root mean squared error (RMS) and best testing result are listed by network type. The back-propagation algorithm adjusts the weights in an attempt to minimize the RMS error. A network is still learning while the RMS is decreasing. The networks in Table 1 did not train well as can be seen from the small change in RMS error. When the number of possible output patterns was large, the RMS was small and remained small, even though the network did not train or test well. Also, when all 37 output patterns were used for some networks, the initial random weights tested the same as and sometimes even better than those of the trained network.

Table 1. Comparison of Network Types and Trainability

Network	A	B	C1	**C2**	C3
Fixation Input Type	number of fixations in grid	percent of fixations in grid	presence of fixations in grid		
Other Inputs	Total fixations, error, gender	Total fixations, error, gender	Total fixations, error, gender	Total fixations, error	Total fixations
Input/Hidden Neurons	16	16	28	26	25
Output Pattern	7 Categories (1-3 or 4-6 arrows w/ distracter type)	7 Categories (1-3 or 4-6 arrows w/ distracter type)	7 Categories (1-3 or 4-6 arrows w/ distracter type)	7 Categories (1-3 or 4-6 arrows w/ distracter type)	7 Categories (1-3 or 4-6 arrows w/ distracter type)
Initial RMS Error (random weights)	.3408	.3436	.323	.3207	.322
Final RMS Error (optimally trained network)	.2864	.2898	.2868	.2905	.2891
Correct Predictions with Test Data	22%	22.6%	22%	**28.7%**	23.6%

The performance of the networks A, B, C1, and C2 for six different output patterns is summarized in Table 2 below. Network C2 tested at rates that are much higher than would be expected by chance. It even correctly predicted 15% of the time with the 37-scene output pattern, which is ten times the probability of randomly selecting a pattern and having that pattern appear from the data.

Table 2. Comparison of Network Performance

Target Output Pattern	N of out-puts	% of Output Types	Prob. of correctly predicting arrow pattern	Prob. of correctly predicting text pattern	% Correct in Test of Optimal Network			
					A	B	C1	**C2**
37 Scenes	37	2.7%	0.15%	0.07%	0%	0%	0%	**15%**
7 Categories 1-3 or 4-6 w/ distracter type	7	14.3%	4.63%	0.39%	22%	22.6%	22%	**28.7%**
3 Categories 1-3 or 4-6 w/o distracter type	3	33.3%	16.22%	0.90%	48%	58%	59%	**67%**
13 Scenes w/o distracter type (distracter as input)	13	7.7%	3.74%	0.21%	2.7%	2.3%	5.5%	**24%**
Distracter type	4	25%	8.11%	0.68%	42%	50%	59%	**62.5%**
Subject as output (scene & distracter type as input)	30	3.3%	N.A.	N.A.	0%	0%	0.4%	**0%**

4. CONCLUSION

The results of this study indicate that it is possible to predict task from eye fixation information, but not 100% of the time. Also, the number of times that individuals fixate on certain locations is less important than whether or not they fixated on them in determining task. Networks that were created using the subject, as opposed to the scene, as the target pattern did not perform well. This implies that there are more similarities among the ways individuals looked at the same scenes than among the fixations of one individual over all scenes. Also, the individual difference of gender was not a useful input in predicting task from eye fixations, and reduced the performance of some network types. Gender was added to test speculation that there are gender differences in spatial ability which may affect how a person performed a visual task. Still, more information is needed to create a network that is better able to predict task, or to prove that individual differences limit the ability of a neural network to generalize from eye movements. Other factors accounted for much of the variation since the best network could predict only 15% of the 37-scene pattern and only 67% of the 3 simple scene categories from the fixation information.

Since the best networks of each phase correctly predicted over 40% of the targets when tested on both the 3-category-without-distracter target patterns and the type of distracter target patterns, the networks may have been responding to the number of arrows or the difficulty level of the distracter types. In other words, the network may have predicted correctly because no distracters required fewer fixations, oblique needed slightly more, and vertical the most fixations. Also, the fact that the networks tested better when determining whether the scene had 1-3 or 4-6 arrows or had text, supports the concept of subitizing. Subitizing is a phenomenon where people usually recognize 1 to 3 objects without counting them, but need to count higher numbers of objects, which may result in different fixation patterns (Peterson and Simon, 2000). Another study that uses fewer scenes and has more participants is needed to confirm the results of this initial study. Neural networks hold promise for generalizing eye movement information for various applications, but more research is needed to isolate the most important inputs to the network.

REFERENCES

California Scientific Software. (1998). *BrainMaker: User's Guide and Reference Manual, 7th Edition.* Nevada City, CA: California Scientific Software.

Crosby, M.E. and Stelovsky, J. (1990). Using enhanced eye monitoring equipment to relate subject differences and presentation material. In R. Groner, G. d'ydevalle, R. Parham (Eds.), *From Eye to Mind: Information Acquisition in Perception, Search and Reading.* 3-22. Amsterdam: Elsevier Science.

Hegarty, M. and Steinhoff, K. (1997). Individual differences in use of diagrams as external memory in mechanical reasoning. *Learning and Individual Differences, 9,* 19-42.

Konar, A. (2000). *Artificial Intelligence and Soft Computing: Behavioral and Cognitive Modeling of the Human Brain.* Boca Raton: CRC Press.

Lawrence, J. (1994). *Introduction to Neural Networks: Design, Theory, and Applications.* Nevada City, CA: California Scientific Software.

Peterson, S.A. and Simon, T.J. (2000). Computational Evidence for the Subitizing Phenomenon as an Emergent Property of the Human Cognitive Architecture. *Cognitive Science, 24,* 93-122.

Viviani, P. (1990). Eye movements in visual search: cognitive, perceptual and motor control aspects. In E. Kowler (Ed.), *Eye movements and their role in visual and cognitive processes.* 353-393. Amsterdam: Elsevier Science.

Student Usage of a Statistical Web-DB

Joan C. Nordbotten and Svein Nordbotten

Department of Information Science, University of Bergen, Bergen, Norway

Abstract
With the multitude of Web databases available today, it important to analyze Web site usage so that improved services can be made. This paper explores the use of log data as an information source for improving user interfaces for official statistical databases. In the paper, student usage characteristics and trends are discussed and compared to those of the general user community. Student use of official statistical databases is rising. Compared to the average user, they show a more explorative behavior, have a higher usage rate, but retrieve less data.

1. Introduction

National Statistical Agencies are increasingly giving free access to Web databases containing national economic and social statistics. The data content, search & retrieval systems, statistical processing programs, and DB usage logs are maintained by the agencies. It is important for the agencies to analyze usage patterns so that improved services can be made.

This study is related to other studies we have made focusing on evaluation of the interface to websites (Nordbotten & Nordbotten, 1999, 2001a and 2001b). The particular case considered is an analysis of *student usage* of a national statistical Web-DB developed and run by Statistics Sweden.

The Statistical Web-DB was first made available on the Internet in 1997. In September 2000, the Web-DB contained some 800 *statistical tables* within 19 topic areas, such as population, health, housing, and labor. In addition to the statistical data, the Web-DB system contains a hierarchical set of *meta-data* describing the tables within each topic area. A special set of usage logs has been created for this database that includes user registration data, data relating users to the metadata and statistical data they retrieve, and the type of data retrieval used: display, download as a text file, and download to a statistical analysis package freely available from the statistical office.

The study is based on 2 samples from the DB usage log taken in September 1999 and September 2000. The data allows both an identification of usage activities as well as a study of their development between the 2 time periods.

337

2. Observations

2.1 Growth in Usage

Users of the study Web-DB must initially register to receive a unique user identification (userId) and give their user category (government, business, education, personal, etc) and geographic location. In Sept. 2000, there were 8316 registered users in 13 user categories. Of these, 497 were educational organizations: university/colleges (349) and high schools/other schools (148). In addition, there were 928 private users, some of whom have given their user category as education. 89% of the users were organizations from which multiple individuals, or secondary users, could be active. Since the userId is associated with organizations, the actual number of individual users is unknown.

Figure 1 shows the growth in the number of educational organizations, universities and high schools, registered from September 1999 to September 2000. It is interesting to see that interest in taking advantage of this source of information is strong in both groups which both increased by a factor of 3. The growth rate for all organizational users for the period was 4.8, indicating that educational organizations have been somewhat slower in making use of this resource. In comparison, individual users had a growth rate of 66 during the same time period, most likely a result on increased Internet activity.

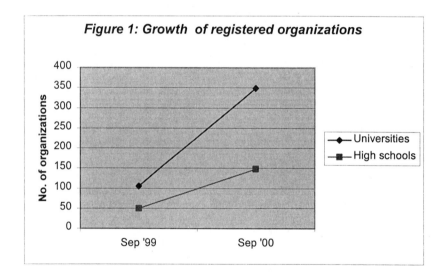

There were 42 universities and 20 high schools, 12.5% of the registered educational organizations, active in September 2000, up from 4 in September 1999, a growth rate of 15.5. The rather low activity level is most likely a reflection of the fact that September is in the beginning of a new school year.

338

2.2. Usage characteristics of the Web-DB

To obtain access to the Web-DB, a user must use the unique identification of his registered organization to authenticate himself. Thereafter each link selection within the database is recorded as a *request* in the DB log. The log entry records the time the request is received, the organization identification (though not the IP address), and the page or service requested. Identification by organization makes it difficult to separate individual *user-sessions* since frequently there were concurrent sessions from the same organization. This is a particular problem for analysis of usage from educational organizations where the number of potential student users is high.

In an initial study of all 13 user categories [Nordbotten, 2001b], a *day-session* was used, defined as all activity from an organization within 1 day. There were 600 active users and 1842 day-sessions in September 2000, giving an average of 3 sessions per user. The 62 active educational organizations had 246 day-sessions, or an average of 4 day-sessions during this period, indicating higher than average activity, presumably from a higher number of individual users.

In order to get a clearer view of individual student access to the statistical data, the current study uses a *topic-session*, defined as the sequence of requests from initiation of a topic search to data retrieval in that topic. A typical topic-session includes an initial request to the DB, followed by several requests for meta-data within the topic of interest and terminates by a request for statistical information in one of the 3 forms described above. As earlier, a topic-session is not equivalent to an individual user session, in as much as a single user may well retrieve data from multiple topic areas in one session.

Using the above definition, there were in all 14.998 topic sessions in September 2000, giving an average of 25 per active user organization. Educational users made 2.916 requests for statistical outputs, 19% of the total, averaging 47 per organization, well above the general average. University students had 1139 topic-sessions, for an average of 27 sessions per active university. High school students initiated 1777 topic sessions averaging nearly 90 per active high school. One explanation of this high activity may be that high school students were working on class assignments, while university students may have pursued personal needs.

Another interesting characteristic is the *output volume* requested. In the log, the volume of statistical output is recorded as rows provided. The average topic-session retrieved about 400 rows of data. University/college students retrieved, on average, 365 rows per topic-session, while high school students retrieved an average of 274 rows per topic-session. The lower than average retrieval rates most likely indicates simpler statistical queries than those used, for example, by government agencies, which retrieved on average over 1200 rows of data.

Data retrieval by the average user was distributed as follows: 68% display, 24% copied to local storage, and 8% use of the free software package. Both student groups retrieved most data to the display screen, 60% and 68% for university and high school students, respectively. High school students used the free statistical package for 14% of their data, while university students used this package for 5% of their data processing. University students copied 35% of their data, most likely for processing in local statistical packages. High school students copied 18% of their data.

The number of requests between an entry into the DB until a request for statistical output reflects how easily the user located desired statistical information. These requests are referred to as *metadata* requests. Measuring this number was difficult due to 2 interrelated reasons: the problem of separating concurrent users from the same registered organization, and the uncertainty associated with determining the start and the end of a session. A useful indicator of the user activities in locating desired information may be the average number of metadata requests per output request, which for all users was 2.6. For university students the average number of metadata requests per output was 2.9 while the average for high school students was 3.0. The slightly higher average may simply reflect relative unfamiliarity with the statistical DB and therefore a more explorative use of the DB.

Figure 2 presents a summary comparison of the log findings for university and high school user organizations.

Active users, Sept. 2000	600	users
Universities:	42	organizations
High schools:	20	organizations
University and high school use	246	day-sessions
Topic-sessions (student-sessions)	2916	sessions
Universities:	1139	sessions
High schools	1777	sessions
Output volume:	903409	rows
Universities:	416689	rows
High schools	486720	rows
Meta-requests per output request	2,96	requests
Universities:	2,90	requests
High schools	3,00	requests

Figure 2: Main characteristics of DB use by students

3. Conclusions

Student use of official statistical databases is rising. Compared to the average user, they show a more explorative behavior, have a higher usage rate, but retrieve less data. Their retrieval preferences are similar to the average user, though university students actually had a higher than average tendency to save/copy the data to their local system, most likely for local processing.

It is well known that analysis of log data alone does not give a full description of site usage. The log data for the Web-DB that we used lacked some information that would have helped to provide a more complete picture for interface evaluation. Particularly:

1. Separation of data for concurrent individual students in the log was difficult because the registered users were organizations with many secondary users (students). One possible solution is to assign a temporary identification to each entrance request, and maintain this for the IP platform until a new entrance request is received. The risk is that the student leaves his computer and a new one takes over without making an entrance request.

2. Linking a chain of requests was uncertain partly because of the reason mentioned in 1 and partly because the log data used did not include a referrer page, i.e. the previous page in a chain of requests. Pattern analysis would be facilitated if this data were available.

3. A drawback of all log data is that the time recorded refers only to request time and not to processing time. For a sequence of requests, we can assume that the processing of one request is finished by the time of the next request. However, the end of the processing of the last request will be undetermined. An obvious solution to this problem is to introduce a log off procedure, but it cannot be enforced since nothing prevents the user from leaving or turning off his computer without logging off.

4. *User* and user *session* are important concepts the definitions of which depend on the purpose of anticipated studies. These concepts should be discussed with designers of log systems to make the log data as useful as possible for later evaluations.

References

Nordbotten, J. and Nordbotten, S. (1999). *Search Patterns in Hypertext Exhibits*. HICSS-32. Proceedings of The Thirty-Second Annual Hawaii International Conference on System Sciences. IEEE. ISBN 0-7695-0001-3.

Nordbotten, S. and Nordbotten, J. (2001a): *Perception of Statistical Presentations Investigated by Means of Internet Experiments*. Hicss-34. Proceedings of the 34th Hawaii International Conference on System Sciences. January 3-6 2001. IEEE. ISBN 0-7695-0981-9.

Nordbotten, S. and Nordbotten, J. (2001b): *A Study of the SSD Web Logs for September 1999 and September 2000* - A Report to Statistics Sweden, Stockholm.

ACTIVITY THEORY AS A BASIS FOR THE STUDY AND REDESIGN OF COMPUTER BASED TASK

Gregory Z. Bedny[1], Waldemar Karwowski[2], Mark H. Seglin[3]

1. Essex County College, Newark, New Jersey
2. Dept. of Industrial Engineering. University of Louisville, Louisville, Kentucky
3. Activity and Human Performance Center of New Jersey, Newark, New Jersey

Abstract:

Human information processing is currently the consensual paradigm for the psychological study of human computer interaction (HCI). While important insights have been obtained from this cognitive psychology and its evaluation of cognitive aspects of the user interface, few professionals find it adequate for the solution of practical problems in HCI. By contrast, activity theory (AT) provides a broad basis for the study of HCI that is sensitive to the real world social and cultural context. Under the rubrics of AT, cognition is an essential component of activity but it is integrated with behavioral and motivational components of human performance. To illustrate the advantages of AT in conceptualizing and in intervening in HCI, we selected a computerized task within an inventory process system of a manufacturing firm that embodies receiving parts, putting them away, storing them, withdrawing and placing them into work-in-process, while, simultaneously tracking their movement and the record of these events. This study is from the perspective of AT.

1. Introduction

AT provides an opportunity to overcome the gap between theoretical research in psychology and practical design in HCI. Transcending exclusive merely the internal mental representation of reality, AT focuses on the task and problems subjects engage, the goals and motivational forces of human activities, subject-object relationship, self-regulation, strategies of performance, mental and practical actions performed by subjects, etc. AT creates a framework for the assessment of computer-based tasks within their respective contexts, providing a non-traditional viewpoint on how practitioners can think about, interpret and comprehend processes associated with HCI. AT attends to socio-cultural contexts, the rules, norms and community in which the activity is embedded. AT differs from Cognitive Psychology in treating cognition as a system of mental actions intimately connected with external behavior (Rubinshtein, 1973). Thus, the cognitive approach be seen as a component of AT with which it should be integrated. We examine computerized task in inventory process from AT to exhibit the fruitfulness of this integration.

2. Systemic-structural analysis and design of activity

One of the strengths of AT is its sixty year old commitment to a systems approach to psychology. We have previously developed a description of activity as a system called "systemic-structural analysis of activity" (Bedny & Karwowski, 2000). System-structural analysis posits a kind of syntax to activity enabling a formulation of activity in terms of "structures. The major units in a functional analysis of activity is a called a function block, a concept that enables approaching activity from the perspective of self-regulation. Another way of studying activity is called a morphological analysis of activity of which the major units of analysis are actions and operations (Bedny & Meister, 1997; Bedny et al., 2000). This work described here is a morphological analysis of computerized inventory process.

Typically, current systemic approaches to describing work tasks concentrate on the systemic description of human-machine interface. What is distinctive about our Self-Regulating Work Activity concept is the formulation of human performance itself as a system that is only possible through AT's development of units of analysis and corresponding methods to guide it. In the morphological description of activity, actions, operations and members of an algorithm supply the conceptual framework for such an analysis. Activity as an object of study embodies multiple distinct aspects requiring a complex system-structural approach to capture the multi-dimensionality of activity during task performance. From this it follows that multiple frames must be employed to characterize a single episode of activity. We adduce four stages of analysis that use diverse procedures to capture the various aspects of the structure of activity – a) qualitative description; b) algorithmic analysis; c) time-structure analysis; d) quantitative analysis. The major importance of the last is the objective evaluation of task complexity (Bedny & Karwowski, 2000). All stages are related to one another in a recursive loop structure in which later stages frequently require revisiting preliminary stages of analysis. Depending on the specific purpose of a study and its object, some stages of analysis and description may be abbreviated or eliminated altogether. In this study of computer based tasks some qualitative analysis and the concomitant algorithmic descriptions of activity are used.

3.   Qualitative and algorithmic stages of analyses

The qualitative stages of analysis include determination of acquired procedures, as well as, levels and steps of analysis appropriate to the purpose. This begins with a general task analysis, transferred into detailed analysis of particular tasks. "Technological analysis" refers to the description of major equipment, tools, raw materials, sequence of basic technological steps, etc. The relationship between computerized and non-computerized work invites examination of the background and training in the use of a computer, subjective relationships to computerization, relative job satisfaction from the computer user, etc. Cognitive task analysis may be included as a step of qualitative analysis. During the qualitative analysis of activity involved in task performance, methods of comparative analysis, such as the contrast between effective and standard performance, error analysis, determination of difficulties and obstacles, etc are deployed.

The next stage of analysis is the algorithmic description of activity including its inner psychological processes. It consists of the subdivision of activity into qualitatively distinct psychological units, called members of the algorithm, and the determination of the logic of their organization and sequence. Each member of the algorithm is limited to between one and three actions as constrained by the limits of capacity of working memory. The rules of combination of actions in time are described by Bedny (Bedny & Meister, 1997). The members of an algorithm are classified according to particular rules. As units of activity, the members of the algorithm are termed "operators" and "logical conditions". Operators consist of actions and operations that transform objects and information such as operators that are implicated in receiving information, analysis and interpretation of a situation, shifting of gears, levers, etc. Logical conditions are related members of the algorithm that determine the logic of selection and realization of different members including decision-making process. Actions as units of analysis constitute, the distinctive features of human algorithm from flow charts widely used to represent human performance.

Each member of the algorithm is designated by special symbols. For example, operators by the symbol "O"and logical conditions by the symbol "l ". All operators involved in reception of information are categorized as afferent operators, and are designated with the superscripts α, as in "O^{α}". If an operator is involved in extracting information from long-term memory, the symbol is μ used as in O^{μ}. The symbol $O^{\mu w}$ is associated with keeping information in working memory, and the symbol O^{ε} is associated with the executive components of activity, such as the movement of a gear. Operators with the symbol O^{ε} designate an efferent operator. In deterministic algorithms, the logical conditions designated with "l " have two values, zero or one. In some cases logical conditions can be combinations of simpler ones. These simple logical conditions are connected through "and", "or", "if-then", etc rules. Complex logical conditions are designated by a capital "L ", while simple logical conditions are designated by a small "l ". Complex logical conditions are built from simpler ones. We propose a method of description not only for deterministic algorithms, but for probabilistic ones as well.

In probabilistic algorithm logical conditions may have two or more outputs with a probability between zero and one. Such logical conditions may be thought of in the following way: Suppose we have logical conditions with three outputs with distinct probabilities of recurrence. In such a case logical condition can be designated as $L_1 \uparrow^{1 (1-3)}$ that possess three or more potential values. In this case we have three versions of output. $\uparrow^{1 (1)}, \uparrow^{1(2)}, \uparrow^{1 (3)}$. Suppose that the first output has the probability 0.2, the second 0.3 and the third 0.5. Knowledge of the probability of the output may be taken into consideration for the study of the probability of performance of different members of algorithms, strategies of performance, calculation of time for performance of the algorithm or components of the algorithm, analysis of errors and evaluation of task complexity. Frequently, in algorithmic description we can use the always-false logical condition designated by the symbol ω. This logical condition is introduced only to make it easier to write. It does not designate real operations performed by the subject. It always defaults to the next member of the algorithm as indicated by the arrow included in the specification of this always-false logical condition.

Algorithms exhibit all of the possible actions and their logical organization and, therefore, constitute a precise description of human performance. The tabular form of the algorithm is formulated in the following way. On the left side of the table there is a column in which we place symbols of algorithms described according to the rules presented above. On the right side, we have a verbal description of the members of the algorithm. The symbols in "l " or "L" for logical conditions in left column include an associated arrow, numbered with a subscript, as $\uparrow^2$. An arrow with the same number, but a reversed position must be presented before another member of the algorithm to which the arrow makes reference, $\downarrow^2$. Thus the syntax of system is based on a semantic denotation of a system of arrows and superscripted numbers. An upward pointing of logical state of simple logical conditions, "l " when, "l"

= 1, that requires skipping the next appearance of the superscripted number with an arrow (e.g. $\uparrow^1\downarrow^1$). A downward pointing arrow with superscript implies that the next denoted operator that follows the logical conditions is to be executed.

4. System-structural analyses of computerized inventory task

An inventory-receiving task was selected as an object of study. Any inventorying process may be presented in terms of three subsystems: 1) stocking; 2) record keeping; 3) work-in-process. The first subsystem refers to the physical movement of items into (In) and out (Out) of stock generating a physical quantity of items on hand. "In" increases stock and "Out" decreases stock. The second sub-system, "work-in-process" (WIP) is a value adding manufacturing process in which diverse raw materials or intermediate products are transformed into ready product. The third sub-system is the record keeping process, that is a complex computerized system that must track all physical movements of parts, purchases, intermediate production, etc. The current study pertains to the first task, called "Inventory receiving task" which entails the reception of parts from different vendors to the re-stocked the warehouse. However, this task should be studied in connection with other tasks because they are interdependent. Figure 1 demonstrates the sequence of tasks before improvement.

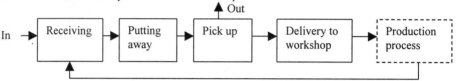

Figure 1. Sequence of the tasks prior to improvement.

The main operation in a receiving task begins with the item being removed from the box and placed in the tote. The tote, thus filled with parts, is placed in "Put-away" area by stock belt. The second task is "Put away" in which the "Put away" operator takes the part from the tote and places on the corresponding shelf. The next task is "Pick-up". The "Pick-up" operator then takes the parts that have been ordered from the shelf and places them into the tote. This tote is later delivered to the shop for production. The pick-up operator also places ready product into tote for sales. Delivering different parts for production is related to delivery task (see Figure 1). This is just a very brief description of the receiving task and its accompanying tasks. The receiver opens the box from the vendor and reads a packing slip. Then the receiver uses a computerized management system. He enters PO (purchase order) number listed on a packing slip and hits an F8 key to check what is still open on the PO. The receiver takes the parts out of the box and compares the order quantity with received quantity. He enters the sequence number of the item on the PO, then changes or confirms the quantity and the price. He assigns allocation if necessary. If allocation is already reserved for the item, the system will select it automatically. All required information is exhibited on the screen. Later this information is printed on the label. As many other tasks that include use of computer, this task is subject to much variation and therefore not readily describable. Accordingly, following preliminary qualitative analysis, an algorithmic analysis is performed, which, in turn, leads back to a more elaborated qualitative analysis. After conducting this analysis and comparing the relationship between this and other listed tasks, the deficiencies of the task design may be uncovered. In this regard, algorithmic analysis may be thought of as a powerful and elaborate tool for the kind of process mapping and job simplification called for by Total Quality Control. While these precise delineations are time consuming, they facilitate the simplification and rationalization of tasks that may be repeated countless times by countless workers. In this they confer the kinds of competitive advantage for which TQM became known in the 1980's.

An algorithmic description of a task is performed with assistance of two tables that contain six pages each. Accordingly, only some fragments of this table are presented below.

Analysis of the established inventory process reveals that the task "delivering to workshop" is possible only after performance of all preliminary tasks (receiving, putting away, and pick-up). However, our study shows that there are some emergency situations in which some parts should be delivered immediately into production. However, according to Figure 1, it may be seen that the existing system cannot perform such a by-pass operation, thereby engendering an inefficient production system with delays in delivery of needed items to production. By the same token, workers are thereby placed into stressed conditions by virtue of the delay in materials for production. In our current example, such emergency situations occur in more than 20% of cases. Based on this data, a different organization of task performance was developed. Under the improved conditions, while the majority of parts may be delivered through the routine five-step process, when required, the receiver may address some parts directly to "delivery to workshop", bypassing the two intermediate tasks.

From this follows that the receiver must assign particular parts in an order according two categories "emergency" or "regular" order. "Emergency" orders are called "work-in-process" (WIP). Receivers must determine through the computer the category of order (work-in-process – Yes/No). Members of an algorithm describe the sequence of actions performed following improvement in this situation (see Fragment 1 of algorithm describing work-in-process). The left column with symbolic description is called the formula of algorithm that is presented in a vertical orientation.

Algorithmic description of activity during performance of WIP. (Fragment 1, after improvement)

O^{α}_{33}	Look at the screen message. (see fig. 2)
$\overset{12}{1\ _{12}\uparrow}$	If screen displays a message, "work-in-process (Y/N)" and the answer is 'Y' go to O^{ε}_{34}, otherwise, go to O^{ε}_{37}
O^{ε}_{34}	Type 'Y', hit "enter" to print out a label, and put label on the part.
O^{α}_{35}	Look at the label to determine which department within the plant the item will be shipped to.
$\overset{13\ (1\text{-}3)}{1\ _{13}\uparrow}$	If it goes to department 1, go to $_1O^{\varepsilon}_{36,}$ if it goes to department 2, go to $_2O^{\varepsilon}_{36,}$ otherwise, go to $_3O^{\varepsilon}_{36.}$
$\overset{13\ (1)}{\downarrow\ _1O^{\varepsilon}_{36}}$	Put the part in box 1.
$\omega_1\uparrow_{\omega 1}$	Always falls logical condition (see O^{α}_{28}).
$\overset{13\ (2)}{\downarrow\ _2O^{\varepsilon}_{36}}$	Put the part in box 2.
$\omega_2\uparrow_{\omega 2}$	Always falls logical condition (see O^{α}_{28}).
$\overset{13\ (3)}{\downarrow\ _2O^{\varepsilon}_{36}}$	Put the part in box 3.
$\omega_3\uparrow_{\omega 3}$	Always falls logical condition (see O^{α}_{28}).
$\overset{12}{\downarrow\ O^{\alpha}_{37}}$.	Check if there is a bin for this item.

It may be seen that this fragment describes activity after improvement when the screen displays "work-in-process" and answer is "Yes (Y)". If according to receiving message, the answer is "No (N)", the receiver bypasses all the members of the algorithm up to O^{α}_{37}. Through precise description of actions performed by users related to "work-in-process", computer programmer can introduce more effective changes in the design of the software providing for an improved performance of the task.

Consider another example. Member of the algorithm $O^{\alpha th}{}_{27}$, l_9, and $O^{th\,\mu}{}_{28}$ (see Fragment 2 of algorithm) performed only before improvement. After improvements they are eliminated. Taking into consideration that this member of the algorithm involves primarily ponderous thinking, mnemonic and decision-making actions, their elimination is a particularly useful example of the reduction of task complexity.

Algorithmic description of activity during task performance. (Fragment 2, before improvement)

*** $O_{27}{}^{\alpha\,th}$	Compare new price with ordered price.
$l_9\uparrow^{9}$	If new price is smaller, go to $^{1}O^{\varepsilon}{}_{31}$. If new price is greater go to $O_{28}{}^{th\,\mu}$
$O_{28}{}^{th\,\mu}$	Mentally calculate the price difference.

This study in general enables significantly enhanced efficiency of the "receiving task". More detailed discussion of this example is beyond the scope of this brief presentation.

5. Conclusion

The current work briefly presents a theoretical concept of the system-structural analysis of activity during the performance of a computer-based task. System-structural studies of activity emphasize multiple approaches to the description of a single object that enables a diverse description of the same object. This requires diverse related stages of analysis of activity.

This study is limited to two stages of analysis, qualitative and algorithmic. In most cases of HCI this is sufficient. It should be noted that in the description of computer based task with multiple approaches to the goal, when a variety of methods are available, probabilistic measures, rather than deterministic measures are used. In computer-mediated tasks, activity is difficult to observe and describe systematically. Users and customers are often at a loss to describe their preferences and requirements. In such circumstances, programmers often develop programs in the absence an adequate comprehension of users' needs. Users frequently reformulate their needs after the system has already been introduced. Software design thus becomes subject to a process of blind trial and error in which the design process must be extensively reiterated. AT permits precise description of a computer mediated tasks. Consequently, evaluation of efficiency can be carried out at an earlier stage of the design process.

References

Bedny, G. Z., Meister, D. (1997). *The Russian theory of activity; Current application to designing and learning.* Mahwah, NJ: Lawrence Erlbaum Associates, Publishers.

Bedny, G. Z., Karwowski, W. (2000). Theoretical and experimental approaches in ergonomic design. Towards a unified theory of ergonomic design. *Proceeding of IEA 2000/HFES Congress,* San Diego, California., v. 5, 197-200.

Bedny, G. Z., Seglin, M. H., Meister, D. (2000). Activity theory: History, research and application. *Theoretical Issues in Ergonomics Science* Taylor & Francis Ltd., 1 (2), 168-206.

Rubinshtein, S. L. (1959). *Principles and directions of developing psychology* Moscow: Academic Science.

Analysis of Visual Search Requirements addressed in current Usability Testing Methodologies for GUI Applications- An Activity Theory Approach

One-Jang Jeng,Ph.D.,   Tirthankar Sengupta

Department of Industrial and Manufacturing Engineering
New Jersey Institute of Technology
NJ, Newark, USA

Abstract

The field of Human Computer Interaction still strives for a generalized model of visual search tasks (icon search, menu search, text search, label search, search through hypertext and feature recognition). The existing models of visual search, in spite of being impressive, are limited under certain perspectives due to lack of generality. The paper tries to provide a holistic approach the modeling of visual search tasks in graphical user interfaces from the Activity Theory (AT) perspective with the aim of rendering a theoretical bridge between HCI and Psychology. Protocol Analysis revealed that complex interaction of behavior, cognition and motor action manifest in these tasks. The results have been analyzed and possible modifications have been identified to provide theoretical support to aid Usability Testing Techniques for GUI applications.

1   Introduction

1.1   Visual Search Research in HCI

The past two decades witnessed immense multi-disciplinary research in cognitive-motor processes, more importantly visual search tasks, in cognitive psychology (Card, Moran and Newell, 1983); user interface design (Shneiderman, 1992), eye movements registration (Russo, 1978) and cognitive modeling using artificial intelligence (Hornof, 1999). Activity theory (AT), developed in Soviet Union from 1930's has evolved to a strong psychological theory and has already been actively involved in HCI. Rubinshtein (1959) introduced the principle of unity of cognition and behavior. This idea was also important in study of the concept of internalization. Vygotsky (1962) defined internalization as the transformation of external performance onto an internal mental plane in social interactions. According to Bedny (Bedny and Meister, 1997) the external action is not transformed into an internal plane but is an important precondition for the formation of internal cognitive actions based on the self-regulation mechanism. Mental actions begin internally with external support. Therefore cognitive psychology, a primary tool for studying human-computer-interactions, possesses one serious drawback - studying cognition in separation from behavior. Cognition and behavior must be studied in unity. This unique approach explains why AT always pays a lot of attention to eye movements during the performance of perceptual and thinking tasks (problem solving)(Bedny and Meister, 1997). Yarbus (1965), the pioneer in this field, proved that eye movements played an essential role in the formation of visual image. Later other scientists demonstrated the importance of eye movements not only in visual search tasks but also in problem solving and thinking tasks. Research was also done in the US to address issues regarding the influence of eye movement on cognitive processes (Russo, 1978). AT embraces cognitive psychology but as a subset of the processes involved in the tasks that conjure up to complex object oriented, goal oriented tasks involved in HCI. This paper identifies the prospect of AT in generating novel models for HCI related tasks such as the visual search, where not only cognitive and motor processes will be involved but also the unconscious and non-verbalized nature of interactions can be studied, thereby providing exact measure of the cognitive complexity involved.

1.2   Activity Theory Approach

AT emerged from the Marxian theory of "dialectical materialism". Founders of AT were Russian scientists Rubenshtein(1959), Vygotsky(1962) and Leont'ev (1978). Basic research in HCI related fields from the AT perspective include information systems, computer mediated communications, HCI, computer science, (Boedker, 1991), AT is defined as a goal directed, artifact mediated set of actions for accomplishing a certain objective. Recent research has defined an activity (Deyatel'nost') as a coherent system of internal mental processes, external behavior and motivation that are combined and directed to achieve conscious goal (Bedny and Meister, 1997). Basically AT promotes system structural approach in the study of human performance. An activity is described as multidimensional system (Bedny, 2000.). Hence Activity can be studied from a cognitive perspective when it deals with process, a morphological analysis when it deals with actions and operations or a functional analysis when its basic unit is a function block.

The task, a fundamental component of AT, can be divided in two basic components- cognitive and motor, which are interdependent and can be further subdivided into cognitive and motor actions. In visual search tasks, cognitive processes constitute the cognitive component and the eye movements constitute the motor component. These processes from the AT point of view are interrelated. This unique feature of AT empowers it to identify a relationship between the cognitive and the motor components of visual search tasks and thereby provides a more general model. This will not only involve the task awareness but also address the phenomena as an interrelation and interaction of the various components (Bedny and Meister, 1997). An experiment was conducted using AT as a theoretical lens to understand a task in which interaction of users interacted with one of the most widely used software, Microsoft Word (© Microsoft Corporation).

2 Methodology

With a verbal protocol approach, speech can emerge as a system of verbal actions, which to some extent correspond to actual performance. The verbal protocol can also be organized as explanation of actions being performed. And these two methods can be combined. In one case subject speak aloud what he did and in another case subject speak aloud how he did. Furthermore AT embraces the fact that verbal speech and thought overlap but do not exactly match each other. Thus verbal protocol in AT should be combined with observation of task performance by different subjects with different backgrounds and different levels of experience with respect to the system. The differences between verbal protocol analysis and real performance can be used as sources of information to find of relationships between conscious and unconscious processes (Bedny, Meister, 1997). This is due to the AT view of verbal protocol as a result of thought and not an exact match of the thought process, which guides the subjects' actual actions. In this study a combination of verbal actions as a system of actions performed by subjects with explanation of the subjects' action has been used. The obtained data were interpreted from a functional analysis point of view. The functional analysis describes the strategies of human performance through analysis of different function blocks in the self-regulation system. (Bedny, *et.al.* 2000). (Refer to the book for a complete illustration of the function blocks). Data obtained from observations and the verbal protocol analysis was assigned to different function blocks (a coordinated system of sub-functions with a specific purpose within the structure of activities). Each function block includes different cognitive processes for achieving a specific purpose in regulation of activity. Function blocks integrate into a self-regulative system with feed forward and feedback connections between them.

The major function blocks in the current study evolved into a goal or goals, past experience, subjectively relevant task conditions, motivation and function blocks involved in evaluative components of the self-evaluation of the task.

2.1 Subjects

Experiments were conducted with three users. All the users were observed and went through a protocol analysis of the task. All of them are regular users of Microsoft Word. However one of them reported superior skills. The second one had a moderate level of experience. The third also was a regular user but was not acquainted with the intricate aspects and were hence considered inexperienced.

2.2 Procedure

The task required the conversion of a piece of text into a table. The desired sequence of intermediate operations is shown in the figure 1. The final product or desired result is shown in figure 2. The optimum steps are shown in figure 1 in a sequence of four steps directed by arrows. An ideal operation involves the use of the following steps.

1. Select the text to be converted by using the mouse. (Step 1 in figure 1)
2. Go to "Table" menu option and then to open the submenu of the "Convert" option.(Step 2)
3. Click to open the "Text to Table option" (step 3)
4. Mark the desired options in the dialog box to obtain the final result.(Step 4).

All the users were experimented at different times so that no user can learn from the other one. The explanation of the task given in 2.1 was not stated but end result or goal (figure 2) was shown to them. The goal of activity was not only stated verbally but also presented in material form (figure 2). The subjects had to achieve the required future state through the set of actions, operations and sub-tasks. They were not only asked to speak aloud what they were doing, but also what they were thinking at each step and why they were doing the particular operation (thinking aloud protocol). Three users each with different levels of proficiency in Microsoft Word performed the task with the protocol analysis. A post-task self analysis was also carried out by the subjects and feedback about the interface was taken. All the subjects used the same computer at different times.

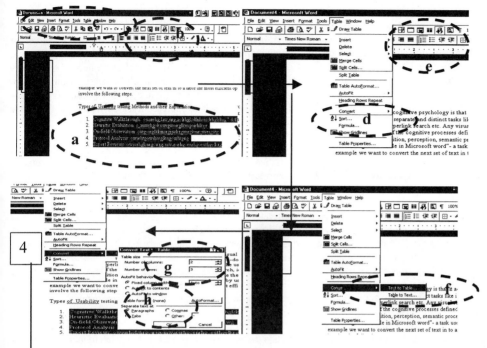

Fig 1: The interaction during the task. The areas where subjects focused most of the times during the task is marked by a dotted circle (Identified using Protocol Analysis).

Cognitive Walkthrough:	Cognitive walkthrough is a review technique where usability specialists
Heuristic Evaluation:	It is a variation of usability inspection where expert evaluators construct
On-field Observation:	Observing users in the field is often the best way to deter mine
Protocol Analysis:	Protocol Analysis is a popular technique used during usability
Questionnaires:	Questionnaires are written lists of questions that you distribute to users.

Cognitive Walkthrough: Cognitive walkthrough is a review technique where usability specialists	Protocol Analysis: Protocol Analysis is a popular technique used during usability
Heuristic Evaluation: It is a variation of usability inspection where expert evaluators construct	Questionnaires: Questionnaires are written lists of questions that you distribute to users.
On-field Observation: Observing users in the field is often the best way to deter mine	

Fig 2: The prescribed product/result   Fig 3: Final product or result more often obtained by subjects.

3   Results and Discussion

Based on the protocol analysis, observation and post-task interview eight areas were identified as areas of the prime focus for the subjects. These areas have been shown in the figure 1 with dotted lines. Each of these areas affects the self-regulation system of the individual and renders one or more function blocks to be more significant than the others during the performance of that task. Problems associated with these areas and their inability to generate effective interactions is explained below.

The text body, which needs to be converted, is identified by "a" in stage 1. This area of interactions on the screen had no significant interaction problem. All the three users quickly identified the strategy to select the text for converting it to a table. Let us consider how the function blocks of experience influences the subjects' strategy formulation, and that of subjectively relevant task conditions and sub-goal formulation matched the requirements of task execution. It is worth mentioning that Subject 3 developed a subjective goal that deviated significantly from the objectively given goal. Subjectively relevant task conditions also differed for him and thus reformulated his goal to produce different and inefficient strategies of execution. For example, he created a separate blank table and put the texts one by one in the respective cells. However, according to instructions he must convert the given text by using the particular software feature which the other two followed. As a result orientational components involved in the strategies became totally inadequate. Reformulation of goal resulted in the change on all the other function blocks of self-regulation. Goal reformulated by subject intermittently, can influence all other mechanisms. This interaction can be viewed as a game of chess where external information becomes an input for developing subjectively relevant task conditions, which are dynamical models of the situation. Subjects can develop totally different mental representations of task from the same screen. The strategy though incorrect from the expectation did meet the requirements but required more time to execute.

Problems were observed in the next stage of task performance when subjects used visual search processes for manipulation of icons, menus and text in combination. Here subjects 2 and 3 were constantly referring to the areas "b" and "c" as the more important components of task at this stage, but the only thing they were needed to identify was the table menu option. As a result subject's image of task developed by the analysis of function block of subjectively relevant task conditions become inadequate at this stage of performance. This produced complex decision-making processes, complicated mental operations and actions, and increased search times. At this stage subjectively relevant task elements were only in the table menu option. However, due to irrelevant task elements the function blocks used up much of the cognitive resources. Complexities of the task were further increased when the subjects clicked a wrong option and were introduced to a completely different situation. This was mostly due to the icons in area "b" and "e", which continuously deviated the subject from obtaining the correct strategy of choosing the "Convert " menu option marked by area "d". This occurred because identificative features of icons matched the image of task worked as a filter for eliminating objectively important elements of the task and extracting irrelevant components of the objectively presented situation. This in turn resulted in the formation of intermittent goals in an incorrect way leading to the generation of negative motivational states. This affected the Goal, motivation and subjectively relevant task conditions and thereby increased the task time. In some cases the subjects rejected the task by opting out. In other cases when subjects overcome negative motivational states and continue to reach the goal the time of task performance was significantly increased. The option of rejecting or not rejecting the task depends on the relationship between the function blocks of sense of the task and difficulty of the task. If the task is perceived as very difficult but not sufficiently significant for subject then they will reject the task performance. Thus the level of motivation, which determines the acceptance or rejection of task, depends on the relationship between the significance and difficulty of task.

As soon as the third stage came up the subjects had no problem in identifying the option as the semantics completely matched the verbal-logical processes- subjects were encountering (area "f"). The function blocks of evaluation and experience rendered a match between the subjective and objective image of the subtask and the execution took no time for opening the dialog box as shown in the fourth stage. Interestingly the inexperienced user could not understand what the two options, "Text to Table" and "Table to Text", in the convert submenu meant, a fact supporting the importance of experience function block.

The fourth stage had most of the problems and even with the experienced user. The dialogue box had given no clue to the subjects in terms of the result. Only the number of rows and columns (area "g") attracted subjects' attention. The second set of choices, a subjectively relevant task component was never paid attention by most of the subjects. On the other hand all the subjects who reached this stage could make out that any option in the area "h" (the dialogue box) would have an effect over the results and thereby had developed an adequate representation of the task. But not how to perform the same? Here discrepancies were between orientational components and executive components of activity. The area "h" does not possess identificative features to let the subject select a correct program of performance. The experienced user tried all the options sequentially to satisfy the criteria developed by evaluation blocks. This reflected a development of trial and error strategies through a sequence of executive and evaluative steps thereby complicating the interactions between the executive and evaluative blocks of self-regulation. As a result the complexity of the task noticeably increased. Subjects 2 and 3 were still using visual processes to search the screen, specifically areas "c", "b" and "e", for identifying options, which were relevant to the task.

From the observations it is clear that at each stage of interaction the function blocks of the self-regulation system either were reconstructed or reoriented in terms of the feedback influences. Cognitive processes were mostly dominated by visual search tasks and a fair amount of time was expended for search of unwanted areas on the screen. This was due to the similar identificative features of the icons and meaning in the text, which rendered an improper state to the users' mental dynamical model of the image of the task. Two inferences come out of these observations. First, in any interaction we cannot define the components of the interaction by simply extracting the elements of the task. These elements include icon search, menu search or text search. Rather, these elements are integrated by function blocks into a structural system, which is relevant to the particular stage of the solution. The more experienced the user is the less the time and resources used on these function blocks for proper execution. A highly efficient user of MS Word uses strategies of performance in which evaluative components of activity significantly were almost automatized. Thus self-regulation of activity achieved a more automated level of functioning. Second, it is better to use an object-action pair for tool tips. Tool tips should be provided for menu options too with the option of activating and deactivating these features. For example the use of tool tip "Columns" in one of the icons confused most of the users. It is used for writing text in two columns. Therefore a better use should be "Columns for Page." Using a tip like "converts the selected text into a table as specified" at the "convert" menu option would definitely be helpful for Subjects 2 and 3. Subject 1, who is not comfortable with it, can deactivate the feature. The basic aim is to guide the user through the process in such a way that he identifies the correct sub-strategies most of the time. This will expedite the process of his learning and the self-regulation function blocks will execute with more efficiency.

4 Conclusion

The functional analysis of HCI related activities reveals the relationship between empirical material and the function blocks of self-regulation, which provides the opportunity to precisely describe users' strategy of performance during HCI. Goal formulation and experience are of prime importance, which necessitate memorization and training as important factors for these tasks. The function blocks of goal and subjectively relevant task conditions function as filters, which determine what information is important for a particular stage of task performance. Thus AT promises to shed light not on the tasks as a separate unit but as a component which is affected by related processes and subject behavior. We have seen usability techniques being criticized for lack of theoretical backup. This may be due to the fact that usability studies are performed on individuals who have unique behavior phenomenon during interaction. Unless we model this behavior the results of psychological research cannot be applied with efficacy to understanding usability problems, which show up during the stage of actual use. It is difficult to chart the myriad of human behavior but it is definitely possible to model the interaction of these simultaneous processes to render usability testing as a useful and cost-effective for the industry. Usability aims to render systems to be a pleasant learning experience for the users. AT aims to expedite the learning process through its holistic approach to the cognitive processes involved. We have only addressed visual search processes but AT can involve all the processes. As more sensory input channels are used the nature of interaction will become more complex and individual models of these tasks may lead to misleading results. Future research is needed for adopting AT for HCI and modifying the established techniques like systemic-structural analysis for efficient implementation.

Acknowledgement:
The authors would like to thank Dr. Gregory Z. Bedny for his support and guidance.

References
Bedny, G.Z., Seglin, M.H., & Meister, D. (1997). Activity Theory: History, Research and Applications. *Theoretical Issues in Ergonomic Science*, VI, 2, 168-206.
Bedny, G. Z. & Meister D. (1997). *The Russian Theory of Activity: Current Applications to Design and Learning*. New Jersey: Lawrence Earlbaum Associates.
Bedny, G. Z.& Meister, D. (1999). Theory of Activity and Situation Awareness: *International Journal of Cognitive Ergonomics*.1999, 3(1), 63-72. New Jersey: Lawrence Erlbaum Associates.
Boedker, S., (1991) *Through the Interface - A Human Activity Approach to User Interface Design*, Lawrence Erlbaum Associates.
Byrne, M.D. (1993). Using icons to find documents: Simplicity is critical. *Proceedings of INTERCHI '93*, New York: ACM, 446-453.
Card, S.K., Moran,T.P., & Newell, A. (1983). *The Psychology of Human-Computer Interaction*. Hillsdale, NJ: Lawrence Erlbaum Associates.
Hornof, A.J. (1999). Computational Models of the Perceptual, Cognitive and Motor Processes involved in the Visual Search of pull-down menus and computer screens. Ph.D. dissertation in Computer Science and Engineering. The University of Michigan, Ann Arbor, Michigan.
Jeffries, R., Miller, J. R., Wharton, C. & Uyeda, K.M. (1991). User Interface Evaluation in the Real World: A comparison of four techniques: *CHI Proceedings*, p.119.
Leont'ev, A.N.(1978). *Activity, Consciousness and Personality*. Moscow: Moscow University Publishers.
Nielsen, J. (1994). *Usability Engineering*. San Francisco, Ca: Morgan Kaufmann Publishers.
Rubenshtein, S. L., (1959). *Principles and Directions of Developing Psychology*. Moscow: Academic Science.
Russo, J. E. (1978). Adaptation of cognitive processes to the eye movement system. In J. W. Sanders, D.F. Fisher, & R.A. Monty (Eds.). Eye Movement and the Higher Psychological Functions. Hillsdale, New Jersey: Lawrence Erlbaum and Associates.
Shneiderman, B. (1992). *Designing the user interface. Strategies for effective human computer interaction*. (second ed.). Reading, Massachussetts: Addison-Wesley.
Vygotsky, L.S.(1962) *Thought and Language*. Cambridge, MA: MIT Press.
Yarbus, A. L. (1965). *Purpose of Eye Movements in Visual Process*. Moscow: Academy of Psychological Science.23.

PROFESSIONALLY IMPORTANT PERSONALITY TRAITS AND PSYCHO-PHYSIOLOGICAL CONDITIONS RELATED TO EFFICIENCY OF ACTIVITIES FACTORS OF SHIP OPERATORS

V.N. Levytskiy[1], Ph.D., M.H. Seglin[2], Ph.D., A. M. Karpukhina[3]

1. Military Merchant Marine Academy, Kiev, Ukraine
2. Activity and Human Performance Centre, Newark, New Jersey
3. Ukraine Academy of Pedagogical Science

Abstract

In this article we describe a method for studying various personality characteristics within the framework of Activity Theory. This study examines psychological features of personality and psychophysiological states of military ship operators who work in computerised ship control systems. Psychophysiological states of operators were examined with multiple methods that include combination of expert evaluation and experimental methods. The latter include measurement of biologically active points of the skin (BAPs), neurophysiological features of personality, the study emotional stability, characteristics of psychological processes, etc. These studies resulted in recommendations for predicting the efficiency of operators in enclosed, computerised ship control environment.

1. Introduction

Activity theory, a psychological approach that comes from the former Soviet Union is emerging as an important new social science paradigm in Western Europe, North America and Australia (Bedny, Meister, 1997; Nardi, 1996;)that through it internationalisation is becoming integrated with certain branches of cognitive psychology (Bedny, Meister, 1999). Historically AT has been closely linked to practical application on the basis of it concept of anthropocentric design. Under this rubric, an individual is considered the subject of an activity, and as the main, key component of the system. This concept has resulted in an intensive search for the most expedient ways of collection, systematisation and analysis of data about individual/operator professional activities in such systems as 'individual – machine – environment', 'individual – computer' etc. AT is attracting particular attention for its utility in the design of digitally mediated environments (Kuuti, 1999). Therefore, we conducted the psychological and psycho-physiological analysis of individual operator activity within this kind of functioning system. Specifically we explored functioning within computer automated ship control systems (CS). In addition to well-known features of operator's activity, there are specific features arising from the specific conditions in which ship operators perform (Levytskiy & Karpukhina, 1979). The widespread use of computers in the ship's CS (so called 'computer supported collaborative work'), leads to psychological changes in the structure and functioning of operator activities. These changes affect cognitive, communicative and personal spheres. They transform executive activity links and the processes of establishing aims, as well as, activities demand-motivational regulation. They necessitate the special study of computer-related communication, so called 'computer – mediated communication' (CMC).

To increase CS efficiency through optimising the dynamic interaction of all ship CS components - - include CMC features -- it is necessary to take ship operator activity features into account. To resolve this problem within the frame of the research area known as 'Human – Computer Interaction' (HCI), making use of the AT approach, we studied the professionally important characteristics (PIC) of ship operator personality (psychomotor and communicative abilities) and the operator's psycho-physiological state (PPS). At the same time, we aimed to build from the analytic study of separate psychological functions, processes and conditions to identify the holistic characteristics of the operator personality, and to understand the personality as a whole (Bedny, Seglin, 1999).

The main task was to undertake the complex research of personality PIC and PPS dynamics of the ship operator in conditions of real activities with both moderate and high emotional and emotional-physical stresses. In addition, we determined the possibilities of using various quick and convenient methods for the quantitative estimation of basic operator personality PIT and his prognostic indices of PPS dynamics during these activities. This should allow us to control and improve the efficiency of the ship operator's activities.

Our research, departs from the functional systems theory as developed by Anokhin (1962) and functional analysis of activity (Bedny, et. al. 2000). Such an approach allows us to carry out the complex study of all functional system mechanisms (including the activity psychic regulation system involved in the operator professional activities). It also enables us to define the amount contributed by each of these systems to the end result of activity – i.e. to its efficiency.

The functional systems of the operator differ according to the nature of the functions being performed – they are psychic, psycho-physiological and physiological. Therefore the condition of the operator as a supersystem can be defined as a psycho-physiological state (PPS). There are numerous physiological, psycho-physiological and psychological indices used for the estimation of the operator's PPS (Nebylitsyn, 1976; Teplov, 1961, etc.)that include diagnostics and patients' condition monitoring, as well as, for the estimation, monitoring and regulation of an organism functional condition of healthy individuals (Karpukhina, 1980; Levitskiy, Megheritsky, 1979). We used the technique derived from of China of electroconductance and resistance of the biologically active points (BAP) of the skin, including the floor of the auricle BAPs (auricular BAPs). Since, these points correspond to the functional condition of the individual organism's most important organs and systems, we consider the practical application of methods using BAP electrical parameters for ship operator's PPS monitoring to be extremely promising.

We combined psychometric indices of various questionnaires subjective estimations of the conditions', corrective trials, tests of attention, thinking, memory etc. Comparison of operators' current PPS indices with the functional norms enables measurement of deviations and assessments of optimisation of the indices to facilitate regulation of an operator's PPS. The ship operator's PPS regulation contributes to stabilisation of the individual's working capacity at optimal level, ensuring the efficiency of the specified activities.

On the basis of the subjective estimation of the significance of personality characteristics considered fundamental to the ship operator in his/her professional activities, we selected the most important qualities. In this process we employed a method of expert assessment consisting of quantification and psychological qualimetry from which we derive quantitative estimation.

2. Experts method of personality evaluation

The main tasks were to determine:
- the correlation between operator's personality traits and efficiency of performance;
- the criteria for predicting ship operators' activity efficiency.

The operators selected from the population were grouped by success level into "more successful" and "less successful" according to an expert by using a five point scale. Determination of underlying psycho-physiological structures were inferred through a factor analysis. The factor analysis was performed both to explore the common factor structure across the two groups as well as to determine differences in the factor structures of "more" or "less" successful groups. Nine common factors were identified and named and interpreted according to the indices that loaded on them. On the basis of this factor analysis (for level of significance $p < 0.01$ and a coefficient of correlation $r > 0.463$ for more successful and $r > 0.435$ for less successful), we obtained the factor image of the indices of basic personality characteristics and ship operator's professional training level.

Activity theory posits a hierarchy of personality attributes that in some way resonates Maslow's well-known hierarchy of motives (Bedny & Seglin, 1999). Under this rubric personality is segmented into four levels: Special attention was given to comparative analysis of integrative indices of personality characteristics as motivation, emotional stability, communicative process quality, strength of neural system, mobility and balance of neural system, etc. There were statistically significant differences in these indices between these two groups. The results of the factor analysis forms a basis for an estimate the lower boundary of psychological traits and professional experience for successful performance of the operators in the ship CS, and also leads to specific ways of improving operators specialist and psychological training.

With the help of regression analysis by the Gauss – Gerdan method an adequate personality model for prediction of the ship operator's success was found. This model is presented by of linear multiple regression equation.

$$Y_{ri} = 0.40 + 0.20 \, X_{wcs} + 0.2X_{ii} + 0.10 \, X_{po} \ldots$$

Where, Y is the integrative index of success of ship operator;

X_{wcs} is the work capacity during emergency;

X_{ii} is the intrinsic interest of the work;

X_{po} is a global characteristic of proactivity.

These three are the broadest, most significant predictors of success. Using correlation analysis those personality traits which are in the most degree connected with the ship operator's activities success were uncovered. By solving this equation we obtained personality resultant indication value, which can be used as an estimation criterion to forecast the ship operator's activities success.

Factors obtained as a result of the regression analysis are the most potent ones in accounting for the variance in the totality of ship operator's measured personal characteristics and they are connected with the operator's activity success. They allowed us not only to estimate personality criteria to forecast ship operator activity efficiency, but, also, to discover how to improve efficiency by influencing such subjective factors of the operator's activities as his/her individual characteristics.

The results of our studies of ship operator's PIC allowed us to recommend quantitative personality criteria to monitor, estimate, forecast and control ship operator's activity efficiency.

3.   Analyses of psycho- physiological state of ship's operators
We studied the ship operator's PPS dynamics during actual working conditions of his/her activities in CS using computers. We monitored the psychological, psycho-physiological and physiological indices of those functional systems, which were most essential to the ship operator's efficient performance (Sapov, Solodkov, 1980; Levytskiy, Karpukhina, 1979). For this purpose, we studied neural features of personality develop by Khilchenko (1978) and organs' functional states by the value of electrical conductivity of the biologically active points (BAP) of a floor of the auricle. In addition to well-known methodologies for hemodynamical indices estimation - auriclediagnostics were included in the group of methodologies.

To assess the PPS dynamics, the indices were registered before and after the operators' day watch. From the whole group of operators participating in the studies, on the basis of expert opinion, two groups also were selected – "more successful" and "less successful",.

The main aims of these studies were to identify:
- the relationship of the PPS dynamics and the operator's efficiency of activities under real conditions of activities while the ship was cruising;
-    informative and prognostic indices of the operator's PPS dynamics;
-    new quick and convenient methods of estimating the dynamics of the ship operator's organism functional condition when subjected to moderate and high emotional-physical stress during ship cruising.

Various mathematical statistical methods including alternative and correlation analysis were applied to the data to evaluate the relationships among variables.

Analysis of the data showed that the most significant differences ($p < 0.001$) in the PPS dynamics of the ship operators before and after watch were for indices of psychic condition. Some psychic processes (thinking, memory), neurodynamics and also between certain indices of cardiovascular and digestive systems functioning. Changes of the value of the integrative index of the floor of the auricle BAP show that this index is a complex parameter characterising the functional shifts in the operator's organism in the process of adaptation.

In addition to comparison of the values of the different ship operator's PPS indices, measured in the general operators' group before and after a strenuous watch lasting 4 hours, an hypothesis was tested concerning the mean values for "more successful" and "less successful" operators' groups before and after watch. A number of indices of psychic conditions and processes, neurodynamics, cardiovascular system and BAP had significant differences before and after operators performed their activities in the ship CS, and these differences were modulated by the operators' groups success level. These indices' values can serve as estimation criteria for forecasting the ship operators' activities success and onset of tiredness, a factor which is especially important in the conditions of long-term cruising.

Correlation analysis of the indices characterising ship operator's PPS dynamics reveal their connection with activities success before and after watch. The indices of thinking activities have the closest connection with the operator's activity success before and after watch. For the group "more successful", it is mainly the indices characterising the operator's organism functional condition (in particular respiratory system and digestive tract) and general organism stress that differentially predict success before watch. For the group " less successful" before watch, it is the indices, characterising mainly psychic conditions and processes (memory, attention, and thinking) that differentially correlate with success. This stresses the necessity of the different emphases in the preparation of operators for performing their activities in the ship CS. Most likely, in the group "more successful" before watch, more attention should be given to the monitoring of the operator's organism functional condition, and in the group "less successful", in addition to the above, attention should be given to psychological preparation, particularly to psychic condition normalisation. After watch for the group "more successful" in comparison with the group "less successful" the highest correlation coefficients were found for those PPS indices which characterise organism adaptive capabilities, its psychological and physiological reserves. Hence all the measures to restore ship operator's working capacity both during short (but especially during long) cruising and during the periods after cruising should be carried out differentially, taking into account operators' individual psychological and psycho-physiological features and their level of professional training.

As the ship operator's PPS dynamics analysis showed, the operator's PPS levels are determined by his/her individuality, psychological and physiological status features. Operator's state depends on internal and external environmental changes connected with ensuring the specified activities efficiency in the ship CS. Therefore, the whole complex of psychological, psycho-physiological and physiological characteristics of the operator's functional systems, ensuring his/her activities efficiency, used in the research, is inadequate for activities efficiency simulation. Hence, on the basis of our research specified aims and tasks, we could resolve only the tasks of selecting and estimating the most informative (in respect to operator's activities efficiency) PPS indices characterising various hierarchic levels of the operator – psychological, psycho-physiological and physiological. This will allow us to monitor, estimate and predict the PPS dynamics in mutual relation to efficiency in the typical operator activity conditions and also to determine the possibilities of regulating PPS, in particular by establishing the relevant professionally important personality characteristics.

Thus, the study carried out confirmed the hypotheses. There is utility for efficiency control of activity in the ship CSs which use computers, selection and quantitative estimation of professionally important personality traits (PIT) and informative prognostic indices for the psycho-physiological condition (PPS) dynamics of an operator performing. The main task has been resolved – the complex examination of personality PIC and PPS dynamics of a ship operator in real activities conditions using a wide range of various high-speed methods for estimating psychological, psycho-physiological and physiological indices.

4. Conclusions:
1. It is possible to construct a model of personality for ship operator activities that predicts success, as well as, obtains the value of the estimation criterion for prediction of ship operator activities success during long cruising. (One of the versions of such model was recommended for practical use in the navy ships.)
2. The quality of communicative processes including computer communication quality is the one of the basic ship operator's professionally important traits, estimation of which requires use of special psychodiagnostic procedure with revealing the level of development in psychomotor and communicative abilities.
3. Of all the ship operator's psycho-physiological condition dynamics indices, the indices of subjective estimation of condition, psychic and neurodynamic processes, gemodynamics and biologically active points (BAP) proved to be the most informative. Some of these indices are estimated with the help of apparatus methodologies (for example, the indices of neurodynamics, gemodynamics, BAP etc.) can find application in ship automated systems of PIT discrete-continuous monitoring and estimation.
4. PIT indices differed in their correlations of activities success with before and after watch for operators groups "more successful" and "less successful". It was shown that for the less successful operators the compensatory-adaptive processes influence in larger measure the PIT psychic component. On the basis of this moderator variable effect an approach was suggested for the preparation of operators for watch and for the restoration of their efficiency after watch. For more successful operators before watch it is necessary to pay more attention to the organism functional condition monitoring, but for less successful ones, in addition to the above, attention should be paid to psychological preparation, especially to lowering anxiety level.

References

Anokhin, P. K. (1962). *The theory functional systems as a prerequisite for the constriction of physiological cybernetics.* Moscow: Academy Science of the USSR.

Bedny, G. Z., Meister, D. (1997). *The Russian Activity Theory: Current application to design and learning.* Mahwah, NJ: Lawrence Erlbaum pulisher.

Bedny, G. Z., Meister, D. (1999) Activity theory and situation awareness. *International Journal of Cognitive Ergonomics,* 3, 63-72.

Bedny, G. Z., Seglin, M. H. (1999). Individual style of activity and adaptation to standard performance requirements. *Human Performance ,* 12, 59-78

Bedny, G. Z., Seglin, M. H. & Meister, D. (2000) Activity theory: History, resarch and application. *Theoretical Issues in Ergonomics Science* 1(2), 168-206.

Karpukhina, A. M. (1980). Biologically active points of skin and problems of regulation of psychophysiological state. *Psychohygiene,* 5 (7), 54-61.

Khilchenko, A. E. (1966). Study of mobility of neural processes in man. *Physiological Journal,* 6, 67-81.

Kuutti, K. (1999). Activity theory as a potential framework for human-computer interaction research. In B. A. Nardy (Ed.) *Context and consciousness. Activity theory and human-computer interaction.* (pp. 17-44). The MIT Press, Cambrige, Massachusetts.

Levytsiy, V. N., Karpukhina, A. M. (1979). *Basics of navy ergonomics* Kiev, Ukraine: Higher Education.

Levytskiy V.N., Mezhiritsky N.I. (1979). For the question of possibility to use the biologically active points' informative parameters for students' functional condition accounting. In A. M.

Karpukhina (Eds.) *Psycho-physiological condition of a man and the skin biologically active points.* (pp. 36-38). Kiev, Ukraine: 1979, pp. 36 – 38. Higher Education.

Nebylitsyn V.N. (1976). *Psycho-physiological examination of individual differences.* – Moscow: Education Publishers.

Sapov I.A., Solodkov A.S. (1980). *Sailors working capacity and functions condition of their organisms.* Leningrad: Medicine Publisher.

356

Theories and Methods in Affective Human Factors Design

Martin G. Helander

School of Mechanical and Production Engineering
Nanyang Technological University
Singapore 639798

ABSTRACT

Users and operators have affective reactions towards artifacts and interfaces. Theses reactions are caused by design parameters either through their perceptual attributes or from the experience in operating or using artifacts/interfaces. A systems framework is proposed to conceptualize Affective Human Factors Design. A variety of customer and operator needs support affective design. Several theories are explained using this framework, including the theory of Flow, Situation Awareness, and Prospect theory.

1. INTRODUCTION

Affective or pleasurable appreciation in design is nothing new. This is what we admit to when we buy clothes, look at beautiful objects, or select a birthday card. But it is fairly a new area in research. And it is fairly new in human factors and in human-computer interaction. Human factors engineering has not dealt with affective issues. In the past there were two sets of dependent variables: those related to human performance (time and error) and those related to physical or psychological pain. We will now consider affect or pleasurable design. The evolution in systems performance measures has taken one step further: from performance and pain to pleasure. But this is a different perspective: It is how the user evaluates – not how to evaluate the user. We are concerned about human actors not human factors.

But to some extent we will still use the same criteria for evaluation – ease of information flow and decision making. And despite the focus on the user we have not escaped the systems problem and use of more complex evaluation criteria. They remain in the background. The most urgent in research is how to address the measurement issues and theory formation:

- How can one measure affective design?

- How can one predict affective design?

- How can one predict user and customer needs for affect?

- What are the analogies with past theories – in motivation, job satisfaction, stress research, information flow, and situation awareness?

- What kind of theories should we develop to obtain a broader understanding of human reactions to affective issues?

1.1 A Systems Model

Theories and methods in affective or pleasure-based design can be approached using a systems model of human information processing as in Figure 1. This is a traditional information processing model, where the system is broken down into three subparts: Environment, Operator/User and Machine. It is a feedback system, and the Machine

357

subsystem, through its feedback, can be considered part of the Environment. The operator input to the machine modifies the machine reaction, and the machine output will then change or upgrade the environment. The operator/user acts on the information in the environment. Information is first perceived and interpreted, decisions are made and finally there is control action. The latter may take the form of manual control of a machine, but it may also be verbal or voice control, which is then interpreted using voice recognition. This model can easily be broadened to include interaction with other people, such as a multi-user scenario.

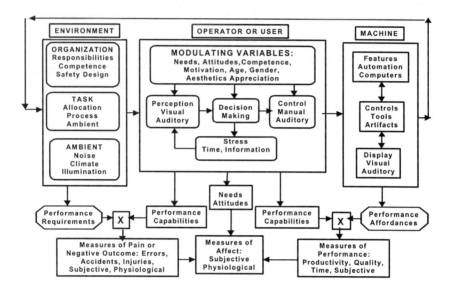

Figure 1. A Systems Model for Appreciation of Affective Design

The design variables in the system are considered independent variables and the measures of systems effectiveness are dependent variables. Thus one may measure Pain or Negative Outcome as indicated in the box in the lower left corner of Figure 1 by using various measures including: operator-user errors, accidents, injuries, physiological (stress) responses, or one may simply ask people by using subjective methods such as questionnaires or surveys. Negative effects come about because the operator is overloaded. One may then compare the performance requirements of the system and the performance capabilities of the operator.

Performance is typically measuring the amount of time it takes to perform a task. The idea is that for a well designed interface it takes less time to perform the task than for a poorly designed interface. One classical example is the time it takes for a telephone directory assistance operator to inform a customer about a telephone number. Through improvements in interface design over the last 20 years this task has now been reduced from about 45 seconds to 18 seconds. Good performance comes about because the system design is so well tuned that it affords good performance. To assess the likelihood for good performance, the affordances of the system must be compared to the performance capabilities of the operator.

Measures of affect or pleasure can be conceptualized accordingly. Users and operators have needs and attitudes with respect to interface design and their own performance (e.g. Khalid, 1999). This can include aesthetics criteria as well as the satisfaction from performing a task well. In figure 1 there are two primary measures of affect: Subjective response or expression and physiological variables. Often users or consumers express their aesthetic preference for objects in words; they will tell sales persons or family members what they like. Sometimes however affective response is influenced by the performance of the user/operator. A task which is so well designed that it affords total concentration may be perceived as very satisfying, see below. An injury caused by poor safety design at the workplace will create aggravation. Thus, as shown in the bottom of Figure 1, the user's experience of good performance as well as the negative outcome or pain, will also influence the affective state of the user or operator.

2. THEORIES AND APPLICATIONS

Below we will mention a few of the theories that have been proposed in recent research. This is by no means a full account, since this paper is limited to a few pages.

Theory of Seductive Interface Design. One important challenge in theory and application is the design of Seductive and Fun interfaces. The main goal is to please the user rather than maximize transactions and productivity. It is of course likely that users will also respond positively towards productivity enhanced measures. Therefore it is necessary to also consider the traditional measures of systems effectiveness.

The research in this area is just about starting. Kahneman (1999) brought psychological theory to "Hedonic Pleasures". The research on Hedonic Values and Seductive interfaces will be a welcome contrast to safety and productivity, which have dominated human factors and ergonomics: from pain and performance to pleasure! Jordan (2000) identified four conceptually distinct types of pleasure that a person can experience when using a product. Physio-pleasure, has to do with the body and the senses, socio-pleasures with inter-personal and social relationships, psycho-pleasure with pleasures to do with the mind (cognitive and emotional), and ideo-pleasure with pleasure to do with values.

Theory of Flow. To create an enjoyable experience in e-commerce, several researchers, particularly in management and business schools, have made reference to Csikzentmihalyi's (1975, 1990) Theory of Flow and Workflow.

Csikzentmihalyi (1975, p. 72) described the flow state accordingly: "players shift into a common mode of experience when they become absorbed by their activity. This mode is characterized by a narrowing of the focus of awareness so that irrelevant perceptions are filtered out, by loss of self consciousness, by a responsiveness to clear goals and unambiguous feedback, and by a sense of control over the environment – it is this common flow experience that people adduce as the main reason for performing an activity."

Trevino and Webster (1992) provided the following definition: "Flow represents the extent to which (a) the individual perceives a sense of control over the interactions with technology, (b) the individual perceives that his or her attention is focused on the interaction, (c) the individual's curiosity is aroused during the interaction, and (d) the individual finds the interaction interesting." The goal is, hence, to design an interface which will improve user attention and make the user feel in total control, and free from distractions from other non-related tasks or distractions due to poor usability. If so, we may have to resort to well established methods in design, such as Usability Assessment, including Task Analysis and Cognitive Walkthrough.

Nel, van Niekerk, Berthorn and Davies (1999) evaluated 20 e-commerce sites. These were classified in two ways: domestic/international and informational/transactional. Thirty-three students were used as experimental subjects. Domestic, informational facilities had better flow than domestic, transactional sites, followed by global, informational and global transactional sites. The authors concluded that the greater flow of informational sites is due to the fact that they require less effort than transactional sites. It is also easier to construct a site for a domestic community than an international community.

Theory of Economic Trade-off and Affect. Theories in economic psychology can be used to predict affect. For example, the prospect theory introduced by Kahneman and Tversky (1982) expresses that the positive personal utility of a gain (i.e. winning $100,000) is much less than the negative sentiment from losing $ 100,000. In other words a person who gains $ 100,000 on the stock market will certainly be pleased, but a person who loses $ 100,000 will feel relatively much worse. Loosing money is terrible; gaining money is OK but nothing we keep being joyful about. Thus affect will often drive investment strategies. A seasoned investor will shy away from risky investments because of the fear of loss.

This theory could be developed for study of consumer strategies in purchasing goods. Some customers are keen to pay a bit more – to buy "quality" goods. In doing so, they have minimized the future possibility that the product will break apart – in effect they have minimized future loss.

Theory of Internet Addiction. The success of Internet auctions is not only a matter of service provided. There is also an element of compulsive behavior, addiction, conditioning and affect. Many users develop addiction to web sites: they develop compulsory behavior in checking the news, financial information, and e-mail. Recent models in psychology may be used as a theoretical foundation to explain why and how users get addicted to the internet (Rabin and O'Donoghue, 1999).

3. DISCUSSION

Theories. The research studies and theories quoted in this paper are mostly from management and business schools. This is because they have taken an early interest in this area.

The Theory of Flow has the goal to design an interface, which will improve user attention and make the user feel in total control. The main goal is to construct a goal-driven task that is free from distractions, such as distractions form other non-related tasks or distractions due to poor usability. There are similarities with the theory of "Situation Awareness" (SA) (Endsley, 1995). A common application in SA has been design of cockpit displays. The information delivery in the cockpit must be so, such that the pilot may anticipate what will happen next, and react quickly. Situation Awareness has been a bit controversial, since it is difficult to define efficient measures of SA. It is likely the Theory of Flow will find it equally difficult to define operational measures. None-the-less SA has not yet been tested out in the context of predicting flow and user delight. New studies are needed here.

Future Development. There are other important challenges. One is the design of Seductive and Fun interfaces (Nielsen, 1996). The research in this area is just about starting. Monk (2000) suggested that a common element associated with enjoying the use of programs is the concept of 'engagement' — the feeling of being within the 'world' of the system. It may seem that "Theory of Flow" or "Situation Awareness" would fit into this sphere of interest. The research on Hedonic Values and Seductive interfaces will be a welcome contrast to safety and productivity, which have dominated human factors and ergonomics.

REFERENCES

Csikzentmihalyi, M. (1975) *Beyond Boredom and Anxiety.* San Fransisco, CA: Jossey-Bass
Csikzentmihalyi, M. (1990) *Flow: The Psychology of Optimal Experience.* New York:Harper Collins
Endsley, M.R. (1995). Toward a theory of situation awareness in dynamic systems. *Human Factors,* 37, 32-42.

Jordan, P.W. (2000) The four pleasures – A framework for pleasures in design. In: P.W. Jordan (Ed.) *Proceedings of Conference on Pleasure Based Human Factors Design,* Groningen, The Netherlands:Philips Design

Kahneman, D. (1999) Objective happiness. In: D. Kahneman, E. Diener and N. Schwarz (Eds.): *Well-Being: Foundations of Hedonic Psychology.* New York: Russell Sage Foundation Press,

Kahneman, D. and Tversky, A. (1979) Prospect Theory: An analysis of decision under risk. *Econometrica,* 263-291.

Khalid, H.M. (1999) Uncovering customer needs for Web-based DIY product design. *Proceedings of QERGO, International Conference on TQM and Human Factors.* pp. 343-348. Linköping University, IKP.

Mandel, N. and Johnson, E. (1999). Constructing Preferences Online: Can Web Pages Change What You Want?" Working Paper, University of Pennsylvania, Philadelphia, PA

Monk, A. (2000) Computers and Fun. In: P.W. Jordan (Ed.) *Proceedings of Conference on Pleasure Based Human Factors Design,* Philips Design, Groningen, The Netherlands.

Nel, D., van Niekerk, R., Berhon, J.-P. and Davies, T. (1999) Going with the Flow: Web Sites and Customer Involvement. *Internet Research: Electronic Networking Applications and Policy, 9,* 109-116.

Nielsen, J. (1996) Seductive interfaces, *www.useit.com/papers/*

Norman, D.A. (1990) *The Design of Everyday Things.* New York: Doubleday/Currency.

Rabin, M. and O'Donoghue, T. (1999) Addiction and Self Control. In: J. Elster (Ed.) *Addiction: Entries and Exits,* New York: Russell Sage Foundation

Trevino, L.K. and Webster, J. (1992) Flow in computer-mediated communication. *Communication Research, 19,* 539-573.

Cross Cultural Issues in Affective Design

Sheau-Farn Max Liang
Honeywell Singapore Laboratory
Honeywell Building, 17 Changi Business Park Central 1,
Singapore 486073
max.liang@honeywell.com

Due to the trend of globalization, the production and consumption of products becomes widespread throughout the world. Consequently, how to design a product/system to fulfill the needs of users in the merging international market becomes a significant issue.

This paper reviews the related literature in usability, affective taxonomy and cultural diversity. A matrix of cross-cultural affective design has been developed. This matrix will continue to evolve as we gain practical experience with our products and their users.

1   Beyond Usability

Among several definitions, the ISO defines the usability as the extent to which a product can be used by specified users to achieve specified goals with effectiveness, efficiency and satisfaction in a specified context of use (ISO 9241-11, 1994). While the effectiveness and efficiency are usually measured in economic and cognitive terms, the satisfaction is related to affective responses of users (Keinonen, 1998). Though many tools are available to measure users' subjective experience with products (Keinonen, 1998), none of these tools pay much attention to users' affective characteristics such as pleasure with products. Furthermore, while the cross cultural issues for the usability of user interface designs have been addressed (e.g., del Galdo and Nielson, 1996; Fernandes, 1995), the impact of users' cultural background on their affective responses to product designs has not been fully analyzed.

To be able to design a product/system with positive affective responses from users, it is necessary to develop a framework for the analysis of users' cultural backgrounds and affective processes during the interaction with a product.

2   Framework of Cross Cultural Affective Design

2.1 Affective Taxonomy

Bloom, et al.'s (1964) affective taxonomy categorized affective domain into five sequential stages from simple to complex. The five categories are list below:

1. Receiving: Being aware of or attending to something in the environment
2. Responding: Showing some new behaviors as a result of experience
3. Valuing: Showing some definite involvement or commitment
4. Organization: Integrating a new value into one's general set of values, giving it some ranking among one's general priorities
5. Characterization by Value: Acting consistently with the new value

This taxonomy was originally used for education objectives. However, these five affective stages are also suitable to be applied in affective design since the purpose of an affective product design is to fulfill users' needs at each stage and facilitate users to establish their values on products/systems.

2.2 Cultural Diversity

According to Webster's dictionary, culture is defined as the total pattern of human behavior and its products embodied in thought, speech, action and artifacts, and dependent upon man's capacity for learning and transmitting knowledge to succeeding generations through the use of tools, language and systems of abstract thought. This definition denotes that the culture is learned and specified to group or category (Hofstede, 1997). Several cultural variables related to attitudes, communication, and time perception have been studied (Hall, 1983; Hofstede, 1997). These variables are listed below:

1. Power Distance: The extent to which the less powerful members of groups or categories expect and accept that power is distributed unequally
2. Individualism/Collectivism: A group or category in which the ties between individuals are loose/tight
3. Masculinity/Femininity: A group or category in which social gender roles are distinct/overlap
4. Uncertainty Avoidance: The extent to which the members of groups or categories feel threatened by uncertain or unknown situations
5. Long-/Short-term Orientation: A group or category in which virtues are oriented towards the future/the past and present
6. Context of Information: The amount of stored/unspoken information in a given communication
7. Monochronic/Polychronic Time: A group or category in which the things are done one-at-a-time/many-at-a-time

The first five dimensions have been used to categorize people into their cultural groups (Hofstede, 1997). In the sense that culture is learned, users' affective experience in the interaction with a product/system can be represented as their affective learning processes for receiving, responding and valuing a new value associated with the product/system. Based on users' existed cultural backgrounds, the affective design principles can be derived at each stage of the affective taxonomy. This matrix for the cross-cultural affective design is presented in Table 1.

Table 1. Matrix of Cross Cultural Affective Design

	Receiving	Responding	Valuing	Organization	Characterization by Value
Power Distance • High • Low					
• Individualism • Collectivism					
• Masculinity • Femininity					
Uncertainty Avoidance • High • Low					
Orientation • Long-term • Short-term					
Context of Information • High • Low					
Time • Polychronic • Monochronic					

2.3 An Example

Two studies on the aesthetics and apparent usability of an ATM user interface design (Kurosu and Kashimura, 1995; Tractinsky, 1997) are reviewed to demonstrate how the proposed matrix might be used for the analysis of the cross-cultural affective design. In the original Japanese user interface design, the status of the ATM system in processing was presented as an image of a lady who bows repeatedly (Kurosu and Kashimura, 1995). However, the

363

corresponding Israeli study stated that this image was totally foreign and changed it to an hourglass instead (Tractinsky, 1997). Based on the results of Hofstede's (1997) national culture dimensions, Japan scored higher in Power Distance and Masculinity than Israel, whereas both countries had similar scores in the other dimensions. Compared to an hourglass, the image of a bowing lady reflects the culture with higher Power Distance and Masculinity. For the stages of the affective taxonomy, both studies can be categorized as the research at the stage of receiving since the aesthetics and apparent usability were the focuses of the studies.

Though this example is a post-hoc analysis, it shows how the proposed matrix might be used as the tool for the analysis of cross-cultural affective design, and how the affective design principles might be derived from the analysis. Further research on the development of cross-cultural affective design principles at each stage of the affective taxonomy is necessary in this relatively new area.

References

Bloom, B. S., Mesia B. B. and Krathwohl, D. R. (1964). *Taxonomy of Educational Objectives: The Affective Domain*, David McKay: New York.

del Galdo, E. M. and Nielson, J. (1996). *International User Interfaces*, Wiley Computer Publishing: New York.

Fernandes, T. (1995). *Global Interface Design: A Guide to Designing International User Interfaces*, AP Professional: MA.

Hall, E. T. (1983). *The Dance of Life: The Other Dimension of Time*, Doubleday: New York.

Hofstede, G. (1997). *Cultures and Organizations: Software of the Mind*, McGraw-Hill: New York.

ISO 9241-11 (1994). ISO DIS 9241-11 Ergonomics requirements for office work with visual display terminals (VDTs):- Part 11: Guidance on usability.

Keinonen, T. (1998). Usability of artefacts. In T. Keinonen (Ed.), *One-dimensional Usability- Influence of Usability on Consumers' Product Preference*. UIAH publication: Helsinki.

Kurosu, M. and Kashimura, K. (1995). Apparent usability vs. inherent usability-Experimental analysis on the determinants of the apparent usability, In *CHI'95 Conference Proceedings*, ACM Press, 292-293.

Tractinsky, N. (1997). Aesthetics and apparent usability: Empirically assessing cultural and methodological issues, In *CHI'97 Conference Proceedings*, ACM Press, 115-122.

Customer needs in Web based interaction: A macro view of usability

Ravindra S. Goonetilleke*, Colleen Duffy* and David Jacques+

* Human Performance Laboratory, Hong Kong University of Science & Technology, Hong Kong
+ Icon Medialab Asia, Hong Kong

Usable Web page design is critical for the acceptance and utility of the Internet. Pages with poor usability can have several negative effects. A recent email posting announced an "exciting, major advance" on a certain web site. In that email, the improvements discussed were described as "more powerful, simpler, faster", "easier-to-use, more logical navigation", "in-depth and in-context information" and so on. Even though all these descriptors seem important to a user, it is not clear as to whether any user can really relate to these descriptors and how related they are to each other. Our study was an attempt to understand this issue. We examined two retail and two airline industry web sites. The effect of having time limits when making transactions was also examined. Hence, there were two independent variables: Industry Type and Time Limit. Sixty-four native Hong Kong subjects aged 18 to 24 years participated in the study. These sixty-four participants were divided into two groups of thirty-two. One of the two groups had an overall time limit (15 minutes to perform all the assigned tasks) while the other group did not. Each participant was asked to use two websites (one from each industry). The analyses on the post-test questionnaires showed that web usability is governed by four important dimensions: Interaction Efficiency, Trust and Safety, Information Content and Access, and Input-Output. These four factors were able to explain most of the variation in the data. This finding has important implications in terms of website design as well as usability testing. The results may be used to simplify usability questionnaires thereby minimizing user-testing time and enhancing the user feedback information.

1. INTRODUCTION

Usable web page design is critical for the acceptance and utility of the Internet. Pages with poor usability can have several negative effects. For example, users can become frustrated from their inability to find information, confusing information displayed, pages under construction, disconnected links, lack of navigation support and so on. As a result of disorganized pages, misleading link names, long pages, and long downloading times, users' time is wasted. All of these factors not only affect the use of a certain site, but, are also largely responsible for much of the unnecessary traffic on the Internet. Poor usability of Web pages may lead to reduced visits to the site, negative feedback from frustrated users, and leave an overall negative image of the site.

Designing usable and exciting pages is not an easy task. There are various guidelines for designing usable interfaces and Web pages (Nielsen, 2000; Johnson 2000; Forsythe, Grose, & Ratner, 1998; Galitz, 1997; del Galdo & Nielsen, 1996; and Fernandes, 1995). These methods are extensive and cover various aspects of interface and Web page design. Nielsen (2000) gives a good review of fundamental errors in web page design. In addition, Fernandes (1995), del Galdo & Nielsen (1996), Lau, Shih, and Goonetilleke (2000), Shih & Goonetilleke (1998,1997), Choong & Salvendy (1998) have studied the impact of culture, language, and locale and proposed guidelines for "global" design.

Even with the use of all such guidelines, web design is almost always subject to usability testing where different characteristics are evaluated using standardized questionnaires. Every time a web site has been upgraded, there may be email announcements with words such as powerful, simpler, faster, easier to use, logical navigation capability, in-depth and in-context content and so on. The question that remains is how important is each one of these characteristics relative to each other in order to make a website exciting to use.

This study attempts to understand the "higher level" factors that are important to users.

365

2. METHODOLOGY

2.1 Stimulus Materials and equipment

Four websites were tested in this study: two retail and the other two related to airlines. The experiment was run using a Compucon computer, NEC multisync A700 monitor, and Genius Net Mouse Pro. The Netscape browser was used for most experiments. On some occasions, the Internet Explorer was used as the browser.

2.2 Participants

Sixty-four Hong Kong Chinese were participants in this study. Their Internet experience and other demographic information were collected at the beginning of the experiment. Participants were paid HKD 75 for their time. The experiment lasted between 1 ½ hours and 2 hours each.

2.3 Experimental Design

There were two independent variables in this study: Industry Type and Time Limit. The two industries were Airline and Retail. The sixty four participants were divided into two groups of thirty-two. One of the two groups had an overall time limit (15 minutes to perform all the assigned tasks) while the other group did not. Each participant in each group was asked to use two websites (one from each industry). The experimental design is shown in Table 1.

Table 1. Experimental Design. N refers to sample size.

		Industry / Site			
		Airlines		Retail	
		"AA"	"AB"	"RA"	"RB"
TIME	Yes	N=16	N=16	N=16	N=16
LIMIT	No	N=16	N=16	N=16	N=16

In order to balance carry-over effects, the order of testing was balanced within each of the two groups. This resulted in eight combinations within each group (AA-RA, AA-RB, AB-RA, AB-RB, RA-AA, RA-AB, RB-AA, RB-AB). Four participants were randomly assigned to each of these combinations.

2.4 Procedure

The experimental procedures were as follows:
1. The participants were briefed about the purpose of the experiment and provided the task instructions.
2. The participant was asked to sign the consent form
3. A pre-test questionnaire was then given to the participant
4. Each participant was asked to perform six tasks on one of the retail sites and four tasks on one of the airline sites.
5. After finishing the tasks, the participant was asked to complete a questionnaire on the usability of the website.
6. The participant was then asked to repeat steps 4 and 5 for the second site.
7. Participants received payment and signed a payment form.

The questionnaire was an attempt to evaluate as much information as possible about each site. A 5-point scale (1=Strongly Agree, 2=Agree, 3=Neutral, 4=Disagree, 5=Strongly Disagree) was used for this purpose and the subjects' opinion was sought with respect to each of the following:

1. The website is easy to use.
2. The way that information presented in the website is easy to read.
3. Each web page contains all the necessary information to do what is required.
4. It is difficult for me to find the information I want.
5. It takes too much time for me to use the website.
6. The sequence of activities required to complete a task follows my expectation.

7. I always receive clear feedback from the system when it completes a requested action (successfully or unsuccessfully).
8. The website is well organized.
9. It is easy for me to navigate through the website.
10. The response times are too slow.
11. It is not easy for me to input (enter) information in the website.
12. I need to be very careful in order to avoid errors.
13. I like to use this website.
14. Conducting transactions on this website is safe and secure.
15. I trust this website completely.
16. I trust this COMPANY/organization completely.

3. RESULTS AND ANALYSIS

The descriptive statistics of the participants is given in Table 2.

Table 2 Descriptive Statistics of participants

Gender	Number	Mean (years)	Std. Dev. (years)	Minimum (years)	Maximum (years)
Female	27	20.63	1.043	19	23
Male	37	20.35	1.274	18	24

Table 3. Analysis of Variance results. The F value and the corresponding probability (p) value (in parenthesis) are shown for each factor.

Question number Variable	Time	Industry	Time*Industry
1-easy to use	0.15 (0.7016)	**26.89 (0.0001)** Airline > Retail	0.04 (0.8480)
2-readability	0.72 (0.3969)	**16.49(0.0001)** Airline > Retail	0.22 (0.6376)
3-contains required info	0.14 (0.7063)	**8.02 (0.0054)** Airline > Retail	0.04(0.8505)
4- difficult to find	1.02 (0.3149)	**17.96 (0.0001)** Retail > Airline	2.00 (0.1602)
5-too much time	0.71 (0.4022)	**17.02 (0.0001)** Retail > Airline	0.17 (0.6835)
6 – expectation	2.01 (0.1590)	**4.29 (0.0404)** Airline > Retail	0.30 (0.5867)
7 – feedback	0.00 (1.0000)	1.39 (0.2400)	1.90 (0.1708)
8 - organization	0.29 (0.5921)	**9.71 (0.0023)** Airline > Retail	0.01 (0.9146)
9 - navigation	0.10 (0.7556)	**14.80 (0.0002)** Airline > Retail	0.27 (0.6040)
10 - response time	1.01 (0.3160)	**31.79 (0.0001)** Retail > Airline	1.46 (0.2293)
11 -inputting	3.14 (0.0789)	3.14 (0.0789)	0.10 (0.7551)
12 - erroneous	1.22 (0.2721)	**16.36 (0.0001)** Retail > Airline	0.00 (1.0000)
13- liking	0.10 (0.7495)	**31.96 (0.0001)** Airline > Retail	0.10 (0.7495)
14 - safe	1.28 (0.2595)	0.40 (0.5303)	0.14 (0.7064)
15- trust website	0.56 (0.4539)	0.10 (0.7480)	0.10 (0.7480)
16- trust company	0.04 (0.8385)	1.04 (0.3091)	0.04 (0.8385)

Due to hypothesized differences between the two industries and among the four web sites, each of the questions in the post-questionnaire was subjected to a 2 (industry type) * 2 (Time Limit/No Time limit) Analysis of Variance. The results are shown in Table 3.

Since the results for many of the variables were similar, we performed a factor analysis to identify potential groups in order to explain the variability in the data. The varimax rotated factor patterns are shown in Table 4. The emergence of 4 important factors (Eigen values greater than 1) is quite significant.

Table 4. Varimax rotated factor patterns (Factor scores greater than 0.5 are shown in bold)

	FACTOR 1	FACTOR 2	FACTOR 3	FACTOR 4
1-easy to use	**0.82**	0.15	0.24	-0.24
2-readability	**0.86**	0.03	0.13	-0.09
3-contains required info	0.33	0.14	**0.70**	-0.11
4- difficult to find	**-0.73**	-0.09	-0.16	0.05
5- too much time	**-0.70**	-0.23	0.08	0.35
6 - expectation	**0.55**	0.03	0.39	0.06
7 - feedback	0.49	0.06	-0.09	-0.15
8 - organization	**0.65**	0.17	0.45	-0.07
9 - navigation	**0.80**	0.16	0.31	-0.13
10 - response time	-0.15	-0.06	-0.04	**0.85**
11 -inputting	-0.08	0.00	**-0.63**	**0.56**
12 - erroneous	-0.40	-0.15	-0.16	**0.53**
13- liking	**0.78**	0.20	0.24	-0.23
14 - safe	0.18	**0.70**	-0.16	-0.26
15- trust website	0.17	**0.86**	0.13	-0.08
16- trust company	0.06	**0.69**	0.44	0.14
Variance Explained	5.02	1.93	1.78	1.71

4. DISCUSSION AND CONCLUSIONS

A significant difference ($p < 0.05$) exists between the Retail and Airline sites for most post-test questionnaire variables. In general, participants favored the retail sites in terms of easy of use, readability, information content, time taken, organization, navigation, response time, general liking and inputting information. For example, more than 75% of the participants felt that the two retail sites were easy to use. However, not more than 30% of the subjects felt that the two airlines sites were easy to use. Interestingly, no significant differences exist between retail and airlines in terms of safety, trusting the company or trusting the website. This is somewhat an indicator that the differences for the usability related variables are primarily as a result of site design, organization and the amount of help provided to guide the user in performing the tasks. It is true that most of the post-test questionnaire variables are somewhat correlated. To avoid repetition and to scale down the problem, it is always useful to categorize all the variables into a set of critical "dimensions".

The factor analysis helped in grouping the variables (Table 4). The variability in the data among the sites can be explained with four basic factors. The emergence of four factors (four dimensions) can give a broad view of a web site (Table 4). Factor 1 loadings are dominated by easy of use, readability, difficulty of finding information, time taken to find information, expectation, organization, navigation and general liking. Factor 2 is dominated by safety, trusting web site and trusting company. Factor 3 by information contained and the information input and lastly Factor 4 by response time, errors, and information input. Based on these variables, Factor 1 can be labeled as "*interaction efficiency*", Factor 2 as "*trust and safety*", Factor 3 as "*information content and access*" and Factor 4 as "*input-output*". Site improvement announcements that we receive through email almost every day capture most of these aspects but give little information towards a coherent model for the web-designer. Even though the Max model proposed by Lynch, Palmiter and Tilt (1999) has built-in the general system characteristics, it is not complete in terms of the interaction or usability as many factors related to interaction have to be deduced based on the block model proposed. As an introductory study exploring the characteristics of two different industry sites, this study has provided valuable insights for usability testing. More work is required not only to identify the problems in site design but also to generate designs based on task analyses and cognitive mappings.

ACKNOWLEDGMENTS

The authors like to thank Richard Ihuel, Paul Liu, Malcolm Otter, Paula Niemi, Peter Lau, Jennifer Lo, Simon Li, and Vico Ho of Icon MediaLab Asia, Hong Kong for their involvement in this study. The assistance of Roger Hung, Anita Wong and Chong Tsz Sum in the preparation work and the experimental investigations is also greatly appreciated.

REFERENCES

Choong, Yee-Yin, & Salvendy, G. (1998). Design of icons for use by Chinese in mainland China, *Interacting with Computers*, 9: 417-430.

del Galdo, E.M. & Nielsen, J. (1996). *International user interfaces*. New York, NY: John Wiley & Sons, Inc.

Fernandes, T. (1995). *Global interface design*. Boston, MA: AP Professional.

Forsythe, C., Grose, E., & Ratner, J. (Eds.) (1998). *Human Factors & Web Development*, Mahwah, NJ: Lawrence Erlbaum Associates, pp. 121-136

Galitz, W.O. (1997). *The essential guide to user interface design*. New York, NY: John Wiley & Sons, Inc.

Johnson, J. (2000). GUI bloopers : don'ts and do's for software developers and Web designers. San Francisco : Morgan Kaufmann Publishers.

Lau, W.C., Shih, H. M., and Goonetilleke, R. S. (2000). Effect of cultural background when searching Chinese menus. In *Proceedings of the 4th APCHI / 6th SEAES Conference 2000 (Ed. K. Y. Lim)*, (pp. 237-243), Amsterdam: Elsevier.

Lynch, G., Palmiter, S., and Tilt, C. (1999). The Max Model: A Standard Web Site User Model. http://zing.ncsl.nist.gov/hfweb/proceedings/lynch/index.html.

Nielsen, J. (2000). Designing web usability. Indianapolis, Ind: New Riders.

Shih, H.M. & Goonetilleke, R.S. (1997). Do existing menu design guidelines work for Chinese? In: G. Salvendy, M.J. Smith, & R.J. Koubek (Eds.), *Proceedings of the 17th International Conference on Human-Computer Interaction*, San Francisco, California, August 24-29, New York, Elsevier, pp. 161-164.

Shih, H.M. & Goonetilleke, R.S. (1998). Effectiveness of menu orientation in Chinese. *Human Factors*, 40(4): 569-576.

Towards Affective Collaborative Design

Halimahtun M. Khalid

Institute of Design and Ergonomics Application
Universiti Malaysia Sarawak
94300 Kota Samarahan, Sarawak
Malaysia

ABSTRACT

This paper addresses human factors issues of collaborative systems that may have implications for design of affective interfaces. A model for understanding human factors at the user, task environment and technology levels is outlined. Findings from the VCODE (Virtual Collaboration in Product Design) project confirmed the need for integrating affective elements into future design of collaborative systems based on user experience of two different VCODE systems.

1. INTRODUCTION

The collaboration marketplace has been evolving over the last 15 years, delivering technologies that enable coordination and information sharing, virtual meetings, and more recently virtual collocation. The promise of these technologies is to improve organisational ability to collaborate, coordinate, and share information in order to facilitate inter- and intra-organisational team work (Khalid, 2000). While these new collaborative media promise to reduce cost and time of information exchange, they have implications on collaborative design processes.

The need to bring together the process of collaboration and computing has led to the development of various collaborative systems. A major purpose of using collaborative systems is to have meaningful interactions with other people. Such interaction richness can be achieved when barriers of space, time and media/document formats are overcome when interacting with others. This includes the ability to talk, see, write and draw in both synchronous and asynchronous manner, access to relevant information, archiving of interactions for future review, and debate of issues on a global basis. However, human-machine interaction could be improved by having collaborative systems naturally adapt to their users, that is, in this case by including emotional communication together with appropriate means of handling affective information. People have strong feelings about if, when, where, and how they want to communicate their emotions (Pichard, 1997). Observations of designers at work reveal both the inherent social processes as well as their communication of emotions (e.g. Cross and Cross, 1995). Although the emotional aspects are not easy to reveal nor to design, they are essential in providing affective interfaces. Such interfaces take into consideration user experience and seek to reduce user frustration during an interaction, besides enabling comfortable communication of user emotion.

Evidently, it is still hard for people to work together through their computers because of the artificial constraints of technology, inadequate interface design, and the poor integration of conventional software with groupware. Also, there is a lack of comprehensive methodologies in computer supported cooperative work (CSCW) for measuring affect as well as for evaluating collaborative interfaces. Testing groupware is also extremely difficult. It demands more than traditional usability studies. To further confound the problem there are no agreed upon measurement metrics for deciding upon the success or failure of groupware. Measuring the end product often shows little difference because people are resilient at working together through even the most limited groupware. This paper raises three issues: what can be learnt from social computing to design affective interfaces for virtual collaboration, how to measure affect in virtual design collaboration, and is user experience necessary in using collaborative systems?

1.1 Mapping social computing to affective computing

Social computing can be defined as the use of computers and computing systems for the purpose of interaction between people for work, entertainment or even simple communication. Claims that social computing systems are able to simulate the face-to-face collaborative environment besides maximising user value by combining real time, interactive and multimedia capabilities are now challenged by Olson and Olson (2000, p. 149). They claimed that emerging collaborative technologies are pragmatically or logically incapable of replicating characteristics of face-to-face human interactions, particularly the space-time contexts in which such interactions take place. These include providing: rapid feedback, multiple channels, personal information, nuanced information, shared local contexts, informal "hall" time before and after, co-reference, individual control, implicit cues and spatiality of reference. Ackerman (2000) further argues that there is an inherent gap between the social requirements of CSCW and its technical mechanisms. The intellectual challenge, therefore, is to explore, understand and resolve this social-technical gap – a fundamental mismatch between what is required socially and what we can do technically. From the perspective of developing an affective interface for collaboration, the social characteristics of virtual collaboration must be identified.

Some key requirements for collaborative systems include: First, the need for a *common ground*. Effective communication between members of a team requires that the communicative exchange takes place with respect to some level of common ground (knowledge that participants have in common, and they are aware that they have it in common). Clark and Brennan (1991) outlined cues that various media provide for people to obtain common ground:

- copresence - same physical environment,
- visibility - visible to each other,
- audibility - speech,
- contemporarily - message received immediately,
- simultaneity - both speakers can send and receive,
- sequentiality - turns cannot get out of sequence,
- reviewability - able to review others' messages and
- revisability - can revise messages before they are sent.

Second, the need for *coupling in work*. Coupling here refers to the extent and kind of communication required by the work. Tightly coupled work strongly depends on the talents of workers, and typically requires frequent complex communication among the group members with short feedback loops and multiple streams of information. Many collaborative design tasks are tightly coupled. Technology, at least today, does not support rapid back and forth in conversation or awareness and repair of ambiguity. Third, the need to be *collaboration* and *technology readiness*. Groupware and remote technologies may be readily introduced in communities and organisations that have a culture of sharing and collaboration. In such contexts, the users would be ready for a technology that may not even deliver its promise.

In a collaborative environment, designers interact by sharing common information and they bring their own personal viewpoints to the integrated vision of the product design process. Functional, aesthetic, environmental and life-cycle issues are each characterised by different viewpoints, goals and constraints that have to be balanced with appropriate tradeoffs. Understanding a group's work practice is inherently more difficult than understanding a single person's work practice. The same individual will relate to others in a groupware context in quite different ways, depending on their personalities, the dynamics of their group, the organisational structures, and their politics.

Workspace awareness brings another dimension to understanding collaborative interactions. It helps people move between individual and shared activities, provides a context in which to interpret other's utterances, allows anticipation of others' actions, and reduces the effort needed to coordinate tasks and resources (Gutwin and Greenberg, 1999). Therefore, electronic virtual workspaces must emulate the affordances of physical workspaces, if they are to support a group's natural ways of working together, such as knowing where others are looking, relating body gestures to items in a workspace, glancing around for awareness, and so forth. For example, gaze awareness is an important aspect of collaborative work over a shared workspace. It informs of a person's focus of attention, and indicates what objects the person attends to in the shared space, and whether both people are looking at the same thing (e.g. Greenberg, 1998). This then relates to the remaining issues concerning measurement and the extent to which user experience is necessary in virtual collaboration. Both issues will be discussed in the context of the Virtual Collaboration in Product Design (VCODE) project.

2. THE VCODE PROJECT

The VCODE project is aimed at evaluating two versions of collaborative systems that support conceptual design by a concurrent engineering team. Generally, the VCODE system comprises an integrated desktop video conferencing system with collaborative multimedia (graphics, text, chat, whiteboard, data sharing), and virtual environment.

Systems. The first VCODE system uses the ISDN-based Intel Proshare conferencing technology (known afterwards as System A) while the second operates on an Internet Protocol-based Mlabs MCS conferencing system (referred to as System B). System A enables free-for-all discussion, while System B allows discussion via turn taking only. System A displays 5 video images of the team members while System B only 3. The virtual environment is a VRML Web-based environment.

Subjects. Two concurrent engineering teams, comprising 5 designers and engineers each, participated in the study. The teams were from two car manufacturing companies, referred afterwards as Team A and Team B. Team A has had prior experience with System B while Team B has not used any of the systems.

Tasks. Subjects performed four tasks, namely: to identify trade-offs in design and negotiate, to negotiate trade-offs and propose conceptual design, to discuss configuration layout of information, communication and entertainment devices in the virtual environment, and to negotiate design details by pointing to features on a whiteboard. Each task was performed for 15 minutes using both VCODE systems.

Environment. Subjects performed the tasks 'remotely' although they were located in the same physical space. The environment simulated the open plan concept of their real physical workspaces.

Method. Subjects were first briefed on the task, then trained on the VCODE systems, followed by performance of two tasks in two experimental sessions. A usability questionnaire was given at the end of each session with 15 minutes rest interval between sessions. The subjects' site behaviour was recorded via a camcorder placed at one side of the subject. The online collaborative behaviour, including verbal communication, screen information and social processes, was recorded directly on to a videocassette recorder.

The experimental design was a repeated measures factorial design with two factors (systems and teams). The data were analysed using 2 x 2 way ANOVA.

Findings and Discussion. The ANOVA tests on subjects' ratings of the systems showed significant differences between the teams in user experience as revealed below:

i. Feelings of fun and free

The collaborative system was viewed as fun to use although there were technological constraints that impeded useful interaction, $F (1,1) = 8.23$, $p<0.05$. Team B preferred System A although their typical collaborative culture prioritise turn-taking which is enabled by System B. Also, this was the first time the team had used the collaborative system, therefore the experience itself in using a new technology and the awkwardness of collaborating electronically may have induced the feeling of fun. Furthermore, this team experienced various technical failures in using System B relative to Team A. It is possible that they tried to overcome the technical failures by having fun.

System A which enables free-for-all discussion is greatly favoured over System B ($F (1,1) =38.12$, $p<.001$). This is because each member feels free to discuss whenever they like without having to wait for their turns as induced by System B. This feeling of 'autonomy' and freedom in system use is crucial in enhancing design creativity.

ii. Usability

A measure of usability is important to ensure successful implementation of technologies. The teams claimed that they will continue to use the systems when provided in their workplace ($F (1,1) = 16.33$, $p< 0.001$). However Team B appears to be more positive relative to Team A, given this is their first experience with the technology.

iii. Task performance

Clearly, the systems supported performance of design tasks significantly, in particular identifying design trade-offs ($F (1,1) =5.59$, $p<0.05$). System A allows better interaction and support over System B. There is also a significant difference between teams in terms of system effectiveness in supporting conceptual design discussion, $F (1,1)= 4.50$, $p<0.05$. Team B feels that they can accomplished the task using the systems. The same team also finds collaborating on the net is helpful in design ($F (1,1)= 14.00$, $p<0.01$).

iv. User frustration

In using collaborative systems to perform conceptual design, the user must be happy with the system. The teams expressed their frustration in terms of: unhappy with the system (F $(1,1 = 8.96$, $p<0.01$); difficult to support task needs (F $(1,1)= 5.04$, $p<0.05$); difficult to share data (F $(1,1) = 16.04$, $p<0.001$); screen cluttered with many windows (F $(1,1)= 10.24$, $p<0.01$); and cannot point to highlight the data (F $(1,1)= 5.56$, $p<0.05$). Team B found System A better due to less technical failures than System B in performing the tasks. Team A seems to find both systems equally difficult in performing all the above tasks despite their prior experience with the system. Both teams preferred a flexible and seamless mode - switching between turn-taking and free-for-all as desired by the task.

From the above, it can be concluded that design of collaborative interfaces must consider user experience. Features to consider that may result in affective user interfaces include:

- free-for-all: enable people to deliberate as and when they like freely
- open-ended: provide multiple ways of combining and recombining system elements, and of introducing new ones
- social engagement: enable people to engage in the activity
- user control: give the user direct access to the activity and its tools
- robust and forgiving: system elements should withstand "misuse" because they will be used in unpredictable ways
- physical/sensory: allow a range of gross and fine interactions
- flexible: permit use by people with range of different skill levels and in different circumstances
- personal: allow the user to personalize the system and its content in a creative way.

3. DESIGN IMPLICATIONS

From the study, we derive some generic implications for design of collaborative interfaces:

- the need to select appropriate collaborative technologies including the base system, multi-user conferencing system and peripherals;
- the need to specify top-down procedures for virtual conceptual design through identification of customer needs, functional requirements and design parameters to performing design analysis and iteration; and
- the need to structure the protocol for team collaboration in terms of rules for communication, rules for negotiation, and rules for information transfer.

Figure 1 summarises the human factors issues of collaborative systems, based on VCODE. Space precludes discussion of the model, but suffice it to say, there are various considerations in designing affective interfaces for virtual collaboration.

4. CONCLUSION

Since team collaboration is a complex process, a variety of analytical approaches will be required in order to be understood fully. This raises greater issues, such as:

- What do we know about having fun in collaboration? Job satisfaction theories such as Maslow's Hierarchy of Needs do not seem to address it. Can Kahneman's theory of hedonic pleasures provide a deeper insight?
- What methods and measurements of user experience? Presently, video-based observational techniques have been used to record social processes. Other established methods include Nagamachi's Kansei engineering and Pichard's wearables for capturing emotions. Are there computerised means for data logging affect directly?
- What are the components of emotion and affect in virtual collaboration? Are they the same for CSCW? How can affective responses be built into groupware?

The above questions beg further human factors research on collaborative systems. Some of these issues will be discussed in the Conference on Affective Human Factors Design (see Helander, Khalid and Tham, 2000).

ACKNOWLEDGEMENTS

The VCODE project is funded by a grant from the Intensified Research in Prioritised Areas (IRPA Grant No. 04-02-09-1401).

REFERENCES

Ackerman, M.S. (2000). The intellectual challenge of CSCW: The gap between social requirements and technical feasibility, *Human-Computer Interaction*, **15**, 179-203.

Clark, H.H. and Brennan, S.E. (1991). Grounding in communication. In L. Resnick, J.M. Levine and S.D. Teasley (eds.), *Perspectives on Socially Shared Cognition*, Washington DC: APA, pp. 127-149.

Cross, N. and Cross, C. (1995). Observations of teamwork and social processes in design, *Design Studies*, **16**, 143-170.

Greenberg, S. (1998). Collaborative interfaces for the Web. In C. Forsythe, E. Grose and J. Ratner (eds.) *Human Factors and Web Development*, LEA Press, pp. 241-254.

Gutwin, C. and Greenberg, S. (1999). The effects of workspace awareness support on the usability of real-time distributed groupware, *ACM Transactions on Computer-Human Interaction*, **6**, 3, 243-281.

Helander, M.G., Khalid, H.M. and Tham, M.P. (2001) *Proceedings of the International Conference on Affective Human Factors Design*, London: Asean Academic Press.

Khalid, H.M. 2000, Human factors of virtual collaboration in product design. In K.Y. Lim (ed.) *Proceedings of 4th APCHI and 6th SEAES Conference*, Amsterdam: Elsevier Science, pp. 25-38.

Olson, G.M and Olson, J.S. (2000). Distance matters, *Human-Computer Interaction*, **15**, 139-178.

Pichard, R. (1997). *Affective Computing*. Cambridge, MA: MIT Press.

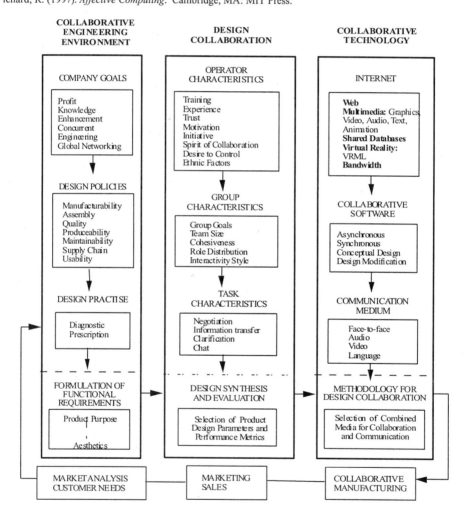

Figure 1. Human Factors of Collaborative Systems

Usability and Beyond: Experiential Aspects of Web Interaction

Peter Jämting and I.C. MariAnne Karlsson

Human Factors Engineering
Chalmers University of Technology
SE-412 96 Göteborg, SWEDEN

ABSTRACT

The aim of this study was to develop and evaluate an instrument for assessing different aspects of human emotional response to website interaction. The underlying assumption was that the design of a web-page (its appearance, usability, and content) influence the individual's emotion response, engagement, and identification. The proposed instrument consists of (*i*) a personal interview, covering the user's emotional state and personality; (*ii*) Hesselgren's Primary Emotion Questionnaire in which initial and overall emotions are addressed; and *(iii)* a main questionnaire assessing e.g. the overall impression of the page and how well it corresponds to the expectations of the user; how well the web page is adapted to the user's personal preferences; as well as some key usability qualities. The results from a user evaluation of three different websites demonstrate a correlation between interest and joy and between interest and liking. A correlation was also found between liking and efficient use. Overall, all five aspects of usability showed a correlation to different aspects of emotion. However, no correlation was found between the subjects' assessment of the appearance of the different web-pages and, e.g., liking or emotional responses. The study indicates that interest but also traditional usability qualities have an impact on emotional response and that emotion is not based merely on 'appearance'. Overall, the study indicates that the instrument has potential but that it needs further refinement in order to be truly useful in addressing human emotional response to web-site interaction.

1. INTRODUCTION

According to ISO, usability is defined as the extent to which particular users can attain particular goals with efficiency and satisfaction in a particular environment (ISO 9241-11). Several methods and tools for evaluating the usability of artefacts have been developed based on this and similar definitions. However, most often the focus has been performance and efficiency while satisfaction has not been a particular theme. Most often satisfaction seems to diminished to the avoidance of physical or cognitive discomfort.

Recently however, the issue of satisfaction and above all 'pleasure' in the use of products have gained attention [e.g. 1]. Cognition is considered to be deeply intertwined with emotion and the term 'seductive interfaces' has been used to describe designs that aim to please and attract users [2]. Actually, a negative reaction to the appearance of, e.g., an interface is considered to create barriers to its successful use and adoption [3].

Thus, satisfaction and pleasure are important aspects of all product design, of interface design, and – not least – web design. However, while the importance of factors beyond usability may have been acknowledged, their application has been noted to be less well understood. The aim of this study was to develop and evaluate an instrument which could be used to identify relevant aspects of human emotional response to website interaction.

2. METHODOLOGY

The aim was approached in an explorative, iterative way: (*i*) A *focus group interview* was completed in order to ensure the relevance of the emotional aspects of web interaction; (*ii*) A *literature study* was completed in an effort to investigate theories and methods presently available; and finally (*iii*) *An instrument* was developed and assessed in a series of *user trials*

3. BUILDING A FRAME OF REFERENCE

3.1. The focus group interview

A focus group interview was conducted with a group of seven persons (four women and three men, their ages ranging from 18 to 60). The participants discussed their individual experiences of different web pages. The group pointed out six aspects as particularly important: Appearance (i.e., colours and layout); Usability (i.e. the efficiency by which information can be found and retrieved); Contents, Emotion, Interest (i.e. that the web page covers the topic searched for); and Target group awareness (i.e. whether the user felt that the web page addressed his or her personal preferences).

3.2. The literature study

The following section summarizes some of the key references in the development of the instrument. The focus has been on theories and methods which have been judged useful in the context of web-interaction.

According to Nielsen [4], usability is a multidimensional property which includes five attributes: Learnability, Efficiency, Memorability, Few Errors and Satisfaction. Satisfaction is defined as the user's experience of the performance of the computer system in comparison with the user's expectations. It is also defined in terms of the 'pleasantness' of using the system. However, the relation between these two dimensions of the concept is not explained.

According to Monö [5], a product communicates a message through the product gestalt (i.e. the totality of colour, material, surface, structure, taste, sound, etc. appearing and functioning as a whole. This message is received, interpreted, and understood (or not) by the user and results in an understanding of the product, in feelings, and emotions. Through the Gestalt, the product describes facts (about the product), expresses properties (of the product), exhorts reactions (from the user), and identifies, e.g., origin and product area. Monö's model has proven satisfactory in formulating and transforming the product's desirable communicative properties into a product shape [e.g. 6].

Hesselgren [7] in exploring the human experience of architecture, divided this experience into four aspects: the practical; the esthetical; the ethical; and the emotional. The instrument developed by Hesselgren based on Plutchick [8] consists of eight questions covering the eight primary emotions: joy, affection, surprise, fear, sorrow, disgust, expectation, and wrath.

Some instruments have been developed specifically for evaluating web-pages. One example of an on-line instrument is W.A.M.M.I (Website Analysis and Measurement Inventory) [9]. The instrument is based on a traditional usability approach. Another instrument is M.U.M.M.S. (Measuring the Usability of Multi-Media Software) [10] which measures the usability of multimedia software. Also this instrument is based on a traditional usability approach but includes the aspect of excitement.

4. THE INSTRUMENT

4.1. Assumptions

The development of the instrument was based on the results from the focus group interview, and the theories found in the literature study. The underlying assumption was that the design of a web-page in terms of appearance, usability, and content influence the individual's emotion response, engagement, and identification.

A fundamental issue was differentiating between a user's statements directed towards the artefact and those directed towards him- or herself: that is, the user can make statements about the web-site (the *object*): "It is.....", and the user can also make statements about him- or herself (the *subject*): "I feel.....". Statements about and directed towards the object would involve the Expression, the Usability and the Content of the web page. Statements about and directed towards the subject would involve the Emotions that the user experience, the Engagement and the Identification. Identification consists of two aspects; Interest and Target group awareness. Usability concerns both the subject and the object since it is, in itself, divided into different aspects.

4.2. The Instrument

The instrument consists of a personal interview and three different questionnaires.

- *The interview* covers the user's initial emotional state and personality (see Figure 1). The questions on emotional state assess the present mood of the subject in order to avoid disturbances due to low hedonic tone. They also map the individual's strategies in dealing with expectations and disappointments. The questions on personality assess the person's leisure interest, colour preferences and interest in, e.g., art, music, and design.
- *Hesselgren Primary Emotion Questionnaire* [7]*(H.P.E.Q)*. In this questionnaire, initial and overall emotions are addressed. The eight questions concern the emotions of joy, liking, irritation, worry, depression, dislike, expectation, and surprise. The questionnaire is given to the subject in the beginning of the evaluation session in order to measure the emotions that the start page evokes, and again after the session assessing the emotional impression of the website.
- The *Main questionnaire (M.*Q.*)* is given to subject after each interaction session. The subject is to assess
    (*i*) the *overall impression* of the page and how well it corresponded to the expectations of the subject; and
    (*ii*) *identification:* how well the web page was adapted to the subject's personal preferences;
    (*iii*) *engagement:* the subject rates the time that he/she would like to spend on the page;
    (*iv*) e*xpression:* the subject is encouraged to pick out a car model which in terms of expression resembles the web page visited. He/She is also asked to motivate the choice; and finally
    (*i*) *usability aspects:* Five questions are posed covering five different aspects: controllability, efficiency, helpfulness, learnability and global usability.

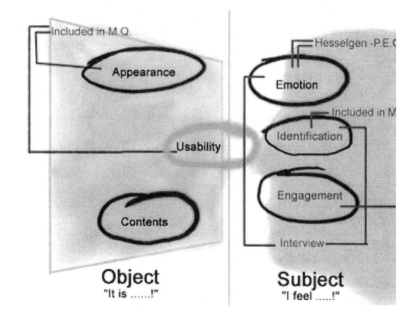

Figure 1.
The instrument

377

5. THE EVALUATION

A series of user trial was carried out in order to evaluate the instrument. After several adjustments, a 'final' evaluation was completed.

5.1. Method
Twenty subjects participated in the evaluation. Criteria for inclusion were age (20-40 years old), gender, and Internet experience (less experienced/ experienced). The pages used in the evaluation were (*i*) the page for Nike's football (soccer) gear [11] and the pages of the Swedish banks (*ii*) SEB [12] and (*iii*) Nordbanken [13] respectively. The bank-on-line pages were to form a basis for a comparison of different web designs while the soccer page was chosen in order to indicate differences in user response due to differences in interest.

After the initial interview, each subject evaluated all three pages in a random order and filled in the H.P.E.Q. and the M.Q. for each page. The interaction time for each web page was limited to ten minutes. Each subject formed their own tasks based on their individual interest in the respective page.

Correlations between e.g. appearance and emotional response were calculated using the Spearman Correlation Coefficient [14].

5.2. Results
The results indicate differences between the subjects' assessment and response to the different designs of the bank sites. However, and somewhat surprising, no correlation was found between the subjects' assessment of the specific *appearance* of the different web-pages and *liking* or between appearance and other emotional responses.

One assumption was that the subjects' approximation of the *time* they wanted to spend on the web page would indicate their *emotional experience*. The study showed a correlation between time and liking (r_s=0.603, $p<0.01$) and between time and expectation (r_s==0.630, $p<0.01$) for one page: www.nikefootball.com. A correlation was also found between time and joy (r_s=0.828, $p<0.001$) for one of the other pages: www.nb.se.

Correlations between *liking* and *efficient use* were found for all three web pages: for Nike (r_s=0.592, $p<0.01$), for NB (r_s=0.665, $p<0.01$) and for SEB (r_s=0.626, $p<0.01$). The results for the Nike page and the SEB page show a negative correlation between dislike and hyperlink predictability (Nike, $r_s = -0.617$, $p<0.01$; SEB, $r_s = -0.726$, $p<0.001$), as well as between dislike and learnability (Nike, $r_s = -0.643$, $p<0.01$, SEB ($r_s = -0.727$, $p<0.001$). Overall, all five aspects of usability demonstrated a correlation with different aspects of emotion.

The rated *interest* in the web page www.nikefootball.com correlated with the experience of *joy* (r_s=0.822, $p<0.001$) and *liking* (r_s=0.756, $p<0.001$). However, the rated interest in the specific web page did not necessarily correlate with an interest in football or vice versa. Instead, interest seemed to be related to whether or not the users felt 'addressed' by the page, felt as though they belonged to the target group (cf. target awareness). Actually, the 'footballs fans' were not appealed by the site and complained about the fact that it did not contain enough information about Nike's products or about the soccer league and the players.

6. DISCUSSION AND CONCLUSIONS

The aim of this project was to develop and evaluate an instrument which could be used to identify and assess aspects of human emotional response to web-site interaction. The assumption was that the design of a web-page in terms of appearance, usability, and content influences the individual's emotion response, engagement, and identification.

It was somewhat surprising that no correlation was found in the evaluation between appearance and liking. One explanation could be that the questions on appearance were not the appropriate ones to elicit this dimension. Consequently, the influence of appearance and expression on emotional response needs to be investigated further. However, the results suggest that appearance and expression are not the only qualities to consider. Actually, there may be other aspects than appearance that are given, as yet, a higher priority by the user. For instance, the results showed that usability in terms of efficient use is connected to a positive emotional experience. The claim by Jordan [1], that usability has become a 'dissatisfier' only in users' evaluation of products, could therefore be debated. At least is usability 'on-the-web' not yet taken for granted.

Furthermore, the study shows that the content of a web page influences the interest of the user and could result in positive emotions. However, there must be a specific match between the user's expectations of the content and the actual content in order to result in a positive emotional experience. Expectation seems to be a key dimension in the users' responses. This aspect needs to be studied further, as should the effect of the users' emotional state before an evaluation session.

Overall, the results indicate a "hierarchy of needs" [cf. 15]), where certain requirements (e.g. for usability and content) may have to be satisfied to a certain level before other aspects of use (e.g. appearance) are considered by the user. However, whatever the 'level of satisfaction', there will always be an emotional response influenced by these, and most probably other qualities. Nevertheless, most existing instruments for measuring the usability of web pages do not consider the emotional aspects of web interaction.

The instrument described is one attempt to address dimensions of web-interaction beyond usability. The instrument shows potential but needs further development. For instance, it is possible that the instrument should be treated as several different instruments which can be used separately or combined in different ways depending upon which dimension is to be investigated. However, a holistic view eliminates the risk of over-emphasizing one dimension and helps to discern relations between different dimensions. The relationship between users and artefacts is complex and built upon several different dimensions, the emotional aspects being one [16]. The challenge for the future is not only to reach an understanding of what these dimensions are and how to assess them. It is to understand how the design of the artefact, the interface, the web-page will influence them. These factors have to be understood if we are to design interfaces and web-sites which attract users and which users can and want to use.

ACKNOWLEDGEMENTS

The authors wish to express their gratitude to Sharon Kühlman-Berenzon, the School of Mathematics and Computing Sciences Mathematics, Chalmers University of Technology.

REFERENCES

[1] Jordan P.W. (2000). *Designing Pleasurable Products.* London: Taylor & Francis
[2] Skelly at http//www.designhappy.com (00-09-15).
[3] Weeimer J. (1995) *Research Techniques in Human Engineering.* Prentice Hall, N.J.
[4] Nielsen J. (1993) *Usability Engineering.* San Diego. Academic Press
[5] Monö R. W. (1997) *Design for product understanding. The aesthetics of design from a semiotic approach.* Stockholm. Liber
[6] Wikström L. (1996) Produktens budskap. Metoder för värdering av produktens semantiska funktioner.
[7] Hesselgren S. (1985) *Om Arkitektur.* Lund. Studentlitteratur
[8] Plutchick, R.(1962). *The Emotions: Facts, Theories and a New Model.* New York: Random House
[9] HFRG Ireland and Nomos Management AB Sweden, (1998) http://www.wammi.com/
[10] http://www.ucc.ie/hfrg/questionnaires/mumms/index.html
[11] http://www.nikefootball.com
[12] http://www.nb.se
[13] http://www.seb.se
[14] Altman D.G. (1991). *Practical Statistics for Medical Research*, Chapman & Hall/CRC
[15] Maslow A.(1970). *Motivation and personality,* 2nd ed., New York: Harper & Row
[16] Karlsson MA (1996). *User requirement elicitation. A framework for the study of the relationship between user and artefact.* Department of Consumer Technology, Chalmers University of Technology, Göteborg.

Designed for Enablement or Disabled by Design?
Choosing the Path to Effective Speech Application Design

Jennifer Lai[a]
Savitha Srinivasan[b]

[a]IBM T.J. Watson Research Center, 30 Saw Mill River Road, Hawthorne, NY 10532

[b]IBM Almaden Research Center, 650 Harry Road, San Jose, Ca 95120

ABSTRACT

Speech technology is often touted as the great equalizer, simplifying computer interfaces and making computers accessible to all. Since speech is a form of communication that we learn early and practice often, many have claimed that speech is by it's vary nature, a totally intuitive way to interact with computers. When computers were first introduced, the required interface was a series of arcane typed commands, known only by a small percentage of the population. With the arrival of the graphical user interface (GUI), computers became more accessible to a larger number of people since commands could be found (at worst) by an exhaustive search through the dropdowns in the menu bar. However with a spoken interface, in many ways we are back to the old model of hidden commands. If you hand somebody a phone and tell him there is a computer on the line, waiting for him to speak, the user is immediately confronted with the question of what is an acceptable command for the system. The job of helping the user past this first hurdle and onto the path towards successful task completion falls to the speech interface designer. Effective speech application design depends on the use of carefully crafted system prompts that cue the user what to say, as well as the use of a grammar designed with an understanding of how users speak in the domain of the application. Design considerations also depend on the type of speech application being created.

1. SPEECH RECOGNITION APPLICATIONS

Traditionally, speech recognition applications may be seen as belonging to one of three categories: dictation systems, navigation or transactional systems and multimedia indexing systems. Each of these application areas offers a set of benefits to the user along with a set of design challenges imposed by the limitations of the technology. Design requirements often vary significantly based on the targeted user group and usage scenario. In the first two categories, speech recognition is used as an input mechanism, occasionally in combination with one or more traditional input modalities such as a keyboard, mouse or telephone keypad. The design issues in this case revolve around maximizing the potential for accurate input and repairing errors easily when they do occur. In the third category, audio-indexing applications, speech recognition can be utilized in two ways. First, it can be used as an input mechanism to recognize the spoken query (as in the first two application categories). Secondly, speech recognition can be used to transcribe the audio contained in the multimedia recordings (e.g. broadcast news or meeting notes). The words in the transcript are used to index into the media so that spoken keywords can serve as search terms to retrieve the relevant audio/video segments. The ultimate goal in audio-indexing applications is to search through an multimedia collection by voice. However, since the speech recognition accuracy is often limited by a multitude of factors such as recording fidelity, speech spontaneity (e.g. umms and ahhs, elongation and shortening of syllables), the audio domain (for example technical versus general), and background noise conditions, the current speech recognition technology is only just beginning to make textual queries on multimedia documents practical. The design of such systems must therefore, support the limits of the technology to yield acceptable accuracy in retrieval while correctly managing user expectations.

2. SPEECH APPLICATION DESIGN

Creating a speech-only interface presents design challenges that are different from those presented in a purely graphical environment. While some basic design rules apply equally to the field of Voice User Interfaces (VUIs) as they do to GUIs (e.g. know your user), there are additional restraints imposed on the speech application designer which stem from the need to work around the limitations of the technology. Designers new to the process of creating speech applications often find themselves at a loss because there are no concrete visuals to sketch. Designing speech interfaces involves understanding conversational styles and the different ways that people use language to communicate. In this paper we will discuss the fundamental issues relating to the practice of sound design in each of these three speech application classes. We will also discuss ways of involving users along the way.

2.1 Dictation Applications

With dictation applications the words spoken by a user are transcribed verbatim into written text. Such applications are used to create text such as personal letters, business correspondence, or even email messages. Usually the user has to be very explicit, specifying all punctuation and capitalization in the dictation. Examples of commercial products for these applications are L&H Dragon Naturally Speaking, and the IBM ViaVoice product. Dictation applications are often referred to as multi-modal applications since they combine the traditional GUI modalities (visual output, mouse and keyboard input) with spoken input and often, spoken output. While many of the technological hurdles in large vocabulary speech recognition have been vanquished (continuous speech, speaker-independent, modeless products are now the norm), using speech to create text can still be a challenging experience. Many users have a hard time getting used to the process of dictating (whether it be to another human or to a machine) and the most successful ones often compose their thoughts before switching the microphone on.

As the user speaks, the text appears on the screen and is available for correction. Correction can take place either with traditional methods such as mouse and keyboard, or with speech. The problem with correcting a speech recognition error with speech, is that it is likely to be wrong the second time as well. This is especially true if the spoken word is not the system's vocabulary. Most speech systems can only "understand" or "hear" words that are contained in the vocabulary. A study that looked at error-correction patterns (Karat, 1999) for these dictation systems found that novice users can get caught is a cascade of errors when correcting an error. Part of the explanation for this is that novices were more likely to use speech to correct the error than experienced users, who tended to fall back on keyboard usage for error correction. Novices also often felt compelled to "correct work as they go" rather than dictating a portion of text and then dealing with corrections after the fact. This further broke up their ability to compose text, and occasionally confounded errors.

When designing a dictation application, especially a niche application (e.g. dictation for radiologists) the designer will do well to first observe the users in the environment where they will be dictating. There are many environmental factors that will affect how well the application will work for the particular user group. In the case of radiologists, ambient noise is a big challenge. Radiologists dictate in a reading room which has several doctors dictating at the same time, support staff moving in and out of the room, as well mechanical noise from large light boxes that display the films needing interpretation. Other user-specific issues are the doctors' willingness and motivation to switch to a speech-based application, as well as levels of computer literacy and acceptance. While a dictation application is best suited to a group of users that is already comfortable dictating (e.g. doctors or lawyers), a study of the design and usage of a dictation application for radiologists (Lai, 1997) found that not all doctors embraced the use of speech technology. Even though the application was successful in decreasing the turnaround time on reports in one hospital from 50 hours to a few minutes, some doctors could not accept the change in roles that it required. Radiologists objected to assuming the additional task of editing and correction that had previously been the responsibility of transcriptionist.

In addition to observing and interviewing users on location prior to the design of the application, field trials are recommended as soon as the dictation application is functional. Speech recognition accuracy (i.e. the number of words that are correctly recognized) can usually be improved by collecting data on location with the user group and refining the data models that drive the recognizor. Domain specific language models, and acoustic models tuned to the circumstances of usage go a long way towards assuring the highest level of accuracy possible. Another potential problem that needs to be evaluated through usability trials is how well users can detect what mode they are in. Dictation applications have two modes. In the first, everything a user says is transcribed verbatim. In the second, speech is used to tell the application what to do (e.g. "open my reports"). It is not uncommon for users to think they are in command mode when they are actually in dictation mode, and vice a versa. The chaos that results from this confusion is easy to imagine; the application jumping from one command to another as it tries to make sense of the stream of dictation.

Field trials with multi-modal applications are helpful because they allow the designer to observe how effective the visual feedback is with regarding to spotting recognition errors when they occur. While it can be cumbersome for the user to correct errors, it can sometimes be worse if the error goes undetected. Take for example a radiology report where the dictated text of "there are no signs of cancer" is instead recognized as "there are signs of cancer" (referred to as a deletion error). The two primary types of recognition problems that need to be communicated to the user are recognition failures and errors. When a recognition error occurs, the system believes that it has understood the speech, but it understands incorrectly. In the case of a recognition failure nothing happens at all. Since feedback of recognition failures is useful, designers should evaluate through user trials if the feedback mechanism they selected is working effectively. If a user asks the system to do something and the system fails to understand the command, feedback should be immediate. Both visual or auditory methods can be used, or some combination of these. While the user will eventually determine that the requested action was not taken, precious time is lost while the user sits there expectantly waiting for the system to respond correctly.

Finally, field trials for dictation applications are useful in determining if the users are having problems with the microphone. Reductions in accuracy due to such problems are very common. While great progress has been made in developing noise-canceling microphones which do a great job of dealing with average levels of background noise, these microphones are still very sensitive to problems related to the position of the microphone in relation to the mouth. If the user is holding the microphone too close to the lips, or too far, or has the microphone angled away from the mouth, there will be a sharp drop in the accuracy rate.

2.2 Navigation or Transactional Applications

Unlike dictation applications, speech is used in transactional (sometimes referred to as command & control) applications to navigate around the application or to conduct a transaction. For example, in this category of applications, speech can be used to purchase stock, reserve an airline itinerary, or transfer bank account balances. It can also be used to follow links on the web or move from application to application on one's desktop. Most often, but not exclusively, this category of speech applications involves the use of a telephone. The user speaks into a phone, the signal is interpreted by a computer (not the phone) and an appropriate response is produced.

For designers, this category of applications often presents some of the greatest challenges. Unlike desktop applications, where the interaction designer has a entire basket of tools at hand for communicating information to the user (e.g. choice of color, font, dialog boxes), with transactional speech applications, the designer is limited to the creation of the system prompts. Careful crafting of the prompts, (i.e. what the system says to the user) is critical to the success of the application. Not only are the prompts required to drive the interaction to a successful conclusion, but the prompts are also the only way to cue the user as to what can be said.

Transactional applications usually rely on the use of a grammar. A grammar is the explicit definition of the phrases and sentences that will be "understood" by the recognizor. If a user speaks outside of the grammar his sentence will be not be "heard" (i.e. the engine will return a failure) or some other sentence will be returned. One that is legal but that the user did not say. The software designer needs to make sure that he or she considers and defines all the possible ways that a person could say things. The other problem with grammars is that if the user pauses for too long while speaking, the sentence will be rejected even if a legal construct was used. Given the need to speak within the grammar, one of the hardest things in the design of transactional speech application is letting users know what they can say.

There are two basic styles of prompts. The first style is often referred to as explicit or directive. In this case, the interaction designer directs the user to say certain words or phrases. For example, "*Say next, delete, or reply.*" Even when the key word "say" is not present in the prompt, explicit prompts tend to elicit a single piece of information from the user, as in the prompt "*call whom?*" or "*what city are you leaving from?*" Systems that use this style of prompts are called directed-dialog systems. Most systems that have been built to date use explicit prompts since this is the easiest way to deal with unpredictable accuracy levels. This type of prompt helps to constrain what the user says and is appropriate to use when either the cost of making an error is high or the recognizor can only handle a small vocabulary robustly. Explicit prompts however, do not make for the most natural type of interaction. On the other end of the spectrum, are the conversational speech systems which use implicit prompts. These prompts provide a more natural interaction. A conversational airline reservation system might start off by asking the user what his travel plans are. The tradeoff, of course, is accuracy. A conversational prompt is more open-ended and therefore the likelihood of an error occurring is higher. For example starting an application with the prompt: "*How can I help you?*" would most likely generate a high degree of errors. This invites people to say just about anything. A more constrained prompt might be "*Which flight are you interested in checking on?* " The system must then be ready to accept phrases such as "*I'm interested in a morning flight from New York to Los Angeles.*" In addition to these two primary categories of prompts there are other important prompting techniques such as tapering prompts or incremental prompts (Yankelovich, 96). In the first technique the designer shortens the prompts over time, assuming that the user needs less information as he becomes more familiar with the interaction, and in the second, additional information is provided as the system deems that it is necessary.

The best way for a designer to determine if she has done a complete job defining the grammar is to spend time listening to the users speak in the domain of the application. Often the application that is being designed already exists in some other form. Perhaps a voice response unit is currently being used, or a portion of a call center application is being automated. Listening in on the calls helps to define the tone of the interaction as well as the vocabulary that is used. Once a trial grammar is in place, the designer should have users test it to see which common constructs have been forgotten. Sometimes, even before the grammar and the speech system are implemented, trials can take place using a Wizard of Oz (WOZ) system. With a WOZ, the users are led to believe (or asked to pretend) that they are interacting with a fully functioning speech system. In reality, the "wizard" is listening to the speech and generating the appropriate system response. When using a WOZ it is important to remember to simulate several speech recognition errors. Since graceful error recovery is an important part of every speech system, this should be a well exercised path in the design !

2.3 Multimedia Indexing Applications

In multimedia indexing applications, speech is used to transcribe words verbatim from an audio file into text. The audio may be part of a video. Subsequently, information retrieval techniques are applied to the transcript to create an index with time offsets into the audio. This enables a user to search a collection of audio/video documents using text keywords. Digital audio and video are becoming increasingly popular such that large collections of multimedia documents can be found in diverse application domains such as the broadcast industry, education, medical imaging, and geographic information systems. Retrieval of unstructured multimedia documents is a challenge today and requires content-based retrieval where the content of the document is examined for the presence or the absence of an object, of words or phrases, or a visual action. Therefore, cataloging and indexing of audio/video has been universally accepted (Wactler, 1999) as a step in the right direction towards enabling intelligent navigation, search, browsing and viewing of digital audio/video.

One approach to multimedia retrieval is to apply image retrieval techniques to key frames extracted from the video. In this approach, an image is posted as a query, and similar images are retrieved. There are two reasons why this approach, in general, has not become popular yet . First, in most practical situations the user does not have such an image handy to formulate the query. Second, the state of the art in content-based image retrieval has not yet reached the semantic level desired by most users. Rather, it is typically done in a feature space, such as color histograms, color layout, color blobs , texture and shapes (Flickner 1995). A more popular approach to video retrieval is to search the audio transcript of the video using the familiar metaphor of free text search (Jones, 1996; Srinivasan, 2000). The indexed transcript provides direct access to the semantic information in the video.

While searching the audio using text keywords proves to be quite efficient, browsing the video is much more time consuming than browsing of text. This is because the user has to play and listen to each of the retrieved videos, one by one, unlike with text where a quick glance at the result page is often sufficient to filter the information. In this case, it is more efficient to browse the visual portion using a video segmentation technique (e.g. a video storyboard). A few pages of storyboard, each showing ten or more key-frames, can cover one hour of video by showing the main visual scenes contained. Therefore, a popular approach to multimedia retrieval is *"Search the speech, browse the video"* where the video and audio are treated as two parallel media streams of information that are related by a common time line. Thus, the audio stream is used for searching and the video stream for quick visual browsing in a complimentary manner to provide the desired video search functionality.

A well known issue in speech indexing is the concept of in-vocabulary terms and out-of-vocabulary terms. This corresponds to the words that can be "understood" or "heard" by speech applications. In-vocabulary terms can be understood by speech systems, and therefore they can be retrieved using keyword search. In contrast, out-of-vocabulary words cannot be "understood" and will be misrecognized instead as a similar sounding in-vocabulary word. This implies that out-of-vocabulary words cannot be retrieved using text keywords. This problem is typically addressed by creating an index of the sub-word or phonetic representations of a word. When presented with a text keyword, it is first translated into its equivalent phonetic representation and the corresponding phonetic index is searched. The accuracy of phoneme recognition however, is limited, particularly in the case of short words (Ng, 1998). Therefore, in practice, combined indexes that comprise of a keyword index and a phonetic index are used to provide the best search performance.

It has been shown that word error rates can vary between 8-15% and 70-85% depending on the domain and tuning of the recognition engine. The 8-15% error rates correspond to standard speech evaluation data and the 70-85% corresponds to "real-world" data such as a one hour documentary and commercials. In general, it has been shown that for an speech recognition error rate of about 30%, a retrieval system can achieve about 80% of the effectiveness of text search engines that operate on perfect text documents (Wactlar, 1999). This has been validated for multimedia collections of a few hundred hours and is as yet unknown for larger document collections.

It is important to have realistic expectations with respect to retrieval performance when speech recognition is used. The user interface design is typically guided by the *"Search the speech, browse the video"* metaphor where the primary search interface is through textual keywords, and browsing of the video is through video segmentation techniques. While the requirements of specific applications vary, our experience indicates that the precision of search results are more important to a user than the recall, i.e. the accuracy of the top-ranked search results is more important than finding every relevant match in the audio. Therefore, the ranking of search results may be biased to address this. In general, since the user does not directly interact with the indexing system using speech input, standard search engine user interfaces are seamlessly applicable to speech indexing interfaces. However, the following design guidelines can result in a more successful speech indexing system: First, since the transcript of the speech is not accurate enough to result in fully readable grammatical text, it is not advisable to display the entire transcript as part of the search results since it can cause a negative impression on the user. Secondly, search interfaces can "guide" the user in the selection of search terms by providing a list of search terms from the transcript.

3. CONCLUSION

Successful speech applications, like other types of applications, need to incorporate both an understanding of the user and the circumstances of use into the design. Each of the categories discussed in this paper emphasizes the need to pay attention to the different characteristics of the interaction when doing system design. The primary difference between speech applications and other applications which use a more mature technology (e.g. keyboard input) is that the designer must take into account the limitations of the technology. Without an understanding of what causes errors in speech, and a focus on both error prevention strategies and graceful error recovery, the resulting application will be unusable at best. For dictation applications, several strategies apply. First of all, study the users in the location where they will be dictating. Pay particular attention to ambient noise levels, and frequency of interruptions. Design the interface to support these aspects of the interaction, as well as supporting the users task requirements (e.g. eyes-free operation). For transactional applications, the designer must understand and reconstruct through the grammar the way the users speak in the domain. Observations, interviews and a wizard of oz setup are all helpful in achieving this goal. Finally, for multimedia indexing applications, the most important design consideration is optimize speech search performance combined with synergistic user interface considerations that further maximize this.

References

1. Flickner, M. et al. (1995), Query by image and video content: The QBIC system, IEEE Computer, Vol 28, No. 5, pp.23-32.

2. Jones, G. J. F., Foote, J. T., Spärck Jones, K., and Young, S. J. (1996), Retrieving Spoken Documents by Combining Multiple Index Sources. In *Proceedings of SIGIR 96, pp. 30-38*, Zurich, Switzerland.

3. Karat, C., Halverson, C., Horn, D., and Karat, J. (1999), Patterns of Entry and Correction in Large Vocabulary Continuous Speech Recognition Systems. In *Proceedings of CHI '99: Human Factors in Computing Systems*, pp. 568-575, Pittsburgh, PA.

4. Lai, J., Vergo, J. (1997), MedSpeak: Report Creation with Continuous Speech Recognition. In *Proceedings of CHI '97: Human Factors in Computing Systems*, Atlanta, GA.

5. Ng, K. and Zue, V. (1998), Phonetic Recognition for Spoken Document Retrieval. In *Proceedings of ICASSP 98*, pp. 325-328.

6. Srinivasan, S. and Petkovic, D. (2000), Phonetic Confusion Matrix Based Spoken Document Retrieval. In Proceedings of SIGIR-2000, Athens, Greece.

7. Wactlar, H., Christel, M., Gong, Y. and Hauptmann, A. (1999), Lessons Learned from Building a Terabyte Digital Video Library. In IEEE Computer.

8. Yankelovich, N. (1996), How do Users Know What to Say? In *ACM Interactions*, Volume 3, Number 6, November/December.

Inclusive Design at the Royal College of Art (RCA)

Roger Coleman, Jeremy Myerson

The Helen Hamlyn Research Centre, Royal College of Art, London – www.hhrc.rca.ac.uk

ABSTRACT

By creating inclusive design exemplars it is possible to envision a user-friendly future, in which design and technology include rather than exclude, offering improved quality of life to people of all ages and capabilities. Delivering this vision requires accessible guidance, design strategies and user-research methodologies that can support and empower designers and managers throughout the development process. The RCA has a longstanding interest in these issues, dating back to the international 'Design for Need' conference in April 1976 and an R&D program which delivered the standard UK hospital bed design. In 1991, the DesignAge program was established at the RCA to 'explore the design implications of ageing populations', and in January 1999, the Helen Hamlyn Research Centre (HHRC) was launched with a broad social mission to explore inclusive design in a period of rapid social and technological change. Key aspects of this work are the involvement of groups of 'critical' users, primarily older and disabled people, and industrial and voluntary sector partners who can put the research and designs into practice.

1. An emergent trend

Many groups are currently excluded from the mainstream design process. This design exclusion takes several forms: older and disabled people suffer from it; so do certain economically vulnerable groups such as manual home-workers, and those marginalised by changing technologies and work practices. In a period of rapid change, bringing people from the margins to the mainstream through inclusive design is important not just from the perspective of social equality but also for business growth through new products and services. In recent years there has been a shift in attitude, away from marginalising disabled and older people as special cases requiring special design solutions, and towards integrating those people in the mainstream of everyday life through a more inclusive approach to the design of buildings, public spaces and more recently products and services (Coleman 1994a, 1997). This has been an accompanied by accessibility legislation and related building codes, and reflected in a growing convergence of consumer demand (and dissatisfaction), governmental concern and business interest around inclusivity. The trend has been driven by awareness of the implications of population ageing; by the challenges of e.g. creating an integrated European Community without sacrificing cultural diversity; and by the militancy of disability organisations. It is not uniform, and there are marked differences between countries, resulting from cultural and historic factors, and in particular from the contributions made by individuals and organizations.

2. Living Longer

Since 1900, UK life expectancy has increased on average by some 2.5 years per decade, one of the great achievements of the 20th Century (Kirkwood 1999). As people age, they become more diverse as individual life-courses give rise to divergent experiences, interests, activities and capabilities. In the context of aging populations, such diversity will increase, in particular as people explore the new possibilities opened up by 25 or more years of life-expectancy (Laslett 1996). With age, people change physically, mentally and psychologically. For most people these changes involve multiple, minor impairments in eyesight, hearing, dexterity, mobility and memory (Haigh 1993). At present, such changes have a significant impact on older people's independence due to an unnecessary mismatch between the designed world and their older capabilities (Laslett 1998). Unprecedented growth in the older age groups will challenge common assumptions both about the nature and spread of disability, and about the level of capabilities that products and services should be designed for if they are to meet the needs of the majority of the population. For example, a wheelchair-friendly world is not necessarily an age-friendly world – although the majority of wheelchair users are old, the great majority of older people are not disabled, and those that are have very different mobility requirements to younger wheelchair users.

By 2020 close to half the adult population of the UK will be over 50 years old, while 20% of the inhabitants of the United States and 25% of those of Japan will be over 65 (Coleman 1993), yet the youth market continues to be looked to as the economic motor for the future. In the UK every additional consumer added to the marketplace over the past century has been an older person. and an opportunity has been missed to develop products and services for the only consumer sector with real growth potential. In the future, mainstream consumer markets will have to concern themselves with the substantial rise in the number of people who are less than able-bodied yet wish to enjoy an active and independent lifestyle. In this context it is increasingly accepted that as many people as possible should have physical and cognitive and intellectual access to the world around them. The problem is that much of the design community has yet to understand what this means, or how to achieve it. The challenge is to create a supportive environment of buildings, products and services and interfaces

(Brouwer-Janse et al 1997) that make it possible for everyone to live independent and fulfilling lives, for as long as possible. The more effectively this can be achieved, the less strain will be placed on social and welfare systems, and the more older people will be encouraged to spend the now considerable wealth they control on the goods and services that deliver independence and quality of life. A further challenge is therefore to develop a consumer offer that closely matches the aspirations of older people.

3. Design for Our Future Selves

The RCA has a long-standing interest in social issues. In 1976 it staged the Design for Need conference, which brought practitioners from around the world to London (Bicknell & McQuiston 1977). Subjects ranged from ecology, environmental policy and the recycling of materials, to self-build housing, workplace design, designing out disability, equipment for emergencies and disasters, and beyond that to design education in developing counties. Two of the closing papers brought the overall thrust of the conference into sharp focus. Mike Cooley talked about 'Design for Social Use', and Victor Papanek's subject was 'Because People Count: Twelve Methodologies for Action'. Both argued for a shift in emphasis from designer to user, and from the individual artefact to the broader social context of design. This concern with the social aspects of design has been further developed through a collaboration between the Helen Hamlyn Foundation (HHF) and the RCA, which built on a seminal exhibition 'New Design for Old' held at the Victoria and Albert Museum in London in 1986 under the aegis of the HHRC (Manley 1986). An international team of designers produced a collection of furniture, clothing, consumer-durables, bathroom and kitchen fittings, door furniture, and personal items, demonstrating that addressing the needs of older people could be a route to innovation, and that lessons learned from one sector of society could be applied to design as a whole.

The next step was DesignAge, an action-research program exploring the implications for design of aging populations. This presented a unique opportunity to take ideas from the margins to the mainstream of design thinking and practice. Because DesignAge was not a teaching program, it could develop as a cross-disciplinary activity engaging with postgraduate students from fashion and textiles to vehicle design, and from photography to industrial design and engineering (Coleman 1994b). The theme of the program 'design for our future selves' was chosen to encourage young designers to engage with ageing as a natural part of the life-course, focusing not on older people *per se*, but on how they saw themselves in the future. The program concentrated on the convergence between social and commercial imperatives; on arguing the case for age-friendly design; and on encouraging industry and the design profession to recognise the opportunities offered by an older consumer market. It has done this by bringing together factual and trends information, along with design relevant tools and guidance (Coleman 1999). The results have included exhibitions, conferences, competitions, new designs and products, publications and a growing body of information about the subject gathered together in a special collection at the RCA.

A central thrust of the program has been to encourage young designers to work directly with older people. This has been achieved through a collaboration with the University of the Third Age (U3A), a fast-growing and self-organised association of retired people. U3A members attend regular User Forums at the RCA, where they meet students, participate in focus groups and other research activities, and discuss consumer issues with professional designers and industry managers (Coleman 1997a, b). Bringing older users into the college gives design students an opportunity to talk through ideas, develop concepts, and later test prototypes and research specific issues of styling, aesthetics and usability. The RCA students and U3A members build up a high degree of trust, and all find the experience interesting and enjoyable. This interaction with older consumers gives students a rapid insight into how to develop appropriate products and services, and into the pitfalls that await them if their approach can be seen to be in the least patronizing. Alongside working with older users, an important strand of the DesignAge program has been a regular competition open to final year MA. The competition began in 1994, and the many entries have served to establish a body of work that gives substance to the concept of 'design for our future selves' (Myerson 1999). The diversity of the competition projects supports this concept by demonstrating how traditional objects can be redesigned, and new concepts developed. Some of the designs represent a sense of freedom and escape, such as yachts or cars or floatation platforms to view the underwater world from. But most home in on the practicalities of restricted lives, with ideas to support household chores, reading, sitting, standing and staying warm, and extending into humanising hospitals and residential care. By no means all of the projects are aimed directly at the mature market, but in tackling such issues as living alone, they bring the age issue into focus.

4. Inclusive Design at the RCA

The DesignAge program is now a core theme of the HHRC, where the research focus is 'inclusive design', a term used to describe a process whereby designers ensure that their products and services address the needs of the widest possible audience. It combines the perspectives of technology push and demographic pull in ways that include the needs of those groups of people in society who are currently excluded from or marginalised by mainstream design practices, due to age or disability or rapidly changing technologies and work patterns. It links directly to the political concept of the inclusive society and its importance is increasingly being recognised not just by Governments as a focus for social equality but by business and industry as a tool for commercial growth (DTI 2000). At the Centre inclusive design is seen as a strategy that can be employed to improve the quality and usability of products, services, buildings and communications, and a key factor is the emphasis placed on working with specific groups facing design exclusion in order to better understand how to overcome it. In 1999/2000, ten new RCA design graduates were teamed up with ten industry partners to undertake a range of collaborative research and development projects responding to social and demographic change. That entailed working closely with the widest spectrum of users to reshape the design process to their needs. Visually impaired travellers journeyed through Heathrow Airport; disabled teleworkers and low-paid pieceworkers described the frustrations of working from home; older people gave feedback on supermarket labelling and domestic appliance instructions; osteoporosis patients and paraplegics tested prototypes; and office workers revealed the limits of balancing life and work. Some of the outcomes of the programme are designed artefacts: new furniture, products and architectural structures. In other cases, design skills have been applied to the research methodology itself, as in a study which equipped flexible 'knowledge workers' with special diaries and cameras to record their working lives. The results point towards the emergence of a new 21st century design paradigm, in which recognition and inclusion of the special and acute needs of different users in the design process will lead towards better solutions for all in society, and broader markets for business. Some examples follow.

4.1   Process to Pleasure – instinctive wayfinding at Heathrow airport

The scale and complexity of the modern airport terminal makes wayfinding for even the most experienced traveller difficult. Add the physical impairments that result from ageing, and airports become an especially daunting prospect for older users. This socially inclusive project teamed two RCA architects Karen Adcock and Carl Turner, with the BAA development group designing Terminal 5 at Heathrow. Its aim was to define and explore a range of inclusive design ideas in order to improve instinctive wayfinding for all, in what will be, subject to the public enquiry, Europe's largest airport terminal. The study investigated current practice in airports around the world. Visually impaired travellers were adopted as a lead user group, and user trials were carried out at Heathrow Terminal 4 in collaboration with London Regional Transport, Royal National Institute for the Blind and the University of the Third Age. From the findings that emerged, the project overlaid a 'sensory map' for T5 onto the existing framework being developed by the BAA team. This formed an alternative reading of the terminal, comprising a series of pleasurable wayfinding elements designed to manage the scale of the building and enhance the process of

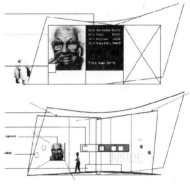

moving through it. Proposals included acoustic arches, tactile pathways, spatial launchpads and a family of furniture, providing information, orientation and comfort for travellers of all ages and abilities. The project will now go into a second year in order to build and test some of the practical recommendations at Heathrow.

4.2   Home Industry – new tools for home-workers

Pieceworkers who work by hand with industrial processes in the home are among the most socially vunerable and economically exploited groups of workers in the UK. A high proportion of those carrying out manual jobs such as sewing, machining, packing and assembly at home, are female and from ethnic minorities. Their work is often repetitive, dirty and hazardous; they lack adequate space and facilities; and they cannot afford to purchase suitable furniture and equipment to support their work. This R&D project by designer Yuko Tsurumaru, involved observation of six homeworking households in South Wales, Gosport and West Yorkshire in order to develop a range of low-tech products that could either be purchased at low cost by the homeworker or supplied free of charge by the employer. Study of homeworkers led to the generation of a series of design proposals targeted at specific user contexts: seating, lighting, trays, bags. Three in particular have been realised: a kitchen workstool in solid beech designed for perching at the kitchen sink to stick and assemble textile samples; a group of table-top organisers in vaccum-formed plastic to assist in small-scale electronic compo-

nent assembly; and, third, a series of stiff paper bags in bright colours to hold and transport rubber trimmings while blending with a domestic interior. Prototypes were tested with users and amendments incorporated as part of an iterative development process. The project demonstrates how simple, inexpensive design can contribute to quality of life for those working with industrial processes and materials in their kitchens, halls and living rooms.

4.3   Playground – inclusive design for disabled teleworkers

Two social trends informed this project. First, the growth in homeworking - more than 30% of the UK workforce will work from home by the year 2006, according to the Henley Centre for Forecasting. Second, the above-average levels of unemployment among Britain's 6.2 million disabled people, many of whom would like to participate in the UK's information technology economy but are prevented from doing so by lack of appropriate tools and support. Working from home is often the only chance for people with disabilities to work at all because the home can incorporate the care and rest facilities they need. But, equally, poor furniture and layout can be a barrier to becoming an effective teleworker. This R&D project by designer Lotta Vaananen, set out to make site-specific design proposals for individual

disabled users as exemplars of an inclusive approach. The habits and needs of disabled teleworkers were investigated through visits and an e-mail questionnaire. From the findings, the designer worked closely with five users in their homes to develop design briefs. Two new furniture products illustrate a move away from hospital-style appliances towards a more playful spirit. Carousel is a large, horse-shoe-shaped desk designed to give wheelchair users the maximum work area from one single point, and with a turntable beneath to simulate the movement of a swivelling office chair when the wheel-chair is driven onto its platform. Swing is a chair for computer users who suffer from severe back pain: it can be hung on pivot points from a support frame from different angles to change the pressure points on the body, and it has a special mattress with ergonomically designed pockets that can be filled with different materials to customise support for the neck and lumbar regions.

4.4   Flexible Workers/Fluid Workscape – supporting work-life balance

Rapid social and technological change is fundamentally redefining patterns of work for many groups of workers, often with unsettling results. For professional 'knowledge workers', work is no longer confined to a geographically-fixed office for set hours but is conducted more or less continuously across a matrix of diverse locations, including the home. Within this more fluid workscape, many people are struggling to find a balance between work and life. The tools and services to support flexible working are inconsistent, and the rituals associated with it are ill-defined. This project set out to develop an understanding of the changing behaviour and needs of increasingly mobile workers, and to propose product and service concepts that aid work-life balance. Twenty-five profes-

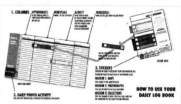

sional 'knowledge workers' in a range of industries were given a specially designed log book and camera and asked to document their work and life activities, workplaces and worktools over a seven-day period through written observation, photography and collecting objects and memorabilia. The results revealed work patterns of increasing complexity and pressure, with more hours, more travel, more weekend working, and more ambiguity between work and home. The findings were mapped against a study of places for flexible working - from rail stations to hotels, libraries and conference centres - and a series of design scenarios populated by identified user types were developed. New product concepts for the 'local juggler, 'highway patroller', 'corporate wanderer' and 'continuous performer' include tools to carry, tools to wear and tools to organise - intelligent luggage, smart clothing and interactive time-planners.

4.5   Impact Wave – clothing that cares and protects

Clothing that provides a level of physical protection while being comfortable to wear is a holy grail in the apparel industry; it has been described as the search for the 'magic t-shirt'. Body impact solutions currently available to the consumer are limited because they are either based on a rigid exterior shell (such as roller blade pads) or some form of foam laminate (ski pant inserts, for example). The former are too inflexible; the latter don't work effectively. This R&D project, by designer Dan Plant, has developed a new flexible system that is incorporated directly into the garment to protect the human body against impacts and abrasions. It comprises two materials combined in multi-layers which stiffen upon impact to provide protection, but flex with the musculature of the body when protection is not required, thus combining safety with comfort. The study looked in particular at applications related to older people. It examined the problems associated with hip joint protection of osteoporosis sufferers and frequent fallers. User groups were undertaken in conjunction with Research into Ageing and AgeNet. Independent technical tests confirmed that the new system, entitled Impact Wave, is up to 10 times more effective on pressure distribution and three times more effective on force than conventional foam and plastic systems with the added benefit of flexibility. The project sought a balance between wearability and compliance to safety standards, and also investigated other potential workwear, leisure and sports markets.

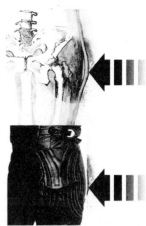

5. Next steps

Until recently, the design community has approached the needs of older and disabled consumers from different specialist perspectives, including rehabilitation design, ergonomics, medical/institutional equipment, and more recently interface design. However, given the trend towards social inclusivity in Europe in particular, there is lacking a coherent approach or methodology that will delivery inclusivity by design. Research suggests that what is needed is a framework for understanding how this can be advanced by countering design exclusion, supported by appropriate design strategies and tools, and a mapping of design-related user research methods. Industrial interest is growing, and the design management community is keen to develop a better understanding of the issues and have access to appropriate design strategies, tools and guidance. In response to this, the authors, along with colleagues from Cambridge University, Central St Martins School of Art and Design, and the UK Design Council, are working on assembling the necessary data and information and preparing it for dissemination to the design community. From this work a broad picture is emerging of what we mean by inclusive design, while case studies from industry are revealing the extent to which it is being put into practice. Some of this work features in a paper presented to this conference by colleagues from the Engineering Design Centre at the University of Cambridge.

REFERENCES:

Bicknell J and McQuiston L. (Eds.) (1977) Design for Need: the social contribution of design. *Conference Proceedings*. London: Pergamon Press & Royal College of Art.

Brouwer-Janse, M.D., Coleman, R., Fulton-Suri, J.L., Fozard, J., de Vries, G. and Yawitz, M. (1997) User interfaces for young and old. *Interactions* 4, 2, 35-46.

Coleman, R. (1993) A demographic overview of the ageing of First World populations. *Applied Ergonomics*, 24, 1, 5-8.

Coleman, R. (1994a) The Case for Inclusive Design – an overview. *Proceedings of the 12 Triennial Congress* (3, 250-252). Toronto: International Ergonomics Association and Human Factors Association of Canada.

Coleman, R. (1994b) Design Research for Our Future Selves. *Royal College of Art Research Papers*, Vol.1, No. 2. London: Royal College of Arts.

Coleman, R. (1997a) Mit alter Menschen arbeiten: ein kollaborative Designprozeß, and Verbesserung der Lebensqualität alteren Menschen durch Design. *Handbuch der Gerontechnik* (II-6.5). Landsberg, Bavaria: Ecomed-Verlag.

Coleman, R. (Ed.) (1997b) *Working Together: a new approach to design*. London: Royal College of Art.

Coleman, R. (1999) Inclusive Design. In P. W. Jordan & W. S. Green (Eds) *Human Factors in Product Design: Current Practice and Future Trends* (pp.159-170). London: Taylor & Francis.

DTI Foresight (2000) *The Age Shift – priorities for action*, report of the Foresight Ageing Population Panel, London Department of Trade & Industry.

Haigh, R. (1993) The Ageing Process: a challenge for design. *Applied Ergonomics*, 24, 1, 9-14.

Kirkwood, T. (1999) *Time of Our Lives: The Science of Human Ageing*. London: Weidenfeld & Nicolson.

Laslett, P. (1996) *A Fresh Map of Life: The Emergence of the Third Age* (2nd ed.). London: Weidenfeld & Nicolson.

Laslett, P. (1998) Design Slippage over the Life-Course. In Graafmans, J. Taipale, V. & Charness, N. (eds.) *Gerontechnology: A Sustainable Investment in the Future*. Studies in Health Technology and Informatics (Vol. 48, pp. 84-92). Amsterdam: IOS Press.

Manley, D. (Ed.) (1986) *New Design for Old*. London: Victoria and Albert Museum and the Helen Hamlyn Foundation.

Myerson, J. (Ed.) (1999) *Design for Our Future Selves: DesignAge Competitions at the Royal College of Art 1994-98*. London: Helen Hamlyn Research Centre.

When Non-Human is Better than Semi-Human: Consistency in Speech Interfaces

Li Gong, Clifford Nass, Caroline Simard

Department of Communication
Stanford University
Stanford, CA 94305-2050
{ligong, nass, csimard}@stanford.edu

Yuri Takhteyev

Department of Computer Science
Stanford University
Stanford, CA 94305
yuri@cs.stanford.edu

ABSTRACT

When the most human-like option cannot be employed for all aspects of an interface, should designers use the human option whenever possible or consistently use the non-human option? The two studies presented here demonstrate that using consistency in an interface is preferable to mixing human and non-human characteristics. The first experiment shows that, in the context of a telephone voice-portal, consistent synthetic speech was liked better, trusted more, and seemed more competent than an approach that juxtaposed recorded natural speech with synthetic speech. The second experiment demonstrates that a synthetic face talking with synthetic speech elicited more intimate disclosure from users than a synthetic face talking with recorded natural speech. Implications for design are discussed.

1. INTRODUCTION

A traditional goal in design is to make interfaces more human-like, on the assumption that anthropomorphic interfaces are more enjoyable and easier to use (Laurel, 1991; Oren, Salomon, Kreitman, & Don, 1997; Stork, 1997). The goal of human similarity has driven research in artificial intelligence (Patterson, 1990), natural language processing (White, 1990), emotion detection and display (Picard, 1997), adaptation (Nass & Moon, 2000), agency (Friedman, 1999), personality (Reeves & Nass, 1996), and vision (Rosenfeld, 1997). Of all these areas, few have received more attention recently than speech output (Lai, 2000; Olive, 1997).

Technologies can produce two types of speech: recorded natural human speech (a human-like option) and computer-synthesized speech (a machine-like option). High-quality recorded human speech, facilitated by new recording and compression technologies and dramatic increases in disk space, reproduces speech that is comparable to speaking with a person in face-to-face conversation (Nass & Gong, 2000). Computer-synthesized speech (also called text-to-speech or TTS), is clearly non-human. TTS produces comprehensible content, but the speech is obviously produced by a machine rather than a person. Naturalness scores for even the best TTS systems are in the poor-to-fair range (Kamm, Walker, & Rabiner, 1997, February), because they exhibit inexplicable pauses, misplaced accents and word emphases, discontinuities between phonemes and syllables, and inconsistent prosody (Nass & Lee, in press).

Although recorded natural speech is superior to TTS in clarity and naturalness, it lacks TTS's flexibility. Specifically, it is very expensive and frequently impossible to have a recorded voice say all possible variations of content. This has led to the situation whereby clearly fixed content, e.g., "Welcome to the message center" is presented in recorded speech, while highly varying content, such as email readers or news readers, speak in pure synthetic speech. In these instances, there is no dispute: Recorded speech is the superior option, but TTS is better than nothing.

The most interesting cases are those in which the nature of the content allows for recorded speech to be regularly used in the first half of a sentence but not in the second half. This occurs in those applications in which the natural sentence structure involves a fixed utterance in the first half of a sentence followed by a varying utterance in the second part of the sentence. Examples include "The most popular book is ...*Harry Potter and the Goblets of Fire,*" "You have received an email from ... *Jane Jones,*" or "The nearest post office is located at ... *52 South Market Street,*" In these instances, the dominant approach is to present the first, fixed part of the sentence in recorded speech, and the second, highly-varying part in TTS. This reflects the goal of making the system as "human as possible." Of course, it is possible to have the TTS voice say both halves of the sentence, providing consistency in exchange for maximizing the quality of each part of the sentence. Is consistency or maximization the most effective approach?

390

A similar question is faced in the pairing of face and speech in talking heads. Synthetic talking heads are hailed as incorporating the advantages and appeals of both face (Laurel, 1991; Massaro, 1997) in a single interface (see, e.g., www.ananova.com; www.extempo.com). Similar to the human and synthetic options in the speech domain, there is also a human and a synthetic option in the face domain. There are enormous constraints in using recorded human faces: memory- and bandwidth-intensity, especially in longer bursts; synchrony between audio and video, and production costs. Hence, recorded human faces are not a viable option at present. However, great strides have been made in the production of clearly synthetic faces, especially with respect to lip-synchronization and emotion manifestation (Massaro, 1997). In fact, lip synchronization has advanced to the point in which synthesized faces can speak with either synthetic speech (the non-human option) or recorded speech (the human option). These voice options again raise the fundamental question: Is it better to maximize the humanness of the voice (recorded speech) when using a synthetic face, or should one instead opt for consistency by having both the face and voice be clearly synthesized?

Unfortunately, the literature on maximization versus consistency is ambiguous. The fundamental rationale behind mixing TTS and natural speech and combining synthetic face and natural speech is *independent maximization*. The idea behind this approach is that humans assess each aspect of an interface independently, and simply apply a weighted average to their assessment of the system. Hence, improving one part of the system, even while holding others constant, should enhance overall quality. Under this framework, juxtaposing recorded speech with TTS and combining a synthetic face with recorded speech is better than limiting the systems to deficient consistency.

An opposing view is that *consistency* between aspects of an interface is more important than independent maximization. That is, instead of a linear addition, the aspects of an interface *interact*. The idea is that rather than an independent response to each aspect of the interface, people respond to the over-all system as a *gestalt*, not analyzing each component for quality. There is a large body of evidence in social psychology that consistency is an important rule governing interpersonal perceptions and interactions (Fiske & Taylor, 1991). For example, people prefer consistency in another person's personality (Hendrick, 1972) and consistency among another's verbal and nonverbal channels (Domangue, 1978). Challenging the independent-maximization approach, the consistency approach argues that mixing human option and non-human option in a single interface presentation is less desirable than consistently using the non-human option for all aspects of the interface. In other words, non-human is better than semi-human when the human option cannot be applied to all aspects of an interface. Hence, following the consistency approach, one should not mix TTS and natural speech; instead, one should use TTS consistently for all content. Similarly, one should pair a synthetic face with TTS instead of with recorded natural speech.

Two studies were conducted to address the debate between the independent-maximization approach and the consistency approach. The first study examined whether to mix natural speech and TTS in a telephone-based portal. The second study examined the pairing of a synthetic face with natural speech or TTS when individuals are asked personal questions.

2. STUDY 1: MIXING TTS AND NATURAL SPEECH[1][2]

A two-condition between-subjects experiment was conducted to test whether one should mix natural speech and TTS or use TTS consistently. The interface was a telephone-based university housing information system which presented users seven sentences for about 2 minutes. A sample sentence is: "If you are a graduate student, your chance of getting your first choice should be at least *five percent*." In the mixed-speech condition, the system delivered the dynamic part of the sentence (italicized) with TTS and the fixed part with recorded natural speech. In the TTS-only condition, the system spoke with TTS for the whole sentence. Twenty-four students were randomly assigned to one of the conditions, with gender balanced in each condition. To avoid potential difficulties in understanding TTS, all participants were native English speakers. The participants received course credit for participating in the study. After using the system, participants filled out a web-based questionnaire.

2.1. Manipulation and Measures

The CSLU Toolkit[3] running on an NT machine was used to run the experiment; a Dialogics board answered the participants' phone calls. The default male voice of the Festival TTS engine in the Toolkit was used for the TTS. An American male graduate student was recorded to provide the recorded natural speech.

---

[1] This research was supported by a grant from the National Science Foundation.

[2] This study was designed and executed by Nass, Simard, and Tahkteyev.

[3] The CSLU Toolkit is downloadable without cost at http://cslu.cse.ogi.edu/toolkit.

Dependent measures were based on items on the web-based questionnaire. The questionnaire asked, "For each word below, please indicate how well it describes the information system you just heard," followed by a list of adjectives. Each adjective was associated with a ten-point, radio button Likert scale anchored by "Describes Very Poorly" and "Describes Very Well."

Responses were combined to create factor scores measuring liking, trust, and perceived competence, three important and distinct aspects of any information system. Factor analysis confirmed that the factor scores were distinct and internally consistent (as determined by eigenvalues and factor loadings). All analyses were based on two-tailed t-tests.

Liking was a factor score of five items: enjoyable, entertaining, friendly, good, and likable ($\lambda = 2.9$; factor loadings ranged from .81 to .95).

Trust was a factor score comprised of three items: realistic, reliable, and trustworthy ($\lambda = 2.3$; factor loadings ranged from .83 to .93).

Perceived competence of the system was a factor score comprised of four items: clever, informative, useful, and well-designed ($\lambda = 2.6$; factor loadings ranged from .82 to .96).

2.2. Results

For all three dependent measures, the TTS-only interface achieved more positive user responses than the mixed-speech interface (see Figure 1). Hence, there was strong evidence for consistency and no evidence for maximization. As predicted by the consistency model, users liked the housing information system using TTS consistently more than the system mixing natural speech and TTS ($t(22) = 2.20$, $p < .05$). Similarly, users trusted the housing information system using TTS consistently more than the system mixing natural speech and TTS ($t(22) = 2.32$, $p < .05$). Finally, the TTS-only system was perceived to be more competent than the mixed-speech system ($t(22 = 3.04$, $p < .01$).

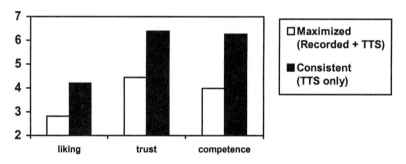

Figure 1: Means for the Housing Information System

2.3. Discussion of Study 1

This study supports the consistency argument by showing that the TTS-only interface achieved more positive user response than the interface mixing natural speech and TTS. The present research suggests that the dominant strategy of mixing recorded speech and TTS may lead to reduced liking, trust, and perceived competence. The results suggest that "doing the best one can at each point in the sentence" is actually inferior to providing a consistent user experience.

3. STUDY 2: SYNTHETIC FACE TALKING WITH NATURAL OR SYNTHETIC SPEECH[4]

The consistency and the independent-maximization approaches also provide differing suggestions in the pairing of synthetic face with either TTS or with recorded natural speech. A simple test of the two face conditions was not appropriate, because we could not determine whether a main effect for type of voice or an interaction between voice

---

[4] This study was designed an executed by Nass and Gong.

392

and face were driving any differences. Hence, we implemented a 2x2 between-subjects experiment that controlled for the main effect of voice: Synthetic face vs. no face [the control] by TTS vs. recorded natural speech.

Participants in the experiment interacted with a computer-based interviewing system that asked questions using one of the four representations: 1) a synthetic face talking with TTS; 2) a synthetic face talking with recorded human speech; 3) TTS without face; and 4) recorded natural speech without face. The talking head or the voice alone was framed as the interviewing agent.

Forty-eight undergraduate students were randomly assigned to condition with gender balanced in each condition. To avoid potential difficulties in understanding TTS, all participants were native English speakers. The participants received course credit for participating in the study.

3.1 Manipulation

The participants interacted with the system on a desktop computer one at a time in a media lab. The CSLU Toolkit was used to create the face and the speech. For the synthetic face, the "Baldi" face provided in the Toolkit was used. The face was placed on the left side of a 43.2-cm diagonal computer screen. The face was 17.8 cm high and 12.5 cm wide. For TTS, the default male voice in the Festival TTS engine in the Toolkit was used. For natural speech, an American male graduate student was recorded. The synthetic face "Baldi" was automatically lip-synchronized with both TTS and recorded natural speech using the Toolkit.

3.2 Measures

The nine-item self-disclosure survey developed by (Moon, 2000) was used in the interviewing task. A sample question is: "What has been the biggest disappointment in your life?" All the questions were open-ended. Participants typed their answers in a text box on the computer screen. Based on participants' responses, two aspects of self-disclosure were captured. The *amount of self-disclosure* was measured as number of words in the participants' responses. The reliability of this index across questions was $\alpha = .85$. The *depth of self-disclosure* was based on intimacy ratings. Two coders, who were blind to the hypothesis, independently rated each of the participants' responses on a five-point Likert-type scale (1 = "low intimacy", 5 = "high intimacy"). The inter-rater reliability was .74; disagreements were resolved by averaging. The reliability of the index across responses was $\alpha = .86$.

3.3 Results

For both the amount and the depth of disclosure, there were significant interaction effects in support of the consistency hypothesis. Participants disclosed more information about themselves when the synthetic face was talking with TTS than talking with natural speech, while the opposite pattern appeared for the no-face control groups $F(1, 44) = 187.9$, $p < .001$ (see Figure 2). Similarly, participants disclosed more intimate information about themselves when the synthetic face was talking with TTS rather than with natural speech, while the opposite pattern was again obtained in the no-face conditions, $F(1, 44) = 13.9$, $p < .001$ (see Figure 3). These results provide additional support for the argument for consistency: When it is the sole medium, recorded speech is desirable , but it becomes less desirable than TTS when combined with a synthetic face. Thus, faces and voices are processed holistically rather than as two separate dimensions.

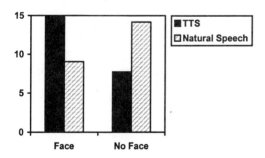

Figure 2: Amount of Disclosure

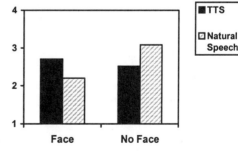

Figure 3: Depth of Disclosure

3.4. Discussion

This study further proves that consistency between aspects of an interface is more important than independent maximization if the maximization causes inconsistency. Synthetic face and synthetic speech are consistent with each other because they are both clearly non-human, while synthetic face and natural human speech are inconsistent.

This study should be extended in two ways. First, user disclosure is a highly attitudinal context because it requires trust from the users. It is important to replicate this study with a cognitive task, such as a comprehension task; in these cases, the focus on cognition might lead to a preference for clarity over comfort. Second, the combination of a clearly human face with both TTS and recorded speech would provide another critical test of the consistency claim.

4. CONCLUSION

While it is tempting to deploy the best technology available, the present research provides a cautionary note. When maximal anthropomorphism on one technological dimension is clearly inconsistent with less anthropomorphism on a different dimension, there are good arguments for constrained consistency. An inconsistent interface which is half human and half non-human is less desirable and effective than an interface which is completely non-human but consistent. As interfaces become more multi-faceted and acquire more capabilities, consistency between different aspects of an interface will increasingly become a critical issue in research and design.

5. REFERENCES

Domangue, B. B. (1978). Decoding effects of cognitive complexity, tolerance of ambiguity, and verbal-nonverbal inconsistency. *Journal of Personality, 46*(3), 519-535.

Fiske, S. T., & Taylor, S. E. (1991). *Social Cognition*. New York, NY: McGraw-Hill, Inc.

Friedman, B. (Ed.). (1999). *Human values and the design of computer technology* (Vol. No. 72). New York: Cambridge University Press/CSLI.

Hendrick, C. (1972). Effects of salience of stimulus inconsistency on impression formation. *Journal of Personality & Social Psychology, 22*(2), 219-222.

Kamm, C., Walker, M., & Rabiner, L. (1997, February). *The role of speech processing in human-computer intelligent communication.* Paper presented at the NSF Workshop on human-centered systems: Information, interactivity, and intelligence.

Lai, J. (2000). Conversational Interfaces. *Communications of the ACM, 43*(9), 24-27.

Laurel, B. (1991). *Computers as theatre*. Reading, Mass.: Addison-Wesley Pub.

Massaro, D. W. (1997). *Perceiving talking faces: From speech perception to a behavioral principle.* Cambridge, MA: MIT Press.

Moon, Y. (2000). Intimate exchanges: Using computers to elicit self-disclosure from consumers. *Journal of Consumer Research, 26*(4), 323-339.

Nass, C., & Gong, L. (2000). Social aspects of speech interfaces from an evolutionary perspective: Experimental research and design implications. *Communications of the ACM.*

Nass, C., & Lee, K. M. (in press). Does computer-synthesized speech manifest personality? Experimental tests of recognition, similarity-attraction, and consistency-attraction. *Journal of Experimental Psychology: Applied.*

Nass, C., & Moon, Y. (2000). Machines and mindlessness: Social responses to computers. *Journal of Social Issues, 56*(1), 81-103.

Olive, J. P. (1997). The talking computer: Text to speech synthesis. In D. Stork (Ed.), *Hal's legacy: 2001's computer as dream and reality* (pp. 101-130). Cambridge, MA: MIT Press.

Oren, T., Salomon, G., Kreitman, K., & Don, A. (1997). Characterizing the Interface. In B. Laurel (Ed.), *The Art of Human-Computer Interface Design.* Reading, MA: Addison-Wesley.

Patterson, D. W. (1990). *Introduction to artificial intelligence and expert systems.* Englewood Cliffs, N.J.: Prentice Hall.

Picard, R. W. (1997). *Affective Computing.* Cambridge, MA: MIT Press.

Reeves, B., & Nass, C. (1996). *The media equation: How people treat computers, television, and new media like real people and places.* New York: Cambridge University Press.

Rosenfeld, A. (1997). Eyes for computers: How HAL could "see". In D. G. Stork (Ed.), *Hal's Legacy: 2001's Computer as Dream and Reality* (pp. 211-235). Cambridge, MA: The MIT Press.

Stork, D. (1997). *HAL's Legacy: 2001's Computer as Dream and Reality.* Cambridge, MA: MIT Press.

White, G. (1990). Natural language understanding and speech recognition. *Communications of the ACM, 33*(8), 72-82.

Wireless Spaces:
enabling technological freedom and creating invisible borders

Ashley Saulsbury
Ansima

ashley@ansima.com

Ing-Marie Jonsson
Dejima Inc.
160 West Santa Clara Street,
San Jose, CA 95113
ingmarie@dejima.com

ABSTRACT

This paper discusses the implications for future wireless networks, supporting middleware, and wireless appliances based on experiences with current and shortly arriving technology. We give a brief overview of wireless networks, and concentrate on the factors and opportunities that will make wireless devices acceptable and thus commercially viable rather than just triumphs of technological innovation. Not all opportunities are to the benefit of the user, and privacy concerns may well affect adoption and usage of new devices. As we shall discuss, there are solutions that can be used either as enabling technologies to empower the user or as disabling technologies to constrain and bind users by creating artificial borders and limiting the scope of the use.

1   INTRODUCTION

Within this paper we concentrate on technology intended for personal use, rather than the infrastructure and devices required for the office or other professional activities. This technology, both fixed and mobile, will be heavily impacted through the use of wireless connectivity, and it is important to understand how this technology can and will be employed both for and against the user. This paper presents conclusions based on observations of market trends and emerging technologies.

1.1   The development of technology appliances

The last two decades have seen the penetration of the computer into the home, both by design and by stealth. Microprocessor technology is so cheap today that most electronic devices contain at least one. As technology has progressed, home computers have become more and more complicated. They now contain a bewildering array of innovations, often hidden behind a non-intuitive user interface (UI) so complicated that only a twelve year old can fully understand it. This is no joke, children accept and learn this UI because they see the computer as a device to explore and master rather than an obstacle to completing the task at hand.

The Internet phenomenon has further expanded the market for the home computer, and as a result highlighted its inadequacies for the job. Ultimately people don't want a computer - it is just a necessary evil to obtain the functionality required, namely; record keeping, a communications tool, and something to entertain at home. Many of these functions would be more accessible if implemented as dedicated devices. The market place has understood this need; next generation WebTVs, TiVos and PlayStations (TiVo, 1999) are just such dedicated terminals - replacing the user-managed combination of the home computer, operating system and application software. These terminals will enable you to browse your bank account on the web, watch movies and shoot aliens, with the more intuitive power button replacing the ctrl-alt-del keyboard salute.

1.2   The introduction of mobile devices

Mobile technology has followed the path of the home computer, only on a more aggressive time-scale. Despite the same development route, mobile technology differs because its interaction model is on a more intimate and personal level. For example, the home computer is for the whole family, but the Palm Pilot is yours alone. This personal aspect allows customization on many levels beyond what we have seen for existing computers.

Today's mobile devices reflect the heritage from their wired counterparts. For example, to provide the same functions of record keeping, entertainment and communications we have the mobile equivalent of the home computer - namely the personal digital assistant (PDA). Manufacturers of these devices are already trying to convince us to purchase and install new applications and additional hardware. However given that we have the opportunity to be continuously connected, can't we skip the "home computer" stage and simply purchase many of these services on the network?

Paralleling the wired world, which favors dedicated appliances, the mobile world can move to a decoupled model of dedicated services accessed through personalized terminals. There are four technologies that need to interoperate to make this a reality, and we describe these briefly below. However, user demands will shape each of these technologies in specific ways and we examine these aspects starting in section 2.

1.3 Wireless Networks

There are three categories of wireless networks divided by their range; wide area networks such as GSM and 3GPP (ETSI, 1992; ETSI, 2000) provide global communication, local area networks such as the 802 family provide communication in the local loop, and finally short range solutions for body area networks and room sized networks area implemented using technologies such as BlueTooth and HomeRF (BlueTooth, 1999; HomeRF, 1998).

The wireless space is implemented as a combination all of these network solutions, with the potential of offering the user anywhere connectivity and true mobility of services. The wireless network solution is almost always combined with a fixed network backbone, responsible for the communication over long distances.

1.4 Middleware

Transparent access to network services is provided through protocols and a layer of supporting functionality. This is also an area offering multiple solutions, ranging from support for general for mobile communication such as Mobile IP (IETF, 1995), to JetSend (JetSend, 1999) for usage in home and office appliances.

Much of this middleware technology is borrowed from the wired Internet, however wireless spaces demand additional or modified features. Aside from plug-and-play features and security issues, wireless middleware will most likely support additional features such as location determination and dynamic name & route determination.

1.5 Access Devices

Formally, an access device provides the method by which a user can interact with services that are provided on the other side of an interconnection network. Such devices must use support from some portion of the middleware to access remote services located across the network. The middleware, in combination with network mechanisms, will enable an access device to be anything from a dumb terminal, to a device that will cache information and services when network connectivity is lost.

1.6 Application services

Services located on the network will provide applications to users. Data storage and retrieval, communication tools, and entertainment options are among the services that can be offered. Such services will make extensive use of middleware connectivity – it is important that services offered by different vendors interact seamlessly.

2. WHAT AND HOW DO CONSUMERS BUY?

Each consumer's purchasing decisions are different, however there are some broad criteria which shape products, and in particular the way those products employ their underlying technology. We examine these in this section, since they will help determine user expectations from technology, how the technology is likely to be used, and the structure of some products and their business models.

2.1 The Price

While the availability of a certain technology can entice consideration, the overriding factor in most consumers' decision to buy is price. Features and options become second order deciding points, although sometimes helping to move someone beyond their originally intended purchase limit. How often have we heard someone ask; I have 3000 to spend on a laptop - which one should I get? A manufacturer's response is to produce the cheapest device required to entice the purchaser's interest, and then introduce a suite of options at minimally priced increments to tease out the actual price for a reasonable machine. Ultimately, while there will always be early adopters for technology, it is availability at the right price to service a perceived need which persuades consumers to buy.

2.2 Form factor and perception

The first stage of the mobile device market concentrated on early adopters of technology. The limiting factors included technology such as the visual interface (screen resolution, color and size), and in the case of 'phones connectivity (availability of service and bandwidth). These technological limitations are currently being resolved, revealing the concerns of the mass market.

To attract the broader market, an important feature, beyond price, is form factor & design - an oft-ignored factor. A classic example is the Apple™ Newton™. It was too small and limited for a home computer, and simply too big to fit a pocket. It had all the features of a Palm Pilot™, but it was too clumsy. The most practical technology, at the right price, and with the right look-and-feel has a stronger attraction than the most advanced technology.

2.2.1 Cultural differences

Cultural aspects also affect the perceived usefulness of design features. For example, in the United States, customers like to have as many options as possible built into their laptops. Screen size should also be as large as possible. This is desirable in a culture where office space is generous, and travel by car is ubiquitous. Japan by contrast is a country that relies on shared work environments and communal transport with many short journeys. This favors lightweight minimalist devices where vital functionality is critically chosen to reduce the weight that has to be carried.

2.2.2   Product and corporate image

Finally, image and perception are the wildcard factors. For example, WindowsCE™, PDAs are based on better technology than their nearest rival, with an equivalent price tag and learning curve. However Microsoft's™ dominant position in the home computer space limits the acceptance of their products by a wary public.

2.3   Personalization

Once a technology has matured, catering to users individuality provides an important differentiator in a market with otherwise similar devices. The purchased appliance becomes an extension of the owner's personality and part of their way of expressing themselves. This personalization is seen today in mobile 'phones with individual ring tones and clip-on colored shells. In the computer industry Apple™ released its iMAC™, and proved that color and style were deciding factors once performance and functionality sufficed. Japan has taken this even further; users now have their 'phones painted to match their nails, hair or clothes. The basic functionality is so ubiquitous that the device is no longer solely bought for its technological benefits, but also for its aesthetic qualities and ability to enhance the owner's image.

3   NEXT WAVE OF WIRELESS DEVICES

We have considered the technological components for wireless mobile devices, and the general criteria by which users can choose to adopt a mobile technology. In this section we consider how these aspects combine to determine future wireless spaces.

3.1   The impact of connectivity

Wireless technology in the mobile 'phone arena is rapidly reaching maturity, while it is just starting to arrive in other fields. New wireless technologies; such as 802.11b, GPRS, and 3G are providing the bandwidth for communication beyond voice and short text messages. This enabling technology will affect many of our everyday practices.

Today we carry our tools with us. For example, while on holiday we take camera and film (or flash RAM). Within the next decade network connectivity will be ubiquitous. Photographs will be delivered directly to storage or printing services from the camera as soon as they are taken (Ricoh 2001). If the camera is lost, stolen or broken you will loose only the appliance, the real valuables, i.e. the 'photos, will be safe. Moreover, reducing the local storage requirement will make the device cheaper. This trend will be true for many service, documentary records will be held in electronic formats - accessible anywhere - physical realizations will be reproductions for viewing purposes.

Manufacturers and service providers are slowly finding the appropriate place for system functionality. For example, Ericsson's T28 World 'Phone includes a voice dialing capability. The handset incorporates sophisticated voice recognition as well as additional memory for storing learned voice patterns. This is an additional benefit for the user as long as there is network service. Other network providers, for example Sprint, offer voice dialing as part of their network service – a design, enabling usage from any 'phone. The advantages are clear; the cost of the recognition hardware is amortized over many users rather than included in the cost of the handset[1]. An extra benefit for the user is that the stored numbers and the voice training are safe even if the handset is replaced, lost or stolen. The Sprint model is still wanting however, because it is a service by the network provider – it becomes a barrier to switching providers. GPRS and 3G 'phones could provide the opportunity for an independent directory service - it remains to be seen whether traditional carriers will allow this to happen.

3.2   The de-coupling of devices, carriers and services

Why carry all your important information with you and risk loosing it? Why pay the expense of storing it locally when it can be retrieved from the network? Why take the time and trouble to manually synchronize data with your home computer when the device can do this automatically when connected to the network? Devices such as Research in Motion's Blackberry (Rim 2001) are a starting step in this direction.

More importantly, why even own the access device? Borrow or rent, for example, a connected camera when going on holiday. Lowering the barrier to purchase (i.e. cheaper device prices), or increasing perceived benefit (e.g. immunity to data loss) increases the likelihood of adoption. However, the user still has to learn how to operate the new device. If the learning curve is too steep this is a barrier to later replacing the technology (a barrier to further consumption). While this may be good for a service provider it may be terrible for a device manufacturer. However, why learn a new user interface when you change your access device? Storing data, including personal preferences,

---

1 Ericsson decided that the incorporation of voice dialing was cheap enough, and gave their handset an advantage in
   markets where the carrier does not provide the feature.

as part of a user profile, in the network would enable personalized UIs to follow the user even when the access device is replaced.

The direction of these trends is toward the separation of functionality between device manufacturers, service providers and communication carriers. This separation will benefit the user, enabling greater competition between providers of each of the three components. Handset manufacturers for GSM phones have already begun, users can already remove the SIM card and switch to another 'phone – the service follows the SIM card. The access device can be selected to suit the occasion, just like a watch or other accessory. The same functionality also enables switching service providers, although this is currently only applicable for selection of communication carriers.

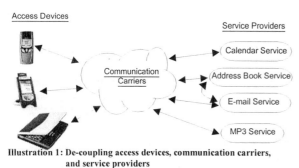

Illustration 1: De-coupling access devices, communication carriers, and service providers

3.4 Disposability

Ultimately personalization and the freedom to switch services at will encourages consumers to upgrade frequently. This leads to a greater need for disposability of access devices. Manufacturers are already working on disposable mobile 'phones. These could ultimately merge with PDAs to become disposable access devices for all types of network services. Aside from the ecological considerations, this model demands turning devices from possessions into consumables.

The economic models of possessions versus consumables are quite different. Except for luxury markets, a possession is sold with a low margin, and the money is made selling the consumables (or service). For example, inkjet printers are cheap, and the profit is made through the price of ink refills. The balance is to keep the consumable price low enough, that the cost of the original equipment is a barrier for abandoning the technology. Therefore consumables aim to maximize margins by minimizing component costs. This reinforces the push to move as many services as possible out of the consumable device and into the communication network.

4 PRIVACY vs. COMMERCE

The difficulty of any network technology is security and privacy. This is even more acute when the user of the network and its services has to rely on the discretion and technology of the various providers.

4.1 Cultural impact

The acceptance of certain network services will be governed by cultural constraints. Convincing people to trust service providers with personal information such as address books and diaries may be difficult, particularly in countries where consumers have weak or non-existent protections from corporations or government. However users can be their own providers for certain services - home servers can be repositories for personal information and the connection to the access device can be encrypted to hide information from the communication carriers en-route. Other providers in the network can offer general and less sensitive services and information.

4.2 Consumer vs. corporate interests

A massive globalization of market places, has been building up since the end of the Second World War. This globalization has generally helped consumers and corporations alike, sometimes to the detriment of local labor markets. However, the arrival of the Internet has provided the consumer with the ability to bypass conventional sales channels, compare prices and purchase across geopolitical boundaries. Corporations use globalization to lower labor costs, but when consumers start using globalization to lower margins, corporations react by re-localizing markets. Starting with the film and music industry's DVD zoning, other industries are following suit. Services and devices can be denied based on where the user is. Equipment can be disabled if bought in a cheap market and imported into an expensive one.

Location capabilities can be provided via, GPS, signal triangulation, or simply nearest base-station identification. However, those capabilities may not be there solely for the user, but also for those collecting information about the user, and enforcing location based rules.

Corporations benefit from knowing their users, collecting data such as, travel patterns, shopping patterns, personal interests, and economical and social status. This has become a geopolitical issue; in some countries, the consumer won't be told, and won't have a choice. In other countries, the consumer may get some theoretical protection - in reality to be ignored, or the culture prevents consumers from voicing their concerns.

5 CONCLUSIONS

Beyond supporting simple person-to-person communication, wireless technology offers the opportunity to relieve users of the burden and worry of precious state. Instead of continually carrying the tools and information they need, they can be held securely, and rapidly retrieved from, reliable but physically remote services. Even the access terminal itself can be a borrowed resource.

There is no all-encompassing technology and wireless spaces will be heterogeneously populated with providers or manufacturers of network connectivity, services, and access devices. Coupling one category to another, like the telephone company providing voice dialing, serves only as a barrier to the user for changing providers. Ideally each of these categories will be decoupled from the others offering consumers the greatest benefit via the opportunity to mix and match. Enforcing this decoupling may require legislative support – difficult when providers span multiple jurisdictions.

Wireless connectivity is fast becoming a commodity in its own right. Users will select communication packages from different carriers based upon their personal needs and preferences. Some providers may themselves resell traffic capacity purchased on a spot market for a given location – mimicking some long-distance telephone providers today.

Many functions and services will migrate from the mobile device and be offered instead over the network as users learn to trust the services and become familiar with the benefits of decoupling from a specific access device – for example, not having to worry about loosing valuable information. Access devices themselves are likely to compete over personalization options such as interaction methods, color style and size. Ultimately the basic functionality becomes ubiquitous, and potentially the device itself becomes disposable.

The key to most of these technological movements is the ability of middleware to facilitate communication and collaboration among user services and applications. Applications need to collaborate on the user's behalf to provide integrated services - for example, as depicted in Illustration 1, your personal address book service co-operates together with your e-mail provider. This will reduce the workload of having to synchronize more than one instance of tools such as calendars and address books.

Importantly the ability of the user to build a personalized interface style and utilize that across many services will be highly valued. Services and access devices can be freely changed with without the user having to learn a new UI each time. This mimics the idea behind glasses, an access device is customized for individual needs and preferences – to be used in all interaction with computer based services and applications.

REFERENCES

BlueTooth, (1999), Bluetooth is a protocol for short-range radio links between mobile PCs, mobile phones and other portable devices. http://www.bluetooth.com/

ETSI (1992), European Telecommunications Standards Institute, Global System for Mobile Communication (GSM), http://www.etsi.org/

ETSI (2000), European Telecommunications Standards Institute, 3rd Generation Partner Project (3GPP), http://www.etsi.org/

IEEE (1980), the Institute of Electrical and Electronics Engineers, (IEEE 802), http://www.ieee.org/index.html

IETF, (1995), Internet Engineering Task Force, Mobile IP, http://www.ietf.org/

JetSend, (1999), A device-to-device communications protocol, http://www.cswl.com/hpjetsend/main.html

HomeRF, (1998), HomeRF is a protocol for high bandwidth communication in the150 ft range, http://www.homerf.org/

RICOH (2001), The RDC-i700 Image Capturing Device http://www.ricoh.co.jp/r_dc/icd/pc/i700/menu.html

RIM (2001): Research in Motion Limited http://www.rim.net

TiVo (1999), A personal TV service that runs on a personal video recorder, http://www.tivo.com/

Context-Aware Mobile Phones:
The difference between pull and push, Restoring the importance of place

Paul J. Rankin

paul.rankin@philips.com, Philips Research Laboratories,
Cross Oak Lane, Redhill, Surrey, RH1 5HA, UK

Abstract

People on the move want information at the right time, in the right place, and personalized for them. This paper describes research and early concept testing towards achieving such relevance on mobile phones, noting the strong influence that the local environment within 10m has on situating human behaviour. Short-range RF beacons therefore offer a powerful way to augment reality, linking a handset to the various virtual associations of a locale, or the place to virtual projections from the user. Processes, which can negotiate sympathetically and confidentially between the current wishes and agenda of the user ('pull') and the range of opportunities generated by their environment ('push') while leaving the user in control, will determine the acceptance of new, location-based mobile services. The solution proposed exploits a set of default profiles, defined by broad user contexts and their calendar, in a two-way negotiation between the external triggers and the actions explicitly initiated by the user, but requires agreement on a common language for describing contexts and services classes. Many research challenges emerge.

1. Ubiquity vs. Situatedness
1.1 Situated behaviour and personalization

Much has been made of the new digital technology's ability to deliver information to a user anywhere and anytime - ubiquity. However, people need structure to help them cope with the information overload of the modern world. Seeing a familiar face or returning to the locale to re-construct an event triggers memory recall. Human behaviour is situated in its current social scene and physical environment, as well as by mood, memories of the past and goals for the future. Our setting instinctively cues our assumptions of the local agenda, social mores, roles or authority, as well as the acceptable behaviour in terms of access rights, ownership or privacy. Just as architects know how to design environments to facilitate certain activities or social interactions, so digital appliances and applications need to sense the user's current context and situation, moderating their behaviour and content offerings in sympathy (Halkia, 2000). Thus 'context sensitivity' is the key to making new technology helpful and easy to use.

Web *browsing* may fit a situation where the user accesses fixed devices such as PC's or set-top boxes, but is the wrong metaphor for accessing the Internet on the move. People then do not want ubiquitous information, rather the right information according to their own personal agenda and mood, at the right time, in the right place, in a form that is short and to the point. The essence is therefore personalized, targeted alerts or searches, falling within this implicit and explicit context of the user, i.e.'pulled push'. The explosion of short messaging on mobile phones and SMS traffic (now globally equivalent to ~10% of all email acts) attests to this model.

Combining knowledge of a user's interest profile and their context (to which their location is a strong indicator) delivers high relevance and therefore perceived value. While web personalization on PC's might be regarded as a luxury, accurate profiling of the individual on the move is fundamental. The counter to this is that mobile portals may gather data throughout the day, not only on whom you call, but also on where you are, what you activate on the web, what alerts you get and when- the 'big brother' spectre. Trusted design and inherently trustworthy technology will therefore discriminate between alternative services and differentiate handset designs.

1.2 Scales, semantics of place and the value-add of location technologies

People are situated on different scales, to which different position-detecting technologies are applicable:
- <1m : Body space (inductive tags)
- <10m : Social context (GPS for latitude/longitude/altitude, RF or IR beacon for logical location)
- <10km, Community (mobile cell id., cellular network time-difference triangulation, e.g. E-OTD)
- <1000km Culture (global internet)

Social behaviour, foreground of attention and so relevancy for push are set on the scale of rooms and buildings, rather than on the scale of street blocks. Locating technologies which only fix position to +/-100m may be useful for finding the nearest Chinese restaurant, but cannot sense whether the user is in a library or the adjacent bar. We postulate that a new class of highly localized and targeted services which rely on more precise positioning will be greatly valued for their currency and relevance, both by users and service providers.

People develop, attach and exchange names for the places they frequent and their habitual routes, viz. 'at home', 'the supermarket', 'on the way to work', 'going swimming'. A private geography of meaning is built up that is much richer than a public map. Relevant research is starting on attaching names to users' haunts and routes, which are automatically learnt (Schmandt et al, 2000). This semantic network, together with that of our daily routines recorded in diaries or agendas echo the way we structure our activities and so the way we mentally classify our personal contexts. In combination then, the user's calendar data, static personal profiling data and geographical semantics offer considerable leverage towards context relevancy.

2. Context-Aware Mobile Devices
2.1 Mobile phones: some market observations

The personal wireless access market is exploding. Shipments were about 650M in 2000 and 1B mobile subscribers worldwide are expected by 2003. By 2002 80% of GSM shipments will be Internet-enabled and 250M users of location-based services are predicted (ARC Group). New technology will lead to new services such as messaging, multimedia, m-commerce (Wireless Strategic Initiative, 2000). Internet access from mobile devices is predicted to be two or three-fold larger than access from fixed devices by 2004.The mobile phone is rapidly becoming *the* personal, intimate device, carried all day long, the sensor of their owner's activity, the holder of personal keys to their profile data and the initiator of secure transactions. Mobile portals moreover have more potential than portals on PC's or set-top boxes as they can track and mediate the user's interface to their physical world, their service opportunities and communication channels as well as just to information access.

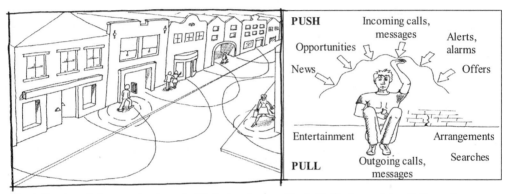

Fig. 1. People and places project 'auras' of their virtual associations, which interact when in proximity

Fig. 2. Unfiltering push is called 'spam', but profiled push is appreciated

2.2 Auras, beacons and the virtual associations of place

By adding fine-grained location-sensing technologies such as GPS or RF beacons, the mobile device can receive alerts of many classes of virtual associations of the locale. These might include local opportunities for services or commercial transactions, located messages, embedded narratives, situated communal activities, social 'wear' or navigation guidance. The interaction with the services themselves may be carried on a different network, e.g. wide-area cellular, rather than over the RF facilitation link. A new world (see Fig. 1) becomes possible: pushed information, commerce and social engagement, hooked onto logical 'hot spots' of activity as defined by beacons

Our vision parallels that of HP's 'Cooltown' (Caswell & Debaty, 2000), where people, places and objects are mirrored by web addresses. These are the virtual associations or opportunities that these three classes of entities can exchange or be examined on. Generalizing, these entities can be seen as short-range broadcasters of 'auras' of identity and opportunity. 'Interaction' thus occurs when such digital auras overlap. Different facets of one entity's aura may be triggered by matches with that of another. This begs work on the affordances of access rights, ownership, copy rights, privacy, authenticity etc. and sophisticated mechanisms for the control of these properties.

3. Proposals to Improve Bluetooth™ for Beacons and Push

Between a beacon and a nearby handset, several RF technologies could be used for the short-range communication such as RFLite, 802.11 wireless LAN or Bluetooth. The latter is particularly interesting because of its commercial momentum and expectations that 40% of GSM mobile phones will be Bluetooth-enabled by 2002. However, the current standard was developed for professional applications and has a number of drawbacks for pushed, opportunistic consumer applications, especially in crowded environments:

- Slow network connection times. The inquiry phase can take up to 60s, but a walking user will pass out of the 10m range of a RF access point in 5-8s.
- Limit on the number of devices in a network (8 in a basic 'piconet')
- Lack of privacy. Unique Bluetooth device identifiers are exchanged before any data is transmitted

For these reasons, we have been developing two approaches to improving this RF technology (Gibson, 2000):

- Bluetooth v1.1-compatible 'Split-beacons' that divide the inquiry/data interaction load across two radios, thereby reducing inquiry times to a few seconds.
- Proposals for connectionless-broadcasting by embedding data in the ID packets of the inquiry scan, eliminating paging and authentication delays. These are backwards compatible with v1.1 and offer very fast information signaling up to 6.4kbytes/s to an unlimited number of anonymous handsets.

The second proposal offers many system uses and opens up a range of new applications. Notice that information about local opportunities and services can be broadcast to users who can remain anonymous at first. (Interestingly, the broadcaster can also remain anonymous.) Graduated, asymmetric disclosure between services and potential customers is thus possible, better fitting the early stages of push (advertising, exploration by the potential customer). Subsequently the user may consent to divulge their details in a transaction, whereupon authentication of the parties may be performed after making a full connection. Connectionless broadcasting over Bluetooth thus offers a way to emulate 'digital auras' for the cheap mass-market.

4. Possible Applications

Beacons which are efficient for pushing offers and pointers to their associated wide-area network services to passing handsets, open up a gamut of applications. A few of over 100 examples which we are studying are:

- Offers and advertisements e.g. to entice customers into a nearby retailer after arousal by their surroundings,
- Public-service broadcasts in crowded places such as train stations, airports or shopping malls
- Located text or voice messages as personal reminders or shared by communities
- Proximity alerts between friends
- Recorded narratives which can be embedded in places and accumulate over time (e.g. as community memory)
- 'Hot badges', which project users' interest profiles or affiliations to seek out others who match (Philips, 2000)
- Places whose media ambience or computing resources automatically respond to the broadcast 'preference auras' emitted by handsets of the users present, i.e. so-called 'ambient intelligence' (Pieper, 1998)

The crucial issue in all these cases is whether the user feels secure, in control and can easily screen or prioritize such opportunity signals according to their own internal context of history, intent, activity and mood (Fig.2)

5. Early Concept Evaluations with Users

To investigate the factors influencing market acceptance and success of the future located-service concepts which we had brainstormed, we set up some 'Wizard of Oz' concept trials in a shopping mall. 12 subjects, aged 16-22 (our lead user group for mobile phone usage models) were taken through a qualitative test. While they walked around a prescribed route, pre-programmed alert sounds, text and graphics on information and offers were triggered on the subject's mobile phone via a signal from another phone used by a 'shadowing' assistant (see Figs. 3,4). This direct experience formed the basis for in-depth discussions, questionnaires on the attractiveness of the service ideas or concerns and participatory design.

In general, reactions to located services in the mall were enthusiastic. The most preferred service examples were practical tools, such as mall guides or travel news, and social tools such as alerts of friends' proximity or located messages. Even specific advertisements were enjoyed as fun. Subjects also helped us generate new service concepts. Concerns were expressed, e.g. on costs when services involved phone calls. Easy user activation of one of a small number of profiles to screen out different categories of pushed alerts by context was thought paramount.

Fig. 3. Entrance to the shopping mall triggers download of changes since last visit

Fig.4. Alerts of 'Virtual Graffiti' left by friends or as reminders to oneself

6. A Mental Model for Context, Push and Pull

A simple mental model for understanding the negotiation process between user-initiated actions (pull) and environmentally-triggered opportunities (push) is being developed. This hinges on the play between two definitions for 'context': one for the external setting of the user, one for their internal agenda. Each context, both external and internal, revolves around three major poles: time, topic (or content) and location (see Fig. 5)

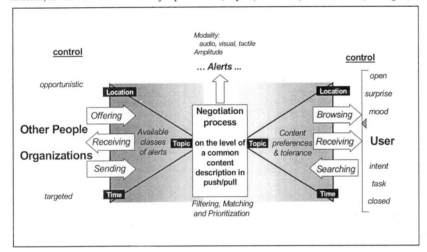

Fig. 5. Services meeting people. A mental model of the negotiation between push and pull

The negotiation process can be initiated on the RHS by the user (pull), or on the LHS by services or other users (push). The locus of control may then pass back and forth. We postulate, as Samulowitz (2000) and Rankin et. al. (1999), a small number of alternate context-specific profiles, plus a static core profile of the user's identity and other stable factors. A default set of profiles may be chosen by the user on a PC beforehand, but we foresee the need for easy adaptation of the filters on the move, e.g. to allow or bar whole categories of alerts. The selection of which of the user's several context-specific profiles is active at any time is done either by the user explicitly, or implicitly (e.g. via their calendar), or may be triggered (with the consent of the user) by their entry to a new type of environment. The negotiation process should control not just the filtering of pushed offers, but the priority of all types of incoming signals and the modalities in which they are announced (c.f. Schmandt, 2000). Finally, we see the need for easy user control of the filter's tolerance to the unforeseen within any one context. This control may be closed down, only allowing strong matches through, or opened up to looser matches when the user has more time on their hands for surprise and exploration (See Fig 5).

7. Research Challenges and Current Work

Manifold challenges are posed by this work, not least of all in the user-interface and interaction design of a new handset to empower the user on the move in this brave new world of pushed opportunities and located social interactions. How can the user be alerted in a way which is 'as easy as looking at a watch' without overly distracting them from their primary task? How can the public, group and private virtual associations of a place be cued and secured? In the design of a beacon infrastructure, how to tradeoff, say efficiency and latency of a beacon-centred system against the trustworthiness of a user-centred system? What parts of the facilitation and interaction stages should be carried on which network, Bluetooth or cellular, and does that depend on the type of the service? How do people think about their 'context', instinctively categorize it, and at what granularity? We are attempting to explore these issues through evolving technology and design research with structured user involvement. Finally, we feel that the next major component to incorporate, after capturing something of the user's profile and sensing their external context, will be their focus of attention, if it can somehow be sensed unobtrusively.

8. Postscript: Mechanisms in a Purely Virtual World

This work is a direct descendant of a vision of a multi-user virtual world by Rankin et al (1998,1999). Many ideas were explored for equipping both abstract representations of both user's presence (a 'StarCursor' avatar) and virtual places ('Content Space') with auras, accompanied by novel mechanisms for a two-way interaction between people and responsive places. In turn, these echoed implementations of projection and awareness of presence, i.e. aura and nimbus, developed by Greenhalgh et al (1995). The StarCursor work attempted to link the user's interest profiles, set according to situation in a representation of their internal world, to alternative views of topic and social properties attached to their local external environment. Projections of personality between the users' StarCursors were automatically matched, controlled by their focus of attention, moderated by the proxemics of their body spaces and contextualized by the environment they shared. This vision of interactions in a virtual world can now begin to be realized in the physical world, through an augmentation of reality using RF auras emitted by beacons or handsets.

Acknowledgements

Many people from different design, human factors and technology disciplines contribute in the team behind this work. Thanks are due for the industry and creativity of: D Bell, R Blake, R Davies, P Fulton, D Gibson, J Griffiths, J Jansen, M Morgan, L Scholten, P Simons, D Walker, J Weston and D Yule.

References

Caswell, D., Debaty, P. (2000). Creating Web Representations for Places, in *Lecture Notes in Computer Science #1927*, Springer-Verlag, 115-126; also http://www.cooltown.hp.com/

Gibson, D. (2000). The Connected Pl@net: The Emergence of Bluetooth[TM] SupraNETS', *Bluetooth Developers' Conference*, San Jose, Dec. 6

Greenhalgh, C., Benford, S. (1995). MASSIVE: A Collaborative Virtual Environment for Teleconferencing, *Trans ACM Computer-Human Interfaces*, 2 (3), 239-261, Sept.

Halkia, M., Streitz, N. (2000). The role of roomware and sensing technology for supporting narratives in ubiquitous computing environments, workshop on *Situated Interaction in Ubiquitous Computing*, ACM CHI Conference, Den Hague, Apr 3; http://www.teco.edu/chi2000ws/proceedings.html

Philips (2000). Visions of The Future. http://www.design.philips.com/vof ; http://www.philips.com/connected

Pieper, R. (1998). From Devices to "Ambient Intelligence": The Transformation of Consumer Electronics, keynote address, *Digital Living Room Conference*, Laguna Beach, June 20

Rankin, P., Spence, R. (1999). A Contrast between Information Navigation and Social Navigation in Virtual Worlds, in A. Munro, K. Hook and D. Benyon (Eds.), *Social Navigation of Information Space*, Springer-Verlag, ISBN 1-85233-090-2, 174-194; Rankin P.et al. (1998). StarCursors in ContentSpace', *Proc. ACM Siggraph*, Orlando, 250

Samulowitz, M. (2000). Designing a Hierarchy of User Models for Context-Aware Applications, workshop on *Situated Interaction in Ubiquitous Computing*, ACM CHI Conference, Den Hague, Apr 3; http://www.teco.edu/chi2000ws/proceedings.html

Schmandt, C., Marmasse, N., Marti, S., Sawhney, N., Wheeler, S.(2000). Everywhere Messaging, *IBM Systems Jnl.*, 39, 3&4, 660-677

Wireless Strategic Initiative, European Commission (2000). The Book of Visions 2000, Visions of the Wireless World, http://www.ist-wsi.org

Natural Interaction Using the Adaptive Agent-Oriented Software Architecture (AAOSA)

Babak Hodjat and Makoto Amamiya

Dejima Inc., 160 W Santa Clara St. #102, San Jose, CA 95113

Kyushu University, Department of Intelligent Systems, Graduate School of Information Science and Electrical Engineering, 6-1 Kasugakoen, Kasuga-shi, Fukuoka 816, Japan

Babak@dejima.com,

amamiya@is.kyushu-u.ac.jp

ABSTRACT

We need to reverse the paradigm in which we contort the expression of our intent to the dictates of the machines functionality and instead turn functionality into an expression of our intent. In this paper we propose an Agent Oriented design for applications where agents interpret the intent of the user and map that onto the applications functionality. Our approach is based on the Adaptive Agent Oriented Software Architecture (AAOSA), a framework providing support for applications structured as communities of collaborative claim-based and message-driven components. An application in AAOSA is designed as a network of agents, where each agent is responsible for all its data structures and operations and takes part in the decision on how to employ them. Therefore, there will only be one agent, or sub-network of agents, responsible for all processes and data structures related to a certain sub-domain. As a result, to add agents to applications simply extends the functionality with only minimal – if any - disruptions to the original behavior of the application. Furthermore, AAOSA uses contextual information to coordinate and resolve contradicting claims, and will in some cases also rely on emergent behavior to achieve the desired result. Based on this approach, AAOSA makes it possible for applications to better incorporate context into the user interactions, thereby making applications more flexible and easy to use.

1. INTRODUCTION

The functionality of most appliances used to be limited, a "home theater," was a TV set with half a dozen broadcast channels, an on/off switch, and a volume control. The interface to the "home theater" was appropriately simple – a remote control with no more buttons than we have fingers on one hand. Today's "home theater" is another story, the TV set alone offers a multitude of choices such as color, contrast, brightness, programmable memory, 500 satellite channels, surround sound and other sound output options and picture in picture. Furthermore, the TV set is just one component in a "home theatre" system that includes a stereo, VCR, and DVD player, each with its own set of functionalities. To keep up, manufacturers try to map this functionality to buttons and other widgets, with the intention of presenting the user with the full scope of what the system can do in one interface.

The evolution of search engines into Internet "portals" has paralleled this tale of functionality sprawl. The problem of finding space for all the buttons and widgets is furthered by the Internet going wireless. Offering access from ever-smaller devices with limited screen real estate. Users are eager to try the new functionality, and they came away disappointed since there is no obvious way to access the promised functionality. The result of this being frustrated users and wasted functionality.

To attack this problem, we need to reverse the paradigm in which we contort the expression of our intent to the dictates of machine functionality and instead turn functionality into an expression of our intent. We need machines that listen to and understand users, and to achieve this we have to rethink the user interface (UI) paradigm. To get a better understanding of what we propose for the new paradigm, let us look at an

Figure 1. The interface should interpret the intent of the user, rather than the user having to learn the interface.

example where we have a machine with a single button. In the old UI paradigm, we would have to know where the button is located, what it will do for us, and how to operate it. This is not in accord with our preferred rule of having the machine listen to us. For our new UI paradigm, the button needs to understand our intent, and map that to what it can do for us. The button would "listen" to us and then decide when it should push itself. This decision could be based on a combination of input modalities, not restricted to speech or text; it may be the status of the machine, our gestures, or us physically pushing the button.

The best interface to machine with a single function is probably a single button. Making the button smarter would probably not be much of an improvement. There is a simple correlation between the operation of the button and our means of operating it. This is the reason why the first remote controls and the first Graphical User Interfaces (GUIs) were simple to use. The complexity of the interface used to be manageable. This is no longer true; applications today have more complex functionality, and even when the functions are simple, the sheer number of them makes them hard to access through conventional interfaces. This is where the new UI paradigm, with buttons that listen, would make a difference. Instead of adding buttons to a UI with limited real estate, this approach allows for scalable UIs by simply having the existing UI listen for the new functionality.

The buttons in the new paradigm still represent a function, however, in addition, they are also responsible for interpreting the intent of the user beyond recognizing the user physically pushing the button. In our approach we represent the functionality of a system, i.e., the buttons, with independent software modules we call *agents*. Our agents are smart enough to coordinate their decisions with one another. Thus, if multiple agents believe that they are the best candidates based on their interpretation of your intent, coordination mechanisms make sure the agents with the strongest case will serve you. There are times when you intend more than one function to be activated, and there are times when the agents simply cannot make a decision based on the available criteria. In these cases, the agents make themselves known by interacting with you to resolve any uncertainty. This is the only time you need to become aware of the existence of the agents. As a side effect, a user will learn the functionality of the system through interactions, and these interactions only occur if the system fails to map intent to functionality.

Using this approach, the buttons disappear; the system is now interpreted in terms of its functionality rather than in terms of its UI. Using the TV set as an example, without the conventional UI you would not have to memorize the TV channel numbers. In our new paradigm, you would not even have to know the network names. The important information would be the type of program *you* would want to see. Leaving it to the agents to map the program types to appropriate channels and network names. This approach of mapping behavior, making claims and collaborating is the essence of the *Adaptive Agent Oriented Software Architecture* (AAOSA), an Agent Oriented Software Engineering Architecture (Iglesias *et al.*, 1998;Jennings *et al.*, 2000)

2. OVERVIEW OF AAOSA

AAOSA provides a programming environment that supports designers and developers by managing the complexities of how to design communication and coordination in a community of agents (Hodjat, Amamiya, 2000). As a direct result of this, agent definitions are declarative rather than procedural. These properties enable designers to focus on the task of modeling the real application instead of spending time on basic support functions.

AAOSA imposes no preferred size on the collaborative agents; rather the granularity of the agent network corresponds to the complexity and inherent parallelism of the application. The divide and conquer approach is used when designing applications; modeling the application in terms of manageable sub-domain elements (i.e., AAOSA Agents). The communication paths between agents will de facto organize the agents in a network or graph, this network can be strictly hierarchical or an arbitrary connected graph. In addition to the architecture of the agent network, the software designer also provides each agent with a set of interpretation policies. These interpretation policies are a rules used to decide when a particular agent should claim responsibility for processing any part or all of the input to the application. The policies are defined in terms of claims, delegations, and disambiguation actions. An agent that claims an input can either process that input, delegate the input to its down-chain agents, or both. Furthermore, the interpretation criterion is not limited to the message content of the input. Other factors such as, process history, probabilities, and outside information (e.g., interaction with other agents) may be used to derive a claim. Interpretation policies are defined using OPAL, a declarative programming language designed as part of AAOSA.

Furthermore, no centralized control is enforced in a network of AAOSA agents. Instead, agents introduce themselves and their abilities to one another at the beginning or during execution. Agents can therefore be added to or removed from the application at runtime.

The AAOSA Software Engineering platform applies powerful Distributed Artificial Intelligence (DAI) techniques (Durfee *et al.*, 1989; Agha *et al.*, 1993; Durfee *et al.*, 1994) implementing the following capabilities:
- Complete encapsulation of responsibilities: Each agent is responsible for all its data structures and operations and the decision to employ them.

- Emergence: When input to a system is unpredictable or restricted, we rely on emergent behavior to achieve the desired output. The emergent behavior is based on agent interactions and is the result of coordination mechanisms in the core of AAOSA.
- Learning: Each agent has learning capabilities to improve its behavior over time.
- Extendibility: The functionality of an agent network is extended by adding agents or by merging with other agent networks. This will cause minimal disruption in the original network behavior.
- Reusability: Agents and agent sub-networks can be reused in different applications and across multiple applications.
- Wrapping legacy code: Existing code can be wrapped and used as an agent in the network.
- Fault tolerance: Removing an agent or a link from the network only reduces the functional scope and does not affect the overall application. Conflicts that occur due to agents misfiring claims are handled by the coordination mechanism in the core system.
- Distributability: This is a result of the system being message-based.
- Multiple entry and exit points: Any agent, regardless of level, accepts multiple simultaneous inputs to the system. In the same manner, output from the system can originate from multiple agents within the agent network.

3. USING AAOSA FOR NATURAL INTERFACES

Creating a grammar is a complicated task, changing grammars based on learning is difficult, and grammars alone are not enough to fulfill our requirements for Natural Interaction Interfaces (NII). A parser can only tell if the input string belongs to the language or not, and our objective is to find the best match for any given input. These and other practical problems using grammatical parsers for natural interaction interfaces prompted AAOSA to implement some improvements (Hodjat et al., 2000). Resulting in a system that can accept non-grammatical input (e.g., "Tea for Jila bring!"), and handle previously un-encountered input (e.g., "Yabadabadee some milk for me!").

The interpretation policies in AAOSA are less restricted than the production rules of a parser. For instance, rather than to require that claims, on which new claims are based, to be in sequence, we only require them to be exclusive. The interpretation policies will determine what the best reduction condition is, and each agent will compute a confidence factor for its claims based on how much its claim differ from the desired one. Using a threshold, claims of higher confidence are used as results. Another difference between the parser approach and our natural interaction system is that context considered in the reductions of a context-sensitive grammar is limited to the input, while in AAOSA the decision to make a claim may be made based on context information not necessarily present in the input. For instance, an agent may decide to make a claim based on the history of successful claims, the status of the semantic domain it is representing, or based on interactions with the user.

Contextual information can also be used during coordination to resolve contradicting claims from agents. Various methods can be deployed to resolve ambiguities. Using the Natural Interaction Interface for A/V equipment, depicted in figure 2, ambiguities occur when more than one agent claims an input or no agents do. In the A/V interface, we have both a DVD and a VCR agent. Each would probably claim the input "play". This would result in an ambiguity in the "A" agent that is trying to decide to which one of its down-chain agents, DVD or VCR, to delegate the input. The information used to resolve this ambiguity is what we call context. What is more important is *how* this context information should be applied. Some examples in this case would be:

Recency: "Did the most recent request that was delegated to Agent "A", belong to the DVD or the to the VCR, and how long ago was this?"

Functional Status: "Is the DVD or VCR playing right now?"

Probability: "Does the user usually mean the VCR when she says, "Play"? The word "usually" here translates into some statistics maintained by the system.

Secondary information: "Is the speaker pointing at the VCR or at the DVD?". This information that could be sent down as part of input from the finger-point agent.

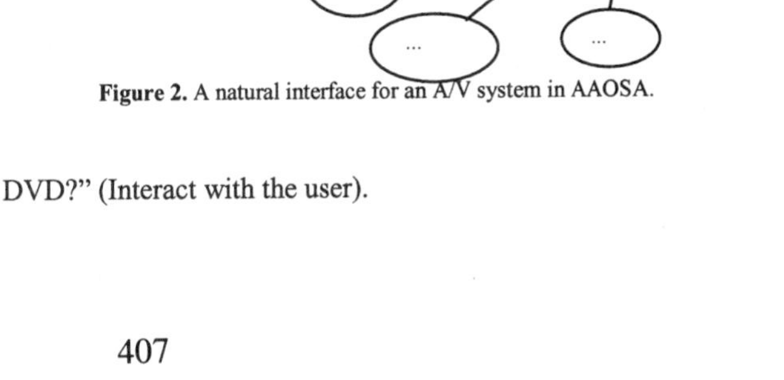

Figure 2. A natural interface for an A/V system in AAOSA.

Interaction: "Do you mean the VCR or the DVD?" (Interact with the user).

Ambiguity: The user may refer to something totally different. The ambiguity occurred in the "A" agent, this is in itself context information to help resolve ambiguities at higher levels in the UI network.

4.   LIMITATIONS OF CURRENT NATURAL INTERACTION METHODS

Word spotting is used in many search engines and computer games. This method is fast and effective in small applications, but not extendable to larger application where users often need to actuate different combinations of functions. The same is true when the same phrase is used to mean different things based on context (e.g., presence of other words, history of the interaction, etc.).

Linguistic methods, in particular grammars, are traditionally used to model the lingual input to natural interaction interfaces. These methods require linguistic skills and the resulting models are difficult to use across applications. The models are language dependant and porting an application from one language to another requires going through all the development steps again.

Statistical methods are usually too inaccurate to be used independent of a language model. These methods try to predict the occurrence of words, and they depend on the availability of a large corpus. It is difficult to tune these methods using thresholds and parameters, and they normally require tweaking by a human expert. Furthermore, the methods are based on the behavior of a specific target audience and have problems adapting to individual users or different audiences.

Semantic frames are used for both linguistic and statistical systems to limit the scope of probable input and simplify the models. This simplification is important when the input modality is at risk of losing accuracy, such as in systems that use speech recognition. A problem here is to help the user navigate through these frames. This is usually accomplished by introducing interactions and dialogs, reducing the system to a semi-menu based interface.

5.   THE EMAIL AND CONTACTS NATURAL INTERACION INTERFACE

We have implemented a NII for an e-mail and contacts application, depicted in figure 3. The NII supports 59 different mailing and contact management functions.

More than 50 agents were used in this application that was designed based on a corpus collected from 30 users. 3000 natural language queries were collected and classified in three categories of simple, compound, and out-of-functionality. The simple corpus consisted of 1414 queries that either invoked a single function in the system or caused a single ambiguity. The compound corpus was 500 queries that invoked more than one function in the system simultaneously, or caused more than one ambiguity. The out-of-functionality corpus was 130 queries that were classified as outside the defined functionality of the application and were tested for the proper ambiguity resolution interaction.

The simple corpus was split at random into

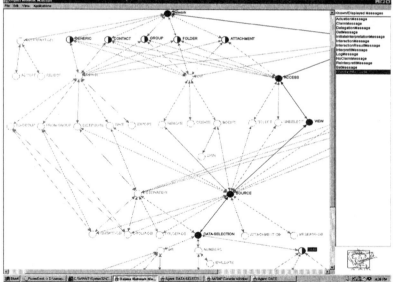

Figure 3. A part of the hyperstructure used for the e-mail and contacts NII.

Simple Queries	Compound Queries	Out-of-functionality Queries
Remove contact with 6504040330	show me John ostrem's	Pass along unread messages to
Open John's contact	address	hank
Delete current message	Get John's info printed	Trash for contact with email
Print messages smaller than 500	Print all the e-mails	address sun
	received today	Forward contact to phone 2222222
	Fax John's message to	Pass along contact to email address
	Albert and print it.	bad.com

Table 1. Sample of the AAOSA e-mail and contacts NLUI corpus.

408

a development corpus and a control group. Development took 10 days for 2 engineers, each working on different agents.

When we tested the system, the overall success rate was 91.7%. The simple corpus test yielded a 98.9% success rate, while the compound corpus test reached a 91.4%. The out-of-functionality test yielded a 51.6% success rate. The control group, set aside from the corpus used when developing the system, yielded an 85% success rate. Listed in Table 1 are some of the sample queries from all three groups, simple, compound, and out-of-functionality queries that were successfully handled by the system.

6. CONCLUSIONS

An application in AAOSA is implemented as a network of agents, where each agent is responsible for all data structures and operations and takes part in the decision to employ them. As a result, there will only be one agent, or sub-network of agents, responsible for all processes and data structures related to a certain sub-domain. Messaging is the only means for communication between agents, which makes it possible to distribute these modules where needed.

Responses from the system may be un-predictable due to the lack of restriction on the input, and in many cases we rely on emergent behavior to achieve desired output. Emergent behavior is a result of the combination heuristics used by the agents. Developers can thus develop an application based on a set of input strings paired with matching outputs and rely on the emergent behavior to provide the best result for previously un-encountered input.

Adding or merging agents with other agent networks extends the functionality of the application and will cause minimal disruption on the original behavior of the application. It is possible to both add and remove agent sub-networks in a NII implemented in AAOSA as the system is running; this will only change the lingual and contextual scope of the application. Agents and agent sub-networks can also be reused in different applications and across multiple applications.

Input to the system is accepted at any level of the agent network, and can be handled by any agent. Each agent also accepts multiple simultaneous inputs. Furthermore, output from the system can originate from multiple agents throughout the application.

AAOSA provides an elegant solution to the complex problem of creating a NII. Ungrammatical and unpredicted input will be accepted by the system, which downgrades gracefully to a dialogue system to get the user to the desired functionality. Different inputs can be used to accomplish the same functional result. Contextual information can be used during coordination to resolve contradicting claims from agents. Interacting with the user can be used during coordination as a means to resolve contradicting claims from agents. This also brings about the means to carry out smart dialogues where users may deviate from answering the system directly by adding extra information or changing the context established in prior interactions of the dialogue.

REFERENCES

Agha, G., Wegner, P., Yonesawa, A. (1993) *Research directions in concurrent object oriented programming*. The MIT Press: Cambridge, MA.

Durfee, E, Rosenschein J. (1994) "Distributed problem solving and multi-agent systems: Comparisons and examples". Proc. of the 13th *International Workshop on Distributed Artificial Intelligence (IWDAI-94)*

Durfee, E, Lesser, V., Corkill D. (1989) "Trends in cooperative distributed problem solving". *IEEE Transactions on Knowledge and Data Engineering, 1(1):63—83.*

Hodjat, B., Amamiya, M. (2000) "Applying the Adaptive Agent Oriented Software Architecture to the Parsing of Context Sensitive Grammars", *IEICE Transaction on Information and System, Vol.E83-D, No.5, pp. 1142-1152.*

Hodjat, B., Savoie, C., Amamiya, M. (1998) "An Adaptive Agent Oriented Software Architecture", Proc. of the *5th Pacific Rim International Conference on Artificial Intelligence (PRICAI '98)*

Hodjat, B., Amamiya,, M., (2000) "Introducing the Adaptive Agent Oriented Software Architecture and its Application in Natural Language User Interfaces", in *Agent-Oriented Software Engineering*, Springer-Verlag,

Iglesias, C.A., Garijo, M., & Gonzalez, J.C. (1998) "A Survey of Agent-Oriented Methodologies". In Proc. of the *Fifth International Workshop on Agent Theories, Architectures, and Languages (ATAL-98).*

Jennings N., Wooldridge, M. (2000) "Agent-Oriented Software Engineering" in *Handbook of Agent Technology* (ed. J. Bradshaw) AAAI/MIT Press.

Simulating user integration during software development

Insa Raue, Christian Leutloff

Department of Computer Science in Mechanical Engineering (HDZ/IMA), University of Technology (RWTH), Dennewartstr. 27, 52068 Aachen, Germany, http://www.hdz-ima.rwth-aachen.de

Abstract

Modern software is being developed in order to fit certain needs of its future users. Therefore it is necessary to involve the users into the software development process especially during the specification and testing phases of the product. But unfortunately there are cases where it is not possible to contact the final users until the software has been finished and delivered to its customers. This paper introduces a successful approach to *simulate* user integration during the software development process.

1. Introduction

In this paper a new approach is presented to integrate aspects of human-computer interaction into the software (SW) development process. The approach is designed for cases when it is not possible to involve the final users into the SW development process before the software has been finished. This approach has been developed and tested in co-operation with a large German software company.

It is very important to integrate the final user into the software development process in order to build the software system with appropriate human-computer interaction. Problems arise, however, if the users cannot be involved into this process, e.g. due to strategic reasons, or because the customers simply do not want any communication between users and software developers. Hence a new approach is suggested in this paper to involve the future users into the software development in a *simulated* process.

On the one hand, the new approach has to ensure that the requirements of future users are collected and incorporated into the software development process. On the other hand, it has to be easy for the software developers to work with the approach in order to ensure that this approach is really *utilised* during the software development process.

2. The Methodology

2.1 The software development process

The current phase models of the software engineering process are frequently based on the *Waterfall Model* suggested by Royce (1970). It can be characterised by the following features. The development process is performed in different sequential steps during the linear non-cyclic process (Figure 1). Users are only integrated in the start phase of the development process. With this linear phase model, the typical steps of a software development process consist of *analysis* (system specification), *design* (system construction), *implementation* (module programming, coding) and *test* (system integration) (Denert 1991; Henning/Kutscha 2000; Balzert 1998).

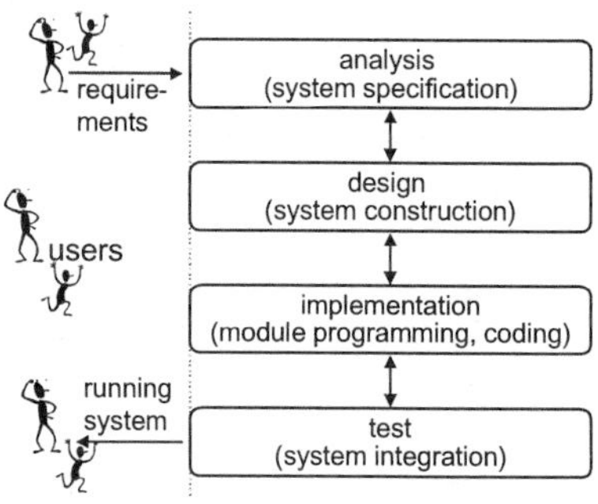

Figure 1: Linear steps of software development (Stoffels 2001; Denert 1991)

2.2 The new approach to user integration

To develop a high-quality software system the context of use must be part of system specification. If the users cannot take part in the specification definition, their work processes can be *simulated* including their utilisation of the software-in-use. In this way the software developers are able to understand the main modes of operation, functions and ways of acting of the users. The approach thus suggested is as follows. The software developers observe the future users at their work and develop concepts as to how to *role-play* these working processes in a laboratory setting. Subsequently this setting integrates *prototypes* of the new software-to-be-developed. Thus they *simulate* the use of their own software prototype within the future work setting including the new software. This laboratory setting uses the concept of *Human-Integrated Simulation* (HIS). The approach can be considered as an integrated form of *role-play* and *prototyping* (Figure 2).

This approach includes four steps (Figure 3).

The *first step* is to become familiar with the present working environment of the users and to analyse their working dialogues. It is necessary to define in detail the problems, to describe the present working environment and its system borders and also the interior world of the software-based dialogues. Overall, this step has the aim to become acquainted with the users' activities and tasks in the present working environment.

In the *second step* the users are described in detail and the scenarios of the future working settings can be written. The scenarios for the *simulation* are designed as if the software system is to be used in *reality*. This includes the description of the different roles of the future users involved.

The *third step* is the simulation with the software developers acting on behalf of the future users. They are playing the roles previously defined. It is important already to set up the completely coded prototype software for the simulation in order to make real working experiences possible. Thus the simulation can come very close to the reality of the working situation of the future users. Two types of roles are to be distributed: The simulators and the observers. The *simulators* pretend to be the real users and *play* them by speaking out loudly what they think about the process, the technical support, the software dialogue and the ergonomics of the interface. The *observers* watch the simulators and write down what they see, hear and think. It might help to video-record the simulation.

During the *fourth step* of reflecting, the simulation participants collect all observations, analyse them and draft the subsequent actions. Thus the analysis of the simulation helps to make decisions about the next steps of software development and improvement.

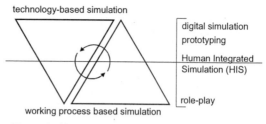

Figure 2: Simulation of software use as a socio technical process (Tschiersch/Kesselmeier

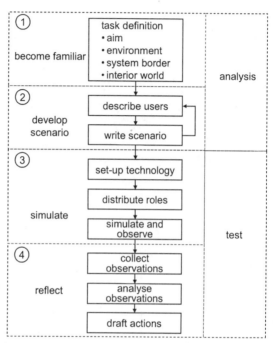

Figure 3: Human-Integrated Simulation of software development

3. The Case study

3.1 The simulation

The approach outlined above was successfully tested within a large-scale software project. Since 1996, several different software companies and up to 100 software developers were involved in this project. The task was to develop completely new software for a new world-wide customer order procedure of a large German car manufacturer (Stoffels, 2001). One particular difficulty was in that the new software was to be integrated into the existing software system of the enterprise. The development included the processes from the creation of a virtual car according to the customer's wishes, up to the production and delivery of the car. The worldwide structure of the global ordering procedure is shown in Figure 4. This ordering process was to be turned upside down by the new software system (Figure 5). Subsequently it is presently changing the manager-controlled production of today into a

411

production strategy of tomorrow which is directly controlled or triggered by customers and sales personnel. Because of its size, the software development project was divided into a series of sub-projects. Each of them were run by 3 to 10 software developers. Thus they produced their software sub-systems or modules in parallel following well-defined strategic steps.

This highly complex project design needed the integration of the future users: the car dealers who are distributed world-wide. It would have been nearly impossible to recruit representatives of these dealers in

Figure 4: Structure of the global ordering procedure

advance to test new software during its prototype phases. Therefore the *Human-Integrated Simulation* was chosen to test certain new software sub-systems and modules during their development process. The first run of this simulation was performed in a 1½-day workshop. The eight participants of the simulation were members of two different SW development teams. Two participants belonged to the team developing certain sub-systems of the new *Global Ordering* system comprising new software-based dialogues between car dealers and central manufacturing plant. This dialogue is to enable the sales personnel to order cars directly from the central manufacturing plant without waiting for the decisions of the ordering administrators who are still centrally located. Two further participants were management staff of the leading software company in the project: the chief designer and one of the project managers. Further participants were software developers who had not been involved into the development of the particular sub-system (or module) to be tested.

During the first half-day, the software developers performed *the first and second steps*. They defined the main scenario of using the new software sub-system including the roles of its different future users (Figure 6). The scenario is simple: Every car dealer around the world has usually at his disposal certain well-defined contingents of different variants of cars to be sold. If a customer wants a specific variant of a car which is not comprised in the contingent of this particular dealer, this dealer negotiates with another dealer in order to get hold of this particular variant for his customer. These two dealers need to involve their *wholesale* dealer into this process of negotiating about the different car variants within their respective contingents, because of the present hierarchical structures of car sales world-wide. The dialogue of these three different dealers is supported by one of the new software sub-systems developed within the Global Ordering system. It is to be tested in this simulation.

The *third and fourth steps* took place on the next day. The participants simulated this scenario by playing the roles of the users corresponding, i.e. the two car dealers and the wholesale dealer. During the simulation, each actor was talking loudly about what they were thinking and experiencing. Therefore it was possible to collect essential information on the new software sub-system being tested, particularly its interfaces. Guiding questions were posed in order to structure the information collected: What features of the new software are useful, which need to be improved? What information and SW functions are missing? How do information and functions need to be presented on the interface to comply with both optimum work processes and SW ergonomics?

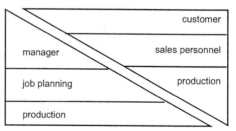

today:

manager-controlled production

Figure 5: Global production control through customers' orders

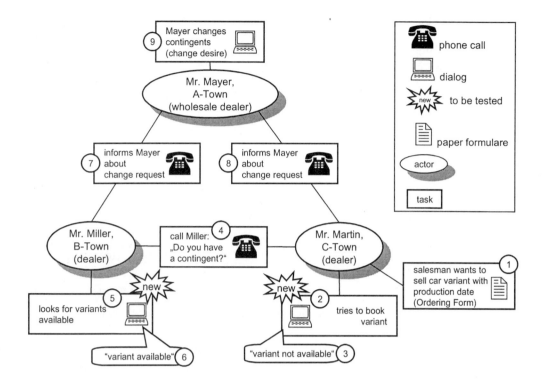

Figure 6: Specific scenario for the simulation of global ordering

3.2 Results

After this simulation, the experiences of the participants were carefully analysed. The result was a collection of clear statements concerning the usability of the new software sub-system. Through the simulation and its analysis, a series of new ideas were produced to improve the human-computer interaction and its user interface. Furthermore the simulation yielded the following two results:

- The in-depth exchange of information and knowledge transfer between the software developers. It included better to understand the tasks which the future user will face when using the new software system.
- The experiences of the software developers to observe the software used in a situation which was very close to reality. This simulation guided the software developers from the *technology-centred* view to the *task-oriented* view of future users.

It has been agreed by the companies involved to add this new simulation approach to the *project guidelines* for software development. Furthermore, it seems to be feasible to use the simulated scenarios in order to train the future users.

4. Integration of the user simulation lab into software development process

The Human-Integrated Simulation is easily integrated into the software development process (Figure 7). The simulation cases are determined and described during the *analysis phase* as they are part of the system specification. Important is that these simulation cases are user-driven and not merely technology-oriented. During the design phase, the simulation cases can be used to verify the optimum design. During the *implementation phase* the developers can use the simulation cases to check that their modules are working properly. The final simulation of the cases is used to *test the whole system*. This seamless integration ensures that the needs of the final users are present during the whole SW development process.

413

5. Conclusions

The user simulation laboratory within the software development process allows to improve the user interface even in cases when it is not possible to get the future users involved. This is achieved through the simulation of the tasks of future users using role-based scenarios. For this task it is necessary to have available within the simulation, the main functional dialogues of the new software system. They need to be linked to the data bases of the manufacturer in order to be used *in real time* during the laboratory simulation. Also there should be enough time

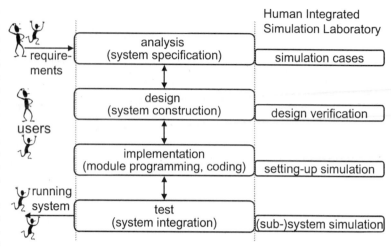

Figure 7: Integrating the simulation approach into the SW development strategy

left after this simulation until the SW distribution date in order to put the results of the simulation into software improvements.

The participants of the simulation should represent different know-how about the SW-based dialogue and its working context. Perhaps half of the software developers should have such expert knowledge (e.g. chief designer, project manager, developers of the particular dialogue) the other participants may be outsiders of this particular part of the SW project.

In this way the Human-Integrated Simulation can gain strong impact on present strategies of software development in the interest of the future users.

References

Balzert, H. (1998). *Lehrbuch der Software-Technik*. Heidelberg, Berlin: Spektrum, Akad. Verlag.

Denert, E. (1991). *Software-Engineering*. Berlin, Heidelberg: Springer-Verlag.

Henning, K. & Kutscha, S. (2000). *Informatik im Maschinenbau*. Band 31, Aachener Reihe Mensch und Technik. Aachen: Wissenschaftsverlag Mainz.

Kesselmeier, H. & Tschiersch, I. (1994). Interaktive Echzeitsimulation –Kombinierte Simulation im Kombinierten Verkehr. In Henning, K. & Harendt, B. & Messelken, M. (Hrsg.) (1994). *Güterverkehr der Zukunft*. Band 5, Aachener Reihe Mensch und Technik. Aachen: Wissenschaftsverlag Mainz.

Royce, W. (1970). Managing the development of large software systems. In *IEEE WESCON*, Aug. 1970, pp. 1-9 (Reprint in Proceedings of the 9th International Conference of Software-Engineering, 1987, Monterey, CA, pp. 328-338).

Stoffels, B. (2000). *Regelkreismodell für die Software-Entwicklung von weltweiten, käufergesteuerten Auftragsabwicklungssystemen in der Automobilindustrie*. Fortschritt-Berichte VDI. Düsseldorf: VDI-Verlag.

Web Document Ranking by Differentiated Expert Group Evaluation

Sea Woo Kim* and Chin-Wan Chung†

*Department of Information and Communication Engineering, †Department of Computer Science
Korea Advanced Institute of Science and Technology
373-1, Kusong-Dong, Yusong-Gu, Taejon 305-701, Korea
seawoo@ngis.kaist.ac.kr, chungcw@islab.kaist.ac.kr

Abstract: This paper introduces a new concept to evaluate web documents through human computer interaction. Together, the developments of the Internet and the popularity of WWW, search engines for web documents have drawn a significant attention. Many search engines have been introduced and developed, but still have difficulty in providing relevant answers to the specific subjects of queries. Currently, a single expert or general users do human evaluation without considering the degree of domain knowledge of evaluators. In this paper, we suggest an automatic creation of a dynamic group of experts among users to evaluate domain-specific documents for the ranking of web documents. The experts have dynamic authority weights depending on their performance of the ranking evaluation. This evaluation by a group of experts provides more accurate search results for users, and reduces navigation efforts more effectively than the evaluation by a single expert.

Keywords: search engine, web document rank, human computer interaction, expert group,

1. Introduction

It is not easy to develop desirable search engine that locate information and tailor it to users' needs in spite of the rapid development of the Internet. The main reason is not due to the lack of data but is related to providing an excess of data with a variety of characteristics and information not specific enough to users. An alternative to the traditional method is to use a ranking measure to improve the quality of search results. Actually, the connectivity of hypertext documents has been found to be a good measure for automatic web citation. This method works on the assumption that a frequently cited document is more popular and important. Many automatic ranking systems have used this citation system to measure the relative importance of web documents. IBM HITS system maintains a hub and an authority score for every document. The hub scores of its parent documents are summed into its authority score. The authority scores of its children documents are summed into its hub score (Gibson 1998) (Kleinberg 1998). By iteration, it determines highly referenced documents. A method called PageRank is suggested to compute a ranking for every web documents based on the web connectivity graph (Brin 1998) with the random walk traversal. It also considers the relative importance by checking ranks of backlink documents, which means that a document is ranked as highly important when the document has backlinks from documents with high authority, such as the Yahoo home page. However, the automatic citation analysis has a limitation in that it does not truly reflect the importance of the varying viewpoints of human evaluation. There are many cases where simple citation counting does not reflect our common sense concept of importance (Brin 1998). Also web documents and academic publication papers are significantly different in terms of the citation analysis. HITS (Kleinberg 1998) and PageRank (Brin 1998) determine the quality or authority of web documents on the basis of the number and quality of the web documents, which are linked to the document. However, the methods for document ranking are static and unable to adapt themselves to a new environment of web patterns.

This paper explores the method of developing a new ranking technique based on human interaction. While the previous approaches have considered topological links of the web, a combination of a broad search of the entire web with domain-specific textual and topological scoring of results is suggested (Aridor 2000). They have knowledge agents

415

to specialize in a specific domain in order to extract the most relevant documents for a given query within a domain of interest. Similarly we have been interested in ranking web documents in a specific domain. In our approach, our search engine goes through the process of a domain-specific web search and shows the list of popular documents for each specialized subject. Each domain is user-defined and can be of any granularity and specialty. The connectivity or pattern of link topology between documents contains implicit information about relative importance of links. Some search engines employ the method based on textual similarity (Aridor 2000)(• • • • • • • • • • • •. Normally they count the frequency of terms in a domain to decide lexical affinities instead of using advanced natural language processing techniques. Yet even then the textual similarity analysis has their own limitations.

We suggest that the importance and level of authority of web documents should be determined by interactions between humans and web documents and domain groups should be responsible for web document ranking for each category. This approach will overcome the disadvantages of the automatic ranking method through incomplete information processing based on citation authority or lexical affinities, which ignore the content of web documents. Currently, we are developing a domain specific search engine that is applicable for a specific item or domain when we want to see the list of top ranked documents for a specialized subject such as recommendable utility software, popular comics, movie lists, and mp3 documents.

This paper is consisted of 4 sections. Section 1 is an introduction, and section 2 is an illustration of methodology of our system. Section 3 is the description of experiment. And section 4 discusses the conclusion and future works.

2. Methodology

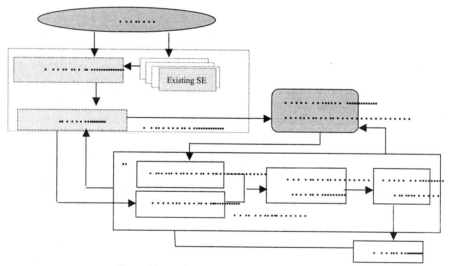

Figure 1 Dynamic Evaluation by Expert Group

A meta-search engine is run to collect addresses of cited web documents from conventional search engines as shown in Figure 1. Addresses of the documents and search engines citing the documents are stored in the document DB. During a query the ranking by the number of citations from a search engine and expert ranking are combined. The combined rank documents are shown to the users and they can be able to visit the web document they decide to explore. These web documents are recorded and monitored. The method of employing an expert group is based on the idea that for a given decision task requiring expert knowledge, many experts may be better than one if their individual judgements are properly combined. In our system, experts decide whether a web document should be classified into a top ranked list for a given category. A simple way is the majority voting system, where each expert has a binary vote for a web document and then the document obtaining equal to or greater than half of the votes are classified into a top ranked list (• • • • • • ••••••

416

1989). The result of the decision for documents is stored along with their addresses. Another possible method we can consider is a weighted linear combination. A weighted linear sum of expert voting results yields the final decision documents. In this paper, we take the adaptive weighted linear combination method, where the individual contributions of members in the expert group are weighted by their judgement performance. The evaluation of all the experts are summed with weighted linear combinations. The expert voting results will dynamically change depending on each expert's performance and credibility. Our approach of expert group decision is similar to a classifier group concept in automatic text categorization (• • ••2000). Their methods use classifiers based on various statistical or learning techniques instead of human interaction and decision.

For each candidate document, experts are required to evaluate the document and make scores. An evaluation score matrix is defined as $X = [\chi_{ij}]$ when the i-th expert evaluates a web document dj with a voting score χ_{ij}. We have a weighted importance or authority over experts for each category. The evaluation score matrix shows a relation between experts and candidate documents. Experts are assigned the same weight at the initial stage, because experts are not differentiated at this stage. The weight is dynamically changed by their activity and feedback from online users about recommended document through voting results. For each web document dj listed as a candidate document, the weighted average score is defined as follows :

$$V(d_j) = \sum_{k=1}^{Ne} r_k \chi_{kj} = \sum_{k=1}^{Ne} \frac{w_k}{\sum_{i=1}^{Ne} w_i} \chi_{kj}$$

where N_e is the number of experts for a given category, and r_k is the relative authority for the k-th expert in the expert pool, and w_k is the weight for the k-th expert member. We suppose w_k should be positive all the time. The weight w_k is a dynamic factor, and it distinguishes bad experts from good experts in terms of their activity and users' voting results.

When some experts show little participation over voting or evaluating incorrectly, their authority weight w_k becomes smaller. For example, when an expert voting score is larger than the weighted average voting score and the average score is smaller than the desirable rank score, the expert is rewarded, or otherwise penalized. Therefore, some experts receive rewards and others receive penalties depending on the weighted average voting score of experts. All experts receive penalties or rewards in our application. It is possible that some experts with too many penalties are excluded from the expert group and new experts are added to the group. However, when users are required to evaluate visited documents, it is burdensome for general users, therefore they will be reluctant to provide feedbacks. In addition, score criteria of general users may be different from those of experts, which will make it difficult to compare numeric scores between experts and general users.

We define an error measure E as a squared sum of differences between desired voting scores and actual voting scores as follows:

$$E = \frac{1}{2} \sum_{j=1}^{n} (V(d_j) - V'(d_j))^2 = \frac{1}{2} \sum_{j=1}^{n} (\sum_{k=1}^{N_e} \frac{w_k}{\sum_{i=1}^{Ne} w_i} \chi_{kj} - V'(d_j)^2)$$

where n is the number of documents evaluated by users, $V'(d_j)$ is the desired voting score for an expert-voting document d_j. We assume this value can be determined by the feedback from general online users. Voting scores of experts should reflect a common idea of users about ranking and satisfy desires of many users to find good information because expert groups are representative of general users and have a good knowledge in a specific domain. Now we use a gradient-descent method over error measure E with respect to a weight w_k and the gradient is given by

$$\frac{\partial E}{\partial w_k} = \sum_{j=1}^{n} [\chi_{kj} - V(d_j)] \frac{\Delta_j}{S}$$

where $S = \sum_{k=1}^{N_e} w_k$ is the sum of weights, and $\Delta_j = [v(d_j) - V'(d_j)]$ is the difference between predicted voting score and users evaluation score for a document d_j.

3. Experiments

We designed a small domain meta-search engine to test our approach, currently only for a movie search engine, which is ultimately aimed to be a general-purpose search engine. We first tested the possibilities to run many expert groups on our search engine through a simulation program. We simulated the dynamic process of web document ranking and creation of expert groups. It also shows the transition of experts through activity of users and the re-grouping process of experts through feedback of users.

We have used a simulation program to run expert groups for web document ranking. However, it is quite different from the real world in many respects. Some experts with good knowledge may have different opinions and views from general users and in reality the access rate of each user is not consistent. The behavior or access pattern of users will vary from category to category. In some categories, it may be difficult to assemble good experts. The activity or interest field of a user does not mean that he/she has expert knowledge.

Figure 2 shows the increase of user's satisfaction as the number of user evaluation increase. The user satisfaction is calculated by the closeness of expert evaluation to the users evaluations. Simulation is held with random expert creation and expert evaluation tendency is also defined as random. The general user evaluates recommended documents and we calculated the difference between them using error measure E. We calculated normalized E and substacts E from 1 for obtaining user satisfaction measure. Experiments were performed 4 times and all of experiments showed the correlation of user satisfaction and user evaluation. In the future, real user data will be applied instead of simulation result.

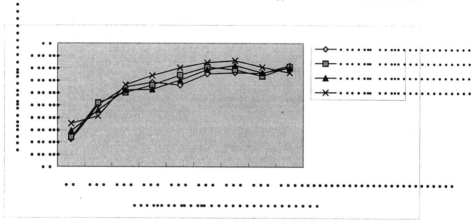

Figure 2. User satisfaction vs User evaluation

4. Conclusion and Future Works

In this paper, we have shown a dynamic evaluation for web document ranking through human interaction. An expert group is automatically formed among users. Each expert has his own authority to evaluate web documents. This authority is dynamically changed using the feedback of users. The expert group evaluation is more attractive compared to other search engines, which can receive more human feedback. Expert groups are formed from the search engine users and each expert evaluates web documents with different authority weights.

Our search engine is distinct in the following aspects. First, experts with domain-knowledge evaluate a web document that leads to a more accurate result. Second, as a meta search engine, our search engine collected URLs from existing search engines such as Yahoo and Altavista. Therefore, URL data size is large enough for providing basic results even when the size of expert recommended documents is small. Third, since domain knowledge is applied in each domain

418

therefore an intelligent search is possible.

Currently, our categorization for all documents is user-defined but not automatic. This text categorization requires much labor cost and time. However, we still need further work to develop a text categorization process to filter out documents irrelevant to a domain with a given query. Also, we will apply our concept to the consumer's evaluation of products in electronic commerce. For example, the consumer will be more informed of good points or weak points of products from an expert's point of view. This information is especially useful when the evaluation size of general customers is not big enough to provide adequate information. Also, other areas requiring domain knowledge will be considered for expanding our system.

References

[1] Aridor, Y., Carmel, D., Lempel, R., Soffer, A. & Maarek, Y., Knowledge Agents on the Web, in Fourth International Workshop on cooperative Information Agents, 2000.

[2] S. Brin & L. Page. The anatomy of a large-scale hypertextual web search engine, in proceedings of the Seventh International World Wide Web Conference, 1998

[3] D. Gibson, J. Kleinberg, & R. Raghavan, Inferring web communities from link topology, in Proceedings of the ACM Conference on Hypertext and Hypermedia, 1998.

[4] J. Kleinberg. Authoritative sources on a hyperlinked environment. In Proceedings of the ACM-SIAM Symposium on Discrete Algorithms, 1998.

[5] R. Liere and P. Tadepalli. Active learning with committees for text categorization. In Proceedings of the AAAI-97, 14th Conference of the American Association for Artificial Intelligence, pages 591-596, 1997.

[6] F. Sebastiani. Machine learning in automated text categorization: a survey. Technical report, IEI-B4-31-1999, Instituto di Elaborazione dell'Informazion, C.N.R., Pisa, IT, 1999.

[7] Maarek, Y&Smdja, F.(1989), Full text indexing based on lexical relations, an application: Software libraries, in Proceedings of SIGIR 89, ACM Press, pp.198-206

[8] C. J. van Rijsbergen, Information Retrieval. Butterworths, London, 1979.

[9] Y. Yan. An evaluation of statistical approaches to text categorization. Information Retrieval, 1(1-2):69-90, 2000.

[10] "Search Engine Review" www.monash.com

[11] "Search Engine Tutorial for Web Designers" www.northernwebs.com

[12] www.dmoz.org

[13] www.henzinger.com/monika/icde

419

THE DEVELOPMENT OF A USABILITY MEASURE FOR FLIGHT MANAGEMENT SYSTEMS

Karen P. Lane* & Don Harris**

* Aerosystems International, West Hendford, Yeovil, Somerset BA20 2AL, UK.
** Human Factors, College of Aeronautics, Cranfield University, Bedford, MK43 0AL, UK.

ABSTRACT

This paper reports on a three-year research programme, which resulted in the development of a series of scales to assess the usability of Flight Management System (FMS) Control and Display Units (CDUs). The methodology adopted is outlined, and evidence of reliability and validity is demonstrated. A content analysis conducted on a large volume of comments collated during the scale development process is also described which provided construct validity evidence for the scales derived and permitted usability recommendations for FMS designers and certification authorities to be made.

1. INTRODUCTION

Usability is "*not a unidimensional product or user characteristic, but emerges as a multidimensional characteristic in the context of users performing tasks with a product in a specific environment*" (Lewis, 1995, p.58). While a number of techniques have been developed over the past fifteen years for assessing usability, none have been developed for, or applied specifically to, the flight deck or its components. For a number of reasons, including, the complex context in which flight crew operate, the usability of flight deck elements is distinct from other professional applications such as personal computers and their related software, or home appliances such as VCRs. It is important, for example, that the automation on board an aircraft is dynamic since changes frequently need to be made very quickly. Further, since the outcome of interactions with FMSs can typically have larger safety implications than those associated with desktop computer interactions, usability can become even more critical.

Aircraft accident rates for the world-wide commercial jet fleet remained at a consistent two to three accidents per million departures over the period 1970 to 1996 (Boeing, 1997). Over the same time period flight decks have been altered greatly, with older mechanical instruments being replaced by electronic displays and computers that can control almost every aspect of a flight (Billings, 1997). The prime purpose of these technological advances has been to increase fuel economy, reduce flight crew workload and increase safety. Advanced, highly automated aircraft can be more productive, reliable and safer than their predecessors when managed appropriately (Wiener, 1988). The introduction of this new automation onboard modern flight decks has not, however, been without problems and its application has also had unintended and unwanted effects (Scerbo, 1996). The result has been that while some pilot errors have been reduced or eliminated other new opportunities for different types of error have been introduced (Woods, Johannesen, Cook & Sarter, 1994). As a result of a number of incidents and accidents that pointed towards difficulties in flight crew interactions with flight deck automation, the Federal Aviation Administration's (FAA) Transport Airplane Directorate commissioned a study to evaluate the flight crew/flight deck automation interfaces across a range of current aircraft. The findings of the research published in the document, *The Interfaces Between Flightcrews and Modern Flight Deck Systems* (FAA, 1996), highlighted a number of vulnerabilities in modern flight decks including flight crew management of automation and situation awareness. In examining why these vulnerabilities existed the research identified a number of interrelated deficiencies in the current aviation system including insufficient criteria, methods, and tools for design, training and evaluation to aid the goal of achieving more human-centred aviation automation.

The FAA Human Factors Team noted that "*Existing methods, data, and tools are inadequate to evaluate and resolve many of the important human performance issues*" (p.3), and subsequently made the recommendation that "*new tools and methods need to be developed and existing ones improved to accompany the process improvements*" (p.4). The research programme described in this paper feeds into this process since a valid and reliable usability measure was developed for assessing the usability of the core flight deck computer, the FMS. Both the European Joint Airworthiness Authorities (JAA) and the US Federal Aviation Administration (FAA) have proposed the introduction of Human Factors Certification Criteria for civil aircraft flight decks. While the regulatory authorities have supported the philosophy behind the human factors certification process, no tools or methods have been proposed as techniques for evaluating the flight deck from a human factors perspective. The tool developed within this research programme could be used in the human factors certification process.

2. OVERVIEW OF RESEARCH PROGRAMME

The research programme described in this paper was conducted in three distinct phases. Phase One involved the development of a reliable and valid FMS CDU usability factor structure. Phase Two focused on validation of the usability factor structure elicited and scale development. With the goal of cross validating the scales developed, a content analysis was conducted on a large volume of qualitative data collected as an adjunct to the scale development process. Each of these three phases will be described in turn.

3. PHASE ONE: DERIVATION OF FMS USABILITY FACTOR STRUCTURE

The first research phase was designed to produce a reliable and valid FMS usability factor structure. The sub-goals of this phase were therefore to:

1. Produce a usability questionnaire that could be distributed internationally to a large sample of line pilots, across a range of fleets and FMS CDUs;
2. Conduct analyses on the data elicited with the goal of determining the factor structure of FMS CDU usability;
3. Examine the internal consistency and validity of the factor structure derived;
4. Generate qualitative data concerning the problems associated with current FMSs, and the improvements pilots would like to see in next generation FMSs.

3.1 Methodology

Following a thorough literature review, the research programme began with a series of in-depth interviews with civil airline pilots to elicit the usability dimensions pertinent to FMS CDUs. From the item pool elicited during the interviews, a self-completion questionnaire was developed which was distributed to a sample of pilots in the US, Canada and the UK.

3.2 Respondents

A total of 252 completed questionnaires were received from respondents working for five airlines across the UK, USA and Canada. Nine (3.6%) respondents were female, while 231 (91.7%) respondents were male. Twelve respondents (4.7%) did not complete that part of the questionnaire. Participants had a mean age of 44 (sd=8), and a mean of 3033 (sd=2762) hours on type. Fifty-five percent of respondents were Captains, while 42% of respondents were First Officers, 3% of respondents did not declare their rank. Respondents in this phase were type-rated on a wide range of air transport aircraft produced by three aircraft manufacturers (labelled Manufacturer X, Manufacturer Y and Manufacturer Z).

3.3 Results

A Principal Components Analysis was conducted on the data to determine the structure underlying the questionnaire items. Ten usability factors were derived labelled Structure & Sequence; Appropriate Functionality; Feedback; Predictability; Information Presentation; Consistency, Flexibility; Error Prevention & Recovery; Processor Speed and Workload.

Cronbach's Alpha analyses demonstrated that the factors had an acceptable level of internal consistency for scale development to proceed ($\alpha > 0.7$). A series of Discriminant Function Analyses, using the factors derived as predictor variables, was used to demonstrate that the factor structure could discriminate between FMS and aircraft types when these variables were analysed together and individually. Evidence of discriminant validity was demonstrated since the factors discriminated between the products of FMS manufacturers A & B. The factors additionally discriminated between the aircraft produced by manufacturers X, Y and Z. In order to assess the predictive power of the usability factor structure elicited, the classification matrices were compared with a chance model using Press's Q statistic and were found to be significant (p<0.01).

4. PHASE TWO – FACTOR STRUCTURE VALIDATION AND SCALE DEVELOPMENT

FMSs were initially designed as a navigation tool for use solely in the Cruise phase of flight when workload levels were low (Billings, 1997). Over time their functionality has been extended throughout all flight phases, from departure through to arrival, but without any significant re-design of the CDU hardware. The purpose of Phase Two of the research programme was to verify the factor structure produced during Phase One, and to produce reliable and valid *scales* rather than factors for assessing FMS usability. Scales had the advantage of providing a simpler, more parsimonious view of FMS CDU usability than the factor structure demonstrated.

4.1 Methodology

This second research phase sought to verify the factor structure produced by applying it to new data and examining its consistency during the Cruise and Descent & Approach phases of flight. Two questionnaire variants were produced, one for the Cruise phase of flight and one for the Descent & Approach phase of flight. The questionnaire instruction paragraph was modified such that participants were asked to rate the usability of their FMS for either the Cruise or Descent & Approach phase of flight only. The usability questionnaire was distributed to a new sample of pilots across five airlines in the UK and Ireland.

4.2 Participants

Of the 356 pilots that participated in this phase of the research, 179 pilots rated FMS usability during the Cruise phase of flight, while 177 pilots rated the FMS usability during the Descent and Approach phase. Fifteen respondents were female (4.2%), while 323 respondents (90.7%) were male. Eighteen respondents (5.1%) did not complete the gender part of the questionnaire. Participants had a mean age of 43 (sd=9), and had a mean of 2777 (sd=2328) hours on type. Sixty percent of respondents were Captains, while 34% of respondents were First officers. Six percent did not complete the "rank" part of the questionnaire. Participants in this phase were type-rated on a wide range of air transport aircraft produced by two aircraft manufacturers (labelled Manufacturer X and Manufacturer Y). No participants type-rated on aircraft produced by manufacturer Z were included in this study, although this was not a problem since analyses conduced during Phase One demonstrated that the usability factors derived could distinguish between all three aircraft manufacturers (X, Y & Z).

4.3 Results

A Confirmatory Factor Analysis (CFA) confirmed the usability factor structure produced in Phase One of the research. Evidence of the stability of the factor structure produced was provided from the fact that the model was valid when applied to the independent data collected for the Cruise and Descent & Approach phases of flight. The CFA model was a good fit for both data sets. Since the factor structure had been shown to be replicable and stable, FMS usability scales could be produced from it with confidence.

In order to examine if the scales still had discriminatory power, a series of DFAs were again conducted. Using these analyses it was demonstrated that the scale scores could again be used to discriminate between FMSs produced by manufacturer X and manufacturer Y. It was also possible to discriminate between aircraft manufacturer X and aircraft manufacturer Y on the basis of the usability scores produced. Further evidence for the validity of the scales was derived from the fact that ratings of the scales were consistent across the research phases. Ratings of usability on the Feedback scale, for example, were consistently more favourable for FMS manufacturer A than for FMS manufacturer B, and were consistently more favourable for manufacturer B than for manufacturer A on the Appropriate Functionality scale. Examining discriminant validity evidence by aircraft type, participants' ratings on the Structure & Sequence scale were consistently more favourable for aircraft manufacturer X than for aircraft manufacturer Y, while ratings on the Information Presentation scale were consistently more favourably for aircraft manufacturer Y than for aircraft manufacturer X. Evidence of criterion validity was demonstrated since it was possible to classify participants' usability ratings by FMS and aircraft manufacturer at a rate which was significantly better than chance.

Ratings of the usability scales did not discriminate between the flight phases rated, which indicated that it was the usability construct, and not, for example, workload which was being measured. The scales were shown to be internally consistent using Cronbach's Alpha analyses. It was concluded that a series of scales to assess the usability of FMS CDUs had been produced and evidence of both reliability and validity was demonstrated. The psychometric properties established during Phases One and Two of the research programme are shown in Table One.

Table One - Psychometric Properties of the Usability Factors & Scales Produced

Property	
Factor structure was internally consistent	✓
Factor structure demonstrated evidence of face validity	✓
Factor structure demonstrated content validity	✓
Factor structure demonstrated criterion validity	✓
Factor structure demonstrated discriminant validity	✓
Usability Factor structure was stable and replicable.	✓
Usability Factor Structure could be generalised across flight phase (Cruise and Descent & Approach).	✓
Usability Scales were internally consistent.	✓
Usability Scales demonstrated evidence of face validity.	✓
Usability Scales demonstrated evidence of content validity.	✓
Usability Scales demonstrated evidence of discriminant validity.	✓
Usability Scales demonstrated evidence of criterion validity.	✓
Usability Scale ratings reflected the FMS and its operational environment and not user characteristics.	✓

5. PHASE 3: CONTENT ANALYSIS

5.1 Purpose

The purpose of this third research phase was to collate the qualitative comments elicited during Phases One and Two, with the goals of (i) identifying problems associated with current FMSs; (ii) making recommendations for improving FMSs; (iii) providing evidence of construct validity for the usability scales developed during Phases One and Two and (iv) aiding interpretation of those scales.

5.2 Procedure

In total 496 discrete comments were derived from the 224 participants who chose to make qualitative comments. After reading each comment a number of times the researcher decided which category the comment was most pertinent to and classified it as such. The allocation of a comment to either the category of "suggested improvements" or "problems encountered" was largely down to way in which the comment was phrased.

Based on this classification a total of 289 of the comments were categorised as referring to issues participants raised concerning their current FMS. A total of 207 comments were categorised as referring to improvements the participants would like to see to future FMSs. A grounded theory approach was adopted in the allocation of comments to category, in that the researcher remained expectation free in the initial stages of research to allow the key issues to emerge via systematic grounded analyses. This approach was adopted so that further evidence of construct validity could be demonstrated.

To ensure intra-rater reliability, the researcher allocated each comment to a category repeatedly until the allocation of comments to categories could be replicated with 95% consistency when the procedure was conducted on sessions a week apart. In order to validate the structure imposed on the data and to determine an inter-rater reliability co-efficient for the ratings made by the researcher, a second human factors expert in the aviation domain also conducted the process of allocating comments to categories. The second researcher was asked to repeat the process conducted by the first researcher until an intra-rater reliability rate of 95% could be established. Analysis of comments was only conducted on those comments on which *both* of the raters agreed on the classification.

5.3 Results

It was concluded that problems associated with current FMSs and suggested improvements for future FMSs were associated with a number of areas which included: Processor Speed, CDU Architecture & Logic, Functional Limitations, Control Devices, Accuracy of parameters (such as wind & fuel), Inadequate Feedback, Error Prevention & Correction, User Guidance & Support, Consistency & Compatibility. The categories of data elicited from this independent study demonstrated the construct validity of the scales produced in the previous two phases. Phase Three results were also consistent with the findings of Phases One and Two in that the number of problem comments associated with particular FMS or aircraft types reflected the usability ratings produced in the earlier phases. Participants type-rated on aircraft with FMSs produced by manufacturer B, for example, made more favourable usability ratings on the Information Presentation scale than those type-rated on aircraft with FMSs produced by manufacturer A. Correspondingly, those pilots type-rated on aircraft with FMSs produced by manufacturer B identified fewer Information Presentation related problems in the content analysis phase than those type-rated on aircraft with FMSs produced by manufacturer A. It was also possible to make usability recommendations for FMS manufacturers and certification authorities on the basis of the study's findings.

6. DISCUSSION & CONCLUSIONS

The product of this research programme was a reliable and valid usability measure that could be used in both the certification and FMS development processes. In Phase One, a reliable and valid FMS usability structure was produced. In Phase Two, it was demonstrated that it was possible to discriminate between FMS and aircraft types on the basis of participants' ratings of the usability scales, but it was not possible to discriminate between user characteristics or flight phases (discriminant validity). In Phase Three, comments made by participants provided evidence of construct validity in that they reflected the content of the scales developed during Phases One and Two.

7. REFERENCES

Billings C. (1997). *Aviation Automation - The Search for a Human-Centred Approach.* Mahwah, NJ: Lawrence Erlbaum.

Boeing Commercial Airplane Group (1997). *Statistical summary of commercial jet airplane accidents, worldwide operations, 1959-1996.* Seattle, WA: Boeing Commercial Airplane Group, Airplane Safety Engineering.

Federal Aviation Administration (1996). Human Factor Team Report on *The Interfaces Between Flightcrews and Modern Flight Deck Systems.* Washington D.C.: Author.

Lewis J. (1995). IBM Computer Usability Satisfaction Questionnaires: Psychometric Evaluation and Instructions for Use. *International Journal of Human Computer Interaction, 7,1,* 57-78.

Scerbo M.W. (1996). Theoretical Perspectives on Adaptive Automation. In, R. Parasuraman & M. Mouloula (Eds.), *Automation Technology and Human Performance* (pp. 37-63). New Jersey: Erlbaum.

Singer G. (1999). Filling the Gaps in the Human Factors Certification Net. In, S. Dekker & E. Hollnagel (Eds.), *Coping with Computers in The Cockpit* (pp. 87-107). Aldershot: Ashgate.

Stanton N. & Baber C. (1996). Factors Affecting the Selection of Methods and Techniques Prior to Conducting a Usability Evaluation. In, P. Jordan, B. Thomas, B. Weerdmeester & I. McClelland (Eds.), *Usability Evaluation in Industry* (pp. 39-57). London: Taylor & Francis.

Wiener E.L. (1988). Cockpit Automation. In, E.L. Wiener & D.C. Nagel (Eds.), *Human Factors in Aviation* (pp. 433-461). San Diego, CA: Academic Press.

Woods D., Johannesen, L., Cook R.I. & Sarter N. (1994). *Behind Human Error. Cognitive Systems, Computers & Hindsight.* State of the Art Report. Dayton, OH: Crew Systems Ergonomic Information & Analysis Center.

An Industrial Case Study of Usability Evaluation in Market-Driven Packaged Software Development

Johan Natt och Dag[1], Björn Regnell[1], Ofelia S. Madsen[2], Aybüke Aurum[3]

[1]Department of Communication Systems, Lund University, Sweden, (johan.nattochdag, bjorn.regnell)@telecom.lth.se
[2]C-Technologies AB, Lund, Sweden, ofelia.madsen@cpen.se
[3]School of Information Systems, Technology and Management, University of New South Wales, Australia, aybuke@unsw.edu.au

Abstract. In market-driven software development it is crucial to produce the best product as quickly as possible in order to reach customer satisfaction. Requirements arrive at a high rate and the main focus tends to be on the functional requirements. The functional requirements are important, but their usefulness relies on their usability, which may be a rewarding competitive means on its own. Existing methods help software development companies to improve the usability of their product. However, companies that have little experience in usability still find them to be difficult to use, unreliable, and expensive. In this study we present results and experiences on conducting two known usability evaluations, using a questionnaire and a heuristic evaluation, at a large software development company. We have found that the two methods complement each other very well, the first giving scientific measures of usability attributes, and the second revealing actual usability deficiencies in the software. Although we did not use any usability experts, evaluations performed by company employees produced valuable results. The company, who had no prior experience in usability evaluation, found the results both useful and meaningful. We can conclude that the evaluators need a brief introduction on usability to receive even better results from the heuristic evaluation, but this may not be required in the initial stages. Much more essential is the support from every level of management. Usability engineering is cost effective and does not require many resources. However, without direct management support, usability engineering efforts will most likely be fruitless.

1. Introduction

When developing packaged software for a market place rather than bespoke software for a specific customer, short time-to-market is very important (Potts, 1995). Packaged software products are delivered in a succession of releases and there is a strong focus on user satisfaction and market share (Regnell, Beremark & Eklund, 1998). Thus, companies tend to put their primary effort into inventing and implementing new functional features that are expected to improve the product. Although developers rely heavily on the number and the existence of new features, usability is recognized as a competitive advantage on its own. Still after many years of usability engineering research, there are many companies that do not approach and explicitly improve usability. Although several methods and techniques exist (Nielsen, 1993) and studies show their cost-effectiveness (Bias & Mayhew, 1994), the seeming difficulty of approaching usability prevents the success of companies. Since usability evaluations of software products are necessary in order to increase user-friendliness (Nielsen 1993), there is a need to put even more focus on usability evaluation methods that are easy to use and adopt and that give fast and appropriate results.

In this paper we present an industrial case study that employs two known usability evaluation methods (Natt och Dag & Madsen, 2000). The study was conducted at Telelogic AB, a large software developing company in Sweden, and the methods were used to evaluate their main product, Telelogic Tau[1], a graphical software development tool aimed for the telecommunications industry. It is shown that the two methods may be used for continuous evaluation of usability without much effort or resources and without any particular experience in usability engineering.

2. Research methodology

Several factors have been taken into consideration when choosing evaluation methods. The methods must be easy to perform and give understandable results that can be utilized in daily work without extraordinary analysis. Furthermore, the methods need to be appropriate for use over and over again, and it must be possible to extend the proficiency in using these methods when experience in usability evaluation within the company increases. If usability experience is lacking, the methods must not be too sensitive to the evaluators' performances.

To obtain satisfactory results that fulfill these requirements, we carefully selected two known usability evaluation methods, a questionnaire and a heuristic evaluation, which give quantitative and qualitative results respectively.

---

1. Telelogic Tau is a registered trademark of Telelogic AB.

2.1 The SUMI Questionnaire

In order to obtain end users' opinions about the software we used a commercially available questionnaire, 'The Software Usability Measurement Inventory' (SUMI) (Kirakowski & Corbett, 1996). SUMI is a standard questionnaire specifically developed and validated to give an accurate indication of which areas of usability should be improved. It is tested in industry and mentioned in ISO-9241 as a method of measuring user satisfaction. The questionnaire consists of 50 statements to which each end-user answers whether he or she *agrees*, *disagrees* or is *undecided*. Only 12 end users are necessary to get adequate precision in the analysis, but it is possible to use fewer users and still obtain useful results.

The questionnaire was sent to 90 selected end users in Europe. Returned answers were sent to the questionnaire designers who statistically analyzed the answers and compared them with the results in a continuously updated standardization database. The database contains the answers from over 1000 SUMI questionnaires used in the evaluations of a wide range of software products. The comparison results in the measurement of five usability aspects:

- *Efficiency* - the degree to which users feel that the software assists them in their work.
- *Affect* - the user's general emotional reaction to the software.
- *Helpfulness* - the degree to which the software is self-explanatory. Adequacy of documentation.
- *Control* - the extent to which the users feel in control of the software.
- *Learnability* - the ease with which the users feel that they have been able to master the system.

Each aspect is represented by 10 statements in the questionnaire and the raw scores for each of the aspects are converted into scales with a mean value of 50 and a standard deviation of 10, with respect to the standardization database. Also, a *global* scale is calculated that is represented by the answers to 25 of the 50 statements that best reveal global usability.

To identify the statements to which the answers differ significantly from the average in the standardization database, *Item Consensual Analysis*, a feature developed by the questionnaire designers, is used. Through comparison between the observed and expected answer patterns, the individual usability problems may be more accurately determined.

2.2 The Heuristic Evaluation

To find usability problems specific to the evaluated software, we used a slightly extended version of a standard *heuristic evaluation* (Nielsen, 1994). In the heuristic evaluation, usability experts go through the interface and inspect the behavior of the software. The behavior is compared to the meaning and purpose of a set of ten guidelines called *heuristics* that focus on the central and most important aspects of usability, such as '*user control and freedom*' and '*flexibility and efficiency in use*'. This enables the evaluator to systematically check the software for usability problems against actual requirements in the specification and given features in the product. The result is a list of identified usability problems that may be used to improve the software.

In the market-driven development organization, there may be little experience in usability and usability experts may not be available. As this was the situation at Telelogic, we used experts on the software from within the company. The number of evaluators needed may then increase, as less usability problems may be found. Experts specializing only in usability tend to find problems mainly related to how easy the system is to operate, whereas domain experts rather find problems related to how well the system responds to its intended behavior. In this sense the usability and the domain experts complement each other, as domain experts, according to Muller, Matheson, Page and Gallup (1998), bring perspectives and knowledge not otherwise available. However, an evaluation is more likely to be conducted in the first place if we initially do not require the involvement of usability experts.

Twelve employees from within the company participated in the study. They were presented with a set of scenarios comprised of different tasks to perform. The method was extended in order to add increased structure to the evaluation and to help the evaluators to stay focused on usability issues. This was accomplished through the introduction of *Usability Breaks* (UB). A UB is a position in the detailed scenario where the evaluator is supposed to (1) stop the execution of the scenario and write down any problems found up to that point together with the associated heuristics, and (2) go through the ten heuristics to identify additional problems with the tasks just performed.

The evaluators were advised to spend 1 uninterrupted hour on finding relevant usability problems. In addition to generating a list of problems, the evaluators were asked to write down during which scenario identified problem were encountered, at what specific UB they were found, the heuristics that apply, the severity of each problem (high, medium or low), and a suggestion of a solution.

3. Results

3.1 The SUMI Questionnaire

Of the 90 questionnaires sent to end users, 62 were properly filled out and returned. The analysis of the returned answers revealed that the evaluated software did not meet the appropriate standards on several aspects of usability. In Figure 1 the medians for each of the six SUMI scales are shown. The median corresponds to the middle value in a list where all the individual scores given by each evaluator have been numerically sorted. The figure also shows error bars for each scale, representing the upper and lower 95% confidence limits. As seen in the figure, all but two of the six SUMI scales are below average. The sub-scale *affect* is the only one that lies above average, indicating that users feel slightly better about this product than they feel in general about software products. The *learnability* sub-scale indicates that the software may be regarded to be as easy or hard to learn as software products are in general.

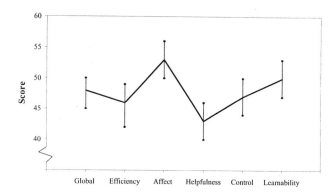

Figure 1. The SUMI satisfaction profile.

The Item Consensual Analysis (see Section 2.1) revealed 9 specific statements that differed significantly from the standardization database (99.99% certain). In Figure 2, the statement that was most likely to differ from the expected is shown. This particular result reveals that the software may not be very helpful. Of the remaining 8 statements that most certainly differed from the standardization database, only 1 statement generated a more positive response than what was expected. As many as 74% of the end users disagreed with the statement '*if the software stops it is not easy to restart it*', indicating that the software *is* easy to restart. Further analysis of the statements revealed several reasons to the low scores in Figure 1.

Statement 28: The software has helped me overcome any problems I have had using it.

	Agree	Undecided	Disagree
Profile	12	19	31
Expected	17.10	30.97	13.93
Chi Square	1.52	4.63	20.93

Figure 2. Results from the Item Consensual Analysis.

3.2 The Heuristic Evaluation

The heuristic evaluation revealed 72 unique usability problems directly related to the software application. A sample usability problem identified by an evaluator is shown in Figure 3. About 20% of the identified problems were considered highly severe, about 65% somewhat severe, and no more than 14% less severe. This indicates that usability needs attention in order to increase the usefulness of the software, which is confirmed by the results from the more reliable SUMI questionnaire evaluation.

Solutions were given to most of the problems but for the 18 that had none a solution may be inferred through the problem descriptions and through the particular UBs in the scenario. The problem descriptions had high enough quality to be used as input into the requirements process.

Figure 4 shows the number of times each of the ten heuristics were used to classify the identified problems. The high use of *flexibility and efficiency in use* indicates that users may get frustrated when using the software. Further analysis of the particular problems related to this heuristic and at which UBs

Problem	How to change a unidirectional channel to a bidirectional channel
Scenario	A
UB	1.3
Heuristic	Flexibility and efficiency in use
Severity	High
Solution	Add a channel symbol to the symbol menu and then add symbols for unidirectional (one in each direction) and bidirectional.

Figure 3. Sample usability problem found by an evaluator.

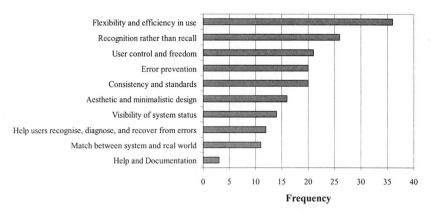

Figure 4. The number of heuristics used to classify the 72 identified usability problems.

the problems were encountered, reveals that there are many tasks that are bothersome to complete due to non-intuitive functionality and a non-supportive graphical interface.

The results in Figure 4 are confirmed by the results from the SUMI questionnaire evaluation. *Efficiency* is identified as a problematic area by both methods and *helpfulness* is related to *recognition rather than recall* and *user control and freedom*. This and the fact that experts on the software were used as evaluators indicate that the heuristic evaluation pinpoints relevant usability problems.

4. Conclusions

In this paper we have presented the results of using two known usability evaluation methods at a market-driven software development company inexperienced in usability. We have found that although experience on usability is lacking, the two methods are easy to use and do not require many resources. The questionnaire is available at a low cost and system experts can easily develop the heuristic evaluation scenarios. Furthermore, the two methods complement each other very well. Both kinds of result were found to be usable and meaningful by the developing company and the generated problem list was particularly welcomed. The results gave them insight into the specific areas that needed improvement and helped them to appreciate the issues to put their effort into.

The selection of evaluators was the most time-consuming task. Mainly, this was because there was little support for usability issues. There was a noteworthy interest from the development department, but we have found that management support on every level in the organization is crucial to effectively get results. Without management support it is not very likely that the results will be used in further development at all. Also, a short 30-minute introduction to the concept of usability will most likely motivate the evaluators to perform even better and be more focused on usability issues in particular.

The initial drawbacks were nevertheless highly compensated by the low cost and the quick and useful results. The estimated costs of applying the methods are shown in Figure 5 (Melchior, Bösser, Meder, Koch & Schnitzler, 1995).

| | Cost (figures in man-days) | | | | | |
Method	Small	Medium	Extensive	Training	Material	Reliability
Heuristic Evaluation	2	4	4	1	None	Medium
SUMI Questionnaire	1	3	3	2	US$500	High

Figure 5. Estimation of cost of applying the two usability evaluation methods.

Currently, we are applying a follow-up study to investigate precisely how the generated problem lists have been used in succeeding releases, what impact on software usability they have had, and to what extent the usability has increased.

Acknowledgements

This work is partly funded by the National Board of Industrial and Technical Development (NUTEK), Sweden, within the REMARKS project (Requirements Engineering for Market-Driven Software Development) grant 1K1P-97-09690.

References

Bias, R. G., & Mayhew, D. J. (1994). Cost-justifying usability. Boston, MA: Academic Press.

Kirakowski, J., & Corbett M. (1996). The software usability measurement inventory: Background and usage. In P. W. Jordan, B. Thomas, B. A. Weerdmeester, & I. L. McClelland (Eds.), Usability Evaluation in Industry (pp. 169-177). London, UK: Taylor & Francis.

Melchior, E.-M., Bösser, T., Meder, S., Koch, A., & Schnitzler, F. (1995). Usability Study: Handbook for practical usability engineering in IE projects. ELPUB 105 10107, Telematics application programme, Information engineering section.

Muller, M. J., Matheson, L. Page, C., & Gallup, R. (1998). Participatory Heuristic Evaluation. Interactions, 5(5), 13-18.

Natt och Dag, J., & Madsen, O. S. (2000). An Industrial Case Study of Usability Evaluation. Lund: Lund University. (ERIC Document Reproduction Service No. LUTEDX (TETS-5390)/1-190/(2000)&local 8)

Nielsen, J. (1993). Usability Engineering. San Francisco, CA: Morgan Kaufmann.

Nielsen, J. (1994). Heuristic Evaluation. J. Nielsen, & R. L. Mack (Eds.), Usability Inspection Methods (pp. 25-61). New York, NY: John Wiley & Sons.

Potts, C. (1995). Invented requirements and imagined customers: Requirements engineering for off-the-shelf software. Proceedings of the Second IEEE International Symposium on Requirements Engineering, pp. 128-130.

Regnell, B., Beremark, P., & Eklundh, O. (1998). A market-driven requirements engineering process: Results from an industrial process improvement programme. Requirements Engineering, 3(2), 121-129.

Supporting Organizational Learning in Usability Engineering: Requirements of Software Development Organizations and Implications for Tool Support

Eduard Metzker
DaimlerChrysler Research and Technology Center, Software Technology Lab,
P.O. Box 2360, D-89013 Ulm, Germany
eduard.metzker@daimlerchrysler.com

Abstract: It is well accepted both by leading software practitioners and academia that structured usability engineering (UE) approaches are required to build interactive systems with high usability. Therefore a number of UE process models have been developed to address the special problems encountered in developing interactive systems. However, surveys indicate that UE methods are still poorly established in industrial software development projects. We present an organizational learning approach that supports the integration, establishment and continuous improvement of UE methods in software development projects.

1 Introduction

UE methodologies such as Usability Engineering (Nielsen 1994; Mayhew 1999), Contextual Design (Beyer and Holtzblatt 1998) or Usage-Centered Design (Constantine and Lockwood 1999) have been successfully applied through various projects. They have proved that they have dramatically been able to improve usability and user acceptance of interactive software systems if they are applied in the correct development context.

On the other hand specific knowledge about exactly how to most efficiently and smoothly integrate UE methods into established software development processes is still missing (Mayhew 1999). It often remains unclear to users of UE methods if and why certain tools and methods are better suited in a certain design context than others (Welie 1999). Little research has been done on integrating methods and tools of UE in the development process and gathering knowledge about interface design in a form that can capture relationships between specific development contexts and applicable methods and tools (Henninger 2000).

One vital step toward bridging this gap is to provide the people involved in UE processes with a tool to effectively support the elicitation and application of these knowledge. Unfortunately, existing tools for supporting UE processes beyond computer-aided design of user interfaces (e.g. (semi)- automatic GUI builders), largely focus on two aspects of the process: supporting the performance and evaluation of usability tests or supporting developers accessing. Most of these approaches assume that there is already a well-established UE process in place that is practiced by the development team and that provides use of the proposed methods and tools. To examine the validity of this assumption, we initiated a survey to find out what kind of development process for interactive systems is practiced by development organizations engaged in engineering interactive systems. Taking the results of this survey into account, we had to revise some of the assumptions about requirements for a tool to support UE processes. These new findings have influenced the development of REUSE, a tool to support organizational learning in UE under the constraints discovered in the survey.

2 Examining UE Process in Practice

To be able to construct a tool for the effective support of UE processes, we needed in-depth knowledge of the future users of such a tool and their requirements. This led to the following central questions:

- What kind of development process for interactive systems is practiced by the development organizations in their projects?
- What typical tasks do the developers have to solve?
- What problems are typical for the development process?
- What kind of support is needed?

2.1   Structure and Performance of the Survey

The survey was elaborated, performed and evaluated in collaboration with industrial psychologists and had the following structure (Wetzenstein and Becker 2000): A questionnaire was used to record both personal data and information on the respondents' professional experience and typical development tasks. A semi-structured interview supplemented by a special set of questions concerning the application of UE activities during the development process as perceived by the respondents formed the core of the survey.

A total of 16 employees from four major companies[i] involved in the development of interactive software systems were selected. The respondents are engaged in developing these systems in projects from diverse domains: military systems, car driver assistance technology or next-generation home entertainment components. The questioning was performed by a single interviewer and the answers were recorded by a second person in a pre-structured protocol document. Each interview took between 90-150 minutes.

The organizations examined are practicing highly diverse individual development processes, however non of the UE development models proposed by (Nielsen 1994; Beyer and Holtzblatt 1998; Constantine and Lockwood 1999; Mayhew 1999) are exactly used.

Furthermore, the persons who are entrusted with the ergonomic analysis and evaluation of interactive systems are primarily the developers of the products. External usability or human factors experts or a separate in-house ergonomics department are seldom available. Furthermore, few of the participants were familiar with basic methods like *user profile analysis* or *cognitive walkthrough*.

The UE methods that are considered to be reasonable to apply by the respondents are often not used for the following interrelated reasons:

- There is no time allocated for UE activities: they are neither integrated in the development process nor in the project schedule.
- Knowledge needed for the performance of UE tasks is not available within the development team.
- The effort for the application of the UE tasks is estimated to be too high because they are regarded as time consuming.

2.2 Survey Conclusions

The results of the survey led to the following conclusions regarding the requirements of a software tool to support the improvement of UE processes:

Requirement 1: Support flexible UE process models

The tool should not force the development organization to adopt a fixed UE process model as the practiced processes are very diverse. Instead, the tool should facilitate a smooth integration of UE activities into the individual software development process practiced by the organization. Turning technology-centered processes into human-centered processes should be seen as a continuous process improvement task where organizations learn which of the methods available best fit in certain development contexts, and where these organizations may gradually adopt new UE methods.

Requirement 2: Support evolutionary development and reuse of UE experience

It was observed that the staff entrusted with ergonomic design and evaluation often lacks a special background in UE methods. Yet, as the need for usability was recognized by the participating organizations, they tend to developed their own in-house usability guidelines and heuristics. Recent research (Billingsley 1995; Weinschenk and Yeo 1995; Rosenbaum, Rohn et al. 2000; Spencer 2000) supports the observation that such usability best practices and heuristics are, in fact, compiled and used by software development organizations. Spencer (Spencer 2000), for example, presents a streamlined cognitive walkthrough method which has been developed to facilitate efficient performance of cognitive walkthroughs under the social constraints of a large software development organization. However, from experiences collected in the field of software engineering (Basili, Caldiera et al. 1994) it must be assumed that, in most cases, best practices like Spencer's are unfortunately not published in either development organizations or the scientific community. They are bound to the people of a certain project or, even worse, to one expert member of this group, making the available body of knowledge hard to access. Similar projects in other departments of the organization usually cannot profit from these experiences. In the worst case, the experiences may leave the organization with the expert when changing jobs. Therefore, the proposed tool should not only support existing human factors methods but also allow the organizations to compile, develop and evolve their own approaches.

Requirement 3: Provide means to trace the application context of UE knowledge

UE methods still have to be regarded as knowledge-intensive. Tools are needed to support developers with the knowledge required to effectively perform UE activities. Furthermore, the tool should enable software development organizations to explore which of the existing methods and process models of UE works best for them in a certain development context and how they can refine and evolve basic methods to make them fit into their particular development context. A dynamic model is needed that allows to keep track of the application context of UE methods.

Requirement 4: Support efficient performance of UE activities

There is a definite need for tool support when it comes to enabling efficient performance of otherwise tedious and time consuming UE activities. Otherwise, these essential activities fall victim to the no-time-to-sharpen-the-saw-because-too-busy-cutting-the-wood syndrome. With respect to requirement 2, this means that the tool has to support the efficient elicitation, organization and reuse of best practices and artifacts relating to UE tasks.

---

[i] DaimlerChrysler Aerospace (DASA) in Ulm, Sony in Fellbach, Grundig in Fuerth and DaimlerChrysler in Sindelfingen (all sites are located in Germany)

3 The REUSE System

To transfer the above requirements into a software tool we have developed the concepts of a *context model* and a set of *USEPACKs* (Usability Engineering Experience Package) with related *context situations*.

In the following sections, we differentiate between two virtual roles played in the utilization of the REUSE system: *readers* who search, explore and apply USEPACKs and *authors* who create, compile and organize USEPACKs.

3.1   The USEPACK Concept

A USEPACK is a semi-formal notation for structuring knowledge relating to UE activities. It encapsulates best practices on how to most effectively perform certain UE activities and includes the related artifacts that facilitate the compliance with the best practice described.

A USEPACK is structured into five sections: The *core information* permits authors to describe the main message of a USEPACK. It is organized according to the pyramid principle for structuring information (Minto 1987). The information first presented to the reader has a low level of complexity, allowing the reader to quickly decide if the USEPACK is worth further exploration. With further reading, the degree of complexity rises, introducing the reader to the experience described. The core information section includes the fields *title*, *keywords*, *abstract*, *description* and *comments*.

The *context situation* describes the development context related to the experience in question. The context situation is generated by using the context model as a template, allowing the authors and readers of USEPACKs to utilize a shared vocabulary for contextualizing and accessing USEPACKs.

A set of *artifacts* facilitates the efficient compliance with the best practice. Artifacts such as templates for usability questionnaires or checklists to perform a user profile analysis allow readers to simplify their work. Artifacts represent an added value to the readers of a USEPACK which helps readers to perform UE tasks more efficiently. A set of *semantic links* pointing to other USEPACKs or external resources which, for example, support or contradict the best practice described. By interlinking USEPACKs and connecting them to external resources, like web pages, a net of related experiences can be created to provide more reliable information than through the isolated sets of USEPACKs. The *administrative data* section is used to store data like the *author(s) of the package*, *the date of creation*, *access rights* and statistical data such as the *number of accesses* to the package and a *user rating*.

3.2   The Context Model Concept

The context model serves as a template to construct the *context situation* for USEPACKs – a semi-formal description of the context in which the information of a USEPACK can be applied. It is organized in a hierarchical structure, divided into sections which contain groups of *context factors*. On the one hand, authors can use the context model to easily construct a description of the context in which the information of a USEPACK can be applied by selecting appropriate context factors from the model. On the other hand, readers can use the context model to specify a context situation which reflects the development context for which they need support in the form of USEPACKs.

Currently our context model contains the following five sections:

The *process context* section for example provides context factors to describe elements of the development process used, such as process phases (e.g. 'User Interface Design') as well as roles (e.g. 'Usability Engineer') and deliverables (e.g. 'UI Styleguide') related to the experience in question. The *project context* section provides context factors to describe project constraints like the size of the development team, budget or project duration which are related to the experience cited. Further sections are provided to capture the domain, technology and quality context of USEPACKs. Users never work directly on the context model, instead they manipulate the model by interacting with components of the REUSE system. These components hide much of the models complexity and provide means for the manipulation of the context model.

3.3   Components of REUSE

The REUSE system *(Repository for Usability Engineering Experience)* provides four components for manipulating USEPACKs, context situations and context models, that are depicted in Figure 1. The *USEPACK editor* component guides authors in creating new USEPACKs and evolving existing packages. To support this task, the editor includes specialized assistants for work on the related sections of a USEPACK.

The *USEPACK explorer* provides different views of the set of USEPACKs available and various filters to facilitate convenient browsing and retrieval of those USEPACKs. The USEPACK explorer facilitates the search for USEPACKs that share certain context factors specified by the reader. An example for one of these filters is the *process context filter*, which builds a graphical representation of the UE process model as defined in the process context section of the context model. By directly selecting process elements in the graphical representation, the reader can easily retrieve the related USEPACKs.

The *context model manager* component enables users to edit and extend the currently used context model with new context factors, e.g. to adapt REUSE to new process models or application domains. By this means, the REUSE system and the underlying models about processes, projects, domains, technologies and quality standards can be evolved in concert with the growing amount of experience to meet the needs of the organization.

The *USEPACK server* displays USEPACKs on a standard web browser and allows convenient browsing of the net of linked USEPACKs.

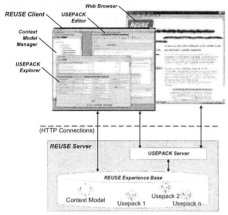

Figure 1 : Architecture of the REUSE System

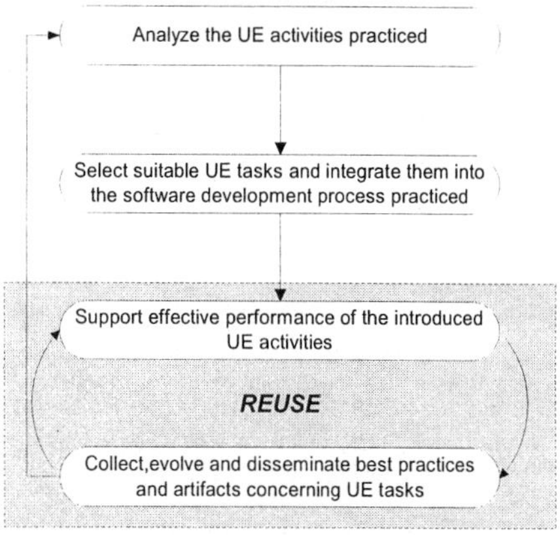

Figure 2 depicted elements:
- Analyze the UE activities practiced
- Select suitable UE tasks and integrate them into the software development process practiced
- Support effective performance of the introduced UE activities
- **REUSE**
- Collect, evolve and disseminate best practices and artifacts concerning UE tasks

Figure 2 : The Experience Based UE Lifecycle

4 Deployment of REUSE in an Experience-Based Improvement Lifecycle

To fully address the requirements discovered in our survey, REUSE has to be tightly integrated in the development lifecycle. Furthermore, establishing UE best practices in technology driven development processes must be seen a continuous process improvement task. We propose a streamlined experience based improvement lifecycle which contains the four steps, depicted in Figure 2. (The shaded rectangle indicates which steps of the improvement process are supported by REUSE.)

Step 1: Analyze the UE activities practiced

The first step comprises an analysis of the practiced UE process and the related UE activities, to elicit when, where and how UE methods are performed within the software development lifecycle in use. This analysis can be performed by using an assessment method like the *usability maturity model* (UMM) assessment questionnaire (Earthy 1999). The deliverable of this step is a documentation of possible improvements of the UE process that is currently used.

Step 2: Select suitable UE tasks and integrate them into the software development process practiced

The results of the first step form the rationale for the selection of UE activities for the improvement of the development process. However, in this step further important factors have to be considered, e.g. the type of system to be developed and project constraints like budget and schedules. The UE activities which have been selected for the improvement of the development process have to be integrated in the software development lifecycle practiced and in the project planning and form the improved development process, supplemented with appropriate UE activities.

Step 3: Support effective performance of the introduced UE activities

At this step resources have already been allocated for UE activities, e.g., a usability engineer was nominated, who is responsible for coordinating and supporting the execution of the various UE activities of the new process. However, the efficiency and impact of the proposed UE methods must be increased by providing information, best practices, tools and reusable process deliverables of past projects (e.g. templates for usability test questionnaires, results of conceptual task analysis or user interface mockups) which facilitate effective performance of the selected UE activities. This set of information should be provided by the usability engineer in form of USEPACKs.

Step 4: Collect, evolve and disseminate best practices and artifacts concerning UE tasks

During the execution of UE activities, experiences and artifacts are generated by the participants of UE activities. Furthermore observations of gaps in the UE process should be documented and used as an input for the next improvement cycle. These best practices and artifacts comprise UE experience and rationale that have to be captured in USEPACKs to facilitate easy reuse in the same or subsequent projects.

Our experience based improvement model contains two cycles: The inner cycle between step 3 and 4 supports the introduction and establishment of UE activities and methods within the practiced software development process. It supports the effective utilization and improvement of UE activities and best practices which are tailored to the needs of the development organization. This cycle is continuously iterated during the development process. The outer cycle which connects step 4 and 1 should be performed at least twice during the development process of a large project or between two subsequent projects as it serves the improvement of the overall UE process. In the outer cycle observations and best practices collected in step 4 should be analyzed and evolved to integrate missing UE activities in the process and remove activities that have failed to prove their utility.

5 Conclusions

We performed a survey to examine the development processes of organizations engaged in the design of interactive software systems and found a high potential for improvement regarding UE activities. The concepts and components implemented in REUSE represent an initial step towards an open tool geared for supporting the improvement of UE processes and to facilitate organizational learning in UE. Finally we outlined how to deploy REUSE in an experience-based improvement lyfecycle.

References

Basili, V. R., G. Caldiera, et al. (1994). Experience Factory,in *Encyclopedia of Software Engineering*. J. J. Marciniak(eds). New York, John Wiley & Sons. **1**: pp.528-532.

Beyer, H. and K. Holtzblatt (1998). *Contextual Design: Defining Customer-Centered Systems*, Morgan Kaufmann.

Billingsley, P. A. (1995). "*Starting from Scratch: Building a Usability Programm at Union Pacific Railroad.*" Interactions **2**(4): pp.27-30.

Constantine, L. L. and L. A. D. Lockwood (1999). *Software for Use: A Practical Guide to the Models and Methods of Usage-Centered Design*, Addison-Wesley.

Earthy, J. (1999). Human Centred Processes, their Maturity and their Improvement, in S. Brewster, A. Cawsey and G. Cockton(eds.), *Proceedings of the IFIP TC.13 International Conference on Human-Computer Interaction*, Edinburgh, UK, British Computer Society, pp.117-118.

Henninger, S. (2000). "*A Methodology and Tools for Applying Context-Specific Usability Guidelines to Interface Design.*" Interacting with Computers **12**(3): pp.225-243.

Mayhew, D. J. (1999). *The Usability Engineering Lifecycle: A Practioner's Handbook for User Interface Design*, Morgan Kaufman.

Minto, B. (1987). *The Pyramid Principle - Logic in Writing and Thinking*. London, Minto International Inc. Nielsen, J. (1994). *Usability Engineering*, Morgan Kaufman Publishers.

Rosenbaum, S., J. A. Rohn, et al. (2000). A Toolkit for Startegic Usability: Results from Workshops, Panels and Surveys, in T. Turner, G. Szwillus, M. Czerwinski and F. Paterno(eds.), *Proceedings of the Conference on Human Factors in Computing Systems*, The Hague, Netherlands, ACM press, pp.337-344.

Spencer, R. (2000). The Streamlined Cognitive Walkthrough Method: Working Around Social Constraints Encountered in a Software Development Company, in T. Turner, G. Szwillus, M. Czerwinski and F. Paterno(eds.), *Proceedings of the CHI2000*, The Hague, ACM Press, pp.353-359.

Weinschenk, S. and S. C. Yeo (1995). *Guidelines for Enterprise-wide GUI design*. New York, Wiley.

Welie, M. v. (1999). Breaking Down Usability, in M. A. Sasse and C. Johnson(eds.), *Proceedings of the IFIP TC.13 International Conference on Human-Computer Interaction*, Endinburgh, UK, IOS Press, pp.613-620.

Wetzenstein, E. and A. Becker (2000). Requirements of Software Developers for a Usability Engineering Environment. Berlin, Artop Institute for Industrial Psychology.

Practicality of Handwriting Japanese Input Interface with and without a Writing Frame

Hiroshi Tanaka and Kazushi Ishigaki

Personal & Service Laboratory, Fujitsu Laboratories Ltd.
64, Nishiwaki, Okubo-cho, Akashi, Hyogo 674-8555, Japan
e-mail: htnk@jp.fujitsu.com, ishigaki.kazu@jp.fujitsu.com

The practicality of using writing frames in a input interface for handwritten Japanese is discussed in this paper. After collecting some questionnaires from subjects of handwriting examinations, we compared the usability of writing frames for inputting handwritten Japanese sentences. According to the comparison of results, input without a writing frame is useful only for inputting handwriting patterns but not very useful for inputting character strings by handwriting recognition. For the practical application of Japanese handwriting recognition without writing frames, it is necessary to find out what kind of field is suitable for non-frame handwriting input, and it is important to improve recognition speed.

1. INTRODUCTION

Since Western language sentences are written word by word, where the words are written separately but the characters within each word are generally connected (Fig. 1 (a)), frame-based input may not feel very natural to users and may not be applicable except in small, mobile computing environments such as PDAs. Conversely, since Japanese sentences are commonly written with separate characters even if the sentences are written cursively (Fig.1 (b)), frame-based input is a natural input method for Japanese language users. In addition, character recognition performance may be better with a frame-based input interface than that on a non-frame interface because use of writing frames might prevent segmentation errors. Primarily for these reasons, most Japanese handwriting systems adopt a frame-based interface.

Although a frame-based interface is a natural input method for Japanese language users, this does not mean a frame-based interface is always better than a non-frame interface. We have received some requests to implement a non-frame interface in our handwriting system[1]. Therefore, we have been investigating different kinds of input interfaces to find one that is appropriate for Japanese handwriting input and to determine the important points of a handwriting input interface that is comfortable for users.

Generally speaking, the handwriting input process is composed of two phases. In the first phase (data entry phase), handwriting data is input into system. In the second phase (recognition phase), a handwriting character recognition engine is used to recognize the input data and convert it into a character code sequence. In this study, frame-based and non-frame input is examined and compared in each phase.

(a) Cursive script (alphabetic characters) (b) Cursive script (Japanese language characters)

Fig.1. Writing in alphabetic and Japanese language character scripts.

2. HANDWRITING RECOGNITION SYSTEM

2.1. Frame-based Character Recognizer

We developed a high-performance handwriting character recognizer based on an integration of online-offline recognizers[1]. We also developed a context processing engine using character bigrams[2] and the user adaptation method [3]. To evaluate the usability of a frame-based interface, we used our character recognizer and context processing engine using only bigrams (no user adaptation).

2.2. Non-frame Character Recognizer

To evaluate non-frame interfaces, a non-frame handwriting recognizer whose performance is good enough for practical application must be used[4][5][6]. We therefore developed a non-frame recognizer based on the character recognizer briefly described in the previous section.

As shown in Fig.2(a), handwriting input is divided into basic segments in the first step of the recognize operation. In the second step, network paths connecting some of the segments are created, and the frame-based character recognizer creates recognition candidates on each path. After the network is created as shown in Fig.2(b), an optimum string is created using a dynamic programming (DP) algorithm. For the evaluation of DP path costs, character recognition scores, and the network path costs calculated using the bigram probability, segment shape, and segment position.

2.3. Recognition Performance

Table 1 shows the recognition performance of each recognizer. The database "kuchibue" is a database of Japanese handwriting collected using a frame-based interface[7]. To evaluate non-frame recognition performance and compare it with the performance of a frame-based recognizer, we used a pseudo non-frame database created from the frame-based database by placing each character pattern in the appropriate sequence in which the gaps between character pattern are the same (Fig.3). We set the gap value d to 4% of the writing frame size s.

(a) Pre-segmentation (b) Optimum path search (DP)

Fig.2. Non-frame handwriting recognizer.

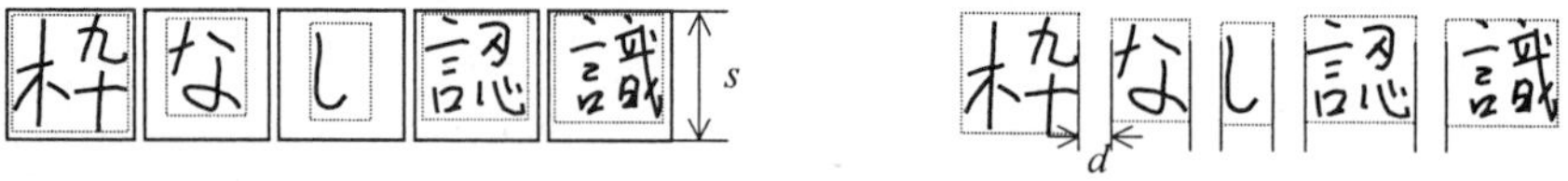

(a) Frame-based database (b) Pseudo no-frame database

Fig.3. Handwriting database.

Table 1. Recognition performance

	Recog. rate	Time (char)
Frame-based	95.1%	13 ms
Non-frame	92.9%	50 ms

Database: HANDS-kuchibue_d-97-06
(120 persons, 11962 patterns/person)
Pentium III 700 MHz (Windows NT4.0)

Table 2. Exp. in using PC (data entry)

Score	1	2	3	4	5
PC.	23	14	33	36	14
pen PC	103	14	2	1	0

Table 3. Exp. in using PC (recognition)

score	1	2	3	4	5
PC	0	0	10	17	3
pen PC	26	3	1	0	0

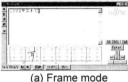 

(a) Frame-based application (b) Non-frame application

Fig.4. Data entry applications.

(a) Frame mode (b) Non-frame mode

Fig.5. Handwriting recognition application.

3. EXAMINATIONS

3.1. Test Application

We created a frame-based data entry test application and a non-frame one (Fig.4), and a handwriting recognition test application that has two input modes (Fig.5). The handwriting recognition application uses our handwriting recognizers (frame-based and non-frame) to recognize input patterns. These applications have an editing interface (edit belt) which makes pen interface more comfortable[8].

In the handwriting recognition application, recognition results are displayed for each characters quickly in the frame-based mode, and are displayed after writing all characters in the non-frame mode.

3.2. Examination Process

The examinations were done with the test applications. In the data entry test, subjects wrote a given set of text strings in the frame-based application, wrote the same text in the non-frame application, and then filled out the questionnaire. In the recognition test, the subjects wrote the same text and input them using handwriting recognition with both input modes. After all of the text was input, the subjects filled out the questionnaire.

3.3. Examination Subjects

The handwriting examinations of the data entry phase were completed by 120 subjects (male=57, female=63). For the recognition phase, the number of subjects was 30 (male=21, female=9). In Tables 2

437

and 3, scores ranging from one to five represent the subjects' individual experience in using PCs (and pen PCs); the number one means "no experiences," two means "less than 1 year," three means "less than 3 years," four means "less than 10 years," and five means "10 years or more."

4. RESULTS

4.1. "Are you satisfied with the recognition performance ?"

Fig.6 shows the subjects' satisfaction with the recognition performance. The scores of 4.1 and 3.2 written beside the figure titles represent the mean opinion score (MOS) which is often used to evaluate speech quality[6]. The MOS of 4.1 for the frame-based recognizer is very high, and the subjects may have been very satisfied with the performance. Comparatively, the MOS of 3.2 for the non-frame recognizer is not very high, but it is higher than the mean score of 3. We think the performance of the non-frame recognizer may not be "adequate," but it may be "acceptable" for our examination.

4.2. "Is Pen Input Useful ?"

Fig.7 shows the answers to the question, "Is pen input useful?" The MOS of 2.9 for the data entry phase and 2.8 for recognition phase are not very bad. This shows that pen input is almost acceptable

4.3. "Which is better, frame-based or non-frame ?"

Fig.8 shows the results of comparing frame-based input and non-frame input. In this estimation, a score of '1' means frame-based input is better, and '5' means non-frame is better. As shown in Fig.8(a), non-frame input is better than frame-based input only for data entry. Otherwise, frame-based input is much better in the character recognition phase (Fig.8(b)). These results indicate that frame-based input is more suitable for text input by handwriting recognition than non-frame input.

4.4. Correlations of Results

This section is about the relationship between these results and the subjects' individual exmerience in using PCs. Table 4 shows the correlations for each result. All correlation values related to C are negative and the absolute values are relatively high. This means that PC users do not prefer to use the non-frame interface. The relationship between A-2 and B shows that users who often use pen PCs do not think the non-frame interface is useful.

4.5. Comments

The comments collected together with the questionnaire give some hints about the reasons for the results. As on answer about pen input usefulness, 37% of the subjects claimed that pen input made their hands tired, and 23% of them said pen input is slow. Otherwise, many subjects claimed that non-frame input is slower than frame-based input by handwriting recognition. They also claimed that frame-based handwriting recognition is more reliable than non-frame recognition. According to these comments, we found three important conditions for a comfortable text input method; non-tiring, quick, and reliable.

5. CONCLUSIONS

We have evaluated the practicality of writing frames for handwritten Japanese sentences on a pen PC by collecting questionnaires and counting the answerson them. One of the results indicates frame-based input is better than using a non-frame interface under the same circumstances as those in our text input examination. Our interpretation of the result is that non-frame text input should not be a replacement of frame-based text input. A non-frame interface is especially suitable for the situations in which a frame-based interface cannot easily be used, such as for a small GUI or on an electronic whiteboard.

438

For practical application of non-frame interfaces, it is more important to improve the response speed of handwriting recognition rather than improve recognition accuracy. We are going to speed up the processing in our non-frame recognizer, and then we intend to evaluate the practicality of non-frame interfaces again.

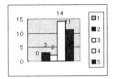
(a) Frame-based (4.1)

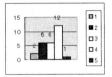
(b) Non-frame (3.2)

Fig.6. Recognition performance

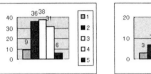
(a) Data entry (2.9)

(b) Recognition (2.8)

Fig.7. Is Pen input useful?

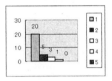
(a) Data entry (3.6)

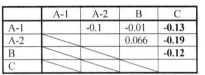
(b) Recognition (1.5)

Fig.8. Frame or non-frame?

Table 4. Correlations for each result

	A-1	A-2	B	C
A-1		-0.1	-0.01	**-0.13**
A-2			0.066	**-0.19**
B				**-0.12**
C				

A-1: use PCs   B: usefulness
A-2: use pen PCs   C: frame / non-frame

REFERENCES

[1] H. Tanaka, K. Ishigaki, K. Nakajima, K. Akiyama, M. Nakagawa, "Hybrid Pen-Input Input Character Recognition System Based on Integration of Online-Offline Recognition," Proc. 5th ICDAR, pp. 209-212 (1999.9).

[2] M. Nakagawa, K. Akiyama, L.V. Tu, A. Homma, T. Higashiyama, "Robust and Highly Customizable Recognition of On-line Handwritten Japanese Characters," Proc. 13th ICPR, Vol. III, pp. 269-273 (1996).

[3] N. Iwayama, K. Ishigaki, "Adaptive Context Processing in On-line Handwritten Character Recognition," Proc. 7th IWFHR, pp. 469-474 (2000.9).

[4] H. Murase, T. Wakahara, M. Umeda, "Online Writing-Box Free Character String Recognition by Candidate Character Lattice Method," Trans. IEICEJ, Vol. J68-D No.4 pp. 765-772 (1985.4)

[5] S. Senda, M. Hamanaka, K. Yamada, "Box-free Online Character Recognition Integrating Confidence Values of Segmentation, Recognition and Language Processing," IEICE Technical Report, PRMU98-138 (1998.12).

[6] T. Fukushima, M. Nakagawa, "On-line Writing Box Free Recognition of Handwritten Text based on Stochastic Models," IEICE Technical Report, PRMU98-139 (1998.12).

[7] M. Nakagawa, T. Higashiyama, Y. Yamanaka, S. Sawada, L.V. Tu, K. Akiyama, "Collection and utilization of on-line handwritten character patterns sampled in a sequence of sentences without any writing instructions," IEICE Technical Report, PRU95-110 (1995.9).

[8] T. Oguni, T. Yoshino, M. Nakagawa, "Demonstration of the IdeaBoard Interface and Applications," Proc. INTERACT 97 (1997.7)

[9] S. Furui, "Digital Speech Processing," Tokai University Press (ISBN4-486-00896-0), pp.130 (1985).

Programming Education on an Electronic Whiteboard Using Pen Interfaces

Taro Ohara, Naoki Kato, Masaki Nakagawa
Dept. of Computer Science, Tokyo Univ. of Agri. & Tech.
Naka-cho 2-24-16, Koganei, Tokyo, 184-8588, Japan
e-mail: sylph@hands.ei.tuat.ac.jp

This paper describes a system to teach programming on an interactive electronic whiteboard that combines the merits of classroom lectures using a black/white board and those of computer processing. Using this system, a teacher can write a program on the board, explain it, make the system recognize it and run the program in front of the class while keeping the attention of the students focused on the board. The system allows input data to be entered by writing input parameters on the board.

1. Introduction

Conventionally, a teacher teaches programming by writing a program on a white/black board and explaining its logic and structure. Students copy the program and its explanation in their notebooks. Then, they are often required to study the program after the class by running it on a computer. Problems may arise at this stage, however, because the program written by the teacher on the board may have contained some errors, the student may make some mistakes in copying the program, and so on.

In order to solve this problem, we have proposed a system for an interactive electronic whiteboard [1]. Using this system, the teacher can verify the correctness of a program that he/she has written immediately, show its execution, and explain how the output is changed when some part of the program is modified, without losing the familiarity and advantages of lectures using chalk and blackboard (Fig. 1).

This paper describes the latest version of our programming education system on an electronic whiteboard system.

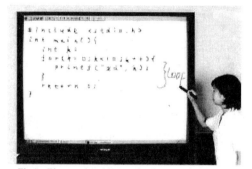

Fig.1   Electoronic whiteboard and
programming education system

2. Design of the Programming Education System

2.1   Design Goals

The specifications of the programming education system are based on the design criteria of the user interface for the electronic whiteboard [2, 3] as follows:
(1) Operability from arbitrary standing position of the user.
(2) Easy operability with a single electronic marker.
(3) Natural extension of the desktop GUI.
(4) Simplicity of displayed contents.
(5) Maximize space for contents while minimizing space for control.

The first of the specifications allows the teacher to operate the board without having to cross the surface or stretch hands from side to side or from edge to edge. The second specification ensures that the teacher can operate the board with a single electronic marker without needing other markers, keyboard, mouse, etc. The third specification ensures consistency with the desktop environment. The fourth criterion allows the teacher to operate the board without confusion so that the students can understand the contents easily. The fifth criterion is to make sure that the surface of the whiteboard is utilized for education as much as possible: buttons, menus, etc. should not hide the contents unless absolutely necessary.

2.2 Functional Design
The following functions are necessary for the programming education system.

(1) Handwritten character recognition
With a marker, writing is the easiest and simplest way to input program text. Without pattern recognition, however, handwriting is just pen-trace patterns and cannot be processed as program text. Therefore, handwritten character recognition [4] is necessary.

To facilitate character recognition, we provide a grid of character input frames to write program text. The character recognition engine accepts the handwritten text, recognizes it within the context of the programming language and outputs a code sequence of program text. We employ a lazy recognition scheme i.e., recognition after all the text is written, because recognizing each character immediately after it is written interrupts the writing of a program. The lazy recognition approach also allows the use of context to help pattern recognition. After program recognition, handwritten character patterns are replaced by font patterns (Fig. 2).

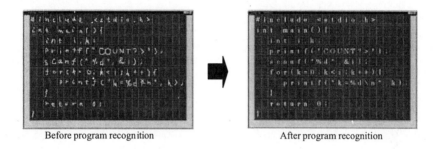

Before program recognition   After program recognition

Fig.2 Handwritten program recognition

(2) Editing
Input, insertion and deletion of program text are the most necessary editing functions. According to the design specifications, the user must be able to perform these operations with a single marker. We implement the input function by allowing the user to write a character pattern in any empty frame. Insertion and deletion can be performed by tapping a marker between lines and dragging right or left (insertion or deletion of character frames within a line), or down or up (insertion or deletion of lines). When a marker is dragged to the right, new frames appear along with the dragging and the user can write characters in them. When the marker is dragged to the left, the characters on the right move over the dragged characters which are deleted (Fig. 3).

Insertion and deletion of lines can be done similarly by shifting lines downward and making new lines or shifting lines upward.

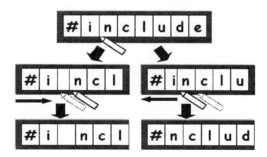

Fig.3   Insertion and deletion of text by sliding character frame

(3) Program execution

A program thus edited can be executed by an interpreter. Using an interpreter rather than a complier, the program execution can be started without the compilation step. The output of the program being interpreted is not displayed directly, but a pipe with the system is created and the output is displayed within the execution window of the system. The input is dealt similarly. Data can be inputted by writing in the execution window and the recognized code is sent to the program through the pipe.

We deliberately avoided the use of a keyboard for input in our system. A keyboard is neither easy to use for a standing teacher nor easy to observe for the students facing the teacher. When the teacher writes some input in the execution window, its recognition result is displayed in a pop-up window. The teacher can confirm the result of recognition, correct it if necessary, and then make the program continue its execution by pushing the execution button (Fig. 4).

(4) Handwriting annotations

A great advantage of using a pen is that one can write almost anything freely. It is very useful for the teacher to be able to write annotations on the program that he/she is explaining instead of merely showing the program. Therefore, our system provides an annotation capability on the source and execution windows with the electronic marker in the same way as on a blackboard (Fig. 5).

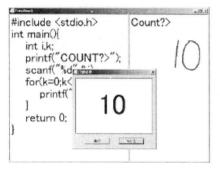

Fig.4 Execution of program

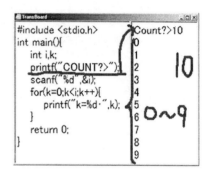

Fig.5 Handwriting annotation

(5) Screen scroll

Since the electronic whiteboard has a limited size, and some programs may be too large to show within the display area, the scroll function is essential. We will describe this in detail in the next section, as it is an issue of user interface.

The standard scroll bar for a window is displayed on the bottom right-hand side of the window. This is convenient for operating in the traditional desktop environment, but on an interactive electronic whiteboard with an electronic marker, it is hard to use. The teacher has to stretch his or her hands from side to side and hide the board by his or her body. This violates one of our basic design goals.

Therefore, scrolling the screen is implemented without using the traditional scroll bar. Scroll area is located around the input area. By touching the marker on any place in this area and dragging it to arbitrary direction, the screen can be scrolled in that direction.

2.3 Design of the User Interface

Design of the user interface is also based on the specifications mentioned earlier in 2.1.

(1) Screen interface

The screen of the programming education system consists of a handwriting input area and a screen scroll area (Fig. 6). The scroll area is located around the input area except for the upper part of the window. When the user taps with the marker somewhere in the scroll area, a tool bar is displayed and operations can be performed from there.

A user can write a program in the handwriting input area.

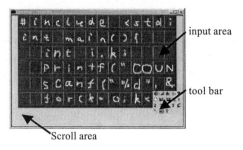

Fig.6   Screen layout of the programming education system

(2) Tool bar
    Since operations in the tool bar are not so frequently performed, the tool bar is displayed only when necessary. When the user taps somewhere in the scroll area with the electronic marker, the tool bar appears there. Moreover, only the buttons used often are displayed, others are hidden. When the functions associated with the hidden buttons are required, the number of buttons displayed can be changed by dragging the scroll area in the tool bar with the marker.

(3) Execution window
    When the teacher is explaining a program, it is useful if he or she can show the source program and its execution results, and annotate them by writing as shown in Fig.5.
    The execution window displays the source program and its execution result within split areas of the window and allows the user to annotate over both the areas. The user can choose either or both areas to be displayed, and can change their proportions of the window as shown in Fig. 7.
    Another alternative is to display the result of program execution in a window separate from the source program, but to allow the user to annotate over two or more windows requires restructuring of our system software for this system.

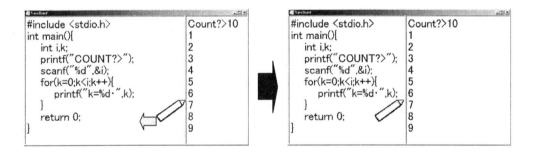

Fig.7   The splitter window displaying the source program and result

(4) Tool bar for the execution window
    The tool bar containing buttons for operations on the execution window appears whenever the execution window is displayed because it occupies only a small area and these operations are used often when the teacher is working on the execution window.

3. Implementation of the Programming Education System
    We have implemented the programming education system on the MS-windows ME using Visual C++ 6.0. Its appearance is shown in Fig. 8.

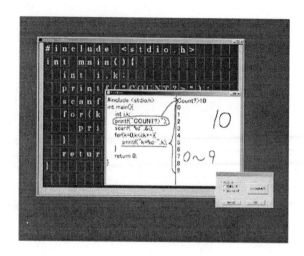

Fig.8 Appearance of the Programming Education System

4. Conclusion

This paper described the design and implementation of our programming education system on an electronic whiteboard system. We are now planning to use the system for teaching a programming course in our university curriculum to evaluate it in actual use.

References

[1] J. Kanda, S. Sawada and M. Nakagawa: "Programming Education System on an Interactive Electronic Whiteboard," Proceedings of the Fourteenth Symposium on Human Interface, Tokyo, pp.601-606 (1998.9)

[2] M. Nakagawa, T. Oguni and T. Yoshino: "Human Interface and Applications on IdeaBoard," Proc. IFIP TC13 Int'l Conf. on Human-Computer Interaction, pp.501-508 (1997.7).

[3] M. Nakagawa, K. Hotta, H. Bandou, T. Oguni, N. Kato and S. Sawada: "A Revised Human Interface and Educational Applications on IdeaBoard," CHI99 Video Proceedings and Video Program and also CHI99 Extended Abstracts pp.15-16 (1999.5).

[4] T. Fukushima and M. Nakagawa: "On-line Writing-box-free Recognition of Handwritten Japanese Text Considering Character Size Variations", Proc. of 15th ICPR, Vol.2, pp.359-363 (2000.9).

Supporting Individual and Cooperative Work Using Scalable Pen Interfaces

Y. Otsuki, H. Bandoh, N. Kato & M. Nakagawa

Dept. of Computer, Information and Communication Sciences
Tokyo Univ. of Agri. & Tech.
Naka-cho 2-24-16, Koganei, Tokyo, 184-8588, Japan
e-mail: yhoko@cc.tuat.ac.jp

Abstract

This paper describes prototyping of a computer-supported system for preparing schools' newspaper, it is a typical example of computer-based environments for supporting creative learning for both individual children and groups of children in primary and secondary schools. Learning is trained through individual creative thinking and communication among individuals. To support this, information processing, more intuitive user interfaces, and seamless coupling of individual work and group work, are all necessary. We have employed pen-based user interfaces for both individual work and group work and designed the groupware so that individual and group works are combined seamlessly. Namely, the materials are prepared by each individual on a display-integrated tablet with an electronic pen, and then they are collected, edited, and merged into the schools' newspaper by electronic markers while being displayed for a group of children on an interactive electronic whiteboard. This paper presents the design, implementation and preliminary evaluation. We hope to clarify the requirements for this type of systems, including their realizations, by the revision at the end of this paper.

1. Introduction

To learn Information Technology (IT) and to utilize IT for learning is indispensable in the information society in the 21 century. There are many problems, however, when children are to use computers. One of the biggest problems is the heavy burden when they have to use a keyboard while thinking or learning. For creative learning, more natural user interfaces are essential so that they are not bothered by the way they interact with computers. We employ pen-interfaces for both individual workspace and cooperative shared workspace. Each child works on a display-integrated tablet with an electronic pen and then collaborates on an electronic whiteboard using electronic markers and erasers. According to the terminology of cognitive psychology, handwriting is so-called "automated" for us since we have been writing by pen since our childhood so that we can concentrate our attention on thinking without being annoyed by how to use a pen. Moreover, we can express our feeling or emotion by our handwriting.

The main reason why we have been working on handwriting-based user interfaces is that thinking is not interrupted by the actions for writing. Thinking and writing form a positive feedback loop system allowing and clarify one's idea. This nature of writing is suited for creative work rather than labor-intensive tasks [1]. Our research also extends to systems using PDAs, desktop tablets, and large electronic whiteboards, which are useful for computerized classroom education [2]. We have been developing several educational applications for the latter system [3].

In primary and secondary education, however, group learning and individual learning are also important as well as classroom learning. Group learning activates discussions among school children, helps themselves share and understand common problems and leads them to the solutions effectively based on their confidential relationship. The research area that group learning belongs to, liberated from physical restrictions and enhanced by computers, is called CSCL (Computer Supported Collaborative Learning).

This paper describes an initial attempt to employ our pen interface resources in order to support creative learning for both individual learning and group learning. Out target is a computer-supported system for preparing school's newspaper in primary and secondary schools. The school's newspaper preparation is a typical example of creative work made by individual children collaborating with each other. Our aim to develop this system is to realize an environment where each individual children and groups of children can creatively learn.

2. Basic Design of the School's Newspaper Preparation System
2.1. Handwriting-based user interface

This system is for school children. We must assume that the users are not well prepared for using a computer and

445

typing a keyboard. Many educational software products require skills for them to use a computer and a keyboard. This is hard even for less creative work. For creative work, this is a much harder problem for children. Therefore, we employ pen interface instead of a keyboard and a mouse. The pen interface can be commonly employed for PDAs, desktop tablets and electronic whiteboards. Thus, it is scalable to the size of the target. It is natural to the users so that creative thinking is not restricted. We expect school children to be able to use it like a sheet of paper with pen and like a blackboard with a piece of chalk.

2.2. Individual work and group work

The process of preparing school's newspaper is divided into individual work and group work. In the individual work, each child collects articles to be used for school's newspaper, and prepares manuscripts for them. In the group work, the group edits the school's newspaper using materials presented by the members. Here, a desktop PC or one of the desktop tablets that the members are working with does not fit the group work. A screen cannot be simultaneously seen from all the members and articles cannot be manipulated by all members. One solution to this problem is to share a large surface virtually and view and access it through each workspace. Another solution is to share the common area physically. We prefer the latter method since the full awareness is shared among all the members: Each child works on a display-integrated tablet with an electronic pen and then collaborate with other children on an electronic whiteboard, which is 70 inches large, using electronic markers and erasers.

3. System for Preparing Schools' Newspaper

This system is composed of desktop environment and electronic whiteboard environment, where the desktop environment with pen input is used by each individual for preparing materials while the electronic whiteboard environment is used for children for collecting, editing and arranging materials into the school's newspaper. Therefore, two functions corresponding to the two environments are necessary that must be seamlessly coherent so that children can use them as if they would prepare materials on a sheet of paper and then discuss them on a blackboard. Fig.1 shows preparation of schools' newspaper using our system.

3.1 Material Preparation Function

The material preparation function is to prepare materials for articles in school's newspaper using a display-integrated tablet with an electronic pen. Fig. 2 shows the screen of this function.

In the material preparation function, there are two modes, one for input handwritten text and the other for correcting segmentation points between characters so that handwritten text can be laid out according to any column width.

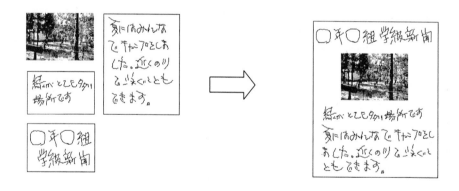

An Individual prepares materials for an article.
(Material Preparation Function)

Group members edit articles using the
materials prepared in the last step.
(Article Editing Function)

Fig.1 Preparation of schools' newspaper on our system.

Fig. 2 The screen of the material preparation function.

Fig. 3 The screen of the article editing function.

In the handwriting input mode, a child can write text on horizontal gridlines displayed on a d isplay-integrated tablet. The user can also erase text, select the thickness of the electronic ink and change its color. Then, the system segments characters using handwriting recognition technology developed at our laboratory [4]. Here, it is important that the system employs handwriting recognition in order to segment handwritten characters rather than replacing handwriting by character font. When the user changes from the handwriting mode to the segmentation modification mode, the system determines character segmentation points automatically. In the segmentation modification mode, the writer can correct mis-recognized character segmentation points.

3.2 Article Editing Function
The article editing function arranges materials that users have prepared using the material preparation function to form a school's newspaper on the electronic whiteboard. The materials include handwritten text, photographs, pictures, figures, voice clips, animations etc. Fig. 3 shows the screen of this function.

This function is used on the electronic whiteboard with surface far bigger than that of tablets. Therefore, it is very important that the user interface doesn't depend on the user's height and standpoint. New scroll bars placed at both sides of the electronic whiteboard solve the problem of the user's height. Fig. 4 shows a general scroll bar and the scroll bar for the electronic whiteboard. The screen can be freely scrolled by tapping the electronic marker anywhere in the scroll bar and dragging it to the direction that the user wants to show.

When the bottom part of the surface is tapped, a menu appears where the user can select functions such as "open a file" or "save to a file " and so on. This arrangement enables even a small child to operate the menu regardless of the standing position. Fig. 5 shows the bottom part after the menu has appeared.

Fig. 4 A general scroll bar (left) and a scroll bar for the electronic whiteboard (right).

Fig.5 The menu bar for the article editing function.

447

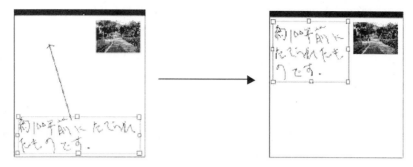

| A handwritten text is specified and moved. | A system judges the column width automatically and handles line-feeds automatically. |

Fig. 6 Automatic line-feed.

When materials are moved, enlarged, reduced or deleted from the newspaper, the column width needed to layout each material is recomputed, and the handwritten text is split into lines (automatic line-feed) at appropriate segmentation points. This is possible because handwritten text has been segmented between characters . Fig. 6 shows how this looks like.

4. Evaluation experiment

We made preliminary evaluations in order to judge the usefulness of each function, clarify problems in the user interface and test whether this system can support creative work. The first evaluation was made to test the material preparation function, and the second evaluation was made to test the entire system for preparing schools' newspaper.

4.1 First evaluation experiment

We invited six children from the second grade to the sixth grade in an elementary school and asked them to use the material preparation function. Fig. 7 shows how they are working using this function.

4.2 Second evaluation experiment

We invited ten children from the first grade to the fifth grade in an elementary school and asked them to use the entire system for preparing schools' newspaper. Fig. 8 shows a scene where they are using the article editing function.

Fig. 7 A snapshot from the first evaluation experiment.   Fig. 8 A snapshot from the second evaluation experiment.

4.3 Opinion of subjects
The school children who used this system gave the following opinions:
- Handwriting is easy to prepare materials.
- Gridlines are useful when inputting sentences.

Moreover, we have observed the following children's behaviors:
- Intervals between handwritten characters get narrower during writing, as shown in Fig. 2, which makes automatic segmentation difficult.
- By passing an electronic marker from a child to another, they can smoothly turn the role of editing to each other.
- Vertical writing is easier than horizontal writing since Japanese children are more familiarized with the former in the Japanese language lessons.
- Automatic character segmentation is not easy for children's handwriting and it takes a certain amount of time to output the result.

The last two problems are being revised. After the revision, usability and effectiveness of the system for making creative learning are evaluated again.

5. Consideration and future work
This paper described prototyping of a computer-supported system for preparing schools' newspaper. It supports creative learning made for individual children and groups in primary and secondary schools.

In the material preparation function, a child can write characters on horizontal gridlines and the system recognizes character segmentation points automatically. However, children's handwriting has narrow intervals between characters. For this reason, automatic segmentation is difficult and correction work by the user is increased. Moreover, since automatic segmentation takes some time, children likely lose their concentrations. For children, lines of character writing boxes rather than just gridlines are much better. They are accustomed to them since they write essays using this type of manuscript papers. Moreover, this dispenses with automatic segmentation and the segmentation modification mode so that children are not annoyed by the delay of processing and they do not have to correct mis-segmentations.

For children, vertical writing is easier than horizontal writing since Japanese children are more familiarized with the former in the Japanese language lessons. The use of this system within a lesson of a primary school is scheduled. We are going to create the input screen for vertical writing so that there is no inconsistency with the language lessons in a primary school.

Acknowledgments
We are thankful to all the people who have cooperated in the evaluation experiment of our system for preparing school's newspaper.

References

1) M. Nakagawa: "Enhancing Handwriting Interfaces," Proc. HCI International '97, Vol. 2, pp.451-454 (1997.8).
2) M. Nakagawa, T. Oguni and T. Yoshino: "Human Interface and Applications on IdeaBoard," Proc. IFIP TC13 Int'l Conf. on Human-Computer Interaction, pp.501-508 (1997.7).
3) H. Bandoh, H. Nemoto, S. Sawada, B. Indurkhya and M. Nakagwa: "Development of Educational Software for Whiteboard Environment in a Classroom," Proc. Of International Workshop on Advanced Learning Technologies, pp.41-44 (2000.12).
4) T. Fukushima and M. Nakagawa: "On-line Writing-box-free Recognition of Handwritten Japanese Text Considering Character Size Variations," Proc. of 15th Int'l Conf, on Pattern Recognition, Vol.2, pp.359-363 (2000.9).

Pen-based Electronic Mail System for the Blind

Nobuo EZAKI Kimiyasu KIYOTA Shinji YAMAMOTO

Toba National College of Maritime Technology, Mie 517-8501, JAPAN, ezaki@toba-cmt.ac.jp
Kumamoto National College of Technology, Kumamoto 861-1102, JAPAN, kkiyota@tc.knct.ac.jp
Toyohashi University of Technology, Aichi 441-8580, JAPAN, yamamoto@parl.tutkie.tut.ac.jp

ABSTRACT

We have developed a pen-based Japanese character input system for the blind (particularly persons with acquired blindness). The user of this system is able to directly input Japanese characters without using a keyboard. This system is composed of a personal computer and a control board with an electric tablet. The blind person is able to get the screen information by using a voice synthesizer. We have investigated the various problems when the blind person edits the document by using this system and solved those problems. From the experimental results, we have confirmed that our proposed system makes easy to input Japanese characters for the novice blind user without the training. We apply to the electronic mail system for the blind person by utilizing this system.

1. INTRODUCTION

In recent years, computer application support for the blind has become an important theme. The reason for this is that a blind Japanese person needs to use many characters of various kinds. There are about 4,000 commonly used characters such as Kanji (Chinese characters), Kana, Katakana, and the Roman alphabet and numerical characters. Braille word processing using an accompanying keyboard is commercially available for blind. However, they have to learn to use the software conversion of Kana to Kanji, which uses a keyboard. This software has to be able to select the correct Kanji character from various candidates of the same Kana-sound (these are called homonyms). Therefore, inputting Japanese characters with a keyboard is quite cumbersome for novice blind users.

As a solution to these problems, we propose an on-line character input system using handwritten character recognition technology instead of a keyboard. Although the user still must select the candidate character for input, the burden on the user is reduced by the development of a high accuracy character recognition algorithm. We have investigated various problems encountered by blind person when they input and edit documents using this system and have solved each of these problems.

2. OUTLINE OF THE SYSTEM
2.1 SYSTEM CONFIGURATION

Our proposed pen-based Japanese character input system is composed of a personal computer with a control board including an electronic tablet as shown in Fig.1. The system automatically starts in the character input mode after the computer boots up. The blind person is able to get screen information using a voice synthesizer. The user inputs characters by using the electronic tablet. The control board, which only has seven buttons, is used for all computer command operations without any use of the keyboard. The control board operation is very

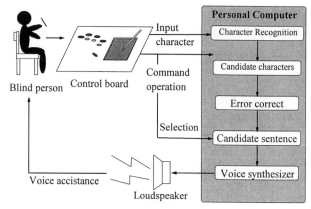

Fig. 1   System configuration

simple and thus the novice blind user is able to operate the system. There are various modes of the system, such as command, character input, file control, editing and mailing modes. The user can easily switch from one mode to another by using the seven buttons on the control board. The voice synthesizer announces when the screen information changes, so the blind user can follow the results of his input. To reiterate, this pen-based system makes it easy for a blind Japanese novice user to input Japanese characters with very little training.

2.2 CONTROL BOARD

The blind person uses tactile sense information effectively. To enable the smooth command input for the blind person, the user's both hands should be fixed in the regular position on the control board. Therefore the control board is designed that the command button is operated by the left hand, and the electric tablet is operated by the right hand, respectively as shown in Fig. 2. Furthermore there is a step in a character input area for the easy description of the blind person. The command buttons of the control board consist of 7 push-type buttons and 1 dial-type button. The push-type buttons are used for the character input control and the command mode change. As these buttons are different size, the user is able to distinguish the option buttons by the size. The dial-type button is used for the cursor movement and the selection of menu options.

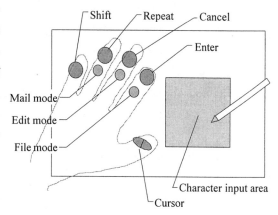

Fig. 2 Control board

3.   CHARACTER RECOGNITION

3.1 RECOGNITION METHOD

A structural analysis method is very useful for the handwritten character recognition. However, as the stroke

451

positions become unstable for the character written by blind person, the above method is no longer useful. We have investigated characteristics of Japanese Kanji characters written by many blind persons. From analysis result, we found the following some stable features [1].

(1)The same blind person can write almost the same stroke shape, stroke number and stroke writing order.

(2)The relative position of the stroke representative points in the partial pattern is stable.

Two kinds of character recognition algorithm; namely, the RDS method and the LSDS method have been proposed for this system by using the above features.

The RDS method is based on the relative direction between two strokes in a writing order. We represent each stroke of the Kanji character by three typical points (a starting point, a middle point and an ending point).

This recognition method is called the LSDS method (the method based on the line segment directions in a stroke) in this paper. A stroke is divided into several line segments by the same segment length. A feature parameter set is expressed as a set of eight direction codes that correspond to each line segment.

The combination of the two methods is desirable to recognize all type of Japanese character written by the blind person, because these methods use different type feature. This method is named the fusion method in this paper.

3.2 ERROR CORRECTION METHOD

From recognition experimental result, it was difficult to distinguish similar characters to use the fusion method that is the recognition algorithm for the one character order. Therefore we adapt the two error correction methods. The first method is by using a tree search algorithm that uses word dictionary and Japanese grammar; namely, Japanese phrase search algorithm based on the Japanese linguistic information. The second method is by using n-gram model, a Japanese phrase is selected by using transition probability and each characters probability. Some candidate Japanese phrases are estimated using these methods.

We examined the recognition test for 11 subjects. The character samples for the recognition are about 250 characters of the illustrative sentence of a letter per one person. From the experimental result, the total average recognition rate was a 90.2% for the fusion method. Furthermore we adapt two error correction methods for the same test samples. The recognition accuracy was improved to a 92.5%. Therefore we confirmed that these methods corrected miss-recognized Japanese characters (Kanji, Hiragana, Katakana, numerals and symbols).

4. DOCUMENT INPUT PROCEDURE

The document input is as following procedure. At first, the blind person writes one character to the electric tablet by using stylus pen. He pushes the [enter] button on the control board. Then, the system begins to recognize the character by using the fusion method. Next the system changes to the character write mode again. The user writes a next character by repeating this procedure. After the one phrase input, the [enter] button is pushed again. Then error correction software starts automatically, and then the voice synthesizer announces the first candidate phrase. If a correct answer is announced, the user pushes the [enter] button of the control board. If a wrong answer is announced, the user pushes the [cursor] button. The system then announces the next candidate phrase. When there is no correct phase in the candidate phrases, the system returns automatically to the re-writing mode.

Hence the cursor movement is a serious problem for the blind person who cannot see the cursor position. So, we regard a Japanese sentence as the character sequence of one dimension. Here the [cursor] button is used for one character movement. It is a basic cursor movement operation in this proposed system. Furthermore we prepare two options of the cursor movement. The cursor movement of one phrase is implemented by pushing the [cursor] button with the [shift] button. To move the cursor position to the beginning/ending of the sentence, the user pushes the [cursor] button and the [escape] button with the [shift] button. Then the system announces the cursor position by the voice synthesizer. Therefore the blind person is able to get the cursor position by above procedure.

5. ELECTRONIC MAIL MODE

In recent years, many blind persons want to use an electronic mail system. However almost all commercial electronic mailing software is not taking use of a blind person into consideration. Therefore we have also added the function of the sending, receiving and managing electronic mail with voice support in our system.

The electronic mail sending method is very simple. After the blind user inputs document, he pushes [Mail mode] button. The system shows the mail mode menu. Then the user selects the send mail mode from the menu. Finally he just only choices an E-mail address from his address book.

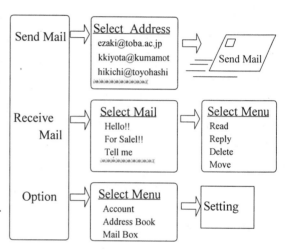

Fig. 3 Electronic mail mode

6. EXPERIMENT

We made experiments to examine the user-friendliness of this system. We measured the time of a keyboard input and a pen input to the same sentences that include 300 characters for the 3 subjects. Total average results of the experiment are shown in Table1. Our experiment showed that the keyboard input method was faster for a user who is skilled in keyboard operation. However, we also confirmed that a novice blind user was able to send an electronic mail using our proposed system without training. Moreover, the voice output time of our system was much more efficient than the commercial Braille word processor. Therefore, it is conceivable that our system would be useful and effective for

Table 1 Result of the sentence input time [S]

		Pen input		Key input	
Input time	Write down	395	Typing	147
	Recognition	10		
	Error correct	51		
Candidates Selection		121		251	
Total time		577		398	

a novice blind person.

7. CONCLUSION

We proposed the pen-based input system for blind persons. This system works to the information devices as the simple Japanese input system for the novice blind person without training. The system also gives a pleasure to write the Japanese character by his hand. We showed that a pen-input method is easier than a key-input method for a novice blind user from the experimental results. We have also applied the electronic mailing mode for the proposed system. When there is an Internet connection for this electronic mailing system, we expect that a novice blind person could be able to communicate with all the people on his address list by using this system.

At present, the prototype is experimentally produced based on the proposed basic concept. Future work is the evaluation experiment by utilizing this system on the school for the blind and the social welfare organization.

REFERENCES

[1] K.Kiyota, T.Sakurai, and S.Yamamoto, "Deformation analysis and classification of on-line handwritten Chinese character for the visually disabled persons." , Trans. IPS Japan, Vol.36.No.3, 1995, pp.636-644.

[2] K.Kiyota, N.Ezaki, T.Yanai, and S.Yamamoto, "A basic design of on-line Japanese input interface for visually disabled person",Vol.J79-A.No.2.1996, pp.310-317.

[3] K.Kiyota, T.Sakurai, and S.Yamamoto, "On-line character recognition for the visually disabled person based on the relative position of stroke representative points.", IEICE Trans. Inf. & Syst..Vol.J80-D-II.No.3.1997, pp.715-723.

[4] K.Kiyota, T.Yanai, N.Ezaki, and S.Yamamoto, "An improvement of on-line Japanese character recognition system for visually disabled persons", Proc. of the 14th International Conference on Pattern Recognition, 1998, Vol.II, pp.1752-1754.

[5] N.Ezaki, T.Hikichi, K.Kiyota , and S.Yamamoto, "A pen-based Japanese character input system for the blind person", Proc. of the 15th International Conference on Pattern Recognition, 2000,Vol.IV, pp.372-375.

User Adaptation in Handwriting Recognition by an Automatic Learning Algorithm

Toshimi Yokota, Soushiro Kuzunuki, Keiko Gunji, Nagaharu Hamada

Hitachi Research Laboratory, Hitachi, Ltd.
Omika-cho 7-1-1, Hitachi-shi, Ibaragi, 319-1292, Japan

ABSTRACT

To improve the on-line handwriting character recognition rate, we developed user adaptation technology that automatically learns character patterns when the recognizer outputs an incorrect result and the user selects any other candidate. This learning algorithm is composed of adding an input pattern to the dictionary and replacing a dictionary pattern with an average of an input pattern and the dictionary pattern.

In evaluation with the kuchibue_d-96-02 database, the recognition rate is improved from 82.0% to 86.7% (4.7% higher) and the dictionary size goes from 373 Kbytes to 409 Kbytes (9.7% larger). Moreover, the correct recognition rate with a small dictionary and the learning algorithm is higher than with a large dictionary without the learning algorithm.

1. INTRODUCTION

A number of on-line handwriting character recognition methods [1,2] are commonly used for data input to small-sized computers or PDAs, because these devices require a natural man-machine-interface such as paper and pencil and a keyboard is not effective. To improve the recognition rate, various deformation patterns must be registered in the dictionary, but this makes the dictionary size larger. When only one person uses the device, only his/her deformation patterns have to be registered so that the dictionary size is smaller. Generally, a user must intentionally register character patterns into the user's dictionary, but this operation is troublesome for her/him. Therefore, we developed user adaptation technology in handwriting recognition which uses an automatic learning algorithm.

2. USER ADAPTATION IN HANDWRITING RECOGNITION

The automatic learning algorithm registers a pattern into the dictionary when the character recognition result is not correct and the user selects one of the other candidate results. For example, suppose the user writes the sign "<", but it is not sharp and long enough, so that recognition system returns the symbol "(" as the result. Then, the user selects the symbol "<" from among the other candidates (Figure 1(1)), and the automatic learning algorithm registers the pair of the user's written sign "<" and the symbol code "<" into the dictionary. The next time the user writes "<" which is not sharp and long, the recognition system correctly returns the symbol "<" for the result. (Figure 1(2))

455

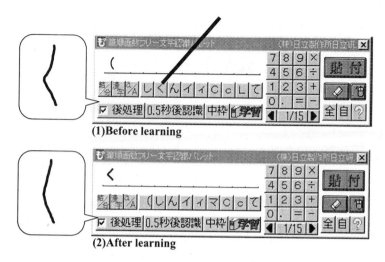

(1)Before learning

(2)After learning

Figure 1: The operation of automatic learning

3. THE AUTOMATIC LEARNING ALGORITHM

This operation is easy and does not annoy a user. However, if an algorithm adds patterns to the dictionary freely and automatically, there is a problem that the dictionary size increases significantly. Therefore, we propose a learning algorithm, which is composed of adding an input pattern to the dictionary and replacing a dictionary pattern with an average of an input pattern and the dictionary pattern. The algorithm restrains the increase of the dictionary size.

Figure 2 shows the algorithm flow.

STEP 1) After the user writes a character pattern and the recognition system returns the result of the first and other candidate codes, if the user selects another candidate code instead of the first, then STEP 2 is invoked for learning the just written pattern.

STEP 2) If the other pattern of the selected candidate code has already been registered in the user's dictionary,

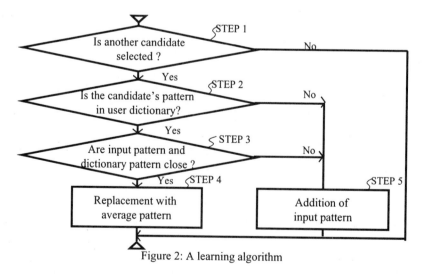

Figure 2: A learning algorithm

then the algorithm goes to STEP 3, otherwise it goes to STEP 5.

STEP 3) The distance of the just written pattern and the already registered dictionary pattern is calculated by the recognition matching function. If the distance is smaller than the threshold value, in other words the two patterns are close, then the algorithm goes to STEP 4, otherwise it goes to STEP 5.

STEP 4) An average pattern of the just written pattern and the already registered dictionary pattern is made, and the dictionary pattern is replaced by the average pattern.

STEP 5) The just written pattern is added to the user's dictionary.

The threshold value in STEP 3 is introduced to avoid making a meaningless average pattern that is from two very different patterns. This algorithm has an effect of restricting the number of registered dictionary patterns.

4. EVALUATION RESULTS

We evaluated the proposed automatic learning algorithm with our character recognition system under the following conditions.

1. The standard dictionaries used have the following two volumes.

(1) Large dictionary: Its size is 373 Kbytes which gives a higher recognition rate because more than one deformation pattern can be registered for each character category in the dictionary.

(2) Small dictionary: Its size is 189 Kbytes. each Kanji category (Chinese character in Japanese writing) has only one pattern registered although other kinds of characters have more than one deformation pattern per each.

2. Data used are three persons' sets inside of the TUAT Nakagawa Lab. HANDS- kuchibue_d-96-02 database [3], which consists of character patterns which are sampled as a sequence of sentences prepared by Tokyo University of Agriculture and Technology. The three sets are mdb0014, mdb0051 and mdb0066, which are representative of data for low/middle/high recognition rates.

3. When 10 candidates contain the correct result for mis-recognition, the automatic learning algorithm registers the pair of the written pattern and the symbol code of the correct candidate into the user dictionary.

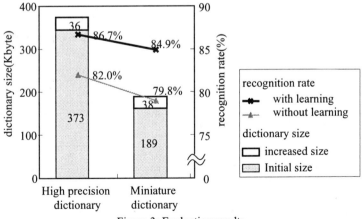

Figure 3: Evaluation results

We compared results with and without the learning algorithm. With the large dictionary, the recognition rate was improved from 82.0% to 86.7% (4.7% higher) and the dictionary size went from 373 Kbytes to 409 Kbytes (9.7% larger). With the small dictionary, the recognition rate was improved from 79.8% to 84.9% (5.1% higher) and the dictionary size went from 189 Kbytes to 227 Kbytes (20.1% larger). Moreover, the correct recognition rate with the small dictionary and the learning algorithm was improved by 2.9% (82.0% -> 84.9%) compared to that of the same recognition system with the large dictionary without the learning algorithm. The former dictionary size was 46Kbyte smaller than the latter.

The kuchibue_d-96-02 database consists of character patterns which are sampled as a sequence of sentences, and we investigated the learning process of the system which recognizes and learns the patterns according to their sequence. Figure 4 shows the averaged recognition rate of up to 1000 patterns in the learning process. Points (x , y) are plotted in the figure, x is the order of the character pattern and y is the averaged recognition rate over the recent 1000 patterns. The recognition rate y is calculated for 1000 samples of x-1000 to x patterns. (In the case that x is smaller than 1000, the recognition rates over 1 to x orderes are plotted.)

After 500 patterns, the rate with learning is larger than that without learning. If a user writes characters in sentences, she/he could feel the adaptation technology effect after inputs of 500 or more characters.

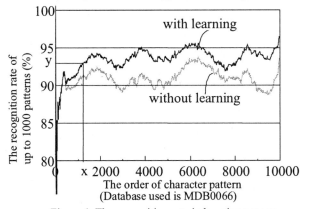

Figure 4: The recognition rate in learning process

5. CONCLUSION

We proposed user adaptation technology in handwriting recognition employing an automatic learning algorithm. In the future, we will modify the technology to decrease user mis-operations.

ACKNOWLEDGEMENT

Prof. Nakagawa at Tokyo University of Agriculture and Technology gave us very valuable advice.

REFERENCES

[1]Nakagawa, M. and Akiyama, K.: A linear-time elastic matching for stroke number free recognition of on-line handwritten characters, Proc. 4th IWFHR, pp38-56(December 1994)

[2]Tappert, T. C., Suen, C. Y. and Wakahara, T.: The state of the art in on-line handwriting recognition, IEEE PAMI, vol.12,no.8,pp787-808(August 1990)

[3] Nakagawa, M., Higashiyama, T., Yamanaka, Y., Sawada, S., Higashigawa, L. and Akiyama, K.: On-line handwritten character pattern database sampled in a sequence of sentences without any writing instructions, Proc. ICDAR-97, pp376-381 (August 1997)

A NEURO-FUZZY ADAPTIVE CONTROL OF INTERACTIVE SYSTEMS

Hans-Günter Lindner
humanIT Human Information Technologies GmbH, GMD TechnoPark
Rathausalle 10, 53754 Sankt Augustin, Germany, lindner@humanit.de

Alexander Nikov, Tzanko Georgiev
Technical University of Sofia, K. Ohridski 8, BG-1000 Sofia, Bulgaria, nikov@tu-sofia.acad.bg

ABSTRACT

The paper presents adaptation and optimization of the interaction structure of interactive systems by application of the control theory principles based on a fuzzy neural network. The interactive system is presented as a control system consisting of two parts: 1) plant of control or controlled plant (model of the real-life process) and 2) neuro-fuzzy-based feedback control algorithm. The last one is detailed presented. A MATLAB-based control structure for adaptive interactive systems is proposed. A MATLAB object-oriented fuzzy neural network is kernel of the control algorithm. By logfile data from a web-based interactive system the neuro-fuzzy-based adaptive control is simulated and studied. The advantages of adaptive control of interactive systems and further developments are discussed.

1. INTRODUCTION

Today's networked world and the decentralization that the Web enables and symbolizes have created information explosion and saturation. This information overload can be reduced by adaptive interactive systems. Usually they have large-scaled very complex structures with high number of unknown or badly estimated parameters. Thus a framework supporting the development of intelligent systems able to learn in dynamic, imprecise, and uncertain environments is needed. This paper presents such framework, showing how it is possible to adapt interactive systems to user combining control theory and neuro-fuzzy-based learning.

The control theory (Vidyasagar, 1995) is appropriate to model the structure and parameters of interactive systems. The adaptive control is a kernel concepts in control theory, which includes on-line estimation of the system parameters and self-tuning of the control algorithm. The modern control theory concepts (Bishop, 1997) are combined with fuzzy sets and neural network ideas and results. Neural networks and fuzzy techniques (Kecman, 2001) are among the most promising approaches to machine learning. The last one is of great importance in many aspects of information technologies, especially human-computer interactions and adaptive interfaces.

The paper is aimed on one hand at application of the control theory principles and of neural network and fuzzy sets concepts, and on the other hand at adaptation and optimization of the system interaction structure, as well as its investigation by simulation.

Main goal of the present work is further decomposition and refinement of the control structure for adaptive interactive systems (Nikov et al., 1999) (cf. Figure 1). In the following its neuro-fuzzy adaptive feedback control is detailed presented and studied by a web-based interactive system.

2. NEURO-FUZZY-BASED ADAPTATION FEEDBACK CONTROL

The synthesis of control algorithms is based on an adequate description of the plant of control (real – life system). Let present the interactive system and its users as a plant of control. For modelling of this plant logfile data from real-life interactive system can be used. This data includes information about initial interaction structure and about user's work with the system. The initial interaction structure can take the form of a selection tree as shown on Figure 3. Detailed description of the control structure for adaptive interactive systems (cf. Figure 1) is given in (Nikov et al., 1999). The neuro-fuzzy-based feedback control contains the following blocks:

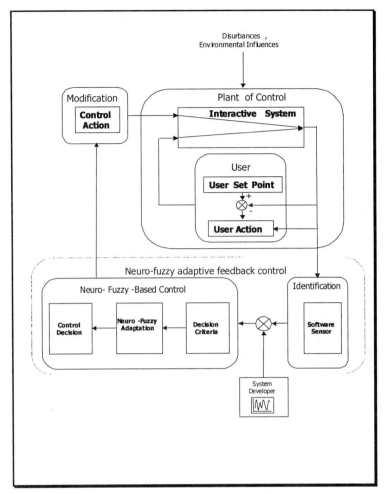

Figure 1. Control structure for adaptive interactive systems.

2.1 Identification

The basic module of the block "Identification" is a "Software Sensor". This block determines the transition matrices $\Omega^m = \left\{ \omega_{ij}^m \right\}$. Each element of these matrices ω_{ij}^m contains the frequencies of transitions between the different interaction points (i and j) during user's work with the interactive system. The following operator presents the software sensor as

$$\Phi_S : L_f \to \Omega^m, \qquad m = 1, 2, \ldots, \tag{1}$$

where L_f is a set of logfile data and $\Omega^m - m$ –the transition matrix.

2.2 Neuro-Fuzzy-Based Adaptive Control Algorithm

The neuro-fuzzy-based control algorithm is presented by the blocks: "Decision Criteria", "Neuro-Fuzzy Adaptation" and "Control Decision". The aim of this control algorithm is to adapt the interactive system (part of the plant) based on user's work with it.

The block "Decision Criteria" builds the training patterns for the block "Neuro-Fuzzy Adaptation". The training patterns include the input and target values for learning of neural network. The operator Φ_P describes the algorithm of training patterns building as follows

$$\Phi_P : \Omega^m \to P_i^m, \tag{2}$$

where P_i^m is training patterns matrix; i–th row of matrix.

The block "Neuro-Fuzzy Adaptation" presents the kernel of adaptive control. Here a fuzzy neural network (Nikov & Georgiev, 1999) is learned by (2). The neural network structure is determined based on user's work with the interactive system (plant).

A fuzzy set P is defined on the space Z as a set of ordered pairs $(z, \mu_P(z))$, where $\mu_P(z)$ denotes the value of the membership function $\mu_P(z): 2^r \to [0,1]$ at a given point $p \in Z$.

The fuzzyfication operator (constructor of the fuzzy set) is presented as

$$f_N(Z): Z \to 2^Z, \tag{3}$$

where 2^Z denotes the set on all subsets of the set Z.

A fuzzy neural network can be defined as

$$Z(i) = R_N \left(f_N (Z(i-1)), f_N (W(i)) \right)$$
$$W(i) = R_L \left(Z(i), W(i), W(i-1) \right) \tag{4}$$

where $Z(i)$ and $Z(i-1)$ are fuzzy sets. $W(i)$ presents network weights on the i–th learning step.

Fuzzy relation R_N describes the activation function of the neurons and R_L is a fuzzy relation describing the learning algorithm.

The learned weights are used on next step of the control algorithm in the module "Control Decision". This module determines a new adapted interactive structure of the interactive system (part of the plant) by an optimisation algorithm.

2.3 Control action
The change of interaction structure is carried out by block "Control Action". This module sets the new adapted interactive structure of the system.

2.4 System Developer
Block "System Developer" models the actions of the developer of system. His/her role is especially important at the development stage of an interactive system. The block has the following functions:
- Supervisor control of the parameters and the rules of the control algorithm and
- Tuning of the control algorithm.

3. MATLAB – BASED SIMULATION
MATLAB is an interactive software product (Pärt-Enander and Sjöberg, 1999), in which a large number of technical and mathematical procedures are available. It supports object-oriented programming enabling the use of classes and objects. The proposed "Neuro-Fuzzy Adaptation Algorithm" is presented and investigated within MATLAB environment SIMULINK (Singh and Agnihotri, 2000). The basic element of the proposed adaptation is the fuzzy neural network. A SIMULINK program is used for this investigation. The SIMULINK diagram of the control system structure is shown on Figure 2.

462

A simulation of the Electronic Funding Information server ELFI [http://www.elfi.ruhr-uni-bochum.de/elfi/] illustrates the neuro-fuzzy-based adaptive control proposed. ELFI provides web-based access to information on research funding. Detailed descriptions of funding programs are maintained in a central database. The user retrieves the information needed from this database using selection trees. When a tree item is selected, appropriate funding information is displayed. All user interactions with ELFI are recorded into a logfile, which provides the basic information for adaptation.

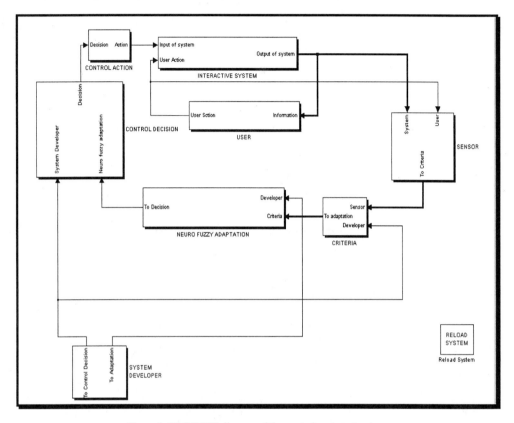

Figure 2. SIMULINK diagram of the control system structure.

The "Interactive System" block is used to present information from logfile derived from user's work with the interactive system. The block "User" presented as a signal noise generator models user actions. Both blocks are an appropriate simulation tool for description of the plant of control. The block "Sensor" determines the transition matrices derived from the logfile. An appropriate set of training patterns is obtained by the block "Criteria". The target of the learning process is determined by "System Developer" block. The fuzzy neural network is learned by the "Neuro Fuzzy Adaptation" block. The interaction structure is optimised by the "Control Decision" block. The "Control Action " block sets the new adapted interaction structure of the interactive system. Based on the learned weights of the fuzzy neural network a new structure of the interactive system is derived. (cf. Figure 4). This structure is obtained by optimisation at maximal four interaction points in a group.

4. CONCLUSIONS
The neuro-fuzzy-based adaptive feedback control of the control structure for adaptive interactive systems (Nikov et al., 1999) is further decomposed and refined. By logfile data from a web-based interactive system the feasibility and functioning of control structure are illustrated and confirmed.

463

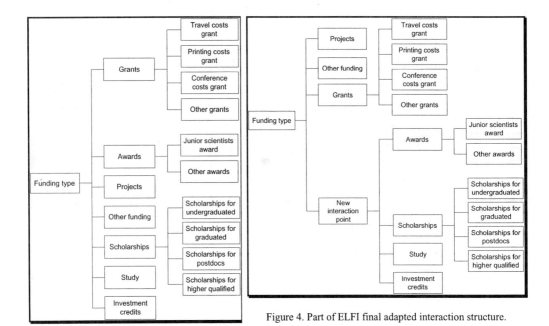

Figure 3. Part of ELFI initial interaction structure.

Figure 4. Part of ELFI final adapted interaction structure.

The advantages of such presentation are: 1) investigation and design of optimal interaction structure of adaptive interactive systems; 2) optimization of parameters of interactive systems; 3) design of stable interactive systems; 4) introduction of novel neuro-fuzzy adaptation approach.

The further developments of the fuzzy adaptive control structure proposed concern:1) further decomposition and refining of control system modules; 2) application to WWW personalization (Fink and Kobsa, 2000), i.e. automatic adaptation of information to the individual user according his/her personal needs, knowledge, preferences or cognitive abilities; 3) support the development of dynamic personalization server (cf. http://www.humanit.de/de/products/dps/index.html).

REFERENCES

Bishop, R. H. "Modern Control Systems Analysis and Design Using Matlab and Simulink". Addison-Wesley, 1997.

Fink, J., and Kobsa, A. "A review and analysis of commercial user modeling servers for personalization on the World Wide Web". User Modeling and User-Adapted Interaction, 2000, 10, 209-249.

Kecman, V. "Learning and Soft Computing: Support Vector Machines, Neural Networks, and Fuzzy Logic Models (Complex Adaptive Systems)". MIT Press, 2001.

Nikov, A. and Georgiev, T. "A fuzzy neural network and its MATLAB simulation". Proc. of ITI99 21 International Conference on Information Technology Interfaces, Pula, Croatia, 1999, 413-418.

Nikov, A., Lindner, H. -G., and Georgiev, T. "A control structure for adaptive interactive systems". In H. -J. Bullinger and J. Ziegler (Eds.) Human-Computer Interaction: Ergonomics and User Interfaces. London: Lawrence Erlbaum Associates, 1999, 351-356.

Pärt-Enander, E. and Sjöberg, A. "The MATLAB 5 Handbook", Addison-Wesley, 1999.

Singh, K. K. and Agnihotri, G. "System Design Through MATLAB, Control Toolbox and SIMULINK". Springer-Verlag, 2000.

Vidyasagar, M. "Nonlinear System Analysis". Prentice - Hall, 1995.

Adaptive tutoring in business education using fuzzy backpropagation approach

Kinshuk
Massey University, Information Systems Department,
Private Bag 11-222, Palmerston North, New Zealand
Tel: +64 6 350 5799 ext 2090, Fax: +64 6 359 5330, kinshuk@massey.ac.nz

Alexander Nikov
Technical University of Sofia, Department of Human Sciences and Design
K. Ohridski 8, BG-1000 Sofia, Bulgaria
Tel: +359 2 9653693, nikov@tu-sofia.acad.bg

Ashok Patel
CAL Research & Software Engineering Centre, De Montfort University, The Gateway
Leicester LE1 9BH, United Kingdom
Tel/Fax: +44 116 257 7193, apatel@dmu.ac.uk

Abstract

To cope with the increasing demand on educational opportunities in current educational environment by constrained resources available to the traditional education sectors, various Intelligent Tutoring Tools were developed for various topics in business education using cognitive apprenticeship framework (Collins et al., 1989). These tutoring tools engage the students in learning by doing (Patel & Kinshuk, 1997) and in addition to their random generator facility, enable a teacher to set problems in a narrative form in a discrete fashion or within a case study. This facilitates the students to face normal real work environment challenges of problem identification and problem definition, which are generally missing in the academic instruction (Stuart, 1997).

To make the systems even more beneficial and adaptive to the student needs, the fuzzy backpropagation (FBP) approach (Stoeva & Nikov, 2000) is used. It is based on a neuro-fuzzy model combining neural networks with fuzzy logic. A detailed analysis of the history of the information the student already looked at establishes the basis of an implicit student model. This model determines the cognitive state (knowledge degree reached) and the learning preferences (e.g. information about student's perceptive capabilities) for each student. In this way by the FBP approach is captured the students problem solving and allocated sub-optimalities of their solutions, i.e. missing conceptions in the student knowledge.

The FBP approach is illustrated by an adaptive tutoring system for business education. The architecture of this system is shown. An example for the marginal costing topic is given. The FBP approach allows easy student modelling and thus adapting the tutoring system to the student. Further details of benefits achieved using this approach are discussed.

1. Importance of Adaptivity

Adaptive systems engage a larger number of interactions without the help of an adapter or another system to change their state or behaviour. Another advantage of adaptive systems is that they are characterised by *design variety* (Benyon, 1993). Instead of the designer trying to get a single solution to a problem, the designer specifies a number of solutions and matches those with the variety and the changeability of the users. Intelligent tutoring systems (ITS) provide individualised instruction for the users, and so such systems must be adaptive.

2. Student Modelling and Adaptation

Student modelling in an intelligent tutoring system (ITS) facilitates the tutoring flexibility as well as individualizing the tutorial discourse. The student knowledge tracing plays a major role in ITS, because it provides appropriate tutoring as well as remediation to create flexibility. The role of student model is essentially to keep the record of the student's understanding as the lesson progresses, and it does so on the basis of student responses (Eklund & Brusilovsky, 1998). The adaptivity helps the educational systems to adjust their interactions based on a changing record of the student knowledge.

In order to create a system that adapts itself to each individual student the system must register each student's action and deduce from that how the student's "state of mind" evolves. Based on this abstraction of the student's state the system can decide how to perform some adaptation. The representation of the student's state of mind is called a *student model*. It contains aspects that are controlled explicitly by the student, such as colour or media preferences, learning style, background knowledge, and other items that can be entered through a questionnaire. The most important part of a *student model* is the information the system maintains about the student's "relation" to the *domain concepts*. Furthermore, the system gathers this information by observing the student's interaction pattern.

3. Fuzzy backpropagation approach for adaptivity

The adaptivity requires some sort of sophisticated mechanism to capture the student interaction and behaviour and analyse it to derive adaptation recommendations based on various pre-defined rules. We have used the fuzzy backpropagation (FBP) approach (Stoeva & Nikov, 2000) for this purpose. FBP approach is based on a neuro-fuzzy model combining neural networks with fuzzy logic. It is an extension of the standard backpropagation algorithm. The student interaction pattern is fed into the FBP as a continuous feed to train the network. The process is not so complicated because the output is already available for verification, but this means that initially the extent of adaptation is rather limited. The sophistication in adaptation increases with time as more and more training data becomes available.

The system tracks the navigation path of the student and keeps the history about the information that has been accessed by the student. This data forms the basis of an implicit student model. With the help of FBP approach, the model determines the cognitive state (knowledge degree reached) and the learning preferences (e.g. information about student's perceptive capabilities) for each student. The FBP approach also helps in identifying sub-optimalities of the students' problem solving and errors in the resulting solutions, i.e. missing conceptions in the student knowledge.

4. System development

A framework for an adaptive intelligent tutoring system is developed, incorporating FBP approach in the student model and inference engine. The design of the system is largely based on the intelligent tutoring tools developed under Byzantium project (Kinshuk, 1996). The prototype is designed with aim to teach *Marginal Costing* for introductory accounting course students. The prototype software is based on inter-relationships of variables in the marginal costing domain and creates an interactive approach for problem setting and problem solving. Figure 1 shows the framework of the system.

The system consists of two main areas: the interactive mode, which allows the student's to practice *Marginal costing*; and the assignment mode, which allows the students to assess the competence achieved in the interactive mode.

The student model of the system contains four main components:
➢ Global preferences of the learner (behavioral component): the global preferences are applied to the whole system.
➢ Specific content presentation related preferences of the learner (behavioral component): it is applied to specific contents.
➢ Domain competence related information about the learner (domain based component)
➢ Student's working history with annotated system feedback (only in problem solving scenarios, for the problems which have sequential processes)

466

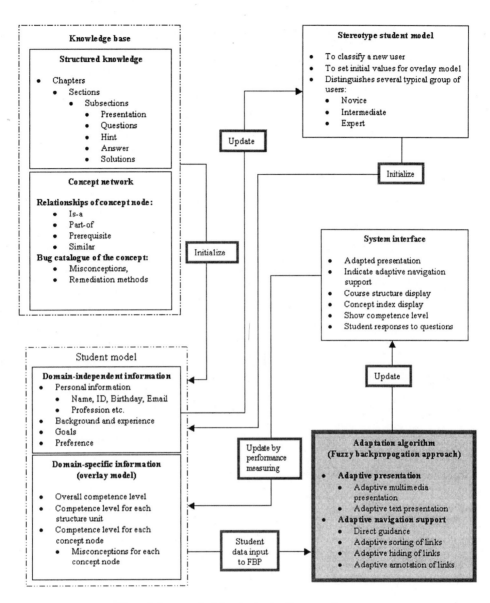

Figure 1. Framework of the adaptive intelligent tutoring system

The student model is used by the system primarily:

i.   to provide adaptive navigation guidance - based on prioritized successors and learner model
ii.   to select coarser/finer granularity of domain content
iii.   to provide context based excursions to other learning units

467

iv.   in making analogies with previously learnt material
v.   in making direct references to previous learnt material
vi.   to provide dynamic messaging and feedback, for example:
   a.   navigation related system messages
   b.   content related system messages
   c.   dynamic progression recommendations based on learner's domain competence and current context

5. Conclusion & Future work

The use of Fuzzy Backpropagation approach to facilitate adaptation to the students in the learning process in intelligent tutoring systems is a novel approach. The project is currently in its early stages, but feedback on the framework design has been quite promising. Since the sophistication in adaptation improves over time (the more training the network receives), the approach seems to be particularly useful for series of systems where interaction patterns observed within one system could be used in other systems. A thorough evaluation will be carried out to confirm these assumptions once the prototype has been developed to a stable phase.

References

Benyon, D., (1993). Adaptive Systems: A Solution to Usability Problems. *User Modeling and User Adapted Interaction*, 3, 65-87.

Collins, A., Brown, J. S. & Newman, S. E. (1989). Cognitive Apprenticeship : Teaching the crafts of reading, writing and mathematics. In Lauren B. Resnick (Ed.) *Knowing, Learning and Instruction*, Hillsdale, N. J.: Lawrence Erlbaum Associates, 453-494.

Eklund, J and Brusilovsky, P. (1998). The value of Adaptivity in Hypermedia Learning Environments: A Short Review of Empirical Evidence. *Proc. of the 2nd Workshop on Adaptive Hypertext and Hypermedia at the 9th ACM Conference on Hypertext and Hypermedia (Hypertext 98)*, Pittsburgh, USA.

Kinshuk (1996). Computer aided learning for entry level Accountancy students. *PhD Thesis*, De Montfort University, England, July 1996.

Patel, A. & Kinshuk (1997). Intelligent Tutoring Tools in a Computer Integrated Learning Environment for introductory numeric disciplines. *Innovations in Education and Training International Journal*, 34 (3), 200-207.

Stoeva, S. & Nikov, A. (2000). A fuzzy backpropagation algorithm. *Fuzzy Sets and Systems*, 112 (1), 27-39.

Stuart, J. A. (1997). A method for teaching problem assessment. *Paper presented at the Frontiers in Education Conference*, Nov. 5-8, Pittsburgh, USA.

AN ALGORITHM AND A SYSTEM FOR
USER INTERFACE ADAPTATION

Alexander Nikov
Technical University of Sofia, K. Ohridski 8, BG-1000 Sofia, Bulgaria, nikov@tu-sofia.acad.bg

Stefka Stoeva
Bulgarian National Library, V. Levski 8, BG-1504 Sofia, Bulgaria, stoeva@gmx.net

Mihail Tzekov
HypoVereinsbank Bulgaria GmbH, Briennerstrasse 52, D-80333 Munich, Germany, m.tzekov@web.de

ABSTRACT
The paper proposes an algorithm that, when applied to sequences of user actions, would allow a user interface to adapt over time to an individual's pattern of use. It is based on a neuro-fuzzy model – the Quick Fuzzy Backpropagation (QuickFBP) algorithm. During the first step of adaptation algorithm the initial interaction structure of the interactive system to be adapted is determined as a tree hierarchy. Then the transition matrices for relevant training patterns are established. For this purpose are used dialogue protocols of particular work sessions with the interactive system from the beginning to completing the selected tasks. To each training pattern corresponds a transition matrix describing the transitions between the interaction points. Further on the basis of the transition matrices a neural network is determined. Then the network weights are trained by the QuickFBP algorithm. These weights are used for construction of an optimal interaction structure with minimum weighted path length. The algorithm proposed is implemented in a software system. The experimental results of the algorithm when applied to real data of the web-based system ELFI are described. The initial and adapted interaction trees of ELFI are shown. The advantages of the algorithm proposed consist in: 1) quasi-unsupervised training of interaction structure weights based on a novel neuro-fuzzy model; 2) creation of optimal interaction structure adapted to user allowing significant quicker access to interaction points; 3) capability to build implicit user model, etc. Possible applications in adaptive web-based systems and especially in personalized e-commerce services are presented.

INTRODUCTION
The adaptive user interfaces (Langley, 1999) improve the ability of software systems to interact with a user by constructing a user model based on partial information and experience with that user. The most promising applications of adaptive interfaces are web-based systems. The growing amount of services and information overload available on Internet makes it very difficult to find and select pieces of information needed by users. Web-based adaptive user interfaces can provide a personalization in user interaction (Fink and Kobsa, 2000; Ardissono et al., 1999) improving user's work in a variety of applications such as e-commerce, distance education or cooperative work.

In recent years, there have been a growing number of applications of machine learning techniques to user-adapted interactions. Machine learning (Hand et al., 2000) supports the acquisition of user models from memorization of user's interaction, which is a difficult problem. Usually the information about the user is limited and it is difficult to infer assumptions about the user. Therefore a leaning algorithm for user interface adaptation combining neural networks and fuzzy logic is proposed.

DESCRIPTION OF THE ALGORITHM

Step 1: Determination of initial interaction structure as tree hierarchy
The initial interaction structure of the interactive system presented as a tree is defined. On Figure 1 is given an example interaction structure shown as a tree hierarchy.

Step 2: Determination of transition matrices

For collection of training patterns with an interactive system are taken the transitions between n interaction points. For this purpose are used the dialogue protocols of particular work sessions from the beginning to completing the selected tasks. To each pattern corresponds a transition matrix. The cells of this matrix consist of the frequencies of

469

transitions between the relevant interaction points. To the pattern with number m, m=1,...,M, corresponds the following transition matrix $U^m = \left\{ f_{ij}^m \right\} i, j = 1,..., N$, where M is the number of all patterns.

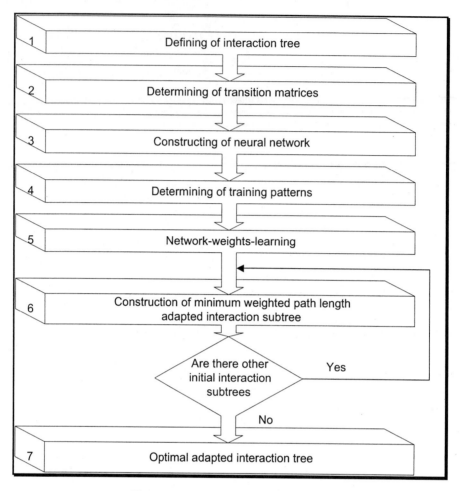

Figure 1. User interface adaptation algorithm.

For generation of the transition matrix $K = \left\{ f_{ij} \right\} i, j = 1,..., N$ the matrices $U^m, m = 1,..., M$, are used. The cells of this matrix contain the sum of frequencies of relevant patterns $f_{ij} = \sum_{l=1}^{M} f_{ij}^l$

Step 3: Construction of neural network

On the basis of transition matrix K a two-layered network structure (cf. Figure 2) is determined. The input neurons $N^{(0)}, N^{(0)} \leq N$, of network layer 0 are relevant to interaction points, which row sum transition frequencies are

470

$\sum_{j=1}^{N} f_{ij} > 0, i = 1,..., N$. The neurons $N^{(1)}, N^{(1)} \leq N$, of network layer 1 are relevant to interaction points,

which column sum transition frequencies are $\sum_{i=1}^{N} f_{ij} > 0, j = 1,..., N$.

The initial weights of layer 1 $w_{ij}^{(1)}$ are normalized to fit into unit interval $[0,1]$

$$w_{ij}^{(1)} = \frac{f_{ij} - f_{\min_K}}{f_{\max_K} - f_{\min_K}},$$

where $f_{\max_K}$ and $f_{\min_K}$ is the maximal resp. minimal frequency element of transition matrix K.

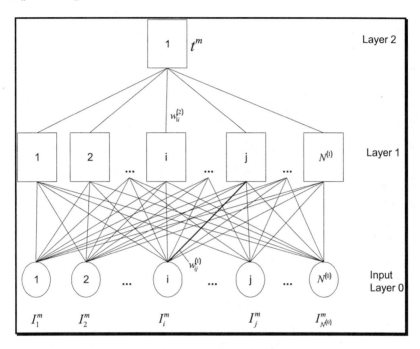

Figure 2: Two layered neural network

The weights at first layer $w_{ij}^{(1)}$ based on transition frequencies between interaction points can be used for forecasting (prediction) of the next user action. The weights $w_{ij}^{(2)}$ at second layer present the weights for interaction points used for constructing the optimal interaction tree.

Step 4: Determining of training patterns for network weights learning
On the base of the transition matrices U^m, m=1,...,M, the matrix $U = \{U_j^m\} j = 1,..., N^{(0)}; m = 1,..., M$ is

determined, where the frequency $U_j^m = \sum_{k=1}^{N} f_{jk}^m$ presents sum of row j of matrix U^m or $U_j^m = f_{jk\_max}^m$ presents

the maximal frequency element in row j of matrix U^m. Then the matrix $T = \{I_j^i\} i = 1,..., N^{(0)}; j = 1,..., M$ of the training patterns is determined. The input values of the neural network for each training pattern with number m

$I^m = \left\{ I_1^m, I_2^m, .., I_{N^{(0)}}^m \right\}$ are calculated by the equation $I_j^m = \dfrac{U_j^m - U_{min}}{U_{max} - U_{min}}$, where U_{min} and U_{max} are the

minimal and the maximal values of matrix U .

The target values t^m are determined in two variants:

1) $t^m = I_{min}^m$, where I_{min}^m is the minimal value of row m of matrix T .

2) $t^m = I_{max}^m$, where I_{max}^m is the maximal value of row m of matrix T .

Step 5: Network-weights learning
The network weights are learned by the fuzzy analogue of backpropagation algorithm - Quick Fuzzy Backpropagation Algorithm (QuickFBP) (Nikov and Stoeva, 2001). It aggregates input values and weights using a fuzzy integral with a psychological background simulating the expert's decision making. According to convergence conditions for QuickFBP algorithm the targets are chosen for maximal profile of input values. Thus quasi-unsupervised learning and building of implicit user model is supported.

Step 6: Optimal interaction subtree
According to functional and ergonomic requirements and constraints the maximal number of arcs r starting from a node in the interaction tree is determined.

The sequential top-down application of an optimization algorithm to each subtree, defined by constraints, generates the optimal tree with minimal weighted path lengths vector (cf. Figures 3 and 4).

Consider a tree T with L leaves. Let $l_1, l_2, ..., l_L$ denote the lengths of the paths from the root to the leaves of T . The l_i 's are called path lengths of the corresponding leaves, and the vector $\lambda = (l_1, l_2, ..., l_L)$ is called the path length vector of T . The path length vector $\lambda = (l_1, l_2, ..., l_L)$ is said to be optimal for the weight vector

$W = (w_1, w_2, ..., w_L)$ if it minimizes the sum $\displaystyle\sum_{i=1}^{L} w_i l_i$.

Step 7: Optimal interaction tree
During this step an interaction tree is generated, which is optimal.

CASE STUDY
The algorithm proposed is implemented in a software system. A case study with the Electronic Research Funding Information server ELFI (http://www.elfi.ruhr-uni-bochum.de/elfi/) illustrates the feasibility of algorithm. The initial and adapted interaction tree of German national research funding organizations as part of ELFI interaction trees is given on Figures 3 and 4. The maximal number of arcs starting from a node in the interaction tree is set to $r = 4$.

CONCLUSIONS
The *advantages* of the algorithm proposed consist in: 1) quasi-unsupervised learning of interaction structure weights based on a novel neuro-fuzzy model; 2) creation of optimal interaction structure adapted to users allowing significant quicker access to interaction points; 3) capability to build implicit user model, etc.

Further developments and applications of algorithm are oriented to:

Adaptive navigation supports the presentation of web links (Brusilovsky, 1996) as: 1) *Direct guidance* where the system decides what is the next "best" node for the user to visit (cf. step 3); 2) *adaptive ordering* sorts all the links of a particular page according to the user model; 3) *hiding* links to irrelevant pages; 4) *adaptive annotation* augments links with a comment with information about the current state of the nodes behind the annotated links.

Personalization of web sites and especially personalized e-commerce web sites (Fink and Kobsa, 2000); support of the development of Dynamic Personalization Server. According to (Abrams et al., 1999) by 2003 about 85 percent

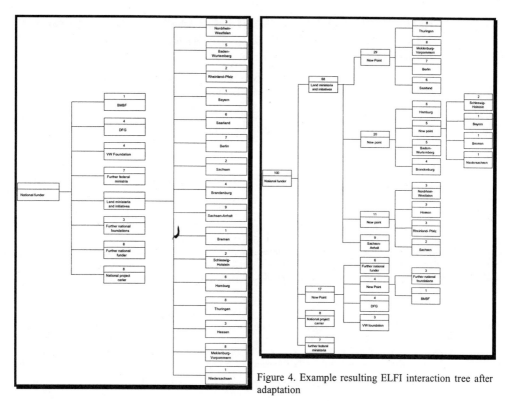

Figure 3. Example initial ELFI interaction tree

Figure 4. Example resulting ELFI interaction tree after adaptation

of all web sites will use some form of personalization. E-commerce web sites offering personalized services convert significantly more visitors into buyers than conventional e-commerce sites (ICONOCAST, 1999).

REFERENCES

Abrams, C., Bernstein, M., deSisto, R., Drobik, A., and Herschel, G. "E-Business: The Business Tsunami". Proceedings of Gartner Group Symposium/ITxpo, Cannes, France, 1999.

Ardissono, L., Barbero, C., Goy, A., and Petrone, G.. "An agent architecture for personalized Web stores". Proceedings of the Third International Conference on Autonomous Agents. Seattle: ACM Press. 1999, 182-189.

Brusilovsky, P. "Methods and techniques of adaptive hypermedia". User Modeling and User-Adapted Interaction, 1996, Vol. 6, No.2-3, 87-129.

Fink, J. and Kobsa, A. "A review and analysis of commercial user modeling servers for personalization on the World Wide Web". User Modeling and User-Adapted Interaction, 2000, 10, 209-249.

Hand, D. J., Mannila, H., and Smyth, P. "Principles of Data Mining (Adaptive Computation and Machine Learning)". MIT Press, 2000.

ICONOCAST. "More concentrated than the leading brand". ICONOCAST. http://www.iconocast.com/icono-archive/icono.102199.html, 1999.

Langley, P. "User modeling in adaptive interfaces". Proceedings of the Seventh International Conference on User Modeling. Banff, Alberta: Springer, 1999, .357-370.

Nikov, A. and Stoeva, S. "Quick fuzzy backpropagation algorithm". Neural Networks, 2001, Vol. 14, No. 2, 231-244.

WEDIS: A TOOLKIT FOR ADAPTABLE CREATION AND SUPPORT OF WORKSHOP PRODUCTION STRUCTURES

Hartmut Enderlein

Chemnitz University of Technology, Erfenschlager Straße 73, D-09107 Chemnitz, Germany
hartmut.enderlein@mb2.tu-chemnitz.de

Alexander Nikov

Technical University of Sofia, K. Ohridski 8, BG-1000 Sofia, Bulgaria, nikov@tu-sofia.acad.bg

Bettina Keil

Chemnitz University of Technology, Erfenschlager Straße 73, D-09107 Chemnitz, Germany
bettina.keil@mb2.tu-chemnitz.de

ABSTRACT

A conception of a disposition toolkit adaptable to customers, customer orders and users is described. It is based on a novel approach integrating fuzzy logic and neural networks for creating adaptable interaction structures. These structures are optimal according to ergonomic requirements. Based on this approach an intelligent adaptation toolkit can be developed and implemented into the new version of an existing disposition software system for creation and support of workshop production structures. Its advantages are presented and discussed.

1. INTRODUCTION

WEDIS is a software system designed to support workshop foremen in their daily dispatching and controlling tasks (Keil, 1998). Principle features of this system are capacity planning, cost accounting and effort feedback and job simulation as such as printing daily workers schedules. WEDIS includes the functions of production planning, of control systems and of dispatching systems. So the foreman is able to handle functions he never had in the past.

It can be used in workshop structures like tool making, mechanical engineering and mounting sections, handling single orders or small series (Keil, 1997). An adaptation of WEDIS to different customers, customer orders and users can significantly improve the user's work and reduce its introduction costs. For this purpose an approach and its relevant intelligent toolkit for adaptation of user interface was developed.

2. DESCRIPTION OF WEDIS

The functional model of WEDIS is shown on Figure 1. The foreman gets custom orders consisting of assemblies which, in turn, consist of several parts. For each part there is a number of working cycles including mounting the parts to assemblies.

The foreman has the task to generate order lists, pieces lists, mounting lists and working cycles. He has to print out all documents, make a daily planning, make sure the daily work and has to feedback the daily actual data.

The daily planning can be automated by using the "capacity planning" tool. The capacity planning process takes into account the workers' skills and the workers and machines availability. The workers fill out the time they need for doing their jobs in their schedules. The foreman collects them at the end of the working day and transfers these data to WEDIS. So an exact cost accounting for each custom order is made. As a result a cost pre-calculation for new, similar orders is possible. Advantage of WEDIS is the integration of the traditionally separated functions "parts list", "working cycle" and "daily schedule" in one interface.

The functional and handling model of WEDIS are interdependent. For defining of a functional model it was necessary to develop the system in a participative way. So could be determined the functions and the handling watched. The handling model was defined by adding events into the functional model and watching on the one hand the user behavior and the other hand the effects on the objects (cf. Figure 2). As first step it was possible to find out typical handlings for the process management and typical reaction in case of disturbances.

Workshop foremen are most likely not familiar with computers. The interface of WEDIS was especially designed for supporting users with limited computer skills. So there are three variants of the system. In order to ensure good user interaction with WEDIS the disposition process was splitted into planning processes and daily work. They are displayed in different colors. So the user is able to recognize the working level and to concentrate to his tasks and not to system handling.

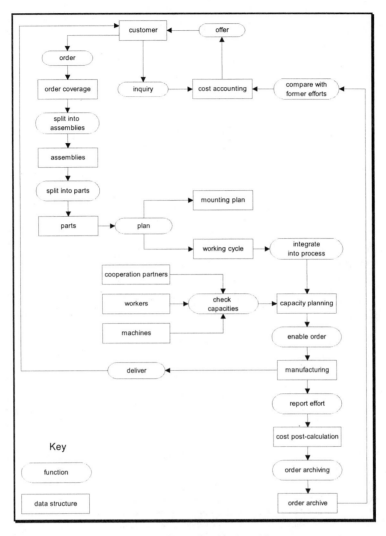

Figure 1. WEDIS functional model

3. INTELLIGENT ADAPTATION TOOLKIT (IAT)
For adapting WEDIS to the customers, customer orders and users the following approach and relevant toolkit are proposed.

3.1 Approach description
For adaptation the recorded logfiles of user dialogue with WEDIS can be used. The interaction points in WEDIS interaction structure should be moved up according to their frequency of use. In this way often-used interaction points are reached through a shorter selection pathway in the newly created interaction structure.

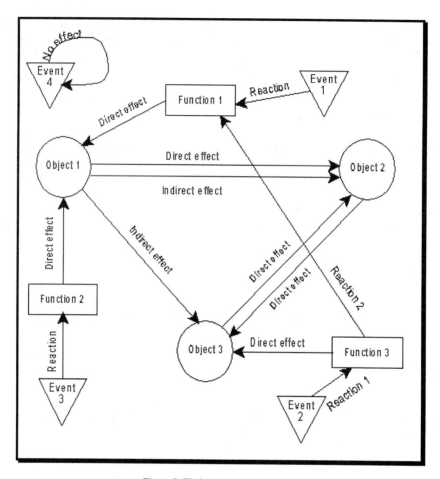

Figure 2. Find out a handling model

For realising this idea the neuro-fuzzy model developed by one of the authors, i.e. the fuzzy backpropagation (FBP) algorithm (Stoeva and Nikov, 2000) is used. It combines fuzzy sets and the adaptable neural networks. The steps of the approach proposed are given in Figure 3.

In step 1 experts define the initial interaction structure of WEDIS presented as a tree hierarchy (cf. Figure 3). It consists of two types of interaction points (IP): 1) dialogue functional interaction points which are connected with the function responsible for manipulation of dialogue objects, e.g. windows, menus (cf. IP1, IP2, IP3, IP7); 2) application functional interaction points (AFIP) which are connected with the functions changing the properties of application objects, e. g. files, text documents, (cf. IP4, IP5, IP6, IP8, IP9, IP10, IP11, IP12, IP13).

On the basis of the logfile data a matrix of transitions between all AFIP pairs is determined (step 1). In case of a menu structure they represent menu action options. The AFIP weights are trained by a modification of FBP algorithm (step 2). Based on ergonomic constraints the maximal number of arcs starting from a tree node can be determined, e.g. the maximal number of options of a menu (ISO 9241-14, 1997). By this way the set of AFIP weights is divided into subsets (step 3). Subsequently for each subset an interaction substructure with optimal weighted path length tree is formed (step 4), where the sum of the weighted path lengths is minimal. The remaining interaction substructures are similarly bottom up constructed to form the overall optimal interaction structure (step 5).

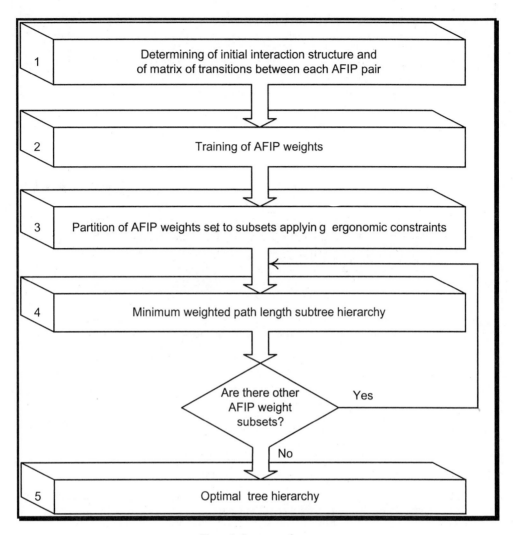

Figure 3. Sequence of steps

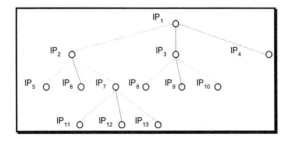

Figure 4. Part of the WEDIS interaction structure

3.2 Intelligent toolkit for the creation of adaptable user interfaces

It is planned to implement the approach proposed in the software toolkit IAT. It should be created with help of the software package MATLAB. After successful testing of the approach the created computer program can be exported as a source code in C++. So the final version of the toolkit can work as module of WEDIS or better as stand-alone system interacting via inter-process communication with WEDIS (cf. Figure 5).

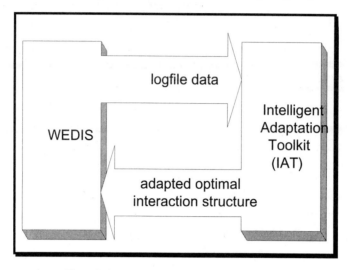

Figure 5. Communication between WEDIS and IAT

4. CONCLUSIONS

A conception of an intelligent adaptable toolkit based on a novel neuro-fuzzy approach is presented. Together with it WEDIS presents a new powerful intelligent disposition toolkit for creation and support of workshop production structures. It can find wide application in workshop structures such as tool making, mechanical engineering, assembly sections, handling single orders or small series, etc. A new object-oriented version of WEDIS suitable for interface adaptation will be developed.

The *advantages* of IAT proposed are: 1) supporting the acquisition of implicit user's knowledge without representation of knowledge in explicit form, i.e. implicit user modelling which is an advantage in comparison to other user modelling approaches; 2) handling logfile data measured in scales of different type and of missing data; 3) rapid search/access time; 4) reduction of introduction costs and training needs. This approach enables: 1) system defaults for different customer groups; 2) system defaults for specific customer orders; 3) system defaults for different level of user experience; 4) individual user adaptation ensuring better user acceptance, etc. IAT can be applied in many other software systems.

Further advantages for WEDIS include: 1) quicker data entering; 2) better user interface by multimedia help system; taking into account recognition aspects and spoken output.

REFERENCES

ISO 9241-14. "Ergonomic requirements for office work with visual display terminals (VDTs) - Part 14: Menu dialogues", 1997.

Keil, B. "Auftragssteuerung leichtgemacht – Werkstattsteuerung fest im Griff. Betriebstechnik aktuell", 1998, 8, 8-10.

Keil, B. "WEDIS - marktorientiertes Produzieren mit Hilfe angepaßter Systeme für Planung und Steuerung". Proceedings of the "Ilmenauer Wirtschaftsforum", Ilmenau, 1997.

Stoeva, S. and Nikov, A. (2000). "A fuzzy backpropagation algorithm", Fuzzy Sets and Systems, 2000, Vol. 112, No. 1, 27-39.

INTERACTION WITH AGENTS SYSTEMS :
PROBLEMATICS AND CLASSIFICATION

Emmanuelle GRISLIN-LE STRUGEON and André PENINOU

L.A.M.I.H., UMR CNRS 8530 - Le Mont Houy, BP 311,
University of Valenciennes, 59304 Valenciennes cedex, France
E-mail : {emmanuelle.grislin, andre.peninou}@univ-valenciennes.fr

ABSTRACT

The development of agent-based approaches, models and techniques, makes it now possible to use them in real applications. Therefore, the use of such systems becomes a topic of interest for the human-computer interaction field. As for us, we propose to study the interactions between human and agents-based systems relatively to a classification, in which the systems are differentiated by the way they make use of agents: a single isolated agent, several independent agents or several cooperating ones. In each of these categories, we attempt to highlight three aspects of interaction: firstly, the interaction functions that are offered to the user, then these functions are connected with the current applications and architectures, and finally the mutual representation that the actors (software agents and persons) have in the interaction.

1 INTRODUCTION

The development of agent-based approaches, models and techniques, makes it now possible to use them in real applications. Indeed, advances in the field now allow to go beyond the research stage. The theory and models proposed for the multi-agent systems and the agents architectures have been developed enough. They can now be used in various situations, even if multi-agent systems design processes and methods are to a large extent still being studied and if real design frameworks do not exist yet, except for very specific areas of systems.

Therefore, the use of such systems becomes a topic of interest for the human-computer interaction field. Previously, our first study concerned the contributions of the multi-agent systems to the adaptive interfaces (Péninou et al. 1999). In this article, we propose to study the interactions between human and agents-based systems relatively to a classification, in which the systems are differentiated by the way they make use of agents: a single isolated agent, several independent agents or several cooperating ones. In each of these categories, we attempt to highlight three aspects of interaction: firstly, the interaction functions that are offered to the user, then these functions are connected with the current applications and architectures, and finally the mutual representation that the actors (software agents and persons) have in the interaction.

2 INTERACTION WITH A SINGLE ISOLATED AGENT: ASSISTING THE USER

2.1 Interaction functions

The concept of agent, seen as an autonomous entity, is often used to study how to help the user. In this case, the objective is not to develop a model or a system from a set of cooperating entities. The objective is to provide the user with a software able of autonomy (or even personality) to help and assist him/her, in an adaptability or adaptivity approach (Kolski et al. 1998). The agent is then responsible of some tasks in an autonomous way relatively to the user. The agents can provide an help to the user while acting as intermediaries between the user and the software. These agents can be set at different levels in the abstraction scale of handled data. Some of them act at the level of the interaction between a user and an operating system (system agents), some others act at the application program level (software agents) and, recently, some agents perform user's tasks in office support software sets (software environment agents).

2.2   Current applications and architectures

Most of the agents of this kind are based on the collaborating interface paradigm as defined by P. Maes (Maes, 1994). It is generally based on modeling the user, or at least some elements that characterize him/her, like preferences, intentions, goals or plans.

The current typical application is web information retrieval. An assistant agent, interface agent or personalized guide, helps the user during web searching. It proposes to the user a personalized informative content that is generally based on a model of his/her preferences. For example, WebMate (Chen, 1998) assists the user during Internet-based information retrieval. It learns the user's profile in order to be able to suggest new keywords to refine the search or to propose a personalized compilation of news. The action of the interface agent Letizia (Lieberman, 1997) personalizes the result of the search and suggests web paths according to the user's topics of interest.

Recognition of intentions, goals or plans is much more difficult to implement. Indeed, it presupposes to have, jointly with the user's actions recording, a deep understanding of the task context and thus a model of activity. However, we can cite the collaboration manager Collagen (Rich, 1998) that makes a recognition of the user's plans, in the aim to make easier the collaboration in a group.

The works quoted above concern agents that are able to provide a content suited to the user. The feedback expected from him/her can be sum up in best cases to a selection of the positive suggestions (well-suited information) or negative ones (non pertinent information). The interactivity is therefore limited, the cooperation is almost nonexistent. An experiment hold by (Payne et al., 2000) attempted to go further. The application was the search of geographical itineraries. Three types of agents have been experimented. They differ from each other by their more or less cooperative nature toward the user, i.e. the way they take into account the user's opinion while solving problems. The results seem to show that cooperating with the user results in a better quality solution but a longer solving duration than when the system is running in a total autonomous and independent way.

2.3   Mutual representations

The internal representation the agent builds about the user has been treated in the previous part. Concerning how the user perceives the agent, some works propose a visual representation of the agent and its activity. This takes sometimes the appearance of an animated character with a more or less anthropomorphic type, like those proposed by (Hayes-Roth et al., 1999). However, this can cause a deception of a user who would expect too high understanding from the software assistant. Note that these two aspects of the mutual representation (from the user toward the agent and reciprocally) are independent issues.

3   INTERACTION WITH SEVERAL INDEPENDENT AGENTS

3.1   Interaction functions

Agents can be used not in an isolated way like previously described but by using the concept of agents groups. The model of the system thus consists in agents interacting with each other. In this context, the concept of agent is above all an abstraction mechanism for system modeling rather than for user-system interaction modeling. The agent approach proposes a new software engineering paradigm for the modeling and the design of distributed applications in which cooperation among agents is a very significant element (Wooldridge et al., 1999). Moreover, in this type of applications, the organizational context holds a major role whose essential characteristics can be represented and handled by agent models.

As regards the interaction with the user, these applications face her/him with an environment in which he/she cooperates with other users in an implicit or explicit way. The applications are various, for example : services recommendation on Internet (Bothorel, 1998), electronic trade (Deschner et al., 1999), workflow (Singh & Huhns, 1999), business process management (Jennings et al., 2000).

3.2   Current applications and architectures

The architecture of independent-agents based systems is often grounded on a "functional" dividing of the tasks. This entails to split the tasks by specialties, the whole coordination being handled by a central agent. The central agent both gives orders to specialists and integrates the results. In such an organization, agents act independently from each other, their interactions being mainly based on data sharing. For example, ACORN (Marsh & Masrour, 1997) is a system of community-based information sharing on Internet between users. The information sharing and exchange is carried out by agents specialized in information retrieval. This system proposes a virtual meeting point between agents : "The Café". In the café, agents can share information about available and relevant documents on the Web.

At the other extreme, interactions between agents can only exist to support the individual goals of each agent, without a real overall goal controlled by the system. That is the case for example in the application developed by Bothorel (1998) of services personalization on Internet by collaborative filtering in virtual communities of interests. In this application of services recommendation between users, each user is represented by a particular agent. By collaborative filtering (exchange of user profiles and local processing), agents cooperate in order to recommend services according to the user profile they deal with.

Lastly, collaborative work support systems require more sophisticated architectures of agents grounded on organizational models (Jennings et al., 2000 ; Singh & Huhns, 1999).

3.3 Mutual representations

Although the system consists of a set of autonomous agents, the explicit interaction between a user and several agents seems rarely developed. Generally, even if the system includes several agents, these ones interact with the user through a single agent often called interface agent. Thus, the user can completely ignore the distributed aspect of the system he/she interacts with. From this point of view, the interaction with such a system is very like the interaction with a unique agent (see section 2). Let us notice that collaborative applications should require additional and specific interaction in order to explain to the user the organizational and cooperation models that agents comply with.

4 INTERACTION WITH MULTI-AGENT SYSTEMS

The third category that we discuss is made of systems in which agents have mutual representations of each other and use a real communication protocol in order to support distribution of tasks and goals. These systems, grounded on collaborating agents, are multi-agent systems (MAS) in the meaning of Ferber (1995), that is agents characterized by some autonomy, cooperation, some self adaptation, and the existence of an overall goal different from the only achievement of local goals of each agent. These systems are designed to solve complex problems by some distribution between agents.

4.1 Interaction functions

From the point of view of the end-user, multi-agent systems supply advanced functionalities for solving specific and complex problems. The multi-expert solving of medical problems, the scheduling of factories, air traffic control, supervision of distribution networks are so many examples of problems where multi-agent concepts can be applied because they can support the distributed nature of these problems (knowledge, space, time, …). Nevertheless, the theory of multi-agent systems was more interested in the resolution of problems by software agents where the user is mainly considered as a "spectator". The shared resolution of problems between multi-agent systems and end-users is a new field opening significant theoretical issues for interaction and a lot of research remains to be done.

4.2 Current applications and architectures

In most of the systems developed to date, interaction with the user is considered as a two part process: an initial parameters setting from the user towards the multi-agent system and the result output from the multi-agent system towards the user, sometimes with some explanations. Because of that, as for the several independent-agents based systems, the proposed solution to support interaction is to implement an "interface agent". This agent is a real agent in the multi-agent system and concentrates all interactions with the user on behalf of all other agents (Avouris & Van Liedekerke, 1993).

For example, the work of Petit-Rozé et al. (2001) proposes a generic architecture of multi-agent systems for personalized information retrieval on distributed and heterogeneous resources of information. The interaction with the users is supported by a set of assistant agents (management of user profiles, results of requests, ...). However each individual user interacts with only one assistant agent which is assigned to her/him during the connecting process. Thus, during interaction, the user can not see the distributed nature of the system itself. The impacts of integrating these special interface agents in the community designed for problem solving (the multi-agent system) is not really studied for the moment.

Let us notice that recent works are addressing this issue of human/multi-agent systems interaction. For example, Hutzler (2000) proposes a new paradigm for the visualization of complex and/or multi-agent systems.

4.3   Mutual representations

During a first study of human/multi-agent systems interaction (Grislin-Le Strugeon & Péninou 1998), we showed that taking into account the user could have a considerable impact on the multi-agent system design. Indeed, with regards to the typical characteristics of multi-agent systems (communication, coordination, cooperation, decision making, ...), the user can hold a separate place or status. When cooperating with the system in order to achieve a specific goal, the user must be able to appreciate and understand the activity of the system at the semantic level that suits his/her task.

When the user carries out some tasks in cooperation with the multi-agent system, he/she has a great priority in the decision-making process. Lastly, the user requires a specific communication protocol, different from the communication protocols that are used among software agents. This protocol has to deal with two levels of communication : interacting about problem solving (task sharing, knowledge sharing), interacting about running tasks (results, explanations).

This study has also revealed two important elements which have to be carefully studied : a) how does the system consider the user, b) how does the user consider the system. From the point of view of the multi-agent system, the user can be considered according to a gradual scale in three different ways. He/she can be regarded as external and thus be ignored. The user can be regarded as part of the environment of the multi-agent system ; for example he/she provides pieces of information. Lastly, he/she can be regarded as an agent of the multi-agent system itself. In this case, the user carries out some particular tasks related to its own competencies and influences the activity of the multi-agent system itself (task allocation, cooperation protocols, and so on).

From the point of view of the user, the multi-agent system can be considered as a black-box or its internal agent structure can be accessible. Moreover, the user can require completely specific information feedback (explanations of the reasoning...). All these elements have an obvious key importance in the design of the user/multi-agent system interaction and in the design and development of the multi-agent system itself, and therefore, on the mutual representations (Grislin-Le Strugeon & Péninou, 1998).

5   CONCLUSION

In this article, we have studied the interaction between humans and agents-based systems. We proposed three classes of systems : i) single isolated agents, ii) several independent agents, iii) several cooperating agents. For each case, we tried to outline the advances of research and the actual limits.

Recent paradigms for distributed systems in software development, and particularly multi-agent approaches, are intended to give dynamics and flexibility to information processing systems. They seem promising for the modeling and development of interactive systems supporting more and more user services. However, we can suppose that a user having at his/her disposal several specialized intelligent assistant agents (internet information retrieval, office assistant, email assistant,...) may expect some consistency among these components during interaction, and may expect that they cooperate: knowledge sharing about the user (his/her preferences,...), joint activity for tasks involving several tools,... This part of the interaction with several agents having to cooperate is still to be studied and tried out.

In another way, the multi-agent support of cooperative information systems and of business process management proposes to model the organization of the work. So, specific research should be done to study the necessary interaction between the user and the organization thus modeled.

REFERENCES

Avouris, N.M. & Van Liedekerke, M.H. (1993). User interface design for cooperating agents in industrial process supervision and control application. Int J. of Man-Machine Studies, vol. 38, pp. 873-890.

Bothorel, C. (1998). Des communautés dynamiques d'agents pour des services de recommandation. In Barthès, J.P., Chevrier, V. & Brassac, C. (Eds). Systèmes Multi-Agents, de l'interaction à la socialité, JFIADSMA'98. Edition Hermès, Paris, pp. 81-98.

Chen, L. & Sycara, K. (1998). WebMate: A Personal Agent for Browsing and Searching. Proceedings of the 2nd International Conference on Autonomous Agents and Multi Agent Systems (AGENT'98), ACM, pp. 132-139.

Deschner, D., Hofmann, O. & Bodendorf., F. (1999). Agent-Based Decentralized Coordination for Electronic Markets. In Bullinger, H-J. & Ziegler, J. (Eds). Human-Computer Interaction : Communication, Cooperation and Application Design (HCI'99). Lawrence Erlbraum Associates Publishers, London, pp. 1152-1156.

Ferber, J. (1995). Multi-Agent Systems : Introduction to Distributed Artificial Intelligence. Addison-Wesley.

Grislin-Le Strugeon, E. & Péninou, A. (1998). Interaction Homme-SMA : réflexions et problématiques de conception. In Barthès, J.P., Chevrier, V. & Brassac, C. (Eds). Systèmes Multi-Agents, de l'interaction à la socialité, JFIADSMA'98. Edition Hermès, Paris, pp. 133-146.

Hayes-Roth, B. & al. (1999). Web guides. IEEE Intelligent Systems, March/April, pp. 23-27.

Hutzler, G. (2000). Du Jardin des Hasards aux Jardins de Données : une approche artistique et multi-agents des interfaces homme / systèmes complexes. Ph. D. Thesis, University of Paris 6, Paris, France.

Jennings, N.R., Faratin, P., Norman, T.J., O'Brien, P., Odgers, B. & Alty, J.L. (2000). Implementing a Business Process Management System using ADEPT: A Real-World Case Study. Int. J. of Applied Artificial Intelligence, 14 (5), pp. 421-465.

Kolski, C. & Le Strugeon, E. (1998). A review of intelligent human-machine interfaces in the light of the ARCH model. Int. J. of Human-Computer Interaction, 10 (3), pp. 193-232.

Lieberman, H. (1997). Autonomous Interface Agents. In Proc. of the Int. Conf. Human Factors in Computing Systems CHI'97, ACM Press.

Maes, P. (1994). Agents that reduce work and information overload. Comm. of ACM, vol. 37, n°7, pp. 31-40.

Marsh, S. & Masrour, Y. (1997). Agent Augmented Community Information - The ACORN Architecture. In Proc. of CASCON'97, Meeting of Minds, november.

Payne, T., Sycara, K., Lewis, M., Lenox, T.L. & Hahn, S. (2000). Varying the User Interaction within Multi-Agent Systems. Proc. of the 4th International Conference on Autonomous Agents (Agents 2000), ACM, pp. 412-418.

Peninou, A., Grislin-Le strugeon, E. & Kolski, C. (1999). Multi-Agent Systems for Adaptative Multi-User Interactive System Design : somes Issues of Research. In H.J. Bullinger, J. Ziegler (Eds.). Human-Computer Interaction : Ergonomics and user interfaces (HCI'99). Lawrence Erlbaum Associates, pp. 326-330.

Petit-Rozé, C., Grislin-Le Strugeon, E., Abed, M. & Uster, G. (2001). Interaction with agent systems for intermodality in transport systems. Proc. of the 9th Int. Conf. on Human-Computer Interaction (HCI'2001).

Rich, C. & Sidner C. (1998). COLLAGEN: a collaboration manager for software interface agents. MERL, Mitsubishi, Report TR-98-21a, March.

Singh, M.P. & Huhns, M.N. (1999). Multiagent Systems for Workflow. Int. J. of Intelligent Systems in Accounting, Finance and Management, Vol 8, John Wiley & Sons Ltd., pp. 105-117.

Wooldridge, M., Jennings, N.R. & Kinny, D. (1999). A methodology for Agent-Oriented Analysis and Design. Proc. of the 3rd Int. Conf. on Autonomous Agents (Agents'99), ACM, pp. 69-76.

The User is an Agent

Guillaume Hutzler

Laboratoire de Méthodes Informatiques (LaMI),
Université Evry-Val d'Essonne, Cours Mgr Roméro, 91025 Evry Cedex, France
hutzler@lami.univ-evry.fr

ABSTRACT

Agents systems are composed of a usually great number of autonomous agents. Due to the autonomy and distribution (physical of functional) of their components, and to the multiple and dynamic interactions between them, agents systems cannot be observed at a given time as a coherent whole. Thus, these systems raise specific issues regarding the design of human-computer interfaces. In this article, we address these issues and give some clues about tractable solutions. The solutions are presented within the *Data Gardens* framework, a generic multiagent platform designed for the real-time visualization of complex systems and the interaction with them.

1. INTRODUCTION

When working with agents systems (Weiss, 1999), users are confronted to the same needs as with any other complex system (Mitchell and Sundström, 1997). On the one hand, users want to understand what is taking place in the system, which requires the design of adequate tangible representations; on the other hand, users want to be able to modify the functioning of the system, which requires the design of adequate interaction protocols. The specificity of agents systems is that they are composed of a usually great number of autonomous agents, which dynamically interact with one another in various different ways, and organize together in evolving structures. The first consequence is that they cannot be observed directly as a coherent whole (Gelernter, 1992). As in the case of natural ecosystems, only small elements of the global picture can be seen at any given time. Furthermore, only local interactions can be observed, leaving the global dynamics out of sight. The second consequence is that interaction protocols must enable actions directed towards a single agent but also potentially actions directed towards groups of agents.

As a result, traditional representation and interaction solutions have to be adapted to the context of agents systems. In this paper, we will successively address these issues, by presenting them in the unifying framework of the *Data Gardens* multi-agent platform. This platform has been designed to allow users to dynamically interact with complex systems. Or, put it more generically, to visualize complex real-time data streams in order to be able to make adequate decisions. We distinguish at least three different contexts for which it may be useful to grasp the complexity of whole systems in a way that is both natural and intuitive:

- In a scientific context, the visualization of a simulated complex system may foster the understanding of its internal dynamics by completing quantitative statistical analysis with qualitative visual analysis. As a result, it can also help design artificial systems with similar global properties.
- In an industrial context, visualization techniques are used as a means to control the functioning of real complex systems such as airplanes or nuclear plants, in order to help operators detect local or global failures. In return, it helps them to make appropriate decisions in the command of the system.
- Finally, dynamic and emergent properties of complex systems may be used in an artistic context as a source of endless novelty. The unpredictability of the system becomes a valuable property that the artist will try to retain, while at the same time constraining the evolution of the system to keep it meaningful.

After the presentation of the *Data Gardens* technical platform itself (in section 2), the paper will be organized by crossing three axis: the first axis corresponds to the functionality (sensible representation in section 3 and interaction in section 4); the second axis corresponds to the context (control/command, analysis/design, art); the third axis corresponds to the type of entities concerned with the representation and/or action (agent, group of agents).

2. THE *DATA GARDENS* PLATFORM

The main idea behind *Data Gardens* is to have societies of agents handle the retrieval and sensible representation of complex data streams. On the one hand, "information agents" retrieve data streams from different distributed sources. On the other hand, a complete society of "representation agents" is used to filter the data and organize it hierarchically. To achieve this, each of the agents is responsible for the representation of some piece of information. Not all of these agents can be represented simultaneously, so they have to "negotiate" with one another in order to decide what information is most important at a given time. As a result, some of the agents will disappear, some will group themselves together, and the remaining agents will find their place in a hierarchical social structure. The resulting visual representation will be driven by two complementary principles: the first one is to translate the social hierarchy of "representation agents" into a graphical hierarchy, most important agents being represented bigger and in a central position in the screen; the second principle is to have perceptual properties of agents reflect their nature, their activity, but also their links with other agents.

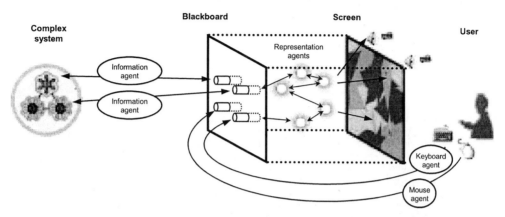

Figure 1. General architecture of the representation system. The information agents are responsible for retrieving the data describing the functioning of some complex system. The representation agents then filter the data and organize it hierarchically before presenting it to the user

Interaction between users and the society of "representation agents" will be obtained by considering each user as a separate source of data. Therefore, an "information agent" will be responsible for the control of a user, examining its actions through the keyboard or mouse, or its physiological reactions through eyes-tracking techniques, or heart rate sensors. In a sense, the user becomes an agent that has the ability to interact with other virtual agents. Therefore, he becomes able to explore an agents system by testing the reaction of the system to given perturbations. Or, he can modify the representation process by modifying the behavior of agents or the priorities attached to the data.

3. VISUALIZATION OF AGENTS SYSTEMS

The issue of visualization strongly depends on the function that is desired. If what we need is to *control* the functioning of a system, then the aim is to give a synthetic view of this system and to highlight dysfunctions as they appear. If what we need is to *understand* the functioning of the system, then the aim is to display the state of the agents composing the system and the interactions between them. These two approaches for the visualization of agents systems are complementary and may be used alternatively when "debugging" the functioning of a system (Giroux et al., 1994; Hart et al., 1997; Ndumu et al., 1999). We will now develop this aspect somewhat further by detailing possible visualization strategies for a "toy" multiagent system of preys and predators. In this system, predators try to trap preys, but a lonely predator can't succeed in it. Thus, when a predator detects a prey, it will call out other predators for help.

3.1 Visualization of agents

The first step in building a visual representation for such a system is to check what elements are important for the understanding of the evolution of the system. Position of preys and predators is naturally of crucial importance and it is equally important to be able to distinguish between the ones and the others. This is what we would call *identification* and *positioning* of agents in the system and it corresponds to figure 2.a. The position in the picture corresponds to the position of the agents in the environment and the shape corresponds to the type of agent (discs for preys, triangles for predators, rectangles for grass patches).

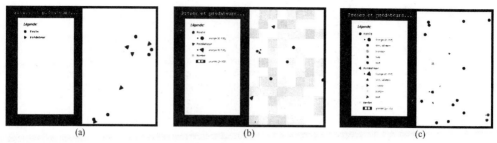

Figure 2. Different steps in the visualization of a prey-predator system. (a) identification and positioning of the agents (b) visualization of the internal state of the agents (c) visualization of the current activity of the agents;

The second step is to identify the *internal state* of the agents. Preys and predators for example are characterized by their level of energy. Preys gain in energy by eating grass, which continually regenerates. Predators gain in energy by catching and eating preys. Figure 2.b shows an example of representation derived from the previous one that takes this parameter into consideration: the color of grass patches and the size of preys and predators are in direct ratio to the level of energy.

The third step is to identify what the agents are presently doing. For preys as for predators, one can distinguish three types of behaviors: rest or random movement, feeding, and hunting behaviors. Colors may be used to differentiate between these various behaviors. Figure 2.c shows an example illustrating this strategy with four colors used (one for each behavior, plus one when the agent is dead).

3.2 Visualization of interactions and groups

More than agents, it is important to visualize interactions between them. In the example, this can be achieved by taking advantage of properties of perception (Arnheim, 1974). By giving identical colors to hunting predators and to fleeing preys, or by orienting the hunting triangles towards the fleeing discs, a link is visually established between the predator end its prey. Movements are also very efficient to display links between the shapes.

Finally, the dynamic creation and disappearance of groups of agents, and more generally global structures, is potentially important to visualize. The creation of hunting groups by predators may thus be visualized by giving a distinctive color to all the agent of a same group. Or we can consider this group as a new entity and replace the individual predators by a single hunting entity synthesizing the information of the different agents.

3.3 Rearranging the visual representation hierarchically

More generally, it may be interesting to completely rearrange the representation in order to display the most important events occurring in the system. To achieve this hierarchical organization of the representation, several strategies can be used.

A first strategy is to design the visual representation of agents so as to have important events become immediately perceptible, for example by using red tints when something important has occurred. Basically, each of the graphical means of visual representation (shape, size, color) can be used to show off specific aspects of the system: a distinct shape, a big size or a contrasted color are different means to enhance the perception of an agent.

A second strategy is to abandon the strict correlation that we used up to now between the position of an agent in the system and its position in the representation, or between agents and their corresponding shapes. In the first case, one can imagine to have the shapes rearranged so that the most important ones are placed in remarkable positions such as the top (figure 3.a) or the center of the screen (3.b). In the second case, the aim is to make the structure of the system more apparent by grouping agents that share common characteristics (figure 3.c).

Position, shape, size and color are different graphical ways to visualize such things as the identity of the agents and their situation within the system, their internal state and their behaviors, the creation and disappearance of interactions and global structures. These graphical solutions are not exclusive of other more classical ones such as text, curves, graphs, but they were useful for the presentation of the example. What is constant however is that agents, interactions and structures must be dynamically displayed so that an interaction becomes possible with a user.

486

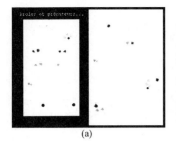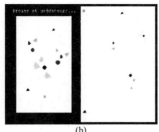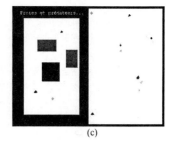

(a)	(b)	(c)

Figure 3. Different forms of hierarchical representation of a prey-predator system. For each picture, the right view corresponds to figure 2.c. The left view corresponds to: (a) most important elements on top (b) most important elements in the center (c) use of "group" entities

4. INTERACTION WITH AGENTS SYSTEMS

To achieve this interaction with a user, the basic idea is to consider that he is an agent in much the same way as other agents. In a sense, this is compliant with the definition of agents as autonomous entities with social abilities, perception and reactivity to their environment, and pro-activeness, i.e. with goal-directed behaviors. Technically speaking, we explained how this might be obtained within our framework. Conceptually, this implies that there should be a common environment comprising both human and artificial agents, and a common interaction space where both types of agents may be in contact.

4.1 A common interaction space

Two elements are required to represent complex systems within the *Data Gardens* platform. Firstly, *information agents* retrieve the necessary data to describe the functioning of the system (and take commands in direction of the system); secondly, *representation agents* make use of the data to build a coherent visual representation of the system. The principle is identical as far as human users are concerned: information agents (mouse and keyboard agents in figure 1) retrieve information about the commands of the user; then, representation agents enable the user to be present within the representation space, next to the representation of the agents system that he wants to interact with. These representation agents are like avatars except that we can have simultaneously several representation agents for a single user.

To take a basic example, one can imagine that the avatar of a user will continually check the information given by the mouse agent and will react accordingly. As a first thing, it will make its position correspond to the position of the mouse, enabling the user to move within the interaction space. Then, he will listen to mouse-click events so as to take predefined actions when the user clicks with the mouse.

4.2 Different interaction strategies

One can imagine several principles of action, directed towards a single agent or towards groups of agents.

 The first possibility is to have a predetermined action directed towards the agent that is closest to the user's avatar. To achieve this, the avatar will have to determine which agent is closest and then interact with it. In the prey-predator system, we may want to kill some of the predators or some of the preys to see how the other population will adapt to this new situation. In this case, we simply need a "killing avatar", whose action will be to kill the agent closest to the mouse location when the user clicks on a button. In the same spirit, we may also desire to catch a prey or predator and see what happens when it comes within reach of an agent of the other population. In this case, what we need is a "grabbing avatar", whose action will be to take control of another agent's position as long as the user holds the button down.

The second possibility is to have a predetermined action directed towards a whole set of agents. For example, we can imagine taking actions that should affect all the agents within a given range. If we want to water the grass, then the grass patches should receive quantities in inverse ratio to their distance from the avatar. What we need is a "watering-pot avatar", whose action is to propagate a "watering-signal" as long as the user holds the button down.

487

All of the examples that we give show reactive modes of interaction because the agents considered have very reactive modes of functioning. The principles however are very generic and may of course be applied to linguistic interactions with cognitive agents as well.

5. CONCLUSION

We tried to address the issue of the interaction with agents system in a generic, yet applied way. To this end, we presented an example that is of course a very simplified vision of the actual complexity of agents systems, but that puts into light some of the basic difficulties raised by these systems. The solutions that we propose are based upon the generic *Data Gardens* framework, more than upon specific visualization or interaction means. In this framework, other visualization means may be used and other interaction modes may be imagined. Furthermore, multiple users may be involved in the interaction with a single agents system, or a single user may be able to interact simultaneously with multiple agents system.

Interestingly enough, this approach may be extended to the design of "classical" computer-human interfaces. Indeed, we talked about *avatars* when we addressed the issue of the interaction with agents. We may have talked of *active tools* instead, which the user could choose from to achieve specific tasks. The difference with classical tools is that these active tools may be assigned specific goals that they will try to reach, responding to users in a way that would be both reactive *and* pro-active. This kind of tools would thus be able to take the current and past contexts into account when acting on behalf of the user, they would be more autonomous in the realization of a task, and they could dynamically adapt when the context evolves. In our opinion, these are properties that traditional computer-human interfaces lack and that should be developed in the future.

REFERENCES

Arnheim R., Art and Visual Perception, Berkeley: University of California Press, 1974.

Boden, M. A., "Agents and Creativity", in Communications of the ACM, July 1994, Vol. 37, n° 7, pp. 117-121.

Chi, E. H., Pitkow, J., Mackinlay, J., Pirolli, P., Gossweiler, R. and Card, S. K., "Visualizing the Evolution of Web Ecologies", in Proceedings of Computer Human Interaction'98, 1998, ACM Press, pp. 400-407.

Davies, N. J., Weeks, R., Revett, M. C., "Information agents for the World Wide Web", in BT Technol. J., October 1996, Vol. 14, N° 4.

Gebhardt, N., "The Alchemy of Ambience", in Proceedings of the 5th International Symposium on Electronic Art, 1994.

Gelernter, D., Mirror Worlds – The Day Software Puts the Universe in a Shoebox... How it will Happen and What it Will Mean, Oxford: Oxford University Press, 1992.

Giroux, S., Pachet, F., and Desbiens, J., "Debugging Multi-Agent Systems: a Distributed Approach to Events Collection and Analysis", in Canadian Workshop on Distributed Artificial Intelligence'94, 1994.

Hart, D., Kraemer, E., Roman, G.-C., "Interactive Visual Exploration of Distributed Computations", in Proceedings of 11th International Parallel Processing Symposium, 1997.

Hutzler, G., Gortais, B., Drogoul, A., "The Garden of Chances: A Visual Ecosystem", in Leonardo, 2000, MIT Press, Vol. 33, Issue 2, pp.101-106.

Ishizaki, S., "Multiagent Model of Dynamic Design - Visualization as an Emergent Behavior of Active Design Agents", in Proceedings of CHI '96, ACM Press, 1996.

Mitchell, C. M. and Sundström G. A., "Human Interaction with Complex Systems: Design Issues and Research Principle", in IEEE Transactions on Systems, Man and Cybernetics, May 1997, Vol. 27, N°3, pp. 265-273.

Moukas, A., "Amalthaea: Information Discovery and Filtering Using a Multiagent Evolving Ecosystem", in Proceedings of PAAM'96, London, 1996.

Ndumu, D. T., Nwana, H. S., Lee, L. C. and Haynes, H. R., "Visualisation and debugging of distributed multi-agent systems", in Applied Artificial Intelligence Journal, 1999, Vol. 13(1), pp. 187-208.

Péninou, A., Grislin-Le Strugeon, E., Kolski, C., "Multi-agent systems for adaptive multi-user interactive systems design: some issues of research", in H.J. Bullinger and J. Ziegler (Eds.) Human-Computer Interaction, Ergonomics and User Interfaces, London: Lawrence Erlbaum Associates, 1999, Vol. 1, pp. 326-330.

Weiss, G. (Ed.), Multiagent Systems: a Modern Approach to Distributed Artificial Intelligence, Cambridge, MA: MIT Press, 1999.

INTERACTION WITH AGENT SYSTEMS FOR TECHNOLOGICAL WATCH

E. ADAM, E. VERGISON*, R. MANDIAU, C. KOLSKI

LAMIH - UMR CNRS 8530, Université de Valenciennes et du Hainaut Cambrésis,
F - 59313 Valenciennes Cedex 9, FRANCE
{emmanuel.adam; rene.mandiau; christophe.kolski}@univ-valenciennes.fr
*SOLVAY Research and Technology, SOLVAY S.A.
Rue de Ransbeek, 310 - B - 1120 Bruxelles, BELGIUM
emmanuel.vergison@solvay.com

ABSTRACT

The boom in Internet technologies and company networks has contributed towards completely changing a good number of habits, which have been well-established in companies for several decades. Documents on paper, exchanged from hand to hand are progressively being replaced by electronic documents often accessible to all the internauts. Enterprises are set off on a race for information: being the first to find the good information becomes an essential objective to competitive enterprises. It is so important to own a fast tool of information search and distribution. Admittedly, tools have already been suggested such as: search engines, meta-engines, tools for automatic search (which set off at determined regular interval), and, more recently, information agent, capable of searching, sorting and filtering information. But teams of actors searching information need teams of agents, which acts in a cooperative way. In this paper, we present a system integrating cooperative information agents for cooperative technological watch. A first prototype has been designed and is currently used in a large company.

INTRODUCTION

Generally, the watch is the faculty that we have to perceive our environment; a kind of state of mind oriented towards exterior. In practical terms, the watch is a technique including three main actions: the information gathering; the analysis and the synthesis of gathered information; the relevant information diffusion.

The term of relevant information means that gathered data have to be subjected to several treatments. So, a first phase of validation consists in checking the validity and the relevance of gathered information, a second one of elaboration and synthesis means extracting the relevant information from the gathered information and grouping it by theme in order to send it to the right person.

This part is essential because even if a piece of information is relevant, it must be diffused to the person that is able to process it. The information must be used as a decision support tool.

There are two stages of evolution of the watch. A first degree of the watch concerns above all papers and data bases: documentation is a general term that groups press, books, studies and data banks together; information watch is characterized by a work specialization of information officers and a best monitoring of information search topic.

A second stage concerns more recent groups: specialized watch; strategic watch; economic intelligence also called technological watch.

The technological watch consists in « the observation and the analysis of the scientific, technical, technological environment, followed by the diffusion well targeted to responsible, of selected and useful information for taking a strategic decision.» (Jakobiak, 1995)

In this paper, we are interested in the technological watch. Thus, the system has to allow the relevant information search and the diffusion of this information to interested persons and principally, to decision-makers.

So, at first, we will describe briefly several existing works on information agents and multiagents systems.

Then, we will propose information agents' architecture for the technological watch (CIACOTEWA for Cooperative Information Agents System for COoperative TEchnological WAtch). And finally, we will present an application of this structure within a cell of watch of a large firm.

2. COOPERATIVE INFORMATION AGENTS SYSTEM FOR COOPERATIVE TECHNOLOGICAL WATCH

A MultiAgent System (MAS) can be defined as a group of agents which interact between themselves directly (indirectly (through of a database). An agent may be defined as an intelligent entity, which is part of a multiagent system (Flores-Mendez, 1999). The space in which the system, and therefore the agent, evolves is called the world. The agent capable of perceiving and modifying its environment (what surrounds it). An agent must be part of a multiagent system it has therefore to have capacities for communication.

As regards the notion of intelligence, the following principle should be underlined: an entity is intelligent if it is capable of learning, that is to say of adapting its knowledge. An agent can therefore be defined as follows: an agent is a adaptive, rational and autonomous entity, capable of communication and action. It may also be adaptive and it adaptivity degree may vary from an agent type to another (Adam, 2000).

An agent generally has acquaintances; these are agents with which it communicates or interacts. Each agent ha elements of knowledge concerning its environment. These elements of knowledge are also called representations (beliefs. An agent has one or several objectives, which are also called goals or desires. According to the importance of th goals, the agent can be required to plan its actions.

According to their goal, MAS have to satisfy certain principles.

2.1 Characteristics of information agent systems

An information multiagent system is composed of *information agents*. Generally these information agents are responsible of: information acquisition and management; information synthesis and presentation; user assistance Information agents can possess more than one of these capacities (Klusch, 2001).

It is often useful to co-ordinate agents' activities through *coordinator agents*. These agents own knowledge on information agents (such as their address or search domain for instance) to which they send requests either in targetec way (if they own knowledge on information agents' competence) or in a general way (by broadcast techniques). Then responsible agents have to gather this information, to check it eventually, to compare it or to filter it.

Most of information multiagent systems are directly in touch with the user, upstream (to integrate new requests) and/or downstream (during the presentation of search results). The principle that recommends the separation of *interface* anc *treatment* parts is also guaranteed by the design of agents based systems. In order to have an interface reactive anc distributed to various users, some Information MultiAgent System (IMAS) propose the use of *interface agents* acting a: an interface between the users and the system.

The terminology is not yet fixed as far as IMAS are concerned (for example, coordinator agents are called AdN (Mediation Agent) in (Camps, 1997) or task agent in (Sycarra, 1997). However, most of IMAS follow the same architecture (cf. figure 1).

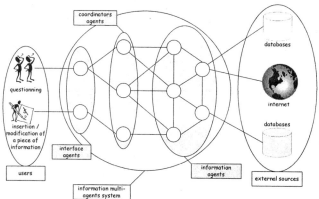

Figure 1. General architecture of a information multiagent system

490

e figure 1 shows an example of IMAS architecture. Actors can interrogate the system about the presence or the ntent of a piece of information thanks to interface agents. These agents can distribute the request among several ordinator agents being in charge of co-ordinate their activities. Information agents can search information on the ernet, in databases or with other information agents (if these information agents own knowledge about data).

en if multiagent systems dedicated to the search on the web, own a similar architecture to the one shown on the figure 1, ents making up them, generally do not co-operate. Moreover, if some of these systems adapt themselves to the user, they : often not able to adapt themselves to Internet network (if a site does not respond any more, the system should be able to d another site for example).

? Proposition for a Cooperative Information Agents System for COoperative TEchnological WAtch (ASCOTEWA)

considering the general study about MAS described before, we defined a MAS whose the goal is to assist hnological watch activities within a team. The CIASCOTEWA architecture (Cooperative Information Agents System COoperative TEchnological WAtch) is based on a set of autonomous entities. Indeed, in this architecture we sociate an IMAS (a CIASTEWA, Cooperative Information Agents System for TEchnological WAtch) for each actor the watch cell.

ch watchman has his/her own sub-system on his/her computer, what allows a larger flexibility at system level than h a centralized system.

leed, integration of a new actor of the technological watch department is realized automatically by his multiagent sub stem, which sends this information to the others CIASTEWA of the system.

e main goal of each sub-system is to search information, gather it, sort it and communicate relevant information.

ch sub-system can be considered as an IMAS, so it must be made up of a *coordinator agent*, an *interface agent* and *ormation agents* whose actions are coordinated, in our case, by an *information responsible agent*.

order to manage all the work, the search (according to requests) and the comeback of results, these agents have to operate through a shared database associated to the user (Shardanand, 1995). This user database is made up of a tabase concerning user requests and a database relating to associated results.

d, in order to have a coherent behavior relative to the others sub systems, each CIASTEWA owns social knowledge lative to the group). This one is composed of: a database containing all the requests of the users of the global system; database containing temporary results of acquaintance of the CIASTEWA; a database containing the keywords that ve to be used by the users to define theirs requests.

, the hierarchy is organized around these bases: the responsible agent is the only one authorized to change their ntents whereas the other agents have only an access for a consultation (except the information responsible agent, ich records the information in the result database).

, four types of agent compose each CIASTEWA :

The interface agent serves as a link between the user and the system. It assists the users to express their requests by allowing the consultation of the base of requests as well as the list of keywords. It allows the interaction between the users and results provided by information agents (i.e. the selection, consultation, deletion, sending of information) or between the different users of the group (for the exchange of information).

The coordinator agent has the mission to coordinate actions of others agents of the CIASTEWA. So, it can be considered as the manager of interface and information search agents. It fixes the communication of information (requests and results) with the other CIASTEWA of the global system.

The information responsible agent executes requests recorded in the request database. For that, it distributes the request to the information agent according to a search strategy. This strategy depends of the user desire. For example, if the user wants to obtain information quickly, the information responsible agent chooses a strategy by *invitation to tender* (where the first results are selected), rather than a strategy by *broadcast* (where the information responsible agent compares all the results).

The information agents have to get information according to the request sent by the information responsible agent and to send them to it.

ter having described the individual role of agents that make up a CIASTEWA, we will now explain the cooperative ctioning within it. For this, we used a method especially suitable when we have to model flows (of communication of data) among entities, where there is a hierarchical capacity between these entities. This method is called

AMOMCASYS (meaning the Adaptable Modeling Method for Complex Administrative Systems, it is complete described in (Adam, 2000)). This method has already been used for the modeling and the design of multi-agent syste for helping actors in human organization (Adam, 1999).

An extract of the treatment model resulted from this method is shown on the figure 2.

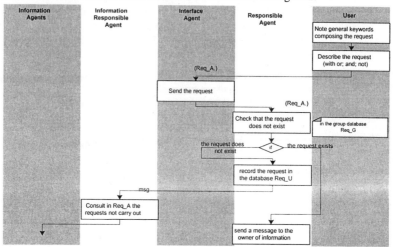

Figure 2. Extract of a cooperative activity of a CIASTEWA

The figure 2 shows an extract of a cooperative activity within a CIASTEWA. The user defines a request to CIASTEWA through the interface agent. This one forwards the request to the responsible agent which checks if t request exists in the requests' group database. If the request does not exist, it is recorded in the user's request databa and a message is sent to the information responsible agent. This message asks it to execute the requests not carry o For that, the agent creates or forwards the request to information agent. If the request exists and has been defined another user, a message is sent to the CIASTEWA of this user in order to get the information.

Thanks to the AMOMCASYS, we have defined other interactions between users and agents of the Cooperati Information Agents System for COoperative TEchnological Watch such as the automatic search of information a their management by each actor (which implies an interaction between users by the sending of mail).

3. RESULTS

A prototype has been developed for a technological watch department of a large company. In this prototype, each ac owns a CIASTEWA, which is able to communicate with other CIASTEWA and which assists communication retrieved information between actors.

In order to rapidly obtain a prototype, all the component of the CIASTEWA have been merged in one agent having th roles (fig. 3): the role of user interface; the role of coordinator and the role of information responsible. The sea engines play the role of information agents queried by the main agent.

The main functionalities have been discussed with actors of the system through models issued from the application the AMOMCASYS method. This allowed us to define the interactions between agents and actors of the team.

The prototype shown in figure 3 has been developed in Visual Basic. Communications between users and betwe agents use mail of actors.

It is currently used in a small industrial team of technological watch composed of three actors. Each agent searct information during the night and compares the results with other agents. It indicates to the user information, which owns in common with others actors (in order to avoid redundant work).

492

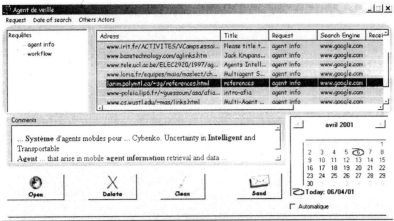

Figure 3. screen copy of the prototype

CONCLUSIONS

anks to the AMOMCASYS method, that have been built to model complex administrative systems (where actors have
ferent levels of responsibility, and who exchange information), we have modeled interactions between actors and
nts in a context of technological watch.

our prototype, which is currently used in a small competitive intelligence service, the functionality proposed in the
ond part of this article has not been totally developed. So, we aims at developing a more complete CIASTEWACO
ng Java language in order to obtain a system that is able to adapt itself to the user (in order to propose new requests
ording to the rejected information for example). We also plan to develop the capacity of the information responsible
nt to choice its search strategy according to user needs.

FERENCES

am E., "Specifications of intelligent human-machine interfaces for helping cooperation in human-organizations", In
H.J. Bullinger, J. Ziegler (Ed.), Ergonomics and user interfaces, vol. 1, London: Lawrence Erlbaum Associates,
1999, pp. 311-315.

am E., Modèle d'organisation multi-agent pour l'aide au travail coopératif dans les processus d'entreprise : application
aux systèmes administratifs complexes (in french), PhD Thesis, Université de Valenciennes et du Hainaut-
Cambrésis, september, 2000.

mps V., Gleizes M.P. "Une technique multi-agent pour rechercher des informations réparties", cinquièmes journées
francophones sur l'Intelligence Artificielle Distribuée & Systèmes Multi-Agents (in french), JFIADSMA'97 (1997),
La Colle sur Loup, Editions Hermès, Paris, 1997, pp. 29-46.

res-Mendez R.A., "Towards the Standardization of Multi-Agent System Architectures: An Overview", In ACM
Crossroads. Special Issue on Intelligent Agents, Association for Computer Machinery, 5.4, 18-24, 1999.

obiak F., L'information scientifique et technique, PUF, 1995.

isch M., "Information agent technology for the Internet: A survey", Data & Knowledge Engineering, Volume 36,
Issue 3, March 2001, pp. 337-372

ardanand U. et Maes P. "Social information filtering: algorithms for automating word of mouth", Conference
proceedings on Human factors in computing systems, ACM Press, May 7 – 11, Denver, CO USA, 1995.

cara K., Decker K., Pannu A. et Williamson M., "Designing Behaviors for Information Agents", Proceedings of the
First International Conference on Autonomous Agents, AGENTS-97, February 5 - 8, Marina del Rey, CA USA,
ACM Press, 1997

INTERACTION WITH AGENT SYSTEMS
FOR INTERMODALITY IN TRANSPORT SYSTEMS

C. Petit-Rozé[*,**] and E. Grislin-Le Strugeon[*]

[*] L.A.M.I.H., Le Mont Houy, BP 311, University of Valenciennes, 59304 Valenciennes cedex, France
{christelle.roze, emmanuelle.grislin}@univ-valenciennes.fr
[**] INRETS – ESTAS, 20, rue Elisée Reclus, 59650 Villeneuve d'Ascq, France.

ABSTRACT

The transport domain is confronted to similar information problems as the other present-day wide data domains. T
aim is to attract the user with a well-customized presentation of the information. In this context, the objective of c
work is to design a personalized information system. The proposed system is based on a multiagent architecture, ma
of four different kinds of agents. Providing personalized information is allowed by the use of profiles, which a
improved with each request of the users.

1 PROBLEMATICS

An information system is intended to supply information to users. However, the data are too often proposed su
as they were found by the system. The user has to find the interesting piece of information in a large amount of suppli
data.

In the transport's domain, the problem is similar. Much information must be spread to the users: the differe
modes of transport, the networks, and the timetables… The information is available on the different web sites of t
transport owners (if they have a web site) or under a paper format. The transport operators attempt to make mc
pleasant the trip of the travelers. But at the present time, each operator keeps and distributes the data with regard to
own transport modes and network. The user has to collect himself/herself the data and to analyze them to choose
better as possible his/her route. So, he/she has to consider various networks and the associated data. That's where t
difficulty lies. It would be well to make easier to the traveler the task of searching for the information.

In view of this assessment, some organizing authorities try to make the transport owners cooperate so as to crea
systems called intermodal information systems that group together the information of several transport owners. Our a
is to put the various transport (passenger land transport) data at public's disposal in a unique system to constitu
intermodal information, which means combining various transport modes from several operators. Our system will
used above all to prepare a trip, which means to give travelers the chronological result of transport modes and th
timetables according to the user's preferences. The goal is to personalize the trip that will be suggested to the user. Th
for a same trip, two users having different profiles may receive two completely distinct route suggestions.

In a software agent based-information system, the interactions between the user and the system get through
software agent. Another software agent has to sort, to classify and to validate the data before proposing them to the use
the goal being to supply "intelligent" information.

Our works concern this type of systems: intelligent information systems able to supply data according to the use
expectations i.e. personalized data.

2 ARCHITECTURE OF THE SYSTEM

2.1 Expectations

What do I expect from an information system on transport that I know it is based on agents and aimed
personalization? First of all, I expect it to deliver at least the basic service of an information system on transport. Then
expect more in reason of the agents' presence in the system. Especially, I would like the system interacts with me ir
natural way: the system should be able to "recognize" me and to provide me exactly with the information I would ha

494

nsidered if I had both enough time and data to do it myself. It should act just like one or more human assistant(s). hat is the general aim of our work on this system and leads the choices made in the architecture of it.

2   Multiagent based architecture

he overall architecture is shown in Figure 1.

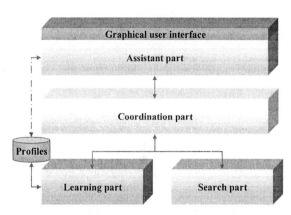

Figure 1: Architecture of the system

The goal of this architecture is to serve as a basis for the setting up of a personalized information system. esigning a personalized system is made easier by the use of intelligent agents. Moreover, to improve the performance ` the system, it is interesting to give specialties to the agents. That's why the above architecture is composed of four rts. The system thus includes two first main parts: the computing part or *coordination part* is responsible for mbining the found data to give a correct answer to the request; the *assistance part* is responsible for communicating formation between the user and the system. The objective is also to give personalized information to the user. So, the stem builds the profile of each user in order to personalize their answers. The third part, called *learning part*, of the stem is responsible for enhancing these profiles. And finally, to allow to the coordinator agent to determine the trip, e *search agents* have to seek data.

3   Characteristics of the agents

The system shown above is based on a multiagent architecture. Each part of it uses agents having specific aracteristics.

The *assistance part* plays as a mediator between the user and the rest of the system. This part is designed to help e users. It is made of the assistant agents. The assistant agent has to interact with the user to recover information and has to help him/her either if the user asks for it, or if the user waits too long before replying. In our application, the sistance of the agents can intervene to search a route, to inquire about a precision... The definition given by Le ardinal and al. (Cf. Ramos, 2000) is similar to the definition that we want to give to the assistant agents. He says that a assistant agent is a constant and close presence founded on principles like the confidence, the communication, the entification and on a common language.

The *learning part* concerns the management of the knowledge about the users. The learning part has to manage e users' profiles. In order to achieve their task, the learning agents have to collect as much information as possible on ch user. These agents have to apply some techniques coming from the statistics, knowledge acquisition (Dieng, 1990), · multiagent learning domains (Vercouter and al., 1998).

The *coordination part* manages the process of request. The coordinator agents know the search agents and their owledge. In this way, they can select a search agent(s) that is well-adapted according to a specific user's request. hen, they can build the result from the information found by the search agent(s) and the user profile given by the arning agent. In brief, the coordinator agent is able to select one or more search agent(s) and to choose the best

495

solution according to the user profile. This solution must be the most relevant for the user. The coordinator agents have the most important role.

The *search part* has to build the result. The search agents have few memories. They know the URL of some databases. They must search the information required by the result. Finally, they return their data to the coordinator agents.

The table 1 sums up the multiagent architecture and describes the characteristics of each kind of agents.

Kind of agents	Type of agent	Characteristics
Assistant agent	Reactive and cognitive	autonomous
Learning agent	Cognitive	autonomous, learning
Coordinator agent	Cognitive	autonomous, cooperative
Search agent	not very cognitive	mobile[1]

Table 1- Characteristics of the agents

3   INTERACTIONS BETWEEN THE USER AND THE SYSTEM

The interaction between the user and the system includes two sides: the acquisition of information from the user and the delivering of information to the user.

3.1   Acquisition of information from the user

So as to answer to the users as well as possible, the system shapes a profile for each of them. This profile is built from the interactions between the user and the software agent. These interactions supply information on the user in a way that can be:

- explicit: the software agent asks the user for a piece of information, a detail;
- implicit: the software agent watches the user's behavior, the requests done and the answers chosen... The agent records the actions and interprets them.

3.1.1   *Explicit acquisition*

Providing information in an explicit way is necessary, even if it is not always pleasant for the user. Indeed, the user may not agree to give somewhat private information, but a basic level of them is yet essential for enabling the personalization system to run correctly. Thus, in our system, the user is lead to identify himself/herself at the time of his/her first use of the system. To register in the system presents several advantages for the users more particularly to receive some new information likely to interest them, and to avoid them to specify again and again the same information, such as search criteria or address, at every request.

As soon as the user decides to register, the assistant agent puts him/her an enrolment form forward. The user must fulfill it. In this form, the user gives some private information (name, first name, address, handicap?) and some transport information (transport season ticket? means of transport to avoid?). From this general information, the learning agent creates a new profile in the profiles' database and returns back a password. This password enables the user to identify for future connections. Matching the general information with the information of some preconceived profile creates the new profile.

At the time of following uses, the information relative to the user is retrieved by the system in order to form the base of the transaction. Beside these data, other explicit information may be asked for. Let's take an example. In our application context, the minimal data required by the system to build an itinerary are represented on the top of Figure (inevitable information). It concerns the information relative to the places and time to leave and arrive; a possible additional piece of information concerns constraints such as heavy or numerous luggage or a baby stroller (on Figure, these information are represented by a dashed line). To enable the system to provide personalized information, we have

---

[1] This characteristic is provided for the definitive version of the system. For the moment, the databases are not already scattered on the Web.

supply it with further information (Cf. part 2 of Figure 2). The user must give some information that can not be duced by the system but that supply some meaningful characteristics, like the existence of a disability and the fact of ·ning a car (on Figure 2, these information are represented by a thin line).

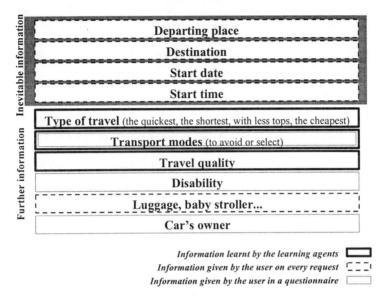

Figure 2: Information about the user

1.2 Implicit acquisition

The implicit acquisition of data is based on inference techniques. The knowledge on the user come from two urces, the stereotypes and the observations of uses.

Using stereotypes is very usual in personalization systems. They are based on a classification of the users into a nited number of categories. Each category includes some characteristics that its members are supposed to share. It ables to anticipate the users' preferences even if some pieces of information remain unknown. For example, the nCall system (Waern et al., 1998) classifies the users according to their job (researcher, editor...). The system livers information related to research events like conferences or workshops. The user's profile includes some sets of ·ywords that are matched to the ones describing the documents in order to infer about their pertinence. The SAIRE stem (Odubiyi et al., 1997) is dedicated to the retrieval of information from NASA and NOAA databases. It fferentiates between three stereotypes (science, general and no-science) according to the user's scientific knowledge ·el. As in ConCall and almost all of the systems, keywords are used. These words can be only listed or organized into ore or less complex frames (semantic nets, for example).

The setting up of our users' models is split into two parts. The first one attempts to define users' categories. rese groups will be used to send half-personalized messages. The second one refines the results of the first step and lapts them to the concerned user. The learning agents enhance the profiles learning new information of the users (on gure 2, the type of learnt information is represented by a thick line). While using the system, the user shows some of s/her needs and preferences[2]. In our case example, some data can be stored about the previous requests of a user: the aces of departing and arriving, the choices made with regard to the kind of transport mode, the duration or its quality.

To sum up, the learning agents have to analyze the user's attitudes, his/her choices to establish connections ·tween the type of travel and the applied criteria.

Our study does not include how the interfaces can be adapted according to the user's preferences and expertise level.

497

3.2 Providing information to the user

The goal of using profiles is to provide personalized information to the user.

The stereotypes are used to remedy the problem of the unknown needs. At the first time the user logs on ♦ system, the learning agent makes the user profile matching with other compatible profiles. Thus it is able to give answer which is somewhat personalized even if it lacks information. This is what we called the half-personalizati step.

Following uses contribute to the enhancement of the user profile; so as to ensure a real personalization. Wher user wants to search the route to go somewhere, he/she logs on the system. He/she enters his/her password and do his/her request, giving either the trip goal, or the exact information. The assistant agent will complete the request necessary. The assistant agent gives the request to a coordinator agent. Then, the coordinator agent processes t demand. It makes a learning agent and search agents intervene. The learning agent builds a message from the us profile and passes on information to the coordinator agent. He adds also some information from the request in the use profile.

Meanwhile, the coordinator agent analyses the request to determine the search agent(s) that will look for t information in the databases. As soon as it receives the information from the search agent(s) and the profile from t learning agent, the coordinator agent chooses the best solution for the user. The result is given to the assistant ager This solution will be shown to the user. If it suits him/her, the result is recorded in the user profile, else the coordina determines another solution without having to re-process the request.

4   DISCUSSION AND CONCLUSION

Personalized information systems are widely developed nowadays. The concepts of agents are often used in tl context. However, we noticed their social aspect remains "underemployed", because many systems are based on one more isolated agents, few are multiagent systems. Our architecture is based on a "really multiple" agents architecture.

Another aspect remains to be wider explored, it is the user-side evaluation of such system. Some questions a not closed yet, such as: what is the limit of the user agreement for answering personal questions? How far and in whi contexts the stereotypes are really pertinent?

One of our following works will be to modify the architecture so that it can manage users in greater numbe The idea is either to supply to an available agent the know-how of the overloaded agents, or to clone a new agent to he the overloaded agents.

REFERENCES

Ramos, M.P. (2000) Structuration et évolution conceptuelles d'un agent assistant personnel dans les domai techniques. Thèse d'université, Compiègne, France.

Dieng, R. (1990) Knowledge acquisition methods and tools. Research Rapport 1319, INRIA.

Vercouter, L., Beaune, P., and Sayettat, C. (1998) Apprentissages dans les SMA. In J.P. Barthès, V. Chevrier and Brassac (Eds.) Systèmes multi-agents de l'interaction à la socialité, JFIADSMA'98. Paris: Hermès, pp. 341-354.

Odubiyi, J., Kocur, D., Weinstein, S., Wakim N., Gokey C. and Graham J. (1997) SAIRE - A scalable agent-bas information retrieval engine. In J.P. Müller (Ed.) Proceedings of the Autonomous Agent conference Agents'97. N York : ACM Press, pp. 292-299.

Waern, A., Tierney, M., Rudstrom, A. and Laaksolahti, J. (1998) ConCall: an information service for researchers bas on EdInfo. Research Report T98:04, Swedish Institute of Computer Science, Human Computer Interaction a Language Engineering Laboratory.

Rozé, C., Grislin-Le Strugeon, E., Abed, M., Uster, G., Kolski, C. (2000) Recherche d'informations personnalisées. Pr of the International Conference NîmesTIC 2000 Ingénierie des Systèmes et NTIC, France, pp. 401-407.

Designing for Affective Interactions

Carson Reynolds and Rosalind W. Picard

MIT Media Laboratory
20 Ames Street, Cambridge, MA 02139-4307
{carsonr,picard}@media.mit.edu

ABSTRACT

An affective human-computer interaction is one in which emotional information is communicated by the user in a natural and comfortable way, recognized by the computer, and used to help improve the interaction. Before the computer or its designer can adapt an interaction to better serve an individual user, feedback from that user must be associated with the actions of the machine: did a specific computer action please or displease the user? Did something in the interaction frustrate the user? One of the essential issues is sensing and recognizing the affective information communicated by the user in a way that is comfortable and reliable. This paper highlights several devices we have built that offer users means of communicating affective information to a computer.

1   INTRODUCTION

There is no one interface that will please all users, no matter how usable or well-designed. Instead, we are in an era where software is adapted to please a user through the effort of the individual user, who must hand-tweak dozens or hundreds of features to her liking. Software that self-adapts to the user is hard to design: the system must obtain a delicate balance between changing in the direction of what the user wants, while providing stability and predictability. A design that changes constantly will be irritating, but so too is one that doesn't change when it should. The hard part is how to know when to change, and how to know if the change has pleased the user or not. The right balance, we argue, is also user-dependent and situation-dependent.

How can a system attain the right balance in the interaction, to please the user? Adopting the theory that successful human-human interactions can be used to inform human-computer interactions (Reeves and Nass, 1996), we suggest addressing the problem of "when and how the computer should adapt" by considering when and how people decide to adapt. Consider, for example, when a person is explaining something to you and you start to look frustrated. In this case the person may carry on for a while, initially ignoring your expression; however, if you continue to look increasingly frustrated, they will eventually pause and (hopefully) politely try to get more information from you as to what is wrong. This is in contrast to most computers today, which don't notice that the user is frustrated, and don't have the skills to respond or to adapt.

Many existing so-called "smart" systems try to fix their behavior to better serve the user. However, in these systems the guiding principle is to automate tasks the user seems to be doing. The presumption is that efficiency is king, and that improving efficiency will please the user. But, when the system is wrong, say, presuming it knows how to fix the user's spelling or how to indent the user's entries, when in fact the user had it right the way he wanted it, then this is a problem where affective communication can help. We suggest that the guiding principle of pleasing the user is greater than the guiding principle of maximal efficiency. Thus, if a user expresses pleasure when the computer fixes his spelling, then the system should continue applying its spell-fixing behavior. If the user expresses annoyance, then the computer should consider asking the user if he would like that behavior turned off or adapted in some other way. Computers that presume to know a user's intentions, no matter how smart the computer is, will sometimes make mistakes that irritate and frustrate users. Thus, if a computer is designed to "fix its own behavior, " and does so in a way that causes the user to indicate increasing frustration, this should be a sign to the system that it needs to take more ameliorative action.

The importance of communicating affective information is often ignored in interaction design. Designers have built systems that try to help a user, but they have not built systems that see if offers of help irritate or please the user. In our minds, this lack of attention to the user's response is disrespectful and short-sighted. How is a system to adapt to a user if it can't even detect if its action is pleasing or displeasing the user? Adapting is not a one-size-fits-all problem. True respect for the user requires adapting to please him or her, and not merely to please some average user that doesn't exist. Even a dog, which presumably doesn't have any sophisticated intelligence about human-human interaction, can tell if it has pleased or displeased its owner. Such affect sensing is foundational to a learning system that interacts with a person– in order to learn, the system must be capable of receiving some valenced feedback – positive or negative – so that its positively-perceived actions can be reinforced and its negatively-perceived actions removed or adapted.

2   BACKGROUND

The idea of using sensors to get positive or negative feedback from an individual is not new. Twenty-five years ago, Sheridan discussed instrumentation of various public speaking locations (classrooms, council meetings, etc.) with switches so that the audience could respond and direct discussion by voting (Sheridan, 1975). However, one of the problems with using switches was that they required active effort on the part of the individual, which distracted them from the first task at hand. If a speaker stops and asks "do you like this?" then flipping a switch "yes" or "no" is natural. However, if your feelings change many times while listening to the speaker, you might prefer to give more continuous feedback, and not have to interrupt your train of thought to do so. Watching somebody's facial expressions subtly change from positive to negative is one way of sensing affective information that is passive – requiring no extra mental effort on the part of the person communicating the emotion.

A number of learning systems, especially robots, are starting to be given the ability to do passive sensing of displays of emotion. The robot Kismet is an example of a computer system that attends to qualities of human speech for indications of approval or disapproval – positive or negative affective information – with the goal of using this information to guide its interaction and help it learn from a person (Breazeal, 2000). By speaking in an approving way at the robot's action, a human "care-giver" serves to reinforce its action. The idea is that affect communication is important for enabling the robot to learn continuously while interacting with a human.

Sensors that detect the expression of emotion can be grouped into two categories: those which passively collect data from the user and those that require the user's intent. WinWhatWhere™ is an example of a passive sensor: it records keystrokes and other data without the user's intent. Passive sensors are convenient, requiring minimal effort on the user's part, but also raise a host of legitimate concerns. When we canvassed non-technical staff around MIT as to whether we could place cameras and microphones or other sensors in their offices or on the surface of their skin to record how they naturally expressed frustration, they all indicated that they would prefer to intentionally express frustration to an interface on the computer, vs. having it sensed passively. Their discomfort with the latter in many cases was because of a feeling of invasion of privacy and loss of control. Thus, although people often use passive methods to sense the affect of another person, we decided to honor users' wishes and try to build means of sensing frustration that were active – requiring them to fill out a web form, adjust a software frustrometer, or click on a specific icon.

In the affective computing group at MIT, we have built a variety of passive and active systems that attempt to gather affective information from a user. Our earliest work focused on wearable computer systems, where there was no traditional keyboard or mouse, but there was contact with the surface of the skin. With these affordances in mind, we tried to develop methods that would passively detect patterns of change in physiology; for example, we built a system that attempted to sense when a user might be frustrated based on measuring changes in skin conductivity and blood volume pressure (Fernandez and Picard, 1998). We are also working on methods that would try to detect such feedback from the face and voice of the user. The rest of this paper focuses on more intentional or active means of sensing, highlighting a few of the prototypes we have built that require no cameras, microphones, or physiological sensing, but that work within the scope of more traditional computer keyboard-monitor-mouse interfaces.

3   DEVICES FOR COMMUNICATING AFFECTIVE FEEDBACK

We have prototyped a number of different designs intended to help people express frustration with their computing systems (Figure 1). The sensors are designed to detect user behaviors that might signal frustration, and to associate the user behavior with the current state of the computer.

Figure 1: Prototypes of tangible feedback devices: pressure-sensing mice and acceleration-sensing voodoo dolls.

Examples of our prototypes include a wireless "voodoo doll" that detects acceleration when picked up and thrown, many different pressure sensitive devices that detect intentional and unintentional muscle tension, and also a microphone which detects a raised voice. Numerous software interfaces ranging from a simple textbox, to thumbs-up/thumbs-down icons, to elaborate graphical displays were also developed.

3.1   Pressure Sensitive Mice

One possibility that emerged early in our inquiry was using force-sensitive resistors and conductive foam in combination with a pointing devices. When coupled with a low cost analog-to-digital conversion (ADC) circuit board, these sensors can be used as inexpensive grip pressure sensors.

Previously, Kirsch had examined using a mouse modeled after Manfred Clynes' "Sentograph" to detect emotional states (Kirsch, 1997). The Squeezemouse (as we call it) focuses less on directional pressure and more on detecting muscle tension and on providing a tangible interface for expressing dissatisfaction. To that end, emphasis was placed upon discovering comfortable locations for pressure sensors.

After building and testing designs in which the entire surface was made to be pressure sensitive, we settled on simpler designs in which only a clearly defined part of the mice was actively sensing. This arrangement had two distinct advantages: it had a fairly linear and predictable response, and it was less prone to being accidentally triggered.

Of course the design of the sensor is

Figure 2: Early Squeezemouse prototype in action.

incomplete without a software user interface. We identified several tasks the interface might perform by conducting informal surveys of users. One thing we learned is that people like to know how data collected about when they are frustrated is used. It is not enough to provide users with tools so that they may easily submit usability bug reports; these tools must also give users information about what information is collected, and when. Furthermore, these interfaces need to allow users to edit content transmitted about the machine state so that any private information may be removed. Consequently, the interface for the squeeze mouse performs a lot of tasks:

- Capture data from analog to digital conversion board.
- Give user feedback from the frustration sensor
- Allow the user to record and transmit usability incidents.
- Allow the user to control exactly what sort of information is reported, without invading privacy.

We settled on an unobtrusive "bulb" that appears on the user's taskbar (a system-wide menu that is available at the bottom of the screen for Windows users). As the user squeezes the mouse harder, the bulb flashes from green to yellow to red (bottom of Figure 3). When a threshold is exceeded, the system records a screenshot, along with a textual dump of the open windows. This is recorded with a running average of the recent pressure on the mouse. When the user clicks on the "bulb" a menu appears allowing the user to edit and send recent frustration events back to usability professionals. A graph can also be displayed for visual feedback.

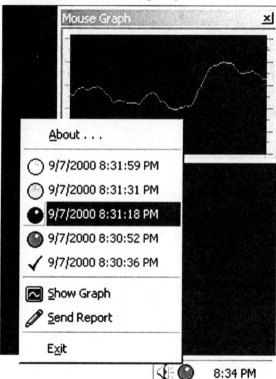

Figure 3: interface for squeezemouse shows level of pressure (graph) and record of events where user triggered feedback, color-coded by severity (red=most severe; green=mildest frustration).

3.2   Frustration Feedback Widgets

Some less elaborate, but quite useful "sensors" we have tested are user interface widgets designed for quickly and easily communicating frustration and usability issues. These have the

501

distinct advantage of not requiring any elaborate hardware, such as special mice and boards. While the widgets don't allow for the same sort of physical expression of frustration, they do allow a greater degree of user control. They do not facilitate passive, unintentional detection.

Figure 4: thumbs-up and thumbs-down icons

One of the first prototypes we tried involved the use of thumbs-up and thumbs-down icons so that the user can register pleasure or displeasure with a particular system (Figure 4). But this early prototype did not provide a mechanism to communicate the severity of the usability incident, only whether a favorable or unfavorable event had occurred. Judith Ramey of the University of Washington had once mentioned that in usability tests, a cardboard "frustrometer" was used to help users express themselves [Ramey, personal communication]. Borrowing from this idea, we developed a software version. This interface allowed for a severity scale to be communicated (Figure 5).

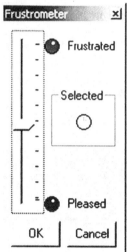

Figure 5: The Frustrometer

4 EVALUATION

As we iterated these interfaces it became increasingly clear that in order to be successful they would have to be considerably more comfortable than the standard mechanism that people use to communicate frustration about their computers: a web form. Consequently, we conducted a pilot study experiments to obtain subjective evaluations of the different sensors and interfaces we've designed. The methodology and preliminary results are sketched out below.

4.1 Methodology

To assess the utility of these sorts of sensors, we designed an experiment to compare two different frustration sensor designs and a more traditional customer-feedback form. After a bit of discussion we agreed that a web feedback form, like those currently in use on many websites represents one commonly used feedback mechanism. Consequently we designed a simple form for the control group to use.

Nine participants solicited from around MIT were asked to fill out a six-page registration sequence to a popular job search site. Additionally, they were asked to send feedback about what they liked and disliked on the website using the Squeezemouse, Frustrometer, or the control interface: a standard web based text form.

What the participants did not know until after the study was that the web pages were designed to be especially frustrating. This was achieved by violating known usability heuristics. Studies have shown that users respond poorly to varied, slow response times (Butler, 1983). Consequently, some pages were made to load especially slowly. To further exacerbate problems, certain long forms were designed so that no matter what sort of information was entered, the form would report errors that needed to be corrected, and forced the user to start filling in the page from scratch (Neilsen, 1999).

After the users completed the registration sequence, they were interviewed and asked to fill out a questionnaire for their condition. After being interviewed and filling out a brief questionnaire, users were debriefed and told of the deception carried out. It was emphasized that the deception was necessary, since it is very difficult to elicit emotional states like frustration if subjects know you are trying to frustrate them.

4.2 Preliminary Results

All participants were asked on the questionnaire about the usability and responsiveness of the registration sequence. The questionnaire presented a seven point scale from (Very Easy) to (Very Hard). For the purposes of this paper, we've chosen to label (Very Easy) as 1 and (Very Hard) as 7. The questions and mean responses (in brackets) are shown below:

How hard was the job registration web form to use?
(Very Easy) ● ● (3.08) ● ● ● ● (Very Hard)

How responsive was the job registration website?
(Very Fast) ● ● ● ● (4.63) ● ● (Very Slow)

Since we were actually interested in the performance of our various frustration feedback sensors, the remainder of the questionnaire dealt more specifically with the sensors. For instance, the participants were asked about how difficult it was to send feedback. The participants were asked about the feedback device for their condition specifically:

- Did you like using the [Web Form, Squeezemouse, Frustrometer] feedback device?
- Did sending feedback interfere with filling out the form?
- How interested are you in using the [Web Form, Squeezemouse, Frustrometer] again?

Each condition had similar questions. For instance, the users of the web feedback form were asked "Did you like using the feedback page?" instead of "Did you like using the Squeezemouse feedback device?" The responses for each condition to these questions are summarized below in Figure 6.

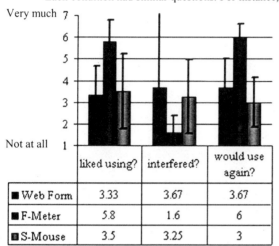

	liked using?	interfered?	would use again?
■ Web Form	3.33	3.67	3.67
■ F-Meter	5.8	1.6	6
▨ S-Mouse	3.5	3.25	3

Figure 6: questionnaire mean results

After performing a single-tailed t-test on this data we found that participants were significantly more likely to say that they would use the Frustrometer again and that they liked using it. So why did the Frustrometer perform better than the Squeezemouse or the web form? We theorize that it is because the Frustrometer was the most accessible, most straightforward interface, and consequently seemed to require less cognitive overhead when reporting feedback. In short: it was less frustrating and distracting than the other options.

Overall users were enthusiastic about being able to send in feedback. One participant noted "I think being able to send feedback while in the middle of a process is cool and sort of prevented me from really losing my temper." Another noted, "The feedback option gave me a sense of power, in the sense that I could complain or compliment about features I dislike or like." Most users seemed to respond positively to the convenience of an accessible and easy to use feedback mechanism: "I liked it being set up such that as soon as I realized there was a problem, I could gripe."

REFERENCES

Butler, T. W. (1983) Computer Response Time and User Performance, in Proceedings of CHI '83.

Breazeal, C. (2000) Sociable Machines: Expressive Social Exchange Between Humans And Robots. PhD Thesis, Massachusetts Institute of Technology.

Fernandez, R. and Picard, R. W. (1998), "Signal Processing for Recognition of Human Frustration," *Proc. IEEE ICASSP '98*, Seattle, WA.

Kirsch, D. (1997) The Sentic Mouse. <http://www.media.mit.edu/affect/AC_research/projects/sentic_mouse.html>.

Neilsen, J. (1999) Top Ten New Mistakes of Web Design. <http://www.useit.com/alertbox/990530.html>

Reeves, B. & Nass, C. (1996). The media equation : how people treat computers, television, and new media like real people and places. Cambridge University Press, New York.

Sheridan, T. (1975) Community Dialog Technology. Proceedings of the IEEE 63, 3, 463-475.

WinWhatWhere Investigator. <http://www.winwhatwhere.com/index.htm>.

Interactive Arts as the Frontier of Future Communication Media
— Learned Lessons from ATR MIC Installations —

Kenji Mase and Ryohei Nakatsu

ATR Media Integration & Communications Research Laboratories
Soraku, Kyoto 619-0288 JAPAN
{mase, nakatsu}@mic.atr.co.jp

ABSTRACT

In this paper, we introduce our research challenges toward realizing future communication media by creating interactive multimedia art installations. Several installations in this direction have already been created and exhibited. These include Interactive Poem, Life Spacies II, Iamascope (interactive kaleidoscope), Tangible Sound, and Augmented Groove.

1. INTRODUCTION

ATR MIC Research has been working to create new communication media and systems by pursuing the science and technologies of vision, graphics, music, virtual reality, interface agents, and social psychology. The evolving communication media and technology will, we envision, facilitate mutual understanding beyond differences in place, time, language and culture. In the course of our research, we have fostered the "Interactive Arts" by providing interactive media artists with a residential laboratory and an open discussion environment for joint/collaborative work with engineering scientists.

Interactive Arts illustrate interesting aspects of future communications. The important characteristics of Interactive Arts include: (i) Bi-directional interaction, (ii) Creation by both artists and audiences, (iii) Active immersion, and (iv) Specialized bandwidth. An active immersive involvement with a successful art piece sometimes gives the sensation of sympathetic communications and realistic experiences.

Many scientists and engineers have dreamed of making intelligent machines since the birth of the first computer. They are interested in realizing artificial intelligence by giving "intelligence" to a mere human-crafted machine. Another dream of some people is to develop a machine that can augment the capability of natural human intelligence. The computer has truly been augmenting human information processing power as an information and networking engine in the Internet age. We can anticipate that in the future we will be using computers that can draw creative and/or Kansei[1] power from humans. Interactive arts are already stimulating human intelligence through their interactions. We should be able to learn how to design future communication systems from interactive art installations.

In the area of human computer interaction and groupware research, it has been considered difficult for computers to join in human communication by facilitating, augmenting, and creating communication itself. Moreover, it has been considered almost impossible to build such a smart computer. Rather, transparent communication systems, such as ClearBoard (Ishii and Kobayashi, 1992), are appreciated because they don't block human creative activities. Today, however, we can see the possibility of a non-transparent system in which a computer actively intervenes and supports the human communication (c.f. Conklin and Begeman, 1988). Computers are becoming much smaller and cheaper relative to computation power and storage space. They are ready to be used exhaustively for at-a-glance trivial purposes. Many creative people have had the experiences of hints for new inventions arising when thoughts or actions are perturbed, when established concept are easily decomposed and associated. The computer is an ideal device to facilitate such perturbation, decomposition and association. We can learn how to support creativity in interface design from interactive arts.

Progress in computing technology in movie, graphics, music, and haptic applications is extending the possibility of computer usage, and it is bringing the art and the technology closer again after science divided them in the Renaissance. Computers have good characteristics of repeatability in operation and performance.

---

[1] "Kansei" is a Japanese word that doesn't have a proper translation in English. It is a kind of integrated sensitivity of feeling, affectiveness, artistry, aesthetics, etc.

Even people without good performing skills can (virtually) play the instrument at its natural speed. As a simple example, the undo function of an electric painting system allows us to recover mis-touches of brush painting. MIDI sequencers allow many novice musicians to compose and listen to their original music without having the performing skills of musical instruments. Multimedia computing technology is bringing artistry into everyday life for everyone.

In this paper, we focus on the above four aspects of future communication media by means of interactive arts. We will introduce five installations created at ATR MIC after discussing each of the four aspect.

2. INTERACTIVE ARTS AND COMMUNICATION

As mentioned in the previous section, Interactive Arts provide the following interesting aspects for future communications.

Bi-directional interaction Its bi-directional interactions with artifacts lead audiences to a better understanding of the installation and the embedded philosophical message of the artists. As we will use some artifact for realizing the future communication media, we can learn the meanings of "understanding" and "message conveying" via interactions with computerized artifact.

Creative Experience Any art piece is the fruit of the creative process of artists. The audience doesn't create the piece in conventional fine arts. However, the Interactive Arts usually asks people to create things during the interaction. At the time of communication, we also need to create the content of communication, such as topic development, dramatic story-telling, and sympathetic expression.

Active Immersion A good Interactive Art installation is well designed to give active immersion with an easy interface to the audience. They are instantly attracted to the artifact and continue to play, exploring the hidden possibilities. When the future communication media technology succeeds, the people are immersed themselves during the communication for deeper understanding.

Specialized Bandwidth Some artists purposefully select the bandwidth of the interaction channel in the design of art. The purpose is to enhance the importance of the chosen communication media. Some artists say they coin the concept first and then choose the available media and technology that is suitable to realize the concept. In any case, such constraints, technological or not, are often effective in providing a creative environment (Finke, Ward and Smith, 1992). The heart of science and technology is to "model the nature", hence we have to select and reduce the bandwidth economically and effectively. Technology may, in the distant future, be able to give the full range of bandwidth for communications. In the meantime, we can learn again from artists how to design the communication system within the focused bandwidth.

3. INSTALLATIONS

Many multimedia content systems have been created at ATR MIC. Some systems were created as pure "Interactive Art" installations by artists. Others were developed as multimedia creation tools or multimedia communication tools by engineers alone or in collaboration with designers. Among these many, we introduce five installations and systems that could be considered as more or less art installations. We discuss how the aspects discussed above are realized in each installation and try to investigate ways to develop future communication media. They use speech, texts, bodies, graspable disks, and un-graspable water as major interaction media of the user side. These are the selected bandwidth for the user's interaction.

3.1 Interactive Poem

"Interactive Poem" (Tosa and Nakatsu, 1997) is a collaborative poem creation environment. A participant and a computer agent exchange short poetic phrases in the installation. A computer agent named "MUSE," which is carefully designed with a face suitable for expressing the emotions of a poetic world, appears on the screen. MUSE utters a short poetic phrase to the participant. The user utters one of the optional phrases displayed on the sub-screen or creates his/her own poetic phrase, all the while listening to MUSE and feeling an impulse to enter the world of the poem. The speech recognition system identifies the uttered phrase and

(a) Interactive Poem   (b) Life Spacies II

(c) Iamascope   (d) Augmented Groove

(e) Tangible Sound 2

Figure 1: ATR Interactive Art Installations
(a) Interactive Poem (©1996-98 Naoko Tosa & Ryohei Nakatsu)
(b) Life Spacies II (©1999 Christa Sommerer & Laurent Mignonneau)
(c) Iamascope (Sidney Fels & Kenji Mase)
(d) Augmented Groove (Ivan Poupyrev et al.)
(e) Tangible Sound 2 (Tomoko Yonezawa & Kenji Mase)

recognizes the emotional element conveyed in pitch, speed, and loudness of speech. Exchanging emotional poetic phrases through this interactive process allows the participant and MUSE to become collaborative poets who generate a new poem and a new poetic world (Fig. 1(a)).

Interactive Poem is a fully bi-directional interaction system where the utterances alternate between a participant and the agent. The user has the freedom to use his/her own phrase and to add any emotional expression to the utterances. The prepared phrases, agent's voice, and visual appearance are well-designed by the artists to provide the active immersion. The length of interaction depends on the prepared phrases and the design of the story.

3.2 Life Spacies II

"Life Spacies II" (Sommerer and Mignonneau, 1999) was originally developed for the ICC Inter Communication Museum in Tokyo as part of the museum's permanent collection. It is an artificial life environment where remotely located visitors on the Internet and the on-site visitors to the installation at the Museum can interact with each other through evolutionary forms and images. Through the Life Spacies II web page, people all over the world interact with the system; by simply typing and sending an email message, one can create one's own artificial creature.

A special text-to-form coding system enables us to use written text as genetic code and to translate it into visual creatures. Form, shape, color, texture and the numbers of bodies and limbs are influenced by the text parameters. The produced creature starts to live and move around in the environment. Depending on the complexity of the written text messages the creature's body design and its ability to move is determined (Fig. 1(b)).

Life Spacies II provides bi-directional interaction with A-Life-like creatures. It also provides an asynchronous communication channel with other visitors through the installation. The texts could be arbitrary. However, many people type meaningful sentences that stimulate creativity. The people are attached to their own creatures from the beginning to enjoy the active immersion of the artwork. As users become familiar with the behavior of the system, they try to extend the life of their creatures.

3.3 Iamascope

"Iamascope" (Fels and Mase, 1999) is an interactive kaleidoscope and a multimedia instrument. The Iamascope installation has been successfully exhibited at many sites throughout the world such as SIGGRAPH97 in L.A., Ars Electronica Center in Linz, Millennium Expo in London, and Kumano Experience Expo in Japan. Using arbitrary images captured by a video camera, it generates kaleidoscopic images that are projected on the wall in front of the performer. As the camera shoots the performer, the generated images consist of the performer, his or her clothes, and anything else in the camera's view. Users can also play music by changing the kaleidoscopic pattern. This is caused by body movements as the system detects such movements to control musical notes and keys. As a result, single or multiple users can draw kaleidoscopic patterns as well as play music by gestures, movements and dances. Figure 1(c) shows the view and the system structure of Iamascope.

Iamascope essentially provides the direct manipulation of images and music. However, the kaleidoscopic symmetrical pattern of the self-image is interacting, feeding back to the players back with well harmonized abstract patterns. Changes in keys, instruments and graphics also give the sensation of bi-directional interaction. People can joyously spend infinitely on generating visual patterns and musical melodies.

3.4 Augmented Groove

"Augmented Groove" (Ivan et al., 2000) is a musical interface that explores use of augmented reality, three-dimensional (3-D) interfaces, and physical, tangible interaction for conducting multimedia musical performance. It was exhibited at SIGGRAPH2000 in New Orleans. Players (users) of the Augmented Groove can play music together simply by picking and manipulating a physical music disk (old LP disks) prepared on a table. The physical motions of the disks are recognized by image processing and interpreted to control musical elements such as timbre, pitch, rhythm, reverb and others. Users can see 3-D virtual graphics attached to the disks whose shapes, color and dynamics reflect aspects of the music controlled by the visitors on the front screen. Several visitors can easily join around the mixing table and play together (Fig. 1(d)).

507

Augmented Groove is also a direct manipulation of graphics and music. It is still more of a music instrument than an interactive art. However, the effective mapping of manipulation to the music parameter control gives easier understanding of the interaction and leads people to enjoy the creative experience and active immersion. Furthermore the graspable disk interface gives the user a strong attachment to the augmented reality world.

3.5 Tangible Sound

"Tangible Sound" (Yonezawa and Mase, 2000) is a musical instrument that uses water as an interaction medium to control the intuitively appealing feeling of musical flow. Performers interact with water flowing from a faucet into four drain funnels. An important aspect of the instrument is that it provides natural tactile feedback when the user touches, scatters and stops the water flow. Water is tangible but not graspable. The spreading water is particularly enjoyable, since it is linked to musical tension. Interacting with water can be a multi-sensorial experience, since it is possible to see, hear and touch the liquid. The user can control harmonies, resonance frequencies, tension, volume, note timing, etc. by controlling the water flow with his or her hands and choosing drain funnels of different heights (Fig. 1(e)).

Tangible Sound is unique because it adopts water as the user interface. The physical characteristic of water provides different sensations and unexpected responses. Water itself is intimate enough to draw the user into playing with it for a long time. Although only a simple control mechanism for musical expression is provided, people begin to explore the various possibility of controls.

4. CONCLUSIONS

The introduced installations and systems are all very well designed to achieve active immersion and creative experience. The former two installations, which were produced by artists, have embedded bi-directional interactions with simple autonomous behavior. The latter three systems were produced by engineers. These have less autonomy but are designed for good direct manipulation of music, graphics, and tactile sensations. These are, even at this moment, providing collaborative/cooperative environments with others in their manipulation and play. For future communication media, bi-directional interaction will be a key factor. Accordingly, agent-oriented technology for multimedia interactions (c.f. Yonezawa *et al.*, 2001), such as embodied entities, music and tangible objects, will become more important.

Acknowledgments

We thank the artists and engineering scientists contributing the installations discussed here. We also thank the people of ATR MIC for their discussion and support of this work.

REFERENCES

Conklin, J. and Begeman, M.L. "gIBIS: A Hypertext Tool for Exploratory Policy Discussion", In *CSCW '88 Proceedings*, 1988, pp. 140–152.

Fels, S. and Mase, K. "Iamascope: A graphical musical instrument", *Computers and Graphics*, 1999, Vol. 23, No. 2, pp. 277–286.

Finke, R. A., Ward, T. B. and Smith, S. M. "Creative Cognition", MIT Press, MA, 1992.

Ishii, H. and Kobayashi, M. "ClearBoard: A Seamless Medium for Shared Drawing and Conversation with Eye Contact", In *Proceedings CHI'92*, 1992, pp. 525–532.

Poupyrev, I. *et al.* "Augmented Groove: Collaborative jamming in Augmented Reality", In *SIGGRAPH 2000 Emerging Technologies*, 2000.

Sommerer C. and Mignonneau, L. "Art as a Living System", in LEONARDO Journal, MIT Press, 1999, Vol. 32, No. 3., pp. 165-173.

Sommerer, C. and Mignonneau, L. "Verbarium and Life Spacies: Creating a Visual Language by Transcoding Text into Form on the Internet", *IEEE Sympo. on Visual Languages (VL'99)*, 1999, pp. 90–95.

Tosa, N. and Nakatsu, R. "Interactive poem", In *Proceedings of the AIMI International Workshop on Kansei -The Technology of Emotion-*, 1997.

Yonezawa, T. and Mase, K. "Tangible Sound: Musical Instrument Using Tangible Fluid Media", In *Proceedings of ICMC2000*, 2000, pp. 551–554.

Yonezawa, T., Clarkson, B., Yasumura. M. and Mase, K. "Context-aware Sensor-Doll as a Music Expression Device", In *Proc. CHI2001 extended abstracts*, 2001, pp. 307–308.

Kansei Interaction in Art and Technology

Haruhiro Katayose[1,] Shigeyuki Hirai[2], Chinatsu Horii[2], Asako Kimura[2], Kosuke Sato[2]

[1]Faculty of Systems Engineering, Wakayama University

[2]Graduate School of Engineering Science, Osaka University

[1]Wakayama 930 JAPAN, [2]Toyonaka Osaka 560 JAPAN

ABSTRACT: Interactive media art formed on media and computer technology realizes kansei communication in terms of man-machine-man or man-machine interface, interaction density of which is rather higher than the other activities engaged in for the technical/availability goals. This paper emphasizes on the importance of direct media manipulation and supplement of insufficient expertise when kansei commutation is assisted by computer technolodies. This paper introduces our concrete activities and systems regarding kansei interaction in art.

1. INTRODUCTION

Kansei is a Japanese term translated into affection, emotion, sentiment, feeling, and atmosphere. It is used as the opposite concept of intelligence. One of the fundamental basis of the intelligence is consistency. On the other hand, kansei is personal and dependent on various circumstances. In Japan, a big national project regarding kansei was executed from both engineering and psychological approach in the end of 1980's. Kansei interaction, derived from Kansei study, especially focuses on the nonverbal communication and intention sharing at the interaction field. Central interest and the technical methodology to realize kansei interaction are similar to those of affective computing[Picard 1997]. From the technical view, computational modeling of emotions, realtime gesture sensing, realtime media-rendering, contents production and its evaluation are central topics.

In the context of media art, interaction technologies plays an important role[Tosa 2000] in the following areas. One is externalization of reflection regarding design process with physical movements. Computer assisted realtime media design and reuse of the trajectory can be compared with invention of paper which has contributed inheritance of literature. The other important point is enhance of intention. Acquirement of expertise is a big gratification in artistic activities. At the same time the hurdle to get expertise can be a factor of giving up the chance of self-expression. Computers can be used as an intention enhancer for those who are willing to express her/his imagination beyond her/his expertise. This is important, especially, in the welfare needs.

In this paper, we are going to introduce our activities regarding Kansei Communications[Katayose 1993] and discuss its possibilities. In the chapter2, a environment of composing media art and two concrete projects will be introduced. In the chapter3, Some systems which complement expertise will be shown. In the chapter4, one approach to evaluate kansei interaction will be introduced.

2. TOWARD NEW INTERACTIVE MEDIA ART

In this chapter, an environment for composing interactive media art will be described. Next, concrete works of computer music and dance stage which were realized on the environment will be introduced.

2.1 ARTISTIC CREATION ENVIRONMENT

The fundamental requirements to externalize reflection regarding design process are gesture sensing and realtime media rendering. Efficient authoring environment is also required to connect gestures and media rendering. We have developed a compact gesture sensor called ATOM[Kanamori 1995] and a framework to write programs of a individual scene called HIAT[Katayose 1997] .

ATOM8 has eight analog input for various transducers and convert the signals into MIDI/serial message (figure 1). The size of ATOM is about one inch cube. The users can choose wireless or wired system in response to artistic requirement. The user can also set communication speed of data output from ATOM in every 10ms.

We use Max, a well-known graphical programming environment for multimedia control, in order to write media contents. Max does not restrict authoring policy. This characteristic of Max sometimes results in the difficulty in implementation of a big pieces. Especially, when artists want to use various sensors which output value data, a heavy data traffic sometimes causes system crash if the program is not carefully tuned. In order to avoid such a trouble, the authors prepared some templates and a methodology called HIATspecified to use ATOM8 on MAX. The methodology regulates the usage of templates, but does not regulate artistic realization. It supports scene transition in blackboard style, easy access to the sensor data, and appropriate inner data control.

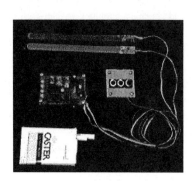 

Figure 1. ATOM8. Left:ATOM8 overview with posture sensors, acceleration transducers, bending treanducers, gyroscope. Right: Wireless ATOM8 Attached to a performer.

2.2 ARTISTIC ACTIVITIES

We have been carrying on two artistic projects using the above environment;Tikukan no uchyu Project and DMI Project.

Tikukan no uchyu Project

Since 1993, we have been producing "Tikukan no uchyu"[Cosmology of bamboo pipes] for the cyber-shakuhachi; a hyper-instrument attached Atom8 to traditional Japanese bamboo flute. The staff are Music/Shakuhachi : Simura Satosi, Video : Masal Ohashi, Engineers : Haruhiro Katayose, Tsutomu Kanamori. "Tikukan no uchyu II" and "Tikukan no uchyu V" was performed at International Computer Music Conference at Hongkong 1995 and at Arhus 1997 respectively.

Tikukan no uchyu has a style that music and video are controlled by the computer-recognized body-actions and skills seen in playing the shakuhachi. The control sources are the triggers which are recognized using sensor-fusion technique, and values which are obtained directly from the motion sensors and time transition. In a religious manner, shakuhachi performance means expression of cosmology or something changing dynamically in the player's mind. Simplicity and complicity live together in shakuhachi performing. The expression of this paradox is the artistic theme of "Tikukan no uchyu." Through the corroborative composition process, cut and trial manner contributed to the design of fundamental part of tone control. For instance, we use "pitch changer" as a rapid grand tone controller. This unusual usage of "pitch changer" has been discovered in a realtime gesture control of sound.

DMI Project

The Dance, Multimedia, Interaction (DMI project) is a project for an interactive multimedia stage featuring dance, which was given the first performance in 1995 November. The title of the piece is "Birth, Evolution, Regeneration and Calm Water." The various gestures of the dancers, including triggers and value information on the stage were detected and used to control sound, CGs, Videos and lights. The artistic concept is harmony between the spiritual and technology. The artistic director is Mariko Takayasu, a choreographer. HIAT played an important role to to realize the stage. Three non-experienced students took charge of three dancers respectively, and wrote programs of the scene where the dancers appeared. The students could write control programs in one month.

Figure 2. Tikukan no uchyu, 1997

Figure 3. DMI Project, 1996

3. MEDIA PARTNER SYSTEM

The activities mentioned above are the examples to assist artistic work for professional users. Another important aspect of computer aided interaction is to enhance intention. In this sense, the usage is for inexperienced people, which leads to the application for handicapped persons. In the following chaptermedia partner systems featuring music applications are introduced.

3.1 PlaytehDE

PlaytheDE is a blues-performance program using 3D movement. It is running on MacOS PC equipped with a video camera. Users can play blues-scaled notes by moving a light pen. Horizontal direction, vertical direction and depth movement, from the camera basis, are assigned to the pitch, volume, density of sound respectively. If using PlaytheDE, inexperienced users who does not know the blues scale can play blues with intuitive gestures. The effect has been verified at the experiment at a music class at a high school. The students group using the PlaytheDE could express the taste of the blues better than the students group using the keyboard.

3.2 TFP

TFP is a music performance system with one finger. One finger tapping controls the tempo and volume of music pieces, maintaining nuance of relative timing and volume of each notes, which are prescribed in a score data. With a additional finger operation, users can control the intervals between notes which should be distinguished from the overall tempo.

After the experiment of the usage of this common system at the primary music education, the following availabilities were reported. The first is the usage as assistance for teachers who have to accompany students. School music teachers are not always trained as they can perform high level musical pieces. TFP can be used to accompany with non-specialty instruments. The second is for expression experience for students. Students can try what they want to express irrespective of their skill in playing the musical instrument using TFP.

Figure 4. PlaytehDE   *Figure 5. Students Enjoying* TFP

The third is gaining true experience of music creation. Music creation process has been long specialized into processes of composition and performance, which should be unitized. The education combining MIDI sequencer and TFP enables the student to experience fundamental music creativity.

3.3 Inspiration

Inspiration is a music system like combining playtheDE and TFP. Inspiration transform the sketchy outline of notes sequence played on the keyboard into the musically correct melody and chord sequence. The users can select the rigidness to comply with the score. The most rigid version behaves like TFP which users can switch melody part and other chord part with sketchy operation.

4. KANSEI INTERACTION AND PHYSIOLOGICAL STUDY

This paper has introduced validity of interactive systems in artistic act ivies. The validity is subjectively understandable. It is, while, difficult to evaluate quantitatively. In this chapter, we would like to describe our activity regarding further evaluation and standpoint toward kansei.

In a field Kansei information processing, main-stream is the realization of computer assisted design system, using keywords of adjectives or reduced parameter space. Most of them utilize multi variate statistical analysis, presupposing the common or generalized kansei information. We are not interested in handling common kansei. We are rather interested in how communicator's kansei and receiver's kansei affect with each other in interaction field, and what are the sources which cause kansei. In this point of view, we have been trying to investigate how the performer's tension or elaboration is conveyed directly to the audience in the performing arts, featuring "Tikukan no uchyu." This conveyance much depends on interaction field; concert type, the audience, place and so on. First, we compared the performer's introspection regarding "tension" with pulse rate known as the index of sympathetic nerve (see Figure 6).

The performer explained that the piece consists of mainly four part; "calm beginning part", "next Growing part", "development part", and the "last calm part." The physiological index showed the different result of performer's evidence. We verified there is not a big correlation between the intense of the performance movement and the physiological indexes[Katayose 2000].

512

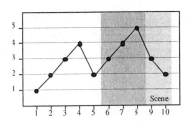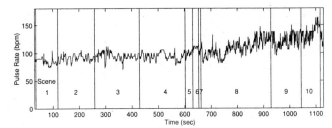

Figure 6. Comparison of introspection and physiological date at the concert. Left: Introspection, Right; pulse Rate

We can set up a hypothesis that a concentration is maintained or still increased at the very last part, where the performer recognizes as the calm part. It is a piece of information to study further kansei interaction.

We are also investigating if audience's physiological data and introspection correlate with those of per former's, inspired by the work done by Senju et al[Senju 1987] .

5. CONCLUSIONS

This paper has presented possibility of kansei communication focusing on artistic activities. Concrete projects regarding professional artistic activities have been shown. Media partner system featuring music has been introduced. We believe the importance and expansion of this field. In future we would like to proceed especially focusing on how to evaluate the the research.

ACKNOWLEDGEMENT

This work is supported in part by Japan Society for the Promotion of Science under grant JSPS-RFTF 99P01404. This study was initiated at L.I.S.T. Authors would like to thank L.I.S.T, Professor Inokuchi and Dr. Shimira for their support and assistance.

REFERENCES

[Katayose 1993] Katayose, H., Kanamori, T., Kamei, K., Nagashima, Y., Sato, K., Inokuchi, S., Simura, S. (1993). Virtual Performer, Proc. Intl. Computer Music Conf., pp.138-145

[Katayose 1997] Katayose, H., Shirakabe, H., Kanamori,T., Inokuchi, S. (1997). A Toolkit for Interactive Digital Art, Proc. Intl. Computer Music Conf, pp.476-478

[Kanamori 1995] Kanamori,T., Katayose, H., Sakaguchi, T., Inokuchi, S. (1995). Sensor Integration for Interactive Digital Art, Proc. ICMC'95. pp.265-268.

[Katayose 2000] Katayose, H., Hirai, S., Kanamori, T., Inokuchi, S. (2000). A Study of Interactive Art -Environment and Physiological Measurement-, Proc. 5th International Conference on Virtual Systems and Multimedia.

[Tosa 2000] Tosa, N. (2000). Expression of emotion, unconsiousness with art and technology, pp.183-202, Affective Minds, Elsevier Science B .V.

[Picard 1997] Picard, R. W. (1997). Affective Computing, MIT Press, Cambridge

[Senju 1987] Senju, M., Ohgushi, K. (1987). How are the players ideas conveyed to audience? , Music Perception, 4, pp.311-323

The Impact of Social Context on Perceptions of Computing Activities

D. Christopher Dryer

Adobe Systems
345 Park Avenue
San Jose CA 95110-2704 U.S.A.
cdryer@adobe.com

Abstract

New wireless, mobile, and collaborative technologies may have an unexpected impact on people's emotions and interpersonal interactions. Social context may be especially important in mediating the emotional impact of computing activities. In this study, participants completed a questionnaire assessing their perceptions of the similarity of the emotional impact of twenty computing activities. The activities described involved a range of social contexts. A multidimensional scaling analysis revealed two dimensions underlying perceptions of these activities. The two dimensions were interpreted as ranging from "public" to "private" and from "work" to "play."

Keywords

Affective computing, collaborative technology, mobile devices, multidimensional scaling, network publishing, social attributions, social context, wireless technology

1. Background

At Adobe Systems, teams are designing the next wave of publishing, called Network Publishing. At the heart of Network Publishing lies the democratization of mass communication. Currently, a few specialists author electronic content using predominantly desktop computers. Their audience typically consumes the content, such as Web pages, also using desktop computers. Nevertheless, the advent of mobile, wireless, and collaborative technologies promises to change the status quo. Designers now envision solutions that allow anyone to publish and consume content anywhere, on any device.

Implicit in this vision is a new interaction paradigm. Traditional interaction technologies were based on a solitary and sedentary model of computing – that is, one person working on a machine in a single location. Network Publishing, however, redefines computing activities as collaborative and mobile. These activities demand a new understanding of users and how they will interact with these technologies. Both research and real-world examples have demonstrated that traditional interaction designs are sub-optimal for collaborative and mobile computing. As noted by researcher Jonathan Grudin (1994), groupware often has fallen short of its promise precisely because designers have failed to account for the new interaction requirements of collaborative systems. Similarly, mobile devices generally have failed whenever designers have employed traditional desktop interaction technologies. These products disappoint users because they have ignored topics such as affective computing (Picard, 1997), group dynamics (Grief, 1988), and social context.

To help ensure the success of Network Publishing solutions, my colleagues and I are studying the social and emotional impact that mobile, wireless, and collaborative technologies can have on social interactions within groups, such as a workgroup or a family. Figure 1 illustrates our model of this impact. This model has four components within a social context: system design, human behavior, social attributions, and interaction outcome. The social context is the environment within which the human-machine interaction occurs, such as in a meeting at work or in a restaurant. Social context can mediate the impact of each of the four components in the model. System design includes specific factors such as the visual appearance of the system, its input and output modes, and its behavior. We propose that system design can impact human behavior, social attributions, and interaction outcome. Social attributions are perceptions and beliefs about other people within a social context. These could be by users of a system, of users of a system, or both. These attributions include perceptions of people's personality as well as perceptions of people being either "in-group" members or "out-group" members. Social attributions can impact both human behavior and interaction outcome. Human behavior describes what people do, in relation to others, as a function of the system. People's familiar ways of interacting might be replaced with unfamiliar ways of interacting, or one person might suddenly gain a social advantage at the expense of others. Human behavior impacts social attributions and interaction outcome. Interaction outcome is a description of what happens when people interact with the system. People might be more or less productive than they were without the system, they might be satisfied or frustrated with the system, and their feelings about other people in the social context may change. Interaction outcome impacts human behavior and social attributions. Over time, interaction outcome also can impact system design, as people adopt socially supportive systems and abandon disruptive systems.

In our previous research (Dryer, Eisbach, and Ark, 1999), we studied the social and emotional impact that mobile devices can have on problem-solving interactions. In one study, we found evidence that people typically have more negative social stereotypes about users of mobile devices than they do about people who do not use mobile devices. In another study, we gave a mobile device (either a laptop computer or a palm-top computer) to one user and had that user engage in a collaborative task with another person (the non-user) who was not using a computing device. We manipulated the design of the device such that the display was either relatively easy to share or relatively difficult to share. (In all cases, however, the users still could share the display.) We found that this manipulation significantly impacted how the non-users felt about the users and how the users felt about the device. In particular, non-users liked the users less in the hard-to-share condition than they did in the easy-to-share condition. Also, users liked the device less in the hard-to-share condition than they did in the easy-to-share condition. The manipulation did not impact how users felt about non-users or how non-users felt about the device.

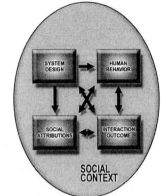

Figure 1: A model of social and emotional impact

This study might suggest that using computing devices in social situation can be impolite, particularly if the device cannot be shared easily. Interestingly, the non-users expressed relative dissatisfaction with the user and not with the device. It is as if the non-users thought it was inappropriate to use a hard-to-share device for a collaborative task. Similarly, the users did not like using the device in that situation, perhaps because they shared the view that such an interaction was inappropriate.

Given these results, my colleagues and I wondered about the social attributions and feelings that people have about various computing activities. In particular, we wondered whether activities might be seen as inappropriate as a function of social context. In earlier research (Dryer, 1998), I have used multidimensional analyses to explore emotional responses in human-machine interactions. I found evidence that interpersonal models of the emotions are well suited to describing emotions in human-machine interactions. In this research, however, I did not examine the impact of the social context in which the human-machine interaction occurs. Anecdotal evidence suggests that the same activity (e.g., "checking email") might have a different social meaning depending on the context (e.g., "in a meeting at work" or "at home").

Therefore, the purpose of the present study was to investigate the psychological structure underlying people's feelings about computing activities in a variety of social contexts. To an extent, questions about Network Publishing defined the domain of the inquiry for this study in two ways. First, the activities of interest are those activities that Network Publishing is intended to support, namely those that involve electronic content consumption, such as reading an electronic report or watching a movie. Second, given the broad scope of Network Publishing, the relevant social contexts are practically anywhere. The forward-looking vision behind Network Publishing sees virtually no limit to potential locations in which electronic media may be consumed. This includes social contexts where many of the activities of interest are simply not possible today, as well as activities in more familiar contexts. In other words, there is a need to assess people's perceptions of activities that they have not yet experienced. Therefore, this is a projective study assessing the interrelationships among people's feelings about possible future experiences with electronic media.

2. Method

2.1 Procedure

To obtain a sample of electronic media activities and social contexts, my colleagues and I reviewed the descriptions of various Network Publishing activities across the industry. Based on this review, we selected five activities (browsing the Web, e-commerce/shopping on-line, reading email, reading a report, and watching a movie) and four social contexts (at home, in a meeting at work, in a restaurant, and in your car). We then combined each of the five activities with each of the four social contexts to get a set of twenty media experiences.

Media experience	Abbreviation
Browse the Web at home	Web/home
Browse Web at a restaurant	Web/rest.
Browse Web in a meeting at work	Web/meet.
Browse Web in your car	Web/car
On-line shopping at home	E-com/home
On-line shopping in a meeting at work	E-com/meet.
On-line shopping in a restaurant	E-com/rest.
On-line shopping in your car	E-com/car
Read a report at home	Report/home
Read a report in a meeting at work	Report/meet.
Read a report in a restaurant	Report/rest.
Read a report in your car	Report/car
Read email at home	Email/ home
Read email in a meeting at work	Email/ meet.
Read email in a restaurant	Email/ rest.
Read email in your car	Email/ car
Watch a movie at home	Movie/ home
Watch a movie in a meeting at work	Movie/ meet.
Watch a movie in a restaurant	Movie/ rest.
Watch a movie in your car	Movie/ car

Table 1: Media experiences

I prepared a questionnaire that presented all possible pair-wise combinations of these twenty items. For each pair, participants were asked to think about how each media experience would feel and then consider the similarities or dissimilarities between those feelings. The participants then rated the similarity or dissimilarity of the pair on a scale from "0" to "5," with "0" being "very dissimilar" and "5" being "very similar." Six people participated in the study. Each was well acquainted with Network Publishing concepts. Two of the participants were women. The participants took approximately 45 to 75 minutes to complete the study.

2.2 Analysis

My colleagues and I analyzed the similarity data using a multidimensional scaling (MDS) procedure. We used MDS to create a solution space in which electronic communication experiences with similar social impact are close together and those with dissimilar impact are distant. The dimensionality of this solution space depends only on the structure of the similarity data. By examining the resulting dimensions, we could map the fundamental differences among these electronic communication experiences. For our analysis, our data were treated as ordinal, the distances as symmetric, and we used a euclidian scaling method, deriving solutions for one through five dimensions.

3. Results

For each of our five MDS solutions, we examined two measures of goodness of fit: Kruskal's stress (s) and the squared correlation of distances (RSQ). Stress values are Kruskal's stress formula 1 and RSQ values are the proportional of variance of the scaled data (disparities) in the matrix which are accounted for by their corresponding distances.

Number of Dimensions	S	RSQ
1	0.42	0.43
2	0.21	0.70
3	0.14	0.80
4	0.10	0.86
5	0.08	0.89

Table 2: Goodness-of-fit.

For both measures of goodness of fit, the greatest change occurred between the one-dimensional solution and the two-dimensional solution, suggesting that two-dimensional solution may be most appropriate. In addition, the two dimensional solution was found to be interpretable. A plot of the experiences in the two-dimensional solution is shown in Figure 2: Derived Stimulus configuration. In this plot, experiences that are closer together are more similar, and experiences that are farther away are less similar.

To interpret the two derived dimensions, I examined how the experiences differed in different regions of the space. On the horizontal dimension, one pole is defined by "watching a movie" and "browsing the Web." Both of these activities tend to involve images and are typically leisure activities. The other pole is defined by "reading a report" and "reading email." These activities are predominately involve text and are often work-related. This dimension seems related to people's personal experience of interacting with different kinds of content. In contrast, the vertical dimension seems related to the social context. One pole of this dimension is defined by "a meeting at work" and "a restaurant" – two very public settings. The other is defined by "home" and "your car," which are more private settings. Thus, one dimension seems to describe a person's feelings about a task separate from the social context, and the other describes a person's feelings about the social context. To highlight the social versus personal distinction of these two dimensions, I have labeled them "Interpersonal Context" and "Intrapersonal Experience."

Figure 2: Derived stimulus configuration

Another way to look at the configuration is to consider the appropriateness of the activities for the social context. This distinction falls out along the diagonal. At one end are activities that carry very little chance of negative social

517

judgment, such as "on-line shopping in your car" and "browsing the Web at home." At the other end are activities that might be very impolite, such as "on-line shopping in a meeting at work" and "browsing the Web in a meeting at work." Similar activities can be associated with very different emotional experiences depending on their social context.

4. Discussion

In this study, participants considered the emotional implications of a range of interactions with electronic media in different social contexts. By assessing the participants' perceptions of the similarities among these experiences, I was able to obtain initial evidence for the fundamental dimensions underlying people's intrapersonal experience and their perceptions of the social context. The intrapersonal experience ranges from work to play activities, and the social context ranges from public to private places.

Another interpretation of the findings focuses on the distinction between activities that are likely to be perceived as impolite and those that are not. One limitation of this study is that most of the activities seem appropriate to the home or car and inappropriate to meeting rooms and restaurants, making it difficult to tell whether the fundamental distinction is socially appropriate to inappropriate, or public to private.

For future work, my colleagues and I would like to address this limitation by finding some activities that might be appropriate to each of the social contexts. We also would like to assess people's perceptions of the activities along the dimensions like propriety, public versus private, work versus play, and emotionally positive versus negative. We could then compare the space defined by these factors to the space generated by the similarity measures to test our interpretation of the psychological structures. Other future work would involve studying and comparing the effects of other system factors, such as how easy it is to share a device, whether activities can involve more than one person, and whether the activities interfere with social interactions. As we uncover the mechanisms that underlie the emotional and social impact of mobile, wireless, and collaborative technologies, we will use this knowledge to inform the design of Network Publishing systems that support positive emotions and social interactions.

Acknowledgements

This paper would not have been possible without the contributions of Philippe Cailloux, Wilson Chan, Sheryl Ehrlich, Darcey Imm, Leah Lin, and Ron Mendoza at Adobe Systems, and Renee Fadiman.

References

Dryer, D.C. 1998. Dominance and valence: A two-factor model for emotion in HCI. *Papers from the AAAI Fall Symposium Series on Affect and Cognition*, AAAI Press, 111-117.

Dryer, D.C., Eisbach, C. and Ark, W. 1999. At what cost pervasive? A social computing view of mobile computing systems. *IBM Systems Journal, 38*, 4, 652-676.

Grief, I. (Ed.) 1988. Computer-Supported Cooperative Work: A Book of Readings. Morgan Kaufmann Publishers, San Francisco, CA.

Grudin, J. 1994. Groupware and social dynamics: Eight challenges for developers. *Communications of the ACM, 37*, 1, 92-105.

Picard, R.W. 1997. *Affective Computing*. MIT Press, Cambridge.

Visualization Techniques Producing Communication in the Creative Community: Self-Propagating Map with an Annual Ring Metaphor

Hisashi Noda, Toshiyuki Asahi

Internet Systems Research Laboratories, NEC Corporation
8916-47, Takayama-Cho, Ikoma, Nara, JAPAN

ABSTRACT

We have proposed a platform for "self-propagating contents " which are open to a wide range of users and are able to stimulate and enhance their creativity through the creation of multimedia content. In this platform, this paper proposes a new space-time visualization technique: an annual ring metaphor. The technique allows users to browse spatial and temporal data simultaneously while feeling the passage of time and grasping the direction of time. This was achieved using a self-propagating map.

1. INTRODUCTION

Although great and rapid advances having been made in hardware and software that support multimedia in the home, office, school, or anywhere, for that matter, in daily life, opportunities are still limited for large populations of users to utilize the multimedia infrastructure fully. For users to receive a great deal of advantages from the coming information-oriented society, they should be able not only to consume information but also to enhance their creativity by producing and distributing information or software contents by themselves.

With this in mind, we have already proposed "self-propagating contents (Asahi, 2000)." The self-propagating contents are open to a wide range of users and are able to stimulate and enhance creativity through the creation of multimedia content, which allows participants to produce yet unknown innovative values as the output of communities. In order to embody this concept, as the first step, a "self-propagating map" is proposed. The self-propagating map enables participants to add highly regional data (from just around their residences, such as an episode concerning narrow lanes, which would never be printed in published maps) with spatial and temporal attributes. Each piece of data will be connected to other pieces according to spatial relationships in the real world. In this platform, this paper proposes a new spatial and temporal visualization technique.

2. PLATFORM FOR SELF-PROPAGATING CONTENTS

The platform for "self-propagating contents (Asahi, 2000)" is described briefly. In the platform, open and flexible human groups are formed with members having different skills, knowledge, viewpoints, etc, enabling them to create multimedia contents with a certain value. One of the models is as follows; many creators provide small pieces of a final output and a producer gathers and edits these pieces into one artifact. The basic system configuration is assumed to include the following sub-systems (Figure 1).

- Producer supporting environment
This sub-system arranges, selects, or edits unit contents contributed from participants so that the output will meet the producer's intent.
- Participants supporting environment
This sub-system encourages participation by providing editing tools and viewers for browsing or appreciating contents edited with the contents engine.
- Contents engine
This sub-system stores contents sent from participants, and also occasionally controls the extent of participant participation according to indications from the contents engine. An important feature that the server should have is flexible distribution control, which will guarantee individual privacy, prevent distributed contents from being illegally copied, impose fees on visitors, and so on.

Assuming that the contents will grow with wide participation, the data is expected to spread temporally and spatially. Therefore, as the first step, we will develop a visualization technique for browsing spatial and temporal data. The visualization technique reported in this paper is placed in the participants supporting environment.

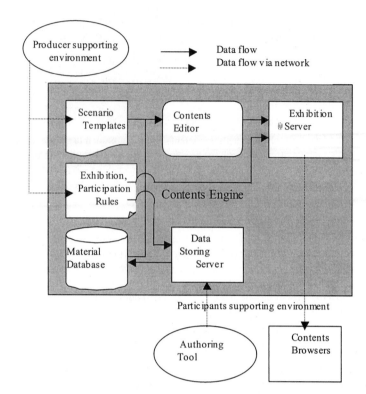

3. ANNUAL RING METAPHOR

This section proposes a new visualization technique for browsing data that have both spatial and temporal attributes. In spatial-temporal visualization on the platform as mentioned above, we assume the following three requirements:

1. Users should be able to view both time and spatial data at once

The existing time-space visualization techniques also deal with this issue. Viewing in both time and space axes may cause users to learn about unexpected relationships between data that they could not find out by merely monitoring one of the axes.

2. Users should be able to feel the passage of time and grasp the proportional length of each period

It is difficult to express a time attribute. In a typical case, it makes no sense to take 2,000 years to express the length of 2,000 years. It is necessary to express the distances and the relationships between the data in a time axis briefly, and to express the length of each period intuitively.

3. Users should be able to use an enjoyable browser

Multimedia contents include artistic photographs, beautiful pictures, realistic videos, attractive animations, and so on. Therefore, aesthetics are important. A bad-looking browser spoils the user's fun. This causes users to get a bad impression of not only the browser but the content as well, and discourages the motivation of participation. Therefore, the browsers themselves should be attractive and enjoyable.

3.1 Annual Ring Metaphor Basics

To meet the above requirements, an annual ring metaphor visualization technique is proposed. An annual ring has the characteristic image of recording the passage of time. Thus, in an annual ring browser, a time axis is assigned in the radius direction and a space axis is assigned in the perimeter direction. In this way, a time axis and a space axis can be expressed at the same time. Figure 2 illustrates the concept of our annual ring metaphor. The data of this example is related to Kyo-kaido. Kyo-Kaido is the traditional road from Kyoto to Osaka. There are various historic spots along Kyo-Kaido.

In the annual ring metaphor, a direction toward the outside means a direction toward the present. Data on the same period is placed in the same ring. A central angle shows the spatial location of the data. For example, in Figure 2, the six stages of Kyo-Kaido ("OTSU," "FUSHIMI," "YODO," etc.) are placed along the perimeter axis on the annual ring browser. As a result, the visual characteristics of an annual ring enable users to instantly grasp the spatial and temporal relationships of the data.

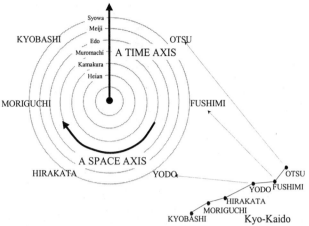

Generally, there is more data in the present than in the past. More data can be indicated toward the outside of the annual ring browser than at the center. Therefore, the browser also has a more effective available display area.

Fig.2. Concept of annual ring metaphor

4. APPLICATION FOR SELF-PROPAGATING MAP

The annual ring metaphor is implemented with a self-propagating map. The self-propagating map is one of the self-propagating contents. Participants will be able to add the highly regional data (from just around their residences, such as an episode concerning narrow lanes, which would never be printed in published maps) with spatial and temporal attributes. Each piece of the data will be connected to other pieces according to spatial relationships in the real world. The details are described as follows.

4.1 Self-propagating Map "KYO-KAIDO"

As a proto-type application, the self–propagating map "KYO-KAIDO" was developed. This is a web page content. It has several scenes related to Kyo-Kaido, which has various historic spots. Figure 3 shows the annual ring browser in this content. If users click on icons representing the data in the annual ring browser, then the scene (essays, animations, videos, photographs, etc.) related to the data is presented. In addition, a morphing technique is used so that users can more easily grasp the spatial location of the data. We use the technique for transforming from a map to an annual ring browser continuously (Figure 4). Users can select arbitrary spots on the map, and the annual ring browser with the space (perimeter) axis that has the selected spots is transformed from the map.

4.2 Reversal Linkage Function

The browser's most important feature is a reversal linkage function. In a typical web page design, creators often put textual or graphical buttons in their pages in order to jump to other pages (i.e., other sites). With self-propagating contents, many cases can be foreseen in which a new participant wants someone using a browser to jump to his/her page. In such cases, the creator can directly stick a button or an icon linked to his/her page on the browser. In other words, participants can add their pages indicating how they are presented in the contents. Of course, a type of filtering function is needed if the reversal linkage is open to the public because people will possibly link unsuitable pages (e.g., malicious or commercial ones) to the contents.

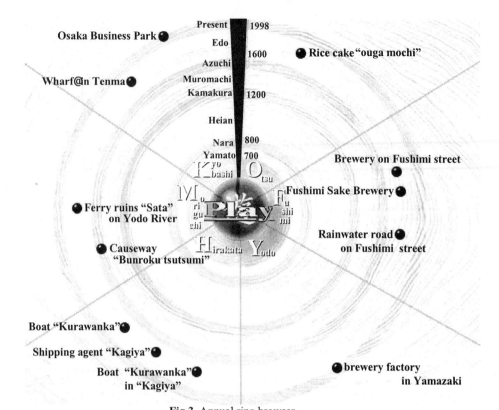

Fig.3. Annual ring browser

Fig.4. Screen shot of transition from map to annual ring browser

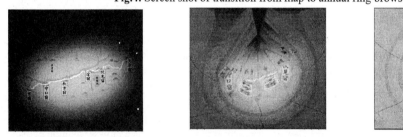

5. RELATED WORK

Many techniques for spatial-temporal visualization have been studied. This section describes some of the differences of conventional techniques from our research. The following approaches are representative of previous research.

(1) Three-dimensional visualization (Kullberg, 1996)

In this technique, a time axis is assigned to one axis in three axes of the three-dimensional space, and the other attributes are assigned the other two axes. However, the three-dimensional space is such that data positioned in the foreground hides data in the background. Considering this issue, we adopted a two-dimensional approach.

(2) Time Tubes (Chi, 1998)

This technique represents the evolution of the Web over long periods of time. The technique can present information structure and time attributes simultaneously. It seems that this technique is not suitable for presenting the proportional length of each period and displaying many temporal data. This is because the aim of visualization is different. The aim of Time Tubes is to aid analyzing; on the other hand, the aim of our visualization is to aid browsing.

(3) Other circular representations

TimeWheel (Chuah, 1997), Circular Histograms (Chuah, 1998), Clan Graph (Terveen, 1998), and so on, are excellent visualization techniques using circular representation. However, we assert that the essence of our research is not circular representation but "metaphor". Our contribution is to find an annual ring "metaphor" that is suitable for space-time visualization.

Furthermore, traditional techniques have no time axis characteristics. Therefore, users cannot feel the passage of time and the proceeding direction of the time axis.

6. FUTURE WORK

- Handling of Two-or Three-Dimensional Space

In this technique, a space is mapped to one dimension. This simplicity may help the users' understanding. However, two- or three-dimensional space can also be handled. We plan to pursue this goal in the future.

- Evaluation

An informal evaluation has already been conducted. We have shown the technique to 70 subjects. 90 % of the subjects understood the metaphor within 10 seconds. However, a formal evaluation of this technique should be executed as the next step. We are considering comparing the technique to existing techniques such as Dynamic Query, three-dimensional techniques, and using a chronology and a map. Such comparisons would be done by subjective evaluations or performance evaluations.

- Innovative Multimedia Contents

We hope to demonstrate that "self-propagating contents" will produce ever-unknown values that will stimulate people's creativity, bring people together, and prompt some to form communities. A self-propagating map is the first step. As the next step, the essential point will be how to control a new participant.

7. CONCLUSIONS

In the platform for "self-propagating contents, we proposed an annual ring metaphor. This technique allows users to browse spatial and temporal data at one time while allowing them to feel the passage of time and grasp the direction of time. The annual ring metaphor is particularly suitable for retrieving aesthetic multimedia data. It was achieved using a self-propagating map. We will deal with the problem of handling two- or three-dimensional space and propose innovative multimedia contents in the future.

REFERENCES

Asahi, T. and Noda, H., "Creative Contents Community: A Multimedia Contents Authoring Environment for New Digital Community," Digital Cities, Lecture Notes in Computer Science, Springer, 2000, pp. 416-426.

Kullberg, R. L., "Dynamic Timelines: Visualizing the History of Photography," Companion of CHI'96.

Chi, Ed H., Pitkow, J., Mackinlay, J., Pirolli, P., Gossweiler, R. and Card, S. K., "Visualizing the Evolution of Web Ecologies," Proc. of CHI'98, 1998, pp. 400-407.

Chuah, M. C. and Eick, S. G., "Managing Software with New Visual Representations," Proc. of InfoVis'97, 1997, pp. 30-37.

Chuah, M. C., "Dynamic Aggregation with Circular Visual Designs," Proc. Of InfoVis'98, 1998, pp. 35-43.

Terveen, L. and Hill, W., "Finding and Visualizing Inter-site Clan Graphs," Proc. of CHI'98, 1998, pp. 448-455.

Collaborative Work Support on Networked Heterogeneous Platforms
— Shared Augmented Interior Design Space —

Koichi Minami[1], Tomi Korpipää[2][3], Tatsuya Shuzui[1], Tomohiro Kuroda[1][2],
Yoshitsugu Manabe[1][2], and Kunihiro Chihara[1][2]
1) Graduate School of Information Science, Nara Institute of Science and Technology
8916-5, Takayama, Ikoma, Nara, 630-0101 JAPAN
2) Nara Research Center, Telecommunications Advancement Organization of JAPAN
8916-19, Takayama, Ikoma, Nara, 630-0101 JAPAN
3) VTT Electronics
P.O. Box 1100, FIN-90571 Oulu, FINLAND
koich-mi@is.aist-nara.ac.jp

ABSTRACT
Some virtual reality (VR) and augmented reality (AR) systems for room remodeling have been proposed. These systems allow users to easily imagine the design of the rooms. However, in foregoing systems, the customer and the designer need to meet somewhere. Moreover, foregoing systems do not support sharing between heterogeneous appropriate platforms based on their situation. The authors proposed a shared AR interior design system using heterogeneous systems including immersive and wearable platforms. This paper introduces the system's overview and technologies to realize the collaborative design space among heterogeneous platforms. The Duplication-Selection-Protocol proposed in this paper allows the users to do comfortable design work between AR and VR platform and the Portable 3D Engine allows a developer to easily configure the system for several platforms.

1. MOTIVATION

In remodeling a room, usually a customer discusses with a designer about the design of his room using a plan map. Some virtual reality (VR) and augmented reality (AR) systems for remodeling rooms have been proposed [1,2]. These systems allow users easily to figure a design of remodeled rooms.

However, foregoing methods require designer and client to meet at a certain place. For example, the customer has to go the designer's office to use such systems. Moreover, foregoing VR and AR interior design systems do not support sharing between heterogeneous appropriate platforms based on their situation. For example, the customer may require seeing the design over the actual room. The designer may require seeing the room with a wide field of view to better understand the proportions and being able to access extensive furniture database in his office.

In this paper, the authors propose shared AR interior design system as an application of collaborative design work support using heterogeneous platforms. The proposed system realizes sharing AR platform and Immersive VR platform such as wearable computer systems and immersive system platforms.

Using the proposed system, the customer can see his room based on the actual room by a see-through HMD. The designer can see the customer's room as immersive image on immersive display in his office.

This paper introduces system's overview and technologies to realize the shared AR collaborative design space on heterogeneous platforms. A proposed smart network protocol "Duplication-Selection-Protocol" for design work allows the users to handle virtual and real objects in AR space in same way and allows to move objects without causing frustration to the users even the user's operation results conflicts. The proposed portable 3D graphics engine is configurable for several types of immersive systems even of significant performance and hardware configuration difference.

2. SHARED INTERIOR DESIGN SYSTEM

The proposed system is shown in Figure 1. Designer is in immersive display system at his office and can see the customer's room as computer graphics. The customer is in his room and wears wearable computer including see-through HMD. The customer can see the room as virtual computer graphics furniture superimposed on the actual room. The designer and the customer can do remodeling the room by adding and moving furniture.

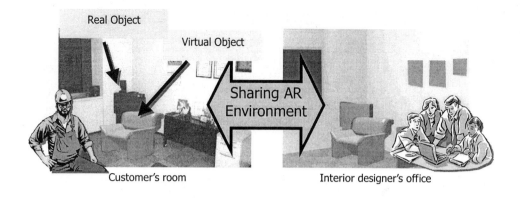

Real Object

Virtual Object

Sharing AR Environment

Customer's room

Interior designer's office

Figure 1 Proposed Shared AR Interior Design System

2.1. NETWORK PROTOCOL

Interior design is usually trial-and-error work. Participants usually repeat moving furniture frequently. In this situation, participant's actions often conflicts with other participants. Moreover, when participants share a space between AR platform and VR platform such as wearable computer and immersive display, the user in VR platform may move a virtual object, which corresponds with a real object under the AR platform. In these cases of conflicts, both spaces have different state of the same object.

To share a virtual space through a network, coherency control between multiple remote places is one of the most important problems [3]. There are two types of coherency control methods: keeping coherency by preventing inconsistency (exclusion control) and by correcting inconsistency after it is detected (lazy consistency).

However, foregoing protocols cannot provide comfortable design work. Methods based on exclusion control require the users to wait until the request become acceptable. Methods based on lazy locking may cause frustration and result in unsightly object jumping, because the method selects only one solution at a time in object handling [4]. Moreover, foregoing protocols do not regard real object handling from a remote place.

The proposed protocol is based on lazy locking. In this protocol, if an inconsistency is detected, for example, if two user's simultaneous object handling results in conflict, this protocol duplicates the object and leaves the users with deciding which object is selected. If the user moves remote real object, this protocol duplicates the object and keeps consistency (Figure 2).

This protocol uses client-server model. A server manages coherency control. AR space is expressed as a tree structure of objects. To detect conflicts, objects in the tree have version number and real-flag that indicates if the object is a real object in either place or not. When the user handles an object, version number of the object is sent to server from his client. The server compares the version number in the message to the number of the object in server's tree. If the version number in the message differs from server's, the server detects a conflict. In case of handling a real object from a remote place, the server checks the real-flag when it receives a message from a client. If the real-flag is set, the server detects a conflict.

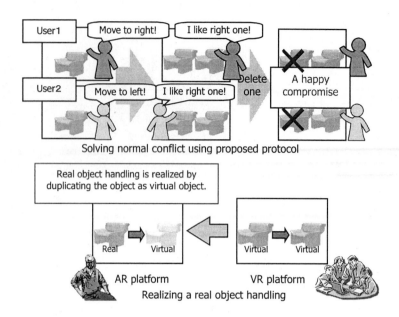

Solving normal conflict using proposed protocol

Realizing a real object handling

Figure 2 Duplication-Selection-Protocol

This protocol was evaluated by a usability test followed by a questionnaire. 5 pairs evaluated the usability of the system using a traditional protocol and the proposed protocol. A protocol that solves conflicts by accepting latest request when multiple messages conflicts was used as foregoing protocol. Using the foregoing protocol, if an user's object handling results in conflicts, the object is jumping. On the proposed protocol, conflict causes object duplication. The evaluation system was run on normal PCs and normal displays.

As results of the evaluation, there were comments that duplicated objects complicate the room when many object handlings result in conflicts. However, most users answered the proposed protocol is somewhat better than the foregoing protocol.

2.2. PORTABLE 3D ENGINE

Nowadays, there are several types of platforms consisting of different kinds of displays, computers and operating systems. The authors have developed interior design system on immersive display platforms, see-through HMD with wearable computer platform and normal display with PC platform. These systems have different hardware configurations. Cylindrical display system (CYLINDRA) is one of the immersive display systems. SGI ONYX2 with two graphics pipes drives that system. Another type of immersive system is a CAVE-like Multi-display system. Multiple networked Windows 95 PCs drive this system, using 2 PCs for each screen for stereo images and a few others for input devices, synchronization and audio. To make a shared application on heterogeneous systems, it must be easily portable to several different platforms instead of having to engineer it separately for each platform.

The authors developed a portable, easily configurable 3D Engine. The engine was developed using OpenGL, GLUT and partly GLX. This engine works on several OS such as Windows, IRIX and Linux and on several display structures (Figure 3). On wearable platform using see-through HMD, the 3D engine uses magnetic sensor for head tracking and input interface.

On multi-display system   On see-through HMD   On cylindrical display system

Figure 3 The Interior Design System on Several Platforms

The 3D Engine has been successfully tested in a number of different configurations, including IRIX 6.5, Linux, and Windows in computers ranging from SGI O2 through several different laptop and desktop PC's to SGI ONYX2. Performance tests were done on office scenes consisting of 1000, 10000 and 100000 polygons. The performance is good as long as the complexity of the scene remains tolerable. The system as such, while usable, is not very user-friendly in very complex design situations, such as designing a large office with hundreds of complex desks and chairs [5].

3. CONCLUSION

The authors proposed a shared AR interior design system on heterogeneous platforms. This paper introduced the overview of the proposed system and technologies to share AR space between wearable computer platform and immersive system platform. The proposed Duplication-Selection-Protocol provides the users with a comfortable design environment even when the operations result in conflicts, and allows users to handle real objects from a remote place. The portable 3D engine allows users to work using several platforms such as wearable computers and immersive display systems.

The proposed application and technologies are expected to assist more practical applications using heterogeneous platforms.

ACKNOWLEDGEMENT

This work is supported in part by CREST of JST (Japan Science• and Technology) and Japan Society for the Promotion of Science under grant JSPS-RFTF99P01404.

REFERENCES

[1] Takehide, T., & Junji, N. Virtual Housing. Nichikagiren, 1996. (in Japanese).

[2] Toshikazu, O., Kiyohide, S., Hiroyuki, Y., & Hideyuki, T. MR Living Room – Considerations on Geometric and Radiometric Registrations. Proceedings of the Virtual Reality Society of Japan Third Annual Conference, 1998, pp.309-312. (in Japanese).

[3] Sandeep, S., & Michael, Z. Networked Virtual Environments -Design and Implementation-. Addison Wesley, 2000.

[4] Jason, L., & Thomas, A, D. CAVERN: A Distributed Architecture for Supporting Scalable Persistence and Interoperability in Collaborative Virtual Environments. In Virtual reality: Research, Development and Applications, 1996, Vol 2.2, pp.217-237.

[5] Tomi, K., Koichi, M., Tomohiro, K., Yoshitsugu, M., & Kunihiro, C. Shared Virtual Reality Interior Design System. The Tenth International Conference on Artificial Reality and Tele-existence, 2000, pp.124-131.

527

Extracting Relations among Files by User's History

Yoshio IWAI, Masashi YAMADA, and Masahiko YACHIDA
Graduate School of Engineering Science, Osaka University
Toyonaka, Osaka 560-8531, JAPAN
iwai@sys.es.osaka-u.ac.jp

ABSTRACT

Computer work such as word processing is a typical operation which users a large number of files. One problem of using computers is an enormous file explosion whose content people cannot remember. In order to reduce this problem, a computer should aid people in finding a target file by showing relations among files. In this work, we propose two methods for finding relations among files from a user's history of his/her computer work. One method is an application of KeyGraph which extracts keywords from documents. Another method is a histogram based algorithm using the command context in the user's history.

1 INTRODUCTION

Computers have come into wide use recently and many software packages have been developed. Most office workers use computers for their operations at the present time. Frequently use of computer generates a large number of files and long term use of computer generates a lot of unwanted or unused files. Such unwanted or unused files should be deleted or moved into other directory for preserving for future use. The need sometimes arise that old files must be referred or reused when we use a computer, but we are not able to access these files because we forgot filenames or we did not put these files in order. If we do not put files in order, we must confirm content of many files for looking up. When we look up files, a tool which helps viewing files reduces our cost to search files.

A great number of documents are converted into electronic form and kept as databases in the present. Full-text searching is mainly used when an user retrieve data from such a database. Full-text searching, however, becomes slow and low accuracy as the content of database increases. Therefore, much work has been done on dividing a large database into some small clusters using keywords contained in documents(Sugimoto, 1998)(Iwayama, 1995). When such a method is applied to personal files, each files are classified into some clusters according to keywords. File relation obtained by such a method is not depended on an user but depended on the content of files. Thus the file relation is not file relation for user, and is simply similarity between files. Moreover, a method using keywords can not find relations of files written in foreign languages and not written in any language like images. It is, therefore, difficult to extract file relations for an user by using keywords.

In this paper we propose two methods for extracting relations among files from user's history. One method uses KeyGraph(Ohsawa, 1999) which extract keywords from documents and the other method uses command relations which is operated by an user.

2 PROBLEM STATEMENT

2.1 Definition of relations among files

Our aim of this research is to extract file relations for an user. We define "file relations for an user" as follows:

1. File A is related to file B if file A is needed to create file B.

2. File A is related to file B if file A is needed to use file B.

3. File A is related to file B if the knowledge of file A affects creation of file B.

We can regard the relation between a source file of computer program like C language and an execution file as an example of type 1. A practical example of type 2 is the relation between an application software and data files (images, binaries, sounds). File A and B is corresponding to data files and an application, respectively. An example of type 3 is that an user duplicate a new file B from other user's file A. Thus "file relation for an user" in this paper has a strong dependence on users, and no dependence on content of files.

2.2 Experimental environment

We use an UNIX OS as a software platform for research. The reasons are: 1) It is easy to keep a log of user's commands by using shells and emacs. 2) It can create a history of each person. 3) Many researcher in our laboratory uses an UNIX OS, so we can keep logs of many user's histories. The second reason is foremost in the above reasons. Single-user system can not create a history of each person and can create only history which contains commands of multi-user. We actually keep a log as user's history: interactive shell command, filenames used by emacs, time-stamps and file paths. Shell command and filenames are easily logged because of instinctive functions of shell and emacs. Time-stamp and file paths are needed for unification of two command histories.

In this work we do not use content of files at all. Our aim is extraction of "file relation for an user", and is not extraction of "similarity of files".

3   EXTRACTION OF FILE RELATIONS BY USING KEYGRAPH

3.1   KeyGraph(Ohsawa, 1999)

KeyGraph was proposed as a method for keyword extraction by using co-occurrence graph connecting with related words. The method adopts a hypothesis that a text is constructed as one stream with some central concepts expressed by keywords. KeyGraph consists of the following three phases:

1. Extraction of basement words
   Frequently used words are extracted as basement words for constructing texts. Basement words are connected with the other basement words (G_a, G_b in fig. 1).

2. Connecting related words
   Calculate similarity between each words in a text and basement words. The word which is related to basement words are connected with basement words (dashed line in fig. 1).

3. Extraction of topic words
   The word which has strongly connected with other words is extracted as the topic word (word 7 in fig. 1).

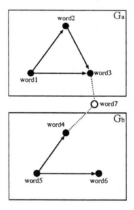

Figure 1: Extraction of basements (G_a, G_b) and topic word (word 7)

3.2   Similarity between user's history and documents

Used file history is created from user's command log in our method. This file history is considered as a history of user's work. A document is consist of some basic concepts and his claim is concluded topic words supported by basic concept words. Similarly, file history can express user's operations to accomplish his purpose. In short, it is reasonable that he will use important files for him during operations. Thus the structure of file history is similar to those of documents.

It is noted that KeyGraph needed punctuation for sentence selection, but file history does not have any punctuation. So we automatically insert a period into a file history when an user logged out. Logout is regarded as a end of today's work. KeyGraph extracts relations of words within the same sentence. As described above, one sentence expresses user's commands from login to logout.

4   EXTRACTION OF FILE RELATIONS BY USING COMMAND RELATIONS

4.1   Command relations

Much work on UNIX system is performed at a CUI terminal because computer experts can do jobs more efficiently by CUI than by GUI. We, therefore, use command history typed on a user's shell and extract file relations from the history. A large number of commands are used on a shell and also used independently, but many related commands might be used. If a sequence of related commands is extracted, the files which related commands used are related with each other. An example follows:

```
% jlatex X.tex
% dvi2ps X.dvi > X.ps
% lpr X.ps
```

'jlatex' is a command for compilation of a source file 'X.tex' written in Japanese LaTeX format to a device independent file 'X.dvi'. 'dvi2ps' is a command for conversion a DVI file 'X.dvi' to a postscript file 'X.ps'. 'lpr' is a command for printing a postscript file 'X.ps'.

'X.tex' is a source file, 'X.dvi' is an intermediate file, and 'X.ps' is a final product file in the above example. This relation is illustrated in fig. 2. 'jlatex', 'dvi2ps', 'lpr' are related with each other and this relation can be mapped into the file relation. It is noted that we do not use prior knowledge of command relation because the command which has the same function does not always have the same name. For example, 'jlatex', 'platex', 'latex209', 'latex2e', 'nttlatex' are so-called 'LaTeX' commands in our UNIX system.

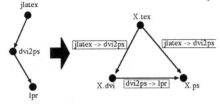

Figure 2: An example of extraction of file relations

4.2 Transition probability among commands

As described in the previous section, we do not use prior knowledge of command relations, so we must extract command relations from used-file history. We use transition probability of commands to extract command relations and we regarded two commands which have a high transition probability as related with each other. The transition probability $P_A(B)$ is defined by the following equation:

$$P_A(B) = N(A, B)/N(A), \qquad (1)$$

where A, B are commands, $N(A)$ is the number of commands in user's history and $N(A, B)$ is the number of commands B which is a command followed by A. The extraction processes of file relation are as follows:

1. Extract shell commands from user's history.

2. Calculate $P_A(B)$ of all A and B.

3. Find the command relation where transition probability is greater than a certain threshold T.

4. Map command relations to file relations.

5 EXPERIMENTAL RESULTS

5.1 Comparison with file relation

We use 11 user's history for experiments in this paper. The whole terms of logging commands are, different for users, from 3 months to 19 months.

We compare with the following 6 graphs in this experiment: 1) file relation graph using command relation, 2) file relation graph using KeyGraph, 3) file relation graph combining the above two methods, 4) file relation graph creating randomly, 5) file relation graph consisted of frequently used files, 6) file relation graph consisted of recently used files. We use four questions for comparison: Q.1) list graphs which contains important files in order, Q.2) list graphs which contains unimportant files in order, Q.3) list graphs which has connections with related files in order, Q.4) list graphs which has connections with no related files in order.

Table 1 shows the number of users answered which graph is the best in each question. From the table, file relation extracted by command relation is not good. The reason is that few relation can be extracted by using command relation. In practice, we can only extract file relations from 6 histories in 29 histories. Fig. 5.1(a) shows file relations of a user by using command relation and fig. 5.1(b) shows file relations of the same user by using

531

KeyGraph. The number of relations extracted by command relation is fewer than those extracted by KeyGraph. As shown in table 1, the graph extracted by combined method is, however, the best result for users. The reason is that command relation extracts a local relation form consequently used commands and KeyGraph extracts a global relation from login to logout. Fig. 5.1(c) is a graph combined with two graphs in fig. 5.1(a) and fig. 5.1(b)..

Table 1: Number of users answered which graph is the maximum in each question.

Graph	Q. 1	Q. 2	Q. 3	Q. 4
1	0	6	2	4
2	9	4	12	3
3	12	5	17	0
4	1	12	0	15
5	4	3	0	6
6	9	3	5	1

5.2   History length adjustification

We examined changes of file relation caused by changes of history length. Fig. 5.2(a) shows a graph generated from the whole history of an user and fig. 5.2(b) shows a graph focused on the file named 'FRS'. The graph in fig. 5.2(a) consists of sub-graphs, but the graph in fig. 5.2(b) consists of one graph centered 'FRS' file which is important for the user. We are likely to think it is the best that an user directly input a period of work, but it is not the best. An user can not remember the dates when he started and finished a job. To aid an user to set a period of work is a good approach to extract file relations.

6   CONCLUSION

We proposed two methods for finding relations among files from a user's history of his/her computer work. One method is an application of KeyGraph which extracts keywords from documents. Another method is a histogram based algorithm using the command context in the user's history.

Moreover, by combining outputs of two method, file relation for each user can be extracted more exactly than by using each single method. We implemented an application for extracting file relations from user's history and it was proved that the application has a high degree of effectiveness of finding file relations from the experimental results.

ACKNOWLEDGMENT

This work is partially supported by the Japan Society for the Promotion of Science under grant JSPS-RFTF 99P01404.

Reference

Sugimoto, M., Hori, K. and Ohsuga, S. (1998) "A System for Visualizing Viewpoints and Its Application to Intelligent Activity Support", IEEE Transactions on Systems, Man, and Cybernetics, Vol.28C, No.1, pp. 124–136.

Iwayama, M. and Tokunaga, T. (1995) "Hierarchical Bayesian Clustering for Automatic Text Classification"; In Proceedings of the International Joint Conference on Artificial Intelligence (IJCAI '95), pp. 1322–1327.

Freeman, E. and Fertig, S. (1995) "Lifestreams: Organizing your Electronic Life", AAAI Fall Symposium: AI Applications in Knowledge Navigation and Retrieval, November, Cambridge.

Ohsawa, Y., Benson, N. E. and Yachida, M. (1999) "KeyGraph: Automatic Indexing by Segmenting and Unifing Co-occurrence Graphs", IEICE, Vol. J82-D-I, No. 2, pp. 391–400.

Ohsawa, Y., Sugawa, A. and Yachida, M. (1998) "Graphical Arrangement of Files on Desktop by KeyGraph"; Tech. report ICS–112–2, pp. 7–12, IPSJ.

Masui, T. and Nakayama, K. (1994) "Repeat and Predict – Two Keys to Efficient Text Editing"; Proceedings of the ACM Conference on Human Factors in Computing Systems (CHI'94), ACM press, pp. 118–123.

Shirai, H. Nishino, J., Okada, T. and Ogura, H. (1999) "An Intrusion Detection Technique Using Characteristics of Command Chains"; IEICE, Vol. J82-A, No. 10, pp. 1602–1611.

Lane, T. and Bradley, C. E. (1999) "Temporal Sequence Learning and Data Reduction for Anomaly Detection", ACM Trans. on Information and System Security, 2(3), pp295–331.

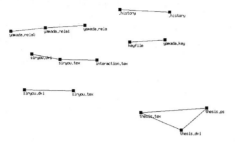

(a) Extracted file relations using command history

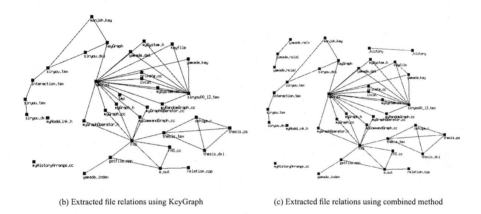

(b) Extracted file relations using KeyGraph

(c) Extracted file relations using combined method

Figure 3: Examples of extraction results

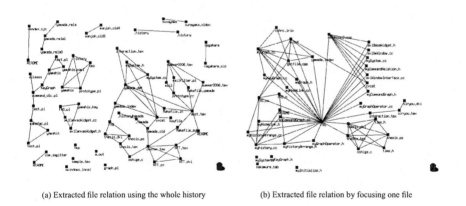

(a) Extracted file relation using the whole history

(b) Extracted file relation by focusing one file

Figure 4: Examples of Extracted file relation

533

Life-like Agent Interface on a User-tracking Active Display

Hiroshi Dohi• • Mitsuru Ishizuka• •

† Graduate School of Frontier Science, University of Tokyo
‡Graduate School of Information Science and Technology, University of Tokyo
7-3-1, Hongo, Bunkyo-ku, Tokyo 113-0033, JAPAN
E-mail: *dohi@miv.t.u-tokyo.ac.jp*

ABSTRACT

We have developed a life-like agent interface with a texture-mapped realistic face, called the *VSA-II (Visual Software Agent-II)*. It significantly improves the life-likeness of the previous VSA implementation, and has been ported to the Windows platform. It interacts with users cooperatively by means of a user-tracking active display, which is a motor drive TFT monitor that automatically turns the screen to the user's face. It can release the user from a seat in front of a computer, which opens new possibilities for human-computer interaction. It also works as a test-bed system for selective interaction; that is, the interaction with the specified user among other users walking around the screen.

1. INTRODUCTION

A life-like agent interface becomes one of the most important issues in multimodal interfaces. An ideal computer interface requires minimal efforts on the user's side. The agent with human appearance would enable a human-like face-to-face communication style; it is an inherently natural and friendly interaction. The human face is one of the most compelling components.

Several research groups have developed a multimodal interface with life-like characters at universities and companies. For example, WebPersona (Rist et al., 1997, André et al., 1998) employs life-like characters for presenting information; cartoon personas and real personas composed of grabbed video material. Microsoft Agent characters (Microsoft, 1998) act as conversational assistants. Some cartoon characters interact with the user via natural-language dialogue. DECface (Waters and Levergood, 1993) uses a texture-mapped 3D face. The face appears to speak by synchronizing lip and displays emotions and moods.

Moreover, several researchers have discussed the effect of displaying a character on a multimodal interface.• • Walker (Walker et al., 1994) reports that, compared to subjects who answered questions presented via text display on a screen, subjects who answered the same questions spoken by a talking face spent more time, but made fewer mistakes. Koda (Koda and Maes, 1996) shows experimental results about the effect of personification. Perceived intelligence of a face is determined not by the agent's appearance but by its competence. It also reports on different results for user groups, who have opposite opinions about personification. Rickenberg (Rickenberg and Reeves, 2000) says that the effects of animated characters are not unilaterally good or bad; they can be either or both.

In general, it has been said that the agent characters should be drawn deliberately as simple cartoon faces, in order not to encourage unwarranted attribution of human-level intelligence (Maes, 1994). The recent rapid progress of computer technology may change this situation. Even characters on the familiar computer game box now have photo-realistic faces.

We have already developed anthropomorphic agent interface systems, and we think realistic faces may have some advantages for the life-like agent interface. We have developed the new life-like agent interface system, called the VSA-II (Visual Software Agent-II). In our VSA-II system, we adopt a texture-mapped 3D face. The advantages of the texture-mapped 3D face as compared to the animated (cartoon) face are:

- The agent is more vivid and realistic.
- The texture is independent of the action control mechanisms.
- We can add some characters without any special artistic skill, even by using our own face.

One of our VSA-II applications is a virtual reception desk. In order to expand the application field of the VSA-II, we have also designed and developed the user-tracking active display. Since the active display turns more than 180 degrees along the user's movement, the character agent can keep eye contact with a user walking around the screen. It would be clear with whom the agent speaks among multiple users. In this paper, we present our life-like agent interface, VSA-II, and the user-tracking active display.

2. USER-TRACKING ACTIVE DISPLAY

2.1 Motor-Drive Active Display

Figure 1. Character face turns.   Figure 2. Screen turns.

It is well known that eye contact plays an important role in every day human-human communication. Eye contact should also be realized between a human and a life-like agent. Our previous VSA system generates a turned facial image within the screen along the user's movement for eye contact (Figure 1). It is plausible if you take a third-person viewpoint and observe the interaction between the user and the agent character. However, the user also watches the agent's face turn away when the user walks sideways. For example, in a theater an audience watches the left (right) side of actor's face when the audience has been seated at the right (left) side of the hall. On the other hand, all audiences in a movie theater watch the same scene at the same time no matter where the seat is.

The Digital Smart Kiosk project (Christian and Avery, 1998) uses the enhanced version of the DECface. It turns the character face in the direction of the person nearest to the screen in radial distance. They report that turning the face about 30% of the actual angular displacement provides a compelling illusion that the face is looking at the user.

The FreeWalk system (Nakanishi et al., 1999) is a 3D virtual meeting space, and the face video image is mapped onto one surface of the 3D pyramid objects. Then the user watches the distorted face. It supports the metaphor of the agent coming (going) to (from) the conversation, rather than becoming a conversation partner.

In our VSA-II system, we have considered another approach. We physically turn the screen in itself along the user's movement by two powerful motors, not the facial image (Figure 2). It can cover more than 180 degrees angular displacement, like a real reception desk at an entrance hall. Although the screen can swing more than 300 degrees, the user-tracking CCD camera restricts the angle. It allows users to walk around the screen.

2.2 Design and Implementation

Figure 3. Front image of active display   Figure 4. Rear image

Figure 3 shows the front image of our active display. The ultrasonic distance meter is attached on top of the screen. Figure 4 shows the rear image. The cylinder like dumbbells is a counter weight. There is a micro CCD camera for capturing a background scene on the tilt motor box.

We have originally designed and built up a new user-tracking active display with reference to our previous prototype system (Dohi and Ishizuka, 1999). The previous system has used a 10.4-inch VGA (640x480) LCD and a pan-tilter for a security camera. One of some vulnerable points is that the LCD size was small and its resolution was insufficient for presenting information. The new active display system loads the Silicon Graphics 1600SW flat panel monitor. It has a 17.3-inch (37cm×24cm), 16:10 wide-format screen, and 1600x1024 high-resolution. Two powerful stepping motors turn this large monitor pan and tilt rapidly and smoothly along the user's movement. The turning speed reaches 180 deg / sec within 0.5 sec; it keeps modest speed for the user's safety.

2.3 User-Tracking Module

The CCD camera with another fast pan-tilter tracks the user's face by the specified color region, and then finds the direction to users. Since the CCD camera moves pan and tilt wide angles, it is difficult to keep flat lighting condition; therefore it may cause wrong results. The ultra-sonic distance meter attached on the screen verifies whether there is a user in front of the screen or not.

The user-tracking module searches the user and controls the direction of the screen to the user. Consequently the ultra-sonic distance meter also directs toward the user's face, and measures the distance between the screen and the user. The measure range is 0.5m ~ 4.0m.

The screen changes the direction randomly to search other candidates in the following cases.
- When it cannot find the user for a certain period (FAILURE), or
- When the user doesn't move for a certain period (WRONG).

The distance information also tells us the user's action; that is, the user approaches, stands in front of the screen, or goes away.

2.4 Towards the Selective Interaction

As compared to current fixed display, this user-tracking active display has some advantages. Current pervasive interfaces implicitly assume that a single user has a seat in front of a computer.

With this active display the screen always directs to the user; therefore the VSA-II character can keep eye contact with the user. The screen turns along the user's movement, therefore the users become aware which user the system tracks; in other words, it would be clear with whom the agent speaks among multiple users. If we can recognize individual users around the screen well, the agent will be intentionally able to specify the user among other users. A selective interaction will be available.

3 VISUAL SOFTWARE AGENT-II

3.1 Texture-mapped 3D Facial Image from a Single Photograph

Figure 5: VSA-II agent character with a live background scene.

Figure 5 shows a screen shot of the VSA-II interface system. On the left is the VSA-II agent character, and on the right is the Internet Explorer. The agent controls the Explorer to present any text and visual information, while she talks with users by synthetic voice. In the previous VSA implementation (Dohi and Ishizuka, 1996, 1997), the agent character has a 3D egg-shaped face. The new implementation, VSA-II, has a character image from the bust up, so the life-likeness is significantly improved.

It is well known that human-like face-to-face style communication gives users great relief. It will reduce the mental barrier for the communication between a human and a computer.

The VSA-II character image is generated from a single full-face photograph. The real photo image is texture-mapped on the 3D facial wire frame model. Since the model is deformable, we can generate in real-time many vivid facial images like wink, speaking, rocking, smile, and so on, even if it is from a still photograph. It is difficult to generate a turned face because we have no materials about the side and rear face. However, the active display compensates this drawback. When the user walks around the screen, the character rocks her head and the screen turns tracking the user at the same time.

It has some advantages to generate the 3D facial image from the single photograph.

- Everyone easily captures his/her own face photograph with a home-use digital still camera.
- It is downloadable through the Internet within a tolerable time period because the character data size is small

Many of the cartoon character agents are appealing for users. The Microsoft agent characters are a good example. They are very attractive. A character can express complex 3D actions although it is the collection of 2D still images. The drawback of the cartoon character agent is that there are less character variations because it is easy to use but is difficult to create a new one. It requires a good taste in a character and a great deal of effort to give the character life-likeness. Another difficulty is that the amount of required data increases in proportion to the number of actions. For examples, the data size of the Microsoft Agent characters are about 1.6 MB ~ 3.3 MB / character. The data size of a single full-face photograph the VSA-II uses is just about 15 KB ~ 60 KB with the JPEG compression, and 3D polygon data is about 16 KB without any compression.

3.2 Live Background Image

A live background image raises the life-likeness of the anthropomorphic agent character. A micro CCD camera for a background scene is attached on top of the screen. We have used the Silicon Graphics Digital Media library for the video capture. The captured image is texture-mapped onto a virtual plane behind the character in real-time, so we can stretch the window. The screen turns along the user's movement, and the background of the character varies simultaneously.

3.3 System Configuration

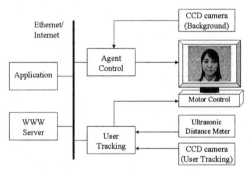

Figure 6. System Configuration

The VSA-II agent control module is implemented on the Silicon Graphics Visual Workstation 320 (Windows 2000, Intel Pentium-III 600 MHz dual CPU). It is written in C language with the OpenGL library. The video stream for the background scene is connected with the on-board video interface. The user-tracking module runs on another Windows computer. Both are connected with each other through the Ethernet.

We can control the agent with a few simple commands from an outer application process. All commands are normal text strings over the TCP/IP protocol. The VSA-II agent face is always rocking and winks, and the screen turns automatically tracking the user, regardless of any commands.

- **SAY** *text*
  The agent speaks the Japanese *text* string with simple lip sync. The Japanese text is usually mixed Kanji characters (ideographs) with Kana characters (phonograms). The system converts any text to the phonogram string by the dictionary, and then the agent speaks it. The control is returned after the speech is complete.
- **OpenURL** *url*
  The system opens the *url* page and displays it in the Internet Explorer. The control is returned immediately.
- **FACE** *action*
  Currently only "smile" is available.
- **LOOK** *pan tilt*
  The agent looks to the direction relative to the screen forcibly. The angles are modest.
- **TURN** *pan tilt*
  It turns both the screen and the user-tracking CCD camera to the direction forcibly. As a result, the system tracks another user.
- **STATUS** *code*
  It returns any status of the system, including the distance to the user.

4 DISCUSSION AND CONCLUSION

In this paper, we describe the life-like agent interface system, Visual Software Agent-II, and the user-tracking active display. The VSA-II significantly improves the life-likeness. Since the texture-mapped 3D face is generated from a single photograph, it is easy to add some characters without any special artistic skill.

We put our VSA-II system on an exhibition as a field test out of our laboratory. There is a narrow aisle between display booths. When the pan-tilter is stopped, it is identical with an ordinary fixed display. Then, many of the guests pass through the familiar sight. However, when the pan-tilter is active, at first the guests become aware that the screen can turn, and next they found which user the agent character tracks. This system is a test-bed and now tracks a single user among many users. If we can recognize the circumstances around the screen well, the interaction with the specified user will be available.

ACKNOWLEDGEMENTS
This work is partly supported by JSPS Grant-in-Aid for Scientific Research No. (C)-12680408.

REFERENCES
André, E., Rist, T. and Müller, J., "WebPersona: a lifelike presentation agent for the World-Wide Web," Knowledge-Based Systems 11, Elsevier, 1998, pp.25-36
Christian, A.D. and Avery, B.L., "Digital Smart Kiosk Project," CHI'98, 1998, pp.155-162
Dohi, H. and Ishizuka, M., "A Visual Software Agent: An Internet-Based Interface Agent with Rocking Realistic Face and Speech Dialog Function," AAAI tech. report 'Internet-Based Information Systems', 1996, pp.35-40
Dohi, H. and Ishizuka, M., "Visual Software Agent: A Realistic Face-to-Face Style Interface connected with WWW/Netscape," IJCAI'97 Workshop on Intelligent Multimodal Systems, 1997, pp.17-22
Dohi, H. and Ishizuka, M., "Visual Software Agent Interface with a User-tracking Flat Panel Display for Continuous Eye Contact," SCI'99/ISAS'99, 1999, pp.242-247
Nakanishi, H., Yoshida, C., Nishimura, T. and Ishida, T., "FreeWalk: A 3D Virtual Space for Casual Meetings," IEEE Multimedia, 6(2), 1999, pp. 20-28
Koda, T. and Maes, P., "Agents with Faces: The Effects of Personification of Agents," 1996, HCI'96
Maes, P., "Agents that Reduce Work and Information Overload," Commun. ACM, 1994, Vol. 37, No. 7, pp. 31-40
Microsoft Corp.: "Developing for Microsoft Agent," Microsoft Press, 1998
Rickenberg, R. and Reeves, B., "The Effects of Animated Characters on Anxiety, Task Performance, and Evaluations of User Interfaces," CHI'2000, 2000, pp. 49-56
Rist. T., André, E. and Müller, J., "Adding Animated Presentation Agents to the Internet," IUI'97, 1997, pp.79-86
Walker, J., Sproull, L. and Subramani, R., "Using a Human Face in an Interface," CHI'94, 1994, pp. 85-91
Waters, K. and Levergood, T.M., "DECface: An Automatic Lip-Synchronization Algorithm for Synthetic Faces," Technical Report CRL 93/4, Cambridge Research Center, Digital Equipment Corporation, 1993
Yang, J., Stiefelhagen, R., Meier, U. and Wawibel, A., "Visual Tracking for Multimodal Human Computer Interaction," CHI'98, 1998, pp. 140-147

Realistic 3D Facial Expression Modeling for Man-Machine Interaction

Yu Zhang, Eric Sung

Intelligent Machine Research Lab, School of Electrical and Electronic Engineering,
Nanyang Technological University, Singapore 639798

Abstract

cial images play an important role in friendly human-
uter interactions as well as in human-human communica-
To express realistic expressions dynamically by computer
ics in a similar manner to that in human being, we
se a fundamentally superior, physically-based approach
nthesize expressions. The development of the 3D facial
l is based on anatomical knowledge. The facial model
porates a physically-based approximation to facial skin and
of anatomically-motivated facial muscles. The skin model
ablished by using a mass-spring system with the nonlinear
gs which simulate the elastic dynamics of a real facial skin.
kinds of facial muscle models are developed to emulate
muscle contraction. Based on Lagrangian dynamics, facial
is deformed as the muscle force is applied on it. The
ing between the muscle actuation and expressions through
dependent AUs results in success to generate flexible and
tic expressions.

Introduction

the present time of advancement not only in machine com-
ion speed but also in machine intelligence, the interaction be-
human being and intelligent machine is getting more inten-
in which computer plays consequently more important role
r daily life. Thus the communication between human being
omputer should obviously become more friendly. It is highly
ted that we shall have one step advance toward the establish-
of heart to heart communication between human being and
uter if the computer would be able to communicate the emo-
nformation with human being. To achieve this goal, it likely
omputer must possess a special interface in order to interact
ly with human being.

insert a human personality in a computer generated inter-
one of the powerful approaches is the modeling of human
expressions . The human face is an extremely visible aspect
erson which, in social interaction, serves the major functions
mmunication in terms of both speech and facial expression.
ological study has indicated facial expressions play a major
n non-verbal communication[8]. However, modeling of the
nd facial expressions still remains a complex task. The face
s from person to person, consisting of a diverse and irregular
ure which makes its representation difficult. This problem is
ounded by the motion of the face which involves complex in-
ions and deformations of the skin and underlying muscle. In
addition, faces are very familiar to us. We have a well-developed sense for distinguishing which behaviors are natural for a face. Consequently, we are likely to notice the smallest deviation from our perception how a face should appear and behave naturally.

With respect to dynamic expression process of facial expressions, there are some research works done in the field of computer graphics. In [5], a real time facial animation system is presented. It uses a simple set of pre-modeled facial expressions to create a wide range of expressions and head movements. Cyriaque Kouadio et al. [1] also animate the facial expressions by morphing the face based upon a bank of 3D facial expressions. These morphing methods are easy to implement and computationally cheap, but usually require the animator to model all the facial expressions before hand. In [2, 9], the face models are classified by defined parameters to describe the attributes of the facial subsections. The operations including interpolation, translation, rotation, and scaling are applied to particular sub-regions of the face to achieve the expression animation. These parameterization methods can create facial images in a short amount of computational time, and can be used for situations which require the real-time performance instead of the accurate facial modeling. In [6], a peudo-muscle based animation approach is proposed. The facial muscle actions are simulated using geometric deformation such as Free Form Deformation but the actural facial tissue dynamics are not implemented. Sakaguchi et al.[11] developed a computer software system in which facial display of animation change its expression continuously to a certain level at a specified time. In this system, the change in facial expression is linear with time and it can not mimic the time-dependent change in human facial expressions.

The purely geometric nature of prior face models mentioned above limits their realism, however, because it ignores the fact that the human face is an elaborate biomechanical system. In this paper, we propose a physically-based model of human face from the anatomical perspective, which has a three dimensional structure of the skin and muscles. The mechanism of generating facial expressions with our model is very close to the actual one in the human face. The skin model is constructed by using the nonlinear spring frames which can simulate the elastic dynamics of real facial skin. Two kinds of muscle models are developed to emulate facial muscle contractions. Action Units in Facial Action Coding System[3] are used to construct the mapping relationship between muscle activation and the facial expressions. When muscles contract, by solving the dynamic equation for each skin node in the facial surface, we can obtain flexible and realistic facial expressions.

AU No.	FACS Name	AU No.	FACS Name
AU1	Inner brow raiser	AU14	Dimpler
AU2	Outer brow raiser	AU15	Lip corner depressor
AU4	Brow lower	AU16	Lower lip depressor
AU5	Upper lid raiser	AU17	Chin raiser
AU6	Cheek raiser	AU20	Lip strecher
AU7	Lid tighter	AU23	Lip tighter
AU9	Nose wrinkler	AU25	Lip part
AU10	Upper lid raiser	AU26	Jaw drops
AU12	Lid corner puller		

Table 1: Action Units employed

Section 2 gives a brief review of the FACS. Section 3 describes the muscle design process in which the distributions of muscular force are modeled. In Section 4, we develop a biomechanical facial skin model, which employs a biphasic strain-stress elastic constitutive relationship to the membrane undergoing deformations. Section 5 illustrates the motion dynamics and numerical simulation of our facial model. The subsequent section demonstrates some experimental results. Finally, the paper closes with concluding remarks.

2 The Facial Action Coding System (FACS)

The simulation of facial expressions requires a mapping of the desired facial expression into facial muscle activation. The Facial Action Coding System (FACS) was developed by Ekman and Friesen [3] for this purpose. Based on photographs of facial expressions, they investigated in fundamental psychological tests the relationship between facial expressions and emotions. The result of this research is a unique categorization of the emotions (happiness, fear, sadness, anger, surprise, and disgust) into primary facial expressions.

The FACS describes the set of all possible basic action units (AUs) performable by the human face. Each AU is minimal facial action that cannot be divided into smaller actions. There are 66 AUs in the FACS. We select 17 units which give strong influence to emotion expression. Table1 shows the 17 AUs which we are actually using out of AUs to make expressions.

3 Facial Muscle Model

The muscles of facial expression are superficial, they are mostly attached to both the skull and the facial tissue. One end of the facial muscle attached to skull is generally considered the origin while the other one is the insertion. Normally, the origin is the fixed point, and the insertion is where the facial muscle performs its action. In human face, a wide range of muscle types exist: rectangular, triangular, linear, sphincter [12]. Two main types of facial muscles are incorporated in our face model. They are linear and sphincter muscles.

3.1 Linear Muscle Model

Linear muscle consists of a bundle of fibers that share a common emergence point in bone and pulls in an angular direction. One of examples is the zygomaticus major which attaches to and

raises the corner of the mouth. Fig. 1 illustrates the linear muscle model with the following definitions

$\mathbf{x}_i$: arbitrary facial skin point

m_j^A : attachment point of linear muscle j at the skull

m_j^I : insertion point of linear muscle j at the facial skin

R_j: the maximal radius of influence

φ_j : the maximal angle of influence

φ_{ji}: the angle between muscle vector and $\mathbf{x}_i$

l_{ji} : the distance between muscle attachment point m_j^A and skin point $\mathbf{x}_i$

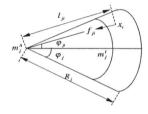

Figure 1: Linear muscle model

On contraction, facial regions close to the skin insertion point of a muscle are affected. The effect of facial muscle contraction is to pull the surface from the area of the muscle insertion point to the muscle attachment point. The muscular influence decreases with both the decreasing of the distance from muscle attached point l and increasing of the angle from muscle vector φ_{ji} . It is assumed that there is zero influence at the point of muscle attachment at the bone (m_j^A) and that maximal influence occurs at the muscle insertion point (m_j^I). Consequently, a fall-off of the muscle force is dissipated through the adjoining tissue in the influence area of the muscle.

l_{ji} is calculated as:

$$l_{ji} = \|m_j^A - \mathbf{x}_i\|$$

l_{ji} and φ_{ji} weight the influence of muscle j at vertex i separately for length factor λ_{ji} and angular factor γ_{ji} :

$$\lambda_{ji} = \frac{l_{ji}}{\|m_j^A - m_j^I\|}$$

$$\gamma_{ji} = \frac{\varphi_{ji}}{\varphi_j}$$

λ_{ji} defines the longitude distance of vertex i to muscle j normalized to values between 0 and 1. A value around 0, or 1 signifies that vertex i lies close to the muscle attachment point , or close to the insertion point, respectively. The influence of the muscle j increases with λ_{ji}. γ_{ji} is defined in [0,1] and represents the latitude distance between vertex i and muscle j. φ_j defines maximal influence angle. Increasing γ_{ji} results in decreasing the influence of muscle j. The muscular force applied at vertex $\mathbf{x}_i$ can be computed as

$$\overrightarrow{f_{ji}} = \alpha_L \Theta_1(\lambda_{ji}) \Theta_2(\gamma_{ji}) \frac{(m_j^A - \mathbf{x}_i)}{\|m_j^A - \mathbf{x}_i\|}$$

uation(4), α_L is the muscular force scaling factor which con-
the magnitude of muscular force. Function Θ_1 scales the mus-
rce according to the length ratio , while Θ_2 scales the muscle
according to the angular ratio γ_{ji} at node $\mathbf{x}_i$. We define

$$\delta_j = \frac{R_j}{\|m_j^A - m_j^I\|} \tag{5}$$

$$\Theta_1(\lambda_{ji}) = \begin{cases} \cos((1 - \lambda_{ji}^{\eta_j}) \cdot \frac{\pi}{2}) & 0 \le \lambda_{ji} \le 1 \\ \cos((\frac{\lambda_{ji}^{\eta_j} - 1}{\delta_j^{\eta_j} - 1}) \cdot \frac{\pi}{2}) & 1 < \lambda_{ji} \le \delta_j \end{cases} \tag{6}$$

$$\Theta_2(\gamma_{ji}) = \cos(\varphi_j \gamma_{ji}) \cos(\gamma_{ji} \cdot \frac{\pi}{2}) \quad 0 \le \gamma_{ji} \le 1 \tag{7}$$

e constant η_j defines the strength of muscle j. A decrease
of η_j increases the muscle influence along the longitude.

Sphincter Muscle Model

like the linear muscle, the sphincter muscle attaches to skin
at the origin and at the insertion, and contracts around a virtual
. An example is the orbicularis oris, which circles the mouth
an pout the lips. Because sphincter muscles do not behave
egular fashion, it can be modeled in elliptical shape and can
mplified to a parametric ellipsoid as shown in Fig. 2. The
tion of the parameters list as follows
: arbitrary facial skin point
epicenter of sphincter muscle influence area
the semimajor axis of sphincter muscle influence area
the semiminor axis of sphincter muscle influence area

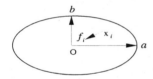

Figure 2: Sphincter muscle model

e muscular force applied at vertex $\mathbf{x}_i$ is computed as

$$\overrightarrow{f_i} = \alpha_s \Theta(r_i) \frac{(o - \mathbf{x}_i)}{\|o - \mathbf{x}_i\|} \tag{8}$$

α_s is the sphincter muscular force scaling factor and

$$\Theta(r_i) = \cos((1 - r_i) \cdot \frac{\pi}{2}) \quad 0 \le r_i \le 1 \tag{9}$$

ation (9)

$$r_i = \frac{\sqrt{y_i^2 a^2 + x_i^2 b^2}}{ab} \tag{10}$$

The Mass-Spring Facial Tissue Model

r synthetic facial tissue is motivated by histology and tis-
iomechanics. Human skin has a nonhomogeneous and non-
pic layered structure consisting of the epidermis, dermis and
dermis. Under low stress, dermal tissue offers very low resis-
to stretch as the collagen fibers begin to uncoil in the direc-
f the strain, but under greater stress the fully uncoiled colla-
bers resist stretch much more markedly [4]. This yields an
ximately biphasic stress-strain curve(Fig. 3).

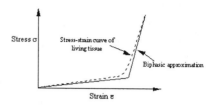

Figure 3: Stress-strain relationship of facial tissue and its biphasic approximation

Our facial model is constructed by polygon meshes(see Fig. 4). It consists of 702 vertices and 1284 polygons, with finer polygons over the highly curved and/or highly articulate regions of face, such as the eyes and mouth, and larger polygons elsewhere, such as cheek and forehead. In order to physically simulate the deforma-

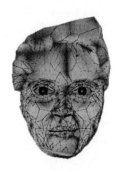

Figure 4: Polygonal face model

tion of the facial skin tissue, we use the mechanical law of particles. The motion of a particle is defined by its physical nature and by the position of other objects and particles in its neighborhood. In our specific case, we only consider the representation of facial surface. The facial surface is composed by a set of particles with mass density m. Their behavior is determined by their interaction with the other particles that define the face surface. In a correspondence with the geometric structure of the 3D face model, each point of the mesh corresponds to a particle in the physical model. To simulate elastic effects of facial skin tissue, we have used concepts from the theory of elasticity [7]. Each mass point is linked to its neighbours by massless springs of natural length non equal to zero. The elastic characteristics of the nonlinear springs is represented by the biphasic curve shown in Fig. 3.

The internal elastic force is the resultant of the tensions of the springs linking $\mathbf{x}_i$ to its neighbours:

$$Q_i = -\sum_{j \in \mathcal{N}_i} k_{ij} [\mathbf{d}_{ij} - d_{ij}^0 \frac{\mathbf{d}_{ij}}{\|\mathbf{d}_{ij}\|}] \tag{11}$$

where:

$\mathcal{N}_i$ is the set regrouping all neighboring mass points that are linked by springs to $\mathbf{x}_i$.

$\mathbf{d}_{ij} = \overrightarrow{\mathbf{x}_i \mathbf{x}_j}$.

d_{ij}^0 is the natural length of the spring linking $\mathbf{x}_i$ and $\mathbf{x}_j$.

k_{ij} is the spring stiffness of the spring linking $\mathbf{x}_i$ and $\mathbf{x}_j$.

$$k_{ij} = \begin{cases} k_L & \varepsilon_j \leq \varepsilon^c \\ k_H & \varepsilon_j > \varepsilon^c \end{cases} \quad (12)$$

The low-strain stiffness k_L is smaller than the high-strain stiffness k_H. Like real skin tissue, the biphasic spring is readily extendible at low strains, but exerts rapidly increasing restoring stresses after exceeding a strain threshold ε^c.

5 Dynamics of the Facial Expression Animation

Based on the FACS, various facial expressions can be created by the combination of the contraction of certain facial muscles. We have chosen 10 types of facial muscles, which are primarily related to facial expressions. Fig.5 illustrates the positions of these muscles. When facial muscles contract, the facial skin points that are in the influence area of the muscle model are displaced to their new positions. As a result, the facial skin points not influenced by the muscle contraction are in an unstable state, and unbalanced elastic forces propagate through the mass-spring system to establish a new equilibrium state.

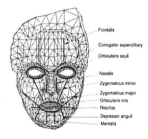

Figure 5: Definition of major facial muscles

Applying external forces to the deformable facial model yields realistic dynamics. In our face model, the net externally applied forces include the muscular force and the gravitational force. Based on the Lagrangian dynamics, the deformable facial model equations of motion can be expressed in 3D vector form by the second -order ordinary differential equation:

$$m_i \frac{\partial^2 \overrightarrow{\mathbf{x}_i}}{\partial t^2} + c_i \frac{\partial \overrightarrow{\mathbf{x}_i}}{\partial t} + Q_i = \sum_{j \in \mathcal{N}_i} f_{ij}(\overrightarrow{\mathbf{x}_i}) + m_i g \quad (13)$$

$$\overrightarrow{\mathbf{x}_i} = [x_i(t), y_i(t), z_i(t)] \quad (14)$$

where m_i is the nodal mass, c_i is the velocity-dependent damping coefficient that controls the rate of dissipation of kinetic energy which eventually brings facial mesh to rest, g is a gravitational constant, Q_i is the total elastic force on node i due to springs connecting it to neighboring nodes, and f_{ji} is the jth muscle's force applied on node i.

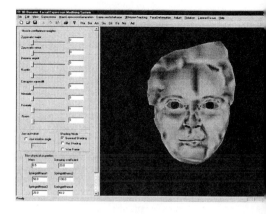

Figure 6: Interface of the facial expression animation sys

To simulate the dynamics of the deformable tissue mesh provide initial positions x_i^0 and velocities v_i^0 for each node numerically integrate the equations of motion forward thr time. In the simulation, we use a second-order Runge-k method [10] to integrate ordinary differential equations.

6 Experimental Results

We have developed a facial modeling system by V C++/OpenGL. It runs on an SGI540, PIII-Xeon 550MHz, 51 memory. Fig. 6 shows the graphical user interface to set the f parameters, such as the combination weight of each facial mu the spring stiffness and damping coefficient of deformable ti the rotation angel of the jaw and the shading modes. By the co panel, we can interactively adjust the values of parameters thr slide-bars or input the value directly. In the experiment, we the physically-based face model to dynamically generate some ical facial expressions. In each expression simulation, we s the major functional muscles based on the AUs of the FACS. 7 shows the dynamic deformation of face model on "Anger". 8 illustrates some primary expressions synthesized by the Gou shaded face model. They are placed side by side with the expressions generated by a man (1st author) for comparison. are also able to model several miscellaneous expressions by li weighted combination of several AUs. Some of them are sh in the Fig. 9. Note that all the synthesized expressions illust are generated dynamically. Regarding the system performance have achieved the simulation of 22 frames per second.

7 Conclusion

In order to realize heart-to-heart communication betweer man being and computer, realistic facial expressions shou synthesized as the communication media. In this paper we proposed a physically-based 3D face model for dynamic ex sion generation. The model has detailed structures of facial t and muscles which are modeled from anatomical perspective cial skin tissue is modeled by a nonlinear spring frame which simulate the elastic biomechanics of real facial skin. Two kin muscle models are developed to simulate real facial muscle traction. Lagrangian mechanics governs the dynamics, dict

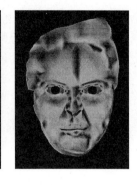
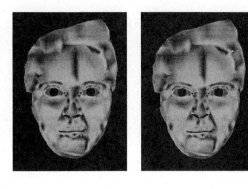

Figure 7: Dynamic change of the face in "Anger"

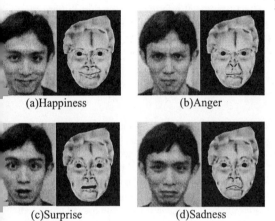

(a)Happiness   (b)Anger

(c)Surprise   (d)Sadness

re 8: Comparison of primary facial expressions gener-
by a man with those synthesized by 3D face model

Figure 9: Some miscellaneous expressions

deformation of facial surface in response to muscle forces. Ex-
nental results show our physically-based model can dynami-
create the flexible and realistic which can be matched to real
an examples.

References

[1] Cyriaque Kouadio, Pierre Poulin, Pierre Lachapelle, "Real-
time facial animation based upon a bank of 3D facial expression," *Proc. IEEE Computer Animation'98,* pages 128-136,
1998.

[2] S. DiPaola, "Extending the range of facial types," *Journal
of Visualization and Computer Animation,* 2(4): 129-131,
1991.

[3] P. Ekman and W. V. Friesen, *Facial Action Coding System
,* Consulting Psychologists Press Inc., 577 College Avenue,
Palo Alto, California 94306, 1978.

[4] Y. Fung, *Biomechanics: Mechanical Properties of Living Tis-
sues,* Springer Verlag, New York, NY, 1993.

[5] Jose Daniel Ramos Wey, Joao Antonio Zuffo, "InterFace:a
real time animation system," *Proc. IEEE Computer Animation'98,* pages 200-207, 1998.

[6] P. Kalra, A. Mangili, N. Magnenat-Thalmann, D. Thalmann,
"Simulaiton of facial muscle actions based on rational free
form deformatios," *Proc. EUROGRAPHICS'92,* pages 59-
69. Cambridge, 1992.

[7] L. D. Landau and E. M. Lifshitz, *Theory of Elasticity ,* Pergamon Press, London, UK, 1959.

[8] K. Mehrabian, J. Ferris, "Inference of attitudes from non-
verbal communication in two channels," *Journal of Consulting Psycology,* 3(31):248-252, 1967.

[9] F. I. Parke, "Parameterized models for facial animation,"
IEEE Computer Graphics and Application, 2(9): 61-68,
November 1982.

[10] W. H. Press, B. P. Fannery, S. A. Teukolsky, and W. T. Ver-
rerling, *Numerical Recipes: The Art of Scientific Computing,*Cambridge University Press, Cambridge, UK, 1986.

[11] T. Sakaguchi et al., "A scenario making tool for face expression animation and real-time motion image playback system," *Technical Report of IEICE,* HC91-57, pages 23-30,
1992.

[12] P. L. Williams, R. Warwick, M. Dyson and L. H. Bannister,
Grey's Anatomy, 37th Edition, Churchill Livingstone, London, 1989.

543

INTEGRATED MULTIMEDIA RECOGNITION AND MOTION SYNTHESIS FOR GENERATION OF 3D SIGN-LANGUAGE ANIMATIONS

Takao Ozawa

Department of Applied Mathematics and Informatics, Ryukoku University
Seta, Ohtsu-City, Siga, Japan   520-2194

ABSTRACT

In this paper we present an integrated text and image analysis followed by a motion synthesis for generating sign-language animations which are played by a puppet in the 3D virtual-reality space. From illustrative texts and pictures(line images) given in sign-language dictionaries we extract movement information, positional information, and shape information of head, hands(wrists), palms, and fingers. Based on the extracted information we determine head and hand movement data, palm directions, and finger/palm shapes. We then compute joint angles of the virtual puppet to generate sign-language animations.

1. INTRODUCTION

Linguistic descriptions of human motions often appear in novels, documents or radio broadcasting. Described human motions are made visual in dramas, plays or movies. Linguistic descriptions in manuals for dance, golf, sign-language, and so on, are especially intended to explain human motions. Unlike coded descriptions such as labanotation(Hutchinson, 1970) linguistic descriptions only are in general insufficient for recognizing the exact human motions, and they are supplemented by illustrative pictures or photographs. Men/women who want to learn dance, golf, or sign-language can somehow understand the motions by integrating information obtained from linguistic descriptions and pictures or photographs. We want to investigate the mechanism of such human multimedia recognition, and one of the objectives of our present research is to analyze and integrate the information expressed in linguistic descriptions and pictures or photographs and then to generate media of quite different nature, that is, sign motions or gestures.

In recent years many sign-language computer-animation systems which translate Japanese(as a natural language) into sign motions have been constructed in Japan(Terauchi and Nishikawa, 1987, Fujishige and Kurokawa, 1997). Practical 2D sign-language animation software packages are now sold in the market. Some of them are used to present notices at public places in sign-language by use of CRT displays, while some others to help people learn Japanese sign-language. In 1997 we built a sign-language animation system, where a puppet in a 3D virtual-reality computer space displays sign motions corresponding to the input Japanese sentences(Tawaraishi, Itani and Ozawa, 1997). Thus another objective of our research is to obtain motional data for the system from the illustrative texts and pictures(line images) given in sign-language dictionaries(Syuwa Communication Kenkyukai, 1992, Zen-Nippon Rohwa Renmei,1997).

2. DATA NEEDED FOR GENERATION OF 3D SIGN-LANGUAGE ANIMATIONS

First we summarize information which should be extracted from texts and pictures.

(1) Frame Information: As is well known, 3D motions of a virtual puppet are achieved by changing its joint angles. Incremental joint angles are obtained by interpolating joint angles for a postures in one frame(scene of animations) and those for a posture in the subsequent frame. For a simple motion the number of frames can be two(the initial frame and the final frame). For a complex motion such as a waving or rotating motion or for a repeated motion the number of frames becomes more than two.

(2) For each frame the following data are need. It is assumed illustrative pictures are drawn on the x-y plane. The z-axis is extended from the back to the front. (2.1) Head movement for calculating the neck joint angles. (2.2) Wrist positions in the 3D space and angles formed between the body and upper arms for calculating the joint angles at the shoulders and at the elbows(Note that these joint angles cannot uniquely be determined from the wrist positions only). (2.3) Directions of palms for calculating the wrist joint angles. (2.4) Finger shapes or direction of fingers for calculating the finger joint angles.

3. MOTION RECOGNITION AND SYNTHESIS

3.1 Recognition Policy

In general illustrative texts contain words describing hand and/or head movements and movement directions (such as "repeatedly move down", indicating wrist movements and directions) together with the relevant hands. In the text analysis such keywords essential in constituting sign motions are detected and the relations among them are examined. The illustrative pictures are line-images showing the upper half of the human body. The hand which plays a significant role in sign motions, is always shown. Thus we try to detect, from the image, the wrist positions(in general the initial positions of movements) and the angles formed between the body and upper arms. The descriptions regarding the positions of body parts together with positional directions("in front of the nose", "from the side of the ear", etc.), if any, in the texts are used to estimate the regions where the search for hands is to be done in the images. Texts often include descriptions for the relation between hands. Such descriptions are utilized to estimate the search region for one hand from that of the other.

Finger shapes are identified as patterns which are determined by combining the results obtained by the searches for them in the texts and images. For a finger shape pattern chosen from the FingerShapeTable a set of finger joint angles is retrieved from FingerAngleTable to generate the finger shape in the 3D space.

The processes of motion recognition and synthesis are illustrated in Figure 1.

3.2 Text Analysis

The major objective of the text analysis is to detect the expressions describing wrist movements and/or fingers shapes. The following tables are prepared for the text analysis

WristMovementTable: a table of verbs for wrist movements(e.g. "ageru(move up)", "sageru(move down)", "hiraku(open)") together with 3D vectors giving the movements.

FingerPalmMovementTable: a table of verbs for finger and/or palm movements which usually indicate finger and/or palm shapes(e.g. "nobasu(stretch)", "mukeru(direct or face)").

HeadMovementTable: a table of verbs for head movements(e.g. "unazuku(nod)", "huru(shake)").

PositionTable: a table of the positions of body parts(e.g. "kao(face)", "hana(nose)") together with their 3D coordinate values in the virtual space. Different values are registered for the right hand and the left hand.

PositionalDirectionTable: a table of positional directions indicating the relation to the positions(e.g. "ue(on)", "sita(under)", "yoko(side)") of body parts. The directions are given as normalized vectors in the virtual 3D space. Different values are registered for the right hand and the left hand.

MovementDirectionTable: a table of the directions of wrist movements(e.g. "ue-e(upward)", "sita-e(downward)") given as normalized vectors in the 3D space. Different values are registered for the right hand and the left hand.

PalmDirectionTable: a table of the directions of palms(e.g. "ue(upward)", "sita(downward)").

FingerShapeTable: a table of Japanese alphabet(hira-kana or kata-kana) and typical finger shapes. For each letter or finger shape a set of finger joint angles is registered in the FingerAngleTable.

MovementModifierTables: tables of modifiers for the movements of wrists, palms and fingers, respectively.

The steps of the text analysis are as follows.

TEXT ANALYSIS(after Morpheme Analysis)

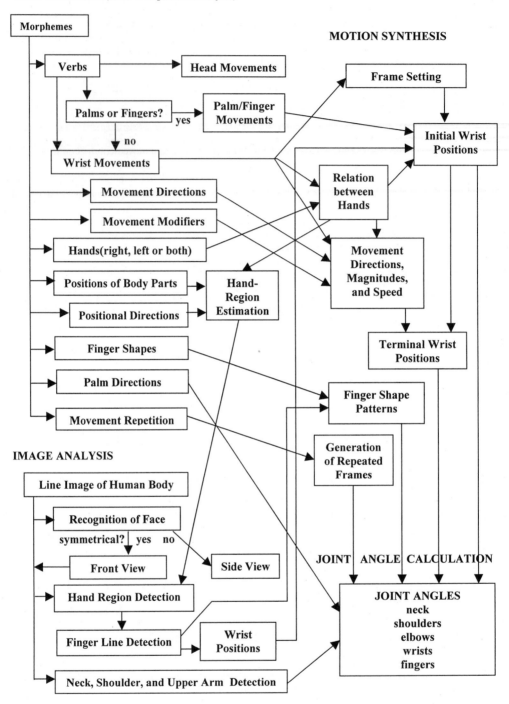

Figure 1. Flow diagram of text analysis, image analysis, and motion synthesis.

(1) The input text is decomposed into a sequence of morphemes. The analysis program "Chasen" developed by Y. Matumoto and his laboratory at the Nara Institute of Science and Technology(NAIST), Japan, is used.

(2) Verbs in the text(Finger/palm shapes are mostly given by verbs in the forms corresponding to the present or past participles in English.) are detected, and checked whether they are among the verbs registered in the movement tables. Certain verbs are listed in both of the tables for wrists and finger/palms. Which of the tables applies is decided by the appearance of "palm" or "finger" which immediately precedes or succeeds the verb.

(3) Appearance of "right hand", "left hand" or "both hands" is detected.

(4) Morphemes indicating the positions of body parts and positional directions are detected. Their relevance to the hands is checked and the 3D data of the positions are used to estimate the hand search regions in the image analysis.

(5) Morphemes indicating finger shapes and palm directions are detected.

(6) Morphemes indicating movement directions and movement modifiers are detected.

3.3 Image Analysis

The major objective of the image analysis is to find wrist positions, upper arm outlines, and finger shapes, if possible. The steps of the image analysis are as follows.

(1) First the face outline is detected. Both front views and side views may appear. Whether the image is a front or side view, is determined by checking the symmetry of the outline showing the face. At present only front views are analyzed in the following. The lines inside the face outline are removed.

(2) Congested lines are used to draw fingers. Thus by using the search region estimation and the information regarding the relation between hands obtained in the text analysis, two most congested rectangular regions are picked up as the hand regions. In each of the regions lines showing fingers are found by using Hough Transform. Based on these lines wrist positions, finger shape patterns, and palm directions(if possible) are determined.

(3) The lines showing the neck are found by searching the lines drawn downward underneath the face. Then the lines showing the shoulders and upper arms are found by tracing oblique lines extended from the bottom of the neck.

3.4 Motion Synthesis

(1) The descriptions in the text for the movements of hands, palms, and fingers appear in sequence. Thus concurrent movements which should constitute a single sign motion are listed in setting up the frames(scenes) of animations.

(2) As for wrist movements the x- and y-coordinate data of the initial positions are obtained from the image. Default z-coordinate values are used unless they are specified by the positional expressions in the text.

(3) The 3D data of the directions and magnitudes of the wrist movements are determined from the 3D data given in the tables corresponding to the verbs, the movement directions, and the movement modifiers specified in the text. By using these data together with the initial wrist positions, the final wrist positions are computed.

(4) The directions of palms and the shapes of fingers are determined as stated in section 3.1.

(5) If an expression indicating repetition of movements exists in the text, the data for the frame are repeated.

4. AN EXAMPLE AND EXPERIMENTAL RESULTS

Example(Figure 2: courtesy of Zen-Nippon Rohwa Renmei, 1997): The following is an illustrative text for "ame(rain)". R and L indicate right-hand and left-hand 3D data, respectively. Figure width=10.0, height=16.0

Karuku yubisaki wo hirogeta ryou   tenohira wo shita ni muke, chiisaku   kurikaeshi   orosu.

(lightly) (fingertips)   (open) | (both)   (palms) (downward) (direct) | (small) (repeatedly) (move down)

*Wrist movements and movement directions: "orosu"[move down] => R=(0.0, -1.0, 0.0), L=(0.0, -1.0, 0.0)

*Relevant hands: "ryou tenohira(both palms)" => both hands

*Wrist movement modifier: "chiisaku(in small magnitude)" => multiplication ratio=2.0

*Palm directions: "shitani muke(direct downward)" => R=(0.0, -1.0, 0.0), L=(0.0, -1.0, 0.0)

*Finger movement(shape): "karuku yubisaki wo hirogeta(slightly open (all) fingertips)" => movement ID=20

*Initial hand position estimation: in front of chest(default): R=(-2.0, -2.5, 1.0), L=(2.0, -2,5, 1.0)

*Initial wrist positions detected from the image: R=(-3.5, -2.0, 1.0), L=(3.5, -2,0, 1.0)

*Final wrist positions by calculation: R=(-3.5, -4.5, 1.0), L=(3.5, -4,5, 1.0)

*Repetition of frames: "kurikaeshi(repeatedly)"
=> generation of two more frames(default)

The first verb "hirogeta" is preceded by "yubisaki (fingertips)". Thus it is judged to be a verb for finger movements(shapes). The second verb "muke" is registered in FingerPalmMovementTable only. The third verb "orosu" is registered in WristMovement-Table only. From the expression "ryou tenohira(both plams)" it is judged that both of the hands are involved in the movement. No expression for the positions of body parts exists, and thus the default hand positions (in front of the chest) are used to estimate the hand regions. The magnitude of the wrist movements is (-1.0)*(2.0) = -2.0 in the y-direction. The y-coordinate value of the final wrist positions is -2.5 - 2.0 = -4.5.

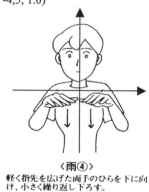

〈雨④〉
軽く指先を広げた両手のひらを下に向
け、小さく繰り返し下ろす。

Figure 2. Illustrative text and image for "rain".

Experimental Results: The analysis results largely depended on the input: if the illustrative text and image are given in the standard forms assumed in the text and image processing, sign motions which fairly agree with human understanding of the illustration, can be obtained. Since our text analysis is based mainly on detecting keywords in the texts, we can obtain desired results from texts which have very poor sentence structures. Verbs in the texts often fail to give exact hand movements, and it is very difficult to construct the details of the movements. Occlusion and absence of colors in the images make the recognition of shapes very difficult.

5. CONCLUDING REMARKS

Apparently our system needs further development to cope with the variety of input texts and images. In order to analyze texts containing simulative expressions it is necessary to implement knowledge-based data into the system. The motion synthesis process needs to include the knowledge of sign-language. We are now trying to detect arrows in the images which indicate the directions of the wrist movements. The detection is often very difficult, because many similar lines exist in the images.

REFERENCES

Hutchinson, A. (1970). Labanotation. New York: Theatre Art Books.

Fujishige, E. & Kurokawa, T. (1997). Japanese Language Processing for Japanese-to-Sign Language Translation. Human Interface, 12, 45-50.

Syuwa Communication Kenkyukai (Ed.). (1992). New Sign-language Dictionary. Tokyo: Tyuoh Hokki Syuppan.

Tawaraishi, Y., Itani, H, & Ozawa, T. (1997). Tools for Constructing 3D-CG of Hand Movements and Their Application to Making Sign-Language Computer-Animations. 1997 IEICE Japan Society Meetings Proc., A-14-2.

Terauchi, M. & Nishikawa, S. (1987). A Case Structure Based Translation System from Japanese Sentences with Analyzed Case Representation into Sign Language. Transactions A, IEICE of Japan, J79-A, 318-328.

Zen-Nippon Rohwa Renmei (Ed.). (1997). Japanese Sign Language Dictionary. Tokyo: Publishing Department of Zen-Nippon Rohwa Renmei.

CRT screens or TFT displays?
A detailed analysis of TFT screens for reading efficiency

Martina Ziefle

Department of Psychology, Technical University (RWTH) of Aachen, Jaegerstr. 17-19, 52056 Aachen, Germany.

ABSTRACT

Two experiments are presented that focus on worker productivity and oculomotor effort in screen reading. A CRT (cathode ray tube) and a TFT (thin flimmer transistor) display were compared regarding visual performance (search speed), oculomotor effort (fixation time and frequency) as well as rated eyestrain. Moreover, it was examined if current electronic displays can match the quality of paper. In experiment 1, the differences between paper, CRTs and TFTs for the above-mentioned measures have been addressed. Visual performance was best on paper, with the fastest search and fixation times and the fewest fixations per line, followed by TFT that was superior to CRT. In experiment 2, two CRTs at different refresh rates (100 and 140 Hz) were compared to TFT. Performance was best for the TFT display, followed by the CRT (with 100 Hz refresh rate) being better than the screen with 140 Hz refresh rate. Thus, with respect to visual performance, a clear order was found in the two experiments, with paper in the first place, followed by the TFT, and CRT. These results confirm that TFT technology might be an important step in screen technology, optimizing visual performance and minimizing eye strain. However, none of the current electronic displays seems to provide conditions under which the performance level of the paper condition was matched.

1 INTRODUCTION

Effects of electronically displayed text on visual performance have been an important research issue for quite some time now. Although there was remarkable technical progress regarding CRT screens (e.g. resolution, contrast, refresh rates), performance is still worse with modern CRTs as compared to hard copies (e.g. Ziefle, 1998, 1999, in press). Moreover, CRT users still report symptoms of eyestrain from more or less extensive on-screen viewing.

Working with TFT screens, some of these problems seem less acute. One of the most obvious advantages of this technology seems that TFT displays do not flicker. Flicker has been reported as one of the main causes of eyestrain and performance decrements observed in CRT screens (e.g. Harwood & Foley, 1987; Jaschinski, Bonacker & Alshuth, 1996; Ziefle, 1999, 2001). Ziefle (1999) showed that visual performance improves with increasing CRT refresh rates (50 up to 100 Hz), suggesting that there is a systematic and monotonic relation between visual performance and CRT refresh rates. However, in a recent publication (Kennedy, Brysbaert & Murray, 1998), saccadic accuracy was found to get worse in refresh rates over 100 Hz, hinting at a curvelinear relationship between visual performance and refresh rates. Apparently, the effects of flicker seem rather complex: On the one hand, information intake is affected by the flicker sensation (perceptual component) in lower (50-100 Hz) frequencies (Bauer, Bonacker & Cavonius, 1983). On the other hand, at high (> 100 Hz) refresh rates intermittent stimulation (physical component) is thought to interfere with saccadic control (Kennedy & Murray, 1993).

From an ergonomic point of view, these findings raise some fundamental questions for computer users: (1) Which screen type (CRT or TFT) leads to a higher productivity? (2) How high is the oculomotor effort in both screen technologies? (3) Do CRT refresh rates above 100 Hz provide for higher visual performance than refresh rates at 100 Hz? (4) Does any current screen match the performance of information intake on paper? In order to answer these questions, two experiments were carried out: Experiment I compares CRT, TFT, and paper and Experiment II compares two CRT conditions (100, 140 Hz), and a TFT screen.

2 EXPERIMENT I

This experiment was conducted to compare three display types: paper, CRT and TFT. Three basic questions were addressed: (1) Which display type provides for the best visual performance (2) Can TFT outperform CRT screens regarding visual performance? (3) Can any of the current electronic displays match the quality of paper?

2.1 Method

Variables: Independent variable is display type: (1) CRT (Elsa Ecomo, 17", 1024 x 768, 100 Hz); (2) TFT (14", 1024 x 768) and (3) paper (oriented vertically). To get a detailed insight into the nature of eyestrain and the underlying processes in information intake, a broad set of measures have been tested. Dependent variables were (1) search time (ms per line), (2) fixation time (the time the eyes need to extract information), (3) fixation frequency per line (the number of fixations that are carried out to scan one line) and (4) the subjective eyestrain in each display (the extent of eyestrain was rated after each condition on a 50-point scale, with 0 = no, 1-10 = very slight, 11-20 = slight, 21-30 = medium, 31-40 = strong and 41-50 = very strong eyestrain. Eyestrain was regarded as a complex sensation of different eye problems, i.e. burning, tearing and aching eyes. Additionally, participants indicated which display the preferred with respect to the given visual comfort.

Task: It is important to choose an experimental task that is able to detect different visual qualities of displays. Moreover, it has to be ensured that the task is visually rather than cognitive strenuous (since the latter could obscure visual effects by different cognitive reading strategies of the participants. Therefore, a continuous visual search task was adopted (e.g. Prinz & Nattkemper, 1986). Participants have to search for predefined target letters ("D" and "Z") scanning through list of letters (distractor letters are F, H, K,L, M, N, V, W, X). They were instructed, to search line for line, from left to right, as in normal reading and to do so as fast and accurately as possible. During searching, horizontal eye movements were recorded. Letter displaying (font, letter size, contrast) was equalized over displays and the room was illuminated by a flicker free bulb.

Participants: 16 participants took part in all three conditions, lasting 50 minutes, each. The presentation order of conditions was balanced over participants. All participants had (corrected to) normal vision (checked by a TITMUS-tester) and wore their corrective lenses during the experiment.

2.2   Results

Mean values were analyzed by a two way (display type as Factor 1, and time on task as Factor 2) repeated measures analysis of variance (ANOVA).

Search performance: On paper, search performance was best with 3800 ms/line. In the CRT, search time was increased by 500 ms per line (4440 ms)and in the TFT by about 270 ms per line (4068 ms) as compared to paper (figure 1). The effects of display type on search time were statistically significant (F (2,30) = 4.1, $p<0.05$). A post hoc contrast testing showed the CRT to be different from hardcopy (t = 3.0; p>0.01), whereas the TFT did not differ significantly from both. Moreover, there was a significant interaction between display type and time on task (Figure 2). After 50 minutes on screen viewing, the performance in the CRT got significantly worse and participants needed 300 ms more to scan one line as compared to the beginning of the search period (F (2,30) = 3.5; $p<0.05$).

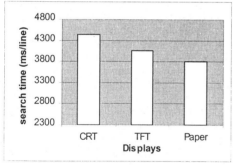

Figure 1: Mean search time (ms/line) for all three displays (CRT, TFT, and paper)

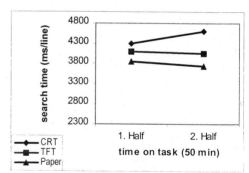

Figure 2: Mean search times (ms/line) for all three displays (CRT, TFT, and paper) as a function of time on task

Eye movements: The same picture was found regarding fixation time and frequency (figure 3 and 4). On paper, mean fixation time was shortest (337 ms) with the fewest number of fixations per line (8.6 fixations). In contrast, fixation time in the CRT was longest with 372 ms and 9.2 fixations per line. The oculomotor effort in the TFT screen was shown to lie between paper and the CRT, with a mean fixation time of 350 ms and 9.6 fixations per line. The effect of display type on fixation time was significant (F (2,30) = 9.6; $p<0.05$). Post hoc contrast showed that paper differed significantly from the CRT (t = 4.3; $p<0.05$) and from the TFT (t = 2.7; $p<0.05$). The difference between CRT and TFT was marginally significant (t = 1.8; $p<0.1$).

Additionally, a marginally significant interaction between display type and time on task was found for fixation times (F (2,30) = 2.8; $p<0.1$). On paper, fixation time significantly decreased after 50 minutes prolonged searching. In contrast, this was not the case in both screen conditions (figure 5). This decrease in fixation time can be regarded as a decrement in the oculomotor effort that reduces over time. Moreover, the advantage of paper was found with respect to the frequency of fixations per line, even if this interaction did not reach statistical significance. Over the search period, the number of fixations that were carried out scanning one line did not change (8.6 fixations in the first, 8.4 saccades in the second half). In other words, the oculomotor effort was kept constant throughout the search period of 50 minutes. In contrast, there was an inverse effect in both electronic displays over time. In the first 25 minutes, CRT and TFT had the same starting point with 9.2 fixations per line. After 50 minutes, however, the oculomotor effort raised with an increment of more than one fixation per line in the CRT and an increment of 0.7 fixation per line in the TFT (figure 6).

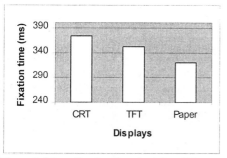

Figure 3: Mean fixation time (ms) for all three displays (CRT, TFT, and paper)

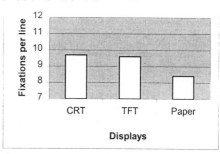

Figure 4: Mean number of fixations per line for all three displays (CRT, TFT, and paper)

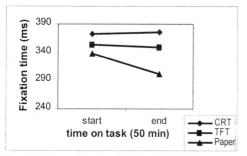

Figure 5: Mean fixation time (ms) for all three displays (CRT, TFT, and paper) as a function of time on task

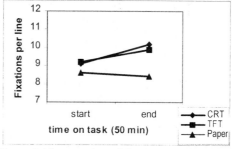

Figure 6: Mean number of fixations per line for all three displays (CRT, TFT, and paper) as function of time on task

Subjective measures: Concerning the ratings of eyestrain and visual comfort, the picture could be completed, even if the differences did not reach statistical significance: the display, that was most preferred (10 out of 16 participants) was paper that was moreover judged to cause the smallest eyestrain. However, CRT and TFT displays did not differ with respect to the preference ratings.

3   EXPERIMENT II
In the second experiment, two CRT screen conditions (refresh rates of 100 Hz and 140 Hz) were compared to TFT. This comparison might help to decide where future research should be directed to: higher refresh rates in CRTs or further development in TFT technology. As in experiment I, the same visual search task was used and the same set of measures have been adopted. In contrast to experiment I, the search period was only 40 minutes for each display.

3.1   Method
Variables: Independent variable is display type: (1) CRT 100 Hz (Iiyama Vision Master Pro, 19", 1024 x 768); (2) CRT 140 Hz and (3) TFT (Iiyama Prolite 39 a, 15.4", 1280 x 1024). Dependent variables were (1) search time (ms per line), (2) fixation time (3) fixation frequency and (4) the subjective visual comfort (preference) and eyestrain.

Participants: 19 participants took part in the three screen conditions, lasting 40 minutes, each. Again, presentation order was balanced over participants that all had (corrected to) normal vision.

3.2   Results

As in Experiment I, mean values were analyzed by a two way repeated measures analysis of variance (display type as Factor 1, and time on task as Factor 2).

Search performance: As can be seen from figure 7, search time was best in the TFT screen (2496 ms per line) and worst in the CRT with 140 Hz (2794 ms per line). Mean search time in the CRT with 100 Hz ranged between the other displays with 2710 ms per line. The TFT search time was better by 7.8% as compared to the CRT with 100 Hz and 10% as compared to the CRT with 140 Hz. Figure 8 shows the interaction between display type and time on task. In the TFT, search time was comparable in the first and second half of the search period, whereas in both CRT conditions search time slightly increased over time. However, search time differences between displays failed to reach statistical significance. Since high standard deviations between participants (32-37%) can be made responsible for that, a post hoc analysis was carried out. In order to reduce between-subject variation, two subgroups were selected: fast readers (n = 6, upper third) and slow readers (n = 6, lower third). In both subgroups, the same pattern of results (TFT > CRT 100 Hz >CRT 140 Hz) was found, thus confirming the validity of results within the total group. For slow readers, search speed was affected significantly by display type (F (2,10) = 4.44; p<0.05), with the TFT to be significantly different from the CRT with 140 Hz (t = 3; p<0.01 and by tendency from the CRT with 100 Hz (t = 1.7; p<0.1). Both CRT screens did not differ significantly from each other.

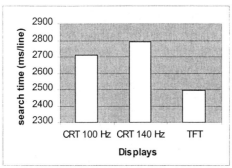

Figure 7: Mean search time (ms/line) for the three displays (CRT 100 Hz, CRT 140 Hz and TFT)

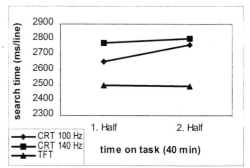

Figure 8: Mean search time (ms/line) for the three displays (CRT 100 Hz, CRT 140 Hz and TFT) as a function of time on task

Eye movements: As can be seen from figure 9, fixation time was shortest in the TFT screen with 247 ms, followed by the CRT with 100 Hz (259 ms) and by the CRT with 140 Hz (268 ms).

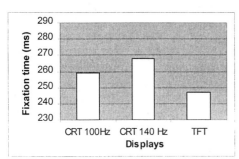

Figure 9: Mean fixation time (ms) for the three displays (CRT 100 Hz, CRT 140 Hz and TFT)

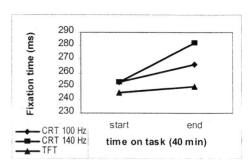

Figure 10: Mean fixation time (ms) for the three displays (CRT 100 Hz, CRT 140 Hz and TFT) as a function of time on task.

The TFT superiority ranged by 11.7% compared to the CRT with 140 Hz and by 9.7% compared to the CRT with 100 Hz. This effect of display type on fixation time was significant (F (2,36) = 8.3, $p<0.05$).

Moreover, fixation time was affected by time on task (F (2,36) = 18.5; $p<0.01$). In other words, fixation time was found to increase over the 40 minutes search period. However, this increment over time differed in its extent between displays, thus resulting in a significant interaction of display type and time on task (F (1,18) = 4.2; $p<0.05$). In the TFT, fixation time did not change over time (1. half: 245 ms; 2. half: 250 ms). In contrast, there was an increment in fixation time over time in both CRT conditions. In the first 20 minutes, mean fixation time is 253 ms in both CRTs, with an increment of 11.5% (282 ms) in the CRT with 140 Hz and an increment of 5% in the CRT with 100 Hz (266 ns) after 40 minutes on screen working. Apparently, the oculomotor effort increased as a function of CRT flicker and time on task. The same result was found in the number of fixation per line (CRT 140 Hz: 10.2, CRT 100 Hz: 9.6 and TFT: 9.3 fixations per line). Even if the results are in the expected direction, no statistically significant effect could be revealed (F (2,36) = 2.34; $p>0.05$).

Subjective measures: Concerning the ratings of eyestrain, the TFT was the display with the smallest amount of eyestrain symptoms, followed by the CRT with 100 Hz and the CRT with 140 Hz.

4   CONCLUSIONS

The focus of the paper was to evaluate current screen types with respect to worker productivity, visual performance, oculomotor effort, and eyestrain. Thus, state-of-the-art CRTs and TFT displays were compared. Moreover, it was tested, if the superiority of paper in text displaying is reached by electronic displays. Comprising the results of both studies, it can be clearly stated that the TFT technology is superior to CRT screens. Visual performance is better by 8-12% as compared to CRT screens (summarized over the different measures). However, participants did not prefer TFT screens over CRT screens as clearly as they preferred paper over electronic displays. One problem seems to be the restricted viewing angle (anisotrope lighting). Also, the not yet so sophisticated anti-aliasing type technologies in TFT screens sometimes make letters appear somewhat "ragged" and "unsharp". Another issue was the comparison of CRTs at different refresh rates at and over 100 Hz. Remarkably, there was an advantage of the 100 Hz screen over the 140 Hz display. This finding might seem odd at first, since it is generally believed that higher refresh rates provide a better productivity. Even if this is true for refresh rates up to 100 Hz, it is clearly not for refresh rates over 100 Hz. Apparently, effects of refresh rates on visual performance seem to follow a curvelinear function, confirming the findings of Kennedy et al. (1998). However, paper is still superior to electronic displays: the gain in visual performance (search speed and oculomotor effort) is 17% compared to the CRT and 10% compared to the TFT. This advantage shows after only 50 minutes of screen viewing and enhances over longer reading periods. Furthermore, paper constantly received the highest preference ratings by the participants.

Further experiments have to evaluate future improvements in TFT-technology. Even with the TFT screens' problems in displaying letters, the present results suggest that this technology is an important step in electronic displaying, optimizing worker productivity and minimizing eyestrain.

REFERENCES:

Bauer, D., Bonacker, M. & Cavonius, C. (1983). Frame repetition rate for flicker-free viewing of bright VDU screens. *Displays, 4*, 31-33.

Harwood, K. & Foley, P. (1987). Temporal resolution: An insight into the video display terminal (VDT) "problem". *Human Factors, 29*, 447-452.

Jaschinski, W., Bonacker, M. & Alshuth, E. (1996). Accomodation, convergence, pupil diameter and eyeblinks at a CRT flickering near fusion limit. *Ergonomics, 39*, 152-164.

Kennedy, A. & Murray, W. (1993, September 16). Pulsation-induced disturbance to eye movement control in the absence of perceived flicker. *Nature, 365*, 213.

Kennedy, A., Brysbaert, M. & Murray, W. S. (1998). The effects of intermittent illumination on a visual inspection task. *Quarterly Journal of Experimental Psychology, 51*, 135-151.

Ziefle, M. (1998). Effects of display resolution on visual performance. *Human Factors, 40*, 54-568.

Ziefle, M. (1999). Bildschirm oder papier - determinanten der leseleistung im medienvergleich [Screen versus paper-determinants of reading performance]. In W. Hacker & M. Rinck (Eds.), *Zukunft gestalten* (pp. 592-604). Lengerich, Germany: Pabst.

Ziefle, M. (2001). Kritische effekte des flimmerns auf die visuelle leistung an selbstleuchtenden medien [Critical effects of visual performance in self-illuminated displays]. *Zeitschrift für Arbeits- und Organisationspsychologie, (45)*, 15-25.

Ziefle, M. (in press). *Visuelle faktoren bei der informationsentnahme am bildschirm* (Visual factors in information processing on CRT screens). Münster, Germany: Waxmann.

Pictures in Mobile Communication

Dr. Pekka Isomursu* and Dr. Minna Mäkäräinen**

*Nokia Corporation, P.O.Box 50, FIN-90571 Oulu, Finland. Pekka.Isomursu@nokia.com
**Solid, Elektroniikkatie 8, FIN-90570 Oulu, Finland. Minna.Makarainen@solidtech.com

ABSTRACT

This paper explores various ways of using pictures in mobile communication. Starting from the general classification of signs by Peirce, we identify different picture classes for mobile communication. For each class we identify the main areas of usage, and discuss the motivation for why the pictures are used in each way. The identified classes and their usage help us to understand the behavior of the user. We can better understand her motivation for purchasing commercial images, using a camera, willingness to pay for sending a picture message etc.

1. INTRODUCTION

The new generation of mobile devices support picture messages which are gaining popularity, especially among young users. Current picture messages are typically symbols, which can be used alone or as an accompaniment to a text message. The most advanced mobile terminals provide the opportunity to send digital images. Further development of current cellular networks and the emergence of the so-called third generation networks will soon enable fast transmission of digital images. This, together with inexpensive digital cameras, will enable new ways of using pictures in mobile communication, provided that the resolution of the terminal screens improve as well.

The use of visual information in controlling the mobile terminal is increasing as well. The user interface (UI) often includes icons with texts to illustrate the function in question. Animations can further advance the icons' clarity. Yet another use for pictures is the personalization of terminals. Several popular services enable the user to download pictures for display on the terminal screen when a certain person calls, or while the phone is in idle mode.

2. ANALYSIS FRAMEWORK

According to Hietala [1] the general classification of signs by Peirce [2] can also be used to classify pictures:
1. Iconic picture, which represents its subject by resemblance (e.g. a photograph on a driver's license).
2. Index picture: the picture and its subject have a commonly known continuity (e.g. hooves print and a horse).
3. Symbolic picture: the meaning of the picture has been agreed and defined by society (e.g. the national flag of a nation represents that nation).

A single picture can be interpreted to belong to any of these categories, depending on the context and cultural knowledge of the interpreter. For example, a picture of the President of Finland can be iconic when it identifies her on her driver's license. The same picture 100 years from now printed in a history book might act as an index to give the reader information about the Finnish society today, when the first female president is in power. Today, school children in Finland see her picture on the classroom wall as an icon of the Finnish nation and society.

As shown later, Peirce's classification does not fit well into certain pictures whose primary nature is to arouse emotional pleasure by being aesthetically pleasant. Whether or not they represent a subject matter, these pictures can be interpreted as constructions of colors and shapes, having no sign value. The purest examples of such pictures are the works of abstract art. The contradiction with Peirce's classification is natural, since Peirce originally classified signs, and by definition a sign always refers to an underlying concept (e.g. [3]). Thus we add a fourth class:
4. Aesthetic picture, whose primary purpose is simply to look pleasant (e.g. abstract art).

We will go on to study the use of these four picture classes within each of the following usage categories:
A. Pictures in user-to-user communication. Visual messages transmitted between users of mobile terminals.
B. Pictures in user-device interaction. Visual information that helps the user in controlling the mobile terminal.
C. Pictures of decorative nature. Decoration to give the terminal a personal touch.

3. PICTURES IN USER-TO-USER COMMUNICATION

ICONIC PICTURES

The iconic use of pictures is evident in scenarios where the mobile terminal replaces personal identification documents, such as a passport or a driver's license. The terminal screen would display the picture and identification information of the person in question, and could be used in a similar way to the documents of today. A picture with identification information, such as name and social security number, can either be displayed on the screen of the person's own device, or it can be sent electronically to the requestor, and displayed on her screen. However, the authenticity of the electronic image has to be confirmed (compare with e.g. UV ink, watermarks, stamps, and special paper in a passport). This digital identity can be secured with a Wireless Identification Module (WIM) in the mobile terminal, and confirmed by connecting to the server of the authority that has certified the user's identity. This authority can be e.g. a government agency for passports, driver's licenses and other official documents, or a commercial company such as a bank or a network operator for credit cards, bank cards and similar.

A picture as personal identification could be replaced by more advanced identification methods, such as fingerprint recognition or eye scanning. However, as these identification methods are technically more complex and typically require expensive technical devices, we do not believe that they will totally replace iconic pictures in personal identification. Additionally, for a human controller (e.g. a customs officer) it is intuitively an assuring method of identification.

Iconic pictures can be used as reminders. Paper format business cards could be replaced by electronic business cards, which include a picture of the card owner so that the receiver can better remember her face in the future. An electronic shopping list with pictures could be used to help remember which particular items to buy. Having a photograph of, for example, a carton of yogurt, would be much more helpful, faster to take and more convenient to read than describing a specific type of yogurt in words to another person going to the shop for you. The pictures can be taken by oneself or from an outside source, e.g. a sales catalogue. Before going shopping, pictures from a catalogue could also give information to the user and help her in the selection process.

As discussed earlier, whether one can trust the picture is a very important question in the case of digital identity. When a picture is used as decoration or as a reminder, securing the picture's authenticity is not an issue. But whenever a picture is in electronic format, people more automatically think of the possibility of picture manipulation. The security aspects of the electronic media are perhaps not yet as advanced as the security features of printed products, and certainly not as well known to the general public. Thus the trust people have in electronic pictures seems to be less than in traditional, printed documents.

INDEX PICTURES

When fast image transmission becomes reality, still pictures and even video images can be used as real-time (or near real-time) electronic postcards, documenting events and activities. In business use, video meetings become common, or maybe just sending a still image of the participants to the other end.

The question of trust is also relevant here. However, the motivation for an individual manipulating a postcard-type index picture clearly differs from the motivation for manipulating an iconic picture used for identification. If postcards are sent to close friends and relatives, there usually is no reason for the sender to manipulate and for the receiver to question the authenticity of the image. However, picture manipulation might become more common for making visual jokes, such as inserting a picture of oneself into an Egyptian sarcophagus and sending that image as a postcard from Egypt. The purpose would not be to cheat the receiver to think that the person actually lay in the sarcophagus of Tutankhamun, but to transfer a visual joke along with the greeting. Another and more serious reason to cheat would be when the sender wishes to deceive the receiver about his/her whereabouts. The existence of companies that provide alibis for people cheating their spouses [4] indicate that this kind of cheating may well happen.

SYMBOLIC PICTURES

Pictures can arouse very strong emotions (e.g. sadness, joy, happiness, even sexual passion) in the viewer. Words, on the other hand, are good for representing intellect [5] [1, pp. 22-23]. Even though actual picture messaging has been introduced only recently in mobile communication, the idea of symbolic picture messages has been around almost as long as text based messaging. Symbolic pictures have been created from textual characters in order to increase the emotional value transmitted with the message. At the beginning of the '80s, when the usage of Internet e-mail grew rapidly, the e-mail culture created the widely used 'smiley' symbol (i.e. the characters :-)) [6] and its numerous variations to express emotions in text messages. In mobile phones, the length of a text message was originally limited to 160 characters. This created a need to write short, economical messages. The original smiley, which requires the space of three characters, can be replaced with the character Ü, which takes up the space of only one character. The same technique of constructing images from textual characters has been widely used for constructing other, often more complicated emotionally meaningful symbols. Figure 1a illustrates a text message showing a picture of a teddy bear. This picture carries a lot more emotional value than the words "teddy bear" do.

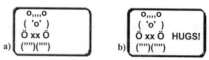

Figure 1. a) A text message consisting of a symbolic picture. b) The same with an added textual message.

Consider the effect of adding the word "HUGS!" next to the teddy bear picture (Figure 1b). The combined message is far more expressive than either of its elements alone. The picture arouses positive, loving emotions, and the text intellectually states that the sender cares for the receiver. The elements of the message in Figure 1b are not independent. This phenomenon is similar to what L. Kuleshov has demonstrated in filmmaking. When he sequentially combined the image of an actor with images of death, a playing child, and a plate of soup, the viewers felt that the same image of the actor expressed sorrow, hunger and affection respectively [7]. When words and images are combined, the image provides emotion to the intellectual message of the word(s).

Recently introduced picture message features in mobile terminals provide a more advanced means for creating pictures: The pictures are not limited to textual characters, but can be constructed as graphical objects. However, the current limitations in display technology limit the use of picture messages to very simple pictures (typically monochromatic, small size and low resolution), and therefore hinder their iconic or index usage. With graphical images, the picture creation process becomes more complicated than with textual characters. This creates an opportunity for commercial services offering symbolic, downloadable pictures. Among others, cellular network operators Orange [8] and Sonera [9] already offer such services.

In the near future digital cameras will become common both as separate items as well as integrated to mobile phones. The users will mainly use their cameras for the creation of index pictures, but we believe that symbolic pictures will still remain popular after this invasion, since they serve a different purpose. A majority of the users would not be able to or would not bother to create symbolic pictures by themselves (as observed by Keskinen [10] from usage of ASCII art by e-mail users). When using symbolic pictures it is more convenient for them to refer to ready-made libraries provided e.g. by the above-mentioned commercial services.

AESTHETIC PICTURES

Abstract, emotionally meaningful pictures are usually used in decoration, but they can also be used as greetings (a.k.a. 'electronic postcards') similarly to index pictures above in Chapter 3. However, we think that the use of aesthetic pictures in communication is rather rare and people prefer using pictures with sign value. This is because the sender and the receiver more easily interpret the emotional message the same way when it is not dependent on the aesthetic preferences of each person. The use of aesthetic pictures in mobile communication today is virtually non-existent, since it is difficult to create works of purely aesthetic value with current screen technology.

4. PICTURES IN USER-DEVICE INTERACTION

Pictures in user-device interaction are used for illustrating the functions provided. These pictures are called icons (not to be confused with the term 'iconic pictures'). The iconic UI follows the graphical UI language introduced in personal computers. An icon usually accompanies textual information, thus giving some extra explanation but not being the only means of control. The importance of icons is emphasized on mobile terminals because their small screens make long textual explanations cumbersome.

ICONIC PICTURES

The use of iconic pictures is currently rare in UIs. However, they can be used as user-specific UI items, such as entries in the phonebook. A photograph of the person in question can be used in identifying the person to be contacted. This feature can be very valuable, for example, for children who cannot yet read.

INDEX PICTURES

For the most part, pictures in a UI are index pictures, since their purpose is to illustrate the function provided. Index pictures provide a causal relationship between the picture and the action to be performed. For example, the picture of a microphone leads the user to think about voice recording, or the picture of a letter and a mailbox refers to sending a written message.

If the UI could be purely iconic, i.e. rely entirely on the use of index pictures, the need for translating the UI texts to different languages would disappear. However, as Hietala [1] states, the interpretation of pictures is culture-specific as well. Although an iconic UI would not remove the need for localizing the UI, it would reduce the amount of localization work. For example, if the UI requires the user to select between "man" and "woman", the localization for each language would require much more work than the localization of the icons representing a man and a woman. People speaking different languages can relate a picture of a woman to a real life phenomenon called woman, provided that the picture shares the visual signs of a woman in that particular culture. The groups of people sharing the same visual background are usually larger than people sharing the same language. While helpful to almost all users, iconic UI can be especially valuable for illiterate people, such as young children or people who have had little or no education.

Replacing all text in the UI with icons would require profound changes in the entire interface structure. Currently used menu-based UIs rely on abstract expressions, which cannot be effectively replaced with icons. For example, an expressive icon for the menu item "Settings" can be very difficult to define. However, even an illogical picture can be easier to memorize than a string of text, especially if the user does not understand the language used. One advantage of words over pictures is that words represent an abstraction of the real life phenomenon, while the picture always defines some specific features of the phenomenon, making it more specific. For example, the word "light" can mean a wide variety of real life phenomena, whereas the picture of a torch specifies the light source. (Notice, however, the earlier discussion of using a picture of a woman as a language-independent sign for "woman".)

SYMBOLIC PICTURES

UIs also utilize symbolic pictures. These pictures have no direct causal relationship with the action, but the relationship has been established by a common agreement. A well-known example is the symbol of a power switch (i.e. ①). Other examples of symbolic pictures are the user-defined symbols known as caller graphics. They are displayed on the screen when a certain person calls. An example of such symbol is given in Figure 2.

Figure 2. An example of caller graphics [9]

AESTHETIC PICTURES

Since the primary purpose of a picture in a UI is to illustrate the action provided, its purpose is never purely aesthetic. However, it still may have an aesthetic dimension as well. For example, although the main purpose of a photograph in a phonebook entry is to provide a recognizable image of the person it represents, most people would prefer the photo also to be aesthetically pleasant.

5. PICTURES OF DECORATIVE NATURE

Westrum [11] divides user needs into functional user needs and emotional user needs. Products must serve the functional needs but that is not all: they can also arouse emotions and provide pleasure. A device is emotionally pleasant when its appearance or use arouses aesthetic pleasure, or if it arouses positive associations. Users seem to get emotional pleasure from having pictures, especially personally selected pictures, in certain situations.

ICONIC PICTURES

Photographs of one's family, pets and other intimate subjects are very popular decorative items. In mobile phones they can be used e.g. as a print in the cover, or displayed on the screen. Note, however, that although the iconic nature of these pictures is vital, they also possess a strong symbolic value to the user. Thus, they can also be interpreted as symbolic pictures: a picture of the wife and kids symbolizes family values, although it is also very important that the wife and kids are clearly recognizable in the picture (= iconic picture).

INDEX PICTURES
Although used primarily for decorative purposes, decorative pictures can also provide some causal relationship between the product features and the picture. For example, the use of colors and shapes can give the user hints on how to hold the device. A mouth painted over the microphone would guide the user to talk in the right direction.

SYMBOLIC PICTURES
Services providing pictures for personalizing the mobile terminal have become very popular during the last couple of years. These pictures are primarily symbolic pictures, and they are chosen on the basis of the personal interests and preferences of the user.

A changeable, personalized cover on a mobile phone helps the user separate her or his own device from others of a similar type, and give the phone a personal touch of its owner. Thus, personalization both helps the owner better identify the product and communicate her interests and image to other people.

To personalize the terminal display, the user can download a picture that can be shown when the phone is not in use. This function is used much for the same purpose as the changeable mobile phone covers: As a means for personal expression and identification. An example of a downloadable symbolic picture is given in Figure 3.

Figure 3. A sample downloadable picture that can be displayed when the phone is not in use [9].

AESTHETIC PICTURES
Decorative pictures do not always have a sign value. Sometimes they are chosen based only on their aesthetic value. For example, abstract shapes and patterns are very popular both in phone covers and downloadable screen pictures. However, sometimes even abstract pictures may, to some extent, have some symbolic value. The user might signal some kind of 'artistic mindset' by choosing an abstract and artistic outlook for her phone.

6. FUTURE WORK

In this paper we have concentrated on still images. However, in our research we have identified scenarios for the use of moving images as well (video calls, video messaging, animations, augmented reality and virtual reality) and we have discovered that the classification of moving images is not natural to the categories introduced in this paper. We intend to study this further in the future. We also intend to study further the identified classification of still pictures, e.g. we plan to analyze how the screen size will affect the use of pictures in the terminal UI.

ACKNOWLEDGEMENT: The writing of this paper was inspired by lectures on Film Studies given by Mr. Pasi Nyyssönen at Oulu University, Department of Art Studies and Anthropology, which is gratefully acknowledged.

REFERENCES
All reference retrievals from the World Wide Web (WWW) were made or verified November 2, 2000.
1. Hietala, V. (1993). Kuvien todellisuus. Helsinki: Gummerus. (in Finnish)
2. Justus Buchler (ed.). (1955). Philosophical writings of Peirce. Dover Publications.
3. de Saussure, F. The Linguistic Sign. In: Innis, R.E. (ed) 1985. Semiotics. Bloomington: Indiana University Press.
4. Alibi co. (2000). WWW: http://www.alibi.co.uk/indexuk.html
5. Freedberg, D. (1989). The Power of Images. The University of Chicago Press, Chicago. 1989
6. Meuronen, P (1999). ASCII taide. WWW: http://media.urova.fi/~pmeurone/ascii/index.html (in Finnish)
7. Pudovkin, V.I.:Film Technique and Film Acting. Ivor Montagu (trans.) 1978. Grove Press, New York.
8. Orange. (2000). WWW: http://www.hutchnet.com.hk/Mobile/orange/o_picture/eng_picture.html
9. Sonera. (2000). WWW: https://www.sonerazed.fi/palvelut/doris
10. Keskinen, M. (1997). Verkon silmä ja korv@: kirjoitus ja puhe kasvokkain sähköpostikommunikaatiossa. In: Mikkonen K., Mäyrä, I. Siivonen, T. (eds.): Koneihminen. Atena. (in Finnish)
11. Westrum, R. (1991). Technologies & Society. The shaping of people and things. CA: Wadsworth Publishing.

MICRO TELEOPERATION SYSTEM CONSIDERING VISUAL REGISTRATION

Akihito Sano, Hideo Fujimoto and Tomoaki Kitagawa

Department of Mechanical Engineering
Nagoya Institute of Technology
Gokiso-cho, Showa-ku, Nagoya 466-8555 JAPAN
{sano, fujimoto}@vier.mech.nitech.ac.jp

Abstract

In this study, a novel micro teleoperation system is discussed. In order to recognize or understand the master-slave manipulators as a tool, their motions should be equal in the 3D space reconstructed by fusing the stereo image. In this paper, a feature-based and adaptive visual servo is proposed as one of the superior methods to realize the visual registration. Especially, the feature-based scheme uses an object image feature directly without computing the object 3D information. The validity of the proposed method is demonstrated by micro teleoperation experiments.

1   INTRODUCTION

It is very important to develop effective and suitable methods to easily execute a micromanipulation and reduce the exhaustion from many hours of operation. Recently, many new potential uses of telerobotics and virtual reality have been explored in medical field [1]- [4].

Taylor et al. [1] have developed a telerobotic assistant for laparoscopic surgery. Sano et al. [2] have proposed a master system with a built-in slave by imaging a kind of compact tool which can be treated easily at the fingertip. In these systems, a part of the transmission and the display of information at high level can be exempt. The reason is that the doctor can operate beside the patient and follow the present operating style.

On the other hand, Green et al. [3] have proposed a telepresence surgery system. Recently, a surgical teleoperator, known as daVinci system, has been designed to provide enhanced dexterity to doctors performing minimally invasive surgical procedures [4]. DaVinci is descended from the Green's system. In these systems, the doctor performs virtually the operation from the surgeon's console.

The development of visual and haptic interfaces, and accurate registration are essential in order to allow a human operator to easily and naturally perform a micro teleoperation. However, these technical subjects are not easy to achieve and have not been seriously considered. Yokokohji et al. [5] have proposed a new concept of visual/haptic interfaces called a WYSIWYF (What You can See Is What You can Feel) display. In this study, the novel micro teleoperation system, named *Micro Dome*, is discussed. In the visual interface, the visual servo scheme is used to get correct visual registration. An alignment of the motion of master to the motion of slave is both visual and spatial.

2   MICRO TELEOPERATION THROUGH MICRO DOME

In this study, the virtual master and real slave system as shown in Fig.1 was developed. The operating environment, named *Micro Dome*, is created. At present, we speed up the development of real spherical display. Micro Dome provides a seamless stereo image and allows the operator to collaborate with the assistant. The parallel mechanism with 3 degrees-of-freedom (3 translations) was adopted as the master-slave manipulators. In the microsurgery, since the range of operation is comparatively small, the application of parallel mechanism that has several superior features such as high accuracy and high

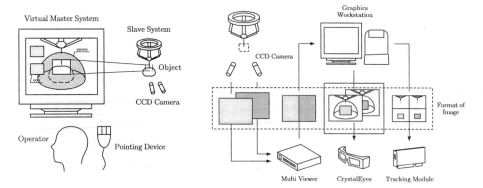

Fig.1: Virtual master and real slave

Fig.2: Flow of image data

stiffness is sufficiently feasible. The CCD cameras (ELMO, UN411) are driven by the step motor in order to look-around the slave environment.

A flow of image data is shown in Fig.2. The video signals from two CCD cameras are sent to the multi-viewer (FOR.A, MV-10D) and are compressed into a video signal. Next, this combined video image is sent to the graphics workstation (O2) and is stored in the memory in a format that is compatible with the texture of Open GL by utilizing the function of video library and DM buffer.

In this system, there are two flows of image data. One is displayed on the monitor of O2 by rendering into the frame buffer (stereo buffer). Another is transferred to the video terminal of O2 by off-screen rendering. The former creates the operational environment in which the operator observes 3D objects by using CrystalEyes2 (Stereo Graphics) and executes the tasks. The real image from the slave site is put on-line in the window that is opened on the virtual dome using the texture mapping procedure. On the other hand, the latter is sent to the color vision module (Fujitsu, TRV-CPW5) and is utilized for tracking and sampling the image feature points which are input signals for the controller of visual servo system as mentioned in the following sections.

3   REGISTRATION BASED ON ADAPTIVE VISUAL SERVO

In order to recognize or understand the master-slave as a tool, their motions should be equal in the 3D space reconstructed by fusing the stereo image as shown in Fig.3. In this study, the visual servo is proposed as one of the superior methods to realize this visual registration. Especially, the feature-based scheme uses the object image feature directly without computing the object 3D position and orientation. Furthermore, by using stereo image data, the tracking in a depth direction (a vertical direction of image plane) is possible.

The image format is divided into four as shown in Figs.2 and 4. The stereo images are arranged on both sides, and the virtual master tool and the real slave tool are arranged on upper side and lower side respectively. As seen from Fig.4, the tips of the slave tool and the tips of the master tool (left and right view) are selected as the image feature points and the references respectively.

Based on a perspective projection, a relationship between the velocity $\dot{\theta}$ (motion) of slave system and the velocity $\dot{\mathbf{x}}$ of image feature can be represented:

$$\dot{\mathbf{x}} = J_{image}J_{robot}\dot{\theta}, \quad J_{image} := \frac{\partial \mathbf{x}}{\partial \mathbf{p}}, \quad J_{robot} := \frac{\partial \mathbf{p}}{\partial \theta} \tag{1}$$

where, θ is a parameter vector which describes the joint variables of the arm. $\mathbf{x}$ is an image feature

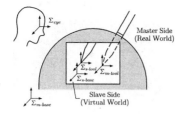
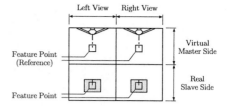

Fig.3: Visual and spatial alignment   Fig.4: Image feature points and references

vector. $\mathbf{p}$ is a vector that describes position and orientation of the endeffector (tip of slave). J_{robot} is a Jacobi matrix of manipulator. Now, $J_{image}J_{robot}$ is replaced by J. J called an image Jacobi matrix or an image Jacobian.

The system containing the CCD cameras is expressed as follows:

$$\dot{\mathbf{x}} = J(\theta)\dot{\theta} \tag{2}$$

In this study, the number of image features is set to be equal to the number of system variables, so that $J(\theta)$ becomes nonsingular. Namely, $\mathbf{x} \in R^n$, $\theta \in R^n$ and $J \in R^{n \times n}$. n is equal to 3. Since $J(\theta)$ is a matrix which is varied by θ, the system is nonlinear.

Now, by introducing a new control input $\mathbf{u} \in R^n$,

$$\mathbf{u} = J(\theta)\dot{\theta} \tag{3}$$

the system can be represented in state space with the state being the image feature vector $\mathbf{u}$ as the following linear system.

$$\dot{\mathbf{x}} = \mathbf{u}, \quad \mathbf{y} = \mathbf{x} \tag{4}$$

In the lower level control, the angular velocity of each joint is controlled so as to track the reference signal that is generated as follows:

$$\dot{\theta} = J(\theta)^{-1}\mathbf{u} \tag{5}$$

The sampling rate is 2048Hz. In the upper level control, the sampling rate in the visual servo is equal to a video rate of 30Hz.

In this study, an on-line estimation of the image Jacobi matrix which has been proposed in the literature [6] is applied. A characteristic of this method is that the Jacobi matrix $\widehat{J}(t)$ is estimated to satisfy the following relationship with respect to the observed $\dot{\mathbf{x}}$ and $\dot{\theta}$.

$$\dot{\mathbf{x}} = \widehat{J}(t)\dot{\theta} \tag{6}$$

Accordingly, the estimation of parameters of image Jacobian is executed in order to track the image feature points into the references.

For the estimation, a least square method with time-varying forgetting factor is applied as follows:

$$\widehat{J}(t) - \widehat{J}(t - \Delta t) = \frac{\left\{ \Delta\mathbf{x}(t) - \widehat{J}(t - \Delta t)\Delta\theta(t) \right\} \Delta\theta(t)^T W(t)}{\rho + \Delta\theta(t)^T W(t)\Delta\theta(t)} \tag{7}$$

where, Δt is a sampling time. $\Delta\theta(t) = \theta(t) - \theta(t - \Delta t)$, $\Delta\mathbf{x}(t) = \mathbf{x}(t) - \mathbf{x}(t - \Delta t)$. $W(t)$ is a weight matrix. $\rho(0 < \rho \leq 1)$ is a forgetting factor.

Since an error due to the on-line estimation is included in the obtained Jacobi matrix, this estimating error is regarded as a multiplicative plant perturbation. Then, the robustly stabilizing controller is designed based on the framework of H_∞ control theory.

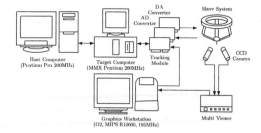

Fig.6: Image data

Fig.5: Experimental system

The design specifications are taken as the following: (1) minimizing the tracking error between the reference $\mathbf{x}_d$ and the observed $\mathbf{x}$, and (2) prespecified saturation limits for applied force $\mathbf{u}$. Now, the transfer function matrix T_{zw} of the closed system is given as follows:

$$\mathbf{z} = T_{zw}\mathbf{w}, \quad T_{zw} = \left[\begin{array}{c} -W_u K(I - PK)^{-1} \\ -W_e(I - PK)^{-1} \end{array} \right] \qquad (8)$$

P denotes the linear plant. The weight functions W_u and W_e were chosen as follows:

$$W_u(s) = \frac{10s}{s + 10^4}I_n, \quad W_e(s) = \frac{10^{-3}}{s + 10^{-4}}I_n \qquad (9)$$

where, the weight W_u is set as a high-pass filter in order to suppress the control inputs $\mathbf{u}$. And the weight W_e is set as a low-pass filter to accomplish the performance specification (1).

The design goal is to find a controller K which stabilizes T_{zw} and keeps the $\mathbf{z}/\mathbf{w}$ transfer function approximately small. The H_∞ design procedure has been carried out utilizing the *MATLAB*.

4 EXPERIMENTAL RESULTS

The experiments of operation through the developed virtual micro dome system are executed in order to examine the validity of proposed adaptive visual servo. Figure 5 illustrates an experimental system that is made up of two subsystems. One subsystem consists of the host computer (Pentium Pro, 200 MHz) and the target computer (MMX Pentium, 200 MHz). The real time OS (Wind River Systems, Inc. VxWorks) is adopted. The second subsystem includes the graphics workstation O2 (MIPS R10000, 195 MHz).

Figure 6 denotes the image data sent to the color tracking module (see Fig.4). Figure 7 shows the snapshot of operational environment that was displayed to the operator. A 3:1 scale factor maps 3cm of master's translation into 1cm of slave's translation. As the results of experiments, when the operator manipulated the master tool by the pointing device, it could be confirmed that the slave tool tracked the master tool in the 3D space reconstructed by fusing the stereo image. Both the master and the slave were observed solidly and were recognized as a tool by the operator. Figure 8 illustrates a step response. In the experiment, each reference in image plane was moved 16 pixels from initial position. As seen from Fig.8, the error converges into zero.

5 CONCLUSIONS

In this study, keeping the medical applications in mind, an interface that gets more accurate visual registration has been developed in order to achieve the effective and suitable micro teleoperation. The results of this study are summarized as follows:

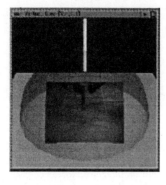

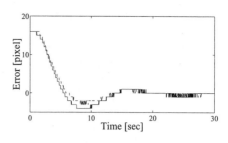

Fig.8: Experimental results (step response)

Fig.7: Snapshot of operational environment

1. The visual servo is proposed as one of the superior schemes to realize the visual registration. In order to recognize the master-slave as a tool, their motions are aligned in the 3D space reconstructed by fusing the stereo images.

2. The nonlinear system was linearized by a nonlinear compensation. And, the adaptive visual servo system consisting of on-line estimator of image Jacobian was designed.

3. In the experiments, it could be confirmed that the slave tool tracked the master tool. Both the master and the slave were observed solidly and were recognized as a tool by the operator.

Our future works are the discussion about haptic registration and the development of real master system. Moreover, we aim to teleoperate from a remote site through a network such as the Internet.

References

[1] R.H. Taylor, J. Funda, et al.: "A Telerobotic Assistant for Laparoscopic Surgery," *IEEE Engineering in Medicine and Biology Magazine*, Vol.14, No.3, pp.279–288, 1995.

[2] A. Sano, H. Fujimoto and K. Kodani: "Development of Combined Master-Slave Tool for Medical Applications," *Proc. of the Japan/USA Symposium on Flexible Automation*, Vol.1, pp.245–250, 1996.

[3] P.S. Green, J.W. Hill, J.F. Jensen, and A. Shah: "Telepresence Surgery," *IEEE Engineering in Medicine and Biology Magazine*, Vol.14, No.3, pp.324–329, 1995.

[4] G.S. Guthart and J.K. Salisbury: "The Intuitive^{TM} Telesurgery System: Overview and Application," *Proc. of the 2000 IEEE Int. Conf. on Robotics & Automation*, pp.618–621, 2000.

[5] Y. Yokokohji, R.L. Hollis and T. Kanade: "What You can See Is What You can Feel -Development of a Visual/Haptic Interface to Virtual Environment-," *Proc of IEEE Virtual Reality Annual Int. Symposium (VRAIS'96)*, pp.46–53, 1996.

[6] K. Hosoda and M. Asada: "Adaptive Visual Servoing Controller with Feedforward Compensator without Knowledge of True Jacobian," *Journal of the Robotics Society of Japan*, Vol.1, pp.313–319, 1996.

Small User Interfaces: Challenges in Designing Holistic Mobile User Experience

Katja Konkka

Nokia, Sinitaival 5, 33720 Tampere, Finland, katja.konkka@nokia.com, (GSM) +358 50 3546980

Abstract

User interface design of small mobile terminals face a challenge how to design user interfaces that support the usage of the mobile Internet and wide variety of services. A further issue is how to make mobile Internet sites and services itself usable. Building a seamless and holistic 'anytime, anywhere' mobile user experience is an issue that has to be taken into consideration when designing mobile terminals and mobile Internet services.

1. Introduction

How about designing user friendly user interfaces (UI) for small mobile terminals that fit to the existing line of products and systems. Lets add to that the challenge of designing UIs that respond to varying user needs and wants, and enable personalisation. Challenging enough? To make it all even more complicated: today the biggest challenge in UI design is not how to design the traditional applications and functionality within the device, but how to design user interfaces supporting the usage of various mobile Internet services. Furthermore, how to make mobile Internet services itself usable via different kinds of mobile terminals.

Generally, providing the Mobile Internet sites and services *(content)*, and manufacturing mobile terminals *(tools for accessing the content)* come from different enterprises. Still, both of them have to work together. The service providers' content and the UIs designed by the terminal manufacturers have to support each other and function coherently. Abundantly this is a technological issue, which will not be discussed in this paper, but it is indeed an UI design issue as well. The service providers give the wheels and the terminal manufacturers should give the vehicle.

The structure of this paper is as follows. Chapter 2 describes briefly what the differences are between designing small mobile terminals and bigger fixed terminals as service platforms. The concept of Mobile Internet, co-operation between service providers and mobile terminal manufacturers, and accessing same content from PCs and mobile terminals are discussed in Chapter 3. Chapter 4 deals with personalising mobile terminals and accessed content. Chapter 5 is about seamless and holistic user experience. And finally, in Chapter 6 there are the conclusions.

2. Mobile Terminals as Mobile Internet and Service Platforms

Designing small mobile terminals that serve as mobile Internet service delivery platforms differs from designing bigger fixed terminals, such as PCs, that serve the same purpose. The role of UI design and usability becomes very relevant when designing small mobile terminals, because the small size and changing content of use make designing challenging.

According to Anne Kaikkonen of Nokia [Kaikkonen 2000] the main differences between mobile terminal and PC as a service platform are the following: firstly, the display in mobile devices is fairly small and there are often problems with contrast. Additionally, the variety of different devices is big. If the difference is not taken into consideration, the service or Internet site that looks good in one terminal may be useless in other. Secondly, the input is slower, and there are lots of variations in the input mechanisms. Thirdly, selection mechanisms vary in different devices e.g. from having one to many soft keys, or cursor movements can be controlled in different ways, e.g. with keys or roller buttons. Fourthly, currently in mobile terminals the connection between the device and service is slower and can be unstable. Fifthly, the amount of information that can be stored in small mobile devices is smaller and dependent on the device. And finally, different WAP browsers have different logic and functionalities. [Kaikkonen 2000]

Additionally, Anne Kaikkonen writes in her article 'Designing usable WAP services' that the context of use of devices and services cannot be predicted when using mobile devices. Thus there may be for example visual and/ or auditory limitations in the usage environment. Also the reason for using the device can be different: Mobile phone is mostly a communication tool and therefore carried around 'all the time'. [Kaikkonen 2000]

3. Mobile Internet and Services

3.1 The Concept of Mobile Internet
The mobile Internet can be provided by Wireless Application Protocol (WAP), which has enabled the development of wireless Internet browsing applications during the past few years, and iMode (in Japan), which also delivers Internet services to mobile terminals. The 3rd generation systems, such as MExE (Mobile Execution Environment), will provide developed platforms for mobile Internet services. The concept mobile Internet also includes, among other things, a possibility to use Internet email, chat, play Internet games, use Internet services, and browse WWW-sites. Most of the mobile Internet services are not necessarily designed specifically for mobile use. Thus, it is not only the terminals that have to be taken into consideration when designing tools for mobile Internet services.

An ideal mobile Internet service for a user means the possibility to access the right kind of content in the right kind of form with a device that supports the usage of the content efficiently. When it comes to WAP and iMode, the industry standardisation work of these technologies has been the way for the mobile terminal manufacturers to effect building seamless mobile Internet services experience.

3.2 Co-operation between Service Providers and Mobile Terminal Manufacturers
The responsibility for designing good and usable mobile Internet sites and services is mainly of service providers, but certainly co-operation between the service providers and mobile terminal manufacturers is one way to make good mobile Internet sites and services.

During the recent years Nokia has co-operated with certain service providers in order to help make good services that are usable via Nokia mobile phones. One example is the Merita WAP bank service: Merita Nordbanken is the biggest bank in Scandinavia. Merita has a long history in developing e-banking solution; for instance it was in front-line in developing banking solutions for Internet and GSM. Merita WAP bank services development started together with Nokia in 1998. The first milestone in service development was to demonstrate the Merita WAP bank service in the Cebit Exhibition in 1999. After that the service ideas were reworked for a commercial version, and launched later in 1999. [Kaikkonen&Törmänen 2000]

Before launching the service paper prototype tests were used to generate the concept. The user interface was designed with Nokia WAP Toolkit, and evaluated with expert users before the Cebit exhibition. After the Cebit exhibition the concepts were reworked, and usability tests were run using real WAP phone prototypes and a server prototype. The project showed that although WAP as a standard was in a quite premature state at the time, it was possible to develop useful mobile services. It was also concluded that a critical aspect in development is to take the usability issues into consideration during the design process from the beginning. [Kaikkonen&Törmänen 2000]

3.3 Accessing Same Content from PCs and Mobile Terminals
The role of a browser UI design, browser meaning a software used to obtain and show web pages with the user's equipment, is important. To give an example: desktop computers can have Windows operating system installed with different kinds of programs, e.g. Netscape browser. Users can download desired content from the Internet. Although somebody else designs the actual content, the Netscape browser's user interface design has its effect on using Internet content as an 'access tool'. In the mobile terminal environment the same applies, the browser of a mobile terminal is a tool for accessing content from service providers. Also because the screen sizes of mobile terminals are smaller than the screen sizes in the desktop devices and the usage content varies, the role of designing a good browser becomes very important. For example there is not much space for menus, thus most of them have to be in sub-menus.

As users might be accessing the same Internet content from their PCs and mobile terminals by using different browsers, and as mentioned earlier all the Internet content desired to be accessed by mobile terminals is not often designed especially for mobile use, an additional challenge occurs: Can we create content regardless of whether it is for the fixed Internet or mobile Internet world?

VTT Information Technology in Finland, a research institute supported by the Finland's government, has implemented a HTML/WML conversion proxy server, through which the users of different WAP phones can access the same web services as they use on their PCs. In their tests, the users have responded positively towards that possibility. [Kaasinen&Kolari&Laakko 2001]

Nokia demonstrated the first XHTML microbrowser on a standard mobile phone in March 2001. XHTML is the language that will be used to create all content regardless of whether it is for the fixed Internet or the mobile terminal world. By narrowing the gap between wired and wireless content, this technology improves the usability of wireless services for consumers. [Nokia 2001]

4. Personalising Terminals and Accessed Content

As there will be more sophisticated terminals and more complicated services available in the near future, the meaning of personalisation increases. Personalisation means that users are able to adjust terminal settings as they like. Just as with an ordinary computer and Windows operating system it may be possible to adjust mobile terminal's screen settings, menus and choose the applications to use. [Konkka 1999]

Furthermore, the personalisation means that users are able to access a wide variety of information and are able to use different services according to the users' personalised needs and wants. In future mobile terminals it should be possible to have a set of service profiles for one user. A user profile may determine services, prices, location specific, or some personal settings for this particular profile. It is possible to have one profile e.g. for work-related calls that the company pays, and second profile for personal calls. Parents may define a third profile for their children allowing calls only to parent's terminals. Hence, the personalisation of a mobile terminal is not just a question of adjusting UI, but adjusting the accessed content by setting user profiles as well.

Additionally, selling products globally to customers of different cultural background brings extra challenges to already challenging product design. The possibility for personalisation might not be enough, if it only allows personalisation within the frames of one cultural area. The logic of product usage and other user interface issues may not always be suitable for all users around the world, not to mention the content accessed via mobile terminals. Still, this doesn't necessarily has to mean localisation to the full extent: The scalability and variability of the whole product to suit different cultures is a basic requirement, instead of designing different products for each culture. This may mean that the UI logic or icons, for example, should be able to vary by modifying the same software in one platform. [Konkka&Koppinen 2000]

5. Building a Seamless and Holistic User Experience

5.1 Holistic and Seamless Mobile User Experience

Now there are more information and communications technologies available than ever in history. All kinds of tools, devices and systems of information and communications technologies – not to forget what the entertainment industry has to offer- have became, and are becoming, available and commonly used. The boundaries between different technologies and devices are breaking down. Instead of isolated pieces of hardware systems, new systems allow multifunctional configurations by connecting software and hardware elements from different devices and systems into one. And, with the mobile terminals this is all put on move. This brings us to the concept 'ubiquitous computing' first coined by Mark Weiser [Weiser 1991]. Ubiquitous computing refers to the proliferation of computing capabilities into everyday devices allowing people to interact with their environment in new ways. The Cooltown concept by Hewlett-Packard [Cooltown 2000] illustrates ubiquitous computing as follows: People can arrive at a room, and with their terminals, browse through on-line representations of the room; what people are, what their backgrounds and interest are, what devices are there, what functions could these devices accomplish, and so on. [Weiser 1991; Cooltown 2000]

In the previous Chapter the seamless mobile Internet service experience was mentioned. But that is not enough. The whole experience of using a mobile terminal should be holistic. A user's mobile terminal should be a part of his/her life when working, playing and keeping in touch with others in a seamless and transparent way. This means taking into consideration all the players in the market of mobile communications and users' lives. Figure 2 outlines these elements: A mobile device should support users' activity practices and communication that vary in context and form. It should also provide pleasant and serendipitous user experiences, and be fully usable regardless of time and place. The style and design of a device should support expressing one's lifestyle and identity. Additionally, a user shouldn't have to think about technologies behind the device, just to get support whenever needed.

Figure 1: Elements of holistic mobile user experience.

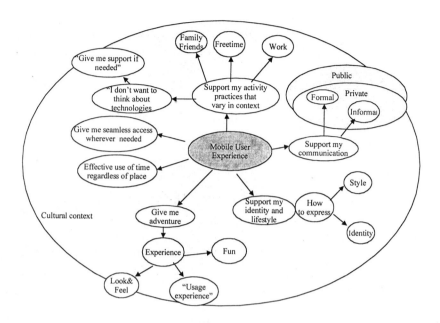

5.2 Methods for Researching End-Users

In order to provide holistic user experience a company has to know its customers. Ethnographic research methods provide a useful way of researching end-user needs and wants that vary in context. As Satu Ruuska [Ruuska 1999] discusses in her paper 'Mobile Communication Devices for International Use – Exploring Cultural Diversity through Contextual Inquiry' the shift from place- and time-dependency to any-time and any-where access to information is creating new challenges in developing personal devices. Developers need to understand users' physical, social and cultural context and the way they are mutually affected by users' activity practices, roles and values. [Ruuska 1999]

Ruuska continues that with traditional market research or usability methods it is very difficult to acquire information on users' activities, roles and culture for product creation purposes. Methodologies that are based on ethnographical research tradition but still are manageable in the business environment are rare. However, some ethnographic research methods, like the Contextual Inquiry [Beyer and Holtzblatt 1998] are providing a structured way to gather contextual information on users activity practices and related cultural aspects. [Ruuska 1999]

6. Conclusions

Designing small mobile user interfaces is a big challenge. Besides the mobile Internet services, and cultural issues there are many other issues challenging the UI design of small mobile terminals: Keys, icons, graphics, sounds, ergonomics, input-methods, out-put methods, look and feel, personalisation and so on. When a user is choosing a mobile terminal in next few years, he will face a wide selection of terminals and services to select from.

Designing mobile devices that serve as mobile Internet delivery platforms is a bigger challenge than designing fixed PC type terminals, because the screen size is small and the usage content varies. Additionally, not all of the mobile Internet services are designed specifically for mobile use.

The whole experience of using a mobile terminal should be seamless and holistic. In order to build such experience the company must know its customers. Ethnographic research methods, complementing traditional market research and usability research methods, provide a useful way of researching user needs that vary in context, and thus help building holistic and seamless 'anytime, anywhere' mobile user experience.

References

[Beyer&Holtzblatt 1998]   Beyer, H. & Holtzblatt, K.: Contextual Design: Defining Customer-Centered Systems. San Francisco: Morgan Kaufmann. 1998.

[Cooltown 2000]   Cooltown – Web Appliances and E-Services. In: http://www.cooltown.hp.com/. 2000.

[Kaasinen&Kolari&Laakko 2001]
Kaasinen, E. & Kolari, J. & Laakko T.: Mobile-Transparent Access to Web Services. Short paper to be published at INTERACT2001, Eighth IFIP TC.13 Conference on Human-Computer Interaction. Tokyo, Japan July 9-13, 2001.

[Kaikkonen 2000]   Kaikkonen, A.: Designing usable WAP services. HCI2001 Proceedings. USA. 2000.

[Kaikkonen&Williams 2000]   Kaikkonen A. & Williams D.: Designing Usable Mobile Services CHI2000 Tutorial Notes, ACM. 2000.

[Kaikkonen&Törmänen 2000]   Kaikkonen A. & Törmänen P.: User Experience in Mobile Banking. HCI2000 Proceedings. USA. 2000.

[Konkka 1999]   Konkka, K.: How people feel mobile communications affects their everyday life. People's reflections in Finland and in the USA. Master of Science Thesis. University of Tampere, Finland: 2000.

[Konkka&Koppinen 2000]   K. Konkka & A. Koppinen, Nokia Mobile Phones, Finland.: Mobile Devices - Exploring Cultural Differences in Separating Professional and Personal Time" International workshop on internationalisation of products & systems 2000; Baltimore, USA. 2000.

[Nokia 2001]   Nokia press release: Nokia demonstrates first XHTML microbrowser on standard mobile phone. Finland. March 21, 2001.

[Ruuska 1999]   Ruuska, S.: Mobile Communication Devices for International Use – Exploring Cultural Diversity through Contextual Inquiry. In IWIPS 1999 Proceedings. USA. 1999.

[Weiser 1991]   Weiser, M.: The Computer for the 21st Century. Scientific American. September 1991.

Using Alternative Realities as Communication Aids in the Participatory Design of Work Environments.

Roy C. Davies, Elisabeth Hornyánszky-Dalholm,
Birgitta Rydberg-Mitchell and Tomas Wikström.

Design@Work, Lund University, Sweden.

roy.c.davies@ieee.org, elisabeth.dalholm@byggfunk.lth.se,
birgitta.mitchell@byggfunk.lth.se, tomas.wikstrom@byggfunk.lth.se

Abstract

Participatory Design is an effective method for capturing tacit knowledge about a workplace and involving workers in the improvement of their own working environment. However, the people from the workplace must also collaborate with designers, ergonomists, builders, architects etc, and thus a common language understandable to all must be found. This is the basis for the Envisionment Workshop – a collection of tools/communication aids/alternative realities which have been carefully assembled and tested over several real design situations to facilitate the participatory design process. The primary tools are full scale modelling, virtual reality, drama and democratic meetings. Our results indicate where in the design process each tool type can be successfully used and how the tools can be combined to lead ideas through from activation and awakening to consolidation to a concrete solution. Nevertheless, the connection between the tools requires further strengthening – this is the basis for our continued research.

Keywords: Participatory Design, Full Scale Modelling, Drama, Communication, Mixed Reality, Ergonomics.

Introduction

Improvement of work environments in which the workers participate in the design process can lead to successful and useful solutions which are better accepted and appreciated than workplace changes performed using non-participative methods (Jensen, 1997; Harrison, 1996).

In Participatory Design (PD), the design team consists of both workers and design experts such as ergonomists, builders, interior decorators and so on. In order that these people with diverse backgrounds can communicate and learn together, common languages must be sought which are, in particular, easy for the workers to use.

These languages can take many forms such as lists of words, sketches on paper, scale models or even computer generated pictures, each with their own capabilities and limitations of expression. In our Envisionment Workshop, we have brought together many such languages (calling them tools) to allow the members of the design team to discuss a variety of aspects of the work environment (Davies *et al*, 2000). The tools we have chosen build on providing an environment in which the workplace can be physically designed and tested in a hands-on and activating fashion.

In particular, we use:

- full-scale modelling where life-size mock-ups of workplaces and tools are built and interacted with using wall, window and door elements as well as cardboard, wood and polystyrene;

- pedagogical drama where the focus is on learning by doing, feeling and experiencing to provide immediate and concrete feedback on social interaction and provide opportunities to test new ways of acting in a more secure environment than reality (Boal, 1985); and

- three dimensional (3D) computerised design and visualisation allowing individuals and groups to design in a more easily modified environment than reality and to communicate design ideas to others.

Usage of these alternative realities is interspersed with democratic meetings, sketching and discussions. In the selection of these tools it has been vital to determine whether each is useful and usable for Participatory Design and how it can best be included in the Envisionment Workshop. Our work, through a number of real design situations involving workplace improvements at companies, has focused on identifying optimal configurations of these tools (both in the outward design and usage protocols) for different design situations, has investigated the usefulness as communication aids in the design process and in particular, has developed and evaluated various ways in which VR can be most effectively used.

This paper briefly describes the work we have performed, some of the design situations we have been involved in, some of the results we have obtained, and our continued research plan.

Communities of Practice

In the Participatory Design of a work environment, there are two principle groups of people – those that come from the workplace being designed (the workplace experts), and those that are helping in the process of design (the PD experts). There may also be others such as management, builders, decorators or the company ergonomist.

Participatory design is a learning process, both for the workplace experts and the PD experts and can be described as a complex interplay of two communities of practice (Wenger, 1998). In the initial stages, the PD experts are apprenticed to the workplace experts to learn the existing work environment. The PD experts start as workplace novices. They must begin by learning about the workplace and act as the humble novices they are in order to win the trust of the employees. The PD experts are also experts in their own fields of ergonomics, design, architecture and so forth. This knowledge is in turn conveyed through co-operation between equals to the workplace experts.

For the worker, tacit knowledge is hard-won and has been acquired through years of training and working in that particular workplace. It is understandably well guarded and the employee needs to know that the PD expert is not going to use this knowledge against them.

During workplace visits, the PD experts talk to the employees, make observations, take notes (though perhaps not overtly), ask leading questions, try to ascertain whether the problems the employees describe are real or symptoms of some larger problem and so on. They must develop a feel for the workplace and the work organisation through experiential and situational learning (Lindberg, 2000). This knowledge is vital to the success of the PD process (Schön, 1983). Furthermore, if the workplace change is being instigated by the management, the feelings of the workers to this change need to also be established, and the management must be included in the following PD exercises to defend the reasons for the changes.

The next stage is for the PD experts to create an atmosphere conducive to workplace analysis and debate. This can be constructed at the workplace or somewhere else; the method chosen for this research being the latter. We have been primarily interested in the structural aspects of a workplace change, and the tools we use reflect this bias. Nevertheless, any physical structure in which people spend time is affected by the activities that occur within and the relationships between the principle actors. Therefore, tools have been chosen that encourage debate about the workplace, work processes and organisation.

The corner-stone tool is full scale modelling within a large full scale laboratory at the university. To this, we have connected the methods of pedagogical drama, virtual reality modelling and democratic meeting techniques. These are part of the workplace of the PD expert, and it is to these premises that the workplace experts are invited. In this situation, they now become the novices and great diplomacy is required to ensure they are not overwhelmed since even in this strange place, the topic of discussion is their workplace.

If the correct atmosphere has been created, if the workplace experts feel the changes are going to be to their benefit and that what they do will have some effect on the changes to be made, the scene is set for a successful participatory design project.

Method

A series of exploratory PD case studies has been performed with the unit of analysis being the individual Envisionment Workshop. Several of the studies have built on the experiences of previous ones, refining a particular combination of tools and usage protocol, whilst others have tested totally novel combinations and, in particular, methods of using VR. For some companies, the design process spanned several studies.

For each study, the fixed parameters were the participants and the workplace problem the participants wished to solve. The participants in a study were chosen by the company, however we had little control over which companies choose to work with us nor who they chose to participate (ideally, the participants should be chosen democratically at the workplace if the process is to be truly participatory).

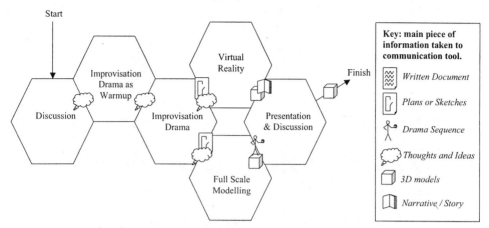

Figure 1: The sequence of events in a typical Envisionment Workshop

In each case, a particular combination of tools and usage protocol was put together as being most appropriate, both to solve the participants' problems and to obtain answers to our research questions. As an example, fig. 1 shows the sequence of events for the second PD session in the design of a new university nearby. The entire session took four hours and involved ten participants. The day began with a general discussion of the current theme – "space for learning" – whilst sitting in the full scale laboratory. This was followed by two drama sessions called "invisible room" improvisation drama. In this exercise, the participants go in one by one to a large empty space and act with imaginary objects within that space. The first time was to practice the method and loosen up the participants, the second to envision a "room for learning" using ideas carried over from the discussion and a rough sketch of the room layout. The group was then split and two subgroups which began the idea consolidation process with either VR or full scale modelling. Finally, the participants came together again to present their ideas to the others and for a final discussion.

The nature of an Envisionment Workshop meant that very little time could be spent by the participants in evaluating the tools – they had a task to do and must take back a concrete solution. This meant that only combinations of tools that we could be fairly certain would work together could be tested in this way, a total failure was not allowed. Furthermore, during a workshop, as much material for later evaluation had to be collected as possible without disrupting the design process.

Another problem is that an Envisionment Workshop may be adapted by the participants as a result of new ideas. A plan is always worked out beforehand (as in fig. 1), however other ideas may spring up taking the workshop in an unexpected direction. Ideally, the workshop should be dynamic, with nothing planned beforehand, just a collection of tools made available and used as appropriate in the discussion phase. In practice, it is deemed wise to have a plan to fall back upon when ideas start to dry up. One benefit, however, of an Envisionment Workshop is that the participants have a concrete, meaningful and engaging task to perform, which is often lacking in experimental situations.

Many companies around the region have taken part in Envisionment Workshops, with projects as small as redesigning a reception area, to the design of a whole new University (fig. 2). Interwoven with these studies have been a number of tests involving students as subjects to investigate new ideas and combinations of technologies, as well as a number of purely technical tests. For more details of these, see Davies *et al*, 2000.

When using each of the communication tools, both workplace and participatory design experts work together. In the case of full scale modelling, the workplace experts are given the main design role, and a task is chosen which has been gleaned from previous methods. Similarly, for design using VR, a task is chosen from previous methods and further worked in the virtual medium. The design process can be said to go through two phases, an idea activation phase and an idea consolidation phase. In the former, discussions, brain-stormings and drama exercises are used to get the participants thinking and to open up as many ideas as possible. These are then sifted through and one or two key ideas taken for consolidation using sketching, VR and full scale modelling. The resulting models are then often tested using role-play (both in VR and the full scale model) and the session is finished by a discussion or debriefing and assembly of material to take back to the workplace.

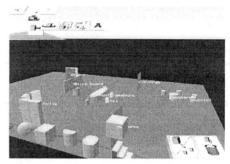

(a) Modelling in the full scale laboratory

(b) Modelling in VR

Figure 2: Participatory Design in full scale and VR

Many other combinations of tools have also been tried (Davies *et al*, 2000). For example, using a VR model of the existing workplace in an immersive VR environment as a basis for initial discussion and idea activation, followed by design sessions in the full scale laboratory.

Results

Some general results from the studies are presented here, however there is insufficient space to go into full detail, the interested reader is referred to Davies *et al*, 2000 and Davies 2001.

The design process has on occasion been split over several Envisionment Workshops over several days. It has been found that a full or half-day intensive workshop, though tiring, produces excellent results if the participants are sufficiently motivated. Furthermore, allowing some time between successive workshops allows ideas to mature which can be further incorporated.

A Participatory Design Virtual Reality tool (the Envisionment Foundry) progressed through several prototypes during the Envisionment Workshops (Davies, 2001), and resulted in the following conclusions (amongst others): A simple interface (see fig 2b) was easily understood by users, but the design process was disrupted by having to share the input devices. This led to frustration and a need for alternative design tools such as pen and paper. The problem was alleviated somewhat by having an expert drive the tool (thus making the interface verbal), but there was still a bottleneck at the interface. Having the VR picture projected on the wall allowed all to see the working design and did facilitate discussion as well as later presentation of the resulting design. Textual labels on objects provided a common language and helped compensate for lack of detail, which was deliberately absent to force a focus on layout rather than details and to simplify the tool.

Multi-user immersive Virtual Reality was found to be useful for initial criticism of an existing work environment and for visualising situations that could not fit in the full scale laboratory.

Full Scale Modelling has been found to be well suited for participatory design as it is a hands-on activating experience which allows full participation by the workplace experts in communication with the design experts.

Free-form drama has been found useful for idea expansion in the early stage of the design process and role-play for testing environments using real work situations.

Discussion and Future Work

The concept of the Envisionment Workshop has worked well in many participatory design situations and the selection of the tools has led to successful designs (as measured by worker-satisfaction). Variations in tool combination and usage have been proposed for re-designing a work environment as well as for designing a new work environment (Davies *et al*, 2000).

However, a number of limitations in the tools have also been noted, particularly, the difficulty for novices to use the VR tool despite a very simple interface and the difficulty of transferring a design from the full scale laboratory to the VR environment.

After analysing these results, it was decided that a number of new research goals be concentrated upon:

1. To develop a system for directly transferring a full scale model into VR without requiring user intervention. Here we have acquired an optical tracking system capable of tracking people and objects in the full scale laboratory and thus making a sophisticated mixed reality system.

2. To simplify and integrate the VR system further into the full scale laboratory. We have also acquired a large VR projection system which will allow VR to be moved out into the full scale laboratory, thus bringing VR into the participatory design work cycle.

3. To determine how best to involve other workers who could not participate directly – here multi-user VR and connection over the internet is being investigated as a potential solution.

4. To investigate the possibilities of moving the Envisionment Workshop out to the workplace. The tracking system could also be used to develop a portable small-scale mixed reality version of the full scale laboratory.

We are currently entering a new stage in our research to investigate these ideas further and working towards a European-wide collaboration.

Acknowledgements

This work has been funded by several groups over the years, in particular, RALF (The Swedish fund for work-life research), NUTEK (The Swedish National Board for Industrial and Technical Development) (both of which are now part of Vinnova – the Swedish Agency for Innovation Systems) and the Knut and Alice Wallenberg Foundation (for equipment funding). All the participants deserve thanks for their patience and interest in being subjects in a research project. Finally, all the other members of Design@Work that are not named in the list of authors that have been part of the team for a while and then moved on are also thanked for their contribution to the total effort.

References

Boal, A., 1985, *The Theatre of the Oppressed*, Theatre Communications Group; ISBN: 0930452496.

Davies, R. C., Brattgård, B., Dalholm, E., Mitchell, B., Wikström, T., Warrén, P., (2000), Combining Virtual Reality with Full Scale Modelling and Drama for the Participatory Design of Work Environments, Publication 70, ISSN: 1104-1080, ISRN LUTMDN/TMAT--1107--SE (in Davies, R. C., (2000), Using Virtual Reality for Participatory Design and Brain Injury Rehabilitation, Doctoral Thesis, Lund University, Sweden, 2000, Publication 69, ISSN: 1104-1080, ISRN LUTMDN/TMAT--1107--SE)

Davies, R. C. (2001), Adapting Virtual Reality for the Participatory Design of Work Environments, *Journal of Computer Supported Cooperative Work*, submitted.

Harrison, S., 1996, The Participatory Design of Work Space, *Proceedings of the Participatory Design Conference*, MIT, Boston, 1996, pp 161-162.

Jensen, P. L., 1997, The Scandinavian Approach in Participatory Ergonomics, *Proceedings of the 13th Triennial Congress of the International Ergonomics Association*, Tampere, Finland, **1**, pp 13-15.

Lindberg, A., 2000, *När goda intentioner möter verkligheten* (When good intentions meet reality), Doctoral Thesis, Institution for Work Sciences, Department of Industrial Production Environments, Luleå University, Sweden, 2000:27, ISSN: 1402-1544, ISRN:LTU-DT--00/27--SE

Schön, D. A., 1983, *The Reflective Practitioner – How Professionals Think in Action*, Basic Books, New York.

Wenger, E., 1998, *Communities of practice: learning, meaning, and identity*, Cambridge University Press, ISBN: 0521663636.

DESIGN, VIRTUAL REALITY AND PEIRCEAN PHENOMENOLOGY

Jean-Marc Orliaguet

Chalmers Medialab, Chalmers University of Technology, 41296 Göteborg, Sweden

ABSTRACT

We suggest an approach to design using virtual reality based on phenomenology, considering the cases of both ordinary and semiotic phenomena. Phenomenology considers all objects present to the mind as 'forms' or 'relational structures' (Marty, 1990). Semiotics studies all possible forms of relations that can connect two objects, when one serves as a representation of the other. As a consequence of defining the reality of an object, either virtual or not, as the agreement by a community on an objective and unique structure associated with that object, we can study the design activity as a dialectic process aiming at instituting objective community-shared structures and the necessary connections between those structures.

1. INTRODUCTION

Design using virtual reality (VR) combines many kinds of experiences apparently belonging to very different categories. Indeed, an ideal form of design occurs when the *mental* object of a designer, externalized in a *material* object, either virtual or not, corresponds to the one of the end-user. A product of design establishes therefore a connection between the realm of ideas and the object world, but to experience the *idea* of an object — either virtual or not — is obviously very different from experiencing the object itself. Yet, the design act combines those two kinds of experiences, and the combination in itself constitutes a third kind of experience.

Our knowledge and experience of virtual objects is so different in a way from a casual everyday experience of 'real' objects that the term 'virtual environments' (VE) was invented, without any quite definite understanding however of what is meant by 'virtual' that would place virtual objects in a category so separate from the category of real-world objects that all 'virtual' experiences would necessitate theories and methods especially invented for that purpose. In any case, a theory focusing on the design process using VR has in one way or another to overcome all the dualisms setting in opposition *virtuality* and *reality*, *mental* and *material*, while avoiding the dangers of reductionism.

Does design using virtual reality necessitate a new theory? On the contrary, we will try to convince the reader that such a theory exists already in the science of phenomenology and semiotics invented by nineteenth-century philosopher C.S. Peirce, and especially in its formalized version (Marty, 1990, 1992). Phenomenology is primarily concerned with describing the direct and immediate experience of what is present to consciousness, refraining from any form of judgment or presupposition on the nature of the experience itself. Peirce defined the phenomenon — or 'phaneron' in his own terms —, as "a proper name to denote the total content of any one consciousness (…), the sum of all we have in mind in any way whatever, regardless of its cognitive value" (Peirce, 1998a, p. 362). Semiotics, based on phenomenology, is the study of how an object can be substituted for another, i.e. when one collective total of objects present to the mind stands for another collective total of objects. The one directly experienced is a representation of the other.

Sartre (1994, p.11) began his essay of phenomenological ontology by commenting on the progress that modern thinking had achieved in overcoming "embarrassing dualisms" by reducing existents, i.e. real objects, to the series of their appearances (the phenomena). By treating objects as appearances, we overcome the duality between 'virtual' and 'real objects'. The 'virtual' partakes in the phenomenon and acquires the same status as any 'real object'. Both terms are not synonymous however. There would be no reason otherwise to study the design process in the particular case of VR; such a theory could be derived directly from any general design theory, so we need to define 'virtual' and 'real' more precisely in order to show their conceptual differences.

The other "embarrassing dualism" that we mentioned earlier was that one of abstract ideas and concrete existents. Here again, the duality is not as strict as it seems when considered from a phenomenological perspective. Every object that can possibly be present to the mind is not restricted to such concrete things as tables, chairs or lines of pixels on a computer screen. Everyone has done the experience of having the presence in consciousness of general ideas, concepts, or abstract forms, in a manner similar to the experience of concrete objects. Both general ideas and existents are thus objects of study for phenomenology and semiotics.

However, before laying down conceptual bases for any design methodology, we need to ask ourselves the question of which, of ideas or existents, predates the other: do ideas originate from existents, or do existents originate from ideas? Functionalism offers an easy solution to the problem. Functionalist principles assert that the form of a concrete object is determined by the function that object is meant to fulfill (Michl, 1995). This approach is well adapted for VR design, provided that the function of every object involved has previously been defined. Since many virtual environments strive to emulate real-world conditions, it is often the case that the form of a virtual object ought to follow from the function of the real-world object from which it is derived. But we see that the functionalist approach is essentially reductionist because it implicitly assumes that every virtual object somehow is the derivative of an already existing real-world object. We must consider the possibility that neither ideas nor existents should have any absolute precedence over the other, which is to say that if the function a thing is not entirely defined from the outset, it must be dialectically constructed.

2. DEFINITIONS AND METHODOLOGY

The elements of a phenomenon are classified on the basis of their form or 'structure' rather than on the basis of their substance (Peirce, 1998a, p.362). Our approach consists in establishing a formal connection between the design process, that produces forms and VR technology that provides the actual 'substrate' into which these forms are embodied. Our theoretical and epistemological bases are the formalization of phenomenology and semiotics done after the works of philosopher C.S. Peirce by R. Marty (1990, 1992). But we may first ask, what is meant by 'real' and what is meant by 'virtual'?

2.1 The reality of the virtual

Conceptually speaking, the term 'virtual' cannot be the opposite of 'real', or the expression 'virtual reality' would be a contradiction in terms.

What is then 'real'? For Peirce (1992) the 'real' is definitely linked to the notion of 'truth': 'Reality' is known at the term of an inquiry aimed at settling a final agreement: "The opinion which is fated to be ultimately agreed to by all who investigate, is what we mean by the truth, and the object represented in this opinion is the real. That is the way I would explain reality" (p.139), or: "'Real' is a word invented in the thirteenth century to signify having Properties, i.e. characters sufficing to identify their subject, and possessing these whether they be anywise attributed to it by any single man or group of men, or not" (1998b, p.434).

And what is 'virtual'? Stemming from Peirce's definition of the term (Baldwin, 1902): "A virtual X (where X is a common noun) is something, not an X, which has the efficiency (virtus) of an X", it is clear that virtual things are virtual not for being unreal but because, they, according to efficiency requirements, fulfill the same function as the actual things that they are substituted for, and this is known *a posteriori*. Imaginary things, on the other hand, like all untested hypotheses are *a priori* 'unreal', existing in the mind only, considered 'untrue' because their existence is dependent, at any given stage of knowledge, on individual minds only. Hence, a virtual thing will be real insofar as it is a true representation of a real thing. But a true representation is more than a graphically accurate representation: the photograph of a piano will never make a virtual piano insofar as a photograph is incapable of playing sounds. The truth of a piano — or the essence of a piano — no matter how large a community agreeing on a final opinion may be defined is clearly to play music. What differentiates something true from something untrue is the existence of essential features or "characters independent of what anybody may think them to be", that any given community of experiencers in some way or another has agreed upon. This posits reality as a social construction where ideas and objects acquire stability through the adherence of the entire community to objective aspects belonging to the structure of objects.

2.2 The essential structure of a virtual object

Insofar as a real thing is defined as having qualities independent of what anybody in particular may think of them, and because, in the same manner, the very way in which those qualities are organized also constitutes a character particular to that thing, it is not just qualities taken separately, but the collective total of qualities and the stability of their structural arrangement that is real. This means that there is in the perceptual configuration of every real object of experience a structure shared by all the members of a community. In Marty's formalization of phenomenology, this common structure is unique; it is named in reference to the Husserlian notion of 'eidos' (Greek: idea, form): the eidetic structure of that object. The eidetic structure of a perceived object is included in every appearance of that object, and in a converse manner, a condition for that object to be present to the mind is that the mind should form its eidetic structure (1990, 1992). The perceptual configuration of an object of experience is described formally by a relational structure incorporating the eidetic structure of the object present to consciousness. While some aspects of this particular structure may in practice be absent from the sensuous field experience, it is often an inference of a psychological nature that in some way or another fills the missing elements by combining elements of 'external' perception with elements of 'internal' perception. W. James (1981, p.747) concisely summed it up this way: "whilst part of what we perceive comes through our senses from the object before us, another part (and it may be the larger part) always comes (…) out of our own head." But in any case it is the *result* of this inference, in the form of a collective total of things present to the mind at the instant of perception, — and not the inference in itself —, that concerns phenomenology.

What is present to the designer's or the user's mind is an object, which formally speaking is a relational structure in which the eidetic structure of the object is included. The designer's task would in principle consist in incorporating into the perceptual form of the concrete object of design the eidetic structure corresponding to the object of his conception. The concrete object formed in this way will be a representation of the mental object conceived by the designer. Now because the essential character of a virtual object is to have the efficiency of the actual object that it is supposed to represent, the eidetic structure of any virtual object needs to incorporate at least the function of the represented object, which we summarize by saying that a virtual object has in essence the function of the object that it represents.

2.3 VR Design and functionalism

What is 'function'? The function of an object is the idea expressing the essential purpose for which that object is designed. 'Form' refers to a structural arrangement, a set of relations according to which the qualities of a any given existing object is organized: i.e. how it must be put together in order to be what it is, which means that the function of an object is potentially contained in the form of that object, but not the opposite. Indeed, if form could be deduced from function, with the implication that there would be such a thing as a function existing prior to form (Michl, 1995), nothing more would be required, in order to design virtual environments, than to identify the function of every object and run a sequence of pre-programmed functionalist precepts.

This approach is highly reductionist, because if 'function' is a first that determines form, then design using virtual reality is fated to simply reproduce or emulate the function of real-world objects, which renders impossible the task of explaining how a new function may arise in a virtual environment before arising in the 'real' world. We see that, by placing function absolutely prior to form, functionalist philosophy applied to virtual reality eventually poses more questions than it answers.

2.4 Dialectic of design

After rejecting the idea that there should be a pre-established function associated with every object, we are led to consider the design activity as a dialectic process where form and function participate as two polarities. The ideal object of the designer posited as universal is the first positive moment of the dialectic. The second and negative moment corresponds the phase of implementation where the idea of the object is confronted with technical limitations of VR technology and has to overcome the resistance to change of the community. A synthesis of both the positive and negative moments in the form of an object of design, combines the abstract idea of an object and its concrete material realization.

In practice, designers tend to create mental objects or 'models' by modifying the structure of an already existing object. If modifications are made on unessential aspects of it, function will be preserved, but if the modifications pertain to eidetic (i.e. essential) structures and provided that the changes are accepted on a community level, it is literally *new* objects and eventually new connections between these objects that are instituted. The design activity can then be regarded as the dialectic process of institution of structures and relations between these structures on the scale of a community (Marty, 1994).

Since there are two kinds of phenomena: those associated with previously memorized experiences (i.e. semiotic phenomena'), and those not associated (i.e. 'ordinary' phenomena) (Marty, 1990, pp. 15-16), the design activity is for 'ordinary' phenomena a process of institution of new *eidetic structures*, and for semiotic phenomena it is a process of institution of *new relations* between these structures.

3. APPLICATION AND RESULTS

3.1 Designing for 'ordinary' phenomena.

'Ordinary' phenomena are, by definition, not associated with memorized experiences. They involve the direct experience of objects in themselves. Nothing is present to the mind that is absent from the realm of direct perception — either internally or externally. The eidetic structure of every object appearing is therefore immediately present in the phenomenon. Hence, designing for ordinary phenomena means for the designer to focus on identifying the eidetic structure of every object that he is working with. It also means to focus on technologies that can generate immediate types of experiences (3D-displays, surround sound systems, force-feedback devices, etc.). The feeling of 'presence' in virtual environments is an instance of this, or the sensation of directly experiencing objects in three dimensions, or the immediate sense of feedback that one experiences with haptic or force-feedback devices, as well as the compulsive idea that forces us to mentally connect things together as if they were connected already, e.g. images and sound. Such experiences can neither be represented without losing their character of immediacy, nor can they be described in simple terms. They must therefore be experienced directly for what they are in themselves.

Now, the total of all immediate experiences of an object contributes to the taking of a habit, the habit of immediately experiencing the object, for the individual as well as for the rest of the community. The dialectic process leading to the formation of eidetic structures is based on the opposition that arises between the particular form of experience of an object by an individual and the experience of the same object by the rest of the community. The level of reality, or 'realness' of an object is a function of the stability of the eidetic structure associated with that object, so the chances of succeeding in modifying that particular structure on the scale of a community are determined by the resistance of the community that may either accept the changes or simply discard them. Hence the designer continuously faces two alternatives: either to preserve the essential (eidetic) structure of that object or to modify some of its aspects while running the risk of failing to communicate the object altogether.

This particular phase in the design process is determined by the technological limitations of the technology. This was particularly the case in the eighties when the rendering power of computers was limited to displaying wire-frame objects. Not surprisingly, objects were designed so as to contain the essential aspects of the object meant to be present to the mind of the user. Wire-frame display techniques were already adequate because they could render most features of the eidetic structure of objects, i.e. vertices and connections between vertices. Yet if too little of the eidetic structure of an object is presented to the end-user, the designer runs the risk of creating ambiguous objects that may be perceived in many very different ways, thereby losing total control on the design process. On the other hand, an excess of realism can never guarantee that an object will not be ambiguously perceived, if by 'realism' is meant to focus on all aspects of an object without making any distinction between what is essential and what is not.

Research themes in relation with the study of 'ordinary' phenomena cover issues of physical presence in VE, immersion, augmented reality, enhanced virtual vision, the use of perspective, perceptual or haptic illusions, etc.

3.2 Designing for semiotic phenomena

By definition, semiotic phenomena are associated with the experience of memorized objects. They involve a) the direct experience of an object of perception, b) the experience of another object that has been memorized, usually absent from the direct field of perception and c) the experience of the mediation linking those two experiences. The object that is directly present to the mind is a *sign* of the object present by mediation, and the connection between them is the *signification,* or *interpretant* of the sign. Designing for semiotic phenomena is very similar to designing for ordinary phenomena, except that, apart from the eidetic structures of objects, it is also the connections between those structures that need to be identified by the designer. A condition, for a law, a rule, or a concept, and more generally for any form of communication within a community to have any efficient effect, is that every individual has assimilated the connections linking the eidetic structures the objects involved, and by 'objects', we mean not only existents but also qualities and general ideas. The designer's task is more complex than in the case of ordinary phenomena. The 'encoding phase' consists in selecting an object whose eidetic structure is at least partially contained or connected by social convention to the eidetic structure of the mental object that he tries to convey. For the end-user, the corresponding 'decoding phase' consists in reconstructing the eidetic structure of the mental object conceived by the designer, by assembling together perceptual structures gathered from previous memorized experiences and connected with the structure of the directly experienced object (nothing is communicable that has not somehow, at least in a partial way, been experienced directly). A condition to succeed for instance in representing the feeling of a three-dimensional object using only two dimensions, is that the experience of seeing in three dimensions has occurred at least once.

The question of how phenomena are formally connected and how community members modify these connections in practice is beyond the scope of this article. But to give an idea, possible research themes in relation with the study of semiotic phenomena are for examples issues of communication and collaboration in virtual environments, the study of social presence in VE as opposed to simply physical presence, the use of metaphors, metonymies in VR, and more generally just any form of representation.

4. CONCLUSION

We cannot claim to have given here more than an overview of Peirce's phenomenology and theory of signs, and the general lines after which we think that it can be applied to design using virtual reality. It is particularly pertinent to the issue that when the level of reality subjectively experienced is a function of the stability of a construction involving the work of community whose members have agreed on some objective structures, then the distinction between real and virtual need not imply that virtual environments should necessarily be derived from 'real' environments. With the expansion of networks and the possibility of reaching towards wider communities, virtual worlds acquire a reality that is determined in the first place by the size of the community but also by the level of agreement between all of its members. To create virtual experiences is therefore an important aspect of the problem, but to understand how these experiences are formally organized and connected together is even more important.

ACKNOWLEDGMENTS

I wish to thank most sincerely for their support Chalmers Medialab and especially research supervisor in *Digital Representation* Sven Andersson.

REFERENCES

Baldwin, J. M. (1902). Virtual. In *Dictionary of philosophy and psychology* (Vol. 2, p. 763). New York: Macmillan.

James, W. (1981). *The principles of psychology* (Vol. 2). Cambridge, MA: Harvard University Press.

Marty, R. (1990). *L'algèbre des signes*. Amsterdam: John Benjamins Publishing Company.

Marty, R. (1992). Foliated semantic networks: Concepts, facts, qualities. In F. Lehmann (Ed.) *Computers and Mathematics with Applications, 23* (6-9). Oxford: Pergamon Press.

Marty, R. (1994). Sémiotique de l'obsolescence des formes. *Design-Recherche, 6*, 31-45. Université Technologique de Compiègne.

Michl, J. (1995). Form follows WHAT? The modernist notion of function as a carte blanche. *1:50 - Magazine of the Faculty of Architecture and Town Planning, 10*, 20-31. Haifa (Israel): Technion Israel Institute of Technology.

Peirce, C. S. (1992). How to make our ideas clear. In N. Houser, & C. Kloesel (Eds.), *The essential Peirce: selected philosophical writings* (Vol. 1, pp. 124-141). Bloomington, IN: Indiana University Press.

Peirce, C. S. (1998a). The basis of pragmaticism in phaneroscopy. In the Peirce Edition Project (Eds.) *The essential Peirce: selected philosophical writings* (Vol. 2, pp. 360-370). Bloomington, IN: Indiana University Press.

Peirce, C. S. (1998b). A neglected argument for the reality of God. In the Peirce Edition Project (Eds.) *The essential Peirce: selected philosophical writings* (Vol. 2, pp. 434-450). Bloomington, IN: Indiana University Press.

Sartre, J. P. (1994). *L'être et le néant. Essai d'ontologie phénoménologique*. Paris: Gallimard.

Computer Games in Architectural Design

Peter Fröst, Michael Johansson and Peter Warrén
Interactive Institute
Space & Virtuality Studio
SE-205 06 Malmö, Sweden
peter.frost@interactiveinstitute.se
www.interactiveinstitute.se

Abstract

The goal of the project presented in this paper is to explore the possibilities to use modern computer games in architectural design and integrate them with easy-to-use modeling tools, in order to facilitate some of the features available in the technically advanced, cheap and wide spread computer games. The project has its point of departure in the observation that several modern real-time computer game engines (Quake, Half-Life etc) are advanced 3D/Virtual Reality simulation tools. They provide the possibility to interact in a multi user environment (over the Internet) and there are good modeling software supporting the games available for free. We have tried the computer game Half-Life integrated with an easy-to-use architectural modeling tool for visualizations, as a multi user environment accessible over Internet and in student projects as a creative environment in architectural design exercises. In this paper are observations and conclusions from different case studies presented. The results are promising and show that modern computer games in combination with other tools can be a very efficient in collaborative architectural design processes.

Key words
Architectural design process, VR computer games, design tool, creative environment.

Background

Tools for architectural design
Our research is aiming at develop and integrate advanced visualization technology into the architectural design process by application of digital tools such as 3D modeling and Virtual Reality. Our focus is primarily on projects where users are actively engaged and a participatory or collaborative approach to design is practiced. A collaborative architectural design framework are presented in the Process Architecture approach by the SPORG Group at MIT (Horgen et al., 1999). Here possibilities to work with a broader sense of design artifacts as muck-ups and design games and the art of using a variety of tools for design has been developed. (Hornyansky-Dahlholm, 1998) at Lund Institute of Technology has described examples of the use of visualization and full scale modeling in participatory architectural design. The usability of Virtual Reality in participatory design has been investigated and developed by Roy Davies and the Design@Work group in Sweden (Davies R, 2000). An outline for www based software for participatory architectural design is outlined in (Cimerman, 2000).

Computer Games in Architectural Design
The rapid technological and artistic development in the game industry compared with the more military/industrial oriented professional VR market is a challenge to research. At Martin Center, Cambridge University (UK), a model of a complete building (10,000 sq.m), built on a computer game Quake platform, has been built and published on the Internet. (Paul Richens, Michael Trinder, 1999). Our group has also followed the development of the VRML standard for communicating 3D models on the Internet. Both these tracks has in more than one way led us to the conviction that there are alternative paths away from the "professional standard" to work with Virtual Reality and 3D as design- and communication tools in architectural design. Modern real-time computer game engines (Quake, Half-Life etc) are advanced 3D/Virtual Reality modeling and representation tools. They provide rich graphical environments where "physical" persons/bodies interact in a multi user environment (over the Internet). They are wide spread, cheap and easy accessible and can be used on a "standard" PC. Advanced modeling tools are available for free. The game technology develops and improves very fast through commercial and non-commercial processes (there is a whole Internet community working hard, sharing information). It is easy to learn and fun to use, that

means stimulates creativity and fast learning. We selected the game Half-Life because it is very open with a SDK available for free. It is very easy to set up and manage your own server. It also supports different kind of level editors such as Worldcraft and Quark, which are both good modeling tools. The method of building libraries with prefabs makes it easy to customize for every session and objects can easily be reshaped and exchanged.

HardHat Designer

Our research goal is to integrate our experiences from user participation in collaborative architectural design using VR Cave technology (Fröst, Warrén, 2000) into a digital modeling and VR-visualization tool based on Half-Life. We have therefore developed a prototype to a quick and easy design tool together with a student in Interaction Technology at Malmö University Collage. It is an extremely "easy to use" digital modeling tool called "HardHat Designer". Here you can move around and build your own environment with elements and furniture on a 2D surface. By a single mouse click the 2D layout will "immediately" be transferred to a lightened 3D/Virtual Reality world in Half-Life. You are only able to build and modify in 2D.

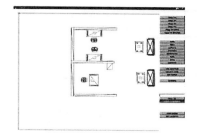

The virtual world in which you build is designed as a laboratory, with fixed ceiling height. Its size can be varied according to task. In the laboratory you have a set of different building elements. You can divide the space with walls in four different lengths plus window and door and you can furnish it with eight items of chairs, sofas, tables

etc. The furniture in this early version of the program was chosen directly out of Half-Life:s standard, rather game-looking library. As a way to present integrated technology in the world, a set of eight prefabricated "placeholders" is available. That is simple geometric figures in different shape and size, from a Palm Pilot to a TV set, which can be used freely. The application is written totally in Java, which we find to be a good tool for rapid development of small systems. The language also offer access to a huge number of different software libraries witch provide the opportunity to, in a simple way, add extra functionality and reuse code.

Experiment cases

Half-Life – a creative environment

Last year (2000) we investigated how you could use Half-Life in communicating certain moods and feelings in architectural design. In a master class in Interaction Design at Malmö University Collage we started our first Half-Life project. The commission was to build a room or a "world" in Half-Life that communicated a certain mode or feeling. After a 2 days "crash course", the students were able to build for 8 – 10 days and construct their own worlds, add textures, interactivity, and lightning. The overall impression by the students was that the editor (Worldcraft 2.2) wasn't hard to learn but had the computers "go mad"

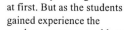

at first. But as the students gained experience the crashes came more seldom and in the end they actually felt that they could work around most of the shortcomings of the editor. They found that when they got a grip of the editor, it was fun an engaging to work with. At the presentation the students where allowed to "play" each others world and make notes of what feeling they experienced inside the other students worlds. To our surprise about 80% of the student manage to communicate their intentions.

This year (2001) the exercise was repeated with the extension that the student had to write a reflective text about their first contact with the Half-Life community. This was part of a second evaluation of the Half-Life environment as a design-tool. The demands for improved graphic capabilities were very clear in this group of students. Also the need of tools for working with more organic shapes was a shortcoming. To compensate for this, some of the students went in another direction by making the environments more interactive and dynamic by using sound, music and opponents to support their "mode". The introduction of movement features as ice-skating, sneaking & spying and the body as DJ clearly points out what architectonic possibilities a virtual environment like Half-Life, which allow the users to experiment with expanded aspects of spatial design and interaction, maintain.

Scenario workshop with HardHat Designer

HardHat Designer was first externally used together with a group of High-tech companies in a scenario workshop with the goal to design different technology integrated workspaces under the theme "Office of the Future". The task was to sketch on technical and spatial solutions, which could support the office work in the future. Participants came from several different invited companies. The participants were divided into three cross groups, which were given the assignment to present a story within a collectively predefined theme, and design a spatial solution for that story. After some introductory discussions, the groups were offered to try the HardHat Designer to visualize their ideas. The actual modeling was circulated within the group so that they all could try "hands-on". The

possibility to immediately view what you had done in 3D/VR was much appreciated and widely used, there was a constant shift between 2D and 3D/VR representations. The 3D/VR mode was used for evaluation of what was built

in 2D but also immediately generated a lot of new ideas, which then was executed in 2D.

The HardHat Designer session lasted for about one hour. After that a presentation was carried out on a large screen display. The persons who made the presentations where placed in front of a large projection of the virtual spaces they just had modeled. They could immediately interact with a Virtual Reality/Half-Life world in scale 1:1 of the scenario they just had designed. They could navigate around freely in the world and show the rest of the participants all the spatial arrangements and where they had placed and integrated the technology in their scenarios.

Design workshop with Multi-user interaction.

A second workshop was performed with a Real-estate Company together with one of its tenants (an R&D Company housed in a traditional office building). After a series of preparing exercises the task here was to use HardHat Designer to model a section of their work environment that had earlier in the workshop been identified as having a need for change. Within this section spatial reorganizations, based in the earlier collaborative discussions, where outlined. Before this second workshop the HardHat Designer where changed and improved in some ways. The surface or "playground" on which you build in 2D could now easily be varied in size up to 80 x 120 meters. You could also display a grid with a ruler. The Half-Life furniture where changed for a set of new more realistic modern furniture.

The participants worked together two persons in each group. They had a hard copy of the actual floor plan as point of departure and started with making a 3D model based on that. The exercise lasted for about two hours. During this time the groups (unskilled in 3D modeling) managed to schematically build workspaces of about 3-400-sq.m, furnish them and in some cases change and even add new rooms. One of the groups executed three different alternatives of the same space. The modeled Half-Life worlds were then displayed on the wall and used at the presentation session to present the proposal from each group about how to change and improve their work environment.

After the workshop the participants and others had access to the worlds on a Half-Life server over the Internet. There the workshop participants, their colleagues who had not attended the workshop, the research team and others could meet, look around and discuss (in Half-Life there is a text chat feature available for simple messages on line) the outcome from the design exercise in a multi user environment.

Conclusions

In our test cases we found that game based VR is a usable tool in architectural design processes: It is possible to use a simple "freeware" game editor and with nearly no computer experience communicate both spatial information and more complicated issues as moods and feelings. A 2D-based rapid modeling tool, HardHat Designer, integrated with a real-time computer game 3D/VR engine, was efficient and easy to use. Totally untrained persons were able to build rather complex furnished and lightened workspaces within short time limits.

Our conclusion is that the way you move around in virtual architectural space is important. The subjective viewpoint and the feeling that you are an actual body in space which are provided in for example Half-Life, are of great importance, especially if you are in a multi user world. The way in which you bump into the environment and a little instability in the first person view (since it tries to simulate the body moving) is also very relevant to the experience of the environment itself.

In the experiment workshops the design tool showed to be effective for expanding ideas and gain a better understanding of the design task. The tool was fun and stimulating to use, promoted innovative thinking and in that way activated the design process. The test cases with HardHat Designer, which the participant made as collaborative design events, expanded and changed the outcome of the exercises. There were observable differences between the list of qualities, which was formulated orally or in written form before the modeling sessions, and the final presented results. Our conclusion is that this is due to the fact that the actual design of the virtual spaces forced the participants to combine different ideas, negotiate and prioritize. In this way the design tool deepened the understanding of the complexity of space and in that way made the result much better. During the relatively short exercises all participating groups succeeded in building rather elaborated spatial configurations, which were innovative and had managed to integrate space and technology in a meaningful way.

The design dialogue was promoted in several ways. By presenting 2D and VR visualizations, the design tool provided a "holding ground and negotiation space" (Hendersson, 1999) which you could argue about and discuss, and in that way invited to a dialogue between the participants. The possibility to display the virtual worlds in real time on a big screen made it possible, during the presentations, to interact with the scenarios and walk around and show the worlds from several different viewpoints. It made the ideas easier to communicate and triggered a lot of comments and reactions from the rest of the participants in the workshops. The multi-user function clearly added communication quality. The possibility to revisit the built worlds in a multi-user environment on the Internet gave a new reflective distance to the outcomes from the engaging design sessions. Now several persons, who had not been involved in the collaborative design work, went into "your" worlds and interacted and made comments to your design. They acted as a second pair of eyes and in that way enhanced the multitude of expressed ideas in the process.

The detailing and visual quality of the 3D/VR worlds generated from HardHat Designer seems to be adequate for the chosen tasks. The few negative comments from users mainly concerned furniture design and lack of daylight. Half-Life seems to be designed primarily for interiors and a "garage" character was apparent in some of the built worlds. Despite this, an important quality in the Half-Life VR environment is the ability to very quickly produce spaces that is artificially lighted. Light has the power to make spatial representations "come alive" without having to overload them with irrelevant detailing. The 3D/VR spaces in our examples were regarded as very understandable and to have character by the participants, although the walls and furniture indeed was very simple.

582

Future work

Our next step is to develop a more integrated application, called "ForeSite Designer", which could be used as an "easy to use" interface to different 3D/VR applications. The goal is to keep the user interface in an extremely simple way, and only add some minor improvements that we identified through our test cases. The need of making quick drawings and in an easy way visualize them in 3D/VR, is something that we will need in the future. The problem right now is that we have to change sketch tools every time we want to use another visualization tool.

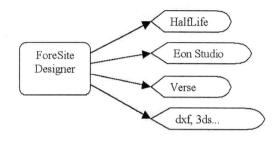

We are also aiming at visually develop the laboratory in which we build our worlds. For example with a tailored library of specially designed parts and furniture, further integrate the multi-user function in the design process and experiment with colors and the level of detail in each scene. In that endeavor we would like to do some testing with non-realistic textures and lightning in different environments, for example design a day lighted laboratory.

Acknowledgements

Johan Torstensson, student in Interaction Technology at Malmö University Collage, Sweden: For developing the HardHat Designer in a student project where Peter Fröst and Peter Warrén where tutors.

References

Cimerman, B. (2000), *Participatory Design in Architecture: can computers help?* In PDC 2000 Proceedings of the Participatory Design Conference, New York.

Davies, R. C. (2000), *Using Virtual Reality for Participatory Design and Brain Injury Rehabilitation*, Doctoral Thesis, Lund University, Sweden

Fröst, P. & Warrén, P. (2000). *Virtual Reality Used in a Collaborative Architectural Design Process*. Proceedings in Information Visualisation 2000, 2000 IEEE Conference on Information Visualization, July 19 – 21 2000, London; England

Henderson, K. (1999), *On Line and On Paper*. Visual Representations, Visual Culture, and Computer Graphics in Design Engineering. The MIT Press, Cambridge, USA

Horgen, T., Kukla, C., Porter, W. L. & Zack, G. (1996) *Staging Documents as the Glue in the Process Design*. The International Working Conference on Integration of Enterprise Information and Process. MIT November 14 – 15 1996.

Hornyansky Dahlholm, E. (1998). *Att forma sitt rum*, Doctoral Thesis, Lund Institute of Technology, Sweden.

Richens, P. & Trinder, M. (1999). *Exploiting the Internet to improve collaboration between users and design team*. The Martin Center for Architectural and Urban Studies, University of Cambridge Department of Architecture, UK.

Latency Compensation in Head Tracked Virtual Environments: Motion Component Influences on Predictor Discriminability

Bernard D. Adelstein, Jae Y. Jung*, Stephen R. Ellis
NASA Ames Research Center, MS 262-2 Moffett Field, CA, 94035
*University of California, Berkeley, CA, 94720

Abstract

Subjective discriminability of artifacts produced by predictive compensation of head tracking time delay in a virtual environment (VE) was examined. Optimizations of a standard Kalman Filter predictor structure according to different cost criteria as well as the amount of compensated time delay both affected this discriminability for a group of 12 observers. New quantitative metrics applied to head motion records from a second group imply that discriminability is influenced both by prediction accuracy and noise and that these metrics can guide the design of improved predictive compensators for head tracked VEs.

1. Introduction

Predictive compensation has long been considered as a means for mitigating the effects of end-to-end (sensor-to-display) latency in virtual environment (VE) systems. (e.g., Liang, Shaw, & Green, 1991; Friedmann, Starner, & Pentland, 1992) This latency, which makes VE images temporally lag the user's input motion, is attributable to the sum of pipeline delays both within and between individual system components such as sensors, the interactive simulation computations, and display rendering. Predictive compensators, operating on current VE sensor measurements of head or body part position and orientation, extrapolate future positions and orientations. Ideally, when extrapolated over the interval matching the actual delay, a *perfect* predictor should cancel out system delay, enabling the temporal consequences of user motion in the VE to be synchronized with the user's actual input, while not introducing additional artifacts (e.g., noise or overshoot). Notwithstanding, we presume that a *practical* predictor need neither remove all latency nor avoid all artifacts, it need only avoid making the user aware of those imperfections.

While visually mediated manual tracking experiments in VEs have demonstrated the potential benefit of predictive compensation for human performance (Wu & Ouhyoung, 1995; Nelson, Hettinger, Haas, Russell, Warm, Dember, & Stoffregen, 1995; So & Griffin, 1996), the direct perceptual impact of such compensation has only recently been examined (Jung, Adelstein, & Ellis, 2000a,b; Adelstein, Jung, & Ellis, 2001). In prior work, the engineering performance of predictors has been evaluated on the basis of simple RMS error metrics (Azuma & Bishop, 1994; Kiruluta, Eizenman, & Pasupathy, 1997; Akatsuka & Bekey, 1998) as well as less rigorously from the cursory appearance of co-plotted motion and prediction traces (e.g., Liang et al., 1991; Friedmann et al., 1992). In fact, minimization of RMS error between the input body part motion and predicted VE output has served as the basis for numerical optimization of predictor parameter sets (Liang et al., 1991; Azuma & Bishop, 1994; Mazuryk & Gervautz, 1995; Kiruluta et al., 1997). However, as can be observed from sample plots in the cited prior studies, these evaluation methods may imply good performance but still fail to completely capture the undesirable noise and overshoot artifacts introduced by prediction. Consequently, we consider these prior metrics to be incomplete indicators of the user's perceptual experience and therefore insufficient for ascertaining the full impact of predictive compensation on human performance in VEs.

The aim of this paper is to use the link between new engineering quantities and subjective evaluation to improve the performance of predictive compensators. To accomplish this, we recap a perceptual experiment on the discriminability of VE predictor artifacts for different predictor parameterizations and different amounts of compensated latency (Jung et al., 2000b; Adelstein et al., 2001). We then introduce quantitative engineering scores derived from head motion recordings that underwent the same predictive compensation used in the initial experiment.

2. Methods

2.1 Predictive Compensation

The predictive compensator used in these studies is based on an Extended Kalman Filter formulation (e.g., Gelb, 1979). Following Azuma and Bishop (1994) and others, the interrelation between motion states is described by a third-order purely kinematic model—i.e., velocity is the derivative of displacement, acceleration the derivative of velocity, and the expected value of jerk (the derivative of acceleration) is zero-valued. The three translational components (x, y, and z) are treated as identical, independent systems of state equations. The rotational components consisting of quaternions for angular displacement and Euler velocities and accelerations are dealt with in unison as an additional single system of state equations. Unlike Azuma and Bishop (1994), we use a computationally more compact and efficient discrete-time formulation with a time-varying state transition matrix for the rotational system of equations. The complete mathematical details of our predictor implementation can be found in (Jung et al., 2000b) and (Adelstein et al., 2001).

Since the state equations of our kinematic model have no numerical weighting terms, this portion of the Kalman Filter (KF) cannot be tuned. This leaves only the standard KF plant and sensor noise covariance matrices as tunable parameters for adjustment of predictor performance. The plant here is the human operator whose stochastic input drives the KF motion states while the sensor is the device measuring the operator's head and body part motion. Because reasonable numerical ranges of VE sensor covariance had little effect on predictor performance, our investigations focused on plant noise parameterization. Plant noise, which at a fundamental level arises from physiological muscle activity and neural signals, is therefore not easy to measure or model analytically. Thus, like Azuma and Bishop (1994) and Liang et al. (1992), we sought to parameterize plant covariance offline by numerically optimizing predictor performance according to desired functional cost criteria.

Since informal analytic observation showed limited predictor sensitivity to the translational plant covariance parameterization, we emphasize in our results the contribution of the rotational system's optimization. This emphasis is also predicated by the obvious dominance of rotation over translation due to the projective geometry for the spatial rendering of head tracked head mounted display (HMD) imagery (Zikan, Curtis, Sowizral, & Janin, 1995), especially given the experimental head motions described below.

2.2 Predictor Optimization

Two optimization cost functions were used in the alternative parameterization of our experimental KF predictor's noise covariance matrices. The first, used by Azuma and Bishop (1994), considers the rotational difference (i.e., the error) between actual and predicted head orientation. The direction-free angular magnitude, or twist, of this spatial rotation error leads to a *twist* cost function that is simply the sum of individual rotation error magnitudes, $\Delta\theta_i$, over the sampled data set ($i = 1,...,N$)

$$J_{twist} = \sqrt{\sum_{i=1}^{N}(\Delta\theta_i)^2}.$$ (1)

While angular error would be expected from projective geometry considerations to be an important indicator for prediction accuracy, the cost function of Eq. (1) does not permit selective weighting of rotation errors in preferred directions. Defining a vector from a point on the HMD to an "object of interest" in the VE enables computation of the translational displacement caused by the rotation error between the predicted and actual location of this rendered object. The resulting cost function is

$$J_{disp} = \sqrt{\sum_{i=1}^{N}\left[\alpha(\Delta x_i)^2 + \beta(\Delta y_i)^2 + \gamma(\Delta z_i)^2\right]}$$ (2)

where α, β, and γ represent weighting factors to selectively penalize error components in a Cartesian coordinate frame fixed to the HMD. For our VE system, the Δy and Δz error components, which lie in the plane of the HMD, are weighted equally. Small twist angle errors, however, produce negligible Δx, making α less important.

Predictive compensation exacerbates the influence of plant and sensor noise on higher frequency VE image jitter, particularly when based on the purely kinematic model described above due to the differentiator action needed to generate velocity and acceleration state estimates from displacement measurements. A cost function, J_{noise}, identical to Eq. (2), by which predictor response to high-pass filtered (> 5 Hz) is minimized, was also formulated. Because an optimization based solely on this noise criterion would yield a predictor with good noise attenuation but poor low-frequency tracking, the hybrid cost function

$$J_{hyb} = A\,J_{disp} + B\,J_{noise}$$ (3)

was introduced, in which the weighting terms A and B allow design trade-offs between the relative importance of tracking accuracy and noise reduction.

Two new predictor parameterizations, termed *twist* and *hybrid*, respectively employing the cost functions in Eqs. (1) and (3), were developed through offline numerical simulations on pre-recorded head motion (~20 s of data sampled at 120 Hz) with the optimizations targeted for a 50 ms look-ahead extrapolation. The hybrid optimization employed two single head motion data sets collected from a single subject (author JYJ): the first containing side-to-side head motion that would later be required of the subjects in the experiment described below, while the second (used for the noise portion of the hybrid design, J_{noise}), was based on a high-pass filtered (2nd order, 0.5 and 5 Hz breakpoints) head motion record while the operator was sitting still and looking straight ahead. Full details on the derivation of the optimization and the implementation and performance of the predictors is available in (Jung et al., 2000b) and (Adelstein et al., 2001)

3. Experiments

3.1 VE System Hardware and Software

The experiment VE and KF predictor software ran on a four CPU (R4400) SGI Onyx workstation with dual-pipeline RealityEngine-2 graphics. The subjects viewed the VE in a Virtual Research V8 HMD. The position and orientation of the subjects' head as well as those of the visually presented target object were measured by separate Polhemus FasTrak instruments, each with a single receiver and single transmitter, and each interfaced to its own Onyx ASO 115.2 KBaud serial port. A variety of software techniques (Jacoby, Adelstein, & Ellis, 1996) enable us to produce fast VEs with low latency (mean ± stdev of 30±5 ms for quaternion reports), high update rates (60 Hz), and reduced temporal variability that make possible the controlled addition of time delays required for our experimental study.

The VE for the experiments consisted solely of the target object, a mid-photopic virtual faceted sphere (i.e., target) in a dark, empty space, as described by Ellis, Young, Adelstein, and Ehrlich (1999a,b). Subjects were seated with the HMD's FasTrak receiver 0.4 m below its transmitter, with the virtual sphere (the so-called "object of interest" used in the hybrid optimization) situated 0.8 m in front of the HMD. Ideally, in the absence of any delay with perfect noise-free measurement, the image of the sphere should move on the HMD's LCD panels in a manner such that it appears to the observer to be fixed in space. However, in the presence of the inevitable delays or other artifacts, the virtual sphere will not appear to be perfectly stationary and may move about the ideal location as the observer's head turns.

3.1 Experiment 1 Protocol

The ultimate goal of our parameterization effort is to generate predictors that remove VE system latency while at the same time not introducing perceptible compensation artifacts. The aim of this experiment is to study user awareness of any artifacts due to the presence of imperfect predictive compensation. The experimental approach is derived from a technique to assess subjective detectability of changes in VE latency (Ellis et al., 1999a,b).

The seated subjects were required to yaw their heads from side-to-side in time to a 80 beat/min metronome (1.5 s per full back-and-forth cycle) while maintaining the virtual sphere in view, which corresponds to a ±15° motion range. Subjects were asked in a two alternative forced choice protocol to judge whether sequentially presented VE conditions were the same or different using any perceivable quality in the appearance of the virtual sphere as they moved their head. The VE could be running either Condition A, at the baseline 30 ms orientation latency without prediction, or Condition B, in which an artificial latency added to the baseline was matched by the predictor's compensation interval.

Each of six equally spaced latency values (16.7 to 100 ms) was blocked into its own randomly ordered set of 20 judgments such that each of the four possible A-B pairings was repeated five times. The proportion of correct discriminations was calculated for each set of 20 responses. The blocking of individual latencies was also randomized. Participants completed all the tests with one predictor before proceeding to the next. The six possible presentation orders for the three predictor parameterizations—*hybrid*, *twist*, and a control condition using the *default* twist-optimized parameters reported by Azuma and Bishop (1994)—were balanced between the 12 participants. The participants were either lab members or paid naive recruits; all had normal or corrected to normal vision and no neuromotor impairments.

3.2 Experiment 2 Protocol

A second group of eleven different participants recruited from among our research colleagues, using the same equipment, were similarly seated and asked to perform the same side-to-side head motion as the previous group. The VE was presented only one time to each participant at the minimum VE system delay with no prediction. A 22.5 s record of each participants' head motion, sampled at 120 Hz, was stored for offline processing to provide objective predictor performance scores for different parameterizations and look-ahead intervals. All participants had normal or corrected to normal vision and no neuromotor impairments.

4. Results

4.1 Discriminability

Fig. 1 shows that the percentage of correct discriminations between the compensated and minimal delay conditions averaged across all twelve subjects grows monotonically with the number of steps of added latency. Essentially, artifact discriminability grows as predictor look-ahead interval is increased. For the balanced proportion of stimulus pairs, 50% correct response would be expected if subjects were guessing randomly. Additionally, discriminability for the hybrid parameterization is lower than for the two parameterizations.

A three-way (latency X predictor type X predictor order) ANOVA was carried out on the proportional responses following the arcsine square root transformation to convert the data to a normal distribution (Sachs, 1984, p. 339). The main effects of added latency ($F = 56.347$; $df = 5,30$; $p < .001$) and predictor type ($F = 11.239$; $df = 2,12$; $p < .002$) on the proportion of correct responses were significant. Neither predictor order nor any of the interactions between the main factors were statistically significant.

586

Post hoc Scheffe contrasts on the transformed-to-normal data indicated that none of the three parameterizations were significantly different from the 50% random guessing level at 16.7 ms of predictive compensation. The upper bound ($p < .1$) for this contrast corresponds to 65% correct response for $N = 12$ subjects. Moreover, the difference in discriminability between the hybrid version and random guessing was also not significant at 33.3. ms (and only marginally significant at 50 ms). In other words, for these particular conditions, prediction artifacts were not specifically discernible. Additional significant ($p < .1$) contrasts between the individual predictors, marked by the heavy vertical bands in Fig. 1, show the hybrid to be less discriminable than the twist parameterization for longer prediction intervals (66.7 to 100 ms).

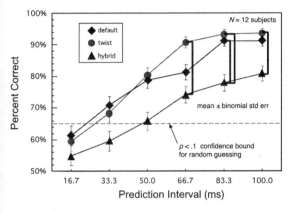

Fig. 1 Percent correct discriminations as a function of predictor parameterization and prediction interval.

4.2 Predicted Head Motion

The second group's sampled head motion records were run through the default, twist and hybrid compensator parameterizations at the same prediction intervals tested in Experiment 1. The predictor's orientation output components for each participant were then evaluated in terms of "complete response" (voluntary motion plus noise) and higher frequency noise resulting from sensor and other system sources. The resulting scores use either the raw twist magnitude directly, calculated similarly to the cost function of Eq. (1), or its decomposition into horizontal and vertical components, as used for Eq. (2).

The "complete response" compares the difference between the input motion (assumed to be undelayed) and the predictor's output when fully compensating an artificially added delay. This first quantity is then normalized by the difference between the input and the uncompensated delay. Thus, "complete response" scores less than unity would indicate that the predictor confers some degree of improvement over the uncompensated latency. The noise scores are computed by high-pass filtering (2nd order, 0.5 and 5 Hz breakpoints) the predictor output for the compensated head motion records and then normalizing by the uncompensated records' high-pass filtered noise. A noise score greater than unity is indicative of the predictor exacerbating high frequency noise, which would be manifest as increased image jitter in the HMD.

The twist and horizontal "complete response" scores averaged across the 11 participants in Fig. 2 show that the prediction error in the twist and horizontal components is less than that for the equivalent uncompensated latency at all prediction intervals. On the other hand, the vertical components were degraded at nearly all prediction intervals, with the hybrid parameterization most adversely affected. It is also noteworthy that the default parameterization fares worst for the twist scores, but best for the horizontal projection of the complete response.

The noise scores in Fig. 2 increase monotonically with prediction interval in all cases, which is not unexpected given the differentiator action of the compensators' KF state equations. In general, the hybrid parameterization's noise grows more slowly with prediction interval and is significantly better in the vertical projection onto the HMD than the other two parameterizations.

5. Discussion

The engineering metrics reported in Fig. 2 confirm that each parameterization obeys the cost criteria which governed its formulation. The parameterization motion data set comprised yaw motion—it essentially had no vertical component other than noise. Thus, the hybrid cost function, which projected noise and prediction errors into HMD coordinates, led to an optimization for voluntary motion in the horizontal component and noise in the vertical. The default and twist parameterizations, both developed for a twist cost function, likewise did best in terms of twist score for the complete response. Because the artifacts associated with the hybrid parameterization proved less discriminable in the subjective experiment, it appears prediction artifact discriminability may be more closely tied to the objective scores in the second experiment, suggesting that the projective decomposition can be useful both for optimization and evaluation purposes.

587

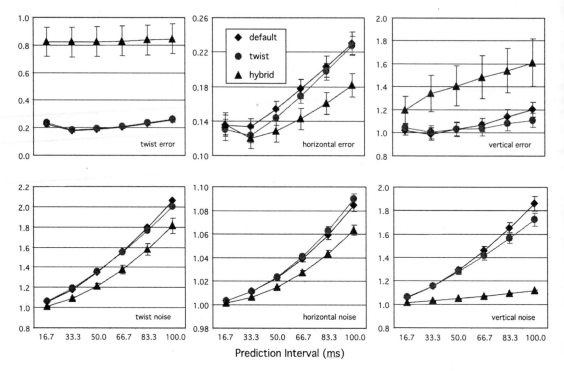

Fig. 2 Predictive compensator effects on twist and on horizontally and vertically projected "complete response" (top row) and high frequency noise (bottom row) motion components as a function of predictor interval. Complete response and noise (mean ± std error for $N = 11$ subjects) are plotted in normalized units described in the text.

References

Adelstein, B., Jung, J., & Ellis, S. (2001). Predictive compensator optimization for head tracking lag in virtual environments. NASA/TM—20001–209627.

Akatsuka, Y., & Bekey, G.A. (1998). Compensation for end to end delays in a VR system. *Proceedings, IEEE Virtual Reality Annual International Symposium*, pp. 156-159.

Azuma, R., & Bishop, G. (1994). Improving static and dynamic registration in an optical see-through display. *Proceedings, SIGGRAPH*, pp. 197-204.

Ellis, S., Young, M., Adelstein, B., & Ehrlich, S. (1999a). Discrimination of changes of latency during voluntary hand movement of virtual objects. *Proceedings, 43rd Annual Meeting Human Factors Ergonomics Society*, Houston TX, pp. 1182-1186.

Ellis, S., Young, M., Adelstein, B., & Ehrlich, S. (1999b). Discrimination of changes of latency during head movement. *Proceedings, HCI'99*, Munich, pp. 1129-1133.

Friedmann, M., Starner, S., & Pentland, A. (1992). Synchronization in virtual realities. *Presence, 1*(1), 139-144.

Gelb, A., ed. (1979). *Applied Optimal Estimation*, 5th ed, Cambridge, MA: MIT Press.

Jacoby, R., Adelstein, B., & Ellis, S. (1996). Improved temporal response in virtual environment hardware and software. *Proceedings, SPIE Conference Stereoscopic Displays and Applications VII, 2653*, pp. 271-284..

Jung, J., Adelstein, B., & Ellis, S. (2000a). Discriminability of Prediction Artifacts in a Time-Delayed Virtual Environment. *Proceedings, 44th Annual Meeting Human Factors Ergonomics Society*, San Diego CA, pp. 499-502.

Jung, J., Adelstein, B., & Ellis, S. (2000b). Predictive compensator optimization for head tracking lag in virtual environments. *Proceedings, IMAGE (Innovative Modeling and Advanced Generation of Environments) 2000*, pp. 123-132.

Kiruluta, A., Eizenman, M., & Pasupathy, S. (1997). Predictive head move movement tracking using a Kalman filter. *IEEE Trans. Systems, Man, Cybernetics—B, 27*(2), 326-331.

Liang, J., Shaw, C., & Green, M. (1991). On temporal-spatial realism in the virtual reality environment. *Proceedings, ACM Symposium on User Interface Software and Technology (UIST'91) 1991*, Hilton Head SC, pp. 19-25.

Mazuryk, T. & Gervautz, M. (1995). Two-step prediction and image deflection for exact head tracking in virtual environments. *Eurographics'95, 14*(3), pp. C-29-42.

Nelson, W.T., Hettinger, L.J., Haas, M.W., Russell, C.A., Warm, J.S., Dember, W.N., & Stoffregen, T.A. (1995). Compensation for the effects of time delay in a helmet-mounted display: perceptual adaptation versus algorithmic prediction. *Proceedings, SPIE Conference Helmet- and Head-Mounted Displays and Symbology Design Requirements II, 2465*, pp. 154-164.

Sachs, L. (1984). *Applied Statistics: A Handbook of Techniques*, New York: Springer-Verlag.

So, R.H.Y., & Griffin, M.J. (1996). Experimental studies of the use of phase lead filters to compensate lags in head-coupled visual displays *IEEE Trans. Systems, Man, Cybernetics—A, 26*(4), 445-454.

Wu, J.-R., & Ouhyoung, M. (1995). A 3D tracking experiment on latency and its compensation methods in virtual environments. *Proceedings, ACM Symposium on User Interface Software and Technology (UIST'95)*, pp. 41-49.

Zikan, K., Curtis, W.D., Sowizral, H.A., & Janin, A.L. (1995). A note on dynamics of human head motions and on predictive filtering of head-set orientations. *Proceedings, SPIE Conference Telemanipulator and Telepresence Technologies, 2351*, pp. 328-336.

DEVELOPMENT OF A COLLABORATIVE VIRTUAL ENVIRONMENT FOR FINITE ELEMENT SIMULATION

M. K. Abdul-Jalil[*] and C.L. Bloebaum[†]

New York State Center of Engineering Design and Industrial Innovation (NYSCEDII)
State University of New York at Buffalo, New York

ABSTRACT

The introduction of Virtual Reality (VR) technology has added a new dimension to the field of scientific visualization. VR brings the sense of realism in visualization technology allowing users to view graphical objects and scientific data similar to the real life. Presently VR is widely used in scientific visualization to improve the human perception and understanding of scientific data. This research extends the VR application one step further into the collaborative application where users in different remote location can interact in a single immersive VR environment. This research is taking advantage of the advent of VR graphics technology and remote connectivity which allow high speed data transfer between distant locations. The environment simulates voice interaction, users representation using avatars and object manipulation. Multiple users can log on at the same time and each will be represented by an avatar. The designers have the ability to view and manipulate a 3D virtual reality model while communicating interactively to each other.

1.0 INTRODUCTION

Scientific problems have become more complex and multidisciplinary in nature. In engineering field for an example, the concept of Concurrent Engineering (CE) is a common scenario where people from multidisciplinary need to work together to produce a common agreement. A complex engineering problem such as the design of an aircraft wing will involve different multidisciplinary teams including, for instance, material technology, manufacturing, design, marketing and finance sectors. During the analysis phase, groups such as aerodynamics, structures, and propulsion, among others, must interact and negotiate in order to complete a complex analysis phase. We may often find these disciplines are geographically separated from each other. An aircraft engine that is attached to a wing might be manufactured in location A while the wing itself is designed and manufactured in location B. In this particular situation, engineers and designers from multidisciplinary divisions need an effective communication tool to achieve an optimum design solution to design problems. However, utilizing visualization techniques is also not an easy task especially when designers are geographically dispersed. One of the primary reasons has been the complexity of the problems worked with. For examples, a huge finite element data with thousands of nodes and element to work with will certainly requires a tremendous amount of computation time. Therefore new computations techniques need to be explored and developed to facilitate fast computation time such as parallel and distributed computing, hierarchical feature extraction and selective visualization. Another important issue is the speed of the connection lines which contributes to the lag time when information is sent to other designers. Eliminating this lag time is important, but another fundamental issue still exists. Users should be able to understand a design representation in a meaningful way especially when dealing with a complex design data or model. Virtual Reality provides users with a new dimension in supplying them with a life-like setup to enable them to interact and manipulate object just like in a real situation. The advent of Internet technology provides another useful means for geographically dispersed designers to communicate with each other.

2.0 RESEARCH MOTIVATION

Many virtual reality technologies have been developed that provide means of interfacing and exchanging data with other programs for seamless transition from visualization to solution and vice-versa. There are a number of CAD packages and modeling language that can be directly used with a lot of virtual reality packages including AutoCAD, OpenGL and Virtual Reality Programming Language (VRML) and Pro/Engineer. The integration of the virtual

---

[*] Research Assistant [†] Professor

reality technology and high-speed data communication has made the concept of collaborative virtual environment an important breakthrough in engineering design. This research is aimed to take these two technologies one step further by bringing designers in different geographic locations into a single immersive Virtual Reality environment via fast networking and Internet connectivity. The environment proposed in this research consists of a virtual room (VRoom) where designers meet and communicate 'virtually'. Each designer can see other participants in avatar form, communicate, and manipulate the design model loaded to the room. The ultimate goal of the CVE research is not merely to reproduce a real face-to-face meeting in every details, but to provide the "next generation" interface for collaborators, at remote locations, to work together in a virtual environment that is seamlessly enhanced by computation and large databases. When designers are tele-immersed, they are able to see and interact with each other and objects in a virtual environment. Their presence will be depicted by life-like representations of themselves (avatars) that are generated by real-time modeling techniques.

Various Collaborative Virtual Environments (CVE) have been developed in the area of distant learning, games and entertainment. A collaborative and completely immersive tennis game[1] has been developed by a group of researchers at MIRALab, University of Geneva. The teleimmersion group at the Electronic Visualization Lab (EVL) at University of Illinois at Chicago has been involved in the development of virtual environment enabling multiple globally situated participants to collaborate over high-speed and high-bandwidth networks connected to heterogeneous supercomputing resources and large data stores. Among the projects developed are CAVE Research Network (CAVERN)[2], Laboratory for Analyzing and Reconstructing Artifacts (LARA), Tele-Immersive Data Explorer (TIDE), Tandem[3], Collaborative Architectural Layout Via Immersive Navigation (CALVIN)[4], Narrative Immersive Constructionist/Collaborative Educational Environments (NICE)[5] and The Round Earth Project[6].

3.0 The Collaborative Virtual Room Environment

The process of simulating a virtual environment is not really complicated and expensive. It requires a good workstation that supports VR software, stereoscopic glasses and a microphone. The combination of stereoscopic mode of the monitor display and glasses creates the sense of depth in a 3D world which simulate the 'immersive' or 'virtual' environment. The CVE developed in this research consist of four major components: The Virtual Room (VRoom), the collaborative module or networking, the FEA module, and the database. The details of every module are explained in the following sections.

3.1.1 The Virtual Meeting Room (VRoom)

The VRoom is the major component of the collaborative virtual environment. It serves as a meeting environment where designers can meet, communicate, analyze design problems in the form of 3D virtual objects, make changes to design parameters and display the effects in a totally immersive environment. The virtual room consists of a room bounded by four invisible walls, a ceiling and a floor. The software used to create the virtual room is WorldToolKit (WTK) [7] which consists of a set of C functions that provide drivers for managing the peripherals used for interacting with the virtual world. The geometries are created using a neutral file format (NFF) which define the coordinate points (vertices) of the object and the polygon connectivity. WTK reads several common graphic file formats including Virtual Reality Modeling Language (VRML), Autodesk DXF, Wavefront OBJ, Autodesk 3D studio mesh, NFF neutral file and Pro/Engineer Render SLP. It also exports to DXF, VRML and NFF file formats. The walk-through simulation is done with the help of a 3D mouse or a regular mouse. There are two distinct virtual rooms developed in this project. The first is with a set of pull-down menus and the other is menuless window-type frame as shown in Figure 1. The one with pull-down menu was designed to be used by the VRoom manager who is in charge of managing the virtual meeting. There is a communication protocol set up to manage the virtual meeting. The manager is responsible to manage the meeting, and to grant permission to save, load, manipulate or make changes to the design model. He or she is the only one who has the access to the database, has the right to grant or deny entry to virtual room and has the right to exploit or change the configuration or setup of models loaded to the virtual room. This is to ensure a virtual meeting run in an orderly manner. The features of the VR menu include the functions to load and save the design model, exit the virtual room, change the scale and position of the model, rotate the model around and load another model. All the controls described are performed using mouse clicks. The menuless virtual room was designed for the remaining participants of the virtual room.

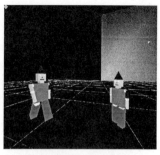 

(a) Participant's window   (b) Manager's window

Figure 1 Sample of VRoom environment with three users logged in at the same time

3.1.2 Networking the Collaborative Environment

The WorldToolKit (WTK) networking capability enables us to build applications that can asynchronously communicate over an Ethernet between several UNIX, PC and IRIX workstations. In this project, up to 16 users can log into a single virtual world where each user can see other people in the world. A Graphical object or avatar representing each user in the simulation is located at the user's viewpoint. WTK network applications share a common API so that a single application can be run on both PC and UNIX platforms without modification. WTK's networking capability is built upon IP (Internet Protocol) and UDP (User Datagram Protocol) guidelines. The database for maintaining the VRoom including the files and images needed for virtual meeting purposes resides on the server machine which is only accessible to the VRoom manager. For network security purposes, only a registered member can get access to the virtual meeting environment. Once connected to VRoom server, a user can work simultaneously or asynchronously with other users. Each user will be represented with an avatar as shown in the Figure 1. Whenever a participant moves around or points at a certain location, the rest or the participants will be able to observe these movements in real time. In addition to that, every user can talk to the room members, point to a position design model, which simulate the situation in real life. The audio capability for VRoom environment was developed using a program called Speak Freely[8] which is currently available at no cost. This is a program that allows two or more people to conduct a real-time voice conference over the Internet or any other TCP/IP network. It supports a variety of compression protocols, such as GSM, ADPCM, LPC, and LPC-10. The program was developed by John Walker, the founder of AutoDesk together with Brian C. Wiles. The program is available for multiplatform including UNIX, Windows 95/98/NT, and Linux.

3.1.3 The Database

One of the major issues in this collaborative virtual environment is real time model update after the perturbation of the original design. A way of sending out the update information in real time has to be found. WorldToolKit package allows graphical objects to be directly read in VRML format from any website. VRML is an ASCII format which is relatively fast to read and save. Therefore in this project the database is placed directly in a web server. Once a model is updated, the rest of the VRoom users are updated in real time. The manager has the capability of accessing the database to manipulate, and allowing changes to be made to the design problem while the rest of the participants only have the capability to communicate to each and to view the design model loaded into the room. This setup is specifically designed in order to make sure the virtual meeting is run in an orderly manner.

3.1.4 Finite Element Analysis (FEA) module

This section aims to illustrate the implementation of a collaborative virtual communication using the VRoom described in the preceding section. A simple cantilever beam problem was chosen as a case study as shown in Figure 2. The initial model was constructed using NFF file format. The most important issue to be solved in a

591

Collaborative Virtual FEA is fast computation to yield real time results. If this is not achieved, the purpose of creating this type of meeting environment is defeated. Conventional FEA package normally takes a considerable

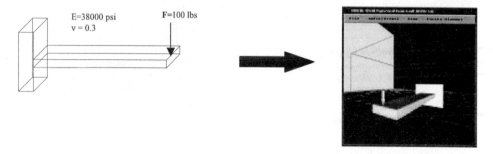

Figure 2 The cantilever beam example problem

amount of time especially when a large scale FEA is conducted. Since the purpose of creating the VRoom is to assist designers in decision-making process, an approximation method with reasonable accuracy can be employed. Therefore the Design Sensitivity Analysis (DSA) technique was chosen to approximate the stress and deflection of the cantilever beam because of its fast approximation capability. The DSA approximates the Von Mises stress or deflection of the perturbed beam based on two original points generated from initial ANSYS runs. In this example, the parameters of the design are mainly the width, length, thickness, force and the material properties. However for the sake of simplicity, only the length and the width are variable. The VRoom manager can vary these two parameters and see the results of the Von Mises stress and deflection as shown in Figure 4. The rest of the users would also be able to see this geometric manipulation and the results in real time.

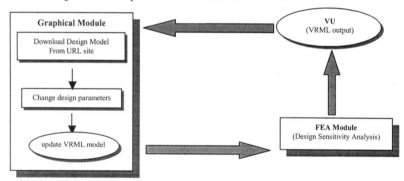

Figure 3 Virtual FEA analysis process

The post-processing of the FEA data is done using a visualization package called VU[13]. VU reads in the output data from FEA module and generate the VRML model of the stress contour and the deflected shape of the beam, which is later, read into the VRoom environment via the web address. The flow of the Virtual FEA process is shown in Figure 3.

4.0 CONCLUSIONS AND FUTURE WORK

The virtual design environment allows designers to gain a unique perspective of a complex design. Through virtual reality techniques, virtual models can be created and manipulated to give designers better insight about engineering problems. With the help of engineering computation methods like finite element analysis, designers are able to discuss real engineering problems in real time giving them an effective means to communicate to each other to facilitate a more effective decision making process. This project is a first step towards the development of a fully collaborative virtual environment for engineering design. Future works will include a more complex engineering problem which will require a more robust database management and powerful computational techniques to produce real-time results.

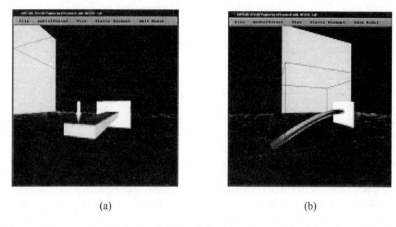

(a)   (b)

Figure 4 The perturbed design model (a) and the deflection and Von Misses stress distribution (b)

ACKNOWLEDGEMENT

The authors wish to acknowledge support of this work under NSF grants DMII 9553210 and DMII 9622314.

REFERENCES

[1] Molet, T., Aubel, A., Capin, T., Carion, S., Lee, E., Thalmann, N., Noser, H., Pandzic, I., Sannier, G., Thalmann, D., "Anyone for Tennis?", *Presence*, Vol.8, No. 2, April 1999.

[2] Johnson, A., Leigh, J., DeFanti, T., Brown, M., Sandin, D., "CAVERN: the CAVE Research Network," *Proceedings of 1st International Symposium on Multimedia Virtual Laboratory*, Tokyo, Japan, March 25, 1998, p.p. 15-27.

[3] B. A. Goldstein, "TANDEM: A Component-Bases Framework for Interactive Collaborative Virtual Reality", *M.Sc. Thesis*, University of Illinois at Chicago, 2000.

[4] Leigh, J., Johnson, A. E., Vasilakis, C., DeFanti, T. A., "CALVIN: an Immersimedia Design Environment Utilizing Heterogeneous Perspectives", *Proc. IEEE International Conference on Multimedia Computing and Systems*, 1996.

[5] Roussos, M., Johnson, A., Moher, T., Leigh, J., Vasilakis, C., and Barnes, C., "Learning and Building Together in an Immersive Virtual World", *Presence*, 8(3), special issue on Virtual Environments and Learning; edited by Winn, William and Moshell, Michale J., MIT Press, pp. 247-263, June 1999.

[6] Johnson, A., Moher, T., Ohlsson, S., Gillingham, M., "The Round Earth Project: Deep Learning in a Collaborative Virtual World", *Proceedings of IEEE VR99*, Houston TX, March 13-17, 1999, p.p. 164-171.

[7] WoldToolKit Reference Manual Release 7.0, Sense8 Corporation, Mill Valley, California, 1997.

[8] Walker, J., Wiles, B. C., Speak Freely 7.1, 20 June 2000, http://www.speakfreely.org.

Issues and non-issues in the production of high-resolution auditory virtual environments.

Fred Wightman[1,2], Doris Kistler[2], and Pavel Zahorik[3]

[1]Department of Psychology and [2]Waisman Center, University of Wisconsin-Madison;
[3]Department of Psychology, University of California-Santa Barbara

The past decade has produced an explosion of interest in and development of virtual environments. Understandably, the emphasis has been in the visual domain. Many would claim that vision is the most important of the five senses, and thus the primary means for interacting with and extracting information about the environment. However, the technical challenges of producing a high-resolution visual virtual environment are enormous, and for this reason current visual virtual environments are still quite limited. The importance of audition, particularly as a complement to vision, is becoming increasingly clear to designers of virtual environments. That "the ears tell the eyes where to look" has been acknowledged in auditory research circles for decades and is now finally becoming accepted in the virtual environment community. The technical obstacles in the auditory domain are fewer, and in fact, it is now possible to produce virtual auditory images that are indistinguishable from real auditory images, something no visual display can match. Unfortunately, although a substantial number of research studies has yielded a solid set of guidelines for producing high quality virtual auditory environments, some of the research findings seem to have been overlooked; unsubstantiated claims based on anecdotal evidence or "conventional wisdom" abound. This paper will attempt to separate fact from fantasy in this area by summarizing the research identifying the real technical and theoretical issues that bear on the ultimate fidelity of a virtual auditory environment. In the process we will also discuss some of what we and others conducting research in this area find to be non-issues. The main topic areas include measurement, processing, and use of head-related transfer functions (especially the need for individualization), head tracking and the impact of head movements, distance perception and the role of reverberation, headphone versus loudspeaker presentation, and the relative importance of interaural time and intensity differences and spectral cues.

1. Introduction The ultimate goal for many virtual environment designers is duplication of some or all of the features of a real environment. Thus, ideally, the virtual environment should be indistinguishable from the real environment. Unfortunately, while great strides have been made, and while technology continues to improve, current virtual environments fall well short of this goal. Even though the focus of development has consistently been on the visual features of the environment, the visual displays used in most virtual environment systems are characterized by poor resolution, long delays, and a narrow field of view, producing an almost cartoon-like representation of a real environment.

The situation is quite different in the auditory domain. The importance of audition, especially as a complement to vision, is becoming clear to virtual environment developers. The technical obstacles to producing a very high quality virtual auditory environment are fewer than in the visual domain, and it is now possible, under laboratory conditions, to produce wide bandwidth virtual auditory images that are indistinguishable from real auditory images (e.g., Zahorik, Wightman, & Kistler, 1995; Hartmann & Wittenburg, 1996; Langendijk & Bronkhorst, 2000). The rules for producing such near-perfect virtual auditory environments and the perceptual consequences of compromises to these rules have been the subjects of numerous research efforts during the past 15 years. Unfortunately, the pace of application of virtual auditory technology (often called 3-D sound technology) has far outstripped the pace of research, to the extent that "conventional wisdom" about the relationships between the technical and perceptual issues has largely replaced rigorous research results.

The purpose of this article is to review of some of the research that addresses important technical and perceptual issues bearing on the ultimate fidelity of a virtual auditory environment. We will also identify some of what we call the "non-issues". The technical issues include the measurement, processing, and use of head-related transfer functions, and the specific means by which the virtual auditory stimuli are presented to listeners (e.g., by headphones and/or by loudspeakers). The main perceptual issues relate to the relative importance of temporal and spectral cues, resolution of positional ambiguities, distance perception, and the role of dynamic cues.

2. Head-Related Transfer Function Issues. In a real environment, sounds reach the eardrums only after substantial modification by the upper torso, head, and pinnae. These acoustical effects must be included for a virtual environment to produce sounds at the eardrums that are identical to the sounds that would have been produced in a real environment. "Head-related transfer functions" (HRTFs) are mathematical functions that describe all the acoustical modifications of a sound from a point in space to a listener's eardrums. Thus, to duplicate at a listener's eardrums the sound produced by a real source at a specific point in space, the left- and right-ear HRTFs for that point in space must be known. Of course, if the real environment to be simulated includes room effects such as reflections from walls, ceiling, and floor, the situation is complicated considerably, since a single source may then produce sound waves that arrive at a listener's ears from several directions. HRTFs can be measured using standard techniques by placing tiny microphones with known characteristics close to a listener's eardrums and measuring the sound produced by a known source at known positions in space (e.g., Wightman & Kistler, 1989; Pralong and Carlile, 1994).

Measurement of HRTFs is a tricky and technically demanding task. One obvious problem is that placing any foreign object in the ear canal involves risk and placing something close to the eardrum involves much greater risk. Although the use of soft silicone probe tubes mitigates the risk considerably, potential changes in the microphone position during measurements, and the requirement that the transfer function of the headphone to be used for virtual stimulus presentation be measured under identical conditions (e.g., Wightman & Kistler, 1989) remain significant problems. There have been attempts to solve each of these problems (Wightman & Kistler, 1989; Pralong & Carlile, 1994; Langendijk & Bronkhorst, 2000), none of which has been universally accepted. The most promising solution appears to be HRTF measurement with a microphone at the entrance to a closed ear canal (Moller, Sorensen, Hammershoi, & Jensen, 1995; Wightman, Kistler, Foster, & Abel, 1995). The argument is that since the ear canal effects are direction-independent, measurement at the ear canal entrance can potentially capture all the "important" directional information in a true HRTF (Moller , et. al. 1995). In fact, perceptual research from our laboratory (Wightman, et. al., 1995) suggests that the resulting virtual auditory environment is nearly identical to that produced from deep-canal HRTF measurements.

The most difficult issue surrounding the use of HRTFs is the fact that they are very individual-specific, and for optimal fidelity the virtual auditory environment to be used by a given listener should be produced using that listener's own HRTFs. If this is not done, the apparent positions of virtual auditory images may be very different from what was intended, with the most frequent error being a "front-back confusion", such that the intended and perceived positions are in opposite hemifields (Wenzel, Arruda, Kistler, & Wightman, 1993; Bronkhorst, 1995). Front-back confusions can be reduced or even eliminated by giving users of a virtual auditory environment access to the same kind of dynamic information that is provided by head movements in a real environment (Perrett & Noble, 1997; Wightman & Kistler, 1999b). Doing so requires coupling of the virtual environment synthesis to head position, which has its own set of challenges and problems. Interesting in this context is the finding that continuous long-term exposure to non-individualized HRTFs apparently leads to nearly complete adaptation, and elimination of the resultant errors (Hofman,Van Riswick, & Van Opstal, 1998). In addition, some recalibration seems to be produced by visual feedback, even without continuous exposure (Zahorik, personal communication). The extent to which this finding may reduce the need for individualized HRTFs in auditory virtual environments (which normally do not lend themselves to continuous long-term exposure) is unclear.

Among the "non-issues" in the HRTF measurement and use are the facts that only the magnitude of the HRTF is important and that only the gross features of the magnitude spectrum appear to be influential. Several published studies confirm that so long as the interaural delay is retained (this can be accomplished with a pure, frequency-independent delay computed from a cross-correlation analysis) the phase of the HRTF is unimportant. For example, minimum-phase approximations to the measured HRTFs produce virtual auditory images that seem nearly identical to those produce with measured HRTFs (Kistler & Wightman, 1992; Kulkarni & Colburn, 1999), a result that is not surprising, since measured HRTFs are essentially minimum-phase anyway. Moreover, linear-phase HRTFs also produce high-fidelity virtual auditory images (Kulkarni & Colburn, 1999; unpublished results from our laboratory), implying that the phase of the HRTF, or, stated differently, the actual shape of the impulse-response representation of the HRTF (i.e., the Head-Related Impulse Response or HRIR) is irrelevant. This result is also not surprising, since HRIRs are short (2-4 ms), very close to the absolute temporal resolving power of the ear (Green, 1971). Studies of the perceptual consequences of smoothing the HRTFs used to synthesize virtual auditory images suggest that only the gross features of the HRTF are influential. Although perfect fidelity requires that there be no smoothing at all (Zahorik, et. al. 1995), the high fidelity of virtual auditory images is resistant even to considerable smoothing, either achieved by shortening the HRIR or by actually smoothing the magnitude spectrum (Kistler & Wightman, 1992; Kulkarni & Colburn, 1998).

Another "non-issue" relates to the problem of interpolating HRTFs. Since a virtual auditory environment should accomodate sounds at any arbitrary point in space, and since HRTFs can be measured only at a finite number of positions, the question of interpolation is a natural one. In the case of HRTFs it is a complex issue, since simple linear interpolation is inappropriate. Linear interpolation of two HRTFs, each of which had a single spectral peak, but at different frequencies, would produce an HRTF with two peaks, not a single peak at the intermediate frequency. Several approaches to this problem have been proposed, from interpolating weights of a Principal Components approximation to the HRTF (Kistler & Wightman, 1992) to the use of spherical basis functions to generate a functional representation of the HRTF space that would allow an analytic solution for estimating an HRTF for any spatial location (Jenison & Fissell, 1996). None of these rather involved approaches is necessary since modern techniques for measuring HRTFs (e.g., Wightman & Kistler, 1999b) allow measurements at a large number of spatial locations, resulting in a relatively dense matrix of measured positions. The change in the HRTF across adjacent measured positions is thus quite small. Linear interpolation then produces virtual images of high enough fidelity that continuous transitions from one position to another are perceptually seamless and smooth.

3. Headphone compensation and loudspeaker presentation issues. An issue of considerable importance, but one which appears not to be widely considered, is the specific way that virtual auditory images are delivered to listeners' ears. There are actually two separate issues here, headphone compensation and fidelity of virtual auditory environments produced by loudspeakers. We will discuss the headphone issue first. Clearly, if the aim is to produce a known waveform at a listener's eardrums, which is the ultimate goal in a virtual auditory environment, the impact of the transducer, which in this case is a headphone, must be measured and eliminated by appropriate compensation. The most straightforward way of doing this is by measuring the response of the headphone at the listener's eardrum using the same techniques and at the same time as the HRTFs are measured (e.g., Wightman & Kistler, 1989; Pralong & Carlile, 1996). In this way, appropriate compensation for the headphone response, which must incorporate listener-specific characteristics, can easily be incorporated into the synthesis algorithms. The headphone response can also be measured with the closed-canal techniques mentioned above; this method is considerably easier. Unfortunately, the response characteristics of even high quality headphones are not only dependent on the specific listener, but also on the specific placement, so some error must be anticipated. For some listeners this placement error can be quite large. Unpublished work from our laboratory shows the standard deviation of repeated headphone response measurements on some listeners to be as great as 10 dB in relevant (5-10 kHz) frequency bands. One might expect that since the headphone response is a non-directional constant its impact on the fidelity of a virtual auditory display would be minimal. However, since it is known that prominent peaks in the spectrum bias apparent position (Middlebrooks, 1992), and it is suspected, but not proven, that prominent notches also have an impact, it is clear that headphone effects must be appropriately compensated for optimal fidelity (Kulkarni & Colburn, 2000).

A high-fidelity virtual auditory environment can be created with either headphones or with loudspeakers. The limitations of the two techniques are related and concern the effect of listener movement. With headphones, a listener's movement causes no change in the sounds presented to the ears unless the listener's position (especially head position) is tracked, and the sounds changed accordingly. The lack of head-coupled changes in the sounds reaching the eardrums is very unnatural and for some listeners causes serious degradation in the realism of the virtual environment. Producing a high fidelity virtual auditory environment with loudspeakers requires that the position and orientation of the listener's head with respect to the loudspeakers be known. This is because the spatial cues produced by the loudspeakers must be in some sense "removed", or cancelled, so that the spatial cues appropriate to the intended virtual source can be delivered to the listener's ears (Moller,1989). Once the listener's head moves, the cancellation process breaks down and the fidelity of the virtual images is compromised. The extent of the compromise for various loudspeaker arrangements and processing schemes is a matter of considerable debate (Bauck & Cooper, 1996; Cooper & Bauck, 1989). It is clear, however, that listener movement is at least as important an issue with loudspeaker presentation as it is with headphone presentation. The solution to both problems will almost certainly require head tracking and subsequent real-time compensation for head position. An additional problem for loudspeaker presentation that could not realistically be solved by head tracking is a result of the effects of echoes and reverberation in the listening room. Unless the room effects are measured at all relevant listener positions, compensation could not be successful.

4. Computing demands - perceptual consequences of compromises. Production of a high-fidelity virtual auditory environment requires substantial computing power. Since important spatial cues exist at high frequencies (5-15 kHz), sampling rates must be in the 40 kHz range. In a typical FIR-filter implementation, each virtual source position requires two HRTF filters (left and right ear) of at least 128 points each, and if room effects are desired, the number of virtual sources is multiplied by the number of echoes added for each source (6 at a minimum for a rectangular room, corresponding to the first-order reflections from the 4 walls, ceiling and floor of the room). If head tracking is implemented, all the filter coefficients (HRTF coefficients, wall reflection coefficients, etc.) must be updated 20 times a second or more. Even with special-purpose hardware and the most current computing platforms, production of a virtual auditory environment of any complexity is a challenge. For this reason, a great deal of research has focused on ways of reducing the computing demands.

In order to assure the utility of a virtual auditory environment produced with limited computing resources it is necessary to know the perceptual consequences of the various compromises that might be considered. There is useful information available on some, but not all, of the relevant issues. One obvious compromise involves reduction of detail in the HRTFs, as discussed above. It is clear that some reduction (smoothing) can be tolerated, but it is also clear that large reductions in computing demands could not be achieved by minor reductions in HRTF complexity. More imaginative solutions could involve, for example, preservation of detail in the HRTF for the ear on the same side of the head as the intended virtual source (the leading ear) and elimination of nearly all detail in the HRTF for the trailing ear (while preserving the interaural time difference). Surprisingly, this rather extreme manipulation has only minor perceptual consequences (Wightman & Kistler, 1999a). It does not even seem to be necessary to preserve the overall interaural intensity difference in such a manipulation. Another possible simplification might be to synthesize the room effects with considerably less HRTF detail than the direct sound wave. There is some evidence that there are virtually no perceptual consequences of this manipulation (Zahorik, Kistler, & Wightman, 1994). It appears that so long as the delays and intensities of the echoes are preserved, their directions (established by the HRTFs) are almost unimportant.

The computational demands required for a high-fidelity auditory virtual environment are determined in large part by the complexity of the environment to be simulated. Complexity, in this case, is measured by the total number of sources to be synthesized. Since each room reflection counts as a source, room effects obviously have a tremendous influence on overall computing demands. Thus, it is important to appreciate the circumstances in which room effects are necessary. There is no doubt that room effects contribute to the overall "realism" of a virtual auditory environment. Although there are few data on this point, there seems to be no disagreement among both scientists and developers that room effects make virtual auditory environments "sound better". The reasons for this are undoubtedly complex, but one is that nearly all everyday listening situations include room effects, and thus simulations with room effects might seem more "natural". The dozens of experiments on the Precedence Effect attest to the fact that our perceptual apparatus seems specifically designed to deal with room effects. Another reason is that the apparent distance of an unfamiliar auditory sound source is cued mainly by reverberation, specifically the ratio of direct to reverberant sound energy (Zahorik, 1998). For this latter reason, incorporating whatever room effects are necessary for accurate distance perception would seem to be a prudent design criterion for virtual auditory environments.

References

Bauck, J. & Cooper, D. H. (1996). Generalized transaural stereo and applications. *Journal of the Audio Engineering Society, 44*, 683-705.

Bronkhorst, A. W. (1995). Localization of real and virtual sound sources. *Journal of the Acoustical Society of America, 98*, 2542-2553.

Bronkhorst, A. W. & Houtgast , T.(1999). Auditory distance perception in rooms. *Nature, 397*, 517-520.

Cooper, D. H. & Bauck, J. L. (1989). Prospects for transaural recording. *Journal of the Audio Engineering Society, 37*, 3-19.

Green, D. M. (1971). Temporal auditory acuity. *Psychological Review, 78,* 540-551.

Hartmann, W. M. & Wittenberg, A. (1996). On the externalization of sound images. *Journal of the Acoustical Society of America,* 99, 1-11.

Hofman, P. M., Van Riswick, J. G. A., & Van Opstal, J. (1998). Relearning sound localization with new ears. *Nature Neuroscience,* 1, 417-421.

Jenison, R. L. & Fissell, K. (1996). A spherical basis function neural network for modeling auditory space. *Neural Computation,* 8, 115-128.

Kistler, D. J. & Wightman, F. L. (1992). A model of head-related transfer functions based on principal components analysis and minimum-phase reconstruction. *Journal of the Acoustical Society of America,* 91, 1637-1647.

Kulkarni, A. & Colburn, H. S. (1998). Role of spectral detail in sound-source localization. *Nature,* 396, 747-749.

Kulkarni, A. & Colburn, H. S. (1999). Sensitivity of human subjects to head-related transfer-function phase spectra. *Journal Acoustical Society of America,* 105, 2821-2840.

Kulkarni, A. & Colburn, H. S. (2000). Variability in the characterization fo the headphone transfer function. *Journal Acoustical Society of America,* 107, 1071-1074.

Langendijk, E. H. A., & Bronkhorst, A. W. (2000). Fidelity of three-dimensional-sound reproduction using a virtual auditory display. *Journal Acoustical Society of America,* 107, 528-537.

Middlebrooks, J. C. (1992). Narrow-band sound localization related to external ear acoustics. *Journal of the Acoustical Society of America,* 92, 2607-2624.

Moller, H. (1989). Reproduction of artificial-head recordings through loudspeakers. *Journal of the Audio Engineering Society.* 37, 30-33.

Moller, H., Sorensen, M. F., Hammershoi, D., & Jensen, C. B. (1995). Head-related transfer functions of human subjects. *Journal of the Audio Engineering Society,* 43, 300-321.

Perrett, S. & Noble, W. (1997). The contribution of head motion cues to localization of low-pass noise. *Perception & Psychophysics,* 59, 1018-1026.

Pralong, D. & Carlile, S. (1994). Measuring the human head-related transfer functions: A novel method for the construction and calibration of a miniature "in- ear" recording system. *Journal of the Acoustical Society of America,* 95, 3435-3444.

Pralong, D. & Carlile, S. (1996). The role of individualized headphone calibration for the generation of high fidelity virtual auditory space. *Journal of the Acoustical Society of America,* 100, 3785-3793.

Wenzel, E. M., Arruda, M., Kistler, D. J., & Wightman, F. L. (1993). Localization using nonindividualized head-related transfer functions. *Journal of the Acoustical Society of America,* 94, 111-123.

Wightman, F. L. & Kistler, D. J. (1989). Headphone simulation of free-field listening I: Stimulus synthesis. *Journal of the Acoustical Society of America,* 85, 858-867.

Wightman, F. L. & Kistler, D. J. (1999a). Sound localization with unilaterally degraded spectral cues. *Journal of the Acoustical Society of America,* 105, 1162.

Wightman, F. L. & Kistler, D. (1999b). Resolution of front-back ambiguity in spatial hearing by listener and source movement. *Journal of the Acoustical Society of America,* 105, 2841-2853.

Wightman, F. L., Kistler, D. J., Foster, S. H., Abel, J. (1995). A comparison of head-related transfer functions measured deep in the ear canal and at the ear canal entrance. *Abstracts of the 18th Midwinter Meeting,* Association for Research in Otolaryngology, 61.

Zahorik, P. (1998). Experiments in auditory distance perception. Doctoral dissertation, University of Wisconsin-Madison.

Zahorik, P., Kistler, D. J., Wightman, F. L. (1994). Sound localization in.varying virtual acoustic environments. In G. Kramer (Ed.), *Proceedings of the 1994 International Conference on Auditory Displays* (pp. 179-186). Santa Fe, NM..

Zahorik, P., Wightman, F. L., & Kistler, D. J. (1995). On the discriminability of virtual and real sound sources. *Proceedings of the ASSP (IEEE) Workshop on applications of Signal Processing to Audio and Acoustics.* New York: IEEE Press.

Rendering Sound Sources In High Fidelity Virtual Auditory Space: Some Spatial Sampling And Psychophysical Factors.

Simon Carlile[1,2,4], Craig Jin[1,3] and Johann Leung[1].

[1]Department of Physiology, [2]Institute for Biomedical Research, [3]Department of Electrical and Information Engineering and [4]Office of the Chief Information Officer, University of Sydney, NSW 2006, Australia

1. Abstract

Auditory localisation performance is an appropriate measure of the fidelity of virtual auditory space. Using this metric, principal component compression of the outer ear filter functions, combined with a spherical spline method is shown to produce a perceptually validated continuous representation of auditory space.

2. Introduction

The fidelity of reproduction is a key issue in the generation of any representation of an object: In a practical sense the fidelity of reproduction is often determined by the use to which the reproduction is to be put. For instance, where the objective is to produce a representation that is as close as possible to a perfect copy of the original object, then, from a perceptual point of view, the limiting factor is the fidelity of the sensory system encoding the representation. On the one hand, when the fidelity of reproduction is low the quality of the resulting perception can be degraded. On the other hand, where fidelity of the representation is higher than the sensory system, then the effort of producing such a high fidelity reproduction is wasted. In the context of virtual displays, properly matching the display and the sensory system requires objective measures of the users psychophysical performance on tasks in real space compared with virtual space. The key issue in evaluating subject performance is the selection of appropriate behavioural measures that relate to the use to which the display is put. In the context of auditory displays, the key perceptual feature that distinguishes virtual auditory space from typical headphone presentation is the precision of the perceived exocentric location of the simulated sources. Over headphones, multiple auditory sources can be discriminated on the basis of differences in the interaural level and timing but the sound sources are still perceived as originating from within the head: i.e. they are lateralized rather than localized. Applying reverberation to the source(s) can produce some measure of spatialisation, however, the resulting perception is of a diffuse rather than precisely localized source. By contrast, a virtual auditory display produces the perception of sound source(s) located precisely in exocentric space. In considering these questions this paper explores (i) the relevant performance characteristic for virtual auditory displays; (ii) the behavioural test that best measures performance; (iii) methods for generating spatially continuous VAS; (iv) experiments evaluating the acoustical distortions (fidelity) of the VAS and (v) evaluation of the psychophysical relevance of acoustical distortions in terms of the user performance measures.

3 Auditory localization performance in free space

Given that the salient perceptual characteristic of a virtual auditory display is the precise localisation of the target, the most relevant behavioural measure is the ability of subjects to localize the source. There is a rich research history examining auditory localisation and a range of methodologies developed (see Middlebrooks & Green 1991 and Carlile 1996 for recent reviews). As with all psychophysical studies, an important characteristic of the experimental design is that it ensures that the performance measured (the dependent variable) is related to the stimulus characteristic that is varied – in this case the location of the source of a sound. Prior knowledge of the potential locations of stimuli has been shown to influence the subjects' responses in sound localisation tasks (see Perrett & Noble 1995, Pralong & Carlile 1996). As a consequence, we have developed an automated stimulus system that allows the placement of a sound source at any position on an imaginary sphere surrounding the subject. A robot arm places the sound stimulus at a randomly chosen location in a darkened anechoic chamber so that the subject also has no prior information as to the potential direction of a sound source. In response to a 150-ms burst of broadband noise (300Hz-16kHz), the subject is trained to turn to face the perceived location of the target and reliably point his or her nose towards the sound source. An electromagnetic-tracking device mounted on the top of the head (Polhemus: IsoTrak) is used to measure the orientation of the head and thus provides an objective measure of the perceived direction.

We have examined performance using a number of statistical measures. The cluster of localisation estimates about an individual target location can be described by the centroid and standard deviation calculated using a Kent distribution (Fisher et al. 1993, Leong & Carlile 1998). The centroid indicates the systematic error in localisation and the dispersion indicates the accuracy. From a study of 19 subjects the systematic errors and the dispersion were smallest for frontal locations close to the audio-visual horizon and largest for locations behind and above the subject (Figure 1: from Carlile et al. 1997). The spherical correlation coefficient (SCC) measures the association between the centroids of the perceived locations and the actual target locations (1 = perfect correlation; 0 = no correlation). The correlation of the data shown in Figure 1 was 0.98 for the data pooled from 19 subjects. The most common type of localisation error is associated with relatively small deviations of the perceived location from the actual location ("local error"). The second type of error typically involves a large error where the perceived location of the target is at a location reflected about the interaural axis. Such errors are often referred to as a front-back confusion or (more properly) cone-of-confusion errors and occur relatively infrequently ($< 4\%$). These have been removed from the data prior to calculating the centroids and the SCC.

In summary, these data provide an insight into the spatial resolving power of the human auditory system for sounds in anechoic space. These also data indicate how accuracy is dependent on the location of the sound. Localisation performance also provides a practical and objective method to explore the range of factors that determine the spatial fidelity of rendered virtual auditory space.

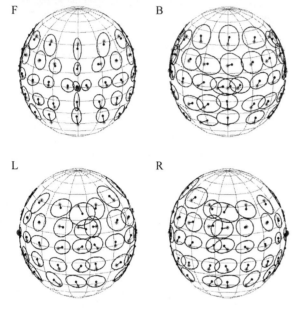

Figure 1: Pooled localisation responses from 19 subjects shown for front (F), back (B), left (L) and right(R) hemispheres. Target locations are shown by the small ''; the centroid of the pooled data by the filled circle. The ellipse = response SD .*

4. Generation of Virtual Auditory Space.

Virtual Auditory Space (VAS) is generated by recording the acoustic filtering properties of the outer ears of individual listeners (the Head Related Transfer Functions: HRTFs) and convolving these with sounds subsequently presented over headphones. The illusion created for the listener is of a sound occurring in extrocentric space at some distance away from the head.

4.1 Recording the Head Related Transfer Function

The Head Related Transfer Function (HRTF) describes the interactions of the sound with the structures of the outer ear. Since the anatomical features of the outer ears are individualized so too are the HRTFs (Jin et al. 2000). In addition, these transformations are highly location dependent (Carlile & Pralong 1994).

A "blocked-ear" recording paradigm is used routinely in out laboratory and involves embedding a small recording microphone in an earplug secured flush with the distal end of the ear canal (Moller et al. 1995 but see also Pralong & Carlile 1994). The recordings are performed inside an anechoic chamber with the subject placed at the center. A speaker, mounted on the robot arm, delivers the stimuli at a radius of one meter from the listener and is able to describe a spherical space with a lower limit of $-50°$ elevation ($0°$: A-V horizon).

The automated procedure results in 400 HRTFs recorded for locations evenly distributed on the sphere, from $-45°$ to $90°$ in elevation. The position of the subject's head was monitored by the magnetic tracking system to ensure head stability throughout the procedure. Impulse responses were measured using Golay code pairs of 1024 bits long sampled at 80kHz (Golay 1961)

4.2 Signal processing of the recorded HRTF filters

The HRTFs are composed of a location independent component (LIC) and a location dependent component. The LIC was estimated by the mean magnitude spectrum of the entire recording and then removed from the recorded HRTF to form the directional transfer functions (DTF). The primary reason for performing such a manipulation is to remove measurement artifacts including speaker and microphone transfer functions and effects due to the precise location of the microphone (Pralong and Carlile, 1994). The desired acoustical stimulus is then convolved with the resulting DTF and presented via in-ear headphones (Etymotic ER2). The frequency responses of the headphones are calibrated to be flat to the eardrum and thereby avoid the need to compensate for the headphone transfer function.

4.3 VAS fidelity measured by localisation performance.

Sound localisation performance for stimuli presented in VAS was assessed in the exactly the same manner as for the free field localisation with the exception that the stimuli were delivered using in-ear headphones. The spatial distribution of localisation errors for sounds located in VAS was very similar to that for sounds presented in the free field (c.f. Figure 1 and Figure 2). On average, dispersion was $1.5°$ greater for stimuli presented in VAS compared to free field. There was a slight increase in the dispersion of localisation estimates for locations behind and also above the subjects when compared to free field localisation.

An increase in the front-back confusion rates (the most prominent form of cone-of-confusion errors) was also seen with average rates rising from around 3-4% in the free field up to 6% for sounds presented in virtual space. The spherical correlation between the perceived and actual target locations (with the cone of confusion data removed) was 0.973.   Furthermore the spherical correlation between the VAS and free field localization was higher still (0.98) indicating that subject biases evident in the free field data were replicated in VAS.

5. Compression of the HRTF filters

Several reports (Kistler & Wightman 1992, Kulkarni & Colburn 1998) have indicated that the HRTF contains a degree of redundancy and several algorithms have been investigated to compress these filters. As well as reducing the storage and computational overheads, HRTF compression is also a first step towards solving the problem of spatial interpolation

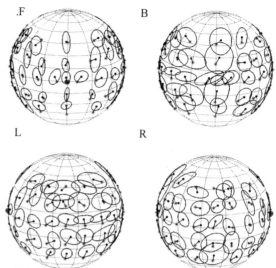

Figure 2: Location estimates pooled for 5 subjects for sounds presented in VAS. All other details as for figure 1

5.1 Principal component analysis

Principal Component Analysis (PCA) is one approach for compressing the HRTFs (Martens 1987, Kistler & Wightman 1992, Chen et al. 1995). The general operation of PCA is based on the decomposition of the covariance of the input matrix via eigen vectors and eigen values. This results in a set of linear basis functions or principal components (PC) and weighting constants which can be ordered by how much variance they account for. Since the PCs are orthogonal, it is possible to reconstruct a reduced fidelity output by linearly expanding only some of the PC dimensions. In addition, the approach is essential non-dimensional as it relies on an analysis of variance. Therefore, different representations of the inputs may have different statistical efficiencies in terms of the distribution of the variance across the PC dimensions. This is explored below.

5.2 Statistical analysis of the PCA compression

In this study, the HRTF filters were represented with various formats in time and frequency domains and the efficiency of the PCA compression was analyzed. In the time domain, unmodified IR filters were used as input (751 taps). In the frequency domain the inputs were represented as linear amplitude, log magnitude and complex value pairs. Since there were 393 unique locations in each set of HRTF recordings, there were 393 dimensions in each input matrix. The results of these analysis using18 sets of HRTF filters are presented in Figure 3. It is clear that representing the HRTF filters in the frequency domain using a linear amplitude format resulted in the highest compression efficiency.

5.3 Fidelity of VAS using varying PC dimensions

The fidelity of VAS generated with compressed HRTF filters was examined using sound localisation accuracy for HRTF filters reconstructed with 5, 10, 20 and 300 PCs. Minimum phase approximation was used to ensure an appropriate magnitude-phase relationship (Oppenheim & Schafer 1989). Three of the 11 original subjects (Section 5.2) participated in this set of auditory localization experiments and their results are shown (Table 1). These data show that VAS using filters with only 10 PCs resulted in performance approaching that for control stimuli (c.f. Section 3). In terms of the original 393 dimensions, the 10 PCs required here to produce high fidelity VAS represents a compression of the HRTFs to less than 2.5% of the original data size.

Table 1: Localisation performance with varying numbers of PC dimensions.

Subiec	Av.	% COC	# of PCs
A	0.89	8.7	5
	0.94	1.8	10
	0.95	0.26	20
	0.95	0.26	300
B	0.92	5.8	5
	0.93	2.6	10
	0.94	2.6	20
	0.94	1.3	300
C	0.85	19.2	5
	0.89	7.0	10
	0.89	5.3	20
	0.89	5.8	300

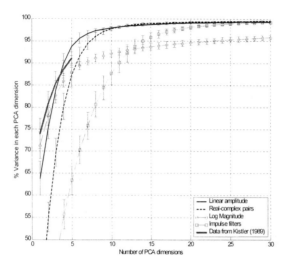

Figure 3 Compression with different input representations

6. Interpolating a continuous VAS

While the real free field is a continuous space, the HRTFs are recorded at discrete locations around the listener. Previous work has looked at the interpolation of HRTFs using different methods (Langerndijk & Bronkhorst 2000, Wenzel & Foster 1993) in both the time and frequency domain. Frequency domain approaches have used straightforward methods such as linear interpolation between nearest neighbors as well as more sophisticated methods such as the Euclidean thin-plate spline (Chen et al. 1995) and the application of a radial basis function neural network (Jenison & Fissell 1996). Generally, the better approaches account for the spherical geometry of the data; however, there are very few studies of psychoacoustical errors associated with HRTF interpolation. One systematic investigation of the psychophysical errors associated with HRTF interpolation unfortunately only examined localization performance at 8 test locations (Langernidijk & Bronkhorst 2000).

6.1 Application of PCA and spherical thin-plate spline

In this section we describe our approach for interpolating and psychophysically validating continuous VAS. As described above, the frequency domain magnitude components of the HRTFs were compressed using PCA to provide a series of PCs and weights as described above. The weights were then interpolated using a spherical thin-plate spline (STPS) according to Wahba (1981). The interaural time delay was estimated using a cross correlation of the impulse response functions for the left and right ears and interpolated using the STPS. The benefits of the STPS are: (1) the approximation is continuous in all directions and is suitable for modeling spherically directional data and (2) the spline is a global approximation incorporating all data around the sphere to provide one interpolation value. The frequency amplitude components of the interpolated HRTFs were reconstructed from the PCs and the interpolated weights. The interaural time delay components were added back into the reconstructed HRTFs as an all-pass delay (see Mehrgardt & Mellert 1977). For the nearest neighbor interpolation, the calculations followed the same procedure outlined above except that the interpolation was based on the 12 nearest neighbors for each location that were weighted inversely as their distance on the sphere.

6.2 Estimation of the magnitude of the spherical errors

HRTF recordings were obtained from 475 locations, 82 of which were selected as a test set and were not used to generate the spherical splines. The magnitude errors for the interpolated HRTFs were calculated for the 82 test locations using estimates derived from the interpolation data sets with a varying number of locations (subsets with 393, 250, 150, 125, 90, 70, 60, 50, 30, 20; subsets covered a range of spatial resolutions varying from about 10° - 45° between neighbors). The root-mean-square error of the magnitude components of the HRTFs was calculated for each test location (Figure 4). Error increased as the number of positions contributing to the spline functions decreased from 393 to around 150 positions and then increased markedly for the smaller sets (Figure 4a,b). The STPS was significantly better than the nearest neighbor interpolation for a given number of locations. A plot of the errors (Figure 4c) demonstrated that the distribution of errors was not uniform throughout space and varied by up to 10dB. This suggests that for sparse data sets the spline fails to accurately model some areas of space where, presumably, the spatially dependent rate of change of the HRTF are relatively higher or have relatively discontinuous changes in the spectral shape.

6.3 Auditory localisation performance

The localization performance of 5 human subjects was measured for noise stimuli presented in VAS rendered using four interpolation sets (250, 150, 50, 20 positions; 10°, 15°, 30°, 45° degrees of resolution). For 150 locations, performance was identical to that using HRTFs measured at the test locations. Although performance was significantly degraded using the sparse sets of 50 and 20 locations, substantial localization capacity was still evident in VAS despite the relatively high levels of acoustic errors in the HRTF estimates (Figure 5). These data indicate that a spherical thin plate spline with as few as 150 recorded filter functions spaced approximately 15 degrees apart was sufficient to achieve localisation performance equivalent to that in the free field

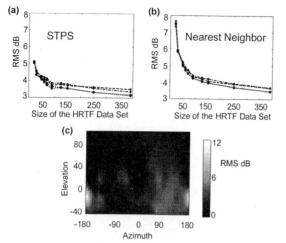

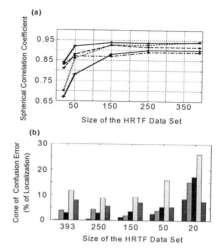

Figure 4: The change in the magnitude of the interpolation errors as a function of the number of measured HRTFs contributing to the spherical spline. (a) the STPS; (b) nearest neighbor interpolation. (c) The distribution of RMS dB error across space for the left ear using STPS for 50 HRTFs

Figure 5: (a) The spherical correlation coefficient is plotted as function of the number of HRTF recording locations contributing to the interpolation model. (b) The % of cone of confusion errors for each subject plotted as a function of interpolation set size.

7 Conclusions

An important finding across these experiments was the robust capacity of the auditory system to support accurate localisation of broadband sounds under conditions of relatively high signal degradation. This suggests that (i) there is a significant amount of perceptually redundant location information contained in broadband sound stimuli and (ii) modern recording and digital reproduction techniques provide a level of resolution that is well beyond the requirement for generating high fidelity virtual space, at least under the stimulus conditions used here. These findings indicate that significant computational savings could be achieved in developing VAS displays that are properly tuned to the perceptual fidelity of the human listener.

References

Carlile S. 1996. The physical and psychophysical basis of sound localization. In*Virtual auditory space: Generation and applications*, ed. S Carlile, Ch 2.:

Carlile S, Leong P, Hyams S. 1997. The nature and distribution of errors in the localization of sounds by humans*Hear Res* 114: 179-96

Carlile S, Pralong D. 1994. The location-dependent nature of perceptually salient features of the human HRTF*J Acoust Soc Am* 95: 3445-59

Chen J, et al.,. 1995. A spatial feature extraction and regularization model for the head-related transfer fucntion.*J. Acoust. Soc. Am.* 97: 439-52

Fisher NI, Lewis T, Embleton BJJ. 1993.*Statistical analysis of spherical data* Cambridge: Cambridge University Press

Golay MJE. 1961. Complementary series.*IRE Trans. Info. Theory* 7: 82-7

Jenison RL, Fissell K. 1996. A sperical basis function neural network for modeling auditory space*Neural computation* 8: 115-28

Jin C, Leong P, Leung J, Corderoy A, Carlile S. 2000.*Enabling individualized virtual auditory space using morphological measurements* International Symposium on Multimedia Information Processing, Sydney

Kistler DJ, Wightman FL. 1992. A model of head-related transfer functions based on principal components analysis and minimum-phase reconstruction. *J. Acoust. Soc. Am.* 91: 1637-47

Kulkarni A, Colburn HS. 1998. Role of spectral detail in sound source localization*Nature* 396: 747-9

Langendijk EHA, Bronkhorst AW. 2000. Fidelity of three dimension sound reproduction using a virtual auditory display*J Acoust Soc Am* 107: 528-37

Leong PHW, Carlile S. 1998. Methods for spherical data analysis and visualisation.*J Neurosci Methods* 80: 191-200

Martens WL. 1987. Principle components analysis and resynthesis of spectral cues to perceived direction. Presented at ICMC

Mehrgardt S, Mellert V. 1977. Transformation characteristics of the external human ear.*J. Acoust. Soc. Am.* 61: 1567-76

Middlebrooks JC. 1992. Narrow-band sound localization related to external ear acoustics.*J. Acoust. Soc. Am.* 92: 2607-24

Middlebrooks JC, Green DM. 1991. Sound localization by human listeners. *Annu. Rev. Psychol.* 42: 135-59

Moller H, Sorensen MF, Hammershoi D, Jensen CB. 1995. Head-related transfer functions of human subjects.*J. Audio Eng. Soc.* 43: 300-21

Oppenheim A, Schafer R. 1989.*Discrete-time signal processing*. New Jersey: Prentice Hall

Perrett S, Noble W. 1995. Available response choices affect localization of sound*Percept and Psychophys* 57: 150-8

Pralong D, Carlile S. 1994. Measuring the human head-related transfer functions: A novel method for the construction and calibration of a miniature "in-ear" recording system.*J Acoust Soc Am* 95: 3435-44

Pralong D, Carlile S. 1996. Generation and validation of virtual auditory space. In*Virtual auditory space: Generation and applications*, ed. S Carlile, Ch 4. Austin: Landes

Wahba G. 1981. Spline interpolation and smoothing on the sphere*AIAM J Sci Statist Comp* 2: 5-16

Wenzel EM, Foster SH. 1993.*Perceptual consequences of interpolating head-related transfer functions during spatial synthesis* Proceedings of the 1993 Workshop on Applications of Signal Processing to Audio and Acoustics, New Paltz, New York

Creating three dimensions in virtual auditory displays[*]

Barbara Shinn-Cunningham

Boston University, Depts. of Cognitive and Neural Systems and Biomedical Engineering
677 Beacon St., Boston, MA 02215

ABSTRACT

In order to create a three-dimensional virtual auditory display, both source direction and source distance must be simulated accurately. Echoes and reverberation provide the most robust cue for source distance and also improve the subjective realism of the display. However, including reverberation in a virtual auditory display can have other important consequences: reducing directional localization accuracy, degrading speech intelligibility, and adding to the computational complexity of the display. While including an accurate room model in a virtual auditory display is important for generating realistic, three-dimensional auditory percepts, the level of detail required in such models is not well understood. This paper reviews the acoustic and perceptual consequences of reverberation in order to elucidate the tradeoffs inherent in including reverberation in a virtual auditory environment.

1. SOUND LOCALIZATION CUES

The main feature distinguishing virtual auditory displays from conventional displays is their ability to simulate the location of an acoustic source. In this section, the basic spatial auditory cues that convey source position are reviewed in order to gain insight into how reverberation influences spatial auditory perception.

The physical cues that determine perceived sound direction have been studied extensively for over a century (for reviews, see Middlebrooks & Green, 1991; Gilkey & Anderson, 1997). Most of these studies were performed in carefully controlled conditions with no echoes or reverberation and focused on directional perception. Results of these anechoic studies identified the main cues that govern directional perception, including differences in the time the sound arrives at the two ears (interaural time differences or ITDs), differences in the level of the sound at the two ears (interaural level differences or ILDs), and spectral shape. ITDs and ILDs vary with the laterality (left/right position) of the source, whereas spectral shape determines the remaining directional dimension (i.e., front/back and left/right; e.g., see Middlebrooks, 1997).

In contrast with directional localization, relatively little is known about how listeners compute source distance. In the absence of reverberation[1], overall level can provide relative distance information (Mershon & King, 1975). However, unless the source is familiar, listeners cannot use overall level to determine the absolute distance (Brungart, 2000). For sources that are within a meter of the listener, ILDs vary with distance as well as direction (Duda & Martens, 1997; Brungart & Rabinowitz, 1999; Shinn-Cunningham, Santarelli & Kopco, 2000b), and appear to help listeners judge distance in anechoic space (Brungart, 1999; Brungart & Durlach, 1999). However, ILD cues are not useful for conveying distance information unless a source is both off the mid-sagittal plane and within a meter of the listener (Shinn-Cunningham et al., 2000b).

The most reliable cue for determining the distance of an unfamiliar source appears to depend upon the presence of reverberation (Mershon & King, 1975; see also Shinn-Cunningham, 2000a). While the direct sound level varies inversely with distance, the energy due to reverberation is roughly independent of distance. Thus, to a first-order approximation, the direct-to-reverberant energy ratio varies with the distance of a source (Bronkhorst & Houtgast, 1999). While many studies show the importance of reverberation for distance perception, there is no adequate model describing how the brain computes source distance from the reverberant signals reaching the ears. Of course, in real environments, reverberation does not just improve distance perception; it influences other aspects of performance as well. The remainder of this paper explores acoustic and perceptual effects of realistic reverberation and discusses the tradeoffs to consider when adding reverberation in virtual auditory environments.

2. ACOUSTIC EFFECTS OF REVERBERATION

Reverberation has a dramatic effect on the signals reaching the ears. Many of these effects are best illustrated by considering the impulse response that describes the signal reaching the ear of the listener when an impulse is played

---

[*] This work was supported by the AFOSR and the Alfred P. Sloan Foundation. N. Kopco and S. Santarelli helped with data collection.

[1] For simplicity, the term "reverberation" is used throughout this paper to refer to both early, discrete echoes and later reflections. In contrast, in much of the literature, "reverberation" refers only to late arriving energy that is the sum of many discrete echoes from all directions (and is essentially diffuse and uncorrelated at the two ears).

at a particular location (relative to the listener) in a room. For sources in anechoic space, these impulse responses are called Head-Related Impulse Responses (HRIRs; in the frequency domain, the filters are called Head-Related Transfer Functions or HRTFs; e.g., see Wightman & Kistler, 1989; Wenzel, 1992; Carlile, 1996). For sources in a room, these impulse responses are the summation of the anechoic HRIR for a source at the corresponding position relative to the listener and later-arriving reverberant energy. In order to illustrate these effects, measurements of the impulse response describing the signals reaching the ears of a listener in the center of an ordinary 18'x10'x12' conference room are shown in the following figures. The sample room is moderately reverberant; when an impulse is played in the room, it takes approximately 450 ms for the energy to drop by 60 dB. While the room in question is not atypical, the listener is always positioned in the center of the room, far from any reflective surfaces, and the sources are at a maximum distance of one meter for all of the measurements shown. These results show what occurs for moderate levels of reverberation. The effects would be much greater for more distant sources, more reverberant rooms, or even different listener positions in the same room.

Figure 1 shows different portions of a time-domain impulse response for the right ear when a source is at directly to the right at a distance of 1 m. The top panel shows a five-ms-long segment containing the direct-sound response (the anechoic HRIR); the central portion shows both the direct sound and some of the early reflections (the first 120 ms of the impulse response); the bottom-most portion shows a close-up (multiplying the y-axis by a factor of 100) of the first 300 ms of the impulse response. Figure 1 illustrates that the reverberation consists both of discrete, early echoes (note the discrete impulses in the middle panel of Figure 1 at times near 11, 17, 38 ms, etc.) and an exponentially-decaying reverberant portion (note the envelope of the impulse response in the bottom of Figure 1).

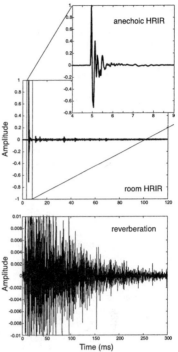

Figure 1: Impulse response to a listener's right ear for a source at 1 m, 90° azimuth in the horizontal plane in a reverberant room.

To compute the total signal at the ears for an arbitrary signal, the impulse response must be convolved with the signal emitted by the distal source (Wightman & Kistler, 1989; Wenzel, 1992; Carlile, 1996). This process can distort and temporally smear the signal reaching the ears. Figure 2 shows the time-domain waveform for a recorded speech utterance (the word "bounce") in its raw form (Figure 2c) and processed through impulse responses to recreate the total signal that would reach the listener's left and right ears (Figures 2a and 2b, respectively) for a source at a distance of 1 m and directly to the right. In the figure, the black lines plot the anechoic HRIR and the gray lines show the reverberant impulse response for the same source position relative to the listener. To ease comparisons, the waveforms are scaled (normalized) so that the maximum amplitude is 1.0 in the anechoic signals.

In anechoic space, the envelope of the waveform reaching the ears is fairly similar to that of the original waveform. Of course, the HRIR processing does cause some spectral changes. For instance, for the right-ear signal (Figure 2b), the sibilant at the end of the utterance (the second energy burst in the unprocessed waveform; i.e., "boun-**CE**") is emphasized relative to the initial energy burst because high frequencies are boosted by the right-ear's HRIR for this source position. Nonetheless, the general structure of the waveform is preserved. Since the direct sound energy to the near (right) ear is much greater than for the far (left) ear, the effect of reverberation is much more pronounced for the left ear signal. For the left ear, the waveform envelope is smeared in time and the modulations in the waveform are reduced (Figure 2a); in the right ear, the waveform envelope is well preserved (Figure 2b). While these graphs only show the distortion of the total waveform envelope, similar modulation distortion occurs for narrow band energy as well (i.e., such as the representation in an auditory nerve).

Figure 3 shows the anechoic and reverberant HRTFs at the left and right ears (left and right columns) for sources both to the right

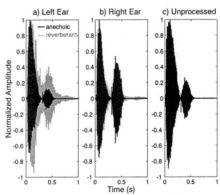

Figure 2: Speech waveform at the ears (anechoic, black, or reverberant, gray) for a source in the horizontal plane (at 1 m, 90° azimuth) and unprocessed waveform.

side (Figure 3a) and straight ahead (Figure 3b) of a listener in the sample room. Transfer functions are shown for both near (15 cm) and relatively distant (1 m) source positions. It is clear that for a source to the right, the energy at the right ear is greater than that at the left (Figure 2a); similarly, the HRTFs for near sources have more energy than those for far sources (compare top and bottom rows in Figures 3a and 3b). As a result, the effect of reverberation varies dramatically with source position. For a near source at 90° azimuth, the right ear reverberant transfer function is essentially identical to the corresponding anechoic HRTF (top right panel, Figure 2a). However, the effect of reverberation on the far (left) ear is quite large for a source at 90° azimuth (left column, Figure 3a).

In anechoic space, HRIRs depend only on the direction and distance of the source relative to the listener. In contrast, nearly every aspect of the reverberant energy varies not only with the position of the source relative to the listener, but also with the position of the listener in the room. The effects of reverberation shown in Figures 1-3 arise when a listener is located in the center of a large room, far from any walls. In such situations, the most obvious effect of reverberation is the introduction of frequency-to-frequency variations in the magnitude (and phase) transfer function compared to the anechoic case. For the far ear, there is also a secondary effect in which notches in the source spectrum are filled by the reverberant energy (e.g., see the notches in the left and right ear anechoic spectra for a 1-m source, bottom row of Figure 3b). However, different effects arise for different listener positions in the room.

We find that in addition to adding frequency-to-frequency fluctuations to the spectral content reaching the ears, reverberation can

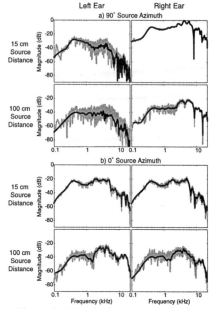

Figure 3: Magnitude spectra of anechoic (black) and reverberant (gray) transfer functions at the two ears for different source positions in a room.

lead to pronounced comb-filtering effects (non-random deviations in the long-term spectrum as a function of frequency) when a listener is close to a wall (Brown, 2000; Kopco & Shinn-Cunningham, 2001). These effects cause larger distortion of the basic cues underlying spatial perception (spectral shape, ITD, and ILD) than those that arise when a listener is relatively far from any large reflective surface. In particular, strong, early reflections can lead to dramatic nulls and peaks in the magnitude spectra, rapid shifts in the phase spectra as a function of frequency, and concomitant distortions of interaural differences. Finally, it is clear that essentially every measurable effect of reverberation depends on the relative energy in the direct and reverberant portions of the HRTFs, which depends on the position of the source relative to the listener, the position of the listener in the room, and the room itself.

3. PERCEPTUAL EFFECTS OF REVERBERATION

Reverberation leads to clear physical effects on the signals reaching the ears. However, when considering whether or how to incorporate reverberation in a virtual environment, the critical question is how reverberation influences perception and performance on different tasks of interest.

Reverberation dramatically improves the subjective realism of virtual auditory displays (e.g., see Durlach, Rigapulos, Pang, Woods, Kulkarni, Colburn & Wenzel, 1992. In many auditory displays that do not include reverberation, sources are often heard in roughly the correct direction, but near or even inside the head. In such anechoic simulations, painstaking care to provide simulations tailored to the individual listener and to compensate for characteristics of the headphone delivery system can ameliorate this "lack of externalization" (e.g., see Wenzel, Arruda, Kistler & Wightman, 1993; Pralong & Carlile, 1996. However, in contrast, good externalization is usually obtained when subjects listen to recordings made from a head in a reverberant setting, even without compensating properly for the headphone characteristic and when the playback is not tailored to the individual listener. Reverberation also provides information about the characteristics of the space itself, conveying information about the room size (e.g., see Bradley & Soulodre, 1995. While realism and environmental awareness are dramatically increased by reverberation, the extent to which these benefits depend on the fine structure of the reverberation has not been quantified. In other words, it may be possible to provide simplified reverberation cues that are less costly to include in a virtual environment but which still convey this information to the listener.

As noted above, distance perception is dramatically improved with the addition of reverberation (Mershon & King, 1975. We find that even when sources are within a meter of the head, where the relative effects of reverberation are small and robust ILDs should provide distance information, subjects are much more accurate at judging source distance in a room than in anechoic space (Santarelli, Kopco & Shinn-Cunningham, 1999a;

Santarelli, Kopco, Shinn-Cunningham & Brungart, 1999b; Shinn-Cunningham, 2000b). In addition, subjects listening to headphone simulations using individualized HRTFs do not accurately perceive source distance despite the large changes in ILDs present in their anechoic HRTFs, but do extremely well at judging distance when presented with reverberant simulations (Shinn-Cunningham, Santarelli & Kopco, 2000a). Further, in reverberant simulations, changes in distance are still perceived accurately for monaural presentations of lateral sources (turning off the far-ear signal), suggesting that the cue provided by reverberation is essentially monaural (Shinn-Cunningham et al., 2000a). While much work remains to determine how source distance is computed from reverberant signals, these results and results from other studies (e.g., Zahorik, Kistler & Wightman, 1994) suggest that simplified simulations of room effects may provide accurate distance information.

Results of previous studies of "the precedence effect," in which directional perception is dominated by the location of an initial source (i.e., the direct sound) and influenced only slightly be later-arriving energy (see Litovsky, Colburn, Yost & Guzman, 1999), suggest that reverberation should have small effect on directional localization accuracy. However, few studies have quantified how realistic room reverberation affects directional hearing. We find that reverberation causes very consistent, albeit small, degradations in directional accuracy compared to performance in anechoic space (Shinn-Cunningham, 2000b). Further, localization accuracy depends on the listener position in a room (Kopco & Shinn-Cunningham, 2001). When a listener is near a wall or in the corner of the room, response variability is greater than when in the center of the room. Based on analysis of the room acoustics, these results are easy to understand; reverberation distorts the basic acoustic cues that convey source direction, and this distortion is greatest when a listener is near a wall (Brown, 2000; Kopco & Shinn-Cunningham, 2001). We also observe that, over time, directional accuracy improves in a reverberant room and (after hours of practice) approaches the accuracy seen in anechoic settings (Santarelli et al., 1999a; Shinn-Cunningham, 2000b). Figure 4 shows this learning for an experiment in which listeners judged the postion of real sources in the room in which reverberant impulse responses were measured. In the figure, the mean left/right localization error (computed based on the difference in ITD caused by a source at the true and response positions, in ms) is shown. The error was computed both for the initial 100 trials (after 200 practice trials in the room, to accustom the listener to the task), and for the final 100 trials of a 1000-trial-long experiment. For each subject, error decreased by the end of the 1000 trials. These results suggest that any detrimental effects of reverberation on directional localization, which are relatively minor at worst, disappear with sufficient training.

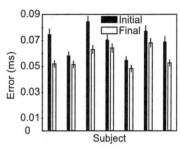

Figure 4: Mean left/right errors at the beginning and end of an experiment run in a reverberant room for each of seven subjects.

Finally, reverberation can interfere with the ability to understand or analyze the content of acoustic sources in the environment (e.g., see Nomura, Miyata & Houtgast, 1991). For instance, one of the most important acoustic signals that humans encounter is speech and much of the information in speech signals is conveyed by amplitude modulations. However, as shown in Figure 2, these modulations are reduced by reverberation. Although moderate amounts of reverberation do not degrade speech intelligibility severely, reverberation can degrade intelligibility. In addition, it is likely that reverberation will degrade signal intelligibility even more when there are competing signals than it will in quiet. Specifically, reverberation decorrelates the signals at the two ears and tends to reduce differences in the level of a signal reaching the two ears. Both of these factors can improve signal intelligibility in the presence of an interfering sound (Zurek, 1993). Thus, we predict that reverberation will have a particularly adverse impact on speech intelligibility in the presence of a masking source, a hypothesis we are currently exploring.

4. DISCUSSION

The computational complexity in simulating realistic reverberant signals is prohibitive. In order to allow real-time, interactive environments to be simulated, many current virtual auditory environments do not include any reverberation. Those that include reverberation often use algorithms that simplify the computations (e.g., by accurately simulating the directional information only for a small number of discrete echoes, and generating decorrelated noise to simulate later-arriving reverberation). The perceptual consequences of these computational simplifications are not well understood. More research is needed to quantify how sensitive the human listener is to these simplifications and to determine how they influence the subjective realism of the display, the ability to judge source distance accurately, and the ability to learn to accurately judge source direction with training.

There are inherent tradeoffs in including reverberation in a virtual environment. As with any complex design decision, the appropriate choice (of whether to include reverberation, how accurately to simulate room acoustics, etc.) depends upon the goals of the display. If the main goal of the auditory display is to provide speech input to the listener, it may be best to exclude any reverberation. If the goal is to provide distance information about arbitrary sound sources, including some form of reverberation is critical; however, one may be able to provide distance

607

information using a very simplified algorithm for generating "reverberation." Along a similar vein, because even moderate levels of reverberation can provide useful distance information and increase the realism of a display (Santarelli et al., 1999a), it may be possible to reduce the energy of simulated echoes and reflections and still provide the perceptual benefits of reverberation while limiting its destructive influence (e.g., on directional localization or on speech intelligibility). Of course, recent work suggests that subjects are sensitive to the agreement of visual and auditory environmental cues (see the article by Gilkey, Simpson, and Weisenberger in this volume); thus, if the main goal is to produce a virtual environment that is subjectively realistic and produces the sense that the listener is truly "present" in the virtual world, the simulation should include the most realistic model of room acoustics that can be integrated into the system.

REFERENCES

Bradley, JS and GA Soulodre (1995). "The influence of late arriving energy on spatial impression." J Acoust Soc Am, 97, 2263-2271.

Bronkhorst, AW and T Houtgast (1999). "Auditory distance perception in rooms." Nature, 397, 517-520.

Brown, TJ (2000). Characterization of Acoustic Head-Related Transfer Functions for Nearby Sources. Electrical Engineering and Computer Science. Cambridge, MA, Massachusetts Institute of Technology.

Brungart, DS (1999). "Auditory localization of nearby sources III: Stimulus." J Acoust Soc Am, 106, 3589-3602.

Brungart, DS (2000). A speech-based auditory distance display. 109th Convention Audio Eng Soc, Los Angeles.

Brungart, DS and NI Durlach (1999). "Auditory localization of nearby sources II: Localization of a broadband source in the near field." J Acoust Soc Am, 106, 1956-1968.

Brungart, DS and WM Rabinowitz (1999). "Auditory localization of nearby sources I: Head-related transfer functions." J Acoust Soc Am, 106, 1465-1479.

Carlile, S (1996). Virtual Auditory Space: Generation and Applications. New York, RG Landes.

Duda, RO and WL Martens (1997). Range-dependence of the HRTF for a spherical head. IEEE ASSP Workshop on Applications of Digital Signal Processing to Audio and Acoustics.

Durlach, NI, A Rigapulos, XD Pang, WS Woods, A Kulkarni, HS Colburn and EM Wenzel (1992). "On the externalization of auditory images." Presence, 1, 251-257.

Gilkey, R and T Anderson (1997). Binaural and Spatial Hearing in Real and Virtual Environments. Hillsdale, New Jersey, Lawrence Erlbaum Associates, Inc.

Kopco, N and BG Shinn-Cunningham (2001). Effect of listener location on localization cues and localization performance in a reverberant room. 24th mid-Winter meeting Assoc Res Otolaryng, St. Petersburg Beach, FL.

Litovsky, RY, HS Colburn, WA Yost and SJ Guzman (1999). "The precedence effect." J Acoust Soc Am, 106, 1633-1654.

Mershon, DH and LE King (1975). "Intensity and reverberation as factors in auditory perception of egocentric distance." Percept Psychophys, 18, 409-415.

Middlebrooks, JC (1997). Spectral shape cues for sound localization. Binaural and Spatial Hearing in Real and Virtual Environments. R. Gilkey and T. Anderson. New York, Erlbaum: 77-98.

Middlebrooks, JC and DM Green (1991). "Sound localization by human listeners." Ann Rev Psych, 42, 135-159.

Nomura, H, H Miyata and T Houtgast (1991). "Speech-intelligibility and subjective MTF under diotic and dichotic listening conditions in reverberant sound fields." Acustica, 73, 200-207.

Pralong, D and S Carlile (1996). "The role of individualized headphone calibration for the generation of high fidelity virtual auditory space." Jornal of the Acoustical Society of America, 100, 3785-3793.

Santarelli, S, N Kopco and BG Shinn-Cunningham (1999a). Localization of near-field sources in a reverberant room. 22nd mid-Winter meeting Assoc Res Otolaryng, St. Petersburg Beach, FL.

Santarelli, S, N Kopco, BG Shinn-Cunningham and DS Brungart (1999b). "Near-field localization in echoic rooms." J Acoust Soc Am, 105, 1024.

Shinn-Cunningham, BG (2000a). Distance cues for virtual auditory space. Proceedings of the IEEE-PCM 2000, Sydney, Australia.

Shinn-Cunningham, BG (2000b). Learning reverberation: Implications for spatial auditory displays. International Conference on Auditory Displays, Atlanta, GA.

Shinn-Cunningham, BG, S Santarelli and N Kopco (2000a). Distance perception of nearby sources in reverberant and anechoic listening conditions: Binaural vs. monaural cues. 23rd mid-Winter meeting Assoc Res Otolaryng, St. Petersburg Beach, FL.

Shinn-Cunningham, BG, S Santarelli and N Kopco (2000b). "Tori of confusion: Binaural localization cues for sources within reach of a listener." J Acoust Soc Am, 107, 1627-1636.

Wenzel, EM (1992). "Localization in virtual acoustic displays." Presence, 1, 80-107.

Wenzel, EM, M Arruda, DJ Kistler and FL Wightman (1993). "Localization using nonindividualized head-related transfer functions." J Acoust Soc Am, 94, 111-123.

Wightman, FL and DJ Kistler (1989). "Headphone simulation of free-field listening. I. Stimulus synthesis." J Acoust Soc Am, 85, 858-867.

Zahorik, P, DJ Kistler and FL Wightman (1994). Sound localization in varying virtual acoustic environments. Second International Conference on Auditory Display, Santa Fe, NM, Santa Fe Institute.

Zurek, PM (1993). Binaural advantages and directional effects in speech intelligibility. Acoustical Factors Affecting Hearing Aid Performance. G. Studebaker and I. Hochberg. Boston, MA, College-Hill Press.

Creating Auditory Presence[‡]

Robert H. Gilkey, Wright State University, Dayton, OH 45435, and Air Force Research Laboratory, WPAFB, 45433
Brian D. Simpson, Wright State University, Dayton, OH 45435 and Veridian Inc., Dayton, 45431
Janet M. Weisenberger, Ohio State University, Columbus, OH 43210

ABSTRACT
The experience in an auditory virtual environment can be extremely compelling, in a way that is rarely, if ever, achieved in a visual virtual environment. We describe experiments that investigate listeners' perceptions of sounds recorded binaurally in rooms of various sizes, as a function of the "listening room," i.e., the room in which these sounds were presented to the subjects. For conditions in which the listening room was a real environment, the sense of *presence* was a function of the similarity of the listening room and the recorded room. However, when the listening room was a visual virtual environment, ratings of presence were not affected by the listening room depicted. The effect of the room match seems to depend on the subjects' prior auditory and visual experience with the listening room. However, neither prior auditory nor visual experience alone is sufficient to explain the effects.

1. INTRODUCTION
1.1 Defining and measuring presence in virtual environments
The concept of "presence" is a major focus in virtual environment research. Presence is related to the user's sense of "being there" in the virtual environment; however, there is no single accepted definition of presence (see, e.g., Ellis, 1996; Heeter, 1992; Held & Durlach, 1992; Sheridan, 1992; Zeltzer, 1992). Some researchers have questioned whether achieving a sense of presence is important for virtual interface design (e.g., Zeltzer, 1992). Indeed, task performance in a virtual environment does not seem to be strongly related to the sense of presence (e.g., Welch, 2000). Slater and Wilbur (1997) argue that task performance improves with increasing user-interface quality, which may also increase the sense of presence. For example, Slater et al. (1996) report that increased presence did not lead to improvements in task performance, but increasing the quality of sensory input increased both the sense of presence and task performance (see also Freeman et al., 1999).

Despite the lack of formal definitions of presence or accepted techniques for its measurement, a number of factors have been identified as likely to increase the feeling of being present in a virtual environment. Slater and Wilbur (1997) define *immersion* as the degree to which a technology can deliver an illusion that is inclusive (other stimulation from the physical world is shut out), extensive (multiple sensory modalities are stimulated), surrounding (a panoramic, wide field of view is presented), and vivid (the display has high resolution, fidelity, and variety of sensory stimulation). Because presence has typically been associated with immersion, factors that increase immersion should increase the sense of presence as well. Of these, Durlach and Mavor (1994) argue that the representation of multiple sensory modalities may be one of the most important determinants of presence.

1.2 Auditory feedback and presence
Gilkey and Weisenberger (1995) described the role of auditory stimulation in determining the sense of presence. Their discussion focused not on virtual environments, but on deafness. Ramsdell (1978) found that World War II veterans who had sustained a sudden profound hearing loss during the war repeatedly described the world around them as "dead" and lacking in movement, and felt that they had lost the feeling of "connectedness," or "coupling" to the world. A number of statements suggested that the world had taken on an unreal quality. From these statements Ramsdell postulated that hearing plays a critical role in establishing the "sense of being part of a living, active world" (p. 503). This role is distinct from the more obvious auditory functions as a communication system and a warning system. He stated that on a more "primitive" level, sound serves simply as the background of everyday life. Background sounds include the incidental acoustic productions of objects in the environment and of ourselves as we interact with these objects (e.g., the ticking of clocks, the dripping of faucets, the impact of footsteps, etc). Ramsdell contended that although we are not necessarily conscious of these sounds, they play a crucial role in maintaining our feeling of "being part of a living world" (p. 501). Gilkey and Weisenberger (1995) used the suddenly-deafened adult as a model system to predict the experience in a virtual environment with limited auditory stimulation, and argued that an inadequate, incomplete, or nonexistent representation of the auditory background would compromise the sense of presence. That is, they noted that for deaf adults, all other sensory modalities are rendered perfectly,

[‡] Work supported by AFOSR F49620-97-1-0231 and F49620-95-1-0106.

with high bandwidth, high-fidelity stimulation; the environment offers an extensive field of view; interactions with the environment produce lawful, orderly, and predictable changes with no unnatural delays and excellent proprioceptive feedback. Nevertheless, deaf adults, *lacking only auditory stimulation*, feel *disconnected* from a seemingly *lifeless* environment. Murray et al. (2000) empirically tested Gilkey and Weisenberger's contention. Subjects were fitted with wax earplugs and sent out into the environment for a 20-min period, during which they conducted typical business (visiting banks, travel agencies, etc.). After returning, they completed questionnaires and reports describing their experiences. Typical responses included feelings of "floating", being a "zombie," being "out of things," and being "more an observer than a participant." Murray et al. concluded that auditory feedback was indeed crucial for establishing a feeling of connectedness to the environment.

In sharp contrast to the experiences of suddenly deafened adults, Gilkey and Weisenberger noted that the experience of a user in an *auditory-only* virtual environment can be extremely compelling, in a way that is rarely, if ever, achieved in a *visual-only* virtual environment. In an auditory virtual environment, users routinely confuse real and virtual stimulation, for example, trying to answer a virtual telephone or failing to respond to a real knock at the door. Moreover, users often feel uncomfortable when a virtual talker violates their personal space; may move to avoid being hit by a virtual auditory object; and sometimes "see" the shadow, or "feel" the breeze, of a virtual auditory object as it passes by. Such observations led us to speculate that auditory input might play a special and critical role in creating presence. Support for this idea comes from Hendrix and Barfield (1996), who found that the use of spatialized sound in a virtual environment increased the sense of presence. Unfortunately, striking experiences with auditory displays are far from universal. Indeed, users are frequently unsatisfied and complain that the sounds appear to be too close to their heads, elevated in front, and of low fidelity. Users experience the display as smaller and less realistic than the real world, something that they are listening to, rather than listening in.

Why are experiences in different auditory displays so different? Individual measurement of the acoustics of the users' head (i.e., individualized head related transfer functions, HRTFs), changing stimulation synchronized to head movements, and wideband stimulation have all been suggested as critical factors. Yet, simple binaural recordings of speech stimuli can make extremely effective auditory displays, even though the HRTFs are not individualized, the stimulation is not affected by head movements, and the bandwidth of the speech is modest.

This paper examines some factors that are often not directly considered, but can nevertheless influence *presence* in auditory-only virtual environments. Specifically, we made binaural recordings in three different-sized rooms and had subjects listen to the recordings while seated in one of the three rooms. In general, the recordings were judged to have greater presence when they were heard in the room where they were recorded. However, results depended on the stimulation conditions and on prior experience with the listening room. Comparison of results across conditions suggests that the role of audition in determining presence may be more complicated than previously concluded.

2. GENERAL METHOD

Binaural recordings were made of "natural" stimuli (the ringing of a telephone, the jingling of keys, and male speech) using the Knowles Electronics Manikin for Acoustic Research. The manikin was positioned in the center of each of three rooms that differed in size ($16m^3$, $60m^3$, and $194m^3$) but were of similar shape and construction. The rooms were largely empty (the large room contained a pillar and some desks stacked up along the wall at one side). The recordings were made at locations surrounding the manikin in azimuth (360°) and ranging from -45° to +45° in elevation at distances of .5m and 1m. The same recordings were used in all experiments.

Each subject participated in only a single experimental session. During each session, the subject was seated in the middle of *one* of the three rooms (i.e., where the manikin had been) and listened to the stimuli through Etymotic ER2 insert earphones (without headphone compensation). A two-interval forced choice task was employed; the same sound source (e.g., the telephone) was presented in both intervals, but the recording room and source location varied across intervals. In each session, 3 blocks of 66 trials were presented in which all 198 possible combinations of sound source, room, and location were presented once. The order of stimulus presentation was randomly varied across subjects. The subjects were instructed to press one of two buttons on a response box after each trial to indicate the interval that contained the most realistic sound (i.e., the sound that "sounded as if it could have come from an object or person in the room with you"). Note that all of the experiments employed a mixed design, such that listening room was a between-subjects factor and recorded room was a within-subjects factor.

3. RESULTS AND DISCUSSION

3.1 Experiment 1

In Experiment 1, 21 subjects listened in the actual rooms where the recordings had been made (7 subjects per room). Verbal instructions were given to each subject while they were seated in the room, providing some exposure to the auditory environment of that room before data collection began. Subjects were told to listen with their eyes open.

Results, averaged across the 7 subjects in each listening room, are shown in Figure 1. The percentage of trials on which the particular recording was chosen as producing the greater sense of presence is plotted as a function of both recording room (bar shade) and listening room (bar cluster). (Note that each cluster of bars totals 150%, because the results of binary comparisons for 3 pairs of conditions are shown.) These data suggest that the listening room has a systematic influence on the subjects' preferences for the recordings. The small-room recordings are most frequently judged as producing the greater presence when heard in the small listening room, are less frequently chosen in the medium-sized listening room, and are least frequently chosen in the largest listening room. The opposite pattern of results was observed for the large-room recordings. The medium-room recordings were chosen less frequently overall (the medium room had an unusually long low-frequency reverberation; we believe that this may have made made the recordings made in that room sound less realistic). Nevertheless, the medium room recordings were chosen more often by listeners in the medium listening room than listeners in the other two rooms. Overall, each recording "sounded best" when heard in the room in which it was made.

Because the subjects had relatively little auditory experience with the room (a few minutes during which the instructions were provided) but substantial visual experience with the room (they had their eyes open throughout the more than 30 min it took to receive the instructions and complete the blocks and breaks), we assumed that the effect was at least partially mediated by visual cues about the listening environment. If so, and if similar effects occur in multisensory virtual environments, then virtual environment designers would need to ensure that there is a *detailed* match between the virtual visual and auditory environments.

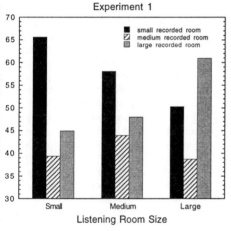 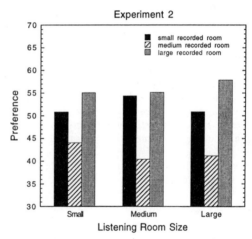

Figures 1 and 2. Percentage of responses indicating preference for a particular recorded room as having the strongest presence, as a function of the listening room. Figure 1 shows data for Experiment 1 (actual listening rooms), and Figure 2 shows data for Experiment 2 (virtual listening rooms).

3.2 Experiment 2

In order to address whether visual cues mediated the perceived realism of acoustic stimuli in Experiment 1 and the implications of such mediation for virtual environments, Experiment 2 replicated Experiment 1 except that subjects sat in a virtual depiction of the one of the 3 listening rooms. The virtual depictions were presented in Wright State University's CAVE™. The CAVE is a 10' x 10' x 10' room composed of rear-projected screens and a top-projected floor. Subjects wear CrystalEyes stereoscopic shutter glasses, yielding images that appear at specified locations in front of or behind the physical walls of the CAVE. Detailed visual representations of the listening rooms were created using Multigen II, including windows, doors, blackboards, electrical outlets, etc. Each of the 20 subjects received instructions in a separate room, inserted the ER2 earphones and donned shutter glasses before being led to the CAVE and seated in the apparent middle of the already presented visual depiction of the listening room (7 subjects in two of the listening rooms; 6 subjects in the third room). The results are plotted in Figure 2 in the same manner as in Figure 1.

There is little impact of the visually-depicted listening room on the subjects' recorded-room preferences (i.e., the three clusters of bars are similar). Indeed, except for the consistently low preference scores for the recordings made in the medium-sized room (also observed in Experiment 1), there seems to be little or no systematic difference among the conditions. These results differ sharply from those for Experiment 1, suggesting two possibilities: 1) the visually-depicted rooms in Experiment 2 did not have the same impact as the actual visual stimulation created by the actual rooms in Experiment 1; or 2) the impact of the listening room, observed in Experiment 1, was not visually mediated. Because the visual depictions used in Experiment 2 were "quite good" and we had little ability to improve them given the technology at our disposal, we designed Experiment 3 to examine how important visual stimulation had been in the actual rooms of Experiment 1. (Note, there were a number of limitations in the CAVE rendering of the rooms; e.g., although in a dynamic virtual environment one tends to "lose track" of the actual CAVE walls and focus instead on the virtual depiction, when viewing these stationary rooms the CAVE walls were clearly noticeable.)

3.3 Experiment 3

In Experiment 3, the subjects' experience with the actual rooms was systematically varied across three conditions. The first two were partial replications of Experiment 1, and the third was a direct replication of Experiment 1. In Experiment 3A, the procedures were identical to those in Experiment 1 except that the subjects were blindfolded and wheeled into the actual listening room in a wheelchair to limit non-auditory experience with the room. In Experiment 3B, the procedures were identical to those of Experiment 1 except that the subjects received experimental instructions in the hallway, inserted the ER2 earphones (which serve as moderately effective earplugs), and were then wheeled into the actual listening room in the wheelchair to limit their auditory experience with the room. In Experiment 3C, all procedures were identical to those of Experiment 1 except that the subjects were seated in the wheelchair in the hallway and wheeled into the actual listening room where they received instructions. A total of 90 subjects participated in Experiment 3 (10 per listening room per condition).

Figure 3 shows the results for Experiment 3A ("only" auditory experience with the listening room), plotted in the same manner as in Figure 1. As can be seen, the room preferences vary with the listening room. However, the pattern of results is not particularly orderly, with, for example, the small room recording receiving its highest ratings in the large room, and the large room recording receiving similar preference ratings in all three rooms. Thus, it appears that eliminating the subjects' visual experience with the room alters the nature of the effects observed in Experiment 1, but does not *eliminate* the listening room dependence. Figure 4 shows results for Experiment 3B ("only" visual experience with the listening room). Although some differences in preference are observed for different listening rooms, they are relatively small and not simply related to the recording-room/listening-room match. Limiting prior auditory experience does severely limit the impact of the listening room on the recorded-room preferences. Interestingly, neither the blindfolded nor the earplugged subjects produced the pattern of results that had been observed in Experiment 1. These results might lead one to question the robustness of the original effects. However, the direct replication of Experiment 1 in Experiment 3C (both auditory and visual experience with the listening room), as shown in Figure 5, produced a nearly identical pattern of results to that observed in Experiment 1, suggesting that the impact of the listening room on subjects' preferences is real and robust.

4. CONCLUSIONS

The results from these various experimental conditions create a somewhat cloudy picture. That is, the results of Experiment 1 and Experiment 3C show a substantial and orderly impact of the listening room on subjects' preferences for different binaural recordings. However, results from Experiment 2 and Experiment 3B suggest that visual stimulation provides at most a modest contribution to this effect. The results of Experiment 3A, on the other hand, suggest that prior auditory experience with the room is not adequate to create the effects observed in Experiment 1 and Experiment 3C. Overall, the results argue that both auditory and visual stimulation are necessary to produce the effects observed in Experiment 1 and Experiment 3C, but that neither auditory nor visual stimulation alone is sufficient to produce these effects. Note, however, that the conclusion that prior auditory experience is not sufficient to explain the effects depends heavily on the data from the 10 subjects in the large listening room under Experiment 3A. Had they shown less preference for the small room and more for the large room (n.b., these are not independent quantities), the results would have agreed with those of Experiment 1.

Taken together, the results suggest that the designers of virtual environments need to be concerned with the detailed match between the auditory and visual scenes. However, further work is necessary to determine specifically how auditory and visual cues must match to preserve the sense of presence.

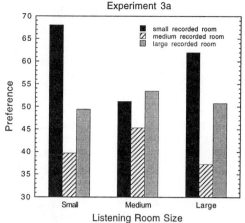

Experiment 3a

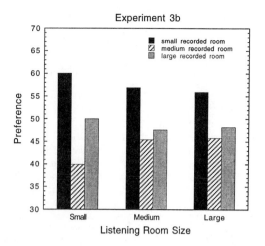

Experiment 3b

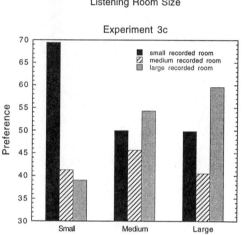

Experiment 3c

Figures 3-5. Percentage of responses indicating preference for a particular recorded room as having the strongest presence, as a function of the listening room. Figure 3 shows data for the auditory-only condition (Exp. 3a); Figure 4 shows data for the visual-only condition (Exp. 3b); and Figure 5 shows data for the auditory and visual replication of Experiment 1 (Exp. 3c).

REFERENCES

Durlach, N. & Mavor, A. (eds.) (1994). *Virtual reality: Scientific and technical challenges.* Washington, DC: National Academy Press.

Ellis, S.R. (1996). Presence of mind: A reaction to Thomas Sheridan's "Further musings on the psychophysics of presence." *Presence: Teleoperators and Virtual Environments 5,* 247-259.

Freeman, J., Avons, S.E., Pearson, D.E., & Ijsselsteijn, W.A. (1999). Effects of sensory information and prior experience on direct subjective ratings of presence. *Presence: Teleoperators and Virtual Environments 8,* 1-13.

Gilkey, R.H., & Weisenberger, J.M. (1995). The sense of presence for the suddenly deafened adult. *Presence: Teleoperators and Virtual Environments 4,* 357-363.

Heeter, C. (1992). Being there: The subjective experience of presence. *Presence: Teleoperators and Virtual Environments 1,* 262-271.

Held, R.M., & Durlach, N.I. (1992). Telepresence. *Presence: Teleoperators and Virtual Environments 1,* 109-112.

Hendrix, C., & Barfield, W. (1996). The sense of presence within auditory environments. *Presence: Teleoperators and Virtual Environments 5,* 290-301.

Murray, C.D., Arnold, P., & Thornton, B. (2000). Presence accompanying induced hearing loss: Implications for immersive virtual environments. *Presence: Teleoperators and Virtual Environments 9, 137-148.*

Ramsdell, D.A. (1978). The psychology of the hard-of-hearing and the deafened adult. In H. Davis & S.R. Silverman (Eds.), Hearing and deafness (4th ed., pp. 499-510). New York: Holt, Rinehart, & Winston.

Sheridan, T.B. (1992). Musings on telepresence and virtual presence. *Presence: Teleoperators and Virtual Environments 2,* 120-125.

Slater, M., Linakis, Usoh, M., & Kooper (1996). Immersion, presence, and performance in virtual environments. In M. Green (ed.), *ACM virtual reality software and technology.* New York: Association for Computing Machinery, pp. 163-172.

Slater, M., & Wilbur, S. (1997). A framework for immersive virtual environments (FIVE): Speculations on the role of presence in virtual environments. *Presence: Teleoperators and Virtual Environments 6,* 603-616.

Welch, R.B. (2000). How can we determine if the sense of presence affects task performance? *Presence: Teleoperators and Virtual Environments 9,* 574.577.

Zeltzer, D. (1992). Autonomy, interaction, and presence. *Presence: Teleoperators and Virtual Environments 1,* 127-132.

613

USING VIRTUAL AUDITORY ENVIRONMENTS TO STUDY SOUND LOCALIZATION

Klaus Hartung

University of Connecticut Health Center, Department of Neuroscience, Farmington, CT, 06030, USA

ABSTRACT
Humans, like most mammals and birds use mainly the signals they receive passively at both eardrums for the localization of sound. The signals of sound sources arriving at the eardrums are distorted in a direction specific manner by the external ears and associated structures due to reflection and diffraction of the incoming waves. If sounds are filtered with filters which simulate theses directional linear distortions and presented over headphones, broadband sounds can be precisely localized to the corresponding position outside of the head. This 'virtual auditory environment' is a powerful tool for behavioral and neurophysiological research in different species. Usually the directional distortions of the external ear are measured with microphones placed in the ear canal. These transfer functions then have to be equalized appropriately for headphone presentation. For some applications the spatial resolution of these transfer functions has to be increased by means of spatial interpolation. In this report, some examples of virtual auditory environments to study sound localization in behavioral and neurophysiological studies will be presented. In order to guarantee a very realistic rendering of the virtual auditory environment the sounds presented over headphones have been compared acoustically, behaviorally and neurophysiologically with sounds emitted in the free field. The results show that the sound spectra at the eardrums are almost identical to the spectra of a real sound source. In a neurophysiological study, it has been shown that the responses of spatially tuned neurons are almost indistinguishable in response to virtual and free field stimulation. The directional cues generated by the external ears are different for different individuals. For research applications it is therefore necessary to measure the transfer functions of each individual and use these individual transfer functions for the generation of auditory virtual environments, otherwise sound sources in the front might be perceived in the back, and/or the spatial pattern of the neuronal responses will be severely distorted. Virtual auditory environments provide a very precise control of all parameters relevant for sound localization. This does not only allow the recreation of natural scenes, but also the creation of complex artificial stimuli in order to study specific mechanisms involved in sound localization.

1. INTRODUCTION
Since beginning of systematic research on the acoustical, psychological and physiological bases of sound localization most studies have concentrated on two cues that arise from the distance of the two ears and head shadowing: interaural time differences (ITD) and interaural intensitive differences (IID). In addition to these cues, important frequency-dependent, directional cues arise from the structure of the external ear (Blauert, 1997 for review). When these external ear based cues are missing or are severely distorted, human listeners typically confuse a sound source from the front to back. Headphone presentation of IID and ITD cues alone results in a hearing event, which appears to be localized inside the head. Changes in the values of the ITDs and IIDs shift the perceived position more to the left or right within the head. Complete, externalized, localization of a sound source requires use of the individualized external ear-based spectral cues. The vast majority of psychophysical and physiological studies on sound localization using headphones have typically only relied on ITD and IID cues, because these parameters are relatively easy to control, and neglected the external ear based spectral cues. The recent advancements in digital signal processing have made it possible to present over headphones sounds that are spectrally shaped to match the effects of directional filtering by the external ears. These direction-dependent filters are usually called 'head-related transfer functions'. Several laboratories have measured HRTFs and presented through headphones, sounds that have been filtered with these transfer functions to create a virtual auditory space (e.g. Wightman and Kistler 1998a 1998b, Brugge et al 1994 Hartung and Sterbing 1997). In this report, some applications of auditory virtual environment in studies to investigate sound localization (Hartung a. Sterbing, 1997, Sterbing and Hartung 1998,1999,2000, Keller et al. 1998, Schlack et al. 2000) using virtual auditory environments are summarized. First, the method for generating a virtual auditory environment will be described. Then, results of comparisons between free field sounds and sounds presented via headphones using HRTF will be shown. Finally, some results and details of the applications of virtual auditory environments in studies about sound localization are shown. These examples demonstrate, that this technique does not only allow to re-create real auditory environment with a very high fidelity, but also provide means to modify the presented auditory cues independently of each other and very precisely in order to study specific mechanisms involved in sound localization. These studies would not be have been possible with classical headphone presentations or free field setups.

2. METHODS

The virtual auditory space is constructed by filtering a source signal with filters called head-related transfer functions (HRTFs) and the resultant signals are presented over headphones. These transfer functions mimic the directional changes in the spectrum of the signal at the eardrums, which are caused by the reflection, diffraction and resonances at the external ears, the head and the shoulders of the listener. For humans, guinea pigs and macaques the physical dimensions of these structures show considerable differences between different individuals. Therefore, it is necessary to measure the necessary transfer functions individually.

2.1 MEASUREMENT OF HRTF

Usually HRTF are measured in anechoic rooms with an array of loudspeakers or a single loudspeaker, which is moved around the subject. Probe or miniature microphones are inserted in the ear canal of the subject. It would be ideal to place the microphone or the tip of a probe directly at the eardrum. This not possible in some cases or would make a measurement extremely difficult. Placing the microphone at the entrance of the ear canal might not capture all directional parameters, because the entrance of the ear canal contributes to the directional filtering of the external ear. Measurements in the ear canal of barn owl's showed, that all directional cues are captures for positions approximately 5 mm and more behind the entrance of the ear canal (Keller et al., 1998). Listening tests of virtual sound localization show that HRTFs, which are measured appr. 5 mm within the ear canal allow authentic reproduction without front-back inversion and inside-head localization, while microphone positions further outside might lead to front/back inversions. The quality of the measurement can be improved, if a microphone-plug combination is used. Then the sound transmission behind the microphone is blocked and no sound pressure minima occur at the position of the microphone (e.g. Møller, 1992). After placing the microphone is placed in the ear canal, for each direction the loudspeaker emits a continuously repeated noise bursts. The sounds are recorded with miniature microphones in the ear canal, averaged in synchrony with the noise bursts. The HRTF is computed by dividing the Fourier transform of this signal with the Fourier transform of the noise burst.

The raw transfer functions cannot be used for the presentation of virtual sound sources. Because of the different shapes and size of ear canal and pinna the influences of the headphones coupled to the ear canal have to be compensated for each subject individually. This is done by measuring the transfer function from the port of the headphone to the port of the microphone in the ear canal when the headphone is placed on the subject. The equalization filters are then designed as FIR filters, using a modified least-square approximation technique leading to a maximal equalization error of less than +/-0.002 dB for all frequencies between 2 and 11 kHz. Small changes in the position of the headphone, which are impossible to notice by visual inspection, might change the transfer function (e.g. Kulkarni and Colburn, 2000). If an exact reproduction is required, the headphone transfer function has to be measured individually (Møller et al. 1995) and the position of the headphone during playback has to match exactly the position during the measurement. In all neurophysiological studies summarized here, the earphones where mounted on a stereotactic device, which held the animal during the HRTF measurement and during the physiological recordings. This allowed a reliable positioning of the headphones and reduced the variability considerably.

2.2 INTERPOLATION OF HRTF

Smooth rendering of dynamic virtual auditory environments (moving sound sources, head movements) or high precision measurements of localization precision require HRTF catalogues with a high spatial resolution (smaller than 5 °) or the application of appropriate interpolation algorithms. Usually, HRTF catalogues are measured with a smaller resolution of 10° to 15° in order to minimize the time required for the measurement. The measurement for one catalogue with this resolution takes approximately one hour. In order to increase the resolution spatial interpolation of the transfer functions has to be applied. Hartung et al. (1998) compared two different algorithms, the "nearest-neighbour" and "spherical-spline" method (e.g. Chen, J. et al.1995). The nearest neighbour method interpolates the missing direction by a weighted summation of the transfer functions of the nearest direction. The weights are reciprocal to the distances. Usually this method is applied on the time domain representation of the HRTF. In the spherical-spline-method not only the nearest directions, but all measured directions are used to calculate the weights of spherical splines. The value at the desired direction can then be calculated by computing the sums of the spherical splines. This method is usually applied independently on the log-magnitude and phase representation of the HRTF. Figure 1 shows the measured HRTF of an artificial head and a typical result for the nearest neighbour and the spherical spline interpolation method. The interpolation is based on a measured catalogue with 15° resolution. The method using spherical splines shows much smaller errors than the nearest-neighbour method. The deviations in the nearest-neighbour method result in changes in the timbre and the perceived direction of the presented sound. In order to get an impression of the distribution of the interpolation errors across space, the mean error for each direction using a third octave band resolution of the HRTF was calculated. Figure 2 shows the distribution of the errors. For the spherical interpolation method the mean errors are relatively small and below 1 dB for most directions. For the nearest neighbour method the mean deviation is for most directions about 1.5 dB and are up to 3.5 dB for some directions. The largest interpolation errors occur for directions of incidence where one ear is completely shadowed by the head (e.g. 105 to 120 deg for the right ear). For these directions the notches in the HRTF between 1.5 kHz and 3 kHz change very rapidly with changing azimuth. No interpolation method can recreate the correct shape of the transfer functions if the HRTF are measured with a 15°- or lower resolution. It is therefore necessary to measure these directions with a higher resolution. The time required for the measurement of a complete catalogue can be kept sufficiently small if the space is sampled with a non-uniform resolution considering the different spatial frequencies across space in the HRTF.

3. ACOUSTICAL AND NEUROPHYSIOLOGICAL COMPARISONS OF FREE-FIELD SOUNDS AND VIRTUAL SOUND SOURCES PRESENTED THROUGH HEADPHONES

The presentation of virtual sound sources, which are filtered through the HRTF and the equalization filter will result in the same signal at the eardrums of a listener as a real world source would create, provided that the position of the headphones is the same during presentation and measurement. This has verified in two studies where virtual sound sources have been used in neurophysiological experiments (Keller et al. 1998, Sterbing and Hartung, 2001). In both studies, the sound spectra at the eardrum created by a real loudspeaker in an anechoic room were compared to the sound spectra at the eardrum that were presented via headphones and filtered with the appropriate HRTF. The sound pressure was measured with a small probe microphone. The tip of the probe was approximately 1 to 2 mm away from the eardrum. Figure 4 shows the results for one direction measured for the studies

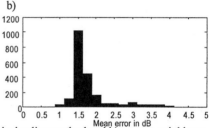

Figure 1: Measured and interpolated HRTF for a direction in the horizontal plane (left ear, 5 Az.)

a)
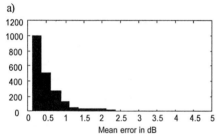
Mean error in dB

b)
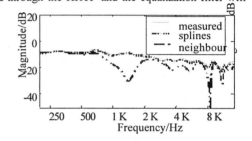
Mean error in dB

Figure 2: Distribution of mean interpolation errors for the a) spherical spline method and b) nearest neighbour method.

with guinea pigs. Only at frequencies above 8 kHz the headphone sound has a slightly higher amplitude than the free field sounds. The differences are typically smaller than +/-2dB between 100 Hz and 10 kHz for guinea pigs. For the measurement in barn owls the differences in the magnitude of the spectra were less than +/- 1dB between 3 kHz and 10 kHz.

For the comparison of the neuronal responses to free field sounds and headphone sounds filtered with HRTF, the HRTFs of barn owls were measured with probe microphones at appr. 2 mm distance from eardrum while the stereotaxic bench and microdrives were present. After equalization these HRTF were used to filter the test noise online with a digital signal processors and then present via earphones to the animal. After headphone presentation, the headphones were removed and the same stimulus was presented via loudspeaker from the same directions. The responses of single neurons in the inferior colliculus to free field and headphone stimulation were compared. The analysis showed that the responses to the free- field stimulus and the virtual sound source were aligned well in space.

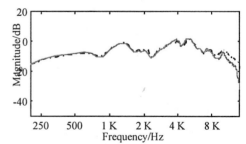

Figure 4: Comparison of the transfer function from a loudspeaker in the free field to the eardrum (grey line) with a corresponding transfer function in the auditory virtual environment (dashed line).

4 APPLICATIONS IN NEUROPHYSIOLOGICAL EXPERIMENTS

Sterbing and Hartung (1999a, 1999b, 2000, 2001) carried out a series of different experiments using virtual sound sources with guinea pigs. For this experiments 122 directions in the upper hemisphere were measured with a spatial resolution of 10° to 15° sampling frequency of 50 kHz. The microphones (Knowles 3046) were positioned a few millimetres inside the entrance to the ear canals of the left and the right ears. The HRTFs were processed using the methods described above. Then the stimuli, white-noise bursts (50 ms duration, 5 ms rise/fall time, 80 dB SPL, 38 dB spectrum level) were filtered off-line with the corresponding transfer functions for all the directions and stored · on hard-disk. During the electrophysiological experiments each virtual direction was pseudorandomly presented 5 times via earphones. It took about 10 minutes to collect the responses for one neuron. The majority of the responding units in the midbrain (ICc, SC) and Cortex (AI) were spatially tuned. The largest subset of the units was activated by virtual sound sources (VSS) positioned in the lower contralateral hemifield. Nevertheless, tuning to high elevation VSS and front or back positions with low, intermediate and high elevation was also measured. Front-back discrimination on a single cell level could be demonstrated. A subset of single units was stimulated with both the correct HRTF and the HRTF of a different guinea pig. In the case of stimulation with the VSS ,of a different individual, they showed either no spatial tuning or altered spatial tuning. In one study (Sterbing and Hartung, 1999a) the correlation of the signals between the two ears was manipulated while presenting unaltered monaural spectral cues and interaural intensitive differences (IID). This manipulation removed interaural time difference cues and tested the processing of spectral cues and IIDs. It was demonstrated that the majority of neurons with high characteristic frequencies did not change their spatial tuning compared to the unmodified signals. This finding allowed identifying the contribution of the different neuronal circuits involved in sound localization and lead to a signal-processing model of sound localization (Hartung and Sterbing, 1999). One advantage of auditory virtual auditory environments is that sound fields with multiple sources can be created relatively easy. In a study aimed to investigate the processing of sounds in a reverberant environment (Sterbing and Hartung, 2000), first, the spatial receptive field of the neurons was measured by testing sound source incidence from the entire upper hemisphere. Then the position of the lagging source (reflection) was kept constant at the neurons′ preferred position, and the leading source was delivered from random locations. Recently a virtual auditory environment has been used to study the auditory responses of cortical multimodal neurons in the area VIP of awake macaques (Schlack et al. 2000). The visual stimuli were projected on a screen. The auditory stimuli were delivered via headphones. In order to match the exact position of the presented visual stimuli, the HRTF were interpolated using the spherical spline method described above. It was shown, that the auditory receptive fields matched the position of the visual receptive fields.

5. CONCLUSIONS

Using virtual sound sources presented through headphones has several advantages compared to loudspeaker presentations. It avoids uncontrolled distortions of the stimulus caused by the setup (e.g. chair, head holder, stereotaxic device, visual projection screens). The behavioral or neuronal response to sounds of the entire auditory space can be tested very fast, compared to free field stimulations where in most cases a loudspeaker has to be moved. It has been shown, that it is possible to present sounds in virtual auditory environment in a very high quality, which make them nearly indistinguishable from free field sounds. In a number of neurophysiological studies, it has been demonstrated that this technique allows a fast and accurate mapping of auditory spatial receptive fields of single neurons and to study the complex mechanisms underlying sound localization. This technique is a very powerful tool to study the various aspects of spatial hearing including the processing of moving sound sources and multimodal interactions.

ACKNOWLEDGEMENTS

This work was carried out while the author was with the Institute of Communication Acoustics at the Ruhr University Bochum, Germany. This work presented here was partly supported by grants form DFG (Ho 450/23-1-347/97) and NATO.

REFERENCES

Blauert, J., "Spatial hearing: the psychophysics of human sound localization," 2nd, rev. edition, MIT Press, Cambridge MA. 1997.

Keller C.H, Hartung, K, and Takahashi, T.R. "Head-related transfer functions of the barn owl: measurement and neural responses," Hear. Res 118. 1998, 13-34

Chen, J. van Veen, B.D. and Hecox, K. E., "A spatial feature extraction and regularization model for the head-related transfer function". J. Acoust. Soc. Am. 97, 1995. 439-452

Hartung, K., Braasch, J. and Sterbing, S.J., "Comparison of different methods for the interpolation of head related transfer functions", AES 16th International Conference on Spatial Sound Reproduction, 1999

Hartung K. and Sterbing, S.J. "Generation of virtual sound sources for the electrophysiological characterization of auditory spatial tuning in the guinea pig,".In Syka, J. (Ed.) Acoustical Signal Processing in the Central Auditory System, Plenum Press, New York, 1997, 407 412.

Hartung, K. and Sterbing, S.J., "A physiology-based computational model for sound localization,". in: Kollmeier, B., Dau, T., Hohmann, V. (eds.) Psychophysics, physiology and models of hearing, World Scientific Publishing, Singapore, 1999, 179-184.

Wightmann, F.L. and Kistler, D.J. "Headphone simulation of free-field listening. I: Stimulus Synthesis," J. Acoust. Soc. Am 85. 1989a, 858-867.

Wightmann F.L. and Kistler, D.J. "Headphone simulation of free-field listening. II:Psychophysical Validation", J. Acoust. Soc. Am 85. 1989b, 868-878.

Brugge J.F., Reale, R.A., Hind, J.E., Chan, J.C.K., Musicant, A.D. and Poon, P.W.F., "Simulation of free-field sound sources and its application to studies of cortical mechanisms of sound localization in the cat," Hear. Res. 1994. 67-84.

Møller, H., "Fundamentals of binaural technology," Applied Acoustics 36, 1992, 171-218.

Møller, H., Hammershøi, D. Jensen, C.B. and Sorensen, M.F., "Transfer characteristics of headphones measured on human ears," J. Aud. Eng. Soc. 43, 1995, 203-217.

Kulkarni, A. Colburn. H.S., "Variability in the characterization of the headphone transfer-function," J. Acoust. Soc. Am. 107, 2000, 1071-1074.

Schlack, A., Sterbing, S., Hartung, K. and Hoffmann ,K.-P. and Bremmer, F., "Spatially congruent auditory and visual responses in macaque area VIP". Soc. Neurosc. Abstr. 26, 2000, 478.

Sterbing, S.J. and Hartung, K "Virtual sound source coding in the inferior colliculus: effects of SPL variation, coherence, stimulus duration and bandwidth," in Kollmeier, B., Dau, T., Hohmann, V. (eds.) Psychophysics, physiology and models of hearing , World Scientific Publishing, Singapore. 1999a

Sterbing, S.J. and Hartung, K. "Space Processing in the Guinea Pig Auditory cortex," Assoc. Res. Otolaryngol., 1999b, Abstr. 153.

Sterbing, S.J and Hartung, K., "The neuronal response to pairs of noise bursts: influence of sound source direction and delay," Assoc. Res. Otolaryngol, 2000, Abstract 4996.

Sterbing, S.J. and Hartung, K., "Virtual sound source coding in the inferior colliculus of the guinea pig," in revision, 2001.

THE ROLE OF SYSTEM LATENCY IN MULTI-SENSORY VIRTUAL DISPLAYS FOR SPACE APPLICATIONS [*]

Elizabeth M. Wenzel

Spatial Auditory Displays Lab, NASA Ames Research Center
Mail Stop 262-2, Moffett Field, CA, 94035-1000, USA

In any virtual environment (VE), system parameters such as the update rate and the total system latency (TSL) may have a significant impact on the smoothness and responsiveness of the dynamic simulation if they exceed perceptual tolerances. The TSL refers to the time elapsed from the transduction of an event or action, such as movement of the head or hand, until the consequences of that action cause the equivalent change in a virtual source/object. The source and magnitude of these latencies may depend upon the context in which the virtual display is used. In space applications, TSLs may include the basic computational overhead in rendering the VE in typical ground-based tasks, as well as the more significant delays (0.4–6 s; Sheridan, 1992) produced by satellite and other communications delays during space-based activities. The presence of latencies creates the potential for a decoupling between the vestibular/kinesthetic sensation of motion and the consequences of motion in the sensory world. Such a decoupling can cause various undesirable perceptual effects, including positional instability of objects or an effective compression of the VE's response to head or hand motion. Recent research has investigated the impact of increasing latency on such measures as localization accuracy for virtual sounds and the ability of viewers to discriminate between different latencies in a visual virtual environment. It appears that subjects are less sensitive to latency in the auditory than in the visual domain, and highly sensitive in the haptic domain.

1. INTRODUCTION

In preparing for and conducting long-term space exploration missions, advanced multimodal displays will be required to enhance cooperative space operations involving a number of computer systems, human participants, and remotely-operated devices. Current interfaces for cooperative tasks such as EVA operations, navigation, control, and scientific visualization often do not allow multi-sensory, human-centered computer interaction. To enable interaction with *multiple* concurrent information streams, it has often been proposed that the human operator be provided with *multiple* display channels—head-tracked, head-mounted stereoscopic video displays, spatial auditory displays, and haptic devices—to provide visual, aural, and force feedback.

The types of tasks that may benefit from multisensory displays can be viewed along a task continuum representing the type of situational awareness (SA) that is required (Figure 1). At one end of the spectrum are tasks that require global SA of a large, spatially distributed task space, such as path planning for a distributed fleet of telerobots during planetary exploration. At the other end are tasks that require localized SA of a small-scale work area, such as tool manipulation during replacement of a malfunctioning electrical component. In between are tasks that may involve both large- and small-scale situational awareness. Such tasks (e.g., guidance of a telerobot over terrain while coordinating with other telerobotic agents and actors in the region) generally require navigation within a larger task-space combined with localized control and manipulation. Corresponding task continua represent the degree of direct engagement required by the actors to do a task and the time-scale required for updating information.

Figure 1 also illustrates continua that represent critical display parameters corresponding to the various types of tasks. For example, localized manipulation tasks that require a high degree of task engagement with immediate information update will require displays that are highly dynamic with tightly-coupled feedback for one or more of the actors involved. For such displays, low latencies, high update rates, and maximum spatial resolution will be required. For global SA tasks, that primarily involve monitoring the status of multiple agents/actors in large-scale regions, the display information may be quasi-static and more loosely coupled to the actions of the actors. Consequently, the constraints on display latency, update rate, and spatial resolution may be relaxed.

Long-duration exploration missions may encompass tasks that represent cooperative tasks at opposite ends of the task-display continua of Figure 1. One example is multi-actor telerobotic path planning in a VE for remote planetary exploration. This task demands large-area (i.e., global) SA and requires deliberate, and therefore slow and loosely-coupled, composition of a potentially lengthy sequence of moves. Latency demands are relaxed because the planned sequence does not need to be evaluated in real-time. A second example is cooperative control of a manipulator arm for on orbit assembly as is currently done on the shuttle and planned for the International Space

Station. Such a real-time task is tightly coupled and highly dynamic, having stringent demands on temporal and spatial fidelity. In order to maintain precise trajectory control, the users must remain highly engaged because the consequences of the manipulation task are immediate. Finally, the manipulation tasks, while requiring a degree of global SA in the initial stages of the trajectory, primarily require local SA in the final part of the task.

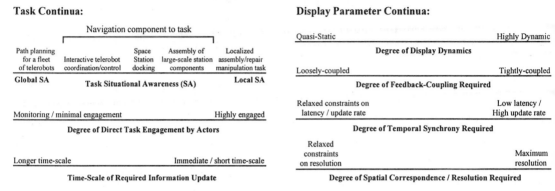

Figure 1. Conceptualization of task/display continua.

2. THE IMPACT OF LATENCY

As Figure 1 suggests, a key experimental issue for the design and deployment of multi-sensory VE displays for NASA missions is the perceptual and performance impact of system and communications latencies. This section reviews recent work on the impact of latency within the sensory modalities.

2.1 Haptic Modality

The haptic modality possesses very little tolerance for latency. Manipulation and control of haptic devices requires a very tightly-coupled feedback system that, unlike vision and hearing, necessarily involves both simultaneous input to (haptic display) and output by (manual control) the human operator (Sheridan, 1992). Research in telerobotic control has shown that latencies on the order of hundreds of ms cannot be tolerated. Such latencies interfere greatly with the ability to control a manual device and, in fact, can be quite dangerous if one is attempting to control a heavy and powerful device such as a robotic arm. Even latencies on the order of tens of ms can be quite problematic (Poulton, 1974) and require that the system compensate for the presence of latency by reducing the bandwidth and gain of the system's response to control inputs. Otherwise haptic devices such as telerobotic arms and virtual haptic interfaces can become unstable, potentially resulting in damage to the device and/or injury to the user. The net perceptual effect of such compensation is to make the interaction appear to be quite sluggish and "squishy" (highly compliant). In general, very little latency can be tolerated in a haptic display without resorting to compensatory filtering by the system, or the use of alternate control strategies such as predictor displays and "move and wait" strategies that reduce the degree of real-time control possible with the device.

2.2 Visual Modality

Much of the prior research on the impact of latency in the visual modality has involved its effect on overall operator performance. For example, the manual control literature (see Sheridan, 1992) indicates that latency has a strong negative impact on the ability to use visual displays and control telerobotic devices. Control accuracy decreases with increasing latency until, at about 300 ms, the user adopts a "move and wait" strategy in order to perform a task, thus largely defeating interactivity in the control interface. Latency has also been shown to interfere with adaptation to positional distortion in visual displays (Held, Efstathiou & Greene, 1966). Ellis, Adelstein, Baumeler, Jense & Jacoby (1999a) have also shown that RMS error in a tracking task in a VE increases with increasing latency, with notable increases in error beginning at latencies of 80 ms. Researchers have also begun to examine users' subjective responses to TSL, particularly in virtual displays. The sense of presence or immersion is reduced by the presence of latency (Welch, Blackmon, Liu, Mellers & Stark, 1996), as is subjective controllability in a tracking task (Ellis, et al., 1999a). Another potential effect of TSL is simulator sickness. As with many types of subjective response, the impact of latency depends upon the nature of the task. Tasks that tend to induce vection appear to be much more susceptible to latency-induced motion sickness than a simple pointing or grasping task. For example, Frank, Casali & Wierwille (1988) showed that visual delays of 170 or 340 ms induced motion sickness when driving a car simulator, while Ellis, et al. (1999a) found no correlation between latency (up to 480 ms) and motion sickness in a 3D pointing and tracking task.

In many of these studies, latency is apparently the most potent factor affecting both the performance and subjective response of the user, dominating other aspects of the virtual display such as update rate and spatial fidelity. One reason that latency may have such a critical effect is that it produces spatial registration errors as well as degradations in display responsiveness. The direction of spatial error produced by TSL will vary depending upon whether the position of the virtual object or the user is considered (Ellis, Adelstein, Jung & Young, 2000). For example, during hand translation of a virtual object, TSL causes a spatial error in the direction opposite to the motion; the "world lags behind" the user's action. Conversely, during head translation while viewing a stationary virtual object, TSL causes a spatial error in the same direction as the movement; the "world moves with" the user.

Since users are apparently quite sensitive to TSL in a visual VE, it is important to know the minimum perceptual tolerance, or just-noticeable difference (jnd), for latency. Such data can define specifications that must be met when designing VE displays. Ellis and his colleagues (Ellis, et al., 2000; Ellis, Young, Adelstein & Ehrlich, 1999b) have collected the first threshold data for latency. Their experiments measured latency discrimination in a simple virtual environment while either manipulating or viewing a 10 cm virtual ball as a function of a paced, stereotyped lateral movement of either the hand or head. A two-alternative forced choice (same-different) task was used in which subjects compared one of three reference latencies (35, 101 or 205 ms) to the reference latency plus an increment (0 to 116.9 ms). For both hand and head motion, hit and false alarm rates were independent of the reference latency, indicating that the data do not follow Weber's Law. Hit rates increased approximately linearly with the incremental latency added to the reference latency, while false alarm rates remained fixed at a value of 0.15-0.20. Comparison of the hit rate data to rates predicted by an equally-likely random guessing strategy indicated that a few subjects could discriminate a 16.7 ms latency difference, while all subjects could discriminate a 33 ms difference. Thus it appears that the jnd for latency in this simple visual task is in the range of 16.7 to 33 ms.

The paced, stereotyped movements used in these studies preclude interpretation that the subjects detected latency *per se*. They may only have responded to the perceived visual displacement of the virtual object with respect to their hand or head. An experiment by Ellis and colleagues is currently underway which examines whether latency itself may be discriminated by requiring subjects to make movements of different speeds during the task.

2.3   Auditory Modality

Relatively little is known regarding the impact of introducing latency in a virtual acoustic environment. One recent study (Sandvad, 1996) investigated the perceptual impact of parameters like system latency, update rate, and spatial resolution during dynamic localization of virtual sources. Update rate and spatial resolution were manipulated by independently changing the parameters of a Polhemus Fastrak, while increased latency was achieved by adding 16.67 ms increments to the minimum latency (29 ms). The subjects' task was to point a toy gun that had a tracking sensor mounted on the handle at the apparent location of an anechoic virtual source. Localization performance was measured by the standard errors of the signed azimuth and elevation components of the pointing response and by the average time between judgements in a block of trials.   It was found that localization performance did not significantly degrade until the system latency increased to 96 ms or the update rate decreased to 10 Hz (compared to best system parameter values of 29 ms and 60 Hz, respectively). Degrading the spatial resolution to 13°, the largest value tested, had little impact on localization error. However, the psychophysical method that was used to measure localization accuracy was self-terminated by the subjects, resulting in trial lengths ranging from about 1.5 to 2.5 s. Such stimuli durations may not have been long enough to allow adequate head-motion sampling by the listeners. Also, the average directions of the pointing responses and the front-back confusion rates (the localization error most affected by enabling head motion) were not reported in Sandvad (1996). The author may have chosen not to report such data because of the large individual differences observed in their data.

Previous studies of localization (Wenzel, 1996; Wightman & Kistler, 1999) have demonstrated that, compared to static localization, enabling head motion dramatically improves localization accuracy of virtual sources synthesized from head-related transfer functions (HRTFs). For example in Wenzel (1996), average front-back confusion rates decreased from about 28% for static localization to about 7% when head motion was enabled for sources synthesized from non-individualized HRTFs. Confusion rates on the order of 5% are typically observed during static localization of real sound sources.

Measurements of the TSL of the system used in Wenzel (1996) indicated a mean of 54.3 ms (+/- 8.8 ms) and minimum and maximum values of 35.4 and 74.6 ms (Wenzel, 1998). Examination of the head motions that listeners used to aid localization in Wenzel (1996) suggests that the angular velocity of some head motions (in particular, left-right yaw) may be as fast as about 175°/s for short time periods (about 1 second).   A maximum TSL of 75 ms could potentially result in short-term under-sampling of relative listener-source motion as well as positional instability of the simulated source. From psychophysical studies of the minimum audible movement angle (Perrott & Musicant, 1977) for real sound sources (listener position fixed), one can infer that the minimum perceptible TSL for a virtual audio system should be no more than about 69 ms for a source velocity of 180°/s. If one assumes that these thresholds are similar for all kinds of relative source-listener motion (e.g., when the source is fixed and the listener is moving), then the latency-induced positional displacement of the simulated sources in Wenzel (1996) may have occasionally exceeded the perceptible threshold. Although listeners did not report obvious instabilities in source position in that study, it is useful to formally investigate the impact of latency in order to characterize the dynamic performance needed in a virtual audio system.

A study by Wenzel (1999) investigated localization of 8-s-duration stimuli in order to provide subjects with substantial opportunity for exploratory head movements. Those data indicated that localization was generally accurate, even with a latency as great as 500 ms. In contrast, Sandvad (1996) observed deleterious effects on localization with latencies as small as 96 ms when using stimuli of shorter duration (~1.5 to 2.5 s). In an effort to investigate stimuli more comparable to Sandvad (1996), a recent study (Wenzel, 2001) repeated the experimental conditions of Wenzel (1999) but with a stimulus duration of 3 s. Five subjects estimated the location of 12 virtual sound sources simulated using individualized HRTFs with latencies of 33.8 (minimum TSL possible), 100.4, 250.4 or 500.3 ms in an absolute judgement paradigm. The virtual sources were fixed in space relative to the listener. The subjects were instructed to imagine an arrow (vector) from the center of their head to the source location. They were also instructed to move/reorient their heads during the localization task while remaining seated. Judgements of the azimuth, elevation and distance of the source (the endpoint of the imaginary vector) were indicated using a graphical response technique. Subjects also rated the perceived latency on each trial.

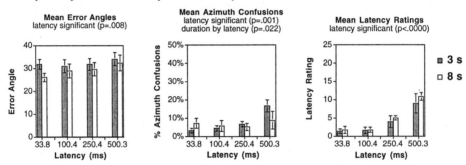

Figure 2. Means for judgement error angles, % azimuth (front-back) confusions, and latency ratings are plotted as a function of latency and stimulus duration. Means were averaged across all applicable positions and subjects. Stimulus duration is a between subjects variable. The error bars represent standard errors for 5 subjects. Significant effects from ANOVAs are indicated. Figure adapted from (Wenzel, 2001).

Comparison of the data for the 3 and 8 s duration stimuli indicates that as a function of latency, localization accuracy is moderately affected by the overall duration of the sound (Figure 2). For example, for the 8-s stimuli, front-back confusions were minimal and increased only slightly with increasing latency. Average error angles (mean of the unsigned angles between the judgement vectors and the vector from the origin to the target location) increased with latency, although the increases were most notable only for the largest latency tested (500 ms). For the 3-s stimuli, the increase in front-back confusions with latency was more pronounced, particularly for the longest latency tested (500 ms). Error angles were higher overall than for the 8-s stimuli, but did not increase with latency. In contrast, mean latency ratings indicated that latency had to be at least 250 ms to be readily perceived for both the 3 and 8-s durations. Thus, there is some suggestion that listeners are less able to compensate for latency with the short duration stimuli, although the effect is not as pronounced as in Sandvad (1996).

The fact that accuracy was generally comparable for the shortest and longest latencies tested in Wenzel (1999, 2001) suggests that listeners are largely able to ignore latency during active localization. Apparently, this is possible even though latencies of this magnitude produce an obvious spatial "slewing" of the sound source such that it is no longer stabilized in space as the head is reoriented. It may be that a localization task *per se* is not the most sensitive test of the impact of latency in a virtual audio system. Other tasks more directly dependent on temporal synchrony, such as tracking an auditory-visual virtual object, may be much more sensitive to latency effects.

3. CONCLUSIONS & FUTURE DIRECTIONS

The data reviewed here have examined perceptual tolerances for latency within each modality. The haptic modality has the most stringent requirements, requiring a very tightly-coupled feedback system that, unlike vision and hearing, necessarily involves both simultaneous input to (haptic display) and output by (manual control) the human operator. The consequences of excessive latency are most serious in this domain and the impact of latency beyond a few tens of ms is both easily detected and highly disruptive to task performance. In the visual domain, the data suggest that latencies of a similar magnitude (16.7 to 35 ms) can be discriminated (Ellis, et al., 1999b; 2000) while somewhat larger latencies (80 or more ms) interfere with operator performance, such as in a visual-manual tracking task (Ellis, et al., 1999a). To date, the detectability of latency in a virtual acoustic display has not been studied, although one can infer velocity-dependent latency thresholds from data on the minimum audible movement angle (Perrott & Musicant, 1977) measured for real sound sources (~70 ms at 180°/s). Localization data for virtual acoustic sources suggests that perceptual tolerances for latency during active localization are much larger, ranging from about 100 ms (Sandvad, 1996) to as much as 500 ms (Wenzel, 1999; 2001), depending upon the nature of the task and the duration of the stimuli. The rating data (Wenzel, 1999; 2001), while a very crude indicator of detectability, suggests that latency must be rather large (250 ms) to be consciously detected by subjects.

Taken together, the data suggest that when implementing multimodal virtual displays for space-based applications in which latency is likely to be present, computational resources should first be devoted to maintaining a tightly-coupled manual control loop. Additional resources could then be devoted to maintaining the visual and then the auditory components of a display. Another way to view this sensory hierarchy is that audition appears to be the more "forgiving" sense in terms of latency requirements. Thus, the auditory channel may actually be a more cost-effective way of providing additional information in a multimodal display. An example of such a situation includes studies currently beginning at NASA Ames that will systematically examine the ability of dynamic auditory cues to enhance subjective perception of haptic "strength" when used in combination with a haptic force-feedback display.

Further work on establishing minimum perceptual tolerances within the modalities in a variety of task contexts is needed, particularly for the auditory domain where discrimination thresholds have yet to be examined. In addition, while clearly a critical issue for multimodal virtual displays, relatively little work has been conducted to examine acceptable tolerances for intermodal temporal and spatial discrepancies. It is likely that the senses will interact when information is combined from two or more modalities in ways not obvious from their individual sensitivities.

REFERENCES

Ellis, S.R., Adelstein, B.D., Baumeler, S., Jense, G.J., & Jacoby, R.H. (1999a). Sensor Spatial Distortion, Visual Latency, & Update Rate Effects on 3D Tracking in Virtual Environments. *Proc. IEEE VR'99*, Houston TX.

Ellis, S. R., Adelstein, B. D., Jung, J., & Young, M.J. (2000). Studies and management of latency in virtual environments. *Proceedings of the 3rd International Conference on Human & Computer*, Aizu, Japan, 291-300.

Ellis, S. R., Young, M.J., Adelstein, B.D., & Ehrlich, S.M. (1999b). Discrimination of changes of latency during voluntary hand movements of virtual objects. *Proceedings of the Human Factors Society*, Houston, TX.

Frank, L.H., Casali, J.G., & Wierwill, W. (1988). Effects of visual display and motion system delays on operator performance and uneasiness in a driving simulator. *Human Factors*, 30, 201-217.

Held, R., Efstathiou, A., & Greene (1966). Adaptation to displaced and delayed visual feedback from the hand, *Journal of Experimental Psychology*, 72, 887-891.

Perrott, D., & Musicant, A. (1977). Minimum audible movement angle as a function of signal frequency and the velocity of the source *Journal of the Acoustical Society of America*, 62, 1463-1466.

Poulton, E.C. (1974). *Tracking Skill and Manual Control.* Academic Press: New York.

Sandvad, J. (1996). Dynamic aspects of auditory virtual environments. *100th Convention of the Audio Engineering Society*, Copenhagen, preprint 4226.

Sheridan, T. B. (1992). *Telerobotics, Automation, and Human Supervisory Control.* MIT Press, Cambridge, MA.

Welch, R. B., Blackmon, T.T., Liu, A., Mellers, B. A., & Stark, L. W. 1996. The effects of pictorial realism, delay of visual feedback, and observer interactivity on the subjective sense of presence. *Presence*, 5, 263-273.

Wenzel, E. M. (1996). What perception implies about implementation of interactive virtual acoustic environments. *101st Convention of the Audio Engineering Society*, Los Angeles, CA, preprint 4353.

Wenzel, E. M. (1998). The impact of system latency on dynamic performance in virtual acoustic environments. *Proceedings of the 16th International Congress on Acoustics & 135th Meeting of the Acoustical Society of America*, Seattle, WA, 2405-2406.

Wenzel, E. M (1999). Effect of increasing system latency on localization of virtual sounds. *Proceedings of the Audio Engineering Society 16th International Conference on Spatial Sound Reproduction*, Rovaniemi, FIN. New York: Audio Engineering Society, 42-50.

Wenzel, E. M (2001). Effect of increasing system latency on localization of virtual sounds with short and long duration. *Proceedings of ICAD 2001, International Conference on Auditory Display*, Espoo, FIN.

Wightman, F. L. & Kistler, D. J. (1999). Resolution of front-back ambiguity in spatial hearing by listener and source movement. *Journal of the Acoustical Society of America*, 105, 2841-2853.

* Work supported by the NASA Aerospace Operations Systems Program and by the Navy (SPAWAR, San Diego).

NEUROREHABILITATION USING 'LEARNING BY IMITATION' IN VIRTUAL ENVIRONMENTS

Maureen K. Holden, Ph.D., P.T.

Department of Brain and Cognitive Sciences, Massachusetts Institute of Technology, Bldg. E25-526b, 45 Carleton Street, Cambridge, MA 02139

ABSTRACT

This paper describes the theoretical background and key features of a novel virtual environment system for motor re-learning that has been developed in our laboratory. The system has been designed to facilitate the retraining of motor control in patients with neurological impairments, such as stroke or acquired brain injury (ABI). It consists of a computer, specially developed software and an electromagnetic motion-tracking device. During training, the arm movements of the patient and a virtual teacher are displayed simultaneously in the virtual environment. The difference between the two trajectories is used to provide the patient with augmented feedback designed to enhance motor learning. Design considerations relevant to Neurorehabilitation patients are discussed. Preliminary results of recent experiments, in which the device has been used to train upper extremity movements in patients with chronic stroke and acquired brain injury, are also summarized.

1. INTRODUCTION

In recent years, there has been great interest in using VE for learning various types of motor and spatial tasks (Darken, Allard and Achille, 1998; Kozak, Hancock and Arthur, 1993). My interest has been to use a virtual environment as a vehicle in which to combine recent discoveries from the field of neuroscience with established principles of motor learning into a new method for assessing and training motor control. I wanted the method to be particularly useful for patients with neurological problems such as stroke or acquired brain injury. In this article I will first review some of the ideas about motor control, neuroplasticity and motor learning that have been incorporated into the design of the system. Next, I will describe some of the key features of the system, and discuss some design considerations relevant to patients. Some preliminary results from patients who have used the system will then be summarized.

1.1 Theoretical Background

Many studies have revealed that the brain seems to specify movement in term of the goals to be accomplished in the external world. How then does the CNS represent these goals and transform them into signals that activate muscles? It appears there are two stages to this process of motor control - planning and execution (Bizzi and Mussa-Ivaldi, 1998). During planning, the neural representation of the goal is developed by CNS; during execution, the motor plan is transformed into neural commands that drive muscles. This hypothesis separates movement kinematics and dynamics.

In this view, movement planning occurs in extrinsic spatial coordinates, *not* in body-centered coordinates. For example, in planning an upper extremity movement, the brain thinks about the trajectory of the hand (limb endpoint) rather than the trajectories of the individual limb segments. This view is supported by psychophysical evidence which has found certain kinematic invariants (approximate straight segments, bell shaped velocity profiles) of the hand path during arm movements (Morasso, 1981). These invariants were not found in the joint movement profiles. These findings were later developed into the theory of "Minimum Jerk" (Flash and Hogan, 1985) and supported with further psychophysical studies (Shadmehr and Mussa-Ivadi, 1994, Wolpert, Ghahramani and Jordan, 1995). Electrophysiological support for the idea has also been provided by recordings from cells in monkey cortical and subcortical areas which show a correlation between the cell's firing pattern and the direction of the hand's path (Georgopoulos, Kettner and Schwartz, 1988).

Once the movement has been planned, the execution phase begins. The end-effector trajectory is tranformed from extrinsic to intrinsic (body-centered) coordinates, and the movement dynamics are generated. How this transformation occurs is the subject of some debate, largely because the solution to this transformation is not unique. That is, any given kinematic trajectory can be achieved by multiple joint movement combinations and/or muscle activation patterns. The brain must find a way to quickly simplify the mechanical complexity of the task, so that the goal can be achieved in a reasonable amount of time. Several theories exist to explain how the nervous system might accomplish this (Hollerbach and Atkeson, 1987; Bizzi et al. 1992; Gandolfo et al., 2000).

Research in robotics has found that artificial neuronal networks exposed to repeated motor commands paired with their sensory consequences are capable of learning fairly complex motor tasks without the need for explicit programming. This learning results from a change in the internal structure of the artificial network, specifically a change in the connectivity among the elements in the network. Based on these results, scientists have proposed a similar process in the CNS. The hypothesis is that learning is the result of repeated exposures to sensory signals coming from the moving limbs as they interact with the environment. The actions produced by the motor areas are initially imprecise, but feedback produces a gradual convergence on the correct solution. Ultimately, this practice leads to an *internal model*, which is embedded in newly formed connections among a group of neurons in the motor areas. The idea is that the brain centers responsible for generating motor commands also serve as sites for storage and retrieval of motor memory. Thus, the brain center that has learned the task becomes the center for expressing the task. There is physiological evidence to support this idea of a neural substrate for internal models (Gandolfo et al., 2000; Grafton et al 1992).

1.2 Implications for Motor Recovery via Plasticity or Relearning:

This cortical link to learning implies a capacity for functional re-organization that may play a significant role in motor recovery following injury to the brain. Because learning and implementation seem to be intermingled in the same cortical areas, it is likely that the mapping of the cortex is probably much more fluid and dynamic than we previously thought. And the remapping seems sensitive to and influenced by practice. In fact, Nudo and colleagues (1996) have found use-dependent alterations of cortical organization in the primary motor cortex (M1) of adult intact primates. Following focal ischemic infarcts, these primates also showed substantial functional reorganization of the motor cortex if they were exposed to a retraining program. However, in the absence of training, such reorganization did not occur. These finding support the idea that the capacity for re-organization is dependent on formation of new synapses, which in turn are dependent on motor training and practice.

The cortical mechanisms which may account for motor learning by imitation of visually observed movement have recently been investigated in both animal and human models (Rizzolatti, Fadiga, Fogassi, et al., 1999; Iacoboni, Woods, Brass, et al., 1999). In the premotor cortex of the monkey (area F5), so called 'mirror' neurons have been discovered. These neurons fire *both* when the monkey *performs* an action and when it *observes* an individual making a similar action. These neurons may represent the output stream of a 'resonance' mechanism that directly maps a pictorial or kinematic description of the observed action onto an internal motor representation of the same action. In humans, a similar mechanism has been identified using functional magnetic resonance imaging (fMRI). This suggests that humans too, may have a mechanism which directly maps visually observed movements onto cells responsible for movement.

2. METHODOLOGY

My goal was to incorporate the discoveries outlined in the Introduction section into the design of a motor re-training system for use in Neurorehabilitation. I wanted the system to serve two purposes: 1) as a research tool, allowing in depth study of motor learning, motor recovery, and the effectiveness of different training techniques.; and 2) as a motor retraining device, useful for therapists treating patients in clinical settings.

2.1 System Description

The VE training system consists of a computer, specially developed software (MIT Patent No. 5,554,033), and an electromagnetic motion tracking device[1]. The tracking device allows the movements of the patient to be

---

[1] Polhemus 3SPACE FASTRAK, Polhemus Inc., PO Box 560, Colchester, VT 05446

monitored and displayed within the context of a virtual environment displayed on the computer. The software provides the tools to create training scenes, and virtual teachers whose trajectories are displayed on the computer screen. A variety of additional features allow the user to receive feedback about his/her performance during practice, and a score following each trial. The score is based upon a comparison of teacher and patient trajectories. Both patient and teacher movements are displayed in real time on the computer screen in the same coordinate frame of reference using 3D graphics. The graphics may be displayed in stereo 3D through the use of head mounted glasses or a stereo screen projection, or as enhanced 2D on a standard desktop or wall-projected display. A subject using the device with a standard desktop display isshown below in Figure 1.

Figure 1. Subject using the VE device

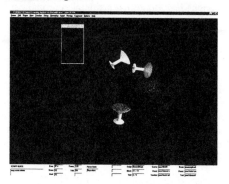

Figure 2. Example of scene used to train a pouring task. Teacher in yellow on right; patient in white to left of teacher. Score and displacement error (red line) are displayed.

Scenes: Training scenes are the 3D 'pictures' (virtual environments) that the patient sees on the computer screen. An example of a scene used to train the functional task of pouring from a cup is shown in Fig. 2. The scenes allow us to create an environmental context with a purpose – one that will elicit a particular movement by providing a goal for that movement. This mimics the way movements are planned in the natural world. The scenes give us a mechanism to control and monitor exactly which movements are being trained during the experimental session. They also provide a vehicle for adjustment of the task difficulty. Scenes also provide a convenient way to set up a particular practice sequence. We have developed a software feature ('Script') that allows us to specify a sequence of scenes (with associated training settings) that will play automatically one after the other, thus simulating a typical rehabilitative therapy session. Both teacher and patient movements may be displayed on the screen as movement of the endpoint only, or as whole arm movement (i.e., the shoulder, elbow, forearm, and wrist movements are rendered in 3-D). In VE the task can be simplified in the early stages of learning, allowing the learner to focus on key elements of the task.

Teacher. The virtual environment provides a mechanism for display of a 'teacher' movement which shows the patient the actual kinematics of hand (end-point) path (i.e., the same way the brain plans the movement). In theory, this should facilitate motor learning by assisting the patient's motor planning process in a natural way. The virtual teacher also affords the patient an opportunity to practice 'learning by imitation'. As discussed in the introduction,, this may facilitate the execution phase of motor control by activating direct links to output neurons associated with task execution. The teacher speed and appearance can be altered in a variety of ways to enhance learning (e.g., whole arm movement may be displayed). The teacher may also be removed from the display so that the subject does not become overly 'dependent' upon guidance to perform the movement. Imitating the teacher in the context of a virtual environment offers some advantages over practice with a real or videotaped teacher. In VE, the patient's and the teacher's limb movements are displayed on the computer screen in real time and in the same coordinate frame of reference. The patient can thus 'get inside' the virtual teacher and imitate the movement exactly. No mental rotations or translations are required. Experimental findings have revealed that cells of the motor areas of primates are involved in mental rotations of the direction of voluntary movements (Georgopoulos et al. 1989). Conceivably, for some patients with cortical motor stroke, the mental rotation of movements required to copy movements demonstrated by the therapist may be difficult.

Augmented Feedback. As described in the Introduction, motor learning occurs when internal models develop as a result of synapse formation following repeated practice of movements that result in at least some success at task completion. It is this trial and error process that we target with our system by providing enhanced feedback about error during practice. This augmented feedback may be applied during the actual movement practice (concomitant), or following completion of the movement (knowledge of results), and is based on a comparison of the patient and teacher trajectories. Presumably, by helping the patient identify the motor strategies that are successful, the development of these internal models can occur more rapidly. And practice, essential to the development of these internal models, seems to be facilitated in VE. This may be because the task practice is experienced by the patients as being 'fun'. In studies on normal subjects, highly augmented and frequent feedback has been found to be effective in improving performance (task 'acquisition') but to lessen task retention and generalization (Schmidt 1991). Thus, our system allows us to vary feedback frequency so that we can provide our subjects with the optimal feedback frequency for learning.

2.2 Design Considerations for Neurorehabilitation

In contrast to healthy subjects, where the focus of VE training is often complex perceptual-motor skills, such as flying a plane, in rehabilitation the goal is usually to re-learn very simple motor tasks. Such tasks as pulling on the sleeve of a jacket must now be practiced in the face of a variety of sensory, motor and cognitive deficits. The exact type of deficit can vary tremendously from patient to patient, even with the same diagnosis. Thus, the system must be flexible, so that the training can easily be adjusted to the needs of different patients. Another major focus is to facilitate transfer of motor learning that occurs in the virtual world to performance in the real world. Flexibility in system design helps us to ensure that the virtual tasks we create resemble the real tasks as much as possible, particularly in terms of the movement required to perform the task.

While very complex, realistic or fanciful graphics and fast-paced game formats are appealing to normal subjects for VE training, they may be overwhelming for patients. When the brain itself has been affected, as in stroke or acquired brain injury (in contrast to disabilities where the brain is intact, like spinal cord injuries), processing of all this information may be too difficult. In contrast, a very simple display, with few movements, may help a patient to focus on the task at hand, and enhance learning.

The issue of possible negative side effects such as cybersickness, altered eye-motor coordination, and postural disequilibrium must be carefully considered. To date, there seem to be no reports of negative side effects using desktop display systems for VE. Most of the problems seem related to the immersive environment created in head mounted displays (HMD) (Stanney, Kennedy, Drexler et al., 1999). It seems obvious that feelings of disorientation, postural disequilibrium, or aftereffects in arm control that produce past-pointing, could be much more dangerous for disabled subjects, in whom a fall and injury could more easily be elicited than in normal subjects. Because trunk and postural control muscles are often adversely affected by neurological impairments, the extra weight imposed on the head by HMD's may be a problem.

Finally, practical issues are important. Keeping costs down, making systems easy to use, and minimizing the amount of equipment attached to the patient, will offer the greatest potential for success and acceptance of VE as a standard treatment procedure in rehabilitation clinics.

Because of many of the factors listed above, we have initially used our system with a desktop display. The display can also be projected on a large screen instead of a monitor, to give a greater feeling of immersion. A significant disadvantage of the desktop display is poor depth visualization. A solution to this problem would be to use a stereo projector - which can project true stereo 3-D onto a wall screen and be seen without glasses or a headset by both the patient and the therapist. Our software has the capability to be used with such a projector, and we plan to try this with patients in the future.

3. RESULTS

Stroke. To date we have used the first prototype of the system to achieve improvements in the reaching ability of 2 stroke patients who were several years post stroke and had shown no improvement in the prior 6 months (Holden, Todorov, and Callahan, et al., 1999). Based on those results, we developed several new features which are in the present system (scoring function, scripts). We have recently begun a study of motor generalization. Results from the first subject in that study are promising in that he improved not only in the task trained in VE, but in related untrained tasks, and on standard clinical measures of motor recovery and function (Holden, Dyar and Callahan, et al., 2000).

627

Acquired Brain Injury. We are currently studying the relearning of a pouring movement in subjects with chronic acquired brain injury (3-18 yr. post injury). To date, we have found improvement for 3 of the 4 subjects in the trained virtual task, on similar and related tasks in the real world, and on standard clinical tests of upper extremity motor recovery and function (Holden, Dettwiler, Dyar, et al., 2001).

3. CONCLUSIONS

Our preliminary results to date indicate that the system shows promise for use in research and clinical applications.

Acknowledgement: This work was supported by the Charles A. Dana Foundation

REFERENCES

Bizzi, E., Hogan, N., Mussa-Ivaldi, S., and Giszter, S. (1992). Does the nervous system use equilibrium- point control to guide single and multiple joint movements? *Behav. Brain Sci. 15*: 603-613.

Bizzi, E., and Mussa-Ivaldi, S. (1998). The acquisition of motor behavior. *Daedalus, 127(2),* 217-232.

Darken, R.R., Allard, T., & Achille, L. B. (1998). Spatial orientation and wayfinding in large-scale virtual spaces: An introduction. *Presence, 7(2),* 101-107.

Flash, T., and Hogan, N. (1985). The coordination of arm movements: An experimentally confirmed mathematical model. *J. Neurosci. 5*: 1688-1703.

Gandolfo, F., Li, C.-S. R., Benda, B., Schioppa, C.P., and Bizzi, E. (2000). Cortical correlates of learning in monkeys adapting to a new dynamical environment. *Proc. Natl. Acad. Sci. 97*: 2259-2263.

Grafton, S., Mazziotta, J., Presty, S., Fristion, K., Frackowiak, R., and Phelps, M. (1992). Functional anatomy of human procedural learning determined with regional cerebral blood flow and PET. *J. Neurosci. 12(7):* 2542-2548.

Georgopoulos, A. P., Kettner, R. E., and Schwartz, A.B. (1988). Primate motor cortex and free arm movements to visual targets in three-dimentional space: I. Coding of the direction of movement by a neuronal population. *J. Neurosci. 8*: 2913-2927.

Georgopoulos, A.P., Lurite, J., Petrides, M., Schwartz, B., and Massey, J. T. (1989). Mental rotation of the neuronal population vector. *Science, 243*: 234-236.

Holden M, Dettwiler, A, Dyar, T, Niemann G, Bizzi E: (2001). Retraining movement in patients with acquired brain injury using a virtual environment. *Proceedings of Medicine Meets Virtual Reality 2001, J.D. Westwood et al. (Eds.),* IOS Press, Amsterdam , pp. 192-198.

Holden, M., Todorov, E., Callahan, J., and Bizzi, E. (1999) Virtual environment training improves motor performance in two patients with stroke: Case report. *Neurology Report 23(2)*, 57-67.

Holden, M.K., Dyar, T., Callahan, J., Schwamm, L., Bizzi, E. (2000). Motor learning and generalization following virtual environment training in a patient with stroke. *Neurol. Report 24 (5),* 170-171.

Hollerbach, J. M., and Atkeson, C. G. (1987). Deducing planning variables from experimental arm trajectories: Pitfalls and possibilities. *Biol. Cybern. 56*: 279-292.

Iacoboni, M., Woods, R.P., Brass, M., Bekkering, H., Massiotta, J.C. and Rizzolatti, G. (1999). Cortical mechanisms of human imitation. *Science 286*: 2526-2528.

Kozak, J.J., Hancock, P.A., Arthur, E.J., & Chrysler, S.T. (1993). Transfer of training from virtual reality. *Ergonomics, 36*, 777-784.

Morasso, P. (1981). Spatial control of arm movements. *Exp. Brain. Res. 42*, 223-227.

Nudo, R.J., Milliken, G.W., Jenkins, W.M., and Merzenich, M.M. (1996) Use-dependent alterations of movement representations in primary motor cortex of adult squirrel monkeys. *Journal of Neuroscience, 16*, 785-807.

Rizzolatti, G., Fadiga, L., Fogassi, L., Gallese, V. (1999). Resonance behaviors and mirror neurons. Arch. Ital. *Biol. 137(2-3):* 85-100.

Schmidt, R.A. (1991) Frequent augmented feedback can degrade learning: Evidence and interpretations. In: GE Stelmach and J Requin, eds., *Tutorials in Motor Neuroscience.* Dordrecht, The Netherlands: Kluwer Academic Publishers, p. 59-75.

Shadmehr,R. & Mussa-Ivaldi, S. (1994). Adaptive representation of dynamics during learning of a motor task. *Journal of Neuroscience 14*, 3208-3224.

Stanney, K.M., Kennedy, R.S., Drexler, J.M., & Harm, D.L. (1999). Motion sickness and proprioceptive after effects following virtual environment exposure. *Applied Ergonomics, 30,* 27-38.

Wolpert, D., Ghahramani, Z. and Jordan, M. (1995). Are arm trajectories planned in kinematic or dynamic coordinates? An adaptation study. *Experimental Brain Research 103,* 460-470.

Using Pinch Gloves™ for both Natural and Abstract Interaction Techniques in Virtual Environments

Doug A. Bowman, Chadwick A. Wingrave, Joshua M. Campbell, and Vinh Q. Ly

Department of Computer Science (0106)
Virginia Tech
Blacksburg, VA 24061 USA
{bowman, cwingrav, jocampbe, vly}@vt.edu

Abstract

Usable three-dimensional (3D) interaction techniques are difficult to design, implement, and evaluate. One reason for this is a poor understanding of the advantages and disadvantages of the wide range of 3D input devices, and of the mapping between input devices and interaction techniques. We present an analysis of Pinch Gloves™ and their use as input devices for virtual environments (VEs). We have developed a number of novel and usable interaction techniques for VEs using the gloves, including a menu system, a technique for text input, and a two-handed navigation technique. User studies have indicated the usability and utility of these techniques.

1 Introduction

The number of input devices available for use in three-dimensional (3D) virtual environments (VEs) is large, and still growing. Despite the proliferation of specialized 3D input devices, it is often difficult for the designers of VE applications to choose an appropriate device for the tasks users must perform, and to develop usable interaction techniques using those devices. The difficulties of 3D interaction (Herndon, van Dam, & Gleicher, 1994) have led to guidelines for the design of 3D interfaces (Gabbard, 1997; Kaur, 1999) and techniques for their evaluation (Bowman, Johnson, & Hodges, 1999; Hix et al., 1999), but there is little work on the design and evaluation of 3D input devices, and on the mappings between input devices, interaction techniques, and applications.

In our research, we have been designing and evaluating 3D interaction techniques using Fakespace Pinch Gloves™, combined with six DOF trackers, as 3D input devices. We had a dual motivation for this work. First, we were dissatisfied with the standard VE input devices (tracked wands, pens, and tablets) because of their poor accuracy and ergonomic problems. Second, we observed that although Pinch Gloves™ were relatively inexpensive and capable of a very large number of discrete inputs, they were not often used in VE applications, except for a few extremely simple techniques. Our goal is to determine whether the gloves can be used in more complex ways for a variety of interaction tasks in VEs, and to develop guidelines for the appropriate and effective use of these input devices.

We begin by summarizing related work, then discuss the characteristics of Pinch Gloves™ that enable their use for novel 3D interaction techniques. Three specific techniques are presented, using the gloves for menus, travel, and text input. We conclude with some generalizations about interaction using the gloves.

2 Related work

There have been a few attempts to organize and understand the design space of input devices for 3D applications, most notably the work of Shumin Zhai (Zhai & Milgram, 1993; Zhai, Milgram, & Buxton, 1996). However, there is still work to be done before we understand how to design interaction techniques that take advantage of the properties of a particular device, and how to choose an appropriate device for a specific application. We are not aware of any empirical studies involving Pinch Gloves™, but they have been used for various interaction techniques (Mapes & Moshell, 1995; Pierce, Stearns, & Pausch, 1999) and applications (Cutler, Frohlich, & Hanrahan, 1997).

Many types of menus have been developed for use in VEs (Angus & Sowizral, 1995; Jacoby & Ellis, 1992; Mine, 1997; Mine, Brooks, & Sequin, 1997), but none of them simultaneously address the issues of efficient, comfortable, and precise selection of a large number of menu items. Similarly, there are many techniques for 3D travel (viewpoint movement) (Bowman, Koller, & Hodges, 1997; Mine, 1995). We present a technique that takes advantage of two-handed interaction and allows users to control velocity. There is little prior work on text input for VEs; this problem has been largely avoided because of its difficulty. Poupyrev developed a system to allow handwritten input in VEs (Poupyrev, Tomokazu, & Weghorst, 1998), but the input was saved only as "digital ink" and not interpreted to allow true text input. Our technique allows intuitive and precise input of actual text, without requiring the user to learn any special key chords.

3 Pinch Gloves™ as a VE input device

Fakespace Pinch Gloves™ are commercial input devices designed for use in VEs. They consist of flexible cloth gloves augmented with conductive cloth sewn into the tips of each of the fingers (figure 1). When two or more pieces of conductive cloth come into contact with one another, a signal is sent back to the host computer indicating which fingers are being "pinched." The gloves also have Velcro on the back of the hand so that a position tracker can be mounted there. They are distinct from *whole-hand* glove input devices, such as the CyberGlove™, which report continuous joint angle information used for gesture or posture recognition. Pinch Gloves™ can be viewed as a choice device with a very large number of possible choices. Practically, however, the gloves have some interesting characteristics that differentiate them from a set of buttons or switches.

The most obvious difference between Pinch Gloves™ and other choice devices is the huge number of possible pinch gestures allowed by the gloves – gestures involving any combination of two to ten fingers all touching one another, plus gestures involving multiple separate but simultaneous pinches (for example, left thumb and index finger and right thumb and index finger pinched separately, but at the same time). This large gesture space alone, however, is not sufficient for usability. Arbitrary combinations of fingers can be mapped to any command, but those gestures will not necessarily provide affordances or memorability. The official web site for the gloves recognizes this fact: "Users can program as many functions as can be remembered" (Fakespace, 2001). In addition, many of the possible gestures are physically implausible. In order to take advantage of the large gesture space, we need to use comfortable pinch gestures that either have natural affordances or whose function the user can easily determine.

The gloves also have desirable ergonomic characteristics. They are very light and flexible, so that they do not cause fatigue or discomfort with extended use. Moreover, the user's hand *becomes* the input device – it is not necessary to hold another device in the hand. Thus, users can change the posture of their hand at any time, unlike pens or joysticks, which might force the user to hold the hand in the same posture continuously. They also support "eyes-off" interaction – the user's proprioceptive senses allow him/her to pinch fingers together easily without looking at them, whereas some other devices may force the user to search for the buttons or controls.

As noted, the gloves can be combined with tracking devices for spatial input. This allows context-sensitive interpretation of pinch gestures. For example, pinching the thumb and index finger while touching a virtual object may mean "select," while the same pinch in empty space may mean "OK."

The use of trackers also means the gloves can support two-handed interaction, known to be a natural way to provide a body-centric frame of reference in 3D space (Hinckley, Pausch, Profitt, Patten, & Kassell, 1997). Of course, two trackers combined with a button device can also allow two-handed interaction, but the gloves give the user flexibility in deciding which hand will initiate the action. One consequence of this is that users can avoid what we have called the "Heisenberg effect" of spatial interaction – the phenomenon that on a tracked device, a discrete input (e.g. button press) will often disturb the position of the tracker. For example, a user wants to select an object using ray casting. She orients the ray so that it intersects the object, but when she presses the button, the force of the button press displaces the ray so that the object is not selected. With the gloves, one hand can be used to set the desired position/orientation, and the other hand can be used to signal the action, avoiding the Heisenberg effect.

Figure 1. User wearing Pinch Gloves™

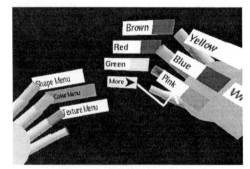

Figure 2. TULIP menu system

Finally, Pinch Gloves™, unlike whole-hand devices, produce discrete input – fingers are either touching or not. Although whole-hand devices can allow the user more natural gestures, it is notoriously difficult to calibrate them and recognize gestures properly (Kessler, Hodges, & Walker, 1995). Pinch Gloves™, because of their discrete nature, can be much more precise.

4 Three-dimensional interaction techniques using Pinch Gloves™

We developed novel 3D interaction techniques based on these properties. Our goal was to "push the envelope" and to go beyond the simple "pinch to grab" metaphor that represents the most common use of the gloves. We also wanted to design techniques that would use natural gestures as well as those where the gesture was more abstract.

4.1 TULIP menu system

Our design of a menu system using the gloves was based on two concepts. First, menu items would be associated with fingers, and would be selected by pinching the finger to the thumb on the same hand (a comfortable and simple gesture). Second, top-level menu items (menu titles) would be associated with the user's non-dominant hand, and second-level items (commands within a menu) would be associated with the dominant hand.

This design is simple to understand, requires no more than two pinches to access a particular menu item, and takes advantage of the natural asymmetry of the two hands (Hinckley et al., 1997). Unfortunately, it also creates an unreasonable limitation of four menus with four items each.

Our solution to this problem is called the TULIP menu system (Bowman & Wingrave, 2001). TULIP stands for "Three-Up, Labels In Palm," meaning that three items are active at one time, while the rest of the menu items are arranged in columns of three along the palm of the user's hand. To access an active item, the user simply pinches the thumb to the appropriate finger. To access other items, the user pinches the thumb to the pinky (the "more" item) until the desired item appears on one of the fingers (figure 2).

This system nicely balances direct selection and visibility of menu items. Given a maximum menu size of N items, it takes no more than $N/3 + 1$ pinches to select any item. Visual cues (highlighting, rotation of the active items, and a clear link between the more item and the next column of items) make it easy for the user to determine which menu is selected, which items are active, and how to make the desired item active.

A usability study showed that this system was more difficult to learn than two other common VE menu implementations, but also that users became efficient relatively quickly, and that the system was significantly more comfortable than the others. To further address the issue of efficiency, we have developed a modified TULIP system for expert users that requires no more than two pinches to select any item, provided that no menu contains more than fifteen items (a reasonable assumption based on interface guidelines). In this technique, to select the fourth item in a menu (such as the item "yellow" in figure 2), the user pinches the *left* ring finger to the right index finger. The left-hand finger determines which *column* of items the user desires, and the right-hand finger determines which item within that column should be selected. This technique requires more practice, but does not require the user to memorize any information, since all items are visible on the palm.

4.2 Virtual keyboard for text input

For the most part, VE applications have avoided the task of entering text information, because no usable text input techniques existed (beyond speech recognition and one-handed chord keyboards, which both require significant user training, and are therefore inappropriate for all but the most frequently used VEs). There are many situations, however, where accurate and efficient text input would increase the utility of VE systems. For example, a user might need to leave an annotation for the designer in an architectural walkthrough application, or enter a filename to which work can be saved after a collaborative design session

Based on these examples, a 3D text input technique would neither require users to type long sentences or paragraphs, nor to approach the speed of typing at a standard keyboard. Our goal was to allow immersed users to be able to enter text without leaving the VE and without a significant amount of training.

We decided to emulate as closely as possible a standard QWERTY keyboard, with which most users are intimately familiar. The basic concept of our virtual keyboard is that a simple pinch between a thumb and finger on the same hand represents a key press by that finger. Thus, on the "home" row of the keyboard, left pinky represents 'a', left ring represents 's', and so on.

We also needed the ability to use the "inner" keys such as 'g' and 'h', and the ability to change rows of the keyboard. We accomplish this through the use of trackers mounted on the gloves. Inner keys are selected by rotating the hand inward. The active row can be changed by moving the hands closer to the body (bottom row) or farther away (top row). Users calibrate the location of the rows before using the system. Still, because the trackers have limited accuracy, fairly large-scale motions are required to change rows or select the inner keys, reducing efficiency.

Special gestures are provided for space (thumb to thumb), backspace (ring to ring), delete all (pinky), and enter (index to index). This set of arbitrary gestures is small enough to be memorized by the user.

Visual feedback is extremely important to make up for the lack of the visual and haptic feedback provided by a physical keyboard. We hypothesized that users would not want to look at their hands while typing, so we attached the feedback objects to the view. These show, for each hand, the location of each character, and the currently active characters (based on hand distance and rotation) via highlighting (figure 3).

In a user study, five users each typed three sentences (six to eight words) using the virtual keyboard. As expected, efficiency was low: users took about three minutes per sentence on average, after five to ten minutes of practice. Most users improved their performance during the study. Much of their time was spent searching for letters (although four users were touch typists) and recovering from errors. However, our goal of minimal training was met, as all users commented on the ease of learning the system. Moreover, one of the designers could type the sentences in an average of 45 seconds each, indicating that the system could be much more efficient with further experience.

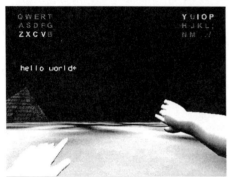

Figure 3. Virtual keyboard technique

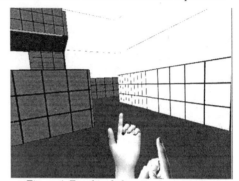

Figure 4. Two-handed navigation technique

4.3 Two-handed navigation
The task of navigation in VEs has been widely studied, including a large number of interaction techniques for viewpoint movement, or *travel*. We wanted to design a travel technique using the gloves that would take advantage of their unique properties, and decided to focus on the gloves' support for two-handed interaction.

Steering techniques for VE travel (Bowman et al., 1997) let the user provide a direction of motion, usually based on the head or hand tracker. With two tracked gloves, we can define the direction of motion based on the vector between the hands. Since both hands are acting as input devices, we can define the forward direction as being toward the hand that produced the pinch gesture (figure 4). Thus, to fly upwards, the user holds his hands one above the other and pinches thumb to index finger on the top hand. Flying back downwards does not require changing the hand positions; rather, the user simply pinches with the lower hand instead.

We enhanced this technique with a velocity control feature (based on the distance between the hands), and a "nudge" feature, allowing the user to move only slightly in a particular direction by using a different pinch gesture.

This technique is quite flexible. In a user study, we found that novice users tend to simulate torso-directed steering by holding one hand close to the body and the other a short distance in front of the body. Expert users have the option to increase their speed for large-scale movements, use both hands to quickly change direction without moving their bodies, and make small movements when it is required by the task at hand. We also found that users' performance on 2D navigation tasks improved significantly after a few minutes of practice, but that it was difficult for users to control direction and remain spatially oriented in 3D navigation.

5 Conclusions and future work
This research has explored the use of Pinch Gloves™ for novel three-dimensional interaction techniques. Our techniques confirm that the gloves can be used for natural gestures, but also show that abstract techniques can take

advantage of the gloves' characteristics for more efficient, usable, or comfortable interaction. In general, we found that abstract techniques require a high degree of visual affordances and feedback, including virtual representations of the hands and indication of the command that will be activated when a pinch gesture is made. We also found that the gloves provide increased comfort over other common devices, and that the large number of possible gestures allows the same technique to be customized for both novice and expert users.

We are continuing to refine the techniques and to develop new ones. For example, we are working on several approaches to numeric input using the gloves, and on lightweight navigation constraints based on glove input. The techniques are also being used in VE applications under development in our laboratory, which will allow us to evaluate them in a setting of realistic use.

Acknowledgements

The authors would like to acknowledge the technical assistance of Drew Kessler and the time and efforts of the subjects in our usability evaluations.

References

Angus, I., & Sowizral, H. (1995). *Embedding the 2D Interaction Metaphor in a Real 3D Virtual Environment.* Proceedings of SPIE, Stereoscopic Displays and Virtual Reality Systems, 282-293.

Bowman, D., Johnson, D., & Hodges, L. (1999). *Testbed Evaluation of VE Interaction Techniques.* Proceedings of the ACM Symposium on Virtual Reality Software and Technology, 26-33.

Bowman, D., & Wingrave, C. (2001). *Design and Evaluation of Menu Systems for Immersive Virtual Environments.* Proceedings of IEEE Virtual Reality, 149-156.

Bowman, D., Koller, D., & Hodges, L. (1997). *Travel in Immersive Virtual Environments: an Evaluation of Viewpoint Motion Control Techniques.* Proceedings of the Virtual Reality Annual International Symposium, 45-52.

Cutler, L., Frohlich, B., & Hanrahan, P. (1997). *Two-Handed Direct Manipulation on the Responsive Workbench.* Proceedings of ACM Symposium on Interactive 3D Graphics, 107-114.

Fakespace, Inc. (2001). PINCH™ Gloves. http://www.fakespacelabs.com/products/pinch.html

Gabbard, J. (1997). *Taxonomy of Usability Characteristics in Virtual Environments.* Unpublished Masters Thesis, Virginia Polytechnic Institute and State University.

Herndon, K., van Dam, A., & Gleicher, M. (1994). The Challenges of 3D Interaction. *SIGCHI Bulletin, 26*(4), 36-43.

Hinckley, K., Pausch, R., Profitt, D., Patten, J., & Kassell, N. (1997). *Cooperative Bimanual Action.* Proceedings of CHI: Human Factors in Computing Systems, 27-34.

Hix, D., Swan, J., Gabbard, J., McGee, M., Durbin, J., & King, T. (1999). *User-Centered Design and Evaluation of a Real-Time Battlefield Visualization Virtual Environment.* Proceedings of IEEE Virtual Reality, 96-103.

Jacoby, R., & Ellis, S. (1992). *Using Virtual Menus in a Virtual Environment.* Proceedings of SPIE: Visual Data Interpretation, 39-48.

Kaur, K. (1999). *Designing Virtual Environments for Usability.* Doctoral Dissertation, University College, London.

Kessler, G., Hodges, L., & Walker, N. (1995). Evaluation of the CyberGlove™ as a Whole Hand Input Device. *ACM Transactions on Computer-Human Interaction, 2*(4), 263-283.

Mapes, D., & Moshell, J. (1995). A Two-Handed Interface for Object Manipulation in Virtual Environments. *Presence: Teleoperators and Virtual Environments, 4*(4), 403-416.

Mine, M. (1995). *Virtual Environment Interaction Techniques* (Technical Report TR95-018): UNC Chapel Hill CS Dept.

Mine, M. (1997). ISAAC: A Meta-CAD System for Virtual Environments. *Computer-Aided Design, 29*(8), 547-553.

Mine, M., Brooks, F., & Sequin, C. (1997). *Moving Objects in Space: Exploiting Proprioception in Virtual Environment Interaction.* Proceedings of ACM SIGGRAPH, 19-26.

Pierce, J., Stearns, B., & Pausch, R. (1999). *Voodoo Dolls: Seamless Interaction at Multiple Scales in Virtual Environments.* Proceedings of the ACM Symposium on Interactive 3D Graphics, 141-146.

Poupyrev, I., Tomokazu, N., & Weghorst, S. (1998). *Virtual Notepad: handwriting in immersive VR.* Proceedings of the IEEE Virtual Reality Annual International Symposium, 126-132.

Zhai, S., & Milgram, P. (1993). *Human Performance Evaluation of Manipulation Schemes in Virtual Environments.* Proceedings of the Virtual Reality Annual International Symposium, 155-161.

Zhai, S., Milgram, P., & Buxton, W. (1996). *The Influence of Muscle Groups on Performance of Multiple Degree-of-Freedom Input.* Proceedings of CHI: Human Factors in Computing Systems, 308-315.

VIRTUAL REALITY AND TRANSFER OF LEARNING

Max M. North, James R. Mathis, Adam Madajewski,
John T. Brown, and Stephanie M. Cupp

Virtual Reality Technology Laboratory
Computer Science and Information Systems Department
Kennesaw State University
1000 Chastain Road, Kennesaw, GA 30144-5591
Max@acm.org

ABSTRACT

This paper reports on the use of virtual reality and its effectiveness on training and attempts to provide evidence that training conducted in virtual reality can be transferred to real-world tasks. The study employed objects, which had to be manipulated and arranged in specific patterns, as stimuli in both physical-world and virtual-world environments. The subjects in the experiment were required to perform the "pick-and-place" operation. Subjects were instructed to move objects, either physically or virtually, in a specified sequence. Subjects were tested based on their speed and the errors they made. The results led to the conclusion that subjects in the Experimental Group did not perform more significantly than the subjects in the Control Group. However the subjects in the Experimental Group did retain their skills learned in virtual reality situation better than the subjects in the Control Group. Thus, virtual reality training was moderately effective, and its transfer to the real-world situation seemed to occur.

1.   INTRODUCTION

The current research in virtual reality provides new insights in training, modeling, simulation, and multisensory human-computer interface (Burg, Hughes, Lisle, Moshell, Carrington and Xin, 1991; Durlach, Aviles, Pew, DiZio and Zeltzer, 1992; North, North and Coble, 1996a). According to many researchers, training accounts for a major portion of the expense in system operation (Kozak, Hancock, Arthur and Chrysler, 1993). The traditional computer-based training programs have provided alternative and improved methods of training; however, virtual reality technology may provide a new opportunity in the area of training (e.g., problem solving). This is a significant area for exploration (North, North and Coble, 1996b). Virtual reality offers the potential for improved training and is currently used as a military and industry training tool (Baum, 1992). It has proven to be cost effective and, in some cases, more effective and efficient than the traditional approach.

According to Bricken (1991), Helsel (1992), and many other researchers, professionals in various educational fields are responding positively to the potential of virtual reality technology as an instructional medium because of what it can offer learners. This reinforces the success virtual reality has experienced as a teaching tool for military, business, education, and other environments. With this unique medium, other environments have benefited when learning to perform complex and dynamic tasks (Alessi, 1988; Mattoon and Thurman 1990; Reigeluth and Schwartz 1989).

Today, it is clear that virtual reality technology is more than the hardware that produces it. Virtual reality is a new paradigm for developing and experiencing interactive and synthetic environments produced through the integration of the computer and the human sensory system (Mattoon and Thurman, 1990; North, North and Coble, 1995, 1996b). Based on Kennesaw State University research and the research of others (e.g., Kozak, Hancock, Arthur and Chrysler, 1993), no empirical, comprehensive, or controlled studies have focused specifically on human factors and issues relevant to training in virtual reality.

The aim of this research project was to design a comprehensive and controlled experiment to provide evidence that training conducted in a virtual environment transferred to training for a real-world task. This work was similar to that of Kozak et al. Detailed examination was made of a classic transfer of a training paradigm in which the perceptual-motor abilities made up the primary component (Holding, 1987). The specific task was termed the "pick-and-place" operation. It was intended that the work of Kozak et al be extended during the first phase of the project and that the research be designed to complement the innovative and leading-edge research undertaken by North, North and Coble, which identified the minimum perceptual factors necessary to create a virtual world and elicited the immersion needed to assist operators in learning a new behavior or skill.

2. METHODOLOGY

2.1 Subjects

Twenty subjects who attended Kennesaw State University in the fall session were selected to participate in this study. The students were self-reported subjects who had not participated in any other virtual reality experiment. After a careful screening, the selected subjects were randomly assigned to two groups - experimental and control. The Experimental Group consisted of four males and six females while the Control Group consisted of five females and five males.

2.2 Virtual Reality Training Scene

By using VREAM virtual reality development software and libraries, the research students developed a scene that consisted of a room in a laboratory similar to the physical room, a virtual chair for the subject to virtually sit on (similar to the physical chair in the laboratory), and a table supporting five empty cans. Textured mapping techniques were employed to add realism to the virtual scene. Also created was a virtual hand with the ability to grasp, pick, and place the virtual cans.

2.3 Apparatus and Task

The virtual reality system for this study consisted of a Pentium-based computer with virtual environment software (VREAM Inc.), a head-mounted display (CyberEye HMD, General Reality Company), an electromagnetic head-tracker (Flock of Birds, Ascension Inc.). A mouse device was used to manipulate, interact with, and navigate the objects within the virtual world.

The subjects in the experiment were required to perform the "pick-and-place" operation. The apparatus needed to perform the task in the real-world environment was: a table, five identical empty aluminum cans, ten paper place-holders, and a digital stopwatch. Participants were seated in front of a white table facing five cans arranged in an arc configuration. Each of the cans was placed on an initial position, marked by a circular place-holder. A second row, the target row, had the same configuration as the initial row of cans, was located two inches behind the cans, and was marked by circular place-holders.

2.4 Procedure

The experiment was administered by four undergraduate computer science and information systems students, and co-researchers under the direct supervision of the principal investigator. They spent several hours learning about the physical and virtual worlds, practicing the manipulation and navigation of objects and performing the intended tasks. In the initial phase, participants also were given an opportunity to become familiar with virtual reality equipment. A virtual hand was created for subjects to practice the "pick-and-place" sequence with virtual cans. Both Experimental and Control groups were involved in this familiarization phase of the study.

Participants were seated at the table so that their target hand (hand with which they performed the operation) was directly in front of the middle can. Although that was the initial position, subjects were free to adjust positions. The facilitator demonstrated to participants how to perform the "pick-and-place" operation. Cans were individually moved from their initial site to a target site. The first can to be moved, can 1, was located to the far right of the subject, and subjects were instructed to move the cans in sequence. Once can 4 was placed on its target position, subjects were to return each can to its initial position, starting with can 4, the last can displaced.

A trial was defined as the time required to move the cans from their initial positions to their target positions and then return the cans to their initial positions. Trials were successive with minimal time between each trial and were recorded despite the misplacement or knocking over of cans. Subjects were instructed to perform the task as quickly and accurately as possible and were assessed based on speed and accuracy. An error was recorded if, when the subject moved a can to a place-holder, a part of the place-holder remained visible. An observer was designated to record errors, and errors were calculated for movement from the initial position to the target position and vice-versa.

Subjects were allowed ten pre-test trials in a real-world environment to determine the most comfortable grip and the most efficient manner for moving the cans. The facilitator gave instructions on when to start each trial. Both the Experimental Group and the Control Group performed the pre-test, which consisted of ten trials of the "pick-and-place" operation in a real-world environment. Although all subjects performed the pre-test, the Control Group was not subjected to the virtual environment and proceeded to the post-test, which consisted of performing the "pick-and-place" operation for ten additional trials in the real-world environment.

Following the pre-test, only the Experimental Group performed the task in a virtual environment. The Experimental Group used a head-mounted display and a mouse for navigational tools in the environment. The environment contained a right hand, a table, cans, and place-holders in an arrangement mimicking the real-world environment. The subject appeared to be seated at the virtual table, and the hand was pre-positioned parallel to the cans, fingers pointing to the right, and palm facing the cans. The facilitator instructed each subject to use the mouse to move the hand toward a can until the can was within reach. The subjects were not to change the hand's orientation. Once the hand was in close proximity to the can, each subject was instructed to "left click and hold" the mouse to grab the can and to release the can when it appeared to be positioned over the target. Subjects were given demonstrations on how to navigate with the hand (i.e., move backward, forward, grasp, etc.). Subjects were then instructed to perform the trial ten times in the virtual environment. Those trials were not timed.

After the Experimental Group implemented the "pick-and-place" operation in the virtual environment, they returned to the real-world environment to perform a post-test consisting of ten additional trials. The experimenter observed the subjects and, using a digital stopwatch, timed all trials performed in the real-world environment. Subjects were allowed to alter the grasp used to pick up cans and to alter the manner of movement from the initial position to the target position. Subjects had experimented with various grasps and movements during the pre-test trials and had determined the most efficient and effective grasps and movements for use. Grasps ranged from clutching the middle of the cans to placing the hand over the top of the can. Movements ranged from lifting the can up from the initial position and placing it on the target position to sliding the cans from the initial position to the target position. Subjects generally seemed to prefer clutching the cans in the middle and picking them up to move them from one place to another. Participants were advised of the time elapsed for each trial, and the cans were reset to the initial position following each trial.

3.   PRELIMINARY RESULTS AND CONCLUSIONS

The purpose of this study, by using a simplistic approach through an experiment, was to investigate the effectiveness of training using virtual environments and to investigate its transfer to a real-world situation. All of the data collected from pre-test and post-test experiments were analyzed by using appropriate statistical procedures.

The results of the t-test indicated that there exists no significant difference between the mean pre- and post-test scores of subjects in the Control Group on their speed (t = 0.97, df = 20, p < 0.01) and error (t = 0.44, df = 20, p < 0.01) of performing the "pick-and-place" operation (Table 1).

TABLE 1. Means (and Standard Deviations) and comparison of mean of Pre-test and Post-test Scores of Experimental and Control Group Subjects on the Speed-and-Error Questionnaire.

	Experimental Group (Received VR Training)						Control Group (Did Not Receive VR Training)					
	Baseline (N=10)		Post-Test (N=10)		Analysis (df=20)		Baseline (N=10)		Post-Test (N=10)		Analysis (df=2)	
Measure	Mean	SD	Mean	SD	t	P	Mean	SD	Mean	SD	t	P
Speed-and-Error Questionnaire												
Speed (10 Trials)	5.33	0.61	5.01	0.85	0.97	<0.01	5.15	0.79	4.88	0.90	0.72	<0.01
Error (10 Trials)	0.51	0.30	0.45	0.31	0.44	<0.01	0.87	0.60	0.67	0.65	0.71	<0.01

The t-value on speed variable for the Control Group was 0.97, which indicated no statistically significant difference at the 0.01 level. That is, there was no statistically significant difference between the mean pre- and post-test scores of the subjects in the Control Group on speed. In addition, the t-value on error variable for Control Group was 0.44, which indicated no statistically significant difference at the 0.01 level. That is, there was no statistically significant difference between the mean pre- and post-test scores of the subjects in the Control Group on error variable.

The results of the t-test indicated that there exists no significant difference between the mean pre- and post-test scores of subjects in the Experimental Group on their speed ($t = 0.72$, $df = 20$, $p < 0.01$) and error ($t = 0.71$, $df = 20$, $p < 0.01$) of performing the "pick-and-place" operation (Table 1). The t-value on speed variable for the Experimental Group was 0.72, which indicated no statistically significant difference at the 0.01 level. That is, there was no statistically significant difference between the mean pre- and post-test scores of the subjects in the Experimental Group on speed. In addition, the t-value on error variable for the Experimental Group was 0.71, which indicated no statistically significant difference at the 0.01 level. That is, there was no statistically significant difference between the mean pre- and post-test scores of the subjects in the Experimental Group on error variable.

Although the results of this experiment did not indicate, as predicated earlier, that the virtual reality training group (Experimental Group) would perform significantly better than the Control Group, the data showed that subjects in the Experimental Group maintained their learning at a higher level than the Control Group. However, both the Control and Experimental groups had lower post-test scores than their pre-test scores in this experiment.

This experiment demonstrated that virtual environment training could be implemented with moderate success. In general, the transfer of learning from the task performed in a virtual environment to the real-world situation assists with retaining the learned task. In addition, the results of this study demonstrated that the virtual environment technology could moderately be a viable tool for training. These conclusions are very limited considering that different tasks may result in different outcomes when utilizing virtual environment technology in a specific training situation. More extensive research needs to be conducted to investigate this hypothesis and these results.

ACKNOWLEDGMENT

This research project is currently sponsored by a grant from the Mentor Protégé Program at Kennesaw State University. Many thanks goes to research students and participants who enthusiastically got involved in this research. Hardware and software for this project was partially supported by grants from Boeing Computer Services and the U.S. Army Research Laboratory where the principal investigator and the virtual reality laboratory were previously housed.

REFERENCES

Alessi, S.M. (1988). Fidelity in the design of instructional simulations. *Journal of Computer- Based Instruction*, 15(2), 40-47.

Baum, D.R., (1992). Virtual training devices: Illusion or reality? Unpublished manuscript, Hughes Training Inc., Minneapolis, MN.

Bricken, M. (1991). Virtual reality learning environments: Potentials and challenges (No. HITL-P-91-5). Seattle, WA: Human Interface Technology Laboratory.

Burg, J., Hughes, C.E., Lisle, C., Moshell, M., Carrington, J., and Xin, L. (1991). Behavioral representation in virtual reality. In D.E. Mullally, M. Petty, and S. Smith (Eds.), *Second Behavioral Representation and Computer Generated Forces Symposium Proceedings*. University of Central Florida, Orlando, Fl.

Durlach, N.I., Aviles, W.A., Pew, R.W., DiZio, P.A., and Zeltzer, D.L. (Eds.), (1992). *Virtual environment technology for training (VETT)*. (BBN Report No. 7661). Cambridge, MA: Bolt Beranek and Newman, Inc.

Helsel, S.K. (1992). Virtual reality and education. *Educational Technology*, 32(5), 38-42.

Holding, D. H. (1987). Concepts in training, In G. Salvendy (Ed.), *Handbook of Human Factors*, (pp. 939-962). John Wiley & Sons: New York.

Kozak, J.J., Hancock, P.A., Arthur, E.J., and Chrysler, S.T. (1993). Transfer of training from virtual reality. *Ergonomics*, 36(7), 777-784.

Mattoon, J.S., and Thurman, R.A. (1990). Microcomputer-based instructional simulation for training research. In D. W. Dalton (Ed.), *Proceedings of the Thirty-Second Annual Conference of the Association for the Development of Computer-Based Instructional Systems* (pp. 366-367). San Diego, CA: ADCIS International.

North, M.M., North, S.M., and Coble, J.R. (1995). Effectiveness of virtual environment desensitization in the treatment of agoraphobia. *International Journal of Virtual Reality,* 1(2), 25-34.

North, M.M., and North, S.M. (1996a). Virtual psychotherapy. *Journal of Medicine and Virtual Reality,* 1(2), 28-32.

North, M.M., North, S.M., and Coble, J.R.(1996b). Application: Psychotherapy, Flight Fear Flees, *CyberEdge Journal,* 6(1), 8-10.

Reigeluth, C.M., and Schwartz, E. (1989). An instructional theory for the design of computer-

Presence in Altered Environments: Changing Parameters & Changing Presence

Dr. Robert C. Allen

University of Central Florida & Naval Air Warfare
Center – Training Systems Division
12350 Research Parkway
Orlando, FL 32826-3276

& Dr. Michael J. Singer

U. S. Army Research Institute for the Behavioral &
Social Sciences, Simulator Systems Research Unit
12350 Research Parkway
Orlando, FL 32826-3276

ABSTRACT. The U. S. Army Research Institute (ARI) has been investigating the use of Virtual Environments (VE) for Dismounted Soldier simulations for several years. One constant issue investigated in the research program has been the issue of Presence. Presence has been defined as the subjective experience of being in one place or environment, even when one is physically situated in another (Witmer & Singer, 1998). In recent research examining performance during a search task with and without a body-matching avatar, no differences were found in the evaluation of presence (Singer, Ehrlich, & Allen, 1998). A recent replication and extension of that research incorporated additional visual display characteristics in a between groups design. The research replicated the previous Virtual Environment (VE) conditions and added expanded display conditions using real world (RW) display mockups, with appropriately reduced resolution and diminished luminance. The four RW conditions were matched to normal helmet mounted displays, doubled the horizontal view, doubled the vertical view, or presented a "normal" field of view. In both the VE and RW conditions, participants were required to perform two tasks in three rooms. In the first room participants were required to follow a marked course as rapidly as possible. In the second and third rooms the participants were required to search for and acquire floor targets in order and drop markers into them. Analysis of the VE presence scores replicated the previous results, finding no difference in rated presence, with the exception of the Naturalness subscale. The Disembodied point of view was rated more natural than the avatar condition. In comparisons between the VE and the matching FOV RW condition, the RW conditions were significantly lower than the VE. However, within only the RW conditions, the Involved/Control subscale was always rated significantly higher than the VE matching condition. It may be that the avatar representation did not provide the immediately obvious information that it was intended to provide, and required longer adaptation times. The Control superiority over a restricted FOV with diminished resolution would seem to also fit this argument, in that hindering or diminishing interactions with the experienced world diminishes the amount of presence one has in the world.

1.0 Introduction

The U. S. Army Research Institute (ARI) has been investigating the use of Virtual Environments (VE) for Dismounted Soldier simulations for several years. One constant issue investigated in the research program has been the issue of Presence. By Presence we mean the subjective experience of being in one place or environment, even when one is physically situated in another (Witmer & Singer, 1998). The efficacy of virtual environments (VEs) has often been linked to the sense of presence reported by users of those VEs. Witmer and Singer argue that presence is a normal awareness phenomenon that requires directed attention and is based in the interaction between sensory stimulation, environmental factors that encourage involvement and enable immersion, and internal tendencies to become involved. This issue is considered as a part of the evaluation of the effectiveness of virtual environment systems. Combined results from previous experiments (Witmer & Singer, 1998) lead to the following conclusions: (1) the Presence Questionnaire (PQ) and Immersive Tendencies Questionnaire (ITQ) are internally consistent measures with high reliability; (2) there is a weak but consistent positive relation between presence and task performance in VEs; (3) individual tendencies as measured by the ITQ predict presence as measured by the PQ; and (4) individuals who report more simulator sickness symptoms in VE report less presence than those who report fewer symptoms. Simulator sickness issues will not be considered in this paper.

As applied to a virtual environment (VE), presence refers to experiencing the computer-generated environment rather than the actual physical locale. This definition provides a common understanding of the concept, but it does not identify the factors influencing presence, nor does it describe the exact nature of the experience. What aspects of the VE or remote environment contribute to the experience of presence? Do individual differences affect how much presence is experienced? What role does immersion, the perception of being enveloped, play in experiencing presence? Does presence result from a simple displacement of attention from the real world to the VE or must one become totally involved in the VE to experience presence? As used here, *attention* includes orienting one's senses toward information sources and selectively processing the available information.

In recent research examining performance during a search task with and without a virtual body representation, no differences were found in the evaluation of presence (Singer, Ehrlich, & Allen, 1998). Although it was not the major focus of the research effort, we had anticipated that not having a visual representation of self would lead to some disengagement from the experience. Two explanations were hypothesized for the search results, one of which also seemed to apply to the presence results. The major hypothesis (that applied to presence) was that the field of view in the Helmet Mounted Display (HMD) effectively cut off any lower Field of View (FOV) one might normally have of one's body. A replication of that research has been conducted incorporating additional and expanded visual display characteristics in a between groups design. The expanded display conditions used a wizard of oz approach consisting of four real world (RW) mockups of HMD displays, with appropriately reduced resolution and diminished luminance. The four RW conditions provided display configurations of 48^0 x 36^0 (horizontal x vertical; called Restricted as it matched normal HMDs), 96^0 x 36^0 (Horizontal Visual Field; HVF), or 48^0 x 72^0 (Lower Visual Field; LVF), and a "normal" field of view group (Control).

2.0 Method

2.1 Participants

Ninety participants were recruited from the University of Central Florida for this study. The participant's age ranged from 18 to 45 with a mean age of 21.04 (SD = 4.64). All participants had normal or corrected to normal vision. Nine females and six males were randomly assigned to each of the six groups.

2.2 Apparatus and Materials

The experiment was conducted at the University of Central Florida, Institute for Simulation and Training (IST) in Orlando, FL, with Visual Systems Laboratory equipment. The Real World (RW) helmet was constructed using a hardhat with a foam rubber lining. A TCM2/50 Electronic Compass Module© (magnetometer/inclinometer) by Precision Navigation™, Inc. was mounted on the hardhat. An adjustable HMD headband was used for mounting the FOV restriction device. Three pairs of plastic welder's goggles, each with cardboard cutouts, were used as masks to restrict the FOV for the RW conditions (48^0x36^0, Restricted, 96^0x36^0, Horizontal Visual Field {HVF}, or 48^0x72^0, Lower Visual Field {LVF}). A wearable computer, constructed by IST, was used for RW data collection, and a hand-held keyboard device - the Twiddler© by HandyKey Corporation - was used to control the computer (start & stop data collection, etc.). The computer and Twiddler© were in a backpack worn by participants, and a fanny pack contained six plastic balls used during the target acquisition task.

The virtual environment was generated through Multigen™ II version 1.5 and specialized software developed by IST. Display generation and data capture was accomplished using a Silicon Graphics™ Onyx RealityEngine2©, with eight 200 MHz, 256 MB RAM, R4400 processors. The VE display device was a Virtual Research Corporation™ V8 Head Mounted Display (HMD). The V8 has a 48^0 x 36^0 FOV with 1820 x 480 color pixels in each lens. Head, shoulder, feet, right arm and right hand motions were registered by six Ascension Flock-of-Birds© sensors and tracked by an Ascension MotionStar© (wired version) with an extended range transmitter.

The Presence Questionnaire (version 3.0) has 32-items, each with a seven-point response scale using three anchors (endpoint anchors and a midpoint). The scoring was conducted using methods recommended by Witmer and Singer (1998). The PQ yields a Total score as well as several sub-scales: Involved/Control, Natural, Interface Quality, Auditory, Haptic, and Resolution. The Involved/Control sub-scale addresses involvement and control issues associated with a given environment, e.g., how well a person could control events or how involved the user was in the experience. The Natural sub-scale measures the naturalness of the interactions the user had with the environment. The Interface Quality sub-scale measures how much the control/display device(s) interfered with a user's experience. The auditory and haptic sub-scales were not included in analyses as this VE configuration did not employ haptic or audio interfaces. The resolution sub-scale is used to measure differences in the perception of the resolution of two or more display devices. Again, this sub-scale is generally not included in an analysis.

2.3 Procedure

The experiment required two groups to perform their tasks in the VE, and four groups to perform the same tasks in the real world. The Real World groups were labeled as Restricted, LVF, HVF, and Control. The two VE groups were labeled Disembodied and Body Model. All groups performed a Guided Movement and Search task in their respective environments.

Each group experienced three phases. During the first, each participant was briefed on the nature of the experiment and signed a consent form. Each participant's eye height, shoulder width, and Inter-Pupiliary distance were then measured and used to adjust the equipment and displays. Visual screening tests were then administered to screen participants for potential visual defects that could affect their performance. These tests included near and far point acuity, stereo and color vision, and a horizontal visual field test.

The VE participants received movement training in two separate practice environments. The first practice environment was simply walking from one empty room to the next without bumping into walls or doorframes. The next practice environment consisted of a series of interconnected rooms that were filled with typical office furniture. A blue line defined a path that the participant was to follow. After a ten minute break, but before reentering the VE, the participant listened to a taped instructional set that explained the practice search task. The task in this environment required the participant to locate two trashcan targets in sequential order and drop a ball inside each. The Body group could always see their virtual body, while the Disembodied group would only see the right hand and fanny pack when they were within two feet of a trashcan.

After completing the last practice task, the participant was given a three minute break. They then heard the instructional set for the Guided Movement task, which required the participant to follow a path defined by the arrows as quickly and accurately as possible, minimizing collisions with boxes, walls, office furniture, or doorframes. After completing this task, the instructional set for the Search tasks of the last two rooms was played.

The procedures for the RW participant differed slightly. After the briefing, the RW participant listened to the taped instructional set for the practice tasks. The participants then enter the practice room and put on the RW equipment. The practice and taped instructions for the experiment were identical to the VE, with minor adaptations to insure that the RW movement matched the constrained VE movement possibilities (no sidestepping, etc.). Once the guided movement task was completed, the procedures for the search task were identical to those used in the VE. After the experiment, the PQ, ITQ, and other questionnaires were completed and the participant was debriefed. All participants were given class credit or payment for his/her time.

3.0 Results

The initial Presence analysis of the VE conditions basically replicated the results obtained by Singer, et al., (1998), finding no difference in rated Presence between a Disembodied Point Of View (POV) and an avatar body representation, with the exception of the Naturalness subscale ($t(84)=2.80$, $p<.006$). The Disembodied POV was rated more natural than the avatar condition in the VE tasks.

In planned comparisons between the VE and Restricted conditions, significant differences were found with all of the PQ scales (excluding the Natural sub-scale), with the Restricted condition being significantly lower than the VE group ratings on those scales. The comparisons found the Involved/Control sub-scale $t(84)=2.74$, $p=0.007$; the Interface Quality sub-scale $t(84)=3.60$, $p=0.001$; the Resolution sub-scale $t(84)=5.23$, $p<0.001$; and the Total score for the PQ $t(84)=3.67$, $p<0.001$ to be significantly lower. (All of these planned comparisons used the average of the VE groups with the exception of the Natural sub-scale, which used the Disembodied group only because that group was significantly higher than the avatar group, see above.)

A series of planned one-way ANOVAs were conducted to test for differences between the RW groups using the PQ scales. While the Involved/Control sub-scale was not significant over the four real world groups (Restricted, Horizontal Visual Field, Lower Visual Field, and Control), the other scales did show significant differences. The Interface Quality sub-scale means differed significantly, $F(3,56)=6.24$, $p=0.001$, and a modified Bonferroni post hoc multiple comparisons procedure, ($\alpha=0.01$) revealed that the Control group rated their experience higher, relative to the Restricted and Lower Visual Field groups. The Natural sub-scale means differed significantly, $F(3,56)=3.97$, $p=0.024$. The Natural sub-scale means were then evaluated with the modified Bonferroni, which revealed that the Restricted group rated their movement as less natural, easy to control, or consistent with the real world, relative to the Control group. The Resolution sub-scale means also differed significantly, $F(3,56)=3.51$, $p=0.021$. When the means were evaluated, the Restricted group's ratings were again lower compared to the Control group. The Total score means differed significantly, $F(3,56)=3.88$, $p=0.014$. The modified Bonferroni revealed that the Restricted group's ratings were lower, compared to the Control group.

A correlational analysis, using the Pearson Product-Moment Correlation procedure, was then conducted between the PQ and the FOV Groups to see whether there was a relationship between a person's experience of presence, as measured by the PQ, and their FOV. Because of the differences found between the RW and VE in the PQ, the correlational analysis was conducted within the RW environment only. In addition, as the total area of regard provided by the LVF and HVF groups is the same (i.e., both the LVF and HVF groups had a total area of regard equaling 3456^0), these two groups were combined. Significant positive correlations were found between several PQ scales and the RW FOV. The PQ Total score and FOV Groups were significantly correlated at $r(60)=0.40$ ($p=0.002$), the Interface Quality sub-scale and FOV Groups at $r(60)=0.49$ ($p<0.001$), the Natural sub-scale and FOV at $r(60)=0.39$ ($p=0.002$), and the Resolution sub-scale and FOV Groups was at $r(60)=0.38$ ($p=0.003$). Since it could be argued that the correlation between FOV and Resolution should not have been conducted with the Control group included (the group had no resolution mask, thus their resolution was different than that of the other groups). To test this, a correlation was conducted without the Control group and the correlation was almost identical and still significant ($r(45)=.33$, $p=.028$).

4.0 Discussion..

The PQ was designed to obtain subjective reactions to factors that might influence a person's sense of presence in an environment, immediately after environmental exposure. In this research, the attempt was made to both replicate previous findings, anchor a real world extension of VE activities to the VE, and investigate possible reactions to currently unavailable display parameters within a similar setting. The discussion will address the results in the same three segments; within the VE, between VE and RW, and within the RW conditions.

The previous research (Singer, et al., 1998) had indicated that there was no reason to expect significant differences in the PQ scales between the Disembodied and Body Model conditions. Finding that the Natural sub-scale ratings differed indicates a complicated response to the VE conditions, since the VE interface devices used by both groups were the same. The non-significant difference on other PQ scales between the two groups indicates that the groups did not differ in the amount of involvement or control, interface quality, resolution, or overall presence experienced. The latter would seem to support the previous findings that a body representation does not increase a person's sense of involvement in a VE experience, relative to a disembodied viewpoint. The Disembodied group rated their movement as more natural, easy to control, or consistent with the real world, relative to the Body Model group. Initially this finding was surprising, however on further reflection it seemed that using the body representation, as constituted for this experiment, may not have been intuitive. Walking in the VE with a Disembodied viewpoint may, in fact, may be more natural given the FOV of the HMD used. The Body Model group was forced to look down at an extreme angle to be able to see (and use) the virtual body representation, something that is not necessary in the natural (real) world. For the Disembodied group, on the other hand, it was irrelevant. During the research, observations indicated that the latter group tended to not look down as long or as often. Therefore the Disembodied groups interaction with the environment, their ability to control movement, and their consistencies with the real world were rated higher (more natural), relative to the Body Model group, because they did not have to adapt to a non-natural feedback stimulus. The body representation may have served to highlight or emphasize the diminished FOV in the HMD, and thus diminish the ratings of naturalness in the environment.

The comparisons between the ratings of the VE and the Restricted FOV RW condition, which matched the VE in FOV, found significant differences between almost all of the PQ scales (excluding the Natural sub-scale), with the Restricted condition being significantly lower than the VE group ratings on those scales. This indicates that the Virtual environment was easier to control, interfered less, made it easier to examine objects, and generally led to a greater overall sense of presence than reality when experienced through a restricted view and decreased resolution mask. The results may be based in the anchoring biases that people brought to the situation. First, expecting that the VE would not be very good, which led to higher ratings for exceeding expectations. Second, people who are used to reality, but then put on a masking helmet, would be expected to rate their experience lower based on normal conditions.

Several differences were found between the RW groups on the PQ scales, as shown in Table 1. While the Involved/Control subscale did not show significant differences for the RW conditions, the results of the Interface Quality sub-scale showed that the Control group rated the interface, devices, and mechanisms associated with their condition as less interfering, relative to the Restricted and LVF groups. The Natural sub-scale results indicated that the Control group rated their movement as more natural or consistent with the real world, relative to the Restricted group. The Resolution results showed that the Control group's ratings were higher in how closely or well they could examine objects, compared to the Restricted group. Finally, the Control group's ratings were higher, in terms of overall sense of presence, compared to the Restricted group.

Table 1.
Presence Analyses for Field of View Conditions.

PQ	VE Disembodied vs. Body Model	Restricted Field of View	Lower Visual Field	Horizontal Visual Field	CONTROL
Involved / Control	ns	ns	ns	ns	ns
Interface Quality	ns	Significantly Less	ns	ns	Significantly Greater
Natural	Disembodied Significantly Greater	Significantly Less	ns	ns	Significantly Greater
Resolution	ns	Significantly Less	ns	ns	Significantly Greater
Total	ns	Significantly Less	ns	ns	Significantly Greater

None of these results are particularly surprising. The LVF mask was rated as being more interfering (Interface Quality sub-scale) than the Control group's "mask", but this may be due to the fact that the former was the most bulky of the three FOV masks, therefore the most uncomfortable to wear. The fact that the Control group's total sense of presence was higher than the Restricted FOV group's may indicate the detrimental effect that the FOV, as provided by the VE HMD used in this experiment, has on presence (perhaps especially when rated by people straight from normal conditions). Prothero and Hoffman (1995) found an increase in the sense of presence when the VE FOV was increased from $60^0 \times 60^0$ to $105^0 \times 40^0$. So, too, does there appear to be an increase in the total sense of presence, as measured by the PQ, when the FOV in the RW is increased from $48^0 \times 36^0$ to a relatively normal FOV and dramatically improved resolution, under similar environmental and task conditions. In addition, increasing the LVF or the HVF by the amounts used in this study did not significantly increase a person's total sense of presence, relative to the Total presence scale for the Restricted group. There may be a point beyond the FOVs of $48^0 \times 72^0$ or $96^0 \times 36^0$ in which a person's sense of presence increases significantly, relative to the sense of presence experienced by the Restricted group, with its $48^0 \times 36^0$ FOV. That point may be somewhere between the FOV's of the LVF and HVF groups in this research, and Prothero and Hoffman's (1995) $105^0 \times 40^0$ FOV.

Within the RW conditions, the Control was always rated significantly higher than the Restricted condition. The avatar used in the VE portion of the experiment seemed to hinder the participants rather than providing intuitive help during the tasks, and was therefore interfered with the presence experience. It may be that the avatar representation did not provide the immediately obvious information that it was intended to provide, and required longer adaptation times. The Control groups superiority over a restricted FOV with diminished resolution would seem to also fit this argument, in that hindering or diminishing interactions with the experienced world diminishes the amount of presence one has in the world. Also as a person's FOV increased, their overall rating that the interface or control devices did not interfere with their experience also increased. As a person's FOV increased, their rating of the resolution (how closely they could examine objects and how well they could examine them from multiple viewpoints) of the RW increased. In addition, as FOV increased, the overall rating of sense of presence also increased.

References

Prothero, J. D., & Hoffman, H. G. (1995). *Widening the field-of-view increases the sense of presence in immersive virtual environments* [On-Line]. Available: http// Hostname: www.hitl.washington.edu Directory: cgi-bin/db.cgi

Singer, M. J., Ehrlich, J. M., & Allen, R. C. (1998). *Effect of a body model on performance in a virtual environment search task* (ARI Technical Report 1087). Alexandria, VA: U. S. Army Research Institute for the Behavioral and Social Sciences.

Witmer, B. G., & Singer, M. J. (1998). Measuring Presence in Virtual Environments: A Presence Questionnaire. *Presence 7*(3), 225-240.

Presence In Distributed Virtual Environments

Patrick M. Commarford, Dr. Michael J. Singer, & Jason P. Kring

U.S. Army Research Institute, Simulator Systems Research Unit, Orlando, Florida, USA
University of Central Florida, Orlando, Florida, USA

Abstract. The U. S. Army is developing distributed interactive simulation (DIS) systems for combat training and military concept development, testing, and evaluation. During an experiment investigating team coordination in distributed mission rehearsal, we investigated changes in response to a Presence Questionnaire (PQ; Witmer & Singer, 1998). The virtual environment (VE) scenario required building search, tool use, and communications protocols while being opposed by computer generated forces. Each participant was trained to standard criteria on all tasks and activities, and then performed 8 missions over several days. The training provided the opportunity to investigate presence responses to simple movement training and complex task training with an automated partner, while using the same VE equipment and comparable graphics. The significant increase in almost all presence scales indicates that a change in complex activities has an effect on presence. We anticipate that teams involved in distributed VE simulations, having more opportunities to interact within the mission, will develop greater levels of presence, and performance achieved by those teams will also be improved.

1. Introduction

The Army Research Institute for the Behavioral and Social Sciences (ARI) at the Simulator Systems Research Unit (SSRU), supported by the University of Central Florida Institute for Simulation and Training (IST), has established a research program in Virtual Environment (VE) technology in order to investigate a wide range of potential applications. The program goals are to "improve the Army's capability to provide effective, low cost training for Special Operation Forces and Dismounted Infantry through the use of VE technology and ICS [Individual Combatant Simulation]" (Knerr et al., 1994, pp.10-12).

The efficacy of VEs has often been linked to the sense of presence reported by users of those systems. Presence is defined as the subjective experience of being in one place or environment, even when one is physically situated in another (Witmer & Singer, 1994; 1998). Witmer and Singer have developed and refined subjective questionnaires that address a person's baseline immersive tendencies (the Immersive Tendencies Questionnaire, ITQ) and responses to the VE situations (the Presence Questionnaire, PQ). The ITQ is divided into three subscales which sum to an ITQ total score. The subscale labels are Involvement, Focus, and Games. The PQ has six subscales (Involved/Control, Natural, Auditory, Haptics, Resolution, and Interface Quality) which also combine to yield a total PQ score. All subscale labels are based on question content.

The concept of presence has been widely discussed (Hoffman, Prothero, Wells, & Groen, 1998; Huang & Alessi, 1999), and several have proposed methods of measuring presence and relating it to possible contributing factors (Nichols, Haldane, & Wilson, 2000; Slater, Steed, McCarthy, and Maringelli 1998; Snow & Williges, 1998). Barfield and Hendrix (1995) have used simple direct questions to show that update rate affects presence. Prothero and Hoffman (1995) have shown that limiting the FOV reduces the amount of presence reported, again using a direct query about the subjective experience of presence. Witmer and Singer (1998) have provided data that supports the concept of presence as a valid construct, as measured by the PQ. The PQ has been shown to relate positively to ITQ scores and to performance.

It is not clear how measures of presence relate to learning and performance in the VE and in the real world (Bailey & Witmer, 1994; Witmer & Singer, 1994;), however, many factors that appear to affect presence are also known to improve learning and performance (Witmer & Singer, 1998). Some situational factors that are believed to increase immersion, such as minimizing outside distractions and increasing active participation through perceived control over events in the environment, may also enhance learning and performance. Other factors may be more internal, such as tendencies toward involvement and selective attention, or familiarity with the task and situation.

During the course of an experiment investigating team coordination in distributed mission rehearsal, we investigated changes in response to the PQ and compared those scores to ITQ scores obtained before VE exposure. Participants in this experiment were trained on all VE equipment and tasks over a 4-hr period in which they were exposed to the VE three or four times. Administrations of the PQ immediately after both simple and more complex training exposures allowed us to track participant involvement in the training using the same VE equipment and comparable graphics, but with increasing task complexity. After training, participants were paired with a partner and completed a series of eight VE missions. PQ measures taken immediately following the first and final VE mission exposure allowed us to track changes in presence in response to scenarios that are similar in complexity as experience increased and performance improved. We were also able to track changes in presence experienced between a complex training session (the final training) and actual mission performance.

The ITQ scale scores were assessed for relationships with the presence scales. The ITQ was expected to correlate positively and more highly with the PQ obtained after the simplest VE situation, as the PQs associated with the more complex VE training and in the actual missions were expected to be influenced by more situational factors. The first hypothesis was:

Hypothesis 1. Higher ITQ scores will be associated with higher simple movement training PQ scores.

Because many of the factors involved in learning and performance logically should increase presence, it would be counter-intuitive if positive relationships between presence and performance or between presence and equipment configurations that increase active participation were not found. In this study, the PQ was administered after several different phases in the experiment and results from the PQ were examined for relationships with task complexity and task experience. One expectation was that scores on the PQ would increase with any change in the VE that increases the amount of interaction required for minimal performance. The initial training focused on learning to walk through a rather simple environment. The final training session focused on the same movement control, while adding equipment operation and team tasks (with an automated partner). This experiment also required repeated team missions, during which the teams were expected to improve in performance. The PQ was administered after the first and last mission, with the expectation that increasing familiarity and capability will support increases in the experience of presence. These expectations are stated in the following hypotheses:

Hypothesis 2. Higher PQ scores will be found after the more complex training situation, relative to the levels found after the simple movement training.

Hypothesis 3. Higher PQ scores will be found after repeated experiences in the VE, relative to the levels found after the initial session.

2. Method

2.1. Participants

Participants were 64 male and female students from the University of Central Florida or were co-op students from one of a number of universities who were working at the Canadian Defence and Civil Institute of Environmental Medicine. All participants were kept unaware of the focus of the research and had normal or corrected-to-normal vision. Participants received monetary compensation for time spent in training and mission rehearsals, and bonuses for completing training and returning for all mission sessions.

2.2. Materials and Equipment

Questionnaires. Questionnaire information was collected using an Access^{tm} database developed by ARI researchers, implemented on a standard Windows95^{tm} platform, and by paper and pencil. In addition to the PQ and ITQ, a biographical questionnaire, the Simulator Sickness Questionnaire (SSQ: Kennedy, Lane, Berbaum, & Lilienthal, 1993), and the NEO Five-Factor Inventory (NEO-FFI^{tm}, Costa & McRae, 1992; copyright by Psychological Assessment Resources, Inc.) were implemented.

Virtual Environment. The VE was rendered at both sites on Silicon Graphics Onyx^{tm} and Reality Engine systems, with Virtual Reality VR8 helmets used for visual display. The participants viewed color, stereoscopic images of the VE. Stereo sound was provided through earphones. The sound included voice communications between each of the participants and the experimenter, and sound effects such as collision noises, doors opening, grenade explosions, and gunfire. MotionStar^{tm} sensors were used for tracking and were placed on each ankle, on the right wrist and elbow, on the harness, and on the helmet. The software was written using Performer, C++, and Java.

Mission Rehearsal VE. The mission rehearsal scenarios were 10-room building VEs laid out along a single corridor approximately four m wide with one 90 degree turn, either to the right or left. The corridors were all scaled to 70 m in length, with the turn at 20, 25, or 30 m. The rooms varied between 5 x 10 and 15 x 10 m in size, with office furniture, home furnishings, warehouse shelving, bookcases, and desks placed in realistic arrangements. Individual rooms ranged from simple to complex with varying numbers of opposing forces and neutrals. The buildings were designed to represent normal offices, a school, a department store, a library, a warehouse, and single story homes. The scenarios were populated with varying numbers of neutral hostages, opposing forces, and gas canisters. Canisters had one of three possible armed states: a) no gas & not armed, b) gas & not armed, and c) gas & armed. The complexity (based on opposing forces and on number and state of canisters) was balanced across scenarios to the greatest extent possible. The order of scenarios was balanced such that no scenario repeatedly followed another, and no scenario started or ended the mission orders more than once.

2.3. Procedures

Before training began, all participants were briefed on the nature and requirements of the training and mission rehearsals. They were also shown a video that demonstrated the equipment, techniques, and mission tasks. After filling out the consent forms, participants started by completing the biographical questionnaire. They then completed the ITQ and an SSQ.

Training. Each participant was trained to criterion on all tasks in a single 4-hr session at least one day prior to the first session of team missions. Participants spent a maximum of 12 accumulated minutes immersed in the environment within a 30-min time period. The participants then had a minimum 30-min recovery time between VE immersions, during which time questionnaires and non-VE training were administered. Participants' physical wellness states were carefully monitored through use of the SSQ, which was administered before and after every VE exposure. During the training session, participants learned to move within the VE, operate all simulated equipment, and follow appropriate communications protocols required in the mission rehearsals. The training concluded with practice on the major coordinated team activities with an automated partner. The PQ was administered after the first (simple movement) and final (complex mission) training VE exposure.

Mission Rehearsals. Following training to criterion, some participants were assigned to a team. Each team completed two sessions during which eight mission rehearsals were performed (four during each session). The two sessions occurred on separate days, with a minimum of 1 day and a maximum of 9 days between sessions. In each mission rehearsal the team moved through a 10-room building searching for and disarming gas canisters, while neutralizing opposing forces and finding hostages. The SSQ was again administered before and after each VE exposure and the PQ was administered after the first and final mission rehearsal. Between each mission, participants engaged in an after action review during which they were required to talk about what they had done wrong and how performance could be improved. After the final administration of the PQ, participants completed the NEO. Singer, Grant, Commarford, and Kring (2001) report contains the analyses of biographical data, SSQ, NEO, ITQ, PQ, communications, and performance.

3. Results

This experiment employed a total of 64 male and female participants, all of whom completed training and some of whom were paired with a teammate and completed all the missions. Thus, virtually all participants completed PQ1 and PQ2, and a subset completed PQ3 and PQ4.

Correlations were conducted between the ITQ scales and both the initial PQ (responses to movement training) and final training PQ (responses to task practice with an automated partner). The only significant correlations between the ITQ scales and the initial PQ scales (using the Bonferroni adjustment on .05 alpha for a family of 28 comparisons) were between the ITQ Focus and the initial PQ Total ($r = .498$, $p < .001$, $N = 59$), PQ Involved/Control ($r = .470$, $p < .001$, $N = 59$), and PQ Resolution ($r = .426$, $p < .001$, $N = 59$). There were no significant correlations between the ITQ and the final training session PQ administration.

A series of planned comparison t-tests were conducted with the total PQ scores for all trained participants over the two training PQ administrations, after the initial movement training (PQ 1) and the final task training with an automated partner (PQ 2). Analyses found significant differences between PQ Total 1 and 2 (t (48) = -4.323, $p <$ .001); PQ Natural 1 and 2 (t (48) = -2.814, $p = .007$); PQ Involved/Control 1 and 2 (t (38) = -2.781, $p = .008$); PQ Auditory 1 and 2 (t (48)= -5.006, $p < .001$); and PQ Haptics 1 and 2 (t (48) = -9.530, $p < .001$). The standard descriptive statistics for the administrations of these scales are presented in Table 1.

Table 1
Presence Questionnaire Subscale Means and Standard Deviations during Training

Presence Questionnaire Scale	Initial Training		Final Training	
	M	SD	M	SD
Total	90.51	12.39	96.73	12.27
Involved/Control	57.90	7.71	60.37	7.81
Natural	13.78	3.25	14.92	3.40
Auditory	9.86	6.22	14.18	5.09
Haptics	3.65	2.83	7.98	3.19

The PQ scales were compared between the last VE training exposure and the first team mission (PQ3) using planned comparisons. All analyses used the Bonferroni adjustment of the alpha level for the family of comparisons. The analyses found significant differences between the PQ Total 2 and 3 (t (39) = 3.390, p = .002), and PQ Involved/Control 2 and 3 (t (39) = 4.424, p < .001). Finally, planned comparisons were also conducted between the PQ administrations after the first (PQ 3) and last (PQ 4) missions conducted by the teams. These analyses found significant differences between the PQ Total for 3 and 4 (t (42) = -3.367, p = .002) and PQ Involved/Control 3 and 4 (t (42) = -3.262, p = .002). The standard descriptive statistics for the PQ scales from these administrations are also presented in Table 2.

Table 2
Presence Questionnaire Subscales

Presence Questionnaire Scale	Final Training		Initial Mission		Final Mission	
	M	SD	M	SD	M	SD
Total	96.73	12.27	91.98	12.61	96.49	11.77
Involved/Control	60.37	7.81	57.0	8.16	59.84	7.85
Natural	14.92	3.40	14.42	3.33	14.74	3.13
Auditory	14.18	5.09	15.0	3.47	14.86	3.50
Haptics	7.98	3.19	7.28	2.21	7.14	2.11

4. Discussion

This research investigated changes in subjective feelings of presence as measured by the PQ. As hypothesized, the ITQ was correlated with the initial PQ responses, and not with the final training session PQ. This supports the argument that internal immersive tendencies relate to feelings of presence in simple environments, but are not predictive of responses to more complex interactions with VE systems. It is suggested that other constructs such as learning, experience with the VE, perceived control, distractions, and performance in the VE account for most of the variance in PQ scores at a higher levels of VE complexity.

PQ scores were higher immediately following mission training with an automated partner than they were following simple movement training, although the same VE equipment was used for each training situation and graphics were similar. These findings indicate that the increased complexity of the final training enhanced the participants' experience of presence. The more complex tasks were likely more engaging and involving for the users, thereby invoking greater feelings of "being there." Also, additional capabilities of object manipulation presumably increased participants' perceived control over the environment, thereby enhancing perceptions of presence. It is also interesting to note, and supportive of the PQ as a valid measure of presence, that Interface Quality and Resolution were the only PQ subscales for which scores did not increase from simple movement training to the more complex mission training.

The PQ scores dropped between the final training session and the first mission session. This unexpected and seemingly counterintuitive finding may be explained by the fact that task difficulty level and responsibility were greatly increased, so much so that participants were often overwhelmed by the initial mission experience. The first mission, for virtually all of the teams, was extremely sloppy and error-ridden. VE equipment was often used improperly and tasks were perceived as extremely difficult. Poor performance during the initial mission rehearsal may have been not only a function of greatly increased task complexity and responsibility, but also of a lapse in time since the last VE exposure and possibly even due to effects of evaluation apprehension. The second PQ was collected to gather responses to a training scenario that occurred after approximately 3 hr of VE and non-VE mission training. However, PQ 3 was administered after the first mission rehearsal, which followed a minimum period of 24 hr without VE exposure or other training. In other words, participants came into the first mission "cold." In addition, the first mission rehearsal was conducted with a human partner, whom each participant had either met briefly or not at all. Research has shown that the presence of additional observers decreases performance on novel or complex tasks (Forsyth, 1999), and this may have contributed to the evaluation apprehension. In addition to its effects on performance, evaluation apprehension could also work as a distracter while immersed in the VE. Distractions imposed by these factors may have worked to decrease perceptions of presence.

As expected, PQ scores were significantly higher for the final mission experience than the initial mission experience. It is likely that as experience and performance increased, participants became more involved in the VE missions and the distractions dissipated. It is also possible that any distractions directly attributable to evaluation apprehension would have disappeared as highly-learned tasks do not appear to suffer performance decrements due to evaluation apprehension, but rather are facilitated by the presence of others (Forsyth, 1999). As participants became more practiced with the VE tasks, their levels of perceived control likely increased, which would presumably enhance learning and increase levels of presence experienced.

This research supports previous findings (Witmer & Singer, 1998) that increased performance and control work to increase subjective feelings of presence. This research also provides further evidence of the validity of a presence construct and of the PQ as a valid indicator of this construct. The next step in VE research may be to investigate the causes of changes found during the training and mission segments of this study. Researchers might overtly manipulate the levels of task difficulty or the amount and intensity of distractions in order to show sensitivity of the PQ measure to task and VE variables.

References

Bailey, J. H., & Witmer, B. G., (1994). Learning and transfer of spatial knowledge in a virtual environment. *Proceedings of the Human Factors & Ergonomics Society 38th Annual Meeting.* Nashville, TN, Human Factors & Ergonomics Society.

Barfield, W. and Hendrix, C. (1995). The effect of update rate on the sense of presence within virtual environments. *Virtual Reality: The Journal of the Virtual Reality Society,* 1(1), 3-16.

Costa, P. T., Jr., & McCrae, R. R. (1992). *NEO-PI-R professional manual: Revised NEO personality inventory (NEO PI-R) and NEO Five-factor Inventory (NEO-FFI).* Odessa, FL: Psychological Assessment Resources.

Forsyth, D. R. (1999). *Group Dynamics* (3rd ed.). Belmont, CA: Wadsworth. (ch. 10).

Huang, M. P., & Alessi, Norman E. (1999). Mental health implications for presence. *CyberPsychology & Behavior,* 2(1), 15-18.

Hoffman, H. G., Prothero, J., Wells, M. J., & Groen, J. (1998). Virtual chess: Meaning enhances users' sense of presence in virtual environments. *International Journal of Human-Computer Interaction, 10(3),* 251-263.

Kennedy, R. S., Lane, N. E., Berbaum, K. S., & Lilienthal, M. G. (1993). A simulator sickness questionnaire (SSQ): A new method for quantifying simulator sickness. *International Journal of Aviation Psychology, 3*(3), 203-220.

Knerr, B. W., Goldberg, S. L., Lampton, D. R., Witmer, B. G., Bliss, J. P., Moshell, F. M. & Blau, B. S. (1994). Research in the use of virtual environment technology to train dismounted soldiers. *Journal of Interactive Instruction Development, 6*(4), 9-20.

Nichols, S., Haldane, C., & Wilson, J. R. (2000). Measurement of presence and its consequences in virtual environments. *International Journal of Human-Computer Studies, 52*(3), 471-491.

Prothero, J. and Hoffman, H. (1995). *Widening the field-of-view increases the sense of presence within immersive virtual environments* (Human Interface Technology Laboratory Tech. Rep. R-95-4). Seattle: University of Washington.

Singer, M. J., Grant, S. G., Commarford, P. M., Kring, J. P., & Zavod, M. (In press). *Team performance in distributed virtual environments* (ARI Technical Report). Orlando, FL: Army Research Institute (Simulator Systems Research Unit).

Slater, M., Steed, A., McCarthy, J., & Maringelli, F. (1998). The influence of body movement on subjective presence in virtual environments. *Human Factors, 40*(3), 469-477.

Snow, M. P., & Williges, R. C. (1998). Empirical models based on free-modulus magnitude estimation of perceived presence in virtual environments. *Human Factors, 40(3),* 386-402.

Witmer, B. G., & Singer, M. J. (1994). *Measuring immersion in virtual environments.* (Technical Report 1014). Alexandria, VA: U. S. Army Research Institute for the Behavioral and Social Sciences, (ADA 286 183).

Witmer, B. G., & Singer, M. J. (1998, June). Measuring presence in virtual environments: A presence questionnaire. *Presence, 7*(3), 225-240.

PERCEPTUAL AND PERFORMANCE ISSUES IN THE INTEGRATION OF VIRTUAL TECHNOLOGIES IN DEMANDING ENVIRONMENTS

Dr. Ellen Haas

U.S. Army Research Laboratory
Aberdeen Proving Ground, MD

The potential exists for the insertion of multiple advanced technologies, including virtual technologies, into future soldier systems such as tanks and helicopters. However, care must be given to the integration of these technologies within these systems. The purpose of the study described in this paper was to determine the effects of the integration of virtual spatial auditory display and automatic speech recognition (ASR) technologies in a tank environment. In the experimental tasks, subjects were instructed to give ASR voice commands and press pushbutton controls in response to simulated monaural or virtual spatial audio radio communications, while simultaneously performing a tank driving task in the presence of tank noise. Independent variables included auditory display configuration (monaural or virtual spatial), driving speed, and subject type (soldier or civilian). Dependent variables included 1) the proportion of radio channel changes correctly recognized by the ASR; 2) the proportion and types of speech errors in radio channel changes; and 3) the number of accurate responses to the radio messages. Results showed that a mean of 95.5% of all subject verbal commands was recognized by the ASR system. ASR phrase recognition and human speech input were not significantly affected by audio display type, driving speed, or subject type. Spatialized radio messages significantly enhanced the ability of the user to understand speech communications from multiple simulated radio channels. Different driving speeds had no significant effect on the listener's ability to process auditory information. In general, ASR and virtual spatial audio displays have significant potential as technologies of interest in noisy, demanding environments in which complex auditory and visual tasks are concurrently performed.

1. INTRODUCTION

The U.S. Army is exploring the insertion of multiple advanced technologies such as virtual spatial audio displays and automatic speech recognition (ASR) into soldier systems such as tanks and helicopters. The purpose of this would be to reduce overload of soldier processing capabilities, to improve soldier mission performance, and to increase situation awareness. Two examples of systems employing multiple virtual technologies are the U.S. Army Virtual Cockpit Optimization Program (VCOP), and the U.S. Army Tank and Automotive Command (TACOM) Crewstation Automated Testbed (CAT).

The VCOP is a new program in which virtual reality is utilized in a helicopter flight system. It is lead by the U.S. Army Aircrew Integrated Systems Program Office at Redstone Arsenal in Huntsville, Alabama. VCOP embraces several independently developed advanced technologies, including a virtual retinal display (VRD) that incorporates a binocular-full-color, high-resolution, high-brightness helmet-mounted display, as well as virtual spatial audio displays, speech recognition, intelligent information management, crew-aided cognitive decision aids, and a tactile vest. Virtual spatial audio displays, also known as three-dimensional audio displays, are used because they allow a listener using earphones to perceive sounds which appear to originate at different azimuths, elevations and distances from locations outside his or her head. Virtual audio displays permit the processing of multiple radio communications, navigation waypoints, and target acquisition to be enhanced because each sound channel is presented at a different spatial location, permitting listeners to selectively attend to one sound at a time (McKinley, Erickson, and D'Angelo, 1995). The purpose of the VCOP program is to employ these multiple technologies to provide the pilot with situational awareness, sensor imagery, flight data and battlefield information in a clear and intuitive manner (Sherman, 2001).

The TACOM CAT program also embraces several independently developed advanced technologies, including virtual spatial audio displays and speech recognition, in a tank environment. As with the VCOP, the purpose of the CAT program is to provide the tank crew with battlefield information in an intuitive manner.

The introduction and use of these advanced technologies, as well as mission demands, have made helicopters and tanks into relatively dynamic, demanding, cognitively complex environments where soldiers must simultaneously monitor multiple visual and audio displays, operate multiple controls, and process large amounts of information in the presence of high levels of noise. Problems may occur when soldiers must perform these tasks at a rapid pace. This could result in information overload, which strains the soldier's mental, visual and audio processing capabilities, and could cause decrements in mission performance and situation awareness.

Researchers have not explored issues dealing with the use of some of these technologies, such as ASR and virtual auditory displays, in demanding environments. For example, no researcher has explored the extent to which virtual spatial audio displays affect the ability of the user to accurately input verbal information into an ASR system. Few if any researchers have investigated the extent to which differing levels of workload affects human performance of both audio display processing and the utilization of ASR control tasks in high noise environments. The use of these advanced technologies raises human factors questions, including the extent to which they can be integrated successfully, separately or together, into demanding crewstation environments.

The goal of this study was to examine the effects of the integration of virtual spatial auditory display and ASR technologies into a noisy tank environment. The first objective of this study was to determine the extent to which the performance of a noise-robust ASR system is affected by high levels of ambient noise and how well the ASR system recognizes voice commands from speakers who experience differing levels of workload. A second objective was to ascertain the extent to which user speech is affected when the subject uses an ASR system in a tank environment containing high noise levels and different levels of workload. A third objective was to determine the effect of virtual spatial audio displays on soldier performance in processing multiple radio communications. A fourth objective was to determine whether the type of incoming auditory information (monaural or virtual spatial), affects user speech input to the ASR system.

2. METHODOLOGY

2.1 Subjects, Facilities and Apparatus

Subjects were six (6) male U.S. Army soldiers and six (6) male civilians, between the ages of 23 and 50 with a mean age of 35 years, stationed or working at Aberdeen Proving Ground, Maryland. All were screened and possessed hearing within thresholds acceptable to the U.S. Army (1991). This experiment was conducted at the U.S. Army Research Laboratory's Hostile Environment Simulator, Human Research and Engineering Directorate, Bldg. 518, Aberdeen Proving Ground, Maryland. Audio display instrumentation included a Tucker-Davis Technologies (TDT) digital signal processing system, a Polhemus Fastrack head tracker, and Sennheiser HD 265 headphones. ASR task instrumentation included a speaker-dependent Verbex portable Speech Commander voice recognition system. Equipment for the driving task included a Silicon Graphics (SGI) Indigo workstation with a 19-inch color monitor, and a Thrustmaster NASCAR Model Pro racing steering wheel with brakes and accelerator.

2.2 Experimental Tasks

The subject was instructed to give ASR voice commands and press pushbutton controls, in response to simulated monaural or virtual spatial audio radio communications. The subject simultaneously performed a simulated tank-driving task. All tasks were performed in the presence of realistic ambient tank noise. Differing levels of workload were manipulated by specifying different speeds of the driving task. Following is a description of the tasks that the subjects performed and the stimuli presented in those tasks.

Radio Communications Task. The subject was instructed to listen to simulated radio communications, and tell the experimenter the origin of a target message using ASR voice commands and pushbutton responses. The pre-recorded radio messages were spoken by individual talkers, presented over the subject's earphones. The messages were presented simultaneously in groups of two, three, or four talkers. Each message consisted of a target call sign, the talker's call sign, and the channel (one, two, three or four) from which the message originated. Only one of the messages was a "target message" which included the listener's call sign ("Zulu two-two"). The subject was instructed to determine which radio channel the target message originated from, push the corresponding button on the four-button pushbutton display labeled with the proper radio channel, and input a verbal command into the ASR microphone using the phrase "Set Radio Channel x," where x indicated the number of the radio channel from which the subject perceived the originating target message. A total of 36 message groups (combinations of two, three or four talkers) were presented to the subject in each of two audio display conditions (monaural or virtual spatial audio). Of these, the subject heard 12 messages with two simultaneous talkers, 12 messages with three simultaneous talkers, and 12 messages with four simultaneous talkers, all presented in random order. This allowed one exposure to every possible combination of talkers (two through four) per session.

The radio messages were presented in two different presentation conditions: 1) Monaural presentation (the condition in existing communication systems), in which speech messages from all simulated radios were routed to both earphones at once, so that the subjects heard all messages in both ears; and 2) Three-dimensional (virtual spatial) presentation, in which radio traffic from the simulated radios one through four was presented from four phantom locations around the subject, at 45°, 90°, 270°, and 315° azimuth, respectively (the position directly in front of the subject was 0° azimuth).

The root mean square (rms) amplitude of each individual radio message measured at the earcup of the subject's headphones was 94 dBA sound pressure level (SPL). These signal levels are intelligible when played in the presence of tank noise, but still provide a safe exposure level. All tasks were performed in the presence of 88 dBA SPL digitized M1A2 interior tank noise, measured immediately outside the subject's headphones. This is a fairly representative sound pressure level for tank noise that is still within the limits of human exposure. All sound and noise levels were within human use hearing conservation guidelines (U.S. Army, 1991).

Driving Task. The subject was instructed to perform a computerized driving task in which he used the Thurstmaster NASCAR steering wheel device to drive on digitized road terrain presented on the SGI Indigo workstation monitor located directly in front of him. The road terrain was a digitized version of the Aberdeen Test Center Course at Churchville, MD. The subject was instructed to drive the course while doing his best to maintain a specific speed (either 15 or 45 mph) while using the pedal accelerator and brake on the floor to maintain speed and control. The subject monitored his speed by means of a speedometer shown on the workstation monitor. Different driving speeds represented differing levels of driving task workload (low and high) and were determined by means of a pilot study before this experiment. Five different versions of the digitized terrain were used in the experiment, including one for training sessions and one for each of the four experimental sessions. The subject performed the driving tasks in the presence of the ambient tank noise while listening to the radio messages.

2.3 Procedure

Prior to the experiment, the subject input voice samples into the speech recognizer, and practiced performing all tasks in the presence of the tank noise. During each experimental session, the subject donned the headphones and the ASR microphone and performed the driving and radio communication tasks at the pre-determined conditions of auditory display and driving speed level, in the presence of tank noise. The radio messages were played at random times, and the subject performed the radio communication, driving and ASR tasks as practiced during the training session. When all 36 radio communication messages had been presented, the experimental session ended.

The subject participated in a total of four experimental sessions on four subsequent days, each session containing a different combination of driving speed and auditory display. In order to eliminate practice or order effects, the levels of each independent variable were counterbalanced so that every combination of every condition was presented in different orders across subjects. Each experimental session lasted approximately 25 minutes. The subject participated in one experimental session per day.

2.4 Experimental Design

A 2 x 2 x 2 mixed factor, repeated measures design was used for data collection and to structure data analysis for the ASR performance and subject speech input. Within-subject variables were auditory display configuration (existing monaural and virtual spatial audio) and driving speed (15 or 45 mph). The between-subject variable was subject type (soldier or civilian).

A 2 x 2 x 2 x 3 mixed factor, repeated measures design was used for data collection and to structure data analysis for subject performance with different audio display configurations. Within-subject variables were auditory display configuration (existing monaural or virtual spatial audio), driving speed (15 or 45 mph), and number of simultaneous talkers (2, 3 or 4). The between-subject variable was subject type (soldier or civilian). All independent variables were fixed effect variables (fixed factors) except for subjects, who were treated as a random effect variable (random factor).

The dependent variables for this experiment included 1) the number of radio channel change phrases correctly recognized by the ASR device (phrase recognition rate), where a whole phrase was considered incorrect if there were one or more recognition errors; 2) the proportion and types of errors in radio channel change phrases incorrectly uttered by the subjects; and 3) the number of times that the subject correctly identified the origin channel of the target message.

3. RESULTS

The different experimental objectives were evaluated by examining the different dependent variables. ASR performance was evaluated by examining ASR phrase recognition rate. Subject ASR input with different levels of workload (different driving speeds) were evaluated by examining the proportion and types of errors in subject phrases. Listener accuracy in processing radio communications with different auditory displays was evaluated by examining the number of times the subject correctly identified the origin of the target message. The extent to which auditory display type affects user speech input into the ASR, was evaluated by examining subject speech input under different audio display conditions. A multivariate analysis of variance (MANOVA) was performed for each dependent variable to determine whether statistically significant differences existed between main effects or interaction effects in the study. Wilk's criterion ($\underline{U}$) was used as the test statistic. Significant effects were explored on a post hoc basis using a Newman-Keuls Sequential Range Test performed at $p \leq 0.05$. The data are described here, with mean scores and standard deviations shown in parentheses (mean; S.D.) in the ensuing text.

3.1 ASR Performance

Each subject input 36 radio communication phrases into the ASR in each experimental session. Of these, the number of phrases correctly identified by the ASR was obtained for each subject. The MANOVA for ASR sentence recognition rate indicated that there were no statistically significant effects for the main effects of auditory display type or driving speed. There were no other significant main effects or interactions, and there was no significant difference between soldiers and civilians. This indicated that different types of auditory display, different driving speeds, and different categories of user (soldier vs. civilian), had no significant effect on ASR sentence recognition rate. The overall mean for ASR performance (collapsed across subjects and conditions) was 34.37 correctly recognized sentences of 36 total sentences per session (standard deviation 2.26), indicating that the Verbex speech recognizer recognized a mean of 95.5% of all subject sentences.

3.2 Subject ASR Input

Each subject input 36 radio communication phrases into the ASR. The proportion of phrases correctly input (recognized by the ASR) was obtained for each subject. Of the 36 phrases input by the subject, the ASR did not recognize a total of 4.5% of all user command sentences. Of these, most (approximately 95%) were attributable to subject speech input errors. The MANOVA for the proportion of sentences incorrectly spoken by the subjects and not recognized by the ASR indicated that no statistically significant effects were found for the main effects of auditory display type or driving speed. There were no other main effects or interactions, and there was no significant difference between soldiers and civilians. This indicated that different types of auditory display, different driving speeds, and different users (soldier vs. civilian), had no significant effect on subject ASR input.

An analysis of types of ASR user input errors indicates that the largest percentage of errors (46%) were run-in phrases, in which subjects uttered two subsequent phrases but did not insert a long enough pause between phrases to allow the recognizer to recognize both as two distinct phrases. Most often, this resulted in neither phrase being recognized by the ASR. Other errors included substitution errors in which subjects used one word for another (17%), insertion errors in which subjects inserted additional illegal words into command phrases (17%), and deletions in which subjects omitted words from command phrases (15%). Mispronounced words consisted of the smallest proportion of errors (5%).

3.3 Accuracy in Processing Radio Communications

The number of correct radio channel identifications in each experimental condition was obtained for each subject. The MANOVA for listener accuracy data, the number of times that the listener correctly identified the origin channel of the target message, indicated that statistically significant effects were found for the main effects of display type ($\underline{U} = 0.432$, $\underline{p} = 0.005$) and number of talkers ($\underline{U} = 0.027$, $\underline{p} = 0.000$) There were no other significant main effects or interactions. Neither driving speed, subject type (soldier or civilian), nor any interaction had any appreciable effect on the subject's ability to process the radio communications.

The data indicate that subjects were able to respond significantly more accurately to radio communications with virtual spatial audio displays (a mean of 18.87 out of 36 total radio communications in each session correctly identified, s.d. = 4.51) than with monaural displays (a mean of 13.29 out of 36 radio communications correctly identified, s.d. = 1.83). The results also showed that as the number of simultaneous talkers increased, the number of accurately identified target messages decreased. As was previously described, the MANOVA for the proportion of phrases incorrectly spoken by the subjects indicated that there were no statistically significant effects for the main effect of auditory display type. Auditory display type had no significant effect on subject speech input.

4. CONCLUSIONS

The first objective of this study was to determine the extent to which high levels of ambient noise affect the performance of a noise-robust ASR system, and how well the ASR system recognizes voice commands by users who experience differing levels of workload. The results indicated that a high mean proportion (95.5%) of all radio communication commands were correctly recognized by the ASR system. This performance was obtained in high level (88 dBA) ambient tank noise. Differing levels of task workload (driving speed) had no effect on recognition performance. These results show that a noise-robust ASR system could be useful in real-life applications in tank and helicopter systems in which soldiers must perform complex concurrent tasks in the presence of high levels of environmental noise. In general, ASR has potential as a technology for "hands-off" command and control tasks such as changing radio frequencies.

The second objective was to ascertain the extent to which user speech is affected when the subject uses an ASR system in a high noise tank environment. The ASR performance data indicated that commands correctly spoken by the subjects were not significantly affected by high levels of ambient noise or different levels of workload, or by concurrent driving tasks. An analysis of user input errors, commands incorrectly spoken by the subjects, showed that the largest proportion, almost half of the user errors, were run-in phrases, in which subjects uttered two subsequent phrases without a pause. This type of error may have been a function of the subject's trying to input more than one verbal command at one time. Subject training and further practice with an ASR system may reduce or eliminate this type of error, as well as the other user input errors described here. However, care also has to be given to the design of the ASR system grammar, to make the words or phrases easy for the user to remember in times of high cognitive demand. In reference to user input, ASR has potential as a technology for "hands-off" command and control tasks in demanding environments.

The third objective was to determine the effect of virtual spatial audio displays on soldier performance in processing multiple radio communications. The results showed that virtual spatial audio radio messages enhanced the ability of the user to understand speech communications from multiple channels. Users were able to identify multiple channel radio communications more accurately with virtual spatial audio than with monaural displays. These results were also found in previous research involving helicopter pilots (Haas, 1998). The virtual spatial audio display was more effective because it provided perceptual cues necessary for the soldier to localize and selectively attend to the messages, even in the presence of concurrent tasks and tank noise.

The fourth objective was to determine whether the type of incoming auditory information affected user speech input into the ASR system. The data indicated that different types of auditory display had no significant effect on subject speech input into the ASR. This indicated that auditory input (monaural or virtual spatial) created no significant cognitive interference with subject input into the ASR. Thus, ASR and virtual spatial auditory displays integrated successfully, separately and together, in a demanding environment with complex, concurrent tasks.

In general, ASR and virtual spatial audio displays have potential as technologies of interest in high noise tank and helicopter environments, such as the VCOP and CAT systems, in which complex tasks are performed.

5. ACKNOWLEDGMENTS

Mr. Scott Dennis, U.S. Army Aircrew Integrated Systems Program Office at Redstone Arsenal in Huntsville, Alabama, provided valuable assistance in writing this manuscript.

6. REFERENCES

Haas, E.C. (1998). Can 3-D auditory warnings enhance helicopter cockpit safety? Proceedings of the Human Factors and Ergonomics Society 41st Annual Meeting, 1117-1121. Santa Monica, CA.

McKinley, R., Erickson, M., and D'Angelo, W. (1995). 3-Dimensional auditory displays: Development, applications and performance. Aviation, Space , and Environmental Medicine, 5, A31-A38.

Sherman, R. (2001). The U.S. Army's Virtual Cockpit. Avionics, March.

U.S. Army. (1991). Hearing Conservation. U.S. Army Pamphlet 40-501.

Effects of vection on the sense of presence in a virtual environment

Martin Olsson[1,3], Kim Vien[1,3], Edith Ng[2], Richard So[2] and Håkan Alm[1]

[1]Division of Industrial Ergonomics, Department of Mechanical Engineering, Linkopings University, Sweden.
[2]Department of Industrial Engineering and Engineering Management, Hong Kong University of Science and Technology, Clear Water Bay, Kowloon, Hong Kong SAR, PRC.
[3]Exchanged students at Hong Kong University of Science and Technology

ABSTRACT

This paper reports the initial data of an investigation on the effects of linear vection on the sense of presence in a Virtual Environment (VE). Twenty-four participants took part in the experiment and the sense of presence was measured by the Presence Questionnaire (Witmer and Singer, 1998) as well as a questionnaire modified from (Slater and Usoh, 1993 and Slater, 1999). Results showed that the lack of vection in a virtual environment do not significantly reduce the rated levels of presence. Significant correlations were found between the results of the modified Slater's questionnaire and some cluster scores of the Presence Questionnaire. Initial results of this experiment are discussed in the context of a VE for skill training.

1.0 INTRODUCTION

Many studies have been conducted to investigate connections between sense of presence, simulator sickness and performance. Among others, there are strong notations that having sense of presence is a necessary condition for better performance. Reversibly, engaging tasks may lead to the fact that users pay more attention to the Virtual Environment (VE) and therefore experience greater sense of presence (Bystrom, Barfield & Hendrix, 1999; Welch, 1999). Studies are also suggesting that persons who experience simulator sickness are distracted from the VE and therefore have less sense of presence (Witmer & Singer, 1996; Nichols, Haldane & Wilson, 1999). Hettinger and Riccio (1992) and McCauley and Sharkey (1992) have attributed the cause of simulator sickness in a VE to the occurrence of vection. According to Hettinger et al. (1990), the term vection was used by Tschermak (1931) to refer to the illusory sensation of self-motion induced in viewing optical flow patterns. Viewing visual representations of motion in any of the linear or rotational axes of the body can induce vection. Hettinger et al. (1990) reported that in a study of simulator sickness, 8 of 10 participants who experienced vection exhibited some form of simulator sickness. Because of the close relationship between the occurrence of vection and simulator sickness, a logical solution to reduce the level of sickness would be to eliminate vection in a Virtual Reality (VR) simulation. However, will the removal of vection also reduce the sense of presence? A review of literature indicates that the relationship between the occurrence of linear vection in a VE and the sense of presence is not known. It is, therefore, the aim of this study to investigate the level of sense of presence in a VE with and without vection.

Witmer and Singer (1998) defined the sense of presence as "the subjective experience of being in one place or environment, even when one is physically situated in another". They also developed a Presence Questionnaire (PQ) to measure the sense of presence in a VE. This PQ consists of 32 questions measuring the qualities of different sensory interfaces (e.g., auditory, haptic, and visual interfaces) as well as the ease of manipulating objects inside the VE. The PQ had 7 cluster scores (involvement / Control, Natural, Interface Quality, Auditory, Haptic, and Resolution) and a total score. In 1999, Slater argued that the PQ was only measuring the qualities of immerson and not the sense of presence. Slater defined the sense of presence as "the belief that they (participants) are in a world other than where their real body are located" (Slater & Usoh, 1993). A comparison between Slater's definition of sense of presence and Witmer and Singer's definition indicates a difference in the use of 'belief' rather than 'subject experience' to describe the sense of presence. Slater proposed to measure the sense of presence by asking participants some open questions. The answers to the open questions would then be analyzed according to the Neuro-Linguistic Programming (NLP) method (www.nlpinfo.com, 2001). In this study, both the PQ proposed by Witmer and Singer (1998) and a short questionnaire derived from Slater and Usoh (1993) and Slater (1999) were used. The short questionnaire consists of five multiple-choice questions and two open questions. This short questionnaire is referred to as the modified Slater's questionnaire in the rest of this paper. This study had two hypotheses:

1. The addition of navigation tours (i.e., vection) will increase the sense of presence because the tours will enable the participants to experience more of the VE.
2. There will be significant correlation relationships among certain PQ cluster scores and the results of the modified Slater's questionnaire.

2.0 METHOD AND DESIGN

Twelve male and twelve female volunteers participated in the study. They were students at the Hong Kong University of Science and Technology and aged between 20 to 24. The experiment had two conditions: (i) VR simulation with linear vection and (ii) VR simulation without linear vection. It is a between-subject design so that each participant will take part in only one condition. Six male and six female participants were randomly assigned to each condition. A few days before the experiment, all participants completed a Motion Sickness Susceptibility Survey (MSSS) questionnaire previously used in a survey of over 500 university students (So *et al.*, 1999). Just before the VR simulation, they were given an Immersive Tendencies Questionnaire (ITQ) developed by Witmer and Singer (1998) and a pre-exposure Simulator Sickness Questionnaire (SSQ, Kennedy *et al.*, 1993) to complete. Then, the participants were instructed to wear a VR4 head-mounted display (HMD). Through the display, they would see a large virtual acoustic room (4.8mx11.7m) with a table in front of them (Figure 1). On the table, there were two speakers, a LCD monitor, some push buttons, some cubes and some cylinders. All of these objects were, of course, 3D images projected through the HMD. During the first two minutes, participants were guided to move their heads to survey the VE systematical. This was to ensure that the participants would be familiar with the use of HMD and the VE. As they moved their heads, the field-of-view changed accordingly. After the first 2 minutes, participants in both conditions were instructed to perform a series of sound localization, visual search, and object manipulation tasks. Instructions were mainly presented on the virtual LCD monitor. At 5, 12, 19, and

26 minutes after the start of the simulation, participants in the condition with vection were given a virtual navigation tour in the fore-and-aft and lateral directions around the room. The speed of travel was 1m/s and the duration of the tour was about 2 minutes. The navigation tour was combined with some visual search tasks. In the condition without vection, participants were just given a visual search task so that they would make similar head movements as those participating the condition with the navigation tours. At the end of each tour or the visual search task, measurements on vection and nausea ratings were taken using a 4-point vection scale and a 7-point nausea scale used in previous studies (e.g., Lo and So, 2001, So and Lo *et al.*, 2001). At the end of the 30-minute simulation, participants were asked to complete a PQ,

Figure 1 A snap shot of the Virtual Environment showing the virtual acoustic room, the virtual LCD monitor and some objects on the virtual table.

the modified Slater questionnaire on presence, and the post-exposure SSQ. The order of presenting the PQ and the modified Slater questionnaire was randomized. The VR simulation was generated by WorldToolKit release 8 running on a SGI OnyxII station. Head and hand positions and orientations were measured by a Polhemus 3SPACE™ tracker system and a CyberGlove™ system.

English was used throughout the whole experiment. It means that all instructions, explanations, and questionnaires were given in English. For some questionnaires, however, Chinese translations were listed side-by-side the English original.

3.0 RESULTS AND DISCUSSION

In this paper, the initial data on the sense of presence of the first 16 participants are partially reported (i.e., 8 participants per condition). The full results can be found in So, Olsson and Vien et al. (2001). The results of the ITQ followed the normal distribution and the results of AONVA indicated that there was no significant difference between the total ITQ scores collected from the two groups of participants ($p>0.2$). This suggests that the two groups of participants did not have significant difference in their rated tendencies to be subjectively involved with an imaginary environment (Witmer and Singer, 1998). The medians of the six PQ cluster scores and the total PQ scores are shown in Table 1. Inspection of the table shows that the various PQ score are similar between the two conditions. Results of Kruskal-Wallis non-parametric test showed no significant difference in all the PQ scores taken between the two conditions (Involvement / Control: $p>0.5$, Natiral: $p>0.15$, Interface Quality: $p>0.8$, Auditory: $p>0.3$, Haptic: $p>0.7$, Resolution: $p>0.8$, and Total: $p>0.4$). The average scores of the five multiple-choice questions of the modified Slater's presence questionnaire were analyzed and no significant difference was found between the scores collected during the two conditions ($p>0.6$). The lack of significant difference between the level of presence collected at the two conditions indicates that the sense of presence was not affected by the removal of vection. Although the result disagree with the hypothesis, it is a useful finding because literature predicts that sickness can be reduced by eliminating the occurrence of vection (Hettinger et al., 1990) and the finding of this study indicates that one can safely do so without affecting the sense of presence.

Table 1   Median PQ cluster scores after a 30-minute VR simulation with and without vection (median data of 8 participants).

Condition	Median						
	Involvement/ Control	Natural	Interface Quality	Auditory	Haptic	Resolution	Total
With vection	50.0	9.0	12.0	14.5	8.0	11.0	97.0
No vection	57.5	12.5	14.0	17.5	9.5	11.0	115.0

A correlation test was conducted between the average scores of the modified Slater's presence questionnaire and the various PQ cluster scores. Since the data collected from the two conditions were not significantly different, they were combined in this correlation test. Results indicated that the average scores of the modified Slater's presence questionnaire was significantly correlated with the Involvement / Control, Natural, and Haptic cluster scores as well as the total PQ scores ($p<0.05$). No significant correlation relationship was found between the average scores of the modified Slater's presence questionnaire and the Interface Quality, Auditory, and Resolution PQ cluster scores, respectively ($p>0.2$).

The significant and lack of significant correlation relationships may reflect the similarity and difference between the PQ and the modified Slater's questionnaire on presence. Further analyses are being conducted and the full results will be reported in So, Olsson and Vien *et al.* (2001).

4.0 CONCLUSION

The removal of virtual navigation tours from a 30-minute Virtual Reality (VR) simulation did not result in significant changes in the rated sense of presence as measured by the Presence Question (PQ, Witmer and Singer, 1998) and a presence questionnaire modified from Slater and Usoh (1993). Since it is expected that the addition of virtual navigation tour will facilitate the occurrence of vection which is associated with symptoms of sickness, the current finding suggests that one could reduce the level of sickness without affecting the level of presence by removing navigation tours from a VR simulation.

Significant correlation relationships were found among the average scores of the modified Slater's questionnaire on presence and the Involvement / Control, Natural, Haptic, and Total cluster scores of the PQ. Further analyses of the complete data set are continuing and the results of the full data set will be published in So, Olsson and Vien *et al.* (2001).

5.0 ACKNOWLEDGEMENT

This study is partly supported by the Earmarked Competitive Grant HKUST6007/98E from the Hong Kong Research Grant Council as well as the Hong Kong / Sweden exchange programme.

6.0 REFERENCES

Bystrom, K.H., Barfield, W. & Hendrix, C. (1999). A Conceptual Model of the Sense of Presence in Virtual Environments. *Presence*, vol. 8, No. 2, 241-244.

Hettinger, L.J., Berbaum, K.S., Kennedy, R.S., Dunlap, W.P. & Nolan, M.D. (1990). Vection and Simulator Sickness. *Military Psychology,* 2(3), 171-181.

Hettinger, L.J. & Riccio, G.E. (1992). Visually Induced Motion Sickness in Virtual Environments. *Presence,* 1(3), 306-310.

Kennedy, R.S., Lane, N.E., Berbaum, K.S. & Lilienthal, M.G. (1993). Simulator Sickness Questionnaire: An Enhanced Method for Quantifying Simulator Sickness. *The International Journal of Aviation Psychology,* 3(3), 203-220.

Lo, W.T. & So, R.H.Y. (2001). Cybersickness in the presence of scene rotational movements along different axes. *Applied Ergonomics,* Vol. 32, No. 1, 1-14.

McCauley, M.E. & Sharkey, T.J. (1992). Cybersickness: Perception of Self-Motion in Virtual Environments. *Presence,* 1(3), 311- 318.

Neuro-Linguistic Programming (2001) Introduction to Neuro-Linguistic Programming. *www.nlpinfo.com.*

Nichols S., Haldane C. & Wilson J.R. (1999). Measurement of presence and its consequences in virtual environments. *International Journal, Human-Computer Studies* (2000) 52, 4 71-491

Slater, M. (1999). Measuring Presence: A Response to the Witmer and Singer Presence Questionnaire. *Presence,* Vol. 8, No. 5, 560-565.

Slater, M. & Usoh, M. (1993). Representations Systems, Perceptual Position, and Presence in Immersive Virtual Environments. *Presence*, Vol. 2, No. 3, 221-233.

So, R.H.Y., Finney, C.M. and Goonetilleke, R.S. (1999) "Motion sickness susceptibility and occurrence in Hong Kong Chinese." Contemporary Ergonomics 1999, Taylor & Francis.

So, R.H.Y., Lo, W.T. & Ho, A.T.K. (2001). Effects of navigation speed on the level of cybersickness caused by an immersive virtual environment. *Human Factors* (in press).

So, R.H.Y., Olsson M., Vien K., Ng E. and Alm H. (2001) Effects of linear vection on rated levels of sickness and presence in a virtual environment (awaiting publication).

Welch, R.B. (1999). How can we determine if the sense of presence affects task performance? *Presence*, Vol. 8, No. 5, 574-577.

Witmer, B.G. & Singer, M.J. (1996). Presence Measures for Virtual Environments: Background and Development. *Draft Ari research note, US Army Research Institute.*

Witmer, B.G. & Singer, M.J. (1998). Measuring Presence in Virtual Environments: A Presence Questionnaire. *Presence,* Vol. 7, No. 3, 225-240.

COMPARING PERCEPTION OF SAFETY IN VIRTUAL AND REAL ENVIRONMENTS

Fang Wu and Vincent G. Duffy
The Hong Kong University of Science and Technology
Department of Industrial Engineering and Engineering Management
Clear Water Bay, Kowloon, Hong Kong SAR, China
Phone: 852-2358-7116, Fax: 852-2358-0062, Email: vduffy@ust.hk

ABSTRACT

Industrial environments can influence perception of safety and thus conditions in the industry can influence workers' safety and manufacturing effectiveness. Though the use of virtual reality may provide some benefits in improving safety and reducing risk in the physical work environment, research on safety perception in a virtual environment has just begun, and the transferability of perception from virtual to real has not yet been well established. This study focuses on comparing perception of safety in a virtual environment with those in the real world. Sequential experimentation techniques are used to build an integrated model of system variables in both environments. To compare the safety perception in both virtual and real environments, two experiments were conducted. The first experiment examined the effects of sound and light on perception of safety in a virtual environment. The second experiment assessed the effects of sound and light on perception of safety in the real. Preliminary results indicate that light and sound have a significant influence the risk perception in real and virtual environments and do not show interactions. One important finding in this study is that light is a transferable factor that does not interact with other variables tested, and light has previously been shown to be an important factor in influencing safety perception and errors in real industrial environments.

1. INTRODUCTION

Identification of a potential hazard is considered an important first stage in safety system implementation (Zimolong, 1997). In this study, a desktop virtual industrial environment can be viewed on the Internet. This desktop platform does not use the head mount display that is commonly used in many virtual environments. Twenty-four subjects were tested using a desktop virtual reality system to determine the influence of factors such as sound, light, distance between machines, presence of a virtual human, and gender on perception of risk which is comprised of perception of hazard, and likelihood of accident. Also, twelve subjects are tested in some of the same virtual conditions as well as in the real environment. Using a technique for combining the data from sequential experiments (Snow and Williges, 1998), a comparison is made for perception of risk between the real and virtual environments.

2. LITERATURE AND HYPOTHESES

Hazard and risk: For the purposes of this study, hazard is considered a set of conditions that can lead to injury or property damage (Laughery and Wogalter, 1997). Risk perception concerns the overall awareness and knowledge regarding the hazards, likelihood and potential outcomes of a situation or a set of circumstances (Laughery and Wogalter, 1997; Zimolong, 1997). Hence, risk perception should be created by some composite measure including measured hazard and likelihood. As a test of face validity, perception of acceptability should show an inverse relationship with hazard.

Hypothesis 1: Acceptability will show an inverse relationship with hazard, while hazard and likelihood will have a high positive relationship that will enable those measures to be combined to create risk.

Light: Illumination can influence industrial performance (Tinker, 1963). Moreover, higher levels of illumination may reduce the demands placed on workers' information processing systems and hence increase spare mental capacity (Boyce, 1973). It is believed that the VR light 'off' condition in a virtual environment will create a difference in perception with regard to the potential hazard to the operator. The VR light 'off' condition does not imply that the screen is completely dark. However the luminance ratios for objects in the foreground and background are changed such that the VR light 'off' appears as a 'poor lighting' condition in the virtual environment (Duffy and Chan, 2000). This condition can occur in the virtual world simply by leaving the default settings, creating a virtual environment in the absence ·of additional virtual light nodes. These light 'nodes' can be placed in the virtual environment at specific locations and intensity. For the remainder of this paper the absence of additional virtual light nodes in the virtual environment will be considered the 'poor lighting' condition in the virtual environment. The environment with additional light nodes will be considered 'good lighting'.

Hypothesis 2: There will be a significant difference in perception of risk for the operator based on the lighting level in the real and virtual environments.

Sound: It has been shown that sound can provide feedback about users actions (Stanney, Salvendy, et al., 1998) and can provide information and understanding about simulation of an industrial environment (Gaver, 1997). However, prolonged noise exposure can be considered a work hazard and limits have been set for permissible noise exposure in the workplace (OSHA, 1983).

Hypothesis 3: There will be a significant difference in perception of risk for the operator based on the sound level in the real and virtual environments.

3. METHOD

The overall purpose of the two experiments was to determine the appropriate methods and the effect of varying virtual industrial parameters on perception of risk in order to determine the transfer of experience from a virtual training condition to a real workplace. A total of 36 volunteer undergraduate students at the Hong Kong University of Science and Technology (excluding engineering students) were recruited after posting the advertisement on the electronic notice board. It took approximately 1 ½ hours for each subject to finish the experiment in both experiments. Each subject was told that the objectives of the experiment included determining the perception of hazard and acceptability. Subjects were also asked to estimate the distances traversed during the walkthrough and distance between machines. For additional details of the computer system requirements and development of software used for testing see Duffy and Chan (2000) or Duffy, Wu and Ng (2000).

3.1 Experiment 1 in a virtual industrial environment

The virtual experiment combined sound, light, human and distance in a 2 (light) X 2 (sound) X 2 (human) X 4 (distance between machines) design. Human is a between-subjects variable. Each subject participated in 16 trials. The two levels of sound were 'no lathe sound during cutting', 35.4 dBA, and 'lathe sound during cutting, 80.4 dBA). The two levels of light were 'good' where the virtual wall measured 51.55 cd/m^2, and 'poor' where the virtual wall measured 5.34 cd/m^2. With and without the virtual human operator constituted the two levels of human. Sound levels in the virtual environment were set and measured prior to the experiment using a Quest 1800 precision impulse integrating sound level meter. In the 'no sound' condition, Bilsom protective sound reduction earmuffs were used. Half of the subjects were shown VR with virtual human operators present in all conditions. The other half were shown VR trials without any virtual human operators present. In each experiment, half of the 12 subjects were tested first in the 'with sound' condition, while the other half were tested in the 'without sound' condition first.

660

A total of 24 subjects participated in the virtual testing only. Throughout these two experiments female and male subjects were counterbalanced.

Participants were shown how to begin a 'walkthrough' in the virtual environment on a predetermined path. Upon completion of the 'walkthrough', two different viewpoints were shown to the participants by the experimenter. One of the two experimental viewpoints shown during virtual testing was 'angle' view. The other was the close up 'straight' view in front of the virtual lathe (and behind the virtual operator in test conditions where the operator was present). Perception data for each of the two viewpoints is later tested to see if it can be combined to create a reliable measure. This is done separately for acceptability, hazard and percent chance of operator error. Each subject was shown series of 16 virtual factory scenarios with different conditions in random order. The subjects were asked to judge the perception of acceptability, hazard and percent chance an operator error from the two viewpoints according to the method described above in the 'subjective rating form' section.

3.2 Experiment 2 in a real industrial environment

The real experiment combined sound and light in a 2 (sound) X 4 (light), within subjects design. There were 2 levels of sound (with ear muff, without ear muff), and 4 levels of light. A total of 12 subjects (6 male and 6 female) participated in the 8 trials in the real environment and 4 selected from the virtual environment. The trials in the VE tested some conditions that were previously tested in experiment 1 (see Snow and Williges, 1998) re: data bridging) in a 2 (light) x 2 (distance between machines) within subjects design. Half participated in real trials first, the other half participated in virtual trials first. The virtual data was used later in an attempt to combine the data of first and second experiments for subsequent comparisons.

During the real testing, a DYNA MYTE 3000 precision CNC Lathe was used to conduct the experiments. According to the experimenter's signals, the technician would set the machine to one of the conditions that had been randomized beforehand. Then subjects were asked to observe the whole machining procedure. Subjects were asked to fill out the subjective rating forms for the three main variables after the machine was turned on before cutting began, and then again after cutting was completed.

4. RESULTS

4.1 Modulus Equalization and correlation

A modified procedure to reduce variability in the fixed-modulus magnitude estimation data arising purely from participants' selection of different moduli and range, called Modulus Equalization, was previously presented by Snow and Williges (1998).

4.2 Experiment 1 in a virtual environment:

Results of hypothesis 1 from a correlation analysis:
- *Results of H1:* The measured variables acceptability and hazard are inversely correlated ($r=-0.42$, $p=0.0001$).

Results of hypotheses 2-3 from a repeated measures ANOVA analysis
- *Results of H2 and H3:* There is a significant difference in perception of risk for the operator based on lighting level, but not sound for the levels used in this experiment.

4.3 Experiment 2 in a real industrial environment:

Two data sets were obtained under the same environmental condition before and after the machine cut (but after the machine was turned on).

Results of hypothesis 1 from a correlation analysis
- *Results of H1:* The measured variables acceptability and hazard are inversely correlated.

Results of hypotheses 2-3 from a repeated measures ANOVA analysis
- *Results of H2 and H3:* There is a significant difference in perception of risk for the operator based on lighting level, and sound for the levels used in this experiment.

5. DISCUSSION

Initially, data were analyzed separately for each of the two experiments. Results of the perception of hazard in virtual and real were later combined into an integrated data set based on the results of the following t-test analyses following the method of Snow and Williges (1998).

Light: The Minolta Chroma meter (CS-100) is commonly used to capture luminance measures from a computer screen. However, in this case, the method of Charness and Dijkstra (1999) is used for capturing the luminance measures in the real environment. In order to compare perception in the real and the virtual environments, measures of light emitted from the wall (luminance) are taken in both the virtual and the real environment using the Minolta Chroma meter (CS-100). In the real environment, the light emitted from the wall is a function of the lighting level in the room set at the four light levels in lux. Preliminary results indicate that risk perception in the VE is greater than in the real.

Sound: Why might sound produce a significant difference in perceived risk in the RE but not in the VE? After comparison of the real and computer-simulated sounds there appears to be difference in the magnitude of sound intensity at some frequencies. Also, it should be noted that the an 80 dBA sound level is permissible for 32 hours by OSHA standards (Sanders and McCormick, 1993). Hence, the subjects estimating the hazard for the virtual operator may not perceive the difference with and without protective earmuffs as being significantly different for good reason. However, an increase to 85 dBA would cut the permissible exposure in half (Sanders and McCormick, 1993). Again, preliminary results indicate that overall risk perception in the VE is greater than in the real.

Comparing virtual and real: It is possible that the subject simply considers the virtual environments shown to be more hazardous than the real. One obvious difference between the two environments is the lack of machine guard in the virtual industrial environment, whereas the real machine has a machine guard covering the chips coming off of the lathe. Though there are differences between the two environments, this is highlighted for discussion purposes. The main purpose of this research is to suggest methods for study of risk perception in virtual environments, and some similarities and differences in the trends and measuring techniques for common parameters in the two environments.

6. CONCLUSIONS

Prior literature has discussed perception of hazard in real workplace. By comparing human perception in a virtual environment with that in the real world, one can determine some similarities and differences for the purpose of understanding the likelihood of transfer of training in the virtual environment. In order to compare the risk perception in both virtual and real environments, two experiments were conducted using the method of sequential experiments. Furthermore, when the results of the two experiments are combined, it is shown that subjects consistently had a higher perception of risk in the virtual environments tested in these experiments than in the RE. Details of the results and a discussion of other relevant variables will be presented at the conference.

7. FUTURE WORK: INDIVIDUAL DIFFERENCES AND PSYCHOPHYSICAL APPROACH

Some future work will be focused on determining the influence of individual differences on perception of hazard and risk. Also, it is expected that perception of hazard under varying light conditions in the virtual environment shows similar characteristics to that in the real, and that there could be Stevens' curves established in the virtual environment. Stevens' (1975), as early as the 1950s, has shown that physical intensity can be translated into psychophysical perception and associated psychophysical scales.

ACKNOWLEDGEMENTS

This study is sponsored, in part, by the Competitive Earmarked Research Grants (CERG) from the Research Grants Council (RGC) of Hong Kong HKUST/CERG 6168/98E and HKUST/CERG 6211/99E.

REFERENCES

Boyce, P. R. (1973). Age, illuminance, visual performance and preference. *Lighting Research and Technology*, 5(3), 125-144.

Charness, N. & Dijkstra, K. (1999). Age, luminance and print legibility in homes, offices, and public places. *Human Factors.* 41, 173-193.

Duffy, V.G. & Chan, H.S. (2000). Effects of virtual lighting on eye fatigue, working paper, *Human Factors and Ergonomics in Manufacturing*, accepted.

Duffy, V.G., Wu, F. & Ng. P.P.W. (2000). Development of an Internet virtual layout system for improving workplace safety, submitted to *Computers in Industry*.

Gaver, W. (1997). Auditory interfaces in M. Helander, T.K. Landauer, and P. Prabhu (Eds.*), Handbook of Human-Computer Interaction*, 2nd completely revised edition, (pp. 1003-1039), Elsevier Science: B.V.

Laughery, K.R. & Wogalter, M.S. (1997). Warnings and risk perception, in G. Salvendy (Ed.), *Handbook of Human Factors and Ergonomics*, 2nd Ed. (pp. 1174-1197), New York: John Wiley & Sons.

Occupational Safety and Health Administration (1983). Occupational noise exposure: Hearing conservation amendment. *Federal Register*, 48, 9738-9783.

Sanders, M.S. & McCormick, E.J. (1993). *Human Factors in Engineering and Design*, 7th edition, McGraw-Hill: New York.

Snow, M. P. & Williges, R. C. (1998). Empirical models based on free-modulus magnitude estimation of perceived presence in virtual environments. *Human Factors.* 40(3), 386-402.

Stanney, K., Salvendy, G., et al. (1998). Aftereffects and sense of presence in virtual environments: formulation of a research and development agenda, *International Journal of Human-Computer Interaction*, 10(2), 135-187.

Stevens, S. S. (1975). *Psychophysics: Introduction to its Perceptual, Neural, and Social Prospects*. New York: Wiley.

Tinker, M. A. (1963). *Legibility of print*. Ames: Iowa State University Press.

Zimolong, B. (1997). Occupational risk management, In G. Salvendy, Ed., *Handbook of Human Factors and Ergonomics*, 2nd Ed., New York: John Wiley & Sons, pp. 989-1020.

IMPACT OF DYNAMIC VIRTUAL AND REAL OBJECTS ON PERCEPTION OF HAZARD AND RISK

Parry P.W. Ng and Vincent G. Duffy

The Hong Kong University of Science and Technology
Department of Industrial Engineering and Engineering Management
Clear Water Bay, Kowloon, Hong Kong SAR, China
Tel: (852) 2358-8237, Fax: (852) 2358-0062, E-mail: parry@ust.hk or vduffy@ust.hk

ABSTRACT

This research is aimed at investigating the effect of environmental factors such as speed of a dynamic virtual object, angle of approach, and distance between subject and virtual dynamic object and light intensity on perception of hazard in the workplace. A series of experiments considering a psychophysical approach and Stevens' Law is conducted. Analysis includes techniques such as *sequential experiments* to bridge data across experiments in virtual and compare to a real industrial environment. The comparison of the analyzed data in virtual and real environments helps to further determine the transferability of performance and perception from virtual reality to reality. Using the preliminary results from the integrated data in the sequential experiments, potential guidelines for using of virtual facility layout in industry will be discussed.

1. INTRODUCTION AND LITERATURE REVIEW

In order to increase productivity and reduce labor cost, most manufacturing companies have adopted the automated systems (for instance, automated guided vehicle (AGV), and robots) for performing material handling and transportation tasks in the production process. Due to the increased application of automated systems in manufacturing industries, safety issues related to speed of the moving AGV, vehicle-to-operator distance, perception of hazard and impact on decision-making must be considered during the design stage of industrial facility layouts. Industrial staff can dive or immerse themselves in the virtual environment and visualize the potential safety problems (Su et. al. 1997). However, first we must understand more about how perception of the virtual relates to reality, and how perception relates to decisions made in an industrial workplace.

With the rapid development of advanced computer technology, the use of Virtual Reality (VR) may provide some benefits in improving safety and reducing risk in the physical work environment (Duffy, et. al. 1997). The actual physical representation of real-world layout for certain tasks was indistinguishable from that of the perception and spatial knowledge acquisition in the virtual world (Arthur, et. al. 1997). Virtual reality can be used in conjunction with simulation to allow the user interact with the environment and have a sense of presence in a dynamic virtual environment (Hollands and Mort, 1995). However, it is not clear to what extent a psychophysical relationship would hold in an environment with moving objects.

2. OBJECTIVES

The objective of this research is to utilize virtual reality techniques to simulate the pre-defined path of the AGV in a virtual environment, and examine how environmental factors influence the perception of hazard and risk in both virtual and real environment. This study is also designed to model an empirical formula for Stevens' psychophysical power function for perception of acceptability and hazard within dynamic virtual work environment. To integrate VR technology in the workplace effectively, it is important to understand the similarities and differences in perception of hazard and risk between virtual and real environments (Duffy and Salvendy, 2000). Therefore, the comparison of analytical data between virtual and real environments can be used to show what extent the virtual experience is transferable to reality.

3. DESCRIPTION OF HYPOTHESIS

According to Stevens' psychophysical power law (Stevens, S.S. 1975), its formula can be written in general form as follows:

$$\psi = k\phi^\beta$$

where ψ = Sensation magnitude
ϕ = Stimulus magnitude
k = Constant
β = Exponent of power function

In this study, as the speed of moving object is held constant, the human perception of acceptability and hazard (i.e. sensation magnitudes) would be expected to follow the relationship based on Stevens' psychophysical power function as distance to a dynamic object (i.e. stimulus magnitude) is decreased in virtual environment. In addition, for a dynamic virtual object with higher speed, the exponent of power function would be expected to increase as it did for comparing the load heaviness between black and white boxes to model the power function (Karwowski, 1996). Figure 1 shows the hypothesized Stevens' psychophysical power relationship in dynamic virtual environment.

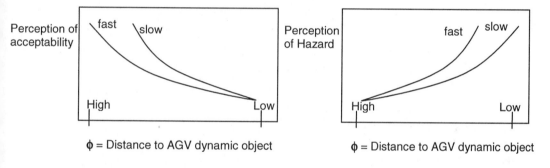

Fig 1. Hypothesized Stevens' psychophysical power relationship in dynamic virtual environment

4. SYSTEM DEVELOPMENT

An Internet-based virtual environment with moving AGV has been developed for testing the above hypotheses. The virtual dynamic AGV object was created according to the real AGV (DAIFUKU FW-10 Forward and backward type) in the manufacturing laboratory at the Hong Kong University of Science and Technology. Its physical dimension of AGV is 1.34m (L) x 0.64m (W) x 0.70m (H). The animation of dynamic AGV moving object can be programmed and simulated by Virtual reality modeling language (VRML). VRML uses the command TimeSensor to define the cycle interval time in seconds to control the time taken for AGV to travel in the virtual environment. By using the PositionInterpolator to define the start and end point of AGV path, the distance traveled of AGV can be predetermined in the program. In the VRML environment, AGV can be controlled (start and stop) through the program (panel of buttons) using an Internet browser with Cosmo player plug-in.

5. METHOD AND PROCEDURE

A series of experiments was conducted to investigate the influence of speed, distance between subject and AGV and angle of approach on human perception of hazard in both virtual and real environments. A comparison is done to help determine the transferability of perception from the virtual environment to reality.

5.1 Method of testing

Forty non-engineering students (20 males and 20 females) were recruited from The Hong Kong University of Science and Technology to participate in the experiment. They are assumed to have no experience to operate the AGV. All subjects were tested in the virtual experiment by using a desktop Pentium III 600 MHz computer and Sony 17" VDT monitor. The whole experiment took approximately an hour. Before performing the experiment, all subjects were required to give their personal information by filling in the informed consent form. The experiment was divided into two sessions: (1) Pretest session and (2) experimental session. In the pretest session, the subjects were instructed to estimate the traversed and interobject distances in real environment.

In the experimental session, the subjects were randomly assigned into four groups based on the subject assignment sheet. Detailed information of four condition groups are shown in Table 1. In each group ten subjects (5 Male and 5 female) participated. The experimental session can be divided into two parts: one is virtual experiment and the other is real experiment. Half of the subjects performed the virtual part of the experiment first. Between real and virtual part, a five minute break was given to the subject. In addition, the virtual experiment was classified into two types: (a) VR-experiment 1 and (b) VR-experiment 2. List of independent variables for virtual and real experiment are tabulated in Table 2.

Condition 1 (5 Male and 5 female) 4 trials in Real experiment first 12 trials in VR-experiment 1	Condition 2 (5 Male and 5 female) 4 trials in Real experiment first 20 trials in VR-experiment 2
Condition 3 (5 Male and 5 female) 12 trials in VR-experiment 1 4 trials in Real experiment	Condition 4 (5 Male and 5 female) (10 subject) 20 trials in VR-experiment 2 4 trials in Real experiment

Table 1. Segment of different experimental conditions for subjects

Experiments (No. of trials)	Independent variables
VR-experiment 1 (12 trials)	• Angle of Approach (30 deg, 90 deg, 120 deg) • Distance to AGV (1m, 2m) • Speed of AGV (0.75 m/s)
VR-experiment 2 (20 trials)	• Speed of AGV (3 m/s, 1.5 m/s 0.75 m/s, 0.5 m/s, 0.375 m/s) • Distance to AGV (2m, 3m) • Angle of approach (90 deg)
Real experiment (4 trials)	• Distance to AGV (1m, 3m) • Angle of approach (30 deg, 90 deg) • Speed of AGV (0.5 m/s)

Table 2. List of independent variables (for 2 virtual and 1 real experiments)

5.2 Experimental procedure

In the real experiment, subjects were instructed to sit on a chair at a preset position. The height of eye level of all subjects was ensured at the standardized level (1.2 meter above floor) and VR light intensity was set to VR light 'on' (see Wu and Duffy, 2001). Before starting the real experiment, the detailed description and objective of the experiment was told to the subjects. Then, the experimenter started to AGV machine. After viewing the AGV motion, the subjects were asked to answer several questions with regard to distance traveled of AGV, speed of AGV, distance to AGV, and perception of acceptability and hazard. The real experiment was finished until all four trials were completed.

In the virtual experiment, subjects were given a brief introduction of the virtual environment by the experimenter. The subjects were needed to sit in front of the Desktop computer. In the first trial, the subjects had chance to become familiar with dynamic moving environment. The subjects were allowed to look around the virtual environment to get some basic understanding of the environment. Experimenter showed the subjects a specific scenario and instructed the subject to view. Then, the subjects were asked to answer the questions regarding perceived distance and speed. Based on the movement of AGV in virtual environment, subjects were needed to judge the mental workload, chance of getting injured to the virtual operator, and perception of acceptability and hazard using fixed magnitude estimation techniques.

6. RESULTS AND DISCUSSION

Detailed results of the current study will be presented at the conference. After applying the six steps' modulus equalization, the perceived acceptability and hazard from each experiment will be transformed to common modulus of that experiment (Snow and Williges, 1998). It is expected that the relationship between the perception of acceptability and distance to AGV would be modeled by the Stevens' psychophysical power function. In addition, more detailed results of exponent of Stevens' psychophysical power function will be calculated. The comparison of the analyzed data in virtual and real environments will help to further determine the transferability of performance and perception from virtual reality to reality. Using the preliminary results from the integrated data in the sequential experiments, potential guidelines of using virtual facility layout in industry will be discussed.

7. FUTURE WORK

It is hoped that the findings of this experiment can be used as the preliminary results of future work. The impact of virtual and real industrial robot on perception of hazard will be investigated. An industrial robot in Robot CAD/CAM laboratory at the Hong Kong University of Science and Technology will be used for testing, and the preliminary virtual industrial robots have been created.

ACKNOWLEDGEMENT

This study was made possible by prior sponsorship of related research projects by the Competitive Earmarked Research Grant RGC/CERG HKUST 6168/98E from the Hong Kong Research Grant Council.

REFERENCES

Arthur, E., Hancock, P.A. & Chrysler S.T. (1997). The perception of spatial layout in real and virtual worlds, *Ergonomics*, 40 (1) 69-77.

Duffy, V.G. & Salvendy G. (2000). Concurrent engineering and virtual reality for human resource planning, *Computer in Industry*, 42, 109-125.

Hollands, R. and N. Mort (1996). A virtual collaborative simulation environment for integrated product and process development, In: B. Delaney, J. Abel and D. Kanecki (Eds), *Proceedings of the 1996 Simulation Muticonference*, April 8-11, New Orleans, Louisiana, pp.23-27.

Karwowski, W., Jamaldin, B., Gaddie, P. and Lee, W. (1996). Linguistic magnitude estimation: a modeling approach for quantifying human perception of load heaviness in manual lifting tasks, *Proceedings of the 4th Pan Pacific Conference on Occupational Ergonomics*, Taipai, Taiwan, November 11-13, (Ergonomics Society of Taiwan, Taiwan, ROC), pp. 380-383.

Snow, M.P. and Williges, R.C. (1998). Empirical models based on free-modulus magnitude estimation of perceived presence in virtual environments, *Human Factors*, 40(3) 386-402.

Steven S.S. (1975). *Psychophysics : Introduction to its Perceptual, Neural, Social Prospects*, Wiley, New York.

Su, C.J., V.G. Duffy, and M.H. Yip, 1997, Toward the virtual factory: a design planning tool for the future. *Proceedings of the 22nd International Conference on Computer & Industrial Engineering*, December 20-22, Cairo, Egypt, pp. 36-39.

Wu, F. and Duffy, V.G. (2001). Comparing Perception of Safety in Real and Virtual Environments. *HCI International 2001, 9th International Conference on Human-Computer Interaction*, New Orleans, Louisiana, USA, August 5-10, 2001, accepted.

DETERMINING THE IMPACT OF TASK COMPLEXITY ON DECISION-MAKING AND RETENTION IN VIRTUAL INDUSTRIAL TRAINING

Gilbert T.C. Leung and Vincent G. Duffy

Department of Industrial Engineering and Engineering Management
The Hong Kong University of Science and Technology
Clear Water Bay, Kowloon, Hong Kong SAR, China
Tel: (852) 2358-8237, Fax: (852) 2358-0062, E-mail: gilbert@ust.hk or vduffy@ust.hk

ABSTRACT

This paper examines the effect of task complexity and experience level on decision-making performance and retention after virtual industrial training. A virtual industrial training system (VITS) is introduced to investigate the relationship between retention, retention interval, risk perception, task complexity, experience level and the presence of a simulated accident. A similar model is developed for mental workload. Training conditions are varied so that the influence of different variables that contribute visual cues, auditory cues and written instructions can be tested. The cues in the virtual training system correspond to real conditions, and are included to provide some warnings that allow decision-making with lower risk and improved performance, even as task complexity increases. Decisions in different machining conditions in the virtual training system cause different machining outcomes such as simulated tool breakage, poor surface finish or good surface finish. This study considers subjects' decisions during a comparable real industrial task as well. Preliminary results regarding the relationships between learning, retention, and task complexity in the virtual and real conditions will be discussed. A discussion of the influence of hazards in the simulation environment on hazard perception and performance in the real task will also be discussed.

1. INTRODUCTION AND LITERATURE REVIEW

A large number of Hong Kong companies experience high turnover and are reluctant to spend a great deal of money on individual training (Duffy and Salvendy, 2000). In many cases training requires substantial downtime. A VITS that does not require machine downtime could possibly be a valuable replacement for traditional training methods (Helander, 1998). The virtual reality environment is relatively less expensive and more flexible (Baum, 1992). In addition, a study of industrial accidents in computer-integrated manufacturing systems shows that the need for additional research with regard to prevention of industrial accidents (Karwowski, 1992; Jarvinen and Karwowski, 1993). As perception of industrial hazards can influence decisions (Zimolong, 1997), comparing real world estimates of psychophysical coefficients to virtual may help predict the transfer of training in order to reduce risk and hazard in a workplace (Duffy, 2000).

This study was divided into three phases. In phase I, an existing VR-CNC training program was modified to allow the user to select different machine settings in PC platform. It was shown that the initial program contributed to significant transfer of training and retention in the procedural task -mill a slot and drill two holes (Duffy, 1998). The knowledge base and decision rules were modified in the VR-CNC training program so that some combinations could provide a good surface finish, while some combinations could provide poor machining performance and relatively unsafe machining conditions. In this study, task complexity is defined as a task that provides multiple paths to obtain desired outcome or multiple possible outcomes (Jacko, 1997). It has been shown that task complexity affects initial performance level and learning rate (Bohlen and Barany, 1966). It has also been shown that recognition of safe machining conditions is improved when the virtual training scenario included a simulated tool breakage (Duffy, 1999).

In phase II hypotheses are derived for predictive models for retention and mental workload from two different mathematical models that consider the importance of task complexity and experience level (Nembhard, 2000; Xie and Salvendy, 2000). It has been shown that experienced participants learn more quickly than inexperienced participants for some industrial tasks (Nembhard, 2000) and that inexperienced can show higher mental workload (Xie and Salvendy, 2000). In this study, decision-making performance is considered when adding risk perception to the models. Subjects try to determine the safe machining condition for varying machining conditions such as spindle speed, workpiece and tool material, feedrate, depth of cut and lubrication. Based on Jacko (1997), it was believed that subjects would find additional complexity of the task when considering workplace safety as well as machining performance. It was believed that risk reduction could be quantified as a function of task complexity and retention (Duffy, 1998).

In phase III, two experiments were conducted in order to examine the impact of retention interval, risk perception, task complexity, experience level and the presence of a simulated accident in training on decision-making performance, retention and mental workload. Previously, retention interval was shown to influence memory for skill and knowledge (Wisher, 1992). In this study, training conditions were varied so that the influence of different variables that contribute visual cues, auditory cues and written instructions could be tested. Given the fact that the auditory interfaces can provide additional information in computer environments (Gaver, 1997), it important to determine whether training related to safety perception and knowledge transfers in tasks where additional complexity of decision making was required.

2. DEVELOPMENT OF HYPOTHESES

This study tested the influence of different task complexity level in the performance of decision-making and retention capacity. The main hypotheses are as follows:

Hypothesis 1. Increased complexity in the combined condition will cause slower reaction time and worse performance in decision-making. This hypothesis is consistent with the information theory in that increased complexity would create a situation that requires increased information needed for decision making.

Hypothesis 2. Experience level, task complexity and retention interval, risk perception and the presence of a simulated tool breakage will influence decision-making performance, retention and mental workload.

Hypothesis 3. Predictive models can be developed for risk reduction as a function of retention interval, experience level, task complexity, risk perception and presence of a simulated accident.

3. METHOD OF TESTING

Decisions in different machining conditions in the virtual training system cause different machining outcomes such as simulated tool breakage, poor surface finish or good surface finish. The cues in the virtual training system correspond to real conditions, and are included to also provide some warnings that allow decision-making with lower risk and improved safety performance, even as task complexity increases. Good machining performance, measured by considering the actual machining finish that would be experienced and decision time, could be achieved in the training condition under any given condition if the correct decision is made. Safe machining performance is considered any decision that does not lead to tool breakage.

Two experiments were designed to test the proposed hypotheses. Experiment 1 was used to ensure that a group of students would perform at a different level when compared with untrained subjects (Shute, *et al.*, 1998). Fifty undergraduates and graduate engineering students at The Hong Kong University of Science and Technology (HKUST) were recruited initially. It was assumed that they have basic CNC machining knowledge. Subjects were asked to run the VR training system for 10 trials and run the real CNC machine for one trial in the 'good' surface finish 'safe' condition. After the training is completed, these subjects would understand the different potential outcomes and would have some understanding of the influence of different machine parameter setting. Subjects in the VR and real environment, having seen the machine settings on paper, are asked to write down the expected outcome for each trial before the VR or real system is run (good finish, poor finish, tool breakage as well as safe/unsafe). Subjects would have the discretion to adjust the machine setting to a safe and good machining condition based on the spindle speed, tool and workpiece material, spindle speed, depth of cut and lubrication.

Experiment 2 was used to test the proposed hypotheses. Eighty subjects would be recruited for conducting this experiment. The forty subjects with best performance were selected from the first experiment. They were regarded as the expert group. Another forty subjects were recruited from the business school at HKUST. It was assumed and confirmed that they did not have any prior CNC machining knowledge. All participants were assigned into five groups. They were G1, G2, G3, G4 and G5 (see Fig. 3), with 16 students in each group. Before the participants started to perform the VR task, they were given a short demonstration.

Again for experiment 2, each subject needed to run the VR training system for 10 trials and run the real CNC machine for one 'good, safe' trial. Every subject was asked to perform a "retention task" immediately after training and in one of two retention intervals. Different task complexity levels would be assigned to different subject group. Half of the subjects would encounter a stimulated accident in the VR-CNC training system. Based on the layout of experiment 2, the impact of different levels of task complexity in decision-making could be tested by comparing performance between groups 1 and 3. In addition, the comparison of performance between groups 1 and 2 would show the influence of retention interval. It is expected that the performance will improve and retention will increase when a simulated tool breakage is present in the VR training scenario (Duffy, 1998). The impact of a simulated accident could be obtained by comparing performance between groups 3 and 4 as well as groups 2 and 5.

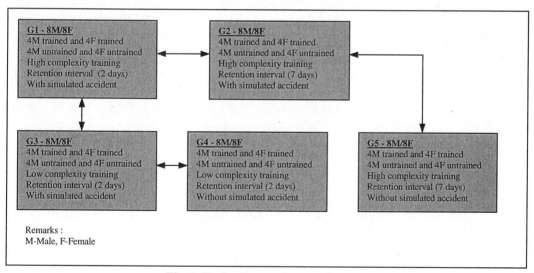

Fig. 1. Design Layout for experiment

4. SYSTEM DEVELOPMENT

The original VR-CNC training program was developed to help train the user for a procedural task (mill a slot and drill two holes) under one machining condition ('good' surface finish and 'safe'). A set of instructions and corrections were included in the form of an Intelligent Pedigogical Agent (IPA) (Ye, *et al.*, 2000). However no decisions were ever made by trainees with regard to whether or not the machining parameters were appropriate for obtaining 'good' surface finish and 'safe' machining conditions. The IPA system gives corrective suggestion back to the user during the VR-CNC training for the procedural task only. It was originally developed using World Tool Kit (WTK) release 7, and operated on a Silicon Graphics SGI Onyx 2 (RE2) workstation. The SGI reality engine II (RE2) uses a 200 MHz Ip 19 processor (Duffy, 1999). Typically, the sale price of this type of machine in Hong Kong is very expensive (around HK$1.1 million). Not many industrial companies can afford this type of machine for training purpose. It is worthwhile to modify the system in a PC window platform. Modifications that allowed consideration of alternate machining conditions and use of the MS-Windows platform for PC-Pentium III with graphics accelerator were done using WTK release 8 for the virtual environment. The additional decision paths and potential outcomes made the training program significantly more complicated. A head mount display was not used since willingness to integrate virtual reality into industrial workplaces would usually not include the willingness of operators to use a head mount display (Duffy, 1999). Some previous studies also mentioned that cyber sickness is generated by factors such as head mount display and speed of moving object. Therefore, the "desktop" VR-CNC training program was adopted in this case.

5. DISCUSSION AND CONCLUSION

This study involves a VR-based CNC (VR-CNC) training program and a real computer numerical control (CNC) machine. Different levels of machining performance and safety occur depending on the machine parameter settings. A simulated industrial accident represents both poorest surface finish and safety conditions and occurs in the virtual environment (VE) with some machine parameter settings. Details of the results of the study, still in progress, will be presented at the conference. Previous literature has been emphatic about the need for industrial engineering and human factors research to develop predictive models that will help product and software developers incorporate research findings at the design stages (Meister, 1996). This study provides an attempt to predict performance based on the retention of knowledge and decision-making prompted by the visual and auditory information provided by virtual training environments in an industrial setting. Retention curves and predictive models of risk reduction should be of particular interest to developers of computer-based multimedia and virtual reality training for tasks with varying levels of complexity in an industrial setting.

ACKNOWLEDGEMENT

This study is sponsored by the Earmarked Competitive Grant RGC/CERG HKUST 6211/99E from the Hong Kong Research Grant Council.

REFERENCES

Baum, D. R. (1992). Virtual training devices: illusion or reality ? Unpublished manuscript. Hughes Training Inc., Minneapolis, MN.

Bohlen, G. A., & Barany, J.W. (1966). A learning curve prediction model for operators performing industrial bench assembly operations. *International Journal of Production Research*, 14, 295-302.

Duffy. V. G. (1998). Modeling risk reduction and industrial performance based on the impact of training in a virtual environment. Research Grants Council. Competitive Earmarked Research Grant Proposal. Project HKUST 6211/99E. The Hong Kong University of Science and Technology, Hong Kong.

Duffy, V.G. (1999). Impact of an intelligent tutor on risk and sound perception in CNC machining. *Proceedings of the IEEE Systems, Man and Cybernetics Conference*, SMC99, October 12-15, 1999, Tokyo, Japan, 1091-1094.

Duffy, V.G. (2000). Toward online virtual safety training in the manufacturing and service industries. *Proceedings of the XIVth Triennial Congress of the International Ergonomics Association and the 44th Annual Meeting of the Human Factors and Ergonomics Society*, HFES/IEA 2000, San Diego, California, USA, July 30-August 4, 2000, 4: 341-344.

Duffy V.G. & Salvendy G. (2000). Concurrent engineering and virtual reality for human resource planning, *Computers in Industry* , 42: 109-125.

Gaver, W. (1997). Auditory interfaces. In: M. Helander, T.K. Landauer, and P. Prabhu (Eds.), *Handbook of Human-Computer Interaction*, 2nd Ed., B.V., Elsevier Science, pp. 1003-1039.

Helander, M.G. (1998). Enhancing design making and communication through virtual environments, Invited seminar presentation. Unpublished notes. The Hong Kong University of Science and Technology, Oct 12, 1998.

Jacko, J.A. (1997). An empirical assessment of task complexity for computerized menu systems. *International Journal of Cognitive Ergonomics*, 1(2) 137-147.

Jarvinen, J. & Karwowski, W. (1993). A questionnaire study of accidents in advanced manufacturing systems. *Proceedings of the Human Factors and Ergonomics Society 37th Annual Meeting 1993 Conference*, pp. 1004-1008.

Karwowski, W. (1992). Worker safety in computer-integrated manufacturing systems: needs for research. In: P. Brodner & W. Karwowski (Eds.). *Ergonomics of Hybrid Automated Systems III*, B.V.: Elsevier Science, pp. 469-480.

Meister, D. (1996). Human factors test and evaluation in the twenty-first century, In: T.G. O'Brien. & S.G. Charlton (Eds.) *Handbook of Human Factors testing and Evaluation*, Mahwah, New Jersey, Lawrence Erlbaum Associates, pp. 313-322.

Nembhard, D.A. (2000). The effects of task complexity and experience on learning and forgetting: A field study. *Human Factors*, 42 (2) 272-286.

Shute, V.J., Gawlick, L.A., & Gluck, K.A. (1998). Effects of practice and learner control on short-and long-term gain and efficiency. *Human Factors*, 40 (2) 296-310.

Wisher, R.A. (1992). The role of complexity on retention of psychomotor and procedural skill, *Proceedings of the Human Factors Society 36th Annual Meeting*, Atlanta, Georgia, Oct 12-16, vol. 2, pp. 1171-1175.

Xie, B. & Salvendy, G. (2000). Prediction of mental workload in single and multiple tasks environment, *International Journal of Cognitive Ergonomics*, 4 (3) 213-242.

Ye, L.L., Duffy, V.G., Yen, B.P.-C., Lin, F. & Su, C.J., 2000, Knowledge modelling methodology for intelligent virtual reality-based industrial training systems. *Asian Journal of Ergonomics*, 1 (1) 73-94.

Zimolong, B. (1997). Occupational risk management. In: G. Salvendy (Ed.). *Handbook of Human Factors and Ergonomics*. New York, John Wiley & Sons, pp. 989-1020.

CONSIDERING INDIVIDUAL DIFFERENCES IN PERCEPTION AND PERFORMANCE IN VIRTUAL ENVIRONMENTS

Colleen M. Duffy and Vincent G. Duffy

Department of Industrial Engineering and Engineering Management
The Hong Kong University of Science and Technology
Clear Water Bay, Kowloon, Hong Kong
phone: 852-2358-7116; fax: 852-2358-0062; email: vduffy@ust.hk

ABSTRACT

In the HCI literature, when considering virtual environments, perception and performance are not commonly discussed in the same study. One of the purposes of this study was to determine the influence of individual differences on perception as well as performance. Data was collected from twenty-four subjects in a virtual industrial environment. From this data, relationships between sense of presence, personal space, immersive tendencies, sense of arousal, mental workload and perception of hazard can be determined. These relationships will be discussed in relation to accuracy of perceived distance in virtual environments. An additional twelve subjects were tested in both the virtual and real environments. Preliminary results of the comparisons between virtual and real will also be shown.

1. INTRODUCTION AND HYPOTHESES

In the HCI literature, when considering virtual environments, perception and performance are not commonly discussed in the same study. One of the purposes of this study is to determine the influence of individual differences on perception as well as performance. One objective is to determine the perception of hazard under varying virtual parameters using an Internet-based industrial environment. The other main objective is to determine the accuracy and performance of distance estimates with varying virtual parameters.

Relationships between sense of presence, personal space, immersive tendencies, sense of arousal, mental workload and perception of hazard can be determined.

Upon inspection of Figure 1, one can see that the virtual industrial environment depicted shows lathe machines and virtual operators in different lighting conditions. The hypotheses are as follows:

Hypothesis 1: It is believed that: Perception of hazard will be greater for subjects that feel the need for additional personal space.

Literature show that some subjects have the desire to have more space in their working environment and that performance will be influenced by that sense of personal space (Hayduk,1978).

Hypothesis 2: It is believed that there will be no significant difference in perceived hazard for subjects that have greater sense of presence or immersive tendencies.

Hypothesis 3: However, it is believed that there will be an interaction between immersive tendencies and lighting, and between immersive tendencies and distance between machines when considering perceived hazard.

In other words, it is believed that subjects who are more likely to feel 'immersed' in the virtual environment are more likely to sense hazard in conditions that are expected to cause differences in perception.

Hypothesis 4: It is believed that immersive tendencies will be correlated with arousal.

Arousal can be measured using a traditional personality indicator.

Hypothesis 5: It is believed that there will be a difference in mental workload based on the presence of a virtual human operator.

These relationships will be discussed in relation to accuracy of perceived distance in virtual environments. Some studies related to immersive tendencies and distance judgements in virtual environments have been done by Witmer (1998).

2. RESEARCH METHODOLOGY AND PROCEDURES

Definitions of the terms above such as sense of presence, personal space, immersive tendencies, sense of arousal, mental workload and perception of hazard have been derived from the literature and fall within the area of perception. Twenty-four subjects were tested on a Pentium 400 computer using a Sony 17" VDT monitor. The subjects participated by viewing the desktop virtual environment using COSMO Player 2.0 on a Netscape 4.5 Internet browser.

Luminance measurements were made in light per unit of projected area Cd/m^2 (Candela per square meter) using a Minolta CS 100 Chromameter. An illustration of front view (close), front view (distant), angle view and luminance levels with VR light on and VR light off are shown in Figure 1. The experiments took approximately 2 hours. Student subjects were given a 10 minute break after ½ of the testing was complete and subjects were compensated for their time. In the lab, with all lights on the illuminance measure was 340 lux. Differences in background-foreground luminance ratios were created by modifying the VR light condition.

A subject was shown the virtual environment by walkthrough on a predetermined path. After the subject had reached the straight (near) viewpoint the subject was asked to record their perception of hazard to the operator and perception of acceptability using a fixed modulus estimation technique (scale of 0-100). Subjects were told that some conditions were 'safe' and some 'unsafe', and 'for the purposes of these trials, a <u>hazard</u> is defined as a condition that can lead to injury or property damage'.

After the subject has turned away from the computer, the scene was modified to 'angle view' and a repeat measure was taken for the purposes of reliability. Two different light levels were shown: vr 'light on', 'light off'. Brightness was set at 31 and contrast was set at 100. The experiments also included asking the subjects about the distance they had traveled during the walkthrough. Additional details of the methodologies can be found in Duffy and Chan (2000). A similar set of questions was asked during the real task.

Details of the methodology for obtaining perception of inter-object and traversed distance are included in Wu, Duffy and Leung (2000). Typically this type of data has not been considered in the literature with regard to individual differences, though it has been mentioned as having the potential to contribute to understanding of the way people perceive virtual environments (Snow and Williges, 1998).

3. RESULTS AND DISCUSSION

Preliminary results of the recent study, still in progress, will be presented. It should be noted, that in studies of this type, sample size is an important consideration. As well, care in designing data collection methods and procedures are important for enabling insights into such individual differences.

4. FUTURE WORK

Additional subjects were tested in both the virtual and real environments. With that, after considering methodologies related to sequential experiments (Snow and Williges, 1998), results of the comparisons between virtual and real based on individual differences can also be shown.

ACKNOWLEDGEMENTS

This study is sponsored by Direct Allocation Grant from Dean of Engineering office at The Hong Kong University of Science and Technology and the Competitive Earmarked Research Grant RGC/CERG HKUST 6168/98E from the Hong Kong Research Grant Council. We would like to thank Annie Fee Fee Ng for her assistance in analyses while on break from Purdue University. Also thanks to Flora Fang Wu for her excellent efforts in system development.

REFERENCES

Duffy, V.G. & Chan, A.H.S. (2000). Effects of virtual lighting on visual performance and eye fatigue, working paper submitted to *Human Factors and Ergonomics in Manufacturing*, accepted.

Duffy, V.G. (2000). Effects of virtual lighting on risk perception and eye fatigue, Manufacturing: Agility and Hybrid Automation III, T. Marek and W. Karwowski (Eds). *7th HAAMAHA Conference, The 7th International Conference on Human Aspects of Advanced Manufacturing: Agility and Hybrid Automation*, Krakow, Poland, August 27-30, 2000, pp. 106-109.

Hayduk, L.A. (1978). Personal space: an evaluative and orienting overview. *Psychological Bulletin.* **85**, 117-134.

Snow, M. P. & Williges, R. C. (1998). Empirical models based on free-modulus magnitude estimation of perceived presence in virtual environments. *Human Factors.* 40(3), 386-402.

Witmer, B. & Kline, P.B. (1998). Judging perceived and traversed distance in virtual environments. *Presence*, 7 (2) 144-167.

Wu, F., Duffy, V.G. & Leung, G.T.C. (2000). A study of inter-object and traversed distance in virtual and real environments, working paper submitted to *International Journal of Cognitive Ergonomics.*

Wu, F. (2000). Developing a methodology for the study of risk perceptions in computer simulated industrial environments. M.Phil. Thesis, The Hong Kong University of Science and Technology, Department of Industrial Engineering and Engineering Management, Advisor, V.G. Duffy.

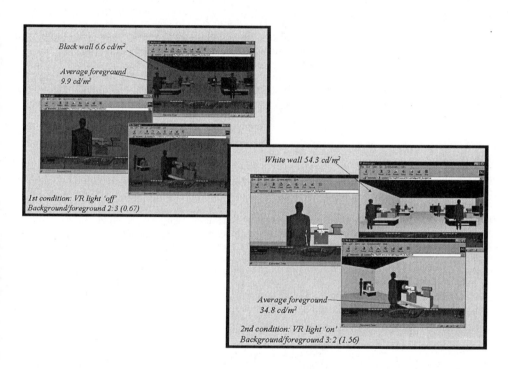

Figure 1. Illustration of front views, angle view and luminance levels with VR light on and VR light off (Duffy, 2000).

Haptic Cueing of a Visual Change-Detection Task: Implications for Multimodal Interfaces

Hong Z. Tan[†], Robert Gray[‡], J. Jay Young[†], and Piti Irawan[†]

[†] Haptic Interface Research Laboratory, Purdue University
1285 Electrical Engineering Building, West Lafayette, IN 47907

[‡] Nissan Cambridge Basic Research
4 Cambridge Center, Cambridge, MA, 02142

ABSTRACT

This study is part of an ongoing program designed to investigate the integration of visual and haptic information in the context of multimodal interfaces. With the current experiments, we study whether haptic cues can be used to redirect spatial attention in a visual task where an observer is asked to detect a change between two scenes. Subjects were asked to look at visual scenes consisting of rectangular horizontal and vertical elements of equal sizes. Their task was to detect an orientation change in one of the elements. Prior to this visual task, the subject was tapped on the back at one of four locations by a vibrotactile stimulator. It was found that reaction time to detect a visual change decreased significantly when the location of the tactor coincided with the quadrant of the visual scene where the changing element occurred. It was also found that reaction time increased when the location of the tactile stimulation did not coincide with the visual quadrant where change occurred. These results have implications for designers of multimodal interfaces where a user can benefit from haptic attentional cues in order to detect and process information in a small area within a large and complex visual display.

1. INTRODUCTION

The growing trend in interface research is towards multimodal human-computer interfaces. This is motivated by the facts that humans naturally employ multimodal information channels for communication, and that multimodal interfaces have been demonstrated to be effective (Oviatt, 1999). Cognitive research has shown that multimodal communication results in increased amount of transmitted information (Miller, 1956). It is well known that a signal with a single varying attribute can at most transmit 2-3 bits to a human observer (for example, we can only identify about 5-7 loudness levels of a fixed-frequency pure tone). However, greater information transmission can be achieved by employing signals with multiple attributes (for example, one can easily identify hundreds of faces at a glance of a person or a photograph, because many facial features contribute to the appearance of a face). This increase in transmitted information can be achieved whether multiple modalities convey different information or encode the same information redundantly (Miller, 1956). Therefore, multimodal interfaces facilitate more natural and efficient human-computer interactions.

One challenge in multimodal interface research is the lack of multimodal interface systems. Robust systems for applications such as speech recognition or gesture interpretation require long-term research and development efforts from a multidisciplinary team of investigators. True multimodal interactions can not take place until problems in each of these application domains are solved. Compared with visual and auditory interfaces, the field of haptic interface research is a less developed yet fast-growing and promising area. For the past several years, we have been developing a tactor-array for the back of a user. We have been studying its effectiveness in conveying directional information for applications such as a haptic navigation guidance system for drivers and blind travelers (Ertan, Lee, Willets, Tan, & Pentland, 1998; Tan, Lim, & Traylor, 2000; Tan, Lu, & Pentland, 1997; Tan & Pentland, 1997, 2001).

Recently, we studied the integration of visual and tactile information about moving objects (Gray & Tan, 2000). In one experiment, we found that tactile pulses simulating motion along the forearm facilitated the speed and accuracy with which subjects discriminated visual targets on the same forearm. In another experiment, we concluded that an approaching visual target's time to contact with the forearm influenced subject's ability to perform tactile discrimination on that forearm. These results demonstrate dynamic links in the spatial mapping between vision and touch. In the current study, we explore this issue further with a paradigm that examined how haptic cueing might affect an observer's visual spatial attention. The long term objective of our research is to investigate the integration of visual and haptic information in the context of cross-modal priming.

In the current study, we investigate whether haptic cues (taps on the back) can be used to redirect spatial attention in a visual task where an observer is asked to detect a change between two scenes. Recent research has shown that attention is required to perceive (even large) changes in a visual scene. This phenomenon, termed "change blindness," occurs in both laboratory (Rensink, O'Regan & Clark; 1997) and real-world (Simons & Levin, 1998) conditions. The proposed explanation for "change blindness" is that we do not form a complete detailed representation of our surroundings. Such a representation occurs only for the small part of the visual field that we are attending. In the typical experimental setup for studying "change blindness", termed the flicker paradigm (Rensink, 2000), two scenes are alternately displayed with a blank inserted between them (to mask motional cues). An observer is asked to respond as soon as a difference between the two scenes is detected. It has been found, using scenes consisting of photographs, that reaction time in such tasks depends on the degree to which the changing element is of interest (i.e., captures the viewer's attention). If attention is the key factor affecting reaction time, then any means of manipulating an observer's attention should affect the reaction time associated with the detection of scene changes. Our experiments are therefore designed to investigate whether such effects can be elicited by drawing an observer's attention to a spatial location via haptic stimulation.

2. METHODOLOGY

2.1 Stimulus

The visual stimuli used in these experiments were based primarily on the flicker paradigm used for the study of "change blindness" (Rensink, 2000). The visual scenes consisted of rectangular elements of equal sizes, but in either horizontal or vertical orientations (Fig. 1). Two scenes, differing only in the orientation of one of the elements, were presented in an alternating order with a blank scene inserted in between. The duration of the two patterned scenes was called the "on time". The duration of the blank scene was called the "off time."

The experimental apparatus for haptic cueing consisted of a 3-by-3 vibrotactile display developed at the Purdue Haptic Interface Research Laboratory. The tactor array is draped over the back of an office chair (Fig. 2). For the experiments reported here, only the four corner tactors (i.e., tactors No. 1, 3, 7, and 9 in Fig. 2) were used. Each tactor could be independently driven by a 60-ms sinusoidal pulse. The frequency of the pulse was between 290-306 Hz (corresponding to the resonant frequencies of the four tactors). The intensity of the vibration was between 26.1-27.9 dB SL (sensation level) under unloaded condition.

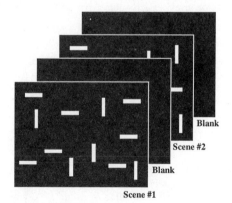

Figure 1. The two visual scenes used in our change-detection experiments (modified from Fig. 2 in Rensink, 2000).

Figure 2. The haptic cueing system. Shown here is a 3-by-3 tactor array draped over the back of a chair.

2.2 Subjects

Ten college students, 5 females and 5 males, participated in the experiment as paid research participants. The average age of the subjects was 21 years. All subjects had normal or corrected vision. They reported no known abnormalities with tactile perception on their back.

679

2.3   Procedures

Before the experiments began, subjects were informed of the nature of the task. Specifically, they were told that they needed to locate a rectangular element on the computer screen that was changing its orientation. Their job was to *locate* and *identify* this element as quickly as possible.

To ensure that the subjects could clearly feel the vibrations presented by the tactor array on their back, and that the subjects could correctly associate each tactor location with the corresponding quadrant on the computer screen, an absolute identification experiment was conducted with the tactor array before each new session. The subject's task was to click on one of the four quadrants on the monitor (represented by four large rectangles) in response to a vibration on the back (e.g., the correct response to a vibration near the right shoulder would be to click on the upper-right quadrant of the monitor). Each subject had to complete one perfect run (i.e., 100% correct) of 60 trials before starting the visual change-detection task. This test with the tactor array was repeated each time the subject left and returned to the chair.

During the visual change-detection task, the subjects were instructed to click the left mouse button as soon as the changing element was found (without moving the cursor over the element). The screen then froze and the color of all elements turned from white to pink. The subjects were required to make a second mouse click with the cursor centered on the element that they perceived to change orientation. The timing of the first mouse click was recorded as the reaction time. The x-y positions of the second click were used to discard trials where the wrong element was identified.

The independent variables employed were the state of the tactors (OFF or ON) and on time (80, 480, and 800 ms). Three 60-trial runs were conducted for each experimental condition and each subject. The order of the eighteen runs (2 tactor state $\times$ 3 on time $\times$ 3 runs) were randomized. For all experimental conditions, off time was fixed at 120 ms. The total number of rectangular elements was fixed at 12 (or equivalently, 3 elements per quadrant). The x-y positions of the elements were chosen randomly within each quadrant with the constraint that the elements never overlapped. For the experiments where tactors were ON (i.e., haptic cueing was present), the interstimulus interval (ISI — the interval from the time the tactor was turned off to the time the first scene was shown on the monitor) was fixed at 50 ms. The percentage of trials with *valid* haptic cues (i.e., the location of the vibrating tactor coincided with the quadrant where the changing element occurred) was fixed at 50%. Our subjects were aware of the fact that the location of the haptic cue may or may not be valid on any particular trial. They were left to decide on their own whether and how they would utilize the information provided by the haptic cues.

Throughout the experiments, subjects were instructed to sit upright with their back pressed against the tactor array. They were instructed not to move their body relative to the chair, or to move the chair relative to the monitor. Headphones were used to block any audible noise from the tactor array. Each subject typically finished all the experiments within 2-3 sessions.

2.4   Data Analysis

The dependent variables were mean reaction time and standard error. For each of the six experimental conditions tested (2 tactor state $\times$ 3 on time), data from all subjects were pooled. Data from the tactor OFF condition served as a baseline measure for reaction time. Data from the tactor ON condition were separated into two subgroups: those with valid haptic cues and those with invalid cues. Mean reaction times for the two groups of trials were computed separately. All error trials (where the subject selected the wrong rectangle element during the second mouse click) were discarded.

3.   RESULTS

In general, our results show that reaction time decreased significantly with valid haptic cues, and increased with invalid haptic cues. For example, results for one subject (S5) are shown in Fig. 3. It can be seen that the average reaction time for each experimental condition increased monotonically with the value of on-time. Compared to baseline measures (i.e., reaction times with no haptic cues, shown as filled diamonds), reaction time decreased with valid haptic cues (filled circles), and increased with invalid haptic cues (filled triangles). This is true for each of the ten subjects tested.

The extent to which valid or invalid haptic cues decreased or increased reaction time varied from subject to subject. For example, shown in Fig. 4 are data from another subject (S9) who has a lower baseline measure than S5 (i.e., faster response without haptic cues). Subject S9 benefited less from valid haptic cues than S5 (i.e., a smaller decrease in reaction time). This subject was also less distracted by invalid haptic cues than S5 (i.e., a smaller increase in reaction time). Subjects S5 and S9 represent the two most extreme observers among the ten subjects tested. We hasten to point out that, despite the obvious differences in the data shown in Fig. 3 and 4, our general

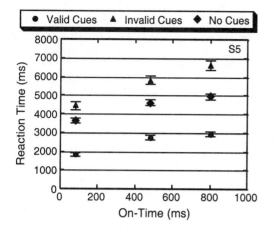

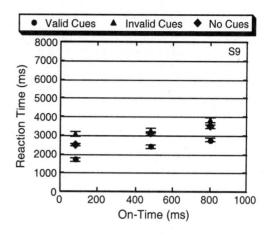

Figure 3. Mean reaction times and standard errors for subject S5.

Figure 4. Mean reaction times and standard errors for subject S9.

conclusion regarding the effect of valid and invalid haptic cues on the reaction time of our visual change-detection task is still valid.

Results averaged over all ten subjects are shown in Fig. 5. As found with individual subject's data, mean reaction time increased as on-time increased, for the valid-cue, invalid-cue and no-cue conditions. Overall, compared with baseline measures, reaction time decreased by 1630 ms (40.6%) with valid haptic cues, and increased by 781 ms (18.9%) with invalid haptic cues. All standard errors are relatively small as compared to the values of reaction time.

One interesting observation from Fig. 3 and 4 is that the datum points for the valid-cue condition for subjects S5 and S9 seem to be quite similar, despite the large differences in reaction time for the invalid-cue and no-cue conditions. To investigate this further, standard deviations of reaction time from data pooled from all ten subjects were computed (Fig. 6). Indeed, it seems that the standard deviations for the valid-cue condition are lower than those for the other two conditions across the three on-time values tested. We therefore conclude, based on the limited data we have collected, that valid haptic cueing reduces the inter-subject variability in the response time for the visual change-detection task employed in this study.

Finally, average number of error trials varied among the subjects tested, with a range of 0-9 per experimental run of 60 trials. Averaged across the subjects, there were fewer than 4 error trials per 60-trial run.

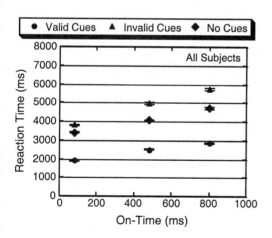

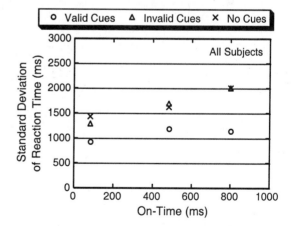

Figure 5. Mean reaction times and standard errors averaged over all ten subjects.

Figure 6. Standard deviations of reaction time from data pooled over all subjects.

4. CONCLUSIONS

In this study, we examined the extent to which haptic spatial cues can speed up or slow down an observer's reaction time to detect a change in a visual scene. Our data suggest that (1) valid haptic cues decrease the reaction time for the detection of a visual change, and (2) invalid haptic cues increase the reaction time (to a lesser degree) for the same task. This general conclusion holds for data from individual subjects as well as for pooled data from all subjects. This conclusion also holds despite the inter-subject differences in their "natural" reaction time (i.e., some subjects tend to react faster than others when no haptic cues are present). Similar results have been reported for visual spatial cueing of a visual change-detection task (Scholl, 2000). Finally, we have some evidence suggesting that valid haptic cues decrease the inter-subject variability in their reaction time.

Our results have implications for designers of multimodal interfaces. In an automobile, for example, a haptic display built into the driver's seat can be useful in alerting the driver of impending collision on one side of the car. In a large and complex visual display for air traffic control, a haptic display used in conjunction with a non-invasive eye-tracking system can remind the operator to look at a neglected area of the visual display, or to pay attention to an area with busy traffic. In general, haptic cueing can provide an effective alternative to visual and auditory cueing for a complex information display.

ACKNOWLEDGEMENT

This work has been partly supported by a gift fund from Nissan Research & Development, Inc., a grant from Honda R&D Americas, Inc., and a National Science Foundation Faculty Early Career Development (CAREER) Award under Grant No. 9984991-IIS. The authors wish to thank Drs. Ron Rensink and Ian Thornton for many insightful discussions on this project.

REFERENCES

Ertan, S., Lee, C., Willets, A., Tan, H. Z., & Pentland, A. (1998). A wearable haptic navigation guidance system, *Digest of the Second International Symposium on Wearable Computers* (pp. 164–165). Piscataway, NJ: IEEE Computer Society.

Gray, R., & Tan, H. Z. (2000). Dynamic spatial mapping between vision and touch. *Abstracts of the Psychonomic Society, 5*, 38.

Miller, G. A. (1956). The magical number seven, plus or minus two: Some limits on our capacity for processing information. *The Psychological Review, 63*(2), 81-97.

Oviatt, S. (1999). Ten myths of multimodal interaction. *Communications of the ACM, 42*(11), 74–81.

Rensink, R. A. (2000). Visual search for change: A probe into the nature of attentional processing. *Visual Cognition, 7*(1-3), 345-376.

Rensink, R. A., O'Regan, J. K., & Clark, J. J. (1997). To see or not to see: The need for attention to perceive changes in scenes. *Psychological Science, 8*(5), 368–373.

Scholl, B. J. (2000). Attenuated change blindness for exogenously attended items in a flicker paradigm. *Visual Cognition, 7*, 377-396.

Simons, D. J., & Levin, D. T. (1998). Failure to detect changes to people during a real-world interaction. *Psychonomic Bulletin and Review*, 5, 644-649.

Tan, H., Lim, A., & Traylor, R. (2000). A psychophysical study of sensory saltation with an open response paradigm. In S. S. Nair (Ed.), *Proceedings of the Ninth (9th) International Symposium on Haptic Interfaces for Virtual Environment and Teleoperator Systems, American Society of Mechanical Engineers Dynamic Systems and Control Division* (Vol. 69-2, pp. 1109–1115). New York: ASME.

Tan, H. Z., Lu, I., & Pentland, A. (1997). The chair as a novel haptic user interface. In M. Turk (Ed.), *Proceedings of the Workshop on Perceptual User Interfaces (PUI'97)* (pp. 56–57). Banff, Alberta, Canada, Oct. 19–21.

Tan, H. Z., & Pentland, A. (1997). Tactual displays for wearable computing. *Personal Technologies, 1*, 225–230.

Tan, H. Z., & Pentland, A. (2001). Tactual Displays for Sensory Substitution and Wearable Computers. In W. Barfield & T. Caudell (Eds.), *Fundamentals of Wearable Computers and Augmented Reality* (Chap. 18, pp. 579–598). Mahwah, NJ: Lawrence Erlbaum Associates.

Virtual Glassboat: For looking under the Ground

Itiro Siio

Faculty of Engineering, Tamagawa University,
6-1-1 Tamagawagakuen, Machidashi, Tokyo 194-8610, Japan,
+81 427 39 8413 siio@acm.org, http://siio.ele.eng.tamagawa.ac.jp/

ABSTRACT

A simple augmented reality system called the Virtual Glassboat is described in this paper. The Virtual Glassboat is a cart mounted computer display which scrolls with the movement of the cart. By pushing the cart over a floor or over the ground, a user can browse the underneath information such as gas or water piping work as if he/she were looking under the ground through the display.

1. INTRODUCTION

In many augmented reality (AR) systems, 3D position trackers or cameras are used to get the exact location of users and objects, and see-through head mounted displays (HMD) are used to provide stereoscopic computer graphics. These devices are usually large, heavy, expensive, and very difficult to calibrate and register [1]. Even when only flat surfaces such as floors or walls are the target to augment the reality, 3D sensors or cameras are usually used to track the user's position.

In most AR applications that browse inside walls, floors or boxes, only 2D position detections and 2D information presentations are necessary. In our previous work [2], we have introduced a simple AR system called the Scroll Browser. This system reveals wiring and pipe-work inside walls on a hand-held display when the user traces the surface with a

Figure 1: The Virtual Glassboat system. The cart mounted display scrolls with the movement of the cart to provide the illusion that the user is looking underground.

2D position detector called a FieldMouse, which is a combination of a bar-code reader and a mechanical mouse. The Virtual Glassboat is an extension of the Scroll Browser to AR applications on floors or on the ground.

2. THE VIRTUAL GLASSBOAT

Figure 1 shows the Virtual Glassboat, a simple AR system that can be used to browse wiring or pipe-work under floors or under the ground. It is a cart-mounted computer system with a face-up display. The user can view graphical information on the display while pushing the cart over a floor or over the ground. The amount of movement and rotation is detected by the computer by monitoring the cart's wheels rotation. It is programmed to scroll out the display in the opposite direction to the movement of the cart. The system thus provides the illusion that the user is looking under the floor or ground through the display frame. The system derives its name from glass-bottom boats used at marine parks to observe creatures underwater.

3. IMPLEMENTATION

We have implemented the first prototype on a commercial cart that has four hard rubber wheels: two steering front wheels and two fixed rear wheels. The position and direction of the cart is detected by measuring the rotation of the fixed rear wheels. As shown in Figure 2, a reflection plate with 32 stripes is attached to each rear wheel (11.25 degrees

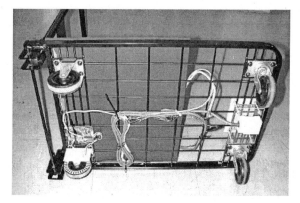

Figure 2: Underside view of the Virtual Glassboat. Optical sensors detect rotation of the rear wheels (left). The RFID reader

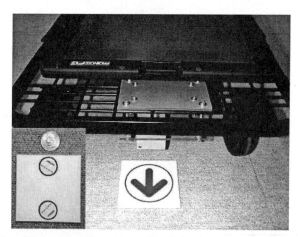

Figure 3: Position and direction of the cart are calibrated by reading the marker plate (center) inlaid with two RFID tags on the underside (left).

resolution) and photo interrupters are installed to measure the rotation. Electronic parts of a mouse are used to interface the photo interrupters with the note book computer on the cart.

In order to calibrate the absolute position and direction of the cart in the real-world, we have made a marker plate inlaid with two RFID (Radio Frequency Identification) tags on the underside and placed it on the floor (Figure 3). When the cart moves over the plate in the direction of the arrow on it, two tags are successively read by the RFID reader mounted on the front of the cart, and the position and direction of the cart are calibrated. In outdoor applications, longer-range RFID readers can be used to detect buried tags under the ground.

A browsing program developed by C language on Linux OS runs in the computer. This program scrolls out the PostScript data on the display on the X Window, calculating the rotation and movement of the cart by reading the information provided by the photo interrupters and the RFID reader. Data that can be displayed as of now are PostScript files made of lines, circles, and letters (Figure 4).

4. APPLICATIONS

The current browser program which displays simple drawing images, can visualize underground piping, wiring and structure. It can therefore be used to locate underground objects to support maintenance or reconstruction work. As RFID stakes have been commercially developed to mark the ground, it will be possible to browse underground objects near the markers by incorporating these stakes into our system.

We have prepared the PostScript data that represents the underground pipeline in front of our faculty building (Figure 5). The sidewalk is about 100m x 20m, and there are 23 manholes (indicated by circles) and underground rainwater and sewage pipe lines. For each manhole, we have installed RFID tags which hold the manhole ID number. By reading one of the RFID tag and moving around the manhole by our system, we can identify and locate the underground pipes.

The Virtual Glassboat can also be used in an interactive art or museum system by displaying virtual objects. For example, we could use it as a virtual glass-bottom boat and chase virtual fish or take walks through coral reef on the floor. We can use it to display multimedia information on a museum floor on which large maps or diagrams are painted.

We have implemented the virtual glass-bottom boat application as shown in Figure 6. Fish animations were prepared on Macromedia Director on Windows 98. Using Director's scripting language Lingo, the rotation and movement of the cart is calculated by reading the cart's photo interrupters which is connected to the mouse port. Fish are programmed to cross the display randomly. The position of the fish is also changed bye the movement of the cart. This gives the user the illusion that he/she is chasing the fish. RFID reader is not used in this application. In the future version with fixed underwater objects like seaweed, rocks, and corals, we should implement RFID tags on the floor.

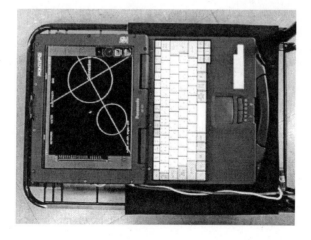

Figure 4: Drawing image is displayed and scrolled.

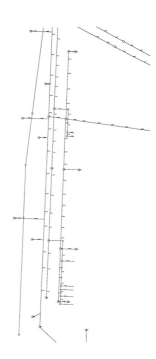

Figure 5: Test field (left) and the PostScript data (right) showing 23 manholes and rainwater/sewage pile lines under the field.

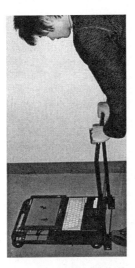

Figure 6: Chasing fish application (left) and fish animation on screen (right).

5. EVALUATIONS

The direction accuracy when rotating the cart 360 degrees at one place is within 2%. The accuracy of the cart movement in straight lines of 1m to 5m is within 3%. Position detection error is accumulated by moving a long distance and by moving with complicated rotations. However, this result shows that our system has sufficient accuracy in actual application, in which the cart position is often calibrated by reading RFID tags.

Smoothness of the floor also affects the accuracy. Especially in the outdoor applications such as the underground piping search in Figure 5, the error increases when running over rough pavement. We suspect the error arises because the rattling wheels introduce malfunction in the photo interrupters. The rotation sensor should be improved for rough surface environment.

On a flat floor, the current prototype provides a convincing illusion that the user is looking beneath the floor. Even on a rough surface, applications that do not require precise positioning, such as the chasing fish application in Figure 6, can be functional.

6. RELATED WORK

As it is mentioned in the introduction, the Virtual Glassboat is an extension of our previous work called Scroll Browser [2].

There are many AR systems that enable to see-through inside a box and a wall. Many of them uses transparent HMDs [3]. In some systems, camera and hand-held display are used to achieve augmented reality [4]. Although the Virtual Glassboat system has no transparent feature, it realized enough augmented reality because the view angle is smaller than HMDs.

Some AR systems are designed for outdoor use [5]. They require large and expensive DGPS (Differential GPS) equipment and a calibration system to operate. Although, the Virtual Glassboat has limitation of 2D display, and works only on flat ground, it operates with simple inexpensive sensors.

REFERENCES

1. Azuma, R. T. A survey of augmented reality. In *Presence: Teleoperators and Virtual Environments*, 6(4), pp. 355–385, MIT Press, August 1997.

2. Siio, I., Masui, T., Fukuchi, K. Real-world Interaction using the FieldMouse In *CHI Letters, Vol.1, Issue 1 (Proceedings of the UIST'99)*, pp.113-119, ACM Press, November 1999.

3. S. Feiner, B. Macintyre and D. Seligmann , Knowledge-based augmented reality, *Communications of the ACM*, pp53-62, vol.36, No. 7, 1993.

4. Jun Rekimoto and Katashi Nagao. The world through computer:. In Proceedings of *the ACM Symposium on User Interface Software and Technology (UIST'95)*, pp. 29-36. ACM Press, November 1995.

5. S. Feiner, B. MacIntyre, T. Hollerer, and T. Webster, A touring machine: Prototyping 3D mobile augmented reality systems for exploring the urban environment. Proc. ISWC '97 (First Int. Symp. on Wearable Computers), October 13-14, 1997, Cambridge, MA.

The Classroom of the Future:
Enhancing Education through Augmented Reality

Jeremy R. Cooperstock
Centre for Intelligent Machines
McGill University
3480 University Street, Montreal QC, H3A 2A7 Canada
jer@cim.mcgill.ca

ABSTRACT

Electronic classrooms offer instructors a variety of multimedia presentation tools such as the VCR, document camera, and computer projection, allowing for the display of video clips, transparencies, and computer generated simulations and animations. Unfortunately, even the most elegant user interfaces still frustrate many would-be users. The technology tends to be underutilized because of the cognitive effort and time its use requires. Worse still, it often distracts the instructor from the primary pedagogical task.

1   BACKGROUND

Information technology promised to empower us and simplify our lives. In reality, we can all attest to the fact that the opposite is true. Modern presentation technology, for example, has made teaching in today's classrooms increasingly complex and daunting. Whereas fifty years ago, the only concern a teacher had was running out of chalk, faculty now struggle to perform relatively simple tasks, such as connecting their computer output to the projector, switching the display to show a video tape, and even turning on the lights! Technology's capacity to improve the teaching and learning experience is evident, but so far, its potential remains largely untapped.

A related concern in the pedagogical context is the effort required to exploit the technology for novel applications, for example, distance or on-line education. The desire to provide lecture content to students who are unable to attend the class in person, as well as to those who wish to review the material at a later time, has been a driving force behind the development of videoconferencing and web-based course delivery mechanisms. Although a number of universities now offer courses on-line, the cost involved in creating high-quality content is enormous. Both videoconferencing and simple videotaping of the lectures require the assistance of a camera operator, sound engineer, and editor. For asynchronous delivery, lecture material, including slides, video clips, and overheads, must be digitized, formatted and collated in the correct sequence before being transferred. Adding any material at a later date, for example, the results of follow-up discussion relating to the lecture, is equally complicated. The low-tech solution, which offers the lecture material by videotape alone, still involves considerable effort to produce and suffers further from a lack of modifiability, a single dimension of access (tape position), and a single camera angle. This prevents random accessibility (e.g. skip to the next slide) and view control (e.g. view the instructor and overhead transparency simultaneously at reasonable resolution), thus limiting the value to the students.

2   AUTOMATED CONTROL

In response to our frustration with this situation, we have augmented our classroom technology with various sensors and computer processing [3]. The room now activates and configures the appropriate equipment in response to instructor activity without the need for manual control (see Figure 1).

For example, when an instructor logs on to the classroom computer, the system infers that a computer-based lecture will be given, automatically turns off the lights, lowers the screen, turns on the projector, and switches the projector to computer input. The simple act of placing an overhead transparency (or other object) on the document viewer causes the slide to be displayed and the room lights adjusted to an appropriate level. Similarly, picking up the electronic whiteboard marker causes a projector "swap." so that the whiteboard surface displays the current slide while the main screen shows the previous slide. Audiovisual sources such as the VCR or laptop computer output are also displayed automatically in response to activation cues (e.g., the play button pressed on the VCR; the laptop connected to a video port). Together, these

688

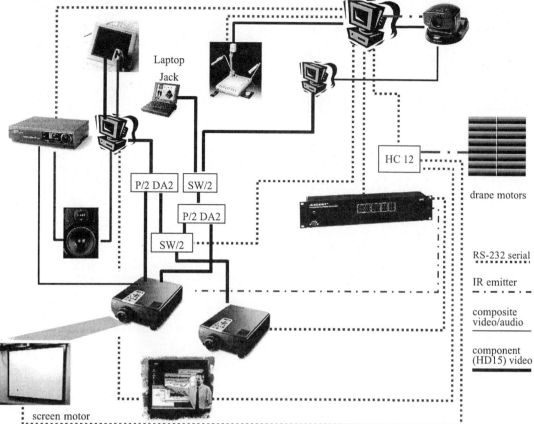

Laptop
Jack

P/2 DA2   SW/2

P/2 DA2

SW/2

HC 12

drape motors

RS-232 serial

IR emitter

composite
video/audio

component
(HD15) video

screen motor

Figure 1. Architecture of our computer-augmented classroom connections. The large black module in the center of the image is the AMX Accent3 controller, which drives various devices under computer control and the HC 12 module is our button-panel unit with microcontroller, pictured separately in Figure 2. The SW/2 units are video switchers, one of them running in an auto-sense mode, such that an active signal on the laptop connection is automatically selected, while the second is driver by computer control. The P/2 DA2 units are video splitters, such that either video signal can be routed to both projectors.

mechanisms assume the role of *skilled operator*, taking responsibility for the low-level control of the technology, thereby freeing the instructor to concentrate on the lecture itself, rather than the user interface.

2.1   Manual Override

Along with such automation, the need for a seamless manual override mechanism becomes paramount. For example, if the instructor raises the lights, the technology must respect that preference. Furthermore, the ability to turn the lights on or off

must not be dependent upon the automatic controller, as it was before this project began.[1]

As a default backup, manual controls for each device (lights, projector, VCR, etc.) should be accessible and functional at all times. Such manual controls serve as basic on/off switches as well as output enable/disable buttons. For example, a single toggle button on the VCR would allow the presenter to select whether or

---

[1] This led to disastrous consequences when the controller became unresponsive, as was the case after any power fluctuations. On at least two occasions, the instructor was unable to control any of the room lights, as no manual override mechanism existed!

Figure 2. Touch-screen interface involving a hierarchical menu structure

Figure 3. The replacement button-panel interface (right) for manual override, providing manual light switches, projector, screen, and drape controls, and a context-sensitive volume dial.

not the video clip being played is projected to the class. By observing the use of these manual override mechanisms, the reactive classroom system can adapt to the preferences of individual users and remember these settings for future use by the same individual. At the end of each lecture, the system resets itself to a default configuration.

Early interviews with instructors revealed that for most users, manual override functions were only required for the room lights and speaker volume, so these were made a top priority. The confusing multi-layered touch-screen menu (Figure 2) was replaced by a simple physical button panel (Figure 3) consisting of six switches for the various banks of lights, another two switches for the projection screen and window blinds, and a volume control knob that adjusts VCR and microphone volume, depending on which is in use.

2.2 Usage Observations

However, advanced users wanted greater control over the selection of projector outputs, for example, display the laptop output on the main screen and the primary computer display (current lecture slide) on the side screen. Unfortunately, our limited deployment of toggle buttons does not permit such flexibility. While the previously described (non-default) configuration is possible, the mechanism by which it is invoked is hardly obvious: picking up the whiteboard marker.[2] A second panel is presently being designed to permit manual input selection for the two projectors.

---

[2] This apparently bizarre mechanism is, of course, related to the projector toggle function, needed when the instructor moves from the digital tablet to the electronic whiteboard.

Interestingly, the one aspect of the current button-panel that failed to achieve improved results over the touch-screen is the volume control. Although the single, context-sensitive control knob is far more accessible than the confusing choice of five independent volume panels (only two of which were actually useful) on the touch-screen, the physical interface of a rotating knob poses problems when used in conjunction with the AMX controller. Since a change in volume is dependent on the response time of the controller, which may be as much as one second when the difference between the current level and the newly selected level is fairly large, users often assume that the controller is not working and start turning the knob faster. The LED directly above the knob, which flashes when the volume is being adjusted, does not, unfortunately, convey sufficient information to the user regarding the state of the system. Once the controller catches up, the user may find that the new volume level is far too low or high. This was not a problem with the touch-screen system since the slow response of the AMX controller could be illustrated graphically by level meters.

3 AUTOMATED LECTURE CAPTURE

In addition to automating device control, the classroom is wired to record a digital version of any presentation, including both the audio and video, as well as the instructor's slides and notes written during the lecture. This recording facility is based on *Eclass*, formerly known as Classroom 2000 [1] a system developed at the Georgia Institute of Technology. Eclass provides a mechanism for the capture, collation, and synchronization of digital ink, written on an electronic whiteboard or tablet, with an audiovisual recording of the class (see Figure 4).

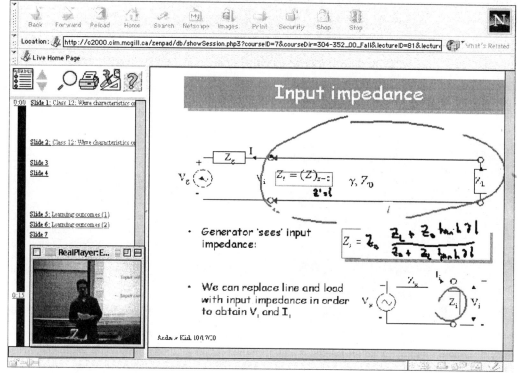

Figure 4. A sample eClass lecture capture being viewed through a web browser and RealPlayer.

At the end of a class, the recorded version of the lecture is then converted into a set of web pages automatically. Each ink stroke written by the professor is linked to the position in the video when that stroke was generated. Students can review the lecture any time after class, randomly accessing portions of the lecture as desired, either from networked university computers or home computers connected by modem.

4 PRESENTER-TRACKING

In order to improve the quality of the video capture, we developed a presenter-tracking algorithm [2] which follows the instructor's movements, even when in front of a projected video screen, thereby obviating the need for a professional cameraman. Device activity, for example, the instructor's use of a pen on the electronic whiteboard, provides additional tracking cues to the camera. Initialization, activation, and recovery from tracking errors are all handled automatically by computer augmentation, allowing

the instructor to remain oblivious to the fact that a recording is being made. The only requirement is that the instructor enters a userid and password to confirm that the lecture should be recorded.

5 FUTURE DIRECTIONS AND CHALLENGES

While interesting in its current implementation, our augmented reality approach to the classroom holds even greater potential when integrated with videoconference technology, in which some, or all of the students are in a remote location from the instructor.

While current videoconference technology has proven to be grossly inadequate for the social demands of effective classroom teaching, we believe that augmented technology may play a role in overcoming this shortcoming. In our envisioned "classroom of the future" scenario, the teaching technology would respond to events in both locations so as to enhance the interaction between instructor

691

and students. For example, a student's hand raised might generate a spatialized background audio cue to draw the instructor's attention toward that student, while a pointing gesture by the instructor toward a remote student could bring about a zoom-in on that student. Figure 5 illustrates a temporal-difference image processing algorithm, used for extraction of the direction of such a pointing gesture. This information, when correlated with the current display, can be used to determine where the remote camera should zoom.

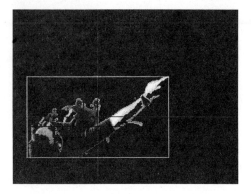

Figure 5. Extraction of a pointing gesture using a temporal difference image processing algorithm.

As a preliminary effort in extending the use of augmented reality to support such interaction, we are developing the infrastructure of a Shared Reality Environment, to provide physically distributed users the sensory experience of being in the same space (see Figure 6).

Figure 6. Video insertion of a remote participant in the Shared Reality Environment. Note that the segmentation algorithm used in this image was unrefined and thus, results in a number of video errors.

A key aspect to allowing for an engaging remote lecture is the use of high-fidelity and low-latency communication of audio and video information, complemented by multichannel, spatialized audio [4], allowing the instructor to capitalize on audio localization abilities for effective interaction with the students. Echo-cancellation, a constant source of headaches for the videoconference technician, becomes an even greater challenge in this context.

ACKNOWLEDGEMENTS

The author would like to thank Aoxiang Xu, Shawn Arseneau, Stephane Doutriaux, and Christian Côté, all of whom contributed valuable components to the technology described in this paper. Special thanks are due to Gregory Abowd and Jason Brotherton of Georgia Tech for their assistance with eClass and allowing us to benefit from their research. Support has come from the Natural Sciences and Engineering Research Council of Canada, Fonds pour la Formation de Chercheurs et l'Aide a la Recherche (FCAR), Petro-Canada, Canarie Inc., and the Canadian Foundation for Innovation. This support is gratefully acknowledged.

REFERENCES

1. Abowd, G., Atkeson, C., Brotherton, J., Enqvist, T., Gulley, P., and Lemon, J. (1998). Investigating the capture, integration and access problem of ubiquitous computing in an educational setting. In Proceedings of Human Factors in Computing Systems CHI '98. ACM Press, New York, pp. 440-447.

2. Arseneau, S. and Cooperstock, J.R. (1999). Presenter Tracking in a Classroom Environment. IEEE Industrial Electronics Conference (IECON'99), Session on Cooperative Environments, Vol. 1, pp. 145-148.

3. Cooperstock, J.R., Tanikoshi, K., Beirne, G., Narine, T., and Buxton, W. (1995). Evolution of a Reactive Environment. Proc. Human Factors in Computing Systems CHI '95, (May 7-11, Denver). ACM Press, New York, pp. 170-177.

4. Xu, A., Woszczyk, W., Settel, Z., Pennycook, B., Rowe, R., Galanter, P., Bary, J., Martin, G., Corey, J., and Cooperstock, J.R. (2000) "Real-Time Streaming of Multichannel Audio Data over Internet." Journal of the Audio Engineering Society, July-August.

PHYSICAL-VIRTUAL KNOWLEDGE WORK ENVIRONMENTS — FIRST STEPS

Thomas Pederson

Department of Computing Science, Umeå University, SE-90187 Umeå, Sweden
+46 90 786 65 48, top@cs.umu.se

ABSTRACT

By examining the common distinction between "physical" and "virtual", from different viewpoints such as users' view and designers' view, and by briefly discussing some development of theoretical and practical tools for integrating the physical with the virtual, this paper presents our first steps towards the goal of designing better knowledge work environments where physical and virtual activities could be performed jointly and with minimal overhead with regard to the gap between the two worlds. Some general problem areas, to be included in the future research agenda, are also identified and briefly discussed.

1. THE PHYSICAL-VIRTUAL ENVIRONMENT GAP

Knowledge work (Drucker, 1973; Kidd, 1994) environments equipped with personal computers tend to create a significant gap between the virtual environment offered by the computer system(s) on the one hand, and the surrounding physical environment on the other (Pederson, 1999). A field study involving 81 professionals categorised as knowledge workers with the aim of investigating the effects of alternating between physical and virtual environments has been performed during 1999. Early analysis supports the belief that the gap is a common and noticeable obstacle in everyday knowledge work (see figure 1 and 2 below) although further empirical investigation is necessary.

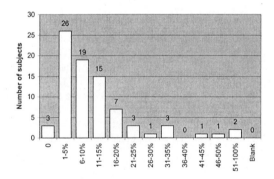

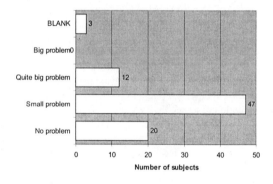

Figure 1: *Perceived proportion of working hours spent on bridging the physical-virtual environment gap.*

Figure 2: *Perceived magnitude of the problem of manually bridging the physical-virtual environment gap.*

2. THE PHYSICAL-VIRTUAL DESIGN PERSPECTIVE

Based on analysis of differences and similarities between physical and virtual environments (e.g., Arias, Eden, & Fischer, 1997; Pederson, 1999), a perspective for design and analysis of integrated physical-virtual environments is derived. The physical-virtual design perspective emphasises a holistic view on the design of knowledge work environments and the objects within them, in order to break loose from traditional distinctions made by designers of software, electronics hardware and architecture (Pederson, 1999). In practice, the physical-virtual design perspective is about categorising physical and virtual objects with the aim to find ways to conceptually and technically link objects

Objects		Physical Environment	Virtual Environment
working objects	raw and refined material building blocks	the wheel of a car, a car itself	embedded software for controlling engine performance of a car
	work process related information entities	paper description how to adjust the breaks on a car	car spare part supplier's contact details on a web page
tools	mechanical	hammer, hydraulics system	resize-box in drawing program (direct manipulation), word processor
	cognitive	pen, paper	calendar with reminder functionality
agents		car mechanic, client	Internet agent searching for best-buy parts on the web
work processes and routines		designed by agents within ("users") and/or outside (organisation managers, *site architects*, other agents)	designed by agents within ("users") and outside (organisation managers, *software designers*, other agents)

Table 1: *An example categorisation of a fictitious environment, a car engineering workshop.*

residing in both worlds to each other. As an illustration, table 1 shows a simple categorisation of objects in a fictitious car engineering workshop. The linking process will not be discussed here, except for the idea itself (see Physical-Virtual Artefacts below), but has been performed rather successfully in the Magic Touch project where objects in a physical office have been linked to virtual objects.

Examining objects within physical and virtual environments, we find that many differences are products of human design activity and not laws of nature. The two environments are usually pure artefacts themselves, designed independently from each other and for different purposes. Thus we assume in our work that most of the differences we identify by observing physical and virtual environments today can be eliminated or synergetically utilised by additional human design activities. It is our belief that this design approach not only has the potential of creating environments in which knowledge workers can act without caring about the physical-virtual environment gap, but that it also leads to the development of new cognitive tools that makes knowledge work more focused, fun and creative.

3. PHYSICAL-VIRTUAL ARTEFACTS (PVAS)

Definition: A *physical-virtual artefact* is an abstract artefact that (1) is instantiated in both the physical and virtual environment, where (2) these instantiations to a large extent utilize the unique affordances and constraints that the two different environments facilitate, and finally (3) where one instantiation of a specific physical-virtual artefact is easily identified if an equivalent instantiation in the other environment is known.

Notation: While PVA refers to both instantiations of a PVA (that is, the PVA as whole), PVA refers to the physical instantiation of a specific PVA and PVA refers to the virtual instantiation of a specific PVA.

An important point is that corresponding PVA and PVA instantiations do not necessarily have to look or behave in a similar fashion. On the contrary, to fully take advantage of the unique affordances and constraints (Norman, 1988) within the physical and virtual environments respectively, they are by necessity different. We believe that the most important characteristic of corresponding PVA instantiations is that they are tightly linked to each other, and that visual and behavioural characteristics similarity is of secondary importance (Pederson, 1999).

3.1 Technological challenges posed by PVAs

3.1.1 *Automatic update of PVA when a PVA has been altered*

Laws of physics constrain the possibilities for a computer system to modify physical artefacts. "Unlike GUI icons, phicons cannot spontaneously disappear or 'dematerialize,' cannot instantly change position or instantly morph into different physical forms..." (Ullmer, 1997). However, some recent research show promising results for particular kinds of physical objects. Certain materials can be reshaped (MEMS, 2000), and paper-like displays painted with ink made up of special chemical substances can change content (EINK, 2000), if suitable electronic signals are sent to them.

3.1.2 Automatic update of PVA when a PVA has been altered

The challenge lies in recognising and digitising the alterations made on the PVA. Also here there is promising development, regarding certain kinds of alterations. Specifically pen-based activities can be tracked using touch sensitive surfaces (e.g. the Crossboard from IBM), or by tracking the movements of the pen itself, e.g. ANOTO (2000).

3.1.3 Automatic revision control as PVAs become consecutively altered

Revision Control is a research area on its own. The only additional challenge is the recognition of PVA alterations and update of PVAs which are problems already mentioned above.

3.2 Theoretical challenges for the Physical-Virtual Design Perspective

3.2.1 Automatic physical-virtual user modelling

In large a classical Artificial Intelligence problem, although interaction viewed from the physical-virtual design perspective has the potential to model "cross-environment" activities without particular attention or bias towards neither environment.

3.2.2 Painting the physical-virtual border

As a perspective encouraging designers to forget about the differences between physical and virtual environments, probably one of it's most important contributions is to give a clear picture of the border which it is trying to erase. What physical phenomena can't be usefully virtualised? What virtual phenomena cannot be given a useful physical representation?

Figure 3: *One of the user's hand holding a paper document.*

3.2.3 The meaning of physical spaces

As indicated by empirical studies in office environments (e.g., Malone, 1983), and as a well-established fact in the area of architecture, much human activity is expressed by moving objects from one place to another. To acknowledge and interpret these "space semantics" (Harrison, & Dourish, 1996) has to be part of the design perspective.

4. THE MAGIC TOUCH PROTOTYPE ENVIRONMENT

The system keeps track of objects based on a wearable, position-tracked RF/ID reader placed on a finger (Pederson, 2000, 2001) (Figure 3). The user of the system can easily assign names and functionality to physical objects and spaces, as well as create PVAs, e.g. link paper documents to Internet URLs. The system maintains a hierarchical representation of the physical environment where each physical space (e.g. book shelf) and object (e.g. coffee cup) is represented digitally (Figure 4). As soon as the user moves a PVA from one place to another, the hierarchical representation is immediately

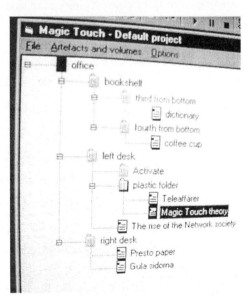

Figure 4: *Spatial relationships between physical spaces and objects visualised as a dynamic hierarchical structure.*

updated. By keeping information about all PVAs in a database, the system fulfils requirement 3 of the PVA definition since this allows users to search for PVA instantiations both in physical and virtual space and inspect them at wish.

4.1   Architecture

A single-user version of the system was first demonstrated in June 2000. At the time of writing, a more flexible client-server based solution is under development. Based on the architecture shown in figure 5, it separates the PVA database from the user interface allowing for a multiplicity of special-purpose user interfaces (of which the hierarchical tree viewer in figure 4 is one), system independency of position tracking and sensor technology, as well as the sharing of PVAs and active volumes among geographically dispersed users.

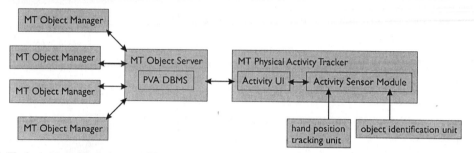

Figure 5: *The basic Magic Touch system architecture consisting of (1) object managers providing special-purpose GUIs, (2) physical activity tracker clients handling input from sensors in a specifical physical environment, and (3) an object server that keeps all authorised clients up-to-date with information about any alteration on Physical-Virtual Artefacts performed by users.*

4.2   Shared active volumes

Within a physical environment facilitated with a Magic Touch system, users can define three-dimensional "active volumes", $\underline{P}VA_{AV}$, in the physical environment that automatically is given a representation $P\underline{V}A_{AV}$ (a folder in the hierarchical structure shown in figure 4). If a specific $\underline{P}VA$ is placed within the $\underline{P}VA_{AV}$, the corresponding $P\underline{V}A$ immediately shows up within the $P\underline{V}A_{AV}$ in the hierarchy. Suppose now that two or more persons have defined identically large active volumes on different geographical locations and that these locations are facilitated with a system such as Magic Touch. Assuming that the system has support for the sharing of active volumes and that the systems have a common database containing information about PVAs (currently under development), as soon as any of the persons would put a $\underline{P}VA$ into "her/his" $\underline{P}VA_{AVshared}$, a corresponding $P\underline{V}A$ would show up in all other instantiations of $P\underline{V}A_{AVshared}$. Interconnected physical-virtual environments allow for more physical interaction among distributed knowledge workers and increases "workspace awareness" (Gutwin, Greenberg, & Roseman, 1996).

5.   MULTI-ENVIRONMENT THINGS (METHINGS)

Space-sharing scenarios such as the active volume-based above has inspired to consider geographically distributed and shared PVAs. Indications suggest that the concept of PVA needs to be generalised in order to encompass the new complexity of these multi-environment things (mEthings).

Networked physical environments has similarities with the concept of ubiquitous computing (Weiser, 1991). However, while ubiquitous computing is something like "access to the virtual world from anywhere in the physical world", networked physical environments is in its extreme "access to the physical world from anywhere in the virtual world".

5.1   General questions regarding mEthings

If the physical-virtual world becomes a giant switchboard of interconnected geographically distributed mEthings sharing complex causal relationships, how do we dare to touch anything? Obviously, it is necessary to make the switchboard transparent in the senses that it should be possible to inspect and "program" by end-users. There might be a trade-off between on the one hand making symbolic manipulation more physical (e.g., Fitzmaurice, & Buxton, 1997; Ishii, & Ullmer, 1997), thereby allowing us to utilise more of our human senses and trained spatial capabilities, and on the other making well-known physical causal relationships less predictable by digitally augmenting things. Another general question regards the problem of privacy and security (Levine, 2000).

6. FUTURE WORK

- To extend and formalise the physical-virtual design perspective into a stable framework for analysis and design of physical-virtual knowledge work environments.
- To finalise the implementation of the client-server based version of Magic Touch and to evaluate the system in general. The Computer-Supported Cooperative Work and Telepresence functions that this version will support, by allowing for the sharing of PVAs and PVA$_{AV}$, are of specific interest.
- To further optimize the Magic Touch wearable device for better ergonomy, as well as more robust identification and positioning mechanisms.

7. ACKNOWLEDGEMENTS

I would like to thank Lars-Erik Janlert, all the students working within the Magic Touch project (in order of appearance): Thomas Saathoff, Fredrik Sandström, Marie Lindström, Susanne Ollander-Åberg, Daniel Granholm, Jonas Hägglund, Peter Landfors, Stefano Salmaso, and Stellan Jonsson. I would also like to thank Anders Broberg, Thomas Johansson and the other colleagues at the Department of Computing Science.

REFERENCES

ANOTO (2000) Anoto digital hand-writing technology. URL: http://www.anoto.com/

Arias, E., Eden, H., & Fischer, G. (1997) Enhancing Communication, Facilitating Shared Understand-ing, and Creating Better Artifacts by Integrating Physical and Computational Media. Designing In-teractive Systems (DIS 97): Processes, Practices, Methods and Techniques Conference Proceed-ings. ACM Press.

Drucker, P. F. (1973). *Managment: Tasks, Responsibility and Practices*. New York: Harper & Row.

EINK (2000) E Ink. URL: http://www.eink.com/

Fitzmaurice, G.W., & Buxton, W. (1997) An Empirical Evaluation of Graspable User Interfaces: towards specialized, space-multiplexed input. In *Proceedings of CHI '97*, ACM Press, 43-50.

Gutwin, C., Greenberg, S., & Roseman, M. (1996) Workspace Awareness in Real-Time Distributed Groupware: Framework, Widgets, and Evaluation. In *Proceedings of HCI'96*, BCS, Springer, 1996.

Harrison S., & Dourish, P. (1996) Re-Place-ing Space: The Roles of Place and Space in Collaborative Environments. In *Proceedings of ACM Conf. Computer-Supported Cooperative Work CSCW'96*, Boston, November 1996.

Ishii, H., and Ullmer, B. (1997) Tangible Bits: Towards Seamless Interfaces between People, Bits and Atoms. In *Proceedings of CHI'97*, ACM Press, 234-241.

Kidd, A. (1994). The Mark are on the Knowledge Worker. In *Proceedings of CHI '94*, Boston, ACM Press.

Levine, R. (2000) The other side of Embedding the Internet. In *ACM Communications*, August 2000.

Malone, T. W. (1983) How Do People Organize Their Desks? Implications for the Design of Office Information Systems. In *ACM Transactions on Office Information Systems*, Vol. 1, No. 1, January 1983, p99-112.

MEMS (2000) Smart Matter Research at PARC. URL: http://www.parc.xerox.com/spl/projects/MEMS/

Norman, D.A. (1988). The Psychology of Everyday Things, Basic Books, N.Y.

Pederson, T. (1999) Physical-Virtual instead of Physical or Virtual - Designing Artefacts for Future Knowledge Work Environments. In *Proceedings of the 8th International Conference on Human-Computer Interaction*, München 22-26th of August, Lawrence Erlbaum Associates. ISBN 0-8058-3392-7.

Pederson, T. (2000) Human Hands as a Link between Physical and Virtual. Position paper in *Proceedings of DARE 2000*, Designing Augmented Reality Environments, ACM, 2000, Elsinore, Denmark.

Pederson, T. (2001) Magic Touch: A Simple Object Location Tracking System Enabling the Development of Physical-Virtual Artefacts in Office Environments. Short paper for the Workshop on Situated Interaction in Ubiquitous Computing, ACM CHI2000. In *Journal of Personal and Ubiquitous Computing*, 5:54-57, February 2001.

Ullmer, B. A. (1997) Models and Mechanisms for Tangible User Interfaces. Master's degree thesis, MIT Media Laboratory.

Weiser, M. (1991) The computer for the 21st century. In *Scientific American*, *265*(3), 94-104.

Dynamic Assignment of Virtual Functions on Physical Objects: Toward the Free Metaphor User Interface

Kazunobu Azuma[1], Hirohiko Mori[2], Makoto Kosugi[1]

[1] Department of Electronic and Computer Engineering, Musashi Institute of Technology
[2] Department of Industrial Engineering, Musashi Institute of Technology
1-28-1, Tamazutsumi, Setagaya, Tokyo, 158-8557, JAPAN
email: linus@leo.bekkoame.ne.jp, mori@ie.musashi-tech.ac.jp, kosugi@cs.musashi-tech.ac.jp

ABSTRACT

In this paper, we describe a way of the intelligent and dynamic assignment of virtual computational functions on physical objects around us as an initial effort to realize a novel concept of Free Metaphor User Interface (FMUI). The major aim of our FMUI project is to try to develop the novel human interface that allows users to control the virtual world with any metaphors which they conceive for manipulating multimedia data. In the Free Metaphor User Interface, users pick up any physical objects around them according to their own metaphor and assign their target file (e.g. a video data file) to it (e.g. a book). The system, then, dynamically constructs the map of the ways to operate the physical object (e.g. turning over a page) to computational functions (e.g. moving forward to the next frame) and allow the user to manipulate virtual media with the physical object. The way of dynamic assignment of computational functions on physical objects described in this paper should be one of the fundamental mechanisms to achieve our goal.

1   INTRODUCTION

Multimedia information is becoming ubiquitous and, in the near future, various types of information, not only in the form of paper document but also multimedia form, will surround us in offices and homes. The current ways of accessing multimedia data, however, is not simple as a way to access paper documents and it is still difficult for novice users to deal with the situation. It can be considered that such difficulties are caused by the mismatch between the Graphical User Interface (GUI) environment and the nature of the multimedia data. In current GUI environment, every bit of information is represented on the two-dimensional display and we manipulate them with two-dimensional input devices such as mouse and keyboard. (In some research projects three-dimensional mouse is developed (Balakrishnan, et. al.,1997.) On the other hand, video data have more than two dimensions, since it has the time dimension. To represent the multimedia data under the GUI environment, it is required to map the time dimension on the two-dimensional space.

To represent information on the time dimension, system designers attempt to compensate for the lack of dimension and to represent multimedia information on the flat space, using metaphors. They naturally make efforts to use good metaphors and to provide good representations. Such efforts are made based on the premise that the metaphors used by the system designers can be shared with all users and the good metaphors for the system designer are also good for all users. From the practical points of view, the premise cannot be upheld because each individual seems to have different metaphors even in the same situation. Therefore, when using a system, a user needs to learn about the designer's model and perform his/her tasks according to the pre-designed manner.

The situation becomes more difficult when dealing with multimedia data. Under the desktop metaphor environment, tasks such as word processing and drawing can be easily represented in the system because all users have experiences in such tasks on the physical desktop and most of them can share common metaphors in them. In case of multimedia information, however, the tasks such as editing the video data are not performed on the physical desktop and, in addition, fewer users have experiences in such tasks. Therefore, in manipulating multimedia data, it is nearly impossible for all users to share the common metaphor because they do not share common experiences. In fact, in our brief interviews to find out about how people found target frames from a video data, one thought about turning pages of a book forward frame by frame, and another thought of pulling at a string to do so. Thus, it is extremely difficult to provide plausible metaphors for manipulating multimedia information for all users.

Therefore, we have started the Free Metaphor User Interface (FMUI) project. The goal of this project is to develop the novel user interface that allows users to control the virtual world with any metaphor that they think of in manipulating multimedia data is proposed. The aim of FMUI is to provide the WYTIHYM (What you think is how you manipulate) environment by inferring each user's metaphor using physical objects. In this paper, we describes the framework of the FMUI and also describe a way of the dynamic assignment, as an initial effort toward our goal, of virtual computational functions on physical objects around us, that is one of the basic mechanism to achieve our goal.

2   THE CONCEPT OF FREE METAPHOR USER INTERFACE (FMUI)

The FMUI is a novel concept of the user interface for manipulating multimedia information. In the environment, users need not follow the metaphor used by the system designer, but can manipulate multimedia authoring systems according to the metaphor that each user thinks of in each task. In the FMUI, all physical objects around us can be used as input devices. When a user has a task of performing and thinking of a metaphor that maps the real world task, the user can select physical objects according to the metaphor. The system, then, dynamically constructs the map in the way to operate physical objects in computational functions and to allow the user to manipulate virtual media with the physical object.

Let us think about the example of picking up some desired frames from a video data file. One user may come up with the idea that turning over pages of a book corresponds to playing forward the video file frame by frame, and he/she picks up a book around them and assigns the target files to the book. Then, the system allows the user to browse the video frames by turning over the pages though the reasoning process where a frame of the video data corresponds to a page of the book. Another user, in the same task, may come up with the metaphor of card cabinet. Then, he/she picks up the physical card cabinet and can browse the video frames by flipping through the cards with fingers.

In this manner, FMUI allow the user to make his/her own working environment according to the mental model with which the user presents his/her own mental images through physical objects to the system and the system reasons the user's mental images with the clues of given physical objects.

3   DYNAMIC ASSIGNMENT OF COMPUTATIONAL FUNCTIONS ON PHYSICAL OBJECTS

In this session, we will explain how to assign computational functions on physical objects by introducing the first prototype system of FMUI that implemented on the LiveDesk[3]. The LiveDesk is a desk which can recognize physical objects and human behavior on the desk and its architecture is similar to the Digital Desk[3]. A digital video camera is mounted above the LiveDesk and is connected to a PC that recognizes physical objects on it, their positions, and actions performed on them. The processed results are projected on the surface of the LiveDesk. Here, in this prototype system, when the user puts some physical objects on the LiveDesk, their graphical images are displayed on the LiveDesk and the user operates the images with the mouse instead of manipulating physical objects. The reason why it is implemented this manner is that it is difficult for a PC to realize the detail human actions on the physical objects in real time.

3.1   The Prototype System

Here, let us assume that the user tries to cut out a part of video data of a video file and he/she thinks of turning over pages of a book for the task of browsing video frames. First, the user brings a real world object (a book in this example) and puts a book on the LiveDesk (Figure 1(a).) The system recognizes the book on the desktop and displays the graphical image of the book (Figure 1(b).) Then, the user selects the icon of the target video file which the user wants to operate, and drags it to the graphical image of the book (Figure 1(c).) At this point, the file is assigned to the book, the system compares the characteristics of the video data with the characteristics of the book, and it determines the map of the computational function and the way the user operates though the reasoning process (Figure 1(d).). In this case, for example, a computational function of forwarding frames corresponds to the task of flipping over the pages of the book. Therefore, the user can browse video frames by turning over the pages of the book (Figure 1(e).).

(a) The book is placed on the LiveDesk.　(b) The characteristics of the book are extracted.　(c) The video data icon is dragged.

(d) The Relational Map between the operations and functions are constructed.　(e) Operate the video by using the image.

Figure 1. Case of Using a Book

Operation on the graphical Image	Operation of the Physical Object	Computational Function
Double-Click the right page.	Leaf through a book forward	Play
Double-Click the left page	Leaf through a book backward	Reverse Play
Click the right page	Turn a page over forward	Fast-forward
Click the left page	Turn a page over backward.	Rewind

Table 1. Map among operations on the graphical image, operations on the physical object (book), and available computational functions

Here, only the relationships between the actions on the graphical images and the ones on the physical objects are fixed in advance. For example, the action of turning over a page is assigned to the action of single click on the right hand page. The obtained relationships, in this example, among the actions on a physical book, the actions on its graphical image, and the available functions are shown in Table 1.In the case, that user thinks of a card cabinet in the same task as above and the results are shown in Figure 2 and Table 2.

In this manner, the FMUI prototype system can infer the user's metaphor on how to perform the task with the computer with the conventional real world behavior. In other words, this process can be considered as that where the system estimates the mental model of the user and links the operation to those occurring with real world objects.

3.2　Linking operations and functions

In the system, there are three types of software objects: one for physical objects, one for data type such as video or audio, and another for computational functions. Among software objects for physical objects, the characteristics of various physical objects, such as types figures, are described. These characteristics can be considered as some factors that are derived from the objects. Figure 3 shows an example of some characteristics of a book. It shows that a book has many two-dimensional pages along the z-axis. Among data type objects, some characteristics of the data files are described. For example, in Figure 4, among video data objects, characteristics such as the array of frames that are two-dimensional pictures are represented in the predicate form. The function object has not only the programs itself but also some rules that determine whether this function can be used in relation to physical objects user picks up and the data type he/she selects.

(a) The file cabinet is placed on the LiveDesk.   (b) The characteristics of the card cabinet are extracted.   (c) Drag of an icon of the video data.

(d) The Relational Map between the operations and functions   (e) Operate the video by using the ima

Figure 2. Case of Card Cabinet

Operation on the graphical Image	Operation of the Physical Object	Computational Function
Click a card	Picking up a card	Move to the frame
Dragging the first card	Picking up card by card forward	Play
Dragging the end card	Picking up card by car backward	Reverse-Play
Dragging a card	Picking up card by card forward from a certain position	Play from a certain position

Table 2. Map among operations on the graphical image, operations on the physical object (card cabinet), and available computational functions

When a user assigns a data file and a physical object, the system compares the characteristics of the data file with the characteristics of the physical object and determines the map of the corresponding elements of the data file and the physical object. For example, in case of a book as the physical object and video data file, a page of the book corresponds to a frame of the video, zaxis to time-axis, and so on. This process is implemented by unification process of the explanation-based generalization (Mitchell, et. al., 1986). Then, this map is applied to the rule part of every function object for the selected data manipulation. If a rule can be applied, the function is added to the list of the executable functions and the system constructs the map of executable functions and the way operations are carried out on the physical objects according to the rule.

4   RELATED WORK

To date, there are various studies about the user interface where physical objects are utilized in the name of Graspable User Interface and Tangible User Interface. For example, in Graspable User Interface (Fitzmaurice, et. al., 1995) and Tangible Bits (Ishii and Ullmer,1997), Phicon, which is the physical version of the icon, is utilized for the user interface. Furthermore, in the research of Triangles (Gorbet, et. al., 1998), the physical objects are used to make a story dynamic by putting them together. MediaBlock (Ullmer and Ishii, 1999) is also a kind of Tangible User Interface for the multimedia presentation and it works as the storage device by linking multimedia data to a block.

These studies are attempts to make the mutual relationship between user and computer by using physical objects, but they are concentrated on the way of using physical objects and the roles of the physical objects are fixed in advance. Our study, by utilizing physical objects as the clue to reasoning user's metaphor and assigning available functions to any physical objects dynamically, aims to allow each user to construct his/her own working environment.

```
physical(book)
sort(frame,t)
frame(flat,dimention(2))
t(axis)
flat(book,page)
dimention_value(book)
square(book,page)
axis(book,Z)
        ….
```

```
sort(frame,t)
frame(flat,dimention(2))
t(axis)
flat (video,frame)
dimention_value(video)
square(video,frame)
axis(video,frame,t)
send(video,frame)
        ….
```

Figure 3. Characteristics of the physical object (Book)   Figure 4. Characteristics of the media (Video)

In research of the visual programming, HI-VISUAL (Hirakawa, et. al.,1990) allows programmers to make computer programs by linking the icons whose figures are physical objects. Though this idea is similar to our study, the kinds of objects used in this system are fixed in advance and the programmer needs to select objects from the prepared list.

5   FUTURE WORK

In this paper, we describe a way of the intelligent and dynamic assignment of virtual computational functions on physical objects around us as an initial effort to realize a novel concept of Free Metaphor User Interface (FMUI). It is a major challenge to completely realize FMUI and this is the initial effort toward it. There are many problems to be resolved.

In the prototype presented here, the system determines the available functions by comparing the characteristics of the media type and physical objects. There are some situations, however, that user needs to link the computational functions to the physical objects directly. The example of this would be the function such as indexing on a frame should be directly assigned with a PostIt. To deal with this issue, we are currently proceeding with a study in the field of drawing software.

Another major issue is how to describe characteristics of each kind of objects. It is hardly possible to describe characteristics of all physical objects around us. In this prototype system, we tried to describe them according to their figures or materials because it would be possible to recognize such characteristics by the image processing techniques in the future. It would still be necessary to build the generic framework for describing them.

REFERENCES

Balakrishnan, R., Baudel, T., Kurtenbach, G., & Fitzmaurice, G.(1997).The Rockin'Mouse: Integral 3D Manipulation on a Plane. In *Proceedings of the ACM CHI '97* (pp.311-318). New York: ACM.

Fitzmaurice, G., Ishii, H., & Buxton, W.(1995). Bricks: Laying the Foundation for Graspable User Interface. In *Proceedings of the ACM CHI* '95(pp.442-449). New York: ACM.

Gorbet, M., Orth, M. & Ishii, H.(1998). Triangles: Tangible Interface for Manipulation and Exploration of Digital Information Topography. . In *Proceedings of the ACM CHI '98*(pp.49-56). New York: ACM.

Hirakawa, M., Tanaka, M., & Ichikawa, T. (1990). An Iconic Programming System, HI-VISUAL. *IEEE Trans. on Software Engineering*, Vol. 16, No. 10, 1178-1184.

Ishii, H. and Ullmer, B., (1997). Tangible Bits: Towards Seamless Interfaces between People, Bits and Atoms, In *Proceedings of the ACM CHI* '97(pp.234-241). New York: ACM.

Mitchell, T. M., Keller, R. M., & Kedar-Cabelli, T. (1986). Explanation-Based Generalization: A Unifying View, *Machine Learning*, vol. 1, no. 1. 47-80.

Mori, H. Kozawa, T., Sasamoto, E., Oku, Y. (1999). A Computer-Augmented Office Environment: Integrating Virtual and Real World Objects and behavior, In H. Bullinger & J. Ziegler (Ed.), *Human – Computer Interaction Vol.2* (pp.1605-1069). Lawrence Erlbaum Publishers.

Newman, W., and Wellner, P.(1992). A Desk Supporting Computer-based Interaction with Paper Documents. In *Proceedings of the ACM CHI* 92 (pp.587-592). New York: ACM.

Ullmer, B., and Ishii, H.(1999). mediaBlocks: Tangible Interfaces for Online Media (video), In *Extended Abstracts of ACM CHI* '99(pp.31-32). New York: ACM.

Development of a Visual Display System for Humanoid Robot Control

Hiroshi HOSHINO, Kenshi SUZUKI, Takashi NISHIYAMA and Kazuya SAWADA

Advanced Technology Research Laboratory, Matsushita Electric Works, Ltd., 1048 Kadoma, Kadoma-City, Osaka 571-8686, JAPAN

ABSTRACT

METI of Japan has launched a national 5-year-project called the Humanoid Robotics Project (HRP) in 1998. In this project, we developed a remote-control cockpit system, which enables an operator to control the humanoid robot at a remote site with sense of presence. This paper describes the development of the visual display system, which is a component of the cockpit system. The developed visual system consists of two subsystems, A Wide Viewing Angle System and a Surrounded Display System. The former captures images around the robot and the latter displays these captured images to a cockpit operator. These subsystems are designed so that an operator can observe the images around the humanoid robot in real time.

1. INTRODUCTION

Nowadays, there is a greater need for automation in the following fields: maintenance of plants and power stations, operation of construction work, supply of aid in case of emergency or disaster, and care of the elderly in the coming aging society. In order to meet these demands, METI, the Ministry of Economy, Trade and Industry established the national 5-year-project of HRP, the Humanoid Robotics Project, in 1998.

In HRP we developed a remote-control cockpit system, which enables an operator to control the humanoid robot as if he/she were at the robot site with sense of presence.

This paper describes the development of a visual display system, which is a component of the remote-control cockpit. The purpose of this visual display system is to represent images around the humanoid robot to a cockpit operator with sense of presence. In other words, a camera system is mounted on the humanoid robot at the remote site and captures the images around it as shown in the right of Fig. 1. The captured images are immediately transmitted to the cockpit, and a display system displays those images to an operator in the cockpit as shown in the left of Fig. 1. Thus, a cockpit operator can perceive the environment around the robot with sense of presence as if the operator were in the remote environment where the humanoid robot was.

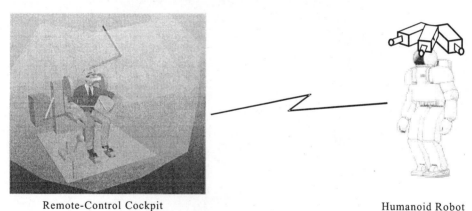

Remote-Control Cockpit   Humanoid Robot

Fig. 1 Concept of a Remote-Control Cockpit System

From the viewpoint of sense of presence we consider the following four factors to be important, a wide field of view, high resolution, binocular stereoscopic display and real-time processing. A wide field of view is necessary so that a cockpit operator can immediately perceive the remote environment around the robot. High resolution is also important so that an operator can grasp the situation in detail. Binocular stereoscopic display enables the operator to precisely perceive the distance and size of target objects in the remote environment. By real-time processing we mean that the system is able to retrieve the images captured by the camera system without any delay and update at a sufficiently fast rate. Real-time processing is very important in order to control the robot smoothly.

In the case of existing systems like CAVE (Cruz et al., 1993) and Semi-Spherical Screen VR System (Shibano et al., 1999) it is difficult to represent the images captured by the camera system in real time for the following two reasons.

1. The source of images is limited to Computer Graphics Workstations.
2. The similarity between the images captured by the camera system and the displayed images to a cockpit operator is not taken into consideration. Therefore it is necessary to process the captured images before displaying to the operator for reducing the incongruity felt by the operator.

Our visual display system is designed such that it possesses all of those four factors, including real-time processing. The developed system consists of two subsystems: A Wide Viewing Angle Camera System and a Surrounded Visual Display System. The two subsystems will be described in detail in the following sections with an experimental example. The paper is organized as follows; Section 2 describes the developed system. The experimental example is shown in section 3 and section 4 concludes this paper.

2. VISUAL DISPLAY SYSTEM

As mentioned in the previous section visual system consists of two subsystems. Let us look into the details of those subsystems.

2.1 A Surrounded Visual Display System

A Surrounded Visual Display System, as shown in Fig. 2, is a multi-screen display system and displays the images around the humanoid robot to a cockpit operator. It has a polyhedral shape, having nine pieces of screens, which are tangential planes to a sphere. An operator looks at these screens at the center of the sphere. The display system has nine pairs of projectors, each pair projecting binocular stereoscopic images onto each screen. Projection light of each projector is circularly polarized by polarization filter.

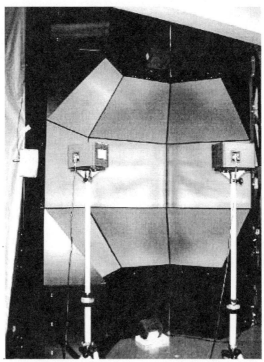

Fig. 2 Photograph of the Developed Surrounded Visual Display System

704

The field of view (FOV) through which an operator looks at each screen is 50° (H) x 39° (V), thus making the FOV of the total display system 150° (H) x 117° (V). The resolution of each projector is 1024 x 768 pixels, producing an angular resolution of 2'56", 0.34 in visual acuity. It is essential that the FOV through which an operator looks at each screen should be equal to that of each camera composing the camera system described in subsection 2.2. By doing so the display system would be to represent the images captured by the camera system in real time.

2.2 A Wide Viewing Angle Camera System

A Wide Viewing Angle Camera System, as shown in Fig. 3, is a multi-camera system, which is mounted on the humanoid robot and captures images around it. In Fig. 3 black cylindrical objects are CCD cameras and the lower glassy objects are reflecting prisms.

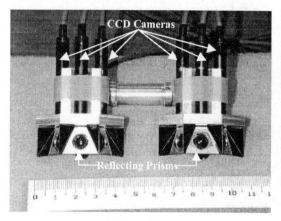 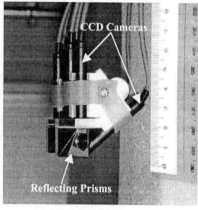

(a) Front View   (b) Side View

Fig. 3 Photograph of the Developed Wide Viewing Angle Camera System

This camera system consists of two sets of four small CCD cameras capturing NTSC formatted images. Analogous to the interval pupil distance (IPD) of humans the distance between the camera sets is 65 mm, thus enabling the system to capture binocular stereoscopic images.

Considering the similarity between the display system and the camera system, it is desirable to use eighteen CCD cameras to capture binocular stereoscopic images in nine directions, but since the payload of the robot is limited, the number of CCD cameras had to be restricted to eight.

Under this constraint, we arranged eight cameras whose captured images are projected onto the four screens as shown in Fig. 4. From the operator's point of view, this screen selection is found to be effective, especially when the humanoid robot walks, enabling the operator to feel sense of presence.

One of the characteristics of our camera system is that it captures a continuous image. It does so by making the center of perspective transformation of each camera optically coincide by reflecting prisms.

The FOV of each camera is 50° (H) x 39° (V), as mentioned in subsection 2.1, thus making that of the total camera system 150° (H) x 78° (V), 19° in the upper vertical and 59° in the lower vertical. The resolution of each camera is 470 x 350 pixels, producing an angular resolution of 6'23", 0.16 in visual acuity. The resolution of the camera system is only one-half of that of the display system because the payload of the robot, as mentioned above, forced us to use NTSC formatted cameras. It is to be noted that cameras with other format with higher resolution such as Hi-Vision cameras are too heavy to be mounted on the robot.

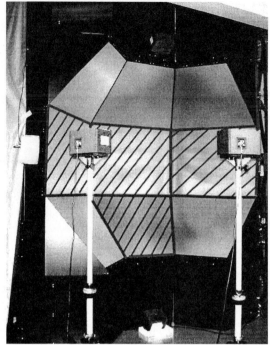

Fig. 4 Pieces of Screens Representing the Captured Images
Hatched screens show the images captured by the camera system.

3. EXPERIMENT

Fig. 5 shows a Wide Viewing Angle Camera System mounted on the humanoid robot. The camera system is mounted so that its height is same as that of a human's eye. Fig. 6 shows a Surrounded Visual Display System representing the images captured by the camera system. Four pieces of screens, as mentioned in section 2, display the images captured by the camera system. Other pieces of screens display CG images in order to assist the operator to control the humanoid robot, for example the operational menu, the map of the remote site where the robot exists, and so on.

4. CONCLUSION

The visual display system described in this paper is well-embedded in the remote-control cockpit in HRP. A cockpit operator is able to control the humanoid robot at the remote site with sense of presence, especially in real time. The validity of the design of the developed system has been verified.

ACKNOWLEDGEMENT
This study is being conducted by Manufacturing Science and Technology Center (MSTC) as part of the Humanoid Robotics Project (HRP) for the New Energy and Industrial Technology Development Organization (NEDO) under the Industrial Science and Technology Frontier Program (ISTF) of the National Institute of Advanced Industrial Science and Technology (AIST) in the Ministry of Economy, Trade and Industry (METI).

REFERENCES
Cruz-Neira, C. et al., "Surrounded-Screen Projection-Based Virtual Reality," Proceeding of SIGGRAPH'93, 1993, pp. 135-142.

Shibano, N. et al., "Development of VR Presentation System with Spherical Screen for Urban Environment Human Media," Transaction of the Virtual Reality Society of Japan, 1999, pp. 549-554 (In Japanese).

Fig. 5 Overview of the Camera System Mounted on the Humanoid Robot

Fig. 6 Experimental Example of the Developed Visual Display System

Virtual Collaborator as Personified Interface Agent for Visualizing Plant Operator's Cognitive Behavior in NPP Plant Control Room

Wei Wu*, Takashi Nakagawa*, Hirotake Ishii**, Hidekazu Yoshikawa**

* Power & Public Utility Systems Dep., Industrial Electronics & Systems Lab., MITSUBISHI Electric Corp.
1-1, Tsukaguchi-Hommachi 8-chome, Amagasaki City, HYOGO, 661-8661,JAPAN
** Graduate School of Energy Science, Kyoto University, Gokasho, Uji City, KYOTO, 611-0011, JAPAN

"Virtual Collaborator" is a sort of virtual robot in 3D virtual reality space, which has human-like appearance and behaves as plant operators do in plant control room; walk, listen, talk, monitor control panels and control the plant through the man-machine interface. A prototype system of such a virtual collaborator has been developed as a virtual operator in a virtual control room of nuclear power plant. It was demonstrated by an example simulation practice for plant emergency situation, that the prototype system by itself could monitor the control panel, diagnose the root cause of plant anomaly, and counteract the transient to safety shutdown state of nuclear power plant in accordance to the operation manual. The implication of this virtual collaborator experiment is that the plant control can be fully automated even in the case of emergent situation so that there would be a possibility that human operator can be excluded from the plant control room, although the above-mentioned experiment was merely a simulation work.

1. Introduction

Recently due to the increased automation by the introduction of modern computer and information technologies, the machinery systems have become so large and complex that the manipulation of the machinery system has become a difficult task for operators. Especially, such tendency is conspicuous in the field of aircraft and power plants. Errors of pilots/operators may give rise to serious accidents. The study on the man-machine interface (MMI) has been extensively conducted to improve the relationship between human and the machinery system.

In this study, authors have proposed the concept of "virtual collaborator (VC)" as an ideal MMI. VC is a sort of virtual robot in 3D virtual reality space, which has human-like appearance and behaves as plant operators do in plant control room; walk, listen, talk, monitor control panels and control the plant through the man-machine interface. A prototype VC system has been developed as a virtual operator in a virtual control room of nuclear power plant(NPP)[1]. The prototype system by itself could monitor the control panel, diagnose the root cause of plant anomaly, and counteract the transient to safety shutdown state of NPP in accordance to the operation manual.

This paper will give a description about the method to visualize operator's internal cognitive behaviors during the plant monitor and control activities by using VC. The visualization method could help to understand real operator's cognitive behaviors and further inspire a new education approach. The configuration of VC is described briefly first, then the visualization method is explained. An example simulation practice for plant emergency situation is conducted to demonstrate the effects of the visualization method. The future application of VC is given in the end.

2. Configuration of VC

VC is configured as a distributed simulation system that consists of (a) an NPP simulator, (b) an MMI simulator, (c) a human model simulator that was constructed by the combination of Petri-net model for the operation procedure and artificial intelligence model of plant diagnosis to simulate plant operators' task for manipulating the plant transient, (d) a human motion simulator and (e) 3D virtual drawing processes that realizes real-time simulation on the network of several workstations. In fact, the three parts of (a), (b) and (c) are the authors' previous product called as simulation based evaluation support system for man-machine interface design (SEAMAID)[2], which is an integrated dynamic simulation system dealing with complex human-machine system interaction in PWR-type NPP. The rest parts of (d) and (e) are a new information 3D visualization mechanism which realizes the SEAMAID simulation as animated behaviors of operator in the main control room of NPP both by 3D virtual reality simulation. The overall software configuration is shown in Fig.1.

SEAMAID has been originally developed as the evaluation system for the design of the control panels in NPP. It consists of three simulators: NPP simulator, MMI simulator, and a human model simulator. NPP simulator is a real time dynamic simulator for an actual PWR-type NPP. It can simulate various kinds of plant anomalies. The MMI simulator is based on an on-line object-oriented database model for the MMI design information of the plant control room. The MMI design information database stores various kinds of information about the equipment in the control room such as layout, shape, location, and CRT menu configuration. MMI simulator is connected to the plant simulator to update the contents of the temporal behavior of instruments on MMI. The human model simulator realizes "intelligent functions" of the virtual operator. It is developed by implementing an extended Reason's memory model[3] into computers as executable programs. The human model simulator has the ability to detect and diagnose an anomaly in NPP and the ability to cope with the anomaly in accordance with the operation manual as well. For the further details of the SEAMAID, please refer to authors' published paper[4] [5]

The human motion simulator generates the human-like motion for the virtual operator during its operation of the virtual control panels. Using the information obtained from the human model simulator such as the virtual operator's action and position, the motion generation process sends the generated motion sequences to the virtual space drawing process. The process can manage a virtual environment in real time, where not only the body motion of the virtual operator but also the various conditions of control room itself can be visualized in 3D virtual space. Figure 2 shows sequential snapshots of the VC's operation in a 3D virtual control room. As for the further detailed information about

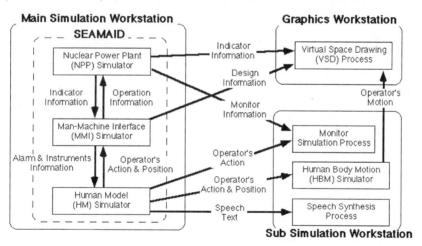

Fig.1 Overall configuration of Virtual Collaborator

Fig.2 Sequential snapshots of the VC's operation in a virtual control room

709

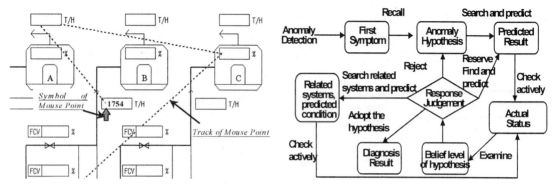

Fig.3 Visualization of MMI operation   Fig.4 Internal Processing for simulating the cognitive

the development of VC, please refer to authors' published paper[1] [6].

3.   Visualization of the Cognitive Behaviors in Anomaly Diagnosis

As shown in Fig.2, operator's external operation behaviors have been visualized with VC. But two kinds of operation behaviors still have not been visualized:(a) Operation behaviors within a CRT interface, (b) Operator's internal cognitive behaviors. Understanding both behaviors would help greatly to make the relationship between human and machinery system better. "Monitor Simulation Process" and "speech Synthesis Process" shown in Fig.1 would take the two types of visualization.

Before explaining the visualization method, the background information about the anomaly diagnosing by VC should be explained first. The anomaly diagnosing behaviors simulated by VC is the behaviors of some NPP engineers in a laboratory experiment conducted by authors before. A CRT interface was provided to show various kinds of information about NPP. Referring to the CRT interface, the experimental subject was asked to detect and diagnose a number of abnormal transients. The CRT interface is a kind of 2D MMI configured by 16 screens to show mimic configuration of NPP subsystems together with the correspondent parameter values. Mouse operation (move & click) is the only required action to operate the CRT interface. Such operations are the external behaviors of operators within the CRT interface and, therefore, should be visualized so that all activities of operator in the control room could be understood well. The concrete visualization method is quite simple by drawing a symbol and lines as shown in Fig.3. The symbol indicates the position of the mouse pointer and the lines indicate the tracks of the mouse movement.

Besides, the more important visualization target is that the internal cognitive behaviors during operator's anomaly diagnosing activities. The developed simulation process of operator's diagnosing process could be illustrated as shown in Fig.4. The diagnosing task sequences are derived from authors' laboratory experimental data. Based on such processing flow, the human model simulator has been validated by comparing the simulation results with the laboratory experimental data[4]. As for the further detailed information about the modeling methods and validation of the operator's anomaly diagnosis, please refer to author's published paper in SMC98 [4].

Here the explanations of each task in the figure are given in the processing sequence.

A)   Detect somewhat abnormal symptom by either alarm or periodic parameter checking. The initial symptom is called as first-symptom.

B)   Recall anomaly hypothesis based on the first-symptom.

C)   Predict the changes in the status of related parameters in accordance with the recalled hypothesis

D)   Verify the predicted changes for the related parameters and compare it with the prediction.

E)   Increase or decrease the confidence level of the hypothesis in accordance with the comparing results.

F)   Based on the changed confidence level, switching process modes among (1) accept the hypothesis, (2)

Table1 Voice-Synthesis for Think Aloud

Situation	Keywords	Examples
Monitoring a subsystem of NPP	Name of the subsystem	"Monitoring" □ ¢ ¢subsystem
Detecting a first-symptom	Name of the first-symptom and the symptom content	□ ¢ ¢is "small" and recall an anomaly hypothesis
Recalling an anomaly hypothesis	Name of the anomaly hypothesis	Maybe it's □ ¢ ¢
Predicting the state of parameters based on the recalled hypothesis	Name of the parameter and its predicted state	In this case, the value of □ ¢ ¢ should be "increasing"
Judging the actual state of the parameter	Name of the parameter, its actual state, and the judgment	Yes, it's "increasing" now Ohh, NO! It's "normal"
Rejecting an anomaly hypothesis	Name of the anomaly hypothesis	The situation is not like to be "□ ¢ ¢
Accepting an anomaly hypothesis	Name of the anomaly hypothesis	"□ ¢ ¢has occurred!
Reconfirming the accepted anomaly hypothesis	Name of subsystem to be reconfirmed	To be sure, check the state of other related subsystems

reject the hypothesis, and (3) check more related parameters. In the case of (2), (3), go back to B), D) separately.

G)   Reconfirm the hypothesis by checking the status of other related subsystems of NPP in the case of (1).

H)   Conclude the diagnosing result and end the simulation.

The above processing is summarized from experimental data of all trials. It represents an overall anomaly diagnosing process observed in the laboratory experiments.

In practice of studying the behaviors of human operators, the think-aloud approach is applied to estimate the internal cognitive behaviors by correlating the speech timing, contents with the external behaviors. In this study, the concept of the think-aloud is utilized conversely. In other words, the voice-synthesis is utilized to generate the speech to express the above internal process explicitly. But, the internal cognitive behaviors of human being are quite complicated to be explained in a number of words. Furthermore, the amount of conscious information came up into human brain could reach a huge number. There is no necessary to express all of the information by voice-synthesis. In this study, the voice-synthesis is made only for several situations in accordance with the above diagnosing task sequence as shown in Tab.1. In each situation, the keywords could be obtained from the internal process of VC so that the synthesis speech contents could be generated. The generated speech contents are in a combined format of two parts: the situation-depend part and the variable part. The situation-depend part indicates the speech situation described in Tab1. The variable part realizes the diversity of human behaviors. Here the diversity means that same speech contents could be expressed in different fashions.

In the verification of the status prediction of related parameters, the emotional condition could be changed in accordance with the comparison results. Especially, the disagreement between the predication and the real status could give operators a stronger impact to his emotional condition. In this study, such emotional effects could also be expressed by including some emotional words before or after the synthesis speech contents, such as "Ohh", "Ohh, No!". This will also increase the personification effects of VC.

4.   Example simulation and Discussion

Based on the above visualization methods, VC was utilized to construct a VR-based education system[6]. Operator's anomaly diagnosis of "Steam Generator Tube Rupture "(SGTR) was visualized in the education system.

From the questionnaires to users of the education system, the effects of the visualization were confirmed such as the user can understand well what the virtual operator thinks in its each activities. The visualization method also provides the function of understanding VC's behaviors online rather than the conventional methods of analyzing the simulation data afterwards to understand how VC diagnosed the anomaly.

The problems remained in the timing of thinking aloud. The MMI operation and thinking aloud were processed in sequence in order to maintain the situation and the contents of thinking aloud. Consequently, the current VC cannot simulate the scene where the operator would think aloud with performing his task in the same time.

The simulation results suggested some requirements for VC in order to enhance the communication function between VC and users. Thinking aloud is a kind of one-direction communication from VC to users. The real meaningful communication requires the information flow from users to VC. Therefore, the further subjects of the study are suggested as recognizing users' voice and facial expression, estimating users' internal condition by the external behaviors (e.g. gesture), and so on. The efforts to realize such functions have been carried out and some achievements have been obtained[7]. Moreover, the feedback effects of such information should be also reflected to change the behaviors of VC as to realize the meaningful communication between users and VC.

Authors think that the further introduction of automation will lead step by step to the stage of full automation where the role of human in the control room will be replaced by computer. It could be imagined that the future plant would be full-automated plant where supervisors in the headquarters could monitor the plant state via reporting from "virtual collaborators" through internet and make consultations with them to manage the plant system such as change of operation mode, need of plant repairs and maintenance, and so on. This is also a plausible image of human-centered automation in future.

References

[1] H. Ishii, W. Wu, D. Li, H. Ando, H. Shimoda, and H. Yoshikawa: A basic Study of Virtual Collaborator –The First Prototype System Integration, Proceedings of The 4th International Symposium on Artificial Life and Robotics, Vol.2, pp.682-685 (1999)

[2] H. Yoshikawa, T. Nakagawa, Y. Nakatani, et al.: (1997), Development of an Analysis Support System for Man-Machine System Design Information, Control Engineering Practice, Vol. 5, No. 3, pp. 417-425.

[3] H. Yoshikawa and K. Furuta, "Human Modeling in Nuclear Engineering", Journal of Atomic Energy Society Japan, Vol. 36, No. 4, 1994, pp.268-278. (in Japanese)

[4] W. Wu and H. Yoshikawa: Study on Developing a Computerized Model of Human Cognitive Behaviors in Monitoring and Diagnosing Plant Transients, Proceedings of 1998 IEEE International Conference on Systems, Man, and Cybernetics: Intelligent Systems For Humans In A Cyberworld , pp.1121-1126 (1998)

[5] T.Nakagawa, Y. Nakatani, H. Yoshikawa, W. Wu, T. Nakagawa, T. Furuta: Simulation-based Evaluation System for Man-Machine Interfaces in Nuclear Power Plants, Proceedings of 1998 IEEE International Conference on Systems, Man, and Cybernetics: Intelligent Systems For Humans In A Cyberworld , pp.1278-1283 (1998)

[6] H. Ishii, W. Wu, D. Li, H. Shimoda, and H. Yoshikawa: Development of a VR-based Experienceable Education System - A CyberWorld of Virtual Operator in Virtual Control Room-. In Proceedings of the 3rd World Multi-conference on Systematic, Cybernetics and Informatics and the 5th International Conference on Information Systems Analysis and Synthesis, pp. 473 478, 1999.

[7] H. Shimoda, H. Ishii, W. Wu, D. Li, T. Nakagawa, and H. Yoshikawa: A Basic Study on Virtual Collaborator as an Innovative Human Machine Interface in Distributed Virtual Environment - The Prototype system and Its Implication for Industrial Application. In Proceedings of 1999 IEEE International Conference on Systems, Man and Cybernetics, pp. V697 V702, 1999.

A SYSTEM FOR SYNTHESIZING HUMAN MOTION IN VIRTUAL ENVIRONMENT

Hirotake Ishii, Kazumasa Sharyo, Daisuke Komaki and Hidekazu Yoshikawa

Graduate School of Energy Science, Kyoto University, Gokasho, Uji, Kyoto, Japan

ABSTRACT

A new designing approach based on object-oriented method is proposed to design a unique but rather complicated virtual environments the salient feature of which is that a human-shaped virtual instructor can behave like a real instructor to educate a trainee about the complicated tasks such as machine maintenance work and plant operation. The advantage of the proposed approach is that it allows the developer to construct various virtual objects for the targeted virtual environment flexibly by combining the components which are independently constructed in advance. In this paper, the elemental technologies used for the object-oriented approach are described such as to realize a virtual instructor, how to design the virtual environments with the object-oriented method and how to configure the system of executing the simulation of the virtual environment.

1. INTRODUCTION

Virtual Reality (VR) technology can be applied for various kinds of industrial and commercial fields such as amusement, education, design support and so on. Especially, various kinds of studies have been made to construct VR-based training environment for machine maintenance work and plant operation (Matsubara, 1997). The VR-based training environment has a number of advantages over the conventional training using real machines or real-size mockups. First, it is safe and economical to conduct training because virtual machines can be used instead of real machines. Second, it is possible to realize instruction functions to support the trainee to learn complicated tasks effectively. Concerning the second advantage, a new training environment has been constructed recently, in which a human-shaped virtual instructor is located to educate the trainee by indicating the trainee's error and demonstrating the complicated tasks (Rickel, 1999). The virtual instructor enables the trainee to learn complicated tasks even if the trainee is alone and a real instructor or the other trainee can not participate in the training. Therefore, such kind of training environment with virtual instructor is very promising as the next generation training environment, but there exists a problem of the workload to construct the training environment for the practical use of real field training.

Then the authors aim at developing a new construction method to reduce the workload for constructing such a next generation training environment. Concretely, the training environment is divided into several subsystems and for each subsystem a new construction method has been developed individually. In this paper, a new construction method is mainly described for human motion synthesizing subsystem which is used to synthesize the virtual instructor's body motion as 3 dimensional computer graphics.

2. ELEMENTAL TECHNOLOGIES TO REALIZE A VIRTUAL INSTRUCTOR

To realize a virtual instructor in virtual environment, 4 kinds of subsystems are necessary to be constructed by the following ways;

Subsystem1 for measuring the trainee's gesture;

> To make it possible for a trainee to manipulate virtual machines, it is necessary to measure the trainee's gesture by using datagloves, polhimus sensors and so on.

Subsystem2 for deciding the virtual instructor's behavior;

> It is necessary to decide the virtual instructor's behavior in according to the states of the virtual machines and the trainee's gesture. For example, if the trainee commits a wrong action, the virtual instructor should point it out and let the trainee correct the error.

Subsystem3 for the management of the virtual environment;

> To make it possible for both trainee and the virtual instructor to manipulate virtual machines just like in a real world, the simulation of the interaction between virtual objects should be based on the related physical laws.

Subsystem4 for synthesizing the virtual instructor's body motion;

> To make it possible for the virtual instructor to show the demonstration of the complicated tasks, it is necessary to synthesize the virtual instructor's body motion as 3 dimensional computer graphics in real time.

In the case of constructing a new training environment for different kinds of training tasks, it is necessary to reconstruct all the subsystems except for subsystem1. So a new method of constructing the virtual environment effectively is required in order to reduce the workload for constructing these subsystems. The authors aim at developing a new construction method with which a training environment for various kinds of training tasks can be constructed easily.

Concretely, for the construction of the subsystem2, the authors have developed a new modeling method for representing the virtual instructor's knowledge for machine maintenance work and plant operation by using Petri net (Endou, 2001). The representation of the virtual instructor's knowledge by using Petri net enables us to construct the virtual instructor's brain visually. For the construction of the subsystem3, the authors have developed the design support system DESCORTE to design the virtual environment in which a trainee can manipulate virtual objects just like in a real world by using human interface devices special for virtual reality (Ishii, 1999). The DESCORTE enables us to design an interactive virtual environment in which physical laws are simulated, only by the guidance of Graphical User Interface (GUI), by setting the information necessary for the training without coding programs. And for the construction of the subsystem4, the authors have developed a human motion synthesizing system AHMSS which is designed based on the idea derived from the concept of affordance, introduced by psychologist James Gibson (Ishii, 2000). The AHMSS enables us to synthesize a various kinds of human body motion easily as 3 dimensional computer graphics. These new construction methods make it possible to reduce the workload for constructing the training environment, but the following problems are remained:

1. The DESCORTE did not take into account of synthesizing the virtual instructor's motion and it is difficult to reuse the information about the virtual environment.
2. The AHMSS did not have the function of making realistic simulation based on physical laws.

Therefore in this study, the authors have developed an object-oriented method for designing virtual environment to realize the remaining three characteristics at the same time:

1. The physical laws can be simulated.
2. The virtual instructor's motion can be synthesized.
3. The information about the virtual environment can be reused to reduce the workload for constructing a new training environment.

In the section 3, the method for synthesizing human body motion is outlined, and in the succeeding sections 4 and 5, the object-oriented method and the system for simulating the virtual environment are described respectively.

3. METHOD FOR SYNTHESIZING HUMAN BODY MOTION

A human has a lot of joints and each joint has the freedom of motion from one to three degrees. So a human has a large number of posture variables. To synthesize the human motion, all of the joint's angles must be specified. Numerous algorithms for synthesizing human motions can be found in literature, but all of them are limited to use for synthesizing a particular motion. For example, the algorithm using 3 dimensional motion capture system is suitable for synthesizing the motion such as walking, gesture. But it is not suitable for synthesizing the motion to manipulate objects with hands. On the other hand, the algorithm of inverse kinematics is suitable for synthesizing the motion to manipulate objects. But this algorithm cannot be applied to the synthesis of the complex motion. Then the conventional method of developing a system using computer animation of virtual humans has been like this; first, what kinds of the virtual human's motion should be synthesized for realizing the system is decided, and then the algorithms and the data for synthesizing those kinds of virtual human's motion are constructed into the system. Although, the motion of the virtual human can be synthesized by such ways, the system is necessary to be reconstructed again when you would like to change the environment.

Then the authors have developed a human motion synthesizing system which is designed based on the idea derived from the concept of affordance. The idea is that the entire algorithms and the information necessary for synthesizing a human motion should be composed in the object database which is an archive for the virtual object's information. This design method makes it possible to use a new algorithm for synthesizing a variety of human motions without reconstructing the human motion synthesizing system. By applying the idea derived from the concept of affordance, it is possible to reduce the workload for constructing the subsystem for synthesizing the virtual instructor's body motion, but this system does not take into account of simulating the physical laws. Therefore in this study, the object-oriented method for designing the virtual environment and the simulation system for simulating the virtual environment have been developed.

4. OBJECT-ORIENTED METHOD FOR DESIGNING THE VIRTUAL ENVIRONMET

In this section, the object-oriented method for designing the virtual environment (OCTAVE) is described. With the OCTAVE method, a virtual environment is constructed by combining plural virtual objects, and each virtual object is constructed by combining plural properties of virtual objects. A virtual instructor is also constructed by combining

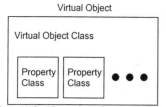

Figure 1 Configuration of a virtual object.

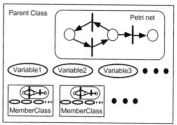

Figure 2 Configuration of a class.

plural virtual objects. Each virtual object and each property of virtual objects corresponds to "virtual object class" and "property class" respectively. Figure1 shows the configuration of a virtual object designed by the OCTAVE method. A virtual object consists of single virtual object class and plural property classes. With the OCTAVE method, as shown in Fig.2, each class consists of the following 3 elements:

1. A Petri-net for representing the discrete states of the class and their transitions,
2. Plural variables for representing the continuous state of the class, and
3. The member classes which add various properties to the parent class.

4.1 Petri-net

With the OCTAVE method, a class has two kinds of the internal state, discrete state and continuous state. The discrete state changes in accordance with the events occurred in the virtual environment. Each discrete state is defined with the process for simulating the property of virtual objects. And when the discrete state of the class changes, the process for simulating the property also changes. For example, when the state of the class for representing the property of the virtual object's movement is "grasped by the other object which has the property 'able to grasp an object'", the process for simulating the virtual object's movement is executed so that the virtual object follows the other object. And when the state of the class which represents the property of the virtual object's movement is "still", the process to fix the virtual object's location and posture is executed. By this way, the transition of the virtual object's state is simulated by changing the process for simulating the virtual object's property.

In the case of constructing a new class, the following informations are mainly defined:

1. The kinds of states the new class can take,
2. The process executed for each state of the new class,
3. How to change the state of the new class in accordance with the occurrence of the events, and
4. The event which should occur next, after the state of the new class changes.

With the OCTAVE method, the above information can be defined by constructing Petri-net. Concretely, one place in the Petri-net corresponds to one state of the class and the existence of a token in the place represents that the class is in the state of the corresponding place. And one transition in the Petri-net corresponds to one event occurred in the virtual environment. With this method, the transition of the class's states can be modeled as the transition of the tokens in the Petri-net.

The important point of the OCTAVE method is that the property of the virtual objects is designed not by indicating the other objects directly, but by indicating the property of the other objects. For example, to design the situation that a pen is located on a desk, the state of the pen is represented not as "located on the desk", but as "located on the object which has the property that some objects can be located on". By this design methodology, it becomes possible to design virtual objects independently of the other virtual objects.

4.2 Variable

With the OCTAVE method, some variables are used to represent the continuous state of the class which can't be represented by the Petri-net alone. For example, the state that location, posture, size of virtual objects would change continuously with time.

4.3 Member class

With the OCTAVE method, a member class is used to add a property to the parent class. Figure 3 shows the relationship between a parent class and a member class. In the Figure 3, the class A, B, C has the property 1, 2, 3 respectively, and the class B and C are the member class of the class A. In this case, the class A has the properties 1, 2, 3 at the same time. By using the relationship between a parent class and a member class in this way, the parent

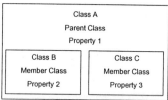

Figure 3 Relationship between a parent class and a member class.

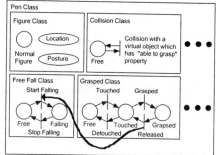

Figure 4 Example of a virtual pen designed

class can be defined by using the definition of the other classes defined in advance. This provision makes it possible to reduce the workload for designing a new class.

4.4 Simulation of interaction between virtual objects

With the conventional method for constructing virtual environments, a virtual object is indicated directly to represent the relationship between two virtual objects. For example, "a pen is located on a desk". But with the OCTAVE method, a property of a virtual object is indicated so that it can make it possible to construct virtual objects independently of the other virtual objects. So, a new function is necessary to simulate the interaction between virtual objects designed with the OCTAVE method. In this study, a particular function is prepared to mediate the information between virtual objects. This function corresponds to "message passing" of the object-oriented concept. For example, to simulate the collision between virtual objects, a particular function is prepared to calculate the location of the virtual objects, detect the collision and inform the occurrence of the collision to the collided virtual objects. With this function, the simulation of the collision between virtual objects can be realized.

4.5 Example

Figure 4 shows an example of a virtual pen designed with the OCTAVE method. The virtual pen consists of Figure Class, Collision Class, Free Fall Class, Grasped Class and so on. The Figure Class adds the property to the Pen Class that the virtual object has shape, location and posture. The Collision Class adds the property that the virtual object collides with the other virtual object. The Free Fall Class adds the property that the virtual object falls. And the Grasped Class adds the property that the virtual object can be grasped by the other object. The variables such as Location, Posture are calculated by the simulation of the virtual environment. For example, when a token exists in the place 'Falling', the variable of the Location is calculated to simulate the falling state of the virtual object. The arc from the transition 'Released' to the transition 'Start Falling' means that after the transition 'Released' fires, the transition 'Start Falling' fires.

4.6 The synthesis of the virtual human's motion

With the OCTAVE method, a virtual instructor is also constructed by combining plural virtual objects, which are constructed by combining plural classes. To make it possible to synthesize the virtual instructor's motion, an "Affordance Class" is prepared to produce the information necessary for synthesizing the virtual instructor's motion. Similarly as the affordance-based method described in the section 3, the Affordance Class is composed in the virtual objects located in the virtual environment. And the information for synthesizing the virtual instructor's motion is transferred from the virtual objects to the virtual instructor by using the message passing described in the subsection 4.4. For example, in the case of constructing a virtual pen, the Affordance Class is prepared so that it would include the information for synthesizing the virtual instructor's motion such as "grasp a pen", "throw a pen". And the Affordance Class is composed in the virtual pen with the other classes described in the subsection 4.5. And the virtual instructor's motion is synthesized by using the information dynamically transfferd from the virtual pen to the virtual instructor.

5. SYSTEM CONFIGURATION OF VIRTUAL ENVIRONMENT SIMULATION

In this study, to make it possible to design the virtual environment in accordance with the OCTAVE method described in the section 4, an original script language has been developed. Figure 5 shows an example of the script language. And the simulation system to simulate the virtual environment in accordance with the script language has been developed. Figure 6 shows the configuration of the simulation system. The simulation system consists of

716

```
#### Free -> Grasp ####
EventNew Grasp_Event
    Variable Instance ClashHand_Instance_Var
    Trigger <(Event)HandFigure_Instance,Clash_Event> ClashHand_Instance_Var
    StateChange Free_State Grasp_State
    Function None , None
        Fire <(Event)HandFigure_Instance,ToNotClash_Event> None , None
        Fire <(Event)BallFigure_Instance,ToNotClash_Event> None , None
        Fire <(Event)HandBound_Instance,Bound_Event> ClashHand_Instance_Var , None
    End_Function
End

#### GetAffordance ####
Event GetAfordanceName_Event GetAfordanceName_Event
    StateChange Free_State Free_State
    Function None , Afordance_Name_Var
        Calc Afordance_Name_Var = <(String)Glasp>
    END_Function
End
```

Figure 5 Example of the script language.

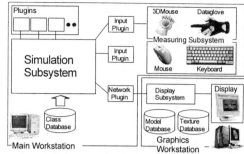

Figure 6 Configuration of the simulation system.

Figure 7 Example snapshots of the virtual environment designed with the OCTAVE method.

Simulation Subsystem, Measuring Subsystem, Display Subsystem and some Plugins. The Simulation Subsystem simulates the virtual environment in accordance with the script language. The Measuring Subsystem measures a trainee's gesture and accepts the indication to the virtual instructor. The Display Subsystem generates 3 dimensional graphics of the virtual environment. And the Plugins are used to add the functions to the simulation system without reconstructing the system. Figure 7 shows the example snapshots of the virtual environment simulation designed and executed by the OCTAVE system.

6.   CONCLUSIONS AND FUTURE WORKS

In this study, the object-oriented method has been proposed to reduce the workload for constructing the virtual interactive environment where a virtual instructor can behave like a real human. Based on the method, the OCTAVE simulation system has been developed to simulate the virtual environment. The proposed method makes it possible to reuse the information about the developed virtual environment, but some problems are still remained to be resolved:

1. The workload for constructing the virtual environment from the beginning is very large.
2. The high level skill is required when constructing the virtual environment according to the proposed method.

Therefore, it is considered that the system with which the virtual environment can be designed through Graphical User Interface with the OCTAVE method should be developed in the future.

REFERENCES

Matsubara, Y., Toihara, S., Tsukinari, Y. and Nagamachi, M., "Virtual Learning Environment for Discovery Learning and Its Application on Operator Training", IEICE TRANSACTIONS on Information and Systems, 1997, Vol. E80-D, No. 2, pp. 176-188.

Rickel, J. and Johnson, W., "Animated Agents for Procedural Training in Virtual Reality: Perception, Cognition, and Motor Control", Applied Artificial Intelligence 13, 1999, pp. 343-382.

Endou, K., Ishii, H. and Yoshikawa H., "A Study on Petri-net based Management Method between Virtual Agent and Real Human in Virtual Collaborative Environment", Human Interface Society 12th Research Meeting, 2001, (To be presented, in Japanese).

Ishii, H., Tezuka, T. and Yoshikawa H., "User-Interface Design of the Support System for Constructing Virtual Environment", Proceedings of the 8th International Conference on Human-Computer Interaction, 1999, Vol. 2, pp. 1182-1186.

Ishii, H., Ichiguchi, N., Komaki, D., Shimoda, H. and Yoshikawa H., "Development of Affordance based Human Motion Synthesizing System", Proceedings of Cognitive Systems Engineering in Process Control 2000, 2000, pp. 241-248.

DEVELOPMENT OF A GROUP TRAINING SYSTEM FOR PLANT WORKERS USING NETWORKED VIRTUAL ENVIRONMENT

Takashi Nagamatsu*, Michiya Yamamoto**, Shigenari Shiba*,
Daiji Iwata**, Hidekazu Yoshikawa**
*Kobe University of Mercantile Marine
**Graduate School of Energy Science, Kyoto University

ABSTRACT

A VR (Virtual Reality)-based cooperative training system is proposed for machine maintenance in nuclear power plant. The NETCOM (NEtworked-vr based Training system for COOperative machine Maintenance) system has been developed on basic software called LASNET (Large Scale Networked virtual EnvironmenT), which can offer flexible CSCW environment for networked virtual environment with multimedia communication between participants. By using the NETCOM, instructor can teach a group of trainees a maintenance procedure how to calibrate the indication of a reactor coolant pump in separate rooms. It was verified by an experiment that the NETCOM system would give a sensation of presence in the course of participated training.

1. INTRODUCTION

Training of machine maintenance work is required to reduce human errors and improve reliability. And it is getting more important especially for large-scale plant such as nuclear power plant. Various mockup machines are used for this kind of training in training facilities. But all such mockups are expensive and there are not many training facilities. Therefore if training can be performed on remote computers, opportunities of training increase and workers would become more skillful.

In the research field of VR, the Networked Virtual Environment (NVE) has been developed extensively. NVE allows multiple users to share virtual space through computer network. The authors developed a base software called LASNET (Large Scale Networked virtual EnvironmenT) (Yamamoto et al., 2000), for constructing NVE aiming at applying for wide use of NVE such as training system, virtual campus, virtual city, etc.

In this study, a training system for machine maintenance in nuclear power plant has been developed using LASNET. Concretely the system provides a training environment of the cooperative machine maintenance work for calibrating the flow rate indication of the reactor coolant pump, which is conducted by several persons in three different rooms: main control room, instrumentation rack room and pump room.

2. LASNET(LArge Scale Networked virtual EnvironmenT)

The authors developed a base software called LASNET. By using the LASNET, a large virtual space can be built by connecting basic units of virtual space which we call "Locale" in order to avoid computational load. Figure 1 shows the overview of LASNET. The LASNET has a server-client configuration. In figure 1, the World Server supervises all the data communication between Locales. Each Locale server can transfer the data with each other through the World Server. It makes the virtual space to be consistent in all the Locales when any object moves between different Locales or the virtual space is expanded by adding a new Locale. The World Server has a special database called World Information Database. It provides the structural data and the geographical data needed for Locales. And it consists of several content files which are described by XML(eXtensible Markup Language).The Locale Server simulates each virtual space by using data of the World Information Database. And the Browsers are the interface for users. The users interact with the virtual space through the Browsers. Interface of the Browser must be designed appropriately according to the application subject. And the World Server is able to add a special user-defined database other than the World Information Database. For example, plant simulator can be added to the World Server and it becomes a part of the LASNET and act as a database. The LASNET is also provided with the function of voice communication. Users can communicate with each other by this voice communication function.

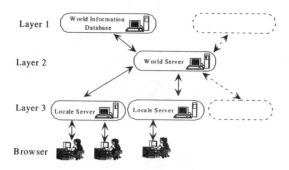

Figure 1. Overview of LASNET

3. GROUP TRAINING SYSTEM FOR PLANT WORKERS

For the purpose of training cooperative maintenance works in nuclear power plant, a prototype system has been constructed on the LASNET. The system is called NETCOM (NEtworked-vr based Training system for CoOperative Machine Maintenance). The objective of NETCOM system is for the efficient training of maintenance workers who are located separately to calibrate the flow rate meters of reactor pump in the three rooms: main control room, instrumentation rack room and pump room. The prototype system has developed by the combination of desktop PCs, because such systemization could be easily realized and widely utilized without any special and expensive devices such as 3D mouse, HMD, etc.

On designing the NETCOM training system, the concept of human-centered design for VR (Gabbard et al., 1999) is applied, and Table 1 shows the key features for the cooperative machine maintenance.

Table 1. Identifying key features of the training system for cooperative machine maintenance

Item	Key features
VE users and User tasks	Users: Trainee, Instructor
	Tasks:
	Moving
	Selecting of the instrument
	Operation of the instrument
	Communication with other users
The Virtual Model	Virtual Plant, Map, Avatar and Machine Maintenance Manual
VE User Interface Input Mechanisms	Mouse, Keyboard, Joystick and Mike
VE User Interface Presentation Components	Display and Speakers

The NETCOM system has developed based on Table 1. As for browser to users, two interfaces have been developed, one for the trainee's interface as shown in Figure 2, and the other for the instructor's interface as shown in Figure 3.

As seen in Figure 2 the trainee's interface has 5 windows: (1) Virtual Model Display Window, (2) Operation Supporting Window, (3) Communication Window, (4) Maintenance Manual Display Window, and (5) Map Display Window. In the Virtual Model Display Window the objects are presented by 3D image. User can move freely by using joystick, keyboard or mouse which user would like best. All the devices have the same function of moving to three directions. User can operate the instruments by using the Operation Supporting Window. When user reaches in the neighborhood of the instrument he would like to operate, the picture of the instrument will appear on this window. And user can operate the instrument by mouse clicking on the 2D picture. Voice communication function is afforded in the NETCOM by using the original function of LASNET, because voice communication by paging or intercom is conventional in the nuclear plant. But, additional function of text chat is also available in NETCOM for auxiliary assistance. The Maintenance Manual Display Window is available for training support. The Map Display Window is the function of LASNET and the window displays the map of whole Locales in the virtual space.

719

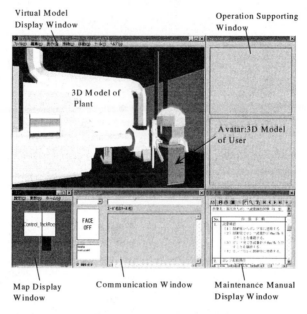

Figure 2. Trainee's interface

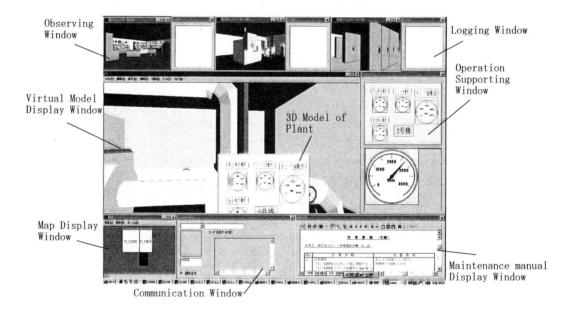

Figure 3. Instructor's interface

In the previous version of NETCOM the instructor used the same interface as the trainee's (Nagamatsu et al., 2000). The result of experimental evaluation showed that specific interface for instructor is necessary. And the requirements for instructor's interface are functions of (1) observing all the rooms and (2) displaying log of trainees' operations.

As shown in Figure 3, the instructor's interface has 11 windows: 5 windows in the middle and lower part have the same function as in the trainee's interface, while the other 6 windows in the upper part of Figure 3 are special to instructor to meet the instructor's requirements. By using the Observing Window, the instructor can watch three rooms to know trainees' behavior in each room. Operation record of the trainee in each room is displayed on each Log Window in text. The instructor can know details and history of trainee's behavior.

The whole NETCOM system configuration is shown in Figure 4. In the NETCOM system, each Locale Server simulates different room, while the simulation of plant instrumentation response that extends over three rooms is generated on the World Server by the help of the World Information Database. Each trainee accesses each Locale of the three rooms through the Browser to join the training together. As for instructor's interface, the Virtual Model Display Window connects to one of the Locales. Instructor's avatar can move to other rooms in virtual space as changing the Locale to connect. Each Observing Window always connects to each Locale sever and get the information of each room. Each Logging Window connects to each Locale sever too.

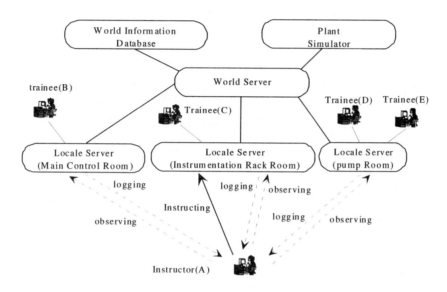

Figure 4. System Configuration of NETCOM

4. EXPERIMENTAL EVALUATION

A scenario-based experiment was conducted to evaluate the effectiveness of the NETCOM system and the usefulness as distributed remote training.

4.1. Method

5 persons participated using a PC, with a display, speakers, a joystick, a mouse and a keyboard. One was an instructor and four was trainees. PCs were connected by LAN, which had enough bandwidth and no noticeable latency. Each trainee was located in different room and couldn't see and hear other partner's doing in real world.
In virtual space, one trainee was in main control room, another trainee was in instrumentation rack room and the other two trainees were in pump room. Instructor moved around in virtual plant.
The Experiment was conducted in order as follows;
(1) Lecture about the content of training and usage of the system,
(2) Filling in a questionnaire sheet about understanding of the procedure of their work,
(3) Training by 4 trainees and instructor to check machines as written in maintenance manual, and
(4) Filling in questionnaires sheet about training and usability.

721

For the subject who played instructor's role, instructor was instructed to observe the trainees, and direct or advise properly when the trainee needed help during the training,

4.2. Results

The result of the questionnaires given to the four trainees is summarized in Table 2. From this questionnaires, it was found that they feel the "sensation of presence" during cooperative machine maintenance work. Almost all the users said that they became to understand the procedure of maintenance work. And the usability of the whole system is good.

Table 2. Score of trainees

Items	0: good, 1:bad Normalized Average score
Sensation of presence for training	0.41
Degree of understanding	0.41
Usability as the whole system	0.25

Some of the instructor's impressions of the system are:
- Additional windows that display the situation of other rooms are convenient.
- I could decide which point to advise to know trainees' location and behavior.

It was found that from the result of the experiment described above, the NETCOM system would be useful for training of plant workers.

5. CONCLUSIONS

In this paper, a VR-based cooperative training system was proposed for machine maintenance in nuclear power plant. The NETCOM system was developed on the basic software called LASNET, which can offer flexible CSCW environment for networked virtual environment with multimedia communication between participants. The interface for cooperative training were constructed based on "Human-centered Design for VR system", and it was verified by an experiment that the NETCOM system would give a sensation of presence in the course of participated training among the users. And the result shows that the system is useful for remote distributed training.

REFERENCES

Gabbard, J. L., Hix, D., & Swan, J.E., II (1999). User-centered design and evaluation of virtual environments. IEEE Computer Graphics and Applications, 19(6), 51-59.

Nagamatsu, T., Iwata, D., Yamamoto, M., Osaka, Y., Matsuzaki, T., & Yoshikawa, H. (2000, November). Interface Design for Networked Virtual Reality based Training of Cooperative Machine Maintenance Work. Paper presented at the Conference of Cognitive Systems Engineering in Process Control, Taejon, Korea.

Yamamoto, M., Iwata, D., Nagamatsu, T., Shimoda, H., & Yoshikawa, H. (2000). A simulation method of distributed virtual environment for training of maintenance work in large-scale plants. Transactions of the Virtual Reality Society of Japan, 5(4), 1103-1112.

Open Simulator: Architecture for Simulating Networked Virtual Environment by Utilizing Online Resources

Michiya Yamamoto*, Yoshihiro Osaka*, Takashi Nagamatsu**, Hirotake Ishii* and Hidekazu Yoshikawa*

*Graduate School of Energy Science, Kyoto University, Kyoto, JAPAN
**Kobe University of Mercantile Marine, Kobe, JAPAN

ABSTRUCT

Rapid and remarkable progress in both computer technology and network technology has opened the way for practical applications of networked virtual environment (NVE) technology. In this study, the authors have proposed a new framework for simulating NVE by utilizing online resources, such as web pages, so that such NVE can be easily created and effectively applied in the area of remote education. This paper describes the framework and its application practice.

1. INTRODUCTION

The Internet has opened tremendous possibilities for remote education, where teachers and students can be communicated with each other through the Internet. The rise of Networked Virtual Environment (NVE) technology has began to enable such teachers to provide multimedia experiential education over the Internet. By utilizing the NVE technology, teachers and students at various locations can participate in a common shared Virtual Environments (VE) over the Internet and interact with VE, which is created by computers. It has begun to be widely utilized for various educational areas, such as language lesson, operation of aviation or power plants, etc. For such NVE, the authors have already developed a NVE-based training system named as NETCOM for the training of cooperative machine maintenance work in nuclear power plants, and confirmed the validity of NVE-based training [Yamamoto, 2000].

From the viewpoint of human computer interaction, visualized 3D images for such NVE-based experiential educations can provide students with precise understanding of the teaching materials. In addition, teachers can virtually create classrooms in 3D virtual space, where students can study together as a group, or move freely from one room to another as in the real world. Therefore, a new style of remote education can be realized by the NVE-based experiential education.

However, the teachers cannot necessarily construct such educational VE as they thought it should be, because it is not easy for them to construct such NVEs by themselves, because they have (1) to construct the environment itself so as to visualize 3D space, (2) to generate dynamic behaviors of various objects in the environment, and (3) to teach students by various teaching materials in the NVE.

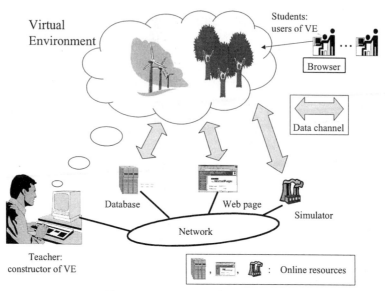

Figure 1: Framework of open simulator.

In this study, the authors propose a new framework of NVE which they call "open simulator" so that teachers can feel free to create and publish their own VEs by focusing on the application area of "experience-centered" type of education. The authors have realized this framework by the following two steps:

(1) Design of "open simulator" framework for simulation of NVEs, and

(2) Implementation of open simulator for supporting users to construct VEs flexibly.

2. FRAMEWORK OF OPEN SIMULATOR

2.1 Backgrounds

The framework of "personal site" has already been proposed for the flexible construction of NVEs [Maeda, 2000], which is accompanied by the progress of WWW technology. Based on the framework, users can easily participate in the construction of their personal NVEs by describing original language "PPML (PalmPlaza Markup Language)", as if they were able to publish web sites by describing web pages by HTML. Virtual environment itself and simple behavior of objects can be described by PPML. The personal site NVE can effectively be utilized in various educational areas, because the professionals of the area can make contribution for the construction of NVEs.

However, there lies the following difficult problem to be overcome when they create various teaching materials: difficulty of integrating existing teaching materials with VEs.

2.2 Framework of open simulator

The authors proposed new architecture for simulation of NVEs as a solution for the aforementioned problem. Figure 1 shows the framework of the NVEs, which is named as "open simulator". It can integrate NVE system with online resources, such as existing simulators or web pages, for the simulation of NVE as teaching materials. The information acquired from the online resources are communicated through the Internet and made effects into NVE.

724

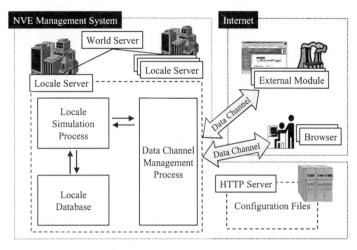

Figure 2: System configuration of MALIONET.

The online resources are called as "external module" and the communication over the Internet is called as "data channel" hereafter. Users can participate in the NVEs by using software named as "Browser".

Users can reuse existing external modules or make external modules by themselves so that they can control the behavior of virtual objects. This architecture of open simulator is also effective for load-balanced computation of large-scale NVE simulations because of the independence of external modules.

3. MALIONET: IMPLIMENTATION OF OPEN SIMULATOR

3.1 Configuration of MALIONET

The authors implemented open simulator to a software system, which is named as MALIONET (Multiple Access of Linked Information Over NETwork). It is composed of locale servers, external modules, data channel and browsers. Figure 2 shows the whole system configuration of MALIONET. A locale server simulates a locale, which is a small unit of VEs proposed by separate server topology [Capin, 1999], for load-balanced computation of the whole VE space.

Locale server is composed of 3 processes: (1) Data channel management process, (2) Locale database, and (3) Locale simulation process. Data channel management process checks whether or not the data channels are linked to the locale. Locale database manages all of the information about all virtual objects. Locale simulation process creates the whole VE by the help of information of locale database. The data from external modules are integrated into NVEs by utilizing these processes.

Data channel management process controls also synchronization between multiple external modules automatically by adding sequential numbers on to the transferred data of data channel. This method is based on optimistic protocols proposed in the area of PDES (Parallel Discrete Event Simulation) [Hybinette, 1999]. Users can construct NVEs without taking into consideration of computational coherency in the synchronization of different external modules.

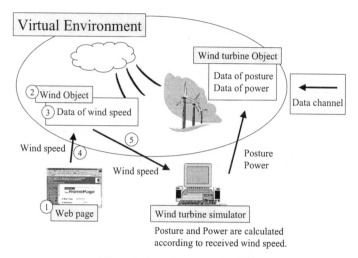

Figure 3: Steps for constructing VE.

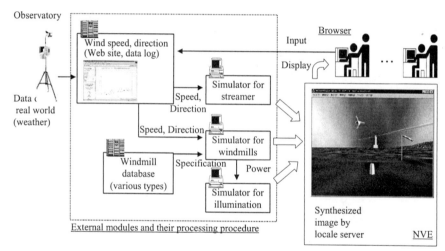

Figure 4: Relationships of multiple external modules for simulation of wind turbine in EEWING.

3.2 How to construct educational VE

When a teacher would like to provide a VE for experiential education, he or she can easily construct the targeted VE. Figure 3 illustrates the steps for constructing VE by the proposed MALIONET with concrete example: (1) Construction of external modules, such as a web page shown as ・ ; (2) Description of configuration files of objects and data in the VE shown as ・ ・and ・ ; (3) Generation of data channel between locale sever and external modules, such as ・ ・and ・ ; and finally (4) Startup of all the system to begin the VE simulation.

MALIONET will enable users to construct NVEs only by describing configuration files and utilizing external modules.

726

4. APPLICATION PRACTICE

The authors configured a VE as application practice by utilizing MALIONET for the purpose of "experientiable teaching" of how the wind power generator works and how the produced electric power would be used. It is a prototype NVE for remote and experiential education and named as EEWING (Experiential learning Environment for WINd Generator). Users can learn with a teacher how the wind generators work and can simulate by themselves whether or not it can supply enough power. We selected this content as teaching materials because it is difficult to acquire the correct knowledge about wind generators, such as types or sizes of generators appropriate for the given speed of wind, or the generated power by the various types of wind generators, etc. with their related quantitative knowledge.

Figure 4 shows the relationships of multiple external modules for the simulation of wind turbine in EEWING. Each external module is shown as a box, and a pair of data channel is shown in a line arrow. As a result of test use of this system, it was confirmed that all of these relationships were correctly managed in real time by data channel management process of MALIONE. For example, the observed wind speed, which will be given in real time on the web site, was inputted correctly into simulators for streamer and windmills.

5. CONCLUDING REMARKS

The authors have proposed a new framework of open simulator for the simulation of NVE. They constructed MALIONET as an implementation of open simulator, and confirmed the validity of the framework by constructing prototype environment EEWING.

The merit of open simulator and MALIONET lies in that the teacher has only to describe configuration files to create NVE and collect proper data from already connected external modules. This is the most significant merit of the open simulator and its implementation of MALIONET, to be compared with the laborious work that the teacher has to create such NVE by programming the whole simulating code of teaching materials and by collecting all the necessary data for executing the simulation.

REFERENCES

Yamamoto M. et al. (2000). Development of Networked Virtual Reality based Training of Cooperative Machine Maintenance Work. *Human Performance, Situation Awareness and Automation: User-Centered Design for the New Millennium, Savannah, U. S. A.* 330-335.

Maeda Y. et al. (2000). PalmPlaza: A Virtual Space Integrating the WWW and Multi-user Communication Platform for Personal Site. *VRSJ Research Report Vol. 4, No. 3.* 13-18.

Capin T. K. et al. (1999). *Avatars in Networked Virtual Environments, John Wiley & Sons, Ltd.*

Hybinette M. and Fujimoto R. (1999). Optimistic Computation in Virtual Environment. *Proceedings of the Virtual Worlds and Simulation Conference (VWSIM '99).* 39-44.

CONFIGURABILITY AND DYNAMIC AUGMENTATION OF TECHNOLOGY RICH ENVIRONMENTS

Thomas Binder & Jörn Messeter

Space & Virtuality Studio
The Interactive Institute
S-205 06 Malmö, Sweden
{Thomas.Binder; Jorn.Messeter}@interactiveinstitute.se

ABSTRACT

The notion of usability has developed since the beginning of the HCI-field from human factors to a concept highly relative to specific use situations and including social aspects concerning communication and collaboration. What comprises usability in concepts based on augmentation technologies? Based on experiences from interaction design in technology rich environments we argue that the concept of augmentation may facilitate a shift in focus within the HCI community away from standardized interface formats and towards a new and diverse world of distributed and dynamic interaction design where the possibilities of configuring interaction for a specific use situation is an important aspect of usability.

1. INTRODUCTION

As the design materials of HCI develop towards ubiquitous computing, the range of potential users and use contexts is rapidly expanding and traditional approaches to interaction design are called into question. In the light of these changes augmentation is in several ways an interesting perspective on computer support. As opposed to the ideas of using computers to automate tasks, the notion of augmentation stress that information technology can be designed for enhancing both the human sensing of the environment and the human capability of intervening in this environment. Similarly augmentation seems to offer a way of avoiding the widespread common-sense duality of the physical and the virtual. Augmented reality (AR) technologies were for long associated with a fairly limited range of human interfaces such as head-mounted displays through which the user could experience visual overlays to the visual appearance of the surroundings. The origin of the representations displayed has in general either been generated offline (as in museum type systems) or they have been the result of complex vision techniques applied to an on-line recorded video image. From a usability or human-centered design perspective these interpretations of augmentation appear to be far too modest and single minded.

2. AUGMENTATION IN PROCESS CONTROL ENVIRONMENTS

The interest in ubiquitous computing beginning with Weiser (Weiser 1991) has mainly been fueled by application examples from environments with a fairly poor penetration of technology such as office or home environments. The attempts to let the human-computer interface transcend established formats of screens and keyboards have been governed by the need particularly in collaborative settings to make information technology blend into the mixed media environments of social interaction (Ishii and Ulmer 1997). While we share this search for new interaction modalities we have found it particularly interesting to study technology-rich environments in which all human activities are highly mediated by technology. The process industry provides an interesting example. The work settings of the process industry are both rich and mature in terms of technology density and represent a very thoroughly constructed setting. In addition the process plant has produced one of the strongest guiding images for the duality between information and actual process: the control room. In later years the focus in the industry has been shifting from well-known automation trends towards a more delicate interest in the interplay between technology and people (Zuboff 1988). To look into interface design for such an environment from the perspective of augmentation along the lines suggested by i.e. Welner et al. (Welner et. al. 1993) invites augmenting the sensing and intervention capabilities of process operators in ways which we believe raise challenging questions as to what, where and how to augment an already highly constructed environment.

2.1 TOWARDS DISTRIBUTED PROCESS CONTROL

In collaboration with process operators at a number of process plants we have worked with the design of mobile interfaces for process control systems. We sought to challenge the strong guiding image of the centralized control room searching for more flexible means for supporting process control work through augmentation. At some plants steps have already been taken towards more distributed process control as PCs in the offices of process operators, and at strategic places in the plant, have substituted the centralized control room. The large geographic area often covered by the plants, further accentuated the tension between centralized control facilities and local needs of access to process information in daily operator work. Ethnographically inspired studies of process control work revealed some characteristics of this tension in the daily work practice.

The importance of physical inspection

Operators spend considerable time making inspection rounds in the plant. During these rounds they can verify the mapping between digital representations of machine components and the physical state of the same components. Discrepancies between readings and actual states are a common source of problems in process control and operators in general learn not to rely on information provided from only one source. Physical inspection allows operators to collect visual, haptic and auditory information not provided by the control system and they are equally attentive to the operation of components as to the quality of processed materials. Presence in the plant is not just a matter of collecting information but also helps operators to establish and maintain a shared understanding of the process. Finally, operators learn about ongoing work from interpreting traces of their colleagues' activities (e.g. tools left for later use or dismounted components) and as they occasionally »bump in« to one another during these rounds they exchange information on current problems. A large part of the work co-ordination is done informally in spontaneous meetings on various locations in the plant.

Dynamics in process control

The general interaction scheme implied by traditional process control systems suppose operators to monitor the system for alarms and, as they appear, take appropriate action in order to bring the system back to the desired state. However, this stands in contrast to research on process operator work practice (e.g. Zuboff 1988) and also to the results from our field study. We found that many alarms are caused by well-known and unproblematic events that do not call for action. Rather, physical inspection allows operators to discover problems before they generate alarms in the system and in general the work strategy of operators is highly proactive. This also reflects the dynamics of process control. The plant is never in any simple sense just up and running. Continuous variations in process input, combined with on-going construction and repair work disrupting normal routines, means that new problem settings constantly evolve. Problems that arise often have consequences further along the production chain calling for collaborative efforts in problem solving. As these situations are approached the exact nature and location of the problem is often not clear. Operators frequently engage in setting up small-scale experiments as a strategy to frame the problem or to test a tentative solution. This may involve the use of temporary instrumentation where readings cannot be provided from the control system. Results from experiments can lead to new experiments and problem solving often evolves as a dialogue with the situation at hand.

Fig. 1. Inspection allows operators to collect information not provided by the control system.

2.2 DISTRIBUTED CONTROL AND AUGMENTATION

The importance of physical presence in the plant together with the proactive work strategies demanded by the dynamics of process control point towards a need for a process control system structure that is distributed rather than centralized. Also, the operators' experimental approach to problem solving requires a highly flexible support where monitoring can be set up based on the nature of the problem. In our view this is a context for human machine interaction where augmentation seems to be a particularly interesting approach allowing access to process information from the control system to naturally »blend in« with the operators interaction with physical machinery. Moving towards distributed control raises at least two issues: the density of access points and the requirements for interaction at these access points.

The density of access points
Placing PCs at strategic places in one plant studied was a step towards more distributed process control but the density of access points was clearly not enough. Operators were frequently forced to switch between interacting with the machine components of the plant and interacting with digital representations of these same components at the nearest PC which disrupted their workflow. Also, the PCs provided the same fixed view of the process as the centralized control system and did not support adaptation of the interface to the problem at hand. The goal of augmented reality systems, as described in most current research, is to enhance the users' view of the real-world by providing context-based information using a mobile device. Ubiquitous access to process information through an AR-system could provide operators with »maximized« density of access points and process information available anywhere in the plant. However, AR-systems are often based on the notion of overlaying the real world with digital information. Such »context-sensitive browsing« of a statically augmented environment does not respond to the process operators need for configuring interaction for the specific problem at hand.

Requirements for interaction – creating and sharing temporary focal sets
The operators' need for flexible access to process information and the issue of distributed control is not simply a dichotomy between *either* centralized control *or* local control facilities in the vicinity of physical inspection points at a particular place in the plant. An operation problem in process control seems only rarely to be localized to a particular geographical location. It is more often related to the malfunction of a combination of components in different parts of the plant, which in a non-trivial way has to be »brought together« at the operators »place of action«. The spatial positioning of these components may be far apart creating a need for remote monitoring. Second, operation problems in process control are dynamic. The set of machine components that needs to be monitored evolve out of the problem situation at hand and reflects the particular perspective on the problem constructed by operators. The components involved in a problem may be described as a set of »focal points« constituting a view of the production system where certain parts of machinery are temporarily connected through a set of casual relationships constructed in a problem framing activity by the operators. We use the term »temporary focal set« to describe a set of focal points associated with a particular problem. The temporary perspective that the focal set represents may change as the operators collect more information about the problem. The established »temporary focal set« is maintained and monitored until the problem is solved. Thus, a core ingredient in operator work at the plant is a constant movement between different temporary focal sets, often set up in parallel, based on evolving potential problem situations that become established, are maintained and then dissolve.

2.3 CONFIGURING TEMPORARY FOCAL SETS

Based on the results from our field studies we entered into a close collaboration with process operators in designing a handheld device for more dynamic process control. The resulting device – the »Personal bucket organizer« or »Pucketizer« (Nilsson et al., 2000) – supports configuration, monitoring and maintenance of temporary focal sets. The Pucketizer provides three functions, which can all be seen as temporary augmentations of the environment providing the operator with an individual view of process information. With the device the operator can establish links to components as he is confronted with them in the plant and lets him organize these into temporary focal sets. This has similarities to the »pick-and drop« functionality suggested by Rekimoto (Rekimoto 1997). The device also facilitates simple monitoring of process information and setting of temporary alarms. Finally the device enables the operator to annotate the environment by leaving audio »notes« on the spot linked to a particular component. Basically the interface concept is an attempt to deal with the same problem of bridging the physical and the virtual world as discussed by Want et al. (Want et al. 1999). However, acknowledging the richness of representations in the process plant we suggest going beyond the duality of digital vs. physical. Rather than distinguishing between representations of plant data in different media we prefer to see the process plant as one large mixed-media interface where operators are dynamically setting up and exploiting augmentations derived from the problem at hand. What we learned from designing for this kind of dynamic augmentation was that there is not in operator terms any fixed spatial organization of the work space to augment, but rather a constantly changing set of spatially floating relations that has to be monitored and evaluated. Furthermore it became clear that establishing a particular view of the process through configuring »the interface«, and exploiting this view to solve operational problems are so closely intertwined that »configuration« can not be meaningfully separated from »use«.

3. STAGING FUTURE AUGMENTED ENVIRONMENTS

The dynamics of process control, where the operators approach to solving a problem evolves as an outcome of a proactive problem framing process rather than as a response to alarms generated by the process control system, is an important aspect of use that in our view has earlier been neglected in designing technology support for this work domain. In designing for this context a user-centered design process with a high degree of user participation therefore seemed a natural starting point. In order for user participation to be successful representations of design ideas are needed that allow users to engage in a dialogue with the designers and the materials of the design situation (Greenbaum & Kyng, 1991). Furthermore, in exploring concepts that transcend traditional formats of human-computer interaction it could be argued that there is a possible tension between the need for visualizing concepts for users early in the process and the amount of resources needed to produce vivid prototypes of augmentation technology. Rapid prototyping of such concepts is difficult to achieve with the tools available, possibly indicating less opportunity for users to participate in the design of augmented environments. The high contingency of process control work leaves the interaction designer with no solid ground for producing use cases to drive conceptual development or prototype testing. As a group we have earlier worked with ethnographic video as a kind of design material in which we capture prototypical work situations that prompts our design work (Binder, 1999). In designing the Pucketizer we expanded this approach by inviting process operators to script possible work scenarios. The scripting was typically made in the plant and in front of a video camera, in order to explore and maintain possible ways of using new devices. An important part of constructing such potential use scenarios is the use of simple mock-ups, as placeholders for new technology and introduced as props in staging future work situations (Brandt and Grunnet 2000). Figure 2 shows a process operator, in dialogue with a member of the design team, using his everyday environment to spur envisionment when confronted with simple mock ups of a new design. In our perspective, adapting techniques for low fidelity prototyping to information technology as a rapidly developing design material seems a promising approach in designing for highly contingent use situations.

Fig. 2. Augmentation provides a good starting point for envisioning new work practices.

4. DESIGNING FOR DYNAMIC AUGMENTATION – CONFIGURING INTERACTIVE SPACES

In our studies the clear mismatch between established interaction design schemes of process control systems, implying an alarm-driven work mode, and the highly proactive work practice of process operators provides a strong argument for pursuing an augmentation perspective in future interface design. The challenge facing us in pursuit of augmentation is however the constant change characterizing such environments as process control. There is no normal state but rather process control work involves constantly shifting between different parallel temporary focal sets constructed from current problems. The spatial distribution of focal points may be large requiring remote monitoring and as problems in one part of a plant often have consequences further down the production chain problem solving often involves coordinating work. In addition to process information from the control system

731

visual, auditory and haptic information from different positions in the plant is often needed. Based on the above we argue that in technology rich and highly constructed environments, using process control work as an example, there is no privileged or »best« spatial position for problem solving and collaboration in process control. The dynamic aspects of problem solving, where experiments are an important part of the problem framing process, create a need for letting the user configure the interaction with the process control system based on the contingencies of the current problem. Therefore, in our view there is no privileged or »best« view of the process.

Beyond process plants

Löwgren (Löwgren 1995) identifies five perspectives on usability that has developed in the HCI-field and still coexist in parallel: general theories for predicting usability based on human factors research; usability engineering relating usability to a specific system and tasks; subjectivity proposing usability as a property of specific use situations; flexibility proposing usability as a property of long term use with continuous adaptation to changing needs; and a slightly more vague perspective of sociality accounting for social aspects of use as collaboration with roots in the CSCW-field. In developing concepts for augmented environments in process control, flexibility and continuous adaptation to specific use situations seem to be the most important aspect of usability. Augmentation however can provide the interesting possibility of designing for configuring interaction for the situation at hand. This will sensitize interaction designers to respond to the dynamic needs of a particular practice rather than searching for general interaction schemes fixing functionality towards an only slowly evolving system of tasks. We would like to stress *configurability* of an augmented environment as an important aspect of quality in use. This can be achieved not through designing generic tools for controlling fixed functionality but through designing supportive interaction technology based on a deep understanding of various practice situations staged in a specific context (Binder 2001). The high contingency of everyday use situations, with no guiding regular use cases, stresses the need for a high degree of user involvement and design representations that support a constructive dialogue between users and designers. In our experience low fidelity mock-ups with little or no functionality inscribed can successfully be used as »props« to »stage« future use situations together with users as a basis for design. This provides possibilities for both grounding design in work practice and stimulating transcendence from existing work conditions to future technological possibilities.

REFERENCES

Binder, T (2001), *Intent, Form and Materiality in the Design of Interaction Technology*, in C Floyd et .al (eds.) Social Thinking Software Practice, MIT Press (forthcoming)

Brandt, Eva and Camilla Grunnet (2000) *Evoking the Future: Drama and Props in User Centered Design*. Full paper presented at PDC00, Nov. 00. New York.

Greenbaum, J. & Kyng, M. (red.) (1991) *Design at Work*. Lawrence Erlbaum Associates. Hillsdale, New Jersey.

Ishii, H. and Ullmer, B (1997). *Tangible Bits: Towards Seamless Interfaces between People, Bits and Atoms*. Proceedings of CHI'97, pp. 234-241.

Löwgren J. (1995) *Perspectives on Usability*. Research Report no. LiTH-IDA-R-95-23, Department of Computer and Information Science , Linköping University, Sweden.

Nilsson J, Sokoler T, Binder T, Wetcke N (2000): *Beyond the control room - mobile devices for spatially distributed interaction on industrial process plants*, Proceedings from Second International Symposium on Handheld and Ubiquitous Computing 2000, pp. 30-45.

Rekimoto, J. (1997) *Pick-and-Drop: A Direct Manipulation Technique for Multiple Computer Environments*. Proceedings of UIST'97, ACM Symposium on User Interface Software and Technology, pp. 31-39, Oct. 1997.

Want, R., Fishkin, K. P., Gujar, A., and Harrison B. L. (1999) *Bridging Physical and Virtual Worlds with Electronic Tags*. Proceedings of CHI'99, pp. 370-377.

Weiser, M. (1991) *The Computer for the 21st Century*. Scientific American, 265 (3), 1991, pp. 94-104.

Wellner, P., Mackay, W., and Gold, R. (1993) *Computer Augmented Environments: Back to the Real World*. Commun. ACM, Vol. 36, No. 7, July 1993.

Zuboff S. (1988) *In the Age of the Smart Machine - the Future of Work and Power*. Heinemann Professional Publishing, Oxford 1988.

An Interface for a Continuously Available, General Purpose, Spatialized Information Space

Rob Kooper and Blair MacIntyre
GVU Center and College of Computing
Georgia Institute of Technology
Atlanta, GA 30332 USA
{kooper,blair}@cc.gatech.edu

Abstract

In this paper we describe an augmented reality (AR) system that acts as a continuously available interface to a spatialized information space based on the World Wide Web. We call such an information space the Real-World Wide Web (RWWW). We present the assumptions we make about the characteristics of such a system, discuss the implications of those assumptions for an AR interface, and describe a RWWW browser we are building.

Keywords: augmented reality, human-computer interaction, adaptive interfaces, mobile computing, wearable computing.

1 Introduction

In our previous work, we have focused on task-specific AR systems (e.g., (Feiner, MacIntyre, Hollerer, & Webster, 1997)). Recently, we have begun exploring how to create AR interfaces for general-purpose 3D information spaces, as suggested by (Spohrer, 1999). In our current research, we are investigating the use of 3D augmented reality (AR) to envelop mobile users in contextually-relevant, spatially-registered information spaces. We envision an enhanced version of the World Wide Web, where each information object (i.e., "web page", containing 2D and 3D visual and auditory information) may have contextual meta-data associated with it, such as a location, person, or activity. This meta-data would be used to decide when and where to present the information objects to users. We refer to such a space as the *Real*-World Wide Web (RWWW), and interfaces to it as RWWW Browsers. In a RWWW Browser, web pages would be registered with real-world locations and presented to the user at the appropriate place and time.

We are particularly interested in RWWW Browsers that can be worn *continuously*, providing users with constant awareness of the information spaces they move through in their daily lives. We are interested in presenting the user not only with information that is relevant to their current location (i.e. restaurants or people near by) but any information that might be pertinent to their context (i.e. to-do items's, calendar entries, messages from friends) (Salber, Dey, & Abowd, 1999; Starner, 1999). Unlike most research in continuously-worn wearable computing, where researchers use 2D heads-up-displays (e.g., (Rhodes, 1997; Starner, 1999)), the ability to place information in 3D around the user raises new opportunities and challenges. In particular, the interface design must balance the conflicting requirements of minimizing the volume of information displayed (to avoid distracting the user and cluttering their visual field) with the need to provide rich context (to capitalize on the users ability to rapidly scan and synthesize data).

2 Background and Related Work

Our work is related to four main areas of research: augmented reality, ambient or peripheral displays, wearable computing and context-aware computing. Our expectation, and that of others (Spohrer, 1999), is that the architecture supporting the general purpose spatialized information space will evolve from the current World Wide Web (WWW), retaining its structure as a loosely organized, heterogeneous collection of information servers. The key difference is that this future WWW will serve up virtual information that contains contextual data describing its association with the real world. Such data might describe a specific geo-spatial location for the information, but might also be more abstract, such as associating the information with a person, time, situation or activity (Abowd, Dey, Brotherton, & Orr, 1999). Because such a system effectively embeds the WWW in the real world, we refer to such a context-based information space as the "Real-World" Wide Web (RWWW).

Our vision of the RWWW is closest to Spohrer's notion of the WorldBoard (Spohrer, 1999). Like Spohrer, our ultimate vision is of a global information space that merges the real and the virtual, contains a variety of content, and is accessible to anyone. Both WorldBoard and the RWWW extend the current WWW infrastructure by enhancing the data provided by existing servers, and by creating new servers that deliver specialized content. Our initial vision differs from that of WorldBoard primarily in the granularity of context that can be associated with information. To simplify the interface, WorldBoard focuses on the use of carefully authored information *channels*. We believe that these simplifications cause WorldBoard to lose much of the character of the WWW that has made it so successful, namely the ability for anyone to author, structure and publish information as they wish; the focus on authored content restricts what can be encountered by the user, and the use of a coarse coordinate system as the only context restricts where and (perhaps more importantly) why information can be encountered.

There have been numerous other AR systems built over the years (e.g, (Feiner, MacIntyre, & Seligmann, 1993)). We are not aware of any other attempts to create a continuously worn, general purpose interface to a spatialized information space such as the RWWW. The most relevant spatialized AR systems to our work are Audio Aura (Mynatt, Back, Want, M., & Ellis, 1998), the Touring Machine (Feiner et al., 1997), and Augmentable Reality (Rekimoto, Ayatsuka, & Hayashi, 1998). Audio Aura is an audio-only AR system designed to provide knowledge workers with a continuous sense of generally relevant, context-sensitive, information (e.g., the state of their E-mail, the activity of their coworkers). The Touring Machine (and other tour systems) exhibits some of the characteristics we desire, but it is designed for a specific kind of data (tours of outdoor sites). In addition, the interface design assumes the user's primary task is taking the tour, and are focussed on the data being presented. Therefore, the Touring Machine interface is not subject to the same constraints we illustrate in Section 3. Rekimoto's Augmentable Reality system shares our goal of placing information throughout the world, but is based on a simple, closed system. The aim of his work was to explore how the spatial attachments could be authored, rather than taking an existing, content-heavy information space and merging it with the world.

Like AR, wearable computing researchers have explored both task specific systems (e.g., (Thompson, Ockerman, Najjar, & Rogers, 1997)) and general purpose, continuous use systems (e.g., (Rhodes, 1997)). However, unlike AR, general purpose, continuous-use systems have received significant attention in the wearable community (e.g., (Starner et al., 1997)). The key difference between AR and typical wearable computers is information spatialization: current wearables are not capable of complex 3D graphics, so they generally use 2D heads-up displays (HUD's) which place the information in a fixed location in 2D. This has the advantage that the user can predict where information will and will not appear, and can adapt accordingly. However, it has the disadvantage that information is limited to a small space and cannot be associated directly with relevant parts of the world. This limits the users ability to take advantage of their spatial memory and perception, and significantly decreases the amount of information that can be displayed.

Finally, our work can be viewed as a context-aware system. It has close similarities to context-aware tour guides (Long, Kooper, Abowd, & Atkeson, 1996), which attempt to present information to the user based on some notion of their current context. Most context aware systems that we know about use hand-held displays, which are suitable because they are aimed at more task-specific, often query-reply, interactions. Unlike these systems, we are interested in presenting the information to the user continuously, and therefore use a see-through, head-worn display. However, many of the issues related to defining, obtaining, and acting on contextual information are common across all of these systems.

3 Assumptions and Design Goals

As the RWWW evolves out of the WWW, it will not suddenly "appear", but will rather be integrated with the existing WWW for the foreseeable future; some WWW pages will have detailed context associated with them, some will have limited contextual information, and some will have none at all. The availability of more complex contextual information will begin to emerge soon, on the tails of initiatives (such as Microsoft's .NET) that are pushing for semantically meaningful structuring of information on the WWW. The corollary to this gradual evolution is that the RWWW will begin to appear "any time now". Given the current interest in creating context-aware applications for mobile phones and palm-sized devices, a rudimentary form of the RWWW is already being developed, with the initial contextual tagging being simple location information. As geo-spatial tagging standards gain acceptance, sites that already allow location-based queries (such as MapQuest and Restaurant.com) will become far more common.

As the RWWW grows, centralized indices (such as google.com), authored portals (such as yahoo.com) and information services (such as Fodors Restaurant Guide) will adapt and allow information to be retrieved automatically, based not only on content, but also on standard contextual cues (such as location or person identity). WorldBoard's idea of channels is closest to today's authored portals, but we believe all three kinds of sites will be important: automatically created indices that crawl the web like today's search engines will be necessary to ensure that as much information as possible is made available to the user, based on their current context, and authored portals and information services will be necessary to ensure authoritative access to common kinds of information.

The implications of these assumptions about the RWWW on any interface to the RWWW are:

- *continuously changing data:* The combination of context awareness and widely distributed servers implies that data may change at any time, and may also be changing continuously.
- *safety:* The distributed authorship of the RWWW implies that both trustworthy and untrustworthy data can be mixed together. Directly displaying unfiltered information from the RWWW may be dangerous (e.g. a malicious person could place a virtual brick wall on a busy highway).
- *heterogeneous data*: The gradual evolution of the RWWW implies that old and new (i.e., tagged and untagged) data will be mixed together for the forseeable future. This implies a need to handle a mixture of location-based, contextually-tagged and untagged data.

As described above, the most significant design choice is that the RWWW Browser is intended to be worn continuously, as in (Starner, 1999). This decision implies two interface characteristics, common to most continuously-available user interfaces:

- *non-interference:* The continuous availability of the system implies that the interface should support continuous awareness of the virtual space without interfering with other tasks the wearer may be engaged in.
- *minimal interaction*: The user should rarely (if ever) be required to interact with the system when they do not want to. While continuously worn systems will require occasional interaction with the user, these interactions should require minimal effort.

Satisfying the first two implications, continuously changing data and safety, is vital for a continuous-use systems; if the system were designed to be worn briefly from time to time, these points would be interesting observations, but might have little practical impact on the interface design.

The fundamental differences in interface design between task-specific and continuous use AR systems are analogous to the differences between applications designed for desktop computers (such as a word processor) and ambient displays (such as the Ambient Room (Wisneski et al., 1998)); it is easy to turn away from a desktop display or iconify the application, but it is a far more significant act for a user to have to leave an Ambient Room if it is interfering with a task they are trying to accomplish. For a RWWW Browser (and the Ambient Room) to be successful, it must present the user with a continuous awareness of their information space, and do it in such a way that does not adversely impact their focal tasks.

Based on the assumption described above, we have come up with a list of design goals for our initial prototype RWWW Browser.

1. The default state of the browser is to display information nodes as simple, consistent icons, instead of presenting the "raw," unfiltered content from the RWWW spatially in the 3D world. This avoids visual confusion, as well as badly timed or maliciously-placed content (*non-interference, minimal interaction, safety, continuously changing data*).
2. The user should be able to organize the information at a coarse level, so they can quickly control what information is displayed (*non-interference, minimal interaction, safety*).
3. Unless the user explicitly requests otherwise, detailed information (especially if it is to remain displayed for any length of time) should be placed in a fixed location relative to the user (such as relative to their head or body). The user can then predict where information will be (*non-interference, minimal interaction*).
4. The user should be able to easily access and dismiss different levels of detail of the information space, ranging from the basic information about a node to the full details of all nodes in the space (*minimal interaction*).
5. The browser should not significantly change the content of the display based on context or location changes. Instead, the browser should peripherally notify the user about potentially significant changes in the information space and provide them with a simple way to explore the new content (*non-interference, safety, continuously changing data*).
6. Non-spatialized data should be displayed at reasonable locations in the environment (*heterogeneous data*).

4 The Initial prototype RWWW Browser

Based on the assumptions and goals stated in the previous section, we have developed an initial interface for our browser that focuses on providing an awareness of the amount of information available, with a coarse indication of

735

the information content (Figure 1). First, web pages are organized by grouping them into related information channels (following (Spohrer, 1999)). The user can control which channels are displayed, giving them fast (but coarse) control of the amount of information in their visual space. Second, unlike previous work where detailed information such as labels are placed in the world (e.g., (Feiner et al., 1997)), web pages are displayed using spatially-located anchors (currently, small "twinkling stars" that are colored to indicate which channel they belong to). This approach gives the user an indication of the amount, location and kind of information available, while minimizing visual clutter. Information that does not have detailed spatial location information, such as a page associated with a time or activity, is automatically positioned in space at a reasonable location by the browser.

Additional information about an anchor is obtained using a simple form of two-level gaze selection, extending the approach we developed in (Feiner et al., 1997) (i.e., the object closest to the center of the screen is selected). The first level (*glance*) immediately displays the title of the selected anchor near the center of the screen, replaces the anchor with a thumbnail rendering of the web page, and begins to draw a red circle clockwise around the thumbnail being glanced at. The second level (*gaze*) is activated by gazing at the same anchor for a few seconds (the time it takes to draw the complete red circle). Gazing causes a yellow circle drawn around the selected thumbnail, and a larger rendering to be placed in a fixed location in the upper left corner of the screen. This approach allows an area to be scanned quickly to obtain more detailed information. Detailed information is always placed at a fixed location (the upper left corner of the display in Figure 1 and Figure 2) so that users can predict where information will appear, and therefore locate it or work around it when necessary. A node remains gaze-selected until a new node is gaze-selected.

To support the situation where a user needs to simultaneously view multiple nodes, or simply wishes to retain a node for future reference, we have added a temporary storage area in the user's body space by allowing the gaze-selected page to be "pushed" to the right around the user's waist. When the user pushes a node, all pages in their body space are rotated to the right (both the saved pages, and the gaze-selected page). To distinguish between nodes that are gaze-selected and those that are saved in the body space, we draw a green circle around the nodes saved in the body space (see Figure 2).

The system alerts the user of significant changes to the set of "relevant information" by flashing the channel indicator along the bottom of the display, but new channels are not displayed without an explicit user action. Currently, our Browser can handle any traditional 2D web page that can be rendered by the Java HTML renderer.

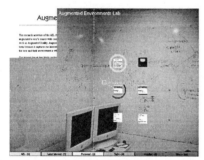

Figure 1 The prototype RWWW browser. The link closest to the center has a text description and small thumbnail (10cm in world coordinates) displayed, connected by a leader line. A red circle is slowly drawn around the thumbnail (over the course of a few seconds). After a short gaze (the time taken to draw the red circle), a larger image of the web page appears in the upper left corner of the screen, and the circle turns yellow to indicate which node is gaze-selected and displayed in the upper left.

Figure 2 The image to the right shows what the user sees when they look down and slightly to the right. The node that is currently gaze-selected (our lab webpage) is to the left, and another node (a homepage) has been saved and is to the right. The small image is a screen shot showing what would be displayed on the HMD if the user looked straight ahead at the information nodes. The green circle indicates the saved node, the yellow circle indicates the gaze-selected node.

5 Discussion and Future Work

In this paper, we list five important issues that must be addressed when creating a continuously worn interface to an information space like the RWWW: non-interference, minimal interaction, safety, continuously changing data, and heterogeneous data. These five issues are then used to develop a list of design goals to which we believe a RWWW browser should adhere. During the design and implementation of our prototype browser, initial tests with other researchers and visitors to our lab aided in the iterative design of the interface, and suggested that our design principles are reasonable.

We believe that coarse organization in channels, as is done currently in the browser, is a useful starting point for organizing the large information space. However, channels alone are not sufficient without the use of contextual information (such as location, time, activity or the identities of nearby people) to prompt the user as to which channels might be most relevant to their current context. We are currently integrating context-sensing infrastructure, developed by others in our research group (Salber et al., 1999), into our prototype to allow us to begin experimenting with the use of context, both to select appropriate channels, and to prioritize the information in the channels. We are also beginning to investigate how knowledge of which physical and virtual objects the user is looking at should affect the display of the information space.

In addition to experimenting with the Browser itself, we are creating a simplified RWWW infrastructure that supports a richer set of content types (such as 3D objects and audio), adds contextual meta-data to each document, and provides the Browser with additional information such as models of the physical environment (to support more controlled spatialized display). These models will not only contain a coarse layout of the room (e.g., the location of walls and windows) but can also contain hints as to where non-spatialized information should be placed.

In this paper, we have discussed our first steps toward implementing a Real-World Wide Web browser. We have developed a set of guiding principles for our experiments, and implemented an initial prototype of a Browser. We hope to take the lessons we have learned, and take the next step toward the eventual deployment of a RWWW Browser.

6 Acknowledgments

This work was supported by Siemens via a GVU Industrial Affiliate Grant, DARPA/ITO under the Information Technology Expeditions, Ubiquitous Computing, Quorum, and PCES programs, ONR under Grant N000140010361, and equipment and software donations from Sun Microsystems and Microsoft.

7 References

Abowd, G. D., Dey, A. K., Brotherton, J., & Orr, R. J. (1999). Context-awareness in Wearable and Ubiquitous Computing. *Virtual Reality Society International Journal, 3*, 200-211.

Feiner, S., MacIntyre, B., Hollerer, T., & Webster, T. (1997). A Touring Machine: Prototyping 3D mobile augmented reality 28 systems for exploring the urban environment. *Personal Technology, 1*(4), 208-217.

Feiner, S., MacIntyre, B., & Seligmann, D. (1993). Knowledge-Based Augmented Reality. *Communications of the ACM, 36*(7), 53-62.

Long, S., Kooper, R., Abowd, G. D., & Atkeson, C. (1996). *Rapid Prototyping of Mobile Context-Aware Applications: The CyberGuide Case Study.* Paper presented at the International Conference on Mobile Computing and Networking (MobiCom'96).

Mynatt, E. D., Back, M. J., Want, W., M., B., & Ellis, J. B. (1998). *Designing Audio Aura.* Paper presented at the ACM Computer Human Interaction Conference (CHI '98).

Rekimoto, J., Ayatsuka, Y., & Hayashi, K. (1998). *Augement-able Reality: Situated Communication through Physical and Digital Spaces.* Paper presented at the International Symposium on Wearable Computers (ISWC '98).

Rhodes, B. J. (1997). *The Wearable Remembrance Agent: A system for Augmented Memory.* Paper presented at the International Symposium on Wearable Computers (ISWC '97).

Salber, D., Dey, A. K., & Abowd, G. D. (1999). *The Context Toolkit: Aiding the Development of Context-Enabled Applications.* Paper presented at the ACM Computer Human Interaction Conference (CHI '99).

Spohrer, J. C. (1999). Information in places. *Systems Journal, 38*(4), 602-628.

Starner, T. (1999). *Wearable Computing and Context Awareness.* Unpublished PhD thesis, MIT, Cambridge, MA.

Starner, T., Mann, S., Rhodes, B., Levine, J., Healey, J., Kirsch, D., Rosalind Picard, & Pentland, A. (1997). Augmented Reality Through Wearable Computing. *Presence: Teleoperators and Virtual Environments, 6*(4), 386-398.

Thompson, C., Ockerman, J. J., Najjar, L., & Rogers, E. (1997). *Factory Automation Support Technology (FAST): A New Paradigm of Continuous Learning and Support Using a Wearable.* Paper presented at the International Symposium on Wearable Computers (ISWC '97).

Wisneski, C., Ishii, H., Dahley, A., Gorbet, M., Brave, S., Ullmer, B., & Yarin, P. (1998). *Ambient Displays: Turning Architectual Space into an Interface between People and Digital Information.* Paper presented at the International Workshop on Cooperative Buildings (CoBuild '98).

AUGMENTED REALITY AND THE DISAPPEARING COMPUTER

Norbert A. Streitz

GMD - German National Research Center for Information Technology
IPSI - Integrated Publication and Information Systems Institute
GMD-IPSI, Dolivostr. 15, D – 64293 Darmstadt, Germany
e-mail: streitz@darmstadt.gmd.de

ABSTRACT

In this paper, I argue for returning to the real world as the starting point for designing future information and collaboration environments and an integration of real and virtual worlds. This is based on the opinion that we have to emphasize again the importance of the rich affordances of the physical, architectural environment. This approach results in augmenting everyday places and objects with information processing capabilities but in a new and unobtrusive way. In the resulting hybrid worlds, the computer as a device will disappear, will be invisible, respectively be in the background, while the functionality will be available in a ubiquitous fashion. I will distinguish between "mental" and "physical" disappearance and discuss a number of resulting design issues and the role of affordances for designing smart artefacts and their user-interfaces. Examples are taken from our work on developing so called "Roomware"-components for the i-LAND environment - an interactive landscape for creativity and innovation - and from a recently launched project "Ambient Agoras: Dynamic Information Clouds in a Hybrid World".

1. INTRODUCTION

The advent of information technology has caused a significant shift: away from real objects in the physical environment as the sources of information towards monitors of desktop computers as *the* interfaces to information. As a consequence, many people consider desktop computers in combination with virtual communities and chat rooms as the future environments for information and communication activities. In contrast to this, I will present an approach that goes beyond the now already more or less traditional desktop computer setting. Instead of Virtual Reality, we argue for Augmented Reality and an integrated design of real and virtual worlds. The resulting hybrid environments provide and require new forms for interacting with information and for cooperation between people. Our application area is the support of collaborative activities in the Workspaces of the Future as fundamental constituents of so called *Cooperative Buildings* (Streitz et al., 1998) where "the world around us is the interface" for people's activities. The idea of going back to reality as the starting point for designing usable work environment, especially also for groups of people, is rooted in our opinion that we have to emphasize again the importance of the rich *affordances* of the physical, architectural environment. Of course, we do not intend to go back to reality as it used to be. Instead, we will be augmenting everyday places and objects with information processing capabilities but in a new and unobtrusive way.

When augmenting the physical environment with computer-based devices providing new functionality, that does not imply to put lots of computers in a demonstrative way all over the place. On the contrary, we want the computer as a device to disappear, respectively be in the background, while the functionality will be available in a ubiquitous fashion (=> ubiquitous computing). This is in line with the earlier proposals of Mark Weiser. In this context, my favorite quote from him is "The most profound technologies are those that disappear. They weave themselves into the fabric of everyday life until they are indistinguishable from it." (Weiser, 1991) And I like to add "... and facilitate a coherent and social experience when interacting and cooperating within the environment by providing appropriate affordances." I take this view and interpret it as the design goal of a two-way augmentation and smooth transition between real and virtual worlds. Combining real and virtual worlds in a computer-augmented environment resulting in hybrid worlds allows to design enabling "interfaces" that build on the *affordances* of everyday reality and virtuality seeking to use the best aspects of both worlds. I will give some examples later on.

My approach is to transform and transcend human-*computer* interaction resulting in rather direct *human-information* interaction and *human-human cooperation* based on *human-artefact* interaction, at the same time making the technology device character of computers disappear.

2. THE DISAPPEARING COMPUTER

In this approach, I distinguish between two types of disappearance of computer devices.

- *physical disappearance* of computer devices is achieved by making the computer-based parts very small so that they can fit in the hand, be interwoven with clothing, be attached to the body or even implanted in the body. In most cases, the computer is integrated in a "shell" of a compound artefact of small dimensions where the features usually associated with a computer are not visible anymore.

- *mental disappearance* (and I will distinguish further between *cognitive* and *emotional* disappearance) of computers is achieved by becoming "invisible" to the "mental eyes" of the users. This can, for example, happen when computer devices are stripped of their usual casing and are embedded in the architectural environment (e.g., walls, doors) or furniture (e.g., tables, desks, chairs) around us, somehow appearing in a new camouflage. The important aspect here is that they are not perceived as computer devices anymore but as embedded elements of augmented artefacts in the environment. Examples are the roomware components we developed (Streitz et al., 1999, 2001).

I distinguish between "cognitive" and "emotional" disappearance in the following way. In the *cognitive disappearance* case, the appearance of the devices is transformed so that they are not perceived as computers anymore but as information appliances for communicative and cooperative situations. The resulting artefacts can be even quite large, as in the example of an interactive wall (our DynaWall – see below - measures 4.40 m wide and 1.10 m high) or an interactive table with a horizontal surface (e.g., our InteracTable - see below) although it is smaller than the wall but still larger than a standard desktop computer display. People perceive more visual stimuli and associations that remind them of available everyday objects than of computer displays and boxes. In these cases, people move the „computer device" character in the background and the functionality in the foreground.

In the *emotional disappearance* case, users are "drawn in" by a high emotional load of the artefact. Take the example of a stone-like artefact (similar to a "Handschmeichler" in German) with a very smooth surface and a round or oval shape that fits the shape of the palm of your hand very nicely. You like to touch and move it in your hand, to play with it, to carry it around in your pocket, etc. Another example would be a game with a very intriguing functionality and easy to handle interaction that guarantees full control over the game. It provides so much fun and excitement that after some time you even forget about the standard desktop display you are using as an output device. In this case, the functionality available, e.g., via a very realistic animation is important for creating this type of disappearance.

3. AFFORDANCES FOR DISAPPEARANCE AND COHERENT EXPERIENCES

Causing computers to disappear is not really our final goal and not only because you might wonder about the answer to the question "how do I interact with an invisible computer?" I like to argue that the notion of disappearance is an important objective on the way to achieve the goal of what we call "coherent experiences" in human-information interaction and human-human cooperation. This becomes obvious when we realize that the visual appearance of the integrated artefact is only one aspect of making computers disappear. It is one design dimension of the affordances of disappearance. The other very important dimension is the type of interaction available. Useful disappearance and finally coherent experience is the result of the combination of macro affordances (physical shape and form factor) and certain micro affordances (e.g., tactile characteristics of the artefact's surface) in combination with the software providing appropriate interaction affordances. In many cases, their design is based on metaphors taken from real world objects and actions in order to be intuitive and thus minimizing the learning effort. In the following, I will present some examples of the Roomware® components we developed at GMD-IPSI as part of the i-LAND environment (Streitz et al, 1998, 1999) in our AMBIENTE-Lab in Darmstadt, Germany. *Roomware®* results from the integration of information technology into room elements as, e.g., walls, doors and furniture. Roomware components are interactive and networked; some of them are mobile due to independent power supply and wireless networks, and are provided with sensing technology.

3.1 The Roomware® Components DynaWall® and InteracTable®

The DynaWall® in our AMBIENTE-Lab is a large interactive wall covering one side of the room completely. The size of 4.50 m (15 ') width and 1.10 m (3' 7") height and the very smooth integration of this very large display (realized by a back projection which is hidden to the user) into the architectural structure creates the impression that you are really writing and interacting with a wall or wallpaper, depending on which metaphor is conveyed, respectively people adopt for themselves. The surface is touch-sensitive so that you can write and interact on it with your bare fingers or with a normal pen (no electronics needed). Several people can write/ interact in parallel in (currently three) different areas of the DynaWall. Beyond these physical affordances, our BEACH software enables very intuitive interaction based on gestures that are reflecting actions with physical objects in the real world (e.g., "take and put", "throw", "shuffle" , ...). When throwing objects (with different accelerations), the speed and thus the flying distance is dependent on the initial momentum provided by the user. People can interact this way immediately after having seen it once.

A similar combination of affordances is provided by the InteracTable®, another roomware component we developed. It has a display size of 65 cm x 115 cm and a diameter of 130 cm (50 "). Beyond the type of interactions available at the DynaWall, it provides additional forms of interaction required by horizontal and round or oval-shaped displays. To this end, we developed in BEACH special gestures for shuffling and rotating individual information objects or groups of objects across the surface so that they orient themselves automatically. This accommodates easy viewing from all perspectives. Furthermore, one can create a second view of an object, shuffle this to the other side so that the opposite team member has the correct view at the same time. Now, everybody can view the same object with the correct perspective in parallel, edit and annotate it.

It is important to note that although the physical appearance of a wall and a table is different the visual appearance of the information objects and the design of interacting with them is done in such a way that it provides a coherent experience. Other examples are the ConnecTables® where moving them together in physical space results in creating a shared workspace between them in the virtual world. More details for interacting with different roomware components are given in (Streitz et al, 2001).

3.2 The Passage mechanism

Passage is a mechanism for establishing relations between physical objects and virtual information structures, thus bridging the border between the real world and the digital, virtual world. So-called *Passengers* (Passage-Objects) enable people to have quick and direct access to a large amount of information and to "carry them around" from one location to another via physical representatives that are acting as physical "bookmarks" into the virtual world. It is no longer necessary to open windows, browse hierarchies of folders, worry about mounted drives, etc. Passage is a concept for ephemeral binding of content to an object. It provides an intuitive way for the "transportation" of information between roomware components, e.g., between offices or to and from meeting rooms.

A Passenger does not have to be a special physical object. Any uniquely detectable physical object may become a Passenger. Since the information structures are not stored on the Passenger itself but only linked to it, people can turn any object into a Passenger: a watch, a ring, a pen, glasses, a wooden block, or other arbitrary objects. The only restriction Passengers have is that they can be identified by the *Bridge* and that they are unique. Passengers are placed on so-called *Bridges*, making their virtual counter parts accessible. With simple gestures the digital information can be assigned to or retrieved from the passenger via the virtual part of the Bridge. The Bridges are integrated in the environment to guarantee ubiquitous and intuitive access to data and information at every location in a building (=> Cooperative Building). For example, a Bridge can be integrated into the table top of an interactive electronic table (=> InteracTable®) in the cafeteria or mounted in front of an interactive electronic wall (=> DynaWall®) in a meeting room. More details can be found in (Konomi et al, 1999).

The preceding examples provide first steps of moving away from human-*computer* interaction and via human-*artefact* interaction towards human-*information* interaction and human-*human* cooperation. But we have also to be aware of the fact that this approach will be only one (important) part of a more comprehensive story. It remains to be seen if taking only metaphors from the real world is the best way for the design of disappearing computers. Similar to the breakdown of the "desktop" metaphor based on icons for folders and trash cans showing us the limitations of that approach, we will witness that a unidimensional approach will cause a lot of problems again. This and other open issues will provide the topics of our new activities described below.

4. CONCLUSIONS AND FUTURE WORK

While the guiding vision for the development of our roomware components was already provided by the framework of *Cooperative Buildings*, the actual development was somehow limited to application scenarios addressing primarily the support of collaborative work in meeting rooms. This limitation was a major motivation for us to extend the scope of the problem space and place more emphasis than before on more comprehensive architectural environments, collections of artefacts and a wide range of activity patterns when interacting with those artefacts and moving in these spaces.

At the beginning of 2001, we started a new project called "Ambient Agoras: Dynamic Information Clouds in a Hybrid World" (www.Ambient-Agoras.org). It is funded by the European Union as part of its proactive initiative "The Disappearing Computer" (www.disappearing-computer.net). The project "Ambient Agoras" aims at providing situated services, place-relevant information, and feeling of the place ('genius loci') to the users, so that they feel at home in the office, by using information technology in an innovative way, e.g., mobile and embedded in the environment. *Ambient Agoras* adds a layer of information-based services to the place, enabling the user to communicate for help, guidance, work, or fun. It integrates information into architecture through smart artefacts, and will especially focus on providing the environment with memory, which will be accessible to users. *Ambient Agoras* will augment reality by providing better affordances to existing places. It aims at turning every place into a social marketplace (= *agora* in Greek) of ideas and information - an *Information Market Place* - where one can interact and cooperate with people. As the title implies, *Ambient Agoras* is a project that brings together technology and people in a particular context. The context and metaphor we will pursue is that of the *Greek Agora*, a place in the center of economic and social activity where information and ideas are exchanged. I will complement my presentation at the conference in August with some results from this project.

ACKNOWLEDGMENTS

I thank all members of the AMBIENTE research division (www.darmstadt.gmd.de/ambiente) for their substantial contributions to various parts of the research described.

REFERENCES

Konomi, S., Müller-Tomfelde, C., Streitz, N. (1999). Passage: Physical Transportation of Digital Information in Cooperative Buildings. In: Streitz, N., Siegel, J., Hartkopf, V., Konomi, S. (Eds.), *Cooperative Buildings - Integrating Information, Organization, and Architecture. Proceedings of CoBuild'99.* LNCS 1670. Heidelberg: Springer, pp. 45 - 54.

Streitz, N., Geißler, J., Holmer, T. (1998). Roomware for Cooperative Buildings: Integrated Design of Architectural Spaces and Information Spaces. In: Streitz, N., Konomi, S., Burkhardt, H. (Eds.), *Cooperative Buildings - Integrating Information, Organization, and Architecture. Proceedings of CoBuild '98*, Darmstadt, Germany, LNCS Vol. 1370, Heidelberg, Germany, Springer, 1998. pp. 4-21.

Streitz, N., Geißler, J., Holmer, T., Konomi, S., Müller-Tomfelde, C., Reischl, W. Rexroth, P., Seitz, P. Steinmetz, R. (1999). i-LAND: an interactive landscape for creativity and innovation. In *Proceedings of CHI'99* (Pittsburgh, May 15-20, 1999) ACM Press. New York, pp. 120-127.

Streitz, N., Tandler, P., Müller-Tomfelde, C., Konomi, S. (2001). Roomware: Towards the Next Generation of Human-Computer Interaction based on an Integrated Design of Real and Virtual Worlds. In: J. A. Carroll (Ed.), *Human-Computer Interaction in the New Millennium.* Addison-Wesley. (to appear in May).

Weiser, M. (1991). The Computer for the 21st Century. *Scientific American*, 1991, 265 (3), pp. 94-104.

Using Augmented Reality to Support Collaboration in an Outdoor Environment

Bruce H. Thomas
Wearable Computer Laboratory
School of Computer and Information Science
University of South Australia
Mawson Lakes, SA 5095, Australia
Bruce.Thomas@UniSA.Edu.Au

ABSTRACT

This paper presents a proposal for an outdoor wearable augmented reality computer system to support collaboration. A key attribute of this system is its ability to use all four time-space configurations for collaboration systems: same time – same place, same time – different place, different time – same place, and different time – different place. The proposed main form of user interaction is the use of hand and head gestures. The seamless movement of information across different display devices, such as head mounted display, PDA, laptop, data walls, and desktop is critical to allow this form of collaboration to integrate traditional work flows. A scenario of a military logistics task is described to illustrate the functionality of this form of collaboration system.

1   INTRODUCTION

Collaboration technology facilitates multi-users to accomplish a large group task. There are a number of ways technology can help these users: combine or merge the work of multi-users, prevent and/or inform users when an item of data is being modified by more than one user, and track the activities of multiple users. One major function of collaborative technology is to help people communicate ideas; collaborative electronic whiteboards are a good example of how collaboration technology may help multi-users communicate, for example the Teamboard system [TeamBoard Inc., 2001]. I believe a wearable computer with an augmented reality(AR) [Azuma, 1997] user interface allows for exciting new collaborative applications to be deployed in an outdoor environment. I refer to these systems as Outdoor Wearable Augmented Reality Collaboration Systems (OWARCS). Like other researchers, [Azuma, 1999], [Feiner et al., 1997], and [Höllerer et al., 1999], I am taking the use of AR from the indoor setting and placing it in the outdoor environment [Thomas et al., 1998, Thomas et al., 2000]. The use of hand-held computing devices communication via a wireless network has been investigated as a means to facilitate collaboration by Fagrell et al. [Fagrell et al., 2000]. Their architecture FieldWise is based on two application domains: first, mobile and distributed service electricians and second, mobile news journalists.

An alternative to pen based computing, wearable computers leave the hands free when the user is not interacting with computer but still allows the user to view data in the privacy of a head mounted display (HMD) [Starner et al., 1995]. A major research issue is the interaction techniques for users to control and manipulate augmented reality information in the field [Thomas et al., 1999]. I propose the use of augmented reality in the field (outdoors) as a fundamental collaboration tool that may be used across a number of application domains, such as maintenance, military, search and rescue, and GIS visualisation. A number of researchers are investigating augmented reality with wearable computers [Billinghurst et al., 1998, Bauer et al., 1999], but I am proposing an overall framework to integrate augmented reality into a traditional work flow.

The paper first presents a scenario of using augmented reality to facilitate communication between a number of people in a logistics framework. A number of interaction techniques are then proposed to map out a direction of investigation into this area of research.

Figure 1: Virtual information tag   Figure 2: Finger framing

2   COLLABORATION

As previously mentioned, collaboration technology facilitates multi-user interactions to achieve a common goal. As with collaborative systems such as distributed white boards and remote video conferencing systems, the main aim of OWARCS is to improve communication between the multiple users to attain their common goal. Overlaying contextually aware information on the physical world is a powerful cuing mechanism to highlight or present relevant information. This ability to view the physical world and augmented virtual information in place between multiple people is the key feature to this form of collaboration technology.

To understand how an OWARCS relates with existing collaboration systems, I use the time - place taxonomy [Ellis et al., 1991]. The time - place taxonomy is defined by the position of the users (same or different) and the time of operation of the collaborative system (same or different). A distinctive quality of activities using an OWARCS is the ability to use all four time-space configurations, where many existing collaboration systems support activities in one or two configurations. An example of how an OWARCS would seamlessly cross these four time-space configurations is a logistics task of supporting an overseas military contingent. [1] I wish to extend the user interfaces for existing outdoor augmented reality systems [Julier et al., 2000].

The scenario starts with an urgent request from an aviation maintenance person for a replacement rotor for a helicopter and to place a virtual marker on the rotor to have the logistics supervisor contact him. The location of the rotor in the warehouse is indicated with augmented reality information in the form of virtual sign posts and virtual line markings on the floor. Someone can quickly find the rotor, and the rotor is moved from a warehouse to the airfield loading dock. An augmented reality information sticker is attached to the container the rotor is placed in stating this is an urgent request. This provides different time - same place configuration for communicating between the person who placed the rotor in the loading dock and the supervisor monitoring the shipment of the rotor. Figure 1 depicts what the supervisor might see in such an augmented display. The annotation is designed to overcome the problem that the box might hidden behind other boxes. This annotation may one or more forms of multimedia information, such as text, line drawings, audio, voice, or digital image. The annotation is registered to the box containing the rotor. The location of the box can be determined through the use of smart sensors or similar technology. The delivery is also recorded in a standard database for information tracking.

At a later time, the supervisor proceeds to check the supplies to be loaded onto the plane. The supervisor reads the virtual note left on the rotor's container: "There are a number of different rotors for the different models of helicopters, please contact ..." He contacts the aviation maintenance person who placed the original order. This information is shown in their HMD and is retrieved through the identification of the smart badge. The aviation maintenance person asks the supervisor to visually inspect the rotor. The supervisor opens the box and shows the maintenance person the rotor via a digital camera mounted on their helmet. This situation is now a same time - different place configuration. The maintenance person views the video on their office workstation while the supervisor concurrently views the rotor through their HMD. The maintenance

---

[1] To make this scenario realistic, I sought advise about military logistics from Dr. Rudi Vernik of the Australian Defence Science and Technology Organisation

744

person indicate where to look via drawing lines over the video image. These drawings can be registered to the rotor's container via simple orientation and position information provided by use of video registration. Fiducial markers can be placed on the container to improve such registration. The maintenance person can direct the supervisor to read off information on an indicated information plate. Both parties can make annotations. The maintenance person can show digital images of similar rotors or they can show a 3D model, for example, to highlight a particular location on the rotor.

Once the two people agree this is the correct rotor, the supervisor places a virtual note on the rotor's container indicating it is an urgent request and has the rotor placed on the plane for shipping. This becomes a different time - different place configuration, as the person in charge at the second airfield will read this note at a later date in a different location. Once the plane lands in the other country, the rotor is placed on a truck with other required items to be sent to the location of the helicopter. While in transit, the truck runs off the road, falls on its side, and dumps its contents onto the side of the road. Many of the containers are damaged. The quartermaster from the helicopter base is contacted and drives out to the site to inspect the rotor and other items. While at the crash site, the quartermaster and truck driver inspect the different containers for damage and reviews each of the manifests as augmented reality information. This becomes a same time - same place configuration. The rotor's container has the urgent virtual information tag on it, and the quartermaster inspects this container first. They find the rotor to be damaged, and the maintenance person contacts the supervisor to get a new rotor ordered. The other items such as clothing are sent on a new truck to the helicopter base.

A key difference with this form of collaboration is the artifact the users are manipulating. This artifact can be characterised by the following features: firstly, it corresponds to the physical world; secondly, the size of the information space reflect physical objects in a large area; and thirdly, the users are able to physically walk within the information space and the physical world simultaneously. This form of collaboration is similar to distributive virtual environment collaboration systems. Both have manipulable 3D models and the position of the users affects their vantage point. The significant differences are that the distances the users are allowed to physically move are larger and there is a one-to-one correspondence with the physical world.

3 INPUT

The use of traditional desktop user input devices is inappropriate in an outdoor environment. New devices and techniques are required to facilitate the mobile nature of the wearable computer platform. For example, how does a user point or select in three space? Using AR in an outdoor setting does not allow for the use of traditional VR six degree of freedom (6DOF) tracking devices. These would be inappropriate for the outdoor setting due to their size and the need for a locationally referenced external source. Body relative 6DOF tracking devices are required to develop pointing devices. There are many issues which make this form of pointing device a difficult problem to solve, but two major issues are as follows: firstly, registration errors will make it difficult for a user to point at or select small details in the augmentation and secondly, pointing and selecting at a distance are a known problem in virtual and augmented reality applications (compounded by the fact the user is outdoors with less than optimal 6DOF tracking of their head and hands).

Therefore, new user interaction techniques are required for an OWARCS, and to state the obvious, the input techniques the users are required to use will have a large impact on the usability of an OWARCS. A key element to the new user interactions is the augmented reality systems have a varying number of coordinate systems (physical world, augmented world, body relative and screen relative) the user must work within. In an outdoor application the registration errors of objects at distance amplifies the differences between the physical and augmented world coordinate systems. In the helicopter rotor example, the problem of selecting the container becomes more difficult the further away the user stands while performing the selection task.

The problem becomes compounded with dynamic elements in the scene, such as moving trucks or people. If objects are moving within either of these coordinate systems (the user moves or an object in the scene moves), the number of coordinate systems the OWARCS has to understand may increase, due to each of the moving objects having their own coordinate systems.

With these problems in mind, I believe the use of hand and head gestures are key to making the collaborative systems usable. In the helicopter rotor example, both users were required to indicate features in the physical world or on the 3D models. Hand and head gestures are an intuitive means of the achieving

Figure 3: Thumb pointing   Figure 4: Laser pointer

this form of interaction. For example, one user may wish to indicate a particular feature by framing the region with their forefinger and thumb (Figure 2), or line of sight to the tip of their thumb (Figure 3), or a laser beam from the tip of their finger (Figure 4). A similar idea in virtual worlds is explored in the paper *Image Plane Interaction Techniques in 3D Immersive Environments* [Pierce et al., 1997]. The region is then highlighted on the desktop display and/or the HMD to provide visual cues to the users. For the HMD users, control of the selected region could be transferred to the head tracking device for gross movements and finer control may be performed by the hands, using a magic lens interaction technique for example. I believe the ability to quickly change input devices and coordinate systems is a key to making this form of interaction feasible.

4   USABILITY OF THE OWARCS

The previously mentioned scenario described a number of different modes of operation to accomplish the given task: single user in the field using a wearable computer; single user in an office using a traditional desktop computer; two users, one in the field - one in an office; two users, both in the field with wearable computers. These different modes of operation suggest the need for seamless movement of information across different display devices. During the operation of the wearable computer, a user may require the display of information on a more standard display device, for reasons of readability of text, details of diagrams, or ability to concentrate on the data.

A HMD may not be the optimal display technology, and as such the transition between the use of the HMD and other devices must be simple to use. I am proposing the use of a vast range of display and computing devices to be deployed. The following are a number of possible examples: 1] Standard desktop computers for office centric work, 2] Wall projectors to produce a data wall for group and ambient interfaces, 3] Notebook computers connected to LAN's or by remote wireless, and 4] Hand-held computers connected by remote wireless. 5] Wearable computers with see-through HMD's connected by remote wireless. Each of these configurations requires a different user interface for the user. The screen size and input device selection is different for each of configuration. The social protocols protecting sensitive information are different for each of the configurations. I propose all of these configurations can use the same information and protocols for information transfer between users and their different information system.

To extend the task of the quartermaster viewing the manifest, the quartermaster could have a hand-held computer (in a similar configuration as described in [Feiner et al., 1997]) to display more detailed information, instead of the manifest displayed as a screen overlay on their HMD. When the quartermaster returns to the truck, an easier interface for ordering the new rotor could be provided by a notebook computer with a wireless communication link. These two forms of computers offer access to more traditional input devices and easier to read display devices. A second option would be to attach these traditional input devices and portable display devices to the wearable computer platform.

As one different example, data walls allow for information displayed on a standard computer screen to be shown in a larger format, for example on the order of 2.0 meters by 1.5 meters. This enables many people to view the information simultaneously. By have a number of people viewing the information, the information can facilitate group activities such as meetings and discussions. Data walls provide a useful ambient interface for a large work area. Ambient interfaces present information to users in such a way as to not dominate their immediate action. The classic example is a person can tell the passage of time from the changing of light levels coming through their office window. This information processing to determine the passage of time is a background activity for the user's cognitive system. By having a data wall present overall contextual information in a large community area (in our example the data wall could be placed in key areas of the warehouse or air-base), users are able understand key overall concepts of the work flow by working near the information, walking past the information, or occasionally popping in to see what is on the wall.

The privacy issues of such a device are clear, but there should be automatic mechanisms to display the appropriate information. The user interface demands are vastly different to a standard desktop configuration. Data walls are information presentation oriented, and input devices are traditionally a video remote control-like device or driven from a standard desktop computer.

5 CONCLUSION

In conclusion this paper presents a proposal for an outdoor wearable augmented reality computer system to support collaboration activities. The signification attributes of the system are as follows: 1] the ability to use all four time-space configurations for collaboration systems (same time – same place, same time – different place, different time – same place, and different time – different place), 2] the use of hand and head gestures as the main form of user interaction, and 3] the seamless movement of information across different display devices (HMD, PDA, laptop, data walls, and desktop). To highlight the effectiveness of such a collaboration system, a scenario of delivering a replacement helicopter rotor was described.

References

[Azuma, 1997] Azuma, R. T. (1997). Survey of augmented reality. *Presence: Teleoperators and Virtual Environments*, 6(4).

[Azuma, 1999] Azuma, R. T. (1999). The challenge of making augmented reality work outdoors. In *First International Symposium on Mixed Reality (ISMR '99*, pages 379–390, Yokohama, Japan. Springer-Verlag.

[Bauer et al., 1999] Bauer, M., Kortuem, G., and Segall, Z. (1999). "Where are you pointing at?" A study of remote collaboration in a wearable videoconference system. In *International Symposium on Wearable Computers*, volume 3, San Francisco, CA, USA. IEEE.

[Billinghurst et al., 1998] Billinghurst, B., Bowskill, J., Jessop, M., and Morphett, J. (1998). A wearable spatial conferencing space. In *Second International Symposium on Wearable Computers*, pages 76–84, Pittsburgh, USA. IEEE.

[Ellis et al., 1991] Ellis, C., Gibbs, S., and Rein, G. (1991). Groupware - some isuues and experiences. *Communications of the ACM*, 34(1):9–28.

[Fagrell et al., 2000] Fagrell, H., Forsberg, K., and Sanneblad, J. (2000). Fieldwise: a mobile knowledge management architecture. In *Proceeding on the ACM 2000 Conference on Computer supported cooperative work*, pages 211–220, Philadelphia, USA.

[Feiner et al., 1997] Feiner, S., MacIntyre, B., Höllerer, T., and Webster, A. (1997). A touring machine: Prototyping 3D mobile augmented reality systems for exploring the urban environment. In *Proc. of ISWC '97*, pages 74–81. IEEE.

[Höllerer et al., 1999] Höllerer, T., Feiner, S., and Pavlik, J. (1999). Situated documentaries: Embbeding multimedia presentations in the real world. In *3nd International Symposium on Wearable Computers*, pages 79–86, San Francisco, CA.

[Julier et al., 2000] Julier, S., Baillot, Y., Lanzagorta, M., Brown, D., and Rosenblum, L. (2000). Bars: Battlefield augmented reality system. In *NATO Symposium on Information Processing Techniques for Military Systems*, Istanbul, Turkey.

[Pierce et al., 1997] Pierce, J., Forsberg, A., M. Conway, S. Hong, R. Z., and Mine, M. (1997). Image plane interaction techniques in 3d immersive environments. In *Symposium on Interactive 3D Graphics*, Providence, RI. ACM.

[Starner et al., 1995] Starner, T., Mann, S., Rhodes, B., Healey, J., Russell, K. B., Levine, J., and Pentland, A. (1995). Wearable computing and augmented reality. Technical Report 355, M.I.T. Media Lab.

[TeamBoard Inc., 2001] TeamBoard Inc. (2001). *TeamBoard*. 300 Hanlan Road, Woodbridge, Ontario, Canada L4L 3P6 www.teamboard.com.

[Thomas et al., 1999] Thomas, B., Grimmer, K., Makovec, D., Zucco, J., and Gunther, B. (1999). Determination of placement of a body-attached mouse as a pointing input device for wearable computers. In *International Symposium on Wearable Computers*, volume 3, pages 193–194, San Francisco, CA, USA. IEEE.

[Thomas et al., 2000] Thomas, B. H., Close, B., Donoghue, J., Squires, J., Bondi, P. D., Morris, M., and Piekarski, W. (2000). Arquake: An outdoor/indoor augmented reality first person application. In *Fourth International Symposium on Wearable Computers*, Atlanta, GA, USA. IEEE.

[Thomas et al., 1998] Thomas, B. H., Demczuk, V., Piekarski, W., Hepworth, D., and Gunther, B. (1998). A wearable computer system with augmented reality to support terrestrial navigation. In *Second International Symposium on Wearable Computers*, pages 168–171, Pittsburgh, USA. IEEE.

Augmented Reality Approaches to Sensory Rehabilitation

Suzanne Weghorst

Human Interface Technology Laboratory
University of Washington

ABSTRACT

The potential for augmented reality (AR) technologies to impact work habits and collaborative work is perhaps most striking for individuals with sensory or perceptual impairments. Commercial display and sensing technologies, in combination with on-board computation capabilities (either in the form of specialized hardware or general-purpose wearable computers), are introducing a new generation of adaptive aids. Spectacles and traditional hearing aids are being replaced by customized and context-sensitive conditional display systems. These technologies will enable much broader access, both to day-to-day interaction and to our increasingly information-based workspace. Two specific examples illustrate many of the technological and human factors challenges presented in the development of these new AR sensory aids. The lessons and technological advances gained from these efforts may have much broader implications for the design and implementation of generic AR systems.

1. Introduction

As interface and computing technologies evolve they may offer hope to people with sensory and neuro-perceptual disorders. Although traditional medical solutions and prosthetic devices continue to advance, wearable computing and personal multi-sensory displays provide a new paradigm for context-sensitive and user-customized "sensory prosthetics". At the Human Interface Technology Lab we are developing several such devices that represent a range of problems in this domain. We present an overview here of two of them that are particularly promising and informative: (1) visual aids based on scanned retinal images, such as the Virtual Retinal Display (VRD), and (2) an augmented reality display that aids Parkinson's Disease patients to overcome frozen gait.

2. Low Vision Aids

While there is evidence of occasional use of natural lenses to focus poor eyesight as early as the 3rd century AD, it was not until roughly 1850 that spectacles designed to improve visual acuity became available to the general public. Today eyeglasses are being at least partially displaced by surgical procedures that augment and correct the optical properties of the eye. Despite these advances, many partially sighted ('low vision') individuals are not able to achieve adequate vision for essential daily tasks. It is estimated that nearly 14 million Americans experience visual problems that impede their enjoyment of everyday living (National Advisory Council, 1998).

Conventional computer displays, for example, pose problems for many people with limited vision. Among the visual challenges presented by conventional displays are insufficient brightness and spatial resolution, flicker (especially with peripheral viewing, a common strategy adopted by many people with macular degeneration), and excessive glare. The deficiencies of conventional visual displays are even more apparent in AR systems, in which computer-generated imagery is superimposed on the user's view of the physical world. Current consumer-level "see-through" head-mounted displays based on reflected LCD panels provide only "ghosty" images that are inadequate in outdoor lighting conditions and are difficult to keep in focus at arbitrary accommmodative distances.

2.1 Scanned Retinal Displays

Scanned retinal displays, such as the Human Interface Technology Lab's virtual retinal display (VRD) technology may provide a significantly better alternative for low vision computer users. In combination with image acquisition technologies and on-board computing, the VRD may also form the basis for a wearable low vision aid. The VRD uses a scanning light beam in place of the conventional image planes of CRTs and LCDs. A very small spot is focused onto the retina and is swept over it rapidly in a raster pattern (see functional schematic in Figure 1). Because the photons are tightly shepherded, this display technique offers exceptional brightness, even at very low light power levels. In addition, the VRD's small 'exit pupil' of collimated light greatly reduces light-scattering and provides for a very wide depth of focus (Kollin, 1993; Tidwell et al, 1995; Johnston and Wiley, 1995).

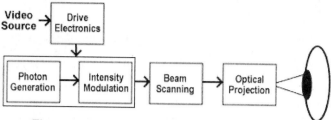

Figure 1. Block diagram of the basic VRD components.

Our prototype VRD projects very low laser power (typically, 50-200 nanowatts), yet is perceived as very bright, colorful, and comfortable to view. Three coherent light sources (red, green, and blue) are combined to provide full RGB color, or one light source can be used for monochrome displays. For each primary color, the current prototype produces a standard VGA resolution image at a refresh rate of 60 Hz.

Although it is being developed as a generic display technology, its unique characteristics make the VRD an exceptionally good image source for people with partial loss of vision, particularly those due to optical anomalies of the eye. Some of the unique aspects of VRD display technology of interest for low vision applications include its:

- very high brightness (at levels unattainable by CRT and LCD),
- collimated light, resulting in less eye strain and accommodative conflict, and
- very small system exit pupil, affording a small entrance pupil into the eye, large depth of focus and greatly reduced glare.

2.2 Low Vision Pilot Studies

Throughout the development of the VRD we have demonstrated the system to a wide variety of people, including many individuals with low vision conditions. Numerous subjective reports of enhanced vision have led us to a more systematic examination of VRD effects on low vision visual perception. The primary objectives of this pilot research have been to determine what types of low vision will benefit most from VRD technology, and to determine if the VRD can be an effective alternative low vision computer interface.

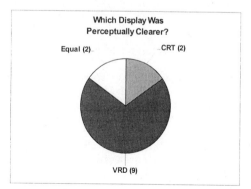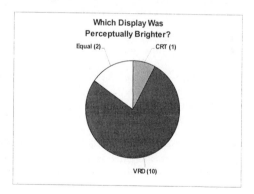

Figure 2. Subjective preference results for VRD vs. CRT display conditions.

In Kleweno's (1999) study thirteen subjects were recruited from the local low vision community. The low vision conditions represented by this group included amblyopia, retinal detachment, diabetic retinopathy, glaucoma, cataracts, nystagmus, aniridia, surface wrinkling retinopathy and strabismus. Using a portable red monochrome version of the VRD, this study compared acuity, reading speed and user preference across four display conditions: (1) standard CRT with white on black contrast, (2) standard CRT with red on black contrast, (3) VRD with red on black contrast with a luminance setting equal to half the measured value of the CRT (white on black contrast), and

749

(4) VRD with red on black contrast with a luminance setting equal to the measured value of the CRT (white on black contrast).

Average reading speed for this pilot subject sample was variable, with no statistically significant increase in average reading speed using the VRD compared with the CRT. However, the relative superiority of the VRD over the CRT display was striking for subjects with optical causes of low vision (e.g., cataracts and corneal aberrations). Results for the subjective comparisons of image quality between VRD and CRT displays were even more compelling (as shown in Figure 2). These pilot low vision subjects clearly preferred the brightness and clarity of the scanned light display over the conventional CRT display.

Although the red monochrome VRD was shown to be effective for several low vision subjects, most subjects indicated that red was clearly not their preferred color for text display. Seibel et al's (2001) study further addressed this issue of optimal text color for a VRD-based low vision AR display. Twelve subjects were recruited, with a wide range of low vision condition, including acromatopsia (lack of color vision), macular degeneration, partial albinism, glaucoma, retinal scarring, and Stargaarts disease (a congenital form of macular degeneration). Of these, eight were able to complete the testing protocol, in which text reading performance was measured for various text colors presented on four different background lighting conditions.

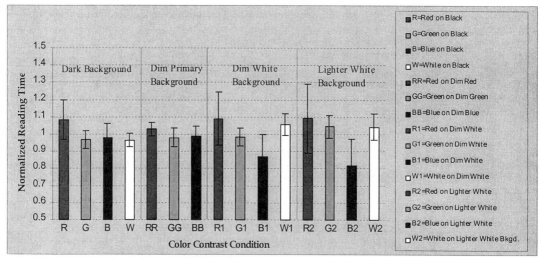

Figure 3. Normalized reading times for all 15 color contrast conditions.

Mean reading times, normalized for each subject within each background condition, are shown in Figure 3. While longer reading times (and thus poorer performance) for red text were expected, the finding that blue text was significantly easier (faster) to read than the other text colors in the ambient lighting conditions was surprising and not previously reported for CRT- and LCD-based low vision reading studies. We are currently systematically exploring these effects to determine how viewing text via the VRD is different from viewing text on conventional displays, and whether the effect will generalize to normally sighted readers with an equivalently challenging task. The overall focus of this research is to determine the optimal image generation parameters and visual aid design specifications for a variety of low vision conditions. Retinal light-scanning technology appears quite promising to (1) optimize low vision access to computing, and (2) form the basis of a wearable low vision aid (in combination with image processing and context-sensitive display).

3. Augmented Perception for Neurological Disorders

Sensory feedback is a vital component of complex integrated motor behaviors, such as walking and talking. In Parkinson's Disease (PD), a degenerative neural disorder, these behaviors are progressively disrupted. While typically characterized as a 'motor disorder' due to failure of the dopaminergic pathways in the basal ganglia, recent evidence suggests that faulty sensory feedback may also play a role in exacerbating these symptoms.

3.1 Parkinson's Disease and *Kinesia Paradoxa*

One of the most debilitating effects of PD is the sudden unpredictable and total inability to take a single step (akinesia). The effects of reduced levels of dopamine on mobility range from the complete inability to initiate ambulation, to small shuffling stutter steps, to normal gait that suddenly freezes (Marsden, 1977). *Kinesia paradoxa* is an interesting phenomenon that has been documented in the PD literature for many decades (Martin, 1967), but has yet to be fully exploited for therapeutic purposes. In PD patients exhibiting akinesia the presence of objects on the ground can often facilitate walking. Typically, *kinesia paradoxa* is demonstrated by placing a series of small objects or markers in the patient's intended path, about one stride length apart. The result can be a dramatic recovery of full stride length walking, but only in the presence of the cues.

3.2 Visual Cueing with AR

Numerous past attempts to develop PD walking aids based on the *kinesia paradoxa* effect have had limited practical applicability. Over the past few years, however, we have developed a number of highly effective AR devices, in which *virtual* visual cues are used to evoke the *kinesia paradoxa* response in place of physical cues (Prothero, 1993; Riess and Weghorst, 1995; Weghorst and Riess, 2001). Using microelectronics and compact head-worn display technologies, these virtual visual cues can be superimposed onto reality, and can be made available on demand in a wearable package. Several working prototypes have been built which demonstrate that virtual visual cues are indeed effective and may have a significant therapeutic impact in the lives of people suffering from PD.

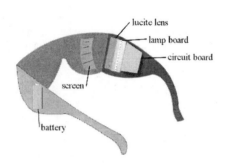

Central Field Cueing Device

Figure 4. Central Field Cueing Device. This implementation uses reflections of elongated "scrolling" LED lamps to project the virtual equivalent of objects onto the real world.

The most cost-effective solution to date uses an array of flashing LEDs, reflected off an optical combiner in the user's field of view, producing enhanced visual flow via "apparent motion" effects (see Figure 4). The efficacy of the LED approach has been demonstrated in a series of trials with experimentally naïve PD subjects.

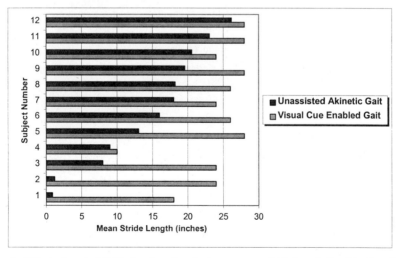

Figure 5. Effect of central field visual cueing device on mean stride length for akinetic subjects.

751

Figure 5 summarizes the results of a series of initial trials with 10 male and 2 female PD subjects, ranging in age from 45 through 74, with subjects ordered by unassisted stride length. Prior to testing with the augmented reality cues, each subject was screened for responsiveness to *kinesia paradoxa* cues using physical markers on the floor. Subjects who exhibited no stride length increase in response to the physical cues were not presented with the virtual cues. It should be noted, however, that less than 10% of the akinetic PD patients tested failed to exhibit *kinesia paradoxa*, and all subjects who exhibited some degree of stride enhancement with physical cues *also* responded positively to the virtual cues. Clearly, the visual cueing effect is extraordinarily robust. The mean stride length ratio (unassisted/cued) for these akinetic subjects ranged from .05 (S1, S2) to .93 (S12), with a mean stride length ratio of .60 across all subjects in this sample.

Since there is a natural "ceiling effect" for stride length (i.e., the maximal normal stride length is relatively fixed for each individual, depending largely on subject height), the benefits of visual gait cueing are proportionate to the degree of akinesia experienced by each subject. The strongest effects were demonstrated for subjects who exhibited freezing or festinating gait (e.g., S1 and S2). It is interesting to note, however, that even mildly akinetic subjects (e.g., S11 and S12) responded positively to the experimental cueing, and that no subjects tested to date have exhibited a *shorter* mean stride length in response to the visual cues. In addition to these quantitative effects, several subjects also exhibited noticeable improvements in gait topology and posture. One subject, for example, showed a marked decrease in her asymmetric foot dragging and shoulder slant while walking using the visual cueing device.

4. Conclusions

The development of small inexpensive microprocessors with low power requirements complements the recent emergence of consumer level head-mounted displays. In addition, the challenge of creating adequate visual displays for wearable applications has engendered new approaches to visual display, including technologies that scan modulated light directly onto the retina. This convergence of enabling technologies bodes well for a new generation of adaptive aids for a wide variety of sensory and neurological disorders. These applications may also have a positive impact on the development of AR for more generic purposes.

REFERENCES

Kleweno, C., Seibel, E., Kloeckner, K., Viirre, E., Furness, T.A. Evaluation of a scanned laser display as an alternative low vision computer interface. *OSA Technical Digest of VSIA Topical Meeting*, 1999.

Johnston, R.S., and Wiley, S.R. Development of a commercial retinal scanning display. *Proc. SPIE. Helmet- and Head-mounted Displays and Symbology Design Requirements II. 2465*: 2-13, 1995.

Kollin, J. A Retinal Display For Virtual-Environment Applications. In *Proceedings of SID International Symposium, Digest Of Technical Papers*, pp. 827, 1993.

Marsden, David C., and J.D. Parkes, Success and problems in long-term levodopa therapy in Parkinson's Disease. *Lancet 12*, 345-349, 1977.

Martin, J.P. *The Basal Ganglia and Posture.* J.B. Lippincott Company, 1967.

National Advisory Council. *Vision Research (1998) A National Plan: 1999-2003*. U.S. Dept. of Health and Human Services (NIH Publication No. 98-4120), 1998.

Prothero, J. *The Treatment of Akinesia Using Virtual Images.* Master's Thesis, University of Washington, 1993.

Riess, T., and Weghorst, S. Augmented reality in the treatment of Parkinson's Disease. *Proceedings of Medicine Meets Virtual Reality 3*, IOS Press, 1995.

Seibel, E.J., Gau, C.-C., McQuaide, S., Weghorst, S.J., Kelly, J.P., and Furness, T.A. Augmented retinal light scanning display for low vision: Effect of text color and background on reading performance *OSA Technical Digest of VSIA Topical Meeting*, 2001.

Tidwell, M., Johnston, R.S., Melville, D. and Furness, T.A. The Virtual Retinal Display - A Retinal Scanning Imaging System. In *Proceedings of Virtual Reality World '95*, pp. 325-333, 1995.

Weghorst, S., and Riess, T. Wearable sensory enhancement aids for Parkinson's Disease. *Medicine Meets Virtual Reality*, Newport Beach, 2001.

Experiential Recording by Wearable Computer

Ryoko Ueoka, Michitaka Hirose

Research Center for Advanced Science and Technology ,University of Tokyo
4-6-1 Meguro-Ku Komaba Tokyo 153-8904 Japan

Abstract

This is a fundamental study on the recording of subjective experience for wearable computer. In this paper, we present five categories which must be included for the recording and recalling of experiences. First, the conscious attention rate during experience is discussed in terms of the results of our experiment. Second the relationship between the conscious attention rate and mental status is discussed in terms of the results of our experiment. Lastly, our preliminary experiment, which involved the recording of experience using sensors covering the five categories, is discussed.

1. Introduction

Today, the wearable computer is becoming a reality due to the significant downsizing of computer devices.(Mann, 1997) It has received a great deal of attention, mainly due to the effectiveness of outdoor information display devices. However, wearable computers also hold great potential for sensing and recording real time information of the immediate surroundings and ambience around the user because they fit the human body closely.

It can be put forth that recording individual experiences is one of the most important and practical applications of wearable computers. The desire to record one's own experiences is so strong that the trend of downsizing recording devices such as cameras and video cameras has been constant. Not only is recorded history worthwhile to allow individuals to recall their experiences, but it will also become a valuable record of the period for the future.

2. Wearable computer for experience recording

First, we define five categories as components of the material of experience. As Figure 1 shows, these parameters, which range from internal to external conditions can be considered as the sources that define human experience. An individual scans these variable information which affect the mind and feelings and lead to the memorization of an event as an experience.

Although "experience" is considered to be very subjective matter, the essence of forming experience is considered to be composed of very objective matter.

By physically wearing computer devices, it becomes possible to obtain various types of information that are synchronous with the behavior of a user. This implies that the wearable computer can be applied as a system to record experience objectively as well as automatically.

The puspose of this study is to make a wearable computer system which records subjective experiences in an objective manner in the future.

Figure1. Five categories as components of the material of experience

Experiences are composed of the events which an individual is exposed to. In general, "experience" is compressed and reformed by the events to which an individual pays attention as an experience memory. In previous experiments, we analyzed the attention rate and the distribution of the attention of the individual using the prototype wearable computer. (Hirose et al,2000, Ueoka et al,2001)

In these experiments, they recorded comments when they encountered an interesting situation or object. In other experiments, they pushed a button when they were paying attention to something. This length of time was then calculated as the total attention time.The rate of the total attention time out of total experience time is calculated as the attention rate.

As Table 1 shows, the results revealed that the average attention rate in experience is approximately 23 %. The distribution of the attention rate is even throughout the total experience. Figure 2 shows the distribution of the attention of one of the users who did the experiment.This indicates that there existed a constantly interesting event which was worth paying attention to during the experience. There was no result in which an user's attention was focused on a single event. This result implied that experience could be reconstructed by compressing whole experience evenly into approximately 23 %.

As an objective method to compress whole experience into approximately 23 %, walking rate can be considered as one of the objective parameters. In a previous experiment, a user's walking rate was calculated by a data taken by a pedometer sensor and vision was also recorded by NTSC(National Television Standard Committee) signal. As an NTSC signal records 30 frames per second, we divided the frame data of the total experience time by the average walking rate; a similar rate with the attention rate was measured .Table 2 presents the results of the experiment.

Table 1. Attention Rate of Wearers

Total Experience Time(M in)	Total Flame	Average Walking Rate(Sec)	Objectively Calculated Rate (%)
8	14,400	0.9	25.6
10	18,000	0.83	20
10	18,000	0.65	19.6
10	18,000	0.85	25.6

Table 2. Calculated Rate by Walking Rate

Total Experience Time (Min)	Total Attention Time(Min)	Attention Rate (%)	Recording Function
14	2.5	17.9	Microphone
12	1.9	15.8	Microphone
10	4	40	Button
10	1.7	17	Button

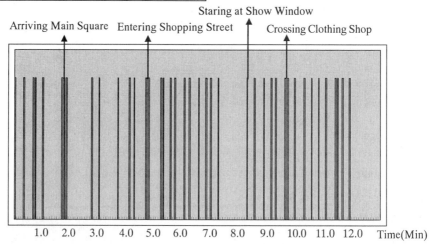

Figure 2. Distribution of the attention

Total Experience Time:12:03Min

Since an experience is considered to be a very subjective record which directly reflects a person's mental status, recording the person's mental status will be useful for recalling more realistic sensations.

We analyzed the correlation between behavior and mental status by recording vision and heartbeat rate variance （variance of R-R intervals of ECG:RRV） which has been reported to be useful as an index of an individual's mental status. (Hirose et al, 1984)

Figure 3 and 4 show the results of RRV value changes during the experience of wearers #1, #2 and #3. Wearers #1and #2 walked around a familiar area. Wearer #3 walked around unfamiliar area.The lower the RRV value, the higher the stress grows. Thus, this result revealed that the wearers experienced constant stress throughout the experiment.

This implied that a person walking outside constantly holds stress. The RRV value of wearers #1 and #2 changed in the middle of the experience. Although the RRV value indicated a relaxed mental status in the middle of experience, the value dropped rapidly. Overlaying recorded video and the RRV value, both wearers' RRV values dropped rapidly 30 or 40 seconds after walking a traffic-jammed road.

As another index of physical changes, we examine the relationship between one's attention and walking pauses.Pausing behavior indicates that there are interesting events taking place near the wearer.

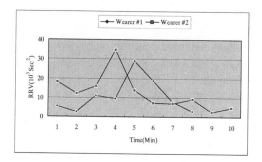

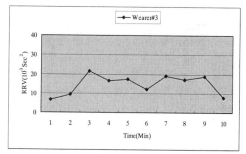

Figure 3. RRV Value of Wearer #1 and #2   Figure 4. RRV Value of Wearer #3

3. Experience Recording Experiment

3.1 Purpose and Method

As a preliminary study of experience recording, we selected several sensors from each of the categories discussed above, and examined the possibility of recording an individual's experience objectively.The selected sensors are listed in Table 3.

The purpose of this experiment is to analyze these objectively recorded data in order to evaluate the effectiveness of reconstructing subjective experience using these records. A user wearing a prototype experience recording system took walks outside while the sensored information was recorded automatically. The sensors in the Emotion category record data every 20 milliseconds and those in the Location and Ambience categories record data every 5 seconds. These data are recorded by a windows machine controlled by a board CPU. Visual and audio data are recorded by a DV recorder.The experiment was performed in two places on different dates by the same person. The system configuration and wearer are shown in Figure 5.

Table 3. Sensors selected for Experience Recording System

Parameter Name	Sensor	Purpose
Emotion	Heart Beat	Recording of RRV as mental index
Emotion	Flat Sensor	Recording of Walking Pause as Physical Index
Location	Grand Positioning System(GPS)	Recording of one's position as location index
Vision	Camera	Recording of one's vision as visual index
Audio	Microphone	Recording of environmental audio as audio index
Ambience	Temprature	Recording of atmosphere as atmospheric index

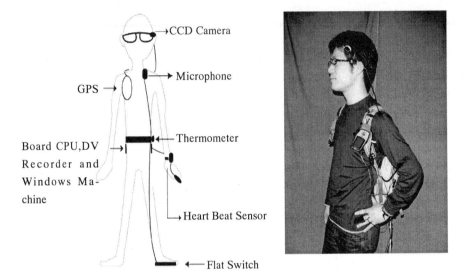

Figure 5. System Configuration and A Wearer

3.2. Results

Figure 6 presents one of the experimental results. The bold line on the map indicates a walking route of the wearer. The result of the data recorded by each sensor is measured by time basis.

4. Conclusion

In this experiment, we evaluated the relationship between objective recorded data and subjective experience. We reached the conclusion that it might be possible to reconstruct subjective experience using objective data set recorded by a wearable computer. In order to realize such a system, we need to increase the number of parameters to enable us to measure more precise physical and environmental status.

References

1.Steve, M.,"Humanistic Intelligence: Wear Comp as a new framework for Intelligent Signal Processing",(Proc. of IEEE) 1998,Volume 86,No.11,Pages 2123-2151

2.Michitaka, H.,Ryoko, U.,Atsushi,H.,Akiyoshi,Y.,"The Study of Personal Experience Record for Wearable Computer Application",(Proc. of the Virtual Reality Society of Japan Fifth Annual Conference) 2000,Pages 389-392

3.Ryoko,U.,Michitaka,H.,Koichi,H.,Atsushi,H.,Akiyoshi,Y.,"Study of Experience Recording and Recalling for Wearable Computer",(Correspondences on Human Interface) 2001,Volume 3,No.1,Pages 13-16

4.Michitraka,H.,Yoshiyuki,N.,Takemochi,I.,"Application for Operational Management using R-R intervals of ECG measured by optical sensor",(Proc of the Japan Society of Mechanical Engineers) 1984,No 844-5,Pages82-84

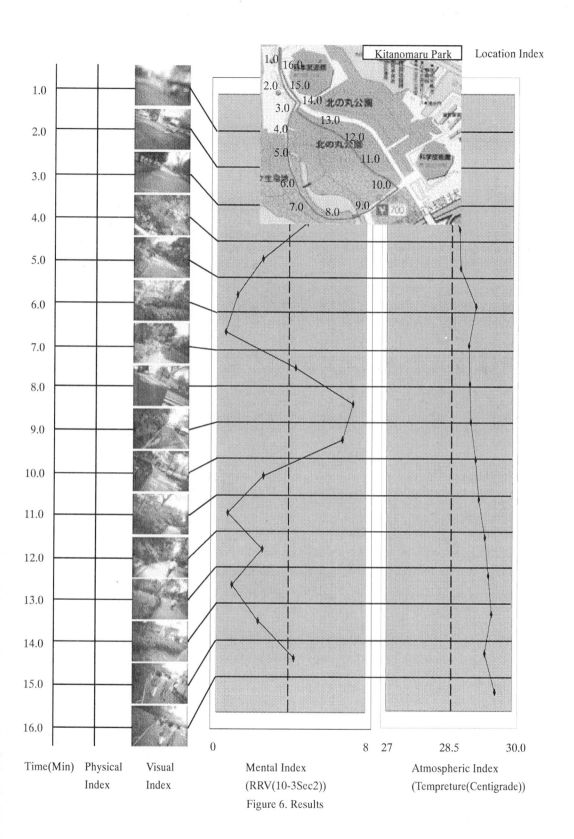

Figure 6. Results

757

VR ON GIGABIT NETWORK: VIDEO AVATAR TECHNOLOGY

Tetsuro Ogi[1,2], Toshio Yamada[1], Michitaka Hirose[2]

[1] Gifu MVL Research Center, TAO
Intelligent Modeling Laboratory, The University of Tokyo
2-11-16, Yayoi, Bunkyo-ku, Tokyo 113-8656, Japan
[2] RCAST, The University of Tokyo
4-6-1, Komaba, Meguro-ku, Tokyo 153-8904, Japan

ABSTRACT

In this study, virtual reality and augmented reality environments generated by immersive projection displays were connected through the Japan Gigabit Network. Particularly, in order to realize a high presence communication in the shared world, video avatar technology was developed. The video avatar was applied to the communication experiments among multiple sites which include virtual reality environments and augmented reality environments. By using this technology, the positional relationship among the users was represented in the three-dimensional virtual world, and the users were able to communicate effectively in the shared world.

1. INTRODUCTION

Advances in the infrastructure of broadband wide area networks have allowed us to transmit a large amount of data between remote places. For example, video image or three-dimensional model data can be sent in real-time through the broadband network. In 1998, Japan Gigabit Network (JGN) was equipped by the Telecommunications Advancement Organization of Japan. JGN is a nationwide optical-fiber network, and it has been used for research and development activities.

In this study, virtual reality environments generated by immersive projection displays were connected through the JGN network to share the virtual world between remote places. Immersive projection displays can generate high presence virtual worlds by projecting high resolution stereo images onto the wide screens. Therefore, in the networked immersive projection displays, remote users can share the virtual world with a high quality of presence as if they are in the same place (Leigh et al., 1997). Particularly, in this study, video avatar technology was developed to realize a high presence communication in the networked immersive environment. Currently, the networked immersive environment is being expanded into the connection among multiple sites which include augmented reality environments as well as the virtual reality environments. Therefore, the video avatar technology was applied to the communication among various sites.

This paper describes the method of generating video avatar, and discuss the communication experiments using the video avatar in the networked virtual world and augmented reality world.

2. VIDEO AVATAR

As for the communication technologies used in the networked environments, video conference system and computer graphics avatar have been used. However, in these technologies, it is difficult to realize a high presence communication in the three-dimensional shared space, because these methods only transmit two-dimensional video images or artificial characters of the users. In this study, video avatar technology which can transmit three-dimensional information about the user using the live video was developed by integrating the concepts of the video conference and the avatar technology (Ogi et al., 2001).

Figure 1 shows the process of making a video avatar. In order to capture the full-length figure of the user in the immersive projection display, a high sensitive camera which has wide viewing angle is necessary. In this method,

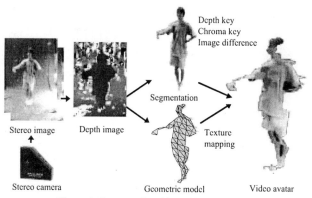

Figure 1. Process of making video avatar

Triclops stereo camera system made by Point Grey Research Inc. was used. Since this camera consists of two pairs of stereo camera modules along horizontal and vertical base lines, it can create an accurate depth image using stereo matching algorithm. The viewing field of this camera is 70 degrees, and the measured depth resolution was about 5.0cm.

Once the depth image is created, the user's figure can be segmented from the background according to the threshold of the depth value. When the blue backdrop is used or the background scene is not changed, chroma key or image difference method can be used in combination with the depth key to segment the clear image of the user. The created depth image can also be used to make a geometric model of the user. Since the three-dimensional position of each pixel in the captured image is determined from the depth data, a geometric model can be made by connecting the pixel positions using a triangular mesh.

Thus, by texture mapping the user's segmented image onto the geometric model, a video-based avatar that has a surface model for the front side is generated. This process was performed in about 10Hz. By transmitting the video avatar mutually between remote places, high presence communication in the shared world can be realized.

3. COMMUNICATION AMONG MULTIPLE SITES

In this study, the video avatar technology was applied to the communication in the networked virtual world among multiple sites. Figure 2 shows the system configuration constructed for the video avatar communication experiment. As for the networked virtual environment, CABIN at the University of Tokyo (Hirose et al., 1999), COSMOS at the Gifu Technoplaza (Yamada et al., 1998) and UNIVERS at the Communications Research Laboratory (Arakawa et al., 1999) were connected through the JGN network. These displays are CAVE-like multiscreen displays, and they can generate highly immersive virtual world by surrounding the user with the stereo images projected on the multiple screens. Therefore, in the networked multiscreen displays, it is expected that the remote users can communicate with a high quality of presence in the shared virtual world.

In the multiple sites communication, users often communicate in various positional relationships such as standing side by side or one behind another as well as meeting face to face. Therefore, in this study, several stereo cameras were placed around the user, and the nearest camera to the other user's viewpoint was selected and used. In addition, the user's position was tracked by the electromagnetic sensor, and the position data was transmitted to the other sites together with the video avatar data. Then, the transmitted video avatar was superimposed at the three-dimensional position on the shared virtual world, and it was used for the communication.

Figure 3 and Figure 4 show the examples of the video avatar communication in the shared virtual world among three sites. In Figure 3, the user in the CABIN is communicating with the video avatars transmitted from the COSMOS and the UNIVERS sites. And Figure 4 shows the communication at the UNIVERS site, where the users at the CABIN site

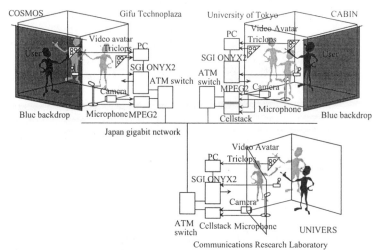

Figure 2. System configuration for video avatar communication

Figure 3. Video avatar communication among three sites (CABIN site)

Figure 4. Video avatar communication among three sites (UNIVERS site)

and the COSMOS site are represented using the video avatar. In these examples, remote users among three sites were able to communicate naturally in the three-dimensional virtual town. Particularly, the positional relationship among the users was effectively represented in the wide area of the virtual town using the video avatar.

4. COMMUNICATION IN AUGMENTED REALITY ENVIRONMENT

Next, the video avatar was applied to the communication in the augmented reality world. The networked immersive environment used in this study also includes an augmented reality environment. In order to generate a high presence augmented reality world, an immersive projection display system using a transparent screen was constructed. The transparent screen is able to superimpose a computer generated three-dimensional image on the real world using a see-through function.

Figure 5 shows the system configuration of the transparent immersive projection display system. In this system, HoloPro screen made by G+B Pronova GmbH was used. This screen is constructed by laminating gelatin film in multi-layer glass, and the screen itself is nearly invisible. When the image is projected from the specific angle of 37 degrees onto the screen, incoming light is redirected in the direction of the viewer. Therefore, the projector can be placed on the floor

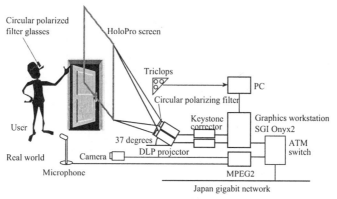

Figure 5. System configuration of transparent immersive projection environment

level hidden from the user's view. When the light goes through the HoloPro screen, the polarization of the light is preserved. Therefore, in this system, the polarizing filter method can be used to generate three-dimensional stereo image.

This display system was used for the video avatar communication. In this method, the stereo image of the video avatar was synthesized in the real world as if it was floating in the air, and the augmented reality world was generated. In this case, since the image projected onto the transparent screen is not seen from the rear side, the stereo camera can be placed behind the screen without disturbing the user's view. By using this system, the remote users can communicate in the actual workplaces such as the design room or the conference room.

Figure 6 shows the example of the video avatar communication in the augmented reality world. In this experiment, the video avatar transmitted from the remote site was projected on the transparent screen, and it was superimposed on the real world. When the augmented reality environments are connected to each other through the network, each user can communicate with the other user in the real world in which he is actually existing.

Figure 7 shows another example of the communication experiment between the virtual world and the augmented reality world. In this experiment, the stereo image of the object put in the real world was also captured by the stereo camera and was transmitted to the virtual world with the video avatar image. Then, the users in the virtual world and the augmented reality world were able to share the same object in the shared world. Thus, by using the video avatar technology, the users among the multiple sites which include the virtual world and the augmented reality world were able to communicate effectively through the network.

Figure 6. Video avatar communication in augmented reality world

Figure 7. Sharing object between virtual and augmented reality environment

761

5. CONCLUSIONS

In this study, in order to realize a high presence communication in the shared world, video avatar technology was developed. By transmitting the video avatar mutually in the networked immersive environments through the broadband network, remote users can communicate with each other. This technology was applied to the communication in the networked immersive world among multiple sites which include virtual reality and the augmented reality environments. In the video avatar communication among multiple sites, the positional relationship among the users was effectively represented. In addition, the users in the virtual world and the real world were also able to communicate using the transparent immersive display.

In this study, though the video avatar data was simply transmitted between each two sites, server technology that controls the transmission of data will be necessary to realize an efficient communication. Future work will also include applying this technology to the practical fields of application, such as the remote education or the collaborative design.

ACKNOWLEDGMENTS

We would like to thank Yoshiki Arakawa, Kenji Suzuki, Makoto Kano, Takashi Imamura and Kazuhiko Hirose for their help in the communication experiments.

REFERENCES

Arakawa, Y., Kakeya, H., Isogai, M., Suzuki, K. and Yamaguchi, F. "Space-shared Communication based on Truly 3D Information Space", Proceedings ICIP'99, 1999.

Hirose, M., Ogi, T., Ishiwata, S. and Yamada, T. "Development and Evaluation of the Immersive Multiscreen Display CABIN", Systems and Computers in Japan, 1999, 30(1), 13-22.

Leigh, J., DeFanti, T.A., Johnson, A.E., Brown, M.D. and Sandin, D.J. "Global Tele-immersion: Better than Being There", Proceedings ICAT'97, 1997, 10-17.

Ogi, T., Yamada, T., Kano, M. and Hirose, M. "Immersive Telecommunication Using Stereo Video Avatar", Proceedings IEEE VR2001, 2001, 45-51.

Yamada, T., Hirose, M. and Iida, Y. "Development of Complete Immersive Display: COSMOS", Proceedings VSMM98, 1998, 522-527.

EMBODIED SPACES: DESIGNING REMOTE COLLABORATION SYSTEMS BASED ON BODY METAPHOR

Hideaki Kuzuoka[a], Keiichi Yamazaki[b], Jun Yamashita[a], Shin'ya Oyama[a], Akiko Yamazaki[c], Hiroshi Kato[d], Hideyuki Suzuki[e], and Hiroyuki Miki[f]

[a]Institute of Engineering Mechanics and Systems, University of Tsukuba, 1-1-1 Tennoudai, Tsukuba, Ibaraki 305-8573, Japan

[b]Faculty of Liberal Arts, Saitama University, 255 Shimo-Ookubo, Urawa, Saitama 338-8570, Japan

[c]Department of System Information, Future University Hakodate, 116-2 Kamedanakano-Cho, Hakodate, Hokkaido 041-8655, Japan

[d]Learning Resources R&D Division, National Institute of Multimedia Education, 2-12 Wakaba, Mihama-ku, Chiba, Chiba 261-0014, Japan

[e]Faculty of Humanities & Social Sciences, Ibaraki University, 2-1-1 Bunkyo, Mito, Ibaraki 310-8512, Japan

[f]IT Laboratory, Oki Electric Ind. Co., Ltd. 550-5 Higashisakawa-Cho, Hachioji, Tokyo 193-8550, Japan

ABSTRACT

In order to design a video communication systems that can embody participants' behaviors, we defined five requirements i.e. 1)arrangement of bodies and tools requirement, 2) orientation of bodies requirement, 3) gestural expression requirement, 4) mutual monitoring requirement, and 5) sequential organization requirement. We named virtual spaces generated by video communication systems that support these requirements as "embodied spaces". In order to realize the embodied spaces, we propose a design guideline named "body metaphor". Then we introduce GestureLaser system and Agora system to show how the guideline can be used to design embodied spaces.

1. INTRODUCTION

CSCW community has been discussing the importance of bodily expressions during communication and various systems that support hand gestures (Tang, 1990), gaze awareness (Ishii, 1991), etc. have been proposed. Most of such systems supports face-to-face discussion and shared drawing. However, some sociologists are interested in human-to-human interaction that requires actions such as pointing or reference, or manipulating objects and artifacts within an environment (Dourish et al., 1992; Goodwin, 1998; Nardi, 1993). Our main interest is to develop remote collaboration systems that support such kind of interaction.

In this paper, we propose a new design principle for remote collaboration systems that consider embodiment of both spatial arrangement and bodily expressions of geographically distributed participants. To be more concrete, "we should arrange a body and a shared artifacts so that participants can monitor each other's body and shared objects."

In the following section we present our design principle, then we introduce two systems to show how our principle can be utilized to design remote collaboration systems.

2. EMBODIED SPACES

Conversational analysis and ethnomethodological studies of workplaces (Heritage, 1997; Goodwin, 1996) are interested in problems of instruction and collaborative work in the real world. Their main interest is to explain how participants organize human actions interactively and sequentially. In his paper, Heath wrote "The emergent and sequential organization of interaction is also relevant to how we might consider the contextual or in situ significance of visual conduct and the physical properties of human environment" (Heath, 1997).

For example, let's think of a case that A asks B to take an object (Figure 1 left). At first, A says to B "take this" and at the same time points to an object (perhaps a book on the table). A's pointing connects with A 's utterance "take this". When B sees A's pointing to the book, B turns his body to the book and says "ok". B's body movement and utterance display his understanding of A's utterance and gesture. After A draws away A's hand which pointed to the book. By withdrawing the hand, A displays A's understanding of B's understanding to all participants, including B (Goodwin, 1998). As shown in this example, participants maintain and reorganize arrangements of bodies and tools interactively to monitor their pointing and other work (Kendon, 1990). According to Goodwin (Goodwin,1996;

Goodwin, 1998), body arrangement of participants is important. For example, when instruction is given, operators move their bodies into appropriate positions to see the shared artifacts. Instructor likewise moves his/her body in such a way that his view of shared artifacts is not obstructed by operators and make sure that the operators are watching his/her gestures while they are given instructions.

However, in video-mediated communication, these kinds of bodily resources easily become disembodied. (Heath et al.,1991; Heath et al.,1992). In order to alleviate this problem, we must design new video-mediated communication systems that can embody participants' behavior (Kato et al., 1997). Based on the above mentioned ethnomethodological studies and our own studies on video mediated communication, we formulate the following five requirements for a system to support remote instruction in the real world.

1. ***Arrangement of bodies and tools requirement***: The arrangement of bodies and tools should be appropriate for monitoring other participants' behavior for both instructors and operators.
2. ***Orientation of bodies requirement***. Each participant should be able to turn his/her body to other participants and shared artifacts.
3. ***Gestural expression requirement***: The instructor must be able to use freely not only verbal expressions, but also body movements and bodily expressions (gestures).
4. ***Mutual monitoring requirement***. Participants should be able to monitor each other's body arrangements, orientations, and gestures mutually.
5. ***Sequential organization requirement***: Sequential and interactive organization of the arrangement of bodies, tools and gestural expression must be possible.

We then named virtual spaces that are generated by video mediated communication systems as "embodied spaces". When participants are co-existing, these requirements, especially mutual monitoring, are naturally realized because an object, A's finger, and B's face are in front of A's face and they are withing A's field of view. Thus A can naturally monitor B's gaze direction and B's actions to an object. In order to satisfy these requirements in remote setting, we attempted to use the ordinary body arrangement as a metaphor for the placement of cameras, the face view, and hand-gesture monitors. We named the metaphor as 'body metaphor'. We believe that the body metahor is one of the effective guidelines that realizes embodied spaces. Due to the limitation of pages, please refer to our previous paper (Kato, 199) for detailed description.

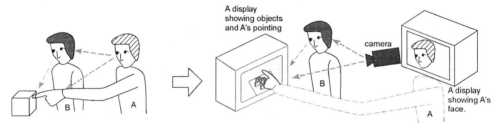

Figure 1. Pointing in case of co-existing situation (left) and pointing using a system based on body metaphor.

Next, we will introduce two systems, "GestureLaser" and "Agora", which were designed based on body metaphor for supporting two different types of remote collaboration.

3. GESTURELASER
The goal of this system is to facilitate remote instruction on physical tasks in the real world, such as repair and maintenance of mechanical devices, laboratory classes, medical treatment.

3.1 System Specification
The GestureLaser is a remote controlled laser pointer developed by the authors. GestureLaser reflects the laser emitter's ray off two orthogonal mirrors into the workspace. One of the features of GestureLaser is that it can be moved like a mouse cursor. As shown in Figure 2, a CCD camera is mounted on GestureLaser. The instructor can monitor not only the position of the laser spot but also the position of the objects and the operator on an image from a CCD camera, and the instructor controls the location of the laser spot with a mouse.

The Second feature of GestureLaser is the fact that it is small and light thus we could mount GestureLaser on a four-wheeled conveyance, which we call GestureLaser Car. This car can move horizontally on the rails, controlled

764

by the instructor's keyboard.

By comparing the figure 3 with the figure 1, it is easily understood that the GestureLaser satisfies the body metaphor when an object, an operator, and the GestureLaser is positioned appropriately.

Figure 2. Overview of GestureLaser system.

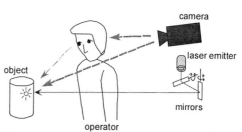

Figure 3. Body metaphor with the GestureLaser.

3.2 Experiment

We performed experiments to see how the requirements of embodied spaces are supported using a GestureLaser and a GestureLaser Car. We analyzed the experiments in the aspect of the body arrangement, the sequentiality, and the gesture expressions (more detail is shown in Yamazaki et al., 1999).

Normally, GestureLaser was installed behind operators. Lower left picture of figure 4 is an image that an instructor monitored on a display, and upper right picture shows the scene that an instructor giving instruction. This figure shows that the instructor could monitor objects, operators, and the laser spot simultaneously. Therefore, it was not only possible for an operator to see an indicated objects and the laser spot, but also an instructor could recognize that an operator was seeing objects and the laser spot. We called this arrangement "the basic arrangement".

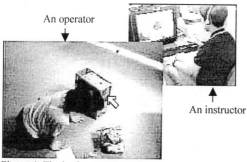

Figure 4. The basic arrangement.

Figure 5. Showing the direction with the laser spot.

Sometimes during the experiment, however, they failed to structure a basic arrangement For instance, from an instructor's point of view, an object was obstructed by an operator or by other objects. In such cases the instructor performed the instruction to restore it to the original state. One of the ways of instruction we observed was that an instructor let an operator structure a basic arrangement by using speech and motion of the laser spot. For example, in order to bring an object to an appropriate position, the instructor said "Turn it to this side" or "Move this" and shows the direction of a movement with the laser spot (Figure 5). Another method was to move the GestureLaser Car and change the instructor's virtual view point. Furthermore, we observed the instance that, when the instructor moved GestureLaser Car, the operator realized the movement and he moved to the position where he didn't obstruct the instructor's field of view. Thus, to some extent, we think that the GestureLaser supports system requirements that were mentioned in section 2.

4. AGORA

4.1 System Specification

Agora is a remote meeting support system (Kuzuoka, 1999). As shown in figure 6, positions of screens and cameras are chosen using the body metaphor of round table meeting of four participants. Two 60-inch screens are settled along two sides of each desk. Waist-up images of remote participants are projected on the L-formed screens by two rear projectors. To maintain eye contact, cameras attached on the screens at eye level capture participants. Cameras are small and flat enough so that they are not too obtrusive for local participants to observe remote participants in the screen. Artifacts and gestures on both desktops are shared by using a similar configuration to Double DigitalDesk (Wellner, 1993).

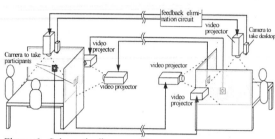

Figure 6. Schematic diagram of Agora system.

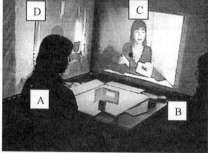

Figure 7. Making eye contact within Agora.

A pair of a video projectors and a camera is mounted above each desk. A camera grabs images from its local desktop and a projector projects down the images from a remote desktop. In this way, waist-up images and hand gestures are projected onto a remote screen and desktop seamlessly. So, participants can distinguish gestures' owner. The arrangement of a desk for shared activity and screens for waist-up images enable mutual monitoring, where a relationship of participants' body and hand, and orientation of participants are reproduced naturally. In other words, body arrangements, orientations, gestural expressions and mutual monitoring of four participants surrounding a desk are reproduced.

4.2 Experiment

We conducted several experiments using Agora. Tasks in the experiments were board games, puzzles, collaborative drawing and a room design task with miniature furniture. The analysis of these experiments showed us that the Agora supports five requirements to some extent. However, we also noticed that the Agora had some problems in comparison with an actual face-to-face conversation. For example, if a remote participant is virtually sitting right in front of a local participant (such as A and C in figure 7) eye contact is maintained. When a remote participant is sitting beside a local paritipant (such as C and B in figure 7), however, eye contact is not maintained due to the Mona Lisa effect.

Figure 7 is a snapshot from our experiment. In this case, participant C is looking at participant B. From A's point of view, it looks that C is looking at B. From B's point of view, on the other hand, C looks as if she is looking at the right side of B. Interestingly, however, when C moved her gaze from the desktop toward B, B realized that C is looking at him and reacted properly.

In CSCW, supporting gaze awareness has been an important issue. Various systems that try to maintain eye contact between more than one participants has been suggested (Okada, 1994). However, it is still difficult for all participants to completely maintain their eye contacts. One solution to this problem may be to use the virtual reality technique. However, the analysis of Agora presented in the previous section suggests that we can think of the issue of eye contact problem from a different point of view. For instance, rather than concentrating on maintaining eye contact perfectly, it may be more important for participants to enable them monitoring changes of postures and gaze and resulting changes of orientations of other participants.

5. CONCLUSIONS

In this paper, we showed requirements to design "embodied spaces". Furthermore, to actually constitute embodied spaces, we proposed the guideline named "body metaphor", i.e. "positions of cameras, monitors, objects, and participants should be defined based on the metaphor of arrangements of body parts of participants and objects for co-existing interaction". Then we described two systems that utilized the guideline.

It should be noted that by considering embodiment of spaces we can analyze effectiveness and problems of a system and compare it with other systems. In other words, the embodiment point of view provides a framework to design and evaluate video mediated communication systems. Our future work is to develop more effective systems while further elucidating the embodiment problem.

ACKNOWLEDGEMENT

The research was supported by Telecommunications Advancement Organization of Japan, Japan Society for the Promotion of Science, Grant-in-Aid for Scientific Research (B), 2000, 12558009, Casio Science Promotion Foundation, Oki Electric Industry Co. Ltd., and Venture Business Laboratory. The experiments with ATM network were supported by Communications Research Laboratory.

REFERENCES

Dourish, P.,Adler, A., Belloti V. and Henderson, A. "Your Place or Mine? Learning from Long-Term Use of Video Communication", *EuroPARK Technical Report EPC-1994-105*,Rank Xerox EuroPARK, UK., 1992.

Goodwin, C. "Professional vision", American Anthropologist 96, 1996, pp.606-33.

Goodwin, C. "Pointing as Situated Practice", presented at the Max Planck Workshop on Pointing Gestures, 1998.

Heath, C., and Luff, P. "Disembodied Conduct: Communication through video in a multi-mediaenvironment", *Proc. of CHI'91*, 1991, pp. 99-103.

Heath, C., and Luff, P. "Media Space and Communicative Asymmetries: Preliminary Observation of Video-Mediated Interaction", Human Computer Interaction, 1992, 7(3), pp.315-346.

Heath, C. "The Analysis of Activities in Face to Face Interaction Using Video", David Silveraman (ed.) Qualitive Sociology, London: Sage, 1997, pp-183-200.

Heritage, J. "Conversational Analysis and Institutional Talk", David Silveraman (ed.), Qualitive Sociology,London:Sage, 1997, pp.161-182.

Ishii, H., Miyake, N. "Toward an Open Shared Workspace: Computer and Video Fusion Approach of TeamworkStation", Communications of the ACM, Vol.34, No.12, 1991, pp.37-50.

Kato, H., Yamazaki, K., Suzuki, H., Kuzuoka, H., Miki, H., Yamazaki, A. "Designing a Video-Mediated Collaboration System Based on a Body Metaphor", Proc. of CSCL'97, 1997, pp. 142-149.

Kendon, A. "Conducting Interaction: Patterns of Behavior in Focused Encounters", Cambridge Univ. Press.

Kuzuoka, H., Yamashita, J., Yamazaki, K., Yamazaki, A. "Agora: A Remote Collaboration System that Enables Mutual Monitoring", Late-breaking Results of CHI'99, 1999, pp.190-191.

Nardi, B., Schwarz, H., Kuchinsky, A., Leichner, R., Whittaker, S., and Sclabasi, R. "Turning Away from Talking Heads: The Use of Video-as-Data in Neurosurgery", Proc. Of INTERCHI'93, 1993, pp.327-334.

Okada, K., Maeda, F., Ichikawa, Y., and Matsushita, Y. "Multiparty Videoconferencing at Virtual Social Distance: MAJIC Design", Proc. of CSCW'94, 1994, pp. 385 – 393.

Tang, J., Minneman, S. "VideoDraw A Video Interface for Collaborative Drawing", Proc. of CHI'90, 1990, pp.313-320.

Wellner, P. "Interacting with the Paper on the DigitalDesk", Communications of the ACM ,Vol. 36, No. 7, 1993, pp.87-96.

Yamazaki, K., Yamazaki, A., Kuzuoka, H., Oyama, S., Kato, H., Suzuki, H., and Miki, H. "GestureLaser and Gesturelaser Car: development of an Embodied Space to Support Remote Instruction," in *Proc. of ECSCW'99*, 1999, pp.239-258.

767

NOVEL HAPTIC TEXTURE INTERFACES

Yasushi Ikei, Naoto Aoki, and Masashi Shiratori

Tokyo Metropolitan Institute of Technology
6-6, Asahigaoka, Hino-shi, Tokyo 191-0065

ABSTRACT

Two models of novel haptic interface are introduced. A projection type texture display was designed to present both tactile (skin) and visual textures with an enhanced sense of presence during user's exploration. A force reflecting type texture display was also developed to provide a force sensation along with a tactile sensation, which enabled three-dimensional surface exploration of a finger accompanied by a skin perception.

1. INTRODUCTION

The authors have developed novel haptic displays which present haptic textures to the user's finger with visual and proprioceptive (force) sensations. The both displays stimulate a skin sensation by a vibratory pin array which renders fine changes on a surface by distributed vibration sources. This technique has been discussed in a previous study on a haptic display—the Texture Display [1]—that is involved in the new display discussed here as a part of it. A problem of the previous display system was that the visual image of an object surface was displayed on a CRT screen deviated from the user's finger position, which reduced a sense of presence. In addition, the proprioception was omitted in the presentation for simplicity of the device. Haptic exploration is evidently concerned with the both two sensations observed along with a skin sensation, therefore the extensions introduced in this study improved the power of expression of the haptic display system significantly.

2. PROJECTION TYPE HAPTIC TEXTURE INTERFACE

A new type haptic display with visual projection was developed by adding a position measurement system and a projection screen to the Texture Display. A new feature of the display is that the user can feel virtual textures at the exact place where the user looks at them. The texture images are projected on to the screen attached to the display's top-plate where a pin-array exists. The user places a fingertip to the pin-array and fix it during haptic exploration on a virtual texture. By moving the entire display with a screen on a table, the user can touch any portion of the projected surface of virtual textures.

Figure 1 shows the photograph of the display system. The display's position is measured by a link mechanism with three joint-angle sensors (potentiometers) which

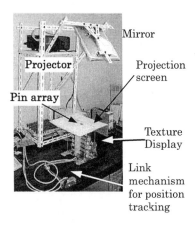

Figure 1. Projection type texture display system

768

provide a two-dimensional position and a rotation angle of the pin array. Figure 2 shows the projection screen and four projected textures on it. These texture images are the surfaces of wallpaper samples for interior decoration. The images are projected vertically by a SVGA projector of 1100 ANSI lumen. The resolution of the image is 73 dpi on the screen. The vibratory pin array that involves fifty wire-pins is at the center of the screen.

The entire system is depicted in Figure 3. A virtual space data is maintained at a host workstation from which a visual data is sent to a PC for visual presentation through TCP/IP. A haptic intensity data is sent to another PC for haptic presentation which returns the finger position in a real space based on link angle data. The PC controls the amplitude of pin vibration according to the relation between a sensation intensity and a pin amplitude.

The system added the measured degree of freedom of the pin array in a two-dimensional space. The previous model measured only x and y positions without a rotation angle, which restricted an arbitrary exploratory motion of the finger on a texture. The new system introduced more natural trace motion for the user to perceive a complicated textures that involved asymmetric elements of bumps. The visual presentation over a pin array screen contributed to enhancing the presence of the bumpy surface.

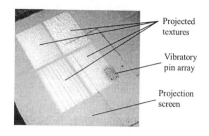

Figure 2. Projected texture images on the display screen

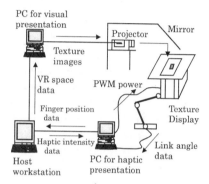

Figure 3. Schematic of the projection type haptic interface system

A subjective evaluation was conducted by using a questionnaire that asked subjects about the difference between this system and the previous model. Ten subjects participated in the evaluation. They reported on the average the significant improvement of presentation quality except for the clarity of a visual image as compared to the CRT screen.

3. FORCE REFLECTION TYPE HAPTIC TEXTURE INTERFACE

3.1 System overview

A pin array type stimulator was attached to the PHANToM haptic display to evoke both skin and force sensations. The PHANToM haptic display provides a force vector at a contact point of a finger to a virtual object's surface. This force constrains the user's finger out of the surface, in addition, it delivers the variation of a surface shape by its change in magnitude and direction based on contact calculation. The variation of a surface is conveyed by vibratory

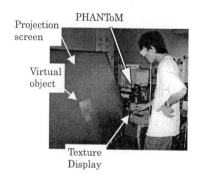

Figure 4. Force reflection type haptic texture interface system

stimulation as well. The user observes the image of a virtual object by using stereo-shutter glasses and examines a surface texture by the index fingertip as shown in Figure 4.

Figure 5 shows a combined mechanism for tactile and force feedback. A pin array of ten pins (two-column and five-row) with a 3-mm spacing was mounted to the stylus of the PHANToM. The user holds a handle placing the index fingertip on the pin array and touches the virtual surfaces. The weight of the pin array display and its handle is compensated by the PHANToM force.

Figure 6 shows the pin array unit made of photo-formed resin. Ten bi-morph like piezoelectric actuators are employed to drive pins independently. The vibration frequency is 250 Hz where the sensitivity of a skin indicates the highest value. The amplitude control of pins permits the user to discriminate at least fifteen levels of a sensation intensity on the average according to a scaling experiment.

The system presents surface textures by vibratory stimulation and a perturbed force. The vibratory stimulation is an approximation of the actual stimulus that occurs when the skin surface is deformed or vibrated in a dynamic contact between the skin and objects. The force conveyed by the system is an approximation as well, since it is a single three-dimensional vector so that it does not produce distribution of varied forces on a skin surface. Thus, the combinatorial use of these two modes of stimulation needs to be properly matched up for producing a touch sensation closer to that from a real contact.

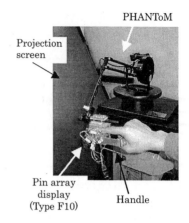

Figure 5. Pin array texture display (Type F10) and the PHANToM

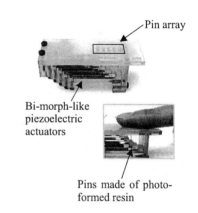

Figure 6. Pin array unit of the Texture Display (Type F10)

3.2 Haptic rendering by tactile and force stimulation

The both modes of stimulation are based on a surface shape data, or a height map of textures. The sensation intensity of vibratory stimulation was calculated in two ways and compared as follows. 1) The two dimensional position of each pin on a surface determined the sensation intensity according to the map when the pin was inside a reference plane on which the texture existed (*2D vib-control*). In this control, the depth a pin intruded into the plane is not considered. 2) Another control determined the intensity in proportion to the depth a pin intruded under the contour of a height map (*3D vib-control*).

The force calculation methods were built in three ways as follows. 1) The direction of force is normal to the reference plane and the magnitude of force is proportional to the depth of intrusion under the contour of a height map (*ref-normal force control*). 2) The direction of force is normal to the contour and the magnitude is proportional to the depth from the reference plane (*shape-normal force control 1*). 3) The direction of force is normal to the contour and the magnitude is proportional to the depth from the contour (*shape-normal force control 2*).

These controls were applied to one-dimensional sinusoidal grating textures with different wave lengths and amplitudes. Eighteen sample textures were arranged in a 3D space for comparison as depicted in Figure 7. The profile of each sample is shown in Figure 8. The subject examined these samples to find which control mode is suitable and its dependency on these features.

Regarding the vibratory stimulation, the *2D vib-control* was more stable providing easily perceivable tactile images regardless of the modes of force reflection. However, it caused somewhat inappropriate sensation since the stimulus intensity was determined only by the two-dimensional position irrespective of the finger force applied to the surface. On the other hand, the presentation of the *3D vib-control* appeared dependent on the mode of force reflection and the texture profiles because the amount of intrusion of a pin under the sinusoid was different among the force modes and contact conditions.

Regarding the force stimulation, the *ref-normal force control* exhibited accurate tracing of the finger to the texture contour as seen in Figure 9a since it did not have the tangential component of force, which enabled a stable trace along the surface shape. Although the shape was clearly perceived, an excessive slippery touch was observed during exploration. This feel increased at the longer wavelengths.

The *shape-normal force control 1* showed a flat finger trajectory (Figure 9b) which did not conform to the surface shape. The vertical component of this trajectory provided no information about the texture, however there was articulate tangential fluctuation of force during a trace which conveyed a clear feel of grating. This tangential force was observed sometimes larger than expected.

The *shape-normal force control 2* showed a fair compliant trajectory to the surface which enabled for the user to perceive the contour shape only by vertical perturbation of finger trajectory. However, the tangential force to detain the finger movement to climb the wall of the curve of a sinusoid was excessively large particularly for textures with long wavelengths. In that case, the stiffness of a surface was observed insufficient which caused a scratchy trajectory sticking at the wall.

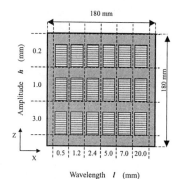

Figure 7. Arrangement of sample textures

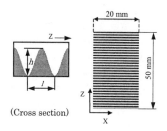

Figure 8. Profile of a texture sample

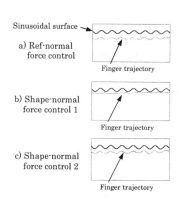

Figure 9. Finger trajectory under three force control modes for the wavelength 7.0 mm and the amplitude 1.0 mm

3.3 Coherency between presentation modes

We adopted the *2D vib-control* for the tactile stimulation since it was stable under the three force control modes. With this *2D vib-control* the force controls were compared in terms of touch feel naturalness regarding the texture samples. Table 1 shows the parameter range in which

Table 1. Textures coherently presented

Force control mode	Wavelength and amplitude $\{l, h\}$
Ref-normal	$\{1.2, 0.2\}, \{2.4, 0.2\}$
Shape-normal 1	$\{1.2, 1.0\}, \{2.4, 1.0\}$
Shape-normal 2	$\{2.4, 0.2\}, \{5.0, 0.2\}$

tactile and force stimulations were observed coherent to haptic sensation. Textures with shorter wavelengths and smaller amplitudes were perceived in a consistent impression generally for all the three force control modes; the tactile sensation and the force sensation were properly fused to form a single haptic event. In the case of textures with a larger wavelength, an improper presentation having a force without a vibratory stimulation occurred when the finger stayed stationary at the bottom of a sine wave. However, if a finger speed is fast enough for such wavelengths, the feel was not improper since it is physically correct that the tactile feedback only takes place when the finger collides at the peak of the waves passing the bottom untouched.

As for the *ref-normal force control*, the presentation was definite for almost all textures; particularly the two stimulations were felt fused at the wavelength around 0.2 mm and the amplitude 0.2 mm. As for the *shape-normal force control 1*, the amplitude of 1.0 mm was appropriate to produce tangential force perturbation. For the *shape-normal force control 2,* the amplitude of 0.2 mm was sufficient since vertical force perturbation was large enough. The long wavelengths were suited to this control mode in that a resistant force and a surface shape were appropriately rendered.

4. CONCLUSIONS

Two models of novel haptic interfaces were developed to improve the presentation quality of the previous model of the Texture Display which conveyed a surface texture sensation by a vibratory pin array. The projection type texture display enabled visual and haptic coordinated presentation and an extended degree of freedom of finger exploration on textures. Although the resolution of visual presentation was still insufficient, presence of virtual textures was significantly increased suggesting an improved resolution will provide much more usable haptic interface for showing details of products being designed. The force reflection type texture interface which delivered both skin and force stimulations could provide a realistic 3D haptic sensation by a novel presentation method which has not been explored. The ranges in which the presentation modes permitted coherent presentation were preliminarily investigated although more extensive observation will be required to clarify the property of this method in detail.

REFERENCES

[1] Y. Ikei, K. Wakamatsu and S. Fukuda, "Texture display for tactile sensation," Advances in Human Factors/Ergonomics, 21B, pp. 961-964, 1997

[2] Y. Ikei, M. Shiratori, S. Fukuda, "Haptic texture presentation in a three-dimensional space," Proceedings of International Conference on Artificial Reality and Tele-existence (ICAT2000), VRSJ, pp. 25-30, 2000

[3] Y. Ikei, M. Yamada, S. Fukuda, "A New Design of Haptic Texture Display - Texture Display2 - and Its Preliminary Evaluation-," Proc. IEEE Virtual Reality 2001, pp. 21-28, 2001

OBJECT MODEL FOR HAPTIC INTERACTION

Koichi Hirota and Michitaka Hirose

Research Center for Advanced Science and Technology, The University of Tokyo
4-6-1, Komaba, Meguro-ku, Tokyo 153-8904, JAPAN
e-mail: hirota@cyber.rcast.u-tokyo.ac.jp

ABSTRACT

Haptic interaction is a topic that has been most intensively studied in the field of virtual reality. We propose a method to simulate real-time haptic interactions based on a physical model. Interaction force in the real world is caused by constraints of the user's motion. A fundamental idea of our approach is to create a spatial map of the variation of such constraints and to compute the interaction force by referring to the map. We propose a fast computation algorithm for real-time and systematic simulations. Through the experiments, we demonstrate that we can apply the method to compute the interaction force while a user is manipulating.

1. INTRODUCTION

The representation of haptic sensation has been an important topic of research in terms of improving the sense of reality in virtual interactions (Burdea, 1996). For the implementation of haptic interaction, not only the haptic devices that transmit force to the user, but also the methods to compute the force that appears in the interaction play an important part (Salisbury et al, 1995).

One of the most typical interactions between the user and the virtual world is object manipulation using the hand or fingers. Various approaches have been proposed to implement object manipulation. We can categorize them into gesture-based, geometric, and physical approaches. Gesture-based manipulation is often employed for the rough handling of objects using glove-type devices. In this approach, the intention of the user is conducted to the system by means of gestures (i.e., the position of hands or fingers); this approach is not suitable for precise manipulation using fingers. In the geometric approach, the contact between fingers and the object is detected, and the behavior of the object is artificially defined based on both the status of geometric contact and the motion of fingers (Kitamura et al, 1999).

Since object manipulation is a physical phenomenon, physical simulation is thought to provide a more realistic result than other approaches. Also, it is a feature of the physical approach that the interaction force is explicitly obtained in the simulation of the behavior of the object. According to knowledge in the field of teleoperation, the sensation of interaction force is essentially important in some kinds of manipulation tasks (Shimoga, 1993). Although there has been an investigation of physical manipulation with interaction force feedback (Yoshikawa and Ueda, 1995), only objects of simple shapes such as spheres were examined.

Simulation of the contact between the user and the object is one of the most basic issues in the implementation of manipulation. Various approaches to detect the contact between solid bodies have been proposed and some of them are sufficiently fast for application to real-time interactions. A problem of these approaches is that they do not provide information on what the relationship between the contacting objects should be. Furthermore, it is difficult to simulate the motion of a solid object while being constrained by the other except in the case where the shapes of those objects are not complex.

In studies of haptic rendering, contact between the user and the object is often modeled as point vs. body contact (i.e., point-based interaction). The constraint-based God-object method (Zilles and Salisbury, 1995) and the 'point-based rendering' method (Ho et al, 1999) are known as typical approaches. In these approaches, it is assumed that the interaction occurs between the representative point, which is called the haptic interface point (HIP), and the object. Since the user's action is directly reflected at the HIP, it is inevitable that the point violates constraints caused by the surface of the object. In the God-object method, another point that indicates the ideal position of the HIP, which is called the ideal haptic interface point (IHIP), is defined, and the constrained motion of the point is computed. In the point-based rendering method, a fast computation algorithm of the idea of the God-object method is proposed. In these approaches, the interaction force is usually defined proportionally to the disparity between the HIP and the IHIP.

There are other cases besides object manipulation where the sensation of interaction force is caused during the spatial motion of the user. The viscous drag that appears when the user (or user's tool) moves in a fluid is a typical example. The friction in some kinds of cutting operations (Tanaka et al, 2000) is also considered as the interaction force that is accompanied by the spatial motion of the user.

We propose a method to simulate real-time haptic interactions based on the physical model. The interaction force in the real world is caused by spatial constraints of the user's motion. A fundamental idea of our approach is to create a spatial map of the variation of such constraints and to compute the interaction force by referring to the map (Hirota et al, 2000). We propose a fast computation algorithm for the real-time and systematic simulations. Moreover, through the experiments, we demonstrate that we can apply the method to compute the interaction force while the user is manipulating objects.

2. SIMULATION OF INTERACTION

The interaction force between the user's finger and the object is derived from the physical constraint on the finger caused by the object. Such physical constraint changes according to the spatial position. For example, the air around the object causes minimal constraint force except in terms of viscous drag, while the material of the object requires a grater force for the finger to enter into or move about the volume. To represent the spatial variation of the material, we divide the space into cells, and define the property of the material on each cell. Several types of model structures are conceivable for dividing the space into cells. The voxel representation (i.e., the array of cubic cells) is a common approach. However, it is a serious problem that the surface of the model becomes jagged. There is another approach in which the space is divided into polyhedra. In this approach, the increase in the number of surface polygons of a polyhedron increases the complexity of the geometric computation. Based on the discussions above, we applied the model that consists of tetrahedral cells (i.e., the polyhedron with a minimum number of surface polygons).

In the simulation using the tetrahedron model, the HIP moves in the tetrahedral cells, and the constrained motion of the IHIP in those cells is computed. The simulation algorithm consists of both the computation of the IHIP's motion in a cell and the transition of the constraint state among cells. The motion of the IHIP in a cell is defined by a 'motion function'. The function determines the position of the IHIP based on the position of the HIP, material in the space, and the constraint state.

Since we assume that the material in each cell is homogeneous, the position of the IHIP computed by the function is valid so long as the IHIP is in an identical cell. However, if the IHIP leaves the cell, we must replace the IHIP on the boundary of the cell, and recompute the IHIP's motion using the material of the neighboring cell. If the IHIP cannot enter the neighboring cell (i.e., the IHIP does not move from the boundary), then we must examine the possibility that the IHIP's movement is constrained on the boundary surface. Similarly, we can think of cases where the IHIP is constrained on an edge or a vertex. To systematically implement the change of the geometric constraint state, we introduce the concept of state transition. In the transition, we assume five states: tetrahedron, surface, edge, vertex constraint states, and outside state.

The above discussion was carried out on the premise that we can experimentally compute the motion of the IHIP under the given condition. The resulting position is determined based on the positions of the HIP and the IHIP, the constraint state, and the material in which the IHIP is constrained.

For the physical simulation of constraints on the IHIP, as well as for the computation of feedback force, we need to define the force acting between the HIP and the IHIP. In our study, we assume that the force is linearly proportional to the disparity between the positions of the HIP and the IHIP. As described above, it was a common approach in previous studies to compute the interaction force in proportion to the disparity between the HIP and the IHIP.

In cases where the IHIP is constrained on a surface or an edge, the motion of the IHIP is limited on these elements. In these cases, the IHIP moves toward a point on a plane or on a line that is closest to the HIP (i.e., the foot of the line perpendicular from the HIP on the plane or the line), rather than toward the HIP. Also, we introduce a flag that indicates whether the IHIP is moving in the model or not. Typically, the flag is needed to simulate the stick-slip transition of the friction state while tracing the surface of an object.

Using these parameters, we develop rules on the motion of the IHIP inside or on the boundary of materials. Namely, we make a function (i.e., the motion function) that defines the resulting position of the IHIP. All tests on the IHIP motion in the state transition are performed using the motion function. By changing the definition of the function, we can represent various constraints on the IHIP.

It is an important feature of the proposed algorithm that the computation time of the state transition is independent of the number of cells and nodes (i.e., points shared by vertices) in the entire model. The computation time depends on the number of cell boundaries that the IHIP passes in a single cycle. In this sense, the computation time increases proportionally to the velocity of the user's finger. In the feedback of interaction force, it is known that the update rate of the interaction force has a serious effect on the reality of the interaction force presented to the user (Shimoga, 1993). In the case of the PHANToM device, it is recommended to keep the haptic update rate at least 1kHz. One of the worst cases of passing a boundary is the edge-vertex-edge transition while tracing a concave shape, and the computation time required for the transition, in the implementation described in the next section, is about 300μs; in a cycle time of 1ms, we can simulate the motion of one IHIP over at least three boundaries.

3. OBJECT MANIPULATION

The friction on the surface has an important effect on the manipulation of objects. One of the most simple approximations of friction is defined by a model in which the coefficients of static and kinetic frictions are defined independent of the velocity. To simulate the gripping force, we define the friction using the model.

It is difficult to compute the distribution of force on the contact area between the finger and the object in real-time. For the precise computation of the distributed force, we need to define the deformable model of the finger and to simulate the physical deformation of the model. On the other hand, the distribution of force is essentially important in manipulation. It is a problem that, in the case of the point contact model, we cannot affect the torque around the normal axis of the contact surface. To avoid this problem, we approximately compute the distribution of force by defining multiple HIPs on the surface of the finger. On each HIP, simulation of the constraint is performed independently.

Based on the force determined in the constraint simulation, the motion of an object is computed using the equations of motion. In our model, we define the shape of an object in a local coordinate system (i.e., object coordinate system), and describe the relationship between the object and world coordinate systems by a transformation matrix. We also define a gravity-center coordinate system, in which the equations of motion of the object are described.

We experimentally implemented the algorithm in a virtual environment. We used a personal computer (PC) (Dual Pentium Pro 200MHz, Windows NT) for all of the computations, and two PHANToM force feedback devices (SensAble Technologies) (Massie, 1996) for haptic feedback. Also, we used the GHOST programming library to control the device. The library enables the execution of haptic and visual processes concurrently. It also schedules the execution of the haptic process every 1ms.

In the experiment, 11 HIPs are defined on each of two fingers (i.e., around the tip of a stylus). Figure 1 (a) shows an example of the force on HIPs while holding a rectangular object, where IHIP positions are indicated by small spheres and the affecting force is visualized by vectors. In the figure, the gravity center of the object is at the center of the rectangle shape, while fingers are holding the object near the edge. Torque to support the object horizontally and tangential force to cancel the gravity are applied to the object by the user at IHIP positions. Although we have 22 HIPs in the simulation, the computation is almost completed within 1ms. This is probably because in the interaction with convex shapes, it rarely occurs that the HIP is constrained on nodes.

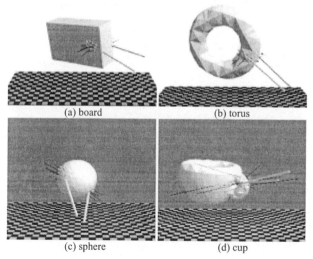

(a) board   (b) torus

(c) sphere   (d) cup

Figure 1: Object Manipulation

Figures 1 (b)-(d) show examples of manipulating objects with more complex shapes. Also in these cases, the computation is executed almost within a cycle time of 1ms.

4. CONCLUSION

In this paper, we proposed a method to physically simulate the interaction with objects in a virtual environment. The interaction force is computed by simulating the constraint caused by the object on HIPs. By introducing physical laws for the simulation of an IHIP's motion, physical forces such as friction is represented.

In the future, we will apply the proposed algorithm for the feedback of tactile sensations on the surface of fingertips. Since the proposed algorithm is capable of computing the slip and stick state on each HIP, it is expected that information on the local friction state can be effectively fed back to the user by integrating tactile devices in the system.

We are also interested in the manipulation operation using an entire hand. For the computation of constraints of an entire hand, we must represent the entire hand shape with a large number of HIPs, and perform the simulation of contacts on all of them. We may have to wait for advances in computing technology before we are able to perform the simulation in real-time.

REFERENCES

Burdea,G., Force & Touch Feedback for Virtual Reality, A Wiley InterScience Publication, New York, 1996.

Salisbury,K., Brock,D., Massie,T., Swarup,N., and Zilles.C., "Haptic Rendering: Programming Touch Interaction with Virtual Objects," Proc. Symp. Interactive 3D Graphics, 1995, pp.123-130.

Kitamura,Y., Higashi,T., Masaki,T., and Kishino,F., "Virtual Chopsticks: Object Manipulation Using Multiple Exact Interactions," Proc. VR99, 1999, pp.198-203.

Shimoga,K., "A Study of Perceptual Feedback Issues in Dexterous Tele-Manipulation: Part I. Finger Force Feedback," Proc. VRAIS'93, 1993, pp.263-270.

Yoshikawa,T., and Ueda,H., "Display of 3-Dimensional Operating Feel of Dynamic Virtual Objects with Frictional Surface," Proc. VSMM'95, 1995, pp.183-188.

Zilles,C.B. and Salisbury,J.K., "A Constraint-Based God-Object Method for Haptic Display," Proc. IROS'95, 1995, pp.145-151.

Ho,C., Basdogan,C., and Srinivasan,M.A., "Efficient Point-Based Rendering Technique for Haptic Display of Virtual Objects," Presence, 1999, Vol.8, No.5, pp.477-491.

Tanaka,A., Hirota,K., and Kaneko,T., "Deforming and Cutting Operation with Force Sensation," J. Robotics and Mechatronics, Fuji Tech Press, 2000, Vol.12, No.3, pp.292-303.

Hirota,K., Hirayama,M., Tanaka,A., Hirose,M., Kaneko,T., "Physically-Based Simulation Of Object Manipulation," DSC-Vol.69-2 (Proc. DSC2000), ASME, 2000, pp.1167-1174.

Massie,T.H, "Initial Haptic Explorations with the Phantom: Virtual Touch Through Point Interaction," Master's thesis, M.I.T., 1996.

Wearable Interface Device

Ryugo KIJIMA

Faculty of Engineering, Gifu University
1-1 Yanagido, GifuCity, Gifu 501-11 Japan
phone: +81-58-293-2759, fax: +81-58-293-3151, e-mail: kijima@info.gifu-u.ac.jp

Abstract:

In this paper, the author proposes a novel configuration using a projection type wearable display, infrared camera with light source, and retro-reflective screen. In this configuration, the image is projected onto the retro-reflective screen. The screen location as well as the finger tip position relative to the user's head is achieved using infrared camera and image processing. These locations are used for the interaction with the displayed object. Owing to the high reflection gain of retro-reflective screens, a small and light weight projector can be used. Also the high contrast between the screen and the other environmental object in the captured infrared image decreases the difficulty of the image processing for the screen and fingertip location. Thus this configuration can offer both the merits of a PDA type and eyeglass type configuration.

1. Introduction

1.1 Third Person Interface

The wearable interface can be classified in terms of person, i.e., the first, second, and third person. The analogy of the mirror could help to explain this notion. First, the full immersion virtual reality (VR) configuration using the closed head mounted display (HMD) or the immersive projection technology such as the CAVE can be categorized as the first person interface, because the user's conciousness enters the virtual world across the mirror (i.e. display). "I" was virtualized as a model in the computer and interact with the virtual object. Next, most of the computer interfaces including PDA, wearable device can be classified as the second person interface. The user focuses the world in the small mirror/display and communicate through the browser, dialogue, etc. Here, "You" means the focused display and input device, through that the communicates.

Now, assuming that there are many fragments of mirror, i.e., many displays, around the user, the user can see many aspects of the world through the fragments of the mirror at the same time, and can move his focus from one to the other freely. Information is not condensed in one monitor as the second person interface, but distributed around the user. "He", "She", "It" corresponds to the each distributed interface. Authors named this type as the third person interface and aim to realize it in this paper.

1.2 Features of Third Person Interface

Ambient Information: In daily life, we are obtaining a large amount of information without noticing it. For example, when we see the tree leaf swinging outside the window, we unconsciously notice that the wind blows. When we go to another town, we feel the level of the activity by seeing the number of people walking around. Such ambient information is important for our life, however, they have been omitted in the context of computation [1]. Since the most general interface, the second person interface is good when our consciousness focuses on it, cannot serve to give the ambient information to the user. As shown in the example above, the third person interface is adequate for this purpose.

Awareness: In the real world, we have different levels of awareness on many events and objects at the same time, and we rapidly change our focus of consciousness from one object to the other [1][2]. The reason why we can quickly give our focus to another object is that we have already been aware of it in the ambiance, and that it is natural and easy to change the target of consciousness in the real world. On the contrary, it is not easy when we are using PC, PDA, namely the second person interface. We need to be notified by sound, blinking graphics at first, and then we need to operate the device, for example, draw the different window to the foreground. The third person interface could enable the natural, easy change of the awareness.

Configuration: The first person interface, such as the Head Mounted Display could deal with the ambiance and enable the easy change of the awareness. However, such configuration is still far from maturity, especially when it is used as the wearable device in our daily life. The largest problem is the sensing of the user's head location. The displayed content is tightly related to the real world and the user's head location, it is necessary to measure it. Usually the magnetic sensor, ultrasonic sensor is used for the VR in the limited laboratory environment, that is difficult to apply to the wearable computing in a larger area of activity. Therefore, another type of configuration is required.

2. Proposed Configuration for Wearable Interace

2.1 Projection Head Mounted Display (PHMD)

The Projection Head Mounted Display (PHMD) is relatively recent technology. The principle of the PHMD was proposed as well as the first prototype by the author in 1995 [3][4]. Afterward, the retro-reflective material was introduced to enable the stereoscopy for the projection display [5] in 1996 and soon the PHMD is used in combination with the retro-reflector [6][7].

The fundamental principle of the PHMD is the optical conjugation between the user's eye and the optical center of the projection. Due to this conjugation, the image is projected from the user's eye position and seen from the same position. This arrangement enables the user to watch the image without distortion when the screen has angle to the line of sight or even the screen is bent, because the viewing transformation is the inverse of the projection transformation and the distortion derived from the screen's shape is canceled.

The author has developed a small eyeglass type projector as shown in Fig. 1. The projection head for the each eye was composed of an LCD panel (Sony LCX009AKB, 0.7 inches, 800x225 pixel), a projection lens (18 mm diameter) and hi-luminance white LED block as the light source. The LED block generated the luminance of approximately 250 millicd/mm2 and was enough for the use within the room, but not for outdoor usage. The weight of this PHMD is about 150 grams.

The PHMD became more practical by the introduction of the retro-reflective material as the screen. This material reflects the incoming light back to the direction of the light source. The image is projected from the user's eye position to the reflector and goes back to the eye position. Almost all the energy from

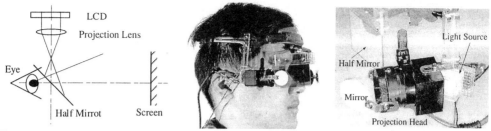

Fig. 1. Projection Head Mounted Display (PHMD) (Left) Principle (Mid) Current Prototype (Right) Projection Head

the projector goes back to the eye and there is only minimal loss. Therefore, small, light weight, dark projector can be applied as the PHMD.

Another merit is the capability of stereoscopy. The image projected from one eye goes back to the same eye and does not reach to another eye. Then each eye of the user can see the different image from the corresponding projector.

The retro-reflective screen is the cloth covered with small glass beads. This is the cheap material that is widely used for the road signs. Many displays can be attached easily in low cost. Also this material can be folded and stored in a pocket like as handkerchief.

2.2. Vision Based Measurement using Infrared Camera

To achieve the relative location of the user's head to the screen, an infrared NTSC CCD camera and the infrared LCDs is attached to the PHMD, and the vision based measurement was applied. When the original shape was known, the screen's location relative to the user's head can be calculated from the image of the screen. The screen served as both the visual screen for the user and the anchor of the vision based measurement at the same time.

Generally speaking, the method to control the environment, other than the image analysis algorithm, is important for the actual application. Since the calculation power of the wearable configuration is limited, the clever way to achieve the clear image of the necessary target is required. The first device is that the LED and camera were optically conjugated with each other. Due to the high gain of retro-reflector, the small, low power LED can serve enough as the bright light source to achieve the clear image of the

screen. The next idea was the differential image calculation. The infrared light was turned on and off synchronized to the frame and the difference of images was calculated. Thus the background noise was decreased largely and the high contrast image of the screen was achieved.

The following is the brief description of the image processing. At first, the edge detection was performed from the captured image and the longest one was regarded as the outline of the screen. After compensating the distortion of the camera, 12 lines at every 15 degrees was circumscribed to this edge. The shape of screen was approximated as a convex polygon with 24 vertices. The number of vertices was

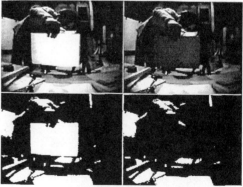

Fig. 2 Differential Image Capture
(Top) Original (Bottom) Binary Image
Infrared Light is (Left) ON (Right) OFF

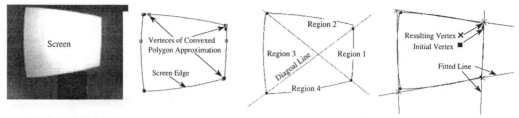

Fig. 3. Detection of Screen Edge and Screen Verteces. (a) Original Image (b) Approximation to Convex Polygon
(b) Shrunken Quadrangle and Division (d) Line Fitting by Mean Suared Error Method

reduced to the quadrangle, referring to the distance between the neighboring vertices and the angle of neighboring edges. Thus the first approximation of the quadrangle was achieved. Next, the image was divided into four regions using by the diagonal lines of this quadrangle. Using the minimum squared error method, four lines were fitted to edges in each region. At last the screen's location relative to the user's head was calculated. This calculation is robust when a part of the screen is hidden by the user's hand, because this is based on the edge information, not the vertices' information.

In addition, when the user places a finger on the screen, it is seen as the dark area in the bright screen. The second order moment of finger image is estimated and fingertip position is calculated. The user can interact with the displayed object with their fingertip.

For example, the user can draw the graphics on the screen with their fingertip. The screen represents the drawing plane in the three dimensional spaces and the user can move it with the other hand, then the spatial object can be drawn.

3. Applications and Demos

Fig. 4 (Top) shows the "virtual window" demonstration. A video camera was placed outside the laboratory building, and displayed on the retro-reflective screen on the wall. According to the user's location, the motion parallax is realized by the simple image based rendering. Fig. 4 (Left) sows the "price tag" application. A small tag of reflector was attached to the product in the shop, and the current price is projected on it. Fig. 4 (Middle) shows the "calendar" application. The calendar itself is the reflector without any print. The date is displayed on it. The user can draw his plan by using the fingertip operation on the calendar. Fig. 4 (Right) shows the "inter-

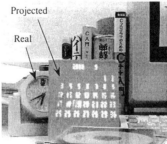

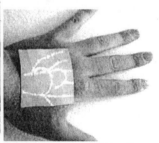

Fig. 4. Applications (Top) Projected Virtual Window on Wall Showing Scene Outside (Left) Virtual Price Tag
Attached To Product (Mid) Projected Calender with Schedule Board Functionality (Right) Visualized Brood Vessel

nal state" application. A reflector is attached to the back of the user's hand. According to the heart beat and blood pressure, the graphics of the blood vessel was animated. These applications are the demonstration of the omni-display environment, the third person interface.

Fig. 5 is the demonstration application to show the active use of the screen location measurement. The sliced image of human brain using the MRI data was displayed on the screen. As the user held the real screen and moved it, the slicing plane in the virtual world reflects the location of the real screen. Thus the user can see various slices of the brain. Also the user can draw the graphical memorandum on the sliced image by their fingertip. The combination of the PHMD, reflector screen and the infrared camera can be applied also as the second person interface in the three dimension space.

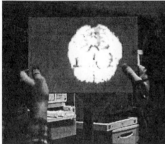

Fig. 5. Application: Interactively Slicing MRI Brain Image with Hand-Held Screen

4. Summary

A novel configuration of wearable display and interface capability was described, to enable the omni-display environment for the third person interface. A small, light weight wearable projector was constructed and used with a retro-reflective cloth screen. This display is combined with the infrared camera and image processing, the movement, location of both of the user's head and fingertip are used for the interaction. Since this configuration is autonomous in terms of sensing, it can be used in larger space of our daily life, not limited to the space of a laboratory.

References

1. Wisneski, C., Ishii, H., Dahley, A., Gorbet. M., Brave., S., Ullmer, B., Yarin, P., Ambient Displays: Turning Architectual Space into a Interface between People and Digital Information, Procs of the 1st Int. Workshop on Cooperative Buildings, pp. 22-32, Springer (1998)
2. Pedersen, E. and Sokoler, T., AROMA: Abstract Representation of Presense Supporting Mutual Awareness. Procs. of CHI '97, pp. 51-58, ACM Press (1997)
3. Kijima, R. and Hirose, M., A Compound Virtual Environment Using the Projective Head Mounted Display Procs of ICAT/VRST '95, pp. 111-121, ACM-SIGCHI (1995)
4. Kijima, R., Ojika, R., Transition between VIrtual Environment and Workstation Environment with Projective Head-Mounted-Display, Procs. of IEEE Virtual Reality Annual International Symposium 1997, pp.130-137, IEEE (1997)
5. Ishikawa, J., 3D B-VISION, Journal of three dimensional image, 10-4, pp.11-14 (1996) (In Japanese)
6. Inami, M., Kawakami, N., Sekiguchi, D, Yanagida, Y., Maeda T., and Tachi, S., Visuo-Haptic Display Using Head-Mounted Projector, Procs of IEEE Virtual Reality 2000 Conference, pp. 233-240, IEEE (2000)
7. Hua, H., Gao, C., Biocca, F., Rolland, J., P., An Ultra-liht and Compact Design and Implementation of Head Mounted Projective Displays, Procs. of IEEE Virtual Reality 2001 Conference, pp.175-182, IEEE (2001)

PROP-BASED INTERACTION IN MIXED REALITY APPLICATIONS

Holger T. Regenbrecht and Michael T. Wagner

DaimlerChrysler AG, Research and Technology
FT3/EV, Virtual Reality Competence Center
P.O.Box 2360, 89013 Ulm / Germany
{Holger.Regenbrecht|Michael.A.Wagner}DaimlerChrysler.Com

ABSTRACT

The use of computer mediated realities will become an integral part of our everyday working life. Our experience with existing Virtual Reality applications however shows that the main obstacle to using these new technologies is their user interface part.

These applications use specific devices and interaction metaphors which can rarely be found in real life. The upcoming technique of Augmented Reality offers the possibility to use well-known metaphors and objects for real-world interaction within the mixed context of reality and virtuality.

The paper presented here shows first results of the use of props to interact within Mixed Reality scenarios. The props are both elements of standard office equipment (e.g. ruler, stapler, etc) as well as custom built objects.

We have designed and evaluated several interface elements to be used for interaction in a Computer Aided Design (CAD) desktop scenario.

A designer uses standard windows applications which are usually involved in todays design processes (e.g. spreadsheets, text processors and internet browsers) while working on a three-dimensional CAD model.

Instead of using only a conventional mouse in front of a computer screen he or she wears a video-through head-mounted display (HMD) and uses probs to interact with both real world objects on his or her desk and computer generated content which is arranged spatially and merged into the real environment.

1   INTRODUCTION

Augmented Reality (AR) attempts to enrich a user's real environment by adding virtual objects (3D models, 2D textures, textual annotations, etc) to it, like described by Azuma (1997) and Milgram et. al. (1994). The goal is to create the impression that the virtual objects are part of the real environment.

The users of the AR system experience the augmented environment through special display devices, which are either worn on the body or are placed in the working environment. From a technical point of view, AR faces three major challenges : (1) to generate a high quality rendering, (2) to precisely register (in position and orientation) the virtual objects with the real environment, and (3) to do so in real-time. Additionally, ergonomics and usability aspects must be taken into account. The AR gear should be comfortable to wear and not too encumbering, especially if it is to be used over prolonged periods of time.

Our general goal is to improve work processes in industrial design and production by introducing novel digital tools. Since it is neither feasible nor desirable to replace existing tools with new ones from one day to the next, it is important to integrate existing tools and work practices with new ones, so as to provide a smooth transition to the new methodology. In the research presented in this article, our focus is on designing the future workplace for office workers in general and for CAD designers in particular. Such a workplace currently consists of a traditional office desk with various desktop utensils, and of a desktop computer operated using keyboard and mouse through a 2D user interface. More specialized CAD workplaces add some limited form of 3D interaction and visualization, based on special 3D input and output.

2   CONCEPT AND IMPLEMENTATION

To fulfill the needs described in the introduction, we have initiated a concept called the MagicDesk, which provides some first tools and metaphors for an almost seamless integration of 2D and 3D information into the real world. Although the concept is based on or limited by the device technology available today, it serves as a testbed for a number of possible application scenarios.

The starting point for the concept is an ordinary CAD desk of an engineer found in most design and development companies. The engineer sits in front of one or two computer screens and operates his or her applications with keyboard and mouse. The main applications are: at least one CAD program, spreadsheet applications, word processor, email program, group- and org-ware, electronic catalogues, and most recently also a web browser.

With The Magic Desk we augment this existing environment by virtual and real elements in two ways: (1) by extending the 2D application space to the whole desk space (instead of limiting it to the monitor), and (2) by placing a 3D object world into the desk space. None of the traditional utensils are removed, only new objects are added to the desk.

Instead of realizing a concept of a projection-based, ubiquitous augmented interface, as for instance introduced by Raskar et. al. (1999), we decided to focus on a head-mounted display (HMD) setup because of the availability of the technology. The HMD can be put on and taken off when convenient and especially if the user wants to work for a longer period of time with the monitor/keyboard/mouse combination, which is the case especially in the early phases of using the MagicDesk. The concept itself described here can be used in almost any configuration and does not depend on any particular HMD model.

The main goals of The Magic Desk are: (a) providing 3D content and interaction in a desktop environment, (b) providing 2D content and interaction in a fairly large space of operation, (c) inclusion of the standard desktop environment, (d) no substitution of 2D workflow and interaction with VR technology, (e) almost seamless integration of real world, 2D computer world, and 3D (VR) world, and (f) natural, intuitive, and consistent interaction techniques within the three domains.

We have implemented The Magic Desk concept on PC platforms running Microsoft Windows 2000 for an easy integration into the working processes of today's CAD engineers.

Although also other techniques were used in the past, the current version of the Magic Desk 6DOF tracking is implemented on top of the ARToolKit library (Billinghurst & Kato, 1999), with some major extensions regarding tracking fidelity and multi-marker tracking. All objects in real space are tracked by colored markers printed out on an ordinary printer. Each object in the real world is tracked by at least two markers for reliable accuracy.

The software architecture and hardware setup of the MagicDesk system is described in detail in Regenbrecht et. al. (2001).

3   VISUALIZATION AND INTERACTION

3.1   2D application space

Besides the two-dimensional working area on the monitor, which is still present, the main applications can be placed in space using a clipboard metaphor (figure 1). Real clipboards hold real 2D Window applications and can be placed on the desk wherever needed. For the sake of easiness, exactly one 2D application (e.g. Microsoft Word) is attached to each clipboard. The clipboard/application can be laid down on the table, can be put on some document holders, can be exchanged with other users, or can be put into the bag or in a drawer. The movement of the applications is therefore tangible in a very easy to use and understandable way (see also Ullmer & Ishii, 1997). Instead of clicking and dragging with a mouse on the (limited) surface of a computer monitor, the user simply grabs the clipboard with the application attached to it and moves it to the desired location.

In addition to the concept of *clipping* 2D windows to the clipboard, 3D contents can also be placed on the real clipboard providing the same tangibility.

Figure 1: A real and augmented clipboards

3.2   3D object space

Analyzing the main tasks of a CAD engineer in using 3D geometrical models, one makes the observation that 3D objects are not usually placed in space in an unconstrained manner. Rather it can be observed that most placement is done by using some sort of planes as a reference. Either the shape of the monitor or the edges of some geometries close to the 3D object to be displayed are used to orient the objects in (quasi-) 3D space. In our desktop setup the table surface serves as such a reference frame. Therefore we have developed a metaphor which utilizes a circle-shaped surface on top of the desk as a reference frame for placing 3D objects on it. This surface, called the cake platter, because it actually is one, can be turned around its upward axis to view the virtual object from different angles (see figure 2). Turning the platter is very easy: simply by turning the outer ring with one or two hands. Since this form of interaction is so natural, novice users intuitively know how to use is right away. The table reference is always consistent.

3.3   Interaction

According to the goals of the concept, 2D interaction techniques should not only be allowed, but rather actively supported by The Magic Desk. Therefore, we support 2D interaction in three different ways: (1) using the standard 2D metaphors with the monitor on the desk, controlling the WIMP-applications with keyboard and mouse, (2) selecting one of the applications running on the clipboards and controlling them in the same way, and (3) interacting with the clipboard applications using a 3D ray cast device. These procedures are integrated into one interface which allows a transition from old to new metaphors. The computer mouse on the desktop can be turned upside down (actually *downside up*) and will become a 3D raycast device, in which case a virtual ray emerges from the front side of the mouse. In this mode, the mouse functions either as a 6DOF device in 3D space, or as a 2D device operating on the surface of a WIMP application. Due to the visual feedback of the interface the user has no difficulty in determining the current mode of operation.

3D operations on the cake platter model can be performed in the 3D mouse mode, where the device is used either as a 3D position or as a 6DOF raycast device. In the 3D position mode only the tip point of the mouse is used for interaction, e.g. for moving 3D points within the virtual model. In the 6DOF mode the ray is used for more remote or more complex operations, like turning the whole model or parts of it. It is also used for all selection tasks in 2D and 3D.

Two special devices are used in the conceptual setup for tasks identified as primary for CAD engineering: clipping planes and lighting. A clipping plane slices the virtual object for better interpretation of the inside structure of the model. The Magic Desk realizes this by providing a rectangular plate with a tangible interface. Moving the plate through the model will clip it at the appropriate location and orientation (figure 2).

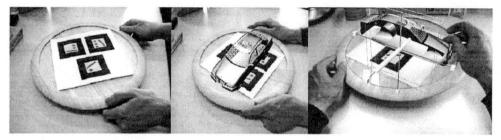

Figure 2: Cake platter in combination with the clipping plane

The lighting concept follows the same simple principle of easiness. A real desk lamp serves as a prop for the virtual light. Moving and turning the real lamp moves and turns a virtual light which illuminates the virtual model on the cake platter (figure 3).

In addition to these two 3D interaction techniques a very basic form of system control, the file or model selection, is implemented using the MagicBook metaphor introduced by Billinghurst et al. (2000). Different 3D models can be chosen out of a book, to be placed on the cake platter. This interface is very natural and intuitive because everyone is used to turn pages in a book. The models within the book are presented in three dimensions and can be viewed from different angles before loading them into the main 3D application space (figure 3).

Figure 3: Real/virtual lamp and real and virtual Magic Book for model selection

More office related props can be illustrated with the following examples. A stapler is used in traditional work to put together some sheets of paper. The same metaphor can be used to append or merge documents. As shown in figure 4, a tracked real stapler functions as a merge tool for electronic documents. Another example is a puncher. In "real life" it is used to prepare sheets for the folder. The virtual equivalent is the archive in elctronic information systems. The puncher in the MagicDesk setup serves as a store tool for the electronic archive on the hard disk (figure 5).

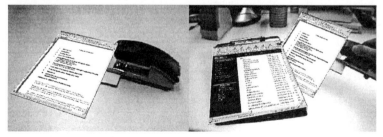

Figure 4: Real stapler merges electronic documents

Very popular in every office environment are PostIt™ notes. They can be placed anywhere by simply temparely glue them to any object, like monitors. Within the MagicDesk context, on the small note sheets are printed seperately identifyable markers. With this it is possible to attach any elctronic document to, either 2D or 3D (figure 5)

Figure 5: Puncher for archive function and PostIt^{TM} note with an electronic document attached

4 CONCLUSION

With this paper we have presented an Augmented Reality system which combines the traditional working process of an CAD engineer with new tangible interfaces for 2D and 3D information. In addition we have introduced some prop-based interaction techniques used within the context of the MagicDesk. At the time of writing of this article no formal user studies have been undertaken. First tests within our laboratories show that the MagicDesk is a very promising concept for bringing AR technology to the desk.

After a positive evaluation of the current implementation we intend to (a) fully integrate the MagicDesk into the Virtual and Augmented Reality System DBView / VPE, (b) add necessary functionality, (c) consider and implement other communication techniques (like CORBA), (d) specialize the system to particular customer needs, (e) add new interaction devices and (f) transfer the whole system to the customer for everyday use.

5 ACKNOWLEDGEMENTS

The MagicDesk was developed together with Gregory Baratoff whom we like to thank very much for his contributions and comments. We would like to thank Steffen Cersowsky and Muriel David for their contributions to the implementation and calibration of the MagicDesk.

REFERENCES

Azuma, R., A Survey of Augmented Reality (1997). *Presence:Teleoperatore and Virtual Environments*, 1997. 6(4): pp. 355-385.

Milgram, P., Takemura, H., Utsumi, A., Kishino, F. (1994). Augmented Reality: A Class of Displays on the Reality-Virtuality Continuum. *Proceedings of Telemanipulator and Telepresence Technologies*. 1994. SPIE.

Ullmer, B., Ishii, H. (1997). The metaDesk: Models and Prototypes for Tangible User Interfaces. *Proceedings of UIST'97*. 1997. ACM. pp. 223-232.

Raskar, R., Welch, G., Cutts, M., Lake, A., Stesin, L., and Fuchs, H. (1998). The Office of the Future: A Unified Approach to Image-Based Modeling and Spatially Immersive Displays. Proceedings of SIGGRAPH 98, Orlando, Florida, July 19-24, 1998

Billinghurst, M., Poupyrev, I., Kato, H., and May, R. (2000). Mixed Realities in Shared Space: An Augmented Reality Interface for Collaborative Computing. Proceedings of ICME 2000, 2000, IEEE. pp. 1641-1644

Billinghurst, M. and Kato, H. (1999). Real world teleconferencing. In Proceedings of CHI '99 Abstracts and Applications, 1999

Regenbrecht, H. ,Baratoff, G., & Wagner, M. (subm.). A tangible AR desktop environment. Submitted for special issue of Computer & Graphics, Elsevier. 2001.

PROJECTION-BASED AUGMENTED ENGINEERING

Oliver Bimber [a] , André Stork [b] and Pedro Branco [c]

[a] Fraunhofer Institute for Computer Graphics, Joachim-Jungius Strasse 11, 18059 Rostock, Germany
+49-(0)381-4024-132, obimber@egd.igd.fhg.de

[b] Fraunhofer Institute for Computer Graphics, Rundeturmstr. 6, 64283 Darmstadt
+49-(0)6151-155-469, stork@igd.fhg.de

[c] Fraunhofer Center for Research in Computer Graphics, 321 S. Main St., Providence, RI 02903, USA
+01-401-453-6363, pbranco@crcg.edu

ABSTRACT

This article presents the early stages of a new projection-based Augmented Reality device. Our prototype represents a seamless combination of a whiteboard and an optical combiner –merging virtual objects with the surrounding real environment. With this, we strive for an efficient and problem specific application of Augmented Reality technology within the engineering domain.

1 INTRODUCTION AND MOTIVATION

Traditional Augmented Reality (AR) displays, such as head-mounted devices entail a number of technological and ergonomic drawbacks. Their limited resolution, the restricted field-of-view, the visual perception issues that are due to the fixed focal length caused by a constant and head-attached image plane, the increased incidence of discomfort provoked by simulation sickness and their cumbersomeness prevent their usage in a number of application areas.

To overcome some of these drawbacks, but also to open new application areas for AR, we propose a projection-based AR concept that combines spatially aligned optical see-through elements (essentially half-silvered mirror beam-splitters) with off-the-shelf projection-based Virtual Reality displays. This concept offers possibilities to combine the advantages of both technologies: the well established projection-based Virtual Reality with the potentials of Augmented Reality.

This paper describes the early stages of a new proof-of-concept prototype that has been developed to extend the scope of projection-based AR towards the engineering domain.

2 RELATED WORK

In contrast to traditional front or rear-projection systems that apply opaque canvases or ground glass screens, transparent projection screens don't block the observer's view to the real environment behind the display surface. Therefore, they can be used as optical combiners that overlay the projected graphics over the simultaneously visible real environment. Pronova's HoloPro system (Pronova, 2001) is such a transparent projection screen. It consists of a multi-layered glass plate that has been laminated with a light-directing holographic film. The holographic elements on this film route the impinging light rays into specific directions, rather than to diffuse them into all directions (as it is the case for traditional projection screens). This results in a viewing volume a 60° horizontal and 20° vertical range in front of the screen, in which the projected images are visible. Regular projectors can be used to rear-project onto a HoloPro screen. However, they have to beam the images from a specific vertical angle (36.4°) to let them appear within the viewing volume. Originally, the HoloPro technique has been developed to support bright projections at daylight and some researchers already begin to adapt this technology for Augmented Reality purposes (Ogi, Yamada, Yamamoto & Hirose, 2001).

However, several drawbacks (mainly due to the applied holographic film) can be related to this technology:

- Limited and restricted viewing area.
- Static and constrained alignment of projector and projection plane (and therefore a no flexibility or mobility).
- Low resolution of the holographic film (the pattern of the holographic elements are well visible on the projection plane).
- Reduced see-through quality due to limited transparency of non-illuminated areas.

Computer augmented whiteboards have been a topic of research over the past years. As traditional whiteboards, a computer augmented or electronic whiteboard presents an environment shared by multiple users, which can be used to discuss, draw or present ideas. We'll analyze the previous work with the focus on the interaction aspects relevant to our system.

The videoWhiteboard (Tang and Minneman, 1991) is one of the early systems that provides tools for shared drawing. A camera captures the drawings and the gestures of each participant on the whiteboard and transmits it to the remote user where it is projected giving the illusion that they are on opposite sides of the same whiteboard. The Clearboard (Ishii and Kobayashi, 1992) is based on the same concept adding gaze awareness by overlapping the image of the user and the drawing on the same display surface. Both, the videoWhiteboard and the Clearboard are suitable for remote conferencing. Another pioneering example of a whiteboard application is the Tivoli (Pedersen, McCall, Mora and Halasz, 1993) system, that has been designed support collaborative meetings. It uses a pen as input device and offers the combination of a traditional graphical user interface that is displayed on Xerox Liveboard and a gesture interface.

A different interaction approach is applied for the BrightBoard (Stafford-Fraser and Robinson, 1996). Here, the user controls the computer by drawing simple sketches or characters on the board. Through image processing the computer distinguishes the sketches from the rest of the drawing. Written words, for instance, can be interpreted as computer commands. Thus, a seamless transition between handwriting and computer control is achieved.

FlatLand (Mynatt, Igarashi, Edwards and LaMarca, 1999) is a system designed to support typical whiteboard applications. A touch-sensitive whiteboard (the SmartBoard^{TM}) accepts normal whiteboard marker input as well as stylus input. Captured strokes are projected onto the board. Gestures serve for managing the displayed visual layout.

HoloWall (Matsushita and Rekimoto, 1997) is a wall-size computer display that allows users to interact with their fingers, hands, bodies, or even with physical objects. Inputs are recognized through infrared lights (IR) and a IR video camera. Since the HoloWall can detect two or more hands (or fingers), multi-hand interaction is possible. Also, user body posture/position or even physical objects that reflect IR light, such as a document showing a 2D-barcode which identifies it, provide an enriched interaction environment.

3 EXPERIMENTAL HARDWARE-SETUPS

Our current prototype (cf. figures 1-3) has the same form-factor as a regular white board. It consists of a mobile Aluminium rack that holds a tiltable wooden frame. The actual white board is replaced by a half-silvered mirror (a 40" x 60" large, and 10mm thick glass pane which has been laminated with a half-silvered mirror foil (3M, 2001)) that simultaneously transmits and reflects light. Thus, we refer to it as "*transflective board*". Technically, the transflective board is used as an optical combiner that reflects stereoscopic 3D graphics off an arbitrary display surface. Using the optical combiner, this graphics (5) is spatially merged with the surrounding real environment. Conceptually however, the transflective board can still be used like a white board –providing the same interaction behaviour. Two different stereo displays have been tested with the transflective board so far:

- A Barco Baron Virtual Table (1) (Barco, 2001) that applies rear-projection and active shuttering using CrystalEyes Shutter glasses (1). In this case, two stereo images are projected sequentially onto the projection surface and shuttering is synchronized via an infrared signal. This is illustrated in figure 1.
- A mobile two-screen front-projection system (6,7) that applies passive shuttering using polarized light. In this case, four beamers (6) project pre-filtered stereo images simultaneously onto the two projection planes (7). The images are separated via polarized glasses (8). This is illustrated in figure 2.

Figure 1: The transflective board: A combination of white board and optical combiner –merging virtual objects with the surrounding real environment. The projection device used for this scenario is a Barco Baron Virtual Table. The dashed lines outline the reflected projection plane.

788

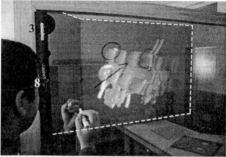

Figure 2: The projection device used for this scenario is a mobile two-screen system. The dashed
arrow illustrate the flow of light. The dashed lines outline the reflected projection planes.

An electromagnetic tracking device (Ascension's Flock of Birds (Ascension, 2001)) is used to support head-tracking and tracking of spatial input devices. In addition, we apply a tool that is usually used to track real markers and cleaning pads on a whiteboard – the Mimio (3,4) (Dunkane, 2001). The Mimio is a hybrid (ultrasonic and infrared) 2D tracking system for planar surfaces which is more precise and less susceptible to distortion than our electromagnetic tracking device. As illustrated in figure 1, its receiver has been attached to a corner of the transflective board.

4   MIXED REALITY TECHNIQUES AND TOOLS

In terms of providing a correct superimposition of the rendered graphics over the real environment, we apply the following techniques (these techniques have been describe in full detail in (Bimber, Encarnação & Branco, 2001):

- *Rendering:* Since the rendered graphics is perceived as reflection by looking at the mirror, several transformations have to be applied to see an unreflected virtual augmentation and to provide correct stereo separation. This is realized by introducing two additional transforms to the transformation pipelines of the rendering framework. A view transformation reflects the viewpoints (the left and right camera positions) that are located on the one mirror side (the side of the stereo display) over the mirror plane. A model transformation reflects the virtual scene that is located on the other side of the mirror (the side of the real environment) vice versa. Since these transformations are affine, they can be simply integrated into a hard- or software-implemented transformation pipeline (such as the one realized by OpenGL (Neider, et al., 1993)). Consequently, only minimal modifications to an existing rendering framework have to be made to support planar mirrors as optical combiners in combination with stereo displays. Furthermore, no additional computational cost is required for these transformations during rendering.
- *Tracking Distortion:* Electromagnetic tracking devices introduce non-linear tracking-distortion and noise over an extensive working volume. We apply smoothening filters to filter high-frequent sub-bands, and positional pre-distortion with the pre-sampled magnetic field of the working volume to minimize non-linear distortion.
- *Optical Distortion:* Optical distortion is caused by the mirror and the projection planes because of refraction and the curvature of the display or the mirror. We have developed several pre-distortion methods that are applied within the 3D spatial space, as well as within the 2D image space.
- *System Delay:* In addition to distortion, end-to-end system delay or lag causes a "swimming effect" (virtual objects appear to float around real objects). To reduce the swimming effect, we apply prediction filters (Kalman filters (Azuma, 1995)) for orientation information and linear prediction for position information.

5   INTERACTION

Indirect and remote interaction with virtual objects through the transflective board, or direct interaction in front of it are possible. A variety of techniques have been implemented and described in more detail in (Bimber, Encarnação & Branco, 2001). For example, a tracked pen is used for direct manipulation in front of the board, as well as for remote interaction behind the board. In addition, a transparent hand-held tablet (which is also tracked) is applied to feature two-handed interaction (direct or remote). Virtual objects can be exchanged between both sides of the board: they can be picked with the pen –either directly or remotely– and can then be pushed, pulled or beamed through the mirror. A virtual laser beam casts from the pen through the mirror to move and place virtual objects behind the board's tangent plane. Additionally, the user's viewpoint can be combined with the pen direction to compute an appropriate selector. The ARToolkit optical tracking system (Kato, Billinghurst, Blanding & May, 1999) is employed to detect different paper markers within the real environment. The markers are used to track real world objects or as placeholder for multi-media information (e.g., images, video or textual information). Since different multi-media contents are attached to specific markers, they can easily be exchanged by replacing the markers, or simply be moved within the real environment. The tracked pen and also the pad feature sketch-based interaction (Bimber, Encarnação, & Stork, 2000): two or three-dimensional freehand sketches are used to build objects by reconstructing their shapes from the sketches, or to interact with the objects (e.g. by sketching assembling steps, etc.)

6   ENGINEERING APPLICATIONS

Today, maintenance and training are the major application fields for Augmented Reality from the industry's point of view (ARVIKA, 2001). Augmented Reality will conquer the earlier phases of the product development process where VR is more heavily used. AR can compensate for the shortcomings of VR, such as fine and precise tactile and haptic feedback. Here, mobility is not a crucial issue. Thus, stationary set-ups that ensure high quality images and a more robust tracking than mobile ones will be preferred for these purposes.

We envision five application scenarios for projection-based AR in the engineering process:

- *Augmented design review:* VR systems are mainly used for design review. But the lack of reality especially concerning haptic and tactile feedback still limits the acceptance of VR as a decision base. This lack can be compensated by full functional rapid prototyping (RP) parts that give the user a real tactile and also aural feedback. To produce many RP parts with, for instance, different colors and materials is cost intensive. The combination of RP parts with virtual overlays that let these parts appear in different colors/materials is a promising 'best-of-two-worlds'-approach.

- *Hybrid assembling:* The feasibility evaluation of the assembling/disassembling processes is a common application of VR whereby the product exists in a digital form, only. But products are frequently altered during their lifecycle. In these situations, the products already exist physically and the need to assemble virtual models into physical mock-ups (PMU) arises. Figure 3 illustrates an example that could be extended towards a hybrid assembling scenario which allows for mounting a virtual engine into a real engine compartment.

- *Hybrid modeling/sketching:* Although CAS and VR systems have changed the styling and design process considerably, clay models still have their place in the design process. During a review of a physical model, requirements to change some of its parts/features appear. Today, these modification requirements are expressed verbally. An Augmented Reality setup would allow to virtually sketch the change onto/over the PMU.

- *Visual inspection of molded parts:* The development of a product comprises the design of tools to manufacture the product parts. Molded parts shrink when they cool down. Thus, the tools are not simply the inverse product geometry. When starting a new production line, the produced parts have to be compared with the intended geometry – the virtual parts. Again, Augmented Reality can give us an efficient to use visual inspection environment.

- *Hybrid ergonomic design:* With virtual humans, ergonomic analysis can be carried out in virtual environments. After having built a virtual model, the manikin is used to evaluate the reachability of surrounding elements. In case of a poor ergonomics, the model has to be changed and the process is repeated. With an Augmented Reality setup the process can be inverted: e.g. a coarse physical muck-up of a cockpit is built and only afterwards virtual buttons are placed in the hybrid environment. This approach leads more directly to an ergonomic design.

Figure 3: The transflective board can be used as a large reach-in system –supporting a direct – manipulative interaction with real and virtual objects. The graphics is projected onto the ceiling.

7   CONCLUSION AND FUTURE WORK

Traditional Augmented Reality displays, such as head-mounted devices entail a number of technological and ergonomic drawbacks that prevent their usage in a many of application areas. With the objectives to overcome some of these drawbacks and to address an efficient and problem specific application of Augmented Reality technology within the engineering domain, we have presented the early stages of a new projection-based Augmented Reality device –the transflective board. Compared to head-attached AR displays, the application of spatial projection displays for Augmented Reality tasks feature an improved ergonomics, a theoretically unlimited field-of-view, a high and scalable resolution, and an easier eye accommodation (Raskar, 1998). Since the technology, as well as the consequential advantages are derived from the well established projection-based VR concept, a seamless combination of VR and AR is also imaginable (some examples are presented in Bimber, Encarnação & Branco, 2001). However, several shortcomings can be related to our approach. Self-reflection of the user in some situations, for example, is one of the drawbacks of the transflective board. Our current approach to address this problem is to integrate the transflective board into engineering-related application scenarios in such a way that these reflections are minimized or avoided. Future research will tackle further technological and usability issues, such as mobility, robust tracking, efficient interaction and the evaluation of implemented scenarios with the industry.

REFERENCES

3M, Corp. (2001) *P-18*. Retrieved from the World Wide Web: http://www.mmm.com/market/construction/html/products/product73_p.html.

ARVIKA – Augmented Reality in Industrial Applications (2001), Retrieved from the World Wide Web: http://www.arvika.de

Ascension Technologies. Corp. (2001*) Flock of Birds*. Retrieved from the World Wide Web: http://www.ascension-tech.com/products/flockofbirds/flockofbirds.htm.

Azuma, R.(1995*) Predictive Tracking for Augmented Reality*. Ph.D. dissertation, University of North Carolina at Chapel Hill, TR95-007.

Barco, Inc. (2001). *Baron*. Retrieved from the World Wide Web: http://www.barco.com/projection_systems/products/product.asp?GenNr=324.

Bimber, O., Encarnação, L.M. and Stork, A. (2000). A Multi-Layered Architecture for Sketch-based Interaction within Three-dimensional Virtual Environments. *Computers and Graphics – The international Journal of Systems and Applications in Computer Graphics*, vol. 24, no. 6, pp. 851-867.

Bimber, O., Encarnação, L.M. and Stork, A. (2001). Seamless integration of virtual reality in habitual workplaces. *To appear in Journal for Industrial Science*, Munich University of Technology, vol. 2.

Bimber, O., Encarnação, L.M. and Branco, P. (2001). The Extended Virtual Table: An Optical Extension for Table-Like Projection Systems. *In revision for Presence: Teleoperators and Virtual Environments*.

Dunkane, Corp. (2001) *Mimio*. Retrieved from the World Wide Web: http://www.dukane.com/AudioVisual/whiteboard.htm.

Ishii, H. and Kobayashi, M. (1992). *ClearBoard: A Seamless Medium for Shared Drawing and Conversation with Eye Contact*. Proceedings of the Conference on Computer Human Interaction CHI'92, pp. 525-532.

Matsushita, N. and Rekimoto, J. (1997). *HoloWall: Designing a Finger, Hand, Body, and Object Sensitive Wall*. In proceedings of UIST'97, 209-210.

Mynatt, E. D., Igarashi, T., Edwards, W. K. and LaMarca, A. (1999). *Flatland: New Dimensions in Office Whiteboards*. In proceedings of the Conference on Computer Human Interaction CHI'99, pp. 346-353.

Neider, J., Davis, T., and Woo, M. (1993) *OpenGL programming Guide*. Addison-Wesley Publ., ISBN 0-201-63274-8.

Ogi, T. Yamada, T. Yamamoto, K. and Hirose, M. (2001). *Invisible Interface for Immersive Virtual World*. To appear at the Immersive Projection Technology Workshop (IPT'01), Stuttgart, Germany.

Pedersen, E. R., McCall, K., Mora, T. P. and Halasz, F. G. (1993). *Tivoli: An Electronic Whiteboard for Informal Workgroup Meetings*. In proceedings of INTERCHI'93, pp. 391-398.

G+B Pronova GmbH. (2001) *HoloPro*. Retrieved from the World Wide Web: http://www.holopro.html.

Raskar, R., Welch, G., and Fuchs, H. (1998). *Spatially Augmented Reality*. In proceedings of First IEEE Workshop on Augmented Reality (IWAR'98), San Francisco, CA.

Stafford-Fraser, Q. and Robinson, P. (1996). BrightBoard: A Video-Augmented Environment. *Proceedings of the Conference on Computer Human Interaction CHI'96*, pp. 134-141.

Tang, J. C. and Minneman, S. L. (1991). *VideoWhiteboard: Video Shadows to Support Remote Collaboration*. Proceedings of the Conference on Computer Human Interaction CHI'91, pp. 157-168.

Issues for Application Development Using Immersive HMD 360 Degree Panoramic Video Environments

Albert A. Rizzo[ab], Ulrich Neuman[a], Tom Pintaric[a] and Martin Norden[a]

[a]Integrated Media Systems Center, University of Southern California, 3740 McClintock Ave, EEB 131, Los Angeles, California, 90089-2561, USA

[b]School of Gerontology, University of Southern California, 3715 McClintock Ave, Los Angeles, California, 90089-0191, USA

Abstract

Panoramic video camera systems may serve as useful tools for creating virtual environment scenarios that would be difficult and/or labor intensive to produce using traditional computer graphic (CG) modeling methods. This paper presents observations from some of our initial research using a 360-degree panoramic video camera. This system captures high-resolution panoramic video (>3Kx480) by employing an array of five video cameras that view real world scenes over a combined 360-degrees of horizontal arc. During playback, users wear a head-mounted display (HMD) and a head-tracking device that allows them to turn their heads freely to observe the desired portions of the panoramic scene. A brief technical description of this system is presented, followed by a detailing of some of the scenarios that we have captured thus far with some discussion of the pragmatic-developer and user-centered issues that we have noted during initial field-testing. While full interaction is limited within these types of panoramic scenarios, initial user feedback has been positive. It is our view that the requirements for certain application areas could be well matched to this form of image capture and delivery. Possibilities for future applications that are well matched to the assets and limitations of panoramic video systems are described along with a discussion of our approach to developing a social phobia exposure-based therapy application with this system.

1. INTRODUCTION

Recent advances in Panoramic Video camera systems have produced new methods for the creation of virtual environments (James, 2001). With these systems, users can playback and observe pictorially accurate 360-degree video scenes of "real world" environments. When delivered via an immersive head mounted display (HMD), an experience of presence within these captured scenarios can be supported in human users. This is in sharp contrast to the constrained delivery and passive viewing of television and video images that have been the primary mode for providing humans with a "virtual eye" into distant times and locations over the last fifty years. During this time, video technology has matured from gray-scale images to big-screen color and digitally processed imagery. Yet one aspect of both the content creation and delivery technology has remained unchanged— *the view is controlled at the source and is identical for all observers*. Along with traditional CG virtual environments (VEs), Panoramic Video (PV) overcomes the passive and structured limitations of how imagery is presented and perceived. The recent convergence of camera, processing, and display technologies make it possible to consider providing each viewer with individual control of their viewing direction. Viewers of PV become virtual participants immersed in the observed scene, creating a new dimension in the way people perceive imagery within these types of VEs. However, as opposed to CG-based VEs, PV has some limitations regarding functional interactivity. Whereas users operating within a CG-based VE scenario are usually capable of both 6DF navigation, and interaction with rendered objects, PV immersion allows mainly for observation of the scene from the fixed location of the camera with varying degrees of orientation control (i.e. pitch, roll and yaw). In spite of this limitation, the goals of certain application areas may be well matched to the assets available with this type of image capture and delivery system. Such application areas may include those that have high requirements for presenting real locations inhabited by real people (e.g. capture of an event of possible future historical significance). As well, alternative methods to create "pseudo-interaction" are also possible, along with the augmentation of panoramic imagery using CG objects that allow for direct interaction. This paper will briefly present the technical details of the 360-degree PV system that our lab has in current use. We will describe the scenarios that we have captured thus far and discuss some of the "pragmatic developer" and user-centered issues relevant for scene capture and delivery of PV. We will then conjecture possible future applications that are well matched to the assets and limitations of PV systems and detail the rationale and approach we have taken to develop a social phobia therapeutic application with this system.

2. BRIEF SYSTEM OVERVIEW AND TECHNICAL DESCRIPTION

Panoramic image acquisition is based on mosaic approaches developed in the context of still imagery. Mosaics are created from multiple overlapping sub-images pieced together to form a high resolution, panoramic, wide field-of-view image. Viewers often dynamically select subsets of the complete panorama for viewing. Several panoramic video systems use single camera images (Nayar, 1997), however, the resolution limits of a single image sensor reduce the quality of the imagery presented to a user. While still image mosaics and panoramas are common, we produce high-resolution panoramic video by employing an array of five video cameras viewing the scene over a combined 360-degrees of horizontal arc. The cameras are arrayed to look at a five-facet pyramid mirror. The images from neighboring cameras overlap slightly to facilitate their merger. The camera controllers are each accessible through a serial port so that a host computer can save and restore camera settings as needed. The complete camera system (Figure 1) is available from Panoram Technologies (Panoram, 2001).

Figure 1. Panoramic Camera

The five camera video streams feed into a digital recording and playback system that we designed and constructed for maintaining precise frame synchronization. All recording and playback is performed at full video (30Hz) frame rates. The five live or recorded video streams are digitized and processed in real time by a computer system. The camera lens distortions and colorimetric variations are corrected by the software application and a complete panoramic image is constructed in memory. With five cameras, this image has over 3000x480 pixels. From the complete image, one or more scaled sub-images are extracted for real-time display in one or more frame buffers and display channels. Figure 2 shows an example of the screen output with a full 360° still image extracted from the video.

The camera system was designed for viewing the images on a desktop monitor. With a software modification provided by FullView.com (FullView, 2001) we were able to create an immersive viewing interface with a SONY Glasstron head-mounted display (HMD). A single window with a resolution of 800x600 is output to the HMD worn by a user. A real-time (inertial-magnetic) orientation tracker (Intersense, 2001) is fixed to the HMD to sense the user's head orientation. The orientation is reported to the viewing application through an IP socket, and the output display window is positioned (to mimic pan and tilt) within the full panoramic image in response to the user's head orientation. View control by head motion is a major contributor to the sense of immersion experienced by the user. It provides the natural viewing control we are accustomed to without any intervening devices or translations.

Figure 2. Still 360-degree PV image extracted from video footage.

3. INITIAL EXPLORATORY FIELD-TESTING

Our initial exploratory field-testing with the PV camera examined performance characteristics under a variety of conditions. The following targeted test environments were chosen that allowed for assessment across a range of lighting, external activity and camera movement conditions. These environments included:

1. An outdoor mall with the camera in a static position in daytime lighting with background structures and moderate human foot traffic, both close-up and at a distance. (*3rd St Promenade in Santa Monica, CA.*)

2. An outdoor ocean pier with the camera in a static position in extremely intense lighting conditions (direct intense late afternoon sunlight with high reflectance off of the ocean) with both long shots of activity on a beach and close-up activity of human foot traffic and amusement park structures on the pier.

3. Interior of an outside facing glass elevator with the camera in a static position and the elevator smoothly rising 15 floors from a low light position (street level shielded) to more intense lighting as the elevator ascended.

4. The camera mounted at the front of a pickup truck bed (near the back of the cab), traveling on a canyon road for 30 minutes at speeds ranging from 0-40 mph under all daylight ranges of lighting (low shaded light to intense direct sun).

5. Same as #4, except at night on a busy well lit street (*Sunset Blvd. In L.A. CA.*), and on a freeway traveling at speeds from 0-60 mph.

6. A University of Southern California Football game within the Los Angeles Coliseum from both static and moving positions in daytime lighting, with extreme close-ups of moving people and massive crowd scenes (40-60 thousand people).

7. An indoor rock concert (*Duran Duran at the Anaheim, CA. House of Blues venue*) from a static position under a variety of extreme lighting conditions in the midst of an active crowd, slightly above average head level with 3D immersive audio recording.

8. A Virtual "Mock-Party" with the camera in a static position in the center of a room with approximately 16 participants. This was our first "scripted" scenario that was shot while systematically directing and controlling the gradual introduction of participants into the scene and orchestrating their proximity and "pseudo-interaction" with the camera. This scenario was created for a therapeutic application designed to address social phobia as outlined in Section 6 below.

4. ISSUES FOR USE OF PANORAMIC VIDEO

Our initial field-testing experience with the capture, production and delivery of PV scenarios, served to provide a range of challenges from both "pragmatic-developer" and user-centered perspectives. We have outlined a series of PV "issues" that are based on our own experience from a developer standpoint and from user feedback provided by approximately 400-500 individuals. The user feedback was obtained informally from persons who experienced various combinations of our scenarios during an exhibit at SIGGRAPH 2000 and from the many demos conducted with visitors to our lab over the last year. This information has helped to guide our PV scenario capture efforts and has provided insight and direction for evolving formal evaluation testing and application development with the PV system. Key issues include:

• The system may present opportunities to capture environments that would be difficult and/or labor intensive to produce using traditional CG methods. This might involve the capture of complex human activity or dynamic environments where realism/fidelity in the primary concern.

• Advance thoughtful consideration of environments where 360 degree viewing would provide a compelling and engaging experience is vital. Perhaps a stroll down Bourbon Street in New Orleans during Mardi Gras would provoke users to become more involved in exploring the 360 degree visual opportunities afforded, as opposed to sitting in the back of a concert hall for a classical concert recital. In the latter situation, the question becomes: How often would one feel compelled to look around at the audience behind them during such a "forward-focused" event?

• The developer is required to take on the role of a "producer/director", as opposed to using the skills required for traditional CG modeling of scenes/people via the use of programming tools. This involves attending to such minutiae as: advance location scouting, getting permission to shoot in some "public" locations, insuring a source of power to run equipment, transporting delicate equipment to environments outside of a lab, scheduling and planning around extreme and changing ambient lighting and weather conditions, working with actors, etc.

• Most users reported a preference for using a HMD for viewing PV scenarios instead of either a single or a three-screen monitor using joystick navigation. This feedback suggests that perhaps too much cognitive "overhead" is required for users to "assemble" a "semi-intuitive" sense of the 360-degree nature of the environment from a long continuous scene "streaming" across a flatscreen. The *naturalistic* integration of body movement with the visual stimuli (particularly with head and neck turning) seems to support the sense of presence that occurred within these scenarios. This may be less of an issue if the camera is moving and providing a forced flythrough effect whereby the user can then select what aspects are of interest in the scenario to focus on. However, in view of the limitations that PV has already (e.g. restricted range of self-controlled navigation and interaction with objects), without the more naturalistic interfacing opportunities afforded by HMD use, it is doubtful that flatscreen PV delivery will get past the novelty level of attraction with users. User preference for flatscreen vs. HMD-delivered PV could be assessed by examining length of self-selected periods of use/exposure by users. Assessment of this issue is planned in our future formal user studies by measuring time that users voluntarily select to spend using PV for both flatscreen and HMD conditions across "camera static" and "camera moving" conditions and how this is related to self-report scores on a presence questionnaire (Witmer & Singer, 1998) and preference rating scales.

• Users rarely self-reported any occurrence of symptoms of cybersickness. Most users sampled the system for between 3 and 10 minutes, but some had actually spent between 10 minutes and one hour in the HMD. This issue will also be formally assessed in our future user studies.

• As would be expected, integration of audio seems to provide for a more compelling user experience. Users were commonly observed to display more positive responses (e.g. *Wow Factor*) to sections where the USC football game footage was accompanied by audio of the stadium full of cheering fans.

• The format that is used to deliver PV scenarios will play a large part in promoting common user consumption. Up until April 2001 we were exclusively delivering the footage via a cumbersome and costly system requiring five separate DVD tape player/recorders. We now have a working prototype for HMD delivery of the scenarios via a media player on a laptop. Eventually, DVD or Internet delivery of these scenarios will be the most usable and "marketable" option for promoting access and adoption.

• The integration of both, branching programs and augmentation of the environment with realistic CG modeled objects, may promote the usefulness, usability and user preference for PV scenarios. This will afford users' with opportunities for independent exploration of larger scale environments and will allow for interaction with specific and relevant objects.

5. POSSIBLE FUTURE END USER APPLICATIONS

Panoramic Video application development decisions require shrewd consideration of the pragmatic match between the assets and limitations that exist with PV scenarios relative to CG approaches and how this relates to potential user preferences. An obvious area could include entertainment and experiential learning applications that allow people to explore environments that are not readily accessible or are dangerous for the user to experience in the real world. One could explore the streets of Paris, dance on the front of the stage at a *Rolling Stones* concert or fly close to an active volcano within the comfort of their own home. A similar approach could be taken for remote real estate inspection and exploration, as has been shown to be useful with more limited still image methodologies. In this possible application, a live "agent" could direct a tour of a building, pointing out features while also allowing the user independent exploration, choice of field of view and control of temporal pacing. The integration of "augmenting" CG avatar/agents that could provide information on requests, keyed with branching programs allowing for independent exploration these environments, might also be an option. The learning of spatial layouts and wayfinding cues through unfamiliar environments could be readily targeted with this system. Indeed, PV systems might have advantages for situations where serious harm reduction could occur via advance navigational training of a dangerous environment (i.e. warfighting, terrorist extraction). As well, PV could be of value in situations where this information serves to inform necessary planning and/or adjustments in travel plans by wheelchair users by providing advance knowledge on inaccessible routes in public buildings. Flythroughs created in interesting compelling environments, in addition to what they may provide for entertainment value, could serve to provide distraction scenarios to promote analgesia or relaxation for persons undergoing painful or anxiety provoking medical procedures. The effectiveness of this approach with traditional CG VR has been demonstrated with burn victims receiving wound care and physical therapy (following skin grafts), and in tolerance studies of experimentally induced pain (Hoffman et al., 2000). These applications are well suited to PV technology and could serve to advance human welfare.

6. SPECIFIC APPLICATON DEVELOPMENT TARGETING SOCIAL PHOBIA TREATMENT

One well matched and potentially cost/effective PV application currently exists in the area of exposure-based habituation therapy for persons with phobias and other forms of anxiety disorders. It is estimated that over 23 million persons in the USA suffer from these debilitating disorders and applications with CG based VR have shown success for treating fear of heights, flying, spiders, and public speaking (Rothbaum & Hodges, 1999). The key to these efficacious approaches is in the systematic and hierarchical presentation to clients, of components of the fear or anxiety producing stimuli or events. Specific to this, we are producing a series of initial feasibility demos addressing social phobia disorder using the PV system delivered with an HMD. Current VR therapy approaches for social phobia disorder using CG are limited by the computationally intensive task of rendering acceptable/believable humans to which the patient is to be gradually exposed. Graded exposure scenarios could be created with the current system, that utilize *real* humans in *real life* situations to create the type of hierarchical stimulus exposure required for this treatment. This could be cost effective to produce and is well matched to the needs of the application area. In addition, this will be the first effort to systematically control "people-exposure" (using actors) within a social environment using our PV camera system. We have already captured "naturalistic" people-laden environments that are being incorporated into this project (Scenarios 1,2,6 and 7, above). To test this application, a collaboration with the USC Psychology Department has been formed to begin initial social phobic user-testing. Within this client/user-centered testing we will be soliciting feedback on a range of issues (i.e. subjective units of discomfort, presence ratings, usability ratings etc.) that will be incorporated in the next iterative design cycle to drive the capture and delivery of relevant stimulus environments.

The initial Virtual "Mock-Party" required the systematic scripting and setting up of a party environment using human actors. Capturing the scenario required that the camera be placed in the middle of a room that was set up to look like a social gathering was about to occur. Actors were introduced into the "party" scenario in a graded systematic fashion. Sixteen attendees entered the room, at a rate of two every 30 seconds, and filled the 360-degree perimeter. The participants interacted with each other as if at a party and then gradually moved closer to the camera at regular intervals to induce a "crowding" effect. At certain predetermined times and distances from the camera, actors then addressed the camera "as if" the camera was the person being treated and asked open-ended questions such as, "Hi, do you know where the bathroom is?" "Hi my name is Jane, what's yours?" or "This sure is an interesting party, huh?", etc. This was followed by pauses of an appropriate length where the actor simply looked at the camera with a neutral expression and then said "Thanks, good to meet you, could you excuse me for a minute" and left the immediate scene. Such "scripting" of events in this scenario (and their eventual evolution) may allow for a form of "pseudo-interaction" to occur. For example, in response to the open-ended questions, the client could practice and "role-play" making appropriate responses with varying levels of guidance by the therapist. Most importantly, clients could be exposed, in a gradual manner, to the anxiety-provoking scenario in a systematic and controllable fashion while other forms of therapeutic strategies are implemented.

If this method shows initial efficacy, then a range of social environments can be produced to target other social scenarios and interactional demands that are relevant for this type of anxiety disorder. Scenarios targeting other phobic stimuli and events (i.e. snakes, heights, closed spaces, etc.) could also be developed for similar therapeutic strategies. Between now and that time, our goals are to produce a variety of environments for these purposes while advancing the enabling technologies for delivery of these scenarios on a standard PC with a low cost HMD. If the technology can be driven to the point where these scenarios are deliverable via DVD or from the Internet, widespread usage will become possible that could be of benefit to persons whose lives are functionally limited by complications resulting from these disorders! Also, pharmaceutical companies might potentially be interested in using these types of scenarios in order to test the efficacy of prescribed medications (i.e., Paxil, Xanax, etc.) under systematically controlled conditions. These sorts of applications are well matched to the technology, could serve to advance human welfare and serve as one type of testbed for testing the usability and compelling nature of Panoramic Video.

REFERENCES

FullView.com Inc. At: www.fullview.com (2001).

Hoffman, H.G., Doctor, J.N., Patterson, D.R., Carrougher, G.J. & Furness, T.A. III. Use of virtual reality for adjunctive treatment of adolescent burn pain during wound care: A case report, Pain, 2000, 85:305-309.

Intersense Inc. At: www.isense.com (2001).

James, M.S. 360-Degree Photography and Video Moving a Step Closer to Consumers. ABC News Report. March 23, 2001. At: http://abcnews.go.com/sections/scitech/CuttingEdge/cuttingedge010323.html

Nayar, S. K. Catadioptric Omnidirectional Camera, Proc. of IEEE Computer Vision and Pattern Recognition (CVPR), June 1997.

Panoram Technologies Inc. At: www.panoramtech.com (2001)

Rothbaum, B.O., & Hodges, L.F. The use of Virtual Reality Exposure in the Treatment of Anxiety Disorders. Behavior Modification, 1999, 23(4), 507-525.

Witmer, B.G., & Singer, M. J. Measuring presence in virtual environments: A Presence Questionnaire. Presence: Teleoperators and Virtual Environments, 1998, 7(3), 225-240.

COLLABORATION WITH TANGIBLE AUGMENTED REALITY INTERFACES.

Mark Billinghurst[a], Hirokazu Kato[b], Ivan Poupyrev[c]

[a] Human Interface Technology Laboratory, University of Washington,

Box 352-142, Seattle, WA 98195, USA

[b] Faculty of Information Sciences, Hiroshima City University,

3-4-1 Ozuka-Higashi, Asaminami-ku, Hiroshima 731-3194, Japan

[c] Interaction Laboratory, Sony Computer Science Laboratory,

3-14-13 Higashi-Gotanda, Tokyo 141-0022, Japan

ABSTRACT

We describe a design approach, Tangible Augmented Reality, for developing interfaces for supporting face-to-face collaboration. By combining Augmented Reality techniques with Tangible User Interface elements, we can create interfaces in which users can interact with spatial data as easy as real objects. Tangible AR interfaces remove the separation between the real and virtual worlds and so enhance natural face-to-face communication. We describe several prototype applications showing how this design approach can be applied in practice.

1   INTRODUCTION

Computers are increasing being used to support communication and collaboration. However, although many researchers have explored how computers can be used to enhance remote collaboration there has been less work in providing support for face-to-face collaboration. This is particularly true for viewing and manipulating spatial data.

Early attempts at supporting face-to-face collaboration were based around computer conference rooms. For example, the Colab room at Xerox (Stefik 1987) used a network of workstations running distributed software applications designed to support brainstorming, document preparation, and other tasks for small group face-to-face meetings. However user's collaborating on separate workstations, even if they are side by side, do not perform as well as if they were huddled around a single machine (Inkpen 1997). This is because the computer introduces an artificial separation between the task space and communication space, preventing people from being aware of the normal communication cues that are used in a face-to-face setting.

One way of removing this artificial seam is by using Augmented Reality (AR) technology to overlaying virtual imagery directly onto the real world. In this way the display space and communication space can become one. In the Studierstube project co-located users can view and manipulate virtual models while seeing each other in the real world, facilitating very natural face to face communication (Schmalsteig 1996). In earlier work we found that users prefer collaboration in an AR setting than an immersive virtual environment and can perform better on some tasks because they can see each other's non-verbal cues (Billinghurst 1997). However, although AR interfaces provide a natural environment for viewing spatial data it is often challenging to interact with and change the virtual content.

Another approach is through Tangible User Interfaces (TUI) (Ishii 1997). In this case physical objects and ambient spaces are used to interact with digital information. Collaborative Tangible User Interface projects have been developed that use digitally enhanced real objects to support face-to-face collaboration or provide a tangible representation of remote collaborators. For example in the Triangles project (Gorbet 1998) several users can gather around a table and play with electronic triangular objects to compose poetry or interactive stories.

797

Tangible interfaces are powerful because the physical objects used in them have properties and physical constraints that restrict how they can be manipulated and so are easy to use. Physical objects are also integral to the face-to-face communication process. Brereton finds that design thinking is influenced by the objects in the environment, and that designers actively seek out physical props to aid in face-to-face collaboration (Brereton 2000). Gav finds that people commonly use the resources of the physical world to establish a socially shared meaning (Gav 1997). Physical objects support collaboration both by their appearance, the physical affordances they have, their use as semantic representations, their spatial relationships, and their ability to help focus attention. Thus TUI combine these properties with an intuitive interface for manipulating digital data.

Although intuitive to use, with TUI interfaces information display can be a challenge. It is difficult to dynamically change an object's physical properties, so most information display is confined to image projection on objects or augmented surfaces. In those Tangible interfaces that use three-dimensional graphics there is also often a disconnect between the task space and display space. For example, in the Triangles work, visual representations of the stories are shown on a separate monitor distinct from the physical interface (Gorbet 98). Presentation and manipulation of 3D virtual objects on projection surfaces is difficult, particularly when trying to support multiple users each with independent viewpoints. Most importantly, because the information display is limited to a projection surface, users are not able to pick virtual images off the surface and manipulate them in 3D space as they would a real object.

To overcome the limitations of the AR and TUI approaches and retain the benefits of physical objects in face-to-face collaboration we have been exploring an interface metaphor we call Tangible Augmented Reality (Tangible AR). In the next section we describe this in more detail and in the remainder of the paper present some Tangible AR interfaces for intuitive face-to-face collaboration.

2   TANGIBLE AUGMENTED REALITY

Tangible Augmented Reality interfaces are those in which each virtual object is registered to a physical object and the user interacts with virtual objects by manipulating the corresponding tangible objects. Thus Tangible AR combines the intuitiveness of Tangible User Interfaces with the enhanced display possibilities afforded by Augmented Reality. In the Tangible AR approach the physical objects and interactions are equally as important as the virtual imagery and provide a very intuitive way to interact with the AR interface. The intimate relationship between the real and virtual also naturally supports face-to-face collaboration.

This approach solves a number of problems. Tangible User Interfaces provide seamless interaction with objects, but may introduce a separation between the interaction space and display space. In contrast most AR interfaces overlay graphics on the real world interaction space and so provide a spatially seamless display. However they often force the user to learn different techniques for manipulating virtual content than from normal physical object manipulation or use a different set of tools for interacting with real and virtual objects.

A Tangible AR interface provides true spatial registration and presentation of 3D virtual objects anywhere in the physical environment, while at the same time allowing users to interact with this virtual content using the same techniques as they would with a real physical object. So an ideal Tangible AR interface facilitates seamless display and interaction. This is achieved by using the design principles learned from TUI interfaces, including:

- The use of physical controllers for manipulating virtual content.
- Support for spatial 3D interaction techniques (such as using object proximity).
- Support for both time-multiplexed and space-multiplexed interaction.
- Support for multi-handed interaction.
- Support for Matching the physical constraints of the object to the requirements of the interaction task.
- The ability to support parallel activity where multiple objects are being manipulated.
- Collaboration between multiple participants

Our central hypothesis is that AR interfaces that follow these design principles will naturally support enhaned face-to-face collaboration. In the next section we describe some interfaces we have developed to explore this.

3   SAMPLE INTERFACES

We have tested the Tangible AR approach in a variety of prototype interfaces, including; Shared Space: A collaborative game designed to be used by complete novices, AR PRISM: An interface for geospatial visualisation, and Tiles: A virtual prototyping application.

All of these projects were built on top of ARToolKit, a software library we have developed for computer-vision based AR applications (Kato 99). ARToolKit tracks camera position and orientation relative to a marked card and so can be used to exactly overlay virtual images on real world objects, making it ideal for developing Tangible AR interfaces.

3.1   Shared Space

The Shared Space interface was a game developed for the Siggraph 99 conference to explore how Tangible AR principles could be applied in an application that could be usable by complete novices. In this experience multiple users stood across a table from one another (fig. 1). On the table were marked cards that could be freely moved and picked up. Each of the users wore a lightweight head mounted display (HMD) with a video camera attached in which they can see a view of the real world. When users look at the cards they see three-dimensional virtual objects exactly overlaid on them. The goal of the game is to collaboratively match up the virtual models, by placing related models side by side. When users get a correct match the pair of models is replaced by an animation of the two objects (fig 2).

Figure 2: Proximity-based Virtual Object Interaction. Placing two related objects side by side plays an animation of the objects interacting.

Figure 1: The Shared Space Siggraph 99 Application

Over 3,500 people have tried the Shared Space game and given us feedback. Users had no difficulty with the interface. They found it natural to pick up and manipulate the physical cards to view the virtual objects from every angle. Since the matches were not obvious some users needed help from others at the table and players would often spontaneously collaborate with strangers who had the matching card they needed. They would pass cards between each other, and collaboratively view virtual objects and completed animations. By combining a tangible object with virtual image we found that even young children could play and enjoy the game. In surveys conducted users felt that they could easily collaborate with the other players and interact with the virtual objects.

3.2   AR PRISM

The AR PRISM interface explored how Tangible AR techniques could be used to support face to face collaboration and object-based interaction with the users environment. The application domain was geographic data visualization and the underlying idea was that multiple users gathered around a real map should be able to see and manipulate three-dimensional virtual geographic information superimposed on the map while seeing each other in the real world. This manipulation should be based on physical markers and tangible interaction techniques. To achieve this the interface combined the ARToolKit tracking technology with an overhead camera that could track object positions and hand gestures relative to a real map. Users stood around the real map looking at it through HMDs with small cameras attached. As the user moved their hand across the surface of the map in their HMD they saw a graphic overlay of longitude and latitude information for the location their hand was pointing at and icons representing data available at that point. If the user wanted to view some of the geological dataset they placed a terrain marker on the map (figure 3) and they would see a virtual terrain model of that location (figure 4). Users could then pick up the models, look at them from any angle, and pass them to each other.

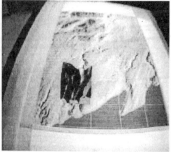

Figure 4: The view inside the Head Mounted Display. As a user places a tile over the real map, a virtual 3D terrain model is shown at that location.

Figure 3: Using AR PRISM

If the user wanted to see a new portion of the virtual terrain they simply moved the marker to a new map location, the system recognized the new location and swaped terrain models. The AR scene and real map were aligned so that features seen on the map could be seen in the three-dimensional virtual model overlaying the same map location. In addition to showing virtual terrain models we also implemented a "Sensor-Data" metapor where other markers could be placed beside the virtual terrain marker to load additional data. For every grid location in the real map there was three-dimensional topographic information available, and for some locations there was also soil, hydrology and well information. So placing a soil tile beside the terrain model would cause the texture on the terrain model to change to show soil type information. Similarly, placing the well marker beside the terrain model would show sub-surface wells.

We are in the process of conducting formal user studies with the AR PRISM interface, although almost one hundred people have tried the software informally. These users generally found the software very easy to use and felt that being able to lift pieces of virtual terrain off the real map was a very compelling experience. Associating a virtual dataset with a physical object meant that this data was as easy to manipulate as the real object. They also liked the ability to see other users in the real world at the same time as seeing the AR model, and felt that this supported natural face-to-face collaboration. However they felt that the HMDs were encumbering and did not provide high enough resolution to clearly read the real map.

3.3 Tiles

Tiles is an AR authoring interface that explored how more complicated functionality could be supported using tangible 3D widgets. The application domain was rapid prototyping for aircraft instrument panels and the interface consisted of a metal whiteboard, a book, and two stacks of magnetic tiles. Using Tiles, one or more users would sit in front of the whiteboard and collaboratively lay out a virtual aircraft instrument panel (Figure 5).

The various tangible elements of the interface served a different purpose. The whiteboard was the working space where users could layout virtual aircraft instruments. The book served as a *menu object*, and users saw a different virtual instrument model on each page. One stack of tiles served as *data tiles* and showed no virtual content until virtual objects were copied onto them. The remaining tiles were *operator tiles* and were used to perform basic operations such as *deletion, copying* and a *help* function. Each of the operations tiles had a different three-dimensional virtual icon to show what their function was (fig 6).

Virtual images appeared attached to the physical objects and could be picked up and looked at from any viewpoint. Interaction between objects was based on physical proximity, however the operation that was invoked by bringing objects next to each other depended on their semantic. For example, touching a data tile

Figure 5: Using the Tiles Interface

that contained a virtual instrument with the trashcan delete tile, removed the virtual instrument. Once virtual instruments have been placed on the data tiles, these can be attached to the whiteboard to layout a prototype virtual instrument panel .

The Tiles interface supported collaboration on a number of different levels. Users could both be wearing HMDs and could see the shared virtual content. If one user didn't have an HMD on he would still be able to reference the virtual models by the physical cards/placeholders there were attached to. There was also an external monitor giving them a view of that the HMD user could see. Finally, the whiteboard can be drawn on so annotations can serve as another tool to help the face-to-face collaboration. This may be especially important for collaboration between users with and without HMDs so people who cannot see the virtual content can still understand what types of models are on the markers.

Trashcan delete widget   Talking head help widget

Figure 6: Widget Representations

4   CONCLUSIONS

In this paper we have advocated a new approach for face-to-face collaboration based on a Tangible Augmented Reality metaphor. As our prototype systems have shown, by combining TUI and AR design principles we can develop interfaces that naturally support face-to-face collaboration and spatial data manipulation. However there is still a large amount of work that needs to be done to compare how collaboration with these types of interfaces is different from more traditional computer supported tools. In addition to running more rigorous user studies, in the future we intend to explore how Tangible AR interfaces can be combined with projective technology, and what other tangible manipulation techniques can be used.

REFERENCES

Billinghurst, M., Weghorst, S., Furness, T. (1998) Shared Space: An Augmented Reality Approach for Computer Supported Cooperative Work. *Virtual Reality : Research, Development and Application*, 1998.

Brereton, M., McGarry, B. (2000) An Observational Study of How Objects Support Engineering Design Thinking and Communication: Implications for the design of tangible media. In *Proceedings of CHI 2000*, April 1-6, 2000, The Hague, The Netherlands. ACM Press, 2000.

Gav, G., Lentini, M. (1997) Use of Communication Resources in a Networked Collaborative Design Environment. http://www.osu.edu/units/jcmc/IMG_JCMC/ResourceUse.html

Gorbert, M. G., Orth, M., Ishii, H. (1998) Triangles: Tangible Interface for Manipulation and Exploration of Digital Information Topography. In *Proceedings of CHI 98*. Los Angeles, CA, USA. ACM Press, 1998, pp. 49-56.

Inkpen, K. (1997) *Adapting the Human Computer Interface to Support Collaborative Learning Environments for Children*. PhD Dissertation, Dept. of Computer Science, University of British Columbia, 1997.

Ishii, H., Ullmer, B. (1997) Tangible Bits: Towards Seamless Interfaces between People, Bits and Atoms. In proceedings of CHI 97, Atlanta, Georgia, USA, ACM Press, 1997, pp. 234-241.

Kato, H., Billinghurst, M., Asano, M., Tachibana, K. (1999) An Augmented Reality System and its Calibration based on Marker Tracking, *Transactions of the Virtual Reality Society of Japan*, Vol.4, No.4, pp.607-616, 1999

Schmalsteig, D., Fuhrmann, A., Szalavari, Z., Gervautz, M., (1996) Studierstube - An Environment for Collaboration in Augmented Reality. In *CVE '96 Workshop Proceedings*, 19-20th September 1996, Nottingham, Great Britain.

Stefik, M., Foster, G., Bobrow, D., Kahn, K., Lanning, S., Suchman, L. (1987) Beyond the Chalkboard: Computer Support for Collaboration and Problem Solving in Meetings. In *Communications of the ACM*, January 1987, Vol 30, no. 1, pp. 32-47.

CONVERGING USER INTERFACE PARADIGMS USING COLLABORATIVE AUGMENTED REALITY

Dieter Schmalstieg [a] & Gerd Hesina [b]

Vienna University of Technology,

[a] Institute of Software Engineering, Karlsplatz 13/188

[b] Institute of Computer Graphics and Algorithms, Karlsplatz 13/186

A-1040 Vienna, Austria - Europe

ABSTRACT

Studierstube is an experimental distributed collaborative augmented reality system, which bridges multiple user interface dimensions. At its core, it uses collaborative augmented reality to incorporate true 3D interaction into a productivity environment. This concept is extended to include multiple users, multiple host platforms, multiple display types, multiple concurrent applications, and a multi-context (i. e., 3D document) interface – into a heterogeneous distributed environment. All this happens almost totally transparent to the application programmer by using a distributed shared scene-graph approach. This paper presents a high-level view of the new *Studierstube* system capabilities.

1   INTRODUCTION

The original *Studierstube* architecture (Schmalstieg et al., 1996; Szalavári et al., 1998) was a collaborative augmented reality system allowing multiple users to gather in a room and experience the sensation of a shared virtual space that can be populated with three-dimensional data. This architecture incorporated simple distribution mechanisms to provide graphics from multiple host computers and shared data from a separate device (tracker) server. It turned out that this approach was insufficient to provide a framework for distributed collaborative augmented reality. An even more limiting factor was that the toolkit allowed to run only a single application at a time. Our efforts towards a follow-up version resulted in support for projection-based platforms (Schmalstieg et al., 1999) and a toolkit for distributed graphics (Hesina et al., 1999).

Our current work on the *Studierstube* project (Schmalstieg et al., 2001) focuses on experimenting with the possibilities of new user interfaces that incorporate collaborative AR. For efficient experimentation, we have implemented a toolkit that generalizes over multiple user interface dimensions, allowing rapid prototyping of different user interface styles. Furthermore we developed some tools to address problems like late-joining users or load-balancing.

2   SYSTEM DESCRIPTION

The system presented in this paper must be understood as an experimental platform for exploring the design space that emerges from bridging multiple user interface dimensions. It can neither compete in maturity and usability with the universally adopted desktop metaphor nor with more streamlined, specialized virtual environment solutions (e. g., CAVEs (Cruz-Neira et al., 1993)). The *Studierstube* user interface spans the following dimensions:

- Multiple users: The system allows multiple users to collaborate. While we are most interested in computer-supported face-to-face collaboration, this definition also encompasses remote collaboration. Collaboration of multiple users implies that the system will typically incorporate *multiple host computers*. However, we also allow multiple users to interface with a single host (e.g. via a large screen display), and a single user to interface with multiple computers at once. On a very fundamental level, this means that we are dealing with a distributed system. It also implies that *multiple types of output devices* such as HMDs, projection-based displays, hand-held displays etc. can be handled and that the system can span *multiple operating systems*.

- Multiple contexts: Contexts are the fundamental units from which the *Studierstube* environment is composed. A context is a union of *data* itself, the data's *representation* and an *application* which operates on the data. Contexts are thus structured along the lines of the model-view-controller (MVC) paradigm known from Smalltalk's windowing system (Goldberg & Robson, 1983): *Studierstube*'s data, representation, and application correspond to MVC's model, view, and controller, respectively. Not surprisingly, this structure makes it straightforward to generalize established properties of 2D user interfaces to three dimensions. Contexts distinguish between two operational modes: master and slave. In master mode a context distributes changes to its graphical database via our distribution toolkit to slaves that are interested in update events. Slaves are more or less passive. That is, slaves update their graphical database only via the aforementioned events or through individual implementation specific features. This approach allows support for local variations (compare MacIntyre & Feiner, 1998).

- Multiple locales: They correspond to coordinate systems in the virtual environment and usually coincide with physical places (such as a lab or conference room, or parts of rooms), but they can also be portable and associated with a user, or used arbitrarily – we even allow (and use) overlapping locales in the same physical space. We define that every display used in a *Studierstube* environment shows the content of exactly one locale. Every context can (but need not) be replicated in every locale; these replicas will be kept synchronized by *Studierstube's* distribution mechanism.

Note that allowing a context to operate in both master and slave mode has implications on how contexts can be distributed: It is not necessary to store all master contexts of a particular type at one host. Some master contexts may reside on one host, some on another host – in that case, there will be corresponding slave contexts at the respective other host, which are also instances of the same application, but initialized to function as slaves. Furthermore contexts are able to swap roles. That is, a master becomes a slave and vice versa. We define this process as migration and distinguish between two different versions:

- *Application migration* allows to transfer running Studierstube applications from one host to another, maintaining intact the state of the user interface as well as the internal state of the application.
- *Activation migration* is the light-weight variant of application migration: an application executes in a distributed system using a replication mechanism, but the responsibility for the computationally critical portion – called activation – of the application can be handed off from host to host without affecting the application or its user(s).

These tools allow us to address the following practical issues in user interface and application management for virtual work environments that are otherwise difficult to resolve:

- *Dynamically changing configurations*: Late joining and early exit of users and their respective hosts can be handled by migrating the set of application instances and their activations for each host.
- *Load balancing*: Applications can be migrated from one host to another if the computational load is too high. This mechanism can also account for situations where it is desirable to support asymmetric configurations, e. g., when a powerful compute server is available to offload less capable workstations, or when not all hosts are able to execute a particular application due to platform or hardware constraints.
- *Utilization of heterogeneous network capacity*: Many distributed virtual environment applications acknowledge variations in network performance by quality of service management and degradation strategies, but make the unrealistic assumption that the networking quality is homogeneous. We observe that small groups of co-located users are also likely to share a high-bandwidth local area network (LAN), while remote collaboration typically relies on a lower performance wide area network (WAN). Activation migration can be used to make the best of both situations simultaneously by allowing finer grained interaction between users sharing a higher bandwidth network.
- *Ubiquitous computing (Weiser, 1991)*: Applications can be made to follow a user across physical locations.

3 CONCLUSIONS

We have presented *Studierstube*, a prototype user interface that uses collaborative augmented reality to bridge multiple user interface dimensions: Multiple users, context, and locales as well as applications, hosts, display platforms, and operating systems. *Studierstube* supports collaborative work by coordinating a heterogeneous distributed system based on a distributed shared scene graph and a 3D interaction toolkit. This architecture allows to combine multiple approaches to user interfaces as needed, so that it becomes easy to create a 3D work environment, which can be personalized, but also lends itself to computer supported cooperative work. Our implementation prototype shows that despite its apparent complexity, such a design approach is principally feasible, although much is left to be desired in terms of quality and maturity of hard- and software.

ACKNOWLEDGEMENTS

This project was sponsored by the Austrian Science Fund *FWF* under contract no. P-12074-MAT. Special thanks to Markus Krutz, Rainer Splechtna, Hermann Wurnig, and Andreas Zajic for their contributions to the implementation, to Anton Fuhrmann for fruitful discussions, to Michael Gervautz for general support, and to M. Eduard Gröller for his spiritual guidance.

REFERENCES

Cruz-Neira C., Sandin D. J., & DeFanti T. (1993). A. Surround-Screen Projection-Based Virtual Reality: The Design and Implementation of the CAVE. In Proceedings of SIGGRAPH '93, 135-142.

Goldberg A., & Robson D. (1983). Smalltalk-80: The language and its implementation. *Addison-Wesley, Reading MA, 1983.*

Hesina G., Schmalstieg D., Fuhrmann A., & Purgathofer W. (1999). Distributed Open Inventor: A Practical Approach to Distributed 3D Graphics. *Proceedings VRST '99,* London, 74-81.

MacIntyre B., & Feiner S. (1998). A Distributed 3D Graphics Library. *Proceedings SIGGRAPH '98,* 361-370.

Schmalstieg D., Fuhrmann A., Szalavari Zs., & Gervautz M. (1996). Studierstube - Collaborative Augmented Reality. *Proceedings Collaborative Virtual Environments '96,* Nottingham, UK.

Schmalstieg D., Encarnação L. M., & Szalavári Zs. (1999). Using Transparent Props For Interaction With The Virtual Table. Proceedings of SIGGRAPH Symposium on Interactive 3D Graphics 1999, 147-154, Atlanta, GI.

Schmalstieg D., Fuhrmann A., Hesina G., Szalavári Zs., Encarnação L. M., Gervautz M., & Purgathofer W. (2001) The Studierstube Augmented Reality Project. To appear in: "Augmented Reality: The Interface is Everywhere", SIGGRAPH 2001 Course Notes, Los Angeles CA, USA, ACM Press.

Szalavári Zs., Fuhrmann A., Schmalstieg D., & Gervautz M. (1998). Studierstube - An Environment for Collaboration in Augmented Reality. *Virtual Reality - Systems, Development and Applications, 3(1), 37-49.*

Weiser M. (1991). The Computer for the twenty-first century. *Scientific American, 94-104.*

Augmented Reality Interface for Electronic Music Performance

Ivan Poupyrev [1], Rodney Berry [2], Mark Billinghurst [3], Hirokazu Kato [4], Keiko Nakao [2], Lewis Baldwin [5], Jun Kurumisawa [2]

[1] Interaction Lab, Sony CSL
3-14-13 Higashi-Gotanda,
Tokyo 141-0022, Japan

[2] ATR MIC Research Laboratories
2-2 Hikaridai, Seika, Souraku-gun
Kyoto 619-02, Japan

[3] University of Washington
[4] Hiroshima City University
[5] Redsmoke Inc.

poup@csl.sony.co.jp, rodney@mic.atr.co.jp, http://www.csl.sony.co.jp/~poup/research/agroove/

ABSTRACT

Starting from the days of the Musical Telegraph, the first electronic instrument, the majority of synthesizers today are still equipped with keyboards, often using the traditional layout of acoustic pianos. The question that many researchers attempt to answer is that of improvement. In this paper we present an Augmented Groove, a novel musical instrument that attempts to depart from traditional approaches to musical performance, i.e. use keyboards, dials or simulated traditional musical controllers. It allows novices to play electronic musical compositions, interactively remixing and modulating their elements, by manipulating simple physical objects.

1. INTRODUCTION

With every technological leap musicians and engineers have been creating new ways to make and play music. This quest has resulted in today's electronic musical instruments, e.g. sound samplers and synthesizers, which opened unprecedented opportunities for musical expression and creativity. Firstly, they allow musicians to synthesize virtually any sound even those that can not be produced naturally, expanding their musical vocabulary. Secondly, musicians can create a composition without just having to use traditional instruments. Today's electronic musical compositions are assembled out of hundreds of looped samples and MIDI sequences, modulated with filters, then sequenced and played back from a computer (Rule, 1999). Therefore, the proficiency in traditional musical instruments is no longer a necessary prerequisite for creative musical expression.

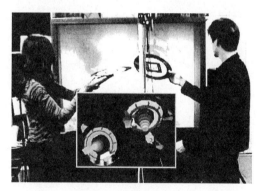

Figure 1: Two users are playing music in Augmented Groove: the insert shows the view on the projector screen

While the way music is *made* has changed significantly, the way it is *played* has not. Most of the synthesizers are still equipped with piano keyboards and arrays of buttons and dials. These controllers are neither easy no enjoyable to use especially for the novice. Currently, we are not aware of any effective musical controllers that would allow the novice to control electronic music as performance, the only option being is simple playback after composition.

This paper presents an Augmented Groove – a musical controller that allows a novice to interactively and collaboratively perform electronic music. We define performance of electronic music as the ability to interactively *remix and modulate elements of musical compositions* created by the composer. The resulted controller is easy and enjoyable to use for novices.

2. RELATED WORK

Recently there has been a significant growth of interest in alternative musical interfaces and controllers (comprehensive surveys in Cutler, Robair, and Bean, 2000; Paradiso, 1997). While some of the new controllers borrow from traditional ways to play music, such as simulating wind instruments or tracking a conductor's baton waved by the user, others completely depart from traditional approaches in search for new methods of artistic expression. Electromagnetic fields, digital pens, touch sensors, data gloves, body posture, soft toys, streaming water and other techniques have been used to interactively control musical performances. Many of these interfaces, e.g. the Digital Baton (Marrin and Paradiso, 1997), allow control of global musical parameters, such as tempo, which inhibits refined musical control, such as adjusting or modulation of single elements of the composition.. At the same time other interfaces, provide control at a too refined level, e.g. control of individual note progression, which makes them as difficult to use as traditional instruments. Finally, many of controllers, e.g. Dance Space , are not designed to be precise and repeatable, focusing instead on "computer-supported improvisation".

3. AUGMENTED GROOVE

In Augmented Groove the musical performance is constructed from a collection of short looped musical phrases, or loops, each carefully composed to fit others so they can be interactively re-mixed. For each individual loop or group of loops a composer can assign filters and effects, to allow the user the ability to modulate musical elements, the range of these modulations, however, is set by the composer to ensure a high quality of performance. Hence, this model targets mostly modern electronic music, which tends to be composed in a very similar fashion (Rule, 1999).

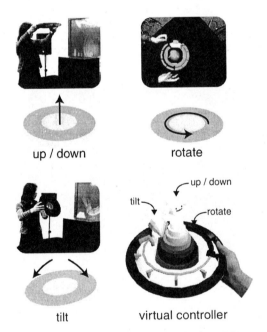

Figure 2: Gestural musical interface in Augmented Groove

During performance[1] the user plays music by physically manipulating simple LP records outfitted with a fiducial markers (Figure 1, 3) and tracked by an overhead camera (Billighurst, Poupyrev, Kato, and May, 2000; Kato and Billinghurst, 1999). To start the phrase playing the user simply flips the record so that the overhead camera can identify its fiducial marker. The system then start playing the corresponding musical sequence. The user can modulate the musical sequence by translating the record up and down, and by rotating and tilting it (Figure 2). The 3D motions of the record are mapped to corresponding modulations, e.g. pitch, distortion, amplitude, filter cut-off frequency, delay mix, etc. More control gestures as well as more complex mappings can be easily added for experienced users. Since the system tracks and recognizes several records at the same time, performers can play several musical phrases simultaneously and collaboratively (Figure 1). Thus, records serve both as physical *musical containers*, grouping together elements of composition, and *tangible 3D controllers*, that allow to interactive modulation, mix and fading between musical phrases.

Augmented Groove is a mixed reality (MR) interface: 3D virtual controllers are overlaid on top of the records providing the user with instant feedback on the state and progression of musical performance. For each control dimension a corresponding graphical element changes depending on the value of the control (Figure 2). For example, as the user raises the record a pyramid in the middle of the overlaid virtual controller (Figure 2) also goes up and when the musical control reaches its limit a small animated character pops up cuing the user. Although virtual controls are not absolutely necessary to control music since the user can hear musical changes, they are important cue for novice users, explaining the interface, visualizing the performance and making the system easier and more enjoyable.

4. INITIAL USER FEEDBACK

Evaluating musical instruments in controlled user studies is difficult (Roads, 1995) because it strongly depends on a user personal preferences. We informally assessed Augmented Groove by collecting informal user feedback during a demonstration at the SIGGRAPH 2000 Emerging Technologies exhibition[1] (Poupyrev, Berry, Kurumisawa, Nakao, Billinghurst, et al., 2000) which allowed visitors to play musical compositions, using three records, i.e. elements of three tracks could be mixed and modulated at the same time (Figure 3).

Figure 3: A Siggraph visitor playing Augmented Groove

We observed that Augmented Groove was enjoyable to use and that visitors were able to effectively control musical performance. Even with three records, there was a large range of possible musical variations, providing room for creativity and exploration: "I can play with this for hours," noted several visitors and a number returned several times with friends and colleagues. We noted that without *any* instructions, visitors had difficulty understanding the interface, however a short explanation was usually sufficient for a user to be able to effectively control a musical

---

[1] Video figures are available at http://www.mic.atr.co.jp/sspace/ or http://www.csl.sony.co.jp/~poup/research/agroove/

performance. Overall, Augmented Groove was highly rated by visitors as fun, enjoyable and creative. The biggest complaint was latency of computer vision tracking system.

5. CONCLUSIONS AND FUTURE WORK

Augmented Groove is a simple, unobtrusive MR interface that allows users to interactively control and modulate electronic musical performance. Although we focused on providing a novice user with an enjoyable and easy to use tool, we have been developing it with a professional electronic musical performer in mind. In future, we a planning to re-design the interface for real musical performances. Technical and interface issues will be addressed, such as lag, physical designs of records, new complex mappings and dynamic gestures.

6. REFERENCES

Billighurst, M., Poupyrev, I., Kato, H., and May, R. (2000). Mixing Realities in Shared Space: An Augmented Reality Interface for Collaborative Computing. In Proceedings of the ICME 2000 (pp. 1641-1644). IEEE.

Cutler, M., Robair, G., and Bean. (2000). The Outer Limits. Electronic Musician, 16(9), 48-71.

Kato, H., and Billinghurst, M. (1999). Marker Tracking and HMD Calibration for a Video-based Augmented Reality Conferencing System, Proc. of 2nd Int. Workshop on Augmented Reality, pp.85-94 (1999). In Proceedings of the 2nd Int. Workshop on Augmented Reality (pp. 85-94).

Marrin, T., and Paradiso, J. (1997). The Digital Baton: a Versatile Performance Instrument. In Proceedings of the ICMC97 (pp. 313-316).

Paradiso, J. (1997). New Ways to Play, Electronic Music Interfaces. IEEE Spectrum (12), 18-30.

Poupyrev, I., Berry, R., Kurumisawa, J., Nakao, K., Billinghurst, M., Airola, C., Kato, H., Yonezawa, T., and Baldwin, L. (2000). Augmented Groove: Collaborative Jamming in Augmented Reality. In Proceedings of the SIGGRAPH'2000 Conference Abstracts and Applications (pp. 77). ACM.

Roads, C. (1995). The computer music tutorial: MIT Press.

Rule, G. (1999). Electro shock! Groundbreakers of synth music. San Francisco: Miller Freeman Books.

SENSORY AUGMENTED WEARABLE COMPUTING AND ITS POTENTIAL FOR HUMAN-COMPUTER INTERACTION

Bernt Schiele

schiele@inf.ethz.ch

Department of Computer Science, ETH Zurich, CH-8092 Zurich, Switzerland

ABSTRACT

The next generation of computers might be literally wearable. Our vision of such a wearable computing device is an intelligent assistant, which is always with you and helps you to solve your every day tasks. Besides size and power, an important challenge is how to interact with wearable computers. An important aspect and unique opportunity of a wearable device is that it can perceive the world from a first-person perspective: a wearable camera can see what you see and a wearable microphone can hear what you hear in order to analyze, model and recognize things and people which are around you. In this paper we argue that a promising direction for interaction is to make the computers more aware of the situation the user is in and to model the user's context. Wearable sensors, such as cameras, mounted to the user's glasses, can recognize what the user is looking at and model what the user is doing.

1. INTRODUCTION

To date, personal computers have not lived up to their name. Most machines sit on the desk and interact with their owners for only a small fraction of the day. Smaller and faster notebook computers have made mobility less of an issue, but the same staid user paradigm persists. Wearable computing hopes to shatter this myth of how a computer should be used. A personal computer should be worn, much as eyeglasses or clothing are worn, and continuously interact with the user based on the context or the situation. With heads-up displays, unobtrusive input devices, personal wireless local area networks, and a host of other context sensing and communication tools, the wearable computer may be able to act as an intelligent assistant. In the near future, the trend-setting professional may wear several small devices, perhaps literally built into their clothes. That way, the person may conveniently check messages, finish a presentation or browse the web while sitting on the subway or waiting in line at a bank. Such wearable devices may enhance the person's memory by providing instant access to important information anytime anywhere. Operating these devices however will be an important issue. Often today's computers require your full attention and both hands to be operated. You have to stop everything you are doing and concentrate on the device [Pentland, 1998]. Using speech for input and output will become more popular but may be quite annoying in many situations. Imagine for example your neighbor on a cross-Atlantic flight constantly talking and chatting with his or her devices.

Although their potential is vast, many of these devices suffer from a common problem: they are mostly oblivious to you and your situation. They don't know what information is relevant to you personally or when it is socially appropriate to "chime in." The goal in solving this problem is to make electronic aids that behave like a well-trained butler or an intelligent assistant. They should be aware of the user's situation and preferences, so they know what actions are appropriate and desirable - a property we call "situation awareness." They should also make relevant information available before the user asks for it and without forcing it on the user - a feature we call "anticipation and availability." An important aspect of a wearable device is that it can perceive the world from a first-person perspective: a wearable camera can see what you see and a wearable microphone can hear what you hear in order to analyze, model and recognize things and people which are around you. A promising direction for interaction with wearable devices is therefore to make the computers more aware of the situation the user is in and to model the user's context. Sensors, such as cameras, mounted to the user's glasses, can recognize what the user is looking at and model what the user is doing. Using sensors of various types, the device can also monitor the user's choices and build a model of his or her preferences. A person may actively train the computer by saying, "Yes, that was a good choice; show me more," or "No, never suggest country music to me." The models can also work solely by statistical means, gradually compiling information about the user's likes and dislikes, and coupling those preferences to the context. For anticipation and availability, the wearable device can take a few key facts about the user's situation to prompt searches through a digital database or the World Wide Web. The information obtained in this manner would then be presented in an accessible, secondary display outside the user's main focus of attention.

The importance of context in communication and interface cannot be overstated. In human-to-human communication contextual information such as physical environment, time of day, mental state, and the model each conversant has of the other participants can be critical in conveying necessary information and mood. Using small body-mounted sensors such as microphones and cameras may enable wearable computers to model and recognize the context of the user and the situation. As processing power increases, a wearable computer can spend more time observing its user to provide serendipitous information, manage interruptions and tasks, and predict future needs without being directly commanded by the user. This contextual information is one way to achieve seamless interaction with the user. We believe that the use of wearable sensors such as head-mounted cameras or wearable microphones combined with software to model and recognize the user's situation and context has the potential to change human-computer interaction fundamentally.

Obviously, a computer interface which uses contextual and situational information to its fullest is more of a long-term goal than what will be addressed in this paper. However, in the following sections we show how computer interfaces may become more contextually aware through machine vision techniques. In this paper we describe two sensory augmented systems. The first system (section 2) uses a head-mounted camera to record and analyze the visual environment of the user as well as to recognize objects the user is looking at. The system can hypothesize which part of the visual environment is interesting to the user and may display information about it when appropriate. The second sensory augmented system (section 3) is a computer vision driven assistant for the real-space game Patrol. The goal of this assistant is to track the wearer's location and current task through computer vision techniques and without off-body infrastructure.

2. RECOGNITION OF OBJECTS USING WEARABLE COMPUTERS

The first example of a sensory augmented computing system is a perceptual remembrance agent, which uses a head-mounted camera to record and analyze the visual environment of the user. In particular a computer vision program recognizes objects in the visual field of view of the user in real-time and displays information the user has associated with them. An application of this sensory augmented wearable system is the museum-gallery guide. A museum is a rich visual environment and is often accompanied with facts and details (from a guide, text or web-page) to be associated with the paintings. For example, as you walk around in a museum you can record video clips of a guide's explanation of the paintings. Such video clips can then be associated with the painting itself so that every time you and the wearable system see the painting again the associated video-clip is replayed. The system has been presented publicly several times including SigGraph 1999 (USA), Darpa Image Understanding Workshop 1998 (USA), Nicograph 1998 (Japan), Heinz-Nixdorf Museum Paderborn Podium 1999 (Germany) and Orbit 2000 (Switzerland) and has been used each time by several hundred people.

An important aspect of the system is that it not only recognizes which painting a user is looking at but also knows how long the user actually looked at it. This piece of information can be used directly in various ways: depending on the duration the user looks at a painting the wearable system may offer to deliver more information about that painting for example by accessing the database of the museum. By assuming that the duration of looking at a painting is correlated with the user's interest and by memorizing which paintings the user looked at, the system may be able to profile the interests of the user. Depending on such profiles the system could then suggest other paintings in the museum. The museum could also attempt to create a database of user-profiles, which could be used to give suggestion to new visitors (depending on their user-profile) or to analyze the organization and effectiveness of a particular exhibition. Even though we have not experimented with the above-mentioned extensions of the system intensively we believe that extensions like these will greatly leverage the usefulness and usability of wearable computing devices.

2.1 Overview of the system

The building blocks of the perceptual remembrance agent are depicted in figure 1. The three main components are the audio-visual associative memory system, the generic object recognition system and the wearable computer interface. See [Schiele et al. 1999] for a detailed description of the system. The audio-visual associative memory system receives object labels along with confidence levels from the object recognition system. If the confidence is high enough, it will retrieve from memory the audio-visual information associated with the object the user is currently looking at and it will overlay this information on the real imagery that the user perceives. This subsystem enables the user to specifically target and shoot footage desired by recording an audio-video clip. After recording such an audio-video clip, the user selects the object that should trigger the clip's playback by directing the head-mounted video camera towards an object of interest and triggering the unit (i.e. pressing a button). The system then instructs the vision module to add the captured image to its database of objects and associate the object's label to its most recent audio-visual clip. Additionally, the user can indicate negative interest in objects, which might get misinterpreted by the vision system as trigger objects (i.e. due to their visual similarity to previously encountered trigger-objects). The primary functionality of the perceptual remembrance agent can be projected on a simple 3-button interface (using for example a wireless 3-button mouse or a simple 3-command speech interface): a record button, an associate button and a garbage button. The record button stores the A/V sequence. The associate button merely makes a connection between the currently viewed visual object and the previously recorded sequence. The garbage button associates the current visual object with a NULL sequence indicating that it should not trigger any play back. Whenever the user is not recording or associating, the system is continuously running in a background mode trying to find objects in the field of view, which have been associated to an A/V sequence. The system acts in consequence as a parallel perceptual remembrance agent that is constantly trying to recognize and explain - by remembering associations - what the user is paying attention to. Figure 1 depicts an example of the process. Here, at an earlier time, an "expert" demonstrated how to program a VCR. The user records the process and then associates the explanation with the image of the VCR body. Thus, whenever the user looks at the VCR he or she automatically sees an animation (overlaid on the left of his field of view) explaining how to use and program the VCR.

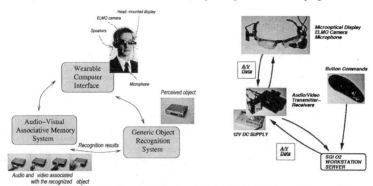

Figure 1: Architecture (left) and hardware (right) of the wearable system

The images sensed by the wearable camera are directly sent to the object recognition system. This system then tries to recognize the objects that the user is looking at. Upon recognition of some object it will send the recognition results (object labels along with confidence levels) to the audio-visual associative memory system. Recognizing objects is one of the most fundamental problems in computer vision and has therefore a long research history. Recognition comes at least in two flavors: "recognition of a cup, a table, a chair" or "recognition of a specific T-shirt, a specific painting". The first case is typically referred to as "classification of objects" whereas the second case is called "identification of objects". Even though humans are very good at classification as well as identification of objects, the classification of objects has proven to be an extremely hard problem by means of machine vision. This is particularly true for unconstrained settings such as using a wearable camera in an arbitrary environment. On the other hand today it is possible to identify objects reliably even with a wearable camera - exactly what is needed for the perceptual remembrance agent.

An important part of the system is the generic object recognition system, which is based on a sound statistical Bayesian framework for object modeling and recognition [Schiele and Crowley 2000]. Objects are represented by multidimensional receptive field histograms of vector responses from local neighborhood operators. A major result of their work is that a statistical representation based on local object descriptors provides a reliable means for the representation and recognition of object appearances. The approach can be used to determine the most probable object, independent of its position, scale and image-plane rotation. The technique is considerably robust to viewpoint changes. The probabilistic recognition algorithm can determine the probability of each object based only on a small portion of the image (15%-30%) and is capable to recognize 100 objects correctly in the presence of viewpoint changes and scale changes. The recognition system runs at approximately 10Hz. In our context this system is used to recognize previously recorded objects and to use the recognized objects as index into the audio-visual memory.

Figure 1 also depicts the major peripherals that are required for the system. Output to the user is rendered via a heads-up display (HUD), which is typically the Micro-Optical clip-on 320x240 VGA device. Attached to the visor is an ELMO video camera, which is aligned as closely as possible with the user's line of sight. Since the user has the option of viewing the world through the camera's point of view, a wide-angle lens was preferred. In addition, a nearby microphone is also incorporated. In the wireless system, audio/video data captured by the system is continuously broadcast using a wireless radio transmitter. The main workstation receives this information, processes the audio and video streams and sends back the processed video and audio for output onto the HUD. All communication occurs via two wireless radio transmitter-receiver pairs (different channels). This wireless transmission connects the user and the wearable system to an SGI O2 workstation where the computer vision algorithm operates. The bi-directional audio-visual connection occurs in real-time (30Hz for the video). The range of the radio frequency (RF) audio-visual link is rated at 900 feet in outdoor conditions but in most indoor and noise conditions we encountered, the user had to remain within 100 feet of the SGI base station. On board batteries permit over 8 hours of continuous operation. All components easily fit into a backpack weighing approximately 5lbs. The PC-based wearable platform, being fully self-contained, allows the user an arbitrary range of movement. However, the rate of processing is slowed down considerably from the SGI due to the reduced computational power of the compact laptop. However, current sub-compact laptops are rapidly approaching the performance levels required for real-time operation and will soon rival the SGI platform.

3. RECOGNITION OF LOCATION AND ACTION OF THE USER WITH WEARABLE CAMERAS

The second sensory augmented wearable computing system is designed for the real-space game called "patrol". Patrol is a game played by MIT students every week in a campus building and provides a scenario to test techniques in less constrained environments. The participants are divided into teams denoted by colored headbands. Each participant starts with a rubber suction dart gun and a number of darts. After proceeding to the second floor to "resurrect" the teams converge on the basement, mezzanine, and first floors to hunt each other. When shot with a dart, the participant removes his headband, and proceeds to the second floor before replacing his headband and returning. While there are no formal goals besides shooting members of other teams, some players emphasize stealth, team play, or holding "territory." Originally, Patrol provided an entertaining way to test the robustness of wearable computing techniques, such as hand tracking for the sign language recognizer [Starner et al. 1998]. However, it quickly became apparent that the gestures and actions in Patrol provided a relatively well-defined language and goal structure in a very harsh "real-life" sensing environment. As such, Patrol became a context-sensing project within itself. Here we shortly discuss some work on determining player location and task using purely on-body sensing.

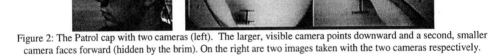

Figure 2: The Patrol cap with two cameras (left). The larger, visible camera points downward and a second, smaller camera faces forward (hidden by the brim). On the right are two images taken with the two cameras respectively.

In this scenario, a two cap-mounted cameras perform sensing. The first camera points downwards to watch the hands and body. The other, smaller camera points forward to observe what the user sees. Each camera is fitted with a wide-angle lens (see Figure 2 for sample images). These cameras are used to determine the location of the player inside the building as well as to model and recognize the action of the player such as aiming, reloading and shooting of the dart guns. This information about the action and location of the player is not only valuable for the wearable computer of the respective player but can be transmitted to other team members. That way other team members are aware of ongoing fights, which team members are involved and how they are positioned to each other. A team strategist may for example deploy this kind of information as appropriate for maintaining territory.

More generally, user location often provides a valuable clue to the user's context. For example, if the user is in his supervisor's office, he is probably in an important meeting and does not want to be interrupted for a phone call or an email unless it is an emergency. By gathering data over many days, the user's motions throughout the day might be modeled. This model may then be used to predict when the user will be in a certain location and for how long. Such information may be invaluable for network caching in the case that the user's wireless network does not provide coverage everywhere on a campus. Several options exist for outdoor positioning such as GPS. However, indoor systems are much less prominent. Active badge systems [Want et al. 1992], [Orwant 1996] and beacon architectures [Long et al. 1996] can trade varying levels of accuracy with the amount of infrastructure that must be installed and maintained. The system described here identifies the user's location solely based on sensing without need for off-body infrastructure.

By identifying the user's current task, the computer can assist actively in that task by displaying timely information or automatically reserving resources that may be needed [Feiner 1993]. However, a wearable computer

812

might also take a more passive role, simply determining the importance of potential interruptions (phone, email, paging, etc.) and presenting the interruption in the most socially graceful manner possible. For example, while driving alone in an automobile, the system might alert the user with a spoken summary. However, during a conversation, the wearable computer may present the name of a potential caller unobtrusively in the user's head-up display.

2.1 Overview of the system

In the Patrol task, location is determined solely based on the images provided by the cameras. Fortunately, the previous known location of the user helps to limit his current possible locations. Location is a first-order Markov process, meaning that the user's next choice of room is limited solely by his current room location. By observing the video stream over several minutes and knowing the physical layout of the building, many possible paths may be hypothesized and the most probable chosen. Prior knowledge about the mean time spent in each area may also be used to weight the probability of a given hypothesis. In order to fully exploit these concepts, a hidden Markov model (HMM) was chosen to represent the environment (see [Rabiner 1989] for HMM implementation details). In the current system, the color variations of three video patches are used to construct a 24-dimensional feature vector in real-time. One patch is taken from the center of the image of the forward-looking camera. A second patch is derived from the downward looking camera in the area just to the front of the player out of range of average hand and foot motion representing the coloration of the floors. Since the nose is always in the same place relative to the downward looking camera, a patch is sampled from the nose. This patch provides hint at the lighting variations of the environment. A 3-state, left-to-right hidden Markov model (HMM) topology is used [Rabiner, 1989]. The patrol environment consists of 14 general areas with two being of very low probability. A grammar limits which area are considered based on the current area. After the model is trained, streams of feature vectors can be submitted to the Viterbi recognizer, which produces the most probable sequence of areas given the model.

In order to determine the user's action the same recognition system as in section 2 is employed. In the context of the Patrol data the system can be used for recognition of image patches, which correspond to particular appearances of a hand, the gun, a portion of an arm, or any part of the background. Feeding the calculated probabilities as feature vectors to a set of hidden Markov models (HMM's) it is possible to recognize different user tasks such as aiming and reloading. Since aiming and shooting are very similar actions, we consider them as the same task. Aiming can be recognized very well, since it is relatively distinctive with respect to reloading and "everything-else". However, reloading and "everything-else" are difficult to distinguish, since the reloading action happens only in a very small region of the image (close to the body) and is sometimes barely visible. See [Starner et al. 1998] for a more detailed description of the system and results.

4. ACKNOLEDGEMENTS

The author would like to express his gratitude to Alex Pentland who provided a very stimulating environment during his stay at the MIT Media Laboratory. In particular the author wants to thank Tony Jebara, Nuria Oliver and Thad Starner, which contributed to the work described in the paper.

REFERENCES

Feiner, S., MacIntryre, B., and Seligmann, D. "Knowledge-Based Augmented Reality", In Communications of the ACM, 36(7):52-62, 1993.

Long, S., Kooper, R., Abowd, G., and Atkeson, C. "Rapid Prototyping of Mobile Context Aware Applications: The Cyperguide Case Study ", In MobiCom, ACM Press, 1996.

Pentland, A. "Wearable Intelligence", In Scientific American presents: Exploring Intelligence, 9(4), 1998.

Rabiner, L. "A Tutorial on Hidden Markov Models", In Proceedings of the IEEE, 77(2):257-286, 1989.

Schiele, B. and Crowley, J.L. "Recognition Without Correspondence using Multidimensional Receptive Field Histograms", In IJCV, International Journal of Computer Vision, 2000, Vol. 26, No. 1, pp. 31-50.

Schiele, B., Oliver, N., Jebara, T., and Pentland, A., "An Interactive Computer Vision System - DyPERS: Dynamic Personal Enhanced Reality System", In ICVS International Conference on Vision Systems, 1999, pp. 51-65.

Starner, T., Schiele, B., and Pentland, A. "Visual Context Awareness in Wearable Computing", In Proceedings of the International Conference on Wearable Computing, 50-57, 1998.

Starner, T., Weaver, J., and Pentland, A. "Real-Time American Sign Language Recognition using Desk and Wearable Computer-Based Video ", In IEEE PAMI, 20(12):1372-1375, 1998.

Want, R. and Hopper, A. "Active Badges and Personal Interactive Computing Objects", In IEEE Trans. on Consumer Electronics, 38(1):10-20, 1992.

Symbiotic Interfaces For Wearable Face Recognition

Bradley A. Singletary and Thad E. Starner

College Of Computing, Georgia Institute of Technology, Atlanta, GA 30332

{bas,thad}@cc.gatech.edu

Abstract

We introduce a wearable face detection method that exploits constraints in face scale and orientation imposed by the proximity of participants in near social interactions. Using this method we describe a wearable system that perceives "social engagement," i.e., when the wearer begins to interact with other individuals. One possible application is improving the interfaces of portable consumer electronics, such as cellular phones, to avoid interrupting the user during face-to-face interactions. Our experimental system proved > 90 % accurate when tested on wearable video data captured at a professional conference. Over three hundred individuals were captured, and the data was separated into independent training and test sets. A goal is to incorporate user interface in mobile machine recognition systems to improve performance. The user may provide real-time feedback to the system or may subtly cue the system through typical daily activities, such as turning to face a speaker, as to when conditions for recognition are favorable.

1   Introduction

In casual social interaction, it is easy to forget the names and identities of those we meet. The consequences can range from the need to be reintroduced to the "opportunity cost" of a missed business contact. At organized social gatherings, such as professional conferences, name tags are used to assist attendees' memories. Recently, electronic name tags have been used to transfer, index, and remember contact information for attendees [Borovoy et al., 1996]. For everyday situations where convention-style name tags are inappropriate, a wearable face recognition system may provide face-name associations and aid in recall of prior interactions with the person standing in front of the wearable user [Farringdon and Oni, 2000, Starner et al., 1997, Brzezowski et al., 1996, Iordanoglou et al., 2000].

Currently, such systems are computationally complex and create a drain on the limited battery resources of a wearable computer. However, when a conversant is socially engaged with the user, a weak constraint may be exploited for face recognition. Specifically, search over scale and orientation may be limited to that typical of the near social interaction distances. Thus, we desire a lightweight system that can detect social engagement and indicate that face recognition is appropriate. Wearable computers must balance their interfaces against human burden. For example, if the wearable computer interrupts its user during a social interaction (e.g. to alert him to a wireless telephone call), the conversation may be disrupted by the intrusion. Detection of social engagement allows for blocking or delaying interruptions appropriately during a conversation.

To visually identify social engagement, we wish to use features endemic of that social process. Eye fixation, patterns of change in head orientation, social conversational distance, and change in visual spatial content may be relevant [Selker et al., 2001, Reeves, 1993, Hall, 1963]. For now, as we are uncertain which features are appropriate for recognition, we induce a set of behaviors to assist the computer. Specifically, the wearer aligns x's on an head-up display with the eyes of the subject to be recognized. As we learn more about the applicability of our method from our sample data set, we will extend our recognition algorithms to include non-induced behaviors.

While there are many face detection, localization, and recognition algorithms in the literature that were considered as potential solutions to our problem [Feraud et al., 2001, Rowley et al., 1998, Schneiderman and Kanade, 2000, Leung et al., 1995], our task is to recognize social engagement in context of human behavior and the environment. Face presence may be one of the most important features, but it is not the only feature useful for segmenting engagement. In examination of 10 standard face databases (> 19,000 images), we found that background contents had little variation. By comparison, scenes obtained from a body-worn camera in everyday life contained highly varied scene backgrounds. In addition to the presence of the face, we would like to exploit the movement of the face with respect to the wearer's camera. Given prior work on the visual modeling of human interaction [Oliver et al., 1998, Ivanov et al., 1999, Moore, 2000, Starner and Pentland, 1998, Starner et al., 1998, Nefian, 1999], we chose hidden Markov Models(HMMs) as the basis of our recognition system.

2 Engagement Dataset

We collected video data from a wearable camera at an academic conference, a setting representative of social interaction of the wearer and new acquaintances. The capture environment was highly unconstrained and ranged from direct sunlight to darkened conference hall. Approximately 300 subjects were captured one or more times over 10 hours. The images in Figure 1 are locations in the video annotated by the wearer to be faces. Our prototype wearable camera video capture system (see Figure 2) consists of: a color camera, an

Figure 1: Representative data set

Figure 2: Marks for user alignment and face capture apparatus

infrared(IR) sensitive black and white camera, a low-power IR illuminator, two digital video(DV) recorder decks, one video character generator, one audio tone generator, a Sony Glasstron head-up display, and four lithium ion camcorder batteries. Output from the cameras is recorded to DV. The HMD augments the user's view with two 'x' characters. The 'x' characters represent known locations for a subject's eyes to appear in the video feed. To capture face data, the wearer of the vest approaches a subject and aligns the person's eyes with the two 'x' characters. The video is then annotated by the user pressing a button which injects an audio tone into the DV stream at the location of the face data.

3 Method

The video data was automatically extracted into 2 second partitions and divided into two classes using frames annotated by the wearer. The two classes were "engagement" and "other". As may be expected, the number of engagement gestures per hour of interaction was much smaller than the number of examples in the garbage class. Since the wearer lined up two x's with the eyes of a viewed subject, the presence of a face could safely be guaranteed to be framed by a 360x360 subregion of the 720x480 DV frame at the annotated locations in the video. Faces present at engagement were large with respect to the subregion. We first convert to greyscale, deinterlace, and correct non-squareness of the image pixels in the subregion. We downsampled the preprocessed region of video to 22x22 images using the linear heat equations to gaussian diffuse each level of the pyramid before subsampling to the next level. Each resulting frame/element in a 2-second gesture example is one 22x22 greyscale subregion (484 element vector). We model the face class

Figure 3: Other and Engagement classes

by a 3 state Left-Right HMM as shown in Figure 3. The other class was much more complex to model and required a 6 state ergodic model to capture the interplay of garbage types of scenes as shown in Figure 3. We plot the mean values of the state output probabilities. The presence of a face seems important for acceptance by the face model. The first state contains a rough face-like blob and is followed by a confused state that likely represents the alignment portion of our gesture. The final state is clearly face-like, with much sharper features than the first state and would be consistent with conversational engagement. Looking at the other class model, we see images that look like horizons and very dark or light scenes. The complexity of the model allowed wider variations in scene without loss in accuracy. Finally, background models could certainly be improved by building location aware models of environment specific features. represented.

4   Results and Evaluation Metrics

Table 1: Accuracy and confusion for engagement detection

experiment	training set	independent test	train confusion, N=843	engagement	other
22x22 video stream	89.71%	90.10%	engagement	82.1%(128)	17.9%(28)
			other	8.6%(63)	91.3%(665)

	test confusion, N=411	engagement	other
	engagement	83.3%(50)	16.7%(10)
	other	8.7%(30)	91.3%(314)

Accuracy results and confusion matrices are shown in Table 1. How effective is leveraging detection of social engagement as compared to continuously running face recognition? If we were to construct a wearable face recognition system using our engagement detector, we would combine the social engagement detector with a scale-tuned localizer and a face recognizer. The cost of the social engagement detector must be sufficiently small to allow for the larger costs of localization and recognition. This is described by the inequality

$$z - R_a * a \geq R_b * b$$

where $z := 1$ is the total resources available, a is the fixed cost of running engagement detection once in sec/frames, b is the fixed cost of running localization and recognition methods once in sec/frames, and R_a and R_b are the rate at which we can supply the respective detectors with frames in frames/sec, respectively.

815

However, R_b has a maximum value determined by either the fraction of false positives U_{fp} multiplied by the maximum input frame rate or the rate at which the user wants to be advised of the identity of a conversant R_{ui}. Thus,

$$R_b * b \geq max\{R_a * U_{fp}, R_{ui}\} * b$$

Note that fixating the camera on a true face could cause up to R_a frames per second to be delivered to the face recognizer. However, we assume that the user does not want to be updated this quickly or repeatedly (i.e. $R_{ui} << R_a$). We also assume that our rate of false positives will almost always be greater than the rate the user wants to be informed, leaving us with

$$1 - R_a * a \geq R_a * U_{fp} * b$$

For comparison purposes, we will assume that the average time per frame of processing for the localization and recognition process can be represented by some multiple of the average detection time (i.e. $b = c * a$). Thus, for a given multiplier c, we can determine the maximum rate of false positives allowable by the face detection process.

$$U_{fp} \leq \frac{1}{R_a * a * c} - \frac{1}{c}$$

Note that if $c \leq 1$, then the localization and recognition process runs faster than the face detection process. This situation would imply that performing face detection separately from face localization and recognition would not save processing time (i.e. localization and recognition should run continually - again, if real-time face recognition is the primary goal).Given a false positive rate U_{fp}, we solve the equation to determine the maximum allowable time for the localization and recognition process as compared to the detection process.Thus, we have a set of heuristics for determining when the separation of face detection and face localization and recognition is profitable.

$$c \leq \frac{1}{R_a * a * U_{fp}} - \frac{1}{U_{fp}}$$

5   Conclusion

Applying the metric from the previous section to our experimental results, we let $U_{fp} = .13$, $R_a = 30$, $a = \frac{1}{60}$ and solving for c we get $c \leq 7.69$. Thus any recognition method used may be up to 7.69 times slower than the engagement detection method and will have a limiting frame rate of about four frames per second. Given that our detection algorithm runs at 30fps, and our knowledge that principal component analysis based face recognition and alignment can run faster than roughly four times a second, we feel that engagement detection can be a successful foundation for wearable face recognition. Post-filtering outputs of detection may help eliminate false positives before recognition [Feraud et al., 2001]. Due to the face-like appearance of the final state of the HMM, it is likely that the output of our method could provide a reasonable first estimate of location to fine grain localization. Other cues including detection of head stillness, eye fixation, and conversational gestures like "hello, my name is ..." will likely reduce false positives[Reeves, 1993, Selker et al., 2001].

We described a platform built to capture video from a wearable user's perspective and detailed a method for efficient engagement detection. We tested our system in a representative scenario and devised a metric for evaluating it's efficacy as part of a face recognition scheme. In doing so, we demonstrated how the design of user interfaces that are aware of social contexts and constraints can positively affect recognition systems on the body. Finally, we have described how the detection of social engagement may be used, in its own right, to improve interfaces on portable consumer devices.

References

[Borovoy et al., 1996] Borovoy, R., McDonald, M., Martin, F., and Resnick, M. (1996). Things that blink: A computationally augmented name tag. *IBM Systems Journal*, 35(3).

[Brzezowski et al., 1996] Brzezowski, S., Dunn, C. M., and Vetter, M. (1996). Integrated portable system for suspect identification and tracking. In DePersia, A. T., Yeager, S., and Ortiz, S., editors, *SPIE:Surveillance and Assessment Technologies for Law Enforcement*.

[Farringdon and Oni, 2000] Farringdon, J. and Oni, V. (2000). Visually augmented memory. In *Fourth International Symposium on Wearable Computers*, Atlanta, GA. IEEE.

[Feraud et al., 2001] Feraud, R., Bernier, O. J., Viallet, J.-E., and Collobert, M. (2001). A fast and accurate face detector based on neural networks. *Pattern Analysis and Machine Intelligence*, 23(1):42–53.

[Hall, 1963] Hall, E. T. (1963). *The Silent Language*. Doubleday.

[Iordanoglou et al., 2000] Iordanoglou, C., Jonsson, K., Kittler, J., and Matas, J. (2000). Wearable face recognition aid. In *Interntional Conference on Acoustics, Speech, and Signal Processing*. IEEE.

[Ivanov et al., 1999] Ivanov, Y., Stauffer, C., Bobic, A., and Grimson, E. (1999). Video surveillance of interactions. In *CVPR Workshop on Visual Surveillance*, Fort Collins, CO. IEEE.

[Leung et al., 1995] Leung, T. K., Burl, M. C., and Perona, P. (1995). Finding faces in cluttered scenes using random labelled graph matching. In *5th Inter. Conference on Computer Vision*.

[Moore, 2000] Moore, D. J. (2000). *Vision-based recognition of actions using context*. PhD thesis, Georgia Institute of Technology, Atlanta, GA.

[Nefian, 1999] Nefian, A. (1999). *A hidden Markov model-based approach for face detection and recognition*. PhD thesis, Georgia Institute of Technology, Atlanta, GA.

[Oliver et al., 1998] Oliver, N., Rosario, B., and Pentland, A. (1998). Statistical modeling of human interactions. In *CVPR Workshop on Interpretation of Visual Motion*, pages 39–46, Santa Barbara, CA. IEEE.

[Reeves, 1993] Reeves, J. (1993). The face of interest. *Motivation and Emotion*, 17(4).

[Rowley et al., 1998] Rowley, H. A., Baluja, S., and Kanade, T. (1998). Neural network-based face detection. *IEEE Transactions on Pattern Analysis and Machine Intelligence*, 20(1).

[Rungsarityotin and Starner, 2000] Rungsarityotin, W. and Starner, T. (2000). Finding location using omni-directional video on a wearable computing platform. In *International Symposium on Wearable Computing*, Atlanta, GA. IEEE.

[Schneiderman and Kanade, 2000] Schneiderman, H. and Kanade, T. (2000). A statistical model for 3d object detection applied to faces and cars. In *Computer Vision and Pattern Recognition*. IEEE.

[Selker et al., 2001] Selker, T., Lockerd, A., and Martinez, J. (2001). Eye-r, a glasses-mounted eye motion detection interface. In *to appear CHI2001*. ACM.

[Starner et al., 1997] Starner, T., Mann, S., Rhodes, B., Levine, J., Healey, J., Kirsch, D., Picard, R. W., and Pentland, A. (1997). Augmented reality through wearable computing. *Presence special issue on Augmented Reality*.

[Starner and Pentland, 1998] Starner, T. and Pentland, A. (1998). Real-time American sign language recognition using desktop and wearable computer based video. *Pattern Analysis and Machine Intelligence*.

[Starner et al., 1998] Starner, T., Schiele, B., and Pentland, A. (1998). Visual contextual awareness in wearable computing. In *International Symposium on Wearable Computing*.

[Sung and Poggio, 1998] Sung, K. K. and Poggio, T. (1998). Example-based learning for view-based human face detection. *Pattern Analysis and Machine Intelligence*, 20(1):39–51.

817

Experiments In Interaction Between Wearable and Environmental Infrastructure Using the Gesture Pendant

Daniel Ashbrook, Jake Auxier, Maribeth Gandy, and Thad Starner
College of Computing and Interactive Media Technology Center
Georgia Institute of Technology
Atlanta, GA 30332-0280 USA
{anjiro, jauxier1, maribeth, thad}@cc.gatech.edu

Abstract

The Gesture Pendant, a computer vision system worn as a piece of jewelry, allows the wearer to control electronic devices in the environment through simple hand gestures. Gestures provide an advantage over traditional device interfaces (such as remote controls) in that they are easily used by all people, including those with such disabilities as loss of vision, motor skills, and mobility. The Gesture Pendant can also be used for medical monitoring—as the user makes gestures, the pendant can analyze the movement of the hands to detect certain tremors, the frequency of which can indicate the progress of some diseases such as Parkinson's.

1 Introduction

The goal of the Gesture Pendant is to allow the wearer to control elements in the home via hand gestures. Devices such as home entertainment equipment and the room lighting can be controlled with simple movements of the hand. Building on previous work that used a wearable camera and computer to recognize American Sign Language (ASL) [Starner et al., 1998], we have created a system that consists of a small camera worn as a part of a necklace or pin. The video from the camera is analyzed and pre–defined gestures are detected, which trigger devices connected to various home automation systems. Thus, the wearer can simply raise or lower a flattened hand to control the light level and can control the volume of the stereo by raising or lowering a pointed finger.

2 Motivation and Interaction

Why do we want to use hand gestures to control home automation systems? Automation offers many benefits to the user, especially the elderly or disabled; however, the interfaces to these systems are generally poor. The most common interface to a home automation system such as X10 is a remote control with small, difficult to push buttons with cryptic text labels that are hard to read. This interface also relies on the person having the remote control with them at all times. Portable touchscreens are emerging as a popular interface, however they present many of the same problems as remotes, with the additional difficulty that the interface is now dynamic and harder to learn. Other interfaces include wall panels, which require the user to go to the panel location to use the system, and phone interfaces, which still require changing location and pressing small buttons.

While speech recognition has long been viewed as the ultimate interface for home automation, there are many problems in this domain. In a house with more than one person, a speech interface could result in a disturbing amount of noise, as all the residents would be constantly talking to the house. Also, if the resident is listening to music or watching a movie, she would have to speak very loudly to avoid being drowned out by the stereo or television. Ambient noise can also cause errors in the speech recognition systems.

One potential problem with using gestures to control devices is the need to contrive enough gestures, or having too many gestures to remember. This, of course, defeats the purpose of having a simple interface. To address this problem, we have experimented with making the pendant system context–sensitive. Listed below are several possible combinations of the pendant and context.

1. **The pendant alone**

   With no context sensitivity, each device must have a distinct gesture for every function the user wishes to control. For example, to adjust the stereo volume, the temperature of the thermostat, and the level of lighting in the room, three different gestures would be required.

2. **The pendant and speech recognition**

   By using speech input, the user can select from a set of devices that all share the same gestures. For example, the user could have a single gesture to control the volume, the thermostat and the light. Before performing the gesture, the user would indicate the desired device by speaking its name. This, of course, would present many of the previously discussed drawbacks of speech recognition, except that the system would be less speech dependent than one which was solely speech–controlled.

3. **The pendant and orientation**

   Another way to select between several devices is to use the physical orientation of the pendant. To choose a target to control, the user faces the desired device. This method of selection can be accomplished by placing the transmitter that sends remote control codes on the pendant rather than near the device being controlled. By limiting the direction of transmission, the signals will only reach the device that the wearer is facing. This method, however, can waste battery power, in that it requires the pendant to send remote control codes for all devices associated with that particular gesture, as it has no way of knowing which device the user is actually attempting to control. One solution for this is to place fiducials on each device, and have the pendant camera recognize what is being controlled. Using fiducials also provides an opportunity for device–specific feedback—if the pendant system knows what device it's facing, it can inform the user by illuminating a light on the device, for example. The biggest downside to using orientation as a source of context is that the user must move from device to device; this could obviously be a problem for the elderly or handicapped.

4. **The pendant and location**

   By using room–level tracking, a set of gestures may be defined on a room–by–room basis. This would require the user to move to the room where the target device is located before performing the gesture. For multiple devices in a room, distinct gestures would be required for each. While this scheme still requires the user to move, it may be reasonable to assume that the wearer will not want to control devices in other rooms.

Such sources of context could be combined in any permutation to create an ideal system for the individual. To date we have tested configurations 1, 2 and part of 3 and are exploring adaptation of our software to recognize fiducials.

3   As an enabling technology

All types of people can use the Gesture Pendant. However, the difficult–to–use interfaces of home automation equipment that make the Gesture Pendant useful for healthy adults make the pendant doubly useful for the elderly. Despite any number of physical impairments that may limit the user's motor skills, mobility, or sight, the Gesture Pendant can still be used, especially for assistive tasks such as opening and closing doors, using appliances, and accessing emergency systems.

The same interface problems are faced by those with disabilities such as cerebral palsy and multiple sclerosis. However, a study has shown that even people with extremely impaired motor skills due to cerebral palsy are able to make between 12 and 27 distinct gestures [Roy et al., 1994], which could be used as input to the gesture pendant. Therefore, the Gesture Pendant can be an interface alternative that can allow people who are unable to use some of the more traditional interfaces to take advantage of the independence that home automation can afford them.

4 Medical Monitoring

As a user makes movements in front of the Gesture Pendant, the system can not only look for specific gestures but can also analyze how the user is moving. Therefore, a second use of the Gesture Pendant is as a monitoring system rather than as an input device—as the user makes a gesture, the pendant can detect the presence and frequency of a hand tremor.

As discussed above, the target populations for the Gesture Pendant are the elderly and disabled. Many of the diseases that these populations suffer have a pathological tremor as a symptom. A pathological tremor is an involuntary, rhythmic, and roughly sinusoidal movement [Elble and Koller, 1990]. These tremors can appear in a patient due to disease, aging, and drug side effects; these tremors can also be a warning sign for emergencies such as insulin shock in a diabetic. Currently, we are interested in recognizing essential tremors (4-12 Hz) and Parkinsonian tremors (3-5 Hz) [Elble and Koller, 1990], since determination of the dominant frequency of the tremor can be helpful in early diagnosis and therapy control of such disorders [Hefter et al., 1989].

The medical monitoring of tremors can serve several purposes. The data can simply be logged over days, weeks, or months for use by the doctor as a diagnostic aid. Upon detecting a tremor or a change in the tremor, the user might be reminded to take medication, or the physician or family members could be notified as appropriate. Tremor sufferers who do not respond to pharmacological treatment can have a deep brain stimulator implanted in their thalamus [Hubble et al., 1996]. This stimulator can help reduce or eliminate the tremors, but the patient must control the device manually. The Gesture Pendant data could be used to provide automatic control of the stimulator. Another area in which tremor detection would be helpful is in drug trials. The subjects involved in these studies must be closely watched for side-effects and the pendant could provide day-to-day monitoring.

5 Gesture Pendant Hardware and Software

The motivation behind the Gesture Pendant calls for a small, lightweight wearable device. At first we considered a hat mount, but concluded that gestures would be too hard to recognize if made in front of the body and difficult to perform if made in front of the hat. Due to the off-the-shelf nature of the components (leading to larger size and heavier weight than ideal), we decided that a pendant form was ideal. Using custom-made parts, the hardware could be shrunk considerably, and other form factors such as a brooch or, assuming sufficient miniaturization, a shirt button or clasp could be possible.

Since the goal of the Gesture Pendant was to detect and analyze gestures quickly and reliably, we decided upon an infrared illumination scheme to make color segmentation less computationally expensive. Since black and white CCD cameras pick up infrared well, we used one with a small form factor (1.3" square) and an infrared-pass filter mounted in front of it. To provide the illumination, we used sixteen near-infrared LEDs in a ring around the camera (Figure 1). The first incarnation had a lens with a roughly 90 degree field of view, but that proved to limit the gesture space too much. A wider angle lens of 160 degrees worked much better, despite the fisheye effect. The design of the pendant is similar to the Toshiba "Motion Processor", which uses a camera and IR LEDs as an input to a desktop gesture–based interaction system [Toshiba, 1998].

The eventual goal is to incorporate all components of the gesture pendant into one wearable device; however, for the sake of rapid prototyping we used a desktop computer to do the bulk of the image processing. This also allowed us to centralize the control system, by using standard home automation devices such as the Slink-E (a computer–controlled universal remote) and X10. To send the video to the desktop, we used a 900 MHz video transmitter/receiver pair. The transmitter is powerful enough that cordless 900 MHz phones do not interfere with it, and the receiver can be tuned to a range of channels to avoid conflicting signals from multiple pendants.

Since one of the groups that we feel could most benefit from the Gesture Pendant is the elderly, it is important to make it as unobtrusive as possible. This means it must be inconspicuous, lightweight, and

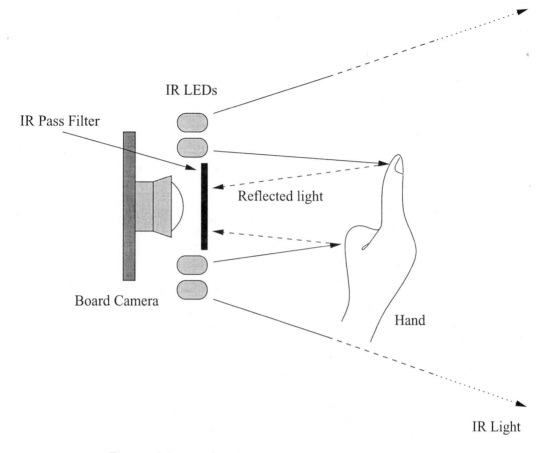

IR LEDs

IR Pass Filter

Reflected light

Board Camera

Hand

IR Light

Figure 1: Side view of pendant with IR reflecting off the user's hand

uncomplicated. Also, since the Gesture Pendant is a wearable device, and one that is constantly in full view, it will be important to make it more attractive. The ring of LEDs makes it appear somewhat jewelry-like, but with a smaller form-factor and some principles of design applied to it, it will become more appealing.

The gesture recognition system incorporates two kinds of gestures: control gestures and user-defined gestures. Control gestures provide continuous control of a device that varies with the magnitude of the gesture, and are recognized with an accuracy of 95%. Control gestures would be useful for devices that provide a range of adjustment, such as the volume of a stereo or the brightness of a light. User defined gestures provide discrete output for a single gesture, and are most useful for on–off tasks such as opening a door. User defined gestures are detected with an accuracy of 97%. The gesture algorithm is described in more detail elsewhere [Starner et al., 2000].

6 Future Work

The current implementation of the Gesture Pendant uses a wireless transmitter to send the video data to a desktop PC where it is analyzed and automation commands are issued. The next step in our work is to place all of the computation onto the body in the form of a wearable computer and eliminate the need for a desktop machine.

The monitoring of tremors and motor skills could be expanded to do more complex analyses of the types of tremors in 3D. For example, Parkinson's sufferers often exhibit a complex "pill rolling" tremor, which could be detected and analyzed. We could also determine more characteristics of the user's motor skills, such as slowness of movement or rigidity, that could indicate the onset of stroke or Parkinson's. We could also design the gestures so that, while they would be used to control devices in the house, they would optimally reveal features of the user's manual dexterity and movement patterns.

Another, more advanced use for the pendant in terms of monitoring would be to observe more about the wearer's activities. For example the pendant could take note of when the user eats a meal or takes medication. It could keep a record of the general activity level of the wearer or notice if she falls down. This would further our goal of providing services for the elderly and disabled that allow them increased independence in the home.

7 Conclusion

We have demonstrated a wearable gesture recognition system that can be used in a variety of lighting conditions to control home automation. Through the use of a variety of contextual cues, the Gesture Pendant can disambiguate the devices under its control and limit the number of gestures necessary for control. We have shown how such a device may have enough merit to be used as a convenience by the elderly but also provide additional functionality as a medical diagnostic.

8 Acknowledgments

Funding for the project was provided in part by the Georgia Tech Broadband Institute, the Aware Home Research Initiative, the Georgia Tech Research Corporation, and the Graphics, Visualization, and Usability Center. Special thanks to Rob Melby and David Minnen for the blob tracking software.

References

[Elble and Koller, 1990] Elble, R. J. and Koller, W. C. (1990). *Tremor.* Johns Hopkins UP, Baltimore, MD.

[Hefter et al., 1989] Hefter, H., Homberg, V., and Freund, H. J. (1989). Quantitative analysis of voluntary and involuntary motor phenomena in parkinson's disease. In Przuntek, H. and Riederer, P., editors, *Early Diagnosis and Preventative Therapy in Parkinson's Disease.* Springer-Verlag Wien, New York, NY.

[Hubble et al., 1996] Hubble, J., Busenbark, K., and Wilkinson, S. (1996). Deep brain stimulation for essential tremor. In *Neurology*, volume 46, pages 1150–1153.

[Roy et al., 1994] Roy, D. M., Panayi, M., Erenshteyn, R., Foulds, R., and Fawcus, R. (1994). Gestural human-machine interaction for people with severe speech and motor impairment due to cerebral palsy. In *Conference on Human Factors in Computing Systems*, Boston, MA.

[Starner et al., 2000] Starner, T., Auxier, J., Ashbrook, D., and Gandy, M. (2000). The gesture pendant: A self-illuminating, wearable, infrared computer vision system for home automation control and medical monitoring. In *IEEE Intl. Symp. on Wearable Computers*, Atlanta, GA.

[Starner et al., 1998] Starner, T., Weaver, J., and Pentland, A. (1998). Real-time American Sign Language recognition using desk and wearable computer-based video. *IEEE Trans. Patt. Analy. and Mach. Intell.*, 20(12).

[Toshiba, 1998] Toshiba (1998). Toshiba's motion processor recognizes gestures in real time. Available at: http://www.toshiba.com/news/980715.htm.

Conversational Speech Recognition for Creating Intelligent Agents on Wearables

Benjamin A. Wong and Thad E. Starner
Future Computing Environments Group,
College of Computing and GVU Center
Georgia Institute of Technology, Atlanta, GA, USA 30332-0280

{bbb, thad}@cc.gatech.edu

Abstract

Paper-based planners and calendar programs run on personal digital assistants are common methods of reminding users of tasks and appointments. Recording an entry using either method requires a certain amount of time. Is it worthwhile (or even possible) to try to create an appointment scheduling system which takes significantly less time and effort? We argue that it is, and we describe one such system currently being built, its potential benefits, and some of the issues it must overcome. This system, called the Calendar Guardian Agent (CGA), runs on a wearable computer to assist in scheduling appointments based on speech captured from a user's everyday conversations.

1   Introduction

There are a variety of methods to remind people of tasks and appointments. These methods include paper-based planners, such as the Franklin Day Planner™, and calendar programs run on personal digital assistants (PDAs), such as the Palm Pilot™. Often these scheduling systems are used during a conversation in which a person is orally scheduling a meeting: "I'll meet you at 3pm tomorrow." However, using these current scheduling systems can incur a significant overhead. There are other methods of scheduling which do not require a large overhead in time, for example, using a human assistant or just trying to memorize an appointment. However those methods have considerable trade-offs in cost and reliability, respectively.

Wearable computing's capability for continuous interaction permits new techniques to facilitate everyday life and social interaction. One of our latest projects, the Calendar Guardian Agent (CGA), is a system to schedule appointments using an agent on a wearable computer which attends to the user's everyday conversations. Although we use techniques similar to a human assistant and memorization to reduce the overhead of scheduling, we expect the costs in terms of monetary expense and unreliability to be reasonable. Note that while portions of the CGA have been completed, the system described in this paper is still under development.

2   Motivation

2.1   Can scheduling be more efficient?

An interesting comparison to make for any scheduling system is to consider that system's convenience versus a hypothetical *human* assistant. For example, a personal secretary can be non-intrusive but always listening, ready to jot down another appointment or remind one of an upcoming event. Contrast this method with the amount of time spent to operate a PDA or paper-based planner and the convenience of real personal assistant becomes clear.

In an informal study of both paper-based planners and PDAs, we found that for appointments that take around 80 seconds to schedule, over three-fourths of the time was spent in the overhead of managing the scheduling systems themselves. We simulated scheduling events by having five pairs of people orally arrange to meet at a specified date six months in the future. The initiator used a script and did not actually spend any time scheduling; the other used their ordinary scheduling system (PDA or paper-based planner). As a baseline we also simulated and timed scheduling events where each party assumed they had a personal assistant listening in, taking care of remembering the details. The average result of that test was around 20 seconds. To calculate a lower-bound for the overhead we found the difference in the time it took to schedule an appointment using a human assistant versus a traditional system. Our study was too crude to show any significant difference between PDAs and paper-based planners. Note that using a human assistant for scheduling achieves a lower overhead but greatly increases the monetary expense.

Once the burden of using a scheduling system rises above a certain threshold, a busy person might resort to memorizing appointments so that she can enter them later when scheduling can be her primary task. This happens despite the unreliability of human memory because of the low cost in time. In general, small increases in system delays can trigger large drops in usability [Rhodes, 2000, Shneiderman, 1998].

2.2 How overhead might be reduced

One of the key features that separates both a personal human assistant and memorization from a PDA or paper-based planner is the speed with which the "knowledge in the head" (to use Don Norman's term, [Norman, 1988]) is transfered to "knowledge in the world". Norman describes "knowledge in the head" as information known by the user, in this case, the appointment details the user knows. Whereas "knowledge in the world" is information that is stored in the environment, i.e., the physical artifact in the world that helps the user to remember. Both PDAs and paper-based planners require the user to explicitly transfer the knowledge from her head to the world. A personal assistant, on the other hand, can *listen in* on the conversation so that ideally nothing more, besides a confirmation, need be done. In the case of memorization of appointments, the transfer is "quick" because it is *postponed* until a more opportune time.

3 Method

3.1 Calendar guardian agent

The techniques of *listening in* and *postponing transfer* for reducing scheduling overhead can be used to describe a new agent for scheduling appointments. This agent, the CGA, will listen to conversations and attempt to assist the user through scheduling the appointment like a human assistant would. If the user is busy, the CGA can be ignored and it will patiently forward the scheduling to a more convenient time as if the user had committed it to memory.

We are currently building a prototype version of the CGA. The initial system has been built around a SaintSong e-PC Pentium III 700MHz brick-computer running UNIX. The system is carried in a vest by the first author through out the day. A MicroOptical heads-up display is used to place a screen in front of the user so that alerts and feedback during scheduling can be received immediately. We found that a noise-cancelling boom microphone, although bulky, is necessary to get good results when using the IBM ViaVoice speech recognition libraries. One important point is that the touchpad for mouse control has been placed very close to the resting position of the user's hand. This means the mouse can be used without any set-up time while sitting, standing, or even walking.

3.2 Dialog tabs—listening and postponing

The graphical portion of the user interface we are building displays the words the user has recently said for visual feedback. The appointment detection module which performs the *listening in* function is not yet complete on our prototype. But once it is, as the user speaks appointment times during normal conversation, a form of dialog box, which we are calling a "dialog tab", will pop up. It is non-modal and appears as a small (two pixel wide) tab on the right side of the heads-up display.

If the user is not paying attention to the screen at the moment, these dialog tabs are too small to be distracting. However, that doesn't mean they are difficult to use. We have taken advantage of Fitts's law by placing the tabs on the edge of the screen rather than the center [Walker and Smelcer, 1990]; if the user wants to click on one it should be easy to do so even in poor motor control situations such as while walking and talking. Additionally, the tabs are stacked in order of arrival and more recent tabs are longer; this make the last uttered appointment especially easy to hit.

Dialogs may be dismissed, without even opening them, by a right click. Hovering over a tab displays the date discussed in the user's conversation. A left click will immediately open the calendar to the mentioned date and time allowing the user to fill-in or correct any fields before committing the appointment to the reminder system. If the user ignores the tabs, they remain on the side of the screen until the user has the time to make scheduling her primary task. In this way, the user can *postpone transfer* of the knowledge in her head that is needed to finalize the appointment. However, unlike plain memorization, the user will be less likely to forget about scheduling appointments deferred in this way because the dialog tabs will jog her memory when she takes a free moment to glance down.

4 Challenges

There are three major issues that must be addressed. Voice recognition can have significant error rates, other's privacy must be respected, and active agents, when poorly implemented, can increase cognitive load.

4.1 Voice recognition accuracy

Current voice recognition techniques, while designed for dictation in an office environment, suffer severe degradation in mobile environments. For example, our system loses accuracy when given the rapid, clipped speech used during conversations. It also suffers from a high number of spurious insertion errors ("false alarms") from ambient noises, such as wind, and can become unusable even with a noise-cancelling microphone. We do not believe there is a single solution to the problem of unreliable voice recognition in an unconstrained environment. Instead we will investigate using several techniques in concert.

If the user is aware that she has just or is just about to orally schedule an appointment she can assist the recognition in the following three ways. First, the user can manually signal the system to "pay closer attention" (that is, to lower the normally high threshold of confidence required to detect a scheduling event). Second, the user has constant visual feedback available of what the voice recognition system thinks she is saying and can repeat misrecognized phrases. Third, the user can adapt her speaking patterns to the grammar she knows has worked with the CGA in the past.

To mitigate the effects of imperfect voice recognition, we are designing the CGA's user interface to have a low cost when mistakes occur. The display should be non-distracting, and it should be easy for the user to dismiss spurious errors or correct entries that are partially right. One of the standard techniques to lower the probability of voice recognition errors is to use language model subsetting [Schmandt, 1994] to restrict the application's domain with a grammar of appointments. Speech recognition software often allows a programmer to examine the "confidence levels" for the individual words recognized. Another standard technique is to set a high threshold on the recognition system to throw out questionable words. A possible path of research is to use a limited version of "topic spotting" so that the recognition system can detect conversations in which it is likely that an appointment will be scheduled, and lower the threshold accordingly.

4.2 Privacy

While some researchers have suggested that wearables can be used to protect a user's privacy [Feiner, 1999], a wearable with recording devices may invade other's privacy [Strubb et al., 1998]. It is important to address these privacy concerns in the design of the system. Even if privacy were not an issue in itself, audio recording of conversations without consent is a legally murky issue. In fact, fourteen states in the United States have laws requiring the consent of all parties in certain situations. Since our system is meant for everyday life, it is almost certain to be used in situations where conversers should have a reasonable expectation of privacy.

The primary method we use to preserve privacy is a "noise-cancelling" microphone which attenuates speech from other people to an essentially inaudible level. We are also currently only saving the text transcript from the voice recognition system; this removes the possibility of later amplifying the audio recording to discern the speech of others. However, if the CGA only has one side of the conversation, some scheduling events can be much more difficult to detect correctly (or at all). We are using two methods to mitigate this. First, the user can assist the CGA by repeating key scheduling times and important points that another speaker has put forth. Results of preliminary tests of this technique are encouraging: since repeating what another person said is a standard social custom for confirming understanding, few people realize that the user may be repeating the appointment for the benefit of the wearable. If we assume the other party also has a networked wearable computer performing speech recognition, the two wearables can negotiate to swap transcripts. The model of swapping transcripts instead of recording other people directly may also improve voice recognition accuracy because each user will have a fixed microphone position and a speech recognition system trained to their voice.

4.3 Cognitive load and Attention

Cognitive load refers to the total amount of mental activity imposed on working (short-term) memory at an instance in time [Sweller, 1994]. The major factor that contributes to cognitive load is the number of elements that need to be attended to. Miller [Miller, 1956], gives the threshold of working memory to be 7 ± 2 items. Because working memory is so limited, the CGA must present an extremely low cognitive-load interface to the user. That is, the CGA must keep out of the user's working memory (as far as possible) even though it is potentially visible during all waking hours of the user's life.

To reduce the cognitive burden our system is designed to be easy to ignore and dismiss. The system will show new information only rarely. Additionally the system uses a "ramping interface" [Rhodes, 2000] in the form of dialog tabs. The dialog tabs show progressively more information to the user as she indicates more interest but always provide an escape mechanism through a right-click. Using a ramping interface also has the benefit of making errors in voice recognition less costly. Since the tabs are so easily dismissed, the user should be able to manage a higher number of false-alarms from the appointment detection module.

5 Related Work

LookOut [Horvitz, 1999] is an agent which parses the text in the body and subject of an email message, identifies a date and time associated with an event and attempts to fill in relevant fields in an appointment book. The system displays its guesses to the user and allows the user to edit its guesses and to save the final result. The LookOut system gives valuable insight into designing an agent interface that can manage uncertainties about the user's goals by collaborating with the user. However, because e-mail messages are already postponable, LookOut does not offer the same benefits as the CGA which works in the domain of oral conversations to help the transfer of knowledge during high cognitive load situations.

Verbmobil [Wahlster, 2000], created by a large consortium from academia and industry, is a speech-to-speech translation system for spontaneous dialogs in mobile situations. Verbmobil can operate in different domains with appointment negotiation being the most relevant to the CGA. Although Verbmobil does not perform any of the scheduling functions of the CGA, it does demonstrate the feasibility of processing spontaneous speech to retrieve appointment scheduling information [Kipp et al., 1999].

CybreMinder [Dey and Abowd, 2000] allows users to create a reminder message to be delivered using a mobile device when an associated situation has been satisfied. A similar project, the Memory Glasses [DeVaul et al., 2000], is an attempt to build a wearable, proactive, context-aware memory aid based on wearable sensors. Both of these projects are good complements to the CGA; they can deliver the appointment reminders which the CGA generates.

6 Conclusion and Future Directions

We have shown an outline for a wearable computer agent we are creating to assist with scheduling appointments. Scheduling, as a task, could be made significantly faster through the use of human memory or a human assistant, but the trade-offs in unreliability or monetary expense are too high. However, we can apply the techniques of "listening in" and "postponing transfer" to create an agent that may strike a balance between the costs.

	Technique	Cost
PDA	—	Time
Human Memory	Listening in	Money
Human CGA	Postponing Both	Reliability Some of each

This proposed system has three major hurdles to clear: voice recognition accuracy, privacy, and added cognitive load. We have described methods for addressing each hurdle. Future work includes completing the appointment detection module and evaluating the system as a whole.

References

[DeVaul et al., 2000] DeVaul, R. W., Clarkson, B., and Pentland, A. (2000). The Memory Glasses: Towards a wearable context aware, situation-appropriate reminder system. In CHI 2000 Workshop on Situated Interaction in Ubiquitous Computing.

[Dey and Abowd, 2000] Dey, A. K. and Abowd, G. D. (2000). Cybreminder: A context-aware system for supporting reminders. In *Proceedings of Second International Symposium on Handheld and Ubiquitous Computing, HUC 2000*, pages 172–186. Springer-Verlag.

[Feiner, 1999] Feiner, S. K. (1999). The importance of being mobile: some social consequences of wearable augmented reality systems. In *(IWAR '99) Proceedings. 2nd IEEE and ACM International Workshop on Augmented Reality*, pages 145–148. IEEE.

[Horvitz, 1999] Horvitz, E. (1999). Principles of mixed-initiative user interfaces. In *Proceedings of CHI'99*, pages 159–166. ACM, Addison-Wesley.

[Kipp et al., 1999] Kipp, M., Alexandersson, J., and Reithinger, N. (1999). Understanding spontaneous negotiation dialogue. In *Proceedings of the IJCAI-99 Workshop on Knowledge and Reasoning in Practical Dialogue Systems*, pages 57–64. International Joint Conference on Artificial Intelligence.

[Miller, 1956] Miller, G. A. (1956). The magical number seven plus or minus two : Some limits on our capacity for processing information. *Psychological Review*, 63:81–97.

[Norman, 1988] Norman, D. (1988). *The Psychology of Everyday Things*. Harpercollins.

[Rhodes, 2000] Rhodes, B. J. (2000). *Just-In-Time Information Retrieval*. PhD thesis, MIT Media Laboratory, Cambridge, MA.

[Schmandt, 1994] Schmandt, C. (1994). *Voice Communication with Computers: Conversational Systems*. Van Nostrand Reinhold.

[Shneiderman, 1998] Shneiderman, B. (1998). *Designing the User Interface: Strategies for Effective Human-Computer Interaction*. Addison-Wesley, Reading, MA, 3rd ed edition.

[Strubb et al., 1998] Strubb, H., Johnson, K., Allen, A., Bellotti, V., and Starner, T. (1998). Privacy, wearable computers, and recording technology. Panel discussion, The Second International Symposium on Wearable Computers, October 19–20, 1998, Pittsburgh, PA.

[Sweller, 1994] Sweller, J. (1994). Cognitive load theory, learning difficulty and instructional design. *Learning and Instruction*, 4:295–312.

[Wahlster, 2000] Wahlster, W., editor (2000). *Verbmobil: Foundations of Speech-to-Speech Translation*. Springer-Verlag. Verbmobil is a speaker-independent and bidirectional speech-to-speech translation system for spontaneous dialogs in mobile situations.

[Walker and Smelcer, 1990] Walker, N. and Smelcer, J. (1990). A comparison of selection time from walking and bar menus. In *Proceedings of CHI'90*, pages 221–225. ACM, Addison-Wesley.

Modeling mobility: Exploring the Design Space for Enabling Seamless Ongoing Interaction for Mobile CSCW

Mikael Wiberg

Department of Informatics
Umeå University, 901 87 Umeå, Sweden
+46 90 786 6115
mwiberg@informatik.umu.se

Abstract

Session management models and technologies to support interaction are typically based on a narrow view of interaction. This view neglects that interaction is ongoing rather than strictly separated as well as it dynamically crosses physical and virtual arenas. In this paper we have theoretically explored the possibilities for and limitations of enabling seamless ongoing interaction across physical and virtual arenas for mobile groups by the use of *interaction objects* and *connection objects*. The exploration was guided by the UMEA approach that has its theoretical roots in Activity Theory. The paper derives some practical requirements for systems intended to support seamless ongoing interaction for mobile groups and discusses their implications for session management model for mobile CSCW systems. Prospects for future work include design and evaluation of RoamWare, a system to support seamless ongoing interaction for mobile groups.

1. Introduction – From one-shot interaction to seamless ongoing interaction

The importance of "mobile meetings" (Bergqvist et al, 1999) to get the work done has recently been argued for in the CSCW (Computer Supported Cooperative Work) literature. Research has explored the nature of such meetings as multi threaded, work related, situated and opportunistic. Research on this topic has also explored how these meetings are initiated and reestablished (Bergqvist et al, 1999). It has also been shown that frequent informal interactions are key to the work of a collaborative organization for knowledge sharing, decision-making and coordination. Spontaneous interactions facilitate frequent exchanges of useful information, and *awareness of ongoing activity* creates shared knowledge and provides a key context for the interactions that occur (Bellotti & Bly,1996). However, so far no one has explored how interaction can be maintained in and between mobile meetings, despite the fact that earlier studies of lightweight interaction (e.g Whittaker et al, 1997) have concluded that current conversational theories so far have focused on *one-shot* interaction, having identifiable beginnings and ends, and that there is a need to extend these theories to also address new problems of *connection, context regeneration* and *conversational tracking* from a seamless ongoing interaction perspective. The issue of *seamlessness* is an important element within the CSCW research area. Ishii and Miyake (1991) defines it as *unobtrusive integration of any noticeable system aspect into the surrounding context*. Borghoff and Schlichter (2000, p. 127) distinguish between several types of seamlessness (i.e. communication media, working mode, phases of the group process, technology, and time). To illustrate the importance of *seamless ongoing interaction*, we will refer to our recent observational study of mobile service engineers at a Swedish telephone operator. During our study we observed that the technicians needed to invent workarounds related to *lack of ongoing interaction* (between them, their clients and the central station staff) (Lindgren & Wiberg, 2000). These problems all lead to disruptions or breakdowns of the work group's interaction and the corresponding need for seamless support for shifting between co-located and dispersed settings. One thing that was clear from the study was that a person never knows if the interaction at a given moment is necessary to continue. This means that the user cannot be expected to maintain all threads of interaction. Instead, the CSCW system must *seamlessly* support the user when the need for interaction occurs.

2. Problems with current session management models from an ongoing interaction perspective

To really understand the underlying model of technology support for enabling *seamless ongoing interaction* there is a need to take a closer look at current session management models in relation to mobile meetings. The term *session management* within CSCW (Computer Supported Cooperative Work) refers to the process of starting, stopping, joining, leaving, and browsing collaborative situations. Currently there are two types of session management models (implicit and explicit) and there are two different models of explicit session management as described below: (1) Initiator-based session management (i.e. through some sequence of dialogs the initiating user invites other users to the collaborative session. Initiator-based session management systems fulfill two goals: to notify users of the existence of the collaboration, and to provide a means for rendezvous with the others in the session.), (2) Joiner-based session management (i.e. the initiating user creates a new session; users must find the session by browsing the list of currently active sessions (or know a priori that the session will be taking place). Once they know that they can attempt to join the session. Joiner-based systems typically only provide rendezvous facilities). However, the explicit models are awkward because they force someone to do a significant amount of overhead work setting up each new session (i.e. not seamless support). Explicit session management seems to be useful in situations where there is a high degree of formality (e.g. board meetings, etc). Explicit session management models do not seam to be the general solution to support mobile meetings. Edwards (1994) has suggested that a more implicit and direct approach would be to have the act of opening the object of the collaboration to provide the potential for collaborative activity itself (i.e. implicit session management). Thus, to collaborate in a meeting, (whether co-located or dispersed) the participants would simply enter the meeting session and automatically be assigned as participants by the meeting support system without any login procedure. Finally, there is a fundamental assumption made in these models that need to be taken care of, i.e. current session management models and technologies to support interaction are typically based on a narrow view of interaction. This view neglects that interaction is ongoing rather than strictly separated as well as it crosses physical (face-to-face) and virtual arenas.

3. Towards seamless ongoing interaction - The UMEA approach

To describe how higher-level user actions, such as participation in series of mobile meetings and computer-mediated interaction, could be supported in computer supported environments Kaptelinin (1997) have developed the UMEA (User-Monitoring Environments for Action) approach. The UMEA approach is theoretically heavily influenced by Activity Theory, and the underlying idea of the UMEA approach is that interactive environments are still application-oriented and that there is a need to support higher-level users actions. UMEA systems are intended to support the accomplishment of higher-level meaningful goals by facilitating reflection, resource integration, portability, and co-operation by focusing on providing users with support for *interaction histories* and *project contexts*. Further, these higher-level user actions belong to *tangled hierarchies* of projects (i.e. activities, such as mobile meetings, performed within one project can have relations to other projects as well). However, since the UMEA approach was originally developed to explore how single user stationary systems should be designed, it is possible to formulate a number of challenges to this approach: (1) Mobile work involves a lot of critical actions outside the computer. Thus these actions must be covered by the UMEA approach. (2) Higher-level user actions and action chains in this case refer to series of connected meetings. (3) Each individual worker has his or her own interaction history containing both physical and virtual sessions of interaction. (4) The higher-level user action involves other persons. These persons past and future actions are important for the continuum of the individuals activities. (5) A project is no longer e.g. the editing of a document across several tools but threads of meetings across physical and virtual arenas. From the tangled hierarchies perspective this also implies that a meeting should be able to mark as belonging to different interaction histories and project contexts. Based on these challenges for the current formulation of the UMEA approach we now explore the possibilities for integrating physical and virtual meetings into seamless ongoing sessions of interaction.

4. Integrating physical and virtual meetings into sessions of seamless ongoing interaction

The term object is used here to represent a mobile meeting (i.e. the interaction object). Each meeting can involve both co-located and dispersed group members. A connection object connects the co-located participants to the session and the line to the remote person indicates that a link is also established to include that particular person in the meeting. Further, each person has his or her own link to the shared connection object that is used for maintaining the session (See figure 1 below):

829

Figure 1. (left) Interaction object model of a mobile meeting and (right) a connection object link.

The term 'interaction object' is used to acknowledge the social part of ongoing interaction (i.e. a focus on mobile meetings, face-to-face interaction and the possibilities for dispersed persons to participate in the interaction) whereas the term 'Connection object' is more technical and is used to establish connections to the participants of a meeting. Connection objects have been used before to establish new sessions (e.g. Kristoffersen, 1998). The circle and the line to the dispersed participant (figure 1 left) symbolizes establish links between participants in a meeting and the connection object. Each connection made by the connection object contains a configuration concerning *interaction distribution, interaction access, and interaction filtering*. This link or connection is described on a technical level of how ongoing interaction could be established and maintained. The interaction distribution channel includes definitions of how interaction at one node (i.e. a group or a person) is mediated through technology to another node. The interaction access channel works the other way around (i.e. how dispersed interaction can be accessed by another node). Finally, the interaction filter determines the flow through the interaction distribution and access channels (i.e. if and when interaction is made available) (see figure 1 right). In pseudo code this can be described as follows:

Interaction distribution	Interaction access	Interaction filtering
While(mobile meeting exist) do Mediate interaction (through Interaction channels configured); // e.g. chat, voice, video, email;	While (mobile meeting exist) do Show available channels; Allow selection of channels (e.g. chat, video, voice); Mediate (interaction);	While (mobile meeting exist) do For all mobile meetings check mobilemeeting[nr].filter; // determine if access and/or distribution is allowed for the particular meetings running;

Of course there are also a lot of social aspects to consider for enabling ongoing interaction for mobile CSCW. Therefore we have developed the model below (figure 2) to illustrate a set of scenarios that allows us to theoretically explore the possibilities for and limitations of ongoing interaction on both a technical and social level of analysis.

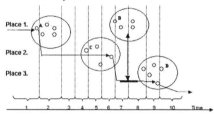

Figure 2. Possible combinations of mobile meetings with face-to-face and remote interaction

The model above illustrates a set of mobile meetings (A, B, C, and D) that happens at different particular places over time. The inner circles represent individuals participating in these meetings. Along the time line a set of numbers and lines has been set to point out some critical phases (1-10) of mobility and ongoing interaction due to the larger work context (i.e action chains). In the model the things happening for a particular individual within a mobile meeting circle is viewed as that particular persons co-located setting, whereas everything else happening is viewed as dispersed settings. According to this, remote colleagues who also might have their own mobile meeting are, from the individuals' viewpoint, dispersed. In the model above the previous presented model of a mobile meeting is complemented by the dimensions suggested by the UMEA approach. More concrete, that

830

is done by connecting the individual's participation (physical and virtual) in several meetings over time (interaction history) depending on the higher-level goals of the user (project context). As figure 2 above illustrates a person enters a mobile meeting physically in phase 1. In phase 2 the person leaves the meeting. However, through the interaction object he/she can still follow the meeting remotely through an appropriate interaction channel (e.g. voice, video, etc). In phase 4 the person in focus enters a new meeting. Meanwhile the meeting continues a new remote meeting is established (phase 6). Of course, the person could remotely follow that meeting as well. However, on a social level that would imply that the interaction focus would be on the remote interaction instead of the surrounding face-to-face interaction. Thus, if that is not appropriate on a social level the interaction filter should be configured to not allow seamless establishment of remote collaboration with another meeting during a present meeting. Another alternative is of course, according to the idea presented by the UMEA approach about human actions as belonging to tangled hierarchies, to link these two meetings together through their interaction object to establish a larger meeting with remote participants for each physical setting. When leaving the current meeting (phase 7) the filter could be set to scan for remote meetings held within a specified work group as to allow for remote collaboration. Still, on a social level, face-to-face interaction is typically priorities so, if another face-to-face meeting starts (phase 8), the user should have the opportunity to leave the remote interaction and, by doing that also distribute a notification to the remote participants about the exit from the meeting. In this way, seamless interaction across physical and virtual meetings could be handled in the same ways as entering and leaving of chat rooms typically represented as: "Person X enters the room", "Person X leaves the room". This seamless integration of the physical meetings into the virtual interaction could also offer similar functionalities as chat rooms (e.g. the possibility the brows the groups interaction history). In this case this functionality could also include face-to-face interaction between some of the participants, remote colleagues, note takings of interest to the rest of the group, etc.

5. Implications for current session management models and preliminary system requirements

Looking at the structure, possibilities and limitations related to enabling seamless ongoing interaction across physical and virtual meetings we have to leave the idea and concept of 'sessions' and current models of session management. If ongoing interaction is to be realized we must first turn away from traditional views of sessions as clearly separated, initiated and finished. As the mobility model illustrates it is not clear when a person can no longer take part in a session or when a session is finished. Further, from a technical viewpoint we have to raise questions concerning how CSCW applications should handle ongoing interaction in the future (if not by starting, stopping, joining, leaving sessions)?

We have, from the above reasoning, derived 5 preliminary and basic requirements for technology to support ongoing interaction for mobile CSCW besides the pseudo code outlined above. These are: 1. *Implicit establishment of meetings* (i.e. a spontaneous meeting should be initiated when it happens for both co-located and dispersed group members). 2. *Fluid maintenance* (i.e. support easy shifting between co-located and dispersed modes of interaction. This requires mobile IT), 3. *Ad hoc meeting distribution* (i.e. when a co-located meeting happens it should also distribute its content to dispersed participants, 4. *Easy access* (i.e. let dispersed participants easily find the meeting and join it whenever they want to) ,and 5. *Meeting priority support* (i.e. support users with visualizations of different threads of meetings to let them easy get an overview of the current state and make it easy for them to make priorities about which meetings to participate in).

Reviewing current support for seamless ongoing interaction across physical and virtual arenas we found no technology that meet these five requirements. There have been some attempts to theoretically inform meeting and interaction support (e.g. Bennett & Karat, 1996; Ljungberg, 1999). However, so far there has been no research attempts to support *seamless ongoing interaction across physical and virtual arenas*. DOLPHIN (Streitz et al, 1994) supports both co-located and dispersed group members during meetings. However, the system does not provide any mobile support, nor does it support seamless shifting between these two settings.

6. Conclusion

In this paper we have theoretically explored the possibilities for and limitations of enabling seamless ongoing interaction across physical and virtual arenas. We have formulated challenges to the UMEA approach to cover CSCW activities and then applied it to explore requirements for how ongoing interaction could be enabled for mobile CSCW. Further, we have discussed the implications of the model for current session management models for CSCW systems. One general conclusion is that, to support mobile CSCW, current session management models need to be extended and move from one-shot interaction to seamless ongoing interaction across physical and virtual arenas.

Future work includes the development of RoamWare, a mobile CSCW system that will meet the five requirements identified. Related to the interaction object model RoamWare will provide *interaction distribution* of interaction objects (mobile meetings), *interaction access* according to browsing of individuals' interaction histories and support for dispersed workers, and finally, *interaction filtering* according to visibility/invisibility on the web, search and visualizations of meetings and their interrelations and multithreadedness according to the ordered and tangled structure of human activities as pointed out by the UMEA approach. In a second phase we plan to focus on filter based notifications of events, and integration with other interaction channels (e.g. voice and video). Finally, we hope that evaluations of RoamWare will give useful insights concerning limitations to ongoing interaction (both technical and social) and the role of face-to-face meetings versus computer mediated remote participation. Further, we hope that the evaluation will provide valuable input to verify the usefulness of session management models for sustained mobile meetings.

7. References

Bellotti, V. & Bly, S. (1996). Walking away from the desktop computer distributed collaboration and mobility in a product design team. In Proceedings of ACM 1996 Conference on Computer Supported Cooperative Work.

Bennett, J., and Karat, J., (1996) Working through meetings: a framework for designing meeting support;; Proceedings of the ACM 1996 conference on Computer supported cooperative work.

Bergqvist, J., Dahlberg, P., Kristoffersen, S., and Ljungberg, F (1999) Moving Out of the Meeting Room: Exploring support for mobile meetings, the European Conference on Computer Supported Cooperative Work, September 1999.

Borghoff, U & Schlichter, J (2000) Computer-supported Cooperative Work Cloth. Springer-Verlag, UK.

Edwards, K., (1994). Session management for collaborative applications; Proceedings of the conference on *Computer supported cooperative work*, 1994, Pages 323 - 330.

Ishii, H., & Miyake, N. (1991) Towards an Open Shared Workspace: Computer and Video Fusion Approach of Teamworkstation. Communications of the ACM, 34:12, 37-50.

Kaptelinin, V. (1997) Supporting Higher-Level User Actions with Representations of Project Context and Interaction Histories, *Working Paper*, Dept. of informatics, Umeå University, ISSN 1401-4580.

Kristoffersen, S. (1998) Developing Collaborative Multimedia: The MEDIATE Toolkit, PhD-thesis, Norwegian Computing Center/Applied Research and Development, Oslo, Norway.

Lindgren, R. & Wiberg, M. (2000) Knowledge Management and Mobility in a Semi-Virtual Organization: Lessons Learned from the Case of Telia Nära. *In Proceedings of HICSS-33*, IEEE-press. Maui, Hawaii, January 4-7, 2000.

Ljungberg, F. (1999) Exploring CSCW mechanisms to realize constant accessibility without inappropriate interaction, Scandinavian Journal of Information Systems, vol 11, p. 115-135.

Streitz, N., Geißler, J., Haake, J., and Hol, J. (1994). DOLPHIN: integrated meeting support across local and remote desktop environments and LiveBoards; Proceedings of the conference on *Computer supported cooperative work*.

Whittaker, S, Swanson, J, Kucan, J & Sidner, C (1997) TeleNotes: Managing lightweight interactions in the desktop, *ACM Trans. Comput.-Hum. Interact.* 4, 2 .

Information Architecture of a Customer Web Application: Blending Content and Transactions

M. W. Vaughan, Ph.D. [a], K. M. Candland, M.A.[a], and A. M. Wichansky, Ph.D. CPE [b]

[a]Usability and Interface Design Department, Oracle Corporation
500 Oracle Parkway, MS 2op10, Redwood Shores, CA, 94065, USA

[b]Advanced User Interface Group, Oracle Corporation
500 Oracle Parkway, MS 2op10, Redwood Shores, CA, 94065, USA

1. ABSTRACT

This case study discusses the methods used, data collected, and results obtained from user activities conducted to design an order management customer application. The application, which blends content and transactions, allows a vendor to provide each of its customers a Web-accessed, secure view of the customer's order and account status. to elicit requirements for the product design, information architecture and high level interaction design of We focus on three levels of design: product definition, information architecture, and high-level interaction design. We discuss the wants and needs analysis, card sorting task, and questionnaire used to collect user data for each of these different levels. We also address the blending of content and transactions at each of these levels.

2. INTRODUCTION

We were approached for help by the Oracle Order Management product development team in January, 2000. They wanted to build an application that would allow a vendor to provide each of its customers a Web-accessed, secure view of the customer's order and account status. We discovered they were venturing into a new area of their domain with little prior knowledge of the user requirements for the application. Specifically, they sought to build: a) a customer-facing product when previously they had built only internally-facing products; b) a product serving two new user populations they had not previously designed for, and c) a web-based application using a new technology paradigm with which they had no prior experience. We offered to help address these issues.

First, we helped them identify and explicitly define their primary user groups. A vendor who installed this application would serve two types of users at a customer's business: professional buyers and accounts payable (AP) representatives. A professional buyer's main function is to negotiate and make purchases for his/her company. An AP representative's main task is to pay a company's bills, i.e., deal with outstanding invoices.

Second, we helped the team identify the design questions that required input from users:

1. What are the types of information and transactions each user group would like to see in such a product?
2. How should we organize the information and functions from an information architecture standpoint?
3. What features do we need to expose at different levels of user interface design?

In working with the team, we realized that we were facing a broad and emerging challenge in web-based design – the need to blend both content and transactions (e.g., Dillon, 2000; Dillon & Vaughan, 1997). This web-based application would require designing for both rich viewing and navigation of information, as well as performing actions on that information. Although more detailed design was needed to specify screen flows and detailed screen designs, in this paper, we will focus on the first three levels mentioned above.

To answer the first question (information types), we conducted a 'wants and needs analysis.' The results helped validate and prioritize existing requirements and identified new features. To answer the second question (organization), we conducted individual card sorting tasks using a set of primary conceptual objects. Based on this information, we were able to extract a generalized conceptual model of the information organization for each user group. For the third question (features), we created a questionnaire designed to capture the frequency with which each group performed certain tasks. The tasks which at least half of the participants performed on a high frequency basis (e.g., daily or weekly) were identified as high priority items to expose in the user interface.

Two user requirements sessions were conducted using all three of these activities – one session with eleven buyers, and another session with twelve accounts payable representatives. Participants were solicited via a recruiting company using a set of screening criteria we had specified. Each session lasted approximately two hours.

3. WANTS AND NEEDS ANALYSIS

For this activity, participants were asked to brainstorm about the content, or information, for an ideal system to help them manage their orders with vendors. They were then asked to specify, individually and on paper, their "top five" pieces of information and indicate why each piece was important. These data were first analyzed for *verbatim* matching and grouped accordingly. Each group was then re-examined for semantic matching (i.e., by meaning) and any re-categorizations were made at this point. Responses were reported in terms of the percentage of participants agreeing that a type of information needed to be included, and was therefore important.

The wants and needs "top five" data analysis resulted in a prioritized list of product content categories and features for each user group. A comparison of the lists revealed that both groups had independently identified four high priority categories (see Table 1). In terms of product design, the lists served to validate and prioritize existing requirements as well as to highlight new features. For example, online ordering was the number one item chosen by 100% of the professional buyers. While this feature had been included in the original product requirements, these data indicated the importance of online ordering, including the buyers' rationale, and provided a strong case for making the feature a higher priority for both user interface design and development efforts. Vendor information was a new content category suggested and ranked highly by both groups. Other requirements were further refined and enhanced, such as shipping information which users suggested should be integrated with third party carrier websites (e.g., FedEx, UPS, etc.) to support more accurate tracking of deliveries. These types of specific, descriptive functional data were used to ensure the utility of the product.

Table 1
Prioritized Content Lists (product content with ~50% or more agreement)

Professional Buyers	Accounts Payable Representatives
1. Online Ordering	1. Shipping Information*
2. Catalog/Product Information*	2. Vendor Information*
3. Price and Availability	3. Account Information
4. Shipping Information*	4. Terms and Conditions*
5. Vendor Information*	5. Buyer Information
6. Terms and Conditions*	6. Purchase Order Number
7. Order Status	7. Invoice Information
8. Create Return Requests	8. Catalog/Product Information*
9. Purchase History	9. Internal Coding Information
10. Outstanding Invoices/Payment History	10. Order Details/Delivery/Taxes
	11. Payment Information
	12. Credit Memos

* Indicates content that appears on both lists.

4. CARD SORTING

For the card sorting task, each participant was given a set of 18 cards where each card contained one name of a conceptual object we thought likely to appear in the product. Any questions about definitions or meaning were clarified, and participants were given blank cards to add any items to their individual piles they thought were missing (these were analyzed seperately from the original set of cards). Each participant was then instructed to sort the cards into groups that were meaningful to him/herself based on his/her work experience.

The card sort data were analyzed as two separate groups, one for professional buyers and one for AP, using cluster analysis software designed for card-based data. The results are displayed as a tree diagram listing all the objects with lines representing the strength of perceived relatedness between the objects (Martin, 1999). Depending on the degree of relatedness, distinct groups and sub-groups of related objects emerge in the tree diagrams.

As would be expected, the diagrams showed similarities and differences in the object groups identified by professional buyers and AP. For example, in Figure 1, both user types grouped 'user profile,' 'logon,' and 'preferences' together, but professional buyers combined this group of objects with another group consisting of 'advertising' and 'news.'

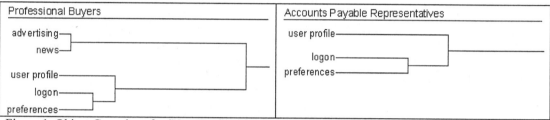

Figure 1. Object Groupings for 'User Profile,' 'Logon,' and 'Preferences'

In Figure 2, The 'order history' object was associated with totally different objects by the different users. Professional buyers placed 'order history' with 'sales orders' and 'shipping' in a sub-group that was combined with 'products' and 'services' while accounts payable representatives put 'order history' in a sub-group with 'account history' that was strongly related to a group of accounting objects including 'credit memos,' 'debit memos,' 'invoices,' and 'payments.' Such differences were consistent with the different tasks performed by the different user types. The 'order history' object is commonly referenced by both user types, but professional buyers put it with other order-related objects because they spend most of their time performing ordering activities while AP representatives placed it with the accounting-oriented objects that they use most often.

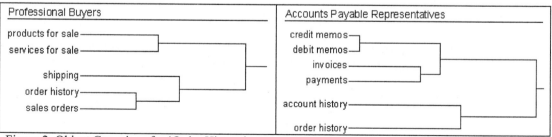

Figure 2. Object Groupings for 'Order History'

The two tree diagrams suggested an information architecture composed of two levels (see Table 2). The first level is a general container for the items at the second level. The second level items are the participants' primary objects. The level one objects that emerged from each tree diagram were very

835

similar conceptually; however, the placement of certain objects within the second level varied by user type. To resolve these differences, we used task frequency data collected from each user type (discussed in the next section). Where objects were associated with tasks that were more commonly performed by one user type, we deferred to their object groups. Based on the card sorting data and the task frequency data we proposed the information architecture shown in Table 2.

Table 2
Information architecture recommended for the application

Level 1 contains…	Level 2
Home	Advertising, News
Purchasing or Orders	Purchase Orders, Order History
Catalog	Products, Services
Accounting or Account	Invoices, Payments, Credit Memos, Debit Memos, Account History
Customer Service	Service Requests, Returns, Repairs, Vendor Contact Information

Using Oracle's user interface design guidelines for browser-based applications, we translated the information architecture (primary groups and their respective objects) into an array of tabs, each of which has its own set of subtabs (see Figure 3 for an example of the Account tab.) Not only do the tabs and subtabs render a visual and textual representation of the information architecture, they afford navigation throughout the application and provide the foundation for high level task flows, both of which are critical to the interaction design of a product.

Figure 3. Account Tab

5. TASK FREQUENCY QUESTIONNAIRE

Finally, we sought to collect information to make informed decisions about what features and functions to expose in the UI. We created a questionnaire listing a set of activities the software was intended to support, and were believed to be performed by accounts payable representatives and buyers. Respondents indicated the frequency with which they performed each activity (e.g., daily, weekly, monthly, yearly or never). The items which at least half of the participants performed on a high frequency basis (e.g., daily or weekly) were identified as high priority items to expose in the user interface.

We found a high degree of overlap between the high frequency tasks and the prioritized content lists resulting from the wants and needs analysis. For example, 'checking item availability' and 'placing a purchase order' were tasks performed by 100% of the buyers on a daily or weekly basis. This was consistent with their request for 'Online ordering' in the Wants & Needs Analysis. This overlap further validated the product's content and feature set planned for development.

The identification of high frequency tasks for each user type also contributed to a number of interaction design decisions. For example, our proposed information architecture had identified primary groups of objects, but provided no direction as to the organization of the objects within each group. In this particular user interface, the objects in a tab are ordered horizontally across a band that hangs below the selected tab. Further, when the user navigates to the tab, the first (leftmost) object/subtab is selected, by default, as indicated by the bold, white text style applied to the subtab label (see Figure 3). The default selection and order of subtabs within a tab, can focus the user's attention on a particular object and reduce the amount of navigation necessary to access objects and tasks.

6. CONCLUSION

At each of the levels of design we faced particular challenges when trying to blend content and transactions. At the product definition level, we helped the product team transform their conception of the application from an information only product, to a information and transaction-oriented product. Based on the user data, we could make concrete suggestions as to what those transactions needed to be.

At the information architecture level, our challenge was to blend the needs of multiple user types. With our user data, we were able to make informed decisions about which user types to focus on for non-overlapping categories of data, such as product availability, as well as how to blend the architecture when users' needs did converge.

For the high level interaction design, the need to carefully integrate content and transactions became even more apparent. For each user type, certain actions were performed on certain information with a high degree of frequency. We chose to segment part of the UI towards buyers and another part towards accounts payable representatives in the form of separate tabs. In addition, some information was highly desired by both groups, such as shipping information. In this case, we created short cuts on the home tab to quickly take either user to the needed information. In follow-up activities, initial screen designs were prototyped in HTML and these same screens were then subject to usability testing.

Overall, we believe that our strategy for combining the need to navigate information as well as act on that information succeeded because we took a holistic, rather than a piece-meal, approach to design. We started from the broadest possible frame of reference, and designed for that case. As transactions and new information are added over time, they can reference the base design and enable the product to grow gracefully. In this way, the product will avoid an erratic development process with features and content added in an *ad hoc* fashion.

REFERENCES

Dillon, A. (2000). Spatial-semantics: How users derive shape from information space. Journal of the American Society of Information Science, 51, 521-528.

Dillon, A. & Vaughan, M. (1997). "It's the journey and the destination" Shape and the emergent property of genre in evaluating digital documents. New Review of Hypermedia and Multimedia, 3, 91-106

Martin, S. (1999). Cluster analysis for web site organization: Using cluster analysis to help meet users' expectations in site structure. internetworking, 2(3). Available at http://www.sandia.gov/itg/newsletter/dec99/cluster_analysis.html

Designing and Evaluating a Web-Based Collaboration Application: A Case Study

Wenli Zhu

Microsoft Corporation, One Microsoft Way, Redmond, WA 98052 USA

ABSTRACT

The Web has evolved from a simple browsing environment to an interactive, goal-oriented, task-driven application and script-enabling platform. More and more Web applications are being developed to allow users to create, edit and manipulate objects in the browser, just like in traditional graphical user interface-based applications. SharePoint™ Team Services from Microsoft® is an example of such a Web application. It is a Web site designed to help a team to collaborate and share information. But it also has commands and dialog-like pages that allow users to modify the content of the Web site in the browser. This paper describes the design and evaluation process of the application and presents results and design solutions in the areas of: Navigation, browsing vs. editing, providing context and feedback, selection model, processes and interacting with other applications. Our findings confirmed and expanded previously identified issues with Web application design, and provided fresh data and design solutions to address these problems.

1. INTRODUCTION

In recent years, more and more Web-based applications have been developed to offer functionalities that go beyond simple browsing. As indicated by Fellenz, Parkkinen & Shubin (1999), there may be several different categories of Web usage: browsing, performing transactions and running applications. Web applications are usually developed out of a desire to reduce the cost of software development and distribution, especially for applications that need to run across multiple hardware and software platforms. As explained by Rice and his colleagues (1996), Web applications allow users to create, edit and manipulate objects in the browser just like in traditional graphical user interface-based (GUI) applications.

While general Web page and Web site design guidelines (such as those identified by Nielsen, 2000) and traditional GUI design guidelines (for example, Shneiderman, 1987) can be applied to Web applications, unique characteristics of Web applications have been suggested and observed (Fellenz et al., 1999; Shubin & Meehan, 1997; Rice et al., 1996), including:

- Navigation: As Shubin & Meehan (1997) pointed out, Web browsers use the page metaphor which is appropriate for browsing Web pages with static text and hyperlinks. The navigation model is simple. Users follow hyperlinks on the page to view more information or use the Back button to retrace steps. With Web applications, however, this navigation model may create problems. When users' main goal is to complete a task in Web applications, links that take people away from the task may be distractive and may result in users taking a detour or losing their work. The Back button could also be problematic. Because the Web browser "caches previously seen pages and allows users to revisit these pages without communicating with the server," as Rice et al. (1996) put it, the browser is essentially allowing users to "travel back in time to an earlier interaction state and attempt to execute the commands as they were presented then." This may result in users executing the same command more than once (ordering more products than intended, for example) or misperceive the Back button as "undo." Some Web applications attempt to solve this problem by opening up a new browser window without the browser controls. However, users may not understand the new window's dependency on the browser and may intentionally kill the browser and thereby inadvertently kill the application.
- Browse vs. edit: Web applications need to support both browsing and editing. Designers of Web applications need to weigh the balance of the two. In the applications that Rice and his colleagues developed (1996), the editing environment was made to look as much like the browsing environment as possible, because they wanted to "minimize the number of different-looking pages to which the user would be exposed." In the end, however, they concluded that it might have been better "not to use the edit-in-place model but rather to use pages just for editing."
- Lack of context or feedback: As Rice et al. (1996) pointed out, "there is a fundamental difference" between the type of interactions supported by the browser and the type of interactions supported by the traditional GUI. In traditional GUI, users select an operand or operands (for example, a paragraph of text) through direct manipulation and then apply an operator by means of a menu selection or a keyboard command. The

operand remains visible during the operation and provides context. After the operation, the operand changes and therefore provides feedback. In Web applications, however, the page that the user is on could be an implicit operand and the links the user chooses will operate on the page. During and after the operation, users may be on different pages and therefore may lose the context of the operation or may not be able to see the feedback.

- Limited interaction model: Compared to traditional GUI applications, the interaction models in Web applications are much more limited. For example, it is hard to implement selection of a single object, menu bars, graying-out options, direct manipulation on objects, right-mouse menus, etc., in a Web application.

- Response time: Slow network connections or limited bandwidth may cause delays in the system response time in Web applications. Navigating from one page to another in a Web application may take longer than moving from a main window to a dialog box in a traditional application.

- Exit: Many Web applications do not have a way to stop "running" without closing the browser. Shubin & Meehan (1997) suggested having a "clearly marked exit" in a Web application to help users remember to finish what they are doing before leaving the application. Interestingly, Rice et al. (1996) confirmed the perceived importance of having a clear exit. They discovered that many of their Web application users had a strong desire to be able to log out of their system. Logging out was not necessary. But users felt a need to log out, even when they were told explicitly that it was not necessary.

2. DESIGN RATIONALE

SharePoint Team Services from Microsoft is a Web-based collaboration application designed to help a team share information. During the design phase, several assumptions and decisions were made (Van Tilburg, 2000):

- It is a Web site. It should look like a Web site. The underlying data structure of the Web site is a list structure, and there are three types of lists that users can use: Documents, discussions and lists. A list that holds documents is called a document library. A list that holds discussions is called a discussion board. A list that holds any other items is called a custom list. Users can create multiple document libraries, discussion boards and lists. Within each list, users can create multiple items. See Figure 1 for an overview of the site structure and Figure 2 for a snapshot of the Web site home page.

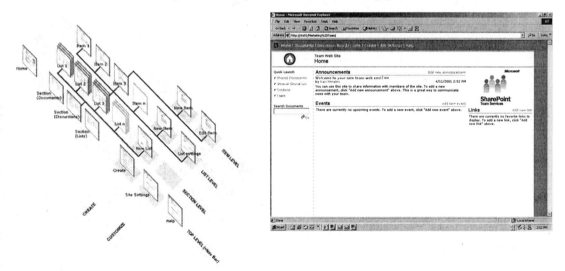

Figure 1. Site Structure (Adapted from Van Tilburg, 2000)   *Figure 2.* Site Home Page

- The design of the Web site is based on assumptions about different user activities and the relative frequencies of these activities. First and foremost, it is a Web site, designed to help people collaborate and share information. It is therefore assumed that users will most often consume information, rather than add or edit information. The Web site is therefore optimized for the browsing behavior. Edit controls are more hidden and only become visible after a deliberate action from the user. Also, no selection model is implemented. To delete an item in a list, users have to choose to "edit" the item first and then delete it.

- There are four basic types of pages: The home page (see Figure 2), the list view page (the page that shows all the items in a list, see Figure 3), the item view page (the page that shows the details of a list item, see Figure 4), the customization summary page (the page that shows all the options for a list) and the customization controls page (see Figure 5). The list view page and the item view page both have editing controls in a special toolbar area above the content area. The customization controls page looks like a traditional GUI dialog, but is not implemented as a pop-up dialog because we want to maintain a consistent "Web" feel of the application.

Figure 3. List View   *Figure 4.* Item View   *Figure 5.* Control Page

- Each page has 4 areas: Navigation bar, title area, left pane and content area. The Web site has one main navigation bar on the top which is always visible. It links to the different sections on the Web site: Documents, discussions, lists, create, site settings and help. On the list view page, the left pane contains links that filter the list (also called views) and on the home page, the left pane contains quick launch links.

3. EVALUATION

A series of usability studies was conducted to understand what potential usability problems users might have. The first study focused on lists, site customization and navigation. The second study focused on "getting started" or "new site creation" experience. The third study focused on document creation, uploading and editing. The fourth study collected quantitative performance measures. The evaluation process was a typical iterative design process, i.e., results from a usability study were used to make design changes and the new design was tested again. A core set of tasks such as those involving users interacting with lists and documents as well as navigation were repeated across all the studies. The following sections briefly review the methodology.

3.1 Subjects

A total of 42 subjects participated in the four usability studies mentioned above. All of them were intermediate to advanced Microsoft Office® users (user experience was classified using a questionnaire developed by Draine and McClintock, 1999). All the subjects were regular Internet users and had intranet in their organizations. All the subjects were working on projects with deadlines and deliverables. Half of the subjects were project leads and the other half were project participants. The profile of the subjects was chosen to represent the target user population of the application. Subjects received free Microsoft software for their participation.

3.2 Tasks

Tasks representative of real-world scenarios were given to the subjects. Most of the tasks were close-ended, i.e., there was one clear outcome to judge whether the task was completed successfully. Some tasks were open-ended, i.e., there was no clear end state that the subjects had to achieve. Tasks were chosen to cover a wide range of possible usage scenarios and feature areas. Repetitive tasks were included, within one usability study and across several usability studies, to compare performance.

3.3 Procedure

Each test session took about 1.5-2 hours. In each test, subjects filled out a background questionnaire and performed the tasks sequentially. Subjects self-reported completion of the tasks. They could also stop at any time.

3.4 Measures

Verbal protocol, success/failure rates, task completion time and errors were collected.

4. RESULTS AND DISCUSSION

Results and observations from the usability studies are summarized below.

4.1 What Worked Well

4.1.1 Separation of commands and content. Throughout the user interface, editing commands are distinguished from the content by: (a) the use of a verb in the command name; (b) the use of a different text color for the command name; (c) the use of icons; and (d) placement of commands in a designated area such as the special toolbar on the list view and item view page with divider lines separating the commands from the content. Results showed that subjects can tell the difference between the commands and the content very well. When subjects needed to accomplish a task using one of the commands, they quickly recognized and used the command. The mean success rate was above 90%, and the mean task completion time was less than one minute. Longer and more specific labels and the combination of icons and labels helped users understand the commands better. For example, a change in the list customization command (from "change list settings" to "change the columns and views") resulted in an increase in success rate (from ~40% to 100%). The results indicated that by separating the commands and the content successfully, the Web site separated editing and browsing successfully.

4.1.2 Navigation and the home page. The content area and the left navigation area got more attention than the top navigation bar. Subjects would look first in the content area and the left navigation area. They would also look there more often. An analysis of link-following patterns on the home page revealed that throughout a test session, 100% of subjects followed the links in the content area whereas 65% followed the links in the left navigation area (the Quick Launch links) and 50% followed the links on the top navigation bar. The results also revealed that subjects tended to use the home page as a "jump board," especially at the beginning, to discover what is available and learn what they can do.

4.2 How Did We Do With Known Problems

4.2.1 The Back button. All the subjects frequently used the Back button. Sometimes, they backed out of an editing page or a customization controls page without confirming the changes or finishing the fields on the page. Most of the time, however, subjects were successful at using the dialog-like pages (for example, the customization controls page shown in Figure 5) to fill out the fields, scroll down to the bottom and click on the buttons. Some subjects used the Back button in the midst of filling out the fields to try to get information from other pages. That resulted in a loss of the data they already entered. It was very rare, however, that subjects used the Back button to get back to an editing page or a customization controls page and execute it again. Subjects mainly used editing commands to open up editing pages and customization controls pages.

4.2.2 Browse vs. edit. In general, subjects did not have a problem separating the two modes. However, some subjects at times tried to directly edit text (of a document or a list item) in the browser. This problem is unique in the Web environment because in traditional GUI applications, there is usually no separation between browsing and editing. For example, in a word processor, users can browse and edit at the same time. Another problem found with editing hints at how much users view the browser and the Web application as separate. In one of the usability studies conducted on an earlier design, subjects had to use the Edit command in the browser (on the browser menu and toolbar) to edit a document. Most of the subjects (90%) failed the task. They typically did not think of using a command in the browser to perform an action in the Web application. To solve this problem, we added a command called "Edit in <application name>" to help users edit a document.

4.2.3 Lack of context or feedback. The fact that some subjects used the Back button to back out of an editing page in order to see the context information on other pages indicated the need of maintaining the context of users' work. Various problems associated with the lack of context or feedback were identified during the iterative design process and design changes were made to address the problems. For example, in earlier designs, users did not get appropriate feedback after adding a new item to a list or adding a new column to a list. Changes were made so that users would be brought to the appropriate pages after executing the editing commands and get feedback on their actions.

4.2.4 Limited interaction model. As mentioned before, no selection model was implemented because we wanted to have a simple design optimized for browsing. This decision made deletion more difficult. Many subjects tried to click or right-click to select a document, a column or a document library. They said that they would really like to be able to directly select an item instead of using the Edit command. They complained that the user interface did not follow the conventional GUI application behavior.

4.2.5 Response time and exit. No problem was found here.

4.3 New Problems Found

4.3.1 Processes and interacting with other applications. For some of the process-oriented or compound tasks, users have to use different pages to accomplish steps in a process or have to jump back and forth between the browser and other applications to finish a process. For example, with an earlier design, when creating a new document library with a custom template and custom columns, users needed to: (a) create a new document library; (b) create a template in an application such as Microsoft Word®; (c) upload the template; and (d) create columns for the document library. When subjects were asked to do this task in a usability study, all of them had difficulties with one step or another. A key problem was that subjects did not understand the steps they had to follow. We did not provide a mechanism (such as a wizard in the traditional GUI) to represent steps taken in a process and walk users through that process. Even if we did, it would have been difficult to lead users through the process of switching back and forth between the browser and another application. In the end, we changed the design so that users no longer needed to know to open an application to create a template. They are now presented with the option to "edit template" which automatically opens up an application for them. The more we can do to help users go through a process, the fewer problems users will have with process-oriented or compound tasks.

4.3.2 Refresh. When subjects used the Back button to go back to a list after customizing it, or when they edited a document in another application and switched back to the browser, they saw the old (cached) page that did not have the changes. None of the subjects understood why. Some thought they forgot to save the changes and tried to repeat the process. Relying on users understanding and remembering to use the Refresh command in the browser was not a good solution. In the end, we implemented automatic refresh in most cases but there were still scenarios where we could not fully solve the problem.

4.3.3 Hyperlink view switcher. As mentioned before, the list view page contains links that filter the list (called views) in the left pane (see Figure 3). For example, when users click on the link labeled "My Tasks," the list will be filtered to show just the tasks "I" (the user) created. Throughout the usability studies, subjects had difficulties understanding how to use these hyperlink view switchers. They all noticed the links but could not understand how the views were defined, how to edit a view or that there was a default view. We do not exactly know why the hyperlink view switchers were so hard to understand, but there may be two possible reasons: views are hard to understand by nature and hyperlinks are not good UI mechanisms to implement views.

5. CONCLUSION

Our experience confirmed many of the problems reported an ' discussed in previous research. Data from our usability studies provided fresh evidence that these problems do exist and must be considered in building Web applications. Our data also suggested new issues that may further complicate the design of such applications. While using the Web as a platform to deliver applications is highly desirable and perhaps even inevitable, it seems the current Web Page metaphor and the application model of the browser will continue to pose unique challenges. New technology and new ways of interacting with the Web, however, will help us find solutions to these challenges.

ACKNOWLEDGMENTS

Martijn Van Tilburg and Mike Morton led the user interface design of SharePoint Team Services and contributed to the evaluation process. The author would like to thank Martijn in particular, for his insights and comments.

REFERENCES

Draine, S. C., & McClintock, M. (1999). Developing a User-Knowledge Assessment Tool for Consumer Software. In the *Proceedings of the Eighth International Conference on Human-Computer Interaction*, Munich, Germany, pp.111-112.

Fellenz, C., Parkkinen, J., & Shubin, H. (1999). Web Navigation: Resolving Conflicts between the Desktop and the Web. *SIGCHI Bulletin, Vol. 31*, pp. 26-28.

Nielsen, J. (2000). *Designing Web Usability.* Indianapolis, IN: New Riders Publishing.

Rice, J., Farquhar, A., Piernot, P., & Gruber, T. (1996). Using the Web Instead of a Window System. In the *Proceedings of ACM CHI 96 Conference on Human Factors in Computing Systems*, Vancouver, BC, pp.103-110.

Shneiderman, B. (1987). *Designing the User Interface: Strategies for Effective Human-Computer Interaction.* MA: Addison-Wesley Publishing.

Shubin, H., & Meehan, M. M. (1997). Navigation in Web Applications. *Interactions, Vol. 60*, pp.13-17.

Van Tilburg, M. (2000). SharePoint User Interface Guidelines. *Technical Report: Microsoft Corporation.* Redmond, WA.

E-commerce User Interface Design for the Web

Lawrence J. Najjar, Ph.D.*

Viant, One Georgia Center, 600 West Peachtree, Suite 2300, Atlanta, GA 30308 USA, lnajjar@viant.com

ABSTRACT

Due to competition and user demands, e-commerce sites must be designed to be very easy to use. This paper provides helpful user interface design suggestions for page format, navigation, catalog, registration, and checkout.

1. INTRODUCTION

According to Forrester Research (Manning, McCarthy, & Souza, 1998), "The rallying cry of the Web will be 'usability, usability, usability.' In a world of dynamic trade -- where information is commoditizing products -- ease-of-use will be the differentiator that drives market share and brand perception." There is strong support for this assertion. Seventy-nine percent of users named easy navigation as the most important characteristic of an e-commerce site (Lake, 2000), and e-commerce sites lose up to 50% of potential online sales because users cannot find what they want (Cohen and Thompson, 1999; Seminerio, 1998). Poor navigation and slow downloads helped cause 83% of shoppers to leave a site (Thompson, 1999) and 78% of shoppers to abandon their online shopping carts (BizRate, 2000).

On the positive side, when we improve the usability of sites, sales can improve. For example, after improving usability, IBM found a 400% increase in sales (Tedeschi, 1999) and Digital Equipment Corporation reported an 80% increase in revenue (Wixon and James, 1992). One study (Creative Good, 2000) estimated that improving the customer experience increases the number of buyers by 40% and increases order size by 10%.

2. DESIGN

Designing the user interface for an e-commerce site is very challenging. E-commerce sites must accommodate nearly all users, include a significant amount of user interactivity, and still be easy to use. To have a significant impact on usability, take special care when designing the overall page format, navigation, catalog, registration, personalization, and checkout.

2.1 Page Format

Design the page size to accommodate America Online (AOL) members. There are over 28 million AOL members (America Online, 2001a), and they buy at three times the rate of other users (Kadison, Weisman, Modahl, Lieu, & Levin, 1998). It is estimated that 70% of the 2000 online holiday spending was by AOL members (America Online, 2001b). AOL members open their Web site windows inside the AOL browser. Without resizing their browser windows, this window-in-window technique restricts the size of AOL members' Web windows to 625 pixels wide by 270 pixels high (America Online). So, to avoid forcing this group of users to scroll their pages horizontally or vertically, design your page to fit within that area.

Since download time is the biggest problem with the Web (Graphics, Visualization, and Usability Center, 1998), design your pages to download quickly. Limit your use of graphics and optimize them for the Web. Do not use plug-ins like Flash (Ragus, 2000b, 2000c; Nielsen, 2000). Do not use Java. Design your pages to download in less than 10 seconds (Miller, 1968; Nielsen, 1994, 1997) over a 56K modem. To do this, limit the size of each page to 40K or less (Lamers, personal communication, February 27, 1996; Sacharow & Mooradian, 1999; Sullivan, 1998).

Except for the product comparison tool, never require the user to scroll horizontally. Avoid forcing the user to scroll vertically on the home page. It is acceptable, however, to put closely related information (e.g., product details, checkout fields) on a vertically scrollable page (Sacharow and Mooradian, 1999). Put important information "above the fold" (above the vertical scroll line).

Format the pages so that user interface elements are in locations that are familiar to users (Bernard, 2001). Put the return-to-Home hyperlink in the top, left corner. To make it serve double-duty as a branding element, use the company's logo as the hyperlink. Put global navigation controls for the major sections of the site across the top of the page (CyberAtlas, 1999). Place along the left side of the page local navigation controls that work inside each major section. Put on the right side or the bottom, left side of each page controls that take the user off the site (e.g., ads). Locate the search entry field near the top of the page, possibly on the left side, below the global navigation controls. Finally, put banner advertisements at the top of the page.

Except for the shopping cart page and checkout pages, put a shopping cart summary (hyperlinked name of each product, quantity, price, subtotal) on each page (e.g., DVDExpress.com) (CyberAtlas, 1999; Ragus, 2000a). Consider putting the shopping cart summary on the left side of each page, below the search entry field and the local navigation controls. Within the shopping cart summary, include links to the shopping cart and checkout.

To easily accommodate registered members, include sign-in entry fields on the home page and a sign-in hyperlink on every page. Provide a welcome notice (e.g., "Welcome back, johndoe@aol.com") inconspicuously near the top of the home page. Be sure to put a "Contact Us" hyperlink on every page, perhaps at the bottom. A great way to get new, loyal users is to provide at the bottom of each page a link to a simple referral form (e.g., ImageExchange.com "Tell a Friend") (Reichheld and Schefter, 2000). Also, provide links to the privacy and security policies at the bottom of every page (Stanley, McCarthy, & Sharrard, 2000).

2.2 Navigation

Design navigation that is simple, intuitive, and obvious. Put the navigation controls in the same locations on each page. Use navigation to tell the user where the user is, how the user got there, and where else the user can go (Fleming, 1998). This is especially helpful to users who arrive at the page not from the home page, but via a search engine or hyperlink. Provide "breadcrumb" navigation on the site. Breadcrumbs are small, hyperlinked page titles at the top of each page, usually above the title of the current page. These hyperlinks show the page titles that the user came through to get to the current page. Like the breadcrumbs dropped by Hansel and Gretel (Grimm and Grimm, 1999), the purpose of breadcrumb navigation controls is to allow the user to easily retrace the user's path.

Limit the number of major sections in the site to about seven. Provide an intuitive name for each major section. Familiar names will make it easy for the user to quickly browse through the site to a desired product. Use these names as the labels on the global navigation controls. Design the navigation so that the user can browse to any product in five clicks or less (Tracy, 2000). It may be better to provide more category names at each level (a broad design) than to provide more levels to click through (a deep design) (Selingo, 2000). Also provide some specialized browse functions to meet specific user needs. These specialized browse functions can include product ensembles (e.g., Guess.com "Styles"), gifts, holiday specials (e.g., Hallmark.com "Gifts"), and discounted products.

Search is an extremely important navigation technique. In one study (Nua, 2001), 77% of the users said they use search to find products. Forty-three percent of the users said that search was the most important online shopping feature. Unfortunately, many e-commerce sites do not design, maintain, or even evaluate their search functions (Hagen, Manning, & Paul, 2000). One evaluation (Hagen et al., 2000) found that on 30 sites the search function failed to meet minimal standards of usefulness or usability. Another study (Nielsen and Tahir, 2001) found that 36% of the time users could not find what they were looking for when they used the search function. To improve search, use meta tag tools (such as Dynabase and Spectra), thesauruses, alternate spellings, and database search engines (such as EasyAsk, Fact City, iPhrase, Mercado, and Requisite Technology) (Guernsey, 2001). For example, the search function on TowerRecords.com allows users to misspell search words (e.g., enter "phl colins" and get "Phil Collins") and to enter hints (e.g., enter "cannibal" and get "Silence of the Lambs"). After adding this powerful search engine, Tower Records doubled the rate at which users made purchases (Guernsey, 2001). Allow users to search by product name, product category, brand, model/item number, and price (Consumer Reports) (e.g., BarnesandNoble.com Bookstore advanced search). Try to get the most relevant hits to appear on the first page of search results. On the search results page, show the search term(s) and allow the user to perform another search, refine the search results, and sort the search results via helpful product attributes such as price or size (e.g., NetGrocer.com).

2.3 Catalog

Never require users to register to see the product catalog. Avoid requiring users to select a city or enter a ZIP/postal code to see a product catalog. Instead, try to let users get directly into the product catalog without performing extra steps. Organize the product catalog the way users expect the products to be organized (e.g., organize clothing by gender). Since users only look at the first two or three pages of a list of products, allow users to shorten a long list of products by using sorting tools (e.g., Webvan.com) (Nielsen and Tahir, 2001). For example, since a user will only be interested in one shoe size, allow the user to shorten a list of shoes by sorting by size.

On each product page, show a small image of the product and provide a link to a larger, close-up image. This technique reduces download times for product pages and increases the chance that a user will buy (eMarketer, 2001). Show only products that are in stock. If you cannot do this, identify the colors, sizes, or models that are out of stock. If the system cannot check a product's inventory status until the user tries to put the product into the shopping cart, tell the user the product is out of stock, when it will be in stock, suggest an alternate product (e.g., same size shirt in different color), and offer to e-mail the user when the product is available. Never allow the user to put an out-of-stock product into the shopping cart. On the product page, provide links to move the product into a wish list and to allow the user to e-mail the page to someone else. If possible, show a shipping cost for the product (e.g., Costco.com).

Since 34% of users say a product comparison tool increases the chances that they will buy from a site (eMarketer, 2001), allow users to select and compare products side-by-side on important, differentiating features (e.g., eBags.com "My Comparison"). Do not limit the number of products that can be compared. Provide links back to each product's detail page. Make it easy for users to remove products from the comparison.

To make the site more helpful, interesting, and "sticky," consider providing expert and customer reviews. Allow users to enter product reviews on the site.

2.4 Registration

The more streamlined the registration process, the more likely users will register and buy (Agrawal, Arjona, and Lemmens, 2001). Even when they are on the registration page, most users will not read or enter information into the registration fields and half the users will leave on each succeeding registration page (Sacharow and Mooradian, 1999). To reduce the number of entries and to make it easier for users to remember their sign-in names, require a user to enter only an e-mail address and a password. Get permission to e-mail notifications (e.g., sales, new products) (Charron, Bass, O'Connor, & Aldort, 1998) and to leave a cookie (e.g., "Remember me when I return. Yes/No"). Provide in the content area obvious links to the privacy and security policies. Also, since users often have trouble entering their e-mail addresses correctly (Rehman, 2000), provide a sample, correct e-mail address ("e.g., johndoe@aol.com").

Gather other user information several ways. Instead of requiring users to register before checking out, let a new user enter shipping and billing information during checkout (e.g., Amazon.com). At the end of checkout, tell the user the benefits of registering (e.g., quicker future checkouts, personalization, order history), ask the user to register, retain the checkout information for registration, and ask the user to provide a password (Nielsen, 1999). To get additional registration information later, present optional, quick, one-question, multiple-choice surveys (e.g., gender, preferred products) when the registered user signs in or revisits the site. Spread out the surveys by several weeks or months.

Allow users to edit the registration and to un-register. Never show the entire credit card number; show only asterisks and the last four digits.

2.5 Checkout

On the shopping cart page, show hyperlinked product names, entry fields with quantities, prices, a drop-down list of shipping choices and costs, an order subtotal (include shipping costs, plus taxes if the user is registered), check boxes to remove products, check boxes to move products into the user's wish list, a button to refresh the cart page, and links to return to the prior shopping page and to check out. Since 56% of users stopped checkout when they saw high shipping costs at the end of checkout (Hill, 2001), it is important to show shipping costs in the shopping cart. Show a subtotal in the shopping cart so users can see this cost estimate before entering their credit card numbers. If a registered user leaves the site without saving the shopping cart or checking out, save the shopping cart contents for up to 90 days (e.g., Amazon.com).

In one study (Rehman, 2000), checkout problems were the primary reason users could not buy from an e-commerce site. Since 27% of users abandoned an order because the site required them to complete cumbersome forms (Sacharow and Mooradian, 1999), do not require the user to register before checking out (Rehman, 2000). Try to put the checkout fields on a single, vertically scrollable page. This will reduce user confusion and will make it easier for the user to change the order. During checkout, provide an unobtrusive link to allow a member to sign in (e.g., "Returning customer? Sign in.") (e.g., BarnesandNoble.com). If the user is a signed-in member, fill in the checkout fields. If the billing address is the same as the shipping address, do not require a new user to enter the same information again in the billing address fields. Instead, allow the user to click a button near the billing address for "Same as shipping address." Refresh the page and show the shipping information in the billing address fields.

When the user makes an error or leaves empty a required field, show the checkout fields page and put an obvious error message at the top (e.g., "We had a problem processing this page. Will you please try to fix the fields that are marked in red?"). Then, adjacent to each field that had an error, use red text to briefly explain what the trouble was and how to fix it.

Make it easy, safe, and reliable to pay. Design checkout to accept several payment methods (e.g., credit cards, checks, debit cards, gift certificates, phone call with credit card number). Twenty percent of users said they stopped an online purchase because they felt the site was not secure (Hill, 2001). So, certify your site's privacy and security through consumer groups such as TRUSTEe or BBBOnLine and display their logos on the checkout page. Provide obvious links that promote consumer protection features such as privacy policy, security protection, a no-questions-asked return policy, delivery guarantees, and customer support e-mail response time guarantees (Agrawal et al., 2001).

After the user enters the checkout information, provide a complete, editable order summary. For example, provide a "Change Order" button that takes the user back to the single, vertically scrollable, editable, checkout information page. After the system accepts the order, provide an order confirmation that includes the order number, instructions for canceling the order, directions for tracking the order and shipment, customer support information (e.g., e-mail, telephone), and, possibly, a promotion (e.g., sale, special deal) (Ragus, 2000a).

For improved security during checkout, if the user is signed in and wants to change the default shipping address, require the user to enter the user's password. This requirement will prevent someone other than the user from making purchases and sending the purchases to a different address.

3. CONCLUSIONS

To be successful, e-commerce user interfaces must be easy to use. The design suggestions described here should make it easier for users to browse, search, register, and check out of an e-commerce site.

REFERENCES

Agrawal, V., Arjona, L. D., & Lemmens, R. (2001). E-performance: The path to rational exuberance. *The McKinsey Quarterly* [On-line], *1*. Available: http://www.mckinsey.com/

America Online. *Webmaster style guide. Recommended page size* [On-line]. Available: http://webmaster.info.aol.com/webstyle/webpages.html

America Online (2001a, January 2). *AOL holiday season shopping reaches $4.6 billion.* America Online press release [On-line]. Available: http://media.aoltimewarner.com/media/cb_press_view.cfm?release_num=50252085

America Online (2001b, March 8). *AOL membership surpasses 28 million milestone.* America Online press release [On-line]. Available: http://media.aoltimewarner.com/media/cb_press_view.cfm?release_num=50252317

Bernard, M. (2001, Winter). Developing schemas for the location of common Web objects. Usability News. Software Usability Research Laboratory, Wichita State University [On-line]. Available: http://wsupsy.psy.twsu.edu/surl/usabilitynews/3W/web_object.htm

BizRate (2000, October 23). *78% of online shoppers abandon shopping carts according to BizRate survey.* BizRate press release [On-line]. Available: http://www.bizrate.com/content/press/release.xpml?rel=88

Charron, C., Bass, B., O'Connor, C., & Aldort, J. (1998, July). Making users pay. Forrester Report [On-line]. Available: http://www.forrester.com/

Cohen, J., & Thompson, M. J. (1999, February). Mass appeal. The Standard [On-line]. Available: http://www.thestandard.com/article/display/0,1151,4927,00.html

Consumer Reports. *E-ratings: Online shopping guide* [On-line]. Available: http://www.consumerreports.org/Special/Samples/Reports/lookfor.htm

Creative Good (2000, June 12). *The dotcom survival guide.* Creative Good [On-line]. Available: http://www.creativegood.com/survival/

CyberAtlas (1999, February 25). *Online stores lacking. E-tailers should follow lead of offline shops* [On-line]. Available: http://cyberatlas.internet.com/market/retailing/taylor.html

eMarketer (2001, March 12). *Turning shoppers on(line).* eMarketer [On-line]. Available: http://www.emarketer.com/estatnews/estats/ecommerce_b2c/20010312_pwc_search_shop.html

Fleming, J. (1998). *Web navigation: Designing the user experience.* Sebastopol, CA: O'Reilly.

Graphics, Visualization, and Usability Center (1998). *GVU's 9th WWW User Survey.* Atlanta: Georgia Institute of Technology, College of Computing, Graphics Visualization, and Usability Center [On-line]. Available: http://www.gvu.gatech.edu/user_surveys/survey-1998-04

Grimm, J., & Grimm, W. (1999). *Complete fairy tales of the brothers Grimm.* Minneapolis: Econo-Clad.

Guernsey, L. (2001, February 28). Revving up the search engines to keep the e-aisles clear. *The New York Times*, p. 10.

Hagen, P. R., Manning, H., & Paul, Y. (2000, June). *Must search stink?* Forrester Report [On-line]. Available: http://www.forrester.com/ER/Research/Report/0,1338,9412,00.html

Hill, A. (2001, February 12). *Top 5 reasons your customers abandon their shopping carts (and what you can do about it).* ZDNet [On-line]. Available: http://www.zdnet.com/ecommerce/stories/main/0,10475,2677306,00.html

Kadison, M. L., Weisman, D. E., Modahl, M., Lieu, K. C., & Levin, K. (1998, April). *On-line research strategies: The look to buy imperative.* Forrester Report [On-line], *1*(1) [On-line]. Available: http://www.forrester.com/

Lake, D. (2000, April 17). *Navigation: An e-commerce must*. The Standard [On-line]. Available: http://www.thestandard.com/research/metrics/display/0,2799,14110,00.html

Manning, H., McCarthy, J. C., & Souza, R. K. (1998, September). *Interactive technology strategies: Why most Web sites fail*. Forrester Report [On-line], *3*(7). Available: http://www.forrester.com/

Miller, R. B. (1968). Response time in man-computer conversational transactions. In *Proceedings of American Federation of Information Processing Societies Fall Joint Computer Conference, 33*, 267-277.

Nielsen, J. (1994). Response times: The three important limits. In J. Nielsen, *Usability Engineering* (pp. 115-163). San Francisco: Morgan Kaufmann. Available: http://www.useit.com/papers/responsetime.html

Nielsen, J. (1997). *The need for speed* [On-line]. Available: http://www.useit.com/alertbox/9703a.html

Nielsen, J. (1999). *Web research: Believe the data* [On-line]. Available: http://www.useit.com/alertbox/99 0711.html

Nielsen, J. (2000). *Flash: 99% bad* [On-line]. Available: http://www.useit.com/alertbox/20001029.html

Nielsen, J., & Tahir, M. (2001, February). Building sites with depth. In *webtechniques* [On-line] *2001*(2). Available: http://www.webtechniques.com/

Nua (2001, March 13). *eMarketer: Web shoppers know what they want*. Nua Internet Surveys [On-line]. Available: http://www.nua.ie/surveys/?f=VS&art_id=905356549&rel=true

Ragus, D. (2000a). *Best practices for designing shopping cart and checkout interfaces* [On-line]. Available: http://www.dack.com/web/shopping_cart.html

Ragus, D. (2000b). *Flash is evil* [On-line]. Available: http://www.dack.com/web/flash_evil.html

Ragus, D. (2000c). Flash vs. HTML: A usability test [On-line]. Available: http://www.dack.com/web/flash Vhtml/

Rehman, A. (2000, October 16). *Effective e-checkout design*. ZDNet/Creative Good [On-line]. Available: http://www.zdnet.com/ecommerce/stories/evaluations/0,10524,2638874-1,00.html

Reichheld, F. F., & Schefter, P. (2000, July-August). E-loyalty: Your secret weapon on the Web. *Harvard Business Review*, 105-113.

Sacharow, A., & Mooradian, M. (1999, March). *Navigation: Toward intuitive movement and improved usability*. Jupiter Communications.

Selingo, J. (2000, August 3). A message to Web designers: If it ain't broke, don't fix it. *New York Times*, p. D11.

Seminerio, M. (1998, September 10). Study: One in three experienced surfers find online shopping difficult. In *Inter@ctive Week* [On-line]. Available: http://www.zdnet.com/intweek/quickpoll/981007/981007b.html

Stanley, J., McCarthy, J. C., & Sharrard, J. (2000, May). *The Internet's privacy migraine* [On-line]. Available: http://www.forrester.com/

Sullivan, T. (1998). *The need for speed. Site optimization strategies*. All Things Web [On-line]. Available: http://www.pantos.org/atw/35305.html

Tedeschi, B. (1999, August 30). Good Web site design can lead to healthy sales. *New York Times e-commerce report* [On-line]. Available: http://www.nytimes.com/library/tech/99/08/cyber/commerce/30commerce.ht ml

Thompson, M. J. (1999, August 9). How to frustrate Web surfers. *Industry Standard* [On-line]. Available: http://www.thestandard.com/metrics/display/0,1283,956,00.html

Tracy, B. (2000, August 16). Easy net navigation is mandatory – Viewpoint: Online users happy to skip frills for meat and potatoes. *Advertising Age*, p. 38.

Wixon, D., & Jones, S. (1992). Usability for fun and profit: A case study of the design of DEC RALLY version 2. Internal report, Digital Equipment Corporation. Cited in Karat, C., A business case approach to usability cost justification. In Bias, R. G., & Mayhew, D. J. (1994). *Cost-justifying usability*. San Diego: Academic Press.

* Lawrence Najjar is an information architect at Viant. He designs digital businesses that are simple and easy to use. Lawrence has 18 years of user interface design experience and holds a Ph.D. in engineering psychology from the Georgia Institute of Technology.

I give special thanks to Laurie Najjar for her helpful editorial assistance.

Design for Better Information Searching

Xiaowen Fang and Shuang Xu
School of Computer Science, Telecommunications,
and Information Systems, DePaul University
243 S. Wabash Avenue, Chicago, IL 60604
Tel: (312) 362-5206   Fax: (312) 362-6116
Email: xfang@cs.depaul.edu

ABSTRACT

Relevance judgment criteria were examined. Based on these criteria, it was proposed that the hyperlink information of a Web page be presented in the abstract to improve the accuracy of relevance judgment. Two search interfaces were designed: one of them adopted the usual interface of the existing Web search engines with no hyperlink information provided in the returned abstracts; another interface provided number of intra-site links and number of inter-site links. A between-subject t-test with 18 subjects was used to test the hypothesis. The independent variable was type of the search interface and dependent variables included users' search performance and users' satisfaction. Results were analyzed and implications to Web search engine designers were presented.

1. INTRODUCTION

Web search engines serve as catalogs of the Web. They index the Web pages by deploying a special computer program called a "spider" or "robot". Users can enter search terms to query against the index database. The search engine returns a list of Web pages along with short descriptions. The returned Web pages are usually ranked based on some relevance measures designed by the search engine, although the ranking is inaccurate in most cases due to the difficulty of defining relevance. The large volume and unstructuredness of the information make it very difficult to efficiently index the Web pages and thus significantly reduce the power of Web search engines in retrieving useful information requested by a user. Inefficient indexing and inaccurate search queries could easily result in millions of hits for a single search query (Fang & Salvendy, 2000). Among the millions of Web pages returned by a Web search tool, many might be totally irrelevant.

Human-beings are not good at dealing with large amount of textual information. Text processing and language comprehension place heavy demands on working memory (Wickens & Carswell, 1997). Because the capacity of working memory is limited to 5 to 9 independent items (Miller, 1956), reading textual information could be inefficient and also time-consuming. In a Web searching process, it is nearly impossible for searchers to review a large number of the returned Web pages. This increases the difficulty of finding useful information when the ranking of the results is not accurate. Based on the latest Web user survey conducted by Georgia Tech GVU center (1998), 73.6% of the users claimed gathering information for personal needs as a primary use of the Web. The same survey also indicated that 45.4% of the users considered finding new information as one of the problems using the Web and 30.0% of the users had difficulties in finding known information such as a Web page.

The objective of this study was to investigate what information should be presented and how Web searching can be improved given richer information in the abstracts.

2. BACKGROUND LITERATURE

Relevance is a fundamental, though not completely understood, concept for documentation, information science, and information retrieval. Studies on relevance fall into following categories (Mizzaro, 1997): foundations, kinds, surrogates, criteria, dynamics, expression, and subjectiveness.

Besides topicality, some other factors are also affecting user's relevance judgment. There are quite a few studies along this line. Taylor (1986) proposed the following six criteria adopted by users in expressing dichotomous relevance judgments: (1) Ease of use, (2) noise reduction, (3) quality, (4) adaptability, (5) time saving, and (6) cost saving. In an empirical study of weather (multimedia) information evaluation, Schamber (1991) identified 10 criteria grouped in three categories mentioned by users: (1) Information (accuracy, currency, specificity, geographic proximity), (2) source (reliability, accessibility, verifiability, through other sources), and (3) presentation (dynamism, presentation quality, clarity). Park (1993) elicited from academic users some criteria affecting relevance judgment, grouped into three categories: (1) "Internal context," containing criteria pertaining to the user's prior experience (for instance, expertise in subject literature, educational background); (2) "external context," concerning the search that is taking place (for instance, purpose of the search, stage of research); and (3) "problem context," representing the motivations and the intended use of the information (for instance, obtaining definitions of something, or frameworks). In another study conducted by Barry (1994), 23 "criterion categories" were found and classified in the following seven "criterion category groups": (1) Information content of the document, (2) user's background/experience, (3) user's beliefs and preferences, (4) other information and sources within the environment, (5) sources of the documents, (6) document as a physical entity, and (7) user's situation. Cool, Belkin, and Kantor (1993) elicited from about 300 subjects 60 criteria underlying document usefulness judgment, grouped in six categories: (1) Topic, (2) content/information, (3) format, (4) presentation, (5) values, and (6) oneself (personal need or use of the document). Thomas (1993) identified 18 factors, grouped in four categories, affecting Ph.D. students relevance judgment in an unfamiliar environment: (1) Information and knowledge sources, (2) feelings of uncertainty, (3) endurance and coordination, and (4) establishing professional relationship. Bruce (1994) individuated some "document characteristics" (author, title, keywords, source of publication, and date of publication) and "information attributes" (accuracy, completeness, content, suggestiveness, timeliness, treatment), that the judges might use when expressing their relevance judgments. Most of these studies are exploratory and preliminary and there is no consensus of the criteria adopted by users. Barry (1994) and Schamber (1994) proposed that there is a finite set of criteria for users in all types of information problem situation.

Because Web searching is in fact an information retrieving process, it is assumed that some of the criteria obtained from information retrieval still apply for Web searching.

3. HYPOTHESIS
Based on above-mentioned relevance judgment criteria, it is proposed that the hyperlink information of a Web page be presented in the abstract to improve the accuracy of relevance judgment. The hyperlink information such as total links, number of intra-site links, and number of inter-site links could indicate the depth and scope of the Web site. One of the 23 "criterion categories" found by Barry (1994) was depth/scope, defined as the extent to which information provided by the document was in-depth and focused. Therefore, users could use the hyperlink information in judging the relevance and thus might gain higher accuracy and efficiency. Hence, it is hypothesized that providing information about hyperlinks in a Web page in the abstract would improve the accuracy of relevance judgment and thus improve user's search performance.

4. METHOD
4.1 Subjects
Eighteen subjects participated in the experiment. A pre-experiment questionnaire was used to determine candidates' knowledge levels of computer use and Web search tools. The questions asked in the questionnaire included questions regarding computer use, library search tool use, and Web search tool use. Table 1 presents the characteristics of the subjects. As shown in Table 1, there were no significant differences between the two groups of subjects in terms of age, computer experience, library search tool experience and Web search tool experience.

Table 1: Subject Characteristics

Variables	Current search interface (n = 9)		Interface with hyperlink information (n = 9)		F	Pr > F
	Mean	Std.	Mean	Std.		
Age	28.9	7.01	23.7	3.12	4.17	0.058
Computer experience score	4.0	0.00	3.8	0.44	2.29	0.15
Library search tool experience score	3.3	0.71	2.7	1.12	2.29	0.15
Web search tool experience score	3.6	1.01	3.6	0.88	0.00	1.00

Note. Scores for computer experience and library search tool experience: 1=less than one year; 2=1-2 years; 3=Over 2 years but less than 5 years; 4=5 years or more. Scores for Web search tool experiences: 1=less than 6 months; 2=Over 6 months but less than 12 months; 3=1-2 years; 4=More than 2 years.

4.2 Task

The task was to find a relevant hub Web site with maximal external links. The subjects were told that they were searching for information for a class project and that they needed to find an informative Web site as the source of information. The objective of the search was to find the relevant hub site, defined as a site that compiles a set of hyperlinks that point to multiple Web sites related to the same topic, with maximal external hyperlinks. The definition of relevance was simplified in this study. Any web site containing a predefined word was considered as a relevant site. Before subjects began the task, they were given a brief on-line instruction that explained the objective (predefined word) of the search task, some beginning relevant keywords, and a general introduction to the related area. After reading the instructions, subjects began the search process and could use any search strategy to find a relevant hub site (i.e., containing the predefined word) with as many external hyperlinks as possible. After subjects entered a Web site, they were asked to use the "Find" function in the browser (an experimental browser developed on the basis of Internet Explorer 4.0) to check whether the predefined word appeared there. If the predefined word did occur in a Web site, this Web site was considered a relevant one. Once subjects found a better relevant hub site, they were told to add the uniform resource locator (URL) address to the bookmark list. The task was performed in a 60-minute period. The task was performed on the Web. All the searches were conducted in the database maintained by the Web search engine, AltaVista. To ensure that subjects searched among the same set of Web sites, a fixed search period, March 1, 1986 to December 1, 2000, was used. The changes that occurred to the Web pages after this period were excluded from the study. The dates in the search period indicated the last dates the Web pages were updated. The fixed search period was set up by software and could not be changed by the subjects. *System response time*, defined as the elapsed time from a mouse click to the end of loading of a homepage, of each Web site was recorded (in one-hundredths of a second) by the computer. A fixed system delay of 16 seconds (Fang & Salvendy, 2000) for each Web site visit was introduced.

4.3 Experiment design

Two search interfaces were designed: one of them adopted the usual interface of the existing Web search engines with no hyperlink information provided in the returned abstracts; another interface provided number of intra-site links and number of inter-site links. A between-subject t-test was used to test the hypothesis. The independent variable was type of the search interface and dependent variables included users' search performance measured by number of external hyperlinks of the best relevant hub Web site identified during one-hour period and users' satisfaction. User's satisfaction was obtained through a satisfaction questionnaire on the scale 1, *lowest satisfaction*, to 7, *highest satisfaction*. There were thirteen questions in total in the questionnaire. Twelve of these questions measured subjects' general satisfaction and were designed on the basis of the job diagnostic survey proposed by Hackman and Oldham (1980). An example of the question regarding general satisfaction is "How much do you agree with the following statement: generally speaking, I am very satisfied with performing the search tasks?" The remaining question measured subjects' satisfaction regarding the abstracts. The question was "How satisfied do you feel with the abstracts provided by the search engine?"

4.4 Procedure

The 18 subjects were randomly divided into two groups. Each participant was asked to sign a consent form before participating in the experiment. After the participant completed a pre-experiment questionnaire, the training session began. The instructions for the search tool that they would use in the experiment were then presented both on the computer screen and on a hard copy. Subjects were asked to read the instructions carefully, and they were allowed to ask questions. One sample search task was displayed in the task window for them to use for practice. They were told to find a relevant hub site with as many external hyperlinks as possible. If a better relevant hub site was found, they were asked to add this Web site to the bookmark list. The training was self-paced. When the participants felt comfortable using the search tool, some questions were asked concerning the training process. After answering these questions, participants were shown the answers. Participants could go back and play with the search tool until they were comfortable with the functions provided by the search tool. The participant could then start to perform a search task. The search behavior of the participant was recorded by the computer. Upon completion of the task, the participants were asked to fill out a satisfaction questionnaire regarding the search tool. The experiment ended then.

5. RESULTS AND DISCUSSION

The intent of the hypothesis was to test the effectiveness of the hyperlink information provided in the abstracts of search results. It was hypothesized that hyperlink information provided in the abstract would improve user's search performance and result in higher satisfaction. The two dependent variables were number of external links of the best relevant hub site and user's satisfaction. The mean values, standard deviations, and test results of each dependent variable for both groups are presented in Table 2.

Table 2 Comparison between Current Search Interface and Interface with Hyperlink Information

Variables	Current search interface (n = 9)		Interface with hyperlink information (n = 9)		Difference in %	F Value	Pr > F
	Mean	Std.	Mean	Std.			
Number of external hyperlinks of the best relevant hub site	211.2	221.82	492.8	272.90	133.3	5.77	0.029
General satisfaction	61	6.06	60.2	8.11	-1.3	0.04	0.84
Satisfaction regarding abstracts	4.3	1.41	5.8	0.71	34.9	6.54	0.022

As shown in Table 2, significant differences were found in number of external hyperlinks of the best relevant hub site identified during 60 minutes [$F = 5.77$, $p = 0.029$] at $\alpha = 0.05$ level. Table 2 also indicates that there was no significant difference in general satisfaction [$F = 0.04$, $p = 0.84$] at $\alpha = 0.05$ level. There were significant differences found in the satisfaction regarding abstracts [$F = 6.54$, $p = 0.022$] at $\alpha = 0.05$ level. Therefore, the hypothesis was supported.

In a study of relevance, Barry (1994) identified depth/scope of the document as one of the criteria that affected human relevance judgment. Bruce (1994) identified some "document characteristics" (author, title, keywords, source of publication, and date of publication) and "information attributes" (accuracy, completeness, content, suggestiveness, timeliness, treatment), that the judges might use when expressing their relevance judgments. Statistical tests indicate that hyperlink information provided in abstracts could significantly improve user's search performance. This result was consistent with previous research findings. In the context of Web searching, hyperlink information could imply the depth, scope, and completeness of the information offered by a Web site. Therefore, this information could be used by users in judging relevance of the Web site without entering it.

Based on Table 2, although no significant differences were found in general satisfaction, the statistical test of the satisfaction scores regarding abstracts indicates that subjects had higher satisfaction regarding the hyperlink information provided in the abstracts. This supports the hypothesis. The higher subjective satisfaction regarding hyperlink information suggests that subjects could have perceived the benefits gained from hyperlink information and could have been willing to use this feature.

6. CONCLUSION
Based on the hypothesis testing, the following conclusion can be drawn:
- Hyperlink information provided in abstracts returned by Web search engines as search results can help users understand the depth and scope of the Web site and thus make an informative relevance judgment.
- Hyperlink information leads to higher user satisfaction.

REFERENCES:
Barry, C. L. (1994). User-defined relevance criteria: an exploratory study. Journal of the American Society for Information Science, 45, 149-159.
Bruce, H. W. (1994). A cognitive view of the situational dynamism of user-centered relevance estimation. Journal of the American Society for Information Science, 45, 142-148.
Cool, C., Belkin, N. J. , & Kantor, P. B. (1993). Characteristics of texts affecting relevance judgments. In M. E. Williams (Ed.), Proceedings of the 14th National Online Meeting (pp. 77-84). Medford, NJ: Learned Information, Inc.

Fang, X. and Salvendy, G. (2000). Keyword Comparison: A User-Centered Feature for Improving Web Search Tools. International Journal of Human Computer Studies, 52, 915-931.

Georgia Tech GVU center (1998). GVU's 10th WWW User Survey. Available online on June 28, 2000 at http://www.gvu.gatech.edu/user_surveys/survey-1998-10/

Hackman, J. R., & Oldham, G. R. (1980). Work redesign. Reading, MA: Addison-Wesley.

Miller, G. A. (1956). The magical number seven plus or minus two: Some limits on our capacity for processing information. Psychological Review, 63, 81–97.

Mizzaro, S. (1997). Relevance: The whole history. Journal of the American Society for Information Science, 48(9), 810-832.

Park, T. K. (1993). The nature of relevance in information retrieval: An empirical study. Library Quarterly, 63, 318-351.

Schamber, L. (1991). Users' criteria for evaluation in a multimedia environment. Proceedings of the 54th Annual Meeting of the American Society for Information Science (pp. 126-133), Medford, NJ: Learned Information.

Schamber, L. (1994). Relevance and information behavior. Annual Review of Information Science and Technology, 29, 3-48.

Taylor, R. S. (1986). Value-added processes in information systems. Norwood, NJ: Ablex Publishing.

Thomas, N. P. (1993). Information seeking and the nature of relevance: Ph.D. student orientation as an exercise in information retrieval. In Proceedings of the 56th Annual Meeting of the American Society for Information Science (pp. 126-130). Medford, NJ: Learned Information, Inc.

Wickens, C. D., & Carswell, C. M. (1997). Information processing. In G. Salvendy (Ed), Handbook of human factors and ergonomics (pp.89–129). New York, NY: John Wiley & Sons.

Increasing Access of Visually Disabled Users to the World Wide Web

Yanxia Yang and Mark Lehto
School of Industrial Engineering, Purdue University, West Lafayette, IN 47907

ABSTRACT

Two experiments examined the use of image processing, as a means of improving image understandability, and, consequently, access to the web by visually disabled users. In Experiment 1, blind subjects interpreted photographic images (typical of those found on the web) more accurately after image processing. Segmentation was somewhat more effective than edge detection. The understandability of less complicated web images, such as symbols, was not significantly increased by either method. In Experiment 2, blind users performed better on web browsing tasks after web pages were simplified using the segmentation method. These positive results are encouraging and support further research on this topic.

1. INTRODUCTION

Web designers have recognized the needs of visually disabled users and have developed various forms of assistive technology (Boyd et al., 1990; Burger & Stoger, 1996). Screen reader software is particularly useful for providing textual information to visually disabled users (Weber, 1996). A screen reader can also provide information about images and hyperlinks by reading their captions. However, the trend toward more graphically and complexly structured web pages is making it increasingly difficult for screen reader technology to inform users of a web page's contents. One major concern is that this approach does not allow users to conveniently (i.e. non-sequentially) access keywords, sentences, and paragraphs of interest while skipping over text they do not care about.

Several approaches have been used to process visual images into a tactile form more easily understood by visually impaired people (Kawai, 1996; Kurze, 1997; Way & Barner, 1997). The latter researchers found that tactile images were better understood by visually impaired subjects after processing using K-means segmentation and Sobel edge detection algorithms. Image processing clearly provides a promising means of providing access to graphical information for visually impaired people. Segmentation and edge detection algorithms are of particular interest due to their relatively small computational requirements. However, not much is known concerning the types of images on the web that should be processed or how this processing should be done. It seems obvious that image processing will be most beneficial for complex images, such as photographs. It also seems reasonable that processing entire web pages may improve browsing performance by blind users, by facilitating non-sequential access to web page elements. These issues are addressed here in two experiments involving blind subjects.

Experiment 1 compares the effect of segmentation and edge detection algorithms on the understanding of typical images, differing in complexity, found on the web. Experiment 2 tests the effect of processing entire web pages on browsing performance.

2. EXPERIMENT ONE

To test how well blind people understand different forms of graphical information, domain dependent images used on the web were acquired, and then simplified using edge detection and segmentation methods. The ability of blind users to understand the images at different simplification levels was measured and the obtained results were analyzed using analysis of variance and least square means tests.

Fifteen categories of frequently occurring domain dependent images were identified for analysis, from 290 randomly selected web pages, depicting groups of people, females, males, automobiles, airplanes, shoes, houses, pots, mailboxes, flowers, phones, TVs, cameras, trees, or dogs. A photograph and an abstract symbol were selected from each category, resulting in 30 domain dependent images to be analyzed. Three raised tactile versions of each domain dependent image were prepared on 8.5 by 11 inch sheets using a Tactile Image Enhancer. One version was the original unprocessed image. The two other versions, respectively, corresponded to images processed using zero-crossing edge detection (Marr & Hildreth, 1980) and modified K-means segmentation (Pappas, 1992) algorithms. Examples of each version of the studied images are shown in Figure 1.

Image Type	Original Image	Simplified by Edge Detection	Simplified by Segmentation
Photograph		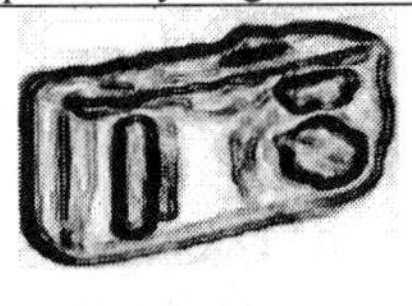	
Symbol			

Figure 1. Example versions of the studied images.

2.1 Procedure

Fifteen blind subjects from Purdue University, the Indiana School for the Blind, and Lafayette, Indiana, participated in the experiment. Two subjects were male, 8 were under 20 years old, 7 were above 30 years old. Nearly all of the subjects was familiar with tactile images. Prior to participating in the experiment, subjects viewed five tactile graphics during a 30 minute training session. The experiment itself was divided into two different sessions. Fifteen images were viewed in each session. The second session was held two to three weeks after the first one. Subjects who viewed photograph images in the first session, viewed symbol images in the second session, and vice versa. Each image within a session was at a different simplification level, five were original images, five were images simplified using edge detection, and five were images simplified using segmentation. This assignment of conditions resulted in three groups of five images, viewed by each subject at one of the simplification levels.

Subjects were limited to two minutes for "viewing" each image. They then answered several questions concerning the image. The first 'open-ended' question asked them to describe the meaning of the image. If they were unable to answer correctly, subjects were asked to choose between the correct answer, a critical confusion close to the correct answer, and two easily ruled out answers[1]. Answers were rated on a four point scale: 1 - incorrect with no understanding; 2 - incorrect, with partial understanding; 3 - correct with the help of a hint; and 4 - correct without a hint. Subjects were then asked to rate their agreement with the two following statements: "The graphic I viewed was very meaningful to me" and "The graphic I viewed was very simple." Ratings were on a seven point Likert scale: 1, *strongly disagree*; 2, *disagree*; 3, *disagree slightly*; 4, *neutral*; 5, *agree slightly*; 6, *agree*; and 7, *strongly agree*. The time needed to complete the experiment was approximately two hours.

2.2 Results

Image identification scores and user rating of image meaning and simplicity were analyzed using ANOVA and least square means analysis. The three independent variables considered in the analysis were Image type, Simplification level, and Subjects. Performance order was randomized rather than included as a factor in the analysis. Because the experiment was constrained by subject and image groupings, a split plot analysis was employed. Image type was tested against the whole plot error with 4 degrees of freedom; simplification level was tested against the subplot error with 64 degrees of freedom.

Table 1 contains the results averaged over all images. Each mean and standard deviation in the table corresponds to 75 observations. Figures 2 and 3 show the distribution of identification accuracy rating scores for photographs and symbols. ANOVA of identification accuracy scores revealed that image type ($F_{(1, 4)}$=51.2, p=0.002) and simplification level had significant effects ($F_{(2, 64)}$=35.5, p<0.0001). The interaction between these variables was also significant ($F_{(2, 64)}$=20, p<0.0001). Least square means analysis revealed that the differences between simplification levels were significant ($F_{(2, 64)}$=54.4, p<0.0001) for photographs, but not for symbols ($F_{(1, 4)}$=1.1, p=0.33). The difference between photographs and symbols was significant for unsimplified images ($F_{(1, 4)}$=131.3, p<0.0001); images simplified with edge detection ($F_{(1, 4)}$=24.3, p<0.0001); and images simplified with segmentation ($F_{(1, 4)}$=8.4, p=0.005). Interestingly, the identification scores for photographs simplified by the segmentation method were not significantly different from those for unsimplified symbols (p =0.17).

---

[1] Hints are often provided when evaluating the ability of blind people to perceive or understand images (Kawai et al, 1990; Way & Barner, 1997).

Table 1. Summary of Means for Experiment One.

Dependent Variables	Simplification level	Photographs		Symbols	
		Means	STD	Means	STD
Identification accuracy	Original	1.7	0.54	3.1	0.73
	Edge Detection	2.6	0.51	3.2	0.57
	Segmentation	2.9	0.57	3.3	0.57
User rating of image meaning	Original	2.3	0.98	4.8	1.30
	Edge Detection	4.3	1.43	5.5	1.16
	Segmentation	4.7	1.28	5.8	1.09
User rating of image simplicity	Original	2.1	0.81	4.9	1.08
	Edge Detection	3.8	1.50	5.5	1.13
	Segmentation	4.7	1.39	5.9	1.04

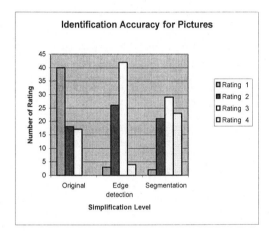

Figure 2. Distribution of rating scores for photograph identification accuracy.

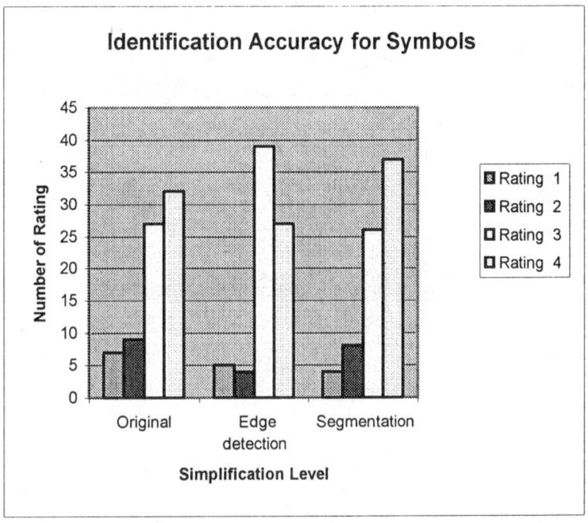

Figure 3. Distribution of rating scores for symbol identification accuracy.

ANOVA of image meaningfulness ratings revealed that image type ($F_{(1, 4)}$=189.9, p=0.0002) and simplification level ($F_{(2, 64)}$=46, p<0.0001) had significant effects. The interaction of these variables was also significant ($F_{(2, 64)}$=9.02, p=0.0004). Least square means analysis revealed that the differences between simplification levels were significant ($F_{(2, 64)}$=47.8, p<0.0001) for both photographs and symbols ($F_{(2, 64)}$=7.2, p=0.0016). The difference between photographs and symbols was significant for original (unsimplified) images, ($F_{(1, 4)}$=88.4, p<0.0001), and for images simplified using edge detection ($F_{(1, 4)}$=21.9, p<0.0001) and segmentation ($F_{(1, 4)}$=14.9, p=0.0003). Photographs simplified by segmentation were not rated significantly different from the original symbols (p=0.8429). For image simplicity ratings, image type ($F_{(1, 4)}$=286.7,<0.0001) and simplification level ($F_{(2, 64)}$=38.5, p<0.0001) had significant effects. The interaction of these variables was also significant ($F_{(2, 64)}$=7.5, p=0.001). Least square means analysis revealed that the differences between simplification levels were significant for photographs ($F_{(2, 64)}$=39.9, p<0.0001) and symbols ($F_{(2, 64)}$=6, p=0.0041). The difference between photographs and symbols was significant for the original images ($F_{(1, 4)}$=88.9, p<0.0001); and for images simplified with edge detection ($F_{(1, 4)}$=31.8, p<0.0001); and by segmentation ($F_{(1, 4)}$=16.9, p=0.0001). Overall, the subjects felt that symbols were simpler than photographs, that images simplified with segmentation method were simpler than images simplified with the edge detection method, and that computer pre-processed images were simpler than the original images.

3. EXPERIMENT TWO

The positive findings of Experiment 1, showing that simplification improved both the identification and understanding of domain dependent graphics in web pages, supported moving on to Experiment 2. Experiment 1 tested individual domain dependent graphics, whereas Experiment 2 tested actual web pages.

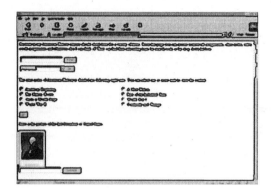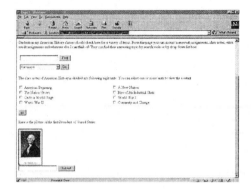

Figure 4. Example web pages in simplified and unsimplified formats.

3.1 Interface Design

As a initial step in developing representative web pages that might be used by blind users, the graphical elements in 50 randomly selected web pages were recorded. The most commonly used elements were data fields (input boxes, text fields) and controls (push buttons, radio buttons, check boxes, list boxes). Each of these domain independent graphic elements were included in the web page design. Domain dependent images were also included, because they were almost always present in the sampled web pages. Task analysis was conducted to identify how experienced blind users currently browse the web. The results of this analysis led us to add several keyboard commands supported by the screen reader software to the web page design. Text captions were provided for every image and input control. Text links, such as "click here," which pose difficulties to visually impaired users were avoided. Frames and tables were not used because they often cause problems when using screen reader programs.

Ten web pages were ultimately developed. Because some of the subjects had only a high school education, web page content was kept very simple. Each web page contained similar graphic elements. Web page content included historical and cultural information that might be relevant in a social studies class; forms for purchasing items over the internet; and routine surveys of consumer preferences. After the web pages were designed, their accessibility was checked using the Bobby application (http://www.cast.org/bobby/). The Bobby reports indicated that each web page met Priority 1 accessibility checkpoints corresponding to the Web Accessibility Initiative (WAI) guidelines recommended in 1999 (http://www.w3.org/TR/WAI-WEBCONTENT/). Two raised tactile versions of each web page screen were prepared on 8.5 by 11 inch sheets using a Tactile Image Enhancer. One version was the original unprocessed image. The other version was processed using the segmentation algorithm of Experiment 1. Figure 4 provides examples of tactile images.

3.2 Methodology

Ten subjects from the original pool of fifteen subjects in Experiment 1 participated in Experiment 2. The task performed by the subjects required them to browse the ten earlier discussed web pages with Internet Explorer to answer three different questions for each web page. The questions asked the subject to find particular graphic elements on the page (i.e. "Please find the input box for customer number."). Subjects located the items using the JAWS screen reader and a raised tactile version of each web page screen. The experiment itself was divided into two different sessions. One group of five subjects used the simplified tactile versions in the first session. After a waiting period of approximately a week, they then used the unsimplified version. The other group used the unprocessed images in their first session, and the processed version in the second. The time subjects spent finding each item was recorded. After completing the tasks for a particular web page, subjects took a short break before starting the next web page. When all ten pages were browsed, they filled out a questionnaire containing 24 items measuring overall satisfaction with the interface (Choong, 1996), and whether subjects felt that simplification helped them identify and locate graphical elements. The browsing task took approximately one hour for each subject.

Before the experiment started, subjects went though a 30 minute training session. During the training session, the experimenter explained the experiment to subjects and instructed them to browse an example web page. Subjects also went through a process in which they touched a raised tactile version of each graphic element while the screen reader spoke its name. The graphic elements were provided on separate sheets. The latter procedure familiarized subjects with the graphical and voice formats of list boxes, data fields, and so forth. Subjects also practiced commonly used keyboard commands for navigating web pages and executing graphical controls.

3.3 Results

Table 2 gives the means and standard deviations of performance time and satisfaction scores for the simplified and unsimplified image conditions. Performance time was analyzed using ANOVA and least square means analysis. The ANOVA revealed that the differences in performance time shown in the table were statistically significant ($F_{(1, 8)}=7.27$, p=0.0273). A paired t-test showed that the difference in satisfaction scores also was statistically significant (p=0.015). An overall Cronbach Alpha (Cronbach & Meehl, 1976) of .71 was calculated for the questionnaire. Interrater reliability was .62 for questions on effectiveness; 0.8 for questions on content; and .48 for questions on presentation.

Table 2. Comparison of Performance Time and Satisfaction Scores.

	Performance Time		Satisfaction Score	
	Mean (sec)	STD	Mean	STD
With Unsimplified Graphics	32.9	16.4	131.4	9.81
With Simplified Graphics	22.9	12.7	144.5	8.55

4. CONCLUSIONS

The ultimate objective of this research is to increase the accessibility of the World Wide Web to visually disabled users through the use of image processing technology. An initial step in this process was to determine how well blind users' understand both simplified and unsimplified graphical information. The results showed, as expected, that blind users found photographs to be more difficult than symbols. They found computer pre-processed images to be more understandable, and providing processed images improved subject performance when browsing the web. Processing photographs often resulted in understanding comparable to that for unprocessed symbols. The segmentation method was found to be a little more effective than the edge detection method. Neither approach increased the understanding of less complicated web images, that is, symbols.

A few specific recommendations can be made for future studies designed to further investigate the hypotheses addressed here. One limitation of this study is that the time blind users required to identify the tactile images in Experiment One was not recorded. The difficulty in finding and recruiting blind subjects also limited the conclusions that could be drawn. Future studies might examine differences between congenitally and late blind subjects. Future work should also expand the pool of sampled images and examine other image processing algorithms to determine which approach is best.

5. REFERENCES

Boyd, L. H., Boyd, W. L. and Vanderheiden, G. C. (1990). The graphical user interface: Crisis, danger, and opportunity. *Journal of Visual Impairment and Blindness*, 84.

Burger, F., Stoger, B. (1996). Access to the WWW for visually impaired people using browsing tools based on GUI: State of the art and prospects. *Interdisciplinary Aspects on Computers Helping People with Special Needs, 5th International Conference*, ICCHP'96, Linz, Austria, 303-310.

Choong, Y. (1996). *Design of computer interfaces for the Chinese population*, Unpublished Doctoral Dissertation, Purdue University, West Lafayette, IN.

Cronbach, L.J. and Meehl, P.E. (1976). *Aptitude and instructional methods—a handbook for research in interaction*, New York.

Kawai, Y.; Tomita, F. (1996). Interactive Tactile Display System - A support system for the visually disabled to recognize 3D objects. *Second Annual ACM Conference on Assistive Technologies*, 45-50.

Kurze, M. and Homes, E. 3D (1996). Concepts by the sighted, the blind and from the computers. *Interdisciplinary Aspects on Computers Helping People with Special Needs, 5th* International Conference, ICCHP'96, Linz, Austria.

Marr, D. and Hildreth, E. (1980). Theory of edge detection. *Proceedings of the Royal Society,* London. B, 207, 187-217.

Pappas, T. N. (1992). An adaptive clustering algorithm for image segmentation. *IEEE Transactions on Signal Processing*, 40, 901-914.

Way, T. P. and Barner, K. E. (1997). *Automatic visual to tactile translation, part II: Evaluation of the tactile image creation system. IEEE Transactions on Rehabilitation Engineering*, 5, p95-105.

Weber, G., Petrie, H. and Mager, R. (1996). Access to MS Windows for blind users. In J. Klaus, E. Auff, W. Kremser and W. L. Zagler (Eds.), *Interdisciplinary Aspects on Computers Helping People with Special Needs*. Vienna: R. Oldenbourg.

Transaction Security in B2C eCommerce: Perceptions and Reality

Jackie Rees

Krannert Graduate School of Management
Purdue University
West Lafayette, IN 47907-1310
Phone: (765) 494-0320
Fax: (765) 494-1526
jrees@mgmt.purdue.edu

Abstract

There are several mechanisms available to ensure acceptable levels of transaction security in a Business-to-Consumer (B2C) eCommerce environment. These security mechanisms are intended to protect both consumer financial information, such as credit card and bank account data, and the financial interests of the company. However, many current and potential eCommerce consumers are uneasy with the perceived lack of transaction security. This study looks at various B2C eCommerce sites and examines the presentation of transaction security information. The findings of this study point to the inadequate presentation of security information as a contributor to consumer concern, rather than the lack of security itself.

Keywords: Information Security, Electronic Commerce, Business to Consumer.

1. Introduction

Initially seen as the promising new model for commerce, Business-to-Consumer (B2C) eCommerce has recently found itself struggling to maintain a presence in many markets. In fact, many companies have either closed down their sites or have consolidated with other stronger players in the market. Many potential factors have been proposed in this current market action. However, one issue that has arisen as a factor in the lack of burgeoning demand for online retail sales is customer service, or the popular perception of a lack thereof. Breaking this factor down further is the lack of a perceived safety in conducting on-line transactions (Lombardi, 1999; Hammonds, 1998). In a brick-and-mortar store, a shopper can typically gauge the "safety" of a transaction through observation or other information channels. If the merchant and/or the shop itself appear less than reputable, the rational customer can choose to use a "safe" form of payment such as cash or can balk and leave the premises. Safety in this context refers to assurances that credit/debit cards are handled appropriately and free from fraudulent use or theft. As more and more consumers begin shopping on-line and on-line revenues comprise a growing share of firms' revenue stream, the ease and safety of transactions have increased. In fact, check and credit card fraud on the Internet comprises a very small amount of all consumer financial transaction fraud and theft (Lombardi, 1999). Despite these improvements, there seems to be resistance to on-line shopping, or at least some hesitation that taints the overall experience for buyers as well as retailers. Therefore, we propose that it is not entirely the security of the system that is at fault but the perception among consumers that accounts for a larger share of the problem. This perception might be due to a lack of information provided to the consumer on a company's eCommerce web site. This paper examines the security statements on the web presences of B2B eCommerce companies. The presence or absence of security-related information is determined and if present, the manner in which the information is presented (placement, graphics, certifications, clarity and level of detail) is analyzed.

1.1 Transaction Security in B2C eCommerce: State of the Industry

Security means many different things to different people. First, we address what is security in this context. Security in eCommerce is the set of processes, both automated and manual that together act to provide assurance as to the integrity of the financial transaction. To the shopper it means the set of controls in place to guarantee the proper exchange of payment for goods or services, free from fraud, abuse, theft or any other civil or criminally liable activities. To the merchant (assuming rationality) it typically means the minimal set of controls necessary to minimize the probability of fraud, abuse, theft or any other civil or criminally liable activities. Next, we discuss the various security mechanisms commercially available intended to protect both merchant and shopper.

There exist various secure payment systems commercially available for the on-line merchant to purchase and implement. One widely used mechanism is SSL (Secure Socket Layer) protocol. The use of SSL allows users to encrypt financial information relating to an order at the client PC (Turban, Lee, King and Chung, 2000). SSL is available in most commercially available and popular Internet browser software, such as Netscape Communicator and Microsoft's Internet Explorer. The shortfall of this setup is that while the transmission of data is encrypted, there are no guarantees once the data are received and stored on the merchant's server. According to recent media reports, the merchant's servers are one of the weak links in financial data assurance. Other protocol systems such as SET (Secure Electronic Transaction) have been proposed but still have the same essential shortcomings of SSL (Turban, et al, (2000).

Other mechanisms exist that can be combined with SSL or SET to provide improved transaction security. VeriSign, Inc. represents what is known as a Certificate Authority (CA). CAs act as a "trusted third party" with authority to oversee a transaction, in a sense verifying and validating the transaction. VeriSign has begun to promote its "brand" as a consumer recognition tool for identifying merchants whose operations conform to VeriSign's expectations of security, analogous to a "Good Housekeeping Seal of Approval" for a particular site (VeriSign, Inc., 2001). Operations such as beyond.com go even further. Beyond.com is itself an e-tailor whose mission is to provide complete on-line store services for other on-line stores (beyond.com, 2001). On-line stores using the services of beyond.com can apply a logo to their website indicating that their site conforms to the standards of beyond.com, an increasingly recognized source of quality in on-line shops. Additionally, existing business verification and validation sources such as the Better Business Bureau have begun "seal of approval" programs for on-line merchants which provide a visible graphic that indicates that the on-line merchant conforms to the certifying organization's guidelines.

Recent media articles indicate that there is a fairly widespread concern among consumers about the security of the online purchasing experience (Lombardi, 1999; Hammonds, 1998). This concern appears to have some validity given the spate of highly publicized "hacks" on companies' servers storing sensitive customer information including credit card data and other personally identifying information. Additionally, given the lack of well-known standards for corporate web page content and design, it might be difficult for most consumers to locate or interpret any security-related information the company has to offer. This leads us into the research question for this study.

1.2 Research Question

Many questions remain unanswered in B2C eCommerce security, including how to create the most secure environment for the least amount of resource while truly enhancing the on-line business experience? The question of how secure are online consumer transactions versus the consumer perception of security is an ever-evolving one. For this research we assume consumers are concerned about security in the on-line environment. We also assume companies engaged in B2C eCommerce activities are operating under bounded rationality and do strive to protect both corporate and consumer assets. If the latter assumption is true and companies are making a reasonable effort to protect consumer information online, are the companies actively raising consumer awareness of these efforts? Therefore, this research will attempt to answer the question of "How is security information on B2C merchant web sites presented?" To help answer this question, we examine several dimensions. Specifically, we examine four factors of security information. These factors are the location of information within the set of pages, the average number of links to find the information, the presentation of information including the presence or absence of graphics and third-party logos and finally the information content which includes clarity and detail.

2. Methodology

The methodology of this preliminary study is based on previously studies performed on the presence and content of information privacy (Culnan, 1999; Goldman, Hudson and Smith, 1999; Landesberg, Levin, Curtin and Lev, 1998). Approximately 85 commercial websites were selected for study. These websites were culled from several popular rankings guides, including the Forrester Group PowerRanking (Forrester, 2001) and ZD Net PC Magazine Top 100 Websites (Willmot, 2000). Only commercial web sites that provided goods and/or services (B2C) directly to consumers were included. Websites that provided intermediary services, such as www.ebay.com and www.priceline.com, were excluded. Three graduate students in MIS completed an instrument for each web site aimed at examining the four factors of security information discussed above. The websites examined can be placed in the following categories: Financial Brokerages, Airlines, Computing Supplies, Apparel, Media (Books, Music and Video), Toys and Games, Health, Flowers, General Merchandise, and Banking. The work was completed over the course of two days and was performed using either Netscape version 4.7 or Internet Explorer 5.0, depending on the preference of the student. The results were tabulated and are presented in Section 3.

3. Results

The first major finding is that the majority of companies sampled are making an effort to include security information on their corporate web sites. Only one site out of the eighty-six sites examined did not provide security information. However twenty-one sites had security information that was difficult to find, meaning not all of the graduate students could find information after a reasonable amount of time. The categories of brokerages (4 out of 9 sites), airlines (3 out of 9 sites), flower retailers (3 out of 7 sites) and banking (3 out of 8 sites) all had a substantial number of sites on which security information was difficult to find or not present. Most security information was available within 2 clicks or links (ranging from 1-3). Many firms locate security information with privacy information. As one of the Fair Information Practice Principles (Laudon, 1996) requires that private information be kept secure, this is a logical grouping. However, the link is often labeled "Privacy Policy" and the user must carefully read through text to find security information. Other locations for security information include "Help Desk" pages (19 out of 86 sites) and "Customer Service" pages (11 out of 86 sites). Other locations included a Frequently Asked Questions (FAQ) page and "Safe Shopping Guarantee" statements. Several sites provided security information at more than one location on their sites. The majority of the firms sampled used only textual information on their security statements (52 out of 86 sites). Most of the remainder participated in third party certification programs (28 out of 86 sites). These firms exclusively used BBB Online, Verisign or TrustE (several firms participated in multiple certification programs). The firms participating in third party programs were almost all from the Computing, Media, Toys and Games and Health categories. Fewer firms used their own graphics in combination with text (8 out of 86 sites). No firms used graphics alone. Font size ranged from small (typically smaller than 10 pt font) to medium (typically 10-14 pt font). Larger fonts were not used according to the students. Wording clarity tended to be average in most cases (one firm was rated as low by all three students and eight were rated as high by all students). Wording detail was similarly rated, with three firms rated by all three students as low and twelve rated by all three students as high.

4. Conclusions

Providing security information is taken very seriously by most merchants in the sample, as measured by the content and design of their security statements. From the merchant point of view, it is advantageous to maximize security investments and to advertise that security for the minimum cost. Many merchants, especially those in the Books, Music and Video category, have found themselves very vulnerable to fraud and theft loss. One interesting result of the study was categorical participation in third party certification programs. Firms in the Computers category most likely have a customer base that is very aware of the implications of security breaches and take extra measures to provide adequate security. Firms in the Healthcare category have been under intense scrutiny regarding privacy concerns and as a result have taken steps to reassure their customers regarding privacy and consequently security of private information. Banking, Brokerage and Airline sites were least likely to participate in any third party certification programs or provide graphics. This could be considered surprising given the dollar volume of transactions processed by these types of firms. In general, having a well-conceived and visible security statement inexpensively build goodwill in a competitive environment.

Some limitations of this study include the sample size and randomness. Only the "top" few sites were used for this study. This allows for bias towards better-developed and maintained sites, as these sites tend to represent the top companies in terms of recognition, hit rates and volume. These sites most likely devote more resources towards design and maintenance. We expect that a larger and more random sample would yield less compliance in terms of providing security information. Also, the measurements for wording clarity and detail need to be more objectively defined to add more insight into the security content. Another issue in most studies of this nature is this information is most likely quite dynamic as the use and acceptance of eCommerce applications grows. Therefore, the information gathered at the time of this study may not be in the same format at a later date.

This study can be characterized as a first step to further examination of security information and how the availability of this information affects online consumer behavior. There is variance in how the information is conveyed to the user. It is logical that different types of companies would organize and present information that is tailored to the needs and interests of their clientele. A future study could examine this tailoring and determine how appropriate this variance is in the attempt to establish well-accepted standards for security information presentation.

References

Beyond.Com. (2001). http://www.beyond.com

Culnan, M. J. (1999). Georgetown Internet Privacy Policy Survey: Final Report. http://www.msb.edu/faculty/culnanm/gippshome.html.

Forrester Research, Inc. (2001). Forrester PowerRankings. http://www.powerrankings.forrester.com./

Goldman, J., Hudson, Z. & Smith, R. M. (2000). Privacy: Report on the Privacy Policies and Practices of Health Web Sites. California HealthCare Foundation.

Hammonds, K. H. (Ed.). (1998, March 16). Online Insecurity. Business Week, 3569, 102.

Landesberg, M. K., Levin, T. M., Curtin, C. G., & Lev, O. (1998). Privacy Online: A Report to Congress. Federal Trade Commission. http://www.ftc.gov/reports/privacy3/toc.htm.

Laudon, K. C. (1996). Markets and Privacy. Communications of the ACM, 39, 92-104.

Lombardi, R. L. (1999). Statistics Show Public's Fears Are Unfounded. Computing Canada, 25, 23-24.

Turban, E., Lee, J., King, D. & Chung, H. M. (2000). Electronic Commerce a Managerial Perspective. Upper Saddle River, NJ: Prentice Hall.

VeriSign, Inc. (2001). VeriSign, Inc. Fact Sheet. http://www.verisign.com.

Willmot, D. (2000). The Top 100 Web Sites. ZDNet PC Magazine. http://www.zdnet.com/pcmag/stories/reviews/0,6755,2394453,00.html#.

THE EFFECT OF POSITIVE AND NEGATIVE INFORMATION IN NOTES TO WEB-BASED FINANCIAL STATEMENTS

Amelia A. Baldwin, Richard B. Dull and Allan W. Graham

Culverhouse School of Accountancy, Culverhouse College of Commerce & Business Administration,
The University of Alabama, Tuscaloosa, AL 35487 USA

School of Accountancy & Legal Studies, College of Business & Behavioral Science, Clemson University,
Clemson, SC 29634 USA

College of Business Administration, University of Rhode Island, Kingston, RI 02881 USA

ABSTRACT

As corporations begin routinely providing financial information on their web sites, users of financial information begin to rely on this delivery method for annual reports. Information producers and users must understand the impact of the report format on users' decisions. Hypertext linking allows information providers to give a direct connection from any specific item to additional relevant information.

Recently, interest has grown among researchers regarding the effect financial information presentation format. Much of this research has included specific information, but not entire sets of financial statements. Hopwood (1996) suggests the need for more research on complete annual reports, not just accounting content of financial statements. Our study extends the concept of format presentation in the era of on-line financial reporting – a concept made possible by the development of the World Wide Web. The current study investigates positive and negative information included in the notes to financial statements presented over the Internet in varied formats. The study examines the differences in decisions, the amount of information used for decisions, and the time to make the decisions, given varied formats and positive and negative information.

1. INTRODUCTION

Today's technological environment allows information providers to make financial information available to decision-makers by a variety of sources and in a variety of formats. Text versions of electronic filings of publicly traded companies have been available through the EDGAR system for some time. Many public companies, using their own web sites, provide current financial statement information in a more esthetically pleasing format, including graphics.

Academic researchers recently started investigating how financial statement information is reported via the Internet. Petravick (1999) reports that 93% of Fortune 150 companies report some type of financial information via the web. Dull (1998) reviewed Russell 1000 companies' web sites and found that 72% of the companies presented some financial information via the web, ranging from quarterly information to complete annual reports. Of these, 60% provided the user with an annual report, most using hypermedia formats.

The increased electronic availability of financial information is consistent with the AICPA's Jenkins Report (1994) focus on the needs of financial statement users. Some users expect companies to provide financial information via the Web. Jensen and Sandlin suggest that accounting researchers should "experiment in the transformation of printed reports into hypermedia databases" (Jensen & Sandlin, 1997, 208).

The Internet's hypertext/hypermedia format may be structured to provide non-linear access to information, allowing flexibility and customization of the search for, and accumulation of information (Kesselman & Trapasso, 1988). Research in the area of hypertext/hypermedia systems suggests that the variety of implementation options, including links, may affect the effectiveness of decisions made using the system (Trumbull, Gay, & Mazur, 1992). Ramarapu, Frolick, Wilkes, and Wetherbe (1997) indicate that certain types of decisions might be influenced by the hypertext format in which the information is presented. Consistent with Hunton and McEwen (1997), Ramarapu et al.

(1997) concluded that a direct method of accessing information resulted in more accurate results for perceptual tasks, and faster decisions for perceptual and analytical tasks. Within an accounting context, Dull, Graham and Baldwin (2000) found that in some cases subjects used more information to analyze financial statements, when links are present.

A steady stream of research has evaluated format differences for decisions based on accounting information. This research has mostly addressed multiple presentation formats for problem solving. Clements and Wolfe (1997) addressed the entire annual report and different formats, but did not investigate decision differences. Hunton and McEwen (1997) looked at sequential versus direct access, but not within the context of hypertext links. The current research extends the Dull, Graham and Baldwin (2000) work, by investigating the effect of positive and negative note information in conjunction within the context of hypertext links. This perspective is important because if the content of the information is significant, and the use of the information is contingent upon the method of access, decision outcomes differ based on the Web format.

Research suggests that decision-making issues relate to information formats, including the links between pages (Kesselman & Trapasso, 1988; Ramarapu et al., 1997; Trumbull et al., 1992). These studies and some in accounting domain (Anderson & Kaplan, 1992; Anderson & Reckers, 1992; Desanctis & Jarvenpaa, 1989) address differences in decisions and/or predictions based on format of information presentation. The creation of a link provides direct access to additional information by electronically directing the user from the current information to additional information. The user has the option to act on this direction by clicking on the link. If the previous research of financial presentation formats hold, one would expect differences in decisions and predictions relative to links.

Within accounting, (Dull et al., 2000) had mixed results when investigating the effect of links on decision-makers. The current study attempts to remove some of the arbitrary differences that may exist when using uncontrolled statements, by using example statements without some of the "noise" of actual statements.

2. METHODOLOGY

The content of the linked/unlinked information may influence the users perceptions of the company and decisions made based on these statement alternatives. In addition to decisions and perceptions, we expect to find differences in the volume of information used to make decisions, and the time used to make decisions. More specifically, the following hypotheses related to the linked/unlinked electronically presented financial statements are investigated in this paper:

Hypothesis 1: Users with links to financial statement notes will use less time in their decision-making process than those users without such links.

Hypothesis 2a: If significant negative information is included in the notes to financial statements, users will have a more negative perception of the company, and decisions or predictions will reflect lower expectations than if the negative information does not exist.

Hypothesis 2b: If significant negative information is included in the notes to *hypertext linked* financial statements, users will have a more negative perception of the company, and decisions or predictions will reflect lower expectations than if the links do not exist.

The research questions were investigated by conducting a laboratory experiment. The financial statements used in the experiment were manipulated for links and the presence of good news or bad news in the notes to the financial statements. The two independent variables (presentation format and good news/bad news) were manipulated by using four web sites that were developed by the researchers. Each site had an introduction page with individual links to each component in an annual report (balance sheet, income statement, statement of cash flows, statement of shareholder equity, auditor report, and each financial statement note). For the experimental company, two of the four sites were non-linked, that is, the individual notes were not directly linked to the financial statements. The other two sites were linked internally (e.g., the balance sheet "inventory" line was linked directly to the inventory note), allowing direct access to the supporting notes. The notes include a good news item for one half of the subjects and a bad news item for the other half. Both the good news item and the bad news item involved the predicted outcome of the probable renewal/loss of a major contract and were reported in the contingencies section of the notes.

2.1   Experimental Task

The experiment was executed using Web-CT, a software tool used primarily for technological support of university courses. The tracking feature documented the information the subjects used while answering the questions posed to them in the experiment: the identity and number of pages accessed, the number of financial statement notes accessed, and the time spent on the total task.

Subjects were advanced undergraduate accounting students at a major East Coast research university. Subjects were randomly assigned to treatment groups. Each student acted in the position of a new credit analyst for a leading regional bank, making the initial credit evaluation for a potential loan client. Each subject was asked to make recommendations on the soundness of the company, predict the next two year's net income, and provide an overall credit rating based on guidelines provided. Differences in the question responses, recommendations, and amount of information used in recommendations were investigated.

2.2   Variables

One independent variable used for the experiment was presentation format. Format for this experiment was defined by whether the statements were directly linked (using hypertext links) to the accompanying notes to the financial statements. The two formats were used to measure the effect of the direct (links to notes provided) versus sequential (links to notes not provided) search of the information. The second independent variable was introduced within each set of statements by providing either a significantly negative or positive financial statement note within the 11 total notes that accompanied the experimental company's financial statements.

Differences among subjects, such as computer experience or course background, that could influence the results, were controlled by the random assignment of subjects to groups. As an additional control, we examined specific potential covariates (all self-reported) that included web experience, comfort with the computer, gender, academic level, and computer experience. The multiple dependent variables included prediction variables, decision variables, the amount of information used, and time.

The decision variables were measured using a seven point Likert-like scale. For example, the question regarding credit rating was measured with one representing "poor credit risk" and seven representing "good credit risk." The amount of information used for the experiment was measured electronically by a count of the pages that the subject accesses during the course of the experiment. The page count was determined by viewing Web-CT's history of pages visited by a particular subject. Duplicate pages visited were eliminated.

The time variable was captured through Web-CT. As a proxy for decision time, the researchers used the total experiment time for each subject. The variable was computed by taking the difference between each subject's first access of the financial statements and the last access.

3.   RESULTS

Based on analyses of the data, support for Hypothesis 1 was found (Table 1). Specifically, those subjects provided links, used significantly less time to complete the task (p=.0182).

With regard to Hypothesis 2a and 2b, the overall MANOVA results for good news/bad news was significant (Table 1). The subjects with bad news gave a significantly lower final credit rating than those with good news (p=.0135) and predicted significantly lower net income in the next two years (p=.007 for year 2, year 1 results are not presented but are significant as well (p=.0516)). The time spent specifically on the note with either good news or bad news (note 7) was examined. Those subjects with bad news spent significantly longer examining the note than those with a good news note (p=.0152). In general, no support was found for the idea that the presence of links influenced the decisions of those viewing either good news or bad news in conjunction with financial statement information.

Table 1

Overall Effects using Manova

Treatment

	F-Value	P-Value	Statistic
Link Style	1.16	.3417	Wilks' Lambda
Good/Bad Note	2.54	.0209	Wilks' Lambda

Two Way Anovas with Link Style and Good/Bad Notes as treatments

Dependent Variable	Model F	Pr > F	Treatments	F-Value	Pr > F		Direction of Effect
Time	3.46	.0382	Link Style	5.92	.0182	**	Linked < Unlinked
			Good/Bad note	.85	.3615		
Time7	3.68	.0314	Link Style	1.28	.2633		
			Good/Bad note	6.26	.0152	**	Good Note < Bad Note
Final Rating	3.70	.0308	Link Style	1.07	.3064		
			Good/Bad note	6.51	.0135	**	Bad Note < Good Note
PredNI02	4.60	.0140	Link Style	1.59	.2118		
			Good/Bad note	7.83	.0070	***	Bad Note < Good Note

Where:
*** Statistically significant difference at p<= .01.
** Statistically significant difference at p<= .05.
* Statistically significant difference at p<= .10.
Time = the time spent by the subject on the experiment
PredNI02 = subject's estimate of the likely net income in the second following year
Time7 = the time spent by the subject viewing Note 7 (which contained either good or bad news).
Final Rating = subject's final credit rating of the company, scale of 1-7 (1 = poor)

4.   CONCLUSIONS

This exploratory research provides information about the environment of increased electronic information dissemination. The Securities and Exchange Commission (1999) categorizes electronic media as a way for companies of any size to remove obstacles relative to providing information to large and small customers. This "leveling of the playing field" should have an impact on the financial markets. It is important to understand the effect of this impact on decision-making. Knowledge concerning this effect is also important as regulators increase their interest in the topic.

The current study is important because disseminating financial information via the Internet is a topic that is growing in importance to the accounting community. Therefore, it is important to understand the complexities of information presentation, and the results that may occur. In light of the current technological environment, research is needed to isolate the causes of differences based on format.
The reporting paradigm appears to be shifting due to the increasing availability of information (Jensen & Sandlin, 1997). Research needs to continue in the area of this new paradigm, to increase the understanding of the implications of the options available to those providing information, and the impact on the ultimate user of the accounting and financial information.
Although the results of this study are mixed, it provides an important step to address the significance of presentation format and decisions made when presenting information over the World Wide Web. Future studies should probably include additional controls, and continue to investigate the affects of format on the decisions of investors.

866

REFERENCES

American Institute of Certified Public Accountants (AICPA). 1994. Improving business reporting -- a customer focus: meeting the information needs of investors and creditors; comprehensive report of the special committee on financial reporting (The Jenkins Report). New York: AICPA.

Anderson, John C., and Steven E. Kaplan. 1992. An investigation of the effect of presentation format on auditors' noninvestigation region judgments. *Advances in Accounting Information Systems* 1:71-88.

Anderson, John C., and Philip M.J. Reckers. 1992. An empirical investigation of the effects of presentation format and personality on auditor's judgment in applying analytical procedures. *Advances in Accounting* 10:19-43.

Clements, C., and C. Wolfe. 1997. An experimental analysis of multimedia annual reports on nonexpert report users. *Advances in Accounting Information Systems* 5:107-136.

DeSanctis, G., and S. L. Jarvenpaa. 1989. Graphical presentation of accounting data for financial forecasting: an experimental investigation. *Accounting, Organizations and Society* 14 (7-8):509-525.

Dull, R.B. 1998. A survey of current, corporate annual reports provided on the World Wide Web, Unpublished manuscript, Indiana University, Indianapolis.

Dull, R.B., and A.W. Graham. 1999. Web-based financial statements: hypertext links and their effect on decisions, Unpublished manuscript, Indiana University, Indianapolis.

Hopwood, A.G. 1996. Introduction. *Accounting, Organizations and Society* 21 (1):55-56.

Hunton, J.E., and R.A. McEwen. 1997. An assessment of the relation between analysts' earnings forecast accuracy, motivational incentives and cognitive information search strategy. *The Accounting Review* 72 (4):497-515.

Jensen, R.E., and P.K. Sandlin. 1997. The paradigm shift: financial reporting will never be the same. *Research on Accounting Ethics* 3:191-209.

Kesselman, M., and L. Trapasso. 1988. Hypertext and the end-user. *Proceedings of the 12th International Online Information Meeting, London*:219-225.

Petravick, Simon. 1999. Online financial reporting. *The CPA Journal* 69 (2):32-36.

Ramarapu, N.K., R.B. Frolick, R.B. Wilkes, and J.A. Wetherbe. 1997. The emergence of hypertext and problem solving: an experimental investigation of accessing and using information from linear versus nonlinear systems. *Decision Sciences* 28 (4):825-849.

Trumbull, D., G. Gay, and J. Mazur. 1992. Students' actual and perceived use of navigational and guidance tools in a hypermedia program. *Journal of Research on Computing in Education* 24 (3):315-327.

US Securities and Exchange Commission (SEC). 1999. *Report to Congress: The impact of recent technological advances on the securities markets* [http://www.sec.gov/news/studies/techrp97.htm]. 1997 [cited July 21 1999].

Adaptive Web interfaces for Electronic Commerce

Nancy J. Lightner, Joyce Jackson

The Darla Moore School of Business, University of South Carolina, 1705 College Street, Columbia, SC 29208

Due to the interactive nature of the Internet, accommodating user's individual preferences is now possible. This paper presents an experiment that tests satisfaction with a Web site that either accommodates or counteracts the visual or verbal information processing tendencies of a user. After simulating an on-line shopping experience, users were asked about their satisfaction with it. Results indicate that regardless of the individual information processing style, the verbal style Web site was preferred. In addition, half of the subjects who experienced the verbal site suggested that pictures would have increased their satisfaction while half of those who experienced the visual site indicated that pictures would have increased their satisfaction.

1. INTRODUCTION

Imagine connecting to a Web site with the intention of buying something. After answering a few questions, the look and feel of the site changes to one that you like better than any site you have visited before. You mark the Uniform Resource Locator as a Favorite, advise your friends to try it, and return again and again to purchase its goods. As a Web e-tailer or an on-line consumer, does this sound too good to be true? New interactive technologies make the extraction and storage of individual information in the Web environment possible. This information may be used to enhance the buying experience for each user. Before we can do this effectively, however, several important areas must be researched. For example, what questions are appropriate to ask as a basis for tailoring the design style of a site to a user? How much more will users like a site that accommodates their style preferences? Will users buy more because they like the site better? This paper focuses on individual on-line preferences, and how providing an interface that adapts to individual characteristics may influence buying satisfaction. The results of the experiment indicate the possible success of an electronic commerce (e-commerce) Web site that accommodates individuals with different information processing styles.

2. BACKGROUND

2.1 Literature Review

Traditional software systems design typically follow the philosophical approach called User-Centered Design (UCD), which focuses on designing from the point of view of the user (Norman and Draper 1986). UCD considers how a user completes a task and interacts with the system that supports the task. This interaction guides the creation of a computer system from the design of the underlying database and system structure to the appearance of the screens. Goodhue and Thompson (1995) formally introduced the concept of integrating user characteristics, the task at hand, and the technology supporting the interaction as the task-technology fit proposition. This proposition posits that a better fit between these three elements will enhance the performance of the user on the task, which is the goal of UCD. An extension of this model includes utilization and elements that influence it (Goodhue 1995). See Figure 1 for a portion of the extended model.

Goodhue's extended model (1995) proposes that User Evaluations of a system act as a surrogate measure for the fit between the task, technology and individual characteristics of the user. How a user rates the system influences the perceived usefulness of the system, which leads to its overall utilization. The concepts of perceived usefulness and utilization are derived from the Technology Acceptance Model (Davis 1989), relative value (Moore and Benbasat 1992), expected consequences of utilization (Fishbein and Ajzen 1975), habit (Bagozzi 1982, Ronis, Yates and Kirscht 1989) and politics (Markus 1983). Since use of a particular Web site, or indeed use of the Web as a buying medium, is completely discretionary, there is interest in creating positive user evaluations and perceived usefulness of an e-commerce Web site.

According to the TTF proposition, the computer interface acts as a facilitator or hindrance to completing a task. In the realm of e-commerce, this interface replaces the bricks and mortar store and must provide at least the same features to be effective. A simple categorization of features of a Web site may be content and style. The Web site should contain expected content, in the form of information and products, and in a style (as reflected in design elements used) that is pleasing and useful to the user.

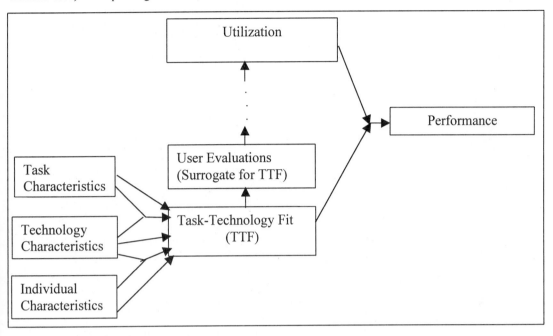

Figure 1: Model of Task-technology Fit and User Evaluations, with Discretionary Utilization (adapted from Goodhue 1995, MIS Quarterly)

This paper focuses on sites with the primary purpose of selling a good or service to an individual. As such, the site should provide information in an entertaining or persuasive manner to entice a purchase. In addition, it should make the buying process fun and efficient for consumers. As a result of display techniques, such as video and font and color variety, Web site designers have creative freedom to fulfill the purpose of the site. The effectiveness of various design techniques, however, has not been rigorously tested in the realm of the Web-based selling. One method of testing effectiveness relies on asking the user to rate their satisfaction with a Web site in a selling environment.

2.2 Hypotheses

Applying the results of a specific individual characteristic scale to a Web design and measuring the impact on the user evaluation in terms of satisfaction results in the following research question: Does tailoring a Web site to an individual style affect the satisfaction with the site?

The results of this research question will indicate the possible success of an e-commerce Web site with individuals having different styles. The hypotheses address satisfaction with a Web site if it is tailored to and if it opposes an individual style. If support is found for these hypotheses, we propose using a brief questionnaire on a user's first visit to a site to determine their individual style preference and altering the site to match that style.

H1: Satisfaction with a Web sites increases if site style matches the results of an individual preference.
H2: Satisfaction with a Web sites decreases if site style opposes the results of an individual preference.

3.0 METHODOLOGY

3.1 Research variables

For testing the hypotheses, an experimental task was designed and conducted to evaluate the impact an individual characteristic has on satisfaction with an e-commerce Web site. The dependent variable of satisfaction was measured using a ten-question survey with each question scored according to a 7-point Likert scale, two of which were reverse coded. Responses were added together to form the satisfaction total. End-user satisfaction is widely used as a surrogate for information technology success. See Mahmood, Burn and Gemoets (2000) for a meta-analysis of its use.

The Verbal/Visual Processing Style measure (Childers, Houston, and Heckler, 1985) was used to determine individual information processing style. This measure is based on Dual Coding Theory (Paivio 1986), which proposes that information coding occurs in one of two subsystems of the brain, either verbal or visual. This theory contends that pictures and words activate independent visual and verbal codes. The measure consists of twenty-two questions and results in two values, one for verbal information processing and one for visual information processing preference. The verbal component has a reported reliability of .81 and the visual component has a reported reliability of .86.

Two Web sites were created for the experimental treatments, one considered verbal and one considered visual. The task that subjects' performed was viewing and interacting with a Web site for the possible purchase of Girl Scout cookies. Girl Scout cookies were selected as the product to purchase since they are familiar and are probably not considered differently by either gender.

3.2 Display Design

Two Web sites of one page each were created to test for site satisfaction. The sites were considered visual and verbal and were designed with the intention of providing equivalent information. The visual site consisted exclusively of design elements that are visual in nature, such as pictures. The site background contained the clover-leaf Girl Scout logo with faces inside. Product descriptions consisted of a picture of the cookie box with two cookies next to it. One of the cookies was broken in half beside the box, to show the outside and inside details of the variety. The pictures were taken from a distance that did not allow subjects to read the words on the box. A drop-down list box containing dashes (-) to indicate None and asterisks (*) to indicate the number of boxes selected to order (one through four). A shopping cart on a button at the bottom of the screen was available for subjects to place their simulated cookie order. See *http://dmsweb.badm.sc.edu/lightner/vis1.htm* to view the actual site.

The verbal Web site used text as the only design element. The background was the words "Girl Scouts of America" repeated over the screen. Product descriptions contained only words. A drop-down list box containing the words None, One, Two, Three and Four was available next to each description to allow a selection of the number of boxes of each cookie type. At the bottom of the verbal site, a 'Buy selection now' link allowed subjects to simulated to place their order. See *http://dmsweb.badm.sc.edu/lightner/verb2.htm* to view the actual site.

3.3 Procedure

Subjects were recruited from various undergraduate and graduate classes in a southeastern university in the United States. After completing an informed consent form, the subjects responded to the verbal/visual processing style questionnaire that measures information processing style. On completion of the questionnaire, subjects were assigned to either the visual or verbal treatment, where they were given brief, written instructions indicating that they were simulating the purchase of Girl Scout cookies on-line. When they finished this simulation, subjects responded to an on-line questionnaire about their satisfaction with the shopping experience. A total of 100 subjects were recruited over a one-week period, which resulted in 97 usable responses. Of the 97, 50 were assigned to the verbal treatment and 47 were assigned to the visual treatment.

4. RESULTS

4.1 Subjects

Of the 97 subjects, sixteen had higher visual processing scores than verbal processing and were considered verbal. Eighty-one of the subjects had higher verbal processing scores. Five subjects had identical verbal and visual scores. Since each subject generated a verbal and a visual score as a result of the verbal/visual processing style questionnaire, a difference in the scores was calculated for each and used in the analysis. The mean difference between verbal and visual scores was 4.77. The range of scores possible on each part of the measure is 11 to 44. Actual test scores ranged from 18 to 34 on the verbal portion and from 13 to 39 on the visual portion. Satisfaction rating scores ranged from 15 to 69 out of a possible 70. The 69 score was reported by a subject given the verbal treatment Web site.

4.2 Data Analysis

An ANOVA was run using the verbal and visual scores, their difference, treatment indication and satisfaction ratings. See Table 1 for a summary of results. In all cases, the verbal treatment was preferred ($\alpha<0.05$) over the visual treatment. These results provide support for Hypothesis I, which indicated that a Web site will be preferred if its style matches that of the individual. However, the results do not support Hypothesis II, which indicated that a Web site would not be preferred if it opposed the style of the individual. Twenty-eight of the 50 subjects assigned to the verbal treatment expressed an interest in seeing pictures of the cookies and 24 of the 47 subjects assigned to the visual treatment expressed an interest in knowing a description of the cookies.

Table 1: Overview of results for Web site satisfaction.

Subjects	N	F value	p-value	R^2 Trt	μ Verbal Trt	μ Visual Trt
All	93	35.84	.0001	.44	49.49	31.70
Visual ss only	16	10.39	.0061	.43	48.89	31.29
Verbal ss only	82	65.65	.0001	.45	49.95	31.79

5. CONCLUSIONS

The results of this experiment indicate that words on an e-commerce Web site result in higher satisfaction than pictures. These findings were consistent, regardless of the information processing style preferred by the subjects. In an environment where Web sites seem to add pictures and graphics for effect, more impact may be achieved by incorporating complete verbal descriptions.

REFERENCES

Bagozzi, R. P. (1982). A Field Investigation of Causal Relations Among Cognitions, Affect, Intentions and Behavior, *Journal of Marketing Research*, *19*, 562-584.

Childers, T. L., Houston, M. J. and Heckler, S. E. (1985). Measurement of Individual Differences in Visual Versus Verbal Information Processing, *Journal of Consumer Research, 12*, 125-134.

Curl, S.S., Olfman, L. and Satzinger, J.W (1998). An Investigation of the Roles of Individual Differences and User Interface on Database Usability. *The DATA BASE for Advances in Information Systems, 29*, 50-65.

Davis, F.D. (1989). Perceived Usefulness, Perceived Ease of Use and User Acceptance of Information Technology. *MIS Quarterly, 13*, 319-340.

Fishbein, M. and Ajzen, I. (1975). *Belief, Attitude, Intentions and Behavior: An Introduction to Theory and Research*, Boston, Massachusetts: Addison-Wesley.

Goodhue, D. L. (1995). Understanding user evaluations of Information Systems. *Management Science, 41*, 1827-1844.

Goodhue, D.L. and Thompson, R.L. (1995). Task-Technology Fit and Individual Performance. *MIS Quarterly*, *19*, 213-236.

Mahmood M.A., Burn J.M. and Gemoets, L.A. (2000). Variables affecting information technology end-user satisfaction: a meta-analysis of the empirical literature. *International Journal of Human-Computer Studies, 52*, 751-771.

Markus, M. L. (1983). Power, Politics and MIS Implementation. *Communications of the ACM, 26*, 430-444.

Moore, G. C. and Benbasat, I. (1992). *An Empirical Examination of a Model of the Factors Affecting Utilization of Information Technology by End Users*. Working Paper, University of British Columbia.

Norman, D. and Draper, S. (1986). *User Centered System Design : New Perspectives on Human-Computer Interaction*, Hillsdale, New Jersey: Lawrence Erlbaum Associates.

Paivio, A. (1986). *Mental Representations: A Dual Coding Approach*. New York: Oxford University Press.

Ronis, D. L., Yates, J. F. and Kirscht, J. P. (1989). *Attitudes, Decisions, and Habits as Determinants of Repeated Behavior*, in Attitude and Structure and Function, Pratkanis, A. R., Breckler, S. and Greenwald, A. G. (editors), Hillsdale, New Jersey: Lawrence Erlbaum Associates.

Validating and Refining User Models of E-Commerce Customers with Usability Test Data

Carl W. Turner

Human Factors Team, State Farm Insurance Companies
Three State Farm Plaza South
Bloomington, IL 61791 USA

Cognitive user models that express customers' "user experience" can help drive the design of e-commerce web sites. Although reliable data on e-commerce customers is often difficult to collect, usability tests are a source of data that can be used to validate and refine predictive user models. A program that combines user modeling and thorough usability testing can influence a company's online enablement strategy and places the human factors analyst at the front of the product development cycle.

1. INTRODUCTION

Human factors analysts understand the processes and techniques that can help drive the design of effective software and e-commerce web interfaces. Frequently, however, their work is performed only during post-design interface testing and evaluation, after most major design decisions have been made. When properly employed, however, cognitive task analysis, user modeling, and user testing can position the human factors analyst to directly influence both corporate enablement strategy and product design and development. This articles discusses a long-term approach for e-commerce design guidance that is built on human factors' traditional strengths of user and task analysis, testing, and product design:

- How user models derived from cognitive task analysis can capture the notion of "user experience."
- How user models can contribute to a company's on-line enablement strategy.
- How high-level models can feed application specific models and prototypes.
- How user models provide the categories for usability test events.
- How usability test data can be used to drive out detail in user models.
- How usability data contributes to company-wide standards and guidelines.

1.1 User Models and Cognitive Task Analysis

User models derived from cognitive task analyses characterize users' behavior and subjective reactions to a system with which they interact. Such analyses are typically performed in preparation for the design of complex systems such as air traffic control centers and nuclear power plant control rooms. User models derived from cognitive task analyses could potentially drive the development of e-commerce sites, but they pose special problems for human factors analysts involved in e-commerce. Unlike the situation found in air traffic control centers, e-commerce tasks are difficult to characterize, the users heterogeneous, and the environments and contexts in which users access Internet sites are uncontrolled. Furthermore, it is difficult to collect data about Internet customers that accurately captures all or even most consumers' behavior. Marketing surveys that aggregate responses to questions do not correlate closely with actual consumer behavior. Clickstream data are available, but are difficult to interpret. If user modeling is to be effective in designing e-commerce sites, there needs to be a dependable way to collect data in order to validate and refine the models.

1.2 Types of Usability Testing

Usability testing is used to best effect in product development when it is conducted iteratively as a part of the user centered design process. There is a wide range of accepted usability test procedures (Rubin, 1994). *Exploratory tests* collect feedback from customers on new product concepts. *Assessment testing* is used to help drive decisions on product design. *Validation testing* is valuable for setting quality baselines and generating historical data. *Comparative tests* are conducted between competing products or between different designs of one's own product. Assessment testing, the most commonly performed type of usability test, is typically a throwaway activity; it works best when designers use it to test hypotheses about prototype designs and obtain direction for redesign. Test materials and procedures vary widely depending on the phase during the development lifecycle the test is run.

1.3 Making Full Use of Usability Testing

Usability testing provides site designers an opportunity to view e-commerce customers under controlled, if artificial, conditions. However, the inherent flexibility of usability testing procedures and the lack of experimental rigor makes it difficult to compare results across tests. There is no experimental control, as it is commonly understood, and the data collected don't add to theory.

Typically, design prototypes are not usability tested against formal models of user behavior. Without a structured model, much of a user's observed behavior is left unexplained or considered uninteresting and not noted during testing. Properly conducted and scored, usability tests can provide a continuous stream of data that can be traced back to one or more models of performance.

2. USER MODELS THAT DRIVE DESIGN

The behavior of e-commerce customers can be modeled at several levels and at different levels of granularity. General models of performance and cognition provide some basic guidelines for design and constrain higher-level, instance-specific models of customers visiting a particular web site. Task-specific models describe the decisions and information required for a user to complete a given task.

2.1 General Models

- Human performance. This model is based on research on perception and memory and includes cognitive limitations. There is a great deal of detailed data available, but is only partially linked to design considerations.
- Page scanning. Based on research on reading and visual search.
- Software interaction. A general model of how users understand how browsers, web sites and software work.
- Meta-cognitive. Recognizes the essential artificiality of the usability test situation. Users often comment on plausibility of the test script, the fidelity of the prototype, and other aspects of the test itself.

2.2 Models that Apply to Specific Instances

- Browser-based applications: task oriented or information seeking. Both types of models are goal based. Task oriented models characterize the customer's goals in terms of procedures. Information seeking models characterize the customer's goals in terms of the site's information architecture.
- Preference for presentation. Individual descriptions of users' preferences for layout, color, font, and graphical appeal.
- Affect. Subjective feelings of trust, security, and confidence in the site or product under consideration.

Used together, cognitive task analyses and usability test data can help produce highly refined user models that can drive the design of complex e-commerce sites. Of course, usability tests are not the only sources of modeling data. Other data may be harvested from experimental studies, surveys, focus groups, and marketing campaigns. Ethnographic and workplace studies are valid sources of behavioral data but time-consuming to collect and interpret. The emphasis in e-commerce customer analysis and modeling efforts should be on behavioral data that can be collected quickly, rather than on self-report data.

3. THE MODELING APPROACH

The modeling approach begins with the assumption that customers rely on a number of mental models when dealing with an e-commerce site. These models may be represented in a number of ways by the modeler; it isn't necessary or even advantageous to select a single representation for all models. A model could start as a simple list of tasks or of the information needed to select a course of action. However, at least one modeling formalism should enable the representation of users' goals and tasks (e.g., John & Kieras, 1994).

Modeling activities start with the analysis and reuse of existing models and data. Literature searches provide a starting point for several types of user models. For example, Byrne et al. (1999) describe a taxonomy of browser tasks that serve as a good starting point for a model of users' understanding of browser based software. As with any complex system, user modeling is never completed, but builds on itself and must be validated continuously.

874

Programmatically, a long-term goal is to create a library of widely applicable, reusable goal-task models ranging from high level, personal goals to task-related practical goals (Turner, 2000).

Low level, task-specific models, such as those for completing an application form, can be derived from task analysis and validated to a high degree of fidelity. Much of the behavior seen in usability testing relates directly to the performance of a given task. Low-level models are traceable to higher-level models, such as those for feelings of trust and security. For high-level models, the analyst is faced with the problem of operationally defining constructs, and deciding what counts as data. Opinions about a site are formed during a task, but users don't reliably offer comments about them. Thus, there is a great deal of skill involved in eliciting useful data about higher-level constructs.

Quantitative user models should 1) account for visible customer behavior, 2) generate predictions of customer behavior, 3) supply direction for design, and 4) capture the notion of the "user experience" with doing business on an e-commerce web site. User experience includes a customer's satisfaction with the content, functionality, ease of use, and utility of doing business on-line with a company's site. User experience also must account for the customer's history with the company in terms of technology, fulfillment, and service.

4. FROM MODELS TO DATA AND BACK

The process of developing models, designing prototypes, and testing end users is an iterative process. Much of the work must go on concurrently in order to meet product deadlines. Figure 1 displays an overview of where modeling and user testing fits within an organizational approach to model-driven development.

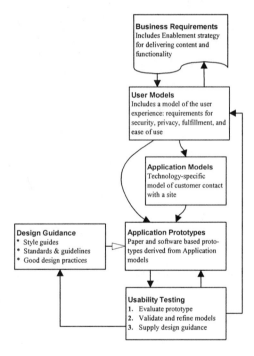

Figure 1. From user models to usability testing.

4.1 Business Requirements and Enablement Strategy

E-commerce companies attempt to define and express their unique value to potential customers. The success of their enterprise depends crucially on the ease with which customers can transact business with the company. Companies must enable current and future customers must be able to learn about and compare their products and services, answer questions and solve problems, and fulfill orders. An enablement strategy describes the company's approach to presenting itself to its customers through its online, phone in, and walk in channels. By defining the customers' information needs and their requirements for security, privacy, and ease of use, detailed user models can directly affect the company's online enablement strategy.

4.2 Modeling the User's Information Needs

In choosing to bank online, a customer weighs the potential advantages in cost and convenience with the potential risk involved. If the customer finds some utility in banking online, then they may gather information on several banks, select a bank, and apply for an account. A high level model of the process is shown in Table 1. As this model is refined and validated, this information informs the choices made on how best to deliver content, present product comparisons, ensure security and trust, and promote good service practices.

The model in Table 1 is not specific to a given technology; it could describe a customer contact through a call center, a visit to a branch, or at a kiosk in a mall. In this example the customer contact is online. This "usage case" form of task analysis emphasizes a customer's decision processes rather than interaction with a specific system (Constantine & Lockwood, 1999). This approach enables the reuse of analyses across different systems and channels.

Table 1. A high-level user model can influence the enablement strategy

Goal: Open and Maintain Online Bank Account(s)	Enablement Strategy: Customer Applications
Assumptions	• Offer multiple contact channels
• Decision made to investigate online banking	• Offer full and fresh content
• Security is a concern but satisfied at time of application	• Enable product and service comparison: Put best features in front of customer
Subgoals	• Offer and define "unique value:" Promote attractive product packages
1. Evaluate banks: Gather information on products, features, and rates	• "Know the customer"
2. Select a bank: Compare and prioritize banks' products and features	• Give feedback in a timely and appropriate manner
3. Apply for account(s): Select a channel and submit application	• Promote feelings of privacy, security, trust
4. Activate account(s): Select payment method and fund account(s)	• Offer service that is easy to find and use
5. Maintain account(s): Fund account(s), check balance(s)	• Build relationship with customer: Show benefits of membership

4.3 Online Account Application Model

Task analyses and reference to high-level user models are employed to generate a technology-specific model of the online application process. The task-specific model in Table 2 is a refinement of part of the higher level goal Open and Maintain Online Bank Account (subgoal 3) shown in Table 1. Note that the process of applying for an account may affect the customer's perception of the advantages and risks of conducting business online with a particular institution. Data on customers' perceptions of products that are still in the design stage are usually only available when usability testing is performed on software prototypes.

In addition to the steps of the process listed in Table 2, cognitive task analysis reveals that customers have expectations about filling out forms online. These expectations derive from previous experience with filling out forms online and on paper, with other companies through traditional channels, with other web sites, with software, and with technology in general. Some expectations include the length of time required to fill out an application, the types and amount of information required to submit a complete application, and how the information will be used. A frequently cited design guideline is, "Don't violate the user's expectations." A well-documented model of the application process makes customers' expectations explicit and serves as a guide to the design of the application process.

4.4 Online Application Prototype

Prototypes are generated from task analyses and existing examples, constrained by company-wide standards and guidelines, industry standards, and product line design guides. User models of customers' perceptions of e-commerce sites act as further constraints on design decisions. As always, good design involves making skilled trade-offs among constraints; it is impossible to optimize all the potential features of an application or a site.

Table 2. User requirements stated in terms of goals

Goal: Apply for Online Account(s)
Assumptions
•   User has collected information and selected a bank
•   User may inquire further about products and features
•   User has expectations for application process: security, ease of use, and feedback
Subgoals
1. Find account application(s): Initiate application and choose product(s)
2. Furnish personal data: Name, address, phone, method of payment
3. Choose feature(s)
4. Submit application and note feedback: who to contact with questions, what steps occur and when
5. Follow up: Sign forms, fund account(s)

4.5 Usability Testing and Categorizing Observations

Task-specific user models, such as those presented in Tables 1 and 2, furnish a description of expected behaviors against which to measure users' observed behavior in usability tests. Each model, as seen in Sections 2.1 and 2.2, supplies the categories to which the observations in usability tests can be assigned. Verbal protocols and observations of significant events during and after testing are traceable to one or more user models.

The impact of usability testing increases as greater use is made of the valuable data generated during testing. In addition to its utility in everyday design activities and model validation, customer-tested workflows, idioms, interactions, and layouts can be captured in corporate and product line design guides. Well-tested design guidance promotes an effective approach to user interface design reuse that is attractive to cost-conscious e-commerce companies.

5. SUMMARY

Implementing a program of user modeling, particularly in its initial phase, is time- and skill intensive. Matching data to models can aid development of e-commerce sites in the long term; however, there is a great deal of overhead involved in order to implement this program. In addition to designers and usability staff, it requires the participation of human factors professionals with a good modeling background. E-commerce companies that seek to understand their customers can benefit in the long term when human factors analysts can better define the "user experience" of transacting business online.

REFERENCES

Byrne, M, John, B., Wehrle, N., & Crow, D. (1999, May). The tangled web we wove: a taskonomy of WWW. In M. Williams, M. Altom, K. Ehrlich, and W. Newman (Eds.), *CHI 99 Conference Proceedings* (pp. 544-551). Pittsburgh, PA: ACM Press.

Constantine, L. & Lockwood, L. (1999). *Software for use: a practical guide to the models and methods of usage centered design.* Pittsburgh, PA: ACM Press.

Cooper, A. (1995). *About face: the essentials of user interface design.* Foster City, CA: IDG Books.

John, B. E., & Kieras, D. (1994). *The GOMS family of analysis techniques: Tools for design and evaluation.* Pittsburgh, PA: School of Computer Science, Carnegie Mellon University.

Rubin, J. (1994). *Handbook of usability testing.* New York: Wiley.

Turner, C. W. (2000, May). A user-centered design methodology for designing intelligent agents. In M. Benedict (Ed.), *Proceedings of the Fifth Annual Conference on Human Interaction with Complex Systems* (pp. 135-139). Urbana-Champaign, IL: HICS.

Decision Support Systems for E-Commerce

Walkyria Rivadeneira and Marc L. Resnick

Industrial and Systems Engineering
Florida International University
Miami, FL 33174

Consumers are shopping on-line in increasing numbers. There has been a corresponding growth in e-commerce sites to meet the needs of these shoppers. However, a challenge that has arisen is that there are too many product choices for consumers to consider fully. The design of product selection decision support systems could alleviate this challenge and streamline e-commerce. These systems can be designed take advantage of the strengths of the Internet while minimize its weaknesses. This research investigates the development of algorithms for the collection of consumer search criteria and presentation of product data to support user needs. An on-line shopping experience was created that simulates realistic e-commerce situations. Two input styles and two output styles that match common consumer decision making strategies were tested. The system that allows users to input data based on importance ratings was superior to one which used absolute minima. Results for output style were inconclusive.

1.0 INTRODUCTION

According to Forrester Research, worldwide Internet retail sales have jumped from a mere $700 million annually in 1996 to over $48 billion in 2000. They had originally predicted that this total would rise to $380 billion by 2003. However, that prediction has recently been reduced. E-marketer predicts that growth in 2001 will be half what is was in 2000. There are several factors that have been used to explain the slowdown in the growth of Internet sales, including poor business models of the companies and privacy issues. However, one of the most important issues seems to be the lack of support on many sites for some basic product filtering tasks.

According to a GVU survey (1999), the four most commonly reported reasons that consumers use the Internet for purchases are for convenience, the availability of vendor information, avoidance of sales pressure, and time. However, just as one of the strengths of the Internet for commerce is the availability of hundreds of vendors for each product, one of its challenges is this same availability. When a retail or business consumer uses the Internet to select a product, he/she needs to search through countless vendors and product choices. Presenting all of them at once would be information overload for the consumer. But if the products are filtered so that only a few are shown, the purchaser must feel confident that the choices are the best ones. An application that would satisfy these two challenges must provide a filtering system that enables the purchaser to input the product characteristics that are most important to his/her search and present the results in a way that can be evaluated easily.

1.1 Consumer Decision Making

Evaluation and selection of products has been investigated by marketing researchers and cognitive engineers for many years. According to Fishbein and Lancaster (1967), consumers define a product by a set of attributes. Consumers differ in their orientation towards these attributes (Swait and Sweeney, 2000). Product selection is the outcome of comparing a personally derived set of criteria. Some criteria may be considered more important than others. The values of all the criteria are integrated and a preference configuration is developed. Based on this configuration the consumer selects a product (Matsatsinis and Samaras, 2000).

Several types of models of this decision making process have been investigated. The ones that seem to describe most consumer purchase situations are the weighted adding and satisficing models. The weighted adding strategy assumes that the consumer can assess the importance of each attribute and assign a subjective value to each possible attribute level. Then the decision maker considers one alternative at a time, examining each of the attributes for that option, multiplying each attribute's subjective value by its importance weight, and summing these resulting values across all of the attributes to obtain an overall value for each option. Then the alternative with the highest value is chosen (Frisch and Clemen, 1994). Weighted adding, however, potentially places great demands on consumers' working memory and computational capabilities and may not be used in many situations. Nevertheless, weighted adding is the decision model that underlies many of the techniques used by market researchers to assess preferences.

Satisficing models propose that consumer decision making is a tradeoff between trying to use as many characteristics as possible to find the best product and a need for efficiency and speed because of time and information processing limitations. The satisficing decision making strategy maintains that consumers consider alternatives sequentially, comparing the value of each attribute to a predetermined minimum level. Only options that meets all minima are considered further. Sometimes the first choice that meets all criteria is selected without consideration of other alternatives. If no alternative passes the cutoffs, more lenient levels can be employed. The extent of processing will vary depending on the exact values of the minima and attribute levels (Simon, 1955).

For a specific purchase, several characteristics are viewed by a particular purchaser as the most critical for that situation. These characteristics change as a function of the consumer and the purchase situation. For each characteristic, there is a minimum level of acceptability and products that do not meet this minimum can be eliminated from consideration immediately in the satisficing strategy but not the weight adding strategy. This filtering effectively reduces the information processing demands by eliminating most product choices and thus the satisficing strategy is easier. Specific comparison of each option is not done until the set of choices has been reduced to a manageable level. Individuals trade imperfect accuracy of their decision in return for a decrease in effort (Bettman, Johnson and Paine, 1990; Johnson and Payne, 1985). Due to this tradeoff, decision makers often select options that are satisfactory but sub optimal when alternatives are numerous and/or difficult to compare (Häubl and Trifts, 2000). This may be descriptive of typical e-commerce purchases where optimality may not be critical.

1.2 Decision Support Systems

One form of managing highly intricate decision environments is to use decision support systems (DSS). A consumer is satisfied with a purchase when he believes that he has thoroughly searched the set of acceptable alternatives without missing an opportunity (Gilovich and Medvec, 1995). When a consumer is using a DSS with an extensive product database, he/she is easily able to locate alternatives that closely match his/her preferences. This increases the likelihood that a set of alternatives that closely matches the consumer's preferences is available. There is thus a smaller likelihood that the consumer will experience a feeling of regret over alternatives that were not considered in the purchase decision. This can lead to greater satisfaction with the decision process. A decision support system that assists purchasers in their product evaluation process by supporting the consumer's decision making process would be more effective than one that provides more information than the consumer can manage or less information than the consumer needs.

1.3 Objective

This study compared the ability of decision support systems based on two competing models of consumer behavior to support the decision making process. One corresponded to the satisficing model of consumer decision making. Another matched the weighted adding strategy, which closely resembles the common Internet practice of providing as much information about all products as possible. Both input styles were matched with alternate output styles that also supported one of these two strategies. The model that matched the product evaluation process of the consumer was expected to be the most effective in supporting product selection, the easiest to use by consumers, and therefore preferred by them.

2.0 METHODS

2.1 Test Materials

2.1.1 Scenario. A product purchase scenario was developed to simulate a realistic purchasing situation. The scenario required the study participant to search for and select a gift for Mothers Day. The study was conducted just prior to Mothers Day to insure that the task was currently relevant to the participants.

2.1.2 Products. A survey was distributed to twenty-five random consumers to collect the important criteria for selecting a Mothers Day gift. Each respondent listed the attributes he/she considered to be important. The nine attributes that were listed most frequently were used in the pilot study. These attributes were the ones used to rate the products and to base the users' search criteria in the experiment.

One hundred products were obtained from popular search engines to collect a range of potential gifts. Actual brand names and prices were used for each product. Each product was scored on a five point Likert scale for each of the nine attributes that were identified in the initial survey. Three individuals performed a preliminary evaluation. Each individual assigned attribute values to the entire set of products. The values were then compared and the final score was obtained by consensus among the group.

2.2 Decision Support Interfaces

2.2.1 Input style. Two input styles were created, corresponding to two common consumer decision making strategies. The weighted-adding input style required participants to rate the level of importance of each attribute on a 5-point Likert scale with verbal anchors of unimportant, neutral, and important and 1, 3, and 5 respectively. The satisficing input style required participants to indicate the minimum acceptance level for each attribute on a 5-point Likert scale with verbal anchors of minimum, average, and maximum and 1, 3, and 5 respectively.

Products in the database were scored according to the decision making algorithm for each strategy. The weighted-adding system calculated a total score for each product by multiplying the product's attribute values with the weighted importance given by the user for each attribute and summing these results across all attributes. The satisficing algorithm eliminated products that did not satisfy all of the minima and calculated a score for each of the remaining products using the sum of the attribute ratings.

2.2.2 Output style. Two output styles were created to support these two decision making strategies. The comparison matrix supported the weighted adding strategy by presenting the top twenty-five products using a picture on the left side of the screen, along with the product's price and scores on each of the attributes (see Figure 1). The recommendation agent supported the satisficing strategy by presenting the top twenty-five products using a picture, a total score, and the price (see Figure 2). Based on these input and output styles, four systems were created, corresponding to a factorial pairing of input and output styles.

2.3 Participants

Seventy-two consumers who were searching for a gift for Mothers Day were identified and invited to participate. There were forty-four males and twenty-eight females. The average age was twenty-seven years. Participants were required to have prior experience shopping online. The participants were recruited in the city of Miami, FL

2.4 Procedure

Before the commencement of the test, an initial screening took place where the prospective participant's eligibility to participate was determined. A pre-test questionnaire was administered to obtain basic demographics and determine if potential participants met the prerequisites of experience with online shopping and a need to acquire a Mother's Day gift. Ineligible prospects were thanked and dismissed.

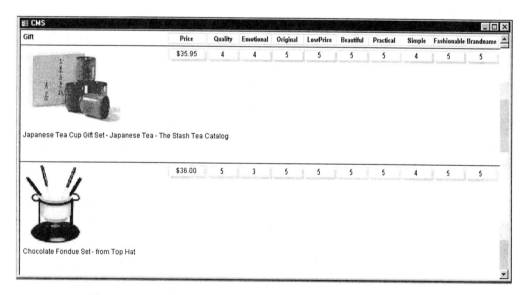

Figure 1. Screen shot of the Comparison Matrix decision support interface

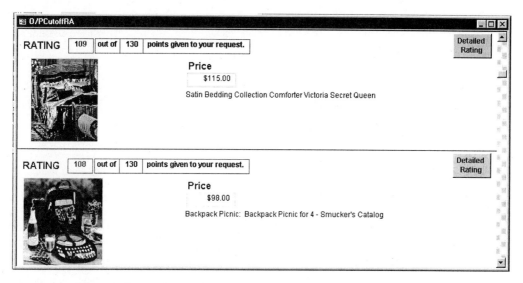

Figure 2. Screen shot of the Recommendation Agent decision support interface

Each participant interacted with two systems, one at a time. The selection of system and the order it was used was balanced to insure that each participant interacted with both decision making styles and that there was no effect of presentation order. The task to search for a Mother's Day gift using one of the decision support interfaces was explained to the participant. The task was subdivided into two subtasks. First, the participant rated the nine attributes according to the input style for that system. Then, the participant evaluated the product information that was presented by the interface and selected a product that best met his/her criteria. The user also had the option to decide that none of the products listed met his/her criteria. A post-test interview was then conducted. Participants rated their satisfaction with the decision support interface. The participant then interacted with the second system using the same procedure. Since time was not a measure collected for the test, the participant was allowed to take as long as he/she required. At no moment was a sense of urgency present, permitting the participant to consider all possible choices before making a decision.

3.0 RESULTS

3.1 User Satisfaction

A two-way Analysis of Variance was used to test the effect of input style and output style on satisfaction with the shopping process. There was a significant difference in reported satisfaction with the input styles (F=7.46 $p<0.05$). The weighted adding input strategy obtained higher ratings of satisfaction than the satisficing strategy, although the magnitude of the difference was small (3.6 compared to 3.2 on the five-point scale). No significant difference was found for the output style or for the interaction between input and output style.

3.2 User Preference

Chi-squared Goodness of Fit tests were used to test the effect of input and output styles on preference for one style over the other. There was a significant difference due to input style (χ^2=40.11, $p<0.05$). The weighted adding input strategy was preferred by 76% of the sample population compared to the satisficing strategy, which was preferred by 24% of the sample population. No significant effect was found for output style.

3.3 Product Choice

Chi-squared Goodness of Fit tests were also used to determine if input or output styles had an effect on whether participants selected one of the top five products recommended by the system. There was no effect of input style on product choice. However, output style did have a significant effect (χ^2=6.224, $p<0.05$). Forty-two percent of the sample population selected a product from among the top-five when using the comparison matrix and twenty-eight percent of the sample population selected a choice among the top-five when using the recommendation agent.

4.0 DISCUSSION

The results revealed some critical insights into the use of shopping assistants and how they should be designed to maximize their benefits to consumers. Widespread availability of electronically provided product information has the potential to significantly affect a consumer's ability to cope with a complex product environment. The presentation of the product information can either facilitate or hinder the decision-making process. This study employed four different decision support systems to test how input and output style affects on-line product search and selection.

User preference was measured because consumers generally revisit web sites that they enjoy using. This is particularly true for service sites such as search engines and shopping assistants. Furthermore, determining which product was truly the "correct" choice is not feasible. The results showed that users preferred the weighted adding input style, but had no preference for output style.

User satisfaction was also compared across all interface formats. User satisfaction has been shown to be critical to generate repeat use of a service site. As with preference, satisfaction data show a superiority of the weighted adding input style but no influence of output style on satisfaction.

The long term success of any service system will also depend on how well it fulfills the needs of the customer. The purpose of shopping assistance systems is to filter the vast number of products available and display only the ones that match the user's needs. The system needs to recommend the product that best fits the needs and expectations of the user. Significantly more participants selected one of the top five products recommended by the comparison matrix compared to the recommendation agent. However, this proportion was less than half for both systems. Few participants in this study selected one of the top recommendations of either style. This may not be the appropriate criterion to use to measure system effectiveness. If the user is satisfied with any of the visible recommendations, he/she can select it for purchase and achieve his/her objective.

This study was a first pass at understanding the complex process of how consumers use product selection decision support systems. Further research is clearly needed. The growth in e-commerce makes the pursuit of this research essential. Search and presentation algorithms that best suit the needs and expectations of users are needed to insure the success of on-line shopping assistants. As the development of electronic agents progresses, these algorithms can be applied to agent design as well.

5.0 REFERENCES

Bettman J.R., Johnson E.J. and Payne J.W. (1990). A componential analysis of cognitive effort in choice. *Organizational Behavior Human Decision Processes*, 45, 111-139.

Fishbein M. and Lancaster (1967). Attitudes and prediction of behavior. In: Fishbein M. (ed). Readings in attitude theory and measurement. New York: Wiley.

Frisch D. and Clemen R.T. (1994). Beyond Expected Utility: Rethinking Behavioral Decision Research. *Psychological Bulletin*, 116 (July), 46-54.

Gilovich T. and Medvec V.H. (1995). The experience of regret: what, when, and why? *Psychological Review*, 102, 379-395.

GVU (1999). GVU Survey Results. Available August 13, 2000 at http://www.searchenginewatch.com/reports/gvu.html.

Häubl G. and Trifts V. (2000). Consumer Decision Making in Online Shopping Environments: The Effects of Interactive Decision Aids. *Marketing Science* 19, 1, 4-21.

Johnson E. J. Payne J.W. (1985). Effort and accuracy in choice. *Management Science*, 31, 394-414.

Matsatsinis N. F. and Samaras A.P. (2000). Brand choice model selection based on consumers' multicriteria preferences and experts' knowledge. *Computers and Operations Research*, 27, 689-707.

PR Newswire (2000). Business-to-Consumer Internet Retail Channel Positioned to Grow Substantially, According to Bank of America Securities Analyst. Feb 4, 2000.

Simon H.A. (1955). A Behavioral Model of Rational Choice. *Quarterly Journal of Economics*, 69, 99-118.

Swait J. and Sweeney J.C. (2000). Perceived value and its impact on choice behavior in a retail setting. *Journal of Retailing and Consumer Services*. 7, 77-88.

Usability Evaluation Methods: What is still missing for the Web?

M. Winckler[a]*, M. Pimenta[b], P. Palanque[a] and C. Farenc[a]

[a]LIHS - Université Toulouse 1
Place Anatole France, 31042 Toulouse Cedex, France.

[b]Instituto de Informática - UFRGS
Caixa Postal 15064, CEP 91501-970, Porto Alegre – RS, Brazil

ABSTRACT

This work presents a study of 23 usability evaluation methods for Web and some tools used to automatic evaluation of Web interfaces. By classifying these methods and tools we have find some missing points for development of new usability methods.

1. INTRODUCTION

When Tim Berners-Lee presented his mockup for the Web in 1989 [1] at the CERN lab, unconsciously he was giving a reason for everyone around the world to use a computer: information supply in large scale and connectivity as never made before. In the last years, the Web is growing exponentially (considering new users and server providers) but, unfortunately, it does not really mean easy-to-use interfaces [2]. In fact, even experienced users are more and more struggling to find the information over the Web due bad design.

Improve the interface design is usually a straightforward process: prototype the interface, identify usability problems and solve them. The problem is that Web presents some special features such as, fast content updating, diversity of technologies to build the interface, cross-platform issues and low budgets for Web projects, etc., that make difficult to apply the traditional usability evaluation methods. In spite of those features, traditional usability methods (such as Heuristic Evaluation and Users Testing) had been used with some changes [3] and new methods [5] (such as evaluation using video-conference) have been proposed. The most recent developments in usability evaluation methods for the Web are concerning with:

- **Low cost methods** to make the evaluation cheap enough to be applied even for small Web project. Gray and Nielsen [4] estimated a budget around US$ 250,00 for each Web page project around the world, so there is not enough money to include any usability evaluation with traditional evaluation methods;
- **Remote evaluation** allows to reach users or evaluators outside of usability labs and in real work conditions taking advantage of distributed nature of Web resources [5];
- **Automated or semi-automated tools** can make easier and faster the evaluation for designers not experts in Computer-Human Interaction - CHI [6]. Tools also can help usability experts collect and analyze special data from user's interaction. Currently, it is available a milestone of methods and tools to accomplish usability evaluation on web interface. Some of them not properly evaluate usability but they can provide some useful information for developers in order to build better Web sites.

Other requirements are not so obvious but not less important are: a) **Real users involvement** in websites projects in a UCD fashion; b) Data from **real user interaction;** and c) **Subjective information** about the

---

* CNPq, Brazilian Council for Research and Development, supports this author.

interface. Nevertheless all effort made still now, the current methods and tools alone not ensure usability in Web projects. There are not graphical designers enough to build all Web interfaces, so people without training on Human-Computer Interaction are making their Web pages and the current usability evaluation methods are too complex to be applied by them.

This work presents a critical discussion about both methods and tools for evaluating Web interfaces highlighting where they are failing to improve usability. Our primary goal is to point out some gaps for further development. Since the main characteristics of each method and tool are also summarized, our secondary goal is to present an up to date state-of-art about usability on the Web.

This work is organized as follow: in the second section we make a clear distinction between Web and WIMP interfaces. In the third section, we present a classification of the analyzed methods considering: user participation, who identify the usability problem, data type collect, degree of automation, experience required for evaluator, information recovered of the evaluation and remote evaluation. The forth section is devoted only for tools for automate usability evaluation. In the fifth we discuss the task of these methods in the design development process for Web interfaces.

2. WEB x WIMP INTERFACES

Since Web have been here we can heard some discussions about differences between Web and Wimp interfaces. We consider the most relevant for our discussion the following ones:

- Users can detour the dialog flow designer for a Web interface. For Wimp, designers have full control over the dialog flow of the interface giving access to presentation units only if some tasks are realized before. For web, browsers facilities such as bookmark, history and direct access to URL let users free to star and skip the dialog flow;
- Presentation units in Wimp, generally, allows much more complex interaction than Web interfaces. We could distinguish two general types of interaction on Web: *hypertext browsing* and *Wimp-objects*. In *hypertext browsing* users can only follow links and fill-in forms. But it is also possible insert *wimp-objects* in Web interfaces that present more complex behavior such as those provide by Java and JavaScript technology. Even when *wimp-objects* are present they affect, generally, the current pages where they are inserted in and, in addition, they also make part of a more complex hypertext systems based in *hypertext browsing*.
- Web interfaces are really easy to build even for non-experienced people. Much more guidance and tools are needed to help those non-experts designer build better interfaces than expert ones. Even expert people could find hard more formal models when modeling very complex systems. So, complex models would be useless without a visual representation.
- Web interfaces are update much more quickly than Wimp interfaces. This has a strong impact over the design once the content of a Web page is strong related with its navigation, so change content or navigation can cause profound damages over general usability when there is no time (or budget) to evaluate daily changes;
- Brower usage is mandatory, they not only show the interface but also include special function that interfere the interface design, for example tools bars supporting navigation and edition (copy, paste, etc.).

3. USABILITY EVALUATION METHODS FOR WEB

We have analyzed 23 usability evaluation methods for Web interfaces that selected from 49 study cases of Web evaluation. Detailed description of this analysis is presented in table 1. Twelve of those methods have been currently used for evaluation of Wimp interfaces. We have classified them according:

a) **User participation**: we distinguish methods that don't require users, such as heuristic evaluation, from others such as user testing (where users are required to use the interface in order to provide data from interaction), and questionnaires (where users only provide comments not using the interface). Finally, if users interact with the interface, what it's the nature of the task realized by them with the interface: normal usage or guided by task;

b) **Who identify the usability problem**: users or evaluators;

c) **What kind of data are collect** for evaluation: directly problem description (as occurs with Heuristic Evaluation method), video tape of test session, verbal user comments, data from questionnaires, log files of user interaction, success observed by evaluators in test sessions, data from simulation, data from inspection. The nature o data collected could be use to determine the complexity of its analysis and quality of problem description.

d) **Degree of automation** of the method: none, capture, analyses or critique;

e) **Level of experience** required for the person dealing the evaluation: few training required; some training or expert in the method;

f) **Remote evaluation**: it is considered if the method could be realized even if users and evaluators can take part in the evaluation in different time and place;

g) **Kind of information recovered**, such as: requirements for interfaces, subjective information (ex; user preferences), natural organization of information, consistence with guidelines (such as ergonomic rules, accessibility, internationalization, etc.), consistence with cognitive models (ex. time of access to information), and malfunction of interface' components (error).

TABLE 1 – Classification of Usability Evaluation Methods.

Method	Interface usage by users			Who identify problem		Data type								Automation type			Evaluator required *	Information recovered **	Remote Evaluation
	Normal	Guided	No direct use	User	Evaluator	prob description	Video tape	User comment	Questionnaire	Log files	Observation	Simulation	Inspection	Capture	analyse	critique			
Card sorting			X		X			X						X			1	1,3	X
Questionnaires			X		X				X					X			2	1, 2	X
User testing in labs	X	X		X	X		X	X		X	X			X			3	1,2,3,6	
Thinking aloud protocol		X			X		X	X			X						3	1,2,3,6	
Icon Intuitiveness			X		X			X									1	1, 3	X
Contextual inquiry	X				X		X	X		X	X			X			3	1,2,3,6	
Log file analysis	X				X					X				X	X		3	3, 6	X
User report CI	X			X		X											1	1-6	X
Remote laboratory testing		X		X	X					X	X			X			3	1,2,3,6	X
Remote formative evaluation		X		X	X	X		X	X	X				X			3	2, 3, 6	X
Inspection using guidelines					X								X				3	4, 5	X
Automatic inspection of structure					X							X				X	3	3	X
Simulation of the interface using user models					X							X				X	3	3	X
Heuristic evaluation					X	X							X				2	3-6	X
Focus Group			X		X	X		X									1	1, 2	X
Merging log files and questionnaires	X				X					X	X			X	X		3	1,2,3,6	X
REmUSINE approach					X							X				X	3	3, 5	X
GOMS approach					X							X				X	3	5	X
Surveys			X		X				X								2	1, 2, 3	X

885

Inspection based on structure visualization				X						X	X	X	3	3	X
Simulation based in content similarity				X						X		X	3	5	X
Automatic visual critic				X						X		X	3	3, 5	X
DomeTree Simulation				X						X		X	3	3, 5	X

* Problem recovered: 1: Requirements for interface, 2: Subjective information (ex. user preferences), 3: Organization of information, 4: Consistence criteria with guidelines (ex. ergonomic guidelines), 5: Consistence with cognitive user models (ex. time access to information), 6: Malfunction of interface's components.

** Evaluator required: 1: few experience required, 2: some training, 3: expert.

4. TOOLS FOR AUTOMATE USABILITY EVALUATION

We also have compared 12 tools for automatic evaluation for Web interfaces. Most of them provide useful help to find problem directly related to HTML code, such as missing alternative tag for images and syntax verifications. Some tools also are able to identify broken links and calculate the downloading time. Brajnik [9] has made a detailed study of 11 tools for automatic usability evaluation considering 44 different items for usability. Our results are similar of that one. The references for tools are presented below:

Link Alarm - http://www.linkalarm.com
CyberSpider - http://www.cyberspyder.com
Set Sim Pro - http://www.northernwebs.com/set/setsimjr.html
Ivry (http://www.crowcastle.net/preston/lvrfy.html
Gif Wizard - http://www.gifwizard.com
WebLint - http://www.ews.uiuc.edu/cgi-bin/weblint
WebTechs – http://www.webtechs.com/
Doctor HTML – http://imagiware.com/RxHTML.cgi
Web Site Garage – http://www.websitegarage.com
WebSat – http://zing.ncsl.nist.gov/WebTools/WebSAT/
HTML Power Analyzer – http://www.tali.com
CSSCheck - http://www.htmlhelp.com/tools/csscheck/

Even though those tools are useful, in the most cases they are not integrated into design tools; typically, they are used during a independent evaluation step. It is not a fault itself, but we think that much more support would be provided for designer if HTML code verification, broken links and so on, are integrated on interface builders fixing problems as soon as they are inserted in the interface. Some Web editors, such as MS FrontPage and Netscape Composer, already give some support and guidance in some level for such problems identified by those tools, for example, inserting size images automatically. Considering this, the usability evaluation step would be mainly centered in solve contextual problems such as *missing information on interface*, *natural organization of information*, *adequacy to cultural aspects*, etc.

5. DISCUSSION ABOUT USABILITY EVALUATION METHODS

We have found that methods traditionally used for evaluation of Wimp interfaces, such as Card Sorting, Questionnaires, User Reporting Problems, Heuristic Evaluation, Focus Group and Surveys methods, are the most frequently reported in the 49 study cases analyzed (41% of all). We suppose that it occurs not only because they are between the less expensive but also because they required few training from evaluators. Easy-to-use methods are important but in fact most of the most methods analyzed are very complex to most people designing Web pages.

The methods based on log file analysis were found to require more experienced evaluators to interpret results. In spite of the diversity of this approach, they still provide few contextual information about usability, and even if there is some tools to automated process, identify the problems is not an easy task. Considering the automation aspect, we agree with the results of Ivory and Hearts [7] about the inexistence about the automatic evaluation based on critique. Most methods cannot be automated and if they support some automation it is strongly based on data capture. It is important to say that automatic methods not are able to capture qualitative information and subjective data such as users preferences. So, methods based on log file analysis, simulation of user interaction and remote user participation, could be used to help designer to manage huge amounts of information.

There is no perfect method that could evaluate all usability aspects of the interface, so more than one must be applied to accurate evaluations. Whatever the method employed, evaluation remains expensive for most projects on Web and inaccessible for most common developers. Tools for design Web interface must provide more guidance and support for designer including automatic code verification of consistence with guidelines as proposed by Scapin et al [11] in order to save time and money of low-budget projects.

Of all methods surveyed, only REmUSINE approach [12] and Automatic inspection of structure [13] are currently using a model of the interface as support to evaluation. Evaluations based on models have been scarcely developed for Web but have been proved their efficacy to help software engineering to build complex and efficient software. Even thought create the interface using models could increase the project complexity, we suggests that is the best approach to integrate user interface requirements in data intensive Web sites.

The advantages of models for design are clear. But it has some important advantages for the evaluation too: automatic verification and correction could be made over the code without design interference; some guidelines could be automatically verified and others, that could not be verified could be informed to users [15]; models give much more elements to deal simulation of the interface.

REFERENCES

1. Berners-Lee, T. Information Management: A proposal. (1999) Available at http://www.w3.org/History/1989/proposal.html
2. Nielsen, J. User Interface Directions for the WWW. *Communications os the ACM*, New York, v.42, n.1, 1999.
3. Nielsen, J.; Sano, D. Sun Web: User Interface Design for Sun Microsystem's Internal Web. *Computer Networks and ISDN Systems*. Amsterdan, v.28, n.1&2. 1995. Available at: http://www.useit.com/papers/sunweb/
4. Gray, W. D.; Nielsen, J. Who you gonna call?: You're on your own something is better than nothing. *IEEE Software*, v.14, n.4, 1997.
5. Hartson, H. R. et. al Remote Evaluation: the network as an extension of Usability Laboratory. *Proceedings... CHI96*, ACM Press, New York, 1996.
6. Scholtz, J.; Laskowski, S.; Downey, L. Developing Usability Tools and Techniques For Designing and Testing Web Sites. *4th Conference on Human Factors & the Web* (1998). Available at: http://www.research.att.com/conf/hfweb/proceedings/scholtz/index.html
7. Ivory; M.; Hearst, M. State of the Art in Automated Usability Evaluation of User Interfaces. Available at: http://www.cs.berkeley.edu/~ivory/research/web/papers/survey/survey.html (2000).
8. Bias, R. Usability Triage for Web Sites. *Proceedings of 6th Conference on Human Factors and Web*. Austin, Texas, 19 June 2000.
9. Brajnik, G. Automatic Web Usability Evaluation. *6th Conference on Human Factors and Web*. Austin, Texas, 19 June 2000.
10. Nakayama, T.; Kato, H.; Yamane, Y. Discovering the Gap Between Web Site Designers Expectations and Users' Behavior. *9th International WWW Conference*, Amsterdam, May 15-19, 2000.
11. Scapin, D.; Leulier, C.; Vanderdonckt, J. Mariage, C.; Bastien, C.; Farenc, C.; Palanque, P. ; Bastide, R. Towards Automated Testing of Web Usability Guidelines. In. *Tools for Working with Guidelines* London Springer; pp. 293-304.
12. Paterno, F.; Ballardin, G. Model-Aided Remote Usability Evaluation. In. *INTERACT'99*, pp.434-442, IOS Press, Edinburgh, September'99. Also available at: http://giove.cnuce.cnr.it/~fabio/interact99.zip
13. Theng, Y. L.; Rigny, C.; Thimbleby, H.; Jones, M. HyperAt: HCI and Web Authoring. HCI'97, Bristol, UK. Available at: http://www.cs.mdx.ac.uk/staffpages/YinLeng/hci97.pdf
14. Automatic Ergonomic Evaluation : What are the Limits ?
15. Farenc, C. ; Liberati, V. ; Barthet, M. In; *CADUI'96 2nd International Workshop on Computer-Aided Design of User Interfaces* - Namur, Belgique, 5-7 juin 1996.

MODELING INTERACTIVE INFORMATION RETRIEVAL (IR): AN INFORMATION SEEKING APPROACH

Amanda Spink

School of Information Sciences and Technology
The Pennsylvania State University
511 Rider I Building, 120 S. Burrowes St.
University Park PA 16802
Tel: (814) 865-4454 Fax: (814) 865-5604
E-mail: spink@ist.psu.edu

Ali R. Hurson

Department of Computer Science and Engineering
The Pennsylvania State University
20 Pond Laboratory, University Park, PA 16802-6106
Phone: (814) 863-1187 Fax:(814) 865-3176
E-mail: hurson@cse.psu.edu

ABSTRACT

Recent research shows that users often engage in successive and multiple information seeking and searching processes on more than one topic at the same time. Successive and multiple topic searching is developing as a major research area that draws together information retrieval (IR) and information seeking studies toward a focus on the broader context of human information behavior. A model of human seeking and searching levels that incorporates successive and multiple topic searching and a theoretical framework for human information coordinating behavior (HICB) are presented. Implications for IR/Web systems design and further research are discussed.

1.   INTRODUCTION

As Web and information retrieval (IR) systems become a major form of information access, human-IR system interaction is being studied in the broader context of their information-seeking behaviors (Spink, 1998). This paper provides a model for information seeking and searching levels (Table 1) that goes beyond the single topic/search approach underlying current assumptions of IR systems design. Recent research shows that: (1) information-seekers often conduct *successive or related searches* over time on the same or evolving topic (Spink, Bateman & Greisdorf, 1999), and (2) humans often conduct information seeking and searching processes on more than one topic at the same time.

This paper will first provide a model that discusses the four levels of information seeking and searching identified in recent studies. Second we identify a range of user behaviors that need to be supported by Web/IR systems and interfaces. The concluding section of the paper discusses the challenge for IR/Web designers and further research issues.

2.   INFORMATION SEEKING AND SEARCHING LEVEL

2.1 Level 1: Single Search on a Single Topic

Current IR/Web systems and interface design is generally based on the assumption that users are engaging in single information seeking processes on single topics when they interact with an IR system. However, recent studies show that users often conduct successive or related searches over time when seeking information on a topic (Spink, 1996; Spink, Bateman & Greisdorf, 1999). A search is defined as all users actions during logging on/logging off period on a database, Web, or IR system. Information-seekers with a topic-at-hand often seek information in stages over extended periods and use a variety of information resources. As time progresses, information-seekers' often search the same or different IR systems for information on the same or evolving topic. As they learn or progress in their work, or as they clarify a topic, or as their situational context changes, users come back to various IR systems for further related searches on the same topic. The process of repeated, related searches over time in relation to a given, possibly evolving, information problem (including changes or shifts in beliefs, cognitive, affective, and/or situational states), is called *a successive searching*.

Table 1. **Model of information seeking and searching levels**

Information Seeking and Searching Level	Number of Search Topics	Numbers of Searches
Level 1: Single search on a single topic	One search topic	One search
Level 2: Successive searching	One search topic	More than one search on a single topic
Level 3: Multiple topic searching	Two or more search topics concurrently	One search on multiple search topics
Level 4: Multiple topic and successive searching	Two or more search topics concurrently	More than one search on multiple topics

2.2   Level 2: Successive Searching

Recent studies show users conduct successive IR searches when seeking information related to a particular topic (Spink, Wilson, Ellis & Ford, 1998; Spink, Bateman & Greisdorf, 1999). Research shows that successive searches are a fundamental aspect of a user's behavior when seeking information related to a topic. Spink, Bateman and Jansen (1999) found that two-thirds of Excite Web users reported a pattern of successive Excite searches on their current topic over time; many reported more than five Excite searches on their topic; and some users reported conducting more than 20 searches on their topic. By user estimates, we find many users repeated Excite searches on the same or evolving topic over time.

Spink, Wilson, Ford, Foster and Ellis (in review) studied searches conducted by librarians with information seekers present. They found that successive mediated IR system searches do not always lead to a decrease in the mean number of items retrieved per search or a greater precision as the number of successive searches increases. Successive searches often provide greater topic clarity for the user, as evidenced by the decline in the percentage of partially relevant items after second searches as information-seekers moved towards more dichotomous relevance judgments over successive searches. Spink, Bateman and Greisdorf (1999) also found a similar result in their study of relevance judgments during mediated successive searching. Previous research by Spink, Greisdorf and Bateman (1998) also found that: (i) partially relevant items decreased over end-user successive searches, and (ii) partially relevant items were significantly linked to changes in the information seeker's understanding of his/her information problem over successive searches. In a study of IR system end-users Vakkari and Hakala (2000) found that partially relevant items increased over successive searches. There is a need for further studies with larger sample sizes to examine these issues.

2.3   Level 3: Multiple Topic Searching

As time progresses, an information-seeker often searches the same or different IR systems for information on more than one topic concurrently. As they learn or progress in their work, or as they clarify these problems and/or questions, or as their situational context changes, users come back to various IR systems for further related searches on more than one topic. The process of searches over time in relation to more than one, possibly evolving, set of topics (including changes or shifts in beliefs, cognitive, affective, and/or situational states), is called *multiple topic information problem seeking and searching processes*.

Results from recent studies of IR system and Web users (Spink, Ozmutlu & Ozmutlu, in review) show that: (1) multiple topic information seeking and searching is a common user behavior, (2) users are conducting information seeking and searching sessions on a continuum from related or unrelated topics, (3) Web or IR multiple topic search sessions are longer than single topic sessions, (4) mean number of topics per Web search session ranged of 1 to more than 10 topics per session with a mean of 2.11 topic changes per session, and (4) many Web search topic changes between hobby and shopping topics.

Multiple topic information seeking and searching is also related to the more micro human cognitive processes of multi-tasking or concurrent information processing. Psychologists have investigated some aspects of multi-tasking/concurrent information processing that occurs over less than second in time. Halford, Maybery and Bain (1986) discuss the two dual-task paradigms and the role of memory during concurrent information processing tasks. Limited research in psychology has focused on information seeking and searching processes.

2.4   Multiple Topic Successive Searching

Our recent research shows that users often work on the resolutions of multiple topics concurrently and conduct successive searches on each topic. Topics may be completely unrelated or related. This leads to some interesting questions for further research. We need to investigate this complex set of behaviors in which different people have multiple information problems concurrently with different levels of information seeking and searching. Such human information behavior leads to a need for users to often coordinate and sustain quite complex information seeking and searching behaviors.

3.   USER INFORMATION BEHAVIORS TO BE SUPPORTED BY IR/WEB SYSTEMS AND INTERFACES

IR/Web interfaces need to adapt to support users engaging in:

- *Single searches*: Single topic
- *Successive searching*: More than one search on a single topic
- *Multiple topic searching*: One search on multiple topics
- *Multiple successive searching*: more than one search on multiple topics

In addition, IR/Web systems and interfaces need to support the following human information behaviors.

3.1   Human Information Coordinating Behavior

Humans' *coordinate* a number of elements, including their cognitive state, level of domain knowledge, and their understanding of their information problem, into a coherent series of activities that may include seeking, searching, interactive browsing and retrieving, and constructing information. The development of our understanding of human information behaviors necessitates a theoretical and empirical explication of the important nature and role of human information coordinating behavior (HICB). There are two levels of HICB's — (1) Information seeking level and (2) Information searching (HCI) level. Humans cognitively coordinate their information seeking level behaviors with their information searching level (human-system interaction) behaviors; including the recognition and making sense of and cognitively articulating an information need or a gap in their knowledge. Human's then coordinate these processes to construct an information-seeking process. Coordination is also related to movement through a human information seeking process. Humans' engage in information coordination in order to move through their information-seeking processes. Part of the information-seeking process is the translation of the information problem into a form that allows them to construct information from texts in the broadest sense and create effective queries. Bringing the elements of their information problem to an effective information seeking and searching process is essential to an effective coordination process.

For example, a human is seeking information on their family history. They enter a library or begin to search the Web. To enable their information-seeking and searching process, i.e., to move forward, they must understand the dimensions of their information problem and coordinate their information seeking and searching processes to the degree that they are able to interact with the functional structure of the library or Web system. The coordination process between information problem and information-seeking/searching process must take place before a human enters a query into the Web or begins to browse the library shelves. The output of the Web search or the books found on the library shelves are coordinated through information feedback by the information-seeker with their information problem through various judgments of the relevance, magnitude and strategic aspects of the information system's output (Spink, 1997).

3.2   Sustaining Human Information Seeking and Searching Processes

At the human cognitive theory and behavior level, we need to help individual humans solve their information problems and sustain their human information behaviors. Sustaining human information behavior processes such as information-seeking processes may include interactions with IR technologies. From childhood humans develop the ability to seek and use information. Within these processes they develop interaction skills to retrieve information from IR systems such as online public access catalogs, the Web and digital libraries. However, research shows that many human interactions with IR systems, such as digital libraries and the Web are short and ineffective (Spink, Wolfram, Jansen & Saracevic, 2001). Humans often do not conduct persistent information seeking and interacting behaviors particularly when interacting with IR systems. We can ask the following questions; how do humans sustain their human information behavior processes and why do they not persist in many instances? How can human information seeking, retrieving and use processes be sustained? Why must they be sustained? Why are the sustaining processes important?

The sustaining aspects are important because human information behaviors, such as information seeking and interactions with IR systems, are complex and cognitively demanding process that must take place over time to be effective. We know that many people browse as an information behavior, preferring to wander and serendipitously meld through information spaces. Browsing can be sustained for longer periods of time than more deliberate searching. However, many tasks require more deliberate searching and more complex interactions with humans and processes or systems. Sustainability includes depth as well as breadth and time. To understand your information problem and articulate and pursue it in sufficient depth and breadth in a manageable timeframe is a key element in human information behavior. Of course, this is based on the assumption that more effective information skills are an increasing crucial element of a successful role in the information or knowledge based economy.

3.3   Coordination Toward Collaboration

IR and HCI are moving toward levels of human and information system collaboration. However, unless a human can effectively coordinate information, collaboration with an information system may be ineffective. Human and information systems must also collaborate to facilitate effective human information coordination. Information coordination goes beyond collaboration – you may collaborate on one level, but not facilitate information coordination.

4.   WEB/IR SYSTEM DESIGN CHALLENGE

Largely, Web/IR systems are currently built following a single search/topic paradigm, i.e., they are designed and operate on the assumption that every search is an unrelated search to any previous, future, or concurrent searches by the user. Research to improve support features for complex information seeking and searching is in its formative stage. We need to further model these complex information seeking and searching processes to examine how successive and multiple topic searches differ from single topic searches. The findings from this research have implications for the design of Web/IR systems. For most users with information problems, seeking information and interacting with IR systems during an information-seeking process will need many forms of support from the IR system during that process. Some key challenges for Web/IR systems and interface designers is to help users keep track of their search processes, coordinate their information problems and sustain their information seeking and searching processes. Some commercial IR systems, such as Dialog (reference), have a "save search" feature based on the assumption that many users come back to the IR system for more than one search on a topic over time. This IR system feature needs to be extended to take account of users successive searches and allow users to store search strategies and results of multiple topic searches for further use or modification. How might multiple topic or successive search sessions be supported by IR systems and interfaces? Does the users stage of an information-seeking process impact the number and performance of concurrent information processes? How might IR systems and interfaces that support successive and multiple information seeking and searching processes be evaluated?

REFERENCES

Halford, G. S., Maybery, M. T., & Bain, J. D. (1986). Capacity limitations in children's reasoning: A dual-task approach. Child Development, 57(3), 616-627.

Spink, A. (1996). A multiple search session model of end-user behavior: An exploratory study. Journal of the American Society for Information Science, 46(8), 603-609.

Spink, A. (1997). A study of interactive feedback during mediated information retrieval. Journal of the American Society for Information Science and Technology, 48(5), 382-394.

Spink, A. (1998). Toward a theoretical framework for information retrieval (IR) within an information seeking context. Proceedings of the 2nd International Information Seeking in Context Conference, August 12-15, 1998. Sheffield, UK: University of Sheffield, Department of Information Studies.

Spink, A., Bateman, J., & Greisdorf, H. (1999). Successive searching behavior during information seeking: an exploratory study. Journal of Information Science, 25(6): 439-449.

Spink, A., Bateman, J., & Jansen, B.J. (1999). Searching Heterogeneous Collections on the Web: A survey of Excite users, Internet Research: Electronic Networking Applications and Policy, 9(2): 117-128.

Spink, A., Greisdorf, H., & Bateman, J. (1998). From highly relevant to not relevant: Examining different regions of relevance. Information Processing and Management, 34(5) (1998), 599-622.

Spink, A., Ozmutlu, H. C., & Ozmutlu, S. (in review). Multiple human information seeking and searching processes. Journal of the American Society for Information Sciences and Technology.

Spink, A., Wilson, T. D., Ellis, D., & Ford, N. (1998). Modeling users' successive searches: A National Science Foundation/British Library Study, D-Lib Magazine, 4(4).

Spink, A., Wilson, T. D., Ford, N., Foster, A., & Ellis, D. (in review). Information seeking and searching. Part 3: Successive searching. Journal of the American Society for Information Sciences and Technology.

Spink, A., Wolfram, D., Jansen, B. J., & Saracevic, T. (2001). Searching the web: The public and their queries. Journal of the American Society for Information Sciences and Technology, 53(2), 226-234.

Vakkari, P., & Hakala, N. (2000). Changes in relevance criteria and problem stages in task performance. Journal of Documentation, 56(5), 540-562.

Using Text Learning to help Web browsing

Dunja Mladenić

J.Stefan Institute, Ljubljana, Slovenia

Carnegie Mellon University, Pittsburgh, PA, USA

Dunja.Mladenic@{ijs.si, cs.cmu.edu}

Abstract

Web browsing is gaining popularity with the growing number of Web users, especially for a casual usage of the Web, when the user does not have a precise query in mind. By observing the user's behavior when browsing, we build a model of promising hyperlinks and use it to highlight hyperlinks on the requested Web pages. In order to do that, we propose text learning methods for handling high dimensional problems (having severals tens of thousands of features) with highly unbalanced class distribution (more than 90% of examples having majority class value). Extensive experimental results were performed on a related problem of modelling Web document content category by using hyperlink to the document. The results show that when modelling by Naive Bayesian classifier, it is highly important how we select the features to be used in the model. Namely, the best performing feature selection in our experiments on Personal WebWatcher data is when the features are scored according to *Odds Ratio* and only a small number of the best features is used for learning.

1 Introduction

Large amount of information available of the Web is attracting many users that are trying to find interesting Web pages. When having an idea about the goal of their search, users typically go to some of the existing search engines and issue a query hoping the target information will be found on some of the top ranked pages. A number of researchers are trying to improve the results of a search engine by addressing the problems such as better query handling by query expansion or stronger query language (eg., request for excluding terms in [AltaVista]), improving search engine's algorithms (eg., hyperlink structure taken into account by [Google]), post processing of the search results by reordering or clustering them (eg., as performed in a research search engine [Manjara] or [Vivisimo]). On the other hand, when the user is not certain what to look for, s/he rather browses the Web by mainly following hyperlinks on the requested Web pages. By processing the requested documents and analyzing their hyperlinks, we can highlight promising hyperlinks and help the user in browsing the Web.

In order to be able to judge the hyperlink we need a model of promising hyperlinks. By promising, we mean here hyperlinks the user is likely to click on, so we want to point them to the user in advance. The model we are using in our approach is based on the content analysis of requested documents and hyperlinks. The methods we use to build this model are text learning methods. The problem of building a model of promising hyperlinks can be seen as a problem of classifying hyperlinks as promising or non-promising. This problem turned out to be non-trivial, by its high dimensionality (ie., a large number of different words that occur in the documents) and a small proportion of positive examples (the proportion of promising hyperlinks among all the hyperlinks is low). To handle this problem, we investigated different ways to reduce a large number of words (perform feature subset selection).

893

2 Domain description

Machine learning problem is here defined as predicting clicked hyperlinks from the set of Web documents visited by the user. This is performed on-line while user is sitting behind some Web browser and waiting for the requested document. Our prototype system named Personal WebWatcher [5] uses text-learning on this problem, learning separate model for each user and highlighting hyperlinks on the requested Web documents. All hyperlinks from the visited documents are used as machine learning examples. Each is assigned one of the two class values: positive (user clicked on the hyperlink) or negative. We use machine learning to model the function $User_{HL} : HyperLink \rightarrow \{pos, neg\}$. Our hope is that this function is also some approximation of interesting hyperlinks (user clicked on hyperlinks that she/he is interested in and skipped all other hyperlinks, that is of course not always true!). We represented each hyperlink as a small document containing underlined words, words in a window around them and words in all the headings above the hyperlink. Our documents are represented as word vectors (using so called bag-of-words representation commonly used in information retrieval) and learning was performed using Naive (simple) Bayesian classifier as commonly used on text data (for overview of text-learning approaches see [6]). For each word position in the document, a feature is defined having a word as its value [4].

Experiments are performed using personal browsing assistant `Personal WebWatcher` that observes users of the Web and suggests pages they might be interested in. It learns user interests from the pages requested by the user. The learned model is used to suggest hyperlinks on new HTML-pages requested by and presented to the user via Web browser that enables connection to "proxy".

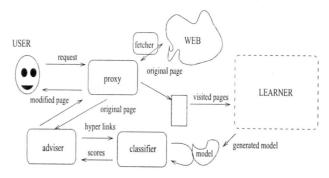

Figure 1: Structure of browsing assistant Personal WebWatcher.

Helping the user browsing the Web is performed here by highlighting interesting hyperlinks on the requested Web documents. We assume that the interesting hyperlinks are the hyperlinks that are highly probable to be clicked by the user. Our problem is defined as predicting clicked hyperlinks from the set of Web documents visited by the user. All hyperlinks that are present on the visited documents are used for constructing machine learning examples.

The data was collected for users participating in the HOMENET project [1]. Results for two users are described in Section 4 with the data characteristics are given in Table 1. For each user approximately 4000 different words occurred in documents resulting here with 4000 features.

3 Feature selection on text data

The usual way of learning on text defines a feature for each word that occurred in the training documents. This can easily result in several tens of thousands of features. Methods for feature subset selection that are used on text are very simple compared to the methods developed in machine

894

Domain (user id.)	Positive class probability	Number of examples	data entropy
usr150101	0.104	2 528	0.480
usr150211	0.044	2 221	0.259
usr150202	0.053	4 798	0.301
usr150502	0.100	2 498	0.468

Table 1: Domain description for data collected from four HomeNet users. It can be seen that we are dealing with unbalanced class distribution, since 10 % or less examples are positive, all other are negative.

learning. Basically, some evaluation function that is applied to a single feature is used. All the features are independently evaluated, a score is assigned to each of them and the features are sorted according to the assigned score. Then, a predefined number of the best features is taken to form the solution feature subset.

Scoring of individual features can be performed using some of the known measures, for instance some measure used in machine learning, such as *Information gain* used in decision tree induction [9]. $InfGain(W) = P(W) \sum_i P(C_i|W) \log \frac{P(C_i|W)}{P(C_i)} + P(\overline{W}) \sum_i P(C_i|\overline{W}) \log \frac{P(C_i|\overline{W})}{P(C_i)}$ In our comparison several feature scoring measures were comapre. Information gain was included as the well known measure successfuly used in some text-learning experiments. Very simple frequency measure proposed in [12] were reported to work well on text data $Freq(W) = TF(W)$. *Odds ratio* is commonly used in information retrieval, where the problem is to rank out documents according to their relevance for the positive class value using occurrence of different words as features [10]. $OddsRatio(W) = \log \frac{P(W|pos)(1-P(W|neg))}{(1-P(W|pos))P(W|neg)}$ Our experiments show that this measure is especially suitable to be used in a combination with the Naive Bayesian classifier for our kind of problems. We propose some variants of *Odds ratio*, to test if the results are sensitive to some modifications in the formula, eg., $FreqLogP(W) = \log \frac{P(W|pos)}{P(W|neg)}$ $ExpP(W) = e^{P(W|pos)-P(W|neg)}$. As a baseline method we used random scoring method defined to score each word by a random number.

Scoring measure	Accuracy					Information score				
best features	10	100	200	500	1000	10	100	200	500	1000
user150101										
ExpP	**94.35**	**94.49**	**94.52**	**94.49**	**94.19**	**0.037**	**0.042**	**0.043**	**0.038**	**0.021**
FreqLogP	94.32	94.49	94.52	94.50	94.18	0.036	0.041	0.042	0.037	0.019
OddsRatio	94.08	94.27	94.04	93.68	93.26	0.030	0.036	0.027	0.016	-0.002
InfGain	94.24	92.62	92.27	92.16	92.09	0.011	-0.064	-0.073	-0.074	-0.071
FreqOddsRatio	92.44	92.35	92.33	92.15	91.87	0.048	-0.045	-0.044	-0.047	-0.06
Freq	91.72	90.85	90.75	90.73	91.74	0.264	-0.242	-0.227	-0.210	-0.197
Random	93.39	93.37	93.37	93.23	92.73	0.005	-0.012	-0.020	-0.035	-0.059
user150211										
ExpP	**96.61**	**96.63**	**96.60**	**96.42**	**95.97**	0	0.001	-0.003	-0.014	-0.039
FreqLogP	96.60	96.62	96.56	96.41	95.97	-0.001	0	-0.005	-0.017	-0.042
OddsRatio	96.61	96.64	96.51	96.22	95.76	0.002	0.005	-0.008	-0.027	-0.055
InfGain	94.31	94.01	93.66	93.29	93.09	-0.226	-0.294	-0.320	-0.334	-0.337
FreqOddsRatio	95.76	95.52	95.43	95.25	94.86	0.130	-0.136	-0.138	-0.145	-0.16
Freq	94.49	92.60	92.05	91.74	91.62	-0.374	-0.40	-0.420	-0.428	-0.427
Random	96.57	996.46	96.38	96.18	95.77	**0.003**	-0.015	-0.035	-0.066	-0.103

Table 2: Comparison of different feature scoring measures giving the average classification accuracy and the average information score of 10 hold-out testing repetitions on the two data sets. The results are given for different number of the best features selected according to each of the scoring measures (10, 100, 200, 500, 1000). The results show here represent a subset of the classification results plotted in Figure 2.

4 Experimental results

We measure classification accuracy, defined as a percent of correctly classified examples and calculated over all classes. We used hold-out testing with 10 repetitions using 30% randomly selected examples as testing examples and reported average value and standard error. Feature selection and learning was performed on training examples only. For each data set we observed the influence of the number of the best features selected for learning to the system performance. Since we have unbalanced class distribution (see Table 1), Classification accuracy can give misleading results. For such domains more appropriate measure is Information score [3]. In the experimental results presented in Figure 2 Classification accuracy and Information score are used to estimate model quality. For both domains the highest Classification accuracy and the highest Information score are achieved by the measures based on Odds ratio: $ExpP, FreqLogP, OddsRatio$ (see Table 2) and Figure 2). For these measures the best vector size is approximately between 60 and 200 best features. This means that the selected feature subset includes just 2% - 5% of all features. The similar reduction (up to 90%) in the number of features used in text-learning was observed in [12]. The other three measures ($InfGain, Freq, FreqOddsRatio$) for most vector sizes perform about the same or even worse than *Random*. Closer look to the words sorted according to Information gain showed that the most best words are characteristic for negative class value (their probability estimated for positive documents is 0). This results were also confirmed by experiments on document categorization into hierarchical structure of Web documents, Yahoo! [7]. This means that in classification, a new positive hyperlink is represented with a word vector almost full of zeros, since it contains very few of the selected best words. In our experiments we didn't remove any common or frequent words. That resulted with html-tags and other common words beeing the most frequent, contributing to the poor performance of the Frequency measure *Freq*. Our explanation for the poor results achieved by the combination of Frequency and Odds ratio *FreqOddsRatio* is that the value of Frequency is standing out in this combination. Odds ratio has most values between 1 and 20 while Frequency has values between 1 and 1000, resulting in their. combination having values between 1 and 2000. In case of the combination of Frequency with logarithm of probability ratio *FreqLogP*, the logarithmic part is standing out and the measure achieves better results.

References

[1] Kraut, R., Scherlis, W., Mukhopadhyay, T., Manning, J., Kiesler, S., The HomeNet Field Trial of Residential Internet Services, *Communications of the ACM* Vol. 39, No. 12, pp.55—63, December 1996.

[2] Joachims, T., A Probabilistic Analysis of the Rocchio Algorithm with TFIDF for Text Categorization, *Proc. of the 14th International Conference on Machine Learning ICML97*, pp. 143—151, 1997.

[3] Kononenko, I. and Bratko, I., Information-Based Evaluation Criterion for Classifier's Performance, *Machine Learning 6*, Kluwer Academic Publishers, 1991.

[4] Mitchell, T.M., Machine Learning, The McGraw-Hill Companies, Inc., 1997.

[5] Mladenić, D., Personal WebWatcher: Implementation and Design, *Technical Report IJS-DP-7472*, October, 1996. http://www-ai.ijs.si/DunjaMladenic/papers/PWW/

[6] Mladenić, D. (1999). Text-learning and related intelligent agents. IEEE EXPERT, Special Issue on Applications of Intelligent Information Retrieval, May-June 1999.

[7] Mladenić, D. & Grobelnik, M. (1999). Feature selection for unbalanced class distribution and Naive Bayes, *Proceedings of the 16th International Conference on Machine Learning ICML-99*, Morgan Kaufmann Publishers, San Francisco, CA. pp. 258-267.

[8] Pazzani, M., Billsus, D., Learning and Revising User Profiles: The Identification of Interesting Web Sites, *Machine Learning 27*, Kluwer Academic Publishers, pp. 313—331, 1997.

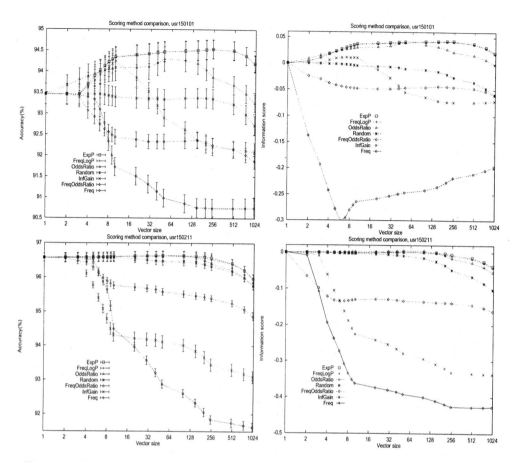

Figure 2: Influence of vector size to Classification accuracy and Information score on data for Home-Net usr150101 (upper) and usr150211 (lower). Notice that curve names are sorted according to the values at the end.

[9] Quinlan, J.R. (1993). Constructing Decision Tree. In *C4.5: Programs for Machine Learning*. Morgan Kaufman Publishers. pp. 17-26.

[10] van Rijsbergen, C.J,. Harper, D.J., Porter, M.F., The selection of good search terms, *Information Processing & Management*, 17, pp.77—91, 1981.

[11] Shaw Jr, W.M., Term-relevance computations and perfect retrieval performance, *Information Processing & Management*, 31(4), pp.491—498, 1995.

[12] Yang, Y., Pedersen, J.O., A Comparative Study on Feature Selection in Text Categorization, *Proc. of the 14th International Conference on Machine Learning ICML97*, pp. 412—420, 1997.

Web Searching Behavior: Selection of Search Terms

Xiaowen Fang
School of Computer Science, Telecommunications,
and Information Systems, DePaul University
243 S. Wabash Avenue, Chicago, IL 60604
Tel: (312) 362-5206   Fax: (312) 362-6116
Email: xfang@cs.depaul.edu

ABSTRACT

This study investigated how users selected search terms and how effective these terms were during a Web searching process. Twenty subjects conducted one search task during one-hour period. The keystroke data was recorded by the computer. The keystroke data included when and which Web sites the subject visited, when and what search terms the subject used, and which Web sites were considered relevant. After the data was analyzed, three primary sources of search terms were identified: (1) the description of the search task, (2) terms derived from the Web pages reviewed by the users, and (3) terms derived from users' domain knowledge. Implications of the findings were discussed.

1. INTRODUCTION

The Web is a loosely organized distributed information base filled with heterogeneous knowledge resources. The information available on the Web has the following characteristics:

- The volume of information is huge.
- The type of information varies widely, from personal Web pages to academic journal articles, from raw scientific data to online shopping.
- The information is unstructured.

In order to retrieve information on the Web, users first need to locate the Web sites containing the information by using the Web search tools.

Web search engines serve as catalogs of the Web. They index the Web pages by deploying a special computer program called "spider" or "robot". The "spider" crawls from site to site on the Web and creates a database that stores indices of Web pages on the entire Web. Users can enter search terms to query against the index database. The search engine returns a list of Web pages along with short descriptions. The returned Web pages are usually ranked based on some relevance measures designed by the search engine although the ranking is inaccurate in most cases due to the difficulty of defining relevance. The large volume and unstructuredness of the information make it very difficult to efficiently index the Web pages and thus significantly reduce the power of Web search engines in retrieving useful information requested by a user. A survey done by Rousseau, Jamieson, Rogers, Mead and Sit (1998) indicated that one of the major problems of using an on-line library system was the difficulty of finding appropriate keywords. Inefficient indexing and inaccurate search queries could easily result in thousands of hits for a single search query (Fang & Salvendy, 2000). Among the thousands of Web pages returned by a Web search tool, many might be totally irrelevant.

In order to design a good search interface that could help users identify useful keywords, designers need to thoroughly understand how users behave, how they select search terms, and how they make relevance judgments. However, research on online search behavior, specially Web searching behavior, is still in its infancy. The objective of this study was to investigate how users select search terms during a Web searching process and how effective the search terms from different sources were.

2. BACKGROUND LITERATURE

Quite a few studies on user search behavior have been undertaken in the context of conventional information retrieval. The following are a few primary studies specifically on search term selection:

- Spink and Saracevic (1997) analyzed the online transaction logs, videotapes, and transcribed dialogue between users and professional intermediaries and identified five sources of search terms: (1) the users' written question statements, (2) terms derived from users' domain knowledge during the interaction, (3) terms extracted from retrieved items as relevance feedback, (4) database thesaurus, and (5) terms derived by intermediaries during the interaction. It was also indicated that search terms from users' written question statements and term relevance feedback were the most productive sources.

- Fidel (1986) conducted a study of search terminology. She derived a decision-tree routine for the selection of search terms from free-text and controlled vocabulary.
- Saracevic, Kantor, Chamis, and Trivison (1988) studied the difference between different experienced searchers in selecting search terms for the same information problem. Considerable differences were found and the average overlap between searchers for the same questions was 27%.
- Hsieh-Yee (1993) indicated that novice and experienced searchers were different in selecting and manipulating search terms for the same information problem.
- Bates, Wilde, and Siegfried (1993) found that search terms selected by humanities scholars were different from the search terms selected by physical scientists.

The context of Web searching is quite different from conventional information retrieval in the following aspects:

- Web searching in most cases involves no intermediaries. Users must interact directly with the Web search tools and usually there are no experienced searchers sitting aside and helping them conduct a search.
- Web is not exactly a database. Information published on the Web is unorganized and unstructured. In other words, it would be a difficult task to efficiently index Web documents.
- Most users are not proficient in using Web search tools. Many of them have never been using any kind of search tool. For these novice users, using the search tools itself would be a challenge without mentioning how difficult for them to construct effective search queries.

Therefore, it is of great importance to examine how users would select search terms in a Web searching context in order to take abovementioned factors into account.

3. METHOD
3.1 Method of data collection

The data used in this analysis were collected during a study on search history of Web searching conducted by Fang (2000). In the following, the data collection process was briefly summarized.

Twenty participants performed a low complexity search task. The objective of the search task was to find as many relevant Web pages as possible during a one-hour period. Relevant pages were defined as any Web pages that contained a predefined word. Before subjects began the task, they were given a brief on-line instruction that explained the objective (predefined word) of the search task, some beginning relevant keywords, and a general introduction to the related area. After reading the instructions, subjects began the search process and could use any search strategy to accomplish the goal of the task. After subjects entered a Web site, they were asked to use the "Find" function in the browser (an experimental browser developed on the basis of Internet Explorer 4.0) to check whether the predefined word appeared there. If the predefined word did occur in a Web site, this Web site was considered a relevant one. Subjects were told to add the uniform resource locator (URL) addresses as bookmarks after they identified the relevant Web sites.

The experiment procedure was as following: each subject was asked to sign a consent form before participating in the experiment. After subjects completed a pre-experiment questionnaire, the training session began. The instructions for the search tool that they would use in the experiment were presented both on the computer screen and on a hard copy. Subjects were asked to read the instructions carefully, and they were allowed to ask questions. One sample search task was displayed in the task window for them to use for practice. They were told to find as many relevant Web sites as possible. When they entered a Web site, they were asked to use the "Find" function in the browser (an experimental browser developed based on Internet Explorer 4.0) to check whether the predefined word appeared in that Web site. If the Web site was relevant, they were asked to add this Web site to the bookmark list. The training was self-paced. When the subjects felt comfortable using the search tool, some questions were asked concerning the training process. After answering these questions, participants were shown the answers. Subjects could go back and play with the search tool until they were comfortable with the functions provided by the search tool. During the task, a brief instruction that explained the objective (predefined word) of the search task, some beginning relevant keywords, and a general introduction to the related area were displayed on the computer screen. The subjects were asked to search as many relevant Web sites as possible in a 60 minute period. After completing the task, subjects were required to complete a questionnaire concerning their satisfaction with the search tool they used.

899

During the experiment, the following data were recorded by computer for analysis:

- A log file. In this file, the URLs of all the Web sites visited by the participant were recorded chronically along with the exact clock time when the Web sites were accessed and length of the network delay.
- All the search queries were recorded chronically along with the exact clock time when they were turned in.
- HTML source of all the relevant Web pages were recorded along with the exact clock time when they were identified.

3.2 Method of data analysis

In order to identify and categorize the sources of search terms, all the search terms used by a participant were first laid out. Then the sources of search terms were determined on the basis of the following rules:

(1) If a search term first appeared in the description of search task, the source of this search term was categorized as search task description;
(2) If a search term first appeared in at least one relevant Web pages identified by the particular participant before it was used in a search query, the source of this search term was categorized as relevant pages;
(3) If a search term appeared neither in the search task description, nor in the relevant Web pages before it was used in a search query, the source of this search term was categorized as domain knowledge.

The log file, all the search queries, and all the recorded relevant Web pages were synchronized and all events were listed chronically. Based on this list, all the identified relevant Web pages were credited to different search queries.

To analyze the effectiveness of search term sources, the first step was to measure the relative retrieval contributions of different search terms. In this analysis, the proportional weighting scheme developed by Spink and Saracevic (1997) was adopted. The basic idea of this scheme was to normalize by weighting the retrieval contribution of all relevant items and not relevant items, taking into account the logical connections between terms in a search query. In this study, only relevant items were considered because irrelevant items were too many to be included in the analysis in the Web searching context. According to this scheme, the total weight for all search terms in a search query in relation to retrieval of a relevant page equals one. This weight is then distributed among search terms based on the given logical combination connecting the terms. The retrieval weight of search terms were calculated based on the following rules:

(1) If only one term in a search query is responsible for the retrieval of a relevant page, its retrieval weight is 1.
(2) If more than one term is responsible for the retrieval of a relevant page, each term received a proportional retrieval weight. If the search query connects two search terms with a logical operator AND, both terms are considered to contribute equally to the relevant retrieval and both receive a proportional retrieval weight of 0.5. Adjacent terms were treated as connected by AND. If three search terms contributed to the retrieval of a relevant page, each search term received a proportional retrieval weight of 0.33.
(3) If two terms are linked by logical operator OR and only one term appeared in the retrieved relevant page, that term was weighted 1. If both terms appeared in the relevant page, each term received 0.5 retrieval weight, as neither term could be excluded as having contribution for the retrieved page.
(4) If a term appeared more than once in the same search query, only ONE occurrence was counted.
(5) A phrase enclosed by quotation marks was considered as one term.

4. RESULTS AND DISCUSSION

Table 1 lists the retrieval weights of different search term sources for all the twenty participants.

Table 1 indicates that all participants of the study had used some search terms from search task description. The percentage retrieval weight of search terms in this category ranges from 9.4% to 100.0%. The average percentage retrieval weight was 62.80% and the standard deviation was 27.291%. In other words, on average search terms from search task description were responsible for retrieving 62.80% of the identified relevant Web pages. This is not a surprising result. Spink and Saracevic (1997) found that that search terms from users' written question statements were one of the most productive sources. The aforementioned result is consistent with the findings of Spink and Saracevic (1997) because the search task description in this study is similar to users' written question statements. In a Web searching process, a user starts a new search when a need for information arises. Obviously, the information need, represented by some kind of question statements, could be a good source of search terms for the initial search query. In this early stage, the user could form some internal representations for the particular search question. As the search process goes on, the user might occasionally revisit the question statements and select terms from them and

900

these terms could be dominant in most search queries without any external help, such as a search intermediary's suggestions.

Table 1 Effectiveness of Different Search Term Sources

Subject No.	Search Task Description		Relevant Pages		Domain Knowledge		Number of Identified Relevant Pages
	Retrieval Weight	%	Retrieval Weight	%	Retrieval Weight	%	
1	10.86	77.6	3.13	22.4	0.00	0.0	14
2	10.67	48.5	0.00	0.0	4.08	18.6	22
3	2.00	66.7	0.00	0.0	0.00	0.0	3
4	13.67	52.6	5.33	20.5	0.00	0.0	26
5	14.00	77.7	0.00	0.0	0.00	0.0	18
6	14.00	56.0	10.50	42.0	0.50	2.0	25
7	1.90	21.1	1.10	12.2	0.00	0.0	9
8	7.00	87.5	0.00	0.0	0.00	0.0	8
9	13.00	92.9	0.00	0.0	1.00	7.1	14
10	6.75	84.4	0.00	0.0	1.25	15.6	8
11	7.50	35.7	0.50	2.4	0.00	0.0	21
12	7.00	100.0	0.00	0.0	0.00	0.0	7
13	3.00	100.0	0.00	0.0	0.00	0.0	3
14	5.10	63.8	2.90	36.3	0.00	0.0	8
15	6.67	39.2	0.00	0.0	2.33	13.7	17
16	7.00	70.0	0.00	0.0	0.00	0.0	10
17	1.00	16.7	1.00	16.7	0.00	0.0	6
18	1.50	9.4	1.50	9.4	0.00	0.0	16
19	19.00	73.1	0.00	0.0	6.00	23.1	26
20	5.00	83.3	0.00	0.0	0.00	0.0	6
Mean		62.80		8.09		4.00	
Std		27.291		13.000		7.411	

It could also be found from Table 1 that 8 out of 20 participants actually selected search terms by reviewing the relevant Web pages identified before. The average percentage retrieval weight of this type of search terms was 8.09% and standard deviation was 13.00%. In other words, on average, search terms from relevant Web pages were responsible for retrieving 8.09% of the identified relevant Web pages. A further correlation analysis indicated that there was a significant correlation ($r = 0.476$ and $p = 0.040$) between the retrieval weight of these search terms and the number of identified relevant Web pages. This means that participants tended to find more relevant Web pages as they selected more search terms from relevant pages. These results are consistent with the findings of a few other studies. In the study undertaken by Spink and Saracevic (1997), term relevance feedback was identified as one of the most productive sources for search terms. Fang and Salvendy (2000) developed a user-centered feature, keyword comparison, which compares relevant Web pages identified by the user and suggests a list of search terms based on frequency. They found in an experimental study that feature keyword comparison improved users' search performance by 77% and satisfaction in using the feature by 35%. All these studies suggest that relevant pages are a good source of search terms.

Among the 20 participants, 6 had used search terms derived from their domain knowledge. The average percentage retrieval weight was 4.00% with a standard deviation of 7.411% for search terms derived from domain knowledge. In other words, on average, search terms derived from domain knowledge were responsible for retrieving 4% of the identified relevant Web pages. This result is consistent with the findings of Spink and Saracevic (1997). They found that one of the five sources of search terms was the user's domain knowledge. It's not surprising that users wouldn't have much domain knowledge when they start a new search on a random field. In this regard, the effectiveness of domain knowledge could be limited.

5. CONCLUSION

In this study, it was investigated that how users selected search terms from different sources in the Web search context with no search intermediaries involved. Based on the analysis, the following sources of search terms were found:

(1)   Search task descriptions were the most productive source of search terms. In this study, search terms selected from search task descriptions were responsible for retrieving 62.8% of the identified relevant Web pages.

(2)   Relevant pages were another important and useful source of search terms. In this study, search terms chosen from relevant Web pages were responsible for retrieving 8.09% of the identified relevant Web pages.

(3)   Domain knowledge was the last source of search terms identified in this study. Search terms derived from domain knowledge were responsible for retrieving 4.0% of the identified relevant Web pages.

However, because of the nature of this study (data were collected in a controlled experiment that was not specifically designed for this study; the definition of the relevance was simplified;), extra cautions should be exerted in generalizing these findings.

REFERENCES:

Bates, M. J., Wilde, D. N., & Siegfried, S. (1993). An analysis of search terminology used by humanities scholars: The Getty Online Searching Project Report Number 1. Library Quartly, 63, 1-39.

Fang, X. (2000). A hierarchical search history for Web searching. International Journal of Human-Computer Interaction, 12, 73-88.

Fang, X. and Salvendy, G. (2000). Keyword Comparison: A User-Centered Feature for Improving Web Search Tools. International Journal of Human Computer Studies, 52, 915-931.

Fidel, R. (1986). Towards expert systems for the selection of search keys. Journal of the American Society for Information Science, 37, 37-44.

Hsieh-Yee, J. (1993). Effect of search experience and subject knowledge on the search tactics of novice and experienced searchers. Journal of the American Society for Information Science, 44, 161-174.

Rousseau, G., Jamieson, B., Rogers, W., Mead, S., & Sit, R. (1998). Assessing the usability of on-line library systems. Behavioral & Information Technology, 17, 274–281.

Saracevic, T., Kantor, P., Chamis, Y., & Trivison, D. (1988). A study of information seeking and retrieving. I. Background and methodology. II. Users, questions and effectiveness. III. Searchers, searches and overlap. Journal of the American Society for Information Science, 33, 161-216.

Spink, A. & Saracevic, T. (1997). Interaction in information retrieval : selection and effectiveness of search terms. Journal of the American Society for Information Science, 48, 741-761.

An Adaptive Agent for Web Exploration Based on Concept Hierarchies

Scott Parent, Bamshad Mobasher[*], Steve Lytinen

School of Computer Science, Telecommunication and Information Systems
DePaul University
243 S. Wabash Ave., Chicago, Illinois, 60604, USA

ABSTRACT

In this paper we present the design of a client-side agent, named ARCH, for assisting users in one of the most difficult information retrieval tasks, i.e., that of formulating an effective search query. In contrast to traditional methods based on relevance feedback, ARCH assists users in query modification *prior* to the search task. The initial user query is (semi-)automatically modified based on the user's interaction with an embedded, but modular, concept hierarchy. This allows for the generation of richer queries than is possible, for example, based on simple query expansion according to lexical variants such as synonyms. In addition, ARCH passively learns a user profile by observing the user's past browsing behavior. The profiles are used to provide additional context to the user's information need represented by the initial query.

1. INTRODUCTION

The World Wide Web is a vast resource of information and services that continues to grow rapidly. The heterogeneity and the lack of structure that permeates much of the ever expanding information sources on the World Wide Web, such as hypertext documents, makes automated discovery, organization, and management of Web-based information difficult. Traditional search and indexing tools of the Internet and the World Wide Web such as Lycos, Alta Vista, Google, and others provide some comfort to users, but they do not generally provide structural information nor categorize, filter, or interpret documents. Often users' poorly designed keyword queries return inconsistent search results, with document referrals that meet the search criteria but are of no interest to the user. Studies have shown that users consistently find the tasks of formulating the right query and managing or filtering the results quite difficult.

In recent years these factors have prompted researchers to design more intelligent tools for information retrieval, such as intelligent Web agents (Boley, et. al. 1999; Bollacker, et. al. 1998; Craven, et. al. 2000; Joachims, et. al. 1997; Lieberman 1997) and mechanisms for incrementally refining users queries (Allan 1996; Eguchi 2000). Traditional approaches to *query reformulation* (also called *query expansion*), generally perform one or both of two tasks: re-weighting the terms in the query and expanding the query using additional terms. These tasks are usually performed using relevance feedback from the search results (Buckley, et. al. 1994) or by expanding the query with lexical variants of keywords such as synonyms (Miller, 1997). Relevance feedback mechanisms allow users to refine search parameters based on explicit judgments on the relevance or non-relevance of results from the initial search. However, these approaches do not allow for the creation of an effective query *prior* to an initial search, and furthermore, they are not adaptive in that they assume user's information needs remain constant at different stages of the search process (Mizzaro, 1997).

In this paper, we describe a client-side agent which is designed to assist users in searching large collections of documents, such as the World Wide Web. The agent utilizes a hierarchically-organized semantic knowledge base in aggregate form, as well as an automatically learned user profile, to enhance user queries. In contrast to traditional relevance feedback mechanisms, our approach allows for semi-automatic reformulation of the user's initial query prior to the search phase. Specifically, our intent is to assist the user in generating richer queries by (a) asking the user to identify portions of the hierarchy which are relevant and irrelevant to the query; (b) using pre-computed term vectors associated with each node in the hierarchy to enhance the original query; and (c) identifying any relevant portions of the user profile to provide additional context for the user's information need. Since there are many hierarchical knowledge bases from which semantically related concepts can be drawn, the design of the system is intentionally modular and not specific to a particular knowledge base. This allows the user to switch among the representations of different domain-specific hierarchies depending on the goals of the search.

---

[*] Please direct correspondence to Bamshad Mobasher (*mobasher@cs.depaul.edu*).

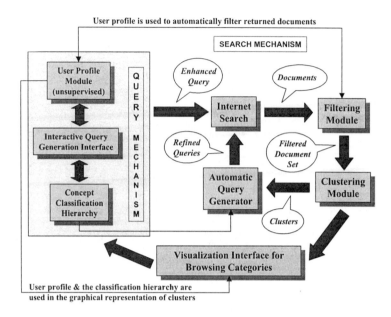

User profile is used to automatically filter returned documents

User profile & the classification hierarchy are used in the graphical representation of clusters

Figure 1. The architecture for ARCH - An Adaptive agent for Retrieval based on Concept Hierarchies

For example, a user interested in performing a general search on the Web can "plug in" the aggregate representation of the Yahoo concept hierarchy. On the other hand, a user interested in searching the Web for specific medical information can, instead, use a medical concept classification hierarchy such as Medical Subject Headings (http://www.nlm.nih.gov/mesh/). It is also possible to use lexical knowlegebases, such as WORDNET (Miller, 1997), in order to capture linguistic relationships among concepts in the search process.

We envision the query enhancement agent to be part of a larger system, as depicted in Figure 1. The system, called ARCH (Adaptive agent for Retrieval based on Concept Hierarchies), combines several capabilities: A highly interactive interface for guiding the user in formulating queries by providing visual tools to graphically view and manipulate the query in the context of the user profile and the concept classification hierarchy; mechanisms for learning a user profile; mechanisms for automatically clustering the search results and automatically generating refined queries according to cluster summaries and the hierarchical knowledge base; and a visual interface for assisting the user in displaying and navigating the results of a query in automatically created semantic categories.

In the rest of this paper we focus on the query generation mechanism for ARCH. We first present our approach for automatically reformulating user's initial query based on the selected portions of the concept hierarchy. We then discuss how the generated query can be further enhanced based on a learned user profile. Throughout the paper, we illustrated the functionality of the query generation mechanism by an example with results based on an experimental data set and the Yahoo concept hierarchy. The clustering and filtering mechanisms of ARCH will be discussed in a future paper.

2. QUERY ENHANCEMENT BASED ON CONCEPT HIERARCHIES

In ARCH, user queries are generated in several stages. In the first stage, the user formulates an initial query, Q_1, which is simply a list of keywords. The system responds to this list by displaying those portions of the hierarchy which are most relevant to the keywords. Once the relevant portions of the hierarchy are displayed, the user is asked to select those categories which are relevant to the intended query, and to "deselect" those categories which are not relevant. The system maintains an aggregate representation of the concept hierarchy. This is to allow for efficient storage and manipulation of the concept hierarchy and to facilitate matching user keywords to nodes in the hierarchy. This aggregate representation is obtained by pre-computing a weighted term vector for each node in the hierarchy which represents the centroid of all documents and subcategories indexed under that node. Specifically, given a node n in the concept hierarchy containing a collection D_n of individual documents and a set S_n of subcategories, the term vector for the node n is computed as follows:

$$T_n = \left(\left(\sum_{d \in D_n} T_d \right) / |D_n| + \sum_{s \in S_n} T_s \right) / (|S_n| + 1),$$

where T_d is the weighted term vector which represents document d (indexed under node n in the hierarchy), and T_s is the term vector which represents subcategory s of node n. Note that the individual documents indexed under n are collectively weighted the same as any of the subcategories under n. Standard information retrieval text preprocessing techniques are performed to initially create a global dictionary of terms from which the features in term vectors are drawn. These techniques include stemming (Porter, 1980) and the removal of stop words. Furthermore, for computing term weights extracted from text we use a standard function of the term frequency and inverse document frequency (tf.idf) as commonly used in information retrieval (Salton & McGill, 1983; Frakes & Baeza-Yates, 1992).

As an example, consider the scenario in which the user may start with a single keyword query "music," using the Yahoo hierarchy. The system may then display an appropriate portion of the hierarchy to the user, containing parents, children and siblings of the node corresponding to the initial query. The user can now select (or deselect) various nodes. In this case we assume that the user is interested in finding information about different types of music, thus selecting the node corresponding to "Genres" in the hierarchy. A portion of the Yahoo hierarchy corresponding to this scenario, as well as the term vector for this node, are depicted in Figure 2. Low scoring terms (stems) appearing in the term vector are not shown in the figure. The term vector which represents the node "Genres" in the Yahoo hierarchy is computed from a combination of those documents indexed under "Genres," as well as the term vectors representing its subcategories (e.g., "Jazz," "Blues," and "New Age").

Once the system has matched the term vectors representing each node in the hierarchy with the list of keywords typed by the user, those nodes which exceed a similarity threshold are displayed to the user, along with other adjacent nodes. An ambiguous keyword might cause the system to display several different portions of the hierarchy. The user is asked to select those categories which are relevant to the intended query, and to deselect those categories which are not relevant. Our approach to the generation of an enhanced query based on nodes in the concept hierarchy is an adaptation of Rocchio's method for relevance feedback (Rocchio, 1971). Using the selected and deselected nodes, the system produces a refined query Q_2, as follows:

$$Q_2 = \alpha \cdot Q_1 + \beta \cdot \sum T_{sel} - \gamma \cdot \sum T_{desel},$$

where each T_{sel} is a term vector for one of the nodes selected by the user, and T_{desel} is a term vector for one of the deselected nodes. The factors α, β, and γ are tuning parameters representing the relative weights associated with each component with the condition that $\alpha + \beta - \gamma = 1$.

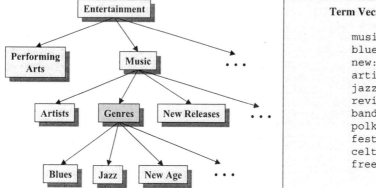

Figure 2. Portion of the Yahoo hierarchy corresponding to the query "music" and the selection of the node "Genres" is depicted on the left. The partial term vector for "Genres" which provides and aggregate representation of the node is given on the right.

905

Continuing with our example, suppose that the user now selects "Jazz" and deselects "Blues." Furthermore, suppose the user also selects "Dixieland Jazz" as a subcategory of "Jazz." This will result in a modified query, more accurately reflecting the user's information need. In particular, the top scoring terms (and their weights) based on the above scenario will be as follows:

> **music:1.00, jazz:0.44, dixieland:0.20, tradition:0.11, band:0.10, inform:0.10**

3. DERIVATION OF USER PROFILES FROM DOCUMENT CLUSTERS

The user profile generation component of ARCH employs several heuristics to automatically (and without user intervention) determine the topics of "interest" to the user. This component works by passively observing the user's browsing behavior over time and collecting and analyzing documents in which the user has shown interest. The heuristics used by the system are based on several factors, including the frequency of visits to a page or a site, the amount of time spent on the page (or related pages within a site), and other actions of the user such as bookmarking a page. Once enough documents have been collected as part of the profile generation, the system clusters the documents into semantically related categories.

As in the case of nodes within the concept hierarchy, each document in the profile is represented as a term vector, with term weights derived using the tf.idf measure. Clustering algorithms, such as k-means, partition a document set into groups of similar documents based on a measure of vector similarity. A common measure used in information retrieval, also employed in ARCH, computes the similarity of two vectors based on the normalized inner product of the those vectors. Our method for the generation of topical profiles is similar to the "concept indexing" method described in Karypis & Han (2000). Individual profiles, each representing a topic category, are computed based on the centroid of the document clusters. Specifically, given the document collection, D and a document cluster $c \subseteq D$, we construct a profile pr_c as a set of term-weight pairs:

$$pr_c = \{\langle t, weight(t, pr_c) \rangle \mid weight(t, pr_c) \geq \mu\},$$

where the significance weight, $weight(t, pr_c)$, of the term t within the profile pr_c is given by

$$weight(t, pr_c) = \frac{1}{|c|} \cdot \sum_{d \in D} w(t, d)$$

and $w(t, d)$ is the weight of term t in the document vector d. The threshold μ is used to filter out insignificant terms within each profile. Each profile, in turn, can be represented as a vector in the original n-dimensional space of terms, where n is the number of unique terms in the global dictionary.

Once the enhance query, Q_2, is derived from the concept hierarchy, each profile can be compared to the query vector for similarity. Those profiles which satisfy a similarity threshold, are then used to further expand the query, resulting in a new query, Q_3. Note that, as in the case of query expansion based on the concept hierarchy, the new query is computed as a weighted sum of the term vector representing Q_2 and the normalized sum of term vectors representing the matching profiles. To continue our example, let us assume that, through her past browsing behavior, the user has expressed interest in a number of documents related to intellectual property and copyright in music (among other topics). Clustering on our experimental data set based on this scenario, resulted in the following matching profile with respect to the enhanced query described earlier:

music:1.00	intellectu:0.48	inform:0.28	peopl:0.20
jazz:0.76	law:0.47	tap:0.21	band:0.20
copyright:0.69	record:0.35	arrang:0.20	musician:0.20
properti:0.60	protect:0.29	product:0.20	author:0.20

This profile was obtained by computing the centroid vector of one of the document clusters produced by the k-means algorithm (in this case those documents relating to intellectual property in music).

906

Finally, the enhanced query vector generated by the system based on concept hierarchy and the user profiles (i.e., Q_3) is as follows:

> **music:1.00, jazz:0.71, band:0.40, copyright:0.22, properti:0.19, intellectu:0.15, law:0.15, artist:0.14, perform:0.14, dixieland:0.13, inform:0.12, protect:0.09, record:0.11, tradition:0.09, musician:0.06**

Performing searches based on this and other queries have shown improved retrieval effectiveness. Due to space constraints, these results are not shown here. A detailed evaluation of the system based on standard measures such as precision and recall will be provided in a future paper.

4. CONCLUSIONS AND OUTLOOK

We have presented the work-in-progress on the design of ARCH, an adaptive agent which can assist users in formulating an effective search query based on a modular concept classification hierarchy and a learned user profile. Preliminary experiments have shown that the agent can substantially improve the effectiveness of information retrieval both in the general context of the Web, as well as for search against domain-specific document indexes. The full system also incorporates mechanisms for categorizing and filtering the search results, and using these categories for performing refined searches in the background. Our future work in this area is focused on a detailed evaluation of the system based on standard information retrieval measures such as precision and recall. We also plan on extending the query generation component by incorporating mechanisms for incremental modification of user profiles and incremental updating of the concept hierarchy based on document categories generated during the search process.

REFERENCES

Allan, J. (1996), "Incremental Relevance Feedback for Information Filtering". In Proceedings of the ACM SIGIR '96, pp. 270-278, ACM Press.

Boley, D., Gini, M., Gross, R., Han, E-H., Hastings, K., Karypis, G., Kumar, V., Mobasher, B., and Moore, J. (1999). "Document Categorization and Query Generation on the World Wide Web Using WebACE". Artificial Intelligence Review, Vol. 13, No. 5-6, pp. 365-391.

Bollacker, K., Lawrence, S., Lee Giles, C. (1998), "CiteSeer: An Autonomous Web Agent for Automatic Retrieval and Identification of Interesting Publications". In Proceeding of the 2nd International Conference on Autonomous Agents, Minneapolis, Minnesota, pp. 116-123, ACM Press.

Buckley, C., Salton, G., and Allan, J. (1994), "The Effect of Adding Relevance Information in a Relevance Feedback Environment". In Proceedings of the ACM SIGIR'94, pp. 292-298, ACM Press.

Craven, M., DiPasquo, D., Freitag, D., McCallum, A., Mitchell, T., Nigam, A., and Slattery, S. (2000), "Learning to Construct Knowledge Bases from the World Wide Web". Artificial Intelligence, 118(1-2), pp. 69-113.

Eguchi, K. (2000), "Incremental Query Expansion Using Local Information of Clusters". In Proceedings of the 4th World Multiconference on Systemics, Cybernetics and Informatics (SCI 2000), Vol. 2, pp.310-316.

Frakes, W. B., and Baeza-Yates, R. (1992). Information Retrieval: Data Structures and Algorithms, Prentice Hall.

Joachims, T., Freitag, D., and Mitchell, T. (1997), "WebWatcher: A Tour Guide for the World Wide Web". In Proceedings of the Fifteenth International Joint Conference on Artificial Intelligence, pages 770-775, Nagoya, Japan, Morgan Kaufmann.

Karypis, G., Han, E-H. (2000), "Concept Indexing: A Fast Dimensionality Reduction Algorithm with Applications to Document Retrieval and Categorization". Technical Report #00-016, Department of Computer Science and Engineering, University of Minnesota.

Lieberman, H (1997), "Autonomous Interface Agents". In Proceedings of the ACM Conference on Computers and Human Interface, CHI-97, Atlanta, Georgia, pp. 67-74, ACM Press.

Miller, G., (1997), "WORDNET: An Online Lexical Database". International Journal of Lexicography, 3 (4).

Mizzaro, S. (1997), "Relevance: The Whole Story". Journal of the American Society of Information Sciences, 48(9), pp. 810-832.

Porter, M. F. (1980), "An Algorithm for Suffix Stripping". Program, 14(3), pp. 130-137.

Rocchio, J. (1971), "Relevance Feedback in Information Retrieval". In Salton, G. (ed.), The SMART Retrieval System: Experiments in Automatic Document Processing, pp. 313-323, Prentice Hall.

Salton, G., and McGill, M. J. (1983). Introduction to Modern Information Retrieval, McGraw-Hill.

INFORMATION AGENTS IN ASTRONOMY

A. Accomazzi, M.J. Kurtz, G. Eichhorn, C.S. Grant, S.S. Murray

Harvard–Smithsonian Center for Astrophysics, 60 Garden Street, Cambridge, MA, 02138 USA

ABSTRACT

Astronomy has probably the most advanced information infrastructure of any science. The overwhelming majority of modern astronomical literature is on–line, is heavily cross–linked and cross–referenced, and encompasses the work of several organizations, worldwide, which collaborate closely. A large fraction of the basic data of astronomy is also on–line, and this fraction is increasing rapidly. An ambitious, international plan to get all important astronomical data on–line, searchable, and cross–linked is now underway. We present the current state of this effort, emphasizing the techniques by which the NASA Astrophysics Data System maintains links to other collaborating institutions and data providers as new information is ingested.

1. INTRODUCTION

The NASA Astrophysics Data System (ADS) has become in the recent years a key component of astronomical research. The ADS provides bibliographic information daily, or near daily, to a majority of astronomical researchers worldwide. A substantial fraction of the ADS functionality and usefulness is made possible by the very existence of a distributed network of resources for astronomical research which has been developed in the last eight years. This new system, still unique amongst the sciences, has been dubbed Urania for the muse of astronomy. Its creation has been possible thanks to the close ties of the major astronomical data groups and to the limited size of the astronomical community.

The Urania system is based on the highest level of data abstraction, astronomical object names and bibliographic articles, rather than the lowest, the actual observed data in archives. This change in the level of abstraction has permitted the creation of a system of extraordinary power, which allows users to search the astronomical literature or object databases and then seamlessly access data catalogs, electronic articles, image and spectra archives by simply following the hyperlinks provided by each search interface. The data centers participating in the network are capable of creating cross–links thanks to the periodic exchange of relational data correlating each other's records via established interfaces and software agents.

In the following sections of this paper we will describe the methodology that the ADS and its collaborators have used to create and maintain such a system. Section 2 describes in more detail the Urania collaboration and its scope; in section 3 we analyze the approach we follow in the creation of bibliographic records from a variety of data sources; section 4 details the procedures that are used to retrieve, archive and process the datasets required in the operation of the ADS; section 5 presents the approach followed in the creation and maintenance of cross–database links; and in section 6 we take a look at what technologies may be used in the future to improve the operation and features of the system.

2. THE ASTRONOMICAL DATA WEB

Conceptually the core of Urania (Boyce, 1996) is a set of distributed cross–indexed lists which maintain concordance of data available at different sites. The primary hubs that form the backbone of the Urania network are the ADS (Kurtz et al., 2000), maintained at the Harvard–Smithsonian Center for Astrophysics in Cambridge, USA, and the SIMBAD database (Wenger et al., 2000), maintained by the Centre de Données astronomiques de Strasbourg (CDS). The ADS maintains an extensive searchable bibliographic database and correlates each bibliographic record with a list of sites which provide data relevant to it. SIMBAD maintains a database of astronomical objects and positions in the sky and correlates each object with the list of articles written about it. SIMBAD also provides a name to object resolver. The possibility for synergy in combining these two data systems is obvious; they have functioned jointly since 1993.

Currently the ADS system (http://adswww.harvard.edu) consists of four semi–autonomous (to the user) abstract services covering Astronomy, Instrumentation, Physics, and Astronomy Preprints. Combined there are more than

2 million abstracts and bibliographic references in the system, and over 1.3 million pages of scanned full–text articles. The Astronomy Service is by far the most advanced, and accounts for 85% of all ADS use. The main ADS search interface allows the user to search by author, title, astronomical object name, abstract text words, or any combination of these (Eichhorn et al., 2000).

The SIMBAD database (http://simbad.u–strasbg.fr) provides identifications, aliases and names of astronomical objects. Any astronomical object name ever mentioned in the literature can be submitted to SIMBAD in order to retrieve basic information known for this object, as well as pointers to complementary data and bibliography. The maintenance of such a system requires a continuous careful cross–identification of objects from catalogues, lists, and journal articles. This ability to gather together any sort of published observational data related to stars or galaxies is the main feature of SIMBAD.

Surrounding the core of Urania, and tightly integrated with it, are many of the most important data resources in astronomy, including the fully electronic journals (currently more than a dozen, accounting for all the major ones), and several data archives which provide access to instrument– and mission–specific datasets. All these groups actively exchange information with the Urania core, point users to it via hyperlinks, and they are pointed to by it. The astronomical journals which are not yet fully electronic, in that they do not support hyperlinked access to the Urania core, also interact with the system. Typically they provide access to page images of the journal, either through PDF files, or bitmaps from the ADS full–text article archive, or both. Bibliographic information is routinely supplied to the ADS, and the SIMBAD librarians routinely include the articles (along with those of the electronic journals) in the object–article concordance database.

An increasing number of data archives are becoming closely connected to the Urania system. For example the Space Telescope Science Institute connects with the ADS bibliographical data via links to papers written about the data in the archive. SIMBAD connects with the High Energy Astrophysics Science Archive Research Center (HEASARC) archive using the position of an object as a search key, HEASARC has an interface which permits several archives to be simultaneously queried (McGlynn and White, 1998), while several archives use the SIMBAD name resolver to permit the use of an object name as a proxy for position on the sky.

3.   CREATION OF BIBLIOGRAPHIC RECORDS

The bibliographic records maintained by the ADS project consist of a corpus of structured documents describing scientific publications (Grant et al., 2000). Each record is assigned a unique identifier in the system and all data gathered about the record are stored in a single text file, named after its identifier. The set of all bibliographic records available to the ADS is currently partitioned into four main data sets. This division of documents into separate groups reflects the discipline–specific nature of the ADS databases, but does not preclude cross–database searches.

ADS continuously receive bibliographic records from a large number of different sources and in a variety of formats (ASCII text, MS Word, TeX, LaTeX, SGML). The ingestion of these records requires a system that can parse, identify, and merge bibliographic data in a reliable way. As more journal publishers and data centers became providers of bibliographic data to the ADS, a unified approach to the creation of bibliographic records became necessary. What makes the management of these records challenging is the fact that ADS often receives data about the same bibliographic entry from different sources, in some cases with incomplete or conflicting information (e.g. ordering or truncation of the author list). Even when the data received is semantically consistent, there may be differences in the way the information has been represented. For instance, while most journal publishers provide us with properly encoded entities for accented characters and mathematical symbols, the legacy data currently found in ADS's databases only contain plain ASCII characters.

To facilitate the ingestion of data in ADS's bibliographic databases, a set of rule–based software procedures have been developed. The process currently used is described by the following sequence of steps:
1. Tokenization: Parsing input data into a memory–resident data structure using procedures which are format– and source–specific. This activity typically involves one or more of the following steps: character set conversion; macro and entity expansion (for TeX and SGML sources); and field–specific formatting (e.g. reduction of author names to a common format).
2. Identification: Computing the unique bibliographic record identifier used by the ADS to refer to this record.
3. Instantiation: Creating a new record for each bibliography formatted according to the ADS "standard" format.

4. Merging: Selecting the best information from the different records available for the same bibliography and combining them into a single entry. This is essentially an activity of intelligent and context–sensitive data fusion, which removes redundancies in the bibliographic records obtained from different sources and creates a "canonical" representation for the record.

Once a bibliographic record has been successfully processed, a corresponding XML document is created in one of ADS's databases. Each bibliographic field (e.g. authors' names or paper title) is represented as an XML element, possibly containing additional sub–elements. Ancillary information about the record is stored as metadata elements within the document. Information about an individual field within the record is stored as attributes of the element representing it. Relationships among fields are expressed as links between the corresponding XML elements.

4.   DATA HARVESTING

Of vital importance to the operation of Urania is the issue of data exchange with collaborators, in particular the capability to efficiently retrieve and process data produced by publishers and other data providers. The task of collecting and entering new bibliographic records in our databases has benefited from three main developments (Accomazzi et al., 2000): the adoption of electronic production systems and relational databases by all collaborating groups; the almost exclusive use of SGML and LaTeX for document formatting; and the pervasive use of the Internet as the medium for data exchange.

The primary means by which users and collaborators submit electronic data to ADS are FTP upload, e–mail, and submission through HTTP requests/responses. While these three mechanisms are conceptually similar (data is sent from a user to a computer server using one of several well–established Internet protocols), the one we have found most amenable to receiving "pushed" data is the e–mail approach. This is primarily due to the fact that modern electronic mail transport and delivery agents offer many of the features necessary to implement reliable data delivery, including content encoding, error handling, data retransmission and acknowledgment. For instance, ADS uses the public domain procmail mail–agent software (http://www.procmail.org) to automatically filter and archive or process any incoming data. Using this paradigm, the email filter allows us to efficiently manage submissions from different collaborators by enforcing authentication of the submitter's email address and proper disposition of the submitted records.

A different approach is used to retrieve data from one or more remote network locations ("data pull"). According to this model, the retrieval is initiated by the receiving side, which simply downloads the data from the remote site and stores it in one or more local files. We have been using this approach for a number of years to retrieve electronic records made available online by many of our collaborators. For instance, the ADS astronomy preprint database is updated every night by a procedure that retrieves the latest submissions of astronomy preprints from the Los Alamos National Laboratory archive, creates a properly formatted copy of them in the ADS database, and then runs an updating procedure that recreates the index files used by the search engine. The pull approach is best used to periodically harvest data that may have changed. By using agents capable of caching and comparing the original timestamps generated by web servers we can avoid retrieving a network resource unless it has been updated, making efficient use of the bandwidth and avoiding re–processing of old data.

5.   CREATION AND MAINTENANCE OF RELATIONAL LINKS

By combining bibliographic data and metadata available from several sources in a single database and by maintaining a list of what properties and resources are available for each bibliography, the ADS system allows users to formulate complex queries such as: "show me the most cited papers ever written about the galaxy M31 which have electronic data tables available and sort them by citation relevance." This query illustrates how knowing whether a particular bibliographic entry possesses a particular property (e.g. whether a data table is available) and its relationship with other bibliographic records (e.g. whether it has been cited by other papers) can be used as a method for selection and ranking of query results. There is no doubt that one of the most important features of Urania's success has been the integration of links between collaborating data providers, which has allowed the implementation of such powerful search interfaces.

In order to allow the maintenance of such an interlinked system, the collaborating data providers need to continuously exchange and update their links to each other. As new data regarding a bibliographic entry or an astronomical object become available, its record must be updated in the relevant databases by merging the new information with the existing one and by updating its relation with respect to other internal and external resources. For instance, when a new paper is published which references an existing article, the record for the latter needs to be updated by establishing a link between the two papers; at the same time, the "citation relevance measure" for the paper, computed as the number of times the paper was cited in the literature, also needs to be updated.

While most publishers of scientific journals have been able to create electronic versions of their journals relatively quickly soon after the explosion in popularity of the web, only a few of them have taken advantage of the new capabilities that the technology has to offer, namely the possibility to create hyperlinks between online documents and related resources. In this respect, electronic publishing in astronomy has always been ahead of its times with the publication by the University of Chicago Press in the summer of 1995 of the electronic version of the Astrophysical Journal Letters which contained hyperlinks from the reference section of articles to bibliographic records in the ADS. Similarly, editors and publishers have now made it their policy to submit electronic versions of data tables appearing in astronomical papers to well−established astronomical data archives, allowing ADS to easily maintain links to these datasets in its bibliographic records. In order to facilitate the creation of such links between data providers, the members of Urania have created a set of HTTP−based interfaces that can be used by collaborators to exchange metadata in an unsupervised fashion. For example, the ADS has made available two different interfaces that allow easy identification of a bibliographic reference: a "reference resolver," which maps plaintext strings into a document identifier, and a "reference verifier," which checks the existence and accuracy of the identifier in the ADS database. Both these services are available as CGI scripts from the ADS web servers (Accomazzi et al. 1999).

While the members of Urania have a close enough relationship which allows the periodic exchange of the relational tables used to create links, in many cases ADS has to use intelligent agents in order to create and maintain links with external data providers. For example, in order to create links to the full−text articles maintained by electronic publishers, we often need to compute the URLs used by the publisher for the full−text paper corresponding to a record in the ADS database from the bibliographic information available to us. Since these URLs are typically specific to each individual publisher (and are not necessarily stable), maintaining these links can often require a substantial effort. One approach that has facilitated this task has been the use of automated procedures to download, parse, and identify the list of online bibliographic resources and relative URLs from each publisher. Once the relevant information has been cross−correlated with the existing records in the ADS databases, links can be created and maintained with confidence.

6. FUTURE DEVELOPMENTS

The ADS and SIMBAD have been successful in creating and maintaining a system which is well−integrated with the electronic publishing community as well as the astronomical data centers. New links are continuously created between collaborating institutions thanks to the use of software agents capable of discovering and cross−correlating new resources as they become available. In order to facilitate the creation and maintenance of such links, members of the Urania collaboration are actively participating in new and promising initiatives.

The Open Archives Initiative (http://www.openarchives.org) develops and promotes interoperability standards that aim to facilitate the efficient dissemination of content. This project was originally developed by members of the electronic preprint community, as a means to enhance access to their data holdings. A steering committee has recently published the specification for the Open Archives Protocol for Metadata Harvesting, designed to foster the exchange of well−structured bibliographic records. The protocol, which is designed around the Dublin Core Metadata Element Set (http://www.dublincore.org), promises to simplify the activities of data exchange among participating institutions through standardization. We expect to make use of this technology in an attempt to create closer ties with data providers in the physics community and with electronic preprint archives.

Within the astronomical community, a more ambitious plan involving the creation of a "Virtual Observatory," is currently being proposed to US and European funding agencies (Brunner et al. 2001). The goal of this project is to foster cooperation among the data centers by providing common front−end services and transparent access to each other's data holdings. This would allow the creation of user interfaces and search tools capable of retrieving data from different archives in a transparent way and presenting them to the user in a consistent and familiar format. Although simple in concept, the implementation of such a system is quite a challenging task given the heterogeneous nature of the data in question and the complexities of the instruments used in the creation of each dataset.

ACKNOWLEDGMENT

Funding for this project has been provided by NASA under NASA Grant NCC5−189.

REFERENCES

1.Accomazzi, A., Eichhorn, G., Kurtz, M. J., Grant, C. S., and Murray, S. S., "The NASA Astrophysics Data System: Architecture," Astronomy & Astrophysics Supplement, 2000, Vol. 143, pp. 85−109.

2.Accomazzi, A., Eichhorn, G., Kurtz, M. J., Grant, C. S., and Murray, S. S., "The ADS Bibliographic Reference Resolver," Astronomical Data Analysis Software and Systems VIII, ASP Conference Series, Vol. 172. Ed. David M. Mehringer, Raymond L. Plante, and Douglas A. Roberts, 1999, pp. 291−294.

3.Boyce, P., "Journals, Data and Abstracts Make an Integrated Electronic Resource," 189th Meeting of the American Astronomical Society, 1996, 189, abstract no. 06.03.

4.Brunner, R. J., Djorgovski, S. G., and Szalay, A. S., "Virtual Observatories of the Future," Astronomical Society of the Pacific Conference Proceedings, 2001, Vol. 225.

5.Eichhorn, G., Kurtz, M. J., Accomazzi, A., Grant, C. S., and Murray, S. S., "The NASA Astrophysics Data System: The search engine and its user interface," Astronomy & Astrophysics Supplement, 2000, Vol.143, pp. 61−83.

6.Grant, C. S., Accomazzi, A., Eichhorn, G., Kurtz, M. J., and Murray, S. S., "The NASA Astrophysics Data System: Data holdings," Astronomy & Astrophysics Supplement, 2000, Vol.143, pp. 111−135.

7.Kurtz, M. J., Eichhorn, G., Accomazzi, A., Grant, C. S., Murray, S. S., and Watson, J. M., "The NASA Astrophysics Data System: Overview," Astronomy & Astrophysics Supplement, 2000, Vol. 143, pp. 41−59.

8.McGlynn, T. and White, N., "Astrobrowse: A Multi−site, Multi−wavelength Service for Locating Astronomical Resources on the Web," Astronomical Data Analysis Software and Systems VII, Astronomical Society of the Pacific Conference Series, 1998, Vol. 145, R. Albrecht, R.N. Hook and H.A. Bushouse, Eds., p.481−484.

9.Wenger, M., Ochsenbein, F., Egret, D., Dubois, P., Bonnarel, F., Borde, S., Genova, F., Jasniewicz, G., Laloe, S., Lesteven, S., and Monier, R., "The SIMBAD astronomical database. The CDS reference database for astronomical objects," Astronomy & Astrophysics Supplement, 2000, Vol. 143, pp. 9−22.

Interactive Visual User Interfaces to Databases

Tugba Taskaya, Pedro Contreras, Tao Feng and Fionn Murtagh
School of Computer Science, Queen's University Belfast
Belfast BT7 1NN, Northern Ireland
f.murtagh@qub.ac.uk

Abstract

We review past achievements, and describe current work, on interactive and responsive visual user interfaces to databases. Such visual user interfaces organize data and information, and also provide interaction with the user. They can be considered as a particular type of agent, helping in the human tasks of information navigating, filtering, seeking, and accessing.

1 Information Clustering and User Interfaces

Information retrieval by means of "semantic road maps" was first detailed in Doyle (1961). The spatial metaphor is a powerful one in human information processing and lends itself well to modern distributed computing environments such as the web. The Kohonen self-organizing feature map (SOM) method is an effective means towards this end of a visual information retrieval user interface.

The Kohonen map is, at heart, k-means clustering with the additional constraint that cluster centers be located on a regular grid (or some other topographic structure) and furthermore their location on the grid be monotonically related to pairwise proximity (Murtagh and Hernández-Pajares, 1995). The nice thing about a regular grid output representation space is that it easily provides a visual user interface. In a web context, it can easily be made interactive and responsive.

Figure 1 shows a visual and interactive user interface map, using a Kohonen self-organizing feature map. Color is related to density of document clusters located at regularly-spaced nodes of the map, and some of these nodes/clusters are annotated. The map is installed as a clickable imagemap, with CGI programs accessing lists of documents and – through further links – in many cases, the full documents. Such maps are maintained for 13000 articles from the *Astrophysical Journal*, 8000 from *Astronomy and Astrophysics*, and over 2000 astronomical catalogs. More information on the design of this visual interface and user assessment can be found in Poinçot et al. (1998, 1999, 2000). For maps in operational use, see:

- A&A: http://simbad.u-strasbg.fr/A+A/map.pl,
- ApJ: http://simbad.u-strasbg.fr/ApJ/map.pl,
- VizieR: http://vizier.u-strasbg.fr/viz-bin/VizieR.

In Guillaume (2000a, 2000b) we developed a Java-based visualization tool for hyperlink-rich data in XML, consisting of astronomers, astronomical object names and article titles. It was open to the possibility of handling other objects (images, tables, etc.). Through weighting, the various types of links could be prioritized. An iterative refinement algorithm was developed to map the nodes (objects) to a regular grid of cells, which as for the Kohonen SOM map, are clickable and provide access to the data represented by the cluster. Figure 2 shows an example for an astronomer (Prof. Jean Heyvaerts, Strasbourg Astronomical Observatory). Given the increasingly central role of XML, the importance of such clustering for data organization and as a basis for knowledge discovery cannot be underestimated. This map is Java-based and client-side.

These new cluster-based visual user interfaces are not unduly computationally demanding, if we assume that they can be set up in advance of use and, if required, periodically updated. Illustrations of processing one million newsgroup messages, and discussion of processing 7 million patent abstracts, can be found on the WebSOM server, http://websom.hut.fi. Further results are available in Oja and Kaski (1999).

913

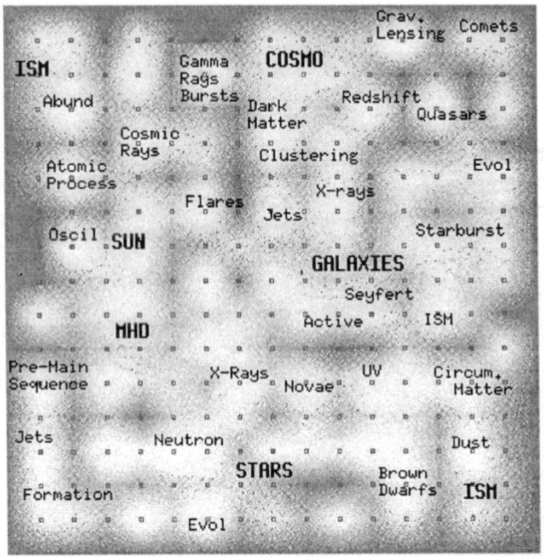

Figure 1: Visual interactive user interface to the journal *Astronomy and Astrophysics* based on 3000 published articles.

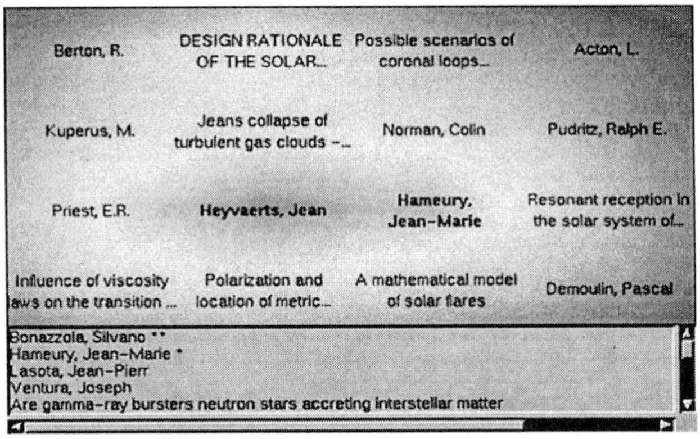

Figure 2: Visual interactive user interface, based on graph edges. Vertices are author names, article titles and (not shown here) astronomical object names. Map for astronomer Jean Heyvaerts.

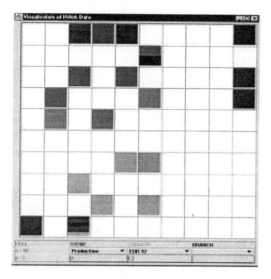

Figure 3: Visual interface to more than 150,000 economic time series. Categories of "countries" and "themes" used.

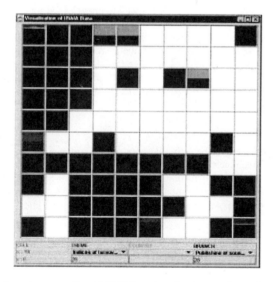

Figure 4: Visual interface to more than 150,000 economic time series. Categories of "countries", "themes" and "branches" used.

2 Input Data for Maps of Information Spaces

We can distinguish between the following types of input for maps of information spaces.

- Keyword-based: the bibliographic maps exemplified in Figure 1 are of this type. The keywords or index terms provide the dimensions of a geometric space in which our objects are located.

- Sparse graph: this was the case for the example discussed in Figure 2. This is highly likely to be the case whenever XML XLink functionality is used as the basis for associations between our objects.

- Dense graph: this is the case for database occupancy, visual user interfaces for which are illustrated in Figures 3 and 4.

In the dense grap case, a convenient way to process the data is to take the dense graph of interdependencies and map or project the objects, using these interdependencies, into a geometric space. We will briefly look at how this may be done.

Principal coordinates analysis is very similar to principal components analysis (PCA). Rather than the usual objects × variables array (e.g., documents crossed by index terms), we are given an objects × objects distance or dependency matrix. A minimal amount of alteration to the approach adopted in PCA allows this type of input data to be handled. Principal coordinates analysis has also been referred to as *classical multidimensional scaling* and *metric scaling* (Torgerson, 1958; see also the short description in Murtagh and Heck, 1987).

We investigated both direct use of principal components analysis, and principal coordinates analysis, on database occupancy data. The principle was the same in both cases: project similarity data into a coordinate space, and use between 2 and 8 best-fitting coordinates (associated with the highest eigenvalues) to characterize the objects. These objects were "countries", economic "themes" and economic "branches". A Kohonen map was then constructed from this data.

Figure 3 considers "theme" and "country" database occupancy frequencies. A color coding is used: themes in red, country in green, and branch in blue. In one grid cell, there can be more than one object. The color intensity varies with density. Also one grid cell is divided into several segments to reflect the different types of object associated with it. In Figure 4, "theme", "country" and "branch" database occupancy frequencies are shown. Clicking on the grid cell give the associated information – a set of economic time series from an OECD and Eurostat (Statistical Office of European Union) database of over 150,000 time series.

3 Conclusion

We have looked at new, closely-related technologies in the area of human-computer interaction. Visual and responsive user interfaces have much promise. The work on visualizing database occupancy which we have described is now being extended to input data derived from relevant documentation. This work is being pursued in the context of the European Fifth Framework IRAIA project, "Getting orientation in complex information spaces as an emergent behavior of autonomous information agents" (http://iraia.diw.de).

References

1. L.B. Doyle, "Semantic road maps for literature searchers", *Journal of the ACM*, 8, 553–578, 1961.

2. D. Guillaume and F. Murtagh, "Clustering of XML documents", *Computer Physics Communications*, 127, 215–227, 2000.

3. D. Guillaume, "Distributed Information Retrieval, Search and Processing in Astronomy", PhD thesis, Université Louis Pasteur, 2000b.

4. F. Murtagh and A. Heck, *Multivariate Data Analysis*, Kluwer, 1987.

5. F. Murtagh and M. Hernández-Pajares, "The Kohonen self-organizing map method: an assessment", *Journal of Classification*, 12, 165–190, 1995.

6. E. Oja and S. Kaski, *Kohonen Maps*, Elsevier, 1999.

7. P. Poinçot, F Murtagh and S Lesteven, "A spatial user interface to the astronomical literature", *Astronomy and Astrophysics Supplement Series*, 130, 183–191, 1998.

8. P. Poinçot, "Classification et recherche d'information bibliographique par l'utilisation des cartes auto-organisatrices, applications en astronomie", PhD thesis, Université Louis Pasteur, 1999.

9. P. Poinçot, F. Murtagh and S. Lesteven, "Maps of information spaces: assessments from astronomy", *Journal of the American Society for Information Science*, 51, 1081–1089, 2000.

10. W.S. Torgerson, *Theory and Methods of Scaling*, Wiley, New York, 1958.

Textual Document Mining Using a Graphical Interface

G. Hubert[1], J. Mothe[1][2],
A. Benammar[1], T. Dkaki[3], B. Dousset[1], S. Karouach[1]

(1) Institut de Recherche en Informatique de Toulouse, 118 Rte de Narbonne, 31062 Toulouse Cedex 4, France.
(2) Institut Universitaire de Formation des Maîtres de Toulouse, 56 Av. de l'URSS, 31400 Toulouse, France.
(3) IUT Strasbourg Sud, Université R. Schuman, 72 route du Rhin – BP315, 67411 Illkirch-Grafenstaden, France
Tel : 05 61 55 87 - Fax : 05 61 55 62 48
Email : {dkaki/dousset/mothe}@irit.fr

ABSTRACT

In this paper we present an approach in order to help the user extracting advanced information from a document set. The approach is based on the application of different mining modules that collaborate in order to provide graphical views to the user. These views correspond to a graphical summarization on document dimension correlations. To proceed, the documents are first represented according to different concept hierarchies that correspond to the document dimensions. Then, the resulting representation are mined using different modules depending on the user's goal.

1. INTRODUCTION

The goal of an Information Retrieval System (IRS) is to retrieve the information that responds to the users' needs from a collection of documents. Most of the IRS displays the results under the form of a list of document references and titles. Then the user browses the returned references and accesses those s/he supposes to be relevant. However, generally the list of returned references is long and it is time consuming to be processed by the users. In addition, the information access (i.e. retrieving the pages that contain the relevant raw information) is not the only goal of the users. The user can be interested in more advanced information as unknown pieces of information and relationships between information. This kind of information (named knowledge) is extracted from the raw information and the user generally discovers it manually while s/he is reading the documents. Some automatic processes have been reported in the literature. One application is the identification of a core set of authors or papers in a given field (White & McCain, 1989) via the cocitation analysis or the determination of the relative authority of Web pages based on the hyper-reference analysis (Kleinberg, 1999). Document classification is another application that can help the user when analyzing a retrieved document set. Post retrieval classification aims at grouping together documents that have been retrieved on the base of the similarity of their content (Hearst, 2000). These applications focuses on a single element of information (document content, citations or hyper-references, authors). However, documents have a wide range of dimensions that can lead to a range of relevant information extracted when analyzing a set of documents (Mothe, 2001). In this paper, we present an approach that takes advantages of different elements of information automatically extracted from the document content. Once extracted, the elements of information are analyzed using different mining components. Each component has its specific aim and the different components collaborate in order to help the user in the information-discovering task.
The paper is organized as follows. Section 2 presents the way documents are represented so that they can be automatically mined. Section 3 and 4 describe the main components of the system and how they co-operate in order to achieve an interactive document analysis.

2. DOCUMENT REPRESENTATION

Information retrieval systems generally represent documents as bags of words/terms, which can be weighted in order to represent the relative importance of those terms. In that case, the terms are automatically extracted from the document contents. Alternatively, a control vocabulary can be used. Terms from the controlled vocabulary are (often manually) assigned to each document. This approach is used in most of the specific digital libraries (Medline, Questel). The same kind of approach is used when documents are assigned to nodes of a concept hierarchy (CH) as it is the case for Web directories such as Yahoo. Whereas these systems use a single hierarchy of terms, our system is based on several ontologies (or CH), each one describing a document dimension. The multidimensionality of the documents is a key point of our approach as it corresponds to the starting point for a data mining process.

In a first phase, the system automatically assigns the documents to the different CH when possible. The nodes to which a document has been assigned correspond to the initial multidimensional document representation. Additionally, we compute a more summarized document representation, which is more adapted to mining processes, under the form of contingency tables.

2.2 Association of documents to the CH

The association of documents to the different CHs is based on Information Extraction principals. When a node from a CH is extracted from the document content, the document is automatically attached to that node. To proceed, templates are defined for each dimension (i.e. for each CH). The extraction process is based on extraction rules, which use syntactic tags from the documents and regular expression matching. Once a syntactic tag has been detected, semantic and filtering functions are used in order to complete the extraction process. Semantic functions take into account synonymy between terms, whereas filtering function eliminate the candidate terms that have been extracted but that do not belong to the corresponding CH.

2.3 Contingency tables

Whereas a bag of words is a representation adapted to information retrieval, it does not fit data mining processes. Contingency tables are a much more adapted representation (Fayyad et al, 1996). Contingency table is a mean to transform non-numerical information into numerical information. A contingency table is obtained by dividing up a population (in our case a set of documents) according to two variables or dimensions, I and J. The columns of the table correspond to the values of the variable J, whereas the lines correspond to the values of I. The intersection Tij of a row i and a column j corresponds to the number of objects in the population for which the variable I has the value i and the variable J has the value j simultaneously. In our approach, a variable corresponds to a CH and the values of the variables are the nodes of the corresponding CH.

A contingency table where I and J correspond to the same variable depicts co-occurrence relationships. For example, a contingency table where I and J are author names (one of the CHs) correspond to co-authoring relationships. When I and J are of different natures, some other relationships are depicted. For example, if I represents the geographic reference of the documents (e.g. the country of the author) and J represents the time (e.g. date of publication), the relative contribution of the countries on the field of the documents is depicted along time. In fact the crossing can involve any kind of information, as soon as it corresponds to a CH. Each contingency table corresponds to a 2-D document representation.

3. MINING COMPONENTS

The mining components all use the representation of the documents resulting from the document analysis (i.e. a 2-D contingency tables, where each dimension corresponds to a CH). The components co-operate in order to mine the corresponding document set. Each component is specialized in a specific mining task. In this section we present the main mining components of the system.

3.1 GeoECD

GeoECD visualizes data that are logically geo-referenced under the form of colored maps. Thus GeoECD can only be activated when the document representation includes a reference to a geographic part of the world. The intensity of the color in the obtained maps reflects the relative contribution of the continents, country or groups of countries to the data. This contribution can be weighted using features related to the countries (e.g. population, GNP, surface). Indeed, the visualized data can be either "absolute" contingencies (e.g. number of foot and mouse disease cases reported) or "relative" contingencies (e.g. number of foot and mouse disease cases reported divided by the total number of animals; the later element being a feature of the countries).

3.2 Clustering

The clustering module groups the data according to the agglomerative hierarchical clustering method. The individuals (rows of a 2-D document representation) are clustered according to the variables (columns of the 2-D document representation). Several modules are available, depending on the method used (single, average or

919

complete linkage). The results are visualized under the form of a dendrogram. This dendrogram can be cut at different levels, depending on the size/consistency of the classes one prefers.

3.3 Factorial analysis

This module mines the data using the correspondence factorial analysis (CFA) defined by (Benzécri, 1973) ; this method belongs to the family of analysis methods that benefits from on the mathematical properties of the Singular Value Decomposition (SVD) of matrices (Eckart & Young, 1936). The general goal of this class of methods is to represent vectors (the individuals) initially represented in an N dimensional space (variables) into a space with a smaller dimension. The CFA uses the SVD, but instead of being based on the absolute object representation (the contingencies), the SVD is based on the object profiles (similar to probabilities or % of contribution). Another specificity of CFA is the distance measure used to compare profiles. The measure that is used ($\chi2$) aims at favoring the specificities instead of the too recurrent phenomena. In addition, $\chi2$ makes it possible to represent the objects and the characteristics in the same reduced space.

To CFA is associated a x-dimensional graphical representation. The first axes resulting from the decomposition (the axes that correspond to the highest values) should be chosen for the representation as they maximize the initial distances between the individuals. In our system, the factorial analysis module uses a 4-D graphical representation.

3.4 LinkMap

LinkMap uses metrics in order to detect links between elements of information and visualizes them in a 4D view. The links are computed using different metrics applied to contingency tables. The gray level used to color the links in the views represents the link's weight: a white link is weak whereas a black one is very strong.

The classification, factorial methods and map representation are not new; classification have largely been used in IRS; factorial methods are used in many other fields that implies the object analysis and maps are similar to the ones used in Geographic Information Systems. The main original point is that these methods are components of an interactive interface and that they can be combined in order to explore a document set.

4. INTERACTIVE MINING

4.1 The document source

The documents used are scientific publications taken from INRA organisms all over the world. The documents are extracted from publications on biomaterials of the Current Contents database. We treat 2454 documents.

4.2 The document representation

2-D document representations are extracted as described in section 2.3. The dimensions as well as the number of different values for each dimension are given in the following Table 1.

Dimension	Number of different values
Author's country : PA	40
Publication source : SO	164
Keywords : KP	923
Publication date : DP	2 (1998 and 1999-2000)

Table 1. Document dimensions

4.3. Example of analysis

In this example, the document set is mined according to the two following dimensions: the country the authors belongs to (PA) and the publication source (SO). Firstly, the classification module is applied in order to group together countries that have the same behaviour with regard to the publication source. Figure 1 displays the resulting dendrogram. The dendrogram can interactively being cut at any level according to the number/size of classes the user prefers.

920

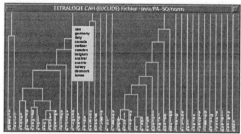

Figure 1. Dendrogram resulting from the classification of countries according to publication source

The result of this classification can be sent to the GeoECD module as shown Figure 2. In that case, to each class resulting from the clustering module is associated a colour.

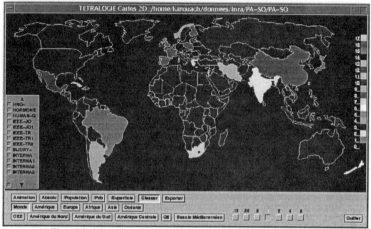

Figure 2: Map visualizing the result of a country classification

Additionally, it is possible to detect what are the countries that have a specific behavior and to extract the reason of these specificities. This is done applying the factorial analysis method on the same document representation. The 4-D representation that results from the factorial analysis is shown Figure 3.

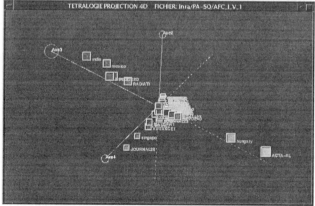

Figure 3. Factorial analysis 4-D representation

921

In Figure 3, the journal ACTA-ALIMENTARIA has a specific behavior (far from the center and close to Hungary) which can be explained by the fact that it is a Hungarian journal.
The real links that exists between the elements of information can be graphically displayed using the linkMap module (see Figure 4).

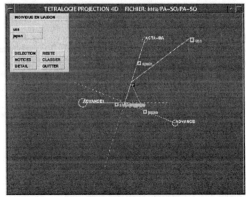

Figure 4: LinkMap results

In this example, we focused on the United Kingdom's case. The gray level shows that there are more common journals between authors of the United Kingdom authors and authors of the USA than between authors of the United Kingdom and authors of Japan.

5.   CONCLUSION

In this paper, we presented an approach to mine a document set according to the different document dimensions. The document dimensions are automatically extracted with regard to different concept hierarchies that correspond to views the user can have on the document set. This initial document representation is then transformed into 2-D document representations (contingency tables) that behave as input of different mining modules. We describe the main mining modules that are provided to the user in order to help him/her mining the document set. We gave an example on how the different modules can co-operate. The two dimensions used in the example correspond respectively to the country of the author and to the publication source. However, they could have been any of the document dimensions. The same kind of analysis would then have been extracted.

Acknowledgement
Research outlined in this paper is part of the IRAIA project supported by the European Commission under the Fifth Framework Programme (IST-1999-10602). However, in this paper, we described our approach which does not necessarily correspond to the IRAIA consortium view.

REFERENCES

Benzécri, J. P. (1973). L'analyse de données. Tome 1 et 2, Dunod Edition.

Eckart, C. & Young, G. (1936). The approximation of one matrix by another of lower rank. *Psychometrika*, Vol.1, pp 211-218.

Fayyad, U. M., Piatetsky-Shapiro, G., Smyth, P., & Uthurusamy, R. (1996). Advances in Knowledge Discovery and Data Mining. *AAAI Press*, ISBN 0-262-56097-6.

Hearst, M. A. (2000). The use of categories and clusters for organizing retrieval results. In *Natural Language Information Retrieval*, Kluwer Academic Press. http://www.sims.berkeley.edu/~hearst.

Kleinberg, J. (1999). Authoritative sources in a hyperlinked environment. *Journal of the ACM*, Vol.46, N.2, pp 212-235.

Mothe, J., Chrisment, C., Dkaki, D., Dousset, B., & Egret, D. (2001, April). Information mining: use of the document dimensions to analyse interactively a document set. *European Colloquium on IR Research (ECIR)*.

White, H. D. & McCain, K. W. (1989). Bibliometrics. In *Annual Review of Information Science and Technology*, volume 24, pages 119-186. Elseiver, Amsterdam.

Users bootstrap searching the Web through interactive agents supporting best practice sharing

Kurt Englmeier[1], Josiane Mothe[2], Bernd Pauer[1],
[1]German Institute for Economic Research (DIW), Königin-Luise-Str. 5, 14195 Berlin,
phone +49 30 89789 367, fax +49 30 89789 200,
{kurt, bernd}@diwsysv.diw-berlin.de
[2]Institut de Recherche en Informatique de Toulouse, 118, route de Narbonne,
31062 Toulouse, phone +33 561 556 322, fax +33 561 556 258,
mothe@irit.fr

Abstract

Dramatic advances in the Internet and software technology have created many new opportunities for information products and services. Unfortunately, the expansion of available information carries with it many associated problems. Information overload is the most prominent one. However, this can be alleviated by a user interface that incorporates a comprehensible view on an information space with personal assistance.

The paper presents a collaboration model that fosters personalized information delivery and experience sharing among users. Information agents resorting to this model observe and analyze users' search and navigation. Unlike in most situations of usage investigation, here the users have full control on the agents' communication behavior and knowledge. The resulting profiles of usage as a special user feedback facilitate the automated induction of accurate interest patterns that leads to a better clarification of use strategies.

1. Introduction

Personalization of user interaction is becoming a more and more prominent marketing instrument. Knowing the user's purchase behavior opens new and more efficient ways in approaching potential consumers. Advertising, product offers, and services tailored to individuals on the basis of knowledge about their preferences promises better positioning among the e-competitors. Personalization in this context means deriving purchase recommendations from the knowledge of who the consumers are, how they behave, and how similar they are to other consumers. Thus, the system's interaction mode has also to capture the user's selections, navigation, and input. This feedback information feeds data mining processes that combine it with already available consumer data which might include histories of purchasing and searching activities as well as demographic and psycho-graphic data. From this structured "knowledge" of consumers salient user profiles are extracted that accurately match products, services and advertisements to individual consumers. Recommendations may be on the basis of the *user's* past activities or on past activities of *other* users with *similar* characteristics.

Personalization outlined so far focuses on **imparting personal data** to devise better and more efficient business strategies. Without any doubt, consumers may benefit from product offers and advertising tailored to their preferences as long as service providers reconcile their system with privacy concerns of the users.

Preserving privacy must be a design objective for the interaction with the user. Profiling must be predictable and controllable to give the users the feeling they are dealing with a system respecting their privacy. Most current systems resemble more spies eavesdropping behind the curtain to siphon personal data to the good of business strategies of a company or the other. Thus, the design approach outlined here puts more emphasis on passing the final control and decision on handling personal data exclusively into the hand of the user.

Sharing personal data takes personalization beyond the scope presented so far focusing on the issue that users often find themselves as lonely information hunters while roaming huge information spaces like the web. Wouldn't it be nice if we had an expert sitting next to us or somebody who recently had the same search problem or, at least, if we could resort to our own search records? "When you look at people dealing with any kind of information system, you realize that each person's decisions – those he or she makes in the course of getting to the right information – are essentially lost to the rest of the world."[1] Navigating the Web should allow people to leave pointers for those who might also navigate along the same paths. This can enhance the users' navigating and browsing by extremely

---

[1] Kantor, P. (1998). Statement quoted in *Wired magazine* 6.03, Mar. 1998, p. 89.

useful information. Such a *collaborative* trait favors equally both users and service providers. This kind of orientation is of crucial importance when it comes to masses of data accessible by all kind of users. However, even for systems drawing on personalization for mainly business purposes the emphasis on collaboration might be pivotal when progress in privacy preservation turns out to be a critical success factor of the e-business application. Users are willing to share personal information if they benefit from recommendations based on shared experiences in browsing and navigating complex data collections *and* if they don't feel themselves eavesdropped by the system.

2. Personal Agency

Tracking individual needs of users leads to personal digital assistants that appear in the system's interaction mode as personal software agents. The approach presented here links the personal agency to the design of user interaction. Information agents resorting to this model observe and analyze the usage of an application, i.e. users' search and navigation in the case of a portal application of an information service. Profiles of usage, reflecting the users' practices, facilitate the automated induction of accurate interest patterns that leads to a better clarification of use strategies.

Thanks to personal agents as assistants, applications with a focus on personalization are in the position to provide information tailored to actual situations and individual user interests. The exchange of usable and useful content is the challenge that defines the broader context of personal agency (Meisel et al. 2000).

3. Semantic Coordinate System

A specific trait in our approach is that personalization as well as searching and navigating takes place in a domain-dependent semantic context. This context can be described by its respective controlled vocabulary organized along a few thesaurus-like morphologies (concept hierarchies). Thus, these hierarchies lend themselves to derive controlled vocabularies. Usually experts develop and maintain these morphologies in order to reflect comprehensively the essentials of a particular thematic area.

As a practical consequence, these concept hierarchies are easy to traverse even for non-experts. It's quite convenient for a variety of users to find significant terms and concepts that fit their information need. It proved to be easier in many cases, just to point to relevant terms instead of finding pertinent concepts in the personal active vocabulary. The inherent information of an hierarchical order showing a concept within its generic concept as well as its subordinal ramifications supports the users' conceptualization of their information need.

Automatically analyzing large amounts of structured and unstructured information benefits from morphologies like thesauri and glossaries. They reflect a comprehensive and structured overview on domain-related knowledge. This enables several expert communities to prepare the basic construction elements for the coordinates of an information space as well as the controlled vocabulary. A notable trait of our approach is presenting document samples in a semantic coordinate system that provides the users with clear orientation during their interaction with the system.

4. A Best Practice Model

What we want to achieve is an approach for information retrieval (IR) applications that fosters best practice (O'Leary 2000) in information search processes. Best practices reflect the knowledge about how users perform successful retrieval activities and attempt to define the best ways of searching large document collections. They are generally based on models that describe and categorize usage scenarios and represent them in explicit specifications of conceptualizations (Gruber 1993).

The controlled vocabulary outlined above endows a best practice model with the necessary common language. In this perspective, best practices can be treated just as a particular group of descriptors. Instead of representing a single document, a part of it, or a group of documents with a significant coherence among them, it stands for an individual or cross-individual scenario of a successful search in the information space. The frequency of a specific navigation scenario among different users distinguishes the successful searches from the less successful ones.

Thus, best practice in searching an information space manifests itself in a set of descriptors that refer to various documents visited in order to satisfy the information need. A best practice can be expressed as the points (documents or navigation steps) on a navigation scenario S visited by a significant number of users. A scenario S can be assumed to be a best practice scenario (*bp*) among various users if every node on this scenario track is visited by at least 95% ($\varepsilon = 0.95$)[2] of all users having passed by its documents.

$$bp = t_1, t_2, ..., t_i, ..., t_m \quad \text{with} \quad \frac{n(t_i)}{n(S)} > \varepsilon \quad ;$$

where

$n(t_i)$ = number of users of node t_i.

$n(S)$ = number of users of the whole scenario.

The system recommends a best practice in an actual situation if an observed scenario O comes close it.

$$bel_{bp} = \frac{n(O)}{n(S)} \text{ with } o_j \in t_j;$$

5. A Model of Personal Agency in IR

If appropriately designed, interface agents support a communication base of high confidence that is indispensable when it comes to transfer personal data to a remote system for analyze purposes. By the interaction with the user, in a step by step manner, a personalized agent may create "knowledge" on the user's retrieval practices. By the collaboration of the different personal interface agents an exchange of "experiences" with different usage scenarios can be realized in order to learn from different profiles. Preserving privacy makes it indispensable to think about ways to make the agents' behavior and its communication act more transparent and modifiable.

5.1. Ontologies

An ontology is a consensual, shared, and formal description of the important concepts in a given domain. It identifies classes of objects that are important in a domain and organizes these classes in a subclass hierarchy. (van Harmelen, Horrocks, 2000) The ontology does not describe necessarily the whole information space together with the interaction model of the application, but rather only those portions that are directly relevant for specific agent tasks. They constitute resources, a particular agent can resort to, and the communication terminology to access instances of the ontology as well as to pass relevant contents to other agents. This means, the ontology describes the objects in this application environment, the relationships between these objects as well as the interfaces to its instances or to the user. The domain dependent part of the ontology's terminology refers to the already mentioned controlled vocabulary. It is indispensable to maintain a semantic context the agents are acting within.

5.2. Communication Language

It has to be pointed out that an agent communication language covers two aspects: the protocol of their communication and the communicative act itself. According to research, a piece of software can be called an agent if it also complies to mental components such as beliefs, capabilities, choices, and commitments. These traits have to be taken into account in the development of agents that can be regarded as a specialization of object-oriented programming (OOP). What is called object in OOP is now an agent. An object's properties represent the mental state of the agent and its methods refer to the communication act such as message passing, response, rejecting, executing commitments, and the like.

Requests to an agent are composed of terms of the controlled vocabulary of the information domain. However, the exchange of concepts based on these terms are wrapped in a protocol. Like KQML the communication language among agents is conceived as both a message format and a message-handling protocol to support run-time "knowledge" sharing. The message itself is opaque to the content it carries. Even though they impart phrases and the like, they rather communicate an attitude about the content (assertion, request, commitment). These performatives (Cohen, Levesque, 1997) define the permissible actions that agents may attempt in communicating with each other.

5.3. Transaction Plan

The capability of agents depends on their ability to process the requests that are directed to them. The agent has to develop an appropriate set of operations. This includes selecting the information sources for the data, the operations

---

[2] This threshold is arbitrary, but a value should be assigned that guarantees the avoidance of ambiguities.

for processing the data, the non-agent processes the data should be delegated to, the sites where the operations will be performed, and the order in which to perform them. The transaction plan helps the agents to orchestrate their operations triggered by the user's request. Moreover, it endows the agent with the capability of producing parallel execution plans, interleaving planning, execution, sensing, and re-planning requests that failed while executing other requests.

Figure 1 shows an example of a transaction plan. For instance, on the user's request, the personal (PA) commits to operate on the actual sequence of query profiles (observed scenario) and follows its plan to elaborate recommendations. The most salient navigation profile may be found, for instance, among the profiles of the user's long-term interests (LTP) or in the remote best practice base. PA may search for similar scenarios in both data sets in parallel or sequentially, according to the user's preferences or its own belief.

```
<pa.recommendation>
    <commit=user.querySequence, user.preferences ?scenarios.similarity > user.preferences.threshold>
        <choice=user.preferences.threshold, self.belief>
            <user.preferences.threshold>
            <ask=user, user.preferences.threshold>
            <subscribe=common.services.preferences.threshold>
        ...
        <choice=scenarios.similarity, self.belief, user.preferences.scenarioSet>
            <user.scenarios.longterm>
                <subscribe=local.services.scenarios.similarity, user.querySequence, user.scenarios.longterm>
                <subscribe.noCommittment -> next.choice>
                    <choice=local.sma.return>
                        <accept, self.belief>
                        <subscribe=local.services.qea, user.querySequence>
                    ...
                    </choice>
                <common.bestPractice>
```

Figure 1: Fragment of the personal agent's transaction plan for retrieving recommendable usage scenarios.

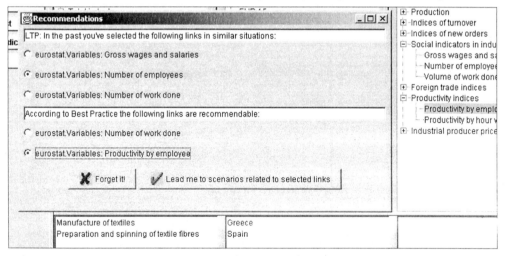

Figure 2: The personal agent gives recommendations with respect to the actual retrieval situation. A suggestion may be derived exclusively from personal records ("In the past you've chosen this direction in similar situations"). In many cases however, it may be pretty helpful for the user if her personal agent also resorts to some common navigation expertise. This recommendations could be like: "The majority of users having navigated so far through the information space chose for the following steps this or that direction, i.e. this or that concept among the related coordinates".

PA subscribes to the service (local or remote) that calculates similarities (SMA) and passes the user's data to a service agent that has committed to PA's request. SMA measures the similarity between an observed scenario and LTPs along its similarity metrics. The precision of the retrieved profiles reflects their respective saliency. Note that both agents "work" exclusively at the user's site.

Only PA knows from its experience with its user which level of precision represents a sufficient similarity. If the results of SMA do not reach this level it probably consults the service board[3] if there are query expansion (QE) services available. If there are, PA passes the results obtained from SMA together with the original query to the respective agent (QEA) for further processing. It's evident that QEA has a decision model allowing it the replacement of terms in a profile without changing its inherent meaning. If the request of PA leads to a number of salient navigation profile, the agent sends a recommendation to the user. It contains useful concept hierarchies to choose together with concepts within the respective hierarchy. In the case that suitable profiles cannot be retrieved locally, PA may resort to remote agents maintaining best practice profiles (BPA), if the user allows the communication and this feature is published on the service board.

6. Conclusions

This paper described important features of an interface agent design with a special emphasis on personalization as a method to share experience among users. The outlined architecture inclines to Knoblock and Ambite's approach (Knoblock, Ambite, 1999) that includes (1) modularity in terms of presenting information sources and agents, (2) extensibility in terms of adding new agents and sources, (3) flexibility in terms of selecting most appropriate sources and operations, (4) efficiency in terms of parallel execution of operations, and (5) adaptability in terms of being able to track semantic discrepancies among usage scenarios.

In our approach we added (6) responsibility to protect user's privacy. All actions of an agent must be predictable and controllable to give users the feeling that the agent's predisposition favors his or her interests over those of the whole user community or the company offering the respective Internet application. The quest in sharing personal data is to develop simple communication conventions that makes an software agent fully controllable by the user while complying with the functional requirements of the remote system it is interacting with. Otherwise personal agency will be rejected like any other malformed metaphor for the design of user interaction.

References

Cohen, P.R.; Levesque, H.J., " Communicative Actions for Artificial Agents", in *Software Agents*, J.M. Bradshaw (ed), AAAI Press, Menlo Park, 1997, p. 420-421.

Gruber, T., "A translational approach to portable ontologies". *Knowledge Acquisition*, Vol. 5, No. 2, 1993, pp. 199-220.

Knoblock, C.A.; Ambite, J.L., "Agents for Information Gathering", in *Software Agents*, J.M. Bradshaw (ed), AAAI Press, Menlo Park, 1997, p. 347-373.

Meisel, J.B.; Sullivan, T.S., "portals: the new media companies". *info – the journal of policy, regulation and strategy for telecommunications information and media*. Vol. 2, no. 5, 2000.

O'Leary, D.E., "Different Firms, Different Ontologies, and No One Best Ontology". *IEEE Intelligent Systems*. Vol. 15, no. 5, 2000, pp. 72-78.

van Harmelen, F.; Horrocks, I., "FAQs on OIL: The Ontology Interference Layer". *IEEE Intelligent Systems & their applications*. Vol. 15, no. 6, 2000, pp. 69-72.

Acknowledgement

Research outlined in this paper is part of the IRAIA project supported by the European Commission under the Fifth Framework Programme (IST-1999-10602). However views expressed herein are ours and do not necessarily correspond to the IRAIA consortium.

---

[3] A service board is an element of the agents' environment containing all services the agents have published or they can subscribe to.

Mining coherence in time series data

George Potamias[1,3] and Vassilis S. Moustakis[1,2]

{moustaki, potamias}@ics.forth.gr

[1] Institute of Computer Science, Foundation for Research and Technology – Hellas (FORTH), Vassilika Vouton, P.O. Box 1385, 71110 Heraklion, Greece

[2] Department of Production and Management Engineering, Technical University of Crete, Chania 73100, Greece

[3] Department of Computer Science, University of Crete, Heraklion 71409, Greece.

Abstract

This paper presents work on modeling coherence between time series data. Work is based on the elaboration of formula that computes the distance between time series. Based on the computed distances, the method exploits the closest neighbor algorithm and leads to the construction of a phylogeny-clustering tree. Using car sales data (available on the www) we demonstrate work done and present preliminary results. We discuss implications of the work performed in correlating time series data with documents including such data.

Introduction

Time series occur in many aspects of economic and social activity. Time series organization of data implies time stamping of individual observations. Observations may represent economic or social activity or even life critical measurements such as those acquired when a patient is under monitoring in a critical care unit. Time series data modeling has been an active area of research in statistics. A variety of models exist, which manifest interest and provide the interested analyst with analysis tools (Box and Jenkins 1976). Expanding interest in data mining and knowledge discovery has contributed to an increase of research awareness in learning using time series data (Morik 2000).

This paper reports preliminary research results and modeling activity in integrating time series data with documents. Motivation originates from the fact that analysts do use time series data to prepare reports. Reports and time series data reside in distributed information archives. There is an emerging need to mine for knowledge over such archives, which contain both reports and data. Specifically, the paper presents a methodology, which investigates similarity between time series data and reports results using exemplar time series drawn from car sales (see legend in Figure 1, in the text). Data-mining methodology is realized by the introduction of a novel algorithmic process and related formulas, for discovering similar and indicative patterns in time-series collections. Final outcome is the clustering of time-series into similar-groups, visualized by the appropriate customization of a phylogeny-based clustering algorithm and tool.

Methodology

Measuring similarity between objects is a crucial issue in many data retrieval and data mining applications. The typical task is to define a function $dist(a,b)$ (or, $sim(a,b)$), between two sequences a and b, which represents how "*similar*" they are to each other. For complex objects, designing such functions, and algorithms to compute them, is by no means trivial. Time series are an important class of complex data objects, and the need for clustering sequential data is profound. While the statistical literature on time series is vast, it has not studied *similarity* notions that would be appropriate for data mining applications.

Our *methodology*, for discovering coherences between time-series is realized by the following steps:

i. *Representation*. A *Piecewise Linear Segmentation* approach is followed. The approach aims to 'weight' the different time-to-time changes (segments) of time-series, based on their '*significance*' according to the whole evolution of the series, i.e. the *intra-significance* of a change. Given a time- series X_a, evolving over a set of k time-points with respective values $y_1, y_2, \ldots y_k$, the *percentage change*, $c_{a,i;i+1}$ is computed by the formula:

$$c_{a,i;i+1} = \frac{|y_{i+1} - y_i|}{y_i} \qquad (1)$$

The significance of a change is decided by the applying a suitable statistical-test. In the current implementation, and in the conducted experiments the *Student's t*-distribution and respective statistics where used.

ii. *Weighting percentage changes.* Besides measuring the intra-significance of changes in a time-series, the respective changes should also be weighted according to the respective time-to-point changes for the whole time-series collection, i.e., the *inter-significance* of a change. The *relative-weight* of a time-series change is computed by the formula,

$$W_{c_{a,i;i+1}} = \frac{c_{a,i;i+1}}{\displaystyle\sum_{\substack{Time\ series \\ in\ \{a,\ b,\ ..\}}} c_{T,i;i+1}}, \qquad (2)$$

iii. *Computing the **distance** between time-series.* Given two time-series X_a and X_b, their distance is computed by the formula:

$$dist\ (X_a, X_b) = \sum_{i=1}^{k-1} |(W_{c_a,i;i+1} \cdot c_{a,i;i+1}) - (W_{c_b,i;i+1} \cdot c_{b,i;i+1})| \qquad (3)$$

✍✍ *Utilization of inter-significance.* If the respective percentage changes are significant then, their weights are computed by formula 2 above. In all other cases, the weights are <u>not</u> taken in consideration, i.e., their values are fixed to 1.

Clustering and Visualization. The computed distances feed an appropriate distance-based clustering algorithm in order to form clusters of similar time-series. The current implementation of our system encompasses the Neighbor-Joining clustering algorithm (Saitou and Nei 1987). The *Neighbor-Joining* algorithm is a phylogeny-based clustering algorithm utilized in biology-systematics and molecular biology. The main utility of the algorithm is that it computes, and outputs a phylogeny-tree that offers a nice and informative *visualization* of the formed clusters.

Experimentation

Six time series data depicting car sales were randomly selected. Data are indicated as TS-1 through TS-6 and are presented in Figure 1. In each time series the horizontal axis presents time periods and the vertical axis sales.

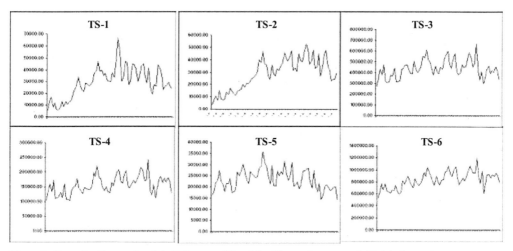

Figure 1: Car sales time series data (*M2-competition*, http://forecasting.cwru.edu/Data/m2comp)

929

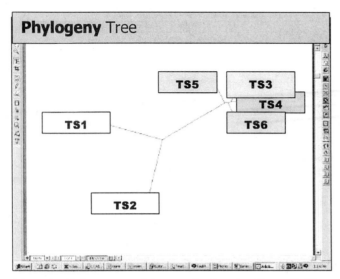

Figure 2: *Phylogeny tree* of the car sales data (see Figure 1). Time series 1 and 2 *(Cheetah and Panter)* are grouped together and isolated by all other time-series. Time series 3 and 4 *(Animal and Bigcat)* are considered similar and grouped together. Time-series 5 and 6 *(Lion and Carrinds)* presents isolated not-coherent objects. Furthermore, time-series 3,4,5, and 6 are isolated from 1 and 2 and could be considered as a sub-group in the hierarchy of clusters.

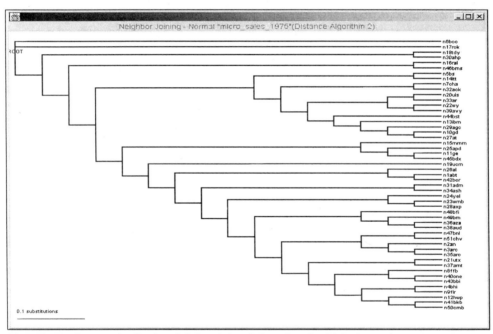

Figure 3: *Dendrogram* visualization of phylogeny-based clustering of multiple time-series

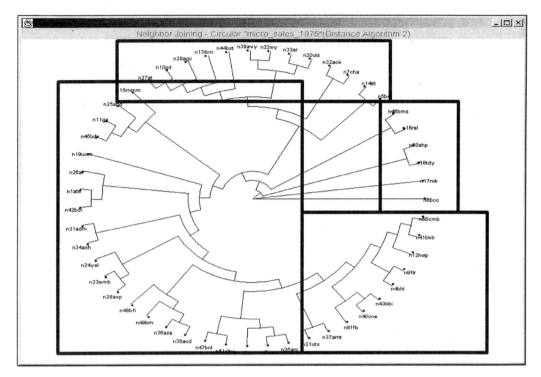

Figure 4: *Circular* **visualization of phylogeny-based clustering of multiple time-series – Identified clusters**

Application of the algorithm, which we present in section on Methodology led to the construction of the phylogeny-clustering tree that we depict in Figure 2. Broader application of the methodology on a wider collection of time series data led to the construction of the phylogeny-clustering tree, which we present in Figure 3. Figure 4 presents an alternative visualization of the phylogeny tree construction. Results point out the feasibility and potential of the selected approach.

Discussion

Preliminary results reported in this paper demonstrate feasibility of an alternative modeling approach to identify similarity and coherence across time series data. Usefulness of the approach in identifying the most relevant time series and corresponding documents (if any) across large (and often distributed) statistical collections is critical. The user tries to identify both data and documents. Modeling coherence across time series, which have been used to prepare documents (such as the documents reporting on industrial or economic activity in a region or across industries) eases retrieval and contributes to the location of the most appropriate information. In other words, modeling coherence provides an alternative indexing approach to a huge collection of both time series and documents.

We are applying this approach in context of a European project titled IRAIA (IRAIA 2001) -- IRAIA stands for *Getting Orientation in Complex Information Spaces as an Emergent Behavior of Autonomous Agents*. Task is to support users to access documents and time series data across repositories that contain thousands of both.

The implemented system offers operations for importing and reading time-series data; computing distances between time-series; clustering of time-series; and visualization of the resulted clustering. The whole system is built in *Java* making it *portable* on various h/w platforms (the java classes of the clustering algorithm were taken from the public-domain library PAL- Phylogeny Analysis Library, http://www.pal-project.org, and were appropriately customized for inclusion in our system).

Furthermore, the approach may be extended to modeling cause-effect relationships between variables, each of which is represented via a different time series.

We maintain that modeling and learning from time series data represents an emerging machine learning domain, which is further motivated by data mining inquiries.

Acknowledgement

Work presented in this paper is partially supported by the European Union, IST Program via the IRAIA project (IST 1999 10602). The partners of the IRAIA project are: DIW (German Institute for Economic Research, D), IFO (Institut fur Wirtschaftsforschung, D), CNRS-UPS (Centre National de la Recherche Scientifique, F), QUB (Queens University of Belfast, UK) and FORTH (Foundation for Research and Technology – Hellas, GR). Responsibility for work presented lies with the authors and not with the IRAIA consortium.

References

Box, G.E.P. and Jenkins G.M. 1976. *Time Series Analysis, Forecasting and Control*. Prentice Hall

IRAIA. Getting Orientation in Complex Information Spaces as an Emergent Behavior of Autonomous Agents. Overview information available at: www.

Morik K. 2000. The Representation Race – Preprocessing for Handling Time Phenomena. In: Mantaras, R.d.L. and Plaza, E., *Proceedings of European Conference on Machine Learning (ECML)*. Springer Verlag

Saitou, N. and Nei, M. 1987. The neighbor-joining method: a new method for reconstructing phylogenetic trees. *Mol. Biol. Evol.* **4** (4), 406–425

How to integrate different text data and fact information.
A conceptual transfer problem in digital libraries and its connection to agent theory.

Jürgen Krause

University of Koblenz-Landau and Social Science Information Center (IZ Bonn),
Lennéstr. 30, D-53113 Bonn, Germany
mailto: Krause@bonn.iz-soz.de; http://www.uni-koblenz.de/~krause/

Today users of scientific information are faced with a highly decentralized, heterogeneous document base with varied content analysis methods. Thus, the main problem to be solved is as follows: users must be supplied with heterogeneous data from different sources, modalities and content analysis processes via a visual user interface without inconsistencies in content analysis, for example, seriously impairing the quality of the search results. The basic ideas of the agent paradigm play an important role in this context. First solutions for transfer modules will be presented from the projects of the research department of the Social Science Information Center in Bonn (Germany).

1   Introduction

Today users of scientific information are faced with a highly decentralized, heterogeneous document base with varied content analysis methods. The traditional providers of information like libraries or information centers have been joined more and more by the scientists themselves, who are developing independent services of varying scope, relevance and type of development in the WWW. Theoretically, groups that have gathered literature or fact information on specialized subjects can crop up anywhere in the world. One consequence of this is the presence of various inconsistencies:

- Relevant, quality controlled data are to be found right next to irrelevant and perhaps demonstrably erroneous data.
- In a system of this kind, descriptor A can assume the most disparate meanings. Even in the narrower context of specialized information, descriptor A, which has been extracted in an intellectually and qualitatively correct manner, and with much care and attention, from a highly relevant document, is not to be compared with a term A that has been provided by automatic indexing in some peripheral area.

Thus, the main problem to be solved is as follows: users must be supplied with heterogeneous data from different sources, modalities and content analysis processes via a visual user interface without inconsistencies in content analysis, for example, seriously impairing the quality of the search results. A scientist who for example is looking for social science information on subject X will not first search the social science literature database SOLIS, and then the library catalogues of the special compilation area of social sciences at the library catalogues and in the WWW before finally trying to find out – each time using different search strategies and systems – whether survey results relating to subject X are available. He wants to phrase his search inquiry in the terminology to which he is accustomed without dealing with the remaining problems.

2   The Problem of Heterogeneity and Vagueness in Information Retrieval

Closer analysis of the problems, which are caused, for example, by the integration of literature databases, library catalogues and WWW sources of scientific institutes, shows that narrow technological concepts, even if they are undoubtedly necessary, are not sufficient on their own (see Krause 1996 for more details). They must be supplemented by new conceptual considerations relating to the treatment of breaks in consistency between the different processes of content analysis. Acceptable solutions are only obtained when both aspects are combined.

933

The IZ group is working on this aspect in Carmen[1]. It is also one central theme of other current projects of the research department of the IZ: ViBSoz[2], Daffodil[3] and ELVIRA[4]. The central features of the chosen approach are transfer components between the different forms of content analysis, which take account of semantic-pragmatic differences and can be – but must not be - modeled as agents (see chapter 3). They interpret the integration between the individual document sets with different content analysis processes (including automatic indexing) on a conceptual basis by cross-referencing the conceptual world of specialist and general thesauri, classifications, etc. In these considerations it does not matter, in principle, whether the semantic-pragmatic differences between two texts, between a text and, for example, the results of a survey in tabular form or a video archive have to be bridged or alternatively between multilingual sources.

2.1   Treatment of vagueness in information retrieval (IR) and the two-step method

The traditional form of vagueness treatment in IR refers to the comparison between query terms and content analysis terms, whereby the document level is regarded as the uniform modeling basis. This is shown most clearly when all documents were automatically indexed. Even if additionally intellectually determined descriptors of a controlled vocabulary were produced, modeling follows this homogeneity demand in principle. The user can either carry out his/her search using one of the two term groups and then base his/her search strategy on the controlled vocabulary or alternatively via the free text terms of automatic indexing using another strategy. If he/she chooses to perform a search via both term areas, differentiation in the match is not distinguished, i.e. it is treated as if the semantic differences in both content analysis methods do not exist.

The problems become even clearer when we use digital libraries and link, for example, a social science literature database such as SOLIS with its own controlled vocabulary (IZ Thesaurus and IZ Classification) with library catalogues, for instance in the ViBSoz Project with the documents of Cologne University Library which is being developed intellectually according to the keyword list of the German Library (DDB) in Frankfurt. A comparison of two such thesauri or classifications reveals that vagueness already occurs at the semantic description level of two document sets to be integrated in the search, and not just in the user query. Terms in a library classification with its controlled vocabulary form a separate description level. It cannot simply be translated on a 1:1 basis into the terms of another classification, e.g. from the area of specialist information systems. The meaning of a descriptor A in the library classification is different from the meaning of the same term in another classification or even in a thesaurus such as the IZ Thesaurus of the SOLIS social science literature database, even if a simply "vague" connection exists. These vagueness relations can be integrated on a non-differentiated way into the modeling of the IR process. In this case vagueness between the user query and documents will be modeled without explicitly treating differences at the document level between two heterogeneously developed document sets.

An alternative "two-step" method plays an important role in the projects of the IZ (ELVIRA, ViBSoz, Daffodil and CARMEN). It is based on the thesis that heterogeneous document sets should first be interlinked through transfer modules (vagueness modeling at the document level) before they are integrated in the superordinate process of vagueness treatment between documents and the query (the traditional IR problem). If, for example, three heterogeneous document sets have to be integrated, transfer modules bilaterally treat the vagueness between the different content analysis methods. The hope behind this form of vagueness treatment, which differs considerably from the procedure used traditionally in IR, is to produce greater flexibility and target accuracy of the overall procedure through separation of the vagueness problem. Different forms of vagueness do not flow uncontrolled into one another, but can be treated close to the causal interface (e.g. the differences between two different thesauri).

---

[1] CARMEN investigates the conceptual problem of heterogeneous data stores on the basis of an exemplary data pool from the contents of major publishers' servers linked to electronic information primarily in the fields of mathematics, physics and social sciences (see Krause/Schwänzl/Plümer 2000).

[2] The "Virtual Social Science Library Project" concentrates on the connection of different library catalogues with the documents of the SOLIS literature database (see Kluck et al. 2000). It is part of the digital library research program of the German Research Association (DFG).

[3] "Distributed Agents for user-friendly Access of Digital Libraries" (see Fuhr et al. 2000))

[4] " Electronic Retrieval and Analysis System for Industry Associations" (funded by the German Federal Ministry of Economic Affairs). ELVIRA started as an online system for time series used by major German industry associations. Text retrieval functionality was added in 1998 to address fact–text integration. The system is now used by companies in more than 350 installations (see Krause/Stempfhuber 2001).

This firstly appears more plausible in cognitive terms and secondly permits the combination of a wide range of modules for treatment of vagueness.

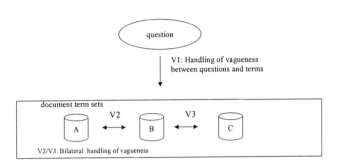

This procedure appears to be highly promising, especially in view of the broad background of empirical knowledge that IR processes differ more in the volume of results than in the quality of the evaluation parameters such as recall and precision. For example, a probabilistic method can be combined with neural transformation modules in the match between the user query and documents. Alternatively, neural transformation modules can also be combined with IR processes based on Boolean algebra. But even if the transformation modules between the heterogeneous document sets use the same similarity function as at the IR level between the user query and documents, the results between the traditional "one-step" and the "two-step" methods proposed here will probably differ.

Within the framework of this approach - used for the first time for text-fact integration in ELVIRA with statistical algorithms - Mandl 2000 examined the question of what neural network methods can be considered for the transfer modules. In this context he encountered transformation networks which were formerly used in the field of modeling of relevance feedback.

2.2   Transformation networks

Transformation networks are a special neural network model which is based on the back propagation approach. Mandl 2000: 124 classifies it as "unsuitable" for IR, but believes it is well-qualified for use for transfer modules between heterogeneous document sets. They have not yet been used to handle explicit transfers between heterogeneous document sets. The transformation network in the context of ViBSoz and CARMEN has a sub-symbolic interpretation layer and is used in conjunction with "Latent Semantic Indexing (LSI)", a condensing method which tries to erase unimportant comparison matrix dimensions which may have an adverse effect on calculation expenditure and similarity calculations. (see Mandl 2000).

Since neural networks generally need more training data than statistical methods, the existence of sufficiently extensive parallel bodies prevents their use in many cases. This fortunately does not apply to the application content of the special virtual libraries (ViBSoz). A comparison of the databases to integrate revealed around 15,000 documents recorded twice, of which around 2,000 were used as test data. The test revealed much better results for the neural network than the baseline of the statistical method with LSI (see Mandl 2000: 206).

3   Transfer modules as agents

At the beginning the treatment of heterogeneity in the IZ projects was intentionally not planned according to the paradigm of agent technology as formulated in the Green Paper on agent technology Odell 2001, in Lieberman 1999 or in the controversy Maes/Shneidermann 1997, although the cross-connections are obvious.

In 1996 when the developments described here commenced, the paradigm of agent technology still appeared to be too imperfect and too closely linked through examples of so-called intelligent agents to advanced attributes such as plan identification. Our prime objective was to solve the problem of heterogeneity using simple methods if possible and complex "intelligent" methods if necessary. The question of whether the developed transfer modules are part of the software agent paradigm can be discussed at three increasingly abstract levels.

3.1 Agent "technology"

The Green Paper Odell 2001: Chapters 4 - 6 tries to define the distinguishing features of agent technology within the meaning of a software development paradigm and software architecture. Typical examples include special agent communication languages, message transportation mechanisms and an agent platform. The transfer modules, which were integrated in DAFFODIL, a multiple agent system for computer science, correspond to this paradigm. The transfer agents are called up in this case during retrieval by the search agent. Carmen and ViBSoz integrate the same modules without showing anything more than non-binding references to agent technology within the meaning of Odell 2001: Chapters 4 – 6.

3.2 Conceptual attributes of agents

Attempts to define agents normally start with generally accepted examples and typical attributes which must or can apply. According to Liebermann 1999: 280, the main characteristic of an information agent is that he "operates in the domain of some external information space, and whose task is to assist the user in accessing ... such information.".

This applies precisely to transfer modules. The user need not personally "translate" a content analysis language into the other language, this takes place in the interaction between several processes on the basis of different knowledge bases created for this purpose, e.g. cross-concordances, co-occurrence networks or neural networks.

Odell 2001 describes the characteristics "autonomous, interactive and adaptive" as obligatory, followed by 15 others of which at least a few should exist. The transfer modules also fulfill these conditions for software agents, but it is noticeable that they are currently restricted to relatively simple basic methods. The fact that characteristics of so-called intelligent agents are ignored more and would only be used after intensive empirical tests depends on the overriding thesis which regards agents as a contrast to direct manipulation techniques of the user.

3.3 Software agents as delegation in contrast to direct manipulation

Maes/Shneidermann 1997 try to basically define software agents at a very high level " ...as a program (or a set of programs) with a specific, understandable competence to which a user can delegate some portion of task". (Wexelblat/Maes 2001: 1).

It is precisely this aspect which is involved in transfer modules. Users will not be forced to use different content development languages in parallel or have to integrate them in their search strategy.

As the discussion within the context of Maes/Shneidermann 1997 shows, Shneidermann's counter thesis is not directed against the above-mentioned basic definition, it is empirically motivated:

"Direct manipulation interfaces are seen as more likely candidates to influence advanced user interfaces than adaptive, autonomous, intelligent agents. User control and responsibility are highly desirable." (Shneidermann 1997: 1)

It is noticeable that Shneidermann always talks about the so-called complex intelligent components, which have hardly been implemented to date in the IZ projects, when he formulates his counterthesis to the agents. He does not see any empirical evidence – and probably does not simply believe – that the postulated concepts such as plan identification will lead to any reasonable results. Contrary to that he sees great opportunities to develop the user-friendliness of information systems through direct manipulation and visualization. We also share this suspicion and believe there is a danger that complex agents will instead counteract the user's wishes, simply because it is very difficult to recognize the user's objectives. The agents must therefore be prevented empirically from making too many practical mistakes in their forecasts. The treatment of heterogeneous data stocks should only include parts of the agent concept, which do not contain the disadvantages directed by Shneidermann against the entire paradigm.

However, this consideration has very little to do with the core of the agent paradigm. In my opinion, it reflects the current state of the art. Dispensing with the term "transfer agents" is therefore more of a precautionary measure in order to avoid unnecessary discussions. The neutral term "transfer modules" does not obscure the sight of the solution of the heterogeneity problem with which we mainly are concerned.

4 Conclusion

The scientific community will not be able to avoid solving the problems of combining heterogeneous document types containing the widest possible range of modalities and medialities conceptually. In this context the basic ideas of the agent technology paradigm will play an important role. Users will want to interlink literature and research project databases with the catalogues of virtual libraries, the WWW homepages of science institutions and fact sources, e.g. data archives with their survey data. In this case integration should not be performed only on a technical level or using solely intellectually created links, as is the case at present. A key role is played here by automatic transfer between different content analysis methods and standardizations of the document sets to be integrated. Based on the initial empirical results, the proposed strategy appears to be highly promising: vagueness problems are not treated non-specifically as a transfer between all documents and the query but will be done cognitively plausible with individual bilateral modules. In this context the neural transformation network model has its place as a transfer module between the terms of different content analysis methods of two document sets.

5 References

Fuhr, N., Gövert, N., & Klas, C.-P. (2000). *An Agent-Based Architecture for Supporting High-Level Search Activities in digital Libraries.* In: Proc. 3rd International conference of Asian Digital Library. Taejon, Korea 2000. p. 247 - 254.

Kluck, M., Krause, J., & Müller, M.; in Kooperation mit Schmiede, R., Wenzel, H., Winkler, S.& Meier, W. (2000). *Virtuelle Fachbibliothek Sozialwissenschaften*: IZ-Arbeitsbericht 19, Bonn. http://www.bonn.iz-soz.de/publications/series/working-papers/index.htm#Virtuelle

Krause, J. (1996). *Informationserschließung und -bereitstellung zwischen Deregulation, Kommerzialisierung und weltweiter Vernetzung.* ("Schalenmodell"). IZ-Arbeitsbericht Nr. 6. Bonn.

Krause, J. & Stempfhuber, M. (ed.) (2001). *Integriertes Retrieval in heterogenen Daten. Text-Fakten-Integration am Beispiel des Verbandinformationssystems ELVIRA* (IZ Forschungsberichte 4). Bonn

Krause, J., Plümer, J. & Schwänzl, R. (2000). *Content Analysis, Retrieval and Metadata: Effective Networking for Mathematics, Physics and Social Sciences*, RC33-Session "New Conceptual Developments in Information Systems and the WWW". In: Proc. Fifth International Conference on Social Science Methodology, October 3 - 6, 2000 organized by - Research Committee on Logic and Methodology (RC33) of the International Sociological Association (ISA), Cologne.

Lieberman, H. (1999). *Personal Assistants for the Web: An MIT Perspective.* In: Matthias Klusch, ed., Intelligent Information Agents, Springer, 1999.

Maes, P. & Shneidermann, B. (1997). *Direct Manipulation Interface Agents: A Debate.* In: Interactions, Vol. IV No. 6. ACM Press.

Mandl, Th. (2000). *Tolerantes Information Retrieval: Neuronale Netze zur Erhöhung der Adaptivität und Flexibilität bei der Informationssuche.* Dissertation. Universität Hildesheim.

Odell, J. (ed.) (2001). Agent Technology. Green Paper of the Agent Platform Special Interest Group. Version 1.0. <http://www.objs.com/agent/index/html [visited: 05.01.2001].

Shneidermann, B. (1997). *Direct Manipulation for Comprehensible, Predictable, and Controllable User Interfaces*, In: Proc. of IUI97, 1997 International Conference on Intelligent User Interfaces, Orlando, FL, January 6 - 9, 1997, pp. 3339. (http://citeseer.nj.nec.com/shneiderman97direct.html)

Wexelblat, A. & Maes, P. (2001). *Issues for software agent UI.* Internet document: http://wex.www.media.mit.edu/people/wex/agent-ui-paper/agent-ui.htm [visited: 05.01.2001].

User Interest Correlation through Web Log Mining

F. Tao, P. Contreras, B. Pauer, T. Taskaya, F. Murtagh

School of Computer Science, Queen's University Belfast; DIW, Berlin

Abstract

Analysing log data can help the information provider understand clients' interests over the information space being served, and adapt it according to users' points of view. This paper describes a novel way of applying data mining techniques to Internet log data in order to find correlated web sections from the user point of view. We explain how data from the log file can be transformed into a set of transactional click-streams and how data mining techniques can be applied on these transactions. A test bed has been developed for transforming web log data and discovering association rules from it. Real log data from a Microsoft web site is used in experiments and evaluation results show that the approach is effective in obtaining useful knowledge of correlated interests at a particular web site. We also describe how we are beginning to mine log data obtained from the IRAIA project, an information retrieval system serving economical data.

1. Objectives

The main objective of web log mining described in this paper is to extract interesting and potentially useful patterns that show users' correlated preferences in accesses to the web pages being served by a particular web server. We investigated various methods of web log mining, most of them providing very primitive mechanisms for reporting statistical facts on accesses, e.g., the number of the accesses to individual files during a period of time, the origin of the users, etc. We believe that by using data mining techniques and systematically analysing the behaviour of past visitors, more sophisticated knowledge of the users' access patterns can be obtained from the web log file. In this paper, we discuss how data in the web log file can be transformed to transactional data and fed to an adapted association rules mining algorithm for correlated pages from users' points of view. There are several kinds of potential usage on the kind of knowledge obtained: From the site-maintaining point of view, it is important reference knowledge for the site administrator in order to maintain a reasonable and well-orientated web layout. Usage information can also be used to directly aid site navigation by hinting at a list of "popular" correlated pages from the current page, thus suggesting new users with the "experience" accumulated from past visitors.

Since the data in the log file does not come ready for mining, we need a data preparation phase where a transaction of user sessions is extracted from the raw web log file. Specially, there are a number of difficulties in cleaning the raw web log file to eliminate irrelevant items, reliably identifying users and user sessions. In this paper, we use a customised transactional model and an algorithm to convert raw web log data into user sessions. Besides this, our contributions include: 1. Development of a data mining test bed with association rules mining currently implemented. 2. Experiments carried out on this test bed, and evaluation on real web log data, shows that the approach is effective in revealing interesting correlated interests from visitors' points of view.

The rest of the paper is organised as follows: section 2 reviews related work on analysing web log data. Section 3 explains the model and algorithm used for user session extraction in the data preparation phase. In section 4, we describe the association rule mining in the context in web log mining, and experiment results and evaluations on real web log data are presented in section 5. Finally in section 6, we describe initial efforts at adapting this approach to mine user behaviour in an economical information retrieval system, where a customised logging mechanism has been designed to collect service usage.

2. Related Work

Research on knowledge discovery from web log data was started originally from web server log analysis tools that provide useful information about user activities and statistical facts on the pages being visited from individual users. The results can provide knowledge such as the number of accesses to individual files during a period of time, the origin of the users, etc. [S01] is a very successful case and more cases are reviewed there. However, these approaches are usually limited in the ability of providing data relationships among the pages, which is often essential when studying visitor correlated interests at the web site.

The idea of applying data mining techniques to web log data was first proposed in [ZXH98], where the authors investigated the probabilities of applying techniques of clustering, predicating and cycle detecting etc. In [TM00], attempts was made to propose a framework of web log mining, where log data is modelled as transactions and an interest context rule is targeted for mining. Chen et al. in [CPY96] introduced the concept of using maximal forward reference in order to break down user sessions into transactions for the mining of traversal patterns. A maximal forward reference is the last page requested by a user before backtracking occurs, where the user requests a page previously viewed during that particular user session. Another system described in [YJMD96] uses the knowledge of user access patterns to dynamically maintain the linkage of the web content.

In [IRAIA], a novel Information Retrieval prototype is being designed. An information space related to economic is labeled according to categories like regions, industries, and variables, and stored in concept hierarchies. Users search or navigate the information space by selecting entries in the concept hierarchy tree for each of the categories. Therefore, the log data of user interaction with the concept hierarchies, and how the entries that they select from the categories to form the queries, can contain useful information on user query patterns when using this IR service.

3. Data Preparation

There are many reasons why we have to prepare the raw log before being able to apply data mining techniques to it. The most essential ones are the incompatibility of data structure and irrelevant information concerning the specific mining task. Therefore, the basic work is to transform the data into a data-mining friendly form and filter out irrelevant information.

By following the HTTP protocol, visitors (by means of the browser client) communicate with a web server. Web log file is a plain text (ASCII) file maintained by the server's logging daemon at the server side. Each HTTP protocol transaction, whether completed or not, is recorded in the log file. Figure 3.1 shows a single HTTP access that is recorded in the log file. A brief description of the fields is listed below. For details, please refer to the HTTP protocol detailed in CERN and NCSA [L97].

```
dejh.ipm.ac.ir - - [08/May/2000:00:47:07 -0700]
"GET /spires/form/hepfnal.html HTTP/1.0" 200
3529
"http://www-
spires.slac.stanford.edu/spires/forms.html"
"Mozilla/4.05 [en] (Win95; I)"
```

Figure 3.1 A single HTTP transaction from the log file.

Transa	Sessions				
E[i]	E[i][1]	E[i][2]	E[i][3]	E[i][4]	E[i][5]
E[j]	E[j][1]	E[j][2]	E[j][3]		
E[k	E[k][1]	E[k][2]	E[k][3]	E[k][4]	
··· ··	··· ···				

Figure 3.3 Transactions and user sessions.

Client: visitor's domain name or IP address that can be resolved to domain name
Auth*: username if registered
Timestamp: Date and time of the access
Request: request method, document path and name, parameters, etc.
Status: status code indicating the result of a request
Cookie*: Crookie ID
Referrer*: previews link address
User agent*: client side browser type

Figure 3.2 HTTP server log structure.
(*:- optional fields)

Unlike the classical basket data mining solutions [AS94], where transaction is defined as a list of itemsets, there is no natural section of a user transaction in the web log data. The web log mechanism simply records every HTTP request when they come from the client side. Sometimes, a single request may trigger other file requests. This, along with the HTTP server's multi-thread and multi-user features, makes user navigational traces nest together in the raw log data. A semantic transaction must be defined with extra effort to meet the demand of mining visitors' correlated interests inside a web site, i.e., adapting them to the association rule mining framework. For this purpose, we define a transaction model with capacity to adjust to session number smoothly according to a time-window parameter. We also define the concept of sessions and discuss how to form transactions from the log entries.

Definition 3.1 Let e∈E, E is the set of all log entries after data cleaning and e is a log entry object with direct attributes such as host, document, timestamp etc., and derived features such as dwell time, interestingness and session ID, etc. (for example, e.interestingness, interestingness∈D.) We denote all these attributes as D and the Domain of D as Dom(D).
A Transaction E[n] can be defined in the following model:

$$E[n] = \{ \ E[n][s] \ | \ E[n][s].host = E[n].host \ AND \ E[n][s+1].timestamp - E[n][s].timestamp < W.\Delta T, \ 1 \leq s \leq |E[n]| \ \}$$

where n is the transaction ID number, s is the session ID number within a specific transaction, ΔT is the time interval configured in the window object W, which will be defined in Definition 2.2. The following illustrates these transaction and session concepts. We can notice that all the log data have been transformed to transactions and sessions, the session is an object with various encapsulated measurements contributing to the user's overall behavior. In another words, the transaction can be regarded as a click-stream generated by a single visitor within a specific time window.

Definition 3.2 A time window W in a transaction set E is a mechanism to categorize sessions into different transactions according to its integrated attribute window-size ΔT, which can be adjusted to simulate the time-span of the browsing transaction semantic. We set it as 20 minutes in our case.

Definition 3.3 Session s = E[n][i] is the ith element in a specific transaction E[n], where $1 \leq i \leq |E[n]|$, s is an object encapsulating all the attributes and features that are possibly derived from either a single log entry or the context entries within a single transaction.

By using the model, we transform the raw log data into a customised database that follows the data format needed in association rule mining. We will explain the association rule mining in the next section. Each entry in the log file can be regarded as a vote from the web site, and a section name is part of the vote to link the visitor's preference to that section. The raw log data is scanned twice: in the first scan, sections, each of which can be a page or a group of pages in a category, are indexed and assigned with IDs. In this step, filtering and counting mechanisms are also involved. Data is read from the raw log file line by line and then fed into a customised filter where irrelevant entries such as icons and auxiliary graphic etc. can be ignored. These requests are often automatically triggered when requesting a content-based page, therefore can not be counted to measure users visiting patterns. For the rest of the data that comes out of the filter, we save them in a hash table where each hash entry is a (key, value) pair. We use the section as the key and the value is an object keeping the section ID and maintaining the section count. At the end of the first scan, irrelevant log entries are ruled out and each of the rest of them is assigned a distinctive ID. We use a hash table to speed up the second scan over the log data where a name and ID correspondence can be easily located.

In the second scan, we maintain a bookmark array to store positions of the file pointers to log entries of the same transaction, i.e., the client addresses are the same and the visiting timestamps fall into the same time window. For each log entry read from the log file, if it is indexed in the hash table, the file pointer is not already recorded in the bookmark array and it also belongs to a same transaction, then we model it as a new session and look up the hash table and use the ID to identify that session. When we finish the whole time window, we get a transaction of sessions represented by section IDs. More features can be obtained and stored for other customized mining tasks (see [TM00] for further information).

4. Justification of Association Rule Mining for Correlated Interest Mining

The association rule mining task was first introduced in [AS94] for mining relationships in large database. In the context of correlated interest mining from web log file, it can be described as follows: let S be the site space represented by a set of pages or sections. And T the transactions, where each transaction represents a click-stream made by a single visitor within a specific time window. A transaction is a set of items representing all the pages requested in a single visit. A set of k items is called a k-itemset. The support of an itemset X, denoted support(X), is the number of transactions in which it occurs as a subset. An itemset is frequent if its support is more than a user-specified minimum support (min_sup) value.

An association rule is in the form of A => B, where A and B are itemset. The support of the rules is given as support(A∪B), and the confidence as support(A∪B)/support(A), i.e., the conditional probability that a transaction contains B, given it contains A. We say A and B are strongly correlated if the confidence is above the pre-specified minimum confidence.

The data-mining task is to generate all association rules in the database, which have a support greater than min_sup and a confidence greater than min_conf. We give below the algorithm in the context of correlated interest mining from web log data. This can be briefly described as a two-step task, as the second step of filtering out weak correlated itemset is less demanding we describe only the first step that finds frequent itemset.

The first iteration computes the set L_1 of frequent 1-itemsets. A subsequent iteration i consists of two phases. First, a set C_i of candidate i-itemset is created by joining the frequent $(i-1)$-itemsets in L_{i-1} found in the previous iteration. This phase is realized by the Apriori-Gen sub-function described below. Next, the database is scanned to determine the support of the candidates in Ci and the non-frequent i-itemsets are ruled out of the candidate set. This process is repeated until no more candidates can be generated.

5. Experiment on Web Log Mining and Evaluation

The approach explained above is used to mine correlated interests from Microsoft web log data provided by UCI Knowledge Discovery Dataset Archive [UCIDA]. The data is transformed and any information that could identify a specific visitor has been pre-processed out. The data records the use of www.microsoft.com by 32710 anonymous transactions. 294 web sections are involved in the log file after filtering.

We implement the transaction model and the mining algorithm in Java and the following screenshot shows the mining result with thresholds of 2% minimum support and 70% minimum confidence.

A	B	Support	Confidence
Windows95 Support	isapi	0.046071537	0.84142935
Knowledge Base, isapi	Support Desktop	0.033231426	0.70860493
Windows 95	Windows Family of OSs	0.032436565	0.91465515
Windows Family of OSs, Windows95 Support	isapi	0.028737389	0.8752327
SiteBuilder Network Membership	Internet Site Construction for D	0.02730052	0.80450445
Free Downloads, Windows95 Support	isapi	0.024640782	0.90561795
Support Desktop, Windows95 Support	isapi	0.024243351	0.8192149
Windows Family of OSs, Internet Explorer	Free Downloads	0.023937633	0.7653959
Free Downloads, Windows95 Support	Windows Family of OSs	0.022439621	0.82471913
Knowledge Base, Windows95 Support	isapi	0.02115561	0.8759494
Windows Family of OSs, Windows95 Support, isapi	Free Downloads	0.020360745	0.70851064
Free Downloads, Windows Family of OSs, Windows95 Support	isapi	0.020360745	0.9073569
Free Downloads, Windows95 Support	Windows Family of OSs, isapi	0.020360745	0.74831456
Free Downloads, Windows95 Support, isapi	Windows Family of OSs	0.020360745	0.82630277

Figure 5.1 correlated interests discovered at threshold of (2%, 70%) as min_sup and min_conf.

The mining result shows that visitors to the Microsoft web site have many correlated interests. After sorting the rules in descending support order, we found that the most frequent correlated interests are "Window 95 support" and

941

"isapi". "isapi". This site proved to be one of Microsoft's main portal URL forwarders where requests are forwarded to the right sections indicated in the URL. Obviously, this rule indicates the knowledge that "Window95 support" takes the largest portion from the requests that were sent to the portal. The second most frequent rule is "Knowledge Base, isapi" => "Support Desktop". This shows that the Desktop support is the highlight in Microsoft Knowledge base from the user point of view. On the other hand, if we sort the rules according to their confidences, we get a list of rules ordered by their correlation strengths. The most strong correlation is between "Windows 95" and Windows Family of OS" (in the third line). This indicates that when "Windows 95" is visited it is very likely (with probability of 91.5%) that the Windows Family of OSs is visited as well in that single transaction.

By adjusting the minimum support and confidence, we can discover more correlated interests with low support but high confidence or reverse. Obviously this knowledge can be essential to understand visitors interests correlation and access pattern in the web site.

6. Conclusions

In this paper, we investigated the possibility of analyzing web log data by using data mining techniques, more specifically, association rule mining for correlated interest patterns that web site visitors left in the web log data. We described a data preparation phase where raw data in the web log file is transformed into transaction forms using a transaction model. We then described the association rule mining model and algorithm in the context of mining for correlated interests. We implemented a test bed and evaluated the approach on real log data obtained from a Microsoft web site. Results show that the approach is effective in discovering visitor correlated interests when browsing a specific web site.

As described in the section of related work, we have started work on integrating a logging mechanism into the IRAIA economic information retrieval system. The logging mechanism will record all the database queries that the users generate by interacting with a GUI based query interface. A similar approach is expected to be applied on the log data, to be mined for correlated interests. Result rules are stored in a knowledge base, which will be used by a hinting mechanism to provide new users with the "experience" of the past users and help them to interact with the query interface of the IR system.

References

[AS94] R. Agrawal and R. Srikant: *Fast Algorithms for Mining Association Rules.* Int. Conference of Very Large Data Bases, Santiago, Chile, September 1994. pp 487-499

[CPY96] M.S. Chen, J.S. Park, P.S.Yu: *Data Mining for Path Traversal Patterns in a Web Environment.* 16th Intl. Conference on Distributed Computing System, 1996. pp385-392

[YJMD96] T.Yan, M. Jacobsen, H. Garcia-Molina, and U. Dayal: *from user access patterns to dynamic hypertest linking.* International World Wide Web Conference, Paris, France, 1996

[TM00] F. Tao, F. Murtagh: *Towards Knowledge Discovery from WWW Log Data.* Intl. Conference on Information Technology: Coding and Computing. Mar. 2000. pp 302-307

[ZXH98] O.R. Zaïane, M. Xin, J.W. Han: *Discovering Web Access Patterns and Trends by Applying OLAP and Data Mining Technology on Web Logs.* ADL 1998. pp19-29

[CT01] P. Contreras, F. Tao: *The Integration of the log mining in the context of IRAIA.* Presented at DIW technical meeting, Berlin, Feb. 2001. http://iraia.diw.de/

[S01] S. Turner. Analog – A Web Log Analyst. http://www.statslab.cam.ac.uk/~sret1/analog/

[L97] A. Luotonen. The common log file format. *http://www.w3.org/pub/WWW/,* 1997
[IRAIA] Getting Orientation in Complex Information Spaces as an Emergent Behavior of Autonomous Information Agents. http://iraia.diw.de/

Constraint-based Immersive Virtual Environment for Supporting Assembly and Maintenance Tasks

Terrence Fernando, Luis Marcelino and Prasad Wimalaratne

Centre for Virtual Environments, University of Salford, M5 4WT, Salford, UK

ABSTRACT

This paper presents a constraint-based virtual environment which is being developed to support the assessment of assembly and maintenance tasks within a CAVE environment. The proposed environment provides an integrated framework for supporting both assembly and disassembly operations. This environment is being evaluated using realistic industrial case studies from the aerospace sector.

1. INTRODUCTION

As specified in *Manufacturing in the Knowledge Driven Economy* (DTI,1999), advances in ICT, global competition and customer demand for more sophisticated products are forcing change in the engineering sectors. Due to global competition, these sectors are now under severe pressure to both reduce the lead time for new products and also improve their quality and customer and market responsiveness. Such pressures have led many to investigate the adoption of concurrent engineering techniques. When using concurrent engineering, specialist knowledge and expertise from the downstream tasks of sequential design, such as manufacturing, assembly and maintenance, are introduced during the early design phases. Practising designers readily accept the principles of concurrent engineering, but there remain significant obstacles to the practice. These obstacles involve the availability of useful, effective design tools to facilitate the principles. Whilst current CAD tools support detailed design of products extremely well, they provide very little support for other life cycle stages such as conceptual design, maintainability, operability, safety and environmental considerations.

The development of a design environment for supporting down stream processes is a challenging research problem requiring an in-depth understanding of issues such as the design process, product life-cycle, simulation techniques, human factors and organisational issues involving expertise from various disciplines. The Advanced Virtual Prototyping Group at the Centre for Virtual Environments at Salford has initiated a research programme to undertake these challenges with several academic and industrial partners. This paper presents the outcome of this research work.

2. SPECIFIC RESEARCH AIMS AND OBJECTIVES

The first phase of this research programme has focused on developing a constraint-based virtual environment within a CAVE environment to assess assembly and maintenance tasks on a virtual prototype. This research is addressing a major question concerning the viability and utility of virtual environments to perform design analysis on a virtual prototype as realistically as with a physical prototype. The specific objectives of this research are to:

- Design and implement a virtual prototyping environment which preserve CAD information to support accurate engineering tasks.

- Design and implement a constraint manager that supports constraint-based interaction between parts to assemble and disassemble parts as intuitively as performed on the physical prototype.

- Design and implement a visual interface which can be configured to exploit the power of immersive display devices (ie. CAVE and Immersive Workbench).

- Design and implement a bi-manual interface to support direct interaction with assembly components to support two-handed assembly and maintenance tasks.

- Conduct user-centered evaluation of the constraint-based virtual environment using a realistic industrial case study.

3. SYSTEM ARCHITECTURE

Figure 1 presents the system architecture of the immersive constraint-based virtual environment which is being developed to investigate the above research objectives. This section discusses the approaches employed in implementing the main modules of this software architecture and current results.

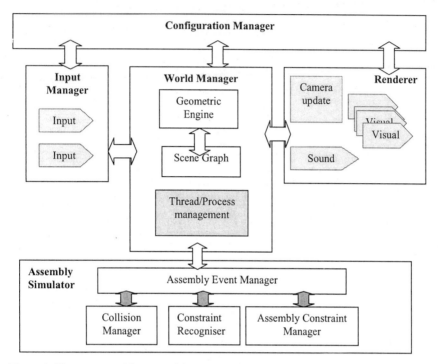

Figure 1: The system architecture of the immersive constraint-based virtual environment.

4. VIRTUAL WORLD MANAGER AND RENDERER

The virtual world has been designed to maintain both geometric and polygonal data. The reason for maintaining geometric data within the virtual world is to facilitate precise collisions between assembly parts and to establish assembly relationships between geometric surfaces. Polygonal data is maintained to visualise the models using standard polygonal rendering techniques. This virtual world representation has been created by integrating the Parasolid geometric kernel and the OpenGL Optimizer graphical toolkit.

The initial loading of the CAD models to the virtual world is performed via the Parasolid geometric kernel. Once the models are loaded, an Optimizer scenegraph is created within the virtual world by extracting polygons for each of the surfaces for each assembly part. The reason for extracting polygons for each surface separately is to provide visual feedback at surface level on collisions. Two way links between the geometric objects and the optimizer objects are maintained to provide a semantically rich model representation within the virtual world.

Figure 2 represents how the virtual world is structured to support the direct interaction between the user and the assembly component with in the CAVE environment. There are significant differences between the interaction performed within a desktop environment, using 2D mouse-like devices, and direct interaction within immersive CAVE environments. In a desktop environment there is no relation between the virtual world space and the real world space occupied by the user. However, in a CAVE environment, the physical CAVE space, in which the user is operating, exists within the virtual world. The user may move this CAVE space within the virtual world space by using a navigational device. In our virtual world representation, we represent the CAVE space as the *user space*. The transformation node at the top of the user space allow the user to position the *user space* anywhere in the virtual world. In this representation, we assume that the user is always within the *user space* and his or his body position are tracked with respect to the centre of the user space. The current *user space* structure maintains the head and the hand positions within the *user space*. The tracker values are assigned directly to the hand and head transformation nodes to maintain their correct position within the virtual world.

The virtual world has two components, the *background world space* and the *assembly component space*. The *background world space* is maintained to provide an visually realistic background environment. The background world space is rendered but not available for interaction. The assembly component space maintain the assemblies which are to be evaluated for assembly and maintainability.

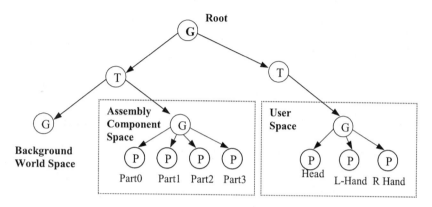

Figure 2 : Virtual World Representation

We have extended the optimizer viewing facilities to render the optimizer scenegraph into different display devices such as a CAVE and Immersive Workbenches either in mono or stereo. The rendering of images on different display walls is performed employing parallel threads to provide real time response.

5. CONSTRAINT-BASED ASSEMBLY AND DISASSEMBLY

The function of the assembly simulator in Figure 1 is to support interactive constraint-based assembly and disassembly operations. It supports real-time simulation of physical constraints within the virtual environment to provide realistic interaction between virtual assembly components. Simulation of an assembly relationship can be considered as a constraint specification and satisfaction problem. For example, alignment of a shaft with a cylindrical hole involves satisfying an axis-alignment constraint. Disassembly operations involve breaking the previously defined constraints by applying an external force.

The previous approaches adopted by the authors and others to simulate the assembly constraints can be found in (Fernando et. al.,1999a; Fernando et.al., 199b). This section provides a brief description of the basic approach and the improvements made to implement a unified framework for supporting both assembly and disassembly operations.

6. BASIC ALGORITHM

Once the assembly parts are loaded into the scenegraph via the CAD interface, the input manager allows the user to grab and manipulate objects in the 3D space. While an object is being manipulated, the position of the moving object is sampled to identify new constraints between the manipulated object and the surrounding objects by the assembly simulator. This is performed in four steps:

Step 1: Detection of collisions between the manipulated object and the surrounding objects. This step is continued until there is a collision.

Step 2: Further testing is made for possible assembly contacts between the surfaces of the collided objects. A constraint is recognised if the geometric surface elements of the collided objects satisfy conditions of a particular constraint type within a predefined tolerance. The current implementation can recognise surface mating conditions such as against, coincidence, concentric, cylindrical fit and spherical fit.

Step 3: When a constraint is recognised, feedback is provided to the user by highlighting the mating surfaces. This newly identified constraint is ignored if the user continues to move the object to invalidate the condition for the constraint.

Step 4: If the user decides to accept the new constraint, the surface description of the mating faces and the type of constraint to be satisfied are sent to the constraint manager. The recognised constraints are satisfied by the constraint manager and the accurate position of the collided assembly part and its allowable rigid body motions are sent back to the scenegraph. This information is used by the scenegraph to define the precise position of the collided assembly part.

The constrained rigid body motions of the assembly parts are used to support realistic manipulations of assemblies without breaking the existing assembly constraints. This is done by converting the 3D manipulation data received from the 3D input device into allowable rigid body motions. A particular manipulation of an assembly model is not allowed if it is not supported by its allowable rigid body motion.

7. EVALUATION RESULTS

The basic algorithm described in section 3.2.1. was evaluated using several industrial case studies to identify the usability of the virtual prototyping environments. These evaluation experiments highlighted the need for improvements in the following areas :

Constraint recognition : The detection of collision at discrete intervals excluded intermediate collisions between objects. This resulted in undetected collisions and constraints. For example, when two parallel faces are coming together, the parallel faces may stay uncollided in one frame and may penetrate in the next frame due to the discrete nature of transformation. As a result, sometimes the users found it difficult to specify against constraints between two assembly parts. Therefore, it was necessary to develop techniques which takes the discrete nature of the 3D input into consideration for performing constraint-based assembly tasks.

Disassembly operations : Once a constraint is recognised and satisfied the user was not able to break that particular constraint in the basic algorithm described in section 3.2.1. As a result, the users were not able to undo an incorrectly specified constraint or to perform a disassembly operations. Some problems encountered due to this limitation is presented below. These examples represent generic problems which need to be supported for performing intuitive assembly and disassembly operations.

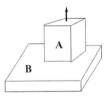

Figure 3 : Breaking an Against Constraint

Figure 4 : Sliding a peg into a hole

Removing a part which lies against another surface : Assume that the user has already established an against constraint to position Block A on top of Block B (Figure 3). Now if the user attempts to move the Block A along the top surface of Block B, this will be supported by the allowbale motion derived by the constraint engine. However, if the user decides to pull the Block A away from Block B, the against constraint needs to be automatically deleted to support this disassembly operation. Therefore further improvements for the basic algorithm was required to distinguish whether the user is attempting a constrained motion or a disassembly operation.

Sliding a peg into a hole : When carrying out assembly operations such as peg and hole type joints, it is necessary for the system to be able to break certain previously established constraints. To illustrate this process, see the scenario shown in Figure 4. Initially, an against constraint is established between the two planar faces. When the cylindrical peg reaches the hole, previously established against constraint needs to be broken. This allows the peg to be slid into the hole. This highlights the need for breaking previously established constraints automatically to allow certain assembly operations. Without this facility, the user will always be forced to establish an alignment constraint first in order to put the cylindrical peg into the hole.

8. IMPROVEMENTS TO THE BASIC ALGORITHM

The improvements made to the basic algorithm is based on attaching positive and negative sensors to geometric surfaces. In this algorithm, positive sensors (bounding boxes) are attached to positive surfaces and negative sensors are attached to holes in each surfaces. The sensors are represented as bounding boxes for each surface on a part. These sensors are used to support the making and breaking of constraints in response to user actions. The following section summarises how these sensors are used to overcome the limitations of the basic algorithms presented in Section 3.2.1.

Constraint Recognition : The sensors act as an "early warning" system to inform the system about surfaces coming together since the sensors have a wider bounding box area. However, although the sensors provide a early warning system, constraints are only recognised whenever there is a collision between two mating surfaces. This has reduced the objects penetration problem. Further techniques such as "ray casting" are being explored to improve constraint recognition.

Constraint Deletion : The assumption made in this algorithm is that a previously recognised constraint is only valid as long as there is a overlap between the two positive sensors attached to the two mating surfaces. A previously recognised constraints is broken if the following conditions are true :

Intersection area between the two positive (mating) sensors becomes null as a result of some relative motion between the two parts.
One of the positive sensors enters a negative sensor area.

These rules allow the user to perform assembly and disassembly operations within a single framework. This algorithm has provided a solution to overcome the limitations of the basic algorithm described in Section 3.2.1.

9. CONCLUSION
The constraint-based immersive environment presented in this paper has been designed and implemented. At present the team is in the process of evaluating this environment using several large industrial case studies from the aerospace industry.

10. REFERENCES
Fernando T., Wimalaratne P., Tan K. and Llewelyn, I. (1999)
    Interactive Product Simulation Environment for Assessing Assembly and Maintainability Tasks, ASME International Mechanical Engineering Congress & Exposition, Symposium on Virtual Environments for manufacturing, Nashville, pages 179 – 189, Nov 1999, ISBN 0-7918-1636-2.

Fernando T., Wimalaratne P. and Tan K. (1999)
    Constraint-based Virtual Environment for Supporting Assembly and Maintainability Tasks , ASME Computers in Engineering International Conference,Virtual Environment Systems, Las Vegas, Sept. 1999, ISBN 079181967-1.

A STUDY OF NAVIGATION STRATEGIES IN SPATIAL-SEMANTIC VISUALISATIONS

Timothy Cribbin and Chaomei Chen

The Vivid Research Centre, Department of Information Systems and Computing,

Brunel University, Uxbridge, United Kingdom. UB8 3PH

ABSTRACT

Visualisations of abstract data are believed to assist the searcher by providing an overview of the semantic structure of a document collection whereby semantically similar items tend to cluster in space. Cribbin and Chen (2001) found that similarity data represented using minimum spanning tree (MST) graphs provided greater levels of support to users when conducting a range of information seeking tasks, in comparison to simple scatter graphs. MST graphs emphasise the most salient relationships between nodes by means of connecting links. This paper is based on the premise that it is the provision of these links that facilitated search performance. Using a combination of visual observations and existing theory, hypotheses predicting navigational strategies afforded by the MST link structure are presented and tested. The utility, in terms of navigational efficiency and retrieval success, of these and other observed strategies is then examined.

1. INTRODUCTION

The representation of unstructured document collections by means of node based graphing methods is an approach that has been widely adopted by the information visualisation community. These visualisation methods aim to support information seeking by providing the user with a spatial map representing an existing model of the semantic structure of a collection. Documents are generally represented as nodes plotted in visual space whereby inter-document proximity implies their degree of similarity.

Visual overviews should guide the information seeker to important information whilst minimising the amount of reading and scanning required. The potential value of overviews, therefore, tends to increase in line with document collection size. Unfortunately as thematic diversity also tends to increase with collection size an overall drop in the accuracy of inter-document proximities can often result. When the reliability of spatial proximity cues are low, the user will be more frequently mislead, reducing their navigational efficiency and causing frustration. Furthermore, large maps can appear quite amorphous and cluttered with few obvious landmarks to guide orientation.

1.1. Using visual links to communicate salient structure

One approach to alleviating this problem is to use other cues to further emphasise the most important or strongest inter-document relationships. Previous empirical work (Hascoet, 1998) has found that users preferred to search using visualisations that made clusters more identifiable and that provided good clues as to which direction to move next once good items had been found. By connecting highly similar documents with visual links, minimum-spanning tree (MST) graphs hold the potential to satisfy both these criteria. MST graphs are always composed of N-1 links, the minimum required to form a continuous tree structure. The example in Figure 1 shows how the branching structure of the MST graph provides additional landmarks and is less cluttered in overall appearance.

1.2. Information seeking performance and visual links

(Cribbin & Chen, 2001) reported results from a study that compared information-seeking performance using a range of graphing algorithms. Principle components analysis (PCA) was used to create a traditional scatter-graph solution from similarity data generated by Latent Semantic Indexing (Deerwester, Dumais, Furnas, Landauer, & Harshman, 1990). MST visualisations were also created from the same data. Participants completed information-seeking tasks, according to a number of given criteria, using each of these visualisations for navigation. Use of MSTs resulted in general and significant performance improvements in comparison to PCA visualisations. This paper examines navigational strategies employed by participants when using the MST visualisation, focusing in particular on the early stages of exploration. It is proposed that the link structures provide a number of navigational cues that, when exploited, lead to superior information-seeking performance.

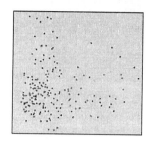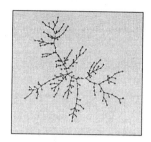

Figure1: PCA and MST graphs representing the same semantic proximity data

1.3. A classification of node types in tree networks

Using MST graphs to represent a document collection provides the user with important cues that are known to support navigation in the real world, most notably pathways, edges and nodes [2] (Lynch, 1960). The continuous link structure provides pathways that support route following and learning. A document radiating a single link clearly indicates the limits of a cluster or pathway. In contrast, a document possessing many links suggests an important document that may form a useful focal point.

In this paper nodes are classified into three types according to the number of direct links to other nodes. An *extremity* possesses a single link, a *thread* node possesses two links and *branch* nodes link to three or more other nodes. From a combination of existing theory and observational evidence acquired from visual replays of task event data a range of hypotheses are put forward.

1.4. Hypotheses and rationale

The analysis is split into two sections. Firstly, an investigation into navigational strategies will test three hypotheses relating to landmark and pathway usage. Branch and extremity nodes are both presented as natural landmark candidates. Secondly, performance benefits in terms of navigation and retrieval performances will be explored in within the context of these strategies.

1.4.1. Using landmarks to guide early orientation

On first presentation, the spatial layout was completely novel to participants. Furthermore, the semantic properties of visible regions were also unknown. When navigating the real world, people tend to identify the most salient landmarks and use them as anchor points from which they can form a working cognitive model of the space (Couclelis, Golledge, Gale, & Tobler, 1987). Our observations also show that landmarks, extremities in particular, seem to be used regularly during the early orientation stage. *H1 therefore predicts that visits to extremity and branch nodes will be comparatively more frequent during the early stages of the task.*

1.4.2. Using pathways to maximise return

The inclusion of a visual link connecting two nodes implies a strong semantic relationship between the two associated documents. Given that participants were aware of this feature it can be expected that, on discovery of a relevant document, users will attempt to capitalise on the pathways radiating from the current node, an activity

---

[2] Lynch use of the term node differs from that of this paper, referring instead to a focal point within a city such as a town square.

949

referred to hereon as chaining. *H2 therefore predicts that discovery of a relevant document will result in an increase in chaining activity.*

1.4.3. Using landmarks to escape from sparse regions

In contrast, users will only spend time browsing information in a particular area if they believe that further useful discoveries are likely and imminent (Chalmers, 2000). Discovering a non-relevant document within a spatial-semantic environment would suggest to the information-seeker that proximal documents are also unlikely to be useful. In such circumstances, it is predicted that users will be more inclined to jump to a different region of the space. Furthermore, they will seek to reorientate themselves by jumping to a landmark node. *H3 predicts that visits to branch and extremity nodes will be comparatively more frequent when the current document node is non-relevant to the task criteria.*

1.4.4. Utility of identified strategies

Correlational analysis will then examine the relationship between the extent to which these strategies were employed and overall task performance. Criteria for success were search recall, precision, general efficiency, lostness and time taken to retrieve the first relevant document. Efficiency is calculated by combining recall and precision using the F-measure formula (see van Rijsbergen, 1979). Lostness is also created with the same formula using a combination of a backtracking measure (time spent re-visiting non-relevant nodes) and the proportion of time spent examining relevant nodes.

2. METHOD

The complete methodology used in this study can be found in a previous paper (Cribbin & Chen, 2001). In brief, four document collections each comprising 200 documents were subjected to Latent Semantic Indexing (LSI) and similarity data computed. Documents were retrieved from the TREC Los Angeles Times database using single keywords. These were alcohol, endanger, gaming, storm. In each case, a set of two-dimensional coordinates and inter-document links were computed using an MST algorithm. The computer display was split into two panes. In the left pane an interactive VRML world, representing the MST model. The right pane displayed full document text when users clicked on a node.

The task examined here required participants to search the visualisation for documents that conformed to a broad set of criteria. For example, for the 'alcohol' collection users were asked to find any document that mentioned an incident of drinking and driving. For each collection approximately 12% (22-26 documents) of all documents were classified as relevant.

Sixteen participants completed the task. All were postgraduate students at Brunel University. Each collection was administered an equal number of times across the sample. Participants received no prior experience of either the spatial structure or the documents contained within. They were instructed to locate and mark all documents relevant to the given criteria. Five minutes was allowed to complete the task although participants could terminate the task earlier if they wished.

3. RESULTS

3.1. Observed uptake of predicted strategies

Early activity was defined as the first five nodes visited by the participant. The proportion of all events (visits, clicks and markings) for both the early and whole task period, was computed by node type. These values were then normalised against chance levels by subtracting the actual proportions of each node type existing within the space.

Figure 2 shows the comparison of early events against events recorded across the whole task period. ANOVA showed a significant quadratic interaction between node type and period, $F(1,15) = 5.01$, $p<.05$. There was a clear bias during early activity towards branch nodes. This was not significantly different from chance, $t(15)= 1.22$ or from the whole task period, $t(15)=1.03$. There was also a tendency for participants to avoid thread nodes and this was significant with respect to chance, $t(15)=1.94$, $p(1) <.05$, and whole period events, $t(15)=2.24$, $p(1) <.025$. Extremity visits occurred at approximately chance level for both the early, $t(15)=.05$, and whole period, $t=1.24$ and there was no significant difference between early and whole period event likelihood, t(15)=.33. H1 is therefore partially supported, although the data suggests that extremities were not viewed as useful landmarks during the early stages of navigation.

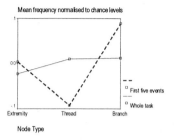

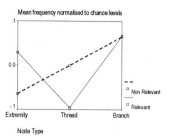

Figure 2: Likelihood of transition to node type as a function of task period

Figure 3: Likelihood of transition to node type as a function of current node relevance

With regards to chaining activity, the likelihood of following a link to an adjacent node was not found to be significantly higher when the source node was relevant, t(15)=.441. H2 is therefore rejected.

To test H3, ANOVA was computed to determine the relative likelihoods of jump destinations from non-relevant documents. Figure 3 shows a linear pattern of means suggesting a positive relationship between the number of radiated links and the attractiveness of a document node. Branch visits were clearly most likely in this context, with the likelihood of a visit being significantly greater than chance, t(15)=1.87, p(1)<.05. Thread visits occurred roughly at chance level and extremity visits occurred at a below chance level, suggesting they were intentionally avoided, although this was not significant, t(15)=1.58.

In comparison, transitions from relevant nodes show the kind of pattern originally expected from non-relevant nodes with means forming a U-shape graph (see Figure 3). The observed difference in distribution was confirmed from a general interaction, F(1,15)=3.33, p<.05. Thread visits fell significantly below the chance level, t(15)=2.64, p(2)<.02. There was an almost identical tendency towards branch nodes visits and a greater than chance tendency towards extremities, although this was not significant, t(15) = .50. At a relative level, extremity visits were more common when the source node was a relevant document. H3 is therefore rejected.

3.2. Utility of observed strategies

The previous analysis showed an overall preference for using branches as landmarks during the early stages of search. Use of this strategy, however, only weakly predicted the time taken to mark the first relevant document (r=-.25, ns). Virtually no relationship was observed between extremity visits and time to first marking was found (r=.09).

Given that a link between two documents implies high similarity, H2 predicted that chaining would be more common from relevant documents. This was not found to be the case. Furthermore, correlations show that it is the degree to which participants chained from non-relevant documents that was most predictive of overall performance. Significant correlations were found for all performance criteria with the exception of time to mark first relevant document. Significant coefficients were in the range r=.52 to .61 (p<.05). Although weakly positive, none of coefficients computed with respect to chaining from relevant documents approached significance.

H3 predicted that jumping to landmarks would be seen as an efficient means of escaping a sparse region of the information space. The strategy of jumping to extremities from non-relevant documents was negatively predictive of performance across all criteria. Participants who used this strategy tended to be less precise with respect to the relevance of documents they marked (r=-.58, p=.02) experienced more lostness (r=-.48, p=.06) and read (clicked on) fewer relevant documents (r=-.45, p=.08).

Despite being a common strategy, the extent to which participants jumped from non-relevant documents to branch nodes did not predict performance in any respect. In contrast, the frequency of transitions to branches from relevant documents was indicative of efficient navigational, but not retrieval performance. For instance the F-measure based just on nodes visited correlated significantly with usage of this strategy (r=.50, p=.05). In contrast, jumping to extremities from relevant documents produced the opposite effect on navigational efficiency (r=-.47, p=.06).

As a final note of interest, the correlational data shows that participants could be split into two more, unexpected styles. It was found in all cases that frequency of transitions to extremities tended to correlate negatively with those to branches (r=-.79, r= -.70, p<.01). This suggests that some users preferred to either pick away at the ends of branches whilst others navigated more extensively around the branch structures. Evidence in support of this assertion comes from negative correlations between thread visits and extremity visits (r=-.55, r=-.51, p<.05). No significant correlations were found between thread and branch visits. Given our data, it seems like the latter strategy was the most successful one.

4. DISCUSSION AND CONCLUSIONS

The analysis of navigation patterns conducted here has provided some interesting insights into the strategies adopted by information seekers when navigating document collections using MST graph visualisations.

When first confronted by a novel document space, users seem to rely on branch, rather than extremity, nodes as landmarks to guide orientation. This strategy, however, did not reliably predict the time taken to identify the first relevant document. When the session data is considered as a whole, however, it seems reliance on branch nodes does become a slightly more successful strategy although this was only in terms of navigational rather than retrieval effectiveness. High reliance on extremity nodes, on the other hand, was generally counter-beneficial to good performance. This was unexpected but may be because searchers did not use extremities in the expected sense. That is, they did not follow up the visit by travelling down the available pathway towards the source branch.

The prediction that chaining activity would be used on discovery of a relevant document as an efficient means of locating others did not seem to be the case. This could be an artifact resulting from the broadness of the relevance criteria given in this task. Clustering of relevant nodes may not have been as coherent as expected. This could have lead to users losing faith in the reliability of the link structure for these purposes. Chaining was, in fact, a more successful strategy when used to navigate from non-relevant documents. Perhaps this is indicative of the utility of taking a generally more methodical approach, particularly when lost. Further analysis is necessary to clarify this issue.

MSTs clearly provide valuable support to the information navigation process. This paper, although only a snapshot, shows that the way these features are exploited is variable and that chosen strategy can have a significant impact on overall performance. Future research and analysis should seek to understand how these features can be best exploited and what further perceptual cues are necessary in order to maximise value of these structures.

5. REFERENCES

Chalmers, R. (2000, 11 November). Surf like a bushman. *New Scientist,* 39-41.

Couclelis, H., Golledge, R., Gale, N., & Tobler, W. (1987). Exploring the anchor-point hypothesis of spatial cognition. *Journal of Environmental Psychology, 7*(2), 99-122.

Cribbin, T., & Chen, C. (2001, January 21-26). *Visual-Spatial Exploration of Thematic Spaces: A Comparative Study of Three Visualisation Models.* Paper presented at Electronic Imaging 2001: Visual Data Exploration and Analysis VIII, San Jose, CA.

Deerwester, S., Dumais, S., Furnas, G., Landauer, T., & Harshman, R. (1990). Indexing by latent semantic analysis. *Journal of the American Society for Information Science, 41*(6), 391-407.

Hascoet, M. (1998, 18 - 23 April). *Analytical versus empirical evaluation of spatial displays.* Paper presented at CHI 98: Human Factors in Computing Systems, Los Angeles, CA.

Lynch, K. (1960). *The Image of the City.* Cambridge, MA: MIT Press.

van Rijsbergen, C. (1979). *Information Retrieval.* London: Butterworths.

Virtuality in Human Supervisory Control

Neville A. Stanton
Department of Design, Brunel University, Egham, Surrey, TW20 OJZ, UK

Anthony D. Roberts, Melanie J. Ashleigh & Francis Xu
Department of Psychology, University of Southampton, SO51 7JH, UK

Virtuality would seem to offer certain advantages for human supervisory control. First, it could provide a physical analogue of the 'real world' environment. Second, it does not require control room engineers to be in the same place as each other In order to assess these effects, a low-fidelity simulation of an energy distribution network was developed. The main aims of the research were to assess some of the psychological concerns associated with virtual environments. First, it may result in the social isolation of the people, it may have dramatic effects upon the nature of the work, and second a direct physical correspondence with the 'real world' may not best support human supervisory control activities. Experimental teams were asked to control an energy distribution network. Measures of team performance, group identity and core job characteristics were taken. In general terms, the results showed that teams working in the same location performed better than team who were remote from one another.

1. Background to the research

With modern technology, it is possible to represent any environment in a similar, yet entirely independent way to the physical world. Furthermore, most organisations appear to be adopting a strategy of increasing centralisation. This has led to considerable re-organisation within these companies and the effect of re-locations has undoubtedly caused some disruption to the workers and their families. The control room environment may become the model for other types of working, for example manufacturing organisations may not need constant presence of workers with increased automation and may opt for remote monitoring at some point in the future. Power generation companies already have remotely monitored gas turbine stations. Therefore, findings from research into human supervisory control will become more widely applicable, beyond the energy distribution companies. In addition, errors in human supervisory control can have potentially disastrous consequences, which can impact upon the lives of many people, beyond those making the errors. This makes human supervisory control an important area of psychological research.

Virtual working environments could negate the need to physically centralise personnel as they could work remotely from each other as well as remotely from the plant they are supervising. The virtual environment could be an analogue of the elements contained in the real world. This representation may have advantages over current mimic displays, as the control room engineer could literally inspect the status of the plant (depending upon the capability of sensor technology) and operate the plant directly. Virtual technologies may also enable the control room engineer to converse directly with other engineers whilst inspecting the same plant. Errors often arise due to misunderstandings over the telephone of what aspect of the plant is being discussed. These errors could potentially be reduced through virtual systems. Virtual systems may introduce new kinds of problems, such overwhelming the operator with information. In short, we might simply be replacing one set of problems with another. It has been suggested that the wealth of information might be more manageable if functional representation techniques are used (Praetorius & Duncan, 1991). This could reduce all system components to six basic functions. Praetorius & Duncan (1991) claim that functional representation reduces the workload of operators and decreases fault diagnosis time considerably. Given that the energy transportation system could be represented both physically and functionally it is worth exploring this further. Whilst virtual environments offer the potential for overcoming physical 'remoteness' there is the potential risk of the social consequences associated with the diffusion of responsibility if the control room engineers are not working in the same physical environment.

A conceptual framework for the research was provided by the 'Levels Of Abstraction Hierarchy' (LOAH) model from Rasmussen, (1986). LOAH model can be used as a hierarchical representation which characterises the different stages of human decision making in supervisory control; describing how the operator converts system goals into control actions whilst interacting with the system. It is also used as a representation of the physical and functional parts of a plant; the systems and sub-components The LOAH hierarchy is divided into five distinct categories: from most concrete level (i.e. physical form) to the most abstract level (functional purpose). The definition of the categories are as follows:

Functional purpose: - The overall meaning of the system and its purpose in the world, e.g. system goals at a high level .

Abstract function: - General and symbolic level of the system, e.g. descriptions in mass or energy terms to convey flow through the system.

Generalised function: - Generalised processes of the system that reflects behavioural structure, e.g. diagram of information flow and feedback loops.

Physical function:- specific processes related to sets of interacting components, e.g. specific sub-systems, such as electrical or mechanical.

Physical form: - Static, spatial, description of specific objects in the system in purely physical terms, e.g. a picture or mimic of the components.

The most persuasive arguments have been made by knowledge theorists (see Goodstien et. al., 1988) and empirical researchers (Vicente, 1997). Vicente in particular, has demonstrated how experimental participants are able to perform process control tasks more effectively if they are presented with both functional and physical information about the system. Rasmussen has argued that this is because people need to work 'top-down' when seeking the purpose of functional requirements and 'bottom-up' when seeking causes of system problems. As systems become more complex and multi-layered, design of interfaces have had to compromise between the physical form and functional purpose. Cognitive control may shift from level to level depending on the situation. In doing so, this may require greater cognitive effort.

This leads us to suppose that there are at least two dimensions of virtuality to be explored: those of virtual team and virtual interfaces. Virtual teams become possible with the advent of technology that enables people to work remotely. Virtual interfaces become possible with the advent of technology that enables designers to realise any environment that they can imagine. Just because these technologies enable new ways of working, it does not necessarily mean that these will lead to any improvements in system performance and worker satisfaction. This research project explicitly addressed these issues.

Specifically the aims of this research were to address the following questions:

- Are there performance gains associated with portraying a human supervisory control environment functionally rather than physically?

- What are the effects of virtuality on the Core Job Characteristics?

- How does virtuality affect team control?

- To what extent is it important to preserve personal identity in a virtual environment?

2. Method

Participants
There were 24 groups of 4 people used in the study, a total of 96 participants. Participants ages were from 19 to 55; a range of 36 years with the mean age being 26. The sample consisted of 74 males and 22 females.

Design

The study aimed to test the effects of location and interface type on teams working in a simulated controlled environment. The study tested between factors using four different conditions, where six teams of four people were asked to perform a simulated task of balancing a gas-network system. They were either working together in the same location (proximal), or working in separate locations (distal), and using either a virtual or abstract interface (24 x 2 x 2). These two independent variables of location and interface type were used to test seven dependent factors. Five of these variables were based on objective behavioural scores taken from the task and two were taken from subjective self-reporting data. The dependent variables were, cost, number of control, team control strategy (Hollnagel, 1993), core Job Characteristics (Hackman & Oldham, 1980), and group identity (Watt & Spears, 1999).

Equipment

Four networked PC's were used for the laboratory-based experiments. Each team member used a PC with either a virtual or abstract interface that represented a geographical area gas network, (e.g. North, South, East or West). Video cameras were used in each laboratory to record behavioural data of team members and allow visual communication across the distal condition. Telephones were used in the distal condition to enable communication amongst the team members. The software used to develop the two interfaces was *World Tool Kit*. The software package *Falcon* was adapted and used to form the link from the server to the four networked machines.

Procedure

Participants were recruited in teams of four. Participants undertook a one-hour training session before performing the task. Initially as a group, participants were given an introduction to the background off the project, followed by a full explanation of the task. Participants were then assigned their individual area names and asked to take their places at the appropriate area gas-network. All groups were asked to perform the same training task which took 20 minutes. After the full training session, the team was asked to carry out the task with no assistance from the researchers. All participants were asked to work together as a team. After completing the main task, participants completed questionnaires in their own time via the computer. They were then paid, asked to sign a receipt, and thanked for their time and participation. The experiment lasted approximately one hour.

Analysis

A variety of statistical techniques were used to analyse the data. Comparison of the experimental groups through the two main independent variables (i.e. proximity and interface) relied upon Analysis of Variance (ANOVA). The exception to this was the analysis of the CONCOM data, for which the Mann Whitney U test was performed. Factor analysis was used to determine the factor structure of the social identity questionnaire. Then ANOVA was conducted on the factor scores.

3. Summary of results

The results are summarised in table 1. In general terms, the proximal condition is favoured over the distal condition and the abstract condition is favoured over the distal condition, although this is not always the case.

Table 1. Summary of statistically significant results

VARIABLE	DIFFERENCE ($p < 0.05$)	FAVOURED COND
CONTROL ACTIONS		
Empty holder	Abstract-distal>others	N/A
COSTS		
Actual minus optimum	Distal>proximal	Proximal
SOCIAL IDENTITY		
Group identity	Abstract>virtual	Abstract
	Proximal>distal	Proximal
	Abstract-distal>abstract-proximal	Abstract-distal
	Virtual-proximal>virtual-distal	
Cohesiveness		Virtual-proximal
CORE JOB CHARACTER		

Skill variety		
Feedback from job	Proximal>distal	Proximal
Feedback from others	Abstract>virtual	Abstract
	Abstract-proximal>abstract-distal	Abstract-proximal
Motivating potential score	Abstract>virtual	Abstract
	Abstract-proximal>abstract-distal	Abstract-proximal
	Proximal>distal	Proximal
CONTEXTUAL CONTROL MODEL		
Opportunistic	Abstract-distal>virtual-distal	Abstract-distal
Tactical	Abstract-proximal>abstract-distal	Abstract-proximal
	Virtual-proximal>abstract-distal	Virtual-proximal
Scrambled	Virtual-distal>abstract-proximal	Abstract-proximal
	Virtual-distal>virtual-proximal	Virtual-proximal

4. Discussion

The findings from this research are interpreted through the research literature. First the reasons why the proximal condition was superior will be discussed under the heading of virtual teams. Second, the differences between the abstract and virtual interfaces will be explored under the heading of virtual interfaces.

Virtual teams

The results suggest that the proximal condition was superior to the distal condition in terms of reduced costs, greater group identity, enhanced motivation and greater tactical control. There is an intuitive appeal that people will perform better if they are located in the same physical proximity because they benefit from mutual stimulation, learning, piggybacking and synergy to produce large numbers of potentially novel and valuable ideas (Valacich et al, 1994). Individuals behaviour may be enhanced by the simple physical presence of others, independent of any informational or interactional influences these others may exert, (Sanders, 1981). A large number of studies have investigated the effects of the physical presence of others on task performance, but they have produced a mass of apparently contradictory results, such that in some studies an individuals performance was enhanced whilst in others it was impaired (for review see Zajonc, 1965; cited in Sanders, 1981). It is likely however, that many of these studies have been confounded by the wide variety of tasks that have been used as part of their methodology. Zajonc (1965; cited in Sanders, 1981), clarified some of this confusion with the proposal that the effect of the presence of others on performance was dependent on the nature of the task. When the task was simple or well learned the presence of others enhanced performance, whilst for complex or novel tasks performance was impaired. In the current study, team co-operation was essential for good performance. Working in the same room as the other team members may be conducive to co-operative tasks particularly as it encourages informal communication.

Virtual interfaces

The findings of the relative superiority of the interfaces were less clear than that of proximity. For control actions and costs there were no differences between the abstract and virtual interface groups. This means that both groups faired as well as each other. Similarly, for the measure of team behaviour using CONCOM, there does not appear to be any differences between the abstract and virtual groups (with the exception that abstract distal group spent more time in the opportunistic control mode than the other three groups). There are differences however, between the interface groups for the psychological measures of social identity and core job characteristics. Despite no performance differences, the psychological differences may be traceable to the nature of the information. Providing the participants with predictive information about the consequences of their actions helps provide feedback from the job. This in turn could enhance feedback from the other team members about progress towards the team goals. This may have, in turn, improved each of the team member's perception of group identity. The interaction between the type of interface and team proximity revealed that feedback (both from the job and other people) was rated as highest in the abstract-proximal group. The goal-oriented displays probably helped promote discussion between team members. It is argued that polygon displays are more easily understood and facilitate decision making (Green et al, 1996), which could have presented the abstract group with an advantage.

5. Main conclusions from the research

In summary, this research project has shown that the proximity of team members is important for improving the cost-effectiveness of system control, helping make control room behaviour more tactical and less scrambled, increasing group identity and intrinsic motivation. Abstract interfaces are associated with higher levels of feedback from the job and others as well as higher levels of social identity. Both abstract and virtual interfaces increase tactical control and reduce scrambled control in proximal settings. From this we may conclude that it is preferable to have control room teams working in the same room and that abstract interfaces can help with some aspects of feedback which may assist in developing group identity.

Acknowledgements

This research project was supported by funding made available through the ESRC's Virtual Society programme (grant reference: L132251038) with additional funds from BG Technology and The National Grid Company.

References

Goodstein, L.P., Andersen, H.B., & Olsen, S.E., (1988*), Tasks, Errors and Mental Models*, Taylor & Francis: London.

Green, C. A.; Logie, R. H.; Gilhooly, K. J.; Ross, D. G. & Ronald, A. (1996) Aberdeen ploygons: computer displays of physiological profiles for intensive care. *Ergonomics*, **39**, (3), 412-4428.

Hackman, J.R. & Oldham, G. (1980) *Work Redesign*. Addison-Wesley; USA

Hollnagel, E. (1993) *Human Reliability Analysis Context and Control.* Academic press: London

Praetorius, N. & Duncan, K.D., (1991). Flow representation plant processes for fault diagnosis. *Behaviour and Information Technology, Vol., 10, No. 1, pp 41-52*

Rasmussen, J. (1986) *Information Processing and Human-Machine-Interaction - An Approach to Cognitive Engineering* (Amsterdam: North Holland).

Sanders, G.S. (1981). Driven by distraction: An integrative review of social facilitation theory and research. *Journal of Experimental Social psychology* **17**, 227-251

Valacich, J.S., George, J.F., Nunamaker, J.F. & Vogel, D.R. (1994). Physical proximity effects on computer mediated group idea generation. *Small Group Research.*, **25** (1) 83-104.

Vicente, K.J. (1997), 'Interface Design: Is it always a good idea to design an interface to match the operator's mental model?' *Ergonomics Australia On-line volume* **22**, (2)

Watt, S. & Spears, R. (1999) personal communication

AUGMENTED REALITY IN MUSEUMS AND ART GALLERIES

C. Baber

Kodak / Royal Academy Educational Technology Research Group, School of Elec. & Elec. Eng., The University of Birmingham, Birmingham. B15 2TT UK

c.baber@bham.ac.uk

Abstract.
In this paper, we report studies into the use of handheld and on-body computing as a means of augmenting media-spaces, such as museums and art galleries. The main focus of the work is on the user-centred design process. Fully functioning prototypes have been developed and are described in this paper. The requirements of the various stakeholders in museums and art galleries are discussed, together with the results of various user trials. The paper concludes with a discussion of augmented reality and information requirements of future media-spaces.

1. AUGMENTING REALITY

Imagine that, when you visited a museum or art gallery, you were accompanied by a Guide who knew about the artifacts on display and who knew what interested you. This Guide could provide an interesting, insightful tour of the artifacts that was tailored especially for you. Imagine further that the Guide was able to use a whole range of different media to present information to you, from annotating the artifacts to showing video clips and playing audio that helped to set the artifacts in context. Imagine that the Guide was not a person but a digital device that you could wear or carry during your visit. Such a device lies at the heart of work into augmenting museums and art galleries. A major European project, completed last year, focused on the provision of audio information to visitors to art galleries, with a view to adapting this information to the needs of different visitors. An interesting aspect of this work was the idea that moving around an art gallery was akin to moving through a hypertext document; as the visitor moved to the next exhibit, so they also moved to the next 'node' in the hypertext (and received information about that exhibit). Recent discussions of this work can be found in Petrelli et al. (2001) and Marti et al. (2001).

The 'digital-guide' scenario raises some of the basic assumptions that underlie the notion of augmented reality (AR). The real world, and the artifacts it contains, can be augmented (or enhanced) through the use of other media. Ideally, this augmentation would be tailored to specific locations or artifacts and adapted to specific individuals. There are many routes via which augmentation can take place; one of the most common is through the super-imposition of visual information onto the real world. For example, the museum visitor could wear a head-mounted display (HMD) that projected an image onto the artifact. Thus, in the work described in this paper, the visitor views the painting with the addition of text and simple marks to highlight points of interest; in effect, the HMD provides a guided tour of the painting by directing the user's gaze to specific points and providing a commentary (via audio or text) or these points. In order to work effectively, an AR system most probably needs to support wearer mobility. This is not to say that 'static' AR is not useful, but that mobile users might require context-relevant information on the move. As far as the domain of museums and art galleries is concerned there are three broad areas of development that are of interest: dynamic labeling; user adaptation; physical hyperspace.

Dynamic Labeling: Milosavljevic et al. (1998) point out that the labels used in museums "...are written according to the assumed knowledge and needs of a single restricted audience model." So, the information used to describe an artifact will only be sufficient for a specific type of visitor (and often for a specific type of visitor query). Thus, the label might contain the name of the artist, the title of the painting and the probable date of completion of the work. Visitors wishing to know more about a painting will need to consult the guidebook or a tour-guide. Consequently, an electronic device that can present information to visitors could be made dynamic in the sense that its content will change for different visitors and can be used to answer a variety of visitor queries. This removes the need for paper labels that detract from the experience and aura of the works of art on display.

User Adaptation: In order for dynamic labels to accommodate a variety of visitors, it is necessary to have some means of adapting to different types of visitor. This requires some form of user model (Petrelli et al., 1999; Sarini and Strapparava, 1998). For instance, Milosavljevic et al. (1998) propose dividing visitors into 'naïve' and 'expert' (depending on their background knowledge), and to allow visitors to select one of several languages. This approach seeks to classify individuals in terms of a set of criteria. An alternative approach would follow from the work of Cooper (1999), and define users in terms of the type of visit that they will make. Falk and Dierking (1992) suggest that it is possible to typify visitors in terms of visit, and we follow this work to propose the following examples: first time visitor; return visitor; systematic route follower; cruiser / browser; artifact spotter; specific artifact searcher. Without going into detail about the defining characteristics of these visit types, it is clear that one might anticipate information requirements to vary across these groups. Current work is being undertaken to survey museum and art gallery visitors to develop this classification scheme.

Physical Hyperspace: Having a system that can determine the visitor's location, and then modify information to suit that location is the central aim of the HIPS project (Not et al., 1997, 1998; Oppermann and Specht, 2000). In this project (and in our work), exhibits are provided with infra-red (IR) transmitters and visitors carry a device that incorporates an IR receiver. When the transmitter and receiver are within range, an audio commentary begins to play. In this work, electronic information is stored in a hypertext format and "...moving around the physical space...the visitor implicitly 'clicks' on meaningful points of the hypertext." (Sarini and Strapparava, 1998). This merging of computer and real worlds has been termed 'physical hyperspace'. The merging could either involve audio information and artifacts (Not et al., 1998), or visual information and artifacts (Baber et al., 1999, 2000).

2. AUGMENTATION OF ARTIFACTS

We have opted for two approaches to the visual augmentation of artifacts, and developed prototypes of each. The approaches share the principle of reflecting a visual image onto the artifact. In one prototype, this reflection is performed using a commercially available HMD, i.e., a MicroOptical HMD (see figure one). Previous work has demonstrated that this approach, when combined with audio commentary, can provide a viable means of presenting information about a painting (Baber et al., 1999; 2000), but that guiding the viewer's attention to specific locations in the painting can affect their ability to recall details of the whole painting. In other words, forcing people to look at certain details seems to impair their ability to see the 'whole' painting. Furthermore, the background knowledge of the viewer has a significant bearing on how well they can relate the information provided to the content of the painting (Baber et al., 2001).

Figure One:
MicroOptical Head-Mounted Display

One of the common complaints against the HMD version was the discomfort caused by wearing such headgear (the discomfort was both physical, as the headset did not comfortably fit all participants, and emotional, as many participants felt that the headset made them look silly). In a second prototype, we sought to move the reflection from the viewer. In this design, the reflection is performed using an angled mirror (see figure two). The well-known visual illusion of Pepper's Ghost was the motivation for the mirror-based system. In this illusion, an image, say of a person sitting in a chair, is projected onto an angled mirror. The viewer sees the reflection and perceives this as being merged with the real objects they are looking at. When the image behaves in an unexpected manner, slightly altering the angle of the mirror makes the image disappear, the viewer could be fooled into thinking they have seen a ghost. By a similar token, our ARDisplay_Case super-imposes a projected image onto an artifact. The prototype provides a simple but highly effective demonstration of the concept, and has allowed us to label a variety of artifacts with information tailored to different groups of users.

Figure Two:
Schematic of ARDisplay_Case
The visitor sees the artifact, together with text and Graphics reflected from the angled mirror.
The displayed information can be selected (for age and language) and navigated using buttons on the front panel.

3. AUGMENTATION OF MUSEUMS

In addition to considering the display of information, our work has focused on the definition of visits, and the use of technology to record the movement of visitors. Using a simple badge system, it is possible to record how long a visitor spends in front of each artifact and the order in which artifacts are visited. This information, stored on the badge that the visitor wears, can be down-loaded at the end of the visit to provide the museum with potentially useful logistically data. Figure three shows a summary of data from visitors in a small museum on The University of Birmingham campus.

Figure Three:
Visitor data

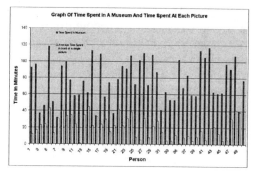

4. AUGMENTATION OF VISITS

Providing visitors with a device that can allow them to receive information about artifacts and record some aspects of their movement, raises the possibility of further augmenting a visit to the museum. In simple terms, this might involve providing visitors with a 'You-are-here' indicator in their display, perhaps through the use of a dot on a map. This was easily implemented in the systems we have developed. At a slightly more sophisticated level, the map display could also include icons for members of the party accompanying the visitor; in this manner, even if the party heads in separate directions, it will be possible to find each other. Finally, given a map display, the visitor could ask for directions to specific location or amenities.

When people leave the museum, they could continue their visit at home through the creation of a personalized web-page. This contains information on the exhibits that they saw, together with links to other information in the museum and announcements of coming events. In the current version, the web-site only contains pages for the exhibits that the visitor 'marked'. The process of marking involves pressing a button on the badge to signify that they have found an artifact of interest; the ID of the artifact is then logged and downloaded at the end of the visit.

5. FUTURE DEVELOPMENTS

The augmentation of museum visits has proved to be a popular research topic in the AR and Context-aware computing communities. This research has, in particular, attended to the complex problem of developing user models that can be used to adapt AR to museum visitors. This work provides useful guidance and models for development. One direction in which we would like to move our work is into the use of models of visits to guide the display of information. However, rather than developing the models on the basis of user profiles, the approach that we are taking is to develop models based on features of context. For example, we know the visitor's location and how long they have spent in front of a painting. Through accelerometers and other sensing devices, we can determine whether the person is sitting or standing, and whether they are actively concentrating on the artifact. Through microphones, we can determine background activity, e.g., is the person speaking to someone? Combining these data from various sensors, allows to predict the most likely features of the context that are affecting the visitor, and then to tailor information displays accordingly.

The work on merging computer-generated images with real objects has raised a number of interesting avenues of further enquiry. The issue of how best of how best to ensure that the image fuses appropriately with the real object has taxed the AR community for several years. There appear to be two main schools of thought: in one, the image and object are fused into a single image, i.e., the image is mixed with a video of the object. This effectively creates a virtual image in which a real object and the computer-generated image are mixed prior to presentation to the person. In the other approach, the object is tracked and the image is fitted onto the location of the object. Both approaches face problems. We have been using the technique developed at HITL, Washington, in which objects are marked using simple geometric shapes and image recognition software identifies the shape and causes a computer-generated image to appear adjacent to the shape. This has proved a simple and reliable means of placing information onto static objects.

REFERENCES

Baber, C., Harris, T. and Harrison, B., 1999, Demonstrating the concept of physical hyperspace in an art gallery, In S. Brewster, A. Cawsey and G. Cockton (eds.) *Human Computer Interaction INTERACT'99 (volume II)*, Swindon: British Computer Society

Baber, C., Bristow, H., Cheng, S-L., Hedley, A., Kuriyama, Y., Lien, M., Pollard, J. and Sorrell, P., 2000, Augmentating MediaSpace: A Socio-Cognitive Engineering Approach, *3rd International Conference on Engineering Psychology and Cognitive Ergonomics*, Edinburgh

Baber, C., Bristow, H., Cheng, S-L., Hedley, A., Kuriyama, Y., Lien, M., Pollard, J. and Sorrell, P., 2001, Augmenting museums and art galleries, *Interact 2001*

Cooper, A., 1999, *The Inmates are Running the Asylum: Why high-Tech Products Drive us Crazy and How to Restore Sanity*, Indianapolis, In.: Sams

Falk, J. and Dierking, L., 1992, *The Museum Experience*, Ann Arbor, MI: Whalesback Books

Marti, P., Gabrielli, F. and Pucci, F., 2001, Situated interaction in art, *Personal and Ubiquitous Computing, 5 (1)* 71-74

Milosavljevic, M., Dale, R., Green, S.J., Paris, C. and Williams, S., 1998, Virtual museums on the information superhighway: prospects and potholes, *Proceedings of the Annual Conference on the International Committee for Documentation of the International Council of Museums (CIDOC'98)*, 10-14 October, Melbourne

Not, E., Petrelli, D., Stock, O., Strappavara, C. and Zancanaro, M., 1997, Person-oriented guided visits in a physical museum, *Proceedings of 4th International Conference on Hypermedia and Interactivity in Museums*, Paris, France. 162-167

Not, E., Petrelli, D., Sarini, M., Stock, O., Strappavara, C. and Zancanaro, M., 1998, Hyper-navigation in the physical space: adapting presentations to the user and to the situational context, *The New Review of Hypermedia and Multimedia, 4*, 33-45

Oberlander, J. and Mellish, C., *Final Report on the ILEX project*, http://cirrus.dai.ed.ac.uk:8000/ilex/final. htm

Oppermann, R. and Specht, M., 2000, A context- sensitive nomadic exhibition guide, *Proceedings of the 2nd International Symposium on Handheld and Ubiquitous Computing (HUC 2000)*, Berlin: Springer-Verlag, 127-142

Petrelli, D., Not, E., Zancanaro, M., Strapparava, C. and Stock, O., 2001, Modelling and adapting to context, *Personal and Ubiquitous Computing, 5 (1)* 20-24

Sarini, M. and Strapparava, C., 1998, Building a user model for a museum exploration and information-providing adaptive system, *Proceedings of the Workshop on Adaptive Hypertext and Hypermedia*, 20-24 June, Pittsburgh, PA

Browsing Patterns in a Virtual Information Space Representation of a Document Database

Collins, J. * & Westerman, S. J. **

*Psychology Group, Neurosciences Research Institute, Aston University, Birmingham, B4 7ET, UK.
**School of Psychology, University of Leeds, Leeds, L52 9JT, UK.

Abstract

This paper reports an experiment in which 'browsing' patterns generated during performance of a task requiring information retrieval from a database of documents were examined. Documents were represented as objects in virtual space, and positioned according to their semantic content. Participants were required to locate all documents relevant to a specific query. Browsing patterns were analysed using an approach in which unique sequences of 'n' documents visited were identified. The frequency of occurrence of these sequences was then represented in a vector for each participant. Cosine coefficients (the angle between vectors in high-dimensional space) were calculated to determine inter-participant browsing pattern similarity. Using this technique, the effects of 'goodness' of the spatial-semantic mapping and individual differences in cognitive abilities were analysed. Results support this technique as a useful method of quantifying browsing behaviour.

1. Introduction

The psychology of the user, particularly with regard to individual differences, is an important issue for consideration with respect to the development of computerised Information Retrieval (IR) systems. This is due to the rapid expansion, and increasing diversity, of the user population, that no longer exclusively comprises experts in computer technology. It can be argued that examination of 'browsing behaviour', i.e. the strategies and patterns of browsing that people employ when searching databases (see Chang & Rice, 1993; Marchionini, 1995), is an important source of information about individual differences in the psychological processes involved in computerised information retrieval. However, while measures of 'recall' and 'precision' are commonly used to analyse search performance (see Harter, 1996, for a discussion of these measures), investigation of browsing strategy is much more difficult. This is largely due to the multidimensional nature of browsing (Chang & Rice, 1993) and the numerous factors both internal and external involved in determining browsing behaviour. For instance the task setting, the system for searching, the task domain, the user, and expected outcomes have been identified as influential factors (see also Marchionini & Shneiderman, 1988). Quantification and comparison of browsing patterns is problematic. Many studies have used verbal protocols as a basis for analysis. However, these generally rely on allocation of individual responses to categories by the experimenter and largely only measure behaviour the individual is consciously aware of. Studies that quantify queries generated, the use of Boolean operators, pages visited etc., while providing valuable information regarding the usability of an IR system, are less informative with respect to specific browsing patterns. Description of sequences of events is required. If fully measurable browsing patterns (in terms of ratio level numerical differences between specific browsing patterns) can be identified this will lead to greater understanding of some of the cognitive issues involved in searching through databases, and provide further insight of users' cognitive maps of the information environment.

Currently most IR systems operate on databases of 'hidden' documents that are retrieved in a list format in response to the input of search words or terms. However, the use of information visualisation techniques and virtual environments has opened up new possibilities for the representation of semantic (contextual) information (see e.g. Cugini, Laskowski & Piatko, 1997; Fox, Frieder, Knepper & Snowberg, 1999; Westerman & Cribbin, 2000). In such environments (i.e. Spatial Data Management Systems - SDMSs) the contents of a database can be represented visually as objects that are located in a multi-dimensional space in which proximity reflects the degree of semantic relatedness between objects. There is evidence to suggest that cognitive processing of semantic inference is spatially structured (Jackendoff, 1983; Gardenfors, 2000). This theory of spatial-semantic processing has been supported by Westerman & Cribbin (2000), who found that performance of a search task using a SDMS improved on several measures with improved 'goodness of fit' between the spatial mapping and the semantic relatedness of represented objects (documents).

The experiment reported here tested a method of sequence analysis ('n-gram' analysis) previously used for Automatic Text Analysis (ATA) (Damashek, 1995) as a means of describing individual differences in browsing behaviour when using a computerised database. This technique involved identifying unique sequences of 'n' documents visited during browsing. For each participant, a vector was calculated reflecting the frequency with which each n-gram occurred. The cosine of the angle between each of these vectors was then calculated and used as an index of browsing pattern similarity between individuals. To test the effectiveness of this measure of browsing strategy, as with Westerman and Cribbin (2000) the effects of manipulations of the 'goodness' of spatial-semantic mapping were examined. It was predicted that browsing strategies would be more similar in conditions where the spatial-semantic mapping was relatively better. Further to this, the predictive capacity of individual differences in cognitive ability was examined. Several predictors of individual differences in information retrieval performance have been identified, such as age (e.g., Westerman et al., 1995) and various cognitive abilities (e.g., Allen, 1992). Westerman & Cribbin (2000) demonstrated a positive effect of spatial visualisation on performance of search tasks in a SDMS. The experiment reported here was designed to investigate whether individual differences in cognitive ability also predict browsing strategy. It was hypothesised that relatively greater similarity of cognitive ability would be associated with relatively greater similarity of browsing patterns.

2. Methodology

The TREC 7 database (see Voorhees & Harmen, 1999) was used to compile a document database for this experiment comprising 100 newspaper articles from the LA Times. Of these, 20 documents were considered by TREC to be relevant to 'risks taken by journalists' (JR), the other 80 were randomly selected and considered irrelevant to JR. Documents were analysed using the 'n-gram' method of ATA (see Damashek, 1995). A list of unique 5-grams (strings five characters long) was compiled for the database, and vectors, each representing a single document, were calculated based on the frequencies with which each 5-gram occurred. A similarity matrix for the document set was generated, using the Vector Space Model (VSM), based on the cosines for all-possible vector pairings. Then a multi-dimensional scaling solution was calculated (stress = .286; r^2 = .485) to map the documents into a three-dimensional virtual environment (VE) in which the documents were represented as objects (spheres). Despite the low r^2 value, as this was the best solution available it was taken to be the 100% condition (i.e. 3D100), in that it accounted for 100% of the semantic variance, notwithstanding random error. Two further three-dimensional VEs were generated by adding random noise to the co-ordinates of the 3D100 condition, thereby varying in the quality of the spatial-semantic mapping, and producing solutions that accounted for approximately 80% (3D80) and 60% (3D60) of the semantic variance accounted for by 3D100 solution. Finally, a two-dimensional virtual environment was created by removing a coordinate from the three-dimensional solution. This accounted for approximately 60% of the variance accounted for by the 3D100 solution (see Figure 1a and 1b for examples of 3D100 and 2D60 environments).

Figure 1a. 3D100 Environment

Figure 1b. 2D60 Environment

2.1 Experimental Procedure

Eighty-four Aston University students (63 women and 21 men) participated in the experiment. The mean age of the sample was 19.68 years (SD=3.57). Participants were semi-randomly assigned to one of four conditions (3D100, 3D80, 3D60, or 2D60). Conditions were matched for scores achieved on measures of Associative Memory (MA), Spatial Working Memory (WM) and Spatial Visualisation (VZ), obtained using psychometric tests from the 'Kit of Factor-Referenced Cognitive Tests' (Ekstrom, French, Harman, & Derman, 1976). A measure of 'speed of comprehension' (SCOLP) was also taken in order to balance the allocation of participants to conditions in terms of reading speed and comprehension (Baddeley, Emslie & Smith, 1992). After completing the psychometric tests, to ensure familiarity with the navigational demands of the task, participants completed a practice task in either a 2D or 3D environment dependent on their environment allocation for the main task. The practice task required the location and selection of a randomly placed red sphere in an environment of green spheres. Participants would have been

964

asked to withdraw if they had not successfully met a minimum performance threshold. In fact, nobody failed to do so.

For the main task, participants were presented with one of the previously described environments and asked to find as many documents as possible relating to a reporter(s)/journalist(s) who faced some sort of personal threat or danger. Pointing at an object using the mouse cursor revealed the associated document. Participants' determined how they searched the environment, what strategies they used, and when they finished based on whether they felt they had found all the relevant documents they were likely to. However, a time frame of between 20 and 45 minutes was stipulated. During completion of the task, the programme recorded information relevant to several measures of performance. These measures included the sequence in which objects were visited (objects were allocated unique numerical identifiers) enabling similarities in browsing patterns to be assessed using an n-gram method similar to that described in 'methodology' for calculating document similarity. Windows 2,3,4, and 5-grams long were passed across the record of objects (documents) visited based on their numerical identifiers for each individual participant. Cosines were then calculated for every pair-wise combination of participants in each of the 'semantic variance' conditions (3D100, 3D80, 3D60, and 2D60), using the VSM as previously described. These were taken to be an index of between-participant similarity in browsing strategy – the larger the cosine between two participants' vectors the more similar the browsing strategies that they employed.

3. Results

To test the effects of the experimental manipulation of semantic variance on between-individual browsing strategies, a 3 (3D100 vs. 3D80 vs. 3D60) x 4 (2- vs. 3- vs. 4- vs. 5gram) ANOVA was calculated in which cosines were the dependent measure (see Figure 2a). There was a main effect of 'semantic variance', $F_{(2,2508)}=79.00$, $p<.001$, with cosines reducing in line with the 'goodness' of the solution. There was a main effect of n-gram length, $F_{(3,2508)}=2712.87$, $p<.001$, with cosine values being greater with shorter n-grams. There was also a significant interaction between 'semantic variance' and n-gram length, $F_{(6,2508)}=7.86$, $p<.001$, such that the difference between 'semantic variance' conditions was more pronounced with shorter n-grams.

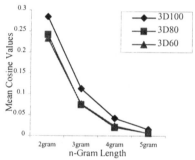

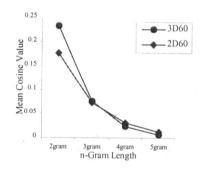

Figure 2a. Mean Cosines Variance by n-Gram   Figure 2b. Mean Cosines Dimension by n-Gram

To test the effects of the experimental manipulation of 'dimensionality' on between-individual browsing strategies, a 2 (2D60 vs. 3D60) x 4 (2- vs. 3- vs. 4- vs. 5gram) ANOVA was calculated in which cosines were the dependent measure (see Figure 2b). There was a main effect of 'dimension', $F_{(1,1672)}=8.22$, $p<.005$, with cosines tending to be greater in the 3D60 condition. There was also a main effect of n-gram length, $F_{(3,1672)}=501.25$, $p<.001$. As with the previous analysis, the magnitude of cosines was inversely related to n-gram length. Finally, there was an interaction between 'dimension' and n-gram length, $F_{(3, 1672)}=14.96$, $p<.001$, such that the cosines were relatively greater for the 3D60 group when calculated with a 2-gram, but fairly similar for both groups when using 3-, 4-, and 5-grams.

For each of the cognitive ability measures (MA, WM, and VZ) difference values were calculated between every unique pair of individuals in each condition (3D100, 3D80, 3D60, and 2D60) resulting in 210 pair-wise difference scores for each ability. These were then correlated with the pair-wise cosine values in each condition, using Spearman's Rho (nonparametric) as histograms showed the distributions in all conditions and for all cognitive ability difference scores to be skewed. As can be seen from Table 1, in the 3D100 condition difference in spatial visualisation scores was negatively correlated with cosines calculated using 2-, 3-, and 4-grams. Pairs of participants with greater differences between their spatial ability scores tended to have less similarity (smaller cosines) in their browsing patterns. Spatial memory was negatively correlated with cosines in the 3D80 condition when browsing pattern cosines were determined with 3-, 4-, and 5-grams, and in the 3D60 condition when browsing patterns were determined using 5-grams.

		n-Gram length			
		2	3	4	5
	MA	0.08	0.14[*]	0.10	0.15[*]
3D100	WM	0.01	-0.00	0.01	0.08
	VZ	-0.24[**]	-0.20[**]	-0.16[*]	-0.05
	MA	0.05	0.11	0.04	0.09
3D80	WM	-0.11	-0.15[*]	-0.23[**]	-0.19[**]
	VZ	0.06	0.09	0.06	0.09
	MA	0.03	0.07	0.13	0.08
3D60	WM	-0.02	-0.10	-0.08	-0.17[*]
	VZ	-0.08	-0.05	-0.04	-0.02
	MA	0.00	0.05	0.07	0.02
2D60	WM	-0.07	-0.01	0.00	0.05
	VZ	-0.02	-0.01	0.00	0.04

Note: [*] $p<0.05$, [**] $p<0.01$: MA = Associative Memory; WM = Spatial Working Memory; VZ = Spatial Visualisation

Table 1. Spearman's Rho Correlations between Cognitive Ability Difference Scores and Cosine Values for each 'semantic variance' Condition and each n-Gram Length.

4. Discussion

The results reported are consistent with the hypothesis that browsing behaviour can be quantified using the n-gram method of analysis. Both quality of spatial-semantic 'mapping' and cognitive ability were predictive of browsing patterns, as assessed using this technique. In terms of the quality of spatial-semantic mapping the results were consistent with the hypothesis that the greater the degree of semantic variance accounted for by the spatial location of objects in a virtual environment (VE) the greater would be the similarity between users' browsing patterns. This does not appear to be a linear effect however, as cosines were very similar for 3D80 and 3D60 solutions. It is perhaps not surprising that there is still much variance unaccounted for. As mentioned above, Chang & Rice (1993) identified many possible determinants of browsing behaviour. However, it may be that stronger results would be obtained if performance were assessed only once the user had the chance to orient themselves in the information space, e.g. once they had located the first relevant document. This may eliminate a source of 'noise' associated with performance.

When n-grams are used to describe browsing patterns, it would seem that relatively short n-grams are more effective. In the analyses presented above, cosine coefficients were inversely related to n-gram length with calculations based on 2-grams accounting for the greatest variance, while very little variance was accounted for by calculations based on 5-grams. Given the number of possible sequences in which documents could be visited it seems reasonable that similarities exist over only a few documents at a time.

With respect to the effects of dimensionality (two vs. three dimensions), there was little difference in cosine values other than those calculated using 2-grams, for which cosines were greater for the 3D60 condition. These results suggest that participants were more consistent in their browsing patterns in the three-dimensional condition. This may be due to the fact that navigation of the three-dimensional environment was relatively more cognitively demanding (see Westerman and Cribbin, 2000) and therefore participants engaged in more planned consideration of browsing behaviour.

Individual differences in cognitive ability were significantly related to browsing behaviour (see Table 1). However, the nature of these effects appears to depend on the amount of semantic variance accounted for by the spatial location of objects in the virtual information space. Spatial visualisation was negatively correlated with cosines calculated using 2-, 3-, and 4-grams in the 3D100 environment. However, spatial memory was negatively associated with cosines based on 3-, 4-, and 5-grams for the 3D80 environment. In 3D60 spatial memory was negatively correlated at 5-gram only. A possible reason for this pattern of results is that spatial visualisation may be highly influential in the acquisition of an individuals' cognitive map when the environment is optimally mapped, but in conditions where spatial mapping is poor spatial working memory may be a more important factor, determining the extent to which the location of objects can be recalled. Browsing performance in the two-dimensional condition may be more random (see above) and therefore not predicted by cognitive ability. This explanation is supported, to some extent, by the findings of Westerman & Cribbin (2000) who suggest that high spatial visualisation allows complex cognitive maps to be generated that can facilitate searching in environments that account for high semantic variance. However, they argue that as the degree of semantic variance accounted for decreases users rely more on search strategies that are either sequential or random (spatial memory was not assessed in this study). It should be noted that, in the present study, all correlations between measures of cognitive ability and browsing performance are fairly weak, the highest being $r = -.24$. However, this may be expected as the use of cosine values for this purpose (i.e. 1 = perfect similarity, 0 = perfect dissimilarity) tends to result in a restriction of range and a skewed distribution. Strong correlations are unlikely under such circumstances.

5. Conclusions

The results from this experiment are encouraging. They support the use of n-grams as a means of quantifying browsing patterns, although it would seem that calculations based on relatively short n-grams should be used. Inter-participant browsing patterns were influenced by experimental manipulation of the 'goodness' of spatial-semantic mapping, and individual differences in cognitive ability also were predictive of browsing behaviour. However, this should be regarded as an exploratory study and further research is needed in this area.

References

Allen, B. (1992). Cognitive Differences in End User Searching of a CD-ROM Index. Proceedings of the Fifteenth Annual International ACM SIGIR conference on Research and development in information retrieval June 21 - 24, 298-309.

Baddeley, A., Emslie, H., & Smith, E. N. (1992). The Speed and Capacity of Language-Processing Test. Bury St. Edmunds: Thames Valley Test Company.

Chang, S. J. & Rice, R. E. (1993). Browsing - A Multidimensional Framework. Annual Review Of Information Science And Technology, 28, 231-276.

Cugini, J., Laskowski, S., & Piatko, C. (1997). Document Clustering in Concept Space: The NIST Information Retrieval Visualization Engine (NIRVE). CODATA Euro-American Workshop on Visualization of Information and Data, Paris France, June 1997.

Damashek, M. (1995). Gauging Similarity with n-Grams: Language-Independent Categorization of Text. Science 267, 843-848.

Ekstrom, R. B., French, J. W., Harman, H. H., & Derman, D. (1976). Manual for kit of factor-referenced cognitive tests. Princeton, NJ., Educational Testing Service.

Fox, K. L., Frieder, O., Knepper, M. M., & Snowberg, E. J. (1999). SENTINEL: A Multiple Engine Information Retrieval and Visualisation System. Journal of the American Society for Information Science 50 (7), 616-625.

Gardenfors, P. (2000). Conceptual Spaces: The Geometry of Thought. Cambridge, Massachusetts: The MIT Press.

Harter, S. P. (1996). Variations in relevance assessments and the measurement of retrieval effectiveness. Journal of the American Society for Information Science, 47, 37-49.

Jackendoff, R. (1983). Semantics and Cognition. Cambridge, Massachusetts: The MIT Press.

Marchionini, G. (1995). Information Seeking in Electronic Environments. Cambridge: Cambridge University Press.

Marchionini, G. & Shneiderman, B. (1988). Finding Facts Vs Browsing Knowledge In Hypertext Systems. Computer, 21, 70-80.

Voorhees, E. M. & Harman, D. (1999). Overview of the Seventh Text REtrieval Conference (TREC-7). NIST Special Publication, 500 - 242.

Westerman, S. J. & Cribbin, T. (2000). Mapping semantic information in virtual space: dimensions, variance and individual differences. International Journal of Human-Computer Studies, 53, 765-787.

Westerman, S.J., Davies, D.R., Glendon, A.I., Stammers, R.B., & Matthews, G. (1995). Age differences in computerised information retrieval. Behaviour and Information Technology, 14, 313-326.

Task Analysis: The Best First Step in User Interface Design

Gwen Pearson, Ph.D.

Phil Weeks, Ph.D.

AT&T Labs
Room D1-3C04
200 Laurel Avenue
Middletown, NJ 07748
USA

AT&T Labs
Room D1-3D26
200 Laurel Avenue
Middletown, NJ 07748
USA

ABSTRACT

This paper describes a large task analysis that we conducted at the AT&T Network Operations Center. The task analysis studied the network manager's job and tasks at the center. The results of the task analysis were used in the design of user interfaces for the network managers. The paper describes the methodology, challenges and benefits of conducting the task analysis. We argue that more projects should conduct a task analysis or workflow analysis as a first step in improving or optimizing user interfaces and processes.

1. INTRODUCTION

A task analysis may be defined as "a description of a job in terms of identifiable units of activities... A task may be defined as a group of related activities directed toward some goal, and often involves people's interaction with equipment, other people and/or media" (McCormick, 1979, p. 92). A workflow analysis includes a broader survey of the inputs and outputs of a particular job. A task and/or workflow analysis is typically the first step in determining how to best deploy user-centered design resources, particularly where process design or user interface design efforts are planned.

Network managers at the Network Operations Center monitor call traffic carried by the AT&T worldwide intelligent networks. We conducted a task analysis of the Network Manager job and tasks. The purpose of performing the task analysis was to understand the Network Manager's job, work processes, tools, and environment with sufficient detail that would allow us to identify aspects of the job and tools that could benefit from process modifications or user interface modifications. By completing this task analysis, we hoped to gain a greater understanding of the Network Manager's job and how the applications and tools help to support their work. Additionally, we hoped to gather information that would help us make informed decisions about detailed user interface design issues.

2. METHOD

2.1 Sampling Strategy

To ensure that we obtained data that were representative of the complete set of network management tasks and users, we sampled from each of the user groups in the center. This approach was based on the assumption that while some common tasks span the groups, many of the tasks performed across the groups vary. Further, the tasks and workload vary across shifts, so we collected data from managers working on each work shift. In addition to the network managers, we had the opportunity to talk with the supervisors of each of the groups.

2.2 Procedures

We used a combination of work observation and structured interviews to collect network management task data, coinciding with selected managers' typical shifts. The one-on-one sessions with network managers were scheduled in advance, and conducted at the user's workstations. The network manager performed typical tasks and was able to walk us through the tasks and answer our questions. In general, the observations and interview sessions took about four hours. Because of the complexity of the job, four hours was about the maximum amount of time we could allow for a session without putting undue burden on both the network manager and the observer. A total of sixteen observation sessions were completed.

A typical interview began with introductions and an explanation for the purpose of the session. Workload permitting, the session began with some background questions, and questions regarding the network manager's preferred computing tool setup at the workstation. Alternatively, if job demands were high, the interview emphasized observation of task performance, asking follow-up questions to clarify activities after the workload had subsided. The ideal interview contained a balance of unobtrusive observation of typical tasks, combined with clarifying questions posed during or after task completion as conditions permitted.

Given the complexity and large number of tasks in the network manager's job, the task analysis focused on a set of benchmark tasks performed by the users. The benchmark tasks included frequently performed network management tasks, as well as tasks that are highly important regardless of their frequency.

2. DATA COLLECTIONS AND ANALYSIS

Data collected from these visits consisted of detailed notes taken by the interviewers during the interviews and discussions with network managers, as well as notes summarizing observations made. Sketches of the console layout and wallboard layouts were also made. Where possible, we collected printouts of the typical computer desktop arrangements of each of the network managers.

The data were analyzed and summarized to produce benchmark task descriptions, user group descriptions, and user profiles. The observations, task descriptions, findings and recommendations of the task analysis were documented for the project team.

3. FINDINGS

An important output obtained from the task analysis was the task descriptions for the benchmark tasks. The task descriptions outlined the detailed series of steps used to perform a given task. The number of steps involved in each of the tasks ranged from a few steps to over thirty steps. This information proved very useful in understanding the complexity involved in carrying out the tasks. Ultimately, this information was used to design new displays that reduced the number of steps and complexity in performing these tasks.

Another important output of the analysis was a set of observations and recommendations derived from common themes in the data. The observations and recommendations were used in the design of new user interfaces. The authors took a holistic approach to the task analysis, which covered the tools, processes and work environment. The project team started several new features and projects as a result of the task analysis findings and recommendations.

4. CHALLENGES OF COMPLETEING A TASK ANALYSIS

There were several challenges and potential obstacles to planning, conducting and completing the task analysis including project schedule pressures, management support, user support, and pressures in the user's work environment. We will discuss each of these challenges and describe how we were able to overcome the potential obstacles.

One challenge to completing the task analysis was getting buy-in from project funders. While the benefits of doing a task analysis may seem academic to those in the Humans Factors community, it still can be a challenge to convince project funders to allocate funds for this type of work. Mainly, this challenge is in convincing the team of the benefits of getting a more direct, detailed level of understanding of the work force's job functions than just working with user managers, user representatives, project managers or developers. However, there is no effective alternative to the direct observation of users to get a complete "user perspective." This more complete understanding leads to better, more integrated and user-friendly system designs. We were able to gain the support for the task analysis by educating both the funder and user organization on the benefits and positive impact of the task analysis.

Attempting to fit a task analysis into an existing project schedule can be a major obstacle to getting support for a task analysis. Project schedule pressures can be unrelenting and driven by forces external to the immediate project team. In addition, human factors engineers are often brought onto a project well into the lifecycle of the project. Usually, a task analysis in not included in the original work plans for a project. Thus, it can be very difficult to fit a task analysis into the time frame of an ongoing project. Obviously, the earlier the human factors team members are brought onto a project, the easier it is to complete a task analysis. The human factors team may have to concretely describe the procedures and what support needed from other team members. Furthermore, human factors team members may have to operate under extreme time pressures. Fortunately, for this project, the project management team understood the benefits of a complete task analysis and from the onset was able to offer us some protection from more immediate time pressures. Yet, even under these almost optimal conditions, the task analysis was competed in parallel with initial interface design work to satisfy project needs.

Once a task analysis has been approved, preparation becomes a critically important task that required more time than we initially anticipated. While it is not trivial to prepare for a task analysis of the type we were conducting, it was important to get started with the interviews as quickly as possible. Depending on the type of task analysis and goals of the task analysis being conducted, it may take more or less time to prepare. In our case, we were interested in gaining a detailed understanding of a set of complex tasks, and to document those finding for other team members. Some of the tasks included in the preparation phase were the identification of benchmark tasks, creation of the structured interview materials, determining the type of data analysis, and scheduling of interviewees. Fortunately, we had the time to complete these tasks. However, other projects may not be able to afford this luxury. In these cases, it's still important to try to include some type of observation of users to understand their job functions before initial designs are set. Smaller, less formal task analysis that may not require full documentation of findings can provide some of the same benefits of a more structured task analysis.

The execution of the task analysis can present additional challenges. A task analysis takes time both for the investigators and for the user community. There may be resistance from the end user community to the task analysis. Some users may resist the task analysis because they feel threatened by the presence of the observer and the observation of their job performance. As a result, it is important to gain the respect and support of the end users. It becomes critical to explain what is being asked of the users, and how the information gathered will be used. Putting the task analysis in the context of the specific project may help in convincing the user that the time they invest will help to create a user-friendly system.

Because we wanted to see the users performing their tasks in the most normal of environments (i.e., not in a staged scenario), we worked with them while they were carrying out their job duties. For our project, this was both a blessing and an obstacle. Due to the intensive nature of the task analysis and the network manager's job, the user's workload increased during the task analysis sessions. As part of our methodology we asked the users to think aloud while working and answer questions we posed, including why they were performing certain actions and the purposes of those actions. Meanwhile, we were busy trying to record this information for later analysis. Needless to say, this very hands-on approach was sometimes tiring for all parties involved. In addition, the center is staffed 24 hours a day, seven days a week. This staffing schedule required us to observe work across all work shifts in order to gain a complete view of the work performed.

Task analysis interviewers and observers have to be able to take advantage of challenging work situations as they present themselves. For example, unusual work or workload events may require that the observers put aside the prearranged plan and simply observe the users in action. When situations such as these arose, we asked if we could remain in the room and observe the users as they managed events happening in the network. We suspended all of our normal interviewing procedures and quietly watched the users. These unusual scenarios provided a different perspective on the work and offered invaluable insights.

Finally, there is the challenge of dealing with the mountains of data often collected during the task analysis. This activity involves analyzing, consolidating and summarizing data into meaningful information useful for the design team. We provided both short and long-term actionable items for the project team.

5. BENEFITS OF DOING A TASK ANALYSIS

While there seem to be many obstacles and challenges to completing a task analysis, the benefits that result from it are well worth the effort. In fact, we argue for the use of Human Factors Best Practices in the entire life cycle of any development project. Some examples of Human Factors Best Practices include task analysis, rapid prototyping, iterative design and usability testing. In this paper, we focus on the benefits of using task analysis methodologies.

In our experience, a task analysis of the scope described in this paper is often not done in the user interface design process. Many of the reasons they are not included are outlined above. However, in the demanding and competitive service industries, companies should strive to include task analyses during the planning and development of products and services.

The benefits of conducting a task analysis include the following:

- Information for detailed user interface design. Ultimately, the purpose of the task analysis is to aid in the design of processes and user interfaces. From the task analysis, we observed first hand the actions performed by users in carrying out job functions. As a result, we had a greater understanding of the steps involved in performing the tasks that would be supported by the user interface designs. We were able to determine ways to make the procedures more efficient and less prone to errors. Thus, the task analysis is a critical first step in designing a user-friendly interface.

- Designs that are better from the beginning. By conducting a task analysis, the first version the user interface design can be much more user-focused than completing the design without clearly defined user needs. Designs based on user needs and an understanding of the users' tasks will result in interfaces that are easier to learn and easier to use.

- A better frame of reference in thinking about design issues. From the task analysis, we had a better understanding of the work environment, which provided a useful frame of reference for thinking about design decisions. From the task analysis, designers have a greater sense of the context of the work and not just the content. During the tasks analysis, we came to understand the full scope of the users' responsibilities, level of job pressures, amount of multitasking, and proportion of down time. The users would not have been able to easily verbalize and describe in sufficient detail some of the contextual information we learned.

- A better understanding of individual differences in performing tasks. During the task analysis we had the opportunity of watching several users perform the same task. We were able to observe the variety of ways a task may be accomplished which helped us to appreciate the potentially wide range of differences among users. While we expected some differences between experts and novices, there were also differences between users with the same skill levels. This broader knowledge of user differences helped us in designing interfaces that fit the objectives of the user's tasks.

- Allows for the opportunity to ask users about what they like and don't like about their systems and processes. We obtained informal information about what end users would change to improve the systems and user interfaces. Over the course of interviewing multiple users, we began to see commonalties and themes in the user comments. We used this information in the design of new tools for the users.

- Identify system performance issues. By observing users interact with the system, we were able to see the limitations of the systems and how the limitations may affect the users in carrying out their duties. This type of information cannot be gained from talking with developers and management alone, and afforded us the opportunity to see performance issues from the user's perspective.

- Identify problems and critical user needs not within the original scope of a project. By taking a holistic approach and examining the user's entire environment in the task analysis, new projects were funded to further improve systems and procedures.

- Better user acceptance of systems. Involving the users in a task analysis allows them to participate in the creation of a solution. From this involvement, users feel more invested and are more likely to support the new or modified systems, which have been designed with their input.

REFERENCES

McCormick, E.J. (1979). Job Analysis: Methods and Applications. AMACOM: New York.

The Comovie Movie Recommender – An Interoperable Community Support Application

Michael Koch, Martin S. Lacher
Technische Universität München, Munich, Germany
[kochm,lacher]@in.tum.de

1. Introduction

Locating relevant information and distributing information to relevant people has always been a challenging task. In the context of the worldwide connection of information resources by the Internet and the increased networking in organizations it has even become more difficult since more (non relevant) information is available and more people are reachable.

Computer based community support systems can provide powerful support in direct exchange of information and in finding people to exchange information with.

In this paper we present a community support application we are currently developing, Comovie, a movie recommender tool. On the example of Comovie we highlight one common problem of todays community support applications: the inability to exchange or reuse information. User profile information and contributed information have to be entered separately for every community support application. So, what one system has learned is not available for other applications. We argue that providing interoperability can solve several problems in community support systems including the problem of 'cold start'.

2. COMMUNITY SUPPORT

In general a *community* is a group of people who share some interest or belong to the same context. The word community itself is derived from the Latin word 'commonis' which means 'in common'. So a community can be seen as a describing identity for a set of people. Mynatt et al. (1997) concretize a bit more:

> `[A community] is a social grouping which exhibit in varying degrees: shared spatial relations, social conventions, a sense of membership and boundaries, and an ongoing rhythm of social interaction.'.

Examples for communities are all students in a university department, all inhabitants in a neighborhood or all people interested in collaborative filtering.

2.1 Basic Support Concepts

The exchange of information in a community can be supported by the following concepts:

- Provide a medium for direct communication and for exchange of comments on objects within the common scope of the community.

- Discover and visualize relationships (membership in the same community, existence of common interests). This can support people in finding possible cooperation partners for direct interaction.

- Use the knowledge about relationships to perform (semi-)automatic filtering and personalization. This helps to reduce the search effort and enables to deal with the information overload.

2.2 Existing Community Support Tools

There are already several applications that are implementing aspects of these basic support concepts. News- and Chat-systems (including different kinds of community networks) provide a place to meet and a communication medium. Buddy systems like ICQ or the AOL Instant Messenger provide detailed awareness information (Michalski 1997). Online communities provide a place to communicate, awareness and a rich functionality for storing and retrieving (community) information. Recommender systems like Movie-Critic (http://www.moviecritic.com/), Knowledge Pump (Glance et al. 1998, Glance et al. 1999) or Jester (Goldberg et al. 1999) do matchmaking on the basis of user profiles and then provide recommendations based on ratings of other community members. Other systems like Referral Web (Kautz et al. 1997) and Yenta (Foner 1997, Foner 1999) focus on expert finding and explicit matchmaking.

3. COMOVIE - A MOVIE RECOMMENDER TOOL

In the Applied Informatics and Collaborative Systems group at Technische Universität München, we have been working on community support for some time now. One of the applications we are building in the project area is the *Comovie* application, a movie recommender system.

3.1 Comovie Basics

The Comovie project was started as a follow-up to the Campiello project (Grasso et al. 1998, Grasso et al. 1999) to explore new user interface ideas for recommender systems. These are applications that recommend items to users using different methods for calculating the recommendations ranging from simple heuristics to automated collaborative filtering.

(Fig. 1. Comovie Web user interface)

Automated collaborative filtering helps users in finding relevant information. The method has received a lot of interest recently since it does not rely on analyzing the content of documents. Instead of trying to understand the content, the method exploits the concept of the 'word of mouth' and tries to find people who have similar opinions and then recommends items that those similar people like (Resnick and Varian 1997, Sarwar et al. 1998). The main idea of the process is that every user rates documents that he has seen and thereby establishes an interest profile. The collaborative filtering system then compares these ratings and determines users with similar interests or tastes by applying different correlation algorithms. Finally, ratings from those similar people are used to generate recommendations for the user.

Based on automated collaborative filtering we have built a community support system for recommending movies. The system collects ratings and free text comments on movies from users and makes these available to other users again. Additionally, the ratings are used to calculate correlations among users and the correlations are used together with the ratings of correlating users to select recommendations.

For building the system we have followed a modular approach. The system consists of

- a database for storing information about the users together with their ratings (user profile database),
- a database for storing information about movies, including free text comments on movies contributed by users,
- a correlator module that reads the ratings from the user profile database and calculates correlations which it stores in the user profile again, and
- a recommender module that takes the correlations and the ratings from the user profile to generate recommendations for a specific user.

Comovie offers a Web based user interface to administer the user profiles, to administer the movie database, to enter ratings and comments, and to request recommendations (see Fig. 1).

3.2 Problems with Collaborative Filtering

Automated collaborative filtering has been used in several other recommender systems like GroupLens, Jester or MovieCritic. The usage and some theoretical work has shown some problems to deal with. These problems are of general interest since they can be found common problems of several community support systems.

First there is the problem of missing explanations. In most of the recommender systems that are based on collaborative filtering, the result of the statistics based recommendation process is just displayed, but it is not explained why the recommender comes to this result (e.g. which ratings were most significant for the correlation that was most significant for the recommendation). As a result current collaborative filtering systems are successful in leisure domains only - when it comes to critical decisions people do not trust the black box approach.

This need for a backing of results can be seen in many other community support systems. Examples emerging are supporting the display of reputation information in online communities and especially in online auctions and opinion sites (Kollock 1999).

Other problems are more related to the correlation and recommendation concepts themselves. The main problem here is 'sparsity'. The term is related to the sparsity of the correlation matrix used in calculating correlations. For a large number of documents of which people do only rate a few there can be too few overlapping recommendations to calculate correlations. Another problem that has to do with the concept itself is the 'cold start' problem. Cold start means that when a person joins the system no profile is available so no recommendations can be made (Resnick et al. 1994, Sarwar et al. 1998).

Current solutions to these problems can be divided into algorithmic approaches and ergonomic approaches (Grasso et al. 1999). Algorithmic approaches focus on providing more sophisticated computation like the introduction of stereotypes or advisors and the introduction of automatic rating agents. Ergonomic approaches focus on enlarging the active usage of the system, either by providing immediate perceived benefit or by integration with (work) practices.

Another solution would be to focus at reuse or exchange of information among community information systems. Feedback and correlations calculated by other community systems could be used by Comovie to overcome the 'cold start' problem. In general, better interoperability and reuse of information among community support applications like recommender systems and online communities could help with several of the existing problems in community support.

3.3 Cobricks – A General Community Support Architecture

In the Cobricks project we are following the approach of better reuse and exchange of information by providing an agent-based community support infrastructure for building community applications.

The basic idea is to modularize applications and to clearly separate user profiles and community information. The application component is further modularized into several agents that allow for reusability in building different applications. In the Comovie application we have for example a community agent keeping track of the community membership, an item agent for storing information about movies and comments contributed by users, a correlator agent for calculating correlations, a collaborative filtering agent for calculating recommendations and a web agent for providing a user interface. This agency communicates with the user interface agent in the user agency which is responsible for storing the user profile and making it available to different applications.

The modularization allows for easily building applications. In addition the reuse of components and the generic data structures and data exchange protocols enable the applications to exchange and reuse information:

- The application can use user profile information from other applications.

- Submitting information (to the item agent) is done in a way and a format similar to other community applications (so it is easy to submit information to more than one application).

- The application is able to exchange information directly with other applications.

All three interoperability issues are of equal importance. The reuse of user profile information allows to tackle different issues of the 'cold start' problem, the exchange of contributed information among community applications and the agent access to community applications help to reach the critical mass for successful community operation and make access to community spaces easier for users.

4. Conclusion and Ongoing Work

In this paper we briefly introduced the Comovie movie recommender system as one example for a community support application. With this application we highlighted some problems quite common to community support applications like the 'cold start' problem where a new user cannot use the recommendation or personalization features. We have proposed the idea of better interoperability among community applications with reusing user profile information and exchanging contributed information to address these problems. The Cobricks agent architecture provides a framework for building such interoperable community applications.

Our current work mainly focuses on implementing different community support platforms that are based on the Cobricks framework and on experimenting with the interoperability features. In addition to the Comovie application we are mainly working with online platforms for different real-world communities like the platform for students and alumni of the Informatics department at Technische Universität München (http://drehscheibe.in.tum.de) or the Entrepreneurship-Portal of Technische Universität München. The work with real-world examples is used to concretize the data and user profile formats and the exchange protocols.

More information on our work can be found in (Koch and Lacher 2000, Koch and Wörndl 2001) and on the Cobricks web site (http://www11.in.tum.de/proj/cobricks/).

References

Foner, L.N.: Yenta: A Multi-Agent, Referral-Based Matchmaking System. In: *First International Conference on Autonomous Agents*, Feb. 1997.

Foner, L.N.: *Political Artifacts and Personal Privacy: The Yenta Multi-Agent Distributed Matchmaking System*. PhD thesis, Massachusets Institute of Technology, Jun. 1999.

Glance, N., Arregui, D., and Dardenne, M.: Knowledge Pump: Supporting the Flow and Use of Knowledge in Networked Organizations. In: Borghoff, U., Pareschi, R. (eds.), *Information Technology for Knowledge Management*, Springer Verlag, Berlin, 1998.

Glance, N., Grasso, A., Borghoff, U.M., Snowdon, D., and Willamowski, J.: Supporting Collaborative Information Activities in Networked Communities. In: Bullinger, H.-J., and Ziegler, J. (eds), *Proceedings HCI International 99*, Vol. 2, Aug. 1999, pp. 422 - 425.

Goldberg, K., Gupta, D., Digiovanni, M., and Narita, H.: Jester 2.0: Evaluation of a New Linear Time Collaborative Filtering Algorithm. In: *Proceedings Intl. ACM SIGIR Conf. on Research and Development in Information Retrieval*, Aug. 1999.

Grasso, A., Koch, M., and Snowdon, D.: Campiello - New user interface approaches for community networks. In: *Proc. Workshop Designing Across Borders: The Community Design of Community Networks*, Seattle, WA, D. Schuler (ed.), Nov. 1998

Grasso, A., Koch, M., and Rancati, A.: Augmenting Recommender Systems by Embedding Interfaces into Practices. In: *Proc. GROUP'99 – Intl. Conf. On Supporting Group Work*, Phoenix, AZ, Nov. 1999

Kautz, H., Selman, B., and Shah, M. Referral Web: combining social networks and collaborative filtering. *Communications of the ACM*, 40(3), Mar. 1997, pp. 63 - 65.

Koch, M., and Lacher, M. S.: Integrating Community Services – A Common Infrastructure Proposal. In: *Proc. Knowledge-Based Intelligent Engineering Systems and Allied Technologies*, Aug. 2000, pp. 56 – 59.

Koch, M., and Wörndl, W.: Community Support and Identity Management. In: *Proc. Europ. Conf. on Computer-Supported Cooperative Work (ECSCW2001)*, Bonn, Germany, Sep. 2001

Kollock, P.: *The Production of Trust in Online Markets*. JAI Press, Greenwich, CT, 1999.

Michalski, J.: Buddy Lists. *Release 1.0*, (6), Jul. 1997.

Mynatt, E. D., Adler, A., Ito, M., and Oday, V.L.: Design for Network Communities. In: *Proc. ACM SIGCHI Conf. on Human Factors in Comp. Systems*, 1997.

Resnick, P., and Varian, H. R.: Recommender Systems. *Communication of the ACM* 3(40), Mar. 1997, pp. 56 – 58.

Resnick, P., Iacovou, N., Suchak, M., Bergstrom, P., and Riedl, J.: GroupLens: An Open Architecture for Collaborative Filtering of Netnews. In: Furuta, R. and Neuwirth, C. M. (eds.), *Proc. Intl. Conf. on Computer-Supported Cooperative Work*, ACM Press, Oct. 1994, pp. 175 – 186.

Sarwar, B. M., Konstan, J. A., Borchers, A., Herlocker, J., Miller, B., and Riedl, J.: Using Filtering Agents to Improve Prediction Quality in the GroupLens Research Collaborative Filtering System. In: *Proc. Conf. Computer Supported Cooperative Work*, pp. 345-354, ACM Press

USABILITY TESTING OF DATA ACCESS TOOLS

John J. Bosley and Frederick G. Conrad

BLS, 2 Massachusetts Ave., N.E., Rm. 4915, Washington, DC 20212

ABSTRACT

Thanks to the World Wide Web (WWW), the public now has access to statistical data produced by many public and private organizations. As a result, there are potential data users who have little experience working with statistical information. Non-expert users find it difficult to cope with the great quantity and variety of data that are available on line, as well as with the specialized technical terms (metadata) used to describe the data. Many producers of statistical information make on-line software tools available to support users' efforts to find the information they want. This paper reviews usability test findings for three tools that the U.S. Bureau of Labor Statistics and the U.S. Bureau of Census offer users. These three tools collectively contain a wide range of features and functions, and the usability tests employed a range of methods to cover various aspects of tool use. The paper employs a model of a generic data access task to integrate the findings and help generalize them. It discusses three types of usability concerns that were evident across the tests. First, a tool's usability depends on including sufficient guidance and instructions for using it. Second, the tool should avoid unnecessary complexity and feature "clutter." Third, usable data access tools must enable users to overcome deficiencies in the way data sets are named and give users some understanding of how statistical databases are organized. The paper provides specific instances of each problem class, and proposes some ways that data access tool designers can avoid or correct these types of usability problems.

1. INTRODUCTION

Citizens can search for and (hopefully) retrieve an enormous amount of statistical data through the World Wide Web (WWW) from many public and private organizations. Data vary widely from site to site in content, quality, timeliness, and many other attributes. Web access opens up these rich data resources to many potential users who may know very little about data, and thus are unprepared to cope with the volume and diversity of the information suddenly available. Several federal agencies in the United States have created software "tools" to help persons trying to locate and use data via the Web.

1.1 Scope of this Paper

This paper reports on usability tests that the authors and colleagues at BLS and the Census Bureau performed to evaluate some of these agencies' data access tools. The goal of the research, of course, was not simply to uncover usability problems but to inform the design of more usable versions of the tools.

We tested the usability of three tools by having a sample of users search for and retrieve statistical data as described in several scenarios.

- A Census Bureau HTML tool enables users to find and download survey data on various social, economic and health topics collected by the Census Bureau and other agencies. This tool had been released for public use prior to the usability tests we conducted. The HTML version that was tested has since been replaced by a Java implementation.
- A BLS Java data query tool that gave users access to more than 50,000 labor force statistical time series collected by BLS. "Time series" contain data sets collected periodically (monthly, quarterly, etc.) and cumulated over many periods. Different series include different measures of labor force characteristics, such as unemployment rate or unemployment level. Different sets of demographic factors such as age, gender, and race characterize different time series as well. The test used a prototype.
- A BLS Java data query tool that gives users access to current hourly wage rates (dollar amounts) for a large number of occupations, occupational classes, and skill levels within specific metropolitan areas. This too was a prototype.

The user interfaces to these three tools were quite different, reflecting the different kinds and organizations of the underlying databases. However, from the user's perspective, the task of finding and accessing data is essentially the same. A common conception of the data users' task can help developers achieve more uniform user interfaces that more closely match the users' understanding of the data search and retrieval process. Users could learn to use such data access tools more easily and transfer more of their data access skills from one tool to the next.

1.2 Conceptual Framework: A Generalized Data Access Task

In general, users gain access to data by selecting characteristics of the available data until just the relevant subset of data has been located, and then by submitting a request for these data. More specifically, users perform three subtasks. (Levi and Conrad, 1999). First, they must specify the attributes of the data of interest with the help of the tool, and send this description to the appropriate database. Ideally, the tool then transparently formulates a query from the user's specifications that the "back end" database software can interpret. The database should send the query results back to the tool, which then displays a set of descriptions (labels, names, etc.) of data elements or datasets that match the user's data specifications.

Second, the user evaluates the candidate descriptions (e.g. data series names) on the basis of how well they fit her or his expectations, and chooses the items that best match the user's concept of the data most relevant to the user's goal(s). None of the descriptions returned may appear sufficiently relevant but if one or more does, the user requests the data thus described. If he or she makes a request, the system retrieves data from the database. Third, the user examines the actual data to determine if they fulfill her goal(s). If unsatisfied with either the candidate descriptions or the actual data, the user may return to the first step to submit a different query. If the user gets no better results, he or she will eventually quit. The tests reported here included little attention to users' evaluations of actual data but future research may focus on tool usability from this standpoint.

2. METHODOLOGY

Two of the three tests covered in this paper were conducted in BLS' usability laboratory, with a browser installed on a standard desktop workstation. The prototype wage locator was tested in the field at a state convention of corporate and union compensation specialists who will be frequent users of the production version of this tool. The prototype was loaded on laptop computers for this remote test.

These tests involved small groups of from 5 to 10 users, although the Census tool was tested repeatedly and so more than 15 users were involved. Except for the wage locator test, participants were recruited from academia, government, the press, and private business. Most were somewhat familiar with using statistical information, but the test team intentionally included some users who were unfamiliar with getting data online using the WWW. Users performed scripted tasks with specific data targets so that the task could be judged on success or failure. In a few cases users were invited to use the tool to look for personally interesting data, so that success or failure in these cases depended on the participant's self-report, not an objective determination.

Although it was usually possible to quantify the users' performance in terms of successful completion of a task script, usability problems were also identified by observing test sessions and asking for user feedback after the sessions. Multiple test observers discussed their perceptions of user problems with each tool to reach consensus on the nature of the problem and its source in tool design. They then communicated a summary of observed problems to tool developers, and in some cases worked with the developers to improve the design.

3. RESULTS

Three major types of problems showed up in all three usability tests. There were numerous additional minor usability problems specific to a given tool, but this section focuses on just the three major types that were evident in all of the tests. These pervasive problem types are:
1.   Insufficient guidance and instructions for using the tool
2.   Complex or "cluttered" screen layout
3.   Deficiencies in data sets labeling and lack of information indicating how data are organized in a database
A brief sub-section is devoted to illustration and discussion of each of these types of problems and to some possible remedies for each.

3.1 Insufficient Guidance or Instructions for Tool Use

The primary interfaces of all three tools gave users too little guidance on how to use the tool, especially top-level orienting guidance. In all three cases, there was ample room on the screen to display additional instructions for use. The interfaces used common and recognizable "widgets" such as drop-down lists, radio buttons, etc, but lacked adequate guidance about the general context within which the tool is intended for use. The interfaces often left users wondering about basic questions like "What sorts of data does this tool help me find?" "How should I describe the data I'm looking for?" or "Where do I start on this interface in order to use the tool effectively?" Sometimes important instructions were provided through a "help" link, but this de-emphasizes the instructionws and makes the user follow the "help" link instead of having immediate access to the information.

Here are some ways to improve the effectiveness of guidance and instructions for using a tool.

- Provide high-level orienting information about the data that the tool works with, such as who collected the data and for what purposes. Don't make users guess or assume that they know much about the data already. Tell them the data topics; if the data are about mortality and age, say so.
- Provide adequate procedural instructions. Something as simple as numbering the necessary steps in a procedural sequence ("1," "2," "3" and so on) takes little space but can greatly increase usability.
- Show as much key guidance on the primary user interface as possible. Provide "help" in small units tied to specific contexts, not in long text segments (Nielsen, 2000). Indicate clearly how to access separate "help" facilities or pages.

3.2 Complex or "Cluttered" Screen Layout

All three tests indicated that data access tools should prominently offer users a small set of "core" features that facilitate quick access to not more than a few data elements. For example, the field usability test of the wage locator tool also showed that a key potential user group—wage analysts—mainly looked for an average hourly wage for just one occupation and one geographic region at a time. The test participants stated that this was not just true for the usability test; it was in fact the way they intended to use the tool in performing actual work. The tool should not display less useful features with equal prominence. Usability testing can also point to unwanted features that can be eliminated to further simplify the screen.

Figure 1 shows the wage locator prototype that was tested. The test results indicated some of its features reduced usability. The list box titled "Select a Level" (lower left) made it possible for users to get a wage rate for different skill levels within an occupation. This field was blank when the application opened. Where a default is lacking users must explore the feature to discover what choices it offers. For this tool, most users did not want to select a level, but rather wanted the "Overall Average Wage" rate. Making that phrase the default choice speeds up data access for many users. Generally defaults should be inclusive, for example, a default for gender data should include both men and women.

The control labeled "Don't know what Level?" actually leads to a "help" screen, and the label should indicate this more explicitly and simply, as discussed in Section 3.1. Finally, the test showed that these users did not need to "Remove" data choices--or indeed to make multiple data choices at all--before retrieving the data. Thus the bottom text box and "Remove" button appear to be unnecessary.

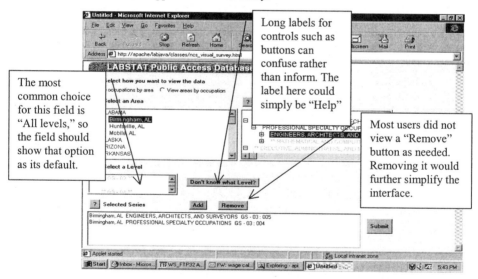

Figure 1—Interface complexity and "clutter"

In Figure 2 in the next section, the different time series labels in the large list box include a common string at the beginning of in every label. This repetition interferes with scanning a list, since all useful distinguishing information is toward the end of the label. It also contributes to making the labels so long that the user must scroll horizontally to view pertinent descriptors, even in a large text box This is a clear usability flaw.

Avoid features that depend on constructing Boolean specifications beyond simple "and-ing"(Sewell and Teitelbaum, 1988). Research shows that more complex Boolean expressions defeat many users (Greene and Devlin, 1990; Spink, Wolfram, Jansen, and Saracevic, 2001). If the tool implicitly performs logical operations, such as automatically "and-ing" individual terms chosen from different lists, users need clear feedback about the result of the operation. It is especially important that the interface shows users whether they are narrowing ("and-ing") or broadening ("or-ing") the set of data specified.

Using design guidelines such as the following can help developers avoid creating tools with this type of problem
- Provide primary access to a simple tool with a small set of features that satisfy the needs of the great majority of users prominently displayed. Give less prominence to rarely-used features, and remove features that are virtually unused
- Avoid asking users for Boolean specifications beyond very simple "and-ing."
- Show defaults in such features as text boxes or pick lists. Use defaults that many users will probably choose.
- Indicate to users that their data specifications will return no results as early as possible.

3.3 Unclear Data Labels and Lack of Cues about Organization of Databases
This problem area has two aspects that are intertwined. (1) Tool users are often expected to understand and use unfamiliar terms from the specialized vocabularies of data or computer experts. (2) Tools fail to give users any cues about how the data are organized in databases. Concerning the first aspect, BLS time series labels often include unfamiliar, specialized terms that make subtle distinctions between data series only expert users will recognize. A prime example is offered by the distinction between data series that include unemployment measures expressed as rates (proportions), and those that are based on unemployment levels (numbers of people). To the non-expert user, the most meaningful part of either data set name is "unemployment." This user may not even notice the technical distinction between "rate" and "level" and may accept and use either, when only one of the two is the appropriate choice for that user. In the time series query tool shown in Figure 2, this vital distinction is made in the top-most text field on the left of the screen. However, the tool treats these defining measures as simply another "characteristic" of series.
Some data labels make distinctions that are understandable but of little value to many users. Data sets may use unusual or irregular groupings of continuous variables like age or income that are of interest to just a few specialists For most users, these irregularities simply adds confusion and unneeded complexity to the data access task.

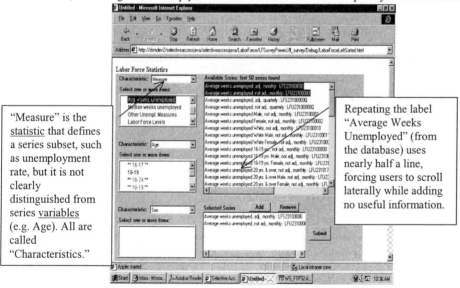

Figure 2. Example—Problems of Data Naming and Organization

Because most existing databases were constructed primarily to serve specialist users inside the producing organization, their structure is opaque to new, relatively unsophisticated users. These users need some cues about the

981

structures of databases whose content is accessible via the Web. The tools under discussion provided users with minimal information about the databases to which they linked. Lacking such cues, users can become disoriented and use the tools less efficiently. For example, instead of simply listing available data series alphabetically, the designer should cluster related series by topic and provide topic labels. This is consistent with Donald Norman's broad design principle to capitalize on "knowledge in the world," i.e. the interface, not just knowledge in the mind of the user (Norman, 1988, pp. 54-80.)

While there is a limit to how much design can overcome poor data labels and provide good clues to database structure, tool developers can take steps such as these to increase usability.

- Provide cues that enable the user to envision data structures; for example, a labeled tree structure
- Provide easy access to definitions of technical terms, through hyperlinks or mouse-overs
- Indicate clearly what data are missing from the data set. "Gray" names of missing elements, or display them in a different color
- Make it easy to (pre)view the results of data specifications, to verify that these results are appropriate and relevant

4. CONCLUSIONS

The three problem types described in this paper grow out of a flawed process for gathering requirements prior to initial tool design. These tests showed clearly that the tools' designers lacked basic information about the kinds of data access tasks many types of users wished to perform, and what features should therefore be embodied in their tool. Instead, expert data analysts employed by the two agencies had largely set the tools' design specifications. Some specifications were based on staff perceptions of the needs of usually equally expert data users outside of government. However, tool designers gathered little or no information on which to base the design of a more widely usable tool from members of the public who can now find agency data on the Web.

The diversity of web users is so great that it is probably impossible to get a complete profile of all possible data users' needs. But the research reported here indicates an urgent need for greater efforts to learn more about what features data access tools should offer to a greater variety of data users. There is every reason to do the best we can, even though we cannot do everything, in gathering more information about the diverse population of users who visit our statistical websites. At a minimum, designers need to know about what kinds of data this diverse user base is most interested in, and how familiar they are with statistical data generally (Hert and Marchionini, 1997).

Frequent usability testing as the tool design evolves can keep the design relevant to user needs. Usability evaluation of a tool when the tool is ready for deployment can increase the chance that users successfully locate and access the data they are seeking. A tool that is hard to use is more likely to lead users to look for data from alternative, possibly less authoritative—but more usable—sources.

REFERENCES

Hert, C. A. and Marchionini, G. (1997). "Seeking Statistical Information in Federal Websites: Users, Tasks, Strategies, and Design Recommendations. A Final Report to the Bureau of Labor Statistics." http://ils.unc.edu/~march/blsreport/mainbls.html

Greene, S. L. and Devlin, S. J. "No Ifs, ANDs, or Ors: A Study of Database Querying," *International Journal of Man-Machine Studies, 32*, 303-326

Levi, M.D. and Conrad, F. G. (1999). "Interacting with Statistics: Report from a Workshop at CHI 99." *SIGCHI Bulletin, 31*, 31-35.

Nielsen, J. (2000). "*Designing Web Usability: The Practice of Simplicity,*" Indianapolis, IN: New Riders Publishing.

Norman, D. (1988). "*The Design of Everyday Things,*" New York: Doubleday/Currency.

Sewell, W. and Teitelbaum, S. "Observations of End-User Online Searching Behavior Over Eleven Years," *Journal of the American Society of Information Science, 37*(4), 234-245.

Spink, A., Wolfram, D., Jansen, M.B.J., and Saracevic, T. (2001). "Searching the Web: The Public and Their Queries," *Journal of the American Society of Information Science and Technology, 52*(3), 226-234.

A Visual Development Environment for Meta-Computing Applications

Pierre BOULET   Jean-Luc DEKEYSER   Florent DEVIN   Philippe MARQUET

Laboratoire d'Informatique Fondamentale de Lille
Université de Lille 1
France

Abstract

Gaspard is a visual programming environment devoted to the development and control of scientific parallel applications.

The two paradigms of parallel programming (task and data parallelism) are mixed in Gaspard: a hierarchy of task graphs operates on array flows. These two levels are mixed in a common metaphor. An application is designed as a printed circuit: the programmer specifies tasks as boards or chips and instantiates tasks by plugging them into slots.

The number crunching applications developped using Gaspard are deployed on metacomputing platforms. The visual specification of the application mapping may be dynamically modified at runtime according to the information provided by Gaspard.

1 Motivation

Gaspard stands for Graphical Array Specification for Parallel and Distributed Computing. Gaspard is a visual programming environment specifically designed for scientific computing. We are particularly interested in time consuming applications such as numeric simulations (such as grand challenge applications). Such applications imply several constraints on the programming environment:

- The users are scientists who do not want to learn a new programming interface or language, hence, the visual environment provides intuitive manipulations.

- Usually, the applications reuse existing code and constantly evolve. Therefore, we use a component-based model where each component can be written in a different programming language and a mixed textual/visual programming environment [6]. All the interactions between the existing codes is automatically handled. This also allows the use of components written by other people as black boxes.

- Usually, numeric simulations need a large computing power. Therefore, they are well-suited to run on a meta-computing system [4] in order to exploit the power of several supercomputers when needed. The parallel activities of the application have to be expressed visually both as task parallelism and data-parallelism. The first prototype runs over CORBA [7]. Other systems are also considered, the underlying low-level system being abstracted.

- The handling of time consuming applications benefits from the overlap of the development and execution phases, because Gaspard programming environment is also an execution environment.

983

Gaspard relies on the previously defined ARRAY-OL model (Section 2). Gaspard applications are designed according to a new visual metaphor (Section 3). The application execution is triggered and monitored in the same environment (Section 4).

2 Application Model

A Gaspard application is a data flow connecting components. These components may be built from lower level (visual or textual) components. The data exchanged between components are parallel data (dynamical multidimensional array of elementary objects). The component graph introduces some task parallelism and the components can be used inside data-parallel iterators [1]. If one uses such a data-parallel iterator, the given component is executed in parallel on all the elements of an array of objects.

The fundamental idea here is that even if the base components may be parallel, the programmer has no need to learn any new parallel programming library or language to intuitively compose its components in a parallel way [5]. Everything is done at the visual level. This allows rapid development on a meta-computing environment.

To build a distributed regular application, Gaspard implements the ARRAY-OL model [3, 2]. ARRAY-OL proposes a two level approach: The first level, which is a global level, defines the task scheduling as a function of dependencies between tasks and arrays. The second level, which is local, details the elementary actions the tasks perform on array elements.

Global Model. The global model looks like the well-known dataflow model: the application is represented as a graph where the node represents the task and the edges define dependencies between tasks. Each edge carries an array. Nevertheless, in the dataflow model, the graph edges carry a continuous token flow and all the tasks are running in parallel; in the ARRAY-OL model, an edge carries an unique array (which may be of infinite size) and each task is triggered only once. A graph node (task) execution produces its output arrays from its input arrays. The task specification and the details of the array element usage are hidden at this specification level.

Array: a Data Structure for Distributed Applications. Distributed applications are organized around a regular and potentially infinite stream of objects. ARRAY-OL captures this stream in arrays with a possible infinite dimension.

Local Model. The local model details the task specification. It defines the access to the objects in the arrays and the computations to be done on those data. The whole task execution is divided into small identical computating units called *Elementary Transformations* (ETs). An ET operates on subsets of the arrays called *patterns*. The output patterns (patterns in the output arrays) are produced by applying an ET on the patterns of the input arrays. So, a task always consists of an iterator constructor; whose iterations are independent.

Fitting and Paving. The patterns are multidimensional arrays. Equidistant elements in a pattern are also equidistant in the array. A pattern may be defined by an origin in the array and a set of vectors (fitting vectors; one vector being associated to each dimension of the pattern).

Two equidistant output patterns are produced by two equidistant input patterns. The array paving with patterns is given by a first pattern in each array and a set of paving vectors.

ET Library or Hierarchical Definition. For each paving iteration, the input patterns are extracted from the input arrays and an ET is applied on these patterns to produce the output patterns. These patterns are then stored in the output arrays.

A library of predefined ETs is available on different computer architectures to process generic data parallel tasks.

A hierarchical extension of ARRAY-OL allows the programmer to define its own ET in ARRAY-OL. Input/output patterns of the first level are considered as arrays on the sub-level of the hierarchy. A new ARRAY-OL global level defines tasks that manipulate these arrays. This hierarchical construction may be applied as many times as necessary.

3 Visual Metaphor

We have chosen the well-known printed circuit metaphor to visualize the previously describe model. The textual components are represented as chips. It is described by a (possibly multithreaded) function supported by the CORBA ORB. The function has to be without side-effect to allow a SPMD execution. The visual components are boards (see the Figure on the next page).

Some links define the board's interface, they are associated to the array-flow. These links can have static or dynamic shapes. The component and its links are "welded" together and so are indissociable.

The chips and boards are plugged into slots, possibly in a recursive way. These slots indicate how data is fed into the plugged components: associations of formal parameters to effective parameters. Especially we can identify two cases: the specification of a data-parallel iterator for a regular application (through a paving/fitting constructor, the plugged component is instantiated in a SPMD model, the execution order is not specified) and the specification of a data-mapping for irregular applications (the origin and the fitting of each pattern in the array are specified). Thus each component can be specified independently of the components it references. Such components are inherently reusable. A component becomes intanciated only when it is plugged into a slot.

Everything except the textual components is visually specified.

Another feature offered to the user is that the Gaspard environment does not impose any programming methodology. Indeed one may adopt a bottom-up or top-down methodology, even a mixture of these two. The environment is strongly typed and the data types are inferred by the environment. When the user acts upon its design, if some new typing information becomes available, it is propagated through the circuit and possible type mismatches are graphically displayed. The types are displayed as colors on the wires between the components, their widths show the rank of the data they carry (the number of dimensions of the arrays).

The inputs and outputs of a component are arrays. These array-component links express the data dependencies used by the compiler to build an execution scheme on our CORBA runtime.

4 Runtime Visualization

As Gaspard is both a programming and execution environment, the user may develop his/her meta-application incrementally and change some components at runtime without having to stop the application. The programming environment is augmented with runtime information on demand when the application is running. Once again, the user has only one environment to be familiar with for both development and monitoring. This greatly facilitates maintenance.

The key concept here is dynamicity which includes three aspects:

• the first is the application's incremental development,

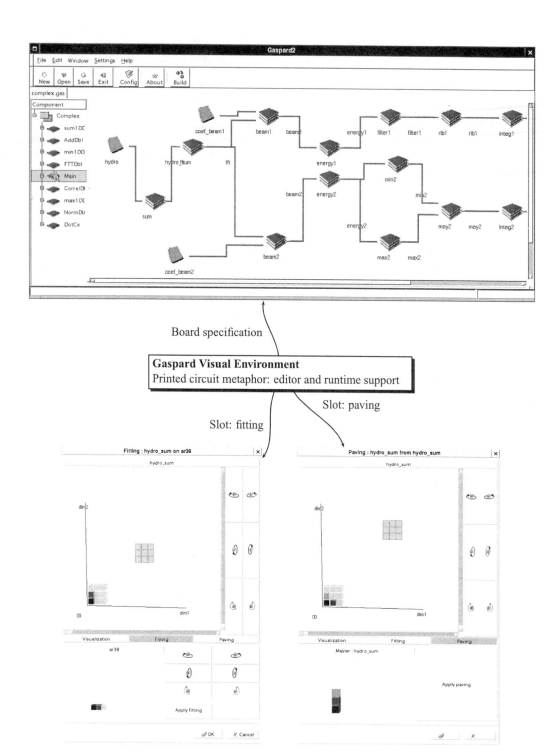

Board specification

Gaspard Visual Environment
Printed circuit metaphor: editor and runtime support

Slot: paving

Slot: fitting

- the second is the ablility to replace, during the execution, a component by another (for example to change the calculation function or to use a more performant or specialized component),

- the third is the migration of a component from one computer-node to another for various reasons such as hardware upgrading or load balancing.

Some work has already been done in building a runtime system that supports these flavors of dynamicity using CORBA [8].

Gaspard proposes a visual way to specify a virtual architecture on which the application is mapped. The mapping is simply done by dragging and dropping. Halos whose colors correspond to the virtual computer-nodes are drawn behind the components. By zooming inside compound components (cards), one may view or modify the precised mapping. A component migration is possible at any time.

The load of the links or the virtual computers can be visualized to highlight load imbalance. These visual indicators help the user in finding a better mapping for his/her application onto the underlying meta-computing system. Indeed, the computation and communication bottlenecks are clearly identified. After a first design of its application, the programmer can then tune it in the same environment without having to rely on external tools as profilers.

5 Perspectives

Visual developpement, mapping and monotoring of scientific applications are integrated in the Gaspard environment. A prototype has validated the programming model, the proposed metaphor and the user-friendliness.

As a consequence of the very time consuming nature of Gaspard applications, the next generation of Gaspard will relies on a new meta-computing platform to benefit of a fault tolerance environment. We are also investigating efficient execution of Gaspard application on SMP (symetric multicomputers) and clusters of SMP. Gaspard will also integrate some semi-automatic mapping stategies to assist the user in the interactive mapping of the application.

References

[1] Pierre Boulet, Jean-Luc Dekeyser, Jean-Luc Levaire, Philippe Marquet, Julien Soula, and Alain Demeure. Visual data-parallel programming for signal processing applications. In *9th Euromicro Workshop on Parallel and Distributed Processing, PDP 2001*, Mantova, Italy, February 2001.

[2] A. Demeure and Y. Del Gallo. An array approach for signal processing design. In *Sophia-Antipolis conference on Micro-Electronics (SAME 98)*, France, Oct. 1998.

[3] A. Demeure, A. Lafage, E. Boutillon, D. Rozzonelli, J.-C. Dufourd, and J.-L. Marro. Array-OL : Proposition d'un formalisme tableau pour le traitement de signal multi-dimensionnel. In *Gretsi*, Juan-Les-Pins, France, Sept. 1995.

[4] *The Grid: Blueprint for a New Computing Infrastructure*, Edited by Ian Foster and Carl Kesselman, Morgan Kaufmann Publishers, July 1998.

[5] Philip Cox, Hugh Glaser and Stuart Maclean. A visual development environment for parallel applications. In IEEE Symposium on Visual Languages, VL'98, Halifax, Nova Scotia, Canada, September 1998.

[6] Martin Erwig and Bernd Meyer. Heterogeneous visual languages - integrating visual and textual programming. In IEEE Symposium on Visual Languages, VL'95, Darmstadt, Germany, 1995.

[7] Common Object Request Broker Architecture. Object Management Group. http://www.corba.org/.

[8] Abdelkader Amar, Pierre Boulet and Jean-Luc Dekeyser. *Assembling Dynamic Components for Metacomputing using CORBA*, Research Report LIFL 2000-10, Laboratoire d'Informatique Fondamentale de Lille, October 2000.

MANAGING HUMAN-CENTERED DESIGN ARTIFACTS IN DISTRIBUTED DEVELOPMENT ENVIRONMENT WITH KNOWLEDGE STORAGE

Marko Nieminen
Email: Marko.Nieminen@hut.fi

Helsinki University of Technology, Department of Computer Science and Engineering, P.O. Box 9600, 02015 TKK
FINLAND

ABSTRACT

In this paper, we present the Knowledge Storage concept and first experiences in applying a prototype implementing this concept it in real settings. Knowledge Storage can be used to store and share human-centered design artifacts in distributed product development environment. It applies concepts from design rationale and provides an initial structure for the management of human-centered development artifacts in distributed development environment. The initial results from the use of the prototype point out several important pitfalls and factors that need to be considered carefully in forthcoming implementations.

1. INTRODUCTION

As human-centered issues gain increasing importance within the boundaries of technical product development, new methods, tools and mechanisms are needed in order to manage all the information that is related to this issue. Methods that address this field have been developed and introduced intensively during the last two decades (see e.g. Muller & al. 1997). Many of the emerged methods do, however, differ remarkably from the technology-oriented methods that have been used traditionally. Technical development work is now affected by methods from humanistic disciplines. As a result, not all aspects of the design can be defined with the same accuracy and determinism, as is the case with technical components. This finding makes it also important to figure out novel ways to manage information in a systematic way that relates to human-centered issues. These requirements make it difficult to construct a human-centered design artifact management system: it needs to have specific features that take into account the special characteristics of human-centered issues and at the same time it must be easily adaptable with other development practices.

Characteristic to contemporary product development work is multi-disciplinary and multi-site working. Multi-disciplinary teams that are located in different geographical locations conduct development work. More often, development groups work beyond organizational boundaries. This makes information management even more challenging. This multidisciplinary multi-site activity sets high demands for the human-centered information management as well. The results from these development activities must be transferred and communicated to all relevant development stakeholders in an understandable form.

In a distributed development environment, also the management of user-centered information suffers from the lack of interaction between project groups. Within a single development project, the findings from user studies may not end up in a product if the required development artifacts are not available (understandable or reachable) by relevant stakeholders. The situation is even more complex in a situation where development groups develop products that are separate from each other from technical standpoint but the audience (user group) is the same. Same users may use different products in the same environment. If one project group has gathered information about its users, this information could well serve other development projects. They could utilize user descriptions, task descriptions, environment descriptions, results of conducted usability tests, user interviews, and user observations that already exist in the development organization.

2. RESEARCH QUESTION

In our study, we wanted to find out, what kind of development mechanism (a concept/tool with accompanying working practices) could support the management of human-centered issues (design artifacts) in a geographically distributed product development.

The underlying reasons for this research question were extracted from our pre-study of six product development companies (Nieminen & Parkkinen 1998) supported with additional interviews from our case organization (in this study). We found out that following topics appear to be problematic for product development practitioners from the standpoint of human-centered development:

988

- *User information does not reach developers well enough.* The organization as a whole may even have information about expected users but there are no well-defined practices that could bring this information to developers' awareness in such a way that it could be included in the product under development.
- *Information is not shared between projects well enough.* Sharing of product and process information between project groups and organisational divisions does not happen as smoothly as it should even when information sharing is excellent inside a single project. Individual project groups work rather independently from each other. Experience about product information, development issues and working practices does not always accumulate in the best possible way. Separate information systems that contain critical development information may make this problem even a more difficult one.
- *Traceability.* It is not easy to go back to history and reconstruct a decision situation: Why did we select this feature? What kind of analysis was in the background of this decision? What was the rationale behind this decision? These kinds of issues are not always documented systematically and therefore the information will remain tacit.

3. BACKGROUND CONCEPTS

The problems mentioned above set the domain of topics that need to be addressed when constructing the mechanism for the management of human-centered issues. Frameworks that provide insights to the problems are knowledge management ("*User information does not reach developers well enough*"), collaborative design ("*Information is not shared between projects well enough*") and design rationale ("*Traceability*"). To an extent, design rationale captures elements from all these domains. All these topics need to be looked from the standpoint of human-centered development activities (as presented e.g. in ISO 13407 and Usability Maturity Model; Earthy 2000).

Nonaka & Takeuchi (1995) present a theory about knowledge creation and knowledge management. This theory provides the overall framework in which the mechanism should operate. According to the theory, knowledge is explicit or tacit. Knowledge gets converted between these two types. When tacit knowledge gets converted to explicit knowledge, externalisation happens. The other three modes of knowledge conversion are socialisation (from tacit knowledge to tacit knowledge), internalisation (from explicit knowledge to tacit knowledge), and combination (from explicit knowledge to explicit knowledge). According to Nonaka & Takeuchi, organisational knowledge creation happens in five phases: I) sharing tacit knowledge (socialisation), II) creating concepts (externalisation), III) justifying concepts (internalisation), IV) building an archetype, and V) cross-levelling knowledge (combination). This process is interactive and the initial information for this process comes from collaborating organisations and from users from the market. The process results in explicit knowledge in the form of advertisements, patents, products and services.

A design rationale system (see Moran & Carroll 1995) may provide support for design, maintenance, learning, and documentation (Lee 1997).

Design support helps designers track the issues and alternatives that are being explored. *Dependency management* can display issues that depend on the current issue or issues that are related to the current issue. More complex systems can actually detect conflicts among various design constraints. *Collaboration and project management* provides the participants of the project with a common and shared workspace. Within this workspace designers can interact with each other. By interacting with each other, designers can form common vocabulary and project memories. This way it is easier to negotiate and reach consensus. *Reuse support* offers designers an index to past knowledge. There may be similar design problems that have been encountered earlier.

Maintenance support is gained through the explanations that have been stored in the previous design decisions.

Learning support. In most advanced systems, both technical systems and humans can learn. A system may be able to suggest more optimal solutions when it detects a non-optimal structure. The most common way to support learning is, however, through knowledge that has been saved in the past to the knowledge base. For instance, new designers can scan through this information and learn from it. A design rationale system supports learning by presenting the designer all design decisions to that are stored in the system. Each piece of information builds up the knowledge base and provides designers with more comprehensive information.

Documentation support. A design rationale system can be used to automatically generate documentation. As documentation is something that reflects the actual events of the development project, a design rationale system gathers documentation in real-time. It has been mentioned that the documentation perspective has been the most successful in the implementation of design rationale systems.

The human-centred aspect to information management (knowledge, artefacts) can be addressed with ISO 13407 and the Usability Maturity Model (Earthy 1999) that follows the standard. In general, the standard points out that developers must be aware of information and artefacts that addresses the context of use (e.g. specification of the range of intended users, tasks and environments), the user and organisational requirements, design solution implementation (with the support of e.g. user interface style guides), and evaluation of the design (e.g. which parts of the system are to be evaluated and how they are to be evaluated; a report of major and minor non-compliances and observations and an overall assessment). Additionally, the various human-centred methods (like some of the ones listed in Muller & al 1997) applied in specific development settings set requirements for the information management structure.

4. THE KNOWLEDGE STORAGE CONCEPT

4.1 Construction of the Knowledge Storage

The development of Knowledge Storage itself contributes to human-centered design: it was designed according to the instructions given by the Contextual Design (Beyer & Holzblatt 1998) method. Contextual Design itself presents a comprehensive set of human-centered development artifacts that are good candidates to be managed with the Knowledge Storage.

The requirements for the prototype were created together with a middle-sized software development company that has several simultaneous development projects going on in different geographical locations. The company has a specific development group that is responsible of human-centered aspects in development work.

4.2 The Knowledge Storage Concept and Meta-information model

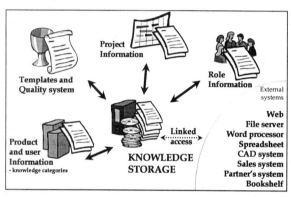

Figure 1. Knowledge Storage contains information about roles (stakeholders of a development project), project (w/phases) and the product (from different aspects via knowledge categories). It also contains various templates (e.g. quality documents, and specification templates) that can be used throughout the development project. The knowledge storage itself contains links with classifying information to various documents that have been created with external development information systems.

Meta-information model. Knowledge Storage consists of four basic information elements that constitute the meta-information model: roles, projects (with phases within the project), templates, and product information (with categories about different aspects describing the product). *Role* captures the different stakeholder groups of product development. An individual participant of a project may act in different roles during a project. *Project* implements basic characteristics of a development project and provides information about the different *phases* of the project. *Templates* provide instructions for practitioners about conducting different activities during a project. Templates may be quality system instructions or files for various office and development applications. *Product information* contains the actual information concerning the details of the product under development. Product information is divided into *knowledge categories* that can be created per project or per organization basis. Human-centered issues form one knowledge category. Knowledge categories can contain information about markets, economic or business issues, customers, users, usability, available and applicable technology, application-specific constraints, product features, and product portfolio.

Context awareness. To make the adding and later searching of information less laborious, the concept (and the prototype) detects the context of the development work. As each developer is required to enter the system with specific *context information* consisting of user's role, current project and phase of the project, all information that is appended or browsed can be stamped with this context information. Having this context information available, it is possible for example to reconstruct the information space that has been used in a specific project. This may be used in another similar project at a later time as a *reference information space*. Should human-centered information be part of this information space, it will also get communicated to the new development group.

The Knowledge Storage prototype has been implemented as a web based application (ASP) and it runs attached to a Web server (MS-IIS).

5. EVALUATION OF THE KNOWLEDGE STORAGE PROTOTYPE

5.1 Methods and Evaluation Environment

Knowledge Storage was tested in a pilot project in real development organization. The evaluation was done with multiple methods: a questionnaire about working conditions, usability testing, server statistics and formal and informal qualitative feedback. The pilot project took place in a software development organization of about 25 persons. There were 11 participants that tested the Knowledge Storage during the evaluation period.

The pilot project was conducted solely in another country and organizational unity than where the prototype was developed. At the end it turned out that the whole testing group was in a single geographical location. The development environment was evaluated, however, to be rather similar in these two locations. There are some common development systems that people from both countries share with each other but because of technical restrictions not all systems have been available for all participants.

The developers of a composite development project were assigned for the initial pilot testing for the system. With "composite project" we mean a project that aims at creating a new product based on parts from other existing or emerging products. Not all participants were directly assigned to the selected development project; there were a few persons who were working full-time for the particular project. An additional half a dozen persons participated in the project on a part-time basis. They worked primarily to other projects and were sort of "associate developers" from the standpoint of the pilot project. This created demands for virtual collaboration throughout the project. Knowledge Storage was expected to support this.

5.2 Evaluation Results

We sent a questionnaire to the possible pilot case participants and received 13 answers. The most important problems mentioned in section "Knowledge management related problems in current work environment" are listed in table 1.

Table 1. The most important problems detected by the questionnaire (N=13; scale 1 = I fully disagree and 5 = I fully agree). The total amount of problem statements in the questionnaire was 20.

Statement	avg	sd
5. There is not enough information about customers and end users	4.7	0.5
2. It is difficult to get a clear picture about what other projects and organisational units do.	4.1	0.5
7. Handling of the feedback from customers is difficult because we have no tools or process for it.	3.9	0.6
9. When a person leaves, too little of his/her knowledge remains in our organisation.	3.9	0.5
13. It is often hard to find answers to specific questions from the development information systems.	3.9	0.8

A usability test with four users was conducted in order to get information about prototype usability. Every test user was given a story to simulate a situation explaining the test tasks. The story specified following issues: the role of test user, the current project, the phase of that project, and the context of tasks. The main results from the usability test were following:

- In general, the user interface was considered easy-to-use. Some confusing concepts, however, existed
- The need for integration of Knowledge Storage and internal development systems is evident, especially with company standard document storage.
- It is crucial to have a "critical mass" of information available in order to be beneficial for the users
- The possibility to upload files to the system is important, not just providing linking paths to the shared files.

The statistics of the usage of the prototype showed that 11 persons from the pilot organization had used the system during the initial piloting. The use of the system had been most intensive at the beginning. During the one-month pilot period, the pilot group had submitted 14 documents that had been indexed with the Knowledge Storage. The documents were from three different development projects and covered various development issues like minutes of a meeting, product descriptions and test results.

We received only a few feedback messages during the beginning period of the pilot case. One of the feedback messages after testing of about three weeks revealed that the system was not fulfilling the expectations that the researchers and developers had thought about. The participating developers felt that they did not have so many documents that they would have needed such indexing and search capabilities within the pilot project.

A final interview was conducted in a teleconference with two representatives of the pilot participants. In addition to the results from the usability test, following issues emerged:

991

- The predefined knowledge categories were not considered suitable despite the fact that they were gathered from the organization's quality system and quality documentation. It was suggested that all stakeholders would have the possibility to add new knowledge categories into the system.
- In order to get realistic results, already a prototype system should deliver key features appropriately
- In the current form, the Knowledge Storage shows all new entries to all persons who log into the system (these are seen in the "Current" page). However, people would like to see only those entries that have been targeted to them. This way they would see only those entries that seem to be of relevance to them at the time. This would decrease the information overflow and ease the focusing to essential issues.
- The Knowledge Storage requires that the users of the system create all indexing information by themselves. The pilot users would have liked the support of an automated indexing system instead of just relying on the "human created indexes".

6. CONCLUSIONS

The answers in formal feedback messages as well as issues presented in the teleconference made it rather clear that the Knowledge Storage in its current form does not fulfill all the expectations and requirements that were set at the beginning of the development project. Despite this fact, the pilot case participants did not feel that the approach has come to a dead end: there are possibilities that Knowledge Storage could support development work in different settings. The initial problems to which Knowledge Storage should provide answers are still there (see table 1). Main problems in the application relate to reasonable amount of data and documents to be managed by the group and the geographical division of development work.

7. DISCUSSION AND FUTURE WORK

Is Knowledge Storage a feasible solution for managing development time information? Can it after all be used to share user related information? Possible causes for the rejection in the pilot case are:
- The group chosen for piloting used only a few documents in their work and they did not see any benefit in using the system for managing these few documents. Similarly, the pilot group was also a small one working in the same office. They did not see any benefit for sharing their knowledge using the new system compared to their current practices.
- The prototype system did not finally include the (project external) human-centered artifacts as it was designed
- The prototype implementation was presented to the pilot group as an index to different *documents*. The possibility of using of the system as a discussion platform or a knowledge gathering tool was not presented to the group well enough. Therefore, this usage did not emerge during the piloting.
- The piloting took place in a "single-project environment" without the need for support for multiple projects.

The prototype and the evaluation provided good insights about possible problems and pitfalls for the upcoming work that aims at the creation of a collaborative virtual desktop for product developers' use.

REFERENCES

Beyer, H. & Holtzblatt, K. "Contextual Design: Defining Customer-Centered Systems". Morgan Kaufmann Publishers. San Francisco. 1998.

Earthy, J. "Usability Maturity Model: Processes" TRUMP version. Version 2.2. Downloadable from http://www.lboro.ac.uk/research/husat/eusc/guides/tr_ump_c.doc

Moran, T.P. & Carroll, J.M. "Overview of Design Rationale" In: Moran, T.P. & Carroll, J.M. (eds.). Design Rationale. Concepts, Techniques, and Use. Lawrence Erlbaum Associates, Publishers. Mahwah, New Jersey. 1996. Pp.1-20.

Muller, M.J., Halswanter, J.H. & Dayton, T., "Participatory Practices in the Software Lifecycle", In Helander, M., Landauer, T.K. & Prabhu, P. (eds.) Handbook of Human-Computer Interaction. Second, completely revised edition. Elsevier science, B.V. Pp. 255-298. Amsterdam, 1997.

Nieminen, M., Parkkinen J. "Usability activities in product development" In: Vink, P., Koningsveld A.P.E., Dhondt, S. (ed.). Human Factors in Organizational Design and Management - VI. Proceedings of the sixth international symposium on human factors in organizational design and management. The Hague, The Netherlands, August 19-22,1998. Pp.433-438. Elsevier / North-Holland 1998.

Nonaka, I. & Takeuchi, H. "The Knowledge-creating Company: How Japanese Companies Create the Dynamics of Innovation". Oxford University Press. New York 1995. p. 284.

Acquiring emotion mappings through the interaction between a user and a life-like agent

YAMADA Seiji
CISS, IGSSE
Tokyo Institute of Technology
yamada@ymd.dis.titech.ac.jp

YAMAGUCHI Tomohiro
Department of Information Science
Nara National College of Technology
yamaguch@info.nara-k.ac.jp

Abstract

This paper describes a human-agent interaction framework in which a user and a life-like agent mutually acquire their emotion mappings through a mutual mind reading game. In these several years, a lot of studies have been done on a life-like agent like a Micro Soft agent, an interface agent, and so on. Through the development of various life-like agents, emotion has been recognized to play an important role in making them believable to a user. For making effective and natural communication between a life-like agent and a human user, they need to identify the other's emotion state from expressions and we call mappings from expressions to emotions *emotion mappings*. If an agent and a user don't obtain these emotion mappings, they can not utilize behaviors which significantly depend on the other's emotion states. We try to formalize the emotion mapping and a human-agent interaction framework in which a user and a life-like agent mutually acquire emotion mappings each other. In our framework, a user plays a mutual mind reading game with a life-like agent and they gradually learn emotion mappings each other through the game.

1 Introduction

In these several years, a lot of studies have been done on a life-like agent like a Micro Soft agent[1], an interface agent[4][5], and so on. A typical life-like agent appears on a Web shopping page and support a user in inputting his/her order and data into the page by speech recognition and synthesis. Through the development of various life-like agents, emotion has been recognized to play a very important role in making them believable to a user[2]. Emotional expressions are also effective to computer-mediated communication between humans[8]. These facts are supported by many psychologists' reports that the emotion significantly influences even rational behaviors as well as instinctive behaviors of a human[7]. Thus some researchers are trying to implement an emotion model on a life-like agent for developing a believable one to a user[2][11]. However there is a significant problem that emotion identification is difficult to both of an user and an agent.

For making the effective communication between a life-like agent and a human user, they need to be able to identify the other's emotion state through the other's expression and we call this task *emotion identification*. If the emotion identification is impossible, they are not able to act human-like behaviors which significantly depend on the other's emotion states. For example, we consider a life-like agent should kindly and carefully behave to a depressed user, and intuitively communicate its computational state to a user through a facial expression. Though emotion identification is always done among human, the emotion identification between a life-like agent and a user becomes far more difficult than between human. Because design of life-like agent's expressions significantly depends on personal preference, culture, and not all the users can understand the emotion by seeing an expression. Consequently a life-like agent and a user need to acquire relation between an expression and an emotion state when they actually encounter. We call such relation *emotion mapping*.

In this paper, we propose a human-agent interaction framework in which a user and a life-like agent mutually acquire emotion mappings each other. In our framework, a user plays a mutual mind reading game with a life-like agent and they gradually learn emotion mappings each other through the game.

First, for describing emotional interactions between a life-like agent and a human, we define an emotion state, primitive emotions, expressions, emotion transition rules and emotion mapping. Next we develop procedures of an agent to learn a user's emotion mappings. Finally, to acquire emotion mappings each other, we develop a mutual mind reading game in which a user and a life-like agent try to recognize the other's emotion state through the other's (facial) expression. They mutually answer their results on the other's emotion, and one having a correct answer gets a score. This game is designed so that a user may enjoy it, and as results, the user's cognitive load is reduced.

Velásquez proposed a emotion model which is based on Minsky's society of mind[6]. His model[11][10] is for generating human-like emotions using a multi-agent system architecture in which each agent corresponds to a primitive emotion like fear, joy and so on. Interactions among agents are defined and emotions are emerged as a result of the interactions. However the purpose of his research is to generate various emotions and moods like a human, and a framework for interaction between an agent and a user like an emotion mapping.

2   Emotion mapping

In this section , we define our framework to deal with emotional interactions between a agent and a human user. First the following primitives are introduced for describing the framework.

- *Emotion state* $S_1, \cdots, S_l$: A variable standing for a state of an agent i's emotion. The next primitive emotion is substituted for this variable.

- *Primitive emotion* $E_i = \{e_1^i, \cdots, e_m^i\}$: E_i is a set of elements of agent i's emotion. We can utilize typical elements like fear, joy and so on.

- *Primitive expression* $X_i = \{x_1^i, \cdots, x_n^i\}$: X_i is a set of primitive expressions. We deal with a single primitive expression, not a combination of plural primitive expressions.

- *Emotion mapping* $M_{j:x \to e}^i = \{x_1 \to e_s, \cdots, x_n \to e_t\}$: This means an agent(or a user) j's one-to-one mapping from a primitive expression to a primitive emotion which was learned by an agent(or a user) i. For simplicity, we assume that a single primitive expression corresponds to a single primitive emotion.

- *Expression mapping* $M_{i:e \to x}^i = \{e_1 rightarrow x_s, \cdots, e_m \to x_t\}$: An agent(or a user) i's one-to-one mapping from a primitive emotion to a primitive expression.

- *Emotion transition rule* $T^a = \{e_t^a \times \bar{e}_t^b \to e_{t+1}^a\}$: An agent(or a user) a's rule determining a transition of an emotion state from a primitive emotion e_t^a to a primitive emotion e_{t+1}^a when the a estimates the other agent(or a user) b's emotion e_t^b from its expression. The $\bar{e}$ is an estimated primitive emotion of e.

Using the above notations, we describe a framework in which a life-like agent a and a user h mutually interact through expressions as shown in Fig.1.

3   Mutual learning of emotion mappings

3.1   What should be learned?

With the framework described in Fig.1, we can define learning of emotion mappings and mutual learning of emotion mappings in the following.

Learning of emotion mappings: An agent(or a human user) acquires emotion mappings $M_{h:x \to e}^a$ (or $M_{a:x \to e}^h$) of a partner.

Mutual learning of emotion mappings: An agent and a human user mutually acquire the other's emotion mappings, $M_{h:x \to e}^a$ and $M_{a:x \to e}^h$.

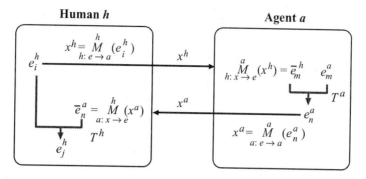

Figure 1 A framework for emotional interactions between a agent and a user.

Since a partner of a user in our framework is a life-like agent which simulates human behaviors, we can assume the following assumptions to restrict our framework for simplifying mutual learning of emotional mappings.

(A1) *Common primitive emotions*: Primitive emotions of a user and an agent are identical, thus $E = E^a = E^h$. Also

(A2) *Partially common emotion transition rules*: Emotion transition rules of a user and an agent are *partially* identical. It is just partially because the emotion transition rules are not so stable depending on situations and contexts.

3.2 Learning in an agent

Because a human user is able to autonomously learn agent's emotion mappings in our framework, we give no restriction to a user within his/her learning. Thus we develop only the learning procedures of an agent.

From an assumption (A1), an agent does not need to acquire user's primitive emotions. Furthermore the user's expressions are observed by an agent with a computer vision system using a CCD camera. Hence if a user's primitive emotion e^h is estimated by a sort of reasoning when a user's expression x^h is observed, an agent acquires an instance of an emotion mapping, $x^h \rightarrow e^h$. After an agent stores sufficient such instances through interactions with a user, it becomes able to estimate a user's primitive emotion from his/her observed primitive expression by a NN(nearest neighbor) method[3].

Thus we need to estimate a user's primitive emotion without his/her observed primitive expression. Under an assumption (A2), we utilize some rules to reason the primitive emotion like the following.

- IF much failure continues THEN a user gets distressed.

- IF a user successively wins THEN he/she feels joy.

4 Mutual mind reading game

This section describes a mutual mind reading game in which a user and a life-like agent try to recognize the partner's emotion state through the other's (facial) expression as the learning task of emotion mappings.

4.1 The Objective of the Mutual Mind Reading Game

The objective of the mutual mind reading game is to collect a set of instances for a NN method both efficiently and broadly for the learning of emotion mapping. An instance is a pair of a estimated primitive emotion and a observed facial expression. In this paper, a game in which a player estimates the partner's emotion state through the facial expression to compete for the accuracy is called a *mind reading game*. "Mutual" means that both the game in which a life-like agent estimates the user's emotion state through his/her facial expression and the game in which the user estimates the agent's emotion state through its facial expression are performed in parallel. The data sampling process for an identification of the user modeling is called enrollment. The main problem of an enrollment is that the user's cognitive load becomes high. To solve this, this game is designed so that a user may enjoy it to play a part in collecting training data actively, and as results, the user's cognitive load becomes low. In general, interactions among multi-agent are classified into two types, competitive or cooperative[9]. A competitive mind reading game is to compete with each other for the accuracy in reading the emotion of an opponent between two players in the game. On the other hand, a cooperative mind reading game is to compete with other pairs for the time to finish reading the emotion of a partner accurately between two players in the game. Note that in both games, an effort to win the game must not give bad influence for sampling the training examples of natural pairs of emotion and expression.

4.2 The Overview of a Cooperative Mind Reading Game

We apply a cooperative mind reading game in which a user and an agent play cooperatively to the learning of emotion mapping.

Assumptions of the game

The primary objective of the game is the emotion identification between a life-like agent and a user. Therefore, both a life-like agent and a user play the game with fixed emotion mapping each other. Besides, users play the game with not to be stretched but to be natural facial expressions as if they behave usual expressions in their daily life.

The rules of the game

1. When a player is able to guess the partner's emotion state through the partner's facial expression, the player says the supposed primitive emotion to the partner.

2. When a player is said a primitive emotion, the player must reply whether "Yes" (the guess is right), "No" (the guess is wrong) or "Near-Miss" (the guess is wrong, but it is close to the answer) as the judgment against the partner's guess.

3. These processes are played until a finish condition of the game is satisfied.

A finish condition of the game

When each two players guess the partner's different primitive emotions for n times of continuation, the game is finished. The frequency of the correct guess of n can be changed according to a degree of difficulty of the game. Note that finally, it is equal to the number of primitive emotions treated in the emotion model.

4.3 Discussions

First, we discuss the assumption of our mind reading game. There are two reasons why fixed emotion mapping is assumed during the game. First reason is that fixed emotion mapping which is equal to that of daily use generates natural facial expressions. Second reason is that fixed emotion mapping makes the learning of opponent mapping easier. Next, we argue about the intention that we used a game for emotion mapping identification. It is not easy for a human to hold various natural emotions in a short time while collecting training examples of facial expressions. For this, the learning of emotion mapping based on the mutual mind reading game is continued until the end condition of the game is satisfied. During the game, various kinds of user's emotions that arise from the result of the guess cause his/her natural facial expression. Finally, we describe convergence conditions of the mutual learning of emotion mapping. The necessary conditions of the mutual learning are the following;

1. Each learning of the opponent emotion mapping is converged.

2. $M_{a:x \to e}^{h}$ which a user learned becomes equal to an inverse mapping of $M_{a:e \to x}^{a}$ of an agent

3. $M_{h:x \to e}^{a}$ which the agent learned becomes equal to an inverse mapping of $M_{h:e \to x}^{h}$ of the user.

$$\underset{a:x \to e}{\overset{h}{M}} = \left(\underset{a:e \to x}{\overset{a}{M}} \right)^{-1} \qquad \underset{h:x \to e}{\overset{a}{M}} = \left(\underset{h:e \to x}{\overset{h}{M}} \right)^{-1}$$

5 Conclusion

This paper describes a human-agent interaction framework in which a user and a life-like agent mutually acquire their emotion mappings through a mutual mind reading game. For describing emotional interactions between a life-like agent and a human user, we defined emotion states, primitive emotions, expressions, emotion transition rules and emotion/expression mappings. Then, to acquire the emotion mapping each other, we developed a mutual mind reading game in which a user and a life-like agent try to recognize the other's emotion state through the other's expression.

Unfortunately the descriptions are conceptual. Thus our system should be fully implemented and we need to verify the feasibility of our framework through various experiments with subjects. We are currently developing a whole framework in which a life-like agent and a human user naturally interact each other.

References

[1] MS agent web page. *http://msdn.microsoft.com/msagent/*.

[2] J. Bates. The role of emotion in believable agents. *Communications of the ACM*, 37(7):122–125, 1994.

[3] B. V. Dasarathy. *Nearest Neighbor (NN) Norms: NN Pattern Classification Techniques*. IEEE Computer Society Press, 1991.

[4] P. Maes. Agents that reduce work and information overload. *Communications of the ACM*, 37(7):30–40, July 1994.

[5] P. Maes and A. Wexelblat. Interface agents. In *Proceedings of the CHI '96 conference companion on Human factors in computing systems: common ground*, page 369, 1996.

[6] M. Minsky. *The Society of Mind*. Simon & Schuster, 1986.

[7] R. W. Picard. *Affective computing*. The MIT Press, 1997.

[8] K. Rivera, N. J. Cooke, and J. A. Bauhs. The effects of emotional icons on remote communication. In *CHI'96 companion*, pages 99–100, 1996.

[9] S. Sen and G. Weiss. Learning in multiagent systems. In G. Weiss, editor, *Multiagent Systems - A Modern Approach to Distributed Artificial Intelligence*, chapter 6, pages 259–298. MIT Press, 1999.

[10] J. D. Velásquez. Modeling emotions and other motivations in synthetic agents. In *Proceedings of the Fourteenth National Conference on Artificial Intelligence*, pages 10–15, 1997.

[11] J. D. Velásquez and P. Maes. Cathexis: A computational model of emotions. In *Proceedings of the First International Conference on Autonomous Agent*, pages 518–519, 1997.

Quantitative evaluation of effect of embodied conversational agents on user decisions.

Kazuhiko Shinozawa, Junji Yamato, Futoshi Naya, and Kiyoshi Kogure

NTT Communication Science Laboratories
2-4 Hikari-dai, Seika-cho
Kyoto, JAPAN 619-0237

{shino, junji, naya, kogure}@cslab.kecl.ntt.co.jp

Abstract

We experimentally investigated the effects of an embodied conversational agent on users' decision-making by using a simple color-name selection task. The experimentation was based on roughly two different situations. First, the subjects selected the name of a color using only a displayed color patch. Second, the other subjects selected it using a displayed color square while hearing an embodied conversational agent's opinion. The agent offered its personal opinion about the color rather than having definite knowledge about the displayed color. In addition, we prepared two types of agent opinions. The results on three conditions indicated that an embodied agent can affect a user's decision and its power of influence changes according to the type of agent recommendation strategy.

1 Introduction

The recent rapid growth in the power of computers has enabled us to develop embodied conversational agents that can interact with their users in a natural and friendly manner by speech recognition, voice synthesis, and the display of actions.

However, even if each communication channel technique were to evolve individually, there is no guarantee that a computer or agent would become a communication partners of a user. What is needed to achieve this partnership? There is no doubt that humans have human partners but they also have other partners as exemplified by some people stating that they always communicate with their dogs, and some people even talking to their dolls. Although an important point of communications is understanding what one's partner wants to say, it is realistically impossible to expect this from a non-human partner. In such case, we can only guess what the partner wants to say. From this point of view, the doll can be a partner.

To confirm that a computer or agent can be a communication partner, investigations are necessary on whether a person can interact with the computer or agent in the same way as the person interacts with other communication partners because the definition of "communication partner" depends on the person who wants to communicate. A represented partner is another human, so we must investigate whether the attitude of a person towards a computer or agent can be made the same as his/her attitude towards another person.

Reeves and Nass gave many examples of people tending to take a humanlike attitude towards media in their book [4]. People can therefore find personality in a computer according to its use of voice and displayed images and can feel a different veracity for the information from the computer than the same information as text. In addition, the ethnicity and personality of an embodied conversational agent were investigated [2].

Takeuchi focused on politeness and showed that people can take a stronger humanlike attitude towards an embodied conversational agent, for example, MS Agent, than the computer only [5].

These researches mainly investigated attitudes towards the computer or agent and little attention was paid to quantitatively measuring and evaluating how an agent's behavior can actually affect its user's decisions. Such an evaluation is important for developing a guidelines to make agent behaviors have more influence on the user's decisions in interactive situations. In this paper, we focused on the effect on user decision-making through recommendations made by the agent.

2 Method

We designed an experimental task to quantitatively measure the effects of an agent on users' decisions. In this task, the subjects saw a color square on a computer display, and were then asked to choose the name of the

Figure 1: Experimental set-up for agent-user interaction

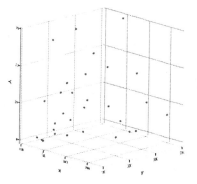

Figure 2: Color Squares

color from two candidates. Most of the color squares and candidates for color names in the experiment are unfamiliar to ordinary people. Each question had no obvious right answer and most of the subjects had no prior reference. We hypothesized that the choices of the subjects might be influenced by the recommendations of an agent.

The experimentation was based on two different situations. First, the subjects selected the name of a color using only the displayed color square (no agent situation). Second, the other subjects selected it using the displayed color square aided by an embodied conversational agent providing a recommendation on one of the two color names. The screen agent offered its personal opinion rather than having definite knowledge about the displayed color. This was to avoid the effect of intelligence because our focus was on the effect of the existence of the agent.

2.1 System

The embodied conversational agent was developed at NTT East R&D Center based on MS Agent. It uses the Japanese speech synthesizer "Fluet" developed by NTT Cyber Space Laboratories [3]. We constructed the experimental system with j-script.

2.1.1 color square

The color square was displayed on a CRT display. All color values were measured by a CRT color analyzer three times a day. The changes in these values were small (less than 10%) for the whole experimentation, and so all subjects saw the same color squares. Figure 2 shows the colors used in this experimentation.

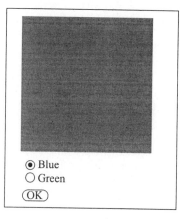

Figure 3: Color Selection

2.1.2 selection method for color name

The subjects selected a color name on a radio button and pushed the OK button by a mouse device.

2.2 Two different recommendations

2.3 Pretest

We carried out two small pretests for the color experiment without an agent before the above experiments and investigated the ratio of color names selected. Each color name was selected from between two choices on a computer display. The selection ratio depended on the order of choices, which is why two pretests were held with the order changed.

Using the obtained results, we prepared two different types of agent speech to investigate which type of speech was more effective in changing the users' decisions.

999

2.4 Experiments

We divided thirty questions into three sections, each section consisting of ten questions. The first and second sections consisted of questions for which the subjects mostly answered one color name. The last section consisted of questions for which they answered the color names almost evenly. In the "Up" type, the speech gradually changed from recommending the less selected color names to the more selected color names. In the "Down" type, it gradually changed in the opposite direction. In the first and second sections, the "Up" type agent and "Down" type agent recommended the opposite color names. In the last section, both agents recommended the same color names.

The subjects were 90 middle-class adults. They ranged in age from 18 to 49. There were 41 men and 49 women. Each experiment had 30 subjects. They were not experts on color names and just ordinary people.

We conducted the following three types of experiments as described above.

Group A. Without an agent.
Group B. With an "Up" type agent.
Group C. With a "Down" type agent.

The subjects received the instruction in these experiments as a color selection task that there was no correct answer in the task and were told that they had to guess the color name. All of them were not told about the agent or given same explanation about the experiments. For groups B and C, the agent appeared and each subject met the agent for first time after hearing the explanation and pushing the START button. This is because we did not want to give any role or authority to the agent beforehand. If the agent were introduced as a communication partner, the subject would find it hard to ignore the agent and the agent would possibly affect the user's decision.

The agent introduced itself and the experiment started. The agent gave recommendations, for example, "This color looks like green to me. What do you think?". We prepared 13 phrases for one color recommendation and had the phrases spoken through a synthesized voice until the subject pushed the OK button; this was to avoid losing the interest of the subject. The agent also displayed behaviors as animations which were presented before the phrases were spoken. If the last phrase was spoken, the order returned to first phrase. However, there were few persons who heard the last phrase.

After the last question, the agent said good bye and disappeared. The subject filled out a questionnaire after the agent disappeared. The items of the questionnaire are listed below.

1. The computer display was easy or difficult to see.

2. The text on the display was easy or difficult to read.

3. The brightness of the display was bright or dark.

4. The computer response was fast or slow.

5. The experiment room was hot or cold.

6. The experiment room was noisy or silent.

7. You were sure of your answers or not.

8. There were many familiar colors or not.

9. There were many familiar names or not.

10. The number of questions was too many or few.

11. The explanation of this experiment was easy or difficult to understand.

12. The time that you spent for this experiment was too long or too short.

Concerning the agent,

13. The agent's phrases were easy or difficult to understand.

14. The agent's motion was natural or not.

15. You could feel familiarity with the agent or not.

16. The agent's voice was easy or difficult to listen to.

17. The agent's voice was too loud or silent.

18. The speed of the agent's utterance was too fast or too slow.

The items for the Group A subjects were items 1 to 12 and for the Group B and Group C subjects they were items 1 to 18. The subjects evaluated these items using five ranks.

3 Result

We estimated the ratios of color names that the agent successfully recommended for each subject and the average ratio in the groups, and then we investigated whether these differences were significant.

There was no agent and no recommendation in Group A, so we focused on the color name recommended in Group B and estimated the ratios of selecting the color name recommended in Group B for Groups A and B, and then we compared the ratios for Groups A and B. The same was done for Group C and A.

The results are shown in Tables 1 and 2. The tables show average selection ratios and p (i.e. p value)

Range	Group A	Group B	p
total(1-30)	0.48	0.54	0.028

Table 1: Comparison Group A with Group B.

Range	Group A	Group C	p
total(1-30)	0.51	0.56	0.035

Table 2: Comparison Group A with Group C.

First, we compared the average ratio of selected color names for all questions between Groups A and B with a t-test. Table 1 shows the average selection ratios in Groups A and B. The difference was significant ($p <$ 0.04). The average in Group B was greater than in Group A. The difference between Groups A and C was also significant ($p < 0.04$) as shown in Table 2. The average in Group C was greater than in Group A.

These results show that the existence of the agent affected users' decisions.

Second, we compared the result among the three groups. However, we could not do this easily because each agent in Groups B and C primarily recommended different color names. However, for questions 21 to 30, the same color names were recommended in Groups B and C, so we could compare the results among the three groups in this range.

The difference between the three groups in this range was significant ($F = 7.820, p < 0.02$) by the analysis of variance (ANOVA). Table 4 shows that the difference between Group A and Group C and between Group B and C were significant from Scheffé's method. These results show that the "Down" type recommendation was more effective than both Groups A and B.

Last, we confirmed equal conditions for all subjects by the questionnaire's results. Each item was assigned a value from 1 to 5 according to the answers of the questionnaire. Figures 4, 5 and 6 show the average values of each item. There are no significant differences between groups by ANOVA.

4 Discussion

Based on these results, we concluded that the agent affected the subjects' decisions and the "Down" type recommendations affected the subjects' decisions more

Range	Group A	Group B	Group C
21-30	0.49	0.49	0.62

Table 3: Comparison between Group A, B, and C.

	p
Group A and B	0.996
Group B and C	0.005
Group A and C	0.001

Table 4: p value between Group A, B and C.

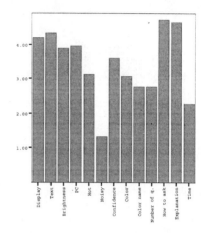

Figure 4: Questionnaire result in Group A

strongly than the "Up" type recommendations.

The results of these experiments provide evidence that an agent can influence a user's behavior. However, the influence on the user's decisions depends on the type of recommendations, that is, "Up" or "Down". This shows that an agent has the power to change users' decisions and only a good design can allow the agent to display its ability in full.

These results remind us of the foot-in-the-door technique and door-in-the-face technique [1]. The foot-in-the-door technique is where a small request that is easy to accept is made first and then the requests slowly change to larger ones. On the contrary, the door-in-the-face technique is where a large request that is difficult to accept is made first and then the requests change to smaller ones gradually.

The "Up" type recommendation corresponds to the foot-in-the-door technique. Each recommendation in the "Up" type changes from a color name that many people do not select to one that many people do select. The "Down" type recommendation corresponds to the door-in-the-face technique because of the same reason. People accept requirements from people that they do not know more easily when they are asked small requirements as opposed to big one. The difference in the results between "Up" and "Down" type recommendations might be able to explain this.

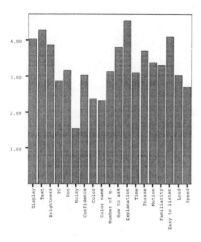

Figure 5: Questionnaire result in Group B

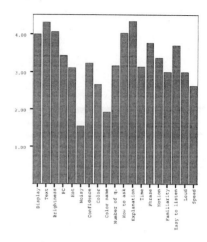

Figure 6: Questionnaire result in Group C

We did not find a significant difference in the subjective familiarity to the agent between Groups B and C from the questionnaire. This suggests that there are few factors from which a user's judgment is changed, and few types of familiarity relationships. It also says that the agent recommendation strategy can be a subtle method for obtaining user confidence and influencing user decisions and behaviors.

The effect of the agent depends on the task. This experimental task was not a serious one, so the subjects could accept the agent's recommendations easily and the agent could have much influence. A more serious task, for example, is the desert survival problem, which is used to investigate human and computer interaction. We are going to conduct the experiment with this task.

Recently, attention has been shifting towards the robot as a communication partner. Both screen agents and robots can be conversational agents. Both of them will be able to synthesize the voice, display facial and bodily expressions, and recognize human speech roughly in the future. The differences between robots and screen agents are that robots exists physically, have three-dimensional volumes, can pay shared attention to physical items with users, and can perform eye contact.

In future work, we will try to investigate the difference between the effect of a screen agent and a real robot from an interactive point of view.

Acknowledgements

The authors thank Syun-ichi Yonemura and Mio Hosoya of the Cyber System Group of NTT-East R&D Center for providing the agent system and the Speech Synthesis Group of NTT Cyber Space Laboratories for providing and fine tuning the speech synthesis system Fluet.

References

[1] Robert A. Baron and Donn Byrne. *Social Psychology*. Allyn and Bacon, 1996.

[2] Clifford Nass, Katherine Isbister, and Eun-Ju Lee. *Truth is Beauty*, chapter 13. MIT Press, 2000.

[3] Mizuno Osamu and Shinya Nakajima. Synthetic speech/sound control language: Mscl. In *3rd ESCA/COCOSDA Proceedings of International Workshop on Speech Synthesis*, pages 21–26, 1998.

[4] Byron Reeves and Clifford Nass. *The Media Equation*. CAMBRIDGE UNIVERSITY PRESS, 1996.

[5] Yugo Takeuchi, Yasuhiro Katagiri, Clifford Nass, and B.J. Fogg. Social response and cultural dependency in human-computer interaction. In *Proceedings of PRICAI*, pages 114–123, 1998.

A DISTANCE LEARNING CASE-STUDY IN TURKEY:
An Example From Istanbul Technical University

Mehmet Mutlu Yenisey

Istanbul Technical University
Dept. of Industrial Engineering
80680 Macka-Istanbul TURKEY
Fax: +90 212 240 72 60
E-mail: yenisey@itu.edu.tr

ABSTRACT

Distance learning, or in other words, distance education had a large amount of improvement in the last century. Four main stages can be described in this development. The first generation started at the end of 19th and the beginning of 20th century. The second generation began at 1970s. In early 1980s, the third generation of distance learning started as a result of new communication technologies like satellite and computer network facilities. Finally, fourth generation started with Internet.

At Istanbul Technical University, studies for distance learning have begun in the mid-1990s. Today, distance learning is being used among three different campuses for undergraduate educational purposes. The distance among these campuses varies from 1 km to 16 kms. Moreover, a satellite-earth communication base has been installed for the distance learning application and there is an exertion for undergraduate students to register and to pursue the courses from University of Virginia. Additionally, Center for Education in Virtual Environment is founded for the coordination and managerial purposes. However, some problems have arisen during its application. These problems are mainly because of the infrastructural deficiencies. Besides, there are some complaints coming from students and, to some extent, from lecturers. These complaints mainly arise from ergonomic conditions in the classroom and deficiencies in the communication system. Additionally, some cultural characteristics are also effective in not being able to achieve the goals desired. In this paper, this distance learning system is discussed and solutions to the problems are proposed after some conclusions are given.

Keywords: Distance learning, Distance education, Internet, World Wide Web, On-campus distance learning, Undergraduate education, Synchronized basis, Asynchronous mode

1. INTRODUCTION

The idea of distance education has existed for many years. With advances in telecommunications and computer technology, new modalities have arisen to enhance the concept of offering an education to anyone, anyplace, at any time (Scollin, P.A., Tello, S.F., 1999). Distance learning, as an important education system, has been in development for more than a century. There are four main stages described for the development. First generation started at the end of 19th and the beginning of 20th century. At that stage, people encountered "correspondence learning". In correspondence learning, some printed materials like textbooks and exercises were being delivered to the students and the assignments and the solutions to the exercises were being mailed back to the instructor.

The second generation began in the 1970s. The main improvement occurred in the delivering medium of course materials. Radio, television, pre-recorded audiotapes and sometimes telephone facilities were used for it. Sometimes audio-conferencing was another part of distance learning. This allowed some interaction between instructor and the students. Hence, the instructor became a part of education environment again.

In early 1980s, the third generation of distance learning started as a result of new communication technologies like satellite and computer network facilities that have emerged. Real-time interaction between instructor and students was obtained by using one of these new technologies. Multimedia and CD-ROM products were also introduced. Instructors and students began to exchange the course materials among themselves on the digital media. These technologies not only allowed learners to interact to instructors, but they also redefined our understanding of the meaning of "distance". Hence, students could be on-campus or off-campus while there was a real-time, two-way interaction.

Finally, the fourth generation has begun with Internet. Especially, new developments on software and hardware technologies allowed the new approaches and produced new paradigm for not only distance learning, but also for entire education systems. The World Wide Web offers educators a new medium to deliver teaching and learning materials. These materials can bring new and exciting ways of learning, and an alternative to traditional teaching techniques. These new techniques can provide solutions to the demands of a changing environment (Allen, 1998).

Information technology can significantly reduce the transactional distance, giving learners access to all the information they require and by allowing them to communicate rapidly, through a variety of media, with teachers and fellow learners. In addition, it reinforces the existing advantages of distance learning by allowing learners to study in a very flexible manner, in the appropriate place, at a speed that suits them, at a lower opportunity cost, and it satisfies the requirements of many adult learners who need further education and training, but who do not wish to have their professional or family lives disrupted by full-time study (Boyle, 1995).

2. CASE OF ISTANBUL TECHNICAL UNIVERSITY

In this section, Istanbul Technical University (ITU) is introduced and Distance Learning activities at ITU are explained.

2.1. About ITU

Istanbul Technical University is one of the major universities in Turkey. ITU was established in 1773, during the time of the Ottoman Sultan Mustafa III as The Royal School of Naval Engineering and its main responsibility was to educate chart masters and ship builders. In 1795, The Royal School of Military Engineering was established to educate the technical staff in the army. In 1847, education in the field of architecture was also introduced. In 1883, it became "Engineering Academy" and the School of Civil Engineering was established to teach essential skills needed in planning and implementing the country's new infrastructure projects. In 1928, the Engineering Academy gained university status and continued to provide education in the fields of engineering and architecture until it was incorporated into ITU in 1944. Finally, in 1946, ITU became an autonomous university, which included the Faculties of Architecture, Civil Engineering, Mechanical Engineering, and Electrical and Electronic Engineering.

Currently, it has (approximately) 16000 undergraduate students, 5000 graduate students, 800 professors, 100 instructors and 1000 research assistants, 11 faculties, 5 institutes and one conservatory. ITU consists of 5 campuses. The distances among these campuses vary between 1 km and 50 kms. ITU is a state university that defines and continues to update methods of engineering and architecture in Turkey. (Official Web Site of ITU, 2001)

2.2. Distance Learning Studies at ITU

Distance Learning studies at ITU can be analyzed within two categories. On-campus Distance Learning activities are those performed among the ITU Campuses. International Distance Learning contains the activities that are performed with other Turkish universities and universities that are abroad. To coordinate the activities, Center for Education in Virtual Environment is established.

2.2.1. On-Campus Distance Learning

Distance learning studies have begun in mid-1990s at ITU. Initially, a distance learning system was installed between two campuses; Macka and Ayazaga -which are about 15 kms away-, in 1997. This system consisted of analogous devices. There is two-way communication. Two distance learning classrooms were built in Ayazaga and

Macka Campuses. It has been used since fall 1998 effectively. Currently, physics and chemistry courses are performed in distance learning medium. It cost about $100,000.- This project was realized by the support of Turkish State Planning Organization and ITU. In December of 2000, third branch of the system was introduced and another distance learning classroom was built at Gumussuyu Campus. It cost about $50,000.- It has a drawing. Radio-link and ATM systems are used for telecommunication. This system works on synchronized basis now.

2.2.2.International Distance Learning

Currently, there is a connection between University of Virginia (UV) and Istanbul Technical University via ISDN. Turkish undergraduate students are able to register to a number of courses at University of Virginia. The courses they have taken are accepted as credit and included within their transcripts. The schematic layout is given in Fig.1.

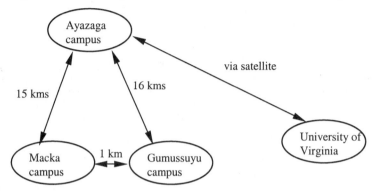

Figure1.Layout of Campus Areas at ITU

Another advanced distance education project was initiated by ITU at the end of 1993 with the support of UNESCO, and aimed at supporting newly established universities, contributing to the education of the universities in the Turkic republics in Central Asia, and in co-operating with the Eastern European universities. Hence, ITU started the process of planning a new multinational educational venture: "The Satellite University of Science and Technology for the Regions of Eastern Europe and Central Asia". The work to realize the project is continuing. $5 million for the first part of the project and $8 million for the second part of the project are needed. It is anticipated that 28 Turkish universities will participate in this project together with (pre-agreed) co-operation of the University of Virginia, USA. A satellite-earth communication base was built and the equipment was installed for this purpose.

2.2.3.Problems

The main problem focused on keeping the system actively. Sometimes there exist quality problems on both video and audio transmission. There is some reaction from students. In this regard, they want to see the lecturer and to make face-to-face communication with him/her. Therefore, they try to sabotage the system by cutting wires, destroying the equipment. It is possible to monitor the classroom during the lecture. This function decreased the sabotage attempts.

2.2.4.Future Plans

At the moment, the system works on synchronized basis. Some applications on asynchronized basis will be considered. Some courses will be prepared by multimedia applications and be installed on the related web site in advance. Those who are registered for that course will follow the lectures on Internet. Some preliminary studies have been completed. VCR records were prepared and were broadcasted on the Web to test. Unfortunately, due to the deficiencies on bandwidth, expected results could not be obtained. But two new centers, Multimedia Center and Software Development Center were established for software and multimedia applications. Meanwhile, the mid-term and final exams of the course titled "BIL101-Introduction to Computers and Information Systems" are performed on the WWW environment. Currently, distance learning is applied to undergraduate education only. The advanced distance education project among ITU, Turkic republics in Central Asia and Eastern European Universities will be completed in near future. This project is really a big one, and requires huge amount of money proportionally to the entire budget of ITU.

3.CONCLUSIONS

Although, there is a lot of investment made for distance learning at ITU, the distance learning is still in the third stage. But, preparations for entering the fourth stage are continuing. The most important and biggest barrier not to achieve the goals desired is the bandwidth of the Internet access in Turkey. But, inter-campus WAN is made of ATM and there is a fiber optic network. Meanwhile, dense distance learning applications will lead to a higher demand for distance learning classrooms and will cause some difficulties in communication infrastructure. At the moment, priority needs to be given to the courses that have some deficiencies regarding classroom and instructor. Before entering the fourth stage, some multimedia applications to support the courses at out-of-course-time could be suggested.

It is necessary for both students and professors to accept and to give importance to the concept of "Distance Learning". Human related factors like emotion and motivation must be considered for applications. Motivation factor is an important issue for a successful system. Motivation of students will lead to decrease negative behaviors like vandalism, and increase interests to the courses. Learning must be transformed from knowledge absorption to knowledge construction by using the tools of new multimedia technologies (Soong, M.H., et al., 2001).

Virtual community has taken its place among communities. Today, people communicate with each other on the virtual environment without meeting themselves physically. Hence, applications on the Web environment gain more importance. Web-based tools provide very good, attractive multimedia possibilities. These tools also have interactivity skills that can help to develop interesting interfaces between humans and computers. In the near future, it is inevitable to use Internet for education (Allen, R., 1998). Additionally, the ability to view content on-demand anytime and anywhere could expand education beyond a primarily synchronous 'live' activity to include more flexible, asynchronous interaction (Bargeron, D., et al., 1999). Asynchronous mode can be easily achieved by Web-based learning.

The trend is toward self-learning and maybe self-teaching systems. We can dream up knowledge-based teaching systems without any professor for the next stage. The system that can analyze the level of learners, can prepare the course accordingly, can trace the improvement of students and can re-organize the course based on feedbacks, and most important of all, -unfortunately- can achieve all these functions without professors like us.

REFERENCES

Allen, R., (1998), The Web: interactive and multimedia education, *Computer Networks and ISDN Systems*, 30, pp.1717-1727

Boyle, R., (1995), Language Teaching at a Distance: From the First Generation Model to the Third, *Systems*, Vol.23, No.3, pp.283-294

Bargeron, D., Gupta, A., Grudin, J., Sanocki, E., (1999), Annotations for streaming video on the Web: system design and usage studies, Computer Networks, 31, pp.1139-1153

Official Web Site of ITU, (2001), http://www.itu.edu.tr/english/default_e.htm

Scollin, P.A., Tello, S.F., (1999), Implementing Distance Learning: Frameworks for Change, *The Internet and Higher Education*, 2(1), pp.11-20

Soong, M.H.B., Chan, H.C., Chua, B.C., Loh, K.F., (2001), Critical success factors for on-line course resources, *Computers & Education*, 36, pp.101-120

REALISING VIRTUAL TRADING:
WHAT PRICE VIRTUAL REALITY?

Soha Maad[a], Meurig Beynon[a], Samir Garbaya[b]

[a]Department of Computer Science, University of Warwick,Coventry CV4 7AL, UK
e-mail: soha@dcs.warwick.ac.uk, wmb@dcs.warwick.ac.uk,
[b]Laboratoire de Robotique de Paris, 10-12 Avenue de l'Europe, 78140 velizy, France
e-mail: garbaya@robot.uvsq.fr

ABSTRACT

Computerisation is about to overtake markets that traditionally depended on physical presence to bring buyers and sellers together in one place. Providing appropriate interfaces for trading environments is a challenging task. In the context of financial trading, the behaviour of different trading parties (investors, brokers and dealers), trading signals, economic and financial indicators, and trading systems, constitutes a complex environment which is difficult to capture in a mathematical model, a computer simulation, or a textual description. This paper discusses the prospects for developing new environments for Virtual Trading that combine Virtual Reality (VR) modelling with a new approach to computer-based modelling that has been developed at the University of Warwick (http://www.dcs.warwick.ac.uk/modelling)

1. INTRODUCTION

Traditional stock exchanges are witnessing major structural changes due to increased competition from alternative trading systems and Electronic Communication Networks and rising investors' demand and financing needs. These structural changes are manifested in the introduction of new trading systems (such as screen based trading), the extension of trading duration, and the opening of new trading channels. The old trading model adopted by traditional exchanges is no longer adequate and new trading models are being introduced, revolutionising old execution, clearing and settlement processes. These developments impact on the behaviour of all market participants (investors, brokers, dealers, and market makers) and are reshaping the financial market microstructure (Harris, 1998) in terms of transaction cost, bid-ask spread, price volatility, trading volume, information effect, and best execution price. The decision process and dealing strategies of market makers are changing, the role of the broker is questioned, and the investor is adopting new trading strategies to boost his profit and gain deeper knowledge of the financial market.

Exploring an ideal trading system minimizing transaction cost and increasing market efficiency is a major concern in the area of financial market microstructure. Interaction in a trading environment is particularly subtle and complex because it combines real-world knowledge and observation with real-time interpretation of abstract numerical data and indicators. Traditional mathematical models are not sufficient for such applications, where human behaviour is of paramount importance. Virtual Reality (Earnshaw et al, 1993), with its orientation towards immersing the human actor in a computer-generated environment, is potentially much better suited to modelling state where human activity is central. VR's capacity to handle objects and their properties, to allow user immersion, and to emulate observation of the real-world using a 3D graphical display, make it an obvious candidate for application in this field.

This paper proposes new principles for the development of environments for virtual trading to deliver VR using an approach to computer-based modelling known as Empirical Modelling (EM). First, the paper overviews EM and VR and their role in constructing environments for virtual financial trading. The paper then discusses the challenges of adopting VR technology to model complicated social environments (such as virtual trading) and proposes the merging of the conceptual framework of the Empirical Modelling approach with the VR design and construction. Second, the paper considers a case study of the Monopoly dealer textual simulation developed by L. Harris. A 2D simulation and a VR scene are constructed and compared with the initial text based simulation. The paper concludes with our findings about the use of VR for modelling a social context such as virtual financial trading.

2. VIRTUAL ENVIRONMENTS FOR FINANCIAL TRADING: THE VR AND EM VISION

2.1 About EM

Empirical Modelling (EM) is a new approach to computer-based modelling that has been developed at the University of Warwick. The Empirical Modelling framework provides a set of principles, techniques, notations, and tools. *Empirical Modelling principles* are based upon *observation, agency, and dependency*. By adopting these principles, EM attempts to represent and analyse systems in a way that can address the complexity of the interaction between programmable components and human agents. The central concepts behind EM are definitive (definition-based) representations of state, and agent-oriented analysis and representation of state-transitions. *Empirical Modelling techniques* involves an analysis that is concerned with explaining a situation with reference to agency and dependency, and the construction of a complementary computer artefact - an *interactive situation model (ISM)* - that metaphorically represents the agency and dependency identified in this process of construing. There is no preconceived systematic process that is followed in analysing and constructing an associated ISM. The modelling activity is open-ended in character, and an ISM typically has a provisional quality that is characteristic of a current - and in general partial and incomplete - explanation of a situation. The special-purpose notation LSD has been introduced to describe agency and dependency between observables. An *LSD account* is a classification of observables from the perspective of an observer, detailing where appropriate: the observables whose values can act as stimuli for an agent (its *oracles*); which can be redefined by the agent in its responses (its *handles*); those observables whose existence is intrinsically associated with the agent (its *states*); those indivisible relationships between observables that are characteristic of the interface between the agent and its environment (its *derivates*); and what privileges an agent has for state-changing action (its *protocol*). The tkeden interpreter is the principal *modelling tool* that has so far been developed: it supports definitive scripts for line drawing and window layout and allows the user to establish dependency relationship between scalars, list and strings using built-in user-defined functions. Dtkeden is a distributed version of tkeden: it allows several modellers to co-operate through communicating definitions and actions within a client-server configuration of tkeden interpreters.

Empirical Modelling emphasizes modelling states and the role of agency in changing state. Agent actions initiate state change. A state is represented in a script of definitions linking observables through dependencies. Agent actions are modelled by redefinitions. In constructing environments for virtual financial trading, EM principles can be useful in construing a situation in the financial market context, and in capturing the state of this situation in a definitive script that can be used to realize and explore different possible construals.

2.2 About VR

Virtual reality tools and technologies supply virtual environments that have key characteristics in common with our physical environment. Viewing and interacting with 3D objects is closer to reality than abstract mathematical and 2D representations of the real world. In that respect virtual reality can potentially serve two objectives: (a) reflecting realism through a closer correspondence with real experience, and (b) extending the power of computer-based technology to better reflect "abstract" experience (interactions concerned with interpretation and manipulation of symbols that have no obvious embodiment e.g. share prices, as contrasted to interaction with physical objects). The main motivation for using VR to achieve objective (a) is cost reduction (e.g. it is cheaper to navigate a virtual environment depicting a physical location such as a theatre, a road, or a market, than to be in the physical location itself), and more scope for flexible interaction (e.g. interacting with a virtual object depicting a car allows more scope for viewing it from different locations and angles). Objective (b) can be better targeted because the available metaphors embrace representations in 3D-space (c.f. visualization of the genome).

Current use of VR is limited to the exploration of a real physical object (e.g. car, cube, molecule, etc..) or a physical location (e.g. shop, theatre, house, forest, etc..). In the course of exploration the user is immersed in the VR scene, and can walkthrough or fly through the scene. The user's body and mind integrate with this scene. This frees the intuition, curiosity and intelligence of the user in exploring the state of the scene. In a real context, agents intervene to change the state of current objects/situations (e.g. heat acts as an agent in expanding metallic objects, a dealer acts as an agent in changing bid/ask quotes and so affects the flow of buyers and sellers). Introducing agency into a VR scene demands abstractions to distinguish user and non-user actions especially when these go beyond simple manipulation of objects by the user hand, or walking through and flying physical locations.

2.3 The motivation for integrating EM with VR

The challenges faced by the use of VR for constructing virtual environments for financial trading are best revealed by drawing a comparison with its use in computer-aided assembly (Garbaya et al, 2000). This comparison reveals a difference in the objective, considerations, approaches, and user role in constructing VR scenes for different contexts.

The main objective in using VR for virtual trading is enhanced cognition of financial markets phenomena; in the case of virtual assembly the main objective is to minimise the need for building physical prototypes. The issues to be considered in applying VR in financial markets and in virtual assembly differ in nature and importance. In virtual assembly, the major concerns are proper 3D picture capturing, conversion, and adding behaviour to objects; in VR for financial trading, they are geometric abstraction of financial concepts, integration with financial database, and distributed interaction. The steps followed to create a VR scene for virtual assembly and for financial markets are different. A linear, preconceived, set of processes can be followed to develop a VR scene for virtual assembly. These can be framed in three stages: defining objects to be assembled, preparing the assembly geometry for visualisation, and adding behaviour to visualised objects. Creating a VR scene for a financial trading context is more complicated and cannot be framed adequately in a pre-conceived way. However, a broad outline can be traced to guide the VR construction process. This involves: identifying entities (both those that admit geometric abstraction and those that have already a well recognised geometric representation) to be included in the VR scene; choosing an appropriate geometric representation for these entities; adding a situated behaviour and visualisation to entities; identifying the external resources (such as databases, files, data feeds, etc.) to be interfaced to the VR scene; and framing the role of the user intervention in the simulation.

Where human intervention is concerned, the user's role in the VR scene is more open-ended in a financial context than in an assembly context. In a VR scene for assembly the immersion of the user is very important. Armed with helmet, gloves, and three-dimensional pointing device (such as 3D mouse and keyboard), the user can manipulate virtual objects with his hands. The user's hands, guided by the user's brain, interact directly with virtual objects. This makes virtual reality environments more appropriate for the assembly task than any alternative technology. Construing financial market phenomena is a function performed by the human brain. The mental model of the designer can be abstracted in a static diagram, a 2D computer artefact, or a VR scene. Geometric objects in the virtual scene might admit no counterpart in the real world - they are purely geometric metaphors. This makes a virtual scene just one of several possible representations. It also motivates a prior situated analysis exploring possible construals pertaining to the social context.

The above comparison highlights the need to support VR technology with principles and techniques to analyse and construe social contexts and to adopt appropriate visualisations for abstract entities (such as financial indicators) that have no real geometric counterpart. Current technologies for Empirical Modelling can help in construing financial situations and in representing state and the analysis of agency in state change, whilst VR offers enhanced visualisation and scope for user immersion and experience of state.

3. CASE STUDY: THE MONOPOLY DEALER SIMULATION

3.1 Description of the case study

Harris's Monopoly Dealer (for more details see http://lharris.usc.edu/trading/#Trading_Game) simulates trading in a dealer market in which there is only one dealer (the user of the simulation model). The user's task (the sole dealer) is to set and adjust bid and ask quotes (raise, lower quotes, or narrow and widen the spread) to maximize his trading profits. The computer model simulates traders arriving at random times to trade with the dealer (user) at his quoted prices. The aim of the simulation is to raise the awareness of its user (playing the role of a dealer) to the trading behaviour of different types of investors (informed/uniformed), and the true value of the security (changing through time and known to informed traders).

3.2 From textual to 2D and VR simulation: the evolving state visibility and state exploration

The Monopoly Dealer is a closed world simulation of a simplified market where there is only one dealer, one share, buyers and sellers, and a time clock. This distances the simulation from reality and limits its scope to convey the true experience of a typical dealer. The simulation uses abstract representations for the buy/sell orders flow, true price, type of investor (informed/uninformed); keyboard press for dealer (user) actions; and mathematical computation of true/realised profit. Simulation results and transaction history are represented by tables. Textual representations are used to display the simulation time, current state, dealer's actions, and warning messages to the dealer.

Many issues surrounding the trading behaviour of buyers and sellers and the strategic decisions of a dealer are abstractly represented in the simulation by a random generation of key variables whose mathematical formulation cannot be determined. These include: the true price of the security; the role of the dealer in the determination of the true price, and transaction price; the type of investor (informed/uninformed); the buyer/seller and transaction flows; and the hidden intentions of the investors to buy or sell. This prompts us to think of different construals, each reflecting a particular scenario, to account for our weakly structured knowledge of the complex trading system. For instance, it might be that the true price is determined by the trading pattern, or that it is influenced in a non-deterministic manner by external events. Rules to govern the interaction are imposed to reflect each possible explanation of the state of observables.

 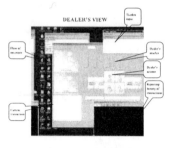 

| 2D simulation (Market View) | 2D simulation (Dealer View) | Original Textual Simulation |

Figure 1. The 2D and textual simulation

A 2D version of the original textual simulation (Figure 1) was developed using the client server architecture of **dtkeden**. The server provides a global market view including the knowledge hidden from market participants (such as the true price, the true position of the dealer, the type of an investor – informed/uniformed), as well as the publicly known information such as the dealer bid/ask/spread quotes, the current status and history of transactions. The client provides a dealer's view that includes the observables that the dealer can view (oracles), such as his position (actual profit and inventory level), the flow of buyers and sellers, and the current status and history of transactions. The dealer's actions (raising/lowering quotes) are also undertaken via the dealer's view and the results of these actions are transmitted to the server. Agents in the model are the dealer, the investor (buyer/seller), and the clock. Many observables are associated with each agent and are classified as oracles, handles, and derivates. An LSD template for the dealer takes the following form:

```
Agent Dealer {
state
    inventory, bid , ask , spread, actual profit, buyers/sellers flow, current status and
    history of transactions, time clock, his estimated true value of the security
Oracles
    flow of orders, order side (buy/sell), order quantity, inventory level, actual profit, his
    estimated true value of security, his knowledge of trader type (informed/uninformed)
Handles
    Bid, ask, spread
Protocols
    if (estimated true price > ask) || (informed trader rush to buy) ) →  raise ask
    if (estimated true price < bid) || (informed trader rush to sell)) →  raise bid
    if (spread is wide)|| (few uninformed traders are trading) → narrow the spread
    if (inventory is approaching the limit of +/-10,000 ) → adjust quotes to attract buy and
    sell orders appropriately
}
```

The state of the model is captured in a script of dependencies such as:

```
informedbuyers_per_time_unit is (true_price-ask)>0 ? rush_rate : normal_rate;
informedsellers_per_time_unit is (bid - true_price)>0? rush_rate : normal_rate;
uninformedbuyers_per_time_unit is (spread<=upperlimit_for_narrowspread)?rush_rate: normal_rate;
uninformedsellers_per_time_unit is (spread<=upperlimit_for_narrowspread)?rush_rate: normal_rate;
screen is(true_price- ask)>0? [buyers_flow, clock, current_transaction, dealer_actions,
dealer_position, dealer_quotes];
```

In constructing a VR scene for the monopoly dealer simulation, the EM analysis was imported. As an additional exercise, we had to find a proper visualization for abstract numeric indicators, agent actions, and the human (user) role in the scene, and to add sound support to produce warning messages to the dealer. The distributed views in the 2D simulation were replaced by a single VR scene including 3 rooms: the dealer action room, the transactions history room, and the hidden knowledge room. Transactions are saved in a file and are visualized in the transaction history room. The set up for the experiment is developed on a Silicon Graphics Machine running Irix6.5, and using Parametric Technology Corporation's VR modelling tool Dvise, and the peripheral includes a 3D mouse as an input device, CrystalEYES glasses for the Stereographic image and 3D auditory feedback. Figure 2 shows snapshots of the VR scene.

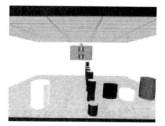

Dealer Actions View   *VR Scene of the Monopoly Dealer*   *Transaction History View*

Figure 2. The VR scene

4. CONCLUSIONS

There is a very significant distinction between VR modelling for areas such as robotics as represented in papers such as (Garbaya et al, 2000), and its application to Virtual Trading. Whilst we can reasonably speak of "using VR to model the reality of a manufacturing assembly process", the reality of the virtual trading environment is an altogether more elusive concept. Where manufacturing assembly deals with objects and actions whose objectivity and real-world authenticity is uncontroversial, virtual trading is a prime example of an activity in which the impact of technology upon human cognition is prominent, and character of its agencies and observables is accordingly hard to capture in objective terms. Empirical Modelling supplies an appropriate framework within which to address the ontological issues raised by such applications of VR (Beynon, 1999). Current research by Cartwright (2001) is aimed at merging EM and VR in a web-based framework. The work carried out for this paper points to the following conclusions:

- A VR scene can help in exploring a particular state in a social context.
- The pre-construction phase for a VR scene can benefit greatly from concepts drawn from the Empirical Modelling literature such as modelling state, state change, and the initiators of state change.
- VR technology needs to be better adapted for the representation of multiple agents acting to change the state and corresponding visualisation in a VR scene.
- The successful application of VR technology in modelling social and data intensive environment relies upon integrating VR with other programming paradigms such as databases and definitive programming.

We propose to apply quantitative and qualitative metrics to our case study to assess the potential benefits of VR in modelling a social context. The profitability of the dealer's position with reference to a particular scenario can be used as a quantitative metric to evaluate our three different simulations. Cognitive Dimensions (Green, 2000) can be used to assess the qualitative aspects of the VR scene. Two dimensions are appropriate: the visibility and the viscosity.

REFERENCES

Beynon, W.M. "Empirical Modelling and the Foundations of Artificial Intelligence, Computation for Metaphors, Analogy and Agents", Lecture Notes in Artificial Intelligence 1562, Springer, 322-364, 1999.

Cartwright, R. I. "Distributed Shape Modelling with EmpiricalHyperFun", Proceedings DALI 2001, to appear.

Earnshaw, R.A., Gigante, M.A., and Jones, H. "Virtual Reality Systems", Academic Press, 1993.

Garbaya, S. and Coiffet, P. "Generating Operation Time from Virtual Assembly Environment", in proceedings of the 7th UK VR-SIG Conference, 19th September, 2000, University of Strathclyde, Glasgow, Scotland

Green, T.R.G. "Instructions and Descriptions: some cognitive aspects of programming and similar activities", in proceedings of working Conference on Advanced Visual Interfaces (AVI 2000). New York: ACM Press, pp21-28, 2000.

Harris, L. "Trading and Exchanges", Draft textbook: December 4, 1998, University of Southern California, Marshall School of Business, p. 1-1.

Test bed System of Image-Based Rendering as VRML Extension

H. Kim, S. Kim, B. Koo, B. Choi and W. Oh

Virtual Reality Center, Electronics and Telecommunications Research Institute,
161 Kajong-dong, Yusong-gu, Taejon, 305-350, Korea

ABSTRACT

The graphics specification for Virtual Reality on internet has been standardizing, with growth of computer hardware technology and dissemination of internet. Virtual Reality Modeling Language 97 (VRML 97) is the last standard graphics specification. Especially, Web3D consortium is promoting the VRML 97 with VRML extension nodes. One of the commonly important requirements for VRML extension is the function of fast and high quality rendering. Image-Based Rendering (IBR) technique is a good solution for that. We present a test bed system of IBR and show its usefulness as VRML extension.

1. INTRODUCTION

The technology of computer makes rapid process and the use of internet also increases rapidly. It gives users hope to see more realistic images on their systems and also requires standardization organizations to establish a standard specification for representing and interacting something. The research of Image-Based Rendering (IBR) techniques is true of the former, and the effort of Web3D consortium is the latter.

Virtual Reality Modeling Language 97 (VRML 97) is the last standard graphics specification approved by International Standards Organization (ISO). Nowadays, The Web3D consortium makes efforts to promote the VRML 97. To do this, the consortium maintains mailing lists for various issues and runs 4 focused teams for better coordination between the consortium and its members. And under the jurisdiction of one or more of the consortium's teams, there are specific working groups to develop standards and technologies, and so forth. The teams review and vote on proposals provided by working groups, following a well-documented, tried-and-true process. The consortium also seeks to build the web3d market by cooperative development and ISO standardization of relevant technologies for real-time, networked 3D graphics. The consortium is actively engaged in the development of standards technologies with other leading organizations such as the World Wide Web Consortium (W3C) for XML and the Motion Pictures Experts Group (MPEG) for MPEG-4 (Web3D consortium, http://www.web3d.org). Therefore, the next generation of VRML standard requires a lot of things including geographical data representation, humanoid animation, multimedia content description, and so on. The method of fast and high quality rendering is one of commonly important requirements, and IBR technique is one of good solutions.

Conventional rendering technique is based on geometric modeling data. So it is subject to modeling scene complexity. But IBR technique is not subject to it as the use of rendered images. Instead, it is affected by the quality and resolution of the images. The conventional technique computes the color values of objects affected by lights every frame and makes scenes of any viewpoint. But IBR technique can make the scenes without computing the color values. IBR technique has been developed from many institutes and university, because it is not depended on scene complexity and so able to keep easily constant update rate. Besides, it is able to make high quality images.

In this paper, our test bed system for using IBR technique is introduced. And we show usefulness of using IBR technique as VRML extension by testing some data. In the next section, related research is briefly mentioned and the third section, our test bed system is introduced. And then, we test rebuilt data acquired by using the system and make conclusion.

2. RELATED RESEARCH

There are many techniques related to IBR. In this section, we briefly introduce some IBR techniques. IBR techniques start to reuse rendered or realistic images. It is very expensive to get or make high quality images. Light Field Rendering technique is an IBR technique. Its basic idea is to generate images from new viewpoints using a two dimensional array of reference images. Given a plan in space, a series of images is taken from different viewpoints on this plane. These images can be treated as an array of rays because for every image, a ray beginning from the viewpoint on the camera plane can be mapped to each pixel on the reference image. With a database of rays for various viewpoints on the plane, new views can be generated (Levoy and Hanrahan, 1996). These new images can give the feeling of three-dimensions as different views are extrapolated from the two-dimensional array of images. Its similar technique is to use Lumigraph (Gortler, Cohen, et. al., 1996)

Using Layer Depth Image (LDI) is another IBR technique. It is a three-dimensional data structure that relates to a particular viewpoint and which samples all the surfaces and their depth values intersected by the ray through that pixel. Each element contains a color, surface normal and depth for surface. Shade, Gortler, et al. suggest two methods for pre-calculating LDIs for synthetic imagery. First, they suggest warping n images rendered from different view points into a single viewpoint. During the warping process if more than one pixel maps into a single LDI pixel then the depth values associated with each source view are compared and enable the layers to be sorted in depth order. An alternative approach that facilitates a more rigorous sampling of the scene is to use a modified ray tracer. This can be done simplistically by initiating a ray for each pixel from the LDI viewpoint and allowing the rays to penetrate the object. Each hit is then recorded as a new depth pixel in the LDI. All of the scene can be considered by pre-calculating six LDIs each of which consists of a 90 frustum centered on the reference viewpoint. (Shade, Gortler, et al., 1998, Alan Watt, 2000).

LDI tree is more advanced techniques than previous one. It has a hierarchy made up different resolution of LDIs, and one reasonable resolution is selected to make new images by selection policies (Chang, Bishop and Lastra, 1999).

When warping any fixed number of source images, some areas of the scene that should be visible in the destination image may not be visible any source image. These holes in the new generated image are usually quite small, and so they can be filled with interpolated color values by using pixel colors in the neighborhood (Jason C. Yang, 2000).

3. TEST BED SYSTEM

We introduce our test bed system of IBR in this section. The system aims to show usefulness by using IBR technique as VRML extension and to provide helpful information. The system consists of three functional parts. One is an analyzer of VRML data, the second is user control subsystem, and the third is a renderer using IBR technique. Figure 1 shows the data process of the test bed system. We use VRML data and render them visually. And any rendered image can be captured with their depth information. Through the user control subsystem, the rendered images are rebuilt and used by the IBR renderer. The IBR renderer makes new images as viewpoint is changed.

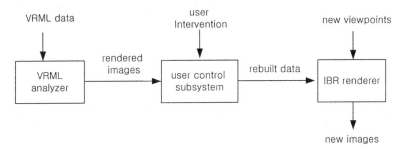

Figure 1. Process of the test bed system

3.1 VRML analyzer

The first one, VRML analyzer, shows graphically VRML data information, for example a hierarchy of the data, viewpoint list, the information of light nodes, and so on. It has functions of VRML data parsing and rendering. And it also provides the polygonal data information and rendering frames per second to help users to analyze current VRML data. After examining the data, we can capture the rendered images for using IBR techniques. The captured information is two separated files, a captured image file and information file. The image file is a general BMP format file with the color RGB value of each pixel, its image size and resolution, and so on. And the other has the normalized depth value of each pixel, related image file name, and the camera information. Figure 2 shows execution examples. At figure 2 (a), its left side window visualizes the hierarchy of VRML nodes. If an item is selected, then it is highlighted on the right view and its information is provided. It is very useful for data editing. Solid mode, wire-frame mode, no texture mode, point mode, etc. are provided as viewing modes and the wire-frame mode is shown at figure2 (b)

| (a) Hierarchy of VRML nodes | (b) Wire-frame view of selected nodes |

Figure 2. VRML analyzer

3.2 User control subsystem

The second one, user control subsystem, supports a good data reconstruction by user intervention. At first, there is a list of captured images at the previous step. We can select or deselect each item as see at (b) of figure 3. The selected images are checked for difference between examination and selection of the image. At (a) and (b) of the figure, the bottom panel of the right side visualizes the viewpoint information of the image with VRML data. It has three view options, x-, y-, z-axis view, which we can select. Figure 3 (a) and (b) are examples of x-axis view and y-axis view, respectively. After we select items, reference image can be made of them for supporting the LDI technique. To make a reference image, we use a dialog window to set the position and orientation, field of view, etc. about any reference camera. Figure 3 (c) shows an execution example of making a reference image. The subsystem gives us useful information in order to make efficient reference images. It can be the more intuitive work to give visual information.

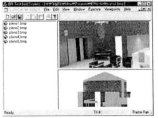  

| (a) Preview of captured images | (b) Selection for reference image | (c) Making a reference image |

Figure 3. User control panel windows

3.3 IBR Renderer

The last one, IBR renderer, visualizes the rebuilt data to check its usefulness. IBR renderer is an image-based renderer. It can be compared with conventional rendering of the previous VRML analyzer. It has similar functionality with that of the analyzer except rendering technique. When mouse moves, a viewpoint changes and the changed viewpoint calls the IBR rendering function. The result is new images made of IBR technique and compared with a current view. Sometimes it has holes because reference images don't have the information. We can cover small holes by using simple heuristic algorithm. But big holes can only be covered with new source images. The type of navigation is usually walk view type. Figure 4 shows the IBR renderer and its result image.

| (a) IBR renderer | (b) Result new image |

Figure 4. IBR Renderer

4. TEST AND RESULTS

The test bed system works on PC environments and is implemented by C++, Microsoft Foundation Class (MFC), and Open Inventor library. An experimental data set has 77,753 triangles and 5,402KB file size, and it is showed at figure 5. At figure 5, (a) is a new image generated from a new viewpoint. The view of the IBR renderer shows at (c). In this figure, the center image generated by IBR technique is the same as (a). And (b) is a captured source image and is used as a texture map data to test simple texture mapping. Figure 5 (c) and (d) can be compared as same viewpoint. The figure (d) is somewhat awkward, because the distance between the viewpoint and each object in the scene is not considered. That is, the pillar's position and width on the right side of each figure (c), (d) are different, even if they have same viewpoint.

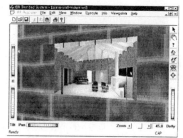

(a) IBR rendered images	(b) A texture map (source image)
(c) IBR renderer view	(d) Simple texture mapping view

Figure 5. IBR vs. simple texture mapping

The table 1 is the comparison of the rendering speed about IBR and simple texture mapping. At the table, the image data size means the size of the reference image data and that of simple texturing map data, exclusive of polygonal data. As we are able to see the result of table 1, the rendering speed of IBR technique is lower than that of simple texture mapping but the technique gives us to feel more cubic than the texture mapping, as figure 5 above.

Table 1. IBR vs. simple texture mapping

	IBR	Simple texture mapping
Image data size	4,767KB	581KB
Rendering speed	About 5~6 frames/sec	About 20 ~ 25 frames/sec

Table 2. IBR vs. conventional rendering

	IBR	Conventional rendering
File size	4,769KB	5,402KB
Rendering speed	About 5~6 frames/sec	About 2 ~ 4 frames/sec

The table 2 is the comparison of the rendering speed between IBR and conventional rendering. The conventional rendering means the rendering that is based on geometry modeling, transformation computation and color shading computation, and so on. In the rendering speed, IBR technique is faster than conventional rendering on special situation, especially, when the size of modeling data is big but the size of rendered range on the screen is rather small and restricted. Because IBR technique uses the information of rendered source images, the new images depend on the quality of the source images and the ranges of camera related with the source images. Therefore, the technique is useful by the use of web portal and background images out of windows. The following figures were rendered images by using amended Blaxxun browser. At figure 6, the building image was rendered by IBR technique but others were rendered by conventional rendering technique. We used Blaxxun3D browser open source and added the functionality of IBR technique to the browser as a VRML extension.

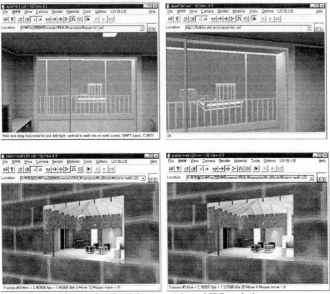

Figure 6. VRML browser adapted IBR technique

5. CONCLUSIONS

The test bed system of IBR and its 3 subparts were introduced. The system consists of a VRML analyzer, user control subsystem and an IBR renderer, and their roles were presented. Using the test bed system, we showed the usefulness of IBR technique, presented the helpful function to reconstruct VRML data, and acquired the new image with high quality. To more easily reconstruct data, visual information and intuitive user interaction interface were adopted, and the rebuilt data for IBR technique were made through trial and error. It was tested to navigate the VRML data and the rebuilt data as a VRML extension by using an amended blaxxun 3D browser. The system worked well but there are improvable points, like adopting various IBR techniques.

REFERENCES

Chang, Bishop and Lastra, "LDI Tree:A Hierarchical Representation for Image-Based Rendering ", Proc. SIGGRAPH 99, 1999

Gortler, Cohen, et. al., "The Lumigraph", Proceeding of SIGGRAPH 96, Aug. 1996, pp. 43 - 54

Levoy and Hanrahan, "Light Field Rendering", Proceeding of SIGGRAPH 96, Aug. 1996, pp.31 - 42

Shade, Gortler, et al., "Layered Depth Images", Proceeding of SIGGRAPH 98, 1998, pp. 231 - 242

Alan Watt: "3D Computer Graphics Third Edition", Addison-Wesley, 2000

Web3D consortium site, "http://www.web3d.org"

Jason C. Yang, "A Light Field Camera For Image Based Rendering", Partial fulfillment of the requirements for the degree of MS at the MIT, June, 2000

Four Ways of 'Being There':
Combined Effects of Immersion and Pictorial Realism
on the Sense of Presence in Virtual Environments

Jan Hofmann[1] and Heiner Bubb[2]

[1] Society and Technology Research Group, DaimlerChrysler AG, D-10559 Berlin, Germany
[2] Chair of Ergonomics, Technische Universität München, D-85147 Garching, Germany

ABSTRACT

The effects of immersion and pictorial realism on the sense of presence in a projection based virtual reality system were investigated. Seventy-seven participants each experienced one of four settings with varied combinations of immersion and pictorial realism. Their sense of presence was measured using a post-exposure questionnaire developed by Schubert, Friedmann, and Regenbrecht (1999). It allows to differentiate between three presence facets (reality appraisal, involvement, and spatial presence). The results show significant effects of both immersion and pictorial realism on two of the presence facets. The influence of immersion proved to be dependent on the experienced degree of pictorial realism. Pictorial realism apparently partially disabled the effects of immersion on presence. Practical implications of these results for the design of virtual reality systems are proposed.

1. INTRODUCTION

In recent years, the sense of presence in virtual environments (VEs) has gained much interest in the virtual reality and broader HCI research communities. The very nature of the cognitive phenomenon of presence, commonly known as the 'sense of being there', is still discussed controversially. Nevertheless it has been suggested – and in some cases shown experimentally – that the degree of presence can be influenced by divers factors.

The objective of this study was to analyse interactions of effects of immersion and pictorial realism (PR) on the sense of presence. The term immersion is used here – as e.g. by Slater and Wilbur (1997) – to describe characteristics of the virtual reality system, the virtual environment, and other factors that enhance the sense of presence. Pictorial realism is usually regarded to be one of these factors (see e.g. Welch et al., 1996). In this study, we varied pictorial realism separately from other immersion factors and thus excluded it from our definition of immersion.

This kind of analysis is particularly useful from a practical point of view. Many researchers suppose effects of presence on the user's behaviour in the VE – e.g. on his or her task performance (see Welch, 1999). The experimental settings of this study can be viewed as 'archetypes' of industrial VEs, providing different combinations of immersion and pictorial realism. Immersion and pictorial realism cause different costs in their generation and usage. Thus, an analysis of the mapping between archetype settings and resulting presence can be valuable: If a certain degree of presence is desirable for an application, effective allocation of resources is facilitated.

2. METHODOLOGY

2.1 Participants, Stimuli, and Apparatus

Seventy-seven people participated in the study (nine female and sixty-eight male). Some of the participants had used virtual reality technologies beforehand, but none of them on a regular basis. They were divided into four separate groups. Each of these groups experienced only one out of four experimental settings. The four settings were designed to provide four different combinations of immersion and pictorial realism (low/low, low/high, high/low, and high/high for immersion and pictorial realism, respectively).

To generate the low and high values of immersion and pictorial realism, we varied several factors simultaneously. The choice of these factors followed a compromise strategy: (1) We looked for factors whose variation was expected to exert a strong impact on the sense of presence. Previous results of divers theoretical and experimental studies were considered to try and predict their impact (e.g. Sheridan, 1992; Steuer, 1995; Welch et al., 1996; Slater and Wilbur, 1997; Lombard and Ditton, 1997; Witmer and Singer, 1998; Regenbrecht, 1999; Regenbrecht, 2000). (2) Factors that would not be varied in a practical application of a system like ours were not considered. E.g., head tracking would not be switched off deliberately, though this might have a profound effect on the sense of presence. Table 1 summarizes the factors chosen and their respective low and high values (see also Figure 1).

Table 1: Immersion and pictorial realism (PR) factors used to influence presence

immersion/PR factor	value for 'low'	value for 'high'
frame rate	appr. 2 – 2.5 frames/s (for each eye)	appr. 11 – 14 frames/s (for each eye)
interactivity	no immersive interaction	object scaling by immersive interaction
duration of exposure	appr. 8, 3, and 7 min (sessions 1 – 3)	appr. 11, 6, and 7 min (sessions 1 – 3)
vividness of scene	no animated objects	animated emergency indicator
mental priming	disillusioning description of experience	illusion-enhancing description
real world ambient light	dim ambient light in real surroundings	only light source was projection
real world background noise	background noise from surroundings	background noises damped
communication w/ instructor	direct addressing of participant	no direct addressing
detailing of 3D model (PR)	most 3D details removed	realistically equipped interior
surface textures (PR)	no textures applied	realistically textured

The virtual scene was the front half of a passenger car interior actually in production. It was based on the original data used in the product development process and was displayed in 1:1 scale. The virtual interior was combined with a real driver's seat and steering wheel to enhance participants' impression of sitting in a vehicle cockpit via haptic sensations.

The virtual cockpit was displayed in a cubic-shaped five-sided back projection system (back projection on ceiling, floor, and three walls of the cube). The length of side of the projection planes was 2.5 m each. The virtual environment ran on an SGI® Onyx 2® graphics engine. The participants' head movements were tracked with a six-DOF tracker (MotionStar® by Ascension®), the left and right eye channels were separated using StereoGraphics® CrystalEyes® shutter glasses. The open side of the projection cube was behind participants' backs.

Figure 1. Views of the virtual cockpits used in the low (left) and high (right) pictorial realism settings. The virtual steering wheel was replaced by a real one.

2.2 Presence measurement

Presence was measured with a post-exposure questionnaire. A questionnaire developed and tested by Schubert, Friedmann, and Regenbrecht in a sequence of studies was chosen (Schubert, Friedmann, and Regenbrecht, 1999; Regenbrecht, 1999). These authors suggested to differentiate between (1) spatial presence (correlated to the 'sense of being there', often considered to be the core concept of presence), (2) involvement, and (3) reality appraisal. The authors claim these to be three different facets of a more general presence construct (see Regenbrecht, 1999). The presence questionnaire by Schubert et al. is designed to measure the three facets separately.

The original version of this questionnaire consists of 14 questions. Their impact on the three presence facets has been proven by factor analysis (Schubert et al., 1999). Minor modifications due to the particular situation of our experiment were necessary, yielding an adapted set of 13 items (seven point Likert scale). The final questionnaire was used twice within each experimental run (versions 'A' and 'B', permuted item order).

2.3 Procedure

All participants experienced three consecutive virtual environment sessions. Before entering the first session, the instructor gave either a detailed description of the enabling technology (disillusioning mental priming, see Table 1) or an enthusiastic description of the VE as a compelling experience (illusion-enhancing mental priming).

In each session, participants were asked to perform a simple task that slightly differed between the three sessions (object comparisons). Participants in the high immersion setting were asked to perform an additional task. This task involved the interactive object scaling mentioned as an immersion factor in Table 1. After completion of their tasks in sessions 1 and 3, participants were asked to fill out presence questionnaires A and B, respectively. While completing the questionnaires they were still sitting in the driver's seat and were still immersed in the virtual cockpit.

3. RESULTS

3.1 Factor analysis of presence questionnaires

The results of questionnaires A and B were subject to separate but equally structured factor analyses (77 cases, all 13 questionnaire items used, main component analysis, Varimax rotation).

The results of questionnaire A were not suitable for factor analysis according to the measure of sampling adequacy criterion (MSA value = 0.763 < 0.8). In addition, the factor loadings could not be interpreted in any sensible way. We thus excluded these results from further analysis. Questionnaire A was filled out only a few minutes after participants' first VE exposure. Probably they could not relate their own impressions to the way the questionnaire requested to express them yet. The results of questionnaire B proved to be suitable for factor analysis (MSA value = 0.809). It yielded three factors with eigenvalues > 1. Numerically, the 13 questionnaire items could unambiguously and completely be assigned to the three factors.

The assigned items allowed a straightforward interpretation of the three factors that corresponded well with that of the original authors (Schubert et al., 1999). We coined the three factors reality appraisal, involvement, and spatial presence after Schubert et al. (1999; they used the term 'realness' for the first factor). Spatial presence is highly correlated to the notion of the 'sense of being there' (see Regenbrecht, 1999). The other two are meant to be self-evident. The factor values of reality appraisal, involvement, and spatial presence for each participant were extracted by regression. They were normalized to a mean of zero and a variance of one.

3.2 Variance analysis of mean presence values

The factor values of the participants of each setting were averaged separately. Figure 1 displays the mean factor values and standard errors of the means. The mean values were subject to separate variance analyses (one independent variable). Due to the high noise expected in presence values, we employed a significance level of 0.1. The results of the variance analyses were the following (see also Figure 2):

Reality appraisal. The variance analyses show a significant dependence of reality appraisal on the immersion setting for high PR (F(1, 34) = 5.13, p = 0.030; higher reality appraisal for high immersion). For low PR no signifi-

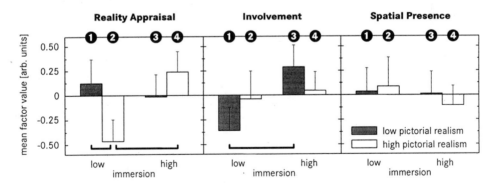

Figure 2. Mean factor values and standard errors of the means for participants' reality appraisal, involvement and spatial presence in the four experimental settings (different combinations of immersion and pictorial realism). The brackets denote significant differences of the means. The numbers refer to Figure 3.

1019

cant influence of the immersion setting was found. In addition, a significant dependence of reality appraisal on PR could be detected (F(1, 31) = 3.04, p = 0.091; for low immersion only). Interestingly, reality appraisal was lower in the higher PR setting. No other effects were significant.

Involvement. The variance analyses show a significant dependence of involvement on the immersion setting for low PR (F(1, 37) = 4.00, p = 0.053; involvement higher for high immersion). For high PR no significant influence of the immersion setting could be detected, and no other effects were significant.

Spatial presence. No significant effects were found.

3.3 Interacting effects of immersion and pictorial realism

The effects of immersion and pictorial realism on the three presence facet can be interpreted in the following way:

(1) Effects on participants' reality appraisal. As expected, lowering the immersion decreased participants' reality appraisal (from setting 4 to 2 in Figure 2). The immersion factor with the strongest influence on reality appraisal was probably the frame rate. In the low immersion setting with low frame rate, the virtual cockpit appeared strongly distorted when users moved their heads. The realistic impression of the high pictorial realism cockpit was partly destroyed. This probably resulted in low reality appraisal.

Participants' reality appraisal values in setting 2 could be interpreted as a reaction to being 'disappointed': Without moving, the cockpit with high pictorial realism looked fascinating; upon head movement, participants were disappointed because the system could not maintain the illusion. Participants reacted to this mismatch with low assessments. This could also explain the significant *increase* of their assessments when *lowering* pictorial realism (from setting 2 to 1). In setting 1, the strong mismatch of high pictorial realism and lacking system power (i.e., low frame rate) that resulted in the disappointment was removed.

But why did the variation of the immersion have no effect on participants' reality appraisal in the low pictorial realism cockpit (settings 3 and 1)?. Probably, these cockpits were just too far from being 'like the real thing' (no colours, no textures applied). Here, the effect of the immersion on participants' reality appraisal was dominated by the low pictorial realism. Their reality appraisal was 'locked' (see the left side of Figure 3).

(2) Effects on participants' involvement. Participants' involvement decreased when immersion was lowered (from setting 3 to 1 in Figure 2). Again, this might be due to the lower frame rate in setting 1. The effect was probably enhanced by the lack of interaction and the other low values of the immersion factors (see Table 1). But immersion affected involvement *only in the low pictorial realism cockpit.* Apparently, participants' attention was strongly captured by the looks of the high pictorial realism cockpit: Their involvement was 'locked' for high pictorial realism (see the left side of Figure 3). In other words, the influence of pictorial realism dominated that of immersion.

(3) Effects on participants' spatial presence. No influence on participants' spatial presence was detected. The systematic effects of immersion and pictorial realism on all three presence facets were overlaid by participants' individual reactions to the offered stimuli. Strong influence of individual user characteristics on the sense of presence has been proposed by various researchers (see e.g. Steuer, 1995; Regenbrecht, 1999). One of these characteristics is the users' willingness to suspend disbelief (Lombard and Ditton, 1997): The willingness to overlook the fact that the experience is mediated by technology. This willingness is *essential* for the development of spatial presence. But it is

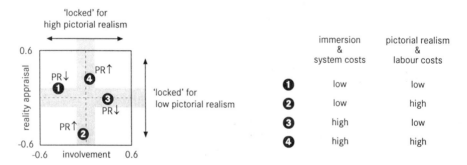

Figure 3. Partial dominance of pictorial realism over immersion regarding their effects on two presence facets (left). The numbers refer to the four experimental settings. The settings are related to general costs involved in their generation and usage on the right.

merely *helpful* for increasing reality appraisal and involvement. Thus, random variations of this willingness among participants have overlaid systematic effects of immersion and pictorial realism on spatial presence most strongly.

3.4 Implications for the practical use of virtual environments

Setting up a virtual environment with high immersion – particularly realizing high frame rates regardless of scene content – usually involves high system costs. On the other hand, achieving high pictorial realism often involves high labour costs: Realistic 3D detailing and textures are not generated automatically in general. Thus, our results might be regarded as a further step towards mapping certain cost types (and preparation time) involved in generating a virtual environment to the achievable effects on the users' sense of presence (see Figure 3).

An example is the following: If for a certain application high reality appraisal is crucial, pictorial realism should apparently be high. But the high labour costs involved in its generation are not well spent if immersion is too low: The mismatch of pictorial realism and system power might lead to a surprisingly low reality appraisal. Low system power, combined with low pictorial realism, might prove to be a better compromise then.

4. CONCLUSIONS

We measured the effects of immersion and pictorial realism on three facets of the sense of presence. Effects of the variation of immersion on reality appraisal and involvement were detected: Both were raised by an increase of the immersion, as common sense would suggest.

But this effect was strongly and *selectively* dominated by the influence of pictorial realism. Pictorial realism 'locked' – depending on its degree – the presence facets at mean values. The influence of immersion was disabled in these cases. We proposed explanations for the effects observed.

The effects of the sense of presence on task performance and other parameters are not well understood yet. *Some* effects are to be expected though. The results of this study help to map efforts or costs involved in setting up a virtual environment to the achievable effects on the user's sense of presence. Thus, we hope our results facilitate the effective allocation of resources in planning and designing virtual environments for practical tasks.

ACKNOWLEDGMENTS

We thank Holger Regenbrecht for valuable advice and the permission to use the questionnaire, Thomas J. Jäger for numerous discussions and for making this study possible, Thorben Deffke and Stefan Thiel for their help in preparing and conducting the experiments, and all our participants.

REFERENCES

Lombard, M., and Ditton, T. (1997). At the Heart of it All: The Concept of Presence. Journal of Computer Mediated Communication [Online], 3(2). Available: http://www. ascusc.org/jcmc/vol3/issue2/lombard.html.

Regenbrecht, H. (1999). Faktoren für Präsenz in Virtueller Architektur. Dissertation, Bauhaus-Universität Weimar.

Regenbrecht, H. (2000). Personal communication.

Schubert, T., Friedmann, F., and Regenbrecht, H. (1999). Decomposing the Sense of Presence: Factor Analytic Insights. Paper presented at the 2nd International Workshop on Presence, University of Essex, 6th and 7th of April, 1999.

Sheridan, T. B. (1992). Musings on Telepresence and Virtual Presence. Presence: Teleoperators and Virtual Environments, 1, 120-125.

Slater, M., and Wilbur, S., (1997). A Framework for Immersive Virtual Environments (FIVE): Speculations on the Role of Presence in Virtual Environments, Presence: Teleoperators and Virtual Environments, 6(6), 603-616.

Steuer, J. (1995). Defining Virtual Reality: Dimensions Determining Telepresence. In F. Biocca & M. R. Levy (Eds.), Communication in the Age of Virtual Reality (pp. 33-56). Hillsdale: Lawrence Erlbaum Associates.

Welch, R. B. (1999). How Can We Determine if the Sense of Presence Affects Task Performance? Presence: Teleoperators and Virtual Environments, 8(5), 574-577.

Welch, R. B., Blackmon, T. T., Liu, A., Mellers, B. A., and Stark, L. W. (1996). The Effects of Pictorial Realism, Delay of Visual Feedback, and Observer Interactivity on the Subjective Sense of Presence. Presence: Teleoperators and Virtual Environments, 5(3), 263-273.

Witmer, B. G., and Singer, M. J. (1998). Measuring Presence in Virtual Environments: A Presence Questionaire. Presence: Teleoperators and Virtual Environments, 7(3), 225-240.

Subjective Intensity Scaling to Vibrotactile Stimulation on the Hand for a Glove System

Seongil Lee and Hyo Sang Lee

School of Systems Management Engineering, Sungkyunkwan University
300 Chunchun-Dong, Suwon, 440-746, Korea
silee@yurim.skku.ac.kr

ABSTRACT

Human responses to vibrotactile stimulation on various parts of the hand were measured using a psychometric scaling method. Eight small DC motors with the diameter of 1 cm were used as contactors in the glove-typed vibrotactile glove system developed in Human Interface Research Laboratory of SKKU. To derive the same level of perceived intensity across the various parts of the hand, each of which has different sensitivity from each other, it is necessary to use different levels of vibration intensity among the loci. In this study, a scheme was developed from measurement of perceived intensity at each hand locus for varied intensity of vibrotactile stimulation above the threshold level. It is expected that the results of the current study would be useful to apply vibrotactile displays to virtual reality or teleoperation.

1. INTRODUCTION

A successful Virtual Reality system or a dexterous hand master should have systems that feedback hand motion and contact information on the hand. Instrumented gloves and exoskeletons such as the Cyber Glove are regarded as the most appropriate system by which the human operator can feel the grasping and manipulating forces exerted either on virtual objects or master arms (Shimoga, 1992). Existing technologies to provide force and touch feedback mostly use actuators and vibrotactile displays. This study presents how vibrotactile display should be used for such a feedback.

For effective use of vibrotactile displays in teleoperation or VR systems, the most important and useful information would be on the intensities far above the threshold. To quantify the perceived intensity of human operator, it is necessary to measure psychophysical responses to mechanical stimulation on the skin. It is already shown that the skin is the most sensitive on the fingertip (Wilska, 1954), and to the vibratory stimulation which has frequencies at around 230 Hz (Verrillo, 1963; Vallbo and Johansson, 1978). These findings, however, of the human responses to the vibrotactile stimulation have been obtained only at threshold levels. For the stimulation above the threshold intensities, it is necessary to measure and scale the perceived intensity using a psychophysical approach (Cholewiak and Collins, 1991).

The subjective scaling of the intensity should be consistent. Once scaled and measured, the subjective intensity of a subject should be reproduced with the same scale with repeated exposure to vibrations.

2. EXPERIMENT

Experiments were conducted to examine the consistency of tactile perception in production and verification of subjective intensity scaling.

2.1 Subjects

A total of 45 college students participated in the experiment. The average age of the subjects was 19.4 years old. Fifteen subjects participated in the first experiment of perceived intensity scaling, and the rest of 30 students participated in the validation experiment.

2.2 Equipment

A vibrotactile glove system with 8 contactors was developed in the Human Interface Research Laboratory in the Sungkyunkwan University. Each contactor is made of a round DC motor with 1-cm diameter, and has the maximum vibratory amplitude of 8 micron. Each contactor is attachable to any locus on the hand using a Velcro tape. The glove is shown in Figure 1.

2.3 Procedures

The test variables were hand locus and stimulation intensity. The dependent variable was perceived intensity of the subject. The eight hand loci investigated in the experiment are shown in Figure 2. The stimulation intensity was varied in ten levels from 10.8 to 22.7 m/s^2. Subjects scaled the perceived intensity in the range from 0 to 7 at each hand locus. The stimulation was turned on until the subjects scaled all the eight loci and finished keying in the perceived intensities. Ten trials were measured for each level of stimulation intensity, and a total of 100 measurement was obtained for each land locus.

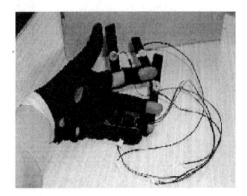

Figure 1. SKKU Glove System

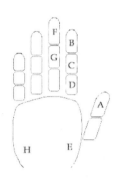

Figure 2. Stimulated loci on the hand

3. RESULTS

Results of the experiments are summarized as follows.

3.1 Perceived Intensity Scaling

Perceived intensity was shown to have a consistent linear relationship with the stimulation intensity as shown in Figure 3(a). Regression analysis for all the test loci presented first-order linear relationships between perceived intensities and stimulation intensities with R^2-values above 0.9.

The results suggest that there should be different schemes to stimulate for each hand locus. The loci on the palm and the thumb, for example, had significantly different slope than other loci on the fingers, and need to be stimulated with stronger vibration to result in the same perceived intensity. On the other hand, the loci on the finger area all showed very similar patterns in the perceived intensity scaling to the vibrotactile stimulation.

Overall, the perceived intensities had a linear relationship with actual stimulation intensities, and showed different sensitivity gradients among hand loci.

3.2 Verification

A follow-up experiment to verify the linear relationship between the stimulation and perceived intensity was conducted. The process used the same psychophysical approach with the reverse protocol. The stimulation intensities were provided according to the stimulation levels of the locus that produced the weakest responses, the H part of the palm. Subjects responded to the vibratory stimuli that were scaled at the same level of perceived intensity with the 7-point subjective rating scale. It was examined whether the subjective rating was consistent to the linear relationship derived in the first experiment.

It seems like that the linear relationship derived in the first experiment was consistent in that the subjects' responses showed a similar pattern in Figure 3(b). However, the thumb and the loci on the palm failed to produce the consistent intensity gradient. Since those loci showed a slow increase in the perceived intensity in the first experiment, they should have shown the similar responses even though the stimulation intensities were scaled up to create high enough responses in the verification test. The loci on the palm actually exceeded the expectation, and showed a lot higher responses. The responses to the lower vibration stimulation intensities in Figure 3(b), which are rather flat, reflects that subjects' responses were inconsistent particularly in lower vibration intensities.

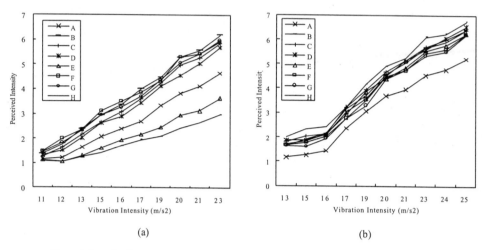

(a)   (b)

Figure 3. Perceived Intensity Responses from the first and verification Experiments

4. CONCLUSIONS

To have an effective feedback system on the human hand using a vibrotactile system, it is necessary to standardize the perceived intensity, not the stimulation intensity, among the loci where the contactors are applied to the skin. The following conclusions can be made from the current study;

- a linear relationship between the perceived and actual vibrotactile intensities
- a differentially and inconsistently scaled sensitivity among various hand loci to the vibrotactile stimulation, particularly in the thumb and palm area from the parts on fingers.
- an overestimation of the perceived intensities on the palm, and an underestimation of the perceived intensities on the fingers than the 1st experiment.

Various virtual reality applications using glove systems must have a feedback system either in the form of force feedback or tactile intensity feedback. The latter mostly uses such systems as vibrotactile or electrocutaneous feedback. Since the hand cannot convey the electric signals effectively, the vibratory form of stimulus signals seems reasonable for use on most glove systems.

This study investigated how the vibrotactile stimulation should be scaled among the hand loci to generate effective and valid feedback. It is expected that the results of the current study would provide a useful information for development of glove-type vibrotactile display systems that can be applied to virtual reality, augmented reality, or teleoperation.

Further studies should make it clear which scheme represents the true psychophysical gradient to the vibrotactile stimulation applied in this study.

REFERENCES

Cholewiak, R. W. and Collins, A., "Sensory and physiological bases of touch," in M. Heller and W. Schiff (eds.), The psychology of touch, Lawrence Erlbaum Associates, 1991.

Shimoga, K.B., "Finger force and touch feedback issues in dexterous telemanipulation," in Proc. 4th Annual Conference on Intelligent Robotic Systems for Space Exploration, New York: Troy, 1992.

Vallbo, A.B. and Johansson, R.S., "The tactile sensory innervation of the glaborous skin of the human hand," in G. Gordon (ed.), Active Touch. New York: Oxford University Press, 1978.

Verrillo, R.T., "Effect of contractor area on the vibrotactile threshold," Journal of the Acoustical Society of America, 1963, Vol. 35, pp. 1962-66.

Wilska, A., "On the vibrational sensitivity in different regions of the body surface," Acta Physiologica Scandinavica, 1954, Vol. 31, pp. 285-289.

Empirical Studies on an Augmented Reality User Interface for a Head Based Virtual Retinal Display

Olaf Oehme, Stefan Wiedenmaier, Ludger Schmidt, Holger Luczak

Institute of Industrial Engineering and Ergonomics, Aachen University of Technology
Bergdriesch 27, D-52062 Aachen, GERMANY
e-mail: o.oehme@iaw.rwth-aachen.de

ABSTRACT

Numerous experimental studies on the design of user interfaces of common desktop monitors exist, whereas empirical studies concerning the design of an Augmented Reality user interface are unknown. Therefore, recommendations relating to the optimal representation size of virtual information for an head mounted display with see through mode, which is used for Augmented Reality, cannot be given at the moment. In order to apply this new technology successfully to an industrial area, such information is urgently necessary. For this reason, three different kinds of displays were tested in this study regarding human information perception. Thus, the smallest target size necessary to provide successful information processing could be determined for different tasks. This study gives exemplary results obtained through tests with the prototype of a "Virtual Retinal Display".

1. INTRODUCTION

Augmented Reality (AR) means the enrichment of the real world with virtual information. Thus, e.g. repair instructions for a machine tool or important installation tips can be directly superimposed on the workers field of view (Luczak et al. 2000, Plapper et al. 1999). An interdisciplinary team of about 20 well-known companies and research centres are cooperating in the ARVIKA project (ARVIKA 2000). In this research project, supported by the German Federal Ministry for Education and Research (BMBF), useful fields of application for this kind of augmented reality technology in an industrial area are explored.

One possibility to overlay the real world with virtual information is to use a head mounted display (HMD). The HMD can be worn just like normal glasses. The information is superimposed on the user's field of view via two integrated LC-Displays. Half-silvered mirrors have to be installed in order to see the real world as well as the virtual information. Thereby, both the see through features of the glasses deteriorate and the size and the weight of the LC-HMD increase (Azuma 1997). Another problem of LC-HMD available on the market today is the limited field of view (FOV) of about 30 degrees. Actions taking place outside the FOV cannot be perceived by the user due to the construction of the HMD. This can evoke the workers' rejection especially in industrial areas. The user only has a limited peripheral view to the real world.

Another possibility of realizing the visual overlay is to use a so-called "Virtual Retinal Display" (VRD). In contrast to the LC-HMD, the VRD addresses the retina directly with a single laser stream of pixels. Thus, a very clear and sharp projection of different information can be ensured (Microvision 2000).

So far, research on see through displays concerning human information perception, task load and possible influences has not been conducted. Especially in the industrial field such investigations are urgently needed. They are necessary in order to provide the developers of augmented reality systems with recommendations for system design and to ensure an easy information access. Results of research on standard computer displays (e.g. Ziefle 1998 a, b) are not transferable to see through displays because the contrast a standard display between the background and the

1026

characters always remains the same whereas the contrast in a merged environment is constantly changing, depending on the users' direction of view.

Therefore, empirical research has been conducted on this subject within the ARVIKA research project. The main question was to find out the differences in human perception using a VRD prototype and other LC-Displays as described above with the aim to find out the smallest and optimum target size for virtual information with an acceptable human error rate.

2. METHODOLOGY

2.1 Independent variables

The three independent variables were the displays, the different tasks and the target size of the virtual information. Four different display types were used, one of them without see through mode for comparison.
Four different tasks were selected from tasks defined in the requirements phase of the ARVIKA project. The participants had to find the displayed virtual information while monitoring a printed paper in the real world. They had to confirm whether the paper contained the displayed information or not (Figure 1).

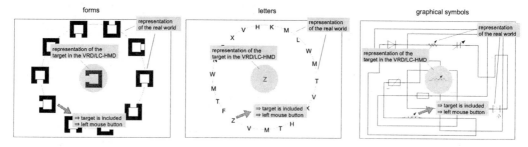

Figure 1: Different tasks for the identification of the virtual display information on printed given data

The first task was to recognise forms (squares which had an opening either at the top, at the bottom, at the left side or at the right side) which was faded into the subjects field of view. In the second task, the participants had to recognise letters; in the third task graphical symbols had to be identified. The fourth and last task took up again the forms from task one, but this time it was animated.
According to the German standard DIN 66234 for the target size of displayed letters on standard computer displays 18 arc minutes were chosen for the smallest target size. The following four target sizes resulted from the duplication of the one before. The representation size was divided into five levels from 18 to 288 arc minutes. This does not apply to the graphical symbols. In a pretest with ten participants it was found that the error rate at 18 arc minutes was to high. When the subjects regarded a symbol or a letter as undetectable, they very likely continued by simply pushing the No-button. Therefore, in this case 36 arc minutes were chosen as smallest target size. Thus, the biggest representation size for graphical symbols were 576 arc minutes.

2.2 Dependent variables

The four dependent variables were reaction time, false response percentage and subjective measures of vision problems and tiredness. The reaction time was measured as the interval between the target onset and the subjects' interaction, which means the confirmation or the negation of the virtual targets' presence on the given printed data described above. The participants estimation concerning visual problems and tiredness was recorded by using a slightly modified category-division technique from Heller (Heller 1981). In a standardised questionnaire the participants were asked to rate their visual problems and tiredness employing on a scale with 50 different levels.

2.3 Apparatus

Three HMD with see through mode were compared. The first display was the Retinal Scanning Display, a monocular prototype of a Virtual Retinal Display (VRD) from Microvision which can be used either on the right or on the left eye. It has a vertical field of view (FOV_V) of 21.37 degrees and the superimposed information appears in monochrome red. The Glasstron from Sony was the second display used (FOV_V = 15.48 degrees, binocular) and the third display used was the binocular i-glasses ProTech from i-O Display Systems (FOV_V = 17.24 degrees).

The displayed information was monochrome red in order to not control the colour perception. All displays had a screen resolution of 640 × 480 pixels and a frequency of 60 Hz. A portable computer (Xybernaut MA IV) were worn on a belt during the whole time of the experiment. To interact – to confirm or to negate – the participants had to push the left or the right mouse button, which is on the portable Xybernaut. The experiment was executed in a darkened room with lightning from a non-flickering lamp with constant current. The subjects were seated on a chair in a fixed position in front of a black background.

2.4 Participants

25 participants were recruited from the ARVIKA project partner Siemens AG from the department Automation and Drives (A&D). All participants were service technicians or service engineers and aged between 21 and 50 years with a mean of 36.9 years (SD = 8.3 years). In the pretest already mentioned before it was found that subjective measures of vision problems and tiredness differed significantly between male and female subjects. Because of the target group of an AR system in the service area being almost exclusively male, it was decided to investigate male subjects only. Participation was voluntary and the subjects were free to finish the experiment at any time.

2.5 Procedure

Before beginning with the experiment, the test persons sight was tested (farsightedness, shortsightedness, stereoscopic sightedness) and the dominant eye was ascertained. After they had been introduced to the technical features they had to identify the virtual display information on printed given data and furthermore to confirm whether they contained this information or not. The independent variables display type and target size were counterbalanced among the test persons. Each subject had to execute the tasks with all displays. Visual problems and tiredness were recorded after each display type.

3. RESULTS

In the ANOVA, significant differences of reaction time and false response percentage among all independent variables (display, task, target size) were observed ($p < 0.01$). The reaction time of the different displays in dependency on the target size is shown in Figure 2.

The Tukey-HSD Post-Hoc-Test showed that the results from the VRD differ significantly from the ones of the other displays ($p < 0.05$). Differences between the two LC-HMD showed no significance ($p > 0.05$). Especially in task "graphical symbols", the VRD proved its advantage compared to the other displays by having the lowest reaction time. These data have to be regarded separately, in order to obtain rules for the design of a VRD interface. In Figure 3, the dependency between reaction time and representation size for the VRD prototype is presented. Here, we only considered the reaction time the test persons needed to answer the task correctly in order to ensure that they did not press any button to continue when they were doubtful.

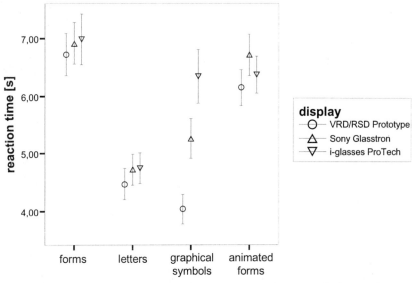

Figure 2: reaction time & 95% confidence interval dependent on display & task

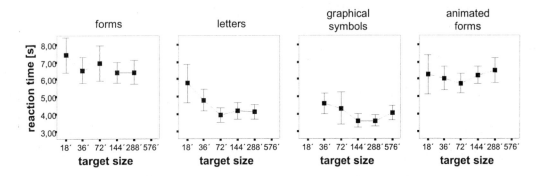

Figure 3: VRD/RSD prototype reaction time & 95% confidential interval dependent on task & target size

In Figure 4, the dependency of the false response percentage on the representation size is shown. It can be observed that the subjects had difficulties to answer correctly when the representation size was too small (see e.g. Figure 3, task "letters"). This is another reason why the reaction time was very slow. From a certain representation size onwards, significant differences concerning false response percentage could not be observed; in turn, an increasing reaction time could be detected. In this case, we assume that this can be explained by an increased degree of occlusion, i.e. a clear sight of the real world is not given. The subjects' ratings concerning visual problems and tiredness were too different to obtain significances.

4. DISCUSSION

Having evaluated the results, certain recommendations for the design of an Augmented Reality User Interface can be given. It can be concluded that different display technologies differ in their usability for different tasks. The VRD has proven useful for complex vectorgraphics like graphical symbols. It has to be taken into consideration that at a certain representation size, e.g. 36 or 72 arc minutes, the error rate cannot be lowered any more significantly, whereas the reaction time can be improved. Furthermore, it can be concluded that letters have to be presented at least with 36 arc minutes, though 72 arc minutes would be better due to the faster reaction time at this size. The best representation size to optimise the reaction time are 144 or 288 arc minutes regarding task "graphical symbols".

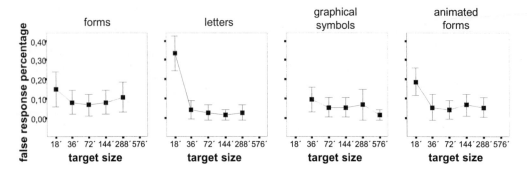

Figure 4: VRD/RSD prototype frp & 95% confidential interval dependent on task & target size

5. ACKNOWLEDGEMENT

The research is funded by the German Federal Ministry for Education and Research (BMB+F) within the research centre 'Augmented Reality in Design, Production and Service' (ARVIKA) under grant no. 01 IL 903 R 4. Dr. Krahl from the department 'Informationstechnik' of DLR deserves additional special acknowledgement for his kind support of the research projects. Furthermore, we want to thank Mr. Bürschgens and Mr. Hamacher for the realisation of the test programs and all partners in the project ARVIKA who supported this research. We would also like to give special gratitude to Wolfgang Wohlgemuth from Siemens AG Automation & Drives for his generous support without which the experiments could not have been executed.

REFERENCES

ARVIKA (2000): Homepage des ARVIKA-Konsortiums: http://www.arvika.de/

Azuma, R. (1997): A Survey of Augmented Reality. In: *Presence: Teleoperators and Virtual Environments 6* (August 1997), No. 4, pp. 355-385

DIN 66234: Norm DIN 66234 Teil 2. *Bildschirmarbeitsplätze; Wahrnehmbarkeit von Zeichen auf Bildschirmen*

Heller, O. (1981): *Theorie und Praxis des Verfahrens der Kategorienunterteilung (KU)*. Würzburg, Würzburger Psychologisches Institut, Lehrstuhl für Allgemeine Psychologie, Forschungsbericht, 1981

Luczak, H.; Wiedenmaier, S.; Oehme, O.; Schlick, C. (2000): Augmented Reality in Design, Production and Service - Requirements and Approach. In: Marek, T.; Karwowski, W.: Manufacturing Agility and Hybrid Automation - III, Proceedings of the HAAMAHA 2000, 2000

Microvision (2000): Homepage Microvision Inc.: http://www.mvis.com/

Plapper, V. ; Wenk, C. ; Weck, M. (1999): Augmented Reality unterstützt den Teleservice. In: *wt Werkstatttechnik* (1999), Nr. 89 H. 6, S.293-294

Ziefle, M. (1998 a): Effects of Display Resolution on Visual Performance. In: *Human Factors* (December 1998), Vol. 40, No. 4, pp. 554-568

Ziefle, M. (1998 b): *Visuelle Faktoren bei der Informationsentnahme am Bildschirm*. Aachen, RWTH, Philosophische Fakultät, Habilitation, 1998

Influence of Different Visualization Techniques of 3D Free-Form Surfaces on Perceptual Performance in a CAVE

Ralf Breining, Wilhelm Bauer, Andrea Gaggioli*, Harald Widlroither

Fraunhofer Institute for Industrial Engineering, Nobelstr. 12, 70569 Stuttgart, Germany
* ATNP-LAB Istituto Auxologico Italiano, Via Spagnoletto, 3, 20149 Milan, Italy

Abstract

Computer-aided design methods are extensively applied in the design of aircraft, automobiles, buildings, computer and many other products. In the next future, CAD methods will be employed in conjunction with 3D immersive displays, which may dramatically improve the possibilities of visualization and interaction offered by common 2D display CAD workstations. Thus, careful evaluation needs to be made as to how the computer-generated object is represented on 3D display. It is commonly recognized that a computer-generated 3D object should be a) an accurate description of the model being designed b) presented in a realistic and integrated format, so that it can be visualized and interpreted without introducing uncertainty regarding to the represented proprieties (Brown, 1995). To ensure that the displayed 3D image will satisfy these requirements, it is important focusing on which perceptual/cognitive operation is performed by the user during the design process. We performed one experiment to examine the effects of different visualization techniques of 3D computer-generated free-form surfaces on depth perception, using Steven's *stimulus magnitude estimation* paradigm. The factors we investigated were presence versus absence of binocular disparity, four different monocular coding techniques (wireframe, flat shading, Gouraud shading, Gouraud shading & normals) and two levels of shape complexity. We found that a stereoscopic image provides subjects with a visual cue that not only enhances perceived relief of 3D forms but also improves accuracy of activities necessary for 3D design, such as depth estimation of free-form surfaces. Furthermore, the nature of the experimental task implies that stereoscopic displays can be most useful when information is presented from a vertical viewpoint.

1. Introduction

The use of conventional computer systems in such complex processes as the design of product shapes requires the user to have a great ability to think in abstract terms. VR systems attempt to reduce this cognitive demand by presenting the user with three-dimensional representation of shapes, tools and workplaces in real time. However, careful evaluation needs to be made as to how the computer-generated object is represented and how the user interact during the design process. The 3D impression should not only contribute to the aesthetical quality of the image, but it should also allow the user to carry out correct inferences about the depth of the real object that the computer generated model represents.

Previous research on human depth perception in 3D environments has addressed two basic questions, that is, which visual cues provide effective depth information and how does the effectiveness of given depth cues changes as a function of the viewing distance. Most studies have focused on subjects ability to estimate egocentric distance and relative distance, while there exists poor research about how depth of single objects is perceived. Real objects have depth and are located phenomenally as well as physically at different distances (Dember & Warm, 1979). That is, objects have an egocentric distance (i.e., the distance of the object from an observer) and a relative distance (i.e., the distance of objects from each other). Objects also have depth in that they are perceived as three-dimensional and some parts of an object look farther away than do other parts (Matlin & Foley, 1992). This type of depth perception can be described as the ability of perceive variations in egocentric distance when looking at different points located on a 3D object.

Depth perception can be studied by manipulating any of the cues to depth (e.g. visual texture, shading, stereo disparity) and determining how perceptual responses are altered. Our experiment has applied this methodology, using Stevens *stimulus magnitude estimation* paradigm. Stereo vision and four monocular coding techniques were investigated. In Stevens original paradigm (1975), the subject is presented a first stimulus of arbitrary intensity and told that its sensation is a particular numerical value (called *modulus*) or is allowed to choose his or her own modulus. Stimuli of different magnitudes are then presented randomly, and the observer is required to assign numerical values to them proportional to their perceived magnitudes. These values then directly provide the scale relating physical magnitude to perceived magnitude. In the modified version of Stevens paradigm subjects are required to report their estimates in a standard metrical unit (i.e. centimetres) and are told the real intensity (i.e. depth) of the modulus. We used this simplified procedure because it was more intuitive. The task consisted in estimating depth of computer generated 3D objects that were observed from a vertical viewpoint. This setting was supposed to stress the importance of the surface features in estimating depth, because in trying to accomplish the task subjects were to find the point of the surface closest to them and then estimate the difference in depth between this point and the object base.

Stereo vision and four monocular coding techniques were investigated. Monocular coding techniques were wireframe (with hidden edges removed), flat shading, Gouraud shading (Foley & van Dam, 1982) and Gouraud shading with surface normals.

2. Methodology

2.1 Design, Subjects, Apparatus and Stimuli

The experiment consisted of two types of object geometry (simple versus complex), four types of graphic images (wireframe, flat shading, Gouraud shading and Gouraud shading with surface normals) and six depth values (15, 30, 45, 60, 75, 90 cm) as completely crossed factors with stereopsis (present or absent). The serial order of depth values was determined by a Latin square arrangement and was the same for all participants. All factors were within subjects.

Twenty-four subjects served as participants in the study (mean age = 29.4 years). All subjects had normal or corrected-to-normal vision and reported no experience with virtual reality, no experience with CAD software and upper-intermediate experience with computer.

3D images were created and displayed on the front wall of a four-walls CAVE capable of both stereoscopic and monoscopic modes. The screen is 300 cm wide and 300 cm high with a resolution of 960 x 960. Stereoscopic condition was created using CrystalEyes time multiplexed LCD shutter glasses. Two basic 3D free-form shapes were created (see Fig.1). The simple one (a) was represented by a medusa-like object, the complex shape (b) was a terrain. Both objects were 50 cm wide. A squared base was added to the scene because during a pilot-experiment subjects reported that stimuli "floated" above the projection plane.

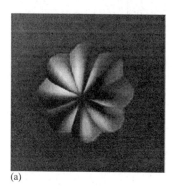 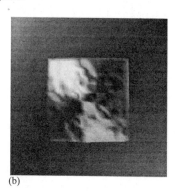

    (a)   (b)

Figure 1a-b. Stimuli used in exp. 1 (both objects are here rendered with Gouraud shading)

2.2 Procedure

Participants were told the purpose of the research and given specific instructions regarding the estimation task. In estimating depths, participants were required to report their estimates in centimetres. In order to facilitate the task, subjects were informed about their egocentric distance from the projection plane and given the edge of the square (100 cm) that represented the base of the 3D objects. Following practice, they estimated depth on each of 48 trials (6 trials for each rendering method; 24 trials for each geometry type) with stereopsis (stereo condition), took five minutes break, then estimated depth to another identical set of 48 trials without stereopsis (mono condition). From the participant's perspective, the depth value that the displayed object had on any given trial was random. In order to control the possible effect of sequence for the rendering conditions, four different sequences were arranged using a 4 x 4 latin square. Subjects were divided in four groups (A,B,C,D) and each group was assigned a different sequence. In each group, half of the subjects estimated depth in stereo first, whereas the other half estimated depth in mono first. Following the experiment, subjects were given a questionnaire, which consisted of information about their background (age, vision, gender, experience with VR, CAD applications and computer).

3. Data Analysis and Results

In order to analyse data, two ANOVAs were performed. The first used depth-difference estimates as the dependent measure and the second used relative errors as the dependent measure for accuracy. Relative error was calculated as follows: Relative error = (Depth estimate - True depth) / True depth.

This represents the percent error in an estimate relative to the true depth difference, with the sign indicating the direction of the error (when the sign is negative, it indicates underestimation; when the sign is positive, it indicates overestimation). Stereopsis, type of geometry, type of rendering and size of depth-difference were the independent variables in both analyses. Fig. 3 illustrates the means of the estimates in stereo and mono condition. The line with crosses represents perfect performance.

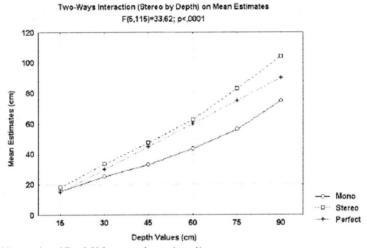

Figure 3. Effect of Stereopsis and Depth Values on estimates (exp. 1).

Fig. 3 shows that observers generally underestimated depth of 3D objects presented in mono, while performance in stereo was clearly more accurate. This is confirmed by the high significant effect of stereopsis in the ANOVA performed on relative error ($F(1,23)=69,65$; $p < 0,000001$). The effect of object complexity on accuracy is reported in Fig. 4. This graph shows that estimates were far more accurate for the simple shape than for the complex shape.

Fig.5 illustrates the effect of graphic coding techniques on accuracy. The main effect is very significant. A Tukey's HSD test was performed to assess pair wise comparisons. It was found that mean relative errors for Gouraud shading differed significantly from each other rendering method, and that there was a significant difference among Wireframe and Normals. Normals and Flat Shading determined more accurate estimates than Wireframe and Gouraud shading and the worst performance was determined by Gouraud shading.

1033

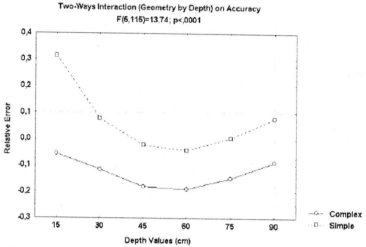

Figure 4. Effect of object complexity on relative error.

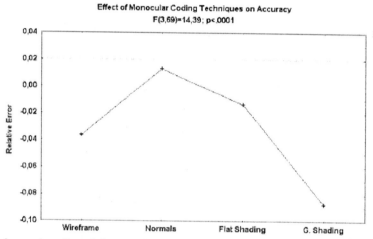

Figure 5. Effect of monocular coding techniques on relative error.

4. Discussion

Estimates and accuracy were significant affected by depth values, stereopsis, object complexity and rendering methods. The highest relative error was determined by the smallest depth value (15 cm). A post-hoc analysis performed on the interaction geometry type by object depth revealed that this effect was principally due to overestimation of the smallest depth value of the simple object (p < 0,01).

Subjects estimated depth of 3D objects more accurately when objects were displayed with binocular disparity than when only monocular cues were provided. Thus, retinal disparity confirms to be a very important depth cue, that provides effective perceptual information not only in estimating egocentric/relative distances but also in estimating distances between parts of an object that lay in different depth positions. Our result agrees with findings of previous studies which have demonstrated that stereopsis can be superior to monocular depth cues under certain circumstances. Schor (1993) indicates that stereopsis is very useful in portraying fine depth intervals (5-10 arc sec) when depth stimuli are presented near the plane of fixation (in this case, the video screen). Other studies indicate that stereopsis can be an effective cue to depth at short distances, (1-2 m) while monocular cues (i.e. texture, perspective etc.) are useful over longer distance (Surdic, 1994). This feature of stereopsis is explained by the fact that this depth cue is based on retinal disparity, which decreases as a function of distance in a non-linear fashion (Surdic, 1994). Our results indicate also that stereoscopic displays can be most useful when information is presented from a vertical viewpoint. This finding is in agreement with results obtained by Davis & Hodges (1995) who have described other conditions where stereopsis is particular effective. The effect of object complexity was highly significant. Complex objects depth was underestimated while simple objects depth - with the exception of the depth value 15 cm - was estimated more accurately.

Another interesting result of this study is that standard CAD monocular coding techniques affect depth estimates in different ways. Depth of wireframe objects was estimated more accurately than depth of Gouraud shaded objects but this result was inverted when normals were added to the surface and when a less realistic shading (flat) was used. This finding can be interpreted through results by Koenderink et al. (1996) who showed that most of the relief of a shape is determined by visual contour and that shading adds very little to the "solidity" of an object if other cues are not available. Todd and Mingolla (1983) performed three experiments to examine the perceptual salience of shading, texture, specular highlights and directions of light sources in providing information about the 3D structure of a cylinder. The results indicated that the shininess of the surface enhanced the perception of the curvature, but had no effect on the perceived direction of illumination, and that shading was generally less effective than the gradient produced by the texture for depicting surface in three dimensions. In our study, texture was not a surface feature but the addition of small normals across smooth-shaded surfaces appeared to be incremental to accuracy.

5. Conclusions

We found that a stereoscopic image provides subjects with a visual cue that not only enhances perceived relief of 3D forms but also improves accuracy of activities necessary for 3D design, such as depth-size estimation of free-form surfaces. Furthermore, the nature of the experimental task implies that stereoscopic displays can be most useful when information is presented from a vertical viewpoint. Another interesting finding of this study is that standard CAD monocular coding techniques affect depth estimates in different ways. In particular, we found that smooth shaded objects are underestimated in depth and that the addition of normals, the use of a less realistic shading (flat shading) and the wireframe coding technique can significantly improve accuracy in estimating.

References

Brown, M.E., Gallimore, J.J. (1995). Visualization of Three-Dimensional Structure During Computer-Aided Design. *International Journal of Human-Computer Interaction*, 7(1) 37-56.

Davis, E.T., Hodges, L.F. (1995). Human Stereopsis, fusion and stereoscopic virtual environments. In W. Barfield and T. Furness (Eds.) *Virtual environments and advanced interface design*. New York: Oxford, 145-174.

Dember, W.N., Warm, J.S. (1979). *Psychology of perception* (2nd ed.). New York: Holt, Rinehart & Winston.

Foley, J., Van Dam, A. (1982). *Fundamentals of Interactive Computer Graphics*. Reading, MA: Addison-Wesley.

Gallimore, J.J., Brown, M.E. (1993). Visualization of 3-D Computer-Aided Design Objects *International Journal of Human-Computer Interaction*, 5 (4), p.361-382.

Koenderink, J.J., Van Doorn, A.J., Christou, C., Lappin, J.S. (1996). Shape constancy in pictorial relief. *Perception*, 25, p. 155-164.

Matlin, M.W., Foley, H.J. (1992). *Sensation and Perception* (3rd ed.). Boston: Allyn and Bacon.

Schor, C. (1993). Spatial constraints of stereopsis in video displays. In S.R. Ellis (Ed.), *Pictorial communication in virtual and real environments* (2nd ed.). London: Taylor and Francis, p. 546-557.

Surdik, R.T., Davis, E.T., King, R.A., Corso, G.M., Shapiro, A., Hodges, L., Elliot, K. (1994). Relevant cues for the visual perception of depth: Is what you see it where it is? *Proceedings of the Human Factors and Ergonomic Society – 38th Annual Meeting*, p. 1305-1309.

Stevens, S. S. (1975). *Psychophysics*. New York: John Wiley.

Todd, J.T., Mingolla, E. (1983). Perception of Surface Curvature and Direction of Illumination From Patterns of Shading. *Journal of Experimental Psychology: Human Perception and Performance*, 9 (4), p. 583-595.

THE VIRTUAL PROCESS VISUALIZATION METHOD
FOR INDUSTRIAL PROCESS CONTROL

Carsten Wittenberg

SIEMENS AG, Corporate Technology - User Interface Design
D – 81730 Munich, Germany, Tel.:+49 89 636 57470
carsten.wittenberg@mchp.siemens.de

ABSTRACT

This paper presents the Virtual Process Visualization (ViProVis) method for supervisory control of technical systems, which contains the three pillars Virtual Process Elements, Goal and State Visualization and Task-oriented Structure. Based on this method, two interfaces for different applications are developed. In the experiments, these graphical user interfaces are be compared to and evaluated against conventional topological interfaces.

1. INTRODUCTION

Complex, dynamic industrial plants are monitored and controlled by human operators in a control room. Increasing requirements regarding quality, economics and ecology, and increasingly efficient and inexpensive automation systems lead to an increasing degree of automation in such industrial plants. As a consequence, the number of operators decreases but the amount of process information to be monitored increases. The operators have to handle a flood of information. Nowadays several processes are managed centrally from one control room with computer-aided control systems. Because of the information preparation required, the operator loses contact with the process. The use of computer monitors means that the flood of process information is represented in a very limited space. However, the ability to identify the relevant process variables from the vast amount of information is important to enable the current situation to be assessed quickly and safely (Johannsen, 1993). A human-machine interface that is adapted to human characteristics can support the operator in this situation. The design of human-machine interfaces has to support human information processing - regarding both the accommodation of information as well as action planning and problem solving.

This contribution presents a human-centered visualization concept called *Virtual Process Visualization*, which contains the three pillars *Virtual Process Elements, Goal and State Visualization* and *Task-oriented Structure* (Wittenberg, 1997; 1998; 1999; 2001). Based on this method, interfaces for two different applications (a distillation column and a central heating system) were prepared and evaluated in experiments.

2. REQUIREMENTS FOR PROCESS VISUALIZATION

Operators monitor and control the technical systems using process visualization. Usually the information is only presented visually on different computer displays. The visualization and the displays are used as a window to the process. The quality of this information presentation determines the amount of support provided to the operator during his daily work. An ideal process visualization can be described as a "transparent window" to the process. Different steps of operators' actions can be described in a simplified manner based on different action models of human behavior (e.g. Rasmussen, 1984; Norman, 1986). The operator perceives process information, interprets this information and determines the state of the (sub)process. The operator compares this actual state with the optimal state – the goal. If there is a relevant deviation between the actual state and the goal the operator determines the actions required. In the final stages the operator plans and carries out the actions required. A user-centered process visualization should support these steps and therefore support the operator.

It is well known from psychology that humans process pictorial information faster and more easily than textual information. These differences are very well described by the so-called *multimodal mode* (Engelkamp, 1991). This model describes different memory areas, which in each case are dependent on the modality of the releasing attraction and possess very different capability characteristics. The advantages of pictorial information - the so-called *picture superiority effect* - were determined from this. Consequently, there should be an advantage in coding process information pictorially to support the operator. Correct operator action planning is based on the mental model. The planning of action steps can be described as a dynamic cognitive simulation of the mental model. This simulation is repeated iteratively with initial parameters changed in each case, in line with simulations of mathematical models on computer systems, until a satisfactory result is achieved. The process visualization has to make the necessary information available, so that the operator is able to form a correct mental model based on system and process identification. Additionally a mental model is usually characterized by a strong figurativeness, so that the mental model should be supported by this kind of pictorial visualization (Dutke, 1994).

In order to make complex technical systems for the operator controllable in all critical and non-critical situations, a function-oriented structuring of the process information is necessary. This enables a transparent representation of the process.

3. VIRTUAL PROCESS VISUALIZATION

The Virtual Process Visualization is a concept which meets the above problems with a special visualization technique familiar from modern computer graphics. Different kinds of relationships (even time-dependencies) can be visualized with this type of visualization. The entire process is represented in a highly pictorial and transparent manner. Based on the superordinate function, subtasks are defined which are linked to appropriate goals. Each task is related to a specific view that includes the process items, the process variables and the relationships between them. A superordinate overall view of the process shows the states of the system and the subtasks. The views of the subtasks show the fulfillment of the goals, the necessary process values, and the statuses of the process items. Because an exclusively quantitative information representation is not possible with pictorial methods, the process values are presented qualitatively. Threshold values and setpoints facilitate the identification of current process conditions. In the concept presented in this paper the threshold values and the setpoints are only visible when the related process values or process items are in abnormal states. The amount of information is thus limited to that required. Different kinds of relationships (even time-dependencies) can be visualized with this type of visualization. This visualization concept is based on three pillars:

- Virtual Process Elements,
- State and Goal Visualization,
- and Task-oriented Structure. These components of the concept will be described in the following sections.

3.1 Virtual Process Elements

The individual process units are implemented as so-called Virtual Process Elements (Wittenberg, 1997). A visual object is developed based on a manifestation typical for the respective group of items. All these graphic objects are developed with respect to a consistent presentation of information and a consistent interaction. Following the idea of direct manipulation, the operators interact with the process items directly. This visualization concept is expected to be efficient with the visualization of process variables and relations between the process items. Visual sizes - such as fill-up levels and flows - are shown directly, others are coded in color and form. Because a quantitative information representation is not exclusively possible with figurative methods, the process values are qualitatively presented. For leading a process the knowledge of the process variables, which are important for the current situation, is indispensable (Färber et al., 1985). However always the absolute numeric value of the process variable is not required. Often only a comparison with setpoints and threshold values or a qualitative representation of the process variable is necessary. For example the absolute level of a buffer is not crucial, but the question in many situations whether the contents are sufficient still for the reaching of the production goal within the next time period. Threshold values and setpoints facilitate the identifier of current process conditions. In the presented visualization technique the threshold values and the setpoints are only visible, if the connected process values or process items are in abnormal states. In this way the number of information is limited to the necessary measure. As an example Figure 1 shows the conventional DIN symbol for a pump and the corresponding Virtual Process Element.

3.2 State and Goal Visualization

Based on the superordinate function, subtasks are defined which are linked to appropriate goals. Each task is related to a specific view that includes the process items, the process variables and the relations between them. A superordinate overall view of the process shows the states of the system and the subtasks. The views of the subtasks show the fulfillment of the goals, the necessary process values, and the statuses of the process items. Because an exclusively quantitative information representation is not possible with pictorial methods, the process values are presented qualitatively. Threshold values and setpoints facilitate the identification of current process conditions. In the concept presented in this paper the threshold values and the setpoints are only visible when the related process values or process items are in abnormal states. The amount of information is thus limited to that required. Figure 2 shows the visualization of the fill-up level in a reservoir. A small line represents the target value (50%). This line is visible only if there is a deviation from the target value. The arrows show the direction of the necessary level change. Like the "target-line" these arrows are again only displayed if there is a deviation. The size of each arrow corresponds to the deviation from the desired fill-up level.

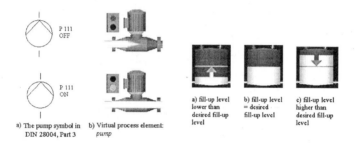

a) The pump symbol in DIN 28004, Part 3

b) Virtual process element: *pump*

a) fill-up level lower than desired fill-up level

b) fill-up level = desired fill-up level

c) fill-up level higher than desired fill-up level

Figure 1: Pump-symbol   Figure 2: Visualization of the actual and desired fill-up level

3.3 Task-oriented Structure

Because the human cognitive capacity is limited the process is structured hierarchically by function to reduce the problem solve space. The basis of this hierarchy is the tasks of the operators, determined in previous analyses, and the statuses of the process items involved that are related to the tasks. A special area in the interface displays the extent to which the tasks have been fulfilled. Each task is linked to goals. The degree of goal fulfillment is represented in the form of colored bars. The color and length of the columns display whether a goal is achieved, still in tolerance, or whether a warning or an alarm is present. The length of the column allows the operator to determine how far the goals have been achieved and when he/she has to intervene (Wittenberg, 1998; 1999; 2001).

4. SAMPLE APPLICATIONS

As a sample application, an interface for the guidance of a distillation process for the separation of a benzene-toluol-mixture is introduced. The mathematical core of the process simulator used was developed as a training simulator for the chemical industry. This simulated chemical process features sufficiently high complexity and demanding technical requirements for the testing of visualization methods (e.g. Wittenberg, 2001). Figure 3 shows a screenshot of the user interface for the distillation column based on the Virtual Process Visualization method.

The second application is a central heating system to maintain the temperature of an apartment complex (SCARLETT; e.g. Inagaki et al., 1998; Moray et al., 2000). Because of the complex relationships between the process variables, controlling this process needs huge manual control effort and is therefore a suitable application for the examination of visualization concepts. Figure 4 shows a screenshot of the energy view on SCARLETT, based on the Virtual Process Visualization method.

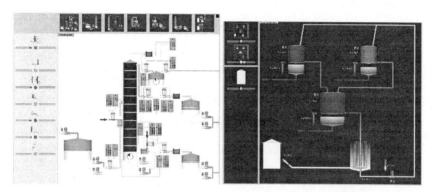

Figure 3: User Interface for a distillation column   Figure 4: User Interface for SCARLETT

5. EXPERIMENTS

In experiments, this visualization method was compared to and evaluated with conventional topological interfaces. These experiments were intended to verify the hypothesis that this visualization is more descriptive, thus making it easier to comprehend the actual process state. The first experiments used the distillation column as an exemplary application. The subjects (German graduate students) had to keep the process in an optimal state although different malfunctions occurred. After each scenario the students had to fill out questionnaires about the scenario, the malfunctions and their subjective impressions. With the questionnaires aspects like general design, navigation, support of the operator performing the tasks, presenting the states and target value etc. were investigated. The Man-Whitney-U test were carried out on the results using the software tool SPSS and showed significant advantages of the Virtual Process Visualization method (Kistner, 1999; Wittenberg, 2001).

The second experiment using the SCARLETT application was designed as a 2x2 matrix. The subjects (Japanese graduate students) were split into in two groups. Each group started with a different interface. After the first stage of the experiment, the type of interface was changed for the second stage. Each stage contained five different scenarios, i.e. different initial values and a different desired temperature in the apartment block. The dependent variables were the mean errors of the temperature of the apartment block, the fill-up levels in the reservoirs, the time to respond to malfunctions and the number of undetected malfunctions. 30 different sets of variables were recorded. The variables were analyzed with the software tool STATISTICA. As expected, differences were found in the time to respond to malfunctions, the number of undetected malfunctions and in the mean error of the apartment block. An ANOVA was carried out, and showed that all these results are significant. There were no significant differences between the two types of interface in the mean errors of the fill-up levels. A very interesting result was that the difference in the mean error of the temperature of the apartment complex depended on the sequence in which the interface types were used. The subjects who started with the interface based on the Virtual Process Visualization method made a significantly lower error (including when they used the conventional interface). In the experiment, two interesting points were found. The first is that there are no significant differences in the mean errors of the fill-up levels of the reservoirs. In discussion with the subjects it emerged that subjects paid the greatest attention to controlling the temperature in the apartment complex. This suggests that the hierarchy of targets was presented clearly. The other interesting fact is that there is a difference depending on the sequence in which the two interface types were used. Subjects who started with the interface based on the Virtual Process Visualization method made significantly less errors than the subjects who started with the conventional interface. This result could prove the hypothesis that a highly pictorial interface presents the process knowledge more clearly and assists in the building of a mental model (see above). The other results (mean error temperature, time to respond to malfunctions, number of undetected malfunctions) also confirm the effectiveness of this visualization concept.

REFERENCES

DIN 28004 Teil 3 (1988): Fließbilder verfahrenstechnischer Anlagen; Graphische Symbole.

Dutke, S. (1994): Mentale Modelle: Konstrukte des Wissens und Verstehens. Göttingen: Verlag für angewandte Psychologie.

Engelkamp, J. (1991): Das menschliche Gedächtnis: Das Erinnern von Sprache, Bildern und Handlungen. 2. Auflage, Göttingen: Hogrefe.

Färber, G., Polke, M., Steusloff, H. (1985): Mensch-Prozeß-Kommunikation. Chemie-Ingenieur-Technik 57 Nr. 4, pp. 307-317.

Inagaki, T., N. Moray, M. Itoh (1998): Trust, Self-Confidence and Authority in Human-Machine Systems. In: Proceedings IFAC Man-Machine Systems, Kyoto, pp. 431-436.

Johannsen, G. (1993): Mensch-Maschine-Systeme. Berlin: Springer-Verlag.

Kistner, J (1999): Experimentelle Untersuchung von Mensch-Prozeß-Schnittstellen. Unpublished Diploma Thesis, Laboratory for Systems Engineering and Human-Machine Systems, University of Kassel.

Moray, N., T. Inagaki, M. Itoh (2000): Adaptive Automation, Trust, and Self-Confidence in Fault Management of Time-Critical Tasks. Journal of Experimental Psychology, 2000, Vol. 6, No. 1, pp 44-58.

Norman, D.A. (1986): Cognitive Engineering. In D.A. Norman, S.W. Draper (Eds.): User centered System Design, Hillsdale New York: Earlbaum, S. 31-62.

Rasmussen, J. (1984): Strategies for state identification and diagnosis in supervisory control tasks, and design of computer-based support systems. In: W.B. Rouse (Ed.): Advances in Man-Machine Systems Research, Volume 1, Greenwich: JAI Press, S. 139-193.

Wittenberg, C. (1997): Unterstützung der menschlichen Informationsaufnahme durch Prozeßvisualisierung mittels virtueller Prozeßelemente. In: K.P. Gärtner (Ed.): Menschliche Zuverlässigkeit, Beanspruchung und benutzerzentrierte Automatisierung, Bonn: Deutsche Gesellschaft für Luft- und Raumfahrt e.V. (DGLR), pp. 183-196.

Wittenberg, C. (1998): Qualitative visualisation of a chemical process using the three-dimensional computer graphics. In: Proceedings of the 17th European Annual Conference on Human Decision Making and Manual Control, France, Valenciennes: LAMIH Laboratoire d'Automatique et de Mécanique Industrielles et Humaines, Université de Valenciennes, pp. 217-226.

Wittenberg, C. (1999): Aufgabenorientierte Visualisierung eines komplexen verfahrenstechnischen Prozesses unter Verwendung dreidimensionaler Computergrafik. In: U. Arend, E. Eberleh, K. Pitschke (Eds.): Software-Ergonomie '99 - Design von Informationswelten, Reports of the German Chapter of the ACM No. 53. Stuttgart: B.G. Teubner, pp. 335-344.

Wittenberg, C (2001): Virtuelle Prozessvisualisierung am Beispiel eines verfahrenstechnischen Prozesses. Fortschritt-Bericht VDI Reihe 22 Mensch-Maschine-Systeme Nr. 5, Düsseldorf: VDI-Verlag.

ACKNOWLEDGEMENT

Part of this project was made possible by the support of the Japanese Ministry of Science, Education, Sports, and Culture MONBU-SHO and the Laboratory for Cognitive System Science (University of Tsukuba, Japan).

CHANGES IN SUBJECTIVE WELL-BEING ASSOCIATED WITH EXPOSURE TO VIRTUAL ENVIRONMENTS (VEs)

Ben D. Lawson, Ph.D.

Spatial Orientation Systems, Naval Aerospace Medical Research Laboratory, 51 Hovey Road, Pensacola FL 32508-1046. blawson@namrl.navy.mil

ABSTRACT

Correctly interpreting the meaning of observed deficits in a Virtual Environment (VE) user's performance or physiological status is difficult without understanding the user's subjective state of well-being. This is a brief review of the adverse changes in subjective well-being elicited by exposure to VEs and other synthetic experiences wherein the usual relation among visual, vestibular, and somesthetic sources of information has been distorted. Symptom prevalence, severity, type, and order are described and tools are recommended for evaluating subjective symptoms and observable signs. Research recommendations are made concerning unknown aspects of VE symptomatology. Gaps in knowledge about symptomatology should be filled to facilitate human-centered design strategies and enhance VE usability.

1. INCIDENCE AND SEVERITY OF DECREASED WELL-BEING IN VIRTUAL ENVIRONMENTS (VEs)

Table 1 summarizes some recent estimates of the proportion of users who suffer from VE-related discomfort. Severity of VE effects can be inferred from the proportion of users who cannot tolerate VE exposure and ask to withdraw from VE experiments.

Table 1. Incidence of adverse reactions to virtual environments.

Source of Estimate	Procedure	% Users with Symptoms	% Withdrawal
Regan & Price, 1994 (experiment)	Stimulus: 20 mins. VE navigation (Provision 200) Sample: n=146	61% of subjects reported adverse symptoms (28% of whom had nausea)	5%
Regan & Ramsey, 1996 (experiment)	Stimulus: same as above Sample: n=20 (placebo + VE)	75% users report adverse symptoms (25% had nausea)	0%
Wilson, 1997 (review)	Stimulus: 20 - 120 mins. in one of three VEs Sample: n=233 in 12 studies	80% report adverse symptoms (symptom type not specified)	5%
Stanney et al., 1998 (review)	Stimulus: 15-60 mins. in various VE studies Sample: ≈ 1000 exposures (personal communication; subjects not specified)	80-95% range for adverse symptoms (symptom type not specified)	5-30%
Howarth & Finch, 1999 (experiment)	Stimulus: VE navigation for 20 mins. via head or hand control (Virtual i-glasses) Sample: n=14 completed both conditions	1. Head control: 100% felt adverse symptoms (64% had mild & 43% moderate nausea) 2. Hand control: 71% adverse symptoms (21% mild & 7% of moderate nausea)	18% during 1st trial (3/17 original subjects; not specified which were in head vs. hand control, but order balanced)
Summary incidence of adverse reactions to VE:		61-100% feel some adverse symptoms; at least 80% estimated in multi-study reviews	0-30% withdrawal; at least 5% in reviews

Displays vary widely among VE manufacturers and within one manufacturer's device over time. Also, individuals vary greatly in their adverse reactions to synthetic experiences. Nevertheless, the incidence and severity of VE side-effects shown in Table 1 seems great enough to warrant concern, especially in light of evidence of additional after-effects such as sensorimotor incoordination and altered gaze control (Durlach & Mavor, 1995), whose relation to signs and symptoms is not fully understood.

2. CHARACTERISTIC ASPECTS OF VE-RELATED SYMPTOMATOLOGY

Table 1 indicates that VE exposure can produce various signs and symptoms; these are elucidated further in Table 2.

Table 2. Most common adverse signs and symptoms elicited by virtual environments.

Source	Most Common Symptoms (No estimates of order of onset reported)
Regan & Price, 1994	Dizziness, stomach awareness, nausea, headache, & eyestrain
Regan & Ramsey, 1996	Dizziness, stomach awareness, nausea, disorientation, headache
Howarth & Finch, 1999	During exposure: nausea was the only symptom the authors mentioned by name. Other adverse symptoms were present, but not listed individually. (After exposure: "car sickness," "discomfort," "hangover," & "feeling vacant")
Kennedy & Stanney, 1997	Most common VE symptom cluster of SSQ = "Disorientation," composed of "difficulty focusing," "nausea," "fullness of head," "blurred vision," "dizzy" (rated eyes open & eyes closed), & "vertigo." Symptoms that only appear in the "Disorientation" cluster & not in any other cluster = "fullness of head," "dizzy," "vertigo."
Summary of common VE symptoms:	Most common symptoms in VEs = dizziness, stomach symptoms, & headache. (Symptoms most unique to VEs = fullness of head, vertigo, & blurred vision)

Post-exposure symptom profiles assessed by the Simulator Sickness Questionnaire (SSQ) indicate that VEs produce more of the "Disorientation" symptom cluster (Kennedy & Stanney, 1997). Three (of the five) descriptors within the Disorientation cluster (viz., "difficulty focussing," "nausea," and "blurred vision") are shared with the other two symptom clusters of the SSQ ("Oculomotor" and "Nausea"), while three descriptors within the Disorientation cluster ("fullness of head," dizzy," and "vertigo") are unique. It would be interesting to determine how individual VE symptoms (rather than clusters) compare to symptoms generated by other synthetic experiences.

3. EARLIEST WARNINGS OF DISCOMFORT: INFERENCES FROM SYNTHETIC ENVIRONMENTS
Most published accounts concerning VEs and simulators report quantitative summaries of total symptom severity before and after exposure without specifying the temporal onset and progression of individual signs and symptoms. Attempts should be made to evaluate symptom onset and progression to see how VEs compare to one another and to other synthetic environments. Such knowledge would aid the development of health and safety guidelines for VEs.

Persons whose jobs will require repeated VE training (e.g., military aviators) should become familiar with their own onset and progression of signs and symptoms, so they can pace themselves accordingly. Early signs and symptoms are also useful pre-nausea criteria for testing adaptation strategies intended to enhance well-being and performance in VEs.

The most thorough elucidation of temporal symptom progression in synthetic environments has come from experimental motion studies wherein the usual relation among visual, vestibular, and somesthetic sources of information has been distorted. Estimates of likely symptom onset and progression from these non-VE studies are shown in Table 3. When reading Table 3, it should be kept in mind that great individual variability exists in the order of symptoms among participants. Fortunately, trained individuals are reliable within themselves during repeated provocative tests (Graybiel & Lackner, 1980; Calkins, Reshcke, Kennedy, & Dunlap, 1987).

The best studies in Table 3 (Reason & Graybiel, 1970b; Miller & Graybiel, 1969) emphasized stomach symptoms due to the highly challenging motion stimuli used, the instructions given to the subjects, and the scoring procedures employed. Although nausea and vomiting are very important aspects of motion discomfort, certain stimuli produce vomiting without marked nausea (Lackner & Graybiel, 1986).

Table 3. Early symptoms and approximate symptom progression during synthetic experiences.

Source	Situation	Individual Signs & Symptoms, in Order of Appearance (Can have overlaps, reversals, or > 1 symptom at a time)
Nieuwenhuijsen, 1958	Voyage in 149 m. ship, New York to Rotterdam with rough seas during part of voyage	Overall, 62% of 193 passengers became sick at sea First indisputable symptoms, in order: General weakness or malaise (32%), headache (29%), slight nausea (23%), dizziness / disorientation (9%), salivation (3%), Cold sweating (2%)
Miller & Graybiel, 1969	The "Coriolis Sickness Susceptibility Index." Test entails moving head & torso while seated & rotating in the earth vertical axis at various speeds in darkness	90% of 250 subjects had stomach symptoms by end Most common effects by "Malaise I" (mild sickness): stomach awareness / discomfort (56%), minimal sweating (14%), subjective warmth or flushing (9%) Most common effects by "Malaise III" (moderate sickness): subjective warmth or flushing (about 70% of subjects); stomach awareness or discomfort (about 60% of subjects) (Also likely: at least minimal cold sweating & minimal pallor)
Reason & Graybiel, 1970a	Moving head repeatedly through 90 dg arcs in four directions while seated (eyes open) inside a rotating enclosure spinning at various speeds in a vertical axis	Overall, 70% of 10 subjects experienced some adverse symptoms during the incremental adaptation protocol Most common first effects noted in the 7 susceptible subjects were stomach awareness (5/7 = 71%), dizziness (2/7 = 29%), pallor (2/7 = 29%) (Note that it was possible for more than one sign or symptom to be reported at first appearance of adverse effects)
Reason & Graybiel, 1970b	"Dial Test" of Kennedy & Graybiel (1965) Entails moving head & torso adjusting dials located around body Subject seated inside a rotating enclosure spinning at various speeds in vertical axis	100% of 41 subjects exhibited adverse signs or symptoms; 96% reported some degradation of overall well-being; 5% vomited (during deceleration at end of dial test) Subjects gave "Well-being ratings, where 0 = "I feel fine," 10 = "I feel awful, as if I am about to vomit." Likely initial effects (at median well-being < 1.5) included minimal dizziness & visible nystagmus Likely effects occurring well prior to the point to vomiting (well-being = 1.6 to 5.0): bodily warmth, moderate dizziness, headache, minimal pallor, minimal cold sweating, minimal stomach awareness, increased salivation
Takahashi et al., 1991	Two hrs. locomotion wearing horizontal-reversing prisms	Earliest symptoms = discomfort (90% of 10 subjects), cold sweating (20%), headache (40%), stomach ache (20%) Most common subsequent symptoms = thirst (60%), nausea (70%), pallor (40%), salivation (40%), & crouching (40%)
Earliest symptoms & typical symptom progression during a synthetic experience		Most common early effects reported: stomach symptoms (5/5 studies), warmth or sweating (4/5 studies), pallor (4/5 studies). Likely temporal progression of the earliest effects (considering all 5 studies): warmth or sweating, dizziness, headache, pallor Likely early progression, considering the two best studies by Miller & Graybiel (1969) and Reason and Graybiel (1970b): warmth or sweating, dizziness, stomach symptoms

A careful study of the onset and progression of individual symptoms can help to distinguish different motion-related syndromes. For example, the fact that drowsiness was not reported as one of the earliest symptoms of motion exposure in Table 3 suggests that "early sopite syndrome" (Graybiel & Knepton, 1976) is not entirely explained by the early stages of "regular" motion sickness observed during sickening motion stimuli. A careful consideration of individual signs and symptoms can suggest ways in which motion syndromes differ from one another and suggest topics for further empirical study.

4. TOOLS FOR MEASURING LOSS OF WELL-BEING DURING OR AFTER VE EXPOSURE

Procedures for obtaining quantitative estimates of signs and symptoms of VE exposure are not uniform across studies. The best criteria for measuring the effects of synthetic experiences derive from multiple symptom checklists initiated in the 1960s by the late Captain Ashton Graybiel (Medical Corps, United States Navy) and developed by a number of researchers during studies of motion discomfort carried out in Pensacola, FL. Two mature forms of the diagnostic criteria for motion sickness emerged from this effort (Miller & Graybiel, 1970a,b; Kennedy, Berbaum, & Lilienthal, 1993). Miller and Graybiel (1970a,b) produced a diagnostic categorization table and a method for scoring signs and symptoms of motion discomfort. Since the Miller and Graybiel diagnostic criteria can be difficult to apply correctly in their original form, a simplified representation is offered in Lawson, Graeber, Muth, & Mead (2001), rendered as a checklist for the scoring of a single motion challenge. The Lawson et al. adaptation of the Miller and Graybiel (1970a,b) criteria is recommended for those investigators committed to using the original Miller and Graybiel criteria, but wishing to limit their metric assumptions.

A similar symptom checklist was developed by Kennedy, Tolhurst, and Graybiel (1965). It eventually emerged as a list of items for measuring general motion discomfort and a smaller subset of items called the Simulator Sickness Questionnaire or SSQ (Kennedy, Berbaum, and Lilienthal, 1993). The SSQ has seen wide use in Navy simulator studies and during other synthetic experiences. The SSQ is recommended for investigators who wish to study VE effects and are not committed to the Miller and Graybiel criteria.

5. GAPS IN KNOWLEDGE ABOUT VE SYMPTOMATOLOGY: TOPICS RECOMMENDED FOR STUDY

Researchers have yet to develop a complete description of the relation between the characteristics of different real or apparent motion stimuli and the particular signs and symptoms elicited by those different stimuli. Important steps have been taken in this direction by Kennedy, Berbaum, and Lilienthal (1993) and Kennedy, Lanham, Drexler, Massey, & Lilienthal (1997), who have identified symptom clusters specific to certain stimuli.

The distinct symptoms of long term exposure to milder stimuli are not well quantified. For example, measuring the sopite syndrome requires looking at greater-than-three factor solutions of the SSQ (Kennedy, Massey, & Lilienthal, 1995) or creating a scale specific to sopite syndrome (Lawson, Kass, Muth, & Summers, 2001).

Graybiel and Lackner (1980) pointed out that the relevant aspect of motion discomfort is dependent upon the situation. For example, during military aviation, rate of adaptation and retention of adaptation to airsickness may be more important than initial susceptibility. A better understanding of these aspects of motion response would prove useful. Subjective, physiological, and performance measures should be employed in concert to achieve this goal.

6. RECOMMENDATIONS

Little is known about the onset and progression VE-induced symptoms. This is because VEs have been introduced very recently and because of the tendency for researchers to publish summary sickness scores in the literature without reporting data concerning the onset, progression, and severity of individual symptoms. As VE technology matures, this reporting trend should be reversed and more individual symptom data presented.

Researchers should continue to explore the relation between signs and symptoms elicited by different VEs (compared to one another) and by other synthetic experiences. The symptoms produced by prolonged or repeated exposure to a given VE should receive as much attention as the symptoms produced by acute exposure. The rate and retention of adaptation to VE side-effects and readaptation after emerging from a VE require further attention as well. Understanding such issues is critical to the design of usable VE displays.

Researchers who neglect to measure a VE user's subjective well-being can be mislead when interpreting the user's physiological status or behavior (Bittner, Gore, & Hooey, 1997). Understanding signs and symptoms will aid the development of rational VE exposure guidelines, warning labels, training syllabi, sickness countermeasures, and sickness predictors (of individual susceptibility before exposure or temporal onset prior to conscious awareness).

REFERENCES

Bittner, A. C., Gore, B. F., & Hooey, B. L. (1997). Meaningful assessments of simulator performance and sickness: can't have one without the other? Proceedings, *Human Factors and Ergonomics 41st Meeting*, Sept. 22-26, Albuquerque, NM.

Calkins, D. S., Reschke, M. F., Kennedy, R. S., & Dunlap, W. P. (1987). Reliability of provocative tests of motion sickness susceptibility. *Aviation, Space, and Environmental Medicine*, 58 (9 Suppl.): A50-54.

Durlach, N. I. & Mavor, A. S. (Eds.). (1995). *Virtual Reality - Scientific and Technological Challenges*. Washington, D.C.: National Academy Press.

Graybiel, A. & Lackner, J. R. (1980). A sudden-stop vestibulovisual test for rapid assessment of motion sickness manifestations. *Aviat. Space Environment. Med*, 51, 21-23.

Graybiel, A. & Knepton, J. (1976). Sopite syndrome: A sometimes sole manifestation of motion sickness. *Aviat. Space Environment. Med*, 47, 873-882.

Howarth, P. A. & Finch, M. (1999). The nauseogenicity of two methods of navigating within a virtual environment. *Applied Ergonomics*, 30, 39-45.

Kennedy, R. S., N.E., Berbaum, K. S., & Lilienthal, M. G. (1993). Simulator sickness questionnaire: An enhanced method for quantifying simulator sickness. *International Journal of Aviation Psychology*, 3, 203-220.

Kennedy, R. S., & Graybiel, A. (1965). *The Dial Test: A Standardized Procedure for the Experimental Production of Canal Sickness Symptomatology in a Rotating Environment*. NSAM-930, NASA Order No. R-93, Pensacola, FL.

Kennedy, R. S., Lanham, D. S., Drexler, J. M., Massey, C. J., & Lilienthal, M. G. (1997). A comparison of cybersickness incidences, symptom profiles, measurement techniques, and suggestions for further research. *Presence*, 6, 638-644.

Kennedy, R. S., Massey, C. J., & Lilienthal, M. G. (1995). Incidences of fatigue and drowsiness reports from three dozen simulators: Relevance for the sopite syndrome. *First Workshop on Simulation and Interaction in Virtual Environments (SIVE '95)*. July 13-15, Iowa City, IA.

Kennedy, R. S. & Stanney, K. M. (1997). Aftereffects of virtual environment exposure: Psychometric issues. In M. J. Smith, G. Salvendy, & R. J. Koubek (Eds.). *Design of Computing Systems: Social and Ergonomic Considerations* (pp.897-900). Amsterdam: Elsevier.

Kennedy, R. S., Tolhurst, G. C., & Graybiel, A. (1965). *The Effects of Visual Deprivation on Adaptation to a Rotating Room*. Naval School of Aviation Medicine. NSAM-918.

Lackner, J. R. & Graybiel, A. (1986). Sudden emesis following parabolic flight maneuvers: Implications for space motion sickness. *Aviat. Space Environment. Med*, 57, 343-347.

Lawson, B. D., Graeber, D. A., Muth, E. R., & Mead, A. M. (2001, in print). Signs and symptoms of human syndromes associated with synthetic experiences. In Stanney, K. (Ed.), *Handbook of Virtual Environments: Design, Implementation, and Applications* (Chapter 30). Mahwah, NJ, Lawrence Erlbaum Assoc., Inc.

Lawson, B. D., Kass, S. J., Muth, E. R., & Sommers, J. (2001). Development of a scale to assess signs and symptoms of sopite syndrome in response to mild or nonsickening motion stimuli. *Proceedings of the 72nd Annual Meeting of the Aerospace Medical Association*, 6-10 May, Reno, NV.

Miller, E. F. & Graybiel, A. (1969). *A Standardized Laboratory Means of Determining Susceptibility to Coriolis (Motion) Sickness* (Tech Report NAMI-1058, 1 p.). Pensacola, FL: Naval Aerospace Medical Institute.

Miller, E F. & Graybiel, A. (1970a). *Comparison of Five Levels of Motion Sickness Severity as the Basis for Grading Susceptibility*. NASA Order R-93. Pensacola, Fla., Naval Aerospace Medical Institute.

Miller, E. F. & Graybiel, A. (1970b). A provocative test for grading susceptibility to motion sickness yielding a single numerical score. *Acta Otolarygologica*, Suppl. 274.

Nieuwenhuijsen, J. H. (1958). *Experimental Investigations on Seasickness*. M. D. Thesis, University of Utrecht, Drukkerij Schotanus & Jens, Utrecht, Netherlands.

Reason, J. T. & Graybiel, A. (1970a). Progressive adaptation to coriolis accelerations associated with 1rpm increments in the velocity of the slow rotation room. *Aerospace Medicine*, 41(1), 73-79.

Reason, J. T. & Graybiel, A. (1970b). Changes in subjective estimates of well-being during the onset and remission of motion sickness symptomatology in the slow rotation room. *Aerospace Medicine*, 41(2), 166-171.

Regan, E. C. & Price, K. R. (1994). The frequency of occurrence and severity of side-effects of immersion virtual reality. *Aviat. Space Environment. Med.*, 65, 527-530.

Regan, E. C. & Ramsey, A. D. (1996). The efficacy of hyoscine hydrobromide in reducing side-effects induced during immersion in virtual reality. *Aviat. Space Environment. Med.*, 67(3), 222-226.

Stanney, K.M., Salvendy, G., Deisigner, J., DiZio, P., Ellis, S., Ellison, E., Fogleman, G., Gallimore, J., Hettinger, L., Kennedy, R., Lackner, J., Lawson, B., Maida, J., Mead, A., Mon-Williams, M., Newman, D., Piantanida, T., Reeves, L., Riedel, O., Singer, M., Stoffregen, T., Wann, J., Welch, R., Wilson, J., Witmer, B. (1998). Aftereffects and sense of presence in virtual environments: Formulation of a research and development agenda. *International Journal of Human-Computer Interaction*, 10(2), 135-187.

Takahashi, M., Saito, A., Okada, Y., Takei, Y., Tomizawa, I., Uyama, K, & Kanzaki, J. (1991). Locomotion and motion sickness during horizontally and vertically reversed vision. *Aviat. Space Environment. Med.*, 62, 136-140.

Wilson, J. R. (1997). Virtual environments and ergonomics: Needs and opportunities. *Ergonomics*, 40, 1057-1077.

THE USE OF REAL OR APPARENT BODY MOTION TO ENHANCE VIRTUAL DISPLAYS

Ben D. Lawson, Ph.D.

Spatial Orientation Systems, Naval Aerospace Medical Research Laboratory, 51 Hovey Road, Pensacola FL 32508-1046. blawson@namrl.navy.mil

ABSTRACT

This paper introduces several strategies for eliciting perceptions of self-motion, self-tilt, or self-entry into abnormal force environments. Discussion centers on the relative merits of real physical acceleration of the user, visually induced illusions of self-motion, and illusions of self-motion induced by locomotion without displacement (as occurs on treadmills). Acceleration perceptions induced by such means can be exploited to enhance the feeling of self-motion or presence within a Virtual Environment (VE).

1. TECHNIQUES FOR ELICITING ACCELERATION PERCEPTIONS

The perception of acceleration includes feelings of self-motion and self-tilt (relative to the upright). The feeling of unusual forces (such as occurs during high performance flight or in non-terrestrial settings) also qualifies as an acceleration perception. The virtual context of the motion perception, not the eliciting stimulus, is what classifies a felt motion as "virtual." Hence, acceleration perceptions in VEs can be elicited by methods involving real physical acceleration of the user or methods wherein felt movement is illusory.

Real motion methods physically accelerate the VE user aboard a moving device such as a centrifuge or allow to him "drive" or actively locomote through a real space that serves as an ambient context for the VE. Illusory motion can be elicited by real physical acceleration (see section 1.1, below). However, the VE literature regarding illusions of self motion usually refers to an illusion of body motion elicited from a stationary user by moving a visual, auditory, or somesthetic surround stimulus relative to the user or by having the user locomote on a treadmill without displacement. (section 1.2, below). Methods for creating the perception of displacement, velocity, acceleration, or abnormal "G-force" upon one's body are introduced in section 1 of this paper. In Section 2, the relative merits of some of the better methods for creating controlled acceleration perceptions are discussed.

1.1 Methods involving acceleration of the user

The most direct way to elicit the perception of dynamically changing body motion through a VE is by real physical motion of the user that is perceived as occurring within the context of the VE. Physical motion through a VE can be achieved "passively" with moving devices such as simulators, centrifuges, or real vehicles. Physical motion can also be achieved by allowing the user to walk through a real environment that serves as an ambient context for the VE. Real motion is an effective and controllable way to create and alter acceleration perceptions quickly. It is a very useful method, provided the monetary cost can be absorbed and the simulations are designed with an understanding of human acceleration sensations (Lawson, Sides, & Hickinbotham, 2001a) and VE side effects (Lawson, Graeber, Muth, & Mead, 2001b).

Physical acceleration of the user is not always perceived veridically; many acceleration-induced illusions are possible. Acceleration can place a "G-force" on the body and thereby elicit the perception of altered body weight or illusory tilt. (Guedry, 1974; Guedry, Richard, Hixson, & Niven, 1973; Lackner & Graybiel, 1980). Such acceleration illusions could enhance the simulation of high performance vehicles or non-terrestrial settings. Other illusions may accompany real motion, such as feeling stationary during prolonged rotation at constant velocity, or the feeling of rotation upon the cessation of prolonged rotation (Guedry, 1974). Thus, real accelerations can produce illusory self-motion as readily as methods that do not involve acceleration. Such illusions can be exploited to enhance the perceived effect of physical acceleration.

1.2 Methods not involving acceleration of the user

There are many ways to elicit acceleration perceptions without physically accelerating the user, most of which are introduced in this section. Acceleration perceptions can be elicited via tilting visual frames of reference, continuous whole-field visual motions, or the aftereffects thereof (Dichgans & Brandt, 1978). This approach requires an understanding of the psychophysics of human perception and the factors that give rise to visually induced discomfort.

Acceleration perceptions can also be elicited via motion of auditory or tactile surrounds or the aftereffects thereof (Lackner, 1977; Lackner & DiZio, 1984). These are promising methods as well, especially when used in conjunction with confirmatory physical motion or visual surround motion. Voluntary tactile/kinesthetic behaviors such as locomotion without displacement, stationary pedaling, or simulated "hand walking" (as well as the aftereffects of these behaviors), are all effective means for eliciting illusory acceleration perceptions (Lackner & DiZio, 1988, 1993.) This is a recommended area for research, albeit the method will be limited to acceleration simulations involving limb movement. Indirect kinesthetic effects can be induced by vibration of skeletal muscles, which elicits a variety of self motion perceptions (Lackner & Levine, 1979; Levine & Lackner, 1979), as can vibration of the cranium (Lackner & Graybiel, 1974).

Finally, several direct methods exist for creating an acceleration perception. Caloric stimulation of the ear canal (National Research Council, 1992) is an effective method and can be applied unilaterally. However, it can be unpleasant, and real-time control of the motion perception is difficult. Low intensity electrical stimulation of the mastoids causes sway (Fernald & Moore, 1966) and shifts the apparent visual vertical (Aarons & Goldenberg, 1964); thus, it is worth exploring as a way to achieve acceleration perceptions.

1.3 Combining different methods for eliciting acceleration perceptions

A controlled feeling of acceleration within a VE can be achieved by combining the more promising stimuli listed above. For example, when motions of the visual field correspond with real motions of the user, feelings of acceleration are enhanced (Boff & Lincoln, 1988). Multi-modal inputs act in concert to create acceleration perceptions in the natural world. Similarly, perceptions of acceleration in VE should be elicited by mutually enhancing combinations of multi-modal stimuli. The best methods will combine real passive or active physical acceleration with visual, auditory surround, or tactile surround motion. Combining different cues to elicit acceleration perceptions is advisable whenever it is feasible.

2. COMPARISON OF DIFFERENT METHODS FOR ELICITING ACCELERATION PERCEPTIONS

The optimal approach to eliciting acceleration perceptions in a VE depends upon the relative importance of the factors that contribute to acceleration perceptions. If virtual displays are to meet the promise of their initial "virtual reality" hype and become fully immersive and effective, a careful consideration of the advantages and disadvantages of different approaches will be necessary. A comparison of three promising acceleration methods is offered in Table 1 (below) to help the reader understand where each method excels and imagine how different methods might complement one another. In Table 1, the highest-ranked method (from column A, B, or C) for each point of comparison (left-most column) is indicated by "↑" for highest ranking on a desirable trait or "↓" for highest ranking on an undesirable trait. In some cases, choosing the highest-ranked method is not applicable or not yet possible.

Table 1. Comparison of three methods for eliciting acceleration perceptions.

Points of Comparison (Relative Pros & Cons)	A) Physical Acceleration Involving Real Displacement (Passively or Actively Produced)	B) Visually Induced Illusory Self Motion without Real Displacement (Vection)	C) Illusory Self Motion via Active Locomotion without Displacement
Flexibility. Range of perceived situations that can be simulated.	↑ Many passive transportation stimuli in 6 degrees-of-freedom can be simulated, as well as active locomotion, e.g., walking through a real space that serves as a context for a virtual space.	Constant velocity and trajectory can be simulated unless visual motion is too brief, fast, or varying. User's movements can delay onset. Does not work in darkness. Illusion varies for different degrees of freedom.	Active locomotion can be simulated without moving through space. Simulations include crawling, walking, running, climbing, peddling, paddling, etc.

Table 1, Continued.	A) Real Acceleration	B) Visually Induced	C) Illusory Locomotion
Isomorphism. Implies: a) stimuli and responses in VE similar to real event; b) absence of VE side effects or aftereffects (unless they are present in the real stimulus).	Nearly isomorphic unless desired movement exceeds available space, in which case simulation quality suffers and head / reaching movements are not isomorphic. Illusions can be exploited at times to enhance the simulation, or real vehicles can be employed instead.	Visual aspect of simple self-motion is reproduced fairly well (e.g., constant velocity). Not isomorphic during changes in simulation velocity or direction, or during the user's own actions (e.g., head movements, reaching behaviors).	Human stride and work during treadmill locomotion are not identical to real locomotion. Still, this remains the best way to simulate active locomotion within a small space.
Compellingness Implies rapid onset of motion perception, perceived self motion without surround motion, and high ratings of "presence" or realism.	↑ Very compelling, especially for vehicle simulation. Space required limits amplitude and duration. When real vehicles are used for simulation, the effect will be very compelling. When space is available, body acceleration via real locomotion is better than using a treadmill.	Very compelling in some circumstances. Not good for changing or rapid virtual motions. Rotational vection is best at constant velocity of surround < 90 dg/s or when axis of vection is aligned with the earth vertical. (Boff & Lincoln, 1988).	Very compelling for simulation of active locomotion in a limited space (especially when coupled with confirming visual information – see column B). Not applicable to passive vehicle motion.
Range of Acceleration Profiles Possible. Range of perceived virtual velocities and accelerations possible.	↑ Many effects possible on moving devices, but duration is limited by the space required. Real vehicles or locomotion also require space, but provide nearly unlimited profiles. In limited spaces, perceived motion can be enhanced by vestibular illusions.	Works best at a moderate, constant velocity (e.g., below 90 degrees/sec for rotational vection). No theoretical limit on duration of stimulus, however.	Limited by the speed, flexibility, and fitness of the user. However, most of the simulations desired will be within normal human abilities.
Temporal Fidelity. How well perception tracks changes in stimulus.	↑ Often rapid onset or recovery of acceleration perception with changes in stimulus. Perception stays in phase with stimulus within normal frequencies of voluntary motion. Vestibular illusions arise during passive motion outside normal range.	Often delayed onset and will not readily follow sudden changes in the velocity or direction of the visual stimulus.	Often delayed onset, but self motion perception stays roughly in phase with stepping. (Stepping direction can be de-coupled from visual motion).
Usefulness for Simulating Altered Gravitoinertial Force Environments	↑ The only method that can create a non-terrestrial force environment or simulate G-forces felt by operators of high performance vehicles.	Can enhance the feeling of falling, but is not ideal for simulating non-terrestrial gravity or motion induced G-forces.	In conjunction with body loading, suspension, or other cutaneous inputs, limited G simulation may be possible.
Behavioral Flexibility. Whether user's activities must be restricted for acceleration perception to be elicited and maintained.	User should be restrained at high G. Head, eye, and limb motions alter perception or produce adverse symptoms. Simulations exploiting real vehicles or real locomotion have the same restrictions as when no simulation is being created.	Works best when the user's head and eyes are still. May not work as well when simulating real situations where field of view or sight of own body is limited (simulation of a view through a periscope).	Works best when the locomotion is at constant velocity and in a predictable direction.

Table 1, Continued.	A) Real Acceleration	B) Visually Induced	C) Illusory Locomotion
Discomfort. How disturbing or "sickening" the method can be. (Lawson et al., 2001b)	↓ Some stimuli can be very disturbing, but discomfort can be decreased by helpful visual or somesthetic inputs (Lawson, et al, 1993, 1997). Drowsiness is one of the most common symptoms. Nausea contributes much to degraded feelings of well being.	Can be fairly disturbing and effects can persist. User closes eyes to avoid stimulus. Cessation of prolonged visual flow is less sickening than cessation of real rotation. "Head symptoms" common.	Few symptoms when walking at a constant speed on a treadmill, but symptoms may arise when walkers are also exposed to discordant visual flow (Durlach & Mavor, 1995).
Other Undesirable Effects. Side and aftereffects besides motion discomfort.	May include balance problems, altered eye-hand coordination, altered vestibulo-ocular and cervical reflexes, and sopite syndrome.	Similar effects as for physical acceleration, but often of milder severity. Effects may persist. Visual problems may arise, such as altered convergence.	Altered limb control and altered gate upon exiting treadmill simulator, possibly other aftereffects as well.
Expense of Building and Maintaining the Display.	↓ A "virtual accelerator" will be expensive. Some moving devices, real vehicles, or real locomotion simulations will be able to take advantage of existing infrastructure.	Cost will tend to be relatively low. Existing computer displays can be leveraged readily. Programming expenses are the main financial limitation.	Costs range from inexpensive (e.g., walking in place or on a treadmill) to expensive (e.g., using sophisticated locomotion devices mounted to each limb).
Technical Difficulty. How difficult the virtual displays are to design, build, and program.	↓ Very difficult approach, but made feasible with knowledge of acceleration psychophysics and by "piggy-backing" virtual displays onto existing motion-capable devices or real vehicles.	Although still an unsolved challenge, the visual motion approach will be relatively easy to design, build, test, and modify.	Not difficult when using simple treadmills to simulate constant velocity, unidirectional walking, but complicated for other behaviors.
Size. Space occupied by visual display.	↓ Large space required for most moving based simulators; very large footprint for centrifuge based devices. Extremely large space needed when using real vehicles or active locomotion through a real space used to mimic the virtual space.	Very small space needed if head mounted, relatively small space needed if visual stimulus surrounds entire body.	Small space for a treadmill or foot mounted device; moderate space for omnidirectional and tiltable treadmill with a suspension/weighting system.
Initial "Market Niche." (Rank-ordering not applicable)	Costly government, sporting, and group entertainment applications. Quality simulation of high performance or active locomotion through a real space mimicking a virtual space. In entertainment, user fees will defray costs.	Low cost single-user applications. However, costly applications will need visual displays of high quality, and hence this method will be employed at many levels in the market.	Moderate cost walking or pedaling in a small space. Higher cost for additional simulations. Portable units could augment training of athletic or dangerous skills.

3. SUMMARY AND RECOMMENDATIONS

Acceleration perceptions can be elicited in VEs by real or illusory motion. Physical motion methods accelerate the VE user aboard a moving device or allow the user to drive or locomote within a real space that serves as a context for the VE. Illusory motion methods induce an illusion of self motion by moving a visual, auditory, or somesthetic surround relative to the user or by having the user locomote without displacement. Inducing acceleration perceptions via physical motion generates a wider range of simulations and is more flexible and compelling. Illusory motion methods require less money and space to implement. Real or illusory motion methods are not mutually exclusive and can be used in combination. Such multi-sensory cueing takes place during a stroll through the real world. A particularly effective way to provide the user with mutually enhancing vestibular, visual, and somesthetic stimuli is to employ physical acceleration stimuli that duplicate the acceleration profiles being simulated, but do so within a real space that serves as an ambient context for the VE (Lawson, et al., 2001a).

REFERENCES

Aarons, L., & Goldenberg, L. (1964). Galvanic stimulation of the vestibular system and perception of the vertical. *Perceptual and Motor Skills*, 19(1), 59-66.

Boff, K. R., & Lincoln, J. E. (Eds.) (1988). *Engineering Data Compendium: Human Perception and Performance*. Wright-Patterson AFB, OH: AAMRL.

Dichgans, J., & Brandt, T. (1978). Visual-vestibular interaction: Effects on self-motion perception and postural control. In R. Held, H. W. Leibowitz, & H. L. Teuber (Eds.), *Handbook of Sensory Physiology* (Vol. 8, pp. 755-804). New York: Springer-Verlag.

Durlach, N. I., & Mavor, A. S. (Eds.). (1995). *Virtual Reality - Scientific and Technological Challenges*. Washington, D.C.: National Academy Press.

Fernald C. D., & Moore, J. W. (1966). Vestibular sway: parameters of the eliciting stimulus. *Psychonomic Science*, 4(2), 55-56.

Guedry, F. E. (1974). Psychophysics of vestibular sensation. In H. H. Kornhuber (Ed.), *Handbook of Sensory Physiology* (pp. 1-154). New York/Heidelberg/Berlin: Springer-Verlag.

Guedry, F. E., Richard, D. G., Hixson, W. C., & Niven, J. I. (1973). Observation on perceived changes in aircraft attitude attending head movements made in a 2-G bank and turn. *Aerospace Medicine*, 44, 477-483.

Lackner, J. R. (1977). Induction of illusory self-rotation and nystagmus by a rotating sound-field. *Aviation, Space, & Environmental Medicine*, 48, 129-131.

Lackner, J. R., & DiZio, P. (1984). Some efferent and somatosensory influences on body orientation and oculomotor control. In L. Spillman & B. R. Wooten (Eds.), *Sensory Experience, Adaptation, and Perception* (pp. 281-301). Clifton, NJ: Erlbaum Associates.

Lackner, J. R., & DiZio, P. (1988). Visual stimulation affects the perception of voluntary leg movements during walking. *Perception*, 17, 71-80.

Lackner, J. R., & DiZio, P. (1993). Spatial stability, voluntary action and causal attribution during self locomotion. *Journal of Vestibular Research*, 3, 15-23.

Lackner, J. R., & Graybiel, A. (1974). Elicitation of vestibular side effects by regional vibration of the head. *Aerospace Medicine*, 45, 1267-1272.

Lackner, J. R., & Graybiel, A. (1980). Visual and postural motion aftereffects following parabolic flight. *Aviation, Space, & Environmental Medicine*, 51, 230-233.

Lackner, J. R., & Levine, M. S. (1979). Changes in apparent body orientation and sensory localization induced by vibration of postural muscles: Vibratory myesthetic illusions. *Aviation, Space, & Environmental Medicine*, 50, 346-354.

Lawson, B. D., Guedry, F. E., Rupert, A. H., and Anderson, A. M. (1993). Attenuating the disorienting effects of head movement during whole-body rotation using a visual reference: further tests of a predictive hypothesis. *Advisory Group for Aerospace Research & Development Proceedings No 541: Virtual Interfaces: Research and Applications*, 18 - 22 October.

Lawson, B. D., Graeber, D. A., Muth, E. R., & Mead, A. M. (2001b, in print). Signs and symptoms of human syndromes associated with synthetic experiences. In Stanney, K. (Ed.), *Handbook of Virtual Environments: Design, Implementation, and Applications* (Chapter 30). Mahwah, New Jersey: Lawrence Erlbaum Assoc., Inc.

Lawson, B. D., Rupert, A. H., Guedry, F. E., Grissett, J. D., & Mead, A. M. (1997). The human-machine interface challenge of using virtual environment (VE) displays aboard centrifuge devices. In M. J. Smith, G. Salvendy, & R. J. (Eds.), Design of Computing Systems: Social and Ergonomic Considerations

Lawson, B. D., Sides, S. A., & Hickinbotham, K. A. (2001a, in print). User requirements for perceiving body acceleration. In Stanney, K. (Ed.), *Handbook of Virtual Environments: Design, Implementation, and Applications* (Chapter 7). Mahwah, New Jersey: Lawrence Erlbaum Assoc., Inc.

Levine, M. S., & Lackner, J. R. (1979). Some sensory and motor factors influencing the control and appreciation of eye and limb position. *Experimental Brain Research*, 36, 275-283.

National Research Council. (1992). Evaluation of tests for vestibular function (Report of the Working Group on Evaluation of Test for Vestibular Function; Committee on Hearing, Bioacoustics, and Biomechanics, and the Commission on Behavioral and Social Sciences and Education). *Aviation, Space, and Environmental Medicine*, 63(2, Suppl.), A1-A34.

HUMAN PERFORMANCE IN VIRTUAL ENVIRONMENTS: EXAMINING USER CONTROL TECHNIQUES

Kay M. Stanney[1], Kelly Kingdon[1], and Robert S. Kennedy[2]

[1]University of Central Florida, Industrial Engineering and Mgmt Systems Dept., Orlando, FL 32816-2450
[2]RSK Assessments, Inc., 1040 Woodcock Road, Suite 227, Orlando, FL 32803-3510 USA

ABSTRACT

When using virtual environments (VEs) for training, designers have to ensure that users can effectively traverse through and interact with the virtual world. For some users this can be an easy task, while for others it can be quite challenging. If basic interaction is cumbersome, then users cannot be expected to achieve effective performance on target tasks. Thus, effective techniques for enhancing human-computer interaction in virtual worlds are needed. This study examined human performance on a set of basic tasks in virtual environments. Tasks included locomotion, object manipulation, and reaction time. Two interaction techniques were evaluated, complete control (6 degrees of freedom [dof]) user movement and streamlined (3 dof) user movement. Performance was measured by time on task, distance moved, and amount of head movement. Results indicated that locomotion and reaction time tasks were performed more efficiently when users had complete control. For tasks that required a change in viewpoint *via only head movement*, complete control was more effective because it allowed users to readily manipulate the view by turning their head as opposed to manipulating with a mouse. However, for tasks that involved *a change in viewpoint via both head movement and position change* (i.e., users 'physically' moved to a new viewpoint), users performed significantly better with streamlined control. For object manipulation tasks, which did not require a change in viewpoint, there were no performance differences between user control conditions. Thus, to enhance human-computer interaction in VEs, providing users with complete control allows for effective performance on both stationary tasks and those requiring head movement only (i.e., body is stationary). With complete control users can adopt a natural mode of interaction, moving their head to update the visual scene as they would to look about the real world, thus eliminating an artificial interaction mode (i.e., the mouse) and allowing for more direct manipulation. With tasks involving both head and body movement, however, user movement control should be streamlined (e.g., by reducing dof) to enhance performance.

1. INTRODUCTION

The efficiency with which one performs within a VE setting is important in both training and entertainment venues. In training scenarios, VEs are usually called upon to replicate real world situations. Here, performance within the VE and transfer of training to the real world are both important indications of learning. In an entertainment venue, performance may affect repeat exposure to the VE and user satisfaction of the system. While many physiological and environmental measures have been taken in VEs to assess participants' well-being and experience, few measures have focused on how humans actually perform within virtual worlds.

By measuring human performance within a VE, one can determine how effectively the system is being used and determine the types of problems users experience while traversing throughout the virtual world. This can be used to perfect the VE design, as well as to compare performance results to real world training situations to ensure the environment is precisely recreated to maximize transfer of training. Furthermore, feedback and knowledge of results can be presented to participants to enhance the training and/or entertainment value of the experience (Lampton, Bliss, & Morris, in press).

One factor thought to affect performance is the amount of control participants have over movement within the virtual environment. A full 6 dof of movement allows the participant to experience x, y, and z translational movement, as well as roll, pitch, and yaw. Traditionally, this full level of control is used in VE systems. Research has shown, however, that 6 dof provides novice users with extraneous movement in the VE, which can actually hinder rather than enhance their experience (Stanney & Hash, 1998). Yet some tasks, such as locomotion may benefit from 6-dof movement, while others requiring movement precision (e.g., object manipulation, reaction time) may be more effectively performed with streamlined movement control. The current study looks at the affects of reducing the degrees of freedom of user control for various types of tasks.

2. METHOD

2.1 Participants

Two hundred and forty participants (141 males, mean age 21.99 [range: 15-45]; 99 females, mean age 20.12 [range: 17-48]) from the University of Central Florida volunteered to participate in this study. Each participant was awarded extra credit for participating, or received a monetary award of $10. All participants had depth perception, were free from seizures, and were in good health when they participated.

2.2 Equipment

A 200Mhz Pentium MMX computer with 64MB of RAM and an Elsa Winner Pro 2000/X with 8 MB RAM graphics board was used to generate the VEs, which were developed using RenderWare software. A Logitech Cordless Mouseman Pro was used as the input device. A helmet-mounted display (Virtual Research V6) equipped with a Virtual iO! Tracker was used to generate the graphics and track the movements of the user.

2.3 Tasks

The tasks examined in this study are a sub-set from the Virtual Environment Performance Assessment Battery (VEPAB). The VEPAB was developed by the U.S. Army Institute to support research on VE training applications (Lampton, Knerr, Goldberg, Bliss, Moshell, & Blau, 1994). The VEPAB tasks include locomotion, object manipulation, tracking, reaction time, and recognition tasks. A sub-set of these tasks were performed in a VE shaped like a maze that consists of 29 rooms and 3 long corridors. The sub-set of tasks evaluated in this study is listed below.

> Locomotion tasks: In this VE there were straightaway, "flying", turning, and "room-to-room" locomotion tasks. Traversal time and efficiency of movement were the performance measures for these tasks.

> Manipulation tasks: For object manipulation tasks, participants used a mouse cursor to interact with and move objects throughout the virtual environment. When the cursor was located over an object, participants pushed the left mouse button to pick-up or "grasp" the object. They could then drag the object to a desired location and release the left mouse button to release the object. Manipulation tasks were performed in both two-dimensional (2D) and three-dimensional (3D) planes. The 3D tasks required participants to move their location within the VE to complete the task, whereas the 2D tasks required no user movement. Manipulation time and efficiency of movement were the performance measures for these tasks.

> Choice reaction time task: For this task, participants entered a room containing a black rotating cube. Selecting the cube made it stop rotating and change to a specific color. The participants then had to turn and look at the wall opposite to where the cube was located and select the cube color from a panel of colors on the wall. Once the participants had selected the correct color from the panel, the cube again turned black and began to rotate. The participants performed this sequence three times. The number of correct responses and time to respond were the performance measures for this task.

2.4 Procedure

Prior to the experimental session, all participants completed an informed consent form and were warned of the possible side effects of VE exposure. Each participant was told that if discomfort became too great during VE exposure, he/she should request termination of the session. Participants were provided with a description of the tasks to be performed, which they read and could ask questions about. Baseline measures of the simulator sickness questionnaire (SSQ, Kennedy, Lane, Berbaum, & Lilienthal, 093), eye-hand coordination, and postural stability were obtained (reports from these measures are not herein discussed). Each participant was randomly assigned to one of the treatment conditions. Experimental conditions were based on amount of user control (complete control - allowed roll, pitch, and yaw, as well as x, y, and z translational movements; streamlined control - allowed for linear movement in the fore–aft [x] and up-down [z] directions, as well as pitch), scene content (simple scenes that used no textures and low ceilings versus complex scenes with textures and high ceilings), and duration of exposure (15, 30, or 45 minutes of exposure). During exposure, the participants maintained a seated position while wearing the HMD and traversed through the maze completing the battery of tasks described above. Performance measures were collected online, and included performance time, an overall score on tasks completed, and total distance (linear and rotational) moved. Immediately following the exposure period, SSQ, eye-hand coordination, and postural stability measures were obtained every 15 minutes for 1-hour post-exposure (reports from these measures are not herein discussed).

2.5 Experimental Design

A 3x2x2 (Duration x User Control x Scene Content) between subjects design was used. Duration conditions were set at 15, 30, or 45 minutes. The two levels of user-initiated control included complete control and streamlined control. The scene content was either simple or complex. The dependent variables related to performance are noted above under Section 2.3 Tasks.

3. RESULTS

Scene complexity was not significant and thus the data were collapsed across these conditions for the analysis of performance. The performance time for the locomotion tasks under each control condition are presented in Figure 1. The results indicated significantly better performance when participants had complete control of movement for the elevator locomotion task ($p < .05$). For this task, those with complete control were 50% faster in their movements than those with streamlined control. For the room-to-room ($p = .057$) and straightaway ($p = .085$) tasks, the effect was not significant but showed a strong trend towards increased performance for the complete control condition. Control condition did not significantly affect the turns ($p = .469$) locomotion task. Similar results were found for the choice reaction time task ($p < .0001$, see Figure 2). For this task, those with complete control were 38.8% faster in their movements than those with streamlined control. Performance distance did not differ significantly with participants' level of control for the locomotion tasks.

Manipulation tasks, where the participant was not required to move to complete the task (i.e., 2D manipulation tasks), showed no significant difference in performance measures between participant control conditions. To complete the 3D manipulation task, participants had to change their viewpoint via both position change and head movement. Here, a significant difference was found in performance distance between the control conditions (see Figure 3), with streamlined control showing significantly ($p < .05$) shorter distances to complete the task. For this manipulation task, those with streamlined control were 25.3% more efficient in their movements than those with complete control. There were no significant differences found in performance time across control conditions for the manipulation tasks.

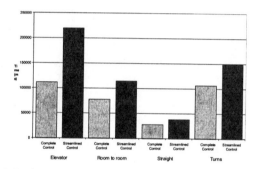

Figure 1. Performance time for locomotion tasks by user control condition.

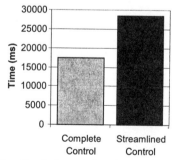

Figure 2. Performance time for choice reaction time task by user control condition.

1053

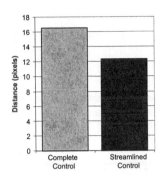

Figure 3. Performance distance for 3D manipulation task by user control condition.

4. DISCUSSION

The locomotion tasks used in this experiment forced users to traverse along relatively set paths throughout the virtual environment. Participants with complete control traveled about the same distance along these paths but in significantly less time compared to those with streamlined control. The choice reaction time task required users to change viewpoint via only head movement. Here too, participants with complete control completed the task in significantly less time than those under the streamlined control condition. Under complete control, participants could readily turn their heads to look towards the desired direction of movement or the desired object. This interaction is similar to real world experiences, where one usually heads in the direction he/she is facing. Under streamlined control, participants had to manipulate the mouse cursor to the left/right to set their heading. Thus, these participants had to map direction in 3D space onto a 2D surface (mouse pad) to effectively move throughout the environment. This mapping into 2D is likely what resulted in the increased performance time for the streamlined condition, although similar distances were covered. The increase in time could be reflective of cognitive processing time to convert heading onto the 2D mouse pad surface.

There were no significant differences found in performance time or object movement distance for stationary (2D) manipulation tasks between the two user-control conditions. This is likely because these tasks are fairly simple to complete and require no participant self-movement. The 3D manipulation tasks required participants to move via head movement as well as change the position of their virtual body. Here, participants performed more efficiently under the streamlined control condition, completing tasks in the same time, but with significantly shorter distances than those under complete user control. For these tasks more than just heading is of importance, users must be able to grab, drag while moving their viewpoint, and manipulate objects in 3D space. Such complex movement within a VE may be more efficient if the degrees of freedom of movement are streamlined such that object and self-movement are more controlled.

5. CONCLUSIONS

Participants performed significantly better on locomotion and choice reaction tasks when provided with complete user control over their movement throughout a virtual environment. Comparable performance on stationary manipulation tasks was maintained regardless of whether user movement control was complete or streamlined. Streamlined control did show an advantage in performance measures for tasks that required a change in viewpoint via both head movement and position change. Thus, to enhance human-computer interaction in VEs, providing users with complete control allows effective performance on both stationary tasks and those requiring head movement (body is stationary). With complete control, users can adopt a more natural mode of interaction, moving their head to update the visual scene as they would to look about the real world. With tasks involving both head and body movement, however, user movement control should be streamlined (e.g., by reducing dof) to enhance performance.

ACKNOWLEDGEMENTS

This material is based upon work supported in part by the National Science Foundation (NSF) under Grant No. IRI-9624968, the Office of Naval Research (ONR) under Grant No. N00014-98-1-0642, and the National Aeronautics and Space Administration (NASA) under Grant No. NAS9-19453. Any opinions, findings, and conclusions or recommendations expressed in this material are those of the authors and do not necessarily reflect the views or the endorsement of the NSF, ONR, or NASA.

REFERENCES

Kennedy, R.S., Lane, N.E., Berbaum, K.S., & Lilienthal, M.G. (1993). Simulator sickness questionnaire: An enhanced method for quantifying simulator sickness. *International Journal of Aviation Psychology*, 3(3), 203-220.

Lampton, D.R., Bliss, J.P., and Morris, C.S. (in press). Human performance measurement in virtual environments. In K.M. Stanney (ed.), *Handbook of Virtual Environments: Design, Implementation, and Applications*, Mahwah, NJ: Lawrence Erlbaum Associates.

Lampton, D.R., Knerr, B.W., Goldberg, S.L., Bliss, J.P., Moshell, J.M., and Blau, B.S. (1994). *The Virtual Environment Performance Assessment Battery (VEPAB): Development and Evaluation*, Army Research Institute, Orlando, FL.

Stanney, K.M. and Hash, P. (1998). Locus of user-initiated control in virtual environments: Influences on cybersickness. *Presence: Teleoperators and Virtual Environments,* 7(5), 447-459.

TOWARD A THEORY OF AFFORDANCE BASED DESIGN OF VIRTUAL ENVIRONMENTS

David C. Gross[1], Kay M. Stanney[2], and Lt. Joseph Cohn[3]

[1]The Boeing Company, M/S JR-80, 499 Boeing Blvd., Huntsville, AL 35824-6402
[2]University of Central Florida, Industrial Engineering and Mgmt Systems Dept., Orlando, FL 32816-2450
[2]Naval Air Warfare Center Training Systems Division, 12350 Research Parkway, Orlando, FL 32826-3275

ABSTRACT

Virtual environment (VE) interface designs are unique because they aim to represent a virtual world, in which users may experience a strong sensation that they are present in, or part of, the computer generated world. This "immersiveness," while generally more engaging and vibrant, brings with it unique design issues that were not of concern with conventional human-computer interaction (HCI). Existing HCI design principles have largely focused on static representations and thus have yet to fully incorporate theories of perception appropriate for the dynamic multimodal interactions inherent to VE interaction. This has led to VE designs whose usability are less than desired, likely because their users cannot readily perceive the actions and functions that can and should be enacted. There is thus a need to integrate a comprehensive theory of perception into VE design. Theories of direct perception, in particular affordance theory, may prove particularly relevant to VE system design because affordance theory explains the interaction of an organism with its environment. As many VEs are intrinsically environmental in that they emulate interaction with a modeled world, examining how an individual interacts with their surroundings when immersed in a VE should prove particularly interesting and informative. Virtual environment design based on affordance theory could help bridge the gap between what HCI provides and VE design needs because affordances purport to explain the communication between objects and observers of an environment. This paper explains why affordances may not be realized in VEs, and provides insights into how three fundamental affordances may be enabled by careful design.

1. INTRODUCTION

A key difference between reality and VEs is that in the real world, an individual generally "knows" what can be done in a natural environment and how to do it. This is often *not* true in HCI, where users often become "lost" within hierarchical menu structures (Sellen & Nicol, 1990), much less specifically in virtual environments. This has led to regular calls in the literature for "natural" or "intuitive" interfaces (Norman, 1988; 1993). Loomis and Blascovoich (1999) state that the ultimate representational system would allow an observer to interact "naturally" with objects and other individuals within a simulated environment, providing an experience indistinguishable from reality. To achieve such systems, the National Research Council's recent survey on VE technology calls for research "on how well the design accounts for human perceptual and cognitive features for human responses" (Durlach & Mavor, 1995, p. 65). One example of such research comes from Dixon, Wraga, Proffitt, and Williams (2000), who have shown that, just as in the real world, observers use their perceived eye-height to judge the size of objects presented in an immersive virtual environment. This is not true in non-immersive (conventional) displays, but is spontaneously evoked in immersive (i.e., helmet-mounted display-based) environments.

A core issue in VE system design is determining how to design a VE in such a way as to enable user's perception, i.e., to understand what can and should be done in the virtual world. This suggests that a productive area of research is the extension and application of psychological theories addressing perception to the design of VEs (Barnard & May, 1999). The question for VE designers is inverted from questions addressed by traditional psychological theories. Instead of understanding how humans perceive what naturally occurs in reality, VE designers need to understand how their designs support and exploit natural perception within the virtual world that they represent. More specifically, the issue for VE design is how the user perceives the objects, properties, and behaviors selected by the VE designer for inclusion in the virtual world. This is a direct result of the fact that the only way for developers to communicate with users is through the artifacts of their design evidenced in the interface (Anders, 1999).

2. BACKGROUND

Computer interface designs that are sensitive to the match between their affordances and their functions should be more usable than affordance insensitive designs. The literature suggests that the particular value of affordance sensitive designs in VEs would be those tasks that correlate well to real world tasks (Norman, 1988; 1993). Such an approach would exploit "natural" affordances (i.e., those experienced by humans operating in the real world) in VE system design. However, the literature is clear that this has yet to be established by empirical study (Eberts; 1994; Okamoto; 1997; St. Amant, 1999). St. Amant (1999) studied the affordances provided by conventional HCIs, but not the challenge of enabling the realization of natural affordances within the computer interface. The aim then should be to develop empirically validated guidelines for enabling the realization of natural affordances within VE interfaces.

2.1 Affordances in Nature

The theory of direct perception purports that humans can (and do) directly perceive what they can do in their environment, or conversely what the environment affords to the human (Gordon, 1989). As part of this theory, affordances are the inherent uses that an object in the world furnishes to their user, for good or ill. As such, VE designs based on affordances could lead to more natural and intuitive interfaces (Gordon, 1989). The following observations regarding the nature of affordances provide insights into how they could be designed into VE interfaces.

Affordances depend on objects in the environment. Self-evidently, different object properties and behaviors result in different affordances. This is the very essence of what it means for objects to be different. A chair is different from a wall because the chair affords sitting, while the wall affords leaning. The organism can perceive the same affordance from multiple objects, but some objects provide a stronger (better fit) affordance than others (St. Amant, 1999). For example, in a room of varying height chairs, the one closest to the observer's knee height will best afford sitting. Designers should thus try to endow objects with properties that are relevant to an individual's purpose within a VE system and thus afford desired skills and appropriate behaviors.

Affordances depend on the organism's action capabilities. Different action capabilities result in different affordances. For example, surfaces that affordance walking to quad-pedal creatures (e.g., a horse) are different than those that afford walking to bipedal creatures (e.g., humans). Animals with relatively rare action capabilities, such as flying, are afforded different actions by their environment than animals that do not possess this capability. Designers should provide affordances that engender appropriate action to attain relevant goal-oriented behavior.

Affordances depend on physical characteristics of the observer. Marik (1987) showed that changes in participants' height effected affordances such as sitting and climbing. This is clear evidence that humans adjust to invariants based on an understanding of their stature in the environment. Further, this experiment showed observers can rapidly and accurately adapt to at least some changes, when they become aware of them. The issue for VE system designers thus becomes whether or not one should somehow represent the physical characteristics of observers in order to engender use of affordances.

Affordances depend on the organism's sources of sensory stimuli. Different senses result in different affordances. Animals such as bats and dolphins, with strong auditory sensory systems, are afforded different actions by their environment than animals without auditory systems, such as snakes. Stephens and Banks (1987) demonstrated that the ability to detect differences in spatial contrast (crucial to object recognition) changes as infants age, suggesting that changes in underlying neural mechanisms underlie the ability to perceive these differences. As humans inherently use multiple senses to perceive affordances, the issue for VE system designers becomes how to develop multimodal representations that elicit such direct perception in virtual worlds.

Affordances depend on integration of multimodal sensory stimuli. Gibson (1979) notes the special cross modal perception between vision and kinesthetics, calling it "visual kinesthesis." Van Der Steen (1996) suggests that a multimodal perception model is needed to describe the dynamics of perceived self-motion in virtual environments. Wertheim (1994) argues that visual-vestibular interaction is crucial for correct perception in ecologically valid environments. One may deduce from research such as Kalawsky (2000) and Marik (1987) that affordances are perceived because of cross modal stimulation, and further that affordances are an effective means of describing how an organism integrates knowledge of its own action capabilities. Virtual environment designers need to understand the types of cross-modal interactions that transpire during direct perception and the possible sensory substitution schemes that can be enacted when a critical sense cannot be represented in a virtual world.

Affordances arise as the organism learns to act within its environment. E.J. Gibson, Riccio, Schmuckler, Stoffregen, Rosenberg, and Taormina (1987) demonstrated that the transversability of different surfaces for infants depended on the mode of locomotion (crawling or walking), and that the pattern of experimentation to learn what the surface afforded differed between crawlers and walkers. Part of learning to act within an environment, is exploration which creates shifting sensory viewpoints. Virtual environment designers thus need to determine the key physiological factors, such as eye-height, that drive direct perception and how to appropriately represent these factors in the virtual world.

2.2 Affordances in VE Design

It is important to note that VEs will present affordances whether or not they are consciously designed. Virtual environments create artificial worlds, whose design defines what the user can do, and the sensory stimuli to be provided. The issue is whether and how a design will enable the realization of particular affordances.

The first observation, that affordances depend on objects in the environment, does not present a particular problem to VE system design. Virtual environment designers, by selecting which objects, properties, and behaviors to represent, can readily choose which affordances a VE will present. The other observations noted above, however, have yet to be fully addressed in current and desired VE designs. First, VE design objectives may include adding atypical action capabilities (such as unpowered human flight), or removing typical ones (such as eating). Second, most VEs provide some but not all of the sensory stimuli of reality. Non-visual mode stimuli in particular are missing in VEs, especially kinesthetic and vestibular stimuli (Loomis & Blascovoich, 1999). As a result, the fact that the realization of affordances depends on integration of multimodal stimuli could prove particularly problematic for perceiving affordances in virtual environments. Finally, the affordances learned in the natural world may not transfer into a virtual world.

The user's internal state, that is the user's beliefs about physical stature, action capabilities, and goals/motivations, available to the user of a VE, does not necessarily correlate to the situation presented by a VE, and this may inhibit the realization of affordances in a synthetic environment. More specifically, as a result of these inaccuracies and absences, a user's ability to correctly perceive available action capabilities (what the user can do in the environment) may be depleted. Furthermore, representations of user characteristics (e.g., size, strength) that would normally be understood through kinesthetic and vestibular stimuli are typically arbitrarily portrayed and unrelated to a user's "real" capability in a virtual world. Due to these shortcomings it is expected that current VE designs will often fail to enable the realization of affordances. Indeed, Stappers (2000) showed that the affordance of passability (fitting through a doorway) was not correctly perceived in a virtual environment. If users are to correctly perceive, operate in, and navigate throughout VEs, it is thus essential for VE designers to address the issue of direct perception. For example, in unimodal (e.g., visual only) representations it may be possible to overlay naturally occurring cross-modal information unto the represented mode. Runeson and Frykholm (1984) demonstrated that observers could accurately estimate the weight of an object simply by viewing a human figure pick the object up – even when the figure was blacked out and illuminated only by small lights attached at critical joints. The visual information successfully provided what would normally be provided via the kinesthetic sense when attempting to pick up the object.

Users of VE systems will naturally learn through their experiences how to operate in and traverse throughout virtual worlds (Pejtersen & Rasmussen, 1997). They will naturally perfect their skills with continued use. Virtual environment designers should thus aim to develop representations that make appropriate actions readily perceivable to support this natural evolution of skills. Table 1 presents key issues that need to be resolved to develop such representations so as to support direct perception in virtual environments.

Table 1. Key issues that need to be resolved to support direct perception in virtual environments.

- How can objects be endowed with properties that are relevant to an individual's purpose within a VE system and thus afford desired skills and appropriate behaviors?
- Which affordances must be provided to engender appropriate action to attain relevant goal-oriented behavior in virtual environments?
- Should VE designers somehow represent the physical characteristics of observers (e.g., eye-height, stature) in order to engender use of affordances?
- How should multimodal representations be designed to elicit direct perception in virtual worlds?
- Which types of cross-modal interactions transpire during direct perception and how should they be portrayed in virtual environments? Is it possible to use sensory substitution when a critical sense cannot be represented in a virtual world?
- Which are the key physiological factors that drive direct perception and how should they be appropriately represented in the virtual world?

2.3 Realizing Key Affordances

Any object, either natural or virtual, offers affordances to its observers. The variety of affordances offered by an object (i.e., the full variety of how humans use objects) is practically infinite. Instead of aiming for realizing a complete set of affordances, designers might better focus on enabling a key subset that would substantially enhance VE interaction. Realizing that task performance in VEs will necessarily involve interaction with objects, three key affordances that form the basis of object interaction include:

1) Passability – an object affords an observer the ability to maneuver between it and other objects close by.
2) Liftability – an object affords to an observer the ability to grasp and move the object, permitting inspection.
3) Catchability –an object affords the ability to be "caught," that is for an observer to pursue and intersect with a moving object.

Passability. Cutting et al. (1999) assert that wayfinding through cluttered environments quickly and safely is a basic and routine task. Warren and Whang (1987) conducted a series of experiments measuring participants' responses to various passageway sizes. They determined that the invariant for free walking passage was the ratio of passageway width to shoulder width. The critical value, i.e. value at which behavior changed from free walking to rotation of shoulders, was 1.3. They discuss the role that the visual and vestibular senses play in perceiving this affordance. Since typical VE designs do not provide vestibular stimulation regarding shoulder width, it is unlikely that users of such VEs would correctly realize the passability affordance, as shown by Stappers (2000). If a VE's design objectives include enabling realization of the passability affordance, it would have to provide some stimulation representing shoulder width, perhaps a visual cue overlaid on the usual ego point of view display.

Liftability. Turvey, Shockley, and Carello (1999) point out that perceiving the affordance of liftability involves interplay between the visual and kinesthetic senses. Since typical VE designs do not provide kinesthetic sensory feedback sufficient for users to understand their own height, reach, and strength, it is unlikely that users will correctly realize the liftability affordance. If a VE's design objectives include enabling realization of liftability, it would have to provide some stimulation representing such information. Bertenthal et al. (1987) suggest that the visual system is sensitive to forces associated with moving objects. Runeson and Frykholm (1981) demonstrated that observers could accurately judge the relative weights of invisible objects when shown a point-light display of a person lifting these objects. Thus, it may be possible to substitute visual information for kinesthetic cues normally conveyed during lifting in order to realize this affordance.

Catchability. Oudejans et al. (1996) investigated the affordance of catchability. They demonstrated that this affordance depends on kinematic body characteristics, in contrast to previously demonstrated affordances, which depend on geometric body characteristics. Participants were readily able to judge their ability to intercept and catch thrown objects in a wide variety of circumstances. Since typical VE designs do not provide kinesthetic sensory feedback sufficient for users to understand their own speed and acceleration capabilities, it is unlikely that users will correctly realize the catchability affordance. If a VE's design objectives include enabling realization of catchability, it would have to provide some stimulation representing such information.

3. CONCLUSION

Virtual environment designers must discover techniques for enabling the realization of natural affordances in VE designs, and to exploit those affordances to increase the usability of these systems. Research needs to determine how affordances are perceived in nature, what is impoverished in VEs that disables the realization of affordances, and how cross modal stimuli can substitute for absent sensory stimuli, thereby enabling the perception of affordances. This paper has discussed three affordances (i.e., passability, liftability, catchability) that may prove influential to VE system design. With current VE technology, it is clear that some form of sensory substitution will be required to enable perception of these affordances.

REFERENCES

Anders, P. (1999). *Envisioning cyberspace: Designing 3D electronics spaces*. New York: McGraw-Hill.

Barnard, P. J. and May, J. (1999). Representing cognitive activity in complex tasks. *Human-Computer Interaction*, 14, 93-159.

Cutting, J. E. (1997). How the eye measures reality and virtual reality. *Behavior Research Methods, Instruments & Computers*, 29(1), 27-36.

Dixon, M. W., Wraga, M., Proffitt, D. R., and Williams, G. C. (2000). Eye height scaling of absolute size in immersive and nonimmersive displays. *Journal of Experimental Psychology*, 26(2), 582-593.

Durlach, N.I. and Mavor, A.S. (Eds.). (1995). *Virtual reality - scientific & technological challenges.* Washington, DC: National Academy Press.

Eberts, R. (1994). *User interface design.* Englewood Cliffs, NJ: Prentice Hall.

Gibson, E. J., Riccio, G., Schmuckler, M. A., Stoffregen, T. A., Rosenburg, D., and Taormina, J. (1987, August). Detection of traversablity of surfaces by crawling & walking infants. *Journal of Experimental Psychology: Human Perception & Performance,* 13(4), 533-544.

Gibson, J. J. (1979/1986). *The ecological approach to visual perception.* Hillsdale, NJ: Lawrence Erbaum Associates.

Gordon, I. E. (1989). *Theories of visual perception.* Chichester, UK: John Wiley & Sons.

Kalawsky, R.S. (2000, March). The validity of presence as a reliable human performance metric in immersive environments. *Presence 2000 -- 3rd International Workshop on Presence,* 27-28 March 2000.

Loomis, J. M. and Blascovoich, J. J. (1999). Immersive virtual environmental technology as a basic research tool in psychology. *Behavior Research Methods, Instruments, & Computers,* 31(4), 557-564.

Marik, L. S. (1987, June). Eyeheight-scaled information about affordances: A study of sitting & stair climbing. *Journal of Experimental Psychology: Human Perception & Performance,* 13(3), 361-370.

Norman, D. A. (1988). *The design of everyday things.* New York: Double Day.

Norman, D. A. (1993). *The things that make us smart.* New York: Addison-Wesley.

Oudejans, R. R. D., Michaels, C. F., Bakker, F. C. & Dolne, M. A. (1996). The relevance of action in perceiving affordances: perception of catchableness of fly balls. *Journal of Experimental Psychology: Human Perception & Performance,* 22(4), 879-892.

Okamoto, A. (1997). Toward rehabilitation cognitive engineering – gap between theory & practice in the human interface of information processing devices for people with disabilities. In M.L. Smith, G. Salvendy, & R.J. Koubek (Eds.), *Design of computing systems: Social & ergonomics considerations* (pp 551-554). New York: Elsevier Science.

Pejtersen, A.M. and Rasmussen, J. (1997). Ecological information systems and support of learning: Coupling work domain information to user characteristics. In M. Helander, T.K. Landauer, and P.V. Prabhu, (Eds.), *Handbook of Human-Computer Interaction* (2nd Edition) (pp. 49-63) Amsterdam: North-Holland.

Runeson, S. and Frykholm, G. (1981). Visual perception of lifted weight. *Journal of Experimental Psychology,* 7(4), 733-740.

St. Amant, R. (1999). User interface affordances in a planning representation. *Human-Computer Interaction,* 14, 317-355.

Sellen, A. and Nicol, A. (1990). Building user-centered on-line help. In B. Laurel (Ed.), *The Art of Human Computer Interface Design* (pp. 143-153). Reading, MA: Addison Wesley.

Stappers, P. J. (in press). Critical ratios as behavioral indices of presence. *Presence: Teleoperators & Virtual Environments.*

Stephens, B. R. and Banks, M. S. (1987, August). Contrast discrimination in infants. *Journal of Experimental Psychology: Human Perception & Performance,* 13(4), 558-565.

Turvey, M.T., Shocklet, K. & Carello, C. (1999). Affordance, proper function, and the physical basis of heaviness. *Cognition,* 73, B17-B26.

Van Der Steen, F. A. M. (1996). Simulating self-motion. *Brain Research Bulletin,* 40(5/6), 473-475.

Warren, W. H., Jr. & Whang, S. (1987). Visual guidance of walking through apertures: Body-scaled information for affordances. *Journal of Experimental Psychology: Human Perception & Performance,* 13(3), 371-383.

Wertheim, A. H. (1994). Motion perception during self-motion: the direct versus inferential controversy revisited. *Behavioral and Brain Sciences,* 17(2), 293-355.

Use of a Motion Experience Questionnaire to Predict Simulator Sickness

Robert S. Kennedy
Norman E. Lane
RSK Assessments, Inc.
1040 Woodcock Road, Suite 227
Orlando, FL 32803

Kay M. Stanney
Susan Lanham *
Kelli Kingdon
University of Central Florida
Industrial Engineering and
Management Systems
4000 Central Florida Blvd.
Orlando, FL 32816-2450

* Now at Fiserv, Inc. CBS International 2601 Technology Drive Orlando, FL 32804-8068

Abstract

There are wide individual differences in susceptibility to sickness in motion environments. Efforts to identify unusually susceptible persons have occurred sporadically for about 50 years.. A Motion History Questionnaire (MHQ) which uses past experiences with and perceptions of potentially provocative situations, has been used in various forms to study airsickness, sea sickness, and rotation effects, usually with useful but modest predictive validity. As part of a larger study of virtual reality (VR) environments, we had the opportunity to develop and validate updated scoring keys for the MHQ on a large sample (around 700 college students). Four composite scores were developed and validated in split samples. The four composites combined correlated .408 and .448 in the two samples, against a criterion of Simulator Sickness (SS) scores obtained after exposure to a helmet-mounted VR display. These validities are highly significant and materially higher than those in previous studies. The paper describes the MHQ, the scoring method, the and the composite development, and discusses MHQ application as a screening tool for identifying individuals who are likely to have sickness problems in training and research applications of simulators and VR systems.

1. Introduction

Efforts to identify individuals who were unusually susceptible to motion sickness started in World War II (Alexander, Cotzin, Hill, Ricciuti, and Wendt, 1945; Birren, 1949), and were renewed in the early 1960's (Kennedy and Graybiel, 1965). The Navy Motion Sickness History Questionnaire (MHQ), intended for pre-exposure use, asked the participant about his or her history in relation to motion sickness, such as whether the participant had become motion sick in a variety of potentially provocative environments. Concurrently another instrument, the Motion Sickness Questionnaire (MSQ) (Kennedy and Graybiel, 1965) was designed for post-exposure use, to assess the participant's symptoms of motion sickness after the experiment. The MSQ has been through several iterations since that time (Kennedy et al., 1984), and a variant, the Simulator Sickness Questionnaire (SSQ) (Lane and Kennedy, 1988) was developed to study the less severe but prevalent version of motion sickness encountered in simulators.

1.1 Previous work with the MHQ

A number of previous studies have examined variants of the MHQ and suggested items indicative of difficulties with motion and motion-related environments. Hutchins and Kennedy (1965) examined the relationship between the MHQ and attrition from flight training. Over six hundred incoming flight students were given the MHQ. Twelve items statistically separated students who successfully completed flight training from those who attrited, including willingness to volunteer for an experiment where others get sick, and a pattern of avoidance of activities such as roller coasters and carnival rides. Lenel, Berbaum, and Kennedy, (1987), studied the MHQ in young adults from a military population. MHQ responses by participants involved in whole-body motion were analyzed against the criterion of subject's vomiting and time under stimulation prior to vomit. Questions about seasickness, feeling better after vomiting, susceptibility to motion sickness, being nauseated in past eight weeks, and self-report of the probability of getting sick were related to presence/absence of vomiting. In 1992, Kennedy, Fowlkes, Berbaum, and Lilienthal looked at the generalizability of the MHQ, originally calibrated against the more provocative environments of air and sea exposure, to the prediction of simulator sickness (SS). Participants (all Navy and Marine Corps pilots) were divided into two groups, differing in the nature and intensity of simulator exposure. All participants were given the MHQ and the SSQ prior to a simulator exercise; the SSQ was given again immediately after the exercise. Nine questions selected from the MHQ related to reported SS, including the extent of experience in ships and aircraft, an individual's perceived susceptibility, and willingness to volunteer for nauseogenic exposures. In addition, MHQ has been predictive of 1) seasickness (Kennedy, Graybiel, McDonough, and Beckwith, 1965); 2) airsickness (Hixson, Guedry, and Lentz, 1984; 3) sickness caused in various laboratory procedures designed to induce motion sickness; and 4) simulator aftereffects (SS) (Kennedy et al., 1984). Typically, while predictive validities for MHQ have been logical, consistent and statistically significant, the validities have not been very large (.20 or less), particularly when exposures are not strongly provocative and the base rate of sickness is relatively low, as in simulator sickness (Kennedy, Dunlap, and Fowlkes, 1990). Other factors such as restriction of range (subjects in these studies were often experienced military personnel), and in some cases relatively small sample sizes, have made the empirical keying of MHQ alternatives to specific environments (such as simulators and virtual reality devices) impractical, and have thus worked to keep the absolute values of MHQ validities relatively modest.

The present study offers an opportunity to perform item analysis of MHQ questions on a large sample and to develop specific keys for using MHQ to predict an individual's likelihood of sickness in a VR environment.

2. Method

Subjects were 766 students at the University of Central Florida in Orlando, ages 18-40. All participants were in good physical health. Participants were compensated for their time and the experiment lasted for four hours. The experiment entailed exposure to scenes with differing content and with varying duration of exposure. Main effects for these variables are reported elsewhere (Stanney, Kingdon, and Kennedy, 2001).

The Motion History Questionnaire used in this study is given in Figure 1. Questions are asked about experiences in environments that sometimes engender motion sickness-like symptoms, about the participant's judged susceptibility to motion sickness, nausea and dizziness, and about likes and dislikes for activities which produce such symptoms in some persons. The MHQ is currently being revised in content and format; the most recent version can be obtained from the authors.

Subjects completed the MHQ prior to the testing period. In general they found the experience challenging and provocative; many (21%) requested to terminate their involvement prior to completion of the experimental regime. Subjects were divided, on the basis of sequential subject numbers, into "Odd" and "Even" samples, each with sample size of 383.

3. The Criterion: Developing MHQ Scoring Composites

The MHQ, as it has evolved through several iterations, has changed slightly in numbers and types of questions and in the way in which questions are posed. The version used in this study had been used in several studies with aviation personnel; although the population of concern in the present study is college students, the same form was used to provide continuity and comparison to previous scoring approaches. As Appendix A shows, there are several different formats of questions (yes/no, always-to-never scales, like/dislike, and symptom lists). To simplify scoring and analysis, all responses were converted to a 1/0 format. For Items 2, 4, 5, 6, 7, and 13, responses of "never" were assigned a value of "0", while all other responses were assigned a value of "1". Items 1, 3, 10, and 14 were not used in the composites developed here. Preference variables (Like, Neutral, Dislike) were coded such that an answer in any column was assigned a "1" and an absence of response was assigned a "0". Symptoms were scored similarly, 1 for presence and 0 for absence of a listed symptom. The resulting dichotomies had both negative and positive natural relationships to the SSQ (high SSQ scores indicate greater symptomatology).

A "rational" approach to determining composites was employed rather than an "analytic" approach. That is, items are selected for the composites based on their content rather than on the statistical properties of the items. Items or item alternatives whose labels suggest that they "go together" tend to be grouped together. Similarly, items were predominantly selected from environments that students were most likely to have experienced. Thus, aircraft, simulators, roller coasters and carnival rides were given preference over items such as swings and trains that are not as common as they once were, over items whose environments were not readily available (gymnastics, cinerama, motorcycles), and over items for which stimuli were likely to be less provocative (merry-go-round, automobiles, hammocks). Further, the "rational" approach does not select or weight items in accordance with sample data. Items are added only with unit weights, and presence of an item in a composite does not depend on item correlations or other statistical information. Composites so constructed are somewhat simpler to score, and much more likely to generalize to other applications and samples outside the present study without coefficient shrinkage.

Accordingly, we identified four likely scoring composites (with the intention that more than one of these might be employed in a given situation). These are (with some tentative labels):

Composite 1: General Susceptibility
Items 2, 5, 6, 7, (-12c), 13. The sum of these variables, possible range –1 to 5.

Composite 2: Stated Dislike of Nauseogenic Environments.
Dislike Aircraft, Simulators, Roller Coasters, Other Carnival Rides. Possible range 0 to 4.

Composite 3: Nausea and Vomiting Experienced during Prior Exposures
Vomiting reported from Aircraft, Simulators, Roller Coasters, Other Carnival Rides (4 variables). Nausea reported from Above exposures (4 variables). Sum of 8 variables. Possible range 0 to 8.

Composite 4: No Reported Symptoms from Exposure
"None" symptoms reported from Aircraft, Roller Coaster, Other Carnival Rides, Long Train/Bus Trips. (Note that Simulators is not weighted here. " No Symptoms" for Simulators had a correlation in both samples opposite to that for other No Symptoms variables, possibly a database error). Sum of 4 variables. Possible range 0 to 4. (Note also that high values on this composite indicate low motion susceptibility).

4. Results and Discussion

Scores on the composites described above were computed for each subject. Correlations of composites with the criterion variable were computed in both the odd and even samples. The criterion was the SSQ score obtained immediately after completion of the experiment, or for quits, the SSQ obtained after they exited the experiment. Table 1 gives the correlations between composites and SSQ for the two samples.

Table 1
Correlations of Composites with SSQ Criterion

Composite	Odds	Evens
1	.302	.352
2	.266	.274
3	.366	.393
4	-.275	-.293

Note: N=383 for each sample

Composite correlations are remarkably consistent across the split samples. They are also much larger than typically encountered in MHQ studies. Composites can also be summed, both to increase validity and to tailor the questionnaire to specific applications. Since the composites are correlated to one another, much of the variance explained by each is multiply determined, and adding composites in any order will eventually provide only modest increases in composite validity. Table 2 shows the validity, and the dropoff in incremental validity, resulting from algebraically summing the composites in the order 1, 2, 3, -4.

Table 2
Correlation of composite sums with SSQ Criterion

Composites Added	Odds	Evens
1	.302	.352
1 + 2	.348	.394
1 + 2 + 3	.400	.442
1 + 2 + 3 - 4	.405	.448

The total validities, .405 and .448, are quite acceptable for MHQ scores based on 22 simply-scored items in a mixed-age, mixed-gender study. It is likely that some useful increments in ability to predict simulator and VR sickness can be obtained by more sophisticated scoring (i.e., using all 5 levels of response to questions about airsickness, motion sickness, and so forth ; using sample-derived integer or least squares weighting, etc). The experiment database supports increased precision, and we intend to pursue additional methods of composite construction in the future. There is a risk associated with more complex scoring, however. The more closely the composite scoring fits the idiosyncrasies of the calibration sample, the lower its likely generalizability to other motion or motion-related environments.

There are some distinct payoffs from being able to estimate with reasonable accuracy who will get sick in, and possibly quit, either studies of virtual reality or actual applications of VR. There is increased interest in training using VR simulations. At least some significant piece of the potential training population (perhaps 25% to 50% depending on the application) may not be able to tolerate the VR training. It may be useful to flag such persons ahead of time for special habituation or incremental adaptation via low-stress exposures. Similarly, experimental studies of VR as a system interface become more difficult when individuals exit the experiment without yielding useful data other than that they left the experiment. It may be of value to know ahead of time which persons are most susceptible, since it may be necessary to design special VR interfaces to reduce the frequency and severity of simulator and VR sickness. MHQ scoring and profiling makes good sense from the standpoint of managing cybersickness. It should eventually be possible with adequate analytic techniques to establish a screening protocol and, with a cost of 5%-10% false positives, correctly identify 50% or more of the persons who are likely to have performance problems or to terminate their exposure..

5. References

Alexander, S. J., Cotzin, M., Hill, C. J., Jr., Ricciuti, E. A., & Wendt, G. R. (1945). Prediction of motion sickness on a vertical accelerator by means of a motion sickness history questionnaire. Journal of Experimental Psychology, 20, 25-30.

Birren, J. E. (1949). Motion sickness: Its psychophysiological aspects. In Panel on Psychology and Physiology (Eds.) A survey report on human factors in undersea warfare (pp. 375-397). Washington D.C.: National Research Council, Committee of Undersea Warfare.

Hixson, W. C., Guedry, F. E., & Lentz, J. M. (1984). Results of a longitudinal study of airsickness during naval flight officer training: Executive summary. (Special Rep. No. 85-2). Pensacola, FL: Naval Aerospace Medical Research Laboratory. (DTIC No. AD A150 887)

Hutchins, C. W., & Kennedy, R. S. (1965). Relationship between past history of motion sickness and attrition from flight training. Aerospace Medicine, 36 (10), 984-987.

Kennedy, R. S., Dunlap, W. P., & Fowlkes, J. E. (1990). Prediction of motion sickness susceptibility: A taxonomy and evaluation of relative predictor potential. In G. H. Crampton (Ed.), Motion and space sickness (pp. 179-215). Boca Raton, FL: CRC Press.

Kennedy, R. S., Fowlkes, J. E., Berbaum, K. S., & Lilienthal, M. G. (1992). Use of a motion sickness history questionnaire for prediction of simulator sickness. Aviation, Space, and Environmental Medicine, 63, 588-93.

Kennedy, R. S., Frank, L. H., McCauley, M. E., Bittner, A. C., Jr. Root, R. W., & Binks, T. A. (1984). Simulator sickness: Reaction to a transformed perceptual world. VI. Preliminary site surveys. AGARD Conference Proceedings No. 372: Motion Sickness: Mechanisms, Prediction, Prevention, and Treatment (pp. 34.1-34.11). Neuilly-Sur-Seine, France: Advisory Group for Aerospace Research and Development.

Kennedy, R. S., Graybiel, R. C., McDonough, R. C., & Beckwith, F. D. (1968). Symptomatology under storm conditions in the North Atlantic in control subjects and in persons with bilateral labyrinthine defects. Acta Otolaryngologica, 66, 533-540.

Kennedy, R. S., & Graybiel, A. (1965). The Dial test: A standardized procedure for the experimental production of canal sickness symptomatology in a rotating environment. (Rep. No. 113, NSAM 930). Pensacola, FL: Naval School of Aerospace Medicine.

Lane, N. E., & Kennedy, R. S. (1988). A new method for quantifying simulator sickness: Development and application of the simulator sickness questionnaire (SSQ). (EOTR 88-7). Orlando, FL: Essex Corporation.

Lenel, J. C., Berbaum, K. S., & Kennedy, R. S., (1987, November). A motion sickness history questionnaire: Scoring key and norms for young adults. (EOTR 87-3). Orlando, FL: Essex Corporation.

Stanney, K. M., Kingdon, K., & Kennedy R. S. (in press). Extreme responses to virtual environment exposure. International Journal of Human-Computer Interaction.

Figure 1. Motion History Questionnaire

MOTION HISTORY QUESTIONNAIRE

Developed by Robert S. Kennedy & colleagues under various projects. For additional information contact:
Robert S. Kennedy, RSK Assessments, Inc.; 1040 Woodcock Road, Suite 227, Orlando, FL 32803 (407) 894-5000

Subject Number: _____   **Date:** _____

1. Approximately how many total flight hours do you have? _____ hours

2. How often would you say you get airsick?
   Always _____ Frequently _____ Sometimes _____ Rarely _____ Never _____

3. a) How many total flight simulator hours? _____ Hours
   b) How often have you been in a virtual reality device? _____ Times _____ Hours

4. How much experience have you had at sea aboard ships or boats?
   Much _____ Some _____ Very Little _____ None _____

5. From your experience at sea, how often would you say you get seasick?
   Always _____ Frequently _____ Sometimes _____ Rarely _____ Never _____

6. Have you ever been motion sick under any conditions other than the ones listed so far?
   No _____ Yes _____ If so, under what conditions?

7. In general, how susceptible to motion sickness are you?
   Extremely _____ Very _____ Moderately _____ Minimally _____ Not at all _____

8. Have you been nauseated FOR ANY REASON during the past eight weeks?
   No _____ Yes _____ If yes, explain _____

9. When you were nauseated for any reason (including flu, alcohol, etc.), did you vomit?
   Only with _____ difficulty _____ with great difficulty _____
   Retch and finally vomited _____

10. If you vomited while experiencing motion sickness, did you:
    a) Feel better and remain so? _____
    b) Feel better temporarily, then vomit again? _____
    c) Feel no better, but not vomit again? _____
    d) Other - specify?

11. If you were in an experiment where 50% of the subjects get sick, what do you think your chances of getting sick would be?
    Almost _____ Probably _____ Almost _____ Probably _____ Certainly _____
    Certainly would _____ would _____ would not _____ would not _____

12. Would you volunteer for an experiment where you knew that: (Please answer all three)
    a) 50% of the subjects did get motion sick? Yes _____ No _____
    b) 75% of the subjects did get motion sick? Yes _____ No _____
    c) 85% of the subjects did get motion sick? Yes _____ No _____

13. Most people experience slight dizziness (not a result of motion) three to five times a year. The past year you have been dizzy:
    More than this _____ The same as _____ Less than _____ Never dizzy _____

14. Have you ever had an ear illness or injury which was accompanied by dizziness and/or nausea?
    Yes _____ No _____

15. Listed below are a number of situations in which some people have reported motion sickness symptoms. In the space provided, check (a) your PREFERENCE for each activity (that is, how much you like to engage in that activity), and (b) any SYMPTOM(s) you may have experienced at any time, past or present.

SITUATIONS	PREFERENCE								SYMPTOMS							
	LIKE EXTREMELY	DO NOT MIND	UNDECIDED	AVERSION *	DISLIKE INTENSELY				AWARENESS OF STOMACH	VAGUE * **	NAUSEA	VOMIT				
Aircraft																
Flight simulator																
Roller Coaster																
Merry-Go-Round																
Other carnival devices																
Automobiles																
Long train or bus trips																
Swings																
Hammocks																
Gymnastic Apparatus																
Roller / Ice Skating																
Elevators																
Cinerama or Wide-Screen Movies																
Motorcycles																

HOW DO DESIGNERS OF WEB SITES TAKE INTO ACCOUNT CONSTRAINTS AND ERGONOMIC CRITERIA?

Nathalie Bonnardel and Aline Chevalier

Research center in Psychology of Cognition, Language and Emotion, University of Provence
29, avenue Robert Schuman, 13621 Aix en Provence, France

ABSTRACT

Due to the extensive growing of the Internet, more and more sites are presented on the World Wide Web (Web). The use of HTML authoring tools (and, especially, WYSIWYG authoring tools) allows not only specialists but also "lay-designers" to design and "self-publish" web sites, either towards professional or personal aims. However, the design of numerous sites presented on the Web can be considered as not satisfying for the users. Towards a better understanding of difficulties encountered by designers of web sites, an experiment was conducted with professional and beginning designers, in order to characterize (1) to which extent they could spontaneously take into account criteria and constraints (especially, related to usability issues), and (2) whether the criteria and constraints spontaneously taken into account were effectively satisfied by the web sites these designers produced. The results allow us to point out cognitive activities for which a support could be specifically useful for designers.

1. INTRODUCTION

The World Wide Web has become a ubiquitous infrastructure, which supports a wide range of services and offers access to an extensive variety of information elements. Consequently, the Web is modifying uses of access to various documents as well as interactions between authors of such documents (especially web sites) and "readers" of these documents or "visitors" of the web sites (see, for instance, Sumner, Buckingham Shum, Wright, Bonnardel, Piolat & Chevalier, 2000). In addition to this side "access to web sites", the Web led to a new type of creative activities: *the design of web sites*. Indeed, the Web allows large organizations, small and medium-sized enterprises as well as individuals from all walks of life to design and "self-publish" their own sites. Such sites may be designed towards various objectives: to present commercial products, to provide visitors with various information elements (e.g., at a large scale or only personal information), to reach educative objectives, etc. Their development does not require a high level of technical expertise, since the use of HTML authoring tools (and, especially, WYSIWYG authoring tools) is not particularly difficult after a training period. Therefore, the design of web sites is not only performed by specialists of new technologies, but also by "lay-designers", i.e. people with little or no formal training in either web site design or its attendant skills (e.g., graphic design or user interface design). However, most of the sites presented on the Web sounds not satisfying for users as well as for specialists in cognitive ergonomics: web sites are frequently difficult to use and they do not meet directly the users' or visitors' objectives or needs (Nielsen, 2000).

From the perspective of improving web sites, numerous research works have been conducted. They are mainly focused on two types of goals:
- analyzing the access to web sites or hypermedia documents (see, for instance, Vora & Helander, 1997; Smith, Newman & Parks, 1997);
- elaborating guidelines and ergonomic criteria specifically adapted to the Web environment (see, for instance, Leulier, Bastien & Scapin, 1998; Nielsen, 2000; Scapin, Leulier, Vanderdonckt, Mariage, Bastien, Farenc, Palanque & Bastide, 2000).

Though such studies are essential, we argue that it is also increasingly important to know more about cognitive processes of web site designers and, especially, to identify difficulties they encounter, which can explain difficultie later encountered by web site users. However, little research is being conducted on designers' cognitive processes though designing web sites appears particularly complex. The complexity of this work is not especially due to the cognitive challenges of authoring in HTML (since HTML authoring tools can be used), but to a variety of tasks necessary to prepare authoring in HTML. For instance, designers have to select relevant information elements to present in the site (according to what will be or could be the users' objectives), to determine how best to present these information elements, and how best to provide navigation and access to them. To perform such tasks, designers have to take into account the viewpoints of different stakeholders, in addition to their own (Bonnardel & Chevalier, 1999; Chevalier & Bonnardel, 2001). Especially, they have to consider the viewpoint of the customer and to take into account potential needs or objectives of web site users. Consequently, much professional design work performed in large organizations is based on cooperation and communication between different stakeholders. On the contrary, individuals and web site designers working in small-sized enterprises mainly work individually. Indeed, while interactions between designers and customers occur at certain stages in the design process, interactions between designers and web site users remain in most cases only "virtual" or imagined. Since such situations are more and more frequent and since the design of web sites is more and more performed by "lay-designers", we are going to present an exploratory story conducted with both beginning designers and professional designers working in small-sized enterprises. In order to contribute to a characterization of difficulties encountered by web site designers, this study aims at determining (1) what is the impact of customers' requirements on the designers' activities, and (2) whether designers "intuitively" or "spontaneously" try to respect and really respect prescribed constraints defined by the customer as well as ergonomic criteria and constraints. Especially, concerning this last type of criteria and constraints, guidelines for the design of web sites have been structured around "Ergonomic Criteria", i.e. well-recognized usability dimensions in human-computer interaction (Scapin & Bastien, 1997; Leulier, Bastien & Scapin, 1998).

2. METHODOLOGY

2.1 Participants

Twelve designers participated in this study:
- 8 "beginning" designers who recently completed a short training course on the design of web sites,
- 4 professionals who have been working, for about 3 years, in small-sized enterprises specialized in the design of web sites, but did not attend a specific and extensive training in cognitive ergonomics, nor in user interface design.

2.2 Experimental task

We asked these participants to design and construct, during 1 hour and a half, the sketch of a web site aiming at presenting new cars for a car dealer. More precisely, to study the role of the customer's requirements on the designers' activities, we provided half of the designers with a "well-defined" schedule of conditions (WSC) comprising numerous requirements, and the other designers with an "ill-defined" schedule of conditions (ISC) comprising only a few information elements. We will refer to this experimental factor as "the level of specification of the schedule of conditions".
In addition, in order to determine which guidelines and ergonomic criteria were spontaneously taken into account, we asked the designers to "think aloud" during their design activities (technique of simultaneous verbalization, see Ericsson & Simon, 1993; Levy, Marek & Lea, 1996).

2.3 Data analysis

The designers' verbalizations were transcribed and separately analyzed by two judges, in order to compare our findings. This analysis comprised two phases aiming at different objectives:
- Quantitatively and qualitatively characterize the criteria and constraints designers spontaneously took into account. More precisely, we distinguished between two types of constraints: "prescribed" constraints, which are derived from the schedule of conditions defined by the customer, and "ergonomic" constraints structured around Ergonomic Criteria (Scapin & Bastien, 1997). During this phase of the analysis, we counted each constraint taken into account, at least, one time by a designer.
- Determine to which extent the produced web sites effectively respected the criteria and constraints identified in the previous analysis phase, whatever their nature (prescribed or ergonomic).

3. RESULTS

3.1 Taking into account prescribed constraints

Designers who had to deal with the well defined schedule of conditions (WSC) took into account, in mean, the same number of prescribed constraints, whatever their level of experience: 7.5 for beginning designers and 7 for professionals (out of a total number of 11 prescribed constraints in the WSC). Moreover, the prescribed constraints taken into account by the designers appeared to be mainly the same (e.g., "to harmonize the colors of the site with the ones of the car dealer's logo").

We also observed that, whatever their level of experience, the designers who had to deal with the ill defined schedule of conditions (ISC) were able to infer between 5 and 6 constraints specified in the WSC but not presented in the ISC.

3.2 Taking into account ergonomic constraints and criteria

Both beginning and professional designers took into account, in mean, respectively, 9 and 10 ergonomic constraints (i.e., constraints structured around Ergonomic Criteria).
However, the level of specification of the schedule of conditions appears to influence the taking into account of ergonomic constraints by beginning designers, which does not seem to be the case for professional designers. Beginning designers took into account less ergonomic constraints when they had to deal with the WSC than when they were provided with the ISC: in mean, respectively, 6.5 vs. 11.5 ergonomic constraints (which corresponds to 36% vs. 63% of the total number of constraints taken into account).
According to the relatively important number of prescribed constraints taken into account by beginning designers provided with the WSC (7.5 constraints out of 11), these designers seem to be focused on prescribed constraints specified in the schedule of conditions. Taking into account such prescribed constraints seems to be performed to the detriment of ergonomic constraints and, thus, of constraints that have been defined in order to facilitate web site users' activities. On the contrary, the fact of being provided with few prescribed constraints (in the case of the ISC) seems to allow beginning designers to take into account more the users' point of view. Indeed, though designers with the ISC were able to infer an important number of constraints, these inferred constraints have a different status than the ones who were explicitly specified in the WSC. Inferred constraints may be considered as less important by the designers (especially beginning designers) than the ones prescribed by the customer (specified in the WSC), though the inferred constraints are the same than the prescribed constraints.

Moreover, the qualitative analysis of ergonomic constraints expressed by the designers shows that only two general criteria were mainly taken into account, whatever the designers' level of experience and the level of specification of the schedule of conditions. These criteria are the following (defined according to Scapin & Bastien, 1997):
- the "prompting", which aims at providing the user with information elements about the state in which he or she is (44% of the total amount of ergonomic constraints taken into account);
- the "workload", both at perceptive and memory levels (27% of the total amount of ergonomic constraints).
Thus, the constraints linked to these two criteria represent about 71% of the total amount of ergonomic constraints taken into account by the designers.

These first results show that the designers spontaneously take into account certain ergonomic constraints, aiming at facilitating the use of web sites, though numerous difficulties of use are encountered in most existing web sites. How can we explain such a discrepancy? In order to bring elements of answer to this question, the web sites produced by the participants in this study were analyzed in order to determine to which extent these sites respected ergonomic constraints as well as constraints linked to the customer, which had been spontaneously taken into account by the designers (comprising both prescribed and inferred constraints).

3.3 Respect of ergonomic criteria and constraints

The results we obtained with beginning designers show that though they were able to spontaneously take into account a relatively important number of ergonomic constraints (in mean, 9 by designer), about 28% of the constraints they previously expressed are not respected in the sketches of web sites they produced, and this occurs whatever the level of specification of the schedule of conditions.

Concerning professionals, the results appear more surprising: though they took into account about 10 ergonomic constraints, whatever the level of specification of the schedule of conditions, we observe differences in the respect of these constraints in the web sites, according to the schedules of conditions. Thus, 60% of the constraints expressed by the professionals who had to deal with the WSC are not respected in the web sites they produced, vs. against only 25% of the constraints mentioned by professionals who were provided with the ISC.

From a qualitative point of view, ergonomic constraints, which are not respected in the web sites, are related to the two general criteria on which the designers focused:
- the "prompting" (e.g., "to put a title on each page of the site", "to have a direct access to the homepage from each page of the site"),
- the "workload" (e.g., "to minimize the scrolling of pages").

1068

Moreover, we identified, in the produced web sites, several ergonomic problems that are not related to ergonomic criteria and constraints mentioned by the designers. These problems are mainly linked to the following criteria: "prompting" (12 types of problems), "compatibility" (4 types of problems), "adaptability" (3 types of problems), "workload" (2 types of problems), "explicit control" (1 type of problems) and "error management" (1 type of problems).

3.4 Respect of prescribed constraints

Whatever their level of experience, the designers who had to deal with the WSC took into account, in mean, the same number of prescribed constraints (see §3.1: 7.5 for beginning designers, and 7 for professionals, out of a total number of 11 prescribed constraints). The analysis of the sketches of web sites shows that these designers concretely respected most of the prescribed constraints they mentioned (6 prescribed constraints were respected in the sites produced by beginning designers, and 5.5 in the ones produced by professionals).

Whatever their level of experience, the designers who had to deal with the ISC appeared to be able to infer between 5 and 6 constraints among the ones specified in the WSC. The analysis of the produced web sites shows that these inferred constraints were concretely respected (in mean, the 5 constraints expressed by beginning designers were respected in the sites they produced, and the 6 constraints expressed by professionals were respected in their sites).

4. DISCUSSION

Though this first study has to be completed by other analyses, it allowed the identification of difficulties encountered by designers of web sites, according to their level of experience in the creation of web sites and according to the level of specification of the schedule of conditions they were provided with. Thus, it appeared that designers, whatever their level of experience and the schedule of conditions, tend to focus on the customer's requirements (i.e. on constraints specified in the WSC or inferred by designers in the case of the ISC). Indeed, the sketches of web sites produced by the designers respected all the constraints (prescribed or inferred) related to the customer. This approach seems to be conducted to the detriment of ergonomic constraints though these constraints were spontaneously taken into account by designers: an important number of mentioned ergonomic constraints are not respected in the produced web sites and, moreover, other types of ergonomic problems were identified in the sites.

These results allow us to point out difficulties encountered by web site designers (and thus to contribute to the understanding of difficulties encountered by web site users or visitors):
- designers spontaneously take into account only certain ergonomic constraints (especially, the ones related to criteria of "prompting" and "workload")
- moreover, they do not apply exhaustively all the ergonomic constraints they take into account.
Different hypotheses can be evoked to explain these behaviours:
- being focused on constraints linked to the customer, designers "forget" some of the ergonomic constraints they previously mentioned and explicitly wished to take into account;
- when they are engaged in developing the web site, they focus on technical aspects of their activities to the detriment of issues linked to the users or visitors (e.g., ease of navigation) — in addition, such usability issues are particularly difficult to consider since the activity of future web users or visitors has to be anticipated;
- though they previously wished to respect certain ergonomic constraints, they encounter difficulties in concretely applying these constraints.

Supplementary studies have still to be conducted in order to bring new elements of explanation and, on these bases, to define ways for supporting designers more adapted to the difficulties they encounter: training in order to learn more about ergonomic criteria and constraints, support for effectively taking into account these criteria and constraints during web site design, and/or support for concretely applying them.

REFERENCES

Bonnardel, N. and Chevalier, A. "La conception de sites Web : Une étude de l'adoption de points de vue". Actes de la Journée satellite de la SELF "Ergonomie et Télécommunications" (Caen, France, 14 septembre), 1999, 83-93.

Chevalier, A. and Bonnardel, N. The role of viewpoints and constraints on the cognitive effort of designers of web sites. In Proceedings of the 7th European Congress of Psychology (London, U.K., July 1-6), 2001, in press.

Ericsson, K.A. and Simon, H.A. "Protocol Analysis : Verbal Reports as Data" (revised edition). Cambridge, MA: MIT Press, 1993.

Leulier, C., Bastien, J.M.C. and Scapin, D.L. "Compilation of ergonomic guidelines for the design and evaluation of Web sites". Commerce & Interaction Report. Rocquencourt, France: Institut National de Recherche en Informatique et en Automatique, 1998.

Levy, C.M., Marek, J.P. and Lea, J. "Concurrent and retrospective protocols in writing research". In G. Rijlaarsdam, H. van den Berg and M. Couzjin (Eds.), Writing Research : Theories, Models and Methodology. Amsterdam: Amsterdam University Press, 1996, 542-556.

Nielsen, J. "Designing Web Usability". Indianapolis: New Riders Publishing, 2000.

Scapin, D.L. and Bastien, J.M.C. "Ergonomic criteria for evaluating the ergonomic quality of interactive systems". Behaviour & Information Technology, 1997, 16, 220-231.

Scapin, D.L., Leulier, C., Vanderdonckt, J., Mariage, C., Bastien, J.M.C., Farenc, C., Palanque, P. & Bastide, R. "A framework for organizing web usability guidelines", In Proceedings of the 6th Conference on Human Factors & the Web (Austin, Texas, June 19th), 2000.

Smith, P.A., Newman, I.A. and Parks, L.M. "Virtual hierarchies and virtual networks: some lessons from hypermedia usability research applied to the World Wide Web", International Journal of Human-Computer Studies, 1997, 47(1), 67-95.

Sumner, T., Buckingham Shum, S., Wright, M., Bonnardel, N., Piolat A. and Chevalier, A. "Redesigning the peer review process : A developmental theory-in-action", In R. Dieng, A. Giboin, G. De Michelis and L. Karsenty, Designing Cooperative Systems: The Use of Theories and Models, Amsterdam: I.O.S. Press, 2000, 19-34.

Vora, P.R. and Helander, M.G. "Hypertext and its implications for the internet". In M. Helander, T.K. Landauer and P. Prabhu (Eds), Handbook of Human-Computer Interaction. New-York: Elsevier Science, 1997, 877-914.

Task-Based Analysis of Internet Search Output Fields

Rebeca Lergier and Marc L. Resnick, Ph.D

Industrial and Systems Engineering
Florida International University
Miami, Florida 33174

The Internet has become a powerful tool for information search and ecommerce. A recent study reported that the Internet and search engines are the most used resource to find information on any topic. However, there are many challenges that reduce the effectiveness of search engines and limit the ability of users to find information. Each directory or search engine operates using a unique search algorithm, ordering scheme, and output style. They provide various fields of the resulting pages, such as title, description, and others. These fields provide information about the resulting link, but there is little empirical evidence to determine which fields help users identify the appropriate result. Some may support the user's search task, but others may simply clutter the interface, or even impede the search process. The purpose of this study was to identify how the most common fields displayed by search engines affect users' decision-making processes, confidence and expectations. Fifteen participants performed four different search tasks using a simulated search engine designed for the study. Each participant reported their pre-click confidence (PCC) in their choice and evaluated each field. Some fields were universally important, whereas others were task specific.

1.0 INTRODUCTION

The Internet has become a powerful tool for information search and ecommerce. There are two main purposes for using the Internet: publishing information in an accessible electronic format and retrieving information published by others (Lightner, Bose and Salvendy, 1996). The Consumer Daily Question Study conducted during Fall 2000 showed that Americans need answers to four questions per day and that the Internet and search engines led the list of information resources used to locate answers to those questions (Sullivan, 2001a). Another study conducted from July 27 to August 1, 2000 by WebTop reported that on average, Americans search the web for information thirteen times per month and 28% search once or more per day. On average, Americans spend 1.5 hours per week searching for information (Sullivan, 2001b). Directories such as Yahoo and search engines such as Google provide the ability to search millions of sites instantly for information. However, there are many challenges that reduce the effectiveness of search engines and thus limit the ability of users to find information.

Each directory or search engine operates using a unique search algorithm, ordering scheme, and output style. By using proprietary formulas or algorithms, they determine which of the millions of pages in their database relate to the topic (Zetter, 2000). Since search engines are very comprehensive, they often include thousands of sites in the list of results. However, these lists often lead users to more confusion and frustration when many of the results are completely irrelevant to the user's search task (GVU, 1999). In the study conducted by WebTop, nearly 88% of users reported at least some level of frustration when getting irrelevant information while searching (Sullivan, 2001b).

Search engines provide various fields of the resulting pages, such as title, description, URL, size, last date modified, and others. But there is little empirical evidence that supports the inclusion of each of these fields. Some may support the user's search task, but others may simply clutter the interface, or even impede the search process. The objective of the search site should be to provide the fields that support effective searching and to eliminate those that slow down or confuse users. When the mechanism for presenting search results is ambiguous, it is left to the judgment of the user to evaluate the quality of the match. At the very least, this adds to the workload of the user. At the worst, it can prevent the user from locating the desired information among the thousands of available hits.

1.1 Background

Several researches have investigated online users' search behavior (Spink, Bateman, and Jansen, 1999; Navarro-Prieto, Scaife, and Rogers, 1999). Spink, Bateman, and Jansen (1999) showed that users tend to perform successive searches for the same search task when their first search does not retrieve a satisfactory set of results. Also, users appear to have very little knowledge about how search engines work and reported having no interest in learning. Users expect search engines to understand their search strategies and automatically create effective queries. Navarro-Prieto, Scaife, and Rogers (1999) investigated the importance of taking into consideration the interaction between the users, the task, and the information presented. The study found that the representations currently used by typical search engines are the cause of multiple problems regarding these interactions. Thatcher (2000) expanded this research to include the effect of user searching experience and found similar results. Another study investigated the effects of Internet experience and domain-specific background knowledge in users' web search behavior (Hölscher and Strube, 2000). They concluded that novices are less flexible in their strategies and return to previous steps of their search process (such as returning to the same page of results) instead of trying another search engine. However, the effects of the inclusion of specific output fields on the selection of a satisfactory result has not been studied.

The search process can be modeled as a problem-solving process in which users must make a series of decisions. Searching is a decision-making process in a natural setting where making a decision is not an end in itself but rather a means to achieve a broader goal (Orasanu and Conolly, 1993). In order to make that decision, users must select from a set of options the alternative that is most likely to lead to successfully achieving the goal (Balasubramanian, Nochur, Henderson, and Kwan, 1999). The searcher evaluates the options in terms of his or her search goals. It is proposed in this study that understanding users' strategies when evaluating search engine results will lead to a better interface design that would help users in their decision making process, increasing the confidence in their decision while reducing frustration.

Several fields of information (such as title) are commonly presented in the results lists of search engines. Typically, users search using certain keyword(s) and would prefer to retrieve only those documents whose title contains those keyword (s) (Shneiderman, Byrd, and Croft, 1997). Description is another common field included for each result. Studies have shown that presenting a field describing the content of the site will increase the probability that the user will select a more relevant result because it provides more information about that web site (Koman, 1998). The file size field provides searchers with a number that would help them approximate the download time. It has been shown that users are more likely to abandon a web site that takes too much time to download in favor of another link that is smaller in size (Wonnacott, 2000). Some search engines provide a field with the date on which the Web page was created or last modified (Hock, 1998). For some search tasks this detail may be relevant for the decision making process and for others it is irrelevant.

1.2 Objective

The purpose of this research was to investigate how users interact with output fields used by search engines to display search results. The objective was to identify how the most common fields used by search engines affect users' decision-making processes, confidence and expectations. The study investigated users' reactions and preferences when interacting with simulated search engine results. Search tasks were designed to simulate the way that actual searches are conducted, particularly the presence of realistic task objectives. By understanding how users of search engines evaluate search results, search engine designers will be able to improve the algorithms that are used to select and order search results as well as improve the usability of the input and output interfaces.

2.0 METHODS

2.1 Participants

Fifteen participants were recruited to participate in this study, 5 females and 10 males. Ages ranged from eighteen to forty. Participants were required to have at least a minimum exposure to the Internet of once per month. At least some self-reported skill with on-line searching was also required to participate in the study. Other demographic characteristics were also recorded to insure a representative sample was achieved.

2.2 Materials
2.2.1 Search Tasks

The task was to search for a web site that best fit the goals of an assigned search task. In order to perform the task, participants were given a simulated search engine output screen containing a list of results for predefined keywords. Four search tasks were created with their associated keywords:

- Find a cat shelter in Miami where you can adopt a cat (cat shelter and Miami)
- Find instructions on how to do the housebreaking training of your puppy dog (dog and house training)
- Find information about damage from the Y2K computer bug after 01/01/00 (Y2K bug and consequences)
- Find an online tutorial for novice investors (online stock tutorial)

2.2.2 Search Engine Models

Search engine models were created to simulate the results page of a search engine. The output was designed to look different than any existing search engine but still contain a subset of the most commonly used fields. This eliminated the potential bias of prior use while testing a controlled set of fields. As with most search engines, the keywords used for the search were included in the output in a text box above the results for each model. A total of 16 models were designed for the test, each one representing a different set of output parameters. Eight of these sets were designed using combinations of the following design parameters: description, URL, category, last modified date, other keywords, similar pages, size and keyword count. These eight combinations are simple models. An example of a simple model is shown in Figure 1. The other eight combinations are the same simple models with a directory added to the output. These are combined models (Figure 2). When interacting with any of the combined models, participants could either select a link from the categories in the directory or go to the list of results. The four search task topics described in Table 1 were used to generate results for each model. For each search task topic, two simple models and two combined models were created, for a total of four models per topic.

2.3 Procedure

Tests were conducted and evaluated in the Usability Lab at Florida International University. All testing sessions were approximately 45 minutes long. A ten-minute training session was used to show participants how to provide verbal protocols.

After the training session, participants were given one of the search tasks described in Table 1. The topic of the search was explained to the participant in detail. He/she was given clear instructions on the objectives of the search task and was invited to ask questions before performing the task. Participants viewed the simulated output for that task and was instructed to either select a link, change the keywords, or select the "next page" option. Participants were asked to think aloud and verbalize their thoughts from the moment the instructions were explained to him/her until the link selection was made. As soon as the link or action was selected, the search task was finished. The test administrator recorded the link selected. Each participant was presented with a total of 4 search models, one for each search task. Participants used two simple and two combined models.

At this point, participants were interviewed in order to understand their decision-making process as well as their confidence about the accuracy of the link selected. Participants were asked to explain why they selected the option as well as to report on a 10-point Likert scale their pre-click confidence (PCC) in that option. Participants were also asked to recall the fields that were present, rate the importance of each one in their decision and to list any fields that would have made their decision easier. This process was repeated for all four search tasks.

3.0 RESULTS

In general, the key factor that determined the PCC of the participants was the inclusion of a description of the link. When a description was present, 76% of the users reported a PCC between 7 and 10 compared to only 40% when description was absent. Additionally, when a description was absent, 78% of the participants specified that one was needed. Link description was particularly important when no directory links were present.

The URL was less important. When it was included, 93% of the participants did not recall seeing it and 20% of them reported that it could be eliminated without a reduction in PCC. Size also did not significantly improve PCC. When it was included, 92% of the participants did not recall seeing it and 51% of them reported that it could be eliminated without a reduction in PCC.

Some characteristics were task dependent. When participants were given the task to search for information on the results of the Y2K computer bug, 67% of the participants reported that the date of posting improved PCC but only 33% did so for the group as a whole.

Some search engines also present advanced features such as the number of matched keywords or links to similar pages. The results show that participants largely ignored these characteristics. When keyword count was included, 90% of the participants did not recall seeing it and 36% reported that it could be eliminated. Similarly, 96% of the participants did not recall seeing the link to similar pages.

The lowest PCC mean value reported was 3.8 for the model that only contained description and size. Ninety-two percent of the participants reported not feeling confident of their decision because of the lack of information provided, specifically date. The topic of the search task for this experiment was "Y2K bug". After participants interacted with the experiment that contained all the fields investigated in this study, they reported a mean PCC value of 8.6. This value was the highest PCC mean value reported among the 16 models. However there were no statistically significant differences between the PCC mean values. Finally, 14 out of the 15 participants reported a favorable opinion of an option to select the fields that would be displayed in the results page.

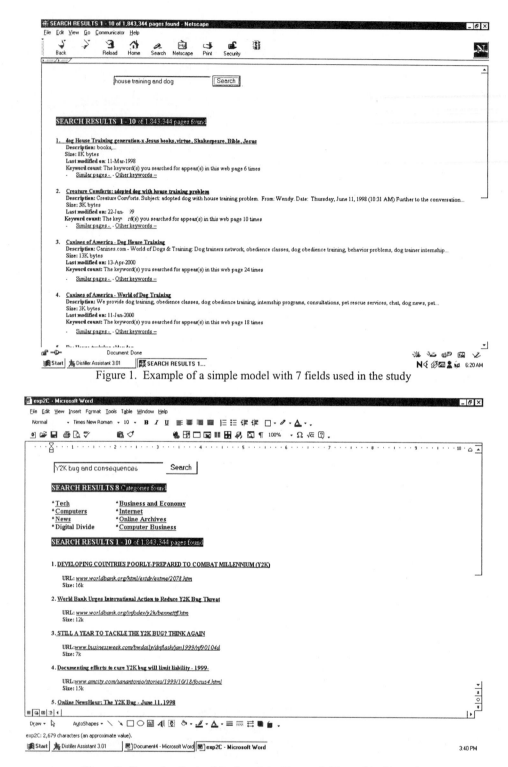

Figure 1. Example of a simple model with 7 fields used in the study

Figure 2. Example of a combined model with two fields used in the study

4.0 DISCUSSION

Research in Usability Engineering and customer service both have found that the ease of use of a web site is one of the main factors in customer satisfaction and repeat visits. For search engines, where there is a limited set of services that are provided, effective usability is even a larger component of customer satisfaction. Surveys have established that the current designs of search engines are frustrating and not satisfying the searching needs of users. Designing search results so that they better support general searching strategies and allow users to customize the search for their individual needs would increase the success of these sites. In order to do this, however, it is essential to understand the evaluation process through which users interpret and select results for further interaction.

This study investigated user behavior and perceptions of search engine results as well as evaluated specific output characteristics. The results support the use of a task specific design to support user behavior and expectations. The presence of task specific fields led to enhanced pre-click confidence and was specifically requested by participants when they were absent. Participants generally did not notice characteristics that were not important for that task. Task description was universally important for all tasks. Task dependent fields included date of posting. Participants did not use advanced features such as keyword matches and similar pages. However, it remains to be seen whether this is because they were not part of the task schema or because users were unfamiliar with how they work or what they represent. Further research is necessary to fully understand user strategies for developing a search objective and parsing search results. However, this study provides needed guidance for the direction in which this research should proceed.

5.0 REFERENCES

Balasubramanian P., Nochur K., Henderson J., and Kwan M. (1999). Managing process knowledge for decision support. *Decision Support Systems,* 27, 145-162.

Graphic, Visualization and Usability Center at the Georgia Institute of Technology (1999). GVU Survey Results. Retrieved August 13, 2000 from http://www.searchenginewatch.com/reports/gvu.html.

Hock R. (1998). A Field Trip. *Online,* May/June, pp 19-22 .

Hölscher C. and Strube G. (2000). Web search behavior of Internet experts and newbies. *Computer Networks,* 33, 337-346.

Koman R. (1998). Helping users find their way by making your site "smelly". Retrieved May 13, 2000 from http://www.webreview.com/pub/98/05/15/feature/index.html.

Lightner N.J., Bose I., and Salvendy G. (1996). What is wrong with the World Wide Web? A diagnosis of some problems and description of some remedies. *Ergonomics,* 39, 995-1004.

Navarro-Prieto R., Scaife M., and Rogers Y. (1999). Cognitive Strategies in Web Searching. *Proceedings of the Human Factors and the Web Conference.* NIST, Washington DC.

Orasanu J. and Connolly T. (1993). The reinvention of decision making. In G.A. Klein, J. Orasanu, J. Calderwood and P. Mac Gregor (eds). Decision Making in Action: Models and Methods. Ablex Publishing, Norwood NJ.

Shneiderman B., Byrd D., and Croft W.B. (1997). Clarifying Search: A User-Interface Framework for Text Searches. *D-Lib Magazine,* January. Retrieved November 16, 2000 from http://www.dlib.org/dlib/january97/retrieval/01shneiderman.html.

Spink A., Bateman J. and Jansen B. (1999). Searching the Web: a survey of EXCITE users. *Internet Research: Electronic Networking Applications and Policy,* 9, 2, 117-128.

Sullivan D. (2001a). Internet top information resource, study finds. The Search Engine Report. Retrieved February 6, 2001 from http://www.searchenginewatch.com/sereport/01/02-keen.html .

Sullivan D. (2001b). WebTop Search Rage Study. The Search Engine Report. Retrieved February 5, 2001 from http://www.searchenginewatch.com/sereport/01/02-searchrage.html.

Thatcher A. (2000). "Experience" and www searching, or the search for the www "expert". *Proceedings of the IEA 2000/HFES 2000 Congress.* Human Factors and Ergonomics Society, Santa Monica, CA.

Wonnacott L (2000). The speed of business: If your pages are slow, your customers will go. InfoWorld.com, Site Savvy. Retrieved October 2, 2000 http://www.infoworld.com/articles/op/xml/00/09/11/000911opsavvy.xml

Zetter K. (2000). How to stop searching and start finding (Internet/Web/Online Service Information). *PC World,* 18, 9, 129.

SIGNS OF TRUST: A SEMIOTIC STUDY OF TRUST FORMATION IN THE WEB

Kristiina Karvonen Jarmo Parkkinen

Department of Computer Science
Helsinki University of Technology
P.0.Box 9700 HUT Finland
Tel. +358 451 60 75

ABSTRACT

E-commerce has been slowed down due to the fact that consumers find Internet an untrustworthy environment, and are not willing to be guinea pigs of a new service and take unnecessary risks by giving their credit card number to an online service they cannot really trust. Yet, if e-commerce is to thrive, creating consumer trust is a necessity – or else there will be no transactions. Not only money, but also private information about an individual customer has high value online: for service providers, it gives invaluable data about customer, about her behaviour, likes and dislikes. This kind of information is key in understanding what will and what will not sell on the Internet. In wrong hands, however, such intimate information may become a powerful weapon against the intimacy, integrity and privacy of a given individual. Trust is needed from the customers in order to be willing to indulge in taking that risk. This paper discusses the notion of trust from a semiotic point of view, seeking to understand and analyse the signs of trustworthiness that the design of a Web site is sending.

1. INTRODUCTION

Making money transactions online is risky business. Even when the service provider is in fact benevolent, this is not enough, for there are also some malicious outsiders that may intervene – the repetitive hacker cases on the Internet have clearly shown us that Internet still remains an insecure place. Giving away private information without serious premeditation clearly still creates a risk to the customer. A lot of work has been done lately to keep hackers out and to ensure the security of online transactions in the near future. Unfortunately, however, making the *technology* work is not enough, but this has to be *communicated* to the consumer in some way as well. The case is especially severe when at current users have grown wary of the online services – and rightly so (Nielsen 1999). To succeed in e-business, the service providers must be able to restore the consumer trust in one way or the other (e.g. Hoffman et.al.1999). They must be able to make their trustworthiness visually manifest somehow.

In the Web, this means that the overall "look" of the Web site of a service provider must give an impression of trustworthiness to the user. Indeed, many studies have shown that "design quality" is one of key ingredients in creating trust in the users towards online services (e.g. Ecommerce Trust Study 1999). Other such ingredients include the use of high-tech features, brand reputation, successful navigation, overall presentation, and user satisfaction in reaching goals. The overall impression of any Web site is, then, created through a multitude of key features. How can we analyse the outcome of these ingredients when put together?

Our answer was to try out a *semiotic* analysis of these Web sites. Semiotics, the study of signs and their use as communicative tools, seemed to us to provide a likely way to grasp the octopus-like character of trust-enhancing complexity with its many tentacles. According to semiotics, seeing is not believing but *interpreting* – a drawing of a tree is interpreted as a picture of a tree, not *seen as* an actual tree (e.g. Nöth 1995; Mirzoeff 1999). These kinds of interpretations are based on agreed rules, mostly unspoken and even unconscious, and vary across different cultures and different times. This means that the interpretation of a given image is likely to change over time, and, like in the case of very old images, such as the cave paintings in Lascaux, France that date more than fifteen thousand years back, the intended meanings of the images may get altogether lost. Visual images succeed or fail to the extent that we can interpret them in a successful way. The same goes for Web imagery – we may interpret it as trustworthy or untrustworthy, for example, based on some key elements in the imagery that contribute to the trustworthiness in some way. We believe that these elements may be found with the help of semiotic tools.

2. METHODOLOGY

As a starting point, we chose the 6 most trusted and the 6 least trusted Web sites from a well-known study by Cheskin Research, the E-Commerce Trust Study, and analysed these sites within a semiotic framework. Respondents of the Cheskin study evaluated the trustworthiness of a total of 102 of some of the highest profile sites on the Internet. Overall, the research showed that the most trusted brands are well-known brands, be it on the Web or in other settings. The sites were:

6 most trusted websites
- http://www.yahoo.com/

- http://www.walmart.com/estore/pages/pg_g1.jsp
- http://home.netscape.com/
- http://www.go.com/
- http://www.blockbuster.com/
- http://www.excite.com/

6 least trusted websites
- http://www.monsterboard.com/
- http://www.spinner.com/index.jhtml
- http://www.cyberkids.com/
- http://www.thewell.com/
- http://www.jennicam.com/
- http://carpoint.msn.com/home/New.asp

A detailed description of the Cheskin study is not outlined here, but instead the outcome of this study is taken at face value. The study can be found at http://www.sapient.com/cheskin/assets/images/etrust.pdf

It is interesting but not surprising that the most trusted sites would be well-known sites. Having trust on a site means having an active relation to the site. The user is in constant contact with the site. Trust is needed to take action, whereas distrust may be passive in nature, and mean withdrawal from action: the site is not used, a purchase is not made, the user clicks her way away from the site. This is why it may be easier for users to list names of trusted sites than distrusted sites. The user is, simply, more familiar with the sites that she has decided to trust enough to use. In practice this means that users may have a list of trusted Web sites (e.g. bookmarks in a browser), but they do not in general have a list of distrusted Web sites.

It is also noteworthy that most online customers are not experts in Web technologies, and are thus not able to evaluate the technical excellence of these sites, nor the lack of it. For most users, the decision to trust or not to trust a Web service is, then, not based on rational evaluations of the actual security provided by that service, but rather that decision is made in an intuitive and spontaneous manner, on the basis of the visual overall impression of that site. Users just do not have enough knowledge nor a proper understanding of how the service works, so what they do is base their initial decision to trust the service based on some visual hints of trustworthiness that the visual design of that site provides. These visual elements are the focus of our semiotic analysis presented here in this paper.

3. TRUSTWORTHY DESIGN

Figures 1-6 present screenshots of the 6 most trusted Web sites. There is a lot of text on the page, and only a few pictures. The general impression is rather peaceful and almost "empty" – the margins are rather wide, and there is a lot of empty space around the text areas. The columns are even and well-organised. The metaphoric model for the sites is clearly that of a newspaper.

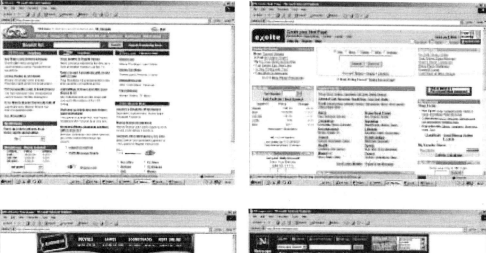

4. UNTRUSTWORTHY DESIGN

Figures 7-12 present the 6 most untrustworthy Web sites. There are a lot cartoon-like and other handdrawn looking images on these pages, and pictures dominate. The overall impression is rather disordered, even chaotic when compared to western-style newspaper layout. The area borders are not clearly defined. In two of the sites the text is included in a single area situated against a darker background, which makes the scene seem to imply the television screen as the metaphorical medium used.

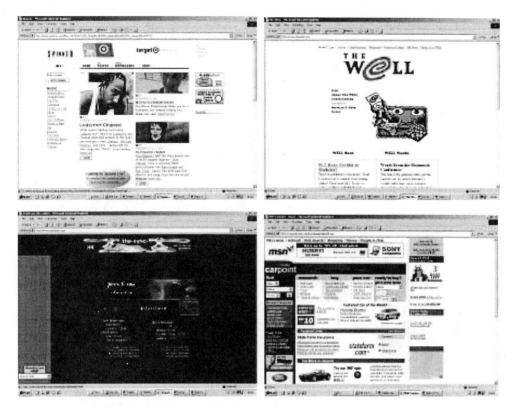

5. CONCLUSIONS

Our analysis shows that the trusted Web sites may be successful for many reasons, and semiotics provides some keys to understand the elements behind it. To start with, the trusted Web sites were more text-based than the distrusted sites. It can be stated that their metaphoric interpretation is the traditional newspaper – a medium that traditionally enjoys a rather high level of trust. The distrusted Web sites were more based on the use of imagery, thus breaking down the possibility to signify any resemblance to a "serious" newspaper layout, but rather referring to a visual metaphor of some kind, e.g. television or even a comic book.

In our previous research, we have noticed that trust may be built differently for users having a different amount of usage experience of e-commerce (Karvonen 1999). This means that the meaning of the visual elements on the Web site that will promote trust in the particular user groups (e.g. novice, intermediary, expert) is likely to differ. Novice users tend to treat each Web page of the site as a singular entity, one image, and only what is clearly visible is counted a s one of its elements. This kind of viewing can be described as *one-dimensional.* Technically experienced users, or expert users, on the contrary, tend to view the Web site as a series of images, or rather as a series of texts, and treat the whole site as one entity, the meaning of which is not all that is visible only, but the visual elements of the site are also treated as signs of underlying elements, technology, and complexity (Karvonen 2000). For the expert users, the site is then essentially a *multi-dimensional* entity.

Novice users – Web site seen as:	Expert users – Web site seen as:
Image	series of images/texts
One-dimensional	multi-dimensional
Visible	hidden/deduced
Entity	Ensemble of entitites

Table 1: Interpretation of a Web site among novice and technical users.

On basis of our analysis of the 6 most and least trusted Websites, we can list the trust-promoting and trust-diminishing elements of these Web sites in *Table 2.*

Trusted	Distrusted
Text-based	Image-based
Empty space as structural element	Empty space as background or as "undefined" space
Strict grouping, visual density	grouping seemingly random
Formal language	Informal language
----	Animations
Structured and linear upper part	Unstructured and nonlinear upper part.
Use of real photos	Use of cartoons

Table 2: Elements of Web design

On basis of our analysis, it seems that the "trusted" elements of the sites were more prominent on the more trusted sites, and less prominent on the less trusted sites. The same is true of the distrusted sites – the distrust elements are at their strongest on the top of the list. There are, then, a lot of sites the design of which is somewhere in the middle, not adhering strongly either to the "trusted design" or the "distrusted design" language. This means that categorising them is not a simple task – no wonder that the customer gets confused.

To promote trust, the site designer should adhere as closely as possible to the presented design rules – it is not enough to follow some and neglect others, for this will not amount to the expected result of trustworthy design. In case of the two sites of the study, the distinction between what is trustworthy and what is not seems indeed to be very small – the Netscape homepage and the msn homepage are very similar to one another, as figures 13 and 14 clearly demonstrate.

Most elements are the same, and the sites seem almost identical. The biggest differences are in that the msn-carpoint site has used a cartoon-like picture of Uncle Sam and its upper part defining line is not as definite as in the Netscape site. The structuring of the latter site is perhaps also somewhat more ordered and newspaper-like. It is likely, however, that actually it is the *contents* of the sites that result in differing opinions about the site: the msn-carpoint site is filled with advertisements, whereas the Netscape-page is more informative in nature. The overall interpretation of a site is, then, always affected, at least to some extent, by the dialogue played between the layout and the content.

6. REFERENCES

ECommerce Trust Study (1999, January). Joint Research Project by Cheskin Research and Studio Archetype/Sapient.
http://www.studioarchetype.com/cheskin

Hoffman, D.L, Novak, T.P., Peralta, M (!999, April).Building Consumer Trust Online. *Commuications of the. ACM* 42, 4 (Apr. 1999), Pages 80 - 85

Karvonen, K (1999). Creating Trust. *Proceedings of the Fourth Nordic Workshop on Secure IT Systems* (Nordsec'99), November 1-2, 1999, Kista, Sweden, pp. 21-36

Karvonen, K (2000). The Beauty of Simplicity. *Proceedings on the conference on universal usability, 2000* on Conference on Universal Usability, 2000, Pages 85 - 90

Mirzoeff, N (1999). An Introduction to Visual Culture. Routledge, London and New York.

Nielsen, J (1999, March 7). Trust of Bust: Communicating Trustworthiness in Web Design. Jacob Nielsen's Alertbox.
http://www.useit.com/alertbox/990307.htm

Nöth, W (1995). Handbook of Semiotics. Indiana University Press, Bloomington and Indianapolis.

AN ANALYSIS OF TECHNOLOGY LEADERS' ATTITUDE TOWARD VIRTUAL CYBERSPACE TECHNOLOGY

Sarah M. North and Max M. North[1]

Clark Atlanta University, Atlanta, Georgia 30314
Computer and Information Science Department
Sarah@acm.org

ABSTRACT

The purpose of this study was to investigate the relationship between selected personal variables and the utilization of virtual cyberspace technology (VCT) as perceived by information technology directors and educational technology leaders. Four personal variables were selected: (1) technology knowledge, (2) technology training, (3) human-interaction skills, and (4) leadership skills. The data was gathered from a two-part questionnaire completed by educational technology leaders and information technology (IT) directors. A total of one hundred fifteen questionnaires were distributed among the fifty-five participating schools and IT organizations. Of this number, sixty-eight properly completed questionnaires were received from thirty-three information technology directors, and thirty-five educational technology leaders. This represented a response rate of approximately 60 percent. The finding indicated that there was a significant relationship among the utilization of VCT; educational technology leaders and IT directors' perception of the utilization of VCT, technology knowledge and technology training. It can also be concluded that, in regard to human-interaction skills, there was no significant relationship between IT directors and educational technology leaders. However, this study indicated that having dynamic leadership skills might increase the utilization of VCT.

1.   INTRODUCTION

Computer technology plays a major role in the United States educational system. It is a dynamic process that continues to change the behavior of instructional leaders and the methods they use to deliver instructional information. In this "information age," it has become increasingly important for school administrators to be skilled in the utilization of modern technology. Computer technology, more than any other form of technology, stands out as a catalyst for revitalizing education (Adams, 1985; Bozeman et al., 1991; Cutts et al., 1993; Goodlad et al., 1986).

The use of computers in education has progressed to the point where, today, nearly all schools depend on computer services for instructional purposes (Watson, 1992; Spuck et al., 1992; Kearsley et al., 1994). The most recent advances made in computer technology have not been introduced extensively into school environments. One such advancement is virtual cyberspace technology. These technologies have been used extensively by industry and the military, for training and the management of everyday activities. Some of the areas of virtual cyberspace technology include: (1) interactive training and electronic videoconferencing, (2) distance learning, (3) electronic brainstorming(EBS), and (4) the creation of virtual reality technology (VRT). According to Sheriden, virtual reality offers a new human-computer interaction paradigm in which users are no longer simply external observers of data or images on a computer screen. Instead, they are active participants within a computer-generated, three-dimensional, virtual world (Sheridan, 1992). These forms of technology have proven to be very effective resource tools because they are efficient and save manpower hours, and they can enhance the work of school systems in the areas of planning, organizing, managing, and academic instruction.

The purpose of this study was to investigate the relationship between selected personal variables and the utilization of virtual cyberspace technology as perceived by IT directors and educational technology leaders. The variables were investigated in terms of the following: (1) technology knowledge, (2) technology training, (3) human-interaction skills, and (4) leadership skills. The results of those interactions were further investigated in terms of the following demographic variables: (1) years of experience, (2) education level, (3) gender, and (4) age.

---

1. The co-author is currently associate professor of computer science and information systems at Kennesaw State University.

2. METHODOLOGY

2.1 Theoretical Framework

Theoretical framework of this research covers the independent, intervening, and dependent variables and the research hypotheses. The focus of this study was to investigate the relationship between selected variables and the utilization of virtual cyberspace technology (VCT) as perceived by IT directors and educational technology leaders in selected public schools and their schools district located in Atlanta, Georgia area, IT directors in colleges, universities and private organization. In fulfilling the purpose of this study, the following null hypotheses were developed and tested:

Hypothesis 1: There is no significant relationship between the IT directors' and educational technology leaders' perception of the utilization of virtual cyberspace technology and their technology knowledge, training, human-interaction skills, and leadership skills.

Hypothesis 2: There is no significant relationship between the IT directors' and educational technology leaders' perception of the utilization of virtual cyberspace technology and their (a) years of experience, (b) level of education, (c) gender, and (d) age.

2.2 Research Design

This research was casual-comparative in nature and quantitative in design. A survey questionnaire, entitled "Factors which Affect the Utilization of Virtual Cyberspace Technology: Study of IT directors' and Educational Technology Leaders' Perceptions" (PETLP)(1) Demographic Instrument, and (2) Utilization Instrument was obtained from the research and development of two sources: Bailey and Lumley (1994), from Kansas State University, and Griffin (1985). By identifying and explaining the relationships between the variables, one can rationalize and have better insight into an understanding of IT directors' and educational technology leaders' perceptions in terms of their utilization of virtual cyberspace technology in public schools and school districts, IT directors in selected colleges, universities and private organization.

2.3 Human Subject Contract

IT directors and educational technology leaders were asked to participate in the study on a voluntary basis. Their anonymity and confidentiality, as they were informed, was ensured. No information was used to evaluate them or for any other purpose other than research. IT directors and educational technology leaders who participated in the study were given a questionnaire with directions for completing the instrument. They were asked to return the questionnaire within one week. No human subject contract was needed since there was no service to be rendered; however, permission to solicit participants' responses was requested from the schools and their school district offices.

2.4 Description of the Instrument

The PETLP-VCT survey instrument used in the study was a modified form of two previously used instruments designed by Griffin (1985) and Bailey and Lumley (1994). Part I, the Administrative Computer Technology Survey (ACTS) instrument, was patterned after an instrument used in a previous study conducted by Griffin (1985). Griffin's survey was designed to investigate the status of computer technology utilization and computer training among educational leaders in Georgia secondary schools. The Bailey instrument was designed to investigate the technology administrator and staff development programs and their utilization of technology to empower leaders for life in the 21st century. The instruments were subjected to a rigorous research and development process by Kenneth Stanage (1996) and Marie Blythe (1996), from Kansas State University. The 1994 Bailey and Lumley instrument was valid and reliable based on the research and development (R&D) process model by Stanage and Blythe. The process included a review of literature regarding the development programs for technology leaders and the development of a prototype through field of testing by experts in the field. The experts included practicing school technology leaders and practicing school administrators. Upon completion of the prototype, a preliminary field test was conducted with eight technology experts (i.e. administrators or technology coordinators who have successfully implemented comprehensive programs). The experts provided the author with suggestions and comments on the content, program model, and usefulness. The authors carefully examined suggestions and comments made by reviewers. The suggestions and comments were incorporated to make the appropriate modifications in the final instrument. Test reliability and content validity were determined by the performance of administrators and technology leaders, participants from the Kansas public schools.

There were two major parts to the PETLP-VCT instrument. Part I, from Griffin, solicited demographic and profile data regarding gender, age, years of experience, educational level, and computer technology training. Part II measured the IT directors' and technology leaders' perceptions on the utilization of virtual cyberspace technology. The questions selected are based on the characteristic variables. The instrument also measured virtual cyberspace technology knowledge, technology training, human-interaction skills, and leadership skills.

The criteria for the development of the PETLP-VCT survey were: (1) Ease of filling out the survey. (2) Clarity and conciseness of the questions and instructions for filling them out. (3) Sufficiency of being directly related to the research questions.

The survey instrument in the form of a questionnaire was divided into two parts. Part I solicited information to develop a profile of the IT directors and educational technology leaders who participated in the study. Part II of the instrument was designed to determine the level of the respondent's utilization of technology. The response mode for Part II was: 1 = Strongly Disagree, 2 = Disagree, 3 = Agree, 4 = Strongly Agree. Data for utilization of the instrument were interpreted in terms of mean scores as follows: 1.00 to 1.50 = Strongly Disagree, 1.51 to 2.50 = Disagree, 2.51 to 3.50 = Agree, and 3.51 to 4.00 = Strongly Agree.

3. RESULTS

An analysis of data gathered from surveys administered to IT directors and educational technology leaders in 55 public schools in two metro Atlanta school districts. The purpose of this study was to investigate selected personal variables and the utilization of virtual cyberspace technology (VCT) as perceived by principals and educational technology leaders. The model presented in Chapter III showed the relationship among the variables. The instrument used to collect the data was a modified version of IT directors' and Educational Technology Leaders' Perceptions of Virtual Cyberspace Technology" (PETLP-VCT), developed by Bailey and Lumley. The research was validated in 1996 by Stanage and Blythe.

A total of 115 fifteen questionnaires were distributed to the IT directors and educational technology leaders in 55 participating schools. Of this number, 68 properly completed questionnaires were received; 33 from IT directors and 35 from educational technology leaders. This represented a response rate of approximately 60 percent.

3.1   Testing the Null Hypotheses

Hypothesis 1: There is no significant relationship between the IT directors' and educational technology leaders' perception of the utilization of virtual cyberspace technology and technology knowledge, technology training, human-interaction skills and leadership skills.

Table 1 presents the correlations that were computed. To test this hypothesis, the Pearson Product-Moment Correlation Coefficient (Pearson r) was used to determine the direction and magnitude of the correlation between their perception of the utilization of VCT and their educational technology knowledge and technology training showed a level of significance of .000, indicated that there was a significant relationship between the two personnel.

Hypothesis 2: There is no significant relationship between the IT directors' and educational technology leaders' perception of the utilization of virtual cyberspace technology and their (a)years of experience, (b)level of education, (c)gender, and (d)age. This was a multi-group correlation of the Pearson Product-Moment Correlation Coefficient (r) was used to determine the IT directors' and educational technology leaders' perception of the utilization of VCT in term of four intervening variables.

To test this hypothesis, the Pearson Product-Moment Correlation Coefficient (Pearson r) was used to determine the direction and magnitude of correlation between the educational technology leaders' perception of the utilization of virtual cyberspace technology and technology knowledge. Table 6 presents the correlations that were computed. Based on the critical level of 0.05, the level of significance was .000, which indicates that the probability of this results occurring by chance is 1 in 1,000.

Table 1. Pearson r for the Relationship Between IT directors and Education Technology Leaders' Perception of the Utilization of VCT and Technology Knowledge

Variable	IT directors			Educational Technology Leaders		
	df	r	Prob. of r	df	r	Prob. of r
IT Knowledge	32	.6735	.000*	34	.7826	.000*
IT Training	32	.5828	.000*	34	.6596	.000*
Human-Interaction Skills	32	-.0217	.905*	34	.2679	.145*
Leadership Skills	32	.2504	.160*	34	.1006	.590*

*P = .05

Table 2. Pearson Product-Moment Correlation Coefficient for the Relationship between IT directors and Educational Technology Leaders' Perception of the Utilization of VCT of Intervening Variables (Years of Experience, Educational Level, Gender, and Age)

Intervening Variables	IT directors			Educational Technology Leaders		
	df	r	Prob. of r	df	r	Prob. of r
Years of Experience						
0	-	-	-	-	-	-
1-5	24	.6421	.000*	19	.6194	.002*
6-10	7	.6258	.049*	7	-.4128	.179
11-15	-	-	-	7	.6872	.044*
15-20	-	-	-	-	-	-
Educational Level						
Bachelor's	-	-	-	11	.5535	.061
Master's	11	.0817	.406	10	.3193	.169
Specialis	12	.6037	.014*	7	.6294	.047*
Doctorate	8	.9021	.000*	4	.7212	.139
Gender						
Male	20	.6004	.003*	9	.6251	.067
Female	13	.6209	.012*	24	.4405	.014*
Age						
20-30	-	-	-	-	-	-
31-40	5	.2303	.355	8	.6704	.034*
41-50	19	.6389	.002*	20	.4003	.040*
51-60	7	.7704	.021*	5	.6835	.102

*Significant beyond the .05

The results in Table 2 demonstrate a highly significant relationship between the IT directors' educational technology leaders' perception of the utilization of the virtual cyberspace technology. The results of the Pearson r correlation coefficient to determine the relationship, which existed between the IT director and technology training and some of the four intervening factors. As stated in the case of the previous hypothesis, and shown on Table 2, there were very similar relationships among the four factors. Once again, data show that the IT directors who had technology training, whether male or female, had obtained the specialist or doctorate degree and had at least 10 years of technology experience. There was a significant degree of correlation in the age range of 41-50 and the age range of 51-60. However, there was no relationship between the 11-15 years and the 15-20 years of experience, bachelor's and master's degree educational level, and the 20-30 and 31-40 age range. An analysis of the data indicates that there was a significant relationship between the independent and dependent variables and the four intervening factors.

4.   CONCLUSIONS

This study revealed that IT directors' and educational technology leaders' perceptions of virtual cyberspace technology utilization were influenced by factors such as technology knowledge and technology training. Human-interaction skills, however, were not affected. The findings of the study also revealed that leadership skills had some influence on the utilization of virtual cyberspace technology. It was also concluded that variables were also affected in terms of (1) years of experience, (2) educational level, (3) gender, and (4) age.

This study found that there was no significant difference in the utilization of virtual cyberspace technology between the IT directors' and educational technology leaders' perceptions. It can also be concluded that technology knowledge, technology training, and leadership skills had a dynamic impact on IT directors' and educational technology leaders' perceptions of the utilization of virtual cyberspace technology.

REFERENCES

Adams, D. "Computer and Teacher Training: A Practical Guide," The Haworth Press, (1985): 11-17.

Bailey, G. D., "Technology Leadership: Understanding technology integration in the 21st century," New York: Scholastic, Inc. Publications, (1994): 1-10.

Bozeman, W. C., and D. W. Spuck, "Technology competence training for educational leaders," Journal of Research on computing in Education, (1991): 23(4),514-529.

Cutts, D. E., Matthew, W. M., Winkle, L., and Nichols, J. L. III, "Administrators Microliteracy: A Challenge for the 80's & 90's," NASSP Bulletin, 66, no. 455, (1993): 53-59.

Goodlad, J. I., and J. F. O'Toole, and L. L., Tyler, "Computers and Information System in Education," New York: Harcourt, Brace, and World, (1986).

Kearsley, G., and Williams, L., "Educational Leadership in the Age of Technology: The New Skills," Educational Technology Publisher Inc., (1994): 5-11.

Sheridan, B., "Musing on telepresence and virtual presence," Presence: Telecoperators and Virtual Environments," 1(1),(1992): 120-126.

Spuck, D. W., Bozeman, W. C., "Training School Administrators in Computer Use," Journal of Research on Computer in Education, 21(2),(1994): 229-239.

Watson, P., "Using the Computer in Education," Englewood Cliffs,NJ: Educational Technology Publishing,(1992).

THE EFFECT OF CHANGES IN INFORMATION ACCESS TIMES ON HYPERTEXT CHOICES

D. Scott McCrickard

Department of Computer Science
Virginia Polytechnic Institute and State University
Blacksburg VA USA 24061-0106
mccricks@cs.vt.edu

ABSTRACT

This research examines ways in which information processing decisions are affected by temporal delays in the acquisition of the information. Specifically, we are interested in examining how browsing behaviors for a collection of hypertext pages differ for various delays in information propogation times. We hope to learn whether the value a user gives a piece of information depends on the amount of time spent obtaining the information, and whether the decision to obtain more information is affected by the time required to obtain it. This paper describes an experiment that examined the reactions of participants to delays in information propagation in a hypertext environment similar to the World Wide Web. The results suggest that rapid propagation times often lead to better performance, but also can lead to poorer path selection.

1 INTRODUCTION

Everyone familiar with the World Wide Web has encountered delays in downloading a page, particularly when network connections are slow. Advances in technology reduce download times; however, even with the availability of rapid connections like cable modems and DSL in the home, over 93 percent of home subscribers still use dialup connections (Nielsen, 1997). Even on fast connections, download delays can occur because of excessive network traffic, slow connections between machines, and other factors. Research into system response times has shown that a response within a second is necessary for the user's train of thought to stay focused (Miller, 1968; Card, Robertson, & Mackinlay, 1991). Since download times for Web accesses typically exceed this time, it is prudent to examine how the delays impact information processing.

Most previous work has only considered accesses to Web pages in isolation, measuring user frustration in accessing individual pages. However, it is rare that users will access only a single page in a session. Typically, users access a series of pages as they browse a Web site and search for information. As they become accustomed to the delay, users may change their browsing strategy to limit the effects of the delay. In fact, it is conceivable that the opposite effect will occur: longer delays may encourage more careful reading of the information and better selection of links, whereas shorter delays would in essence attach lower worth to the information and result in poor decisions in selecting links to follow.

This paper examines ways in which information processing decisions are affected by temporal delays in the acquisition of information. Specifically, we are interested in learning whether the value a user gives a piece of information depends on the amount of time spent obtaining the information, and whether the decision to obtain more information is affected by the time required to obtain it. We plan to examine this question in a hypertext environment similar to the World Wide Web.

The hypertext model provides the perfect environment for incomplete information processing. A hypertext link is an embedded access point to another document that is activated by clicking on a highlighted word or phrase in the current document. The very nature of this link suggests that a user will move on to the next document before completely processing the information in the current one. If the costs (in time) of the current document as well as of obtaining a new document are low, people might follow a link without completely exploring the current one. On the other hand, if the costs are higher, more time should be spent on evaluating the current information and in choosing more information.

The remainder of this paper describes a small pilot study that examined people's reactions to delays in information propagation in a hypertext environment, specifically the time they spent on a given page and the number of choices required to answer questions.

2 RELATED WORK

Effects of system delays on user performance have been a topic of interest since the development of timesharing systems (Miller, 1968; Card et al., 1991; Shneiderman, 1998). In general, response times should be as fast as possible. Consider the research area of graphical systems, in particular virtual reality (VR) systems, where expensive graphical calculations often result in slow display rates that have the potential to affect the user experience. For example, Mackenzie and Ware (MacKenzie & Ware, 1993) looked at the effects of delays between movements and results when users are trying to acquire a target and found that even a 75 millisecond lag time resulted in performance degradation.

A related area of interest emerged with the advent of the Web and the propogation delays experienced in downloading information. Jacob Nielsen argues that the design of pages that require scrolling should be avoided, as should the design of high graphics sites (Nielsen, 1994). Reasons given are that Web users want their information quickly and compactly, ideally within two seconds, and they are unlikely to spend more than 10 seconds waiting for a page to load. Chris Johnson reports that as retrieval delays increase, measured user value for the information decreases (Johnson, 1997). However, this result was shown for a single download with a fairly long delay that could be over 70 seconds (like those experienced when downloading video clips). We suspect that shorter delays of up to only 16 seconds (like those for pages with limited graphics) will result in different behavior, and that regular and repeated delays in download time will lead to changes in user strategy that could result in greater value attached to Web pages with longer download times.

Several research efforts have examined the effects of system delays on the strategies of users. Two groups of researchers examined strategies of participants performing data entry tasks (Teal & Rudnicky, 1992; O'Donnell & Draper, 1996). They found that participants chose strategies based on system delay, with the choice related to the cost in time of the strategy. We observed in our experiment the coping mechanisms participants used when presented with long delays.

3 EXPERIMENT

The experiment asked participants to answer a series of five questions using information found in a hypertext environment. Each question required the participant to traverse through the environment until a solution is found, then to enter the information into an entry box. The hypertext environment is similar to that used in most Web browsers like Netscape Navigator and Microsoft Internet Explorer. Links to other pages are indicated by underlined and colored text. The participant clicks on a link to load the corresponding page and can move between previously viewed pages with "back" and "forward" buttons. Generally, solutions can be found by following a number of different paths.

The pages in the environment are simplified linked text-only versions of hypertext pages found on the Web. Each question has a different set of hypertext pages associated with it. The participants were asked to find pieces of information within the environment. We simulated a delay in the loading time of each page, as if the page were being downloaded from a remote site. The experiment employed a between-groups design, with the participants divided into three groups. One group always had a delay of 2 seconds, another 8 seconds, and the third 16 seconds. We measured the amount of time required and the number of links traversed to find the solution, and we observed the users and noted outward frustration that they experienced with the delays.

Fifteen participants completed the experiment, five for each of the page-loading delays. The participants were computer science graduate students with extensive Web browsing and searching experience. By using experienced participants, we hoped to minimize the learning curve effects for our hypertext interface. The nature of the tasks and a description of the interface were explained to them before they began. To minimize the time required to read and process the question, the participants first read the question, then pressed a button to load the first page and begin the timed search task. The participants were told that time is an important factor. The system recorded the time of every user action (scrollbar clicks and drags, entries

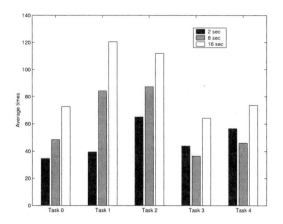

Figure 1: Average times spent finding and entering the correct answer for each of the five tasks (labeled 0-4).

in the entry box, and link traversals). This information, combined with observations of the participants interacting with the environment, helped reveal the browsing patterns of the participants.

We expected that the time to obtain some piece of information will be in direct proportion to the time taken to process the information. That is, the greater the time required to obtain a page of information, the greater the time a user will spend evaluating that information before obtaining more information. There are two reasons we expected this. First, since the participants invest more time in obtaining the information, they should value it more and thus study it more closely. Second, since obtaining more information is costly, the participants should spend more time choosing the next link from the information that they already have.

4   QUESTIONS AND ANSWERS

This section describes each task used in the experiment, outlining why it was included and discussing reactions from the participants to the questions. We selected questions of increasing difficulty with the expectation that the early rounds would provide the opportunity to get acclimated to the experimental setup.

The first question asks how many runs the Atlanta Braves scored in a baseball game several weeks ago. That game featured the last big offensive output by the Braves, so it is mentioned at least in passing on many of the pages. The majority of the participants in each group found the correct answer in the first visited link. We hoped that this task would provide participants with the opportunity to get accustomed to the environment and the questions.

The second question asks how many books author Isaac Asimov wrote and published. The hypertext environment is patterned after a Web-based FAQ (Frequently Asked Questions) list on Asimov. None of the questions in the FAQ list exactly matched the one to be answered, and in fact only two links contained the correct answer. As a result, the participants visited many more links to find the correct solution. Surprisingly, more links were visited on average for larger delays, perhaps because the participants did not yet have a feel for the amount of time it would take to load a page.

The third question asks the participant to find the year in which work on Mount Rushmore was begun. The pages for this question are very verbose, requiring a participant to scan a lot of information to find the link that will lead to the answer. As we anticipated, the participants with smaller delays were more likely to click the first link that seemed at all relevant, since the cost of loading a page is fairly small.

The fourth question asks for the number of Western films that feature John Wayne. There are a number of promising links in the visible portion of the page, however, none of these links lead directly to the solution. The participants had to traverse through several pages to find the answer. By this point, it seemed that the participants with 8- and 16-second page loading delays altered their strategy to scan the entire page to try to identify the best link. Most of the participants with 2-second delays seemed to choose the best link visible without scrolling while most with the 8- and 16-second delay scrolled down in an attempt to find a better link. This resulted in a smaller mean time to find the solution for the participants with an 8-second

1088

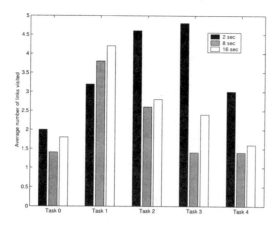

Figure 2: Average number of links visited in finding the correct answer for each of the five tasks.

delay than for those with a 2-second delay.

The fifth question asks for the length of the Mississippi River. Similar to the prior set of pages, this set is verbose and can be difficult to navigate. Again we saw that the participants with a 8-second delay complete the tasks more rapidly than those with a 2-second delay. While the mean times for the last two tasks were lower for the 8-second delay compared to the 2-second delay, the participants with a 16-second delay had a significantly higher average, probably in part because they were showing signs of frustration and boredom at the large loading delays (see Section 5 for details on this phenomenon).

Note from Figures 1 and 2, the participants with 2-second delays found the answers the quickest in the first task, even though they visited slightly more links. By the later rounds, the gap in links visited for the participants with 2-second delays was much greater, so much so that the average times for those with 8-second delays were slightly smaller than those with only a 2-second delay.

5 RESULTS

In analyzing the results, analyses of variance (ANOVAs) were performed to check for statistical significance among different conditions of the experiment. If the ANOVA revealed a significant difference, pairwise t-tests were performed to determine which conditions differed.

As expected, the participants with the shorter delay required less time to find the answers to the questions, with mean times per question of 47.5, 60.5, and 88.6 seconds, respectively ($F(2, 72) = 7.20$, $MSE = 11044$, $p < .01$). The group with a 16-second delay consistently performed the most poorly throughout the rounds ($p < .001$, $p < .01$). The participants in this group typically would mutter and groan after clicking a link, and their attention would wander in that they would look around the room or down at the floor. This seems to be in line with results from Web studies that show that people rarely wait more than 10 seconds for a page to load before hitting the stop button and looking for the information elsewhere. The expectation has been established that hypertext information should be obtained in under 10 seconds, and longer delays are not readily tolerated.

However, the difference between groups with 2-second and 8-second delays were only marginally significant ($p < .09$). In fact, when only the last three rounds were considered (when one would expect that participants were acclimated to the delay), there was not a significant difference in the times to find the solution ($p = .43$). Given that the means are approximately equal (54.5 and 56.6, respectively), there is not even a trend to suggest that a difference exists, a surprising result given the difference in the delays. This suggests that participants with a longer delay perhaps are considering the information they obtain more carefully and making more intelligent decisions.

One expectation was that participants with a longer delay would choose the links they follow more carefully, resulting in smaller numbers of links followed for participants with longer delays. While the mean number of links followed per browsing task was smaller for the groups with longer delays (3.5, 2.1, and

2.7 respectively), the difference was not significant ($F(2,72) = 1.63$, $MSE = 12.04$, $p = .20$). This may be because the tasks only required the participants to follow a small number of links to find the solution. Perhaps more complex browsing environments would result in more significant differences.

The results of this study suggest that rapid propagation times do not necessarily translate into faster times on tasks in which users must browse for information. It may be more difficult to find information in a minimalist Web site without graphical cues to aid in navigation than a more user-friendly site with graphical navigation aids, even though pages in the latter may take longer to download.

Observing the participants revealed their frustrations in long delays. Two of the participants, both with the 16-second delay, quit the experiment before answering all of the questions. (Their results are not included in the statistics.) They claimed that they were bored with waiting and suggested that perhaps the delay was longer than we had claimed. This result is in line with one noted by Johnson (Johnson, 1997), who predicted that at some point in time, users of a remote information retrieval system would "run out of time" and quit their download. For those participants who did endure the longer delays, many were not completely focused on the task. One participant played with the window while waiting for the page to load, others looked around, several verbally harassed the proctor for building an experiment that was so slow. The participants seem to have an expectation that a hypertext page should load in under 10 seconds, as noted in (Nielsen, 1994). Participants approach the problem differently depending on the delay, similar to the strategies observed in (O'Donnell & Draper, 1996). The participants with a 2-second delay use a rapid-fire selection method, those with an 8-second delay use the delay to think and plan, and those with a 16-second delay become annoyed and distracted.

6 CONCLUSIONS

The results of our study show that people spend more time examining information when its cost (in time) was higher. We speculate that this is because they spent more time obtaining the data and because they would have to spend more time obtaining new data. It may be that (up to a point) the participants use the page loading delay to think about the information they had previously viewed and to plan their future strategy. One might think that it would take longer to find a piece of data in a hypertext environment when there is a larger delay and a larger information processing time. However, this is not always the case. In this experiment, better decisions were made when more time was spent analyzing the data, resulting in less overall time spent on the search.

References

Card, S. K., Robertson, G. G., & Mackinlay, J. D. (1991). The information visualizer: An information workspace. In *Proceedings of CHI 1991* (pp. 181–188). New Orleans, LA.

Johnson, C. (1997). What's the web worth? The impact of retrieval delays on the value of distributed information. In *Workshop on Time and the Web*. Staffordshire, England.

MacKenzie, I. S., & Ware, C. (1993). Lag as a determinant of human performance in interactive systems. In *Proceedings of INTERCHI 1993* (pp. 488–493). Amsterdam, the Netherlands.

Miller, R. B. (1968). Response time in man-computer conversational transactions. In *Proceedings of the AFIPS Fall Joint Computer Conference* (Vol. 33, pp. 267–277). Washington, DC.

Nielsen, J. (1994). *Usability engineering.* Boston, MA: AP Professional.

Nielsen, J. (1997). The need for speed. *Alertbox.* (Updated version: August 2000)

O'Donnell, P., & Draper, S. (1996). How machine delays change user strategies. *SIGCHI Bulletin, 28*(2), 39–42.

Shneiderman, B. (1998). Response time and display rate. In *Designing the user interface: Strategies for effective human-computer interaction* (3rd ed., pp. 351–370). Addison-Wesley.

Teal, S. L., & Rudnicky, A. I. (1992). A performance model of system delay and user strategy selection. In *Proceedings of CHI 1992* (pp. 295–305). Monterey, CA.

MATCHING NAVIGATIONAL AIDS TO THE TASK

Stephania Padovani and **Mark W. Lansdale**

Cognitive Ergonomics Research Group, Department of Human Sciences, Loughborough University
Loughborough, Leicestershire, United Kingdom, LE11 3TU
S.Padovani@lboro.ac.uk; M.W.Lansdale@lboro.ac.uk

ABSTRACT

In this paper we describe a series of experiments conducted with two different types of navigational aids (a global site map and a bookmark tool) in different contexts. We investigated the effectiveness of these navigational aids for different types of tasks, within Web sites based on different metaphors and under time pressure. The results highlight the importance of considering task aspects when designing and implementing navigational aids within Web sites since these factors strongly influence users' navigation strategies, navigation performance and understanding of the Web site structure.

1. INTRODUCTION

Many researchers have investigated navigation problems in hypertext systems. One of the major problems, which is also the most frequently mentioned in the literature is that of disorientation. Disorientation has been described as a subjective feeling of being "lost" for not having a clear knowledge of the relationships within the system. It impairs users' performance by reducing their ability to generate new routes within the system of retrieve already traced routes, therefore compromising both information-seeking and information-retrieval. The most recent solution to tackle the disorientation problem recommends the implementation of supplementary navigation tools that provide both guidance and a more direct access to the information contained in the hypertext nodes (Kim & Hirtle, 1995; McDonald & Stevenson, 1998; Nielsen, 2000). These supplementary navigation tools are usually referred to as navigational aids.

Our research is mainly concerned with the psychological principles underlying the utilisation of navigational aids. A major aim of this paper is to demonstrate that the effectiveness of navigational aids is highly task and context-specific but in a principled way. Their utility varies according to the characteristics of the information space being navigated, the nature of the task being performed, and the conditions under which this task is performed.

2. METHODOLOGY

2.1. Experimental design

This study comprised twelve experimental conditions manipulating four variables in a between- and within-subjects design. The three between-subjects factors were:

- *type of navigational aid available* (each group of participants had either a global site map available as a separate screen within the Web site, a bookmark tool available as a pull-down menu from the Internet browser or no navigational aids);

- *time constraint* (some groups of participants had to finish the task as fast as they could whereas other groups were allowed as much time as needed to complete the task and to try to learn how to move around the Web site);

- *Web site context* (two isomorphic Web sites with respect to structure were built, the first of which utilised a spatial metaphor in which each Web site node was a room in a house and the map depicted the house plan, and the second utilised a social metaphor in which each node was a student's web page and the map abstractly depicted the Web site network).

The within-subjects factor was *type of task to be performed* (the first stage of the task involved finding specific target screens - information-seeking whereas the second stage of the task involved retrieving those specific target screens - information retrieval).

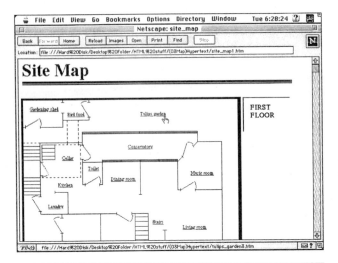

Site Map

Figure 1:
The upper part of the site map available from the spatial-metaphor Web site seen within the browser's window. Users can access any screen by clicking on its name on the site map. Once a screen is visited, its link on the site map fades.

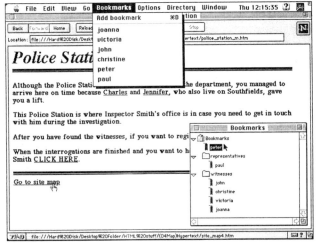

Figure 2:
The bookmark tool available from the Internet browser with a list of nodes added by a participant and these nodes being grouped into user-defined categories within the bookmark organiser window.

2.2. Participants

One hundred and seventy student volunteers from different departments at Loughborough University served as participants. Ten students took part in each of the two initial non-navigational-aid conditions with the spatial-metaphor Web site and fifteen in each of the ten other conditions. All participants had previous experience in navigating the Internet but not all of them had utilised navigational aids before. Participants were tested individually after receiving verbal instructions on how to use the Web site navigation system.

2.3. Generic Task

The experimental task consisted of an information-seeking stage followed by an information-retrieval stage. On the spatial-metaphor Web site, participants were required to find rooms with flowers on the first part of the task and water these flowers on the second part of the task. On the social-metaphor Web site, participants were required to identify witnesses among the students on the first part of the task and contact these witnesses for the interrogation on the second part of the task.

2.4. Procedure

Each experimental session involved performing the experimental task using the Web site, sketching a map of the site and answering a questionnaire. First of all, each participant received instructions on how to use the Web site navigation system, including the navigational aid available. Then, they performed the experimental task, which comprised an information-seeking sub-task followed by an information-retrieval sub-task. Immediately after completing the task, participants were required to sketch a site map from memory without looking back at the Web site. Finally, participants filled a questionnaire on their navigation and memory strategies, navigation difficulties, benefits of the navigational aid, difficulty to sketch the map and accuracy of the sketch produced.

2.5. Data collection and analysis

All navigations were automatically logged during the experimental sessions. The participants' navigations were analysed qualitatively in terms of their overall navigation strategies and quantitatively through performance measures under the following categories:

- *time measures* (indicate how long the participants took to complete the task);

- *navigation pattern measures* (describe the participants' navigation paths);

- *efficiency measures* (show how close the participants' navigation paths were to the optimal navigation path);

- *use of supplementary navigation tools* (characterise the utilisation of the back button, site map and bookmark);

- *task accuracy measures* (measure the proportion of target screens achieved during the realisation of the task).

The map sketches were analysed in terms of the criteria utilised to organise the information nodes, the type of representation chosen to depict the Web site structure and accuracy (number of nodes recalled and correctly placed on the map). These measures altogether allowed us to quantitatively analyse the participants' navigations, obtain an overview of their navigation strategies, and evaluate their understanding of the Web site structure. The questionnaire results, in turn, served as the participants' self-reports of their navigation and memory strategies. They were compared and contrasted with the experimental results.

3. SUMMARY OF RESULTS AND DISCUSSION

This section concentrates only on the most relevant findings of this study. Because of space restrictions we will not describe the CHI square calculations performed with the navigation strategies results and the factorial ANOVA undertaken with the navigation performance and sketch measures. It is important to mention, however, that all findings reported are based on significant results obtained in the statistical analysis.

3.1. Navigation strategies

The navigation strategies classification developed in this study considers how the users traversed the Web site structure and which navigation tools they utilised to do so. Some examples of the main categories within the proposed classification are:

- *end-of-branch strategy* (the user follows a branch until its end before starting to pursue a new branch);

- *meeting-point strategy* (the user accesses all immediately adjacent screens from a singles screen – the 'meeting-point' – before moving to another 'meeting-point');

- *navigational aid – node – navigational aid strategy* (the user selects a screen through the navigational aid and returns to the navigational aid to access the next screen).

A comparison between the navigation strategies adopted by the participants within the various experimental conditions revealed a significant effect of the type of navigational aid, type of metaphor and time pressure.

The role of the metaphor: For example, when no navigational aids were available, participants who navigated the non-spatial-metaphor Web site adopted more thorough and systematic strategies (such as end-of-branch) than participants who navigated the spatial-metaphor Web site. These results indicate that a rather arbitrary and unfamiliar Web site structure leads its users to opt for navigation strategies that are very "tied" to the Web site structure itself when they lack the support of navigational aids.

The role of the navigational aid: In contrast, when a site map is available, the majority of the participants navigating both versions of the Web site adopt hybrid strategies on the first part of the task that include both patterns based on pure node-to-node navigation and the utilisation of the site map. Compared to participants without navigational aids (who were more strategic during the first part of the task because they were guiding themselves by the colour of the links), site map users were more strategic on the second part of the task when they utilised the site map as their main access tool either to go straight to all targets or to search again for missing targets. The introduction of a site map also cancelled the effect of the metaphor upon the participants' navigation strategies.

The availability of a bookmark tool also leads its users to adopt hybrid strategies on the first part of the task (gradually increasing the utilisation of the bookmark list as "key-screens" are added) and single strategies on the second part of the task, bookmark users' navigation strategies were not affected by the type of metaphor upon which the Web site they navigated was based.

The effect of time pressure: When time pressure is imposed, the participants within all experimental conditions tend to adopt more hybrid strategies or to navigate at random, without any explicit navigation strategy. This indicates that time pressure impairs the development of organised navigation strategies even when navigational aids are made available.

3.2. Navigation performance

The role of the metaphor: When users do not have any navigational aid available, their navigation performance is strongly affected by the type of metaphor upon which the Web site is based. When compared to those participants who navigated the social-metaphor Web site, participants who navigated the spatial-metaphor Web site utilised less the back button, revisited fewer unnecessary screens, were more efficient and achieved a higher proportion of targets during information-retrieval. These results suggest that the spatial metaphor improves the participants' ability to retrace their paths to the targets or to infer new paths that avoid screens unnecessary for the task at hand.

The role of the navigational aid: On the information-seeking part of the task, map users and bookmark users who navigated the spatial-metaphor Web site took fewer steps and completed the sub-tasks faster than participants who did not use any navigational aid. In contrast, when navigating the social-metaphor Web site, bookmark users were significantly slower and took more steps to complete the information seeking stage of the task when compared to their map and non-aid counterparts. These results suggest that a site map provides a better support for information-seeking tasks regardless of the spatial metaphor being used. This point is corroborated by the fact that site map users under both metaphor conditions used their navigational aid significantly more frequently for information seeking than bookmark users.

On the information retrieval part of the task, the results are quite different. In this case, bookmark users were faster, more efficient and accessed a higher proportion of screens using the bookmark list when compared to map users or participants under the non-aid conditions. This indicates that the bookmark tool provides a better support for information-retrieval tasks.

The effect of time pressure: Our results have shown that the bookmark tool was significantly more vulnerable to time pressure than the site map. For example, bookmark users who navigated the spatial-metaphor Web site when put under time pressure no longer used their bookmarks as "anchors" to help them with information seeking. Also, the majority of these time-pressured bookmark users forgot to bookmark targets during the information seeking stage and were, therefore, less efficient than their non time-pressured counterparts on information retrieval. One reason for this might be that bookmark use requires a deeper level of processing since it actually involves selecting specific nodes according to their importance to the task and then adding the nodes to a list in order to be able to make use of the bookmark list later. Conversely, a site map may well be regarded as a pre-structured list of shortcuts since its users do not have any participation in the storage or organisation of the information nodes available in the site map.

3.3. Learning of the Web site structure

In terms of learning of the Web site structure, map users who navigated the spatial-metaphor Web site and were not under time pressure could produce significantly more accurate sketches than time-pressured map users and users who navigated the same Web site without navigational aids. This difference disappears when map users navigate a non-spatial metaphor Web site. In the latter case, not even allowing users more time to learn about the information space can improve their understanding of the Web site structure. These results might indicate that a site map can act more efficiently as a memory aid when it represents a structure based on a spatial metaphor.

On the other hand, bookmark users' site map sketches were significantly less accurate than those produced by map users or even by participants who did not have any navigational aid available; revealing that they did not learn as much about the Web site structure. A reason for this might be that the bookmark tool actually removes the memory requirement and, therefore, tends to impoverish the user's learning of the Web site structure. Hence, Web sites requiring users to understand their structure should not use bookmarks alone.

3.4. Questionnaire results

Utilisation of the navigational aid for orientation purposes: The results of the questionnaire indicated that the bookmark tool was used exclusively to access the targets screens involved in the experimental task more directly, whereas the site map was also used for orientation purposes. Although both site map users and bookmark users felt better oriented than those users who navigated without any navigational aids, when enquired about how they guided themselves through the Web site to avoid getting lost, the majority of site map users mentioned that they utilised the site map for this purpose, whereas bookmark users reported on using the 'back button' or the colour of the links instead of the bookmark tool.

Utilisation of the navigational aid to support memory strategies: The site map and the bookmark tool as well as improving navigation performance and orientation (in the case of the site map), acted as memory aids during the realisation of the experimental task. On the information-seeking part of the task, for example, site map users memorised the position of the targets on the map instead of trying to remember their names. Bookmark users at this stage did not develop any memory strategies because they could store the targets in the bookmark list. On the information-retrieval part of the task, whereas site map users still needed to remember the targets positions on the map and count them, bookmark users only needed to select them from the bookmark list.

Benefits brought by the navigational aids: Both site map users and bookmark users pointed out some benefits of utilising these navigational aids. Users stated that the navigational aids helped them to complete the experimental task faster and more accurately. However, when asked about the their sense of orientation and comprehension of the Web site structure, whereas the majority of the site map users considered that the site map helped with this respect, only a few bookmark users believed that their orientation and comprehension of the Web site structure was enhanced by the bookmark tool.

4. CONCLUSIONS AND FUTURE WORK

This study aimed to investigate the effectiveness of a global site map and a bookmark tool within different contexts. Our results have shown that a global site map was more helpful for information-seeking tasks whereas a bookmark tool provided a better support for information-retrieval tasks both within a spatial- and a social-metaphor Web site.

Another important difference was that the global site map improved the users' understanding of the Web site structure whereas the bookmark tool actually impoverished it by completely removing the requirement for remembering paths or even the names of the screens involved on the task. On the other hand, the bookmark tool acted as memory aid both for information-seeking and information-retrieval tasks, whereas the site map supported the users' memory strategies only partially during the information-seeking stage of the task.

These findings suggest that the two navigational aids could complement one another if utilised together for tasks that combine information-seeking and information-retrieval and also involve learning the Web site structure. Current research is aimed at investigating how efficient users are at combining the utilisation of these tools and identifying their utility for different types of tasks.

REFERENCES

Kim, H., & Hirtle, S. C. (1995). Spatial metaphors and disorientation in hypertext browsing. *Behaviour and Information Technology*, 14(4), 239-250.

McDonald, S., & Stevenson, R. J. (1998). Effects of Text Structure and Prior Knowledge of the Learner on Navigation in Hypertext. *Human Factors*, 40(1), 18-27.

Nielsen, J. (2000). The user controls navigation. In Weiss, S & Eberhardt, J (Eds.) *Designing Web usability.* (pp. 214 – 222). Indiana, United States of America: New Riders Publishing.

The influence of language proficiency on web site usage with bilingual users

William G. Hayward* & Kwok-Kit Tong[†]

*Department of Psychology, Chinese University of Hong Kong, Shatin, NT, Hong Kong
(william-hayward@cuhk.edu.hk); [†]Department of Management, City University of Hong Kong, Kowloon
Tong, Kowloon, Hong Kong (mgkktong@cityu.edu.hk)

In this study, we examine Chinese-English bilinguals' performance at completing tasks on web sites with identical Chinese and English versions. Participants were required to perform two tasks with each web site, and the time necessary for completing the tasks was recorded. Despite participants' competency with English, tasks that were completed in Chinese were performed more quickly, and participants judged that they were easier. Crucially, this advantage for Chinese web sites was not simply due to a reduced fluency with English; when English comprehension ability was included as a covariate, performance on the Chinese web sites remained superior to the English sites. We argue that the choice of language for a web site should not simply be due to a threshold of competency among users, but instead should reflect users' preferences for their native language.

1. Introduction

Recent years have seen an increased concern with issues of usability in web site design (e.g., Spool et al, 1999; Nielsen, 2000). One report on web site usability problems in the 2000 Christmas season estimated that global business may have lost $14 billion in sales because of customers' difficulties with using their sites (Rehman, 2000). Clearly, the World Wide Web's usefulness as a medium of commerce and communication is dependent upon users being able to complete transactions; if they cannot, the Web loses its reason for existence.

In concerns about usability, one issue often neglected is the language used for the web site. It is generally assumed that the language chosen should be that of the users, so that English is used for American web sites whereas Chinese is used for web sites in China. But in the increasingly internationalised Web, many web site operators will hope to attract visitors from around the world who may not be fluent with the operator's language. In these situations, how should the operator determine the language or languages for the site?

Typically, usability experts suggest using the language with which users are most proficient. This guideline sounds intuitively obvious, but there are at least two reasons to question its validity. First, if the users of a web site speak different languages, there may be no single language which is the language of highest proficiency of all users. In this situation, different versions of the web site will need to be created if all users are to be able to browse it using their most proficient language. Of course, there is always the possibility that a user who speaks a different language from those available for the web site will attempt to use the site; in this case, she must use a language which is not the one with which she is most proficient. Thus, in the extreme, there may be no way to ensure that all users experience the web site in their preferred language.

Second, it is not clear that proficiency should be the only basis upon which the language of a site should be chosen. The Web is primarily an English-language environment, often requiring English-language input, and thus for contextual reasons users may come to expect to browse web sites in English, even if they are more proficient at another language. In addition, it may be detrimental to a web site to use different languages for different users. It may be impossible to create texts in different languages that express the exact same concepts, and thus translation from one language to another may result in the creation of multiple, distinct messages. Equally, it may be important for branding purposes that a company is perceived as belonging to a particular geographic or social region (e.g., inner city, United States), and in that case translation of its web site may reduce perceptions of that location. In this latter case, one could make a strong case that, as far as usability is concerned, any language is suitable for a web site, as long as users have some threshold of proficiency.

Third, people may perceive that the context of a web site is richer and more updated when it is presented in the original language. A web site is the medium for a stream of continuously updated information. However, updating different language versions of a large web site may be a difficult, if not impossible, process. People may therefore end up believing that the "main" web site contains more information than versions in different languages.

1096

Hence, people may actually prefer using the English version of a web site, particularly if it is the site of a foreign-based organization.

These issues are particularly relevant in technologically advanced, bilingual societies such as Hong Kong. Here, the native language (spoken form: Cantonese; written form: complex Chinese) is used most widely in everyday life, but much education and business is conducted using English. In order to be fully functioning members of the workforce, people must have a reasonable degree of proficiency in both languages. If an American company is preparing a web site for the use of Hong Kong citizens, which language(s) should they choose? If the company chooses to use English and Chinese, it risks diluting and compromising its marketing message. On the other hand, if it uses only Chinese, it risks altering the perception of itself as an American company, and if it uses only English, it may alienate users who are not as proficient in English.

To investigate these issues, we recruited 64 Chinese-English bilinguals, and had them perform two tasks at each of eight web sites. Each web site had identical Chinese and English versions, and participants were assigned at random to either the Chinese versions or the English versions. Completion times were recorded, and a questionnaire was administered to determine participants' subjective evaluations of task difficulty. Of interest was whether there would be differences in participants' abilities to use the different versions of the web sites, and also whether any differences were due to a lessened proficiency with English as compared to Chinese.

2. Method

2.1. Participants.
Sixty-four participants were recruited, all of whom were students at the Chinese University of Hong Kong (CUHK). CUHK is a bilingual teaching institution in which students must be able take classes in both English and Chinese. All participants reported subjectively that they were more proficient in Chinese than English. They were paid HK$50 (US$6.40) for participation in the study.

2.2 Stimulus Materials.
Eight web sites were chosen, each of which had versions in English and complex Chinese that were, in other respects, essentially identical. We attempted to get a variety of economic sectors (banking, retail, travel, food, government), and a mixture of local Hong Kong and foreign organisations. The sites were viewed on 15" monitors with PC-compatible computers running Netscape Navigator.

We also designed a questionnaire to determine participants' subjective evaluations of the difficulty of the tasks. Nine questions asked about the difficulty of each task and each site in general, and whether they would like to have used Chinese or English to do each task. Note that the questionnaire also contained additional items that tapped participants' judgments about the web site, and the operator of the web site. These are reported elsewhere (Tong & Hayward, 2001).

Each participant was also given a test of English written comprehension (Martin, 1970). Our sample score was normally distributed, Kolmogorov-Smirnov $Z = 1.04$, *ns*.

2.3 Procedure.
Participants were required to perform two tasks on each of eight web sites. On each site, one task involved the retrieval of information (e.g., the price of a particular item), whereas the other task required interaction with the site (e.g., buy a retail item). Completion time was recorded. At the conclusion of each web site, participants were given the questionnaire which asked them about their ability to find information on the site.

Participants were randomly assigned to a language group; one group of participants used web sites written in Chinese, and the other group used sites in English. The order with which web sites were seen was counterbalanced across participants.

3. Results

Figure 1 shows the raw completion times. Participants were faster at completing web sites when they could use Chinese and they were faster at performing the information search tasks. Neither of these results is surprising, as participants verbally reported that they preferred to use Chinese, and the interactive tasks simply required a greater degree of participation than the information-search tasks. In order to normalise for differences between the two tests when conducting the statistical analyses, completion times were converted to standard scores. Standard scores convert the mean of the original distribution to zero and the standard deviation of the original distribution to units equal to 1. Note that normalizing the data in this way gives each task an identical mean (of 0), and thus only language differences are examined by the statistical analyses.

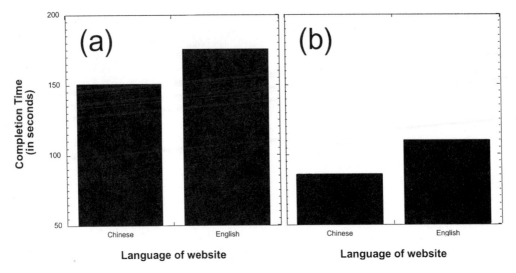

Figure 1. Mean completion times (in seconds) for (a) interactive task and (b) information retrieval task.

An ANOVA on the normalized data showed that there was a significant main effect for Language, $F(1,62)=14.19$, $p<.01$. Although important, of primary interest was the role of English proficiency on performance differences between English and Chinese. To examine this effect, we took subjects' proficiency scores and entered them into an analysis of covariance (ANCOVA), where language was the main factor and English proficiency was a covariate. Proficiency was a statistically significant factor, $F(1,60)=8.72$, $p<.01$, but when differences associated with English proficiency were covaried out of the analysis, the language effect remained statistically significant, $F(1,60)=10.93$, $p<.01$. Thus, although some differences in the data may be due to English proficiency, the increased ability with using Chinese language web sites is not solely due to that factor. There are other reasons that our users were better at performing tasks with Chinese-language than English-language sites than simply their lower proficiency in English.

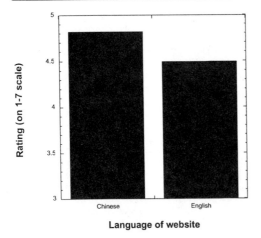

Figure 2. Mean ratings of task difficulty (on a seven-point scale).

1098

It is possible that when measuring differences in completion times we could find statistically significant differences that were so small they were essentially meaningless. To ensure that this was not the case, we asked participants, after each web site, to rate their subjective perceptions of the ease of the task (eg. "Did you find the interactive/information-search task easy to do or difficult to do?", "Do you find this web site well organized or disorganized?"). Participants rated their responses on a seven-point scale. During the test, direction of responses (whether high scores represented positive or negative attributes) was varied; for purposes of analysis, some questions were reversed so that high scores always represented positive judgments. Mean responses are shown in Figure 2. Participants who used the sites in Chinese rated them as easier to use than those who used them in English, as shown by a significant main effect in a one-way ANOVA, $F(1,62)=4.89$, $p<.05$. Thus, participants were sensitive to differences in ease of use uncovered by the completion times. We again performed an ANCOVA with proficiency scores as the covariate and language as the independent variable, and found that the independent variable was still marginally significant even after taking differences in English proficiency into account, $F(1,60)=3.55$, $p<.065$.

4. Conclusions

Two main conclusions can be drawn from this study. First, participants completed tasks more quickly when they were able to use their native language. Although not entirely surprising, it should be noted that these users were required to have proficiency in English to the standard of taking university classes. As such, they are certainly fluent in English, and probably fairly typical of many communities in the world where English is not the native language but is heavily used. When designing web sites for such communities, it is tempting to use English as the language for the site because all prospective users will speak it with a reasonable standard. These results suggest that even an ability that allows for tertiary education will result in inferior web performance than if one was able to use one's native language. We found that using English slowed performance by about 25% relative to Chinese, which would likely translate in terms of usability to considerable frustration with the web site. One language, it appears, does not fit all.

Second, this finding assumes more significance because it was not solely due to problems with English fluency. Even though our participants were bilingual, they were generally much more confident using Chinese than using English. As such, one explanation for the increased ability with Chinese is simply that it reflects differences in fluency. The ANCOVA suggests that this was not the case. Even when taking differences in English fluency into account, participants still showed significant differences in the use of English and Chinese web sites, and their subjective impressions of their interactions with these sites showed a marginal effect. The logical extension of this result is that bilingual users who are equally fluent in two languages may still prefer to use their native language for web site tasks. Although we do not have data to explicitly support this position, we have enough evidence to show that using language proficiency in *any* respect to determine the language of a web site may be flawed. Rather, a designer should attempt to use the native language of users of a web site if (s)he wants users to interact as easily as possible with the web site.

In summary, although language proficiency may influence web browsing ability to some extent, it is not the only determinant of language preference. One major implication of this result is that decisions of which language to use for a web site cannot be made simply on the basis of proficiency. It may be the case that native languages are almost always preferred, regardless of proficiency with a second language (such as English). These issues will be the focus of future research.

Acknowledgments

This research was supported by the Hong Kong office of Icon Medialab Asia, and we would like to thank David Jacques, Malcolm Otter, and Richard Ihuel for assistance in conducting the study.

References

Martin, M. H. (1970). *Listening and comprehending: Audial Comprehension Tests Oral and Written.* London: MacMillan Education Limited.

Nielsen, J. (2000). *Designing Web Usability.* Indianapolis, Indiana: New Riders.

Rehman, A. (2000, October). *Holiday 2000 e-commerce: Avoiding $14 billion in "silent losses."* New York: Creative Good, Inc. (www.creativegood.com).

Spool, J. M. et al. (1999). *Web site Usability: A designers guide.* San Francisco, CA: Morgan Kaufmann.

Tong, K. K., & Hayward, W. G. (2001). *Speaking the right language in web site design: Effects of usability on attitudes.* Manuscript submitted for publication.

VISUALISING AND ANIMATING VISUAL INFORMATION FORAGING IN CONTEXT

Chaomei Chen, Timothy Cribbin

The VIVID Research Centre
Department of Information Systems and Computing
Brunel University
Uxbridge UB8 3PH, UK
Tel: +44 1895 203080
E-mail: {chaomei.chen, timothy.cribbin}@brunel.ac.uk

ABSTRACT

Optimal information foraging provides a potentially useful framework for modelling, analysing, and interpreting search strategies of users through a spatial-semantic interface. Improving the understanding of behavioural patterns of users in such environments has implications for the design and refinement of a range of user interfaces. In this article, we outline the role of optimal information foraging in the study of visual information retrieval and how one may use visualisation and animation techniques to put behavioural patterns in context. Behavioural patterns of information foraging in an information space are visualised and animated to aid further in-depth analysis of search strategies.

1. INTRODUCTION

According to the optimal information foraging theory (Pirolli & Card., 1995), searching for information is, in many ways, very much like hunting. The most fundamentally common strategy between the two focuses on the trade-off between the profitability of a target and the risk involved in achieving the goal at hand (Figure 1). When we search the Internet, we may consider the downloading time as a factor if we are using a modem and the size of the file to be downloaded is huge.

Figure 1. Information foraging focuses on risks as well as the profitability of a target.

The key to the analysis of an optimal information foraging process is therefore similar to the analysis of a traditional food foraging process several thousands of years ago. It is the judgement of the profitability versus the risk that determines one's search strategy, search trails, and other behavioural patterns. When we deal with an information space, we can estimate the profitability of a document in terms of its intrinsic values, such as relevancy to our tasks at hand. And we can estimate the risk of locating and retrieval a document in terms of our subscription fees, per item purchase fees, waiting time, or phone bills for Internet connections.

If we conceptualise a visual information retrieval process as information foraging, we may gain additional insights into users' search strategies and their behavioural patterns in the context of spatially and semantically organised information spaces. For example, one can estimate the profitability with the proportion of relevant documents in a specific area of an information space divided by the time it will take to read all the documents within this area. In their study of the Scatter/Gatherer system, Pirolli and Card found that even a simplified model of information foraging shows how users' search strategies can be influenced. Their study suggests that users are likely to search widely in an information space if the query is simple, and more focused if the query is harder (1995). According to the profitability principle, harder queries entail higher cost to resolve and the profitability of each document is relatively low. In general, users must decide whether or not to pursue a given document on the course of navigation based on the likelihood profitability of the document.

The optimal information foraging theory provides a useful framework for the modelling and analysis of behavioural patterns involving visual navigation. In particular, users are concerned about how to maximise the potential profitability, where to search, and how long to search in a specific area.

In this article, we present a conceptual framework for modelling behavioural patterns in visual information foraging. In particular, this visual information foraging takes place in a spatial-semantic interface of a thematic information space. This framework constitutes the optimal information foraging theory, Hidden Markov Models, spatial-semantic interfaces, and a taxonomy of visual navigation. In this article, we focus on the role of information foraging in our approach. A related study focusing on Hidden Markov Models is reported in (Chen, Cribbin, Kuljis, & Macredie, 2000). The overall approach is illustrated through an example in which visual navigation data were drawn from an information retrieval experiment. Finally, implications of this approach for understanding users' navigation strategies are discussed.

2. INFORMATION FORAGING IN VISUAL NAVIGATION

In order to study users' information foraging behaviour, we constructed four thematic spaces based on news articles from the Los Angeles Times newspaper retrieved from the Text Retrieval Conference (TREC) test data. Each thematic space contains the top 200 news articles retrieved through a single keyword query to the document collection. The four keywords used were *alcohol*, *endangered*, *game*, and *storm*. Corresponding spaces were named accordingly by these keywords.

The user interface for each thematic space is designed based on a popular organisation principle, which is closely connected to Gestalt laws in cognitive psychology. Gestalt laws are several principles developed in earlier 20th century concerning how people perceive patterns. For example, the proximity principle says that people tend to see natural groupings purely based on proximity relationships displayed (Figure 2). In this article, we focus on searching behaviours associated with spatial-semantic interfaces designed based on relevant Gestalt laws. We expect that users' search behaviours should reflect this spatial-semantic feature.

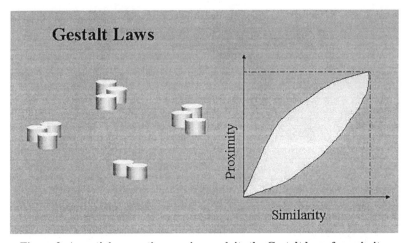

Figure 2. A spatial-semantic mapping exploits the Gestalt law of proximity.

For each thematic space, we generated a document-to-document similarity matrix using Latent Semantic Indexing (LSI) (Chen & Czerwinski, 1998; Deerwester, Dumais, Landauer, Furnas, & Harshman, 1990). A few types of spatial-semantic interfaces were produced, including Pathfinder networks (PF) and minimum spanning trees (MST) (see Figure 3). In MST-based visualisation, $N-1$ explicit links connect all the documents together. Users can see these links on their screen. In PF-based visualisation, additional explicit links are allowed as long as the triangular inequality condition is satisfied. Detailed descriptions of the use of these techniques for information visualisation can be found in (Chen, 1999).

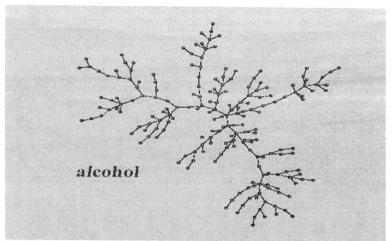

Figure 3: A spatial-semantic interface of the ALCOHOL thematic space.

Users in this study are self-selected students at Brunel University. They performed a series of tasks in each thematic space through spatial-semantic interfaces. Usage data was logged to a computer file in each session, including the occurrence of an event, the time stamp of the event, and the target document on which the event takes place.

The design of the tasks follows Shneiderman's mantra: overview, zoom, filter, details on demand, which highlights users' cognitive needs at various strategic stages in visual information retrieval. At the top level with Task A, users need to locate and mark documents relevant to a search topic (Figure 3). For example, users were asked to locate and mark any documents that mention an incident of drink driving. 20~25 documents were judged as relevant by experts for TREC conferences. Task B is more specific than Task A, and so on. We expect that users would narrow down the scope of their search from Task A through Task D and that this should be evident in their trails of navigation.

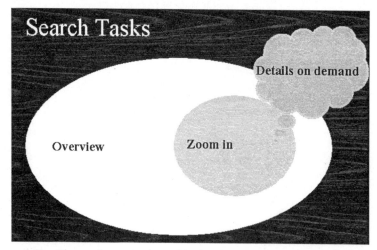

Figure 4. Shneiderman's mantra is used in the characterisation of interactive tasks.

The screen display is designed as follows. Documents in red are not relevant to the search tasks. The course of navigation appears as dotted yellow links. Relevancy judgements made by experts are provided in the TREC test data. Documents relevant to the original search are marked with a bright yellow dot in the centre. If the user marks a document as relevant in a search session, this document will be coloured in blue. Upon the user visits a document, a dark circle is drawn around the current document. The time spent on a document is denoted by a growing green belt until the user leaves the document. If the user comes back to a previously visited document, we will see a new layer of dark circle and an additional layer of green belt will start to be drawn. One can choose to carry these discs grown from one task into the next task and a red disc indicates how long the user has spent on it in the previous task.

Based on the above reasoning, we expect to observe the following patterns concerning users' navigation strategies:

1. Spatial-semantic models may reduce the time spent on examining a cluster of documents if the spatial-semantic mapping preserves the latent semantic structure.
2. Spatial-semantic models may mislead information foragers to over-estimate the profitability of a cluster of documents if the quality of clustering is low.
3. Once users locate a relevant document in a spatial-semantic model, they tend to switch to local search.
4. If we use the radius of disc to denote the time spent on a document, the majority of large discs should fall in the target area in the thematic spaces.
5. Discs of subsequent tasks are likely to be embedded in discs of preceding tasks.

4. RESULTS

Figure 5 shows user *jbr*'s navigation trail for Task A in the ALCOHOL space, who performed the best in this group. Task A corresponds to the initial overview task in Shneiderman's taxonomy. Users must locate clusters of relevant documents in the map. Subsequent tasks are increasingly focused.

As shown in the trajectory map, user *jbr* started from the node 57 and moved downwards along the branch. Then the trajectory jumped to node 105 and followed the long spine of the graph. Finally, the user reached the area where relevant documents are located. We found an interesting trajectory pattern - once the user locates a relevant document, he tends to explore documents in the immediate neighbouring area. This confirmed our expectation. The frequency of long-range jumps across the space decreased as the user became familiar with the structure of the space. The trajectory eventually settled to some fine-grained local search within an area where the majority relevant documents are placed, and it didn't move away from that area ever since, which was also what we expected.

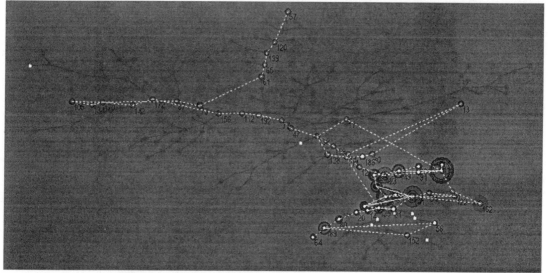

Figure 5: User *jbr*'s trails in searching the ALCOHOL space (Task A).

In the trajectory replay, the time spent on a document is animated as the radius of a green disc growing outward from where the document is located. This design allows us to find out whether the majority of large green discs appear in areas with a high density of relevant documents, and whether areas with a low density of relevant documents will only have sporadic passing navigation trails.

5. DISCUSSION AND CONCLUSION

This is an integrated approach to the study of behavioural semantics. The thematic space was exposed to users for the first time in Task A. Apart from the structural model, no navigation cues were readily available to users. Users must first locate areas in the thematic space where they can find documents relevant to the task. The optimal information foraging theory provides an appropriate description of this type of processes.

Our framework allows us to introduce descriptive and normative modelling techniques. We have conducted an in-depth study of visual information foraging strategies utilising Hidden Markov Models (HMMs) (Chen et al., 2000). The visual inspection of information foraging trails is encouraging. Animated trails and optimal paths generated by

HMMs have revealed many insights into how users were dealing with the tasks and what are the prevailing characteristics and patterns. Replay and animate HMM-paths over actual trails allow us to compare transition patterns in the same context.

In conclusion, many of our expectations have been confirmed in the visualisation and animation of trails of information foragers in thematic spaces. The task we have studies is a global information foraging in nature. The initial result is promising, especially with the facilities to animate user trails within the thematic spaces. Future studies should expand the scope of tasks to cover a fuller range of information foraging activities. Visual-spatial interfaces should be carefully designed for future studies so that fundamental issues can be addressed.

ACKNOWLEDGEMENTS
The work was supported in part by the Council for Museums, Archives and Library in the UK (RE/089).

REFERENCES
Chen, C. (1999). *Information Visualisation and Virtual Environments*. London: Springer-Verlag London.

Chen, C., Cribbin, T., Kuljis, J., & Macredie, R. (2000). Footprints of information foragers: Behaviour semantics of visual exploration. *(Submitted for Publication)*.

Chen, C., & Czerwinski, M. (1998, June 1998). *From latent semantics to spatial hypertext: An integrated approach.* Paper presented at the 9th ACM Conference on Hypertext and Hypermedia (Hypertext '98), Pittsburgh, PA.

Deerwester, S., Dumais, S. T., Landauer, T. K., Furnas, G. W., & Harshman, R. A. (1990). Indexing by Latent Semantic Analysis. *Journal of the American Society for Information Science, 41*(6), 391-407.

Pirolli, P., & Card., S. K. (1995). *Information foraging in information access environments.* Paper presented at the Conference on Human Factors in Computing (CHI '95), Denver, CO.

How rich is the World-Wide-Web?

Dr. Heike Franz

EDS
Bertha-Von-Suttner Str. 12
65189 WIESBADEN / GERMANY
Heike.Franz@eds.com

Abstract

This paper investigates the Internet and its place in IT Research. First information design on the web is discussed to outline the unique properties of the web. Then, established theories such as Media or Information Richness (MRT) (Daft and Lengel, 1984), Social Presence Theory (Short, Williams and Christie, 1976) and Cues-Filtered-Out-Theory (Sproull and Kiesler, 1986, 1991) are deliberated in the context of the Internet. These theories form a good basis for research, but need to be empirically validated. Empirical validation however can be challenging due to the far reaching impacts and the dynamic nature of the Internet. It is suggested to extend the area of research from IT and HCI to the domains sociology, psychology and anthropology.

1. Introduction

This paper investigates the Internet and its place in IT Research. The primary question is: is it possible to use established theories from Computer Mediated Communication (CMC) to explain observations of how people interact with the "net" and how user-friendly they perceive the interaction to be.

First I define and delineate some of the key terms of this paper. Then I describe information design or architecture, how information is organized in more traditional media and how the information architecture differs on the World-Wide-Web (WWW). Hence this part focuses on aspects of Graphical User Interface (GUI), how the interface is intended to work and how it is actually perceived by users.

Having this basis established, Media or Information Richness (Daft and Lengel, 1984), Social Presence Theory (Short, Williams and Christie, 1976) and Cues-Filtered-Out-Theory (Sproull and Kiesler, 1986, 1991) will be reviewed. I shall discuss how these established theories can be interpreted when applying them in the context of the Internet.

At this point in time (fall 2000) I make the assumption that most user access the Internet via PC and laptops, thus I do not yet cover the increasing area of handheld devices and mobile phones (WAP, IMTS)

2. The Internet

The Internet and the World-Wide-Web are omnipresent in our lives. Everybody talks about this, technophobes are familiar with it, but what exactly is it? Internet, also called the Net, is a generic term for over 50,000 different data networks around the world. Developing quietly and unobtrusively for nearly 30 years, the Internet is clearly the most widely discussed IT topic in the latter part of the 1990s and in the beginning of the new millennium. Furthermore, it is rapidly becoming one of the most important information technologies (Schneiderman, 1997), although having no particular owner. As such it is more a concept rather than a piece of property. Everybody can contribute and every day new knowledge is added, by individuals as well as by organizations, for private reasons as well as for business and commercial purpose. These contributions form an ever-growing collection of hyperlinked documents, presenting the information in text, graphical, animation and other formats. The accumulation of all this information is what we call the World-Wide-Web, the Web or WWW. Thus, the Internet can be seen as the network that supports communication, whereas the WWW represents the information created by the public and accessible to the public via exactly this network. Often the terms Internet and World-Wide-Web are used concurrently, as will be done here in this paper.

3. Information Design on the Web

This section discusses structure, format and complexity of information on the Internet.

Chapter 3.1 "Information Structure" focuses on information hierarchy and covers hypertext, navigation and links.

"Information Format" then investigates text, graphics, colour, font, and multimedia aspects.

The section "Information Complexity" brings it together: here I discuss the impact of information format and structure on cognitive information processing or in other words: can we handle all the information we were asking for.

3.1. Information Structure

When discussing information structure I like to use the term "Information architecture". Architecture usually is related to a physical structure, in a building, for example. This physical structure is visible and usually we are familiar with this type of structure. Buildings have been around for a while. Web site composition is distinct: information is organized differently, it is intangible and often not in line with the user's mental model. This becomes more noticeable when we compare the structures to more traditional media, such as books, journals or any other printed tangible media, but also e-mail. Electronic information systems in general and e-mail in particular have been more widely used in organizations since the late 1980s, thus it has been around for a while and has lost the touch of the unfamiliar for most of the working population. The web is more complex and less harmonious.

Königer and Janowitz (1995) analyzed this topic of information structure in depth and propose to classify information along four dimensions. These are:

Selection: to decide which type and amount of information is needed and required and select it accordingly. Management Information Systems, for example, have to select the type of information needed by managers to make decisions for the organization. Newspages are intended to inform a broad audience about the latest news, news not yet in today's newspaper.

On the Web it is no longer necessary to restrict information or to select important information. The Web contains tens of millions of documents and the number is growing exponentially. Consequently, a particular topic for a Web page certainly has been covered, discussed, been analyzed, referenced or mentioned elsewhere. Thus, we can make use of hyperlinks and connect the site to chunks of information that are related, relevant or of interest for this particular page.

The potential implications are twofold: the information provider can offer a far more varied picture and discuss a topic from various perspectives. It is up to the user of this information or the visitor of the page whether to make use of all the details. The negative side is that unlimited information can possibly be linked. Often hyperlinks are created haphazardly and randomly and this is the first step for the visitor to get lost.

Time: to identify and elucidate the life cycle of information. An annual report indicates by its name to which time period it relates. Without this information it is difficult to assess for the user whether or not it is relevant. One of the benefits of the Internet as an information source is its actuality and relevance. Some pages update their news hourly; stock rates have the latest information. Well-designed pages make this clear, but reality shows that this often is not the case. Sometimes flashing signs and banners signal changes, but these changes can be months old.

Hierarchy: to establish order in terms of detail. On the Internet hierarchy is of pivotal importance. In more traditional media, hierarchy is established by using headings, different font types, paragraphs or other design tools. The user thus knows whether he is looking at a summary point or details. The situation is different on-line. In genuine textual format, for example in e-mail applications using plain text, font type, colour and other formatting tools are restricted. Thus the writer does not have many options to express the hierarchy of his information. If he or she, at the same time, has poor writing skills, this can exacerbate the problem. For website design we have more tools to determine hierarchy. Visually, information hierarchy can be expressed by the proper use of colour, white space, scale and contrast (Pirouz, 1997, p. 42). However, empirical research results to date are conflicting. White space for example, recommended by various authors as a means of structuring information, was found to hinder finding information. Users also rated sites with more white space lower than cluttered pages (Spool et al. 1999). Yet, empirical research in this area is still in its infancy and we have to interpret the results with care. To establish information hierarchy in text-only documents and make a Web site user-friendly, document designers and/or webmasters have some more work to do when creating and designing this particular page. A bad option is to simply convert an existing text document into HTML (hypertext mark-up language) and upload the document, although the metaphoric representation of traditional media can be a good starting point. A further point to consider is how the information is used. Will it be viewed on-line? Will the user need a printed hardcopy? Depending on these and other issues, information must be organized into subdocuments, categories or manageable chunks. The categories have to be meaningful and should be in logical sequence (Schriver, 1997). Ideally, each chunk of information or node (see next section) should be self-contained. Finally, these self-contained chunks are linked via hyperlinks. When deciding how much information to put on a web page, designers have two options: one is to create lots of short pages with few links and no need to scroll down. The drawback? This strategy forces users to visit more pages. The second option is to create fewer and longer pages with more links that make users scroll to see what lies "below the fold". Although many popular books recommend on the first strategy, Spool et al. (1999) found out that this is not necessarily the case. It is however essential to make it evident and clear to the user which of both strategies has been selected, else the user might leave the site before coming close to the information he is looking for. Web surfers are typically impatient.

Sequence brings information in a certain order along one defined continuum, for example alphabetical order or decimal numbering system. Sequence of information has a new meaning in the Web context likewise: one of the characteristics of WWW is the use of hyperlinks and hypertext to connect nodes and subdocuments. Nodes or subdocuments can contain a combination of text, graphics and/or other form of data. Nodes are connected to each other by electronic cross-references, called links. A link anchor (< biblio >) presents these links. Hypertext is a relatively new and highly non-linear way of structuring information (Rosenfeld and Morville, 1998), thus contradicts the concept of sequence. Hypertext encourages readers to move from topic to topic rapidly and nonsequentially (Berk and Devlin, 1991). This is in stark contrast to traditional engagement with written prose or information in general: people engage with prose linearly, one word at a time, in sequences that make clauses (Schriver, 1997). Books are read from page one, most of the time in a sequential order. Thus, hypertext readers must assume a much more active role than readers of printed text (Berk and Devlin, 1991). Physical attributes such as the length of a text vanishes when the information is concealed behind a screen. The view is limited and restricted by the screen size; there is always the feeling of not getting the whole picture. There might be more outside the immediate screen display or "below the fold", but it is intangible and invisible. The reader can not grasp all elements and the on-line display can be different from the mental model he applies for this type of information. The result of the non-linear structure is the "Lost in Cyberspace" phenomenon. It occurs when a reader has been following a long chain of links and suddenly finds that he has lost his chain of thought. Certainly there are GUI design options to help or hinder the users finding their way.

3.2. Information Format

In addition to the more complex information structure, Web sites make extensive use of eye-catching pictures, lively images and elegant styles. Technology makes it possible to create fancy design and as an old saying goes: a picture says more than a thousand words. Presenting market shares for a product in a pie chart, for example, requires less cognitive effort to process the information than presenting it in a table of numbers. With the capability to combine information, words, photos, sounds, video snippets and graph, the Web designer is enabled to publish not just plain text and numbers, but beautifully illustrated page layouts. Well applied and considered carefully, this takes care of individual preferences in terms of presentation format. If applied inappropriately, this variety fails its purpose: Schriver (1997) made the observation that people don't accept pretty pictures and numerous links, if the associated prose is low in information. Visitors to the site are disappointed when they do not find the details of information they are expecting. The positive side in regard to variety of information format is the diversity of format choices. If designed thoughtfully and with a particular user group in mind, it is possible to create pages that make it easier to absorb and process the information.

At the same time this might be a risk: people can be overwhelmed, when they see the same piece of information in different formats. The same link can, for example, be displayed as an icon or as a hypertext-line and it might be confusing if this is not clear to the user (Spool, 1999). Hence, a variety in formats can confound rather than increase understanding; a thought that was brought up by Engdahl (1993) even before the Internet became mainstream. This in turn leads to the issue of information complexity.

3.3. Information Complexity

Information complexity is one potential variable that impacts the interpretation of data. Although a preferred variable in empirical research, the definition of complexity is not consistent, omitted or very general as in Engdahls (1993) definition ("quantity of cues"). Information complexity is found to be an important concept with multiple dimensions and correlations to the variables Information Overload, Media Richness Theory and Social Presence Theory (Franz, 1999). These constructs are relevant in the context of this paper likewise, but how does the medium "Web" impact complexity? The points discussed so far, information structure, diversity of format choices and hypertext certainly can contribute to complexity. A further argument to consider is non-verbal cues, which are present in face-to-face communication. In text-only CMC it can be argued that non-verbal cues are filtered, a topic that I will elaborate in a further section. As such, information complexity is reduced, because the likelihood of conflicting cues is reduced. Stohl and Redding (1987) interpret the usage of CMC as a sampling method or coping mechanism to deal with overload. They argue that electronic messaging is low in "Social Presence" (as defined by Short, Williams and Christie, 1976) and richness. The lack of social presence and richness reduces the number of social cues and multiple levels of meaning to be processed. The argument can be challenged. If the context of information or communication is omitted, interpretation of information is rather more difficult. This becomes notably important in today's multi-national or global marketplace, when members of different ethnic cultures communicate. Is it easier for an American to communicate with a British person when they speak face-to-face or when they communicate via E-Mail? Applying these findings to the Web, various hypotheses are possible. Technology and options in terms of information format can compensate for non-verbal cues filtered out, provide context to the user and thus help process information. The alternative hypothesis is that, as a result of the same features, information becomes more complex, thus the user feels overwhelmed, overloaded and lost in Cyberspace. The integration of information, for example, might be lost if the user followed numerous links and no longer sees the context of the node he is currently visiting.

These hypotheses will be further elaborated in the context of information or media richness, which follows next.

4. Information Richness and the Web

For the unprepared reader this heading might look funny, since the web is information and as such it is rich in information, right? Here in this section I step back some years and have a closer look at some of the most influential, dominating and most discussed theories in the domain of CMC. For a long time the theory had its proponents and it opponents, often the discussions were getting rather emotional. Still, this theory contributed to a better understanding and in the context of the Internet some arguments and hypotheses have to be re-evaluated. In this chapter I discuss Daft and Lengel's (1984) Media Richness Theory, Social Presence Theory (Short, Williams and Christie, 1976) as well as Sproull and Kiesler's Cue-Filtered-Out Concept (1991). One can see that these theories had their heyday in the pre-Internet area, yet it is attempted to use them as a framework. All three theories can be considered to be closely linked and are often discussed together.

4.1. Media or Information Richness Theory

One of the most influential and for years dominating theories in this area is the work by Daft and Lengel (1984) on media or information richness.[1] As a criterion to define richness of a medium, Daft and Lengel (1984) use the term information capacity. Information capacity (of the medium) is determined by feedback immediacy (interactivity), cue transmission capacity (words, voice inflection, body language), ability to convey personal feelings and emotions in a message and language variety. Daft and

Lengel (1984) suggest that media can be ranked according to their richness and they define this property as the ability of information to change understanding within a certain time interval. Face-to-face is considered to be the richest medium, and written communication the leanest. MRT is a strong and established foundation for discussion and conclusions can be drawn when applying this theory to the Web. For example, e-mail or text-only CMC is considered to be rather lean in a continuum of various media. When compared to e-mail the Web is without doubt much richer: interactivity or immediacy of feedback is higher than in e-mail, although this can be an automated process. An example is the automatic confirmation sent out by amazon.com, after a book order has been placed.

Cue transmission capacity is higher than in text-only CMC likewise. As was discussed earlier, numerous presentation formats are possible here and are indeed the fundamental characteristics on the Web. Photos can be scanned and displayed, video clips can be reviewed for product information. Graphical features such as banners can be added to introduce motion, animation, feelings and emotions. Theoretically, even body language and voice can be transmitted, when using web cam tools. The ability to convey personal feelings and emotions is possible, albeit not at the same extent as in face-to-face communication. Often it is rather mechanical, a programmed reaction generated by the application. Does this trick the user and does he perceive a real person to be present?

4.2. *Social Presence Theory*

In their Social Presence Theory, Short, Williams and Christie (1976) propose that those media that transmit greater amounts of varied information simultaneously (including non-verbal information) possess greater Social Presence. They are friendlier, more emotional, and more personal and make users to perceive others as being psychologically present.

The result of low social presence can be a problem in interpreting information, a phenomenon related to non-verbal cues filtered out in electronic media. I elaborate Social Presence a bit further to make evident that even with the communication partner perceived as being "present", it is different. People try to make sense of a phenomenon based on previous experience, and this forms a mental model of the transaction. They might adapt behaviour from interactions in other modes to the Net interaction. In a human-to-human sales situations, for example, the shopper is allowed to browse, examine items, and check prices without giving any personal information to the store's employees. Only after the shopper has decided to purchase do they give the salesperson their name, address, and credit card number. One web site required users to fill out a form with personal data before giving them access to the part of the site containing service and price information. This tactic proved to be a deal-breaker, driving users away from the site (Spool, 1999). Is there a lack of Social Presence and does the user have a problem giving information to an invisible person? Certainly his mental model of a physical sales process plays an important role in this observation.

4.3. *Cues-Filtered-Out Theory*

When non-verbal information and hence social information is filtered or reduced, certain effects are hypothesized. Less social information can equalize or obscure status differences (Sproull and Kiesler, 1991) and thus can make users evaluate and interpret information based on facts and not on where it is coming from. On the negative side, and this might depend on the users' attitude, frame of reference and prior experience, the interpretation of the information can be constrained. This constraint is a result of the missing context and source of information (Walther, 1992). How reliable, for example, is medical advice on the Web? To a non-expert the advice might sound very professional, but what is the formal acceptance of the author of this information? Engdahl (1993) offers a different interpretation of non-verbal cues in the communication or information retrieval process. He argues that cues must be congruent or they confound rather than increase understanding. Typically, when more varied types of simultaneous information cues are transmitted, the number of cues is higher. A higher number of cues can increase the likelihood of incongruity, can make the situation more complex and result in difficulties in understanding of the information transmitted (Engdahl, 1993).

In summary, the first school of thought represents the cuelessness-theories and proposes obstacles in information interpretation due to a lack of cues. It can be assumed that this assumption is no longer valid for the Web; it was argued that the increased capabilities to display information act as a coping mechanism for non-verbal cues filtered out. The second group suggests the same as a result of too many and possibly conflicting cues, which increase the complexity of information. This observation might be valid in the context of Web applications.

Traditionally MRT, Social Presence and Cues-Filtered-Out were considered to be strongly correlated. If a medium is high in richness it is said to be high in social presence, because fewer cues are filtered out.

5. Conclusion

Summarizing it can be said that traditional established theories are applicable in the context of the Internet as a framework, yet there are some weaknesses:

It is difficult to test them empirically, because Internet users are not a homogenous group and demographic differences can have a significant impact of a typical transaction. Further the possible applications within the Internet are too numerous.

It is even questionable whether to see the net as a communication platform or rather a completely new model of doing business and leading our lives. This in turn would suggest extending the area of research from IT and HCI to the domains sociology, psychology or even anthropology.

References

[1] Berk, E. & Devlin, J. 1991. *"What Is Hypertext"*. Berk, E. & Devlin, J. (Ed). "Hypertext/Hypermedia Handbook". New York, NY 10020: McGraw Hill PublishingCompany, Inc.

[2] Daft, R.L. & Lengel, R.H. 1984. "Information richness: A new approach to managerial behavior and organizational design". *Research in Organizational Behavior*, vol. 6, p.191 - 233.

[3] Franz, H. 1999. "The Impact of CMC on Information Overload in Distributed Teams". *Proceedings of the Thirty-First Annual Hawaii International Conference on Systems Sciences.*

[4] Engdahl, R.A. (1993). "COMPLEXITY AND RICHNESS: Inversely Related Information Structure Variables With Implications for Improved Communications and Research on Problem Formulation". *Organizational Development Journal*, Vol. 11, No. 3, pp.19 -

[5] Königer, P. & Janowitz, K. 1995. "Drowning in Information, but Thirsty for Knowledge". *International Journal of Information Management*, vol. 15, no. 1, p.5 - 16.

[6] Pirouz, R., developed with Weinman, L. 1997. *"Click Here"*. Indianapolis, IN 46290: New Riders Publishing.

[7] Rosenfeld, L. & Morville, P. 1998. *"Information Architecture for the World Wide Web"*. Sebastopol, CA: O'Reilly & Associates.

[8] Schriver, K.A. 1997. *"Dynamics in Document Design - Creating Texts for Readers"*. New York: Wiley Computer Publishing - John Wiley & Sons, Inc.

[9] Schneiderman, B. 1998. *"Designing the User Interface - Strategies for Effective Human-Computer Interaction"*. Reading, Massachusetts: Addison-Wesley Longman Inc.

[10] Short, J.; Williams, E. & Christie, B. 1976. *"The Social Psychology of Telecommunications"*. London: Wiley.

[11] Sproull, L. & Kiesler, S. 1986. "REDUCING SOCIAL CONTEXT CUES: ELECTRONIC MAIL IN ORGANIZATIONAL COMMUNICATION". *Management Science*, vol. 32, no. 11, p.1492 - 1512.

[12] Spool, J.M.; Scanlon, T.; Schroeder, W.; Snyder, C. & DeAngelo, T. 1999. "Web Site Usability - A Designer's Guide". San Francisco, CA., Morgan Kaufmann Publishers, Inc.

[13] Sproull, L. & Kiesler, S. 1991. *"Connections - New ways of working in the networked organization"*. Cambridge, Massachusetts: The MIT Press.

[14] Stohl, C. & Redding, C.W. 1987. *"Messages and Message Exchange Processes"*. Jablin, Frederick M., Putnam, Linda L., Roberts, Karlene H. & Porter, Lyman W. (Ed). Handbook of Organizational Communication - An Interdisciplinary Perspective. Newbury Park, Beverly Hills, London, New Delhi: SAGE Publications.

1109

Panchronic for Active Media

Francisco V. Cipolla-Ficarra

Human-Computer Interaction Lab
F&F Multimedia Communic@tions Corp.
Cas. Pos. 60 - 24100 Bergamo, ITALY
Email: f_ficarra@libero.it

Abstract

We introduce "panchronic" in order to increase the communication quality in hypermedia/multimedia system. Our goal contained three objectives. The first one, to introduce a new attribute or criteria of the quality: panchronic. Panchronic is a new metric for MECEM (Metrics for the Communication Evaluation in Multimedia). The second, to present an index similar to a chart with all elements and criterium that we considered necessary into a first stage in order to evaluate the panchronic of hypermedia system "off-line" and "on-line" –this chart is constantly updated. The third, to present the first results of the evaluation with 20 commercial CD-ROMs.

1. Introduction

Based on digital technology, multimedia/hypermedia "on-line" and "off-line" integrates static/passive media (i.e. formatted date, text strings, images, graphics, maps, etc.) and active/dynamic media (sound, video and animation). We use active media's to create more attractive, useful, and efficient multimedia interfaces. For example, animation can help cut through the complexity of an interface. Furthermore, show what can or can not be done, guide a user as to what to do or not do, review what has been done. Animation can help us review the past, understand the present, and describe the future. In brief, active media's helps navigate around the multimedia/hypermedia systems [1].

From its origin, semiotics has always been related to advances in technology in order to analyze their contents [2-5]. In other words, decrease the noises into the process of communication between the multimedia systems and the user.

The analyse of a layout, a dynamics, a content, a structure, facilitates the acceptance of multimedia/hypermedia systems [5]. Thus, a primary one-to-one relationship exists between content and structure. However, between the four basic components there is a bi-directional relationship and each of their constituent parts often function as small units or subclasses which are themselves interrelated. We are currently working on four new sub-dimensions inter-related with the previous ones. These are: panchronic ("*pan*" prefix derived from Greek for all or every and "*chronos*" time) and orientation, in other words, access times to information nodes, resources of acceleration in the interaction, synchronization, and principal elements for basic and efficient navigation (inter-related with the dynamics aspect); transformation (inter-related with the structure aspect); isotopy lines (inter-related with the content aspect); and realism, a tendency toward the simulation of the reality (inter-related with the layout aspect). Graphically, this could be represented as:

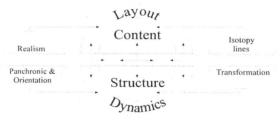

Fig. 1: Panchronic area

The evaluation panchronic should be reconsidered form the users viewpoint, so that they can be made more efficient [6-8]. We assume that the most important point in creating easily navigable multimedia system, because, panchronic is the synchronization between sound (music, voice, etc.), video and animation.

2. MECEM and Panchronic

Metrics for the Communication Evaluation in Multimedia (MECEM) introduces a new vision of the content evaluation for multimedia system: communication. A user interface can be defined as the connection between a person and the screen display of a computer system [9]. If we try evaluate the interface aspect of the screen, three perspectives on screen information can be distinguished. The first perspective is concerned with universality. The second deals with the overall framework of the information on the screen, which is called inference. The third perspective focuses upon empathy. The interest reader can refer to the bibliography in [10].

The attributes presented are compatible and complementary with other evaluation method: MEHEM, Methodology for Heuristic Evaluation in Multimedia [11] [12]. In the empirical evaluation, detected principal failings are registered in the form of a list. This list increases the good organization of the empirical evaluation in terms of results and reduced costs. This methodology, attributes and interactive systems analyst are very important and essential for communication "off-line" and "on-line", principally Internet and intranet.

Consistent use of communication in multimedia will not emerge until the field is more mature. The maturation process requires consolidation of existing development methods and tool integration mechanisms, quantitative data on best practices, and widely accepted agreement on the intellectual content of the field of software engineering, supported by both academic and professional programs for practicing software developers. For example, in heuristic analyses of hypermedia systems it is habitual to refer to studies as being carried out by groups of people, yet no mention is ever made of the problems that arise when working in a team and how they can be solved. These problems, which include the leadership, roles and make-up of the members of the team, the number of members and the formation of subgroups, task delimitation, communication networks (i.e., star, chain or circle) and environmental conditions, to name but a few, can ultimately affect the results of the research.

Just as, for example, computer science has the systems analyst (whether functional of managerial), who does not need other analysts to carry out his work, it would be advantageous to introduce the figure of the interactive systems analyst, whose knowledge and/or experience come from a background of social sciences and formal sciences, but not mathematical, physical, chemical, industrial engineers, etc. (they do not have a formation or knowledge in social communication). We do not need mathematicians or astronomers to design the structure of an E-commerce multimedia system, for example.

Such an analyst would be able to carry out the analysis of these systems and reach reliable results unaided, without the need for a test group. Group evaluation at a later stage might be useful to reinforce the technique employed. Today, we need multimedia/hypermedia systems analyst and not computer specialists oriented to the communication (the meaning is very different) [13].

Panchronic is a new attribute for quality in hypermedia system and we will elaborate a metric for the evaluation of the interactive communication. Evidently, communication is not synonym of the usability. Today may be the usability has a second place in the human-computer interaction, for example.

Usabilty metrics are expensive and are a poor use of typically scarce usability resources [14]. It is common for usability to double as a result of a redesign. In fact, whenever you redesign a website that was created without a systematic usability process, you can often improve measured usability even more. However, the first numbers you should focus on are those in your budget. Only when those figures are sufficiently large should you make metrics a part of your usability improvement strategy [14].

MECEM has proved the reliability of the obtained results and has cut down costs. The costs cut is due to the fact that it does not require:

- A group of evaluators or a sample of users who interact with the system at the moment of the evaluation.
- A laboratory to carry out the evaluation.
- Equipment and maintenance of specialized devices.
- Previous training of the users and eventual aids of the evaluators.

The profession of evaluator or heuristic analyzer has been included in the worldwide spread multimedia systems. It is a new profession, which requires specialists with knowledge and experiences in the field of the multimedia/hypemedia systems. It is located on the borderline between formal science and factic science [15].

The intersection between MECEM and MEHEM offers a useful tool to facilite the utilization of information technology for the design and development of hypmerdia/multimedia systems for the support of interactive communication.

3. Panchronic Evaluation

We have elaborated an index similar to a chart with all elements and criterium that we considered necessary into a first stage in order to evaluate the panchronic of hypermedia system "off-line" and "on-line" –this chart is constantly updated.

ELEMENTS AND/OR CRITERIA CONSIDERED NECESSARY FOR PANCHRONIC EVALUATION	CODE
• Synchronization between: audio and images.	SYI
• Synchronization between: audio and texts.	SYT
• Synchronization between: images and next link/frame.	SIN
• Synchronization between: audio and next link/frame.	SAN
• Stop audio and go to next link/frame.	SPO
• Stop animation and go to next link/frame.	SPN
• Stop video and go to next link/frame.	SPV
• Play animation and go to next animation (two or more animations simultaneously).	PLA
• Play video and go to next video (two or more animations simultaneously).	PLV
• Play sound and go to next sound (two or more animations simultaneously).	PLS

We present a little random set of the CD-ROMs (plays and educational titles) for the panchronic evaluation – alphabetical order. Besides, we can see the "binary" results, that is to say, the multimedia system has or has not panchronism (yes = YP, not = NP, and more or less = ML).

TITLE	RESULT	TITLE	RESULT
1. A bug's life	NP	11. MovieTalk –Il tenente Colombo	ML
2. Art vision –Arte visión	ML	12. NHL 2000	YP
3. Chicken Run	ML	13. Nox	YP
4. French language	YP	14. Pingu	YP
5. Galleria d'arte: Renoir	YP	15. The Feeble files	NP
6. Gli etruschi	NP	16. The Secret of Monkey Island	ML
7. Great atlas of the world	ML	17. The Simpsons Virtual Springfield	NP
8. I learn how to read	YP	18. Tomb Raider II	ML
9. Lemmings revolution	YP	19. Top of the pops	YP
10. Mission Van Gogh	NP	20. Zipi and Zape	YP

The marketing of these commercial multimedia applications "guarantees" hours of entertainment, education and information. However, the content, the dynamism and structure of these CD-ROMs are found to be incomplete when we apply MEHEM and MECEM. For example:

- Textual reading. In the reading of the text, pauses and diction should be taken into account. Important mistakes have been observed in the correlation between what is heard and what is written. Furthermore, there are serious errors in syntax and spelling when the user changes languages (43 % of cases).

- In the synchronisation between audio and video in this type of CD-ROM it is "normal" to find numerous mistakes.

At present, several firms dedicated to the international production and distribution of multimedia systems for young people in the design stage use children or young people (relatives of the designer) to know their opinions. These young users are neither chosen at random nor inserted in evaluation laboratories, thereby damaging the objectivity of the results obtained, since the basic rules of heuristic evaluation are not respected. Finally, the CD-ROM "Pingu and their friends" is a good example for panchronic evaluation (see Aneex 1).

4. Conclusion

The main advantage of the MECEM is that the heuristic evaluator can evaluate the system. Later on, users can check the results obtained in the evaluation. The verification with the proposed method does not increase the quality of the results. The purpose of the verification is to confirm the detected failings. The verification process consist in giving out a set of tasks to each of the user starting from the failings that the heuristic evaluator has detected. This verification must be carried out in a laboratory. However, the cost of the verification of failings can be high because of the infrastructure and staff that are necessary to carry out experiments with users.

When one considers the user's age it is observed that an infant user needs more interaction than an adult one. A multimedia system which content may be a scientific museum requires a longer time for observation and reading, for which reason the user's navigation activity and panchronic decreases. In the multimedia systems designed for entertainment, the adult user prefers a content with a single plot, while the younger ones need more than one plot. The plot refers to a story or tale which content is divided into three basic parts, such as introduction, development and ending. Simultaneously, the synchronization between audio, text and images are essentials for quality of the interaction.

This new attribute panchronic is very important and essential for multimedia communication "off-line" and "on-line". Obviously, we need distinghuish between usability and communication. Today, the communication includes the usability.

Acknowledgments

The author would like to thank Spanish, Portuguese and Italian distributors and producers of CD-ROMs: Contacto Visual (Viana do Castelo –Portugal), DeAgostini (Milano –Italy), Electronic Arts (Madrid –Spain), Microsoft (Madrid –Spain), Nex Media (Badajoz –Spain), RCS (Milano –Italy), UbiSoft (Milano / Sant Cugat –Italy and Spain), and Zeta Multimedia (Barcelona –Spain).

References

[1] Lennon, J. "Hypermedia Systems and Applications: World Wide Web and Beyond". Springer-Verlag (Berlin, 1997).

[2] Eco, U. "Semiotica". Bompiadi (Milano, 1994).

[3] Nöth, W. "Handbook of Semiotics". Indiana University Press (Bloomington, 1995).

[4] Jorna, R., Van Heusden, B. "Semiotics of the user interface". Semiotica, 109, 3-4 (1996), 237-250.

[5] Cipolla-Ficarra, F. "A Method that Improves the Design of Hypermedia: Semiotics". International Workshop on Hypermedia Design (IWHD'95), Springer-Verlag (Montpellier, 1995), pp. 249-250.

[6] Jeong, T. et al. "A Pre-scheduling Mechanism for Multimedia Presentation Synchronization". Proc. IEEE Multimedia Systems '97 (Ottawa, 1997), pp. 379-386.

[7] Steinmetz, R. "Human Perception of Jitter and Media Synchronization Model for Multimedia System". IEEE Journal on Select Areas in Communications. Vol. 14, 1 (January 1996).

[8] Herng-Yow, C., et al. "An RTP-based Synchronized Hypermedia Live Lecture System for Distance Education". Proc. ACM Multimedia '99 (Orlando, 1999), pp. 91-99.

[9] Shneiderman, B. "Designing the User Interface". Addison-Wesley (New York, 1997).

[10] Cipolla-Ficarra, F. "MECEM: Metrics for the Communications Evaluation in Multimedia". International Conference on Information Systems Analysis and Synthesis – ISAS'00 (Orlando, 2000), pp. 23-31.

[11] Cipolla-Ficarra, F. "MEHEM: A Methodology for Heuristic Evaluation in Multimedia". Proceed. Sixth International Conference on Distributed Multimedia Systems (DMS'99), IFIP, Elsevier (Aizu, 1999).

[12] Cipolla-Ficarra, F. "MEHEM for the Representative Evaluation of the Quality in Multimedia System". Proc. International Conference on Information Systems Analysis and Synthesis - ISAS '99 (Orlando, 1999), pp. 31-36.

[13] Cipolla-Ficarra, F. "Multimedia Evaluation: Isomorphism and Narration". Proc. International Conference on Information Systems Analysis and Synthesis – ISAS '98 (Orlando, 1998).

[14] Nielsen, J. "Usability Metrics". www.useit.com (January 2001).

[15] Cipolla-Ficarra, F. "Evaluation of Multimedia Components". Proc. IEEE Multimedia Systems '97 (Ottawa, 1997), pp. 557-564.

[16] BBC Multimedia and Zeta Multimedia CD-ROM. "Pingu y sus amigos" –Pingu and their friends (1999).

Annex 1: Panchronic for Active Media

This is a good example for panchronic evaluation.

Promoting On-Line Customer Confidence Through Page Design

Raquel Montania and Marc L. Resnick, Ph.D

Industrial and Systems Engineering
Florida International University
Miami, Florida 33174

For consumers to shift purchases on-line, it is imperative that the benefits significantly outweigh potential risks. There is evidence that customer service, product quality, and privacy protection are the biggest customer concerns on line. The purpose of this research was to find out if design elements of the user interface of an e-commerce site affect these concerns. The specific objective was to determine if the presence and prominence of customer service links, reputation managers, and privacy policy links on the web site of an unknown Internet retailer are believed by customers, generate positive customer perceptions, and improve the likelihood of purchase from the site. Sixty-four experienced Internet users were recruited for this study, which consisted of an on-line shopping experience in which the user had to decide from which of four web sites of unknown Internet flower retailers to purchase a gift for a sick friend. Participants were more likely to select the sites that had highly prominent customer service links, privacy policy statement links and reputation managers. Expectations of customer service, product quality and privacy protection were higher for these sites.

1.0 INTRODUCTION

1.1 Customer Relationship Management

Customer Relationship Management (CRM) is the use of information to better satisfy customers' needs and offer personalized service. CRM components include customer service, relationship managers, automated sales and marketing, and other enhancement services. The end result is a customer that feels confident that his/her needs in terms of quality, reliability, customer service, privacy, trust and others will be met.

Customer service is an essential ingredient of commercial transactions. A study by Servicesoft Technologies (1999) revealed that 87% of online shoppers will abandon a merchant's website and click to a competitor's site if they experience poor customer service. Conversely, 79% said they have increased their patronage and spending at a website when customer service was favorable (E-marketer, 2000). Trust in customer service is particularly important for e-commerce because of the risk customers are exposed to when they have no physical evidence of vendors' presence or quality (Gefen, 2000). Selnes and Gonhaug (2000) found that vendor reliability has a strong effect on satisfaction and subsequently the buyer's desire to continue the relationship with and talk favorably about the company. Lack of reliability creates negative emotions and negative affect from the customer toward the company and subsequently reduces the motivation to be loyal.

Reputation managers are systems that provide customer or independent feedback about products or services (Nielsen, 1999b). The feedback can be published as text (testimonials) or as ratings (4 out of 5 stars). It is also possible for companies to provide their own ratings of products and services, however the credibility of these ratings has been questioned (Montania and Resnick, 2001). Reputation managers have become popular in e-commerce because consumers can't touch the merchandise, they can only see a representation (Sisson, 1999).

A survey by the Georgia Tech Research Corp (1998) found that the protection of personal information was one of Internet users chief concerns. However, CRM is enhanced when the Internet retailer can create a detailed picture of each customer based on personal information and navigation behavior (Walsh and Godfrey, 2000). A balance must be established between vendors' collection of personal information and their use of it. This balance is often outlined in the site's privacy policy.

1.2 CRM and the Customer Interface

It takes time for customers to appreciate the CRM initiatives of an Internet retailer (Nielsen, 1999a). When a potential customer visits a site for the first time, his/her initial perceptions of that company are based on features that are visible on the company's main pages. These initial impressions drive the development of that customer's relationship with the company. A potentially critical factor is the visibility of reliable customer support. Without the assurance of quality customer service, consumers may not develop sufficient trust. However, the ability of a company to promote perceptions of customer support through the design of its site is unknown. Ratings and testimonials seem to affect customers indirectly by suggesting that other customers have successfully interacted with the company. The presence of this information may lead to increased perceptions of trust and reliability. Knowledge of what kinds of personal information a company collects and how it is used is generally contained in a privacy policy. The visibility of this policy may determine how much confidence customers have in its validity.

1.3 Objective

The objective of this study was to determine how visible customer support, reputation management, and privacy policy features on the web site of an unknown Internet retailer affect perceptions of trust, reliability, and privacy protection. The presence of these CRM features was expected to enhance these perceptions and thus lead to increased intentions to purchase products from the company.

2.0 METHODS

2.1 Participants

Sixty-four participants were recruited from the Miami metropolitan area to participate in this study. There were twenty-seven females and thirty-four males. Ages ranged from eighteen to fifty-three with a mean of twenty-seven. Participants were required to have at least moderate experience with the Internet and seventy-eight percent reported on-line shopping experience.

2.2 Materials

One hundred twenty-eight fictitious e-commerce stores were created. Each store was represented by the front page of its Get Well department and included a set of floral bouquets and related gifts. Each participant interacted with four sites. The only material differences among the sites corresponded to the presence and prominence of links to various channels of customer service (CS), ratings and testimonials (RT) attributed to previous customers, and a privacy policy (PP). Superficial differences in appearance such as color scheme and company names were created to insure that they were perceived as different companies. However, key information such as price levels and the number and variety of products available were consistently maintained. A full factorial combination of all variables was used to create eight basic models. Black and white screen shots of two contrasting models are shown in Figures 1 and 2 respectively. Each model would have appeared in one of four different color schemes depending on the participant.

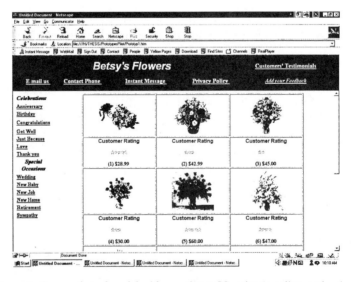

Figure 1. Screen shot of model with prominent CS, privacy policy, and ratings

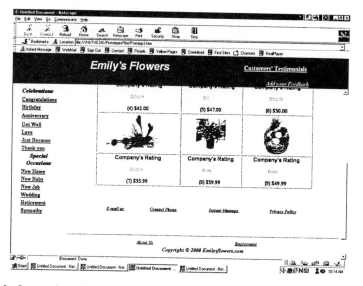

Figure 2. Screen shot of model with low prominent CS and privacy policy and no ratings.

A scenario was described to each participant in which he/she had just heard that a friend was in the hospital. The task was to purchase a floral gift bouquet to be delivered to the hospital that day. The participant was informed that an Internet search for local flower delivery companies that offered same day delivery had retrieved four sites. The **Get Well** pages of each site were opened but minimized and visible only through the toolbar at the bottom of the Windows interface. The order from right to left in which the company icons were displayed on the toolbar was randomized for each participant. Participants were instructed to view each of the company main pages in any order for as long as they wanted. Their task was to select the company from which they would purchase the gift bouquet.

2.3 Procedure

On arrival, participants were informed that the performance of new web sites was being tested to determine whether they met the needs of users such as themselves. The evaluator prepared the computer before each task and provided instructions. Participants were told that no external assistance would be provided. After they selected the company from which they would purchase, they responded to a brief questionnaire in which they rated their perceptions of the customer service, products, and protection of personal information that the company they chose would provide using a 5-point Likert scale where 1 indicated complete dissatisfaction and 5 indicated complete satisfaction. Then they were asked to review the other sites and rate them using these same parameters. Finally they were asked about the criteria they used to decide and some general questions about customer service and privacy policies.

3.0 RESULTS

In order to verify that none of the superficial design details significantly affected the participants' responses, the effects of site name, color scheme and product list were measured using Pearson Chi-Squared tests for each one. None of these had a significant effect on the participants' site selection. Due to the cautionary results of a pilot test, two-way ANOVAs were used to verify that there was no interaction between color and any of the study's dependent variables. These tests found no significance for color or any interactions with color.

3.1 Preference and site choice

A three-way ANOVA was used to evaluate the effects of each of the three design parameters on participants' preference ratings for each site. A significant effect was identified for all three variables (CS: $F=7.845$, $p<0.05$; RT: $F=21.497$, $p<0.05$; PP: $F=5.215$, $p<0.05$) but there were no significant interactions. Binomial tests

1117

were used to determine the effects of these preferences on site choice. All three binomial tests were significant (all p<0.005). Figure 3 shows the number of participants who chose to purchase from each of the site models.

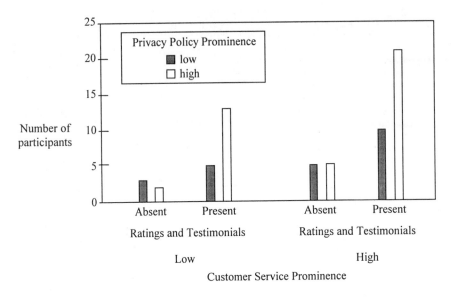

Figure 3. Number of participants choosing each model

3.2 Ratings of Customer Service, Product Quality and Privacy Protection

Results from additional three-way ANOVAs indicated that the expectations of service were also affected by each of the design parameters. Expectations of customer service were significantly increased by the prominence of customer service links ($F=28.511$, $p<0.05$). Likert scale ratings of customer service expectations increased from 3.0 to 3.8 when links to email, telephone and Instant Messaging contacts with customer service were made prominent. Product quality perceptions were significantly increased by the presence of customer ratings and testimonials ($F=45.019$, $p<0.05$). Likert scale ratings of perceived product quality increased from 2.7 to 3.5 when customer ratings and testimonials were added. Expectations of privacy protection were significantly increased by the prominence of privacy policy links ($F=16.143$, $p<0.05$). Likert scale ratings of privacy protection expectations increased from 2.7 to 3.4 when links to the privacy policy were made prominent. Table 1 summarizes these effects. In addition, perceptions of privacy protection were also increased by the presence of customer ratings and testimonials ($F=5.13$, p ,0.05). Likert scale ratings of perceived product quality increased from 2.9 to 3.2 when links to the privacy policy were made prominent.

Table 1. Summary of Likert scale perceptions for each design parameter

Perception type	Design Feature Present/Prominent	Design Feature Absent/Hidden
Customer service expectations	3.8	3.0
Perceptions of product quality	3.5	2.7
Privacy protection expectations	3.4	2.7

4.0 DISCUSSION

4.1 Customer Service

Participants had higher expectations of customer service for sites that had prominent customer service links. This was true even though this feature provided no explicit evidence of the quality of the customer service that would be provided. It appears that customers' perceptions of service are influenced by the design of the links to customer service. One possible explanation for this is that participants assumed that if a company provides prominent links, they are not concerned about customers using these services. Alternately, there may be a simple connection between the saliency of a link and users memory of it, and this increased likelihood of recall affected their perceptions.

4.2 Privacy Policy

Participants also had higher expectations of privacy protection for sites that had prominent links to their privacy policy. As with customer service, the prominence of the link adds no evidence to its quality. No participant attempted to click on the link and read the policy. Again, it could be a determination that companies that choose to make such links prominent must not be concerned with the information being accessed, or a relationship between saliency, recall, and perception.

4.3 Reputation Managers

The presence of reputation managers such as customer ratings and testimonials does provide real information that can be used by consumers to enhance their decision making. However, there is some question as to the validity of these ratings and consumers trust in them (Montania and Resnick, 2001). This study provides evidence that reputation managers do affect consumer choice, even if consumers report a lack of trust in them. An unconscious contribution of this information to perceptions of product quality can have the same effect as a conscious one.

4.4 Conclusion

This study investigated whether the presence of links to customer service, privacy policies, and reputation managers can affect consumer choice despite the presence of no real or confirmed information. Not only did these design parameters affect choice, they also contributed to increases in expectations of customer service, privacy protection and product quality. Studies of human decision making have found that people minimize effort by incorporating cues from the environment as short cuts about hidden information. This appears to be a strong example of such behavior.

5.0 REFERENCES

E-marketer (2000). Statistics. *Emarketer.com*, October 14, 2000. Available at: http://www.emarketer.com/estats.

Gefen D. (2000). E-commerce: the role of familiarity and trust. *Omega The International Journal of Management Science*, 28, 725-737.

Georgia Tech Research Corporation (1998). Graphics, Visualization, and Usability Center's World Wide Web User Survey-1998. Available at: http://www.gvu.gatech.edu/user_surveys/survey-1998-10/graphs/graphs.html

Montania R. and Resnick M.L. The effects of visible customer support in retail ecommerce. To appear in *Proceedings of the Industrial Engineering and Management Systems Conference*. Institute of Industrial Engineers. 2001.

Nielsen J. (1999a). Trust or Bust: Communicating Trustworthiness in Web Design. *Alertbox*, March 1999. Available at: http://www.useit.com/alertbox/990307.html.

Nielsen J (1999b). Reputation Managers are Happening. *Alertbox*, September 1999. Available at: http://www.useit.com/alertbox/990905.html.

Selnes F. and Gonhaug K. (2000). Effects of Supplier Reliability and Benevolence in Business Marketing, *Journal of Business Research*, 49, 259-271.

Servisoft Technology (1998). Servicesoft Survey Confirms Gap Between Web-Based Service Offerings and Customer Demand. *Servicesoft.com Press Release*. June 14, 1999. Available at: http://www.servicesoft.com

Sisson D. (1999). Ecommerce – Schemas & Concept. *Philosophe.com*, October 15, 1999. Available at: http://www.philosophe.com/commerce/schemas.html.

Walsh J. and Godfrey S. (2000). The Internet: A new Era in Customer Service, *European Management Journal*, 18, 1, 85-91.

NAVIGABILITY IN SEARCH SITES OF BRAZILIAN INTERNET

Marcelo Nunes Medeia, Leonardo Viana, Vitor Malheiros Ehmman,Cláudia Renata Mont'Alvão, M.Sc.

UNICARIOCA – Centro Universitário Carioca - Industrial Design Department
Av. Paulo de Frontin, 568, Rio Comprido, Rio de Janeiro, RJ, Brazil – Zip code 20261-243
Phone: + (5521) 502 1001 Fax: + (5521) 502 4172, 539 7573
Email: medeia@ medlei.com.br, leoviana@ajato.com.br, cmontalvao@pobox.com

ABSTRACT

This paper describes the basic characteristics of the interaction between the user and some of the available search sites in Brazilian Internet. These sites were analyzed where was possible to find out their principal informational and technical problems that, for several reasons, end for hindering the users' interaction with the same ones. The results obtained with the accomplishment of a pre-test indicates the existence of the problems delimited in the research.

1. THE PROBLEM

The number of Internet users is increasing daily in a very fast way since the 90's. These users access the *World Wide Web* aiming different objectives. Anyway, many times they need to search information that is available within thousands of *Home Pages* and *Web Sites* in the whole WWW. This necessity broke out the creation of the *Search Sites*.

Search sites are web sites where the main proposal is to offer the user faster and more efficient ways of search at Internet. These sites use "search mechanisms" created from the programming resources named CGI (*Common Gateway Interface*). There is two main distinct search mechanism:

a) *Search in catalogues* – where the user can search using keywords or choosing from a subject index;
b) *Search tools* - where just search for keyword is possible.

Once this type of *Search in catalogues* can search only in this own databank, the *Search Tools* search all Internet environment. It is also possible for the user search using *Advanced search* functions (that has a lot of different names, as "*Search option*", "*Refining your search*") when the user can customize the word or sentence that he/she is looking for.

1.1. Search Catalogues

These kind of search site offer two different options for the user's search: *Keyword Search* and *Subject Index*. Once searching using *Keyword Search* the user types the word(s) or phrase that must be find in a proper case, and then start the search (usually clicking a "Search" or "Enter" button). The mechanism will locate in its databank all the requested information and will present a result list for the user, in the way:
Home Page/Web Site Title: short description of the page's content.

If the user click twice in that underlined title, the user will open the *home page* or *web site*. These underlined titles are named *hyperlinks* (or *links*).

The second possibility offered by the *Search Catalogues* is the *Subject Index* that works as an encyclopedia index. It is a structured index, where the user just chooses one of the listed topics (clicking it) and other sub-indexes are opened. Anytime the user opens a subject or a subdivision a new screen appear, as that shown when he/she search for words (*Home Page/Web Site* Title: short description of the page's content). The layout mechanism of a *Keyword Search* is shown in Figure 1.

1.2.   Search Tools

This kind of search mechanism does not offer the option for searching using the *Subject Index*, just *Keyword Search* . Although, while the *Search Catalogues* are limited in search their our databank, the Search Tools search in all

Internet the word(s) or phrase typed by the user. So that, the search is more comprehend – but the results are presented in the same way as the first one.

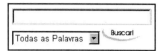 

Figure 1 - Examples of two Search mechanisms that the button (Busca) begins the search.

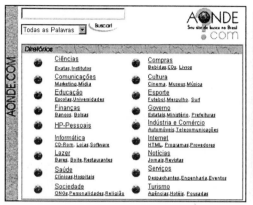 

Figure 2 - Examples of 2 Brazilian search sites using *Search Catalogue* (left) and *Search Tool* (right).

The *Advanced search* are options which the user can restrict the search result, as:
- search in a specific place (ex., country, state, city, etc.);
- search a specific file type (ex., sound, image, text, presentation etc.);
- search a specific phrase (ex., complete phrase, 2 specific words, etc.);

These offered options presented by the *Advanced search* allow the user a search optimization. Nevertheless, if each search site has its own *Advanced search* option, using a specific terminology, the user become lost, and avoid using this tool.

2. WEBSITES, HYPERTEXTS AND NAVIGATIONAL PROBLEMS

Most of the problems detected in search sites are related to how the information is presented. From the point of the system view, some characteristics like displays quality; number, layout and location of each window; hypertext information structure and navigation tools can influency the interactions (Moraes and Padovani, 1998)

Considering the *Advanced search* facility and efficiency if compared to Search tool, it is believed that is primordial that Search sites let the users know the main procedures, how to use it, in its first page. And later, some hyperlinks that leads to more complete information. Anyway, these sites believe that, if the Search tool doesn't work, the user will look for other way on search the information he/she want. In most of the cases, it doesn't happen, once the user are more interested in complete his/her search than decipher the tool enigmas. According Moraes & Padovani (1998), the selective attention is an adaptive mechanism allow us concentrate our cognitive resources in just one and unique channel while monitoring our environment, perceiving that signals that are priority. It is important to use highlights our visual clues in navigation tools to attract the user, supporting the selective attention mechanism during a search task in a hypertextual system.

The other option shown by the studied search sites, *Subject Index*, presents serious problems. The main cause is the absence of analogy between the methodology used in classifying the subject. Learning is word that includes a variety of phenomena, since memorization until the complex comprehension of rules and acquisition of new mental and motor skills. Analogy can make easier this learning process – if the user can relate this new information to another that he already know, it will become easier learn and remember this new information (Moraes & Padovani, 1998).

Standardized index and a map listing the contents of each index subject could decrease the number of steps performed by the user and also comprehend the hypertextual structure where he/she is navigating.

5. THE EXPLORATORY RESEARCH

5.1. Hypothesis

As hypothesis the research group believe that a non-standardization and the deficiencies in how to proceed an *Advanced search* has influence in the users' difficulties while using the search sites. And more, the time spent by the user while looking for an information is enormous, what generates expenses with telephonic services and the Internet provider. It happens because it is necessary a lot of steps (pages that must be opened using the hyperlinks) until the user find the desired page if he/she is using the *Subject Index*, or the infinity of phrases that the user may test using the *Keyword Search*.

For this explanatory research 2 available sites in Brazilian Internet were analyzed *Cadê?* and *Yahoo!Brasil*.

5.2. The sites used in this research

The *Cadê?* search site was created in 1995 by KD Systems and, nowadays is the most used search site in Brazilian Internet. It contents about 150,000 site addresses in its databank and 300,000 visitors per day. *Yahoo!* is a very well known search site in the world, and includes different language versions for Americas, Europe, Asia and Pacific countries. *Yahoo!Brasil* is a Brazilian Portuguese version available since 1999, and is one of the most important search site in Brazilian Internet. You can see the logotypes and complete Internet address of each site in Figure 3.

www.cade.com.br

www.yahoo.com.br

Figure 3 – *Cadê?* (left) and *Yahoo!Brasil* (right) logotypes and addresses

It was tested two types of search that are experienced by the users: *Keyword search* and *Subject Index*.

The tasks that must be performed by the volunteers were subdivided in 2 themes, as shown below.
Theme 1: Open the official site of Porto Alegre City Hall.
This search must be performed in both *Cadê?* and *Yahoo!Brasil* search sites, using *Keyword search*. The goals were open the site at www.portoalegre.rs.gov.br just searching in both sites and after receive instructions, using *Advanced search*. After some preliminary tests, we could agree that 10 steps were enough to finish the task. The search results in each site was:
I. *Cadê?*:
- the keyword **prefeitura** (city hall) returned 366 occurrences and it was not possible find the site after 10 steps;
- the keywords **porto alegre** returned 2156 occurrences and it was not possible again find the site after 10 steps;
- the *Advanced search* using the phrase **prefeitura municipal de porto alegre"**, or using **prefeitura and porto and alegre** returned just 1 occurrence when the site can be opened in 2 steps.

II. *Yahoo!Brasil*:
- the keyword **prefeitura** returned 84 occurrences and it was possible find the site after 2 steps if the user choose the right hyperlink;
- the keywords **porto alegre** returned 69 84 occurrences and it was possible find the site after 2 steps if the user choose the right hyperlink;
- the *Advanced search* using **prefeitura + porto + alegre** returned 2 occurrence when the site can be opened in 2 steps (or 4 steps if the subject choose the wrong hyperlink).

1122

Theme 2: Open the official IBM Brasil website (in Portuguese).
This search must be performed in both *Cadê?* and *Yahoo!Brasil* search sites, using *Subject Index*. The goal was open the site www.ibm.com.br. After some preliminary tests, we could agree that 15 steps were enough to finish the task. The search results in each site was

1. *Cadê?*:
   - it was necessary 3 steps:
   > INFORMÁTICA > Empresas ..... web site IBM
2. *Yahoo!Brasil*:
   - it was necessary 5 steps:
   > INFORMÁTICA e INTERNET > Computadores Pessoais > IBM > Páginas Corporativas: IBM@ ... IBM website
   - it was necessary 6 steps:
   > INFORMÁTICA e INTERNET > Empresas > Hardware > Fabricantes > IBM ..... web site IBM
   - or 7 steps:
   > NEGÓCIOS e ECONOMIA> Empresas > Informática > Hardware > Sistemas > Fabricantes .... IBM website
   Note: the symbol(>) means a step inside the Subject Index and (...) to an external link.

5.3. Research tools and subjects

During this exploratory research were used: a) 2 PC computers using Windows 98 Operational System, Internet Explorer navigator, fax modem 3.600 Kbps; b) Instructions for the subjects about the experiment; c) Questionnaire to take note of some characteristics of each subject; d) 2 Detailed instructions containing how to proceed an *Advanced search*; e) Datasheet to register the user's performance during the experiment. All of the 30 tested subjects have already used some search sites, but are not professionals or experts in Web design or development. It was a random sample with no structured profile of gender, age, background and formal education.

5.4. The sites used in this research

A pre-test was conducted which some problems where detected:

a) the browser highlights the links used by the volunteer. During the pre-test was observed that some subjects follow these highlights, when performing the given test. So, some changes in browser's configurations were necessary to solve this problem;

b) once beginning the search in a determined kind of tool, using the second one the subject starts to develop his/her own strategy avoiding to repeat the wrong procedures. During the exploratory research this problem was solved when half of the volunteers began the task using the *Keyword search*, and the other half, *Subject Index*.

6. RESULTS AND DISCUSSION

6.1. Questionnaire – the user profile

The questionnaire answered by the subjects has, as objective knows a little bit more about them. Ii was possible to notice that is not frequent the use search site (80% uses less than once a week). Most of them know that exist an *Advanced search* tool, but once they cannot remember how to use it (because differs from one search site to another) they prefer not using.

Other interesting result from the questionnaire was that 30% noticed that there are some problems in search sites – ergonomic and navigational problems that entail consequences as irritation and stress once they cannot find what they are looking for. The obtained results are shown in Table 1.

Table 1 - Subjects Profile

Gender	Male		Female	
	80%		20%	
Age	12 to 18 y.o.	18 to 25 y.o.	35 to 50 y.o.	50 or more y.o.
	70%	10%	10%	e 10%
Academic Background	High School (Incomplete)	High School	Graduation Course (Incomplete)	Graduation Course (Complete)
	70%	0%	20%	10%
How frequent do you use search sites?	Less than once a week	Once a week		Once a day
	80%	10%		10%
Which search site do you use?	Just *Cadê?*	*Cadê?* and *Altavista*		*Yahoo!Brasil* and *Cadê?*
	80%	10%		10%
All of the subjects use Keyword Search because it is easier and/ or faster.				
30% use Subject index, and 70% do not use this option because.				
30% use *Advanced search*, 10% do not know this tool, 60% know the tool but don't use this option because are not used to or forget how it works				
70% of the subjects agree that search sites has problems – they can not find what they are looking for.				

6.2. Results

The only task that was completely performed was Keyword search using instructions on how proceed and *Advanced search*. Its true that this task was easier than the first one (Keyword search) but we could also attest that the necessary number of steps for performing the two tasks also decreases.

The task that presented worst performance was Subject index using *Yahoo!Brasil*. We believe that this fact is related with the number of branches shown for each topic. When the volunteer started the search in a wrong way he/she got lost, and was necessary more steps to complete the task. It is important to mention that the users with previous experience of search sites and *Advanced search* were better succeeded.

As conclusion form this pre-test we can affirm that there is no kind of standards in these studied search sites. A same kind of search can be easier performed in one site than the other one. The Keyword search is more efficient when using *Yahoo!Brasil*. On the other hand, its *Advanced search* seems more confusing for the user than Cadê? Search site. When comparing the results Subject index is a more simple structure shown by Cadê? Is better than the Yahoo!Brasil complex one. Aiming corroborate the initial hypothesis, the experiment must include, in its next phase, with a new and stratified sample randomly selected from subjects that did not participate in this pre-test.

REFERENCES

Moraes, A. et al (1994). "Ergonomia e Interação Homem-Computador, Usabilidade de Interfaces: A Construção de uma Linha de Pesquisa". In Anais P&D Design 98, vol.2 Rio de Janeiro: AEnD-BR, 1998, 38-48.

Moraes, Anamaria de, PADOVANI, Stephania et al. *"A Cognição Humana e o Processo de Navegação em Sistemas Hipertextuais"*. In: Anais P&D Design 98, vol. 2. Rio de Janeiro: AEnD-BR, 1998, 68-77.

Perfetti, C. "Text and hypertext". In Hypertext and Cognition. New Jersey: Lawrence Erlbaum Associates Publishers, 1996, 157- 62

Development and evaluation of an information retrieval system for user groups and the WWW

MASE Motohiro and YAMADA Seiji
CISS, IGSSE, Tokyo Institute of Technology
E-mail:{mase, yamada}@ymd.dis.titech.ac.jp

Abstract

In this paper, we propose a framework for searching information through both the WWW and a human group. Though the information retrieval using a search engine in the WWW is very useful, we cannot acquire local information owned by a person and not explicitly described in text. A user knows neither where target information is in the WWW nor who knows in a human group. Thus we integrate the information retrieval in the WWW with that in a human group, develop heterogeneous resource information search HERIS as a multi-agent system.

1 Introduction

Information retrieval with the WWW as information resources has spread. The WWW includes huge Web pages and is updating all over the world. Thus the WWW is very useful as information resource. However, it is difficult for a user to investigate the location of information because huge information is ubiquitous disorderly. Generally, we use a search engine for information retrieval in the WWW.

Though the information retrieval with a search engine in the WWW is very useful, we cannot investigate the specific information like configuration of LAN in a laboratory and is personal knowledge like memo of application installation. Such information is not opened to public in the WWW in spite of being useful for other person. We called this information *closed information* contrasted with *open information* like the Web pages in the WWW. Only a part of personal knowledge is opened in the WWW, the most of closed information is not described in a document. To acquire closed information, asking a person who has the information is best. Moreover, results of a search engine often includes non-relevant Web pages, because filtering in accordance with user's intention is not enough. On the other hand, closed information is considered to be filtered by person in advance. Hence searching closed information through a human group is no less important than information retrieval in the WWW. Since judging where target information in the WWW and who knows the target information is very difficult for user, we propose **HERIS**(HEterogeneous Resource Information Search): a framework for searching information through both the WWW and a human group and indicating adequate information resource which should be accessed.

In the matter of sharing personal knowledge on Web pages, sharing URLs in a bookmark file has been studied. Mori and Yamada proposed Bookmark Agent[3]: a multi-agent system consisting of an agent which constructs a profile from a user's bookmark file by analyzing the HTML files. An agent broadcasts a query from user to other agents, and shows the Web pages related to the query to a user. Unfortunately their system deals with only bookmark files, and does not have a mechanism for connection among users.

2 System overview

Fig.1 shows the system overview of HERIS. HERIS is an information retrieval framework that searches both of the WWW and a human group. We build this framework as a multi-agent system consisting of user agents and SE(search engine) agents, and use contract net protocol[1][7] to communication among the agents.

A user agent and a SE agent cover a user and a search engine. Each of the agents manages a user profile and a SE profile. A profile consists of two types of sub-profiles: a knowledge profile and a resource profile.

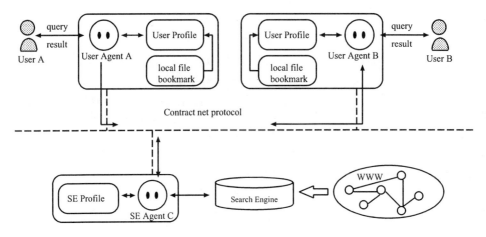

Figure 1: System overview

A knowledge profile contains weighted terms related to knowledge of a user or a search engine. A resource profile contains other properties of an information resource.

A user gives a query HERIS, a user agent that received a query from a user becomes a manager agent and all other agents become contractor agent. A manager agent broadcasts a task announcement message including the received query to contractor agents. Next, contractor agents compute similarity between the query in the task announcement message and their profiles, and reply bid messages including the similarity and resource profiles. After investigating the bid messages, a manager agent selects adequate contractors that win a bid, and send the query as a client message to the selected contractor agents. As a result, a manager agent receives a result message including retrieval information from the selected contractor agents, and integrates the messages. Finally, a user is shown a list of adequate Web pages and person who can answer the query with audio/video applications, and selects information resource from the list.

3   Construction of profile

3.1   User profile

A user profile consists of a knowledge profile and a resource profile.

Knowledge profile   A knowledge profile of a user profile stands for what a user know. We consider that user's information consists of two types of the information. The first type is described document like personal memorandum on application install, LaTeXsource file of a paper and so on. The second type is owned by a person and is not explicitly described in documents. We think that the Web pages that a user looks at well usually is the second type of information. Thus HERIS builds a knowledge profile by investigating user's local files and Web pages stored in a bookmark file that a user selects. The following shows the procedure of constructing a knowledge profile.

1. HERIS shows windows for a user to select the local files and Web pages. By selecting files registered into a profile from a local file and Web pages, a user can restrict to reference from other users.

   If local files selected by a user include LaTeXsource files, they are translated into HTML files using latex2html command. The set of these preprocessing local file is D_l. HERIS actually fetches the selected Web page on-line. The set of HTML files is D_h. HERIS sets D_l and D_h to a document set D in all.

Table 1: Weights of HTML tags

tag	weight	tag	weight	tag	weight
`<title>`	10	`<meta>`	10	`<h1>`	10
`<h2>`	8	`<h3>`	6	`<h4>`	4
`<h5>`	2	`<h6>`	1	`<big>`	2
`<em>`	2	non tag	1		

2. HERIS uses TFIDF[6] and the weighting using the structure of HTML tag for extraction of term and weight from D. In extracting terms and frequency from D, if a document is a HTML file, frequency is weighted according as tag. A set of tags and weights is shown in Table.1. Terms and weights are computed by applying TFIDF.

3. HERIS sums the weight for each term in all documents. All set of term and weight is a knowledge profile of a user profile. Finally, HERIS shows a profile editor window. A user can add/delete terms and edit the weights in this window.

Resource profile A resource profile consists of three elements, presence, cognitive load and social relation. Presence shows whether user is at his/her desk. Cognitive load shows user's busyness. Social relation is computed with ontology among users in their group.

3.2 Search Engine profile

A search engine profile consists of a knowledge profile and a resource profile. A knowledge profile is built by the method used in MetaWeaver[4]. A resource profile has an element, network load. Sending ping command to a search engine and measuring response time investigate network load.

4 Selecting adequate information resources

4.1 The computation of the similarity between the query and profile

We applied a vector-space model[5] to calculation of a similarity between a query and a profile. A similarity is computed as a cosine coefficient of document vector of a query and a profile. When a similarity exceeds a threshold, a bid message is transmitted to a manager agent.

4.2 Selecting the contractor agent

A manager agent selects adequate contractor agents as follows. A manager computes evaluations for contractor agents with weighted linear sum of a similarity and elements of a resource profile, chooses higher rank of contractor agents with high evaluation and requests them to send their result messages.

After receiving the result messages, a manager agent integrates the messages and indicates a result like Fig.2. A list of users who can answer the query is displayed on the left-hand side of this window, and a hit list of Web pages is displayed on the right-hand side. When a user selects a person from the list, if a selected person accepts a user's request, a user can talk with the person with audio/video applications. Fig3 shows the communication established between the users.

5 Experimental evaluation

We conducted two types of experiments for evaluating HERIS. For the following experiments, we used eight user agents and a single SE agent for Google. We implemented HERIS using Ruby and GTK+, and used vat and vic[2] for audio/video communication between users.

1127

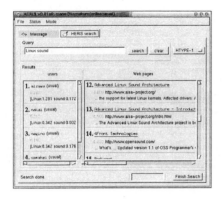

Figure 2: Indication of retrieved results

Figure 3: Communication among users

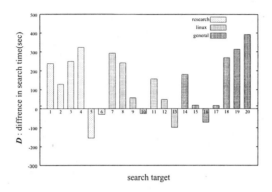

Figure 4: Difference in search time

5.1 Comparison with a search engine

We compared HERIS with a search engine Google. We used two targets for each domain, "research", "linux" and "general", and investigated the difference in search time until relevant information was found. The results are shown in Fig.4. The equation for computing D(difference in search time) is D = (search time of a search engine) - (search time of HERIS). The search time of HERIS is less than that of a search engine, if D is positive value. We see from Fig.4 that it was fifteen searches that the search time of HERIS is less than that of a search engine among twenty searches, and a feature by domains is not seen. We consider that the difference in search time is due to the difference in features of human and search engine as the information resource. A result of a search engine often includes many non-relevant Web pages because it is difficult for users to describe intention in adequate query. Thus finding relevant page takes much time.

However when asking a person directly, a user can tell his/her intention in a conversation with a person who may have the target information. Therefore a user can find the target information easily. Moreover a user may receive advice about more adequate query filtered by a person.

5.2 The feature of search with HERIS

The feature of HERIS is to search the target information through both the WWW and persons. We investigated situations in which person would be accessed. Also we recorded the following logs for every search with HERIS. The recorded logs are "query", "time accessed to the information resource", "order accessed to the information resource" and "comment about acquired information and situation". From analyzing the recorded logs, three types of search process were found. At first, without browsing Web pages at all, a user accessed a person first. Next, although a user browsed Web pages, since the target information was not found or a part of the information was found, a user accessed a person at last. Finally a user accessed a person while browsing Web pages, acquired an adequate query from a person, and searched with a new query again.

As an absorbing instance, some persons searched with a query of the higher concept, without a direct query related to target information. For instance, when a user searched what is sound driver in VineLinux, a user used a query that was not "VineLinux, sound driver" but "VineLinux", and asked a person indicated by system. In this case, user may ask a person who does not have target information. However, if a person asked by a user has target information, a user may be able to retrieve the information that is not registered in a knowledge profile and not described in a document. The same query in search with a search engine causes useless search results against user's intention. This technique of search is characteristic in HERIS searching for a human group.

6   Conclusion

We proposed HERIS: an information retrieval framework that searches open information and closed information through both the WWW and a human group. We developed the information retrieval system as a multi-agent system consisting of user agents and SE agents. Profiles for each person and each search engine are built, and the system selectively searched information resources consisting the WWW and a human group. As a result, a user acquires the results of hit lists from the WWW and direct audio/video connections to a person who can answer the query. It was found through experimental evaluation that the information retrieval using a human group as information resources is valid and a search with HERIS has a character different from conventional search engines.

References

[1] R. Davis and R. G. Smith. Negotiation as a metaphor for distributed problem solving. AI Memo 624, Artificial Intelligence Laboratory, Massachusetts Institute of Technology, May 1981.

[2] S. McCanne and V. Jacobson. vic: A flexible framework for packet video. In *The Third ACM International Multimedia Conference and Exhibition (MULTIMEDIA '95)*, pages 511–522, New York, November 1996. ACM Press.

[3] M.Mori and S.Yamada. Bookmark-agent: Information sharing of urls. In *The 8th International World Wide Web Conference(WWW-8)*, pages 70–71, 1999.

[4] M. Mori and S. Yamada. Adjusting to specialties of search engines using metaweaver. In *WebNet 2000 World Conference on the WWW and Internet*, pages 408–412, 2000.

[5] G. Salton. *Automatic Text Processing*. Addison-Wesley, 1989.

[6] G. Salton and M. J. McGill. *Introduction to Modern Information Retrieval*. McGraw-Hill, Tokio, 1983.

[7] R. G. Smith. The contract net protocol: High-level communication and control in a distributed problem solver. In *IEEE Transaction on Computers*, number 12 in C-29, pages 1104–1113, 1980.

USER APPROPRIATE PLAN RECOGNITION
FOR ADAPTIVE INTERFACES

Marc Hofmann and Manfred Lang

Institute for Human-Machine Communication
Technical University of Munich, D-80290 Munich, Germany
{hofmann, lang}@ei.tum.de

ABSTRACT

Adaptive user interfaces are often based on plan recognizers, which consider just optimal action sequences to reach a goal. The algorithm we present is an approach towards user appropriate plan recognition, i.e. it stays abreast of the fact that in complex domains users often behave sub-optimal to achieve their goal because of a lack of knowledge about the domain and about its commands. Furthermore our algorithm is able to deal with goals with various ways to achieve.

1. INTRODUCTION

User assistance systems usually have a component, which models the user's potential goal for controlling an adaptive interface. In this paper we describe a plan recognizer to infer the user's goal regarding previously observed user actions. Our domain is a Unix file system with a standard Unix shell for entering commands. We refer to files and directories as *objects* the user may manipulate by *operators*. Operators can be interpreted as sub-plans; usually they are Unix commands or groups of Unix commands. For plan recognition we interpret a Unix plan as a vector of operators for manipulating a number of objects:

$$plan = operator(\ object\ ) \tag{1}$$

As mathematical basis for the plan recognition algorithm we make use of Bayesian belief networks [Pea88]. Charniak and Goldman first have used Belief networks for plan recognition [Cha92]. The main feature of our plan recognizer is its ability to exploit optimal and also sub-optimal user behaviour, i.e. the plan recognition process is user appropriate. Hence our networks differ in the topologies and probability tables from the networks of the authors mentioned. We introduced a hierarchical structuring of plan networks in four layers for an user adequate representation of plans.

2. METHODOLOGY

2.1 Requirements

To gather knowledge about typical user acting in the Unix environment and to gather training data for the plan recognizer, experimental subjects have been given various Unix tasks. Analyzing the resulting plans lead to conclusions, which lead to a number of requirements for our plan recognizer.

In our test plans, all users acted sub-optimal, i.e. they made use of commands, which are irrelevant for achieving the goal. Nevertheless these actions can be interpreted as characteristic action patterns for certain plans, which can be ascribed to a lack of knowledge of the domain and its commands. Wrong usage of commands and mistyping are also frequently made mistakes. Furthermore each user seems to have an individual approach to plan completion.

Finally our goal was to build a plan recognizer that stays abreast of imperfect acting of users and which also exploits information apart from the optimal sequence of actions for classification of plans. For an appropriate weighting of actions and sub-plans, we decided for a probabilistic mathematical fundament, namely Bayesian belief networks.

2.2 Basic structure

For the basic structure of our algorithm, refer to Fig. 1. The user has a certain goal he wants to achieve. Therefore he needs a solution for his problem, a sequence of actions leading to that goal. This sequence is defined as *plan*. The task of the plan recognition system is inferring the user's plan by means of previously observed actions. Therefore the plan recognizer is fitted with a plan library, a database containing all potential plans to reason about. For each

1130

plan a *plan model* is generated as basis for the plan evaluation. As this plan model is a Bayesian belief network, we refer to the networks as *plan model networks*. With every new user action all potential plans of the plan library have to be evaluated, given the current and previous user actions as observations. For each plan hypothesis an evaluation measure EM is calculated, which reflects the belief that the observed actions are part of that plan. The plan with the maximum evaluation measure is the result of the plan recognition process.

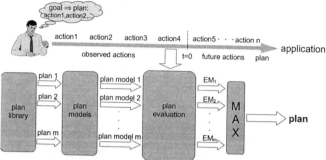

Figure 1: Basic structure of the plan recognition system

2.3 Plan model networks

When using Bayesian belief networks the main task is to find an adequate topology for the network and choosing the proper conditional probability tables for modeling a certain problem. A plan model network is structured hierarchically in four layers, which will now be described in detail.

Plan hypothesis layer

The plan hypothesis layer is the top layer of a plan model network. It allows direct inference of the belief that the observed user actions are part of that plan. Therefore the confidence of a plan is modeled by one discrete, Boolean state variable. At the beginning it is assigned a neutral probability distribution as a priori probability P(plan), i.e. both states are equally likely. We refer to this node as *plan node*. As we interpret a plan as the manipulation of objects, we represent each object to manipulate by a Boolean state variable, which is linked with the plan node by an arc. Moreover the object nodes are linked among each other. Fig. 2 shows the topology of the network.

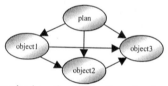

Figure 2: Topology of the plan hypothesis layer for a plan for manipulation of three objects

The structure of the belief network and appropriate conditional probability tables of the object nodes enable modeling the following logical AND-function:

$$plan = object1 \wedge object2 \wedge object3 \dots \tag{2}$$

Equ. 1 reflects the fact that a plan is only completed if the user manipulated all relevant objects in the way the plan is meant for. The a posteriori probability of the plan node P(plan | object1, object2, ...) allows inferring the completion of the plan only if the corresponding objects are manipulated correctly and completely.

Object layer

The object layer models how objects are manipulated. It provides one Boolean node for each operator an object has to be manipulated with. For the conditional probabilities we again have a logical AND-function of the object node with *operator nodes* as parameters. In analogy to the plan hypothesis layer the manipulation of one object is only completed if all necessary operators have been used on that object. This part of the network reflects optimal behaviour, only the essential operators to achieve a goal:

1131

$$object = operator1 \wedge operator2 \wedge ... \tag{3}$$

Now operator nodes, which reflect sub-optimal actions, but nevertheless for that plan characteristic action patterns, are linked to the object node. We refer to these operators as *optional operators*, which may support the belief in a particular plan when observed, but the object may be also manipulated correctly and completely without using these operators. For training the conditional probabilities we gathered training data by giving a number of Unix tasks to various experimental subjects. The emerging plans are used for weighting the contribution of optional operators to plan completion. This is a task of training with complete knowledge, so the probabilities can be determined according to the frequencies of occurrence of optional operators. For weighting the contribution in that way that observing an optional operator supports the belief in a plan, only conditional probability values between 0.5 and 1 are of interest. Therefore we map the division of the number of optional operators observed in the training data ($n_{opt\,op}$) and the number of interesting manipulations of a particular object (n_{obj}) on the range from 0.5 to 1 for the "yes"-state. This results in the following equations:

$$P(opt\,operator = y \mid object = y) = \frac{1}{2}\left(1 + \frac{n_{opt\,op}}{n_{obj}}\right) \qquad P(opt\,operator = n \mid object = y) = \frac{1}{2}\left(1 - \frac{n_{opt\,op}}{n_{obj}}\right) \tag{4}$$

The conditional probability values reflecting the contribution of an optional operator to other plans are chosen neutral, because we treat each plan individually and independent of other plans. This fact and equ. 3 enable even very rarely observed optional operators to contribute to the belief in a plan.

Fig. 3 shows the topology of a typical object layer consisting of two operators and *n* optional operators.

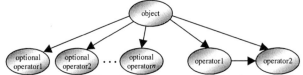

Figure 3: Topology of the object layer with two operators and *n* optional operators

Operator layer

The operator layer models the way an operator is created by a number of user actions. It provides one Boolean node for representing an operator and one Boolean for each user action. This layer also combines optimal and sub-optimal usage of actions. Fig. 4 shows the topology with *n* optional actions and a few actions for modeling optimal behaviour. As the structure shows, it's possible to model different approaches to that operator. Hence synonymous actions with different commands, different options and parameters, but with equal effect can be modeled. In Fig.4 the combination of action1 and action2 has the same effect as action3. In the case of two or more actions to create the operator, the structure and the conditional probabilities are chosen according to a logical AND.

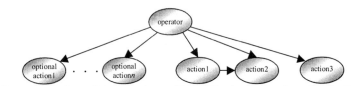

Figure 4: Topology of the action layer with *m* optimal actions and *n* optional actions

The conditional probabilities of the optional action nodes are chosen in analogy to the object layer.

Action layer

The action layer consists of nodes directly representing user actions. Each node is related to various mapping information to ensure the action is mapped on the right node as new information. For modeling an action it is decomposed according to its syntax. Fig. 5 pictures the topology of an action with the syntax pattern <command> <options> <object1> <object2>. The decomposition enables the plan recognizer to deal with mistyping, i.e. in case of a wrong syntax component we only map the observation of right components on the action layer's nodes. The arcs and conditional probabilities again are chosen according to the following logical AND-functions.

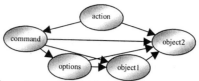

Figure 5: Topology of the action layer for an action with the syntax <command> <options> <object1> <object2>

The decomposition of actions does not apply to *optional action* in order to put not too much emphasis on them. Hence each optional action is represented by one single node.

2.4. Plan evaluation

After creating a plan model network for each plan of the plan library, the plan evaluation process can start. Fig. 6 pictures an excerpt of a plan model network, which takes a number of optional operators and optional actions into account. Below the structure user actions and their corresponding directory information are listed. Every new user action is compared with the commands, directories and objects, which are assigned to nodes. If the action matches with a node, that state variable is given the state "yes", i.e. it is instantiated and represents the observation of that user action. This information is propagated through the whole network to support the belief in that plan. User actions that are not represented by any nodes are not considered for the plan evaluation.

Commands for changing the directory ("cd") are treated in a special way. Not the command itself will be mapped on the plan model network, but the current directory. If the user changes into a certain directory, the state variable representing that directory will be instantiated. Leaving that directory results in some kind of backtracking, the instantiation of the corresponding state variable will be revoked.

Plan: „**Decompress and print all files of the directory /d1/d2**"

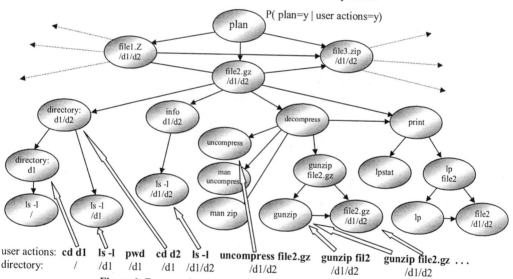

Figure 6: Excerpt of a plan model network and a stream of user actions

If an object has been manipulated completely by using the correct operators, the statistical dependency between the plan node and optional operators or actions for manipulating that object, is blocked, as the object node is instantiated. That means further mapping will not affect the belief in that plan. In this case we make use of the Bayesian belief network's phenomena of "d-separation" [1].

All plan hypotheses have to be treated in the same way. To compare each plan hypothesis with each other, we first calculate the belief in a plan, given the observations of previous user actions. Inferring the belief *Bel(plan)* for k observed actions can be done according to the following equation:

$$Bel(plan = y) = P(plan = y \mid action1 = y, action2 = y, ..., actionk = y) \tag{5}$$

The belief has to be calculated for all plans to reason about. To be able to compare all plans, we use the belief *Bel(plan)* to determine the percentage of how complete a plan is modeled by the observed actions. As plans may consist of different numbers of objects to manipulate, we have to multiply the number of objects n_o of the plan and divide it by the maximum number of objects n_{max} a plan can have. The result is the following equation for the evaluation measure *EM*:

$$EM = (Bel(plan = y) - 0.5)\frac{n_o}{n_{max}} = (P(plan = y \mid action1 = y, action2 = y, ..., actionk = y) - 0.5)\frac{n_o}{n_{max}} \tag{6}$$

The evaluation measures of all plans have to be calculated. The plan with the maximal evaluation measure is the plan to decide for.

3. RESULTS

Evaluating a plan recognizer quantitatively is always a crucial task, because the recognition rate heavily depends on the degree of completion of the plan. As the main feature of our algorithm is the user appropriate plan evaluation, we tested the algorithm as the main component of a user assistance system [Hof01]. A number of experimental subjects with different Unix-skills have been given a number of tasks with the assistance system offering partial task completion on the basis of the plan recognizer's output. The plan library consisted of 20 plans. Figure 7 proves the acceptance of the whole assistance system with the plan recognizer as the centra/l component. The target group, users with little or medium Unix experience judged the assistance system it to be helpful. It has been expected that Unix experts didn't accept the assistance as the system hasn't been created to cope with their needs.

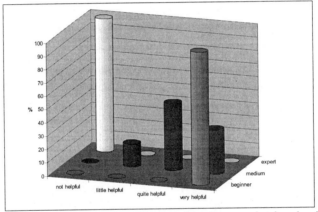

Figure7: Result of an investigation on the acceptance of the plan recognizer based assistance system

4. CONCLUSIONS

Our plan recognizer proved to work well, especially its main feature, the ability to make use of optimal as well as sub-optimal user acting proved to be the key for plan recognition for user assistance systems. Future work will be left to the automatic and dynamic generation of the plan library.

REFERENCES

[Pea88] Pearl, J.: "Probabilistic Reasoning in Intelligent Systems: Networks of Plausible Inference", Morgan Kaufmann, California, 1988

[Cha92] Charniak, E. and Goldman, R.B.: "A Bayesian Model of Plan Recognition", Artificial Intelligence, 64(1), 1992, 53-79

[Hof01] Hofmann, M. and Lang, M.: "A Dialog Model for Offering Task Completion for complex Domains", Poster Proceeding HCII 2001 (New Orleans, Lousisans, USA), (this conference)

Interactive Discovery for Acquiring Trends of Web Information

Wataru Sunayama and Masahiko Yachida

Graduate School of Engineering Science, Osaka University
1-3 Machikaneyama, Toyonaka, Osaka 560-8531 Japan

Abstract

Social activities are divided into two types. One is a creative activity by the combination of the existing object, and another is an imitative activity by which the created matters settles. Since a creation is realized by the combination of existing knowledge and information, people cannot create new things without thinking about previous works and their proper combinations. This paper proposes a framework for interactive discovery. The framework consists of User, Search System, Data Mining System and Interface for supplying knowledge. Namely, this proposal is also a creation which is realized by the combination. However, this does not mean Web Minings but means a practical use of extracted knowledge. Such a system will be very important for this new century.

1 Introduction

Needless to say, the Internet has become active in recent years. Now that the growth of technologies for the Internet are remarkable, a social activity without the Internet may be disappeared. The Internet users want to acquire information what they didn't know. However, they are hard to seek information by proper keywords as inputs of search engines. Because they don't have knowledge related to unknown domain or new things. Therefore, this paper proposes a framework which aids search processes by the interaction between a user and a search system. Namely, this framework will:

1) Make a user interest concrete.

2) Aid a user to acquire relational information.

3) Aid a user to discover unknown relationships among search keywords.

In this paper, along with these points 1) and 2), the point 3) is the most notable. This framework supports discovery of unknown relations, viewpoints and combinations of information concealed in WWW database. By this discovery of new combinations, one will be able to find a new theme of study or will be able to create a hot-selling product.

As for this combinations, social activities are divided into two types. One is a creative activity by the combination of the existing object, and another is an imitative activity by which the created matters settles. As human begins have a nature of "Tire", people always seek novel things. Therefore, it can be said that the society is kept by mutually creations. The information in WWW is unknown for a user but known for the author of the Web page. Therefore, a new idea comes from a combination of known ideas. A brand new combination may has a brand new viewpoint. However, this number of combination will be so enormous that a person cannot match by hand.

A person need to aware something as a concrete word for a creation. After the awareness, he/she examines the word by some relational evidences or previous works. If the results of examinations are good, the thing is more widely examined to know the possibility of application and expansion. The keywords of this creative process are "Word" and "Examination". An abstract mind is stimulated by words and becomes a concrete mind. Unless a person knows the word, he/she cannot examine and realize the idea. Along with this, examination is necessary for making an idea concrete and for expanding the idea.

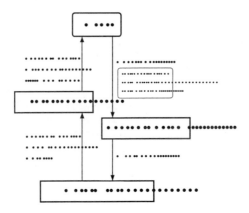

Figure 1: A framework for interactive discovery

In this study, the framework provides users with keywords as chances that can be viewpoints, and users can discover new relationships from WWW information through the sequence of the Web search. The framework proposed in the next section regards those keywords as important.

2 Framework for Interactive Discovery

A framework for creative activities is shown as the Figure 1. This framework consists of User, Search System, Data Mining System and Interface for supplying knowledge. A search system in the framework needs search keywords as user's input, and outputs arranged relational keywords in a two-dimensional interface after a Data mining processes for extracting useful information. Users make his/her own interest concrete, acquire relational information and do creative activities with a discovery of unknown relationships by the repetition of a search and a supply of keywords. Relational keywords are extracted from Web pages matched with user's each search keyword. In the interface, the system supplies not only keywords but also some additional information such as a summary of output Web pages. In the rest of this section, each of them is defined and explained.

2.1 User discovers a viewpoint of the combination

The framework which provides users with keywords as chances, which can be viewpoints, is proposed. As for this viewpoint, an agreement or a disagreement is needed for two keywords which are combined. In short words, to discover this viewpoint is a discovery for creation. This may be a kind of discovery by co-occurrence[Langley 87]. For example, two companies will be merged by a viewpoint that a company must survive or must pursue profits. Therefore, this framework aims at constructing a new system which aids users to discover new combinations by suggestions of viewpoints representing an agreement or a disagreement. However, nothing will come without inputs corresponding to user's mind or interests. So users use Internet search systems with holding latent or vague motivations in their mind.

2.2 Support System for Search Systems

Three creative purposes of using search engines, as in the Figure 1, are investigation, verification and extension. As these keywords are already described in 1, Investigation and Verification aim at focusing information, and Extension aims at extending information related to a search keyword. Therefore, a support system for search engine is surely not a search engine but is a contrivance to support above purposes. The contrivance is realized by Data Mining and an interface for displaying knowledge.

2.3 Data Mining from Web pages

Data Mining[Fayyad 96] is a methodology to seek useful rules and knowledge from enormous data wearhouse. Some of them are derived from association rules[Agrawal 94] and conditional probabilities defined as co-occurrances of data. In the Data Mining module of the framework, some relational keywords of search keywords are extracted from current Web pages. The features of Web database are as follows;

1. Enormous:Not all data can be in use.

2. Dynamic:The data is always changing.

3. Heterogeneous:A data includes some topics and viewpoints.

The most important point is how to restrict data for a data-mining, such as pages in specific domains, pages retrieved by a keyword(including a specific keyword), its freshness and so on. Along with these, the same things are applicable to a single Web page. That is, a constructor must think how to divide and how to interpret a page to extract essences corresponding to each user desires.

2.4 Interface for Knowledge Refinement

It is important for a user to understand relationships between search keywords and existing Web information. A two-dimensional search interface is needed to know tendencies of Web information, for making a concrete search condition and for getting an idea of a new topic. Therefore, relational keywords, as outputs of this framework, are arranged neatly in a two-dimensional search interface. Some of relational keywords are related to multiple search keywords, so each keyword will be classified in groups of the n-th power of 2 when n is the number of search keywords. Because each keyword is related to a search keyword or not. If the number of search keywords is more than 4, search keywords are currently clustered in 3 groups in order to arrange keywords distinguished in a two-dimensional plane. This clustering is executed automatically by using co-occurances of relational keywords appearing in real Web pages.

Practically, though keywords are ultimate summarized information of Web pages, those are fragments of sentences. Some users may want a summarized sentence that is chained by keywords, because a word has various meanings. Therefore, some complementary information will also be useful. The interface for interactive discovery needs some components which make up each loss occured by the restriction of the data.

3 Experimental System

Currently, though the prototype system is under construction, each module have already worked separately [Sunayama 99, Sunayama 00]. User's potential interests are inferred by relationships among search keywords the user input. A search system is useful for not only seeking Web pages including avalable information, but also understanding a keyword how it is related to other topics. So, each extent of a search keyword may have common topics. These common topics can be a user's potential interest.

Now, one of the Data Mining methods to extract relational keywords is to select keywords commonly appeared in Web pages including a search keyword[Sunayama 99]. A two-dimensional search interface have already appeared in [Sunayama 00]. Relational keywords were arranged in two-dimensional interface, and users could make out the relationship between search keywords and relational keywords easily.

In the Figure 2, search query was "Woods AND Golf AND Ichiro AND Baseball AND Sasaki", and relational keywords are arranged. (All search keywords and relational keywords are translated from Japanese to English. The database is about 14000 Web pages are downloaded from WWW related to Japanese public entertainment and sports by January, 2001.) By this search, no Web page was hit with this search query because golf and baseball are ordinarily inconsistent. However, relational keywords related to each search keyword is certainly exist in WWW. Therefore, by using such information, the relationship is emerged by a data mining method for discovering a viewpoint.

In the interface, Search keywords were clustered by common relational keywords, the pairs ("Woods" and "Golf") and ("Sasaki" and "Baseball") were clustered in the same category. Some keywords arranging in the interface will be hard to explain why those keywords are output as relational keywords. Such keywords

1137

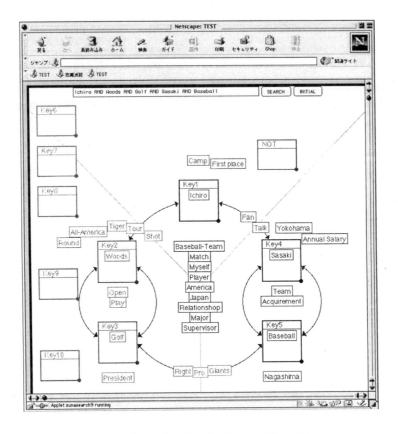

Figure 2: Search Interface for Discover Viewpoints

will treated as unknown or new common points of search keywords. Then, the user will examine the details of the keywords why the keywords have output. For example, as "Giants" related to Baseball and Golf, input a new search query "Baseball AND Golf AND Giants". The user had known the fact that Giants baseball players are likely to play golf. After this process, the interface became as Figure 3. The user could examine the details of the facts by using these appearing keywords. In general, the keywords can be start points of the examining process to find a new research topic and to find a strategy of administration.

This cycle of search and acquiring information will give birth to new ideas. This system supplies users with chances for knowing trends of the world because a user will be in the state that the user only knows unknown viewpoints of the combination.

4 Conclusion

This paper proposed a framework for interactive discovery for creative activities. Creative activities are necessary for our usual life and the affluent society. Especially, it is effective for people to imply new viewpoints of the combination which has never been thought out. Certain symbolic words are useful for users to concrete one's idea, and to know relational knowledge and to expand their ideas. May this new century will be a creative century!

1138

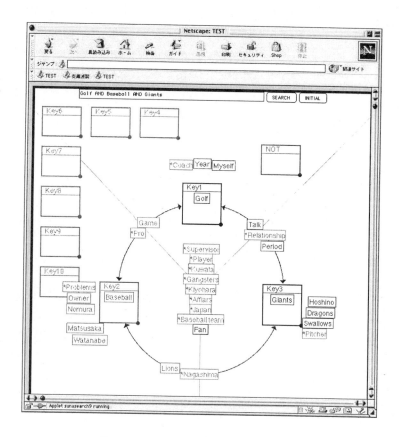

Figure 3: Search Interface for Examination

References

[Agrawal 94] Agrawal, R. and Srikant, R.: Fast algorithms for mining association rules, In Proc. of the 20th VLDB Conference, pp.487 – 499, (1994).

[Fayyad 96] Fayyad, U., Piatetsky-S., G., and Smyth, P.: "From Data Mining to Knowledge Discovery in Databases", *AI magazine*, Vol.17, No.3, pp.37 – 54, (1996).

[Langley 87] Langley, P., Simon, H.A., et al:
"Scientific Discovery - computational Explorations of the Creative Processes", *MIT Press*, (1987).

[Sunayama 99] Sunayama, W., Osawa, Y. and Yachida, M.: "Computer Aided Discovery of User's Hidden Interest for Query Restructuring", *in Proc. International Conference on Discovery Science'99*, pp.68 – 79, (1999).

[Sunayama 00] Sunayama, W. and Yachida, M.: "An Aiding System for Internet Surfings by Associations: Get One through Chances", *IEEE Proc. of the fourth International Conference on Knowledge-based Intelligent Information Engineering Systems (KES 2000)*, pp.808 – 811, (2000).

Assistance of Web browsing by indicating the future Web pages

NAGINO Norikatsu   YAMADA Seiji

CISS, IGSSE, Tokyo Institute of Technology

{nagino,yamada}@ymd.dis.titech.ac.jp

Abstract

This paper describes a novel method to assist a user in gathering interested Web pages on the WWW by narrowing a search space using sequences of the user's current browsed pages. Various methods of gathering interested Web pages for a user on the WWW have been proposed. Those approaches can't make an agreeable response to user's interest shift dynamically. We think those approaches don't try to understand which links will be selected and what kind of pages will be interested by a user in the past and the future.

To cope with this problem, rules to select a link which a user wants see are introduced in our system, and the rules called "**Navigation Rule**". Each rule has a class of Web page types in the condition part and a link-type in the action part. Also they are weighted for indicating the preference. Every time user selects a link to see the next page, the weights of matched rules are increased. The rule having the highest weight value was applied and some matched Web pages are fetched by Web robots again and again. We called the search strategy "**Navigation Search**". Web robots don't gather some useless Web pages and the search space will be narrow using the **Navigation Search**. Also we describe the way to provide an interface which display Web pages gathered by Web robots. A user can understand that a user will be reach what kind of Web pages with this interface. Also using the interface, a user also be able to select some links to interest Web pages directly without crawl all path.

1 Introduction

The accessible information through the Internet is increasing explosively as the WWW becomes widespread. In this situation, the WWW is very useful for a user who wants to gather interesting information. However there is a significant issue that a user does not know where the information exists. A practical and simple solution of the problem is to use a search engine like MetaCrawler, AltaVista, YaHoo with the interesting information as a query. The search engine provides a list of relevant Web pages to a user. Unfortunately, since a database of a search engine is very huge and adequate filtering is hard, many Web pages including irrelevant ones may be indicated.

To cope with this problem, various methods of gathering interested Web pages for a user using Web robots on the WWW have been proposed. WebWatcher[2] and Letizia[4] are able to indicate the Web pages that a user wants to see next. Using browsing history, they learn to predict useful Web pages for a user. However these systems do not consider sequences of browsed Web pages. FishSearch[1] and InfoSpiders[6] are distributed online search algorithms based on technique of artificial life for gathering relevance information. These approaches can't make an agreeable response sufficiently to user's browsing shift dynamically. PWM[8] is search algorithm for gathering interest Web pages with user's anytime controlling Web robots. In this system, gathered Web pages was divided into clusters, and user selects can select clusters about which he/her wants know more. Web robots refer the selected cluster, therefore search space for Web robots was narrowed. But the search space was not well narrowed because a user doesn't interest all Web pages in a selected cluster. The method with such traditional search strategies can't navigate sufficiently for a user.

In this paper, in order to give some rules valid weights and narrow a search space, we extend the condition part of the rules to a sequence of classes of Web page types. It is important to support a user in his/her browsing task that a search space for Web robot is narrowd without gathering useless Web pages. We propose the search strategy called "**Navigation Search**" based on some rules called "**Navigation Rules**". Supporting to reach to his/her interested Web pages efficiently by gathering more deep Web pages and displaying Web pages gathered by Web robots intelligible is very important too. We also describe a method to provide interface display Web pages gathered by Web robots at browsing task. Using this interface, a user will be able to modify crawling direction of Web robots easily.

2 Search strategy

2.1 Traditional strategies

The following main search strategies are available for gathering Web pages.

- *Breadth-first search*[7]
  Web robots use this search strategy when they gather Web pages for database used by some search engines. By using this search strategy, subjects of search Web pages is spread, and gathered Web pages will be uniform in subjects. In addition, probability that Web robots access to particular Web sites concentrically will be decreased.

- *Depth-first search*[7]
  To search some relevance Web pages, this search strategy look like browsing strategy by internet users. The behavior based on the assumption that some pages liked from a page similar to the source of page. This assumption was declare in ARACHNID[5].

- *Strategy using artificial life technique*
  Agents have energy, which is gained from relevant Web pages and lost from irrelevant pages. Agents having high energy can reproduce themselves and others having low-energy may die. Relevant Web pages is identified with this algorithm.

To navigate for a user with a current seeing Web page, agents have to gather relevance Web pages in a short term. Furthermore, as a user can predict the kind of Web pages linked a page partly, agents have to deepen the search space in the term. If above search strategies apply to agents for user's navigation, some problems will appear. Using breadth-first search, important Web pages aren't sufficiently gathered. Although, a search space for agents will be larger, and we can't reach to the depth Web pages. Using depth-first search, as user's interest changes frequently, very simillar Web pages are gathered in a few subjects. Using some techniques based on artificial life, it is very difficult to reflect user's interest to Web robots. PWM[8] integrates breadth-first search and depth-first search in order to gather Web pages interest for user's efficiently. Web robots search Web pages based on breadth-first search in PWM to gather Web pages in various subject. In order to make narrow the subject of gathered Web pages for a user, the Web pages will be divided into some clusters with SOM[3] and clusters are displayed in a window as a 2D map. A user can select clusters about which he/her wants know more, and Web robots gather Web pages that will be placed the cluster. The way to select a cluster seems like depth-first search. But a user isn't interested in all Web pages in the cluster which was selected, because a search space was not well narrowed.

For navigate to a user, we have to need novel search strategies instead of the one integrated above search strategies. We propose a novel search strategy called "**Navigation Search**" based on using rules in this paper.

2.2 Navigation Search

Many navigation systems compare keywords in a current Web page or weighted keywords in his/her profiles with keywords in the next linked Web page. But we may not gather important Web pages because of differences between Web pages in meaning. In such a case, it is important to understand correctly at the meaning of this document. On the other hand, links are selected based on the kind of Web pages in our method. Our rules are represented as the following

$$\textbf{if } p_t \in C_j^p \textbf{ then select } l_t \left(L \cap \in C_k^l \right)$$

The t is a sequential number that describe the frequency of Web pages p selection by a user, and it is increased incrementally. The $p_{t=0}$ is current Web page a user seeing. C^p means a set of Web pages in the Web page class, C^l means a set of links in the link class, and l means a link. The L is a set of links in current Web page. The j and k is labels of the Web page classes. If some rules apply to some Web pages in browsing history and the current Web page, the next link in a current Web page is determined. Example of Web page classes C^p are shown as the following.

- *"Link Page"* : Web pages in this class include many links to others. The threshold of number of links is set beforehand.

- *"Image Page"* : Web pages in this class include many images. The threshold of number of links is set beforehand.

- *"English and Japanese Page"* : Web pages in this class written in English and Japanese half and half.

- *"Page of K"* : Web pages in this class include all keywords in one's set K.

Example of link classes C^l are shown as the following.

- *"links near an image"* : Links in this class be placed near an images.

- *"next link"* : Links in this class are the first link of the other links which aren't selected by a user yet.

These rules are used by Web robots for gathering the next Web page. By applying rules to browsing history and predict Web pages again and again, Web robots can crawl automatically. The systems using this rules will not be adapted for users' interest shift dynamically. Because the systems can't understand the kind of Web pages the user want at that time. Using these rules, many links matched these rules are selected, and search space of Web robots will be spread.

Then in order to give all rules the valid weights and narrow a search space, we extend the condition part of the rules to a sequence of Web page classes as the following.

$$\text{if } p_{t-i} \in C^p_x \wedge \cdots \wedge p_t \in C^p_z \text{ then select } l_t \in \left(L \cap C^l_k \right)$$

We called this rules "**Navigation Rules**". At the time of applying a rule, compare the condition part of the **Navigation Rule** with the sequence of user's browsed Web pages. Furthermore by compare the sequence include Web pages correspond to selected links with **Navigation Rules** again and again, Web robots gather Web pages based on the kind of Web pages a user want. If many **Navigation Rules** was matched the sequence, most weighted rule is applied at first. Weights of **Navigation Rules** are modified with feedback from a user as shown later, and we expect the weights make values relative to probability of behavior the user will do. To also approach a human behavior, the action part of rules include not only link types but also actions about 'back' and 'forward' for functions of Web browser. We called this search strategy "**Navigation Search**".

3 System's Overview

The system's overview with **Navigation Rules** for gathering Web pages shown in Fig1. The system always observe user's browsing task. **Navigation Rules** are compare with sequence of browsed Web pages each user's selecting of links, and selected links are written in the working memory. Web robots always gather Web pages from the WWW based on condition of working memory asynchronously with user's browsing task. List of links for gathering in working memory is also modified by user feedback. Gathered Web pages by Web robots based on matched rules provide to a user as an interface for understanding the kind of Web pages gathered easily. This interface makes a user to access interested Web pages directly.

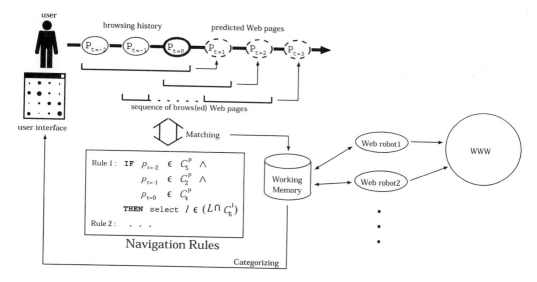

Figure 1: System's Overview

4   Adjusting weights of a rule by user feedback

In order to give all **Navigation Rules** the valid weights, weights of rules applied frequently make be high. User feedback for learning based on rules doesn't require explicit evaluation of Web pages, and it performed naturally. It is important to be decreased a load of evaluation of gathered Web pages. If links which predicted with our method are selected by a user in his/her browsing, weights of the **Navigation Rules** which determined selection of those links are increased. The weights of the other **Navigation Rules** are decreased. Some rules which value decrease under the threshold are eliminated for the cost of calculate and validity of the **Navigation Rules**. The weight values of many **Navigation Rules** which have a single Web page class in the precondition tend to be, high because such **Navigation Rules** match to browsing history frequently. Therefore **Navigation Rules** weight with value relative to the length of Web page classes in the precondition. We can expect to adapt our system for a user in changing user's interest frequently. Furthermore gathering useless Web pages will be decreased because **Navigation Rules** have valid weights, and Web robots can reach more deep level Web pages.

5   User Interface

User interface shown in the Fig1 display Web pages gathered by Web robots visually. The window of a user interface is displayed in the side of the Web browser window during browsing, and the contents indicated the interface window is updated whenever a user select a new link or a user is only looking the Web page. When some links are recommended to a user graphically by modifying original Web page contents like any other systems, if the recommendation is not valid, it is obstacle to a user. Our interface is not prevent user's browsing task, and it can use whenever the user want use. Predicted Web pages in this interface display that a user want what kind of Web pages and will reach what kind of Web pages.

There are two types of Web pages when it was displayed we think; the Web page types in condition parts of **Navigation Rules** and subjects of Web pages. For example, some Web page clusters are displayed for each subjects at the first, and Web pages displayed for each class of condition part in **Navigation Rules** ordering by the weights. Web pages are divided into some clusters based on SOM[3]. Clusters are displayed in an interface window with some characteristic keywords for the clusters. If a user select a cluster, Web pages in the cluster ware displayed for each Web page classes with the rate that the Web pages locate into the cluster. A user can understand predicted Web pages easily using by this interface. If a user find an interested Web page, he/she can see the Web page directly by selecting the link of the Web page.

6 Conclusions

In this paper, we defined the rules called "**Navigation Rules**" using sequence of browsed Web pages instead of only access log of browsing, and we proposed the way to predict links and actions which the user will do. And we described the way to gather Web pages for navigation by Web robots based on **Navigation Rules** called "**Navigation Search**", and then described the way to construct a user interface that can use whenever a user want. The characteristic of our interface is that a user can give feedback without prevent his/her browsing tasks. Our system can navigate a user effectively by narrowing a search space for Web robots and gathering Web pages based on the user's interest.

References

[1] P. De Bra and R. Post. Information Retrieval in the World-Wide Web: Making Client-based Searching Feasible. *Computer Networks and ISDN Systems*, 27(2):183–192, 1994.

[2] T. Joachims, D. Freitag, and T. Mitchell. WebWatcher: A Tour Guide for the World Wide Web. In *Proceedings of the Fifteenth International Joint Conference on Artificial Intelligence*, pages 770–775, 1997.

[3] T. Kohonen. *Self-Organizing Maps*. Springer, Berlin, Heidelberg, 1995. Second Extended Edition 1997.

[4] H. Lieberman. Letizia: An Agent That Assists Web Browsing. In *Proceedings of the Fourteenth International Joint Conference on Artificial Intelligence*, pages 924–929, 1995.

[5] F. Menczer. ARACHNID: Adaptive Retrieval Agents Choosing Heuristic Neighborhoods for Information Discovery. In *Machine Learning: Proceedings of the Fourteenth International Conference*, pages 227–235, July 1997.

[6] F. Menczer, R. K. Belew, and W. Willuhn. Artificial Life Applied to Adaptive Information Agents. In *Working Notes of the AAAI Symposium on Information Gathering from Distributed, Heterogeneous Databases*. AI Press, 1995.

[7] S. Russell and P. Norvig. *Artificial Intelligence –A Modern Approach–*. Prentice-Hall, 1995.

[8] S. Yamada and N. Nagino. Constructing a Personal Web Map with Anytime-Control of Web Robots. In *Conference on Cooperative Information Systems*, pages 140–145, 1999.

EVALUATING VISUALIZATIONS:
A METHOD FOR COMPARING 2D MAPS

Thomas Mandl & Maximilian Eibl

Information Science - University of Hildesheim - Germany
Social Science Information Centre - Berlin - Germany

Two-dimensional maps are a valuable interface element for the visualization of information retrieval results. Various methods exist for the creation of these maps. This article shows, that these methods may create very different maps in which objects may have different distances from each other. The evaluation method is based on the users perspective. These results show that the mapping method has to be chosen very carefully and different methods should be tested for an application.

1. INTRODUCTION

Visualization is generally regarded as a good method to support user interaction. Concerning information retrieval (IR) this support can be applied at different stages of the retrieval process (Eibl 2000). As long as Boolean operators are used a visualization can simplify the construction of queries at an early stage. Other visualizations operate at a later stage and try to explain the result set.

Evaluations of the visualizations are rare. Usually a general advantage of a visual display over a textual is assumed. In respect to the theories of perception and cognition this assumption seems to be plausible. Still, this does not already rectify a visualization. Today, several different approaches employing visualization exist to each stage of the IR-process. But there are no evaluations comparing two concurring visualizations. This article introduces a method to compare two-dimensional map layouts.

2. VISUALIZATION IN INFORMATION RETRIEVAL: TWO DIMENSIONAL MAPS

Most of today's visualizations emphasize the result set. Here, two different approaches can be identified. First the usage of the metaphor of information space and second the visualization of the search criteria within the documents. Visualizations of the information space usually resemble star fields. Search criteria are represented by bigger stars spread out on the screen. Documents are represented by smaller stars positioned between the search criteria according to their respective relevance. That way, patterns in the document space emerge. This model is based on the concept of one way attraction: the documents are attracted by the corresponding search terms but not vice versa.

Another way visualizing the result set is the use of maps. Their aim is twofold: First, they try to reduce complexity by reducing high-dimensional data to a spatially two-dimensional display. Second, they enable the user to browse the document set rather than to search. The basic idea of maps is to express semantic relations by spatial distance: closely related documents are presented next to each other. Non-related documents are presented remote from each other.

Two-dimensional displays are based on mathematical reduction methods which compress a large vector space of terms into only two dimensions. Most often, the Kohonen Self Organizing Map (SOM, cf. Kohonen 1998) is used. Other methods are latent semantic indexing (LSI, cf. Berry et al. 1995) or factor analysis.

Two-dimensional document maps have attracted considerable attention in recent years. Meanwhile, commercial implementations can be used in the internet[1]. Especially two implementations which are

---

[1] http://www.newsmaps.com/   http://www.cartia.com/

based on the Kohonen Self-Organizing Map (SOM) and which are accessible over the internet made this interaction technique more popular. Chen et al. 1996 applied the SOM to internet documents derived from the Yahoo entertainment section. Due to the large number of documents, the visualization is organized in layers. Therefore, the user needs to choose from a map of labeled document clusters in the first steps. In the last interaction step, a cluster of documents which can be selected is visualized.

A even more recent example of a large scale implementation is WEBSOM[2] (Kohonen 1998) which includes over one million documents, in this case contributions to news groups. WEBSOM also uses a layered structure to overcome the problem of visualizing large amounts of documents.

However, the layered structure introduces the necessity to navigate between different maps. This makes the direct interaction with the intuitive two-dimensional map far more complicated. Therefore, the directness and the associate navigation unfold their advantages best for a small amount of objects which fits on one map. Consequently, SOM have also been used to the display the result set of a retrieval system (cf. Roussinov et al. 1999).

The developers of two-dimensional display usually do not give a rationale for their choice for a mathematical method. It remains unclear why SOM or LSI is being used and whether a specific method has an advantage for the domain.

3. EVALUATION OF VISUALIZATIONS IN INFORMATION RETRIEVAL

Though there are many approaches in document retrieval using visualization only few evaluations are conducted. Appropriate quantitative evaluation methods for evaluations have not been established yet. Most formal studies rely on the standard information retrieval measures recall and precision.

Eibl (2000) reports on 25 visualizations but can make out only two formal evaluation studies: SENTINEL (Knepper et al. 1998) and J24 (Odgen et al. 1998). Both were tested during the 7th Text Retrieval Conference (Trec-7). Both Systems could not prove to be superior to textual search systems. SENTINEL lead to worse results than most comparative systems. Nevertheless, in Trec-7 SENTINEL lead to better results than the previous year in Trec-6 when it did not use a visualization. But it can hardly be attributed to the added visualization because to many changes in the entire retrieval engine were made. J24 was compared to ZPRISE but could not outperform it. Eibl (2000) conducts a user test demonstrating the superiority of the visualization DEViD over the systems freeWAIS and Messenger.

Other user tests were conducted by Swan et al. (1998) and Chen et al. (1996). Chen et al. (1996) conducted a user test for a SOM and the qualitative results were promising. During Trec-6 Swan et al. (1998) compared the systems AspInquery and AspInquery Plus to ZPRISE. AspInquery and AspInquery Plus are based on the Inquery search engine and offer two additional user interfaces: a aspect window for aspect oriented retrieval called AspInquery and a 3D visualization called AspInquery Plus. Concerning recall, ZPRISE outperformed AspInquery and AspInquery outperformed ZPRISE.

4. EVALUATION METHOD

Since usually no reason for the choice of a specific two-dimensional mapping method is given, the question remains, whether these methods lead to different maps at all. Here we want to introduce a possible method of evaluating different mapping processes.

The users point of view is central for the evaluation of visualizations as well as of information retrieval systems. A formal evaluation in information retrieval requires a ranked list of objects which is not provided by the two-dimensional map. In order to evaluate two-dimensional displays with information retrieval methodology, one could map the query into the two-dimensional space and calculate the distance to the documents. By using the distance as a ranking criterion, a ranked list could be obtained. However, the advantage of a document map lies in the possibility of associative browsing through interesting areas of documents. That means, that a user most likely does not evaluate the documents sequentially according to their exact distance from a query.

---

[2] http://websom.hut.fi/websom

Another possibility is based on the perspective of the user while browsing. The closer semantically similar documents are grouped the more distinct and useful a map is. The user will look at documents close to the one he focused. Therefore, the evaluation method uses one document as the starting point and calculates the Euclidian distance to all others. By sequentially using all documents as starting point, a similarity matrix is obtained for each mapping method. The ranked lists can be compared in order to determine the degree of correlation between the methods.

This resembles the comparison of two similarity matrices row by row. The correlation between two ranked lists was measured using the following coefficient:

$$\text{Spearman:} \quad r = 1 - \frac{6 \sum_{i=1}^{n} d_i^2}{n(n^2 - 1)} \qquad \text{(Hartung 1984)}$$

The Spearman correlation measure were calculated for each row. The average of all rows was then calculated as the correlation between the matrices.

The Spearman coefficient has already been applied for the evaluation of IR systems in Mandl (1998) in order to measure the differences between similarity matrices derived by different algorithms. Figure 1 outlines the process of comparing the two maps. The steps in figure 1 need to be carried out for each document.

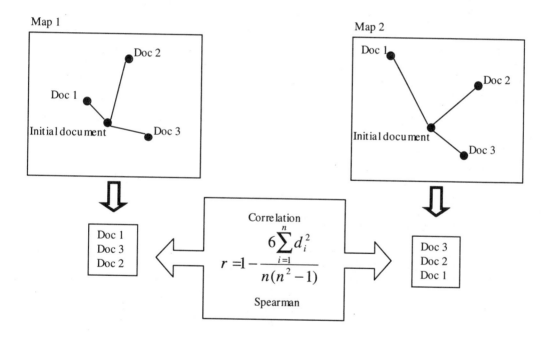

Figure 1: The evaluation process for one document

5. EXPERIMENTAL DESIGN AND RESULTS

The data chosen for the experiment consists of 1000 documents from the database of the Social Science Information Centre in Bonn, Germany. The documents are indexed intellectually. This data is now part of the international initiative for the evaluation of multilingual information retrieval systems CLEF[3] (Cross Language Evaluation Forum). CLEF provides a test bed for multilingual retrieval systems including documents, queries and relevance assessment.

From this document set, two two-dimensional displays were constructed. One is based on latent semantic indexing (LSI, using software provided by Berry et al. 1993) and the other on the Kohonen Self-Organizing map (SOM). The evaluation method showed that there is indeed no correlation between the two two-dimensional maps an therefore between the two methods. For a subset of 1000 documents the correlation was only 0.053 on a scale from 0 for no correlation at all to 1 for identical maps. Two-dimensional maps are very efficient for smaller sets. Therefore, the correlation for sets of 10 and 100 documents was also calculated. From the set of 1000 documents we extracted 100 sets with 10 documents and 10 with 100 documents. Although there are some sets with a higher correlation, the average of all sets lies in the same order of magnitude as the correlation for the set of 1000 documents.

Table 1: Results

Number of documents	1000	100 (average of 10 sets)	10 (average of 100 sets)	Highest correlation for a set of 10
Correlation	0.053	0.037	0.013	0.34

6. SUMMARY

The best evaluation of a two-dimensional map is a user test. However, the value of browsing techniques are often hard to assess in quantitative user tests. The chosen method evaluates the distance between document therefore taking into account the users perspective during browsing. The results clearly show that the methods for creating a two-dimensional display lead to very different maps. Therefore, the methods have to compared for the document set and the proper method needs to be chosen carefully. This result also shows that quantitative evaluation methods for visualizations need to be developed and refined.

7. REFERENCES

Berry, M., Do, Th., O'Brien, G., Krishna, V., & Varadhan, S. (1993): *SVDPACKC (Version 1.0) User's Guide*. Technical report. Computer Science Department. University of Tennessee in Knoxville, USA. Retrieved November 30, 2000, from the World Wide Web: http://www.netlib.org/svdpack/svdpackc.tgz

Berry, M., Dumais, S., & Letsche, T. (1995): *Computational Methods for Intelligent Information Access*. In: Proc. ACM Supercomputing '95. San Diego, CA. pp. 1-38.

Chen, H., Schuffels, Ch., & Orwig, R. (1996): *Internet Categorization and Search: A Self-Organizing Approach*. In: Journal of Visual Communication and Image Representation. 7(1). pp. 88-101.

Eibl, M. (2000). *Visualisierung im Document Retrieval. Theoretische und praktische Zusammenführung von Softwareergonomie und Graphik Design*. IZ-Forschungsbericht 3, Informationszentrum Sozialwissenschaften, Bonn. (includes CD-ROM with English summary and animation)

Hartung, J. (1984): *Lehr- und Handbuch der angewandten Statistik*. München, Wien.

---

[3] http://www.clef-campaign.org

Kohonen, T. (1998): *Self-organization of Very Large Document Collections: State of the art.* In: Niklasson, L., Bodén, M., Ziemke, T. (eds.): Proceedings of ICANN '98, 8th International Conference on Artificial Neural Networks, Springer: London. vol. 1, pp. 65-74.

Knepper, M., Killiam, R., Fox, K., Frieder, O. (1997). *Information Retrieval and Visualization using SENTINEL.* In: Proceedings of the 7th Text Retrieval Conference Trec-7, S.393-397.

Mandl, Th. (1998): *Learning Similarity Functions in Information Retrieval.* In: Zimmermann, H.J. (ed.): EUFIT '98. 6th European Congress on Intelligent Techniques and Soft Computing. Aachen, Germany, 8.-10.9.1998. pp. 771-775.

Mandl, Th. (2000): *Tolerant Information Retrieval with Backpropagation Networks.* In: Neural Computing & Applications. Special Issue on Neural Computing in Human-Computer Interaction. Vol. 9 (4). pp. 280-289.

Odgen, W., Davis, M., & Rice, S. (1998). *Document thumbnail visualizations for rapid relevance judgements: When do they pay off?* In: Proceedings of the 7th Text Retrieval Conference Trec-7, S.599-612.

Roussinov, D., Tolle, K., Ramsey, M., & Chen, H. (1999): *Interactive Internet search through automatic clustering: an empirical study.* In: Proceedings on the 22nd annual international ACM SIGIR conference on Research and development in information retrieval pp. 289-290

Swan, R., Allan, J., & Byrd, D. (1998): *Evaluating a Visual Retrieval Interface: AspInquery at TREC-6.* In: CHI `98 Workshop on Innovation and Evaluation in Information Exploration Interfaces.Los Angeles, April 1998. Retrieved November 30, 2000, from the World Wide Web: http://www.fxpal.com/CHI98IE/submissions/long/swan/index.htm

Visual Information Retrieval for the WWW

Harald Reiterer, Thomas M. Mann, Gabriela Mußler

Computer and Information Science, University of Konstanz, Germany

Abstract

In this paper we present the conception and the evaluation of a visual information retrieval system for the Web. Our work has been motivated by the lack of good user interfaces assisting the user in searching the Web. The selected visualisations and the reasons why they have been chosen are explained in detail. An evaluation of these visualisations as an add-on to the traditional result list is presented.

1 Introduction

Some of the main challenges of the Web are problems related to the user and his interaction with the retrieval system. There are basically two problems: *how to specify a query* and *how to interpret the answer provided by the system*. Surveys have shown that users have problems with the current paradigm of information retrieval systems for Web search simply presenting a long list of results (Zamir, Etzioni 1998). These long lists of results are not very intuitive for finding the most relevant documents in the result set.

The above empirical findings motivated us to develop a new type of user interface for Web retrieval that supports the user in the information seeking process by providing selected visualisations in addition to the traditional result list. Systems combining the functionality of retrieval systems with the possibilities of information visualisation systems are called *visual information retrieval systems*. An important aspect of visual information retrieval systems is their possibility to visualise a great variety of document characteristics allowing the user to choose the most appropriate for his task.

This paper presents our main design ideas developing a visual information seeking system called INSYDER[1]. In chapter 2 we discuss, with the focus on the visualisations, the new features of the system. Chapter 3 presents our synchronised visualisation approach of Web search results and the results of a summative evaluation. Conclusions and an outlook are given in chapter 4.

2 INSYDER - a Visual Information Retrieval System

During the development of the INSYDER system it was not intended to develop new visual metaphors supporting the retrieval process. The main idea was to select existing visualisations for text documents and to combine them in a new way. Nowadays there are a lot of visualisations of search results in document retrieval systems available (Hearst 1999). Our selection of existing visualisations was based on the assumption to find expressive visualisations keeping in mind the target user group (business analysts), their typical tasks (to find business data on the Web), their technical environment (typical desktop PC), and the type of data to be visualised (text documents). The primary challenge from our point of view was the intelligent combination of the selected visualisation supporting different views on the retrieved document set and the documents itself. The primary idea was to present additional information about the retrieved documents to the user in a way that is intuitive, fast to interpret and able to scale to large document sets.

Another important difference of our INSYDER system compared to existing retrieval systems for the Web is the comprehensive visual support of different steps of the information seeking process. The visual views used in INSYDER support the interaction of the user with the system during the formulation of the query (e.g. visualisation of related terms of the query terms with a graph), during the review of the search results (e.g. visualisation of different document attributes like date, size, or relevance with a scatter plot or visualisation of the distribution of the relevance of the query terms inside a document with a TileBar), and during the refinement of the query (e.g. visualisation of new query terms based on a relevance feedback inside the graph representing the query terms).

The retrieval aspects of the visual information seeking system INSYDER have not been in the primary research focus. Nevertheless the system offers some retrieval features that are not very common in today's Web search engines (Reiterer et al. 2000).

It is for sure not new to combine visualisations and information retrieval aspects, but nowadays systems which do a dynamic search with a document attribute generation and the different visualisations of these attributes and docu-

---

[1] INSYDER (INternet SYstème DE Recherche) was funded by a grant from the European Union, ESPRIT project number 29232.

ment inherent data are new. Our approach aimed at getting the highest added-value for the user combining comp o-
nents like dynamic search, visualisation of the query and different visualisations of the results and information retrieval techniques (e.g. query expansion, relevance feedback).

3 Visualisations supporting the Information Retrieval Process

3.1 Visual Query Formulation

From the literature it is well known that users have problems formulating their information need (Pollock, Hockley
1997). This led to the demand of methods to overcome the problem of lacking knowledge to formulate queries. The
idea of the visual query formulation is to help users to specify their information need more precisely using query
expansion techniques and visualisation. The query expansion is implemented using a knowledge base, which is built
upon terms and related concepts. Users can benefit from using it in two ways: either by changing their query terms
leading to a more precise result set or by expanding their original query with additional terms from the knowledge
base, which will result in a broadened result set, which could be much more satisfying, too. And as a side effect using terms from the knowledge base, spelling mistakes can be minimised, too. We propose a visual query, which will
show the user related terms for his query (Figure 1), taking into account other successful solutions and ideas from
automatic query expansion and query visualisation, e.g. (Voorhees, Harmann 1998), (Zizi, Beaudouin-Lafon 1994).
As an intuitive way to express the relation of terms, we propose to use a graph for their visualisations. The entry
point for the visualisations is the query entered by the user (e.g. *WWW visualisation*). The original and the resulting
related terms are then presented in a graph and tree view. The graph view represents the terms with their "near" related terms. E.g. *WWW* is expanded to *Web, Internet, Media,* etc.. (see Figure 1). The tree view resembles the overall
term space and therefor contains all terms related to a distinct term. These distinction is made to keep the graph view
as easy as possible to survey. The broader related terms are displayed in the graph view using a hypertext metaphor
for navigation: Clicking on a term (e.g. *Internet*) will show all other related terms for this distinctive term, depending on the number of all the related terms (e.g. *Hypertext, Usenet* etc.), it may come up as a circle with the expanded
term as a centre (Figure 2). The two views are synchronised, which means that a term selection in the graph view
will select the equivalent in the tree view and vice versa. If the graph view does not contain the selected term from
the tree view, a new graph is created. Following the hypertext metaphor the visual query provides also a history
function, so that the user can keep track of different graphs.

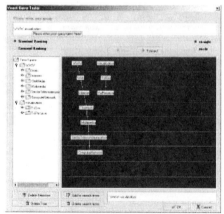

Figure 1: Visual query

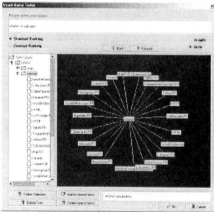

Figure 2: Examination of a term

The retrieval system uses two approaches to rank documents. The standard ranking takes all terms into account for
its ranking, while the concept ranking takes the concepts from the knowledge base. These concepts can be weighted
by the user to express his information need more precisely. In the visual query we therefore have also two ways
foreseen. The standard follows the description above. The concept ranking (by selecting the appropriate radio button), will expand the graph view in two ways. Using the context menu the user can select the weight of each concept. A '+' sign is used as a marker for the importance, which will be shown above that term. According to the type
of ranking terms are added to the search terms by using the appropriate button. The search terms are displayed in an
own highlighted text entry field next to the button. Deleting the original search terms there (e.g. to take more precise

1151

terms), the user can see them still in the upper entry field.

If the user uses the relevance feedback option of the system, the process is basically the same. The difference is that instead of the user the system provides the entry point terms for the graph and term space.

3.2 Visualisation of Search Results

The main idea behind our visual information retrieval approach is to present additional information about retrieved documents to the user in a way that is intuitive, fast to interpret and which is able to scale large document sets. One important feature is the possibility to group documents that share similar attributes. We have used two different approaches depending on the additional information presented to the user:

- Predefined document attributes: E.g. title, URL, server type, size, document type, date, language, relevance. The primary visual structures to show the predefined documents attributes are the Scatterplot (a similar idea could be find in Ahlberg et al. 1994) and the Result Table (Figure 6).
- Query terms` distribution: This shows how the retrieved documents related to each of the terms are used in the query. The primary visual structures to show the query terms` distribution are the Bargraph (a similar idea could be find in Veerasamy et al. 1995), the TileBar (Hearst 1995) and the Stacked Column.

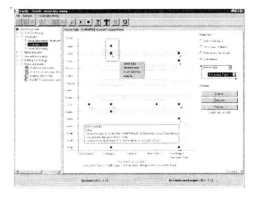

Figure 3: Scatterplot

Figure 4: Barchart

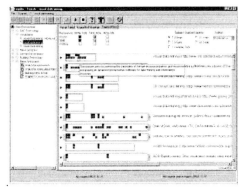

Figure 5: TileBars

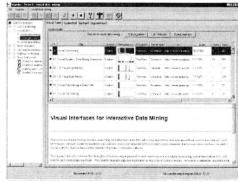

Figure 6: Result table

The visual mappings of web documents we have chosen are text in 1D: Bargraph (Figure 4), TileBars (Figure 5), Stacked Column and text in 2D: Scatterplot (Figure 3). This final selection of the visual structures was based on a field study, an extensive study of the state of the art in visualising text documents and the design goal to orientate our visual structures as far as possible on typical business graphics. The field study shows that all users have a good understanding of this kind of graphics and use them during their daily work (e.g. in spreadsheet programs). Similar conclusions, mainly based on an overview of the research done in the area of visualisation of search results in document retrieval systems, can be found in (Zamir 1998).

Another important design decision was to use a *synchronised multiple view* approach. It offers the user the possibility to choose the most appropriate visualisation view for his current demand or individual preferences. Our approach has similarities with the idea of "Multiple Coordinated Views" with "Snap-Together Visualisation (STV)" (North et al. 1999), e.g. offering the user coordinated views for exploring information.

3.3 Evaluation

The primary goal of the summative evaluation was to measure the added value of our visualisations in terms of effectiveness (accuracy and completeness with which users achieve task goals), efficiency (the task time users spent to achieve task goals), and subjective satisfaction (positive or negative attitudes toward the use of the visualisation) as dependent variables for reviewing Web search results. Knowing advantages of the multiple view approach documented in user studies (North et al. 1999), we didn't intend to measure the effects of having Scatterplot, Bargraph and TileBar/Stacked Column (also called SegmentView) *instead* of the List and Table. We wanted to see the added value of having these visualisations *in addition* to the Table and List.

From the factors influencing the design of a visual structure (Mann, Reiterer 2000) we decided to vary *target user group, type* and *number of data*, and *task* to be done. These have been determined as the independent variables. *Technical environment* and *training* was identical for all tests. The test setting covered all combinations of the different kinds of information seeking tasks (specific and extended fact finding), different kinds of users (beginners and experts), amount of results (30, 500), number of keywords of each query (1,3,8) and the chosen combinations of different visualisations.

A short entry questionnaire was used to record demographic data of each user. Then each user got a standardised system demo using a predefined ScreenCam recording presenting each visualisation. After that each user had about 10 minutes to get familiar with the system and to ask questions if he had problems using it (learning period). The users were then asked to answer the 12 test task questions as quick as possible. During the tasks the users were requested to "think aloud" to enable the evaluation team to understand and record their current actions. After accomplishing the test tasks the users had to answer a questionnaire of 30 questions regarding their subjective satisfaction and to suggest improvements of the system.

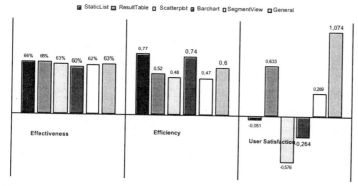

Figure 7: Evaluation results

Added values of the visualisations: In most test cases the users made use of the visualisations (using only the visualisation or using it in combination with the Result Table to answer the test questions). From this we conclude that the majority of the users expected an added value of the visualisations.

Effectiveness: The effectiveness of the visualisation is measured with the help of the degree of fulfilling the test tasks. E.g., if 8 out of the 12 tasks were solved, the effectiveness is 66,6% out of a maximum of 100%. As it can be seen in the left part of Figure 7 (Effectiveness) there was no significant advantage of using a specific visualisation combination. All visualisations performed nearly as good as the static list, which was used for reference purposes.

Efficiency: The efficiency of the visualisations has been defined as the effectiveness divided by the time the test persons needed to fulfil a test task. As no absolute minimum or best time exists for this test setting, the values derived are only comparable to each other. In Figure 7 (middle part) it can be seen that the Barchart combination performed second of all visualisation combinations. If we take into account that the Static List is something familiar to the user (well known from search engines), the Barchart has an outstanding role. Surprisingly it performs worst when looking at the effectiveness, but as the values are in a small interval, we do not give too much strength to this effect. Also the fact that the subjects often used the Scatterplot combination, and therefore probably expected a high added value from using it, but had in realty a low value in effectiveness and efficiency can be taken as a hint that training effects

1153

could have a high influence on the results. This will have to be evaluated in a next step.

User Satisfaction: The user satisfaction is derived from the final questionnaire based on a Likert-scale (-2 to +2). Therefore positive and negative values occurred. For the user satisfaction an overall value has been calculated summarizing a number of questions. Figure 7 shows that this general impression of the visualisation was satisfying. This means that the majority of the test persons thought that none of the visualisations are dispensable. They also had the impression that the visualisations helped them to solve a task. The subjective impression of the Scatterplot was the worst. Users might have performed better, if they would have had more training time for the use of the Scatterplot and by performing better, it is likely that they have a more positive attitude towards a distinctive visualisation. Interestingly, most of the test persons were in a better mood after using INSYDER (positive mood before the test 92,5%, after the test 97,5%).

Influence of target user group, type and number of data, type of task. The numbers of documents, the numbers of keywords, the type of users, and the task type have shown to influence the efficiency of the visualisations.

4 Conclusion and Outlook

The results of the evaluation of our visual information retrieval system for the Web have motivated us to go ahead. Our main design ideas for the development of a visual information retrieval system for searching the Web have been successful. Most of the users make use of our synchronised multiple visual views and regarded them a nice enabling technology to find the most relevant documents in the search result. The evaluation results have shown that effectiveness and efficiency do not really increase when using visualisations, but the motivation and the subjective satisfaction do. We assume that more training time is needed to use the system effectively and efficiently.

Throughout the ideas presented above we are still working on the enhancement of the overall system. This includes the visualisations of the search results, developing specific filter functions supporting dynamic queries in combination with our visualisations, the visualisation algorithms and particularly the user interface of the whole application.

References

(Ahlberg et al. 1994) C. Ahlberg; B. Shneiderman. Visual Information Seeking: Tight Coupling of Dynamic Query Filters with Starfield Displays. In: Proc. ACM CHI'94 pp. 313-317.

(Hearst 1995) M. A. Hearst. TileBars: Visualization of Term Distribution Information in Full Text Information Access. In: Proc. ACM CHI'95; 59-66, 1995.

(Hearst 1999) M. A. Hearst. User interfaces and visualization. Modern Information Retrieval. R. Baeza-Yates and B. Ribeiro-Neto (eds.). Addison-Wesley (New York): 257-323, 1999.

(Mann, Reiterer 2000) T. M. Mann, H. Reiterer. Evaluation of different Visualizations of WWW Search Results. Proc. Eleventh International Workshops on Database and Expert Systems Applications (DEXA 2000). Con-ference: Greenwich, UK, September 4-8, 2000 (IEEE Computer Society).

(North et al. 1999) C. North; B. Shneiderman. Snap-Together Visualization: Coordinating Multiple Views to Explore. University of Maryland, technical report CS-TR-4020 June 1999.

(Pollock, Hockley 1997) A. Pollock and A. Hockley. What's Wrong with Internet Searching. D-Lib Magazine, 1997, http://www.dlib.org/ dlib/march97/bt/03pollock.html [1999-02-01].

(Reiterer et al. 2000) H. Reiterer, G. Mußler, T. M. Mann and S. Handschuh: INSYDER - An Information Assistant for Business Intelligence. Proceedings of the annual International ACM SIGIR Con-ference on Research and Development in Information Retrieval SIGIR '00, Athens 24-28 July 2000.

(Veerasamy et al. 1995) A. Veerasamy; S. B. Navathe. Querying, Navigating and Visualizing a Digital Library Catalog. In: Proc. DL'95. http://www.csdl.tamu.edu/DL95/papers/veerasamy/veerasamy.html [1999-03-24]

(Voorhees, Harman 1998) E. M. Voorhees and D. K. Harman (eds.): NIST Special Publication 500-242: The Seventh Text Retrieval Conference (TREC-7) Gaithersburg, Maryland (Government Printing Office (GPO)) 1998. http://trec.nist.gov/pubs/trec7/t7_proceedings.html [1999-12-20].

(Zamir, Etzioni 1998) O. Zamir and O. Etzioni. Web Document Clustering: A Feasibility Demonstration. SIGIR 1998. http://zhadum.cs. washington.edu/zamir/sigir98.ps [1999-03-23].

(Zamir 1998) O. Zamir. Visualization of Search Results in Document Retrieval Systems. General Examination's Paper, University of Washington, http://www.cs.washington.edu/homes/zamir/papers/gen.doc [2000-09-13]

(Zizi; Beaudouin-Lafon 1994) M. Zizi; M. Beaudouin-Lafon. Accessing Hyperdocuments through Interactive Dynamic Maps. Conference on Hypertext and Hypermedia Proceedings of the 1994 ACM European confer-ence on Hypermedia technology. 126-134.

Toward A Human-Web Interface

Kang Zhang
Department of Computer Science
University of Texas at Dallas
Richardson, TX 75083-0688, USA
kzhang@utdallas.edu

Jiannong Cao
Department of Computing
Hong Kong Polytechnic University
Hung Hom, Hong Kong, China
csjcao@comp.polyu.edu.hk

[Abstract] *This paper presents a general framework that addresses the issues in Web site design and navigation through an integrated graphical approach. The framework uses a simple and intuitive graph formalism throughout the development life cycle so that a single mental map is maintained for Web designers and users. The paper describes the framework in the context of a Web tool, known as WebStar, that has been developed for a pilot study.*

1 Introduction

The current development of Web sites with complex interconnections of large number of Web pages has been largely an ad hoc process. There has been no commonly accepted methodology, which supports ease of design, navigation, and maintenance of sophisticated Web sites. As the number of Web sites is increasing in an exponential order, with the huge information space provided by the Web, users become increasingly confused when they navigate a growing number of Web sites; finding the right information also takes longer time. The problems are partially due to the unstructured nature of the current organisation of Web sites. One well-known problem is *lost in hyperspace* [2] where the user becomes disoriented during the process of navigation. For example, in most of the existing Web browsers, the continuous process of jumping from one location to another can easily confuse the user. The main reason for this is that the user does not know the current context of space with respect to the overall information space.

Attempts have been made to develop tools and facilities to assist users to overcome the above problem. Most of the tools are designed only for one stage of Web design, navigation, and maintenance, such as FrontPage for Web design; browsers and lenses [5][7][9], history lists [3], bookmarks and filters [11] for browsing and navigation. WebOFDAV [5] also tries to maintain the user's mental map during navigation and help the user in keeping focus on his/her main goals. Yet, these approaches do not support the integration of Web site design, navigation, and maintenance.

In order to improve the design, maintenance, and navigation of WWW, better tools that enforce structure in the design phase, while supporting fully integrated maintenance and navigation capabilities, are urgently needed. We believe that a complicated Web system can be made more structured and navigated more easily through graphical visualisation and graphical operations. More importantly, maintaining a uniform view throughout the design, navigation, and maintenance cycle can reduce considerable development effort and enhance the navigation efficiency. The goal of the work reported in this paper is to propose an integrated view throughout the Web development cycle. The paper presents an integrated tool framework that supports Web design, maintenance, and navigation.

2 The Human-Web Interface

Different from most existing work, the work discussed in this paper focuses on a uniform view of the design, maintenance, and navigation of the Web. We call such a uniform view the *Human-Web Interface* (HWI), which can be facilitated in a simple framework. The HWI framework shown in Figure 1 consists of the support for three major activities: Web site design, navigation and browsing, and maintenance and updating. The front-end of the user interface consists of a *Graph Editor and Navigator* (GEN) for Web site construction and navigation, *Filters* and a Web browser. This combined front-end forms the human-web interface (HWI).

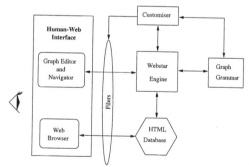

Figure 1: The HWI framework

The Web designer uses the *Graph Editor* of GEN to design and construct Web sites as graphs to be transformed and processed by the *Site Generation Engine*. The Engine performs grammatical check of the constructed graphs according to the predefined Web graph grammar [13], transforms the validated graphs into inter-related HTML files, and generates an internal data structure for debugging and maintenance purposes. The HTML files and the data structure are stored into the *HTML Database*.

The *Filters* lying between the GEN and the back-end allow Web pages to be classified and Web sites to be structurally organised and viewed according to various criteria. For example, a *domain filter* allows only the pages of a particular domain to be navigated and displayed. Filtering

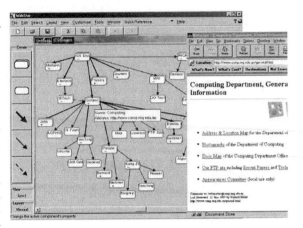

Figure 2: Navigation on a Web graph in the HWI

rules can be defined on various structures, including graph structure, Web context, and document structure [6]. Section 4 discusses information filtering in more details.

To compare HWI with traditional HCI (human-computer interface), we consider the following three aspects: the main functionality of the interface, the device for which the interface is suitable and designed, and the target of the communication that the interface facilitates.

- **Functionality**: The major role of a HWI is to act as a window to the world while a HCI could just be for a standalone computer. Therefore, the main objective of a HWI is to facilitate information gathering and retrieval while that of a HCI is to facilitate operations on a computer.
- **Communication target**: Related to the above difference, the communication target of a HWI is human while that of a HCI is machine. The human-to-human communication through HWIs may be direct such as in a net-meeting, or indirect such as in usual Web browsing. To support indirect human-to-human communications in various professional domains, we need commonly understandable and agreed communication protocols. The XML standard [12] has been motivated precisely for this reason.
- **Device**: A HWI could be installed not only on a computer, but also on a PDA (portable digital appliance), a mobile phone, or a television set. In the latter case, the HWI needs no Web design function and thus would not be equipped with a graph editor and Web site generation engine. The display could also be much more simplified. For a PDA for example, the display may only include clickable texts and running texts for navigation and browsing.

The proposed HWI framework provides a webmaster to effectively design and maintain a Web site with the uniform graphical view and allows users to navigate Web sites graphically by direct manipulation and information filtering as desired. A webmaster designs and generates a Web site by drawing the Web graph that conceptually represents the site structure. Navigation and maintenance of the Web site are performed on the same Web graph by the user.

During design or navigation, the user can click on any graph node to enter directly into the page symbolised by the node without going through all the intermediate pages. This direct access method via a Web graph is much more efficient than linear access method in conventional browsers.

We have designed an experimental tool, called WebStar, that realises the above HWI functionality through visual Web programming for design and Web visualisation for navigation. Figure 2 depicts a snapshot of the WebStar HWI when navigating the Computing Department of Macquarie University by a simple click on the node "Comput". The home page shown on the right-hand side represents the node "Comput" in the navigation window on the left-hand side.

3   Web Graphs and Design Notations

Visual structures and relationships are much easier to reason about than similar linguistically described structures. This is why designs in many application domains have been conducted on graphical representations. Using visual programming techniques to graphically design Web sites and Web pages will obviously enable more visual artists and other non-computing professionals to develop their own Web sites easily. The main philosophy behind the HWI framework is its consistent visual approach to Web design, navigation, and maintenance. This section introduces the concept of Web graphs and their notations and describes the graph operations.

A graph G (N, E) consists of a finite set N whose members are called *nodes* and a finite set E whose members are called *edges*. An edge is an ordered pair of nodes in N. A node of a graph G1 can itself be another graph G2, which is called a *sub-graph* of G1. The properties of a graph may be inherited by its sub-graphs. We regard the organisation of a Web site of any size as a graph, known as *Web graph*. A node in a Web graph represents a Web page, and an edge represents a link from one page to another. The World Wide Web is certainly the largest Web graph that is still expanding all the time. For scalability and convenience of design and navigation, we define a special class of nodes, called *group*. A group represents a set of pages that are connected to a common parent page, and share the same set of attributes.

The *distance* between a pair of nodes, node A and node B, is defined as the number of intermediate nodes along the shortest path between A and B (including B). A sub-graph of graph G consisting of a node A and all such nodes in G that have a distance of N or shorter from A is called *A's level-N sub-graph of G*, or simply *A's level-N sub-graph*.

A graph class provides the general common properties that dictate whether certain operations are applicable to the corresponding graph objects. A Web graph can be constructed using a combination of tools: a graphical editor for constructing a Web graph at the high level and a Web page tool for constructing Web pages at the lower level. The graphical editor supports two-dimensional construction of Web graphs with direct manipulation.

3.1 Graph Operations

At the design level, the graph-oriented operations for editing, navigation, and maintenance are categorised into the following groups:

- *Graph editing*: this category is for the construction and editing of a Web graph, whose nodes are associated with Web pages or are lower level Web graphs, i.e. sub-graphs;
- *Sub-graph generation*: these operations derive sub-graphs according to some classification or filtering criteria. Sub-graphs are easier to navigate and maintain than a full graph. Categorising sub-graphs also removes unrelated information in the display and during processing.
- *Query*: these operations provide information about the graph, such as the number of nodes in a graph, whether an edge exists between two nodes, and how a node is reached from another node. Such information is useful for the Web design and maintenance.
- *Time and version control*: graphs may change over a period and may reach a particular state at a predetermined time. A graph from an early state may be partially reused and incrementally updated for a later graph. Graph states are represented through versioning.
- *Viewing*: these operations include applying various filters in selecting certain categories of sub-Webs, graph updating due to a change of any graph element, and collapsing of a sub-graph into a graph node for reducing clustering.

Since graph objects may carry different meanings under different states, the semantics of a graph operation will depend on the context defined by the Web designer. Graph operations can be implemented according to their contexts but all provide the same interface to the designer.

3.2 Visual Notations

The WebStar Human-Web Interface uses a small number of simple notations, as shown on the left hand side of the screen in Figure 2, to design and visualise the components of a Web graph.

- The round-cornered rectangle denotes a Node that represents a Web page. The label is used to identify the node and the page.
- The stacked double-node denotes a Group, representing a group of Web pages that are combined together either due to their commonality or for the brevity of viewing. It is like a Web template or class, which can be used to generate similarly structured-pages. A Group also has a label that identifies a specific class of pages. This notation also enforces the consistency in the pages belonging to one Group.
- The thin arrow denotes an Edge that represents a Web hyperlink. This is the most common link seen in Web pages.
- The thick arrow is called a Gedge, short for Group Edge, which represents an edge coming out of or entering a Group. The difference between an Edge and a Gedge is that a Gedge connects to a Group and thus refers to all the Nodes belonging to the Group (some kind of inheritance). For example, if there is a Gedge connecting a Group A to a Node D and Node B is a member of A, then B is also connected to D. More

importantly, the Nodes connecting by Gedges to a Group share a common set of characteristics and attributes. This is useful in generating consistent look-and-feel pages.

- The broken-line arrow is called a **Hedge**, short for Hidden Edge. To support scalable visualisation, WebStar allows a number of nodes to be collapsed into a single node, which is called a *super-node* and represented as a Group graphically. This operation achieves the zoom-out effect. A super-node can be expanded to allow its contents to be viewed or modified. The expansion achieves zoom-in. The zoom-in and zoom-out effects help designers to view their designs easily. Collapsing and expanding could be performed in various levels or depths. It may be defined as either a connection between pages of different domains, or as a connection between a collapsed node and its neighbouring node. When generated automatically by WebStar, a Hedge indicates the existence of a connection (in form of hyperlink)

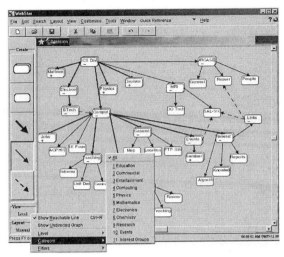

Figure 3: Selecting a filter during navigation

between collapsed nodes. The designer may use Hedge at the design stage. In this case, a Hedge denotes either an Edge or a Gedge between two nodes and the designer has not yet made a decision in an early stage of design.

4 Information Filtering

The *lost in hyperspace* [2] problem occurs in Web navigation results from the user's lack of knowledge of the overall structure of the information space, mainly due to the unstructured Web organisation. Organising Web sites structurally and navigating them graphically provides a solution to the problem. By presenting a map of the underlying Web sites interconnections, the HWI framework provides the user a sense of direction as to where s/he is in the overall information space. It also gives the user some guideline of what other related information is available and how to access it. During navigation, the graph of a Web site being viewed is displayed in such a manner that unnecessary information has been filtered out and the node representing the current page is highlighted in the graph.

The WebStar HWI supports several modes of displaying, including *Level* view, *Domain* view, *Category* view, *Pattern* view, and *Constraint* view. The Level view allows the user to choose the level of Web page pointers to display, i.e. a given level of linked pages in the Web graph relative to a given node. The Domain view shows the pages of a given application domain. If the Web designer has classified all the pages according to some application criteria, the user can choose Category view to see a given class of pages. The Pattern view allows the viewer to see some common patterns in a Web graph. Finally, the Constraint view shows all the pages that satisfy a given set of constraints.

The views described above are implemented by various filters as described below. Web designers can design or customise their own filters to suit their specific application purposes.

- **Level filter**. If the user selects a particular node, say node A in a Web graph, and specifies a level N, the system will display A's level-N sub-graph. The Level filter uses hierarchical nature of the Web information space to set the levels. Only information of the given level will be shown on the view. "Collapse" and "expand" operations use the Level filter to decide whether a particular page will be shown.
- **Domain filter.** The domain-related pages can be displayed if the user provides several keywords of an application domain.
- **Category filter**. The Category filter allows user to view the Web pages that belong to certain categories. The user could choose categories to be viewed from a list of available categories. This list of categories is gathered when the Web designer adds new pages into the Web graph and categorises them when filling the property form.
- **Pattern filter**. Groups of Web pages with certain patterns can be filtered and shown in the view. A typical use of this filter is when the user attempts to remove redundant pages by comparing similar sub-graph

1158

patterns. The filter accepts a sub-graph (perhaps generated using another filter) from the user and finds the matching patterns to be displayed. The matching process can be slow for a large Web graph.

- **Constraint filter**. Web pages that satisfy certain constraints can be displayed in the view. Examples of constraints are file size, file attributes, file's creation dates, etc. For instance, the user may wish to view all the pages under a certain file size, or view only text contents when a PDA is used for navigation.

Figure 3 shows the selection of one of the filtering options (three types of filters have been implemented at this stage). It is possible for the user to combine several filters in obtaining desired information. One example of combined filtering effect is when the Category and Level filters are both selected. The display will show only the page of the selected category but within the given distance from a particular page. The user can also set the order of applications of the filters when multiple filters are applied at the same time. Filtering process may override the default properties used in the system. Preference in execution of the filtering operations can also be applied.

5 Conclusion and Future Work

This paper has presented a visual approach to Web site design, navigation, and maintenance. It advocates the integration of the tools and consistency of the views for all activities, ranging from Web page and Web site design, navigation and browsing, to Web system maintenance, while preserving the same mental map for both the Web designer and the Web user throughout these activities. We have implemented a prototype of WebStar in Java, which is capable of generating Web sites from Web graphs drawn on the WebStar Graph Editor. Most of the presented features have been implemented. We plan to adapt an existing layout algorithm to support more pleasant viewing during navigation and maintenance.

This work has opened up many opportunities for further investigation. Our future work will include:

- Extending WebStar with CSCW capability such that more than one person may be involved in the Web design at the same time. We will incorporate and adapt some recently developed techniques in concurrent graph editing [1].
- Security features will be built into the framework so that different groups of people may access different parts of a Web site.
- We will also conduct empirical studies in order to evaluate the usability of the HWI in real world applications.

6 References

[1] D. Chen and C. Sun, A Distributed Algorithm for Graphic Object Replication in Real-Time Group Editors, *Proceedings of ACM Conference on Supporting Group Work*, Phoenix, Arizona, USA, Nov. 1999. 121-130,

[2] D. W. Edwards and L. Hardman, Lost in Hyperspace: Cognitive Mapping and Navigation in Hypertext Environment, *Intellect Books*, Oxford, 1989.

[3] E. Frecon and G. Smith, WebPATH – A Three Dimensional Web History, *IEEE Symposium on Information Visualization*, N. Carolina, October, 1998, *http://davinci.infomatik.uni-kl.de/vis98/archive/fp/papers/webpath.html*.

[4] S. Hinton, From Home Page to Home Site: Effective Web Resource Discovery at the ANU, *Proceeding the 7th International World Wide Web Conference*, 14-18 April 1998, Brisbane, Australia.

[5] M. L. Huang, P. Eades, and R.F. Cohen, WebOFDAV - Navigating and Visualising the Web On-line with Animated Context Swapping, *Proceedings of 7th International World Wide Web Conference (WWW7)*, Brisbane, 14-18 April 1998.

[6] M. L. Huang, et al. Dynamic Web Navigation with Information Filtering and Animated Visual Display, *Proceedings of Asia Pacific Web Conference (APWeb98)*, Beijing, China, 27-30 September, 1998, 63-72.

[7] W. Lai, et al. Web Graph Displays by Defining Visible and Invisible Subsets, *Proceedings of 5th Australian World Wide Web Conference (AusWeb99)*, Ballina, Australia, 17-20 April, 1999. .

[8] Y.S. Maarek and I.Z.B. Shaul, WebCutter: A System for Dynamic and Tailorable Site Mapping, *Proceedings of 6th International World Wide Web Conference*, 1997, 713-722.

[9] D. C. Muchaluat, R. F. Rodrigues, and L. F. G. Soares, WWW Fisheye-View Graphical Browser, *Proceedings of IEEE of the 1998 Multimedia Modelling*, 1998.

[10] D. Siegel, *Creating Killer Web Sites: The Art of Third Generation Site Design, 2nd ed.*, October 1997, Hayden Books, ISBN: 1568304331.

[11] M. B. Spring, E. Morse, and M. Heo, Multi Level Navigation of a Document Space, *http://www.iis.pitt.edu/~spring/mlnds/mlnds/mlnds.html*.

[12] W3C, Extensible Markup Language (XML) 1.0, *http://www.w3.org/TR/REC-xml.html*, October 2000.

[13] D.Q. Zhang and K. Zhang, Reserved Graph Grammar: A Specification Tool for Diagrammatic VPLs, *Proceedings of 1997 IEEE Symposium on Visual Languages (VL'97)*, Capri, Italy, 23-26 September 1997, 284-291.

iScape: A Collaborative Memory Palace for Digital Library Search Results

Katy Börner

SLIS, Indiana University, 10th Street & Jordan Avenue, Bloomington, IN 47405, USA
Email: katy@indiana.edu

ABSTRACT

Massive amounts of data are available in today's Digital Libraries (DLs). The challenge is to find relevant information quickly and easily, and to use it effectively. A standard way to access DLs is a text-based query issued by a single user. Typically, the query results in a potentially very long ordered list of matching documents making it hard for users to find what they are looking for.

This paper presents iScape, a shared virtual desktop world dedicated to the collaborative exploration and management of information. Data mining and information visualization techniques are applied to extract and visualize semantic relationships in search results. A three-dimensional (3-D) online browser system is exploited to facilitate complex and sophisticated human-computer and human-human interaction.

Informal user studies have been conducted to compare the iScape world with a text-based, a 2-D visual Web interface, and a 3-D non-collaborative CAVE interface. We conclude with a discussion.

1. INTRODUCTION

Large amounts of human knowledge are available online - in form of texts and images but also as audio files, 3-D models, video files etc. Given the complexity and the amount of digital data it seems to be advantageous to exploit spatial metaphors to visualize and access information. At the same time, a growing number of projects are collaborative efforts that bring people with different skills and expertise together. Domain experts are often spread out in space and time zones and consultation and collaboration has to proceed remotely instead of face-to-face. Required are user interfaces that can access multi-modal data, are available on a standard PC and at any time, and that support collaborative information access and information management efficiently and intuitively.

In this paper we present research on iScape (Information Landscape), which is part of the *LVis (Digital Library Visualizer)* project (Börner et al., 2000). iScape is a world in Active Worlds (AW) *Educational Universe*, a special universe with an educational focus. It aims at the support of the navigation through complex information spaces by mapping data stored in digital libraries onto an 'information landscape' which can then be explored by human users in a natural manner. In particular, iScape displays retrieval results in a multi-modal, multi-user, navigable, virtual desktop virtual world in 3-D, which is interconnected with standard web pages. Documents are laid out according to their semantic relationships and can be navigated collaboratively. Document full texts, images or even videos can be displayed in the 2-D web interface on demand. Users can change the spatial arrangement of retrieved documents and annotate documents ultimately transforming the world in a 'collaborative memory palace'.[1]

The subsequent section presents an overview of research on the visualization of search results. The design and the current capabilities of iScape are explained in Section 3. First results of an usability study comparing the efficiency and accuracy of the iScape environment with a text-based, and a 2-D visual Web interface as well as a 3-D non-collaborative CAVE interface are reported in section 4. We conclude with a discussion of the work.

2. VISUALIZATION OF SEARCH RESULTS

The majority of today's search engines confront their users with long lists of rank-ordered documents. The examination of those documents can be very time consuming and the dosument of real interest might be hidden deep inside the list.

---

[1] Memory palaces refer to highly evolved mnemonic structures. They have been developed in classical Greek culture to manage and recite great quantities of information. Basically, a memory palace is a non-linear storage system or random access memory that is responsive to the user's position in an imagined space.

Peter Anders (1998) argues on p. 34, that '*The memory palace could resurface as a model for future collective memory allowing users to navigate stored information in an intuitive spatial manner*' and that '*... cyberspace will evolve to be an important extension of our mental processes*' allowing us to '*... create interactive mnemonic structures to be shared and passed from one generation to the next.*'

Visualizations are used to help understand the search result. For example, Spoerri's (1993) *InfoCrystal* or *TileBars* (Hearst, 1995) visualize the document-query relevance.

Recently, several approaches have been developed that cluster and label search results to provide users with a general overview of the retrieval result and an easier access to relevant information (Hearst, 1999). An example is the *scatter/gather* algorithm developed by Cutting et al. (1996). Clustering can be performed over the entire collection in advance reducing the time spent at retrieval time. However, post-retrieval document clustering has been shown to produce superior results (Hearst, 1999, p. 272) because the clusters are tailored to the retrieved document set. Labels for clusters need to be concise and accurate such that users can browse efficiently.

Other research utilizes visualizations to support query formulation, the selection of an information source, or to keep track of the search progress (Hearst, 1999, p. 257). For example, Ahlberg & Shneiderman (1994) established a set of general visual information seeking principles comprising *Dynamic Query Filters* (query parameters are rapidly adjusted with sliders, buttons, maps, etc.), *Starfield displays* (two-dimensional scatterplots to structure result sets and zooming to reduce clutter), and *Tight Coupling* (interrelating query components to preserve display invariants and to support progressive refinement combined with an emphasis on using search output to foster search input).

Another line of research exploits spatial, real world metaphors to represent information. Examples are the *Data Mountain* by Czerwinski et al. (1999) or *StarWalker* by Chen & Carr (1999). The former enables users to manually organize a relatively small information space of personal bookmarks. The latter uses automatic data mining and artificial intelligence techniques to display citation networks of large documents sets in 3-D. *StarWalker* uses Blaxxun's community platform http://www.blaxxun.com/products/community_platform to display documents by spheres that are connected by citation links. Clicking on a sphere displays the original full text version at ACM's website in the web browser frame. Multiple users can visit this space together and communicate via the build in chat facility. However, to our knowledge users cannot change the semantic layout of documents or annotate them.

Librarea is a world in the main *Active Worlds Universe* in which real librarians can create functional, information-rich environments, meet with other librarians from around the world, create a work of art, etc. (http://librarians.about.com/careers/librarians/). However, *Librarea* does not apply data mining or information visualization techniques to ease the access and manipulation of information.

The *LVis (Digital Library Visualizer)* comes with two interfaces: a 2-D Java applet that can be used on a desktop computer (Börner, 2000) as well as a 3-D immersive environment (Börner et al., 2000) for the CAVE (Cruz-Neira et al., 1993). However, only a small number of documents can be visualized on a standard screen without overlap and the CAVE interface exploits 3-D but is a very limited resource. In addition, the CAVE requires users to use a special input device such as the joystick like 'wand' which takes practice to learn.

To our knowledge there exists no semantically organized visualization of search results that can be collaboratively explored, modified, and annotated in 3-D and is accessible via a standard desktop computer.

3. THE iSCAPE WORLD

The iScape world is a multi-modal, multi-user, collaborative virtual environment in 3-D that is interconnected with standard web pages. It was created using *the 3-D Virtual Reality Chat & Design Tool* by Active Worlds (AW) http://www.activeworlds.com/. Figure 1 shows the AW interface. In contains four main windows: a list of worlds on the left, a 3-D graphics window in the middle showing the iScape world, a Web browser window left, and a chat window. At the top are a menu bar and a toolbar for avatar actions.

Entering iScape, users can explore different search results from the Dido Image Bank at the Department of the History of Art, Indiana University http://www.dlib.indiana.edu/ collections/dido/. Dido stores about 9,500 digitized images from the Fine Arts Slide Library collection of over 320,000 images. Latent Semantic Analysis (Landauer et al., 1998) as well as clustering techniques were applied to extract salient semantic structures and citation patterns automatically. A Force Directed Placement algorithm (Battista et al., 1994) was used to spatially visualize co-citation patterns and semantic similarity networks of retrieved images for interactive exploration. The final spatial layout of images corresponds to their semantic similarity. Similar images are placed close to one another. Dissimilar images are further apart. Details are reported elsewhere (Börner et al., 2000; Börner, 2000).

Users of iScape are represented by avatars (see fig. 3, left). They can collaboratively navigate in 3-D, move their mouse pointer over an object to bring up its description (see Fig. 1), click on 3-D objects to display the corresponding web page in the right Web frame of the AW browser, or teleport to other places. The web browser maintains a history of visited places and web pages so that the user can return to previous locations.

1161

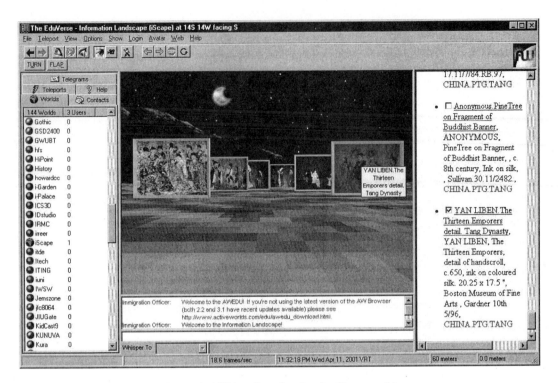

Figure 1: AW interface showing the iScape world

Besides exploring the space collaboratively, users can select and move images thus changing the semantic structure of documents to resemble and communicate their experience and interest. In addition, they can annotate images.

4. INFORMAL USER STUDY

Cugini et al. (1999) report an extensive usability study in which they evaluated text-based, 2-D and 3-D visual interfaces to search results employing a document three-keyword axes display to visualize query relevance. In particular they showed how visualizations might lead to either increased or decreased cognitive load.

So far a set of informal user studies has been conducted to compare the efficiency and accuracy of the iScape environment with Dido's original text-based interface see fig. 2, left, the 2-D LVis Java applet Web interface depicted in fig. 2 right, and the 3-D non-collaborative LVis CAVE interface shown in fig. 3, right. Note that all four interfaces depicted in fig. 2 and 3 display the same search result, namely *chinese paintings from the 5th Dynasty period* (keyword equals china.ptg.tang) retrieved from Dido.

Inspired by the study by Cugini et al. (1999) we tried to make the visual appearance of the four interfaces as similar as possible, preserving as much of the functionality as possible. In particular, the background and floor of iScape's AW interface and the LVis CAVE interface have been changed into the same yellow as the Dido text-based interface & the LVis 2-D Java applet. The exact same spatial layout of images was used for the 2-D and the 3-D interfaces. The proportion of images to the human user in the CAVE interface and images presented in AW to the users' avatar are identical.

Each subject was confronted with two interfaces and had to solve a set of retrieval tasks such as:
(1) Find an image given part of its textual description.
(2) Find an image given its image.
(3) Find two images that are visually similar to an image, and
(4) Find all images by the same artist.

1162

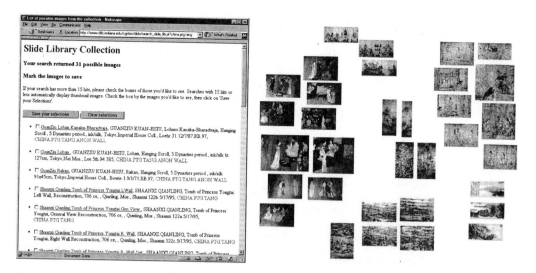

Figure 2: Dido text-based interface and the LVis 2-D Java applet

In general, retrieval over textual image descriptions was superior in response time and accuracy for task 1 and 4. As for task 2 and 3, the 2-D Java interface was superior over the 3-D interfaces if the number of images was small and images did not overlap.

Figure 3: iScape AW interface and the LVis CAVE interface

For larger numbers of images, users exploited 3-D navigation to quickly get different vantage points on image sets. The bird's eye view was used frequently. Response time and task accuracy have been better for the AW browser than for the CAVE interface which is very like due to the fact that users interact in AW via a standard mouse and keyboard while they are required to learn an unfamiliar interface in the CAVE. Like Cugini et al. (1999) we found that users are typically familiar with text-like operations such as scrolling and selecting but they need time to learn how to use graphical interfaces. Also, it took some practice in 3-D navigation and object selection before they could exploit spatial metaphors in 3-D. The overall user satisfaction with the 2-D Java and the 3-D interfaces was higher than for the text-based interface. Detailed results of this usability study are forthcoming.

5. DISCUSSION

This paper introduced iScape, a virtual desktop world for the display of retrieval results. Users of iScape can collaboratively experience the semantic relationships between documents or access concrete images/documents.

Besides, they can manipulate the semantic structure of documents to resemble and communicate their experience and interest.

We believe that the computational power and high-end interface technology available today should be exploited to build DL interfaces that are easier to use and interact with and that assist users in the selection, navigation and exploitation of information.

Still, our knowledge about the strengths and limitations of 3-D, immersive (desktop) worlds is very limited. Detailed usability studies are necessary to provide guidance on the selection of appropriate interfaces and visualizations for specific user (groups), tasks, and domains.

6. ACKNOWLEDGEMENTS

This work would not have been possible without ActiveWorld's generous free hosting of the iScape world, Mandee Tatum's and Lucrezia Borgia's continuous support, and the active research environment in EduVerse. We are grateful to Eileen Fry from Indiana University for her insightful comments on this research as well as ongoing discussions concerning the Dido Image Bank. Maggie Swan provided insightful comments on an earlier version of this paper. The SVDPACK by M. Berry was used for computing the singular value decomposition. The research is supported by a High Performance Network Applications grant of Indiana University, Bloomington.

REFERENCES

Ahlberg, C. & Shneiderman, B. (1994) Visual Information Seeking: Tight Coupling of Dynamic Query Filters with Starfield Displays, Proceedings of CHI'94 Conference on Human Factors in Computing Systems, ACM press, pp. 313-480.

Anders, P. (1998) Envisioning Cyberspace: Designing 3-D Electronic Spaces, McGraw-Hill Professional Publishing.

Battista, G., Eades, P., Tamassia, R. & Tollis, I.G. (1994) Algorithms for drawing graphs: An annotated bibliography. Computational Geometry: Theory and Applications, 4 (5), pp. 235-282.

Börner, K., Dillon, A. & Dolinsky, M. (2000) LVis - Digital Library Visualizer. Information Visualisation 2000, Symposium on Digital Libraries, London, England, 19 -21 July, pp. 77-81. (http://ella.slis.indiana.edu/~katy/IV-LVis/)

Börner, K. (2000) Extracting and visualizing semantic structures in retrieval results for browsing. ACM Digital Libraries, San Antonio, Texas, June 2-7, pp. 234-235. (http://ella.slis.indiana.edu/~katy/DL00/)

Chen, C. & Carr, L. (1999) Trialblazing the literature of hypertext: Author co-citation analysis (1989-1998). Proceedings of the 10th ACM Conference on Hypertext.

Cruz-Neira, C., Sandin, D. J. and DeFanti, T. A. (1993) Surround-screen projection-based virtual reality: The design and implementation of the CAVE, in J. T. Kajiya (ed.), Computer Graphics (Proceedings of SIGGRAPH 93), Vol. 27, Springer Verlag, pp. 135-142.

Cutting, D., Karger, D., Pedersen, J. & Tukey, John W. (1992) Scatter/Gather: A Cluster-based Approach to Browsing Large Document Collections, Proceedings of the 15th Annual International ACM/SIGIR Conference, Copenhagen.

Hearst, M. (1995) TileBars: Visualization of Term Distribution Information in Full Text Information Access, Proceedings of the ACM SIGCHI Conference on Human Factors in Computing Systems (CHI), pp. 59-66.

Hearst, M. (1999) User Interfaces and Visualization. In Modern Information Retrieval, Ricardo Baeza-Yates & Berthier Ribeiro-Neto, chapter 10, Addison-Wesley.

Landauer, T. K., Foltz, P. W., & Laham, D. (1998) Introduction to Latent Semantic Analysis. Discourse Processes, 25, 259-284.

Sebrechts, M. M., Cugini, J. V., Laskowski, S. J., Vasilakis J. & Miller, M. S. (1999) Visualization of search results: A comparative evaluation of text, 2-D, and 3-D interfaces. Proceedings on the 22nd Annual International ACM SIGIR Conference on Research and Development in Information Retrieval, pp. 3-10.

Spoerri, A. (1993). InfoCrystal: A Visual Tool for Information Retrieval. In Proc. Visualization'93, pages 150-157, San Jose, California (1993). IEEE Computer Society.

The Indiana University Department of the History of Art Dido Image Bank, http://www.dlib.indiana.edu/collections/dido/

USER BEHAVIOR IN HYPERTEXT BASED TEACHING SYSTEMS

Anja Naumann[1], Jacqueline Waniek[1], Josef F. Krems[1] & Diana Hudson-Ettle[2]

[1]Dept. of Psychology & [2]Dept. of English Language and Linguistics,
Chemnitz University of Technology, 09107 Chemnitz, Germany
anja.naumann@phil.tu-chemnitz.de

ABSTRACT

The aim of this study was to identify the characteristics of an internet based teaching system developed for second language acquisition (specifically of English grammar) with the intention of developing design principles which would improve navigation and knowledge acquisition. Several studies have shown that the advantages and disadvantages of hypertext depend on the complexity of the set task (reading or information seeking). The present study was conducted on the basis of these findings to determine whether similar differences occur in the behavior of users approaching second language acquisition via hypertext. Furthermore, user behavior was expected to vary according to different user profiles, e.g. computer literacy and previous knowledge. The results show that in didactic hypertexts the complexity of the task leads to differing results in knowledge acquisition and navigational behavior. In addition, orientation problems known to be typical of hypertext are also evident in both kinds of task. Previous knowledge appears to exert a greater influence on knowledge acquisition than personal features, such as experience with hypertext.

1 INTRODUCTION

Studies comparing hypertext and linear text predominantly report more orientation and navigational problems (Ohler & Nieding, 2000) and less knowledge acquisition for hypertext than for linear text (Chen & Rada, 1996; Gerdes, 1997). Furthermore, the quality of interaction with hypertext seems to depend on the complexity of the task (Chen & Rada, 1996).

A first, preliminary study (Naumann, Waniek, & Krems, 2001) compared electronically prepared exploratory linear texts and hypertexts with regard to knowledge acquisition and processing load. As expected, when reading the linear text, participants acquired more knowledge than when reading the hypertext version. The hypothesis that more orientation problems occur in a hypertext environment than with linear text was also supported. Furthermore, the results recorded for information retrieval, that is the number of correct answers, the length of answering time, and knowledge acquisition, did not differ whether obtained via linear text or hypertext. In addition, fewer orientation problems were reported for the hypertext version than for the linear text. The results show that an improved hypertext design is essential for users reading a text with scant previous knowledge. On the other hand, hypertext seems to be at least as suitable as linear text, or even more so, for the purpose of information retrieval. The advantages and disadvantages of hypertext are thus seen to be dependent on the task.

A follow-up study was undertaken to determine whether there was evidence of differences in user behavior regarding didactic texts which could be seen - as in the exploratory texts - to be dependent on the complexity of the task. Furthermore, user behavior was expected to vary according to different user profiles, e.g. computer literacy and previous knowledge, e.g. level of education, experience with computers, and previous knowledge.

2 METHOD

1.1 Participants

30 students from the English department (from 1st to 12th semester) with a medium computer literacy participated in this experiment. 23 of the students were female and 7 were male, and the average age was 22 years (SD = 3.7).

1.2 Material and design

The Chemnitz Internet Grammar is a hypertext learning environment which is being developed by the English Department at Chemnitz University of Technology (see: http://www.tu-chemnitz.de/phil/InternetGrammar). The deductive learning component of The Chemnitz Internet Grammar consists of three modules: explanations, examples, and exercises (for more background information see Schmied, 1999). The three parts are variously linked. Screenshots of a sample page of each part and the relevant contents page are shown in figure 1.

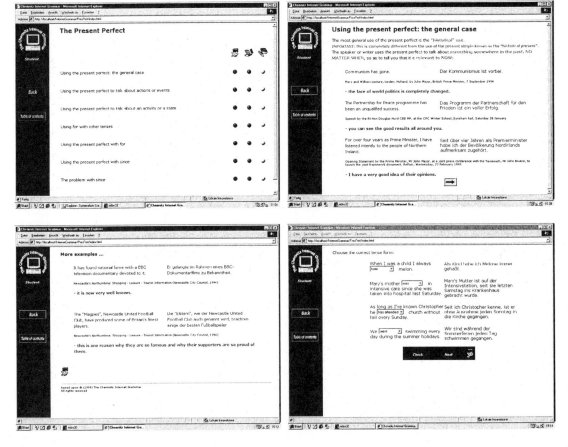

Figure 1. Contents, explanation , example, and exercise pages of a section of the Chemnitz Internet Grammar (see: http://www.tu-chemnitz.de/phil/InternetGrammar)

Two chapters (Present Perfect and Present Continuous) of the original Grammar were adapted for laboratory use to fulfill the preconditions for the experiment. Each chapter consisted of approximately 50 screen pages without scroll bars. Navigation was possible by clicking embedded links and symbols and using the back button and the table of contents button in the left frame.

The data collected were the number of correct answers in knowledge tests on the Present Perfect or Present Continuous before and after the task, demographic data, reported orientation and navigation problems and difficulties mentioned pertaining to the text and text design (information acquired by means of questionnaires). Furthermore, logfiles of navigation behavior were recorded on the participant's computer. These logfiles list the links and buttons used, reading time, visited pages, and the length of time spent on these pages. In addition, for the Present Perfect, the number of correct answers and eye movements were recorded.

2.3 Procedures

Each participant read one chapter (Present Continuous) of the Internet Grammar on the screen in the first session and answered questions with the help of another chapter (Present Perfect) in the second session. There was an interval of at least three days between each session.

Before reading the text on the one hand or answering questions with the help of the text on the other, participants were asked to fill in a questionnaire requiring demographic information, details of computer and Internet experience and an assessment of their attitude towards computers etc. An additional questionnaire on orientation and navigational problems, difficulties concerning the text content and text design was given to each participant on completion of the task.

During a brief introductory phase after filling in the questionnaire, participants were given basic instructions on how to navigate through the text. The eye tracking system was then calibrated and the actual experiment was implemented. The task in the first session was to read the chapter on the Present Continuous. The task in the second session was to answer 22 questions with the help of the chapter on the Present Perfect. Preceding and following each task the knowledge of the participants was tested by a gap-filling exercise in a text with 37 gaps.

3 RESULTS

By reading through the Internet Grammar texts (the Present Continuous section), participants acquired more knowledge than by trying to answer questions with the help of the Internet Grammar (Present Perfect). Participants with more previous knowledge acquired less additional knowledge than participants with less previous knowledge ($r = -.62$, $p<.01$).

After *reading* the chapter Present Continuous, participants scored significantly higher ($M = 28.4$, $SD = 2.8$) in the knowledge test than they did before the reading ($M = 26.5$, $SD = 3.1$; $t(28) = -3,15$, $p<.01$). This is shown in figure 2. Participants who clicked more often on the contents page ($r = -.41$, $p<.01$) and stayed there longer ($r = -.43$, $p<.05$), acquired less knowledge. Orientation seems to distract from the actual learning process. Furthermore, participants who reported that they liked working with computers attempted more exercises within the text. Participants who found the text interesting rated the text as less demanding. No interaction was found between knowledge acquisition and computer literacy, frequency of internet use, reported orientation problems, total reading time, number of pages visited, number of attempted exercises, rating of the text and other navigational behavior.

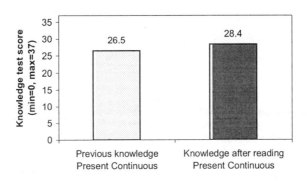

Figure 2. Knowledge before and after reading the chapter Present Continuous of the Internet Grammar.

The group with the the highest scores in knowledge acquisition (more than 4 points, n = 8) had significantly less previous knowledge (M = 23.1, SD = 1.9) than the group with the lowest scores (less than 1 point, n = 9; M = 8.6, SD = 2.7, F(16)=23.1, p<.01). This group with the highest scores clicked significantly less frequently (M = 5.9, SD = 4.7) on the contents page than the group with the lowest scores (M = 13, SD = 7.1; F(15) = 5.2, p<.05) and the group with the highest scores spent less time on this page in all (M = 21.7s, SD = 12.2) than the group with the lowest scores (M = 45.2s, SD = 21.9; F(15 = 6.5, p<.05).

In *answering questions* with the help of the explanation pages of the section on the Present Perfect in the Grammar, participants did not score significantly higher in the knowledge test than they had done before the task. Participants with more previous knowledge had higher results for knowledge after the task (r = .61, p<.01) and answered more questions correctly (r = .45, p<.05). Furthermore, participants who spent a longer period on explanation pages (r = .58, p<.05) and more time on the visited pages as a whole (r = .45, p<.05), acquired more knowledge. However, participants who acquired more knowledge, answered fewer questions correctly (r = .44, p< .05) in the same total processing time.

The group of participants with the highest scores in knowledge acquisition spent a significantly longer period of time on the example pages (M = 1418.4s, SD = 461.5) than the group with the lowest scores for knowledge acquisition (M = 980.9, SD = 268.2; F(14) = 9.7, p< .05). This group with the highest scores had also fewer orientation problems (M = 2.0, SD = 0.9) than the group with the lowest scores (M = 2.8, SD = 0.8; F(16) = 5.7, p< .05). Whereas knowledge acquisition increases with the time spent on pages read, the number of correct answers decreases. No interaction was found between knowledge acquisition and computer literacy, frequency of Internet use, time for answering the questions, number of correct answers, number of performed exercises, rating of the text, and other navigational behavior.

4 CONCLUSIONS

In the task of *reading* through a chapter of the Internet Grammar, the requirements of orientation appear to distract from actual knowledge acquisition. Participants who clicked more often on the contents page and stayed there longer, acquired less knowledge. In order to improve knowledge acquisition in this hypertext learning environment, the development of orientation support and navigational aids is deemed essential.

In *information retrieval*, a long search for information, especially in the explanations section, goes in line with more knowledge acquisition but not with a more successful answering of the questions. Knowledge acquisition increases with the time spent reading the pages visited. That is to say, not finding information appears to interact with incidental knowledge acquisition. Furthermore, reported orientation problems are associated with a low degree of knowledge acquisition. This supports the contention that there is a need to provide assistance for orientation and navigation in hypertext.

To summarize, it can be claimed that differences in knowledge acquisition and navigational behavior related to the complexity of the task are also to be found in the processing of didactic hypertext. This supports the results of previous studies on other text types. In addition, the orientation problems typically associated with hypertext are encountered in the two kinds of task investigated.

Previous knowledge seems to exert considerable influence on the amount of knowledge acquisition, whereas personal characteristics like experience with hypertext or level of education do not seem to be very relevant. Learners with little previous knowledge appear to benefit more from the text than learners with more previous knowledge.

Based on these results and on a forthcoming analysis of navigational paths and eye movement of different user groups, design principles for user oriented structural aids will be developed for future use in the Internet Grammar and in similar applications.

ACKNOWLEDGEMENTS

We would like to thank the German Science Council DFG, which provided support for this research (Grant: BO 929/13-1).

REFERENCES

Chen, C. & Rada, R. (1996). Interacting with Hypertext. A Meta-Analysis of Experimental Studies. *Human Computer Interaction, 11,* 125-156.

Gerdes, H. (1997). *Lernen mit Text und Hypertext [Learning with Text and Hypertext].* Lengerich: Pabst.

Naumann, A., Waniek, J. & Krems, J.F. (2001). Knowledge acquisition, navigation and eye movements from text and hypertext. In: U.-D. Reips & M. Bosnjak (Eds.). *Dimensions of Internet Science.* Berlin: Pabst Science Publishers.

Ohler, P. & Nieding, G. (2000). Kognitive Modellierung der Textverarbeitung und der Informationssuche im WWW [Cognitive Modeling of text processing and information seeking in the WWW]. In: Batinic, B. (Ed.). *Internet für Psychologen* (2nd rev. ed.). Göttingen: Hogrefe.

Schmied, Josef (1999) Applying Contrastive Corpora in Modern Contrastive Grammars: The Chemnitz Internet Grammar of English. In: Hasselgard, H. & S.Oksefjell (Eds.). *Out of Corpora Studies in honour of Stig Johansson* (pp. 21-30). Amsterdam: Rodopi.

Mutual Evaluation and Situation Map to Encourage the Participation in Web-based Conferencing

Yu Shibuya and Yoshihiro Tsujino

Kyoto Institute of Technology
Matsugasaki, Sakyo-ku, Kyoto 606-8585 JAPAN

ABSTRACT

The web-based discussion system, which embodies the situation map, has been constructed and evaluated experimentally. In the situation map, several kinds of useful information for participants are visualized. They are the mutual evaluation of the opinion, the activity level of it, and so on. Comparing with text-based representation, the situation map was useful and helpful for participants to find designated opinions. From the subjective evaluation, it was found that the mutual evaluation and the activity level of the opinion are important factor to select the interesting opinion.

1. INTRODUCTION

Many on-line interaction environments have been established and used for a lot of purposes. We have been developing web-based discussion systems and using them for the conferencing in our community and classroom (Shibuya, 1997; Choui, 1998). One of the issues of our research is to find suitable methods for encouraging participation. Here, the word "participation" means our member's reading/writing opinions. If people can find interesting opinions easily, they might be encouraged to participate to the conferencing.

In order to characterize the opinion, two kinds of information are introduced in this paper. The first is the mutual evaluation among participants, that is, they evaluate their opinions each other. The second is the activity level of the opinion which represents how often read or referred recently.

Furthermore, it is important to show such information in suitable way. There are many researches in information visualization. For example, *PeopleGarden* is a graphical representation of users based on their past interactions (Xiong, 1999). We introduce the situation map which is a visualization of various information in conferencing. If there are a large number of opinions, it is difficult for participants to understand the current situation or to find the interesting opinion through the text-based representation. However, it might be easy to do so by well-designed visualization.

2. WEB-BASED DISCUSSION SYSTEM

2.1 System Configuration

The system configuration of our web-based discussion system is shown in Figure 1. The system consists of a server and many client computers. They are connected via computer networks. The server is used for both web and database server. Participants can access the server using their web browser on the client from anywhere at anytime.

If participants can find interesting opinions easily, they might be encouraged to read or make their comments on such opinions. We focused on the two factors of the opinion. They are the score of the mutual evaluation and the level of the activity.

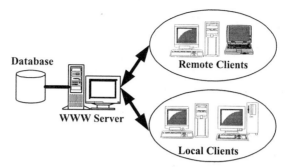

Figure 1. System configuration of the web-based discussion system.

2.2   Mutual Evaluation

Evaluations of opinions will help participants to find interesting opinions. It might be possible that particular staffs are assigned for the evaluations. However, it is difficult for them to evaluate the large number of opinions quickly. Furthermore, evaluations may be toward the direction depending on their taste. In order to overcome these problems, the mutual evaluation is introduced in this paper. That is, participants evaluate each opinion and give the score to it.

Each opinion is evaluated with its originality, applicability, and readability by participants who have read it. The range of score is from 1 to 5 and 5 is the best. If the score is simply added up to the total, a new good opinion is underestimated because only a few participants have read it. On the other hand, an old ordinary opinion is overestimated because many participants have read it. However, if the average score is used, some unique opinions might be underestimated. In this paper, each score is cubed and added up to the total score of the opinion in order to avoid the underestimation, so that we should be able to pick up as many interesting opinions as possible.

In our system, participants can evaluate the opinion only before writing their response to it, because they should read each opinion carefully and try to understand what is mentioned in it. The total score is stored for every opinion and used for the situation map as mentioned below.

2.3   Activity Level

Each opinion has its life span in network conferencing. An opinion is born when an author posts it. If there is no access to the opinion for a long period, it might be regarded as dead. That is, the opinion life span is from its birth to death. In other words, the opinion is active when it lives. In this paper, the activity level is introduced for characterizing the opinion.

When the opinion is born, its activity level is set to a predefined value. The activity level decreases as the time past. If there is a read or reference of the opinion, a predefined amount of value is added to its activity level.

In Figure 2, a typical life span of an opinion is shown. As shown in Figure 2, this opinion was referred 5 times and a comment was written about it. In this case, the amount of increased level by receiving a comment is larger than that by reference.

It is easy to understand that the high activity level does not always indicate the high quality of the opinion. For example, the opinion, which has just been born, has the moderate activity level but there is no evaluation whether the opinion is interesting or not. On the other hand, if a good opinion has not been accessed for a long period, it might not be interesting to current participants.

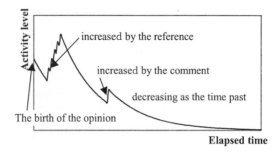

Figure 2. A typical life span of an opinion.

The interesting opinion to participants should have both high activity level and good mutual evaluation. In order to encourage the participation in web-based conferencing, it is important that interesting opinions should be found easily. For such purpose, the situation map is introduced in this paper.

2.4 Situation Map

The situation map is used for visualizing both the mutual evaluation and the activity of the opinion. This map is used to find interesting opinions among lots of opinions. In Figure 3, a situation map is shown. In this figure, each opinion is drawn as a rectangle on the map. The vertical axis denotes the total score of the mutual evaluation. The size of the rectangle denotes the level of the activity. The bigger rectangle is, the higher activity level is.

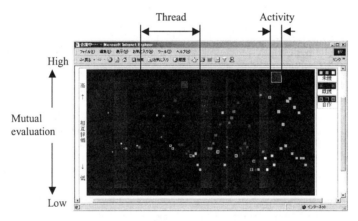

Figure 3. Situation Map

An opinion belongs to a thread of the discussion. Each colored vertical band represents each thread respectively. In the band, each opinion is located left to right in the order of its birth and the horizontal distance among them is fixed. So, the width of the band represents the amount of the opinions in it. Three colors are used to distinguish each band and its opinions. At this moment, pink, yellowish green, and light blue are used.
It is useful for participants to distinguish the unread opinions from the already read opinions. For such purpose,

the brightness of the color of unread opinions is set to higher value than that of others. Furthermore, the participant is not interested in reading his/her own opinions. In order to distinguish his/her own opinions from others, the border of the rectangle is drawn in bright color.

When participants point a rectangle by the mouse, the title of the opinion will be pop-upped. This title is helpful for the participant to predict the content of the opinion without opening it.

The situation map might also be helpful for the new comers or participants who have been absent for a long time. They can find a current active thread by checking the width of the band and the size of the rectangle. They can also find interesting opinions which are located higher region of the map and drawn in bigger rectangle. So, they can easily catch up the mainstream.

3. EXPERIMENT

In order to clarify the usefulness of the situation map, it is compared with the text-based representation as shown in Figure 4. Following 6 tasks are used for experiment.

(1) Find an opinion which has the highest score in mutual evaluation.
(2) Find an opinion which has the highest level of activity.
(3) Find an opinion which belongs to the largest thread and has the highest score of mutual evaluation in the thread.
(4) Find an opinion which belongs to the largest thread and has the highest level of activity in the thread.
(5) Find an opinion which has the highest score of mutual evaluation among unread opinions.
(6) Find an opinion which has the highest level of activity among unread opinions.

Figure 4. A text-based representation of the conferencing

There were 11 subjects and 95 opinions had been written for this experiment. Each subject was asked to select the designated opinion correctly and as soon as possible. Selection time was measured and used for evaluation. After above tasks, subjects were asked to answer the questionnaire about the situation map.

In the questionnaire, subjects were asked whether the following factors were important or not in selecting the opinion. They are the score of mutual evaluation, the activity level, thread representation, the number of opinions in the thread, distinction between unread opinions and read ones, and distinction between

1173

participant's own opinions and others. Subjects were also asked the suitability of the visualization of these factors.

4. RESULTS

From the experiments, the average time of selection is shown in Table 1. As shown in Table 1, the average selection time using situation map is significantly shorter than that using text-based representation in every task ($p < 0.05$).

Table 1. The average time of selection.

Task no.	Average selection time [sec]	
	Situation map	Text-based representation
(1)	9.97	13.67
(2)	9.14	16.08
(3)	9.71	18.53
(4)	10.21	29.13
(5)	11.11	27.60
(6)	13.21	35.63

From the subjective evaluation, the score of the mutual evaluation and the level of the activity of the opinion are important factor to select the interesting opinion. They thought the thread which has a lot of opinions might be interesting discussion. Furthermore, they would like to select unread opinions than others. In our situation map, the score of mutual evaluation and the distinction between unread opinions and others were well represented. However, visualization of both the activity level and the number of opinion didn't get so good evaluation from subjects. We should improve it in next step while the further experiments are needed to evaluate our situation map.

5. CONCLUSION

The mutual evaluation and the activity level of the opinion might encourage the participation in conferencing. The situation map, which represents various information of the conferencing, was introduced to the web-based discussion system and evaluated experimentally. Comparing with text-based representation, the situation map was useful and helpful for participants to find designated opinions. From the subjective evaluation, it was found that the mutual evaluation and the activity level of the opinion are important factor to select the interesting opinion. However, further experiments are needed to evaluate the system.

6. REFERENCES

Choui, M., Kuwana, G., Shibuya, Y., & Tamura, H. (1998). "Network Support for Imaginative Communication in Technological Systems and Advanced Societies", Proc. the 7th IFAC Symposium on Man-Machine Systems, pp. 585 - 590.

Shibuya, Y., Kuwana, G., & Tamura, H. (1997). "Teaching Information Literacy in Technical Courses Using WWW", Advances in Human Factors / Ergonomics, vol. 21B, pp. 723 - 726.

Xiong, R. and Donath, J. (1999). "PeopleGarden: Creating Data Portraits for Users", Proc. the 12th annual ACM symposium on User interface software and technology, pp. 37 - 44.

MURBANDY: A Dynamic One-Screen User Interface to GIS Data

Maximilian Stempfhuber[*] and Bernd Hermes[1]
Luca Demicheli and Carlo Lavalle[2]

[1]Social Sciences Information Centre (IZ), Lennéstr. 30, 53113 Bonn, Germany
[2]European Commission, DG Joint Research Centre, Space Applications Institute,
Strategies and Systems for Space Applications (SSSA) Unit, TP 261, Ispra (VA), I-21020, Italy
[*]Corresponding author: st@bonn.iz-soz.de

To support the user in his information seeking task, graphical user interfaces with direct manipulation have proven to be superior to command line interfaces with formal query languages – at least for the non-expert user. Dynamic queries and query previews further enhance usability by tightly integrating query formulation and a calculated preview of the results the user can expect. We combined both with dynamic screen layout and a Visual Formalism, resulting in a highly flexible status display that integrates query formulation and review of results into a single screen. In addition, it visualises dependencies between query attributes at the most detailed level, making them available to the user for query refinement and interactive exploration of results.

1 Introduction

Graphical user interfaces (GUI) for information systems allow users to interactively specify their queries without the need to learn formal query languages. These languages, like SQL, have shown to be difficult to comprehend, mostly because the Boolean operators' semantics (AND, OR and NOT) do not exactly match their semantics in natural language (Greene et al. 1990, Hertzum&Frøkjær 1996). Although GUIs use direct manipulation or form-fillin instead of typing at a command line, they are often based on the Boolean retrieval model too. It is the case with form-based screen designs, where search attributes (e.g. title, keyword, author or publication year) are implicitly connected with the AND operator to reduce the result set. Multiple values within the fields are normally OR-ed to let the user specify alternatives.

To close the gap between query formulation and result display, which are often located at separate screens, *dynamic queries* (Ahlberg et al. 1992) and *query previews* (Doan et al. 1996) combine both at the spatial and temporal level. The Dynamic HomeFinder (Williamson&Shneiderman 1992) uses direct manipulation of search attributes and adequate visualisations of the result set, which allows fast and reversible operations whose impact on the query result is immediately visible. To improve query performance in networked information systems, where long response times can reduce the benefits of dynamic queries, query previews (Greene et al. 1999) use metadata to calculate the size of the result set in advance. Now the user is able to explore the data and refine his query before it is sent to the server. At the same time he gets valuable information about how search attributes influence each other.

2 Dynamic screen layout and tight coupling

While the systems mentioned above have proven to be superior to command line interfaces, they sometimes limit the complexity of queries. This is the case if more than one value per search attribute can be specified. Not all controls allow multiple selections or they fail to adequately display options and selected values if their number is large. A status display to reduce short term memory load is often missing due to space restrictions on the screen. If present, it doesn't always allow interaction, but forces the user to return to some earlier screen for query refinement.

In addition, the user may not always be able to comprehend how the underlying logic works and influences the result set. While search attributes are normally combined with AND to reduce the result set, multiple values of one attribute are (at least in fact retrieval) combined with OR and would otherwise result in zero-hits queries. Query previews often only display the total of the results, failing to visualize dependencies in the data and successful combinations of search terms, or to let the user control the query at this fine-grained level.

To solve this problems, we propose tight coupling between the controls which represent the search attributes and a dynamically growing status display (figure 1). The general idea was introduced with the ELVIRA information system (Stempfhuber 1999) and was very well accepted by the users. The selection lists in the lower area of the screen let the user select rather than enter the search terms (recognition vs. recall) while the status display above dynamically grows and shrinks according to the number of entries selected in the lists below. The status display is enhanced with means to directly enter or delete search terms, so that there is no need to re-locate an entry in the selection list to de-select it. Experienced users can use the entry field in the status display to type in numeric shortcuts or search terms, which are then located in the selection list with the help of a thesaurus. The status display is still visible while the results are displayed. The user is able to compare query and result within one screen, change the query with the means of the status display and send it to the server again.

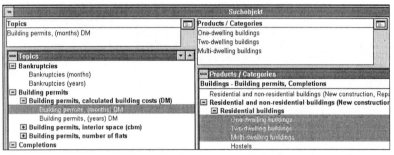

Figure 1: Dynamic layout and tight coupling between selection lists and status display

3 Visual Formalisms

The proposed solution and the systems mentioned so far fail to visualize dependencies between query attributes – or their values – and the elements of the result set. Because multiple values of an attribute are often combined with the Boolean OR operator and the attributes themselves are combined with AND, query previews give only the total of records in the result set, but may fail to visualize which *combinations* of search values appear in the result set.

Visual Formalisms (Nardi&Zarmer 1993) are generic visualizations like maps, tables or charts, which have to be adapted to a certain domain when used. They use humans' information processing capabilities – like detecting patterns from spatial arrangements – and visualise dependencies between data. They are also well known through frequent use and do – in contrast to metaphors – not require explicit transfer of knowledge between domains. What makes them useful in computer interfaces is that they are interactive, being output and input at the same time.

Attribute	B1	B2	B3	
A1	☒		☐	Σ
A2	☐	☒		
A3	☒	☐	☒	
	Σ			

Figure 2: Visual formalism for displaying dependencies between attributes

Our visualization uses a table – a Visual Formalism – where each axis shows one search attribute (figure 2). The table dynamically grows or shrinks depending on the number of values specified for each attribute. The cells of the table contain one of three possible values:

- *Empty cell*: no dependency between the corresponding attribute values.
- *Unmarked checkbox*: dependency in the data which the user can activate to include it in the query. (mark the checkbox by clicking the mouse)
- *Marked checkbox*: activated dependency which adds records to the result set.

1176

The number of records resulting from an activated cell can be displayed within the cell or as sum for rows and columns. This gives the user an idea how the result set will grow or shrink when activating / de-activating a cell. Figure 3 shows this for a document retrieval system, where the two attributes to combine are the search terms and the databases to saerch (adapted from Stempfhuber 2001). In the example the user has excluded the term 'legislation' from the database FORIS.

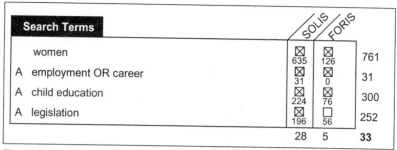

Figure 3: Visualising dependencies between search terms and databases

4 The MURBANDY user interface

In the project MURBANDY (Monitoring Urban Dynamics, Lavalle et al. 2000), methods for monitoring urban dynamics of European cities are developed and indicators are created, which make these dynamics and the influence on cities' peripherals understandable. The MURBANDY WWW user interface combines dynamic screen layout and our Visual Formalism into a one-screen solution for information search, visualization and exploration.

In the initial screen (figure 4), selection lists for land use classes and cities are tightly coupled with a status display. The status display contains in the beginning only an empty field, to let experienced users type in cities or land use classes. It grows as values are directly entered or selected in the lists, and shrinks when they are deleted or de-selected. Usage of screen real estate is constantly optimised and the load on short term memory is reduced. Beginners and experts are supported equally well through the selection lists (recognition vs. recall) and the entry field in the status display, where even hierarchy numbers from the land use classification can be entered as a shortcut. Smooth transition from novice to expert is therefore supported within one single user interface.

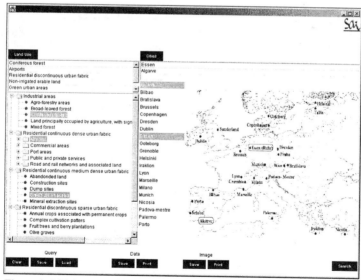

Figure 4: Initial screen with adapted selection lists and status display

1177

After executing the query, the status displays for both search attributes are transformed into a tabular query preview (figure 5). While the search attribute 'land use' stays in place, the attribute 'cities' was rotated and became the column heading of the preview table. By clicking the mouse on the labels 'land use' and 'cities', the selection lists can be opened again to modify the query. But since the table is interactive, additional entries can be directly typed in or deleted. This allows reformulation or refinement of the query within the result screen. The status display here is the constant part of the user interface, giving the user the impression that he remains on the same screen for query formulation and result display.

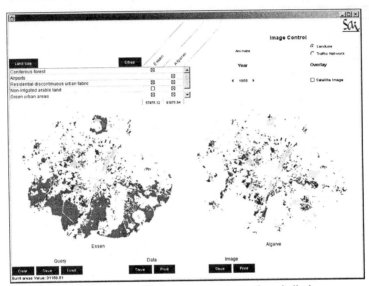

Figure 5: Integration of status display, query preview and result display

The tabular query preview lets the user specify the query at a more detailed level than it is possible with the selections lists and the status display in the initial screen layout (figure 4). There, all values of one attribute are combined with OR and the two attributes are combined with AND, giving the Cartesian product as a *possible* result set. Adapting the values of the attributes dynamically as query formulation proceeds (cf. Doan et al. 1996, Greene et al. 1999), will not always give precise information on existing combinations. As soon as two values are selected in one attribute (e.g. 'airports' and 'coniferous forests') dependencies to the reduced list of values of the other attribute (e.g. 'Essen' and 'Algarve') are not clear (both land use classes for both cities?).

The query preview exactly shows the dependencies between the selected values of both attributes, clearly stating which combination is available in the database. Because it is tightly coupled with the visualization of the results – the coloured maps of the selected cities – it can simultaneously be used for exploration. De-selecting a land use class in the preview table will instantly hide it in the corresponding map. Certain combinations of land use classes and cities can be temporarily excluded from the query without returning to some previous screen.

Conclusion

We presented a new combination of dynamic queries and query previews, which is enhanced with dynamic screen layout and a Visual Formalism. The integrated design results in a one-screen user interface where query formulation and display of results happens within one window. Important parts of the screen, like the status display, stay at fixed locations and give the overall impression of a static layout. Dynamically resizing or hiding controls gives the needed flexibility for displaying changing amounts of information.

The Visual Formalism of a table was used to visualize dependencies in the underlying data. It exactly shows, which combinations of attribute values will lead to records in the result set. With this information, the user can modify his query or explore the results at the most detailed level.

1178

We already implemented two systems using our tabular query preview. One is the MURBANDY WWW user interface, which will shortly be available to the public (http://murbandy.sai.jrc.it). The other is an interface prototype for text retrieval in heterogeneous databases. Both have undergone heuristic evaluation with domain experts with very promising results. The systems are currently being evaluated with non-expert users in controlled user tests. The focus will be on how comprehensible the tabular query preview is, and if the underlying principles are successfully transferred between domains and parts of larger systems by the users.

References

Ahlberg, C., Williamson, C., & Shneiderman, B. (1992). *Dynamic Queries for Information Exploration: An Implementation and Evaluation.* In: Proceedings of CHI'92 Conference on Human Factors in Computing Systems, Monterey, CA United States, May 3-7, 1992, pp. 619-626.

Doan, K., Plaisant, C., & Shneiderman, B. (1996). *Query previews in networked information systems.* In: Proceedings of the Third Forum on Research and Technology Advances in Digital Libraries, ADL '96, Washington, DC, May 13-15, 1996, pp. 120-129.

Greene, S. L., Devlin, S. J., Cannata, P. E., & Gomez, L. M. (1990). *No Ifs, ANDs, or Ors: A study of database querying.* In: International Journal of Man-Machine Studies. Vol. 32, pp. 303-326.

Greene, S., Tanin, E., Plaisant, C., Shneiderman, B., Olsen, L., Major, G., & Johns, S. (1999). T*he end of zero-hit queries: query previews for NASA's Global Change Master Directory.* In: International Journal on Digital Libraries. Nr.2, pp. 79-90.

Hertzum, M., & Frøkjær, E. (1996). *Browsing and querying in online documentation: a study of user interfaces and the interaction process.* In: ACM Transactions on Computer-Human Interaction. Vol. 3 (2), pp. 136-161.

Lavalle C., Demicheli L., Casals C. P., Turchini M., Niederhuber M., & McCormick N. (2000). *Murbandy / Moland.* Technical Report European Commission Euroreport. (In press).

Nardi, B. A., & Zarmer, C. L. (1993). *Beyond Models and Metaphors: Visual Formalisms in User Interface Design.* In: Journal of Visual Languages and Computing. Nr. 4, pp. 5-33.

Stempfhuber, M. (1999). *Dynamic spatial layout in graphical user interfaces.* In: Bullinger, H-J.; Ziegler, J. (eds.). Human-Computer Interaction. Communication, Cooperation, and Application Design. Proceedings of HCI International '99, Munich, Germany, August 22-26, 1999. Vol. 2, pp. 137-141.

Stempfhuber, M. (2001). *ODIN. Objektorientierte Dynamische Benutzungsoberflächen. Behandlung struktureller und semantischer Heterogenität in Informationssystemen mit den Mitteln der Softwareergonomie.* Dissertation im Fach Informatik an der Universität Koblenz-Landau (to appear).

Williamson, C., & Shneiderman, B. (1992). *The Dynamic HomeFinder: Evaluating Dynamic Queries in a Real-Estate Information Exploration System.* In: Proceedings of the Fifteenth Annual International ACM SIGIR Conference on Research and Development in Information Retrieval, 1992, pp. 338-346.

The Relation of Usability and Branding of Financial Web-Sites
an integration of qualitative and quantitative methods

Sabrina Duda, Michael Schiessl & Rico Fischer

eye square GmbH
Sabrina Duda & Michael Schiessl
Hagenauer Str.14
10435 Berlin
Germany

Abstract

We conducted a usability and branding study for a financial web-site.The aism of the study were to evaluate the usability of the web-site, and to evaluate the influence of the web-site on the image of the company.

Image test design followed a pre-post-design ratio. Attitudes were indexed with two different measurements. Mind mapping for assessing individual constructs and indexing explicit attitudes and the Implicit Association Test (IAT, Greenwald, 1989) to index the influence on implicit attitudes.

Usability was measured with a scenario based testing of the web-site. The subjects should fulfill certain tasks and think aloud. They had to rate their subjective feeling of how good they managed the tasks. The experimenter rated the quality of completing the tasks as well. After the site testing subjects got a questionnaire with usability questions. Main results demonstrate the need of a multi-method approach in investigating the image of a brand.

1 Introduction

During the last years traditional banking had to face a dramatic change. More and more customers choose the option of online-banking from home than going to the actual counter them self. That means that the appearance of a financial company and its own presentation on a web-site becomes more important in terms of customer binding and acceptance than ever before. One factor of the general acceptance of a web-site is based on its usability. In this study we did not only focus on usability in terms of „ease of use", moreover, we raised the question what impact the web-site-design might have on the image of the brand. The main question was therefore, how does the interaction with a web site influence the image of the brand?

2 Methodology

2.1 Subjects

Ten participants took part in this study. Five out of ten were customers of the company. Participants had to fill out a questionnaire concerning personal data including pre-experience in financial questions, internet experience, education and so on. All together we can state that the participants were well educated (9 had a university degree) and well experienced with computers and internet (daily use). All of them described their knowledge about stocks and funds as good.

2.2 Procedure

To measure the image of the company we used a pre-post-test design. We started with the image-pre-testing using a Mind Mapping procedure as an explicit index of attitudes and the Implicit Association Test (IAT) as an implicit measurement of attitudes towards the web-site of the company.

Mind Mapping describes a procedure which enables the experimenter to find out about an individual construct a participant creates towards a company. Our participants were asked to write down all their associations concerning the brand on little paper cards. Afterwards they had to put them in an individually chosen order around the logo of the company. The Mind Mapping procedure reveals a highly intentional and controlled judgment of the brand and therefore, it represents an explicit index of attitudes.

As a second method to investigate the image of the brand the Implicit Association Test was used. Implicit associations can be understood as automatic and unconscious information processing by the low level channels of the brain. Stimuli of the environment receive a fast and pre-attentive processing which is the basis of further attending. These kind of processes reflect our first impression we have of our surroundings, people or the web-site of a company. Metaphorically spoken, one might argue, that decisions made up on implicit associations are based on our intuitions.

The IAT-method is based on two general dimensions. An evaluation dimension and a target dimension. The target dimension contained the logo of the company which had to be assessed in terms of three evaluation dimensions (table 1) included in this study. Figure 1 shows a sample of the IAT-procedure.

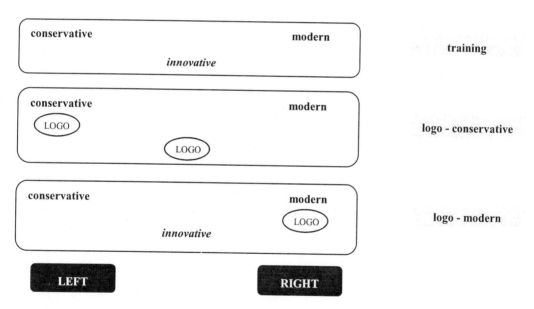

Figure 1. Example of an IAT-session using the evaluation dimension „conservative vs. modern". In a training session words appeared in the lower middle of the screen and had to be categorized whether they described a more conservative term (left button) or a modern term (right button). In the first experimental trial words as well as the logo of the company could appear in the lower middle of the screen. The logo was combined with the response button of the conservative words (left). That means in addition to conservative words the left button had to be pressed in response to the logo. In a second experimental trial the response to the logo was combined with the

response button for modern word (right). In both sessions reaction times were measured. The mean reaction times were afterwards compared with each other. The difference described the implicit association of the brand in terms of the evaluation dimension.

In this presentation a dummy-logo is used because we were not allowed to insert the logo of the investigated company.

1	conservative / modern	outdated, yesterdays, obsolete, old contemporary, innovative, up to date, trend
2	self / others	I, myself, self, mine foreign, others, they, those
3	success / failure	open, nice, pretty, healthy dumb, evil, ugly, dirty

Table 1. Evaluation dimensions and wordlists used in the IAT.

All participants performed on the Mind Mapping Test and the IAT in the image pre-test session. This was followed by the usability testing in which participants had to explore the web-sites of the company in solving 13 user tasks. These tasks were designed to investigate the content of the web-sites. Examples were finding an appropriate investment fund, getting informed about financial questions, job offers, help functions and so on. Participants were video taped and asked to report their thoughts and impressions (thinking aloud protocols). After each task they had to rate on a five point scale how well they performed on it (1-very well … 5 very bad). The experimenter rated the quality of completing the tasks as well. This took all together 40-60 minutes.

A post-image testing took place after this session, involving the same experimental design of the IAT as before. In terms of the created mind map (cards in distance to the logo) participants got the chance to change positions or add new cards with associations. In addition participants had to fill out an explicit usability and image questionnaire (Usability@Metrics).

3 Results

3.1 Usability
Because our main focus involved the question about a potential change of the image after dealing with the web-site, only general usability results are reported.

Participants described their performance on the tasks generally as good. The gave positive feedback about the "look and feel" of the sites. Usability problems occurred when they had to deal with the "deeper" content. Participants reported difficulties when looking for specific funds or stocks. Especially topics concerning the potency of the company (funds, stocks, financial questions) lead to increased usability problems.

3.2 Image
In the pre-test the method of Mind Mapping revealed three main dimensions that people associated with the brand: A strong connection to the mother company, seriousness and safety, performance and potency of the company. In the post-test four participants took the opportunity to change their original map into the direction of more innovation. In addition, four testers put more value on autonomy and independence of the mother company. Only two testers reduced the potency of the company.

As stated above the IAT measures more the implicit associations. Figure 2 shows the result of the IAT in dependence of the three evaluation dimensions in the pre- and post-test session.

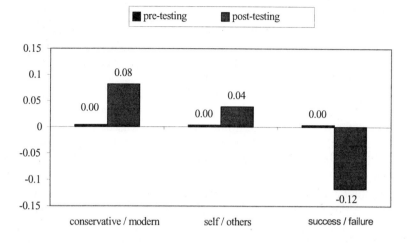

Figure 2. Results of the IAT in the pre- and post-testing. The chart shows the z-transformed differences in reaction times between the categorization tasks for each evaluation dimension. Positive scores indicate the availability of positive parts of the concept.

The results clearly demonstrate that participants did not have any implicit and automatic associations regarding the brand before dealing with the web-sites. Therefore, it can be concluded that participants do not have an established visual binding towards the brand.

In the post-testing session, results indicate a slight shift in the direction "modern" and a tendency towards the own self concept. A more strong effect is revealed in the dimension "success / failure". After performing the usability tasks participants seem to have more implicit associations in the direction of reduced potency of the brand.

4 Discussion

The aim of this study was to demonstrate how the usability and the image of a brand are interconnected. We used a web-site of a financial company to elaborate this question. Image was tested with two different methods. The Mind Mapping procedure representing an explicit index of attitudes and the Implicit Association Test (IAT) representing an implicit index. Whereas the Mind Mapping procedure is a controlled and intentional process of presenting attitudes and opinions, the IAT is an unbiased method which allows to show attitudes beyond the so called "cognitive scissor". The results are quite interesting in terms of these two approaches.

The main result of the Mind Mapping procedure in the pre-post-test design is an intentional positive change in attitudes towards the brand in terms of more innovation and more autonomy. Only two out of ten participants reduce the experienced potency of the company after using the web-sites. On the other hand, the usability tests show significant problems of participants performing on sites containing financial information (stocks, funds), a topic which actually defines the strength and the power of a financial company.

The results of the IAT make the connection between usability and image of a brand more evident. After participants showed no specific representations of the brand in the pre-test, results of the post-test also demonstrate a positive change of the implicit associations of participants following the use of the web-sites in direction "modern". We argue, that this might be based on the generally positive judgment of the participants in terms of the usability dimension "look and feel". The usability problems experienced in dealing with the "serious" contents of the financial sites (stocks, funds) reflect the drastically decrease in the associated potency and power of the brand (see figure 2). Although, participants did not show this directly in the "open" Mind Mapping method, the results of the IAT support this argumentation.

The question now, why participants did not report the decrease in potency of the brand while implicitly having such associations, is one that can not be answered within the frame of this study. One assumption could be that a controlled and intended behavior is much more complex than a simple implicit, automatic association. Implicit means in this term that participants don't have (yet) cognitive access to these representations which in fact lead further action.

Nevertheless, the study provides clear evidence of the importance of a multi-method approach investigating the image of a brand. Only the use of an implicit and explicit index of attitudes allowed us to get a more differentiated view on the complexity of the connection between the usability of a web-site and the image of a brand.

Based on the results presented in this study we would assume that although innovation, autonomy and self concept go congruent with a general satisfaction, but that the critical factor underlying the intention of buying a product (in our case a fund) is related with the anticipated and associated strength and potency of the brand. Moreover, this associated potency is directly related to usability.

LOG FILES ANALYSIS TO MEASURE THE UTILITY OF AN INTRANET

Stéphane Vokar, Céline Mariage, Jean Vanderdonckt

Université catholique de Louvain – Institut d'Administration et de Gestion
Place des Doyens, 1 - B-1348 Louvain-la-Neuve, Belgium
{vokar, mariage,vanderdonckt}@qant.ucl.ac.be – http://www.qant.ucl.ac.be/membres/jv

1. INTRODUCTION

A user interface (UI) evaluation should assess two aspects of the studied interface: its utility and its usability (Farenc,1995). Utility is related to the adequacy that should exist between functions provided by a UI and the user's task (Farenc,1995). Usability evaluation is an assessment of the conformity between a system's performance and its desired performance (Whitefield,1991). An interactive system combines a user and a computer engaged upon some task within an environment (Whitefield,1991). So to perform usability evaluation of Web applications, the human computer interaction has to be assessed. Multiple methods exist to conduct web site evaluation (Ivory,2000). One of them is the "log files analysis" method which consists in recording and analyzing user actions by observing log files saved on a particular machine. This method measures the utilization of the site in terms of different quantitative metrics, such as frequency of use or navigation traffic. Attempting to evaluate the utility and the usability of a web site just by looking at log files produced by the system poses a series of challenges due to the differences between utilization on the one hand, and utility and usability on the other hand. Many reasons impede the evaluator to obtain an accurate UI evaluation: the recorded actions are often from server side, not client side; data collected can be falsified by any utilization of dynamic IP addresses, proxy servers and cache of browsers. This method is heavily based on the definition of user sessions, determined by a predefined time. After an inactivity of this elapsed time, another session is assumed to begin without necessarily meaning another session, not reflecting the reality.

We present some interpretation of results obtained by the log files analysis on a particular web site, an Intranet, and propose inferences made from collected data to provide preliminary results to assess the utility of the related UI. Conducting a log files analysis typically consists of performing five activities (Drott, 1998):
1. *Collecting*: a software captures and gathers data into a data base for every usage from any user.
2. *Analysis*: a software organizes and structures collected log data to produce an analytical report of a site usage, along with graphics.
3. *Visualization*: a software graphically and interactively present log data resulting from the analysis to facilitate the manipulation of high level data and its understanding.
4. *Critique*: a software attempts to critique the results provided by an analysis and to identify potential utility and usability flaws.
5. *Remediation*: a software suggests to repair utility and usability flaws identified in the previous step either in an automated way or a computer-aided manner.

As many tools exist either on the market or in the research/development community, it is important to classify the scope of these tools. For this purpose, we introduce a design space defined along three dimensions:
1. The coverage of evaluation steps: this dimension precises to what extent the concerned tool is able to support the five activities as identified above;
2. The localization: this dimension informs the physical location where the software is installed and running; it can be on client-side, on server-side, or hybrid (shared by both a client and a server)
3. The periodicity of data analysis: this dimension specifies the frequency of log data analysis as a consequence of the previous collecting and organizing activities; this can be in real time (as soon as there is a visitor, the analysis is updated accordingly), periodic (when the evaluator programs some interval to launch the analysis), or aperiodic (when the analysis is launched on-demand).

This design space may raise an important series of various configurations for log files analysis. Therefore, we preferred to summarize them into four main categories. Fig. 1 graphically depicts one of the classical log analyzer categories, the one which collects and analyses data on the server, allows visualization on both the client and the server, but typically does not support activities beyond visualization. WebTrends (http://www. Webtrends.com) is considered as a representative example.

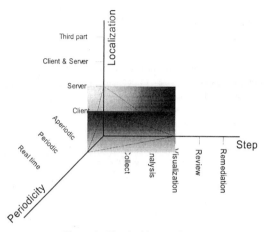

Figure 1. Classical log analyser

2. METHOD

Our usability evaluation concerns an Intranet application of a big Belgian Bank Group. This one is composed of those five topics: What's new?, Current news, Practical guide (phone book, meal, permanent information), sites of various bank departments (training, human resources, societies, year 2000, documentary CD-ROM), access to some applications (e.g., stock exchange rate, net banking, and press report). Most of web pages are available in two of three languages. They are composed of three frames: a first on the left which contain the table of content, a second on the top which is a tool bar designed to make easier the navigation (changing language, coming back to the homepage, coming back to upper level, search, help), and the late which is the document. All Web pages are generated dynamically by ASP technology. This Intranet is located on a local area network composed of personal computers with Windows NT 4.0 as operating system. IP addresses are assigned dynamically by the server (protocol DHCP), but they remain fixed at least two weeks. If a computer has not been used during this period, his IP address is removed, and it will receive another at the time of the next network connection.

The tool used in this evaluation is WebTrends Log Analyzer 3.6. It produces eight types of metrics: general statistics (number of hits, number of page views, number and average length of users sessions, number of users), resources accessed (most/least requested pages, top entry/exit requests, top paths through site, single access pages, most downloaded files), adds (number of mouse clicks on each add viewed), visitors and demographics (top users, most active areas), activity statistics (activity level by time), technical statistics (errors), referrers and keywords (top referring sites, top search phrases), browsers and platforms (most used browsers and platform).

To perform a usability evaluation of this application, we used log files analysis method, with WebTrends. This kind of method presents some limitations. Recorded actions in log can be inexact with the utilization of dynamic IP addresses, proxy servers and cache of browsers. Moreover, this method uses user sessions. A user session is the set of actions (all hits) performed by one user. The session terminates when user is inactive during a predetermined period of time. This concept is an approximation of the reality assessable with difficulty. If the idle period is fixed to 10 minutes, a user visiting a site twice during this period, the system count only one session, instead of two. We base our evaluation on the following hypothesis

- Users are not used to read the screen. They prefer print the documents of interest. We fixed the idle period at one minute.
- IP addresses are considered like fixed.
- The majority of the users have their own computer. To each client computer corresponds to a user.
- Most of web pages are generated dynamically, so we could consider the browser cache influence negligible.

1186

3. RESULTS AND DISCUSSION

The general statistics about the utilization of the site we obtained are presented in Fig 2.

General statistics	
Date & Time This Report was Generated	Tuesday April 04, 2000 - 12:37:05
Timeframe	01.02.00 00:00:17 - 29.02.00 23:59:58
Number of Hits for Home Page	337,587
Number of Successful Hits for Entire Site	6,339,898
Number of Page Views (Impressions)	1,096,477
Number of User Sessions	536,303
User Sessions from Belgium	0%
International User Sessions	0%
User Sessions of Unknown Origin	100%
Average Number of Hits Per Day	218,617
Average Number of Page Views Per Day	37,809
Average Number of User Sessions Per Day	18,493
Average User Session Length	00:00:27
Number of Unique Users	4,722
Number of Users Who Visited Once	143
Number of Users Who Visited More Than Once	4,579

Fig 2 General Statistics

We did not utilize the concept of Hit in our inferences. Indeed, a hit is not really representative of the user action as it cannot indicate how much pages a user is visualizing (Fuller,1996). A hit is a request of a file sent to the server by the browser of the client computer. This file is an element of the html page. This element can be a graphic. So if a page contains 10 graphics, the visualization of the page needs one request for the page and 10 more for the graphics. For this reason, we utilize the Page Views concept, which indicates the number of pages impressed. Moreover, the collected data present some incorrectness because of the User Session concept, an approximation of the reality, as indicated earlier. With those metrics we could determine the proportion of users among the employees and the frequency of use of the application. We computed the following ratios:

- Users proportion = $\dfrac{\text{Number of unique users}}{\text{Number of potential users}} = \dfrac{4722}{5000} = 94\%$

- Average number of visit by a regular user per day = $\dfrac{\text{Average number of user sessions per day}}{\text{Number of user who visited more than once}} = \dfrac{18493}{4579} \cong 4$ (1)

- Average number of pages views by a regular user per day = $\dfrac{\text{Average number of pages views per day}}{\text{Number of user who visited more than once}} = \dfrac{37809}{4579} \cong 8$ (2)

- Average number of pages views by a regular user per visit = $\dfrac{(2)}{(1)} = \dfrac{8}{4} = 2$ (3)

Approximately 94% of the Intranet potential user have visited it during the month we made this usability evaluation (February 2000). The average user of this Intranet visits it a little more than four times per day, and visualizes about eight pages, which made roughly two pages per visit. The following table (figure 3) presents the most requested pages groups. This concept regroups all the pages of the same subject content. We reorganized the results in function of user sessions instead of hits (figure 4).

Most requested pages groups			
Group name	hits	% hits to all groups	Sessions
1 Bank sites	379,343	24.01%	90,066
2 Practical guide	367,877	23.29%	197,765
3 Phone book (Practical guide)	206,464	13.07%	153,834
4 Human resources (Bank site)	114,836	7.27%	50,502
5 IS Sites (Bank site)	111,387	7.05%	12,771
6 Meal (Practical guide)	85,744	5.42%	30,299
7 Stock exchange rates	64,549	4.08%	59,589
8 Royal sport and cultural circle (Bank site)	59,848	3.78%	11,985
9 Current news	57,024	3.61%	39,431
10 Documentary resources (Bank site)	33,761	2.13%	4,258

Fig 4 Most requested pages groups

	Group name	% user session to all group
1	Phone book (Practical guide)	28,68%
2	Stock exchange rates	11,11%
3	Human resources (Bank site)	9,42%
4	Current news	7,35%
5	Meal (Practical guide)	5,65%
6	IS Sites (Bank site)	2,38 %
7	Royal sport and cultural circle (Bank site)	2,22%
8	Documentary resources (Bank site)	0,08%

Figure 4. Most requested pages groups.

So we can observe the user preferences in terms of content. The summary of activity for report period make an idea about Intranet usage. The same critics about hit and user session concepts can be applied in this case.

Summary of Activity for Report Period	
Average Number of *Users* per day on Weekdays	25,407
Average Number of *Hits* per day on Weekdays	300,803
Average Number of *Users* for the entire Weekend	685
Average Number of *Hits* for the entire Weekend	5,755
Most Active Day of the Week	Tue
Least Active Day of the Week	Sun
Most Active Day Ever	February 01, 2000
Number of Hits on Most Active Day	345,027
Least Active Day Ever	February 13, 2000
Number of Hits on Least Active Day	1,791
Most Active Hour of the Day	09:00-09:59
Least Active Hour of the Day	02:00-02:59

Figure 5. Activity Statistics for Report Period.

- Average number of visit by a regular user per day =

$$\frac{\text{Average number of user sessions per day during the week}}{\text{Number of user who visited more than once}}=\frac{25407}{4579}\cong 5 \qquad (1)$$

- Aveiage number of pages views by a regular user per day =

$$\frac{\frac{\text{Avergage number of pages views per day}}{21}}{\text{Number of user who visited more than once}}=\frac{\frac{1096477}{21}}{4579}\cong 11 \qquad (2)$$

- Average number of pages views by a regulai user per visit = $\frac{(2)}{(1)}=\frac{11}{5}=2$ (3)

Figure 6. Inferences from Activity Statistics.

The average user of this Intranet visits it between five and six times a day, and visualizes eleven web pages that is to say about two web pages per visit. The most active period is Tuesday between nine and ten o'clock. It's the day where the most employees are present and the time where they come at a work. Visiting the Intranet seems to be one of their first work of the day. We can then thinking that the bank's employees as an important working tool consider this Intranet.

4. CONCLUSION

When we use a log analyser we have to be careful in the interpretation of the results. In this study, the used tool, WebTrends bases his analysis on hits and user sessions. The first concept is totally useless because a hit is just a request of a file to a server. This file is most of time a part of a web page. So it is better to base the analysis on to the page view concept that represents the pages visualized by users. The second concept is an approximation of the reality, because WebTrends cannot identify exactly the end of a user session. A user session is the set of all actions performed by a determined user.

We have identified five types of configurations. Unfortunately, any is the best. Either the tools provides a large amount of statistics well synthesized, which do not correspond rigorously to the reality because they use session concept and collect only data onto the server side, either they provide a large scale of accurate measures from the client and server side, but without analysis method. Moreover, most of the tools can't generally provide reviews and remediation in function of the analysis results. In reason of those limits, a log analyzer like WebTrends is not enough to achieve a utility and/or a usability study of a web site. It can however constitute a first step in the evaluation process to identify easily, quickly and cheaply potential problems to investigate later, with other methods, like user testing. Relating to utility, we have identified the proportion of users, the contents they prefers, and their usage frequency. This permitted us to assume that the Intranet is considered as an important working tool by the bank's employees. We can't however assert more, because a log analyzer is not appropriated to assess the satisfaction degree of users. For it, it's better to use another method, like questionnaires for example.

REFERENCES

Farenc, Ch., Palanque, P., and Vanderdonckt J., "User Interface Evaluation: is it Ever Usable?", In Proceedings of 6th International Conference on Human-Computer Interaction HCI International'95, Advances in Human Factors/Ergonomics Series, Vol. 20B Symbiosis of Human and Artifact: Human and Social Aspects of Human-Computer Interaction, Elsevier Science B.V., Amsterdam, pp. 329-334.

Ivory, M., "State of the Art in Automated Usability Evaluation of User Interfaces", Working Paper, Univ. Berkeley, 2000. Accessible at http://www.cs.berkeley.edu/~ivory/research/web/papers/survey/survey.html

Whitefield, J., Wilson, F., and Dowell, J., "A framework for human factors evaluation", Behaviour and Information Technology, January 1991, Vol. 10, No. 1, pp.65-79.

Drott, M.C., "Using Web Server Logs to Improve Site Design", In Proceedings of 16th ACM Conference on Systems Documentation SYSDOC'98, ACM Press, New York, pp. 43-50.

Fuller, R. and de Graaff, J.J., "Measuring User Motivation from Server Log Files", Proceedings of 2nd Interactional Conference on Human Factors and the Web HFWeb'96. Accessible at http://www.microsoft.com/usability/webconf.htm

Scapin D., Leulier C., Vanderdonckt J., Mariage C., Bastien Ch., Farenc Ch., Palanque Ph., Bastide R. (2000), A Framework for Structuring Web Design Guidelines, Proceedings of 6th International Conference on Human Factors and the Web HFWeb'2000 (Austin, 19 June 2000), Ph. Kortum & E. Kudzinger (éds.), University of Texas. Accessible at http://www.tri.sbc.com/hfweb/scapin/Scapin.html

Too Many Tools – Overtooling The Web
A Usability Study on Internet Viewbars

Anke Ahrend, Sabrina Duda

eye square, Hagenauer Str. 14, 10435 Berlin, Germany

ABSTRACT

In this paper we will show how new tools on the Web, i.e., viewbars, are accepted by the users and what features of such tools on the Web become increasingly important. We present the results of a usability and image testing revealing how simplicity affects the user's experience with benchmarking aspects being of importance. This paper also gives distinctive recommendations with regard to viewbars in general and the prototype viewbar.

1   INTRODUCTION

The Web is flooded with various tools entailing different features. This usability study focuses on viewbars revealing usability problems as well as on user acceptance/ preferences. The increasing glut of these tools makes necessary personalization approaches with the users becoming more and more demanding with regard to tailored-to-fit solutions. Testing such small tools quite simply provides information on these aspects. The prototype's level of functionality was reduced, i.e.: horizontal prototyping. A horizontal prototype is a simulation (Life et al. 1990) of the interface where no real work can be performed. In a Web example, this means that a user can execute all navigation and search commands but without retrieving any real documents as a result of these commands. Horizontal prototyping makes it possible to test the entire user interface. Advantages are: Fast implementation with the use of various prototyping and screen design tools; it can be used to assess how well the entire interface hangs together and feels as a whole. Fake data and other content is used for demonstration (Nielsen, 1997).

2   USABILITY AND IMAGE TESTING: COMPARING A PROTOTYPE VIEWBAR TO TWO VIEWBARS ALREADY ONLINE

In this test, two viewbars already on-line - FairAd and alladvantage (http://www.fairad.com, http://www.alladvantage.com) - were compared to one prototype viewbar, Maxamyzer. Scenario-based testing of the viewbars was the means of measurement for usability; i.e., different tasks had to be fulfilled by the subjects while thinking aloud, the commentator rating the quality of the tasks' completion at the same time. Right after having tested each viewbar, the testers had to fill out a questionnaire regarding usability and image. Afterwards, questions focussing on user and consumer behavior as well as their expectations on viewbars including various aspects (Internet shopping, personalisation, surfing for money/incentives, responding to market research questions and other questions, etc.) led to a short discussion with the commentator. As the prototype showed a new composition of such a tool on the Web, the study also included usability aspects (use of the mouse, direct manipulation, drag & drop) as well as user acceptance of the prototype's unique design.

Below, the three different viewbar screenshots (Fig. 1 - 5) and their structures are listed:

Figure 1: FairAd

Figure 2: alladvantage

Figure 3: Maxamyzer

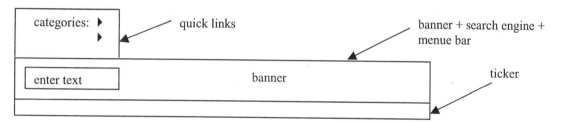

Figure 4: Structure of common viewbars

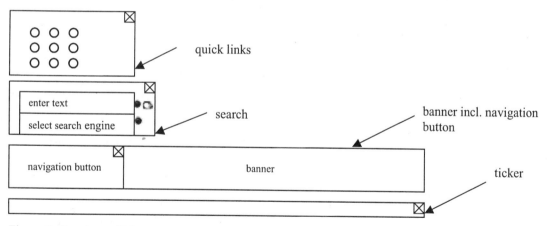

Figure 5: Structure of Maxamyzer prototype

The most important features and components of viewbars currently available on the Internet are: search engines, quick access to pre-defined Web-pages (personalization possible), news tickers, surfing for money, marketing banners with on-line marketing surveys. Those viewbars generally consist of one banner including all the features mentioned above with no provision of modularity. In comparison to these viewbars, the protoype shows a quite different structure: the viewbar consists of different modules (banner, search engine, quick links, news ticker) whereas the different modules can be activated as many times as desired (resulting in, e.g., one banner, two search engines, and four quick links on the screen) and each module be placed on the screen according to one's own option, plus the user may choose from different designs ("Green, Air, Metal, Plastic, Chocolate").

The 20 participants were moderate computer and Internet literate, 16 of which had Internet access at home and were of higher education, the average age being 31. Most users (17) had never made use of viewbars before but were familiar with the term "surfing for money" and marketing banners on the Internet. All of them were interested in new tools on the Web showing different features.

The users' tasks included activating the tool, dragging it over the screen (drag&drop), having the stock quotes displayed, using a specific search engine plus typing in a particular word, and deactivating the viewbar.

The three viewbars were tested in alternating order.

2.1   Results of the usability and image testing

Table 1 shows the **rating of the different tasks' completion**.
Ratings: 1 – 5: (completion of tasks: 1 = excellent, 5 = very bad)

Tasks	FairAd	alladvantage	Maxamyzer
Activating the viewbar	1.5	2.2	3.2
Dragging the viewbar over the screen	4.1	3.5	2.4
Having stocks displayed	3.5	1.4	4.0
Using the search engine	4.7	5.0	2.1
Deactivating the viewbar	4.8	4.7	1.5

Table 1: Tasks and relevant ratings

Table 1 ratings provide the following averages (1 = excellent, 5 = very bad):
FairAd: 3.72, alladvantage: 3.36, Maxamyzer: 2.64.

Several **usability problems** regarding the Maxaymzer prototype were revealed by user performance. These usability problems are shown in Table 2:

Type of usability problem	no. of users facing problems	no. of users without problems
Control via right mouse button	16	4
Identification of navigation button	18	2
Differentiation banner/ navigation button	17	3
Activating other modules	17	3
Changing design	15	5
Selecting search engine	5	15
Identifying lens	15	5

Table 2: Usability problems of Maxamyzer prototype

The modularity was appreciated by most of the users (12) as common viewbars were considered not to be enough flexible, whereas a modular tool including drag&drop features gave the users the feeling of completely controlling the tool according to their personal and specific desires. The missing menue bar of the prototype created problems for 18 users (they could not find it at all), 9 of which clearly requested a menue bar.

Search engine and quick links were requested and necessary for all users. 19 of 20 participants requested personalized links, i.e. having their 10 - 20 most visited and favorite sites included in the quick links. The search engine's symbols (i.e., arrows, lens) were criticized as they were way too small.

Changing the design from, e.g., "Green" to "Chocolate" was considered by most of the users (17) as "nice to have" and was rated with 2.6.

Fun of use and support for surfing the Web was not realized by all the users immediately. As soon as they noticed that such a tool with an integrated search engine makes obsolete invoking a search engine's home page, and that the search results are displayed instantly, the acceptance rose by 80%.

Based on these results, Table 3 states the overall users' rating.
Rating: 1 – 5 (1 = like it very well, 5 = not at all)

Viewbar	Rating
FairAd	3.8
alladvantage	2.0
Maxamyzer	3.0

Table 3: Overall users' rating

Benchmarking:

Surfing for money:
Younger users were more familiar with and attracted by this feature; 10 users decided for reimbursement, e.g. as an equivalent for their phone bill.

Marketing banners:
They were of no great importance to the users. 13 participants wouldn't answer to questions displayed on a banner. If the answers were given instantly, 15 users would participate in benchmarking, the participation depending on the users' respective interests.

Name of the tool:
The rating of 4.2 (1 = excellent, 5 = very bad) showed that the name "Maxamyzer" did not meet the users' expectations with regard to such a tool and only 3 users associated the name to the feature of the viewbar of letting the tool grow by activating different modules.

3 RECOMMENDATIONS AND CONCLUSION

Regarding viewbars in general:
The study showed that the users preferred conservative and "easy-to-grasp" tools. The design should be simple yet pleasing. Personalization in fact plays an important role for Web users according to their increasing demands toward Web applications.

Ease of use was the most important dimension for satisfaction. The meaning of such tools was not understood and/or accepted by all the testers, according to their statements mainly due to German Internet user and consumer behavior (e.g., why shop on the Internet when the next store is so close, etc.). Nevertheless, seeing the advantages of such a tool led to a much higher acceptance.

Regarding the prototype Maxamyzer:

The majority of the users (15) would prefer having the possibility of changing the size of the viewbar. Also, a dynamic tool that could be placed in the back- or foreground as desired was requested (17 users) pursuing a different layer UI approach. Another feature asked for by most of the participants (16) was the packetizing of the different modules with one click making the structure even more flexible.

As users are opt to use devices as known; control via right mouse button is critical; double clicking as an action as well as cognition is preferred; the navigation button should be made clear for identification (icon, labelling) or a menue bar should be included. In addition, a help function would be of use as well.

The arrows and the lens to the right of the search engine are definitely too small; consequently, their size shall be changed.

The name Maxamyzer does not provide for associating it with the tool and its features, and should, therefore, be thought over.

Modularity and personalization play an increasing role for tools on the Web as the users' demands are increasing and such features can be easily adopted to those requirements and result in greater user acceptance and satisfaction.

REFERENCES

http://www.fairad.com

http://www.alladvantage.com

Life, M.A., Narborough-Hall, C.S., and Hamilton, W.I., eds. 1990. *Simulation and the User Interface.* Taylor and Francis, London, UK, ISBN: 0-8506-6803-4

Nielsen, J. Usability Engineering. *The Computer Science and Engineering Handbook, 1450,* CRC Press 1997, ISBN 0-8493-2909-4.

A CORPUS OF DESIGN GUIDELINES
FOR ELECTRONIC COMMERCE WEB SITES

Costin Pribeanu, Céline Mariage and Jean Vanderdonckt

National Institute for Research and Development in Informatics, Bd Averescu 8-10 – R-71316 Buccharest, Romania
Pribeanu@acm.org – http://www.ici.ro/chi-romania/pribeanu/index.htm
Université catholique de Louvain, Place des Doyens, 1 – B-1348 Louvain-la-Neuve, Belgium
Vanderdonckt@qant.ucl.ac.be – http://www.qant.ucl.ac.be/membres/jv

ABSTRACT

Designing and evaluating the usability of electronic commerce web sites is a rather complex activity that involves design knowledge. One particular form of this knowledge consists of design guidelines, which can ensure some minimal form of usability. However, these guidelines are widespread throughout the literature and expressed in various forms. This paper attempts to provide a framework for structuring guidelines, exemplifying it for electronic commerce web sites.

1. INTRODUCTION

A design guideline for an interactive system consists of a widely accepted principle that ensures some form of usability of the user interface (UI) of this system. For more than fifteen years, guidelines helped designers to improve the UIs by relying on the statement of these guidelines. However, guidelines suffer from drawbacks such as (Bastien & Scapin, 1993): widespreading throughout literature, inconsistent vocabularies, homogeneousness of sources and disciplines, potential conflict, and lack of structure. In this paper, we attempt to improve the structure of design guidelines by reporting on a template for structuring guidelines and by exemplifying it on electronic commerce web sites. It is expected that guidelines structured according to this approach will be more easily communicated to designers, understood by developers, and propagated throughout the interested organizations.

2. TEMPLATE FOR STRUCTURING GUIDELINES

The proposed template for structuring guidelines consists of a list of attributes (table 1) based on a general guideline model refined by trial and error (Vanderdonckt, 1999) and expanded to consider modern guidelines. In this table, the rows in gray are not implemented in our current database, but are encoded in the field Comment. Fields marked with an asterisk means the following: in order to ensure some consistency over the guideline collection some explanations, procedures and a list of possible values (where appropriate) are given in this guide. A subset of these attributes is defined afterwards. An entity-relationship model was defined in order to implement a guidelines base (figure 1).

The guideline identifier is an alphanumeric identifier denoting the position of the guideline in the section guide. It is not known at the moment when guidelines are collected. However, if the collection is structured in sections, the number of the guideline within the section should be provided.

The guideline title is a brief and representative sentence using an active verb. It is strongly recommended to be as short as possible provided that the guideline meaning is preserved. Examples of titles are: *Provide on-line shopping carts..*

The statement describes the principle that should be followed when designing the interface, for instance: *Provide means to create on-line shopping carts that help users to keep track of what they already bought.*

For each guideline a list of bibliographic references quoting the guideline should be provided. The reference specifies the author name and the year, for instance: [IBM00-2] IBM ease of use: "Web design guidelines" section: e-commerce topics. Product information. http://www-3.ibm.com/ibm/easy/eou_ext.nsf/Publish/615

No	Attribute name	Definition	Status
1	ID_Guideline	an alphanumeric identifier denoting the position of the guideline in the all library; is an AutoNumber	simple mandatory
2	Title	a brief and representative sentence written in a simple sentence style: {[condition(s)], {imperative verb	{subject, active verb}, object complement}}; must appear on one screen line	simple mandatory
3	Statement	a complete statement written in a natural language (preferentially in the same title sentence style) describing the guideline, in particular how to apply it	simple mandatory
4	Rationale	one or many sentences explaining why the guideline is important	simple facultative
5	Exception	one or many sentences explaining possible exception cases with reference to negative or positive examples	multiple facultative
6	Comment	in his field can appear any other information about the guideline, which is not covered by the template fields, in particular information from the entire ERA schema; temporally the information in grey from this template might appear	simple facultative
7	Reference	a list of bibliographic references quoting the guideline (author(s), year, title)	multiple facultative
8	Criteria	criteria respected by the guideline; use elementary criteria's; preferentially one ergonomic criteria might be chose	multiple facultative*
9	Linguistic Level	linguistic level in which is situated the guideline	simple facultative*
10	Example	example depicting a UI violating the guideline: textual explanation; picture; reference; original web address; local directory / file the example is positive or negative (Ill_type), normal or exceptional (Ill_Statut)	Multiple facultative
11	Relationship	relationship established with other guideline according to a link typology to be specified: the guideline source and the guideline destination of the relationship	multiple facultative
13	GlossaryWord	words defined for the indexation of the guideline; supertype of Eval_Method, Criteria and Ling_Level	multiple facultative
14	EvalMethod	methods for the evaluation of the guideline	multiple facultative
15	UI type	the user interface type for which the guideline is valid: web, textual, graphical, vocal, tactile, virtual; can be more than one	---
16	Development phase	a set of phases to whom it might concern: requirements specification, design, implementation, evaluation, training (education) ; can be more than one	---
17	Activity domain	the activity domains the interface are targeted to and which may benefit from the guideline; can be more than one	---
18	Task	User's tasks that benefit from the guideline; tasks are meaningful in the activity domain; can be more than one	---

Table 1. Template for structuring design guidelines.

1196

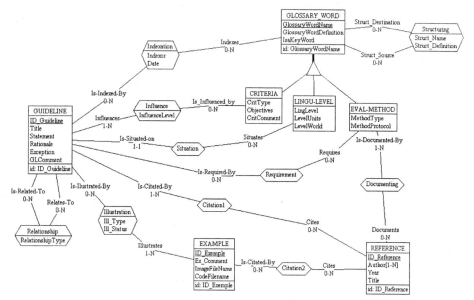

Figure 1. Entity-relationship model of the guidelines base.

According to Nielsen (1986), the linguistic level to which the guideline is applied can be *one of* the following: goal, pragmatic, semantic, syntactic, lexical, alphabetical, physical. The linguistic level refers to levels of decomposition of a task. Examples of linguistic levels are given in table 2.

N°	Linguistic level	Example
7	Goal	visit a shelf in a real store
6	Pragmatic	visiting a product category page
5	Semantic	select a combo box that links to the desired the product category page
4	Syntactic	display the list, select the item
3	Lexical	items in the combo box, label "Select a category", a button
2	alphabetic	a colour, the shape of an object
1	physical	moving the mouse

Table 2. Linguistic levels of a guideline.

The complete list of ergonomic criteria adopted for this template is based on the report of Bastien & Scapin (1993). Additionally an elementary criterion – adaptability - was added to adaptability criteria group. Also, the last criteria – compatibility – was expanded in three elementary criteria: task compatibility, user compatibility, environment compatibility. Root ergonomic criteria are: guidance, workload, explicit control, adaptability, error management, consistency, significance of codes, and compatibility. Only elementary criteria should be used. There are 21 elementary criteria in this set. For example, *Significance of codes* concerns the relationship between a term and /or a sign and its reference. Codes and names are significant to the users when there is a strong semantic relationship between the codes and the items or actions they refer to.

The rationale consists of one-two phrases justifying the guideline. For example, <u>Guideline</u>: Use hypertext to structure the content space into a starting page that provides an overview and several secondary pages that each focus on a specific topic. <u>Rationale</u>: The goal is to allow users to avoid wasting time on those subtopics that don't concern them.

A negative example is a reference to a UI violating the guideline. A positive example is a reference to a UI respecting the guideline. For each example, we have:
- a textual description that explains why or how the guideline was violated; if possible indicate the right solution;
- a picture that illustrates and documents the explanation;

1197

- the reference where the example was found;
- original web address;
- a local directory / file name where the example is saved for a future use.

Both the original location and the saved page should be kept because the original site could change in time thus leaving the web address and/or the example obsolete.

Guideline statement: Preserve previous attribute values as selected by the user

Negative example: The Bluefly e-store does not preserve the size setting when choosing a product to examine it in more detail. Original address:

http://www.bluefly.com/list/catg.asp?zone=mens&sort=fresh&brand=&catg=ls&min=0&max=0&color=&listview=brand&sid=V0U9VC68P1S92LTA00A3HJBDN59B2EBA&mapsz=M

Local address:

D:\Guidelines\Examples\bluefly\list\catg.asp

1) The size as previously selected by the user

2) The user selects this products to examine it

3) The selection is lost so the user has to redo it. It might not observe !

Positive example. Each image must have alternative text.

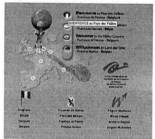

This invitation web page related to the linguistic choice has got alternative text One ore more cases when the guideline does not apply, with negative and positive examples.

A set of relationships established with other guidelines according to a link topology. It can be one or many of the following: related-to, is the inverse of, inherits from, is inherited by, generalises, specialises, precedes, follows, updates, precises, is a, is comprised in, belongs to, includes, is included in, is a user of, is the owner of, has property, is an example of, is provided by, uses, is used by, is a requirement for, is alternative to, is contrary with, is conflicting with, updates / is outdated by,...

The user interface type for which the guideline is valid. It can be one or many of the following: Graphical User Interface, Character User Interfaces, web, virtual reality, tactile, audio,...

A set of phases to whom it might concern. It can be one or more of the following: requirements specification, design, implementation, evaluation, training, education.

The activity domains the interface are targeted to and which may benefit from the guideline. It can be one or more of the following: e-commerce, e-banking, e-documents, CAD,...

The user's tasks that benefit from the guideline. The tasks are meaningful in the activity domain. It can be can be more than one task for which the guideline applies. For an e-commerce web sites the task could be classified within following categories: *user support* (i.e., knowing the site, understanding return policies, understanding shipping conditions, contacting, looking for help), *product navigation* (choosing a store, choosing a product category, choosing a brand or a designer, specifying a display, product examination, detailed product examination), *product picking* (add to wish list, examining the wish list, add to shopping cart, examining the shopping cart), and *checking out* (checking the order, specify payment details, tracking an order).

Under the keywords section are classified other attribute values that might be relevant for the guideline organisation.

5. Conclusion

In this paper, we introduced a template for structuring design guidelines for the UI of interactive computer-based systems. This template is currently being used to input design guidelines into knowledge bases of the MetroWeb project. This software is a Web-based tool for working with guidelines: any user may input a guideline, submit it for revision to developers or human factors expert, and validate it. Once guidelines are accepted in a knowledge base, they can be accessed through the web.

For instance, a user interested to know usability guidelines applicable to mobile devices, such as WML-compliant cellular phones, may access the guideline base for mobiles devices. In this base, the user can select any subset of guidelines and retrieve them from the base to form his/her own working base. Examples can then be added, along with annotations. Guidelines for which some possibility for automated or computer-aided evaluation are considered for an extension of this tool.

REFERENCES

Bastien, C.J.M. and Scapin, D. "Ergonomic Criteria for the Evaluation of Human Computer Interfaces", Technical Report No.156, INRIA, Rocquencourt, 1993.

Nielsen, J. "A virtual Protocol Model for Computer-Human Interaction", International Journal of Man-Machine Studies, 1986, Vol. 24, No. 3, pp.301-312.

Vanderdonckt, J. "Development Milestones Towards a Tool for Working with Guidelines", Interacting with Computers, 1999, Vol.12, No.2, pp.81-118.

Getting off to a good start: the tricky business of the discovery and design phase

Volker Frank
Paul Osburn

Vertigo Software, Inc.
51 Washington Avenue
Point Richmond, CA 94801
www.VertigoSoftware.com

Abstract

The author's, a designer and an engineer, introduce the seven-step process for an interdisciplinary team that addresses several common problems in the development of web applications and outlines the best practice at Vertigo Software. Application development for the Internet is usually subject to severe constraints, most notably time, which often leads to communication break-downs within the team. The result often is software that is difficult to use, does not meet user needs, is troubled by bugs and performance problems, which leads to delays in the launch date. Paradoxically, the delays are too often caused by the rush to start code development early in an effort to meet aggressive deadlines. The seven-step process outlined below for the discovery and design phase of application development projects suggests ways to minimize communication problems between designers and engineers through interdisciplinary teamwork and provides a solid foundation for clear priorities, reliable planning, and solutions that are user-friendly and meets their needs. Best of all, in the experience of the authors the time spent up-front not writing code leads to higher efficiency during the development and deployment phase and allows the team to meet aggressive deadlines.

1. Introduction

Developing applications for the Internet introduces several new issues to the software development process that is often reflected in the quality of the solutions. The first version of many products are often unusable, have serious performance problems, and are plain no fun. While research and requirements gathering before starting any development has gotten more and more friends in the world of desktop applications over the years, in the high-pressure environment of Internet applications, it is still a tough sell. The time -to-market pressure is overbearing, as the first-mover advantage is a one-time chance to leap ahead of the looming competition. Additionally, it is so seductive to rely on the "quick-and-easy" update possibilities of incremental release cycles, which suggests that you can always fix problems after the initial launch. In the eyes of the entrepreneurial spirit, both in start-ups and divisions within corporations, analysis and planning activities are often seen as delay tactics borrowed from another time that are inappropriate in the current competitive environment. Consequently, we have in some respects moved back to the stone age of software development, where technology rules and design is reduced to lipstick in the service of "brand building". The analysis and user-centered design process put forth suggests seven effective steps to avoid the following common problems in the development of web applications:

- Limited time to create extensive documentation that nobody studies as closely as necessary and the resulting break downs in the communication between designers and engineers
- Poor understanding of scope that gets in the way of setting meaningful priorities
- Feature creep during the development phase that can only be accommodated by compromising the integrity of the solution, introduces disproportionate number of bugs and often leads to delays in the deployment of applications
- Unmet user needs
- Disproportionate training efforts that force users to adapt to poor software, which translates immediately into failure for applications that are accessible by the public (Application Service Providers).

1200

2.   The Process

The Vertigo Software Engagement Framework relies on seven key steps and deliverables to lay the solid foundation of the implementation and deployment phases. Integral to the success of this process is the interplay of roles that designers and engineers play in the development of these deliverables and who leads the effort in which step to ensure the effective knowledge transfer between team members.

2.1.   The Participatory Design Session

In this session the lead designer and engineer meet with a group of representatives for users, marketing and product strategy, and, if applicable, the IT organization, support, service and internal consulting. In the authors' experience, these sessions are most successful if the group does not exceed seven people who are primary stakeholders in the success of the product and have the decision making power to resolve the often conflicting interests. Generally, these sessions require the commitment of all participants for 2 to 4 days of ideally uninterrupted work time. To avoid interruptions it helps to meet away from the office. It is also important to stress the need for consistent participation so that every member of the team understands all the previous decisions made and the facilitator can cultivate a constructive and creative work atmosphere.

In this stage of the process the designers lead the effort to ensure they uncover all the information they need for the user and task analysis and the conceptual design, such as roles, workflow, mental models, information architecture, competitive information, market analysis, etc. Special attention has to be given to the changes that the new system introduces to the workflow, roles and responsibilities. Users are generally good at describing what they currently do but require coaching to envision the new system that addresses their pain points and introduces efficiencies.

The lead engineer clarifies basic system and technical requirements that gives him a high-level understanding of the implementation constraints. Based on this understanding, s/he plays primarily an observing and supporting role to answer any technical questions that may arise or caution participants, when ideas start to drift too far out of the realm of the feasible. However, this is primarily a time for brainstorming and envisioning the ideal world, so the veto of feasibility is used judiciously.

2.2.   User and Task Analysis

The design team creates user profiles by attributes, such as roles, responsibilities, educational backgrounds, computer literacy, and domain expertise. For each role, the document captures use case scenarios and lists the individual tasks within a use case together with the goals and decision making strategies. Additionally, the document captures significant environmental factors, such as workflow, physical constraints, work culture, etc. [1]. The document is reviewed with all participants of the design session to refine the shared understanding and add the necessary detail.

In this document the design team outlines the structure of the requirements gathering process by identifying the use cases and tasks that are the quasi table of contents for the requirements.

2.3.   Technical Exploration and Prototypes

At the same time as the design team is working on the user and task analysis, the engineering team explores any technical risks that arise from suggestions of the participatory design session. The resulting technical proof-of-concept prototypes allow the engineers to feed information back into the design process. They also enable the team to manage the overall scope and identify feasible design solutions.

2.4.   The Design Narrative

The design narrative consists of schematic representations of the UI screens and their descriptions that capture any functionality not reflected in the static screen representations. They explain strategies for decision making and the accomplishment of users' goals. The development of the conceptual design is an interdisciplinary task, where designers and engineers collaborate to arrive at an overall solution that is feasible within the constraints and meets the users' needs. The collaboration is a principle way in which designers and engineers get on the same page.

The schematic screen representations are an excellent tool to gather feedback in form of a paper prototype from users outside the core team in addition to the review by the participants of the design session. The conceptual designs often require several iterations. While the first iteration is well received because they so clearly signal progress, the fidelity of the representation does not change from one iteration to the next, which makes them very prone to acceptance by executive decision before all the kinks get worked out and all the feedback can be integrated. At this stage client education and expectation management often becomes critical for the success of the deliverable.

2.5. Requirements Gathering

Contrary to conventional wisdom, it is our experience that the requirements are best gathered by the engineering team. They reference the task analysis document for the high-level structure of the requirements by use cases [2] and compare those with the results of the design narrative. This allows the engineers to wrap their heads around all the suggestions, which often uncovers oversights, corner cases and exceptions that have not been adequately addressed. This step leverages the difference in mind set and approach between designers and engineers. It also give the engineers control over the amount of documentation, i.e., granularity of the requirements, they need for the implementation phase and hence avoids excesses and busy-work or gaps in critical areas.
In addition to the description of the requirements, the engineers also capture software specifications that describe more specific implementation strategies for later phases, which allows them to think through technical issues in more detail early on.

2.6. Architecture Design

Based on the requirements and software specifications, the engineering team can now develop the software architecture, database schemas, system specifications for the deployment environment and any APIs for adjacent systems. The resulting document will provide a solid foundation for estimating the overall project effort and the schedule. Now is an important checkpoint to determine the priorities for the development effort and work together with the client to develop strategies for release cycles that meet aggressive timelines and take marketing and user needs into consideration.

2.7. Test Plan

Based on the requirements captured by the engineering team, the design team identifies the necessary test scenarios that the software has to meet for the first release. This step fulfills two important functions. The design team is forced to double-check the requirements and discovers any misunderstandings or ambiguities before any code has been written. Secondly, the test scenarios are invaluable for the engineers during development to determine, if their code meets both the specified requirements and user's tasks, as expressed in the test scenarios.

3. Conclusion

The development of software for the Internet happens in an aggressive and highly constrained environment, which has been aggravated by the current economic climate. Under these circumstances the demands on the design process to balance user needs, marketing strategies, and feasible technical solutions have dramatically increased. To meet those demands Vertigo Software has developed an interdisciplinary engagement framework for the discovery and design phase that has a good track record. It is based on seven steps and deliverables that facilitate efficient knowledge transfer between subject matter experts and users, and the design and engineering team with several checks and balances that allows the teams to identify misunderstandings and oversights early on in the process. The process allows for open-ended brainstorming in the beginning and then condenses and prioritizes the design and proposed functionality in iterative steps that keeps all stakeholders in the loop, provides opportunities for user feedback, and ensures a feasible solution: A solution with a predictable outcome that enjoys high user acceptance rates and stays within the business parameters of people responsible for the bottom line.

References

[1] Hugh Beyer, Karen Holtzblatt (1998), *Contextual Design: Defining Customer-Centered Systems*, San Francisco: Morgan Kaufmann
[2] Suzanne Robertson, James Robertson (1999), *Mastering the Requirements Process,* New York: ACM Press

User expectations and iterative design: a case study

Clemens Lutsch M.A.[1]
Dipl.-Ing. (FH) Bernhard Ertle[2]

IconMedialab AG, Office Munich, Infanteriestr. 19 / Geb. 3, 80797 Muenchen, Germany
[1]clemens.lutsch@iconmedialab.de
[2]bernhard.ertle@iconmedialab.de

Abstract

Projects in the web-environment tend to be overruled by the term "user-experience". Clients and producers often hope to cover both usability and suitability for the business with this term. This does not work. It is essential to address user expectations to identify the user's goals. After that, usability specialists are able to adopt these goals in an iterative design-process to build a system with specific qualities.

Using an ecommerce-case of an sports-goods online-shop, the differentiation of specific user-expectations and the transformation into a system can be presented.

1. Introduction

The actual diversity of Websites is immense and an phenomenon of a growing online-universe can be observed, determined by content, features and commerce.

Companies that are about to join this stampede are challenged with the question how to work out a proper positon to be recongnizable, remarkable and therefore… rememberable. They very often decide to surrender to a term called "user experience". But this is similar to the idea to build a house by moving in first… without knowing in what exactly you are moving in.

In terms of site-development, this means that during project planning nice visuals, text-tonalities and high-end features take precedence over concepts and the development. As a result, suitability to fit business and user's goals are neglected.

Assuming the formula "the more my system provides, the more success I will have" is not valid any longer (in fact, it was never valid anyhow ;-) we will have to look into processes, qualities and actions that allow the developer of an online system to include what is needed by the ultimative validation instance: the user.

Using the standarized human-centred design process ISO 13407, user expectations are integrated as an essential determinant of development activities.

In this article the major activities in establishing of human-centred design goals are adressed and shown by using an ecommerce-case of an online-shop.

2. User expectations

First, a proper definition of the term "user expectation" must be provided: A user expectation is a statement of a user that affects functionality, content or performance of an interactive system. It adresses both quantitave and qualitative issues of functionalities and content.

According to the human-centred design process, it is relevant to understand the context of use of the system we are about to build. That is, the characteristics of the users, tasks and organizational issues that will affect the system. After the context of use is defined, usability specialists are about to face the challenge to identify the user's expectations. There is no way to do this except to gather input directly from the defined set of users. This can be achieved by using interviews, surveys or focus groups. It is essential that this input is documented in a way that allows the usability specialist to derive user expectations from. The following list of characteristics has to be met to model user expectations out of various sources of user input.

1. Separate wishes from ideas
2. Reformulate every wish to "I will…"
3. Evaluate the "I will…" statement against context of use
    a. In case of a contradiction: skip statement
    b. In case of non-conflict: establish "I will…" statement as an user expectation
4. Identify home of expectations (user group profile)
5. Set up criteria for effectiveness, efficiency and satisfaction of user expectations

Explanation of the steps:

Separate wishes from ideas: Out of the sum of user statements, the usability specialist identifies those statements that can be connected to a preferred quality of a service that is expected. These statements differ from other input that is often communicated with less emotion, indifferently or as unreflected repetitions.

Reformulate every wish to "I will...": This is obvious.

Evaluate the "I will..." statement against context of use: If the user statement would jeopardize the context of use, it has to be reconsidered if this risk is worth taking. That means potential redefinition of the business concept, scope and/or creative strategies of the system. In most cases, the statements that challenge the context of use will not be addressed further.

Identify home of expectations (user group profile): The usability specialist links the expectation to the user group that produced the original statement. The user group shall be one of those that were identified during the specification of the context of use. Therefore every user expectation can be traced to specified user group(s). This means, that every user group "owns" some expectations. This is even more important if the system is build for very different user groups. The assignment of expectations to different user groups also provides a nice overview of the differenciation of the system that is to be adressed.

Set up criteria for effectiveness, efficiency and satisfaction of user expectations: This activity leads back to standarized criteria of an interactive system (refer to ISO 9241 part 11). This has to be done in order to make measurements during development (or prototyping) of the system and of course after deployment.

3. Usability goals

Regarding the "big picture" of a project, it would not be very efficient to stick only with "I will..." statements. Documented Usability Goals contain information about specific characteristics of the system and a validation plan that describes the means of evaluating ergonomic quality and which ensures a human-centred design process. Both, usability goals and the validation plan occupy the position of a sensible quality statement because they formulate specific performances of the system which is to be built.

A usability goals document includes also an indentification of stages in the specification and development of the system that are subject to evaluation, the identification of responsible persons (Usability evaluators), defining prototype-level (paper/graphic/functional etc.) and prototype techniques, identification of issues on user-tests (recruiting, time, amount of test-persons) and the reporting of evaluations. In addition, the follow up of recommendations during the ongoing process (applied iterative design) is included.

Sometimes, usability goals have to be defended against clients and/or against technological determinants. It is important, that usability goals have to be regarded with the same priority as business goals and technological issues. If the system does not meet the user's expectations, then the user will not be satisfied, the user will not use the system, the user will not revisit the system, the client will not gain good reputation... investments will be wasted. Another path to oblivion is to set up client requirements as usability goals. As stated above, user input is to be gathered and evaluated against client input (requirements). In best case, user input refines the client perspective (the "added value" to a system: the user involvement). In worst case, user input contradicts client statements: the concept of the client is wrong, again a kind of refinement has to be done.

The following qualities have to be met in order to detail usability goals:
1. Don't specify business goals as usability goals
2. Don't deal with visual (creative) design-specifications
3. Assure adherence to established standards and/or guidelines (ISO, W3C, WAI...)
4. Don't use emotional statements to identify goals ("the site will be a great tool to...blabla")
5. Don't put more than one goal in a usability goal-statement

Fig. 1: Visualization of the modeling of usability goals and the integration into a human-centred design process:

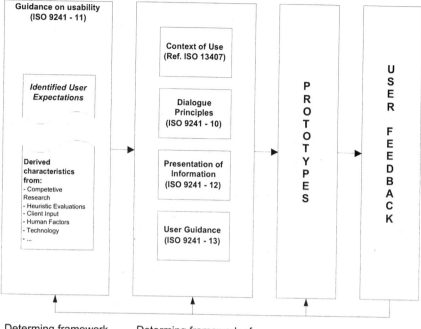

Guidance on usability
(ISO 9241 - 11)

Identified User Expectations

Derived characteristics from:
- Competetive Research
- Heuristic Evaluations
- Client Input
- Human Factors
- Technology
- ...

Context of Use
(Ref. ISO 13407)

Dialogue Principles
(ISO 9241 - 10)

Presentation of Information
(ISO 9241 - 12)

User Guidance
(ISO 9241 - 13)

P R O T O T Y P E S

U S E R   F E E D B A C K

Determining framework that specifies the characteristics and the qualities of a system that is to be developed.

Determining framework of guidelines that outlines development issues.

4. A case-study: Navigate according to user's expectations

The following part describes the process of integration of user expectations into an ecommerce-project. The client was a sports-goods-seller (established in the order-by-mail and brick-and-mortar worlds).

Ecommerce in Germany is a little bit tricky, since the Germans do not very much use credit-cards. In addition, German users are very concerned about Internet security. Because of these reservations, every quality that supports trust-building is welcomed by aware clients. Usability is one of the keys to create user loyalty and therefore the client asked to include usability qualities into the redesign of their online-shop.

Taking in account the actual offline-order catalogue, the client stated that people will search by product category. But the client itself regards (and wants to be regarded) the shop as the "house-of-brands". Our usability team was able to offer additional input to validate correct strategies for the information retrieval. We conducted ad-hoc surveys (short questionnaires in the brick-and-mortar shops) with internet-users to profile potential users (demographical data and preferences in sports, brands and qualities).

In addition, focus-groups were conducted to separate the "wishes" from other, more indifferent statements.

The expectations varied very much according to gender, age and preferred sports. Underlying was the categorization of products. The choice was: cut all other ways to provide product-categorization as the smallest point of agreement, or provide multiple entries to address as many user preferences as possible. We decided to opt for the second choice and leave it to the user to choose the way s/he wants to browse for products.

The major categories that were identified: Sport-Context, Brand, Sports and Product-Category

1. Sport-Context
    All kind of sportal activities that are related to each other regarding where or when they take place. Winter sports, Racketsports, Sport at the beaches etc.
2. Brand
    Sport-goods filtered by brand
3. Sports
    Products filtered by the type of sport, eg. Soccer, Trekking, Tabletennis, Jogging, Boxing etc.
4. Product-Category
    Products, clustered by type, eg. food, equipment, clothes, shoes, glasses, bags etc.

The strategic conclusion to ensure both client needs and user expectation was to adopt the product-categories as underlying paradigm. The major categories are implemented as filters, that provide a specific view of the categories. Does the online-shop not offer a shoe of a specific brand was selected by the user, the shoe category is not visible.

In addition, efficient and adequate information on products (Table 1), high security (Table 2), advices, payment by invoice and more qualified statements on preferences have been gathered.

Table 1: Requested information on sport-goods

	age-group (years)	frequence	% of age-group
valid	<=20	38	67,9
	21-30	56	87,5
	31-40	35	**94,6**
	>=41	9	90,0
	sum	138	

Table 2: The site should provide highest security

	age-group (years	frequence	% of age-group
valid	<=20	30	53,6
	21-30	37	**57,8**
	31-40	19	51,4
	>=41	2	20,0
	sum	88	

Fig. 2: Modelling Navigation-Paradigm using a a usability goal deducted from user expectations

Client statement: Users will serach by product-category.
User expectations: I will look for products according to my activities
                         I will also look for products of specific brands
                         I will of course look for product-categories (using product-families)
Usability goal (work paradigm): The user can choose different ways to access a product

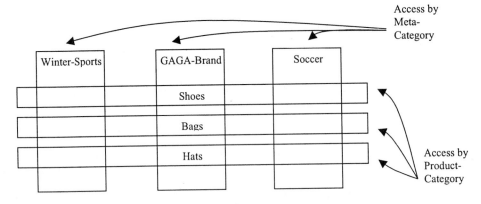

5. Conclusion

Addressing user input not only as "input" but utilize this information as expectation-statement can support the desgin process of an interactive system in an effective way. Through identification of user expectations and the additional set up of usability goals, usability specialists are able to provide measurable criteria of a system-quality. The follow up of the development of the system becomes more transparent both to the client and development-staff. The ecommerce-case showed the flexibility of the process that implements user expectations in a highly emotional environment (trust, security, market-oriented emotions). A work-paradigm that treats user expectations as quality statements is supporting the human-centred design and the role of usabiliy specialists in a multi-disciplinary environment.

The essential precondition to adopt these quality-statements is the dedication to usability principles, awareness of human factors and adherence to established ergonomic guidelines. This basis is, unfortunately, still under heavy development in most companies. User expectations for themselves don't make a system more usable.

References

Eberleh et al.: Einführung in die Software-Ergonomie. 2nd Edition, Berlin: de Gruyter 1994

International Organization for Standardization: ISO 9241-10: Ergonomic requirements for office work with visual display terminals (VDTs) – Part 10: Dialogue Principles, Geneva 1996

International Organization for Standardization: ISO 9241-11: Ergonomic requirements for office work with visual display terminals (VDTs) – Part 11: Guidance on usability, Geneva 1998

International Organization for Standardization: ISO 9241-12: Ergonomic requirements for office work with visual display terminals (VDTs) – Part 11: Presentation of information, Geneva 1997

International Organization for Standardization: ISO 9241-13: Ergonomic requirements for office work with visual display terminals (VDTs) – Part 13: User guidance, Geneva 1998

International Organization for Standardization: ISO 13407: Human-centred design processes for interactive systems, Geneva 1999

Kids´ Space

Michael Schiessl, Sabrina Duda & Rico Fischer

eye square GmbH
Michael Schiessl & Sabrina Duda
Hagenauer Str. 14
10435 Berlin
Germany

Abstract

In this study we evaluated virtual environments for children. We compared two virtual web applications: 4kidz and KinderCampus. The aim of the study was to evaluate children's acceptance of the system and the usability of the virtual environment. Usability was measured with a scenario based testing of the website. The kids should fulfil a variety of tasks. While exploring the web site the attention of the kids was measured by counting the eye-fixation time and rate on the screen. Concerning usability KinderCampus which could be characterized as a very intuitive virtual environment showed the best results. The navigation of the system is based on a space metaphor. Even illiterate children at the age of four were able to use the system and showed great pleasure while exploring the virtual environment. 4kidz caused many usability problems. Contrary to the other systems the navigation concept was not based on a metaphor concept. Especially the younger kids where not able to use the system and therefore, got frustrated. Results highlight the need for navigation concepts that follow a consistent spatial metaphor. Results also showed that the system has to give a very fast and clear feedback, otherwise kids loose track very easily.

1 Introduction

In the last years many new virtual environments for children have been developed. All of them have the intention to be a safe and funny place for kids on the web. But how does a virtual environment have to look like in order to achieve a high acceptance from children. This study was conducted to demonstrate the important role of the structure of an interface and its relation to usability and acceptance by children. For this reason we tested two virtual web applications: 4kidz and KinderCampus. The basic concept of the providers will be briefly explained.

Figure 1. Loading pages of KinderCampus (left) and 4kidz (right). While the instruction of the 4kidz-page only indicates a delay, KinderCampus offers a "loading-game" to keep children busy for the mean time.

Figure 1 shows the loading pages of the two virtual web applications. Both designed a comic space ship which will be used as a medium to get to different topics of the sites. This is meant to transfer the concept of a world of its own, away from reality – a kids universe with lots of interesting things to explore. Figure 2 and 3 clearly demonstrate the different strategies used within the system.

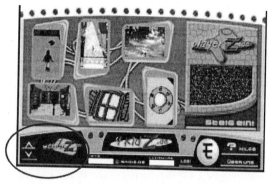

Figure 2. Throughout the whole system Kinder-Campus continues the "space metaphor". Different planets circle in space and the space ship called cab is used to virtually transport children to the topics of interest. Once a kid is "on" a planet, comic characters do funny things if the mouse arrow is pointed at them, indicating that something will happen when clicked on it. This kind of presentation reflects an outside perspective.

Figure 3. 4kidz focuses more on a "game-boy" or desktop metaphor. The buttons up and down in the lower left can be used to switch between topics which are indicated beside the buttons. The windows connected through wires and cables offer direct access to the different topics. If one would want to continue the space concept then this game-boy or desktop metaphor would reflect an inside view of the space ship.

To summarize the two providers we state that KinderCampus realizes a consequent space metaphor in which navigation is supported by stimuli features (i.e. moving objects) whereas 4kidz uses a "game-boy" strategy in which symbols as navigation tools play a more important role.

2 Methodology

2.1 Subjects
Thirty children from 4 to 12 years (mean: 8.9) participated in this study. Results showed that the internet competence of average German kids is very high. Even the 4 year old kids where able to use a mouse and generally got along with the websites.

2.2 Procedure
To answer the questions how children are able to use the two virtual application and which impact this has on the general acceptance we applied qualitative methods and the Semantic Differential Scale. In the usability testing the two different providers were balanced within the subjects. A scenario based testing required subjects to perform on a number of KinderCampus- or 4kidz-relevant tasks. (KinderCampus: i.e. How many phones are in the city hall? Send an email! Go into the zoo! 4kidz: i.e. Go to the skateboard-game! Send an email! and so on) While solving the tasks performance was observed and recorded. In addition, quality of performance was rated on a scale from 1 (solved tasks without any problems) to 5 (needed lots of help to solve the task).

The attention was also recorded as a function of fixation time and rates on the screen and the times of looking away. Afterwards participants were interviewed about their likings and problems concerning their performance on the two web applications (i.e. Which site did you like better and why? Which age group do you think this site is suited best for?).

A normally used standard method of thinking aloud while performing the tasks could not be included in this study. We found that children less than 7 years were not able to succeed in this method.

3 Results

Results indicate a clear preference of the KinderCampus web-site. Nine out of ten children who made preference statements preferred KinderCampus to 4kidz. Surprisingly, these preferences for KinderCampus were not related to age. Only one child had no preference. In addition, participants were asked to give a total score on a scale from 1 to 5 (1 equals very good, 2 equals very bad) concerning their general judgement of the provider. KinderCampus was scored with 1.6 and 4kidz with 2.1. The rating of the performance quality (usability tasks) on an equivalent scale (1-5) revealed an advantage of the KinderCampus sites with a 1.7 compared to 4kidz with 3.5. That means, usability tasks were much easier to solve when using the KinderCampus sites. 4kidz sites revealed more usability problems. Not all children could use the navigation system 4kidz as efficiently as the one of KinderCampus. They described it as more difficult and complicated In addition, concerning 4kidz especially younger children reported problems with the ambiguity of symbols that lead to mistakes.

4 Discussion

The aim of this study was to investigate the acceptance and the usability of two virtual environments which were based on two different interface structures.

KinderCampus represents a strict "space-concept" in which navigation is primarily guided by intuition. Different application levels are indicated by planets that move through space. This "space-metaphor" makes it especially for young children easy to shift levels without the necessity of creating a mental representation of the system structure. KinderCampus made thoughtfully use of psychological knowledge about capturing visual attention in the design of the virtual environment. Links that indicate a gate to a different level contain so called "invitation-characteristics". Animations, pop-out effects, faces and animals are known in their ability of capturing visual attention in a bottom up manner. Kids are therefore, invited to attend to those stimuli and often times just naturally try to click on them. Once they did so, the figure or the face does funny things and leads to a different level. Kids learn to use these stimuli as navigation tools. That is what we described as navigation by intuition. It is not surprising that only basic mouse-competence is needed to perform in such an environment.

The usability results of our study confirm these assumptions and demonstrate how easy it is even for illiterate children to navigate in this kind of "space-metaphor" structure. The ease of use is connected to a low frustration rate and leads to a enormously high preference of this environment compared to the competitive provider.

4kidz on the other side focuses more on a "game-boy" metaphor. The structure of the system is less intuitive and, moreover, different levels and applications are placed in a hierarchical order. Navigation tools carry a more symbolic character than self-inviting features (see also figure 3).

Therefore, attention is not as much captured by physical features - it has to be directed intentionally towards the symbols and can be described as a controlled top-down process. This leads to the problem that children must understand, that the symbols represent actions or commands similar to street signs. They have to encode the signs and associate them with their anticipated goals. In order to navigate successfully within the virtual environment of 4kidz it can be summarized that children must have created a mental representation of the system structure. It can be assumed that especially younger children will have problems with this.

Our results show a tendency that is consistent with this hypothesis. Especially younger children report complexity and difficultness of the 4kidz sites as negative. All children show a less accurate performance on the usability tasks compared to KinderCampus. It is interesting to note, that even older children which seem to have less problems with the navigation prefer the KinderCampus site.

To summarize the results of this study we can state that an interface using an ecological context (i.e. "space metaphor") and additionally involves links that are action coherent, as described above, lead to a better task performance and acceptance than interfaces using a hierarchical structure. It can be assumed that the younger the children the more intuitive the navigation within the system should be.

SOME PRELIMINARY INVESTIGATION ABOUT THE ORGANIZATION OF USER INTERFACE DESIGN GUIDELINES

Jean Vanderdonckt

Université catholique de Louvain, Place des Doyens, 1 – B-1348 Louvain-la-Neuve, Belgium
Vanderdonckt@qant.ucl.ac.be – http://www.qant.ucl.ac.be/membres/jv

ABSTRACT

Several tools for working with guidelines have today appeared, either in the commercial market or in the do-main of research and development. Since these software frequently manipulate guidelines during many development steps of a user interface of an interactive application, they can overthrow any approach followed to develop this application. They also raise the fundamental question of to what extend can we trust in these software. To answer this question, we introduce five development milestones through which we need to go to get such a tool for working with guidelines.

1. INTRODUCTION

Ensuring the ergonomic quality of user interfaces (UIs) is become one of the major worrying in developing interactive application since more than a decade. To characterize this, utility and usability are often distinguished. Utility translates the appropriateness of a UI with respect to the functional goals of an interactive application; in this way, it remains user independent. As regards for the next, usability translates the appropriateness of this UI with respect to the operational user's goals ; it therefore heavily depends on the user. The UI of a word processor could be considered as useful in providing a facility for computer-aided creation of a table of contents of a given document; the usability of this UI will depend on the easiness with which a user will be able to carry out this task, without being blocked by an error, but being supported to fix it.

The debate of interactive applications which are functionally rich (thus, useful), but operationally poor (thus, unusable) is still open:

- Intensive, experienced users tend to prefer a useful interactive application, even if not usable rather than a usable interactive application which is not very useful.
- Episodic, novice users reject an interactive application as soon as it is unusable, even if not very useful; at the limit, they prefer an application which is a little useful, but usable.

To study, express and insure the usability of a UI, several disciplines can help every person who is responsible for developing the UI : participatory design, cognitive psychology, contextual inquiry, software ergonomics,... In this last, many methods have already proved their positive impact on the usability of a UI: evaluation methods with or without users, heuristic evaluation, usability testing as well as any ergonomic approach grounded on guidelines. We generally define a *guideline* by a design and/or evaluation principle to be observed in order to get and/or guarantee the usability of a UI for a given interactive task to be carried out by a given user population in a given context. In the rest of this text, we assume a general method for including guidelines throughout the development life cycle of a UI for a particular interactive application. Guidelines are mostly used during

- The specification phase: a set of guidelines is delimited as requirements for the future UI.
- The design phase: guidelines are exploited in order to decide an appropriate value for each design option by considering the context (which includes the interactive task, the user population and the working environment in which the users are carrying out their task).
- The prototyping phase: guidelines are exploited to obtain as soon as possible a static or working UI prototype that can be showed, tested and evaluated.
- The programming phase: guidelines are gathered to guide, orient, precise, insure a UI development within the developing environment for the targeted computing platform.
- The evaluation phase: the resulting UI is evaluated towards guidelines which are often those which have been selected in previous phases.

1212

- The documentation and certification phase: guidelines which have been manipulated in previous phases are instructed in a documentation for documenting an interactive application for communication, reuse, maintenance or commercial promotion purposes; this phase typically consists in exposing the UI towards these guidelines.

Aside from these situations, guidelines also reclaimed a new land: the land of teaching UI design based on guidelines throughout the life cycle. This paper is the editorial of a special issue completely devoted to tools for working with guidelines and underlying activities: searching for guidelines, collecting, working, writing, propagating, using, applying, verifying and, more recently, teaching and learning. These activities are progressively better understood and structured into methods; they are today manipulated by a series of persons who often cumulate roles: project leader, methodologist, designer, analyst, developer, evaluator, human factors specialist, psychologist, teacher, computer scientist.

To support these people in their activities, research and development commercialized since 1989 with NaviText SAM (Perlman, 1989) a wide range of tools for working with guidelines. From this production and this commercialization, a profound overthrow emerged in the current practice of using guidelines as observed in several questions raised at three levels:

1. At the level of the transformation source:

- To what extend can we thrust in guidelines played in an ergonomic approach?
- Where guidelines can be located in existing methodologies for developing interactive applications, especially during the above phases and how?
- What are the exact roles played by these people who manipulate guidelines in the quoted activities and how do they insert themselves in the conjunctive texture of a development team?

2. At the level of the transformation target, which is often transmitting the same questions at the software level, of its contents and its usage :

- To what extend can we thrust in guidelines incorporated into software tools? More particularly, if the confidence level of a guideline is given acceptable at the beginning, what is the resulting confidence level after transformation?
- Where guidelines can be located in existing methodologies for developing interactive applications, especially during the above phases and how? How are these locations redefined after their usage in a software tool? More particularly, to what extend could a person be assisted in an ergonomic approach by such a software tool? To what extend could we delegate a portion of the activities to these software tools and how?
- What are the exact roles played by people who are using software tools for working with guidelines and how do they insert themselves in the conjunctive texture of a development team? More critically, if we consider a person who recently learned software ergonomics and who is using such tools, is this person able to assume the same roles of a person who acquired extensive experience in human factors, but who is not using such tools? For instance, could a human factors expert be substituted by an inexperienced designer assisted with software tools for working with guidelines? Could this expert be also assisted by these tools?

3. At the level of the transformation itself, from source to target:

- How can we manage to develop an efficient and valid software tool?
- On which guidelines can we ground this tool and how can they be incorporated in the software tool?
- What are both negative and positive consequences of this transformation at the different steps of this transformation?

To raise these question, to interrogate contents and practices at these three levels already induces a new light to shed on UI usability resulting from the new approach and their collective and symbolic stakes. If we are not able to today – would we be ever?– to provide an appropriate answer to all the questions, we are nevertheless able to bring some answer elements by examining the steps which transform the old approach, based on a manual guidelines usage, to a new approach, based on software tools for working with guidelines. For this purpose, we introduce five development milestones through which such a tool should go (figure 1). Each milestone is placed as output of the five steps of this complete transformation:

1. The guidelines collecting is aimed at collecting, gathering, merging, compiling guidelines coming from world-wide available ergonomic sources into initial guidelines (milestone #1).

2. The guidelines organization is aimed at sorting, classifying, unifying these initial guidelines into organized available guidelines (milestone #2).
3. The incorporation of organized guidelines into a methodological approach is aimed at locating, stating precisely, circumscribing, specifying the intervention points of organized guidelines subsets into a methodology for developing an interactive application grounded on these guidelines (milestone #3).
4. The operationalization of approach located guidelines is aimed at transforming, reexpressing, reformulating them according to an operational and internal representation which executable, manipulable by a finite state machine (milestone #4).
5. The usage of operational guidelines is aimed at examining, analyzing, developing various ways to record, consult, access guidelines from a tool for working with guidelines (milestone #5).

This tool is the conclusion of the complete transformation and this milestone is the ultimate milestone of the whole process for developing such a tool.

2. SOME RESULTS AND DISCUSSION

Two major problems encountered in organization are:

1. Guidelines should all be translated into some canonical form. Since initial guidelines are written with vocabulary and conventions that are local in the different disciplines, they all vary along the axes identified above and require a constant translation into a common format.
2. Guidelines are insufficiently classified, a much regretted flaw, which is unfortunately aggravated by an even greater range of classifications:
   - By ergonomic criteria, as in Scapin's ergonomic guide.
   - By usability factor, as in ISO 9241 standard.
   - By interaction style, as in Mayhew's guide.
   - By widget, as in IBM CUA style guide.
   - According to an object-oriented model on input/output.
   - By linguistic level, as in Marcus's « User Interface Standard Manual ».
   - By importance level, as in Banks's standard.
   - By type of widget, as in Farenc's ERGOVAL automatic evaluation tool (Farenc, 1999).

Every attribute of the general guidelines model could potentially become an organizing category at a high or intermediate level. Unfortunately, no known study underpins any preference for a particular classification. Taxonomical approaches remains the most frequently used support for classification. However, a particular organization (for instance, in order to certify a standard compliance) that suits one goal can become totally inappropriate for another (for instance, in order to learn about usability from the standard).

In operationalization, guidelines integrated into methodologies are used manually. Such guidelines have often provided for development teams, but have tended to be ignored, underestimated and underused for several reasons:

1. First, the current form of the guidelines themselves, as already noted, implies a difficulty of interpretation, their decontextualisation limits application, variations in importance and appropriate target development role.
2. Second, there are reasons related to the onerous and inefficient demands of guideline management in their current form. The huge amount of guidelines exceeds human mental capability: the paper format turns developers into medieval scholars. Only a part of interviewed designers looked for guidelines related to a particular problem, they infringed on average 11% of guidelines because they did not know how to access them.

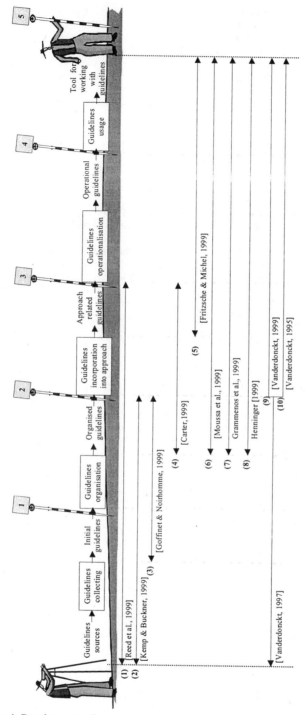

Figure 1. Development milestones towards a tool for working with guidelines.

Today, we do not know a lot about software ergonomics as it is since the weight, the priority, the relevance of guidelines is varying and hard to assess. Similarly, we do not know a lot about the sufficiency and the validity of guidelines, though many authors claim their necessity. We even do not know more about meta-ergonomics, that is the usability of guidelines themselves when used during development to insure UI usability. We finally do not know anything at all on the potential degradation of guidelines undergone by all transformations (for instance, what is the impact of replacement and reduction guidelines?). At most, we are able to locate where these degradations occur and to explain them. Thus, all we can do is to very moderately believe in used guidelines. Finally, the knowledge we possess about the usability of software tools for working with guidelines is also non existent. And yet, it seems fundamental when we are talking about UI usability that the UI of the software tool for working with guidelines itself is usable in their contexts. The scientific community may benefit from an empirical study of the usability of such tools to assess whether they are usable. The poor status of our knowledge may partially explain why it is impossible for us to develop this milestone further.

7. CONCLUSION

After going through the five milestones (from guidelines collecting to their usage), we are able to count up all the problems we encountered: huge amount of guidelines, guidelines increasing with time, guidelines wide spreading, guidelines variation in function of their discipline, required translation of guidelines, guidelines variation in function of their foundation, contextual independence of guidelines, guidelines variation in function of their contents, guidelines variation in function of their presentation, guidelines variation in function of their detail level, guidelines variation in function of their complexity, guidelines variation in function of their importance level, guidelines variation in function of their scope, guidelines variation in function of their development phase, guidelines variation in function of their target person, necessity and insufficiency of guidelines, conflict between guidelines, vanishing of connections between guidelines, contextual independence of guidelines, reduction of guidelines, independence of guidelines with respect to users, insufficient guidelines illustration, insufficient guidelines references, insufficient guidelines classification, insufficient operationalization of guidelines, multiple reference of guidelines, heavy guidelines management,...

REFERENCES

Carter, J., "Incorporating standards and guidelines in an approach that balances usability concerns for developers and end users", Interacting with Computers, Vol. 12, No. 2, pp. 179-206

Farenc, Ch., Palanque, Ph., and Bastide, R., "Embedding Ergonomic Rules as Generic Requirements in the Development Process of Interactive Software", In Proceedings of 7th IFIP Conference on Human-Computer Interaction Interact'99, IOS Press, 1999. Accessible at http://lis.univ-tlse1.fr/~farenc/papers/interact-99.ps

Fritzsche, H., Michel, T., "Formalization and Proof of Design Guidelines within the Scope of Testing Formally Specified Electronic Product Catalogues", Interacting with Computers, 2000, Vol. 12, No. 3, pp. 209-223.

Goffinet, L., Noirhomme-Fraiture, M., Automatic cross-referencing of HCI guidelines by statistical methods, Interacting with Computers, 1999, Vol. 12, No. 2, pp. 161-177.

Grammenos, D., Akoumianakis, D., Stephanidis, C., "Integrated support for working with guidelines: the Sherlock guide-line management system", Interacting with Computers, Vol. 12, No. 3, 2000, pp. 281-311.

Henninger, S., "A methodology and tools for applying context-specific usability guidelines to interface design", Interacting with Computers, Vol. 12, No. 3, 2000, pp. 225-243.

Kemp, B., Buckner, K., "A taxonomy of design guidance for hypermedia design", Interacting with Computers, 1999, Vol. 12, No. 2, pp. 143-160.

Moussa, F., Kolski, Ch., Riahi, M., "A model based approach to semi-automated generation for process control interactive applications", Interacting with Computers, Vol. 12, No. 3, 2000, pp. 245-279.

Perlman, G., "Asynchronous Design/Evaluation Methods for Hypertext Technology Development", In Proceedings of ACM Conference on Hypertext Systems Hypertext'89, New York: ACM Press, 1989, 61–81.

Reed, P., Holdaway, K., Isensee, S., Buie, E., Fox, J., Williams, J., and Lund, A., "User interface guidelines and standards: progress, issues, and prospects", Interacting with Computers, 1999, Vol. 12, No. 2, pp. 119-142.

Vanderdonckt, J., "Accessing Guidelines Information with Sierra", In Proceedings of Interact'95, pp. 311–316.

Vanderdonckt, J., "Conception assistée de la présentation d'une interface homme-machine ergonomique pour une application de gestion hautement interactive", Fac. Univ. Notre-Dame de la Paix, Namur, July 9, 1997.

Vanderdonckt, J., "Development Milestones Towards a Tool for Working with Guidelines", Interacting with Computers, 1999, Vol 12, No. 2, pp. 81-118.

USER INTERFACE AND OTHER FEATURES OF A VIDEOTELEPHONE SET AIMED AT THE ELDERLY - a pilot assessment of depth and breadth trade-off of screen menus

Seppo Väyrynen[1], Minna Törmänen[1], Vesa Tornberg[2] & Tatu Prykäri[2]
1)   University of Oulu, Work Science Laboratory, FIN-90014 Oulun yliopisto, Finland
2) ˙ Navicre Ltd, Oulu, Finland

ABSTRACT
The videotelephone set studied here is a prototype developed during the multimedia Home Aid Communication System project. The main objective of the project was to create new product and service concepts and also to develop a system which incorporates social and health care services, commercial services and other daily services and contents. The target group consisted of elderly and disabled people, who, with the support of ICT-based services, would be able to live independently and safely in their own homes. The technology used in the project is based on the computer, phone and videoconferencing technologies. The videotelephone set comprises the basic PC technology with a 15-inch touchscreen-only UI and a codec card providing a rapid video and voice connection via, for instance, ISDN. This paper mainly focuses on determining some preliminary basic solutions of the menu structure on the screen. The pilot trials with elderly subjects (n=7) showed the advantage of breadth over depth. This was revealed by comparing two menu-tree forms, both of which were used to search the same services as goal tasks.

1.   INTRODUCTION

The number of elderly individuals among citizens (and customers) is increasing in industrialised societies. The changing needs, varying capabilities and impaired mobility tend to confine the elderly and most of their activities to their homes. The rising costs of care in hospitals and other service institutions and the increasing proportion of elderly people out of the total population make it vital to ensure independent living of the aged in their domestic environment. And their own wishes and mental models of daily living are compatible with this aim. The decreased need and possibility to resort to physical mobility can obviously be enhanced/substituted by "e-mobility" using Information and Communication Technology (ICT).

The use of video telephony (VTP) to replicate face-to-face contact is one form of multimedia communication technology (Väyrynen&Pulli 2000). Via VTP, many actions, objects and places can be seen or shown along with the participants. Telepresence is also possible as one form of telecontact. For the elderly, VTP may turn out to be an essential (assistive) technology for everyday life. They can communicate and be served remotely using VTP. This will guarantee and prolong the autonomous and full life of the elderly.

The multimedia Home Aid Communication System (mmHACS) is a 3-year project, the main objective of which is to create new technical products. Further, it also aims to provide a whole system of social and health care services, e-commerce, e-banking, and other services linked to hobbies, family contacts, entertainment, shopping, parish life, etc. There is a strong emphasis in the project on (1) user studies, (2) concept creation and relevant communication about this aspect, and finally, (3) usability studies with prototypes (Väyrynen et al.1999, Ikonen et al. 2000). The overall aims of the ongoing project are:
-    to observe and analyse thoroughly the needs, the users and the organisational context (user study),
-    to choose, develop iteratively and design in participatory ways ICT & service concepts and to further experiment with them to establish systems applicable to care and living,

-   to demonstrate concepts in collaboration with equipment manufacturers, telecom operators, service providers and organisational / individual users.

In accordance with the literature (Sanders & McCormick 1993, Shneiderman 1998), our background knowledge and experiences of mmHACS emphasise the importance of the user interface (UI) and adequate knowledge of the characteristics of elderly users. The latter author reviewed the role of the menu structure, though not especially in software used by the elderly. The aim of this paper was hence:
-   to gather basic performance and preference data of iteratively and participatorily developed alternatives of menu layouts of various services on the touchscreen of a VTP device for the elderly, and especially,
-   to gain firsthand experimental evidence of how to design an optimal menu structure.

Thus, two designs with various combinations of depth, i.e. the number of levels in the hierarchy, and breadth, i.e. the number of items (service options) per level, were compared in the context of real aged end-users and the typical services needed by them. The two designs were implemented on the UI of a VTP set of Videra Ltd, a member of the mmHACS consortium.

2. METHODOLOGY

The comparison was based on three variables: (a) the "extent" of the interactive tasks, here the number of pushes on touchscreen buttons, (b) the operating time, s, needed to navigate from the start after the instruction until the service is found, and (c) subjective satisfaction and replies to the questions posed by the researchers. The following two menu-tree forms were experimentally studied:
-   design 5^2, 5^n items at each level (n = 1,2; with an increasing degree of speciality)(Fig.1)
-   design 3^3, 3^n items at each level (n = 1,2,3; with an increasing degree of speciality)(Fig.2)

The subjects (n=7) were 60-80 years old, and they were carrying out a user trial linked with their daily living, with both menu-tree forms applied in a random order. The subjects used each design on a different day. The trials were video-recorded and analysed later. Each trial comprised service-finding tasks, which involved a need to seek goals. An example of the instruction to find a goal used in the trials was: "Find contact to Pharmacy." The following 10 typical daily living services were used as tasks: (1) TV, (2) Shop, (3) Calendar, (4) Booking, (5) Phone number of a friend, (6) Taxi, (7) Pharmacy, (8) Phone number inquiry, (9) Lotto, (10) Library.

NOTEBOOK	HEALTH	SERVICES	PHONE CALLS	ENTERTAINMENT
Reminder	Doctor	Home Care	Numbers	TV
Calendar	Emergency Duty	Shop	Sister	Library
Photograph album	Booking	Bank	Friend	Lotto
Word processor	Pharmacy	Taxi	Friend	Chatline
Booking	Health visitor	Church	Phone number inquiry	www

Fig.1 Design 5^2 included 25 services grouped as 5 items on 2 levels.

The distribution of (a) and (b) measures by service tasks and by design were tabulated, and the raw data were processed further based the "relative score" method recommended by Nielsen (2001). Here, the corresponding ratio was called Comparative Performance Percentage (CPP), which is a ratio for comparing two designs. CPP was calculated as a geometric mean separately for both (a) and (b) by this equation.

$$CPP = \sqrt[n]{X_1 \cdot X_2 \cdot \ldots \cdot X_n}$$

n= number of tasks=10, in which for each task separately,

$$X(a)_{1,2,...,n} = \frac{pushes\ in\ design\ 3^3}{pushes\ in\ design\ 5^2} \cdot 100\%, \ and \ X(b)_{1,2,...,n} = \frac{time\ in\ design\ 3^3}{time\ in\ design\ 5^2} \cdot 100\%$$

This kind of formulation of the equation shows, with the value of CPP minus 100%, how much better/worse design 5^2 is compared to design 3^3.

SERVICES	PHONE CALLS	FREE TIME
HEALTH	DIAL PHONE NUMBER	NOTEBOOK
Booking		Calendar
Pharmacy		Reminder
Doctor		Photograph album
ERRANDS	PHONEBOOK	HOBBIES
Bank	Sister	Chatline
Taxi	Friend	Library
Shop	Friend	Church
HOME CARE	INFORMATION SERVICE	ENTERTAINMENT
Cleaning	Phone number inquiry	TV
Health visitor	Taxi	WWW
Meal service	Telephone advice service	Lotto

Fig 2. Design 3^3 included 25 services grouped as 3 items on 3 levels.

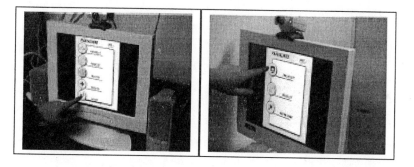

Fig 3. User trials of the UI prototypes. The elderly subjects used both design 5^2, which is presented on the left, and design 3^3 shown on the right.

3. RESULTS

The results clearly show design 5^2 to be superior to design 3^3, and this trend is very consistent as evaluated by both task and subject. (Tables 1 and 2).

Table 1. Trial performance medians (min and max) based on each subject's raw data when carrying out the tasks. The data are shown as both the number of pushes on the touchscreen (a) and the operating time needed to find each service (b).

Task	(a) Pushes (median(min-max))		(b) time, s (median (min-max))	
	5^2	3^3	5^2	3^3
1	2 (2-12)	7 (3-14)	34 (18-145)	58 (19-192)
2	2 (2-4)	3 (3-5)	15 (7-71)	15 (9-61)
3	2 (2-4)	3 (3-10)	11(9-22)	42 (21-123)
4	4 (2-8)	7 (3-13)	30 (5-58)	69 (21-136)
5	2 (2-2)	3 (3-6)	12 (7-30)	26 (17-55)
6	4 (2-8)	5 (3-16)	40 (6-64)	42 (8-143)
7	2 (2-4)	3 (3-15)	13 (5-27)	19 (12-110)
8	2 (2-4)	3 (3-11)	11 (6-53)	30 (9-73)
9	2 (2-8)	9 (3-13)	21 (10-46)	66 (18-90)
10	4 (4-10)	9 (5-18)	27 (20-45)	45 (9-104)

Table 2. Calculated values of Comparative Performance Percentage (CPP) for both (a) and (b) variables by subjects.

Subject	CPP	
	(a) %	(b) %
1	207	209
2	149	197
3	176	245
4	160	120
5	175	238
6	173	137
7	247	267
Mean	184	202

Examples of the (c) expressions of subjective satisfaction in debriefings:
" This (design 5^2) is quite easy"
" This (design 5^2) is considerably easier than the other."
" I live alone, so it might be nice to call my children by the videotelephone because I could see them."

4. CONCLUSION

The previous results reported in the literature (e.g. Kiger 1984, Shneiderman 1998, Goldstein et al. 1999) suggest that, as far as tree-structured menus are concerned, breadth should be preferred over depth in design. This pilot study, with a special emphasis on a VTP set aimed at the elderly, confirmed this general rule. More detailed rules for applications designed for elderly users warrant further research.

In this trial, the subjects did not have much experience of computers, and they can therefore be considered novices in terms of menu structure. In design 3^3, the third level was difficult to find. Design 5^2 was more effective for use by the elderly, and they also reported more subjective satisfaction with design 5^2. The number of pushes and the total time showed design 5^2 to be 84% and 102% better in performance, respectively. But it naturally remains to be unverified whether 5^2 is good enough or whether even better designs can be found. The navigation problem seems to increase in severity as the depth of the hierarchy increases. Generally, one possible solution could be menu maps, which are sometimes used on web sites (Shneiderman 1998). The capacity of the working memory limits the number of menus. The maximum number of items that can be held in the working memory is 7±2 (cf. Sanders & McCormick 1993). Design 5^2 is well within these limits. The results of menu-tree designs indicate that the breadth of the menu aimed at the elderly should be maximized within the above limits, whenever possible, to avoid an excessive increase in the number of levels of the menu-tree.

The obvious aspects that should be focused on later in the project include: tree menu or network menu, menu-tree design on a small screen as in mobile phones or PDAs (Goldstein et. al.1999), cyclic vs. acyclic menus, classification (grouping of services in categories), the role of mental models and cognitive capacity, etc. Menu design could be used to enhance the possibilities of the elderly to maintain the VTP set under their intellectual control. In addition, iterative improvements will be achieved through further trials with more design alternatives, more users, and more evaluation feedback based on both performance and preferences.

REFERENCES

Goldstein, M., Anneroth, M., Book, R. (1999), Usability evaluation of a High-fidelity smart phone prototype: task navigation depth affects effectiveness In: Bullinger, H-J, Ziegler, J. (Eds.) *Human-Computer Interaction: Communication, Cooperation and Application Design. The 8th Int. Conference on HCI, Munich, August 22-26, Vol. 2*, pp. 38-42. Mahwah, NJ: Lawrence Erlbaum Associates, Publishers.

Ikonen, V., Väyrynen, S., Tornberg, V. (2000). New technology in health and social services - ethics and ergonomics in videotelephone use among workers and clients. In: Fostervold, K.I., Endestad, T. (Eds.) *Ved Inngangen til Cyberspace-ergonomisk tenkninig inn i et nytt årtusen. Proceedings fra Nordiska Ergonomisälskapets Årkonferanse 2000. NES med NEF*, pp. 27-30.

Kiger, J. I. (1984). The depth/breadth trade-off in the design of menu-driven user interfaces. *International Journal of Man-Machine Studies, Vol.20*, pp. 201-213.

Nielsen, J. (2001). Usability Metrics. In: http://www.useit.com/alertbox/20010121.html

Sanders, M., McCormick, E. (1993). *Human Factors in Engineering and Design, 7th edition.* New York: McCraw-Hill.

Shneiderman, B. (1998) *Design the User Interface: Strategies for Effective Human-Computer Interaction, 3rd edition.* Reading: Addison-Wesley.

Väyrynen, S. , Pulli, P. (2000).Video Telephony. In: W. Karwowski (Ed.) *Int. Encyclopedia of Ergonomics and Human Factors, Vol. I*, London: Taylor & Francis. pp. 757-764.

Väyrynen, S. Tornberg, V., Kirvesoja, H. (1999) Ergonomic Approach to Customised Development of Videotelephony Application. In: Bullinger, H-J, Ziegler, J. (Eds.) *Human-Computer Interaction: Communication, Cooperation and Application Design. The 8th Int. Conference on HCI, Munich, August 22-26, Vol.2*, pp. 261-265. Mahwah, NJ: Lawrence Erlbaum Associates, Publishers.

THE USABILITY OF ON-LINE SHOP IN CHINA

Xiaolan Fu[a], Guomei Zhou[a] and Ming-Po Tham[b]

[a]Institute of Psychology, Chinese Academy of Sciences, Beijing 100101, P.R. China
[b]Human Interface Lab, Motorola Electronics Pte. Ltd, 02-02 Techpoint, 10 Ang Mo Kio Street 65, Singapore 569059

ABSTRACT

Web usability is the focus of much attention in current HCI research and is now widely recognized to be an important ingredient contributing to the success of e-businesses. The present study used psychological method to compare the usability of the Web sites of two very well established and popular on-line retail stores in China. In this study, 40 Chinese undergraduate students, half males and half females, participated in a two-session experiment in the laboratory. In the first session, the subjects made simulated purchases of two items in each of the two on-line shops to familiarize themselves with the two shops. In the second session, subjects were given a questionnaire including a series of questions related to the usability of the Web sites of the two on-line shops. The results showed that there are significant differences between the two on-line shops with respect to customer's satisfaction and preferences and their ranking of the various aspects of the usability of the shops' Web sites. The differences distinguishing these two on-line shops and the factors which Chinese e-shoppers deemed important to usability and customer satisfaction are summarized and discussed in the paper.

1. INTRODUCTION

Namely, an on-line shop is a virtual shop founded in e-media space and based on the Internet. In mainland China, on-line shop is still at its trial stage. Nevertheless, the number of on-line buyers is expected to increase as the number of Internet users increases. According to the reports of China Network Information Center (CNNIC, 2000a, 2001), the number of Internet users had increased from 8.9 million of 1999 to 22.5 million of 2000, and the percentage of on-line buyers had increased from 8.79% of 1999 to 31.67% of 2000 in China. Chinese Internet users are also optimistic to the development of on-line shopping: 81.71% thought that China would develop large-scale on-line e-commerce within 5 years, and 61.41% chose on-line shopping as the most promising on-line enterprise (CNNIC, 2001).

Nevertheless, other data paint a more sobering picture of e-business. Many studies have found that up to 60-75 percent of e-shopping transactions were interrupted, and e-tailers were having difficulty in building loyal customer bases as on-line customers turned back to traditional brick-and-mortar shops or print catalogues to make their purchases (E-Commerce Times, 1999a, 1999b, 1999c; ZDNET, 2000; Zona Research, 1999). Similar observations were made in Asian cities. For instance, in an on-line "survival test" staged by the Chinese government in three major Chinese cities in the latter half of 1999, all the participants expressed negative views of the adequacy of China's e-commerce industry (Agence France Press, 2000).

Obviously, any on-line shops have three main purposes: to persuade customers into buying their stuffs, to build tight relations with customers, and to attract more potential customers. One important way to help realize shopping behavior is to deeply understand the properties of e-media space and the interaction between customers and e-media

1222

space. The usability of on-line shop addresses the relationship between an on-line shop and its customers. Usability is the quality of a system that makes it easy to learn, easy to use, easy to remember, error tolerant, and subjectively pleasing (Usability First, 2001). The usability of on-line shop is important because in all cases, lack of usability can cost time and effort, and can greatly determine success or failure of an on-line-shop. We can learn to be better on-line shop designers by learning design principles and design guidelines. But even the most insightful designer can only create a highly-usable on-line shop through a process that involves getting information from actual users. Web design criteria should focus on what users are trying to do, where they are trying to go rather than on how the technology works (Nielsen, 2000).

Usability is one of the focuses of the Human-Computer Interaction field. However, there are few usability studies on Chinese Web sites, especially on Chinese on-line shops. The present study used psychological method to compare the usability of the Web sites of two very well established and popular on-line retail stores in China. The merchandises sold by these two on-line shops include a large variety of home appliances, electronic equipment, mobile telephones, IT products, flowers, gifts, postcards, books, tickets, etc. The main objective of this study is to investigate Chinese users' preference for the interface design of on-line shop.

According to our demographic comparison between on-line buyers and browsers on the data provided by CNNIC from the results of the survey conducted in June of 2000 (CNNIC, 2000b), over 50% of the buyers attained bachelor degree or above; 22.7% of the buyers are in computer industry or IT industry and 12% of the buyers are in the area of science research and education; and largest proportion (21.4%) of the buyers reside in Beijing. These findings make us believe that college students should be the most probable potential on-line buyers in the near future. Thus, the detailed objective of this study was as following: (1) To investigate the preference of Chinese college students in Beijing to Chinese on-line shop design; (2) To provide the design guidelines for an ideal interface to facilitate the potential on-line shopping in China.

2. METHOD

2.1 Subjects

40 undergraduate students from colleges in Beijing were as participants, half male and half female, age 18-23 (average=20.28). Their WWW experiences were considered: 21 of them spent less than 5 hours per week on browsing general Web sites; Other 19 students spent at least 5 hours per week on browsing general Web sites.

2.2 Material

A questionnaire includes background, feeling on shopping on-line, browsing behavior, form design, content design, and general evaluation. A six-point scale was used (0: the lowest degree; 5: the highest degree).

2.3 Procedure

Within-subjects design was used in this study. All subjects participated in a two-session experiment in the laboratory. In the first session, the subjects made simulated purchases of two items in each of the two on-line shops to familiarize themselves with the two shops. The browsing sequence of the two on-line shops and participants' genders were balanced. In the second session, subjects were given a questionnaire including a series of questions related to the usability of the two on-line shops. Subjects were allowed to revisit the two on-line shops when completing the questionnaire.

3. RESULTS

3.1 A comparison between the two on-line shops

The results showed that there are significant differences between the two on-line shops with respect to participants' satisfaction and preferences and their ranking of the various aspects of the usability of the shops (Table 1), though there is no significant difference between two shops in participants' preferences to color using and to style of writing. Overall, Chinese undergraduates preferred Web A than Web B.

Table 1   A comparison between Web A and Web B*

Aspects	Web A	Web B	Significant level of the difference
General evaluation on Web site design	3.53	3.13	p=0.020
Homepage design is helpful to understand the whole structure of Web site	3.43	2.82	p=0.016
Preference for the Web site	3.33	2.82	p=0.011
Confidence in buying something back	3.33	2.85	p=0.026
Satisfaction with shopping process	3.30	2.70	p=0.003
Familiarity with the Web site	3.25	2.76	p=0.000
Preference for interface design	3.22	2.75	p=0.055
Satisfaction with whole layout	2.66	2.29	p=0.022
Quantity of advertisements	2.58	2.21	p=0.019
preference for style of writing	3.23	3.07	p = 0.354
preference for color using	2.94	3.16	p = 0.393

* A six-point scale was used (0: the lowest degree; 5: the highest degree).

3.1.1 The different characteristics of the two on-line shops

Sixty percent of participants thought Web A's homepage design was more helpful to hold its whole structure because Web A had following characteristics: concise, top navigation, and detailed categories of merchandise in center of the page.

Fifty-five percent of participants preferred Web B in its color using because Web B was greenness, lively, bright, color singleness, identical predominant color, gentle, and light background. But other thirty-five of participants preferred Web A in its color using because it was staid, in good taste, and substantial. One participant liked warm color because it was winter when he participated in this study.

Some participants (47.5%) preferred Web A in its style of writing because it was concise and comprehensive, clear at a glance, objective, and attractive. But other participants (27.5%) preferred Web B in its style of writing because it is humor, informal, agile, flexible, and close to life.

3.1.2 The different behaviors in the two on-line shops

The percentage of participants who would search first when they wanted to buy specific things on Web A is 42.5%, less than 55% on Web B. One main reason might be that Web A had a detailed category in center of its homepage; while the category on Web B was in left bottom corner, and not detailed enough, so that more participants would search first on Web B. However, the percentages of participants who would stroll first when they did not want to buy specific things on both Web A and Web B are 65%.

The percentage of participants who would read the introduction of on-line shopping or the introduction of the Web site is 35% on Web A, less than 45% on Web B. The reasons given by participants include: Web A had clear structure and detailed category, and Web A was same to other Web sites, so that how to do shopping was clear at a glance; However, Web B was an auction Web site, participants were not familiar with it, and its page design was not good enough, so that more participants would read the introduction of Web B.

The percentage of participants who would pay their attention to advertisements is 75% on Web A, more than 50% on Web B. One main reason might be that the quantity of advertisements on Web A was more than on Web B. The reason given by participants is that Web B had fancy icons similar to advertisements. Two more other reasons of overlooking advertisements include that they were not interested in the content of the advertisements and that they thought clicking advertisement might slower their speed or waste their time.

Among the participants who reported they paid attention to advertisements on Web A and on Web B, 37% and 30% reported that they clicked advertisements, respectively. The reasons of clicking advertisements include that participants were interested in the content of the advertisements and that they saw the word "free" there.

3.2 Browsing behaviors and preferences

Some questions in the questionnaire are related to the participants' general browsing behaviors and their preferences to the interface design of Web site.

3.2.1 General browsing behaviors

Sixty percent of participants reported that they paid their first attention to content, but the others paid their first attention to navigation. Sixty-five percent of participants reported that they tended to browse a pop-up window, while the others tended to close it immediately.

The sequence of browsing a screen reported by participants is: top left corner (45% of participants' first choice) or center (45% of participants' first choice), then top right corner (10% of participants' first choice), bottom left corner, and finally bottom right corner.

Sixty percent of participants reported that they opened 3-4 windows at the same time, 25% opened 5-9 windows, and 2.5% opened 10 or more windows. Participants reported that they tended to forget the content of the windows when they opened 3-4 windows or more.

3.2.2 Preferences to the form design of Web site

The sequence of convenience of navigation locations is: top (45% of participants' first choice), left, right, center, and finally bottom. Seventy-five percent of participants preferred scrolling in one page to linking to other pages, but the others preferred linking to scrolling. 72.5% of participants thought that a search engine in every page was necessary, but the others did not think it was necessary.

The importance of characteristics of color using in turn is: comfortable feel (70% of participants' first choice), colors match fit, identical predominant color, color layout, and color contrast. Participants' preference for predominant color is: blue (37.5% of participants' first choice), and then green (25% of participants' first choice). Participants' preference for predominant font is: Song (37.5% of participants' first choice), then Kai, Ad font, Li, and finally boldface;

Participants' preference for the link color before linking is: blue (37.5% of participants' first choice), and then purple; and their preference for the link color after linking is: red (37.5% of participants' first choice), then orange, blue, and purple. Obviously, the web is becoming a genre with its own established conventions. Other studies have shown that blue-underlined text is the most reliable indicator of links and provide the most click-throughs. Using another color drastically reduces click-throughs.

3.2.3 Preferences to the content design of Web site

The photo of the commodity (40% of participants' first choice), its function (27.5% of participants' first choice), and its price (15% of participants' first choice) were important information for participants to know about a commodity. The price of the commodity (32.5% of participants' first choice), its function (27.5% of participants' first choice), and its photo (20% of participants' first choice) were important information for them to buy a commodity. Books (22.5%

of participants' first choice), tickets (22% of participants' first choice), computer (15% of participants' first choice), and gifts (12.5% of participants' first choice) were most suitable to be sold on-line.

3.2.4 Others

Participants preferred buying through bid (2.62), and preferred public sale (2.93). Participants agreed that the price on-line was lower than the price in traditional marketplaces (3.09). Participants liked personalization (4.01). And finally, participants preferred paying cash. The results of their preferred payment mode are: pay cash when receiving goods (82.5%), pay on-line (7.5%), remit by post office (5%), and others (5%).

4. CONCLUSIONS

In on-line shop design, there is a constant tension between wanting to control the way a page looks and allowing users to set their own preferences. Designers would want to shape the overall look of a page. However, users should not be prevented from customizing certain elements for themselves. In fact, users can set their own preferences for: text link colors (visited and unvisited), background color, navigation place, font, et al.

In this study, our main findings are as follow:

(1) Generally, a good Chinese on-line shop should be a web site with the following characteristics: detailed category of commodity, search engine in every page, navigation located the top of the page, concise interface, harmonized color, light background, with C2C bargaining and secondhand commodity, simple registration and shopping procedure.

(2) Overall, more Chinese undergraduates paid their first attention to content than to navigation; They preferred browsing a pop-up window to closing it immediately; They preferred scrolling in one page to linking to other pages.

(3) Chinese undergraduates' preferred predominant color is blue, preferred predominant font is Song, and preferred payment mode is to pay cash when receiving goods. They thought that the photo, function, and price of commodity are important information for them to know about a commodity and to buy a commodity.

REFERENCES

Agence France Press. (2000). China's E-Commerce Fails Test – And Brave Volunteers. September 7, 1999.

CNNIC. (2000a). http://www.cnnic.net.cn/develst/cnnic2000.shtml.

CNNIC. (2000b). http://www.cnnic.net.cn/develst/cnnic200007.shtml.

CNNIC. (2001). http://www.cnnic.net.cn/develst/cnnic200101.shtml.

E-Commerce Times. (1999a). Message to E-Tailers: Deliver Better Service or Become E-Toast. September 13, 1999.

E-Commerce Times. (1999b). E-Tailers Waking Up to Customer Service Needs. October 19, 1999.

E-Commerce Times. (1999c). 75 Percent of E-Business Destined to Fail. November 12, 1999.

Nielsen, J. (2000). Designing Web Usability. Indianapolis: New Riders publishing.

Usability First. (2001). http://www.usabilityfirst.com/intro/index.txl.

ZDNET. (2000). Shoppers of the Web Unite: User Experience and Ecommerce. February 24, 2000

Zona Research. (1999). Shop Until You Drop? A Glimpse in Internet Shopping Success.

ACKNOWLEDGEMENTS

This project was funded by Motorola Electronics Pte. Ltd. The authors wish to thank Mr. Yuming Xuan of Institute of Psychology, Chinese Academy of Sciences, and Mr. Jianwen Zhou of Human Interface Lab, Motorola Electronics Pte. Ltd, for their valuable suggestions and comments made in this study.

A Study of Evaluating E-commerce Web from the Viewpoint of Usability Engineering
A Case Study of Ticket-selling System of Taiwan Railway Web

*Shing-Sheng Guan

*Department of Visual Communication, National Yunlin University of Science and Technology

Abstract

This study is based on the methods of usability engineering including Ameliorative Coaching and performance measurement methods and using the ticket-selling system of Taiwan Railway Web (TRW) as a case to evaluate the usability of E-commerce. The. This study has three purposes: 1.) Investigating the usability problems of TRW. 2) If a manual was provided in the web will affect the users operation performance. 3) According to the investigation results, some suggestions and guidelines were provided and they should be good tools for designers in planning a web of E-commerce.

Keywords: Taiwan Railway Web☐E-commerce☐Ameliorative Coaching Method☐Usability engineering

1. Introduction

Much research data reveal about 33% to 66% of Web shoppers are forced to cancel their bargains due to the operation-unfriendly (Chang, 1999). Therefore, some consumer websites help to solve those operational inconveniences by providing on-line consultation. We can see many e-commerce websites such as http://bid.com.tw, http://www.amazon.com, http://www.coolbid.com and so on also provide on-line consultation system like "New Driver on Road" or "Help on Line". The secondary public transportation in Taiwan is the railway whose quantity of transport (24.7%) is a little bit lesser than that of the highway (69.9%)(MOTC, 1995). Although possessing indefinite business potential, the railway websites could not provide on-line consultation in this regard. Whether or not there are problems on error operation or user-unfriendly is a worthy issue to be discussed. Therefore, taking the ticket-selling system of Taiwan Railway web as an example, the author tries to evaluate the usability of e-commerce websites by means of the methods of usability engineering, including Ameliorative Coaching Method and Performance Measurement☐Nielsen, 1993☐.

The purposes of this study are as follows: First of all, discuss the usability question within the ticket-selling system of Taiwan Railway Web. Secondly, does the presence or absence of on-line auxiliary instruction affect system operation? Finally, provide related suggestion and instructional principles as the reference for website designers.

2. Research Methods

This study makes evaluations not only on searching function and ordering-ticket function within the ticket-selling system of Taiwan Railway Web, but also on the results of error ratio and subjective opinion of operation, designing and the whole satisfaction, so as to know the problem of usability and provide amelioration and designing suggestion.

According to Yam's investigation on website users (Chang, 1999) and the list of passenger levels taken from Railway Bureau (MOTC, 1995), the college student is the most important subject in terms of surfing Internet or traveling by train. Consequently, the author selects 32 college students, including 8 female students and 24 male students, as our subjects to carry out usability test. The 32 students were divided into two groups, i.e. novice user and semi-novice user, basing on the degree of their understanding about the ticket-selling system of Taiwan Railway Web. The so-called "novice user" is the one who surfs on Internet frequently, but has never paid a visit to the ticket-selling system of Taiwan Railway Web; while "semi-novice user" is the one who surfs on Internet frequently and has ever used the system, but only limited to searching not ordering any tickets from it. Basically, each group has 16 subjects; we then re-divide each group into two subgroups based on two main features—whether or not having searching experience and reading instruction. Totally there are four groups; subjects without searching experience and reading instruction are belonging to group one (G1); those without searching experience but have reading instruction are group two (G2); those who have searching experience but without reading instruction are group three (G3); those who both have searching experience and reading instruction are group four (G4).

The purpose of making instruction manual is to explore whether or not the existing of on-line auxiliary instruction would affect the operation? Whether or not the provision of on-line consultation will help the users out when encountering operational problem? The content of the instruction manual includes standard operational processes of searching and ordering tickets, the most common-used functions in the ticket-selling system of Taiwan Railway Web.

The method for testing usability in this study includes Ameliorative Coaching Method and Performance Measurement. The Ameliorative Coaching Method combines Coaching Method (Mack & Burdett, 1992), Thinking Aloud 〔Nielsen, 1993〕, and Retrospective Testing (Hewett & Scott, 1987). The advantages of Ameliorative Coaching Method include saving more experimental time than Retrospective Testing, observing users' operational processes and understanding users' real thought. The disadvantages are the experimenters have to stay beside the subjects for note taking and provide oral instruction and assistance while the subjects encounter difficulties. Such method is strenuous and time consuming for the experimenters. As for Performance Measurement, it is used for evaluating the operative error ratio and the feeling of operation, designing and the subjective satisfaction on Taiwan Railway Web after their operation. The error ratio is taken from the average frequency on error, which made by subjects when operating each typical task. As regards the degree of operation, designing and the subjective satisfaction, we apply Likert's Scale method, which involves seven levels (from worst to best).

3. Typical Task

The most common-used functions in the ticket-selling system of Taiwan Railway Web are searching and ordering tickets. The Focus groups established four typical tasks and standard operational processes as the sequence of the ticket-selling system of Taiwan Railway Web:

Typical task one—looking up the timetable of the train: look up the trains that head for Hsinchu from Tainan between 10:00 a.m. and 5:00 p.m. 1. Click "Time table and ticket price. 2. Click "Line" you want. 3. Select the departure station under the icon "From". 4. Select the arrived station under the icon "To". 5. Key in "Time". 6. Click "Start" or "Arrive". 7. Click "Numbered express" or "Express, ordinary". 8. Click "Query".

Typical task two—order one-way ticket based on train number: order one-way ticket with train number 1003 from Taipei to Taichung on January 24th. 1. Click "order ticket using train number". 2. Key in "passport number". 3. Select "original station". 4. Select "destination station". 5. Select "departure date". 6. Key in "order ticket number". 7. Key in "train No. code". 8. Click "Start to Order".

Typical task three—order ticket using train type: order two 自強號 tickets from Kaohsiung to Tao-yuan at 7:34 a.m. on January 23rd. 1. Click "order ticket using train type". 2. Key in "passport number". 3. Select "original station". 4. Select "destination station". 5. Select "departure date". 6. Key in "order ticket number". 7.Select "train type". 8. Select "start time". 9. Select "due time". 10. Click "order ticket number".

Typical task four—ordering ticket using train number or train type: order one-way 自強號 ticket with train number 1019 at 2:30 p.m.

We can select any one of the two modes to order ticket. One of them, which includes eight steps (i.e. typical task two), is ordering ticket by using train number; the other of them, which includes ten steps (i.e. typical task three), is ordering ticket by using train type. These typical tasks are designed to understand which kind of task the subjects will choose to finish ordering ticket if the subjects are provided with complete train information and have experienced the previous three typical tasks.

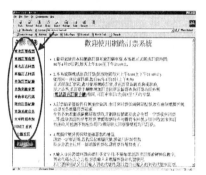

Figure 1 The category of ticket-selling system of TRW   Figure 2 The layout of ticket-selling system of TRW

4. Result and Analysis

The result of this study contains two parts: First of all, examine whether there are differences among error ratio, design, operation and the subjective satisfaction. Secondly, analyze the error ratio on each typical task, and explore the problem with its order of severity on usability of each typical task for the convenience of an important basis of re-designing.

4.1 Error ratio, design, operation and the subjective satisfaction

Apply One way ANOVA with LSD (Least-Significant Difference) method to carry out multiple comparisons with the value generated from the error ratio and subjects' subjective evaluation, then to explore whether there is an obvious difference among error ratio, design, operation and the subjective satisfaction.

1. In the analysis of error ratio, the result reveals that the error ratio of G1 is higher than that of other three groups; on the contrary, the error ratio of G4 is the lowest among the four groups.

Table 1- The analysis among four groups

Error Ratio	Design	Operation	Satisfaction
X4	X1 X3	X1	X1
X3 X2	X3 X2 X4	X3 X2	X2 X3 X4
X1		X2 X4	

X1: the average score of group one (G1); X2: the average score of group two (G2); X3: the average score of group three (G3); X4: the average score of group four (G4).

2. In the analysis of design, G1 got the lowest score of design among four groups; on the contrary, G4 got the highest one.

3. In the analysis of operation, G1 obviously got the lowest score in terms of operation than other groups. Moreover, the score of G2 is higher than G1 (X1<X2).   This kind of result is the same as G3 and G4.

4. In the evaluation of the subjective satisfaction, the result shows that there is only one difference between G1 and the others.   Obviously, the subjective satisfaction of G2 is higher than G1; however, there is no clear distinction between G3 and G4.

4.2 The error ratio analysis of the typical tasks

Find out every possible usability question and its error ratio based on four typical tasks. We could take those data as an important reference for the redesigning in the future.   The error ratio analyses are as Table2:

Table 2. The error ratio analysis of the typical tasks

I. Time table and ticket price	◻◻◻
1◻Click◻Time table and ticket price◻	31.25%
2◻Select◻which line◻	9.38%
3◻Select◻From◻station	6.25%
4◻Select◻To◻station	12.50%
5◻Key in◻Time◻	6.25%
6◻Select◻Start◻◻◻Arrive◻	15.63%
II. Order ticket using train number	
1◻Click◻Order ticket using train number◻	18.75%
III. Order ticket using train type & duration time	
1◻Click◻Order ticket using train type & duration time◻	6.25%
8◻Select◻Start time◻	59.38%
9◻Select◻Due time◻	53.13%
IV◻Order ticket using train number (68.75%)	
1◻Select◻Order ticket using train number◻	15.63%
7◻Key in◻Train number code◻	3.13%
IV◻Order ticket using train type & duration time (31.25%)	
1◻Select◻Order ticket using train type & duration time◻	9.38%
9◻Select◻Due time◻	3.13%

The experimental results are summarized as follows: 1.) Observers who hadn't contacted TRW and didn't use the manual had the highest operation errors. Observers who had contacted TRW and used the manual had the lowest operation errors. 2.) While the observers having the same pioneer experience, the operation manuals were provided or not could influence the operation performance. 3.) Observers who hadn't contacted TRW and didn't use the manual had the lowest subjective satisfaction. Observers who had contacted TRW and used the manual had the highest subjective satisfaction.

5. Discussion and Suggestion

From the analysis of error ratio among four typical tasks, this study finds out some problems about usability within operational interface. Also, it provides ameliorative method and designing suggestion as the basis for website designers. The four important points are:

1. The interface displays messages hierarchically; the functions for query and ordering ticket are basically placed on the first level, then, more sub-functions are placed on the following levels.

2. The most common-used functions should be placed on the top of the other ones; for example, the icon of "timetable and ticket price" should be placed on the top of the others ones.

3. There should have consistency between time of the original station or destination station and time of start or arrive. For example, the corresponding relationship should exist between starting time and arriving time.

4. The differences between the original system of TRW and the suggestion of this study are as follows:

(1) Original system applied linear placement on all names of train stations, no matter it is a big station or a small one.

(2) The suggestion of this study is to adopt Formosa profile, boldface and enlarge the written character, color the station name of municipality, or list all stations hierarchically from big one to small one.

From the usability engineering point, four guidelines were summarized for designers in planning a web of TRW and E-commerce in this study. The guidelines are summarized as follows: 1) A good web of E-commerce should provide "On-line manuals" to help users to sort out their operational problems. 2) Except emphasizing the web's style or the other visual effects, a good web should concentrate on the usability o to decreases the users' operational errors. 3) The information are displayed at web should be keep their consistency to avoid the users getting cognitive confuse. 4) The design of functional commands for web shouldn't be complicated or unnecessary and the meaning of the icons shouldn't be too similar.

6. Reference

□□□□□□□□□□□□□□(1999)□□□□□□□□□□□□□□□□□□□□

□□□□□□□□□□□□□(1999)□□□□□□□□□□□□□□□□□□□□□□□

Hewett T.T., Scott S. (1987). The use of thinking-out-loud and protocol analysis in development of a process model of interactive database search. IFIP INTERACT'87 Second Intl. Conf. Human-Computer Interaction, pp.51-56.

Mack R.L., Burdett J.M. (1992). When novices elicit knowledge□Question-asking in designing, evaluating and learning to use software, In The Psychology of Expertise□Cognitive Research and Empirical AI. Springer-Verlag (pp.245-268). New York.

Nielsen J. (1993). Usability Engineering. (pp.192-197) New York□AP Professional.

Institute of Transportation MOTC□(1995)□□□□□□□□□[online]□ Available: http://www.iot.gov.tw/chinese/ency/ency.htm. (Jan 23, 2001)

The User Interface of an Engineering Data Management System on WWW

Ching-Chih Hsu and Sheue-Ling Hwang*

United Microelectronics Corporation, Science Based Industrial Park, Hsin-Chu Taiwan
*Department of Industrial Engineering & Engineering Management,
National Tsing Hua University, Hsin-Chu Taiwan

ABSTRACT

The combination of engineering data management system (EDMS) and WWW offers an integrated management system and extends the area of data transmission. Since the interface design may influence the efficiency of the system and user performance, this research intends to evaluate and modify an EDMS on WWW according to principles of the human-computer interface. Following the literature review and the analysis of the user requirement, the interface of a current system was evaluated and modified by applying VBS(Visual Basic Script) and Active X to raise the system efficiency. Then, an experiment was conducted to verify the effects of the improvement. The outcomes of the experiment were analyzed and discussed.

1. INTRODUCTION

In the product development process, large amount of engineering data are produced and need to be processed. In order to reduce the delays and costs caused by inefficient engineering data management, many companies start to consider an engineering data management system (EDMS) (Peng and Trappey, 1996). Therefore, an EDMS is a management system of handling product data and process with the help of computer technology, dealing with the problem of the traditional document process and management, such as product design, analysis, manufacture, product management, quality management, and of other support administration (Peng, 1997).

In addition, each department of the company uses computer technology to achieve the best performance of itself, but ignoring the integrated performance of the company, inducing the problem of the inner and outer communication. More seriously, these engineering data cannot be intercommunicated and integrated, inducing the mistake and confusion of data. Therefore, how to construct an integrated engineering data management system is important （Sigal et al., 1995; Wang and Fulton, 1994; Urban et al., 1994; Peng and Trappey, 1996）. The integrated engineering data management system should provide a complete management system to enterprises by combining the web technology, database technology, data translation in application software, and development technology of user interface (Chen, 1997).

The previous papers of engineering data management and product data management can be divided into two approaches. The first approach mainly devoted to discuss the development structure and the function of an EDMS. The second approach tried to use case studies to learn how to implement an EDMS in the real task. Besides, some papers of the second approach also considered about human-computer interface. For example, Peng and Trappey (1996) have developed and implemented the CAD-integrated engineering-data-management systems for spring design, including product model definition, CAD-database communication, and human-machine interface. Chen (1997) also investigated human-computer interface design for an engineering data management system. On the other hand, some of the second approach started to combine EDMS & WWW. For instance, Peng (1997) constructed a STEP based engineering data management system on web site, and using CGI to design web pages of the system. But very few papers have developed the EMDS on web site and also thought about human-computer interface on design. Though Wang (1997) has completed a WWW interface design for an O-O engineering DBMS, many problems are still worth investigating further. For example, using more powerful tools (e.g. VB5/ActiveX) to design web pages on WWW, allocating the proper work distribution of EDMS between client site and server site on web. This study will modify Peng's engineering data management system on web site and try to meet the requirement of human-computer interaction. Furthermore, the work distribution of EDMS between client site and server site on web will be discussed.

2. DATA BASE and WWW

The WWW provides a unique opportunity for the development of database applications and also offers tremendous potential for rapid development of mechanical products to meet global competition (Buccigrossi. et al., 1996). The WWW is a global, interactive, dynamic, cross-platform, distributed, graphical, and hypertext information system on the Internet. The main property of WWW technology is that WWW clients can works on different hardware platforms with different software data standards, so the user can retrieve and access information with most existing platforms and terminal types. (Nye, 1996) There were some advantages of applying the web technology such as quick-design, easy to be integrated, new trend, and commercial potential. However, the disadvantages of applying the web technology might exit, such as security problem, low transfer speed, confusing layer, and immature interface design languages (Wang,1997).

Two-tier database-oriented client-server had been used to construct an EDMS before the appearance of the WWW. This architecture presents the user interface and information on the client machine, and manages data on the server. It has some advantages such as simplicity, and some natural constituencies, but it cannot deliver the network-centric, collaborative, and scalable applications envisioned in most reengineering plans. In addition, the biggest problem of two-tier structure is "client fat". In other words, in two-tier structure, the workload of the client is heavy enough to reduce the efficiency of the execution. So the three-tier architecture has been advocated. The three-tier architecture enlarges the flexibility of the client-server architecture, but it needs a standard interface between the software parts. Luckily, the WWW appeared timely and HTTP became the standard interface between the client and the server (Robertson, 1995). From the database perspective, the combination of database and WWW can offer relational data manipulation, high-speed search capabilities, and industrial-grade data input and retrieval. From the Web perspective, this combination offers user friendliness, cross-platform compatibility, and high-speed prototyping capabilities (Whetzel, 1996).

3. METHODOLOGY

3.1 Visual Basic Script and Active X

Visual Basic is a powerful language for designing the interface of windows, but this language did not support the Web page design in the past. Now, this is not a problem, because VBScript and ActiveX have had the capacity for the Web page design. VBScript is written in the HTML, which can control the Web pages and can increase the function of Web pages. ActiveX is the architecture including VBScript, ActiveX documents, ActiveX controls and other technology. In other words, VBScript is the controller of the program, and ActiveX is one of the components of VBScript. The new architecture can make the Web pages livelier.

According to the previous description, HTML is the basis of the web design, CGI can be used to access data bases, and VBScript and ActiveX can make some works to be executed in the browser, so these four tools will be applied in this study.

3.2 User requirements analysis

When designing a web page, the first thing is to know the users' intention of using this interface. The designers have to understand what users want to do, what information users want to transmit, and what users' purposes are. Thus the user requirements and system functions have to be analyzed.

This engineering data management system was constructed on the WWW, so the potential users of this system are very wide, complex, and almost no limit. Thus, the designer has to consider the requirements of the different users. Usually, the users can be divided into three groups. The first group are the general users who only browse but do nothing. The second group is the suppliers from cooperative manufactures, satellite manufactures, and other related firms. The users from manufactures and firms can use this system to query, transmit, or receive engineering data. The last group is the inner staffs who can access data, update data, and maintain the system. Actually, the second and the third groups are the expected users of the system in this study, and the general users seldom enter into this system. Thus, designing the interface of this system, the users are focused on the ones from the manufactures and firms. Since the executive efficiency is very important for the users who use the EDMS frequently, the transmission speed has to be considered. For security, the users have to input the user name and password before executing some tasks. In order to protect the data integrity from destruction, the users can view the data but cannot change the real data of the server.

3.3 Evaluation of a current system

An Object-orient engineering data management system developed by Peng (1997) contain seven subsystems which are engineering drawing management system, technical document management system, project management system, material specification management system, product structure management, engineering change management system, and basic system management. In order to understand the interface design problems of this system, a checklist is developed for the web pages evaluation. Most of the items in this checklist are from Chang's checklist (Chang, 1996) and ISO 9241. Some other items are from the literature related to the web page design guideline.

From the results of the checklist evaluation and the user analysis, there are some weaknesses in the original system such as 1) Unnecessary icons, 2) Too many layers, 3) Weak links, 4) Insufficient information, and 5) Error check.

The functions of this system do not need to be changed, but the tasks of these functions and the workflow can be modified. For example, when the users try to inquire the product structure of one product, the users must process six steps as shown in Figure 1. There are too many layers in this process. Because the users only inquire the product structure, the product data list may not be shown. The inquiry method can be prearranged previously. If the users want to change the inquiry method, there are different buttons to use in the web page. So step 4 and step 5 can be skipped, and the procedure becomes 4 steps.

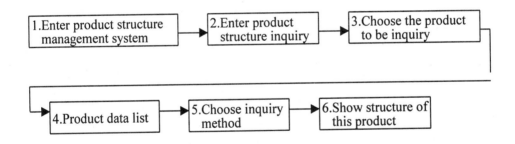

Figure 1 Procedure of Product Structure Inquiry

Using frames and a homepage to rearrange the links, the operative layers can be decreased, and the unclear icons are eliminated to speed up the transmission time. In addition, applying VBS and Active X, the system efficiency can be improved.

3.4 Experiment

In order to verify the effects of the improved system, an experiment was conducted. The original and the

1234

improved interface design are compared in operating time and the users' subjective responses.

This experiment included three parts, and each experimental part had two different interface designs, the original and the improved ones. In order to balance the effect of the subject's individual difference, each subject participated in both systems.

Twenty students of Industrial Engineering at National Tsing Hua University participated in the experiment with payment. Their ages were from 22 to 30 with the average age 24.3 years old. All subjects had the experiences of browsing on WWW.

4. RESULT

The results revealed that there were significant differences on performance between two systems ($p < .01$). According to the mean of the operating time (Table 1) and t tests, it is shown that using " frame" and homepage to construct the links can help the users to link to the web pages easily and quickly. As to the standard deviation of the measures (Table 1), the standard deviations of the improved system were smaller than the original system in experimental parts A and B. It shows that the improved interface can provide consistent operating method for different users. In addition, the range of the operating time in the original system is larger than the improved system (Table 2). In experimental part A and B, the performance of the original system may change for different users and questions, but the improved system decreases the difference.

Additionally, the results of subjective evaluation showed that there were no significant differences between two systems. Thus, the elimination of the unclear icons may speed up the transmission time without decreasing the user's satisfaction.

In this study, VBS and Active X were used successfully for the web pages design, and the server load of the original system was decreased.

Table 1 Analysis of Objective Data

Part		Total Average	Standard Deviation
A	The original system	1.3783 (min)	0.2521
	The improved system	1.1689 (min)	0.2041
B	The original system	0.285 (min)	0.045
	The improved system	0.193 (min)	0.034

Table 2   Range of operating time

	The original system	The improved system
Experiment A	1.0608 (min)	0.7133 (min)
Experiment B	0.146 (min)	0.131 (min)

ACKNOWEDGEMENTS

This study was supported by National Science Council under contract number NSC87-2212-E007-002.

REFERENCES

Buccigrossi, R., Crowley, A., and Turner, D. (1996). "A comprehensive system to develop secure Web accessible databases." WebNet96 - World Conference of the Web Society. Proceedings, 70-75.

Chang, J. M. (1996) "Development of HCI design tools". Master's Thesis, National Tsing-Hua University.

Chen, H.-C. (1997). " Human-computer interface design for an engineering data management system." Master's Thesis, National Tsing-Hua University, Hsin-Chu, Taiwan.

Lynn, N. (1996). "Keep your eyes peeled as the web engulfs client/server computing", Communications Week, September 18, 21.

Peng, T -K. and Trappey, A J C. (1996). "CAD-integrated engineering-data-management system for spring design", Robotics & Computer-Integrated Manufacturing. 12(3), 271-281.

Peng, T -K, (1997). "STEP based engineering data management system", Ph.D. Dissertation, National Tsing-Hua University, Hsin-Chu, Taiwan.

Robertson, B. (1995). "Client/server the standard way?" Network Computing, January 15,149.

Sigal, J., Foley, D., and Kempfer, L. (1995) "Hands-on product data management", Computer Aided Engineering, 4(2), 42-48.

Tibbetts, J., and Bernstein, B. (1996). "Great tools, bad architecture." Information Week, June 3, 128.

Urban, S. D., Shah, J. J., Liu, H., Roger, M. T., Jeon, D. K., Ravi, P., and Bliznakov, P., (1994) "A heterogeneous, active database architecture for engineering data management", International Journal of Computer Integrated Manufacturing, 7(5), 276-293.

Wang, C -C, (1997). "A WWW interface design for Object-oriented engineering DBMS", Master's Thesis, National Tsing-Hua University, Hsin-Chu, Taiwan.

Wang, C.-Y. and Fulton, R. E. (1994). "Information system design for optical fiber manufacturing using an object-oriented approach", International Journal of Computer Integrated Manufacturing, 7(1), 61-73.

Whetzel, J. K.(1996). "Integrating the World Wide Web and database technology", AT&T Tech. J. (USA), March-April, 75(2), 38-46.

Internationalization and localization: Designing for Asian Users

Pei-Luen Patrick Rau, Shueu-Farn Max Liang*

Department of Management Information Systems, Chung Yuan Christian University, Chunli 320, Taiwan

*Honeywell Singapore Laboratory, Singapore 486073

Abstract

The objective of this study is to implement Usability Engineering into every phase of the Web site development process, on the basis of Internationalization and Localization perspectives for honeywell.com/your home. The first step was to develop the usage scenarios. Three usability specialists in Taiwan carried out one heuristic evaluation session for the current honeywell.com/your home Website. The usability problems were analyzed and possible solutions to these problems were discussed. In the next phase, cluster analysis was utilized to test the current information architecture. Based on the results, suggestions for improving the Web site usability were provided. The results demonstrate the User-Centered Design (UCD) approach, and stress international and local issues in Web site development to Web site designers.

1. Introduction

The Honeywell Inc. Web site is utilized to demonstrate products and services. The honeywell.com your home Web site (http://content.honeywell.com/yourhome/) is aimed at introducing home and building control products, such as centralized home control, humidity control, thermostats, etc. To design a usable honeywell.com/your home Web site, usability specialists in Taiwan and Singapore conducted a series of tests that involved developing scenarios, conducting heuristic evaluations for the current services and a card sorting experiment.

The content of the honeywell.com your home Web site consists of products and services, concerns, home advice, your home 2000, etc. The two channels, products and services, and the concerns, are considered the main content of this Web site. There are four major service concerns in the honeywell.com/your home Web site: Comfort, Convenience, Safety, and Health. Products are classified into these four concerns.

Aykin (1998, 1999) developed design guidelines for Internationalization and Localization. Aykin stressed the importance of Internationalization and Localization in Web sites. Honold (1999) collected publications about cross-cultural Usability Engineering from 1975 to 1998. Honold classified these papers into three main phases and concluded that the research findings must be deeply integrated into the product development process.

2. Scenario development

The goal of scenario development is to accommodate the honewell.com/your home Web site to the users.

Daily life scenarios are helpful for designers to define the target audience and to characterize what happens when users perform typical tasks. Three usability specialists participated in the development of the four scenarios.

The four scenarios were developed to help usability specialists and Web site developers to define their users, users' tasks and goals for the Web site. The tasks for the target audiences include:

- Browsing
- Searching
- Reading

The goals of the target audiences include:

- Learning about products
- Comparing Honeywell products with others in the retail market
- Looking for services provided by Honeywell

The results were provided to the design team at Honeywell and utilized for the heuristics evaluation and prototype development.

3. Heuristic Evaluation

The evaluation procedure includes five steps:

1. The usability specialists went through the honewell.com/your home Web site at least twice.
2. The usability specialists discussed the heuristics.
3. The usability specialists discussed typical usage scenarios for honewell.com/your home, which were developed in the early phase.
4. The usability specialists evaluated honewell.com/your home separately for one-and-half hours. The usability problems found with references to the violated usability principles were recorded.
5. The usability specialists discussed the usability problems that were found and suggested improvement solutions.

The four sets of heuristics (Garzotto et al., 1995; Nielsen 1993, 1994, 2000) were categorized into thirteen heuristics for the three usability specialists to evaluate the Web site. The thirteen heuristics were developed based on the heuristics for evaluating hypertext systems by Garzotto et al. (1995) and Nielsen (1993, 1994, 2000). The frequencies of usability problems found in the honeywell.com/your home Web site are shown with the heuristics in Table 1.

As shown in Table 1, many usability problems were found with richness, consistency, readability and self-evidence heuristics. Thirteen problems were found with the richness, aesthetic and minimalist design. The other three heuristics, readability, consistency, self-evidence and the match between the system and the real world were also associated with most of the usability problems. These results imply that the content of the Web site and the presentation of the content may be not satisfactory to users in Taiwan.

Only a few usability problems were found with the heuristics concerning user operation and tasks, such as predictability, ease (of use), and internal locus of control support. This result implies that the Web site is not difficult

1238

to browse. The likelihood that users will become disoriented while browsing the current Web site is expected to be low. According to the results, usability specialists suggested improving the Web site content and its presentation, particularly about products and services, without changing the other parts of the Web site.

Table 1. Examples of Heuristics and Frequencies of Usability Problems

Heuristics	Usability Problems
• Richness: the abundance of information items and ways to reach (Garzotto *et al.*, 1995) • Aesthetic and minimalist design (Neilsen, 2000)	13
• Ease: information accessibility and how easy to grasp operations (Garzotto *et al.*, 1995) • Match between system and the real world (Nielsen, 2000)	1
• Consistency: conceptually similar items in a similar fashion and different items differently (Garzotto *et al.*, 1995) • Consistency and standards (Nielsen, 1993, 1994, 2000)	6
• Self evidence: how well users guess the meaning and the purpose of content or navigational element (Garzotto *et al.*, 1995)	5
• Readability: the overall feeling about an application's validity (Garzotto *et al.*, 1995)	8
• Support internal locus of control (Garzotto *et al.*, 1995) • Visibility of system status (Nielsen, 2000)	2

4. Card Sorting

Card sorting is a technique for discovering the mapping between the user conceptual model and the information displayed in the interface (Nielsen, 1993). The User Interface Architecture and Design group at IBM successfully utilized the card sorting technique for designing the overall structure of the IBM Web site (Lisle, Dong, and Isensee, 1998). Knowledge elicitation techniques such as verbal reports, clustering methods, scaling methods, or experimental simulation are often used to measure or evaluate an individual's knowledge (Benysh, Koubek, and Calvez, 1993). The outcome from card sorting is used for clustering analysis or multidimensional scaling. Clustering analysis is one alternative method to multidimensional scaling. The former emphasizes the categories in the data, and the later emphasizes the dimensional and spatial structure of the data (Eberts, 1994). Clustering analysis was used to group 271 functions in a system according to the ratings of experienced users (Tullis, 1985). Clustering analysis can be used in Web site design, clustering related items or links in a Web page to show their meaningful relationships.

The card sorting technique was utilized to retrieve subjective judgments on the similarity of the honewell.com/your home content. Clustering analysis was then conducted to construct the hierarchical taxonomy. The material, participants and procedure are discussed in the following:

• Material: 29 items were chosen to represent the content of honewell.com/your home. The items were products and concerns, categorized into four major concerns. All of the items were printed on cards in English. A dictionary and instructors were available if participants had any question about the meaning of the items.

1239

- Participants: 16 undergraduate and graduate students in Department of Management Information Systems, . CYCU.
- Procedure:
  1. A deck of index cards with the 29 items displayed was prepared. The cards were randomly ordered and provided to the participants.
  2. Each participant was tested in an individual session, to assure that all of the responses were not influenced by the other participants.
  3. The participants were required to arrange the index cards into groups according to the logical relationships.
  4. The participants were instructed to combine their initial groupings into larger clusters.
  5. Participants were also asked to name the groups and explain their naming criterion while all the cards were grouped into a single cluster.
  6. The card sorting results were recorded and used for cluster analysis.

The test results from card sorting were collected for the cluster analysis. The cluster analysis results showed that four clusters were derived according to the card sorting results. There are four concerns in the current honeywell.com/your home Web site: Comfort: Convenience, Safety and Health. The comfort concern was identified by the participants as Cluster 1 and the second half of Cluster 4, without three items. The convenience concern was also identified as Cluster 2, without two products, Web-enabled Home Controller Gateway and Wireless Devices.

Two out of the three Comfort concern items absent in the original category were categorized as a Convenience concern. This result indicates that these two concerns were conceptually close to one another for the participants. The third concern, Safety, and the fourth concern, Health, were grouped together most of the time. These two concerns were not as close as the Convenience concern to the Comfort concern. Further analysis is required to review the four-concern concept and to develop the information architecture for the honeywell.com/your home content.

5. Conclusions

These results demonstrated the User-Centered Design approach and stress international and local issues to Web site designers. Scenario development helps usability specialists and Web site developers to define their users, user tasks and the goals of the Web site. By defining the target audience for the honeywell/your home Web site in Asia, Web site developers are able to identify local design issues. The heuristic evaluation results suggested improvement for the Web site content and its presentation, particularly about products and services, without changing the other parts of the Web site. The card sorting results, which reviewed the four major products and service areas, provided options for future information architecture development for this Web site.

Acknowledgment

This research is collaborated with and supported by Human Factors program at Honeywell Singapore

1240

Laboratory. The authors would also like to express their gratitude to National Science Foundation in Taiwan (NSC 89-2213-E-033-066) for their support.

References

Aykin, N. (1998). Internationalization and Localization of the Web Sites, *Workshop presented at the Asia-Pacific Computer Human Interaction Conference (APCHI)*, July 1998.

Aykin, N. (1999). Internationalization and Localization of the Web Sites. *Proceedings of the 8th International Conference on Human-Computer Interaction (HCI International '99)*, Munich, Germany, August 22-27, 1999, 1, pp. 1218-1222.

Benysh, D. V., Koubek, R. J., and Calvez, Z. (1993). A comparative review of knowledge structure measurement techniques for interface design. *International Journal of Human-Computer Interaction*, 5(3), pp. 211-237.

Eberts, R. (1994). *User Interface Design.* NJ: Prentice-Hall.

Garzotto, F., Mainetti, L., and Paolini, P. (1995). Hypermedia design, analysis, and evaluation issues *Communication of ACM*, 38, pp. 74-86.

Honold, P. (1999). Cross-cultural Usability Engineering: development and state of art. *Proceedings of the 8th International Conference on Human-Computer Interaction (HCI International '99)*, Munich, Germany, August 22-27, 1999, 1, pp. 1232-1236.

Lisle, L., Dong, J., and Isensee, S. (1998). Case study of development of ease of use web site. *Proceedings of the 4th Conference on Human Factors on the Web,*
http://www.research.att.com/conf/hfweb/proceedings/lisle/index.html.

Nielsen, J. (1993). *Usability Engineering.* AP Professional, Academic Press, San Diego, USA.

Nielsen, J. (1994). Heuristic evaluation. In Nielsen, J., and Mack, R. L. (eds), *Usability Inspection Methods.* New York, NY: John Wiley & Sons, pp. 25-62.

Nielsen, J. (2000). *Ten Usability Heuristics.* http://www.useit.com/papers/heuristic/heuristic_list.html.

Tullis, T.S. (1985). Designing a menu-based interface to an operating system. *Proceedings of the ACM CHI'85 Conference.* San Francisco, CA, 14-18 April, pp. 79-84.

Cultural Differences in Web Development

Pei-Luen Patrick Rau, *Xiaowen Fang

Department of Management Information Systems, Chung Yuan Christian University, Chunli 320, Taiwan

*School of Computer Science, Telecommunications and Information Systems, DePaul University, USA

ABSTRACT

This study investigated the effects of cultural differences in the development of web sites. A search query tracker was developed to record each participant's searching behavior on the Web. The information tracker was installed in each participant's computer. In Phase I of the study, participants in Taiwan were asked to use Yahoo! Taiwan to conduct a set of search tasks. In Phase II, participants in US will be asked to use Yahoo! to carry out the same search tasks. The layout of Yahoo! Taiwan is the same as the layout of Yahoo!, and the search directory of Yahoo! Chinese has been translated from the English directory of Yahoo!. The information tracker recorded the search keywords and all the directories clicked during the test period. An evaluation was carried out by the participants in the second stage of the experiment. The participants were asked to fill out the satisfaction questionnaire for the search tasks. Then the participants were invited to join a discussion session for problems and comments of searching tasks during the experiment.

1. INTRODUCTION

The Web provides a rich base of information organized in hypertexts. The main purposes for using the Web (Lightner, Bose, & Salvendy, 1996) include publishing information in an electronic format for worldwide attention and retrieving information about a particular subject or topic. To retrieve information on the Web, users need to connect to the Web sites containing the information. The huge, increasing number of Web sites and the unstructured data on the Web often make it difficult for users to search for specific information.

To retrieve information on the Web, users often utilize search tools such as Web directories and search engines. Web directories are organized as a hierarchy of subject categories. Usually, search engines can provide updated information on the Web better than Web directories. The problems of using search engines are that they often return either zero hits or thousands of hits to a query and the heterogeneous search strategies and various interfaces often frustrate users (Sullivan, 1997).

Fang and Salvendy (1999) developed a template for search queries based on user-centered design principles. The results from this experiment showed that the user-centered template design improved users' search performance by 70% as compared to the current search engines. Fang (2000) developed a hierarchical search history for Web searching to assist users in controlling a Web search process. The results from this experiment suggested that with a highly complex task, search history improved users' search performance by 124% and user satisfaction by 42.5%,

1242

compared to the current search engine process.

The growth in Web use and the fact that many Web sites are developed for international audiences imply that cultural effects are very important to Web site development. Many researches on Web site development have explored the impact of cultural effects in the context of a single country. However, researches examining the cultural effects on Web usability are relatively few. The goal of this study is to investigate the effects of cultural differences between Chinese and American in developing searching tools for Web sites. We selected Chinese users in Taiwan for this study because of the relatively high popularity of the Internet in Taiwan. Forty percent of the households in Taiwan are connected to the Internet. Taiwan is ranked as fourth in the world, next to Denmark, USA, and Singapore, in the percentage of households connected to the Internet (Netvalue, 2001). The national language in Taiwan is Mandarin Chinese. Chinese in Taiwan use traditional Chinese characters, unlike the simplified Chinese characters in China.

2. PHASE 1

This research consists of two phases. In phase one, an experiment was conducted by Chinese participants in Taiwan. To examine cultural effects on the development of Web search tools, the same experiment was conducted by American participants in the US.

2.1 Experimental design

A search query and search history tracker was developed to record each participant's results for assigned search tasks on the Web. The tracker was installed in each participant's computer. During the experiment, participants were allowed to use only the tracker program to perform the search tasks on Yahoo! Chinese. The layout of Yahoo! Chinese is the same as the layout of Yahoo!, and the Web directory of Yahoo! Chinese is translated from the English directory of Yahoo!. The presentations of the search results for Yahoo! and Yahoo! Chinese are also the same except for the differences in language. The tracker program recorded the search keywords, URLs, and all of the subjects in the Web directory selected. A satisfaction questionnaire was designed to investigate the participants' degree of satisfaction with the Web search results. The questionnaire consisted of sixteen questions. Fifteen questions were designed on a scale of 1 (lowest satisfaction) to 7 (highest satisfaction). The last question was an open question asking for suggestions and comments from the participant.

2.2 Tasks

Before the participants began the tasks, they were given brief online instructions that explained the objective of the search tasks and a general introduction to the related area. After reading the instructions, participants began the search process and could use any search strategy to find relevant Web pages. Ten search tasks were assigned to the participants written in their native languages:

1. Find a few Web sites of manufacturers of wearable computers
2. Find a few Web sites about the Linux system
3. Find a Web site of a Taiwan high school

4. Find some information about the Chicago Transit Authority (CTA) – Chicago's public transportation system

5. Find the invoice price of a 2001 Honda Accord

6. Find some commercial Web sites that sell online shopping cart software

7. Find some Web sites about the actress Julia Roberts

8. Find some Web sites that can help you look for new jobs

9. Find an online travel agency to book air tickets

10. Find today's top stories

2.3 Participants

Eight participants in Taiwan volunteered to take part in this experiment. The 8 participants in Taiwan were undergraduate students in the Department of Management Information Systems, CYCU.

2.4 Procedure

Each participant was asked to complete a background questionnaire before the experiment. Each participant was given instructions for using the tracker program presented both on the computer screen and on a hard copy. When the participants felt comfortable using the tracker program, they were asked to conduct the search tasks. After completing the tasks, participants were required to complete a questionnaire concerning their satisfaction.

2.5. Results and Discussion

The results for the degree of satisfaction were obtained through the satisfaction questionnaire (Table 1).

Table 1. Results for the Degree of Satisfaction

	Question	Mean	SD
1	Generally speaking, I am very satisfied with performing the search tasks.	3.63	1.60
2	It is easy to find information in this portal site.	3.88	1.64
3	The site is very interesting to me.	3.88	1.13
4	Information in this portal site is well organized.	4.13	1.25
5	I often felt lost when searching for information in this site.	3.75	1.49
6	I like this site very much.	3.75	1.28
7	Information provided by this site is of high quality.	4.13	1.64
8	I would use this site to search information regularly.	3.75	1.67
9	The categorization is easy to follow.	4.25	1.49
10	I would recommend this site to my friends.	3.88	1.46
11	I am generally satisfied with the kind of work I did in the search tasks.	3.50	1.60
12	How satisfied do you feel with the amount of independent thought and action you could exercise in the search tasks?	3.38	0.92
13	How satisfied do you feel with the worthwhile accomplishment you got from performing the search tasks?	3.25	1.58
14	How satisfied do you feel with the amount of challenge in the search tasks?	3.63	1.19
15	How satisfied do you feel with the level of mental effort required to perform the search tasks?	4.00	1.20

As shown in the Table, there were only four questions (4, 7, 9, and 15) that received a response of 4 points or above from participants in Taiwan. The fact that the responses to question 4 (Information on this portal site is well organized.), 7 (Information provided by this site is of high quality.), and 9 (The categorization is easy to follow.) were all above 4 points, which implied that the participants in Taiwan were somewhat used to the Web category design and the Yahoo! Chinese search engine results.

The tracker program recorded every search step made by each participant. The participants could make more than one attempt for each task. The numbers of successful steps and the total number of attempts for the search tasks are presented in Table 2. As shown in the table, it took from 2.4 to 4.5 steps for participants in Taiwan to conduct the search tasks successfully. Task 3 (Find a Web site of a Taiwan high school), 4 (Find some information about Chicago Transit Authority (CTA) – Chicago's public transportation system), 5 (Find the invoice price of 2001 Honda Accord), and 9 (Find an on-line travel agency to book air tickets) were relatively difficult, in terms of the number of steps, for participants in Taiwan.

Table 2 Numbers of Steps for Search Tasks

Tasks	Successful Trials		Total Trials	
	Mean	SD	Mean	SD
1	3.0	1.86	2.4	1.38
2	2.6	1.32	2.5	1.34
3	3.5	1.92	3.8	2.09
4	3.3	1.50	3.1	1.56
5	4.5	5.68	3.7	4.12
6	3.0	1.46	2.7	1.27
7	3.0	2.26	3.0	2.21
8	2.6	0.70	2.6	0.67
9	3.4	0.84	3.2	0.95
10	2.4	1.24	2.3	1.14

3. PHASE TWO

The same experiment was conducted by American participants in the US. The eight participants in the US were students in the School of Computer Science, Telecommunications and Information Systems, DePaul University. The experimental design, search tasks, and procedure for this experiment were the same as described previously. The results in Phase two will be compared with the results in Phase one, to investigate the effects of cultural differences in developing searching tools for Web sites.

ACKNOWLEDGMENT

The authors would also like to express their gratitude to National Science Foundation in Taiwan (NSC89-2213-E-033-066) for their support.

REFERENCES

Choong, Y.Y. and Salvendy, G. (1998). Design of icons for use by Chinese in mainland China, *Interacting with Computers*, 9, 417-430.

Choong, Y.Y. and Salvendy, G. (1999). Implications for Design of Computer Interfaces for Chinese Users in Mainland China, *International Journal of Human Computer Interaction*, 11(1), 29-46.

Fang X. and Salvendy, G. (1999). Templates for search queries: A user-centered feature for improving Web search tools, *International Journal of Human Computer Interaction*, 11(4), 301-315.

Fang X. (2000). A hierarchical search history for Web searching, *International Journal of Human Computer Interaction*, 12(1), 73-88.

Lightner, N.J., Bose, I., & Salvendy, G. (1996). What is wrong with the World Wide Web?: A diagnosis of some problems and prescription of some remedies. *Ergonomics*, 39, 995-1004.

Netvalue (2001). *Global snapshot: going online across the world* [Online]. Available:

http://www.netvalue.com/corp/presse/index_frame.htm?fichier=cp0025.htm

Sullivan, D. (1997). *The major search engines* [Online]. Available: http://www.searchenginewatch.com/major.htm

Why Do Experts Predict False Alarms? An Empirical Investigation into the Validity of Expert Evaluations of Instructional Multimedia Software

Maia Dimitrova, Helen Sharp and Stephanie Wilson

Centre for HCI Design, City University, London EC1V 0HB, United Kingdom
+44 207 477 8993, +44 207 477 8481, +44 207 477 8152
maia@soi.city.ac.uk, hsharp@soi.city.ac.uk, steph@soi.city.ac.uk

Abstract: During the past few years, methods for expert-based usability evaluation of Instructional Multimedia software have been emerging, however there is very little empirical evidence of their validity, i.e. do these methods identify issues that would impact users and the achievement of their tasks or not. This paper presents an empirical study which measures and compares the validity of three expert-based evaluation methods. The results show that between 28% and 50% of the usability problems predicted by experts were invalid. We also reveal some of the reasons behind the invalid predictions and discuss the implications for the usability evaluation process.

1. Introduction

A number of expert-based methods for evaluating Instructional Multimedia (IMM) software have been developed during the past few years (e.g. Barker and King 1993, Heller and Martin 1999), however there is little empirical evidence of their validity. There have been some attempts to critically review different evaluation techniques for instructional software (e.g. Reiser and Kegelmann 1994, Tergan 1998). These critiques, however, do not provide empirical evidence of the number and the nature of the problems predicted. Thus, no definitive conclusions can be drawn regarding the validity of the techniques.

We define an expert-based evaluation method for IMM to be *valid* if the predictions made highlight issues that will hinder the way they interact with and learn from the multimedia application if they are not rectified. When measuring the validity of a method it is important to establish which predicted problems are valid and separate them from false predictions, which identify issues that will not have any effect on the users' interaction or the achievement of their learning tasks. It is also essential to identify the reasons why experts identify false alarms so that measures can be taken to improve the validity of evaluation methods (EMs). Validity is typically measured as the ratio of 'real' problems experienced by users during user tests to the number of problems predicted by experts (John and Marks 1997). We do not consider this approach to be adequate when evaluating IMM applications, as it may lead to a number of issues being inappropriately classed as false alarms. The reason for this is that experts typically uncover problems regarding the suitability of the media to represent the content, the completeness of the material, and the effectiveness of the instructional approach. Such issues can have an effect on the users' knowledge construction processes, however the users may not be aware of these effects, nor have they the necessary knowledge to identify such issues. Thus, by comparing all expert-predicted problems to those identified by users, such issues may inappropriately be counted as false alarms. We believe that independent instructional specialists, rather than users, should determine whether such problems are valid or not.

In this paper we report an empirical study which investigates the validity of three expert-based usability evaluation methods for IMM – one cognitive walkthrough, one checklist and one taxonomy-based approach. We present the results of evaluations made using each method, and establish the proportion of valid predictions to false alarms. We then discuss the reasons why the evaluators made invalid predictions.

2. Expert-Based Evaluation Methods

2.1 Multimedia Cognitive Walkthrough (MMCW) (Faraday and Sutcliffe 1997)
This approach concentrates on cognitive aspects of using multimedia (MM) presentations. It is designed to review how effective a MM user interface is in supporting key cognitive processes, such as attention to and comprehension

of information presented concurrently in multiple media. The method is intended to aid evaluators in reviewing MM presentations segment-by-segment and assessing their effectiveness in supporting users' cognitive processes during their interaction with the multimedia presentation. To be able to do that, the evaluators need to follow three consecutive steps: evaluate the attentional design first, then the combination of visual and audio media, and finally the appropriateness of the media selected. In each step they are guided by a set of guidelines.

2.2 Interactive Multimedia Checklist (IMMC) (Barker and King 1993)
This method uses a checklist of ninety questions divided into twelve categories, each of which embodies essential principles of good design of IMM applications. The evaluation categories cover issues of user engagement, interactivity, tailorability, appropriateness of the media mix, mode and style of interaction, and quality of the end-user interface. The authors suggest that experts, such as multimedia designers and instructional specialists, review the design of the IMM for the presence of certain design principles and good design features.

2.3 Multimedia Taxonomy (MMT) (Heller and Martin 1999)
The Multimedia Taxonomy is a three-dimensional categorisation of multimedia issues which is intended to provide a framework for analysing MM applications. The first dimension represents different *media types*, such as text, graphics and sound. The *expression* dimension consists of different modes of expressing such media, i.e. as an elaboration, representation or abstraction. Finally, the *context* dimension comprises six categories, including target audience, content matter, interactivity, quality, and aesthetics. The taxonomy contains 120 cells, bounded by those three dimensions. For each cell of the taxonomy evaluators can ask questions which they feel are relevant to the particular media type or aspect of design.

3. Study Background

3.1 The Instructional Multimedia Software
One section of a multimedia learning environment for studying Mathematics was evaluated. The section covers the principles of exponential functions and their graphs. A series of 23 screens present the Maths content in textual, graphical and animation formats. Once a topic is selected, the student may navigate within it either linearly or non-linearly. Interactive quiz-like tests are also included.

3.2 The Evaluators
Ten professionals with expertise recommended by the authors of each method took part in the evaluations. Two multimedia designers with 2 and 4 years of experience respectively, a mathematician with software design experience and a lecturer in Maths used the Multimedia Taxonomy. Two multimedia designers with 2 and 9 years of experience respectively and two lecturers in Maths used the Interactive Multimedia Checklist. Finally, the Multimedia Cognitive Walkthrough was applied by an HCI specialist, with 5 years of usability engineering experience and a multimedia designer with 2.5 years of experience. The authors of the MMCW do not recommend the method to be used by subject matter experts as it concentrates on low-level multimedia design issues.

3.3 Expert Evaluations
The evaluators individually reviewed the IMM application applying one of the methods to identify potential usability and learning problems. The evaluators were asked to write evaluation reports describing the problems they found and specifying the evaluation criterion which led them to the problem. The latter was necessary to provide a clear indication as to whether the problem was identified using the EM or the experts' own judgement.

3.4 User Tests
In order to collect data about the actual occurrence of the predicted problems with users, an empirical usability study was performed with four representative students. Pre-exposure tests were administered to establish the students' prior knowledge of the material and their computer literacy. The students were then given four tasks to perform with the IMM application, which consisted of learning about different principles of exponential graphs. During the study the students were asked to think aloud while performing the tasks. All user sessions were video- and audio-recorded. After the students had completed the tasks, they were interviewed by the experimenter to identify what their attitude was towards different aspects of the application, such as the media design, the navigation aids, and the learning

support provided. At the end, comprehension tests were administered to reveal the knowledge students gained while working with the software. The material covered by the students was divided into 20 knowledge propositions, each of which was tested in the comprehension tests.

4. Results from the Expert Evaluations

The original problem sets produced by the experts were refined to combine the same comments made by reviewers using the same method and to exclude errors due to evaluators' unfamiliarity with the rest of the application. A total of 191 unique problem descriptions were generated by the experts, which were divided into problems whose detection evaluators attributed to the EM and those identified by the experts' own personal judgement. The total number of problems generated is shown in Table 1.

Table 1: Problems Predicted by the Experts Using Each Evaluation Method

Method	Total number of problems	Number using EM	Percentage using EM	Average per expert using EM	Number from own judgement	Percentage from own judgement	Average per expert from own judgement
MMCW	34	19	56%	9.5	15	44%	7.5
IMMC	88	70	80%	17.5	18	20%	4.5
MMT	69	46	67%	11.5	23	33%	5.75

As can be seen from the data in Table 1, in total 135 problems were predicted using the EMs. The evaluators using the Interactive Multimedia Checklist identified the most problems attributed to the EM. The experts using the Multimedia Taxonomy identified fewer problems. Finally, in the case of the Multimedia Cognitive Walkthrough the least number of problems were identified. Although these figures provide an indication of the predictive power of each EM, what is really important is to find out how many of the problems predicted by the evaluators are *valid* and how many are *false alarms*. In the following sections we validate the predicted problems using user test data and the opinion of two independent reviewers.

5. Validation of Predicted Problems

5.1 User Test Results
The videotape data from the user tests was analysed to identify problems using a set of nine criteria, such as 'the learner articulated a goal but cannot succeed in achieving it without external help from the experimenter'. 51 unique problems were found to match the criteria. The comprehension test results showed that the students had problems understanding certain principles of exponential graphs. In total the students had difficulties comprehending 13 of the 20 knowledge propositions. Thus, as a result of all user tests we found that the students encountered 64 problems in total, i.e. 51 usability and 13 comprehension problems.

5.2 Validation Using User Test Results
Before comparing the predicted and the user problems, the predicted problem sets had to be refined to remove the issues that users could not be expected to identify. These fell into three categories. The first category included problems concerning the accuracy of the Maths content and the notation used, the second one comprised issues regarding the adequacy of different assessment techniques, and the third one represented issues concerning whether expert system facilities are required to support learners. At the end, 44 problems were excluded of the 135 problems predicted using the EMs, and 91 remained to be validated by the user test data.

To be able to match the expert and user problem sets, six matching rules were established. For instance, problems were matched if both problem descriptions described the same user behaviour or the same fault with the same design feature. As a result of the problem matching we found that 58 out of the selected 91 predicted problems were actually experienced by the users. This means that 33 problems were not experienced by the users and can be classed as false alarms.

5.3 Independent Expert Validation

One instructional expert and one instructional designer, both familiar with the design of the Maths software, reviewed the 44 problems not validated with the user test data. They were asked to point out any issues the considered invalid in terms of their appropriateness for the application and whether they were likely to affect users' interaction and comprehension of the material. As a result of their analysis it was established that 20 false usability problems were identified during the IMMC evaluations, two false alarms were predicted during the MMT evaluations and none during the MMCW.

Table 2 presents a summary of the validity calculations for each method based on the two steps of validation.

Table 2: Validity Calculations for Each Evaluation Method

Method	Total problems using the EM	Total valid problems	% Valid problems	Number false alarms by users	Number false alarms by experts	Total false alarms	% False alarms
MMCW	19	12	63%	7	0	7	37%
IMMC	70	35	50%	15	20	35	50%
MMT	46	33	72%	11	2	13	28%

The false alarms figures revealed in the last two columns of Table 2 give cause for concern. As comprehensive validations of predicted problems are not possible during real software development, there is a danger that developers could spend valuable time and effort redesigning software in response to expert evaluations, without making any improvement to the usability and with a risk of introducing new problems. To improve expert evaluation methods, understanding of the reasons for generating false alarms is needed and measures for improvement should be taken. In the following section we investigate the origin of the false predictions made during the expert evaluations.

6. Analysis of Invalid Predictions

6.1 MMCW

All but one false alarm were identified by the same evaluator. This evaluator had some multimedia design and testing experience, but they had not used any formal evaluation methods before, and did not have sound cognitive psychology knowledge. This evaluator seem to have interpreted the guidelines in a general sense, rather than in their specific meaning within the context of cognitive processing of MM presentations. For instance, he applied the term 'focus' in its broad sense rather than to signify directing user's attention to different audio and visual media. The same evaluator also did not seem to be able to determine easily which guidelines were applicable to the IMM software and tried to apply most of them. As a result, he generated nearly three times as many problems than the more experienced evaluator (16 and 6 respectively), however 38% of them were classified as false alarms. This evidence suggests that evaluators with limited cognitive psychology knowledge and limited usability evaluation experience require more training and further explanation of the method to be able to apply it effectively.

One false alarm was generated by the more experienced evaluator, who used a media selection guideline to make a recommendation about the media resources used. The suggestion, however, was considered inappropriate by the two independent reviewers. The false alarm was perhaps raised because the Maths content does not fit well in any of the information types suggested in the method, and thus the media selection guidelines, based on these information types, cannot be used successfully for Maths material.

6.2 IMMC

A major factor found to influence the validity of the predictions was the nature of the checklist questions. We distinguished between two different kinds of question, which differ in the level of scientific knowledge and the level of subjective judgement required to answer them. One kind simply asks the evaluators to check for the presence of a good design feature, e.g. "does the product monitor user performance?". Such questions require a simple 'yes' or 'no' reply, and the level of subjectivity and the level of specialised knowledge required to answer them are low. Most of the replies to such questions raised valid problems, however some were classed as false alarms by the independent reviewers. This was because some good design features not present in the application were considered irrelevant or inapplicable to the IMM application. As the evaluators who took part in the study were not familiar with the design rationale of the Maths software, they could not make such decisions. Thus, this method is more suitable for use by evaluators who have sufficient knowledge of the objectives and the design rationale of the application to be evaluated. Another set of questions require the evaluators to make predictions regarding how an aspect of the IMM will influence the users, their motivation, attitude and behaviour. An example of such question is: "can the user identify with the goals and objectives and build their own personal plan for achievement?". Such questions require pedagogical knowledge and a fair amount of subjective judgement on the part of the evaluators, who did not constitute typical users and thus would have had difficulties predicting users' attitude and behaviour with accuracy. The data shows that 75% of the false alarms by users were incurred when answering this type of questions.

6.3 MMT

Evaluators using the MMT are given freedom to comment on aspects of the MM interface they feel are appropriate and to provide their own judgement. We found that the most false alarms were generated by the least experienced MM designer: 8 out of 13 false alarms, or 62%. On the other hand, the expert who generated the most number of comments (18 in total), of which only one was classed as a false alarm, had considerable software design and evaluation experience, and knowledge in Maths. Thus, it appears that the validity as well as the thoroughness of the predictions made using the Multimedia Taxonomy will greatly depend on the evaluator's own expertise and experience in usability evaluation. Thus, such taxonomies are more suited to be used by more experienced evaluators.

7. Conclusions

The results presented in this paper reveal that the predictions made using the three evaluation methods are not as effective as usability practitioners would like them to be. Although a considerable amount of the problems predicted were valid, between 28% and 50% of them were found to be false alarms as they described issues which did not have any effect on the users or their performance. Thus, rectifying such issues is unlikely to result in any improvement to the usability and the user acceptability of the multimedia application, but will only waste valuable resources.

In this paper we also suggest possible reasons for the generation of these false alarms, which range from the evaluators' experience, through to the applicability and clarity of the evaluation criteria, to the amount of expert judgement required to apply the evaluation methods. In order to improve the validity of existing methods, a number of improvements could be made, including improving the clarity of the evaluation criteria and providing more detail to help less experienced evaluators. New EMs should aim to combine the expertise of different specialists in analysing specific design issues of learning with multimedia, and support them in inferring the likely effects of the MM design on users and their learning processes.

8. References

Barker, P. and King, T. (1993). Evaluating Interactive Multimedia Courseware – A Methodology. *Computers in Education*, 21 (4), pp. 307-319.

Faraday, P. and Sutcliffe, A. (1997). Evaluating Multimedia Presentations. *The New Review of Hypermedia and Multimedia,* 3, pp. 7-37.

Heller, R.S. and Martin, C.D. (1999). Multimedia Taxonomy for Design and Evaluation. In B. Furht (ed.): *Handbook of Multimedia Computing*, pp. 3- 16, CRC Press.

John, B.E. and Marks, (1997). S.J. Tracking the Effectiveness of Usability Evaluation Methods. *Behaviour and Information Technology*, 16 (4/5), pp. 188-202.

Reiser, R.A. and Kegelmann, H.W. (1994). Evaluating Instructional Software: A review and Critique of Current Methods. *Educational Technology Research and Development*, 42 (3), pp. 63-69.

Tergan, S. (1998). Checklists for the Evaluation of Educational Software: Critical Review and Prospects. *Innovations in Education and Training International*, 35 (1), pp. 9-20.

Wireless Interaction for Large Screen Displays[1]

Wyatt D. Bora, Peter A. Jedrysik, Jason A. Moore, Terrance A. Stedman

Air Force Research Laboratory/IFSB
525 Brooks Road
Rome NY 13441-4505

Abstract— The increasingly complex battlefield environment drives the requirement for the presentation and interactive control of the endless stream of information arriving from a diverse collection of sensors deployed on a variety of platforms. Without the benefit of extensive data fusion and correlation to present a true picture of the battlespace from all information sources, the situational awareness picture is, at best, fragmented. Collaboration and interaction is also needed for operators within a command center and among remote geographic locations. The need to display and manipulate real-time multimedia data in a battlefield operations command center is critical to the Joint Commander directing air, land, naval and space assets. The Interactive DataWall being developed by the Advanced Displays and Intelligent Interfaces (ADII) technology team of the Air Force Research Laboratory's Information Directorate (AFRL/IF) in Rome, New York is a strong contender for solving the information management problems facing the 21st century military commander. It provides an ultra high-resolution large screen display with multi-modal, wireless interaction. Commercial off-the-shelf technology has been combined with specialized hardware and software developed in-house to provide a unique capability for multimedia data display and control.

1. Introduction

The problem is the inability to effectively display and manipulate large amounts of real-time, multimedia data in a Command and Control (C2) environment. Migration to electronic media for mission planning is progressing. However, conventional media such as large paper maps with acetate overlays and Plexiglas boards for grease pencil annotation are still the norm. The transition has been delayed since new methods of operation are often slow to be accepted, the use of individual workstations diminishes the *big picture* in a C2 environment, and an intuitive, unencumbered means of Human Computer Interaction (HCI) is limited.

The approach is to utilize commercial off-the-shelf (COTS) products to the greatest extent to create a very high-resolution, tiled wall display. It is usable by both operators who are in close proximity to the screen, and by commanders who typically work at a distance. Multiple users can directly manipulate the wall display using speech recognition and wireless pointing devices for unencumbered interaction.

The uniqueness of the AFRL/IF Interactive DataWall is twofold. Its enhanced computer display capability tiles the output of multiple computer displays into a single workspace. The current configuration has a total display resolution of approximately 3.9 million pixels over a 12' x 3' screen area. Its enhanced HCI capability allows for wireless interaction with the display. Speaker independent speech recognition and conventional mouse functionality via camera-tracked laser pointers have been incorporated.

2. High Resolution Tiled Display

A tiled display consists of *n* x *m* distinct display devices, such as video projectors, each displaying a portion of an entire screen area. The Interactive DataWall's display consists of three horizontally tiled video projectors producing a very high-resolution, near seamless, large screen display (Fig. 1). Properly balanced color and brightness, and proper alignment of the display devices is critical to minimize any distractions caused by the seams between image tiles. Variations in chromaticity and luminosity among tiles can cause inconsistency in color and brightness across the screen. Vertical or horizontal disparity of the tiled images can cause segmentation of objects at the seams. Gaps between image tiles can cause discontinuity of the imagery.

The large screen display provides a global view of the information space, and allows the users to make comparisons and find relationships between items. It also makes collaboration within a localized working environment much more effective.

2.1 Window Management

The SGI version of the Interactive DataWall uses X-META-X, a COTS software package from X-Software. X-META-X provides a meta window manager functionality, merging the separate X Window managers of the individual display drivers. X-META-X provides two significant features. It allows windows to be repositioned anywhere on the tiled display and lets them be resized to the maximum size of the total display area.

PC DataWall implementations (see section 4) running Microsoft Windows NT use video card drivers that support multi-headed configurations. The Windows 2000 Operating System has native support for multiple displays so special video drivers are not required for tiling.

Figure 1 AFRL/IF's Interactive DataWall

2.2 LIVE VIDEO FEED CAPABILITY

An important source of information for C2 is live and recorded video from the battlefield. Intelligence, Surveillance and Reconnaissance video from manned and unmanned air vehicles, weapons systems, and ground forces can give the commander an up-to-date view of the battlespace to assess enemy forces and battle damage. Teleconferencing with other geographically dispersed mission planners, or receiving weather information via television broadcasts are also examples of useful video data that can be displayed.

Initial video capability of early versions of the Interactive DataWall consisted of low quality, looped MPEG clips stored as files on the display computer. Playback of this type of video consumes processing resources as the MPEG file is decoded. Clips of any substantial length can occupy large amounts of disk space. The desire was to display high-quality, real-time video information on the DataWall without significant processing impact. A hardware component called the SuperView 1000 by RGB Spectrum was incorporated into the DataWall architecture to provide this capability. It allows up to four real-time, full-color video windows to be displayed simultaneously. The video windows are overlaid on an individual display tile of the DataWall. Four live video feeds from Digital Satellite System (DSS) dishes provide the inputs to the SuperView to demonstrate the capability. However, any NTSC video source can be fed to the SuperView and displayed on the DataWall, including video from cameras and VCRs. Software was developed to provide window frames for the video windows to facilitate their manipulation. Each video window can be independently positioned and resized. The low-overhead window framing software is the only processing required to add this live video capability to the DataWall.

3. WIRELESS INTERACTION

The purpose of the Interactive DataWall is much more than providing a high-resolution summary device. It embraces the notion of a truly interactive environment that avoids the tethers of conventional human-computer interfaces such as keyboards and mice while maintaining the user's ability to use traditional input devices. The Interactive DataWall is an environment that allows unencumbered user interaction from various locations near and far. Speech recognition is supported via wireless microphones and 900Mhz cordless

telephones. Conventional mouse functionality is achieved using camera-tracked laser pointers.

3.1 CAMERA TRACKED LASER POINTER

To provide a wireless mouse-like mode of input, a camera-tracked red laser pointer is used. Three video cameras equipped with red filters are positioned behind the screen on each video projector. The cameras' views of the screen are dark fields until the laser dot comes into a camera's field of view. The live video from the three cameras is processed and the data is subsequently sent to the display computer for proper positioning of the cursor. It allows all the functionality of a conventional mouse including: dragging/dropping windows, resizing windows, and interacting with graphical user interface (GUI) widgets such as buttons and scroll bars.

PC/Video Capture Card Implementation— The original laser pointer tracking implementation consisted of three Personal Computers (PCs) equipped with video capture cards. These computers provided a frame by frame screen capture for each video camera positioned behind the screen. The frames were analyzed using ADII developed software on the PCs and subsequently transmitted to the display computer for cursor positioning.

Although quite effective, the process suffers several limitations. First, the system cannot operate in real time. Second, both resolution and update rate are constrained by the processing power available. Lastly, the cost/benefit ratio was difficult to justify.

Custom Laser Pointer Tracking Hardware—To address the limitations of the PC implementation, the ADII team invented specialized hardware to track the laser pointer. It functions in near real time, with readily expandable resolution, and at a small fraction of the cost.

The device combines a microcontroller, video processing logic, a pair of counters, and real time control logic, which together track the laser dot image. The process involves synchronizing the counters to follow the camera video. At the point in time when the camera "sees" the dot, the counters will contain values representative of the dot's position on the display surface. These values are then passed on to the display computer in near real time via serial cable to position the cursor.

In order to evaluate the effectiveness of the device, a side by side comparison between it and the PC implementation was conducted. The exercise confirmed several significant advantages of the laser pointer tracking hardware: better detection rate, faster detections, increased resolution and much lower cost. The complete cost for the tracking hardware unit is just $450.

Although the device was initially used to track the position of a laser pointer's dot on a rear projection video display, it can be easily adapted to scan a pointer image on any projection (rear or front) or non-projection direct view display, including printed material such as a paper map. The technique also lends itself well to multiple panel displays as both the cost and complexity of the device scale linearly.

Laser Pointer Tracking Software— The laser pointer tracking hardware detects a spike in the NTSC video signal from the cameras. It informs the display computer of the laser dot's location through a serial port when the signal surpasses a customizable threshold. The tracking hardware analyzes 512 by 480 pixels at 60 Hz.

Before using the laser pointer with the Interactive DataWall the operator must first perform a calibration process. This involves the detection of a number of initialization points and the generation of a lookup table. This table contains the mappings of each camera pixel to a corresponding pixel on the displayed image.

The custom tracking hardware detection method has a lower resolution than the display with which it typically interfaces. Therefore, it is difficult to correctly register the laser dot's location in reference to the displayed image. Several programs using different algorithms were developed to handle the interpolation and extrapolation.

Ideally, the camera's view of the projected image would resemble Figure 2. The black border is the total area that the camera can see and the gray region represents the projected image. In reality, the camera can see more than just the projected image and

Figure 2

cannot be perfectly placed in the center of it, which yields the resultant image in Figure 3. This is an exaggeration to facilitate explanation of the correction algorithm. The user is then instructed to point the laser at each corner of the projected image

Figure 3

Figure 4

giving the information shown in Figure 4. From that information the intersection points of the line segments, (P1, P2) with (P3, P4) and (P1, P4) with (P2, P3) is calculated (Fig. 5). Using these points, a table is created that maps a single camera location to a specific screen location. Since there is a resolution difference in the capabilities of the camera and projector, the algorithm relies on percentages for making the calculations.

Figure 6 shows the method to obtain a specific location (point D1) in the projected image. The dashed lines are projected from the intersection points N1 and N2. The intersection of the dashed lines and line

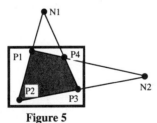

Figure 5

Figure 6

segments (P1, P2) and (P2, P3) are percentages of the overall length of these segments. For this example, assume the intersection points are (25%, 50%). If the projected image resolution is 800 by 600, that will yield a screen coordinate of (200, 300). This operation is continued until all camera locations are exhausted. In the first implementation of the DataWall this algorithm worked exceptionally well. However, due to the short distance between the cameras and the projected image on the Portable DataWall (see Section 4), the required wide-angle camera lens introduced barrel distortion.

Thus the camera sees an image similar to that in Figure 7 as opposed to that shown in Figure 3. This caused the correction algorithm to be accurate in the corners but inaccurate elsewhere. Algorithms exist that can provide the terms of a polynomial to fit the

Figure 7

barrel distortion exactly. These algorithms were tested and found to converge but ran too long to be usable. A simplified approach was tested and proved to be successful. Instead of providing just four data points per screen, the user provides nine data points. This produces four quadrants in which to perform the original linear based algorithm (Fig. 8). This technique can be expanded to handle other lens anomalies by adding more control points. While this is not a perfect solution, it performed the task required with minimal computation time and complexity.

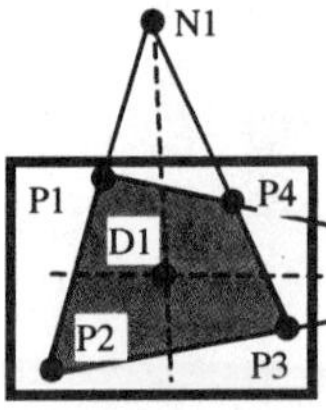

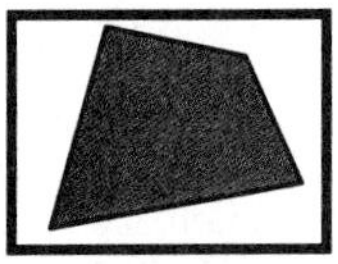

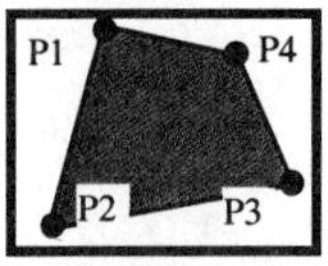

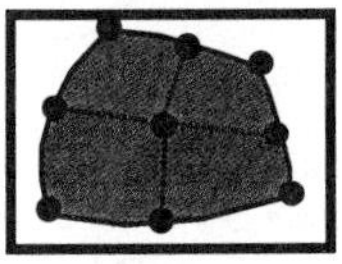

Figure 8

Once the calibration is complete and the table generated, the runtime software can be used to control the system mouse. A mouse resource window that models a three-button mouse appears after executing the runtime program. This window is borderless and resembles a computer mouse (Fig. 9). It allows the user to select the way in which the laser pointer is to be used. If there are no buttons

Figure 9 Mouse Resource Window

selected, the system mouse cursor simply follows the operator's laser dot. If the top left button is selected, the runtime software will generate a left mouse button down event the first time the laser dot is seen. A mouse button up event will be generated when the laser is turned off. The other two top buttons operate similarly, enabling the user to perform operations requiring middle and right button events. Two additional buttons at the bottom of the mouse resource window make distant interaction with the DataWall much easier. The lower left button emulates a single left mouse button click on release. When this button is selected the runtime software will **not** generate a left mouse button down event the first time the laser dot is seen, but simply moves the cursor. When the laser is turned off (released), the left mouse button down event is generated followed by a mouse button up event. This allows users at a distance to more easily click GUI buttons by first positioning the cursor then releasing. The lower right button emulates a double left mouse button click on release. This mouse emulation application allows operators to wirelessly manipulate any GUI.

3.2 CONTINUOUS SPEECH RECOGNITION

Speech recognition is another wireless interaction method supported on the DataWall. Using a COTS continuous speech recognition system called Nuance 7.0 from Nuance Communications, operators are able to interact with applications using a wireless microphone or 900Mhz cordless telephone. In order to voice enable an application, a grammar set is defined to trigger the actions desired. A very important capability of the Nuance system is that it is speaker independent. Any user can immediately interact via voice without having to train the system for his/her voice. Furthermore, the Nuance product supports accent and dialect recognition as well as international support for over 20 languages.

Users are able to manipulate windows and interact with applications using a combination of voice commands and laser pointer input. For example, by saying the command, "Minimize Window" the window manager will cause the active window, selected by the laser pointer, to be minimized just as if it had been performed through standard input devices. Another useful set of voice commands involves associating a phonetic letter with a window. For example, by uttering "Assign Alpha to Window," that window can then be called to the foreground by the command, "Get Me Alpha" or maximized by "Full Screen Alpha." Application hotkey sequences can also be triggered by voice. For example, on a PC, "Go to End" will cause Ctrl+End to be sent to the window with focus, causing the cursor to go to the end of a Microsoft document. While a useful command set has been identified, specific voice command grammars can easily be tailored to the end user's needs and preferences.

4. DATAWALL IMPLEMENTATIONS

AFRL/IF has successfully implemented a number of interactive DataWalls each consisting of three horizontally tiled video projectors. The first implementation, which serves as a development system, uses LCD projectors each displaying 1280 x 1024 pixels for a total display resolution of 3840 x 1024 pixels across a screen area 12' x 3' (Fig. 1). This far exceeds the state-of-the-art in single element display systems. A SGI Onyx workstation with three Reality Engines drives the display.

In the interest of developing a less costly alternative to the SGI-based DataWall, a PC-based version was successfully implemented to support the Joint Battlespace Infosphere program. Again three LCD projectors were utilized to achieve the same total display resolution of 3840 x 1024 pixels. A quad processor PC running Microsoft Windows 2000 drives the display. It has the same wireless interaction capabilities as the SGI-based DataWall.

To provide a deployable version of the Interactive DataWall to support the testbed for the forward deployable element of the Configurable Aerospace Command Center Integrated Technology Thrust Program, a Sun workstation and PC-based Deployable Interactive DataWall (DID) was implemented (Fig. 10).

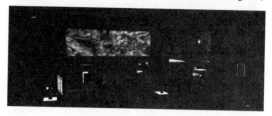

Figure 10 Model of the DID

This display can be switched between a Sun Enterprise 3500 or a dual processor PC. It is housed in an extensively modified Air Force S-530 A/G Standard Rigid Walled shelter, with its own Tactical Generator Set and Environmental Control Unit. Due to the unique, short-throw, rear-projection requirements, three LCD projectors with special short-throw lenses are used. In this configuration, each projector displays 1024 x 768 pixels for a total display resolution of 3072 x 768 across a screen area 9' x 2¼'. It should be noted that even in this most reduced resolution configuration, the total display resolution still exceeds the state-of-the-art in single element display systems. The DID is also equipped with hardware to support window encapsulated live NTSC video feeds. This allows the commander and staff to view and manipulate information from such sources as real-time surveillance video, satellite broadcasts, and VCR mission playbacks.

A Portable Interactive DataWall (PID) designed to be disassembled, transported, and reassembled quickly has also been developed. Unlike the DID, this version is intended for indoor use and easy transport. The DID requires a forklift and flatbed truck to transport it, while the PID was designed to be easily rolled to a location by two people. It is PC-based and fits through a conventional 3' doorway. Its footprint is 9¼'W x 5'H x 2¼'D while folded, and 9¼'W x 6¼'H x 4½'D extended. It is also a 1 x 3 LCD projector configuration. Each high-resolution projector displays 1280 x 1024 pixels for a total display resolution of 3840 x 1024 across a screen area 9' x 2¼' (Fig. 11). Projectors and cameras with wide-angle lenses are used to accommodate the extremely short-throw requirements. New versions of the PID have been developed that come apart in 3 sections to facilitate transport.

Figure 11 Portable Interactive DataWall

5. CONCLUSIONS AND FUTURE WORK

The ability to support multiple *simultaneous* users is a limitation of the Interactive DataWall. Although multiple users can interact with the display through speech and laser pointer, they cannot simultaneously. The problem with the current laser pointer-tracking method is the inability to differentiate multiple red laser pointers on the screen.

If the multiple pointer-tracking problem is solved, there is still the issue of the window manager supporting multiple cursors. Currently only one cursor is supported under the operating systems used in all versions of the DataWall. To assure all legacy applications can be supported on the Interactive DataWall without modification has precluded the development of a custom solution to the window manager and/or operating system. Hopefully the increasing popularity of tiled displays will motivate the commercial sector to develop a solution.

Multiple simultaneous speaker recognition is also desired. In order to support multiple speakers, a Dialogic PCI telephony card can be used to provide four separately addressable audio input/output channels to a PC system running Nuance 7.0. A wireless headset and telephone unit is connected to a line-simulator to give the telephony card the correct signal levels. A potential problem with multiple speakers in close proximity is a situation where one user's microphone picks up another user's voice. This difficulty may be overcome by using highly directional microphones. Another possible solution to this audio interference is to use Nuance Verifier 2.0, which allows identification of the speaker registered to a particular microphone. Of course, once the multiple pointer issue is solved, the problem of associating a user's voice with a particular pointer must still be resolved.

The Interactive DataWall has been demonstrated to a number of visitors at AFRL/IF at Rome Research Site. The Deployable and Portable Interactive DataWalls have been deployed and demonstrated to a number of off site conferences and military exercises. They have been extremely well received, so the ADII team is confident that research efforts are headed in the right direction for next generation Command and Control systems. Although a considerable amount of research and development remains, extremely capable systems have been created, integrating a significant amount of commercially available hardware and software.

REFERENCES

[1] G. A. Alphonse and J. Lubin, "Psychophysical Requirements for Tiled Large Screen Displays," *SPIE Vol. 1664 High-Resolution Displays and Projection Systems*, 1992, pp.230-240.

[2] P. Jedrysik, J. Moore, M. Brykowytch, and R. Sweed, "The Interactive DataWall," *DoD C4ISR Cooperative Research Program 1999 Command and Control Research and Technology Symposium*, 1999, pp.955-970.

[3] P. Jedrysik, J. Moore, T. Stedman, and R. Sweed, "Interactive Displays for Command and Control," *2000 IEEE Aerospace Conference*, 2000.

The AOF (Authoring on the Fly) system as an example for efficient and comfortable browsing and access of multimedia data

Wolfgang Hürst, Rainer Müller

Institut für Informatik, Universität Freiburg,
Georges-Köhler-Allee, D-79110 Freiburg, Germany
Email: {huerst, rmueller }@informatik.uni-freiburg.de

ABSTRACT

Multimedia is more and more becoming a ubiquitous part of our life, because of the tremendous advancements in recording, storing, and processing, which have been achieved over the last couple of years. In the (not so far) future, we will be confronted with innumerable amounts of this kind of data. As a consequence, users need to be able to easily access the data, to browse through it, to filter and retrieve required information, and so on. In this paper we will illustrate this scenario using an example application from the teaching and learning scenario. We will introduce the AOF system for automatic presentation recording, which was developed by our research group and which can be used to support and enlarge traditional learning. When using recorded presentations for teaching at our university, we identified several requirements and features which should be provided to achieve a high usability. We will show the user interaction functionality which we integrated in the AOF system to fulfill these needs. Techniques ranging from an appropriate presentation of retrieval results to functions for easy and comfortable navigation in a multimedia file will be discussed. While originally being developed for the teaching and learning scenario, the solutions we propose are useful for any application where intensive user interaction with multimedia data is required.

1. INTRODUCTION AND MOTIVATION

Authoring on the Fly (AOF) [1,2] is an advanced system for automatic recording of multimedia presentations. It is used by a variety of universities and institutions [3,4] to record and preserve talks and complete lectures for further (offline) usage. With AOF, presentations are recorded 'on the fly', i.e., automatically in the background while the lecturer is giving a talk in a traditional way. There is no need for manual recording or post-processing. All data streams which occur in the lecture room are recorded with an optimal granularity. The resulting multimedia document consists of several media streams (e.g., audio with the voice of the lecturer, a whiteboard stream containing the used slides and handwritten annotations, as well as an optional video image of the lecturer and/or the audience). A combined synchronized replay of the single media streams is guaranteed through our synchronization model [5]. That way, a high quality multimedia document is automatically produced, which preserves the live-character and classroom experience in the best possible way.

Using the AOF system at our university for a couple of years now, we got a lot of feedback from the students and gained experience with using this type of multimedia data for teaching and learning. The possibility to repeat and re-listen to a (recorded) lecture, e.g., when doing homework or when preparing for an exam, doubtlessly offers great advantages to the students. On the other hand, they are confronted with a tremendous amount of information. They are not only faced with textual information (text books, scientific papers, handwritten notes, slide printouts from a lecture, etc.) as in the past, but with multimedia data (audio, video, computer animations, etc.) as well. As a consequence, students must learn how to deal with this 'information overflow', how to find and how to identify valuable information, how to filter important from irrelevant parts, and so on. They need a possibility to interact and 'work' with the material. For example, when replaying a recorded lecture, it should be possible to quickly skip something you already know and to jump to the next interesting part. In the same way, users want to go back to a specific position to repeat a part which they did not understand completely. For a system designer, this means that comfortable browsing mechanisms and navigation functionality must be integrated in the system. It is important for the usability to enable the users to quickly glance over a the data, to randomly scroll through a document, to arbitrarily repeat any part, and so on.

Additionally, with a growing amount of data being available – recorded lectures as well as other kinds of multimedia documents – it becomes important to offer some sort of search functionality. Query processing and retrieval mechanisms are normally used, when searching (very) large databases. Browsing, on the other hand, can be very useful for searching smaller document sets. This is not only true for multimedia retrieval, but for any kind of information search. For example, when searching for one out of thousands of books in a library, one would usually go to the library catalogue. When looking for a useful book in a bookshelf containing less than fifty books, browsing the shelve seems to be more appropriate. In practice, often a combination of both approaches is applied: After placing a query, the retrieved subset is browsed by the user to find the results which are relevant to the particular needs. This second step is very important. When placing a query to a huge data collection, the returned (sub)set of documents can get very large. However, not all of the retrieved documents are really relevant to a particular user – on the one hand, because of imperfect index calculation and/or retrieval algorithms, and, on the other hand, because of ambiguous user queries or different intentions of the users (which can not be identified through the according query). Hence the user must filter the data to identify the documents (or the parts of a document) that are really relevant to his/her particular needs.

Summing up it may be said that browsing and interaction functionality is an important feature when using recorded presentations for offline learning for two reasons: First, to fulfill the user's request for easy and intuitive navigation mechanisms, and second, to assist and help the user when searching for information within the recorded data. In the following we will discuss some issues in browsing and navigating multimedia data using the AOF system and the teaching and learning scenario as an example. After discussing some related work in Section 2.1, we will illustrate in Section 2.2 how the presentation of a retrieval result can support user browsing. Section 2.3 addresses the question, what features can be offered to easily navigate in the data and to interact with it. We use the teaching and learning scenario to illustrate our approaches and solutions, however, the basic ideas can be applied to other multimedia applications as well. AOF documents are a good, representative example for multimedia documents in general because they consist of several media streams (audio, video, whiteboard annotations, etc.) with different characteristics. Additionally, the AOF replay model [5] is an open model, i.e., new, additional media types can be added very easily. The teaching and learning scenario is a good application area to study interaction with multimedia data because of the high requirements requested by the users.

2. BROWSING AND NAVIGATION OF AOF DOCUMENTS

2.1 Related Work

Browsing multimedia is a difficult, non-trivial problem. Discrete data, such as (short) text documents, images, and even Web pages, can usually be browsed very easily by just looking at them. This is not true for multimedia data whose content is changing dynamically over time (e.g., audio, video). In addition to the continuous nature of the data, they are often rather long and lack the existence of assisting information, such as punctuation marks, paragraphs, headlines, or tables of contents. Most techniques used for browsing multimedia data try to overcome these problems (a) by doing a time compressed replay, i.e., by viewing the document at a higher speed or frame rate, and (b) by extracting particular, representative information from a document, which is useful to get an idea of the according content. This can be some meta information, as well as a specific part of the document. For example, when retrieving video clips, some key frames could be extracted. In [6], meaningful frames are automatically identified and presented to the user in a 'filmstrip' view. The continuous video stream is transformed into a discrete stream of still images, which are easier to browse. The common fast-forward and backward function known from VCRs is an example for time compressed replay. The SpeechSkimmer [7], a system for browsing speech recordings, is a good example where both techniques are applied – information extraction as well as time compression. Important fragments of the audio file are identified (for example, the beginning of a new topic) by detection of unvoiced segments or pauses of the speech signal as well as by examination of pitch frequencies. Additionally, the user can not only jump between those outstanding parts during replay, but listen to a time-compressed version of the document. Various techniques, such as replay at a higher frame rate or shortening of unvoiced parts are applied. Replaying speech at a higher speed rate still allows users to classify (not necessarily understand) the content of a speech recording. Continuous speech remains comprehensible even if replayed up to twice normal speed (compare [7]).

As already discussed in the previous section, browsing is often used as a second step of a search process after placing a query and retrieving some (probably) relevant documents. Some approaches try to support this browsing or filtering process of the user by extracting particular, meaningful information from the data. This additional (application specific) information is presented along with the relevance values calculated by the retrieval algorithms. For example, when retrieving stories from a news database, the question when and where a particular event from the news story took place can be very important for a user. Therefore, [8] introduced so called visual maps and timeline presentations for news retrieval. The retrieved news stories are not presented in a ranked list according to their relevance, but instead along a timeline or on a topological map. In [9], TileBars are introduced to represent the result of a query request when retrieving longer text documents. This special representation clearly indicates the position and context of a relevant part of a retrieved document, an information which is very important and helpful when retrieving long text documents

2.2 Presentation of Retrieval Results in AOF

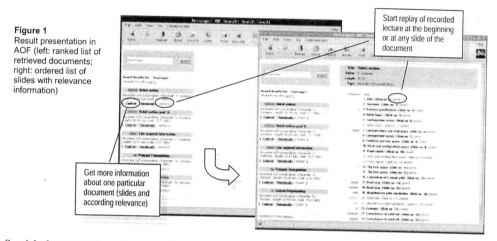

Figure 1
Result presentation in AOF (left: ranked list of retrieved documents; right: ordered list of slides with relevance information)

Start replay of recorded lecture at the beginning or at any slide of the document

Get more information about one particular document (slides and according relevance)

Special characteristics when retrieving recorded presentations. When retrieving recorded presentations, the information *when* and *in which context* a particular part of a document appears can be very important for a user. For example, the definition or explanation of a term is most likely to be found near the first appearance of this term in the document, since new terms are usually introduced before they are used on a regular basis. Thus, a slide at the beginning of a document might be more relevant to a particular user than one in the middle or at the end, even if given a lower relevance by the retrieval algorithms. On the other hand, users interested in a detailed discussion about the search term, might be more interested in a part of the document where the according search term occurs on more consecutive slides, even if each of these slides has a lower relevance.

Additionally, users generally have very different intentions when questioning a retrieval engine, even when using the same words in their query. For example, entering a specific term could represent the wish to retrieve an exhaustive answer, i.e., all documents from the database with any (even a very small) relevance to this particular query term. However, another user entering the same search term might just want to look up its exact definition thus being interested not in all documents with a certain relevance but just in a very particular position of one document.

The AOF retrieval interface. AOF considers the characteristics specified above by presenting retrieval results in a special way. After placing a query to the AOF search engine, the user gets a list of documents, which is ordered according to their relevance (Fig. 1, left). A user interested in full document retrieval can start replay immediately. Users who are looking for a more detailed information of a particular document can select a link which brings up a presentation of the content of the document, i.e., a list of all used slides (Fig. 1, right). Each slide has an associated relevance value, indicating the relevance of the according part of the document. The slides are not ordered according to these relevance values, but according to the position at which they appeared during the lecture. By preserving the original order of the slides, it is clearly indicated when and in which context a particular slide appeared. This information would be completely lost when presenting the slides as a ranked list.

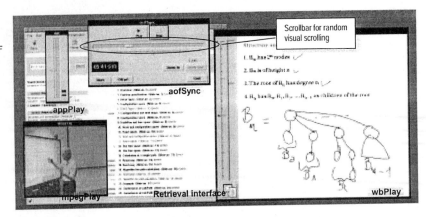

Figure 2
Viewer for AOF documents.

2.3 Browsing, Navigation, and Interaction in AOF

Presenting retrieval results in an appropriate way is a first step to support users in browsing and filtering the according documents. Additional browsing functionality is offered in AOF once the viewer is started. Figure 2 shows a snapshot of the AOF viewer. *aofSync*, the main module, replays the audio stream and guarantees a synchronized replay of all other media streams. Each media type is replayed through its own so called helper application (*wbPlay* for the whiteboard and annotation stream, *appPlay* for external applications shown during the lecture, and *mpegPlay* for the recorded video).

Before starting regular replay, a user can browse the slides as well as the external applications, using the helpers *wbPlay* or *appPlay*, respectively. Browsing the snapshots of the slides can also be done by selecting the according link in the overview generated by the retrieval engine. Snapshot based browsing of the slides offers a good way to get a first idea of the content of the overall document. Additionally, quick access to every position within the document is possible in real-time – even during replay – because our synchronization model provides real-time random access [5]. This random accessibility can be used, e.g., to browse all relevant parts of a document step by step by selecting the according links in the overview shown by the retrieval interface. This offers a convenient way to skip non-relevant parts of the lecture during replay.

In addition to be able to jump to some outstanding positions of the document during replay, a more granular browsing is required, especially when the document contains a lot of dynamic information, such as handwritten annotations. In Section 1 we illustrated that students need a flexible and easy way to interact with the data, e.g., they want to be able to repeat or skip random parts of a lecture. For this reason, we introduced a technique called *random visible scrolling* for browsing any kind of visual data whose content is changing over time. Using the scrollbar contained in *aofSync* (compare Fig. 2), the user can visually scroll through the document in real-time, i.e., move along the timeline - independent of speed and direction - while all visual data is instantly displayed. Thus, random visual scrolling enables a use-controlled time-compressed replay. Evaluations with users of the system confirmed that random visual scrolling is a very useful technique for two tasks: Getting a quick overview of the content, as well as localizing a particular part of a document where some specific information can be found. Additionally, we identified that users often remember specific situations or contents in a visual way. For example, when students use the recorded lectures to do their homework, situations such as "I remember there was a little blue graphic at the bottom of the slide, when the lecturer made some comments on this particular topic. I want to find that." happen very often. Random visual scrolling offers a very convenient and intuitive way to deal with such situations.

3. SUMMARY AND FUTURE WORK

Fig. 3 summarizes the features offered by our system for browsing and navigating the documents. Users can browse the data to get an overview of the content or to find a particular information. They can skip a topic and jump to the next one or they easily can repeat any random part. All of these features are particular useful when searching for some information. Using the functionality shown in Fig. 3 from the top to the bottom, the user gets more and more detail about the document and its content. However, each level of browsing can be skipped as well as repeated (indicated by the grey arrow on the right side). Each user is able to chose which level is the most useful for the

1260

actual situation. Easy and comfortable interaction with the data is provided. The techniques in (a) extract some important information from the data which is representative for the content, e.g., start-points of parts which are relevant to some query, or snapshots of the slides. Random visible scrolling (b) enables a time-compressed replay of the (visual) data streams. Offering user-controlled time-compressed replay, instead of relying on time-compressed replay at a fixed speed, is one of the key features of our system.

The viewer with the according browsing functionality was developed some time ago and is in use on a regular basis now. The query processing and retrieval engine was just added to the system and will be evaluated in a class next term. Our future work goes into the direction of extracting more meta information, for example, automatically finding content related clusters of a lecture to provide the user with a more detailed segmentation of the document. Additionally, it would be interesting to examine the speech browsing methods proposed

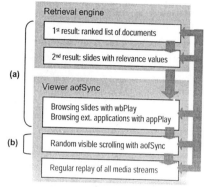

Figure 3 Retrieval and browsing functionality offered by AOF.

in [7] in combination with our techniques for visual browsing, i.e., combining random visual with random audible scrolling. Another interesting question is if and how the techniques we developed can be applied to other applications and scenarios. While originally being developed for the teaching and learning scenario, features such as random visible scrolling can also be useful in tasks such as news retrieval – or any other application where retrieval and browsing of continuous media streams is important.

ACKNOWLEDGEMENTS

This work is part of the project *"Generation of Multimedia Documents on the Fly"*, which is supported by the German Research Foundation (Deutsche Forschungsgemeinschaft DFG) as part of V3D2.

REFERENCES

[1] Müller, R., Ottmann, T. (2000) The "Authoring on the Fly" System for Automated Recording and Replay of (Tele)presentations, *ACM Multimedia Systems Journal*, Vol. 8/3, 2000.

[2] Authoring on the Fly (AOF) – Homepage: http://ad.informatik.uni-freiburg.de/mmgroup/aof/

[3] Authoring on the Fly – Users: http://ad.informatik.uni-freiburg.de/mmgroup/aof/users/

[4] VIROR – Virtuelle Hochschule Oberrhein (Virtual University Oberrhein): http://www.viror.de

[5] Hürst, W., Müller, R. (1999) A Synchronization Model for Recorded Presentations and its Relevance for Information Retrieval, *Proceedings of ACM Multimedia '99*, 7th ACM International Multimedia Conference, 1999.

[6] Hauptmann, A.G., Witbrock, M.J., Christel, M.G. (1997) Artificial Intelligence Techniques in the Interface to a Digital Video Library, *Proceedings of the ACM CHI '97*, ACM Conference on Human Factors in Computing Systems, 1997.

[7] Arons, B. (1997) *SpeechSkimmer*: A System for Interactively Skimming Recorded Speech, *ACM Transactions on Computer-Human Interaction*, Vol 4/1, 1997.

[8] Christel, M.G. (1999) Visual Digests for News Video Libraries, *Proceedings of ACM Multimedia '99*, 7th ACM International Multimedia Conference, 1999.

[9] Hearst, M.A. (1995) *TileBars*: Visualization of Term Distribution Information in Full Text Information Access, *Proceedings of ACM CHI '95*, ACM Conference on Human Factors in Computing Systems, 1995.

1261

Combining Speech with Sound to Communicate Information in a Multimedia Stock Control System

D. Rigas[1], H. Yu[2], D. Memery[1], D. Howden[1]

[1]School of Informatics, University of Bradford, Bradford BD7 1DP, United Kingdom
Tel.: +44 (0) 1274 235131 Fax.: +44 (0) 1274 233920 E-Mail: D.Rigas@bradford.ac.uk

[2]School of Computing and Engineering, University of Exeter, Exeter EX4 4QF, United Kingdom
Tel.: +44 (0) 1392 264685 Fax.: +44 (0) 1392 217965 E-Mail: H.Yu@exeter.ac.uk

ABSTRACT

This paper describes two experiments using speech and sound to communicate information to users in a multimedia stock control system. The first experiment used sound only and the second experiment tested combinations of simultaneous presentation of speech and sound. Various types of information (e.g., types of stock, windows, numbers) were communicated using this approach. Results indicated sound on its own communicated information successfully and also the simultaneous presentation of sound and speech increased the load of information that can be communicated successfully. A number of empirically derived design guidelines are also discussed. These guidelines concentrate in the design of messages that combine different types of auditory stimuli.

1. INTRODUCTION

The use of auditory stimuli in multimedia systems and user interfaces has attracted significant research attention in recent years. There are many different types of auditory stimuli, which can be used to communicate information. Some of these types include structured musical stimuli or earcons, synthesised or recorded speech, and environmental sounds or auditory icons. For instance, designers could use structured musical stimuli at the basic level with single notes or a short series of single notes to communicate simple events. If more complex musical structures (e.g., particular rhythm or harmony) are considered then more information can be communicated.

A number of experimental applications using one or another type of auditory stimuli have been developed and empirical investigations still continue. However, there is a limited set of practical design guidelines to help developers take advantage of the auditory medium as a whole by combining different types of auditory stimuli. The synergy of simultaneously utilising different auditory media offers a promising case for unique, creative, and usable multimedia systems or auditory interfaces of the future.

2. RELEVANT WORK

Auditory feedback can be generally divided into synthesised speech, environmental sounds (or auditory icons) and structured musical sounds (or earcons). Musical sound is a rich medium containing numerous structures introduced by musicians over many years of human evolution. Also, given that we live in an age where multimedia systems are fully capable of producing musical sounds relatively easily and effortlessly, the use of structured musical stimuli in interfaces is currently at a relatively low level. The auditory channel, as a whole, has been neglected in the development of user-interfaces, possibly because there is very little known about how humans understand and process auditory stimuli. It is not intuitively obvious how to use musical structures in interface design. Current user interfaces focus heavily on visual interaction. The consequence of this is that user interfaces have become more and more visually crowded as the user's needs to interact with the computer increase. This creates considerable difficulties for blind users. Sound has already been utilised successfully to communicate graphical information to blind users (Rigas and Alty 1997). There are also other examples of user interfaces that accommodate special needs (Edwards 1995). Other experiments in the literature suggest that the auditory channel provides alternative ways of conveying information to general user interfaces (Brewster 1994).

Sound has also been used to communicate the execution of programs (Rigas and Alty 1998) and the contents of complex databases (Rigas et al. 1997). Other work includes the CAITLIN tool. The tool assists novice programmers in debugging activities by communicating statements and structures (e.g., loops, selection) within programs (Vickers and Alty 1996). Furthermore, auditory stimuli were successfully used in multimedia applications that communicated spatial information (Rigas et al. 1999) and information related to mobile telephony (Rigas et al. 2000).

This paper describes two sets of experiments in which complex structured musical stimuli and speech were simultaneously used to communicate information. The platform for the experiments was a prototype of a multimedia stock control system (see Figure 1). These experiments have provided an experimentally derived overall viewpoint for the simultaneous use of speech and structured musical sound. These results aim to help audiovisual or multimedia user interface designers or developers, who wish to incorporate auditory media in their designs.

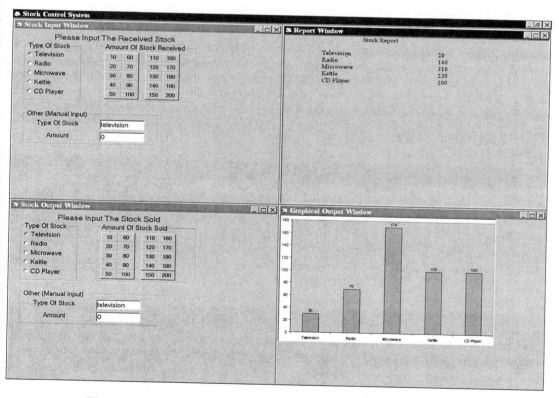

Figure 1: The visual interface of the multimedia stock control system.

3. EXPERIMENTAL PLATFORM AND METHODOLOGY

A prototype multimedia stock control system (see Figure 1) was developed to provide a platform to investigate the simultaneous use of synthesised speech and structured musical stimuli in one communication metaphor. The system has the provision of using other types of auditory stimuli such as environmental sounds (i.e., auditory icons) or special sound effects. The information communicated by the system was in the problem domain of a stock control application. Two experiments were performed using this experimental platform.

The sample consisted of 32 people. All subjects were students of computer science related courses. The age range was between 21 to 40. The musical knowledge of all subjects was assessed via a questionnaire. The experimental session involved two stages. Subjects were presented with sound stimuli only in the first stage and with simultaneous use of speech and sound in the second stage. The stimuli in both experiments were presented once for each individual trial. Thus, one auditory message was presented to subjects for each message. Subjects recorded their interpretation on an answering sheet. At the end of the experiments, all subjects were requested to answer a post-experimental short questionnaire in order to collect user feedback.

3.1 Experiments Using Structured Musical Sound

Four distinctive rhythms were utilised to communicate the four windows provided by the experimental system. Subjects were presented with the auditory stimuli once and 1 minute was given to all subjects in order to answer the relevant question. Each answering sheet had 16 questions in total. On overall, 489 out of 512 questions about auditory messages presented were answered correctly. More specifically, 21 subjects (or 66% of the sample) answered correctly all 16 questions in the answering sheet, 7 subjects answered correctly 15 questions, 2 subjects answered correctly 14 questions, 1 subject answered correctly 12 questions, and 1 subject answered correctly 8 questions. The first rhythm, which communicated the window 'stock received', was recognised correctly by 96.8% of the sample (or 124 answers out of 128 questions). The second rhythm, which communicated the window 'stock sold', was recognised correctly by 97.6% of the sample (or 125 correct answers out of 128 questions). The third rhythm, which communicated the window 'report', was recognised correctly by 91.4% of the sample (or 117 correct answers out of 128 questions). The window 'report' was confused with window 'graphics' 11 times out of 128 presentations to subjects. The fourth rhythm, which communicated the window 'graphics', was recognised correctly by 94.5% of the sample (or 121 correct answers out of 128 questions).

On overall, the above results indicate that subjects could easily interpret correctly the information communicated by the four rhythms. Designers could take a real advantage of the auditory medium if the structured musical stimuli is successfully combined and used with other types of auditory stimuli such as synthesised speech. The experiment described below addresses some of these issues by investigating the simultaneous use of synthesised speech with the rhythms tested above in the first experiment.

3.2 Experiments with Simultaneous Use of Speech and Sound

This experiment aimed to test the simultaneous use of synthesised speech and structured musical stimuli to communicate information to users. The auditory stimuli consisted of 640 synthesised speech messages to communicate numerical values, 640 synthesised speech messages to communicate types of stock in words (e.g., radio), and either rhythm 1 or rhythm 2 as explained in the previous experiment. The role of rhythm was to communicate that the stock was either 'sold' or 'received' and it was continuously presented as the synthesised speech messages were communicated. On overall, 73% of the speech messages that communicated numerical values were correctly recognised, 89% of the speech messages that communicated type of stock in words were also recognised successfully, and 93% of the stock status 'received' or 'stock' were recognised successfully.

Subjects were also requested to answer a short questionnaire after the experiment. 75% of the sample believed that musical sound and speech could be easily recognised and interpreted by users when they are presented in one single auditory message. 78% believed that the simultaneous use of speech and sound can successfully communicate more information than using any of them on their own. Furthermore, 97% believed that the visual presentation of the windows for a stock control application could benefit by the complementary and simultaneous use of speech and sound.

4. DISCUSSION

These results indicate that speech complemented with non-speech sound can communicate successfully information in multimedia systems. A part of the overall message (information) is communicated using speech and another part using sound. This is particularly useful for user interface or multimedia designers who wish to use this part of media. The experimental results and observations suggest the following guidelines for auditory user interface or multimedia designers:

- Structured musical stimuli should be low base (avoid low frequency notes).

- Speech messages should be short, sharp, and clear.

- Provide mechanisms for user feedback or confirmation before the user interface proceeds to any further actions.

- Use different types of synthesised speech voices according to the musical sounds.

- Provide a review mechanism for the auditory message. The use should be able to pause, stop, review or forward auditory messages.

- Place the synthesised voice message in the fore ground and other non-speech stimuli at the background of the auditory message.

Experimental results indicated that subjects' concentration level, interest to interpret and perceptual context contributed to successful perception. Users had to maintain high concentration levels in order to interpret the whole of the auditory message. Subjects needed to highly concentrate particularly when the auditory message combined speech and non-speech sound.

5. CONCLUSION

Results of this study indicated that speech and sound could be successfully combined to communicate information. The simultaneous use of speech and sound did not confuse subjects who on overall successfully perceived both types of stimuli. It is believed that auditory messages could utilise different types of auditory stimuli to communicate information. For example, these types of stimuli could include the use of musical sound and speech but also environmental sounds and special sound effects. A further set of experiments is under way to investigate how successfully these different types of auditory stimuli could be simultaneously combined in one communication metaphor. This approach of integration could have particular application to a wide range of applications, which range from audio-visual or multimedia systems to interfaces for mobile telephony.

REFERENCES

Brewster, S. A. (1994). Providing a Structured Method for Integrating Non-Speech Audio into Human-Computer Interfaces. PhD Thesis, University of York, England, UK.
Edwards, A. D. N. (ed.) (1995). Extra-Ordinary Human-Computer Interaction. Cambridge: University Press.

Rigas, D., Alty, J. L. & Long, F. W. (1997). Can Music Support Interfaces to Complex Databases? In EUROMICRO-97, New Frontiers of Information Technology, pp. 78-84. IEEE Computer Society Press.

Rigas, D. & Alty J. L. (1997). The Use of Music in a Graphical Interface for the Visually Impaired. In Howard, S. Hammond, J. & Lindgaard K. (Eds.): *Human-Computer Interaction, INTERACT'97: Proc. of the 5th IFP Conference on Human-Computer Interaction*, pp. 228-235. Sydney, Australia: Chapman and Hall.

Rigas, D. & Alty, J. L. (1998). Using Sound to Communicate Program Execution. In EUROMICRO-98, Engineering Systems and Software for the Next Decade, pp. 625-632, IEEE Computer Society Press.

Rigas, D. (1996). Guidelines for Auditory Interface Design: An empirical Investigation. PhD Thesis. Loughborough University, England, UK.

Rigas, D., Memery, D., Hopwood, D. and Rodriques, M.A. (2000): "Empirically Derived Issues in Auditory Information Processing for Mobile Telephony", in *IEEE International Conference on Information Technology: Coding and Computing, IEEE Computer Society, ITCC - 2000, pp. 462-469*, Las Vegas, USA.

Rigas, D., Hopwood, D. and Memery, D. (1999): "Communicating Spatial Information via a Multimedia Interface", in *EEE Proceedings of the 25th Euromicro - 1999 Conference, IEEE Computer Society, pp. 398-405*, Milan, Italy.

Vickers, P. & Alty, J. L. (1996). CAITLIN: A Musical Program Auralisation Tool to Assist Novice Programmers with Debugging. In Frysinger, S. & Kramer, G. (eds.), Proc. of ICAD'96, pp. 17-23.

North American Videotex:
Bringing Computer Networks to the Public Too Soon

Eric Lee, Ph.D.

Saint Mary's University, 923 Robie Street, Halifax, Nova Scotia, B3H 3C3
elee@stmarys.ca

ABSTRACT

Telidon, a videotex precursor to the Internet, was an early attempt (early 1980's) to synthesize computers and telecommunications in the provision of interactive information retrieval services, teleshopping, and other services directed at all members of society as potential users in their homes or businesses. Employing a Telidon terminal, a keypad, and their home television as an information display, users accessed services and databases using telephone lines or cable to retrieve information stored in centralized computers. From the earliest stages of development, Telidon was heralded as a revolution because it would provide ordinary people in their homes with access to the information of the entire world. Over 30 different studies were undertaken by researchers in psychology, computer science, and library science on a wide variety of applied psychological and human factors issues in the design and development of the Telidon system. These included design and testing of alternative information search systems for users, interface design, reading text using screens rather than printed text, the role of images and pictures in communication, and social implications of introducing computer services to people in their homes. Very early in the development of the Telidon system, human factors research identified several key weaknesses which ultimately contributed to the failure of field trials of the Telidon system. For example, Telidon users were forced to search for information using hierarchical menu structures that were tedious, time-consuming, and error-prone despite numerous studies suggesting that hybrid information retrieval systems offering several alternative methods of searching for information (e.g., menus and keywords) were markedly superior. Implications for human factors practitioners are discussed.

1. HISTORY OF TELIDON

In 1978, the British launched a commercial videotex system called *Prestel*. This first videotex system delivered pages of text from a central computer to terminals in homes where the information was displayed on the TV set. Not wishing to be left behind in the race to develop the next generation of communications technology, the Canadian Department of Communications asked a team of technical experts led by Herb Bown at the Department's Communications Research Centre (CRC) to put together a quick simulation of Prestel to show to senior management. Bown recognized at once that his team's previous research could be used to provide a more powerful competitor to the British system. Bown's team had been working on image modeling protocols called *picture descriptor instructions (PDI)* that allowed aesthetically-pleasing images to be compressed into a few hundred bytes. They quickly developed the technology to store and transmit videotex pages with graphical images of much higher quality than were possible using the British system (which was primarily a text delivery system with some relatively crude graphics). Given the limited bandwidths available at the time, this Canadian system, called *Telidon*, was a significant improvement in videotex technology.

Technically, there was a cost associated with this improved graphic capability. The terminal in the home required to control the display of information of the home TV required considerable computing power to render the images that were modeled with PDIs. At that time, a computer with a Z80 CPU and 64 kilobytes of RAM cost over $2000. Bown argued that the cost of computing and memory was irrelevant in the long term. Home terminals with sufficient computing power and memory would be inexpensive within a few years.

1.1. THE TELIDON PROJECT

Deputy Minister Bernard Ostry and Assistant Deputy Minster Douglas Parkhill orchestrated the launch of the *Telidon* program, Canada's homegrown videotex system. The Canadian Cabinet approved funding amounting to $9,000,000(CDN) in 1979. This initial money was expended in developing the technology and evaluating it in a series of field trials. A further $50,000,000(CDN) was allocated in 1982 to extend the Telidon technology into marketing tests and establish a viable videotex industry in Canada.

Late in 1979 the objectives for a national {Telidon} system were proclaimed. The objectives of this national videotex system were technical, economic, and social. Technically, Telidon was supposed to rival the British and French videotex systems, particularly for the US market. As well, it had to provide flexible overall system design to incorporate possible future technological developments. Economically, Telidon had to achieve the fastest possible expansion of services and systems without prejudice to other social priorities while retaining adequate Canadian control and ownership. Social aims included protection of the right of access to information, privacy, and freedom of speech.

1.2. THE HYPE

The Telidon project generated considerable publicity in Canada. The launch of the project was accompanied by a flurry of press releases that were picked up by the media. In 1979, microcomputers were just being introduced. The IBM PC would not be developed for another three years. Everyone was wondering what these tiny computers would do for them, so Telidon was promoted as an application that would put computers in people's homes, attached to their television sets, providing a service that they would use every day.

Key people traveled around Canada and internationally, delivering public lectures and speaking to the press. Along with the descriptions of the technical superiority of the Telidon technology over the character-oriented graphic schemes of Prestel and the French Minitel systems; the experts described the expected impact of this new technology. It was those descriptions that caught people's attention. One poster from 1978 proclaimed that "[Videotex} can tell you everything this poster does without wasting ink or killing trees. [Videotex] is a new two-way TV system and can display, at your command, words and pictures on tour home TV screen. [Videotex will] be linked to data banks all over the world. And, if you ask me a question, I'll answer at the speed of light." A pamphlet from this era (1980) proclaimed that "Interactive television is here!" It went on to say "Videotex is a colour TV with a keypad, an open door to the world's information, recreation, education, a threat to the TV broadcast industry, a home computer "for the people," a communications protocol, and a highly flexible system with a range of uses and a range of terminals." Even as early as 1979, many experts realized that Telidon could have extensive social impacts if adoption was widespread. Expectations included: "Everyman his own publisher, universal access to knowledge, work from home, decentralization, and greater freedom of choice."

Bold words, even in this era of the World Wide Web, but truly startling in 1979 when only a few hobbyists owned micro-computers and only a small percentage of them were equipped with a 300 baud modem. But these words tell a fascinating story.

First, there was considerable confusion about what the technology actually was. The public presentation of videotex placed emphasis on the relatively trivial idea that the home television would be used as the display device. The technology was often called "Interactive television," even though it had none of the characteristics of television, lacking animation, audio, and photographic realism. There were two reasons for this emphasis. First, the intent was to enhance the technology so that it would eventually include these characteristics. In fact, photographic realism was built into the PDI technology from the beginning, but was almost never used because it was too slow to be transmitted by the leading edge 1200 baud modems that were used in the field trials. Second, it was believed that a public not yet familiar with computing technology would understand the television metaphor much better. Third, the program wanted to give the impression, particularly to decision-makers, that videotex was going to be as important in Canada as television.

Second, there was a presumption that Telidon was going to be universal. All of the objectives that were discussed publicly were about how a videotex service was going to be established in every home without disrupting existing social values.

Third, there was a much greater vision of the range of services than most people associated with videotex. Though the primary emphasis was on information retrieval, there was an expectation that people would also want to use the system for messaging, electronic transactions, and downloading software.

1.3. FIELD TRIALS

Once the project was established, considerable effort was devoted to collaborating with industry. Most of the telephone companies in Canada sponsored field trials of the technology. Some broadcasters, cable companies, a major newspaper chain, many post-secondary educational institutions, were also participants in field trials. In the field trials, Telidon was used to provide weather information to farmers in the prairies, news and sports information to homes in major cities, educational information in schools, and mass transit schedules in shopping centres.

Ultimately, information providers wanted to provide three classes of services: information access (via the keypad and home TV), transaction services, and other. Information access services included public information and closed user groups. Expected transaction services included simple messaging between users, telex/twx connections, conversational mode between users, and simple data input (e.g., to join a book club). Miscellaneous services included computational capabilities, games, and telesoftware.

The Government of Canada contributed substantial resources to the field trials, but, in fact, the collaborators provided the bulk of the funding themselves. Canadian industry bought into Telidon in a big way.

The objective of the field trials was to provide a stepping stone to the establishment of a viable videotex industry in Canada. Companies that had developed expertise and were bringing a valuable product to the public in the field trials expected to start charging customers and generate a profit after the field trials ended.

1.4. THE END

In 1985, when market studies showed that the Telidon systems were not going to generate profits, cabinet terminated further funding for the project, and the Telidon project ended. Within two years, most of the key members of the Telidon team had left the government to work in industry. Though government funding ended in 1985, efforts to develop a videotex system continued in industry for a few more years. For example, a couple of years later, Bell Canada launched another videotex system called Alex. It, too, failed to achieve profitability.

Over a decade later, the World Wide Web has swept across North America like wildfire. Few remember that 10-15 years earlier Telidon attempted to accomplish a similar effect by providing home and business users with access to and graphics. Today, Telidon is listed on the Dead Media web site www.deadmedia.org).

2. HUMAN FACTORS ISSUES

1.1. TELIDON BEHAVIOURAL RESEARCH

Telidon is important to us because human factors figured prominently in its development. At the launch of Telidon, the Department of Communications already had a small human factors research group called the *Behavioural Research Group*, which consisted of three experimental psychologists and a research assistant. With the launch of the Telidon project, this group convinced management that many social and human factors issues would have to be considered if Telidon were to succeed. The group was renamed the *Behavioural Research Division* and expanded to include about 6-8 psychologists and post-doctoral fellows and a similar number of research assistants. In addition, they were given research labs, equipment, and a much larger budget.

Not all human factors work was conducted in-house. The Director of Behavioural Research made an explicit decision to develop more human factors research capabilities outside the government, primarily in the universities. Funding was broken into small research contracts and spread across as many different Canadian university researchers as possible. Some funding was used to support a joint meeting in Canada between behavioural researchers in Canada who were working on Telidon and behavioural researchers in France who were working on Minitel. A few others from outside the project were invited to that meeting. Other funding was used to support a small conference on Telidon Behavioural Research for government researchers and invited researchers from across Canada.

Research results were not only published in peer-reviewed journals, but interim results were rapidly published in-house by the federal government and distributed at government expense to ensure immediate access to timely information within Canada. Seven edited books containing journal-like articles were produced during the life of the program. Researchers from all around the world, as well as participants in the field trials, requested thousands of copies of these government research reports.

1.2. TELIDON-RELATED HUMAN FACTORS RESEARCH

Behavioural research on Telidon focussed on five general areas:

(1) navigation through information databases,
(2) interface issues,
(3) the role of pictures and graphics in Telidon,
(4) office and business issues,
(5) social issues, and

(6) applications.

Characteristic of this research was that it was never restricted to examination of the technology as it existed at any given time. Every effort was made to explore ways of improving the Telidon system.

An obvious area for research was the apparent difficulty typical users might experience in trying to find information using the system initially provided by the engineers for retrieving information from the database. Rather than embedded hypertext links and keyword searches supplemented by the restricted use of menus, it was expected that Telidon navigation would proceed through a hierarchy of menus. Moreover, all pages of information in the database were numbered according to this hierarchical tree structure. Researchers addressed a number of questions appropriate to the web today. How often would people select the wrong menu item? Would the probability of selecting the wrong menu item depend on the level of the menu in the hierarchy? How often would people find the information they wanted? How many items should be listed on one menu? How soon would people give up if the information were not available? Would people browse through menus if they did not have an explicit question? Would people be able to construct their own private menus? Were there alternative methods of searching for target information?

Another area for research was the keypad. What keys were required on a keypad? How should the keys be arranged? How large should the keys be, and how should they be spaced? Would people want a full keyboard? Should keyboard keys be arranged in an alphabetic, QWERTY, or in Dvorak layout for untrained users?

Telidon required research on reading text from a low-resolution television screen. What font should be used? How many pixels were required to represent a character: 5x7 or 7x9? Should lower-case characters be displayed in a small font? Would ascenders and descenders that extended above or below the line improve the readability of a low-resolution font? How should characters be spaced? How could accents be placed on characters, especially if the font only included upper-case characters?

Given the centrality of graphics and images to Telidon (in contrast to the more text-oriented European competitor's systems), many researchers studied problems and issues in this area. How important were graphics? How would information providers use graphics? Could graphics be used to replace text? Could graphic icons reduce the speed of menu selection? How could people create their own graphics with a minimum of training? Should graphics be displayed as they were downloaded, or downloaded in the background and revealed all at once? How complex should graphics be, considering that complexity increased download times?

Telidon required research on applications. How would Telidon be used? Were there business uses of Telidon? What kind of games would be most popular? Would people use transactional services? Would messaging be an important application or not? Were there medical applications, especially to remote areas? Could speech therapy be delivered by a minimally trained medical assistant using Telidon? Could people use Telidon to work from home? Was the technology sufficiently powerful to be used to deliver educational materials?

The Telidon field trials raised a host of questions. Which services were most popular? How were people affected by long download times? How long would people spend reading each page? And the big one, would people pay for the service?

Finally, there were broad social questions. How could equitable access be ensured? Would this reduce people's privacy? If anyone could publish, how could objectionable material be policed? If objectionable material was prohibited, how could free speech be guaranteed? Would people living with disabilities benefit from the technology? Would Canada be able to maintain its cultural sovereignty if the technology became popular in the United States? Who would own information? Would French culture be enhanced or eroded?

To answer all of these questions, the Division of Behavioural Research used a number of different methodologies, including laboratory studies with both non-functional prototypes and working systems, guidelines drawn from the literature on human factors, protocol analysis, questionnaire-based surveys, field trial analyses, interviews with experts, and meta-analyses of the collaborator's field trial reports.

3. BENEFITS OF THE BEHAVIOURAL RESEARCH

The benefits of this work are obvious.

First, we addressed many human factors problems that bedevil the web and the field of human-computer interaction to this day. Other laboratories have independently replicated and expanded on most of our results. The results have generalized to other technologies and other situations. This provided Canada with a body of human factors results that provided a useful foundation for later work on computers and the web.

Second, we stimulated interest in human factors research in Canada. A number of established Canadian experts in human factors will admit that their first attempts to conduct human factors research started with Telidon. Younger researchers may not realize that their mentors in graduate school had their roots in the Telidon project.

Third, we developed expertise in a wide range of methodologies. Because the Telidon project was so ambitious, beginning with the development of a specific technology, and ending with national scale field trials, there was ample opportunity to use most of the different kinds of human factors methodologies that had been developed by 1980, and even to invent a few of our own.

4. REASONS FOR THE DEMISE OF TELIDON

Analysts and researchers have proposed many alternative reasons for the ultimate failure of Telidon to catch on at the time. The most commonly mentioned problems included:

(1) the lack of useful information (available for retrieval during the field trials),
(2) the high cost of terminals (because demand never snowballed),
(3) information and services developed for by different field trial operators were not shared (each operator developed their own database of information),
(4) a technology in search of a market or application, and
(5) the regulatory environment which required all information provided by field trial operators had to be in both English and French (effectively doubling development costs).

However, three human factors problems have also been cited as major contributors to the expiration of Telidon:

(1) user friendliness of the system was low,
(2) Telidon graphics, while superior to the text-oriented videotex systems of other countries, was a poor substitute for full pictorial and video representation, and
(3) the hierarchical tree structure of menus used to retrieve target information and services was highly cumbersome, time-consuming, and even irritating.

From the ashes of disaster into a human factors boom: The legacy of large databases

Gitte Lindgaard

Carleton University, 1125 Colonel By Drive, Ottawa, Ontario K1S 5B6, Canada

ABSTRACT

The project started as an ergonomic review of customer service operators' jobs in the Australian national telecommunications carrier. The computer system had been purchased without adequate understanding of local technology- and user requirements. Consequently, system adaptations were ongoing for several years from before cutover commenced and well into the development of the fourth generation of the GUI front end.
The user interface evaluation identified serious usability problems with screen design, system workflow, and navigation, which severely compromised operators' performance. However, despite these findings and alarming symptoms of stress-related problems among operators, management refused to take ownership of the problems until we presented these in terms of losses to the organization based on simple conversions of data into dollars. The effect of these figures allowed the Human Factors (HF) Team to devise and roll out a comprehensive HF program in which Human Factors activities were integrated into all systems development procedures.

1. INTRODUCTION

The implementation of a new application into an existing network of systems supporting a diverse range of real-time core functions in a large organization remains a huge challenge. Even today after nearly 30 years of progressive automation in the telecommunications industry, it rarely follows a smooth, untroubled path. The larger the network of people, systems, and functions an application will support, the more difficult it is to predict what, when, where, and how problems are likely to occur. However, precisely because problems are to be expected in the immediate post-implementation phase, one would expect implementation to be guided by thorough, flexible plans and strategies for dealing with the process and the ways it will be monitored. For customer support systems, one would expect plans to include predictions of ways staff will be affected, training and support, the timeliness of these relative to implementation, and an estimation of the time it will take to bring performance up to pre-implementation levels.

In the case discussed here none of these precautions had been taken. The organizational chaos following implementation was exacerbated by quick-fix remedies that the organization was forced to put into place, which eventually gave way to a coherent long-term strategic approach. While this chaos was extremely traumatic for the staff who were affected by the system, it turned out to be the turning point for the HF Team who had been looking for an opportunity to move from a reactive into a pro-active role influencing the design and selection of systems.

1.1 The system and its implementation

The application was one of the critical core systems containing details of all services and equipment associated with every telephone number connected to the Australian national network. Details include customer name, address, type of equipment, number of lines and sockets on the premises, as well as information about the relative location of cables and exchanges. It also provides access to technicians', engineers', and cable assigners' work schedules, and it provides information to the white pages system, the billing system and many others that are an integral part of the customer service network.

Despite warnings in the literature of the need to take people and the impact on their jobs into account when implementing a new application (Furnas, 2000; Greif, 1991), no analysis of people's needs, their jobs or workflow had been undertaken prior to purchasing this application. It was bought from another telecommunications carrier under the assumption that similarity in services implies similarity in work practices. The implementation plan focused entirely on technical requirements while totally ignoring the human and organizational context in which the system would be operating. During system cutover, several major problems became evident. Technically, the system had not been designed to support several thousand operators as it was expected to do here. This slowed down the system response time to such a degree that it often took over six minutes to display the next screen in a transaction; system downtimes were also frequent and lengthy. Operators thus never knew whether it was down or merely slow. In an effort still to service customers well, operators often took transaction details by hand and entered them into the system later. Predictably, this led to a huge backlog of paperwork causing further service delays. Customers who were neglected in this way re-joined the calling queues in which a waiting time of up to 40 minutes was common. The fact that telecommunications services were a monopoly at the time is also important. 'Subscribers' were not regarded as 'customers', and the notion of 'customer service' was quite foreign to the organization. The operators' 'Statement of Duties and clearly emphasized technological over human needs, demanding adequate knowledge of the equipment but no specified personal qualities or interpersonal skills. It was simply not recognized that these operators were the 'front door' of the organization.

2. THE ERGONOMIC REVIEW

2.1 Operators' jobs

One major problem for operators was that the customer database was not always in place by the time the system was cut over. Existing customer records could thus often not be retrieved. Because of an unforgiving, cryptic interactive dialogue, users were uncertain as to whether the customer record was missing from the database, or whether the search string was incorrect. For example, an entry for someone named 'Bill Owen Smith' could have been entered as 'Smith, William', 'Smith, B', 'Smith,W', 'Smith,W.O', 'Smith,William Owen', and so on. The system could only recognize an exact match. In addition, the PABX used by the operators was unable to drop waiting calls even when the caller had hung up before being attended to. Together with a host of usability problems, these contributed to significant stress levels among the operators. Specifically, absenteeism increased by more than 200% during the nine-month study. Staff turnover rates escalated to 120%, there were two suicide attempts, three nervous breakdowns, and several operators went on long-term stress leave. Yet, senior management refused to take ownership of these problems or acknowledge a link of these to the introduction of the system.

2.2 Operator training

The training program provided for new operators was based on a very narrow range of tasks, limited to transactions supported by the application. During the study, however, it became obvious that job demands were quite complex and versatile, requiring skills that went far beyond the technical competence taught in training. For example, more than 40% of all calls taken during the study were inquiries, very few of which required interaction with the application. A tendency to use jargon and neglect giving feedback to a customer waiting online was noted: customers were left unattended for up to 35 minutes during which time the operator diligently pursued the query. The operators were unaware of differences in the perception of time between the waiting and the serving party. Information on the voluminous, complex organizational policies and procedures is available in several other computer systems, but the operators were not trained to use these, and nor were they encouraged to pass on calls to their supervisor. Instead, they chose to give incorrect information to customers rather than risk 'looking silly'. No training was offered on stress management, yet every operator encounters difficult, rude and offensive customers in their daily routine. The impact of these kinds of situations on operators' self esteem was not assessed, but it cannot have helped people who were already seriously intimidated by the system.

2.3 Usability evaluation

Operators performed 10 different transactions such as establishing or disconnecting telephone lines, changing the billing address, the phone number, or the status of an existing number (silent/public). Each transaction comprised a set of screens ranging from four to 13. Screens were presented in a fixed sequence, and the system would allow the next screen to be displayed only when the presently displayed screen had been completed and all the data had been recognized and accepted. Yet, acceptance of data by the system did not necessarily ensure that the entry was also correct, as several alternatives were often acceptable. Screens could only be traversed in one direction, so if a transaction was rejected, the user would have to start it over. Not surprisingly, operators spent between 30 and 60 minutes daily correcting data entry errors. If a transaction was accepted at the operator level, it could still be rejected at another level. Rejections at this level amounted to 30% of all transactions. Another group of people employed specifically for this purpose processed these. The application was found to have serious usability problems, including:

- Navigation: No visual distinction between different areas of the system to signal different transactions; no online help; no indication of where in a transaction the user currently is, or of how many screens are left in the transaction.
- Screen design and layout: Cluttered, disorganized; too many alignment points to allow effective visual scanning; no indication of which fields are mandatory and which are discretionary; no visual difference between captions and field entries; no indication of field length or the type of input required.
- Terminology: jargon and incomprehensible codes that are unrelated to user actions or transactions prevail.
- Feedback: the user is only told whether a record entered is acceptable at the end of a transaction. If it is not, the user is not told where the problem is or what should be done about it; error messages are uninformative and unspecific.
- Redundancies: the user is often required to enter certain items several times on different screens. For example, in one transaction, the customer's name is entered four times. Many items apparently serve maintenance staff rather than the user, but it is not clear which items the user can safely ignore.
- User control: once the user exits from a screen, the only way to get back to it is to loop through the entire transaction, as there is no provision for backtracking. Once an order has been submitted, there is no mechanism for retrieving or removing it.

The findings from the usability evaluation as well as from observations confirming the existence of these serious problems were presented to management together with a three-phase strategy for fixing the interface. However, even with this evidence management still refused to recognize the extent of the problems or take for solving them. It was not until we, in our frustration at having witnessed the problematic effects of the system on staff and customers, decided to calculate the cost of the problems to the organization. The formula took into consideration the time wasted by activities that should be unnecessary during transactions, the communication bill clocked up by operators calling someone else in the organization to obtain information that should be at their fingertips, training costs, the cost of others who were inconvenienced by system flaws, the cost of rectifying the 30% rejections.

An activity analysis (Wilson, 1995) was conducted to quantify the operator requirements during customer calls. A telephone receiver was plugged into the operator's PABX terminal to enable the researcher overhearing conversations with customers. Seated behind the operator, it was also possible to observe his/her actions and the computer responses. Each incoming call was timed from pick-up till hang-up, and every type of action was noted according to a previously generated action-taxonomy.

From a sample of several thousand calls, the average amount of time spent on each activity performed during a transaction was calculated, initially, to understand the composition of activities during typical calls. This analysis showed that operators spent roughly 22% of their time on activities other than interacting with customers or the computer, all of which could readily be eliminated or at least dramatically reduced. These observations showed that operators spent some 85 minutes, on average, every day on such extraneous activities (paperwork, walkabouts, phone calls). The cost associated with operator time lost by locating information, which should be at their fingertips amounted to some $5.6Mill. The telephone bill for internal calls came to $1.5Mill per annum, training costs to $10.6 Mill, and so on. The savings that could easily be made amounted to $30Mill. When these figures were presented to management, implementation of the suggested solution for this application commenced immediately. This was precisely the opportunity the Human Factors team had been hoping for to roll out its program.

A three-phase program was undertaken to bring the system up-to-date. First, the screens were cleaned up by reducing the alignment points to facilitate visual scanning; captions were presented in capital letters with field entries in mixed case; redundant information and unnecessary codes of no relevance to the operator were removed, and each screen was given a unique title related to the transaction. This enabled a reduction of screens, in one instance from 13 down to four to complete a transaction. In the second phase, the system was integrated with relevant parts of the billing system. In the third phase, a GUI interface was designed for this and the billing system simultaneously, relying on a user task approach.

3. THE HUMAN FACTORS STRATEGY

In order to integrate HF requirements into development processes such that they cannot be creatively interpreted or ignored, our experience suggests that four criteria must be satisfied: (1) know what HF can offer developers without retarding the development process; (2) devise and apply product styles and provide tools to assist adherence to this; (3) formalize organizational obligation to adhere to the style; (4) tie quality control processes to project funding.

Our goal was very clear: we wanted HF to be an accepted, integral part of product development. It was necessary first to identify those areas in which we could score a few 'quick wins' to produce tangible results. In management terms, this means revenue generation and/or cost savings. We systematically reviewed data from recent and current projects (user documentation, GUI, IVR) to identify time wastage's that could easily be turned into savings. User documentation was important to secure funding for the Human Factors project because it is easy to standardize with electronic templates, because costs are well documented, and predicted savings can very quickly be verified. It was also an easy sell because most senior managers had experienced difficulties as product users. IVR systems were another area in which it is equally easy to show the need for standardization across products and implement a corporate image through consistency in style.

3.1 Positioning Human Factors

Human Factors expertise had traditionally been sought mainly to put out fires that occurred so late in the development cycle that serious problems could not be rectified. The Human factors team was thus working in a reactive mode in which no one was obliged to request or use our expertise. Our efforts had little effect on the end products. In the first instance, we collected all the tools, techniques, methods and HF principles that we had successfully applied in a coherent toolkit, the Human Factors Kit (HFK), specifying when, where, and how each could be applied in the design and development cycle. The package included tools for capturing user requirements, performing task analyses, drafting Test & Evaluation Plans, performing evaluations, and so forth. Since the number of development projects by far exceeded the number of HF experts, suitable training courses, workshops and seminars were developed for staff with different responsibilities related directly or indirectly to product positioning, development and sales.

A corporate 'look and feel' was first specified for GUIs, user documentation, and voice-based interfaces. Some 38 individuals representing different areas of the organization were invited to the relevant Style Guide committees to draft a corporate style outlining the appearance and behaviour of future products. This was done to ensure wide ownership of the end product. Highly experienced style guide consultants with intimate knowledge of national and international standards as well as the relevant platform Style Guides led the committees. The process was advertised widely with an open invitation to review the first draft. One aim of this process was to promote ownership of the end product through wide participation, which, in turn, would be an effective way to recruit HF advocates in the organization. Reviewers' comments and suggestions were incorporated in the next version, which was printed and packaged with the development support tools into the HFK. It may seem wasteful to print materials with such a short life cycle. However, it was done to create a sense of excitement, to obtain wide internal publicity with the introduction of the HFK ensuring that everyone knew about it, and that it was strongly endorsed by senior management. A sample of generic voice- and GUI interfaces were then produced to show the recommended style in a tangible fashion and to support developers in their efforts to apply them.

Shortly after the introduction, the Style Guides were made available online to facilitate efficient updating and distribution of the documents. The Human Factors Team remains the focal point to which all user feedback is directed. However, the original steering committees 'owns' the Style Guides which are updated whenever there are enough comments to warrant a review meeting. Urgent issues were settled by the HF team immediately, but were presented to the committee for approval. To obtain documents, users register online, and once registered, they are automatically informed of document updates via email. Downloads are handled overnight to avoid overloading the email system. This communication process enables continuing iterations and avoids confusion about document versions.

At the same time as the HF strategy was rolled out, a project development review process was introduced to improve project overview and management. Within this process, all development projects are reviewed at certain points during the design and development cycle. Project funding is tied to this process such that permission to proceed is given for a single step at a time, until the next review point. The introduction of this process was very timely for translating the HF policy into definite requirements and deliverables that could readily be embedded within the project development methodology. The review process enabled us to specify HF goals and tools for achieving this at every point of the process, much as described recently by Vredenburg (2001), to ensure that products would meet users' needs and reach high usability standards. So, for example, it is mandatory for all projects to complete a User Needs Analysis (UNA) which comprises user profiles, task analyses and a 'high level' user interface to be designed in collaboration with users and other stakeholders very early in the development process. In the next phase, quantitative usability goals are specified on the basis of the outcomes of the UNA, and the usability goals are linked with usability dimensions, benchmark tasks, and so forth, within the Usability Test & Evaluation Plan (Lindgaard, 1994; 1995). One major advantage of the review process is that it is neither owned nor controlled or enforced by the Human Factors Team.

In the early days, specific training courses, workshops, seminars and presentations were prepared to assist project teams meeting the HF requirements specified in the review process because the HF team was just not big enough to be involved in every development project. Seminars introducing the concept of HF to senior management were conducted regularly in all major Australian cities, and training is offered to project teams as per need. The work of the HF team was re-structured allowing individual team members to work simultaneously with several large development projects. A user representative was assigned to work full-time in development projects exceeding $100,000. Special training was offered for these usability advocates who worked closely with the HF team member who assumed responsibility for the relevant project. More recently, large project teams tend to employ Human Factors specialists rather than user representatives, but contact with the Human Factors Team is maintained.

As the story shows, the success of the HF strategy was a mixture of opportune events, good luck, and above all, of the HF Team being prepared. The strategy was ready to roll the moment the opportunity arose, complete with cost estimates, facilitators and committees lined up, and HFK implementation time lines. Had we not been ready at that time, I am not convinced that the necessary funding and support would have been made available.

ACKNOWLEDGEMENTS

This research was sponsored by Telecom Australia while I was head of the Human Factors Team. I would like to thank my colleague Ian Milburn for his role in this study, Joe Havloujian, Josephine Chessari, and Liz Bednall from the Human Factors Team, and all the committees, reviewers, and facilitators who helped make the HFK happen.

REFERENCES

Furnas, G.W. (2001). Future design mindful of the MoRAS, **Human-Computer Interaction, 15 (2-3), 205-261.**
Greif, S. (1991). Organizational issues and task analysis. In B. Shackel & S.J. Richardson (Eds.), **Human factors for informatics usability** (pp. 247-266). Cambridge, UK: Cambridge University Press.
Lindgaard, G. (1993). Widening the usability horizon beyond HCI: A navel-gazing path to improved customer services. **Proceedings 14th. International Symposium Human Factors in Telecommunications**. Darmstadt, 29-38.
Lindgaard, G. (1994). **Usability testing and system evaluation: A guide for designing useful computer systems**. London, UK: Chapman & Hall.
Vredenburg, K. (2001). IBM Industry, **Interactions, March-April, 41-44.**

Political, Social and Commercial Problems of Rolling Out ISDN in the UK

Prof. John M Griffiths

Department of Electronic Engineering, Queen Mary, University of London, Mile End Road, London, E1 4NS, UK
j.m.griffiths@elec.qmw.ac.uk

ABSTRACT: *The ISDN was seen by technologists as a universal service which would sell itself by its obvious benefits. That this did not happen can be traced to the fact that ISDN design embraced the technology it was seeking to supersede rather than current users needs. Thus a whole raft of unnecessary complexity was added which was of no user benefit, slowed application development and raised regulatory difficulties.*

1. Introduction

Until the early 1960s telephone conversations were carried by systems in which the variations in air pressure due to the arriving sound waves at the microphone were converted to corresponding variations in the current flowing in a circuit. At the distant end these current variations were converted back into sound pressure variations by a small loudspeaker held close to the ear. The electric current and sound pressures varied in a similar manner and hence were *analogous* to each other and were therefore known as *analog systems*. The principle sounds simple but has many hidden difficulties as the current in the wires can be modified by internal effects in the equipment (distortion) and external effects (interference). Thus the current variations at the far end of the connection are different to those which were launched from the start of the connection and the added distortion and interference cannot be distinguished by the telephone network from the original signal, but is passed on to the recipient who then finds it difficult to understand what the distant speaker is saying. As distances increased the problem became severe. Local telephone calls were not a problem but the creation of intercontinental connections was more of an art than a science. In 1937 A H Reeves invented a system whereby, instead of transmitting the analog current in its original form, the magnitude of the current was measured many times a second and the numbers representing each measurement were transmitted instead. In the case of telephony the measurements were made 8000 times a second and the numbers used were in the range +2047 to -2047. The system is called Pulse Code Modulation (PCM) and any book on telecommunications will give the details. At the far end the reverse process converts the measurements back to a varying current. As might be imagined the electronics required is complex and it was not until the 1960s that the arrival of the transistor (and, subsequently, the integrated circuit) made the system practical.

The advantage of PCM is that it can be made immune to distortion. Providing the numbers are accurately transmitted from end to end the reconstituted signal is independent of the transmitted distance. In practical systems the numbers will have to be reconstituted along the path as they are transmitted, but, providing this is done before there is too much distortion and interference, no harm will be done. By way of demonstration the picture to the far

24   *24*   *24*

left illustrates an American Quarter Dollar. I have attempted to get its size correct but the various reproduction stages that this paper has been through mean that this will not be true. However I have measured its diameter in mm and this is noted below it. We then subject the image to distortion, just as would happen to the waveform of a current travelling along a wire. The information relating to the diameter from measurement in the image was not reliable initially but after distortion is now meaningless. Nevertheless the number is still legible, and can be copied here as 24 for onward transmission. The image of the Quarter is therefore less informative than the number. Thus PCM through its use of numbers alone was able to remove the barriers to long distance telecommunications and make intercontinental telephony the commonplace it is today. It should be mentioned here the limitation of PCM; in the above example the diameter of the Quarter is only specified to 1 mm whereas the diameter can be measured to the

1277

accuracy is of the measuring equipment. Similarly in transmitting a series of numbers representing the speech the accuracy is finite. The error introduced by rounding the measurement is known as "quantising distortion" but can be made as small as necessary and the range mentioned above (+2047 to -2047) has proved to be more than adequate for telephony and, in practice, a non-linear compression process is used to reduce the range to 256 values which can be represented and transmitted as 8 binary digits.

PCM is now universally employed between your local exchange and the distant local exchange to which your correspondent is attached. Used thus PCM has few human factor implications; the fact that the distortion is no longer dependent on the distance is hardly a problem. A small difficulty arises on very long connections; inevitably at the distant end of a connection some of your conversation will be echoed back to you. In the old days the transmission in each direction was so poor on such long connections that these echoes were reduced to negligible levels by the time they returned. With PCM there is no loss and distant echoes are a nuisance, but can be controlled.

2. The ISDN

The above application of PCM only extended between local exchanges. The conversion of speech to numerical or "digital" form was a major breakthrough for long distance transmission but it was argued in the 1970s that further benefits would accrue if the digital connection could be extended to the user. There were two main strings to the argument.

- there would be some small further improvement in speech quality through avoiding the analog connection from telephone to exchange. This could be linked with a more comprehensive signalling system so that the setting up of services such as conference calls, call transfer, and so on, could be simplified
- there was a demand for data services which could be met using the same equipment. Speech was being transmitted as 8 bit 'octets' or 'bytes' at the rate of 8000 octets/second corresponding to a bit rate of 64 kbit/s. This could be used as a data transmission system linking user terminals to mainframe computers. Remember that in the 1970s we were a long way from the commonplace of the computer at home.

The resulting service was to be known as the 'Integrated Services Digital Network' or ISDN for short. There were technical problems associated with providing this service[1] but they are not of concern here. I was employed by BT and involved in ISDN development from about 1976 until 1994 and therefore accept some responsibility for some of the decisions made. The title originated in 1971 in an international (CCITT) committee meeting. It must be remembered that in those days data modems operated at 1.2kbit/s and that the step to 64 kbit/s seemed so enormous that it was felt that it would solve all conceivable needs.

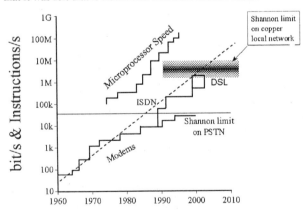

The figure (left) shows how the speed of connections provided by data modems have increased over the years, under the influence of user demands and technology development. In the 1970s the introduction of ISDN was expected to take place around 1980. If that had happened it would have represented a major step in telecommunications service. As things turned out the ISDN service can be seen more as part of natural evolution rather than a landmark change. In fact looking at the dotted trend line in the figure the ISDN was clearly late. It is my contention that the relegation of ISDN to an evolutionary step rather than a major innovation, was due to the lack of consideration of the user.

3. ISDN Design

As mentioned above, ISDN was seen as an opportunity to converge telephony and data into a single telecommunications access. The often used comparison was with the power supply. We have power sockets on the

1278

wall into which we can plug a wide range of appliances. The comparison was more persuasive in the UK where the 13 amp 230 volts supply interface means that the same sockets can be used for shavers, kettles and all the other portable items found around the house. The argument was that ISDN would simply offer, through a single socket type, all the telecommunications that a user would need for all types of service. In fact the final ISDN interface performance fell short in many ways; in particular it was neither simple nor was it adequate.

3.1 The Interface Adequacy Problem

As I mentioned above speech is encoded into 64 kbit/s channels and the switched network was configured on this basis. This does not mean that data rates are limited to 64 kbit/s as perfectly practical methods of aggregating several 64kbit/s channels are in use. The ISDN connection between the user and the local exchange uses the existing copper pair which had been provided for telephony, and in the studies made during the 1970s and 1980s it was concluded that two 64 kbit/s channels could be supported on the pair. I have no problem with this conclusion as the technology of the day meant that to achieve more was impractical. What was unfortunate was that this limit was enshrined in the interface. An ISDN interface supports two 64 kbit/s only and cannot be extended; no studies were made of potential user needs although the marketing fraternity in the mid 1980s made it clear that this was a serious problem. It was accepted that the initial offering would only carry two channels but future technology might well offer greater capacity. The advent of DSL on copper pairs demonstrates the truth of this and could offer 10 - 20 channels; optical fibres could offer thousands of channels.* If the ISDN interface had been extensible then its existence could have embraced the new technology.

A further inadequacy of the interface relates to the intercommunication between terminals. The interface was designed so that, just like powers sockets, multiple sockets could simply be connected in parallel and terminals plugged into them would operate automatically. However terminals could not intercommunicate. This was at a time when Local Area Networks (LANs) had become commonplace and their uptake had demonstrated that users needed this facility. Some proposals were made to provide interconnection but they were not incorporated. Of course it was possible to make a call from one terminal to another via the ISDN but this had financial implications and used up both channels immediately. The problem was that the interface was based on a sort of updated telephone connection which allowed a telephone and facsimile machine to co-exist, but had not surveyed what users would really like.

3.2 The Complexity Problem

Returning to the power supply analogy, the power supply interface is simple and people are widely capable of wiring a plug. Even the provision of house wiring is widely understood. The interface to the ISDN is complex. Most of the 280 pages of my book [1] on the ISDN are concerned with describing the interface and the protocols associated with it, and that description is by no means complete; the full specification is about 3 times as long. Partly this is justified but much of it provides functions which are of no interest or application to the user.

The difficulty lies in the term "Service" included in the title. To a user a service might be to speak to someone, to see someone or to exchange textual or graphic information with another user or a database. In the case of ISDN a service included emulating the various data services that were extant in the 1980s including various rates:
0.6 kbit/s V.6 and X.1, 1.2 kbit/s V.6, 2.4 kbit/s V.6 and X.1, 3.6 kbit/s V.6, 4.8 kbit/s V.6 and X.1, 7.2 bit/s V.6, 8 kbit/s I.460, 9.6 kbit/s V.6 and X.1, 14.4 kbit/s V.6, 16 kbit/s I.420, 32 kbit/s I.460, 48 kbit/s V.6 and X.1, 56 kbit/s V.6, 64 kbit/s X.1. In addition there were facilities to emulate the following modems; V.21, V.22, V.22 bis, V.23, V.26, V.26 bis, V.26 ter, V.27, V.27 bis, V.27 ter, V.29, V.32, V.35

You may argue that this is of little interest to you, and that is really the point; none of these facilities are of any interest to a user. I could have filled many pages of this paper with ISDN specifications which you as a user would have found of absolutely no interest whatsoever and even I, as a technologist, could not advance a reason why you should be interested, or why it would be useful! People want to transfer information in some way, generally as quickly as possible. They have no interest in how it was done before unless they wish to continue to use some existing investment, such as a piece of equipment, which is a subject we will come to later.

---

* I know that an alternative interface of 23 or 30 channels is specified but this has other limitations which really prevent its being a contender.

3.3 The Application Problem

Another difficulty with the ISDN was what it was going to be used for. To technologists like me, the fact they we could see many applications for it suggested that it would be widely used. We were therefore somewhat surprised that marketing people were confused as to its launch. Their problem was that ISDN did not meet the usual criteria in any way, as it was:

- Unfocused in Usage

  The marketing ideal is the 'killer application' which may be carefully targeted at a specific range of specific customers who can then be provided with appropriate support. However, the whole concept of the basic ISDN service and its interface is to offer a wide range of services from a wide range of products which can be 'mixed and matched' on the same basic access. It is therefore difficult to make a clear benefit statement particularly as salespeople have to spread their expertise over a wide range of applications. Experience has shown that early customers actually installed ISDN with only a single initial application in mind.

- Unfocused in Location

  One of the virtues of the ISDN is its ubiquity, arising from the fact that it is based on the telephone network. However, the result is a very thin initial spread of support from marketing, sales, operational and technical activities, and application support.

- Competing with Other Networks

  ISDN is an alternative to leased lines and other specialist networks. There is no benefit to the administration or users if the result is simply a transfer of traffic from one network to another. To be useful, a complementary role for ISDN needed to be established.

- Unclear on Customer Benefits

  A 64 kbit/s switched service was only of potential use to a customer until terminals and applications were actually available. But there was a chicken and egg problem. Terminal apparatus suppliers would not produce products until the network was available and customer demand was clear; administrations were reluctant to launch the network until customers (and terminals) were available.

The solution to these problems was to engineer a 'killer application' by taking a reasonably well founded product and pushing it in a form which could be understood by all. The product chosen in the UK was 'Leased Line Backup'. In the UK digital private circuits are widely used, but if they fail it could take considerable time to repair them. Equipment had been developed to automatically transfer a connection to the ISDN in the event of such a failure. This equipment was sold in a box with a postcard, the completion and posting of which would result in ISDN lines being installed at the right places. A further advantage of this choice of application is that the users would be company telecommunications managers who were to a degree 'technically literate' and hence able to see through some of the technology fog and appreciate its benefit to them.
Nowadays the same approach is still taken but the 'killer application' has moved on to internet access. Even with modems approaching 56 kbit/s rates, the fast call set-up of the ISDN and the possibility of aggregating the two channels to make 128 kbit/s gives a significant advantage

3.4 The Transition Problem

A further problem arose from the complexity of the ISDN service. Much play was made of how different and improved the ISDN service was compared with the ordinary telephone service. However this had an unforeseen consequence.
One of the problems of selling ISDN was that its interface was incompatible with the standard telephone, answering machine and group 3 fax machine. Thus a customer would have to replace all his existing equipment on moving to the ISDN, even if the old equipment were working perfectly. The obvious solution was to incorporate a digital-analog conversion unit in the ISDN customer termination. Unfortunately our emphasis of the degree to which ISDN was better and different to the conventional telephone had made its mark, and the UK regulator (OFTEL) indicated that incorporating a digital-analog conversion in the network box was unacceptable as such a conversion would be a

1280

value added service and thus must exist in a region which was open to competition. Had we emphasised the commonality of the services provided by the telephone and ISDN, just regarding the ISDN as an improvement on the other, then this is probably an interpretation which would not have arisen. After about 7 years OFTEL was prevailed upon to accept that a telephone interface could be presented via an ISDN connection. This is now sold by BT as 'Highway' products. However I am convinced that had the technical hyperbole of the ISDN been avoided, and the service to user been emphasised, then the difficulty would never have arisen, to the benefit of user and operator.

3.4 The Present and the Future

At present the ISDN service is rolling out well. The figure shows how the numbers are increasing fast. It is interesting to see that the rate of increase is similar across many countries and that the installed figures are essentially dependent on the time at which the service was launched, and that this depended on when a reasonable case could be made to people that the product was of value to them.

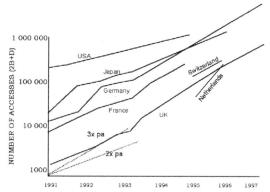

Some countries did not have a significant digital leased line network and for them the killer application was obvious and early.

As for the future, the 2 channel ISDN service cannot expand into a multi-channel service as its interface is set in stone. We are getting local access transmission systems, collectively known as xDSL, which would allow 10 - 20 channels but they will have to look elsewhere for interfacing to the user. Broadband ISDN has been discussed for many years, based on Asynchronous Transfer Mode (ATM) technology but Internetworking Protocol (IP) has much to commend it, although it will probably have to adopt some of the ATM features to allow the expansion to a global high-speed service.

4. Conclusions

The ISDN has essentially emerged as a technology which was designed to consolidate the telecommunications technologies which went before it rather than the future needs of the user. Hence it was encumbered by many unnecessary features which in no way contribute to its usefulness, but make it difficult to use. On the other hand features which customers would have found useful such as flexibility and extensibility were not available. It is easy to say that the needs of the user should be taken into account at all stages but this is easier said than done. As a technologist I recall many times seeking guidance from academic human factors and marketing people on what was needed but with little response. It is much easier to analyse with hindsight than to synthesise at the time. Perhaps the most important thing is to keep it simple. The several hundred pages of ISDN specification almost inevitably must contain people-unfriendly stuff.
How did the camel-that-should-have-been-horse arise? One major problem is the CCITT negotiation realities of the time. In practice consensus was required in committees and the easiest way of getting that was to accept every proposal that does not technically conflict with anything else. The early history of IP in no way involved such committees. The sheer simplicity of IP has little impact on people, and applications can be built on this neutral foundation as required.

Reference

[1] J M Griffiths "ISDN Explained: Worldwide Network and Applications Technology" 3rd Edition, published by John Wiley and Sons Ltd. ISBN 0 471 97905 8

Next Big Things that Were Not

Thomas E. Whalen, Ph.D.

Communications Research Centre, 3701 Carling Avenue, Ottawa, Canada, K2H 8S2
Thom.Whalen@CRC.ca

ABSTRACT

Many technologies that were widely touted as the Next Big Thing turned out to be less than spectacular successes. These projects generally start with a clear vision, are centrally-managed, well-funded, and highly-publicized. Because they promise so much and raise unrealistically high expectations, most of them are unable to satisfy the expectations that they created. Typically, one of the problems endemic to these projects is that they intermix the technology and the application of the technology. It is important to keep these concepts distinct because technologies invariably become obsolete sooner or later, but applications may remain important indefinitely.

Human interface design is prominent in these projects because the success of the project can be limited either by a failure to satisfy an important need or by a badly-designed user interface. Therefore, these projects can benefit from the involvement of human factors experts. In return, human interface experts can benefit from their involvement in these projects if they recognize their role, do not promise an unrealistic degree of success, and use the opportunity to increase their own expertise. Their expertise will be valuable for longer if they work at the level of the application than at the level of the technology.

1. NEXT BIG THINGS (NBTs)

The past few decades have seen some amazing technological successes. Thirty years ago, computing meant lumbering mainframes with insatiable appetites for resources. Today the average person owns and uses desktop, laptop, and now, palm-sized computers. Twenty years ago, people had to conduct their personal business at their local bank branch during "banking hours" between 10 AM and 3 PM. Today, money can be whisked instantly from their account with the touch of a few buttons at any hour of the day from anywhere in the world. Ten years ago, digital communication networks were the private demesne of elite researchers. Today, elementary school students use the Internet for their homework assignments.

Yet, there have been numerous other technologies that were promised to be the Next Big Thing (NBT) that would revolutionize our lives; technologies that do not spring to mind so readily when we think about important technological advances. Remember Teletext? People were going to get access to news, weather, and sports by decoding digital information stored in the vertical-blanking interval of their analog television signal. Remember Fifth Generation Computing? Parallel computers that executed declarative programming languages natively were going to revolutionize everything computational. Remember Logo? Children would routinely learn to think recursively in elementary school by programming electronic turtles. Remember video-on-demand? There would be no reason to rent videotapes because people would be able to select movies from giant libraries stored on centralized video servers. Remember the information superhighway? A unified digital network would supercede the Internet. Remember information push? There would be no need to surf the Web because information would be delivered automatically to your desktop computer.

1.1. BIG VISION

All of these NBTs, and many others not listed here, have similar characteristics[1]. First, all began with clear, well-articulated vision. White papers described the technology, its expected impact, and the development process that would be required to bring them to completion.

---

[1] The critical reader will note that this paper offers many bald-faced assertions without proof of their truth. The experienced reader will be able to offer a counter-example from his or her own experience that will contradict one assertion or another. The only defense offered to these criticisms is that the assertions offered in this paper should be considered as general hypotheses that will not necessarily apply in every instance. It is only hoped that these assertions will "ring true" as being consistent with most readers' experiences in most cases.

1.2. BIG MONEY

Second, they were all big projects. Governments and industry spent hundreds of millions of dollars on most of these projects; even back in the seventies when a hundred million dollars was still a lot of money. And they involved a lot of people. All money goes to someone somewhere. It is impossible to spend a large amount of money without touching a great number of people.

1.3. BIG EXCITEMENT

Third, they were all exciting projects. They were announced with loud fanfares and widely reported in the popular media. Informed members of the public had heard about them. Popular books that described the technologies and their expected impacts were sold in shopping malls to the general public. Experts scrambled to be affiliated with the projects; or scrambled to become famous nay-sayers.

1.4. STRONG LEADERSHIP

Fourth, most of them had strong, centralized leadership. An organization – sometimes a single company, sometimes a government department, sometimes a consortium formed specifically for the project – was identified as the leader of the project. That organization had control over the design of the technology; was the official spokesperson for the media; and controlled the largest single source of funding. This control was not necessarily absolute. Other organizations might have their own sources of funding; be willing to speak to the media; or design non-compliant technologies. But invariably, the lead organization had a much greater impact in each of these areas than all the others combined. Within the lead organization, one person typically led the project. That was usually not the person who had invented the technology, nor even the person who first conceptualized the project. Typically, it was a charismatic personality who committed him or herself to the vision, attracted resources, managed the project at the highest levels, and championed the project to the media.

1.5. LIMITED SUCCESS

Fifth, none of these projects had the impact that was predicted at the outset. Because each began with a clear vision, each had clearly-defined objectives; generally to be widely adopted in the marketplace within a few years. None of them even came close. Though most of them were not abandoned completely, they were usually removed from centre stage within their lead organization; and lost the attention of senior management. Most often, a small core group of people who were loath to give up continued to dedicate time to the technology in hopes of making some kind of breakthrough that would revitalize the project.

2. THE BIGGER THEY ARE, THE HARDER THEY FALL

So many NBTs follow similar paths in their development that one is compelled to find a reason for the pattern. The underlying reason is obvious. It is human nature to want to be part of an important, successful project and to not want to be associated with a failure. Furthermore, the higher a person is placed in the organizational hierarchy, the more he or she will be driven to want to be associated with big, successful projects. If an idea looks like it will be successful when it is first proposed, senior management will be quick to buy into it. These people are in a position to re-deploy staff and re-direct resources. Four groups emerge within the organization: the core development group that first proposed the project and believes in it; senior managers of the project who promote it wildly, beyond the expectations of the core development group; staff who have been re-purposed, and who may not be as committed as they appear; and a large group of outsiders at all levels in the organization who are not part of the project and who have lost resources to it.

Once these four groups are in place, the evolution of the project is inevitable. The core development group and senior management reinforce each other. The core development group is not business-oriented and believes the business case that senior management is making. This project is bigger and more important than they ever realized and they are at the centre of it. On the other side, senior management is not paying sufficiently close attention to the limitations of the technology. They are led to believe that the core development group will solve any technical limitations that arise. The re-deployed staff members have little choice but to believe the claims of senior management and the core development group and quickly learn to recite the official catechism. The outsiders mutter among themselves, but know better than to complain publicly.

Because the core development group already has basic technology in place for demonstrations, initial success comes easily. But senior management expectations are unrealistic. Some limitations are encountered that cannot be overcome. Technology does not scale upward, or is too expensive for the marketplace, or does not address user needs, or requires the cooperation of other industry sectors that have not bought in, or requires a critical mass of users that is never reached. Usually, there is more than one limiting problem.

When it becomes obvious that expectations are not going to be fulfilled, the disenfranchised outsiders raise their voices. Uncommitted staff members realize that they are going to be associated with a failure if they do not quickly abandon the project. In public, senior managers boldly declare the project a success, regardless of the perceptions of the uninformed; in private, they blame the core development group and scramble to find a safe haven. They cannot afford to be caught without a chair when the music stops. The core development group blames everyone for their failure to understand the technology. Those with important skills are re-deployed back to the core business of the organization; often disguising and re-purposing the NBT technology rather than abandoning it.

A year later, few people are willing to admit in public that they ever worked on the NBT.

3. THE BENEFITS OF NEXT BIG THINGS THAT WERE NOT

The eagerness of people to abandon an NBT that did not fulfill expectations masks the importance of these projects. Let us ask a simple question: Where would we be without the NBTs of decades past? Would we have broadband to the home if we had never developed ISDN? Would we have intelligent agents if we had never developed expert systems? Would we have the Web if we had never developed videotex? Would we have desktop video if we had never developed videoconferencing rooms? Maybe not.

NBTs provide at least four important benefits, even when they do not meet their expectations.

3.1. NBTs Provide Intermediate Technologies

First, they provide intermediate technologies. New technologies are seldom cut from whole cloth. Most often, they are a variation on an earlier technology. NBTs usually require the development of several technologies. Though people often only look at an NBT as the application of a single core technology, typically there are a number of other additional technologies that have to be developed. Human interface technologies are often prominent among these supporting technologies. If an organization spends a hundred million dollars developing new technologies, it is almost guaranteed that that they will be the stepping-stones to the next generation.

3.2. NBTs Provide Expertise

Second, NBTs provide a large pool of expertise for future projects. Very often, close examination reveals that the experts who are developing current leading edge products apprenticed on a recent NBT. The demand for resources created by NBTs stimulates a surge in hiring; the need for new skills created by NBTs stimulates a surge in training; and the collapse of the NBT releases these newly trained experts back into the workforce.

3.3. NBTs Provide Information about Needs

Third, NBTs provide valuable information about the need for new products. Often NBTs collapse from a failure of the technology to meet user needs. In these cases, organizations see that the need exists and get a good estimate of its magnitude. At the end of the project, they know what technology would be required to meet the need, and how inexpensive that technology would have to be. Other times, NBTs collapse because there was no user need at all. In that case, management has also learned something valuable.

3.4. NBTs Stimulate Inter-sector Collaboration

Fourth, the high public profiles of NBTs stimulate different industry sectors to collaborate. Inter-sector collaboration is especially obvious when content-based industries, like the music and movie industries, are stimulated to collaborate with high-technology industries. But close examination will reveal much less obvious collaborations, such as manufacturing and resource industries that have entered into collaborations with content or technology-based industries in response to an NBT.

4. THE ANATOMY OF A NEXT BIG THING

An NBT can be viewed as an attempt to develop a ubiquitous application for a new technology. While simplistic, this conceptualization highlights the two critical aspects of an NBT: the technology and the application. Basically, the application is what people will be doing and the technology is the way people will do it. Telemedicine is an application; doctors will practice medicine at a distance. Satellites, local area networks, and the Internet are the technologies that have been used at various times to enable the telemedicine application. Similarly, hypertext is an application and videotex, teletext, and the World Wide Web are hypertext technologies. Management information systems are applications and RDBMS and MIS software are the technologies.

4.1. TECHNOLOGIES ARE DIFFERENT THAN APPLICATIONS

Usually, the application and technology are confused in everyone's mind; confusion that is often exacerbated by managers who give both the application and the technology the same name. When terms like, *videotex, MIS, network computer, HDTV*, and *electronic cash* are used by both experts and the mass media alike, they sometimes refer to the application and sometimes to the technology.

This is a dangerous confusion because the application and the technology are different things that have different characteristics and different weaknesses.

4.2. TECHNOLOGIES BECOME OBSOLETE

The biggest difference between an application and a technology is that technologies always become obsolete. A new technology that displaces an old technology is called a *disruptive technology* (Christiansen, 1997). Some technologies, like paper books, are very resistant to disruptive technologies because they are simple, cheap, and powerful. Other technologies, like eight-track audiotape, are very vulnerable because a newer technology, like the cassette audiotape, is superior in almost every way. But even the most venerable technologies will become obsolete eventually, maybe not disappearing completely, but, like the paper scroll, being relegated to a minor role in society.

In contrast, applications never become obsolete. Applications, like telemedicine, remain as a goal even after project after project fails to fulfil the promise because the technologies used are too expensive and insufficiently powerful to satisfy the need. Thus, there is no such thing as a *disruptive application*. This provides a way to distinguish between an application and a technology. Technologies are vulnerable to obsolesce and applications are not.

4.3. APPLICATIONS MAY BE UNNEEDED

Applications can be unneeded. For example, in the late '90s, several organizations produced *information push* technologies that were intended to replace Web surfing. Rather than having to search the Web every day for the same kind of content, like weather or sports scores, that information would be sent to a person's computer automatically so that current information would always be available. The technology that was developed worked well, but the project fizzled. It is clear in hindsight that, even though this appeared to be a good application, there was insufficient need to sustain it.

In contrast, technologies are never unneeded. Rather, any current technology can be re-targeted toward another application. Though there was insufficient need to support the use of information push technologies for delivering current event information to every computer screen, delivering software updates automatically may well be a needed application for this technology.

This model suggests that applications are persistent and technologies are not. This often leads to a misperception that organizations that pursue the same application that has failed before are failing to learn from history. Everyone has heard the nay-sayers criticize companies for attempting to develop a picturephone one more time, even though every picturephone project ever attempted before has failed. But these companies are not flogging a dead horse; rather, they are creating a new project by attempting to apply a new technology to an application that is still needed. If they fail, it is either because not enough information about the parameters of the need was available from the previous attempt, or because the technology was mis-represented as finally being able to fit within the parameters of the need.

5. THE ROLE OF HUMAN INTERFACE RESEARCH AND DESIGN

Because NBTs always include an application, there is always a human interface; and, because the human interface is prominent, the human interface design is often the most prominent single supporting technology.

NBTs fail to meet over-ambitious expectations because they encounter at least one limitation, be it technological, economic, needs-based, or something else. Because the human interface is prominent, when an NBT fails to meet expectations, the question about the possible failure of the human interface always arises. In some cases, such as the picturephone developed by Bell Laboratories in the 1960s and '70s, there were clear human factors problems. Although the human factors problems were not the only, or even the

1285

major, limitation on the success of the technology, they were prominent enough to spur many organizations to pay far closer attention to human factors design than they had done previously – to the long-term benefit of the organizations and their customers.

Conversely, because the end of an NBT is precipitated by an encounter with some kind of fundamental limitation, even the best human interface design cannot ensure the success of the NBT. No single aspect of the NBT can be executed well enough to make the NBT overcome the limiting factors that it will encounter in other areas.

A less prominent, but more important and more pervasive use of human factors in NBTs is in needs analysis. Many NBTs would never be launched if organizations conducted an accurate analysis of the need for the technology or application before they made their first announcement to the media. And at the other end, organizations would gain much more benefit from their NBTs if they collected a complete body of data about the usage of the NBT during the project and built a proper model of user needs afterwards.

However, these needs analyses are seldom done. NBTs almost never begin with a proper needs analysis because both the core development group and senior management want a big, exciting project; neither wants to be told that their idea will not fly just as they are taking off. There should never be an NBT in which lack of user need was the critical limitation if a proper needs analysis was conducted beforehand because the project should never be launched.

Historically, once NBTs were launched, good usage data was seldom conducted during the project because everyone was working too hard to overcome limitations as they were encountered. This is less true in recent projects because there is increasing awareness that usage data has long-term value. But even when good usage data is collected, effort is seldom devoted to constructing an accurate model of user demand; because, once the NBT has been abandoned, there are no human factors people left to work on it.

6.   GENERAL LESSONS FOR HUMAN INTERFACE RESEARCH AND DESIGN

What then, should human interface experts do about NBTs?

First, be aware of the characteristics of NBTs so that you know if you are becoming involved in one, as opposed to simply being part of the core business of your employer.

Second, avoid involvement in an NBT? Certainly not! NBTs are where the big action is. They are where resources are spent freely and where senior management is determined to use those resources to push back the leading edge of technology.

Third, do not promise that you will make the NBT successful. No single technology, not even one as important as human interface design, makes an NBT successful. Rather, remember that you are working to ensure that the human interface is not the limitation that will cause the NBT to fail to meet expectations.

Fourth, use the opportunity to develop your own expertise. For human interface experts, the most important expertise that you can develop is to gain experience with new human interface design and evaluation methodologies. You will never have a better opportunity to increase your expertise; and, when the NBT collapses, the value of your expertise to the organization will determine if they reassign you to another project or if another organization will be eager to hire you away.

Fifth, if you want to have a lasting impact on the state of the art, work at the level of the application. Remember that technologies become obsolete, but applications do not. Any substantive expertise that you develop at the level of the application will be important for longer than expertise at the level of the technology.

Sixth, plan to survive the collapse of the NBT. Some technologies do fulfil expectations and become the next big thing, but this is very rare. Your career plan should not depend on the success of the NBT. Make certain that the work that you do can be described on your resume without explicit reference to the NBT. Watch for opportunities to leave the NBT project before the final collapse.

Seventh, have fun. Do not lose sleep over the likely collapse of the NBT because you are in no position to save it from itself.

REFERENCE

Christensen, Clayton M., The Innovator's Dilemma: When New Technologies Cause Great Firms to Fail, Harvard Business School Press, 1997.

HUMAN-COMPUTER INTERACTION SYSTEM DESIGN

Dr. J.E.L. Hollis, Mr. H.W. Choo#, Mr. S. Takel* and Mr. J. Morrison*

RMCS, Shrivenham, UK. *DERA, Bincleaves. #Choo is now at the MOD, Singapore.

ABSTRACT

A user-based approach to HCI system design is described. A good "requirements specification" for HCI design is difficult to achieve without effective user input. This input is best if the user is closely involved at an early stage, preferably the design initiation stage, of the HCI system. A good methodology is essential in order to extract the information for, and reaction to, a design. By using a mixture of soft system techniques, a general methodology for HCI system design is achieved. A particular HCI design, a multi-sensor unmanned underwater vehicle platform sponsored by the Ministry of Defence, illustrates the user-based design process along with the issues of VV & A.

INTRODUCTION

The work described here was sponsored by the Maritime Mine Weapons & Countermeasures Department (MMWCD), Vehicle Technology Group, at DERA, Bincleaves. A variety of sensors for the detection and identification of under water mines is currently being researched by MMWCD. Future, Unmanned Underwater Vehicles (UUV's) are to be fitted with a suite of such sensors to carry out underwater surveillance missions. The UUV is being designed for real-time communication with an operator on a sea surface mother platform (near term), and also for complete autonomous operation (long term). A high degree of data fusion is required to present key information to the user from the multiple information streams from the sensor suite. Both sensor and vehicle control systems are required for the platform. The task was to develop a user-approved, man-machine-interface (MMI) for the interchange of the required information between console and rig for the UUV system. The ergonomics of MMI was important in order to ensure relevant information (and only the relevant information) reached the operator and reached him at the right time. The success of missions is often highly dependent on such information management.

The aim was a HCI Requirements Specification & a Prototype UI for a multi-sensor UUV with full (VV&A).

DESIGN METHODOLOGY

A good methodology and suitable techniques are essential to achieve an effective, efficient soft system design. Many techniques have been put forward over the years for HCI design [1-16]. Here Soft Systems Methodology (SSM) [13] and the RESPECT[1] Technique [14] have been integrated to produce an effective design method. This method is exampled in this paper by its application to a new UUV system for MMWCD.

RESPECT is a technique for assisting soft system design by providing a framework under which to work.

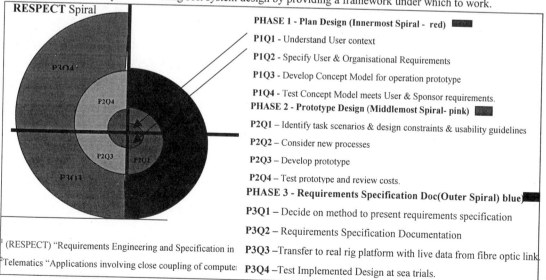

RESPECT Spiral

PHASE 1 - Plan Design (Innermost Spiral - red)

P1Q1 - Understand User context

P1Q2 - Specify User & Organisational Requirements

P1Q3 - Develop Concept Model for operation prototype

P1Q4 - Test Concept Model meets User & Sponsor requirements.

PHASE 2 - Prototype Design (Middlemost Spiral- pink)

P2Q1 – Identify task scenarios & design constraints & usability guidelines

P2Q2 – Consider new processes

P2Q3 – Develop prototype

P2Q4 – Test prototype and review costs.

PHASE 3 - Requirements Specification Doc(Outer Spiral) blue

P3Q1 – Decide on method to present requirements specification

P3Q2 – Requirements Specification Documentation

P3Q3 –Transfer to real rig platform with live data from fibre optic link.

P3Q4 –Test Implemented Design at sea trials.

[1] (RESPECT) "Requirements Engineering and Specification in
[2]Telematics "Applications involving close coupling of compute

The European Usability Support Centres (EUSC) under the European Commission is promoting RESPECT adoption. The objective of RESPECT is to provide a technique for user-centred requirements specification via a systematic framework for extracting and specifying the real user requirements from sponsors and end users. The system design is greatly enhanced by having end-user's in the loop at an early stage and involving them throughout the design.

The framework defined in RESPECT is a spiral process based upon the design cycle as detailed in the draft user-centred design standard ISO 13407 (1997b). RESPECT provides a broad framework for Requirements Engineering that aids meeting the users needs more closely, achieving quality and meeting overall objectives. Neither functionality nor usability is given priority. They are subservient to the objective of providing a system that enables the designer to meet the user goals in the real world. RESPECT focuses on the importance of obtaining a complete understanding of users' needs, and validating emerging requirements against potential real world scenarios of usage.

The RESPECT framework is shown diagrammatically in Figure 1. In this a spiral framework acts through 4 activity quadrants, namely Context, Requirements, Design and Test. The spiral, cycles around these 4 activities, in 3 phases. Namely, Phase 1, the Planning of Design via the User-based Context and Design Concept phase, Phase 2, the Prototype Design and user Testing phase, and Phase 3, the User Requirements Formal Specification & Full Documentation phase.

SSM allows the capture of a systemic view of the problem situation using Rich Pictures,

- The mission statement and definition of system's objectives using Root Definition (RD),
- The capture and organisation of a systemic account of purposeful system activities (i.e. what activities are required & their relationships) required to meet the system's objective using a task Conceptual Model CM),
- The definition of performance indicators that will lead to the identification of actions for improvement using Measurement of Performance (MOP) parameters,
- A constant reminder of the attributes of the project with the CATWOE definition, and
- A constant reminder of what the system is about via the PQR (What, How, Why) Structure basis.

The framework of SSM enables systems thinking, which allows continuous learning and reflection of the problem situation (i.e. the project and its status) through an iterative cycle of real and conceptual world comparison. This allows the use of SSM in the front-end analysis and throughout the ensuing life cycle of a product.

The purpose of SSM is not on finding a solution to a specified problem, but rather, it is to understand the situation in which a perceived problem is thought to lie. The situation is understood by engaging the stakeholder in an iterative learning cycle using system concepts to reflect and elaborate perceptions of the real world, and to identify actions to improve situations. It can be viewed as an inquiry system based on a Human Activity System, using it to apply to human-computer systems engineering, focusing on purpose, people, constraints and the 'world view' of a human-machine system. This enables the designer to understand the organisation and the context of use of the system that is being developed

PHASE 1 – Plan

The aim of Phase 1 is to capture the user and sponsor views and to identify and analyse the task processes necessary. This is the first cycle phase (P1) round the spiral RESPECT framework. The activities within each quadrant (Q1-4) were:

P1.Q1 User Context
- Summarise project
- Identify stakeholders and Users
- Specify User desired system characteristics
- Describe technical environment
- Describe physical environment
- Describe social and organisational environment.

P1.Q2 Capture Goals & Tasks
- Identify User goals & tasks
- Review current processes in Users environment
- Review other similar systems & products

P1.Q3 Concept Design
- Produce Design ideas
- Produce a Concept Model.

P1.Q4 Test / Review
- Undertake expert (i.e. User) review.
- Collate revised ideas & suggestions.

HCI Requirements Specification and UI Prototype Design for the DERA new UUV system

The HCI Requirements Specification and UI design for the DERA UUV system were developed in accordance with the spiral framework defined in RESPECT using SSM.

P1Q1 - Phase 1, Context

In this quadrant, the project background information and the system's desired functionality were gathered.

A user and stakeholder analysis was then performed in order to identify the range of different user groups and their characteristics. A description of the required environments for each user group was produced. The needs of each user group were captured and highlighted.

The information leading to the required products was obtained through interviewing the Project Sponsor and the Operational Users teams. A semi-structured interviewing technique was adopted for this purpose. The advantage of this technique was the combination of both structured and unstructured (flexible) interviewing techniques [9-Preece], allowing the elicitation of specific information and at the same time allowing the exploration of hidden issues under a free information exchange.

Before the conduct of (any) interview, an initial briefing was given to set the context of the interview. A general introduction to background issues and main goals to be achieved out of the interview were presented. This was to ensure that the interviewees understand the objectives of the interview.

An important aspect of this phase is the importance of projecting the interviewer's objective such that the audience (i.e. the interviewees) are convinced. This allows confidence to be built-up between the parties and thus encourages focused and barrier free discussion.

P1Q2 - Phase 1, Requirement

This stage identified User's goals and the features required of the system.

The goals for the new system were identified and the current processes were reviewed. Similar systems or products on the market that are relevant to this development were reviewed. The possible functions or features that should be included or excluded from the new system were captured. Also ideas for overcoming identified problems were detailed.

The information for the above products was obtained through interviewing the Project Sponsor and the Operational Users. The users involved in this project were from the School of Maritime Operation, HMS Dryad. The operational users provided access to the current system, i.e. Naval Autonomous Intelligent Console (NAUTIC) and to the current simulators used for training operators. This allowed a good understanding of the current system. The operations, tasks and procedures of the current system were further understood by running through the operation of the simulator system. The benefit of using the training simulator was that it allowed the repeated execution of any scenario until the operation and procedure was fully understood by the designer. In the process of researching and understanding the current system, SSM was used as an inquiry system. A primary task conceptual model (CM) with its Roots Definition (RD) and Measures Of Performance (MOP) were developed to capture the current operation of the system. The CM was developed in conjunction with the users and clarification of any ambiguous information was thus obtained immediately. Having understood the current system, the differences between the activities of the current system and that required of the new system were identified. The goals and features for the new system were thus developed. Mutual confidence was gained by talking the same language as the users (learning their terminology). Further enhanced by demonstrating professionalism through using a methodology and preparing thoroughly for the sessions. SSM was also used during the discussions with the users. SSM's framework allowed structuring of the discussions. This allowed the discussion to stay in focus and at the same time explore hidden issues.

P1Q3 - Phase 1, Conceptual Model for System

The perceived HCI was a computer-based, graphical driven, highly intelligent, multi-screen display and control interface that could provide an effective UI for the user to operate the UUV. It was to cover the extensive functionality that the multi-sensor, data fusion, controlled of the UUV platform being developed. The user console was envisaged as a set of easily accessible screens, which would allow sensor control, data display and intelligent data compilation for assisting user assimilation of intelligence gathered by the sensors as well as providing direct control of the UUV rig. The multi-screens should allow relevant displays at all times, with ultimate User control of both the sensors and the platform motion during mission conduct. Further multi-screen designs allow mission planning through to debriefs.

An Object-Oriented (OO) Paradigm [11 - IBM] was used for the conceptual design. The necessary object views for the top-level objects were identified. Paper prototypes were developed to visualise how the objects would look on screens and to test User reaction. Ideas and design concepts that were identified in the above activities were also iterative reflected in the paper prototypes. Analogies with the current system were used extensively in the development of th paper facades. This allowed the potential users to quickly grasp the design and understand implications. During a pap design of Phase 1, the top-level objects and top-level user tasks for the new system were identified.

A Hierarchical Task Analysis (HTA) was carried out at this stage and an Interface State Transition Diagram (ISTD) wa developed to show how the UI would handle the task interaction captured in HTA.

The synergetic combination of paper prototype with the HTA and ISTD, demonstrates their use in a UI development environment. These combined techniques provided an effective representation of the UI for the user and sponsor.

P1Q4 - Phase 1, Test of the conceptual model of UI

The design ideas and concept model were tested by an "experts" (Users) review of the User goals. The Users ra scenarios against the paper prototypes supported by ISTD and HTA. Ideas and concepts that were feasible were kept th others were rejected. The model was also evaluated against the original Sponsor and Designer's ideas. This is a goo example of how the RESPECT method can work in practice.

PHASE 2 – UI Prototyping and User-based Testing

This is the second cycle around the spiral RESPECT framework. The aim of Phase 2 was to identify new processes ar functions for the UI of the UUV system, as well as developing a prototype system with a simulation produced by usin Visual Basic. An assessment was made to ensure that the design ideas and concepts from Phase 1 formed a sound bas for development of a VB UI prototype that simulated the real rig. User confidence in the new UI concepts was built up

P2Q1 - Phase 2, Context

Task scenarios that would be used as a way of testing the system design were identified in the first quadrant of th phase. Each scenario represented a user goal within a particular context of use and at least one scenario was produc for each user goal. General usability goals to be achieved for each scenario were identified.

The scenarios that represented common or important task situations were first identified. The generation of scenari encouraged the user to consider all possibilities of the user tasks and at the same time helped to identify usability target Scenarios were generated with the help of the paper prototypes and early UI screen façade prototypes. As use interacted with the UI screen prototypes, users were prompted to consider all the possible scenarios that could happe The benefit of using prototypes (even simulated prototypes) to generate scenarios was that it gave something physic for the user to look at rather than relying on pure imagination. Furthermore, the prototype was also verified along wi the identified scenarios at the same time. It also allowed gathering of information that was initially left out in Phase The general usability goals and guidelines for the UI were identified from the list of identified scenarios. This w achieved by identifying key design and performance characteristics of the UI that the users deemed essential for th successful conduct of a particular scenario. The usability goals produced a checklist for the UI should conform to.

P2Q2 - Phase 2, Detailing Requirements

For each scenario, a set of steps representing each interactive process was developed. At the same time, a list potential functions and features to support these processes was also documented.

The needs so defined were expanded. A set of steps was listed to demonstrate how the system should work. Along wi each step, potential functions and features required from the system were identified. These steps were also used update the HTA developed in Phase 1. HTA allowed capturing of tasks that needed to be performed by the user in a to down hierarchical and logical sequence. This ensured that all top-level tasks and their sub-tasks were considered. Th interaction between the Prototype Façade, HTA, ISTD and System Model was so developed.

P2Q3 - Phase 2, Prototype UI Design

The simulated interactive façade was developed using Visual Basic. Backdrops with active windows, updated wi stored rig sensor data (simulating the real fibre optic link) were used. Control buttons and display windows enabled th sensor interaction and rig control. The design of the façades was based on the paper prototypes developed in Phase 1.

The prototype was developed on a Pentium 200Hz MMX, 800x600 resolution, and laptop. Developing the prototype the laptop allowed the prototype to be easily brought to the user's site (HMS Dryad) and to the sponsor's site (DER Weymouth). The final design screen size should be based on the display sizing guidelines specified in Def-Std 00-25.

P2Q4 - Phase 2, Testing

Walkthrough techniques were used to test the UI prototype. Iterations of the UI design & walkthroughs were done.

P3 - Phase 3 – User Requirements Documentation and system implementation and User guide

The aim of Phase 3 was to consolidate all information gathered from the previous phases in order that a definitive user-oriented HCI "Requirements Specification" could be produced. The final multi-screen HCI was serviced by the fibre optic data link from the sensors that supply data to the controls.

Item	Product	Description
3.1	General system characteristics	Specified the system characteristics identified during concept stage.
3.2	Organisation structure	Specified the intended organisation structure of the system.
3.3	Task scenarios and interaction steps	Specified the interactive steps required to carry out key tasks defined by task scenarios.
3.4	Technical environment	Specified hardware and software characteristics and user physical interfaces of the system.
3.5	System functions and features	Specified the system functions & features identified during the analysis and concept stage.
3.6	User interface design	Specified the details of user interface that the system would offer.
3.7	User Support	Specified the support needed for user, e.g. documentation, training,
3.8	Physical environment	Specified the physical environment that was identified in the analysis and concept stage.
3.9	Social and organisation environment	Specified the organisational requirement identified during the analysis and concept stage.
3.10	Standards and style guides to apply	Specified the standards & guidelines for the system to comply with.
3.11	Test plan	Testing the feasibility of the system during implementation phase.
3.12	Implementation plan	Provided a guideline of the implementation of the system.

Conclusions

The HCI Requirements Specification and the UI design for the UUV were developed successfully using the integration of the RESPECT framework with the supporting methods of Soft System Methodology, Object-Oriented Paradigm, Interface State Transition Diagrams and Hierarchical Task Analysis that we have successfully embedded together.

It was demonstrated in this work how the requirements specification for, and the design of, a UI evolved systematically by using the RESPECT spiral framework. Unlike in purely iterative framework methodologies, which iterate until the requirements are met, the evolutionary characteristics of the spiral framework of RESPECT allows the design to be evolved as information and knowledge are gained from each iteration. This makes the RESPECT spiral framework a truly user-centred approach and User Centred Design is fundamental for achieving high usability in a product.

The façade developed using Visual Basic dramatically improved the outcome of this work. Firstly, it provided the Users & Sponsor with a realistic representation of the final system. Secondly, it instilled such confidence in the User & Sponsor during the UI development, even though the prototype was just a simulation. Another benefit of using Visual Basic was that it dramatically reduced the turn-around time between iterations. Changes made to the façade could be easily done on-line. This definitely impressed the Users & Sponsor.

The combined use of prototyping techniques, user models, conceptual models and system models proved an effective aid for communication with users and the system sponsors. These techniques allowed the demonstration of the UI Design and System Behaviour without developing an actual real physical system connected to a real rig.

RESPECT captures the best practices of contemporary design techniques. It has all the properties of a good methodology. Highly recommended for the development of Requirements Specifications and as a Design Aid.

1. Booth P., An Introduction to Human-Computer Interaction, LEA.

2. Defence Standard 00-25, Human Factors for Designers of Equip.

3. Shelley A. (DRA) & Nicholson L. (Thorn EMI), Designing For Effectiveness, Battlefield Systems Internat. Conf 94 Proceedings.

4. Shneiderman B., Designing the User Interface- Strategy for Effective Human-Computer Interaction (3rd Ed), PH.

5. Jahnsen, Kramarics & Langset, Human-Computer Interfaces for Fast Patrol Boats, FFI Report.

6. Hix D. & Rex Hartson H., Developing User Interfaces Ensuring Usability through Product & Process, Wiley, 1993.

7. Maddix F., Human Computer Interaction – Theory and Practice, Ellis Horwood Series.

8. Preece J., Human-Computer Interaction, Addison Wesley.

9. Dix A., et al, Human Computer Interaction, 2nd Edition. Prentice Hall.

10. Object-Oriented Interface Design, 1st Edition IBM Common User Access Guidelines.

11. Shackel B, HUSAT Research Centre, UK. Human Factors & Usability in Human-Computer Interaction, edited by Jenny Preece-Wiley

12. Checkland P. & Scholes J., Soft Systems Methodology in Action-Wiley

13. Maguire M., RESPECT User Requirements Framework Handbook, European Usability Support Centres, HUSAT Research Institute.

14. Wards P. T. and Mellor S. J., Structure Development for Real-time Systems, Volume 1,2 & 3,Yourdon Press.

15. Morrison J. and Takel S., Very shallow and surf zone MCM reconnaissance, DERA Bincleaves.

Effects of Computer Freezes on Physiological Measures

Kaoru Suzuki

College of Engineering, Hosei University, 3-7-2 Kajinocho, Kogani-city, Tokyo
184-8584 Japan
E-mail: suzuki@ergo.is.hosei.ac.jp

ABSTRACT

This study investigated the effects of computer freezes on physiological measures which would be expected to reflect internal changes of the body. In this study, heart rate, heart rate variability index, the %LF, galvanic skin reaction, and skin surface temperature of the subjects entering a manuscript using the word processing program were measured. Changes in some of the indices, which were observed when the computer was frozen by the experimenter, indicated that the freezing gave a certain extent of the mental impact on the subjects.

1. Introduction

Since personal computers have been made more complicated to "improve" their function, sudden freezing, i.e. the computer does not accept keyboard inputs, mouse movements or mouse clicks, has become a common malfunction. However, even though users are becoming accustomed to them, the freezing might not be good for their "hearts".

This study investigated the effects of computer freezes on physiological measures which would be expected to reflect internal changes of the body, p articularly changes in autonomic nervous system (ANS) activity.

2. Measures

Many measures may be used to observe the internal changes in the human body. In this study, heart rate (HR), heart rate variability (HRV) index, the %LF (described later), galvanic skin reaction (GSR), and skin surface temperature were measured.

The HR, HRV index, and the %LF can be calculated from a series of inter-beat interval values. An inter-beat interval is equal to an R-R interval represented on the electrocardiogram (ECG) and the ECG signal can be obtained from electrodes attached to the chest. The ECG signal was converted into a pulse train, each of the pulses corresponding to an R-wave represented on the ECG. By clocking inter-pulse intervals, a series of inter-beat intervals was recorded on a personal computer. Based on the recording, the HR and the HRV (Suzuki and Hayashi 1993, not variance) index were calculated. (Although the HRV index used for this study was essentially the same as that described in the literature, a small modification was applied to the method of calculating the index. The index was calculated as a summation of two spectral power values, which were normalised and corresponded to the two lowest spectral bands.) In addition to the instantaneous HR and the HRV index, the %LF was also calculated as the intensity of the low frequency component (0.05-0.15Hz), which was relative to all one (0.05-0.4Hz) of the heart rate variability (Suzuki, 2000).

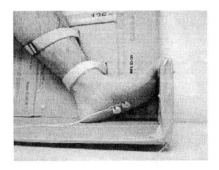

Fig. 1 Electrodes attached to the arch.

Fig. 2 A dialog box with an OK button.

There are several indices related to the galvanic skin reaction. In this study, the skin resistance reflex (SRR) was employed. The SRR is defined as the change in the resistance between two electrodes attached to the skin. The resistance is changed by perspiration and the SRR, usually measured at the palm, is often used to detect emotional disturbance. The electrodes were attached to the arch of a foot (Fig. 1), since both hands were used for keying. The distance between the electrodes was 3 cm. They were covered with plastic caps, which were attached to the skin with instant glue to ensure proper contact. The leg was loosely fixed to a board (Fig. 1), since placing the foot on the floor would change the contact resistance. The time constant to obtain the SRR, or to detect a change in the resistance, was 0.33 second. The SRR was recorded on paper with a pen recorder.

Most of the peripheral blood vessels tighten with an increase in sympathetic nervous system activity, which should decrease skin surface temperature. A thermograph camera (TH3108ME, manufactured by NEC san-ei Instruments Ltd.) was set above a CRT display located in front of the subject. The output was recorded on a personal computer at even intervals.

3. Software

A software program, which was to be run on a computer used by a subject, was developed that enabled the experimenter to remotely freeze the computer or to restart the operating system. In the frozen state, the target computer will not accept keyboard inputs, mouse movements or mouse clicks. A thawing command from the remote computer (or depressing a special key combination on the target computer) released the frozen state. The command to restart the operating system caused the target computer to display a dialog window with an OK button (Fig. 2). The dialog window imitated the genuine one used to signal the occurrence of a violation and termination of a program. When the subject depressed the OK button, the operating system was restarted without, however, warning the subject to save opened files. Of course, unsaved files would be lost. (If the subject did not depress the OK button and ignored the dialog window, he could continue the task without restarting the operating system.)

4. Experiment

The subjects, eight male university students (age: 21-22), did not know the aim of the experiment. Each subject was asked to enter, at their own pace, a manuscript of approximately 1000 Japanese characters, which would normally require approximately 30 minutes to enter by an ordinary subject who was slightly familiar with the word processing program. The characters were copied from a typed manuscript. The task was performed in a sound attenuation chamber to avoid any interactions with the experimenter. The chamber was temperature controlled, since changes in temperature would affect the GSR. When a subject had entered approximately 95% of the manuscript, the experimenter sent a command to freeze the computer. After three more minutes, the experimenter sent a command to restart the operating system. Three minutes after the second command, the experiment was terminated.

5. Results and Discussions

The measures, except temperature, were calculated for each of the three sections, i.e. from three minutes before freezing to freezing, from freezing to restarting, and from restarting to three minutes after restarting. The HR was calculated by dividing the number of beats within a section by three minutes. Since a temporal window is required to calculate the %LF, its length was set to three minutes. Because the HRV index is obtained for each beat, the maximum value within each section was calculated as the most important one. The number of the SRR pulses was counted for every section. The maximum pulse height, within all three sections, of the SRR pulse relative to the base line was examined, and those pulses whose heights were above 10 percent of the maximum were counted. The temperature was measured every minute at the subject's forehead (between the eyebrows).

Figure 3 shows the mean value (averaged over the subjects) and confidence intervals of the HR. Although the HR varied with individuals, which made the confidence intervals greater, the averaged difference within a subject among the sections might become significant. Results of the analysis of variance (ANOVA) showed that the differences in the HR among the sections approached significance ($p < 0.10$).

Figure 4 shows the mean value and confidence intervals of the %LF. Results of the ANOVA showed that the difference in the %LF among the sections was not significant. However, since four out of eight subjects showed a similar tendency, results of the Friedman test were significant ($p < 0.05$). Although the maximum HRV index showed a tendency that was similar to the %LF, it was not significant.

Figure 5 shows the mean value and confidence intervals of the SRR pulses per minute. Results of the ANOVA showed that the differences in the SRR frequency among the sections approached significance ($p < 0.10$). If one subject who showed a quite different tendency from the others is omitted from the analysis, the differences become significant ($p < 0.05$).

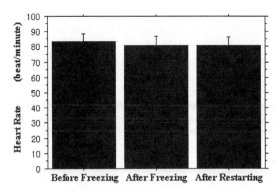

Fig. 3 Changes of the HR.

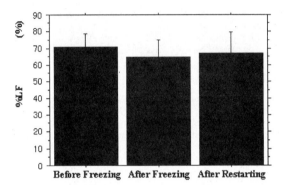

Fig. 4 Changes of the %LF.

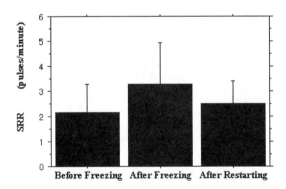

Fig. 5 Changes of the SRR frequency.

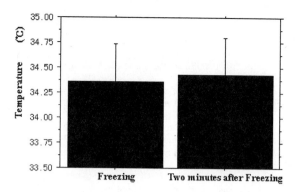

Fig. 6 Difference in the temperature.

Figure 6 shows the comparison of the mean temperature (averaged over the subjects, but excluding one subject who looked away from the camera when the temperature was sampled) between the time when the computer was frozen and two minutes after. If only these two points are taken into account, the difference in the temperature is significant (p<0.01).

Generally, emotional disturbances can result in excessive mental strain, which in turn causes sympathetic attenuation. This interpretation may be applied to the results of this experiment since, on average, the HR (Fig. 3) and the %LF (Fig. 4) decreased after the computer froze. Immediately before the freeze, the subjects are assumed to have been concentrating their attention on typing, since the task was nearing completion. In such circumstances, the sympathetic nervous system's activity should be relatively high. Because most of the peripheral blood vessels become tighter with an increase in sympathetic nervous system activity, the skin surface temperature should decrease Hence, an increase in the surface temperature (Fig. 6) can result from the sympathetic attenuation caused by the emotional disturbance or loss of concentration on the typing task.

6. Conclusions

The results of this experiment imply that computer "freezes" are not good for computer users, even those familiar with computers and their operating systems. Since the subjects in this experiment did not show remarkable responses after restarting of the operating system, it was impossible to consider about suitableness of a dialog window telling occurrence of fatal errors. Although the results did not achieve high statistical significance, the trend was in the direction that would indicate greater emotional disturbance when the computer froze. An additional factor is that the task was not one that the subjects were personally invested in, such as a deadline for an article. In such cases, the frustration level with computer freezing might be expected to be greater. Further studies will be required to investigate the mental impact of computer malfunctions on computer users.

REFERENCES

Suzuki, K. & Hayashi, Y. (1993). On a simple and effective method to analyze heart rate variability. In Smith, M. & Salvendy, G. (Eds.): *Human Computer Interactions: Applications and Case Studies (Vol. 1), Proc. 5th Int. Conference on Human-Computer Interaction* (HCI International '93, Orlando, USA, August 8-13, 1993), pp. 914-919. Amsterdam: Elsevier.

Suzuki, K. (2000). The sympathetic nervous system symptoms as reations against motion sickness. In Zheng, X. & Zhang, Y. (Eds.): *Proceedings of IEEE-EMBS Asia-Pacific Conference on Biomedical Engineering (Vol. 1)* (IEEE-EMBS Asia-Pcific Conference on Biomedical Engineering, Hangzhou, China, September 2-28, 2000), pp. 169-170. Beijing: World Publishing Corporation International Academic Publishers.

A Study of Ergonomic Requirements
for Japanese Character Sizes and Matrixes on Computer Displays

Ryoji YOSHITAKE and Motoyuki SATOH

Human Factors, Yamato Laboratory, IBM Japan Ltd.
1623-14, Shimotsuruma, Yamato-shi, Kanagawa, 242-8502 Japan

ABSTRACT

The present study was carried out to find an optimum character size and character matrix on computer displays for Japanese characters. At first a field study was performed to assess the actual conditions under which computer users do their daily work. The field study revealed that smaller fonts than are ergonomically recommended were frequently used. Secondly, a human factors' experiment was carried out to find an optimum size. Three kinds of displays with different pixel densities were used, because we predicted that pixel size would affect the users' preferences for character appearance. The results showed that the preferred character height decreased as the pixel pitch of the display increased and the pixels became smaller. The preferred character matrix, however, was almost constant. This demonstrated that the character matrix was given a much higher priority than character height, because the matrix controls such factors as the optimum stroke width and the general appearance of a font. We suggest new ergonomic requirements for character height and matrix size for Japanese characters based on the field study and the experiment. The proposed character height that is the minimum for readability is 20 minutes of arc, and the preferred range is 25 to 35 minutes of arc. Regarding the character matrix, we recommended that 11 x 11 be the minimum, and more than 15 x 15 is preferable.

1. INTRODUCTION

It is known that a most suitable character size does exist when reading sentences and characters on computer displays, although it varies slightly depending on the purpose and the context of use. Therefore, international ergonomic standards for computer displays such as ISO 9241-3 (1992) and ISO/FDIS 13406-2 (2000) specify appropriate character sizes to be used. Regarding the readability on computer displays, not only the character size but also the character matrix is a key factor. Japanese Kanji characters require more pixels than alphanumeric characters, because they consist of many strokes (graphic element). As a result, the current ergonomic requirements for the character sizes and matrixes for Japanese characters call for significantly larger characters and more pixels in each kanji character matrix than are used of the alphanumeric characters. These requirements seem to be based on the studies performed more than fifteen years ago (Yoroizawa and Inoue, 1984).

In the past few years the quality of computer displays is very much improved, as measured by higher pixel densities, focus improvements, and more colors or shades of gray. Moreover, computer users can easily change the sizes of characters on the current displays at any time. Due to these changes in the computer usage environments, it seems that the preferred sizes and matrixes of Japanese characters have decreased relative to the required ranges described in the current ergonomic standards. The authors started their investigation with the purpose of verifying whether the values recommended in those ergonomic standards are still appropriate for users. First we performed a field study to assess the actual conditions under which computer users do their daily work. Next, a human factors' experiment was carried out to find the appropriate character sizes and character matrixes for Japanese characters.

Table 1. Character height requirements from ISO/FDIS 13406-2

	Unit	Minimum for legibility		Preferred range for legibility	
		Latin origin	Asian	Latin origin	Asian
Character height	minutes of arc	16	25	20 – 22	30 – 35

Table 2. Character format requirements from ISO/FDIS 13406-2

	Unit	Minimum used for numeric and upper-case-only presentations	Minimum Used font size	Minimum for reading for context or if legibility is important	Preferred font size
		Latin origin	Asian	Latin origin	Asian
Character matrix (width to height)	pixels	5 x 7	15 x 16	7 x 9	24 x 24

Note) The character format requirements for Asian characters are applicable for fixed sized font only.

2.   CURRENT ERGONOMIC STANDARD REQUIREMENTS
   Since ergonomic standards for characters displayed on computer display already exist, the current requirements related to this study are shown in Tables 1 and 2.

3.   FIELD STUDY
3.1   Methods
   The daily working conditions of fifty computer users working at a development office were investigated. An investigator visited their cubicles one by one, and examined the settings of their computers. The examined items included:
   Type of display, size of active area of display, number of pixels (e.g.1024 x 768), type of operating systems, the fonts and their sizes selected in display properties, frequently used applications (e.g. email program and word processor), fonts and sizes used by each application, display tilt angle, typical viewing distance, typical line of sight angle, typical angle of view, and user characteristics such as visual acuity, age, and frequency of computer work.
   Since the users were using various kinds of applications, only word processors and email programs were focused on in this study, because all of the users are familiar with those two. The character sizes and matrixes used can be determined from the collected data. After all the checks had been finished, the users' satisfaction with the fonts and sizes used were assessed by interview.

3.2   Results and Discussions
   The subjects who participated in this study were 11 female and 39 male Japanese between 23 and 54 years of age, working daily with visual displays for at least 2 hours per day.
   The averages and standard deviations (S.D.) of the character heights and character matrixes collected in this study are shown in Table 3. Histograms for character height and character matrix appear in Figures 1 and 2. It was found that 56 percent of the users were usually using font heights of 25 minutes of arc or less for their word processor applications. For email programs, 78 percent of the users had font heights of 25 minutes of arc or less. Regarding character matrixes, half of the users used a font with a height of 13 pixels for word processing, and about 80 percents of the email programs were configured with a height of 11 pixels. This result comes from the default setting for each application. The heights of 13 and 11 pixels were the default settings for the word processing and email programs, respectively. What is most important is the users' satisfaction with those settings. Over 92 percent of the users participating in this study answered that they were satisfied with most of the settings on the font and size. Only 8 percent of the users felt the fonts used were too small. However, the users who answered they were satisfied might simply be accepting the default settings. Since the results obtained from this study did not provide strong indications about the optimal conditions, an experiment was performed to find the most preferred character sizes and matrixes.

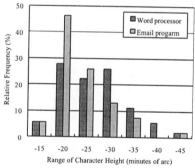

Fig. 1. Histogram of character height used

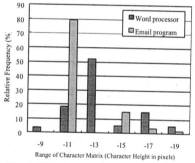

Fig. 2. Histogram of character matrix used

Table 3. Averages of character heights and character matrixes used in daily computer work

Applications	Character height (minutes of arc)		Character matrix (character height in pixels)	
	Average	S.D.	Average	S.D.
Word processor	24.0	6.3	13.5	2.3
Email program	21.2	5.8	11.9	1.9

4. HUMAN FACTORS' EXPERIMENT

4.1 Methods and Procedure

Three kinds of liquid crystal displays (LCD) with different pixel densities were used in this experiment, because we predicted that pixel size would affect the users' preferences on character appearance. The pixel pitches and densities of the displays used in this experiment are shown in Table 4. Eight files using with different font sizes were prepared. Table 5 shows the experimental conditions for character matrixes and character heights (minutes of arc) of each display when the viewing distance is 500 mm. The other factors such as line spacing were proportionally the same among the 8 files. The text itself was quoted from an article in a Japanese newspaper. Two sets of files with the typefaces "MS Mincho" and "MS Gothic" were prepared to investigate the effect of the fonts.

Table 4. Pixel pitch and density of LCDs used in this experiment

	$L_{display}$	$M_{display}$	$H_{display}$
Pixel pitch (mm)	0.307	0.264	0.213
Pixel density (ppi)	83	96	119

Table 5. Experimental conditions of the presented fonts (when viewing distance is 500 mm)

Character matrix (character height in pixels)	Character height (minutes of arc)			Stroke width (% of character height)	
	$L_{display}$	$M_{display}$	$H_{display}$	Mincho	Gothic
21	44.3	38.1	30.8	4.8	9.6 (double strokes)
19	40.1	34.5	27.8	5.3	5.3
17	35.9	30.9	24.9	5.9	5.9
15	31.7	27.2	22.0	6.7	6.7
13	27.4	23.6	19.0	7.7	7.7
11	23.2	20.0	16.1	9.1	9.1
9	19.0	16.3	13.2	11.1	11.1
7	14.8	12.7	10.3	14.3	14.3

The subjects who participated were 8 females and 19 males, working daily with visual displays for at least 2 hours per day. The mean age of the subjects was 36 years (range: 24–54). They all had near visual acuity, either natural or corrected, of at least 0.8 for both eyes.

The experiment was conducted in a human factors' laboratory. The three displays were arranged on a large table side by side. A chin rest was used to fix the viewing distance at 500 mm and insure the angle of view was perpendicular to the center of each display. The room illuminance was adjusted about 300 lx on each display surface. All of the texts were presented in the positive mode (dark characters and light backgrounds), and the background luminances of all displays were about 50 cd/m².

The subjects entered the test room one by one. Before starting the test, the subject was asked to fill in a background questionnaire, and his/her visual acuity was checked with an "ErgoVision" tester. After the test, instructions were provided, and the subject was asked to rate the presented text by using 5 or 7 grade scales on 7 characteristics:

Q1) How do you feel about the character size?
Q2) How do you feel about the stroke width?
Q3) How do you feel about the line spacing?
Q4) How do you feel about the character spacing?
Q5) How do you feel about the smoothness of the characters?
Q6) How do you feel about the character legibility?
Q7) How do you feel about the readability of the text?

The total number of conditions presented to a subject was 48 (3 displays, 8 character matrix sizes, 2 font types). The presentation order for each subject was changed to reduce viewing-sequence-related effects.

4.2 Results and Discussions

The focus of this paper is the rating score on readability (Q7), because "Ease of reading" was one of the most important factors and it provided us with some valuable suggestions. Figures 3 and 4 show the average rating scores on readability as a function of character height. Figures 5 and 6 illustrate the average rating score on readability as a function of the character matrix. The most preferred character height was different when the pixel pitch of the displays was changed. Regarding the character matrixes, on the other hand, the shift of the most preferred character matrix was very small, even if the pixel density of the displays was changed. This result indicated that character matrix was given a much higher priority than character height.

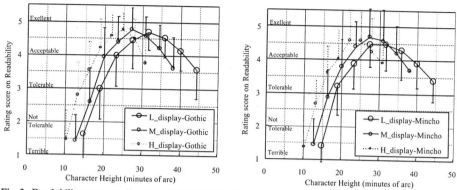

Fig. 3. Readability score as a function of CH (Gothic)   Fig. 4. Readability score as a function of CH (Mincho)

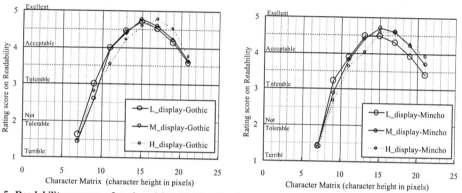

Fig. 5. Readability score as a function of CM (Gothic)   Fig. 6. Readability score as a function of CM (Mincho)

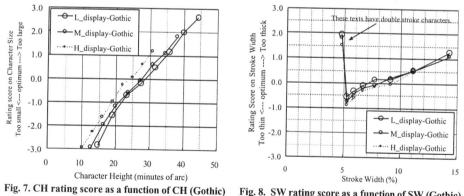

Fig. 7. CH rating score as a function of CH (Gothic)   Fig. 8. SW rating score as a function of SW (Gothic)

Fig. 7 shows the average rating scores on character size as a function of character height for the Gothic font. It shows the same tendency as the readability results. The average rating scores on stroke width as a function of stroke width (%) are shown in Fig. 8. This shows that there was little difference in the preferred stroke width among the three displays. The Mincho font had the same tendency as Gothic. Although it is possible to increase the character height by increasing the number of pixels, the stroke width becomes relatively too thin if the width of the stroke is only one pixel. Not only does this effect degrades the appearance of the font, but also it causes a visible contrast loss. This negative effect seems to be especially strong for elderly people who have lower contrast sensitivity (Kubota et al. 1999).

To determine the ergonomic requirements, quadratic regression equations for Figures 3, 4, 5 and 6 were used. Table 6 shows the peak values of each rating score and the values when the rating score is 3.5 as an index of the acceptable lower limit on those regression curves. For character height, the preferred size was reduced as the pixel size of the display was reduced. However, there must be a lower limit. Since the acceptable limit of $L_{display}$ is about 21 minutes of arc, we would like to propose 20 min of arc as the minimum ergonomic requirement for readability. The preferred range for the ergonomic standard seems to be 25 to 35 minutes of arc for Japanese characters. Regarding character matrixes, we propose that the minimum size is 11 x 11, and the preferred size is larger than 15 x 15. Since some Kanji characters have many strokes, it is impossible to precisely represent the shape even if the character matrix is 15 x 15. From the readability point of view, however, it is optimum because the stroke width is suitable and the appearance of the font is good. If higher density displays succeed in the market in the future, these requirements might have to be rewritten. As another approach, this concern could be resolved by using anti-aliased fonts (Yoshitake, 1996).

Table 6. Peak values and y=3.5 values of rating scores on the quadratic regression curves.

	Font type	Preferable value (y:peak)			Acceptable lower limit (y=3.5)		
		$L_{display}$	$M_{display}$	$H_{display}$	$L_{display}$	$M_{display}$	$H_{display}$
Character Height	Gothic	32.9	28.6	23.9	21.3	18.8	15.6
(minutes of arc)	Mincho	32.4	28.7	24.1	21.5	18.8	15.9
Character Matrix	Gothic	15.5	15.8	16.3	10.1	10.4	10.6
(CH in pixels)	Mincho	15.3	15.8	16.5	10.2	10.4	10.9

5. CONCLUSIONS

We suggested ergonomic minimums for character heights and character matrixes for Japanese characters, based on a field study and a human factors experiment. The proposed character height is that the minimum for readability is 20 minutes of arc, and that the preferred range is 25 to 35 minutes of arc. For the character matrix, we recommend that 11 x 11 is the minimum, and more than 15 x 15 is preferred.

ACKNOWLEDGMENT

The authors express appreciation to Ms. Chinatsu Saitoh who contributed to the conduct of the human factors experiment.

REFERENCES

ISO 9241-3: Ergonomic requirements for office work with visual display terminals (VDTs) – Part 3: Visual display requirements. 1992.

ISO/FDIS 13406-2: Ergonomic requirements for work with visual displays based on flat panels – Part 2: Ergonomic requirements for flat panel displays. 2000.

Yoroizawa, I. and Inoue, M. "Legibility Evaluation of Dotted Matrix Displayed Japanese Characters on VDTs Under Varying Dot Matrix Counts", *The Journal of the Institute of Television Engineers of Japan*. 1984, Vol. 38, No. 3, pp. 237-244 (in Japanese).

Kubota, S., Matsudo, K. and Marumoto, K. "Preferred Image Characteristics of Transmissive Liquid Crystal Displays for Older", *The Journal of the Institute of Image Information and Television Engineers*. 1999, Vol. 52, No. 9, pp. 153-160 (in Japanese).

Yoshitake, R. "Japanese Anti-Aliased Font Evaluation". *Hito to shisutemu*.1996, Vol. 2, No. 1, pp. 3-10 (in Japanese).

A NEW INDEX FOR CHARACTERIZING THE RESPIRATORY HEAERT RATE VARIABILITY

Hirohisa Mizuta*,Yasutake Aoki*,and Kazuo Yana**

* College of Engineering, Hosei University, 3-7-2 Kajino-cho, Koganei City
Tokyo 184-8584 Japan
** Hosei University Research Institute, California
800 Airport blvd. #504, Burlingame, CA 94010 U.S.A.
yana@k.hosei.ac.jp

ABSTRACT

This paper proposes to utilize a new index called degree of system nonlinearity $d.n.$ for characterizing the respiratory heart rate variability. The degree of system nonlinearity is defined as the fraction of the output power which is associated with the input signal but cannot be described by its linear transformation. An adaptive method to estimate the index is proposed. The method has been applied to the simultaneously recorded heart rate fluctuations and the instantaneous lung volume. The analysis showed that the index takes higher value in upright posture than in supine posture. The index will be a new index to characterize the autonomic nervous activity behind the heart rate fluctuations.

1 INTRODUCTION

Heart rate variability (HRV) analysis has been widely accepted as a mean to access the autonomic nervous activity. The spectral analysis has been dominated for the characterization of the HRV and widely utilized in practice. For more precise characterization, HRV has been related to other physiologic quantities such as respiration or blood pressure. In this paper we would like to propose a new index for characterizing the HRV associated with respiration. It is well known that the respiration affects the HRV. Here, we assume the HRV as output and the instantaneous lung volume (ILV) as input signals. Then an index called the degree of system nonlinearity ($d.n.$) is defined to measure how much nonlinear the system is. The index takes value 0 when the system is linear and takes 1 the maximum value when output is uncorrelated while they are deterministically associated. Simultaneous recording of both heart rate and instantaneous lung volume has been made and the index was estimated for healty male subjects both in supine and upright posture.

2 METHODOLOGY

2.1 Definition

Suppose we observe the input and output stationary time series $x[n]$ and $y[n]$ of unknown target system described as

$$y[n] = f[X[n]] + \eta[n] \tag{1}$$

Here, $X[n]$ is a transversal type input vector including a unity element to describe the system output bias, *i.e.*

$$X[n] = [1, x[n], x[n-1], \cdots, x[n-L+1]]^T \tag{2}$$

$f[.]$ is a nonlinear function characterizing the system. η_n is an unobservable exogenous output noise which is statistically independent from input signal. Now, we introduce an associated linear system of the target system as,

$$y[n] = W^T X[n] + e[n; W] \tag{3}$$

Here, W is the vector $[w_0, w_1, \cdots, w_L]^T$ of the filter coefficients. Optimal filter coefficients W^* minimizing the mean squares error $E[e^2[n; W]]$. is as well known, the solution of the normal equation. The quantity

$$\sigma^2_{e[n;W^\bullet]} - \sigma^2_{\eta[n]} \qquad (4)$$

is regarded as the excess output signal power which cannot be described by the associated optimal linear system. The degree of system nonlinearity $d.n.$ is defined by normalizing the quantity as,

$$d.n. = \frac{\sigma^2_{e[n;W^\bullet]} - \sigma^2_{\eta[n]}}{\sigma^2_{y[n]} - \sigma^2_{\eta[n]}} \qquad (5)$$

The $d.n.$ has the following properties which are desirable as an index of system nonlinearity.
(P1)$d.n.$ takes the value between 0 and 1.
(P2)$d.n. = 0$ when the system is linear.
(P3)$d.n. = 1$ when the output signal is uncorrelated with input while they are deterministically associated by $f(.)$.
(P4) $d.n. = a(0 \leq a \leq 1)$ when 100 % of output signal power which is originated from the input signal cannot be accounted by the linear transformation of the input signal.

2.2 Adaptive estimation of $d.n.$
A recursive estimation method of $d.n.$ defined by (5) will be described in this section. For a time series $z[n]$ we first define

$$S_z[n; l] = \sum_{k=1}^{n} \lambda^{n-k} z^l[n], 0 < \lambda < 1, l = 1, 2, \cdots \qquad (6)$$

Then,

$$E[S_z[n; l]] = \frac{1 - \lambda^n}{1 - \lambda} E[z^l[n]] \qquad (7)$$

Hence, for n large, $(1 - \lambda)S_z[n; l]$ can be regarded as an estimate of $E[z^l[n]]$. Thus, the variance of $z[n]$ can be estimated by the following recursive formula.

$$S_z[n; l] = \lambda S_z[n - 1; l] + z^l[n] \qquad (8)$$
$$\sigma^2_{z[n]} = (1 - \lambda)(S_z[n; 2] - S_z^2[n; 1]) \qquad (9)$$

Considering the data windowing effect by λ in (6), above estimate can be utilized as a running estimate of the $\sigma^2_{z[n]}$ in nonstationary case. Now, $\sigma^2_{y[n]}$ in the definition of $d.n.$ in (5), can be simply estimated by (8)(9) substituting $z[n]$ in (8) by $y[n]$. $\sigma^2_{e[n;W^\bullet]}$ in (5) is similarly estimated by substituting $z[n]$ in (8) by the residual time series $e[n; W]$ of the optimal linear system. The residual $e[n; W]$ is obtained by (3) where W is updated by a standard recursive algorithm such as RLS or LMS . In estimating $\sigma^2_{\eta[n]}$ a parametric system model $f(X[n]; \Theta)$ which is capable of expressing a large class of nonlinear system is introduced. Volterra system expansion or multi-layer perceptron may be a candidate for such a parametric model. Here we adopt Volterra system expansion because the standard RLS or LMS algorithm, suited for adaptive processing, can be applied for the estimation of system parameter vector Θ introducing an extended input vector consists of input polynomials. Assuming that the parametric model is good enough to simulate the target system, the residual $\eta[n; \Theta]$ of the parametric system model is regarded as an estimate of $\eta[n]$. Hence we may obtain a recursive estimate of $\sigma^2_{\eta[n]}$ by (8)(9) substituting $z[n]$ in (8) by $\eta[n; \Theta]$. The estimation accuracy of $\sigma^2_{\eta[n]}$ depends on how well the system model approximates the target system. However even if we adopt lower order Volterra system expansion which does not fully capture the system characteristics, (5) could still be an index of the degree of the system nonlinearity condition to the specific nonlinear system model. We may call the estimate of $d.n.$ using the estimate of $\eta[n; \Theta]$ the conditional system nonlinearity.

3 RESULTS
Simultaneous recordings of EKG and ILV have been made for 14 healthy male subjects. The data length

for each subject was 5 minutes. The instantaneous heart rate (IHR) has been constructed from the EKG signal sampled at the rate of 200Hz. IHR has been resampled at 4Hz yielding HRV signal. ILV were also sampled at 200Hz and resampled at 4Hz synchronized with HRV. Mean estimated values of $d.n.$ were 0.227 for supine and 0.450 for upright posture. Standard errors were 0.08 and 0.192 respectively. Pared t test showed significant difference between mean $d.n.$ values with $p < 0.001$. For above estimation Volterra system expansion of the order 3 were used for nonlinear output prediction.

4 CONCLUSIONS

A new index called the degree of system nonlinearity $d.n.$ for characterizing the transfer characteristics from ILV to HRV has been introduced. An adaptive method to estimate the index from observed input and output time series has been also proposed which will be useful for the real practice. Data analysis showed the significant difference in $d.n.$ between supine and upright posture. This finding implies that the index may reflect the sympathetic and parasympathetic balance although further physiologic interpretation of this finding is necessary.

REFERENCES

S. Akselrod, D. Gordon, F.A. Ubel, D.C. Shannon, A.C. Barger, and R.J. Cohen, "Power spectrum analysis of heart rate fluctuation: A quantitative probe of beat-to-beat cardiovascular control," *Science*, vol. 213, pp.220-222, 1981.

R.D. Berger, S. Akselrod, D. Gordon, and R.J. Cohen, "An efficient algorithm for spectral analysis of heart rate variability," *IEEE Trans. BME,*vol.33, pp.900-904, 1986.

M.L. Appel, R.D. Berger, J.P. Saul, K.M. Smith, and R.J. Cohen, "Beat to beat variabaility in cardiovascular variables: Noise or Music?," *J. Amer. Coll. Cardiol,*vol.14, pp.1139-1148,1989.

J.P. Saul, R.D. Berger, P. Albrecht and R.J. Cohen, "Transfer function analysis of autonomic regulation: II Respiratory sinus arrhythmia," *Amer. J. Physiol.*, vol. 256, pp. H153-H161, 1989.

K. Yana, J.P. Saul, R.D. Berger, M.H. Perrott, and R.J. Cohen, "A time domain approach for the fluctuation analysis of heart rate related to instantaneous lung volume,"*IEEE Trans. BME*, vol.40, pp.74-81, 1993.

H. Yoshida, M. Komai and K. Yana, "An index of system nonlinearity and its estimation," *Proc. International Joint Conference on Neural Networks*, pp. 2021-2024, 1993.

T.J. Mullen, M.L. Apell, R. Mukkamala, J.M. Mathias, and R.J. Cohen, "System identification of closed-loop cardiovascular control: effects of posture and autonomic blockade," *Am. J. Physiol.*, vol. 272, pp. H448-461, 1997.

H. Mizuta and K. Yana, "Heart rate signal decomposition," *Method Inform. Med.*,vol 39, pp.200-203, 2000.

T. Yanai, Y. Yamamoto, N. Kishi and K.Yana, "Evaluation and monitoring the accumulated mental work load using heart rate variability," *Proc. IEEE-EMBS Asia-Pacific Conf. Bio. Med. Eng.*,pp.167-168,2000.

K. Kotani, I Hidaka, Y. Yamamoto, and S. Ozono, "Analysis of respiratory sinus arrhythmia with respect to respiratory phase," *Methods Inform. Med.*,vol. 39, pp.153-156, 2000.

Evaluation of keyboards for Personal Computers
-Performance, subjective evaluation and EMG study-

Shin'ichi FUKUZUMI* and Masayuki KOBAYASHI**

*Consumer PC Division, NEC Corporation
1-10, Nisshin-cho, Fuchu, Tokyo, 183-8501, Japan
E-mail: s-fukuzumi@aj.jp.nec.com

**NEC Design, Ltd.
20-36, 2-chome, Takanawa, Minato-ku, Tokyo 108-0074, Japan
E-mail: mas-kob@design.nec.co.jp

Abstract

Three kinds of QWERTY layout keyboards (traditional liner type, natural design type and ergonomic design type) and two kinds of M-system layout keyboards (liner type and ergonomic design type) were evaluated in point of performance, subjective evaluation and EMG (Electromyogram). We found that traditional linear type QWERTY layout keyboard was good for performance, and about ergonomic design type, both a QWERTY layout and an M-system layout keyboard were evaluated lower than any other keyboards in the point of subjective evaluation and fatigue. We also found that keyboards layout can be evaluated by EMG. As the problems of these two keyboards have become clear, we propose to improve usability of these keyboards on the basis of this result.

1. Introduction

Many study about keyboard layout, keyboard profile and so on have been carried out because a keyboard is one of important input devices for VDT (Visual Display Terminals) works (Nakaseko, 1986 and Yoshitake, 1995). A traditional typewriter keyboard layout is applied to a VDT's. However, this layout may not necessarily be best for using Asian characters. ISO 9241-4 (1998) presents the ergonomic requirements about keyboards. However, a requirement about a layout of Asian character is not described. This time, we evaluated different layout keyboards during the input of Japanese sentences. The objective of this study is to clarify any problems about the layouts. We will make use of this result in develop a suitable keyboard for users.

2. Experiments

2.1 Keyboards

In this study, three kinds of QWERTY layout keyboards (called a traditional liner type (KB-A), a natural design type (KB-B) and an ergonomic design type (KB-C)) and two kinds of M system layout keyboards (called a liner type (KB-D) and an ergonomic design type (KB-E)) were evaluated in point of performance, subjective evaluation and change in physiological data. In these five keyboards, a traditional liner type is a generally used keyboard, a natural design type is Microsoft-made and the others are developed in our company. In this, "M" of M system layout means the initials of developer's name in our company. Figures 1 to 5 show the pictures of these keyboards.

Figure 1. QWERTY liner type keyboard (KB-A)

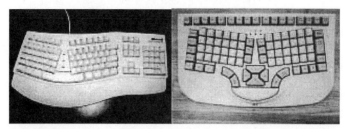

Figure 2: QWERTY natural design type
Keyboard (KB-B)

Figure 3: QWERTY ergonomic design type
keyboard (KB-C)

Figure 4: M-liner type
Keyboard (KB-D)

Figure 5: M-ergonomic design type
keyboard (KB-E)

2.2 Evaluation methods

An experimental task was to input Japanese sentences for about 15 minutes while a participant shows Japanese manuscripts. As for performance, an input velocity and an error rate were measured. As for subjective evaluation, a five-point numerical scale was used for evaluating general, any keys, a palm rest, home positions, and physiological fatigue. Each participant's internal report was also collected. An index of objective fatigue evaluation, EMG (Electromyography) for arms was measured. EMG was led from a muscle of extensor carpi radialis longus and a muscle of flexor carpi radialis for each arm. EMG data was applied to FFT. Figure 6 shows the positions of wearing the electrodes for measuring EMG.

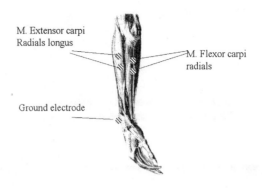

M. Extensor carpi
Radials longus

M. Flexor carpi
radials

Ground electrode

Figure 6: Positions of wearing the electrodes for measuring EMG

2.3 Participants

Three male and three female who usually use QWERTY layout keyboards and two male and three female who usually use M system layout keyboards were participate in the experiment.

2.4 Procedure

In this experiment, participants who usually use QWERTY layout keyboards evaluates three type of QWERTY keyboards (KB-A, KB-B, and KB-C), and who usually use M system layout keyboards evaluates two type of M system layout keyboards (KB-D and KB-E), to avoid the influence of experience.

Participants try to input Japanese sentences according to sample sentences using each keyboard. For each task, an input velocity (character/min) and an error rate (%) are measured. After the task, participants check a five-point numerical scale with 1 being "Poor" and 5 being "Excellent" for key typing.

During the task, EMG is measured. Five seconds EMG data in the early task and in the late task were applied to FFT, its sampling frequency was 1 KHz. The distribution of frequent power is known to shift to lower frequent bands due to muscle fatigue and frequent bands under 40 Hz is known to be less related to fatigue (Gilmore, et al., 1985). From these, the content rate of FFT power value in EMG frequency bands from 70 Hz to 160 Hz was calculated. The change in the ratio of the content rate in the early task to that of in the late task was defined as an index of fatigue. This ratio value means that fatigue level is higher according to the increase of the value.

3. Results and Discussions

3.1 Performance and subjective evaluation results

Table 1 shows the result of subjective evaluation. These data are the average values of all participants for each keyboard type. In this table, the average values were calculated using data obtained from five-point numerical scale except "Input velocity" and "Error rate".

The results of QWERTY layout keyboards, for an input velocity, the KB-C and the KB-B were significantly slower than the KB-A. These significant levels were 1% and 5%, respectively. For an error rate, there were no significant differences among these keyboards. As for subjective evaluation, the KB-A got higher evaluation than the others. The results of M system layout keyboards, there were no significant differences between the KB-D and the KB-E about an input velocity and an error rate. As for subjective evaluation, the KB-D got higher evaluation than the KB-E. Two kinds of the ergonomic design type keyboards got higher scores about "cursor key", "palm rest" and "home position". From these, we found that the KB-C and the KB-E were usable for users in point of them.

Table 1: Performance and subjective evaluation result for both types of keyboards

	QWERTY layout			M system layout	
	KB-A	KB-B	KB-C	KB-D	KB-E
Input velocity (Character/min)	51	36.2	42.7	44.7	34.3
Error rate (%)	6	6	7	6	6
Total evaluation	3.5	2	2.7	4	2.6
Input characters	4	1.8	2.7	3.8	3.6
Input symbols	3.5	2	2.3	2.6	2.6
Editing	2.8	2.7	2.2	3.8	2.8
Cursor	2.7	1.7	2.3	3	3.2
Space key	3.2	2.2	3.8	3	3
Return key	3.3	2.8	3	4	3.2
Palm rest	2.8	2.8	3	3.3	3.4
Home position	3.3	2.3	2.7	2.6	3
Load to one's hand	2.8	2.3	3	4	1.8

3.2 EMG results

Figure 7 to 10 shows the results that the change in the ratio of the content rate of FFT power obtained from EMG data in the early task to that of in the late task. In these figures, the ratio value means that fatigue level is higher according to the increase of the value.

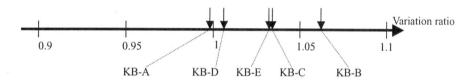

Figure 7: The results of the content rate of FFT power (M. extensor carpi radialis longus of right arm)

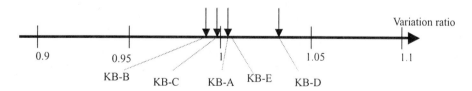

Figure 8: The results of the content rate of FFT power (M. flexor carpi radialis of right arm)

1307

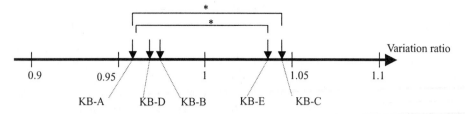

Figure 9: The results of the content rate of FFT power (M. extensor carpi radialis longus of left arm)

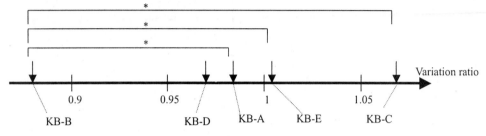

Figure 10: The results of the content rate of FFT power (M. flexor carpi radialis of left arm)

From the results of EMG analysis, as for two measurement positions of right arm, there was no significant difference among the keyboards. However, as for left arm, two kinds of the ergonomic design type keyboards were found to increase the content rate value significantly. Considering with Table 1, about two kinds of the ergonomic design type keyboards, the design of the cursor key, the shift key by thumbs and the home positions key was not accepted for the participants. These evaluation results are similar to the results of subjective and performance evaluation. Therefore, EMG is a useful index for keyboard layout evaluation.

4. Conclusions

We found that traditional linear type keyboard was good for performance, and the liner type keyboard developed in our company was also good for users. However, both QWERTY layout and M system layout ergonomic design keyboards were evaluated lower than any other keyboards. As these two keyboards have got good evaluation in point of "cursor key", "palm rest" and "home position", we propose to improve their usability on the basis of this result.

References

Gilmore, L. D. and De Luca, C. J.: Muscle fatigue monitor (MFM): Second generation, IEEE Transactions on Biomedical Engineering, Vol. BME-32, No. 1, pp75-78, 1985.

ISO9241-4: Ergonomic requirements for office work with visual display terminals (VDTs)- Part4: Keyboard requirements, International Organization of Standardization, 1998.

Nakaseko, M.: Ergonomic design of Keyboards, Japanese journal of Ergonomics, Vol. 22, No. 2, pp53-61, (in Japanese), 1986.

Yoshitake, R.: Relationship between Key Space and User Performance on Reduced Keyboards, Applied Human Scienece, Journal of Physiological Anthropology, Vol. 14, No. 4, pp287-292, 1995.

On the usability of user manuals

Lutz Groh and Martin Böcker

Siemens Information and Communications Mobile ICM MP TI 3
Haidenauplatz 1, D-81667 Munich, Fed. Rep. of Germany
+49 89 722 61719, +49 89 722 36961
lutz.groh@mch.siemens.de, martin.boecker@mch.siemens.de

ABSTRACT

At Siemens, we have made it a tradition to ensure the quality of our products by paying particular attention to the quality of both the product itself and of the user manual. This paper discusses the role of the user manual in today's information and communication products. Reasons for investing in good user manuals are discussed and different options for the quality assurance of user manuals are presented. Quality control measures include usability tests of user manuals and an example of such a test is briefly described. The general background to the work reported is in portable and fixed-network phones. Lessons learnt, however, apply not just to telephones but information and communication products in general.

1   The role of user manuals in today's information and communication products

1.1   Current challenges in user interface design

Even though user manuals are often assigned a low priority by manufacturers, manuals are perceived by most users as being an integral part of "the product", playing an important role (a) when first familiarising with the product and its options and (b) for troubleshooting.

It is a frequently uttered common-sense comment that user documentations are superfluous if the user interface (UI) of a product is sufficiently self-explanatory. Since this goal is hardly ever achieved, the documentation has to support the user wherever controls and indications as well as user procedures for a particular feature are not immediately obvious (at least to the novice user). User manuals more and more frequently compensate for shortcomings of those user interfaces that follow current trends which have the potential of making the product less "self-explanatory":

Miniaturisation: The current trend towards miniaturising consumer products requesting the smallest possible realisation of a device leaves designers limited room for the user controls and indications – the minimum size of which is more and more frequently dictated by human physiology than by technological progress.

Featurism: A further trend in information and communication products concerns the number of supported features. „Featurism" refers to the implementation of the largest possible number of features in a device regardless of whether they make sense for a given device (Lindgaard, 1993). Since the majority of users only use a small percentage of all the features implemented in a complex device, the UI designer has to make an educated guess (or, preferably, a decision based on empirical user requirements) on how to implement features that are less likely to be used. These are usually buried deep in the menu trees and options lists and difficult to find without documentation.

User interfaces of „hybrid" devices: The cross-breeding of devices from different traditions (e.g. a mobile phone with an MP3-player) confounds UI elements from different traditions (functions like "Play", "Pause" and "Record" cannot be mapped easily on the 12-key keypad of a phone) requiring support for the user through documentation.

International User Interfaces: Many products are not produced with a localised variety but only as one global version. The UI for such global products often employ symbols which have to be learnt.

These trends lead to products being not fully self-explanatory thus presenting the need for supporting documentation.

1.2   Reasons for investing in the development of good user manuals

Additional sources of motivation for investing effort into user manual design include:

Legal requirements: In many countries, user documentation (in the local languages) is a legal or regulatory requirement.

Follow-on costs: The user of a novel device may first try to find out by himself how the device is operated. If he fails, he may consult the user manual. If the manual does not easily provide the required information, the user may have to resort to calling the service hotline (provided the manufacturer operates a service call-centre). Since the costs to the manufacturer of one call to the service hotline may destroy the profit margin for that particular exemplar of the product, and since waiting time in the call queue further lowers the customer satisfaction, most manufacturers are highly motivated to design the user manual in such a way as to help the user and making the call to the service hotline superfluous.

The "message" of the user manual: The user manual plays a role in the overall perception of the product's usability. It is an expression of a company's external presentation (e.g. as high-tech, state-of-the-art, and user-centred).

It has to be noticed, however, that strong cultural differences exist between the readiness to read a user manual or to consult it at least (see Honold, 1999).

2   Assuring quality in user manuals

2.1   Paper-based manuals

A user manual is of (sufficiently) high quality, if a user finds a correct and complete description of the functionality of a product in a form and language that he finds easy to understand. In particular:

Correct and complete description: An explanation that does not assume much prior knowledge, describing (preferably only) the delivered product, providing a complete description without being too repetitive.

Structure: A user manual taking the form of a small booklet possibly does not require a formal structure. User manuals of a dozen pages or more should include a table of contents and ideally also a subject index. Correct cross references are also very helpful. Distinct function groups (e.g. set-up and installation procedures) are often explained in separate booklets ("Quick Start", "First steps"). However, too many different bits of paper ("confetti") can be highly confusing.

Layout: Conventions for good page layout and typography should be followed.

Language and style: The user manual should be written in the local language(s) (e.g. French and Flemish for customers in Belgium). Of further importance is the choice of linguistic style: a product targeted primarily at young people may be documented in a youthful style that older users may find irritating (this includes the choice of informal personal pronouns).

Terminology: In many areas of information and communication technology there is a co-existence of two terminologies, namely the language of the "expert" and the language of the novice or low-level user. For example, the expert language includes acronyms for device features that a novice is unlikely to know (e.g. CCBS for "Completion of calls to busy subscribers"). The user manual should reflect the terminology used in the user interface, while giving the expert a chance to find his terminology as well, e.g. in the subject index which ideally includes synonymous terms with appropriate references.

Tables and figures: These have the potential of representing a large quantity of information using less space than needed for the representation of the same information in text (provided that the meaning of rows and columns is made clear to the reader). Similarly, figures can express complex (e.g. spatial) information in a simple and economic way.

2.2   Electronic user manuals

These include information provided on local, self-contained media (e.g. CD-ROMs delivered with the product), pre-installed information (e.g. help systems in PC-software) and information made available on the internet.

Interactive user manuals (e.g. on CD-ROM or in the internet) offer the possibility to a potential buyer to get to know the new product in a playful way even before purchasing it - he receives a help for deciding whether he likes and understands the product. The manufacturer can combine user documentation with advertising elements, product promotions and direct ordering. User manuals on the web can be easily updated and be made available worldwide in all languages provided by the manufacturer (e.g. an Italian user living in Sweden may prefer the Italian to the Swedish documentation).

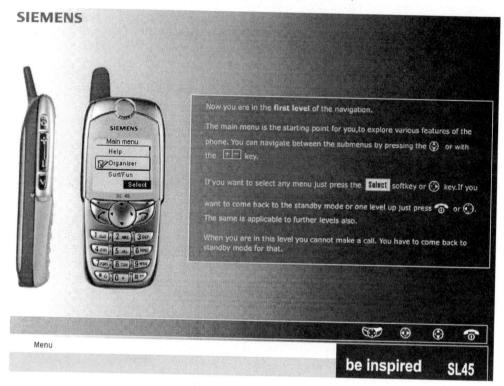

Fig. 1: Example of an interactive user manual for a mobile telephone (www.my-siemens.com).

The example in Fig. 1 illustrates the benefits of an interactive approach guiding the user with accompanying text while allowing him to explore the product by himself.

Integrated help-systems: These are help-systems directly implemented into the product (e.g. in PC software). The information is made available easily, however, in many cases it cannot be updated (e.g. help systems in telephones and PDAs). Furthermore, reading pages of information on the screen can be tiring, which is why they are often printed out by the user. If no additional paper-based manual is provided, the costs of manual printing is shifted over to the buyer.

Acoustic User Manuals: Spoken user manuals on CD or cassette appeal to audio-oriented users: they receive a spoken explanation of how the new device works. The usability of spoken documentation depends inter alia on the verbal comprehensibility, the reading speed and the control options of the medium. Speech is also a suitable medium for partially sighted and blind users provided that no visual references ("Press the red key") are made. Acoustic instructions made available as MP3-files on the internet can also be updated easily.

2.3 Further quality assurance measures

The following measures are further options for assuring the quality of user documentation:
External expertise: Experts (communications scientists, linguists, ergonomists and Human Factors experts) assess the quality of a document applying formal and textual suggestions for improvement.
External editors: Editors check the documentation according to linguistic and formal criteria (spelling, punctuation, correctness of the table of contents, of the subject index, of cross references and of the numbering of tables and figures).
Usability-Tests: The usability of the user manual in the context of the corresponding product is being tested in the usability lab with real users (see Section 3).

3 Usability testing user manuals

3.1 Motivation for conducting usability tests of user manuals

Usability tests are in many cases the most important source of insight into the quality of a user manual. Many manufacturers limit usability testing to tests of the product itself, often taking the form of quick-and-dirty tests conducted with few subjects many of whom are colleagues. At Siemens, we have extended usability testing to include tests of the user manual itself. After two years of systematic testing, the overall conclusion is that testing the usability of the user manuals proved vital for the identification of shortcomings both in the manual itself as well as those in the user interface of the product. Conducted at an early stage in the development process, the results of the test can be used for changes to the manual prior to printing or for preparing a second edition for later printing.

3.2 Methodological considerations

Essentially, all standard procedures used for professional usability-testing also apply for the testing of user manuals. The following aspects are particularly important:
Test subjects: The choice of subjects is critical for achieving meaningful results: they should be external (i.e. they should not be employees of the manufacturer), they should belong to the target group of the product, and they should not be used for more than one manual test (to avoid breeding test experts).
Test material: The manual to be tested should be at least a stable draft; it is pointless to test a "building site" of a manual with large gaps. Test scenarios and test tasks are usually chosen to include essential features, new features and those the authors feel unsure about.
Data capture: Questionnaires and interviews capture standard usability criteria like efficiency, effectiveness and satisfaction. Observations of users struggling with a particular feature point to the need of re-writing the manual. Keeping the data points constant across tests allows for the documentation of changes between documentation for the own products and of differences to those of competitor products.

3.3   Example of a usability test of manuals for cordless phones

A conventional, booklet-type of user manual of an existing Siemens product (Gigaset cordless phone) was tested against an alternative and cheaper approach, namely the a large A3-format sheet folded three times yielding 16 pages, employed by a competitor for a cordless phone as well.

During the test with 24 external test subjects, ten test tasks had to be performed by the subjects who were asked to first read the relevant section in the user manual. The tasks included setting up the phone for its first use, making an outgoing call, changing the ringer melody, registering an additional handset and setting up a conference call. 12 subjects used the booklet-type of manual and 12 the folded sheet (each with the respective product).

One general trend that emerged from the test was that some tasks were more difficult to solve than others, regardless of the manual type. E.g. registering an additional handset is a complex procedure involving many steps whereas accepting a call is simple. In many instances, the sheet-type manual received less favourable ratings in terms of being more ambiguous, less easy to understand, and more cumbersome. These ratings may be more due to level of detail and choice of style and not dependent on the paper format.

Asked directly for a judgment on the external appearance of the manuals using semantic differentials, the differences are more obvious: the booklet was rated more clearly arranged, modern, nice and appealing than its sheet-style counterpart (see Fig. 2). Asked for their satisfaction

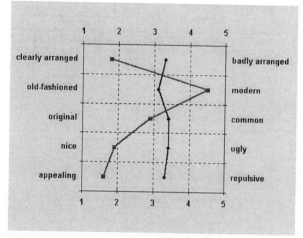

Fig. 2: Semantic differential ratings of the two manual types booklet (squares) vs. folded sheet (diamonds).

with the respective user guides using a 7-point smiley scale (ranging from 1 "sad" to 7 "happy"), the results favoured again the booklet (6.1) over the leaflet (4.2). The results of this usability test provided a solid basis for a decision for future Siemens user guides and is a good example for the role of usability testing for assuring the quality and usability of user manuals.

Literature

Honold, P. (1999). Learning how to use a cellular phone: Comparison between German and Chinese users. In: *Technical Communication*, Second Quarter 1999, p. 196-205.

Lindgaard, G. (1993). WOW – 568 Smart Features on your PABX: What really determines the uptake of technology? *Human Factors in Telecommunications*. Darmstadt, HFT. p. 29-38.

Semantic Lenses:
Exploring Large Information Spaces More Efficiently

Christian Rathke, Markus Alexander Wischy, Jürgen Ziegler

C. Rathke, HBI Stuttgart, Germany, rathke@hbi-stuttgart.de
M.A. Wischy, CT SE 2, Siemens AG, Munich, Germany, markus.wischy@mchp.siemens.de
J. Ziegler, Fraunhofer IAO, Stuttgart, Germany, juergen.ziegler@iao.fhg.de

ABSTRACT

The amount of directly available information is growing exponentially every year. Finding something is becoming increasingly difficult. Classic search engines usually return several hundreds or thousands of hits. Browsing the result list has become awkward and does not allow getting a good overview of what has been found. This paper proposes the use of semantic lenses, a special form of magic lenses [Bier et. al, 1993], in order to explore large information spaces more efficiently. The goal is to increase the flow of information between user and information system by dense visualization and high interaction rate. The undertaken usability tests indicate that Semantic Lenses improve the search time in information spaces for complex queries and repeated location of items.

1. THE SEARCH PROCESS

The search process is characterized by three separate activities of the information system: formulation of the search request, presentation of the results, and visualization of the details. After the user has formulated the request, the system generates a list of matching topics, which may then be looked at in detail. Usually, each of these activities uses their own screen. Navigation becomes complicated if a request must be reformulated. But this happens quite often: the desired result is rarely found in one step. Rather, users incrementally narrow in on their search targets. The user's task of the search process thus involves "narrowing" and "exploring" as recurring tasks: after the user has reformulated ("narrowed") a query, he explores the information space of the result. The exploration tasks involves high interaction between the result view and the detail view. After the user has gathered enough information about the result space, he can then again reformulate the query, explore the result space and repeat these steps until he has found the correct results.

The solution proposed here for faster searches is to enhance the information flow between user and information system: the user learns more about the information space in shorter time. This can be achieved by two paradigms: first by transferring more information per interaction and second by increasing the number of interactions per time slice.

According to these paradigms, we have developed Semantic Lenses as a special form of magic lenses to display a high-density overview of the results and at the same time gain instant access to the interesting details.

2. SEMANTIC LENSES

Semantic Lenses are interaction devices for dynamic and continually adaptive visualizations of information. They exploit the intrinsic capabilities of the human visual system to track moving objects within their contexts. Instead of a distinct separation of visual impressions, dynamic visualizations are continually modified by, e.g., spatial transaction, rotation or zooming. Complex information structures are experienced in a simulated world of abstract information objects or visual representations of real objects. The benefits of Virtual Reality are applied to abstract information spaces.

On a technical level, Semantic Lenses are graphical interaction devices. They are usually composed of a frame and at least one filter. The filter implements a mode and the frame separates the device from its surrounding environment. When the filter is inactive the lens becomes transparent (Fig. 1).

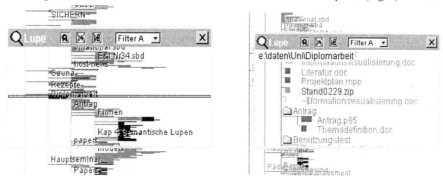

Figure 1: Example of an inactive and an active Semantic Lens

The frame provides space for other interaction elements such as additional filters or filter modificators. The size of the lens may be adapted as well, using the frame. The selected filter is activated as soon as the mouse is moved inside the lens. The lens may then be moved around by simple dragging. If the mouse is inside the lens (as opposed to the frame area) the representation generated by the filter is continually updated. This generates an effect similar to a physical magnifying glass.

3. USABILTIY TESTS

To validate and wage the efficiency of the Semantic Lens usability tests in the context of an example application have been carried out. The example application is a file browser with a similar purpose as the Windows Explorer. File systems can become very large and finding files in such a structure is a common problem. Also, with the Windows Explorer there exists an established tool to which the test results can be related. The file system overview is provided by a hierarchical graph with directories as intermediate nodes and files as the terminal nodes. In this example the entire file hierarchy consisting of several hundreds of directories and files is displayed on a single screen. The Semantic Lens is used to magnify the partial structure under its focus and at the same time to display the file names and some of their properties such as file type, size and creation date.

The usability tests were carried out in a lab equipped with video recording facilities. The main test evaluated the behavior of 10 subjects carrying out various predefined tasks with a total time of approximately 60 minutes for each subject. After the main test, questionnaires were used to collect impressions and suggestions from the subjects.

Figure 2 shows average navigation times with standard deviation for three successive navigation tasks for the same target. It compares the performance of using the lens with using the Windows Explorer for a given file with complete path description (e.g. \dir-a\subdir\file-c). The users navigated to the same file four times, changing from lens to Explorer after each iteration. The first three searches were done in a row, the fourth after a 15 minute break.

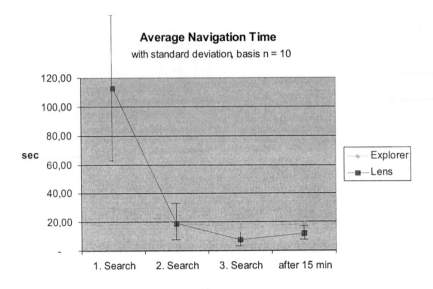

Average Navigation Time

with standard deviation, basis n = 10

Figure 2: Average navigation times for three selection tasks

The figure shows that the first search is significantly slower with the Semantic Lens. The Information space was unknown to the user and the free interaction of the lens didn't naturally direct the search along the directory path like the Explorer. The second task, finding the file again, is not significantly slower or faster than the explorer. This shows that the users can reuse their knowledge about the information space and can select the file much faster than on the first search. On the third search the lens is significantly faster than the Explorer. This is true also for the forth search. These findings show that spatial visualization, direct access and fast exploration lead to faster access for repeated selections. Another effect is visible: spatial information can be remembered better than textual information (e.g. file paths), even after a break of 15 minutes, the navigation time was significantly faster than the Windows Explorer.

In the questionnaire the users rated the Semantic Lens as effective (0.25), efficient (0.20) and easy to learn (0.55, averages on a five scale Likely-Scale). This shows a positive impression of functionality and usability of the Semantic Lens. The complete results of the usability tests can be found in [Wischy, 2000]. The users were students and interns at the Fraunhofer Research Lab. The users differentiated in age (21 to 32), sex (three female, seven male) and GUI experience (2 – 10 years). No correlation between these variables and the tested variables were visible. The reliability of the results is 90% that the search times vary maximal amount of 20% [Nielsen, 1993]. This is sufficient for the statements made previously. Though the test only shows the efficiency of young, untrained users. Preliminary test data of trained users of semantic lenses show that the lens leads to faster access even at the second search and the difference between explorer and lens in the first search disappears.

4. RELATED WORK

Magic Lenses, a "See through Interface" was first introduced by Mauren Stone and other members of Xerox PARC [Bier et. al, 1993]. The lens metaphor was applied for graphic and text processing. The semantic function of the filter was noted in [Fishkin, Stone, 1995], where an application of querying a database was introduced. Other applications of the lens metaphor were presented in [Viega et. al, 1996] (3D Graphics) and [Hudson et al., 1997] (Graphical Debugging).

None of the above undertook usability tests to acquire empirical data for the effectiveness of the lens metaphor. No application of Semantic Lenses for file or web searches is known to the authors.

5. SUMMARY

Overall, the positive results show the potential usefulness of Semantic Lenses for exploring large information structures. The usability tests carried out are the first ones providing empirical data for the practical usage of Semantic Lenses. The evaluation of the questionnaires has lead to a variety of improvements over the original design. First trials have been undertaken to use the Semantic Lens concept in web searches (e.g. search in the Yahoo directory) and further applications of Semantic Lenses will be evaluated.

REFERENCES

[Bier et. al, 1993] Toolglass and Magic Lenses: The See Through Interface, Proceedings SIGGRAPH 1993, Eric A. Bier, Maureen C. Stone, Ken Pier, William Buxton and Tony D. DeRose

[Fishkin, Stone, 1995] Enhanced Dynamic Queries via Movable Filter, Proceedings of SIGCHI 1995, Ken Fishkin and Maureen C. Stone

[Hudson et al., 1997] Debugging Lenses: A New Class of Transparent Tools for User Interface Debugging, UIST 1997, Scott E. Hudson, Roy Rodenstein and Ian Smith

[Nielson, 1993] Usability Engineering, Academic Press, San Diego, USA, 1993

[Stone, Fishkin, Bier, 1994] The movable filter as a User Interface Tool, Proceedings of SIGCHI 1994, Maureen C. Stone, Ken Fishkin, Eric A. Bier

[Viega et. al, 1996] 3D Magic Lenses, UIST 1996, J. Viega, M. J. Conway, G. Williams and R. Rausch

[Wischy, 2000] Semantische Lupen: Ein Dialogwerkzeug zur Exploration von großen Informationsräumen, Diploma Thesis, University of Stuttgart, Markus Alexander Wischy, May 2000

USING EYE TRACKING DATA TO INDICATE
TEAM SITUATION AWARENESS

Henning Boje Andersen, Christian Rud Pedersen and Hans H.K. Andersen

Risø National Laboratory, Systems Analysis Department, DK-4000 Roskilde, Denmark

ABSTRACT

This paper proposes that Team Situation Awareness be defined in terms of operators' mutual knowledge of a range of task features - namely mutual knowledge of task distribution, goals and shared task parameters. Mutual knowledge, which is distinguished from shared knowledge and common knowledge, involves higher order intentional states in the form of A's beliefs about B's beliefs about A's beliefs etc. Finally, we suggest that Eye Tracking data be gathered and analysed in observational and experimental studies to indicate operators' perception of their teammates' actions and activities including their gaze.

1. INTRODUCTION

The notion of Situation Awareness (SA) was originally introduced and designed in order to capture - indirectly - a broad type of cognitive failures. Thus, while it is difficult to define the range of cognitive functions that need to be sustained if SA is to be maintained, it is easier to characterise what is common - in cognitive-environment terms - to all situations in which an operator will loose SA. So the label is a highly useful and easily understandable way of referring to what happens in technical work settings when an otherwise skilled human operator gets out of tune with the system to be controlled or the surroundings in which the system (say, an aircraft) operates.

However, most technical work settings involve *teamwork* and typically require co-ordination between teams. So, humans working in crews or teams will co-ordinate their performance between themselves in order to achieve shared work-oriented goals. People working together in a technical setting (cockpit, operating theatre, ATC tower, ship's bridge, etc.) will co-ordinate their response in either an explicit mode (planning, discussing action options) or implicitly by listening to and observing or just catching a glimpse of each other's activities. In this paper we are mainly focused on the implicit co-ordination and its role in establishing and maintaining so-called *Team Situation Awareness* (TSA).

There are some rather distinct lines of background research that we recommend be brought to bear on TSA[1]. These lines of research cover

    (a) the notion(s) of mutual knowledge (belief) as introduced and applied in linguistics (pragmatics - see e.g., Clark, 1986) and philosophy of language (Grice, 1957) and later applied in Artificial Intelligence and game theory.

    (b) the basic notions of intentional relations including beliefs as applied in developmental psychology and ethology (e.g., Barresi & Moore, 1996).

Finally, we suggest that methods for studying co-ordination in terms of gaze and gestures as well as speech derived from ethnographic studies within CSCW (computer supported co-operative work) be applied in addition to traditionally ET data analysis and that theories and results about gaze recognition and social cognition be used as inspirational sources as well.

1. TEAM SITUATION AWARENESS

Before seeking to charactise *team* situation awareness let us recall that, following Endsley (1995), the individual situation awareness of an operator is usually defined in terms of an operator's three-fold accomplishment of three generic tasks: (1) picking up perceptual cues of the system to be controlled; (2) integrating these cues into a coherent and valid dynamic model; and (3) predicting future states. Definitions roughly along these lines (with which we largely concur - but confer the additions to follow) will emphasise that SA consists in maintaining and updating a

---

[1] For reasons of space we cannot here list nor discuss the many relevant references to these background lines of research. However, extensive references are provided in the technical reports by the authors: see e.g., Andersen 1998; 2001; and Pedersen et al., 2001).

coherent representation of the situation that is sufficiently comprehensive and valid to allow the operator to meet current goals.

Looking at the individual operator, this type of characterisation of SA may be expanded and elaborated along the lines mentioned. But if the operator is working in a setting where he or she need to know and make assumptions about the beliefs and awareness of other operators (say, about his or her co-pilot; or about the surgical or the anesthetic team), the characterisation is fundamentally inadequate. For in this type of case, the "situation" of which a competent operator needs to maintain awareness will include not just the system to be controlled but also - and essentially so - the knowledge and awareness of his or her team mates. It hardly needs arguing that the resources required to solve the tasks at hand will involve teammates' knowledge and awareness and priorities; nor that awareness of the "situation" cannot exclude these.

Following the suggestion by a number of authors, we distinguish between an operator's long-term and his or her short-term (situation bound) knowledge of his or her domain and work setting. Thus, Cooke et al. (2000) distinguish between what they call an operator's *mental model* and his or her *situational model.* Whichever label we use, this distinction is clearly relevant and needed. In addition, however, for teamwork to succeed, team members will have formed models of their fellow teammates - so, we may talk about a *teammate model.*

However, we need to apply the same distinction between a long-term and situationally determined model to an operator's conception about and expectations vis-à-vis his or her teammate. An operator will have some generic expectations concerning the knowledge, competence, work goals, practices, norms etc. that a fellow team member will have and follow (this is part of the professional culture) even before they have met. For instance, pilots in larger airlines will meet and fly together for a few days after which they are not liable to work together for several months or even years. These generic expectations correspond to a non-individualised teammate model and are not specific to any given day or situation. Then, for any given work session the operator will form a specific "situational team-mate model" - that is, a dynamic model of his or her fellow team member's current awareness including priorities and possibly his or her shortcomings and strengths. (It goes without saying that a team member's expectations about his or her colleague's competence and norms will be individualised and much richer if they have worked together for some time.)

2. TEAM SITUATION AWARENESS AS MUTUAL KNOWLEDGE
In this section we will argue, very briefly, that TSA needs to be defined in terms of higher-order representations and intentional states (knowledge, beliefs, trust, goals etc.) mutually held by team members. (For details see Andersen, 2001, and Pedersen et al., 2001). Consider the following descriptions of intentional states: (a) the nurse notices the CO_2 level rising; (b) the doctor does not consider that the CO_2 level is abnormal; (c) the nurse thinks the doctor has noticed that the CO_2 level is abnormal, but is not sure; (d) the doctor notices nurses fidgeting with anesthesia monitor and realises she is calling his attention to the CO_2 display (sub-part of the monitor display).

Clearly, statements (a) and (b) refer to first-order intentional states, that is, a subject's state of belief (or non-belief or awareness vs. lack of awareness) , whereas (c) refers to a second-order intentional state - the nurse's belief about the doctor' beliefs. Finally, (d) may be construed as a third-order or even higher-order state. So, using subscript numbers to denote level of intentional state, lets us imagine, for instance, that in the scenario the doctor is extremely busy and the nurse thinks he might be annoyed if alerted in a direct way to a parameter he has already noticed: "I am afraid$_3$ that, if he has already noticed the CO_2, he may think$_2$ that I have little trust$_1$ in him [i.e., believe$_1$ him to be inattentive]".

The CSCW literature that involves ethnographical methods contains somewhat similar observations about the achievement of team members' mutual awareness - and recognition of lack of mutual awareness - of system states. For instance, Heath and Luff (2000) point out in their well-known study of co-ordination in the London Underground Control Rooms that the "mutual availability of the various information allows personnel to presuppose that information available to one is available to all; a presupposition which is dependent upon the systematic ways in which the individuals monitor and participate in each other's actions and activities....For example, a glance towards the fixed line diagram, a gesture towards the radio phone ... can ...provide resources through which a colleague can recognise the actions and activities of another." (ibid., p.121).

On the classical definition of mutual knowledge, two persons, A and B mutually know that p if: (i) A and B both know$_1$ that p; (ii) both know$_2$ that the other knows$_1$ that p; (iii) both know$_3$ that (ii) obtains; and so on.

Clearly, in complex real time domains, team members cannot share all situational knowledge but they must distribute attentional resources. Therefore, it would be misguided to try and define TSA in terms of (just) mutual knowledge of situational parameters. Our proposed characterisation of TSA is therefore (in very brief terms) that it is a necessary (not a sufficient!) condition for team-mates A and B to maintain TSA that they have mutual knowledge of each other's basic professional competence, of their *de facto* shared and *de facto* distributed tasks ("who is monitoring what and who is taking care of which task?"), of the values and significance (interpretation) of parameters within shared tasks, of situational priorities and finally, that each team-mate succeeds in maintaining SA of his or her task domain[2]. In the following sections we describe a proposed technique for assessing mutual awareness of system parameters and we touch briefly on results from a pilot study of operators' use of gaze and visual orientation to co-ordinate and inform each other of actions and concerns.

3. MEASURING TEAM SITUATION AWARENESS AS MUTUAL KNOWLEDGE

Based on the above conception of the cognitive co-ordination involved in real time team work, we have devised a short battery of questions (ATSA: Assessment of Team Situation Awareness) designed to elicit a subjects' estimates of his or her team-mate's situation awareness and view of task allocation in addition to the subjects' own first-order knowledge of significant system parameters and their trends. The brief questionnaire is intended for use (and has been applied in pilot studies of anesthesia simulations involving medical Crew [Team]Resource Management training) during pre-determined interruptions of audio/video prompted *debriefing* sessions immediately following an experimental trial or a training session (conf. Hansen, 1991, for descriptions of this debriefing technique).

The appendix contains the ATSA questionnaire used to query subjects, at pre-determined probes, about their own judgement about individual parameters and their estimate of their teammate's judgements. Results of using the questionnaire during pilot trials (se section 5) involving anesthesia simulations were largely positive. Trainees need about 5 min. of familiarisation with the questionnaire prior to the debriefing session. Debriefing is conducted after the exercise has taken place and the above-mentioned estimates of parameters are elicited as video/audio cued recall.

4. GAZE AND EYE MOVEMENTS SERVING AS CUES TO TEAM-MATE ATTENTION

As alluded to above in connection with the Heath and Luff study of the London Undergrond controllers, operators achieve the greater part of their co-ordination through implicit means. They rely on shared visual and auditive cues in shared work space (shared in the sense that they are - and are known by team members to be - readily available to both team members); they pick up their fellow team members' direction of attention (or lack of directed attention) by noticing gaze direction and they let each other know what they are attending to by direction of head and of gaze.

The ability to shift our attention in the direction towards which another person's eyes are turned seems to be an innate competence. For instance, Hood et al. (1998) report that "infants as young as 3 months attend in the same direction as the eyes of a digitized adult face". Similarly, Langton et al. (2000) note that the structure of the eyes of humans "provides us with a particularly powerful signal to the direction of another person's gaze", and they point out that gaze direction "is analysed rapidly and automatically, and is able to trigger reflexive shifts of an observer's visual attention" (ibid., p. 50).

5. PILOT STUDY OF TEAM CO-ORDINATION AND GAZE

The authors conducted an observational pilot study of anesthetists' performance and visual behaviours during critical patient scenarios in a comprehensive anesthesia simulator.[3] The pilot study was made in order to (1) test the usability and face validity of the Assessment of Team Situation Awareness questionnaire (see appendix) and (2) collect and analyse Eye Tracking (ET) data from realistic, critical scenarios.

---

[2] Several additional conditions - notably integration of information - need to be added to characterise TSA. See Andersen, 2001 for further discussion.

[3] We are grateful for support and generous advice from the Danish Institute of Medical Simulation at the Herlev University Hospital (Copenhagen). Our special thanks go to Anne Lippert, MD, Dept of Anesthesiology, Herlev University Hospital.

But the study was not a clinical trial, there was no dependent/independent variables distinction, and most importantly, the number of trials (i.e.,2) and subjects (i.e.,2+2) was too small to permit generalisations and inferences)[4]. Yet, its results may serve to illustrate the use of ET data to indicate co-ordination mechanisms and phases at which TSA breaks down. During each scenario, the doctor and the nurse wore eye tracker helmets, and recordings of their visual behaviours were supplemented with one video recording of overall operation scenery and one video track of the monitoring screen.

The study revealed that the doctors and nurses, while rarely focusing on each other's gaze, at crucial points seemed to follow each other's line of gaze. There was a preponderance of simultaneous dwells on areas of interest (patient, monitor) though a consistent division of labour was also observed. In addition, the subjects' chief implicit mode of acquiring awareness of their teammate's activities was simply to visually sample what the hands of their teammate were doing. E.g., during one very hectic episode of one of the scenarios when the doctor was busy and highly concentrated administering IV infusion while verbalising his worries and hypotheses to the surgeon and the nurse, the nurse urgently needed confirmation that the drug she had in her hands was the intended one. So she waved the drug label in front of the doctor's field of vision and he nodded. Finally, subjects were observed to perform an additional visual check on the monitoring apparatus whenever their teammate announced a slightly deviant or unexpected values. (Confer Pedersen et al., 2001, for further details).

REFERENCES

Andersen, H.B. (2001) "Team Situation Awareness. Definitions and Measures. Technical Report R-1259, Risø National Laboratory, 4000 Roskilde Denmark. 2001.

Andersen, H.B., Garde, H., Andersen, V. (1998). "MMS: An electronic message management system for emergency response". IEEE Trans. Eng. Manag., 45, 132-140.

Andersen, H.H.K. and Hauland, G. (2000): "Measuring Team Situation Awareness of Reactor Operators During Normal Operation: a Technical Pilot Study. In: Proceedings of the First Human Performance, Situation Awareness and Automation Conference, Georgia, Oct. 15-19, 268-273.

Barresi, J. & Moore, C. (1996). Intentional relations and social understanding. Behavioral and Brain Sciences, 19(1), 107-122.

Clark, H.H. (1996). Using Language. Cambridge: Cambridge University Press.

Cooke, N. J., Salas, E., Cannon-Bowers, J. A., and Stout, R. J. (2000). "Measuring Team Knowledge". Human Factors, 42, (1), 151-173.

Endsley, M. (1995). "Toward a Theory of Situation Awareness in Dynamic Systems", Human Factors, 37, (1), 32-64,

Grice, P. (1957). "Meaning". Philosophical Review, 66. 377-388.

Hansen, J. P. (1991). "The Use of Eye Mark Recordings to Support Verbal Retrospection in Software Testing". Acta Pcychologica. 76, 31 – 49.

Heath, C. and Luff, P. (2000). Technology in Action. Cambridge: Cambridge University Press.

Hood, B.M., Willen, J.D. and Driver, J. (1998). "Adult Eyes Trigger Shifts of Visual Attention in Infants". Psych. Sci., 9, (2), 131-134

Langton, S.H.R., Watt R.J. and Bruce, V. (2000). "Do the Eyes Have It? Cues to the Direction of Social Attention". Trends in Cognitive Sciences, 4, (2), 50-59.

Pedersen, C.R., Andersen, H.H.K. and Andersen, H.B. (2001) "Team Situation Awareness and Gaze". Technical Report R-1269, Risø National Laboratory, 4000 Roskilde Denmark. 2001.

---

[4] Two sessions were conducted, the sessions being different both in terms of the scenario (script) used and trainee team. Each team consisted of a physician in training (2rd or 4th year of specialising) and an experienced anaethetist nurse (5 or 8 years of anaesthesia experience). Each session lasted 35-45 minutes including (20-30 minutes excluding) pre-operative and peri-operative procedures. Both scenarios required tight team collaboration when the critical symptoms were introduced (one scenario involved a surgeon-induced vein puncture, the other involved a severe allergic reaction approaching an anaphylactic shock).

Appendix: Assessment of Team Situation Awareness (ATSA) Debriefing Form [parameters exemplified from anaesthesiology]

Subject name / no.:	Session no.:	Team no.:	Interruption no.:

Please estimate your and your colleague's workload **right now**?	**Very high** (am using all my attention on my tasks; will refuse interruption by external calls)	**Somewhat high** (have few resources to spare; would be reluctant to accept external calls)	**Medium** (am occupied but do not feel any great load; will accept simple external calls)	**Somewhat low** (am slightly occupied; external calls be welcome)	**Very low** (merely monitoring a normal, non-complicated anaesthesia; external calls welcome)
Your own workload - mark one →					
Your colleague's workload - mark one →					

		Pulse	Systo-lic	Dia-stolic	Oxyg.-satu-ration	CO_2
Your own estimate of parameters	Current trend - within the last couple of minutes the parameter has been rising, level or falling (please insert ↑, →, or ↓):					
	Write your estimate of the value NOW of this parameter:					
	Indicate your confidence in your above estimate on a scale from 1 to 10 (1 = entirely uncertain, 10 = entirely certain)					
Your estimate of your colleague's knowledge of parameters	Please enter: **1** (= correct) or **2** (=possibly wrong but not critically so) or **3** (= possibly critically wrong)					
Task allocation	*Whose* task is it to monitor the parameter in this phase? Please enter **M**, **B** or **C** ("primarily **m**ine", "shared=**b**oth of us" "primarily my **c**olleague's"):					
	Do you believe your colleague will agree with your view of task distribution? Write **Y**(es) or **N**(o) or "**–**" (= don't know)					

Combined Analysis of Verbal Protocols and Eye Movements

John Paulin Hansen°, Gunnar Hauland† & Henning Boje Andersen†

°IT-university of Copenhagen, †Systems Analysis Department, Risø National Laboratory, Roskilde, °†Denmark.

ABSTRACT

This paper reviews and illustrates advantages of combining verbal protocols and eye movement recordings in cognitive task analysis. Combining these two types of data may be done in various ways, as we illustrate by examples from ten years of research. The main focus of interest in most applications of cognitive analysis is the subject's use of meaningful information. This suggests that "seen" and "unseen" and thus frequency of dwells on specific targets become important variables. Eye movements and verbal protocols will often complement each other. Ambiguities in utterances can be clarified by examining the eye movements, and the subject himself may explain the rationale behind scanning strategies. Practical advice on how to facilitate the encoding of dwells on areas of interest is provided. Eye movement recording are expected to become widespread within cognitive ergonomics, as new and improved tracking systems are introduced.

1. INTRODUCTION

Cognitive processes play a crucial role in the study of human behaviour within ergonomics. Typical examples of cognitive processes are problem solving strategies, co-operative procedures, reasoning, and users' mental models. Yet, while ergonomists must pay careful attention to cognitive processes, they also realise that these processes cannot themselves be observed directly: they are important but are in principle hidden, that is they cannot possibly be observed directly. Therefore, in research as well as in ordinary life we infer the occurrence of cognitive processes by means of a variety of indicators. They include subjects' verbalisations, their overt actions, and their eye movements. To study cognitive processes, the ergonomist must use these derived data to describe these hidden or unobservable processes. In the context of ergonomics, the subjects' cognitive process are often related to their work scenario. Therefore, it will nearly always be highly relevant to include significant work environment conditions in the overall data set selected, e.g. system status. This paper will describe, from a practitioner's point of view, how the different types of data referred to above may be collected and integrated.

2. VERBAL PROTOCOLS

Language provides the most important window into human cognitive processes. Since these processes are not themselves the subject of direct observation, it is obvious for the investigator to ask the subject to simply state what he is thinking. A verbal protocol is intended to contain aspects and components of thought processes which the subject can observe in his own mind. There are 4 different types of verbal protocols, depending on their use and the manner in which they are collected: in an *interruptive protocol* a task is stopped (interrupted), and the subject is asked to state what his thoughts are about the task or problem he is facing; when the subject has verbalised his thoughts, the task is continued. A *retrospective protocol* collects the subject's responses to questions about the problem solving at the end (in retrospect) of the task. While differing in their time span, interruptive and retrospective protocols are similar in that they both require the subject to report on the mental processes he has just been going through. So in this sense they are both retrospective, and they are thus similar to other verbal data collection formats such as questionnaires and interviews. What discriminates the interruptive from the retrospective protocols are therefore differences in instructions and procedures.

A way of asking a subject "what are you thinking" is simply to ask him to "think aloud" during the process of solving a problem. This method is called *concurrent verbal protocol* because it is intended to be a simultaneous trace of an ongoing cognitive process. A concurrent verbal protocol is fundamentally different from retrospective protocols, introspection, and other types of verbal data because it is an attempt to trace cognitive processes while they are being performed. The concurrent protocol is a transcription of ongoing verbalisation (e.g. recorded on audio or video tape or hard disc), and when it is transcribed, it is usually time-tagged and broken down into columns of statements following a serial timeline of verbalisations. Closely similar to concurrent protocols are recordings and transcriptions of two or several subjects' spontaneous verbalisations during their activities under study. Such protocols, which we suggest be called *naturalistic protocols*, are not usually distinguished from concurrent (think aloud) protocols, but it is nevertheless useful to differentiate between these two types. Thus, naturalistic protocols will typically be collected with no prior instruction by the ergonomist to the subjects since their very purpose is to collect exactly and faithfully what subjects utter during specific working situations. Protocols of this type are in fact similar to what linguists refer to as corpora, i.e. transcriptions of naturalistic samples of speech. A transcription of the tape of a cockpit voice recorder, for example, will provide an analyst with the "naturalistic" communication within the crew and between the crew and others such as cabin attendants, ATC etc.

3. ANALYSIS OF EYE MOVEMENT DATA

A cognitive eye movement analysis will involve an inference from eye movement data to the nature and content of a subject's thinking. But there is a long way from sampling the eyeball position relative to the scull and its surroundings to a semantic analysis of the content of thought.

First of all, there is no standard way of defining a single fixation operationally. Karsh & Breitenbach [1] have defined a fixation as a cluster of minimum data points (e.g., 100 milliseconds) whose deviations from the centroid is less than the maximum resolution set. They believe that the specific minimal duration and the fixation range (the resolution set) must be tailored to the specific needs of the researcher. We have observed that variations of just 25 milliseconds in the definition of a fixation can result in qualitatively different fixation patterns from the same data.

Secondly, there is no simple relationship between the length of fixations and the amount of information obtained. Moray and Rotenberg [2] found that instruments were fixated more frequently after a plant failure, but that dwell times were unchanged. Russo and Rosen [3] found few differences in dwell time between conditions in which different diagnostic strategies were used. Weber and Andersen [4] found that the variation in terms of frequency of dwells and distribution of dwells on targets (instruments and runway) was greater among subjects than among task scenarios, and thus their experiment showed that the variation among pilots for any specific aircraft take-off scenario is greater than the variation for any given pilot across normal and abnormal take-off scenarios.

Thirdly, there are different psychological theories about the relationship between eye movements and cognitive processes. A so-called "strong eye-mind assumption" [5] considers that a fixation will continue until all cognitive processes activated by the fixated word have been completed thus leaving out all other forms of cognition during fixation. This immediacy hypothesis does not hold because one can delay encoding of contents of words in a sentence (as well as of e.g. display elements). The assumption of an identity between attention and gaze direction, a so-called "identity hypothesis", does not hold either because one can process information presented outside fovea.

To overcome these problems, the visual unit of analysis can be defined on a much lower resolution than single fixations of e.g. 100-250 milliseconds. Instead of "one single fixation", this unit of analysis can be referred to as "a dwell" [7]. The main interest in a cognitive gazeline analysis is the subject's use of meaningful information. This suggests that "seen" and "unseen" and thus frequency of dwells on specific targets become important variables. Since a dwell may consist of several single fixations, the target is an area rather than the exact location of a gazeline. To extract information from a visual target representing e.g. a cockpit instrument, the fovea may be moved around on this instrument thus moving the line of gaze within such a target. We shall use the term "area of interest" (AOI) for a meaningful location of dwells.

3.1 Encoding areas of interest

When encoding eye movement recordings, the AOIs that has been established through the task analysis are used as categories. Advanced systems with combined head tracking and eye movement recording make it possible to do this encoding automatically as head tracking allows the system to correct for head movements. Thus, fully automated analyses of visual behaviour are possible for static areas of interest. Such static areas can be graphically represented in a model of the real scene: instead of having a Point-Of-Gaze (POG) marker superimposed on a video image of the scene, the POG marker may be displayed in a 3D model and logged as hits on AOIs [6].

If the AOI is dynamic or moving, or if the content of an AOI on a display is changing, then additional data (capture of video frames or log of display changes) is needed in order to perform automated analyses. In some field settings, for example, it may not be possible to get a log from the system. Thus, the eye movement analyses will be partly automated and partly manual. Furthermore, some analyses also require a more explorative approach, i.e. it may not be possible or desirable to define all AOIs a priori, which is often necessary for automated procedures. Thus, there are several analyses of visual behaviour that may include manual analyses: moving (natural) targets, AOIs that change during dwell time, and explorative analyses. For this, the video image of the scene with a superimposed POG marker is needed. Some eye tracking systems allow a scene video add-on so that the superimposed POG follows the calibration of the integrated eye head calibration.

Manual encoding of eye movement data is feasible at the low level of resolution represented by the AOI approach. Depending on the definition (e.g. type of AOI categorisation, AOI size and number of AOIs), a manual analysis of dwells on AOI can be reliably performed with 1:2 the normal playback speed. In our experience, a 1:1 playback speed can reveal a reliable dwell frequency distribution but with fewer hits for each AOI category [7].

A useful manual scoring support is to designate a button to each of the AOI categories on a concept keyboard with a simple graphical overlay of the AOIs. When a particular button is pressed during the review, its category label may then be written into the database in time synchronicity with the other events stored. The advantage of using such a board is firstly, that the spatial layout of AOIs support memory in a better way than letter or number codes and secondly, that the problem of head down during data coding can be avoided because the AOI representations on the concept keyboard can be seen through the peripheral vision. Ideally, two human factors specialists should carry out their scoring of the same episodes independently with identical sets of categories to control for inter-subjective differences in the scoring.

4. COMBINING VERBAL PROTOCOLS AND AREAS OF INTERESTS

Besides establishing the relation between task specific cognitive processes and the areas of interest, it is crucial to know how the subject actually interprets the information seen. An additional data source is needed to decide this. In our opinion, concurrent verbal protocols or subjects' verbalisations during retrospections of their own eye movements provide important indications.

The use of joint methodologies is strongly advocated because eye movement recording and verbal protocols complement each other. For instance, Hauland and Hallbert [7] have been able to solve partly the problem of incomplete concurrent verbal protocols. By examining the dwells of plant operators, they could identify (disambiguate) the meaning of 64% of the unclear statements from the verbal protocols. For instance, the operators gave statements like: "What is the situation here? It looks normal". By examining the eye movement recording, it was possible to establish the exact reference of "It", cf. Table 1.

TIME	ALARM	AOI	VERBAL PROTOCOL
07'00"	HL:LoLpreszr		"Low level in pressurizer, of course"
07'10"			"It is not getting enough water"
07'20"	HL:circuit radiation		"Oh, yes- we've got a leakage here"
07'30"			"Between the primary and secondary circuit"
07'50"		YA00	"Then I suppose the best thing to do is to shut it down"
08'00"		YB00	"What is the situation here? It looks normal"
08'10"		YD00	"Secondary circuit radiation"
08'20"		YA00	"It should have been possible to detect something on the primary side"
08'30"		YB00	"...that one of them is not as they should be."

Table 1. Example of a joint verbal protocol and AOI encoding. The AOI column refers to process states represented on a control panel.

4.1 Combining concurrent verbal protocols and AOIs: an example from ship navigation

Hansen and Itoh [8] conducted a task analysis of a simulated ship navigation task using a combination of concurrent verbal protocols, eye movement recordings, and external video recordings. First, the cognitive activities were categorized in "decision ladder" terms [9]. On basis of the recordings, the navigation task was then divided into operating modes and monitoring modes. Each operation sequence begins with a scanning of the surrounding environment ("Observe & Identify"), immediately after the detection of an action cue ("Activate"), and it terminates with the actual manipulation of a control device ("Execute") often with a shortcut from "identification" to "procedure formulation". In contrast, the monitoring mode is defined as a sequence of cognitive processes without a final manipulation of a control device; the navigator concludes that no actions need to be taken. Examples of a monitoring mode identified from the verbal protocols were: *"I have started to look at the radar. ... I have a nice approach as far as I can see"* and *" Now I have a slight starboard turn. It is my intention is to go more to the starboard. So, that is OK".*

Data collected made it possible to specify typical task sequences and calculate the amount of time spent on each task element. On this basis, a cognitive model of ship navigation behaving compatible to a real human navigator under various environmental conditions has been developed [10].

4.2 Combining retrospective protocols and AOIs: an example from a usability study

Hansen [11] compared text editor users' retrospective comments, which they provided when they were watching a video recording and a recording of their eye movements. Users were significantly more problem-orientated while retrospecting their eye movements compared to retrospection of a standard video recording. They also verbalized their visual strategies more often during the retrospection of the eye movement recording. Some of the eye movements would not have been intelligible if the user had not explained them: *"Here comes the one I need. Even though I have found the command I need, I just read through the others to be sure..."* or *" "...I was only looking at the words at the top and thinking. Then it might be that my eye are moving, but really I don't think I used it for anything"* (ibid.).

4.3 Combining naturalistic protocols, retrospective protocols and AOIs: an example from anaesthesia

As mentioned above, retrospective operator comments may be elicited by asking operators to comment on an audio/video recording of their own activities during a work scenario they have just been involved in. Their comments, which may be cued by additional recording of their eye fixations (in addition to the a/v recording of the entire work scene) will typically be a highly useful and efficient means of getting data about the "meaning" of their own overt actions (including their visual behaviours), their reasons for doing what they did, as well as explanations of the reference and sense of their, possibly highly truncated, utterances contained in the naturalistic protocol. For instance, in a recent study [12] the authors applied the interruptive technique *after* the actual work session (an emergency anaesthesia scenario) in order not to disturb the performance of the doctor and nurse during the scenario itself. The interruptive protocol was gathered at four pre-determined segments of the scenario; that is, the operators (doctor and nurse) were presented with the a/v recording of the scenario and recording of their own visual behaviours. Then, at each of the pre-determined segments, the recording was stopped (screens blanked out) and subjects were asked to indicate their own estimate of a few important parameters (e.g., pulse rate, CO_2 saturation) and their estimate of their team mate's estimate and their view of current task allocation. In this way, subjects' awareness of system parameters as well as their ability to correctly predict the awareness of their fellow team member (shared and mutual knowledge) was collected.

5. PRACTICAL CONCERNS WHEN COMBINING VERBAL PROTOCOLS AND AOIs.

The recordings of verbal utterances and eye movements should be strictly synchronised – the specific precision required being determined by the analysis task at hand. The log of activation of controls and the log of environmental parameter values should receive the same time tags as the other data sources, but in practice they will be restricted to the timing precision and resolution of the simulator or the equipment that produces the log. It is both effective and useful to have a reference signal emitted by the simulator or task environment (a lamp that goes on for instance) and to have this event recorded simultaneously on the eye movement recording and in the log.

Some practical considerations when recording eye movements are important too. If the visual work space is limited and fixed (e.g. only a single computer monitor) and the operator is constantly located in front of it, remote tracking systems have a great advantage by the fact the subject need not wear any apparatus. But if the operator is moving around or paying attention to several independent information sources, the system has to be head-mounted.

A head-mounted system should be low in weight (less than 600 grams) if the subject is to wear it for more than a few minutes. Some systems are likely to slip if the subject coughs or nods vigorously, and this may then require a disruptive and annoying re-calibration.

The precision required of the system will in general depend on the nature of the subject's tasks or the experiment performed. In general, accuracy of eye tracking systems should be expected to vary from user to user. But systems with a precision greater than 1 degree over a viewing angle of +/- 30 degrees horizontally and +/- 20 degrees vertically are to be preferred.

For most systems, it is not a problem when subjects wear glasses. In some cases, the glasses may have to be tilted slightly to get the IR reflections away from the pupil. Similarly, contact lenses normally do not cause problems. Equally, room lighting should not affect the system. But most IR systems are quite sensitive to direct sunlight and incandescent light sources with a large amount of infra-red light. The sampling rates needed are also dependent on the type of experiment or observation. In most cases, ergonomic analyses of dwells may be performed well at sampling rates around 25 - 50 Hz.

The ease and duration of the calibration process is very important. An experienced eye tracker operator will be able to calibrate adult subjects within 3 to 5 minutes. If the calibration session takes too long (say, more than 10 minutes), the subject may get tired and loose his concentration. If the system has to be re-calibrated repeatedly, most experiments and observations will be seriously disrupted. Some systems require the subject to wear bundles of wires and may restrict his normal viewing angle. These systems of course have to be used with great precaution in experiments or observations where safety is an important issue, e.g. when a subject drives a car in real traffic or administers anaesthesia during a real operation. The final choice of an eye tracking system should also consider whether pupil size data or eye blink data are provided.

Even systems that satisfy the criteria listed above may be rather troublesome to use [13]. However, recent developments in tracking technologies show promising performance under extreme conditions with sunlight and vibrations. The precision of the new systems may not be as high as 1 degree, but for dwell analysis on AOI's, a precision of e.g. 3 degrees will be acceptable in a broad range of real-life task scenarios.

6. CONCLUSIONS

Based on our decade long experience with collecting and analysing eye movements in combination with subjects' verbal protocols, the authors are convinced that this method is a highly useful and sometimes unique source of data for cognitive task analysis. At the same time, while data from eye tracking alone or from verbal protocols alone may be analysed separately, we believe that for the *interpretation* of these data sources it is useful and uniquely informative to apply an analysis that uses information from both sources.

REFERENCES

[1] Karsh, R., & Breitenbach, F. W. (1983). Looking at looking: The Amorphous Fixation Measure. In R. Groner, C. Menz, D. F. Fisher, & R. A. Monty (Eds.), *Eye Movements and Phychological Functions: International Views.* (pp. 53-64). Hillsdale, N.J.: Lawrence Erlbaum Press.

[2] Moray, N., Rotenberg, I. (1989): Fault Management in Process Control: Eye Movements and Action. *Ergonomics; 32,* (11): 1319-1342.

[3] Russo, J.E., Rosen, L.D., (1975): An Eye Fixation Analysis of Multialternative Choice. *Memory and Cognition; 3:* 267-276.

[4] Weber, S. & Andersen, H.B. (2001): Pilots' eye fixations during normal and abnormal take-off scenarios. *HCI International 2001.* Aug. 5-10, 2001, New Orleans, LA, USA.

[5] Just, M.A. and Carpenter, P.A. (1980): A Theory of Reading: From Eye Fixations to Comprehension. *Psychological Review, Volume 84, Number 4,* pp. 329-354.

[6] Andersen, H.H.K. & Hauland, G. (2000): Measuring Team Situation Awareness of Reactor Operators During Normal Operation: a Technical Pilot Study. In: Kaber, D.B. and Endsley, M.R.; Human Performance, Situation Awareness & Automation: User-Centered Design for the New Millenium. *Proceedings of the First Human Performance, Situation Awareness and Automation Conference,* Georgia, Oct. 15-19, pp 268-273.

[7] Hauland, G. & Hallbert, B. (1995): Relations between visual activity and verbalised problem solving: a preliminary study. In: Norros L. (Ed.): *5th European conference on cognitive science approaches to process control,* Espoo, Finland, pp. 99-110.

[8] Hansen, J. P., & Itoh, K. (1995). Building a cognitive model of dynamic ship navigation on basis of verbal protocols and eye-movement data. In: Norros, L. (Ed.): *5th European conference on cognitive science approaches to process control,* Espoo, Finland, pp. 325-337

[9] Rasmussen, J. (1986). *Information processing and human-machine interaction: An approach to cognitive engineering.* New York: North Holland.

[10] Itoh, K., Yamaguchi, T., Hansen, J.P. & Nielsen, F.R. (2001): Risk Analysis of Ship Navigation by Use of Cognitive Simulation. *Cognition, Technology & Work 3,* pp. 4-21

[11] Hansen, J. P. (1991): The use of eye mark recordings to support verbal retrospection in software testing. *Acta Pcychologica.* 76, p. 31 – 49.

[12] Andersen, H. B., Pedersen, C..R. & Andersen, H.H.K. (2001) Using eye tracking data to indicate team situation awareness. *HCI International 2001,* Aug. 5-10, 2001, New Orleans, LA, USA

[13] Hansen, J. P., Hansen, D.W & Johansen, A.S. (2001): Bringing Eye Gaze Interaction Back to Basics. *Proceedings of 1st International Conference on Universal Access in Human-Computer Interaction,* New Orleans.

Eye-Tracking Applications to Design of New Train Interface for the Japanese High-speed Railway

Kenji ITOH*, Masahiro ARIMOTO** and Yasuhiko AKACHI**

*Tokyo Institute of Technology, Tokyo, Japan
**Central Japan Railway Company, Nagoya, Japan

ABSTRACT

The present paper presents an eye-tracking project on new interface design of the bullet train cockpit that adapts to the improved train control system, ATC (Automatic Train Control). We specifically report an eye-tracking application to analysing train drivers' learning processes to a new bullet train interface. eye-tracking data were analysed applying a normative approach because of the task's well-defined property. In this approach, we introduced a principle of "*Gaze Relevance*" that is closely connected with the fundamental concept of safe and stable operations of the bullet train. The gaze relevance is involved by three metrics to uncover train drivers' information acquisition for a specific activity: *gaze economy*, *gaze redundancy* and *gaze robustness*. Based on application results of the proposed approach to train drivers' eye-tracking data in simulator experiments, we discuss their adaptation to a new train system and its interface.

1. INTRODUCTION: DRIVING TASK OF BULLET TRAIN

IT (information technology) applications enable train operators to perform more efficient and adaptive control in the high-speed railway ever than before. A Japanese high-speed train, "Tokaido Shinkansen", is running between Tokyo and Osaka (ca. 550 km distance) at the maximum speed of 270 km/h for two hours and a half. For the high-speed transportation, safe and highly reliable operations of the train are required, and the ATC (Automatic Train Control) system contributes to transportation these requirements as their technological background. In this control system, the upper speed boundary - which is primarily based on the radius of track and distance to a train running ahead or to the next station - is constantly displayed as a traffic signal in a speedometer. When the running speed exceeds the ATC signal, the train is automatically braked to reduce its speed to the upper limit.

In the current ATC system, the signal is changed in the discrete levels, e.g., 30, 70, 170, 220, 255 and 270 km/h. An advanced ATC system is planned to introduce in the next few years. In the new system, the ATC signal is adaptively changed continuously, not in the discrete level, by taking into account a train's braking performance. This allows a train driver to perform more flexible and effective control of the bullet train for the stable operation.

In the normal situation, a driver is required to operate a bullet train following the planned time schedule for all stations to be passed or stopped at. For this purpose, he sets several checkpoints between successive two stations and there he checks the current driving states in progress. Based on the states and much other information, he decides the running speed to the next station, and adjust it with the acceleration lever according to the current state of slope of the track and so forth. During this driving process he is continuously monitoring outside scene and various information sources such as the speed and the ATC signal indicators in the instrument console to maintain his situation awareness and anticipate events and states going on. As can be seen from this task description, under the normal condition, the driving task is performed in skill-based manner (Rasmussen, 1986) and its quality and efficiency highly depend on the driver's visual monitoring and attention allocation within relevant information sources in the train interface and outside environment. For such task characteristics, the eye-tracking technique is useful to analyse train drivers' cognitive processes toward various ergonomic purposes such as interface design, job redesign, and design of a training programme and of an operation procedure (Itoh et al., 1998; and 2000).

In this paper, we mention an eye-tracking application to cognitive task analysis of the train driver's learning with the new interface. As an analysis framework employing eye-tracking technique, we present a principle of "*Gaze Relevance*" that is involved drivers' information acquisition strategies from the aspect of efficient and reliable attention. Train drivers' learning processes are speculated based on the application results of this analysis framework to their eye-tracking data recorded in two-day experimental sessions.

2. ANALYSIS FRAMEWORK OF BULLET TRAIN OPERATIONS

2.1 Principle of Gaze Relevance

In the normal driving situation, a task structure can be described clearly in advance: task goal, constraints, working procedure and human mental processes. A *normative approach*, which prescribes how a system and/or an operator should behave, is suitable to employ for this category of well-defined task. As a process/activity-based analysis from a normative point of view, we propose a principle of "*gaze relevance*" as a framework for analysing eye-tracking data by comparing with an ideal eye-gaze sequence for each specific activity or process in the task.

The concept of gaze relevance comprises three subordinate metrics: gaze economy, gaze redundancy and gaze robustness. The *gaze economy* is referred to as a metric of how economically an operator acquires information required for a specific activity. This can be paraphrased as how promptly the required information acquisition is completed after an activity cue is provided to an operator. The other two metrics are relating to reliable activity ensured by multiple or redundant inspections, not only a single input of the required information. The *gaze redundancy* is a metric on which the information relevant to an activity is examined multiple times not only from different information sources but also from the same information sources. This metric can be evaluated by the number of fixations or fixation duration at

1328

relevant information within a certain time interval after an activity cue is provided. Relevant information pieces only from different information sources are counted as robust information pieces to ensure more reliable activities for the *gaze robustness*. This metric is particularly important to perform reliable operations even with a uncertain system environment by malfunction or faulty system components.

2.2 Operation Model and Hierarchical Information Description

With the introduction of the new ATC system, there exists a major change in the driver's operations when the bullet train is under control of the ATC system and immediately before this situation. A cognitive model for the changed operation is depicted in Figure 1, applying the scheme of ITM (Information Transition Modelling) (Itoh, 1998). As can be seen in this figure, the driving operations consist of the following seven activities: (1) anticipation of the forthcoming ATC-control area, in which the ATC signal is reducing continuously, (2) confirmation of entering the ATC control area, (3) anticipation of the forthcoming ATC braking area, in which the train speed reducing with the ATC brake control, (4) confirmation of applying the ATC brake, (5) anticipation of making free from the ATC brake, (6) confirmation of non-ATC brake control, and (7) confirmation of change to a higher ATC signal.

A driver is required to acquire corresponding information piece(s) to each activity in. For example, for Activity 3, he judges whether an ATC brake will be soon applied based on the information on estimated time or distance to the ATC braking area. This information can be directly acquired or generated alternatively by one or more information pieces displayed in the instrumental console. Such a relation between information pieces on the interface and the required information for each activity is represented hierarchically as shown in Figure 2. In this figure, information pieces laid at Level 0 are the ones directly required to perform the activity. Some of Level 0 information pieces are

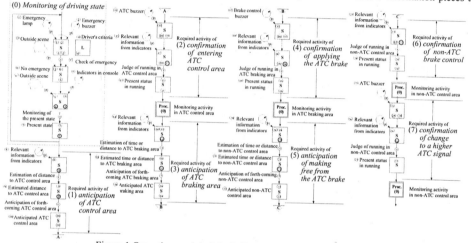

Figure 1 Operation model of the bullet train under control of the ATC system

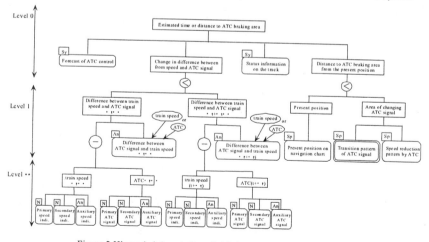

Figure 2 Hierarchal description of driving information for Activity 3

themselves displayed in the train interface and the others are generated by one or multiple Level 1 information pieces. As such, Level 2 information pieces are subordinate ones immediately to Level 1. In this hierarchical description, an information piece displayed in the interface is described in a node with round-shaped corners while one depicted in a rectangle node is generated by the driver's mental operation. For information pieces presented in the interface, this figure also describes their displaying form: analogue (abbreviated as "An" in Figure 2), digital or numerical (N), verbal (V), symbolic (Sy), temporal (T), and spatial (Sp) representation. Through this information generation process, we assume that the driver performs only simple arithmetic and relational operations to produce a higher level of information piece. For example, the information piece on "the difference between the train speed and the ATC signal", which is laid at Level 1 in Figure 2, is generated by an arithmetic operation, subtraction, from two lower level information pieces, "the train speed" and "the ATC signal". The latter two information pieces are laid at Level 2, and these can be obtained from either one of three ATC/speed indicators

2.3 Calculation of Gaze Relevance Metrics

A schematic example of gaze relevance metrics calculation is shown in Figure 3. For gaze economy, a calculation scope is spanned from the start of cue input for a particular activity to its completion, i.e., at the moment when enough information is acquired to achieve the activity. The gaze economy is defined as percentage fixation(s) at information piece(s) required for an activity over total fixations during this interval. Applying the number-based procedure, the gaze economy ratio is calculated as the number of fixations at required information divided by the total number of fixations to complete the activity. In the example shown in Figure 3, only the third fixation in the sequence, Gaze No.3, looking at "Forecast of ATC control", was required enough to achieve this activity, i.e., for the first inspection, while Gazes No.1 to 3 were performed during this interval. Accordingly, the ratio is calculated as 1/3=33%. In calculating a duration-based ratio, data on the gaze duration are applied to instead of those of the number of fixations. In this example, the duration-based gaze economy ratio is obtained by 0.20/(0.52+0.32+0.20)= 19%.

To produce gaze redundancy and gaze robustness ratios, we need to define the relevance level, i.e., how much time an activity is repeated reliably as multiple inspections. When the relevance level is set at 2, the calculation scope is ranged from the cue input for the activity to the moment when the second inspection is completed, and similarly to the end of the third inspection in the relevance level of 3. The gaze redundancy ratio is calculated by dividing all the fixations at relevant information by the total fixations in the corresponding calculation space. In calculating the gaze robustness, we define a robust information piece as only the first fixation at each relevant information piece in its calculation scope. A gaze robustness ratio can be obtained by dividing a set of robust information by a set of the total fixations. Supposing the relevance level of 3 in the example of Figure 3, the last fixation at "Auxiliary speed indicator", Gaze No. 8, was the second watch in the calculation scope, and therefore it was removed from a set of relevant information to form a set of robust information.

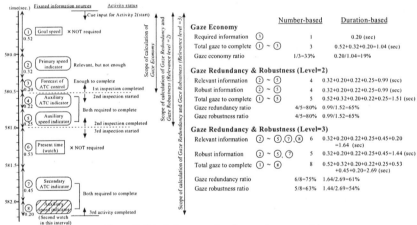

Figure 3 Calculation example of gaze relevance metrics

3. EXPERIMENTS

The bullet train company usually provides drivers with a two-day training programme so that they can well adapt to a new train system and its interface. Following this standard of training systems, drivers' learning processes with the new interface were examined by use of eye-tracking data during two-day experimental trials. Six bullet train drivers participated in the experiment, in which they performed a series of driving sessions with a NAC eye-mark recorder (Model 8) using a bullet train simulator. They ranged from age 25 to 45 years (averaged 35.3 years) with 2-19 year experiences of bullet train operations (averaged 9.2 years). All the participants had normal vision without correction.

The simulator handled the dynamic behaviour of the bullet train with a VCR-projected screen approximately 1.5 metres in front of the driver. The simulator equipped the identical interface to that in the actual train cockpit except for an instrument console designed for the new ATC system. The instrument console examined in the experiment had two major displays, a speedometer and a navigation display, as shown in Figure 4. The speedometer included most information pieces indispensable for driving operations of the bullet train, e.g., speed indicators, ATC signal indicators, and several emergency lamps. The navigation display presents a driver higher abstraction level and detailed information on driving states which helps him to maintain his situation awareness, e.g., ATC signal transition chart and forecast of speed reduction pattern and distance and time to the next station.

A task performed in the experiment was driving operations in the normal situation, viz., to operate the bullet train to follow the planned time schedule as precisely as possible for the same driving area – ca. 70 km distance taking 15-20 minutes – with various driving scenarios. Each scenario included descriptions of a planned time schedule, thus allowance time to the next station, position of slow-down area or no such an area, congestion state on the track, and whether to be passed or stopped at the next station. We selected seven typical scenario for experimental sessions: four standard scenarios having moderate time allowance (15 seconds – one minute), two longer allowance scenarios (3-4 minutes), and two stressful scenarios running on the congested track.

During a two-day experiment - taking about eight hours a day - each subject was instructed on the new ATC system and its interface design as well as the experimental procedure for approximately one hour. Then he performed two training sessions without an eye-mark recorder to learn the new interface. During these sessions, the experimenters provided the subject with suggestions, e.g., on how to operate new interfaces and how the train behaved in the new ATC system, when he needed. After the training sessions, he performed 13 experimental sessions in total, during which his eye-movement data were recorded. As a standard procedure of the experiment, he performed a block of two or three sessions successively for about one hour. After each block of experimental sessions, he took a one hour break, during which he was checked by the experimenters how well he could adapt to the new interface and was provided with the instruction on the interface repeatedly. Then, he returned to the next block of experimental sessions. He performed five experimental sessions in the first day, and the rest of experimental sessions were performed on the next day. After the subject completed the experimental sessions, we obtained his subjective preferences on the new interface design and his comments on how easily he adapted to the new interface both using questionnaire and by a debriefing process with replaying eye-tracking video.

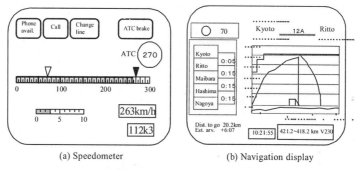

(a) Speedometer   (b) Navigation display

Figure 4 Designed interface of bullet train

4. RESULTS

Transitions of the gaze relevance metrics with sessions are shown applying the duration-based calculation for two subjects in Figure 5. In this figure, averaged ratios over Activities 1 through 6 are plotted with sessions for the three metrics in the relevance level of 2 for gaze redundancy and gaze robustness. In several sessions, all the activities, 1 through 6, were not performed during an entire period, and such sessions were excluded from graph plotting in this figure. As for the gaze economy, Subject 2 (S2) employed a slightly more efficient gaze strategy than Subject 1 (S1). From the transition of this metric with sessions, S1, who had the shortest professional experience of the subjects, does not seem to have learned with the new interface until Session 5. After this session, however, this metric was improved with session, particularly the ratio in the last session was very high despite using the most difficult scenario. For the gaze economy of S2's, who had the longest experience, a changing point exists betweens Sessions 3 and 4. Until Session 3, its ratio was about 10% lower than the later sessions, where the ratio was almost constant at higher level. This may indicate that this subject could adapt well to the new interface after the fourth trial. For example, the gaze economy is increased by 15% after his leaning, comparing by the same difficult scenario (Sessions 3 and 13).

Similar patterns can be seen for S1's transitions of the gaze redundancy and gaze robustness, with the latter metric about 10 % lower than the former. Like his gaze economy, the ratios of these two metrics in the first half were changed up and down and then improved in the later sessions. Integrating this result with that of gaze economy, this subject may have needed more learning sessions to adapt to the new interface. In contrast, S2's gaze redundancy was constant at higher rate having a little deviation. Regarding his gaze robustness, it seemed to be decreased until Session 3, and then to be increased after this session. All these three metrics can be influenced not only by the learning level but also by other individual and task factors. The scenario used in Sessions 3 and 13 was the most difficult while those of Sessions 1 and 2 were the easiest to follow the planned time schedule. This may have affected to Session 3 as decrement ratio of

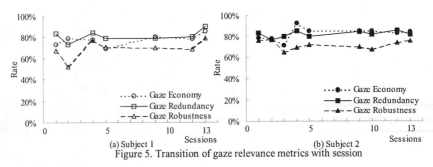

(a) Subject 1

(b) Subject 2

Figure 5. Transition of gaze relevance metrics with session

the gaze robustness. As his learning effect, the gaze robustness ratio was improved with sessions using the same scenarios, e.g., Sessions 3 and 13; Sessions 4 and 12.

In the metrics calculation mentioned in this section so far, the duration-based procedure was applied to the two subjects' eye-tracking data. To discuss compatibility between the number- and the duration-based calculations, we performed the correlation analysis using data samples of combination of subjects, sessions and activities. As a result, highly significant correlations were obtained in the three gaze relevance metrics: $r=0.926$ (gaze economy), 0.932 (gaze redundancy), and 0.964 (gaze robustness). This may suggest to have possibility to apply the number-based procedure, which is more time-saving, to eye-tracking data with enough reliability in analysing human cognitive performance.

5. CONCLUSION

In the present paper, we proposed a principle of "Gaze Relevance" and its calculation procedures for analysing a well-defined cognitive task, following a normative approach. This principle was applied to analysis of the bullet train drivers' learning processes with the new ATC adapted interface for managing its effective replacement from the present system, including design of a training programme. In addition to this analysis, the drivers' learning processes were also uncovered from macro-scopic viewpoint by using eye-movement data during an entire task though we did not mention it. As one such analysis, transition networks of attention allocation during an entire task were generated to represent the driver's overall performance in state monitoring process. As a performance index on stability of attention allocation, entropy was calculated based on the transition network. Also, a distribution of eye-gaze duration allowed us to speculate some characteristics of a particular information piece such as its degrees of importance, usefulness, ease of use and visual complexity in the instrument console.

The present project had two other different objectives on installing the new interface to the bullet train. For these objectives, we also conducted a series of eye-tracking experiments to evaluate alternative interfaces. As another IT application to the interface design, we examined effects of a higher level driving information, e.g., the optimal speed which guides the train to reach the next station in time, not only on operating performance but also on other human characteristics such as vigilance, workload, fatigue and subjective preferences. The proposed principle of gaze relevance and its analysis procedure were also applied to these project objectives, and several useful suggestions were obtained for design of the new bullet train interface.

In a future project, we will tackle with another category of bullet train operations, namely so-called *ill-structured* tasks. A typical example of this category is operations in an emergency situation such as treatment of control system's sudden malfunction or fault in the train cockpit. In this situation, task goal as well as its constraints can neither be defined in advance. Nor its procedure exists before such a event takes place. There may be many possibilities of actions to treat with this situation and its outcome may depend on which action sequence is taken. For these characteristics, we plan to apply a *formative approach* (Vicente, 1999), which describes requirements that must be satisfied so that a new train interface and a driver can behave in a desired way, to cognitive task analysis of operations in abnormal situations.

ACKNOWLEDGEMENTS

We would like to acknowledge Shuji Komatsu, Norihiro Shimizu and Toshio Itonaga, Central Japan Railway Company, for their corporation in this project. We also thank Hirotaka Aoki, Hironao Tanaka, Takashi Haraguchi and Akira Ishii, Tokyo Institute of Technology, for their assistance in the experiments.

REFERENCES

Itoh, K. (1998). A Cognitive Modelling Scheme for Task Analysis and Its Application to Computer Editing Task, *International Journal of Cognitive Ergonomics*, Vol. 2, No. 3, pp.269-295

Itoh, K., Hansen, J.P. and Nielsen, F.R. (1998). Cognitive Modelling of a Ship Navigator Based on Protocols and Eye-movement Analysis, *Le Travail Humain*, Vol. 61, No. 2, pp. 99-127.

Itoh, K., Tanaka, H. and Seki, M. (2000). Eye-movement Analysis of Track Monitoring Patterns of Night Train Operators: Effects of Geographic Knowledge and Fatigue. *Proceedings of the 14th Triennial Congress of the International Ergonomics Association*, IEA 2000. Vol. 4, pp.360-363, San Diego, July-August.

Vicente, K.J. (1999). *Cognitive Work Analysis: Toward Safe, Productive and Healthy Computer-Based Work*. Lawrence Erlbaum Associates, Mahwah, NJ.

Empirical Evaluation of a Novel Gaze-Controlled Zooming Interface

Marc Pomplun, Nada Ivanovic, Eyal M. Reingold, and Jiye Shen

Department of Psychology, University of Toronto, 100 St. George St., Toronto, Canada M5S 3G3

ABSTRACT

In the present paper, we present a novel gaze-controlled interface. It allows the user to magnify and inspect any part of an image by just looking at the part in question and subsequently shifting gaze to another window. No manual input is required to control this process. The interface was empirically evaluated in a multi-session experiment employing a comparative visual search tasks that required several steps of zooming in and out of a search display. Each participant's performance was assessed separately for using gaze control and using a mouse as the input device and compared between conditions and across sessions. The results demonstrate that participants' performance with the gaze-control interface is quite comparable with a standard mouse input device and that using the gaze-control interface can be learned very quickly.

1. INTRODUCTION

When we interact with computers today, we typically feed input into the computer with our hands controlling a mouse or keyboard, while we employ our eyes to examine the computer's output on a screen. Intuitively, it seems beneficial to this process to also use the eyes as input devices - by measuring eye movements indicating shifts of attention – and thus eliminate the motor control loop for arm and hand. On the one hand, eye movements can indeed be used as a fast and direct means of controlling a computer, but on the other hand, they are not completely under conscious control, so when using gaze-control interfaces we may sometimes trigger actions unintentionally (the "Midas-Touch Problem"). It is a crucial point in the design of gaze-controlled interfaces to minimize such undesirable effects.

Several different kinds of gaze-controlled interfaces have been developed for handicapped persons (e.g., Frey, White & Hutchinson, 1990; Levine, 1981; Parker & Mercer, 1987; Spaepen & Wouters, 1989). The most widely employed paradigm in this field of research is "typing by eye", which enables the user to type text by fixating and thereby "pressing" keys on a virtual keyboard displayed on a computer screen (e.g. Stampe & Reingold, 1995). Although gaze-controlled interfaces are particularly useful for handicapped users, their applicability can be extended to facilitate human-computer interaction in normal population as well. A promising candidate for gaze control is one of the classic interface functions, namely zooming in and out to view an image (for instance, a map, diagram, or camera picture) at different resolutions.

Goldberg and Schryver (1995) proposed a gaze-controlled zooming interface analyzing the spatial and temporal distribution of preceding fixations to determine whether the user intends to zoom in, zoom out, or keep the current zoom level. Their methodology first collects samples of eye-gaze locations looking at the stimuli just prior to the user's intent to zoom, which are subsequently broken into temporal snapshots and connected into a minimum spanning tree, and then clustered according to user-defined parameters. A multiple discriminant analysis that uses cluster size, gaze position and pupil size statistics is then performed to formulate optimal rules for assigning observations into zoom-in, zoom-out or no-zoom conditions. Goldberg and Schryver did not report the results of an actual implementation of their proposal, but such an interface would inevitably have two drawbacks: First, the analysis of fixation patterns would require a sufficient number of fixations to determine the user intent. Since people make about two to four fixations per second, there would be a dragging delay of at least one or two seconds between a change in the user's intent and the system's response. Second, given the variability of eye-movement patterns across different situations and users, the system would perform a certain proportion of misinterpretations, causing actions that are unintended by the users who lose control over the system.

When implementing a different kind of interface, Jacob (1991) avoided such problems by dividing the computer screen into two windows presented side by side. Window A was a geographic display of ships, and window B showed some information about one of the ships, namely the last one the user had looked at in window A. This way, users could select one of the ships in window A by means of eye fixations and then read information about that ship in window B. Originally, Jacob (1991) predicted that computer commands triggered by eye movements would be difficult for subjects to get accustomed to, since eye movements are naturally used to screen the environment and are relatively hard to control. However, the results emphasized the advantage of using eye movements to activate system commands, in comparison to the more standard, button-press mouse device.

In the present paper, we used a similar two-window design to build a convenient and reliable gaze-controlled zooming interface. Our implementation employs two square windows *A* and *B* of the same size. Window *A* on the left shows the whole image. When users look at window *A*, a square, highlighted selection marker follows their eye movements through the picture. As soon as users switch their gaze to window *B* on the right ("zoom in"), it displays a magnified part of the image – the last one that was selected in window *A* by looking at it for at least 120 ms. As long as users inspect window *B*, the screen does not change, i.e., the selection marker in window *A* remains in the position corresponding to the image part shown in window *B* (see Figure 1). This enables users to switch back to window *A* ("zoom out") without losing their bearings. The interface only provides two magnification levels. More levels would require either more windows or gaze-triggered zoom selection buttons at the expense of resolution or efficiency respectively. However, our interface allows quick and reliable zooming operations that are suitable for a variety of tasks, given an appropriate choice of magnification factor between the two windows.

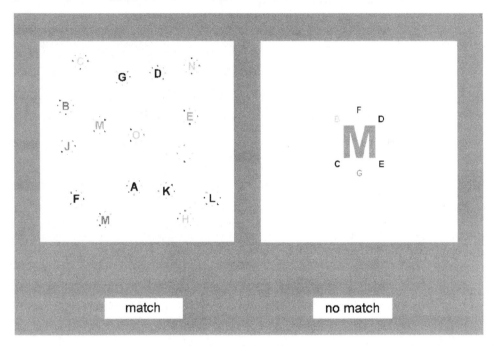

Figure 1: Layout of the zooming interface. While users inspect the left window showing an overview image, a highlighted square is continuously centered on their gaze position. As soon as the gaze switches to the right window, the area highlighted prior to the eye movement is shown magnified in the right image. The two fields labeled "match" and "no match" can be triggered by looking at them for a minimum duration and were used to register participants' response in the experimental task.

In order to evaluate the efficiency of gaze control with regard to our interface, we had participants perform in a dual-scale variant of comparative visual search (see Pomplun, Reingold & Shen, in press; Pomplun, Sichelschmidt, Wagner, Clermont, Rickheit & Ritter, 2001) requiring multiple zoom-in and zoom-out operations in each trial. The use of gaze control was compared to the use of a standard input device - a computer mouse – in the same task. To investigate practice effects, participants' performance was assessed in multiple sessions for each of the two control modes. Since all participants were used to controlling computers with a mouse, we expected only small practice effects for the mouse condition, but large effects for the gaze condition.

2. METHOD

2.1 Participants

Four undergraduate students from University of Toronto and York University participated in the study. All of them had normal or corrected-to-normal visual acuity and had no color vision defects. They were aware of the fact that the purpose of the study was to compare two different user interfaces.

2.2 Apparatus

Eye movements were recorded with the SR Research Ltd. EyeLink system, which is a video-based eye tracker operating at a sampling rate of 250 Hz (4 ms temporal resolution) and measures a participant's gaze position with an average error of less than 0.5 degrees of visual angle after a 9-point calibration at the beginning of the experiment. Before each trial, a 1-point re-calibration was performed to compensate for possible shifts of the system's headset. By default, only the participant's dominant eye was tracked in our study. The EyeLink system uses an Ethernet link between the eye tracker and display computers for real-time saccade and gaze-position data transfer. Stimuli were presented on a 19-inch Samsung SyncMaster 900P monitor with a refresh rate of 120 Hz and a screen resolution of 800 by 600 pixels. A gaze-contingent marker was implemented, which followed the participant's eye movements with an average delay of 14 ms.

2.3 Stimuli

On each trial, two square windows with a length of 10° appeared to the left and right of the center of the screen (see Figure 1). While the right window B was initially blank, the left window A contained a random distribution of 16 large letters with diameters of about 0.5°. All of these letters were different and had distinct colors, except for one pair of identical letters that also had the same color. Each of the large letters was surrounded by a set of eight circularly arranged small letters A to H (diameter 0.07°) in distinct colors. The positions of the small letters A to H and their colors were randomized for each large letter and across trials. Participants had to decide whether the two identical large letters shared an identical small letter in the same color - for example, a green B - which was the case in 50% of the trials. The task required zooming operations, because due to the screen resolution of 800 by 600 pixels, the small letters were illegible in window A. In order to magnify part of the search display, participants could move their gaze to the right window B. During such a saccade, a four-fold magnification of the area selected during the last fixation in window A was painted into window B. Thus, in window B, the large letters had diameters of 2° and the small ones diameters of 0.28 degrees, which made them clearly legible. After participants switched back to window A, window B was blanked again. Since it was virtually impossible to memorize all letter-color combinations at a time, many steps of zooming in and out were required to complete the task.

2.4 Procedure

Participants performed in two experimental conditions: In the "gaze" condition, participants controlled the computer with their gaze as described above, and in the "mouse" condition, they controlled the selection marker in window A with a computer mouse and pressed a mouse button to have the selected area magnified and displayed in window B. Once they determined the match or non-match, participants selected a corresponding key located in the bottom of the screen and made their response by clicking on it (mouse condition) or fixating on it for at least 500 ms (gaze condition). Each participant completed six sessions in intervals of approximately 48 hours. Each session consisted of a block of mouse trials and a block of gaze trials, each of which included 50 trials with short breaks after every tenth trial. The order of experimental conditions was counterbalanced across participants. Response times, error rates, and the number of magnifications per trial were taken as efficiency measures and compared across the two experimental conditions and across sessions.

3. RESULTS AND DISCUSSION

Each participant's response time, error rate, and number of magnifications (zoom-in operations) were recorded as efficiency measures across the two interface conditions and the six sessions (see Figure 2). To reduce noise, only data of the "no match" trials with correct response were considered, because in these trials, participants had to search through all eight small letters surrounding each of the matching large letters. Due to the small number of participants, no statistical analysis of the data was conducted.

Generally speaking, the results did not reflect strong practice effects on either of the two input modalities. Participant #1 exhibited faster response times in the gaze condition in the later sessions, but her error rate increased correspondingly, suggesting a potential speed-accuracy trade-off in performance. Interestingly, participant #2 presumably had short-term practice effects in the gaze condition. She produced longer response times and more magnifications in sessions 1, 3, and 5, in which the gaze condition trials preceded the mouse condition trials, than in sessions 2, 4, and 6, in which the mouse condition trials preceded the gaze condition trials, leading to a zigzag pattern in the corresponding diagrams. For participant #3, the only clear indication of a practice effect was a very slight negative slope in the number of magnifications along the six sessions in both conditions. Finally, participant

#4 demonstrated a considerable, almost linear practice effect across sessions resulting in approximately the same efficiency gain in both conditions. His error rate and number of magnifications, however, did not vary considerably over time. Taken together, very minimal practice effects were observed and were mostly similar for the gaze and the mouse interfaces. This indicates that practice effects were largely attributable to task demands rather than to differences between the two interfaces.

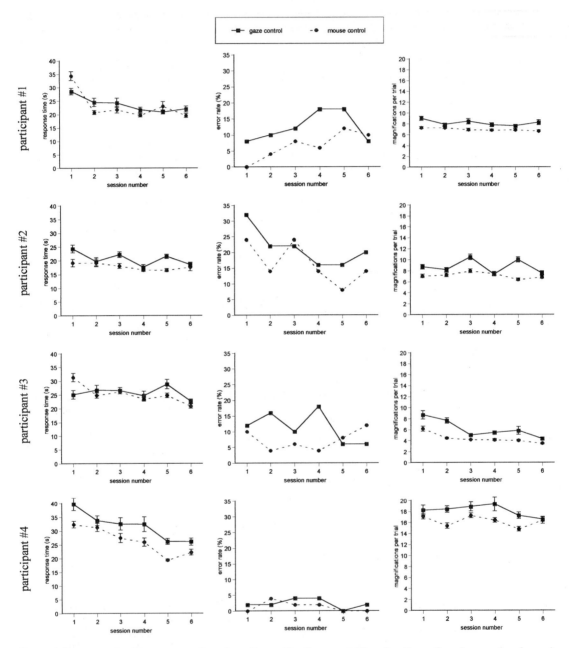

Figure 2: Response time, error rate, and number of magnifications per trial as functions of session number for each of the four participants and each interface condition (gaze vs. mouse). Error bars indicate standard errors within each session.

Comparing the absolute efficiency for gaze and mouse control, we find that, on average, more magnifications per trial were triggered in the gaze condition (10.28) than in the mouse condition (8.69). There are two potential factors contributing to this difference: First, in some trials, the gaze-position measurement may have been imprecise due to headset shifts, so the interface did not magnify the letter that the participant looked at, but instead magnified a neighboring item. Second, similar to the Midas-Touch Problem, when inspecting a particular letter in window *A* and intending to magnify it, participants may unintentionally have looked at another letter before switching to window *B*, so the wrong letter was magnified. When designing the gaze interface, we empirically tested a range of duration thresholds for triggering the magnification function in order to find the best compromise for precise control and quick responsiveness. The minimum dwell time of 120 ms was determined to provide both features appropriately, but obviously cannot completely eliminate unintentional selections.

The sporadic occurrence of unintentional selections may have contributed to the finding that using the gaze interface led to more incorrect responses (11.83%) than using the mouse interface (7.92%). Since all items in the comparative search task looked somewhat similar – large letters surrounded by small ones – in some cases participants probably did not notice when an unintended item was magnified and consequently produced an incorrect response. Conceivably, this might not happen so easily for different stimuli, for example, when applying the zooming interface to pictures or video streams showing real-world scenes, and therefore should not be considered a general shortcoming of the interface.

Finally, response times were slightly longer in the gaze condition (25.44 s) than in the mouse condition (23.19 s). This difference was smaller than 10%, demonstrating that, despite occasional unintentional selections for magnification, using the unfamiliar gaze interface and wearing the eye tracker headset did not considerably slow down participants' task performance as compared to the familiar mouse interface. Given that practice effects in the current experiment were small and similar across conditions, it seems that gaze control can be considered a valid alternative to standard control methods in suitably designed interfaces such as the one described here.

The findings of the current study suggest that the potential of gaze-controlled interfaces has not yet been fully realized. Contrary to Jacob (1991), who predicted that computer commands triggered by eye movements would be difficult for subjects to get accustomed to, and Goldberg and Schryver (1995), who suggested that a gaze-control interface is an inappropriate substitution for a mouse device, our interface shows that eye movements can in fact substitute manual input. Further empirical research is necessary to develop algorithms that can determine the users' intent from their spatiotemporal eye-movement patterns more reliably and hence minimize the Midas-Touch Problem. Appropriately designed gaze-controlled interfaces could then more and more replace conventional interfaces in areas of application where hands-free interaction with machines is desirable. Moreover, advanced gaze-controlled interfaces for handicapped people could considerably improve many aspects of their lives.

REFERENCES

Frey, L. A., White, P. K. & Hutchinson, T. E. (1990). Eye-gaze word processing. *IEEE Transactions on Systems, Man, and Cybernetics*, 20(4), 944-950.

Goldberg, J.H. & Schryver, J.C. (1995). Eye-gaze determination of user intent at the computer interface. In J.M. Findlay, R. Walker & R.W. Kentridge (Eds.), *Eye Movement Research: Mechanisms, Processes, and Applications*. Amsterdam: Elsevier.

Jacob, R.J.K. (1991). The use of eye movements in human-computer interaction techniques: What you look at is what you get. *ACM Transactions on Information Systems, 9* (3), 152-169.

Levine, J. L. (1981). *An Eye-Controlled Computer*. Yorktown Heights: IBM Thomas J. Watson Research Center, Res. Rep. RC-8857.

Parker, J. L. & Mercer, R. B. (1987). The ocular communicator: a device to enable persons with severe physical disabilities to communicate using eye movements. *Exceptional Child*, 34(3), 221-226.

Pomplun, M., Reingold, E.M. & Shen, J. (2001, in press). Investigating the visual span in comparative search: The effects of task difficulty and divided attention. *Cognition*.

Pomplun, M., Sichelschmidt, L., Wagner, K., Clermont, T., Rickheit, G. & Ritter, H. (2001). Comparative visual search: A difference that makes a difference. *Cognitive Science, 25* (1), 3-36.

Spaepen, A. J. & Wouters, M. (1989). Using an eye-mark recorder for alternative communication. In A. M. Tjoa, H. Reiterer and R. Wagner. (Eds.), *Computers for Handicapped Persons*. Vienna: R. Oldenbourg

Stampe, D.M. & Reingold, E.M. (1995). Selection by looking: A novel computer interface and its application to psychological research. In J.M. Findlay, R. Walker & R.W. Kentridge (Eds.), *Eye Movement Research: Mechanisms, Processes, and Applications*. Amsterdam: Elsevier.

An Assessment of a Live-Video Gaze-Contingent Multi-Resolutional Display

Eyal M. Reingold[*], Lester C. Loschky[+], Dave M. Stampe[*] and Jiye Shen[*]

[*]Department of Psychology, University of Toronto,
100 St. George Street, Toronto, Ontario, Canada M5S 3G3; E-mail: reingold@psych.utoronto.ca
[+]Department of Psychology, University of Illinois at Urbana-Champaign & Beckman Institute,
405 N Mathews Ave., Urbana, IL 61801; E-mail: loschky@uiuc.edu

ABSTRACT

This study investigated the effects of gaze-contingent multi-resolutional displays (GCMRDs) on target acquisition in live video. In GCMRDs, an area centered on gaze position is displayed in high-resolution, while outside regions are presented in lower resolution resulting in substantial video compression. Participants had to detect targets in video clips shot from a helicopter flying over landscapes. Small moving targets in the form of hollow circles, colored to match the background (but with increased luminance), were added to the clips. The task for the participants was to visually acquire (fixate) the target as rapidly as possible, and track it until the video clip ended. We manipulated the size of the high-resolution circular area (window). All other regions were blurred. There were four display conditions: 1) blurred - no window, 2) small window (1.5° radius), 3) large window (3° radius), and 4) nonblurred - all high-resolution. The results of the experiment were rather counter-intuitive. Whereas performance in the nonblurred condition was clearly the best, the blurred condition resulted in faster acquisition times than either of the window conditions. These results are explained with reference to visual attention and their implications for GCMRDs are discussed.

1. INTRODUCTION

Advances in computer imaging technology are revolutionizing applications for defense, medicine, industry, communications, education, and entertainment. They have made possible very realistic and immersive flight simulators for pilots, allowing them to train and practice difficult maneuvers without risk to people or equipment (Hughes, Brooks, Graham, Sheen, & Dickens, 1982), allowed face-to-face communication of people thousands of kilometers apart (Agostinho Pavao, Carvalho, & Pacheco da Rocha, 1998), and even enabled a surgeon to perform surgical operations on a patient a continent away (Rovetta et al., 1993). Much more is possible, but there is a major hurdle to overcome. Many of these applications require computationally intensive imagery, which is either beyond the reach of current technology, or is too expensive. Such a problem exists in intensive graphics applications when the computational requirements of generating images exceeds the available resources for image generation, or in communication when the bandwidth requirements of transmitting images (usually expressed in bits per second - bps) exceeds the carrying capacity of the communication link (henceforth referred to as the bandwidth problem). Conventional compression techniques are not sufficient to perform the enormous reductions of bandwidth required for the most demanding display and imaging applications.

Interestingly, the human visual system solves a similar problem of providing high-resolution, wide field of view vision using a finite amount of brain resources and limited capacity communication links (the optic nerves). The solution implemented in the visual system involves varying the acuity of the retina, so that only a very small region of the retina, the fovea, possesses high acuity vision, and using eye and head movements to position the fovea on areas of interest in the visual scene. Gaze-contingent multi-resolutional displays (GCMRDs) exploit these properties of the human visual system in attempting to solve the bandwidth problem (Reingold, Loschky, Stampe, & McConkie, 2001). Gaze-contingent displays are important extensions of gaze-monitoring (eye plus head tracking) systems, and function by modifying a displayed image as rapidly as possible to reflect changes in the participant's gaze position. Application of gaze-contingent displays for image compression involves transmitting high-resolution data only for the area near the user's point of gaze. The remainder of the image may then be sent in lower resolution, with a large savings in image bandwidth. Because the acuity of the eye decreases with distance from the fovea (and thus distance from the point of gaze) (e.g., Thibos, 1998), it should be possible to discard large amounts of detail from the peripheral image areas without causing performance reductions (Loschky & McConkie, 2000). Thus, GCMRDs distribute display resolution according to the spatially variant acuity of the retina, in order to obtain substantial savings in resources without perceptibly degrading the image. GCMRDs monitor or predict gaze position (the location the observer is looking as a combination of head and eye position), and center the highest resolution region of the image at that position. GCMRDs have been implemented or proposed across a wide range of demanding display and imaging applications such as flight simulators, teleoperation and remote vision, teleconferencing, telemedicine and medical training simulations (Reingold et al., 2001).

At present, most of the implemented GCMRDs have employed bi-resolutional displays. Such displays have two levels of resolution: a high-resolution area centered on the user's point of gaze, or area-of-interest (AOI), and lower resolution everywhere else. When such AOI displays have been used in flight simulators, a common observation has been that users reportedly prefer larger AOIs, because with smaller AOIs, the edges are more visible (e.g., Turner, 1984). Note that if the edges of the AOI are salient, they could interfere with performance of peripheral tasks. Furthermore, studies have demonstrated that another factor, referred to as foveal load, could interfere with performance of peripheral detection tasks (e.g., Crundall, Underwood, & Chapman, 1999; Ikeda & Takeuchi, 1975; Mackworth, 1965; Pomplun, Reingold & Shen, in press; Williams, 1985; 1989; see Williams, 1988, for a review). For example, Holmes, Cohen, Haith, and Morrison (1977) demonstrated that the mere presence of a foveal item that subjects were instructed to ignore resulted in poorer peripheral task performance (see also Ikeda & Takeuchi, 1975; Mackworth, 1965). The authors interpreted this finding as a general interference effect; the foveal item draws the attention of the observer and thus interferes with processing of other stimuli in the visual field. Given that this decline in peripheral task performance was sometimes found to be greater for targets at larger eccentricities (Mackworth, 1965; Williams, 1985), it was argued that the foveal load reduced the useful field of view, leading to the controversial term 'tunnel vision' (see Williams, 1988). Accordingly, the main goal of the present study was to investigate if AOIs may similarly interfere with the performance of a peripheral moving target acquisition and tracking task.

2. METHODOLOGY

2.1 Participants

18 paid participants were included in the study (ages 19 to 35). All participants had normal or corrected-to-normal vision.

2.2 Stimuli, Design and Task

Stimuli were full-color video clips shot from a helicopter flying over landscapes (desert and canyon). Scenes with the horizon visible were preferred as this type of shot has a wide range of visual flow velocities and directions. The clips were approximately 3 sec long each, shown at a rate of 30 frames-per-second at a resolution of 320 x 240 pixels. The average luminance was about 60 fL. For each clip a hollow ring target 6 pixels in diameter was added. This was implemented by designating the target's positions at the start and end of the clip, and interpolating its location on intermediate frames. Targets were visible and moving throughout the clip from beginning to end. Target motion was at a constant speed, averaging 8 degrees per second across clips. This movement was designed to mimic the apparent direction of the visual flow in the images, which was predominantly top-to-bottom. There were four directions of target motion: vertically down the left side, vertically down the right side, diagonally down and to the left, and diagonally down and to the right. Visually, the target motion and size approximated the perceived motion of a missile as viewed from a cockpit. A relatively large 30° field of view enhanced the sense of self-motion. Target color was selected by the average color of the background on which it appeared. The luminance was raised by 40 to 80%. This coloring technique was selected to make target search difficult in static scenes. The blurred videoclips were produced using a process equivalent to a Gaussian low-pass filter of 1.0 cycles/degree. Both the target and the background were blurred. The blurred target was still discriminable from the background, although sometimes only by its motion.

Resolution-defined windows (AOIs) were created by combining blurred and nonblurred versions of the same clip, running simultaneously and synchronized in time. The non-blurred version of the videoclip was displayed inside the window, and the blurred version of the videoclip was displayed outside the window. We manipulated the size of the high-resolution circular window, with all other regions being blurred. There were four display conditions: 1) blurred - no window, 2) small window (1.5° radius), 3) large window (3° radius), and 4) nonblurred - all high-resolution. All windows were centered at the participant's gaze position, as measured by the EyeLink gaze tracking system. The edges of the window remained sharp; there was no blending region between the window and the background.

The task for the participants was to acquire (look directly at) a target as rapidly as possible, and track it until the video clip ended. No other response was needed. A trial sequence began with a fixation dot on a blank screen. The participant fixated the dot, and the experimenter initiated the trial when the gaze cursor stabilized. The fixation dot disappeared, and after approximately 0.5 sec the videoclip started. When the videoclip ended, the fixation dot reappeared, and the next trial began. Participants received a practice block of 8 trials, followed by 4 experimental blocks of 16 trials each, for a total of 64 trials per subject. Each block contained 4 trials for each of the 4 display conditions (blurred, small window, large window, nonblurred). In addition to measuring target acquisition time, subjective impressions of image quality were collected.

2.3 Apparatus and Measurement

The SR Research Ltd. EyeLink gaze tracking system used in this research has high spatial and temporal resolutions. The three cameras on the EyeLink headband allow simultaneous tracking of both eyes and of head position, computing true gaze position with unrestrained head motion. By default, only the participant's dominant eye was tracked in our studies.

The eye tracker host computer, a 100 MHz Pentium PC, supplied gaze position information to the display computer. The display computer was a 100 MHz 486-DX PC, that controlled stimulus presentation, integrated incoming video signals, and displayed blurred video as background imagery and normal resolution video as a window (AOI) at the participant's point of gaze. The gaze-contingent multi-resolutional images were displayed on a 17" monitor in front of the participant. The display was positioned at a viewing distance of 60 cm so that the total field of view was 30° (horizontal) x 22.5° (vertical). The total system display update delay (the time taken from an eye movement to a change in the display) was on average 21 ms.

A 9-point calibration was performed at the start of the experiment followed by a 9-point calibration accuracy test. Calibration was repeated if the error at any point was more than 1°, or if the average error for all points was greater than 0.5°. Before each trial, a black fixation target was presented at the center of the display. The participant fixated this target and the gaze position measured during this fixation was used to correct any post-calibration drift errors. Throughout each trial, the experimenter was able to view on a separate monitor the target path, overlaid with a cursor corresponding to real-time gaze position. If the experimenter judged that gaze-tracking accuracy had declined, the experimenter initiated a full calibration before the next trial. However, this occurred very infrequently.

3. RESULTS

Subjectively, participants were quite satisfied with the quality of the window and display. The window's coupling to their eye movements seemed natural and transparent, but the edges of the window were perceptible even in the large window condition. In several instances during practice trials, the participants complained that the window had "stopped moving"--they had inadvertently started to watch the window rather than the target. In other words, they thought they were following rather than moving the window.

There were 3 dependent measures quantifying performance: 1) Latency to move (defined as the time from the start of the video clip until the first eye movement (saccade); 2) Acquisition time, defined as the time until gaze position was within 2° of the target; and 3) Average error (distance between gaze position and target) during the first 1000 ms. As shown in Table 1 and Figure 1, while performance in the nonblurred condition was clearly the best, the blurred condition resulted in better performance than either of the window conditions. Performance did not significantly differ across the 2 window conditions. This pattern was evident for all 3 dependent measures. The results of the experiment were rather counter-intuitive as more visual information resulted in poorer detection performance.

Table 1

Mean Target acquisition time, latency to move, and tracking error (see Figure 1).

	Blurred	Small Window	Large Window	Nonblurred
Target acquisition time (ms)	466	529	554	402
Latency to move (ms)	272	342	318	248
Error (deg) during the 1st sec.	6.84	7.72	7.53	6.2

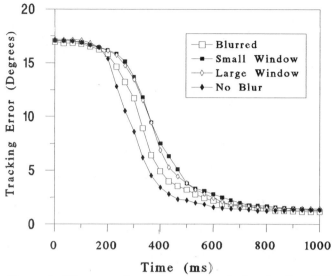

Figure 1. Distance of gaze from target during the first 1000 ms of viewing the clip.

4. CONCLUSIONS

Our results are quite thought provoking, and are in need of further study. Within the context of the current study, it is clear that the bi-resolutional displays (i.e., gaze contingent window conditions) led to inferior performance compared to the all low-resolution condition (i.e., blured condition). It appears that our AOI conditions may have captured attention at the fovea, and in this way made it more difficult for viewers to attend to the peripheral target. As explained in the introduction, our effect is similar to other findings of interference with performance on peripheral detection tasks as a function of increased foveal load (e.g., Crundall et al., 1999; Holmes et al., 1977; Ikeda & Takeuchi, 1975; Mackworth, 1965; Pomplun et al., in press; Williams, 1985; 1988, 1989). This could have occurred due to the salience of the edge between the high- and low-resolution areas. Alternatively, it may have occurred due to increased salience of the AOI in comparison to the lower resolution surround. For example, Loschky and McConkie (2000) used gaze-contingent bi-resolutional displays and found that viewers' eyes traveled shorter distances as the high resolution AOI became smaller and peripheral resolution became lower. They explained this in terms of a reduction in the salience of potential eye movement targets in the low-resolution periphery. Another way of saying this is that as the visual periphery was blurred, potential eye movement targets in the high-resolution AOI became relatively more salient.

Nevertheless, we expect that these results might change if we vary our display conditions, tasks, and stimuli. For example, if we enlarged our high-resolution AOI, at some point our effect would disappear (since the largest possible AOI was our all high-resolution window condition). It would be interesting, therefore, to investigate the relationship between our effect and the size of the high-resolution AOI. Similarly, one could ask the same question regarding the degree of peripheral image blurring. Again, the lowest level of peripheral blur was in our all-high-resolution condition. Thus, we can imagine a response surface in which both high-resolution AOI size and peripheral resolution are varied, and speed of target acquisition rises and falls as a function of each.

We also wonder whether the effect we have found might be mediated by the sharpness of the edge between the high and low-resolution regions. For example, what would happen to this effect if the edge were blended, as is commonly done in flight simulators (Dalton & Deering, 1989; Tong & Fisher, 1984; Warner, Serfoss, & Hubbard, 1993)? Or, if the image was smoothly degraded from the center of the image outwards, would the effect disappear? A recent study by Loschky and colleagues (Loschky, McConkie, Yang, & Miller, 2001) found that saccade lengths were still shortened in the highly blurred conditions even when there was no edge. Thus, if these two effects, shorter saccades due to peripheral blur, and longer acquisition times in our current study's window conditions, come from the same source, namely greater foveal salience, we would expect that window edge is not a major factor. However, this is clearly an open question.

An interesting question, is regarding the relationship between our current effect and the task and stimuli we used. That is, we expect that under conditions in which it is important to identify a target, or to discriminate it from non-targets, the window conditions would undoubtedly be superior to the blurred condition where no details could be seen. Therefore, the effect we found in this experiment may be specific to the task of identifying simple moving targets, or more generally, attentional capture by peripheral motion cues.

In sum, this effect, delayed acquisition of peripheral moving targets in an AOI condition relative to an all low-resolution condition, is certainly in need of further study. However, it does show that there are important unanswered questions regarding the human factors of gaze-contingent multi-resolutional displays.

REFERENCES

Agostinho Pavao, J., Carvalho, P., & Pacheco da Rocha, N. (1998). Perspectives on the implementation of a videotelephony based DAVIC system. Paper presented at the 9th Mediterranean Electrotechnical Conference (MELECON 98).

Crundall, D.E., Underwood, G., & Chapman, P.R. (1999). Driving experience and the functional field of view. *Perception, 28,* 1075-1087.

Dalton, N. M., & Deering, C. S. (1989). Photo based image generator (for helmet laser projector). Proceedings of the SPIE: The International Society for Optical Engineering, 1116, 61-75.

Holmes, D. L., Cohen, K. M., Haith, M. M., & Morrison, F. J. (1977). Peripheral visual processing. *Perception and Psychophysics, 22,* 571-577.

Hughes, R., Brooks, R., Graham, D., Sheen, R., & Dickens, T. (1982). *Tactical ground attack: On the transfer of training from flight simulator to operational red flag range exercise.* Paper presented at the Proceedings of the Human Factors Society - 26th Annual Meeting.

Ikeda, M., & Takeuchi, T. (1975). Influence of foveal load on the functional visual field. *Perception & Psychophysics, 18,* 255-260.

Loschky, L. C., & McConkie, G. W. (2000, November 6-8, 2000). User performance with gaze contingent multiresolutional displays. Paper presented at the Eye Tracking Research & Applications Symposium 2000, Palm Beach, FL.

Loschky, L. C., McConkie, G. W., Yang, J., & Miller, M. E. (2001). Perceptual effects of a gaze-contingent multi-resolution display based on a model of visual sensitivity. In P. N. Rose (Ed.), Proceedings of the Fifth Annual Federated Laboratory Symposium on Advanced Displays and Interactive Displays (pp. 53-58). College Park, MD: Army Research Laboratories.

Mackworth, N.H. (1965). Visual noise causes tunnel vision. *Psychonomic Science, 3,* 67-68.

Pomplun, M., Reingold, E. M., & Shen, J. (in press). Investigating the visual span in comparative search: The effects of task difficulty and divided attention. *Cognition.*

Reingold, E. M., Loschky, L. C., Stampe, D. M., & McConkie, G. W. (2001). Multi-resolution gaze-contingent displays: An integrative review. Manuscript submitted for publication, University of Toronto.

Rovetta, A., Sala, R., Cosmi, F., Wen, X., Sabbadini, D., Milanesi, S., Togno, A., Angelini, L., & Bejczy, A. (1993). Telerobotics surgery in a transatlantic experiment: Application in laparoscopy. Proceedings of the SPIE: The International Society for Optical Engineering, 2057, 337-344.

Thibos, L. N. (1998). Acuity perimetry and the sampling theory of visual resolution. *Optometry & Vision Science, 75(6),* 399-406.

Tong, H. M., & Fisher, R. A. (1984, May 30-June1, 1984). Progress Report on an Eye-Slaved Area-of-Interest Visual Display, Report No. AFHRL-TR-84-36. Paper presented at the Proceedings of 1984 Image Conference III, Phoenix, AR.

Turner, J. A. (1984). Evaluation of an eye-slaved area-of-interest display for tactical combat simulation. In *Proceedings of the 6th Interservice/Industry Training Equipment Conference and Exhibition* (pp. 75-86).

Warner, H. D., Serfoss, G. L., & Hubbard, D. C. (1993). Effects of area-of-interest display characteristics of visual search performance and head movements in simulated low-level flight. Final technical report (Sep 89-Jul 92) (AL-TR-1993-0023). Williams Air Force Base, AZ: Armstrong Laboratory.

Williams, L.J. (1985). Tunnel vision induced by a foveal load manipulation. *Human Factors, 27,* 221-227.

Williams, L.J. (1988). Tunnel vision or general interference? Cognitive load and attentional bias are both important. *American Journal of Psychology, 101,* 171-191.

Williams, L.J. (1989). Foveal load affects the functional field of view. *Human Performance, 2,* 1-28.

Pilots' Eye Fixations During Normal and Abnormal Take-Off Scenarios

Steen Weber and Henning Boje Andersen
Risø National Laboratory
DK-4000 Roskilde, Denmark

Abstract: An observational study was conducted to examine pilots' reactions and eye fixations during widely differing take-off scenarios. Eye tracking data were collected for ten crews (twenty pilots) across six different scenarios (60 trials) yielding data for both pilot flying (PF) and pilot not flying (PNF) for each trial. It was found that pilots' eye fixations, both when regarded in terms of frequencies and total time and when regarded in terms of transitions of focus, were largely uniform across the quite different take-off scenarios. No significant difference was found between normal and abnormal scenarios or between scenarios ending in successful take-off and in RTO (rejected take-off). On the other hand, large differences were found between individual pilots for each scenario and across scenarios.

1 Introduction

The experiment reported here was planned and carried out as a collaborative effort between Aerospatiale (Toulouse, France) and Risø National Laboratory (Roskilde, Denmark). The purpose of the experiment was to gather information about line pilots' response and visual orientation behaviour during normal and abnormal take-off scenarios. A part of the experiment that will not be reported here was designed to assess pilot behaviours under non-standard cockpit configurations. A detailed analysis of results from this part has been completed (Bove & Andersen, to appear), but the data and results described in this paper concern standard cockpit configuration only.

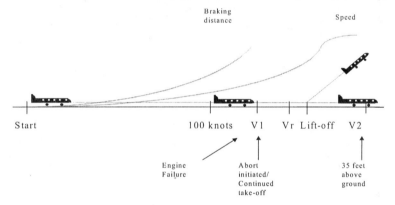

Figure 1: The main phases of the take-off. As the figure shows the braking-distance increases rapidly relative to the increase in speed. The V1-speed is determined as the latest point at which a pilot can abort the take-off in the case of an engine failure (occurring shortly before V1). Dependent on whether the take-off is aborted or continued at V1, the plane will either stop at the end of the runway or reach a height of 35 feet and the speed V2 at the end of the runway.

It is part of standard operating procedures that a take-off may be rejected before V1 has been reached. Indeed, the speed of V1 is defined to mean the speed below which the aircraft can be brought to a stop on the runway by normal braking power and at which the aircraft is capable of successful take-off with loss of power of one engine (Schwab, 1989). Pilots know that if they intiate an RTO (rejected take-off) after V1 has been achieved they may overrun the runway. Indeed, a number of accidents have occurred because the RTO decision was made too late or not at all. Indeed, about 12% of all accidents occurs during take-offs (Person, 1989). It is therefore of some interest to investigate to what extent pilots' attention and visual orientations are changed when symptoms of substandard aircraft performance appear.

1.1   Materials

The experiment was conducted during 1997 with the use of a fixed-base research simulator at Aerospatiale configured for the Airbus A340, a four-engine, heavy passenger jet, the largest of the Airbus family. Data collection equipment is described below.

1.2   Subjects

Twenty pilots (ten crews) were recruited for the experiment. All pilots except one were Airbus 320/330/340 pilots - 15 line pilots, 4 instructors (one pilot was undergoing Airbus training). Mean age of subjects was 43 (24-57). The pilots were introduced to the simulator and were allowed two or three take-offs and some flight to familiarise them with the simulator before the experimental trials were started. The subjects who had the role of PF (or PNF) had the same role throughout the trials. The PF was seated in the left seat, except for one crew.

1.3   Scenarios

The scenarios used for the experiment consisted in one normal take-off scenario and five abnormal scenarios: Hanging brakes; Engine fire; Reduced thrust and engine fire near V1; Wrong weight of aircraft; Interruption by ATC.

1.4   Experimental design

The total experimental design included the comparison between cockpit configurations, as noted above, and was planned as a within-group (cross-over) design. The part of the experiment reported here was planned as an *observational study* the target of which was the distribution of visual attention by Pilots Flying (PFs) and Pilots Not Flying (PNFs). Hence, being observational in nature, this part of the experiment was not designed to measure the impact of independent variables (e.g., the parameters of the different scenarios) on independent variables, viz., decisions (continued take-off or Rejected Take-Offs / RTOs) and outcomes (GO-decisions and successful take-offs; speed at which RTO is initiated or distance to end of runway). Altogether, the data reported here are based on 59 trials (118 Eye tracking recordings), each of the ten crews taking each of the six scenarios once. (One of the 60 trials had to be excluded due to problems with data collection).

The scenarios were unknown to the crews, who were informed only about the total number of trials and that "some" of these would be abnormal. The order of the scenarios was randomly distributed across the crews, except that the first trial always started with scenario A (the normal scenario). After the sequence of six trials, each crew was debriefed (lasting about 30 minutes).

1.5   Data

Simulator data were logged and included speed, position on runway, position of throttles, as well as scenario-dependent data (weight, temperature, abnormal parameters - e.g., engine fire, retarded accelleration - unknown to subjects). Each pilot wore an eye tracker (ASL 4000SL) and a/v recordings were made on separate tapes. Eye fixation data have been recorded with the use of MacShapa and analysed with the aid of tools built at Risø.

We used six areas of interest or visual targets for scoring the eye tracking data: **PFD** (primary flight display; **ND** (navigational display); **Runway**; **Engines**; **ECAM** ("Electronic Centralised Aircraft Monitoring System"); "**other**" (target other than the five above).

The analysis of eye fixations and their correlation with simulator data covered data from the take-off phase between 80 knots and up to V1 or to RTO and stop on runway (no RTOs were made after V1).

2   Results

In the following we present eye fixation results in terms of: (A) pilots' distribution of visual attention by accumulated *time* of fixation and by *frequency* of fixation; (B) scan patterns or frequencies of focus transitions from targets to targets for PFs and PNFs; and (C) outcomes (RTOs, speed) for the six scenarios.

2.1   Times and frequencies of eye fixations between 80 knots and V1 or RTO.

We first describe the distribution of visual focus in terms of times and frequencies. Naturally, the PFs look at the runway and the PNFs look at the instruments, as anyone would expect. See figure 1 which illustrates the distribution of attention by time and frequency for each of the pilots across all scenarios and all crews. In general, minor differences can be found between distributions along the time and the frequency scales; therefore, in the following we do not quote results separately from the two scales. It should be added, however, that the mean dwell time (the mean time of a fixation on a given target) varies considerably - see table 1. As shown in figure 2, the PFs regularly and briefly look at the PFD (primary flight display).

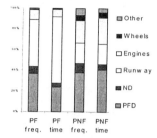

Target	PF	PNF
PFD (primary flight display)	0.79	1.14
ND (navigational display)	0.80	0.56
Runway	1.74	1.10
Eng. (Engines)	0.74	1.25
ECAM ("wheels")	0.77	1.06
Other	0.47	0.62

Figure 2: Distribution of visual attention by time and frequency for each pilot position across all scenarios and all crews.

Table 1: Mean fixation (dwell) time per pilot per target

In the following figures, we illustrate relations by showing data from PNF alone due to space restrictions. In figure 3 is shown, for all PNFs and for each of the scenarios the distribution of visual attention in terms of frequency, while figure 4 illustrates the distribution for all scenarios across crews.

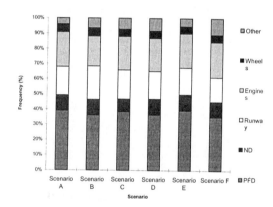

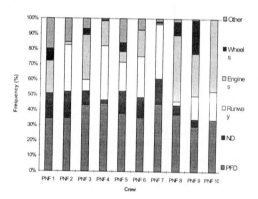

Figure 3: Distribution of the frequency of visual attention in terms of frequency for all PNF's for each of the six scenarios

Figure 4: Distribution of the frequency of visual attention of each individual PNF for all scenarios

It should be noted that there is a *high degree of uniformity across the rather different scenarios* whereas pilots vary widely among themselves for any scenario or set of scenarios. Similarly, when one looks at any specific crew, there is an individual "attention distribution profile" for each pilot which varies less across scenarios than, for any scenario, across pilots.

2.2   *Times and frequencies of eye fixations between RTO and halt*

Out of the 59 trials, 21 ended as RTOs and 38 as GO trials (compare section below on 'Outcomes'). Similar results as above concerning variation also obtains with respect to RTO trials: There is little variation among the scenarios in terms of the crews' distribution of visual attention; but there is, in contrast, a fairly pronounced variation among crews across the scenarios. Figure 5 shows the distribution in terms of frequency for all PNFs for each scenario involving RTO trials. Similarly, in figure 6 is shown the distribution of attention for individual PNFs across all RTO trials.

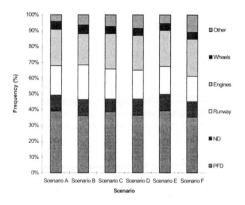
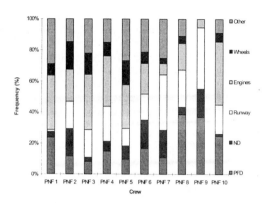

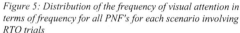

Figure 5: Distribution of the frequency of visual attention in terms of frequency for all PNF's for each scenario involving RTO trials

Figure 6: Distribution of the frequency of visual attention in terms of frequency of attention of individual PNF's across all RTO trials

2.3 Eye focus transitions between 80 knots and V1 or RTO.

Scan patterns of PF and PNF were calculated in terms of the number of focus transitions from any given target to and from any other target (Runway, PFD, ND, Eng., Wheels, "other"). We have calculated the frequency of all chains of length 1, 2, ..5. A chain of length 1 is a "landing" on a target, a chain of length 2 goes from x to y or y to x; a chain of length 3 goes from y to x to z (z possibly equal to x) etc. In figure 7 is illustrated all chains of length 2 by frequency for the PF and PNF across all scenarios.

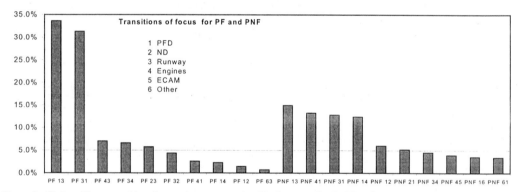

Figure 7: Chains of focus transitions from and to targets ordered by frequency for the PF and PNF across all scenarios.

A more perspicuous illustration is provided in figure 8, which shows, for both pilots, their three most frequent focus transitions. While there are obvious differences between the two pilots, no salient differences were found in focus transitions between scenarios. More significantly, no difference was found in eye focus transitions between trials ending with a successful take-off and trials ending in an RTO. As before, there were rather large differences among individual PFs and PNFs.

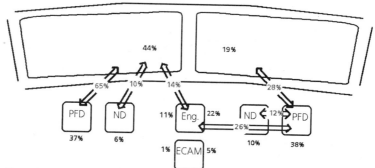

Figure.8: Numbers on arrows indicate percentage of transitions to and from targets, Numbers at target represents total percentage of fixations

2.4 Eye focus transitions between RTO and halt

While there are differences (as expected) of focus transitions between successful take-offs and RTOs as measured from RTO to halt, there are no noticeable differences between scenarios. Again, pronounced differences were found among crews.

2.5 Outcomes

Out of the 59 trials, 21 ended in RTO and 38 in successful GOs. The two engine fire scenarios ended in RTOs for most crews, while three crews continued in the face of the ATC scenario. See table 2.

Finally, no order effect was detected: the distribution of eye fixations and the relative frequency of focus transitions remained the same across the individual trials of the crews.

Scenario	A	B	C	D	E	F	sum
RTO	0	0	8	10	0	3	21
GO	10	9	2	0	10	7	38

Table 2: Distribution of GO and RTO decisions

3 Discussion

It was expected that there would be individual differences among PFs and among PNFs - and the nature and size of these differences were in line with results of the study of Racca (1995):. However, it was not expected that there would be so relatively very minor differences across the scenarios in terms of eye fixations and focus transitions. There was, as described, very little variation across trials, and the normal trial, A, was not significantly different from the non-normal trials. Equally, it was somewhat unexpected that the trials leading to an RTO were not significantly different from trials not ending in an RTO in terms of eye fixations or scan transitions.

References

Aarons, R. N. (1988). Takeoff management breakthrough. Business & Commercial Aviation. February 1988, 85-88.

Bove, T & Andersen, H.B. (to appear). The effect of an advisory system on pilots go/no-go decision during take-off. Reliability Engineering and System Safety, to appear.

Person, L. H. (1989). To Go - Or Not to Go: Situation Awareness on Takeoff. Flight Safety Foundation - Flight Safety Digest - October 1989, 35-42.

Racca, E. (1995): Visual perception, pilot workload and resources management strategies. Proc. of the 8th Int. Symp. Aviation Psych., April 24-27 1995, 25-30.

Schwab, A. (1989): V-speeds can vary. Professional Pilot. November 1989, 82-86.

A Spacing Method for Readable Arrangement of Characters

Masazumi MIYOSHI[†], Yoshifumi SHIMOSHIO[†], Hiroaki KOGA[†], Michinari SHIMODA[†]
and Keiichi UCHIMURA[‡]

[†] Kumamoto National College of Technology,
2659-2 Suya Nishigoshi-machi, Kumamoto 861-1102, Japan
and
[‡] Kumamoto University,
2-39-1 Kurokami Kumamoto-shi Kumamoto 860-8555, Japan

ABSTRACT

This paper proposes a quantitative spacing method for readable arrangement of characters within a specified area like a signboard, headline, etc. Characters written on a signboard are arranged into a sequence with proper spacing. If the arrangement of characters on the signboard is a sentence, it will be desirable that the arrangement of characters is easy to read. To obtain an index of spacing for the readable arrangement of characters, the relations between spacing and readability are investigated. Two kinds of quantitative indices of spacing for Japanese *Kana* characters and English alphabet letters are obtained.

1. INTRODUCTION

In typography it is well known that spacing is important for the readability of sentences. By the way, a sentence is a set of words expressing a statement, and the way of spacing varies every writing system of the language. In the English writing system, a word is represented by a combination of the English alphabet letters and the spaces between words are wider than the gap spaces between letters so as to be readable. Besides, the Japanese writing system uses two kinds of character sets called *Kanji* and *Kana*, and the way of spacing varies to each character set. The *Kanji* character set, which has been introduced into Japan from China, is one of ideograms. Therefore, a *Kanji* character corresponds to a word. There are about 2,000 regular *Kanji* characters nowadays. On the other hand, the *Kana* character set, which has been developed from the *Kanji* character in Japan, is one of phonograms and consists of about 50 *Kana* characters. Since the *Kana* character is a phonogram, a word is basically represented by a combination of characters as well as English alphabet letters. Most of Japanese sentences are composed of both the *Kanji* characters and the *Kana* characters. When the Japanese sentence contains a lot of *Kanji* characters, each *Kanji* character leads the corresponding word. Therefore, it is not so difficult to read the sentence, even if the characters are arranged at regular intervals. But it is difficult for children to read it, because there are characters of many kinds in the *Kanji* character set. Then, a sentence for children contains a lot of the *Kana* characters. However, the sentence in which the *Kana* characters are arranged at regular intervals is not readable, because the boundary between words is not clear. For this reason, the Japanese sentences which a lot of the *Kana* characters are contained need proper spacing as well as English sentences need spaces between words. Thus, Japanese sentences written by the *Kana* characters have the similar way of spacing to English sentences. For this reason, this paper deals with the English alphabet letters and the Japanese *Kana* characters, and the relation between the spacing and the readability is investigated. From now, for convenience, both letters and characters are simply described as characters, except for a few cases where they need to be distinguished.

Now, this paper proposes a spacing method, in which the spacing is quantitatively evaluated, for readable arrangement of characters. In order to evaluate the spacing quantitatively, the spaces between characters are represented by means of a potential field like an electric field. The potential field is based on the theory of the induction field in vision (Yokose, 1966). Concerning this theory, it was found that a visual distance between characters has relation to a potential value in the potential field induced by the characters, and that the relation is little influenced by the character shapes (Miyoshi et al, 1999; Miyoshi et al, 2000). This is useful in quantifying a visual distance between characters without taking account of the character shapes. The proposed method in this paper uses the potential value as a quantitative index of the spacing for the readable arrangement of characters. The readable arrangement of characters is obtained by experiments.

2. MEASURE OF THE DISTANCE BETWEEN CHARACTERS

This section describes a measure of the distance between characters using a potential at space between

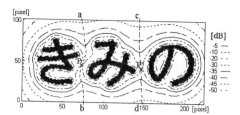

Fig.1 Contour map of the potential distribution induced by three Japanese *Kana* characters

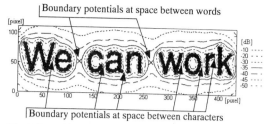

Fig.2 Quantities of the distance between characters and the distance between words

characters. Yokose supposed that a certain effect spreads around a figure and the effect generates an induction field on retina during looking at the figure (Yokose, 1966). Then, he derived an empirical formula for the potential in the field from psychological experiments by means of the light stimulus threshold. This is well known as the theory of the induction field in vision. Figure 1 is an example of the potential distribution induced by three characters. The potential value is evaluated by means of a geometric calculation on the character arrangement (Miyoshi et al, 1999; Miyoshi et al, 2000). Figure 1 shows the potential values around the characters with a contour map. It is shown that the potential at a place close to each character is high and the potential at a far place from any character is low as well as an electric field. Besides, a potential valley appears nearby a boundary between adjacent characters. There is the point P, shown in Fig.1, where the potential is the maximum on the line a-b along the valley between two characters "き(ki)" and "み(mi)". We define a boundary potential between the two characters as the potential at the point P. In the same way, the potential at the point P', shown in Fig.1, is the boundary potential between two characters "み(mi)" and "の(no)", too. Thus, there is a boundary potential every space between adjacent characters. The boundary potential has a characteristic that the more two characters approach together, the more the boundary potential between them increases. That is, the boundary potential is in inverse proportional to the distance between characters. This paper uses the boundary potential as a measure of the distance between characters.

3. EXPERIMENTS IN THE READABLE ARRANGEMENT OF CHARACTERS

This section describes experiments to obtain an index of spacing for the readable arrangement of characters. Now, the English spacing usually segments a sentence every word, while there are several ways of segmenting in Japanese. In this paper, the Japanese spacing segments a sentence as a word or a word plus a few characters is a unit, and for convenience the unit is simply called a word as well as the English segmentation.

3.1 Procedure

In experiments, the readability of a simple sentence, which is printed with single line and lateral typesetting, is investigated using a questionnaire. The spacing varies with both the distance between characters and the distance between words. These distances are quantitatively evaluated by the boundary potentials as shown in Fig.2. The experiment procedures are as follows.

(1) Japanese samples: The three kinds of sentences (see Fig.3) are used as the Japanese samples. These are chosen from sentences in a newspaper, magazine, etc. The character font of all samples is 24-point font size Gothic. The distances between characters vary with four steps at the boundary potential of -24dB, -30dB, -33dB and -36dB. The distances between words vary with seven or eight steps every the distance between characters. Therefore, the thirty-one samples with various spacing are made every sentence, and a total of 93 Japanese samples are prepared.

(2) English samples: The three kinds of sentences (see Fig.6) are used as the English samples. These are chosen from the titles of song and have the same number of characters but different number of words. The character font of all samples is 24-point font size Helvetica. The distances between characters vary with four steps at the boundary potential of -10dB, -20dB, -30dB and -36dB. The distances between words vary with six or seven steps every the distance between characters. Therefore, the twenty-six samples with various spacing are made every sentence, and a total of 78 English samples are prepared.

(3) Survey: The readability is surveyed using a questionnaire to each sample. The questioning is the following: looking at the sample one by one, and answer the readability of the sample sentence. The answer is a choice among three which are readable, unreadable and neither. The samples are shown at random and are looked at a distance of 50 cm front. The experiment environment is a lighted classroom usually. The subjects are students aged from nineteen to twenty-three. The numbers of the subjects for the Japanese and the English are 17 and 25 persons, respectively.

(4) Readability: The answered choices which are readable, unreadable and neither are numerically evaluated at 1, 0

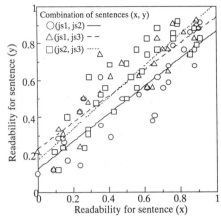

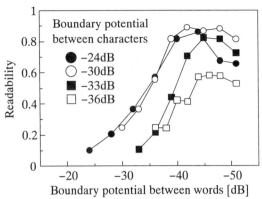

Fig.4 Readability to the distance between words in Japanese sentence

Fig.3 Comparison of readability variations between Japanese sentences of the following three kinds:
(js1)"しぜんに そだてたい たくましさ",
(js2)"しんせいきを になう こどもたち",
(js3)"きみの こいびとは いるかな".

and 0.5, respectively, and then the mean values are calculated every sample. The readability of a sample is defined as the mean value to the sample. Therefore, the readability of the sample which all subjects answer "readable" comes to 1. Conversely, the readability of the sample which all subjects answer "unreadable" comes to 0.

3.2 Experiment for the Japanese sentences

First, the variations of readability about the spacing among the three kinds of sentences are compared. Figure 3 shows the correlations of readability between the different sentences when the spacing changes variously. Note that the spacing is the same for each symbol in Fig.3. The correlation coefficients for each combination in Fig.3 are

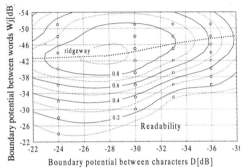

Fig.5 Contour map of readability to varying the distance between characters and the distance between words in Japanese sentences. (Dot line on map shows the ridge-way of readability)

0.90 for js1-js2, 0.91 for js1-js3 and 0.90 for js2-js3. It is shown that the relation between spacing and readability is little influenced by the kinds of sentences. Next, it is shown how the readability varies with spacing. Figure 4 shows that the variation of the readability, which shows the average value of the three kinds of sentences, to the distances between words on each distance between characters. It is shown that the readability proportionally increases when the distance between words expands. The two variations of readability at the boundary potentials between characters of −24dB and −30dB are almost the same, while the variations of readability at −33dB and −36dB shift to the lower boundary potential as the distance between words expands more. Figure 5 shows how the readability is related to both the distance between characters and the distance between words, in 3-dimensional contour map. There is a ridge-way of the readability, as shown by the dot line. This paper uses the ridge-way to determine the optimum distance between words at the arbitrary distance between characters, and derives the quantitative index of spacing for readable arrangement of characters from the ridge-way. Approximating the projection line of ridge-way shown in Fig.5 to a quadratic, the optimum distance W_j[dB] between words at the distance D[dB] between characters is expressed by the following equation:

$$W_j = -0.0166D^2 - 0.620D - 48.1 \qquad (1).$$

Consequently, this equation is the index of spacing for readable arrangement of characters in Japanese.

3.3 Experiment for the English sentences

In the same way as the Japanese sentences, first the variations of readability about the spacing among the three kinds of sentences are compared. Figure 6 shows the correlations of readability between the different sentences when the spacing changes variously. The correlation coefficients for each combination are 0.97 for es1-es2, 0.96 for es1-es3 and 0.95 for es2-es3. It is shown that the relation between spacing and readability is little influenced by the

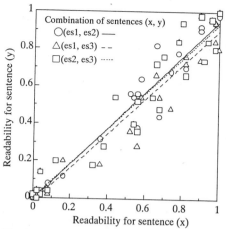

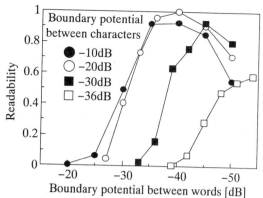

Fig.7 Readability to the distance between words in English sentence

Fig.6 Comparison of readability variations between English sentences of the following three kinds:

(es1)"We can work it out",

(es2)"Eight days a week",

(es3)"All together now".

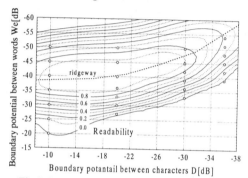

Fig.8 Contour map of readability to varying the distance between characters and the distance between words in English sentences. (Dot line on map shows the ridge-way of readability)

kinds of sentences. Next, Figure 7 shows the variations of the readability as well as Fig.4. It is shown that the readability proportionally increases when the distance between words expands. The two variations of readability at the boundary potentials between characters of –10dB and –20dB are almost the same, while the variations of readability at –30dB and –36dB proportionally shift to the lower boundary potential as the distance between words expands more. These are similar to the Japanese spacing mentioned above. Figure 8 shows how the readability relates to both the distance between characters and the distance between words, in 3-dimensional contour map. There is a ridge-way of the readability, as shown by the dot line. In the same way, approximating the projection line of ridge-way shown in Fig.8 to a quadratic, the optimum distance W_e[dB] between words at the distance D[dB] between characters is expressed by the following equation:

$$W_e = -0.0285D^2 - 0.658D - 42.2 \qquad (2).$$

Consequently, this equation is the index of spacing for readable arrangement of characters in English.

4. COHESIVENESS WITHIN A WORD

It is considered that a written word is a cohesive set of characters. This section defines cohesiveness within a word as a degree of the character cohesion within a word, and describes experiments in the relation between the cohesiveness within a word and the distance between characters.

4.1 Procedure of experiment

The cohesiveness within a word is surveyed using a questionnaire. The procedures are as follows.

(1) Samples: Word samples for questionnaire are cut out from the sentence samples used for the previous experiments. Hence, the distance between characters and the character font are the same as the sentence samples. The words used in experiment are the eight kinds of Japanese words (しぜん, そだてたい, たくましさ, しんせいき, になう, こどもたち, きみ こいびと) and the eleven kinds of English words (We, can, work, it, out, Eight, days, week, All, together, now). Since the distance between characters is changed with four steps every word, the total of 32 Japanese word samples and 44 English word samples are prepared.

(2) Survey: In questionnaire, the definition of the cohesiveness within a word is notified in advance. The questioning is the following: looking at the sample one by one, and answer the cohesiveness of the sample. The answer is a

choice among three which are cohesive, non-cohesive and neither. The samples are shown at random and the other conditions are the same as the experiments in the readability. The numbers of the subjects for the Japanese and the English are 15 and 16 persons, respectively.

(3) Cohesiveness: The answered choices which are cohesive, non-cohesive and neither are numerically evaluated at 1, 0 and 0.5, respectively, and then the cohesiveness is defined as the mean values calculated every sample, as well as the readability.

4.2 Experiment Result

Figure 9 shows how the cohesiveness within a word varies with the distance between characters. Error bars in the figure show the degree of variation with the kinds of word. It is found that the cohesiveness within a word depends on the distance between characters and it is little influenced by the

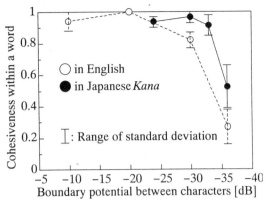

Fig.9 Cohesiveness within a word to the distance between characters, in English and Japanese.

kinds of word. Besides, as the distance between characters expands, the cohesiveness starts to decrease rapidly at the boundary potential between characters of –25dB in English and –33dB in Japanese each.

4.3 Cohesiveness within a word and Spacing

Figure 9 shows that the cohesiveness within a word is still high despite varying the distance between characters when the distance between characters is not so wide, that is the boundary potential is more than –25dB in English and –33 dB in Japanese each. On the other hand, both Fig.4 and Fig.7 show that the readability of a sentence depends on only the distance between words when the distance between characters is not so wide. Consequently, the above suggests that the character arrangement within a word looks like a lump because the cohesiveness within a word is high when the distance between characters is not so wide. Conversely, when the distance between characters is wide, the distance between words must expand according to the distance between characters to give a readable arrangement of characters because the cohesiveness within a word decreases.

5. CONCLUSION

This paper proposes a quantitative spacing method for readable arrangement of characters, and investigates the relation between the readability and the spacing on sentences in Japanese and in English, each. In experiments, the following are found.

(1) When the distance between characters is not so wide, in which the boundary potentials between characters are at least –30dB in Japanese and –20dB in English, the readability of a sentence depends on only the distance between words.

(2) When the distance between characters is wide, the readability of a sentence depends on both the distance between characters and the distance between words.

(3) The cohesiveness within a word depends on the distance between characters and it is little influenced by the kinds of word.

As a result, two kinds of equations, which are used for the Japanese *Kana* characters and the English alphabet letters, are obtained as the quantitative indices of spacing for readable arrangement of characters.

REFERENCES

Yokose Z. (1966). Pattern recognition and reading of letters by machine. The Japanese Journal of Ergonomics, Vol.2, No.3, pp.10-16

Miyoshi M, Shimoshio Y, Koga H, Ideguchi K. (1999). Character Arrangement Design with *Kansei* Information by Using the Theory of Induction Field on Retina, The Transaction of IEIC, VOL.J82-A, No.9, pp.1465-1473.

Miyoshi M, Shimoshio Y, Koga H, Ideguchi K, Uchimura K. (2000). A Quantitative Evaluation Method of Spacing for The Character Arrangement with Human *Kansei*. 4th World Muticonference on Systemics, Cybernetics and Informatics, IIIS, SCI2000, Proceedings, vol.10, pp.87-91.

Eye-tracking analysis of priming effects of pre-watched television commercials on reading patterns of printed advertisements

Hirotaka Aoki and Kenji Itoh

Department of Industrial Engineering and Management, Tokyo Institute of Technology, 2-12-1, Oh-okayama, Meguro-ku, Tokyo 152-8852, Japan

ABSTRACT

This paper discusses the influences of pre-watched television commercials (CM's) on readers' behaviour during reading of newspaper advertisements based on eye movement data. To analyse the influences, we classify components arranged in newspaper advertisements into eight categories in terms of their contents and information types. Next we calculate the following two indices for any one of the classified categories: The fixation durations and the number of fixations. These indices seems to represent readers' information processing to newspaper advertisements. The influences are analysed mainly based on these indices that can be generated by the eye movement data combined with the classification of the arranged components. A series of experiments was carried out with twenty female subjects by use of an ASL 4000 eye tracking system. Each subject participated in three experimental sessions performed for three consecutive days. In each session, a subject was asked to read eight newspaper advertisements. In the second and third sessions, each subject watched sixteen CM's, two of which included the same brands advertised in the newspaper advertisements, before reading the newspaper advertisements. Based on data obtained in the experiments, we will discuss the aforementioned influences. In addition, we will also discuss relationships of individual attributes with the influences.

1. INTRODUCTION

With the advance of Information Technology, various new advertising media (such as World Wide Web) are generated increasingly. In addition, the number of television channels increases suddenly along with the beginning of a digital broadcasting using broadcasting/communication satellites. Therefore, the environments around advertising business are changing rapidly. Reflecting these backgrounds, "media planning" has been receiving increasing attention. The term "media planning" indicates the function of deciding the best combination of the advertising media by which maximum effects can be obtained in limited costs. To achieve effective media planning, it is of great importance to analyse interactions as well as qualitative differences among different advertising media on advertising effects. For instance, it is necessary to clarify such things as follows: What kind of interaction exists between advertisements in papers, magazines, television, and WWW, etc., what features does each of these media have, and so forth.

Considering these background issues, we especially focus on influences of watching television commercial (CM) on patterns of reading newspaper advertisements. The choice of focus is motivated by the following reasons: Newspaper advertisements are read actively by consumers. Therefore, it is recognized as one of advertising media which persuade consumer to realize the advertised products/services. On the other hand, CM's are viewed passively. On that account, it seems to be one of advertising media which direct consumers' attention towards information relating to products/services. It can be said that both of these roles are typical functions of advertisements. Thus, it is useful to analyse the existing influences between these two media on advertising effectiveness.

Many studies have been performed to evaluate the effects of advertisements, mainly in the field of marketing and consumer research. For this purpose, a number of studies employ an approach that examines a consumer's memory traces for the advertised products using questionnaires (Pieters and Bijmolt, 1997). Recent studies have concentrated on a consumer's cognitive and attitudinal aspects on the basis of theories of human information processing (MacInnis and Jaworski, 1989). These aspects are of great importance in designing advertisements to draw a consumer's attention to the advertisements. However, it is usually very difficult to analyse a consumer's cognitive/perceptual attitude or behaviour since the cognitive processes of the product information are automated unconsciously.

Eye-tracking analysis seems to have a potential ability for investigating human cognitive/perceptual processes that are performed unconsciously in the field of transportation human factors (e.g., Itoh et al, 1998) and so forth. Some researchers have showed the eye tracking technique to be a useful methodology for issues especially relevant to consumers' behaviour performed unconsciously (e.g., Pieters et al., 1999; Aoki and Itoh, 2000). In the light of this fact, eye tracking seems to have potential as methodology for analysing consumers' information processing of advertisements.

The objective of the present paper is to analyse influences of pre-watched CM's on the readers' behaviour during reading of newspaper advertisements employing eye-tracking technique. In this study, we concentrate on aspects of readers' information acquisition from advertisements.

2. EVALUATION OF READERS' INFORMATION ACQUISITION

Considering the role of CM's, we expect such influences of viewing of CM's on reading patterns of newspaper advertisements as described below: CM's attract viewers' attention towards advertised brands. Therefore, pre-watched CM's causes the viewers to move their attention from information of brand name, catch copy and other information to that of explanation about benefit during reading of newspaper advertisements, and to exert more information of the brand's benefits from the newspaper advertisement. In addition, we assume that the strength of the CM's influence varies depending on viewers' attitudes to the advertised brand.

In order to estimate the readers' information processing to newspaper advertisements, we calculate the number of fixations and the fixation durations using readers' eye movement data. The reasons for this calculation are as follows: Eye movement data can generally be classified into saccades and fixations. Saccades are rapid movements of the eyes, and the sensitivity to visual input is suppressed during these eye movements. Between the saccades, eyes remain relatively still during fixations. It is often said that information is mainly acquired during fixations, and the link between attentions and fixation locations is tight. In addition, it is also said that a single fixation may correspond to a chunk especially during viewing of pictorial information (Loftus, 1972). According to these assumptions, it is assumed that the number of fixation represents the amount of information extracted by a reader. Fixation duration is another index relating to the depth level of information processing.

In addition, components arranged in newspaper advertisements are classified into the following two categories in terms of representational form of information: Textual and pictorial information. These two categories are further divided into sub-categories in terms of their involving contents in the advertisement as follows: Brand name, explanation about benefit, catch copy, and others. An example of the classification is described in Table. 1. This classification enables us to combine eye movement data with advertising contents. The aforementioned indices are calculated for any one of the classified sub-categories. Based on these indices, we evaluate readers' information processing to each sub-category in order to analyse influences of pre-watched CM's on newspaper readings.

Table 1. Classification of components arranged in advertisements

Type of information	Type of content	Typical examples
Pictorial	Brand name	Logos
	Explanation about benefit	Sentences on the benefits/utilities of users
	Catch copy	Decorated phrases which attract readers' attention
	Others	Other pictorial objects
Textual	Brand name	Brand name
	Explanation about benefit	Figures representing situation in which the advertised service/product is used
	Catch copy	Pictorial objects which attract readers' attention(e.g. celebrities)
	Others	Other textual objects

3. EXPERIMENTS

A series of experiments was carried out with twenty female subjects. In these experiments, their eye movement data were recorded during reading of newspaper advertisements using an ASL 4000 eye-tracking system. In this study, we perform three experimental sessions for each subject. In each session, a subject read eight newspaper advertisements once. Three of the shown newspaper advertisements were used for analysis. Two of them were selected from financial services (Ad 1, Ad 2), and the other one was from jewelries (Ad 3). The reasons for this selection are described below. The contents of Ad 1's CM are almost similar to those of Ad 1. On the other hand, the contents of Ad 2's CM except brand name do not coincide with those of Ad 2. Ad 3 does not have its CM. Each subject's eye movement data was recorded during all of her reading time (including time for reading newspaper articles).

All the subjects were homeworkers. Each of them has much interest in financial services. Their ages ranged from thirty-three to forty-nine years. They were paid about twenty dollars per hour for their participation. Before experimental sessions, self-report including attitudes towards advertisements, preferences/interests for services/products advertised in the analytic newspaper advertisements, and so forth were collected from all the subjects. Because of technical problems relating to the eye-tracking system, thirteen of participating subjects' data were selected for analysis.

Each subject's individual attributes are briefly described bellow: Subjects 1, 2, 6, 9, and 11 had known the brand/service name advertised in Ad 1. Subjects 2, 4, 6, 7, and 11 had known those advertised in Ad 2. Subjects 4 and 11 like the brand advertised in Ad 1. However, subject 11 does not like the brand of Ad 2. The rest of subjects' preferences towards brands in Ad 1~3 are neutral. As to interests in purchasing/using the brands, subjects 2, 4, 9, and 11 are interested in purchasing/using the brand in Ad 1. Only subject 11 has no interest in purchasing/using the service advertised in Ad 2. The rest of subjects' level of interests is neutral.

In the first session, each subject read eight newspaper advertisements including Ad 1~3. On the next day, the second session was carried out. At first, each subject watched sixteen CM's including two CM's adverting brands of Ad 1 and Ad 2. The CM's were inserted in a talk program, which was broadcasted within a few week of its broadcast debut. After that, the same eight newspaper advertisements were exposed to each subject. The third session was also performed on the next day. In this session, the procedure was almost the same as that in the second session. The difference from the second session is that each subject asked to join questionnaire test after reading newspaper advertisements. The questionnaire includes questions about individual attributes (e.g. degree of comprehension for advertised contents). The subject was not informed of the purpose of the experiments, but provided only with explanation of the experimental procedures and the eye-tracking system. Thus, all the subjects did not always watch all of the exhibited advertisements in these experimental sessions.

4. RESULTS AND DISCUSSIONS

4.1 Transcription of Reading Sequences

Subjects' eye movement data for each advertisement were transcribed in time-line with type of fixated components, as shown example transcriptions in Figure 1. In this figure, advertisements' components are arranged along with the vertical axis. The upper area in each time-line represents type of content, and the lower represents type of information. Using these time-line transcriptions, we can obtain easy comprehension of positions of fixations, fixation durations, content and information of fixated components, and so forth. For example, it can be easily understood the followings: In the second session, S4 watched a component, which shows brand name in textual style, included in Ad 2 for ten sec., and moved her gaze toward another textual component referring to catch copy.

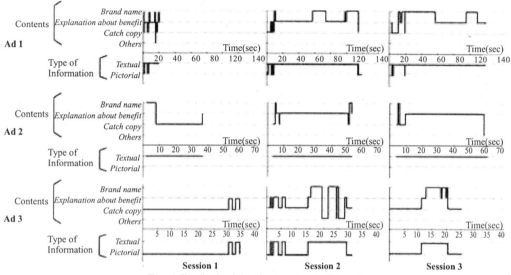

Figure 1. Time-line description of reading sequences (S4's case)

4.2 Influence of Pre-watched Television Commercials

Using the aforementioned transcription, we calculated the fixation durations and the number of fixations in any one of classified components' eight sub-categories for each session. To identify the tendency of both indices' transition between sessions, we defined the following three transition patterns: Ascending, constant, and descending. These transition patterns are distinguished using a half of the indices' standard deviations for each component's sub-category as a basis. Table 2 indicates the number of transition patterns for all the combination of advertisements and the subjects. As can be seen from this table, a conspicuous difference in tendency of the number of fixations for Ad 1 can be identified in Constant (1st to 2nd session) –Constant (2nd to 3rd) case. For Ad 2 (with its brand's CM having different contents from Ad 2) and Ad 3(without its brand's CM), it seems that the numbers of fixations in components classified into explanation about benefit are much larger than that for Ad 1(with its brand's CM having similar contents with Ad 1). It is also observed that the numbers of fixations for each session are almost constant for Ad 2 and Ad 3. However, tendency of the index's transition for Ad 1 varies depending on the individuals. From this result, we can deduce that readers' information extraction from components showing explanation about benefit may be influenced by the pre-watched CM including similar contents with the newspaper advertisement. We will investigate such influences observed in the experiments in detail.

In Ad 1, seven kinds of benefits which the advertised brand has are described. These benefits present advantages of the brand advertised in Ad 1. To analyse the observed influences in Ad 1, we calculate the rate of reading of explanation about benefit for all the combination of the seven benefits, sessions, and subjects. The rate can be calculated as follows: Number of characters which a reader surely read divided by total number of printed characters. The characters read by a reader are identified by the following two criteria: Duration for which she fixates them (more/less than 100 msec.) and continuity of her sequence of fixations during reading. Table 3 shows the rate of reading of explanation about benefit calculated using S1, S3, and S7's data. Each of these subjects' data has typical tendency of influences observed in the experiments. S1's transition of number of fixations for Ad 1 can be classified into Ascending (1st to 2nd session)-Descending (2nd to 3rd) case. In the first session, S1 read textual components regarding benefit 6 and 7(see Table 3). After watching the CM, she paid her attention towards benefit 1~3, and 5 as well as benefit 6, and did not move her gaze at components relating to benefit 7. In this case, it is conjectured that the CM played a role as a cue in her selection of exhibited information. Such conjecture can also be adapted to S4, S8, and S11. On the other hand, S3, whose pattern of transition of number of fixations is Ascending (1st to 2nd session)-Ascending (2nd to 3rd), read no words before she watched the CM. In the second session, S3 paid a little attention to all the components regarding eight kinds of benefit evenly. At last, S3 concentrated her gaze towards benefit 2 and 3. According to this result, it seems that the CM activated S3's intention to pay attention to its brand. Besides, S7's case can be considered as follows: S7's transition pattern of number of fixations is characterized as Ascending (1st to 2nd session)-Constant (2nd to 3rd) pattern. S7 read almost all the textual components regarding benefits carefully in the first session. In the second session, S7 chose components regarding benefit 1~5, and she read some/all parts of components regarding benefits again in the third session. S13 shares the same kind of tendency. It can be said that some parts of the tendency observed in S7 and S13 are owing to the pre-watched CM. However, it seems to be difficult to identify whether the observed tendency is primarily due to the pre-watched CM in this case.

In summary, these results may suggest that the CM, which is exhibiting similar contents to those of its newspaper advertisement, promotes most readers' positive attitudes towards its newspaper advertisement. However, it is obvious that other factors (such as high/low involvement, product/service related knowledge, and so forth) as well as the pre-watched CM affect the readers' behaviour during reading of newspaper advertisement. Considering these issues, we will discuss relationships between subjects' attributes and their recorded eye movement data in the next sub-section.

Table 2. Tendency of indices' transition between sessions

Tendency Transition from 1st to 2nd session	2nd to 3rd session	Ad	Whole advertisement Fixation duration	Num. of fixations	Brand name Fixation duration	Num. of fixations	Explanation about benefit Fixation duration	Num. of fixations	Catch copy Fixation duration	Num. of fixations	Others Fixation duration	Num. of fixations
Ascending	Ascending	Ad 1	1	2	0	0	1	1	0	0	0	1
		Ad 2	0	0	0	0	0	0	1	0	0	0
		Ad 3	1	0	0	0	0	0	0	0	0	0
	Constant	Ad 1	0	0	0	2	0	2	1	0	1	0
		Ad 2	1	3	0	4	1	0	0	2	1	0
		Ad 3	1	0	1	1	1	0	0	0	0	0
	Descending	Ad 1	4	5	2	3	4	4	1	4	2	3
		Ad 2	4	0	5	1	4	0	4	3	2	3
		Ad 3	4	1	3	1	4	0	4	4	2	4
Constant	Ascending	Ad 1	2	1	3	2	0	2	3	2	2	1
		Ad 2	1	0	0	1	2	0	1	1	4	1
		Ad 3	2	0	0	0	2	0	1	0	2	1
	Constant	Ad 1	3	3	4	2	3	2	1	2	7	1
		Ad 2	4	9	4	6	2	12	2	5	6	3
		Ad 3	3	4	4	5	4	12	2	1	6	1
	Descending	Ad 1	2	0	0	0	3	0	1	0	0	1
		Ad 2	0	0	2	0	0	0	0	0	0	1
		Ad 3	0	4	0	2	0	0	0	2	1	1
Descending	Ascending	Ad 1	1	2	2	3	2	2	1	2	0	1
		Ad 2	2	1	0	1	2	0	2	1	0	2
		Ad 3	1	0	3	1	1	0	2	1	2	2
	Constant	Ad 1	0	0	1	1	0	0	5	3	1	5
		Ad 2	1	0	2	0	1	1	3	1	0	2
		Ad 3	0	4	2	3	0	1	1	2	0	2
	Descending	Ad 1	0	0	1	0	0	0	0	0	0	0
		Ad 2	0	0	0	0	1	0	0	0	0	1
		Ad 3	1	0	0	0	1	0	3	3	0	2

Numbers represent those of cases observed in the experiments over all the subjects

Table 3. Rate of reading of explanation about benefits in textual

Content	S1 1st	2nd	3rd	S3 1st	2nd	3rd	S7 1st	2nd	3rd
Benefit 1	0	0.3636	1	0	0.2192	0.2192	1.0000	0.2192	1.0000
Benefit 2	0	0.7043	0.6451	0	0.1881	1.0000	1.0000	1.0000	0.8118
Benefit 3	0	0.3064	0.5107	0	0.1881	1.0000	1.0000	0.1881	1.0000
Benefit 4	0	0	0	0	0.3030	0	1.0000	1.0000	0.6969
Benefit 5	0	0.1566	0.4666	0	0.0833	0	0.4666	0.3833	0.3833
Benefit 6	0.1757	0.1747	0.1747	0	0.1747	0	1.0000	0	1.0000
Benefit 7	0.6551	0	0	0	0.6551	0.6551	0.2068	0	0.2068

4.3 Relationship of Readers Attributes with Indices' Transition

In this sub-section, we discuss the observed patterns of indices' transition from the viewpoint of relationship with individual attributes. Table 4 indicates the number of observed transition patterns of number of fixations on whole of Ad

2 falling into three categories in terms of subjects' daily attitudes towards newspaper advertisements. As can be seen from this table, the remarkable difference among these three categories (pay-attention/neutral/not-pay-attention) is that the number of Ascending (1st to 2nd session)- Constant (2nd to 3rd) transition cases seems to be large for subjects classified into pay-attention group. Ad 2' contents are different from those of the exposed CM except for brand name. These subjects seem to have acquired more amount of information from Ad 2 after they watched its CM. This result may suggest that CM influences people who pay attention to advertisements in print media to extract product/service information from printed advertisements.

In addition, Table 5 shows the number of transition patterns of the fixation durations for brand name in Ad 2 falling into three categories in terms of individual perceived importance of CM's as referring sources to decide to purchase financial services. The conspicuous tendency of transition patterns is as follows: Fixation durations for brand name of subjects perceiving CM's as important seem to increase after exposition of the CM. On the other hand, it seems that those of subjects perceiving CM's as not important tend to decrease or keep values of the index after exposition of the CM. It is difficult to put an interpretation on this observed result. One reason may be as described below. Subjects classified into Descending (1st to 2nd session)-Constant (2nd to 3rd) case considered the CM as an effective referring source. In addition, they are more interested in services advertised in the CM than those in Ad 2. Therefore, the subjects decreased their level of information processing to Ad 2 after they watched the CM.

Table 4. Tendency of transition of number of fixations on whole of Ad 2

Transition from		Daily attitudes to printed advertisement		
1st to 2nd session	2nd to 3rd session	Pay attention	Neutral	Not pay attention
Ascending	Ascending	0	0	0
	Constant	3	0	0
	Descending	0	0	0
Constant	Ascending	0	0	0
	Constant	3	4	2
	Descending	0	0	0
Descending	Ascending	0	0	1
	Constant	0	0	0
	Descending	0	0	0

Table 5. Tendency of transition of fixation duration on brand name in Ad 2.

Transition from		Level of importance of CM's as referring sources to decide to purchase financial services		
1st to 2nd session	2nd to 3rd session	Important	Neutral	Not important
Ascending	Ascending	0	0	0
	Constant	0	0	0
	Descending	0	3	2
Constant	Ascending	0	0	0
	Constant	1	3	0
	Descending	1	1	0
Descending	Ascending	0	0	0
	Constant	2	0	0
	Descending	0	0	0

5. CONCLUSION

In the present paper, influences of pre-watched CM's on patterns of reading newspaper advertisements were analysed by use of eye movement data. The analysis was based on two indices, i.e., the fixation duration and the number of fixations, relating to readers' states of information acquisition. These indices were calculated by use of eye movement data and classification of the arranged components in advertisements. Using these indices, we analysed the influences of pre-watched CM's in terms of these indices' transition between sessions. According to these analysis results, we could obtain some implications. For example, it could be conjectured that the CM including similar contents with a newspaper advertisement influences readers' information extraction from components showing explanation about benefit.

We plan to continue analysing data obtained in the experiments in more detail. Especially, we will bring out the types of interaction between newspaper advertisements and CM's. These analysis results will allow us to obtain some implications for achieving effective media planning.

ACKNOWLEDGEMENTS

This research was partly supported by Grant-in-Aid for Encouragement of Young Scientists, No. 12780329, the Japan Society for the Promotion of Science. We would like to acknowledge Yuichiro Kato and other staffs, Daiko Advertising Inc., for their corporation in this project. Maho Kasai, FCB Worldwide Japan, provided useful ideas to this project. Special thanks to Motokazu Andoh and Tomohisa Takeo, Tokyo Institute of Technology, for their assistance in the experiments and data analysis.

REFERENCES

Aoki, H. and Itoh, K. (2000). Analysis of cognitive attitudes to television commercials on the basis of eye tracking data, *Proceedings of the 14th Triennial Congress of the International Ergonomics Association*, *1*, 38-41, San Diego, USA, July-August.
Itoh, K., Hansen, J.P. and Nielsen, F. R. (1998). Cognitive modelling of a ship navigator based on protocols and eye-movement analysis, *Le Travail Humain*, *61(2)*, 99-127.
Loftus, G. R. (1972). Eye fixations and recognition memory for pictures, *Cognitive Psychology*, *3*, 525-551.
MacInnis, D. J. and Jaworski, B. J. (1989). Information processing from advertisements: Toward an integrative framework, *Journal of Marketing*, *53(4)*, 1-23.
Pieters, R. and Bijmolt, T. (1997). Consumer memory for television advertising: A field study of duration, serial position, and competition effects, *Journal of Consumer Research*, vol. 23, no. 4.
Pieters, R., Rosbergen, E. and Wedel, M. (1999). Visual attention to repeated print advertising: A test of scanpath theory, *Journal of Marketing Research*, *36(4)*, 424-438.

Building Bridges from Theory to Practice

Susan Dumais and Mary Czerwinski

Microsoft Research, One Microsoft Way, Redmond, WA 98052

ABSTRACT

The bridges from basic cognitive theory to applied human-computer interaction practice are neither as strong nor as complete as we would like to see. This paper outlines some areas in which the two disciples complement each other. We also describe several challenges such as designing from first principles, generating useful results and guidelines, and moving professionally between the two disciplines. Although much progress has been made in the last few decades, we believe that both disciplines can benefit from a closer partnership, and we hope that this panel begins to build some of these bridges.

1. INTRODUCTION

Cognitive psychologists who work in the software industry typically find themselves designing and evaluating complex software systems to aid humans in a wide range of problem domains, like word processing, interpersonal communications, information access, finance, remote meeting support, air traffic control, or even gaming situations. In these domains, the technologies and the users' tasks are in a constant state of flux, evolution and co-evolution. Cognitive psychologists working in human-computer interaction (HCI) design may try to start from first principles developing these systems, but they often encounter novel usage scenarios for which no guidance is available. For this reason, we believe that there is not as much application of theories, models, and specific findings from basic psychological research to user interface (UI) design as one would hope. However, several analysis techniques and some guidelines generated from the literature are useful. In this paper we outline some efforts in HCI research from our own industrial research experience, demonstrating at what points in the design cycle HCI practitioners typically draw from the cognitive literature. We also point out areas where no bridges exist. It is our goal to highlight opportunities for the two disciplines to work together to the benefit of both.

1.1 Building Real Bridges: The Millennium Bridge Example

A recent example of bridge building provides a nice example of the complex interactions between design, empirical evaluation, and iterative refinement. The Millennium Bridge was central London's first new bridge in over 100 years. The design was selected from hundreds of entries and built by leading architects and engineering firms. The stainless steel and aluminum pedestrian bridge formed a "blade of light" over the Thames River linking St. Paul's Cathedral to the new Tate Modern Art Gallery. On June 10, 2000, the bridge opened with much fanfare. Only a few hours later, however, the bridge closed briefly because it swayed violently in the wind under the weight of the many pedestrians who crossed it on opening day. Two days later on June 12, 2000, it closed semi-permanently for analysis and repairs. The bridge remains closed today, almost a year later.

What happened? All suspension bridges move to some degree, and walking adds additional motion. Normal walking pace is about two steps a second, so a vertical force of 2 hertz is produced. Military commanders have long told their troops to break step when crossing bridges to avoid these vertical movements. Designers know this, standards require accommodating for it, and the Millennium Bridge engineers designed and tested for it. This worked fine. However, there is also a smaller horizontal movement. As we walk, one foot pushes left and the other pushes right, so there is an additional 1 hertz horizontal movement. When large numbers of people walked across the Millennium Bridge, an unintentional synchrony of stepping occurred. As the bridge moved slightly in the horizontal direction, people instinctively adjusted their steps in synchrony with the movement of the bridge. When large numbers of people do this at the same time the movements in the bridge become quite noticeable. The designers (and standards) missed the horizontal synchrony in walking, what they now call lock-in.

Despite a good deal of experience building suspension bridges, standards that have been developed and improved over time, and complex mathematical models to predict and test how the bridge would behave, this one did not quite work in the real world. Why? Actual behavior was outside expectations -- of standards, models and practice. Engineers are now revising the design with new simulation models to examine ways to dampen the lock-in effect. Hopefully the visibility of this failure will also lead to changes in standards and practice as well.

1.2 Building Bridges in HCI Design

Much the same cycle of initial design from principles and practice, usability testing, careful failure analysis, and iterative design happens all the time in building complex human-computer systems. One might even argue that designing HCI applications is much more complex that building bridges because of the many perceptual and cognitive processes involved. Regardless of where one stands on the relative complexity of design in these two arenas, the design cycle is much the same. Contributions from psychology can and often do influence all aspects of the HCI design cycle, some more successfully than others.

Throughout our discussion we provide examples from information retrieval and notification management. The former deals with how best to organize search results for efficient and accurate analysis, the latter focuses on the design of intelligent notification systems to convey important information so that users can efficiently make use of it.

2. BASIC RESEARCH IN COGNITIVE PSYCHOLOGY AND HCI DESIGN

Just as in bridge building, we cannot currently design complex HCI systems from first principles. Useful principles can be drawn from the sub-domains of sensation, perception, attention, memory, and decision-making to guide us on issues surrounding screen layout, information grouping, menu length, depth and breadth. Guidelines exist for how to use color in user interface design, how to use animation and shading, or even what parameters influence immersion in virtual worlds. Designers also often borrow ideas from best practices, such as very successful products that have been iteratively refined and accepted broadly in the marketplace. However, there will always be technology-usage scenarios for which the basic research simply does not exist to guide us during design. The complexity, learning and interaction effects seen in HCI usage scenarios are daunting. In addition, HCI systems are used by a very wide range of users for a large variety of tasks. Finally, information itself is an extremely complex material to work with, which makes the design task all the more challenging.

So while there are a large number of areas within psychology from which HCI design can draw, theory falls far short of covering the entire design life cycle. We begin by examining the areas of psychology that have proven most useful to user interface design and describe what seems to work well during the design process and what areas could be improved.

2.1 Experimental Methods

Many of the methods that psychologists use during their academic careers are very useful to the HCI practitioners and researchers. This is often the first way in which a psychologist in industry can add value to a product team. For example, the team may want to design an educational computer application for children, but the programmers, designers and technical writers may not understand how to measure the performance of users of the software (small children), or may not understand the various developmental stages that need to be considered in the user interface design. The psychologist can work with the team to develop metrics of ease of use and satisfaction, including learning, efficiency and engagement. The psychologist will not only know the proper methods to use in the study of children's interaction with the early prototype of the system, but also which statistics to use in the analysis and how to interpret and communicate the findings from the study effectively. These are all skills that the programmers and other team members do not typically have, and they are therefore considered valuable by the team.

These same skills are useful in an industrial research environment as well, where new technologies are developed. The inventors have ideas about the benefits of a particular idea, but they typically have little experience in designing the right experiments and tasks to study the questions of interest. In our work we have used visual search tasks, dual tasks, reaction time and accuracy studies, deadline procedures, memory methods like the cued recall task and others to explore new technologies and interaction techniques. Problems with the proposed new technology are almost always identified, as rarely is a new design flawless in its first stages. And, iterative design-test-redesign works to improve the existing technology implementation from that point forward.

The downside of using traditional experimental designs and tasks in HCI work is that factorial designs with tight control and many subjects are simply not feasible given the time and resource constraints faced by design professionals. In addition, most real world designs consist of many variables and studying all possible combinations just is not possible (or perhaps appropriate). Finally, as we saw in the bridge example, it is often the unanticipated uses of a system that are the most problematic, and these by their very nature are difficult to bring into the lab ahead of time.   To understand some of the issues, think about how you would go about designing and evaluating a new voice-input word processor over a six-month period across multiple users, applications and languages.

One might conjecture that important independent variables could be studied in isolation with the best of those being combined into a final design. In reality, this rarely works in HCI design. We have witnessed teams that have iteratively and carefully tested individual features of a software design until they were perfected only to see interactions and tradeoffs appear when all of the features were united in the final stages of design or when the system was used in the real world. Let us take a simple example from information retrieval. Highlighting the user's query terms in the listing of results allows for more rapid identification of matching sections within documents. Semantic processing of query terms allows users to find more relevant information (e.g., normalization and spelling correction so that 'Susan Dumais" matches "Susan T. Dumais" or "Dr. Sue Dumais", or synonym expansion so that "HCI" matches "human-computer interaction" and perhaps even "computer-human interaction"). How do you combine these two features of highlighting and expansion? The more complex the processing that goes on, the harder it is to know what to highlight in the actual user interface to the information retrieval system.

In addition to experimental methods, practitioners use a wide range of observational techniques and heuristics (e.g., field studies, contextual inquiry, heuristic evaluation, cognitive walkthrough, rapid prototyping, questionnaires, focus groups, personas and scenarios, competitive benchmarks tests, usage log collection, etc.) to better understand system usage and to inform design. These techniques are often borrowed from anthropology or sociology, and rarely would a system design be complete without these research techniques and tools. Often a qualitative description of the jobs people are trying to do, or how a system is used in the field by real people doing their real work is much more valuable than a quantitative lab study of one small system component. Observational techniques, task analysis, and related skills are much used in the practice of HCI but little represented in most cognitive psychology curricula. This is an area in which psychologists find their background and training lacking when first introduced to applied work. In addition, the methods themselves are not necessarily as well honed or useful as they could be in influencing user interface design.   More research at the basic level is needed to better understand and extend these field methods.

An important but often overlooked contribution is that psychologists often act as user advocates in informal but important ways. Being sensitive to individual differences, including the important realization that product managers and programmers are not "typical" users (for most systems) is a real contribution. Simply getting programmers to acknowledge that the user knows best will improve design.

Evaluating and iterating on existing interfaces is only one aspect of the HCI design task. Generating a good design in the first place or generating alternatives given initial user experiences is equally important, and we now turn to that problem.

2.2 Specific Findings and Guidelines
There are areas in which basic cognitive psychology research has helped the domain of HCI by providing specific findings and guidelines for design. Some of the best-known and often-cited examples of the applicability of results from basic psychology are the rather low-level perceptual-motor findings that have been used quite effectively in HCI design. For example, Fitt's Law and Hick's Law have been used to design and evaluate input devices for years. The Power Law of Practice, and the known limits on auditory and visual perception have often been leveraged in the design of interactive systems. Many other results are seen in guidelines for good control layout, the use of color and highlighting, depth and breadth tradeoffs in menu design, and many general abstractions have made their way into "libraries" of parameters describing typical response times, for example. Using findings and guidelines like these allow designers to start with a good initial design, or prevent silly mistakes, but it does not guarantee a good system when all the variables are combined together into one design.

Even though specific findings and guidelines have proven useful in some cases, there also exist many problems in their use, which limits their effectiveness. Guidelines are often misused or misinterpreted. For example, using the rule of thumb of having only 7+-2 items in a menu often leads to poor design. Or applying visual search guidelines without taking into account characteristics of the items such as their discriminability, cohesiveness, or labels can also be problematic. In addition, guidelines are often written at too abstract a level to help with specific designs, or alternatively they are too specific for a given usage context. A designer often finds it very difficult to combine all of the recommendations from specific findings or guidelines without knowledge of the costs, benefits or tradeoffs of doing so.

An example from our work in managing interruptions illustrates some of these issues. The problem area of interruption while multitasking on the computer is one where much of what we have learned from basic psychological research has been applicable in system design. For example, 100 years of attention research has taught us about limited cognitive resources, and the costs of task switching and time-sharing even across multiple perceptual channels. But how does one go about designing an effective interface for alerts when the user may benefit from the information? And how does one design an interface that allows the user to easily get back to the primary task after a disruptive notification?

In practice, specific findings guide our intuitions that flashing, moving, abrupt onset and loud audio heralds will attract attention. But how much attraction is too much? How does this change over time? Do users habituate to the alerts? How does the relevance of the incoming messages affect task switching and disruption? Some studies in the literature suggested relevance was influential, while others found that surface similarity was more important in terms of task influence. All of these studies used tasks that were not representative of typical computer tasks (i.e., they were too simple, demanded equal distribution of attention, etc.). We ran our own studies and showed the benefits of only displaying relevant notifications to the current task in order to mitigate deleterious effects (Czerwinski, Cutrell & Horvitz, 2000).

When, as part of the same notification system design, we attempted to design a "reminder" cue that was visual and could help the user reinstate the original task context after an incoming notification, specific findings were again not useful. We found through our own lab studies that using a "visual marker" as a spatial placeholder to get users back to a point in a primary task was not enough. Instead, users needed a "cognitive marker" that provided more of the contextual, semantic cues related to a primary task (Cutrell, Czerwinski & Horvitz, 2001). No specific findings existed in the literature for exactly what visual anchors could be used in a display to help users tap into their mental representations of a task, or how many visual retrieval cues would be needed to reinstate a computing task context most efficiently. Again, prototypes needed to be designed and tested.

In both the design of notifications, and the corresponding reminder system, paradigms from very basic cognitive research had to be utilized in new lab studies to examine the broader usage context and the specific user interface design chosen for the target task domains.

2.3 Analytical Models
Cognitive architectures and analytical models are sometimes used to evaluate and guide designs. GOMS, first proposed by Card, Moran & Newell (1983), is certainly among the most influential models in HCI. GOMS uses a general characterization of basic human information processing, and the GOMS technique (Goals, Operators, Methods and Selection rules) for task analysis. Some successful applications of GOMS have been reported (Gray et al., 1993; John & Kieras, 1996). GOMS is most useful in evaluating skilled error-free performance in designs that have already been specified in detail, and can be costly in terms of time or evaluator training. In one of our experiences using GOMS to evaluate two user interfaces for trip planning, the GOMS model was not all that helpful. Using the GOMS analysis the two systems were equally usable and efficient. A subsequent laboratory study, however, showed that one design was more successful overall.

Several more general cognitive architectures have been proposed as well, including EPIC, ACT and SOAR. HCI design has been a fertile testing ground for these efforts since complex tasks are a challenge for any such architecture. Pirolli (1999) provides a nice review of the characteristics of these architectures and issues involved in applying them to HCI design.

2.4 Theory
Most psychological theories are descriptive, not prescriptive, and are tested using simple paradigms that may or may not scale up to real world HCI scenarios. Both of these factors limit their applicability. For example, many memory studies examine pairs of words, and attention studies examine the ability to quickly detect the onset of a letter either with or without a predictive cue. The point is not that these paradigms have not proven to be valuable as basic researchers refine and evolve their theories. Insofar as these basic behaviors emulate components of complex task behaviors, they are extremely useful to practitioners as well as the basic cognitive scientist. The point is more that most psychological theories are

developed by examining isolated phenomena in carefully controlled lab settings, with little or no guidance about complex interactions and tradeoffs, which are critical for design.

We have also observed that the details of cognitive theories often do not matter in practice. In system design, does it really matter whether the "bottleneck" in attention is early or late? Not really, designing to support attention as a critical resource is what matters if the system is to be usable. But, few theories guide designers in how to do this. Wickens' Multiple Resource Theory (Wickens & Carson, 1995) is a good attempt toward this end. Does it matter if visual processing is serial or parallel? Again, not really, as simply knowing that visual scanning slopes are often linear with an increase in set size is what matters for most cases. So, from a practical perspective, much of what is enthusiastically debated in basic cognitive research has little practical implication to HCI designers trying to build useful, efficient interaction systems. Another point to reiterate here is that theories from cognitive psychology are often not predictive enough during the initial stages of design, and not effective for articulating tradeoffs later during design.

HCI is complex, dependent on many unpredictable variables, and ever changing during interaction, and this probably is not going to change much over the next decade or so. Humans are complex, the tasks they perform and the information they work with are equally complex, the systems they interact with are varied, and the combined design problem (which is what HCI is about) is daunting. Cognitive theory and empirical results are currently not up to the challenge.

3. CONCLUSION

We opened this paper with the premise that the design and evaluation of HCI systems is challenging. We further argue that basic results and theory can provide reasonable starting places for design, although perhaps experience in an application domain is as good. Theories and findings from psychology are not, however, as useful as one might hope. Getting the design right from the start is very hard even with these starting points, so any real world design will involve an ongoing cycle of design, evaluation, failure analysis, and redesign.

We argue that there is a need to train psychology students to think about which experimental results are relevant, what the effect size is (rather than what the significance level is), and how to analyze costs, benefits, and tradeoffs in a multi-disciplinary, fast-paced product development environment. Iterative design is a valuable tool in improving designs. But it is not sufficient to catch all the problems with use of technology in the real world or to adequately guide new design. We need a broader range of observational and analytical techniques than we are used to. Theories need to be extended to address cognition in complex real world scenarios and be more prescriptive to help in design. For psychology to be the "mother of invention" in HCI design, we need to begin building some of these bridges.

4. REFERENCES

Card, S. K., Moran, T. P., & Newell, A. (1983). *The Psychology of Human-Computer Interaction*. Lawrence Erlbaum.

Cutrell, E., Czerwinski, M. & Horvitz, E. (2001). Notification, disruption and memory: Effects of messaging interruptions on memory and performance. To appear in *Proceedings of Interact 2001*, Tokyo.

Czerwinski, M., Cutrell, E. & Horvitz, E. (2000). Instant messaging: effects of relevance and time. In S. Turner, P. Turner (Eds), *People and Computers XIV: Proceedings of HCI 2000, Vol. 2*, British Computer Society, 71-76.

Gray, W. D., John, B. E., & Atwood, M. E. (1993). Project Ernestine: A validation of GOMS for prediction and explanation of real-world task performance. *Human-Computer Interaction, 8*, pp. 237-209.

John, B. E., & Kieras, D. E. (1996). Using GOMS for user interface design and evaluation: Which technique? *ACM Transactions on Computer-Human Interaction, 3*, 287-319.

Pirolli, P. (1999). Cognitive and engineering models and cognitive architectures in human-computer interaction. In F. T. Durso (Ed.) *Handbook of Applied Cognition*, John Wiley and Sons, pp. 443-477.

Wickens, C. D. & Carswell, C. M. (1995). The proximity compatibility principle: Its psychological foundations and relevance to display design. *Human Factors, 34*, 473-494.

The Effectiveness of Visual vs. Auditory Cues in Visual Search Performance: Implications for the Design of Virtual Environments.

Thomas. Z. Strybel and Diane L. Guettler

Department of Psychology, California State University Long Beach, Long Beach, CA 90840

ABSTRACT

In virtual environments, cuing the user to the location of important events may be necessary, and researchers are evaluating the effectiveness of auditory and visual cues in this regard. Although stimulus-driven visual cues produce fastest search times, the inability to ignore such cues could be disastrous in some work environments. Therefore, we evaluated the effectiveness of auditory spatial cues in orienting attention when stimulus-driven visual cues were also present. When the visual cue validity was 90%, invalid auditory spatial cues reduced search times. When the auditory cue validity was 90%, invalid visual cues increased search times. When neither cue was reliable (validity=50%), participants usually searched the visual cue area before the auditory cue.

1. INTRODUCTION.

Virtual environments can enhance the quality of human-computer interaction, particularly when they are applied to complex environments. A virtual environment is a computer-based generation of a natural or abstract environment; the user is immersed in a real or artificial world, and interacts directly with components of this world (Bullinger, Bauer, & Braun, 1997). For immersion, the virtual environment should have an extended visual field of view (often a full 360°), with the information displayed a function of the current position and orientation of the user. One particularly promising application of virtual environments is the aerospace cockpit. A virtual environment could provide configural displays about the status of the aircraft and the immediate environment outside the aircraft. Head Mounted Displays (HMDs), currently used in rotocraft cockpits, are examples of primitive virtual environments. With HMDs, flight and sensor data are projected on a lens attached to the pilot's helmet. Some information displayed depends on the pilot's current head position.

Performance costs can offset the potential benefits of virtual environments, however. If the information available to the user depends on current eye and head position, the user will miss critical information when he or she is not oriented properly. In cockpit applications, failure to detect and respond to critical information could be disastrous. Designers could reduce these consequences with mulitsensory displays that present information via the auditory and tactual, as well as visual, modalities. Of course, proper utilization of these capabilities places added burdens on the designers of the displays because the choices for display location, modality and format are expanded. For usable virtual environments, designers should incorporate the advantages and disadvantages of different sense modalities. For example, although auditory spatial resolution is poorer than visual spatial resolution, the auditory system has a greater field of view. This omnidirectional property of audition means that objects can be heard anywhere in the environment without the need for repositioning the auditory apparatus. Designers must also consider how information from each modality affects the user's selection of relevant sources from the environment. Research in this area, known as selective attention, has been focused primarily on the effectiveness of visual cues in covert orienting (the eyes and head remain stationary). The effectiveness of visual and mulitsensory cues in overt orienting (information selected by repositioning the eyes and head) has received less attention although overt orienting is multimodal in nature. For example, the time required to position the eyes on a visual target is reduced when a spatially-coincident auditory signal accompanies the target (e.g., Frens, Van Opstal, & Van Der Willigen, 1995). These mulitsensory effects are usually attributed to the role of the superior colliculus in regulating orienting behaviors. In the deep layers of this structure, multimodal maps of space have been observed and these direct the orienting behaviors of the organism. The multimodal maps are based on a single frame of reference, meaning that spatial maps specific to individual sensory systems have been transformed into a common reference frame in the superior colliculus (e.g., Stein & Meredith, 1993).

These features of the orienting system suggest that cueing critical information in virtual environments could be enhanced with auditory spatial cues because head and eye positioning are faster with multimodal targets, and the auditory system is omnidirectional. Many behavioral studies are consistent with this observation. Auditory spatial cues substantially reduce the time required to locate and identify a visual target in a large search field. The benefits of auditory spatial cues depend on the characteristics of both the visual search task and the auditory cue. The important visual factors are target distance, target contrast, and the number of nontarget "distractors" in the local (area immediately surrounding the target) and global (remaining area) search field. The important auditory factors are cue precision and amplitude (e.g., Rudmann & Strybel, 1999; Strybel et al., 1995; Strybel, Vu, & Castagna-Osorio, 2000). Although providing auditory spatial cues to the target's location can improve search performance, complex environments might have multiple targets and cues, and the user must respond first to the most critical events. It is important, therefore, that designers know if the user can control his/her search strategy in the presence of multiple cues and targets. In selective attention research, a distinction is made between stimulus-driven and goal-directed cues (e.g., Egeth & Yantis, 1997). Stimulus-driven cues, such as abrupt visual onsets or feature singletons, capture attention; the user has difficulty ignoring them. The intentions of the observer control goal-directed cues. Although stimulus-driven cues produced the faster search times, the inability to ignore the cue could be disastrous in complex environments. Whether auditory spatial cues are stimulus-driven or goal-directed is debatable. With covert orienting, Spence and Driver (1997) showed that auditory cues to both auditory and visual targets exhibited stimulus-driven properties. However, visual cues appeared stimulus-driven only to visual targets. With overt orienting, Rudmann and Strybel (1999) showed that search times to displaced audio cues were sometimes longer than when no cue was present. Strybel Vu and Castagna-Osorio (2000) obtained similar results when the cue was presented in background noise. Fujawa and Strybel (1997) showed that uninformative, high-intensity auditory cues interfered with visual search performance. Although these findings suggest that auditory spatial cues are stimulus-driven, there has not been a direct comparison of visual and auditory spatial cues for overt orienting. In the present experiment we evaluated the stimulus-driven properties of auditory spatial cues when visual stimulus-driven cues were also present. We also varied the validity of the cues, or the percentage of trials on which the cue accurately signaled the target. For high validity conditions (90%), we investigated the extent to which an invalid cue could be ignored. For uninformative validity conditions (50%), we determined which cue the user would prefer.

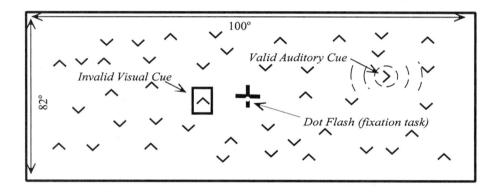

Figure 1. Search field and stimuli for a valid auditory cue and invalid visual cue that is closer to the fixation cross than the target (not to scale).

2. METHOD.

    2.1 Participants. Six participants with normal hearing and normal or corrected-to-normal vision were tested. Two had participated in previous experiments on auditory spatial processing. Participants were paid $10.00/hour for their participation.

    2.2 Apparatus. The experiment was conducted in large (approximately 3 m x 4.3 m) dimly-lit, semianechoic room. All surfaces of the room were covered by Martek 10.16-cm acoustic foam, except for a small window separating the test room from the control room. The participant was seated 1.3m in the front and center of a large (100° horizontal by 82° vertical) screen in the middle of the room. The search field was projected through the window onto the screen by two projection panels and overhead projectors in the control room. Visual stimuli (targets, distractors, visual cues, and fixation cross) were white symbols projected on a dark background with a contrast of 73%. Auditory cues were provided by 45 Blaupunkt 7.6-cm loudspeakers positioned behind the screen in eight concentric circular rings with radii between 12° and 45°. The number of speakers in each ring varied between two and eight. The acoustic cue was high-pass noise (lower frequency cutoff = 2000 Hz.) at 70 dB A-weighted, pulsed at 10 Hz. A microprocessor with Tucker-Davis Technologies' programmable modules controlled all visual and auditory stimulus generation, trial sequencing, and response collection.

2.3 Procedure. The stimulus arrangement is shown in Figure 1. Participants were instructed to locate and identify a visual target (left or right pointing arrow) in a distractor field (up/down facing arrows) by pushing one of two buttons attached to a hand-held box. Targets were presented at one of three distances from fixation, 18°, 30° or 42°. The density of the distractors was either 16% or 32%. Each participant was tested in five cue/validity conditions. In two of these conditions, only one cue was available on each trial (validity = 100%). These conditions were considered baseline conditions because they provided a measure of the effectiveness of the auditory and visual cue in signaling target location when presented alone. In the remaining three conditions, two cues were provided on each trial, one auditory and one visual. These conditions varied the percentage of trials on which the cues were valid (90% auditory-10% visual, 90% visual-10% auditory, and 50% visual-50% auditory). For the 90% validity conditions, the participants were instructed to ignore the invalid cue because it rarely signaled the target. For the 50% validity condition, the participants were told that neither cue was reliable and there was no best strategy. A valid auditory cue was the noise presented from a loudspeaker directly behind the target. A valid visual cue was a rectangle surrounding the target. An invalid cue was either the sound or rectangle presented at a distractor in the opposite hemifield. The distance of the invalid cue, relative to the target was also varied (closer, equidistant or farther from fixation). For example, Figure 1 presents a valid auditory cue in the right hemifield and invalid visual cue in the left hemifield. The invalid visual cue is closer to the fixation cross relative to the target.

The specific sequence of events on each trial was as follows. At the beginning of the trial, a fixation cross and pattern of "X"s corresponding to the location of targets and distractors was projected. These remained on the screen for a variable foreperiod (1000 - 2000 ms). To ensure that the participant remained fixated on the cross until the trial began, a fixation task was presented in the last 100 ms of the foreperiod. A filled circle was briefly flashed either above or below the center of the cross, as shown in Figure 1. At the end of the foreperiod the cross was removed and the targets, distractors and cues were presented. The distractors were displayed by removing the upper/lower half of each distractor "X." The target was displayed by removing the left/right half of the target "X." The visual and auditory cues were also turned on. Because the visual target and distractors were already on the screen, the visual cue was seen as an abrupt onset. The participant located and identified the target direction by pushing the corresponding response button. After the target response, the participant indicated whether the fixation flash occurred on the top or bottom half of the cross. The participant was told to respond to the target as quickly as possible, but accuracy was more important than speed. The fixation response was not timed.

Each participant ran three sessions per cue-validity condition. Within a session, ten trials at each of three target distances, two distractor densities and three invalid cue relative locations, were presented in random order, making a total of 180 trials per session. The number of valid and invalid trials was determined by the validity condition. Three sessions were run at one condition, before going to the next, with the order of conditions randomized. The first trial block in each condition was considered practice, and was not included in subsequent analysis.

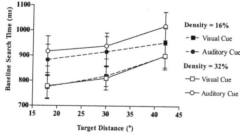

Figure 2: Baseline Condition. Mean search times as a function of target distance and distractor density for auditory and visual cues presented alone.

3. RESULTS

Mean search times were computed for each subject at each condition after incorrect-response trials and trials producing reaction times more than two standard deviations from the mean were omitted. We analyzed the validity conditions separately because in each validity condition, the number of valid trials varied. For the baseline conditions (validity = 100%), a three-way, Cue Modality x Distractor Density x Target Distance repeated measures analysis of variance was performed on the mean search times for each subject. Significant Cue x Distance and Cue x Distractor Density interactions were obtained [$F(1,5) = 19.78$; $p = 0.007$; $F(2,10) = 4.35$; $p = .04$, respectively]. These interactions are illustrated in Figure 2. The visual cue produced faster overall search times and was unaffected by distractor density.

Distractor density affected search times with the auditory cue, with 32% density search times being on the average 40 ms longer than 16% density. Search times with both visual and auditory cues were affected by distance, although the effect of distance was more pronounced in the visual cue condition. In Figure 2 it also appears that the effect of distance in the auditory modality was lessened at 16% distractor density. The three-way interaction of cue modality, distractor density and distance was only marginally significant, however.

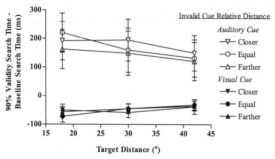

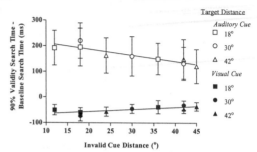

Figure 3A: Validity = 90%. Change in search times relative to baseline as a function of target distance and invalid cue relative distance.

Figure 3B: Validity=90%. Change in search times relative to baseline as a function of invalid cue and target distance from fixation.

At 90% validity, we subtracted the baseline search times from the valid cue search times for each subject, because of the relative advantage of the visual cue, and the difference in the number of valid and invalid trials. Significant Cue x Target Distance [$F(2, 10) = 4.33$; $p = .04$] and Cue x Target Distance x Invalid Cue Relative Distance [$F(4,20) = 4.20$; $p = .01$] interactions were obtained, as shown in Figure 3A. Most surprising was the effect of invalid auditory cues on search times with the valid visual cues: search times were shorter than with the visual cue presented alone. This benefit of invalid auditory cues was reduced with target distance. Search times with valid auditory cues, on the other hand, increased by 100 - 300 ms over the auditory baseline condition. The interference was highest at the shortest target distance, and decreased with distance. A consistent effect of the invalid cue relative distance was not evident in Figure 3A, so we replotted these data in Figure 3B to account for the actual distance of the invalid cue from fixation. From this figure it appears that the benefits of the invalid auditory cue to searching with the valid visual cue, and the costs of the invalid visual cue to searching with the valid auditory cue are linearly related to the distance of the invalid cue from fixation. The absolute value of the slope of the valid auditory function (2.4 ms/°) is more than double the slope for the visual function (.82 ms/°).

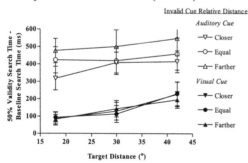

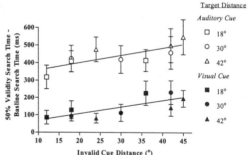

Figure 4A: Validity=50%. Change in search time relative to baseline as a function of target distance and invalid cue relative distance.

Figure 4B: Validity=50%. Change in search times relative to baseline as a function of invalid cue and target distance from fixation.

Figure 4 presents the change in search times for the 50% validity condition. Significant main effects of distance [$F(2,10) = 13.36$; $p = .001$] and invalid cue relative distance [$F(2,10) = 3.84$; $p = .05$] were obtained. A significant Cue x Invalid Cue Relative Distance interaction was also obtained [$F(2,10) = 11.67$; $p = .002$]. As shown in Figure 4A, search times in each modality increased relative to the baseline search condition; the cost was greatest for valid auditory cues. The relative distance of the invalid cue affected search time for valid auditory cues most. We obtained the greatest increase in search times when the invalid visual cues were farther from fixation. When we replotted the search times against the actual invalid cue distance in Figure 4B, most of the search time variability was accounted for. Here, the slopes are nearly identical (4.2 ms/° and 3.8 ms/° for auditory and visual cues respectively.)

4. DISCUSSION.

We obtained no adverse effects of invalid auditory cues when the validity of the visual cues was 90%. Instead, the auditory cues reduced the time to locate and identify the target, contrary to previous experiments on auditory spatial cueing. Rudmann and Strybel (1999) showed that search times with auditory cues that were displaced from the target were longer than when no cue was provided in some conditions. Spence and Driver (1997) showed that auditory cues can be stimulus-driven in that they increased reaction time when inaccurate. We believe the difference in experimental outcomes between our findings and previous research is due to either the absence or inadequacy of the visual cue in past works. No visual cue was provided with the auditory cue in Rudmann and Strybel, although the pattern of distractors sometimes cued the local target area. Spence and Driver tested each cue in separate sessions; the cues were never presented simultaneously. Our results at 90% validity are consistent with Frens et al.'s (1995) findings regarding eye movements to multimodal stimuli. The latency of the eye movements to high intensity auditory and visual stimuli were reduced by approximately 50 ms, relative to the saccadic latencies to the visual stimuli in isolation. This effect diminished as the stimuli were spatially separated. Note in Figure 3B that the benefits obtained here are diminishing slightly with increasing distance, as predicted by Frens et al.

When the validity of each cue was 50%, our participants usually searched the visual cue before the auditory cue because the cost of the invalid cue was less, as shown in Figures 4A and 4B. With both cues the cost of the invalid cue increased with its distance. We believe this is due to the time required to move from an invalid cue to a valid cue. As the distance of the invalid and valid cue increased, it took longer to reorient to the cue in the opposite hemifield. Moreover, given that the visual cue produced faster search times in isolation (as shown in Figure 2), participants probably oriented to the visual cue more quickly after searching an invalid auditory cue than they oriented to the auditory cue after searching an invalid visual cue.

In summary, our results suggest that stimulus-driven visual cues interfere more with auditory cues even when users know the auditory cues are valid. Users prefer visual cues to auditory cues when neither cue is reliable. Designers of mulitsensory environments could probably provide auditory spatial cues without concern about visual cue interference. On the other hand, designers should be careful about using auditory spatial cues to signal the most critical events, because visual onsets may interfere with them. These recommendations are limited, however, to the specific stimulus characteristics tested here. From Figure 2, it appears that our visual cue was more conspicuous and accurate than the auditory cue. Degrading either the contrast or the accuracy of the visual cue might produce a greater effect of auditory spatial cues. Suggesting this possibility, Frens et al. (1995) showed that saccade latencies to low-intensity visual cues increased considerably, but saccade latencies to low intensity auditory cues were not affected.

REFERENCES.

Bullinger, H., J., Bauer, W., and Braun, M. (1997).Virtual Environments. In G. Salvendy (ed.) Handbook of Human Factors and Ergonomics 2nd Edition (pp. 1725-1760). New York: Wiley.

Egeth, H.,E., and Yantis, S. (1997). Visual attention: control, representation, and time course. Annual Review of Psychology, 48. 269-297.

Frens, M., A., Van Opstal, A., J., and Van der Willigen, R., F. (1995). Spatial and temporal factors determine auditory-visual interactions in human saccadic eye movements. Perception and Psychophysics, 57, 802-816.

Fujawa, G., and Strybel, T., Z. (1997). Auditory spatial facilitation with valid and invalid cues. Proceedings of the Human Factors and Ergonomics Society 41st Annual Meeting, (pp. 556-560). Santa Monica, Ca: Human Factors and Ergonomics Society.

Stein, B., E., and Meredith, M., A. (1993). The Merging of the Senses. Cambridge, MA: MIT Press.

Rudmann, D., and Strybel, T., Z. (1999). Auditory spatial facilitation of visual search performance: Effect of local vs global visual distractor density. Human Factors, 41, 146 - 160.

Spence, C., and Driver, J. (1997). Audiovisual links in exogenous covert spatial orienting. Perception and Psychophysics, 59, 1-22.

Strybel, T., Z., Boucher, J., Fujawa, G., and Volp, C. (1995). Auditory spatial cueing in visual search tasks: effects of amplitude, contrast and duration. Proceedings of the Human Factors and Ergonomics Society 39th Annual Meeting, (pp. 109-114). Santa Monica, Ca: Human Factors and Ergonomics Society.

Strybel, T., Z., Vu, K.-P., L., and Castagna-Osorio, J., T. (2000). Auditory spatial cueing of visual search in the presence of noise (Abstract), Proceedings of the Human Factors and Ergonomics Society 44th Annual Meeting. Santa Monica, Ca: Human Factors and Ergonomics Society.

Stimulus-Response Compatibility in Interface Design

Kim-Phuong L. Vu and Robert W. Proctor

Purdue University, Department of Psychological Sciences, West Lafayette, IN 47907

ABSTRACT

Principles of stimulus-response compatibility have been incorporated into many design guidelines because maintaining compatibility leads to better performance. Although designers may be generally familiar with basic compatibility principles, they likely are not familiar with more complex ones. Payne (1995) found that naïve subjects were not very accurate at predicting performance for different stimulus-response mappings. These results suggest that designers may not be able to predict whether a particular display-control configuration will lead to better performance than another. The present study verified Payne's findings, but showed that subjects' estimates of performance can be improved with little practice with the different stimulus-response mappings.

1. INTRODUCTION

Stimulus-response compatibility (SRC) has been a topic of concern in human factors and ergonomics since the field's earliest days. SRC refers to the fact that people respond more quickly and accurately with some mappings of stimuli to responses than with others (Proctor & Reeve, 1990). Thus, SRC effects have been incorporated into many design principles for human-computer interaction (see Proctor & Van Zandt, 1994). Because SRC effects are robust, general rules of compatibility have been adapted to the design of interfaces, and guidelines recommend maintaining compatibility whenever possible (Andre & Wickens, 1990).

Spatial SRC effects range from fairly intuitive relations to more complicated ones. When the stimuli are lights presented in left and right locations and responses are left and right keypresses, the compatible mapping of left stimulus to left response and right stimulus to right response yields better performance than the alternative incompatible mapping. The outcome of this task is intuitive, and designers would likely choose to use the compatible mapping whenever possible. However, compatibility effects for many display-control relations are not so straightforward. For example, when stimuli and responses vary along both horizontal and vertical dimensions, compatibility may be maintained for both, one, or neither of the dimensions. Not surprisingly, performance is best when both dimensions are mapped compatibly and worst when both are mapped incompatibly. However, when compatibility can be maintained for only one dimension, it is not obvious whether maintaining compatibility for the horizontal or vertical dimension is more important. Andre and Wickens (1990) recommended maintaining compatibility along the horizontal dimension over the vertical one because results obtained by Nicoletti and Umiltà (1985) showed that the horizontal dimension is dominant, a phenomenon known as right-left prevalence. However, recent studies we conducted demonstrate that right-left prevalence only occurs for situations in which the environment provides a salient frame of reference for the horizontal dimension. Right-left prevalence can be reversed to top-bottom prevalence if the display or control configuration makes the vertical dimension more salient (Vu & Proctor, in press). Thus, for more complex relationships between the stimulus-response (S-R) mappings, compatibility effects are not easily predicted.

A way to determine whether a display-control configuration is optimal is to obtain performance measures with different configurations. The main concern of usability testing is to detect oversights in the design of a product that would affect its usefulness. Usability testing has the goal of determining whether a product is easy to use, effective, and preferred by users (Neilsen, 1997). Usability is evaluated by observing users perform tasks with a product or different versions of the product. However, because usability testing is expensive and time-consuming, designers can only observe a few participants. Thus, when designers want a large-scale evaluation of user preference for a product, they may choose to use questionnaires. In general, users' judgments are relatively reliable, especially when the users are familiar with the task. For example, a recent study of metacognition, or people's knowledge of their own cognitive processes, showed that individual ratings of their expertise were better predictors of their performance on word-processing tasks than were their estimates of how frequently they use word-processing applications (Vu, Hanley, Strybel, & Proctor, 2000). However, user preference and performance are not always correlated. Users may indicate that they prefer a design that would lead to poorer performance (Bailey, 1993).

Establishing the relation that exists between users' metacognitive judgments and actual performance with respect to SRC effects is important because a product design is influenced by the designers' choices as well as by users' preferences. Payne (1995) conducted a study asking naïve subjects to rate the usability of different S-R configurations that differed in terms of spatial mappings. He showed subjects illustrations of a four-choice task with four different S-R mapping conditions (see Figure 1). The difference between the four conditions is whether the inner and outer S-R pairs are mapped compatibly or incompatibly. For Condition 1 (incompatible condition), both the inner and outer pairs are mapped incompatibly. For Conditions 2 and 3 (the mixed conditions), the outer pairs are mapped compatibly and the inner pairs incompatibly, or vice versa. For Condition 4 (compatible condition), both the inner and outer pairs are mapped compatibly. With these conditions, performance is best for the compatible condition because the stimuli are directly mapped to their corresponding responses. It is second best for the incompatible condition because there is a systematic relation between the S-R pairings that allows subjects to adopt a "respond opposite" rule. Performance is worst for the mixed conditions because two mappings are in effect and subjects must decide which mapping is appropriate on the current trial before responding (Duncan, 1977). The slowing of responses with mixed mappings is called a mixing cost and is a common finding (Proctor & Vu, in press). The results of Payne's study showed that subjects correctly predicted that performance would be best for the compatible condition, but incorrectly predicted that performance would be better in the mixed conditions than in the incompatible condition. However, the illustrations of the tasks that Payne used in his study to present the different S-R mappings may have biased his results. That is, looking at Figure 1, the diagram of arrows connecting the stimuli to their assigned responses appears more complex for the incompatible condition than for the mixed conditions. Thus, subjective judgments of difficulty may have been based on the display complexity. It is possible that subjects can predict performance better for the different mapping conditions if the illustrations of the tasks look equally simple or complex, or if the individual S-R pairing for each mapping condition is presented separately.

The purpose of the present study was to evaluate whether Payne's (1995) results are obtained using different types of diagrams for the mapping conditions. We also included scenarios that depict basic compatibility effects for two-choice tasks to see whether subjects can predict performance for easier tasks. In addition to obtaining subjective judgments of SRC, we evaluated users' judgments before and after they performed with the specific mapping conditions. Evaluation of pre- and post-tests allowed comparison of judgments of the relative compatibility of tasks when subjects are familiar with the tasks in comparison to when they are not. Results from our study provide insight regarding the relationships between usability judgments of SRC and actual performance.

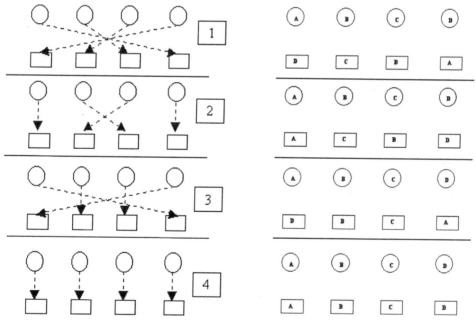

Figure 1. Illustration of the four-choice task used Payne's (1995) study and Experiment 1. Arrows connected the stimuli to their assigned responses.

Figure 2. Illustration of the four-choice task used in Experiment 1. Letters were used to designate stimulus positions and their assigned responses.

2. EXPERIMENT 1

Experiment 1 was designed to assess whether subjects can accurately predict actual performance for different S-R configurations. Subjects' performance estimates for two-choice compatibility tasks were obtained in addition to those for the four-choice tasks used in Payne's (1995) study. The two-choice task consisted of compatible and incompatible mappings of left and right stimuli to left and right response keys, with the stimulus location information conveyed in three different modes. For physical positions, the stimuli were circles in left and right positions. For arrow directions and location words, the stimuli were left- or right-pointing arrows and the words "Left" or "Right" placed in the center of the display. After completing the ratings for the two-choice task, subjects were given the four-choice tasks used in Payne's study. As mentioned in the introduction, the illustration of the four compatibility tasks that Payne used may have biased his results because the configuration of the arrows connecting each stimulus to its assigned response varied in complexity across the different conditions. To rule out this perceptual bias, we used four different types of illustrations to depict the four mapping conditions. Two of the displays used arrows to connect the stimuli to their assigned responses and two used letters to designate the mappings (see Figures 1 and 2). The illustrations for the four conditions and the individual S-R pairings were presented simultaneously or on separate pages (each condition was presented on a different page with each S-R pairing illustrated separately). If subjects rated the incompatible condition as more difficult than the mixed condition because its illustration looks more complex, presenting each mapping condition on a different page with the individual S-R pairings illustrated separately may reduce this perceptual bias.

2.1 Method

2.1.1 Subjects. Forty subjects were recruited from Purdue University's Introductory Psychology subject pool. They received partial course credit for their participation.

2.1.2 Questionnaire. The questionnaire was divided into two parts. The first consisted of three basic two-choice compatibility tasks in which subjects were instructed to estimate their mean reaction time (RT) with each S-R mapping and the percentage of errors (PEs) they would make. Subjects completed three tasks that differed only in terms of the mode by which the stimulus-location information was conveyed: Physical positions, arrow directions, or location words. Each task consisted of a compatible and incompatible mapping of left and right stimuli mapped to left and right response keys. The scale for RT estimates ranged from 250 ms to 500 ms in 25 ms increments. The scale for PE estimates ranged from 0% to 2.25% in 0.25% increments. The order of the tasks was counterbalanced between subjects.

The second part consisted of the four-choice task used in Payne's (1995) study, in which the inner or outer S-R pairs are mapped compatibly or incompatibly. There were four different displays, to each of which 10 subjects were randomly assigned. The first display was analogous to that used by Payne in which arrows connected the stimuli to their assigned responses. The second display also used arrows to connect stimuli to their assigned responses, but each condition was presented on a different page with each S-R pairing presented individually on the page. The third and fourth displays used letters to designate the stimulus positions and their assigned responses. The conditions and S-R mappings were presented simultaneously in the third condition and separately in the fourth condition. The RT scale ranged from 350 ms to 600 ms in 25 ms increments, and the PE scale ranged from 0% to 2.25% in 0.25% increments. The increased scale for RT was designed to reflect the range for actual performance. The order in which the conditions were presented was randomized.

2.2 Results

2.2.1 Two-choice tasks. Subjects' estimates of RT and PE were submitted to separate 3 (Stimulus Mode: physical locations, arrow directions, or location words) x 2 (Compatibility: compatible or incompatible) ANOVAs. For both RT and PE, the main effect of stimulus mode was significant, $Fs(2,78) > 5.85$, $ps < .004$. Subjects estimated that performance would be better when the stimuli were physical locations ($Ms = 322$ ms and 0.71%) than when they were location words ($Ms = 335$ ms and 0.85%) or arrow directions ($Ms = 347$ ms and 0.87%). There was also a main effect of compatibility for both RT and PE, $Fs(1,39) > 91.00$, $ps < .001$. Subjects estimated an 80 ms and 0.52% advantage for the compatible mapping over the incompatible one. The interaction of stimulus mode and compatibility was not significant for either measure, $Fs < 1.5$.

2.2.2 Four-choice tasks. Subjects' estimates of RT and PE were submitted to separate 2 (Outer Compatibility: compatible or incompatible mapping for the outer S-R pairs) x 2 (Inner Compatibility: compatible or incompatible mapping for the inner S-R pairs) x 2 (Display: arrows or letters) x 2 (Format: simultaneous or separate) ANOVAs. Inner and outer compatibility were within-subjects factors, and display and format were between-subjects factors. The two-way interaction of outer and inner compatibility was significant for both the RT and PE data, $Fs(1,36) > 9.94$, $ps < .003$. Subjects estimated that performance would be best for the compatible condition ($Ms = 373$ ms and 0.24%), intermediate for the mixed conditions ($Ms = 453$ ms and 0.98%), and worst for the incompatible condition, ($Ms = 491$ ms and 1.27%). Format and Display did not interact significantly with inner and outer compatibility, $Fs(1,36) > 3.22$, $ps > .08$, although there was a trend to reduce the RT estimate for the incompatible condition to the same level of difficulty as that for the mixed conditions when each condition and the individual S-R pairings were presented separately.

2.3 Discussion

Subjects were able to correctly predict the typical spatial compatibility effect obtained in a two-choice task. However, they were not able to predict the different magnitudes of compatibility effects obtained with different stimulus modes. Actual performance with these tasks shows that the compatibility effect is largest for physical locations, intermediate for arrow directions, and smallest for words, and that the overall time to respond to these stimuli is usually shortest for physical locations, intermediate for arrows, and longest for words (Proctor & Vu, in press). However, subjects estimated the compatibility effect to be equivalent for all stimulus modes, and that it would take longer to respond overall to arrows than words.

For the four-choice task, the results replicate Payne's (1995) results. Subjects predicted that performance would be best for the compatible condition, intermediate for the mixed conditions, and worst for the incompatible condition. These predictions do not correspond with actual performance, which is usually for the incompatible condition than the mixed conditions. Using letters or arrows to designate the S-R pairs did not produce different results, but presenting each condition and S-R pair separately tended to reduce the RT estimates for the incompatible condition.

3. EXPERIMENT 2

The results of Experiment 1 indicate that subjects were able to accurately predict a compatibility effect for two-choice tasks. However, they were not able to predict the magnitude of compatibility effects across stimulus modes or actual performance on four-choice tasks involving more complicated S-R mappings. Although this finding is not surprising, it has important implications. As Payne (1995) indicated, many designers are not aware of compatibility relationships and may use less than optimal designs because they cannot predict performance accurately. The literature on metacognition, although not directly linked to human-computer interaction, indicates that subjects are fairly accurate about predicting their performance, especially if they have some experience with the task (see, e.g., Vu et al., 2000). Although designers may not have ample time to investigate performance with a wide variety of display and control configurations, they should be able to devote limited time to evaluating several configurations. The purpose of Experiment 2 was to determine whether limited exposure to different S-R mappings could alter subjects' initial estimates of performance to correspond more closely with actual performance.

3.1 Method

Twenty-eight subjects were given the same questionnaire as in Experiment 1. After completing the questionnaire, they performed 40 trials with each individual task, and estimates of RT and PE were obtained after each task was completed. For the four-choice task, only the illustration used by Payne (1995) was included because the display format manipulation had little effect in Experiment 1.

3.2 Results

3.2.1 Two-choice tasks. Subjects' estimates of RT and PE were submitted to separate 2 (Condition: pre-test and post-test) x 3 (Stimulus Mode: physical locations, arrow directions, or location words) x 2 (Compatibility: compatible or incompatible) ANOVAs. For both RT and PE, the main effect of stimulus mode was significant, $Fs(2,54) > 7.68$, $ps < .001$. Subjects estimated that performance was best for physical locations ($Ms = 309$ ms and 0.51%), followed by arrow directions ($Ms = 327$ ms and 0.65%) and location words ($Ms = 336$ ms and 0.69%). There was also a main effect of compatibility for both RT and PE, $Fs(1,27) > 42.00$, $ps < .001$. Subjects estimated a 41 ms and 0.34% advantage for the compatible mapping over the incompatible one.

Most important, condition interacted significantly with stimulus mode for RT and PE, $Fs(2,54) > 6.63$, $ps < .033$. For the pre-test, subjects estimated that performance was best for physical locations ($Ms = 316$ ms and 0.68%) followed by location words ($Ms = 334$ ms and 0.72%) and arrow directions ($Ms = 342$ ms and 0.87%). However, in the post-test, the order for arrows and words were switched, yielding estimates of 303 ms and 0.34% for physical locations, 313 and 0.48% for arrow directions, and 339 ms and 0.67% for location words. These latter estimates correctly reflect the relative ordering for actual performance.

3.2.2 Four-choice task. Subjects' estimates of RT and PE were submitted to separate 2 (Condition: pre-test or post-test) x 2 (Outer Compatibility: compatible or incompatible) x 2 (Inner Compatibility: compatible or incompatible) ANOVAs. The interaction of outer and inner compatibility was significant for both RT and PE, $Fs(1,27) > 36.1$, $ps < .001$. Subjects estimated that performance would be best for the compatible condition ($Ms = 393$ ms and 0.34%), intermediate for mixed Condition 2 ($Ms = 445$ ms and 0.79%), and worst for mixed Condition 3 ($Ms = 464$ ms and 0.98%) and the incompatible condition ($Ms = 462$ ms and 0.92%).

More importantly, this interaction was modified by a significant three-way interaction of Condition x Outer Compatibility x Inner Compatibility for RT and PE, $Fs(1,27) > 13.75$, $ps < .001$. In the pre-test, subjects estimated that performance would be best for the compatible condition ($Ms = 388$ ms and 0.36%), intermediate for the mixed conditions ($Ms = 449$ ms and 0.92%), and worst for the incompatible condition, ($Ms = 491$ ms and 1.29%). However, in the post-test, subjects estimated that performance would be best for the compatible condition ($Ms = 398$ ms and 0.31%), intermediate for the incompatible condition, ($Ms = 433$ ms and 0.55%), and worst for the mixed conditions ($Ms = 460$ ms and 0.84%). These latter estimates reflect the ordering of actual performance.

3.3 Discussion

The results of Experiment 2 show that subjects were able to adjust their initial estimates of performance on a given task to reflect that of actual performance after little practice with the task. For the two-choice task, subjects were able to adjust their estimates of the overall ordering of RT and PE for the three stimulus modes. However, they were still not able to predict the relative magnitudes of the compatibility effects for each stimulus mode. For the four-choice task, the results of the pre-test replicate those of Payne (1995) and Experiment 1, in which subjects estimated that performance would be worst for the incompatible condition than the mixed conditions, even though the reverse is true for actual performance. However, subjects were able to correctly adjust their estimates after 40 trials of practice with each mapping. In the post-test, subjects correctly indicated that the incompatible condition yielded better performance than the mixed conditions.

4. CONCLUSION

Taken together, the results of Experiments 1 and 2 indicate that naïve judgments of performance for different S-R configurations are not particularly accurate. For two- and four-choice tasks, subjects know that performance is best when spatial compatibility is maintained. However, their judgments do not accurately reflect the way in which performance varies as a function of different stimulus modes or of mixed versus incompatible S-R mappings. These results are consistent with those of Payne (1995) and show that his findings were not a consequence of how the S-R mappings were illustrated. However, Experiment 2 showed that after limited practice with the individual S-R mappings, subjects adjusted their initial judgments of performance to more accurately match actual performance. Thus, minimal experience with alternative configurations is sufficient to improve relative estimates of performance.

5. REFERENCES

Andre, A. D., & Wickens, C. D. (1990). Display-control compatibility in the cockpit: Guidelines for display layout analysis. *Technical Report: NASA Ames Research Center*. Moffett Field, CA.

Bailey, R. W. (1993). Performance versus preference. In the *Proceedings of the Human Factors and Ergonomics Society 37th Annual Meeting* (pp. 282-286). Santa Monica, CA: Human Factors and Ergonomics Society.

Duncan, J. (1977). Response selection rules in spatial choice reaction tasks. In S. Dornic (Ed.), *Attention and performance VI* (pp. 49-61). Hillsdale, NJ: Erlbaum.

Nicoletti, R., & Umiltà, C. (1985). Responding with hand and foot: The right/left prevalence in spatial compatibility is still present. *Perception & Psychophysics, 38*, 211-216.

Nielsen, J. (1997). Usability testing. In G. Salvendy (Ed), *Handbook of human factors and ergonomics* (2nd ed.; pp. 1,543-1,568). New York: Wiley.

Payne, S. J. (1995). Naïve judgments of stimulus-response compatibility. *Human Factors, 37*, 495-506.

Proctor, R. W., & Reeve, T. G. (Eds.) (1990). *Stimulus-response compatibility: An integrated perspective*. Amsterdam: North-Holland.

Proctor, R. W., & Van Zandt, T. (1994). *Human factors in simple and complex systems*. Boston, MA: Allyn and Bacon.

Proctor, R. W., & Vu, K.-P. L. (in press). Eliminating, magnifying, and reversing spatial compatibility effects with mixed location-relevant and irrelevant trials. In W. Prinz and B. Hommel (Eds.) *Common mechanisms in perception and action: Attention and performance, Vol. XIX:*. Oxford: Oxford University Press.

Vu, K.-P. L., Hanley, G. L., Strybel, T. Z., & Proctor, R. W. (2000). Metacognitive processes in human-computer interaction: Self-assessments of knowledge as predictors of computer expertise. *International Journal of Human-Computer Interaction, 12*, 43-71.

Vu, K.-P. L., & Proctor, R. W. (in press). Vertical versus horizontal spatial compatibility: The role of salience in two-dimensional compatibility effects. *Journal of Experimental Psychology: Human Perception and Performance*.

Cognition in a Dynamic Environment

Francis T. Durso, Jerry M. Crutchfield, & Peter J. Batsakes

University of Oklahoma, Department of Psychology, Norman, OK 73019

ABSTRACT

The human-technical system of air traffic control allowed us to determine the extent to which performance deficits, one caused by an automation failure and one caused by an individual difference in working memory span, could be attributed to a diminished understanding of the situation, that is a diminished situation awareness (SA). Results suggested that the automation failure decrement was due, at least in part, to SA. The superior performance of high span operators, however, did not seem to be due to better SA. Design efforts to improve SA would be appropriate to remedy consequences of the automation failure, but might be misguided in the case of working memory span.

1. INTRODUCTION

Each year, industrial workers are asked to perform increasingly complex cognitive tasks. Understanding how operators control and monitor complex, dynamic environments, especially interactive safety-critical ones, requires an appreciation of both the operator and the accompanying automated arsenal. For example, air traffic controllers are highly trained, highly selected individuals with a variety of cognitive skills and abilities who use a computer augmented radar display and other automation to accomplish the safe and expeditious movement of the nation's air traffic. Clearly, the controller, the automation, and their interaction are critical to the success of such a venture.

Cognition in a dynamic environment is often studied under the rubric of situation awareness (SA; Durso & Gronlund, 1999). Although SA research is conducted primarily in the aviation domain (e.g., Endsley & Garland, 2000), the work is relevant to other domains, such as the development of SA aids in computer supported cooperative work. Awareness services (e.g., Walker, 1998) reflects HCI's recognition that intelligent aids could assist remote collaborating teams by supplying the members with SA about the task and the other members of the team. Researchers have proposed a number of methods to study SA, including subjective measures (the participant judges SA), implicit performance (the participant detects a problem), and situation queries (the participant answers questions).

We have been employing one version of the query approach that we refer to as Situation Present Assessment Method (SPAM; Durso et al., 1998). The name implies that the situation remains available to the operator throughout the assessment rather than being removed (cf. Endsley, 1995), and the SPAM acronym reminds us that we are presenting participants unsolicited requests for information, and thus may influence normal processing as do most on-line measures of cognition. In SPAM, we assume, as cognitive psychology did 40 years ago, that latencies can reveal information about how operators understand. For information easily accessible in memory or from the environment, latencies should be small. For information not understood by the operator, search through memory or the environment should result in increased latencies. SPAM acknowledges that knowing how to find something in the environment is a relevant part of SA; it does not have a memory component; and it measures SA when it succeeds, rather than by inferring SA through failures. In SPAM, the operator first indicates a readiness to accept a question, is given the question, and then answers the question. SPAM also allows us to disentangle the time required to prepare for a question, that could be affected by other task relevant activities, work load, and so on; from the cognitive activity association with understanding and SA.

In the current study, we report preliminary data directed at determining the extent to which performance detriments caused by psychological and technological changes can be attributed to a diminishment in SA. We hope to cause the performance deficit of technology by introducing an automation failure (e.g., Parasuraman, Mouloua, & Molloy, 1994) into a low-fidelity air traffic control (ATC) simulation. This automation failure is interesting for ATC researchers because it has features similar to the redundant backup system currently in place. We hope to produce the psychological-dependent performance deficit by comparing participants with a large working memory span to those with a small one. This choice is interesting because working memory has been assumed to be an important part of situation awareness, and because we have discovered in previous work (Crutchfield, Durso, & Bleckley, 1999) that low span operators perform more poorly than high span operators.

2. METHODS
2.1 Participants

Participants were screened based on a test of working memory, Operation Span (OSPAN) (La Pointe & Engle, 1990). Participants scoring within national high (18+; 19 - 31) and low (≤10; 2 - 10) quartiles (Engle, Kane, & Tuholski, 1999) were given the opportunity to continue to participate.

2.2 Materials

The OSPAN test presents participants with a series of simple mathematical equations, each followed by a one-syllable word. The participant's task is to read the equation out loud, evaluate the equation, and say the word. After several equation-word pairings (3 to 6), the participant is asked for perfect serial recall of the words. Each participant received multiple trials yielding scores that could range from 0 to 60.

The performance task was the air traffic scenarios test (ATST). Our ATST was a modification of the Federal Aviation Administration's ATST. The ATST forms part of the battery being developed to select controllers. It presents participants with a simulated radar air space complete with 2 airports, 4 jetway exits, and a variable number of moving aircraft. Participants can tell aircraft identification, speed, destination, and altitude from data blocks presented near the radar blip. The task is made more complex by the fact that airports portrayed in the airspace will change required landing directions and that 10% of all control actions given to aircraft will result in a readback error. Participants were told to listen for readback errors and to correct them by pressing a "repeat" button and reissuing the control instructions.

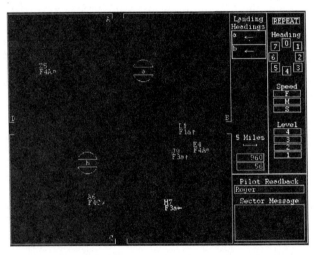

In the field, air traffic control specialists use paper flight progress strips, in part, as a redundant backup to their automated technologies. The ATST was supplemented to include paper flight progress strips, similar to real flight progress strips but adjusted for this simulation. They included all the same information (including destination) that the aircraft datablocks included, and thus they served as a redundant backup to the datablock automation. Our participants were instructed to place a check mark by all aircraft designations on the flight progress strips as they accepted each aircraft into their airspace. The ATST screen appears in the adjacent figure.

The SPAM queries consisted of two sets of six air-traffic related questions. The questions were specific to each 16-minute scenario and of the six questions, three were related to the automation manipulation and will be analyzed here (e.g., At what final speed should aircraft K0 be?, What is the destination of aircraft U3?). In SPAM, the participant is warned that a question is in queue, to which the participant responds "ready." The time to respond "ready" is used as a measure of workload. The question is then asked and the time to answer the question is used as a measure of understanding or SA. The questions were presented orally at random times while all information normally available to the participant as they controlled traffic remained so.

2.3 Procedure

After screening for high and low OSPAN individuals, participants were taught how to control aircraft. Participants performed 5, 16 min practice scenarios. Each scenario began with five activated aircraft with a total of 27 aircraft appearing from start to finish.

Two test sessions, each containing two 16 min scenarios, followed practice. One scenario in each test session was designated an automation failure. The order and scenario used as the automation failure was counterbalanced across participants. In the automation failure condition, aircraft datablocks did not include aircraft destination information, forcing participants to rely on the redundant backup system. Participants had to find designated aircraft destinations by reading flight progress strips adjacent to the radar display.

The first test session was used to collect performance data during the automation failure and normal condition. In this session the participants were only given the task of controlling traffic. The second test session was used to collect information about situation awareness. In this session participants had to respond to SPAM questions while engaged in traffic control.

3. RESULTS

In all analyses, we submitted the dependent variables to 2 x 2, Automation (Normal, Failure) x Span (High, Low) analysis of variance. Given the preliminary nature of the work, we held outcomes to an alpha of .10. Pairwise tests are one-tailed t statistics.

3.1 Performance measures

We combined the performance measures collected during the final training session into two dependent variables: Traffic errors and Clearance delay. Traffic errors included failures to maintain separation standards between aircraft, sending aircraft to the wrong destination, transitioning aircraft illegally, and so on. Clearance delay was the cumulative time required to accept an aircraft from the simulation, transferring control of the aircraft to the operator.

Figure 1 shows the cumulative time to issue clearances on incoming aircraft. The measure increases as a function of the number of aircraft and the time required to accept each aircraft. Analysis of Clearance Delay yielded an effect of Automation. Both high and low span operators were hurt by automation failure, $F(1,22)=16.14$, p=.001.

Figure 1: Clearance Delay	Figure 2: ATC Errors

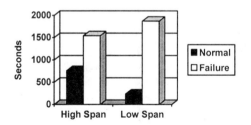

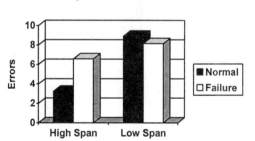

High span operators made fewer traffic errors than did low span operators, $F(1,22) = 1.98$, p = .07 This is an important demonstration of the impact of the individual difference variable on performance and is consistent with earlier work (Crutchfield, et al., 1999). The effect of automation was not reliable.

3.2 Situation awareness measurement

It is tempting to assume that the differences observed in performance could be attributed to differences in SA. However, without a direct test of SA such an assumption would be unwarranted. This section attempts to determine if our two variables produced their deficits via diminished SA.

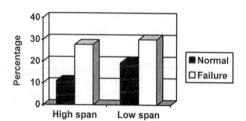

Figure 3: SPAM Errors

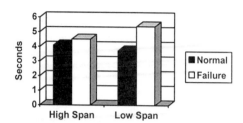

Figure 4: Answer Time

First, as evident in Figure 3, total errors were not especially low. Even in the conditions when automation was present, the high span error rate was 11% and the low span was 19%. Automation failure increased error rates for the high span operators and the low span operators, $F(1,22) = 6.71$, $p = .017$. Thus, we have evidence that that automation failure hurt the ability of operators to understand the traffic situation. However, there was no difference between the high and low span operators. Figure 4 shows the time required to answer SPAM questions after the participant indicated "ready." We see again a main effect of automation, $F(1,22) = 3.09$, $p = .09$, indicating that operators took longer to answer the SPAM questions during the failure condition. And again, there were no differences that could be attributed to the difference in working memory span. Overall, it seems clear that SA is implicated in the detrimental effects of automation failure. However, SA does not seem to be the reason that high span individuals made fewer ATC errors than low span individuals.

If SA does not account for the tendency for high span operators to make fewer ATC errors than do low spans, what might account for the individual differences? In an effort to find an explanation, we looked at the time individuals took to indicate that they were ready to attempt a question. These ready times appear in Figure 5.

Figure 5: "Ready" responses

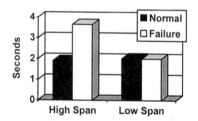

Figure 5 suggests that high span operators, but not low span operators, were hurt by the automation failure. It even appears in the automation failure condition that the low spans are outperforming the high spans. In fact, this turns out to be more a strategy difference than any real difference in workload. When only those items that ultimately will be answered correctly are considered, both span groups are hurt by automation failure, but neither significantly so.

It seems from these data that low span operators are less able to evaluate their workload. Even though low span operators make more SPAM errors under automation failure (Figure 3), they do not adjust the time to accept questions. High spans experiencing an automation failure, on the other hand, take longer before they are willing to attempt a question. Perhaps it is not that high and low span operators differ in their understanding of the situation but rather in their management of it.

4. CONCLUSIONS

The automation clearly had detrimental effects. In particular, it increased the time to grant clearances to newly arrived aircraft. SPAM results suggested that SA was implicated in the impact of automation failure. It was also the case that individual differences in working memory as measured by OSPAN affected performance: High span operators made fewer ATC errors than did low span operators. However, SA did not seem to be related to this performance difference. This is important for a number of reasons. First, it illustrates that assuming SA underlies differences in observed performance is not always prudent. Second, many models of SA have assumed that working memory is a critical component of it, yet here neither SPAM errors nor SPAM times confirmed this. Instead it may be the case that high and low span individuals differ in the management of resources. Our low span participants were especially poor under automation failure at determining when they would accept a question; in fact, they accepted questions as rapidly under automation failure as they did under automation.

In summary, both the psychological variable of working memory and the technological variable of automation failure had effects on the operators' control of simulated air traffic. In our simulation, which forced controllers to rely on a redundant backup in the event of a radar failure, automation influenced understanding. However, not all differences in performance will have SA as an underlying cause. For example, our working memory effect did not. In other human-technical systems, it may prove to be the case that performance is affected by SA or by other cognitive procedures, like the management of resources.

REFERENCES

Crutchfield, J., Durso, F. T., & Bleckley, M. K. (2000). Predictors of air traffic control success. *American Psychological Society,* Miami Beach.

Durso, F. T., & Gronlund, S. (1999). Situation awareness. In F. Durso (Ed.), *The Handbook of Applied Cognition,* Chicester: Wiley, pp. 284-314.

Durso, F. T., Hackworth, C., Truitt, T. R., Crutchfield, J., Nikolic, D., & Manning, C. A. (1998). Situation awareness as a predictor of performance in en route air traffic controllers, *Air Traffic Control Quarterly,* 6, 1-20.

Endsley, M. R. (1995). Measurement of situation awareness in dynamic systems. *Human Factors,37,* 65-84.

Endsley, M. R. & Garland, D. J. (Eds.)(2000). *Situation Awareness Analysis and Measurement.* New Jersey: Lawrence Erlbaum Associates.

Engle, R. W., Kane, M. J., & Tuhulski, S. W. (1999). Individual differences in working memory capacity and what they tell us about controlled attention, general fluid intelligence, and functions of the prefrontal cortex. In A. Miyake & P. Shah (Eds.) *Models of Working Memory: Mechanisms of Active Maintenance and Executive Control,* New York: Cambridge University Press, pp. 102-134.

La Pointe, L. B., & Engle, R. W. (1990). Simple and complex word spans as measures of working memory capacity. *Journal of Experimental Psychology: Learning, Memory, and Cognition, 16,* 1118-1133.

Parasuraman, R., Mouloua, M, & Molloy, R. (1994). Monitoring automation failures in human-machine systems. In M. Mouloua and R. Parasuraman (Eds.), *Hum,an Performance in Automated Systems: Current Research and Trends,* New Jersey: Erlbaum, pp. 45-49.

Walker, W. F. (1998). "Rapid Prototyping of Awareness Services Using a Shared Information Server." SIGCHI Bulletin 30(2): 95-101.

Keeping Up with Busy Schedules:
Using Personal Data Assistants (PDAs) Effectively

Douglas Herrmann and Carol Yoder

Indiana State University
Psychology Department
Root Hall
Terre Haute, IN 47809

ABSTRACT

Personal data assistants can remind people effectively of intentions that they would forget otherwise. These devices eliminate the need for people to remember intentions over long time periods. People need only remember the intention after the warning signal reminds them to execute the intention. The warning signal of the PDA "externalizes" the intention, thereby ensuring that the intention is activated in conscious memory at or near the appropriate time. It seems likely that as the designs of PDA render them more user friendly, their use will become more common.

1. INTRODUCTION

Keeping on top of today's busy schedules requires prospective memory, that is, memory for the things we need to do (Kvavilashvili, 1992; Meacham & Leiman, 1982). The most difficult aspect of executing intentions is that they must be completed at a certain time or within a certain interval (Brandimonte, Einstein, & McDaniel, 1996; Harris, 1984). If a person remembers to perform an act but does not perform it at the appointed time, the person's memory is nevertheless said to have failed.

The consequences of forgetting an appointment or a chore can be onerous. Professionally, failures to meet certain appointments or execute certain chores can irritate a boss and even put someone's employment at risk. Some professions require carrying out so many intentions that doing so is stressful, e.g., air controllers continually consult prospective memory as they monitor locations of aircraft (Vortac, Edwards, & Manning, 1995). Personally, failures to remember appointments or doing chores can get others upset. Sometimes such failures hinder or even destroy relationships (Herrmann, 1996).

Because of the negative consequences of forgetting appointments and chores, people cue themselves to remember with a variety of cueing devices. Cueing practices have been used for centuries. In this century, a variety of products have been developed to remind people of their intentions (Petro, Herrmann, Burrows, & Moore, 1991). Such products range from a conspicuous calendar to personal data assistants (PDAs) that trigger alarms on or about the time of intention execution (Herrmann, Yoder, Wells, & Raybeck, 1996). This presentation reviews recent research by us and by others that indicate some of the variables that influence the efficacy of PDAs.

1.1 Operation of a PDA as a Reminding Device. PDAs deliver a warning signal and exhibit on a screen a written directive of behavior to be executed when the signal is presented. Each intention involves a retention interval that spans the time from when the intention was formed to the time the intended act is to be carried out (Wilkins & Baddeley, 1978). A response is judged as successful if it occurs in an expected interval of time, bounded by a lower and an upper time limit. Responses prior to this window are early; responses after this window are late (Harris, 1984). These temporal properties of intentions are presented below in Table 1.

When a person prepares a PDA to remind oneself to perform an intended act, he or she typically sets the PDA to provide a warning signal to go off shortly before a scheduled appointment (Herrmann, Brubaker, Yoder, Sheets, & Tio, 1999). On hearing the signal (a relatively high pitched beep), the user knows that the time has come to perform the intended act. The length of time that occurs between the sounding of a signal and the time when the intended response should be executed is called the anticipatory interval because the warning signal anticipates the time that an action is to be executed.

Table 1
Temporal Properties of an Intention Cued by PDA

Retention window - the duration of time after an intention has been formed up to the time that a response is to be made.

Anticipatory interval - the duration in time that begins just after a reminder alarm has been set off and ends at the response time that the response is to be made.

Response window - the interval of time in which a response is regarded as being on time.

The length of a retention interval is made less important by PDAs because a person does not have to worry about an intention until the warning signal alerts a person that the anticipatory interval has begun (Meacham & Kushner, 1980). The duration of the anticipatory interval is critical to correct responding (Ceci, Baker, & Bronfenbrenner, 1988). A person may respond too soon if the anticipatory interval is too early relative to the intended time of a response as some people may forget to respond at all. If the anticipatory interval is made too brief after the warning signal, a person may be caught off guard and not have sufficient time to respond. Alternatively, an optimal anticipatory interval enhances the likelihood of appropriate responding (Herrmann et al., 1999).

The width of the response has an important influence on responding also. This refers to the idea that acceptable responses must occur within a certain time interval or response window. Smaller windows are generally more difficult for responding than large windows because it is harder to respond within a small window than a large window. If a person is expected to arrive for an appointment at a specific time, not too early and not too late, failure is more likely to occur. Alternatively, a response interval that is too wide can lead to failure as well because a person will lose track of the time for responding.

2. USEFULNESS OF PDAs FOR REMEMBERING APPOINTMENTS

2.1 Likelihood a Person Will Make Use of a PDA. An important factor that influences the usefulness of PDAs is obviously people's inclination to use these devices. Our research indicates that only a small percent of the population actually use these devices on a regular basis, perhaps about 5% of those people in management or executive positions (Herrmann et al., 1999). However, McCurne (2000) surveyed 26 top level executives and found that 50% of them used PDAs. More people purchase PDAs than use them. Like the VCR, many people want to own a PDA but after owning it they are put off by how much they have to learn in order to use it. Presently, it seems likely that most people will never make use of a PDA, even people busy with active schedules who could benefit from using one. People who do not use a PDA remember their intentions in other ways, such as with a planner, a calendar, post-it notes, and memo pads. Our research indicates that people often do not adopt the use of a PDA because they are accustomed to using other reminding methods (Herrmann et al., 1999).

2.2 Facilitation of Remembering After being Reminded by a PDA. Probably the most important skill for PDA users to learn is to remain aware of an intention after a warning signal has been sounded. Since the signal precedes the time of action by a brief anticipatory interval, typically minutes or seconds, it may seem surprising that people would ever forget to carry out the intention but most users do sometimes. Memory retention is one issue, but remembering may also be influenced by motivation concerning the prospective memory task. If a task is complicated, aversive, or less important, people may be more likely to forget. Alternatively, characteristics of users that affect motivation, such as anxiety or depression, may also affect remembering (MacLeod, 1999).

Several subskills have to be mastered if a person is to continue to use a PDA (Andrzejewski, Moore, Corvette, & Herrmann, 1991; Herrmann et al., 1999). These subskills include learning how to: enter scheduling information on the reminder's keyboard; enter data accurately while engaging in concurrent tasks (such as engaging in conversation); schedule effectively, so as to not overwhelm oneself with setting too many signals; judge the optimum time for warning signals to precede an intention (which should consider the broader context of that day's activities). Other mechanical constraints are also critical, such as remembering to carry the PDA everywhere (or at least have it accessible and within hearing range of the PDA's signal) and remembering to check the electrical supply for the PDA.

2.3 Users of PDAs. In brief, the individual who is most likely to use a PDA regularly is someone whose work and home impose many demands on him or her, who likes gadgets, who is easily fatigued, who tends to forget appointments, is ambitious, and can afford a PDA. Middle and upperclass individuals are more likely to use PDAs. Similarly, men are more likely to use PDAs than are women.

Life style is also an important factor in use of PDAs. While highly organized individuals remember intentions more than less organized individuals, some highly organized people may not feel the need for a PDA. On the other hand, people with an ambitious cognitive style may make more effective use of a PDA than those who are laidback (Searleman & Gaydusek, 1996). More effective use of a PDA may also occur when it is used by people who are more oriented to the environment than to their thoughts (Wichman & Oyasato, 1983). Lifestyle appears to be pertinent to PDA use. While individuals with a relaxed lifestyle will forget intended acts less often than individuals with a hectic schedule of appointments, relaxed people may benefit more from a PDA although they may not see a PDA as valuable.

2.4 Nonusers of PDA. First and foremost, many people will probably never use a PDA because these people are technophobic. They find the use of devices like PDAs makes them feel anxious. However, many people who are not technophobic have other reasons for not using a PDA. PDAs typically cost substantially more than paper appointment books and planners. More important, many people who avoid PDAs do not believe that these devices will simplify their lives. These people do not like having to push buttons, carry the device with them, and other tasks required to use PDAs. They find the multiple steps required to use a PDA difficult to learn and, even after learning, to be too time consuming to execute. As a result, they are reluctant to invest the effort and time necessary to learn how to use these devices. They find that appointment books or planners can be updated and changed by writing in them more easily than information can be inputted on tiny keypads or by writing on a PDA's screen in computer-readable script.

Another major concern of PDA nonusers is a fear of lost data. PDAs can be destroyed more easily than appointment books and planners. Stepping on an appointment book or planner will usually not render either one useless; stepping on a PDA can render it useless. PDAs that are not charged in sufficient time may lose all of their information. Nonusers fear not being able to access one's schedule, which may be an unlikely event but not an unrealistic concern. Many PDAs now can be hardwired to a cradle that transfers a person's schedule to a computer, providing a back up record of what is on the PDA should the PDA be damaged or lose its charge. Nevertheless, nonusers fear they will be unable to transfer information prior to damage or loss of charge. When nonusers consider the various ways that information may be lost with a PDA, it appears to them that managing an appointment book or planner is more under individual control than is a PDA.

A PDA nonuser is accustomed to an appointment book or a planner and usually has not experienced a trial run with a PDA. Nonusers are accustomed to reading and rereading their appointment books and planners wherever they go. They are not accustomed to auditory warning that remind one of impending responsibilities. They may even find the warning signals as irritating. As a consequence, the nonuser does not get the experience with a PDA that would enable them to better appreciate the benefits that PDAs might provide them. Thus, the PDA nonuser view PDAs as adding more complexity to their lives.

3. CONVERTING THE NONUSER OF PDAs

Our research indicates that experience alone with the use of a PDA will not necessarily convert the PDA nonuser into a user. After a group of graduate students used a PDA for over two months, they said that the PDA was useful and that it kept them from forgetting some appointments. However, only 15% of them said that they would continue to use a PDA in the future. They preferred the use of their appointment book or planner to the PDA.

However, there are other factors that make a person a user or a nonuser. A person who has an above average number of obligations will consider the use of a PDA. People who tire more easily than others and who are forgetful tend to find the PDA worth using. And, obviously, people who can afford the cost of a PDA are going to be more likely to purchase a PDA.

3.1 How PDAs May be Designed to Win Over Nonusers. One way that nonusers may come to perceive PDAs in a more positive light may be to create simple function devices. For example, devices may be devoted just to scheduling and remembering intentions. This would eliminate the number of buttons that confront potential users, making the device easier to use. A second way to render these devices more appealing would be build them out of nearly indestructible material. Also, the device would become more competitive with the appointment book and planner if the device contained a backup system that would preclude data loss from loss of energy. Third, stores selling PDAs should encourage customers to check back in after a period of time to help them with any problems they may have encountered. Additionally, customers might be given an 800-phone number where expert help could be obtained.

4. ACKNOWLEDGMENT

We thank our students, Mike Sarapata and Ayako Sakuragi, and our colleagues, Brad Brubaker, Nancy Rankin, and Judy Swez for helping us understand the issues discussed here.

5. REFERENCES

Andrzejewski, S. J., Moore, C. M., Corvette, M., & Herrmann, D. (1991). Prospective memory skill. Bulletin of the Psychonomic Society, 29, 304-306.

Brandimonte, M., Einstein, G. O., & McDaniel, M. A. (Eds.), (1996). Prospective memory theory and applications. Mahwah, NJ: Lawrence Erlbaum Associates.

Ceci, S. J., Baker, J. G., & Bronfenbrenner, U. (1988). Prospective remembering, temporal calibration, and context. In M. M. Gruneburg, P. E. Morris, & R. N. Sykes (Eds.), Practical aspects of memory: Current research and issues (pp. 360-365). Chichester, England: Wiley.

Harris, J. E. (1984). Remembering to do things: A forgotten topic. In J. E. Harris & P. E. Morris (Eds.), Everyday memory, actions and absent-mindedness (pp. 71-92). New York: Academic Press.

Herrmann, D. (1996). Improving prospective memory. In M. Brandimonte, G. Einstein, & M. McDaniel (Eds.). Prospective memory: Theory and applications. Mahwah, NJ: Erlbaum.

Herrmann, D., Yoder, C., Wells, J., & Raybeck, D. (1996). Portable electronic scheduling/reminding systems. Cognitive Technology, 1, 36-44.

Herrmann, D., Brubaker, B. Yoder, C., Sheets, V., & Tio, A. (1999). Devices that remind. In F. Durso (Ed.). Handbook of applied cognitive psychology. Mahwah, NJ: Erlbaum.

Kvavilashvili, L. (1992). Remembering intentions: A critical review of existing experimental paradigms. Applied Cognitive Psychology, 6, 507-524.

MacLeod, A. (1999). Prospective cognitions. In T. Dalgleish & M. Power (Eds.). Handbook of cognition and emotion. New York: Wiley and Sons.

McCune, J. C. (1999). Technology giveth and taketh away. Management Review, 88, 7.

Meacham, J. A., & Kushner, S. (1980). Anxiety, prospective remembering and performance of planned actions. Journal of General Psychology, 103, 203-209.

Meacham, J. A., & Leiman, B. (1982). Remembering to perform future actions. In U. Neisser (Ed.), Memory observed: Remembering in natural contexts (pp. 327-336). San Francisco: Freeman.

Petro, S., Herrmann, D., Burrows, D., & Moore, C. (1991). Usefulness of commercial memory aids as a function of age. International Journal of Aging and Human Development, 33, 295-309.

Searleman, A., & Gaydusek, K. A. (1996). Relationship between prospective memory ability and selective personality variables. In D. Herrmann, C. McEvoy, C. Herzog, P. Hertel, & M. Johnson (Eds.), Basic and applied memory (Vol. 2). Mawah: Erlbaum.

Wichman, H., & Oyasato, A. (1983). Effects of locus of control and task complexity on prospective remembering. Human Factors, 25, 583-591.

Wilkins, A. J., & Baddeley, A. D. (1978). Remembering to recall in everyday life: An approach to absentmindedness. Practical aspects of memory (pp. 27-34). London: Academic Press.

Vortac, O. U., Edwards, M. B., & Manning, C. A. (1995). Functions of external cues in prospective memory. Memory, 3, 201-219.

Automating Measurement of Team Cognition through Analysis of Communication Data

Preston A. Kiekel[1], Nancy J. Cooke[1], Peter W. Foltz[1], and Steven M. Shope[2]

1. Department of Psychology, New Mexico State University, Las Cruces, NM 88003
2. Sandia Research Corporation, 4200 Research Drive, MSC3-ARP PO Box 30001, Las Cruces, NM 88003

ABSTRACT

In this paper we propose a general methodological approach for semi-automatically assessing team cognition using communication data. The approach rests on four premises: 1) analyzing communication data is a means of assessing team cognition, 2) both substantive content and physical quantity measures of communication are needed, 3) sequential flow methods are especially helpful to effectively make use of communication data, and 4) analysis of both data types can be automated with contemporary tools, both statically and sequentially. We begin by illustrating the first three points. Next, we briefly review commonly employed methods of communication analysis, and note the difficulties of analyzing such data. We then suggest that appropriate automatic methods for analyzing communication data are becoming increasingly available, and we give some examples of our approach. Finally, we conclude by discussing the implications of this approach, especially for team training and groupware design.

1. INTRODUCTION: A HOLISTIC DEFINITION OF TEAM COGNITION

For the purposes of this paper, any small group of people collaborating on a task constitutes a team. Though teams have been a ubiquitous component of most organizations for some time, research on team cognition is relatively new. Team cognition can be defined as the team analog of individual cognition, i.e., the thoughts and knowledge of the team. This is distinct from "social cognition," which revolves around individuals' cognitions about other people. Certainly, social cognition is relevant to team cognition, but team cognition is also the interaction of team members as they collaborate. As such, it must be somewhat distinct from individual cognition, which has a physical mechanism for storage and processing. The thoughts and knowledge of a team are not contained in a single brain, yet there is an emergent property when a group of people collaborate (Steiner, 1972).

Usually, this emergent property is treated as the sum of individual cognition, and this has value. However, the fact that current measures of team cognition focus on the similarities among the individuals' cognition (Langan-Fox, Code, & Langfield-Smith, 2000), makes a paradigmatic assertion that the tasks in which we are most interested are those in which there is little individuation of subtasks. Some teams require specialization of individuals, such as surgical teams. For such teams, should our definition of the team's cognition still focus only on similarities among individual's cognition (Cooke, Salas, Cannon-Bowers, & Stout, 2000)? Just as an individual has contradictory and dynamic thoughts, so individual team members have different ideas and knowledge regarding a task. So it is perfectly reasonable to define team cognition at the holistic level, where the team is the unit, rather than at the level of averaged individual cognition.

If teams are to be the unit of analysis for this holistic definition, we will need to measure behaviors exhibited by the team as a whole. Just as we use think aloud protocols to examine what an individual is saying to herself when she thinks, so we can use the communication inherent in the collaboratory process to assess team cognition in a holistic way. In this respect, team cognition is easier to measure than individual cognition; teams need not interrupt their process in order to "think-aloud," since they are always "thinking aloud" in some sense. With a newly formed team, team cognition begins as the sum of individual cognition. Then, as the team "thinks" (interacts), dynamic changes occur in the "team mind" as a natural result of the interaction. Effects of this process on performance depend on the type of task (Steiner, 1972). Thus, an analysis of team communication provides a window through which to view team cognition.

2. TYPES OF COMMUNICATION DATA AND COMMONLY EMPLOYED METHODS

Communication measures can be characterized as "physical" (i.e., relatively low-level measures such as duration of speech, e.g., Watt & VanLear, 1996) and "content" (what is actually being said). Both can be useful in characterizing team cognition. For instance, major drops in communication frequency over time may be driven by implicit learning of the task, leading to less need for explicit discussion. Likewise, content data might show that a team talks about mechanics of the software early in the session, but later only talks about more substantive issues.

Though physical measures can often be taken in real time by human or machine recorders, content measures are typically taken on transcripts made from taped interaction. After transcription, a coding scheme is employed

1382

that classifies the utterances into meaningful categories (Emmert, 1989). Analyses can be either time-static (e.g. code counts), or sequential. Codes can themselves be sequences, such as "argument followed by name-calling."

Statements-- the basic verbal units teams use to define a common frame of reference-- are themselves driven by previous statements. Thus, we cannot readily isolate a statement without its context. This gives an advantage to sequential analyses over static. However, both types of analysis are important for assessing team cognition. As an example of a useful static analysis, it would be quite helpful to know that 98% of a team's communication is argument and 0% is planning. The approach advocated here includes both physical and content data, and emphasizes sequential flow analysis over static measures. This yields a 2*2 table of possible data types (see Table 1).

Hand coding can be time consuming, for both content and physical measures. For physical hand coding, speech duration can be captured by recording onset/offset times, at the rate of one hour per hour of interaction. For content coding, Emmert (1989, p. 244) cites a study that required 28 hours of transcription and encoding per hour of tape. We propose more automatic methods to circumvent such time requirements, and also to increase reliability.

Table 1. Types communication analysis, with an example of each.

	Static	Sequential
Content	Number of arguments	Number of arguments followed by insults
Physical	Total seconds spoken	Number of Person A speaking after Person B

3. AUTOMATIC ANALYSIS OF COMMUNICATION DATA

To collect physical data, we developed software that records quantity of verbal communication as a K^2*N communication log matrix of dichotomous values. K is the number of team members, and N is the number of time intervals (e.g., seconds) across which the communication spans. All possible pairs of the K speakers account for the K^2 columns. At each time interval, a measure is automatically taken of which team members are talking, and to whom. This creates the N rows in the communication matrix. The result enables rapid analyses of sequential flow.

To code content data, we use Latent Semantic Analysis (LSA) (Landauer, Foltz, & Laham, 1998). LSA is a computational linguistic technique which can measure the semantic similarity between strings of text. Its "knowledge" of the language is based on a semantic model of domain knowledge acquired through training on a corpus of domain-relevant text. Through a statistical analysis of how words occur across contexts (e.g., paragraphs), LSA generates a high-dimensional semantic space, in which each original word, as well as larger units of text (utterances, paragraphs, documents), are represented as vectors in the space. The derived vectors for words and utterances can be correlated by taking the cosine between their vectors. This permits the matching of strings based on semantic relatedness, rather than just direct keyword overlap.

LSA can be applied to communication data in a variety of ways, with differing degrees of automation. A completely automatic means of categorizing content is to use LSA to generate a correlation matrix of every utterance with every other utterance, then cluster highly correlated utterances. Each original utterance can be classified according to the cluster of which it is a part. A less automatic method is to develop a coding scheme, and create text strings that are prototypically representative of each coding category. Then, each utterance can be classified according to which of the prototypical categories it correlates with the best. Even less automatically, raters can code a subset of the dialogues, and use LSA to compare each new utterance to one of the pre-classified utterances. LSA can further be used to automatically analyze the coherence, quality and amount of information flowing between speakers, based on analyzing utterances. This permits a wide range of measures that can be correlated to team performance.

4. EXAMPLES

Our approach incorporates several sequential data analysis methods to further refine the data, both in content and in physical dimensions of communication. Tools include lag sequential analysis (Bakeman & Gottman, 1997), graphical display methods, ARIMA models (Suen & Ary, 1989) and fourier analysis (Vallacher & Nowak, 1994), network models such as PRONET (Cooke, Neville, & Rowe, 1996), etc. For instance, we have analyzed team discourse using communication log data, LSA applied to the transcribed dialogue, and PRONET as a graphical, sequential data compression tool. The sequel describes an example, using LSA to assess communication content.

4.1 Content data: static and sequential analyses

Dyads collaborated for one hour to write an essay on censorship and pornography. We used LSA to correlate each utterance with every other utterance. The correlation matrix was then submitted to factor analysis and cluster analysis to give a notion of what sentences were related. We boiled an hour's worth of discussion down to 13 statement types. In a sense, showing that these specific statements were typical of the communication is a static analysis of content. Further analysis would include counts of how frequently each statement type was uttered, by whom, how much information in multidimensional LSA space is conveyed by each speaker, etc.

We went on to find which statements tend to follow each other by using PRONET (Cooke, Neville, & Rowe, 1996). PRONET is a sequential analysis that relies on the network modeling tool, Pathfinder (Schvaneveldt, 1990). Transition probability matrices among a set of nodes are input to the Pathfinder algorithm, and a network representation of prominent pairwise connections is generated. Here, each statement was coded by classifying it as an instance of a statement type. Statement types were nodes in a lag-1 transition matrix.

Figure 1 shows the output. One possible interpretation is that this dyad's "thought processes" are focused on retaining agreeability. Strong assertions, such as the PornBad statement type, tend to be followed by weaker, more clearly acceptable statements (e.g., "children need to be protected from pornography"), or by clarifying, opinion seeking statements (e.g., "what do you think?"). Note also that specific questions tend to precede questions regarding punctuation, implying that more specific clarifications tend to motivate the team to write.

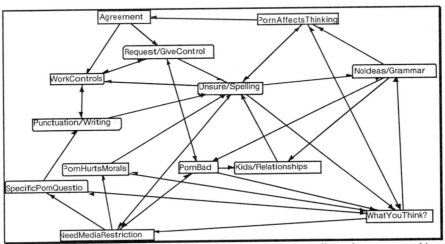

Figure 1. PRONET representation of LSA-classified statement types during a discussion on censorship and pornography. Nodes are statement types. Arrows indicate prominent lag-1 transitions.

Examination of the network helps to create interesting hypotheses about the team's process. It is important to validate the interpretations implied by this type of analysis with other measures. In the next example, the team process and cognition uncovered by the methods we describe was validated by human observers viewing videotaped dialogue. In addition, converging validity was determined against other measures of team cognition, and predictive validity was assessed using measures of team process, performance, and situation awareness.

4.2 Physical data: sequential analysis

This time we use a sequential analysis of physical data taken from the communication log. Each of the three team members had a specialized role: AVO, PLO, or DEMPC. Teams flew a simulated plane for 10 missions. Six events were defined, one for each team member beginning or ending a speech sequence. These events were treated as nodes in a lag-1 transition probability matrix, fed into PRONET (see Figure 2 for output).

Networks for missions 1, 2, and 4 are identical. DEMPC appears to be the focal point between AVO and PLO. At mission 5 this flow pattern is interrupted, and appears relatively chaotic. In particular, the connections between PLO and DEMPC are gone. In subsequent networks, there is never again a completed trapezoid between PLO and DEMPC. In fact, the two are never connected at all, except at mission 9. We chose this team because the PLO and DEMPC had a fight during mission 5. What we find interesting here is that, using the PRONET method, with only lag-1 transitions for this example, we can clearly portray this team's pattern of interaction. Information flow is severely hampered between two team members because of a conflict in team process, and after that, the team's pattern is altered. In essence, the way the team "thinks" is changed by this shock to the system.

4.3 Physical data: static analysis

We conclude with an example of a semi-static analysis of physical data taken from the same team's communication log output. Our goal was to identify the pattern of communication dominance among team members. We separated the mission into segments of one minute duration, and formed a multinomial model of how much each team member spoke during that minute. These models were then used to generate expected values for every other minute, whose observed deviation was tested with a χ^2. We test every minute against every other minute for similarities in the dominance pattern. Those segments whose differences could not be detected with a χ^2 were pooled to form a new model. The process of pooling and testing was continued until all "similar" minutes were pooled, and only "distinct" dominance patterns remained. For this static analysis example, we retained counts of distinct patterns the team exhibited during each mission. In addition, flow between minutes could be examined by sequential methods.

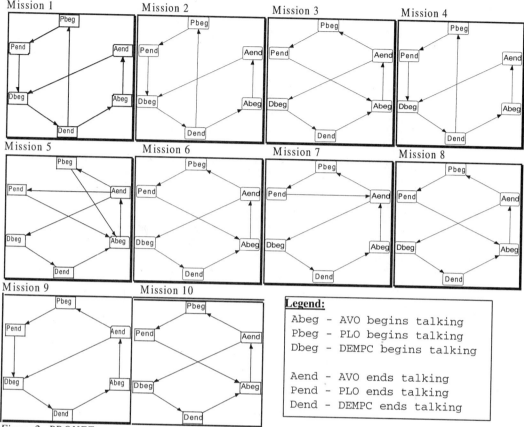

Figure 2. PRONET representation of shift in team communication pattern across 10 missions.

The number of statistically distinct communication patterns in any mission is a measure of communication stability for that mission. We used this to estimate how well the team had an established process for the passing of knowledge. The more patterns exhibited during a mission, the less stable the team's communication, and so the less stable we predict their team knowledge to be.

We plotted reverse-scaled z-scores of the number of distinct patterns and distinct patterns per minute against z-scores for performance (see Figure 3; performance data were missing for the 10th mission) and situation awareness (not shown). Except at mission 1, where communication patterns were stable in spite of presumably low team knowledge, general shifts in performance and situation awareness correspond to the number of distinct patterns.

5. CONCLUSION AND IMPLICATIONS

A general principle of our approach is to take communication data and reduce it to its core components. Then look for shifts in the communication patterns. Next, tie these pattern shifts to known events and internal team trends. This process is essentially an application of the dynamical systems paradigm (e.g., Vallacher & Nowak, 1994). Employed in an exploratory way, it can lead to general predictive principles of team cognition. These principles can then be applied to other teams by relying solely on the communication data to assess team cognition.

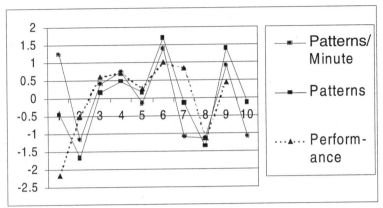

Figure 3. Z scores for number of communication patterns per mission and performance scores across 10 missions.

The efficiency of these methods has not been explicitly computed at this time. But we know that time spent recording who is talking to whom at what time, is eliminated by use of the communication logger. Furthermore, time spent coding text can been eliminated by the use of LSA. With text-based computer mediated communication, even transcription becomes unnecessary.

There are a number of implications that automated, sequential measurement tools have for design and for cognitive modeling. First, these methods will allow for rapid and on-line assessment of team cognition. This is important for evaluating and designing training programs and system interfaces. Moreover, poor process detection can automatically trigger interventions on the part of the system. Second, computerized tools such as PRONET and LSA allow the assignment of numeric values to variables in models of team cognition, facilitating the development of relatively specific predictive models.

6. REFERENCES

Bakeman, R., & Gottman, J. M. (1997). *Observing interaction: An introduction to sequential analysis* (2nd ed.). Cambridge: Cambridge University Press.

Cooke, N. J., Neville, K. J., & Rowe, A. L. (1996). Procedural network representations of sequential data. *Human-Computer Interaction, 11,* 29-68.

Cooke, N. J., Salas, E., Cannon-Bowers, J. A., & Stout, R. (2000). Measuring team knowledge. *Human Factors, 42,* 151-173.

Emmert, B. J. L. (1989). Interaction analysis. In P. Emmert and L. L. Barker (Eds.), *Measurement of Communication Behavior*, pp. 218-248. New York: Longman.

Landauer, T. K., Foltz, P. W., & Laham, D. (1998). An introduction to Latent Semantic Analysis. *Discourse Processes, 25(2 & 3),* 259-284.

Langan-Fox, J., Code, S., & Langfield-Smith, K. (2000). Team mental models: Techniques, methods, and analytic approaches. *Human Factors, 42,* 242-271.

Schvaneveldt, R. W. (Ed.) (1990). *Pathfinder associative networks: Studies in knowledge organization.* Norwood, NJ: Ablex.

Steiner, I. D. (1972). *Group processes and productivity.* New York: Academic Press.

Suen, H. K. & Ary, D. (1989). *Analyzing quantitative behavioral observation data.* Hillsdale, NJ: Lawrence Erlbaum Associates.

Vallacher, R. R., & Nowak, A. (Eds.) (1994). *Dynamical systems in social psychology.* San Diego: Academic Press.

Watt, J. H. & VanLear, C. A. (Eds.) (1996). *Dynamic patterns in communication processes.* Thousand Oaks, CA: Sage.

7. ACKNOWLEDGEMENTS

This work was supported by ONR Grant No. N00014-00-1-0818.

Strategies for Designing Usable Interfaces for Internet applications

Nancy J. Lightner

The Darla Moore School of Business, University of South Carolina, 1705 College Street, Columbia, SC 29208

ABSTRACT: Preferences for certain characteristics of an on-line shopping experience may be related to demographic data.   This paper discusses the characteristics, demographic data and interface preferences by demographic group. The results of an on-line survey indicate that overall, respondents are satisfied with their on-line shopping experiences. Males and females have similar preferences, ranking security as the most important characteristic. Higher satisfaction ratings were reported by respondents with more education, higher income or higher age. Although speed of the site received consistently low ratings, differences in preferences in other categories between demographic groups were found.

1. INTRODUCTION

The Internet represents ubiquitous computing, with access to an audience all the time. In addition, it is available for many purposes, from communicating with friends to collaborating on research problems with international implications. In the world of commerce, it is the desire of companies to sell as many products as they can. With the opportunity to reach potential customers, companies seem to have posted Web sites without considering questions such as "How do I present my product so people will buy it?" and "What is most important to my customers on a Web site?"

One popular saying is "you can't please everyone". In marketing, this saying translates into the targeting of customers through branding, advertising campaigns and product design. The concept behind identifying a target market is that individuals of certain demographic characteristics have similar likes, dislikes and preferences. The discussion presented here centers around targeting consumers on the Internet according to demographic data. Since Web sites are currently the only interface to the Internet, designing those sites according to target market preferences should enhance consumers' shopping experience, and perhaps, consequently, motivate them to purchase and repurchase from the Web site.

2. METHODOLOGY

2.1 Research variables

A survey was developed and implemented in an attempt to provide direction on how to best accommodate users of various demographic groups by designing with the variables of speed, sensory impact, security, information quality and information quantity in mind.

We selected speed, sensory impact, security, information quality and information quantity as the aspects of an electronic commerce (e-commerce) experience that is determined by the style and content of a Web site and can be manipulated by site designers and developers. These variables were selected where other research areas provide direction. In addition, popular press about e-commerce drove the selection of these variables.

The first variable of concern when defining an e-commerce experience is speed.   Since the commercialization of the Internet, the speed of access is of concern to users (Lightner, Bose & Salvendy 1996). We are in a society that places great importance on time. Speed is determined by the selection of graphics, techniques that include movement and by the navigational design of a site. Providing tools such as one-click shopping may enhance the speed of the shopping experience.

The discipline of Marketing focuses on capturing and holding the attention of consumers so that they eventually purchase products. The *Handbook of Marketing Scales* (1999), a compilation of measures to use to predict consumer behavior, includes measures of Optimal Stimulation levels and Exploratory Buying Behavior among other characteristics. These measures are reported to relate to a wide range of consumer-related behaviors. As a composite, some of the measures that address stimulus needs are categorized as 'sensory impact' in our study. Sensory impact is accomplished with stylistic elements such as color and movement. Examples of content related to sensory impact are product descriptions that focus on the experience of the product rather than product specifications.

Security on the Internet is of concern to users, as credit card and identity theft seem to threaten every on-line purchase (Easson 2000, Ilioudis & Pagalos 2000). A secure site is currently conveyed by prominently displaying a symbol such as a lock on the site and by having the uniform resource locator (URL) begin with 'https://'. Another method of achieving a perception of security may be to display a rate of secure transactions achieved to date.

Information Quality is an input variable used by Delone and McLean (1992) in their model of determining information systems success. Information Quality in our study is defined as the perception that the information contained in the site was true. This is related to vendor reputation. New vendors may choose to convey information quality by linking to reputable sites or displaying seals of quality.

Information Quantity means how much information is available about the products and the process of using the on-line system. Providing such information may occur by using pictures, words or a combination of the two. In marketing, a preference for information seeking is measured by the Exploratory Information Seeking questionnaire (Bearden and Netemeyer 1999), which indicates finding information for fun.

2.2 Survey

An on-line survey was developed containing twenty-five questions and an area available for comments. See *http://dmsweb.badm.sc.edu/lightner/survey.htm* to view the complete survey. Questions one through ten asked for basic demographic information such as age, sex, education, family size, income and amount of on-line purchases made. Question eleven asked respondents to rate their satisfaction with making purchases on-line. Questions twelve through twenty-five asked respondents to rate the importance of Web site characteristics according to a 7-point Likert scale ranging from 'very strongly disagree' to 'very strongly agree'. Each of the five site characteristics (speed, sensory impact, information quantity, information quality, security) was represented by two questions apiece. The four remaining questions asked about other aspects of e-commerce, such as the ability for comparison shopping, the use of a shopping cart, vendor reputation and price.

Students at three universities were asked to participate by submitting a survey. In addition, several listserv groups consisting of Information Systems and Human-Computer Interaction academics and professionals were solicited for participation. Total data collection occurred in two weeks time, from February 20 to March 7, 2001.

3. RESULTS

See Table 1 for an overview of the results from the survey (N=305). Since two questions were included in each characteristic value, reliability ratings were calculated and are reported in the third column of Table 1. Findings indicate that overall satisfaction with on-line shopping is high (μ=5.25 where 5='agree'). The three most important features for a Web site are security, information quantity and information quality. Evaluation of the results according to various demographic categories indicates that males and females have similar expectations of on-line experiences. They also report similar satisfaction with those experiences. Respondents possessing more education report more satisfaction, as do those with higher income. Older respondents (>20 years) tended to report higher satisfaction with on-line shopping than did younger (<20 years) respondents.

Table 1: Survey results for Web site characteristics

Characteristic	Mean / St. Dev.	α
Security	6.07/ 1.12	.77
Information Quantity	6.06/ 1.03	.63
Information Quality	6.05/ 1.05	.71
Speed	5.75/ 1.05	.54
Sensory Impact	4.72/ 1.30	.49

The speed of the Web site was indicated as fourth in importance of these characteristics, while sensory impact was last. When demographic preferences were analyzed, speed consistently received similar ratings from all demographic groups. The other characteristics had significantly different ($\alpha<0.05$) ratings between demographic groups. Older respondents preferred information quality and quantity while younger respondents preferred more sensory impact. Both age groups expressed the same preference for security. Educational levels impacted ratings for all categories except speed, with those not having college degrees preferring more in all other categories. Those with incomes of less than $60,000 per year preferred more information quantity and more sensory impact.

4. CONCLUSIONS

The results of this survey indicate that preferences for style and content considerations on a Web site are related to demographic data. With this information, Web site designers may enhance the on-line shopping experience for their target audience.

REFERENCES

Beardon, W. O. & Netemeyer, R. G. (1999) *Handbook of Marketing Scales*, second edition, Sage Publications, Inc., Thousand Oaks, CA.

Delone, W.H. & McLean, E.R. (1992) "Information Systems Success: The Quest for the Dependent Variable", *Information Systems Research, 3*, 60-95.

Easson, W. (2000) "Computer viruses, worms and Internet security", *Veterinary Record, 147*, 252-252.

Ilioudis, C. & Pangalos, G. (2000) "Development of an Internet Security Policy for health care establishments" *Medical Informatics and the Internet in Medicine, 25*, 265-273.

Lightner, N.J., Bose, I. & Salvendy, G. (1996) "What is Wrong with the World Wide Web?: A diagnosis of problems and prescription of remedies" *Ergonomics, 39*, 995-1004.

Enabling Universal Access -
Minimum Requirements for Content Preparation

Harald Weber[1] and Constantine Stephanidis[1,2]

[1]Institute of Computer Science (ICS)
Foundation for Research and Technology-Hellas (FORTH)
Science and Technology Park of Crete
Heraklion, Crete, GR-71110, Greece
{harald, cs}@ics.forth.gr

[2]Department of Computer Science, University of Crete

ABSTRACT

The objective of the paper is to describe minimum requirements for the preparation of accessible content in Information Society Technologies. The aim is to ensure that all citizens can access the content anywhere, at any time and through any available device. Based on the assumption that this paradigm can be met by systems capable of adaptation to the diversity of users, contexts of use, and access devices, the paper elaborates on three different areas: (i) provision of information in different modalities, (ii) adaptation policies, and (iii) design and evaluation of systems capable of adaptation.

1. INTRODUCTION

Economies and societies are undergoing a remarkable change of paradigm regarding the use of computing technologies, which is characterized by multiple dimensions of diversity in the categories of users, the computing devices with different interaction mechanisms and modalities, and the usage patterns and contexts (Stephanidis & Emiliani, 1999). A key element elaborated for coping with diversity is the one of adaptation (Stephanidis, 2001), i.e. of software products that automatically adapt their interactive behaviour according to the needs and requirements of users, to the context of use and to the technology adopted.

Adaptation is required to support optimal fit between the users, the access devices, the tasks at hand, and the contexts of use. To achieve this fit, adaptation spans from simple modifications in the interface or in the content presentation to complex changes, e.g., in the overall interaction metaphor or in the modality of presentation. However, adaptation creates a certain overhead. User interfaces that are capable of adapting to the users, access devices, or contexts of use are based on a more complex software architecture, while digital content needs to be enhanced through metadata, which defines the way content is to be generated and to be presented to the user. This paper focuses on the content aspect, while user interface adaptation is described elsewhere (Stephanidis, 2001a).

2. UNIVERSAL ACCESS

The overall aim of adaptation in the context of this paper is to provide *universal access*, which is defined as the conscious and systematic effort to proactively apply principles, methods and tools of universal design, in order to develop Information Society Technologies, which are accessible and usable by all citizens (Stephanidis et al., 1998b).

Adaptation can be separated into two different system behaviours (Stephanidis, 2001b): *adaptability* and *adaptivity*. *Adaptability* denotes the ability of a system to pre-configure its human-computer interaction according to knowledge, which is available *at the time of system start-up*. This knowledge might, for instance, encompass information on the different devices used for accessing content, known user characteristics, preferred usage patterns, or the actual context of use (e.g., physical environment, user's task, user's current activity). *Adaptivity* as the second instantiation of adaptation updates the interface and the content presentation *during use* according to changes of user characteristics (e.g., improvements in skills to perform a certain task), of the access device (e.g., for continuous interaction), of preferred or appropriate usage patterns (e.g., due to shift of user's attention), or of the usage context (e.g., in ubiquitous use).

Universal Access has been successfully applied in the context of Human-Computer Interaction, where it refers to efforts to practically apply principles, methods and tools of universal design in order to develop high quality user interfaces. These user interfaces are accessible and usable by a diverse user population with different abilities, skills, requirements and preferences, in a variety of contexts of use, and through a variety of different technologies (figure 1; see also Stephanidis (2001a)).

However, adaptation at the user interface level does not seem to be sufficient for addressing the needs of the users. For example, it has been shown that content adaptation is an effective tool to support learners in performing a learning task (e.g., (Milosavljevic, 1997)), or users to navigate in hypertext structures (e.g., (Brusilovsky, 1996), (Fink, Kobsa & Nill, 1997)). Also, providing access to information for users with intellectual disabilities cannot be achieved by interface adaptation alone: It requires efforts in the area of content preparation. This paper will describe the minimum requirements for content preparation to provide universal access.

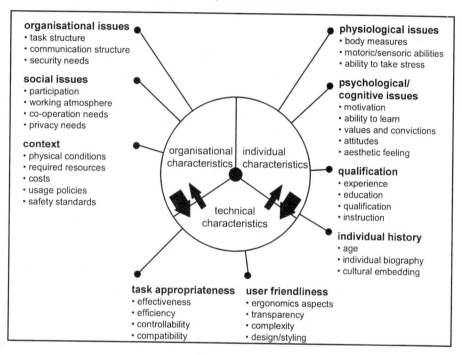

Figure 1: Diversity of users, contexts of use, and access technologies (Weber, 1999)

3. MINIMUM REQUIREMENTS FOR CONTENT PREPARATION

Content preparation intends to support diverse objectives, e.g., search support, maintenance, or information presentation. The focus of this paper is to describe minimum requirements for content preparation as to ensure that all citizens can access the content anywhere, at any time and through any available device, i.e., that they enjoy universal access. The paper elaborates on three different areas: (i) provision of information in different modalities, (ii) adaptation policies, and (iii) design and evaluation of adaptable and adaptive systems.

3.1 Provision of different modalities of presentation

Redundant presentation modalities facilitate access to information for highly diverse users, in different contexts of use, using various access devices. For instance, mobile phone users can control in which modality an incoming call is communicated (e.g., by sound, display, vibration), how certain callers can be identified (e.g., display of name or number, specific sounds associated with callers, different vibration patterns), or how certain modality attributes can be adjusted (e.g., contrast of display, background light, loudness) to optimally suit the current context of use or the preferences of the user. Another example, which currently is under discussion in Europe to provide participation for all in the knowledge society (*eEurope 2002* initiative of the European Commission (2000)), refers to access to web-based information. If web pages are properly designed (e.g., by following the W3C-WAI guidelines of the World Wide Web Consortium (2001)), they are accessible by users with diverse sensory abilities (e.g., blind or visually impaired users), but also through different access devices and software applications (e.g., non-visual browsers like the AVANTI web browser, see (Stephanidis et al., 1998a)).

To minimise efforts for content preparation, the provision of content in one modality and the availability of mechanisms for automatic transformation in all other required modalities would be desirable. Although much research work is ongoing in the field of automatic transformation across different modalities (e.g., graphic to text (Mittal et al., 1998)) or within modalities (e.g., machine translation (Dorr, Jordan & Benoit, 1999)) is often not possible or the resulting quality is not satisfactory. Therefore, the need to provide redundant information arises. This, however, introduces well-known problems of redundancy to content preparation, namely overhead (e.g., with regard to storage space, bandwidth, human efforts), proneness to errors (e.g., incorrect translation) and inconsistencies (e.g., time delays between updates of corresponding information segments). Additionally, specific modalities exhibit unique characteristics, which might not be available in other modalities. The visual modality, for instance, facilitates almost parallel perception of spatially dispersed information, supports the concept of "overview", and allows for high data rates. Similar characteristics cannot easily be replaced by other modalities, e.g., audio or tactile presentations, which are more constrained and mainly provide sequential processing at medium data rates.

Research into approaches to replace one modality by another or to combine certain aspects of different modalities in one presentation will also affect content preparation. Provided that a minimum requirement for adaptation is to be able to present content in different modalities without loss of essential information, knowledge about the quality of expressiveness of corresponding information segments (i.e., segments prepared for different modalities) needs to be conveyed to the adaptation system. Content preparation has to provide this "quality information" concerning the segments to allow for an informed decision-making process, performed by an adaptation unit. In those cases where full equality of information across modalities is not given or possible, combinations of suitable modalities might succeed. However, as long as research on the combination of modalities to optimally support intended impact on the user is not consolidated, content preparation needs to include complementarity information about information segments.

3.2 Provision of input to adaptation policies

Adaptation behaviour might range from complete user-initiated adaptation, where the user decides on the time, extent and kind of adaptation, to complete system-initiated adaptation, where the same tasks are performed by the system (Dieterich et al., 1993). While user-initiated adaptation requires the users to know that they can adapt the system and how this can be achieved, system-initiated adaptation suffers from a lack of control by the users. Systems capable of (self-)adaptation usually combine characteristics of both approaches. Adaptation implies change, and frequent or major changes, e.g., in the user interface or in the modality of presentation, might burden or confuse users and lead to decreased usability. Therefore, the conditions ("thresholds") that have to be met to initiate adaptations must recognise and respect user's preferences. These conditions are defined in adaptation policies, which are responsible for the "when", "how", and to "what extent" adaptation has to be applied.

Adaptation policies may depend on the purpose of an application or on a usage situation. For instance, the purpose of educational software is to make a user familiar with specific educational content. Although other quality attributes are also of importance, effectiveness of educational software is of highest priority. Depending on the specific user, adaptations might be necessary to convey this content in the most effective way. However, if the usage situation changes, e.g., to the preparation for an examination, effectiveness as a primary aim might be replaced by efficiency, and the same content should be presented to the same user differently (e.g., as summaries or as questions for self-assessment).

Consequently, adaptation policies themselves should be adaptable to the goals of the users. Therefore, it is necessary to provide information concerning how a certain information segment can contribute to specific goals. This requires metadata to be created at content preparation time, defining suitable "views" to this information.

3.3 Design and Evaluation

Traditional quality attributes of interfaces, which are addressed, e.g., in evaluation tools to assess the usability of software, need to be re-defined in the context of self-adapting software. Consistency, for instance, as one of the main design objectives to support usability must not be restricted to the user interface (Grudin, 1989) or to the interface including documentation, on-line help and on-line or video-taped tutorials (Perlman, 1989). New devices might automatically adapt the modality of presentation to the context of use, might provide continuous interaction across different devices (Paramythis et al., 2001), or might be shared by several users at the same time (e.g., CSCW). Consequently, under this new paradigm, consistency must span across different devices, modalities, and contexts of use.

However, the task to define consistency in a way that includes a complete change, e.g., of the underlying interaction metaphor, is not solved yet. The new challenges posed by the changing paradigm have a direct influence on design and evaluation. First, designers get little support in terms of guidelines to properly design adaptable and adaptive systems (e.g., with regard to consistency). Much in the same way, content preparation is missing concrete guidelines about how to transform information segments into different modalities, and how to define suitable combinations of different modalities. Equally important for design and development is the second aspect, namely evaluation. Evaluation of these systems cannot be operationalised at this point of time. Most existing evaluation tools do not consider the dynamic components of adaptation (including adaptation policies), which are not covered by traditional usability, accessibility or other quality characteristics. Although the objective of a "high quality of interaction" is desirable, currently there are no appropriate tools available to assess or measure it in systems capable of adaptation.

4. CONCLUSIONS

Content in Information Society Technologies is required to be accessible by anyone, anywhere and at anytime, and this implies considerable requirements for its preparation. As a first step, semantic redundancy should be introduced along different modalities. Whenever full redundancy is not possible (due to the use of modalities with different powers of expressiveness), content preparation has to provide "links" (metadata) to other content segments of an information entity that complete the modality-constraints. Design needs to ensure that the combination of entities in the different modalities:

(i)   is accessible for the user (taking the respective user characteristics, tasks, context of use and device characteristics into consideration),

(ii) represents the full "view" on the content (without omissions), and

(iii) is equally efficient, usable, and acceptable in all presentation alternatives.

This, however, requires new design and evaluation approaches that cover whole interaction episodes including the adaptation process (i.e., its quality with respect to necessity, appropriateness, and effectiveness) as well as user-, context-, device- and task-specific aspects of the interaction.

REFERENCES

Brusilovsky, P. (1996). "Methods and techniques of adaptive hypermedia". *User Modeling and User-Adapted Interaction, 6,* 87-129.

Dieterich, H., Malinowski, U., Kühme, Th., Schneider-Hufschmidt. M. (1993). "State of the Art in Adaptive User Interfaces". In M. Schneider-Hufschmidt, T. Kühme & U. Malinowski, *Adaptive User Interfaces.* (pp. 13-48). Amsterdam: Elsevier Science.

Dorr, B. J., Jordan, P. W., & Benoit, J. W. (1999). "A Survey of Current Paradigms in Machine Translation". In M. V. Zelkowitz (Ed.), *Advances in Computers, 49*. New York: Academic Press.

European Commission. (2000). "eEurope 2002 – An Information Society for All – Draft Action Plan". Prepared by the European Commission for the European Council in Feira. 19 – 20 June 2000. Brussels.

Fink, J., Kobsa, A., & Nill, A. (1997). "Adaptable and adaptive information access for all users, including the disabled and elderly". In A. Jameson, C. Paris & C. Tasso (Eds.), Proceedings of the 6th International Conference on User Modeling. (pp. 171-173). Wien: Springer-Verlag.

Grudin, J. (1989). "The Case Against User Interface Consistency". *Communications of the ACM, 32*, 1164-1173.

Milosavljevic, M. (1997). "Augmenting the user's knowledge via comparison". In A. Jameson, C. Paris & C. Tasso (Eds.), Proceedings of the 6th International Conference on User Modeling. (pp. 119-130). Wien: Springer-Verlag.

Mittal, V. O., Moore, J. D., Carenini, G., & Roth, S. (1998). "Describing complex charts in natural language; a caption generation system". *Computational Linguistics*, 34, 431–468.

Perlman, G. (1989). "Coordinating consistency of user interfaces, code, online help, and documentation with multilingual / multitarget software specification". In J. Nielsen (Ed.), *Coordinating User Interface Consistency*. (pp. 35-55). Boston: Academic Press.

Paramythis, A., Leidermann, F., Weber, H., & Stephanidis, C. (2001). "Continuity through User Interface Adaptation: a perspective on Universal Access" (submitted to HCII 2001).

Stephanidis, C., Paramythis, A., Akoumianakis, D., & Sfyrakis, M. (1998a). "Self-Adapting Web-based Systems: Towards Universal Accessibility". In C. Stephanidis & A. Waern (Eds.), Proceedings of the 4th ERCIM Workshop on "User Interfaces for All". Stockholm, Sweden. [On-Line]. Available at: http://ui4all.ics.forth.gr/UI4ALL-98/stephanidis2.pdf

Stephanidis, C., et al. (1998b). "Toward an Information Society for All: An International R&D Agenda". *International Journal of Human-Computer Interaction, 10*, 107-134.

Stephanidis, C., & Emiliani, P.L. (1999). "Connecting to the Information Society: a European Perspective". *Technology and Disability Journal, 10*, 21-44.

Stephanidis, C. (Ed.) (2001a). *"User Interfaces for All - concepts, methods and tools"*. Mahwah, NJ: Lawrence Erlbaum Associates.

Stephanidis, C. (2001b). "Adaptive techniques for Universal Access". *User Modelling and User Adapted Interaction International Journal,* 10th Anniversary Issue, *11* (to appear).

Weber, H. (1999). "Design for all - a sketch of challenges for HCI designers". In H.-J Bullinger & J. Ziegler (Eds.), *Human-Computer Interaction: Communications, Cooperation, and Application Design*, Volume 2 of the Proceedings of the 8th International Conference on Human-Computer Interaction. (pp. 777-781). Mahwah: Lawrence Erlbaum Associates.

World Wide Web Consortium (W3C) Web Accessibility Initiative (WAI). http://www.w3c.org/wai

Industry Design Practices: Differences in the Approach to Design Conceptualization

Kay M. Stanney

University of Central Florida, Industrial Engineering and Mgmt Systems Dept., Orlando, FL 32816-2450

ABSTRACT

This study identified three main focuses of design conceptualization adopted by industry design companies. Some companies spend considerable effort in information design, where the focus is on exploration and discovery. Others focus on more traditional human-computer interface design concerns, such as the sensory and perceptual aspects of system or product usage. Still others focus on interaction design, where the concern is determining user action strategies, designing product features, and specifying the discourse between user and system or product. Beyond these differences in design conceptualization, companies also differ in their *focus* during this process. There is a clear distinction among companies in the emphasis they place on understanding the user and usage context, the propensity to be document driven, the priority given to following a formal development process, and the tendency to jump in and start prototyping.

1. INTRODUCTION

Content development is the design and construction of the objects, actions, interaction techniques, and other artifacts of human-computer interaction (Isdale, Fencott, Heim, & Daly, in press). Content design often evolves via process- or document-driven approaches, and relies on such techniques as heuristic evaluations, cognitive walkthroughs, checklists, and guidelines / standards to ensure usability (Mayhew, 1999; Nielsen, 1993). Yet little is known about how well these approaches and expert evaluations relate to actual use of products and services (Green, Kanis, & Vermeeren, 1997). Other techniques, such as scenario-based design (Carroll, 1995), may provide useful information in laboratories or controlled settings but often break down when utilized in more ecologically valid (Gibson, 1966) work settings. While this technique infuses real-world context and constraints into the design process, it does so only for those contexts (i.e., scenarios) that can be "explicitly envision(ed) and document(ed)" (Carroll, 2000, p. 44). Thus, situated and impromptu utilities of a product or service are often overlooked by scenario-based design, as are sectors of the user population (e.g., those with functional limitations or disabilities) for which scenarios are not envisioned or developed.

More recently, content design practices have begun to incorporate a greater emphasis on studying and understanding the semantics of work environments (Vicente, 1999), often through contextual inquires, such as ethnographic studies (Nardi, 1997; Takahashi, 1998) or focus groups (Kuhn, 2001). Through participant-observation practices, efforts are made to understand the tasks, work practices, artifacts, and environment of which a service system or product will become a part (Stanney, Maxey, & Salvendy, 1997; Stanney, Smith, Carayon, & Salvendy, in press). This is often achieved by content designers immersing themselves in the target environment, thereby becoming accustomed to and familiar with the actual experiences and constraints of target users. Yet, while the benefits of such user experience characterization are espoused (Takahashi, 1998), in two recent reviews (*Interactions* 7[2] and 8[2]) of industry design practices, only about half the companies reported actually using such techniques.

2. CHARACTERIZING THE USER EXPERIENCE

The traditional development lifecycle must be expanded to address characterization of the user experience. While there are many variants of this waterfall style lifecycle (e.g., Leslie, 1986; Parkin, 1980), traditionally the process commences with some form of problem/opportunity identification and definition (Stanney, Maxey, & Salvendy, 1997). In contrast, when designing based on the user experience, the process commences with exploratory rather than definitional activities, where designers are focused on discovery and invention (Bear, Teasley, & Carroll, 2001). Through contextual inquiries, competitive analyses, and prior art research, designers focus on the actual needs and intents of target users. They identify users' goals, desires, and behavioral patterns, usage context, and competitors, as well as technology barriers (Crow & Yang, 2001). Informed design decisions can then be made based on observations of the real behaviors of target users (Armitage, 2001). Stanney et al. (1997) identified key factors that should be elicited during exploratory activities. These factors include:

- target users' capabilities and limitations (both cognitive and physical);
- user requirements (e.g., situated roles and responsibilities);
- organizational factors (e.g., interaction with customer base, social issues);
- task requirements (e.g., cooperative task activities; critical inputs and outputs to a service system);
- equipment/system specification (e.g., artifacts that support task activities); and
- environmental factors (e.g., informal practices) that the service environment or product supports.

Through the familiarity gained by this involvement, designers can develop systems whose format and layout of the content presentation is appropriate for target user groups. Yet eliciting this information can be a time-consuming and costly process. For example, MONKEYmedia reports spending a third of its project development time and budget on design analysis (Bear et al., 2001).

3. DESIGN CONCEPTUALIZATION PROCESS

Based on the *Interactions* (7[2] and 8[2]) industry design practice profiles, the real difference in design philosophy tends to be at which phase companies enter the design conceptualization process, which refers to the activities that transpire prior to coding or bending metal. The design conceptualization process can be divided into three phases (Bear et al., 2001) (see Figure 1). Information design is the first phase, which involves discovery and invention. Through contextual inquiries designers try to incite new design concepts and develop a thorough understanding of users and real-world implications of use. IDEO calls this characterization of the user experience and focus on accommodation of existing activity patterns "transaction engineering" (Stillion, 2000, p. 33). Through the information derived from this phase designers can establish the social circumstances (i.e., the often unconsidered but essential conditions that must be in place to enable use) within which the service system or product must survive, thereby "empowering" users to utilize the service or product within its intended context of use (Henderson, 2000). Companies who are involved in the information phase can be further differentiated based on how they conduct this phase. Some companies address this phase via market surveys, customer feedback, focus groups, and other such "removed" analyses, while others become involved in actual on-site field studies and participant-observer activities.

The information design phase is followed by the interface design phase, which considers more traditional human-computer interface design concerns such as the sensory and perceptual aspects of system or product usage. For companies who commence design conceptualization at this phase, interface design emanates from the ideas of visual, graphics, or interface designers, often evolving via brainstorming sessions focused on reaching consensus, decision-making, and problem solving (Kuhn, 2001). The designers often represent the users' perspective, using their knowledge of user profiles, tasks, task flows, display principles, and human information processing to derive design concepts. The design outcomes from this phase generally involve specification of mental models (i.e., high-level organization), navigation, and layout.

Finally, the interaction design phase focuses on user action, product features, and the discourse between user and system or product. Companies who enter the design conceptualization process at this phase often use existing documentation on users, gathered through marketing studies, usability data, and customer feedback, to drive the design of user actions and features. The focus is on interaction style, prototyping, and evaluation. The design outcomes from this phase generally include low-level organization, interface objects and actions, and input/output sequences.

It is important to note that the specification in Figure 1 of which phase companies enter the design conceptualization process is based on self-report of design philosophy and design process (see *Interactions* 7[2] and 8[2]) and thus is open to interpretation. Even if one were to take issue with the exact placement of the phase at which these companies enter the process, however, there is a clear distinction among these companies (again based on the self-reports) in the *focus* of the design conceptualization process. Whereas some companies place great emphasis on understanding the user and usage context (e.g., IDEO, MONKEYmedia), others focus on jumping right into design generation (e.g., Quicktime Apple, Play). More specifically, there were five main focuses identified from the *Interactions* (7[2] and 8[2]) industry design practice profiles, including user research driven, market research driven, document driven, process driven companies, and development driven (see Table 1).

3.1 User Research Driven

Companies who are user research driven seek to identify workflows, artifacts used in work processes, and work constraints, such as deadlines, dependencies on coworkers, noise, and distractions. This information is

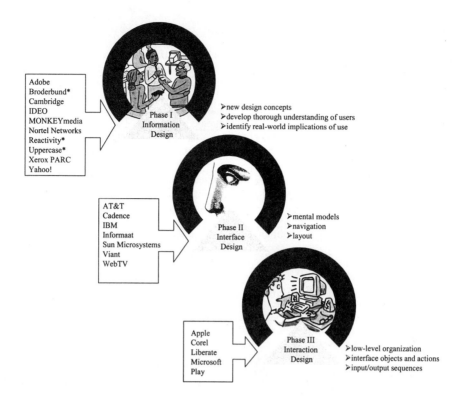

Adobe
Broderbund*
Cambridge
IDEO
MONKEYmedia
Nortel Networks
Reactivity*
Uppercase*
Xerox PARC
Yahoo!

Phase I
Information
Design

➤new design concepts
➤develop thorough understanding of users
➤identify real-world implications of use

AT&T
Cadence
IBM
Informaat
Sun Microsystems
Viant
WebTV

Phase II
Interface
Design

➤mental models
➤navigation
➤layout

Apple
Corel
Liberate
Microsoft
Play

Phase III
Interaction
Design

➤low-level organization
➤interface objects and actions
➤input/output sequences

Figure 1. Phase at which companies report entering the design conceptualization process.
* Information design based on marketing studies rather than observation.

primarily used to develop a thorough understanding of target user groups, as well as to identify latent user needs, innovative product concepts, and shortcomings in current work practices that can be supported by redesigned or new products and services. As IDEO's Danny Stillion (2000, p. 32) suggests, with this approach "A combination of user observations and technology research often reveals paths to innovation and inspires the design process."

3.2 Market Research Driven

Market research driven companies focus on customer feedback analysis, focus group input, and other market information sources to direct design. Market research results are generally provided to design groups via documents that describe market needs, potential market size, and features that could enhance existing products, as well as provide customer definitions (Ahn, 2000). With this approach, the initial design conceptualization is often done by market research firms. Some companies, such as Uppercase and Reactivity, are both market driven and process driven.

3.3 Document Driven

With this approach, key documents are prepared by different groups (e.g., marketing, engineering, interface design) and drive the design and development processes. The documents that drive the user interface design describe in detail user requirements, visual appearance, and interaction strategies (Dykstra-Erickson & Hoddie, 2000). As Tom Spine (2000, pp. 70-71) of Sun Microsystems suggests, "The design needs to be in the project schedule. It needs to have clearly identified inputs, usually a functional specification. And it needs to have a clearly identified deliverable, usually a user interface design specification. Like other project documents, the design specification is subject to review and revision." With this approach initial design ideas are often generated via sticky notes, design notes, and affinity lists of brainstorming ideas, which evolve into the user specification document. Some companies, such as Xerox Parc (Russell, 2000), develop a "design brief," which is a document that follows the design throughout the development lifecycle, specifying the design problem, approach, resources, competition, advantage/ideas, and delivery.

Table 1. Differences in design focus during the design conceptualization process.

Design Focus	Results	Representative Companies
user research driven	characterize users and usage context, identify innovative product concepts and shortcomings for targeting new products and services	Adobe, IDEO, MONKEYmedia, Nortel Networks, Yahoo!
market research driven	describe market needs, potential market size, and features that could enhance existing products, as well as provide customer definitions	Broderbund, Reactivity, Uppercase
document driven	documents that provide details on user requirements, visual appearance, and interaction strategies	Quicktime Apple, Sun Microsystems, Xerox Parc
process driven	via formal process design feature sets, map feature lists to user interface specifics and interaction strategies	AT&T, Cadence, Cambridge Technology Partners, Corel, IBM, Informaat, Microsoft, Reactivity, Uppercase, Viant
development driven	design concepts, prototypes, drawings	Liberate, Play, WebTV

3.4 Process Driven

Process driven companies follow a formalized design process, from functional requirements definition, identification and characterization of users tasks, to iterative design, development, and testing. For example, Corel's process has four primary stages: business plan stage, preliminary development stage, project development stage, and post-development stage (Deevey, 2000). Whereas, Uppercase's (Tarlin, Nielsen, & D'Arlch, 2000) process has five phases: conceptualization, investigation, specification, design, and validation. Through such processes, feature sets, user interface designs, and interaction strategies are designed. As David Cortright (2000, p. 39) of Microsoft indicates, "For any feature design, the process always begins with three components: a goal (or goals), justification data, and user scenarios... Using this information, the program manager works with the feature team (developers and testers), as well as other resources throughout the company ... to come up with a design proposal." Thus, the design evolves through the formal stages of the development lifecycle, with interface designers contributing to the process at each stage.

3.5 Development Driven

Some companies focus on generation of design concepts and rapid prototyping right from the start. With this approach, designers are encouraged to try various ideas using quick prototyping tools and obtain feedback from product team members or users (Hartford, 2000). Hartford (2000, p. 51) indicates that "programmers and designers are encouraged to simply try various ideas and methods via rough, quick approximations." WebTV often obtains initial design concepts from a user interface engineer or graphics designer, from which they follow with iterative user testing of the design (Darnell, 2000). These companies focus more on building the product than following a specified process or writing a design specification (Palmer, Fulker, Liston, Misconish, & Arnold, 2000). The general sentiment of these companies tends to be that too much formalism in process or documentation stifles creativity.

4. CONCLUSIONS

Expanding the design conceptualization process to include information design changes the focus of early design efforts. Rather than experts driving design via brainstorming techniques, in which efforts are made to reach consensus, make decisions, or solve problems, during information design the goal of designers is to explore, investigate, and understand users as they go about their activities in the actual context of use. The outcome of information design is reach qualitative data that can inspire unique design ideas, provide a deep understanding of users' needs, and identify real-world implications that can make or break the success of a service system or product in the marketplace. Beyond their approach to design conceptualization, industry design practices differ in their *focus* during this process. There is a clear distinction among companies in the emphasis they place on understanding the user and usage context, the propensity to be document driven, the priority given to following a formal development process, and the tendency to jump in and start prototyping. Understanding differences in the approach to design conceptualization can assist companies in allocating design resources to resolve current product shortcomings. If innovation and creativity are lacking, more time may need to be spent in the information design phase. If products are difficult to understand, then the interface design phase may need to be extended. Finally, if products are difficult to interact with, interaction design may need more focus.

5. REFERENCES

Ahn, E. (2000). Broderbund. *Interactions*, 7(2), 16-19.

Armitage, J. (2001). Viant. *Interactions*, 8(2), 75-79.

Bear, E.G., Teasley, B., and Carroll, L.P. (2001). MONKEYmedia. *Interactions*, 8(2), 63-69.

Crow, D. and Yang, M. (2001). Reactivity. *Interactions*, 8(2), 71-74.

Carroll, J.M. (Ed.) (1995). *Scenario-based Design: Envisioning Work and Technology in System Development*. New York: John Wiley.

Carroll, J.M. (2000). Five reasons for scenario-based design. *Interacting with Computers*, 13(1), 43-60.

Cortright, D. (2000). Microsoft Corporation. *Interactions*, 7(2), 39-40.

Darnell, E. (2000). WebTV Networks, Inc. *Interactions*, 7(2), 77-81.

Deevey, K. (2000). Corel Corporation. *Interactions*, 7(2), 27-31.

Dykstra-Erickson, E. and Hoddie, P. (2000). Quicktime Apple. *Interactions*, 7(2),11-15.

Gibson, J. J. (1966). *The senses considered as perceptual system*. Boston: Houghton-Mifflin.

Green, W.S., Kanis, H., and Vermeeren, A.P.O.S. (1997). Tuning the design of everyday products to cognitive and physical activities of users. In S.A. Robertson (Ed.), *Contemporary Ergonomics* (pp. 175-180). London: Taylor & Francis.

Hartford, S. (2000). Play, Inc. *Interactions*, 7(2), 50-53.

Henderson, A. (2000). Design of what? Six dimensions of activity (Part 1 of 2). *Interactions*, 7(5), 17-22.

Isdale, J., Fencott, C., Heim, M., and Daly, L. (in press). Content Design for Virtual Environments. In K.M. Stanney (Ed.), *Handbook of Virtual Environments: Design, Implementation, and Applications*. Maywah, NJ: Lawrence Erlbaum Associates.

Kuhn, K. (2001). Problems and benefits of requirements gathering with focus groups: A case study. *International Journal of Human-Computer Interaction*, 12(3&4), 309-325.

Leslie, R.E. (1986). *Systems Analysis and Design: Methods and Invention*. Englewood Cliffs, NJ: Prentice Hall.

Mayhew, D. J. (1999). *The Usability Engineering Lifecycle: A Practitioner's Handbook for User Interface Design*. San Francisco, CA: Morgan Kaufmann.

Nardi, B.A. (1997). The use of ethnographic methods in design and evaluation. In M. Helander, T.K. Landauer, and P.V. Prabhu (Eds.), *Handbook of Human-Computer Interaction* (2nd Edition) (pp. 361-366). Amsterdam: North-Holland.

Nielsen, J. (1993). *Usability Engineering*. Boston: Academic Press Professional.

Palmer, J., Fulker, J., Liston, A., Misconish, D., and Arnold, P. (2000). Liberate Technologies. *Interactions*, 7(2), 36-38.

Parkin, A. (1980). *Systems Analysis*. Cambridge, MA: Winthrop.

Russell, D.M. (2000). Xerox Palo Alto Research Center. *Interactions*, 7(2), 82-86.

Spine, T. (2000). Sun Microsystems, Inc. *Interactions*, 7(2), 70-72.

Stanney, K.M., Maxey, J., & Salvendy, G. (1997). Socially-centered design. In G. Salvendy (Ed.), *Handbook of Human Factors and Ergonomics* (2nd Edition) (pp. 637-656). New York: John Wiley.

Stanney, K.M., Smith, M.J., Carayon, P., and Salvendy, G. (in press). Human-computer interaction. In G. Salvendy (Ed.), *Handbook of Industrial Engineering* (3rd Edition). New York: John Wiley.

Stillion, D. (2000). IDEO Product Development, *Interactions*, 7(2), 32-35.

Takahashi, D. (1998). Technology Companies Turn to Ethnographers, Psychiatrists. *Wall Street Journal*, 10/27/98.

Tarlin, E., Nielsen, P. and D'Arlch, C. (2000). Uppercase, Inc., *Interactions*, 7(2), 73-76.

Vicente, K.J. (1999). *Cognitive Work Analysis: Towards Safe, Productive, and Healthy Computer-Based Work*. Mahwah, NJ: Lawrence Erlbaum Associates.

Bringing System Requirements into Focus:
The Power of Knowledge Structures

Sherrie P. Gott, PhD
University of Texas Health Science Center at San Antonio

ABSTRACT

The essential objective of any body of information, regardless of the communication medium used, is to connect with the audience – the user, the student, the consumer, or whomever. Through advances in modern cognitive psychology, we now know that to optimize the engagement of the user/student with knowledge-based automated systems, developers must consider the knowledge structures that users bring for interacting with the system. Specifically, learners will have key mental models (or mental representations) in a given domain that they use in structuring and integrating new knowledge. Some of these models have even been found to have high generalizability across domains of learning. There are methods now available for eliciting such cognitive requirements, which have been found to be particularly effective in developing the various components of tutoring systems, including tutoring content, on-line help, student evaluation, and documentation. There are some critical guidelines to follow in eliciting the cognitive requirements for knowledge-based systems, and those guidelines will be explicated in this paper. Essentially, the guidelines and associated knowledge-elicitation methods are grounded in studies of expertise and various cognitive task analysis techniques coming out of the cognitive sciences.

1. Introduction and Background

School-based learning has been described as the acquisition of theoretical knowledge and the development of general competency skills (Resnick, 1987). By contrast, practice-oriented instructional programs are directed at some specific criterion performance, such as computer programming or electronics troubleshooting or acute care nursing, not at general academic proficiencies. As a consequence, the level of achievement desired is beyond the initial stages of learning, where factual knowledge bases are constructed. Rather, emphasis is on the later stages of skill acquisition, where knowledge is proceduralized and procedures are in turn smoothed out via practice.

Because the targeted level of achievement in practice-centered training is a specific performance, the nature of the targeted performance has considerable influence on the design of instruction and associated assessment. The more overt and observable the elements of the performance, the more it lends itself to traditional forms of apprenticeship training and assessment. For example, a carpenter's apprentice could generally learn a great deal by following a conventional apprentice regimen of observing the master, executing a task with support and critique from the master, and then continuing to practice with the master's support, eventually becoming autonomous. Similarly, assessment could be accomplished by evaluating observable behaviors and products of the behaviors, using standards provided by expert craftspeople. However, practical learning experiences for modern workplace tasks are not so easily handled. Performances are now far more mental and less physical. Internalized cognitive structures and processes have replaced the external behavioral elements of work performances of an earlier time. Because so much of "the action" is now internalized and unobservable, learning through observation has been significantly hampered. No longer is it effective to focus on overt behaviors and observable end products. Rather, practical learning experiences and assessments for modern work environments must be targeted at the internalized cognitive concepts and processes that lie behind expertise.

Knowledge-based instructional systems designed to facilitate practice-centered training have now been developed and evaluated, yielding encouraging results to support their efficacy as tools to enhance the complex mental performances that dominate modern workplaces (Collins, Brown, & Newman, 1989; Gott & Lesgold, 2000; Lajoie & Lesgold, 1989; Lesgold & Katz, 1992). These systems are often termed intelligent tutoring systems (ITSs), and the effective ones have been grounded in Piagetian notions of schemata (or knowledge structures). Pedagogically, they have followed Dewey and the recent constructivist movement where learning-by-doing, apprenticeship-based experiences are emphasized (Gott, Lesgold, & Kane, 1996; Perkins, 1991). In a manner similar to traditional apprenticeships, ITS instruction follows the pattern of modeling the desired cognitive performance, then supporting the novice performer as s/he attempts in a series of successive approximations to reproduce the expert performance across different contexts and situations until smooth autonomy prevails.

The content for effective ITSs is derived through the use of cognitive task analysis (or knowledge elicitation) methods that can elicit and make observable the internalized processes and knowledge structures of performers in a domain (Gott, 1989; Gott & Morgan, 2000; Hall, Gott, & Pokorny, 1995; Means & Gott, 1988). Once identified and documented, the knowledge structures and processes can precisely inform instructional content, learning activities, and assessment, ultimately functioning to bring the system requirements of intelligent tutors into sharp focus (Glaser, Lesgold, & Gott, 1991). The goal of this presentation is to explicate some of the critical guidelines for conducting cognitive task analysis studies in order to optimize knowledge elicitation aimed at complex workplace performances.

2. Cognitive Task Analysis Guidelines

2.1. **Identify ill-structured tasks as the targets of instruction.** In any domain it is the non-routine, ill-structured tasks – i.e., the complex problem solving tasks requiring critical thinking skills – that pose the greatest difficulty in modern workplaces. While advanced technologies have tended to automate certain functions in the work environment, machines are able to perform only routine, well-structured tasks, where it is possible to pre-specify the exact steps to take to attain the desired solution state. Conversely, ill-structured tasks are dynamic and multivariate, and thus solution steps cannot be easily specified in advance to enable automation. They often require the performer to bring to bear multiple types of knowledge and skill, as well as to think on multiple levels in the service of a variety of goals (Simon, 1973). Examples of ill-structured, non-deterministic tasks include medical diagnosis, electronic troubleshooting, and computer programming. Cognitive scientists have studied these domains intensively. *Tasks that require performers to think on multiple levels in dynamic task environments, deploying multifaceted skill and knowledge components in the process, are quintessential ill-structured tasks.*

2.2. **Work with domain experts to specify particularly difficult cases or scenarios based on the identified tasks.** The cases, and their variants, eventually become the content for the instructional system. For example, in the aircraft electronics domain, an ill-structured task might be stated at a very general level as "Diagnose and repair a malfunctioning antenna on the jet aircraft." A case-based scenario for this task would identify both the specific cause of the malfunction, e.g., "Loss of digital signal to the antenna's XYZ circuit board due to a faulty ABC switch," as well as the presenting symptoms, i.e., what the initial symptoms of a faulty ABC switch would be. In developing instructional content, it is very useful to specify cases that can be altered at a later time to become less difficult. Variants of the original scenarios provide learning-by-doing practice opportunities for students on progressively more difficult cases. For example, a variation of the case of the "loss of digital signal to the XYZ circuit board" could be digital signal loss due to another (less complex) component on the signal path.

Once the particular cases have been developed, domain experts participate in a workshop setting to specify all the possible paths that workers might explore in the solution search process. The specification of anticipated paths and outcomes constitutes the "problem space" for a given scenario. Every piece of equipment used in the search for a solution is also specified as part of the problem space. Once the problem spaces have been anticipated, it is possible for experts to engage in simulated problem solving sessions, where one expert poses a particular case to a colleague (fellow expert) who does not know the source of the problem in advance.

2.3. **Capture expert performances on scenarios derived from ill-structured tasks at a fine-grained level of detail.** In a given domain, the thinking that experts apply to ill-structured tasks differs in fundamental ways from the thinking of performers at less-advanced stages of development. Experts typically deploy diverse types of knowledge and skill, which interact in the problem solving (or critical thinking) process. A taxonomy that has been useful in characterizing the various types of knowledge used by domain experts on complex problems is illustrated in Figure 1.

This figure shows an abstracted version of an empirically derived cognitive model of technical performance. The model highlights the important interplay among (a) strategy, (b) tactics (procedures), and (c) conceptual (or system) knowledge. The Strategic Knowledge component sits on top of and controls the two remaining interactive components – Procedures (tactics or operations) and System Knowledge. This model posits

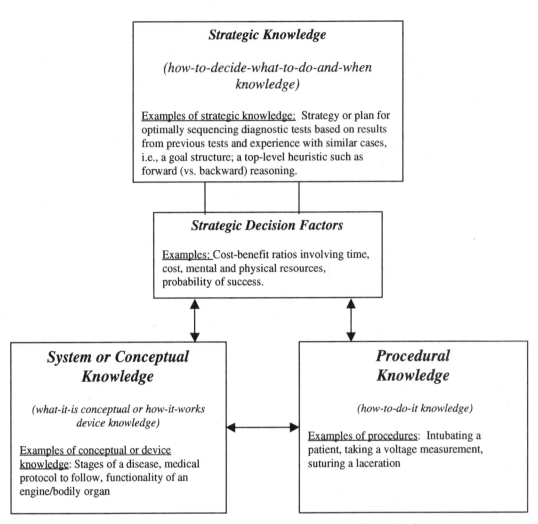

Figure 1. Abstract Cognitive Model of Complex Problem Solving Performance

that a top-level plan or strategy deploys pieces of system knowledge and procedural subroutines as needed and as driven by strategic decision factors such as time, effort, payoff, and resource efficiency. Problem solving on ill-structured tasks is thereby represented as multilevel, complex decision making, which involves choices among various top-level and intermediate-level strategies, concepts, and operations.

In effect, detailed cognitive task analyses instantiate this model for particular scenarios to reveal how experts integrate conceptual, procedural, and strategic knowledge into the diverse mental models needed in a domain (Gott & Lesgold, 2000; Gott & Morgan, 2000; Hall, Gott, & Pokorny, 1995; Means & Gott, 1988). Interestingly, these mental models often cannot be found in any formal curriculum, meaning they may have never been formally taught, but rather derived experientially by experts over time. Such expert performance models can in turn provide improved targets for instruction and learning assessment (among other applications), because they are very precise, situated characterizations of the most advanced levels of performance being demonstrated in the workplace (Glaser, Lesgold, & Gott, 1991).

Having expert performance models as targets for instruction is necessary but not sufficient for intelligent tutoring systems, however. The expert models only specify the final goal state for learners. Cognitive performance models at novice and intermediate levels are also needed so that a skill acquisition trajectory can be mapped out to guide instruction from the student's initial learning state to the goal state.

2.4. Capture performances on the same scenarios from non-experts to reveal the normal learning progression in the domain as well as the common barriers to expertise. The acquisition of complex skills occurs incrementally, in successive approximations of mature practice; therefore, the sequencing of instructional events is critical since it must promote the maturation process. A guiding principle of intelligent tutoring systems has been to decompose the targeted knowledge/skill base and reorganize it to fit learning. With most instruction, the reverse is true, that is, learning is expected to conform to the way knowledge is organized in some external curriculum (or system). The fit of knowledge to learning depends upon careful instructional sequencing so that the learner is always building on the foundation of prior knowledge. Cognitive task analysis studies that capture performances from a range of performers in a domain provide the skeleton for a learning trajectory to decompose the targeted knowledge into a natural instructional progression.

In addition to capturing the normal skill progression in a domain, cognitive task analyses can reveal important expert-novice skill differences that can further enhance and inform instruction. For example, research into the nature of expertise has consistently revealed differences between expert and novice knowledge structures that recur across domains. A common finding is that novice mental models are typically organized around the *surface-level* features of objects or scenarios in a field of practice, while experts have *deep-structure* models (Chi, Glaser, & Farr, 1988). Because of their limited knowledge structures, the reasoning of novices is restricted to the superficial dimensions of a problem while experts can reason about a problem at deeper, functional levels of understanding. Further, on diagnostic tasks, where hypothesis-testing is demanded, studies have shown that experts distinguish themselves from novices by engaging in forward reasoning, i.e., reasoning forward from what is known about a medical condition to some unknown diagnosis (Groen & Patel, 1988). Novices, by comparison, often engage in backward reasoning, where they fix on a diagnosis early on and then conduct tests (test hypotheses) to confirm their premature diagnosis. When they encounter any disconfirming evidence, they either ignore it or misinterpret it as support for their initial diagnosis. Intelligent tutoring systems can be designed to detect novice misconceptions and faulty reasoning such as this as well as to provide appropriate remediation – *if* cognitive task analysis studies are conducted to reveal such phenomena.

2.5. Develop the student evaluation component for an ITS based on standards articulated by experts in evaluating a full range of performances on the system's problem set. To be complete, every instructional system must incorporate a mechanism for assessing the learner's progress through the curriculum. Cognitive task analysis studies offer an intriguing approach to developing a student evaluation component. Research in the domain of electronic troubleshooting (Gott & Lesgold, 2000) has demonstrated the feasibility of eliciting standards to evaluate student performance from domain experts by having them independently review the solutions generated by novice, intermediate-level, and other expert performers on a common set of scenarios. Of course, students can be evaluated on obvious standards such as successfully finding the source of a malfunction within a certain amount of time; however, competent performance is characterized not only by solution success but, more importantly, the manner in which the solution process unfolds. More specifically, experts define success in terms of problem solving logic and resource efficiency. For example, diagnostic experts invoke standards such as "thoroughly test all suspect components (or sub-systems) before replacing or repairing a component part." The premature swapping of parts is penalized by experts who evaluate performances, even if the premature swap is a fortuitous one that solves the problem. Another example from diagnostic domains concerns the manner in which performers use the vast array of diagnostic tests available to them. A targeted, resource-efficient approach is valued by experts over a less-selective, shotgun approach, where all manner of tests are conducted to "see what turns up."

Studies where expert evaluative judgments have been elicited as the standards for student assessment have shown that the standards can often be traced to the Strategic Knowledge component of expertise (see Figure 1.). It is that component that accounts for the goal structure and other strategic processes responsible for the organization, coherence, and general execution of a performance. Strategic knowledge has been called "the glue" of cognition, but often it is *tacit knowledge*, meaning it is not routinely made explicit by experts. If it remains tacit in a community of practice, it is extremely time-consuming to acquire through lengthy experiential learning. Alternatively, if it is made explicit and observable via a cognitive task analysis, it becomes knowable, i.e., an instructional objective having high payoff.

1403

3. Summation.

Detailed cognitive task analysis studies enable learning environments where students construct understanding in authentic contexts, rooted in the needs of practice. Cognitive performance models reveal the knowledge structures of both expert and non-expert performers as they solve ill-structured problems. These structures in turn provide both the targets and learning pathways for knowledge-based instructional systems. Carefully sequenced learning activities are achieved by capturing performances at different stages of the skill acquisition process. As a tool for practice-oriented instruction, cognitive task analysis derives perhaps its greatest utility from eliciting knowledge that is directly tied to its uses in the world. Learning thereby occurs at the highest level of authenticity, and typically tacit knowledge such as goals, strategies, and assumptions is made explicit and knowable by students. In sum, knowledge structures create a sharp, detailed blueprint for instructional system requirements.

4. References.

Chi, M.T.H., Glaser, R., & Farr, M.J. (1988). The nature of expertise. Hillsdale, NJ: Erlbaum

Collins, A., Brown, J.S., & Newman, S. (1989) Cognitive apprenticeship: Teaching the crafts of reading, writing, and mathematics. In L. B. Resnick (Ed.), Knowing, learning, and instruction: Essays in honor of Robert Glaser. (pp.453-494). Hillsdale, NJ: Erlbaum.

Glaser, R., Lesgold, A.M., & Gott. S.P. (1991) Implications of cognitive psychology for measuring job performance. In A.K. Wigdor & B.F. Green, Jr. (Eds.), Performance assessment for the workplace. Vol. II (pp. 1-26). Washington, D.C.: National Academy of Sciences.

Gott, S.P. (1989). Apprenticeship instruction for real world tasks: The coordination of procedures, mental models, and strategies. In E.Z. Rothkopf (Ed.), Review of research in education, Vol. XV (pp. 97-169). Washington, D.C.: American Educational Research Assn.

Gott, S. P., & Lesgold, A.M. (2000). Competence in the workplace: How cognitive performance models and situated instruction can accelerate skill acquisition. In R. Glaser (Ed.), Advances in instructional psychology, Vol VI, Hillsdale, NJ: Erlbaum.

Gott, S.P., Lesgold, A., & Kane, R. S. (1996) Tutoring for transfer of technical competence. In B.G. Wilson (Ed.), Constructivist learning environments: Case studies in instructional design. (pp. 33-48). Englewood Cliffs NJ: Educational Technology Publications.

Gott, S.P. & Morgan, S. (2000). Front end analysis: From unimpressive beginnings to recent theory-based advances. In S. Tobias & J. D. Fletcher (Eds.). Training and retraining: A handbook for business, industry, government, and the military. New York.: Macmillan.

Groen, G.J., & Patel, V.L. (1988). The relationship between comprehension and reasoning in medical expertise. In M.T.H. Chi, R. Glaser, & M.J. Farr (Eds.), The nature of expertise. Hillsdale, NJ: Erlbaum.

Hall, E.P., Gott, S.P., & Pokorny, R.A. (1995). A procedural guide to cognitive task analysis: The PARI methodology. (AFHRL-TR-95-108). Brooks AFB TX: Armstrong Laboratory.

Lajoie, S., & Lesgold, A. (1989). Apprenticeship training in the workplace: Computer coached practice environment as a new form of apprenticeship. Machine-Mediated Learning, 3, 7-28.

Lesgold, A., & Katz, S. (1992). Models of cognition and educational technologies: Implications for medical training. In D.A. Evans & V.L. Patel (Eds.), Advanced models of cognition for medical training and practice (pp. 255-264). NATO ASI Series F, Vol 97. Berlin: Springer-Verlag.

Means, B. & Gott, S.P. (1988). Cognitive task analysis as a basis for tutor development: Articulating abstract knowledge representations. In J. Psotka, D. Massey, and S. Mutter (Eds.), Intelligent tutoring systems: Lessons learned. Hillsdale, NJ: Erlbaum.

Perkins, D.N. (1991, May). Technology meets constructivism: Do they make a marriage? Educational Technology, 31(5), 18-23.

Resnick, L.B. (1987). Learning in school and out. Presidential address at the Annual Conference of the American Educational Research Association: Washington, DC.

Simon, H.A. (1973). The structure of ill-structured problems. Artificial Intelligence, 4, 181-201.

A Meaning Processing Approach to Analysis and Design

John M. Flach

Psychology Department, Wright State University, Dayton, OH 45435

ABSTRACT

Classically, cognitive psychology has treated "meaning" as if it was a product of information processing. That is, "meaning" has been considered to be the result of "interpretation" of a symbol. One consequence of this perspective is that the environment has been treated as a symbol – (i.e., as if its structure was arbitrary). However, researchers are now beginning to appreciate that environments are not arbitrary symbolic structures, rather they are ecologies that impose lawful constraints on performance and that have non-arbitrary consequences. A meaning processing approach starts with the assumption that the structure and consequences of work ecologies can be the stimulus for cognitive processes. From this perspective, meaning is considered to be the raw material of information processing. This paper considers some implications of a meaning processing perspective for analysis and design.

1. EXPERTISE, MEANING, & DESIGN

From the earliest work on expertise (e.g., de Groot, 1965) it has been apparent that experts see the objects of their domain in terms of meaningful patterns or chunks. This ability allows experts to process information more efficiently (e.g., to hold more information in working memory) and it helps experts to "zero-in" on solutions in situations where novices are faced with a confusing set of ambiguous alternatives. So, it would seem natural for someone interested in expertise to address the question of meaning: what constitutes a meaningful pattern?

There seems to be two important dimensions to the question of meaning. One dimension is the subjective dimension. That is, meaning is an *interpretation*. It is an attribute of a person's awareness. For example, people may have different opinions about what constitutes a "safe" following distance when driving on a highway. The different subjective opinions might reflect differences in skill, experience, and attitudes toward risk – a mother and her teenage son may have different criteria for what they judge to be a "safe" following distance.

A second dimension to meaning is the objective dimension. That is, meaning is the *significance* relative to some event or situation. To address this objective dimension in the driving example, one would like to know whether there are objective criteria for a "safe" following distance. In order to come up with objective criteria for a safe distance (e.g., minimum stopping distance) one might consider the control dynamics of the car (e.g., handling properties, braking performance), the properties of the road surface (e.g., wet or dry), the states of the various cars (e.g., relative positions and velocities), and the skill of the drivers (e.g., reaction time). Here meaning tends to be grounded in the lawful constraints on a dynamical control system.

Classically, the subjective dimension of meaning has been the domain of cognitive psychology. For cognitive psychology, "meaning" has typically been treated as a subjective product of information processing. For the most part, the objective dimension of meaning has been ignored by cognitive psychology. The objective dimension of meaning typically falls within the domain of engineering. Engineers consider physical laws and principles to derive the criteria and boundaries for safe operation. However, if a dimension cannot be specified and measured objectively, it is rarely considered in an engineering analysis.

Returning to the question of expertise, both subjective and objective dimensions of meaning are important. That is, the meaningful patterns that experts "see" are not merely subjective constructions of mind, but rather reflect objective properties of the domain of expertise. In other words, experts would be expected to have good situation awareness, in that their awareness or interpretation of a situation would generally reflect the objective constraints of the situation. Thus, an expert driver's judgment about a "safe" following distance would be expected to conform closely to the objective criterion for this boundary. The research literature supports this. For example, with training in tracking-tasks humans tend to converge on solutions that are typical of "optimal" control systems (e.g., Pew & Baron, 1978). The fact that experts typically generate objectively good options as the first alternatives considered is also consistent with the idea that the patterns that they see are both "subjectively" meaningful to the expert and "objectively" meaningful with respect to the domain constraints. The interdependence between the objective and subjective dimensions of meaning is illustrated clearly in Vicente and Wang's (1998) analysis of the expertise literature.

A premise of this paper will be that an objective for design of human-machine systems is to facilitate and support expertise. That is, a well-designed system should allow the users of that system to function as experts. In other words, a good design will shape the user's interpretation of "what is subjectively meaningful" so that it is

1405

congruent with the objective constraints that determine "what is objectively meaningful." One path to this goal, is "ecological interface design" or "EID." This approach, first proposed by Rasmussen and Vicente (1989), attempts to make the objective constraints of a work domain visible to humans by making these constraints explicit properties of the interface representation. This is typically accomplished using configural graphic displays in which geometric properties (e.g., colinearity, closure, shape, and symmetry) are mapped to the work domain constraints or domain semantics (e.g., see Bennett & Flach, 1992). However, Tanabe (2001) has recently observed that EID needs to be complimented by ecological approaches to instruction and training. An ecological approach to instruction and training would focus on the development of "internal representations" that were congruent with the objective domain constraints, as reflected in the ecological representations.

2. WHAT'S NEW?

At one level, the idea of designing interfaces, instructions, and training around the objectively meaningful aspects of a work domain seems to be simple, common sense. How else would you do it? In this section we will consider how it is typically done. We will argue that the issue of "meaning" is typically side-stepped in the design process where syntactical constraints (e.g., rules and procedures) are typically substituted in place of semantic constraints.

The standard approach to the design and analysis of work can be traced back to the ideas of scientific management typically associated with Taylor and Gilbreth. The premise of this approach was that it would be possible to use experimental methods to identify the "one best way" (typically defined as satisfying a quality goal while minimizing time per unit of product). Scientific management involved "task analysis" (e.g., time and motion studies) to identify the "one best way" and then "work design" to provide the tools, training, instructions, and incentives so that workers would follow the "one best way." This approach led naturally to the design of assembly lines where work was designed to involve monotonous repetition of well-defined procedures. The scientific management philosophy was easily generalized to cognitive work. It simply required that task analysis be expanded to consider not only the manual activities of workers, but also the cognitive or information processes involved in work (e.g., Fleishman & Quintance, 1983). However, the focus remained on the "one best way." That is, management defined the procedures, and the job of workers was to follow the procedures as trained and instructed by management.

Following this scientific management philosophy, work design typically gets divided between the domain engineers and the human engineers. The domain engineers set the procedures. These procedures are generally based on an analysis of the objective constraints within a domain. They might reflect a path through the domain constraints that is optimal relative to some criteria (e.g., minimizing time, minimizing waste, minimizing cost) or the path might be chosen to satisfy safety goals (e.g., a path that keeps the system well within the bounds of safe operation). Thus, the choice of procedure is based on an analysis of the objective aspects of meaning. However, this analysis of meaning generally remains with the domain engineers. The only thing handed over the wall to the human engineers is the procedures.

The human engineers then have the task of insuring that the humans will follow the procedures defined by the domain engineers. And as noted above, they are typically asked to do this with little or no explicit information about the objective constraints that motivated the choice of one set of procedures or another. So, it is typical that the human engineers who design the interfaces, instructions, and training systems have little domain knowledge. Thus, their design efforts are organized around the procedures, not the constraints from which those procedures were derived. For example, the displays are designed with the explicit goal to help the operator see where (s)he is in the procedural sequence – to recognize when a step is completed and to specify the next step in the sequence contingent on the outcome of the previous step. The displays are designed to lead the users through a step by step process and to alert or warn the users whenever they deviate from the procedure. Similarly, the instructions and training systems are designed to "teach" the procedures. For cognitive work, these procedures are typically presented as a set of rules (if this, then that, etc.). The question of "why this procedure" is considered to be a problem for the domain engineers, not a problem for the operators or for the human engineers. Theirs is not to reason why, but to follow the procedures as specified. The result is typically a very efficient, but very brittle system – an automaton.

3. COGNITIVE SYSTEMS ENGINEERING

Limitations of the scientific management philosophy began to become apparent in the analysis of safety for complex processes like those involved in nuclear power production (e.g., Rasmussen, 1986; Ramussen, Pejtersen, & Goodstein, 1994). Displays and training designed around procedures worked adequately for routine operations under normal conditions (e.g., routine startup). However, when operators were asked to diagnose faults, problems arose. For a complex system it is impossible for domain engineers to anticipate all possible faults. It is impossible to design procedures for every fault. On the other hand, the displays and training of operators has focused on procedures. Thus, operators typically don't have either the knowledge or the display representations that might support reasoning with respect to the deep structure of the systems that they are operating. When unexpected events force the system away from normal trajectories, operators can sometimes become completely lost without resources to help find solutions. Concern for how to better support operators in fault diagnosis led to the development of Cognitive Systems Engineering and to the concept of EID.

Cognitive systems engineering is an attempt to design systems to support meaning processing. That is, it is an attempt to bridge the gap that has classically existed between the domain engineers and the human engineers. Cognitive systems engineers work in concert with domain engineers to identify the objective constraints within a work domain. They work in concert with human engineers to design interfaces, instructions, and training systems that explicitly represent those constraints in order to facilitate the operators' understanding of the process. For example, EID attempts to make the objective constraints (or deep structure) of the process visible in the interface. In applying EID to power plant control rooms, Yamaguchi and Tanabe (2000) have designed displays that reflect the basic principles of thermodynamic systems (e.g., mass and energy balances), the functional relations among subsystems (e.g., heat exchange, pumps), and the physical connections among the components. This representation has been designed with the explicit goal of helping operators to understand the constraints that govern the processes being controlled. In a complimentary way, cognitive systems engineers design training systems around first principles and design instructions around operating goals. The assumption of this approach is that an operator who understands the objective constraints and the operating goals for the system will be better able to follow procedures for routine situations, and possibly will be better prepared to "invent" procedures for non-routine events, such as faults.

To avoid the "brittleness" associated with rule based automatons that breakdown whenever they are forced beyond the assumptions underlying their programs – it is important to provide the human operators in complex systems with the knowledge and the tools that they need to adapt procedures to contingencies that were not (and possibly could not have been) anticipated by the domain engineers. The scientific management philosophy tended to constrain operators to follow pre-determined paths. Cognitive engineering, on the other hand, tends to empower operators to explore the work domain more freely. The Cognitive engineering approach does not require that design to support procedural compliance be abandoned all together. For high-risk systems, like nuclear power plants, procedural compliance may be an important aspect of managing safety. However, it is clear that procedural compliance will never be sufficient to insure safety in complex systems! It is essential that designers provide operators the support that they need to go beyond procedures to address meaning, when required! The system must be given the capacity to adapt to contingencies not anticipated in the procedures.

4. AN ECOLOGICAL PERSPECTIVE

The cognitive systems engineering perspective on work is complimented by an ecological approach to cognition and human performance. Just as cognitive systems engineering attempts to bridge the gap between domain engineering and human engineering, ecological psychology attempts to bridge the gap between objective and subjective poles of meaning. For example, Gibson's construct of affordance was explicitly formulated to help bridge the gap between the two dimensions of meaning:

The notion of affordance implies a new theory of meaning and a new way of bridging the gap between mind and matter. To say that an affordance is meaningful is not to say that it is "mental." To say that it is "physical" is not to imply that it is meaningless. The dualism of mental vs. physical ceases to be compulsory. One does not have to believe in a separate realm of mind to speak of meaning, and one does not have to embrace materialism to recognize the necessity of physical stimuli for perception (Gibson, 1982, p. 409).

Figure 1 attempts to illustrate an ecological perspective on meaning. Meaning is viewed in terms of three classes of constraints: constraints on goals (values); constraints on action, and constraints on information. Each of these constraints is a joint function of the human actor and the environment. Within the ecology, the "objects of study" are these constraints and the relations among these constraints. Thus, an ecological perspective blurs the distinction between actor and environment (mind and matter; user and work domain). The focus is shifted to the mutual constraints that bound the functional coupling between animal and environment – these constraints determine what should be "meaningful." For an ecological approach to human performance, these meaningful aspects of an ecology are considered to be the raw materials for cognition – **not** the products of cognition. Thus, cognition is visualized as "meaning processing" as opposed to "information processing." Classically, the question of cognition has been to discover, "Why people think about (see) the world as they do?" For ecological psychology, the question has been altered to, "How are people able to think about (see) the world as it is?" [This is a paraphrase of Mace's (1977) contrast between classical and direct approaches to perception (p. 63 – 62)]. In order to address

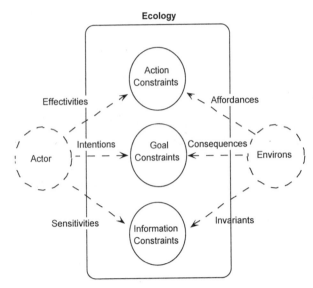

Figure 1 . Within an ecology the actor and environment fade into the background and the mutual constraints on action, goals (values), and information emerge as the objects of interest.

this second question, it is essential to have a theory about how the world "is." Thus, again it is essential to bridge the gap between subjective and objective properties of meaning. Design for meaning processing, then means to develop interfaces, training, and instructions that help people to "see" or "think about" the world as it "is."

5. ECOLOGICAL TASK ANALYSIS

Classically, task analysis has been framed around the activities (both manual and cognitive) of workers (either in terms of actual or required activities) (Fleishman & Quaintance, 1984). This is perfectly reasonable when work is defined in terms of procedures. However, when work involves problem solving (e.g., fault diagnosis) or requires creative solutions (e.g., command and control) an analysis organized around procedures will not be adequate. In terms of Simon's (1981) parable of the ant on the beach, classical task analysis has attempted to describe the ant's path as a sequence of actions.

A meaning processing approach requires that the task analysis shift from describing the ant's path to describing the "beach." That is, an ecological task analysis must describe the work landscape. In other words, it must describe the constraints (goals, actions, and information) that shape the ant's path (see Vicente, 1999 for a more thorough discussion of this metaphor and the lessons for cognitive work analysis). As Flach and Dominguez (1995) have argued this requires a shift from focus on the "user" to focus on "use." Flach (2000) outlines some of the implications for the process of task analysis. For example, he observes that for knowledge elicitation it is not sufficient to interview "operators" to get a better understanding of their mental models. It is also important to interview other domain experts (e.g., domain engineers and system designers) in an attempt to separate the "subjective constraints in a mental model" from the "objective constraints in the work domain." Flach also observes with respect to table-top analysis that it is not sufficient to read the operating and procedure manuals. It is also important to study the domain texts and documents on the history and evolution of the work domain. The significance of a historical perspective for understanding cognitive work is nicely illustrated by Hutchins' (1995) analysis of ship navigation. Separating the subjective from the objective is impossible from any single perspective – discovering the invariant properties of a work domain requires multiple perspectives. Thus, different users, domain engineers, system designers, managers, procedural manuals, domain texts, historical documents each offer the potential for unique and valuable insights on "the way the world is."

6. DESIGNING TO SUPPORT MEANING PROCESSING

To the extent that the objectively meaningful constraints within a work domain can be discovered through ecological task analysis – the goal of design is then to make those meaningful constraints visible to operators. As with any design question, there is no simple recipe for doing this. There are many ways to do it wrong and only a few ways to do it correctly. The best guide to doing it right may be examples of past successes. I recommend two examples that might help guide people who what to design systems with the goal of supporting meaning processing. One example is the graphical displays that are being developed by the Japanese Atomic Energy Research Institute (JAERI) for control of a nuclear reactor (Yamaguchi & Tanabe, 2000). This graphical display reflects a detailed ecological task analysis guided by Rasmussen's abstraction hierarchy framework. In this case, a rich variety of detailed quantitative models derived from first principles were available from the domain engineers. The abstraction hierarchy provided a framework for relating the different quantitative models and submodels into a coherent functional model of the process (that spans multiple levels of abstraction and decomposition). The design challenge then was to develop a visualization that captured the functional model. The graphical formats used in the interface were inspired by Vicente's (1992) work with the DURESS microworld and Beltracchi's (1987, 1989) ideas for displaying information based on the Rankine Cycle.

A second example, is the BOOK HOUSE interface for libraries (Pejtersen, 1989). Unlike the nuclear domain, there are no quantitative models for processes associated with "finding something interesting to read." Science has not provided us with a set of first principles for the dynamic of "interesting." So, the ecological task analysis for the BOOK HOUSE focused on the strategies used by the general public and by expert librarians. This reflects the assumption that "expert" librarians are people who are well-tuned to the landscape of this work domain. Note that the intended users in this case are the general public, not the librarians. However, the librarians were considered the experts on this domain. That is, the assumption is that the strategies that skilled librarians have evolved to assist people to find something interesting are intelligent adaptations to the objective constraints of this domain. The BOOKHOUSE then used a 3D spatial metaphor to arrange virtual tools designed to support the search strategies that librarians found to be most successful.

7. SUMMARY

The JAERI control room and the BOOKHOUSE examples are radically different both in terms of the nature of the ecological task analyses and in terms of the graphical design solutions. However, these differences help to highlight the central issues for a meaning processing approach. At a deep level, both projects reflect a commitment to discovering meaning as the focus of the task analysis and to representing meaning as the focus of the graphical design. This is the heart of a meaning processing approach – to identify the objectively meaningful constraints of situations and to shape people's awareness to reflect those constraints, so that the people are able to behave intelligently as they navigate through the work domain. In sum, the challenge of a meaning processing approach for analysis of work is to shift attention from the surface features of work (e.g., procedures) to the deep structure (e.g., constraints on safe operation). The challenge for design is to make that deep structure explicit to workers through interfaces, instruction, and training.

8. REFERENCES

Beltracchi, L. (1987). A direct manipulation interface for water-based Rankine Cycle heat engines. IEEE Transactions on Systems, Man, and Cybernetics, SMC-17, 478-487.

Beltracchi, L. (1989). Energy, mass, model based displays and memory recall. IEEE Transactions on Nuclear Science, 36, 1367-1382.

Bennett, K. & Flach, J.M. (1992). Graphical Displays: Implications for divided attention, focused attention, and problem solving. Human Factors, 34, 513-533.

de Groot, A.D. (1965). Thought and choice in chess. The Hague, the Netherlands: Mouton.

Flach, J.M. (2000). Discovering situated meaning: An ecological approach to task analysis. In J.M. Schraagen, S.F. Chipman, & V.L. Shalin (Eds.) Cognitive task analysis. (pp. 87 – 100). Mahwah, NJ: Erlbaum.

Flach, J.M. & Dominguez, C.O. (1995). Use-centered design. Ergonomics in Design, July, 19 - 24.

Fleishman, E.A. & Quaintance, M.K. (1984). Taxonomies of human performance. Orlando, FL: Academic Press.

Gibson, J.J. (1982). The affordances of the environment. In E. Reed & R. Jones (Eds.) Reasons for realism. (pp. 408-410). Hillsdale, NJ: Erlbaum.

Hutchins, E. (1995). Cognition in the wild. Cambridge, MA: MIT Press.

Mace, W.M. (1977). James J. Gibson's strategy for perceiving: Ask not what's inside your head, but what your head's inside of. In R. Shaw & J. Bransford (Eds.) Perceiving, acting, and knowing: Toward an ecological psychology. (pp. 43 – 65). Hillsdale, NJ: Erlbaum.

Pew, R.W. & Baron, S. (1978). The components of an information processing theory of skilled performance based on an optimal control perspective. In G. E. Stelmach (Ed.). Information processing in motor control and learning. (pp. 71 – 78). New York: Academic Press.

Pjetersen, A.M. (1989). The BOOK HOUSE: Modeling users' needs and search strategies as a basis for system design. Riso-M-2794. Roskilde: Riso National Laboratory.

Rasmussen, J. (1986). Information processing and human-machine interaction. New York: North Holland.

Rasmussen, J., Pejtersen, A.M., & Goodstein, L.P. (1994). Cognitive engineering. New York: Wiley.

Rasmussen, J. & Vicente, K.J. (1989). Coping with system error through interface design: Implications for ecological interface design. International Journal of Man-Machine Studies, 31, 517-534.

Simon, H.A. (1981). The sciences of the artificial (2nd ed.). Cambridge, MA: MIT Press.

Tanabe, F. (2001). Ecological interface design. Presentation at the JAERI Workshop on Ecological Interface Design. Tokai, Japan. (March, 2001).

Vicente, K.J. (1992). Memory recall in a process control system: A measure of expertise and display effectiveness. Memory & Cognition, 20, 356-373.

Vicente, K.J. (1999). Cognitive work analysis. Mahwah, NJ: Erlbaum.

Vicente, K.J. & Wang, J.H. (1998). An ecological theory of expertise effects in memory recall. Psychological Review, 105, 33-57.

Yamaguchi, Y. & Tanabe, F. (2000). Creation of interface system for nuclear reactor operation. Proceedings of the IEA 2000/HFES 2000 Congress, 571 – 574. Santa Monica, CA: The Human Factors and Ergonomics Society.

I didn't do it: Accidents of automation

Mark S. Young and Neville A. Stanton

Department of Design
Brunel University
Runnymede Campus
Englefield Green
Egham
Surrey TW20 0JZ
England

1. ABSTRACT

The present paper is concerned with accidents involving the interaction of humans with automated systems. Experience of automation in aviation is used to postulate the potential benefits and pitfalls of transferring such technology road vehicles. A brief review of the literature on human error is presented. Theories of accident causation are covered, with a focus on psychological problems relevant to the operation of automated systems. The paper then turns to the application of this knowledge to some notable case studies in aviation. As automation in road vehicles becomes more prevalent, such problems are likely to transfer to cars and their drivers. These problems are discussed with respect to issues of automation design and driver training. The paper concludes with a discussion about the future of accident research, contrasting past problems in aviation with future concerns about vehicle automation.

2. INTRODUCTION

Accidents and disasters, particularly those involving complex technological systems, are becoming an all-too common feature of the daily news. Industries such as aviation, rail, and nuclear power have all come under the health and safety spotlight lately. There have been a number of high-profile crashes involving advanced passenger aircraft, the so-called 'glass cockpits'. These aircraft are highly automated, and employ numerous computer processors in an effort to assist the pilots with the task of flying. Despite – or perhaps because of – their complexity, glass cockpits have presented brand new problems for researchers in human factors and design. Where there were errors of perception or action, there are now deeper cognitive problems of attention and mental workload. Recent advances in vehicle technology mean that the humble road car will soon become a similarly complex piece of equipment as a jet airliner. As such, we may expect to see a new breed of automation-related accidents on the roads. Before hypothesising about these future problems, though, it is necessary to review the academic literature on accident causation and look at previous problems with aviation automation.

3. ACCIDENT CAUSATION

There is a considerable body of literature devoted to an understanding of accident causation (e.g., Cox & Cox, 1996; Glendon & McKenna, 1995; Reason, 1990; 1997). Models of accident causation can be broadly classified into sequence models, influence models, and multi-causality models. These models differ in terms of their subjects of focus. Sequence models imply accidents are the result of a single causal chain in a linear sequence of events. The influence models represent the other extreme, where anything (e.g., the task, the environment, the interface) is thought to influence human behaviour. Finally, multi-causality models integrate sequence and influence models, by considering the interaction between people and events that produce accidents.

A seminal example of a generic theory of accident causation is Reason's (1990) resident pathogen metaphor for latent failures in technological systems. In his discussion of accident causation, Reason identifies various human contributions to the breakdown of complex systems. These include failures in decision-making at management levels, as well as unsafe acts by operators associated with the accident.

Although the models of Reason (1990) and others can usefully be applied in hindsight to explain disasters, there is increasing disillusionment in the human factors community of this 'autopsy' approach to preventing accidents (e.g.,

Wickens et al., 1998). In light of this, it may be preferable to address accident research within a general 'systems' paradigm (von Bertalanffy, 1950). A system is defined as a *"...set of interrelated elements, each of which is related directly or indirectly to every other element..."* (Ackoff & Emery, 1972). This emphasises the point that system elements cannot be construed as separate, self-contained, elements. All elements of the system are interrelated and interconnected. Therefore, changes in one element will have an effect upon the entire system. As Woods (1987) writes, accidents are rarely a result of purely technical failure, nor are they just a result of human failure. In particular, the greatest insight into performance problems can be found at the boundary between the social and technical system components (Lockett & Spear, 1980). This implies that maintaining quality two-way communications between human and machine is imperative for safe operations.

4. ACCIDENTS OF AUTOMATION

It should be stated at this point that – whilst this review may have pessimistic overtones – automation does have its advantages. Indeed, glass cockpit aircraft are statistically safer than conventional aircraft (Byrne, 1996). However, although it can yield benefits of improved performance and efficiency, automation can bring with it a whole new set of difficulties.

One of the most prevalent problems with automation, particularly in aviation, is mode errors (Stanton & Marsden, 1996). The simplest example of a mode error is attempting to set the time on a digital clock, when the clock is actually in alarm mode. System functionality and flexibility can increase with the introduction of modes. Complex, event-driven systems may change modes without input from, or feedback to, the operator. However, this can cause confusion and increased cognitive demand as the user tries to keep track of mode transitions and the system state (Sarter & Woods, 1995). Consequently, 'automation surprises' may occur, in which the system behaves according to specifications, yet this is quite different to that which the operator expects or desires. Due to a lack of communication from the automation to the pilot, a loss of control and crash can occur under conditions which the pilot should ordinarily be able to cope.

Automation surprises are often determined to be the cause of aviation accidents involving modern 'glass cockpit' aircraft (e.g., Learmount, 1994; Sedbon & Learmount, 1993). Many incidents are due to a lack of feedback from the system, to the extent that even pilots experienced with automation are sometimes surprised (Hughes, 1995; Hughes & Dornheim, 1995). Feedback was also implicated when it was found that mode confusions are often only detected by observing the system response, rather than the automation displays (Palmer, 1995). Byrne (1996) recommended that instead of fighting the computer, pilots should occasionally switch it off and look out of the window.

A tangible example of this is provided by Beaty (1995), who describes an accident at an air display in France on 26th June 1988. An Airbus A320, one of the new generation of 'fly-by-wire' aircraft, was making a low pass with the gear extended. Unfortunately, one of the features of the A320 was an ability for automatic mode transition. With the aircraft at a low altitude and the gear down, the automated systems assumed that the pilot was attempting to land. It therefore switched into landing mode and throttled back the engines. The pilot, not being completely familiar with the operation of the aircraft type, did not realise exactly what was happening. At the end of the pass, the aircraft failed to gain height and crashed into trees at the end of the runway. The pilot was effectively fighting the automation, attempting to pull up when the computer was determined to land. Three of the 136 people on board were killed.

Another classic example of a mode confusion, also involving an A320, ended in tragedy on 20th January 1992 near Strasbourg (Sedbon & Learmount, 1993). In this case, the descent mode was confused on the basis of a decimal point on the instrument display. Instead of entering a flight path angle of 3.3°, the crew did not notice the absence of the decimal point (the sole indicator the flight management system's mode status), and entered a vertical speed of 3300 feet per minute. As vertical navigation was left entirely to the automatic system, the crew ignored all other clues about the abnormally high descent rate, and the consequence was a 'controlled flight into terrain'. Only six people survived out of the 93 on board.

Pilots generally enjoy flying automated aircraft, however they are aware of the drawbacks in performance when resuming manual control (James et al., 1991). Over the next 15 years, technology will make automatic controls and collision avoidance commonplace in cars, with similar potential pitfalls (Walker, Stanton & Young, in press). Stanton & Marsden (1996) summarised the advantages and disadvantages of aviation automation, and applied the knowledge proactively to future systems in road vehicles. Problems may include technical reliability, skills maintenance, and equipment designs which induce errors (such as the mode confusions observed in aviation). For instance, overdependence on automated vehicle systems has been related to skill degradation or inattention (Hoedemaeker, 1999), and this in turn could result in more serious consequences in the event of system failure.

There has already been a small amount of empirical work directed at automated vehicle systems, mostly investigating the effects of longitudinal control systems such as Adaptive Cruise Control (ACC). In one study, Nilsson (1995) investigated the effects of ACC in critical situations. Of those who crashed, participants were four times more likely to have been using ACC than driving manually. Nilsson attributed this to the expectations that drivers had about ACC – the ACC behaviour did not correspond with manual driving control.

Stanton et al. (1997) explored the effects of ACC failure on driver performance. Participants were required to follow a lead vehicle with ACC engaged. At a predetermined point, the ACC system would fail to detect the lead vehicle braking, necessitating participant intervention to avoid a collision. It was found that one-third of all participants collided with the lead vehicle when ACC failed.

A major problem for future automated vehicles will lay in counteracting the inevitable deskilling of drivers. Whereas airline pilots are highly trained individuals, there is a much wider degree of variability in skill amongst the driver population. Less experienced drivers have been shown to be less effective in responding to emergency situations involving automation than their skilled counterparts (Young & Stanton, in press). Ironically, more automation will result in fewer skilled drivers, forcing higher investments in either retraining or design.

5. IMPLICATIONS

Where technological progress is concerned, the question most often asked is 'when will it happen?' Rather than thinking about what *can* be done, though human factors has often argued whether it *should* be done, and if so, how we might best go about it (Harris & Smith, 1997; Parasuraman, 1987; Wiener & Curry, 1980; Young & Stanton, 1997). Areas where future technology can potentially assist drivers have been derived from accident causation data (Broughton & Markey, 1996) and an analysis of driver support needs (Hancock et al., 1996; Owens et al., 1993). For instance, the argument for ACC and other longitudinal support devices is based on the fact that over a quarter of all road traffic accidents are due to rear-end collisions (Gilling, 1997). Intelligent vehicle highway systems therefore have the potential to improve safety and traffic flow on future roads (Biesterbos & Zijderhand, 1995; Hancock & Parasuraman, 1992). However, they also have the potential to overload and confuse the driver if they are not designed appropriately (Verwey, 1993). This has led to the call for human-centred design in future vehicle systems (Hancock et al., 1996; Hancock & Verwey, 1997; Owens et al., 1993), and greater attempts to support drivers rather than replace them. Stanton & Marsden (1996) review the arguments for allocating system functions between humans and automation, reiterating the need for "consideration and involvement of the driver in the automation of tasks" (p. 47).

The systems view offers a useful perspective for considering design issues. Rather than see the user and the technology as separate entities, they are actually part of a single interactive system, and the goal should be to optimise the performance of that system as a whole (Singleton, 1989). This involves exploiting the capabilities of each component of the system, and being aware of their limitations. Particularly in the case of driving, it must be realised that humans are actually very capable of performing the task, and any additional devices should be problem-driven rather than technology for its own sake.

A problem-focused approach (cf. Owens et al., 1993) might use the available technology to provide a different solution. For instance, if the problem of rear-end collisions is so significant, it follows that drivers have some difficulty perceiving the closing speed of leading traffic. ACC radar sensors could be used to provide drivers with information about the relative (or actual) speed of the lead vehicle, perhaps warning them if this crossed some threshold. This is more in line with a driver assistance philosophy, and resembles some ideas about future collision warning systems (Broughton & Markey, 1996; Gilling, 1997; Janssen & Nilsson, 1993). This would also solve the problems of automation failure, as now the system simply provides them with extra information about the task they normally perform.

Some advocate a preference for support systems rather than full automatic control, as this fosters human strengths while compensating for weaknesses (Grote et al., 1995). This is illustrated by the positions of two major aircraft manufacturers, who have ventured further into automated territory in recent years. Airbus use a 'hard protection' system in their A320 and A340 series, employing automation to prevent error, and hence it can override the pilot. Boeing, on the other hand, opted for 'soft protection' in their 777 aircraft, using automation as a tool to aid pilots, and not giving it the authority to override pilot control (Hughes & Dornheim, 1995). It is also interesting to note that McDonnell Douglas sit somewhere in between these extremes, using a soft protection system but giving the automatic controllers more authority for fuel, electrical, hydraulic, and pneumatic systems.

Any hybrid automated/human system must exhibit superior performance than the human alone, thus favouring the primary task (Hancock & Parasuraman, 1992; Hancock et al., 1996). This may often mean implementing low-technology solutions, or possibly not using the full potential of the computer in favour of optimising human performance (Hancock et al., 1996; Owens et al., 1993). Ideally, technological support systems should act like a driving instructor in the passenger seat – subtle enough so as not to cause interference, but accessible enough so as to provide assistance when needed.

It would be a shame if the unique skills and flexibility of human operators were to become redundant with the advent of automated systems. This is not to say that technology should not be used, but that more thought should be given to its appropriate implementation. Human capabilities have proved time and time again to be crucial in critical situations. Take the example of United Airlines Flight 232, on 20th July 1989 (Faith, 1996). The DC-10 aircraft had suffered an explosive failure of its number two engine (the DC-10 has three engines, one on each wing and one mounted on the tail structure; it was the latter engine which failed), which in the process had destroyed the hydraulics to the control surfaces. Unable to fly the aircraft by conventional means, the crew enlisted the help of another pilot who happened to be on board, and they managed to fly to an airport at Sioux City, Iowa, purely by using the balance of the two remaining engines. The landing was not ideal, and unfortunately 112 lives were lost. However, if it was not for the ingenuity and skill of the pilots, the 184 survivors would certainly have died too.

It is in situations such as these that the value of human input is realised. The keepers of future technologies should also realise this, and instead of trying to design humans out of systems, they should integrate the human fully and nurture their abilities. If we do not learn from the aviation example, miscommunications will result in cries of "I didn't do it!" from the drivers of future vehicles. To appreciate the impact of automation, we need a complete knowledge of the relative abilities of human and machine, and the way in which they work together. Neither component of the system is infallible, but by exploiting the strengths of each, the system as a whole really can be greater than the sum of its parts.

6. REFERENCES

Ackoff, R. L. & Emery, F. E. (1972). On Purposeful Systems. London: Tavistock.

Beaty, D. (1995). The naked pilot: The human factor in aircraft accidents. Shrewsbury: Airlife Publishing Ltd.

von Bertalanffy, L. (1950). The theory of open systems in physics and biology. Science, 13, 23-29.

Biesterbos, J. W. M., & Zijderhand, F. (1995). SOCRATES: A dynamic car navigation, driver information and fleet management system. Philips Journal of Research, 48(4), 299-313.

Broughton, J., & Markey, K. A. (1996). In-car equipment to help drivers avoid accidents. (TRL Report no. 198). Crowthorne, Berkshire: Transport Research Laboratory.

Byrne, G. (1996). High wire balancing act. Electronic Telegraph [On-line], 552.

Cox, S. & Cox, T. (1993). Safety, Systems and People. Oxford: Butterworth-Heinemann.

Faith, N. (1996). Black box: the air crash detectives - why air safety is no accident. London: Boxtree.

Gilling, S. P. (1997). Collision avoidance, driver support and safety intervention systems. Journal of Navigation, 50(1), 27-32.

Glendon, A. I. & McKenna, E. F. (1995). Human Safety and Risk Management. London: Chapman & Hall.

Grote, G., Weik, S., Wafler, T., & Zolch, M. (1995). Criteria for the complementary allocation of functions in automated work systems and their use in simultaneous engineering projects. International Journal of Industrial Ergonomics, 16(4-6), 367-382.

Hancock, P. A., & Parasuraman, R. (1992). Human factors and safety in the design of Intelligent Vehicle-Highway Systems (IVHS). Journal of Safety Research, 23(4), 181-198.

Hancock, P. A., Parasuraman, R., & Byrne, E. A. (1996). Driver-centred issues in advanced automation for motor vehicles. In R. Parasuraman & M. Mouloua (Eds.), Automation and human performance: Theory and applications. (pp. 337-364). Mahwah, NJ: Lawrence Erlbaum Associates.

Hancock, P. A., & Verwey, W. B. (1997). Fatigue, workload and adaptive driver systems. Accident Analysis and Prevention, 29(4), 495-506.

Harris, D. & Smith, F. J. (1997). What can be done versus what should be done: a critical evaluation of the transfer of human engineering solutions between application domains. In D. Harris (Ed.), Engineering Psychology and Cognitive Ergonomics Volume 1: Transportation Systems (pp. 339-346). Aldershot: Ashgate.

Hoedemaeker, M. (1999, November 11). Cruise control reduces traffic jams. De Telegraaf.

Hughes, D. (1995, January 30). Incidents reveal mode confusion. Aviation Week and Space Technology, 56.

Hughes, D., & Dornheim, M. A. (1995, January 30). Accidents direct focus on cockpit automation. Aviation Week and Space Technology, 52-54.

James, M., McClumpha, A., Green, R., Wilson, P., & Belyarin, A. (1991). Pilot attitudes to flight deck automation. Human Factors on Advanced Flight Decks One Day Conference (pp. 3.1-3.8). London: The Royal Aeronautical Society.

Janssen, W., & Nilsson, L. (1993). Behavioural effects of driver support. In A. M. Parkes & S. Franzen (Eds.), Driving future vehicles. (pp. 147-155). London: Taylor & Francis.

Learmount, D. (1994, May 11). Airbus points to pilots in Nagoya crash. Flight International, 5.

Lockett, M. & Spear, R. (1980). Organisations as Systems. Milton Keynes: The Open University Press.

Nilsson, L. (1995). Safety effects of adaptive cruise control in critical traffic situations. Proceedings of the second world congress on intelligent transport systems: Vol. 3 (pp. 1254-1259).

Owens, D. A., Helmers, G., & Sivak, M. (1993). Intelligent Vehicle Highway Systems: a call for user-centred design. Ergonomics, 36(4), 363-369.

Palmer, E. (1995). 'Oops, it didn't arm.' - A case study of two automation surprises. 8th International Symposium on Aviation Psychology. Columbus, Ohio: Ohio State University.

Parasuraman, R. (1987). Human-computer monitoring. Human Factors, 29, 695-706.

Reason, J. (1990). Human Error. Cambridge: Cambridge University Press.

Reason, J. (1997). Managing the Risks of Organisational Accidents. Aldershot: Ashgate.

Sarter, N. B., & Woods, D. D. (1995). How in the world did we ever get into that mode? Mode error and awareness in supervisory control. Human Factors, 37(1), 5-19.

Sedbon, G., & Learmount, D. (1993, December 22). Training 'inadequate' says A320 crash report. Flight International, 11.

Singleton, W. T. (1989). The mind at work: psychological ergonomics. Cambridge: Cambridge University Press.

Stanton, N. A., & Marsden, P. (1996). From fly-by-wire to drive-by-wire: safety implications of automation in vehicles. Safety Science, 24(1), 35-49.

Stanton, N. A., Young, M., & McCaulder, B. (1997). Drive-by-wire: The case of driver workload and reclaiming control with adaptive cruise control. Safety Science, 27(2/3), 149-159.

Verwey, W. B. (1993). How can we prevent overload of the driver? In A. M. Parkes & S. Franzen (Eds.), Driving future vehicles. (pp. 235-244). London: Taylor & Francis.

Walker, G. H., Stanton, N. A. & Young, M. S. (in press). Where is computing driving cars? International Journal of Cognitive Ergonomics.

Wickens, C. D., Gordon, S. E. & Liu, Y. (1998). An introduction to human factors engineering. New York: Longman.

Wiener, E. L., & Curry, R. E. (1980). Flight-deck automation: promises and problems. Ergonomics, 23(10), 995-1011.

Woods, D. D. (1987). Technology Alone is not Enough: reducing the potential for disaster in risky technologies. Human Reliability in Nuclear Power. London: IBC.

Young, M. S., & Stanton, N. A. (1997). Automotive automation: Investigating the impact on drivers' mental workload. International Journal of Cognitive Ergonomics, 1(4), 325-336.

Young, M. S. & Stanton, N. A. (in press). Size matters. The role of attentional capacity in explaining the effects of mental underload on performance. In D. Harris (Ed.), Engineering Psychology and Cognitive Ergonomics Volume 5/6. Aldershot: Ashgate.

AIR-TO-GROUND, GROUND-TO-AIR: WARFARE OR TECHNOLOGY TRANSFER?

Don Harris

Human Factors Group, Cranfield University, Cranfield, Bedford MK43 0AL, UK.

ABSTRACT

This paper presents a conceptual framework (the Five 'M's model) for use in evaluating the likely success of technology transfer between application domains. The model is based upon a sociotechnical system approach to the study of contemporary ergonomics. The model is illustrated with reference to the transfer of interface technologies, initially from the aerospace sector to the automotive sector but now, after further development, subsequently back to the aerospace sector once more.

1 TECHNOLOGY TRANSFER

Technology transfer is becoming an issue of increasing importance. Great efforts are being made concerning the transfer of technology from military applications into those in the civilian arena (*e.g.* the transfer of vision enhancement systems from military aircraft initially into civilian aircraft and subsequently into road vehicles). Similarly, technology that was previously only found in commercial products is now finding its way into consumer products. However, not all transfers of technology are successful and in some instances the unsuccessful or inappropriate transfers of technology may have profound safety implications. There is no method to assess the likely success of a transfer of technology before the process is complete. Furthermore, the human interface with technology tends to be evaluated solely in terms of its usability and functionality, however, when transferring technology from one application domain to another the wider context needs to be assessed. This paper proposes an evaluative framework that explains why technology transfer may (or may not) be safe and/or successful.

1.1 Technology and the transfer of technology

The stereotypical notion of technology is of a 'thing', a complex mechanical or electronic device. However, 'technology' is better characterized as an approach. Technology is the application of scientific principles to solve practical problems. Levin (1996) described technology as having three facets: material artifacts; the use of these artifacts to pursue a goal; and the knowledge to use them.

Transfer of technology will *often* involve the cross-fertilization of engineering principles between application areas but will *always* involve an operator, therefore it will *always* involve ergonomics (Levin, 1996). The human is usually the component that makes or breaks a successful transfer of technology. Shahnavaz (1984) labeled technology that was not ergonomically re-designed to the requirements of its new environment 'transplanted technology' and observed that such transfers were 'doomed to failure'. Noy (1997) suggested governments should have the dual responsibility of developing technologies that enhance safety but discouraging those that compromise safety. This applies not just to new technologies but also their transfer. Many technology transfers occur between safety critical industries. However, it is not just the equipment involved *per se* that compromises safety, it is how it is used on a day-to-day basis. Current ergonomic evaluations are poor at predicting and evaluating this latter aspect. The modern discipline of ergonomics is concerned with the study and optimization of work systems as a whole, not just the human-machine interface. Nevertheless, the prospective evaluation of new products within a wider sociotechnical system context is not usually considered. Abuse and misuse design scenarios, secondary usage scenarios, third party user scenarios and 'worst case' design scenarios (Hasdogan, 1996) are critical in the design of human-machine systems, as there is little evidence to suggest that equipment will always be operated in the manner prescribed. This must change if technology transfers are to be successful. A framework is required to predict the likelihood of success or identify potential pitfalls in the transfer of technology from one application domain to another within this wider context.

1.2 The Five 'M's: a sociotechnical system model of contemporary ergonomics

Ergonomics is no longer just about the optimal integration of user (*huMan*) and *Machine* to perform a task (or *Mission*) within constraints imposed by the physical environment (*Medium*). This approach has been extended to incorporate the societal aspect of the *Medium*, another component of the task environment. Furthermore, in the case of commercial products, *Management* structures also need consideration. This *Five 'M's* system approach, described in Harris & Smith (1997) is an extended and adapted form of the work of Miller (1983). The *huMan* aspect of the Five 'M's Model encompasses such issues as the size, shape, personality, capabilities and training of the end users. Taking a traditional user-centered design approach, the human component of the system is the ultimate design forcing function. The *Machine* component of the model is the system with which operators interact. Taking a user-centered design approach, Ergonomists of the 1970s and early 1980s would simply be concerned with the human-machine interface. With the sociotechnical system approach, though, Ergonomists are now equally as concerned with how a new machine will affect working practices as a whole. Within the Five 'M's Model, a *Mission* is undertaken when the *huMan* and *Machine* components come together to perform a prescribed function. Within commercial spheres the *Management* will task this. It is usually the *Machine* and *Mission* components on which designers fixate. Despite the advances made in socially-centered design in the 1990s (*e.g.* Stanney, Maxey & Salvendy, 1997) the role of ergonomics in product development is still often confined to the user interface. In commercial products, the *Management* has a key role in dictating performance standards, either through the selection and training of personnel or through the standards it imposes for the technical performance of equipment. However, these requirements are secondary to those imposed by the physical or societal aspects of the *Medium*. *Management* must work within these physical constraints and also ensure their operations are conducted within the formal frameworks demanded by society (*e.g.* the legislative framework; the level of safety required, the minimum standard of user competence, or recently the product's environmental impact) and also the less formally specified rules and norms of society. The inter-relationships between the Five 'M's can be described with reference to a simple diagram (figure 1).

The Five 'M's model can be used to provide predictions about the likely success of technology transfer from one domain to another. Put quite simply, the more characteristics described within the Five 'M's the donor and recipient application areas have in common, the more likely the technology will transfer successfully. The idea that the more aspects the donor and recipient applications have in common, the higher the likelihood of success is not new (*e.g.* Cardwell, 1994). However, until now, a framework for evaluating the likely success of technology transfer has not been proposed.

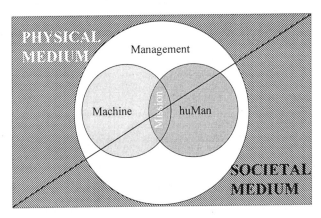

Figure 1   Representation of the inter-relationship between the meta-components of the Five 'Ms' model.

2   CASE STUDY – VISION ENHANCEMENT SYSTEMS

Night vision systems were originally developed by the military for use in performing night attacks from low-flying aircraft. A variety of techniques have been used, including image intensifiers and the use of cameras operating in either the near- or far-infrared part of the spectrum. As this technology matured and reduced in size, weight and cost it rapidly found its way into a variety of other applications, such as tanks and night vision scopes for rifles. More recently, this technology has found its way into motor vehicles. General Motors has developed a passive night vision system operating in the far-infrared that projects a non-collimated image onto a small area at the bottom of the windshield. This is being offered as an optional extra in its Cadillac brand of cars. This system has between three-to-five times the range of low-beam headlights and double the range of high-beam headlights. At 60 mph headlights provide the driver with about 3.5 seconds of vision of the road ahead. With this passive, far-infrared based system the driver will have up to 15 seconds to react on a straight road. Furthermore, the system helps drivers see beyond the glare from the headlights of oncoming vehicles. Some European manufacturers are taking a slightly different approach, using an active, near-infrared source to illuminate the road ahead and them projecting a collimated image onto a head-up display (HUD) immediately in front of the driver's eyes. These systems are likely to be introduced onto the market in the next few years. However, the key point is that with a significant reduction in the cost of the technology *per* se and the further savings possible through mass production these night vision systems will be available to drivers at a price premium of hundreds, rather than tens of thousand, of dollars. Furthermore, with this massive decrease in cost over the years it is now possible to re-introduce the technology back into the aerospace sector but this time into the General Aviation (GA) sector. The question that this possibility raises, though, is concerned with if this would be desirable.

GA is the highest risk of all the aviation activities. In 1997 in the USA the accident rate for GA aircraft was approximately 7.05 per 100,000 hours compared to the corresponding rate for commercial aircraft, which was 0.291 per 100,000 hours (NTSB, 2000). Of the 1,848 accidents that occurred 11.4% occurred in twilight or at night. While this may seem to be a relatively small percentage, it should be borne in mind that the vast majority of GA activity occurs during the day. Considerably less than 5% of GA flying is undertaken at night suggesting that the accident rate at least doubles during the hours of darkness. Even taxiing an aircraft at night, especially on an unfamiliar airfield is a relatively hazardous undertaking. Providing the pilots with better night vision will reduce this risk, helping avoid collisions between taxiing aircraft, collisions with ground vehicles and reducing the likelihood of runway incursions. These low cost night vision technologies developed for road cars offer a potential safety benefit for the GA pilot. The technology offers the ability to display onto a HUD, with near-daylight clarity, a view of the surrounding environment, for example a perspective view of the runway on approach and landing. It will also be possible to display information such as the optimum approach path onto the HUD in real time. Situation awareness may be further enhanced through the provision of altitude, attitude and airspeed data and also when combined with data from a global positioning system receiver, navigational data. However, all these potential benefits need to be evaluated within the larger sociotechnical system context before their true net safety benefit can be evaluated.

2.1 Technology transfer: Air-to-ground

Firstly, it is worth considering the transfer of night vision technology from its original military applications to that of the road environment. There are some parallels between the physical *Medium* aspects of the military pilot's and the car driver's task. Driving is a 'head-up and eyeballs-out' task, somewhat akin to flying a fast military jet at 200 feet through mountainous terrain. A car driver on a busy multi-carriageway road may be a second away from the vehicle in front and less than one-half of a second away from a vehicle to the side. A pilot may be a similar time gap away from the sides of the valley. Both the pilot and the driver need to spot unexpected events in their surrounding environment as quickly as possible, be they pedestrians or the Enemy. In both cases the system operator is actively in the control and monitoring loop. However, in terms of the other 'M's, this is about as far as the similarities between the two applications of night vision technology goes.

In terms of the *huMan* aspect of the system the contrasts between the car driver and the military pilot are quite stark. The fast jet pilot undergoes a rigorous selection and training process over an extended period of months and years. Very few candidates who start the process of becoming a fast jet pilot actually achieve their ultimate aim. The pilots are qualified on each particular type of aircraft and specific training is given for each type of mission and for each individual aircraft system. Training is recurrent and pilots are required to demonstrate their competency every few months. Once a car driver has obtained a license, subject to making neither major nor frequent transgressions of the rules of the road, they essentially have a license for life. Drivers are permitted to drive any motor vehicle that falls within the broad categories covered by their license, anywhere, at any time of the day and in any conditions. Pilots are subject to much tighter regulation. For example, in civil aviation specific ratings are required to fly at night, in poor visibility or in certain categories of airspace. The legislation framed by the aviation authorities (an aspect of the societal *Medium*) dictates not just the person requirements, but also the 'fit' between the person and the *Machine* to perform an assigned *Mission*. Furthermore, the *Missions* undertaken by military pilots and car drivers are quite different. The pilot is undertaking a high jeopardy risk *Mission* that is not optional. Most journeys undertaken by car drivers do not involve such risk and are mostly optional! In terms of the *Machine* component, aircraft are excellent examples of commercial products whereas cars are the archetypal consumer product. As Cushman & Rosenberg (1991) note, the human engineering requirements for these two categories of product are very different.

The crux of the argument is that while there may apparently be a good reason for installing night vision systems into motor vehicles when the physical *Medium* alone is considered, further examination of the other aspects of the wider system suggest that this may not be such a good idea. ISO design standards (an aspect of the societal *Medium*) are being developed for in-vehicle systems but these standards are advisory rather than mandatory. Furthermore, they can only specify the design of the system (the *Machine*). They cannot regulate for its appropriate usage, which is very much a product of the *huMan* component (Hasdogan, 1996). Mobile telephones should not be used in a moving motor vehicle; cruise control systems should not be used in heavy traffic. Are night vision systems in motor vehicles another technology that may be either used inappropriately by drivers through a lack of knowledge about the system's capability and operation, or perhaps deliberately used improperly? There is already evidence to suggest that drivers drive faster in restricted visibility when using such systems (Stanton & Pinto, 2000). It could be argued that while the driver of a vehicle equipped with a night vision system is driving at the same or a lower level of risk, the level of risk for all road users has actually increased (see Wilde, 1982).

There is also the issue of the role of *Management*. *Management*, in a complex, safety critical system, plays a central role in maintaining its safety. It is also the link between the standard of performance required from the *huMan* and the *Machine* to perform the *Mission* and the rules and norms of society (the societal *Medium*). *Management* ensures that systems are maintained to the standards required and the *huMan* and the *Machine* are fit for service on a day-to-day basis. This role is entirely missing in the privately owned motor vehicle. Furthermore, unless specifically prohibited by the national design, construction and use regulations for motor vehicles, it is permissible to fit anything in a car. Contrast this with a commercial or military aircraft where the manufacturer or operator may be required to demonstrate that any modification does not compromise flight safety.

2.2 Technology transfer: Ground-to-air

Many of the same concerns discussed previously are also applicable when technology is being transferred in the opposite direction, as in the case of motor vehicle night vision technology being transferred back to the air into a GA aircraft. Using Cushman & Rosenberg's (1991) product classification scheme, GA aircraft can best be classified as consumer products. As has been alluded to previously, the human engineering requirements for consumer products are very different to those required by commercial products. Unlike a professional pilot but in some ways similar to a car driver, the GA pilot requires no specific type rating to fly a single-engined light aircraft. As long as s/he also holds a night rating and the aircraft is suitably equipped, that pilot can undertake a flight at night. Under current aviation regulations there would be no training, licensing or type rating requirement for a GA pilot to operate an aircraft with a HUD-mounted night vision system. Indeed there would not even be a formal requirement for them to even read the manual! In terms of the Five 'M's model, the *huMan/Machine/Mission* 'fit' is not specified in any way by the regulations (societal *Medium*) and similar to the case of the motor vehicle, the *Management* part of the system is missing. Furthermore, from the point of view of the airworthiness authorities, the certification criteria for a night vision system implemented on a HUD in a GA aircraft are unclear. Indeed, it is unlikely that any of the current certification criteria will be applicable, with the exception of those associated with crashworthiness.

3   LESSONS LEARNED?

Any potential disbenefits accruing from the use a HUD-based night vision system in a GA aircraft need to be identified and assessed *before* the system is finally cleared for use. These must be assessed in a wider, socio-technical system context. It would be unwise to implement the complex symbology from a military or commercial aircraft onto the HUD of a GA aircraft. A new symbol set must be developed which is immediately intuitive. It is required to be operated by the critical end user (the low-hours, night rated private pilot) with no training but it must not impose either additional workload nor induce error. At the same time, the symbology should enhance the accuracy with which the pilot is able to conduct all aspects of the flight (be it an approach to landing or a cross-country navigational exercise) and increase his/her awareness of the flight situation. However, before undertaking detailed human-centered design, the wider implications of the introduction of such systems into GA aircraft should be considered.

The arguments in favor of HUDs for the display of information (as an adjunct to their night vision function) center around the assumption that they allow the parallel monitoring of the HUD symbology and the external environment. However, on occasions this has been demonstrated not to be true. Paradoxically, the design feature claimed to confer the major benefits of a HUD (that the information display and real world spatially overlap) can lead to interference between these sources of information. Information from one source captures the pilot's attention thereby stopping information from the other source reaching the pilot's conscious awareness. Furthermore, the use of conformal symbology on the HUD (symbology that bears a direct relationship with some feature of the external environment) is also likely to obscure objects in the environment (Wickens & Long, 1995). Attention may sometimes be commandeered by this object at the expense of other items, either on the HUD or in the real world. This includes objects at the same location, such as a runway obstruction, as was demonstrated by Martin-Emerson & Wickens (1997). Alternatively, the pilot's attention may be captured by a single element on the display that is being used for the task at hand ('coning' of attention). Although a different attentional mechanism is at work here, the result is the same. The HUD symbology captures the pilot's attention. Furthermore, attentional capture is most likely when workload is high, for example, during approach and landing. One potential advantage of the HUD is that it increases the possibility that changes in the environment will be noticed when compared to the use of head-down displays. This is true if the pilot is monitoring for an expected event, however this has been found not to be the case if the event is unexpected.

As Stuart, McAnally & Meehan (2001) note 'More ominously, pilots in general aviation are unlikely to receive specific training in the use of HUDs, and therefore may be more susceptible to attentional capture.' It is essential that a night vision system enhances the safety of *all* users of the airspace, not just the safety of the pilot of the aircraft so equipped. The same applies to road vehicles: the technology must promote the safety of *all* road users. The results described above throw some doubt about the suitability of night vision systems for road vehicles and by inference, also suggest that the same problems may be evident when transferring the technology back to the GA sector. It is possible that without a system-wide evaluation of the potential effects of the transfer of this technology (in either direction) a new safety problem may be introduced into the air or onto the roads rather than achieving a net safety benefit.

REFERENCES

Cardwell, D. (1994) *Fontana history of technology.* London: Fontana Press.
Cushman, W.H. & Rosenberg, D. (1991). *Human Factors in Product Design.* Amsterdam: Elsevier Science publishers B.V.
Harris, D. & Smith, F.J. (1997). What can be done versus what should be done: a critical evaluation of the transfer of human engineering solutions between application domains. In D. Harris (Ed.) *Engineering Psychology and Cognitive Ergonomics (Volume 1: Transportation Systems)* pp. 339-346. Aldershot: Ashgate.
Hasdogan, G. (1996). The role of user models in product design for assessment of user needs. *Design Studies 17,* 19-34.
Levin, M. (1996). Technology transfer in organisational development: an investigation into the relationship between technology transfer and organisational change. *International Journal of Technology Management, 2,* 297-308.
Martin-Emerson, R. and Wickens, C.D. (1997). Superimposition, symbology, visual attention and the head-up display. *Human Factors, 39,* 581-601.
Miller, C.O. (1983) A comparison of military and civil aviation safety. *Proceedings of the Airline Pilots Association Symposium.* Washington, D.C.: ALPA.
National Transportation Safety Board (2000). *Annual Review of Aircraft Accident Data – US General Aviation Calendar Year 1997.* Report NTSB/ARG-00/01. Washington D.C.: NTSB.
Noy, Y.I. (1997). *Ergonomics and safety of intelligent driver interfaces.* Mahwah: Lawrence Erlbaum Associates.
Shahnavaz, H. (1984). The importance of ergonomics in developing countries. In, H. Shahnavaz & H. Babri (Eds), *Ergonomics in developing Countries* (pp. 4-27). Luleå: Luleå University Press.
Stanney, K.M. Maxey, J. & Salvendy, G. (1997). Socially Centered Design. In, G. Salvendy (Ed.) *Handbook of Human Factors and Ergonomics (2nd Edition)* pp. 637-656. New York: John Wiley.
Stanton N.A. & Pinto, M. (2000). Behavioural compensation by drivers of a simulator when using a vision enhancement system. *Ergonomics, 43,* 1359-1370.
Stuart, G.W., McAnally, K.I. and Meehan, J.W. (2001). Head-up displays and visual attention: integrating data and theory. *Human Factors and Aerospace Safety, 1,* 103-124.
Wilde, G.J.S. (1982). The theory of risk homeostasis: implications for safety and health. *Risk Analysis, 2,* 209-225.
Wickens, C.D. and Long, J. (1995). Object vs space-based models of visual attention: Implications for the design of head-up displays. *Journal of Experimental Psychology: Applied, 1,* 179-193.

From Fly-By-Wire to Drive-By-Wire

Neville A. Stanton and Mark S. Young
Department of Design, Brunel University
Egham, Surrey, TW20 0JZ

Whether we like it or not, automation is gradually taking over the driver's role. Full vehicle automation is predicted to be on British roads by 2030 (Walker, Stanton & Young, 2000). Whilst it is accepted that some drivers will still want to control their vehicles manually, many may be pleased to relinquish the role to automatic systems. Many of the computing technologies have been grounded in aviation systems (Stanton & Marsden, 1996), and technologies like adaptive cruise control are taking over from the driver already (Richardson, Barber, King, Hoare & Cooper, 1997). Whilst at present the technologies are working independently of each other, full integration of these systems will make an autonomous vehicle a commercial reality. Whilst the engineering of these vehicle's seems on track, the understanding of the interaction between the computing, the vehicles behaviour and the drivers reactions seem much less clear. Basic Human Factors research into the driver's behaviour when supervising an autonomous system is required. Whilst there is a tradition of research in human supervisory control in static technologies spanning some two decades (Reason, 1990; Hollnagel, 1993; Stanton, 1996), much less is known in transportation systems. What is known from automation in aviation might give some cause for concern (Billings, 1993; Stanton & Marsden, 1996). The workload benefits of autopilot systems have been welcomed, but incidents associated with the reduction in the pilot's situational awareness have led to a more cautious acceptance (Beaty, 1995).

1. Automation in Aviation

Stanton & Marsden (1996) noted that automation has been implicated in a number of fatal aviation accidents. Root cause analysis by accident investigators have identified psychological factors such as boredom and inattention under conditions of low workload, cognitive strain under conditions of very high workload, failure of automated systems to meet pilots' expectations, and over-reliance on the technology. There is no reason to believe that these factors are only in the domain of airborne automation and, if poorly designed, they could transfer to automation in ground transportation.

In aviation, two different automation philosophies have emerged: hard automation (the Airbus philosophy) and soft automation (the Boeing philosophy). Airbus believe that the automation technology exists to prevent the pilot from inadvertently exceeding safety limits. *Airbus fly-by-wire aircraft have hard speed envelope protection features that will prevent the pilot from stalling the aircraft in most circumstances and from pulling more than 2.5g even in an emergency* (Hughes & Dorheim, 1995). Boeing, on the other hand, believe that automation is a tool to serve the pilots, and the pilots have full authority to override the automated systems if they want to. The pilots have access to the full performance envelope and will not be overridden by the automated systems. *A soft protection scheme requires the pilot to apply more force on the yoke when exceeding 3.5 degree of bank and when pulling the yoke back as the aircraft decellerates below the minimum maneuver speed* (Hughes & Dorheim, 1995). Naturally, both Airbus and Boeing think that their scheme is best (Hughes & Dorheim, 1995).

2. Four Problems

In anticipating the degree of vehicle automation that might become standard issue in 2030, it is wise to consider other transportation domains where automation dominates. Stanton & Marsden (1996) identified 4 principal negative outcomes in their discussion of lessons learnt from fly-by-wire technologies. These were: shortfalls in expected benefits, problems with equipment reliability, problems with skills maintenance and error inducing designs. Given, as we suggested at the beginning of the paper, many of the concepts for vehicle automation owe much to automation in aviation, much could be learned from the experiences of automation in aviation.

Shortfalls in expected benefits

A major problem associated with automated aids arises when the system in question fails to deliver the expected benefits. Performance shortfalls can take a number of forms. For example, one common problems is that automated systems are frequently less reliable than anticipated when introduced into the operational arena. They can also sometimes prove more costly to operate than originally envisaged by the design teams. In yet other situations, automation can have detrimental effects on human performance due to increases (or reductions) in the amounts of information which must be monitored and processed by the user.

To pursue this latter theme one stage further, there is now good evidence available to suggest that automation in aviation has occurred quite rapidly in areas of work where pilot workload demands are already quite low, for example, routine in-flight operations. Automation here has led to increased boredom of flight crews. Conversely, the allocation to automation in areas with inherently high pilot work rates, for example, take-off and landing, can contribute greatly to cognitive strain and team stress due to the need to process ever increasing amounts of information (Billings, 1993; Weiner, 1985; 1989). Indeed, there are several well documented case histories in which automation induced cognitive stress contributed to the occurrence of a serious accident.

Inattention to flight instruments was cited as a probable cause of an accident involving an Eastern Air Lines L-1011 at Miami, Florida on the 29th December, 1972. The crash was thought to have occurred following an accidental autopilot disconnect which went undetected for a considerable amount of time. The crew also failed to notice an unexpected descent in sufficient time to prevent impact with the ground in the Florida Everglades. The accident report noted that the three crew members plus an additional jumpseat occupant were preoccupied with the diagnosis of a minor aircraft malfunction at the time the accident occurred.

Cognitive strain was identified as a factor in an accident which occurred at Boston's Logan Airport in 1973. In this incident, a Delta Air Lines DC 31 struck the seawall bounding the runway killing all 89 persons on board. The cockpit voice recorder indicated that the crew had been experiencing difficulty with the Sperry Flight Director while attempting an unstabilised approach in rapidly changing meteorological conditions. The accident report concluded that the accumulation of minor discrepancies deteriorated in the absence of positive flight management in a relatively high risk manoeuvre. Specifically, the crew were preoccupied with the information being presented by the flight director to the detriment of paying attention to altitude, heading and airspeed control.

A less dramatic example of an instance where automation has failed to meet prior expectation can be illustrated with reference to the case of Ground Proximity Warning Systems (GPWS) which produced a high level of spurious alarms when first introduced into the cockpit environment. The experience of GWPS is similar to many other instances where the implementation of first-generation automation has had detrimental effects on the performance of flight crew due to problems inherent in the prototype design, for example, Traffic Collision Avoidance Systems such as TCAS-II (cf, Billings, 1993).

Equipment Reliability
The question of equipment reliability is clearly an important consideration in the automobile automation debate. Equipment reliability appears to significantly affect human performance in a number of circumstances.

Perhaps the most obvious way that the reliability of automation might effect the quality of human performance arises when the automated system in question consistently malfunctions. In this case, one would expect that the user would, over time, lose confidence in the device to the extent that they prefer to operate in the manual mode wherever possible. Such a hypothesis is supported in a number of scientific publications , conference proceedings and incident reports (see for example, Weiner and Curry, 1980) and need not be considered further in detail here.

Similarly, loss of faith in the reliability of automated aids will occur where devices are prone to faults of an intermittent nature. Intermittent failures in automated aids are potentially more serious for human cognition because they can frequently go undetected for long periods of time only to manifest themselves at a critical phase of operation. Witness the case of Delta flight 1141 which crashed shortly after take-off from Dallas Fort Worth Airport in 1988. The accident was attributed in part to an intermittent fault in the aircraft's take-off warning system which should have alerted the flight crew to the fact that the aircraft was wrongly configured for the operation being performed (National Transport Safety Board, 1989).

Perhaps the most surprising way in which the reliability of automation can cause serious problems for users comes not from system deficiencies, but rather from equipment which has a well proven reliability record accumulated over many years of operation. In this situation, flight crews often come to over depend on automated aids when they are operating in conditions beyond the limits of their designs. Billings (1993) has discussed this aspect of automation at length and suggested that there are many examples where:

> "...automated systems, originally installed as backup devices have become de facto primary alerting devices after periods of dependable service. These devices were originally prescribed as a "second line of defence" to warn pilots when they had missed a procedure or checklist item. Altitude warning devices and configuration warning devices are prime examples"

Over-reliance of technology was a factor in an incident involving a China Airlines B747-SP which occurred 300 miles north-west of San Francisco in February, 1989. Towards the end of an uneventful flight, the aircraft suffered an in-flight disturbance at 41,000 feet following loss of power to its Number 4 engine. The aircraft, which was flying on autopilot at the time, rolled to the right during attempts by the crew to relight the engine, following which, it subsequently entered into an uncontrolled descent. The crew was unable to restore stable flight until the aircraft had descended to an altitude of 9,500 feet, by which time it had exceeded its maximum operating speed and had sustained considerable damage. In conducting its enquiry, the NSTB concluded that a major feature of this incident was the crew's over dependence on the autopilot during the attempt to relight the malfunctioning engine, and that the automated device had effectively masked the onset of the loss of control of the aircraft.

A similar conclusion was obtained for another incident which in this case involved a Scandinavian Airline DC-10-30. In this incident, the aircraft overshot the runway at JFK Airport, New York by some 4700 feet. The pilot was, however, able to bring the plane to a halt in water some 600 feet beyond the runway's end. A few passengers sustained minor injuries during the evacuation of the aircraft. The enquiry noted that again the crew had placed too much reliance upon the Autothrottle Speed Control System while attempting to land. It was also noted that use of the autothrottle system was not a mandatory requirement for a landing of the type being performed.

Training and Skills Maintenance
A third way in which automation has been found to have detrimental effects on the quality of human operator performance concerns the knock-on effects which automated aids can have on the knowledge and skills of an individual. The tendency for humans to rapidly lose task related knowledge and skills in partially automated environments is a well documented psychological phenomenon. In the aviation domain the accident involving the collision between two B747's at Tenerife appears to be particularly relevant. In this accident, a highly experienced KLM Training Officer with considerable operational experience, failed to ensure that adequate runway clearance had been given prior to commencing take-off. The findings of the Spanish Commission set up to investigate this incident part attributed causality to the fact that the KLM pilot had insufficient recent experience of route flying with the 747.

While the problems of deskilling are well known, much less understood are the strategies whereby the knowledge and skills possessed by an individual can be developed or maintained such that they can regain control of the system in the event of a malfunction. Barley (1990) has suggested that flight crews often have to deal with the problem of skill maintenance by periodically disengaging the automated systems to refresh their flying skills and/or relieve the boredom of a long-haul operation. One would expect, however, that more effective methods of refresher training could be implemented to ensure the retention and development of automated tasks which rely on human intervention following failure of the technology.

Despite the assumption that skills can be developed through standard proficiency training programmes, there are many examples which can be taken from accident and near-miss reports, which indicate that the human in an automated environment only rarely receives adequate training and exposure to manual task performance. The poor quality of Air Traffic Controller training, for example, was cited as an important factor in two aircraft separation incidents which were investigated at Atlanta Hartsfield Airport on the 10th July, 1980. The investigators concluded that the collisions were the result of inept traffic handling on the part of controllers, and that the ineptitude was due in part to the inadequacies in training, procedural deficiencies, and the poor design of the physical layout of the control room.

Similar criticisms have been made in relation to the standards of preliminary and refresher training received by flight crews, and more than one accident has been attributed in part to mistaken actions made by trainee officers flying unfamiliar aircraft. An example, here is provided by the case of the Indian Airlines A320 (a reduction from an aircrew of 3 to 2 persons accompanied the introduction of automation into the airbus) which crashed short of the runway at Bangalore on February 2nd, 1990 killing 94 of the 146 persons on board. In this incident the primary cause was attributed to the failure of the trainee pilot to disengage the flight director which was operating in an incorrect mode, and the failure of the crew to be alert to the problem in sufficient time to prevent the accident. All members of the flight crew were killed in the accident which may, in retrospect, have been prevented by more effective training in the use of fly-by-wire technology.

Error Inducing Equipment Designs
It has already been suggested that many prototype automated systems are introduced with inherent design flaws which can compromise the effectiveness of the human-machine combination. In the majority of cases, residual design faults are rapidly identified in the operational arena and rectified in second generation technology. In some cases, however, identification of system shortcomings leads not to redesign, but rather to an engineering fix in which a system is, to a greater or lesser extent, patched up.

In their account of cockpit automation, for example, Boehm-Davies *et al* (1983) discuss a case in which a proposal to rectify problems inherent within air traffic control-flight crew voice transmissions by means of a CRT cockpit data link would increase the propensity of the flight crew to make reading errors, rather than the errors of hearing which appeared to be occurring at that time. Furthermore, they suggested that the adoption of such methods of communication would have the effect of depriving flight crews of important information regarding the location of other aircraft within the vicinity. One possible consequence of such a transition could be an increase in the number of air traffic separation incidents.

The experience with Inertial Navigation Systems (INS) would seem to offer a more concrete example of automation with a propensity to induce (or indeed amplify) pilot errors. In this system, developed for flight management purposes, pilots are required to enter way-point co-ordinates by means of a computer console. Incorrect data entry can have catastrophic consequences. It is now widely believed that the aberrant flight of the Korean Air Lines B-747, which was destroyed by air-to-air missiles over Soviet airspace in 1983, was due to the incorrect entry of one or two waypoints into the INS prior to its departure from Anchorage. In less dramatic fashion, the near collision over the Atlantic between a Delta Air Lines L-1011 and a Continental Airlines B-747, was also attributed to incorrect waypoint entry in this case in the Delta aircraft. At the time of the incident the L-1011 had strayed some 60 miles away from its assigned oceanic route.

3. Automation in Automobiles

In learning the lessons from automation in aviation we may anticipate at least 4 potential types of problem for automation in automobiles: (i) shortfalls in expected benefits, (ii) problems with equipment reliability, (iii) training and skills maintenance, and (iv) error inducing equipment designs. Conclusions for these are drawn.

(i) Automatic systems seem to have shortfalls in expected benefits when introduced into the operational arena. In terms of vehicular automation, this could mean that they turn out to be less reliable (e.g. the collision avoidance system fails to detect approaching object), more costly (e.g. the automated systems add substantially to the purchase price of the vehicle) and have an adverse impact upon human performance (e.g. automation seems to make the easy tasks boring and the difficult tasks even more difficult).

(ii) Automatic systems can have problems related to equipment reliability. In terms of vehicular automation, this could mean that drivers lose their trust in the automated systems (e.g. the driver prefers to choose the manual alternative), intermittent faults could go undetected until the context becomes critical (e.g. the failure reveals itself immediately prior to the vehicle impacting at high speed into another vehicle) and the driver becomes so dependent upon the automated systems that they operate them beyond design limits (e.g. invoking Automatic Intelligent Cruise Control in non-motorway situations).

(iii) Automatic systems seem to lead to problems related to training and skills maintenance. In terms of vehicular automation, this could mean that driving skills could be stripped away through lack of practice by automation being in control. This is likely to make the driver even more dependent upon the automated systems. If drivers are not performing a function, how can they be expected to take it over adequately when the automated systems fail to cope?

(iv) Finally, automatic systems seem to induce errors in users. In terms of vehicular automation, this could mean that design flaws lead to driver errors when interacting with the automated systems, for example specifying the wrong target speed and distance with Automatic Intelligent Cruise Control. Of particular concern is the possible introduction of mode errors (i.e. the driver believes the system to be in one mode when it is actually in another). Mode errors are most likely when controls have more than one function and the mode the system is not transparent.

From our analysis of automation in the context of aviation, we see the need for caution in the pursuit of automation of driver functions. This need for caution is also voiced by pilots in their own domain in discussing the A320 Airbus, as the following quote indicates:

"I love this aeroplane, I love the power and the wing, and I love this stuff [pointing to the high-technology control panels] *but I've never been so busy in my life...and someday it* [automation] *is going to bite me"* (Anon, 1991)

This observation makes two problems with automation very clear: the problem of increased workload and the anticipated problem of lack of co-ordination. We propose that allocation of function needs to explicitly examine co-ordination and co-operation between human and automated sub-systems if the problems of automation in aviation are not to be replicated through automation in automobiles. Automation can have beneficial effects upon system performance, but automation for its own sake can have unforeseen risks which can only be determined by a structured evaluation.

References

Barley, S. (1990) The Final Call: Air Disasters...When will they ever learn?. London: Sinclair-Stevenson.

Beaty, D. (1995) The Naked Pilot. Airlife: Shewsbury.

Billings, C. (1993) Aviation Automation. Lawrence Erlbaum: New Jersey.

Boehm-Davies, D.A.; Curry, R.E.; Wiener, E.L. & Harrison, R.L., (1983) Human factors of flight deck automation. Ergonomics. 26: 953-961.

Hollnagel, E. (1993) Context and Control. Academic Press: New York.

Hughes, D. & Dorheim, M. A. (1995) Accidents direct focus on cockpit automation. Aviation Week & Space Technology, January 30, 52-54.

Reason, J. 1990) Human Error. Cambridge University Press: Cambridge.

Richardson, M.; Barber, P.; King, P.; Hoare, E. & Cooper, D. (1997) Longitudinal driver support systems. Proceedings of AutoTech I.Mech.E.: London, 87-97.

Stanton, N. A. (1996) Human Factors in Nuclear Safety. Taylor & Francis: London.

Stanton, N. A. & Marsden, P. (1996) From fly-by-wire to drive-by-wire: safety implications of automation in vehicles. Safety Science, 24, (1) 35-49.

Stanton, N. A. & Young, M. (1998) Vehicle automation and driving performance. Ergonomics. 41 (7), 1014-1028.

Walker, G.; Stanton, N. A. & Young, M. S. (2000) Where is technology driving cars? A technology trajectory of computing in vehicles. Paper submitted to the International Journal of Human-Computer Interaction.

Weiner, E.L., (1985) Cockpit automation: In need of a philosophy. SAE Technical Report, 851956.

Weiner, E.L., (1989) Human factors of advanced ("glass cockpit") transport aircraft. NASA CR-177528, Coral Gables, Florida: University of Miami.

Weiner, E.L., and Curry, R.F., (1980) Flight deck automation: Promises and problems. NASA TM-81206. Moffet Field: CA.

USING AEROSPACE TECHNOLOGY TO IMPROVE OBSTACLE DETECTION UNDER ADVERSE ENVIRONMENTAL CONDITIONS FOR CAR DRIVERS

A. Amditis[1], L. Andreone[2], E. Bekiaris[1]

1. Institute of Communication and Computer Systems (ICCS) / National Technical University of Athens
9, Iroon Politechniou str., 15773 Zografou , Athens, Greece
angelos@esd.ece.ntua.gr

2. Fiat Research Centre, Strada Torino 50, 10043 Orbassano (TO), Italy
l.andreone@crf.it

Abstract

Microwave radar and far infrared sensor technologies have been used in aerospace applications for decades but both their technical limitations and high prices have prohibited their use in the automotive field. Instead, in this field low-cost and performance sensors were used. Thus, the typical automotive radar sensor is limited to the detection of moving targets and cannot recognise stationary obstacles. The relevant vision enhancement systems on the other hand, had poor behaviour in low visibility cases. Road safety and efficiency have requested the introduction of such improved environmental knowledge acquisition and leaded to the development of the first generation of ADAS (Advanced Driver Assistance Systems). As most fatal accidents occur at night or are influenced by rain, ice and fog, an improved obstacle detection system over the next 200 meters, has been estimated to lead to an accident reduction rate of up to 25% and a significant reduction in traffic congestions due to poor visibility conditions. It seems that the latest developments in the aerospace field regarding such sensors open now a window for their automotive application. Low cost microwave radars have emerged at the 77 GHz frequency, able to be seen 200 m in front, whereas far infrared sensor cost has been reduced by up to 30%, substituting rare materials used before (i.e. Germanium) by new materials (i.e. TEX glass) of lower cost and using more compact and less costly microbolometer sensors. In the described application, two such sensors (a microwave radar and a far infrared) of aerospace origin are being redesigned and integrated to meet the relevant automotive application requirements (in terms of range, volume, accuracy, cost, etc.). Even most important, emphasis is given on designing a completely new HMI, so that any car driver can use it. Indeed this constitutes an equal challenge as the technical one, since the airplane pilot is trained to translate and understand the complex sensors output, while the car driver is not. To make things worse, reaction margins of car drivers are usually smaller than airplane Pilots. For this application, an innovative HMI is designed and tested, in order to present the two sensors fused output in an understandable way to the driver, without requesting that he/she will take his/her eyes off the road. A number of image presentation techniques and use of obstacle symbols/metaphors will be comparatively tested in laboratory by 3 car manufacturers and several research institutes Europe-wide. The results of this work are expected to bring a new dimension not only in automotive obstacle detection systems technology but also to their HMI concept and relevant user's acceptance.

1 Introduction

Microwave radar and far infrared sensor technologies have been used in aerospace applications for decades but both their technical limitations and high prices have prohibited their use in the automotive field. Instead, in this field low-cost and performance sensors were used. Such systems were firstly introduced for defence applications allowing pilots to have a clear view of the traffic and tactical scenario in front them even in large distances and in adverse weather conditions. Soon similar systems

1425

were introduced also to the commercial aircrafts for a number of tasks improving significantly safety and efficiency. Such applications include:

- Weather radars – Microwave radars are used for many years now for detecting weather phenomena in front of the aircraft allowing the pilot to have a clear image of the situation in front of him.
- Traffic and obstacles detection systems. Radars are also used of course for detecting the traffic around the aircraft in order to avoid possible collisions between aircrafts. In the recent years IR sensors are also used for similar purposes covering mostly small distances, night and adverse weather conditions cases. In these cases IR sensors work better that the microwave radars.
- Safe approach, landing and taxi of an aircraft in near-zero visibility. In attempting to create a system that will assist the pilots on those tasks, an accurate "picture" of the outside world is required. This include the detection of terrain and obstacles For many years military aircraft have used Forward Looking Infra Red (FLIR) sensors superimposed on a Head Up Display (HUD) (or Head Mounted Display (HMD)) to create what is known as an Enhanced Vision System (EVS). This type of design can greatly increase the view outside the aircraft when flying at night and in poor visibility. Unfortunately, the FLIR is susceptible to certain densities of fog that render the sensor inoperative. To overcome this problem, the use of Millimetre Wave Radar (MMW) has been considered. The result being that the two types of sensor are complimentary to each other and so, the current state-of-the-art systems are attempting to merge the two different sets of data for common display to the pilot

In all the above cases the two sensors were handled separately in terms of HMI and fusion of their data and findings. Only during the last few years engineers start to work seriously in merging the functions of these two sensors starting from the defence applications.

One of the interesting projects having to do with the possible merging of radar and infrared sensor data is currently undertaken for aircraft application by the Braunschweiger Institut fuer Flugmachanik of DLR to develop a system to ensure a safe landing in reduced visibility conditions caused by fog, rain and snow. According to the forecasts at least five years will be needed until this system will be in use (1).

Since the mid 80's several efforts have been undertaken to transfer these technologies also to the automotive applications field, in order to develop anti-collision driver support systems, able to provide an effective support to vehicle drivers, especially in cases of reduced visibility, Indeed, around 37% of road accidents with injury (2,3,4,5,6,7) occur in conditions of limited visibility, like darkness and fog. Moreover, accident statistics at EU level (2) show that numbers and causes of vision-related accidents are comparable in the different EU countries. The worse environmental conditions in the North are compensated by the better ability of local drivers to cope with them; thus, leading to a problem of truly pan-European dimension calling for common technological action within Europe and beyond.

2 The Performance Levels and Problems of Stand-Alone Sensors

Currently far infrared and radar sensors have been separately employed in various prototypes, to support the driver in different reduced visibility conditions. However, mature as they may be in technological terms, they have not been successful yet to present the complete traffic picture to the driver, especially under particularly adverse environmental conditions.

The current lack of success in the technology transfer from the aerospace field to the automotive one is due to the very different requirements of the automotive applications, in terms of scenario complexity, visibility range and human machine interaction requirements (car drivers are far less trained and skilled in the use of information devices than airplane pilots).

A number of different technological approaches that have been used so far and their limitations and shortcomings are summarised below.

- Ultrasound techniques, useful at very short distances since the technology is strongly dependent on propagation medium variations.
- CCD / CMOS sensors and image processing devices, active in the visibility range, but not offering a sensible benefit in any reduced visibility condition.
- Near infrared sensors (laser radar, infrared sensors) which do not offer a sensible benefit in fog.
- Far infrared sensors (sensitive in the range of 8-14 µm), providing thermal images of the scenario independently from any light conditions. The enhanced visibility in fog and heavy rain condition is dependent on droplet dimension.
- Microwave range sensors (radar), operating properly in any poor visibility condition, providing the detection of objects in the road ahead. Information from stationary objects require powerful signal processing to be extracted; their functionality is limited in complex scenario like urban traffic.

If one focuses on the benefits and shortcomings of radar and infrared sensors however, one can remark that the two technologies are fully complementary, as shown in the following Figure (a bullet denotes appropriate functioning a question mar a specific problem).

Sensor Type	Adverse Weather Condition				Traffic scenarios				Information Presentation
	Fog	Snow	Night	Heavy rain	Urban	Number of lanes	Traffic Density	Day or night	
Anticollision Radar	•	•	•	•	? [3]	•	•	•	? [4]
Far IR Sensor	? [1]	? [2]	•	? [2]	•	•	•	•	•
EUCLIDE System	•	•	•	•	•	•	•	•	•

Figure :Strengths and weaknesses of anticollision radar and far IR sensors, when used in automotive applications

1. The performance is depends on the type of fog
2. Decrease of performance in heavy snow and rain
3. Object detection problems in complex traffic scenario like in city and inside tunnels
4. No enhanced perception of the external environment, only warnings

3 EUCLIDE integrated approach

The main limitation of state of the art microwave radar sensors is related to the detection of objects not belonging to the road, like bridges, to the difficulty to extract road geometry, and to the relatively rough classification of objects types. However, this information can be extracted by image processing of infrared sensor data. The far infrared detector provides an enhanced image of the road ahead (up to 200 m, depending on the selected optical field of view) without the need of external illumination. The development of appropriate real time image processing algorithms will allow the extraction of essential features to support the radar object tracking. This will be particularly important for overhead object alarm suppression at long ranges.

On the other hand, really heavy rain and snow can cause a decrease in the infrared image quality, since big droplets are seen as "spots" of different temperature, thus decreasing image quality. In fog the behaviour of the infrared sensor depends on fog droplet dimension, so that the advantage (in terms of increased visibility) could be decreased in presence of particular type of fog made of droplets of big dimension. Fusing the infrared sensor data with radar-recognised objects in such cases may however significantly decrease the droplets problems, by filtering those thermal images that do not correspond to a radar-recognised object.

In addition, the driver may not easily interpret the thermal images of an infrared sensor not may he/she gain an enhanced traffic conditions perception by discrete visual or audio warnings on obstacles ahead from a radar sensor. When however merging the two data sources, a much more intuitive and continuous traffic scene feedback may be provided to the driver.

For the above reasons, EUCLIDE Consortium decided to merge a state of the art radar and infrared sensors, both in terms of data fusion and user interface to the driver, to support the driver in r educed visibility, due to night and adverse weather conditions but also to warn the driver even in good visibility, when dangerous situations occur, thus addressing driver's distraction too.

3 EUCLIDE system specifications

The EUCLIDE vision enhancement system will offer the following innovative functionalities to the driver: the possibility to distinguish obstacles from what is outside and above the road, to identify the type of obstacle and to give the driver an enhanced perception of the road ahead. To achieve this goal, the following developments are planned:

- The 77 GHz scanning radar, developed for collision warning and avoidance, will be optimised, to achieve improved objects handling and classification, road estimation and prediction and a variety of modifications to the tracking subsystem.
- The high-resolution far infrared camera will be miniaturised in terms of camera and lenses electronics and will be equipped with a real time image processing board.
- A novel automotive sensors data fusion software will be developed, based upon relevant military technology, for plot level sensor fusion.

However, the greatest innovation will rest with the new systems human machine interaction, that will feature an efficient and effective way to combine different warning signals to support the driver by providing the right help only when needed (both in conditions of reduced visibility, and in good visibility to minimise driver's distraction). The aim is to develop an HMI effective and easy to use, open and interoperable to be adopted in on-vehicle advanced driver assistance system functions, avoiding overload of information, and characterised by an open architecture for future expansions.

The permanent thermal image display of the infrared camera will be replaced by a reduced optical information for driver support and driver warning. The warning and support strategies will be managed in an innovative way with optical and acoustic media. The human machine interface development will be carried out and continuously verified in close interaction between the partners for the HMI design and those for the HMI technical and ergonomic evaluation.

The technological step forward is to be found both in the definition of driver warning strategies and in the design and development of a head-up display based human machine interface to be applied both for images and symbols, with the following advantages.

- Simplification of the set-up and electronic control. This will be reached by avoiding the use of 2 optical elements for image projection and magnification (such set up would be critical especially in terms of on-vehicle integration and tolerances).
- System component cost reduction
- Component miniaturisation .
- Use of mass production components for cost effectiveness to allow a reasonable time to market
- The usability to display essential on-vehicle information.

- The open on-vehicle architecture for the applicability to essential on-vehicle information, like vehicle warning failure, and other on-vehicle warning support system
- Optimisation in terms of binocular vision adaptability.

4 Conclusions

EUCLIDE Consortium aims to merge a microwave radar and a far infrared sensor, initially developed for and widely used in aerospace applications and later adapted for the automotive environment, to achieve the required functionality improvements (in terms of obstacle recognition and localisation, rain/.fog droplets filtering and user interface optimisation), to support car drivers under all types of adverse visibility scenarios (i.e. heavy rain, fog, night).

Although the state of the art emanates from the aerospace technological field, the automotive environment poses much higher requirements on obstacles localisation precision and user interface design. Therefore, an extensive sensors improvement, data fusion and user interface re-design needs to take place the so developed sensors will be installed on a highly dynamic driving simulator (of DaimlerChrysler in Berlin) and two cars (a city car of Fiat and a luxury car of Volvo). A series of simulator and test track tests will be performed, estimating the new systems reliability, efficiency, usability and user acceptance under various parameters, such as: type of road, traffic conditions, road illumination, meteorological conditions, light conditions and vehicle speed range.

The results of the project may however be fed back to the aerospace sectors, since merging of such sensors is currently the state of the art within this application field too; leading to aerospace systems that are more accurate and easier to be operated by the pilots.

5 References

1. Bild der Wissenschaft 2/2000
2. Promote Road Safety in Europe COM(97) 131 fin.
3. Rapport Annuel: Sécurité routière 1996. (Institut Belge pour la Sécurité Routière)
4. Federal Statistics Institute, Offenbach, Germany, 1996
5. Great Britain - Transport Statistics Report. (Department of Transport, 1996)
6. Northern Ireland - Road Traffic Accident Statistics: Annual Report, 1996. (Central Statistics Unit)
7. Sweden - Swedish National Road Administration, 1996

Cognitive Flight-Deck Automation for Military Aircraft Mission Management

Axel Schulte, Peter Stütz, and Wolfgang Klöckner

ESG Elektroniksystem- und Logistik-GmbH, Experimental Avionics,
P.O. Box 80 05 69, 81605 Munich, Germany. E-mail: aschulte@esg-gmbh.de

This paper describes an approach to flight-deck automation in the field of tactical combat mission management. A concept for a functional prototype system, the so-called *Tactical Mission Management System* (TMM), will be given. The TMM has been implemented as a functional prototype in the Mission Avionics Experimental Cockpit (MAXC), a development flight simulator at ESG and evaluated with German Air Force pilots as subjects in simulator trials. Therefore, the TMM has been compared with a reference cockpit avionics configuration in terms of task performance, workload, situation awareness and operator acceptance. After giving a brief overview on the system concepts his paper reports on the experimental design and results of the simulator trial campaign.

1. INTRODUCTION

Performing military combat missions in an uncertain dynamic tactical environment, presents a potentially intolerable workload for the crew. Therefore, research and development activities are conducted in order to assist flight crews while performing the tasks with respect to a safe and successful mission completion. The functions required to provide intelligible interaction and efficient use of such a system are derived from a concept taking into consideration the process of human information processing. Such cognitive automation concepts have already been proved successful in other crew assistant programmes for transport aircraft (Walsdorf & Onken, 1998). They are incorporated in the concept and extended towards an application for the air-to-ground attack role. The approach aims at the provision of crew assistant functions which focus upon the monitoring and if necessary the retrieval of the integrity of superior goals such as safety, combat survival and mission accomplishment. In order to realise this approach, a way had to be found how to deal with human goal knowledge in machine systems.

The TMM facilitates functions such as tactical situation and threat analysis as well as more advanced cognitive modules such as conflict management. The system is able to continuously keep up with the change of situation parameters gathered by on-board sensors and data-link communication, in order to provide situational adapted assistance. It incorporates the functionality to detect conflicts in mission progress and finally supports the crew by autonomously generating suggestions for conflict resolution. The system performs its tasks in parallel to the flight deck crew and, therefore, behaves like an additional electronic crew member.

2. SYSTEM DESIGN CONCEPT

The following sections discuss the approach to cognitive flight-deck automation on the basis of a co-operative automation philosophy. A simple model of the human problem solving strategy based upon Rasmussen's work (1986) is used to identify shortfalls in automation. A model of superior goals and the deduction of resulting tasks is given.

2.1 Co-operative automation

The variety of tasks to be performed by the pilot during a tactical flight mission result in a workload on all work process levels (see Figure 1), ranging from skill-based manipulatory control (bottom) through rule-based system interaction (middle) up to general knowledge-based problem solving tasks (top). Conventional automation traditionally focuses on relieving the crews from exhausting routine actions, thereby being granted full autonomy in certain well defined areas. Expanding this strategy of automation into task domains primarily subjected to rule- and knowledge-based crew action, leads to severe problems in the area of man-machine interaction. Significant for this kind of development are very complex avionics structures and functions taking over full autonomy for comprehensive parts of the flight, while reducing the pilot to a mere solver of abnormal situations (Billings, 1997). Therefore, several authors, e.g. (Walsdorf & Onken,

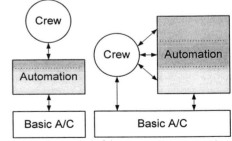

Figure 1: Conventional / co-operative automation

1998) demand new progressive methods when it comes to expand automation into all aspects of flight and mission management. One promising way to proceed is the concept of an automated system acting in a co-operative relation rather than separating the crew from the basic aircraft systems (see Figure 1). Being well aware of the complex task to perform, the crew interprets the output of the automation system as the recommendation of an additional electronic crewmember. The decision to accept or reject the machine's advice is reserved for the crew. Proceeding this way, the crew is kept continuously in the decision loop and is able to employ the full strength of human performance. At the same time the crew takes advantage of the particular strengths and abilities of the system.

2.2 Human problem-solving strategy

Besides an effective means of communication, a prerequisite to establish this partnership relation between man and machine is the implementation of transparent functional behaviour within the automation system. In order to make the machine output easy to comprehend and evaluate, and to establish a close-partner work relationship, both crew and machine have to reason from the same principles. Thus, the analogue problem-solving strategies and mechanisms have to be implemented in the automated system in a similar way to that which can be found in the human counterpart. This concept is the core element of cognitive automation.

Figure 2 depicts the elementary steps of human problem solving (Rasmussen, 1986), which are to be transferred equally into the machine system. On the lowest level a state-oriented acquisition of environmental signals is conducted and direct manipulatory output is generated. Problems demanding a certain amount of data abstraction and knowledge transfer typically cannot be solved on this level. Therefore, further data interpretation is necessary, taking into account superior context knowledge. A task-related aggregation and fusion of information can be derived as a result. On the structure-oriented level the data so-gained is further processed using additional rule and knowledge bases in order to reach a more profound problem diagnosis and a spectrum of possible solutions. Again, back on the context-level, a decision on how to proceed is found by the use of planning and forward-simulation results. Then, the derived solution is passed to the state-level for execution, thus ensuring successful problem solving under consideration of all relevant circumstances and all available information. Simple automation implementations within clearly-cut task domains such as auto-pilot or flight director systems usually can be seen as immediate instantiations of the processing steps on the state-level, which are directly connected through functional relations. Autonomous planning functions are state-of-the-art in today's Flight Management Systems. Assisting the crew also on the higher levels of their problem solving tasks, some abstract and therefore more versatile task knowledge has to be made available to the machine.

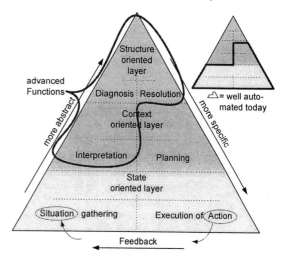

Figure 2: Human cognitive problem solving behaviour

2.3 Functional breakdown

In order to derive a functional breakdown of the TMM the generic processing steps are translated into specific functions. On the state-oriented layer situational parameters such as aircraft sensor and system signals, data-link information such as tactical elements and mission order, and on-board database entries e.g. terrain elevation data are to be considered. These data are analysed in order to identify context-specific features concerning pilot behaviour, aircraft movement and the external tactical situation. The monitoring of the threat accumulation along the planned trajectory is a typical example for a context-spanning analysis. The determination of the current phase of flight is essential for the model-based prediction of expected crew actions required to cope with the mission constraints. It considers all relevant task domains such as flight guidance along the pre-planned track, systems operation, weapon deployment, as the human operator does. It computes the current tasks and task parameters relevant for the crew in the present situational context. The module *conflict detection and resolution* builds up a hierarchy of general goals to be followed throughout the mission, such as flight safety, threat avoidance and mission accomplishment. Utilising the results of the tactical situation interpretation and the monitoring of the flight situation-dependent tasks, the system

1431

figures out violations of these goals. After negotiation with the pilot, a proposal how to resolve the conflict will be passed to appropriate machine agents for conflict resolution. Implementing this human-like goal-task-model ensures machine problem solving strategies which are easy to anticipate for the pilot. *Planning* is the most important conflict solving agent activity. The tactical mission management system offers a fully autonomous mission and route planning capability to the crew, including terminal operations planning, transit flight planning, tactical low-level flight trajectory planning and the use of attack procedure templates. Feedback is obtained in two ways; externally, due to manipulatory action of the crew and, thereby, alteration of the situation; and internally by the continuous re-consideration of the planning result in terms of goal integrity. Finally, the Tactical Mission Management System provides an appropriate *man-machine interface* on the flight deck, in order to manage the information flow and the crew interactions. The main components are an advanced primary flight display utilising a perspective 3D synthetic vision symbology, a tactical mission management and navigation interface, and speech synthesis.

3. SYSTEM EVALUATION

In order to prove the approach, the MMS has been implemented as a functional prototype and integrated in the flight and scenario simulation environment. It has undergone an experimental evaluation with operational personnel in spring 2001. The following sections give details on the experimental design and the evaluation results.

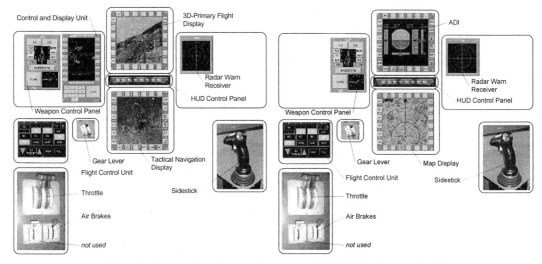

Figure 3: Compared cockpit configurations: TMM (left), reference (right)

3.1. Experimental design

For the evaluation of the Tactical Mission Management System a comparative study was chosen. Two different simulator set-ups were configured, on the one hand representing the basic functions of a reference combat aircraft cockpit (e.g. Tornado) and on the other hand demonstrating the MMS functions and displays (Figure 3). Following the experimental procedures the pilots had to perform a dedicated test mission with each of the cockpit con-

figurations. Figure 4 shows the phases of the test mission located in the south-west regions of Germany i.e. (1) tactical transit, (2) low-level ingress, (3) attack, (4) low-level egress and (5) tactical transit. During the low-level phase the mission was supported by computer-generated units such as SEAD-forces for suppression of enemy air defence and AWACS. Using the TMM-configuration the aircraft was participant of a tactical data-link network providing data on other participants and surveillance information. During the mission the tactical situation (i.e. hostile SAM sites) was supposed to change several times forcing the pilots

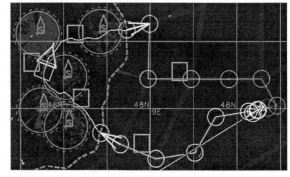

Figure 4: Scenario of the test mission for system evaluation

to react accordingly (e.g. route adaptation, re-planning, threat avoidance), thereby workload being imposed on the operator. The subjects were four German Air Force pilots (partly flight instructors) from the Fighter Bomber Wing 34, Memmingen at an age of 30 to 38 years. Their flight experience ranged from a total of 900 to 3000 flying hours on Tornado and other NATO combat aircraft. During a one day familiarization phase the pilots had the opportunity to train the handling of the simulator and the interaction with the TMM before the test mission had to be performed.

3.2. Evaluation results

The main scope of the evaluation of the Tactical Mission Management System was to account for improvements in comparison to the reference system in terms of system performance, pilot performance, pilot's situation awareness, workload and subjective ratings. The following sub-sections report on the specific results. Table 1 provides a collection of criteria concerning flight safety and mission accomplishment. The results make evident that the pilots performed notedly better with the TMM than without. The assessment of ground collisions and the frequency of AGL minima violations make clear that the TMM caused a significant risk reduction by a better ground separation. Due to a better threat avoidance with the TMM the SAM shots could be reduced. Performance criteria such as meeting the Time-over-Target (TOT), hitting the target or reaching the destination could be fulfilled by all pilots quite well. Only pilot 4 did not reach the destination with the reference system due to a flame-out condition. In general it was found that fuel consumption could be decreased with the TMM during the mission. Another observation was made concerning the number of violations of the Airspace Co-ordination Order (ACO) routing, which could be totally eliminated by the use of the TMM. So, the risk of being hit by friendly fire could be minimized.

	Pilot 1		Pilot 2		Pilot 3		Pilot 4	
	REF	TMM	REF	TMM	REF	TMM	REF	TMM
Ground collisions [#]	1	-	-	-	1	-	1	-
50 ft AGL violations [#]	5	-	0	-	3	-	9	1
150 ft AGL violations [#]	39	2	17	5	19	5	36	23
SAM shots [#]	-	-	-	-	1	-	1	-
ΔTOT [s]	L 8.0	E 0.4	E 0.3	E 2.1	E 2.3	E 0.2	L 3.8	L 2.1
Target hit	OK	OK	OK	OK	OK	OK	OK	OK
Destination reached	OK	OK	OK	OK	OK	OK	no	(OK)
Fuel on Board @ Touch down [%]	4.4	14.3	9.1	13.8	4.3	12.5	0.0	8.8
ACO violations	5	-	5	-	5	-	8	-
Mission accomplished	no	OK	OK	OK	no	OK	no	(OK)

Table 1: Global criteria of flight safety and mission accomplishment

One of the most important features of the TMM is the pilot assistance in optimizing a threat minimal route under a dynamically changing hostile threat theatre. Figure 5 shows the total and mean threat exposure computed with an underlying worst-case scenario. Comparing the threat exposure of a direct routing (1st column) with the result of the low-level route planner of the TMM (3rd column) makes the advantage obvious. The total threat accumulation could be decreased from about 8500 to 5700 %km. Due to the longer way, the effect on the relative threat exposure is even more noticeable (55 down to 30%). The columns 2 and 4 in Figure 5 show the threat values of the actual flown trajectories, again under consideration of a worst-case scenario. It is obvious that the co-operation between the system and the pilots yields another improvement in terms of threat avoidance due to synergetic effects. It should be emphasized that the improvement of threat avoidance with the TMM could be achieved in combination with a much better ground separation (Table 1). The TMM was designed to reduce the operator's workload by providing functions to support a better situation awareness and particular automation functions. During the experiments measurements of situation awareness and workload were conducted. Therefore, the experiment was stopped at dedicated points of time in order to perform the NASA Task Load Index and the Situation Awareness Global Assessment Technique (SAGAT) (Endsley, 1988). The evaluations were conducted four times each experimental run (reference system and TMM). The measuring points were the task situations 1 (Transit Ingress), 2 (Low-level Ingress), 4

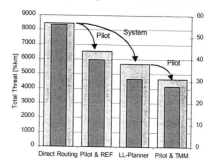

Figure 5: Threat exposure

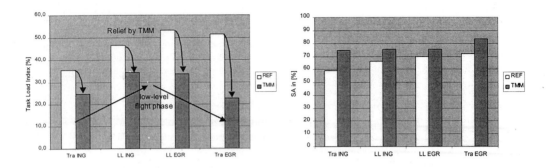

Figure 7: NASA TLX (left) and SAGAT (right) results over mission phases

(Low-level Egress) and 5 (Transit Egress) according to Figure 4. Figure 6 shows the results of the assessments averaged over the four subjects. Concerning the NASA TLX rating it was found that the overall workload could be reduced massively by use of the TMM with an expected slight increase of workload during the low-level phases of the mission. The situation awareness assessment was based upon the evaluation of a total of 26 multiple-choice questions concerning situational features. The right diagram in Figure 6 shows the weighted results. An increase in situation awareness of about 10 to 15%, in particular during the early mission phases, can be observed.

Finally, Figure 7 shows some selected but representative results of subjective ratings given by the pilots by the end of the two day evaluation period. The subjects fully agreed with the hypothesis that the TMM provides a better *big picture* in terms of global situation awareness. TMM is qualified to increase mission efficiency according to the pilots. The operators regarded the presented technology of Tactical Mission Management and crew assistance to be absolutely necessary, suited and adequate.

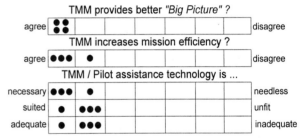

Figure 6: Subjective ratings (selection)

4. CONCLUSIONS

After having gathered years of experience in cognitive flight-deck automation and crew assistance, this paper presents the results of an exceedingly successful effort in the field of military combat aircraft. The functional breakdown of the *Tactical Mission Management System* has been derived from a model of human information processing, in order to approach a co-operative automation principle. A laboratory prototype system has been evaluated with fighter-bomber pilots in simulator trials. Besides the fact that the system was very well appreciated by the pilots, objective measures evidence a significant increase in performance in terms of threat avoidance and mission efficiency. These results could be achieved in conjunction with a noticeable reduction of the terrain collision risk and the operator's workload. Therefore, the system approach is highly recommended for application in the advancement and automation of combat aircraft.

REFERENCES

Billings, C.E. (1997), *Aviation Automation. The Search for a Human-Centered Approach.* Lawrence Erlbaum Associates: Mahwah, NJ.

Endsley, M.R. (1988), Situation Awareness Global Assessment Technique (SAGAT), National Aerospace and Electronics Conference.

Rasmussen, J. (1986), *Information Processing and Human-Machine Interaction. An Approach to Cognitive Engineering.* North-Holland.

Walsdorf, A. & Onken, R. (1998), 'Intelligent Crew Assistant for Military Transport Aircraft', *Sensor Data Fusion and Integration of Human Element.* NATO RTA SCI. Ottawa, Canada.

Speech Recognition in the Joint Air Operations Center – A Human-Centered Approach

D. T. Williamson[a] and T. P. Barry[b]

[a]Human Effectiveness Directorate, Air Force Research Laboratory,
2255 H Street, Wright-Patterson AFB, OH 45433-7022

[b]Sytronics, Inc.,
4433 Dayton Xenia Rd, Bldg 1, Dayton, OH 45432

ABSTRACT

This paper discusses the design and implementation of a prototype speech recognition interface to the Theater Air Planning (TAP) module of Theater Battle Management Core Systems (TBMCS). This effort supported an initiative at the Command and Control Battlelab at Hurlburt Field, FL to assess the operational benefits of speech recognition for data entry applications in a Joint Air Operations Center environment. A human-centered approach to the speech interface design resulted in the highly successful "TAPTalk" speech system which allowed operators the ability to verbally build the daily air tasking order as much as 40% faster than with conventional keyboard and mouse interfaces. Results from several operational assessments, including the recently completed Joint Expeditionary Forces Experiment 2000 (JEFX 2000), will be discussed along with plans to transition TAPTalk to the numbered Air Forces for operational use, are discussed.

1. INTRODUCTION

The latest generation of military command and control (C2) software applications provide planners in the Joint Air Operations Center (JAOC) with powerful tools to plan and execute the daily air tasking order (ATO). With this power comes a new level of complexity that pushes the limits of menu-based graphical user interfaces. Becoming proficient in the use of each application takes considerable time. This is a significant problem in today's JAOC environment where many operator positions are filled with "augmentees", military personnel with little or no operational planning background. Methods for improving the learning curve and increasing the efficiency of data entry are needed. The idea of using speech recognition to augment the conventional keyboard and mouse with spoken commands was proposed as a Kenney Battlelab Initiative for assessment by the Command and Control Battlelab (C2B) at Hurlburt Field, FL.

The primary objective of this initiative was to determine if speech recognition had reached a level of maturity for it to be considered a viable input technology for C2 software applications. In order to evaluate the concept, more specific objectives were defined to evaluate the overall goal. The following objectives were defined for the C2B operational assessment of the prototype speech recognition interface to the Theater Battle Management Core Systems (TBMCS) Theater Air Planning (TAP) module, which is responsible for the generation of the ATO in the JAOC:

1) Determine the impact of speech recognition on the completion times for standardized sets of TAP ATO planning tasks.
2) Determine the intuitiveness or naturalness of speech as an operator interface and the extent to which speech recognition enhances the mission planning process.

To accomplish this assessment, the C2B requested the help of the Air Force Research Laboratory's Human Effectiveness Directorate, AFRL/HECA, based on their years of experience in developing and evaluating speech recognition interfaces in various military applications [1] [2] [3] [4]. This paper discusses the design, implementation and evaluation of the speech recognition interface to the TAP module, known as TAPTalk.

2. SPEECH INTERFACE DESIGN

In designing the speech interface, focus was placed on TAP's Planning functions. In particular, three areas were investigated: menu navigation, data entry, and database query. Each of these areas is discussed below.

2.1. Menu Navigation

For menu navigation, the goal was to provide short goal-oriented commands that would take the operator directly to the desired function, rather than provide a menu-speak interface that required the verbal clicking of each menu in the same sequence as was required with the mouse. A good example of this direct access capability was in Flow Planning. To perform this function with the mouse, the operator had to click through six menu items on two separate windows.

To perform this function with speech, the operator simply spoke the command *"Plan air location flow"* or *"Plan air alert flow"*, resulting in a faster and more direct method for initiating a particular planning function within TAP. In addition, the use of synonyms was used often to allow for command flexibility. In the above example, *"air location"* and *"air alert"* were synonymous and could be spoken interchangeably.

2.2. Data Entry

In designing the data entry functions, a convention was used that required the operator to enter data by speaking the name of the data field followed by its value. This avoided any cognitive load associated with remembering the various field names, since each field was clearly labeled on the screen. For example, "Type A/C" is a screen label corresponding to the aircraft type data field to be populated with a verbal command such as *"Type Aircraft F16C"*.

One exception to this rule was in the assignment of a target to a particular military unit. The operator had the option of using this powerful command to rapidly assign a specific target (identified by an alphanumeric id) to one of the available aircraft units during ground target mission planning. As an example, the command *"Assign B00101 to 48th Fighter Wing"* specified that the 48th Fighter Wing be being assigned to the target identified as B00101. This verbal command took the place of over twenty separate manual actions needed to accomplish the same task. With just a few additional commands, such as entering a time schedule window and mission type, an entire strike mission could be planned in a matter of seconds. This showed the unique ability of speech to perform a compound data entry task with one short utterance.

2.3. Database Query

Another key feature that was designed into the TAPTalk interface was the ability to perform database queries. While not exercised extensively during the operational assessment, this capability could potentially be invaluable in the JAOC by allowing operators to gain rapid access to data during planning sessions. One example of such a query was suggested by an operator during subject matter expert interviews conducted early in the development of TAPTalk. During resource allocation in ground target mission planning, the operator often has to select a particular standard conventional load (SCL) that can have a cryptic identifier such as 2A65X2M84LB10. Operators often require additional information about this SCL and will query the database to view its configuration details, requiring over 30 mouse and keyboard actions to accomplish. To perform the same query with speech, the user simply clicks on the particular SCL in the Resources area and says, *"What's this SCL"*. The query is then performed with the resulting component list that makes up the SCL displayed on the screen. This is just one example of how speech could be used to rapidly gain access to desired information during the planning process.

3. SPEECH SYSTEM SELECTION AND IMPLEMENTATION

There were a number of commercial speech recognition products considered for interfacing to the TAP software. Three classes of speech recognition systems were available: large vocabulary speech-to-text dictation systems, small to mid-sized vocabulary industrial systems, and mid to large vocabulary telephony systems. In selecting the appropriate system for integrating into TAP, several factors were considered. First, it had to function in a Solaris/UNIX environment to be compatible with equipment in place in the JAOC. In addition, the ideal product would not have a requirement for operators to enroll or train the system before use. Of the commercial speaker independent, UNIX-based systems available, Nuance 6 from Nuance Communications, Inc. was selected. Nuance provided a highly scaleable client/server architecture with an excellent set of development tools. Also, Nuance's natural language processing feature became indispensable in creating a highly flexible speech interface. This specific feature is discussed in more detail in the following section.

3.1 Natural Language Slot Processing

To extract the semantics or meaning from speech commands into a form that can be easily interpreted by the interface software, Nuance uses a natural language slot concept. The following is an excerpt from the TAPTalk grammar to illustrate this slot filling approach:

Command
(request [(?air refueling) tanker (a r)] support)

Natural Language Slots
{<command_id window>
 <window_id request_refuel>}

In this example the following speech commands are possible:

"request air refueling support"
"request refueling support"
"request tanker support"
"request a r support"

However, the only important information the speech interface is concerned with is the resultant values in the natural language slots, *command_id* and *window_id*. The *command_id* slot value is *window* indicating that a window is being requested. The *window_id* slot contains the specific window being requested, in this case *request_refuel*. This makes it very easy to interpret meaning from the commands, especially when there is considerable input flexibility. Using a strict command parsing approach would result in having to consider all possible alternatives, making the software development task more difficult. With natural language slots, the task is much easier since there is only one simplified interpretation possible.

3.2 TAPTalk Command Interpretation

Once the slots are filled and sent to the speech interface software, this information needs to be packaged and sent to TAP to change a menu, enter a data item or perform a query. The user interface to TAP is a custom fourth generation language (4GL) developed by Lockheed Martin to handle the building of the graphical user interface. A messaging system was devised that formats the slot information into a packet of data that it sends to TAP for processing by the 4GL. This same messaging system was also used to send commands back to Nuance to change grammars, keeping it in sync in case the user manually selects a different window with the keyboard or mouse.

3.3 Push-to-Talk Implementation

During an early implementation of the TAPTalk prototype, the mute switch on the microphone cord was used as a sort of push-to-talk switch. This had one significant drawback. If the user paused too long in the middle of an utterance, the system would time-out and often reject the entire utterance, requiring the user to repeat the command. What was needed was an external switch that could signal to the speech software when to start and stop recognizing the command sequence and thus allow the user to naturally pause during the push-to-talk event without the threat of the system timing out. This was implemented in a later version as a foot switch interfaced with the RS-232 serial port of the Sun workstation. This foot switch had the added benefit of freeing the operator's hands to hold planning worksheets containing the information needed to plan the missions.

3.4 PronTool Implementation

Another human-centered design feature built into TAPTalk is a Pronunciation Tool or PronTool which allows the user to enter synonyms or refine pronunciations for certain grammar elements such as base names, aircraft, locations, units, or targets. For example, if the base KFFO is in the database, the default pronunciation is "K F F O" or "kilo foxtrot foxtrot oscar". Using the PronTool the user can easily add "Wright Patterson" as an alternate phrase with the resulting "KFFO" forwarded to the interface software for data entry. This feature greatly enhances the users' satisfaction by enabling the tailoring of these elements.

4. ASSESSMENT RESULTS

A series of operational assessments was performed to determine the effectiveness of the TAPTalk speech interface. Of primary interest was how much speech recognition reduced the data entry time required to build an ATO. Of secondary interest was how accurate the Nuance engine was in recognizing various users with different dialects and no significant prior experience using speech recognition systems. Results of these assessments are presented in the following sections.

4.1 Operational Assessments at C2 Battlelab

The first series of assessments was performed at the C2 Battlelab Initiative Facility at Hurlburt Field, FL. Sixteen male subjects participated in the two-week assessment of the TAPTalk interface. Eight subjects from the local Command and Control Training and Innovation Group (C2TIG) participated in the preliminary evaluation. This group represented a wide range of planning experience, from none at all to over ten years of JAOC experience. Eight different subjects, military planners from various numbered Air Forces, Navy and Marine Corp participated during the second week, the formal assessment period. Planning experience for this group ranged from less than a year to more than ten years of JAOC planning experience. Prior to the operational assessment, representative aircraft mission requirements were mapped into mission planning worksheet packages. During the assessment, each subject entered the information from the assigned mission worksheets into TAP using the workstation keyboard and mouse augmented with TAPTalk or, for comparison, using only the standard keyboard and mouse interface. The worksheet packages were sized to be completed in 90 minutes or less. Introductory briefings on the features of speech recognition were conducted prior to each phase of the assessment. During the first 2 days of each assessment phase, subjects were provided both conventional TAP training and TAPTalk speech recognition training. Subjects were then provided an opportunity to become familiar with the use of TAP and the speech recognition software. Practice sessions on mission planning were conducted using training worksheets and sample speech recognition scripts.

After the two-day TAP and TAPTalk orientation and practice period, the three-day data collection session began. During each assessment session, C2B data collectors monitored each subject's progress and the accuracy of their mission planning. All subjects' verbal commands were logged in computer files for later analysis. During both weeks, the subjects were given exit surveys after completing both their first and last pair of planning sessions. Results from these initial assessments revealed that speech reduced the data entry time by about 10.6% with a recognition accuracy of 97.4%. Subjective questionnaire data showed a strong preference for the speech interface.

4.2 Operational Assessment at JEFX2000

Based on the lessons learned from the previous assessments, several design changes were made to the TAPTalk system to significantly improve the data entry process. The most important changes were the implementation of multi-field commands and the hardware foot switch for push-to-talk operation. Multi-field commands allow the operator to speak several data entry field commands in a single utterance. The foot switch implementation has already been addressed in Section 3.3 and greatly improved input flexibility over the microphone mute switch. Both of these improvements were evaluated in a follow-on C2B initiative during the Joint Expeditionary Forces Experiment 2000 (JEFX2000) exercise conducted at Hurlburt Field.

During JEFX2000, planners in the Master Air Attack Planning cell routinely exercised the TAPTalk system to build the daily ATO. During a formal technical assessment, the difference in data entry time with TAPTalk was over 30% faster than manual entry. Speech recognition accuracy was measured at over 98% during this assessment. Survey data collected during JEFX2000 indicated strong user preference of the TAPTalk system and was therefore recommended for fielding as soon as possible.

5. DISCUSSION

The operational assessments showed that speech recognition was sufficiently mature to function reliably in the Joint Air Operations Center environment. When the proper commands were used, the Nuance speech recognition software achieved greater than 97% word accuracy. Speech recognition proved extremely useful for entering data, navigating menus and locating infrequently used information. The operational assessments showed that speech recognition could significantly reduce the time required for ATO development. In addition, the operator training required to become proficient with the speaker-independent speech recognition software was minimal.

Subjects noted that anything which reduces the frustration level for augmentees would result in a higher quality product and a reduction in fatigue levels. While it is possible to plan missions using only a keyboard and mouse, when implemented effectively, speech recognition can significantly enhance the use of the TAP module. These enhancements include shortening the time required to develop an ATO, reducing the training required for new augmentees, and assisting experienced operators to more effectively utilize the lesser used, but powerful query capabilities of the TAP application.

Based on the success of the TAPTalk assessments and strong user support, the TAPTalk interface has been recommended for operational use. The TAPTalk system has evolved into a highly efficient alternative method for developing the ATO. Design iterations have been made incorporating user feedback and results from the speech logs captured during actual use. These speech logs were an invaluable tool for refining the vocabulary and grammar structures to more accurately represent the planners' preferred commands. Plans are currently underway to fully incorporate TAPTalk into the fielded version of TBMCS.

REFERENCES

1. Barbato, G. J. "Integrating Voice Recognition and Automatic Target Cueing to Improve Aircrew-System Collaboration for Air-to-Ground Attack", In *Proceedings of the Research and Technology Organization Panel: Sensor Data Fusion and Integration of the Human Element*; RTO-MP-12 (pp. 24-1 to 24-11) published February 1999. System Concepts and Integration (SCI) Symposium, Ottawa, Canada, 14-17 September 1998.
2. Williamson, D. T. "Flight Test Results of the AFTI/F-16 Voice Interactive Avionics Program", In *Proceedings of the Voice I/O Systems Applications Conference* (pp. 335-345). Alexandria, VA, October 1987.
3. Williamson, D. T., Barry, T. P., and Liggett, K. K. (1996). Flight test performance optimization of ITT VRS-1290 speech recognition system. In *Audio Effectiveness in Aviation: Proceedings of the Aerospace Medical Panel Symposium*; AGARD-CP-596 (pp. 35-1-35-6) published June 1997. Copenhagen, Denmark, October 1996.
4. Williamson, D. T., and McDowell, J. W. "The implementation of a voice actuated radio management system in a C-135 aircraft", In *Proceedings of AVIOS '86 Voice I/O Systems Applications Conference* (pp. 144-151). Alexandria, VA: American Voice Input/Output Society, September 1986.

DESIGNING DYNAMIC HUMAN-MACHINE TASK ALLOCATION IN AIR TRAFFIC CONTROL: LESSONS DRAWN FROM A MULTIDISCIPLINARY COLLABORATION

Serge DEBERNARD, Jean-Michel HOC

LAMIH, UMR CNRS 8530
Université de Valenciennes et du Hainaut-Cambrésis
Le Mont Houy, 59313 Valenciennes Cedex 9, France
Email : serge.debernard@univ-valenciennes.fr, jean-michel.hoc@univ-valenciennes.fr

ABSTRACT: This paper deals with Human Machine Cooperation and more specifically with Dynamic Allocation of Tasks in the Air Traffic Control domain. It presents previous studies in this area and the main results obtained. These results have led us to design a new assistance to air traffic controllers based on the principle of task delegation and the implementation of a Common Work Space for improving human machine cooperative activities.

1. INTRODUCTION

For a complex system, the choice of the appropriate automation level is a difficult step, where the human needs must be considered precisely. Indeed, if too complete an automation can affect the human abilities, an incomplete automation can be not sufficient, firstly for coming up to the human's assistance needs, and secondly for keeping the reliability and the productivity of the system (Garland, 1991). Automation can be achieved in accordance with two main lines. The first one consists in the definition of function allocation between the human(s) operator(s) and a control system, this allocation being static, (see Hollnagel, 2000). The second one, which can be complementary to the first one, focuses on Human-Machine Cooperation (HMC).

The work presented in this paper is coming from a reflection and a cooperation between two research teams in engineering sciences and cognitive sciences. After a short presentation of the Dynamic Task Allocation (DTA) principles, this paper presents the main results of previous experimental investigation in the domain of Air Traffic Control (ATC). These results lead to the design of a new assistance to air traffic controllers based on the principle of task delegation and on the implementation of a Common Work Space (CWS).

2. HUMAN-MACHINE COOPERATION

The HMC is a research domain which is of interest of several disciplines; a part of them are interested in the description and the understanding of the psychological mechanisms underlying cooperative activities; the other are interested in the definition of tools and of human-computer interfaces for supporting these cooperative activities. So, defining exactly what is HMC remains a difficult task. From Hoc (in press), *"Two agents are in a cooperative situation if they meet two minimal conditions:*
- *Each one strives towards goals and can* interfere *with the others on goals, resources, procedures, etc.*
- *Each one tries to manage the interference to* facilitate *the individual activities and/or the common task when it exists.*
The symmetric nature of this definition can be only partly satisfied".

In this definition, the notion of goal does not refer to the global goal to reach when supervising and/or controlling a process, but to the goal for achieving a particular task. The word « interference » refers to the normal interaction between the activities of several agents, but also to conflicts between the agents, on the results of their activities, or on the means for achieving their tasks.

It is clear that in HMC, it is necessary to improve the interactions between the agents that allow the correct fulfillment of their tasks, and to minimize the possible conflicts. Nevertheless, this definition shows the limits of HMC, especially if the "machine" or computer must facilitate the human activities, because it will be necessary to prevent or detect conflicts with the human operator.

Royer (1994) specifies that a system is more cooperative when this cooperation integrates several levels, these levels being perception, analysis, decision and action. So, cooperation is not only a coordination of actions between several agents, but depends also on the merging of perceptions, on the confrontation of situation analyses, and on the convergence of decisions.

Schmidt (1994) presents different forms of cooperation between agents, which allow to define cooperation more precisely on the basis of type of interaction between them:

- In *Augmentative* cooperation, the agents have the same abilities, but their capacities are insufficient for achieving all the tasks to be performed alone. So, these tasks must be shared between them.
- In the *debative* form, the agents have also the same abilities, but they compare their results for obtaining the best solution.
- In the last mode, called *integrative* cooperation, the two agents have not the same abilities, and it is necessary to integrate their contribution for achieving a task.

In all these definition of cooperation, several notions appear: roles, tasks, functions and activities. The literature is not always very clear on their meanings. Before presenting the concepts of DTA, we propose the next definition for avoiding any misinterpretation.

- The notion of role corresponds to the responsibility of an agent on a part of work that he/she must achieve. We will see latter in this paper that, in the ATC domain, two human operators can have different roles in the task although they have the same abilities.
- A task is a work to perform, and corresponds to one or several goals to reach and that, under constraints. These goals can be "external"; the human operator must infer them from the situation and from the global goal given by the system designer. They can be also "internal", when a human operator defines him/herself its own goals for reaching some other ones. Constraints are also external and internal: external when a human operator, for example, must follow a procedure for performing a task; the internal constraints correspond to the operator's resource as, for example, his/her own abilities or his/her own temporal availability.
- We consider that activities and functions are at the same level of abstraction. These (human) activities / (artificial) functions correspond to an agent's "production". That concerns information elaboration, diagnosis, schematic and precise decision making, and solution implementation.

Our research concerns the DTA that is initially an augmentative form of cooperation. DTA consists in assisting the human operator, firstly with the integration of an automated system that is able to perform some tasks, and secondly by allocating the tasks between each agent in a dynamic way. So, the automated system must integrate all the functions that are necessary for performing a task entirely, from information elaboration to solution implementation, (Debernard et al., 1990). An optimal DTA aims at finding a task sharing that optimizes the process' performance and that takes into account the two decision-makers' abilities and capacities. So, it is necessary to define the set of the shareable tasks in real time that corresponds to the intersection between the set of tasks belonging to the human abilities, and the set of tasks within the assistance tool abilities. In DTA, the allocation of the shareable tasks is performed by a module called allocator, which the main function is to inform each decision maker about who performs which task and how. This allocator is controlled by a dispatchor. This function can be performed by an automated system, or by the human operator him/herself, (Millot, 1988) :

- In explicit DTA, the dispatchor function is achieved by the human operator with a specialized interface. So, the human operator allocates the shareable tasks with his/her own criteria.
- In implicit DTA, the dispatchor function is achieved by an automated system which performs the allocation in accordance with an algorithm and with criteria defined by the designer. These criteria can require to measure or estimate some human-machine system parameters like the human workload, the system performance, etc.

The implicit DTA has the ergonomic advantage to discharge the human operator from the management of the allocation. Nevertheless, this mode is more complex to implement than the explicit mode, because the choice of algorithm and criteria requires a serious study for integrating the human operator correctly in the human-machine system. To avoid this difficulty, it is possible to adopt an intermediate mode called assisted explicit DTA. In this mode, the initial allocation is performed by the automated system, but the human operator can modify it if he/she disagrees with the proposal for any reason.

We have implemented and evaluated these different DTA modes in the ATC domain. These studies are presented in the next section.

3. DYNAMIC TASK ALLOCATION IN AIR TRAFFIC CONTROL

The French ATC is a public service in charge of the flight security, the regulation of air traffic and the flight economy for each aircraft that crosses the French airspace. The security is of course the main objective and consists in detecting and preventing the collision between the aircraft. These situations are called conflicts and controllers must avoid them.

Our studies concern the "en route" control more particularly, in which the air space is divided into several sectors. Each of them are supervised by two controllers who have the same qualification. These human operators have a radar screen that displays the position of each aircraft in the sector in real time, and have some paper strips that contain the flight plan of each aircraft, i.e. the flight goal (trajectory and flight level) to reach. These two information means constitute a work space that is shared between the two controllers and that allows them to cooperate, (Rognin, Blanquart, 2001).

The first controller has a tactical role and is called radar controller (RC). He/she must supervise the traffic in order to detect conflict between aircraft and then resolves them by modifying the initial trajectory of one or several aircraft by sending a verbal instruction (heading, flight level, etc.) to pilots. He/she must also integrate the aircraft entering the sector. The second controller has a strategic role and is called planning controller (PC). This role consists in avoiding an overload of the tactical level, i.e. an overload of the RC. So the PC must predict the future traffic that the RC will manage, and negotiate the enter and exit aircraft flight level with the adjacent sector PC, in order that RC will not have too much conflict to resolve. So, PC performs "traffic filtering", which has nevertheless some limits because it is impossible to increase the number of aircraft thoughtlessly, and changing aircraft flight level is not a economical solution.

So as to face the air traffic increase, one solution consists in giving an active assistance to RC in order to increase the tactical level capacity. Now we present our previous studies realized in collaboration with CENA (French acronym for French Research Center on ATC). These studies aim at integrating a DTA in order to keep the controllers "in the loop".

In a first study called SPECTRA V1 (Debernard, 1993), only the tactical level was concerned: a DTA has been implemented between the RC and an automated conflict resolution system called SAINTEX. Two modes of DTA have been evaluated: the implicit mode and the explicit mode, where RC performed himself the task allocation. Professional controllers (certified controllers) have participated in the experiments on a software platform that integrates a realistic air traffic simulator. SAINTEX (Angerand, Lejeannic, 1992) is able to resolve conflicts between only two aircraft (called binary conflict); it has its own strategy and it is fully autonomous, i.e. it performs completely the resolution task, from the detection to the implementation of instructions, including rerouting.

The results obtained with SPECTRA V1 (Debernard & al., 1992) have shown that the implicit mode allowed to obtain a better security and a better PC's workload regulation than explicit mode. In fact, allocating a tactical task (conflict resolution) and, at the same time, a strategic task (conflict allocation) to RC causes an increase of his/her workload and a decrease in performance; the same results have been found in the domain of aircraft piloting (Jentsh et al., 1999). Nevertheless, the implicit mode was rejected by controllers; they preferred to keep the full control of the situation. Indeed, the SAINTEX strategy, based on the notion of binary conflict, could lead to some decisional conflict between the two decision-makers. In some situations, controllers did not understand the situation as SAINTEX did, because they considered that a third or fourth aircraft participated in the conflict.

Keeping the advantages of implicit mode for the tactical level, but with a human control of the allocation, was the research goal of SPECTRA V2, (Crévits & al., 1993; Lemoine & al., 1996). In this second study, the two levels was concerned: At the tactical level, a DTA has been implemented with an explicit and an assisted explicit mode; at the strategic level, an assistance to air traffic filtering has been implemented too, in order to help PC to manage the two resources of the tactical level – RC and SAINTEX. In the explicit mode, the two controllers can control the allocator whereas, in the assisted explicit mode, only PC can control it. A cognitive analysis of human activities had shown that in the explicit and the assisted explicit modes, the controllers had a more anticipative behavior than in a situation without assistance: They planned their own activities better and were less subjected to the traffic events, (Hoc, Lemoine, 1998). This result has shown indirectly that assistance tools allowed controllers to work "correctly" by giving some time for better preparing a global plan for controlling the situation. On the other hand, the assistance tools facilitated the human-human cooperation: They allowed controllers to build a Common Frame of Reference (COFOR) more easily by reducing elaboration activities. Nevertheless, the same problem appears: The SAINTEX' strategy can hamper controllers because they don't share the same representation of a conflict. Furthermore, in the assisted explicit mode, a complacency phenomenon (Smith & al. 1997) led RCs to relax their monitoring of the overall traffic, because they were less implicated in the allocation than the explicit mode. To solve these problems, a new project called AMANDA (Automation & MAN-machine Delegation of Action) is presently studied.

4. AMANDA PROJECT

In this project, two main ideas are studied:

- The principle of task delegation by controllers to a new system called STAR (French acronym for tactical system for assistance to resolution).
- The principle of a Common Work Space (CWS) which allows controllers and STAR, firstly to communicate together, and secondly to share the same representation of the problems, i.e. the conflicts.

As was the case of the SPECTRA studies, we try to help controllers by allocating some tasks to an assistance tool in accordance with an augmentative cooperation, but the new system will not have its own strategy for resolving a conflict. STAR will take the strategy given by controllers into account, in accordance with an integrative cooperation, and then will try to transform this strategy in a new trajectory that allows to resolve the conflict. If the controller agree with the proposed solution, he/she could allocate the task to STAR. Nevertheless, if the strategy given by controllers do not allow STAR to calculate a "good" trajectory because another aircraft is present (interfering aircraft), STAR integrates this aircraft into the problem and informs controllers in accordance with a debative cooperation. So, the controllers could negotiate with STAR by introducing another strategy or constraints. This kind of cooperation is called task delegation: It is really a task (a goal to achieve under constraints) that is explicitly allocated to the assistance tool, but this allocation integrates some constraints from controllers that are presented in our case under a strategy form. So, the assistance tool does not perform the task of conflict resolution entirely as it was the case with SAINTEX.

The controller's activities and the function of STAR will be integrated within a shared problem representation. It is possible to define a problem as a particularly situation that requires some activities (or functions) from an agent to clarify their characteristics, to define the strategy, then the solution, and to implement this solution. The Rasmussen's problem solving model (1980) allows us to define the contents of the CWS, from the information produced by these activities, (Lemoine, Debernard, 2001). Several disciplines have already considered the Common Work Space idea:

- In the case of Air Traffic Control, Bentley & al. (1992) have presented a shared work space that provides an adapted presentation of air traffic to different users, on different machines, to make the use of shared entities easier.
- Decortis and Pavard (1994) have defined the shared cognitive environment as a set of facts and hypotheses that are a subset of each agent's cognitive universe.
- Hoc (in press) has introduced the notion of Common Frame of References (COFOR), which is a set of information built and maintained by the cooperating agents. These information refer to their situation representation, but also to their goals, plans and resources.

A first experiment designed in the AMANDA project, where two RCs had to cooperate on the same traffic, has shown the prominence of cooperative activities that allow to build and maintain the COFOR, (Carlier, Hoc, 1999). For HMC, the computerized implementation of a COFOR (i.e., the CWS) should lead to a "richer" control environment because it allows to confront the agents' representations. Cooperation must be built around this work space, which is common for all the agents (in our project, the two controllers and STAR). Each decision maker supplies this space in accordance with their own competencies and with the initial function allocation made by the designer. Then, each agent can peruse information that is presented on the CWS, to check them (mutual control activity) or to negotiate if their representations are incompatible (Lemoine & al., 2001).

The next table shows the contents of our CWS. Each line refers to an activity, and for each activity are shown:
- The generic information produced by the activity.
- The effective contents of the CWS: A problem is a set of conflicting aircraft called "cluster"; a strategy is modeled as one or several "directives" (for example, "turn AFR365 behind AAL347"); a solution is a delayed instruction (for example, "turn AFR365 30° to the left at 22:35").
- The last column shows the agent that performs the activity/function and provide CWS with information.

5. CONCLUSION

This paper has presented the concepts of DTA in the ATC domain. Some previous studies have shown the interest of a CWS in HMC, which will allow a better cooperation between human operators and assistance tool. The project AMANDA has been presented. The contents of the CWS that AMANDA integrates has been defined in order to implement the principle of task delegation between controllers and the assistance tool STAR. Now, some experiments must be achieved in order to evaluate these principles. Especially, we will pay attention to the effective use of the assistance tool, to the expected decrease in human-machine decisional conflict, to the expected extinction of the complacency effect, and to the possible disturbances in subsymbolic (automated) activities on the radar display, produced by intensification of symbolic (attentional) activities through human-machine symbolic communication on conflict patterns.

Activities / Functions	Information	Contents of CWS	Allocation
Elaboration of inform.	Initial information	none	
Diagnosis	Problems	Cluster Creation Cluster Updating	PC, (RC) PC, RC, (STAR if incompatible representation)
Schematic decision making	Strategies for resolving a problem	Directives	PC, RC
Precise decision making	Solutions	Delayed instruction	PC, RC, STAR from a directive
Implementation of solutions	Solutions implemented	Instruction	PC, RC, (STAR if the cluster is delegated to it)
Allocation of tasks	*Allocations*	*Delegation*	*CR, (CO)*

REFERENCES

ANGERAND L., LE JEANNIC H. "Bilan du projet SAINTEX" [Results of the SAINTEX project]. Note R92009, Centre d'Etudes de la Navigation Aérienne ; Décembre 1992.

BENTLEY, R., RODDEN, T., SAWYER, P., SOMMERVILLE, I. "An architecture for tailoring cooperative multi-user display". Proceedings of CSCW'92. November1992

CARLIER X., HOC J-M. "Role of a common frame of reference in cognitive cooperation: sharing tasks in air-traffic control". In proceedings of CSAPC'99. pp. 67-72. Presses Universitaires de Valenciennes. 1999

DEBERNARD S. "Contribution à la Répartition Dynamique de Tâches entre opérateur et système automatisé : application au contrôle du trafic aérien" [Contribution to the DTA between a human operator and a computer system: application to ATC]. Thèse de Doctorat ; Université de Valenciennes. 28 Janvier 1993

DEBERNARD S., VANDERHAEGEN F., DEBBACHE N., MILLOT P. "Dynamic task allocation between controller and AI systems in air-traffic control". 9th European Annual Conference on Human Decision Making and Manual Control. Ispra, Italy. 1990.

DEBERNARD S., VANDERHAEGEN F., MILLOT P. "An experimental investigation of dynamic allocation of tasks between air traffic controller and A.I. system". 5th IFAC/IFIP/IFORS/IEA Symposium on Analysis, Design and Evaluation of Man-Machine Systems. The Hague, The Netherlands. June 9-11, 1992

DECORTIS F., PAVARD B. "Communication et coopération : de la théorie des actes de langage à l'approche ethnométhodologique" [Communication and cooperation: from the speech acts theory to the ethnomethodological approach]. In Pavard B. (Ed.), Systèmes Coopératifs de la modélisation à la conception. Toulouse, France, Octarès .1994

GARLAND D.J. "Automated Systems : the human factor". NATO ASI series. Vol. **F73**. Automation and Systems Issues in Air Traffic Control. pp 209-215. 1991.

HOC J-M. "Towards a cognitive approach to human-machine cooperation in dynamic situations". Int. J. Human-Computer Studies (2001) **54**. In press.

HOC J-M., LEMOINE M-P. "Cognitive evaluation of human-human and human-machine cooperation modes in air traffic control". The International Journal of Aviation Psychology ; Vol. **8**, N°1, pp. 1-32, 1998

HOLLNAGEL E. "Principles for modeling function allocation". Int. J. Human-Computer Studies. **52**, pp. 253-265, 2000

LEMOINE M-P., DEBERNARD S. " A Common Work Space For Human-Machine Cooperation In Air Traffic Control". Paper accepted for Control Engineering and Practice. 2001

LEMOINE M-P., DEBERNARD S., CREVITS I., MILLOT P. " Cooperation between humans and machines : first results of an experimentation of a multi-level cooperative organisation in air traffic control". Cooperative Supported Cooperative Work : the journal of collaborative Computing ; Vol. **5**, N°2-3, pp. 229-321 ; December 1996

MILLOT P. "Supervision des procédés automatisés" [Supervision of automated process]. Edition HERMES. Paris. Décembre ; 1988

ROGNIN L., BLANQUART J-P. "Human communication, mutual awareness and system dependability. Lessons learnt from air-traffic control fiels studies". Reliability Engineering and System Safety. **71**, 327-336, 2001

SCHMIDT K. "Cooperative work and its articulation :requirements for computer support". Le Travail Humain, tome 57, N°**4**, pp.345-366 ; 1994

SMITH P.J., McCOY E., LAYTON C. "Brittleness in the design of cooperative problem solving-systems: the effects on user performance". IEEE Transactions on Systems, Man and Cybernetics. Part A: Systems and Humans. **27**, pp. 360-371

DEVELOPMENT OF A REAL-TIME THINKING STATE MONITORING SYSTEM FOR PLANT OPERATION

Hirokazu Nishitani, Yuh Yamashita and Taketoshi Kurooka

Graduate School of Information Science, Nara Institute of Science & Technology

8916-5 Takayama, Ikoma, Nara 630-0101, JAPAN

ABSTRACT

We have been studying the use of multiple channel electroencephalogram (EEG) data to infer a human's thinking state. As a result, off-line thinking state estimation has been confirmed to be effective in experimental studies on simulator training during malfunctions and mathematics problem solving. In this research, we developed a real-time system that monitors a human's thinking state on the basis of off-line results and evaluated its effectiveness experimentally using mathematics problem solving.

1. INTRODUCTION

Recently, many applications using EEG pattern recognition methods have been studied [e.g. 1]. Among them, a global pattern of EEG data from multiple channels was correlated to a functional state. For example, a simple method to identify an emotional state with EEG data was developed using a linear regression model by Musha et al. [2]. They defined a pattern vector of cross-correlation coefficients between 10 electrodes in three types of bands and the emotional state with four orthogonal components: anger/stress, sadness, joy and relaxation. However, this definition of the emotional state is not suitable for directly analyzing a plant operator's functional state in an emergency because it is difficult to consider why such an emotional state was evoked. To analyze the functional state of a plant operator, it is desired to know the operator's thinking state rather than his/her emotional state.

In the previous study, we defined three basic modes of logical thinking: confidence (mode A), inference (mode B) and confusion (mode C) [3]. This classification was validated by cluster analysis [4]. We also conducted experiments to obtain EEG data while a subject was solving mathematics problems [5]. By taking an off-line approach, we built a linear regression model or an artificial neural network (ANN) model that correlates pattern vectors called CC vectors calculated from EEG data with associated thinking states. It was experimentally confirmed that these models can estimate human functional states in logical thinking jobs [6]. To date, the estimation method has been used in the off-line mode. However, real-time estimation will extend the application of this method. In this paper, we develop a prototype of a real-time thinking state monitoring (RTSM) system and consider a conception of plant operations with the RTSM system.

2. OFF-LINE ESTIMATION

An industrial simulator training system of plant operations was used for the experiments [3]. The system was composed of the plant control system called DCS and a boiler plant simulator for training. The boiler plant simulator was operated during 10 kinds of malfunctions in Table 1. The experimental procedure is summarized as follows:
 (1) The subject is trained under four kinds of fundamental malfunctions to have the basic skills of boiler plant operation and fault diagnosis.
(2) Experiments evoking the basic thinking state modes are performed to build an estimation model.
(3) Meaningful time periods, indicating when each of the basic modes is evoked, are selected with reference to observational data, and an estimation model is built with the EEG data in these periods.

(4) Both an ANN model and a linear model are built using the obtained EEG data.

Other experiments were performed to evaluate the thinking state estimation results. The malfunctions used for the model evaluation are also shown in Table 1. The experimental procedure is summarized as follows:

(1) The operator's thinking state during a malfunction is estimated by the constructed ANN and linear models. In addition, the operator's behavior is represented with a flow graph from observational data obtained from video images, along with audio recording and an interview.

(2) The thinking state estimation results by the two models are compared with the operator's behavior, and the correspondence is examined in detail.

The estimation results from the ANN model and linear model and the observational data were in good agreement. In the ANN model, the dominant mode shows a value close to 1 and an indifferent mode shows a value close to 0. Therefore, it is possible to judge the dominant mode automatically, which is an advantage when the ANN model is used in real-time monitoring.

Table 1   Malfunctions for model building and evaluation

Malfunctions	Subject Training	Mode-Evoking	Evaluation
Failure of deaerator feed water valve	√		
Clogged high-pressure strainer of fuel system	√		
Breakage of drum discharge valve (discharge of steam)	√	√ Mode A	
Misindication of boiler level meter (step change)		√ Mode B	
Misindication of steam flow rate (sinusoidal change)		√ Mode C	
Breakage of fuel heater pipe	√		√
Sticking of pressure control valve of heavy oil pump			√
Misindication of steam flow rate (step change)			√
Sticking of feed water valve			√
Misindication of boiler level meter (sinusoidal change)			√

3. REAL-TIME ESTIMATION SYSTEM

3.1 System architecture

We developed a prototype RTSM system, and its configuration is shown in Fig. 1. The system consists of four sets of EEG measurements with up to 14 electrodes, four amplifiers for EEG data, a signal processor box, a video camera, and a host PC. The host PC controls the signal processor box function. The PC can also calculate the thinking states of at most four subjects and displays these results. Both EEG data and control signals transfer through a SCSI interface between the host PC and the signal processing box. Video images are taken by the video camera and transferred to the host PC through an IEEE1394 interface. The host PC captures the video images and displays them on the monitor along with other information.

3.2 System functions

The host PC sends the processor box a control signal to start measuring the EEG data. EEG data is measured at a sampling rate of 100 Hz and stored in a buffer in the processor box. The host PC receives this EEG data every 5.12 seconds. In the host PC, the received EEG data is immediately transformed to CC vectors through the FFT (fast Fourier Transform). From the CC vectors, the intensity of each thinking state mode is calculated by the ANN model, and three mode intensities are displayed on the PC monitor. Figure 2 shows the graphical user interface on the monitor of the host PC. The upper right window is the system's controller. Through this window, the user can start and stop monitoring and set the monitoring conditions. In the lower right window, the video images of the subject are displayed. In the left window, the level of each mode intensity is displayed as a bar chart every 5.12 seconds. These three graphs are regarded as trend graphs of the thinking state of the operator. The time courses of multiple channel raw EEG data are also displayed in this window.

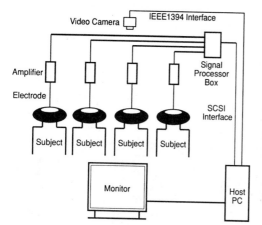

Figure 1. Architecture of real-time
thinking state monitoring system

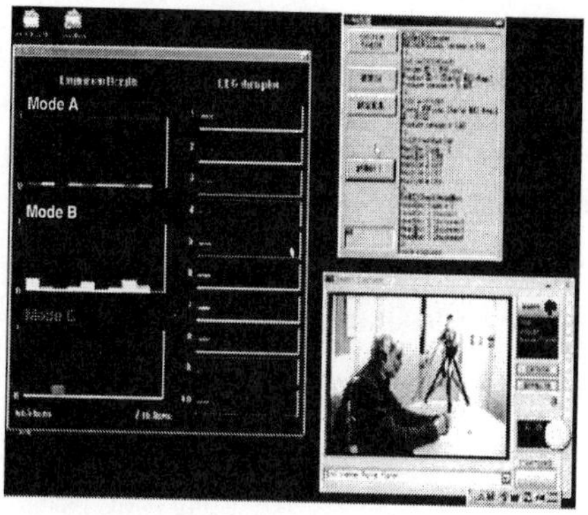

Figure 2. Display on PC monitor

3.3 Model accuracy

It is difficult to build a robust model that can be applied for anyone at any time. It is necessary to check whether the model is appropriate for the present real-time environment. As preparation for real-time monitoring, therefore, the accuracy of the estimation model should be known. We evaluate the accuracy as follows:

(1) Half of the CC vectors obtained in model building experiments are chosen at random, and a model is build with these CC vectors.

(2) The rest of the CC vectors are inputted into the model and the classification outputs are evaluated. The number of correct classifications is counted when the output intensity of each mode is greater than or equal to 0.5.

3.4 Evaluation test of the RTSM system

We tested our prototype RTSM system in an experiment. Rather than a plant operation, the solving process of mathematics problems was analyzed. This was because mathematics problem solving, like plant operations, requires logical thinking but is easier for analyzing a subject's thinking process with his/her answer sheets. In this study, therefore, we tested the RTSM system with mathematics problems such as factorization and simultaneous equations.

The experimental procedure was as follows:

Step 1: Mode evoking experiments are carried out for some mathematics problems.

Step 2: With the EEG data obtained in step 1, meaningful time periods are segmented with reference to observational data based on answer sheets and interviews. The CC vectors corresponding to three basic modes are obtained.

Step 3: By off-line calculations, an ANN model is built with half of the obtained CC vectors.

Step 4: The ANN model is evaluated with the rest of the CC vectors. If the classification accuracy is satisfied, the model is applied to real-time monitoring.

Step 5: Thinking-state monitoring proceeds with the RTSM system.

3.5 Test results

The RTSM preparation (steps 1-4) was done for about three hours. The resultant classification accuracy is shown in Table 2. Table 2 shows that the model produces a high level of correct classifications, and the possibility of misclassifications is small. Accordingly, we can apply the model for real-time monitoring. Then, the thinking state was monitored while the subject was solving five mathematics problems in about 15 minutes. In various situations, the thinking state mode coinciding with the subject's thinking process was indicated. A typical example is shown in Fig.3. This figure shows the display on the RTSM monitor when the subject was at a loss for how to proceed with a task using his tactics and so adopted another solving technique without confidence. Immediately after the subject changed his approach, the thinking state turned from Mode C to Mode B on the thinking state output window.

Table 2 Classification accuracy
of a built model

Test data	CC vector for Mode A (142 data)	CC vector for Mode B (132 data)	CC vector for Mode C (149 data)
$y_A \geq 0.5$	120 data	12 data	4 data
$y_B \geq 0.5$	4 data	96 data	0 data
$y_C \geq 0.5$	11 data	7 data	73 data

where $y = [y_A, y_B, y_C]^T$

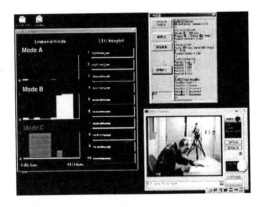

Figure 3. Display on PC monitor
from mode C to mode B

3.6 Utilization of RTSM in plant operations

Figure 4 shows a conception of plant operations with RTSM. In conventional plant operations, only the plant state monitoring and its feedback to operations are considered. In plant operations with RTSM, operator thinking state monitoring and its feedback to operations can also be considered. Namely, we can consider the entire plant operation system, including the plant operators, and devote a lot of attention to helping the total system fulfill its function.

4. CONCLUSIONS

We have developed a prototype real-time thinking state monitoring system for plant operations. First, both an ANN model and a linear model were applied as estimation models and experimentally evaluated by off-line analysis. The estimation results showed that the ANN model has some advantages over the linear model for real-time use. Second, a prototype RTSM system with the ANN model was developed and evaluated experimentally. The thinking state of the subject was monitored in real time while the subject solved mathematics problems. In various situations, the thinking state mode coinciding with the subject's thinking process was indicated. Third, a conception of plant operations with RTSM was also proposed.

It is important to be able to adjust the thinking state estimation model to accommodate day-to-day variations. We are now evaluating a model-adjustment technique and considering the use of other physiological measurements as supplemental model inputs. The RTSM system is expected to be used in many areas to help humans operate large and complicated systems.

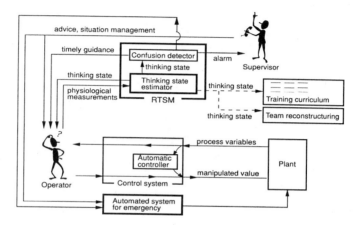

Figure 4. Conception of plant operation with real-time thinking state monitoring

ACKNOWLEDGEMENTS

The authors gratefully acknowledge the partial financial support from the Japan Society for the Promotion of Science under research for the future program (JSPS-RFTF96R14301) and by a Grant-in-Aid for Scientific Research (B) (#12480098).

REFERENCES

1. Freeman, F.G., P.J. Mikulka, L.J. Prinzel and M. W. Scerbo; Biological Psychology, 50, 61-76, 1999
2. Musha, T., Y. Terasaki, H. A. Haque and G.A. Ivanitsk; Artificial Life and Robotics, 1, 15-19, 1997
3. Kurooka, T. , M. Kisa, Y. Yamashita and H. Nishitani; Proceedings of 7th IFAC/IFIP/IFORS/IEA Symposium on Analysis, Design and Evaluation of Man-Machine Systems (MMS98), 539-544, Kyoto-Japan, Sept.16-18, 1998
4. Kurooka, T., Y. Yamashita and H. Nishitani; Proceedings of 15th World Congress of International Measurement Confederation (IMEKO-XV), Paper 18.2.3, Osaka-Japan, Jun.13-18, 1999
5. Kurooka, T., Y. Yamashita and H. Nishitani; Proceedings of IEEE International Conference on Systems, Man, and Cybernetics (SMC'99), II.407-412, Tokyo-Japan, Oct.12-15, 1999
6. Kurooka, T., Y. Yamashita and H. Nishitani; Computers and Chemical Engineering, 24, 551-556, 2000

Case Study: Incident Reporting in a Neonatal Intensive Care Unit

Daniela K. Busse

Department of Computing Science, Glasgow University, 17 Lilybank Gardens, Glasgow G20 8QQ, Scotland, UK

ABSTRACT

The occurrence of medical adverse events is a growing cause for concern worldwide. Critical incident reporting schemes have recently been suggested as an effective means to tackle the problem of medical adverse events. There are few comprehensive frameworks that accommodate the specific requirements of local settings as well as generic issues in incident reporting. The local setting radically influences a scheme's successful implementation and maintenance. Major issues that impact the overall success of incident reporting schemes concern the format of data collection, and especially, a meaningful data analysis. This paper reports on the introduction of a critical incident reporting scheme to a neonatal intensive care unit (NICU). Issues concerning the implementation and maintenance of the reporting scheme are discussed. Incident analysis is described in terms of the process and the results of incident categorization. The implications of such a viewpoint are considered.

1. INTRODUCTION

Adverse events in medicine have become a major public concern in recent years, after hospital tragedies such as the Bristol Baby Case in the UK (Kennedy et al. 2000). This spurned scientific studies into the nature of medical adverse events, their incidence, and whether their characteristics contain any clue as to how they can be remedied (Leape, 1994). Both the UK and US national governments have acknowledged the issue and have called into being organizations such as the US National Patient Safety Foundation, and the UK National Patient Safety Agency. The need to reduce 'human error' in medicine has thus been publicly recognized, and there is an urgent demand to implement safety measure that can tackle the problem effectively.

Critical incident reporting schemes have been cited as a major safety tool to combat human error and adverse events in medicine. These kind of schemes are established in safety-critical domains such as aviation and process control as complementary safety management tools. Incident reporting schemes facilitate the collection and archiving of data concerning critical incidents, or near-misses, that have occurred in those industries. Typically, a distinction can be drawn between standardised, national schemes of broad scope but less depth, and smaller, local schemes. Local schemes permit close scrutiny of in situ adverse events. Their implementation can also be fine-tuned to the local culture and conditions, a prerequisite for successful delivery of the scheme. This is at the expense of the benefits of statistical evaluation of the collected data that a more extensive, standardised data set offers. Both types of scheme operate by presenting employees with a data collection form, which prompts for a number of charactistics of the problem description. The nature of the form varies, but there are some fundamentals common to most schemes: general circumstances of the adverse events need first to be established (e.g. at what time did the incident occur?), as well as general facts about the reporters themselves (e.g. how experienced they are). Typcially, this is followed by asking for a narrative description of the incident, and includes questions about the presumed 'causes' and contributing factors, and how the incident was detected. This is question is not typically included, but is clearly valuable for future incident avoidance. This data can then be used to instigate further, more in-depth investigation of the incident. It can also be categorised and archived for statistical purposes, if the data set permits. Major national safety schemes are NASA's Aviation Safety Reporting System (ASRS), and its UK equivalent CHIRP (Confidential Human Factors Incident Reporting Programme). An example for a local incident reporting scheme that has been implemented in healthcare and maintained for over 10 years is described in (Busse and Wright, 2000).

This paper offers a report that touches on the prerequisites to successfully running an incident reporting scheme. The implementation that is described here attempted to localize an incident reporting scheme by taking contextual factors such as safety culture into account. It also had staff actively participate in the conception of the scheme and the design of the reporting form. This is argued to be crucial in achieving long term staff participation, and is argued to optimize the scheme's efficiency. The scheme that is described here also attempted to incorporate standardization issues that are posed by schemes implemented on a grander, e.g. national, scale. It thus presents a matrix of issues that still need to be addressed in any safety-critical domain that employs critical incident reporting schemes as part of ongoing safety management.

2. INCIDENT REPORTING

The cost of adverse events is high; not only in human suffering, but also in compensation claims and the need for prolonged treatment of afflicted patients. In 1990, the Harvard Medical Practice Study (Leape, 1994) investigated the occurrence of patient injury caused by treatment - so-called adverse events. It found that nearly 4% of patients suffered an injury that prolonged their hospital stay or resulted in measurable disability. Leape (1994) pointed out that, if these rates are typical of the US, then 180000 people die each year partly as a result of iatrogenic ('doctor-caused') injury. In the UK, a Department of Health report (2000) revealed that as many as 850,000 adverse incidents are happening in UK hospitals each year. This, in terms of litigation and the extra care needed by victims, added up to a £2bn bill. (BBC News 02/15/2001). There is an urgent need to make patient safety one of the highest priorities. The BBC also cites fears that the medical community is "complacent" about the toll of accidents, and notes that to date the National Health Service "did not even collect figures on the number of medical accidents".

Until recently, evidence of medical incidents (or near-misses) was mostly anecdotal. In the case their existence was acknowledged, the data typically did not leave a hospital's boundaries. This lack of distribution of incident data can lead to the replication of similar, preventable incidents across hospitals. This not only concerns e.g. faulty or badly designed equipment which might lead to deadly consequences. It also concerns badly designed work procedures that might be in place in hospitals across the country, the safety threat of which might only be recognised locally and sporadically. Similarly, it concers drugs that might have similar sounding names, but that have very different effects on a patient's condition. In order to prevent incidents needlessly repeating themselves, incident data must be recorded, analysed, and then made available for distribution. In industries such as process control for chemical plants and power stations, incident reporting schemes have often been used as 'early warning schemes'. This has yet to translate to the medical domain.

Introducing incident reporting into hospital wards is the first step towards recording information on incidents' nature and frequency. Summary data can then be used for trend analysis to identify systematic sources of error, and to prompt more in-depth analysis of potential causes. However, in order to to base valid and relevant conclusions on this frequency counts, the classification of incidents clearly needs to be *meaningful*. For instance, in current incident studies, most of the incidents' precursors are perceived to be 'human error' (e.g. Runciman et al., 1993) if it does not fit a category such as 'equipment failure'. Such categorisation might provide an initial filtering of immediately attributable equipment faults, but does not tell us much about how to prevent future instances of such 'human error'. Furthermore, as soon as poor equipment design is considered as an instance of equipment failure, there is no telling as to what constitutes error-inducing design (such as similary named drug containers) and what constitutes human error (mistaking the drug containers). This clearly poses a very real problem for the validity of incident data and its analysis. However, this subjectivity in the incident classification process, and the resulting spurious precision of trend analyses based on the data, is not sufficiently recognized as would seem necessary for any wider distribution of incident data beyond the local setting.

There have been some notable attempts at creating grounded and relevant categorizations schemes in the recent past. An influential example is reported in Runciman et al. (1993) who studied anaesthesia incidents in Australian hospitals as part of the Australian Incident Monitoring Study (AIMS). This study was subsequently extended to also investigate intensive care unit (ICU) incidents. The AIMS categorization scheme presents an integrated summary of previous categorization schems, and has had substantial impact on future ones. For the study reported in this paper, the AIMS-ICU categorization scheme was utilized to analyze incident data that had been collected in a neonatal intensive care unit. In the following sections, the outcome of this process is reported.

3   THE NICU INCIDENT REPORTING SCHEME

In the Neonatal Intensive Care Unit (NICU) in which the study took place, current safety management includesd informal checks, communication and consultation with fellow members of staff, morbidity and mortality meetings, and an adverse events reporting scheme, which addresses incidents that in fact resulted in harm to the patient or to staff and that legally require investigation. Near-miss adverse events that do not require legal investigation, but that could also lead to harm to the patient or staff, were dealt with on a local and immediate basis. They were not

SYSTEM-BASED FACTORS		HUMAN FACTORS	
Physical environment / infrastructure	#	**Knowledge-based error**	#
Lack of space / room		Lack or faulty knowledge..............................	1
Lack of facility		Error of:	
Excessive noise		Judgement..	1
High unit activity level......................................		Problem recognition / anticipation	
	4	Diagnosis	
Staff mealtime		Treatment decision	
Handover / ward round		Use of investigation procedures	
Lack of support staff		Timing of investigation procedures	
		Omitting intended treatment...........................	1
		Incorrect charting..	1
Equipment (including monitors)		Incorrect prescription	
Unavailable equipment		Incorrect interpretation of information	
Inadequate equipment		Information not sought	
Poor design..	4	Information not available	
Poor maintenance			
Equipment failure.......................................	3		
Inadequate inservice		**Rule-based error**	
		Patient assessment inadequate	
		Patient preparation inadequate	
Work Practices / Policies / Protocols		Failure to check equipment	
Communication problem.............................	1	Misuse of equipment	
Inadequate assistance		Unfamiliar equipment	
Lack of supervision		Unfamiliar environment	
Inadequate training		Unfamiliar patient	
Inadequate protocol		Failure to follow protocol...............................	2
Insufficient staff.......................................	1	Labelling error	
Unable to contact staff		Calculation error	
Inapprop. staff / patient allocation..................	1		
		Skill-based error	
		Distraction / inattention.................................	1
		Fatigue	
Other System-based Factors:		Haste	
		Stress...	1
_____			
		Technical error	
		Fault of technique	
		Inexperience..	1
		Uncooperative patient....................................	1
		Difficult patient body habitus...........................	1
		Patient physiological factors	
		Other Human Factors:	
		_____	

Figure 1 Incident Categorisation and Analysis

documented or kept track of, and distribution of known sources of error in the system was at best infrequent. There was demand to complement the existing safety management measures with a critical incident reporting scheme.

3.1. Critical Incident Definition

A 'critical incident' was here defined as follows: "A critical incident is an occurrence that might have led (or did lead) – if not discovered in time - to an undesirable outcome. Complications that occur despite normal management are not critical incidents." Staff that participated in the study were also asked to fill in an incident reporting form "if in doubt". This reflected the intention to collect rich, qualitative data, rather than data that would be fit for statistical analysis.

3.2. Set Up

A 'critical incident' thus include near-misses as well as actual adverse events. Reporting schemes involve staff reporting critical incidents using the provided reporting forms on a voluntary and anonymous basis. The incident reports are regularly analysed and categorised. The main aim of the analysis is to identify factors contributing to the causation of incidents that may be rectified. Accordingly, similar incidents are hoped to be avoided in future.

3.3. Incident Reporting Form

The incident form was developed iteratively, and evaluated by means of a questionnaire survey of the unit staff (Busse, 2000). The current form covers the following questions: the first section asked for a "description of what happened"; 'Drug Confusion Error' is treated as a category distinct from other types of critical incident on the form, due to its known frequency (Bogner, 1990). This separate treatment allowed for more specific data to be gathered on drug errors. Other questions related to what factors contributed to the incident, and which factors minimized it. The next section covers details on the circumstances: which procedure was being carried out, which monitoring was being used, and which equipment failed (if any). A question on the presence of alarms was added to the form, since it was felt that incidents discovered through alarm sounding fell in a sufficiently distinct category of incident circumstances. This was then validated by the data that was collected. One section of questions touched on the personnel that was involved in the incident and its detection. The data was collected on an anonymous basis, so the reporting staff was not asked to provide contact details. However, experience levels and job titles were covered in the personnel section. Another section noted the estimated and actual outcome of the incident to the patient (or in terms of other costs). The final section provided an open-ended question regarding suggestions for improvements by the reporting staff - future prevention of similar incident being the primary goal of the incident reporting scheme.

3.4. Incident Analysis

The AIMS-ICU analysis scheme was used. AIMS used a reporting form that consisted primarily of given categories which were to be ticked off by the reporting staff (see figure 1). This way of recording staff reports could potentially lead to decreased analysis time (since the reporters essentially did the categorization themselves). It could also be argued that this decreased the degree of indirection in the analysis process – categorization based on fairly subjective interpretation of gathered data could be replaced by the reporter's own interpretation of the actual events. In the NICU study, the form that was used was specifically developed to address the local needs (see section 3.4. above). However, the collected data was subsequently analysed by assigning it to categories as listed in the AIMS. The breakdown of the results is shown in the following section.

3.2. RESULTS

As can be seen in Figure 1, 14 causes of incidents were classified as system factors, whereas only 11 where classified has human factors. This is in contrast to findings in comparable studies, where up to 80% of causal factors were classified as human error. The most commonly attributed subgroup of causal factors was 'equipment' with 7 occurrences, followed by 'physical environment/infrastructure' (4) and 'knowledge-based errors' (4). Furthermore, the categories 'rule-based errors', 'skill-based errors', 'technical errors', and 'work pratices' were also all represented in the results. All causal factors could be categorised, and 'other system factors' and 'other human factors' were not assigned in this study.

However, it was found that for most incidents, multiple categorizations were necessary. There were mostly several causal factors per incident, for instance the categorizations listed above show at least two categorisations of causal factors per incident. The combination of factors proved to provide a more meaningful picture of the incident's causation and its potential future prevention, than single categorizations. This confirms previous findings (Busse and Wright, 2000). The most striking finding in this study was arguably the difficulty of arriving at a *meaningful* classification of incidents. The nature of a *meaningful* classification has not yet been discussed, let alone been operationalized, in the current discourse on incident reporting. Early work in the process control domain covered substantial ground in delineating a meaningful analysis of Human Error (Rasmussen, 1982), but such work has still be addressed in incident categorization and analysis.

4. CONCLUSIONS

The NICU incident reporting scheme succeeded in achieving staff engagement and participatory design of the reporting form. Most importantly, it impacted the overall safety culture in the unit, by raising awareness of clinical incidents, and raising the belief that the occurrence of 'mistakes' can be dealt with in a constructive way by highe level management (rather than following a 'punitive perfection model' (Leape, 1994). Equipment failures could be followed up by either contacting the manufacturer directly, and also for instance by being able to refer to the incidents as evidence of insufficient design of specific devices. This data provides the basis to pass on valuable lessons to other NICUs that e.g. use similar devices.

The UK Department of Health (2000) stated as their main conclusion to the Chief Medical Officer's report: "We believe that, if the NHS is successfully to modernise its approach to learning from failure, there are four key areas that must be addressed. In summary, the NHS needs to develop:

> ➤ unified mechanisms for reporting and analysis when things go wrong;
> ➤ a more open culture, in which errors or service failures can be reported and discussed;
> ➤ mechanisms for ensuring that, where lessons are identified, the necessary changes are put into practice;
> ➤ a much wider appreciation of the value of the system approach in preventing, analysing and learning from errors."

Additional to this, however, it needs to be stressed that the incorporation of the specific local conditions and requirements are necessary for the successful maintenance of an incident reporting scheme. Furthermore, if a unified mechanism for analysis is determined, it needs to take into account that the classification scheme should not only aid causal factor 'counting', but also their meaningful analysis.

REFERENCES

BBC News, "Medical accidents - unstoppable?". http://news.bbc.co.uk/hi/english/health/newsid_1171000/ 11717 12.stm, Feb 15 2001.

Bogner, M. S., (ed.) "Human Error in Medicine", Lawrence Erlbaum Associates, 1994.

Busse, D.K. "Staff Perception of Critical Incident Reporting as Part of Clinical Safety Management", in Proceedings of the PsyPag Annual Conference, Glasgow University, 2000

Busse, D.K. and Wright, D. "Classification and Analysis of Incidents in Complex, Medical Environments", in Johnsons, C.W. (ed.) Special Edition of Topics in Health Information Management on Human Error and Clinical Systems; THIM Vol 20 Issue 4.

Department of Health, UK, Chief Medical Officer. "An Organisation with a Memory". Crown Copyright 2000

Kennedy, I., Howard, R., Jarman, B., and Maclean, M. "The Inquiry into the management of care of children receiving complex heart surgery at The Bristol Royal Infirmary. Interim Report: Removal and retention of human material", Crown Copyright, UK, 2000.

Leape, L. L. "Error in Medicine." JAMA: The Journal of the American Medical Association 272(23): 1851-57, 1994.

Rasmussen, J. "Human Error: a Taxonomy for Describing Human Malfunction in Industrial Installations." Journal of Occupational Accidents 4: 311-333, 1982

Runciman, W. B., A. Sellen, et al. "Errors, Incidents, and Accidents in Anaesthetic Practice." Anaesthesia and Intensive Care 21(5): 506-519, 1993.

Predicting Situation Awareness Errors using Cognitive Modeling

Troy D. Kelley
Debra J. Patton
Laurel Allender

Army Research Laboratory
AMSRL-HR-SE
APG, MD, 21005-5425

ABSTRACT

The U.S. Army Research Laboratory's Human Research and Engineering Directorate has developed a series of computational cognitive models within the Atomic Components of Thought-Rational (ACT-R) (Anderson & Lebiere, 1998) cognitive architecture to attempt to predict errors made by soldiers on navigation-related tasks while they wore helmet-mounted displays (HMDs). The study used 12 infantry soldiers who were required to perform a series of navigational tasks while wearing HMDs. During the exercise, the soldiers were asked a series of probe questions pertaining to information that had been displayed on the HMD. An hypothesis was developed about the causes of errors that soldiers made in response to the probe questions. The hypothesis was used as the basis for ACT-R models which, in turn, were evaluated against the data. The modeling effort here was not a typical example of "curve fitting," in which modelers examine data that they hope to later match by developing a cognitive model. Instead, these models were purely predictive from the beginning. The error data were not examined until after the initial models were completed. Results indicated that, after some adjustment, the computational cognitive models were able to predict the likelihood of soldiers answering a probe question correctly or incorrectly, based on the activation levels associated with the multiple memory chunks pertaining to a given question. Future directions of predictive cognitive models for interface design are discussed.

1. INTRODUCTION

The use of computational cognitive architectures for the investigation and prediction of human error is a new and important development for human-system design. In the past, traditional error analysis has taken the form of error classification or the development of error taxonomies (Norman, 1981; Rassmussen 1982; Reason, 1990). While error taxonomies can be useful for the identification of general types or classes of error, taxonomies do not necessarily provide the explanatory and predictive power required to understand the underlying cognitive processes or to prevent error occurrence. Error classification alone does not provide the ability to examine "what-if" scenarios in the way that a cognitive architecture would.

Examination of leading cognitive architectures such as ACT-R and SOAR (Newell, 1990) reveals that their particular strengths are in the description of skill acquisition rather than the prediction of error. In the case of ACT-R, this largely attributable to its long history in the field of intelligent tutoring systems (Anderson, 1993). Because of this, some researchers have called for entirely new cognitive architectures to be developed, which would allow more complete modeling of errors within complex dynamic environments (Grant, 1996). While new architectures for error prediction may be useful in the future, existing architectures have a wealth of support and research that easily justifies using them for current error prediction modeling. Furthermore, the perceived limitation in error prediction with ACT-R, for example, perhaps lies more of how it has been used rather than a limitation of the architecture itself.

ACT-R was used for this study. It is freely available for government and academic research from Carnegie Mellon University. It is a symbolic, production system architecture, capable of low-level representations of memory structures. ACT-R is implemented in the common LISP programming language as a collection of LISP functions and subroutines, which can be accessed by the cognitive modeler. For this project, we used Macintosh Common LISP and ACT-R 4.0 running on a G3 Apple Macintosh computer running system 8.5.1.

The data used for development of the cognitive models reported here were collected by members of the Human Research and Engineering Laboratory's Soldier Performance Division as part of a study entitled "A Comparison of Soldier Performance Using Current Land Navigation Equipment With Information Integrated on a

Helmet-Mounted Display" (Glumm et al., 1998). As part of the study, 12 male infantry soldiers performed a series of navigation tasks while wearing helmet-mounted displays (HMDs). During the navigation exercise, soldiers were asked a series of probe questions at pre-determined coordinates along the path. The soldiers were queried about their positions with respect to various objectives (e.g., targets, way points). The probe question technique was used as a measure of situation awareness, operationally defined as a correct response. Each question was phrased to elicit either a "yes" or "no" response. Mistakes made in response to the probe questions were the errors that were modeled for this effort.

2 METHODOLOGY

Each participant navigated a densely wooded path while wearing an HMD. The total length of the path was 3 kilometers and consisted of four segments (or legs) of different lengths that intersected five way points: The lengths of the path legs were 550, 700, 850, and 900 meters. The terrain was flat with elevation contours of 2 to 3 feet. The ground was covered with fallen trees and branches which, in some areas, were concealed by grass approximately 8 inches tall. There were some small streams along the path and marshy areas with standing water. Except for a few short, muddy sections of path that lacked ground cover, the hardy grasses and vegetation that grow in this area tend to recover quickly from footsteps, revealing little evidence of previous subjects.

A total of 20 probe questions was asked during the entire course. The first five probe questions are shown in the first column of Table 1. All the information necessary to answer the questions was available on the HMD. There were four different HMD screens (target, way point, enemy, and path), each of which could be accessed by the soldier via a small, belt-mounted keyboard. Participants were able to access any HMD screen at any time, except that immediately after a probe question, the HMD was blanked until the question was answered. For the overall study, subjects traversed the course in both HMD and non-HMD conditions; however, only the HMD data were used for the modeling effort. Data were collected by use of a computer linked to the HMD. The data collected were time stamped to indicate, among other things, the time when any of the four screens were selected for viewing and the time when each probe question was presented. In this way, the amount of time between the presentation of a given, relevant screen and the probe question could be calculated, as well as the number and timing of any other, irrelevant screens selected for viewing in the interim. (Test participants, of course, had no way of knowing what the next probe question would be, and therefore had no way of knowing which screens would be relevant and which irrelevant. They had been instructed to access the four screens at will in order to help them with their navigation-related tasks.) These were important variables that were used to build the ACT-R model.

Table 1.

First Five Probe Questions Given to Each Soldier During the Scenario and the Hypothesized Memory Retrievals Needed for Each Question as Represented in the Final Models

Probe Question	Memory Retrievals
1) Are you within 50 meters of your next target?	Target Memory Chunk Pace Count Memory Chunk
2) Are there friendly units only to the left of your path?	Friendly Memory Chunk Direction Memory Chunk Unit Memory Chunk
3) Are you within 100 meters of the next way point?	Way point Memory Chunk Pace Count Memory Chunk
4) Is there an enemy unit within 200 meters of your left?	Enemy Memory Chunk Direction Memory Chunk Unit Memory Chunk
5) Are you within 50 meters of your next target?	Target Memory Chunk Pace Count Memory Chunk

2.1 Cognitive Model Concept

The goal of the model was to predict when errors would be made by soldiers in response to probe questions. The model development was predicated on the development of an hypothesis of why errors were made.

The hypothesis was relatively simple: memory decay for the memory chunks needed to answer the probe questions correctly would lead to errors. The more time available for memory decay and the more competing or similar memory chunks, the more likely an error. The probe questions all pertained to information that could be displayed on the HMD. The HMD was blanked during the presentation of the probe questions; thus, information was not immediately available and had to be recalled from memory. Also, various other events, both external (e.g., objects along the path, an irrelevant HMD screen) and internal (e.g., rehearsing or checking information that was not related to the probe question) may have occurred between the viewing of the relevant HMD and when the probe question was asked. Finally, participants may or may not have actually viewed an HMD screen that was relevant to answering a given probe question, since the participant had to choose which HMD screen to view at any given time.

In summary, our hypothesis regarding error generation was based on the decay of a soldier's memory for the information that had been displayed on the HMD. More specifically, the likelihood of a soldier producing an error on a probe question would depend on the amount of decay for each separate memory chunk required to answer the probe question correctly. ACT-R assigns memory chunks an activation level, and this level can be lowered in the model through various means, one of them being time-based decay. In the end, this was a good hypothesis for ACT-R to evaluate, since one of the strengths of ACT-R is its computational simulation of memory decay.

For this study, seven subjects with similar experimental conditions were modeled, and five subjects with dissimilar experimental conditions were dropped. Because of the complexity and overall duration of the navigational task to be modeled, each subject and each question were modeled separately. This reduced the complexity of the modeling effort; however, it also reduced the amount of predictive power that the models would be able to achieve. For example, it eliminated the possibility of evaluating confusion across probe questions.

Each model (which consisted of one subject answering one probe question) was generated from a model template. The model template incorporated general aspects of the navigational task, but not the specific sequence of events that a given subject experienced during his scenario; this information was inserted later. The general model template included walking, checking the display, listening to messages, and responding to probe questions. One note here is that the model assumes nearly perfect perception of incoming information. In other words, if the model presents a screen, it assumes that all the displayed information was viewed and entered into memory perfectly, with no errors in the encoding process. The difference in each model was the precise sequence of events modeled prior to each probe question. In other words, even with the probe questions held constant, depending on their strategies or even small variations in the outside world, different subjects went through slightly different sequences of events prior to answering the question (e.g., precisely when each HMD screen was accessed).

A Macintosh HyperCard ™ stack was created to read the raw data files and then generate the unique LISP code that represented each soldier's unique experience or sequence of events. The unique LISP code was then inserted with ACT-R productions.

The modeling effort here was very extensive, and in many ways, atypical from more traditional cognitive modeling. In all, 2 million lines of ACT-R code were written, compiled, run, and analyzed. Additionally, the modeling effort was challenging because prediction of behavior required a standard set of default parameters to be chosen that would hopefully be representative of the end behavior. Specifically, ACT-R has several parameters that are typically set by their being mapped through iteration to real-world data. Because our models were predictive, we had to make estimates for the default values of many ACT-R parameters. In some cases, this was relatively easy, since a few ACT-R parameters have "unofficial" default values; however, this was difficult for other parameters where default parameters, even "unofficial" ones, do not exist.

2.2 Model Development

In order to begin building the predictive models, first the experimental protocol and the HMD displays were reviewed. The information available to help answer each one of the 20 probe questions was examined. The time when each HMD screen was accessed and each probe question was asked were entered into individual subject's models. The initial assessment, then, was to determine which was the single most important memory chunk for each question and to return or retrieve that single chunk from the model's declarative memory. The average activation levels of the single memory chunk associated with each question from this initial set of ACT-R models were compared to the experimental data; however, the results were not significant.

It was noted, even in the beginning, however, that answering probe questions could involve the retrieval of multiple memory chunks, each with its own decay, and therefore potentially multiple conflicts or partial matches with existing memory chunks (see the second column in Table 1). As a verification, a correlation between errors and these estimates of the raw number of memory chunk retrievals needed for each probe question was computed, and that was significant $r(19) = .43$, $p < .05$. In other words, a probe question that we had hypothesized would need four memory chunks was more likely to be answered incorrectly than a question needing only one memory chunk.

Following these initial analyses, the models were revised to accommodate multiple memory chunks being retrieved. Also, the initial analysis revealed that we had a restriction of range problem with the activation levels of the memory chunks. To counter this limitation, certain ACT-R parameters were changed in order to increase the variance among activation levels. Third, following a consultation with a military subject matter expert (SME), it became clear that an important variable had been omitted from the initial models. Infantry soldiers typically keep track of something called a "pace count." This is a count of the number of steps they have taken while walking which then enables a soldier to calculate the total distance traveled in meters. Infantry soldiers can then use this information to estimate how far certain distances are by determining how far they have walked. This memory chunk, which is constantly revised, was then added to the ACT-R models. The final models contained memory retrieval productions for multiple memory elements and the pace count.

In general, there were memory chunks for targets and way points. Targets could be either friendly or enemy and they were designed so that they could be confused with each other by via ACT-R's partial matching feature. The direction memory chunk was treated as a separate memory chunk and not as a slot belonging to another memory chunk. This allowed for retrieval of the direction chunk separately from targets or way points. The pace count memory chunk was constantly updated by the simulation; this was designed to simulate the soldier keeping track of his approximate distance traveled as a function of his footsteps.

3 RESULTS

The ACT-R models of each question for each subject (a total of 140 models) were each run 40 times. For each subject, the activation values of each memory chunk across the 40 runs was averaged. Then the activation values of all the memory chunks retrieved for each question were averaged. In other words, the activations from the 40 runs were averaged across runs, across memory chunks, and across subjects to yield a single activation level for each question. In this way, the resulting single activation level per question could be correlated with the percent errors for each question. This analysis yielded a significant correlation. Our hypothesis was that lower activation levels would lead to a higher likelihood of making an error. This enabled use of a one-tailed Pearson's Product Moment correlation coefficient between the average activation value and the percentage of errors for each question $r(19) = -.43$, $p < .03$.

4 CONCLUSIONS

Error prediction using cognitive architectures is a new and exciting application of computational cognitive modeling, but it requires further application and refinement to show its viability to the system design process. In this study, we were able to make general predictions concerning the error rates of probe questions and the subsequent effects on situation awareness using cognitive modeling. The models had to be "tweaked" during the development process, albeit only slightly. It was an adjustment nonetheless. It was also unclear precisely how much benefit the cognitive modeling effort added over the initial (non-modeling) estimates of error generation. One could only speculate that real benefits of modeling would be incurred during the system modification phase, which would allow for errors to be tested on future designs by the use of previously developed models. The benefits of such an iterative error modeling process to system design could be significant; therefore, such efforts should continue to be pursued by the human factors community.

REFERENCES

Anderson, J. (1993). Rules of the Mind. Hillsdale, NJ: Earlbaum.

Anderson, J. R., and Lebiere, C. (1998). Atomic components of Thought. Hillsdale, NJ: Earlbaum.

Grant, S. (1996). SimulACRUM: A cognitive architecture for modeling complex task performance and error. In: Proc. 8th European Conference on Cognitive Ergonomics. (ECCE 8) Granada, Spain, 97-102.

Glumm, M. M., Marshak, W. P., Branscome, T. A., McWesler, M., Patton, D.J., Mullins, L.L., (1998). A Comparison of Soldier Performance Using Current Land Navigation Equipment with Information Integrated on a

<u>Helmet Mounted Display</u>. (Army Research Laboratory publication No. ARL-TR-1604). Aberdeen Proving Ground, Aberdeen, MD.

Newell, A. (1990). <u>Unified Theories of Cognition</u>. Cambridge: University Press.

Norman, D.A. (1981). Categorization of Action Slips. <u>Psychological Review, 88</u> (1), 1-15

Rasmussen, J. (1982). Human Errors – A Taxonomy for Describing Human Malfunction in Industrial Installations. <u>Journal of Occupational Accidents, 4</u> (2-4), 311-335.

Reason, J. (1990). <u>Human Error</u>. Cambridge: University Press.

Author Note

The author would like to acknowledge the contribution of Nancy Grugle (ngrugle@vt.edu) to this effort. Correspondence concerning this article should be addressed to Troy Kelley, Army Research Laboratory, Human Research and Engineering Directorate, AMSRL-HR-SE, APG, MD, 21005-5425. tkelley@arl.army.mil

ALARM RESOLUTION TRAINING AND JOB AID FOR EXPLOSIVES DETECTION SYSTEM OPERATORS

Melissa W. Dixon, Ph.D.

Federal Aviation Administration
Aviation Security Research and Development, AAR-510
William J. Hughes Technical Center, Bldg 315
Atlantic City International Airport, NJ 08405

ABSTRACT

The present paper presents the results of an evaluation of previous and current alarm resolution performance of Explosives Detection System (EDS) operators in United States airports. The evaluation was conducted in three phases, with a small sample. Phase I involved the data analysis of existing performance data for operators, for both field tests conducted by the Federal Aviation Administration and on-line testing software. Phase II evaluated the training program that was recently revised for new hires. While Human Factors Engineers (HFE's) evaluated the course in its entirety, from the slides and operator manuals to the alarm resolution protocols and written exams, only the results of the in-class alarm resolution exercises are discussed. In Phase III, HFE's conducted a field evaluation of the performance of newly certified operators, as well as operators trained with the previous training. HFE's used real bags, of which half contained a simulant improvised explosive device and half contained a non-threat alarm object. The results showed a trend in which operators trained with the new training performed somewhat better at detecting threats than did operators trained with the previous training program. Because the performance data of EDS operators is sensitive security information, only the general results of the evaluation could be discussed in this forum.

1. INTRODUCION

The effectiveness of the national civil aviation security system is highly dependent upon people, especially those who screen passenger baggage. It is the mission of the Federal Aviation Administration's (FAA) Aviation Security Human Factors Program to evaluate and improve human performance within the civil aviation security system. As part of this mission, the Human Factors Program conducted an evaluation of a revised training program designed to enhance the performance of Explosives Detection System (EDS) operators. The present paper presents a summary of the major results of this effort. However, due to the sensitive nature of this information, detailed information about operator performance cannot be included in a publication of this sort.

The CTX 5500 DS™ (CTX) is an EDS manufactured by InVision Technologies, Incorporated. The device uses Computed Tomography (CT) technology to scan the checked baggage of airline passengers for the presence of explosive materials. Because this technology is expensive and the effective throughput is slow relative to the number of bags that people check with the airlines for cargo, only a small percentage of these bags are scanned.

When a bag enters the first scanning chamber, the CTX takes a scan projection X-ray image. The computer then analyzes this image for potential objects that require CT slices for further inspection. The bag is transferred to a second chamber, in which a rotating X-ray source takes "slices" of the identified objects of interest. Finally, computer algorithms analyze each "slice" and compare the data against a database of known explosive materials. If the acquired data matches the information in the database, the system alerts the operator of potential explosive threats by displaying the X-ray of the entire bag on one monitor and the CT slices of the alarm areas on another monitor. Once the CTX system alerts the operator of a potential explosive, the operator must resolve whether the alarm is an actual threat or a false alarm. Operators base their decisions on a well-defined alarm resolution protocol.

Their final decision is either to reject or clear a bag. A "clear" bag is one that the operator deems to be free of a potential IED, and a "reject" bag is one that is deemed to have a potential IED and requires further investigation. The information that is presented by the CTX to operators is complex and the alarm resolution protocol consists of a multi-step decision-making process.

Potential CTX trainees must have previous experience as airport screeners for X-ray machines at the airport checkpoint. All trainees participate in a two-week training course and pass an Operator Qualifying Test (OQT) before they can operate the CTX as a certified operator. In the past, trainees were not permitted to takes notes on the alarm resolution protocol due to its sensitive nature. They instead were on their own to remember the steps for effectively resolving alarms.

It is important to keep in mind that, aside from being presented with an occasional FAA test bag, every alarm that an operator sees is created by an ordinary non-threat item in the bag. However, the operator must be prepared to recognize a real Improvised Explosive Device (IED) should one be present in a passenger's bag. This fact alone may increase the likelihood that operators will bypass and eventually forget the correct steps to follow because they know that the probability of seeing a real threat is virtually zero. In an attempt to maintain the vigilance of screeners, the Aviation Security Human Factors Program, in conjunction with X-ray vendors, developed Threat Image Projection (TIP) software. TIP allows screeners to be trained on IED's while providing a measure of on-line performance for security companies, airlines, and the FAA.

On X-ray machines, TIP superimposes, at a pseudo-random interval, a stored X-ray image of a threat item onto the X-ray image of a real passenger's bag. The TIP image database consists of IED's, knives, guns, and other hazardous materials. When TIP is in operation, the screeners press a designated button each time they believe that they see a threat item. When the operators press the button and a TIP image is present, they are congratulated for finding a TIP threat and the threat is highlighted. The TIP image is then removed so that the operators can look at the real bag image that was occluded by it. When operators press the button and no TIP is present, they are told that no TIP image was projected and they should follow their normal procedures if they believe a threat is present. Operators are also informed when they miss a TIP image.

TIP on the CTX is slightly different than on X-ray. On the CTX, TIP images consist of stored images of complete bags, not just of a threat in isolation. TIP bags can include IED's or non-threat alarm objects that resemble the characteristics of explosive material. When operators make their final decision for the bag and it is a TIP, they are informed that the bag was a TIP and then the image of the real bag inside the scanning chamber is shown. However, they are not given detailed feedback. The TIP system records the operator's decision, as well as other operational information, for each real and TIP bag displayed.

Despite training and the presence of TIP on all CTX machines, previous operator performance on covert FAA field agent tests has been less than ideal. To help operators avoid using incorrect alarm resolution procedures, the FAA Security Equipment Integrated Product Team has recently revised the CTX training course to include a new alarm resolution protocol and a job aid. Because of the sensitive nature of the protocol, the job aid includes only key questions to trigger recall of the steps during training and in the field. The goal of the effort reported here was to evaluate the performance of new trainees during the revised training as well as in the field.

2. METHOD

The evaluation of the revised training was conducted in three phases. Phase I evaluated the field performance of operators trained with the previous course. Phase II evaluated the quality and effectiveness of the revised training materials, with a focus on the performance of operators using the revised alarm resolution protocol and job performance aid (JPA). Phase III tested operators' ability to identify IED's, as well as their knowledge and use of the alarm resolution protocol once they were certified operators.

2.1 Phase I: Operator Performance with Previous Training

In Phase I, human factors engineers (HFE's) evaluated two main measures of performance for operators who were trained with the previous training program, Threat Image Projection (TIP) and FAA field agent test results. FAA field agent tests consisted of simulated explosives inserted into FAA test agent bags. The FAA field agents make arrangements with the airline ticket agents to tag the bag like any other checked bag that requires scanning by the CTX. In this way, the operator is unaware that they are being tested. The agent announces that a bag is an FAA test once the operator has finished processing it. HFE's calculated the probabilities of detection and false alarm for FAA field agent tests conducted at 25 airports between September 1997 and October 2000.

Phase I also incorporated an analysis of TIP data. HFE's calculated the probabilities of detection and false alarm for TIP data collected from 8 airports between January and July 2000.

2.2 Phase II: Training Evaluation

During Phase II, HFE's observed and evaluated the 2-week revised CTX training course at a large airport in the United States. Seven trainees were enrolled in the training. The CTX training package includes nine training modules or lessons, written exams and quizzes, an operator's guide, the alarm resolution job aid, and on-the-job training with a CTX simulator. HFE's evaluated the course materials for the following eight fundamental human factors criteria for training: completeness, clarity, conciseness, consistency, compactness, currency, construction, and communication. HFE's also evaluated the performance of trainees on the in-class exams, practice alarm resolution, and the FAA OQT. This paper will focus only on the in-class alarm resolution performance of trainees for images on the simulator, which was conducted on days 8, 9, and 10.

2.3 Phase III: Operator Performance after Revised Training

During Phase III, HFE's returned to the airport visited in Phase II to test both newly certified and existing CTX operators. The five newly certified operators had worked on the CTX for approximately 2 weeks, whereas the three existing operators had worked on the CTX for 14 months. HFE's administered written tests of operator knowledge with respect to the alarm resolution protocol steps, as well as a test of alarm resolution performance using 20 FAA test bags. Ten of the FAA bags contained an IED and ten contained a non-threat item that produced a false alarm on the CTX. HFE's observed operators' use of the alarm resolution protocol with regular stream of commerce bags in addition to the FAA test bags. Following these tests, HFE's downloaded decision data that the machine recorded for real and TIP bags. HFE's evaluated the performance of operators on the written tests, FAA test bags, TIP images, and protocol use for real bags. This paper focuses only on the alarm resolution results for FAA test bags and TIP images.

3. RESULTS

The following sections present the overall results of the three phases of this effort. Because information about the ability of CTX operators to detect IED's is security sensitive information, detailed information about operator performance cannot be disclosed in this forum. Requests to the author for information regarding the details of these results will be reviewed and approved on an individual basis.

3.1 Phase I: Operator Performance with Previous Training

Because the FAA field agent tests take into account operator performance using the CTX monitors as well as their hand searches of rejected bags, there were three possible outcomes: Pass (correct detection of the IED on the CTX monitor), Fail (incorrectly clearing a bag with an IED), Pass CTX-Fail Search (correct detection on the CTX monitor but failure to find the IED during hand search). Because the primary interest of this data assessment was to determine operator performance on the CTX and not on hand search, the Pass and Pass CTX-Fail Search results were summed to obtain an estimate of passing rates just on the CTX machine (Pass CTX). Of all the tests attempted by field agents, only 86.26% were conducted successfully. An example of an unsuccessful test could be a machine failure or a field agent being recognized by the operator. Because an operator can only reject or clear a bag on the CTX, 50% represents chance performance for correctly detecting an FAA IED bag. The results showed that operators performed above chance for the field agent test bags.

The mean time to resolve TIP bags was 34.05 seconds, with 74.48% of bags containing only 1 threat. For TIP images, operators obtained a relatively high probability of detection (P_d) but their probability of a false alarm (P_{fa}) was also somewhat high. Note that the P_d for TIP images was substantially higher than that obtained for FAA field agent bags. Although this could be due to true ability to detect threats, an alternative argument is that it could be due to operator awareness of when a TIP image is presented. In fact, some operators have reported that they know when a TIP image is presented on the CTX, especially if they have seen the bag enter the chamber but a bag with different characteristics shows up on the CTX monitor. They have also reported that there may be a slight delay before a TIP image is presented relative to a real bag image. Because of the ratio of IED images relative to non-threat images in the TIP database, there is a high probability that operators will correctly detect an IED TIP if they reject the bag. Therefore, if operators know that they are being tested and they know that failure to detect an IED is worse than incorrectly rejecting a non-threat bag (i.e., a false alarm), then they may reject all test bags.

3.2 Phase II: Training Evaluation

Although an HFE evaluated all of the relevant training materials, only the data for the in-class alarm resolution procedures are presented here. On day 8, each student processed only 2 bag images. On days 9 and 10, each student processed 5 bags. Because operators used the job performance aid throughout this training exercise, these results serve as a small-scale validation of the revised protocol. Overall, the P_d was very high and the P_{fa} was very low. All trainees passed the final exam for the course, but only 5 of the 7 trainees passed the OQT and became certified CTX operators.

3.3 Phase III: Operator Performance with Revised Training

Operators processed 20 FAA test bags each, for a total of 160 trials. However, 7 (4.38%) of the trials did not produce an alarm after three attempts and were declared "no test" bags. Operators who completed the revised training course earned a higher P_d and slightly higher P_{fa} compared to operators who completed the previous training course. However, separate Chi-square analyses on the relationship between operator performance and type of training for IED bags and non-threat alarm bags did not yield significant relationships; IED bags, $\chi^2 (1) = 1.72, p = 0.19$, and non-threat bags, $\chi^2 (1) = .12, p = 0.73$. Had more data been gathered for operators, the detection rates might have shown a relationship to the training received. The detection rates on the IED test bags for operators who received the previous training are similar to the detection rates recorded for FAA field agent tests at the same airport, which is interesting considering that FAA field agent tests are covert tests.

The overall mean time to resolve an FAA test bag was 79.59 seconds, with no differences between operators trained with the previous training (78.78 seconds) and operators trained with the revised training (80.09 seconds). The mean number of slices that operators took per bag was 17.48. Operators trained with the new training took significantly more slices (19.95) than did operators who received the previous training (13.56), $F(1, 151) = 4.84, p = 0.03$. This is important because the previous protocol instructed operators take additional slices of any threat that they felt required further investigation. The revised protocol, as one of the first steps, requires operators to take additional slices of all threats identified by the CTX, regardless of whether the operator at first glance believes that the object is suspicious. It is unclear whether this difference contributed to the higher detection rates for newly certified operators.

HFE's also analyzed the performance of operators for TIP images. While newly certified operators performed better than previous operators on the FAA bags, the opposite was true for TIP images. A Chi-square analysis showed a significant relationship between operator performance on IED TIP images and type of training, $\chi^2 (1) = 13.24, p < .001$. This relationship was not observed for non-threat TIP images, $\chi^2 (1) = 1.48, p = 0.22$. Because the newly certified operators had only operated the machine for two full weeks prior to the testing, they had only one-twelfth the TIP data compared to existing operators. Based on this difference, strong conclusions cannot be made about TIP performance and differences due to training. Moreover, existing operators had much more experience with the TIP system, and as previously discussed, there is some indication that cues might exist when a TIP image is presented. New operators may not have learned these cues yet, so their data may be more representative of true performance.

The mean time to resolve a TIP alarm bag was 29.99 seconds. Operators trained with the revised training had significantly longer processing times (53.89) than did operators trained with the previous training (28.12), $F(1, 151) = 13.70, p < .001$. Note that new operators increased their processing times by about 50% for FAA bags compared to TIP bags, while previous operators almost tripled their processing times for FAA bags. The mean number of slices that operators took per bag was 3.99, which is one-fourth the number of keystrokes made for FAA test bags. Operators trained with the new training took significantly more slices (8.91) than did operators who received the previous training (3.60), $F(1, 151) = 10.49, p = .001$. These differences in processing times, keystrokes, and slices taken may indicate that the presence of an observer and the atmosphere of a test situation may have introduced some bias and modified true screener behavior. They may have been more likely to use the protocol that they were taught in class or to look at the images more closely.

It is also important to mention here that although operators were given copies of the job performance aid when they finished training, none of the newly certified operators used it on the job. One person indicated that she regularly studies the job performance aid, but not while she is working.

4. CONCLUSIONS

The results of Phase I showed that CTX operators trained with the previous training program showed less than ideal detection rates for FAA field agent test bags. While their performance was quite good for TIP images, they showed moderate false alarm rates. This difference between performance on FAA field agent test and TIP tests might result from operator awareness of when a TIP image is presented. This is further supported by operators' high false alarm rate for TIP images. If operators suspect that an image is a TIP image, then they may be more likely to respond by rejecting the bag, which increases their probability of detecting an IED. Phase II showed that when operators use the job performance aid, they produce acceptable detection and false alarm rates. Finally, Phase III demonstrated a trend, whereby newly certified operators obtained slightly better detection rates for the FAA bags than did existing operators. The fact that newly trained operators took more slices, but processed bags in the same amount of time, lends support to the argument that the revised training produces better prepared CTX operators. The opposite result was obtained for the TIP bags, with existing operators detecting more TIP IED images than newly certified operators. This may have been due to an awareness of the presence of a TIP image by existing operators, which is supported by their high false alarm rate for TIP images relative to FAA test bags. A follow-on evaluation of the performance of newly certified screeners on TIP and FAA tests would provide important information about changes in detection rates over time.

Axiomatic And Pragmatic Approaches To Modelling Of Cognition

Erik Hollnagel

CSELAB, Institute of Computer and Information Science, University of Linköping,
SE-581 83 Linköping, Sweden, erik.hollnagel@ida.liu.se

Abstract. The modelling of cognition is a central issue in the study of human-machine systems. A distinction can be made between an axiomatic and a pragmatic definition of cognition. In the former case, cognition refers to a set of internal functions that humans have, and that possibly may be shared by a limited number of sophisticated technological artefacts. In the latter aces, cognition refers to a quality of any system that has certain performance characteristics, notably that it is able to maintain control of a process. This anchors the definition on the characteristics of the tasks and of system performance, rather than on the possible constituents of the mind and explanations of plausible mechanisms for the same. The two positions lead to very different approaches to the modelling of cognition, and it is argued that the pragmatic approach is preferable in terms of both functional and structural simplicity and predictive power.

1. WHAT IS COGNITION?

Since the early 1990s it has been *comme il faut* to use the adjective "cognitive" to adorn otherwise dour technical terminology. This has led to interesting neologisms such as cognitive tools, cognitive models, cognitive task analysis, cognitive ergonomics, cognitive structures, cognitive processes, cognitive engineering, cognitive design, etc. The development has left a great many people somewhat puzzled, not least those who are struggling with the practical problems of human-technology interaction. Once reason for their bewilderment is that it is uncertain what the meaning of "cognitive" is in the first place; another reason it that is often not clear whether this development denotes a change of terminology or a change of substance. It does little to improve the situation that the cognoscenti use the term with the tacit understanding that everybody knows what it means, and that there therefore is no need to define it properly.

Aside from the mock definition that cognition is that which goes on "between the ears", the term has broadly speaking been used to describe the psychological processes involved in the acquisition, organisation and use of knowledge – emphasising the rational rather than the emotional characteristics of human behaviour. Etymologically, the term it is derived from the Latin word *cognoscere*: to learn, which in turn is based on *gnoscere*: to know. A Pickwickian approach would therefore define cognitive tasks as those tasks that require or include cognition. This easily leads to an axiomatic position, which starts from the fact that humans are cognitive beings (or that humans have cognition), that human performance therefore has a cognitive component, and that models of human performance for that reason must be models of human cognition.

However, following the same line of reasoning one could also argue that human actions are driven by motives and have emotions, hence that human performance has a motivational and an emotional component. This would, indeed, be fully justified but would also pose serious problems for information processing theories and models of human-machine interaction and human behaviour. Although it may possibly suffice to describe how emotions affect cognition, and thereby remain on the *terra firma* of human information processing (Mandler, 1975), it should not be ruled out that complementary theories of motivation and emotion may be possible and necessary. While it is evidently true that humans have cognition, the axiomatic position also makes it difficult to extend the notion of cognition to other entities, such as technological artefacts and organisations. (The problems would be even worse for motives and emotions.) The axiomatic position also begs the question of what cognition is, assuming that it is more than an epiphenomenon of information processing.

2.   LEVELS OF MODEL DESCRIPTION

Among the many uses of the term cognition, cognitive modelling is one of the earliest and most widespread. Accepting the common notion that a model is a condensed description of the essential features of an object or reference system in some respect, it is possible to distinguish among several dissimilar levels of modelling:

◆   **The level of performance**, meaning the characteristic overt behaviour of a person – or more generally of a system. This can focus on specific performance aspects, such as failures or "errors" (Reason, 1990), or more generally the orderliness of performance (Hollnagel, 1998).

◆   **The level of internal functions** that constitute the commonly accepted basis of performance or actions. These refer either to what we know from introspection (e.g., attention, planning, decisions, problem solving) or to what we can infer with reasonable confidence from systematic observations and experiments (e.g. decision heuristics, forgetting, capacity limitations). The internal functions are typically described in terms of information processes, although information processing and cognition are not strictly synonymous.

◆   **The level of (hypothetical) brain functions** that correspond to established mental faculties or capacities, but described at the level of cell assemblies rather than at the level of neurones. This can be supplemented by descriptions at the level of molecular function that focus on properties of individual neurones, cell components, and even further down to the level of chromosomes and genes.

Without in any way being exhaustive, the three levels do represent significantly different types of modelling and strongly suggest that a description that only involves one level may be incomplete. In practice the different levels complement each other and serve to reduce possibly ambiguous descriptions and explanations. It is also common to rely on a single level of description where the choice is dictated by the purpose of modelling, for instance one of the following:

•   A **scientific** purpose, to aid understanding of something (a phenomena, a system). In this case a model explains the phenomenon in question and provides an account of the mechanisms or functions (the causal or functional architecture) that underlie the phenomenon. (This, of course, begs the question of what an explanation is since it can clearly not be the model in itself.)

•   An **engineering** purpose, for instance to construct a representation of a phenomenon or system that can be used to **calculate** or **predict** how system events will develop. Here the model is a correct translation of essential functional relationships and dependencies into a form, which enables controlled manipulation of the independent parameters (including the environment). An engineering model does not necessarily constitute an explanation, but can be a black box, a finite state automaton, or a convenient representation of input-output correlations. Common examples are neural networks, mathematical models (differential equations), linear regression models, supervisory control models, Kalman filters, etc.

- A **cybernetic** purpose, in the sense of providing the representation necessary to control a system. This usage is based on the Law of Requisite Variety, which can be expressed as saying that a regulator of a system must be a model of that system (Conant & Ashy, 1970). Although control implies a certain amount of prediction, the need for precision and details is quite different from the engineering use of models.

The axiomatic approach to modelling of cognition takes for granted that the model must serve a scientific purpose, and that a description at the level of internal functions therefore is the most appropriate. It is, however, important that the intended usage of a model is explicitly acknowledged, both because there is a need to communicate with others and because of the implications for how the results from the modelling can be applied. The industrial need of models is not just to be able to understand something better, but rather to find efficient – or at least workable – solutions to practical problems. The level of model description should therefore correspond to the level of practical problem issues and in that respect an emphasis on cognition, as the internal functions of the human mind, may not be the best choice.

3. COGNITION AND CONTROL

While it is a truism that human behaviour cannot be explained without understanding what goes on in the mind, this does not imply that there is only one way of describing it. Indeed, it is also a truism that what goes on inside a system cannot be described separately from what goes on outside, and *vice versa*. The mistake of the axiomatic approach has been to confuse the fact that there are general characteristics of cognition (common traits, common functions, features across situations) with the assumption that cognition is an internal and context free process. The alternative is to accept that the general characteristics, hence cognitive functions and the orderliness of performance, simply reflect the relative constancy of the environment, to paraphrase Simon (1970).

The pragmatic approach to modelling of cognition does not view cognition as a set of functions or processes in their own right, but rather as a way of explaining how people – and systems – are able to maintain control of what they do. The important issue is therefore not what cognition **is** but what cognition **does**. To say that a system is cognitive (rather than to say that it has cognition) means that it is able to do certain things, specifically that it is able efficiently to achieve particular purposes. In consequence of that, modelling should be of the system's performance, i.e., of cognition and context together rather than of cognition alone. The basic premise is that cognition is always embedded in a context, which includes demands and resources, the physical working environment, tasks, goals, organisations, the social setting, etc. In order to be able to understand how a human-machine ensemble can do meaningful work, it is therefore necessary that modelling occur at the level of meaningful system behaviour, rather at the level of the underlying processes.

One place where this need becomes abundantly clear is in the tight coupling between human performance and time. As pointed out by Decortis & De Keyser (1988) time has always been treated in a perfunctory manner in the fields of human-machine action and human-computer interaction. Despite the fact that everything we do takes place in time, that everything we do has a certain duration and therefore uses time, and that everything we do needs coordination with what others do and when they do it, time is conspicuous by its absence from the information processing models. The rough effects of time are recognised in relation to specific issues such as short-term retention and workload, but the fact that a major part of our reasoning and planning is about temporal relations is difficult to reconcile with the axiomatic approach to modelling and to the representation of cognition as internal functions in the mind.

3.1 Time, Control, And Cognition

In the pragmatic approach to modelling of cognition, the emphasis is on a system's ability to maintain control and this also forms the basis for the definition of cognition. One consequence of that is that time is fundamental to the modelling approach, since actions always take place in time and must consider the dynamics of the process that is being controlled. (Note that the term process is used in the widest possible sense, to mean anything that changes appreciably relative to the time period of description.) This is clearly seen from the paradigmatic representation of the pragmatic approach, which is the cyclical model shown in Figure 1. As shown here, at lest four types of time are important for how a system controls a process. These are:

- The time needed to evaluate events (T_E). Control requires feedback, and the controlling system must therefore spend some time to evaluate the information it receives or selects. Regardless of whether this information is the expected outcome of previous actions or the consequence of unexpected events, the evaluation takes some time – particularly in the latter case. The quality of the evaluation furthermore depends on the time that can be spent on it.

- The time needed to chose an action (T_S). The effectiveness of control depends on choosing actions that efficiently achieve intended goals. The actions are chosen on the basis of the current understanding (called the construct in Figure 1) and the choosing itself is a process that takes time, particularly on higher levels of control (Hollnagel, 2001).

- The time needed to carry out the chosen action (T_P). This can be understood in two different ways, either as the time that it actually takes to do something or as the time that is assumed to be needed to do something (the expected performance time). While the former may vary in ways that cannot always be foreseen, the latter should be a matter of making an estimation, based on experience. In practice, this is often more difficult than it sounds, and estimates of the time needed to do something may be quite imprecise.

- Finally, the time that is available for the controlling system either to evaluate, choose, or to carry out a chosen action. This time may vary considerably depending on the nature of the process (e.g. the difference between flying an airplane and controlling the flow from a dam), and very often the controlling system has little possibility of increasing it.

The short treatment given here only serves to indicate how time can be represented in the pragmatic approach to modelling of cognition. In relation to the line of arguments presented here, it is noticeable that the cyclical model does not refer to cognition as information processing, but instead focuses on certain characteristics of the performance of a controlling system. The pragmatic approach to modelling of cognition does not make any assumptions about the nature of internal functions, and indeed does not describe internal functions at all. It is, of course, inevitable that the description of performance at some time will require a description of plausible "mechanisms", but when that happens the choice should be determined by the principle of parsimony rather than the axioms of human information processing (e.g., Bye et al., 1999). This relieves the modellers of the need to account for phenomena that are artefacts of the model, heeding the advice of Broadbent (1980) that research should not start with a model of man, but rather with the practical problems.

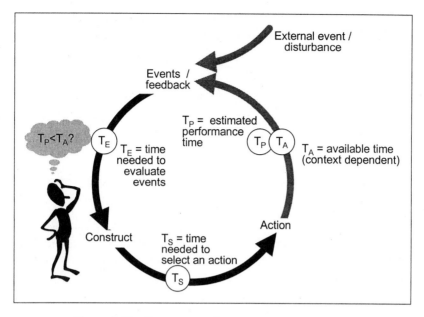

Figure 1: Significant types of time in the cyclical model.

For the practitioner, engineering and cybernetic models may be adequate to provide the required solutions. But even for the scientific use of models, it may be worthwhile to consider whether it would not be better if the modelling of cognition started top-down from the problems that need answers, rather than bottom-up from the traditional assortment of information processes and cognitive structures.

4. REFERENCES

Broadbent, D. E. (1980). The minimization of models. In A. J. Chapman & D. M. Jones (Eds.), *Models of man*. Leicester: The British Psychological Society.

Bye, A. Hollnagel, E. & Brendeford, T. S. (1999). Human-machine function allocation: A functional modelling approach. *Reliability Engineering and Systems Safety, 64*, 291-300.

Conant, R. C. & Ashby, W. R. (1970). Every good regulator of a system must be a model of that system. *International Journal of Systems Science, 1*(2), 89-97.

Decortis, F. & De Keyser, V. (1988). *Time: The Cinderella of man-machine interaction*. IFAC/IFIP/IEA/IFORS Conference on Man-Machine Systems, 14 - 16 June 1988, Oulu, Finland.

Hollnagel, E. (1998). Context, cognition, and control. In Y. Waern (Ed.). *Co-operation in process management – Cognition and information technology*. London: Taylor & Francis.

Hollnagel, E. (2001). IEEE

Mandler, G. (1975). *Mind and emotion*. New York: Wiley.

Reason, J. T. (1990). *Human error*. Cambridge, U.K.: Cambridge University Press.

Simon, H. A. (1972). *The sciences of the artificial*. Cambridge, MA.: The M. I. T. Press.

Human Model Simulation of Plant Anomaly Diagnosis (HUMOS-PAD) to Estimate Time Cognitive Reliability Curve for HRA/PSA Practice

Wei Wu*,Takashi Nakagawa*, Hidekazu Yoshikawa**

*Power & Public Utility Systems Dep., Industrial Electronics & Systems Lab., MITSUBISHI Electric Corp.
1-1, Tsukaguchi-Hommachi 8-chome, Amagasaki City, HYOGO, 661-8661,JAPAN
** Graduate School of Energy Science, Kyoto University, Gokasho, Uji City, KYOTO, 611-0011, JAPAN

This work is related with human model development for real-time simulation of nuclear power plant (NPP) operator's cognitive behavior to diagnose plant anomaly through the control interface. The employed method for HUMOS-PAD is largely based on blackboard architecture to configure human information processing mechanism on computer. The information processing is modeled as a number of data-processing subtasks in human working memory (WM) and long-term memory (LTM). Efforts were made to model the inherent variety and diversity in human behaviors. The simulation results demonstrated that the developed model not only can simulate individual operator's cognitive behaviors well, but also can give good statistical analysis results for human cognitive reliability analysis (HRA) in probabilistic safety analysis (PSA).

1   Introduction

NPP system is one of the "human in the loop" systems, which consists of human operators, machinery system, and man-machine interface (MMI). Though the reliability of the machinery system has been enhanced much, there remain further efforts of human factors, towards the enhancement of the safety and reliability of the system as a whole. So far, carrying out large-scale experiments is the main approach to deal with human factors in human machine interaction[1]. The experimental data and the data analysis results do provide valuable information for the MMI design, evaluation, HRA and so on. But, not only vast time and enormous cost are required, but also the applicability of the experimental results is limited within the tested scope[1,2].

A new approach of using human model simulation has been put forward[2,3] by modeling the real skilled operators behaviors, called as human model here. The merit of the human model approach is its much more flexibility and applicability by modifying MMI simulator, or human model, or NPP simulator in accordance with the required situation. Also such numerical simulation experiments could be repeated with very cheap cost compared with the large-scale experiments.

Needless to say, the important subject of human model approach is how well the model could simulate human behaviors. Especially, with respect to the application of such approach to HRA as related with PSA for NPP, it has been pointed out that the model should represent well versatile human behaviors on monitoring and controlling the process plant with various environmental effects surrounding human's tasks being taken into consideration[3]. This paper describes a fundamental study conducted to model operators' internal cognitive behaviors during detecting and diagnosing an abnormal transient in NPP. It aims at making the application level of human model closer to the level where the analysis and the evaluation of NPP safety and reliability could be conducted by means of human model simulation. The developed system is called as human model simulation of plant anomaly diagnosis (HUMOS-PAD).

2   Laboratory Experiment

A laboratory experiment was conducted to collect data for modeling operator's behaviors in detecting and diagnosing abnormal transients. The laboratory experiment itself was conducted for the purpose of obtaining data on cognitive behaviors of operators who would detect an abnormal transient and then identify the root cause through a MMI of NPP simulator. The MMI used in the experiment is a kind of CRT-based interface configured with 16 screens where various kinds of information were presented to show the dynamic status of NPP emulated by a PWR simulator. Alarm sounds were not used intentionally and very limited warning messages were presented on the CRT interface. Subsequently, subjects had to detect the occurrence of an abnormal transient by himself rather than by depending on alarms. Three engineers participated in the laboratory experiments as the subjects. They either had engaged in

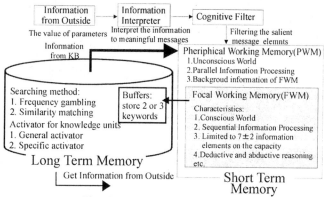

Information from Outside → Information Interpreter → Cognitive Filter	

The value of parameters | Interpret the information to meaningful messages | Filtering the salient message elemnts

Information from KB

Pheriphical Working Memory(PWM)
1.Unconscious World
2.Parallel Information Processing
3.Background information of FWM

Searching method:
1. Frequency gambling
2. Similarity matching

Activator for knowledge units
1. General activator
2. Specific activator

Buffers: store 2 or 3 keywords

Focal Working Memory(FWM)

Characteristics:
1.Conscious World
2. Sequential Information Processing
3. Limited to 7±2 information elements on the capacity
4.Deductive and abductive reasoning etc.

Long Term Memory

Get Information from Outside

Short Term Memory

Fig.1 Human modeling framework

Tab.1 Variable factors in detecting and diagnosing an abnormal transient

Anomaly Detection	Anomaly Diagnosis
Variation degree of parameters for judging something wrong	Hypothesis recalling
Parameter reference frequency	Sequence of examining hypothesis
Parameter reference sequence	Selecting display type
	Confidence level of making decision

designing control system and training simulator of PWR plant, or had experiences of being an instructor in a PWR operator-training center. Subjects' behaviors in the laboratory experiment could be divided into two phases: monitor phase and diagnosis phase since the subject were asked to push a button located in every screen to indicate explicitly that he detected something wrong and push another one to indicate that he got the root cause. Experimental data were analyzed separately for the two phases so that the analysis results could provide a basis for the human modeling development. The attention of data analysis is concentrated on both the common tendency in the behaviors of all experimental subjects and the differences in the individual cases. The common tendency was utilized to clarify the fundamental mechanism of the cognitive behaviors and suggest modeling framework as described later. The individual characteristics were utilized to suggest how to tune the developed model to simulate the diversity and variety in the cognitive behaviors. Table1 summarizes the variable factors in detecting and diagnosing abnormal transients.

3   Human Model Development

Authors have made efforts to develop a computer simulation model based on the experimental data and data analysis results. Figure1 shows the human modeling framework. It is basically the expanded framework of J. Reason's memory model[4,5] by adding the deductive and abductive reasoning mechanisms that were not considered in the original model. There are five components in the human modeling framework. Among them, short-term memory (STM) and long-term memory (LTM) are core parts. The deductive reasoning and abductive reasoning are performed in STM to formulate hypotheses and evaluate them. During the information processing in FWM, necessary information could be retrieved from LTM by sending cues from STM. LTM represents human knowledge database including various kinds of his own experiences, and learned theoretic knowledge. And there are two ways to search LTM by the cues sent from FWM: similarity matching and frequency gambling.

There are two types of information to be processed: one is in STM and the other one is in LTM. The knowledge model of authors' human model is constructed by modeling such two types of information. As for the former, working memory element (WME) is introduced to model the information elements in STM processing. While, as for the latter, knowledge database (KDB) that has a graphical network structure is devised for operators' knowledge and experience about diagnosing abnormal transients.

Data structure of a WME is as shown in Table 2. *Category* has several types corresponding to several processing contexts in the information processing. *Processing State* describes whether the WME had been processed by STM processing. *Processing Priority* defines the order of the processing sequence. And *Impression Index* shows preservation time of a WME in STM. All the data are defined by PWM information processing.

Only the WME whose *Processing Priority* is higher and *Processing State* is not "Yes" will be transferred into FWM. Information processing in FWM is mainly conducted to define the cues for searching KDB in accordance with *Category* and *Contents* of WME, e.g., to verify hypothesis, or to make final diagnosis decision. The total number of WME and the number of WME whose *Processing State* is not "Yes" can be considered as an evaluation-index of human mental workload in cognitive information processing. Also the cognitive processing procedures could be

1471

Tab.2 Data Structure of WME

Number	No.XXX
Category	Alarm, hypothesis, value prediction, trend prediction, value state, trend state, diagnosis results
Content	Defined in accordance with Category of WME
Processing State	Yes or No or Reserved or Verifying
Processing Priority	Number from 0 to 2
Impression Index	Initial value=7, will be decreased by 1 per second if Processing State is set to Yes

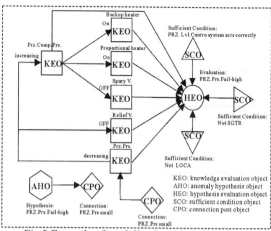

Fig.2 Example of modeling knowledge database

clarified by tracing *Contents* of WME,.

KDB is the model of LTM that stores various experiences and various kinds of knowledge. The configuration of LTM is assumed as a sort of network structure that maintains systemically the relationships, processing procedures of the knowledge and experiences. In this study, operators' experiences and knowledge about diagnosing abnormal transients are divided into two groups: knowledge about plant control system, and experiences or knowledge about accidents in NPP. Furthermore, the concept of *knowledge module* is defined to represent sets of knowledge, e.g., knowledge about LOCA, SGTR, and Pressure Control System. As an example, figure2 shows the model of knowledge module of "Pressurizer Pressure Control System". The model is constructed based on the logical relationship of each components of the pressure control system. While the knowledge module about accident describes the experiences or knowledge about what symptoms will be observed if such accident occurs.

As seen from Fig.2, the knowledge network database is constructed by five types of objects and pointing arrows between them. The pointing arrows between the knowledge objects represent the cause-and-effect relationship between them. The concrete contents of the relationship are described in the format of IF-THEN rules in each object. "Knowledge Element Object (KEO)" is the fundamental elements for constructing the network-configuration knowledge database. It stores the information related to the parameter's steady value, location, rules of cause-effect relationship with other KEOs and so on. "Anomaly Hypothesis Object (AHO)" is the model of anomaly hypothesis. AHO stores the prediction about the status of plant parameters in the format of "declare type" rules as shown in Fig2. "Hypothesis Examination Object (HEO)" is the model of knowledge utilized to examine the hypothesis represented by AHO. The hypothesis examination was modeled as an accumulation process of the belief on the correctness of the hypothesis (conviction or negation). Such subjective belief is represented in quantity called here as *Confidence Score*. We think that when a certain level is reached by accumulating *Confidence Score*, subject will do the decision of accepting or rejecting a hypothesis. Here such belief level is called *Confidence Level*. The *Confidence Level* could be different case by case. For the cautious person, it may be a high, while for others it may be low. Operators will get more parameters' information related to the hypothesis before *Confidence Level* is reached. "Sufficient Condition Object (SCO)" is utilized to model the re-confirmation activities in the cognitive behaviors of anomaly diagnosis that were observed frequently in the case of cautious operators. KEOs are also called as necessary conditions compared with SCO. That means all symptoms represented by KEOs should be checked during examining a hypothesis. Subsequently, *Confidence Level* of an abnormal hypothesis would be increased or decreased during the verification of necessary and sufficient conditions of the correspondent HEO. Finally, "Connection Post Object (CPO)" is just an object defined as such "the connection posts are identical if the names of connection posts are same". CPO could be utilized to model the connection of different knowledge modules. It is very useful in modeling the formulation of abnormal hypotheses over different knowledge modules.

Combined with WME and the network-structured KDB model, Fig.3 shows the configuration and inter-relationships of the various kinds of cognitive information processing on the basis of the human modeling framework. In PWM processing, four tasks are performed: making WME based on various information, setting the

1472

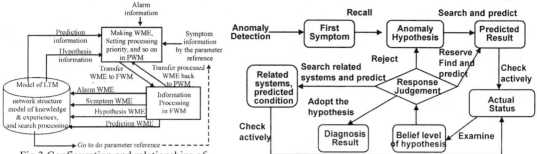

Fig.3 Configuration and relationships of
information processing tasks

Fig. 4 Anomaly diagnosis procedures

processing priority of WME, transferring WME to FWM, and changing the impression index of WME. While in FWM processing, the tasks are performed based on the category of WME: (1) "alarm": putting keywords based on WME into LTM to search the parameter corresponding to the alarm, (2) "value/trend prediction": putting keywords based on WME into LTM to search the location of the parameter to be checked next and then go to do aggressive reference, (3) "value/trend state": putting keywords based on WME into LTM to find a new hypothesis to be verified, or to change CS of the hypothesis that is verified at that moment, (4) "hypothesis": deciding whether to accept the hypothesis or to deny it, or putting keywords based on WME into LTM to search a new parameter to be checked next for accumulating more value/trend state information, and so on. As for the LTM processing, the main task is to search the database according to the keywords. Such task includes searching the parameters to be checked next, searching hypothesis to be verified, searching rules to change confidence score of hypothesis, and so on.

Figure 4 shows the executing sequence of the information processing tasks in Fig.3, as explained below.

A) Detect somewhat abnormal symptom by either alarm or periodic parameter checking.

B) Recall anomaly hypothesis based on the first-symptom that is the initial symptom detected.

C) Predict the changes in the status of related parameters in accordance with the recalled hypothesis

D) Verify the predicted changes for the related parameters and compare it with the prediction.

E) Increase or decrease the confidence level of the hypothesis in accordance with the comparing results.

F) Based on the changed confidence level, switching process modes among (1) accept the hypothesis, (2) reject the hypothesis, and (3) check more related parameters. In the case of (2), (3), go back to B), D) separately.

G) Reconfirm the hypothesis by checking the status of other related subsystems of NPP in the case of (1).

H) Conclude the diagnosing result and end the simulation.

The above processing sequence is summarized from experimental data of all trials. It represents an overall anomaly diagnosing process observed in the laboratory experiments. The individual characteristics of experimental subjects could be simulated with changing setting parameters to realize the effects of the variable factors summarized in Tab.1.

4   Validation and Application of HUMOS-PAD

HUMOS-PAD has been utilized to conduct numerical experiments. The validation of HUMOS-PAD is conducted by comparing the numerical experimental results with the laboratory experimental results. The comparisons were made on several aspects such as anomaly detection time, first symptom, the relationship between the first symptom and first hypothesis, detailed diagnosing procedures[6]. Figure5 shows a comparison of detailed diagnosing procedure between "Subject I" and HUMOS-PAD. The operation activities were divided into several phases that are explained in Tab.3. The diagnosing patterns were simulated well by HUMOS-PAD from the inter-comparison results. Defenses in the detailed parameter reference are considered as the inherent variety in human behaviors. The other validation results also turned out that not only the common tendency of cognitive behaviors could be simulated well, the individual characteristics were reflected well also by changing the setting parameters in the common modeling framework of HUMOS-PAD.

Authors have further applied HUMOS-PAD to estimate time-cognitive reliability curves that reflect an important human error probability parameter used in NPP HRA/PSA[7]. Conducting a sort of Monte Carlo simulation by using HUMOS-PAD with changing the input parameters values related with the variability of human information processing at man-machine interface, the TCR curves were estimated by numerical experiments. Figure 6 shows the

Operation Sequency of Subject A in Simulation Trial No.10		Correspondent Relation	Human Model Simulation	
Time (sec.)	Action		Time (sec.)	Action
82	check "PRZ. Prs." and detected the anomaly	A	119	check "PRZ. Prs."
86	switch to "PRZ. Prs. Control System" screen		125	detected the abnormal transient
88	watch trend graph of "PRZ.Prs." and "PRZ. Comp-Pres."	B	129	watch the variation-trend of "PRZ Prs."
102	switch back to "Summary" screen		137	watch the variation-trend of "PRZ. Comp-Pres."
108	switch to "FW system", and then to "RMS"		142	check "Prop-heater"
110	watch radiation monitor of CV-Gas-MNT, CV-Dust-MNT, SG-Blowdown-MNT, and		146	check "Backup-heater A1" and "Backup-heater A2"
114	switch back to "FW system", then to "Summary" screen	C	150	watch the variation-trend of "CV-Gas-MNT"
119	check "PRZ. Prs." and check "PRZ. Lvl."		165	watch the variation-trend of "CV-Dust-MNT"
125	switch to "PRZ. Lvl Control System" screen		173	check "PRZ. Lvl."
126	watch trend graph "PRZ. lvl" and "CVCS-IN flw."	F	176	watch the variation-trend of "PRZ. Lvl."
130	switch back to "Summary" screen, and then to "PRZ. Prs Control System" screen		182	watch the variation-trend of "PRZ.Lvl."
136	watch trend graph of "PRZ.Prs." and "PRZ. Comp-Pres."		193	watch the variation-trend of "SG-Blowdown-MNT"
140	check "PRZ. Comp-Pres."		199	watch the variation-trend of "Condenser-Gas-MNT"
141	check "PRZ. Prs."	E	206	check "A-SG Lvl."
147	switch back to "Summary" screen, then to "PRZ. System" screen		208	check "FW flw."
155	check "Prop-heater" and "Backup-heater A1"		211	watch the variation-trend of "FW flw."
162	switch back to "Summary" screen		231	watch the variation-trend of "PRZ. Comp-Pres."
163	check "CVCS-In flw." and "CVCS-out flw."	F	253	check "CVCS-In flw."
174	Check A- B- C-SG Lvl and Prz.			Warning "PRZ. Lvl Low!"
186	switch back to "FW system" screen, then to "RMS" screen	D	255	check "PRZ. Lvl."
190	watch radiation monitors		259	check "CVCS-IN Cont. V"
191	push "Identified" Button	G	273	Output diagnosis result and terminate simulation

Tab.3 Activities Phases

Phases	Operation Activities
A	Detected the occurrence of the anomaly by checking PRZ.prs.
B	Examining "PRZ.Prs Control System Failure"
C	Examining LOCA
D	Examining SGTR
E	Re-confirming "PRZ.prs Control System"
F	Re-confirming "PRZ.lvl Control System"
G	Terminating diagnosing process

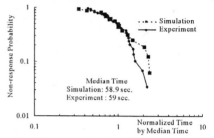

Fig.5 Comparison of operation sequence between simulation and experimental data in the case of diagnosing LOCA

Fig.6 Inter-comparison of TCR curves of LOCA between simulation and experiments

comparison of TCR curve in the case of diagnosing LOCA between the estimated one by HUMOS-PAD and the one obtained from the laboratory experimental data. Both the shape of the curve and the median time utilized for the normalization are agreed well with each other.

It could be concluded that the developed HUMOS-PAD model generally agreed with the experimental observations. From these results of inter-comparison of human model simulation with those of experimental data, it has been suggested that TCR curves required by HRA/PSA could be estimated by numerical simulation experiment using plant operator model, rather than by conducting large-scale experiment. Other studies[8] such education system, distributed simulation system have utilized HUMOS-PAD and obtained good results. Authors would like to expect wider applicability of the model thanks due to its much more flexibility.

References

[1] A.J.Spurgin, et al., EPRI NP-6937, Vol.2, 1990.

[2] H.Yoshikawa, et al., "Development of an Analysis Support System for Man-Machine System Design Information", Control Engineering Practice, Vol.5, No.3, 1997, pp417-425

[3] B.Kirwan: "The Requirements of Cognitive Simulations for Human Reliability and Probabilistic Safety Assessment", Proc. of CSEPC96, 1996, pp292-299.

[4] J.Reason, "Human Error", Cambridge University Press, 1990.

[5] H.Yoshikawa, K.Furuta: "Human Modeling in Nuclear Engineering", Journal of Atomic Energy Society, Japan. Vol.36, No.4, pp268-278, 1994. (in Japanese)

[6] W.Wu: " Study on Modeling Nuclear Power Plant Operator's Cognitive Behaviors at Man-Machine Interface and Its Experimental Validation". Dissertation submitted for the doctor degree in Kyoto University. Dec. 1999.

[7] W. Wu, H. Yoshikawa, T. Nakagawa, A. Kameda, M. Fumizawa: Human Model Simulation of Plant Anomaly Diagnosis (HUMOS-PAD) to Estimate Time Cognitive Reliability Curve for HRA/PSA Practice, PSAM5-Probabilistic Safety Assessment and Management, Proceedings of PSAM5, Vol.2, pp.1001-1007, 2000

[8] H. Ishii, W. Wu, D. Li, H. Shimoda, and H. Yoshikawa: Development of a VR-based Experienceable Education System - A CyberWorld of Virtual Operator in Virtual Control Room-. In Proceedings of the 3rd World Multi-conference on SCI & ISAS, pp. 473 478, 1999.

Integration of a Driver Cognitive Model into a Traffic Micro-Simulation Tool

Delphine Delorme

California PATH – University of California at Berkeley
Richmond Field Station, Bldg. 452
1357 S. 46th Street
Richmond, CA 94804-4648
Tel: (510) 231 9455 - Fax: (510) 231 9565
delorme@nt.path.berkeley.edu

Abstract: This paper discusses the fusion of human factor and physics approaches for the development and integration of a driver cognitive model into a traffic simulation tool. This synthesis is discussed via the presentation of the model and the development of the modules comprised within this architecture. These modules are in charge of simulating the perceptive and cognitive aspects involved for driving and also vehicles dynamics. A simulation case illustrates the merge between these approaches for a car following situation where a distraction of the visual attention justifies a late braking.

1. INTRODUCTION

Driver's models are usually produced in two fields: physics and human sciences (Human Factor and Psychology applied to driving). The motivations underlying their production range from traffic phenomena understanding, design of advanced vehicle control and safety system (AVCSS) and understanding driver's risk taking, reactions and errors. Therefore, the type of model produced varies in terms of formalism, implementation and format of the result produced. The main difference between physics and psychology is in the manner the driver is represented. For physicists, the emphasis is on the vehicle, its acceleration, trajectory and so on. For the human factor or psychologist, the focus is on the driver's perceptive and cognitive capacities.

The aim of the integration of a cognitive driver model into SmartAHS (Automated Highway System), a traffic micro-simulation tool is twofold. First, it addresses the understanding and reproduction of human control characteristic resulting in traffic phenomena, such as shock waves, and therefore provides insight for countermeasure design. Second, it permits testing the benefit of driving assistance devices through comparison of driver-simulated traffic versus semi-automatic or fully automatic vehicles flows. These comparisons are realized in term of throughput and traffic steadiness.

SmartAHS simulates traffic constituted of fully automated vehicles. Therefore, the integration of a driver model in such a perspective is to replace the vehicle "machine" control by a "human" control. However, the human control is expected to operate actions more complex than maintaining a speed or keeping a vehicle on a track. Indeed, the goal is to integrate human behavior through the simulation of the perceptive and cognitive aspects underlying this behavior. Consequently, it attempts to "couple" the physicist and human sciences approach in terms of formalism and variables: the result of a simulated driver decision has to be translated in terms of acceleration of a vehicle.

We choose to adapt COSMODRIVE (Tattegrain-Veste, Bellet, Pauzie, Chapon 1996), an already existing driver model, to SmartAHS. COSMODRIVE describes the cognitive aspects underlying the driving activity around the classical hierarchy, strategic, tactical and operational. It also includes several other modules in order to integrate perception, cognitive resources management and emergency management. Its modular aspect allows a step-by-step development.

The driver model development currently focuses on highway driving and on the processing of a driving situation in progress, as one of the first goals of this model is to address systems such as Automatic Cruise Control (ACC). In this paper, we will present the architecture of the model, then the type of simulation it can produce and we will conclude with the presentation of the next steps thought for the model development.

2.   MODEL DEVELOPMENT

The architecture of the model presented in Figure 1 concentrates on the implemented modules, perception, tactical, operational and execution and the interactions between these modules. The perception module receives information from the road environment. So far the main function of this module is to provide the perception of range and range rage with a leading vehicle. This perception is realized through a model describing the scaling of relative velocity between vehicles (Hoffman & Mortimer 1996). The distribution of visual attention is also managed at this level. The visual attention is allocated to the front of the vehicle by default (perception of range and range rate) and diverted to other locations (side, dasboard, mirrors and others) by two means: if there is a need to check information, if driver's attention is attracted (e.g. a ringing car phone). The visual attention cannot be distracted from the front for more than 2 sec. The values used for the glance duration on the different locations considered are extracted from several studies dedicated to the analysis of eye movement in Human Factors (Bhises et al. 1986, Rockwell 1988, Wierville & Tijerina 1998)[1].

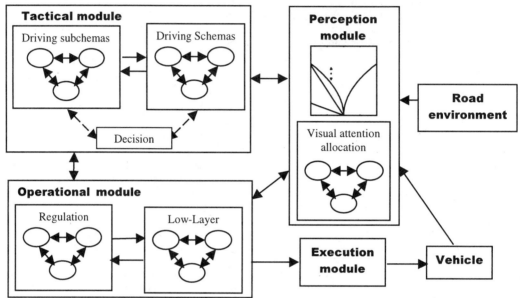

Figure 1: Architecture of the implemented driver model

The tactical module is in charge of processing a driving scene in progress. To do so, it is necessary to reproduce part of the driver knowledge, a categorization process, allowing transitions between the different schemas present in this knowledge and a decision process for adapting the behavior to the changes of the driving situation. Hence, we first divided highway driving in two broad categories: itinerary management, with entering, exiting or changing of highway, where some physical constraints exist from the environment and allow to create some absolute zone, and what we call "highway cruising", where the main constraint is represented by the surrounding traffic. We then focused on highway cruising and considered that in this mode, drivers categorize the situation in progress as a function of their goal and that the different goals correspond to the maneuvers he can operate (Dubois & Fleury 1993). The knowledge present at this level is described within a schema.

The maneuvers or schemas are the following: driving alone, pursuit, following a car and overtaking. Each schema represents a different goal related to traffic management. For example, while driving alone, the goal is to maintain a desired velocity and the trajectory, in pursuit, the goal is to choose between following and overtaking, in following, the goal is to regulate the gap with the leading vehicle etc. The transitions between the schemas are operated based on the information provided about the leading vehicle by the perception module and the value of the current velocity versus the desired velocity. The choice of overtaking can also be taken when the driver is on the following schema if

---

[1] See Song & Delorme 2000 for a more in-depth description of the perception module

too many violations of the desired gap occur. The schemas are represented using the formalism of finite state machine.

Final state machines were also used for representing the subschema. A subschema represents the different actions and regulation to be undertaken for the realization of the maneuver described by a schema. For example, in the following schema, we consider that drivers will prefer to be in one out of three different following zones (Ohta 1996) depending on their goal and on traffic density. These zones are defined by the time gap with the leading vehicle, the zones and thresholds are: dangerous (0-0.6 sec), critical (0.6-1.1 sec), and comfortable (1.1-1.7 sec).

The operational module task is to receive signals and commands from the tactical module and translate them to throttle, steering and braking input for the actuators on the vehicle. For this purpose it utilizes several continuous time control laws that make use of the information provided by the perception module to calculate the actuator inputs required for a particular maneuver. For instance, if the driver wants to maintain the desired velocity when driving alone, the operational module involves a control law for preferred velocity tracking as well as vehicle dynamics and human-factors considerations. The operational module is activated by receiving a message from the tactical module and generates the corresponding control input to the vehicle.

3. SIMULATION CASE

The simulation case presented here illustrates the consequences of visual distraction on car-following regulation. The scenario we conceived consisted in creating a distraction that would occur simultaneously with a hard braking from the leading vehicle (deceleration a –0.39g). Figure 2 represents the difference between a simulation where the driver is not distracted (attentive) and a simulation where the driver is distracted for 2 seconds.

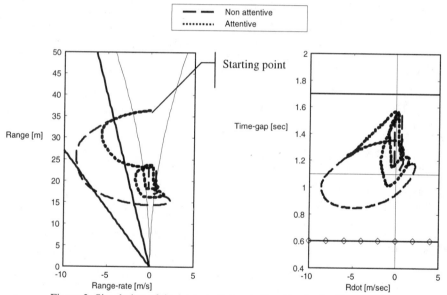

Figure 2: Simulation of the impact of visual allocation for braking reactions

The graph on the left of Figure 2 represents the variation of the range between two vehicles. The thin solid lines represent the driver's perception limits. Within these lines, drivers perceive that the relative velocity between the two cars is null, on the left side, driver perceives he is closing with the leading vehicle, on the right side, he perceives that the gap with leading vehicle is opening. The two others solid thick lines represent normal Time to Collision (TTC) and TTC for hard braking. The dotted lines represent the range and range rate with a leading vehicle. At the starting point, the leading vehicle is at a distance of 36 meters and the relative velocity between the two vehicles is null. The graph on the right represents the same data, here the time gap replaces the range and illustrate that the driver was initially in the comfortable following zone.

This simulation shows that the attentive driver reduces his speed as soon as he crosses the normal TTC line while the non-attentive driver will go to the hard braking TTC, as he does not have the information that he is closing with the leading vehicle. The graph on the right also shows that the attentive driver has a better regulation of his time gap, as he almost does not enter in the critical zone. This figure illustrates how physicist and human sciences can get fused for describing driver behavior.

The following schema has been calibrated with data provided by the University of Michigan, Transportation Research Institute. It was also used in another PATH research program oriented toward the evaluation of ACC and Cooperative ACC versus drivers' throughput (VanderWerf et al. 2000).

4. CONCLUSIONS

The current state of development allows the simulation of most driving situation encountered while driving on a highway. The calibration of part of the model and its use for other project indicate the suitability of this approach for supporting the design and evaluation of AVCSS. The perceptive and cognitive aspects of the model are still being developed. Our current work addresses the development and integration of a mental representation, which will allow the simulation of traffic situation more complex, like the one involved in the itinerary management. The other aspects we plan on developing are the calibration and validation of the model. In this aim, we will proceed at a naturalistic data collection with an instrumented car.

REFERENCES

Bhise, V.D., Forbes, L.M., Farber, E.I., (1986) Driver behavioural data and considerations in evaluating in-vehicle control and displays. Paper presented at the Transportation research board, National academy of Sciences, 65th Annual Meeting, Washington D.C.

Dubois D. and Fleury D. (1993) From classifications to cognitive categorization: the example of the road lexicon. 4th Conference of the International Federation of Classification Societies Paris September

Hoffmann E.R. and Mortimer R.G. (1996) Scaling of relative velocity between vehicles. Accident Analysis and Prevention, 28-4, 415-421.

Ohta H. (1993) Individual differences in driving distance headway. In A.G. Gale (Eds) Vision in Vehicles IV, (pp. 91-100) North Holland, Amsterdam.

Rockwell, T. H., (1988) Spare visual capacity in driving-revisited: new empirical results for an old idea In A.G. Gale (Eds)Vision in vehicles II, (pp. 317-324) North Holland, Amsterdam.

Song B. and Delorme D. (2000) Human driver model for SmartAHS based on cognitive and control approaches. Proceedings of the 10th ITSA Annual Meeting, Boston

Tattegrain-Veste H., Bellet T., Pauzie A., Chapon A. (1996) Computational driver model in transport engineering: COSMODRIVE. Transportation Research Record, 1550, 1-7.

VanderWerf, J., Shladover, S., Krishnan, H., Kourjanskaia, N. and Miller M. (2000). "Modeling the Effects of Driver Control Assistance Systems on Traffic", Paper No. 01-3475, presented at 80th Annual Meeting of Transportation Research Board, Washington DC

Wierwille W. W., Tijerina L. (1998) Modeling the relationship between driver in-vehicle visual demands and accident occurrence In A.G. Gale (Eds) Vision in vehicles VI. (pp. 233-243) North Holland, Amsterdam.

Acknowledgment: The author would like to thank Bongsob Song, Joel VanderWerf, Thierry Bellet and Hélène Tattegrain-Veste for their help in developing the ideas presented in this paper.

This research was supported by a grant from California DOT, Caltrans, (MOU 369).

MODELING AND PREDICTION OF HUMAN DRIVER BEHAVIOR

Andrew Liu and Dario Salvucci*

MIT Man Vehicle Laboratory, Rm 37-219, 77 Massachusetts Ave., Cambridge, MA 02139
*Nissan Cambridge Basic Research, 4 Cambridge Center, Cambridge, MA 02142

ABSTRACT

Knowledge of the current and future driving context could facilitate the interaction between human driver and advanced driver assistance systems. A driver's intended actions (the future context) can be inferred from a number of sources, including the driver's current control actions, their visual scanning behavior, and the traffic environment surrounding them. In an approach similar to hidden Markov models, the intended actions (e.g., to turn or change lanes) are modeled as a sequence of internal mental states, each with a characteristic pattern of behavior and environmental state. By observing the temporal patterns of these features, it is possible to determine which action the drivers are beginning or intending to execute. This approach has been successfully demonstrated in a variety of simulated driving conditions for a wide range of driver actions including emergency maneuvers. In these studies, only the control actions of the driver (i.e., steering and acceleration actions) were used to infer the driver's state. We are presently exploring the use of the driver's visual scanning behavior as another source of information about the driver's state. Visual scanning behavior offers the additional advantage of prediction of driver actions since scanning generally takes place in areas ahead of the current car position.

1. INTRODUCTION

Our interest in modeling the behavior of humans, specifically human drivers, stems from our desire to improve the interaction with various types of automated systems. If these systems could recognize the human's behavior and anticipate future actions, it could adjust its behavior to better suit the needs of the human operator. Driving behavior is the observable result of multiple levels of information processing and motor control (Michon, 1985; Boer, 1998) and it is becoming clear that models that capture both high-level cognitive processing and low-level operational control are needed. For practical applications, these models must capture the behavior of the overall population but also have facilities to adapt to a particular person or driver.

In this paper, we discuss an approach to modeling driver behavior in which we assume that the driver has a large number of internal cognitive states, each with its own associated control behavior and interstate transition probabilities (Pentland & Liu, 1999). The states of the model can be organized in a hierarchy to describe both short-term behaviors, such as passing or turning, and long-term behaviors, such as lane keeping. The control behaviors can be thought of as typical "operational-level" processes in the widely-adopted Michon (1985) driving model and the transition matrices as the cognitive components working at the "tactical-level" of driving. We will also examine whether driver eye movements, which have also been described with stochastic models, can be used within this human behavior modeling framework.

2. MARKOV DYNAMIC MODELS

To implement such an approach, we must make observations of the driver's state, decide which model applies to the current state, and make a response based on that model. But the internal states of the driver are not directly observable, thus we must use an indirect estimation process on the observed behavior (e.g., steering or braking behavior in the case of driving). We have adapted the expectation-maximization methods developed for use with hidden Markov Models (HMMs) to perform this estimation task. Traditional HMMs used to model human behaviors such as speech (Rabiner & Juang, 1986) or gestures (Pentland, 1996) do not capture properties such as smoothness or continuity in their statistical framework. The small-scale structure of human behavior needs to be described by a set of dynamic models that are coupled together into a Markov chain. Unlike a simple multiple dynamic model approach in which the states have a fixed likelihood at each time step, the likelihood of any state in our Markov dynamic model (MDM) makes use of the estimate of the current internal state to adjust the transition probabilities.

In summary, the MDM describes how a set of dynamic processes must be controlled in order to generate the observed behavior, rather than describing the signal itself.

Since human behavior often changes over time due to greater experience or age effects, any model of the driver must also be capable of learning or adapting to these changes. For example, it is well known that driving behavior changes as drivers gain more experience. Thus it is necessary for the MDMs to be able to capture these changes over time. This is theoretically simple to implement since any classification of new behavior can be used as additional evidence to re-estimate the parameters of the model. Thus, MDMs should be capable of matching the behavioral changes in the driver over time, provided that the changes come about slowly.

The MDM approach shares some similarities with modeling approaches based on cognitive architectures, such as ACT-R (Anderson & Lebiere, 1998), and specific models developed within these frameworks, such as the ACT-R driver model (Salvucci, Boer, & Liu, in press). Both approaches are computational models capable of predicting driver behavior. The approaches integrate both low-level perceptual and motor processes with high-level cognitive processes. Typically, the cognitive architecture relies on a central cognitive processor to maintain the mental model and direct the execution of lower-level behaviors. In MDMs, the cognitive process is captured in the transition probabilities connecting the individual models of a particular driving action. In cognitive architectures, the lower level processes are often encapsulated in a set of production rules which could capture the dynamics of the system specified by the dynamic models used in MDMs. Both approaches incorporate some form of learning into the model. This is done explicitly in cognitive architectures either with new rules, or implicitly by the strengthening or weakening of rules based on their usage. Similarly, MDMs can implicitly learn by continuously re-estimating their parameters using their own classifications of data. However, the approaches also have significant differences as well. Primarily, the functions of the models are quite different. The cognitive architectures have been used to simulate human behavior and predict behavior in novel conditions. The MDMs are to be used in real-time applications performing fast recognition of the driver state or prediction of behavior in the near-term. In addition, MDMs do not have the formalization of the cognitive processes such as with cognitive architectures, which have been developed from a theory of human cognition and perceptual-motor behavior. Instead, MDMs capture the cognitive processes within a probabilistic framework connecting the hidden cognitive states. There is never an explicit "model" of the cognitive processes at work.

3. EXPERIMENTAL IMPLEMENTATIONS

We have used the MDM approach in several experiments to test whether human driver behavior can be accurately recognized with MDMs in real-time and to determine what behavior must be observed to infer the driver's cognitive state. One study in a driving simulator (Pentland & Liu, 1999) looked at a number of general driving actions such as stopping, turning, changing lanes, and passing. The driving environment was an urban setting with autonomous traffic, buildings, and lane markings. Subjects performed various driving actions when prompted by text commands on the screen. The commands were generally completed in 5-10 seconds depending on the complexity of the action and surrounding environment. Their control of steering angle, steering velocity, car velocity, and car acceleration was recorded at 10Hz. The dynamic models used were specific to the simulated car (Nissan 240SX). Using three-state models estimated from the collective data (8 subjects, 20 minute sessions, about 600 different actions), we examined the ability of the models to classify the actions that occurred 2 seconds following the presentation of the text commands. This point generally occurs before any large changes in car position, velocity or acceleration occur. The mean recognition accuracy over all actions was 95% ± 3.1%. In comparison, Bayesian classification of the actual physical control data (measured accelerator, brake and steering wheel positions) were not statistically different from chance performance. The reason is that the pattern of brake taps, steering wheel movements and accelerator motions occur over a range of time scales and vary seemingly randomly, as the pattern depends on microevents in the environment, shifts in driver attention, and other disturbances. Only when these data are integrated by the dynamic models to obtain the state variables of the car motion do we begin to see the desired human control behavior.

Other variants of this approach have also been used with success for other driving maneuvers. Kuge et al (2000) constructed a HMM-based driver behavior recognition model to characterize emergency lane changes, normal lane changes, and lane keeping. The models were estimated from steering angle and steering angle velocity information only. The models were trained and tested on data collected in a driving simulator. The accuracy of continuous recognition of emergency lane changes with these models was 98% with a misclassification rate of 0% for normal lane changes and 0.3% for lane keeping. Again, correct recognition tended to occur within 0.5 – 0.7 seconds after the on-screen cue to begin the action.

Oliver (2000) developed a set of MDMs based on actual on-road driving data and examined the recognition performance of models estimated from various sets of vehicle (e.g., velocity, steering angle) and environmental (e.g., lateral lane position or relative position of other vehicles) parameters. She found that recognition accuracy was greatly improved when both vehicle and environmental data were used. Recognition with only vehicle data was quite good for certain maneuvers such as stopping, starting, and passing but poor for lane changes. The addition of environmental data improved the overall recognition performance since it provides additional context for discriminating between similar maneuvers such as passing and lane changes. Similar to the other studies described here, the models, on average, were able to recognize the maneuvers 1 second before any significant change (20%) in the car or contextual signals. Recognition performance was somewhat lower overall than the performance found in the previously described simulator studies. This is not surprising since there are many factors in the real world that could influence the execution of the driving action.

4. DRIVER EYE MOVEMENTS

Eye movements are another source of information about the cognitive state of a person. There are numerous examples illustrating changes in eye fixation patterns with different mental states (e.g., Yarbus, 1967; Stark & Ellis, 1981). This is not really surprising as the context of the task determines the salience of different features in the visual scene. It is important to note that patterns of eye movements are somewhat idiosyncratic— that is, the actual gaze patterns may be quite different from person to person — but it seems that the fixation locations in the scene are generally similar. Driver eye movements also exhibit different patterns of fixation behavior for different driving tasks such as lane keeping (e.g., Mourant & Rockwell, 1971), curve negotiation (e.g., Land, 1994), and car following (e.g., Veltri, 1995). Very often the gaze location tends to be ahead of the car by 2-5 seconds, thus providing for the possibility of actually predicting upcoming actions. Eye movements also change with different tasks involving instruments inside the car, such as with a moving map display system (Antin, 1990).

Stark and Ellis (1981) were one of the first to model eye movements as a Markov process, showing that eye movements were well modeled by a first-order Markov process. The location of a current fixation is dependent only on the previous location. Thus the pattern of eye fixations is captured in the set of conditional probability matrices which can be empirically measured. For a first order Markov process, an m x m matrix, M_1, contains the probabilities of a fixation in Region R_i being followed by a fixation in region R_j. If only a few entries in a row have high transition probability, then there is a high probability of passing through a cyclic sequence. This approach provides a quantitative method for modeling eye movements and statistically differentiating between models.

5. EXPERIMENTAL ANALYSIS OF EYE MOVEMENTS

The Markov analysis was applied to driver eye movements in the context of lane keeping/curve negotiation and car following to determine if characteristic patterns in driver behavior could be identified (Liu, Veltri, and Pentland, 1998). The underlying assumption was that driving could be considered as a combination of "basis" or one-action situations and that the Markov transition matrix for driving over a period of time could be predicted by a linear combination of the Markov transition matrices characterizing those basic situations. A first-order Markov analysis of the data was performed on gaze data collected in a simulated single lane roadway. The results of that analysis found statistically different transition matrices characterizing the different driving tasks, suggesting that the additive model of driver fixation behavior was reasonable. To restate this in terms of our MDM framework, the driver must transition between the two cognitive states since the control of eye movements by multiple cognitive processes can only occur serially. The observed gaze data merely reflect the information acquisition needs of these processes.

Oliver (2000) also incorporated gaze data into the MDMs used in her experiment and found that they generally slightly improved the recognition performance of the models for lane changing. The gaze data were obtained with video post-processing and only incorporated information about the driver's glances to the various mirrors in the car or whether the driver was looking to the right or left. Considering how drivers systematically view the external scene, recognition performance could probably be significantly improved by improving the resolution of the gaze data such that features of the external visual scene (e.g., tangent points in curves or near and far points ahead of the vehicle) could be identified. We hope to be able to perform such an analysis on data from a recent empirical study of driver behavior in a simulated highway environment (Salvucci & Liu, in preparation). We captured steering data and eye movement behavior during normal lane keeping and curve negotiation as well as lane changing. Since the experiment was run in a simulator it is relatively easy to recover the gaze locations in terms of visual scene features. Our preliminary analysis of the lane changing data show that significant changes in the predominant gaze location on the roadway correlate with the onset of the maneuver. Furthermore, there is an increase in the frequency of gazes to the inside mirror approximately one second before the lane change. We hope that these changes can be captured in the MDM to improve the speed and accuracy of recognizing lane changes.

6. CONCLUSIONS

In this paper, we have described an approach for modeling the behavior of drivers such that we can also perform real-time recognition. It is hoped that these models will be useful in improving the interaction between humans and intelligent automated systems. We briefly described the similarities and differences in our efforts to use cognitive architectures to model human driver behavior. Finally, we described past and current efforts to implement our MDM framework in simulated driving tasks and discussed the possibilities of using gaze information to further improve the recognition performance of the MDMs.

7. ACKNOWLEDGMENTS

We thank Prof. Sandy Pentland for his guidance with HMMs and in developing the modeling approach.

REFERENCES

Anderson, J. R., and Lebiere, C. (1998). *The atomic components of thought.* Hillsdale, NJ: Erlbaum.

Antin, J.F., Dingus, T.A., Hulse, M.C., and Wierwille, W.W. (1990). An evaluation of the effectiveness and efficiency of an automobile moving-map navigation display. *International Journal of Man-Machine Studies*, 33, 581-594.

Boer, E.R, and Hoedemaeker, M. (1998). Modeling driver behavior with different degrees of automation: A hierarchical framework of interacting mental models. In *Proceedings of the 17th European Annual Conference on Human Decision Making and Manual Control*, Valenciennes, France.

Kuge, N., Yamamura, T., Shimoyama, O., and Liu, A. (2000). A driver behavior recognition method based on a driver model framework. In *Proceedings of the 2000 SAE World Congress*, Detroit, MI.

Land, M.F., and Lee, D.N. (1994). Where we look when we steer. *Nature*, 369, 742-744.

Liu, A. (1998). What the driver's eye tells the car's brain. In: Underwood, G., (Ed.), *Eye Guidance in Reading and Scene Perception* (pp. 431-452). Oxford: Elsevier Science.

Liu, A., and Pentland, A. (1997). Towards real-time recognition of driver intentions. In *Proceedings of the IEEE Intelligent Transportation Systems Conference* (pp. 236-241). Boston, MA.

Liu, A., Veltri, L., and Pentland, A.P. (1998). "Modelling changes in eye fixation patterns while driving." In A.G. Gale et al., (Eds.), *Vision in Vehicles VI* (pp. 13-20). Amsterdam: Elsevier.

Michon, J. A. (1985). A critical view of driver behaviour models: What do we know, what should we do?, In: Evans, L. & Schwing, R.C., (Eds.), *Human Behaviour and Traffic Safety*. New York: Plenum Press.

Mourant, R.R. and Rockwell, T.H. (1970). Mapping eye movement patterns to the visual scene in deriving: an exploratory study. *Human Factors*, 12(1), 81-87.

Oliver, N. (2000). Towards perceptual intelligence: Statistical modeling of human individual and interactive behaviors. Doctoral Thesis, Program in Media Arts and Sciences, Massachusetts Institute of Technology, Cambridge, MA.

Pentland, A., and Liu, A. (1999). Modeling and prediction of human behavior. *Neural Computation*, 11, 229-242.

Salvucci, D.D., Boer, E.R., and Liu, A. (in press). Towards an integrated model of driver behavior in a cognitive architecture. *Transportation Research Record*.

Salvucci, D.D., and Liu, A. (in preparation). Control and monitoring during highway driving.

Stark, L. and Ellis, S.R. (1981). Scanpaths revisited: Cognitive models direct active looking. In: D.F. Fisher et al. (Eds.), *Eye Movements: Cognition and Visual Perception* (pp. 193-226). Hillsborough, NJ: Erlbaum.

Veltri, L. (1995). Modelling eye movements in driving. Master's Thesis, Dept. of Electrical Engineering & Computer Science, Masachusetts Institute of Technology, Cambridge, MA.

Yarbus, A. (1967). *Eye Movements and Vision*. New York: Plenum Press.

Applications of cognitive architectures: Limits and potential

Christian Lebiere[1] and Dieter Wallach[2]

[1] Human-Computer Interaction Institute, Carnegie Mellon University, 5000 Forbes Avenue, Pittsburgh, PA 15213
[2] Department of Psychology, Basel University, Bernoullistr. 16, 4056 Basel

ABSTRACT

We describe the principles of cognitive architectures and their potential for applications to the modeling of human-computer interaction in complex dynamic environments. By providing a precise computational account of complex interactions, they go beyond existing techniques in providing a detailed understanding of the impact of system design on human cognition. We introduce the ACT-R hybrid cognitive architecture (Anderson & Lebiere, 1998), emphasize the fundamental properties of its subsymbolic level, and briefly describe its application to the modeling of a synthetic air traffic control task.

1. INTRODUCTION

As the field of Human-Computer Interaction (HCI) matures, there is increasing demand for formal methods to systematize the application of empirical results in cognitive psychology to practical problems in the field, including user-interface design and usability evaluation. Cognitive modeling attempts to provide that answer by providing computational models that closely approximate the workings of human cognition. It can be viewed both as a research tool for theory development and as an engineering tool for applying cognitive theory to real-world problems (Gray & Altmann, in press). While most of the past cognitive modeling work in HCI was largely based on general-purpose programming languages, there have been more and more approaches in recent years that rest on *cognitive architectures* as their theoretical basis. Gray, Young and Kirschenbaum (1997, p. 302) regard cognitive architectures as the most important contribution to a theoretical foundation for HCI "since the publication of *The Psychology of Human-Computer-Interaction* (Card, Moran, and Newell, 1983)". Given that the publication of the aforementioned book is occasionally credited with the creation of the field of Human-Computer Interaction, the role of cognitive architectures in HCI is obviously endorsed by a renowned legacy. Gray et al. speculate that architectures are in the process of becoming the preferred route for bringing cognitive theory to bear on HCI. Judging from the recent literature, this view is clearly supported: While applications of architectures to HCI issues are illustrated in only 11 articles published in the ten-year period from 1983-1993 (Gray et al., 1997), more than three times as many have been published in the past five years.

2. COGNITIVE ARCHITECTURES AS A THEORETICAL FOUNDATION FOR HCI

What is a cognitive architecture? A cognitive architecture embodies a comprehensive scientific hypothesis about the structures and mechanisms of the cognitive system that can be regarded as (relatively) constant over time and independent of a task (Howes & Young, 1997). On the one hand, a cognitive architecture provides an integrative theoretical framework for explaining and predicting human behavior in a wide range of tasks. On the other hand, architectures are theoretically justified, implemented software systems that allow for the modeling of different phenomena. By postulating a core system of theoretically motivated constructs, cognitive architectures constrain our theorizing and provide an empirically supported vehicle for understanding human behavior. Grounding a model in the structures and mechanisms of a cognitive architecture removes unprincipled degrees of freedom often associated with models that postulate different sets of mechanisms for each new task studied (Remington, Shafto, & Seifert, 1992).

While early approaches of specifying an architecture concentrated on higher-level cognition, most of such *unified theories of cognition* turned out to be "brains in a box" that were typically ignorant with regard to the modeling of perception and motor behavior. In contrast to the dynamically changing environments involving multiple simultaneous goals typically found in complex event-based systems, early modeling approaches were often concerned with human behavior in simple, static reaction time experiments. Consequently, the resulting models were of only limited interest for the modeling of interactive tasks in HCI research. More recently, embodied architectures, incorporating theories of human perceptual-motor capabilities, have widened the field by allowing the construction of generative models that directly interact with dynamic, event-based applications in the same principled manner as human subjects. These models are concerned with the orchestration of different aspects of cognition with processes of perception and have demonstrated how the structures and mechanisms of an architecture can guide theorizing in HCI, helping to close the gap between cognition and overt behavior.

In this paper we will concentrate on a discussion of the ACT-R cognitive architecture and its relevance for HCI research. As a recent survey of more than 100 published papers has shown[1], models developed in the ACT-R architecture have been successfully applied to an extraordinary wide body of empirical data, covering research on as diverse areas as perception and attention, learning and memory, problem solving and decision making, spatial reasoning, dynamic system control, use and design of artifacts, game playing, scientific discovery, language processing, emotion and cognitive development.

3. THE ACT-R COGNITIVE ARCHITECTURE

The ACT-R architecture can be divided into two layers: a *cognition layer* that handles all aspects of higher level cognition and a *perceptual/motor layer* that mediates the interaction between cognition and the external environment. The cognition layer is based on a production system that distinguishes between a procedural and a declarative memory. Procedural knowledge is encoded in procedural memory using modular *condition-action-rules* (productions) to represent actions to be taken when certain conditions are met. So-called *chunks* are used to store simple facts in declarative memory. Sub-symbolic activation processes control which productions are used and which chunks are accessed. Productions request the retrieval of a chunk from declarative memory, with the chunk's activation level determining both its probability of retrieval and its retrieval time. A hierarchical representation of goals is used to control information processing where exactly one goal is designated to be active at each instance of time. ACT-R includes learning mechanisms for the acquisition of new productions and chunks as well as the adaptive tuning of subsymbolic quantities associated with the symbolic knowledge units. ACT-R's cognitive layer is augmented by a perceptual/motor system that allows the architecture to directly interact with the external environment.

Models created within the ACT-R framework are *generative* and *reactive:* the model actually *generates* actions using the knowledge encoded in its chunks and productions and it *reacts* to events initiated by the task requirements (Gray et al., 1997). Thus, an ACT-R model can act on an interface just like a human user by moving a mouse, clicking on buttons on the screen, operating a joystick or typing on a keyboard. ACT-R can move its attention to certain locations on the screen, encode the object that is present there and can track the trajectory of objects in a dynamic environment. Using its audition and speech components an ACT-R model can simulate the hearing of stimuli and/or instructions and respond to them with simple phrases. The formulation of an interactive cognitive model enforces a strict necessity to construct a truly *complete* model, encompassing all steps from the perception of an environment, to deciding for appropriate actions and executing them physically. In this process of creating an interactive model, important questions about the interplay of cognition, perception and action necessary to successfully master a task are uncovered.

From the perspective of ACT-R, applications in HCI provide a strict and demanding real-world test bed for the scope of its underlying theory. Cognitive models go beyond task analysis in grounding each step in the theory of the architecture, providing accurate predictions and detailed explanations for the behavior observed. They embody the cognitive implications of a given interface design in an inspectable representation that identifies the declarative and procedural knowledge necessary to operate a device and predicts the performance on new systems. By generating the behavior they purport to explain, ACT-R models can be used as analytical surrogates for empirical data, permitting an efficacy analysis of different design alternatives and thereby narrowing the design space for future redesigns. Running a model on an interface provides a stringent test of the sufficiency of its underlying assumptions that contrasts with the tedious and error-prone hand-simulation of descriptive models.

4. THE ROLE OF GOMS

Before the advent of cognitive architectures, the GOMS analysis technique (John & Kieras, 1996) was regarded as a prime candidate to provide a theoretical foundation for HCI. The GOMS methodology is based on the idea that performing a task can be usefully analyzed by decomposing it into the Goals, Operators, Methods and Selection rules required for its completion. While goals describe what a user wants to achieve, operators refer to the elementary actions that a user can perform to operate a device. Methods describe operator sequences that accomplish a goal. If several methods are available in a certain context, selection rules pick a method to accomplish a goal. GOMS techniques make a priori quantitative predictions about human behavior and have been used in the design of real-world systems, leading to the creation of better systems or the cancellation of inferior ones (Gray, John, & Atwood, 1993).

However, GOMS models have fundamental limitations in their attempt to approximate human cognition. They are designed to capture expert performance and cannot realistically account for the full learning curve from beginner to expert level. GOMS models are developed to represent routine cognitive skills and are thus not suitable to the representation of problem solving and other high-level cognitive activities. Moreover, all parameters used to

---

[1] See http://act.psy.cmu.edu

quantify performance in GOMS models are constant, making it impossible to represent interactions that change over time. Although even early studies (Mack, Lewis, & Caroll, 1983) showed that new users spent up to a third of their time in error recovery, modeling non-routine problem solving or error-prone behavior seems to be beyond the scope of GOMS. Finally, a GOMS analysis focuses on the procedural knowledge required to operate a device. GOMS is not intended to represent with any precision the declarative knowledge that constitutes a significant part of the user's knowledge of computing environment, and indeed is the basis for most procedural knowledge.

5.   PROPERTIES AND APPLICATIONS

At the symbolic level, ACT-R is relatively similar to GOMS: goals are a central organizational principle for both, GOMS operators correspond to the basic actions available to ACT-R, methods correspond to sequences of production firings and selection is performed in ACT-R by the conflict resolution process that chooses between competing productions. Our behavior, however, does not typically follow neat deterministic rules but is instead constantly evolving, erratic but robust. This is probably because the everyday human environment is also dynamic and approximate, constantly changing and very seldom presenting us with the same situation twice. The rational analysis underlying the ACT-R theory (Anderson, 1990) assumes that human cognition survives and thrives in such environments by using statistical learning mechanisms to continuously tune itself to the structure of the environment. Unlike GOMS, ACT-R is a hybrid system that provides a rich subsymbolic layer, enabling the structured cognition of the symbolic layer to adapt itself to complex dynamic environments.

A cognitive architecture that would model human cognition operating in such environments needs to meet at least three specific criteria. To adapt to changes in the environment, it needs *learning mechanisms* that tune the performance of the architecture to adapt and operate optimally in its environment, and they must do so in real time, without the often exorbitant number of training cycles required by connectionist models. As previously mentioned, ACT-R has learning mechanisms that create declarative chunks to encode new pieces of information, that reinforce the activation of chunks with practice and decay it with disuse to make the most frequently and recently accessed information more available, that proceduralize frequently performed series of actions to make them more efficient, and that tune the utility of the resulting productions to reflect their history of success in order to select the most efficient strategies. The second criterion is *stochasticity*. Human behavior is generally hard to predict, resulting in perfect performance one moment and sudden error the next. From an evolutionary perspective, stochasticity in behavior is a desirable characteristic, both to fool opponents and to explore all the possible ways to solve a problem. The third criterion is *generalization*, i.e. the ability to apply existing knowledge to new, slightly different situations in order to provide robust behavior. Again this is an essential quality in unpredictable natural environments where nothing is ever exactly the same. The PIPES model (Lerch, Gonzalez, & Lebiere, 1999) is an instance of an ACT-R model of human interaction with a complex dynamic environment that fulfills all three criteria. This dynamic decision-making task, adapted from an actual mail-sorting facility, consists of scheduling a finite number of resources to solve competing requests in a constantly changing real-time environment. While solving the problem perfectly is computationally very difficult, humans can quickly learn to solve it near-optimally. The ACT-R model learns instances of problem solving that it then generalizes to new but similar situations, while the stochastic nature of the model enables it to explore promising new solution paths. Adaptivity, stochasticity and generalization are fundamental features of human cognition that will arise in any setting, but especially in complex dynamic environments.

There is every indication that artificial computing environments that have been recently developed share many of the fundamental statistical features of our original cognitive environment (Anderson & Schooler, 1991, Hubermann, Pirolli, Pitkow, & Lukose, 1998; West & Salk, 1987). Moreover, new environments are being designed that explicitly incorporate some of the outside world's features. For example, many software systems now include pull-down menus that dynamically reorganize themselves depending upon their history of use. Just as repeatedly picking fruit or hunting game in the same location ultimately leads to depletion and requires adapting by finding new territories, in such a system repeatedly using a menu item leads to its migration and requires finding its new location in the menu. Fortunately, human cognition is well suited to operating in such environments, but if one is to accurately model it, say to guide the design of such systems, one cannot rely on tools that provide only precise, deterministic and unchanging behavior. An instance of the adaptivity of human cognition to complex environments is the ACT-R model of the Kanfer-Ackermann air traffic control task (Lee & Anderson, in press), which shows that controllers optimize their behavior in large part by learning which parts of a complex interface contain relevant information and should be paid attention to and which do not and can be ignored, thus saving attentional resources.

We illustrate the need for adaptive, stochastic and robust cognition by applying ACT-R to the task of modeling human interaction with a synthetic air traffic control simulation, which provides a challenging, complex and dynamically changing environment (Lebiere, Anderson, & Bothell, 2001). The AMBR modeling comparison is an

1486

Air Force program designed to showcase the state-of-the-art in cognitive modeling by applying competing architectures to a common, challenging task. The display of the AMBR ATC task is shown in Figure 1 (a). It includes a radar screen as well as a number of text windows in which messages appear from neighboring controllers requesting action. Failure to perform those actions in a timely fashion results in penalty points. Average performance in terms of penalty points for subjects and model is displayed in Figure 1(b). The left three bars correspond to a condition in which aircraft requesting action were color-coded on the radar screen, and the right three bars correspond to the standard condition in which scanning of text windows to determine the action requested was necessary. Within each condition, the three bars correspond to increasing simulation speed. While the model was quite simple, featuring only 5 declarative chunks and 36 simple production rules, it easily captured the subjects performance, especially the positive impact of interface support (color-coding of aircraft) and the negative impact of the rate of change of the environment (simulation speed).

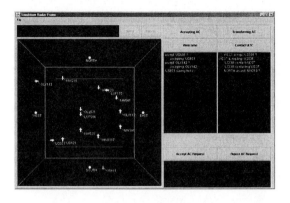
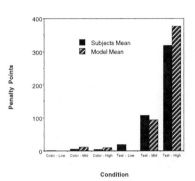

Figure 1: AMBR ATC display (a) and mean performance for subjects and model in various conditions (b)

Subject performance in this task is highly variable. Figure 2a shows that the model nicely captures the range of variation in subject performance without any changes in the model but instead by relying on the natural stochasticity of the architecture's subsymbolic level, amplified by the interaction with the dynamic environment. Figure 2(b) shows that the model captures the spectrum of errors committed by the subjects, not by being engineered in the model but instead as emergent properties of the constraints on the cognitive power of the architecture.

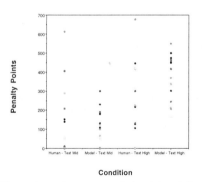
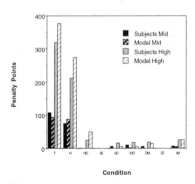

Figure 2: performance variation (a) and distribution of errors (b) for subjects and model in the 2 hardest conditions

The model was also validated by accounting for other measures of behavior such as response times, selection probabilities and self-reports of cognitive workload. Though ACT-R is goal-directed, the model accounts for the subjects multi-tasking behavior through interruptions triggered by detection of message onsets at the perceptual level. This provides an example of how applications to demanding tasks force architectural changes. The next phase of the AMBR modeling comparison will increase the difficulty of the task by adding a subtask requiring category learning, which will exercise the architecture's learning mechanisms and its capacity for generalization.

One limit to the precise modeling of complex tasks is the type of data available. Aggregate data such as average performance or response time provides a useful constraint, but if the various strategies at the lower level, such as perceptual search strategies, are to be elucidated then finer-grained data at a similar scale is necessary, such as eye-tracking protocols.

6. CONCLUSION

HCI and cognitive architectures live in a *mutually* beneficial relationship. Cognitive architectures provide a comprehensive, detailed, and quantitative theory of human cognition and performance that has matured to be applicable to dynamic real-world tasks. While such applications promise to be informative about the cognitive consequences of design decisions, they might, also pinpoint potential theoretical gaps or shortcomings in an architecture's assumptions about the human cognitive system. Although limitations remain in the use of cognitive architectures, such as the lack of model reuse and of precise fine-grained data to guide model development, we claim that they provide the *via regia* to arrive at a deeper understanding of complex interactive behavior.

REFERENCES

Anderson, J.R., "The adaptive character of thought". Hillsdale, NJ: Erlbaum, 1990.

Anderson, J.R., & Schooler, L.J., "Reflections of the environment in memory". Psychological Science, 1991, 2, 269-408.

Anderson, J.R. & Lebiere, C., "The Atomic Components of Thought", Hillsdale, NJ: Erlbaum, 1998.

Card, S., Moran, T. & Newell, A., "The Psychology of Human-Computer Interaction", Hillsdale, NJ: Erlbaum, 1983.

Ehret, B. D., "Learning where to look: The acquisition of location knowledge in display-based interaction", Dissertation Abstracts International: Section B: the Sciences & Engineering, 2000, 60(10-B), 5239.

Gray, W. D., & Altmann, E. M., "Cognitive modeling and human-computer interaction:, In W. Karwowski (Ed.), International encyclopedia of ergonomics and human factors. New York: Taylor & Francis, Ltd., in press.

Gray, W.D., Young, R.M. & Kirschenbaum, S.S., "Introduction to architectures and human-computer interaction." Human-Computer Interaction, 1997, 12, 301-309.

Gray, W.D., John, B.E. & Atwood, M., "Project Ernestine: A validation of GOMS for prediction and explanation of real-world performance", Human-Computer Interaction, 1993, 8, 237-309.

Howes, A. & Young, R.M., "The role of cognitive architectures in modeling the user: SOAR'S learning mechanism", Human-Computer Interaction, 1997, 12, 345-389.

Huberman, B.A., Pirolli, P., Pitkow, J., & Lukose, R.J., "Strong Regularities in World Wide Web Surfing". Science, 1998, 280: 95-97.

John, B. E., & Kieras, D. E., "Using GOMS for user interface design and evaluation: Which technique?" ACM Transactions on Computer-Human Interaction, 1996, 3, 287-319.

Lebiere, C, Anderson, J. R., & Bothell, D., "Multi-tasking and cognitive workload in an ACT-R model of a simplified air traffic control task", In the Proceedings of the Tenth Conference on Computer-Generated Forces and Behavior Representation. Norfolk, Va, 2001.

Lee, F. J., & Anderson, J. R., "Does learning of a complex task have to be complex?", A study in learning decomposition. *Cognitive Psychology*, in press.

Lerch, F. J., Gonzalez, C., & Lebiere, C., "Learning under high cognitive workload", In Proceedings of the Twenty-first Conference of the Cognitive Science Society, pp. 302-307. Mahwah, NJ: Erlbaum: 1999.

Mack, R.L., Lewis, C.H. & Carroll, J.M., "Learning to use office systems: problems and prospects", Assoc. Comput. Mach. Trans. Off. Inf. Syst., 1983, 1, 254-271.

Remington, R.W., Shafto, M.G. & Seifert, C.M., "How human is Soar?", Behavioral and Brain Sciences, 1992, 15, 455-456.

West, B.J., & Salk, J., "Complexity, organization and uncertainty", *European Journal of Operational Research, 1987, 30,* 117-128.

Evolution of socio-cognitive models supporting the co-adaptation of people and technology

Guy A. Boy

European Institute of Cognitive Sciences and Engineering (EURISCO),
4 avenue Edouard Belin, 31400 Toulouse, France

Abstract

Socio-technical systems of our post-industrial era embed their own internal cognitive mechanisms and behavior. New information technology has induced new practices and human roles. Human-centered automation is not new. During the XXth century, automation technology and resulting practice have evolved using the same underlying models of cognition. These models have emerged incrementally for rationalization purposes. Recently, the concept of affordance popped up as a necessary cognition model to explain human-objet relations that are either natural or learnt. Situated patterns or cognitive automatisms are constructed from training and routine practice. Conversely, technology incrementally incorporates situated patterns as external augmentations of people's initial capacities. The cognitive function paradigm has been developed to help analyze the way cognition is distributed among various human and machine agents. Finding the right balance between externalizing cognition into tools and forming human situated patterns by training is the key issue. For that matter, dynamic and incremental ethnomethodological approaches can bring a lot to human-centered design.

An interpretation of the automation evolution

Bernard Ziegler, a former Vice-President of Airbus Industrie, made the following observations and requirements from his experience as a test pilot and distinguished engineer: "the machine that we will be handling will become increasingly automated; we must therefore learn to work as a team with automation; a robot is not a leader, in the strategic sense of the term, but a remarkable operator; humans will never be perfect operators, even if they indisputably have the capabilities to be leaders; strategy is in the pilot's domain, but not necessarily tactics; the pilot must understand why the automaton does something, and the necessary details of how; it must be possible for the pilot to immediately replace the automaton, but only if he has the capability and can do better; whenever humans take control, the robot must be eliminated; the pilot must be able to trust automation; acknowledge that it is not human nature to fly; it follows that a thinking process is required to situate oneself, and in the end, as humiliating as it may be, the only way to insure safety is to use protective barriers." (Ziegler, 1996).

Ziegler's very high level observations and requirements come from his very rich experience. Cognitive science could benefit from them by proposing appropriate models of cognition that would rationalize this experience. Rasmussen's model has been extensively used over the last decade to explain the behavior of a human operator controlling a complex dynamic system (Rasmussen, 1986). This model is organized into three levels of behavior: *skill*, *rule* and *knowledge* (Figure 1).

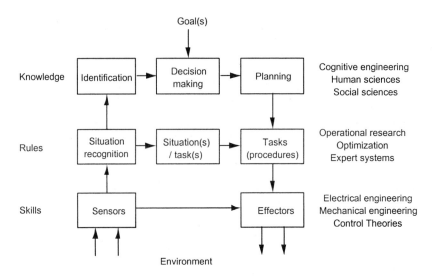

Figure 1. Rasmussen's model, the evolution of automation
and the emergence of contributing disciplines.

Historically, automation of complex dynamic systems, aircraft in particular, have led to the transfer of human operators' skills (e.g., performing a tracking task) to the machine. Autopilots have been in charge of simple tracking tasks since the 1930s. This kind of automation was made possible using concepts and tools from electrical engineering, mechanical engineering and control theories, such as mechanical regulators, *proportional-integral-derivative* controllers (PID), Laplace functions and stochastic filtering. Autopilots were deeply refined and rationalized during the 1960s and the 1970s. Human skill models were based on quasi-linear models' functions and optimal control models. Human engineering specialists have developed quantitative models to describe and predict human control performance and workload at Rasmussen's skill level. They have been successfully applied to study a wide range of problems in the aviation domain such as handling qualities, display and control system design and analysis, and simulator design and use.

The second automation revolution took place when the rule-based level was transferred to the machine. In aviation, a second layer was put on top of the autopilot to take care of navigation. The *flight management system* (FMS) was designed and implemented to provide set points for the autopilot. A database is now available onboard with a large variety of routes that cover most of the flights in a specific geographical sector. Pilots need to program the FMS by recalling routes from the database and eventually customize them for a specific flight. Once they have programmed the FMS, the aircraft is "capable of flying by itself" under certain conditions, i.e., the FMS is in charge of the navigation task in pre-programmed situations.

Today, human factors research and practice have evolved towards *cognitive engineering*, and *hermeneutics*, because the control of highly automated systems does not require the same abilities and requirements as traditional tools. Human operators mostly work at Rasmussen's knowledge-based level. Basic operations are delegated to the machine, and humans progressively become managers of (networked) cognitive systems. Humans need to identify a situation when there is no pattern-matching (situation recognition) at the rule-based level, to decide according to specified (or sometimes unspecified) goals, and to plan a series of tasks. These are typical strategic activities. Some people are good at strategic activities, others prefer to execute what they are told to do. In any case, the control of cognitive systems requires strategic training. For example, using the Web has totally transferred the shopping task to Rasmussen's knowledge-based level. The selection of food items is made using virtual objects. The delivery is planned with respect to the customer's schedule and the nature of the food.

Technology has always shaped the way people interact with the world. Conversely, interacting with the world has direct impact on how technology evolves. Rationalization of experience feedback influences the development of theories that make new artifacts emerge. In a technology-driven society, this goes the other way around, i.e., the use of artifacts make new practices and new jobs emerge, as the film technology did for example. The twentieth century was rich in technology innovation and development. The speed of evolution of technology and resulting practices is very sensitive to economical impacts. In some cases, when economical benefits were not obvious a priori, but the evolution of human kind was at stake, technological advances were decided at the political level, such as designing and developing a technology that enables a man to walk on the moon. Today following these grandiose projects, we realize that human-centered automation, and more generally human-centered design, is not effectively taken into account at the political level yet. A new paradigm needs to be found to better understand the balance between human and machine cognition.

The cognitive function paradigm

The concept of cognitive function helps analyze how people (human *agents*) and information-intensive systems may interact. A cognitive function can be interpreted in the mathematical sense or in the teleological sense. The former interpretation leads to the definition of an application transforming an input into an output. The input is usually a *required task* to be performed. The output is the result of the execution of the task. We usually say that a human uses a cognitive function that produces an *activity* or an effective task. The latter interpretation leads to the definition of three attributes of a cognitive function: a *role*, e.g., the role of a postman is to deliver letters; a *context* of validity, e.g., the context of validity of the above role is defined by a time period that is the business hours and a specific working uniform, for example; a set of *resources*, e.g., the resources necessary to execute the function are a bicycle, a big bag and a delivery procedure. A resource is a cognitive function itself. In a big town, the chief postman delegates the letter delivery task to other postmen. Note that when the postman returns home (different context) the above described cognitive function is no longer active, and he activates different cognitive functions such as father and husband (different roles) using different resources such as affection for his family and helping in homework activities.

Some cognitive functions are learned or artificially constructed and are called cognitive artifacts. Other cognitive functions are innate. Information technology also has affordances that need to be found in tools that enable people to: generate information, i.e., tools that make information explicit to others; maintain information awareness, i.e., tools that enable people to be aware that appropriate information exists somewhere; access information, i.e., tools that enable people to access appropriate information at the right time in the right format; understand information, i.e., tools that enable people to understand information chunks.

People develop appropriate cognitive functions to speed up, and increase both comfort and safety of their job, i.e., the tasks that they usually perform. These cognitive functions can be *soft-coded* or *hard-coded*. When they are soft-coded, they usually appear in the form of procedures or know-how stored in their long-term memory. When they are hard-coded, they usually appear in the form of interface devices or manuals that guide users in their job. In both cases, cognitive functions can be either *implicit* or *explicit*. When they are implicit, they belong to what is usually called *individual expertise*. When they are explicit, they belong to what is usually called *sharable knowledge*. Sometimes, cognitive functions remain implicit for a long time before becoming explicit and easily sharable. When a cognitive function is persistent, it can be elicited and transferred to a machine that will perform it for its user. This transfer process is commonly called *automation*. Various levels of automation can be implemented with respect to both human factors involved in the execution of the task and the technological limitations.

Conventional design is based on goal-driven methods, i.e., designers start with an overall goal in mind and attempt to decompose this goal into sub-goals until basic actions can be derived and effectively performed. Goal-driven approaches to design are strongly anchored in industry since they lead to manageable and explainable products. Resulting products are usually technology-centered. In many cases, they are easy to maintain also. However, goal-driven design approaches ("I want to do this!") do not handle end-users requirements well. At the other end of the spectrum, event-driven approaches ("If this happens, what should I do?") to design tend to foster participatory design and use of experience feedback data. Since design is intrinsically iterative, event-driven approaches to design can be very time-consuming and very unstable. I introduced the *Cognitive Function Analysis* (CFA) approach that supports event-driven human-centered design (Boy, 1998). CFA can be seen as cognitive function allocation. CFA proposes to handle these issues in two necessary and complementary ways: categorization of experience feedback cases into cognitive functions that may be re-used in design; use of an integrated methodology based on the use of active design documents that enable design teams to implement participatory design.

Embedding cognition into tools: Looking for the right balance

Nobody questions the use of the clock today: the clock's cognitive function is to provide the time to its user. Its context of validity is determined by several parameters such as the working autonomy of the internal mechanism or the lifetime of the battery. Its resources include, for instance, a battery, the ability of its user to adjust time when necessary or to change the battery. Note that the user is also a resource for the clock's cognitive function.

Today, external cognitive functions have become more complex than the clock. People delegate to these external cognitive functions some actions that they used to perform before. They have to plan, monitor, negotiate, supervise, communicate, cooperate and coordinate with composite artifacts, e.g., a travel agent needs to work with a composite world-wide travel system (i.e., a composite artifact that embeds many external cognitive functions). This system is a network of a large number of computer systems. The travel agent learns about information traffic jams in the network, local crashes and tricks for booking a trip using a cheaper carrier, for example. These are cognitive functions that are relevant to his or her job. They are valid in time-specific contexts such as "during a holiday period", or "the Paris Orly airport is always very busy on Monday mornings". Travel systems have taken this into account for a long time with pricing. Cheap flights are usually available during the day, not in the morning or the evening. Today, such systems learn very fast, i.e., both human and external cognitive functions adapt very fast. The travel agent needs to assimilate and accommodate more cognitive functions to handle the increasing number of options.

The integration of new information technology in the life support systems, such as human/organizational learning and human-centered design, should enhance its role of preserving cultural heritage, improving knowledge transfer, social integration, and safety. Human and organizational learning tries to adapt people to the world. I propose five attributes of learning that come from my own experience: motivation; learning how to learn; efficiency; allowing for errors in order to learn from them; memory retention. Consequently, dynamic and incremental ethnomethodological approaches can bring a lot to human-centered design. Human-centered design tries to adapt technology to people. This co-adaptation process should contribute to: the development of autonomy and subsequent mandatory co-ordination, as well as collective and individual learning; the removal of barriers caused by social or geographical isolation; the openness to the external world and facilitation of synergy with local resources. I strongly believe that good individual and collective human knowledge management starts at school and is a life-long experience. Therefore, socio-cognitive models that are learned in the first place strongly determine the way people adapt to technology and their ability to construct new socio-cognitive models. Our children will really live in and develop the information-based society because they will have grown up in the emerging virtual world of videogames and Internet technology.

References

Boy, G.A. (1998). *Cognitive function analysis*. Ablex, Stamford, CT.

Rasmussen, J. (1986). *Information Processing and Human-Machine Interaction - An Approach to Cognitive Engineering*. North Holland Series in System Science and Engineering, A.P. Sage, Ed.

Ziegler, B. (1996). *The flight control loop*. Invited speech. Final RoHMI Network meeting. EURISCO, Toulouse, France, September 28-30.

EVALUATION OF THE HMI SUGGESTED BY SIMULATIONS OF OPERATOR TEAM BEHAVIOR AND ITS APPLICATIONS

Ken'ichi Takano, Toshiaki Sano and Kunihide Sasou

Human Factors Research Center,
Central Research Institute of Electric Power Industry
2-11-1, Iwadokita, Komae-shi, Tokyo 201-8511 JAPAN
Email:takano@criepi.denken.or.jp

ABSTRACT

In a large and complex system, such as a nuclear power plant, it would be valuable to evaluate the adaptability and usability of a Human Machine Interface (HMI). Among various ways of evaluation, simulation is a useful and effective means in the designing stage. CRIEPI has developed the Man Machine Simulator (MMS), by which standard operator team behavior including strategic, tactical and cognitive behavior in coping with an anomaly even caused by multiple malfunctions can be reasonably simulated. The fidelity was confirmed by comparing the simulation results with experimental ones and by conducting an expert judge. This paper describes the evaluation of the typical HMIs, i.e. 1st generation and 2nd generation control panel, in domestic utilities by using MMS as a tool. The evaluation indicator is a summation of delayed operations and omitted operations deviating from standard operating procedure. Total walking distance, and total number of utterances and operations was adopted as indicators of the HMI usability. The parameters changed in the simulations were, the speed of individual operator's behavior (individual characteristics), the cooperation style within team (team characteristics), and the type of the HMI (interface). The obtained results suggested us that omission of critical operations that could affect plant dynamics into further deterioration were not observed if priority of operations could be kept and the influence of individual characteristics was larger than those of other parameter change. However, it was also suggested that the 2nd generation panel was not so improved because of longer walking distance due to wider size of 2nd generation panel than that of 1st generation panel. Other than evaluating HMIs, the MMS can be applicable to the operator training and support to demonstrate standard operators' behavior in coping with an anomaly in which procedures would not have been stipulated previously.

1. INTRODUCTION

Superiority of human machine interfaces (HMIs) become crucial in carrying out the critical operations under severe time pressure. Especially in coping with multiple malfunctions occurring at a large and complex system, sequential operations demand high workload to each operator in a team. Then, adaptability and usability of HMIs should be taken into consideration so as to avoid a disaster. As an evaluation of HMIs consists of multiple factors, usually experimental method was adopted by preparing a few types of the mock-up control panels and actual operator teams. In this method, evaluation was owing to operators' subjective score. However, it took much resource if several types of mock-up control panels and several sets of actual operator team should be prepared. In this point of view, simulation is effective means in the designing stage. To evaluate HMIs by simulations, user should request concurrent superiority of the specific HMI for wide variety of malfunctions and also demand the reality and fidelity of the simulation itself. Authors have been making intensive effort to develop the Man Machine Simulator (MMS), by which standard operator team behavior including strategic, tactical and cognitive behavior in coping with an anomaly even caused by multiple malfunctions could be reasonably simulated based on a fully knowledge base architecture (Takano, Sunaoshi and Sasou, 1999). The fidelity was confirmed either by comparing the simulation results with experimental ones or by conducting an expert judge. This paper describes the evaluation process and results of the typical HMIs, that includes the 1st generation and 2nd generation control panel utilized in domestic nuclear power plants by using MMS as a evaluation tool. Followings are presented here as a simulation results and discussion: 1) Influences to the performance observed in changing individual characteristics; 2) Influences to the performance in changing cooperative style within a team; 3)Evaluation of the difference between 1st generation and 2nd generation control panel.

2. OUTLINE OF TECHNICAL BASIS OF THE SIMULATION

This simulation system mainly composed of following four modules:1) Plant module: 1100MW commercial level of BWR plant dynamics calculating program, introduced from "GSE", a Swedish company (plant simulator); 2) Human machine interface modules: modeling the 1st generation of the main control panel (BWR Training Center Co.Ltd., BTC No.3 simulator) and the 2nd generation panel (BTC No.4 simulator), on which all indicators, switches and display were 3-dimensionally coordinate in virtual space; 3) Operating team modules: composed of four MCR operators and one field operator model each has detection, sensory buffer, thinking, knowledge base, working ,memory, action and utterance micro models, respectively; 4) Human-human interface modules: this model controls communication and role sharing among operating team.

Controls of operating team behavior is totally based on a mental model, that is composed of a set of rules and knowledge base by which an instance for a specific anomaly can be created (Takano, Sasou and Yoshimura, 1995). This mental model was supposed to be created in his mind in the beginning of event by synthesizing both current information on plant status and accumulated knowledge. The functions of this mental model are: 1) listing possible causes, the ways to identify these causes, and causal remedies diagnosis; 2) taking immediate response on identifying the annunciator indication (alarm); and 3) envisioning future scenarios and taking preventive measures along with anticipated scenario prognosis. And, it was also implied in the previous work that the mental model could explain the operational sequence seen in the experiments. In order to deal with an anomaly caused by multiple malfunctions, the operator can create multiple mental models corresponding to each of them.

3. SIMULATION CONDITIONS

3.1 Individual Characteristics

In the simulation, content of knowledge base was fixed as the same for all operators. The other primary characteristics influencing to human performance suppose to be necessary time to execute actions, utterances, and cognitive process (thinking cycle time). This simulation model has a function to vary the time required to execute above. Standard necessary time for each action and utterance was estimated based on the MTM: Measurement Time Methods, (Mundel, 1982) and standard time for thinking cycle time was quoted from the reference (Card, 1983). Then there were 4 condition set as shown in Table 1.

Table 1. Four of simulation condition set as the individual characteristics: execution time required.
(Standard time is set as 1.0, the other is magnifying factor to standard condition)

Condition	Action	Utterance	Thinking cycle time
Condition 1 (quick)	0.5	0.5	0.5
Condition 2 (standard)	1.0	1.0	1.0
Condition 3 (slow)	1.5	1.0	2.0
Condition 4 (very slow)	2.0	1.5	4.0

3.2 Team Characteristics

The variety of the styles of team cooperation also was changed as following: 1) strict role sharing case (each operator covering fixed territory):*condition TC1* as a standard; 2) flexible role sharing (only monitoring and checking read being exchangeable):*condition TC2*; 3) flexible role sharing (every role being exchangeable): *condition TC3*.

3.3 Human Machine Interface

Adopted HMIs for making an evaluation were 1st generation (HMI 1)and 2nd generation control panel (HMI 2)as shown in Fig. 1, in which modeled face configuration and plane figure was dedicated.

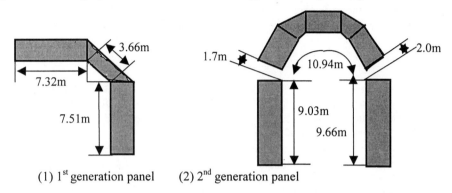

(1) 1st generation panel   (2) 2nd generation panel

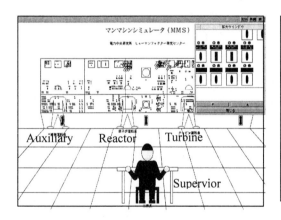

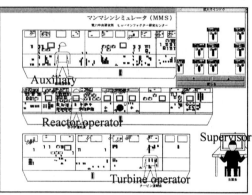

(3) 1st generation panel   (4) 2nd generation panel

Fig.1 Plane figures of the control panel and modeled face layouts of the HMIs evaluated

4. INFLUENCE TO THE PERFORMANCE

Total ten cases of anomalies, all of which were multiple malfunctions, were applied to the simulation for evaluating the layout of HMIs, and the influence due to the individual and team characteristics. For example, representative one is "Total loss of feedwater supply + failure of high pressure core injection starting up", and "Two of circulation pumps trip + Radiation high at main steam pipe + loss of grand steam". Simulations were executed of total 240 cases (ID 4cases×TC 3cases×HMIs 2cases×anomalies 10cases) for evaluating the influence to the plant and operators' performance.

4.1 Possibility to Induce an Unexpected Scenario

These simulations were modeled to urge operators to practice an operation based on the priority of its importance. In other words, an operator should select the most important operation if there were plural operations to be requested on the operator at the same time. From the results obtained by the simulations, There was no case found in which it went to an unexpected way of further deterioration in spite of the event sequence being each different. However, as a matter of course, omissions and delays in executing operations were observed in each case. This suggested us that most of all protection not to lead to severe accident could be available by the interlocks that were actuated automatically when necessary. Operators' commissions to the plant were composed of necessary actions in order to protect primary equipments and components. However, in fact, there are critical operations not to bear an unexpected bad transient. Thus, if operators could execute the critical operations due to the priority, there was little possibility to come out to a severe accident. Spare time would be prepared for avoiding severe accident.

4.2 Omissions and Delays of Operations to Protect Equipments

Typical examples are shown in Fig.2 that gives influence from defined conditions of the individual characteristics, the team characteristics and HMIs. Obtained results summarized as following:
1) Influence from the change of the individual characteristics was the largest of all, and the omissions were found in executing operations of lower priority;
2) There observed several cases that operations which were implemented by operator not in charge in TC2 and TC3 were effective and valuable. However, in the most of cases, walking distance and utterances were significantly increased in TC2 and TC3. Eventually, the operations were omitted or delayed in TC2 or3.
3) The effect of HMIs change from HMI 1 to HMI 2 was suggested to be worse rather than little. The reason is, mentioned later, the walking distance of HMI 2 became longer than HMI 1 due to the dimension shown in Fig.1.

This simulation results show that the layout and proper task sharing is crucial, especially the size of control panel is a key issue.

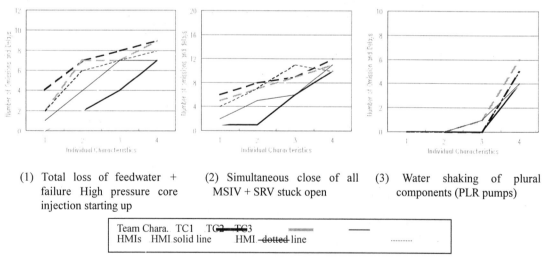

(1) Total loss of feedwater + failure High pressure core injection starting up

(2) Simultaneous close of all MSIV + SRV stuck open

(3) Water shaking of plural components (PLR pumps)

| Team Chara. TC1 | TC2 TC3 | ▬▬▬▬ | ——— |
| HMIs HMI solid line | HMI dotted line | | ············ |

Fig.2 Influence to the numbers of omissions and delays observed in the simulations when every conditions were changed for three anomalies.

Obtained results concerning walking distance, the number of utterance and operations were analyzed to find following:

1) The walking distance in HMI 2 was quite longer than that of HMI 1.
2) According to the time consumption due to necessary walking time, the number of operations and utterances were a little reduced. But the reduction was not so significant and prominent.
3) The influence of the individual characteristics to walking distance and utterances was identified, but it was not so larger than that of the number of omissions and delays.

This results shows us that the improvement from HMI 1 to HMI 2 might not so effective, then designer must pay attention to the size of the control panel.

5. FURTHER APPLICATION OF THE MMS

At present, the developed MMS can be applicable to the operator training and support to demonstrate standard operators' behavior in coping with an anomaly in which procedures would not have been stipulated previously. As shown in Fig.3, this training system can offer the total and concurrent understanding of strategic, tactical and procedure level dealing to unexpected emergencies based on the universal mental model.

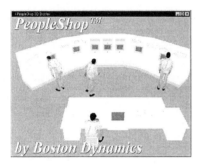 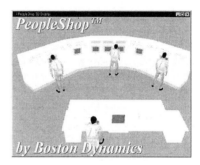

Fig.3. Training system: An example of displays both team behavior and the control panel

6. CONCLUSION

This paper describes the evaluation of HMIs by simulation using the MMS as a tool. The simulations were carried out to ten of the multiple malfunctions, and gain the effective suggestions. This MMS can be applied to an Operator Training System using Virtual Reality in order to give a set of systematic and concurrent training in front of the control panel of the tailored nuclear power plant to each operator.

References

Takano,K., Sunaoshi,W., Sasou, K.(1999). Intellectual Simulation of Operating team Behavior in Coping with Anomalies Occurring at Commercial Nuclear Power Plants. *Human Computer Interaction: Communication, Cooperation, and Application Design volume 2* (Eds. Bullinger and Ziegler). Lawrence Erlbaum Associates, Publishers, Mahwah, New Jersey, USA. pp.1201-1205.

Takano, K. Sasou, K., Yoshimura, S. (1995). Structure of operators' mental models in coping with anomalies in nuclear power plants. *Int. J Hum-Comp Stud* 47, pp.767-789.

Mundel, ME. (1982). Productivety measurement and improvement. In. Salvendy G, Ed. Industrial Engineering. John Wiley & Sons, New York.

Card, SK (1983). The psychology of human computer interaction, Lawrence Erlbaum Associates, Hilslade, NJ,.

Experimental Study on Expert Behavior for Situation Comprehension

Makoto Takahashi*, Daisuke Karikawa*, Akira Ishibashi** and Masaharu Kitamura*

* Department of Quantum Science and Energy Engineering, Tohoku University,
Aramaki-Aza-Aoba 01, Aoba-ku, Sendai, 980-8579, JAPAN
** Human Factor Research of Japan, 3-4-8 Nishi-Shinbashi, Minato-Ku, 105-0003, JAPAN

ABSTRACT

The cognitive behavior of experienced pilot has been studied with the emphasis on the cognitive resource management under insufficient information conditions. A PC-based flight simulator has been utilized to simulate the task of approach phase of the flight by Boeing 767-300 type aircraft. Two subjects, one is expert pilot and the other is novice, were instructed to perform the task with degraded information condition, such as lack of True Air Speed, Altitude, Pitch Angle, etc. The subjects performance has been evaluated based on the integrated deviation from the desired path. The results show that the expert pilot could compensate for the lack of necessary information for the approach phase by using the mental model of the aircraft behavior developed through preceding experiences.

1. INTRODUCTION

Most of the complex systems, such as nuclear power plant, chemical plant, aircraft are supposed to be operated by highly experienced experts. Thus, a human-machine interface, which is a key component to realize higher level of safety, should be designed to be consistent with expert cognitive behavior in order to avoid causing human errors. Although the mechanism of cognitive skill acquisition in relatively simple task domain have been widely studied (J.R.Anderson Ed., 1981, M.K.Singley and J.R.Anderson Ed., 1989), it is quite difficult to clarify specific behavior of experts facing large-scale, complex systems because in most of the cases the expertise can be hardly modeled explicitly. The study of expertise is one of the main research topics in the theory of Naturalistic Decision Making (NDM), which has drawn considerable attention in the field of cognitive work analysis (C.E.Zsambok and G.Klein Ed.,1997). Although the characteristics of expert behavior has been examined in variety of practical task domains such as fire fighting, nuclear power plant operations, etc. in the context of NDM, the process of decision-making has been studied dominantly.

In the present study, the cognitive behavior of an experienced pilot and a novice subject during the final approach phase has been studied comparatively with the emphasis on the dynamic situation comprehension under the conditions in which some of the important information is not available. The purpose of the present study is to validate a working hypothesis, that is; " The expert can keep the performance level even when some of information resources are missing by using the mental model of the system." A clear insight to the mechanism of compensation for lacking information and the required mental model are most useful in establishing a clear guideline for better design of HMI for expert pilot of an aircraft and other artifacts as well.

2. METHOD

2.1 Experimental setups

The PC-based flight simulator (Microsoft Flight Simulator 98) has been utilized to simulate the task of final approach phase of the flight by Boeing 767-300 type aircraft. In order to enhance the reality of the simulation, some add-on software listed in Table 1 have been jointly utilized with the original simulation software. Fig.1 shows the hardware configuration of the experimental system. Two displays showing same graphics image are prepared. One is for a subject (Pilot Flying ; PF) and the other is for the experimenter acting as Pilot Not Flying (PNF). A control wheel and throttle are prepared for the subject to control the aircraft. Although PF and PNF should check their operation each other in the practical line operations, PNF gave no advice to PF and did no voluntary actions in this study. PNF in this experiment performs only two tasks as follows;

- Operation of flaps, lading gears and instruments according to PF's order.

Table1 Add on Software List

Software	Sales Agency	Uses
Microsoft Flight Simulator 98	Microsoft Inc.	Basic simulation
Photo Realistic Scenery	TWILIGHT EXPRESS Inc.	Detailed outside sight
United Airlines Boeig 767-300	free software	Simulating flight dynamics of B767-300
Boeing757/767 Panel v7.0 for Fs98	free software	More precise cockpit instruments for B767-300
FS98 Boeing767 Sound Package v2.0	free software	Engine sound, Environmental sound
Flight Data Recorder version 6.1.0	free software	Data recoding for analysis

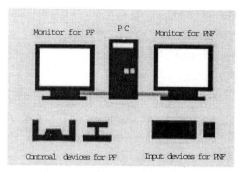

Fig.1 The Hardware Configuration of
The Experimental System

Fig.2 The Experimental Task Scenario

- Reading out warnings on Engine and Crew Alerting System (EICAS)

2.2 Subjects and task scenario

Two subjects, a highly experienced pilot (Total flight hour: 20000hrs. Flight on B767: 3000hrs. Subject-A) and a novice (Subject-B), participated in the simulator experiments. The subject-A had about five hours of practice for simulator, which gave him ample time for adapting to the simulation. The subject-B was well accustomed to the simulation environments although he has no real experience of aircraft operation. The reality of the simulator has been verified by the Subject-A. It has been confirmed that the reality of the simulation is precise enough as long as the tasks performed in the experiments are concerned. Although the simplified tasks have been used in related studies of aviation (A.J.Masalonis et.al., 1997, S.L.Chappell et.al., 1997), the reality of the contents of the task has been considered as most important factor in the present study.

The experimental task scenario starts when the B767 aircraft is at 19 mile south-east of Osaka Itami Airport at the altitude of 3500 ft. as shown in Fig.2. The series of tasks instructed to the subjects are:

1. Maintain altitude at 3500 ft. until glide slope capture
2. Capture glide slope and descend following the glide slope.
3. Reducing speed by flap maneuver and engine power control.

All autopilot systems have been disengaged.

2.3 Interfaces

The subjects were instructed to perform the task under degraded information condition, such as lack of true air speed, altitude, pitch angle, etc. The important point in this experiment is how the subject can compensate for the loss of important information to complete the landing task. The five types of interface have been utilized;

Fig.3 The Example Image of Type 4 Display

Type1. Normal Interface (All indicators/signals are available.)
Type2. True air speed (TAS) is not available.
Type3. Altitude is not available.
Type4. Pitch angle is not available.
Type5. Outside sight is not available.

Fig.3 shows the example image of Type 4 display.

2.4 Evaluation measures
The operational performance of each subject has been evaluated according to the following measures recorded digitally during the task execution:

- Lateral location
  The deviation from the ideal flight path has been taken as an index of aircraft lateral behavior.
- True air speed
  The smoothness of the deceleration process has been evaluated. When any abnormal acceleration or deceleration resulting in stall is observed, the performance is considered to be degraded.
- Altitude
  The integrated deviation from the ideal glide slope profile has been taken as an index of aircraft vertical behavior.
- Pitch angle
  The control of pitch angle is quite important to perform stabilized landing. The pitch angle is controlled both by engine power and by control wheel. The variation of pitch angle has been taken as an index of stability of the vertical maneuver. If the pitch angle shows abnormal and/or sudden change, it is considered that the performance is degraded.
- Bank angle
  The bank angle should be zero in this scenario. Thus the averaged bank angle per second has been taken as an index of aircraft stability.
- Heading
  The variation of the heading angle has been taken as an index of stability of the lateral maneuver.

Interviews have been performed to clarify the subjects' behavior and the way of compensation for the lacked information.

1500

3. RESULTS

The results of the cases when pitch angle is not available are only shown in the following as the representative of other cases. Fig.4 shows the time trend of pitch angle both for normal interface and for Type 4 interface when the indicator of pitch angle has not been provided. In case of subject-A, the significant difference has not been observed. On the contrary, in case of subject-B, significant deviations are observed when using Type4 interface. As the results of interview after the simulation trials, the following differences in strategy for aircraft maneuver have been clarified for each subject.

<Subject A>
- It is difficult to know how the pitch angle changed when adjusting the control wheels.
- Too much attention has been paid to the vertical speed meter in order to compensate for the information concerning pitch angle. This results in poor attention allocation to the power and speed control.
- Pitch control tends to fall behind.

The subject-A tried to deal with these problems by the strategies shown below;
- The pitch angle has been estimated based on the vertical speed, and the bank angle has been estimated based on fluctuation of the heading angle.
- Keep the control of pitch angle as small as possible.
- There is a relationship among pitch angle, engine power and vertical speed. By using this relationship, the pitch angle can be estimated qualitatively if the engine power is fixed. The pitch angle response to the modifying maneuver has been estimated based on this relationship.

<Subject B>
- Both pitch angle and bank angle are hardly estimated.
- Taking too much time on the control of bank angle, which results in the poor attention on pitch angle control.
- No operational resources can be allocated to other parameters such as air speed, even if the deviation was found.

The subject-B could do nothing to deal with this situation.

When the attitude director indicator is unavailable, subject-A could compensate the lack of pitch angle indicator by the vertical speed and the degradation of the performance was not significant, while subject-B (novice) showed significant deviation of pitch angle and bank angle. From the results of the interviews and the recorded measures, it has been shown that the expert can estimate the pitch angle information from various available information sources.

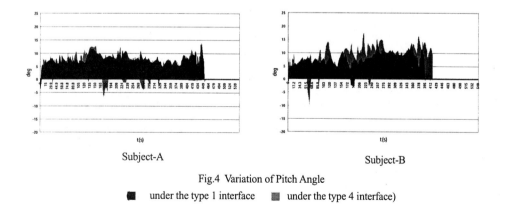

Subject-A   Subject-B

Fig.4 Variation of Pitch Angle
(■ under the type 1 interface   ▨ under the type 4 interface)

1501

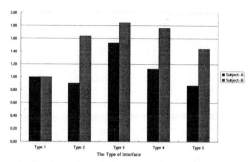

Fig.5 Integration Deviation from Graide Slope (Normalized by the value of Type 1)

Fig.5 shows the performance measure of deviation from the ideal glide slope for all types of interfaces. It also shows significant degradation in case of subject-B, while the degradation was not significant in case of subject-A. Although the directional information for keeping the correct glide slope path has been provided, subject-B could not allocate the attention resources to this information, which resulted in the degraded performance in terms of altitude control. This result implies that the cognitive resources used for the compensation for the masked information could be controlled properly and the interference effect among competing control goals could be kept smaller in case of expert pilot. It has also been that the degree of degradation varied depending on the kind of lacking information. This fact implies that the value of information can be different according to the kind of tasks and facing situations.

The expert behavior facing high workload situation is characterized by the well-developed mental model of the aircraft, which can be used to estimate the lacking information. This mental model enables expert pilot to allocate cognitive resources to other monitoring task to maintain the performance level.

Although the subject-A shows smaller degradation in performance measures, it should be pointed out that the some slip type error has been observed in the landing procedures when he facing the type 4 interface. (He failed to give an order to PNF to set the flap to landing position.) This type of error has never been observed in case that all information is available. This fact implies that the possibility of committing errors becomes higher because of larger cognitive burden even for the expert.

4. CONCLUSION

In the present study, the cognitive behavior of experienced pilot has been studied with the emphasis on the situation comprehension under insufficient information conditions. As the results of the comparison between expert and novice subjects, it has been shown that the expert behavior is characterized by the appropriate cognitive resource allocation and the mental model of the aircraft developed through the long time experiences. Although further experiments including autopilot operation and more complex maneuver are on the way, authors believe that the results of the present study can provide important information for the design of the HMI for well-experienced experts.

REFERENCES

John R. Anderson Ed., "Cognitive Skills and Their Acquisition", Lawrence Erlbaum Associates, 1981

Mark K. Singley, John R. Anderson Ed., "The Transfer of Cognitive skills", Harvard University Press, 1989

Caroline E. Zsambok, Gary Klein Ed. "Naturalistic Decision Making", Klein Associates Inc., 1997

Anthony J. Masalonis et.al. "Instrument Failure Detection And Workload In Simulated General Aviation Flight During Manual And Automated Lateral Tracking", Ninth International Symposium On Aviation Psychology, Volume 1, pp.780-785, 1997

Sheryl L. Chappell, "The Effects Of Experience And Automation On Failure Detection", Ninth International Symposium On Aviation Psychology, Volume 1, pp.773-779, 1997

Ontology Processing for Technical Information Retrieval

Takuya OGURE[*1]   Keiichi NAKATA[*2]   Kazuo FURUTA[*2]

[*1] QUEST, The Univ. of Tokyo
[*2] Graduate School of Frontier Sciences, The Univ. of Tokyo

Abstract

In general, much smaller number of concepts is used in technical documents than in general documents. From this observation, we propose a novel IR method based on the vector model by exploiting ontologies, i.e., systemized concepts that cover specific knowledge in a domain. We have developed a GUI tool to support management of ontologies, and are planning an experiment to evaluate the proposed IR method.

1. Introduction

The characteristic feature of technical documents such as design specification or maintenance guidance is that their contents are limited in scope and terms used are relatively poor in variety. This leads to our observation that there is a possibility of taking advantage of a new method of information retrieval (IR) specifically for technical documents.

Conventionally, Full Text Retrieval[1] (FTR) by word matching is widely used, as well as indexing by local rules, such as author's affiliation or published date. Such local indexing rules are hardly available when the documents are accumulated in large scale. By contrast FTR is one of successful methods for a large-scale document-base management because of its effectiveness and simplicity. However, the effectiveness of FTR degrades in some cases such as follows:

1. When query words include synonymy or polysemy.
2. When the query cannot be described precisely by keywords.

The latter case corresponds to the situation in which the user doesn't know what s/he actually

1503

needs, e.g., s/he is searching for a new idea or concepts. FTR is definitely effective in most cases, but another IR method is desired not as the replacement of FTR but as its complement.

2. Overview of the Proposed IR Method

2.1. Ontology

When we use the Internet web portal site such as "Yahoo" or "AltaVista", there exist two types of site searching, namely keyword matching and category index selection. The category index can be seen as a systemized conceptual index that reflects concepts entries that users tend to be interested. The conceptual index is powerful for IR in the ideal situation in which these concepts cover all of users' interests and all documents are properly related with concept nodes. This is naturally difficult to achieve for general documents but the situation is somewhat different for technical documents because its contents are highly limited as well as its concepts.

A systemized set of concepts that covers specific knowledge in a domain is called an "Ontology" [2]. Usually an ontology has concept nodes linked to each other through some relations (Fig. 1). Constructing an ontology brings a lot of benefits not only to IR but also to inference, e.g., a medical expert system[2].

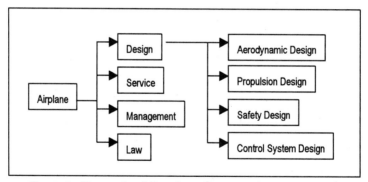

Fig 1. An Example of Ontology (in case of a tree graph.)

2.2. IR Model

Most FTR usually uses a Boolean Model (BM), in which query words are given as conjunctions and disjunctions (e.g., advanced search mode of "AltaVista"). Generally in BM the user puts some feature items of document in an AND/OR combination to retrieve desirable documents. On the other hand in Vector Model[3] (VM) documents are represented as a surrogate of feature vector in which the strength of pre-defined features about the document is listed as vector elements.

User's queries are converted to similar vectors so that the closeness can be compared by calculating their inner product or cosine value. This method is more flexible than BM, which is rather discrete.

The feature vector elements are decided by the IR system designer, for example they are set as the keywords frequency of a document. Here we propose a new feature vector definition in which the vector elements correspond to ontology elements. This feature vector should represent the concepts described in technical document properly if a well-defined domain ontology is provided. Here onwards we refer to this particular feature vector as "ontology vector".

2.3. Processing of Ontology Vector

To determine the values of the document's ontology vector, it is desirable to adopt the direct judgment by human experts who have read the documents, though it is difficult to keep its consistency and carry it out through a large number of documents with vector elements. We assume that the ontology vector of the document can be generated by the summation of vectors of the terms that appear in the document. This idea can be implemented as follows.

1. Prepare the term list concerned with the target technical domain.
2. Check which element in the ontology should be associated with each term. An ontology vector of every term can be obtained from this association.
3. Extract terms that exist in the prepared list from a document.
4. Sum up the ontology vectors of these terms to get a document's ontology vector. (Fig. 2)

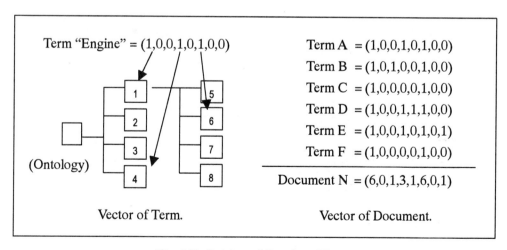

Fig. 2 Definition of Ontology Vector

Users' query should be given a similar ontology vector, by selecting ontology element list or converting terms selected from the term list to be compared with the documents' ontology vector. In this way the system can share the human's comprehension as an ontology and process it to retrieve the required documents.

3.Management of Ontology and Ontology Vector

Actual technical ontology we can use is usually incomplete, so it must be improved constantly upon usage. According to the changes in ontology, ontology vectors of terms need to be modified manually, in addition to the continuing enrichment of term list. Generally speaking a thesaurus such as the ontology vector dictionary would introduce problems when applied to the maintenance of term lists. In order to support the management of ontologies and ontology vectors, we have developed a GUI tool shown in Fig. 3. This tool can only deal with ontologies in a tree structure, something like a category tree, and the main characteristics of this tool is that not terms but ontology nodes keep association between terms and ontology nodes. With this approach, the tool supports the following operations graphically.

1. Associate terms with ontology nodes by dragging a term icon onto an ontology icon.
2. Create a new ontology node for branching a concept or adding sub-concepts. Only the terms in the parent node of the newly created concept have to be re-checked.

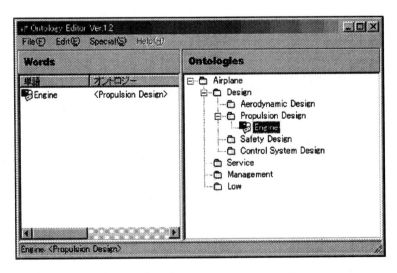

Fig.3 Ontology management tool

3. Move an ontology node to another place. Associated terms under the branch are moved according to the movement with associations preserved.
4. Delete terms or ontology nodes. Associations are discarded.
5. Add new terms with an automatic redundancy check.

4. Future plan

We are planning an experiment to evaluate the effectiveness of this IR method by using a "test collection", i.e., a set of document data, query, and the correct data set generated by hand. We have chosen articles on artificial intelligence from the Japanese test collection[4] so that we can construct specific domain ontology for the retrieval task. We will compare this IR method with other methods by comparing precision and recall rates after preparing the ontology and ontology vector of terms using the tool described above.

5. Conclusion

We proposed a novel IR method for technical domain knowledge that utilizes ontology of the domain. It is expected that this method is powerful if a well-defined ontology is given, that is expressive enough to include most concepts that users consider in the domain. This assumption is impractical for general documents but is reasonable for technical documents. We adopted the vector model for elastic matching, naming the feature vector as "ontology vector".

In order to evaluate this method, we are planning to compare its performance with other methods using a Japanese test collection. The ontology management tool has so far been developed and an ontology for the test collection is currently being constructed.

References

[1] Chuichi Kikuchi: A Fast Full-Text Search Method for Japanese Text Database; Journal of the Institute of Electronics, Information and Communication Engineers. **J75-D-1 No.9**. 836/846 (1992).In Japanese.

[2] G. Van Heijst et al.: Using explicit ontologies in KBS development; Int. J. Human-Computer Studies. **45**. 182/292 (1997).

[3] Scott Deerwester et al.: Indexing by Latent Semantic Analysis; Journal of the American Society for Information Science.**41-6**. 391/407 (1990).

[4] NII-NACSIS Test Collection for IR Systems (http://research.nii.ac.jp/ntcir/).

CREATIVITY OR DIVERSITY IN COMMAND AND CONTROL ENVIRONMENTS

Mats Persson[a] and Björn Johansson[b]

[a]Department of Operational Studies, Swedish National Defence College,
P.O. Box 27805, SE-115 93 Stockholm, Sweden
[b]Department of Computer and Information Science, Linköpings Universitet,
SE-581 83 Linköping, Sweden

ABSTRACT

Within areas of crisis management new visions are expressed on how technology could facilitate the work for those exerting command and control. In this paper we present results from a pilot study of decision teams working together in a laboratory setting of an envisioned command and control environment. One of the prevailing assumptions within most new visions is that having access to a common and dynamically updated representation of the outer world will facilitate the decision makers' work analysing data from different sensors in real or near real time. Furthermore, novel ways of presenting data on shared displays is supposed to support the decision makers' creativity when handling unanticipated situations. In turn creativity should result in better and faster decisions. The results presented are obtained from audio and video recordings and indicate that the creative discussions among the subjects that have near real-time information do take place, but also that they seem to focus merely on the current situation. Thus, their work in comparison with teams that are provided with delayed, but human-filtered data about the situation, seems to become more reactive and that the capability to anticipate events and consequences from made decisions decreases.

1. INTRODUCTION

In a Swedish research project called ROLF 2010 (a Swedish acronym that stands for joint mobile command and control function for the year 2010), efforts are made to develop a command and control environment aimed for the future. The work in the project is based on an overall vision, which describes how Sweden's national defence and rescue services command and control units may appear within 10-15 years' time. The vision is founded on the present tasks assigned to the Swedish Total Defence, which includes both armed as well as civil defence. Development of the concept rests upon several assumptions, where the most apparent are; (1) that technology will make it possible to considerable reduce the size of a staff unit; (2) that the seating arrangement will facilitate communication among staff members; and (3) that use of all data available from a vast number of different sensors about current situation together with advanced technical means to present data will contribute to and facilitate the staff members' awareness of the situation (Sundin & Friman, 1998, 2000). Hence, by sitting round a table with common artefacts presenting shared data, and at the same time be able to monitor each interlocutor's eyes and interactions is thought to be one of the best ways to start a creative discussion with a common focus (Artman & Persson, 2000). In turn it is supposed to support co-operation in order to efficiently handle a vast flow of data, as well as facilitating problem solving and the capability of making better decisions.

Much of the functionality that is proposed to be appropriate by future decision makers derive from results from current research and advances within various signal-processing techniques together with available and predicted computer power, which have given rise to a number of novel human-computer interaction modalities. Traditionally, organisations as e.g. the rescue services or the Military have been hierarchical structured, where the command and control function has been located in the uppermost hierarchical level. Today arguments can be found within those communities that current hierarchical organisations are considered too rigid for being able to act and react on situations in future and highly dynamic environments. As an option of handling the dynamics, so called "network organisations" have been proposed as a solution. The structure of network organisations is considered to be a possible answer to several problems of exerting command and control among military communities (e.g., see Alberts, Gartska, & Stein, 2000; Brehmer & Sundin, 2000; Cebrowski & Garstka, 1998). Implementation of such an organisational structure imply that traditional hierarchical levels of command could be flattened out, reduced or maybe completely removed. Reducing levels of command is considered advantageous and necessary to shorten the time for reaction upon changes in the environment, since data processing within every level of command is considered time consuming and delaying the possibility to take action. Furthermore, it is assumed that as long as a decision-maker is provided with enough data presented in an understandable way he or she will be able to make "optimal" decisions. As a consequence larger amount of data has to be handled by the commanders within network organisations in comparison with traditional hierarchical organisations. Thus, greater demands will be made on mission controls as the ROLF 2010 staff unit, since "…it will require that the commander and his staff will be able to handle greater amounts of information and greater complexity than before" (Brehmer & Sundin, 2000). One reason why larger amount of data has to be handled could depend on that the organisational filters that are present in hierarchical organisations are reduced or completely removed within network organisations. Hence, there are clearly some very significant differences between the envisioned systems and the traditional ones (see table 1). Although these differences never have been tested, they are to a large extent assumed to be valid among many western countries today where developmental projects aimed to create new C2-systems are, or is being, started.

Table 1:   Characteristics of C2-systems compared to envisioned C2-systems.

"Traditional" C2-Systems	Envisioned C2-Systems
• Organised in hierarchies • Information distributed over a variety of systems, analogue and digital. Most common medium is text- or verbal communication. • Data is seldom retrieved directly from the sensor by the decision-maker. It is rather filtered through the chain of command by humans that interpret it and aggregates it in a fashion that they assume will fit the recipient. • Presentation of data is handled "on spot", meaning that the user of the data organises it him/her self, normally on flip-boards or paper-maps. The delay between sensor registration and presentation depends greatly on the organisational "distance" between the sensor and the receiver	• Organised in networks. • All information is distributed to all nodes in the system. Anyone can access data in the system. • Powerful sensors support the system and feed the organisation with detailed information. • Data is mostly retrieved directly from sensors. Filtering or aggregation is done by automation. • Presentation is done via computer-systems. Most data is presented in dynamic digital maps. The time between data retrieval and presentation is near real-time. • It is possible to communicate with anyone in the organisation, meaning that messages do not have to be mediated via different levels in the organisation.

1.1   Problem

In this paper we present results from experiments made, where the aim was to examine differences in performance and behaviour between decision teams presented data that is filtered by the organisation (as in hierarchies) with decision teams that were presented data that was unfiltered (as proposed for network organisations). Our hypothesis when conducting the experiment was that there were a difference in performance between teams supported with unfiltered data presented directly from the field, and teams working in a more "traditional" way presented filtered data. Our second research question concern whether shared, but unfiltered, data presented for the decision team supports creativity in the way that have been predicted by the visionaries. As an example, is it so that the staff to plan future actions uses the time gained from not having to organise data? Will it effect the division of labour within the staff?

2.   METHODOLOGY

In order to examine these two questions, we have tested the existing command and control laboratory at the Swedish National Defence College with its current ROLF 2010 set-up. To make it possible to test teams in dynamic decision-making, we have used a micro-world called C3-fire, which provides an excellent support for quantitative data retrieval where operational, collaborative and personal work information is logged in the session server (Granlund, 1997; Johansson, Granlund, & Wærn, 2000). The C3-fire micro-world generates a task environment in which team members co-operate to extinguish a forest fire. The simulation includes the forest fire, houses, and different kinds of vegetation, computer-simulated agents such as reconnaissance personnel and fire-fighting units. The fire and the fire-fighting organisation could be considered as complex and dynamic systems that change autonomously, as well as a consequence of actions made on them. In the C3-fire micro-world, the subjects are to take to roles of a staff commanding two ground-chiefs. The ground-chiefs are the persons who are in direct command of the fire-brigades used two extinguish the fires in the simulation. The staff consists of four persons. Two staff-officers who communicate with the outer world (the ground-chiefs) and two who serve as the commanding officer and assistant. These four persons decide together on where the ground-chiefs are to move their fire-fighting units. The subjects used in the study are a total of randomly picked professional military officers at the rank of captain. They where randomly assigned to one of the two conditions. A total of 60 subjects divided into ten teams where tested.

Two sets of data are collected; video-recordings and log-files from the micro-world. This combination makes it possible to observe effects on team behaviour and creativity, as well as to judge their performance in the simulated environment by quantitative measurement. In all sessions the subjects are audio and video recorded from two angels. All quantitative data were analysed and evaluated. Regarding the qualitative data, we have focused our analysis of video and audio recordings on the third trial, which makes over five hours of analysed material in total.

2.1 Design and procedure

The design is a two (unfiltered versus filtered) by three (trials) design with repeated measures for the trials only. There are two independent variables, namely the updating of the map. Either the map is updated directly (unfiltered information) from the ground-chief (who controls the fire brigades), which creates an illusion of automatic updating within the staff (the ground-chief is the sensor) or manual updating, which is based on e-mail messages from the ground-chiefs (filtered information). All the trials were exactly the same with the exception that the starting-point for the fire scenario was rotated and mirrored between the trials, which the subjects did not now.

All teams were informed about the nature of their task and the roles within the team. They were also able to choose which roles to take themselves. The following roles are available; two ground-chiefs (who control the fire brigades), two communication officers, one commander and one assistant (who is responsible for the documentation of the actions taken). Then they were given a fifteen-minute training session to acquaintance themselves with the C3-fire environment. After the training session, the subjects got five minutes to discuss within the team before the first trial started. Each trial lasted 30 minutes and was followed by a five-minute evaluation session where the trial was replayed for the subjects 10 times the normal speed.

3. RESULTS

The task of dynamic decision-making also challenges the researcher, since such a task even in the laboratory is partly dependent on uncertain aspects like luck and chance. Therefore, we believe that it is important to consider both process and outcome when studying this area. There is a clear risk that results only based on performance in dynamic decision-making has very low explanatory value. During our analysis of the video-recorded material, we have tried to focus on events specific to the conditions rather than incidents unique to a certain team. Below, we first present the statistical findings in terms of performance and then the main observations from the video-recordings.

Since the C3-fire micro-world is divided into 400 cells that theoretically could burn down, performance is calculated in terms of burned down cells in the simulation. The fewer cells that are burned down during a trial, the better. As seen in the result from the log-files presented in figure 1 there is no significant statistical difference between the unfiltered data versus filtered in terms of performance. Accordingly, our hypothesis has to be rejected. However, after examining the video and audio recordings from the experiments, we have found some interesting differences between the two conditions.

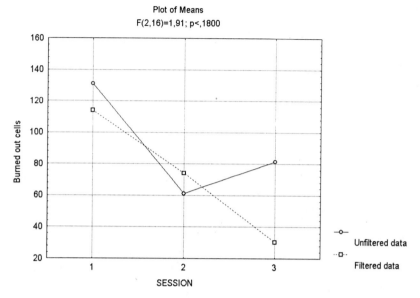

Plot of Means
F(2,16)=1,91; p<,1800

Figure 1: Means of number of burned out cells for the two conditions per trial.

It seems like the shortened delay in information presentation has an impact on the planning that the team conducts. Generally, in both conditions the teams make plans in advance, which they probably have intended to carry through. In the unfiltered condition however, they often seem to loose focus of their original plan and start to perform actions that effect events in the near future. In the filtered condition, the teams more often follow their original plans to the end. The division of labour within the staffs may be pre-arranged by the experiment leaders to some extent, but the real division of labour, meaning who takes part in which activity clearly differs between the two conditions. In the filtered conditions there is a distinct division between the commander and his assistant and the two staff-officers. If we would take an analogy to an ensemble, the commander is in this case clearly the conductor. The work-procedure has some distinct differences, especially concerning the communicative aspects. In the filtered condition, a typical conversation in the staff-room may sound like this:

```
Staff-officer: Fire in B6 (reads incoming mail out loud)
Assistant: Fire in B6 (types event into database, updates map with information)
Commander: Move car two to B6
Staff-officer: Car two to B6
```

Excerpt 1: Excerpt taken from a session provided with filtered data

What we can see above is that data in this condition is treated on an "as is" basis. It is distributed into the staff in the same fashion as it was filtered. The procedure of handling new information does not differ greatly from the traditional approach in staff-work as seen in the field. In contrast, the unfiltered condition actually seems to support a more "creative" environment. There is a much more non-distinct division of labour in that condition. In this case, the staff-officers also take the roles of decision-makers and engage in the activity by the updated map. It seems like the seating arrangement suggested in the ROLF 2010-writings (see above) actually has the desired effect. Typically, the staff-members in this condition gather around the representation where the unfiltered information is presented and discusses it. They "discover" new events in the environment they are to control. These "discoveries" may also be the reason to their somewhat reactive behaviour, since every new event creates a discussion, which in turn usually leads to a new decision.

4. CONCLUSIONS

These results show that there is no clear effect of the conditions in terms of performance. The results obtained from this simulated task rather indicate the opposite. The difference may not be "visible" to the observer until the results are at hand. However, we shall not neglect the possibility that the unorthodox way to receive unfiltered data confused the subjects (whom where all professionals). This could mean that the assumed improvement in performance is achievable with proper (more) training.

A natural question to rise in this context is whether a "creative" environment really is desired? If we approach command and control work from a traditional view, there has always been a clear division of labour within the staff. This might be because humans need an orderly-arranged work procedure to handle exceptional events and uncertainty. A creative environment might be better suited for less time-critical tasks. It can be argued that the time used from an actual event to a decision is longer in the unfiltered condition because of the time consumed discussing the "raw" data. What we mean by this is that although the filtered data reaches the staff later than it would have done unfiltered, time is saved due to the fact that the information is already interpreted and aggregated. Also, the team as a unit opens up for different interpretations which demands time before a consensus is reached, rather than getting pre-filtered data. Is it not so that the unfiltered data merely supports discussion about what is going on in the observable world? This is not necessarily the same thing as creativity. We could also question whether it is possible to make filters that can organise data in a way that actually is an improvement to the human filters used today. In a network organisation like the ones proposed in the envisioned systems, the notion of 'command' and 'control' is more unclear than in an hierarchical system, which also suggests that it will be much harder to predict the way in which data has to be filtered to fit a specific person in a specific situation.

We suggest that more longitude studies are conducted and that more complex organisational structures than the one presented in this paper should be tested. Such studies could answer two critical questions that have risen from our results: Is a network organisation superior to the traditional hierarchical, and will training of teams using unfiltered data improve their performance beyond the possible limit of filtered?

REFERENCES

Alberts, D. S., Gartska, J. J., & Stein, F. P. (2000). *Network Centric Warfare: Developing and Leveraging Information Superiority*. Washington, DC: National Defense University Press.

Artman, H., & Persson, M. (2000). Old Practices - New Technology: Observation of how established practices meet new technology. In R. Dieng & A. Giboin & L. Karsenty & G. De Michelis (Eds.), *Designing Cooperative systems: The Use of Theories and Models - Proc. of the 5th Int. Conf. on the Design of Cooperative Systems (COOP'2000)* (Vol. 58, pp. 35-49). Amsterdam: IOS Press.

Brehmer, B., & Sundin, C. (2000). Command and Control in Network-Centric Warfare. In C. Sundin & H. Friman (Eds.), *Rolf 2010 The Way Ahead and The First Step: A Collection of Research Papers* (pp. 45-54). Stockholm: Elanders Gotab.

Cebrowski, A. K., & Garstka, J. J. (1998). Network-Centric Warfare: Its Origin and Future. *U. S. Naval Institute Proceedings, 124*(1), 28-35.

Granlund, R. (1997). Microworld Systems for Emergency Training. In Y. Waern (Ed.), *Co-operative Process Managment: Cognition and Information Technology*. London: Francis and Taylor.

Johansson, B., Granlund, R., & Wærn, Y. (2000). *The communicative aspects of distributed dynamic decision making in the ROLF environment*. In proceedings of the 5th conference on Natural Decision Making, May 26-28, Stockholm.

Sundin, C., & Friman, H. (Eds.). (1998). *Rolf 2010 - A Mobile Joint Command and Control Concept*. Stockholm: Elanders Gotab.

Sundin, C., & Friman, H. (Eds.). (2000). *Rolf 2010 The Way Ahead and The First Step: A Collection of Research Papers*. Stockholm: Elanders Gotab.

LEVELS OF AUTOMATION IN EMERGENCY OPERATING PROCEDURES FOR A LARGE-COMPLEX SYSTEM: PROBABILISTIC ANALYSIS ON HUMAN-AUTOMATION COLLABORATION

Hiroshi Furukawa*, Yuji Niwa** and Toshiyuki Inagaki*

* Institute of Information Sciences and TARA Center, University of Tsukuba, Tsukuba, 305-8573, Japan
** Institute of Nuclear Technology, Institute of Nuclear Safety System, Inc., Mihama-cho, 919-12, Japan

ABSTRACT

For appropriate operations of large and complex human-machine systems, tasks must be allocated among human operators and automated systems according to the situations of the tasks. Nevertheless, basic methods or techniques, especially quantitative ones, are not mature enough to be used as rational tools for human-machine design. This problem can be defined as 'determination of a proper level of automation.' A goal of this research is to establish a theoretical framework for quantitative evaluation of automation levels, which can be used in real domains. A method using discrete event simulation and quantitative indexes is developed and applied to a particular artifact: a nuclear power plant.

1. INTRODUCTION

Benefits of automated systems can be clearly enumerated in comparison with human operators, such as capability of extraordinary precision in control, lower error rate, higher processing capability, and extension of operator's perceptual and cognitive capabilities (Sheridan & Parasuraman, 2000; Wickens & Hollands, 2000). Those are benefits of the systems themselves as opened-loop systems, but the benefits and costs must be evaluated on total human-machine systems which are closed-loop systems of target artifacts, the automated systems, and human operators. Many accidents and incidents revealed that the discrepancies were not trivial matters. Details of the costs and benefits of automation are delineated comprehensively in Parasuraman & Riley (1997) and Wickens & Hollands (2000). Interactions between the constituents are dynamic processes in most cases. It suggests that proper schemes for task allocation among human operators and automated systems can be different for different situations. The necessity of dynamic task allocation is indicated through many researches (Scerbo, 1996; Inagaki, 1998). Because timing of reactions by human operators or automated systems is one of crucial points in evaluating the appropriateness of automation schemes, the evaluation must be practiced in quantitative manner. We must have quantitative data for what may happen if one scheme has been adopted for an automated task. Nevertheless, the basic methods or techniques, especially quantitative ones, are not mature enough to be used as rational tools for human-machine design.

Authors are developing a theoretical framework for quantitative evaluation of task allocation schemes (Furukawa, Inagaki, & Niwa, 2000). The framework is based on a method for behavior estimation of human-machine system, using cognitive task simulation with probabilistic models. A target usage of this method is preliminary evaluation of designs at initial phase of designing human-machine systems. The schemes considered in this research are based on "levels of automation" (Sheridan, 1999). To describe a problem of task allocation between human and automation, Sheridan suggested ten levels of automation (Level 1-10) for four stages of a process in operations: Information Acquisition (Task-A), Information Integration (-I), Decision (-D), and Control (-C). At designing phase of systems, designer must select a proper level for each of four stages in every process. Alternative schemes are defined by combination of levels of automation and objective tasks. The number of schemes is so many that evaluation tests with human subjects are difficult to perform. This approach may identify worse cases which must be avoided, and discover effective parameters that change the behaviors of human-machine systems.

This paper delineates results of the application of the proposed method to actual tasks during an emergency situation at a nuclear power plant (NPP) with typical quantitative indexes for the evaluation, i.e., expected values (Sheridan & Parasuraman, 2000). The indexes are defined on each of *execution time of a process* and *cognitive workload indexes of a human operator* which are acquired by the simulations, taking failure rates of target systems, human error rates, and error rates of automated systems into account. The results show facts that advantages of automation levels depend on situations, and the target domain is too complex to be evaluated by qualitative methods.

2. METHOD

This section elucidates our method and an actual task for NPP during malfunction.

2.1 Simulations with Cognitive Task Networks

A simulation code for estimating behaviors of human operators and automated systems is based on cognitive task networks (Laughery, 1999). Through cognitive task analysis, processes in an operation are resolved into sets of cognitive tasks and paths. Each task is described as a model with time and cognitive resources which are required to execute the task. The execution time is simulated in a probabilistic manner based on normal distribution model (its mean time and standard deviation, SD, are assumed). The cognitive resources are simulated with the Multiple Resource Model proposed by Wickens (Wickens & Yeh, 1986), where five types of resources are defined: visual, auditory, cognitive, motor, and speech. Indexes describing cognitive workload, which are necessary to use the resources, are assigned in a model using data from a reported database (Micro Analysis and Design, 1997). Branching conditions in paths are described in a probabilistic or logical manner. Task allocations to agents (operators or automated systems) are fixed based on the adopted levels of automation. This simulator is implemented using WinCrew (Micro Analysis and Design, 1997), a discrete event simulation-modeling tool. Monte Carlo simulations using this model will provide quantitative time data on performance of each task (succeeded/failed, execution time, etc.) and total momentary workload index of every operator.

2.2 Cognitive Tasks during Malfunction of NPP

The proposed method was applied to an Emergency Operating Procedure (EOP) of a Steam Generator Tube Rupture (SGTR) failure at a pressurized water reactor. SGTR failure is worth investigating, because this is one of assumed failure modes in safety judgment for permission of NPP construction, and it is expected that cognitive workload of operators at the mode is higher than at other failure modes. The EOP has seven steps, and each has several small steps inside them (fictitious data for this research). A single step in Step 3 is modeled and evaluated, which is a task 'cutting coolant flow of a particular pipe.' It seems that operations of NPP are highly automated, but the truth is not. Because their safety is indispensable, the large-complex systems are designed in conservative manners. At Step 3-1-1, an operator orders a Human Interactive Computer (HIC) to close a valve. The HIC send the signal to a Task Interactive Computer (TIC), and the TIC manipulates the valve automatically. After that, the operator must confirm that the valve is closed completely. If the operator inferred that the task completes, he/she tries to perform a next step, Step 3-2-1. If not, the operator must contact a field operator and direct him/her to close the valve manually (Step 3-1-2), which takes much longer time than Step 3-1-1, and then must perform Step 3-2-1. After execution of Step 3-2-1, if the operator finds out that Step 3-1-1 was not completed and he/she did not execute Step 3-1-2, he/she must execute Step 3-1-2 and Step 3-2-1 again. This is the risk in estimating current situation. The malfunction in the valve controlling system is reported as one of dominant failures in this EOP (Bertucio & Julius, 1990).

2.3 Models with Different Levels of Automation

A type of the target task of automating is Task-C. A set of Task-A, -I and -D before the Task-C is recognized as a stage where a human operator is trying to comprehend state of the system. The scheme of automation in this evaluation is that HIC must confirm the result of Step 3-1-1 (Task-A, -I and -D), and must perform only Step 3-2-1 (Task-C) or both Step 3-1-2 (Task-C) and Step 3-2-1 depending on result of the confirmation. A human operator must have correct state comprehension (Task-A, -I and -D), and is asked to perform actions (Task-C) according to an adopted level of automation.

This research evaluates three levels of automation (Levels 5, 6 and 7) among ten defined by Sheridan (Sheridan, 1999). The levels are selected under two conditions: HIC has an ability to perform Task-C (viz., higher than Level 4) and an operator can recognize it if the HIC executed an action (viz., lower than Level 8).

> The computer suggests one action alternative, and
> > Level 5: executes that suggestion if the human approves.
> > Level 6: allows the human a restricted time to veto before automatic execution.
> > Level 7: executes automatically, then necessarily informs the human.

A cognitive task network for this automation scheme is shown in Figure 1, where Level 5 is adopted. A-AID-* indicates that this task is a set of Task-A, -I and -D (AID) and is allocated to an automated system, H-C-* indicates Task-C allocated to a Human operator. *-C-1 (with mean 450 sec, SD 150 sec) is Task-C of Step 3-1-2, and *-C-2

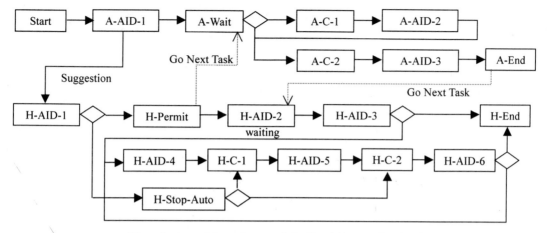

Figure 1: A cognitive task network for Step3 (Automation Level 5).

(with mean 10 sec, SD 1 sec) is that of Step 3-2-1. Conditions in branches are defined with three Rs: a result of Step 3-1-1 (R1: Succeeded/Failed), a result of situation identification by HIC (R2: S/F), and a result of situation identification by a human operator (R3: S/F). When the operator agrees with the suggestion by HIC, he/she approves (H-Permit) and tries to comprehend situation (H-AID-2). When the operator does not agree with it, he/she can stop the HIC (H-Stop-Auto) and perform the task manually (from H-C-1 or H-C-2). After HIC executed actions, the operator must identify the situation (H-AID-3). If R1=F and R2=S, he/she must perform C-1 and C-2. If R1=S and R2=F, he/she can consummate this process, because the valve has been closed completely. C-1 was not necessary to perform.

2.4 Expected-Value Analysis
In this research, several indexes are selected to evaluate the levels of automation for the target task, i.e., expected values of execution time and several cognitive workload indexes. The Monte Carlo simulations are executed under eight (=2^3) different states, corresponding to combinations of the results [R1 R2 R3] (SSS, SSF, ..., FFF). In Level 7, because a human operator does not do tasks of state comprehension, there are only four states ([R1 R2]: SS, SF, FS, FF). The probability of occurrence of each state can be defined by failure rate of the valve, error rate of HIC for state comprehension, and human error rate for state comprehension. The expected values are calculated as weighted averages of data by the probability of occurrence. The dedicated simulation for each state can provide data which are useful to see details of the dynamic processes, even the case where the probability of occurrence is very small. The results of a detailed analysis on each of the states are reported in Furukawa, Inagaki, and Niwa (2000).

3. RESULTS
This section describes main results of the simulations with three types of automation schemes (Level 5, 6, 7). Ten thousand trials of Monte Carlo simulations were performed under each of eight states. The failure rate of closing the valve was set as 3.100E-3/demand (fictitious data), based on fault-tree analysis using a generic database for component failures (Eide & Calley, 1993). The error rate of HIC and human error rate for state comprehension were 0.01 (Low) or 0.05 (High) tentatively. The expected values were given for four different situations at Level 5 and 6; [HIC error rate, human error rate] = ([Low, Low], [Low, High], [High, Low], [High, High]), and for two situations at Level 7; [HIC error rate] = ([Low], [High]).

3.1 Expected-Value Analysis on Execution Time
Table 1 shows expected values of execution times at each level of automation. When the error rate of HIC for state comprehension is low, the result is: Level 6 < Level 7 < Level 5, where A < B denotes that expected value of execution time at level B is longer than at level A. Thus, Level 6 is the best under this condition, whether the human error rate is Low or High. It seems incomprehensible that the expected value at Level 7 with a low HIC error rate is worse than at Level 6 with a higher human error rate. Main cause of this result is a correcting action of the human operator against a situation where the HIC estimated that TIC had failed to close the valve automatically

Table 1: Expected values of execution times at each level of automation.

Error Rate for State Comprehension		Level of Automation		
HIC	Operator	5	6	7
Low	Low	56.08	39.14	40.92
Low	High	74.91	39.54	
High	Low	56.27	40.01	58.91
High	High	75.07	41.1	

Table 2: Expected values of time integration of total momentary workload index.

Error Rate for State Comprehension		Level of Automation		
HIC	Operator	5	6	7
Low	Low	390.39	315.75	171.26
Low	High	649.96	326.77	
High	Low	390.64	322.43	172.09
High	High	645.23	338.33	

Table 3: Expected values of time averages of total momentary workload index.

Error Rate for State Comprehension		Level of Automation		
HIC	Operator	5	6	7
Low	Low	6.40	8.08	4.82
Low	High	6.67	8.10	
High	Low	6.40	8.07	4.64
High	High	6.65	8.09	

Table 4: Expected values of total time where workload index is larger than 6.4.

Error Rate for State Comprehension		Level of Automation		
HIC	Operator	5	6	7
Low	Low	48.58	31.63	22.52
Low	High	67.41	32.03	
High	Low	48.77	32.51	22.59
High	High	67.57	33.59	

Table 5: Expected values of total time where workload index is larger than 7.6.

Error Rate for State Comprehension		Level of Automation		
HIC	Operator	5	6	7
Low	Low	43.38	31.31	22.52
Low	High	61.76	31.66	
High	Low	43.19	31.60	22.59
High	High	61.56	32.65	

Table 6: Expected values of total time where workload index is larger than 13.7.

Error Rate for State Comprehension		Level of Automation		
HIC	Operator	5	6	7
Low	Low	4.50	0.03	0.01
Low	High	23.68	1.35	
High	Low	4.37	0.09	0.07
High	High	22.78	1.35	

and tried to execute Step 3-1-2 (manual closing) by mistake. The operator could modify the situation in Level 5 and 6 in the early part of the process, but not in Level 7. The execution time of t is much longer than that of Step 3-1-1, and the mistake by TIC causes serious damage to total execution time. The advantage of human operators is significant when the error rate of HIC for state comprehension is high. The difference of the execution time between at Level 6 and 7 is approximately 20 seconds, while nearly 1 second when the HIC error rate is low. The total result is: Level 6 < Level 5 (Human error rate is Low) < Level 7 < Level 5 (Human error rate is High).

Regardless of the error rates of HIC and human operators, Level 6 is the best to minimize the expected value of execution time. This result shows that the most proper level of automation can be identified using the index; an expected value of a total execution time of a target task. It is not true that higher-level automation can always abridge the execution time of processes. Advantages of higher-level automation must be discussed in their relations with characteristics of tasks adopted. In this case, it might be difficult to estimate the result of this evaluation delineated in this section using any qualitative analysis methods.

3.2 Expected-Value Analysis on Cognitive Workload

Table 2 shows expected values of time integration of total momentary workload index at each trial (= sums of [a workload index multiplied by its time length]). Expected values of time averages of the workload indexes (= [the time integration] divided by [execution times of the trials]) are in Table 3. Tables 4, 5 and 6 are expected values of total time where cognitive workload indexes of human operators are larger than 6.4, 7.6, 13.7, respectively. The levels are the typical levels of workload in this process and were selected to see total time with low, medium and high levels of workload. These three types of indexes present different aspects of states of human cognitive workload at the operation. The first and second types of indexes (time integration and time average) indicate general aspects of cognitive workload, and can be used for macroscopic analysis on automation levels. The third type of indexes (total times) depicts much-detailed characteristics of the workload. These indexes draw an inner structure of time-varying cognitive workload during an operation with an automation level.

These results show: (1) The results of evaluation with the first and third types of indexes, i.e., time integration and total times, are the same: Level 7 < Level 6 < Level 5. The result gave good agreement with the general understandings of the automation levels concerned in this research. (2) The result of evaluation with the second type of index, i.e., time average, is: Level 7 < Level 5 < Level 6. The Total workload (time integration) at Level 5 is larger than at Level 6, but the average is smaller. The reason is that the execution time at Level 6 is much shorter than at Level 5. The result of the expected-value analysis on execution time claims that Level 6 is most proper as the automation scheme for the target task. However, it should be examined whether there are severe ill effects of using Level 6 on cognitive workload of human operators. The result (1) says that even though workload at Level 6 is much better than Level 5, the workload is much higher than Level 7. Fatigue of human operators is a real issue. Additionally, time average at Level 6 is the worst in the three levels, as described in the result (2). In this context, an important issue is an amount of room between the real workload and limits of cognitive resources of human operators. It must be noted that there is possibility that performance of human operators will be corrupted because of high cognitive workload at operations.

4. CONCLUSIONS

This paper delineates a theoretical framework for quantitative evaluation of automation levels based on a probabilistic simulation method to estimate behaviors of human operators and automated systems. Several expected-value indexes for the evaluation are defined, based on possibilities of occurrence of the states concerned. The method was applied to a nuclear power plant for evaluation of task achievements during abnormal operating conditions. The results of these evaluations for NPP show that advantages of levels of automation must be discussed in their relations with characteristics of tasks adopted. They also show that the proposed method can provide effective knowledge for improvement of automated systems designs. The next step of this research is to extend the proposed method into a complete framework for determination of proper levels of automation. One of them is implementation of an error model, where effects of cognitive workload are considered.

ACKNOWLEDGEMENT
This work has been partially supported by Grants-in-Aid for Scientific Research 12680366 and 11780369 of the Japanese Ministry of Education, Culture, Sports, Science and Technology.

REFERENCES
Bertucio, R. C., & Julius, J. A. (1990). *Analysis of Core Damage Frequency: Surry Unit 1 Internal Events.* NUREG/CR-4550, Vol.3, Rev.1, Part 1.
Eide, S. A., & Calley, N. B. (1993). Generic component failure data base. *Proceedings of PSA'93* (pp.1175-1182).
Furukawa, H., Inagaki, T., & Niwa, Y., (2000). Operator's situation awareness under different levels of automation; evaluations through probabilistic human cognitive simulations. *Proceedings of 2000 IEEE International Conference on Systems, Man, and Cybernetics, Tennessee* (pp.1319-1324).
Inagaki, T. (1998). Situation-adaptive autonomy: Trading control of authority in human-machine systems. *Proceedings of Third Automation Technology and Human Performance Conference* (pp.154-158).
Laughery, K. R. (1999). Modeling human performance during system design. In E. Salas (Eds.), *Human /Technology Interaction in Complex Systems, 9* (pp.147-174). Stamford: JAI Press.
Micro Analysis and Design (1997). *User's manual of WinCrew: Windows-based workload and task analysis tool.*
Parasuraman, R., & Riley, V. (1997). Humans and automations: use, misuse, disuse, abuse. *Human Factors, 39,* 230-253.
Scerbo, M. W. (1996). Theoretical perspectives on adaptive automation. In R. Parasuraman and M. Mouloua (Eds.), Automation and Human Performance (pp.37-63). New Jersey: Lawrence Erlbaum Associates.
Sheridan, T. B. (1999). Human supervisory control. In A. P. Sage and W. B. Rouse (Eds.), *Handbook of Systems Engineering and Management* (pp. 591-628). New York: John Willey & Sons.
Sheridan, T. B., & Parasuraman, R. (2000). Human versus automation in responding to failures: An expected-value analysis. *Human Factors, 42,* 403-407.
Wickens, C. D., & Hollands, J. G. (2000). Complex systems, process control, and automation. *Engineering Psychology and Human Performance* (pp.513-556). New Jersey: Prentice-Hall.
Wickens, C. D., & Yeh, Y.-Y. (1986). A multiple resource model of workload prediction and assessment. *Proceedings of the IEEE Conference on Systems, Man, and Cybernetics* (pp.1044-1048).

Performance recovery and goal conflict

Magnhild Kaarstad

Institutt for energiteknikk,
OECD Halden Reactor Project
P.O. Box 173
No-1751 Halden

ABSTRACT

This paper will present an experiment performed in the Halden Man Machine Laboratory (HAMMLAB) in order to investigate the impact of conflict between a safety goal and a productivity goal on performance recovery among nuclear power plant operators. A second objective was to develop a tool for studying recovery actions, and a third objective was to find some aspects or factors that are important for recovery.

Sixteen licensed operators from the Loviisa plant participated in the experiment. All participated in 4 different experimental conditions. Goal conflict did not have the expected impact on recovery for these operators. Operators performed better when they are focused on a safety goal. The tool for studying recovery actions worked successfully. Some important aspects with the nature of recovery actions were uncovered.

1. INTRODUCTION

For many years, developments in safety policies have been made in the context of suppressing or preventing human error. The "zero accident" policy has long been interpreted as a "zero error" policy. This exclusive reliance on the "error suppression" approach has recently been questioned by various researchers (e.g., Frese, 1991, Wioland and Amalberti, 1998). Now, there is a shift in safety paradigms towards finding solutions that may prevent the consequences of human error by providing opportunities for recovery (Kontogiannis, 1999). Research on recovery can provide valuable input to the design of error-tolerant systems, and may in the long run give input to training programmes and to establish teamwork styles or strategies that seem to be efficient for recovery.

1.1. The error handling process

Studies in error recovery have tended to distinguish three processes in error handling or error recovery, namely: (1) *error detection* - realising that an error is about to occur or suspecting that an error has occurred; (2) *error explanation* or localisation - explaining why an error occurred; and (3) *error correction/ recovery* - which refers to knowing how to undo the effects of the error and to achieve the desired state. (e.g., Sellen, 1994).

A joint experiment was performed between the Human Cantered Automation (HCA) project and the Performance Recovery project. In the performance recovery project, the focus of attention was on recovery. The experimenter set an error, and operators were told that they, or one of their crewmembers, pretendably had made an error. The operators' task was to recover from the unsatisfactory process condition. The advantage with this way of studying recovery is that the same error is "made" by all operators, and it is easy to compare results.

1.2. Control of complex systems

Performance recovery is in this context defined as the process of regaining control of an operator-plant system that is not in an acceptable situation. People's ability to control complex systems is important in considerations of safety for two reasons. First, many accidents are due to failures to achieve and maintain control over some system, and second, the very task of achieving safety in an organisation can be seen as a matter of controlling a complex system.

Control theory specifies the general prerequisites for control. These prerequisites are the same, regardless of whether we consider a system that is controlled manually, or one that is controlled automatically. There are four of them (Brehmer, 1993): clear goals, observability, action possibilities, and a model of the system.

1518

All of these prerequisites are problematic in safety work. In this experiment the focus of attention is on 'Clear goals'. The purpose is to investigate how the processes of regaining control / recover from an unsatisfactory process condition will be influenced by introducing a goal conflict in some experimental conditions.

Locke, Smith, Erez, Chah and Schaffer (1994) conducted two studies on intra-individual goal conflict. In both studies conflict was negatively related to at least one performance outcome. Weber (1999) used existing data from a previous HAMMLAB experiment, to investigate whether goal conflicts affected performance among NPP operators. There where no clear results in this study. However, in phases of scenarios where safety was considered as the most important goal, the performance increased.

1.3. General hypotheses and investigation areas

The three main objectives in this study, were as follows:
1. One of the main issues in this experiment was to analyse the impact of a safety-production goal conflict on performance recovery. Safety and production seem to be the most important goals of operators. These different goals may sometimes work in the same direction, but there are also situations, in which it is impossible to follow two goals at the same time. In this experiment, goal conflict was manipulated experimentally. It is hypothesised that operators who are faced with a goal conflict will have poorer and a more time-consuming performance recovery than operators who experience no such conflict.
2. The second aspect in this experiment was to test a system for online self-rating of recovery performance. The rating was performed online by the operators in seven periods on a scale ranging from 1 to 10 (1 = no recovery, 10 = full recovery) every 1.5 minutes. The operator's online self-rating of recovery performance was compared with a post hoc process expert rating.
3. The third objective was to seek to uncover, through a factor-rating questionnaire, what factors help operators to regain system control. After the scenario, operators were asked to evaluate how important 10 different factors were when their recovery rating decreased or increased. Of special interest was to identify what factors the operators explained as most important when the curve crossed into the area for acceptable or full recovery.

2. METHODOLOGY

2.1. Research facility

The experiment was performed in HAMMLAB, in a research simulator that simulates the pressurised water reactor from the Loviisa plant in Finland.

2.2. Participants

Eight crews of 16 licensed operators from Loviisa participated in this experiment.

2.3. Scenarios

A process expert developed the scenarios according to the following requirements: One of the scenarios should be safety oriented and the other production oriented. The recovery task should be possible to perform within a relatively short time-frame (within 10 minutes). It was also important that the error should be of such a nature that it would not be possible with just one action (e.g., scram) to recover it, so that some variation in recovery performance could be found.

2.4. Instructions
Just before the simulator started running, operators were given a description of a situation their plant pretendably was in. One instruction led the operators into a safety-oriented mode, and the other instruction led the operators into a production-oriented mode.

1519

Instruction focused on Safety
There has recently been a serious incident at your plant with a safety inspectorate investigation. The management has therefore stressed the importance that the plant is run very safe.

Instruction focused on Production
Due to a thunderstorm, other electrical suppliers are disconnected from the grid. Your plant is still up running, and has been asked to produce electricity until the other electrical suppliers are up and running again.

2.5. Experimental design

A 2*2*7 (instruction*scenario*period) repeated measures design was used to test the hypothesis concerning action recovery. That is, by combining the two independent variables 'scenario-type' and 'instruction', conflict/no-conflict conditions were produced, since their goal-orientation would either be in conflict (safety-production) or not in conflict (safety-safety or production-production). The relationship between instruction and scenario-type is shown in table 1. The grey areas indicate the conflict conditions, while the white areas represent the no-conflict conditions.

Table 1 The two independent manipulations, instruction and scenario type

	Scenario type	
Instruction	*Production oriented*	*Safety oriented*
Production oriented (p)	No conflict	Goal conflict
Safety oriented (s)	Goal conflict	No conflict,

3. RESULTS

No main effect of conflict on time to recover successfully was found, $F (1,15) = 1.13, p < .72$ ($\hat{\omega}^2$, .0). As shown in figure 1, the operators' recovery curves in the conflict and the no-conflict situations are almost identical.

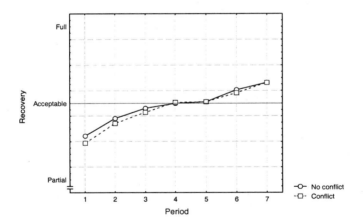

Figure 1: Conflict and no-conflict recoveries throughout the scenario.

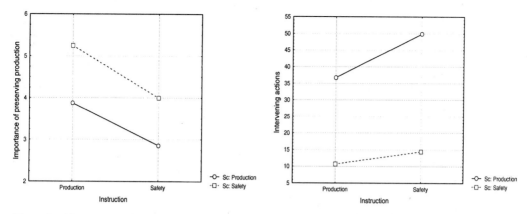

Figure 2: a) Importance of preserving production is affected by both scenario and instruction (left). b) The effect of instruction and scenario on intervening operator actions (right).

A main effect of scenario was found in respect to the operators' judgement of the importance of preserving production, $F(1,15) = 17.03$, $p < 0.001$ ($\hat{\omega}^2$, .20), showing that preserving production was perceived as being more important in the safety scenario, than in the production scenario (Figure 2a). This raises doubt about whether the scenarios exhibited the desired production-safety distinction. The instruction, however, seems to have had the planned effect, i.e., the importance of preserving production was higher when the production instruction was given than when the safety instruction was given, $F(1,15) = 4.25$, $p < 0.06$ ($\hat{\omega}^2$, .05) (Figure 2a).

A general operator performance score was obtained. The data did not support the assumption that operators perform better in the no-conflict situation, $F(1,7) = 1.98$, $p < .21$ ($\hat{\omega}^2$, .03). However, Performance was better when a safety instruction was given than a production instruction, $F(1,7) = 12.34$, $p < .01$ ($\hat{\omega}^2$, .26) (Figure 2b).

There are several alternative explanations to these findings. The most obvious, is that the operators who participated in this study are not affected by conflicting goals. They perform their job safely and routinously despite of different surrounding conditions.

Another explanation for the finding is related to the scenarios chosen in this study. As figure 2a indicates, operators report a significant higher emphasis on maintaining production in the safety-oriented scenario than in the production-oriented scenario. This result may indicate that the operationalisation of scenario in this experiment was not efficient enough, and that the type of scenarios selected is the reason for the lacking result of the effect of goal conflict on recovery. However, if a "clean" safety scenario was made, there would have been a risk that operators chose to scram the reactor in a safety instruction situation. This would have resulted in no interesting findings regarding the recovery process. The aim in this experiment was to create scenarios where it would be possible to see some variation in performance recovery.

Yet another possible explanation to the finding is that the Performance Recovery experiment was conducted in the end of the HCA experiment. Different operator actions during the HCA scenarios may have affected the recovery experiment, in that the starting point for this experiment became very different for the various experimental conditions.

However, an interesting result of this experiment is that when operators are presented with the production-oriented instruction, their performance deteriorates. This finding is supported by Weber (1999), who found that in phases of scenarios were safety was considered most important, the performance increased. Operators act more safe and efficient when they are instructed to be focused on safety. This finding will be of interest for safety work.

1521

Operators did not report any trouble in filling in their recovery curve online, and a similar curve was drawn online by the operators and post hoc by the process expert. Differences and similarities in scoring between operators and the process expert will be analysed in more detail later.

Factors that were most often reported for performing an acceptable recovery, were understanding of the situation, the process development and information from process displays. For reaching a full recovery, information from and discussion with crewmember were considered very important by the operators.

4. CONCLUSIONS

This experiment showed that goal conflict did not have the expected effect on recovery. It may be too early to conclude in general, as different countries and different industries sure have different safety cultures. It is, however, important to note, that a production oriented instruction seem to deteriorate performance.

Some factors of importance for recovery were identified, and an efficient tool for studying recovery actions was developed in this experiment. This tool will be used in future studies where the purpose is to study other aspects of recovery.

REFERENCES

Brehmer, B. (1993). Cognitive Aspects of Safety. In Wilpert, B., and Qvale, T.: (Eds), *Reliability and safety in Hazardous Work Systems. Approaches to Analysis and Design,* Lawrence Erlbaum Associates, Hillsdale, 23-41

Frese, M., (1991). Error management or error prevention: two strategies to deal with errors in software design. In Bullinger, H.J. (Ed.), *Human aspects in Computing: Design and use of interactive systems.* Elsevier, Amsterdam, 776-782.

Kontogiannis, T. (1999). User strategies in recovering from errors in man-machine systems. *Safety Sience 32, 46-68.*

Locke, E.A., Smith, K.G., Erez, M., Chah, D.O., and Schaffer, A. (1994). The effects of intra-individual goal conflict on performance. *Journal of Management, 20:* 67-91.

Sellen, A.J. (1994). Detection of everyday errors. *Applied Psychology: An International Review 43:* 475-498.

Weber, M. (1999). *Goal Conflict in the Process Control of a Nuclear Power Plant. Effects on Operator Performance and Implications for its Measurement together with Plant Performance.* Diploma Thesis, University of Bern, Switzerland.

Wioland, L., Amalberti, R. (1998). Human error management: towards an ecological safety model - a case study in an air traffic control microworld, *9th European Conference on Cognitive Ergonomics, Ireland, August.*

Cognitive gearing: a measure of system effectiveness?

Carole DB Deighton and Iain S MacLeod

Aerosystems International Ltd
West Hendford
Yeovil
Somerset BA20 2AL
United Kingdom
e-mail: carole.deighton@aeroint.com

ABSTRACT

Supervisory control and monitoring tasks characterise human work in many complex safety critical environments. The ability to maintain a high level of alertness, situation awareness and optimal performance, by all operators and teams, particularly during repetitive or prolonged operations is questionable. Human operators manage their time and resources according to the perceived requirements of the task and their assessment of the operating environment. Cognitive Gearing is a facet of expertise that allows an individual to allocate their cognitive resources appropriately with regard to their environment and following the correct detection and evaluation of overt and covert signals and events. In this paper it is argued that system effectiveness measures should assess the extent that a system can support cognitive gearing and provide sufficient and timely cues to enable the appropriate allocation of cognitive resources by the single user and the team.

1. INTRODUCTION

Supervisory control and monitoring tasks characterise human work in many complex safety critical environments. For example, military and civil, land, sea and air transportation systems are a synthesis of distributed teams of human operators monitoring and controlling sub-systems to achieve a common goal. The ability to maintain a high level of alertness, situational awareness and optimal performance during such operations, by all human operators, particularly during repetitive or prolonged operations, is questionable. Human operators of complex systems manage their workload, attenuate their level of performance and situation awareness according to skills, expertise, the *perceived* requirements of the task, and their assessment of the operating environment. An appreciation of the level of cognitive resources that should be allocated to a task at a given moment in time is an expertise developed by the expert user. Systems are not necessarily designed for operation by a homogenous group of expert users and the design needs to accommodate the skills, capabilities and biases of novice users. Cognitive gearing is a facet of expertise that allows an individual to allocate their cognitive resources appropriately with regard to their environment and following the correct detection and evaluation of overt and covert signals and events.

In this paper it is argued that system effectiveness measures should assess the extent that a system can accommodate variations in human attention, i.e. support cognitive gearing and provide sufficient and timely cues to enable the appropriate allocation of cognitive resources by the single user or team. A model summarising the main components of a system effectiveness methodology, which incorporates cognitive gearing and influencing factors, is given in figure 1. Section 2 describes each component of figure 1 and section 3 elaborates the case for using state assessment methods to indicate system effectiveness. The argument is summarised in section 4 along with the recommendations of the paper.

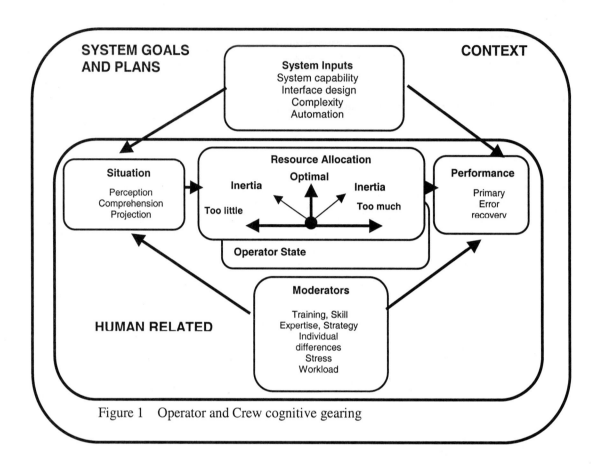

Figure 1   Operator and Crew cognitive gearing

2.   A MODEL OF COGNITIVE GEARING

2.1   Cognitive Gearing

The notion of human-system resource management encompasses the skill of 'cognitive gearing' characterised by the experts' ability to reallocate resources appropriately following the correct detection and evaluation of overt and covert signals.

The centre of figure 1 provides an abstract representation of the range of cognitive demand that may be experienced by the operator. The initial state for resource allocation may be work underload, overload or a point between these two extremes. The actual operator state will be influenced by a multitude of factors; figure 1 highlights the importance of operator Situation Assessment and moderators associated with skill, expertise, training, work strategy, individual differences and operator stress.

In both the underload and overload scenarios the operator is required to rapidly assess the system state using cues originating from other team members, instrumentation and the external environment. The underload state is characterised by boredom, fatigue, distraction, frustration and under-arousal. The operator needs to overcome the inertia created by these states and to 'gear-up' psychologically and physiologically to the work demands (Dyer-Smith and Wesson, 1995). Lag in the system, as the operator prepares to respond to the system input, has implications for system safety, particularly when the system events are unanticipated. Conversely, in the overload state the operator may be required to attend to information from diverse sources for short or longer periods. In this scenario the operator may be fatigued from too much to do, over-focus on system inputs, delay, restructure or abandon tasks. Unexpected system inputs require the operator to reassign cognitive resources, decreasing attention in some areas and increasing resources in other areas. Cognitive gearing is relevant in this scenario with an emphasis on the appropriate reassignment of resources.

2.2 Factors influencing Cognitive Gearing

A variety of factors may impact the capability of the operator or crew to cognitively gear-up and respond promptly and appropriately to system demands. Figure 1 summaries many of the factors which encompass components of the human, the system and associated interactions and the operational context.

2.2.1 System Input and Situation Assessment

The lag in the human system and operators' subsequent ability to respond to system inputs effectively will be influenced by their Situation Awareness (SA), which in turn is influenced by the efficacy of system design and the operators' current cognitive state. The topic of SA has been research extensively and figure 1 recognises the proposed three levels of SA: perception of elements in current situation (level 1); comprehension of current situation (level 2); and projection of future states (level 3) (Endsley, 1995). It can also be argued that SA is an integral part of complex open skills (MacLeod, Taylor and Davies, 1995).

Supervisory control and monitoring systems embedded within safety critical systems have the potential to detach the human operator from essential cues that are indicative of a need to reallocate cognitive resources or 'gear-up'. Understanding the type and form of cues that must be provided by hi-technology systems to support cognitive gearing is essential. Philosophies on the design of automation exist (Weiner and Woods, 1995). The extent that these frameworks encourage system designs that support cognitive gearing is unknown.

2.2.2 Performance

Cognitive Gearing is an intervening variable influencing successful error recovery and the re-acquisition of SA. The human operator and team that are adept at optimising their resources will be most effective at responding to unexpected system inputs during normal and emergency operations. System feedback or Knowledge of Performance (KOR) Results provides a direct indication of the optimal allocation of cognitive resources, at a particular moment in time and this is beneficial to system effectiveness.

2.2.3 Moderating Factors

Moderating factors associated with the operating style, behaviour and strategies adopted by the operator will impact cognitive gearing. Studies of personality types most prone to work underload and which can readily adapt to changing system demands have been conducted (Dyer-Smith et al, op cit, Farmer and

Sundberg, 1986). The importance of Cognitive Gearing to modern industry is emphasised by the work conducted by Dyer-Smith et al into the development of computer based selection tools to measure this phenomenon. Operator and team selection and training programmes to promote effective behavioural strategies for cognitive gearing can be envisaged. The provision of activities simply to provide stimulation for bored operators is considered counterproductive, particular if such tasks detach the operator from the 'control-loop'.

3   EVALUATION METHODOLOGY

Given the adaptability of the human operator and team and the dynamic nature of work, it is argued that system effectiveness should be measured according to the capability of the system to support cognitive gearing. The human factors' community is recognising this change of emphasis in system evaluation. The statement of a particular workload utilisation along with the assumption that a constant SA is a mark of good system design is considered limited. Moreover, concentrating on the development of systems and procedures to detect system failures, without equivalent effort placed upon the behaviours required to recover the system, is regarded as restrictive.

So what methods can be used to assess cognitive gearing? Indicators of cognitive gearing, such as changes in Operator State from boredom to absorbed in a task, provide an important method for evaluating system effectiveness. The reconsideration of operator and team state assessment methodologies as a core approach to the assessment of system effectiveness is advocated. This position is supported by earlier studies conducted by Comstock, Harris and Pope (1988), who concluded a need to focus on operator state assessment and particular hazardous sates associated with work underload. More recently, Braby (1993), developed a state assessment technique which assessed operator work according to twenty-two states ranging from boredom and fatigue through to activation and challenged (as shown in figure 2).

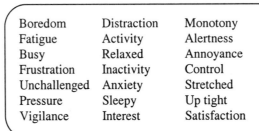

Boredom	Distraction	Monotony
Fatigue	Activity	Alertness
Busy	Relaxed	Annoyance
Frustration	Inactivity	Control
Unchallenged	Anxiety	Stretched
Pressure	Sleepy	Up tight
Vigilance	Interest	Satisfaction

Figure 2 States arising during work underload and overload

4   FINAL NOTES

A considerable amount of research has been conducted within the Human Factors community to develop subjective, objective and performance indicators of system effectiveness. These methods which originate predominately from the literature on workload and SA enable the production of numerical data to develop profiles of work and the identification of peaks and troughs. In this paper the measurement of operator and team state was advocated as an alternative methodology for assessing system effectiveness. The benefit of such an approach being the development of profiles of states that were indicative of the operators' or teams' readiness to respond to inputs from the system. A change in emphasis within the psychological and physiological domains to compile a battery of techniques that focus on the detection of operator changes in state is advocated. In many respects the foundation for these studies is evident within the literature on stress and performance. System design philosophies, particularly those which have been developed to control the introduction of automation within safety critical industries should be reconsidered to determine their validity with respect to the promotion of Cognitive Gearing.

It is proposed that, with further development of this work, points along the abstract scale ranging from underload and overload could be profiled with typical operator states. The delay between changes in operator state along the continuum would then provide a robust indication of the capability of the human to reallocate cognitive resources in response to system demands. Knowledge of the hazardous nature of remaining in a given state would have to be considered.

The briefly outlined approach moves away from the unresolved debate of what constitutes an acceptable magnitude of workload or workload profile reported by the traditional multidimensional workload rating scales. Moreover, this method, supported by physiological correlates provides an objective approach to workload assessment among complex teams where task interruption for data collection is unacceptable.

REFERENCES

Braby, C.D., (1993). The development of a subjective measure of work underload. In, R.S. Jensen and L.A. Rakovan (Eds.), *Seventh International Symposium on Aviation Psychology, Columbus Ohio, April 26-29.*

Comstock, J.R., Harris, R.L., and Pope, A.T. (1988). Physiological assessment of task underload. *Space Operations Automation and Robotics (SOAR 88) Workshop, Dayton Ohio.*

Dyer-Smith, M and Wesson, D.A (1995). Boredom and Expert Error. In, S.A Robertson (Eds.), *Contemporary Ergonomics pp 56-61.* London: Taylor and Francis.

Endsley, M. (1995). Toward a theory of situation awareness in dynamic systems. *Human Factors* 37(1): pp 32-64.

Farmer, R. and Sundberg, N.D. (1986). Boredom Proneness. The development and correlates of a new scale. *Journal of Personality Assessment,* 50(1), pp 4-17.

MacLeod, I.D., Taylor, R.M., and Davies, C.L. (1995). Perspectives on the Appreciation of Team Situational Awareness. In, D.J. Garland and M.R. Endsley (Eds), *Experimental Analysis and Measurement of Situation Awareness. pp 305-312.* Florida: Embry Riddle Aeronautical University Press.

Weiner, E.L. and Woods, D.D. (1995). The focus of automation; which tasks should be automated? *Presented at the Industrial Summer School on Human-Centred Automation.* France: Saint-Lary, Pyrenees.

BASIC PRINCIPLES TO DESIGN AN ADVANCED HUMAN ADAPTED SUPERVISORY SYSTEM

Bernard RIERA and Slim TRIKI

Department of Man-Machine systems, LAMIH, University of Valenciennes, Le Mont Houy, BP 311
59300 Valenciennes, FRANCE

ABSTRACT

In this paper we propose some specifications for an advanced human adapted supervisory system (AHASS) integrating representation characteristics of the production system, such as functional, structural and behavioral aspects based on cognitive engineering models, with the use of advanced algorithms of detection and location. The challenge is never to cross the boarder between support and assistantship. To reach this goal, in this paper we propose some basic principles and we develop the idea to design high level information useful for human monitoring and diagnosis tasks.

1. INTRODUCTION

Supervision of highly automated processes is an interdisciplinary research area. Knowledge's in the fields of automation, process knowledge, machine engineering, "work post" ergonomics, cognitive ergonomics, working psychology, sociology and so on are necessary to design efficient supervisory systems. This is because supervision is an activity in which man, despite the increasing automation of the last years, is still present. Our research concerns monitoring tasks and diagnosis tasks in continuous processes. In this paper we propose some specifications for an advanced human adapted supervisory system (AHASS) integrating representation characteristics of the production system, such as functional, structural and behavioral aspects based on cognitive engineering models, with the use of advanced algorithms of detection and location. The main idea is to design a supervisory system well balanced between human and technical aspects. Indeed, man machine system centered approaches can deal to another extreme like purely technical approaches. The challenge is never to cross the boarder between support and assistantship. To reach this goal, in this paper we propose some basic principles and we develop the idea to design high level information useful for human monitoring and diagnosis tasks.

2. HUMAN SUPERVISORY TASKS

This paper deals with the "real time" supervision of the process. As a consequence, human supervisory tasks are at a medium level between strategic functions (planning for instance) and tactic functions linked to the command part of the production system. Human operator (HO) receives continuously continuous (measures) and discrete (states and alarms) information coming from the supervisory system and acts on it sometimes. Supervisory tasks necessitate high level of knowledge and can be shared in three classes : command tasks, monitoring tasks and diagnosis tasks.

2.1. Command tasks

HO can act in different ways on the process, by modifying set points (tuning for instance) or change of mode (manual for instance). HO acts on the process during specific phases of the process in normal (like the start up or the shut down) or abnormal running (failures, perturbations). In this last case, when it is possible HO has to correct or to compensate the failure.

2.2. Monitoring tasks

HO supervises the process, filters available information in order to evaluate process state and its future evolution. During this activity, HO is an observer. Indeed, operator looks at process information without acting to modify the current state. The main preliminary goal of supervisory tasks is to find an abnormal running. The secondary goal is to update the operator's mental model of the process in order to take the good decisions when a problem occurs. This one can be detected by the mean of an alarm (passive detection by the mean of a discrete information) or by his own observation of process running (active detection). In supervisory control, HO is in a certain way, out of the loop of control-command because he (or she) does not get enough information all the time about what happens in the process. HO sometimes receives either no information or too little information to maintain a good picture of system state

2.3. Diagnosis tasks

After detecting a problem, HO has to diagnose the process state. This consists in determining the primary causes but also the effects on the process. First of all, human operator has to check the validity of the problem. A large-scale system is supervised by the mean of numerous sensors. Some of them are used for feedback control; others exist only to inform the operator about the process. In the first case, a sensor failure will involve bad effects on the process (disturbance with process effects). It is not true in the second case (disturbance without process effects). After, if it is a disturbance with process effects, diagnosis is to foresee if the process will be able to reach its steady state. Indeed, automatic control can compensate and stabilize the process. If it is not the case, human operator has to act on the process. One can notice that the earlier the detection is made, the easier bad consequences could be avoided. The capacities of abstraction and the fact that people may take decisions even in the case of great uncertainties explains why it is not possible to withdraw human being from supervisory loop.

3. COGNITIVE OPERATOR MODEL REQUIREMENTS

Models of cognitive control must include higher-level analytical problem solving at one extreme and also the control of action at the other. That can be illustrated by the mean of a discussion about level of automation. Level of automation has consequences on the operator's information requirement. When a function is performed by an automatic system, the operator does not need the same information as if he (or she) has to carry out it. Bye (Bye, 1999) indicates that HO requires less information in the case of monitoring. Consequently, automation may result in a reduction of information, which over time can deal to bad consequences on the operator process knowledge and thereby make them less able to cope with unexpected events. The notion of feed-back is here fundamental. Cognitive operator models like proposed by Neisser (Neisser, 1976) (the perceptual circle) explicitly show the coupling between past and future events. Indeed feedback information modifies current understanding of the situation, which control the goals having to be performed by the mean of actions, which supply information. These cyclical models like COCOM (Hollnagel, 1998) enable to explain several problems encountered by HO. One complementary description of cognitive activity is to use « situation awareness » (SA) and « decision making » (DM) concepts. (Endsley, 1998) defines SA as being « the perception of the elements in the environment, within a volume of time and space, the comprehension of their meaning and the projection of their status in the near future ». DM can be defined as the « choice of actions having to be performed ». Boarders between SA and DM are not completely defined. The idea we will keep in mind is that HO builds a picture of production system by the mean of available information. The SA directs considerably DM. We notice the importance of perception. Today, supervisory systems do not support an efficient human perception of the process adapted to monitoring and diagnosis tasks. As a consequence, supervisory tools today available do not facilitate SA and DM.

4. TECHNICAL IMPROVEMENTS

In this paper, algorithms based on a real time observation of the process will be studied in adopting following definitions: failure is different from breakdown because there is not a "stop" of the function but only a malfunction.

Detection: the detection stage is the automatic decision, which consists in determining if a failure has appeared. **Location:** this word (as diagnosis) has multiple definitions. We consider that location consists in determining the origin of the failure and more particularly the first measured variable which is characteristic of the origin of the problem. The words « structural location » are used when the physical component, source of the problem is determined. **Isolation:** This consists in identifying and characterizing the failure. That means to determine what the cause of the failure is. For that, a model of the failures has to be available. Algorithms involved in these three decision activities are called «Fault Detection and Isolation algorithms» in Automatic control. Usually, people involved in supervisory control (cognitive or technical engineers) use different words « supervisory systems », « alarms systems » or « diagnosis systems » with different meanings. In all cases, all the decisional activities can be included, from the detection to the reconfiguration. A supervisory system has to include at least these two activities. Algorithms must enable HO or automatic system to take the right decisions and actions. It is interesting to notice that HO is not taking into account in the design of these algorithms. However, it is not possible to withdraw completely HO from supervisory loop. As a consequence, even if automatic control becomes more and more robust to disturbances and FDI algorithms efficient, HO is still the last decider in the process. The principle of methods based on a normal running of the process is to perform redundancy analysis (Franck, 1995). It is in fact a generalization of material redundancy. An analytical redundancy relation is deduced from a model, between variables of which measurements are available. All these methods are very attractive in a theoretical point of view. However, they require identification of reliable dynamical models very difficult to obtain for large-scale systems. In addition, very often, mathematical models are based on the state representation which is not may be well adapted to human adapted supervisory applications. At least, it is important to notice all these methods have been designed in order to improve the passive detection. There are also other methods using analytical algorithms but based on different models of the process much more explicit than state representation. These models are better adapted to support man-machine cooperation, because they have a power of explanation. It is the case for instance of causal graph which is a graphical representation composed of nodes characterizing variables and oriented relations between nodes representing cause to effect relations. The simplest causal graph is the signed oriented graph. The causal graph is not a new concept. It has been widely used in the preliminary design of chemical processes and control structure design for industrial processes (Biegler, 1997). When knowledge is available, it is useful to model more accurately the relations between variables. In this case, causal graph can support detection and location algorithms based on a generation of residues (Evsukoff, 1998), which can easily be used to justify results to Human operator.

5. BASIC PRINCIPLES TO DESIGN AN ADVANCED HUMAN ADAPTED SUPERVISORY SYSTEM (AHASS)

The description of a complex system stated so far is a structural one, which can be viewed as the organization of the system in space (Modarres, 1996). It corresponds to a picture of the system showing the whole system. This structural perspective can not described time-based properties of complex systems. The system organization in time can be described with functional and behavioral models. For several authors (Modarres, 1996), the functional hierarchy is the central backbone of the system model describing the system's state-time behaviour. Behavioral hierarchy shows the ways that functions are realized. In addition, structural hierarchy shows system parts that perform functions. Consequently supervisory tools, in order to favor an active perception of process state, must enable HO to see the process according to different points of view: structural, functional and behavioral. This is reached via : design of supervisory tools dedicated to each task, integration of high level information coming from advanced supervisory algorithms, and explanation capabilities in order to improve human machine interactions (HMI).

5.1. Dedicated supervisory tools

The first part of the paper has shown that supervisory tasks have different objectives and require different kinds of information. Monitoring requires a global view of the process. On the other hand, diagnosis requires multiple points of view at different levels of abstraction. As a consequence, an AHASS must propose advanced displays dedicated to monitoring and diagnosis based on these different points of view. There is a great deal of research concerning new modes of representation e.g. ecological interfaces (Vicente, 1995), functional interfaces, Mass Data Display (MDD) (Beuthel, 1995). Especially, the MDD idea is very interesting. MDD concept appeared during the year's 80ths, the basic concept is the presentation on a single display of all the data allowing an operator to supervise a process. Each variable is symbolically represented by at least one of its intrinsic characteristic.

5.2. High-level information

Today supervisory systems offer real time measurements and trends of variables. With regard of the numerous variables, probability that HO looks at the right variable at the right moment is low. In fact, HO monitors the process by the mean of alarms (it is to say discrete information) that are activated when a threshold is triggered. These thresholds are like tolerance width which are difficult to design because if they are too narrow, there will be a lot of false alarms and if the tolerance width is large, the detection will be late. It will be much more efficient to HO to work with for each variable a target value (TV). The target value is easy to define for a regulated variable in a steady state, it is the set point. In addition, even in large-scale process, it exists behavioral models of normal running of some parts of the process (used for FDI algorithms). These models can supply in real time several target values of process variables. In our approach, advanced supervisory systems have to be designed to give useful information about process state to HO. New kinds of discrete and continuous high-level information can be thought of doing to reach this goal (cf. figure 1). Advanced discrete information (ADI) are in fact advanced alarms. That means they are based on the comparison between the measure and a value coming from a dynamic model of the process. As seen previously, if the model is right, the triggering of the advanced alarms will be better and more sensitive to actual process failures. But, all the problems of passive detection are still present. Advanced continuous information (ACI) are based on real time measurements but give to HO, synthetic information that would have required a cognitive processing to be humanly obtained. In other words, ACI use the good capabilities of perception of human being. Different kinds of ACI can be designed. For instance, a first classification deals with the use or not of a dynamic process model. In the first case, ACI can be based on the residue, which is the difference between the variable value and its normal value coming from a model, in the second case, directly on the measure. The second classification depends on the signal processing performed : real time or along a temporal window. In this last case, the signal identification can for instance determine the shape of the signal : step, ramp, first order, second order, ... We argue in this paper that ACI must improve active detection. It is important to notice that integration of continuous information in a supervisory system can considerably modify the HO's work usually based on measurements and alarms.

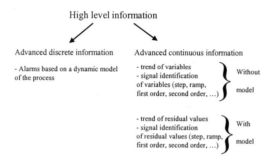

Figure 1. High-level information (Riera, 2001)

6. AN EXAMPLE OF ACI : ECOLOGICAL INTERFACE FOR ANALYSIS OF TRENDS

Today, the display of trends in SCADA system is very simple. The current time is at the right of the graph and there is a scrolling on the left depending on the sampling time defined during the design of the supervisory system. The scales are also fixed and there is seldom possibility of changing it. Human Operator uses trend displays to anticipate, to detect a problem and to make a diagnosis. When several alarm occurs, HO has to find the variable source of the problem. For that, he or she has to compare the different trends and uses the causality principle to determine the cause which has involved effects. In addition, HO compares the shape of the trends. Today with the trend interfaces, HO has to find by himself all these information. The idea we have tested consists of displaying the trends in a new way, focusing on the changes of deviation. The algorithm (not developed in this paper) is based on a segmentation of the graph. To illustrate the concept, figure 2 shows a trend. This graph is composed of 864 measurements with a fixed sampling time. If we look at the graph, it is possible to analyze roughly the behavior of the variable. However, with regard to the scale in the Y axis, there are some parts of the graph which necessitate to change the scales in order to be analyzed. That takes time and requires a will from the operator. An ecological interface (it is to say a direct perception interface) for trend is proposed figure 4. The graph is divided in pieces of straight lines. From the 864 points, the algorithm has extracted 32 points characterizing the set of measures. The time is presented as a bar-graph and the rectangles symbolize for one segment the minimum value and the maximum value. Inside the rectangle, the trend for the segment is represented. In addition, a code of color characterizes the trend : no change, step, slow up trend, up trend, fast up trend, slow down trend, down trend, fast down trend. Hence, with this interface it is possible to see all the variations of the graph. For instance, inside the circle, one can notice that the part is in fact an oscillation with a low amplitude.

7. CONCLUSIONS

Improvements of supervisory systems require to have a global approach, integrating new algorithms, new displays and taking into account human behavior by the mean of the use of cognitive models. It is very surprising to note that supervisory tools have not evolved for 20 years.

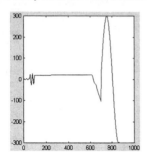
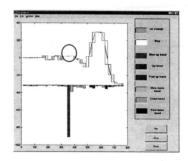

Figure 2. Pedagogical example   Figure 3. Ecological interface for trend display

REFERENCES

Beuthel, C. "Advantages of mass-data-display in process S & C". IFAC analysis, design and evaluation of man-machine systems, MIT Cambridge, 1995, USA, pp.439-444.

Biegler, L.T. and Grossman, I.E. "Systematic methods of chemical process design". Prentice Hall, 1997, New Jersey.

Bye, A., Hollnagel, E. and Brendeford T. S., "Human-machine function allocation : a functional modelling approach", In Reliability Engineering and System Safety, Elsevier, N° 64, 1999, 291-300.

Endsley, M. "Design and evaluation for situation awareness enhancement", Proceedings of the Human actors Society 32nd Annual meeting. 97-101. 1998, Santa Monica, CA: Human Factors Society.

Evsukoff, A., Gentil, S. and Montmain, J. "Fuzzy reasoning in co-operative supervision systems", In Control Engineering Practice, 2000, N°8, Pergamon Press, pp 389-407.

Frank, P.M. "Fault Diagnosis in Dynamic Systems Using Analytical and Knowledge-based Redundancy - A Survey and Some New Results". Automatica (pp.459-474), 1995, Pergamon Press, Vol. 26, N°3.

Hollnagel, E. "Cognitive Reliability and Error Analysis Method (CREAM)", Oxford, 1998, Elsevier.

Modarres, M. "Functional Modeling for Integration of Human-Software-Hardware in Complex Physical Systems". Fourth International Workshop on Functional Modeling of CTS, 1996; Athens, Greece, pp. 75-100.

Neisser, U. "Cognition and reality", San Francisco, WH Freeman, 1976.

Riera, B. "Specifications, Design and Evaluation of an Advanced Human-Adapted Supervisory System", In Cognition, Technology & Work Springer-Verlag London Limited, N°3, 2001, 53-65.

Vicente, K.J., Christoffersen, K., Pereklita, A. "Supporting Operator Problem solving Through Ecological Interface Design". IEEE Transactions on Systems, Man, and Cybernetics, SMC 25, 1995, N°4, pp.529-545.

An Integrated Framework for Tactical Human-Machine Systems Engineering

Arne Worm
Swedish Defence Research Agency
Man-System Interaction Division
P.O. Box 1165, SE-581 11 Linkoping, Sweden
E-mail: arne.worm@foi.se

ABSTRACT

Based on research in the emergency management and military domains this paper reports on research and development of theories, methods and tools for modeling, analysis and accident prevention in precarious time-critical air traffic control, process control, emergency response and military operations. We performed identification, modeling, and synthesis of Tactical Joint Cognitive Systems, and their inherent command, control, and intelligence activities. We found significant relations between workload, time pressure, catecholamine levels in saliva samples, and cognitive complexity.

1. INTRODUCTION

In air traffic control, military operations, emergency management and many industrial processes mission performance relies increasingly on distributed organizations and system architectures to attain high safety and effectiveness without risking excessive resource depletion. The nature of such complex dynamic processes and operations are high-risk activities, where human and artificial team members together perform tasks requiring extreme mobility, efficiency, agility and endurance. These systems incorporate numerous team players, widely distributed across the whole theatre of operations. They can operate autonomously for certain time periods and in specific areas, but primarily they are forced to co-ordinate their actions very accurately with one another. Tactical commanders and operators frequently encounter severe threats and critical demands are imposed on cognitive capacity and reaction time. In the future they will be making decisions in situations with operational and system characteristics that are highly dynamic and non-linear, i.e. small actions or decisions may have serious and irreversible consequences for the entire mission.

1.1. Theory

Our underlying principle was integration of well-established scientific disciplines into a pioneering research direction, Action Control Theory (ACT), based on the assumption that operational feedback, stress measures, process performance measures and outcome performance measures can be integrated, and that information in Command and Control systems must meet rigorous demands regarding:
- Reliability: How do I know that this information is true?
- Relevance: What is the use and purpose of this information?
- Availability: How do I access this information, and what are my sources?
- Diagnosticity: Does this information represent what I want to know?
- Complexity: Can this information be integrated to a reasonable and comprehensible entirety?

1.2. Models

The point of departure in this ACT-based systems modeling was the definition of a Tactical Joint Cognitive System (TJCS). The concepts of a Tactical Joint Cognitive System are depicted in Figure 1.

Figure 1. The *Tactical Joint Cognitive System.*

A Tactical Joint Cognitive System is an aggregate of one or several instances of four principal sub-system classes:
1. *Technological Systems*, for example vehicles, intelligence acquisition systems, communication systems, sensor systems, life support systems, including the system operators.
2. *Command and Control Systems*, consisting of an information exchange and command framework, built up by technological systems and directly involved decision-makers.
3. *Support Systems*, comprising staff functions, logistic functions, decision support functions, organisational structures, and various kinds of service support.
4. *Tactical Teams*, composed and defined according to (Salas et al., 1992) as: "Two or more people who interact, dynamically, interdependently, and adaptively toward a common and valued goal/objective/mission, who have been assigned specific roles or functions to perform, and who have a limited life-span of membership."

The next step was integration of these concepts into a Tactical Action COntrol Model (TACOM). The principal components of the TACOM are the Mission Environment, the Tactical Joint Cognitive System, the Situation Assessment function, and the Cognitive Action Control function, derived primarily from the work of Brehmer (1992), Klein (1993) and Worm (2000). From the TACOM the Mission Execution and Control Model (MECOM) is constructed. The MECOM consists of one or several TACOMs extended with control theoretic components, to handle system disturbances, model error, and to allow an adaptive and balanced mix of feedforward and feedback control. The last step in the model formation process was combining and aggregation of several MECOMs into unilevel and multilevel MECOMs, respectively.

2. METHODOLOGY

To build, test and validate the theories and models, a comprehensive and novel approach was considered necessary. We developed a set of methods and tools for modeling, analysis, and evaluation of tactical forces and their abilities at the lower organizational levels: Mission-critical skills of individual operators and teams, commander mission resource management, and overall unit performance.

2.1. The TRIDENT Method Package

The primary objective of TRIDENT was to develop a coherent and straightforward package of theoretically sound and empirically validated methods and techniques for human-machine systems analysis in the setting of tactical mission scenarios. The components of TRIDENT are described in Worm (2000) and are summarised below:

- Using the Action Control Theory (ACT) Framework (Worm, 2000) for conceptual modelling of dynamic, complex tactical systems and processes, of their states and state transitions.
- Identification of mission and unit state variables, and of action control and decision making mechanisms for process regulation (Worm, 2000).
- Mission Efficiency Analysis (Worm, 2000) of fully manned and equipped units executing full-scale tactical missions in an authentic environment.
- Measuring information distribution and communication effectiveness (Worm, 2000).
- Measuring workload by means of the NASA Task Load Index (Hart & Staveland, 1988).
- Assessing team member psychosocial mood by means of the Mood Adjective CheckList (MACL, Sjöberg et al., 1979).
- Assessing situation awareness (Endsley, 1995) as a function of mission-critical information complexity (Svensson et al., 1993)
- Measuring level and mode of cognitive, context-dependant control of the team members, and identifying what decision strategies were utilised by the team and team members.
- Applying reliability and error analysis methods for investigating failure causes both in retrospect and for prediction (Hollnagel, 1998).
- Validating identified constructs and measuring their influence using advanced data analytic procedures.

2.2. Studies

Numerous battle management and emergency response studies have been carried out to test, refine and augment the modelling, measurement, data collection and analysis concepts of TRIDENT. Implementing these ideas for tactical mission analysis in potentially dangerous, stressful and cognitively complex environments showed to be very effective, with high acceptance among the subjects; all of them were trained and skilled professionals performing their daily tasks in their accustomed work environment. However, it is occasionally claimed that reliability and validity of subjective workload ratings are insufficient. Inspired by the results of Svensson et al. (1993), who studied workload and performance in military aviation, Zeier, (1994) who studied workload and stress reactions in air traffic controllers, and Holmboe et al. (1975), who studied military personnel performing exhausting battle training, we designed a study in order to elucidate to what extent hormonal physiological stress indications are linked to the rating, observation and data collection methods normally used in TRIDENT to assess workload and tactical performance. The details of the studies are described in Worm (2000).

3. RESULTS

From the studies we could identify a number of particularly interesting causes of mission failure or poor performance. The predominant error modes were:

- Timing of movement and of tactical unit engagement.
- Speed of movement or manoeuvre, which is especially important in the initial phase of engagement.
- Selection of wrong object. The environments of ground warfare or emergencies offer many opportunities for choosing wrong objects, in navigation, in engagements, or in visual contact.

After a retrospective cognitive reliability and error analysis (Hollnagel, 1998) we found that mission failure or poor performance in every case could be attributed to:

- Slow or even collapsed organizational response.
- Ambiguous, missing or insufficiently disseminated, communicated and presented information.
- Equipment malfunction, e.g. power failure or projectile/missile impact.
- Personal factors: inexperience, lack of team training etc.

The results from the hormonal response study suggest three potentially significant mechanisms influencing how the team is able to execute mission control, which consequently also influences mission efficiency:

- Time-dependant filtering functions like defence and coping mechanisms according to the cognitive Activation Theory of Stress (Eriksen et al.; 1999, Levine & Ursin, 1991).
- Performance limiting factors due to specific mission and task situation factors and resource requirements (Reason, 1997; Hollnagel, 1998; Worm, 2000).
- Balance between feedforward and feedback in mission-critical action control (Reason, 1997; Worm, 2000).

4. CONCLUSIONS

Our theoretical achievements were a complicated and arduous venture, in that we have constantly striven for empirical evidence. Nevertheless we feel that we have reached a scientific breakthrough. Implementing our ideas in tactical mission analysis in potentially dangerous, stressful and cognitively complex environments was effective and very rewarding. Using the ACT / TRIDENT approach as an advanced systems engineering support will facilitate:

1. Identification of limiting factors of a specific individual, unit, system, procedure or mission.
2. Assessment of the magnitude of influence of these factors on overall tactical performance.
3. Generation and implementation of measures to support, control and improve insufficient capabilities and contribute to successful accomplishment of future missions.
4. Methodological support in future integrated C3I systems analysis, development and evaluation.
5. Improving training programs for tactical decision making and resource management.

REFERENCES

Brehmer, B. (1992). Dynamic decision making: Human control of complex systems. Acta Psychologica, 81, pp. 211-241.

Endsley, M. R. (1995) Towards a theory for situation awareness in dynamic systems. Human Factors, 37, pp. 32-64.

Eriksen, H. R., Olff, M., Murison, R., & Ursin, H. (1999). The time dimension in stress responses: relevance for survival and health. Psychiatry Research, 85, pp. 39-50.

Hart, S. G., & Staveland, L. E. (1988). Development of a multi-dimensional workload rating scale: Results of empirical and theoretical research. In P. A. Hancock & N. Meshkati (Eds.), Human Mental Workload. Amsterdam, The Netherlands: Elsevier Science B.V.

Hollnagel, E. (1998). Cognitive Reliability and Error Analysis Method (CREAM). Amsterdam, The Netherlands: Elsevier Science B.V.

Holmboe, J., Bell, H., & Norman, N. (1975). Urinary Excretion of Catecholamines and Steroids in Military Cadets Exposed to Prolonged Stress. Försvarsmedicin, 11, p. 183.

Klein, G. A. (1993). A recognition-primed decision (RPD) model of rapid decision making. In G. A. Klein, J. Orasanu, R. Calderwood, & C. E. Zsambok (Eds.), Decision Making in Action: Models and Methods. Norwood, NJ: Ablex.

Levine, S., & Ursin, H. (1991). What is stress? In M. R. Brown, G- F. Koob, & C. Rivier (Eds.), Stress – Neurobiology and Neuroendocrinology. Marcel Dekker, New York, pp. 3-21.

Reason, J. (1997). Managing the Risks of Organizational Accidents. Ashgate.

Salas, E., Dickinson, T. L., Converse, S. A., & Tannenbaum, S. I. (1992). Toward an understanding of team performance and training. In R. W. Swezey and E. Salas (Eds.), Teams: Their training and performance, pp. 3-30 Norwood, NJ: Ablex Publishing Corporation.

Sjöberg, L., Svensson, E., & Persson, L.-O. (1979). The measurement of mood. Scandinavian Journal of Psychology, 20, pp. 1-18.

Svensson, E., Angelborg-Thandertz, M., & Sjöberg, L. (1993). Mission Challenge, Mental Workload and Performance in Military Aviation. Aviation, Space and Environmental Medicine, Nov., pp. 985-991.

Worm, A. (2000). On control and Interaction in Complex Distributed Systems and Environments. Linkoping Studies in Science and Technology, Dissertation No. 664, Linkoping University, Linkoping, Sweden. ISBN 91-7219-899-0.

Zeier, H. (1994). Workload and psychophysiological stress reactions in air traffic controllers. Ergonomics, 37, pp.525-539.

The Galvactivator: A glove that senses and communicates skin conductivity

Rosalind W. Picard and Jocelyn Scheirer

MIT Media Laboratory
20 Ames Street, Cambridge, MA 02139-4307
{picard,rise}@media.mit.edu

ABSTRACT

The galvactivator is a glove-like wearable device that senses the wearer's skin conductivity and maps its values to a bright LED display, making the skin conductivity level visible. Increases in skin conductivity tend to be good indicators of physiological arousal --- causing the galvactivator display to glow brightly. The new form factor of this sensor frees the wearer from the traditional requirement of being tethered to a rack of equipment; thus, the device facilitates study of the skin conductivity response in everyday settings. We recently built and distributed over 1000 galvactivators to audience members at a daylong symposium. To explore the communication potential of this device, we collected and analyzed the aggregate brightness levels emitted by the devices using a video camera focused on the audience. We found that the brightness tended to be higher at the beginning of presentations and during interactive sessions, and lower during segments when a speaker spoke for long periods of time. We also collected anecdotes from participants about their interpersonal uses of the device. This paper describes the construction of the galvactivator, our experiments with the large audience, and several other potentially useful applications ranging from facilitation of conversation between two people, to new ways of aiding autistic children.

1   INTRODUCTION

The skin conductivity response, also known as the electrodermal response, is the phenomenon during which the skin momentarily becomes a better conductor of electricity when either external or internal stimuli occur that are physiologically arousing. Arousal is a broad term referring to overall activation, and is widely considered one of two main dimensions of an emotional response. Measuring arousal is therefore not the same as measuring emotion, but is an important component of it. Arousal is not only an indicator of emotional activation; it is also a strong predictor of two important aspects of cognition: attention and memory (Reeves and Nass, 1996). Highly arousing events tend to attract attention and be more memorable than low-arousing events. We are interested in devices that help people communicate emotional information as part of our work in affective computing (Picard, 1997).

Skin conductivity is not a new signal to study. The empirical study of electrical changes in human skin was described more than 100 years ago by Vigouroux (Vigouroux, 1879; Vigouroux, 1888) and by Fere (1988). Since that time, the skin conductivity response has been widely studied in psychophysiology research (e.g., Boucsein, 1992). Skin conductivity is one of many signals commonly included in lie detection tests. It is also often included in studies of cognitive workload and stress.

Skin conductivity is sensitive to many different stimuli; thus, it is often hard to determine what caused a particular skin conductivity response. Typically, events of a novel, significant, or intense nature trigger a sudden increase in skin conductivity. Arousal level tends to be low when a person is sleeping, and high in activated states such as rage or mental workload. When you engage in a mental workload task, such as solving a series of math problems (even if not particularly hard), the level will tend to climb rapidly and then gradually decline. In general, the commencement of a new engaging experience tends to cause the skin conductivity to respond with this characteristic behavior.

Because many different kinds of events can elevate skin conductivity (strong emotion, a startling event, pain, exercise, deep breathing, a demanding task, etc.) it is impossible for an outsider to tell what made your sensor glow unless several potentially confounding factors are controlled. For example, if you remain seated in a comfortable chair in an auditorium, where temperature and humidity are fairly constant, and where your physical activity level does not change, and wear the sensor in the way that it was designed to be worn, then changes in the illumination about your baseline are likely to be meaningful indicators of changes related to psychological arousal.

Much has been learned about emotional, cognitive and physical correlates of the skin conductivity response. However, most research has been conducted on subjects in an unfamiliar laboratory setting, wired to a rack of equipment, who are told to act natural. Our design of an inexpensive glove-like wearable sensor enables the sensing to take place with people who are going about their daily activities. The use of an LED to communicate the signal makes it easy for the wearer to see the signal – there is no need to use a keyboard or computer; the changing level can be easily interpreted by children or adults. The only limitations on the wearer's activity are that the glove is not designed to be used under water. The glove has successfully endured a trip through a washing machine, and worked fine once washed, but it is not designed to work while immersed. The galvactivator attempts, through its small size, inexpensive components, and wearability, to bring the sensing of skin conductivity out of the laboratory, and into a naturalistic setting that encourages people to play with the sensor and to learn more about their skin conductivity response.

Because the LED is very bright, and appears on the back of the hand, the device can also easily communicate the wearer's skin conductivity response to others. The wearer is free to hide this signal (by hiding the back of the hand, or by dialing the LED off) or to show it to others. We think it is important that the wearer maintain control over the communication of the signal, especially since the signal is one that is not usually communicated to others in an ongoing way. We describe below an experiment where a large audience was given the opportunity to communicate this signal to the performers on stage, as well as to other individuals through more personal interactions.

2 APPARATUS

The galvactivator hardware consists of a small printed circuit board with analog circuitry that amplifies the skin conductivity signal, and maps it to a super-bright LED (LRI, Inc.). The electrodes are standard nickel-plated clothing snaps. The circuit is powered by two 3V-lithium ion batteries. The board and electrodes are embedded in a neoprene glove that slips over the hand and is secured with a velcro strap. The board can be easily pulled out to replace the batteries if needed. The back of the glove allows access to a thumb-wheel potentiometer (for setting individual baselines), and houses a star-shaped PVC window which serves to help diffuse the directional beam of the LED. Adjacent to the thumb-wheel is a small jack where the actual signal can be read out and recorded by a computer if desired; however, we designed the glove primarily to be worn without attachment to any other devices.

Figure 1: Galvactivator circuit board, Galvactivator on a hand.

The glove form-factor was designed to maximize comfort and minimize motion artifacts with a snug fit. To this end, the placement of electrodes differs slightly from traditional placement options (see Figure 2), but the signal from the galvactivator correlates with a traditional skin conductivity sensor at $p < .001$. Because of time and budget limitations, we designed the galvactivator as a one-size-fits-all device to be worn on the left hand. Consequently, some individuals required small modifications in order for it to fit correctly (i.e., small slits made in the finger hole for larger hands).

Figure 2: Traditional Placements (left, middle) and our placement (right).

Once the subject has set his proper baseline (established by resting for 5 minutes, then dialing the wheel until the light is dim), the signal is mapped linearly to the brightness of the LED. The display is sensitive enough to display both tonic (low frequency or long-term) and phasic (high frequency or short-term) components of the skin conductivity signal. However, reading the brightness of an LED is not as accurate as reading a numerical value output on a graph. The device was not designed for scientific recording and analysis of detailed changes in skin conductivity, but was designed to make it easy for the wearer and for those he communicates with, to easily see the patterns of change.

3   LARGE AUDIENCE USE OF THE GALVACTIVATOR

We conducted a large-scale test of the device, distributing 1200 galvactivators to attendees at the SENS*BLES symposium at MIT in October 1999. Audience members received their galvactivators mid-morning. They were given a short demonstration of how to properly put on the device. Then they were led through five minutes of a shared experience while wearing the device (described below) and were free to wear (or not wear) the device the rest of the day and take it home with them afterward, to keep.

Figure 3 illustrates the aggregate brightness output recorded from a section of a couple hundred people in the audience before, during, and after the shared experience. The graph shows the overall brightness of the video signal as measured by a camera positioned in the upper left of the auditorium. The auditorium was held to a constant darkness level so that the primary changes in brightness would be due to the collective galvactivator LED's. In Figure 3, we first see a calibration step where the audience was asked to hold their left hands so that the camera could see the LED, in order for us to obtain an estimate of the maximum brightness value that could be obtained (the devices were set to maximum brightness initially.) Then the audience was asked to turn their LED's away from the camera's view, to obtain an estimate of minimum brightness. Afterward, the audience was told about what would happen next (which elicited laughter as well as a heightened sense of expectation.) Then they were told to get out a balloon that had been given to them earlier, and they were led into a 30-second relaxation period, followed by instructions to dial the thumb-wheel so that their light was dim. (In theory, the wearer should rest at least five minutes before dialing the light dim; however, given time restrictions and the difficulty of anyone relaxing in the presence of 1200 people when they know they are part of a live experiment, we decided that 30 seconds was going to be acceptable.) Next, they were instructed to blow up their balloon to popping. The deep breathing combined with the hundreds of loud noise bursts and anticipation of their own balloon bursting, brought most people to their individual maximum value. (In Figure 3, this peak is somewhat less than the peak of everyone pointing the device, dialed to maximum brightness, directly at the camera, since not everyone's LED would be perfectly visible to the camera; however the peak is substantially higher than the preceding valley, where everyone set their baseline after the brief rest period. Episodes of clapping tended to raise the hands, pointing more LED's directly toward the camera, increasing the brightness the camera detected.

Continuous video of the audience was collected throughout the day, as well as video of the stage. Analysis of the videotapes shows significant areas of high and low brightness as correlated with stage events. The brightness levels of the video do not perfectly reflect the output of the galvactivators; for example, someone might hide his signal so that only he could see it, and in some cases there were camera flashes, despite that we prohibited photographers from using flashes while we were recording. Nonetheless, the brightness values that were measured appeared to be meaningful. In general, we found that with the arrival of a new speaker or performer on stage, that brightness increased (with audience anticipation as well as with their clapping.) Elicitation of laughter and of questions from the audience also tended to make the brightness increase. Live demonstrations also led to brighter values than did segments where a speaker just spoke with PowerPoint slides. These are all reasonable reactions, given what is known about skin conductivity and psychological arousal.

As well as acting as a device for mass communication, the galvactivator can be used to learn about one's own bodily responses in a myriad of situations–interacting with a computer, reading a book, engaging in conversation with others, and so forth. We collected many anecdotes from symposium participants, for example:

- *"It made me more aware of how I was feeling"*
- *"I noticed that the galvactivator lights up every time I laugh…"*
- *"We had a series of brainstorms the following day, and whenever an idea was sparked, before it was voiced, the device glowed."*
- *"It was a bit disheartening to see the audience go dim as I talked to them."*

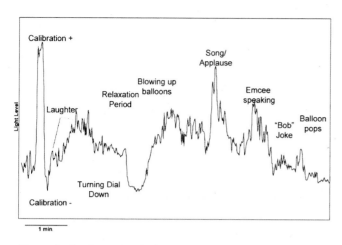

Figure 3: Brightness level of sensors from a segment of the audience.

Several people were observed having fun trying to make each other light up. Because feeling self-conscious or embarrassed can also raise skin conductivity, the device often led to conversations which included admissions of such feelings. Because the skin conductivity response is not usually visible to others, it raises many questions regarding the psychology of communicating this signal, and what impact this can have on inter-personal relationships. These issues are being explored in ongoing research with teens and their peers and parents.

4   POSSIBLE APPLICATIONS

A comfortable wearable skin-conductivity sensor has many potential applications. In addition to the mass-audience communication described above, one can imagine stadiums full of sports fans lighting up in the color of their team when the game gets exciting. One can also imagine its use among smaller audiences. One case we have explored is use in classrooms. Tenth and eleventh grade classes were given galvactivators to wear. After setting proper baselines, the students wore the gloves during ordinary classroom activity. The brightness was observed to be very low when they were instructed to engage in their regular reading task. It increased when the students were asked to write about something that interested them in their journals. A few students commented that wearing the device all day helped them gain insight into their personal learning style, including which types of activity engaged them best.

For communication purposes, the device affords new opportunities in nonverbal expression; there is anecdotal evidence that the galvactivator can change the course of person-to-person conversation. One student, who was wearing the device when she got in a fight with her parents, reported that after her mother saw the device glow, she asked the daughter about it, and they were then able to communicate more openly about the daughter's feelings.

We have also employed the skin conductivity signal as one of several signals used by a computer in trying to discern the affective state of the user. Skin conductivity tends to increase with anger and frustration; together with other signals from the user, it can be analyzed by the computer to try to infer how the interaction is proceeding (Scheirer et al.; Picard et al.).

Certain groups of people display characteristic skin conductivity patterns. For example, autistic children's skin conductivity tends to fluctuate between dramatic highs and lows, consistent with the difficulty these children experience in self-regulation, affecting their attentiveness and ability to interact with other people and things. A wearable skin conductivity sensor, coupled with a "squeeze vest" appears to be able to help the child, based on the principle that touching and squeezing a child from the sides tends to calm the child and bring their skin conductivity back into a normal range (work of Carlos Elguero of Univ. Mass, Amherst.)

Skin conductivity also tends to increase with pain and with stress, suggesting other potential therapeutic uses of the device where it might assist patients in helping manage these conditions. In particular, for patients who are nonverbal, changes in the LED brightness might help communicate to loved ones and to caregivers an aspect of what the wearer is feeling. Seeing a non-verbal friend "light up" when they recognize you can be very gratifying if they have no other means of acknowledging your presence.

5 CONCLUSIONS

The galvactivator is a new wearable device that senses the skin conductivity response – a correlate of physiological arousal – and makes the signal visible through a glowing LED. It takes a signal that has traditionally been studied in stringent laboratory settings and makes it available for study in naturalistic, real world environments. New modes of communication with the device have been demonstrated in both large audience settings and in smaller interpersonal communications. Collaborations using the device are currently being explored with educators, clinicians, and industry.

For more details about how the device is constructed, and for a list of frequently asked questions about skin conductivity, there is more information online at http://www.media.mit.edu/galvactivator.

ACKNOWLEDGMENTS

We would like to thank Nancy Tilbury and Jonny Farringdon of Philips Design & Research for their assistance with the design of the glove. Their aids in the UK included: Kelly Barrett, Philippa Wagner and Andrew Moore at Philips, Tirath Dillon and Nav Dillon at ACE Sportswear, Richard Extell and Kevin Niblett at GS and John McLeod at Micro Thermal Systems. Several colleagues at the MIT Media Lab gave valuable help during various stages of circuit design, prototype design and production of 1500 gloves: Dana Kirsch, Dan Overholt, Ted Selker, Joe Paradiso, Carson Reynolds, Matt Norwood, Henry Wu, Blake Brasher, Pamela Mukerji, Martin Howard, Joey Richards, Josh Strickon, and Eric Scheirer. We also thank George Northover and Scott Cohen at Accutek and David Allen at LRI.

REFERENCES

Boucsein, W. (1992). *Electrodermal Activity*, Plenum Press, New York.

Fere, C. (1988). Note on changes in electrical resistance under the effect of sensory stimulation and emotion. *Comptes Rendus des Seances de la Societe de Biologie* (Series 9), 5, 217-219.

Picard, R. W. (1997). *Affective Computing*, MIT Press, Cambridge, MA.

Picard, R. W., Vyzas, E. and Healey, J. (to appear) Toward Machine Emotional Intelligence: Analysis of Affective Physiological State. *IEEE Transactions on Pattern Analysis and Machine Intelligence*.

Reeves, B. & Nass, C. (1996). *The media equation : how people treat computers, television, and new media like real people and places*. Cambridge University Press, New York.

Scheirer, J., Fernandez, R., Klein, J., and Picard, R. W. (to appear). Frustrating the User on Purpose: A Step toward Building an Affective Computer. *Interacting with Computers*.

Vigouroux, R. (1879). Sur le role de la resistance electrique des tissus dans le'electrodiagnostic. *Comptes Rendus Societe de Biologie* (Series 6), 31, 336-339.

Vigouroux, R. (1888). The electrical resistance considered as a clinical sign. *Progres Medicale*, 3, 87-89.

Homepages with Emotions:

Quantitative Relations Between Homepage Design Factors and Emotions

Jooeun Lee, Jinwoo Kim, Dongseong Choe

Human Computer Interaction Lab, Yonsei University

Seoul, 120-749, Korea

jinwoo@yonsei.ac.kr

ABSTRACT

Emotional aspects of homepages are becoming more important as people spend more time in cyberspace. This research aims at identifying quantitative relationships between key design factors and generic emotional dimensions so that we may develop homepages with target emotions more efficiently. In order to achieve this goal, we conducted three related studies. In the first study, we identified thirteen emotional dimensions that people usually feel from diverse homepages. In the second study, we identified the key design factors that professional designers frequently use in developing various homepages. Finally, in the third study, we identified the quantitative relationships between the key design factors and the thirteen emotional dimensions. This paper concludes with implications and limitations of the study results.

1. INTRODUCTION

As the Internet spread widely into our society, cyberspace has become an important part of our daily life (Hefley and Morris, 1995). Accordingly, homepages, which are the central locations of cyberspace, have become more important than ever (Vora, 1998). When people first started to use the Internet, homepages tended to focus on providing basic functions, such as search engines or directories. However, as users gradually got used to these basic functions, cognitively convenient functions were not sufficient in satisfying user needs (Picard and Andrew, 1998). Especially when homepages are used for electronic commerce, the emotional aspects of homepages have gained increased importance (Kim and Moon, 1998). Accordingly, site developers strive to design homepages that can evoke specific emotions, so that it fits well with the overall site strategy (Schenkman and Jonsson, 2000). However, most prior studies in web site design have only been concentrated on the functional usability of homepages (e.g., Fuccella, 1997). Although several studies have dealt with the emotional aspects of physical product and system design (Picard and Andrew, 1998; Schenkman and Jonsson, 2000), few researches have focused on the emotional aspects of homepage design. The primary goal of this research is to identify the design factors that induce homepage visitors to feel certain target emotions. In order to achieve the primary goal, we assume that human emotions have several dimensions (Ortony, Clore, and Collins, 1988). Therefore, we first examined what kinds of emotions people usually feel when interacting with various homepages. Then, we examined what were the important design factors that might be related to those emotions. Finally, we investigated the relationship between the generic emotional dimensions and the key design factors to provide guidelines for effectively designing homepages with target emotion.

2. STUDY 1: DIMENSIONS OF EMOTION FOR HOMEPAGES

In the first study, we conducted an extensive search for emotional adjectives that had been used in previous studies of emotions in web interface (Kim and Moon, 1998). Consequently, 278 emotional adjectives were identified as expressing generic emotions. These included basic emotional adjectives such as happy and sad, as well as more complex emotions such as modern and elegant.

Then, twelve expert designers were recruited for selecting a set of sample homepages. Each of them was asked to bring four homepages that they thought were distinctively different from other homepages, resulting in a total of forty-eight pages collected, which were then classified into clusters according to the emotion that they induced. This classification resulted in twelve clusters and one homepage from each cluster was selected as the most representative within each cluster.

Using the twelve sample homepages and 278 emotional adjectives, a survey was conducted in order to identify the basic dimensions of emotion. Four hundred eighteen respondents participated in the survey, and each examined a random set of four of the twelve sample homepages. Respondents were asked to watch each of the four homepages for three minutes and then to answer a questionnaire consisting of the 278 emotional adjectives.

Cluster analysis was performed to estimate the similarities among the emotional adjectives and hierarchical diagrams were constructed to indicate these similarities. Factor analysis was performed to determine representative emotional adjectives for each category. We finally selected thirty adjectives based on the factor loadings in each category, which were then verified based on their Cronbach Alpha values. The thirteen basic dimensions of emotions, their representative adjectives, and the Cronbach alpha coefficients are presented in Table 1.

Table 1. Emotional Dimensions and Representative adjectives

Emotional dimension		Representative adjectives			Cronbach Alpha
E1	Light	Faint	Bright	Fresh	0.6982
E2	Tense	Tense	Sharp		0.6390
E3	Strong	Strong	Powerful		0.7670
E4	Static	Neat	Balanced	Calm	0.6793
E5	Deluxe	Deluxe	Elegant	Valuable	0.7916
E6	Popular	Popular	Familiar		0.6815
E7	Adorable	Adorable	Cute		0.8778
E8	Colorful	Colorful	Vibrant	Sexy	0.6410
E9	Simple	Concise	Simple		0.7813
E10	Classical	Classical	Conventional		0.6787
E11	Futuristic	Futuristic	Surreal		0.3436
E12	Fantastic	Mystic	Vague		0.7472
E13	Hopeful	Reliable	Hopeful		0.7837

3. STUDY 2: KEY DESIGN FACTORS

The second study was conducted to identify the key design factors that were frequently used to express the thirteen emotions that were found in the first study. A total of thirty professional designers with more than three years of experience in homepage design participated in the experiment. The subjects were given a two-page description about a homepage a client requested. The descriptions were exactly the same except for the emotional adjectives that were used to describe the client's preference about the requested homepage. Subjects were given two hours in the experiment and the entire design session was recorded with video cameras. Then, they were allowed two to three days to collect appropriate images and design materials. They were asked to keep a self-report design log during this period. Finally, the designers returned for an additional two hours to implement their homepage using computers. This session was again recorded using video cameras. Consequently, fifty-two homepages were developed in total – four sample homepages for each of the thirteen emotional dimensions that had been identified in the first study.

Protocol analysis was conducted with the data from the lab sessions and content analysis was performed with the interim self-report design logs. The results of the analyses suggest that the design factors that professional designers frequently use can be classified into three categories: objects, background, and the relationship between objects and background. In terms of objects, title, menu and main images are most frequently used in homepage design. In terms of background, layout, color and texture are found to be used most often by the professional designers. Finally, in terms of the relationship between object and background, designers usually consider the color relation between objects and background in three aspects: hue, brightness, and saturation.

4. STUDY 3: QUANTITATIVE RELATIONS BETWEEN EMOTIOS AND DESIGN FACTORS.

The third study was conducted to investigate the quantitative relations between the emotional dimensions identified in the first study and key design factors that were found in the second study.

A total of five hundreds fifteen subjects participated in the third study. The subjects were asked to view each of the fifty-two homepages that had been developed in the second study. For each homepage, they were asked to answer thirty questions related to the representative emotional adjectives that had been identified in the first study. Regression analyses were conducted for each of the thirteen emotional dimensions.

The results of the regression analyses are presented in Table 2 below. The first column of Table 2 indicates the thirteen emotional dimensions identified in the first study. The second column of Table 2 presents regression equations in which key design factors that are found to have significant impacts on the corresponding emotional dimensions are described in an abbreviated form. The number stands for a beta coefficient of the selected design factor, and the following initial stands for the design factors. For example, t stands for title, m for menu, i for main image, b for background, rbt for the relationship between background and title, rbm for the relationship between background and menu, and rbi between background and main image. The next initial following a dash (_) stands for specific features of selected design factors. For example, t stands for texture, o stands for layout, h stands for hue, *s* stands for saturation and *b* for brightness. Finally, an index number is attached at the end if there are more than one design features for the selected design factors. Therefore, m_h_1 stands for the first hue of the menu bar if it consists of more than one hue.

Table 2. Regression Equations and Adjusted R-Square for Thirteen Emotional Dimensions

Emotional Dimension	Regression Equation	Adjusted R-Square
(E1) Light	$E1 = -0.429\ m\_h + 0.282\ m\_b - 0.165\ t\_h + 0.386\ i\_o + 0.474\ i\_b + 0.924\ b\_b + 0.159\ rbt\_h + 0.587\ rbi\_b$	0.909
(E2) Tense	$E2 = 0.246\ m\_t + 0.441\ m\_h\_1 + 0.410\ m\_h\_2 + 0.435\ t\_o - 0.303\ t\_h\_1 + 0.24\ t\_h\_2 - 0.118\ i\_h + 0.670\ i\_b + 0.248\ b\_o + 0.721b\_h$	0.958
(E3) Strong	$E3 = 0.447\ m\_b + 0.446\ t\_o + 0.345\ i\_t + 0.333\ b\_o$	0.657
(E4) Static	$E4 = 0.516\ m\_o - 0.384\ m\_h - 0.426\ t\_h + 0.570\ i\_t\_1 - 0.478\ i\_t\_2 - 0.355\ i\_h\_1 - 0.175\ i\_h\_2 + 0.265\ i\_b - 0.672\ b\_t$	0.936
(E5) Deluxe	$E5 = -0.372\ t\_h + 0.578\ b\_t$	0.495
(E6) Popular	$E6 = 0.459\ m\_t - 0.363\ m\_h - 0.264\ t\_h\_1 - 0.218\ t\_h\_2 - 0.227\ t\_h\_3 - 0.702\ i\_h\_1 - 0.128\ i\_h\_2 + 0.834\ b\_b - 0.844\ rbm\_b + 1.262\ rbi\_b$	0.932
(E7) Adorable	$E7 = 0.209\ m\_o - 0.354\ m\_h + 0.400\ i\_o + 0.631\ i\_b + 0.271\ b\_h + 0.305\ rbt\_h$	0.797
(E8) Colorful	$E8 = 0.550\ m\_o + 0.273\ i\_s - 0.311\ b\_h + 0.472\ rbm\_h$	0.766
(E9) Simple	$E9 = 0.298\ m\_h + 0.414\ m\_b + 0.339\ t\_o + 0.534\ i\_h + 0.255\ i\_b + 0.379\ b\_i - 0.468\ b\_h$	0.623
(E10) Classical	$E10 = 9.245\ t\_s + 0.499\ i\_o + 0.254\ i\_h - 0.352\ b\_h\_1 + 0.252\ b\_h\_2 + 0.336\ rbt\_h$	0.424
(E11) Futuristic	$E11 = 0.457\ i\_o + 0.536\ i\_i - 0.481\ i\_h\_1 - 0.173\ i\_h\_2 + 0.457\ b\_o - 0.456\ b\_h + 0.158\ rbt\_h + 0.390\ rbt\_b$	0.859
(E12) Fantastic	$E12 = 0.343\ m\_o + 0.617\ m\_b - 0.278\ t\_h\_1 + 0.212\ t\_h\_2 + 0.621\ i\_h + 0.405\ b\_h$	0.741
(E13) Hopeful	$E13 = 0.421\ m\_o - 0.348\ m\_h + 0.342\ b\_b$	0.395

For example, people tend to feel lighter (E1) when the color of a menu bar is bright (m_b), but neither blue nor purple tone (m_h). Also, a green tone of a title (t_h) decreases the light emotion, while an ellipse outline (i_o) and bright color (i_b) of a main image increase the light emotion. People also tend to feel light when background is in bright color (b_b), when background and title are contrasted in hue (rbt_h), and when the brightness of background and main image are not contrasted (rbi_b). Finally, the last column of Table 2 indicates adjusted R squares of the corresponding equations. The results in Figure 2 suggest that design factors are in fact related to each of the thirteen emotional dimensions.

5. CONCLUSIONS AND DISCUSSIONS

This paper reports the results of three studies that investigate the design of homepages with target emotions. In the first study, we conducted a survey that helped us to identify thirteen emotional dimensions. Key design factors for emotional homepages were identified in the second study. In the third study, we conducted an experiment in which the quantitative relations between the key design factors and the thirteen emotional dimensions were identified. The results clearly indicate that several design factors are significantly important in inducing certain emotions while people interact with homepages.

This study has several limitations. First, even though we can say which design factors are important to elicit certain feelings, we cannot explain why they are important. Further analyses of verbal protocols of the professional designers during the second study are underway to answer this question. Second, the emotional adjectives were asked in the domestic languages and participants in the three studies were all local residents. In order for the study results to be generalized, further studies should be conducted in different cultures and nations.

ACKNOWLEDGEMENT

We would like to thank the 'Korea Science and Engineering Foundation' for providing us with a generous research grant (Grant #2000-2-0132).). Special thanks to Donguk Lee, Jungpil Hahn, Sungjoon Park, Kyunguk Park and Sui Park at Human Computer Interaction Lab at Yonsei University, and President Choe Jaehack and Director Seok YoonChan at Hihome, Co.

REFERENCES

Hefley, B. and Morris, S. J, "An Introduction to the Internet and the World Wide Web", *Proceedings of the CHI '95 Human Factors in Computing Systems*, May 7-11, Denver, CO, 1995, http://www.acm.org/sigchi/chi95/Electronic/documnts/tutos/wh_bdy.htm.

Kim, J. and Moon, J. Designing towards emotional usability in customer interface- trustworthiness of cyber-banking system interfaces. *Interacting with Computers*, 10(1), pp. 1-29, 1998.

Schenkman, B. and Jonsson, F., Aesthetics and preferences of web page, *Behaviour & Information Technology*, 19(4), pp. 367-377, 2000

Picard W. R, and Stern A., Panel on Affect and Emotion in the User Interface, *Proceedings of the 1998 International Conference on Intelligent User Interfaces (IUI '98)*, January 6-9, San Francisco, CA, pp. 91-94, 1998

Fuccella, J., Using User Centered Design Methods to Create and Design Usable Web Sites, *Proceedings of the 15th Annual International Conference on Computer Documentation (SIGDOC '97)*, October 19-22, Snowbird, UT, pp. 69-77, 1997

Ortony, A., Clore, G., and Collins, A., *The Cognitive Structure of Emotions*, Cambridge University Press, Cambridge, 1988.

Vora R. P, "Designing for the Web: A Survey", *Interactions*, 5(3), pp. 13-30, 1998.

Trust in the Online Environment

Cynthia L. Corritore, College of Business, Creighton University, Omaha, NE 68178, cindy@creighton.edu
Beverly Kracher, College of Business, Creighton University, Omaha, NE 68178, bkracher@creighton.edu
Susan Wiedenbeck, College of IST, Drexel University, Philadelphia, PA 19104, susan.wiedenbeck@drexel.edu

ABSTRACT

Trust has been identified as one of the most, if not the most, important element in electronic commerce and online education (Cheskin/Sapient, 2000). However, while trust has been widely studied in other fields and in the offline environment, its' study in an online context is just beginning. In this paper we review the key work done on trust and identify five essential items to guide development of a general model of online trust.

1.0 INTRODUCTION

Trust has been a current topic of research in many fields since the 1950's. Each field has produced its own set of concepts, definitions, and findings. The outcome is a multi-dimensional family of trust concepts, each with a unique focus, which can provide a foundation for the study of trust in an *online* environment. This paper takes a first step in a dialog about online trust, based on a multidisciplinary review of trust research and an identification of the essential elements required to develop a model of online trust. We begin by reviewing offline trust literature from several different fields, then outline five essential categories of information needed to develop a general model of online trust. Last, we propose topics for research in online trust.

2.0 REVIEW OF THE ONLINE TRUST LITERATURE

2.1 Psychology

The research on trust in the field of Psychology research has focused on examining the characteristics of trust and how individual traits affect trusting behavior. It embodies a large variety of approaches to defining and examining trust with little interaction between the approaches. Psychologists maintain that trust is a basic feature of all social situations that demand cooperation and interdependence, and is vital to social life and personality development (Kee and Knox, 1970; Rotter, 1971). The research tends to take one of two approaches: one that focuses on individual personality differences and developmental factors in early and later life which lead to a 'trusting' personality (Erickson, 1963) or an examination of trust in interpersonal relationships, transactions and cooperative ventures (Deutsch, 1962). No definitive link between individual personality differences and trusting behavior has been identified.

Interpersonal trust has been studied to a great extent in the context of mixed-motive games such as the Prisoners Dilemma (Kee and Knox, 1970; Deutsch, 1962; Rotter, 1971). In this approach, trust is inferred from cooperative behavior. In such research, Deutsch (1962) found that communication appeared to promote trust. Effective communication contained: 1) intention and expectations, 2) an outline of the basic features of the trusting relationship, 3) an indication of how violations will be treated, and 4) an outline of a method of absolution from violations of trust. Shared, common goals and mutual benefits, as well as control over another's behavior, and the ability to predict behavior and perceive trustees intentions and motivations also identified as important to trust (Deutsch, 1962; Lewicki and Bunker, 1995). Experience, expectations, and confidence in another's abilities were also found to be key in trust decisions (Deutsch, 1962; Lewicki and Bunker, 1995). These factors can be grouped as, structural factors (ie. incentives, power, communication), situational factors (ie. degree of control in a given situation), or dispositional factors (ie. motivational orientation, personality, attitudes) (Kee and Knox, 1970).

Deutsch (Deutsch, 1962) identified two types of trust: interpersonal and mutual. More recently, Shapiro, Sheppard, and Cheraskin (1992) proposed three types of trust: calculative (deterrence-based), knowledge-based (enough information available to predict behavior), and identification-based (internalized others desires and intentions). Lewicki and Bunker (1995) extended this model by proposing that the three types of trust are actually phases.

2.2 Sociology

Sociologists believe that trust is a deep assumption for modern society and essential for social interaction (Giddens, 1990; Buskens, 1998). For example, Misztal (1996) sees trust as social capital, used to improve efficiency of society by facilitating coordinated actions. Others have studied trust in the context of games and define it implicitly as cooperation (Snijders and Keren, 1999). Most sociologists also agree that trust is a means to decrease social complexity. This was illustrated by Garfinkel's (1963) classic experiments in which the removal of apparently inconsequential features of everyday conversation significantly disrupted the social interaction subjects 'trusted' to be in place. Generally sociologists distinguish between personal, or face-to-face, trust versus trust in some type of societal structure (Barber, 1983; Giddens, 1990). Social perception, the ability to read others intentions and inclinations, the stakes involved and their importance is also important (Barber, 1983; Snijders and Keren, 1999). Giddons (1990) identified bias in the interpretation of social events and a cognitive inertia to preserve trust by individuals. Other trust factors include situational context and the variability of trust depending on the context (Lewis and Weigert, 1985; Buskens, 1998).

Barber (1983) proposed that trust has three dimensions: continuity of natural order, technical competence of actors in roles, and fiduciary obligations of actors (putting others interests ahead of ones own). Trust has also been defined in terms of risk-taking decisions in social contexts (Lewis and Weigert, 1985, Misztal, 1996) while game theorists view trust as a rational process and that cooperation in a two-person gaming situation defines trust (Buskens, 1998).

2.3 Philosophy

Baier (1986) is the grandmother of trust, or more precisely, entrusting. Her focus is interpersonal trust, though in the field of philosophy trust is studied at all levels including personal, interpersonal, organizational and social. Baier defines trust as "letting other persons take care of something the truster cares about." Philosophers debate the value of trust. Some argue that is it virtuous and intrinsically valuable (Baier, 1986; Brenkart, 1998). Koehn (1996) finds only instrumental value in trust. She proposes a dialogical test for ethically good (authentic) trust where trustors and trustees are willing to openly dialog about each other's intentions and interests. All philosophers seem to agree that the definition of trust includes a trustee in a state of vulnerability and a perception that an other is trustworthy. However, further clarification of the psychological state of the trustee is disputed and is the basis of several taxonomies of trust. Brenkart (1998) identifies three types of definitions of trust, where trust is an attitude, a prediction or a voluntary action of entrusting. There is also conscious or unconscious trust (Baier, 1986; Koehn, 1996). Flores and Solomon (1998) distinguish simple trust, blind trust, basic trust, authentic trust and articulate trust. While simple trust is given in a situation, authentic and articulate trust are created over time and in relationships (Flores and Solomon, 1998; Baier, 1986).

2.4 Human-Computer Interaction

In the domain of human-computer interaction trust is relatively new topic. Fogg and Tseng (1999) argue that the trustworthiness of a computer is a key element of computer credibility, along with computer expertise, i.e. the computer's perceived knowledge and skill. Several HCI studies of electronic commerce interfaces have begun to identify how the interface affects trust. Kim and Moon (1997) found that the manipulation of visual elements, such as use of color and clipart, can influence the user's perception of trustworthiness of an electronic commerce interface. Further work by Lee, Kim, and Moon (2000) indicates that the factors positively related to trust in an electronic commerce interface include: provision of comprehensive information, perception of shared values between the electronic commerce store and the user, perception of frequent, high quality communication, and internet store specificity. In addition, it was found that the level of involvement with the product mediates the effect of these factors on trust. In a large survey Fogg, Marshall, Laraki, Osipovich, Varma, Fang, Paul, Rangnekar, Shon, Swani, and Treinen (2001) found that people perceive cues of trustworthiness on web sites. They suggest that it is important for web sites to convey honesty and lack of bias via the interface. Fogg, Marshall, Kameda, Solomon, Rangnekar, Boyd, and Brown (2001) also found that the author's photograph increases the trustworthiness of an online article, while banner ads affect perceptions of the trustworthiness of a web site's content in different ways depending on what is being advertised. Other studies focusing on computer-mediated communication (Bos, Gergle, Olson, and Olson, 2001) have found that partners playing a game using face-to-face and video channels of communication establish trust and cooperation quickly, while those using text-only fail to establish trust and cooperation. However, if participants met face-to-face before engaging in text-only communication they did show trust in the subsequent communication (Zheng, Bos, Olson, and Olson, 2001).

2.5 Marketing

Recent work in marketing focuses not on discrete transactions but rather on relationship marketing, i.e., establishing and maintaining relational exchanges of long duration. A frequently quoted *definition* of trust is "a willingness to rely on an exchange partner in whom one has confidence" (Moorman, Zaltman, and Deshpande, 1993). The vendor's credibility (expertise) and benevolence (positive intentions toward the buyer) have been identified as facets underlying trust (Ganesan, 1994). The *antecedents* of trust in marketing relationships are vendor reputation (Ganesan, 1994), perception that the vendor organization has made investments on the buyer's behalf, e.g. customization of products (Ganesan, 1994), large size of the vendor organization (Doney and Cannon, 1997), and high communication and shared values between the vendor and buyer (Shelby and Hunt, 1994). *Outcomes* of trust are the establishment of long-term exchange relationships (Ganesan, 1994), reduction of uncertainty in relationships, cooperation, and amicable resolution of disputes (Shelby and Hunt, 1994).

Recently, the concept of relationship marketing has been expanded to database marketing in which the seller organization establishes a long-term, trusting relationship with individual consumers, unmediated by a salesperson, e.g. direct marketing by phone or over the Internet. Milne and Boza (1999) argue that successful database marketing depends on trust, and that trust is engendered by treating the customer's personal information fairly. Milne and Boza's study indicates that the database marketing industry has low trust based on the practices of consumer profiling and sharing information with third parties without permission. However, it seems clear that trust relies on more than fair treatment of customer information. Other researchers propose to gain trust in Web marketing by taking advantage of the Web's ability to add value by increasing useful information available to consumers (Urban, Sultan, and Qualls, 1998). Other approaches to trust-building that need to be investigated are integration of the distinct organizations involved in a purchase and using brands effectively as a symbol to evoke trust on the Web.

2.6 Management

Management scholars argue that trust enables us to live in risky situations (Bhattacharya, Devinney and Pillutla 1998), reduces fear (; Mayer, Davis and Schoorman, 1995; Hwang and Burgers, 1997) and transaction costs (Wicks, Berman and Jones, 1999). The assumption by management theorists is that individual people, organizations, and systems can be trusted. Organizational theorists think about trust at the micro level (for example, trust in workgroups) or meso level (trust in one's organization) (Hosmer, 1995). Strategists discuss trust towards various stakeholders (Wicks et al, 1999), or inter-organizational trust (Dodgson, 1993). Recently, Handy (1995) has discussed trust in virtual organizations, and Meyerson, Weick and Kramer (1996) have focused on trust in temporary systems, distinguishing swift and slow trust. Management literature is replete with hypotheses about the causes of trust. Repeated interaction, alignment of interests, shared identity and predictability are some of the identified causes of specific types of trust (Bhattacharya et al., 1998). Other researchers focus on the causes of trust in general including the trustee's integrity, competence, loyalty, consistency and openness (Whitney, 1994), the trustee's reliability and fairness (Whitney, 1994) and effective communication between the trustor and the trustee (Dodgson, 1993).

2.7 E-Commerce

Trust research in the area of electric commerce (ecommerce) is unique for several reasons. First, the field has focused entirely on online rather than offline trust (Jarvenpaa, Tractinsky, and Saarinen, 1999; Marcella, 1999, Sisson, 2000). Second, ecommerce trust research includes discussions about whether technology itself is a proper object of trust (Javenpaa, et al., 1999; Marcella, 1999). Third, much of the ecommerce literature focuses on trustworthiness rather than trust. Ecommerce theorists are interested in the question of how web designers and developers can create sites to which users will return which they will find 'trustworthy'.

In the ecommerce literature, the Cheskin/Sapient report (1999) is repeatedly referenced when designating ways to create trustworthiness. The Cheskin/Sapient report is somewhat controversial since it appears to confuse trust and trustworthiness (Sisson, 2000). Nevertheless, the report is one of the first of its kind and will have to be considered by anyone in the ecommerce trust field. The Cheskin/Sapient report identifies six building blocks of trustworthiness, namely, seals of approval, branding, fulfillment, navigation, presentation and technology. Three of these six (navigation, branding, fulfillment) are key. When effective navigation is combined with either a well-known brand or effective fulfillment, it appears that there is a strong likelihood that the site will be perceived as trustworthy. The Cheskin/Sapient report shows how the six building blocks of trustworthiness can be divided into 28 specific ways to establish trustworthiness. For example, effective navigation is communicated in three ways, namely, through navigation clarity (navigation and content terminology are distinct), access (navigation system is consistent and easy to find), and reinforcement (prompts, instructions, etc. on how to use the navigation).

3.0 TOWARD A MODEL OF TRUST

Our purpose in conducting a multidisciplinary review of the trust literature was to ultimately synthesize a general model of trust. Since little research exists on trust in an online environment, this initial trust model would be built on concepts and research from the offline environment with subsequent testing in an online environment. We have identified five basic groups of information that define the elements required for a general model of online trust. They are: 1) the object(s) of trust, 2) the antecedents of trust, 3) the consequences of trust, 4) the types of trust, and 5) the stages of trust.

By objects of trust we mean the target of the trust: what the trustor is trusting. This can range from a micro to a macro level. For example, at the macro level trust can be placed in the entire online environment. In contrast, at the micro level the trustee can include individuals, entities (eg. computers), or specific websites. In-between fall groups, organizations, brands, and third party authorities. A model of trust must identify which of these objects it addresses. Next are the antecedents of trust. These are things that precede trust and can either provide a suitable environment for trust or be causative. Antecedents can be either personal or external/environmental. Personal include factors such as personality traits and predispositions, cultural expectations, experience, comfort, and psychological phenomena such as conformational bias. Experience addresses previous knowledge of and experience with the trustee as well as familiarity with the domain. The external and environmental factors include the notion of risk, vulnerability, imperfect knowledge and uncertainty, a perceived lack of control, issues of privacy, the inability to be able to completely monitor the trustee, and the predictability of the trustee. The quality of the relationship management, issues of user interface design, and third party authorities (such as VeriTrust) are also considered external.

The actual consequences of trust include a variety of factors, some focused on individuals and others with a wider relevance. Some examples include satisfaction, a reduction of fear and complexity, and an improvement in efficiency. More general consequences include decreasing perceived risk and engagement/disengagement behaviors such as increased customer loyalty, repeated business, and increased communication and cooperation. The fourth set of factors are the types of trust. We maintain that there are many types of trust, most of which can be conceptualized as a continuum. Some examples are: 1) fiduciary (duty) vs. self-interest, 2) initial vs. mature, and 3) swift vs. long-term. Other types of trust exist which do not lend themselves to a continuum, such as deterrence- vs. knowledge- vs. identification – based trust. Last are the stages of trust. Many different stages of trust have been proposed. For example, trust can be seen as movement through a series of stages beginning with a widespread distrust, through local trust and local cooperation, to eventual 'universal' trust in a particular context. Trust has also been described as moving through a calculative phase, to a knowledge-based phase, and finally to an identification phase. However, all stages of trust share a reference to a long-term progressive development.

4.0 FUTURE RESEARCH

We identify the first step in the examination of online trust to be the development of a general model of trust. While this is a challenging step, we believe that it is necessary in order to guide subsequent research on the topic of online trust. With a general model in place, individual parts of the model can then be empirically tested. Even more importantly, the model can be tested and refined in specific contexts. We identify three primary contexts: transactional (e-commerce), informational (sites that focus primarily on information provision), and entertainment. Another related area for research is examination of distrust – how is trust lost in an online environment and how can it be recovered?

REFERENCES

Baier, A. (1986). Trust and antitrust. *Ethics,* 96, 231-260.

Barber, Bernard. (1983). *The Logic and Limits of Trust.* New Brunswick, New Jersey: Rutgers University Press.

Bhattacharya, R., Devinney, T.M., and Pillutla, M.M. (1998). A formal model of trust based on outcomes. *Academy of Management Review* 23(3), 459-472.

Bos, N., Gergle, D., Olson, J..S., and Olson, G.M. (2001). Being there versus seeing there: Trust via video. *CHI 2001 Extended Abstracts* (pp. 291-292). New York: ACM.

Brenkert, G.G. (1998). Trust, morality and international business. *Business Ethics Quarterly*, 8(2), 293-317.

Buskens, V. (1998). The social structure of trust. *Social Networks*, 20(3), 265-289.

Cheskin Research and Studio Archetype/Sapient. (1999). Ecommerce Trust Study. http://www.sapient.com/cheskin. Accessed 5/9/2000.

Deutsch, M. (1962). Cooperation and Trust: Some theoretical notes. *Nebraska Symposium on Motivation*, 10, 275-318.

Dodgson, M. (1993). Learning trust, and technolgical collaboration. *Human Relations*, 46(1), 77-95.

Doney, P.M. and Cannon, J.P. (1997). An examination of the nature of trust in buyer-seller relationships. *Journal of Marketing*, 61, 35-51.

Erickson, E.G. (1963). *Childhood and Society*. New York: W.W. Norton.

Fogg, B.J., Marshall, J., Kameda, T., Solomon, J., Rangnekar, A., Boyd, J., and Brown, B. (2001). Web credibility research: A method for online experiments and early study results. *CHI 2001 Extended Abstracts* (pp. 295-296). New York: ACM.

Fogg, B.J., Marshall, J., Laraki, O., Osipovich, A., Varma, C., Fang, N., Paul, J., Rangnekar, A., Shon, J., Swani, P., and Treinen, M. (2001). What makes web sites credible? A report on a large quantitative study. *CHI 2001 Conference Proceedings* (pp. 61-68). New York: ACM.

Fogg, B.J. and Tseng, H. (1999). The elements of computer credibility. In M.G. Williams, M.W. Altom, K. Ehrlich, and W. Newman (Eds.), *CHI 99 Conference Proceedings* (pp. 80-86). New York: ACM.

Fox, A. (1974). *Beyond contract: Work, Power, and Trust Relations*. London: Faber and Faber Limited.

Ganesan, S. (1994). Determinants of long-term orientation in buyer-seller relationships. *Journal of Marketing*, 58, 1-19.

Garfinkel, H. (1963). A conception of, and experiments with, 'Trust' as a condition of stable concerted actions. In O.J. Harvey (Ed.), *Motivation and Social Interaction: Cognitive Determinants*.

Giddens, A. (1990). *The Consequences of Modernity*. Cambridge, U.K.: Polity Press.

Handy, C. (1995). Trust and the virtual organization. *Harvard Business Review*, 73(3), May-June, 40-50.

Hosmer, L.T. (1995). Trust: The connecting link between organizational theory and philosophical ethics. *Academy of Management Review*, 20(2), 379-401.

Jarvenpaa, S.L., Tractinsky, N., and Saarinen, L. (1999). Consumer trust in an Internet store: A cross-cultural validation. *Journal of Computer-Mediated Communication*, 5(2). http://www.ascusc.org/jcmc/vol5/issue2. Accessed 10/19/00.

Kee, H.W. and Knox, R.E. (1970). Conceptual and methodological considerations in the study of trust and suspicion. *Conflict Resolution*, 14(3), 357-366.

Koehn, D. (1996). Should we trust in trust? *American Business Law Journal*. 34(2), 183-203.

Lee, J., Kim, J., and Moon, J.Y. (2000). What makes Internet users visit cyber stores again? Key design factors for customer loyalty. *CHI 2000 Conference Proceedings* (pp. 305-312). NY: ACM,

Lewicki, R.J. and Bunker, B.B. (1995). Trust in relationships: A model of development and decline. In B.B Bunker and J.Z. Zubin (Eds.), *Conflict, Cooperation, and Justice: Essays Inspired by the Work of Morton Deutsch*. San Francisco: Jossey-Bass.

Lewis, D. and Weigert, A. (1985). Trust as a social reality. *Social Forces,* 63 (4), 967-985.

Marcella, A.J. (1999). *Establishing Trust in Virtual Markets*. Altamonte Springs, FL: The Institute of Internal Auditors.

Mayer, R.C., Davis, J.H., and Schoorman, F.D. (1995). An integrative model of organizational trust. *Academy of Managaement Review*, 20(3), 709-734.

Meyerson, D., Weick, K.E. and Kramer, R.M. (1996). Swift trust and temporary groups. In R.M Kramer and T.R. Tyler (Eds.), *Trust in Organizations: Frontiers of Theory and Research* (pp.166-195). London: Sage Publications.

Milne, George R. and Maria-Eugenia Boza. (1998). Trust and concern in consumers' perceptions of marketing information management practices. *Journal of Interactive Marketing.* 13, (1), 5-24.

Misztal, B.A. (1996). *Trust in Modern Societies: The Search for the Bases of Social Order*. New York: Polity Press.

Moorman, C., Deshpande, R., and Zaltman, G. (1993). Factors affecting trust in market research relationships. *Journal of Marketing* 57(1), 81-101.

Morgan, R.M. and Hunt, S.D. (1994). The commitment-trust theory of relationship marketing. *Journal of Marketing,* 58, 20-38.

Rotter, J.B. (1971). Generalized expectancies for interpersonal trust. *American Psychologist*, 26, 443-452.

Sisson, D. (2000). *Ecommerce*. http://www.philosophe.com/commerce/trust.html. Accessed 5/17/2000.

Snijders, C. and Keren, G. (1999). Determinants of trust. In D.V. Budescu, I. Erev, and Zwick, R. (Eds.). *Games and Human Behavior: Essays in Honor of Amnon Rapoport* (355-383). Mahwah, NJ: Lawrence Erlbaum.

Urban, G.L., Sultan, F., and Qualls, W. (1998). *Trust Based Marketing on the Internet*. White paper, W.P. 4035-98, Cambridge, MA: Massachusetts Institute of Technology, Alfred P. Sloan School of Management.

Whitney, J.O. (1994). *The Trust Factor: Liberating Profits and Restoring Corporate Vitality*. New York: McGraw-Hill.

Wicks, A.C., Berman, S.L., and Jones, T.M. (1999). The structure of optimal trust: Moral and strategic implications. *The Academy of Management Review*, 24(1), 99-116.

Zheng, J., Bos, N., Olson, J.S., and Olson, G.M. (2001). Trust without touch: Jump-start trust with social chat. *CHI 2001 Extended Abstracts* (pp. 293-294). New York: ACM.

Intrinsic Motivation, Ease of Use and Usefulness Perceptions as Mediators in Computer Learning

Susan Wiedenbeck[1] and Sid Davis[2]

[1]College of IST, Drexel University, Philadelphia, PA 19104 USA, susan.wiedenbeck@drexel.edu
[2]College of IST, University of Nebraska, Omaha, NE, 68182 USA, sidneydavis@unomail.unomaha.edu

This study examines the roles of the interaction style and the learner's prior exposure to other interaction styles in promoting task performance, *mediated* by the intrinsic motivation of the learning environment and users' perceptions of the usefulness of the software and of their ability to use the software successfully. As expected, the results confirm the importance of directness in the interaction style. However, they also indicate that intrinsic motivation has a strong effect on performance via its effect on perceived ease of use. This suggests that software designers should not only give special attention to creating software that promotes interface *directness*, but that also promotes *intrinsic motivation*.

1. INTRODUCTION

Research shows that interaction style has a strong impact on initial software learning and that learning a new interaction style is affected by prior experience with other interaction styles (Benbasat and Todd, 1993; Davis and Bostrom, 1993; Davis and Wiedenbeck, 1998; Wiedenbeck, 1999). Also, research has begun to study the effects of behavioral variables on learners. Such behavioral variables include users' motivation and perceptions about their ability to use software effectively to achieve desired outcomes. These variables appear to have a strong impact on software learning and may mediate the effects of other training variables, such as type of training and interaction style. In this research we study the effect of interaction style and the user's prior familiarity with other interaction styles mediated by three factors: the intrinsic motivation, or engagement, of the learning environment, the user's perception of the ease of use of the software, and the user's perception of the usefulness of the software. The contribution of this research is that it integrates theories of skill development, motivation, and self-efficacy to suggest *how* certain interaction styles and prior experience lead to effective learning. We address four general questions: 1) Does interaction style affect intrinsic motivation and ease of use perceptions, 2) Does prior experience with another interaction style affect intrinsic motivation and ease of use, 3) Does prior experience with a similar interaction style have a greater effect on intrinsic motivation and ease of use than experience with a different style, 4) Are intrinsic motivation, ease of use, and usefulness perceptions related to performance?

2. THEORETICAL BACKGROUND AND MODEL

Ausubel (1961) describes a theory of cognitive skill development known as assimilation theory. In this theory there are two kinds of learning: meaningful learning and rote learning. Meaningful learning occurs when the learner works with the material to draw connections between the material being learned and related knowledge in long-term memory, known as the assimilative context. This results in a deep understand of the material learned, unlike rote learning which is based on memorization without making meaningful connections. In computer learning, some interaction styles may promote meaningful learning. Hutchins, Hollan, and Norman (1985) argue that direct manipulation provides an environment in which the distance between the user's intentions and the actions needed to carry them out is small. An interface metaphor increases directness by allowing the user to carry out actions on objects in a well-understood semantic context. Furthermore, given the visibility of actions and objects, there are few syntactic barriers to achieving user goals. Extending this line of thinking, we may argue that there is a continuum of interaction styles. Direct manipulation gives the greatest support to meaningful learning. Menu-based interaction has many of the same qualities as direct manipulation in that it makes actions and objects visible, but lacks icons that seek to evoke a metaphor in a concrete, visual way. Command-based interaction does not present an interface metaphor or make actions and object visible and thus supports meaningful learning least well. In our study we compare learning of a more direct interaction style (menu) to a less direct style (command) in two contexts: initial use by novice users with no prior experience and subsequent use by users who have gained a prior assimilative context from exposure to direct manipulation software. The assimilative context resulting from prior exposure to direct manipulation software is expected to aid performance of subsequent users.

Bandura's Social Cognitive Theory (1986) identifies two critical cognitive factors that affect people's behavior and interactions with the environment: self-efficacy and outcome expectations. Self-efficacy is defined as the individual's belief about his or her ability to successfully carry out given behaviors. Outcome expectations are the individual's expectations about whether performing a given behavior will have positive results. In the computer domain, Compeau and Higgins have shown that self-efficacy beliefs and outcome expectations have a direct effect on computer performance (1995), and that self-efficacy also has an indirect effect on performance by modifying the individual's outcome expectations. Davis's (1989) Technology Acceptance Model has a close relationship to Social Cognitive Theory. Davis developed two constructs, perceived ease of use (PEU) and perceived usefulness (PU), that map respectively to self-efficacy beliefs and outcome expectations. PEU is defined as the extent to which the user perceives software to be easy to use. PU is defined as the user's perception of the extent to which software will aid job performance. Davis found that features of the software have a strong effect of PEU and PU, and that in turn PEU and PU affect later intentions to use the software. As in Compeau and Higgins' work, PEU affected intentions to use software both directly and indirectly by modifying PU (i.e., poor perceptions of ease of use lowered users' expectations of positive outcomes from using the software). In our study we use PEU and PU as measures of self-efficacy and outcome expectations respectively. We measure the PEU and PU of first-time novice users of menu and command-based software and also of users of menu and command-based software who had prior exposure to direct manipulation software. PEU and PU are expected to be greater for the menu than for the command-based software because of the menu software's greater directness. PEU and PU are expected to be greater in subsequent-use than in first-time use because in subsequent-use the learner has a strong direct manipulation assimilative context that will lead to perceptions of greater efficacy and more positive outcomes.

A theory of intrinsic motivation arises from the work of Csikszentmihali (1975) and Malone and Lepper (1987). Intrinsic motivation is a drive from inside the self to carry out an activity whose reward comes from the enjoyment of the activity itself, rather than from an outside source (Csikszentmihali, 1975). A person can become so involved in an activity that is highly intrinsically motivating that they lose track of time and events. This state is referred to as a flow experience (Csikszentmihali, 1975). Factors that make an activity intrinsically motivating are: a level of *challenge* (Csikszentmihali, 1975) in which there is a match between the task and the person's skills, *curiosity* (Malone and Lepper, 1987) which arises from a reasonable level of complexity and incongruities between what the person expects and what actually occurs in the activity, *control* (Lepper and Greene, 1978) in which the person is given a sense of control over the choices made in the activity, and *fantasy* (Malone and Lepper, 1987) which provides metaphors and analogies to help the person better understand the activity. In the computer domain, research by Webster and Martocchio (1992) and Venkatesh (1999) has shown that intrinsically motivating training environments can lead to better outcomes. Although Hutchins et al. (1985) do not use the term "intrinsic motivation," the component of *engagement* in their theory of direct manipulation is clearly related to the intrinsic motivation construct. They argue that engagement arises from the feeling of working directly on objects from the world rather than on surrogates and that this feeling of directness is evoked by interface metaphors. Following this line of reasoning, we argue that interaction styles can be classified on a continuum of engagement. Direct manipulation with its visual metaphor and direct action on objects would be most engaging, followed by menu, which also uses direct action on objects and may evoke a metaphor via the presentation and organization of textual menus. Command would be is least likely to evoke engagement in the learner because of the indirection of actions and the lack of metaphor. In our study we compare learning of a more engaging style (menu) to a less engaging style (command) in the two contexts of first-time use by novice users with no prior experience, and subsequent use by users who have gained prior experience from exposure to direct manipulation software. It is expected that the more engaging style will increase ease of use perceptions, which will in turn increase perceived usefulness and task performance. It is also expected that higher engagement will occur in subsequent use than in first-time use because of the familiarity gained from prior exposure to the direct manipulation software.

3. METHOLOLOGY

The participants were 173 undergraduate students selected on the basis of having little prior knowledge of computers or word processing. The average age of the participants was 21 years. There were 106 males and 67 females.

Word processing was used as the application domain for the study. We used commercial word processing packages in order to do our testing in a complex software environment. In particular, we wanted systems with many choices and multiple ways to achieve a goal. We tried to choose software that was as purely menu or command-based as possible in commercial software. The menu style software was a fully menu-based version of Word running on a PC. This version uses only textual menus for actions and does not contain iconic direct manipulation features. Arrow keys are used to select text on the screen, not a mouse. The command-based software was a full screen editor, Vi, with a powerful text formatting program, TROFF. In this system, the interaction is wholly via commands. The text of a document and document formatting commands are placed in a file created with the editor. Then the file is processed by the formatting program to produce a file containing the formatted document, ready to print. This software ran on a mainframe and was accessed using a personal computer as a terminal with the mouse removed. The participants did not seem to be aware of differences in hardware extraneous to the experiment (e.g., the command users believed they were using a personal computer not a mainframe). The menu and command-based applications ran under different operating systems, but the participants never interacted with the operating system. The two word processing applications differed in their advanced functionality, but both offered the basic functions used in this experiment. Given the differences between the two word processing applications, this should be considered a wholistic comparison, as described by Whiteside, Jones, Levy, and Wixon (1985). As in their work, the largest sources of variance were controlled (task set, training materials, time to train, time to perform tasks). Among the remaining differences, the interaction style was overwhelmingly the most salient.

Half of the participants were assigned to the first-time use condition. These participants were given no prior exposure to other interaction styles. Half of these participants were assigned to the menu-based software and half to the command-based. They were trained in a two-hour session. The first 55 minutes consisted of hands-on practice with basic word processing tasks guided by a self-study training manual. At the end of the training period, participants filled out three instruments PEU (Davis, 1989), PU (Davis, 1989), and Intensity of Flow, an instrument measuring engagement, or flow (Webster, 1989). Following the questionnaires, the participants carried out a series of evaluation tasks for 35 minutes without the aid of the training manual. These evaluation tasks were scored for correctness and used as a measure of performance.

The other half of the participants was assigned to the subsequent-use condition. These individuals participated in two sessions three days apart. In the first session, an assimilative context was created by exposing participants through hands-on practice to basic word processing tasks in a direct manipulation style interface. In the second session, the participants were trained to use the menu-based or command-based software, in exactly the same manner as the first-time use group. The objective was to determine how prior exposure to a direct manipulation assimilative context for word processing would affect the perceptions and performance of participants using the menu and command-based applications.

4. RESULTS

The basic results are shown in Figure 1. The path coefficients are shown for each path with the significance of the coefficients indicated. The findings indicate that all paths are significant, as predicted, with three exceptions: the paths from training condition-to-PEU and from PU-to-task-performance are not significant. The path from training-condition-to-intensity-of-flow is significant, but the results are in the opposite direction of the anticipated result, that is, participants in the first-time use condition experienced greater intensity of flow than subsequent-use participants.

Figure 1. Research Model with Results[†]

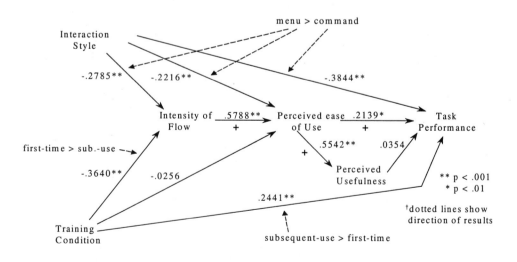

1555

We performed two analyses of the interaction of training condition and interaction style. First, we held the training condition constant while varying the levels of the interaction style variable. Three differences emerged. The paths from interaction-style-to-intensity-of-flow and from interaction-style-to-perceived-ease-of-use are significant in the subsequent-use model but not in the first-time use model. Also, the path from perceived-ease-of-use to task performance is significant in the subsequent use but not the first-time use model, a difference between conditions we had not anticipated.

Our second interaction analysis involved holding the interaction style constant while varying the levels of the training condition variable. This analysis showed the performance across training conditions for each interface. We found that the menu-based group had significantly higher scores on task performance in the subsequent use condition than in the first-time use condition. However, the command-based participants scored about the same in both first-time and subsequent use conditions.

5. DISCUSSION

Menu-based interaction leads to better task performance than command-based. Our path models show that this better performance comes from two sources: a direct effect of interaction style on task performance and an indirect effect of increased intensity of flow (engagement) and perceived ease of use. The direct effect can by explained by the directness component of Hutchins et al.'s (1985) theory of direct manipulation. The indirect effect can be explained by the second component of their theory, engagement. While several studies have addressed the effects of directness, this study begins to explain the mechanisms of *engagement*.

In terms of the interface factor, our results show that for learners using a menu or command-based word processing application for the first time there are no differences in intensity of flow (engagement) or in their ease of use perceptions. First-time users across interfaces reported relatively high levels of flow and PEU. However, for participants in the subsequent-use condition, using the menu or command-based word processor *after* experience with a direct manipulation word processor, there are differences. Command users experienced less intensity of flow and less perceived ease of use than menu users. These interactions of training condition and interaction style may be interpreted in terms of the assimilative context. First-time users are complete novices and have little relevant assimilative context for evaluating any interaction style. Their perceptions of flow and ease of use are largely insensitive to the interaction style, i.e., all interaction styles evoke similar levels of flow and ease of use. Subsequent-users have an assimilative context by which to evaluate the interaction style, i.e. their prior experience with a direct manipulation style. Given this direct manipulation assimilative context, they judge the menu style more similar to that context and evaluate it more highly than the command style in engagement and perceived ease of use. In effect, their experience with direct manipulation becomes the standard by which they evaluate all subsequent interfaces.

In terms of the training condition factor, our results show that subsequent-users achieved overall higher task performance scores than first-time users. However, the interaction analysis indicates that this is attributable to the menu group. The performance of participants in the command style did not differ between the training conditions. This seems to indicate that there was not sufficient similarity between the direct manipulation assimilative context and the command-based software for assimilation to occur. For the behavioral variables, we had expected both intensity of flow and perceived ease of use to be greater given prior experience. However, for perceived ease of use there was no difference between the first-time and subsequent-use conditions, while intensity of flow was greater in first-time use (see Figure 1). These results are not explained by the interaction of training condition and interaction style; if interaction style is held constant and training condition allowed to vary there is no difference between first-time and subsequent-use in either interaction style. The reason for these results is not certain. With respect to intensity of flow, the similarities between the prior direct manipulation style and the subsequent menu style may have been too small to evoke challenge, which is a necessary component of engagement. On the other hand, the differences between the prior direct manipulations style and the subsequent command style may have created too much challenge, too little control, and little curiosity. With respect to perceived ease of use, the lack of difference could be explained by the fact that participants were administered the PEU immediately after the training period but before the evaluation tasks, where they actually tried out their skills independently. Learners in both first-time and subsequent use learned the software during the training period by using a hands-on training manual that guided them step-by-step through operations. The supports given in the manual may have made first-time and subsequent use participants perceive the software at that time as equally easy to use, whether they had prior experience with the direct manipulation software or not. A useful manipulation would have been to administer the PEU again after the evaluation tasks.

We verified a positive relationship between perceived ease of use and perceived usefulness, as expected form the literature (Davis, 1989), but did not find the expected effect of perceived usefulness on performance. This implies that the participants in our experiment did not have a clear idea of the benefits of word processing software or how they would use the application in the future.

6. CONCLUSION

Two basic conclusions support past research: 1) a more direct style of interaction aids performance and 2) prior experience aids performance in cases where the interaction styles have recognizable similarities. Going beyond earlier results, this study shows the importance of engagement, which affects performance through its effect on perceived ease of use. This suggests that designers should attempt to create software that promotes engagement as well as directness. Another finding with implications for design and training is that first-time users experienced the same levels of engagement regardless of the software used. However, subsequent users' assessments appear to be influenced by the assimilative context they have gained from prior experience. Designers need to be aware of the learners' likely exposure to different interaction styles to successfully design engaging software. For complete novices, it may be difficult to improve learning through engagement.

REFERENCES

Ausubel, D. P. (1963). *The Psychology of Meaningful Verbal Learning*. New York: Grune and Stratton.

Bandura, A. (1986). *Social Foundations of Thought and Action: A Social Cognitive Theory*. Englewood Cliffs, NJ: Prentice-Hall.

Benbasat, I. and Todd, P. (1993). An experimental investigation of interface design alternatives; icon vs. text and direct manipulation vs. menus. *International Journal of Man-Machine Studies,* 38, 369-402.

Compeau, D. R. and Higgins, C. A. (1995). Application of social cognitive theory to training for computer skills. *Information Systems Research*, 6(2), 118-143.

Csikszentmihalyi, M. (1975). *Beyond Boredom and Anxiety*. San Francisco: Jossey Bass.

Davis, F. D. (1989). Perceived usefulness, perceived ease of use, and user acceptance of information technology. *MIS Quarterly*, 13(3), 319-340.

Davis, S. A. and Bostrom, R. P. (1993). Training end users: an experimental investigation of the roles of the computer interface and training methods, *MIS Quarterly,* 17, 61-85.

Davis, S. and Wiedenbeck, S. (1998). The effect of interaction style and training method on end user learning of software packages, *Interacting With Computers*.

Hutchins, E. L., Hollan, J. D. and Norman, D. A. (1985). Direct manipulation interfaces. *Human-Computer Interaction*, 1, 311-338.

Lepper, M. R. and Greene, D. (1978). *The Hidden Costs of Reward*. Hillsdale, NJ: Lawrence Erlbaum Associates.

Malone, T. W. and Lepper, M. R. (1987). Making learning fun: a taxonomy of intrinsic motivations for learning. In R. E. Snow and M. J. Farr (Eds.), *Aptitude, Learning, and Instruction,* (pp. 223-253). Hillsdale, NJ: Lawrence Erlbaum Associates.

Venkatesh, V. (1999). Creation of favorable user perceptions: Exploring the role of intrinsic motivation. *MIS Quarterly*, 23(2), 239-260.

Webster, J. (1989). *Playfulness and Computers at Work*. Unpublished doctoral dissertation, New York University, New York, NY.

Webster, J. and Martocchio, J. J. (1992). Microcomputer playfulness: development of a measure with workplace implications. *MIS Quarterly*, 16(2), 201-226.

Whiteside, J., Jones, S., Levy, P. S., and Wixon, D. (1985). User performance with command, menu, and iconic interfaces. In Borman, L. and Curtis, B., eds., *CHI'85 Proceedings: Human Factors in Computing Systems*. New York: ACM, 185-191.

Wiedenbeck, S. (1999). The use of icons and labels in an end user application program: an empirical study of learning and retention. *Behaviour and Information Technology*, 18(2), 68-82.

Cross-Cultural Studies of the Computers are Social Actors Paradigm:

The Case of Reciprocity

Yasuhiro Katagiri

ATR MI&C
2-2 Hikaridai, Seika-cho, Soraku-gun
Kyoto, 619-0288 Japan
katagiri@mic.atr.co.jp

Clifford Nass

Department of Communication
Stanford University
Stanford, CA 94305-2050
nass@stanford.edu

Yugo Takeuchi

ATR MI&C
2-2 Hikaridai, Seika-cho, Soraku-gun
Kyoto, 619-0288 Japan
yugo@mic.atr.co.jp

ABSTRACT

A number of studies have demonstrated that people treat computers socially. However, all of these studies have been performed in the United States and assume social rules derived from U.S. culture. The present research demonstrates, in the context of a comparative study of the United States and Japan, that individuals apply cultural norms derived from their unique cultures to determine their responses to computers. In Study 1, U.S. and Japanese participants interact with either a helpful or unhelpful computer. They are then asked to perform a task by either the computer they worked with or an identical computer on the other side of the room. U.S. participants exhibited both behavioral and attitudinal reciprocity and (some) retaliation effects, responding more favorably to the first computer when it helped and more favorably to the second computer when the first was unhelpful, consistent with an individualist culture. Japanese participants exhibited similar attitudinal patterns, but no behavioral differences. A follow-up study in Japan demonstrated that Japanese participants acted consistent with a collectivist culture, treating the second computer differently only when it was a different brand.

1. INTRODUCTION

Over the past ten years, Nass and colleagues (see (Reeves & Nass, 1996) and (Nass & Moon, 2000) for reviews) have performed a series of studies demonstrating that individuals will respond socially to computers. They have shown that people will be polite to a computer (Nass, Moon, & Carney, 1999) and will treat technologies as a teammate (Nass, Fogg, & Moon, 1996), a specialist (Nass, Reeves, & Leshner, 1996) and a flatterer (Fogg & Nass, 1997). They have also demonstrated that people will assign a gender (Lee, Nass, & Brave, 2000; Nass, Moon, & Green, 1997) and a personality (Isbister & Nass, 2000; Nass & Lee, in press; Nass, Moon, Fogg, & Reeves, 1995) to computers.

It is important to understand how this wide range of social responses to computers is demonstrated. In all of these experiments, the researchers employ a particular paradigm:

1) Go to the social-psychological literature and select a finding describing how humans treat other people.

2) Re-write the statement of the theory so that it refers to how humans treat *computers* (rather than other people).

3) Design and run an experiment that demonstrates that the re-written social rule is correct. That is, have participants interact with a computer and demonstrate that attitudes and behaviors that are observed match what would be observed if two people were interacting.

Using this model, the Computers are Social Actors (CASA) paradigm has been verified with respect to over 100 social rules (see (Reeves & Nass, 1996) and (Nass & Moon, 2000) for reviews).

While this approach seems culturally independent, there are actually two implicit biases in the research. First, virtually all of the research which the paradigm draws on was done in the United States with U.S. participants. While the importance of this limitation is rarely made explicit in the human-human research (there is a tradition for

U.S. researchers to simply assume that anything discovered applies across all cultures (Kitayama & Markus, 1992), it is likely that at least some discoveries are culturally-dependent. Second, all of the CASA studies were done with U.S. participants. While this was consistent with the source of the research, it leaves the CASA paradigm open to the same criticism as most other psychological research: The discovery that people respond socially to computers may not be applicable to other cultures, even if the rest of the psychological literature is applicable.

There is a second set of motivations for comparing individuals from different cultures. There has been a growing interest in international issues in interfaces (del Galdo & Nielsen, 1997; Fernandes, 1995; Khaslavsky, 1998). There are a number of reasons for this. First, as hardware prices have dropped, more and more countries can have a critical mass of machines that warrants attention by the software industry. Second, the growth of the Internet and the Web have greatly facilitated cross-national interactions via computers, interactions that both blur and highlight cultural differences. Third, with the advent of more richly social interfaces that include voices, characters, and natural language, cultural and cross-national differences move to the fore. Finally, the general societal concern with a Euro-centric approach to research has influenced the HCI community along with all other academic communities.

Despite the interest in cross-cultural differences in interfaces, there has been little systematic research. Much of the literature is filled with charming anecdotes, such as a character's gesture that is innocuous in one country but obscene in another, or a translation mistake that offends users. A second set of activities uses "deep description" and ethnographic methods to illustrate the most unique aspects of a particular culture's responses to a particular interface. While these methods provide provocative insights into cultural differences, they provide warnings rather than a systematic understanding of how to design for different cultures.

How can one address these deficiencies in the psychological, CASA, and human-computer interaction literatures? The most interesting and challenging approach is to identify a psychological finding concerning relationships between humans that leads to different predictions in different cultures, that is, a study in which an experiment with two people gives or would give different results when performed in the two cultures. One then performs the experiment replacing one of the people with a computer. If CASA works cross-culturally, participants in each culture should respond to the computer precisely as they would if they encountered another person from that culture.

The present studies adopt this approach, testing whether the CASA theory is robust across cultures. We begin with the concept of "reciprocity," and discuss that the Japanese view norms of reciprocity differently than do Americans. We then present two experiments that demonstrate that U.S. and Japanese participants exhibit responses to computers that are consistent with their cultural norms.

2. RECIPROCITY

Reciprocity, defined as "the rule that people should help those who help them" (Gouldner, 1960, p. 173), is one of the most fundamental norms of human behavior: Every culture trains their members to observe the rule of reciprocity (Cialdini, 1993; Gouldner, 1960; Greenberg, 1980). The rule of reciprocity is also very powerful (Cialdini, 1993): "Those who receive a favor or benefit seem to be absolutely obligated to return it in some way" (Fogg, 1997). Among humans, the core mechanism driving reciprocity seems to be a feeling of indebtedness (Fogg, 1997)

The critical issue in reciprocity research (and to a lesser extent, retaliation research) is how broad the obligation to reciprocity extends. The United States is an individualistic culture (Triandis, 1990). In these cultures, obligations are very narrow: individuals reciprocate *only* to the person that helped them. Conversely, Japan is a collectivist culture (Nakane, 1970), in which people generally associate with others on a group-oriented basis, and people's behaviors are strongly influenced by the considerations of one's affiliating groups (Nakane, 1970; Triandis, 1990). Thus, the obligation, and hence the reciprocation, will be directed to people in the helper's group as well as to the person himself or herself.

To make this concrete, imagine that Person A helps Person B. After that interaction, either Person A or Person C (a member of Person A's group) asks for help. In the United States, when Person A helped, Person B would help Person A much more than Person C. In Japan, however, the responses would be different. If Person A helped, we would find no significant difference between B's response to Person A and Person C, because B would feel a similar, collectivist obligation to both. To demonstrate an effect for the Japanese, we would have to have a condition in which the comparison was between person A and a person D that was in a different group than A; in that case, Person B would have no obligation to D and would thus be more helpful to A than D.

The above description provides guidance in how to test our hypotheses. First of all, we would expect a two-way interaction between country and A vs. C (a person in the same group as A), resulting from the reciprocity in the U.S. and a lack of reciprocity in Japan. We could then follow up this study with a study in Japan that compared B's responses to A, C, and D, the person in a different group; we would predict no difference between A and C, while both A and C would obtain more help than D.

2.1. Reciprocity and Computers

While the above seems logical and clear when A, C, and D are people, this analysis seems absolutely absurd when applied to computers. It should be obvious that users should not feel moral obligations to a computer: If *Computer* A helps a user, we would not expect a user to be more helpful to *Computer* A than to a different but physically identical *Computer* C that asked for help. Computers do not have distinct selves (but see (Nass et al., 1999)), so it is ludicrous to respond differently to one computer than another.

However, in a remarkable study, Fogg (see (Fogg, 1997) and (Fogg & Nass, 1997)) demonstrated that U.S participants do in fact respond to computers guided by the rules of reciprocity in an individualistic culture (see the left half of Figures 1 through 5 below). Specifically, U.S. users provided significantly more help to a computer that helped them than a different but identical computer that asked for help.

In the present paper, we first present a replication of Fogg's study in Japan. We determine whether Japanese participants will *not* exhibit reciprocity with respect to computers when the computers are identical, consistent with a collectivist culture. We then present a follow-up study that represents a critical test of collectivism with respect to computers via a three condition experiment: Computer A asks for help, an identical Computer C asks for help, or Computer D, a *different model*, asks for help.

3. EXPERIMENT 1: COMPARISON OF U.S. AND JAPANESE RECIPROCITY TO SIMILAR COMPUTERS

The experiment was a 2 (country) x 2 (same computer/different computer) between-participants design. Country was a quasi-experimental factor; the other factor was between-subjects. Participants in both countries ($\underline{N}$ = 22 in each country) were randomly assigned to condition.

In essence, the experiment consisted of two tasks. The first task created a situation in which the computer helped the subject. In this case, the computer performed an ostensible web search and consistently provided the information that the subject needed.

The second task created a situation in which the subject could help the computer. In this case, the subjects could provide information to the computer on how they perceived colors on the screen in an ostensible effort to help the computer create a color palette to match human perception. Half the subjects worked with the same computer as before, and half worked with a different—but identical—computer. This second task allowed subjects to choose how much—or how little—help to provide the computer, by allowing them to perform as many color comparisons as they wished. After completing the second task, participants filled out an attitudinal questionnaire (for full details of the experiment, see (Fogg, 1997)).

All aspects of the experiment were identical in the U.S. and Japan, with the exception that all stimulus materials and questionnaires were translated into the native language of the two countries.

3.1. Results

As can be seen in Figure 1, United States participants performed significantly more color comparisons when asked by the helpful computer they had worked with than when asked by a different computer, $t(20)$ = 4.90, p < .001, reflecting the norms of an individualistic culture. Conversely, in the collectivist culture of Japan, there was no significant different in the number of color comparisons, $t(20)$ = 0.33, p > .74. In the 2 x 2 analysis, there was a significant interaction between country and which computer asked for help, F(1,40) = 4.18, p < .04, demonstrating the clear cultural difference between the two countries.

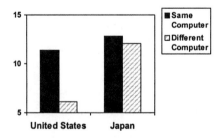

Figure 1: Number of Color Comparisons

4. EXPERIMENT 2: JAPANESE RESPONSES TO DIFFERENT BRANDS OF COMPUTERS

While the interaction in Experiment 1 provides suggestive evidence that the norms of a collectivist culture are applied to computers, a direct test of this claim requires us to distinguish the computers by "group," and to demonstrate that when the second computer belongs to a different "group" than the first, we see less helpful behavior directed toward the second computer.

Of course, computers do not belong to "groups" in the same ways that people belong to groups. The closest analog is brand; computers of the same model and brand might belong to the same "group," while computers from different brands might belong to different "groups."

To address this possibility, we replicated Experiment 1 in Japan but added a third condition, one in which the two computers were of different brands. Specifically, for half of the participants in this condition, they used an IBM PC for the search task (which instantiated helpfulness) and a Mac for the second task; the other half of participants used a Mac for the search task and an IBM PC for the second task, thereby balancing for any "brand" effects. In the other two conditions, half of the participants used (one or two) PCs and the other half of the participants used (one or two) Macs. This allowed us to control for any idiosyncratic differences with respect to brand.

4.1. Results

Participants ($\underline{N}$ = 80) were randomly assigned to one of the three conditions. Consistent with Japanese participants responding with collectivist norms with respect to brand, there was a significant difference between conditions, $F(2,77) = 6.6, p < .01$ (see Figure 2). Post-hoc comparisons (Tukey h-s-d) indicated that different-brand participants exhibited significantly less color comparisons than either the one-computer ($p < .01$) or two-computer ($p < .01$) participants. Consistent with the previous experiment, there were no differences between the one-computer and two-computer participants ($p > .5$).

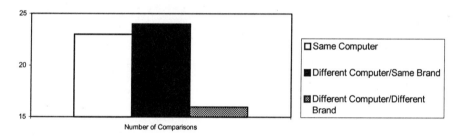

5. DISCUSSION

Historically, concern with internationalization of interfaces has focused on problems of translation (del Galdo & Nielsen, 1997; Watzlawick, 1976) or idiosyncratic cultural rules (e.g., an owl character is wise in the United States but evil in Germany). While these problems are important and certainly not solved, the present research suggests that culture is a much broader and deeper determinant of user attitudes and behaviors. When people apply social rules to guide their responses to technology, those responses are not just simplistic applications of social universals; instead, people automatically and unconsciously search for analogues between human characteristics and technological characteristics in order to guide their behaviors. Users in the United States, among the most individualistic of cultures (Boorstin, 1973), look for ways to individuate computers by viewing each machine as autonomous. Japanese participants, similarly, extend their core notion of group to a brand, the closest analogue in the technology world. Designers should be careful to determine how characteristics of technology will implicate particular, culturally-dependent social rules.

This research extends the Computers are Social Actors in two ways. First, it is one of the first studies to demonstrate behavioral effects, as distinct from the attitudinal effects present in previous studies. Second, this study demonstrates that computers are *culturally-embedded* Social Actors. That is, computers are treated using the norms for treating people *within the culture*. It is possible that if a computer was clearly marked as from a particular culture, for example, English-language computers in non-English-speaking countries, users would treat the computer as a foreigner rather than as a member of the culture.

The current research also demonstrates the importance of modesty in drawing conclusions from experiments done within a particular culture. This means not only that one must test interfaces in multiple cultures; it also means that one cannot blithely apply *psychological* studies performed in one country to design the interface in another.

In sum, the present research provides a cautionary note that creating the same application for different cultures requires more than translating words and pictures; it also requires a sensitivity to the entire range of social and cultural norms.

REFERENCES

Boorstin, D. J. (1973). *The Americans, the democratic experience* (1st ed.). New York: Random House.

Cialdini, R. B. (1993). *Influence: Science and Practice* (3rd ed.). New York: Harper Collins.

del Galdo, E., & Nielsen, J. (1997). *International User Interfaces*. New York: Wiley and Sons.

Fernandes, T. (1995). *Global Interface Design*. San Diego: AP Professional.

Fogg, B. J. (1997). *Charismatic Computers: Creating More Likable and Persuasive Interactive Technologies By Leveraging Principles from Social Psychology* (Unpublished doctoral dissertation): Stanford Unviersity.

Fogg, B. J., & Nass, C. (1997). *Do users reciprocate to computers?* Paper presented at the ACM CHI.

Fogg, B. J., & Nass, C. (1997). Silicon sycophants: The effects of computers that flatter. *International Journal of Human-Computer Studies, 46*(5), 551-561.

Gouldner, A. W. (1960). The norm of reciprocity: A preliminary statement. *American Sociological Review, 25,* 161-178.

Greenberg, M. S. (1980). A theory of indebtedness. In K. Gergen & M. S. Greenberg & R. H. Willis (Eds.), *Social Exchange: Advances in theory and research* (pp. 3-26). New York: Plenum.

Isbister, K., & Nass, C. (2000). Consistency of personality in interactive characters: Verbal cues, non-verbal cues, and user characteristics. *International Journal of Human-Computer Interaction, 53*(1), 251-267.

Khaslavsky, J. (1998). *Integrating culture into interface design*. Paper presented at the CHI Conference, New York.

Kitayama, S., & Markus, H. R. (Eds.). (1992). *Emotion and Culture: Empirical Studies of Mutual Influence.* Washington, DC: American Psychological Association.

Lee, E.-J., Nass, C., & Brave, S. (2000). *Can computer-generated speech have gender? An experimental test of gender stereotypes.* Paper presented at the CHI 2000, The Hague, The Netherlands.

Nakane, J. (1970). *Japanese Society*. Berkeley, CA: University of California Press.

Nass, C., Fogg, B. J., & Moon, Y. (1996). Can computers be teammates? *International Journal of Human-Computer Studies, 45*(6), 669-678.

Nass, C., & Lee, K. M. (in press). Does computer-synthesized speech manifest personality? Experimental tests of recognition, similarity-attraction, and consistency-attraction. *Journal of Experimental Psychology: Applied.*

Nass, C., & Moon, Y. (2000). Machines and mindlessness: Social responses to computers. *Journal of Social Issues, 56*(1), 81-103.

Nass, C., Moon, Y., & Carney, P. (1999). Are people polite to computers? Responses to computer-based interviewing systems. *Journal of Applied Social Psychology, 29*(5), 1093-1110.

Nass, C., Moon, Y., Fogg, B. J., & Reeves, B. (1995). Can computer personalities be human personalities? *International Journal of Human-Computer Studies, 43*(2), 223-239.

Nass, C., Moon, Y., & Green, N. (1997). Are machines gender-neutral? Gender-stereotypic responses to computers with voices. *Journal of Applied Social Psychology, 27*(10), 864-876.

Nass, C., Reeves, B., & Leshner, G. (1996). Technology and roles: A tale of two TVs. *Journal of Communication, 46*(2), 121-128.

Reeves, B., & Nass, C. (1996). *The media equation: How people treat computers, television, and new media like real people and places*. New York: Cambridge University Press.

Triandis, H. C. (1990). Cross-Cultural Studies of Individualism and Collectivism. In R. A. Dienstbier (Ed.), *Current theory and research on motivation.*: University of Nebraska Press.

Watzlawick, P. (1976). *How Real Is Real*. New York: Vintage.